THE
BUILDING ESTIMATOR'S
REFERENCE BOOK

**A Reference Book Setting Forth Detailed
Procedures And Cost Guidelines For Those
Engaged in Estimating Building Trades**

30th Edition, 1st Printing

Editor-In-Chief
Victor P Ticola

Technical Editor
Jerrold Ratner, CPE, CCP

Associate Editors
Peter F Ticola

Published By
FRANK R. WALKER COMPANY

Eugene R. Callahan, Chairman

700 Springer Drive
Lombard, Illinois 60148
(800) 458-3737
www.frankrwalker.com

We will be grateful to readers of this volume who will kindly call our attention to any errors, typographical or otherwise, discovered herein. We also invite constructive criticism and suggestions that will assist in making future editions more complete and useful.

FRANK R. WALKER COMPANY
Publishers of
Walker's Practical Accounting and Cost Keeping for Contractors
Walker's Insulation Techniques and Estimating Handbook
Walker's Quantity Survey & Basic Construction Estimating
F.R. Walker's Remodeling Reference Book
Walker's Manual for Construction Cost Estimating
Walker's Pocket Estimator
Walker's Practical® Forms for Contractors

ISBN: 978-0-911592-30-6
Library of Congress Catalog Card Number 15-23586

Printed in the U.S.A.

IMPORTANT

How To Use This Book

In a book of estimating cost data that is used in all parts of the U.S. and in many foreign countries, it is impossible to quote material prices and labor costs that apply in every locality. To make the best use of a reference like this one, it takes some computation on your part.

All labor cost tables give the quantity of work that a worker or crew should perform per hour or per day, together with the number of hours required to complete a certain unit of work. We have taken this data and multiplied by wage rates that are the national average to determine labor unit prices for a given item.

The wage rates used in the labor cost tables found throughout this book are U.S. averages. Each rate is an average of the total wage and benefits that resulted from collective bargaining for that trade throughout the U.S. The rate shown may or may not reflect wage and benefit rates in your area.

You will want to insert your local wage scales and material prices into the cost tables to ensure accurate estimates. A standard method of tabulation has been used throughout this book, an example of which is given below:

When framing and placing wood floor joists up to 2" x 8" (50 x 200 mm) in buildings of regular construction, a carpenter should frame and place 550-600 b.f. (1.30-1.42 cu.m) per 8-hr. day at the following labor cost per 1,000 b.f. (2.36 cu.m):

	Hours	Rate	Total	Rate	Total
Carpenter	13.9	$....	$....	$28.25	$392.68
Labor	4.5	....	....	22.25	100.13
			$....		$492.81

The blank spaces are for you to insert your local wage rates and extend the prices as shown below:

	Hours	Rate	Total	Rate	Total
Carpenter	13.9	$21.50	$298.85	$28.25	$392.68
Labor	4.5	17.25	77.63	22.25	100.13
			$376.48		$492.81

Note that there is no factor included for overhead and profit on any of the costs in this book, because this is a variable item that each contractor must establish for each individual project.

Because of wide variance in state sales tax laws, this cost has not been included, and the user must apply the prevailing rate to obtain the total material cost.

To accommodate the maximum number of contractors, the insurance and taxes applicable to labor costs have not been included in any of the labor unit costs contained in this book, because some contractors prefer to incorporate these costs as part of overhead, while others maintain them as direct costs.

Using the Metric System

In many ways metric is more coherent and efficient than the standard English (Imperial) inch-pound system. The metric system avoids the need for much tedious conversion. For example, lineal measurements are expressed in the meter and its decimal multiples. Compared with all the various units of measure—inches, feet, yards—in the English system, conversions in metric are accomplished by simply moving a decimal point to the left or right.

Rather than create two separate editions of the same book, one metric and the other inch-pound, we feel that our users are better served by this dual English-Metric guide. All contractors, even ones who do mostly government work, will continue to do a substantial amount of work in the traditional system.

It is difficult to make a comprehensive and general statement about the conversion of data in this book to metric, but for the most part, it has been what is called "soft" converted. Nothing changes about a material item itself. An inch-pound measurement is simply expressed in metric. For example, 2" x 4" lumber is 50 x 100 mm. For lineal dimensions the key conversions are 1 inch = 25 mm and 1 foot = 300 mm.

The metric conversion of data is this book was only possible with the able assistance of Mr. Suresh Ceyyur.

TABLE OF CONTENTS

2. SITE WORK
CSI DIVISION 2

3. CONCRETE
CSI DIVISION 3

4. MASONRY
CSI DIVISION 4

5. METALS
CSI DIVISION 5

16. ELECTRICAL

17. MENSURATION

01   GENERAL REQUIREMENTS

A building project is the effort of many parties. Traditionally, the primary parties are the owner, the architect/engineer, and the contractor. But the owner will also have as consultants a lawyer for legal matters; a realtor for land purchase and building management; and accountants and financiers, who provide short-term cash for building construction and long-term mortgage when the project is complete

The Role of the Architect

The obvious role of an architect is to create and design the working drawings for a client. Even though this is a very simple view. The architect's role begins with visualizing the needs of the client from conception to completions. Success in the architect's role depends on this ability to work with the client, the various contractors, a multitude of engineering firms, and manufacturing companies. The technical knowhow to incorporate the vast amount of materials and methods of construction and their impact on the environment places a tremendous responsibility on the architect.

The architect brings to the project several more in-house and outside consultants as well: structural and mechanical engineers, site planners, fire protection engineers, interior space planners, environmental engineers and landscape architects, to name but a few.

The architect also coordinates with government agencies that control and monitor local zoning, building codes, and city planning. Even though normally it is the contractor—not the architect—who takes out a building permit, the latter must assure that based on the plans and specifications for the project, permits can be issued. The architect also relies on the cooperation of many suppliers and manufacturers whose products are designed into a structure.

The Role of the Estimator

Estimators are the most important person in a contractor's organization, because it is up to them to prepare estimates that will make the contractor money. The most efficient project organization or the best

1

purchasing department in existence cannot make money on contracts taken below cost.

The estimator should be practical and must possess a thorough understanding of job conditions, of how the work will be carried on in the field, and of the operations necessary to assemble the materials and put them in place in the structure. The estimator must be able to visualize and must be able to take a set of drawings and develop a mental picture of the building. The estimator must know every branch of the work to be handled, the materials required, and the labor operations necessary to convert a pile of gravel, sand, brick, cement, lumber, steel, and glass into a completed building. The job superintendent or foreman may be able to take a set of drawings and lay the entire job out in their mind and picture the building progressing from the foundation to the finished structure, but unless they possess the knowledge necessary to compute the quantity of materials and the labor cost of putting them in place, they will never be able to prepare an accurate estimate of cost. The following requisites are essential for the making of a good estimator:

1. Ability to read and measure plans
2. Possess a basic knowledge of math.
3. Able to visualize the project from the drawings.
4. Have a working knowledge of materials and methods of construction.
5. Possess the know how to develop unit cost for materials and labor.
6. Possess an intimate knowledge of labor performances and operations and be able to convert them into costs in dollars and cents. This requires a working knowledge of what a worker will do under given conditions and is obtained only by a careful study of different jobs completed under varying conditions. This involves time and motion study knowledge.
7. Have an amount of common sense to determine the project requirements and their costs – at a future point in time.

The Role of the Contractor

On the contractor's team, there are various subcontractors and the producers and suppliers of the equipment, materials, services, appliances, tools, and machinery called for by the architect and required to complete the job. The contractor must also cooperate with governing officials and must take out all building permits and conduct their operation without violating local laws that govern everything from blocking traffic to waking up the neighbors. And then there are the trade unions, the local utilities, and the ever-necessary bankers.

When land becomes scarce, money rates increase, and ecological concerns grow, the exclusive club of owner-architect/engineer-contractor is forced to let in others, and the triangle becomes more of a circle. On large projects today, the owner may be the large landowner who wishes to develop

2

their holdings, a government that is working to rebuild the inner city, or a large investment company that needs to keep its investments active. The architect's concern may no longer be the individual building but an entire complex, a neighborhood, or a whole new town. And the general contractor may find themselves in demand less for craftsmanship abilities than for managerial capabilities. But no matter how small or how complex the project, the owner has a need, the architect/engineer develops that need into satisfactory plans, and the contractor interprets the plans into an actual building.

Erecting a building is a complex undertaking, and seldom is one firm capable of doing all phases of the work. Yet the owner or developer usually prefers to let one contract and make one firm responsible for the completion of the project. That firm is then known as the General Contractor (General or GC), Prime Contractor (Prime or PC), or Construction Manager (CM), who usually assumes this role when the owner asks for firm price bids. It is they who will compile all material, labor and service, and other costs called for on the architect's drawings and specifications. Where the contractor's own firm is not able to do work called for, they will turn to other contractors who have expertise in those fields. These are the Subcontractors (or subs), and if the project is awarded to the general, the subs will enter into contracts with the general and be responsible to the general rather than directly to the owner. The general contractor will ask for bids from several subcontractors in each trade. This enables the general contractor to put together the best price and what they believe to be the best team to perform the work.

The amount of work performed by specialty subcontractors varies to suit each project's needs. This is determined by the general contractor's in-house capabilities and the project scope. This major decision is not predicated solely upon economic factors. Each time some portion of the work is assigned to a subcontractor, there is introduced into the construction team another member, whose strengths and weaknesses translate directly into the total performance plan. The general who attempts to "broker" a project by maximum use of subcontractors in order to limit financial responsibility will find that other hazards have been created. There will be a considerable loss in coordination and production, because the general has relinquished the right and ability of direct control in the assignment of personnel, materials, and equipment. They might be subjected to a subcontractor who does not have their same dedication to the project.

Initially, it may appear that the large dollar value assigned for subcontractor performance insulates the general from financial loss on that portion of the project, but this is usually found to be false economy. Upon signing the subcontract document, the general contractor has inherited any faults and deficiencies that exist in the subcontractor's company. The extent of these faults becomes apparent during the life of the contract, and they adversely affect the project schedule, subjecting the general contractor to delay claims from other subcontractors as well as from the owner. In addition, the general contractors own direct costs will increase in direct proportion to delayed completion.

In the most severe instances, the general may be forced to terminate the subcontract. There may well be excess costs incurred due to additional procurement costs and associated delays. It is an accepted practice to employ certain subcontractors who possess expertise in a particular discipline. However, they must be carefully selected on the basis of financial capacity, technical ability and the proven ability to perform consistent with the requirements of the contract documents. This policy can ensure a successful project, a credit to the owner, the A/E, and the contracting team.

The average percent of work performed by subcontractors for a general contractor cannot be precisely determined, but surveys conducted by the Associated General Contractors of America indicate from 40% to 70%.

The general contractor must retain the financial ability to perform the work volume they have under contract, including allowances for unanticipated delays in receiving payments for work performed. Such delays seem to be increasing at an alarming rate, placing an additional burden on the financial resources of the general. There must be in-house personnel who are experienced in the particular type of projects in which the firm is engaged. Of equal importance, there must be adequate managerial skills to coordinate the field construction activities and the administrative functions into a total effort. In private work or negotiated bidding the general contractor is normally expected to provide many services to the owner and the architect, before the contract award as well as during construction. The general contractor's knowledge and contribution may be the deciding factor in receiving a contract award or in their expanding a project under construction.

How much of the contract will be sublet will vary with each general contractor. The American Subcontractor's Association claims that 90% of the work force in the building construction industry is employed by subcontractors. It is certainly quite possible for a subcontractor to erect a greater portion of the project than the general, and in some cases, the general is forced to sublet work. Some manufacturers insist that their products be installed by contractors licensed to them. For example, in curtain wall construction the specifications will call for the supplier to install the product in order to put the responsibility for any corrective work on one source. Obviously, if anything like 90% of the contract is to be sublet to others, the general contractor becomes more of a construction manager and must have the managerial skills to administer the work, schedule and coordinate all the firms involved so that the project proceeds without delay.

In addition to building and managerial skills, the contractor must be financially responsible. Not only must they meet their own payroll and overhead costs, but they also must pay their subcontractors and suppliers in advance of being reimbursed, hopefully with some profit, by the owner. Usually the general contractor will put in a request for payment each month for that work completed the month before. These requests may have to be accompanied by <u>waivers of lien</u> for the amount requested from each sub and supplier. In submitting such a waiver, the sub gives up their right to file a claim against the property (this is a general statement the contractor needs to check with each state law). To obtain a waiver, the general will have had to pay the sub or supplier at least the major portion of the payment request amount. Usually there is a <u>retainage</u> of at least 5% to 10% by the general contractor to the subs and in turn by the owner to the general contractor. Retainage is not completely released until the project is substantially complete. It has become common practice to reduce the retainage by 50% when the project is 50% satisfactorily completed.

A contractor whose financial affairs are in good order and who has a contract with a responsible client, who in turn has secured adequate construction and mortgage loans, should have little trouble in securing a loan from their bank to finance the month-to-month payments. The contract will serve as the loan security. But before quoting a final price for a project, the contractor should determine the length of the period they must finance, estimate the amount involved, and add the cost of the money to the job cost. It should not have to come out of the contractor's profit.

The successful general contractor is one who has a broad knowledge of building construction. Their initial contact with a prospective client may be about site selection, sources of construction finance, or the advantages of one type of construction over another. The general must develop contacts with architects, engineers, bankers, realtors, mortgage brokers, soil engineers, and the local enforcers of building and zoning codes. Their knowledge in these allied fields may be the deciding factor in them being considered for the job.

Most municipalities and many states require subcontractors and general contractors to be licensed. There is no general rule governing licensing in all jurisdictions. A state license may not fully qualify a contractor to practice in a municipality within that state. Many suburban areas will require additional fees from out-of-town contractors. Some states, for public work projects, may give in-state contractors a preference over out-of-state contractors. One must always inquire about the laws governing the site of construction.

Often a license or permit bond is required guaranteeing a public body that the contractor will comply with all applicable statutes and ordinances. If land development is part of the contract, subdivision bonds may be required guaranteeing proper installation of roads, sewers and other utilities, today even extending to provision of schools and park systems.

To operate as a union shop or merit/open shop firm is another decision. If selecting to operate as a union shop contractor, one must enter into proper relationships with the unions. The unions will have established wages and fringe benefits that must be met. To operate as a merit/open shop a source for obtaining skilled labor must be determined.

Finally, one must also consider the type of business relationship that defines the firm; that is, whether to practice as an individual, a partnership, or a corporation. Consult your lawyer and banker.

Construction Management

In Construction Management (CM), a general contractor or engineering company enters into a contract with the owner prior to the bidding period and acts in a managerial and advisory role. Bid packages for the project are usually taken under the construction manager's supervision. These are typically separate contracts for different trades, phases or other areas of the project. These contracts are customarily solicited, received and awarded on behalf of the owner by the CM firm.

The construction manager is paid their fees by the owner. This fee is separate from other job expenses. CM fees are determined in one of several ways:

1. A stated number of dollars plus expenses. The expense items would include the hourly wages paid the principals billed at an agreed upon fixed rate, plus employees' time billed at an agreed upon multiplier to cover overhead and profit, plus a multiple of expense for any outside consultants.
2. Same as above but without a fixed fee, all billing being on a multiple of direct expense.
3. An agreed upon fixed fee.
4. A percentage of the total construction cost.

Services rendered vary, but generally the CM is hired early in the planning and will work with the architect and the owner in setting up job budgets, bidding and construction schedules, site development, and selecting building systems. They will participate in value engineering studies. Not only will they determine what material or procedure is least costly, but which ones will give optimum quality within the owner's budget. The CM will work with the architect/engineer in arranging specifications to include the proper extent of work for each phase to be bid. They will prepare prequalification criteria for the bidders. They will conduct pre-bid conferences to acquaint the bidders with the extent of work covered under the various contracts, let them know what facilities are available to them at the job site and what they must furnish themselves, the project schedule, and special procedures to be followed.

When the bids are received, the CM will evaluate them and recommend that one be accepted or that the phase should be modified and rebid.

When the construction begins, the CM will coordinate all work on the site to coincide with the agreed to schedules, budgets, and quality standards. The CM will recommend corrective procedures and job changes that they feel are necessary, but the decisions are subject to the approval of the

GENERAL REQUIREMENTS

architect and owner. The CM will maintain an office at the job site with sufficient personnel to inspect the work, record job progress, keep up-to-date records of drawing and specification changes, shop drawings, samples, operating manuals and other records which will be delivered to the owner upon completion of the project. A CM will set up and maintain the project accounting system. They will provide reports on the project cost status, payment status, and an analysis of each contract and the project cash flow. The construction management approach has gained a wider acceptance in recent times, especially in federally sponsored construction.

It frees the contractor from many of the risks of a lump sum contract. But if it limits their losses, it also limits their profits, because any job savings that the CM's expertise may bring about accrue to the owner.

The CM format is not attractive to all owners, especially not to the ones who desire to minimize their direct activity with the project during the construction period. Many owners find it more desirable to employ the services of a general contractor on a well-defined, firm price contract basis.

Bidding for a Contract

Construction contracts are usually awarded in one of two ways- competitive bidding or negotiation. Competitive bidding is the method used for most public contracts and quite a number of private contracts. It is most effective when complete working drawings and specifications are available and when contractors are screened so all who bid are of the same caliber. Bids based on sketchy plans and vague specifications can be a waste of time. There is typically a bidder that will underbid the knowledgeable contractors. One of the most valuable contacts an aspiring contractor can make is to get on the bidding list of an architect who produces detailed drawings and tight specifications and insists on their being carried out. The architect has already spent many hours working with the owner to make most of the critical decisions, and by the time bids are taken, the contractor's responsibilities are clearly set forth. The contractor can proceed, if awarded the contract, knowing that the finished project will most probably satisfy the owner, and satisfied owners are one of the most valuable assets a contractor can have.

In private work the architect, often with the owner's help, will limit the list of bidders. This is a prequalification process. With known, prequalified bidders it is possible to evaluate the competition and know that the lowest bidder will be awarded the contract. Public work contracting is usually open to any qualified bidder. The qualifications are generally concerned more with financial status than reputation for quality. In order to propose a low bid against unknown competition, a good contractor must be certain their organization can bring to the project unique experience and efficiencies. Otherwise, they can be underbid by those desperate enough for work to bid without profit, or under cost, just to get a job.

Private work is announced in several ways. It can be listed in trade publications, at builders' exchanges or plan rooms and via electronic newsletters, etc. A real estate agent or a banker can often steer a contractor to jobs. Public work is almost always publicly advertised. Often, to get on a bidding list, a contractor must be prequalified. There are various ways this is done, by organizations, governmental bodies, and banks, but in general, the information that must be submitted will follow that contained in AIA Document A305 Contractor's Qualifications Statement (AIA form A305 Contractor's Qualification Statement). This requires that a firm—whether corporation, partnership, or individual—provide details on: length of time in business; names of principals; percent of work performed by own staff; record of and reasons for any past failures to complete a job; major projects now under construction; major projects completed in past five years; a record of experience of firm employees; bank references; trade references; name of bonding company; and a statement of financial conditions including current assets, liabilities, and the auditor's name. However, often just being a member in good standing of a local contractor's association and a bank reference will satisfy an inquiry.

Once on a bidding list, a contractor will receive an <u>Invitation to Bid or Bid Notice</u> for each prospective job. This form will include such information as job location, type of proposal, the number of sets of drawings and specifications to be furnished to bidders, what to do should discrepancies or errors be found in drawings and specifications, bulletins to be issued later, when and how bids will be opened, and when and where bids are due. A proposal sheet is usually attached so only the blanks need be filled out, but sometimes the contractor must furnish their own.

Often the invitation to bid is accompanied by an <u>Instruction to Bidders</u> further defining the job restrictions such as completion dates, milestone dates, visiting job site, special conditions, etc.

Bid deposits may be required, the amount being either a lump sum or a percentage of the proposal. The <u>bid bond</u> guarantees that the bidding contractor, if awarded the contract, will enter into the contract and furnish a <u>performance and payment bond,</u> if required. If they do not honor their bid, they forfeit the amount of the bond. Or, it may be the difference between their bid and the amount the owner can contract the work for with another contractor. Bid bonds—and the later performance, material, labor, maintenance, completion, supply, and subcontractor bonds—are often encountered in public work, but may not be required in private work, where the contractor's reputation is deemed sufficient and the cost of bonds unwarranted. Bid bond costs are customarily minimal, if any, and borne by the contractor. Performance and payment bonds costs are also typically covered by the contractors and/or subcontractors and should be included in the bid amount. Unless noted otherwise, bonds are not usually paid for by the owner directly. Bonds are further discussed under "Definitions" in this chapter.

GENERAL REQUIREMENTS
Negotiating a Contract

The other type of contract, one that is <u>negotiated</u>, allows the owner control over the selection of the general contractor. It is often used on complex work, for example, in remodeling jobs where the space is occupied and must be done piecemeal; interior work where strict control of subcontractors is desired; contracts in which mechanical equipment is a major factor and the owner may prefer one manufacturer over another; work which must proceed in great haste where the architect may not be able to produce complete plans in the time allotted and work will be let in sections (i.e., foundation work, then steel framing, then masonry). Often the contractor will proceed as with bid jobs in submitting a lump sum price, usually on the safe side and known as an <u>upset</u> or <u>guaranteed maximum price (GMP)</u>. The GMP will include overhead and profit. However it may be included as a firm, fixed price or expressed as a percentage of the job total.

A variation of the negotiated contract is fast track, design build <u>construction</u>. In this arrangement the project may be started before all the plans are fully developed. Each phase of the job, such as foundation, masonry, carpentry, etc., is bid separately, just before the phase is required to be installed. Some advantages and disadvantages are:

1. The project can be started earlier, rather than wait until all the design details are settled. Cost may be expended on items that have to be redone or features are designed "larger" than needed.
2. The bidders need not add contingencies to cover unknowns such as wage and material cost rises or adverse job conditions. The owner bears the risk of having costs increase beyond what was budgeted.
3. The bidding time is increased from three or four weeks to several months. If coordination is not done properly between bid packages, work will be left out of the bid scopes.

As the phases are bid, the successful subcontractors may be assigned to a general contractor in the same manner as a lump-sum contract; or the general may act in the role of a project manager, in which case each subcontractor for each phase will have a direct contract with the owner. This variation is referred to as <u>multiple bidding</u>.

Contract Documents

Once it is determined which contracting firm is to do the job, a formal contract will be drawn up. The Contract Documents usually should include the Owner-Contractor Agreement; the General Conditions of the Contract; Supplementary Conditions of the Contract (if any); the Working Drawings, giving all sheet numbers with revisions; Specifications, giving page numbers; and Addenda or Bulletins issued prior to contract. After the contract is signed, any changes are added as Modifications to Contract and would take the form of written amendments, Change Orders or architectural interpretations.

The Frank R. Walker Company has a wide range of forms available for the contractor's use. The forms are categorized by the following headings: Contract & Bid Forms (including Contract Change Order, Contractor's Sworn Statement, Proposal, Purchase Order, Sub-Contract Agreements, Telephone Bid, Transmittal Letter, and Waiver of Lien), General Estimate Forms, Job Estimating Forms, Quantity Take-Off Sheets, Recapitulation Sheets, and Time & Labor Distribution Forms (including Daily Construction & Material Reports, Daily Time, Material, and Labor Distribution Sheets, Extra Work Order, Record of Employment and Earnings, and Weekly Time Cards). The category links to the forms can be viewed at http://www.frankrwalker.com/c-3-forms.aspx.

The Owner-Contractor Agreement is a legal matter, but for many jobs standard forms are used. If the architect furnishes the Owner-Contractor Agreement, they will likely use one printed by the American Institute of Architects (AIA), Washington, DC. These documents are:

A101 Standard Form of Agreement Between Owner and Contractor where the Basis of Payment is a Stipulated Sum. This is a standard form of agreement between owner and contractor for use where the basis of payment is a stipulated sum (fixed price). The A101 document adopts by reference and is designed for use with A201, General Conditions of the Contract for Construction, thus providing an integrated pair of legal documents. When used together, they are appropriate for most projects. For projects of limited scope, however, use of A107 might be considered.

A132 Standard Form of Agreement Between Owner and Contractor where the basis of payment is a Stipulated Sum, Construction Manager-Adviser Edition. A132 is a standard form of agreement between owner and contractor for use on projects where the basis of payment is a stipulated sum (fixed price), and where, in addition to the contractor and the architect, a construction manager assists the owner in an advisory capacity during design and construction. The document has been prepared for use with A232, General Conditions of the Contract for Construction, Construction Manager-Adviser Edition. This integrated set of documents is appropriate for use on projects where the Construction Manager only serves in the capacity of an adviser to the owner, rather than as constructor (the latter relationship being represented in documents A133 and A134). A132 is suitable for projects where the cost of construction has been predetermined, either by bidding or by negotiation.

A105 Standard Form of Agreement Between Owner and Contractor for a Small Project, and A205 General Conditions of the Contract for Construction of a Small Project. A105 is for use on a project that is modest in size and brief in duration, and where payment to the contractor is based on a stipulated sum (fixed price**A107 Standard Form of Agreement Between Owner and Contractor for Construction Projects of Limited Scope Where the Basis of Payment is a Stipulated Sum.** A107 is intended for use where the basis of payment is a stipulated sum (fixed price).

GENERAL REQUIREMENTS

The document contains abbreviated general conditions derived from A201 and is appropriate for construction projects of limited scope not requiring the complexity and length of the combination of documents A101 and A201, General Conditions of the Contract for Construction. A107 is appropriate for use when the owner and contractor have established a prior working relationship (e.g., a previous project of like or similar nature), or where the project is relatively simple in detail or short in duration.

A107 Abbreviated Standard Form of Agreement Between Owner and Contractor for Construction Projects of Limited Scope Where the Basis of Payment is a Stipulated Sum. A107 is intended for use where the basis of payment is a stipulated sum (fixed price). The document contains abbreviated general conditions derived from A201™-1997 and is appropriate for construction projects of limited scope not requiring the complexity and length of the combination of documents A101™-1997 and A201™-1997, General Conditions of the Contract for Construction. A107 is appropriate for use when the owner and contractor have established a prior working relationship (e.g., a previous project of like or similar nature), or where the project is relatively simple in detail or short in duration.

A102 Standard Form of Agreement Between Owner and Contractor Where the Basis of Payment is the Cost of the Work Plus a Fee with a Negotiated Guaranteed Maximum Price. This standard form of agreement between owner and contractor is appropriate for use on most projects requiring a negotiated guaranteed maximum price, when the basis of payment to the contractor is the cost of the work plus a fee.

A103 Standard Form of Agreement Between Owner and Contractor Where the Basis of Payment is the Cost of the Work Plus a Fee without a Guaranteed Maximum Price. The A103 is appropriate for use on projects when the basis of payment to the contractor is the cost of the work plus a fee, and the cost is not fully known when construction begins.

A133 Standard Form of Agreement Between Owner and Construction Manager as Constructor where the basis of payment is the Cost of the Work Plus a Fee with a Guaranteed Maximum Price. This document is intended for use on projects where a construction manager, in addition to serving as the advisor to the owner, assumes financial responsibility for construction of the project. The construction manager provides the owner with a guaranteed maximum price proposal, which the owner may accept, reject, or negotiate.

A134 Standard Form of Agreement Between Owner and Construction Manager Where the Construction Manager is also the Constructor and Where the Basis of Payment is the Cost Plus a Fee and there is no Guarantee of Cost. Similar to A133, this CM-constructor agreement is also intended for use when the owner seeks a construction manager who will take on responsibility for providing the means and methods of construction. However, in A134 the construction manager does not provide a Guaranteed Maximum Price (GMP).

A141 Agreement Between Owner and Design-Builder. A141 obligates the Design-Builder to execute fully the Work required by the Design-Build Documents, which include A141 with its attached exhibits, the project criteria and the design-builder's proposal, including any revisions to those documents accepted by the owner, supplementary and other conditions, addenda and modifications. The Agreement requires the parties to select the payment type from three choices: (1) Stipulated Sum, (2) Cost of the Work Plus Design-Builder's Fee, and (3) Cost of the Work Plus Design-Builder's Fee with a Guaranteed Maximum Price. A141 with its attached exhibits forms the nucleus of the Design-Build Contract. Because A141 includes its own Terms and Conditions, it does not use A201.

A142 Agreement Between Design-Builder and Contractor. A142 obligates the contractor to perform the work in accordance with the Contract Documents, which include A142 with its attached exhibits, supplementary and other conditions, drawings, specifications, addenda, and modifications. Like A141, A142 requires the parties to select the payment type from three choices. (1) Stipulated Sum, (2) Cost of the Work Plus Design-Builder's Fee, and (3) Cost of the Work Plus Design-Builder's Fee with a Guaranteed Maximum Price.

The AIA forms are widely accepted within the industry as contract standards between the owner, architect and contractor. The AIA has for ease of understanding there documents separated the documents into series. It is suggested that the owner, contractor, architect and engineer become familiar with all AIA forms and use the appropriate form or forms for a specific project. Listed below are the documents by Series.

A – Series: Owner – Contractor Documents:

A101-2007 Standard Form of Agreement Between Owner and Contractor where the Basis of Payment is a Stipulated Sum

A102-2007 Standard Form of Agreement Between Owner and Contractor Where the Basis of Payment is the Cost of the Work Plus a Fee with a Negotiated Guaranteed Maximum Price

A103-2007 Standard Form of Agreement Between Owner and Contractor Where the Basis of Payment is the Cost of the Work Plus a Fee without a Guaranteed Maximum Price

A105-2007 Standard Form of Agreement Between Owner and Contractor for a Residential or Small Commercial Project

A107-2007 Abbreviated Standard Form of Agreement Between Owner and Contractor for Construction Projects of Limited Scope

A132-2009 Standard Form of Agreement Between Owner and Contractor where the basis of payment is a Stipulated Sum, Construction Manager-Adviser Edition

A133-2009 Standard Form of Agreement Between Owner and Construction Manager as Constructor where the basis of payment is the Cost of the Work Plus a Fee with a Guaranteed Maximum PriceA134-2009 Standard Form of Agreement Between Owner and Construction Manager as

GENERAL REQUIREMENTS

Constructor where the basis of payment is the Cost of the Work Plus a Fee without a Guarantee Maximum Price

A141-2004, Agreement Between Owner and Design-Builder

A142-2004, Agreement Between Design-Builder and Contractor

A151-2007 Standard Form of Agreement Between Owner and Vendor for Furniture, Furnishings and Equipment where the basis of payment is a Stipulated Sum

A232-2009 General Conditions of the Contract for Construction, Construction Manager-Adviser Edition

A201-2007 General Conditions of the Contract for Construction

A251-2007 General Conditions of the Contract for Furniture, Furnishings, and Equipment

A305-1986 Contractor's Qualification Statement

A310-2010 Bid Bond

A312-2010 Performance Bond and Payment Bond

A401-2007 Standard Form of Agreement Between Contractor and Subcontractor

A503-2007 Guide for Supplementary Conditions

A533-2007 Guide for Supplementary Conditions, Construction Manager-Adviser Edition

A701-1997 Instructions to Bidders

A751-2007 Invitation and Instructions for Quotation for Furniture, Furnishings and Equipment

B – Series: Owner – Architect Documents:

B101-2007 Abbreviated Standard Form of Agreement Between Owner and Architect

B105-2007 Standard Form of Agreement Between Owner and Architect for a Small Project

B107-2010 Standard Form of Agreement Between Owner and Architect for Limited Architectural Services for Housing Projects

B108-2009 Standard Form of Agreement Between Owner and Architect for a Publicly-Funded or Insured Project

B141-1997 Standard Form of Agreement Between Owner and Architect with Standard Form of Architect's Services

B141CMa-1992 Standard Form of Agreement Between Owner and Architect, Construction Manager-Adviser Edition

B142-2004, Agreement Between Owner and Consultant where the Owner Contemplates using the Design-Build Method of Project Delivery

B143-2004, Agreement Between Design-Builder and Architect

B144ARCH-CM-1993 Standard Form of Amendment to the Agreement Between Owner and Architect Where the Architect Provides Construction Management Services as an Adviser to the Owner

B151-2007 Standard Form of Agreement Between Owner and Architect for Furniture, Furnishings and Equipment Design Services

B152-2007 Standard Form of Agreement Between Owner and Architect for Architectural Interior Design Services

B161-2002 Standard Form of Agreement Between Client and Consultant for use where the Project is located outside the United States

B162-2002 Abbreviated Standard Form of Agreement Between Client and Consultant for use where the Project is located outside the United States

B204-2007 Standard Form of Architect's Services: Value Analysis, for use where the Owner Employs a Value Analysis Consultant

B205-2007 Standard Form of Architect's Services: Historic Preservation

B206-2007 Standard Form of Architect's Services: Security Evaluation and Planning

B207-2008 Duties, Responsibilities and Limitations of Authority of the Architect's Project RepresentativeB210-2007 Standard Form of Architect's Services: Facility Support

B211-2007 Standard Form of Architect's Services: Commissioning

B214-2007 Standard Form of Architect's Services: LEED® Certification

B305-1993 Architect's Qualification Statement

B503-2007 Guide for Amendments to AIA Owner-Architect Agreements

B727-1988 Standard Form of Agreement Between Owner and Architect for Special Services

C – Series: Architect – Consultants Documents:
C101-1993 Joint Venture Agreement for Professional Services

C132-2009 Standard Form of Agreement Between Owner and Construction Manager as Adviser

C401-2007 Standard Form of Agreement Between Architect and Consultant

C727-1992 Standard Form of Agreement Between Architect and Consultant for Special Services

D – Series: Architect – Industry Documents:
D101-1995 Methods of Calculating Areas and Volumes of Buildings

D200-1995 Project Checklist

G – Series: Architect – Industry Documents:
G601-1994 Request for Proposal-Land Survey

G602-1993 Request for Proposal-Geotechnical Services

G605-2000 Notification of Amendment to the Professional Services Agreement

G606-2000 Amendment to the Professional Services Agreement

G612-2001 Owner's Instructions to the Architect Regarding the Construction Contract, Insurance and Bonds, and Bidding Procedures

G701-2001 Change Order

GENERAL REQUIREMENTS

G701CMa-1992 Change Order, Construction Manager-Adviser Edition

G702-1992 Application and Certificate for Payment

G703-1992 Continuation Sheet

G704-2000 Certificate of Substantial Completion

G704CMa-1992 Certificate of Substantial Completion, Construction Manager-Adviser Edition

G704/DB-2004, Acknowledgement of Substantial Completion of a Design-Build Project

G705-2001 List of Subcontractors

G706-1994 Contractor's Affidavit of Payment of Debts and Claims

G706A-1994 Contractor's Affidavit of Release of Liens

G707-1994 Consent of Surety to Final Payment

G707A-1994 Consent of Surety to Final Reduction in or Partial Release of Retainage

G709-2001 Work Changes Proposal Request

G710-1992 Architect's Supplemental Instructions

G711-1972 Architect's Field Report

G712-1972 Shop Drawing and Sample Record

G714-2007 Construction Change Directive

G714CMa-1992 Construction Change Directive, Construction Manager-Adviser Edition

G715-1991 Supplemental Attachment for Accord Certificate of Insurance 25-S

G716-2004 Request for Information

G722CMa-1992 Project Application and Project Certificate for Payment, Construction Manager-Adviser Edition

G732-2009 Application and Certificate for Payment, Construction Manager-Adviser Edition

G737-2009 Project Application Summary, Construction Manager-Adviser Edition

G803-2007 Amendment to the Consultant Services Agreement

G804-2001 Register of Bid Documents

G806-2001 Project Parameters Worksheet

G807-2001 Project Team Directory

G808-2001 Project Data

G808A Construction Classification Worksheet

G809-2001 Project Abstract

G810-2001 Transmittal Letter

Setting Up the Job

Having been awarded the contract, the general contractor must then proceed to sign up the subcontractors. The standard AIA A401 Standard Form of Agreement Between Contractor and Subcontractor can be used for this purpose.

15

This document is intended for use in establishing the contractual relationship between the contractor and subcontractor. It spells out the responsibilities of both parties and lists their respective obligations, which are written to parallel A201, General Conditions of the Contract for Construction. Blank spaces are provided where the parties can supplement the details of their agreement. A401 may be modified for use as an agreement between the subcontractor and a sub-subcontractor. Many general or prime contractors will have their own contract documents for these uses. As with any contract make sure it is reviewed, and modified if necessary, before signing.

Owner-Contractor Agreement calls for the successful bidder to submit a list of subcontractors for approval prior to award of contract, and the owner and architect may object at that time to any firm on the list. If there is an objection, the contractor has two alternatives: submit an acceptable substitute with the change in cost, if any, in which case the owner may at their discretion accept or reject the bid; or the contractor may withdraw the bid without penalties. Additional expenses occasioned by owner's or architect's rejection of subcontractors after award of contract will be added by change order to the contract sum and paid for by the owner. There is conflict on this issue, and extreme caution is to be observed. The contractor must also submit a progress schedule and a schedule of values based on the subcontracts. The AIA agreement further sets forth the contents of agreements between subcontractors and the general contractor and the method of paying the subcontractors. AIA Document A401 may be used, or the shorter Frank R. Walker Forms #133 or #144 at http://www.frankrwalker.com/c-7-contract-bid-forms.aspx. The general contractor, with the cooperation of the subs, architect, and owner, will file for and usually pay for all permits. Often it is necessary to have selected certain subcontractors before filing for permit, because the permit may require the license number of certain trades. The Supplementary General Conditions will usually set forth the initial site requirements—office, telephone, toilets, water, light, heat, barricades, temporary partitions—that the general contractor must provide. The prime may also need to make the arrangements with local utilities.

Carrying Out the Work

The administration of the contract is clearly set forth in the AIA standard contractor agreements. The general contractor is responsible for executing the work in strict accordance with the contract documents and for turning the completed project over to the owner in full conformance with them. The prime contractor must provide the project supervision necessary to check all labor, materials, and procedures to accomplish a successful project.

The architect serves in an advisory capacity to the contractor and the owner, interpreting drawings and specifications, approving shop drawings and samples, checking conformity of the job with contract documents,

16

making periodic inspections to determine progress and approving payments to the contractors, and approving final acceptance of the contract. The architect will also have consultants visit the site at certain intervals.

Owner's may want more supervision than the architect provides in the standard contract. They may specify additional time on the project or hire a separate Owner's Representative. In addition, public building inspectors will probably visit the site periodically.

The owner has another option, and that is to have a Construction Manager (CM) whose cost for their services can vary from 1.5% to 3% of the construction cost. This percentage will vary according to specifics of the project. Due to market and project conditions large fluctuations can be encountered. The CM must be an unbiased entity working for the owner who will coordinate the work between the owner, architect, engineer and the contractor. If not this can lead to having severe problems in performing the work. The CM will be responsible for daily logs, schedule adherence and monitoring of payments and assurance that the project is being built in accordance with the plans and specifications. Usually the architect, contractor and construction manager have separate contracts with the owner.

Getting Paid

Applications for payment are usually made monthly. The Frank R. Walker Form #591 at http://www.frankrwalker.com/c-7-contract-bid-forms.aspx is an excellent one for this. It provides a column for the list of subcontractor names, what work the contract is for, the total amount of each contract, the amount previously requested, and finally the balance to complete. The bottom of the form provides a summary of total amounts, less percentages retained, and a space for notarizing. Waivers of lien showing that the contractor has paid for all materials, labor, and subcontractor billings are usually required to accompany the request for payment.

The architect then checks the application, and if it agrees with what they feel is proportionate to the amount of the contract, they will issue a certificate of payment. This decision may be reviewed by either the company writing the performance bond or the company supplying the interim financing. All of this takes time for the contractor to prepare, the architect to check and approve, and the owner to process, so the contractor must keep all billings and payments current. With the high cost of money today, short term loans to tide a contractor over for a month because they didn't get around to filing a proper or complete request for payment can eat away at profits as surely as having to correct mistakes on the job.

When standard AIA contract forms are used, then the *AIA application for payment* will also be used to retain the same document format. Upon final completion and acceptance, the contractor should be prepared to file final waivers along with any written guarantees, certificates of inspections, operating instructions, etc., that may have been called for.

Definitions

Standard contract forms use certain terms that have special meanings within the profession. Some of these are explained below:

Addenda: Modifications to the contract documents issued by the architect/engineer's office during the bidding period. Sometimes referred to as "bulletins". The addendum can be written, graphic, or both.

Alternates: Additions or subtractions to a contract sum for substitutions asked for by the architect, which the contractor must submit with the proposal.

Approved Equal: A contract term that allows the contractor to substitute another product for one specified by the architect, providing it is "equal" in all respects. Sometimes, the "or equal" must be accompanied by a proposal with a list of substitutions. The owner or the architect has the final say as to what is equal.

Arbitration: A method of settling a dispute whereby parties in disagreement submit their differences to a neutral third party. It is an alternative to formal litigation.

Bid: A proposal stating the sum for which the contractor will complete a project.

Bid Bond: A bond furnished by the contractor that guarantees they will sign a contract on the basis of their bid.

Cash Allowance: Sums set forth by the architect for specific items that the contractor must include in the proposal. Upon completion of a job, the contractor must make an accounting of these sums, returning any unspent amounts to the owner. Often used for buying hardware, face brick, light fixtures, and other special items.

Certificate of Occupancy: Issued by a government authority, authorizing occupation of a building.

Change Order: A written authorization by the architect to change the basic contract, plans, or specifications because of additions or deductions from it.

Contract Time: The period of time established in the contract to complete the contract or specific tasks. It can be expressed in calendar days, weeks, months, or final completion date.

Cost Breakdown (or Schedule of Values): A schedule prepared by the contractor showing how the contract sum is divided among the various divisions of the work, which is filed before work is started. The architect will use this as a basis for approving applications for payment and waivers. It should be accurate and kept up to date, reflecting any changes in the contract.

Extras: Construction costs outside the original contract amount. A change order usually is for additional work, or reduction of work scope.

Final Acceptance: When the owner accepts a project as complete.

Payment and Performance Bond: A bond assuring that the contractor will complete a project in accordance with contract documents including discharging all financial obligations.

18

GENERAL REQUIREMENTS

Letter of Intent: Written prior to drawing up complete contract documents to tell the contractor a formal agreement is forthcoming. Often this authorizes certain preliminary work such as filing for building permits and site clearance.

Liens: Legal claims against an owner for labor and materials for construction of a project.

Liquidated Damages: A sum established by the owner, and stipulated in the contract as a charge to the contractor for damages suffered by the owner because of the contractor's failure to fulfill the contract obligations.

Maintenance Bond: Guarantees owner that contractor will rectify defects in workmanship or materials reported within a specified period following final acceptance of the work under contract. Often this is included as part of the Performance Bond and usually runs for one year. It may also be issued for longer periods covering specific phases of the work, such as roofing, curtain wall, paving, etc.

Punch List: A list, established by the owner or the owner's representative, of work yet to be done or to be corrected by the contractor.

Retainage: A percentage set forth in the Owner-Contractor Agreement to be withheld from each payment to the contractor. Usually 10%, reduced to a total of 5% upon substantial completion.

Separate Contract: A contract let by the owner directly to a contractor other than the general contractor. Usually for special equipment, landscaping, interior furnishings, etc.

Shop Drawings: Drawings prepared at the contractor's expense showing how special items shall be fabricated or installed. These usually are prepared by the subcontractors and are standard for such items as structural and ornamental steel, reinforcement steel, cut stone, millwork, partitions, door frames, fire protection, electrical fixtures, HVAC, etc. Shop drawings must be submitted to the architect or engineer for approval.

Subcontractor Bonds: A performance bond given by a subcontractor to the general contractor guaranteeing performance of contract and payment of labor and material bills.

Substantial Completion: When the project reaches a state where sufficiently complete to allow the owner to occupy it but not necessarily finally accept it.

Superintendent: Contractor's representative at the job site. This individual should not to be confused with the architect's superintendent (administration) or owner's clerk of the works.

Supplier: Sometimes referred to as a vendor. An organization that supplies materials, fabrication, or equipment for the project.

Supply Affidavit (Bond): Written notarized statement to the owner by the manufacturer that the materials delivered to the project comply with contract documents.

Unit Prices: Amounts asked for on the proposal for furnishing materials per unit of measurement. That is, cost of concrete per cu. yd. (cu. m, m³), floor tile per s.f. (sq. m, m²), or steel per ton (kilogram, kg).

Upset Price: An amount agreed to by the contractor as a maximum cost to perform a specific project. Used on Cost Plus a Fee jobs.

Warranty: A guarantee by the contractor or a manufacturer covering quality, workmanship, and performance for a specified period.

Value Engineering

A primary objective of management is to achieve the maximum value for every expenditure. Many cost control programs have been devised to assist in this quest. One of the best methods to control the cost of construction is to use professional estimators to assist in the budgeting and pre-construction phase of the project. Whether they are retained by the owner or employed by contractors, estimators have the necessary skill sets to make sure the project comes in close to the budgeted amount. A more formal method used to reduce costs on projects is known as *Value Engineering* (VE). Too often the cost reduction process or value engineering implemented on a project is simply cutting scope to reduce the costs. This is not value engineering.

The value methodology is based on identification of function. In essence, it asks, What does the client really need in order to accomplish an objective? This is more difficult to quantify than it might appear. In order to be sure that the basic function is uncovered and clearly defined, a multidisciplinary team is assembled. The team should be composed of persons who can address all sides of an issue and usually includes the following:

1. certified value specialist
2. architect
3. structural engineer
4. mechanical engineer
5. electrical engineer
6. cost estimator

In the VE process, a specific time is allotted for study. A typical project might require one week for the *pre-workshop phase*, one week for the *workshop phase*, and another week for the *post-workshop phase*.

During the pre-workshop phase, the design documents are studied. When practical, a site visit is made. If a cost estimate exists, and if in the opinion of the team it does not represent the true cost of the design as it then exists, it is modified. If there is no cost estimate, then one is produced by the team cost estimator.

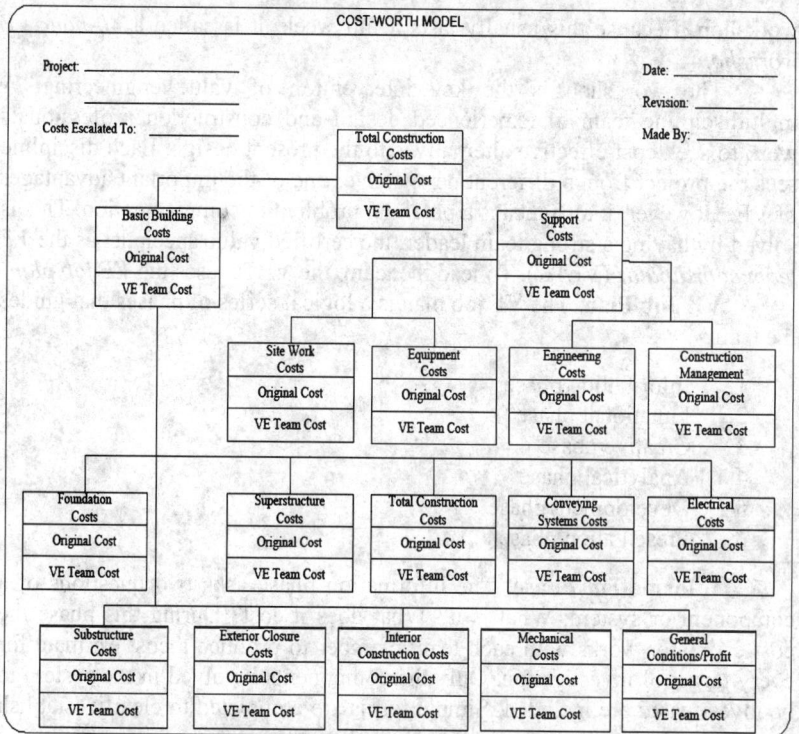

COST-WORTH MODEL

Courtesy of James Hudson Associates
Spotsylvania, VA

Cost/Worth Model. A *cost model* is produced by the cost estimator and the certified value specialist. This gives high cost visibility to the functions that compose the total project. Sometimes these models follow the CSI format—site, foundations, superstructure, etc. At other times, they might highlight design features, such as penthouse, annex, or parking. The purpose of the cost model is to graphically show how the costs are distributed by function.

The team next introduces a subjective factor called *worth*. This factor may modify the value of each cost by providing a comparative cost based on the experience of the team. For example, the team may perceive that the cost shown for HVAC in the cost model is much higher than for comparable projects. They determine what they consider the lowest cost possible to accomplish the basic function.

The cost model is then modified by giving each entry a worth, and it is then a *cost/worth model*, which will enable the team to see where the cost exceeds the worth by a significant amount. These become *targets* for the VE workshop.

VE Workshop. After a thorough review of the design documents, a site visit, and the preparation of a cost/worth model, the VE team convenes a workshop. Because this usually lasts a full week, it is called a *40-hour VE workshop*.

The workshop is the key intervention of value engineering. A multidiscipline team of experienced design and construction professionals work to seek cost effective alternatives to the present design. Each discipline sees the project from a different perspective, one of the important advantages of VE. However, it also means a potential problem in communication. This is solved by having a strong team leader, the certified value specialist as the *VE team coordinator* (VETC). To lead the team, the VETC uses the *VE job plan*.

VE Job Plan. The VE job plan is a logical series of phases that guides the team:

1. Information phase
2. Functional phase
3. Creative phase
4. Analytical phase
5. Development phase
6. Presentation phase

Information Phase. The information phase asks two questions of a component or system: What is it? What does it cost? During this phase the cost estimator works with each team member to produce a cost estimate for every target of investigation. Only the components involved in the system to be investigated are included. Simple forms are completed to clearly establish the present design for each target to be challenged.

Functional Phase. Now that the team has picked its targets and determined the costs, the functions are carefully analyzed. The questions asked in this phase are: What does the system do? What must the system do? The differences between answers to these two questions lead the team to fresh solutions that eliminate the functions that may be desirable but are not essential to the design.

Creative Phase. Each function is addressed to seek cost effective alternatives. Often this can be done by discussions among the members, and an effective technique is "brainstorming", where ideas are thrown out by any member as they occur, by association, contiguous thoughts, opposite concepts or the like.

What is sought in this phase is a great number of rough ideas, not engineered solutions, and the reason being that an idea advanced by one member automatically triggers other ideas from other members of the team. Creativity seems to work best when negative expressions are not allowed to surface. The subconscious mind is at its best with positive direction, and it is this subconscious resource that we are trying to tap.

Analytical Phase. The long list of ideas generated during the creative phase is examined. Similar ideas are grouped so that a comparative analysis can be made to determine those ideas that must be discarded and those that show merit and should be developed. Several techniques are used to assist in this comparison. The experience of the team enables them to grade ideas with reasonable accuracy. The forms keep the team on track and expedite the comparative analysis.

Development Phase. One or two alternatives are selected by the team for development into the preferred solution. Closer cost estimates are made to compare with the present design.

The team must be aware of its responsibility. As professionals they are expected to propose concepts, systems, and components that fulfill the basic functions required to properly satisfy the project. This is not a cheapening process! Anyone can use lower cost materials to reduce the cost. The goal of value engineering is to propose alternatives that completely fulfill the needed functions.

Often the savings are in life cycle cost (LCC). The alternative may have the same or higher initial cost but may have significant cost savings over the life of the facility, product, etc. These LCC are highly desirable for most clients, though in some projects the "first costs" are the primary consideration. The VE team seeks this type of preliminary guidance during the pre-workshop phase to ascertain the real interest of the decision makers in LCC.

It is the duty of the VE team to make very specific proposals. The designers do not need a "laundry list" of good ideas. Only a clear, well-defined alternative can be of use in modifying the design. The proposal must be clearly written, with sketches, to convey the concept.

Presentation Phase. Alternatives are useless unless they are implemented. It is imperative that the VE team prepare a written report, complete with all of the back-up worksheets prepared during the workshop, so that the decision makers fully understand the reasoning. In addition, whenever possible a verbal presentation should be given.

While the project is fresh in their minds, members of the VE team can give a complete oral recitation of their findings, proposals, and emotional reaction to the present design and to the proposed design. Often, on the last day of a five-day workshop, the owner, designer, user, and other decision makers are asked to attend a presentation. During the presentation, each team member is given the opportunity to express without reservations the feelings that underlie the proposals. This might include minority opinions on some of the proposals. A properly led workshop seeks truth, not spectacular cost savings. This usually results in many cost effective alternatives, but it might also reveal serious deficiencies in the design. Great care is taken to divulge such findings with diplomacy and compassion.

Effectiveness of Value Engineering. VE has been used in construction for many years. When properly conducted, it has produced average savings of more than 10% of the estimated cost without sacrificing quality. VE has proven effective in controlling the cost of design and construction and is being used widely by several federal government agencies.

CONSTRUCTION FINANCE

Before getting into the details of estimating the materials and labor that go into a project, let us take a quick look at the "material" that makes it all possible—money.

The Cost of Money

Let's say a house that cost $75,000 ten years ago now costs $250,000. But add finance charges, based on a 20% down payment and a mortgage for twenty-five years, with interest charges of 7% ten years ago and 7% today. The house really cost $143,700 ten years ago and costs $479,000 today. The increase in the initial cost of the house in ten years was 333%, but the increase in the cost of financing was 333%. . Obviously, the point is that money must be used as efficiently as any other commodity. And by efficiently we mean not only spent wisely but carefully planned for, shopped for, drawn upon only at the exact time it is needed, and paid for on time.

Our concern here is not with the money that the contractor needs to run their own business affairs, but the money to pay for the project itself. This breaks down into two types of financing—immediate cash to pay for the labor and materials as they are placed on the job, and once the project is completed, a loan against the project with repayment scheduled over many years. These are known as the short term or the construction loan and the long term or mortgage loan.

Project financing is not usually the direct concern of the contractor. It is the owner's or project developer's responsibility to secure both the short and long term loans. But it is essential that a contractor understand all phases of the project financing and their relationship to it.

First, few projects today are paid for from funds the owner already has accrued. They must find someone willing to loan the money to make the project possible, and that someone must have enough faith in the project to be willing to wait for repayment of the loan from money generated by the success of the project. In order to "sell" the project to a loan source, the owner must prepare a complete analysis of all aspects, including reliable cost projections, and will often turn to a contractor for this information. It is important that the contractor understands the commitment associated with such an estimate, what the lender "sees" is what the contractor later "gets" if the project loan is approved.

Second, the construction loan is the source from which the contractor will be paid, and they will have to work closely with the lender and the job inspectors to get their money on time. If the contractor is inefficient in compiling the requests for payment, the worker's and suppliers will go unpaid unless the contractor borrows the money until the next period the lender makes the payouts. The interest the contractor will have to pay will cut into the profit as surely as if they had underestimated the materials for the job.

24

Third, the owner may be very naive in construction finance. They may build only one project in their entire life. The local bank may be the only financial contact. The local banks are an excellent starting place to seek construction funds; they know the owner and the community and the owner's best interests are presumably the bank's best interests. But the bank may have other business loans with the owner, and they may not be willing or even legally able, to make an additional loan of the size, at the rate, and for the length of time the owner requires. Thus the project may die in the bank's board room. The owner may lack the time and the sophistication to search further, and so the contractor loses the job as surely as if they had been underbid by a competitor. But there are other money sources besides banks, and an aggressive contractor will develop these other contacts and steer the owner to them to salvage a good job that otherwise might never get off the drawing board.

Fourth, as we mentioned at the very beginning of this chapter, the triangle of owner-architect-contractor is nowadays often more of a circle, with each overlapping the other's territory. The contractor might also be the owner, either as the project developer or as the official landlord who then leases the project with perhaps an option to buy. Such arrangements can be very profitable and can be used in slack times to keep a contractor's force busy. But a good contractor does not always make a good landlord or even a good client for their company.

Sources of Money

The main sources of money are:

1. Commercial Banks - short and intermediate term financing
2. Savings and Loans - small residential loans
3. Insurance Companies - restricted long term loans
4. Pension and Trust Funds - long term loans. These funds are often administered by commercial banks.
5. Real Estate Investment Trusts (REITS) - long and short term loans, land loans, second mortgages, sell and leaseback deals.
6. Governmental Agencies - FHA and VA loans for residential work (guarantee loans)
7. Government Sponsored Bonds - industrial revenue bonds for projects that increase employment. Local governments issue tax-exempt bonds secured by the project income.
8. Mortgage Banking Firms - long and short term loans, land loans, second mortgages, sell and leaseback deals.

Each source has an area it is most interested in, but they compete to a degree. Many projects will draw on two or more sources. Short term loans are for three years or less. Long term loans are seldom for less than ten years and may run up to thirty years.

One thing all loan sources have in common is that they are "selling" their money, not "buying" your project. When the project is presented to them, they not only want to know the physical details but how the project will generate the money to pay back the loan. The essence of a good presentation is the proof that the loan can be repaid.

Mortgage Loans

While it is the construction, or short term, loan that is drawn on first, and land and "front money" loans may precede it, the mortgage, or long term, loan is shopped for first. Once the owner has the mortgage loan lined up, the construction loan is more easily secured, because it will eventually be paid by the mortgage loan when the project is completed. Further, the interest charged for the mortgage loan may decide whether the project can generate the income to make the necessary payments. If one fails to get a commitment for the mortgage interest and interest rates soar, the debt service may take all the projected profits.

Small projects can usually be financed both long and short term at local banks and savings and loan associations. These institutions make no direct charge for reviewing proposals and can usually give a yes or no answer in a few days, so one can shop several sources at the same time with no obligation. Once a loan is approved and accepted, there are certain charges called closing costs, and a part of these costs, known as points, origination fees, or the discount, cover the cost of setting up the loan. These points are a one-time charge made as a percentage of the loan and are deducted from the amount of the loan. If the points are 2%, then the amount of two dollars is deducted from every hundred dollars of the loan. The usual range of points is from 1% to 3%, but in states where there are legal limits set on the interest that may be charged, points have been quoted as high as 7% as a way to get around the usury laws. Such points are common to all loan sources and are one of the important items to shop for.

Mortgage Banker

For larger projects and projects that generate income, one should check out all the money sources. It isn't likely that even the most gregarious contractor knows all the institutional sources, or even if they did, could keep abreast of their current interests and capabilities. One must consult a specialist, and that specialist is the mortgage banker, as we shall call them here to distinguish them from a commercial banker.

The mortgage broker is an intermediary between borrowers and lenders, although their firm may have some funds of its own to invest. They are a "matchmaker" who brings parties together. To do this they must carefully evaluate both sides, shopping for the best terms for the owner and for projects most compatible with the portfolio of the lender. For consummating the transaction, they charge a flat fee of around 1% to 2% of all the loans placed. This fee is in addition to the usual closing costs that are

26

charged by the lender. Both expenses are paid by the borrower. Many lenders require a 1% fee at the time the loan application is filled. This fee is refundable if the lender rejects the loan but becomes earned once the lender approves the application and is credited to the fee due the lender at the time of settlement.

There are other benefits in working with a mortgage broker. They can advise on the techniques of financing as well as the sources. There are numerous options available, and the owner will want to review them all with their accountant and lawyer and select what may suit their particular situation best. Decisions to be made would include the desired length of the mortgage; whether to select a loan at lower rates but "locked in" for many years or one at higher rates but with prepayment provisions; what personality—that is, items not part of the building itself such as equipment, carpet, drapes and the like—should be included in the mortgage; what owner assets should be pledged; what existing property should be made part of the new mortgage (wrap-around mortgage); what "kickers" might be offered, such as sharing income or equity with the lender if certain income or worth is exceeded; what advantages second and third mortgages might have; what method of repayment of the mortgage loan should be adopted (usually a constant covering interest and principal); payments of interest only, with the principal paid when the note is due; a constant payment of principal with interest figured on the declining balance; no payments on either interest or principal until the note is due but with the interest compounding; or a combination of the preceding with refinancing at certain periods to fit the owner's needs. There are no fixed answers that cover all situations. These decisions can contribute as much to the success of the project as decisions about room arrangements or types of building materials.

Having determined the best package for the owner, the mortgage banker will then shop various money sources for the best deal. Each source will be looking for certain features, and the final deal probably will be a compromise.

The mortgage banker's opinions are useful in other ways. Having reviewed numerous similar projects at the preliminary stage, they can often suggest refinements in the physical arrangements, as it relates to potential income, that they have seen work out well in other situations. And finally, in accepting or refusing a project, they give their opinion as to its probable success. If two or three mortgage bankers turn a project down when they are placing others, the owner should review the figures. The risk elements must be greater than those in the normal market. The mortgage broker need not be just a necessary evil. They can be a creative part of the building team.

Selling the Lender

Whether one works directly with a loan organization or through a broker, one must be prepared to make an effective presentation of the project.

While each type of project will emphasize different elements, all should include the following:

1. Physical analysis of the project including a surveyor's plot plan, an architect's sketch plans and specifications, an artist's rendering of the developed site, and a contractor's estimate of the cost;
2. Market feasibility study;
3. Financial analysis of the ownership including personal statements;
4. Financial analysis of the project.
5. Draft Environmental Impact Study (DEIS).

A great deal of thought must go into the presentation of the project. Once the project is approved, it is difficult to make major changes. With the use of computers presentations it's easy to "overkill" the presentation. A too glamorous presentation may backfire. It is utility not beauty that sells the lender. An impressive entrance framed by a tree lined drive circling a large fountain may be just an artist's window dressing and will be the first thing abandoned when the final budgets are reviewed. But that may be the feature a lender remembers and insists be built. Another lender, with their eye on the cash flow, will see such embellishments as so much maintenance expense and as a detriment rather than asset.

In addition to the contractor's estimate, the lender will make one also and may ask the borrower to have an independent certified estimator make up a third. This is usually done by a member of the American Institute of Real Estate Appraisers, whose members carry the title MAI (Member Appraisal Institute), or senior member of an appraisal society such as the American Society of Appraisers. The certified estimate is made by a certified professional estimator.

The owner's financial statement may be prepared by their certified public accountant, but often the lender will have their own standard form to fill out. The proposed contractor may also have to file a financial statement. Some loans will only be made if performance bonds are posted. Before presenting a proposal an owner will want to check with their lawyer and accountant as to the best form of ownership for the project-individual, corporate, partnership or LLC. Some buildings are owned "at arm's length". A third party actually owns the building, and the promoter of the project becomes the tenant.

The market feasibility study shows the compatibility of the project with the surroundings, means of public access, the effect on the environment, and compliance with zoning regulations. Similar successful projects in the area should be noted, and the continuing need for additional facilities proven. And finally, the adaptability to another occupancy and resale value should be evaluated.

The most carefully scrutinized section of the presentation will probably be the projected value of the project based on income, or whether the project will generate enough income to repay the loan on time with

28

interest. What is included here will vary with the type of project. Where rents are involved, rent rolls should be projected. The lender in such cases may stipulate that a certain rent level be reached, usually around 80% of the projection, before the mortgage becomes final. If not, the mortgage will be for a lower amount. This is called setting a floor for the mortgage. The spread may be as much as 20%, and if the owner does not reach the agreed upon rent level, they must find another source to cover this amount. This will be discussed later.

The Commitment

Once a project is accepted by a lending organization, a commitment paper or contract is issued stating that the lender will lend a certain amount of money for a certain length of time at a certain rate of interest, providing the borrower fulfills certain stated conditions and closes the loan by a certain date. This commitment, if the owner accepts, will become the basis for the first mortgage on the project. There may be other mortgages either on the project as a whole, or on some phase of it, such as furnishings, equipment, or the land itself. But the holder of the first mortgage takes precedence over all other claims except taxes, and has the authority to take possession of the project upon proven default and to sell it and deduct the amount due from the selling price, which is usually determined by a public auction. If any excess cash remains, it then goes to the holder of the second mortgage, then to the third, if any, and so on, with the original developer the last to receive any distribution. Holders of mortgages other than the first cannot instigate the sale of the project. They are subordinate to the first mortgage holder and therefore carry a higher risk and charge a higher rate of interest.

The commitment will be quite specific and should not be thought of as a preliminary proposal, but rather, as the basis of a contract that one must live with for many years. This is why the presentation of the project must be carefully thought out and based on realistic goals.

Short Term Loans

Having received a commitment, the owner must then work fast to finalize the plans. The commitment will carry a deadline date for acceptance that may be only a couple of weeks away. It will also carry an expiration date by which time all closing documents shall be submitted to the loan company so that the final mortgage loan may go into effect. If this date is not met, the owner forfeits an agreed upon deposit, which is usually around 1% of the mortgage amount and is deposited with the lender's agent upon accepting the commitment. One must then proceed immediately to secure a construction loan, gap financing (if it is needed), and "front money".

The construction loan is a short term loan to cover the building costs during the erection of the project. With the mortgage loan commitment in hand, the construction loan repayment is guaranteed providing that the project is completed in a satisfactory manner, so it is not necessary to sell the feasibility of the project all over again. But the construction loan lender does take the risk that the owner or the contractor will not get in financial difficulties during the construction period, or that the projected budget will not prove inadequate. The lender may then have to complete the project themselves. Because the mortgage amount may be less than the stated amount if certain income levels are not reached, the construction loan will be based on the mortgage "floor" rather than the "ceiling" figure, which may deduct as much as 20% from the amount needed to complete the construction. Further, the construction loan is seldom for more than 80% of the floor figure, and like all loans, is subject to points and other closing costs, so perhaps another 2% will be deducted from the loan.

While shopping around might improve one's position, the construction loan will still fall short of the amount needed to complete the project. The missing funds must be supplied by the owner and represent their *equity* in the project. If the borrower can pledge sufficient liquid assets, that are cash, stocks, bonds, or other unencumbered property, they might be able to raise the additional funds from the same source as the construction loan. If not, they will have to take in a partner or borrow from those who take high risks. One such source is those who loan the difference between the floor and the ceiling of the mortgage. This is known as *gap financing*. They will issue a commitment to pay this difference if the income generated by the project is less than that specified to qualify for the full or ceiling amount. To obtain such a commitment one must pay in advance a flat fee, usually around 5% of the amount to be loaned. If the project reaches the income level to qualify for the full mortgage, and the gap loan is not needed, the fee is not refundable. If it is needed, the loan becomes a second mortgage against the project and is repaid with interest.

In setting up the project budget, there are other costs to be met besides construction costs. First, the land must be obtained. While this is sometimes leased or paid for under a mortgage arrangement, the first mortgage commitment may ask that the land title be unencumbered. In addition to land costs, there may be land development expenses such as surveys, soil investigation borings, utility extensions and connections, and access roads or sidings. There also may be purchasing expenses, including attorney and real estate agent fees, taxes and permits, and zoning fees. If the zoning for the land is to be changed, it can be both time-consuming and expensive.

And the architect must be paid. The fee will run from around 4% for a large industrial plant up to 15% for a fine residence or interior remodeling job. Some of this fee will not become due until the job is completed.

When the loan commitments are accepted, they may require a good faith deposit of around 1% of the loan. Consulting fees must be paid to accountants and lawyers. At loan closings legal and recording fees and transaction taxes may be charged, and if asked for, a performance bond obtained.

30

GENERAL REQUIREMENTS

When the work is started, there is always a lapse of time between paying for permits, materials, and salaries and being reimbursed for these items from the construction loan. Usually, each request for payment from the construction loan must be accompanied by waivers of lien. In actual practice the contractor may finance this period rather than the owner, but the lender may still demand that the owner show they have the money to cover these start-up costs.

How much front money, or equity, an owner will need to launch the project will vary with the type of project, the money market, and the owner's reputation. It is often said that an owner with a proven need, a piece of property free of debt, and an architect's set of plans can obtain all the financing they will need. We have used the term "owner" in our preceding discussions as if the one who builds the project were the one who will occupy it. Today even large corporations may choose to be tenants rather than owners and will rely on developers to put the project together. The contractor may be asked to be the developer or at least to become a partner in it. Often the owner of the real estate and the architect will be included as limited partners, their limits of both liability and equity being proportional to the cost of the services or land contributed by them. When the construction loan is paid off by the mortgage loan, they are also often bought out and the ownership reverts to the general partner. However, some mortgage commitments guard against such arrangements by adding a penalty fee if there is any change of ownership within a stipulated time.

Rates of interest on construction loans vary from week to week and community to community but are usually tied to the prime rate, which is the interest charged by commercial banks to their preferred customers for short term loans. Currently construction loans are quoted at 2-1/2% to 4% over the prime. High risk loans will have considerably higher rates.

It is not our intention to encourage the contractor to become the project entrepreneur. But with the cost of money tied so closely to the success of the project, from an income-producing standpoint a contractor must know all the procedures in order to follow the time and budget schedules, to process monthly draws on time, and to turn the completed project over to the owner in such condition as to be completely acceptable to the mortgage lender.

Interim Financing

This particular type of financing has nothing to do with paying for the construction costs of a project. Nor does it have to do with long-term mortgage financing, which is repaid over an extended period from the proceeds of the revenue generated from the project.

The definition of the word interim is provisional, intervening, interlude or relating to a time period. This is the type of financial assistance required by virtually all contractors, regardless of size, for short-term

31

working capital to compensate for unanticipated delays in receiving payments from owners for work performed and properly invoiced.

The source is usually a commercial bank that is responsive to the needs of the contractor. Not all banks are receptive to this type of business, and the contractor must establish banking relations with a bank that has personnel who possess intimate knowledge of the financial structure found in the construction industry. Once an acceptable financial institution has been selected, then the contractor should identify a particular person within the bank with whom a personal relationship can be established.

Next, it is the obligation of the contractor to prove, to the complete satisfaction of the bank, the credibility of the technical and financial assets of the company. This can be accomplished by furnishing certified (CPA) financial statements of the company along with a written detailed resume of the qualifications of each person directing the destiny of the company. Progress, at this point, should place the contractor in a position to obtain from the bank a general commitment as to the limit and terms under which they would participate in granting short term loans (usually 30 to 90 days). The ideal situation is to obtain from the bank a letter of credit outlining its position so that at any time during the next 12 months, a loan could be arranged without further review from a committee. This avoids days, or even weeks, of delay from the time the loan application is presented until the proceeds are available.

At the time a loan request is made, the position of the contractor can be significantly enhanced by presentation of a cash flow analysis of all current projects, projecting anticipated expenditures and income on a monthly basis so that the amount of money required and the availability of sufficient income to repay the loan can be clearly defined.

Expect the interest on this type of loan to be from 1% to 6% above the current prime rate. Banks consider this high risk business. If the current prime rate is 8% and the loan is assigned 3% over prime, this means the contractor is paying 11% for the use of the money. Obviously, a contractor is forced to impose restrictions on the use of such costly money. It cannot be used as continuous working capital, which at best has an earning capacity of about 10%. It has to be justified by large discounts. For example, prompt payment of equipment invoices would affect an early delivery date that would equate with a time and cost saving justifying the interest rate.

The availability of interim financing provides a measure of security for the contractor, who is assured that external financial assistance, is available to augment working capital during periods of depressed cash flow.

DRAWINGS AND SPECIFICATIONS

The owner's dreams and the architect/engineer's visions must eventually be put down in a form the contractor can use as a basis for bidding the project and for constructing in enduring materials. This form has developed into the dual one of *working drawings* and *specifications*.

GENERAL REQUIREMENTS

Basically, the drawings present the design, the location, and the dimensions of a project. The specifications give the quality required. In the past drawings were done in pencil or ink on tracing paper or cloth, which is easily reproducible, and prints were delivered to the contractor.

Now the drawings are made on the computer using CADD (Computer Aided Design and Drafting) technology to create 2D or 3D drawings. The architect retains the original drawings and CADD files in their office. The specifications, except on very small jobs for which they may be incorporated on the drawings, are produced separately. Both are made available to each contractor bidding the job in either electronic or hard copy formats. The electronic format is becoming more prevalent in the industry and has enabled estimators to utilize onscreen based estimating software.

The bidding process usually occurs in a relatively short time frame which forces the contractor to quickly interpret the bidding documents and form a realistic and complete bid. This chapter presents an outline of what information a contractor should expect to find in a bid set and where it can be found.

The Working Drawings

Arrangement. A typical complete project will find several drawing sheets, usually all of uniform size, bound into one job set. The first sheets typically include the site location, the complete job title, index of the drawings, symbols, definitions, and architect/engineers of record.

Following the general information sheets are the civil drawings. The typical order for civil drawings is site survey, pavings, grading, utilities and then details. This will be followed by the sheets showing architectural details, usually in the following order: basement plan, first floor plan, upper floor plans, exterior elevations, sections, interior elevations and details.

The architectural sheets are followed by the structural set including footing and foundation plans, basement framing plans, floor framing plans, roof framing plans, and then the structural details. These are followed by the heating and ventilating plans; then the plumbing plans, and finally, the electrical plans, each starting with the lowest floor and working upward.

Each sheet should have a title block in the lower right-hand corner with the sheet number (usually "G-" for general information, "C-" for civil, "A-" for architectural, "S-" for structural, "M-" for heating and ventilating, "P-" for plumbing, "FP-" for fire protection, "E-" for electrical, "I-" for instrumentation and "T-" for technology); the number of sheets in each set (i.e., A-1 of 7); the date made plus each date it has been revised, and the initials of the person or persons who drew and approved the sheet.

Revisions are sometimes listed outside the title block. Some architects, rather than revising a drawing once it is printed, will issue supplementary drawings which must be attached to the appropriate set as they are issued.

Type of Drawings. Most working drawings for building construction are based on *orthographic projection*, which is a parallel projection to a plane by lines perpendicular to the plane. In this way, all dimensions will be true. If the plane is horizontal, the projection is a plan; if vertical, it is an elevation for outside the building, or a sectional elevation if through the building.

The only descriptive drawing that presents a building as the eye sees it is the perspective. But perspectives are used mainly to study the building and present it to the client in an easily understood form. A perspective is seldom useful for presenting information on working drawings.

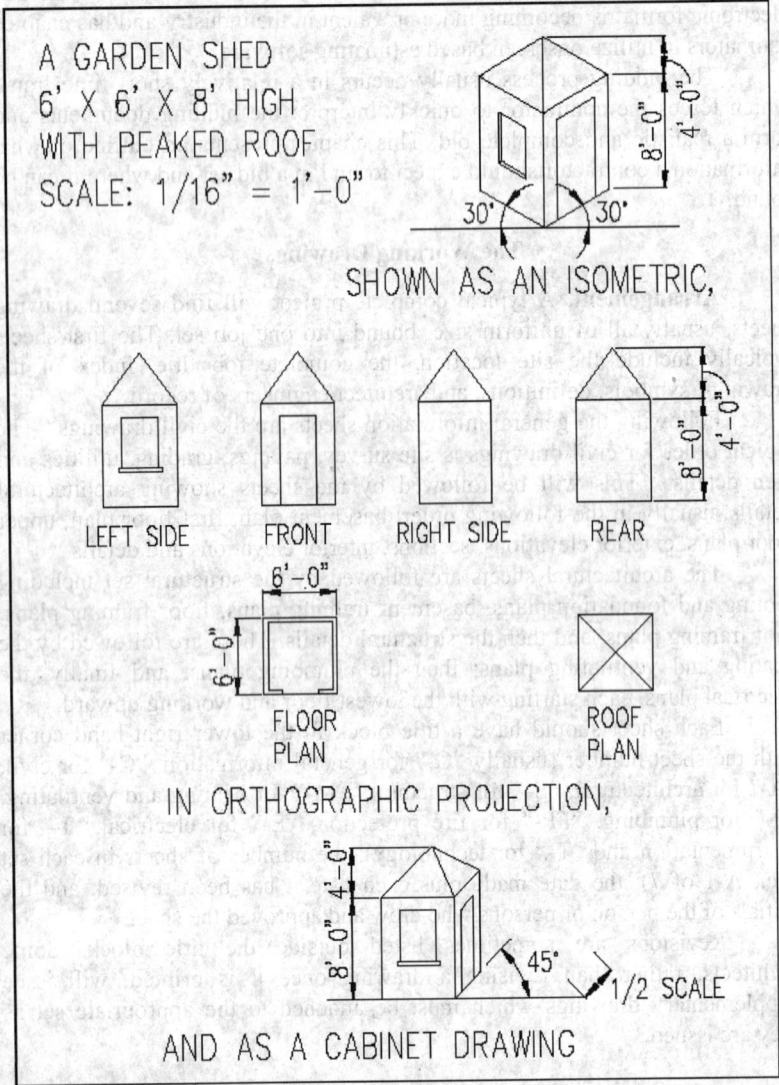

A GARDEN SHED
6' X 6' X 8' HIGH
WITH PEAKED ROOF
SCALE: 1/16" = 1'-0"

SHOWN AS AN ISOMETRIC,

LEFT SIDE   FRONT   RIGHT SIDE   REAR

FLOOR PLAN   ROOF PLAN

IN ORTHOGRAPHIC PROJECTION,

1/2 SCALE

AND AS A CABINET DRAWING

GENERAL REQUIREMENTS

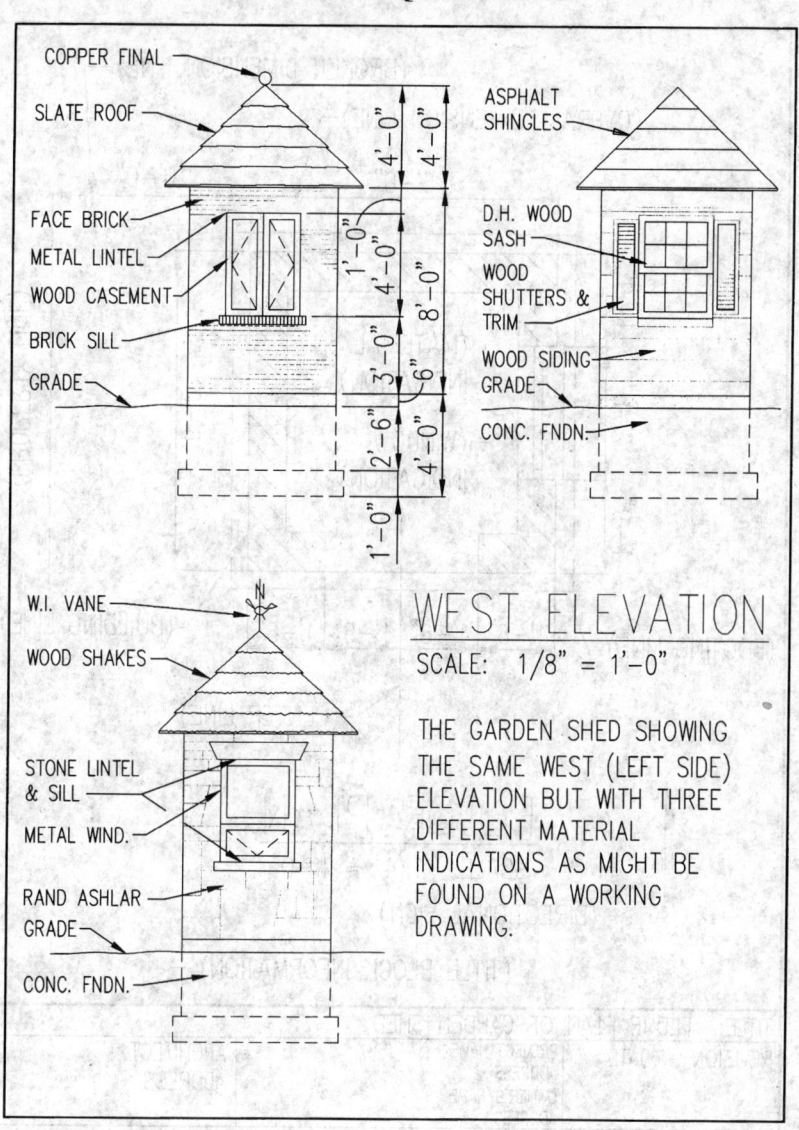

COPPER FINAL
SLATE ROOF
FACE BRICK
METAL LINTEL
WOOD CASEMENT
BRICK SILL
GRADE

4'-0" 4'-0"
1'-0"
4'-0" 8'-0"
3'-0" 3'-0" 6"
2'-6" 4'-0"
1'-0"

ASPHALT SHINGLES
D.H. WOOD SASH
WOOD SHUTTERS & TRIM
WOOD SIDING
GRADE
CONC. FNDN.

W.I. VANE
WOOD SHAKES
STONE LINTEL & SILL
METAL WIND.
RAND ASHLAR
GRADE
CONC. FNDN.

WEST ELEVATION
SCALE: 1/8" = 1'-0"

THE GARDEN SHED SHOWING THE SAME WEST (LEFT SIDE) ELEVATION BUT WITH THREE DIFFERENT MATERIAL INDICATIONS AS MIGHT BE FOUND ON A WORKING DRAWING.

35

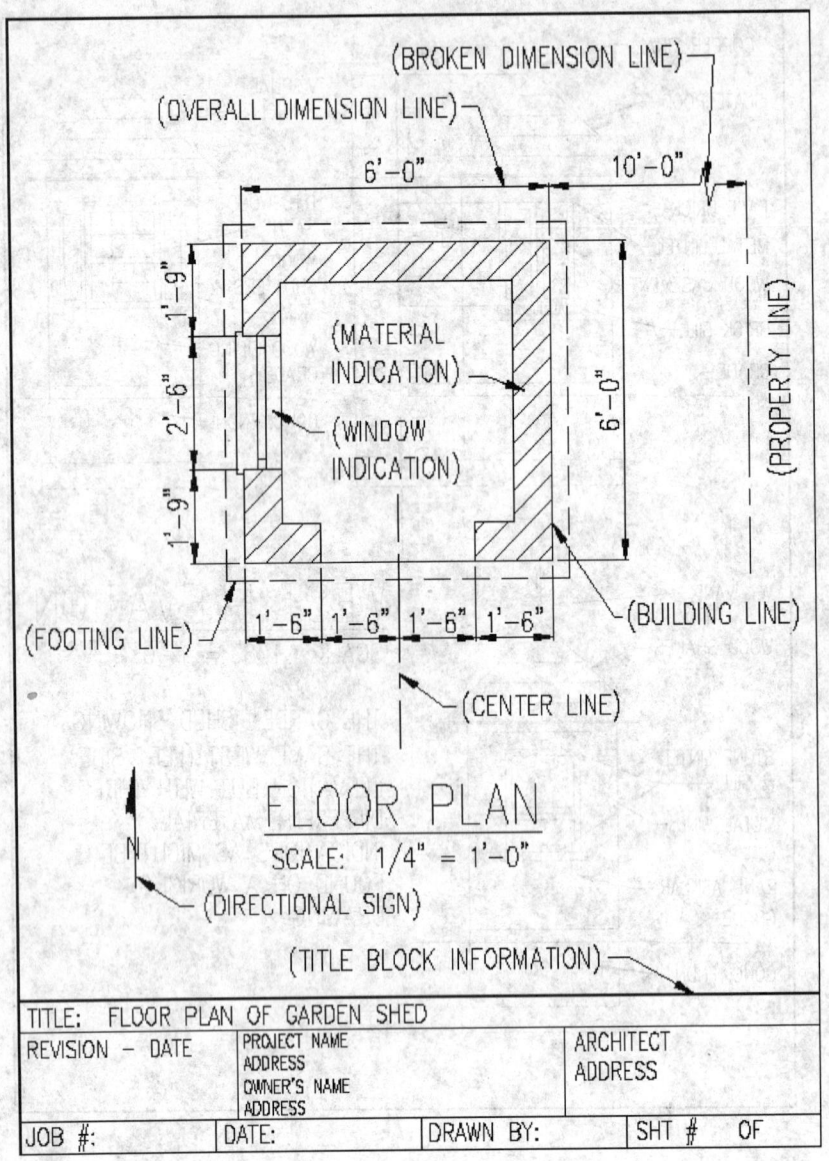

(BROKEN DIMENSION LINE)

(OVERALL DIMENSION LINE)

6'-0"   10'-0"

1'-9"

(MATERIAL INDICATION)

(WINDOW INDICATION)

2'-6"

6'-0"

(PROPERTY LINE)

1'-9"

(BUILDING LINE)

(FOOTING LINE)   1'-6" 1'-6" 1'-6" 1'-6"

(CENTER LINE)

FLOOR PLAN
SCALE: 1/4" = 1'-0"

N

(DIRECTIONAL SIGN)

(TITLE BLOCK INFORMATION)

TITLE: FLOOR PLAN OF GARDEN SHED			
REVISION – DATE	PROJECT NAME ADDRESS OWNER'S NAME ADDRESS		ARCHITECT ADDRESS
JOB #:	DATE:	DRAWN BY:	SHT # OF

However, other pictorial presentations are helpful to the builder. Two of these are the isometric and the cabinet projection. Isometrics are drawings in which all horizontal and vertical lines have a true length, and those lines parallel on the object are also parallel on the drawing. Vertical lines are vertical, but horizontal lines are set at 30 or -30 degrees.

Isometrics are often used for piping diagrams and to show complicated intersections, such as on sloped roofs. Cabinet drawings are those with the front face shown in true shape and size, as if it was an orthographic projection, but they simulate a perspective. The sides are shown receding at 45 degrees and at 1/2 scale (1:20). Variations of this are oblique drawings, where the angle and side scale may be anything that best shows the object, and cavalier drawings, where the side scale is at the same scale as the front. Cabinet drawings get their name from the fact that they are often used for cabinet work.

Scale. The architect's scale, with the inch divided into 1/4, 1/8, 1/16, 1/32, is standard for building construction in the United States. The engineer's scale, with the inch divided into tenths, is sometimes used in structural work or on site plans. However, it is advantageous to have architectural, structural, and mechanical drawings all at the same scale for a job.

The metric scale is divided into centimeters and millimeters, 2.54 centimeters equaling one inch. It is the common scale throughout the world. (For ease of conversion it is accepted practice to use 25 mm to the 1 inch or 300 mm = 12 inches).

The SI metric scale is based on a decimal with the meter representing a unit of 1.0. The Metric Conversion Act of 1975, as amended by the Omnibus Trade and Competitiveness Act of 1988, establishes the modern metric system (System International, S.I.) as the preferred system of measurement in the United States. Even though S.I. is the preferred system, it is typically used only on government projects.

Plans and elevations are usually drawn at no less than 1/8" = 1'-0" (1:100); plot plans at 1/16" = 1'-0" (1:200). Complicated plan areas, such as toilet rooms, are usually repeated at 1/4" (1:50) to 1/2" (1:20) scale. Wall sections and cabinet work must show more detail and are from 3/4" (1:200) to 1-1/2" (1:10) scale. Certain areas of sections will be further detailed at 3" (1:5) scale, while moldings and decorative items might even be drawn at half to full scale.

Unit	Abbreviation	Length, Meters	Approximate U.S. equivalent
kilometer	km	1,000	0.62 miles
hectometer	h	100	109.36 yards
decameter	dkm	10	32.81 feet
meter	m	1	39.37 inches
decimeter	dm	0.1	3.94 inches
centimeter	cm	0.01	0.39 inches
millimeter	mm	0.001	0.04 inches

Note: For this book the National Institute of Building Sciences (NIBS), Committee for Metric in Construction recommendation of soft conversion metric that is: 1" = 25 mm or 1'- 0" = 300 mm.

Dimensioning. While scales should be clearly identified for each drawing, either in the title block or beneath each individual layout, contractors should verify the scale based on known dimensions. A contractor should never rely on the stated scale since this scale could be inaccurate for any number of reasons.

Notes on Drawings. The drawings should not be cluttered with notes, which should be in the specifications, but architects will vary on this. All notes on the drawings need to be read since the architects will place information that can and will affect the total cost of the project. Sometimes the notes are used to explain what needs to be included on a specific item that cannot be represented in the drawing. Specification should govern, and all notes on drawings should be checked against the specifications and cross-referenced. If a contractor finds discrepancies during bidding, it should be reported to the architect in writing and the architect asked for a written statement, known as an *Addendum*, to clarify it.

Finish Schedules. Schedules may appear either as part of the drawings or the specifications. Room finish schedules are often on the drawings next to the floor plan to which they refer. Schedules for lintels, columns, footings, doors, and frames are usually on the drawings, while hardware schedules are usually in the specifications. Fixture schedules for mechanical and electrical work are usually found on the drawings but may also be found in the specifications.

Reproduction. *Blueprints* show the work to be performed on the project. In the past, the ammonia copying process was used which produced white lines on a blue background or a black, blue, or maroon line on a white background. Today, many architects and contractors have their own printing equipment for small scale projects. Typically, architects will either rely on either supplying electronic files or use a printing company to distribute the plans. Small prints may be run off on one of the many types of office duplicating machines.

GENERAL REQUIREMENTS

In the past, the process to reduce or enlarge a drawing utilized a *Photostat*, which is a photographic process that produces negatives as positives. There were also processes that transfer blueprints to cloth drawings or to *sepia prints*, which could be altered, added to, and printed just like an original tracing. Since a majority of drawings are now electronic, the documents can easily be reduced or enlarged using computer software. The ease of changing documents makes it necessary to carefully review each new issue of drawings for changes. While the architects customarily will note changes as notes or "clouds" on the drawing where changes were made, the contractor should review all areas of the drawings to confirm the changes were identified. Changes to the drawings can substantially affect the amount of work the contractor must perform. **Interpreting Lines.** The architect will use various lines in making the drawings. Thick solid lines are used to outline the edges of the plans and elevations. Lines consisting of long and short dashes in sequence denote center lines. Short dashes usually indicate hidden lines but are also used on windows to show where they are hinged if they are operable. Thin lines are used for dimensions. These terminate in arrows to show the limit of the given dimensions. Repeat marks or the notation "do" indicates the repetition of a given dimension. Dimension lines are often broken by a zigzag which shows the object in the drawing has been shortened.

Symbols

Material indications and items related to plumbing, heating and ventilation, fire protection and electrical work have been standardized. Typically a symbol schedule is included on the drawings. The following are many of the usual symbols a contractor may encounter, but always check with the plans and specifications for building materials, because symbols will vary between design firms and from project to project.

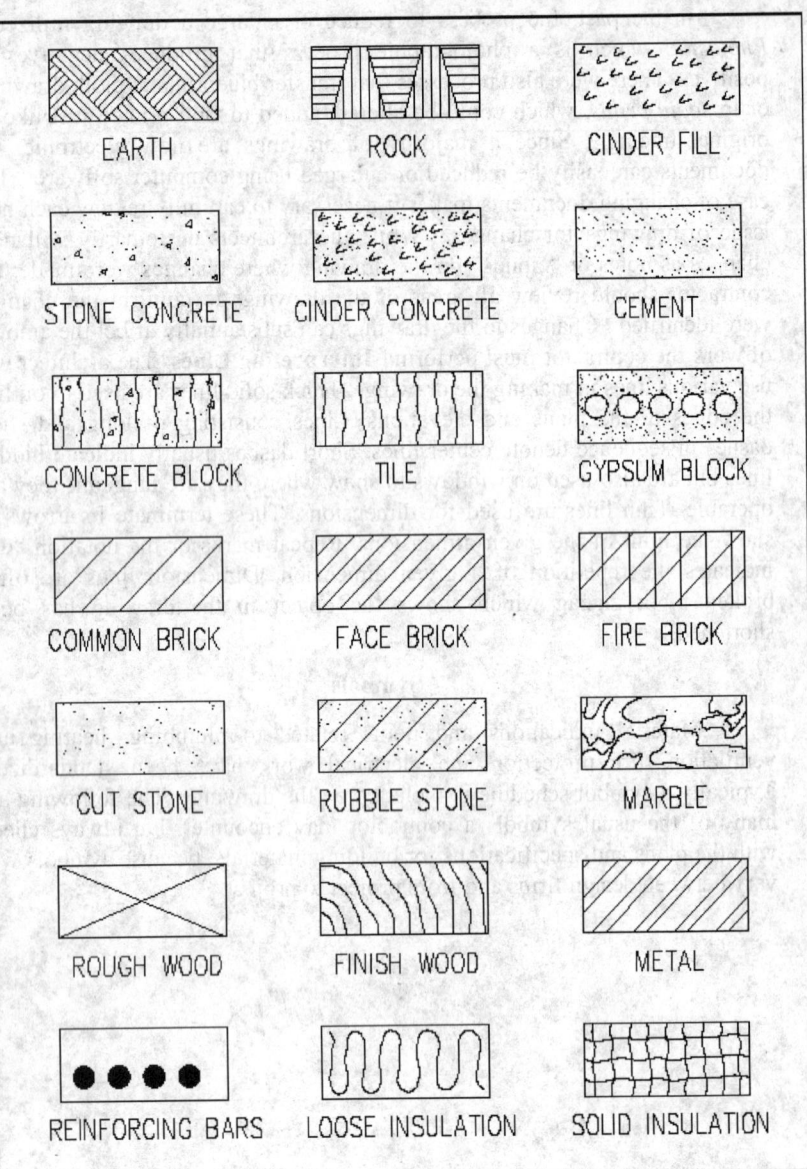

EARTH	ROCK	CINDER FILL
STONE CONCRETE	CINDER CONCRETE	CEMENT
CONCRETE BLOCK	TILE	GYPSUM BLOCK
COMMON BRICK	FACE BRICK	FIRE BRICK
CUT STONE	RUBBLE STONE	MARBLE
ROUGH WOOD	FINISH WOOD	METAL
REINFORCING BARS	LOOSE INSULATION	SOLID INSULATION

GENERAL OUTLETS:

CEILING	WALL	
◯	─◯	Fixture Outlet
Ⓑ	─Ⓑ	Blanked Outlet
Ⓒ	─Ⓒ	Clock Outlet
Ⓓ		Drop Cord Outlet
Ⓕ	─Ⓕ	Fan Outlet
Ⓙ	─Ⓙ	Junction Box
Ⓡ	─Ⓡ	Recessed Outlet
Ⓢ	─Ⓢ	Pull Switch
Ⓧ	─Ⓧ	Exit

Surface Fluorescent

R   Recessed Fluorescent

Bare Lamp Fluorescent

CONVENIENCE OUTLETS:

	Duplex
WP	Waterproof
R	Radio
	Range
●	Floor
▲	Special Purpose

SWITCH OUTLETS:

S	Single Pole Switch
S_2	Double Pole Switch
S_3	Three Way Switch
S_D	Automatic Door Switch
S_E	Electrolier Switch
S_K	Key Operated Switch
S_{CB}	Circuit Breaker
S_{MC}	Momentary Contact
S_{RC}	Remote Control
S_{WP}	Weatherproof Switch
S_F	Fused Switch

PANELS & CIRCUITS:

	Lighting Panel
	Power Panel
	Two Wire Branch Circuit
	Three Wire Branch Circuit
	Under Floor Duct & Box
G	Generator
M	Motor
	Controller

42

SIGNALING SYSTEMS

⊡	Push Button
☐	Buzzer
☐	Bell
CH	Chime
◇	Annunciator
⊡	Electric Door Opener
◁	Telephone Switchboard
◁	Interconnecting Telephone
◀	Outside Telephone
F	Fire Alarm Bell
F	Fire Alarm Station
⬛	City Fire Alarm
W	Watchman's Station
H	Horn
N	Nurse's Signal Plug
R	Radio Outlet
TV	Television Outlet
☐	Interconnection Box

PIPE FITTING SYMBOLS:

———————	Low Pressure Steam
—/——/——/—	Medium Pressure Steam
—//——//——//—	High Pressure Steam
— — — — — —	Return Line
—O——O——O—	Condensate Line
— —·— —·— —	Make Up Water
— — — — — —	Air Relief Line
—— FOF ——	Fuel Oil Flow
—— FOR ——	Fuel Oil Return
—— FOV ——	Fuel Oil Vent
—— A ——	Compressed Air
—— RL ——	Refrigerant Liquid
—— RD ——	Refrigerant Discharge
—— C ——	Condenser Water Flow
—— CR ——	Condenser Water Return
—— CH ——	Chilled Water
—— CHR ——	Chilled Water Return
—— H ——	Humidification Line
—— D ——	Drain Line
—— B ——	Brine Supply

HEATING SYMBOLS:

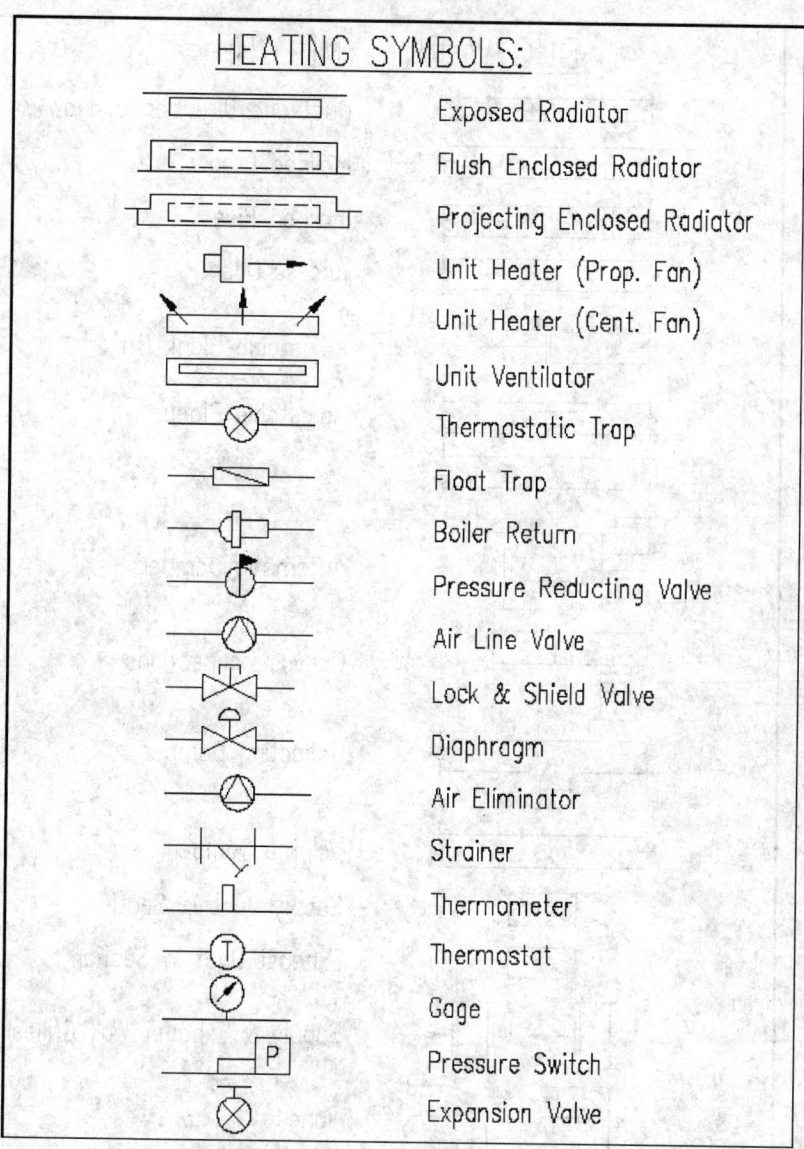

Exposed Radiator

Flush Enclosed Radiator

Projecting Enclosed Radiator

Unit Heater (Prop. Fan)

Unit Heater (Cent. Fan)

Unit Ventilator

Thermostatic Trap

Float Trap

Boiler Return

Pressure Reducing Valve

Air Line Valve

Lock & Shield Valve

Diaphragm

Air Eliminator

Strainer

Thermometer

Thermostat

Gage

Pressure Switch

Expansion Valve

DUCTWORK SYMBOLS:

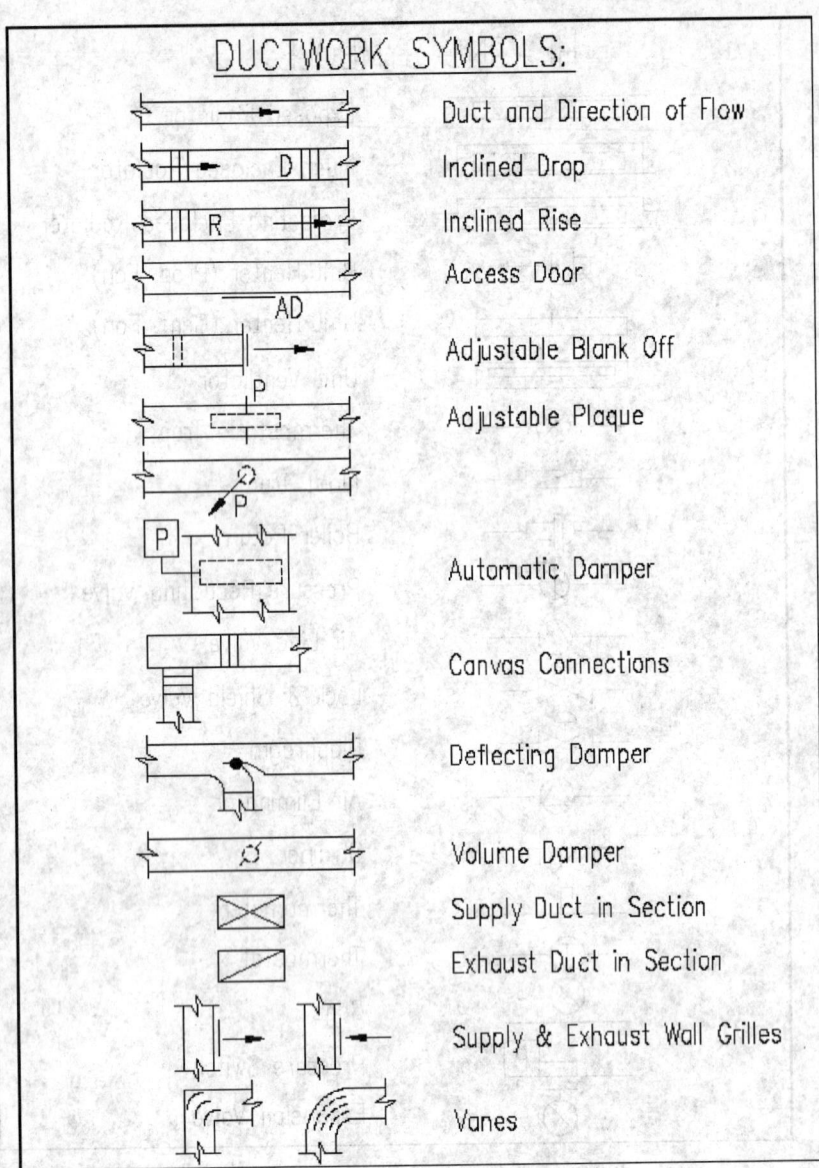

Duct and Direction of Flow

Inclined Drop

Inclined Rise

Access Door

Adjustable Blank Off

Adjustable Plaque

Automatic Damper

Canvas Connections

Deflecting Damper

Volume Damper

Supply Duct in Section

Exhaust Duct in Section

Supply & Exhaust Wall Grilles

Vanes

GENERAL REQUIREMENTS

PIPING SYMBOLS:

—‖—	Flanged Joint
—┼—	Screwed Joint
—C—	Bell & Spigot Joint
—✕—	Welded Joint
—○—	Soldered Joint
⊙‖	Flanged Elbow—Up
○‖	Flanged Elbow—Down
‖⊙‖	Flanged Tee—Up
‖○‖	Flanged Tee—Down
‖▷‖	Flanged Reducer—Concentric
‖◁‖	Flanged Reducer—Eccentric
‖⋈‖	Flanged Gate Valve
‖●‖	Flanged Globe Valve
►◁‖	Flanged Check Valve
‖☐‖	Flanged Cock Valve
‖S‖	Flanged Safety Valve
‖☐‖	Flanged Expansion Joint
▭	Reducing Flange
‖‖	Flanged Union
‖---‖	Flanged Sleeve

47

Abbreviations. Each project will have project specific abbreviations listed on the drawings. It is important to read the abbreviations since they can vary from project to project. The following is a list of standard abbreviations used and accepted within the construction industry. The list is alphabetized by abbreviation.

A	ampere; air; area
AB	anchor bolt
ABAND	abandoned
ABT	about
ABUT	abutment
ABV	above
AC	alternating current; acoustical
A/C	air conditioning
ACC	access
ACI	American Concrete Institute
ACFL	access floor
ACPL	acoustical plaster
ACR	acrylic plastic
ACS	access
ACT	acoustical tile
ADA	American with Disabilities Act of 1992
AD	area drain; access door
ADD	addendum; additional
ADH	adhesive
ADJ	adjacent; adjustable; adjust
ADMIN	administration
AFF	above finished floor
AGA	American Gas Association
AGG	aggregate
AIA	American Institute of Architects
AL	aluminum
ALM	alarm
ALT	alternate; altitude
AMP	amperage; ampere
ANCH	anchor; anchorage
ANOD	anodized
AP	access panel
APPD	approved

APPROX	approximate
APT	Apartment
APX	approximate
ARCH	architect; architectural
ASB	asbestos
ASC	above suspended ceiling
ASPH	asphalt
ASSY	assembly
AT	asphalt tile
AUTO	automatic
AUX	auxiliary
AVG	average
AWG	American Wire Gauge
AX	axis
B	boiler; bathroom; bidet
BAL	balance
BBD	bulletin board
BBL	barrel
BC	bottom of curb; broom closet
BD	board; blow down (pipe)
BDL	bundle
BEL	below
BET	between
BF	board foot; back face
BH	baseboard heater
BIT	bituminous
BJT	bed joint
BL	building line
BLDG	building
BLK	block
BLKG	blocking
BLW	below
BLW FL	below floor
BM	bench mark; beam
BOT	bottom
BP	back plaster; base plate; bearing pile; blue print

48

BPL .... bearing plate
BRG .... bearing
BRK .... brick
BRKR .... breaker
BRZ .... bronze
BS .... both sides
BSMT .... basement
BT .... bent
BTM .... bottom
BTU .... British Thermal Unit
BTUH .... British Thermal Units per hour
BU .... built up
BUR .... built up roof
BVL .... beveled
BW .... both ways

C .... course; curb; Celsius
C to C .... center to center
C & G .... curb and gutter
CA .... compressed air
CAB .... cabinet
CAD .... cadmium; computer-aided drafting
CANTIL .... cantilever
CAP .... capacity
CB .... catch basin
CC .... cubic centimeter
CBL .... Concrete Block
CBLSTN .... cobblestone
CCF .... hundred cubic feet
CCTV .... closed circuit TV
CEM .... cement
CENT .... centrifugal
CER .... ceramic
CF .... cubic foot; cellular floor
CFL .... counterflashing
CFM .... cubic foot per minute
CFS .... cubic foot per second
CG .... corner guard
CHAM .... chamfer
CHBD .... chalkboard
CHEM .... chemical
CHR .... chilled water return
CHT .... ceiling height

CI .... cast iron
CIP .... cast in place; cast-iron pipe
CIPC .... cast-in-place concrete
CIR .... circle
CIRC .... circ; circumference
CJ .... control joint
CK .... caulk
CLG .... ceiling
CLL .... contract limit line
CLS .... closure
CLR .... clear
CLS .... closure
CM .... centimeter
CMP .... corrugated metal pipe
CMT .... ceramic mosaic tile
CMU .... concrete masonry unit
COAX .... Coaxial
COL .... column
COMB .... combination; combustion
COMP .... compression
COMPO .... composition
COMPT .... compartment
CONC .... concrete
CONST .... construction
CONT .... continuous; control
CONTR .... contract; contractor
COORD .... coordinate
COP .... Copper
CORR .... corrugated
CPE .... certified professional estimator
CPM .... critical path method
CPR .... copper
CPT .... carpet
CR .... chromium
CRG .... cross grain
CRS .... course
CS .... countersink; combined sewer; cast steel; cast stone
CSMT .... casement
CT .... ceramic tile; cork tile
CTR .... counter

49

CTSK	countersunk screw	DP	dampproofing; dew point; distribution panel
CTWT	counter weight		
CU	cubic	DPR	damper
CU. FT.	cubic foot	DR	door; drive; dining room
CULV	culvert	DRB	drainboard
CU. M.	cubic meter	DS	downspout
CU.YD.	cubic yard	DT	drain tile
CV	check valve	DTA	dovetail anchor
CWT	hundred pounds	DTL	detail
CY	cubic yard	DTS	dovetail anchor slot
		DUP	duplicate
D	drain; depth; discharge	DW	dumbwaiter; distilled water
d	penny (nail size, i.e., 10d = 10 penny nail)		
		DWG	drawing
DA	doubleacting	DWL	dowel
DB	direct burial; decibel	DWR	drawer
DBL	double	DS	downspout
DC	direct current		
DEFL	deflect; deflection	E	east; enamel
DEG	degree	EB	expansion bolt
DEM	demolish	E to E	end to end
DEMOB	demobilization	EE	each end
DEP	depressed	EF	each face
DEPT	department	EJ	expansion joint
DET	detail	EL	elevation; elevator
DF	drinking fountain	ELECT	electric
DH	double hung	ELEV	elevator; elevation
DHW	domestic hot water	EM	expanded metal
DI	drop inlet; ductile iron	EMER	emergency
DIA	diameter	EMH	electrical manhole
DIAG	diagonal	EMT	electrical metallic conduit; thin wall conduit
DIAM	diameter		
DIFF	diffuser	ENC	enclose
DIM	dimension	ENG	engine; engineer
DIP	ductile iron pipe	ENGR	engineer
DIR	direction	ENT	entrance
DISC	disconnect	ENTR	entrance
DISCH	discharge	EP	electrical panelboard; explosion proof; edge of pavement
DIV	division		
DK	deck		
DL	dead load	EQ	equal
DMH	drop manhole	EQPT	equipment
DML	demolition; demolish	EQUIP	equipment
DMT	demountable	ERP	emergency receptacle panel
DN	down		

50

ESC .... escalator
EST .... estimate
EW .... each way
EWC .... electric water cooler
EWH .... electric water heater
EXC .... excavation; excavate
EXCA .... excavate
EXG .... existing
EXH .... exhaust
EXP .... exposed; expansion
EXS .... extra strong
EXT .... exterior; extinguish

F .... female; fill: fuse
f .... Fahrenheit degree
FA .... fire alarm; fresh air
FAB .... fabricate
FACP .... fire alarm control panel
FAS .... fasten
FB .... face brick
FBD .... fiberboard
FBO .... furnished by others
FBRK .... fire brick
FC .... foot-candle
FD .... floor drain; fire damper
FDN .... foundation
FDR .... feeder
FE .... fire extinguisher
FEC .... fire extinguisher cabinet
FF .... factory finish; far face;
          finish floor
FFE .... finished floor elevation
FFL .... finished floor line
FGL .... fiberglass
FH .... fire hydrant
FHMS .... flathead machine screw
FHR .... fire hose rack
FHS .... fire hose station
FHWS .... flathead wood screw
FIG .... figure
FIN .... finish
FJT .... flush joint
FL .... floor; fire line
FLCO .... floor cleanout
FLG .... flashing; flange; flooring

FLR .... floor
FLSH .... flashing
FLUOR .... fluorescent
FLX .... flexible
FN .... fence
FND .... foundation
FO .... fiber optics; finished
          opening
FOB .... free on board
FOC .... face of concrete
FOF .... face of finish
FOM .... face of masonry
FOS .... face of studs
FPRF .... fireproof
FPL .... fireplace
FPM .... foot (feet) per minute
FR .... frame
FRA .... fresh air
FRC .... fire-resistant coating
FRG .... forged
FRT .... fire-retardant
FS .... full size; far side; federal
          standards
FT .... foot; feet
FTG .... footing
FT. LB. .... foot-pound
F to F .... face to face
FUR .... furred
FURN .... furnish; furniture
FUT .... future
FVC .... fire valve cabinet
FW .... fire water

G .... gauge; gas; gas main or
          service; girder
g .... gram
GA .... gauge, gage
GAL .... gallon
GALV .... galvanized
GB .... grab bar
GC .... general contractor
GCMU .... glazed concrete masonry
          units
GD .... grade
GEN .... generator; general

GF	ground face	HPS	high pressure steam
GI	galvanized iron	HR	hour
GKT	gasket	HT	height
GL	glass	HTG	heating
GLB	glass block	HV	high voltage; heating/ventilating unit
GLF	glass fiber	HVAC	heating/ventilating and air conditioning
GLZ	glazing		
GND	ground	HW	hot water; heavy wall
GOVT	government	HWD	hardwood
GP	galvanized pipe	HWH	hot water heater
GPD	gallons per day	HWR	hot water return
GPH	gallons per hour	HWS	hot water supply
GPL	gypsum lath	HX	hexagonal
GPM	gallons per minute	HYD	hydrant; hydraulic
GPPL	gypsum plaster	HZ	hertz
GPT	gypsum tile		
GR	grade; granite	ID	inside diameter; identification
GRN	granite		
GSS	galvanized steel sheet	ILK	interlock
GST	glazed structural tile	IN	inch
GT	grout	INCIN	incinerator
GV	galvanized	INCL	inclusive; incline
GVL	gravel	INSUL	insulation
GYP	gypsum	INSC	insulating concrete
GYP BD	Gypsum Board	INSF	insulating fill
		INSTL	install
H	height; high	INSUL	insulation; insulated; insulator; insulating
HB	hose bib		
HBD	hardboard	INT	interior
HC	handicapped; hollow core; high capacity	INTM	intermediate
		INV	invert
HD	heavy duty	IP	iron pipe
HDR	header	IPS	iron pipe size
HDWE	hardware	IR	irrigation
HES	high early-strength cement	IW	indirect waste
HH	handhole	J	joist
HK	hook	J-BOX	junction box
HM	hollow metal	JB	junction box
HOR	horizontal	JC	janitor's closet
HORIZ	horizontal	JCT	junction
HOSP	hospital	JF	joint filler
HP	horse power; high pressure; high point	JT	joint
HPR	high pressure return		

GENERAL REQUIREMENTS

K	Kilopound (1,000 pounds); Kelvin
KCP	Keene's cement plaster
KG	kilogram
KIT	kitchen
KLF	kips per lineal foot
KIP	kilopound (1,000 pounds)
KIT	kitchen
KM	kilometer
KO	knockout
KPL	kickplate
KSF	kips per square foot
KSI	kips per square inch
KW	kilowatt
KWH	kilowatt hour
L	length; left; liter; long
LA	Landscape Architect; lighting arrester
LAB	laboratory; labor
LAD	ladder
LAM	laminated
LAT	Lateral
LAV	lavatory
LB	lag bolt; pound (weight)
LBL	label
LBS	pounds
LBS/HR	pounds per hour
LBS/LF	pounds per lineal foot
LBS/SqIn	pounds per square inch
LC	light control
LCD	liquid crystal diode
LF	lineal foot; lineal feet
LG	long; length
LH	left hand
LIQ	liquid
LL	live load
LLH	long leg horizontal
LLV	long leg vertical
LMS	limestone
LN	length
LOC	locate; local
LONG	longitudinal

LP	lightproof; light pole; lighting panel; low point
LS	lump sum; limestone
LT	light
LTL	lintel
LV	low voltage
LVR	louver
LW	lightweight
LWC	lightweight concrete
M	meter; thousand (brick, etc)
MACH	machine
MAINT	maintenance
MAS	masonry
MATL	material
MAX	maximum
MB	machine bolt; mail box
MBF	thousand board feet
MBH	thousand BTU per hour
MBR	member
MC	medicine cabinet
MCC	motor control center
ME	mechanical engineer
MECH	mechanical
MED	medium
MET	metal
MEZZ	mezzanine
MFD	metal floor deck
MFG	manufacture
MGD	million gallons per day
MH	manhole
MHW	mean high water
MI	mile; malleable iron
MID	middle
MIN	minimum
MIR	mirror
MISC	miscellaneous
MK	mark
ML&P	metal lath and plaster
MJ	mechanical joint
MK	mark
MLD	moulding
MLW	mean low water
MM	millimeter

MMB	membrane	OF	outside face
MO	month; masonry opening	OH	overhead
MOB	mobilization	OHD	overhead door
MOD	modular	OHMS	oval-head machine screw
MON	monument	OHWS	oval-head wood screw
MOV	movable	OI	ornamental iron
MPA	megapascals	OJ	open-web joist
MPH	miles per hour	OP	opaque
MR	mop receptor	OPNG	opening
MRB	marble	OPH	opposite hand
MRD	metal roof deck	OPP	opposite
MSL	mean sea level	OPS	opposite surface
MTD	mounted	OZ	ounce
MTFR	metal furring		
M^3	cubic meter	P	pole; power; pump; page; pitch
MTHR	metal threshold	PA	pascal
MTL	material; metal	PAR	parallel
M^2	square meter	PB	panic bar; push button; pull box
MULL	mullion	PBD	particle board
MWK	millwork	PCC	precast concrete
MWP	membrane waterproofing	PCF	pounds per cubic foot
		PCPL	Portland cement plaster
N	north	PCMP	Perforated Corrugated Metal Pipe
NA	not applicable	PE	porcelain enamel; professional engineer
NAT	natural	PED	pedestal
NEC	national electrical code	PEJ	premolded expansion joint
NF	near face	PERF	perforated
NI	nickel	PERI	perimeter
NIC	not in contract	PFB	prefabricated
NMT	non-metallic	PFL	pounds per lineal foot
NO	number	PFN	prefinished
NOM	nominal	PG	plate glass; pressure gauge
NR	noise reduction	PKG	parking
NRC	noise reduction coefficient	PL	plate or property line
NS	near side	PLAM	plastic laminate
NTS	not to scale	PLAS	plaster
		PLUMB	plumbing
O	oxygen	PLYD	plywood
O to O	out to out		
OA	overall; outside air		
O & P	overhead and profit		
OBS	obscure		
OC	on center		
OD	outside diameter		

54

GENERAL REQUIREMENTS

PLYWD	plywood	RD	roof drain
PMP	Perforated Metal Pipe	RE	reinforced
PNL	panel	REF	reference
PNT	painted	REG	register; regular
PR	pair	REINF	reinforcement
PRC	precast reinforced concrete	REM	remove
		REQD	required
PREFAB	prefabricated	RES	resilient
PRESS	pressure	RET	return
PRF	preformed	REV	revision or revised
PROV	provide	RF	roof
PSC	prestressed concrete	RFG	roofing
PSF	pounds per square foot	RFH	roof hatch
PSI	pounds per square inch	RFL	reflected
PSIG	pounds per square inch gauge	RH	right hand
		RL	railing; roof ladder
PT	point; point of tangent; potential transformer	RM	room
		RND	round
PTC	post-tensioned concrete	RO	rough opening (RGH OPNG)
PTD	paper towel dispenser; painted	ROW	right of way
PTN	partition	RPM	rotations per minute
PTR	paper towel receptor	RR	railroad
PVG	paving	RT	rubber tile
PVC	polyvinyl chloride	RVS	reverse
PVMT	pavement	RVT	rivet
PWR	power	R/W	right of way
		RWC	rainwater conductor
QUAL	quality		
QT	quarry tile	S	south; sanitary sewer; supply; sign
QTR	Quarter	S4S	Surfaced four sides
QTY	Quantity	SAN	sanitary
		SBSTA	substation
R	riser; radius; relay panel	SC	solid core
RA	return air; registered architect	SCH	schedule
RAD	radius	SCHED	schedule
RB	rubber base	SCR	screen
RBL	rubble stone	SCT	structural clay tile
RBT	rabbet; rebate; rubber tile	SD	storm drain
RBR	rubber	SE	structural engineer
RCF	raised computer floor	SEC	section; second
RCP	reinforced concrete pipe; reflected ceiling plan	SF	square foot
		SFGL	safety glass
RECP	receptacle	SH	shelf

55

SHO	shoring	SYS	system
SHT	sheet		
SHTH	sheathing	T	tread; top; tangent; telephone
SIM	similar		
SK	sketch; sink	T & B	top and bottom
SKL	skylight	T & G	tongue and groove
SL	sleeve; slope; Siamese line	TB	towel bar
		TC	terra cotta; top of curb
SMH	sanitary manhole	TCP	terra cotta pipe
SNT	sealant	TEL	telephone
SP	soundproof; soil pipe	TEMP	temperature
SPC	spacer	TER	Terrazzo
SPEC	specification	TERM	terminal
SPK	speaker	THHN	nylon jacked wire
SPL	special	THK	thickness
SPLR	sprinkler	THR	threshold
SQ	square; 100 square feet	THRU	through
SQ FT	square foot; square feet	THW	insulated strand wire
SQ M	square meter	THWN	nylon jacketed wire
SQ YD	square yard	TKBD	tackboard
SS	sanitary sewer; stainless steel	TKS	tackstrip
		TLT	toilet
SSK	service sink	TOC	top of concrete
SST	stainless steel	TOL	tolerance
ST	steel; storm sewer; straight	TOT	total
		TP	top of pavement
STA	station	TPC	top of pile cap
STD	standard	TPD	toilet paper dispenser
STG	seating; storage	TPTN	toilet partition
STL	steel	TR	transom
STO	storage	T/R	top of rail
STR	structural	TRANS	transformer
STRL	structural	TRK	track
STY	story	TST	top of steel
SUBS	subcontractors	TS	top of stone
SUR	surface	TSL	top of slab
SUSP	suspended	TV	television
SVCE	service	TW	top of wall; tempered water
SW	switch		
S/W	sidewalk	TYP	typical
SWGR	switchgear	TZ	terrazzo
SY	square yard		
SYM	system	UC	undercut
SYMM	symmetrical	U/C	under construction
SYN	synthetic		

GENERAL REQUIREMENTS

UCI	uniform construction index	W	west; wide; watt; water main; wire
UG	underground	W/	with
UH	unit heater	W/O	without
UL	underwriters laboratory	WB	wood base
UNEXC	unexcavated	WC	water closet
UNF	unfinished	WD	wood
UNIF	uniform	WE	water elevation
UNO	unless otherwise noted	WF	wide flange
UR	urinal	WG	wired glass; water gage
USG	United States Gauge; United States Gypsum Company	WH	wall hung
		WHB	wheel bumper
		WI	wrought iron
USGS	United States Geological Survey	WIN	window
		WM	wire mesh; water meter
USS	United States Standard	WO	without
UTIL	utility	WP	waterproofing
		WPT	working point
V	voltage; volt; vent; valve	WR	water repellent
VA	Volt Ampere	WS	waterstop; weather-stripping
VAC	vacuum		
VAR	varnish	WSCT	wainscot
VAT	vinyl asbestos tile	WST	waste
VB	vapor barrier or vinyl base	WT	watertight; weight
		WTR	water
VC	vertical curve	WTW	wall to wall
VCP	vitrified clay pipe	WV	water valve
VEL	velocity	WWF	welded wire fabric
VENT	ventilate	XARM	cross arm
VERT	vertical	XFR	transfer
VF	vinyl fabric	XFMR	transformer
VG	vertical grain	XH	extra heavy
VIB	vibration	XHD	extra heavy duty
VIN	vinyl		
VJ	v-joint	Y	wye
VLF	vertical lineal feet	YD	yard
VNR	veneer	YHYD	yard hydrant
VOL	volume	YI	yard inlet
VRM	vermiculite	YR	year
VT	vinyl tile		

Specifications

Arrangement. The Uniform System developed by the Construction Specifications Institute (CSI) has become the accepted industry standard for setting up specifications. This system not only sets forth an arrangement for

57

specifications, but it is equally applicable to data filing (most of the manufacturers' catalogs carry the CSI symbol, a rectangle enclosing three ellipses, and the Uniform System number) and to cost accounting systems.

This format is ideal for developing estimates and estimating check lists, and later on in this chapter, when we discuss setting up the estimate, we will refer the reader back to the list below for this purpose.

The CSI format is divided into four parts:

1. Bidding Requirements
2. Contract Forms
3. General Conditions
4. Specifications

The first three parts—bidding requirements, contract forms and general conditions—are not specifications sections. The CSI Masterformat™ refers to these as documents rather than sections. Prior to 2004 the CSI was divided into Divisions 1 through 16 which contain the specifications for work items as listed below. In 2004 the CSI approved a new MasterFormat™ which is being slowly accepted by government agencies as well as private industry. This conversion from the original 16 divisions to 49 divisions will take time and money. The 16 division format is still prevalent and therefore, this book is still using the pre 2004 16 divisions for this edition. The new MasterFormat™2004 edition is also listed after the 16 divisions for your use and information.

BIDDING REQUIREMENTS, CONTRACT FORMS AND CONDITIONS OF THE CONTRACT

00010	**Pre-Bid Information**
00020	Invitation to Bid
00030	Advertisement for Bids
00040	Prequalification Forms
00100	**Instruction to Bidders**
00120	Supplementary Instruction to Bidders
00130	Pre-Bid Conference
00200	**Information Available to Bidders**
00210	Preliminary Schedules
00220	Geotechnical Data
00230	Existing Conditions
00240	Project Financial Information
00300	**Bid Forms**
00400	**Supplements to Bid Forms**
00410	Bid Security Forms
00420	Bidder Qualification Forms
00430	Subcontractors List
00440	Substitution List

GENERAL REQUIREMENTS

00450	Equipment Suppliers List
00460	List of Alternates/Alternatives
00470	List of Estimated Quantities
00480	Noncollusion Affidavit
00500	**Agreement Forms**
00600	**Bonds and Certificates**
00610	Performance Bonds
00620	Payment Bonds
00630	Warranty Bonds
00640	Maintenance Bond
00650	Certificate of Insurance
00660	Certificates of Compliance
00700	**General Conditions**
00800	**Supplemental Conditions**
00810	Modification to General Conditions
00820	Additional Articles
00830	Wage Determination Schedule
00900	**Addenda**

This system is the generally accepted form for specifications today. This book is organized according to the system. What follows is a complete listing of the 16 CSI divisions and their subdivisions:

01	**GENERAL REQUIREMENTS**
01010	**Summary Of Work**
01020	**Allowances**
01021	Cash Allowances
01024	Quantity Allowances
01025	**Measurement of Payment**
01030	**Alternates/Alternatives**
01035	**Modifications Procedures**
01040	**Coordination**
01041	Project Coordination
01042	Mechanical and Electrical Coordination
01043	Job Site Administration
01045	Cutting and Patching
01050	**Field Engineering**
01060	**Regulatory Requirements**
01070	**Identification Systems**
01090	**References**
01091	Reference Standards
01092	Abbreviations
01093	Symbols
01094	Definitions
01100	**Special Project Procedures**
01200	**Project Meetings**

01210	Preconstruction Conferences
01220	Progress Meetings
01245	Installation Meetings
01300	**Submittals**
01310	Progress Schedules
01320	Progress Reports
01330	Survey and Layout Data
01340	Shop Drawings, Product Data and Samples
01360	Quality Control Submittals
01380	Construction Photographs
01400	**Quality Control**
01410	Testing Laboratory Services
01420	Inspection Services
01425	Field Samples
01430	Mock-ups
01440	Contractor's Quality Control
01445	Manufacturer's Field Services
01500	**Construction Facilities and Temporary Controls**
01505	Mobilization
01510	Temporary Utilities
01520	Temporary Construction
01525	Construction Aids
01530	Barriers and Enclosures
01540	Security
01550	Access Roads and Parking Areas
01560	Temporary Controls
01570	Traffic Regulation
01580	Project Identification and Signs
01590	Field Offices and Sheds
01600	**Material and Equipment**
01610	Delivery, Storage and Handling
01620	Installation Standards
01630	Product Options and Substitutions
01650	**Facility Startup/Commissioning**
01655	Starting of Systems
01660	Testing, Adjusting and Balancing of Systems
01670	Systems Demonstrations
01700	**Contract Closeout**
01710	Final Cleaning
01720	Project Record Documents
01730	Operation and Maintenance Data
01740	Warranties and Bonds
01750	Spare Parts and Maintenance Materials
01760	Warranty Inspections
01800	**Maintenance**

GENERAL REQUIREMENTS

02	**SITE WORK**
02010	**Subsurface Investigation**
02012	Standard Penetration Tests
02016	Seismic Investigation
02050	**Demolition**
02060	Building Demolition
02070	Selective Demolition
02075	Concrete Removal
02080	Hazardous Material Abatement
02100	**Site Preparation**
02110	Site Clearing
02115	Selective Clearing
02120	Structure Moving
02140	**Dewatering**
02142	Sand Drains
02144	Well Points
02146	French Drains
02148	Relief Wells
02150	**Shoring and Underpinning**
02152	Shoring
02153	Needle Beams
02154	Grillage
02156	Underpinning
02158	Slabjacking
02160	**Excavation Support Systems**
02162	Cribbing and Walers
02164	Soil and Rock Anchors
02166	Ground Freezing
02167	Reinforced Earth
02168	Slurry Wall Construction
02170	**Cofferdams**
02172	Double Wall Cofferdams
02174	Cellular Cofferdams
02176	Piling with Intermediate Lagging
02178	Sheet Piling Cofferdams
02200	**Earthwork**
02210	Grading
02220	Excavating, Backfilling and Compacting
02230	Base Courses
02240	Soil Stabilization
02250	Vibro-Flotation
02270	Slope Protection and Erosion Control
02280	Soil Treatment

03250	Concrete Accessories

03300	**Cast-in-place Concrete**
03310	Structural Concrete
03330	Architectural Concrete
03340	Low Density Concrete
03345	Concrete Finishing
03350	Concrete Finishes
03360	Specially Placed Concrete
03365	Post-Tension Concrete
03370	**Concrete Curing**
03400	**Precast Concrete**
03410	Structural Precast Concrete-Plant Cast
03420	Structural Precast Post-Tensioned Concrete-Plant Cast
03430	Structural Precast Concrete-Site Cast
03450	Architectural Precast Concrete-Plant Cast
03460	Architectural Precast Concrete-Site Cast
03470	Tilt-up Precast Concrete
03480	Precast Concrete Specialties
03500	**Cementitious Decks and Toppings**
03510	Gypsum Concrete
03520	Insulating Concrete Deck
03530	Cementitious Wood Fiber Systems
03540	Composite Concrete and Insulation Decks
03550	Concrete Toppings
03600	**Grout**
03700	**Concrete Restoration and Cleaning**
03710	Concrete Cleaning
03720	Concrete Resurfacing
03730	Concrete Rehabilitation
03800	**Mass Concrete**

04	**MASONRY**
04100	**Mortar and Masonry Grout**
04150	**Masonry Accessories**
04200	**Unit Masonry**
04210	Clay Unit Masonry
04220	Concrete Unit Masonry
04230	Reinforced Unit Masonry
04235	Pre-assembled Masonry Panel Systems
04240	Non-reinforced Masonry Systems
04270	Glass Unit Masonry
04280	Gypsum Unit Masonry
04290	Adobe Unit Masonry
04400	**Stone**
04410	Rough Stone

64

04420	Cut Stone
04440	Flagstone
04450	Stone Veneer
04455	Marble
04460	Limestone
04465	Granite
04470	Sandstone
04475	Slate
04500	**Masonry Restoration and Cleaning**
04510	Masonry Cleaning
04520	Masonry Restoration
04550	**Refractories**
04555	Flue Lines
04560	Combustion Chambers
04565	Firebrick
04570	Castable Refractories
04600	**Corrosion Resistant Masonry**
04605	Chemical Resistant Brick
04610	Vitrified Clay Liner Plates
04700	**Simulated Masonry**
04710	Simulated Stone
04720	Cast Stone
05	**METALS**
05010	**Metal Materials**
05030	**Metal Coatings**
05050	**Metal Fastening**
05100	**Structural Metal Framing**
05120	Structural Steel
05140	Structural Aluminum
05140	Steel Wire Rope
05160	Framing Systems
05200	**Metal Joists**
05210	Steel Joists
05250	Aluminum Joists
05260	Composite Joist System
05300	**Metal Decking**
05310	Steel Deck
05320	Raceway Deck Systems
05330	Aluminum Deck
05400	**Cold Formed Metal**
05410	Load-Bearing Metal Stud Systems
05420	Cold-Formed Metal Joist Systems
05430	Slotted-Channel Framing Systems
05450	Metal Support Systems
05500	**Metal Fabrications**

66

06195	Prefabricated Wood Beams and Joists
06200	**Finish Carpentry**
06220	Millwork
06240	Laminates
06250	Prefinished Wood Paneling
06255	Prefinished Hardwood Paneling
06260	Board Paneling
06300	**Wood Treatment**
06310	Preservative Treatment
06320	Fire Retardant Treatment
06330	Insect Treatment
06400	**Architectural Woodwork**
06410	Custom Casework
06420	Panelwork
06430	Stairwork and Handrails
06440	Wood Ornaments
06450	Standing and Running Trim
06460	Exterior Frames
06470	Screens, Blinds and Shutters
06480	Custom Wood Turning
06500	**Structural Plastics**
06600	**Plastic Fabrications**
06610	Glass Fiber and Resin Fabrications
06620	Cast Plastic Fabrications
06630	Historic Plastic Reproductions
06650	**Solid Polymer Fabrications**
07	**THERMAL and MOISTURE PROTECTION**
07100	**Waterproofing**
07110	Sheet Membrane Waterproofing
07120	Fluid Applied Waterproofing
07125	Sheet Metal Waterproofing
07130	Bentonite Waterproofing
07140	Metal Oxide Waterproofing
07145	Cementitious Waterproofing
07150	**Dampproofing**
07160	Bituminous Dampproofing
07175	Cementitious Dampproofing
07180	**Water Repellents**
07190	**Vapor Retarders**
07195	**Air Barriers**
07200	**Insulation**
07210	Building Insulation
07220	Roof and Deck Insulation
07240	**Exterior Insulation and Finish System**
07250	**Fireproofing**

08110	Steel Doors and Frames
08120	Aluminum Doors and Frames
08130	Stainless Steel Doors and Frames
08140	Bronze Doors and Frames
08200	**Wood and Plastic Doors**
08210	Wood Doors
08220	Plastic Doors
08250	**Door Opening Assemblies**
08255	Packaged Steel Door Assemblies
08260	Packaged Wood Door Assemblies
08265	Packaged Plastic Door Assemblies
08300	**Special Doors**
08305	Access Doors
08310	Sliding Doors and Grilles
08315	Pressure Resistant Doors
08320	Security Doors
08325	Cold Storage Doors
08330	Coiling Doors and Grilles
08350	Folding Doors and Grilles
08355	Chain Closures
08360	Section Overhead Doors
08365	Vertical Lift Doors
08370	Industrial Doors
08375	Hanger Doors
08380	Traffic Doors
08385	Sound Control Doors
08390	Storm Doors
08395	Screen Doors
08400	**Entrances and Storefronts**
08410	Aluminum Entrances and Storefronts
08420	Steel Entrances and Storefronts
08430	Stainless Steel Entrances and Storefronts
08440	Bronze Entrances and Storefronts
08450	All-Glass Entrances
08460	Automatic Entrance Doors
08470	Revolving Entrance Doors
08480	Balance Entrance Doors
08490	Sliding Storefronts
08500	**Metal Windows**
08510	Steel Windows
08520	Aluminum Windows
08530	Stainless Steel Windows
08540	Bronze Windows
08600	**Wood and Plastic Windows**
08610	Wood Windows

08630	Plastic Windows
08650	**Special Windows**
08655	Roof Windows
08660	Security Windows and Screens
08665	Pass Windows
08670	Storm Windows
08700	**Hardware**
08710	Door Hardware
08740	Electro-Mechanical Hardware
08760	Window Hardware
08770	Door and Window Accessories
08800	**Glazing**
08810	Glass
08840	Plastic Glazing
08850	Glazing Accessories
08900	**Glazed Curtain Walls**
08910	Glazed Steel Curtain Walls
08920	Glazed Aluminum Curtain Walls
08930	Glazed Stainless Steel Curtain Walls
08940	Glazed Bronze Curtain Walls
08950	Translucent Wall and Skylight Systems
08960	Sloped Glazing Systems
08970	Structural Glass Curtain Walls
09	**FINISHES**
09000	**Exterior Insulation Finish System (EIFS)**
09100	**Metal Supports**
09110	Non-loadbearing Wall Framing Systems
09120	Ceiling Suspension Systems
09130	Acoustical Suspension Systems
09200	**Lath and Plaster**
09205	Furring and Lathing
09210	Gypsum Plaster
09215	Veneer Plaster
09220	Portland Cement Plaster
09225	Adobe Finish
09230	Plaster Fabrications
09250	**Gypsum Board**
09260	Gypsum Board Systems
09270	Gypsum Board Accessories
09300	**Tile**
09310	Ceramic Tile
09320	Thin Brick Tile
09330	Quarry Tile
09340	Paver Tile

09350	Glass Mosaics
09360	Plastic Tile
09370	Metal Tile
09380	Cut Natural Stone Tile
09400	**Terrazzo**
09410	Portland Cement Terrazzo
09420	Precast Terrazzo
09430	Conductive Terrazzo
09440	Plastic Matrix Terrazzo
09450	**Stone Facing**
09500	**Acoustical Treatment**
09510	Acoustical Ceilings
09520	Acoustical Wall Treatment
09525	Acoustical Space Units
09530	Acoustical Insulation and Barriers
09540	**Special Wall Surfaces**
09545	**Special Ceiling Surfaces**
09550	**Wood Flooring**
09560	Wood Strip Flooring
09565	Wood Block Flooring
09570	Wood Parquet Flooring
09580	Wood Composition Flooring
09590	Resilient Wood Floor System
09600	**Stone Flooring**
09610	Flagstone Flooring
09615	Marble Flooring
09620	Granite Flooring
09625	Slate Flooring
09630	**Unit Masonry Flooring**
09635	Brick Flooring
09640	Pressed Concrete Unit Flooring
09650	**Resilient Flooring**
09660	Resilient Tile Flooring
09665	Resilient Sheet Flooring
09670	Fluid Applied Resilient Flooring
09675	Static Control Resilient Flooring
09678	Resilient Base and Accessories
09680	**Carpeting**
09682	Carpet Cushion
09685	Sheet Carpet
09690	Carpet Tile
09695	Wall Carpet
09698	Indoor/Outdoor Carpet
09700	**Special Flooring**
09705	Resinous Flooring

71

09710	Magnesium Oxychloride Floors
09720	Epoxy Marble-Chip Flooring
09725	Seamless Quartz Flooring
09730	Elastomeric Liquid Flooring
09750	Mastic Fills
09755	Plastic Laminate Flooring
09760	Asphalt Plank Flooring
09780	**Floor Treatment**
09785	Metallic Type Static Disseminating and Spark Resistant Finish
09790	Slip Resistant Finishes
09800	**Special Coatings**
09810	Abrasion Resistant Coatings
09815	High Build Glazed Coatings
09820	Cementitious Coatings
09830	Elastomeric Coatings
09835	Textured Plastic Coatings
09840	Fire-resistant Paints
09845	Intumescent Paints
09850	Chemical Resistant Coatings
09860	Graffiti Resistant Coatings
09870	Cost Systems for Steel
09880	Protective Coatings for Concrete
09900	**Painting**
09910	Exterior Painting
09920	Interior Painting
09930	Transparent Finishes
09950	**Wall Covering**
09955	Vinyl Coated Fabric Wall Covering
09960	Vinyl Wall Coating
09965	Cork Wall Coating
09970	Wallpaper
09975	Wall Fabrics
09980	Flexible Wood Sheets and Veneers
10	**SPECIALTIES**
10100	**Visual Display Boards**
10110	Chalkboards
10115	Markerboards
10120	Tackboards
10130	Operable Board Units
10140	Display Tracking System
10145	Visual Aid Board Units
10150	**Compartments and Cubicles**
10160	Metal Toilet Compartments
10165	Plastic Laminated Toilet Compartments

72

GENERAL REQUIREMENTS

10170	Plastic Toilet Compartments
10175	Particleboard Toilet Compartments
10180	Stone Toilet Compartments
10185	Shower and Dressing Compartments
10190	Cubicles
10200	**Louvers and Vents**
10210	Metal Wall Louvers
10220	Louvered Equipment Enclosures
10225	Metal Door Louvers
10230	Metal Vents
10240	**Grilles and Screens**
10250	**Service Wall Systems**
10260	**Wall and Corner Guards**
10270	**Access Flooring**
10272	Rigid Grid Access Floor Systems
10274	Snap-on Stringer Access Floor Systems
10276	Stringerless Access Floor Systems
10290	**Pest Control**
10292	Rodent Control
10294	Insect Control
10296	Bird Control
10300	**Fireplaces and Stoves**
10305	Manufactured Fireplaces
10310	Fireplace Specialties and Accessories
10320	Stoves
10340	**Manufactured Exterior Specialties**
10342	Steeples
10344	Spires
10346	Cupolas
10348	Weathervanes
10350	**Flagpoles**
10352	Ground Set Flagpoles
10354	Wall Mounted Flagpoles
10356	Automatic Flagpoles
10358	Nautical Flagpoles
10400	**Identifying Devices**
10410	Directories
10415	Bulletin Boards
10420	Plaques
10430	Exterior Signs
10440	Interior Signs
10450	**Pedestrian Control Devices**
10452	Portable Posts and Railings
10454	Rotary Gates
10456	Turnstiles

10458	Detection Specialties
10500	**Lockers**
10505	Metal Lockers
10510	Wood Lockers
10515	Coin-operated Lockers
10518	Glass Lockers
10520	**Fire Protection Specialties**
10522	Fire Extinguishers, Cabinets and Accessories
10526	Fire Blankets and Cabinets
10528	Wheeled Fire Extinguisher Units
10530	**Protective Covers**
10532	Walkway Covers
10534	Car Shelters
10536	Awnings
10538	Canopies
10550	**Postal Specialties**
10551	Mail Chutes
10552	Mail Boxes
10554	Collection Boxes
10556	Central Mail Delivery Boxes
10600	**Partitions**
10605	Wire Mesh Partitions
10610	Folding Gates
10615	Demountable Partitions
10630	Portable Partitions, Screens and Panels
10650	**Operable Partitions**
10652	Folding Panel Partitions
10655	Accordion Folding Partitions
10660	Sliding Partitions
10665	Coiling Partitions
10670	**Storage Shelving**
10675	Metal Storage Shelving
10680	Storage and Shelving Systems
10683	Mobile Storage System
10685	Wire Shelving
10688	Prefabricated Wood Storage Shelving
10700	**Exterior Protection Devices for Openings**
10705	Exterior Sun Control Devices
10710	Exterior Shutters
10715	Storm Panels
10750	**Telephone Specialties**
10755	Telephone Enclosures
10760	Telephone Directory Units
10765	Telephone Shelves

GENERAL REQUIREMENTS

10800	**Toilet and Bath Accessories**
10810	Toilet Accessories
10820	Bath Accessories
10880	**Scales**
10900	**Wardrobe and Closet Specialties**
11	**EQUIPMENT**
11010	**Maintenance Equipment**
11012	Vacuum Cleaning Systems
11014	Window Washing Systems
11016	Floor and Wall Cleaning Equipment
11018	Housekeeping Carts
11020	**Security and Vault Equipment**
11022	Vault Doors and Day Gates
11024	Security and Emergency Systems
11026	Safes
11028	Safe Deposit Boxes
11030	**Teller and Service Equipment**
11032	Service and Teller Window Units
11034	Package Transfer Units
11036	Automatic Banking Systems
11038	Teller Equipment Systems
11040	**Ecclesiastical Equipment**
11042	Baptisteries
11044	Chancel Fittings
11050	**Library Equipment**
11052	Book Theft Protection Equipment
11054	Library Stack Systems
11056	Study Carrels
11058	Book Depositories
11060	**Theater and Stage Equipment**
11062	Stage Curtains
11064	Rigging Systems and Controls
11066	Acoustical Shell Systems
11068	Folding and Portable Stages
11170	**Instrumental Equipment**
11072	Organs
11074	Carillons
11076	Bells
11080	**Registration Equipment**
11090	**Checkroom Equipment**
11100	**Mercantile Equipment**
11102	Barber and Beauty Shop Equipment
11104	Cash Registers and Checking Equipment
11106	Display Cases
11108	Food Processing Equipment

11110	**Commercial Laundry and Dry Cleaning Equipment**
11112	Washers and Extractors
11114	Dry Cleaning Equipment
11116	Drying and Conditioning Equipment
11118	Finishing Equipment
11120	**Vending Equipment**
11122	Money Changing Equipment
11124	Vending Machines
11130	**Audio-visual Equipment**
11132	Projection Screens
11134	Projectors
11136	Learning Laboratories
11140	**Vehicle Service Equipment**
11142	Vehicle Washing Equipment
11144	Fuel Dispensing Equipment
11146	Lubrication Equipment
11150	**Parking Control Equipment**
11152	Parking Gates
11154	Ticket Dispensers
11156	Key and Card Control Units
11158	Coin Machine Units
11160	**Loading Dock Equipment**
11161	Dock Levelers
11162	Dock Lifts
11163	Portable Ramps, Bridges and Platforms
11164	Dock Seals and Shelters
11165	Dock Bumpers
11170	**Solid Waste Handling Equipment**
11171	Packaged Incinerators
11172	Waste Compactors
11173	Bins
11174	Pulping Machines and Systems
11175	Chutes and Collectors
11176	Pneumatic Waste Systems
11190	**Detention Equipment**
11200	**Water Supply and Treatment Equipment**
11210	Pumps
11220	Mixers and Flocculators
11225	Clarifiers
11230	Water Aeration Equipment
11240	Chemical Feeding Equipment
11250	Water Softening Equipment
11260	Disinfectant Feed Equipment
11270	Fluoridation Equipment
11280	**Hydraulic Gates and Valves**

11285	Hydraulic Gates
11295	Hydraulic Valves
11300	**Fluid Waste Treatment and Disposal Equipment**
11302	Oil/Waste Separators
11304	Sewage Ejectors
11306	Packaged Pump Stations
11310	Sewage and Sludge Pumps
11320	Grit Collecting Equipment
11330	Screening and Grinding Equipment
11335	Sedimentation Tank Equipment
11340	Scum Removal Equipment
11345	Chemical Equipment
11350	Sludge Handling and Treatment Equipment
11360	Filter Press Equipment
11365	Trickling Filter Equipment
11370	Compressors
11375	Aeration Equipment
11380	Sludge Digestion Equipment
11385	Digester Mixing Equipment
11390	Package Sewage Treatment Plants
11400	**Food Service Equipment**
11405	Service Storage Equipment
11410	Food Preparation Equipment
11415	Food Delivery Carts and Conveyors
11420	Food Cooking Equipment
11425	Hood and Ventilation Systems
11430	Food Dispensing Equipment
11435	Ice Machines
11440	Cleaning and Disposal Equipment
11445	Bar and Soda Fountain Equipment
11450	**Residential Equipment**
11452	Residential Appliances
11454	Built-in Ironing Boards
11458	Disappearing Stairs
11460	**Unit Kitchens**
11470	**Darkroom Equipment**
11472	Transfer Cabinets
11474	Darkroom Processing Equipment
11476	Revolving Darkroom Doors
11480	**Athletic, Recreational and Therapeutic Equipment**
11482	Scoreboards
11484	Backstops
11486	Gym Dividers
11488	Bowling Alleys
11490	Gymnasium Equipment

11492	Exercise Equipment
11494	Therapy Equipment
11496	Shooting Ranges
11500	**Industrial and Process Equipment**
11600	**Laboratory Equipment**
11650	**Planetarium Equipment**
11660	**Observatory Equipment**
11680	**Office Equipment**
11700	**Medical Equipment**
11710	Medical Sterilization Equipment
11720	Examination and Treatment Equipment
11730	Patient Care Equipment
11740	Dental Equipment
11750	Optical Equipment
11760	Operating Room Equipment
11770	Radiology Equipment
11780	**Mortuary Equipment**
11850	**Navigation Equipment**
11870	**Agricultural Equipment**
12	**FURNISHINGS**
12050	**Fabrics**
12100	**Artwork**
12110	Murals
12120	Wall Decorations
12140	Sculpture
12160	Ecclesiastical Artwork
12170	Stained Glass Work
12300	**Manufactured Casework**
12301	Metal Casework
12302	Wood Casework
12304	Plastic Laminated Faced Casework
12345	Laboratory Casework
12350	Medical Casework
12360	Educational Casework
12370	Residential Casework
12380	Specialty Casework
12500	**Window Treatment**
12510	Blinds
12515	Interior Shutters
12520	Shades
12525	Solar Control Film
12530	Curtain Hardware
12540	Curtains
12600	**Furniture and Accessories**
12605	Portable Screens

78

GENERAL REQUIREMENTS

GENERAL REQUIREMENTS

13500	**Recording Instrumentation**
13510	Stress Instrumentation
13515	Seismic Instrumentation
13520	Meteorological Instrumentation
13550	**Transportation Control Instrumentation**
13560	Airport Control Instrumentation
13570	Railroad Control Instrumentation
13580	Subway Control Instrumentation
13590	Transit Vehicle Control Instrumentation
13600	**Solar Energy Systems**
13610	Solar Flat Plate Collectors
13620	Solar Concentrating Collectors
13625	Solar Vacuum Tube Collectors
13630	Solar Collector Components
13640	Packaged Solar Systems
13650	Photovoltaic Collectors
13700	**Wind Energy Systems**
13750	**Cogeneration Systems**
13800	**Building Automation Systems**
13810	Energy Monitoring and Control Systems
13815	Environmental Control Systems
13820	Communications Systems
13825	Security Systems
13830	Clock Control Systems
13835	Elevator Monitoring and Control Systems
13840	Escalators and Moving Walks Monitoring and Control Systems
13845	Alarm and Detection Systems
13850	Door Control Systems
13900	**Fire Suppression and Supervisory Systems**
13950	**Special Security Construction**
14	**CONVEYING SYSTEMS**
14100	**Dumbwaiters**
14110	Manual Dumbwaiters
14120	Electric Dumbwaiters
14140	Hydraulic Dumbwaiters
14200	**Elevators**
14210	Electric Traction Elevators
14240	Hydraulic Elevators
14300	**Escalators and Moving Walks**
14310	Escalators
14320	Moving Walks
14400	**Lifts**
14410	People Lifts
14420	Wheelchair Lifts

GENERAL REQUIREMENTS

15250	Mechanical Insulation
15260	Piping Insulation
15280	Equipment Insulation
15290	Ductwork Insulation
15300	**Fire Protection**
15310	Fire Protection Piping
15320	Fire Pumps
15330	Wet Pipe Sprinkler Systems
15335	Dry Pipe Sprinkler Systems
15340	Pre-action Sprinkler Systems
15345	Combination Dry Pipe and Pre-action Sprinkler Systems
15350	Deluge Sprinkler Systems
15355	Foam Extinguishing Systems
15360	Carbon Dioxide Extinguishing Systems
15365	Halon Agent Extinguishing Systems
15370	Dry Chemical Extinguishing Systems
15375	Standpipe and Hose Systems
15400	**Plumbing**
15410	Plumbing Piping
15430	Plumbing Specialties
15440	Plumbing Fixtures
15450	Plumbing Equipment
15475	Pool and Fountain Equipment
15480	Special Systems
15500	**Heating, Ventilation and Air Conditioning (HVAC)**
15510	Hydronic Piping
15515	Hydronic Specialties
15520	Steam and Steam Condensate Piping
15525	Steam and Steam Condensate Specialties
15530	Refrigerant Piping
15535	Refrigerant Specialties
15540	HVAC Pumps
15545	Chemical Water Treatment
15550	**Heat Generation**
15555	Boilers
15570	Boiler Accessories
15575	Breechings, Chimneys and Stacks
15580	Feedwater Equipment
15590	Fuel Handling Systems
15610	Furnaces
15620	Fuel-fired Heaters
15650	**Refrigeration**
15655	Refrigeration Compressors
15670	Condensing Units

GENERAL REQUIREMENTS

16140	Wiring Devices
16150	Manufactured Wiring Systems
16160	Cabinets and Enclosures
16190	Supporting Devices
16195	Electrical Identification
16200	**Power Generation – Built-up Systems**
16210	Generator
16250	Generator Controls
16290	Generator Grounding
16300	**Medium Voltage Distribution**
16310	Medium Voltage Substations
16320	Medium Voltage Transformers
16330	Medium Voltage Power Factor Correction
16340	Medium Voltage Insulators and Lighting Arrestors
16345	Medium Voltage Switchboards
16350	Medium Voltage Circuit Breakers
16355	Medium Voltage Reclosers
16360	Medium Voltage Interrupter Switches
16365	Medium Voltage Fuses
16370	Medium Voltage Overhead Power Distribution
16375	Medium Voltage Underground Power Distribution
16380	Medium Voltage Converters
16390	Medium Voltage Primary Grounding
16400	**Service and Distribution**
16410	Power Factor Correction
16415	Voltage Regulators
16420	Service Entrance
16425	Switchboards
16430	Metering
16435	Converters
16440	Disconnect Switches
16445	Peak Load Controllers
16450	Secondary Grounding
16460	Transformers
16465	Bus Duct
16470	Panelboards
16475	Overcurrent Protective Devices
16480	Motor Control
16485	Contactors
16490	Switches
16500	**Lighting**
16501	Lamps
16502	Luminaire Accessories
16510	Interior Luminaries
16520	Exterior Luminaries

16535	Emergency Lighting
16545	Underwater Lighting
16580	Theatrical Lighting
16600	**Special Systems**
16610	Uninterruptible Power Supply Systems
16620	Package Engine Generator Systems
16630	Battery Power Systems
16640	Cathodic Protection
16650	Electromagnetic Shielding Systems
16670	Lightning Protection Systems
16680	Unit Power Conditioners
16700	**Communications**
16720	Alarm and Detection Systems
16730	Clock and Program Systems
16740	Voice and Data Systems
16770	Public Address and Music Systems
16780	Television Systems
16785	Satellite Earth Station Systems
16790	Microwave Systems
16850	**Electrical Resistance Heating**
16855	Electric Heating Cables and Mats
16880	Electric Radiant Heaters
16900	**Controls**
16910	Electrical Systems Control
16915	Lighting Control Systems
16920	Environmental Systems Control
16930	Building Systems Control
16940	Instrumentation
16950	**Testing**
16960	Electrical Equipment Testing
16970	Electrical System Startup/Commissioning
16980	Demonstration of Electrical Equipment

CSI MasterFormat™ 2004 edition by division numbers and titles

Procurement and Contracting- Requirements Group:
Div.01 Procurement and Contracting Requirements

Specifications Group:
General Requirements Subgroup:
Div.01 General Requirements
Facility Construction Subgroup:
Div.02 Existing Conditions
Div.03 Concrete
Div.04 Masonry
Div.05 Metals
Div.06 Wood, Plastics, and Composites

GENERAL REQUIREMENTS

Div.49 Reserved for Future Expansion

Subdivision. Each section of each division of the specifications is divided into the Scope or Work Included, Materials listing the quality of all materials to be furnished, and finally, Fabrication and Erection, defining the quality of workmanship desired.

Scope includes items to be furnished and installed, items to be furnished but installed by others, items furnished by others and installed under the section, and finally, related work in other divisions of the specifications. Scope will also list some general requirements such as those for shop drawings, samples, tests, methods of delivery, and storage that may be required. No matter how detailed the scope section, it will not include each and every item. The drawings must always be checked against scope, which often carries the phrase "including, but not limited to, the following".

Materials will list the materials to be used in one of several ways, often found in combination. The closed specification will list a single trade name, and the specified product that must be furnished. The contractor's option specification (or bidder's choice) lists more than one trade name, and the contractor may choose from those listed. The substitute bid specification might list a choice of trade names and the bid must be based on one of the products included, but the contractor is allowed to suggest a substitute at the time of submitting a bid, naming the amount they would subtract for using the alternate product.

A variation is the product approval specification which asks the contractor to submit any substitutions prior to submitting a bid. If the architect approves the substitute, it will be put in an addenda sent to all contractors. This "or approved equal" type specification is the most common. It provides the widest competition and rewards the contractor who knows their products, prices, and sources.

However, it is the architect who decides whether the product is equal. If the proposed substitution is not stated as approved in an addendum, then the alternate material should not be used in the estimate.

The product description specification is used to describe bulk materials and some manufacturers' articles, the latter particularly in government work. The performance specification describes not the material but what work is required to produce strength, mechanical ability, or similar measurable results.

Specifications based on reference standards refer to those which are published by industry associations and organizations. They are encountered often, are general in nature, and usually apply only in part to the specific job. A contractor should build a library of these published specifications.

Following is a list of construction associations, many of which offer commonly referred standards. In addition, many architects will refer to the published specifications of a named manufacturer. These are readily available and generally are included in installation procedures.

GENERAL REQUIREMENTS

Fabrication and Erection will set performance standards for both shop and field work and will include, if required, sections on protection, cleaning, guarantees, warranties, maintenance, and operating instructions. The specifications of trade organizations and manufacturers' installation instructions are often made a part of this section.

Acoustical Society of America (ASA)
Melville, NY 11747
www.acousticalsociety.org

Air Conditioning Contractors of America (ACCA)
Arlington, VA 22206
www.acca.org

Air Conditioning, Heating, and Refrigeration Institute (AHRI)
Arlington, VA 22201
www.ahrinet.org

Air Diffusion Council
Schaumburg, IL 60195
www.flexibleduct.org

Air Movement and Control Association (AMCA)
Arlington Heights, IL 60004
www.amca.org

The Aluminum Association (AA)
Arlington, VA 22209
www.aluminum.org

American Arbitration Association (AAA)
New York, NY 10019
www.adr.org

American Association of Cost Engineers International (AACE)
Morgantown, WV 26505
www.aacei.org

American Association of State Highway and Transportation Officials (AASHTO)
Washington, DC 20001
www.transportation.org

American Concrete Institute (ACI)
Farmington Hills, MI 48331
www.concrete.org

American Concrete Pavement Association (ACPA)
Rosemont, IL 60018
www.pavement.com

American Concrete Pumping Association (ACPA)
Lewis Center, OH 43035
www.concretepumpers.com

American Council of Engineering Companies (ACEC)
Washington, DC 20005
www.acec.org

American Fence Association (AFA)
Glen Ellyn, IL 60137
www.americanfenceassociation.com

American Forest & Paper Association (AF&PA)
Washington, DC 20036
www.afandpa.org

The American Institute of Architects (AIA)
Washington, DC 20006
www.aia.org

American Institute of Building Design (AIBD)
Washington, DC 20045
www.aibd.org

American Institute of Steel Construction (AISC)
Chicago, IL 60601
www.aisc.org

American Institute of Timber Construction (AITC)
Centennial, CO 80112
www.aitc-glulam.org

American Insurance Association (AIA)
Washington, DC 20037
www.aiadc.org

American Iron and Steel Institute (AISI)
Washington, DC 20001
www.steel.org

American National Standards Institute (ANSI)
Washington, DC 20036
www.ansi.org

American Plywood Association (APA/EWTA)
(Engineered Wood Technology Association)
Tacoma, WA 98466
www.apawood.org

GENERAL REQUIREMENTS

American Road & Transportation Builders Association (ARTBA)
Washington, DC 20007
www.artba.org

American Society of Civil Engineers (ASCE)
Reston, VA 20191
www.asce.org

American Society of Heating, Refrigeration and Air Conditioning Engineers (ASHRAE)
Atlanta, GA 30329
www.ashrae.org

American Society of Landscape Architects (ASLA)
Washington, DC 20001
www.asla.org

American Society of Mechanical Engineers (ASME)
New York, NY 10016
www.asme.org

American Society of Plumbing Engineers (ASPE)
Des Plaines, IL 60018
www.aspe.org

American Society of Professional Estimators (ASPE)
Nashville, TN 37214
www.aspenational.org

The American Society of Safety Engineers (ASSE)
Des Plaines, IL 60018
www.asse.org

American Society for Testing and Materials (ASTM)
West Conshohocken, PA 19428
www.astm.org

American Subcontractors Association (ASA)
Alexandria, VA 22314
www.asaonline.com

American Water Works Association (AWWA)
Denver, CO 80235
www.awwa.org

American Welding Society (AWS)
Miami, FL 33126
www.aws.org

American Wood Council (AWC)
Leesburg, VA 20175
www.awc.org

American Architectural Manufacturers Association (AAMA)
Schaumburg, IL 60173
www.aamanet.org

American Coatings Association (ACA)
Washington, DC 20005
www.paint.org

Architectural Precast Association (APA)
Fort Myers, FL 33919
www.archprecast.org

Architectural Woodwork Institute (AWI)
Potomac, VA 20165
www.awinet.org

Asphalt Institute (AI)
Lexington, KY 40511
www.asphaltinstitute.org

Associated Air Balance Council (AABC)
Washington, DC 20005
http://www.aabc.com

Associated General Contractors of America (AGC)
Arlington, VA 22201
www.agc.org

Association of Home Appliance Manufacturers (AHAM)
Washington, DC 20036
www.aham.org

Association of the Wall and Ceiling Industry (AWCI)
Falls Church, VA 22046
www.awci.org

The Association of Union Constructors (TAUC)
Arlington, VA 22209
www.tauc.org

Barre Granite Association (BGA)
Barre, VT 05641
www.barregranite.org

GENERAL REQUIREMENTS

Builders Hardware Manufacturers Association (BHMA)
New York, NY 10017
www.buildershardware.com

Brick Industry Association (BIA)
Reston, VA 20191
www.bia.org

Building Stone Institute (BSI)
Chestertown, NY 12817
www.buildingstoneinstitute.org

Cast Iron Soil Pipe Institute (CIPI)
Atlanta, GA 30316
www.cispi.org

Ceilings and Interior Systems Construction Association (CISCA)
St. Charles, IL 60174
www.cisca.org

Construction Industry Institute (CII)
Austin, TX 78759
www.construction-institute.org

Construction Innovation Forum (CIF)
Walbridge, OH 43465
www.cif.org

Construction Management Association of America (CMAA)
McLean, VA 22102
www.cmaanet.org

California Redwood Association (CRA)
Pleasant Hill, CA 94523
www.calredwood.org

Carpet and Rug Institute (CRI)
Dalton, GA 30722
www.carpet-rug.org

Concrete Reinforcing Steel Institute (CRSI)
Schaumburg, IL 60173
www.crsi.org

Construction Safety Council (CSC)
Hillside, IL 60162
www.buildsafe.org

93

Construction Specifications Institute (CSI)
Alexandria, VA 22314
www.csinet.org

Design-Build Institute of America (DBIA)
Washington, DC 20004
www.dbia.org

Door and Hardware Institute (DHI)
Chantilly, VA 20151
www.dhi.org

Expanded Shale, Clay and Slate Institute (ESCSI)
Chicago, IL 60601
www.escsi.org

Finishing Contractors Association (FCA)
Bethesda, MD 20814
www.finishingcontractors.org

Gypsum Association (GA)
Hyattsville, MD 20782
www.gypsum.org

Hardwood, Plywood and Veneer Association (HPVA)
Reston, VA 20190
www.hpva.org

Illuminating Engineering Society (IES)
New York, NY 10005
www.ies.org

Indiana Limestone Institute of America (ILIA)
Bedford, IN 47421
www.iliai.com

Institute of Electrical and Electronics Engineers, Inc. (IEEE)
Piscataway, NJ 08855
www.ieee.org

International Code Council (ICC)
Washington, DC 20001
www.iccsafe.org

International Masonry Institute (IMI)
Annapolis, MD 21401
www.imiweb.org

GENERAL REQUIREMENTS

Kitchen Cabinet Manufacturers Association (KCMA)
Reston, VA 20191
www.kcma.org

Maple Flooring Manufacturers Association (MFMA)
Deerfield, IL 60015
www.maplefloor.org

Material Handling Industry of America (MHIA)
Charlotte, NC 28217
www.mhia.org

Mechanical Contractor Association of America, Inc. (MCAA)
Rockville, MD 20850
www.mcaa.org

Metal Building Manufacturers Association (MBMA)
Cleveland, OH 44115
www.mbma.com

Molding and Millwork Producers Association (MMPA)
Woodland, CA 95695
www.wmmpa.com

National Asphalt Pavement Association (NAPA)
Lanham, MD 20706
www.asphaltpavement.org

National Association of Elevator Contractors (NAEC)
Conyers, GA 30012
www.naec.org

National Association of Home Builders (NAHB)
Washington, DC 20005
www.nahb.org

National Association of Surface Finishing (NASF)
Washington, DC 20005
www.nasf.org

National Association of Women in Construction (NAWIC)
Fort Worth, TX 76104
www.nawic.org

National Building Granite Quarries Association, Inc. (NBGQA)
Washington, DC 20005
www.nbgqa.com

National Concrete Masonry Association (NCMA)
Herndon, VA 20171
www.ncma.org

National Council of Acoustical Consultants (NCAC)
Indianapolis, IN 46268
www.ncac.com

National Demolition Association (NDA)
Doylestown, PA 18901
www.demolitionassociation.com

National Electrical Contractors Association (NECA)
Bethesda, MD 20814
www.necanet.org

National Electrical Manufacturers Association (NEMA)
Rosslyn, VA 22209
www.nema.org

National Fire Protection Association
Quincy, MA 02169
www.nfpa.org

National Glass Association (NGA)
Vienna, VA 22182
www.glass.org

National Hardwood Lumber Association (NHLA)
Memphis, TN 38134
www.nhla.com

National Institute of Building Sciences (NIBS)
Washington, DC 20005
www.nibs.org

National Lime Association (NLA)
Arlington, VA 22203
www.lime.org

National Precast Concrete Association (NPCA)
Carmel, IN 46032
www.precast.org

National Roofing Contractors Association (NRCA)
Rosemont, IL 60018
www.nrca.net

GENERAL REQUIREMENTS

National Roof Deck Contractors Association (NRDCA)
Westford, MA 01886
www.nrdca.org

National Safety Council (NSC)
Itasca, IL 60143
www.nsc.org

National School Supply and Equipment Association (NSSEA)
Silver Spring, MD 20910
www.nssea.org

National Society of Professional Engineers (NSPE)
Alexandria, VA 22314
www.nspe.org

National Stone, Sand and Gravel Association (NSSGA)
Alexandria, VA 22314
www.nssga.org

National Swimming Pool Foundation (NSPF)
Colorado Springs, CO 80919
www.nspf.com

National Terrazzo and Mosaic Association (NTMA)
Fredericksburg, TX 78624
www.ntma.com

National Wood Flooring Association (NWFA)
Chesterfield, MO 63005
www.nwfa.org

North American Contractors Association (NACA)
Greensboro, NC 27404
www.infonaca.com

Painting and Decorating Contractors of America (PDCA)
St. Louis, MO 63146
www.pdca.org

Perlite Institute, Inc.
Harrisburg, PA 17110
www.perlite.org

The Plastics Industry Trade Association (SPI)
Washington, DC 20006
www.plasticsindustry.org

Plastics Pipe Institute (PPI)
Irving, TX 75062
www.plasticpipe.org

Plumbing and Drainage Institute
North Andover, MA 01845
www.pdionline.org

Porcelain Enamel Institute (PEI)
Norcross, GA 30010
www.porcelainenamel.com

Portland Cement Association (PCA)
Skokie, IL 60077
www.cement.org

Precast/Prestressed Concrete Institute (PCI)
Chicago, IL 60606
www.pci.org

Professional Construction Estimators Association of America (PCEA)
Charlotte, NC 28216
www.pcea.org

Professional Landcare Network (PLANET)
Herndon, VA 20170
www.landcarenetwork.org

Professional Women in Construction (PWC)
New York, NY 10022
www.pwcusa.org

Resilient Floor Covering Institute (RFCI)
La Grange, GA 30240
www.rfci.com

SAVE International (SAVE)
Dayton, OH 45402
www.value-eng.com

Sheet Metal and Air Conditioning Contractors National Association
(SMACNA)
Chantilly, VA 20151
www.smacna.org

The Society of Cost Estimating and Analysis (SCEA)
Vienna, VA 22182
www.sceaonline.com

GENERAL REQUIREMENTS

Society for Protective Coatings (SSPC)
Pittsburgh, PA 15222
www.sspc.org

Southern Forest Products Association (SFPA)
Kenner, LA 70065
www.sfpa.org

Southern Pine Inspection Bureau (SPIB)
Pensacola, FL 32524
www.spib.org

Stained Glass Association of America (SGAA)
Raytown, MO 64133
www.stainedglass.org

Steel Deck Institute (SDI)
Fox River Grove, IL 60021
www.sdi.org

Steel Joist Institute (SJI)
Florence, SC 29501
www.steeljoist.org

Steel Window Institute (SWI)
Cleveland, OH 44115
www.steelwindows.com

Structural Building Components Association (SBCA)
Madison, WI 53719
www.sbcindustry.com

Tile Council of North America (TCNA)
Anderson, SC 29625
www.tileusa.com

Tilt-Up Concrete Association (TCA)
Mount Vernon, IA 52314
www.tilt-up.org

Truss Plate Institute (TPI)
Alexandria, VA 22314
www.tpinst.org

Underwriters Laboratories, Inc. (UL)
Camas, WA 98607
www.ul.com

U.S. Metric Association, Inc. (USMA)
North Ridge, CA 91325
www.metric.org

Vermiculite Association (TVA)
Hampshire, UK GU35 9LU
vermiculite.org

Water Systems Council (WSC)
Washington, DC 20007
www.watersystemscouncil.org

West Coast Lumber Inspection Bureau (WCLIB)
Portland, OR 97281
www.wclib.org

Western Wood Products Association (WWPA)
Portland, OR 97204
www.wwpa.org

Wire Reinforcement Institute (WRI)
Hartford, CT 06103
www.wirereinforcementinstitute.org

Wood Truss Council of America (WTCA)
Madison, WI 53711
www.woodtruss.com

Alternates. At the time of bidding, the architect often wants to investigate the cost, either upward or downward, of a substitute product, material, or procedure, or the cost of adding or subtracting work from the contract. Such requests for alternate prices are usually noted at the end of the section of the specifications to which they refer, and are repeated at the end of Division 1 of the specifications and on the proposal form. On the typical lump sum proposal form, the Alternate follows the Statement of the Lump Sum price in a form such as the following:

If the following Alternates are accepted,

	Add:	Deduct:
Alternate No. 1	$___	$___
Alternate No. 2	$___	$___

A request for alternates complicates bidding. Sometimes, the extent of the alternate is unclear. Since the architect will accept the price change for an alternate with the same responsibility as attached to the lump sum figure, the contractor must be certain of all the possible contingencies involved should the change be accepted.

Often an alternate is put only under the one section that makes the basic substitution, but other subcontractors' work is also affected by the change. It is the contractor's responsibility to foresee this. As an example, under Carpentry, an architect might include:

"Alternate No. 1-For substituting one 3'-0"x6'-8"x1-3/4" (900 x 2 000 x 44 mm) door with frame for the two 3'-0"x6'-8"x1-3/4" (900 x 2 000 x 44 mm) doors and frame shown between the living and dining rooms, deduct $____."

While the door cost will be less, the wall construction will be more. Yet there might be no mention of this under the division for the latter.

Cash Allowances. Sometimes, the architect does not have a final decision from the owner on certain items. Rather than leave them out of the lump sum proposal, the architect will state a definite budget amount in the specification that is to be included in the bid. This is often encountered with the selection of brick, hardware, and other finish items. As an example:

"Facing brick shall be included under an allowance of $150 per M, f.o.b. job site."

Every bidder will include the same material cost, rather than just guessing what type of face brick the owner may choose and using that cost. Cash allowances are almost always only for the cost of material and applicable taxes delivered to the site. Labor to set and overhead and profit on the material should be included in the lump sum price unless specifically noted otherwise.

Upon completion of the job, the actual cost of the items furnished under Allowance shall be tabulated with any savings credited to the owner, any overruns added to the contract cost.

Unit Prices. Where quantity of materials is in doubt, but quality is known, the specification may ask for unit prices. As opposed to allowances, unit prices are set by the bidder, not the architect; they are firm and binding on the bidder, the lump sum price being adjusted up or down, only according to variation in the quantity actually used on the job.

Unit Prices are included on the proposal form for each material per unit of measure. For example, unit prices are often asked for concrete per square yard (square meter), piling per lineal foot (meter), partition block per square foot (millimeters squared), etc. These prices should be complete with all costs, profit, and overhead included.

Unit Prices must be quoted with care, even if the amount of work covered by them is small, because they will be compared directly with those submitted by other bidders. The owner may expect the same unit prices to be applicable on work that is added or deducted from the contract scope. Added work unit prices should be quoted at a different price than deducted scope unit prices.

Addenda

Few plans and specifications are perfect when they leave the architect's office, and contractors bidding the job should call to the architect's attention all discrepancies they note. These, plus changes the architect and owner may wish to make after the plans and specifications have been issued but before bids are turned in, are incorporated in the Addenda. This usually takes the form of a printed sheet or sheets the same size as the specification to which it should be attached. Addenda eventually become a part of the Contract Documents.

Even changes to the drawings often can be described and included this way. But sometimes additional drawings or sketches must be issued, or a drawing is reissued with a revision date.

Each addendum is given a number, and the changes it covers are listed in the same order as the divisions in the specification, with drawing changes last. Usually, addenda are acknowledged on the proposal form in a space provided for it. The single addendum or several addenda are also included in the Owner-Contractor Agreement.

Upon receipt of an addendum, the contractor should mark their set of the specifications and drawings with appropriate notes, or with a stamp on each affected page saying "See Addendum #_____." The Addendum will usually identify the change by division and page number and a reference to location on the page. As an example:

3:01-Scope: Add item "h) Remove existing wire fence along north edge of property."
6:02-Materials: Delete paragraph referring to concrete masonry lintels and add "...lintels in block work shall be units of sizes and shapes shown on the lintel schedule."
Dwg. Sheet A-2: Move rolling shutter door from center of Bay C-D to just north of Column B.
Dwg. Sheet A-4; delete ladder to roof.

An alternate method of incorporating addendum or amendments to the bid documents is to post the changes to the documents. This is done by adding the changes directly to the affected documents. The old sections are marked as void. The new sections are then attached by taping, gluing or stapling to the original document.

Change Orders

Change orders are modifications issued after the contract is signed. Errors on drawings and specifications are usually uncovered during bidding. If these errors or omissions are not corrected prior to the bid, the contractor should note them as exceptions in their bid documents. A contractor who submits a bid without noting these problems may very well end up having to pay for the items anyway to complete the job or face a lengthy lawsuit. Public work contracts typically do not allow for bid exceptions so all problems must be identified and resolved prior to submitting the bid.

Some change orders are necessary due to job conditions, changes required when plans are submitted for permits, substitutions made to satisfy

102

GENERAL REQUIREMENTS

the owner or mortgage holder, products not being available, or because as the job progresses, certain changes for the benefit of the job become obvious.

The contractor must cooperate in making these changes. This is set forth in Article 12 of the General Conditions. Sometimes no cost is involved. Most often, an extra is called for. This may be handled by submitting a firm price change, or by order to proceed on a time and material basis. The change order is issued by the architect but signed by the owner and contractor. The format of the change order will follow closely that of the addendum, except that costs of changes and tabulation of the new contract amount are added. The Frank R. Walker Form L-101 is a typical contract change order format and can be purchased at http://www.frankrwalker.com/c-7-contract-bid-forms.aspx.

Summary

As can be imagined, what appears on drawings will sometimes differ with what is stated in specifications. There is no legal precedent that one is more binding than the other, although some specifications will state which is to be followed in case of inconsistency.

In general, sizes, quantities, design, and location will be best taken from the drawings. Materials, quality, procedures, and general requirements are most reliably defined in the specifications. Never rely solely on one or the other. A drafts-persons delineation of a stone wall in elevation may not show the coursing called for in the specifications at all, while the specifier might not make special mention of a stone sill condition carefully detailed on the drawings. One must always complement the other.

SETTING UP THE ESTIMATE

Estimating is one of the most important aspects of a building contractor's business. In nearly every instance it is necessary for a contractor either to estimate the cost of the work, or to bid in competition with others before a contract for the work is awarded. If a contractor's bid is too low and a contract is awarded on the basis of it, the contractor often must complete the work without profit and sometimes at a loss in time and money.

A contractor must know the size, complexity, and quality of work that their firm can efficiently handle. Exceeding these limits often leads to a poor contract, which can be costly not only monetarily but also in terms of dissatisfied customers and loss of future opportunities for bidding.

Most contractors fall into categories as to the type of work which their organizations can handle efficiently, expeditiously, and profitably. A heavy construction contractor specializing in structures such as airport runways, highways, bridge abutments, etc., usually should forego bidding on such work as churches, banks, hospitals, or modern office buildings. A contractor whose organization and personnel are geared to a particular category of construction can find it difficult if they stray from their specialty.

Many of the smaller contractors in business today are former employees of contracting firms who have gone into business for themselves. Many of these people doubtless were exceptional mechanics, foremen or superintendents, and had a thorough understanding of how the work should be carried out in the field. Sometimes they lack business ability and training to estimate costs accurately. Some of these contractors will make a reasonable profit on their work and remain in a similar sized business. Others will learn the importance of estimating and maintaining accurate cost records to grow their business. Unfortunately many more of them will fail and eventually go back to work for someone else.

There are various reasons for these failures, but probably the most common one is the inability of the person estimating costs to come up with realistic and profitable estimates. The procedures involved in preparing an accurate, detailed cost estimate necessary for a realistic bid are relatively simple. Yet almost all instances of failure of contracting firms are the result of poor estimating practices, resulting in the submission of unsound and unprofitable bids. There is no substitute for complete and detailed material take-off and pricing and for cross-checking estimated costs with actual costs so that adjustments in pricing can be made for future bidding.

Another common reason for failure in the contracting business is overexpansion. Again, the contractor must know the capabilities of their office and supervisory staff and how much they can adequately handle at any one time. The successful contractor does not take on more work than their staff can take care of.

Bonding and insurance companies have certain minimum requirements that contractors applying for surety bonds or contractual liability insurance must meet. Some of the most important considerations that such companies make before bonding an applicant are:

1. The technical ability of the contractor applying for a bond, whether or not they understands the business thoroughly, and their capabilities in the preparation of detailed cost estimates;
2. The reputation of the contractor for honesty and their standing among those with whom they does business;
3. The financial ability of the contractor, and whether or not they have sufficient capital at their disposal to carry on the business, purchase materials and supplies, meet their payrolls and current expenses, etc.

Bonding companies seldom get the opportunity of bonding the best contracting firms. On public work projects the bidding is usually open to all bidders and in some instances as many as twenty or more bids may be received on one job. In most cases, the lowest responsible bidder is awarded the contract. The low bidder often may be the poorest risk because of substandard estimating procedures, lack of technical ability to perform the work as specified, and/or a lack of financial responsibility. On privately owned projects, the risk to the bonding company may not be as great, because the bidding list can be controlled. Only contractors with the proper credentials are usually asked to submit proposals.

Architectural and engineering firms that handle work for private owners are often asked to select contractors to bid on a project. They will attempt to choose bidders whom they know can perform the work equally well and will carry out the intent of the drawings and specifications in the best interest of the owner.

Others besides the contractor attach importance to a correct and detailed estimate. The success of a contractor depends as much on a well prepared estimate as it does on the manner in which they purchase materials, lets the subcontracts, or carries out the work on the job.

The Detailed Estimate

The skill level of the estimator determines the reliability of the estimate. If similarly skilled estimators are preparing estimates on the same project, they will attain similar results. If one of the bids is substantially low or exceptionally high, there may be many reasons for this. One of them may well be the skill level of the estimator. Other reasons could include the approach taken in determining how to perform the work, preferential pricing from major subcontractors and suppliers or economic conditions of the company. While no two estimators will have the same cost, they will have similar estimated costs based on their comparable experiences. The following preconstruction estimates are prepared by contractors, construction management firms, architects, engineers and independent cost estimators. The skill level and professional quality of the estimates prepared by these firms and individuals can vary widely. Estimates prepared by estimating professionals are generally the best quality. Construction estimates prepared by inexperienced estimators or non-estimating personnel will have less reliable results.

The estimating of a project should include the amount of detail represented in the plans and specifications. The basic estimate line item detail consists of a quantity, a unit hour, total hours, labor dollars, unit material and total material costs. Other items to consider are equipment unit cost and total equipment costs. If an item of work is to be subcontractor performed work item there will be a subcontractor price. For example, it might take 8 hours to set 100 lineal feet (30 meters) of 1/2" (12.50 mm) rigid conduit. The material will be priced separately. Using line item details such as these will provide the most reliable estimate results. There are variations on this where the unit hours may be crew rate based production hours. This is the crew method. For example, a concrete pouring crew of 8 workers will place 100 cubic yards (76.46 cu. meters) of concrete per day. Then again there may be labor unit cost rates based upon historical labor dollars if a contractor is only working in a local area. This method can be the least reliable method of estimating line item costs.

Each contractor needs to develop their own unit labor hour and costs from historical cost records that reflect their own company's particular experience and expertise. This expertise can vary widely from company to company and it is up to the estimator to develop the estimates reflecting this cost history. In the final analysis of estimates and bids, they should be accepted as representing the projected final costs for performing certain prescribed work within a stipulated time schedule, taking into consideration all of the variables that a prudent contractor can anticipate during a construction contract based on information available at the time of bidding. To prove the validity of the bid, one must consider the extreme risks that the contractor is willing to assume by furnishing payment and performance bonds and by entering into a legal contract dedicating the technical and financial resources of the firm to a successful completion of the contract in hopes of realizing a modest profit.

Estimating can be a time consuming and costly operation. Under no circumstances should an estimator or contractor produce a less than totally professional estimate that will stand the test of proving each element contained in the estimate. In line with this, if there is not enough time to correctly prepare the estimate, the contractor should give serious consideration to not bidding the project. Mistakes in the estimate/bid preparation period will have long lasting consequences. These consequences include missing out on a contract because the estimate prices developed were too high. It also includes losing money because the estimated costs were too low.

Estimate Types

There are several types of cost estimates that should be required before the bid estimate is needed. These estimates are to make sure the proper budget is developed and the budget is maintained during the design period. In many situations the contractors or estimators will be called in to prepare these estimates. Some owners rely upon construction managers to provide these as part of the preconstruction services. Some rely upon the architects or engineers to provide the estimating prior to the bid period. These estimates may be performed by or in conjunction with the owner's staff, contractors, subcontractors, professional estimators, or the estimating section of the engineering or architectural firm involved on the project. Each type of cost estimate should serve the purpose for which it was intended.

In building construction there are several significant estimating milestones. These are commonly in conjunction with the design milestones of the project. The first estimate is usually identified as the Budget/Feasibility Estimate. The next is the Schematic Design Milestone Estimate followed by the Design Development Milestone Estimate. After these estimates is the Construction Documents Milestone Estimate. These are applicable to all types of projects. An effective cost control estimating package on a design-bid-build project will differ substantially from a design build project.

GENERAL REQUIREMENTS

Two major cost societies have identified these milestones as the Levels of the Estimate. These are the American Society of Professional Estimators (ASPE) and Association for the Advancement of Cost Engineering International (AACEI). ASPE identifies six major milestones and AACE five Classes of Estimates. ASPE uses the common terms above for identification while AACEI uses Class 1 through 5 as their identification system. Others in construction have used different terms but are not as common as these two societies.

In addition to these milestones there may be cost studies (estimates) prepared in between or during these milestones to keep the project budget on track or for other purposes. Some of these studies may be for value analysis to reduce the costs of the project. If this is the case, to be highly effective these studies should be prepared prior to the Design Development Milestone. One of the reasons for this is to keep any redesign efforts to a minimum so the project schedule can be maintained.

Budget/Feasibility Estimate. The budget or feasibility estimate, once it has been developed, is effectively cast in stone. No matter what happens throughout the life of the project, the costs stated in the budget estimate are what everyone will quote until the project is completed. If the budget estimate does not accurately represent the work to be performed, then it might mean one of two things.

If costs have been overstated the project might be eliminated before it gets started. On the other extreme if the project starts with insufficient funding, the work might have to stop in midstream, or it will be necessary to borrow funds from other sources, delaying the start of other work.

While the budget estimate does not reduce the importance of the other types of cost estimates, it is usually the governing cost estimate of the project. At this stage the project either gets the green light or dies in committee. The subsequent estimates in the preconstruction period are used to confirm the project budget is being maintained or if there are variances that they are acceptable.

In order to minimize budget problems this estimate, along with the entire series of estimates, needs to identify the scope of the project in as much detail as practical. An estimate that is prepared by using allowances is no better than a guess. Defining the scope is a straightforward process. The estimator needs to work with the design team and owner to determine this scope. Additionally the estimator must have good historical cost information. The scope outline for a building can follow the customary divisions and subdivisions in format. Depending upon the building type and scope the budget estimate can be one page with a few line items to several pages with numerous line items.

When preparing the budget on a building project it is imperative to know the end use of the building. The costs to build a retail strip center are vastly different from a high rise office building. Similarly a hospital has different costs than a museum.

The estimate pricing format needs to avoid the use of lump sum, all inclusive allowances for pricing. By using the building area, wall areas, site size and other quantities, the estimate scope will be defined, even at this early stage. While unit hours and other itemized costs may not be practical at this time, historical unit costs are certainly appropriate. When using these unit costs care needs to be used in their application. If the costs are from a different type of project they may require substantial modification for use on the current project.

The importance of an estimate review cannot be emphasized enough at this milestone. The scope of the budget estimate must be confirmed by the owner as representative of their needs and requirements on the project. A guide to consider in developing and reviewing the budget is if you do not see a line item – it is not in the budget.

Estimate reviews may present other obstacles to having successful budgets developed by the contractor. There may be "pressure" to change the estimated budget amount. This can be due to a wide variety of reasons including corporate politics. If changes are made to the budget for invalid reasons the ones to suffer will be the estimators that relented and the project.

Schematic Design milestone Estimate - 3% to 5% Overall Design Completeness. At this stage the estimator has preliminary scope and drawing information. This will typically include floor plans and elevations. There may be outline specifications available. If the specifications are not available there may be a Program Statement for the building. A Program Statement is a document that identifies the needs and requirements on the project. In many cases it identifies the uses for each room and may have preliminary scopes for interior and exterior finishes.

Using the information from the Budget/Feasibility cost estimate, the estimator also uses this new information, along with discussions with the owner's staff and the A/E, to expand to this level of estimating.

The format of this estimate needs to conform to the format of the previous estimate. This enables the estimator to verify the scope is proper for the project. It also enables the estimator to readily identify the scope and cost changes that will occur from the previous estimate to this estimate. As a part of the identification process, the reason for the change needs to be identified as this information is presented to the building owner.

While the previous estimate may have relied upon general or parametric quantities and unit prices, it is now appropriate to use comprehensive information for each line item of the estimate.

The quantity takeoff or item list generated in the budget estimate needs to be developed to a greater extent in this and subsequent estimates. Individual items of work are broken down, where possible, into units of cost for labor, equipment, material, and subcontractor. While it may be easy to use unit prices and general quantities at this estimate milestone, they are to be avoided as much as possible,

108

GENERAL REQUIREMENTS

An example of the level of detail required at this milestone can be illustrated with concrete pad footings in the budget estimate. The budget estimate would have a line item of: Pad Footings - 200 Cubic Yards with a unit price that included labor, material and equipment costs associated with the pad footings. The Schematic Design Milestone Estimate should have quantities developed for these Pad Footings with line items such as: Excavation (Cu Yds), Fine Grade (Sq Ft), Formwork (Sq Ft), Rebar (Lbs), Concrete (Cu Yds), Anchor Bolts (Each), Finish (Sq Ft) and Cure & Protect (Sq Ft). These quantities will have unit & total hours, hourly & total labor dollars, and unit & total material dollars. Equipment (not small tools) required for the line item may be included as a part of the line item with a unit & total equipment dollar value. It may also be listed as a separate line item with the equipment identified by type and size with a number of operating hours. These hours would then have an hourly and total dollar value.

Prior to finalizing this or any other estimate the estimator should conduct a site investigation. The basic determinations on how the work will be performed are reviewed and refined or revised as required.

Design Development Milestone Estimate - 35% to 50% Overall Design Completeness. This cost estimate is an expansion on the detail of the previous Schematic Design Milestone cost estimate. The same principals apply to this as mentioned for the previous estimate. **70% to 98% Construction Documents Milestone Estimate(s).** At the 70% to 98% design completion milestone, at least one estimate is required prior to issuing documents to the contractors for bidding. However, depending upon the project conditions, two or more estimates may be required in this design stage. Projects with problems meeting the budget or having design completeness issues may require these additional estimates. Again these estimates are prepared on a highly detailed basis with itemized quantities for virtually all of the project scope. This estimate uses the same format as the previous type of cost estimate.

Construction Documents Milestone and Bid Cost Estimate - 100% Overall Design Completeness. These are two different estimates prepared at the same time, by separate entities. The first is the Construction Documents Milestone Estimate. It is prepared by the Owner's Representative (Construction Manager, Architect, Engineer or Estimator). This estimate is used to evaluate the bids being submitted by the contractors. While this estimate can rely upon vendor quotes for materials, this estimate must be prepared independently of any bidder on the project. This is the only method to determine if the bids being submitted are reasonable (not too low or too high). It is also used to verify the control budget on the project for payment and billing purposes.

The Bid Cost Estimate is prepared by contractors and subcontractors submitting bids on the project. The prime contractors will prepare detailed estimates for their own self-performed work scope and receive bids from

109

subcontractors on their scopes of work. The prime contractor will then assemble the bid into the lowest practical price. This is then submitted as the bid cost estimate.

Both of these estimates are prepared in the greatest amount of detail practical. In general terms this mean 65% to 75% of the identified general trade items can be purchased right from the estimate sheets. The mechanical and electrical detail should be such that 90 to 95% of the listed items can be purchased directly from the estimate sheets.

The use of this detailed estimate for the contractor extends far beyond being the basis of a bid. It serves as the guide for the purchase of materials and the award of subcontracts. It also provides the information for the establishment of a comprehensive cost control system so that project progress can be monitored to determine if the estimated material and the labor costs are consistent with the actual costs incurred during performance of the contract.

Post-Bid Pre-Award Requirements.

Review of Unit Price Estimates. In examining unit price contracts, the estimator examines the total price as well as the individual unit quantities and prices. Significant variations from the estimate, or from other bidders, should be further investigated by checking on estimated quantities, computations, and basic assumptions about construction methods. Where the bid of the apparent responsible low bidder is within an acceptable range of the estimate, the estimator should focus on the unit prices in order to identify imbalances in unit vs. time, "front loading", or in creating a "windfall" where there might be significant variations in the actual quantities of a particular item of work.

Review of Lump Sum Bids. In reviewing bids on lump sum contracts, the estimator should initially determine whether the apparent low bid varies significantly from their estimate, or from the other bidders. The estimator then prepares for the contractor's submission of the detailed breakdown, which is compared with the detailed estimate to verify the estimate and to identify possible imbalances. Where there is a significant variation between bids and the estimate, or where there are wide variations among the bidders, the estimator attempts to determine the cause.

Bid/Estimate Reviews - It is normal practice for an estimator working on behalf of the project owner to review bids on a project and to compare them with the estimate of the work. The estimator attempts, where possible, to identify variations from the estimate and variations among the bidders. Significant differences are investigated to check the estimating system and to identify possible alternative methods of performing the work. The usual output of such a review would be a recommendation to the owner to either award a contract or not award one.

110

GENERAL REQUIREMENTS
Estimate Formats

Building construction is based upon trades such as general, mechanical, electrical and other specialty subcontractors performing certain scopes of work on a project. These scopes generally follow what is commonly known as the Construction Specifications Institute divisions of work. These are typically in a 16 division (pre-2004) format or 49+ divisions (post-2004) format. Another format is known as UniFormat. UniFormat does not follow the trade breakdowns. This organization follows major components of a building such as basement, shell and superstructure.

When preparing estimates for a project in the pre-construction period and at bid time, the best format is one that provides the best level of information for making cost decisions.

Pre-construction estimates often use historical cost information generated by trade (CSI 16/49) contracts. The use the CSI 16/49 format makes it simpler to research and apply historical costs. It makes it easier track costs from the budget through the bidding stage. The CSI 16/49 format lends itself to cost effective estimating services in the preconstruction period.

At bid time the estimate/bid format needs to be kept as simple as possible. A bid with few alternates will encourage bidders to bid on a project. A bid with few breakouts will do the same. More alternates and more breakouts not only hinder getting good bidders, they can lead to mistakes in bids.

In some cases unit prices and alternates are unavoidable. Sometimes they are warranted. A unit price may be required when soil conditions preclude having firm prices for caissons and piles. One or two clear cut alternates may be appropriate due to the project conditions. In any case, unit prices and alternates should be minimized.

Estimate Check Lists and Practices

Estimates are done under the pressure of a deadline. Even in the best of circumstances errors will crop up. When using pre-packaged software, custom spreadsheets or even paper and pencil, another person should track the numbers to make sure they were done correctly. Errors may not be obvious with software. A lot of people think if it looks good it must be good. It is not unusual to receive a bid from a subcontractor with an obvious error in the figures. This is an indication that the bid was not checked prior to tendering, and it could severely affect the company offering the bid and the general contractor using the bid as part of the general estimate.

Check your profit. How much of this is certain, and how much contingent on optimism? Set forth a definite percentage for contingencies and try to hold to it. Keep the profit apart. If there is a questionable item, alert the owner and architect in writing, and then if possible, qualify your bid accordingly. Some proposals provide a space for this. Give a firm price on what is clear. Set up an allowance to cover what is in doubt with a proposal, stating what is to be charged against the allowance and why.

Finally, make certain all items are covered. Check not only the drawings and specifications, but also any bulletins addenda. An addendum changing wood paneling from full height to wainscot in an office means a savings in millwork. But now, of course, the drywall above needs to be taped and finished for painting. So both the drywall and painting contractors must be alerted, even though the addendum only refers to a change in the millwork.

For a comprehensive check list of direct scope estimate items, based on the Construction Specifications Institute (CSI) 16/49 divisions, refer to the Specifications section earlier in this chapter.

The following listing is intended to cover every type of structure from the smallest residence to the largest skyscraper, commercial or industrial plant. Compare it with your estimate before submitting your bid, because it might be the means of including some items that would otherwise be omitted from your estimate. This may require a few minutes extra time, but always remember, it is much better to be safe than sorry. Forgetting an item has cost many a contractor their profit. Also remember, item checklists are there to prevent repeat mistakes!

A master checklist for every estimator should include the Bid Document Inventory, Estimating Assignment, Direct Estimate, Wage Rate Development, Bid Document Reviews, Takeoff General Practices, Labor Hour General Practices, Material Pricing General Practices, Subcontractor Dealings General practices, Formatting the Estimate/Bid, Estimate Reviews along with Bid Statistics/Evaluations.

The Bid Document Inventory Checklist is used to make sure that all of the bid documents are present, the correct revision and legible. It does not matter if they are being used electronically or on a hardcopy basis. Check the drawings to verify all of them are included with the correct revision number. It is not unheard of to having missing drawings and incorrect revision numbers. Make sure the specifications are complete with all sections and pages. There may be missing or illegible sheets/pages no matter what format. Notify the issuer immediately if there are any discrepancies. The same holds true for each addenda or amendment to the bid documents. Additionally the estimator must mark the documents are void or superseded so they are not used incorrectly in the preparation of the estimate/bid.

The Estimating Assignment Checklist is developed to prepare the estimate in an efficient manner. It does not matter if it is one estimator or several estimators; the estimate preparation time has to meet the bid deadline. A standard checklist having the activity required with the beginning and end times is a good start. Activities would include items such as Decide to Bid, Order Plans and Specifications, Determine Subcontracting Areas and Send Requests for Quotes, Takeoff Materials, Contact Material Vendors, Develop Wage Rates, etc., ending with Submit Bid on the bid day.

The Direct Estimate Checklist is developed to make sure the entire scope of work is bid. In this the scope packages are developed for the subcontractors. Records are kept as to which subcontractors are responsible for bidding each package. These are monitored during the bid period to make sure the subcontractors are bidding or not bidding and to respond to questions that develop during the bid period. These are coordinated with the self-performed estimate scope by the prime contractor. The same thing is done for material and equipment suppliers on the project as the quantities are developed and finalized.

The Wage Rate Development Checklist is used to develop new or confirm existing hourly rates for the craft and staff labor on the project. Craft and staff hourly rates include three main components. These are wages, fringe benefits and payroll burdens.

The craft rates will include the base wage per hour. It may also need to consider pay items such as overtime, shift differential pay, high pay and/or toxic pay. These are premiums to the base wage. Sometimes overtime is used as an incentive to attract craft labor to a project. All pay items are considered taxable.

The staff rates will include the base wages per hour and/or per week or month, depending upon the pay classification. Management and supervisory personnel are generally paid on a weekly or month salary basis. Others are paid on an hourly rate for their services. Overtime is generally applicable for the hourly rate staff and not the management and supervisory staff.

Fringe benefits may vary from craft to craft and definitely will vary from open shop (non-union) to union agreement fringe benefits. Fringe benefits will include items such as Health and Welfare, Vacation and Holiday, Pensions, Apprenticeships, legal and other items. The fringe benefits may also include travel pay and/or subsistence for distant projects. Sometimes subsistence is used by contractors as a method to attract craft labor to a project, similar to overtime. Generally not all fringe benefits are taxable. Some fringe benefits, such as Vacation and Holiday, will be taxable.

Fringe benefits for staff can be similar to the craft. They may include sick and vacation time, holidays, pension or contributions such as to 401k accounts. Other fringe benefits may include a vehicle, cellular phone, or expense account. These fringe benefits may or may not be taxable events.

The third portion of the craft and staff hourly rates is the payroll burden. The payroll burdens are the taxes and insurances mandated by local, state or federal law. A portion of it is also from company operations. These are applied to the wages and/or fringe benefits. Some are applied to all wages or to only the straight time portion if overtime is involved. Some are applied to the first x dollars of annual income, not the total annual income. Payroll burdens include Social Security (FICA) tax, Medicare insurance, Federal Unemployment Insurance (FUTA), State Unemployment Insurance (SUI), Worker's Compensation Insurance (WC) and company insurance coverage for items such as property loss and property damage.

The best solution for determining the fringe benefits and payroll burdens applicable to each craft and staff rate category is to consult with the accountant, insurance and legal advisors for your firm. The following briefly covers some of the burdens associated with hourly rates:

The Federal Insurance Contributions Act, F.I.C.A. (Social Security), requires contractors to pay a tax on all wages. For 2013 the employee's share of this tax is 6.20% on the first $113,700 of taxable wages paid to each employee for the year (2013). Medicare basic hospital insurance 1.45% on all wages is deducted. These are deducted from the employee gross wage amounts.

The employer is also obliged to contribute 6.20% from each employee's salary or wages (capped to first $113,700). This cost must be included in the hourly rate. In addition to the social security tax, the employer must also contribute to the Medicare tax at a rate of 1.45% on wages or salary.

Regular FICA	6.20% of wages up to $113,700 annually
Medicare Tax	1.45% of wages (no limit)
Total	7.65%

Federal Unemployment Tax ACT (FUTA). This tax applies to the first $7,000 of wages paid each employee during the calendar year 2013. The rate is 6.0% but a credit of up to a maximum 5.4% of total wages for contributions paid into State Unemployment Funds for a total federal tax rate of 0.6%. Federal Unemployment Tax is imposed on employers and must not be deducted from wages of employees.

State Unemployment Tax. This amount (percentage) will vary from state to state and companies past record of claims. Note above that a credit is allowed against the federal rate.

Worker's Compensation. This insurance provides the contractor's employees with the benefits specified by the various state and federal compensation laws in the event of injury or death arising out of the employment. Most compensation acts contain at least some provisions specifying benefits for the most common occupational diseases.

Rates vary widely among the states, craft labor and staff labor categories. In states where medical benefits are limited, it may be advisable to carry full or extra-legal medical coverage.

Contractor's Public Liability Insurance protects against liability for bodily injury to a member of the public arising out of the contractor's operations and against property damage liability. It can be had under either a schedule general liability form policy or a comprehensive general liability form.

Under the schedule form, the several hazards to be insured are specifically named, and the coverage may be applied to specified locations and operations of the insured contractor. The comprehensive form is broader

114

in nature and covers all the usual hazards of a general liability nature other than those specifically excluded.

Insurance companies are almost always willing to expand coverage conditions to cover excluded risks or to amend with broader terms the definition of exclusions. The broad form comprehensive general liability endorsement is an example of a large package of additional coverage.

Property Damage Insurance. The contractor should make a careful study of what their responsibilities are in regard to damage to other property and to the work completed so that their policies will mesh with those carried by the owner.

Takeoff General Practices must be established to minimize any problems with defining the work. Items should be listed or identified with enough information to clearly represent the work. It does not do any good to list an item as Concrete – 10 cubic yards. The proper way to list the materials is to have enough detail to quote or purchase the materials. This minimizes the time it takes to price out the work and the opportunity to make a mistake.

In addition to this the takeoff must consider many elements besides a simple listing of the materials. As it progresses were any interference uncovered. Are there conflicts between the drawings and specifications? Were any areas showing scope duplications or omissions? How did the plan views compare to the details and sections? Have there been any allowances in the quantities? Have questions been submitted? Have answers been received? Have adjustments been made due to the answers?

Labor Hour General Practices must be followed for each estimate. First priority is to use historical information whenever possible. This information must be adjusted to match the current project conditions. Comparing the current project to previous projects is important in this adjustment. Labor for items in a power plant is not comparable to labor in a strip center. Not only should there be specific comparisons to other jobs but compare to the overall company average or range of labor hours encountered previously. If no labor history is available use a crew analysis to determine the hours required for the item of work.

Material Pricing General Practices involves dealing with vendors. Identify the materials required and provide the listing of materials to them for pricing in the estimate. The general preference is for vendors to provide a price for each line item rather than quoting all of the items on a lumps sum or firm price basis. This should help reduce the material costs on the project since the vendor will not be responsible for quantity variances. Others will argue against this because they are fearful their quantities will be released to the competition due to unethical vendors.

Material listings need to be released to vendors as soon as possible in the bid period. Oftentimes vendors will provide contractors with standard prices to be used in bids with each bid having different adjustment factors. Many mechanical and electrical contractors use pricing services of this type for their bulk materials. When vendor quotes are received they need to be documented if not in writing. A vendor quote should be documented containing the firm, date, representative and scope of the quote. It should note if the taxes are included or excluded. Notes should be made concerning the delivery time and if there are extra charges for getting the material and/or equipment delivered to the project location. If delivery is to a warehouse the estimate must include the time and equipment to get the materials to the project site.

Subcontractor Dealings Practices cover dealing with subcontractors in an aboveboard ethical manner. First and foremost is that their pricing should never be released to their competitors. This is considered bid shopping and while not illegal is highly unethical in the construction industry. Bid shopping is one of the fastest ways to ruin your reputation in the business. It will also preclude you from receiving the best prices on future projects from the subcontractors and material vendors.

In dealing with subcontractors be clear on what is needed and when it is needed. Any questions on scope need to be addressed during the bid period so problems can be resolved. Scope needs to be defined in advance of the bid day. This will allow their bid day price to be evaluated with a minimal opportunity for mistakes.

Formatting The Bid to match the bid requirements is imperative. Should the owner request a lump sum, firm price for the entire project, the bid must be formatted to match this requirement. Should the format have intermediate breakout prices to arrive at the total, the estimate must be formatted to match this requirement. Whatever format is requested, the estimate needs to be prepared to match in order to minimize mistakes in transferring the amounts to the bid form. That being said, owners and architects will sometimes request unreasonable amounts of detail in the bid format. Discussion should be held with them early in the bid period to minimize the number of bid categories or items in the bid form. The more bid items, the more likely an error.

Estimate Reviews need to be done prior to submitting the bid. These reviews are not just to be done in the few minutes prior to submitting the bid either. These reviews need to be done at least the day before the bid will be submitted. So the contractor's staff can focus on the bid day assembly of the complete bid.

The estimate reviews need to make sure all "plug" amounts have be identified and eliminated or identified for closing out on bid day. Plug numbers are amounts put in an estimate that are allowances until the actual numbers are determined. If the plug amounts are still being used, the adjustments can be done through bid adjustment sheets (spreadsheets or packaged software). A review of the numbers should also be implemented to check for transpositions and related errors. Even though computers and estimating software packages are a great help in estimating, they do not recognize if the numbers are right or wrong, they just process them fast!

GENERAL REQUIREMENTS

Review the numbers to be sure they are being summarized or carried properly on the summary sheet. Sometimes inserting a line in a spreadsheet will cause problems on the summary sheet or give an incorrect total.

The Bid Review Checklist is done to make sure the bid documents are being submitted in accordance with the bid request. Make sure all of the exceptions or qualifications, if any, are sated and complete. Review what steps were taken to minimize or eliminate the exceptions. This can be through the questions having been answered or other measures. If the bid does not allow exceptions what impact will this have on the bid amount? Can the exception be identified in dollar terms? Is the exception so much that not listing it will pose a serious risk to your company?

Also review the bid documents to make sure there is the right amount of time to complete the work. A preliminary schedule should be prepared to validate the time required. Is the duration reasonable? Can there be improvements made to the schedule or can the schedule slip without major financial penalties? Will there be any special equipment or manpower requirements that will pose a problem? What about long lead times on major equipment?

All of the above should be developed over a period of time based on the actual operations of each company. Many of the items will be common to all contactors. Some will be specific to a few contractors. It will be a very rare item that is specific to only one company.

Office Overhead Expense

The item of overhead expense is neglected by many contractors, but it is of such importance that every contractor should give it careful consideration. Every contractor has a certain fixed expense that must be paid regardless of the amount of work done or contracts received, and these items should be charged to office overhead. This is sometimes referred to as General and Administrative (G&A) costs.

G & A includes but not limited to such items as office rent, fuel, lights, telephone and telegraph, stationery, office supplies, advertising, trade journals and magazines, donations, legal and accounting expenses not directly chargeable to any one job, fire and liability insurance for the office, club and association dues, office employees, such as bookkeeper, administrative assistants, clerks, estimators, janitor services and salaries of executives, along with travel and entertainment expenses of good will nature and depreciation on office furnishings and equipment. These items should be estimated for a year and then reduced to a certain percentage of the total annual business, as follows:

Description	Rate
Non-reimbursable Salaries*	$70,000.00
Employee Benefits*	$21,000.00
Legal & Accounting Fees	$5,000.00
Insurance*	$10,000.00
Office Rent	$5,000.00
Office Utilities	$3,000.00
Office Supplies	$2,000.00
Job Procurement & Promotion	$7,000.00
Depreciation*	$2,000.00
Maintenance*	$2,000.00
Job Management* (incl. transportation)	$5,000.00
Education/Regulations	$2,000.00
Miscellaneous	$6,000.00
Total	$140,000.00

*Indirect expenses only, that is, those not reimbursed
by fees from individual projects.*

The list includes many more expenses than a small contractor will have and considerably less than those carried by larger contractors, but it serves as an example of the items that should be included as overhead or general expense. For example, suppose you have an overhead expense of $300,000 per year, and do an annual business approximating $3,000,000. On this basis your overhead or general expense would average 10.00% of the total annual business.

Overhead may run 6.00% to 15% for smaller firms. Larger firms may have overheads that are as little as 1% to 2% of the annual volume. Even if a contractor cannot readily determine their annual overhead costs, they should not be ignored. Every contractor can approximate their overhead based on their average yearly business by meeting with their accountant periodically during the year.

As the workload changes during the year, so does the overhead rate applicable to bidding new work. Many smaller contractors do not maintain an office and perform the greater part of their office duties themselves, so only such items of expense as they actually incur should be included.

Proposals for jobs taken on a time and material basis and provisions for extra work on jobs taken on a stipulated sum basis may require the contractor to state the percentage to be added for overhead, plus that for profit. In a job bid from five medium-sized contractors, these varied as follows: 8%+4%, 8%+7%, two at 10%+5%, and one at 10%+10%. While overhead percentage varied only from 8.00% to 10.00%, profit varied from 4.00% to 15.00%.

Whether these same percentages are used in figuring competitive bids is, of course, each contractor's own determination. But certainly, some addition to cover office overhead should be made, unless the organization is so successful that the year's overhead has already been absorbed by other jobs, or the work is of such nature that it is primarily subcontracted and office expenses are minor.

All manufacturers, jobbers, and retailers add their overhead expense to the cost of their products when computing the selling price, and there is no reason why contractors should not do the same, always keeping their eye on the competition.

Office Furniture and Equipment. All expenditures for office furniture, typewriters, computers, calculating machines, filing cabinets, etc., should be kept as a separate account and a certain percentage of this amount charged off each year as depreciation. The depreciation, however, should be charged as a part of the overhead expense of conducting the business, as the cash value of the equipment decreases each year.

In a completely efficient office, construction equipment would be constantly utilized on the job and would be charged as a job cost, not office overhead. However, there are always slack periods when equipment is idle, or it might have been bought for a specific job but still have a useful life. It can be depreciated and charged to overhead until it can be put to use again. The Internal Revenue Department publishes fixed schedules of allowed depreciation.

Some contractors set up a "rental" cost for equipment owned by the firm but chargeable to individual jobs. Today the trend is to rent tools and equipment from outside firms specializing in this, which simplifies bookkeeping-all costs are directly charged to the job and no depreciation need be carried-and frees the contractor's capital for more lucrative investment. A discussion about equipment is included later in this chapter.

Insurance. Insurance costs directly chargeable to office overhead would include fire and theft for office property, liability for office premises, workmen's compensation for office employees, and automobile insurance for trucks and cars owned by the firm.

Liability insurance is a must in today's construction market. This is often included as an office overhead, a necessary cost of doing business. However, if the policy has rather low limits, the contractor will find most specifications will require additional coverage, and this would be a job expense rather than overhead expense.

Surety bonds, performance bonds, completion insurance, property insurance for work under construction, and worker's compensation for on-job employees can be charged directly to the job costs and are discussed under project indirect costs.

119

Project Indirect Costs

In addition to home office overhead expenses there are project related indirect costs. These indirect costs are incurred in support of performing the work. Basic project indirect costs to consider in each estimate include Mobilization, Project Staff, Temporary Buildings & Construction, Temporary Utilities, Construction Equipment, Project Office Expense, Small Tools & Consumables, Weather Protection, Outside Services, Insurance, Taxes & Bonds, Home Office Support, Demobilization, Escalation (Cost Growth) and Finance Expense.

When estimating these expenses the estimator must know what is being provided by the owner, or if the project includes a Construction Manager, and what is being provided by them. When determining what will be provided by others, determine if what is being provided will be adequate for the project. If not, the added costs need to be included as part of the project indirect costs scope of the estimate.

Each job will entail certain expenses that are chargeable to the job as a whole. These are set forth in the General Conditions (the AIA articles usually made a part of the contract) and the Supplementary Conditions and/or General Requirement Section of the Specifications. These will include such items as, resident engineer's office and supplies, contractor's supervision, temporary buildings, enclosures, barricades, offices, toilets, utilities, protection, clean-up, permits, surveys, photographs, tools and equipment, quality assurance/quality control (QA/QC), insurance and benefits, sales taxes, surety bonds, warranties and training of new equipment.

Sometimes, these are covered only in a general way, and it is up to the contractor to visit the site to determine what is needed. Often the unions and local laws may be more demanding than the architect's specifications, and the contractor must know what items they are liable for.

Job overhead and office overhead are best kept separately, although occasionally employees' insurance, tools and equipment, and other items may be divided between office expense and job expense.

Many contractors figure a certain percentage of the cost of the job to take care of all overhead expense and plant charges, but this is not the correct method to use and does not give the contractor or estimator a reliable basis for computing the cost of these expenses. Each item entering into the cost of job overhead expense should be estimated the same as any other branch of work.

Mobilization: Mobilization is the moving of personnel, material and/or equipment to the project site in order to perform the work. This may be as little as bringing in gang boxes of tools or a skid loader. It takes time and equipment to get these items to the site. These costs need to be included as a part of the project costs.

120

GENERAL REQUIREMENTS

Project Staff: The project staff may include part-time or full time allocations of personnel on the project. If a project manager or superintendent is required, their time needs to be estimated and included in the estimate. In general this category is for management or other support personnel on a direct hire basis.

The proper method of determining these costs is to estimate the salaries of the various personnel required at a certain price per week or per month for the entire estimated time they will be required on the job.

Temporary Buildings & Construction: If a project office is required, this is where the cost needs to be included. If there are temporary storage needs, such as with cargo trailers their costs need to be included here.

This category should include temporary fences, bridges or platforms, temporary sheds and storerooms, temporary enclosures over hoisting engines, concrete mixers, temporary runways, ladders and stairs in the building before the permanent stairs are installed, temporary dustproof partitions to isolate work areas from building space still occupied, and temporary closures for doors and windows before the permanent doors and windows are in place.

It should also include costs for protecting or repairing areas damaged by the new construction. There may be damage to sidewalks, pavements, and landscaping that cannot be prevented. Protection of windows or foundations and walls maybe required. Newly installed work such as wood flooring may need protection from other work activities.

Temporary Utilities: This category includes items of cost for electricity, water, sanitary facilities, telephones and other related items. Not only does the weekly or monthly cost need to be estimated, but any hookup and disconnect fees associated with the utilities should be included.

Practically every job must have some provision for sanitary facilities. Local codes usually govern what must be supplied. Chemical portable toilets usually suffice. On large jobs, however, a more elaborate setup is usually specified requiring a complete water supply, drain, vent, and fixture installation. This work is sometimes included in the plumbing specifications.

In the case of a remodeling job where permission is granted to use the existing facilities, an item should be included in the estimate for periodic and final cleaning of the toilet room and its appurtenances.

On most many projects water for construction purposes must be provided by the prime contractor for all trades. This item can become a high-cost operation, especially in barren, undeveloped localities where either a well must be drilled or water must be hauled to the job. When a water main is available, a plumber must be hired to tap it, install water meter, and run a water line to strategic locations on the job. This work is sometimes included in the plumbing specifications. Also included under this heading are the charges made by the local water department for the volume of water used. Often hydrants can be used at a set fee from the municipality.

The temporary power can vary, from a small charge for tapping into existing adjacent power lines and running entrance cable into a meter setup 4from which extension cords may be run, to complex installations requiring long run pole lines, transformers, and switch gear, such as for electric mixers, hoists, and other heavy duty electric power equipment. This work is sometimes included in the electrical specifications. Also included in this item are the power company charges for energy consumed and rental of transformers, etc.

Construction Equipment: A portion of the construction equipment can be considered direct cost. This would include backhoes excavating for footings, cranes for hoisting and lifting materials, along with other similar items. However, some of the construction equipment on a project does not get estimated as a direct cost of the work. This may include flatbed trucks for material deliveries, pickup trucks for superintendents, a warehouse forklift and so on. These costs need to be estimated and recovered for the project in this category.

Equipment ownership or rental is a consideration in each bid. If the contractor owns the equipment, it may be depreciated for tax purposes. However, if the inflation rate is high, the depreciation allowed by the Internal Revenue Service might not be adequate to cover replacement costs, and the tax savings might not be so attractive. On the other hand, a high inflation rate might make it possible to buy equipment, use it, and resell it for what one paid for it.

The advantages in renting or leasing include:

1. Eliminates tying up capital or borrowing against bank credit for equipment that most probably will not return as much money as if those assets were applied toward financing another job or taking advantage of cash discounts;
2. Offers flexibility in choosing equipment; one can try one product on one job and another on a different job, until one finds just the right piece of equipment for each situation, or if the item does not perform to expectations, it can be returned;
3. Eliminates personal property taxes on the equipment;
4. Eliminates maintenance costs and the expense of training personnel to handle those tasks and minimizes possible losses due to obsolescence and vandalism;
5. Eliminates costs of shipping, handling and storage, and tying up trucks and drivers to move this equipment, as this will be done by the leaser.

The question then becomes whether to rent or to buy and depreciate plant equipment. In a large community where several rental sources are available, renting can be preferable, leaving the contractor's money for other forms of investment, such as building up capital in order to take on larger jobs or investing in land on which a contractor cannot only hope to make a profit on resale but also can insist their firm be used as contractor for any development.

If the contractor does elect to purchase equipment, they should set up separate records for each item. Be sure to consider the Internal Revenue Service's depreciation schedule allowed, the probable life, probable maintenance and storage cost, actual cost of transporting the item to and from the job, and the costs of operating the equipment on the job. Enter an hourly rate chargeable for the use of the item on each job estimate.

Charges should be sufficient to replace the item when it is worn out, as well as to cover interest charges if a loan is taken out to purchase it and a separate fund, often known as a sinking fund, set aside to cover future replacement. These charges should not be carried as profit. If the equipment is idle for long periods, the costs could be transferred to general overhead, but this is rather dodging the issue. If the equipment costs are accurately charged against each job, a contractor can easily tell whether the costs of ownership exceed current rental costs, always keeping in mind that the funds tied up in plant ownership could be making money elsewhere. Plant and equipment costs are discussed in greater detail later on in this chapter, under "Tool and Equipment".

Project Office Expense: this category is used to cover the site office expenses associated with the project. There can be paper goods, postage, copies, and other office supplies. There may be a need for computers, software and support.

Small Tools & Consumables: Each craft requires small tools and consumables for the performance of the work. For a carpenter it may include their tools and for laborers it may include rags and cleaners. Rather than identifying each specific item, costs for these items should be tracked by craft hours. The number of hours in the direct estimate can then be used to develop a total cost for both of these categories.

Weather Protection: Weather protection required on a project is dependent upon the stage of completion and the time of year. Placing concrete by not only require heated concrete provisions in the direct cost, but may require the use of unit heaters and tarps. The cost of the unit heaters, fuel and traps, along with the labor to setup, maintain and dismantle them would be estimated in this section. Cold is not the only element to consider. Heat, rain, winds, and weather factors need their own protection considerations.

Outside Services: The use of outside testing services for material testing, non-destructive examinations, or even an Authorized Inspector (boiler work) would be included as costs in this section. If the contractor was responsible for site security these costs would be included in this section too.

Site surveys are also included in this section. It may also be necessary Where it is necessary to have photographs and video tapes taken at intervals to show job progress, the cost should be included in this section. At times the contractor together with the architect can take and log the photos at pre-determined times or when special conditions arise. Even if they are not required by contract, photographs should be taken for job progress purposes and marketing uses.

Insurance, Taxes & Bonds: The hourly labor rates for the project should include the customary labor burdens for taxes and insurance. The insurance costs to include here may include insurance policies such as excess liability insurance, select equipment floaters and other special insurance requirements.

Special insurance policy costs should be included in this section. These include the following types of insurance coverage.

There are two common builders' risk forms that may be written. The first is **Builders' Risk Reporting Form**. This form affords coverage for the contractor from the time material is placed on the job site until the building is completed. Under this form the contractor reports to the insurance company the actual value of the building at the end of each month as construction progresses and pays a premium each month on the additional values that go into the building. The other is **Completed Value Builders' Risk Form**. This form of coverage is normally preferred by a contractor, because it gives 100% coverage without the necessity of keeping records as to the value of the building as it progresses or reporting at the end of each month.

This form is written for the completed value of the building. The normal rate charged is about 55% of the rate on the basis that at the beginning of the job there is little value and at the end you have the full value. Normally the premium under the two types of policies is remarkably close.

The interests of a general contractor can be covered and they may exclude or include the interests of subcontractors who also may be employed in the construction of the building.

Temporary structures, tools, and equipment that belong to the contractor and are on the building site may also be included in the coverage for the hazards insured.

Today, many contracts that require the contractor to provide the coverage call for coverage on all risk of loss basis, naming the owner, lender, and numerous other parties, as well as the contractor, as insured's in the policy.

In addition, many contractors require the inclusion of specified exposures related to the project. Delay in opening and bank financing changes are two of the more frequently seen "soft cost" coverage requirements.

124

GENERAL REQUIREMENTS

On occasion, contracts with clauses that are not contained in the insurance section of the contract but that are found in other sections, such as force majeure, risk of loss, indemnification, and work place safety, leave the contractor with liabilities that may or may not be covered by insurance. For example, the insurance section might specify limits of liability to be purchased for policies provided by the contractor, but another section of the contract or specification, contractual indemnification or risk loss, may specifically state that the contractor's liability is full and not limited to or by any insurance requirements otherwise stated in the contract.

Costs of such insurance vary by location, type of construction, and available public fire protection and may be obtained from your local insurance representative.

Contractor's Equipment Floater: Many contractors possess many thousands of dollars in contracting machinery and equipment. To properly protect these values, such equipment and machinery is usually insured under a contractor's equipment floater policy.

Insurance under this form is quite flexible so that many situations may be insured to fit the contractor's particular needs. The insurance is normally written on a named peril policy which includes loss or damage by fire or lightning, explosion, windstorm, flood, earthquake, collapse of bridges and culverts, theft, landslide, collision, and upset. Or if desired, a policy may be issued protecting from practically all risks of physical loss or damage. Such contracts usually write a deductible clause.

Installation Floater: The contractor who specializes in plumbing, heating, air conditioning, or electrical equipment may find an installation floater more suited to their needs. This type of policy may be written for an individual job or may be written to cover all the jobs they may secure during a year's time.

Coverage may be on a so-called "named peril" policy, usually including the hazards of transportation of material to the job site and covering fire and lightning, windstorm and hail, explosion, riot and civil commotion, smoke damage, and aircraft and vehicle damage at the job site. It may also include coverage for vandalism and theft if desired.

If provision is not made in the specifications that fire insurance will be carried by the owner during construction, it is important that the contractor protect their interests by carrying some type of builders' risk insurance. This is normally written on a standard fire policy together with an extended coverage endorsement that extends the fire and lightning provisions of the standard fire policy to include windstorm, hail, explosion, riot, smoke damage, and damage from aircraft and vehicles. An additional endorsement, vandalism and malicious mischief, may be obtained. Some companies afford a broader coverage of an all risk nature although there are a number of exclusions

Taxes may include sales taxes, property taxes or other requirements determined by the contract provisions. Most states have sales or use taxes applicable to building construction as well as to other commodities. This tax varies with the different states, and there are different methods of levying it. In some states, the tax is levied on all permanent building materials and is paid by the contractor when they pay for their materials. In other states the tax is applicable to the entire cost of materials entering into the construction of the building and is paid by the contractor directly to the state. In either case, the contractor should add this cost to the estimate/bid amount. Otherwise, it will make a big dent in estimated profits. Note that some jobs will be tax exempt. However, it is up to the client to establish this tax exemption, and the contractor should qualify their bid accordingly.

Bond costs may include fidelity bonds for employees, installation warranty bonds and material supply bonds in the case of major equipment or long lead items. The payment and performance bond is customarily added after all costs are determined. Building permits would also be included in this section of the indirect costs. Check with local governing agencies for costs in the area where you are building for all construction permits required including possible utility permits.

Bonds are usually issued to cover a single job and are applicable to each project. Some cover only a single trade and may be included in a subcontract sum, such as warranties or roofing bonds. Bonds taken out by the general contractor are usually carried in this category.

Bonds are usually issued by insurance companies or surety companies specializing in such bonds. A division of the U.S. Department of Labor, Small Business Administration, has been set up to provide surety bonding for small and minority contractors. Surety bonds are credit guarantees of contractors and are issued after careful scrutiny by underwriters of a contractor's financial statements, skills, work in progress, past work, and general reputation. Bonds are not purchased by contractors, but rather, issued for the payment of a fee based on contract cost. Some of the types of bonds encountered are discussed below.

Bid Bonds. Bids are invited by advertisements, and the bidder may have to submit with the bid a certified check, usually for 5% of the bid, or a bid bond, usually for 10% of the bid. This requirement is made to insure that bids will be submitted only by qualified contractors who will not only enter into the contract but qualify for performance and payment bonds.

The bid bond guarantees that as the contract is awarded, the principal will, within a specified time, sign the contract and furnish the required performance and payment bonds. If the principal fails to furnish the required performance bond, there will be a default under the bid bond, and the damage will be the difference between the principal's bid and that of the next highest bidder, not exceeding the penalty of the bid bond.

126

GENERAL REQUIREMENTS

Surety companies generally follow the practice of authorizing a bid bond only after the performance bond on the contract has been underwritten and approved, so there are few defaults under bid bonds.

Performance and Payment Bonds. These are required on a vast majority of contracts. Some owners and most public work require that each contractor to whom a contract is awarded shall furnish bonds which contain two obligations: a performance bond to indemnify the owner against loss resulting from the failure of the contractor to complete the work in accordance with the plans and specifications; and a payment bond to guarantee payment of all bills incurred by the contractor for labor or materials for the work.

The federal government, under the Miller act, requires that a contractor furnish two separate bonds, one for the performance and one for the payment of labor and materials. Some states follow the same procedure.

The risk of the surety under the construction contract bond is if the contractor fails in the performance of their obligation to complete the job, or to pay the bills for labor or materials. The surety assumes the responsibility for completing the project and paying the bills if the contractor fails.

The underwriting skill is designed to weigh the hazards involved against the contractor's qualifications. If the contractor's ability is deemed sufficient to warrant the assumption that they will be able to complete the work and pay the bills, the bond is issued. If not, it is declined.

The general practice is to underwrite each bond on an individual basis. The contractor cannot assume that every request will be favorably received by underwriters. The contractor's volume of work that is still outstanding at the time of the bond request is a key factor.

Some of the most common causes of failure on the part of construction contractors are abnormal labor costs, fire or other loss not adequately covered by insurance, unforeseen restriction or withdrawal of credit, weather or subsurface conditions that increase the cost but not the contract price, over extension beyond the contractor's financial capacity, unavailability of materials, failure of subcontractors, death or disability of key men in the contractor's organization, a poor accident record, diversion of funds from one job to pay off overdue accounts on other jobs, inexperience in the type of work involved, or dishonesty on the part of the contractor's employees.

Costs of performance bonds depend on amount of contract and length of coverage, as well as the contractor's reputation and whether the coverage is for a portion or complete cost of work.

Maintenance Bonds. This bond may be required to guarantee an owner that the general contractor agrees to correct all defects of workmanship and materials for a specified time following completion. Often the above performance bond can be arranged to cover this period which usually lasts one to three years.

License or Permit Bonds. Bonds provided by a contractor to a public body, guaranteeing compliance with local codes and ordinances. If the contractor regularly operates within an area requiring such bonds, this cost should be carried under office overhead, because it is a normal cost of doing business. But if it is required for a specific job, it can be charged directly as a job expense.

Supply and Subcontractor Bonds. Guarantees from manufacturers and installers for specific items and workmanship furnished, the cost of which is usually borne by them and included in the subcontract price for that trade rather than in overhead.

Home Office Support: Even though a project may have part-time or even full time project personnel, there is oftentimes support from the home office. A part of this is the percentage determined to be applicable to all projects that was covered in the home office overheads. Another part is the specific support needed based upon the contract terms and conditions. The key to estimating the correct costs is not to duplicate the general home office overhead and the project specific home office support requirements.

Demobilization: The costs for demobilization on a project do not automatically include all of the mobilization costs – only in reverse. Since the new project would have the costs to mobilize the items, they should not be estimated twice. This is also a part of contracting risk in that there is another project than can use the same items. Demobilization should include costs for final job cleanup, some relocation items and other costs related to the specific project closeout. There may even be an offsetting credit for salvage items that will assist in keeping the estimated cost down.

Escalation (Cost Growth): Escalation is the increase in prices (cost growth) from the time the project bids to when the work is performed. This cost growth can occur in wages, materials and equipment. Wages can increase if a project will span across a wage increase period, even if an open shop concern. Long lead materials and equipment may increase, or even decrease, from the time the quotes were received to the time the materials are ordered. This increase in costs needs to be determined, even if no money is added to the estimate for these events. It may or may not come out of the profit amount by the end of the project.

Finance Expenses: Finance expense is incurred on every project. It is simple. The contractor must work a period of time before a billing can be submitted. The architect then has several days to approve the pay request. The owner then takes time to review and pay for the first period of work. Even then the payment may have retainage withheld. Meanwhile the contractor is working (spending money) and self-financing the costs on the project. This is called the progress payment delay lag. The contractor must either borrow money (pay interest) or take money from an account (lose interest) to pay these costs. This is the finance expense amount and can be calculated by building a progress schedule with preliminary costs. As a general rule the contractor should try to be cash positive not much later than 30% though the project. If the finance expense is not added to the bid, the amount should be calculated because it will be coming out of the profit amount in the end.

Profit

The amount of profit to be added to the estimated cost of the work is a question that a contractor must answer individually for each bid. There is no set amount that can be added. It all depends on local conditions, competition, cost of money, and how badly the job is wanted.

On small jobs, alterations, remodeling, and similar work, a contractor is justified in adding 20% to 30% profit to the actual cost, but must ask themselves whether they can actually obtain this amount.

On new work, where it is possible to estimate cost with a fair degree of accuracy, a contractor is entitled to 10% to 15% on the actual cost of the work (job overhead included in the actual cost of the job), but it is safe to say that competitive figures submitted for many jobs show a 5% instead of a 10% to 15%. A contractor is entitled to a fair profit of 10% profit, but getting it is another matter.

Another topic of considerable discussion is whether or not a contractor should add a certain percentage to the estimate to take care of contingencies. The answer is both yes and no. Yes, because it is not necessary to be quite so careful in the preparation of the estimate, and if any items have been overlooked, there is a fund to take care of omissions. But a contractor who is not careful in the preparation of the estimates never knows how much should be added to provide for forgotten items or contingencies. No, because if you add a percentage for contingencies, it is quite possible that one or more of your competitors has not, and another job is lost.

The only safe way is to be as careful as possible in listing all items from the plans and specifications. Include them all, and it will not be necessary to add for contingencies. This includes everything to be furnished and all labor costs, including job overhead expense and a percentage for profit.

An item often overlooked, but which can become a major expense, is the finishing up of a job and making corrections after the job is completed. Often, there is confusion on the job at the end and it is difficult to say who chipped the paint, scratched the door, broke the window, stole the light bulb, broke the door handle, or backed into the fencing. Most of the trades are on the job doing touch-up work, the owner is moving in, the decorator is hanging drapes, and children are running wildly through the corridors. But the general contractor gets the complaint first, and in the interest of having a satisfied client, will probably assume the responsibility for correcting this. In a warehouse, such problems may be minimal; in a fine residence, you might be called back daily for weeks.

Other factors an estimator must evaluate in setting up a bid are the weather conditions that are apt to prevail at the time of construction; possible rises in labor rates and the possibility of strikes if labor contracts expire during the construction period; material shortages and strike actions at the manufacturers level; the money supply and its cost; and public relations with the community in which the project is located and possible job stoppage or slowdowns because of environmental or safety concerns which the drawings and specifications fail to cover.

Estimate Check List

Estimates are done under pressure, on a deadline. Every estimator has a check list of items typically included in their estimates. The format for the check list can be in the CSI structure or in a detail format of what is usually included. The estimate over time becomes the checklist used repeatedly.

Finally, make certain all items are covered. Check not only the drawings and specifications, but also any bulletins addenda. An addendum changing an item, such as asphalt shingles to concrete tile, ½ drywall ceilings to 5/8" drywall ceilings, etc. can be costly if the estimator didn't catch the change. Using a check list every time helps the estimator eliminate errors. Forgetting an item has cost many a contractor his profit.

Determining Costs

The detailed estimate is a projection into the future of the anticipated costs of a project. It reflects all of the known conditions at the time of preparation. It is important to maintain a cost system that uses as a base each of the components contained in the estimated cost. This is necessary in order to compare estimated costs with actual costs incurred during construction. If the estimated costs are realistic and accurate, they will produce the anticipated profit.

Equally important, the periodic review of actual cost versus estimated cost will serve as a data base that can be adjusted, either by increasing or decreasing the unit costs, so that future estimates will be produced more competitively and profitable.

Cost control systems can take many forms with the use of computers. The system can be relatively simple and still produce accurate information, or they can be designed in an elaborate and detailed format to provide information relating to costs for every item of material and labor productivity for the entire project. The use of computers in estimating, and computer cost management tied to the estimate facilitate in this review.

The selection of the system is one of personal choice, but it must be remembered that any cost system requires considerable expenditure of time in accumulating the material and labor records, in distributing to the proper categories of work, and then, in calculating the costs of each and recording this information on progressive cost record forms so that the actual costs can be monitored while the work is being performed. When this information is available, it is then possible, should an overrun appear, to review the situation to determine if a remedy is possible, such as adjusting crew balance, pursuing another method of construction, or allocating additional supervision to a specific operation. The system selected should be one that serves the needs of the company. Anything in excess of this is a burden and wasteful in both time and cost.

CONSTRUCTION SCHEDULING

In order to complete a construction project within the estimated cost, the contractor must ensure that individual work activities are completed within the time allowed by the cost estimate.

There are three traditional methods of construction scheduling:

1. Critical Path Method (CPM)
2. Bar (or Gantt) Charts
3. Program Evaluation and Review Technique (PERT)

The CPM method involves identifying individual activities within a project, determining the relationships among the activities, and presenting the information in a graphical network diagram.

The Bar Chart method of scheduling uses bars to show the start and end dates and duration of individual activities, but they do not show the relationship among the activities.

The PERT method of scheduling is another network diagraming technique for the scheduling of research and development projects.

CPM and PERT differ in several ways especially concerning the estimation of activity time durations. CPM is widely used in the construction industry and relies on past experiences to estimate the duration of activities. PERT uses three time estimates (optimistic time, most likely time, and pessimistic time) to calculate the expected time based on probability and statics. PERT offers the contractor another method when asked to create a schedule for a project that hasn't any historical experience. With either method, the highest sum of activity durations that form a continuous chain of sequential activities through the planned project, allowing contingency time for weather and other delaying factors, is the scheduled duration of the entire project. Because the critical path method (CPM) is one of the most widely used planning and scheduling tools, we will use it as the basis for our discussion of planning.

Planning

Planning for a construction project involves the understanding of the project's contract requirements, deliverables, the means and methods of construction and defining how the project would be constructed. Even the most simple construction projects would require some measure of planning. The estimator would usually be the first to consider all of the planning elements of the project in preparation for the estimate. The estimate for the general requirements and site works are the direct result of this planning effort. In small projects the estimator/project manager may also be the planner for the project. In more complex projects, planning involves more individuals, including the scheduler, site manager, and resident engineer. In these complex projects early planning is very important and may start as early as the conceptual design phase and before finalization of design to ensure constructability of the project.

Project Scope and Work Breakdown Structure. Clear definition of the project scope and identification of the project deliverables is vital for the proper planning of a construction project. Project scope is defined in the contract documents such as terms and conditions, drawings, and specifications. In addition to the contract documents, planning requires a clear understanding of the site conditions, site access for equipment and materials, local labor markets, and available equipment, and all the other factors that can affect construction. The schedule needs to be coordinated with the estimating and job costing portions of the contractor's operations.

By understanding the scope, the project can be broken down to smaller components. Each component can be further broken down into even smaller components. The result of this process is called the Work Breakdown Structure (WBS), which is defined by the Project Management Institute as an "oriented grouping of project elements that organizes and defines the total work scope of the project. Each descending level represents an increasingly detailed definition of the project. Experienced planners typically develop a WBS based on work area and type of work. The lowest level of detail on the WBS is called a Work Package. The following figure shows an example of a WBS for a construction project.

Construction Means and Methods

Construction means and methods basically answer the question of how the project is going to be constructed, what equipment will be used, and what technologies will be adopted. It identifies what materials to use, and what are the labor needs. Defining the construction means and methods require the careful examination of the following factors:

GENERAL REQUIREMENTS

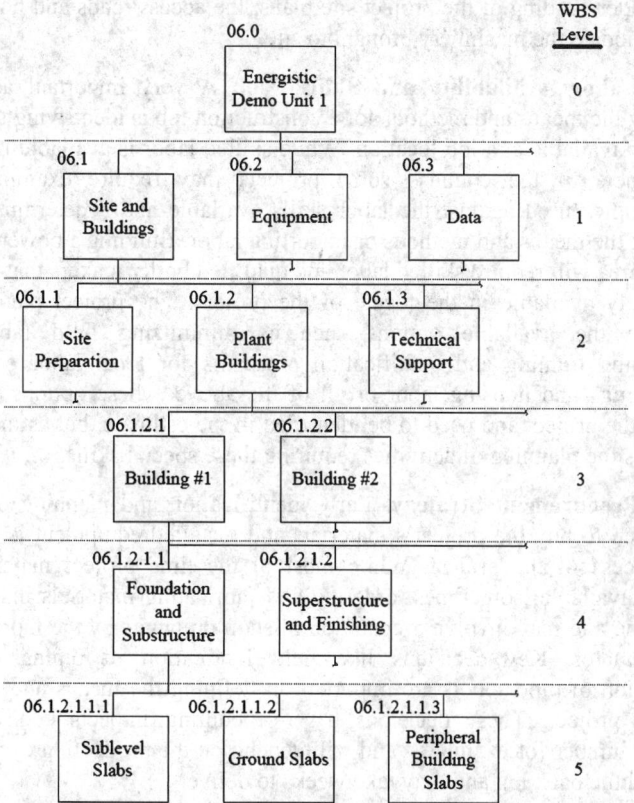

Energistic Demo Unit 1 — 06.0 — WBS Level 0

06.1 Site and Buildings / 06.2 Equipment / 06.3 Data — Level 1

06.1.1 Site Preparation / 06.1.2 Plant Buildings / 06.1.3 Technical Support — Level 2

06.1.2.1 Building #1 / 06.1.2.2 Building #2 — Level 3

06.1.2.1.1 Foundation and Substructure / 06.1.2.1.2 Superstructure and Finishing — Level 4

06.1.2.1.1.1 Sublevel Slabs / 06.1.2.1.1.2 Ground Slabs / 06.1.2.1.1.3 Peripheral Building Slabs — Level 5

Drawings and Specifications. Drawings show the location of various features on the site and illustrate details about the building (e.g., construction details). For specific trades, drawings show the height, position and bulk of the building. Specifications set out the minimum standards or codes necessary to obtain relevant sizes for all structural components. Specifications may also detail the fixtures and fittings. Understanding the drawings and specification is the first step in planning how the project is going to be constructed. The drawings and specifications will define the materials used. The specifications often identify the methods used in construction, e.g. pre-cast concrete instead of cast in place.

Technology. The size of the project and its location needs to be evaluated to determine the technologies and methods used. Questions facing the planners would be typically of the sort: Does the amount of excavation require just a bobcat for cut and fill or will we need larger bulldozers and loaders? Are we hauling dirt off site or will we be piling it for backfilling later? Do the quantities of concrete justify a batch plant on site? Do we have room for the batch plant? All these questions need to be addressed with a

133

clear understanding of the project site plans, the access roads and how can it accommodate the machinery brought to site.

Labor Availability and Skills Pool. A very important aspect of defining the means and methods for a construction job is identifying the skills of labor available at a job location. While skilled labor is available in almost all corners of the country, some projects may require examining the technologies used against the labor skills available and a determination of revising the means and methods or importing labor. Building a power plant in a rural area will require skilled labor and qualified boiler makers that may not be readily available in the locale of the project. The project planner will evaluate the available options such as minimizing field fabrication, developing training and qualification programs for local labor or maybe "importing" and housing labor or all of the above. These options will add cost to the project and need to be identified in the estimate, but as important, it affects the planning of activities requiring these special skills.

Procurement Strategy. Early identification and planning of where and when to buy the project's materials and installed equipment is vital for the success of the project. While most of the time project materials are readily available, sometimes, ordering and fabricating materials may take a long time and may involve preparation of shop drawings by the fabricator or subcontractor. Key decisions like field fabrication of piping or shop fabrication of pipe spools are important in defining the means and methods for the project. These decisions have scheduling impacts—delivering a house's lumber for example—and will depend on the production cycle in the mill and the back log and may take weeks to deliver.

Institutional Constraints. In addition to the above factors to be considered in planning a project, there are institutional constraints on how, when and what to build. These are usually governed by the local authorities in simple projects but may require also state and federal approvals. These constraints range from a simple permit application to a municipality and its planning board to an elaborate environmental process that involves the community feedback, public hearings and approvals from several agencies. Acquiring all the required permits for the project may put constraints on the project. Typical examples of these constraints are: limited or no nightshifts, limit on size of trucks and equipment coming in and out of the site, dust control etc. Permits may take a long time and needs to be incorporated in the project plan.

Project Phasing and Staging. On larger projects, especially if construction occurs within the confines of a site that is used by others (highway spur, or road construction on a used road), the need to phase and stage the project is very important to the successful planning of the project. The terms phasing and staging are used interchangeably and sometimes confusingly. For simplicity, phasing will be the different distinct phases of

134

the project usually associated with a new type of work beginning each new phase.

In our road construction example an addition of new high speed left lane may have the following phases: Phase I is to construct a bypass by modifying the right shoulder to accommodate traffic. Phase II is to put jersey barriers to isolate left lane for construction and construct new lane and left shoulder in the median. Phase III would be to divert traffic to the newly constructed lane and shoulder and begin demolition (or repair) of the old lanes and construction of the new pavement. Phase IV would be removal of barriers and final road markings etc. There is usually more than one way to phase a project, and that brings us to staging.

Staging is identifying for each phase how equipment and materials will be brought in and out of site during each phase. Staging and phasing are best represented on the site plans marking the location of site facilities, equipment, storage and site circulation, and areas of construction in each phase. Staging and phasing diagrams will mark areas of temporary construction needed during each phase. Each phase would be assigned a duration and the sum of durations add up to the project duration. This duration is usually estimated based on experience, and is the first step in scheduling the project. It is important to note that Phasing and staging should respect and follow the WBS for the project. If the WBS did not accommodate the phasing and staging of the project the WBS will need to be revised. It is important to also note that the phasing and staging have to take into account the design considerations like construction joints, different elevations etc.

Weather Considerations. Weather conditions play an important role in the successful completion of construction projects. Weather impacts construction in many different ways and all should be considered thoroughly. The following are examples of weather conditions that impact the construction planning and scheduling.

Severe cold will impact earthworks, and most contractors will avoid beginning earthworks in the winter if the soil freezes. Severe cold temperatures will also impact concrete cure time and therefore the schedule. Concrete additives to improve concrete performance in different weather conditions may be expensive; therefore, the schedule impact of concrete cure times should be evaluated against the cost impacts.

Heavy rainfall may require site erosion and sediment control during construction and may interrupt work. This is of particular importance in road works and earthworks where construction disturbs the natural surface.

High wind will impact high rise construction and heavy crane lifts.

Severe hot temperatures will impact sensitivity of measuring equipment, and leveling and measurements and laying out activities will have to be scheduled for early morning.

Scheduling

Scheduling Methods/Tools. The use of bar charts started during the industrial revolution of the late 1800s. An early industrial engineer named Gantt developed these charts to improve factory efficiency. Bar charts are often called "Gantt Charts."

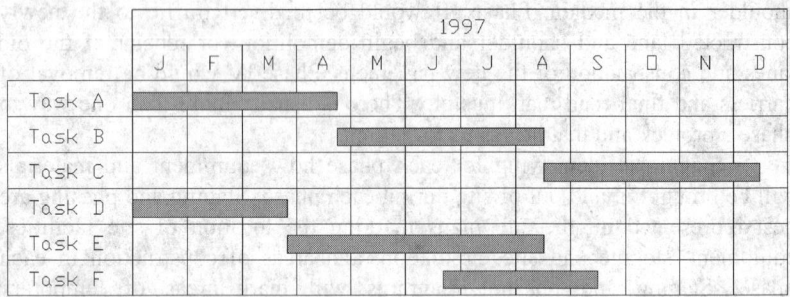

Critical Path Method (CPM) for project scheduling began in the 1950s in two parallel applications. The U.S. Navy developed the Project Evaluation and Review Technique (PERT) to develop the schedule for the construction of its Polaris Program. This technique used probabilistic durations for activities. At the same time, DuPont was developing the Critical Path Method for its construction projects. The CPM used deterministic duration for activities. CPM became the most common tool for scheduling construction activities.

There are two methods for CPM calculations, arrow diagramming and precedence diagramming. In the arrow diagramming method, project activities are shown as arrows. Circles at the beginning and end of activities are called nodes. Pairs of nodes or letters are used to identify each activity.

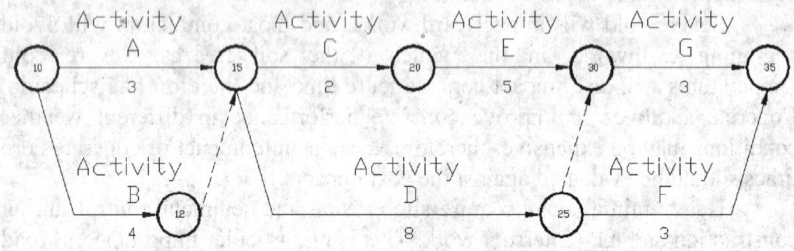

Arrow Diagramming

In the precedence diagram, activities and their durations are shown "on the node." Sequence between tasks is shown with arrows between related activities.

GENERAL REQUIREMENTS

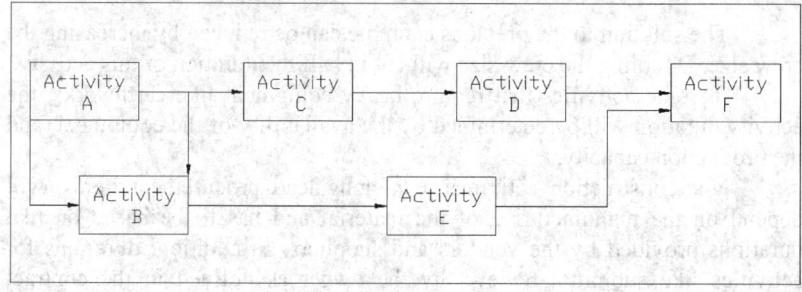

Precedence Diagramming

Precedence diagramming is capable of representing activities that start or end in parallel with other activities. Precedence diagramming became the dominant method for CPM calculation and will be described in more detail later.

Identification of Activities. An activity is any significant unit of work within the WBS' work package. There is no one "right" way to define activities for a given project. Specific activities are created as the amount of work of each type and in each area are estimated. It is important to note that the definition of activities depends on the estimate and should be representative of the estimate details. Some of the typical criteria used to identify activities are:

1. Construction type activities should not be longer than thirty days and not less than three days duration,
2. Include activities for the owner, designer and subcontractor work not just the prime contractor work,
3. Add activities for material and equipment,
4. Include milestone activities for all specified customer due dates and
5. Add activities for client tasks, such as submittal approvals, that could delay the project.

Activity Durations. The activity duration is the time it takes for an activity to be completed, given the planned amount of material, labor and equipment. While there are different factors that affect the duration of an activity, the most important factor in determining the construction activity duration is the quantities to be installed, and the crew size available.

Example:
Activity description; Above Grade CMU 12' wall (quantity 10,800 units) assume a crew of 9 laborers calculate the duration for this activity.
Reference section "04220 Concrete Masonry Units" in this book, the production of one laborer per 8 hour day is 110 -130 units a day.
Solution:
Average production per laborer = (110+130)/2 = 120 Units/Day
Crew production per day = 120x9= 1080 units per day
Duration = quantity/crew production per day = 10800/1080= 10 days.

The solution to the previous example can be reduced by increasing the crew size. Doubling the crew size will cut in half the duration of this activity.

In some activities that require heavy equipment like earthworks, the activity duration will be determined by the availability of the equipment, and the production capacity.

Non-construction activities, e.g., long lead procurement items, will depend on the manufacturing of the material and has to be based on real durations provided by the vendors and suppliers. In addition, durations for activities like submittal reviews by the owner as defined in the contract documents should be represented on the schedule accurately.

CPM calculations. The purpose of scheduling is to calculate the project completion date and the dates for completion of project activities and intermediate milestones. The CPM method will calculate two sets of dates for each activity the Early Start-Early Finish and Late Start-Late Finish. As stated earlier precedence diagramming (also called network diagram) graphically represents the relationships between the project activities. For every project, there is one Start activity and one Finish activity.

Early Start (ES) of an activity is the earliest time the activity can possibly start, allowing for the time required to complete preceding activities.

Early Finish (EF) of an activity is the earliest possible time by which an activity can be completed; it is determined by adding the activity duration to its early start time. (ES + Duration = EF)

Late Finish (LF) of an activity is the very latest it can finish and still allow the project to be completed on time.

Late Start (LS) of an activity is the very latest it can be started and still allow the project to be completed on time. It is determined by subtracting the activity duration from the LF time. (LF – Duration = LS)

Forward Pass. The forward pass calculated the Early Start (ES) and Early Finish (EF) of each activity, using the formula (ES + Duration = EF). The calculation proceeds from project start to project completion, from left to right. Each activity ES starts as soon as the last of its predecessor activities is finished. If two activities converge at the same node, then the path with the greatest duration is the Early Finish time.

Backward Pass. The Backward Pass calculates the Late Finish (LF) and Late Start (LS) of each activity. The calculation proceeds from project finish to project start, working from right to left. The LS is calculated by subtracting the activity duration from the LF of its earliest successors. The Backward Pass with the smallest duration is the Late Start time.

138

GENERAL REQUIREMENTS

All project activities are related to one another by one of the following relationship types:

Finish to Start or FS: This is the most common relationship between two activities and indicates that a successor activity cannot start until its predecessor activity is finished.

Start to Start or SS: In this relationship, the successor activity can start after the start of the predecessor activity.

Finish to Finish or FF: In this relationship, the successor activity cannot finish until the predecessor activity is finished.

The project activities should also contain the following information:
Manpower: The number and craft types anticipated for the work. This will aid in manpower forecasting.

Labor Hours and Labor Dollars: The total craft hours and labor dollars for the work. This will assist in forecasting the cash flow and anticipated costs.

Material Quantity and Dollars: By including the material quantities and dollars the contractor can use these to assist in payment for material and cash flow forecasting.

Equipment Hours and Dollars: These will assist in equipment and cash flow forecasting.

Network Logic Diagram: Network logic diagram represents all the relationships between all the project activities. Every logic diagram should have one start activity, such as "Notice to Proceed", and one finish activity, such as "Project Acceptance", to properly perform the scheduling calculations.

Free Float is the amount of time an activity can be delayed without affecting the early start of the following activity, and also affecting any other activities in the network.

Total Float is the amount of time an activity can be delayed without delaying the end date of the project, and is defined as the difference between the LS and ES of an activity. Activities with 0 Total float are critical activities and are said to fall on the Critical Path.

Using the arrow diagram shown on page 136, complete a Forward Pass to calculate the ES and EF times of each activity to determine the finish time of the project. Next, complete a Backward Pass to calculate the LS and LF times of each activity. Remember, the forward path with the greatest number of days and the backward path with the least number of days will identify our critical path. The diagram should look as follows:

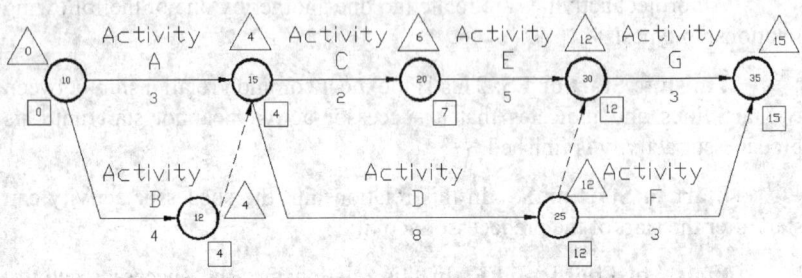

Arrow Diagramming – Forward/Backward Pass

Next we can use this updated arrow diagram to complete our time analysis table below.

Time Analysis Table								
Activity	Nodes	Duration	ES	EF	LS	LF	Free Float	Total Float
A	10-15	3						
B	10-12	4						
-	12-15	0						
C	15-20	2						
D	15-25	8						
E	20-30	5						
-	25-30	0						
F	25-35	3						
G	30-35	3						

In order to fill out the time analysis table, we look at the values we got from our forward and backward passes through the arrow diagram above. The forward pass has times shown with a triangle shape and the backward pass has times shown with a square shape. In the table, we have identified the node's starting and ending point. The ES column will correspond to the starting node value on the forward pass. The EF column is per our equation on page 139. The LF column will correspond to our ending node of our backward pass. The LS column is also per our equation on page 138. The free float column and the total float column is the difference between our ES and LS values. The table below is now complete.

Time Analysis Table								
Activity	Nodes	Duration	ES	EF	LS	LF	Free Float	Total Float
A	10-15	3	0	3	1	4	1	1
B	10-12	4	0	4	0	4	0	0
-	12-15	0	4	4	4	4	0	0
C	15-20	2	4	6	5	7	1	1
D	15-25	8	4	12	4	12	0	0
E	20-30	5	6	11	7	12	1	1
-	25-30	0	12	12	12	12	0	0
F	25-35	3	12	15	12	15	0	0
G	30-35	3	12	15	12	15	0	0

It should be noted that most commercial software today already calculate these values and generate a table similar to this one.

GENERAL REQUIREMENTS

Critical Activity is an activity with **zero float.**

Contingency Time. Planning and scheduling of a project must include consideration of the impact of weather and other potential delaying factors. The development of individual activity durations is based on the concept that the actual productivity for each activity can be monitored and controlled to that duration. The best and most consistent method of planning for potential and anticipated extra days, for which contract time extensions will not be allowed, is to include time contingencies as a separate and specific consideration. The contractor should review a particular path of activities that is likely to be delayed, review the overall duration of that path, anticipate the time impact, and include a time contingency as a separate activity at the end of the path. This process places float on the critical path to accommodate nonproductive project time and assures that the scheduled start date for activities occurring later in the project are more realistic.

Resource Constraints. Construction project resources can be summed in the following main categories:

1. Labor.
2. Materials and installed Equipment.
3. Installation Equipment.
4. Subcontracts.
5. Small Tools and consumables.

The construction schedule provides the means to plan ahead the resource requirements for the project. Some of these resources drive the activity durations, while others drive when activities can start. Labor and installation equipment would drive the activity duration. In the example provided earlier the duration was driven by the availability and size of crews, and labor production rates. Similarly the availability of concrete pumps and the number of concrete trucks that could be delivered to site in a day would drive the duration of a concrete pour activity. A more complicated example of resource constraints is when different concurrent activities compete on the utilization and availability of a limited number of shared equipment like cranes, man-lifts, welding machines etc. CPM scheduling provides tools for "resource loading" of schedule activities to identify equipment and labor utilization and means to optimize resource usage. (Resource loading and Resource leveling are subjects for further study by the reader). In addition to driving the activity durations some resources would drive when an activity can start. One example would be the long lead delivery materials, like special curtain-walls, which may require approval of detailed workshop drawing prior to fabrication and for which there is a long backlog of orders. If not planned correctly, the delivery of the curtain wall would drive when the installation would take place, even if the predecessor construction activities are completed.

Monitoring/Updating. Once the project schedule is completed and agreed on, this schedule becomes the "Baseline" schedule against which all progress will be measured. The baseline schedule is frozen and saved and a copy of the baseline would be used for monitoring and updating. Progress should be monitored periodically, ideally weekly or monthly. The update period is the time period between the previous update and the current update. All activities scheduled during the update period should be monitored and progress is measured. Once the progress is updated an evaluation of the remaining works need to take place to take into account changes in construction progress and all update information is incorporated in the schedule update. This entire process of monitoring and updating should take place promptly and on a timely basis agreed with the entire project team. The monitoring and updating process is detailed below.

Measuring Progress

Activities Started and Finished. All activities that are scheduled to start or finish during the update period are identified and the actual start and finish dates recorded. Attention should be paid to activities that were supposed to start but did not, as well as activities that did not finish on time. Activities that did not start on time are delayed by predecessor activities and the full impact of this delay needs to be understood. If the activity is on the critical path, this delay will impact the project completion date unless mitigation efforts are addressed. Similarly activities on the critical path that did not finish on time will impact the project completion date.

Activities Percent Complete. Activity percent complete need to be identified for started activities or activities continuing through the update period. The percent complete of an activity is either a function of the quantity of the activity e.g. feet of pipe erected. Percent complete could also be a function of the key driving resource like equipment hours, or labor crews, or number of blocks placed etc.

Productivity. Productivity is the measure of the crew's ability to meet estimated production rates for unit quantities. If the estimate's production rate for pipe installation is 4Ft/Labor hour, and the actual production rate is 3Ft/hr., the productivity is 75%. Productivity measurement is a good indicator for the crew's capability to finish the remaining works in the scheduled duration.

Remaining Durations. Based on percent complete and productivity, the remaining duration on an activity could be assessed.

Example:
Activity: install chilled water pipe
Quantity: 1600 ft (From estimate)
Production rate: 4 ft/hr (from estimate)
Crew size 5 pipefitters (40 hrs/day)

GENERAL REQUIREMENTS

Duration: 10 days (40hrsX4ft/hr)

Percent Complete: 20% and Productivity: 80%

Solution:

Installed quantity= percent complete X estimate quantity = 0.2X1600= 320 ft

Remaining quantity= 1600- 320=1280 ft

Remaining production rate = 0.8X4 = 3.2 ft/hr

Production per day = 3.2 X40= 128 ft/day

Remaining duration = remaining quantity / production per day
$$= 1280/128 = 10 \text{ days}$$

In this example the new estimated the total duration for the activity was driven by the productivity and is 12 days instead of 10.

Review of Remaining Works

Identification of Remaining Works. When updating activities' remaining durations, it is also important to review all the remaining works to ascertain that all remaining scope is covered. Sometimes changes in construction means and method may require adding or re-sequencing activities to complete remaining scope.

Learning Curves. In evaluating the remaining works it would be important to factor in the fact that production rates at the beginning of a process could be improved as the construction crews get more familiar with the work. In the example above, we assumed that the productivity for the pipe installation will continue at 80%. It could be fair to consider that since this activity only started, that the crew's learning of the process could improve productivity.

New or Unforeseen Activities. Remaining works could change by changing project scope, or by unforeseen activities or events, and new activities may need to be added. A heavy snow fall may interrupt site works and require adding an activity for snow removal prior to resuming the scheduled works is only one example.

Changes in Means and Methods. Site conditions may require change of construction means and methods and revision of the remaining works to reflect new conditions. An example of change in means and methods is changing a trench excavation from open cut in a normally dry site conditions to bracing and shoring after a major rainfall that deemed the soil unstable for open cut excavation. Adding shoring and bracing will require additional durations and may require change in excavation and earth moving equipment all impacting the schedule for the remaining works.

Schedule Updates

Updating the Schedule Activities. Once all of the above factors are considered the schedule is ready for update. Most scheduling is performed

143

using scheduling software, which is becoming more user friendly. Before updating the schedule, a backup copy of the previous update should be maintained. Updating the schedule involves moving the data date to reflect the current update period and incorporating progress and remaining works.

Update Logic. The logical sequencing of events may require updating depending on the review of the remaining works.

Run Scheduling Calculations. Some software will automatically recalculate the schedule while others will require the scheduler to deliberately initiate the schedule calculation. In either case, the schedule update should be reviewed after the schedule calculation to ensure all updates have been incorporated accurately.

Reporting. Once the schedule is updated, communicating the new schedule to the project team is very important for the success of the project and meeting the schedule dated. Most software allows presenting the project information in many different ways and reports could be customized to the audience needs. Main reporting types are discussed below.

Summary schedules and Milestones. Summary schedules and milestone schedules provide a top level summary of the projects main activities and key milestone important to the project. These schedules should be easy to read and usually fits in one page. These schedules are very useful in communicating project goals to management, clients, or for high level coordination.

Look-Ahead Schedules. Look-ahead schedules are very important for managing project activities, and are usually 3-week, or 90 day look-ahead. The three week look-ahead filters only those activities that are occurring in the upcoming three weeks and should provide all detailed activity information, it is mostly used in weekly construction coordinating meetings. The 90-day look ahead schedules provide a wider window to keep the project focused on the near future needs. The 90- day look ahead is more common on larger projects.

Detailed Schedules, Grouping and Sorting of Activities. In larger projects, schedule details may require different sorting and grouping requirements for different trades, or locations. In this case schedule reports grouped by discipline, or location may be required to facilitate communicating schedule information on as needed basis.

Critical Path Analysis. Critical Path was defined earlier as activities with 0 float. An analysis of all activities on the critical path is very important to focus the project team on the activities that will impact the project duration. In some situations, analysis of near critical activities is also important as these activities may easily slip into the critical path if they are delayed. Depending on the size and duration of the project near critical

144

GENERAL REQUIREMENTS

activities could be activities that have floats ranging from one week to a month. The Critical Path will almost certainly change from one update to the next. It rarely stays the same and if it does, the update may be flawed.

Progress Meetings/Delays Mitigation. Progress meetings are necessary to conduct on a periodic basis to ensure that the schedule reporting is communicated to project team, and may be necessary to initiate mitigation action plans to reduce potentials for delays. Progress meetings also helps keep the project team focused on the schedule and the project completion dates.

CONSTRUCTION EQUIPMENT

The estimator must have the ability to match the construction equipment to the job conditions. The construction equipment costs will be carried in the direct and/or indirect costs of the estimate. Equipment is must be matched to meet the project requirements. For instance when excavating the footings for a foundation the use of a backhoe or excavator with dump trucks is more appropriate than the use of scrapers. It is also important to use the right size of equipment for the project. The excavation of the footings may require the use of a backhoe or excavator with a one cubic yard bucket, not an excavator with a three cubic yard bucket.

It is also imperative to match the size of the equipment to the other equipment involved in the work operation. If you have an excavator that can operate at 80 cubic yards per hour, the most efficient use of dump trucks for hauling is to have the right number and capacity of trucks to haul eighty cubic yards per hour. This will minimize the downtime of the excavator and standby time of the dump trucks so the best possible costs can be met on the project. If the wrong size or type of equipment is used on the project, it will only lead to inefficiencies that increase the cost of the work.

Determining Hourly Rate

Whether the equipment is rented or owned, the estimator should charge for it at a fixed hourly rate. This rate would include the basic rental plus delivery, setup, and dismantling charges. It would then be divided by the working hours that the equipment stayed at the site. In some cases one might elect to charge only for the hours that the equipment was in use at the job site. The remainder of the time would then be charged to indirect costs at a standby rate which excludes all operating expenses. Operating costs would be added as a separate item only for the actual in-use hours.

Equipment rental rates vary greatly throughout the country, depending on local practices and conditions. They might also vary according to how the equipment is used, such as for hard service in rocky soil or for use in sand.

It is standard practice to base rates upon one shift of 8 hours per day, 40 hours per week, or 176 hours per month for a 30 consecutive day period. In some urban areas, union building trades work 7 hours per day, 35 hours per week. Rental equipment will be invoiced to the contractor based on 40 hours per week, 176 hours per month regardless of the shorten daily work day. Equipment rental/lease rates are negotiable. If it is for one day, the previous will hold true. However many rental/lease practices discount rental rates. Paying the rate for three days will get the equipment for five days. Paying for three weeks can get the equipment for the month, etc.

The leaser usually bears the cost of repairs due to normal wear and tear on non-tractor equipment. On rubber-tired hauling and tractor equipment, the lessee may be required to bear all costs for maintenance and repair.

The equipment rented should be delivered to the lessee in good operating condition and should be returned to the lessor in the same condition as delivered, less normal wear.

Rates used herein are f.o.b. the lessor's warehouse or shipping point. On short term rentals or leases, the lessee usually pays all freight or delivery charges from shipping point to destination and return. Lessee also pays all charges for unloading, assembling, dismantling, and loading where required. On longer term rentals or leases the lessor may include all of these costs in the rental/lease rate.

Rentals are subject to the terms and conditions of the leaser's rental contract. Rates do not include contractor's insurance, license fees, sales taxes, or use taxes. They also do not include the operating expenses, unless specifically identified in the agreement terms and conditions.

Where the contractor owns the equipment, the estimator must arrive at an hourly cost to substitute for the basic rental. To do so, they must consider the following:

    a) Original capital investment
    b) Anticipated life
    c) Depreciation
    d) Estimate of maintenance
    e) Estimate of taxes, insurance, interest, and storage

Items c), d), and e) are expressed in percentages of the original capital investment, item a). The estimator will need the cooperation of the firm's accountant to determine the hourly cost figure, as well as some knowledge of the possible future use of the equipment. The capital investment cost should include all original delivery, test, and start-up charges. The expected life may be based on Internal Revenue tables. It is usually entered as 2000 hours per year (40 hour week, 50 week year).

Depreciation cost is the original capital investment less any resale or scrap value. This is often so small that depreciation on all but highway trucks may be figured at 100%. However, in times of high inflation, for equipment held only a short time, the inflation rate may approximate the depreciation rate. The equipment would appear as having zero depreciation.

146

GENERAL REQUIREMENTS

Maintenance allowances will vary from 40% to 100% of the investment cost, but most equipment will fall in the 65% to 75% bracket. Those items falling below 50% would include air tools, gas engines, small motors, mortar mixers, vibrators, drills, screw jacks, small power tools, towers, small tractors, trailers, and trucks. Those items for which over 75% of investment cost should be used would include bituminous equipment, hoppers, crushers, derricks, electric generators, hydraulic jacks, pumps, and rollers. Taxes, insurance, and interest percentages should be worked out with the accountant. Ten percent used to be an acceptable figure for these items, but with today's rates 15% would probably be more realistic, and 3% is generally added for storage.

A formula to arrive at this entry would be:

$$\frac{p* (n + 1)}{2} + p** \text{ (useful life)} = \text{total percentage}$$

p* is percentage for taxes, insurance, and interest.
p** is percentage for storage.

If the useful life is five years, p* is 15% and p** is 3%, then the total percentage to be entered under item e) would be:

$$\frac{15(5+1)}{2} + 3x5 = 45 + 15 = 60\%$$

Once items a) through e) are determined, the hourly rate can be figured as follows:

$$\frac{\text{Capital investment}}{\text{Anticipated life in hours}} \text{ x sum of \% for items c), d), and e)}$$

As an example, assume the contractor has purchased a $200,000 piece of equipment. The following is a reasonable cost projection:

a) Capital investment $200,000.00
b) Anticipated life 5 yrs at 2000 hrs   10,000 hrs.
c) Depreciation $10,000 scrap value   95%
d) Maintenance   70%
e) Taxes, insurance, interest & storage   60%

Hourly Use Charge =
$200,000 ÷ 10,000 hrs. = $20.00 per hr.
$20.00 x (95% + 70% + 60%) = $45.00 per hr.
$20 x (0.95+0.70+0.60) = $45.00 per hour

Many equipment manufacturers will provide calculations to be used for their equipment. Check what is available online from the manufacturer of your equipment. Care must be used in adapting their calculations to your practices. Oftentimes the equipment manufacturer calculations are based upon ideal conditions – which are rarely encountered.

02 SITE WORK

02010 SUBSURFACE EXPLORATION

When preparing estimates on any structures involving foundations or earthwork, there is a need to obtain all information available concerning the geotechnical conditions and characteristics of the soils to be encountered in the subsurface. Frequently some of this information is made available on the working drawings as an aid to the contractors, but complete reliance on such limited information places the contractor at risk, especially when the drawings provide for a disclaimer of responsibility by the architect/engineer for the accuracy or completeness of the subsurface information. A site investigation is mandatory, because many conditions can be determined through adequate visual observations of the particular site involved and adjacent property. There can be discovery of previous stream beds, evidence of prior dumping of debris, existence of abandoned mining operations, existence of flood plains, appearance of rock outcroppings, overhead structures and wires including many other situations that would strongly influence the construction means and methods and the associated costing involved.

In urban areas records are usually available from utility companies and highway authorities concerning subsurface conditions affecting their installations. All 50 states and the District of Columbia plus Canada have incorporated laws normally called "Call Before You Dig" telephone number. Anyone, including home owners, is required to notify this agency before they dig. The time to call prior to digging will vary with each location; the telephone number is normally toll free. If a contractor digs before properly notifying the local agency the contractor is responsible for all repairs if required and possibly subject to a fine. It must be recognized that any penetration of the ground surface has attendant risk, and the accumulation of

149

knowledge of the conditions to be encountered during excavation will tend to control the risk to a reasonable level thus avoiding exposure to a catastrophic situation.

Excavating test pits or caissons produces accurate information in cohesive soils, insofar as this method permits visual inspection of the soil in place, and samples removed for laboratory tests may be carved out of the pit walls in a relatively undisturbed condition. Costs of this work are high, however, and will be about the same as excavating costs for comparable pits and caissons.

However, most subsoil investigations require studies of the soil at greater depths than can be economically reached by test pits. For these projects, various methods of soil boring and sampling are used. A relatively small diameter hole is sunk into the ground from which soil samples are obtained.

Hand Auger Boring Method. The hand auger boring method, satisfactory for highway explorations at shallow depths, can be used only for preliminary investigations for foundations. Auger borings are used in cohesive soils and cohesion less soils above ground water. Depths up to 20 feet (6 meters) may be reached. Soil samples obtained through this method are disturbed to such an extent that little or no information is furnished about the character of the soil in its natural state. Where the auger hole is filled with water, sample disturbance is particularly acute. Auger holes are most useful for ground water determination.

Hand auger borings generally cost from $20.00 to $50.00 per lin. ft. ($65.61 to $164.004 per m) depending on the soil, the amount of boring necessary, and the diameter of the hole, which will vary from 4" to 8" (100 to 200 mm).

Split Spoon Sampling. A split barrel sampler known as a split spoon is frequently used for obtaining representative samples. This method is described in detail in *ASTM Method D-1586 Standard Test Method for Penetration Test and Split-Barrel Sampling of Soils.*

In securing split spoon samples, the drill hole is opened and cleaned to the desired sampling level by drilling bits and wash water, or alternatively, the hole may be drilled with a regular power auger. Drill casing or drilling mud is used when it is necessary to prevent the soil above the sampling level from closing the drill hole. A drive pipe is attached to the upper end of the drill rod and a slip jar weight placed on the drive pipe.

The sampling spoon is then driven into the soil by repeated blows of a drop hammer. The number of blows required to drive the spoon are recorded in the logs. The spoon is then brought to the surface, and the sample is removed, classified, and placed in a glass jar.

The ASTM method specifies a 2-inch (50 mm) OD x 1-3/8 inch (33 mm) ID split spoon driven 18" to 24" (450 to 600 mm) by a 140 lb. (64 kg) weight falling 30 inches (750 mm). Various other sizes of split spoons and

150

drop weights and lengths of drops are employed. The split spoon method is most useful in granular or dense soils.

Thin-Wall Tube Sampling. The method that yields the least disturbed samples is thin-wall tube sampling. This method is described in detail in *ASTM Method D-1587. Standard Practice for Thin-Walled Tube Sampling of Soils for Geotechnical Purposes.*

In this method the hole is opened and cleaned as in the split spoon sampling method, but a thin-wall tube, usually with an outside diameter of 2", 3", or 5" (50, 75 or 125 mm) is then attached to the end of the drill rod and lowered to the bottom of the boring. The tube is then pushed into the soil by a hydraulic piston arrangement with a rapid and continuous motion.

The thin-wall tube is then raised to the surface, cleaned, labeled, and sealed to prevent loss of moisture. It is then taken to the laboratory for testing. Here it is usually cut into short lengths, and the sample in each length is ejected and tested.

Samples are normally taken at 5-foot (1500 mm or 1.5 m) intervals or every time a soil change is expected, determined by watching the overflowing wash water and feeling the resistance to penetration of the drill bit.

Cost of drilling and sampling is usually between $12.00 and $35.00 per lin. ft. ($40.00-$115.00per m) depending on the depth, diameter, and difficulty of drilling, plus mobilization and demobilization charges for equipment and workers.

Laboratory testing, for density, moisture content, compression, or mechanical analysis, is performed at an additional cost, depending on what tests are needed. This practice is limited to soils that can be penetrated by the thin-walled tube. This sampling method is not recommended for sampling soils containing gravel or larger size soil particles cemented or very hard soils. Other soil samplers may be used for sampling these soil types.

Core Borings. In addition to soil samples, information is frequently wanted on the nature or thickness of underlying rock. Core borings are made for this purpose with rotating coring tools such as diamond drills, shot drills, or carbide bits. Continuous samples are generally recovered for examination and further testing. *ASTM Standard D-2113, Standard Practice for Rock Core Drilling and Sampling of Rock for Site Investigation.* Diamond Core Drilling for Site Investigation, should be called for in performing core borings.

The cost of making core borings will depend upon the location of the work, quantity, and the difficulty encountered in getting to the holes.

The drilling costs per lin. ft. (300 mm) should be estimated at from $30.00 to $60.00 ($99.00-198.00 per m), with the lower costs on jobs of larger size. To the above prices, add Mobilization and Demobilization charges for equipment and workers and all required testing charges.

Other Methods. There are many methods used to investigate soil conditions for foundation purposes. Some are reliable and sufficient to use for design purposes while others produce inadequate or dangerously deceptive data. In problem soils the gathering of the basic information under controlled conditions is important, but to obtain definitive and factual information there is a need for professional engineering interpretation. In such instances the best approach is to obtain the services of a soils or geotechnical engineer with the laboratory facilities to perform a complete subsurface investigation and provide a written report of the findings and recommendations with a geotechnical profile for remedy of the problems to be encountered.

Engineering and laboratory services are for the most part inexpensive but can become expensive based on accessibility, weather, depth of penetration, water, type(s) of soil, boulders, cobblers, and rock. However, when balanced against the magnitude of potential loss and redesign that could be encountered without this advance knowledge, the cost is justified.

Engineering costs vary widely by geographic location and services required. There usually is enough competition to allow the contractor to solicit proposals from several geotechnical engineering firms, who provide services that are tailored to the specific project.

Seismic resistivity, vane shear tests, cone penetration test, and nuclear methods are more frequently being used in subsurface investigations, performed by specially qualified seismic engineers having the necessary equipment and regularly engaged in this type of work. Cost estimates will be tendered to the contractor predicated upon the services required.

02100 CLEARING

Clearing and grubbing of a construction site by the removal of trees and vegetation has become costly because of federal, state and local government environmental protection regulations. Most areas prohibit the burning of trees. If the wood does not have a market value, the use of chipping machines is employed. This method requires additional personnel to cut the trees into manageable lengths, remove the branches, and transport them to the chipping machine. Some of the cost can be recovered by selling the by-product to nurseries and landscape companies to use as mulch. Stump disposal is very expensive since stumps cannot be buried at the project and have to be transported to a distant acceptable disposal area.

SITE WORK

Clearing, Cutting & Chipping		
Per Acre		
Light	6 " diameter	$1,650 - $2,090
Medium	12 " diameter	$2,750 - $3,300
Heavy	16 " diameter	$3,850 - $4,950
Per Sq. Meter		
Light	152 mm diameter	$ 0.41 - $ 0.52
Medium	305 mm diameter	$ 0.68 - $ 0.82
Heavy	406 mm diameter	$ 0.95 - $ 1.22
Grub Stump Plus Haul Charge, per ea. . . . .$30.00 - $55.00		

If chipping is not required, average prices can be reduced by 25%.

Selective Cutting Down and Removing Trees

The cost of cutting down and removing trees will vary with their size and the method used in cutting and disposing of them.

The following production times are based on a crew of 3 or 4 workers, cutting down trees by hand, removing branches, and cutting the trunk into 4'- 0" (1.2 m) lengths. Labor costs of digging down around roots and removing a stump are given extra. It is assumed the limited quantity would not require chipping.

Cutting - Removing Trees by Hand			
Approx. Diam. of Tree	Approx. Height of Tree	Labor Hours	Removing Stump, in Hrs
8 " - 12 "	20 ' - 25 '	9 - 12	8
14 " - 18 "	30 ' - 40 '	12 - 16	10
20 " - 24 "	45 ' - 50 '	20 - 24	12
203 mm - 305 mm	6.1 M - 7.6 M	9 - 12	8
356 mm - 457 mm	9.1 M - 12.2 M	12 - 16	10
508 mm - 610 mm	13.7 M - 15.2 M	20 - 24	12

If necessary to cut trunks and stumps into small pieces and load into trucks, add extra for this work. Add also for removal from the site.

Today's projects benefit from the many improvements in portable saws—gasoline, air, electric and hydraulic. Trees are removed with power chain saws speeding the work. A chain saw will cut within 2 to 4 inches (50 to 100 mm) of the ground, but if the stump must be removed, it is usually done by hand or "stump" grinding machine.

Chain saws are gas- , electric- , or air-powered; 12", 24", 36", 48" (300, 600, 900, 1200 mm) and larger; and rent for about $40.00 per day.

Labor required using a chain saw, with 3 to 4 workers in the crew cutting trees, removing branches, and cutting the trees into 2'-0" (600 mm) lengths, should average as follows:

Cutting - Removing Trees with Power Saws		
Approx. Diam. of Tree	Approx. Height of Tree	Labor Hours
8 " - 12 "	20 ' - 25 '	2.25 - 3
14 " - 18 "	30 ' - 40 '	3 - 4
20 " - 24 "	45 ' - 50 '	5 - 6
30 " - 36 "	50 ' - 55 '	10 - 16
40 " - 48 "	55 ' - 65 '	18 - 24
203 mm - 305 mm	6.1 M - 7.6 M	6.10 - 7.62
356 mm - 457 mm	9.1 M - 12.2 M	9.14 - 12.19
508 mm - 610 mm	13.7 M - 15.2 M	13.72 - 15.24
762 mm - 914 mm	15.2 M - 16.8 M	15.24 - 16.76
1016 mm - 1219 mm	16.8 M - 19.8 M	16.76 - 19.81

A faster and more economical method of removing trees is by using a bulldozer or loader, provided it does not cause damage to adjacent trees.

Trees up to 15 inches (375 mm) diameter or trees such as pine, which have small root systems, can be rapidly uprooted by raising the dozer blade or shovel bucket to maximum height and pushing against the tree. It may be necessary to strike the tree a few times to loosen it.

Trees with diameters of 24 inches (600 mm) or more, or trees like elms, maples, and oaks, which have heavy, wide-flaring root systems, usually require root cutting before they can be pushed over. First make a trenchcut with the dozer blade or shovel bucket across the roots on the side opposite the direction in which the tree must be felled. Backfill the trench and form a ramp up to the trunk so the machine can strike as high on the tree as possible. Unless the tree is exceptionally large or well-rooted, it can now be pushed over. If not, try cutting the roots on each side of the tree perpendicular to the first cut, and if necessary, the roots on the side toward which it will be felled.

Once the tree is felled, it can be pushed or dragged to a disposal area where it can be sawed into lengths for hauling or burning (where permitted). *A Word of Caution* - Always watch for falling limbs. In the case of pines, if the whipping action is too great, the top may snap off and kick back on the machine or operator. Also, as the tree begins to fall, the machine must be backed up quickly so that the upturning root system will not come up under the machine, possibly damaging the unit or hanging it up.

154

SITE WORK

An experienced operator, using a medium to large tractor, 80-HP (60 kg) and up if crawler type, 160-HP (120 kg) and up if rubber-tired, with ground conditions affording good traction, and no problems regarding steep slopes, should fell trees at the following rates:

Removing Trees by Machine		
Approximate Diam. of Tree	Approximate Height of Tree	Minutes Req'd for Falling
8 " - 12 "	20 ' - 25 '	5 to 10
15 " - 24 "	30 ' - 50 '	15 to 20
27 " - 36 "	55 ' - 75 '	25 to 30
203 mm - 305 mm	6.10 M - 7.62 M	5 to 10
381 mm - 610 mm	9.14 M - 15.24 M	15 to 20
686 mm - 914 mm	16.76 M - 22.86 M	25 to 30

Add time for pushing felled trees to disposal area and for disposing of them or chipping them.

Stripping & Storing Topsoil

Following the site clearing operation, the stripping and storing of topsoil, if available, will be the next phase. In the interest of economy, the recovery of the maximum quantity of existing topsoil is important.

However, any area cleared of trees and stumps will have a loss factor of approximately 15% to 50% depending on the number and size of stumps removed. The remaining topsoil should be stripped, cleaned of foreign materials, stored in a convenient location, and protected from erosion or contamination.

The depth of the topsoil influences the production rate. Less quantity would be available in a given area if it were 4" (100 mm) in depth as opposed to 6" (150 mm) and yet the same area would have to be stripped by the dozer. The selection of the storage area is important. It should not only provide protection for the topsoil material, but it also should minimize the haul time involved in the storing and redistribution of the material to the final location.

Strip and Store Topsoil		
Description	Per Cu. Yd.	
Dozer 200 Hp, 4" - 6" Depth, 100' haul-average conditions	$1.14	- $1.42
Dozer 200 Hp, 4" - 6" Depth, 300' haul-average conditions	$1.52	- $1.89
Dozer 400 Hp, 4" - 6" Depth, 100' haul-average conditions	$0.95	- $1.18
Dozer 400 Hp, 4" - 6" Depth, 300' haul-average conditions	$1.52	- $1.89
Description	Per Cu. Meter	
Dozer 150 kW, 100mm - 150mm Depth, 30.48 M haul-average cond.	$1.49	- $1.86
Dozer 150 kW, 100mm - 150mm Depth, 91.44 M haul-average cond.	$1.98	- $2.48
Dozer 300 kW, 100mm - 150mm Depth, 30.48 M haul-average cond.	$1.24	- $1.55
Dozer 300 kW, 100mm - 150mm Depth, 91.44 M haul-average cond.	$1.98	- $2.48

Note: Erosion control and protection are additional

If adverse conditions are present increase unit price by 20% to 33%. Price range depends on terrain and capacity of equipment.

The redistribution of topsoil usually requires some preparation of the ground prior to placing the topsoil.

Construction activities cause rutting, depressions, and general unevenness of the site area, and to avoid excessive quantities of topsoil being used, the site should be graded to the designed contours less the depth of the topsoil. Any preparatory work such as scarifying, raking, or rolling should be accomplished before the introduction of the topsoil. Requirements for soil conditioning, fertilizer, seeding etc., become cost items in addition to the basic spreading cost. When specifications have rigid conditions concerning the quality of topsoil, it may be necessary to submit samples to a laboratory for analysis to establish compliance.

Off-site purchase of topsoil is very expensive and can vary from $20.00 to $60.00 per cu. yd. ($26.15-78.47 per cu. m) depending on the quality and availability.

156

Spread and Prepare Topsoil		
Description	Unit	Average Cost
Scarify subsoil-average conditions	Sq.Ft.	$0.03
Spread, 300' with dozer, Average conditions	Cu.Yd.	$1.88 - $3.13
Hand dress topsoil, ($20.00 average hourly rate)	Cu.Yd.	$3.75 - $5.00
Rolling, push roller	Sq.Ft.	$0.06 - $0.09
Description	Unit	Average Cost
Scarify subsoil-average cond.	Sq.M.	$0.32
Spread, 91.44 M with dozer, Average conditions	Cu.M	$2.46 - $4.09
Hand dress topsoil, ($20.00 average hourly rate)	Cu.M	$4.90 - $6.54
Rolling, push roller	Sq.M.	$0.65 - $0.97

Caution is to be exercised in determining if the specified depth of topsoil is loose or compacted depth. Depending on the characteristics of the material involved, topsoil will compact from 33% to 50%. This could mean to obtain 4" (100 mm) compacted depth would require 6" (150 mm) of loose material. An estimator not taking this into consideration would discover a 50% shortage from the estimated quantities.

If topsoil is to be placed in a series of small restricted areas or at the perimeter of buildings, walks, drives, or other obstructions, additional hand labor will be required to place, grade, and hand compact the material.

It is absolutely essential that the estimator become familiar with local laws regarding disposal. Silt fence for containment during stripping and spreading operations costs about $0.75 to $3.00 per lin. ft. (300 mm). *If salt hay is used*, costs vary greatly depending on the availability of material, usually between $1.00 and $6.00 per lin. ft. (300 mm).

02060 DEMOLITION

Demolition can be divided into several main types - demolition of entire buildings or structures, controlled removals for historic preservation work, general building demolition of items such as doors and frames, windows, floors, ceilings, and partitions and controlled blasting.

In most areas of the country, there are specialty contractors whose sole business is the demolition of entire buildings and structures. These contractors will use a wide range of heavy equipment, even explosives. Controlled blasting with its many laws and regulations governing explosives

157

it is advised a specialty contractor with experience be consulted for the controlled demolition work. It is best to solicit a quotation for demolition. However, the following per cubic foot (cubic meter) unit prices will provide some budget guidelines in the absence of quotes:

General Demolition		
Building Type	Unit Price per Cu.Ft.	Unit Price per Cu.M.
Low Rise-Steel Frame	$0.54	$19.07
Low Rise-Concrete Frame	$0.72	$25.43
Low Rise-Masonry Wall Bearing	$0.63	$22.25

Demolition associated with historical preservation work is highly specialized, and since each project has its own particular set of intricacies, it is extremely difficult to produce accurate historical cost data that can then be saved for use in future estimates.

Of less intricate nature than the controlled demolition of historical preservation works, but closely related, is general building demolition. It is this type of work that we will address for the various types and items of demolition and their appropriate labor production tables and costs.

The estimator must always take into consideration factors such as working space, is the building occupied or not, method and route in removing debris to the trash receptacle or truck, special equipment requirements, temporary bracing and shoring, dust and security control partitions, length of the haul to the dump site, and special use permits, EPA permits, and dump fees.

Demolition Using Compressed Air Tools

Compressed air is used extensively in all classes of construction work including such operations as concrete breaking, masonry chipping, excavating, compacting earth, and driving sheet pilings. To reduce the overall operating cost (ownership or rental rates), it is always wise to use an air compressor that is capable of handling at least two pneumatic tools. The use of two tools lowers the rental rate per hour for the machine down to a level half of what it would be if only one operator was using the machine to do an operation. For example, if you had 1,000 sq. ft. (90 sq.m) of concrete sidewalk to demolish, using one compressor with two hammers and two workers would require the same total number of labor hours as one worker. However, your equipment rental hours would be half of that required if you used only one worker on the air compressor.

The following chart shows the classification of work to be demolished by an air hammer, the daily production of work per crew of two laborers at 8 hours each, and the total number of manhours required per 100 sq. ft. (10 sq.m) or per lin. ft. (mm), whichever is applicable. Keep in mind that the

SITE WORK

production rates shown in the following table only include the material break-up, not transport/removal to the trash receptacle.

Crew Daily Production			
Description	Crew Daily Production	Man-hours per 100 Sq.Ft. (9.29 sq.m)	Man-hours per 100 Lin.Ft. (30.48 lin.m)
Wall Footing			
2' wide x 1' thick	125 Lin.Ft	......	12.80
3' wide x 1'-6" thick	65 Lin.Ft	......	24.62
Slab on Grade - 4"	450 Sq.Ft.	3.56	......
Slab on Grade - 8"	250 Sq.Ft.	6.40	......
.61M wide x .30M thick	38.1 M	......	12.80
.91M wide x .46M thick	19.8 M	......	24.62
Slab on Grade - 100mm	41.8 Sq.M	3.56	......
Slab on Grade - 200mm	23.22 Sq.Ft.	6.40	......
Concrete Foundation Walls Lightly Reinforced			
6" Thick	200 Sq.Ft.	8.00	......
8" Thick	160 Sq.Ft.	10.00	......
10" Thick	135 Sq.Ft.	11.85	......
12" Thick	110 Sq.Ft.	14.55	......
Header Curb	225 Lin.Ft.	......	7.11
Curb & Gutter	175 Lin.Ft.	......	9.14
150 mm Thick	18.58 Sq.M	8.00	......
200 mm Thick	14.85 Sq.M	10.00	......
250 mm Thick	12.54 Sq.M	11.85	......
300 mm Thick	10.22 Sq.M	14.55	......
Header Curb	68.58 Lin.M	......	7.11
Curb & Gutter	53.34 Lin.M	......	9.14

Partition Removal

The following labor production table gives the daily crew (2 laborers, 8 hours each) output and the manhours required per 100 s.f. (9.3 cu.m) for the demolition of various types of partitions. The labor production hours do not include the transport/removal of demolished materials to the trash receptacle.

Partitions 10' High (3000 mm)		
Type of Partition	Daily Production Sq. Ft.	Man-hours per 100 Sq.Ft.
Metal Stud & Drywall	850	1.88
Plaster & Lath on Wood Stud	650	2.46
4" Concrete Masonry Units	650	2.46
8" Concrete Masonry Units	500	3.20
12" Concrete Masonry Units	400	4.00
4" Brick	500	3.20
Type of Partition	Daily Production Sq. Meter	Man-hours per 10 Sq.Meters
Metal Stud & Drywall	79	2.02
Plaster & Lath on Wood Stud	60	2.65
100mm Concrete Masonry Units	60	2.65
200mm Concrete Masonry Units	46	3.44
300mm Concrete Masonry Units	37	4.31
100mm Brick	46	3.44
Partitions 15' High (4500 mm)		
Type of Partition	Daily Production Sq. Ft.	Man-hours per 100 Sq.Ft.
Metal Stud & Drywall	800	2.00
Plaster & Lath on Wood Stud	600	2.67
4" Concrete Masonry Units	600	2.67
8" Concrete Masonry Units	450	3.56
12" Concrete Masonry Units	375	4.27
4" Brick	450	3.56

Partitions 15' High (4500 mm) (continued)		
Type of Partition	Daily Production Sq. Meter	Man-hours per 10 Sq.Meters
Metal Stud & Drywall	74	2.15
Plaster & Lath on Wood Stud	56	2.87
100mm Concrete Masonry Units	56	2.87
200mm Concrete Masonry Units	42	3.83
300mm Concrete Masonry Units	35	4.60
100mm Brick	42	3.83

Door & Frame Removal

The following labor production rates are for the removal of doors and frames from walls to be demolished. Keep in mind that if a frame is to be removed from a wall that is to remain in place, more care must be used to prevent needless damage and costly repairs to the wall. As a guide the hours shown in the following table could be doubled for the removal of metal door frames from masonry walls that will remain in place. The production times shown in the table do not include any transport of the demolished materials to the trash receptacle.

Door and Frame Removal	
Description	Man-hours per unit
Single wood door & frame in stud partition	1.00
Pair - wood doors & frame in stud partition	1.25
Single door and frame in masonry partition	1.25
Pair doors and frame in masonry partition	1.50
Single hollow metal door and frame in stud partition	1.50
Pair hollow metal doors & frame in masonry partition	2.50
Single hollow metal door & frame in masonry partition	2.25

Window Removal

The following table shows various types of window materials and the required man-hours per unit for their removal, including removal of trim. The production times shown in the table do not include transport of the demolished materials to the trash receptacle.

161

Window Removal			
Description	Window Size		Man-hours per unit
	English	Metric	
Wood and Vinyl up to	3 ' x 5 '	900 mm x 1500 mm	0.75
Wood and Vinyl up to	5 ' x 6 '	1500 mm x 1800 mm	1.25
Aluminum up to	3 ' x 5 '	900 mm x 1500 mm	1.25
Aluminum up to	5 ' x 6 '	1500 mm x 1800 mm	1.75
Steel up to	3 ' x 5 '	900 mm x 1500 mm	2.00
Steel up to	5 ' x 6 '	1500 mm x 1800 mm	2.50

Floor Finishes Removal

The following table shows various types of finish flooring materials and the required manhours per 100 sq.ft. (9.29 sq. m) for their removal. The production times shown in the table do not include disposal transport of the demolished.

Floor Finishes Removal		
Type of Flooring	Man-Hours per 100 sq.ft.	Man-Hours per 10 sq.m
Vinyl composition tile	2.00	2.15
Wood parquet	2.50	2.69
Wood strip	2.75	2.96
Carpet & pad - tack strip	0.40	0.43
Carpet - glued down	1.00	1.08
Ceramic tile	2.50	2.69

Ceiling Finishes Removal

The following table shows various types of ceiling finish materials and the required man-hours per 100 sq. ft. (10 sq. m) for their removal. The production times shown in the table do not include transport of the demolished materials to the trash receptacle.

SITE WORK

Ceiling Removal		
Type of Material	Man-Hours per 100 sq.ft.	Man-Hours per 10 sq.m
Drywall	1.00	1.08
Suspended plaster	3.00	3.23
Suspended acoustical tile and grid	1.25	1.35
Plywood	1.00	1.08

If the ceiling finishes are more than 10' (3 m) above the finished floor, add a 10% labor factor multiplier for every 2' (0.60 m) increment over 10' (3 m).

Wall Finish Removal

The following table shows various types of wall finish materials and the required manhours per 100 sq. ft. (10 sq. m) for their removal. The production times shown in the table do not include transport of the demolished materials to the trash receptacle.

Wall Removal		
Type of Material	Man-Hours per 100 sq.ft.	Man-Hours per 10 sq.m
Drywall	0.60	0.65
Plaster	1.00	1.08
Wood paneling	0.85	0.91
Ceramic tile	2.25	2.42

Roofing Material Removal

The following table shows various types of roofing materials and the required man-hours per 100 sq. ft. (10 sq. m) of their removal. The production times shown in the table do not include transport of the demolished materials to the trash receptacle.

Roofing Materials Removed		
Description	Man-hours per 100 Sq.Ft.	Man-hours per 10 Sq.m
Built up roofing-4 ply w/gravel	3.50	3.77
Rigid insulation board	0.75	0.81
Shingles - composition strip	0.90	0.97
Shingles - wood	1.25	1.35
Shingles - slate	1.30	1.40
Sheathing - plywood	0.95	1.02
Sheathing - 1" (25mm) plank	1.10	1.18
Sheathing - 2" (50mm) plank	1.50	1.61

Siding Removal

The following table shows various types of siding materials and the required manhours per 100 sq. ft. (10 sq. m) for their removal. The production times shown in the table do not include transport of the demolished materials to the trash receptacle.

Siding Removal		
Description	Man-hours per 100 Sq.Ft.	Man-hours per 10 Sq.m
Metal - horizontal	2.75	2.96
Metal - vertical	3.00	3.23
Wood - horizontal	3.00	3.23
Wood - vertical	3.25	3.50
Shingle - wood	3.50	3.77
Plywood	1.50	1.61
Vinyl Siding	2.75	2.96

The reader should be aware that the above productions do not take into account hazardous materials such as lead, asbestos and petroleum base products. The reader is referred to the Hazardous section of this text.

Disposal

The disposal of construction material takes many shapes today. Asphalt, concrete, steel, glass etc may need to be separated and disposed at separate dumps or may be required to be recycled at a recycling center. The cost of a construction dumpster will vary and the estimator must for each estimate call and get a quote. Generally a 30 Cu.Yd. (23 Cu. Meter) dumpster will cost about $700.00 per load.

164

SITE WORK

Some owners want to be LEED-certified (Leadership in Energy and Environmental Design) part of which requires the designer to use recycled and salvaged materials. Therefore, the estimator is cautioned that he or she needs to review the drawings and specifications carefully and be aware of all Federal, state and local requirements as to what materials are being removed and where and how they are disposed of.

Core Drilling

One of the quickest cleanest, vibration free and tidy methods to cut a hole in a concrete slab or wall is by core drilling. The core drill rig uses a circular-shaped diamond drill bit that is operated by a large drilling machine mounted on a stand. The diamond drill requires continuous water circulated into the core bit to provide lubrication and cooling. The small amount of water should not be a problem. The amount of water used is controllable, either by damming or suctioning, to prevent damage to surrounding finishes.

Core drill bits come in diameters from 1/2" to 24" (13 to 610 mm) in size and the depth of cut allowed by the average coring bit is about 12" (300 mm). However, special bits are made to allow boring through columns or walls up to 48" (1220mm) in thickness or more.

Core drilling is usually performed by specialty contractors (Semcor Inc. Keyport, NJ - Mr. Wayne Camp) acting as subcontractors whose only business is to provide this service to general contractors. Prior to quoting a project it is best to obtain a firm price quote from a core drilling company. For estimating core drilling is usually priced by the unit inch, which is the diameter in inches (mm) x the depth in inches (mm). For example a hole 4" (100 mm) in diameter and 10" (250 mm) deep would be equal to 40 unit inches. Currently the going rate per unit inch is approximately $1.00 to $1.50 depending on the following: a) Quantity of holes to be drilled, b) location, c) set-up time required and d) mobilization. Horizontal holes will cost about 10% to 15% more than a vertical hole. If reinforcement steel is to be encountered, add 50% to 100% more.

Cutting Concrete with Portable Concrete Saw
(Flat Saw/Walk Behind Saw)

Many states specify that control joints in highway paving be cut with a concrete saw, rather than produced by forming. The advantages of this method are improved ridability of joint; less sealing material required to seal joint; better aggregate interlock at joint; and no spalling or cracking at joint, which reduces maintenance costs.

This technique is also widely used for cutting floors and paving for repair work, trenches for electric conduit, pipe lines, and sewers, and also for scoring before breaking out concrete for machinery bases and other foundation work.

Target saws from Electrolux located in Kansas City and Diamond Products located in Ohio are a good example. They are available in electric, hydraulic, diesel, and gasoline models. The most popular models are the 18 Hp model, the 37 Hp model, and the 65 Hp model. Approximate prices range from $3,000.00 for the basic model manually propelled to $15,000.00 for the self propelled 65 hp unit. The water pump and spotlight are optional features along with larger blade guard sizes. Many accessories are available on all of these units including a propane option for inside cutting when the proper ventilation is available. The larger 65 Hp engine is generally used for sawing contraction joints in airports, highways, and turnpikes. The smaller units are widely used in patching, trenching, and plant maintenance work. Electrically powered models are available from 3 Hp and up.

Two types of blades are used in cutting concrete with concrete saws as follows: 1) Green-Con wet-dry blades for cutting green concrete 2) diamond blades, of various specifications, for cutting cured or old concrete.

The cost of cutting concrete with a concrete saw varies considerably, depending on age of concrete, size of rebar, type of aggregate, blade specification used, accessibility of work, skill of operator, etc.

Assuming proper blade specification is used for the cutting job, green concrete 24 to 48 hours old may be cut 1" deep at the rate of 12 to 14 l.f. (3.6-4.2 m) per minute and 3" (75 mm) deep at the rate of 6 to 8 l.f. (1.8-2.4 m) per minute. These rates cover only actual cutting time. Overall production will be reduced considerably depending on length of each individual cut, number of machine moves, changes in direction of cuts, etc. Blade wear for a Green-Con blade of proper specification should cost from $0.01-0.02/l.f. (0.033-0.066/m) of cut 1" (25 mm) deep. For a 2" (50 mm) cut, blade cost should be $0.015-0.03/l.f.; 3" (75 mm) cut, $0.02-0.04 per l.f. (0.06-0.13 per m).

When cutting old or cured concrete, using a diamond blade of correct specification, cutting time may be double the amount given above for green concrete and cost of blade wear will probably run 5 to 6 times that given above. Inside cutting of a concrete slab on grade (trench) with rebar may be cut 6" deep at the rate of 40-50 l.f. (12.2-15.3 m) per hour. Trenches 6" deep x 12" wide are typically cut into 2' x 2' pieces for easy removal. Outside cutting of a concrete slab such as bridges and piers 8"–10" deep with rebar may be cut at the rate of 65–75 l.f. (19.9-22.3 m) per hour. Typically bridges and pier decks are cut into 6' x 8' panels and loaded onto truck beds with the aid of a crane or excavator. Roadway concrete with minimal rebar may be cut 8"–10" deep at a rate of 100–150 l.f. (30.5-45.8 m) per hour. Asphalt roadway may be cut 4"-6" deep at a rate of 200–225 l.f. (60.1-68.6 m) per hour. Blade wear considering proper specification when cutting 8-10" deep concrete should cost between $0.05-$0.06 /inch foot. It is best to obtain a firm price quote for any concrete cutting that is required for a project, but if a quick budget estimate is needed, flat sawing concrete is usually priced by the inch foot. For example a slab that is 6" deep would be priced at $0.20-$0.60/inch foot or $1.20-$3.60/l.f. Asphalt 6" deep would be priced at $0.10-$0.20/inch foot or $0.60-$1.20/l.f.

166

SITE WORK

Cutting Concrete Door Openings with Wall Saw

Typical door openings are 3' x 7' x 12"-16" deep (.91M x 2.13M x 300 mm – 400 mm). To cut these openings a wall saw that operates on a track is utilized. Wall saws are available from Diamond Products located in Ohio, Electrolux located in Kansas City and many other manufacturers. Costs for the wall saw units range between $10,000.00-$15,000.00 plus a power unit to run the saw that ranges between $20,000.00 and 30,000.00. These units are typically powered by hydraulic, diesel and electric. The openings are cut with continuous water which can be controlled by damming or with the use of a vacuum. Over cutting is generally allowed and there are options using a chain saw and core drilling the corners when over cuts are not allowed, but will not be discussed at this time. Diamond blades used for this application are specifically manufactured for wall saws with diameters between 18"–72" (450 mm – 1800mm). The average wall saw blade is 30" (750 mm). The cost of cutting concrete with a wall saw varies considerably, depending on age of concrete, size of rebar, type of aggregate, blade specification used, accessibility of work, skill of operator, etc. Wall sawing is usually performed by specialty subcontractors as Semcor Mr. Wayne Camp) whose only business is to provide this service to general contractors. It is best to obtain a firm price quote from any concrete cutting company for cutting that is required for a project, but if a quick budget estimate is needed, wall sawing is usually priced by the inch foot. For example a wall that is 12" (300 mm) deep would be priced at $3.00-$4.00/ inch foot ($9.84 - $13.12/25mm-foot) or $36.00-$48.00/ l.f.($118.11 - $157.48 / LM)

Assuming proper blade specification is used for the cutting job, the cutting rate will be 4–5 l.f. (1.22–1.55 m) per hour. Blade wear considering proper specification should cost between $ 0.15-$0.20 /inch foot. Typically a 30" blade will cut 12" deep into a wall at a blade cost of approximately $900.00 and last 4,500 inch feet.

When the depth of a wall exceeds 24" (600 mm) and cannot be cut from both sides then other options is available including the use of a wire saw. This saw utilizes pulleys, a main frame and wire with diamond beads. 2" (50 mm) holes are drilled into the concrete to sling the diamond wire. The power unit can be electric, hydraulic, gas or diesel.

02200 EARTHWORK

It is not difficult to compute the number of cubic yards of earth to be excavated or removed from the premises on a specific project, but it is quite another matter to estimate the actual cost of performing this work, because of the many factors and possible unknown items entering into the cost. This includes such items as type of soil, whether or not water will be encountered, pumping, whether the banks of the excavated portion will be self-supporting or whether it will be necessary to brace and sheet them, disposal of excavated earth, length of haul to dump, type of material, etc.

The prudent estimator will examine the equipment that is available when preparing the cost estimate. The accurate choice of equipment at the bidding stage will determine the final costs. Type of equipment selected, as well as the size of equipment, should be based on several factors-subsurface information, site conditions, volume of material to be moved, and availability of specific equipment.

Estimating Quantities of Excavation. Excavating is measured by the cubic yard, cu.yd. (cubic meter, cu.m, m3). One cubic yard contains 27 cu. ft. (3' x 3' x 3')

General or Mass Excavation. When computing the volume or cubic contents of any basement or portion to be excavated, take all measurements from the outside face of wall footings, add 6" to 1'-0" (150 to 300 mm) on each side to allow for placing and removing forms, and then allow for sloping the sides of the excavation so they will stand reasonably stable without being supported by bracing or sheet piling. The amount of slope necessary to provide a "safe" hole will vary with the kind of soil to be excavated, depth of excavation, etc.

On construction projects the Federal Occupational Safety and Health Administration (OSHA), as well as state and local agencies, require for excavations of five feet deep and over soil support of sheeting or some other type. If none is used, then the side banks of the excavation shall be sloped back to the soils natural angle of repose so that the excavation sides will remain stable.

Digging in previously undisturbed material and assuming that no water or unstable conditions exist, most estimators, when taking off quantities for excavation, use a 1:1 slope; that is, one horizontal to one vertical, for sand and gravel; a 1:2 slope for ordinary clay and a 1:3 or 1:4 slope for stiff clay.

In some cases job conditions may not permit the sloping of banks and then it is usually necessary to sheet and brace the banks to prevent accidents that result in damage to adjoining property, or worse, worker injury or death.

On jobs with column footings that project beyond the wall footings, additional cuts must be made into the banks at the column locations to accommodate the additional width of the footings. The volume of this cut is computed by taking the column footing length plus 1'-0" or 2'-0" (300 or 600 mm) for work space plus "layback" of banks on both ends and multiplying by the amount of projection beyond the wall footing to obtain the area of the additional cut, which is then multiplied by the depth of the cut to obtain the volume of additional excavation required.

The depth of the general cut is usually taken from the underside of the topsoil, previously removed, to the underside of the floor slab. Where sand, gravel, or other fill material is to be placed under the concrete floor, the depth should be taken to the underside of the fill material.

Example: Assume you are figuring the general or mass excavation for a building 100'-0" x 50'-0" (30 x 15 m) at grade. The bottom edge of the

168

excavation is to be 2'- 0" (600 mm) from the edge of footing(s). Also assume the depth of the excavation to be 8'-0" (2.4 m) and the soil will stand at a slope of 1:1, one horizontal to one vertical. The calculation for excavation volume shall be figured as follows:

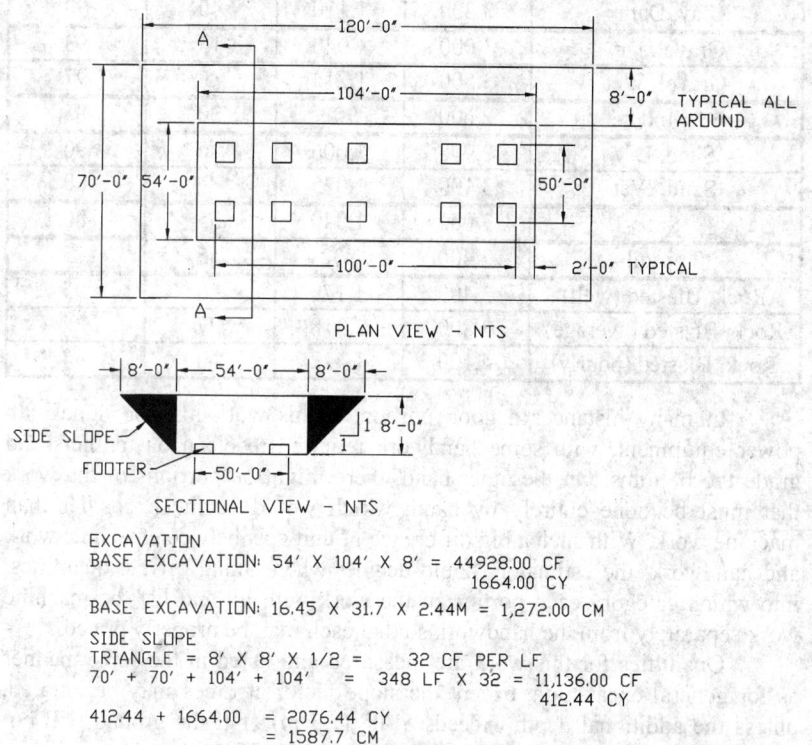

PLAN VIEW – NTS

SECTIONAL VIEW – NTS

```
EXCAVATION
BASE EXCAVATION: 54' X 104' X 8' = 44928.00 CF
                                    1664.00 CY

BASE EXCAVATION: 16.45 X 31.7 X 2.44M = 1,272.00 CM

SIDE SLOPE
TRIANGLE = 8' X 8' X 1/2 =    32 CF PER LF
70' + 70' + 104' + 104'  =   348 LF X 32 = 11,136.00 CF
                                             412.44 CY

412.44 + 1664.00  = 2076.44 CY
                  = 1587.7 CM
```

In some cases a portion of or the entire building site may be high and require a grading cut to produce the required contours. In these instances the grading cut should be figured first and the general or mass excavation computed afterwards taking the depth from the underside of the grading cut to the bottom of the floor slab or fill under the floor.

Trench and Pit Excavation. To complete the building excavation quantity take-off, there are usually additional items of excavation to figure for footings, foundations, pits, trenches, etc., which extend below the level of the general or mass excavation, or which occur in unexcavated areas. This kind of excavation is always estimated separately from the general work as it is usually a more expensive operation.

169

Typical Examples of Swell Factors				
Material	Weight per BCY (lbs)	Weight per BCM (kg)	Percentage Swell	Original Density (%)
Clay, Dry	2,900	1,006	35	90
Clay, Damp	3,200	1,110	40	90
Gravel, Dry	3,000	1,040	15	93
Gravel, Wet	3,500	1,214	5	97
Mud	2,600	902	20	85
Sand, Dry	2,900	1,006	10	90
Sand, Wet	3,100	1,075	5	90
Silt	3,000	1,040	35	80
Topsoil	2,400	832	55	75
Rock, Blasted (well)	3,400	1,179	20	95
Rock, Blasted (average)	3,800	1,318	27	0
Rock, Blasted (poorly)	4,100	1,422	40	30

In many instances a good portion of this work may be done with power equipment, with some handwork required to clean out corners and grade the bottoms. On the other hand, there might be portions of the work that must be done entirely by hand, which would be much costlier than machine work. With such a big difference in unit cost between machine work and handwork, the estimator should decide, when taking off the quantities, into which category each portion of the work will fall and list the machine work separately from the handwork so that each may be properly priced.

Quantities for this work are measured and listed in the same manner as for general excavation, except the slope factor at times may be ignored, unless the additional depth exceeds 5'-0" (1.5 m) or if the ground will not stand safely for this depth with vertical banks. Under these conditions, allowances for sloping the banks should be made in the same manner as for general excavation, or allowances for soil support, such as sheeting and bracing, need to be included.

Remember that the plan dimensions shown and used to compute the amount of material to be excavated are sometimes referred to as pit measured or **bank cubic yards** (cubic meters); that is, the material as it rests in its natural state. After the material is removed, and either stored or loaded into trucks, it swells. The same material is now referred to as **"loose cubic yards (cubic meters)"**. The swell factors will vary with the type of materials being excavated.

Example: Dry gravel has a swell factor of 15%. If the number of bank cubic yards of dry gravel is 350 cubic yards (268 CM), then the number of loose cubic yards would equal 350 x 1.15 or 402.5 LCY (307.7 LCM).

The best percentages that an estimator can apply to current jobs is from the knowledge from past jobs. When the actual figures are not available, then the table above will serve as a substitute until the estimator can develop some history of jobs. If only swell applies to the earthmoving operation, then the measured quantity times the percentage of swell becomes the **swollen, loose, or trucked cubic yards** to be removed.

For example, how much dry clay would have to be trucked out of a hole that is 5' deep by 50' long and 30' wide (1.5m x 15.2m x 9.1m)? Since volume is equal to length (in feet) times width (in feet) times depth (in feet) all divided by 27, this would equal the measured quantity or bank cubic yards. Thus 5' x 50' x 30' ÷ 27 = 277.77 bank cubic yards (212.4 BCM). Using our swell factor for dry clay, we would have swollen or loose cubic yards = 277.77 x 1.35% = 374.99 or 375 SCY (286.7 SCM).

Backfill. After all excavating items have been figured, the amount of backfill should be computed. An easy method is to compute the displacement volume of the construction that is to be built within the limits of the excavation; that is, the volume of footings, piers, the basement volume figured from the underside of the fill under the floor to the elevation of the top of the general cut, etc., and deducting this volume from the sum of the general or mass excavation and the trench and pit excavation. The remainder is the volume of backfill required.

Another method of figuring backfill, usually more difficult, is to compute the volume from the actual dimensions of the spaces requiring backfill.

Some specifications require interior backfills to be made with selected materials, such as sand, gravel, bankrun gravel, etc., while the exterior backfill may be the excavated material, if suitable and approved. In this case it is necessary to keep the two items separate as the interior backfill material will probably have to be purchased and brought in from outside sources.

Backfill very often is specified to be compacted by mechanical tampers and a degree of compaction is also specified. In these instances the backfill operations must be considered very carefully, because this is expensive work.

When calculations for the backfill are complete, the estimator should consider the **compaction factor**. This is the measured difference between the truck, loose cubic yardage, delivered quantity and the final in-place measured quantity. Backfill material purchased in loose cubic yards (cubic meters) after compaction will not yield the same in-place quantity.

If both swell and compaction apply to the earthmoving operation, then the swollen amount to be moved is calculated by the following formula:

$$SCY = \frac{(MCY \text{ or } BCY) \times \text{Rate of Swell}}{1.0 - (\text{Eng. \%} - \text{O.D.\%})}$$

$$SCM = \frac{(MCM \text{ or } BCM) \times \text{Rate of Swell}}{1.0 - (\text{Eng. \%} - \text{O.D.\%})}$$

where SCY (SCM) is swollen cubic yards (swollen cubic meters), MCY (MCM) is measured cubic yards (measured cubic meters), BCY (BCM) is bank cubic yards (bank cubic meters), Eng.% is the percentage or compaction specified by the engineer, and O.D.% is from the table.

Many times the specifications will call for an Engineer's percentage or a compaction percentage. The formula above uses Engineer's percentage specifications minus the Original Density percentage.

For example, if you need to determine the number of SCY (SCM) of wet sand needed to fill 8" under a 180' x 60' (203mm x 54.9m x slab fully compacted (or 100% Engineer's %), then you would have MCY (MCM) = 180' x 60' x 0.67' ÷ 27 = 268 BCY (204.9 BCM). Using the equation above, we can now substitute and solve for SCY (SCM).

$$SCY = \frac{268 \times 1.05}{1.0 - (100\% - 90\%)} = 312.67 \text{ SCY (239 SCM)}$$

The estimator must remember that the plan dimensions shown and used to compute the amount of material to be excavated are sometimes referred to as **pit measured or bank cubic yards**; that is, the material as it rests in its natural state. For example, never disturbed or compacted by man, the natural density of most clay is 90%, which means it contains 10% voids by volume. When the excavation is disturbed, cut or moved, the soil changes from bank cubic yards to swollen or loose cubic yards. The swell percentage for clay is 40%. Therefore, to determine the number of swollen cubic yards, we multiply the number of bank cubic yards by the percentage of swell.

In a fully compacted state, the soil loses all prior swell and the original percentage of voids. Therefore to determine the number of swollen cubic yards needed for compaction, we divide, using the formula above.

The engineer will indicate the percentage of compaction on the drawings or in the specifications, such as 100% Standard Proctor, 98% Stand. Proc., 100% Modified Proctor, or 95% Mod. Proc., etc. The whole point of soil compaction is to increase soil density and load bearing capacity of the soil by adding water to the soil.

The standard Proctor test known as "T-99" is usually specified for fill material designated for use under building slabs, sidewalks, and under grassy areas in utility trenches. The modified Proctor test known as "T-180" is usually specified for fill materials designated for used under federal state highways, airport runways, and concrete drives. As stated above, to determine the number of swollen cubic yards needed to compact an excavation, we need to know the rate of swell percentage, the compaction percentage and the original density percentage of the soil.

As an example, to find the number of swollen cubic yards needed to compact a trench 4' wide by 6' deep and 210' long (1.2m x 1.8m x 64m) in dry sand with a 98% Std. Proc. specification, we would consult the table and formula above. From the table we see that the swell percentage for dry sand

172

SITE WORK

is 15% and the original density is 90% (10% voids). Therefore swollen cubic yards would equal (4 x 6 x 210 ÷ 27 x 1.15) ÷ (1 – (98% – 90%)) = 233.33 SCY (178.4 SCM).

Therefore, additional yardage (meters) will be required. For example, assume a hole to be backfilled that is 9'0" x 9'0" x 3'0" (2.7 x 2.7 x 0.9 m) deep; measured in place, 9 bank cu. yds. (6.5 cu.m) of material will be required. If the hole is backfilled with dry gravel, then 9 x 1.15 = 10.35 loose cubic yards. After compaction, assuming a compaction loss of 15%. Therefore, 10.35 ÷ 0.85 = 12.18 compacted cu. yds. (8.1 cu.m).

Estimating Trenches. To calculate the number of swollen cubic yards (cubic meters) of excavation needed, we must determine the square footage (square meters) of the proposed trench cross-section multiplied the lineal footage (lineal meters) of the trench divided by 27 to get cubic yards (cubic meters). Remember, adjust this quantity for swell when excavating and compacting during backfill.

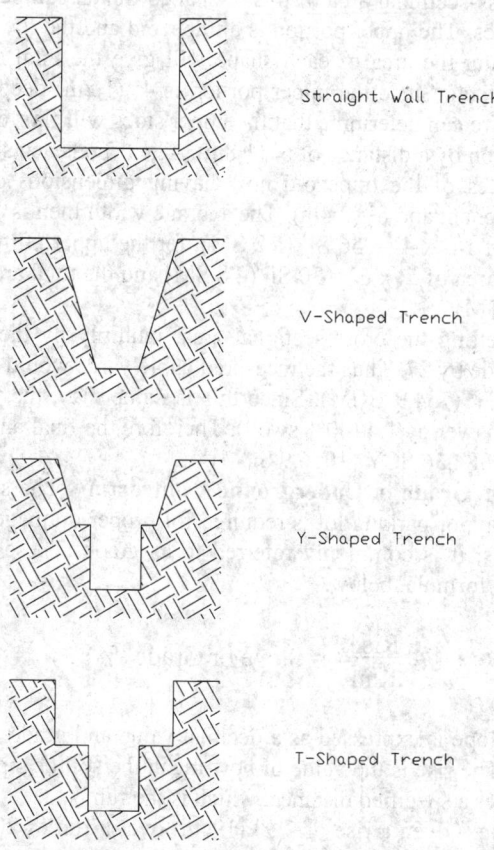

Straight Wall Trench

V-Shaped Trench

Y-Shaped Trench

T-Shaped Trench

Four typical trench cross-sections

173

Example: Find the total swollen cubic yards of damp clay to be removed if the trench is 310' (94.5m) long and using the figure below.

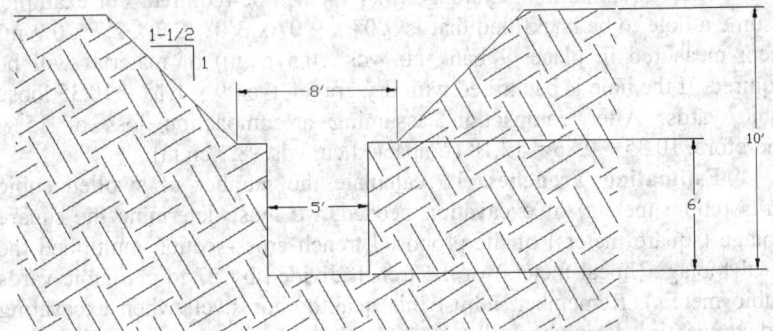

The cross-sectional area of this Y-shaped trench can be split into two geometric shapes. The upper portion is a trapezoid and the lower portion is a rectangle. Adding the area of each shape will give the total cross-sectional area for this trench. Since the upper portion is 4' (1.2m) deep and the sides slope at 1½:1, we can determine that the upper sides will flair out from our 8' (2.4m) dimension by a distance of 6' (1.8m), 4' x 1.5 on each side. This gives the parallel sides of the trapezoid now having dimensions of 20' (6.1m), mouth of the trench, and 8' (2.4m). The average width then is 14' (4.3m) and gives an area of 14' x 4' = 56 SF (5.2 SM), for the upper portion. The lower portion has an area of 5' x 6' = 30 SF (2.8 SM) and the total area will be 56 + 30 = 86 SF (8 SM).

After getting the cross-sectional area, multiply by the length of the trench and divide by 27. Thus the measured cubic yards would be 86 x 310' ÷ 27 = 987.4 BCY (754.9 BCM). Since this is damp clay, the final step is to multiply this answer by 1.4 (40% swell). Therefore the final answer would be 987.4 x 1.4 = 1382.36 SCY (1056.9 SCM).

Slope or Grade in Underground Contracting. The slope of a pipe, trench, driveway, or parking lot is required for proper drainage. When this is applied to soils, it is commonly referred to as grade. These are calculated using the same formula, below.

$$\frac{\text{Rise'}}{\text{Run'}} = \text{Slope or Grade (\%)}$$

Here, slope is expressed as a decimal value and grade is expressed as a percentage. The rise is the value of how much the soil or pipe will increase or decrease over a specified distance, which is the run.

Example: Given a rise of 7½" (19cm) over a run of 200' (61m), find the slope or grade. The first step is to convert the rise to feet so that the units

174

of measure are the same. Thus 7½" is 7.5 ÷ 12 = 0.63'. Now using the equation above, we 0.63' ÷ 200' = 0.003 slope or 0.3% grade.

Example: A driveway with a grade equal to 0.4% and a run of 280' (85.3m), would have a rise of how many inches (centimeters)? Using the formula above, we can cross-multiply to get 280' x 0.4% = 1.12' (.341m) for the rise. Now to convert this to inches, 1.12' x 12 = 13.44" or about 13-7/16" (34.1cm).

Excavation and Fill for New Site Grades. Excavation and filling for new site grades, including cuts and fills for drives, walks, and parking areas, vary from simple to complex according to the ground conditions. Estimating quantities for this work may be rather difficult and some surveying or engineering knowledge will greatly help.

There are a number of methods to determine volumes of cut and fill required. The method that is most adaptable to the particular problem will be chosen by the estimator after considering the problem, with the information given and the accuracy desired.

Slope Factor Table. When a plan-view of sloping banks or areas are shown or scaled on the drawings and a design slope is specified, use the multipliers in the table below to determine the actual sloping area.

Slope Ratio Horiz. Run : Vert. Rise	Multiply the Measured Flat Length (or Area) by
1/2 : 1	2.236
3/4 : 1	1.667
1 : 1	1.414
1 1/2 : 1	1.202
2 : 1	1.118
2 1/2 : 1	1.077
3 : 1	1.054
3 1/2 : 1	1.040
4 : 1	1.031
4 1/2 : 1	1.024
5 : 1	1.020
5 1/2 : 1	1.016
6 : 1	1.014

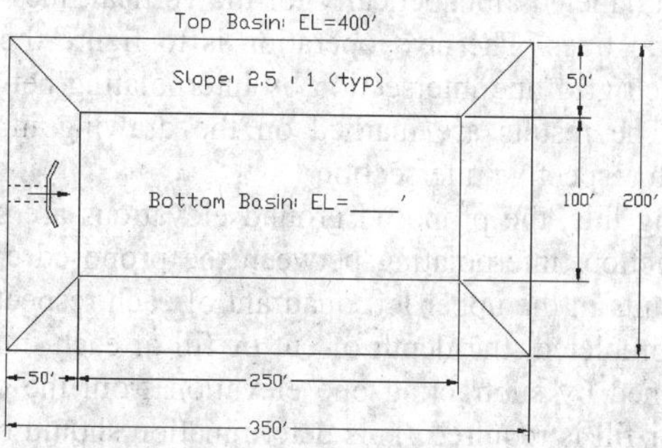
Top Basin: EL=400'
Slope: 2.5 : 1 (typ)
Bottom Basin: EL=____'
50'
100'
200'
50'
250'
350'

Catch Basin

In order to estimate the amount of stone needed to line the catch basin in the figure shown above using 0.25 tons of stone per SY (0.271 metric tons per SM), the first step is to determine the total "flat" square foot (SM) area of the four sloping sides of the retention basin. This would be 2 x 50' (15.2m) wide x 200' (61m) = 20,000 SF (1,858 SM) and 2 x 50' (15.2m) wide x 250' (76.2m) = 25,000 SF (2,323 SM) or a total of 45,000 SF (4,181 SM).

Once the total flat area is determined, adjust the total flat area using the appropriate slope factor from the table above. With a 2½ : 1 slope, the multiplier is 1.077. Thus 45,000 SF x 1.077 ÷ 9 = 5,385 SY (4,503 SM). Next calculate the quantity of stone required by multiplying by the factor given. 5,385 SY x 0.25 = 1,346.25 tons (1,221.3 metric tons).

It is also possible to calculate the elevation at the bottom of the catch basin using the slope. A proportion needs to be set calculating the vertical rise/fall. This is 2.5 ÷ 1 = 50' (15.2m) horizontal ÷ ??? vertical. Thus 50' ÷ 2.5 = 20' (6.1m) vertical rise/fall. Now the bottom elevation would be 400' (121.9m) – 20' (6.1m) = 380' (115.8m).

Cross Sectional Method

One of the easiest and most frequently used methods of computing grading cuts and fills when the plot plan shows both original and proposed contours is that of gridding or dividing up the area to be graded into squares, rectangles, triangles, or combinations of these figures, of regular and convenient dimensions, thus forming a series of truncated prisms. The depths of cut or fill at the corners of these prisms are computed and the volume of each prism may be determined by averaging these depths and multiplying the average by the cross sectional area of the prism. The results of these computations are totaled, keeping the cuts separate from the fills, and the total excavation and/or fill required is obtained.

Example: Assume the contour sketch shown on the next page to be a 400' x 200' (120 x 60 m) portion of a site grading plan showing both present and proposed contour lines. Since both the present and proposed surfaces have fairly regular slopes, as indicated by the even spacing of contour lines, a grid spacing of 50 feet in both directions has been chosen and grid lines drawn as shown, labeled alphabetically for the vertical lines and numerically for the horizontal lines. The next operation is to figure the present ground elevation for each grid line intersection by interpolating between the present contour lines. The results are marked on the drawing in the upper right quadrant of each respective intersection.

Following this, the proposed ground elevations are similarly figured for each intersection, interpolating between the proposed contour lines and entering the results in the upper left quadrant of each respective intersection. When this is completed, the depth of cut or fill at each intersection can be readily determined by subtracting one elevation from the other and noting whether a cut or fill is required. This determination should be made for each intersection and the results entered in the proper quadrant, using the lower left for cuts and the lower right for fills.

When all intersections have a depth value entered for either cut or fill, the plan should be examined closely to see where the grading operation will change from cut to fill or vice versa. In the sketch it will be seen that advancing from left to right on horizontal grid line "1" a change from fill to cut occurs between vertical grid lines "D" and "E" and by interpolating between the fill and cut depths the zero point "a" can be established approximately 36 feet to the right of intersection "D-1."

Similarly, zero points "b, d, e, and g" are established on the horizontal grid lines and zero points "c and f" on the vertical grid lines. These points are then connected by a line "a, b, c, d, e, f, g" and for all practical purposes it may be assumed that all points on this line are at the proposed elevation and require neither cut nor fill. In this case the line forms the boundary between cuts and fills. The change from cut to fill and vice versa could, of course, occur many times in actual problems depending on the irregularity of the present and proposed contour lines.

After the previously described operations have been completed, the actual volume of cut and/or fill required can be computed.

The volume of cut or fill in each prism in the grid system is obtained by averaging the corner depth values of each prism and multiplying by its cross sectional area. Thus, for the square prism "A-1, B-1, B-2, A-2" the volume of fill would be 1/4 x (2.4' + 1.6' + 1.6' + 2.5') x 50.0' x 50.0' or 5,063 cu. ft. (143 cu.m). All grid prisms are similarly treated, keeping cut and fill results separated, except those prisms through which the zero line passes. So, for square prism "D-1, E-1, E-2, D-2" it will be seen that this volume has been divided into two trapezoidal prisms "D-1, a, b, D-2" and "a, E-1, E-2, b". To compute the respective fill and cut volumes of these prisms they must be divided into triangular prisms first and then the corner depths may be averaged and multiplied by their respective cross sectional areas. Thus for the triangular prism "D-1, a, b" the volume of fill would be 1/3 x (0.5' + 0.0' + 0.0') x 1/2 x 50.0' x 36.0' or 150 cu. ft. (4.25 cu.m). Similarly, the volume of fill for the triangular prism "D-1, b, D-2" would be 1/3 x (0.5' + 0.0' + 0.3') x 1/2 x 50.0' x 21.0' or 140 cu. ft. (4 cu.m). The volume of cut required for the trapezoidal prism "a, E-1, E-2, b" is figured in the same manner. In the case of square cross section "C-2, D-2, D-3, C-3" it will be seen that the zero line crosses one corner of the square and divides it into a triangle "c, D-3, d" and a five-sided figure "C-2, D-2, c, d, C-3". The five-sided figure should be further divided into three triangles "C-2, D-2, c", "C-2, c, C-3" and "C-3, c, d" and the volumes of fill for each triangular prism would be as follows:

1/3 x (0.7' + 0.3' + 0.0') x 1/2 x 50.0' x 30.0' or 250 c.f. (7 cu.m)
1/3 x (0.7' + 0.0' + 0.5') x 1/2 x 50.0' x 50.0' or 500 c.f. (14 cu.m)
1/3 x (0.5' + 0.0' + 0.0') x 1/2 x 36.0' x 20.0' or 60 c.f. (1.7 cu.m)

When there are several prisms with the same cross section to be either cut or filled the volume of such can be computed as one solid by assembling them as follows. Multiply each corner depth by the number of typical prisms in which it occurs and then add these results and divide by 4. This is then multiplied by the cross sectional area of one prism. For example the volume bounded by A-1, D-1, D-2, C-2, C-4, B-4, B-5, A-5, A-1 can be obtained by one computation because it is composed of a series of prisms having the same cross section. In the summation of depths, those at A-1, D-1, D-2, C-4, B-5 and A-5 are taken but once, those at B-1, C-1, C-3, A-4, A-3 and A-2 are multiplied by 2, at C-2 and B-4 the depths are multiplied by 3 and at B-2 and B-3 they are multiplied by 4. All of these values are then totaled and divided by 4 and then multiplied by the cross sectional area of one prism giving the volume of fill required.

When a volume has been computed for each grid division of the site, either individually or collectively, it is only a matter of summing up the cuts and fills to obtain the total excavation or fill required to bring the site to the proposed grades.

End Area Method

It often happens that a cut and fill operation is to be performed over a long and narrow area such as for roads, levees, ditches, etc. In this case the usual method of computing the earthwork quantities is by means of end areas.

In jobs of this sort a profile plan and a topographic plan is generally furnished to the contractor as part of the bidding documents. Using this information, sections are taken perpendicular to the longitudinal centerline of the work, usually at each 100-foot (30-meter) station, at each radical change in either present or proposed grades and where the cut changes to fill or vice versa. These sections are plotted to scale showing the original grades as obtained from the topographic plan and the proposed grades as obtained from the profile plan and design details of the surface required. After this has been done, it will be apparent that areas are enclosed by the plotted lines representing the present and proposed grades at each section and these areas should be computed.

To obtain the volume of cut and fill between each section, the usual and easiest method is to average the areas of two adjacent sections and multiply by the distance between them, keeping the cuts and fills separate. This method is not quite accurate, but the results are on the safe side and in practically every case are close enough for estimating purposes.

Example: Assume the following sketch to be a 300-foot (90-meter) portion of the topographic plan and profile plan of a proposed road location. The topographic plan shows the undulations of the present ground surface by the contours and the location of the road by the center and shoulder lines. The profile plan shows the relation between the present ground surface and the proposed road bed grade at the centerline of the road. Also assume the road bed is to be 20.0 feet wide and the shoulder slopes are to be 1-1/2 : 1, that is 1-1/2 horizontal to 1 vertical.

To compute the volumes of earthwork involved in this proposed grading operation, by means of the end area method, sections must be taken perpendicular to the center-line of the proposed road and the present and proposed grade lines plotted. In this case sections have been taken and plotted, as illustrated in the sketch, at even stations 15+00, 16+00, 17+00, and 18+00. Sections have also been taken at stations 15+07, 15+37, 16+50 and 16+65 because at these points the cut or fill at a shoulder of the proposed road is zero. A section is taken at station 17+65 because at this point the present grade makes a radical change in slope.

In plotting the sections any convenient scale may be used and if cross section paper is used the job will be much easier.

If a greater degree of accuracy is required the prismoidal formula must be employed which necessitates figuring the middle area between two adjacent sections. This middle area is obtained by averaging the corresponding dimensions of both end areas and computing an area from the average dimensions; it should not be taken as the average of the two end areas. To obtain the volume between adjacent sections using the prismoidal formula, multiply the length between sections by one-sixth of the sum of the two end areas and 4 times the middle area combined.

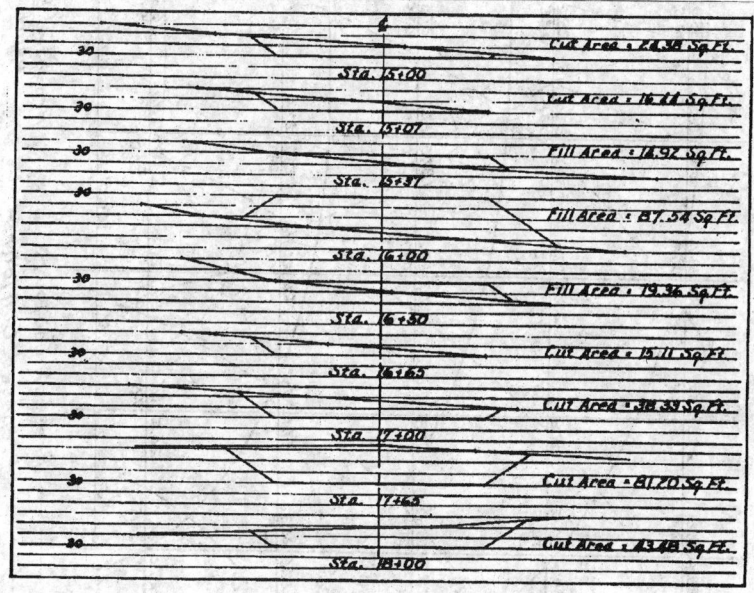

Method of Plotting End Areas

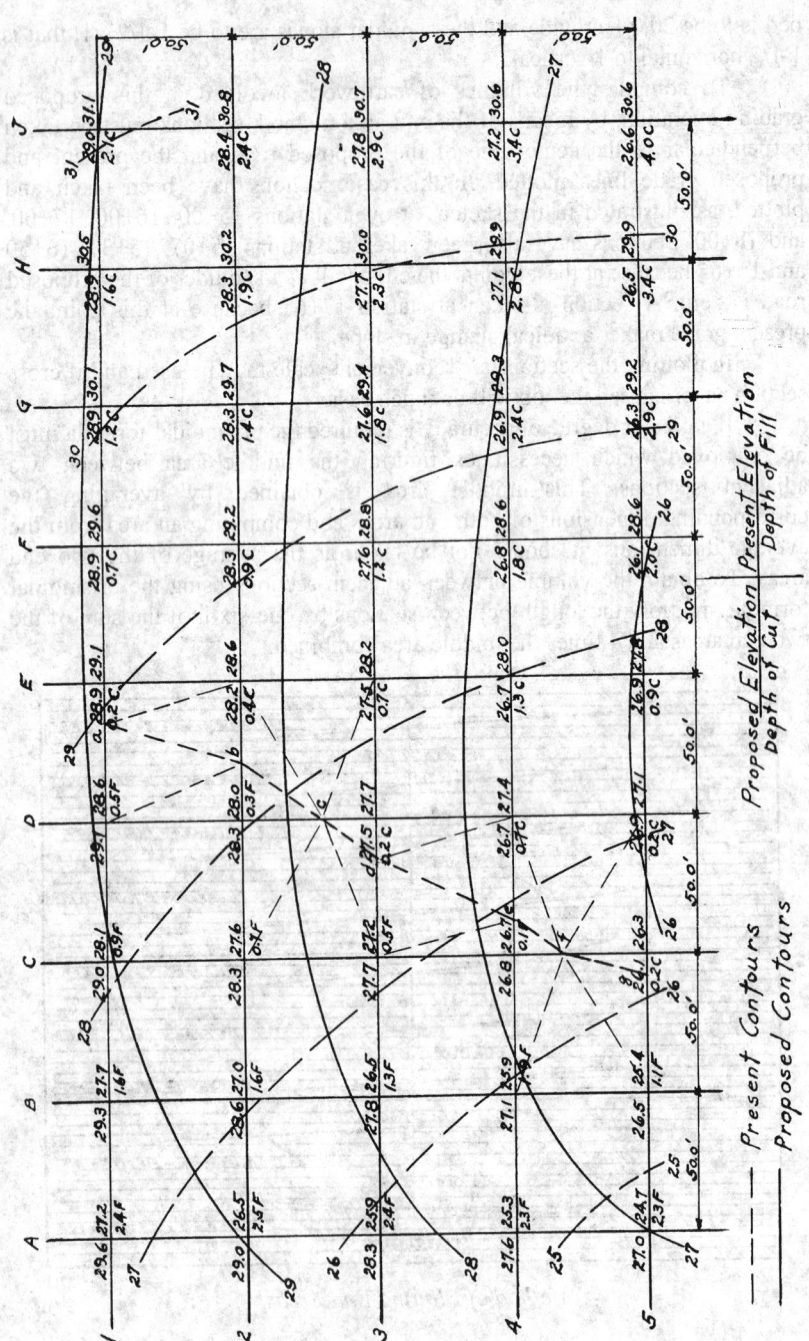

Method of Cross Sectioning a Contour Plan

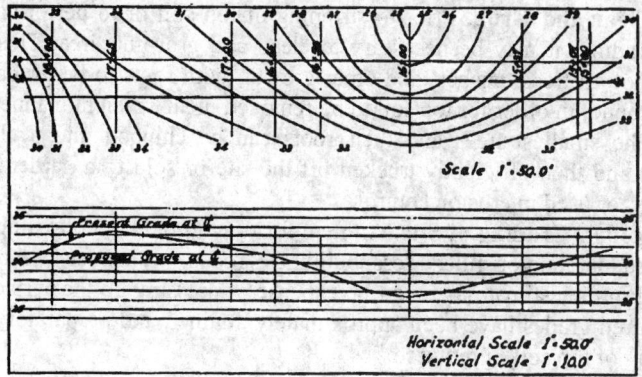

Topographic and Profile Plans

Most specifications call for the fills to be placed in layers and then specify some degree of compaction to be attained. This should be considered by the estimator, because additional equipment such as sheepsfoot rollers, pneumatic rollers, or tandem rollers will be required plus labor for the operation.

As each section is plotted, the area enclosed by the present and proposed grade lines is computed and the result tabulated. Final computations are made as follows:

Station	Dist. Between Stations, in Ft.	Actual Cut Ft.	Avg Cut Sq. Ft.	Actual Fill Sq. Ft.	Avg Fill Sq. Ft.	Cu.Ft. Cut	Cu. Ft. Fill
15+00		24.38		-			
	7		20.41		-	142.87	-
15+07		16.44		-			
	30		8.22		7.46	246.60	223.80
15+37		-		14.92			
	63		-		51.23	-	3,227.49
16+00		-		87.54			
	50		-		53.45	-	2,672.50
16+50		-		19.36			
	15		7.56		9.68	113.40	145.20
16+65		15.11		-			
	35		26.72		-	935.20	-
17+00		38.33		-			
	65		59.77		-	3,884.73	-
17+65		81.20		-			
	35		62.34		-	2,181.90	-
18+00		43.48					
Totals	300					7,504.70	6,268.99

Clear and Grub. Prior to stripping of top soil or to performance of rough grading, it may be necessary to clean and grub the area. This is the removal of all trees, bushes, and other growth, with roots to be grubbed out of the ground, in order for top soil to be removed in an efficient manner.

The smallest trees and their roots can be chipped into a chipping machine and then efficiently trucked off the site or sold to a garden supply company, or used in erosion control.

Rough Grading of Site. An item for rough grading, consisting of the total area cut and/or filled, less buildings, roads, walks, etc., should be included in the estimate to cover the expense of dressing up the surface after the required grades have been approximately attained and preparatory to the spreading of top soil, if required.

Spreading Top Soil. In most instances the specifications call for spreading top soil over site areas not covered by other construction, usually to a depth of 6 inches.

Finish Grading of Site. Fine grading of top soil, usually including a hand raking operation, is generally called for and should be listed in the estimate as an item, the quantity being the area used in computing the volume of top soil to be spread.

Other Grading. Another grading item to be considered is grading the subgrade for floor slabs, walks, paving, etc. Some contractors also include an item for grading bottoms of footings, when they are to be dug by machine, and include an amount for labor squaring and cleaning the holes.

Disposal of Excess or Borrow. After all excavation, backfill, grading cuts and fills, etc., have been computed and listed in the estimate, the total cuts and the total fills should be compared to determine whether there will be an excess of materials to dispose of or a deficiency of materials to be made up by borrowing or purchasing from outside sources. The results of this comparison should be listed in the estimate. The top soil comparison should be kept separate from the other materials, because the cost of any top soil that must be purchased usually is much more expensive than ordinary fill material. Also, any excess of top soil may find a ready market and be sold, reducing the overall cost of the earthwork. The difference between the cuts and fills will give the bank or compacted volume required. In case of a deficiency of materials, this compacted volume quantity must be adjusted for shrinkage due to compaction, because the material will probably be bought by loose measure. In the same manner all excess bank measurement volumes must be adjusted for swell due to bulking to obtain the true volume that must be handled in the disposal operation. The amount of increase for gravels and sand is from 5% to 12% and for clays and loams from 10% to 30%.

Pumping or De-watering. Finally, careful consideration should be given to the probability of pumping or de-watering operations. In some parts of the country this is no problem whatsoever, quite the reverse being the case. However, in most localities ground or climatic conditions make it necessary to include pumping or de-watering as an item in the estimate.

This item is quite variable, being affected by the season of the year as well as the locale. If only rainwater run-off is expected, an allowance for a few small pumps may be sufficient. If, however, the job is in close proximity to a body of water or springs are present, it might be necessary to de-water part or all of the operational area by means of a wellpoint or deep well system, in which case the contractor should consult a company that specializes in this work. (See the section on wellpoint systems in this chapter.)

Estimating the Labor Cost of Excavation. Before an excavating estimate is made up or a bid submitted, the estimator should visit the site and ascertain the exact conditions surrounding the work to be performed.

The natural grade of the lot should be noted to compute the depth of the excavation from grade to underside of floor slab; the nature of the soil should be determined, whether sand, loam, clay, or rock, and whether or not water is likely to be encountered. Water usually means an additional cost-pumping out the hole day and night until the foundations are completed. Rock usually means blasting, with a greatly increased cost for drilling and blasting, then excavating and loading the rock.

The safest method of determining the kind of soil is to dig a test hole the approximate depth of the basement, or take a set of borings. From this one can obtain a fairly accurate idea of the conditions to be encountered.

In most areas the soil strata, as well as the ground-water table, is fairly consistent, and its character may be ascertained by talking to contractors or engineers who have worked in the area for some time. In most larger jobs, the plans will show the results of test borings ordered by the owner or architect/engineer.

If the lot is low and the basement does not cover the entire lot, part of the excavated soil may be used for filling. On the other hand, if the natural grade of the lot is above the sidewalk or established grade, it may be necessary to remove a certain amount of soil from the entire lot.

In large structures having basements and sub-basements, the excavation may extend 40 to 50 feet (12 to 15 m) below grade. This requires the construction of steep runways, haul roads, or ramps so that trucks may get in and out of the hole, and an extra truck or bulldozer may be required to help the loaded trucks out of the hole. These items must all be taken into consideration when pricing any excavating job.

If it is necessary to remove the excavated soil from the premises, locate the nearest available landfill, and ascertain just how the soil will be handled at the landfill (for example, whether it is necessary to spread it).

National and local environmental protection agency rules may require the contractor to test material to be excavated and disposed of, or even to be placed back into the same area, after excavation. Local EPA rules at times require local disposal site operators not to accept material without a certification from the contractor stating that no contaminants are in the material. Materials excavated that have contaminants must be disposed of at designated disposal sites. The types and amounts of contaminants will determine the amount of additional costs.

These are all items that affect the cost and should be given careful consideration by the estimator.

In most localities there are contractors who specialize in excavation and are familiar with the kind of soil encountered in different parts of the area. These contractors usually quote a price per cu. yd. (cu. m) for all of the general excavation, or they may quote a lump sum for the entire job.

Trench, pier, and all hand excavating are usually performed by the general contractor.

METHODS OF EXCAVATING

Excavating costs will vary with the kind of soil and the method used in removing and hauling to a final destination.

Except on very small jobs, practically all excavation today is performed by power operated equipment-power shovels, cranes with dragline or clamshell buckets, bulldozers, tractor excavators, front end loaders, scrapers, and trenching machines. Grading, spreading, and backfilling are performed by bulldozers, tractor excavators, and front end loaders.

In shallow excavations, where the excavated earth may be spread over the site not more than 50 feet (15 m) from the excavation, bulldozers or loaders (a tractor equipped with a shovel that may be elevated to load directly into trucks or over the sides of the excavation) are frequently used. This type of excavator is also used for excavating basements, loading directly into trucks.

What was at one time referred to as a back-digger power shovel is now on tracks or rubber tire and commonly called a backhoe, which is extensively used for basement and trenching excavation. It is equipped with a bucket that pulls or crowds toward the machine as it excavates, which is always located on the bank above the excavation, working from the sides and ends of the excavation.

In large excavations, the use of a regular backhoes or hydraulic front shovels are the most satisfactory and economical method to use. The backhoe will excavate and load the trucks as fast as they can be handled in and out of the hole and will enable the job to proceed much faster.

If it is desirable to keep the backhoe on the bank above the excavation, but the depth is greater than the hoe can reach efficiently, a crane equipped with a dragline bucket or clam bucket is used.

For individual pier footings, small pits, and hard-to-reach spots, a crane equipped with a clamshell bucket or a small combination backhoe/loader either on rubber tire or skid mounted loader can be used.

On medium to large grading jobs, self-propelled or tractor-hauled scrapers are used quite often and perform cutting and filling operations creditably.

SITE WORK

Quantities and production times given in the tables below are average. Experienced excavators will perform more work than given, but an inexperienced worker will perform less, so average conditions have been used.

Hydraulic power excavating equipment has become very extensive and is used almost 100 percent on the excavation machines.

HAND EXCAVATION

Most construction projects require extensive use of hand labor to accomplish the selective types of excavation which cannot be performed by machine excavators.

In the excavation of footings, it is frequently found to be cost effective to combine the use of machines for the bulk excavation and then to "hand dress" the bottoms, sides, and corners to produce the precise horizontal and vertical planes required. Similarly, this is true in finalizing the subgrade elevation under slabs on grade, walks, and patio slabs.

The quantities and production rates given in the following table are for average conditions.

Hand Excavation		
Material & Operation	Average No. Cu. Yd. per 8-Hr Day	Man-hours per Cu. Yd.
Sandy loam		
Small footings, hand	6	1.33
Trenches 3'-4' deep, hand	6	1.33
Piers 6' deep, hand 1 lift	5	1.6
Piers 12' deep, hand 2 lifts	3	2.66
Heavy soil and clay		
Small footings, hand	5	1.60
Trenches 3'-4' deep, hand	5	1.60
Piers 6' deep, hand, 1 lift	4	2.00
Piers 12' deep, hand, 2 lifts	2	4.00
Excavation with 1/2-Cu.Yd. backhoe, casting to side		
Trench footing	300	......
Pier footing 6' deep	200	......
Pier footing 12' deep	150	......

Hand Excavation (continued)		
Material & Operation	Average No. Cu.M 8-Hr Day	Man-hours per Cu. M
Sandy loam		
Small footings, hand	4.59	1.02
Trenches .9-1.2 m deep, hand	4.59	1.02
Piers 1.8 m deep, hand 1 lift	3.82	1.22
Piers 3.7 m deep, hand 2 lifts	2.29	2.03
Heavy soil and clay		
Small footings, hand	3.82	1.22
Trenches .91M-1.22 m deep, hand	3.82	1.22
Piers 1.83 M deep, hand, 1 lift	3.06	1.53
Piers 3.6 M deep, hand, 2 lifts	1.53	3.06
Excavation with 0.40 Cu. M. backhoe, casting to side		
Trench footing	229.38	......
Pier footing 1.83 M deep	152.92	......
Pier footing 3.66 M deep	114.69	......

Hand Trimming Trench, Piers And Slabs After Machine Excavation		
Description	Sq.Ft. per 8- Hour day	Sq.Ft. per Hour
Bottom and side areas	240	30
Fine grade under slabs	600	75
Description	Sq.Meter per 8- Hour day	Sq.Meter per Hour
Bottom and side areas	22.30	2.79
Fine grade under slabs	55.74	6.97

Backfilling By Hand With Compaction Material From Stockpile		
Ordinary Light Soils		
Description	Cu. Yd. per 8-Hr Day	Man-hours per Cu. Yd.
Soil distribution laborer	12.00	0.67
Compactor operator	12.00	0.67
Description	Cu. Meters per 8-Hr Day	Man-hours per Cu. M.
Soil distribution laborer	9.18	0.51
Compactor operator	9.18	0.51

No allowance for compactor equipment cost.

MACHINE EXPLORATION

Excavating Using Hydraulic Backhoe Tractor (Crawler) 1.0 Cu.Yd. (.76 Cu.Meter) Bucket		
Type of Material	Avg. No. Cu.Yd per 8-hour day	Backhoe & Operator hours per 100 Cu.Yds.
Sandy Clay/Moist Loam	720	1.11
Gravel/Sand	690	1.16
Clay, Hard	570	1.40
Rock, Well Blasted	450	1.78
Rock, Poorly Blasted	300	2.67
Type of Material	Avg. No. Cu.M per 8-hour day	Backhoe & Operator hours per 100 Cu.M
Sandy Clay/Moist Loam	551	0.00
Gravel/Sand	528	0.00
Clay, Hard	436	1.45
Rock, Well Blasted	344	2.33
Rock, Poorly Blasted	229	3.49

Above production is based on the following: project conditions,
average to above average; operator, average; obstructions, none;
swing to 60 degrees or 1/4 turn. Material placed into waiting trucks.

Excavating Using Hydraulic Backhoe Tractor (Crawler) 1 1/4 Cu.Yd. (.96 Cu.Meter) Bucket		
Type of Material	Avg. No. Cu.Yd per 8-hour day	Backhoe & Operator hours per 100 Cu.Yds.
Sandy Clay/Moist Loam	900	0.89
Gravel/Sand	840	0.95
Clay, Hard	570	1.40
Rock, Well Blasted	450	1.78
Rock, Poorly Blasted	300	2.67
Type of Material	Avg. No. Cu.M per 8-hour day	Backhoe & Operator hours per 100 Cu.M
Sandy Clay/Moist Loam	688	1.16
Gravel/Sand	642	1.25
Clay, Hard	436	1.84
Rock, Well Blasted	344	2.33
Rock, Poorly Blasted	229	3.49

Above production is based on the following: project conditions, average to above average; operator, average; obstructions, none; swing to 60 degrees or 1/4 turn. Material placed into waiting trucks.

Excavating Using Hydraulic Backhoe Tractor (Crawler) 1 1/2 Cu.Yd. (1.15 Cu.Meter) Bucket		
Type of Material	Avg. No. Cu.Yd per 8-hour day	Backhoe & Operator hours per 100 Cu.Yds.
Sandy Clay/Moist Loam	1080	0.74
Gravel/Sand	1020	0.78
Clay, Hard	870	0.92
Rock, Well Blasted	660	1.21
Rock, Poorly Blasted	300	2.67

Excavating Using Hydraulic Backhoe Tractor (Crawler) 1 1/2 Cu.Yd. (1.15 Cu.Meter) Bucket		
Type of Material	Avg. No. Cu.M per 8-hour day	Backhoe & Operator hours per 100 Cu.M
Sandy Clay/Moist Loam	826	0.97
Gravel/Sand	780	1.03
Clay, Hard	665	1.20
Rock, Well Blasted	505	1.59
Rock, Poorly Blasted	229	3.49

Above production is based on the following: project conditions, average to above average; operator, average; obstructions, none; swing to 60 degrees or1/4 turn. Material placed into waiting trucks.

Excavating Using Hydraulic Backhoe Tractor (Crawler) 1 3/4 Cu.Yd. (1.34 Cu.Meter) Bucket		
Type of Material	Avg. No. Cu.Yd per 8-hour day	Backhoe & Operator hours per 100 Cu.Yds.
Sandy Clay/Moist Loam	1260	0.63
Gravel/Sand	1200	0.67
Clay, Hard	1020	0.78
Rock, Well Blasted	750	1.07
Rock, Poorly Blasted	510	1.57
Type of Material	Avg. No. Cu.M per 8-hour day	Backhoe & Operator hours per 100 Cu.M
Sandy Clay/Moist Loam	963	0.83
Gravel/Sand	918	0.87
Clay, Hard	780	1.03
Rock, Well Blasted	573	1.40
Rock, Poorly Blasted	390	2.05

Above production is based on the following: project conditions, average to above average; operator, average; obstructions, none; swing to 60 degrees or1/4 turn. Material placed into waiting trucks.

Excavating Using Hydraulic Backhoe Tractor (Crawler) 2.0 Cu.Yd. (1.53 Cu.Meter) Bucket		
Type of Material	Avg. No. Cu.Yd per 8-hour day	Backhoe & Operator hours per 100 Cu.Yds.
Sandy Clay/Moist Loam	1440	0.56
Gravel/Sand	1380	0.58
Clay, Hard	1140	0.70
Rock, Well Blasted	870	0.92
Rock, Poorly Blasted	720	1.11
Type of Material	Avg. No. Cu.M per 8-hour day	Backhoe & Operator hours per 100 Cu.M
Sandy Clay/Moist Loam	1101	0.73
Gravel/Sand	1055	0.76
Clay, Hard	872	0.92
Rock, Well Blasted	665	1.20
Rock, Poorly Blasted	551	1.45

Above production is based on the following: project conditions, average to above average; operator, average; obstructions, none; swing to 60 degrees or1/4 turn. Material placed into waiting trucks.

Excavating Using Hydraulic Backhoe Tractor (Crawler) 3.0 Cu.Yd. (2.29 Cu.Meter) Bucket		
Type of Material	Avg. No. Cu.Yd per 8-hour day	Backhoe & Operator hours per 100 Cu.Yds.
Sandy Clay/Moist Loam	1800	0.44
Gravel/Sand	1710	0.47
Clay, Hard	1440	0.56
Rock, Well Blasted	1080	0.74
Rock, Poorly Blasted	720	1.11

Excavating Using Hydraulic Backhoe Tractor (Crawler) 3.0 Cu.Yd. (2.29 Cu.Meter) Bucket		
Type of Material	Avg. No. Cu.M per 8-hour day	Backhoe & Operator hours per 100 Cu.M
Sandy Clay/Moist Loam	1376	0.58
Gravel/Sand	1307	0.61
Clay, Hard	1101	0.73
Rock, Well Blasted	826	0.97
Rock, Poorly Blasted	551	1.45

Above production is based on the following: project conditions, average to above average; operator, average; obstructions, none; swing to 60 degrees or1/4 turn. Material placed into waiting trucks.

ESTIMATING THE COST OF OWNING AND OPERATING CONSTRUCTION EQUIPMENT

The following information on the costs of owning and operating power shovels, hoes, cranes, truck cranes, tractors, and scrapers will give a general idea of the costs involved in making up an estimate for this type of work.

The equipment purchase prices given are only a relative guideline to be used in the tables. What is most importance is the format for determining these costs, not the price of the equipment listed, which will vary greatly depending on locality, method of purchase, number of pieces bought at one time, etc.

The price f.o.b. factory should cover the complete machine with all variable equipment and accessories, such as various attachments, light plant, magnet generators, and clamshell bucket.

Total Investment or Cost of Equipment. Depreciation, interest, taxes, and insurance are directly related to the initial investment in construction equipment. In addition, certain other costs may be estimated by their normal relationship to this figure. Therefore, the proper determination of the total cost is a basic requirement.

Approximate Prices of Power Shovels, Draglines, Clamshells, and Lift Cranes

The following are approximate prices, f.o.b. factory, and will give the estimator some idea of the investment required in shovels and cranes of various sizes:

191

Backhoe Crawler		
Bucket Size of Hoe		Cost
3/4 Cu. Yd.	0.57 Cu. M.	$ 130,000
1 Cu. Yd.	0.76 Cu. M.	$ 145,000
1 1/4 Cu. Yd.	0.96 Cu. M.	$ 200,000
1 1/2 Cu. Yd.	1.15 Cu. M.	$ 250,000
2 Cu. Yd.	1.53 Cu. M.	$ 275,000
2 1/2 Cu. Yd.	1.91 Cu. M.	$ 450,000
3 Cu. Yd.	2.29 Cu. M.	$ 525,000
4 1/2 Cu. Yd.	3.44 Cu. M.	$ 750,000

Lift Cranes - Crawler Mounted			
Lift Capacity			Cost
Ton	Kips	Metric tonne	
40.0	80.0	36.29	$ 390,000
50.0	100.0	45.36	$ 400,000
70.0	140.0	63.50	$ 491,000
100.0	200.0	90.72	$ 800,000
150.0	300.0	136.08	$ 1,200,000
200.0	400.0	181.44	$ 1,350,000
400.0	800.0	362.87	$ 2,200,000

All Terrain Hydraulic Cranes			
Lift Capacity			Cost
Ton	Kips	Metric tonne	
20.0	40.0	18.14	$ 295,000
40.0	80.0	36.29	$ 521,530
70.0	140.0	63.50	$ 875,000
100.0	200.0	90.72	$ 1,160,000
150.0	300.0	136.08	$ 1,600,000
200.0	400.0	181.44	$ 1,820,000
300.0	600.0	272.15	$ 2,500,000
500.0	1000.0	453.59	$ 2,600,000

SITE WORK

Approximate Rental Costs for Excavating Equipment with Operator				
Description	Metric	Day	Week	Month
Backhoe Crawler Mounted with One Operator				
1/2 Cu.Yd.	0.38 Cu. M.	$758	$3,567	$14,871
1.0 Cu.Yd.	0.76 Cu. M.	$793	$3,721	$15,492
2.0 Cu.Yd.	1.53 Cu. M.	$1,143	$5,301	$21,807
3.0 Cu.Yd.	2.29 Cu. M.	$1,850	$8,484	$34,542
3 1/2 Cu.Yd.	2.68 Cu. M.	$1,943	$8,903	$36,216
Compactor (with operator)				
1000 Lb Ram	453.6 Kg	$350	$1,727	$7,512
1000 lb. Vibrator	453.6 Kg	$353	$1,741	$7,584
5000 lb. Vibrator	2,268.0 Kg	$528	$2,528	$11,511
2000 Lb	907.2 Kg	$359	$1,766	$7,667
Crawler Mounted Lattice Boom Crane (with operator plus oiler)				
45 Ton	40.82 M. Ton	$1,096	$5,188	$21,756
50 Ton	45.36 M. Ton	$1,098	$5,198	$21,796
75 Ton	68.04 M. Ton	$1,201	$5,659	$23,642
110 Ton	99.79 M. Ton	$1,395	$6,536	$27,148
200 Ton	181.44 M. Ton	$1,878	$8,711	$35,848
Crane-Cable Truck (with operator plus oiler)				
30 Ton	27.22 M. Ton	$1,253	$5,896	$24,588
70 Ton	63.50 M. Ton	$1,573	$7,337	$30,353
90 Ton	81.65 M. Ton	$1,603	$7,416	$30,989
115.2 Ton	104.51 M. Ton	$1,727	$8,028	$33,118
Crane-Hydraulic Truck (with operator plus oiler)				
9.7 Ton	8.80 M. Ton	$1,081	$5,120	$21,486
44.7 Ton	40.55 M. Ton	$1,434	$6,711	$27,847
72.3 Ton	65.59 M. Ton	$1,840	$8,539	$35,160
108.3 Ton	98.25 M. Ton	$1,982	$9,179	$37,721
Drill (with laborer)				
35 Lb, Drill (light)	15.88 Kg	$217	$1,083	$4,752
80 Lb, Drill (Heavy)	36.29 Kg	$218	$1,085	$4,759

Approximate Rental Costs for Excavating Equipment with Operator				
Description	Metric	Day	Week	Month
Scraper (with operator)				
11.0 Cu.Yd.	8.41 Cu. M.	$866	$4,054	$16,821
23.0 Cu.Yd.	17.59 Cu. M.	$1,355	$6,256	$25,628
Tractor - Dozer (with operator)				
105 Hp	78.3 Kw	$683	$3,226	$13,508
180 Hp	134.3 Kw	$995	$4,632	$19,132
Loader (with operator)				
1 1/2 Cu.Yd.	1.15 Cu. M.	$532	$2,548	$10,796
2 1/4 Cu.Yd.	1.72 Cu. M.	$588	$2,801	$11,810
Concrete Pump (with operator)				
75 Cu.Yd per Hr	57.35 Cu. M.	$661	$3,128	$13,118
117 Cu.Yd per Hr	89.46 Cu. M.	$774	$3,639	$15,159

The above costs do not include move in/out charges, permits, daily fuel or other daily operating expenses.

Basic Hydraulic Backhoe Unit with Attachments. Naturally, the ideal situation would be to have the most suitable and economical backhoe, crane, or dragline for each job, but this is seldom practical or profitable. More frequently, the contractor must take into consideration the handling of a wide range of work with varying conditions, requiring more or less frequent and easy conversion (for example, backhoes with attachments convert to drill machines and forklifts, cranes to draglines clamshell operations and pile driving) by changing attachments on the basic unit.

Power cranes and backhoes are designed to accommodate a wide range of front-end attachments and working tools to meet the various job requirements. The attachments are classified into three basic groups: the crane boom, and the hoe boom. These attachments generally are interchangeable in the field.

Since the basic purpose of the backhoe boom is for backhoe work, the crane-type boom serves for all other materials handling tools, such as dragline buckets, clamshell buckets, orange peel buckets, hooks, hook blocks, tongs, grabs, clamps, grapples, concrete buckets, skull crackers, and pile driving leads.

Crane type booms can be extended for extremely high lifts by inserting standard sections in the center of the boom and by boom tip extensions called jibs. These jibs are used primarily to extend horizontal reach with boom raised vertically close to maximum elevation. They also provide an increase in vertical lifting range. Since the jib adds weight at the

SITE WORK

outer lifting radius and extends the working range, it has a limiting effect on lifting capacity.

Average Useful Life of Hoes and Cranes. A reasonable allowance for the exhaustion, wear and tear of property used in business, including a reasonable allowance for obsolescence, is permitted in figuring depreciation on your equipment.

Average Useful Life of Hydraulic Backhoes				
Bucket, Cu.Yd.	Bucket, Cu. M.	hp	Years	Hours
3/8 - 3/4	0.29 - 0.57	110	7	7,900
1 - 1 1/2	0.76 - 1.15	138	7	7,900
2 - 2 1/2	1.53 - 1.91	222	6	9,210

Average Useful Life of Draglines, Clamshells & Cranes				
Bucket, Cu.Yd.	Bucket,Cu. M.	hp	Years	Hours
3/8 - 3/4	0.29 - 0.57	154	9	14,940
1 - 1 1/2	0.76 - 1.15	195	9	14,940
2 - 3	1.53 - 2.29	230	9	14,940

The grouping shown for lifting cranes is a separate grouping and does not necessarily relate to the machines by size in cu. yds. For example, many 3/4 cu. yd. excavators would have a greater rating than 5 tons and might be rated in tons to fall in a different group than they do as excavators.

Depreciation. The straight line method of figuring depreciation is used. This method is in general adequate, but if it is desired to take changing prices for equipment into account in estimating, an annual appraisal to determine a current value should be considered.

Therefore, for the years and hours used, the following percentage of depreciation for the various groups is used:

Years	Hours	% per year	% per Hour
5	10,000	20.00%	0.01000%
6	12,000	16.67%	0.00833%
8	16,000	12.50%	0.00625%
9	18,000	11.11%	0.00555%
10	20,000	10.00%	0.00500%
12	24,000	8.30%	0.00416%

Interest, Taxes and Insurance. Interest, taxes, insurance, and storage are usually charged at 15% of the average investment, of which 10% is interest (to be adjusted for current rates) and 5% covers taxes, insurance, storage, and incidentals.

195

Average investment must be established, based on the number of years used for depreciation. Since the first year is considered as 100% and since the investment is considered at the beginning of each year, the method of calculating average investment for a 5-year depreciation period is as follows:

1st year	100% of total investment
2nd year	80% of total investment
3rd year	60% of total investment
4th year	40% of total investment
5th year	20% of total investment
	300% of total investment

Average investment equals 300% divided by 5 years equals 60% of total investment. The percent of Total Investment to use for Average Investment for 5, 6, 8, 9, 10, and 12 year periods are as follows:

No. Years Depreciation Percent Average Investment per Year

5	60%
6	58.33%
8	56.25%
9	55.56%
10	55%
12	54.17%

Therefore, the percentage of Total Investment per year and per hour to use for interest, insurance, taxes, and storage, based on 20% of average investment for the groups considered here, is as follows:

Interest, Insurance, Taxes, Storage Based on 20 Percent Average Investment per Year				
Average Investment	Years	Percent of Total Investment		
		Hours	Per Year	Per Hour
60.00%	5	10,000	12.000%	0.006000%
58.33%	6	12,000	11.666%	0.005833%
56.25%	8	16,000	11.250%	0.005625%
55.56%	9	18,000	11.112%	0.005556%
55.00%	10	20,000	11.000%	0.005500%
54.17%	12	24,000	10.834%	0.005417%

Fixed Costs, Repairs, Maintenance, and Supplies. These are difficult to estimate. The figures used are not based on actual records but are believed to be representative. In the case of backhoes, this cost is based on 100% of the total investment spread over the life of the machine and includes repairs, replacement parts and repair labor for normal operation, and such

196

items as labor, parts, supplies, bucket teeth, oil and grease for lubricating machinery, as well as seasonal overhaul. It does not include engine fuel or lubricating oil.

For draglines and clamshells, 80% of the total investment spread over the economic life of the machine is a fair average cost of these items.

In the case of lifting cranes, 60% of the total investment spread over the economic life of the machinery may be used.

The owner of an excavator should anticipate three things in regard to repairs and maintenance. First, the average figures used presume proper maintenance. Without this, repair bills will greatly exceed the estimates given here. Second, it is assumed the equipment will be used within its specified rating. Third, these expenses are not uniform each year.

Maintenance is then further defined in two major categories. The first is normal maintenance as a daily operating expense, which includes regular servicing on the job, such as fuel, changing oil, filters, greasing, brakes and hydraulic fluid levels, minor tune-ups and the like.

The second category is defined as major maintenance, which requires that the equipment be idle for an extended period while repairs are made in the shop or field. An example of major maintenance would be the undercarriage of a dozer.

Provision should be made for a periodic general overhaul, and it should be expected that both the amount of repairs required and the losses caused by shutdowns will increase as the equipment becomes older.

Maintenance and Supplies, Including Labor Associated with Shovels and Backhoes				
Size Cu.Yd.	Years	Percent of Total Investment		
		Hours	Per Year	Per Hour
3/8 - 3/4 Cu. Yd. (0.29 - 0.57 Cu. M.)	5	10,000	20.00%	0.01000%
1 - 1 1/2 Cu. Yd. (0.76 - 1.15 Cu. M.)	6	12,000	16.67%	0.00833%
2 - 3 1/2 Cu. Yd. (1.53 - 2.68 Cu. M.)	8	16,000	12.50%	0.00625%
4 1/2 Cu. Yd. (3.44 Cu. M.)	10	20,000	10.00%	0.00500%

Based on 100% of total investment spread over life of machine.

For example, assume a 1.0-c.y. (0.76 cm) shovel cost $68,000. Repairs, maintenance, and supplies equal $68,000.00 x .00833% = $5.66 per hr.

Operating Costs, Engine Fuel, and Lubricating Oil. A formula for estimating the approximate amount of diesel fuel consumption for equipment of this type is as follows:

$$\frac{\text{BHP x Factor x lbs. fuel per HP hr.}}{\text{Weight of fuel per gallon (liter)}}$$

= Gallons (Liters) per Hour, where

BHP = Brake HP of engine-or rated HP
Factor = Factor for this use 50 to 60%
Diesel = 0.5 lbs. per brake horsepower hour.
Diesel = 7.3 lbs. per gallon (U.S.)

From this formula we obtain an approximate diesel fuel consumption of 0.034 to 0.041 gals. (0.13-0.15 liters) per horsepower hour (based on 50% to 60% factor respectively), and suggest using an average of about 0.040 gal. per horsepower hour for normal operating conditions.

Backhoes and shovels normally will consume a greater amount of fuel than the other types of machines considered here. Therefore the larger consumption rate indicated should be used for estimating fuel for shovels and the smaller rate for draglines and clamshells. Machines used for lifting crane service only usually operate intermittently and fuel consumption is difficult to estimate.

Example: A 100-horsepower engine in shovel service is estimated to consume per hour: 100 x 0.040 = 4.0 gals. (15 liters) of diesel fuel per hr. Lubricating oil is considered with fuel, because it varies with the size and type of engine. It usually includes a complete change every 100 hours plus make-up oil between changes. Allow 15% of fuel costs.

Actual fuel consumption should be measured in the field. However, it this is not practical consumption can be estimated when the machine application is known.

Labor Operating Costs. The labor rates, as well as the number in the operating crew, vary in different parts of the country and on different jobs. The contractor must estimate this cost. Costs related to crew costs, besides rates of pay, are employer contributions for Federal Insurance Contribution Act, Workmen's Compensation Insurance, Unemployment Compensation, Overtime, Paid Holidays, and Contractor's Contributions to Union Welfare and Pension Funds.

Dragline Yardages. As in the case of any excavation machine, dragline yardage production is affected by the type of material to be excavated, the depth of cut, the swing before unloading, the type of

198

Hourly Short Boom Dragline Output in Cubic Yards									
Bucket Capacity - Cu. Yd	3/8	1/2	3/4	1	1 1/4	1 1/2	1 3/4	2	2 1/2
Class of Material									
Light Sandy Clay	70	95	130	160	195	220	245	265	305
Sand and Gravel	65	90	125	155	185	210	235	255	295
Common Earth	55	75	105	135	165	190	210	230	265
Clay, Hard and Tough	35	55	90	110	135	160	18	195	230
Clay, Wet and Sticky	20	30	55	75	95	110	130	145	175

Hourly Short Boom Dragline Output in Cubic Meters									
Bucket Capacity - Cu. Meter	0.29	0.38	0.57	0.76	0.96	1.15	1.34	1.53	1.91
Class of Material									
Light Sandy Clay	53.5	72.6	99.4	122.3	149.1	168.2	187.3	202.6	233.2
Sand and Gravel	49.7	68.8	95.6	118.5	141.5	160.6	179.7	195.0	225.6
Common Earth	42.1	57.3	80.3	103.2	126.2	145.3	160.6	175.9	202.6
Clay, Hard and Tough	26.8	42.1	68.8	84.1	103.2	122.3	13.8	149.1	175.9
Clay, Wet and Sticky	15.3	22.9	42.1	57.3	72.6	84.1	99.4	110.9	133.8

* The dragline is working in the optimum depth of cust for maximum efficiency.
* The dragline is working a full 60 minutes each hour with no delays.
* The dragline is making a 90° swing before unloading.
* The bucket loads are being dumped into "properly sized" hauling units.
* The proper type bucket is being used for the job.
* The dragline is being used within the working radius recommended by the manufacturer for machine stability.

199

unloading (unloaded into hauling units, cast onto spoil banks, etc.), the degree of continuity in the operation, etc.

The table on the opposite page gives an approximate idea of the difference in maximum production possible subject to conditions listed.

How Degrees of Swing and Depth of Cut Affect Dragline Yardage Output. The following table gives the effect of depth of cut and angle of swing on dragline yardage production. Variations in degrees of swing, in particular, have a marked effect on production and this is important to keep in mind when laying out a job as the shorter the swing the more the yardage.

Table Giving Effect of Depth of Cut in Feet and Meters and Angle of Swing on Dragline Output									
Depth of Cut in % of Optimum	Angle of Swing in Degrees								Meters
	30	45	60	75	90	120	150	180	
20 '	1.06	0.99	0.94	0.90	0.87	0.81	0.75	0.70	6.10
40 '	1.17	1.08	1.02	0.97	0.93	0.85	0.78	0.72	12.19
60 '	1.24	1.13	1.06	1.01	0.97	0.88	0.80	0.74	18.29
80 '	1.29	1.17	1.09	1.04	0.99	0.90	0.82	0.76	24.38
100 '	1.32	1.19	1.11	1.05	1.00	0.91	0.83	0.77	30.48
120 '	1.29	1.17	1.09	1.03	0.98	0.90	0.82	0.76	36.58
140 '	1.25	1.14	1.06	1.00	0.96	0.88	0.81	0.75	42.67
160 '	1.20	1.10	1.02	0.97	0.93	0.85	0.79	0.73	48.77
180 '	1.15	1.05	0.98	0.94	0.90	0.82	0.76	0.71	54.86
200 '	1.10	1.00	0.94	0.90	0.87	0.79	0.73	0.69	60.96

Clamshell Production. The clamshell excavator is not to be considered a high production machine but rather a machine to be used where the work is beyond the scope of other types of equipment. For example, a condition which usually requires a clamshell is where digging is vertical or practically straight down as in digging pier holes or shafts. Digging in trenches that are sheathed and cross-braced generally calls for a clamshell, because the vertical action of the bucket enables it to be worked through the cross-bracing. Jobs requiring accurate dumping or disposal of materials are usually clamshell jobs. Also, for high dumping jobs, whether it be charging a bin, building a stockpile, or wherever the material must be dumped well above the machine level, the clamshell is well adapted. In general, the clamshell can operate vertically and dig or spot dump below, at, or above the level of the machine.

200

SITE WORK

Hourly Short Boom Clamshell Handling Capacity, Cubic Yards									
Bucket Capacity - Cu. Yd	3/8	1/2	3/4	1	1 1/4	1 1/2	1 3/4	2	2 1/2
Class of Material									
Moist Loam or Sandy Clay	50	65	95	120	140	155	170	190	225
Sand and Gravel	45	60	85	110	130	140	160	175	205
Common Earth	40	55	70	95	115	125	145	160	185

Hourly Short Boom Clamshell Handling Capacity, Cubic Meters									
Bucket Capacity - Cu. Meter	0.29	0.38	0.57	0.76	0.96	1.15	1.34	1.53	1.91
Class of Material									
Moist Loam or Sandy Clay	38.23	49.70	72.64	91.75	107.04	118.51	129.98	145.27	172.04
Sand and Gravel	34.41	45.88	64.99	84.11	99.40	107.04	122.34	133.81	156.74
Common Earth	30.58	42.05	53.52	72.64	87.93	95.58	110.87	122.34	141.45

It must be thoroughly understood that the above production is based on the most ideal job and management conditions.

201

The clamshell is only effective where the materials to be handled are relatively soft or loose. Conditions are so variable on clamshell operation that a table showing typical production has little value.

However, in an effort to establish some point to work from, the following table has been prepared showing maximum production to be expected from clamshell excavators operating under the following conditions: The clamshell is engaged in open digging such as a basement or large footing, with the permissible cut at least a full bucket depth.

1. The depth of the excavation is not more than 20'-0" (6.09 m).
2. The quantities are in terms of bank measure.
3. There is no wasted time, 60 minutes of digging each hour.
4. The clamshell is making a 90° swing before unloading.
5. The bucket loads are being dumped into "properly sized" hauling units.
6. The clamshell is being used within the working radius recommended by the manufacturer for machine stability.

How Degrees of Swing Affect Clamshell Yardage. The output of a clamshell operating at a steady pace is affected by the swing before unloading in the same manner as a dragline. The following table gives the effect of angle of swing on clamshell yardage production.

Angle of Swing in Degrees							
30°	45°	60°	75°	90°	120°	150°	180°
1.32	1.19	1.11	1.05	1.00	0.91	0.83	0.77

Job and Management Factors. Ideal conditions seldom, if ever, exist in the field, and it is necessary to rationalize these figures by some other factor to compensate for the fact that actual job conditions differ widely from the perfect conditions assumed so far.

There are two sets of factors on every job that have a great deal to do with the output of equipment on the job. They are the job factor and the management factor.

Job factors are the physical conditions pertaining to a specific job that affect the production rate, other than the class of material to be handled. These may be divided into three general headings:

1. topography and the dimensions of the work, which include the depth of cut and whether the work will require much moving within the cut or from one cut to another;
2. surface and weather conditions, which include in some cases the difference between summer and winter work, and also the question of drainage of either surface or underground water;
3. specification requirements that control the manner in which the work must be handled, or indicate the sequence of various operations, and

202

which also show the amount of bank sloping or cutting close to finished line and grade that is required.

The above conditions are all inherent in the job itself and can be taken into account by the contractor in making up the bid. Another group of factors which affect output are the management factors. These cover conditions which pertain to the efficiency of the operation and thus affect output. These are grouped under the head of the management factor because, in general, they are made up of items which management can determine and control. These include:

1. selection, training, and direction of personnel; the quantities given in the tables indicate what can be reasonably expected by an experienced operator who is willing to work; the availability of trained workers and the incentive to produce must be considered;

2. selection, care, and repair of equipment; the quality of the inspection and preventive maintenance program can have much to do with the lost time on the job;

3. planning, laying out the job, supervising, and coordinating the operations is a very important factor in increasing efficiency; providing the right number and size of hauling units is one of the first things to show its effect.

Foremanship and supervision are important. When everything is working smoothly, there is added incentive for high production rates.

Because both job and management factors must be considered jointly, that is, the output yardages must be modified by both a job and a management factor, the following table consolidates these factors.

To estimate the yardage, select the proper combined job and management factor from the following table. First, select a job factor classification of excellent, good, fair or poor, whichever applies to your job conditions and follow straight across the management factor classification of excellent, good, fair or poor, whichever applies to the management conditions. The resulting figure is the combination job management factor by which the yardage figures from the basic output tables (as modified by the conversion factors for depth of cut and swing) must be multiplied to give the yardage factor that can reasonably be expected under the job management conditions that will exist on the job.

Usually, pit and quarry and heavy construction jobs will fall into groups 1, 2, or 3, whereas highway grading will usually fall into groups 2, 3, or 4 because of the variations in cuts, required machine travel, close cuts, etc.

Combination Job and Management Factors				
	Management Factors			
Job Factors	1	2	3	4
	Excellent	Good	Fair	Poor
1. Excellent	0.84	0.81	0.76	0.70
2. Good	0.78	0.75	0.71	0.65
3. Fair	0.72	0.69	0.65	0.60
4. Poor	0.63	0.61	0.57	0.52

Dragline Excavating

The cost of dragline excavating will vary in the same manner as shovel excavating, according to the size of the job, size of the machine, and the capacity and speed of the hauling units. But the hauling hazards are reduced substantially by the trucks not having to be loaded in the hole, eliminating the necessity of trucking up a ramp.

As may be determined by comparing the basic output tables for shovels and draglines, the dragline has about 75% to 80% of the basic output capacity of shovels. However, the dragline definitely has a place in excavating for construction work by virtue of its generous digging reach, its ability to dig under extremely wet conditions (it can stand on the top on dry, firm footing), and the easier haul with no ramps to climb, all conditions that under certain circumstances would make shovel digging very expensive if not totally impossible.

In excavating for building construction, the dragline is affected by job and management factors the same as shovels and the same basis for computing daily output may be used as follows.

Basis for Computing Daily Output of Dragline Excavator. The following is based on a dragline having a 1 cu. yd. (0.76 cu.m) bucket.

Based on a Dragline Having a 1 Cu. Yd. (0.76 Cu.M) Bucket.		
Bucket capacity	1 Cu.Yd	0.76 Cu. M.
Bucket efficiency, average material	90.0%	90.0%
Load carried	0.9 Cu.Yd.	0.69 Cu. M.
Average operating cycle	20 Seconds	
Cycles per hour	150 Cycles	
Cu. Yds. per hour, operating 100% time	162 Cu. Yd.	123.87 Cu. M.

Based on a Dragline Having a 1 Cu. Yd. (0.76 Cu.M) Bucket.		
Operating efficiency in basement excavation due to machine delays, traffic delays to hauling units and clean-up work, approx.	50% to 60 %	50% to 60 %
Average output per 8-hr. day	540- 650 Cu.Yds.	412.9-497.0 Cu. M.
Operating efficiency on heavy construction, roads, dams, open cut excavation, etc., including machine delays and clean-up work approximately 66.67 Average.	66.67%	66.67%
Average output per 8-hr. day	720 Cu.Yds.	550.51 Cu. M.

Backhoe Excavating

The backhoe is a cross between the dragline and the shovel, which is why it is the equipment of choice today. It incorporates some of the characteristics of each and overcomes some of the limitations of each. The backhoe is primarily a unit for digging below machine level, but it will dig harder material than the clamshell or dragline, because the weight of the boom itself may be used to force the dipper into the material. It is, however, limited in digging by the length of the boom and stick. The backhoe dipper can be controlled more accurately than the dragline bucket and is better suited to close-limit work.

In building construction the backhoe is used to dig trenches, footings, and basements. On small residence basements, it offers many advantages. It digs straight, vertical side walls (in soil which will stand); it cuts a level floor; it trims corners neatly and squarely; it can dig sewer and waterline trenches; it always works from the top on dry, safe ground; and it reduces hand trim to a minimum.

The typical process of digging a basement with a backhoe is to dig a trench around the four sides of the basement, scooping out the center as you go.

On small work it is necessary to have low-cost, simple and easy means of moving the backhoe from job to job. This is best done by using a single-purpose trailer which can be loaded or unloaded in 15 to 20 minutes. The cost of moving the backhoe from job to job will vary with the distance between jobs.

On time studies made on several small basement jobs, where the basements contained from 275 to 350 cu. yds. (210-268 cu.m) of excavation, a backhoe excavator equipped with a 3/4-cu. yd. (0.57-cu.m) bucket averaged from 66 to 88 cu. yds. (50-67 cu.m) per hour.

An example of the cost of excavating a small basement containing approximately 350 cu. yds. (268 cu.m) of excavation, with the soil placed around the excavation, is as follows:

Backhoe Excavating for Small Basement			
Description	Hours	Rate	Total
Move-In Charges	......	$ 110.00	$ 110.00
Excavating			
Backhoe Operator	5.00	$ 37.89	$ 189.45
Labor, clean-up	5.00	$ 26.16	$ 130.80
Backhoe charge	5.00	$ 87.00	$ 435.00
Total Direct Cost			$ 865.25
Cost per Cu. Yd			$ 2.47
Cost per Cu.M			$ 2.86

The above costs do not include removal of excavated earth from site or backfilling.

Backhoe Excavating Trench (Courtesy Caterpillar, Inc.)

Data on Backhoe Costs and Operation

Description	Size of Backhoe in Cu. Yds.						
	3/4	1	1 1/2	2	3 1/4	4 1/4	5 1/2
CY per Hr, based on 100% Time	135	180	270	360	585	765	950
CY per Hr, 100% Time and 90% Efficiency	122	162	243	324	527	689	855
CY per Hr, 100% Time and 66.7% Efficiency	90	120	180	240	390	510	633
CY per Hr, 100% Time and 50% Efficiency	68	90	135	180	293	383	475
CY per 8-Hr. Day, 66.7% Efficiency	720	960	1,440	1,920	3,120	4,080	5,064
CY per 8-Hr. Day, 50% Efficiency	544	720	1,080	1,440	2,344	3,064	3,800
Avg Cost Backhoe, Including Freight and Unloading	$128,000	$145,000	$264,000	$300,000	$525,000	$820,000	$855,000
Depreciation, Percent Per Year	12.20%	13.90%	20.80%	22.60%	56.60%	67.80%	74.20%
Depreciation per Hr, @ 2,000 Hours per Year	$7.81	$13.44	$36.61	$45.20	$198.10	$370.64	$422.94
Interest, Taxes, Insurance, 20% per Year (2,000 Hours per year)	$12.80	$14.50	$26.40	$30.00	$52.50	$82.00	$85.50
Hourly Consumption of Fuel, (Gal per Hr)	3.20	4.00	6.80	9.90	12.90	19.00	22.70
Fuel Cost per Hour*	$12.80	$16.00	$27.20	$39.60	$51.60	$76.00	$90.80
Engine Lub.Oil Cost of Fuel	$1.90	$2.40	$3.90	$5.70	$7.33	$11.00	$13.00
Cost per Hr., Repairs, Maintenance & Supplies	$2.10	$2.40	$3.40	$4.40	$7.10	$10.90	$12.00
Total Hourly Cost, Not Including Labor, Supervision, Workman's Compensation, Unemployment, Social Security, etc.	$40.61	$52.74	$104.31	$134.80	$329.53	$569.54	$646.94

*Fuel costs based on diesel fuel at $4 per gal.

Data on Backhoe Costs and Operation - Metric							
Description	Size of Backhoe in Cu. Meters						
	0.57	0.76	1.15	1.53	2.48	3.25	4.21
Cu.M. per Hr, based on 100% Time	103	138	206	275	447	585	726
Cu.M. per Hr, 100% Time and 90% Efficiency	93	124	186	248	403	526	654
Cu.M. per Hr, 100% Time and 66.7% Efficiency	69	92	138	184	298	390	484
Cu.M. per Hr, 100% Time and 50% Efficiency	52	69	103	138	224	292	363
Cu.M.Y per 8-Hr. Day, 66.7% Efficiency	551	734	1,101	1,468	2,386	3,120	3,874
Cu.M. per 8-Hr. Day, 50% Efficiency	413	551	826	1,101	1,789	2,340	2,905
Avg Cost Backhoe, Including Freight and Unloading	$128,000	$145,000	$264,000	$300,000	$525,000	$820,000	$855,000
Depreciation, Percent Per Year	12.20%	13.90%	20.80%	22.60%	56.60%	67.80%	74.20%
Depreciation per Hr, @ 2,000 Hours per Year	$7.81	$13.44	$36.61	$45.20	$198.10	$370.64	$422.94
Interest, Taxes, Insurance, 20% per Year (2,000 Hours per year)	$12.80	$14.50	$26.40	$30.00	$52.50	$82.00	$85.50
Hourly Consumption of Fuel, (Liters per Hr)	12.11	15.14	25.74	37.48	48.83	71.92	85.93
Fuel Cost per Hour*	$12.84	$16.05	$27.29	$39.72	$51.76	$76.24	$91.08
Engine Lub.Oil Cost of Fuel	$1.90	$2.40	$3.90	$5.70	$7.33	$11.00	$13.00
Cost per Hr., Repairs, Maintenance & Supplies	$2.10	$2.40	$3.40	$4.40	$7.10	$10.90	$12.00
Total Hourly Cost, Not Including Labor, Supervision, Workman's Compensation, Unemployment, Social Security, etc.	$49.56	$63.93	$123.33	$162.50	$365.62	$622.70	$710.45

*Fuel costs based on diesel fuel at $1.06 per liter.

On large projects a combination of machines together with a backhoe with a 4-5 Cu.Yd (3.05 – 3.82 CuM) bucket can be very cost effective. With extended boom backhoes can reach a digging depth to 30' (9.1 Meters).

The costs in the previous two tables do not include removal of excavated earth from site or backfilling.

Truck and trailer time to deliver the hoe and removing same at completion of job will vary with length of haul. In medium to large work at moderate depths, hoe production exceeds that of shovels which are being use almost exclusively more and more for mining operations and very large projects. Output falls off considerably backhoes as greater depths are required, this is with all types of excavation equipment.

For the smaller machines, the size backhoe is usually chosen for the width trench it will cut, but for the larger sizes, the capacity of the machine is chosen on a production basis.

Data on typical backhoe bucket dimensions and weight is given in the following table.

Dipper Bucket Outside Total Width Added Weight		
Bucket Capacity Cu. Yd.	Bucket Outside Width Inches	Total Width Added for 2 Side Cutters- Inches
3/8	20 " - 24 "	4 " - 6 "
1/2	24 " - 28 "	4 " - 9 "
3/4	28 " - 39 "	4 " - 9 "
1	33 " - 45 "	4 " - 9 "
1 1/4	39 " - 45 "	4 " - 9 "
1 1/2	39 " - 45 "	4 " - 9 "
Bucket Capacity Cu. Meter	Bucket Outside Width mm	Total Width Added for 2 Side Cutters- mm
0.29	500 - 600 mm	100 - 150 mm
0.38	600 - 700 mm	100 - 225 mm
0.57	700 - 975 mm	100 - 225 mm
0.76	825 - 1125 mm	100 - 225 mm
0.96	975 - 1125 mm	100 - 225 mm
1.15	975 - 1125 mm	100 - 225 mm

Production for Backhoe Excavating							
Volume of Excavation in Job							
Cu. Yds	1,000	1,500	2,000	2,500	3,000	4,000	5,000
Cu. Meters	765	1,147	1,529	1,912	2,294	3,058	3,823
Actual Operating Time of Shovel							
3/8 Cu.Yd. (0.28 Cu. M.) **Backhoe** with an Average Daily Output of 300 Cu Yds (230 Cu.M.)							
Foreman*	27	40	53	67	80	107	133
Shovel Operator	27	40	53	67	80	107	133
1/2 Cu.Yd. (0.38 Cu. M.) **Backhoe** with an Average Daily Output of 400 Cu Yds (306 Cu.M.)							
Foreman*	20	30	40	50	60	80	100
Shovel Operator	20	30	40	50	60	80	100
3/4 Cu.Yd. (0.57 Cu. M.) **Backhoe** with an Average Daily Output of 600 Cu Yds (460 Cu.M.)							
Foreman*	13	20	27	33	40	53	67
Shovel Operator	13	20	27	33	40	53	67
1.0 Cu.Yd. (0.76 Cu. M.) **Backhoe** with an Average Daily Output of 720 Cu Yds (550 Cu.M.)							
Foreman*	11	17	22	28	33	44	56
Shovel Operator	11	17	22	28	33	44	56

For all of above, foreman or Laborer in charge of project, handling and expediting trucks, etc.

Production for Dragline Excavating Using a 1/2 Cu.Yd. (0.76 Cu. M.) Dragline based on an Average Daily Out Put of 560 Cu Yds (428 Cu.M.)							
Volume of Excavation in Job							
Cu. Yds	1,000	1,500	2,000	2,500	3,000	4,000	5,000
Cu. Meters	765	1,147	1,529	1,912	2,294	3,058	3,823
Actual Operating Time of Shovel							
Foreman*	14	21	29	36	43	57	71
Shovel Operator	14	21	29	36	43	57	71

For all of above, foreman or Laborer in charge of project, handling and expediting trucks, etc.

SITE WORK

Sample Excavating Estimate

The following example is based on a basement containing 5,000 cu. yds. (3823 cu. m) excavation, average conditions, using a 1-1/4 c.y. (0.96 cu. m) backhoe, loading into trucks, based on an average of 900 cu. yds. (688 cu. m) per 8-hr. day.

Add cost of trucks for removing excavated soil, overhead expense, and profit. The above method for estimating shovel excavation can also be used for hoe, dragline, and clamshell excavation by merely substituting the proper production values and equipment charges for those pertaining to shovels.

Sample Excavating Estimate using 1 1/4 Cu YD (.96 Cu. Meter) Backhoe			
Description	Hours	Rate	Total
Bringing Shovel to Job	1	$500.00	$500.00
Excavating			
Foreman	48.00	$28.16	$1,351.68
Backhoe Operator	48.00	$37.89	$1,818.72
Backhoe Costs (rental)	48.00	$55.00	$2,640.00
Total Direct Cost - 900 Cu. Yd.			$6,310.40
Cost per cu. Yd			$7.01
Cost per cu. Meter			$9.17

Note: Add trucking costs if required to haul excess material off Project

Hauling Excavated Materials in Trucks. Where an excess of excavated materials over and above the amount required for backfill exists, it will have to be removed from the job by trucks.

The usual procedure followed in these cases is to determine beforehand the approximate quantity of material required for backfill, etc., and this amount subtracted from the estimated excavation quantity will give the volume of excavated material to be hauled away. The excess material should be loaded and hauled as the excavation progresses so that no or minimum rehandling is necessary.

Types of Trucks. Truck hauling units are mainly of two types: conventional highway hauling units and off-the-highway heavy duty hauling equipment.

The conventional highway hauling unit is the most frequently used. For the average contractor it is the most economical type of truck to use. It is adaptable to practically every job and may be used for other hauling purposes as well.

For the contractor specializing in large earth moving projects, the high production, heavy duty, off-the-highway hauling unit is the most economical equipment to use, because it is specifically designed for this purpose. Not limited by state highway regulations as to weight and size, it has a high ratio of net weight to payload, the additional weight being found in its more rugged construction, such as heavier frame, axles, body, hoist, and engine.

Combination Backhoe/Loader (Courtesy Caterpillar, Inc.)

Hourly Cost of Ownership and Operation of Trucks. To estimate hauling costs, the contractor must know the equipment costs to own and operate. This can be determined by the same method previously described for excavating equipment.

Attention is called to the manner in which tire costs (original, replacement, and repair) are handled. Recognizing the fact that the economic life of tires will not be as long as the life of the vehicle, the original value of the tires is deducted from the total price before computing the hourly charge for depreciation and is then treated as an operating cost.

The hourly cost of tire replacement and tire repairs varies greatly according to operating conditions, road surfaces, and tire loadings. While tire costs in ordinary trucking operations normally are figured on a mileage basis, in construction work, with its rough roads, short hauls, and relatively low speeds (below 30 mph), they are figured on an hourly basis and are computed by dividing the replacement cost by the estimated tire life in hours. The average life expectancy of truck tires in construction work is from 3,000 to 3,500 hours. Tire repairs should run about 15% of the hourly tire cost.

The following tables give data on various sizes of rear dump trucks for both conventional highway hauling units and heavy duty off-highway equipment.

Table Showing How Hourly Cost of Ownership and Operation May Be Derived for Conventional Highway Rear Dump Trucks Used in Construction Work Based on Economic Life of 3 Years of 2,000 Hours Each in Tons

	5	12	15	20	25
Capacity in Tons	5	12	15	20	25
Capacity in Cu. Yds., Struck	4	6	10	15	20
Approximate Delivered Price	$47,000	$48,000	$136,000	$176,000	$200,000
Original Tire Value	$2,350	$2,400	$6,800	$8,800	$10,000
Depreciation, .0167% of Total Investment Less Tires	$7.46	$7.62	$21.58	$27.92	$31.73
Interest, Taxes and Insurance, 0.0067% of Total Investment	$3.31	$3.38	$9.57	$12.38	$14.07
Hourly Tire Replacement Cost, Based on 3,000-Hr Life	$0.78	$0.80	$2.27	$2.93	$3.33
Hourly Cost of Tire Repairs, 15% of Hourly Tire Replacement	$0.12	$0.12	$0.34	$0.44	$0.50
Maintenance and Repairs (.004% of Total Investment)	$2.00	$2.00	$6.00	$7.00	$8.00
Fuel Cost per Hour, 0.03 Gal. per Hp	$20.40	$25.20	$30.60	$39.00	$39.60
Oil and Grease 15% of Fuel	$3.06	$3.78	$4.59	$5.85	$5.94
Total Hourly cost, Not Including Labor, Supervision, Unemployment, Social Security. Overhead Expense and Profit	$37.13	$42.89	$74.94	$95.53	$103.17

Table Showing How Hourly Cost of Ownership and Operation May Be Derived for Conventional Highway Rear Dump Trucks Used in Construction Work Based on Economic Life of 3 Years of 2,000 Hours Each in Metric Tons					
Capacity in Metric Tons	4.54	10.89	13.61	18.14	22.68
Capacity in Cu. Meters, Struck	3.06	4.59	7.65	11.47	15.29
Approximate Delivered Price	$47,000	$48,000	$136,000	$176,000	$200,000
Original Tire Value	$2,350	$2,400	$6,800	$8,800	$10,000
Depreciation, .0167% of Total Investment Less Tires	$7.46	$7.62	$21.58	$27.92	$31.73
Interest, Taxes and Insurance, 0.0067% of Total Investment	$3.31	$3.38	$9.57	$12.38	$14.07
Hourly Tire Replacement Cost, Based on 3,000-Hr Life	$0.78	$0.80	$2.27	$2.93	$3.33
Hourly Cost of Tire Repairs, 15% of Hourly Tire Replacement	$0.12	$0.12	$0.34	$0.44	$0.50
Maintenance and Repairs, .004% of Total Investment	$2.00	$2.00	$6.00	$7.00	$8.00
Fuel Cost per Hour, .15 liters per kW	$20.40	$25.20	$30.60	$39.00	$39.60
Oil and Grease 15% of Fuel	$3.06	$3.78	$4.59	$5.85	$5.94
Total Hourly cost, Not Including Labor, Supervision, Unemployment, Social Security. Overhead Expense and Profit	$37.13	$42.89	$74.94	$95.53	$103.17

Table Showing How Hourly Cost of Ownership and Operation May Be Derived for Off-Highway Rear Dump Trucks Based on an Economic Life of 3 Years of 2,000 Hours Each

	18	22	35	50	60
Capacity in Tons	18	22	35	50	60
Capacity in Cu. Yds., Struck	13	16	25	38	44
Horsepower	180	260	405	641	720
Approximate Delivered Price	$225,000	$321,000	$375,000	$400,000	$563,000
Original Tire Value	$11,250	$16,050	$18,750	$20,000	$28,150
Depreciation, 0.01% of Total Investments less Tires	$21.38	$30.50	$35.63	$38.00	$53.49
Interest, Taxes and Insurance, .0067% of Total Investment	$15.83	$22.58	$26.38	$28.14	$39.61
Hourly Tire Replacement Cost Based on 3,000 Hours	$3.75	$5.35	$6.25	$6.67	$9.38
Hourly Cost of Tire Repairs, 15% of Hourly Tire Replacement Cost	$0.56	$0.80	$0.94	$1.00	$1.41
Maintenance and Repairs, .004% of total Investment	$1.42	$2.02	$2.36	$2.52	$3.55
Fuel Cost per Hour, $0.025 Gal per Hp	$18.00	$26.00	$40.50	$64.10	$72.00
Oil and Grease, 15% of Fuel	$2.70	$3.90	$6.08	$9.62	$10.80
Total Hourly Cost	$63.63	$91.15	$118.13	$150.04	$190.23

Table Showing How Hourly Cost of Ownership and Operation May Be Derived for Off-Highway Rear Dump Trucks Based on an Economic Life of 3 Years of 2,000 Hours Each

Capacity in Metric Ton	16.33	19.96	31.75	45.36	54.43
Capacity in Cu. Meter, Struck	9.94	12.23	19.12	29.05	33.64
Kilowatts	134	194	302	478	537
Approximate Delivered Price	$225,000	$321,000	$375,000	$400,000	$563,000
Original Tire Value	$11,250	$16,050	$18,750	$20,000	$28,150
Depreciation, 0.01% of Total Investments less Tires	$21.38	$30.50	$35.63	$38.00	$53.49
Interest, Taxes and Insurance, .0067% of Total Investment	$15.83	$22.58	$26.38	$28.14	$39.61
Hourly Tire Replacement Cost Based on 3,000 Hours	$3.75	$5.35	$6.25	$6.67	$9.38
Hourly Cost of Tire Repairs, 15% of Hourly Tire Replacement Cost	$0.56	$0.80	$0.94	$1.00	$1.41
Maintenance and Repairs, .004% of total Investment	$1.42	$2.02	$2.36	$2.52	$3.55
Fuel Cost per Hour, $0.13 Liters per kW	$18.00	$26.00	$40.50	$64.10	$72.00
Oil and Grease, 15% of Fuel	$2.70	$3.90	$6.08	$9.62	$10.80
Total Hourly Cost	$63.63	$91.15	$118.13	$150.04	$190.23

*The foregoing tables are for the purpose of showing the items that must be taken into consideration in figuring hourly costs but do not necessarily reflect current prices of equipment, tires and fuel.

Power Shovel Excavating (Courtesy Caterpillar, Inc.)

Cost of Hauling Excavated Material in Trucks. Hauling costs vary with the capacities of the trucks and time of the hauling cycle.

Truck capacity, for proper sizing, should be at least 4 times the dipper or bucket capacity of the excavator. This is important, because the efficiency of the excavator is seriously affected by undersized hauling units due to the increased hazards of truck delays.

The table on the next page gives theoretical spotting time cycles for various sizes of backhoe excavators working at 100% efficiency and loading trucks of 4 times bucket size capacity.

For dragline operation the above time values may be increased from 20% to 30%.

The hauling time cycle consists of several operations-spotting of truck under excavator; loading truck by excavator; traveling to dumping area; dumping; returning to excavator.

Some of these operations, such as spotting, loading and dumping, can be minimized by good management and supervision. Traveling time, however, will vary with the haul distance, traffic conditions, road conditions, etc., and must be carefully analyzed for each job.

When the hauling time cycle has been estimated, the number of loads each truck is able to haul per day can be determined and multiplying this by the capacity of the truck will give the theoretical daily haul of each truck in cu. yds., assuming 100% job efficiency. This amount should be multiplied by the expected job efficiency factor to obtain the net yardage hauled. Divide the daily truck cost by the net yardage hauled and the result will be the hauling cost per cu. yd.

The number of trucks required to keep the excavator going can also be determined by dividing the hauling time cycle by the loading time

Haul Units Needed to Spot Under Backhoe per Hour in medium Digging

Size of Bucket Cu.Yd.	Minimum Haul Unit Capacity at 4 Times Bucket Size Cu. Yd.	Approx. Backhoe Cycle in Seconds 90° Swing No Delay	Loading Time 4 - Bucket Truck in Seconds	Time Spotting Cycle for Steady Oper. in Minutes	Number of Spots Required per Hour for Steady Operation	Size of Bucket Cu. Meter	Minimum Haul Unit Capacity at 4 Times Bucket Size Cu. M.
3/8	1 1/2	19	76	1.26	48	0.29	1.15
1/2	2	19	76	1.26	48	0.38	1.53
3/4	3	20	80	1.33	45	0.57	2.29
1	4	21	84	1.4	43	0.76	3.06
1 1/4	5	21	84	1.4	43	0.96	3.82
1 1/2	6	23	92	1.53	39	1.15	4.59
2	8	25	100	1.66	36	1.53	6.12
2 1/2	10	26	104	1.73	35	1.91	7.65

cycle using 100% excavator efficiency as a basis. This is done to be sure that a hauling unit is under the excavator at all times. While an excavating job may actually work out to a 50% or 60% efficiency, there usually are times on the job when the excavator is working at full capacity. To take advantage of this, sufficient, properly-sized hauling units must be on the job. In large operations many contractors maintain standby units to step in when one of the hauling fleet trucks breaks down or is otherwise unusually delayed.

In sizing trucks to conform to dipper or bucket capacities, the struck capacity of each is used, assuming that the heaping of loads on both trucks and dippers or buckets will offset each other and produce bank measure quantities.

However, all estimating should use actual job data rather than theoretical figures. Some materials do not follow the usual swell ratios and job efficiency ratings vary widely. Following book figures to the letter in excavating, or any other branch of construction work, can be disastrous. Book values, at best, can only give what can be done under a definite set of conditions. The contractor must exercise his judgment in estimating this work and base his figure on his own particular set of conditions, i.e., equipment, supervision, operators, local materials and local weather.

Tractors and Scrapers

The tractor is probably the most used piece of earth moving equipment on the construction job. It can perform a wide variety of operations, and equipped with the proper accessories, it is a high production and efficient tool for such jobs as site clearing, grading, shallow excavating, etc. In addition the tractor is often called upon to act as the "strong man" in freeing stuck trucks, hauling or dragging equipment, sheds, etc., into place, miscellaneous hoisting using a snatch block and cable, etc. In general, the tractor is almost indispensable on the job.

Equipped with bulldozer apparatus, the tractor, either rubber-tired or crawler type, has demonstrated its ability to do a wider variety of jobs than any other earth moving tool.

The bulldozer usually is employed to open up the job, which may consist of anything from clearing brush and vegetation to the removal of average size trees, up to 30 inches (750 mm) in diameter, stumps, rocks, paving, old floor slabs, old foundations, etc.

Stripping and stockpiling top soil on small quantity jobs with a haul distance from 200 to 400 feet (61 to 122 m) or less is most efficiently done by bulldozer. On larger jobs this is done more economically by scrapers.

On small quantity, shallow excavation jobs with a short haul, tractor dozers are moderately efficient.

Crawler type bulldozers are reasonably economical on pushing distances up to 200 feet (61 meters). Under the same conditions the rubber tired dozer's economical push distance is about 400 feet (122 meters).

Most backfilling jobs fall to the lot of the bulldozer, which outperforms all other equipment.

Tractors also perform a valuable service when they are called upon to provide the additional power for push loading scrapers. On a production scraper job, dozers helping to load these units more than pay for themselves.

On residence and other small jobs, a loader usually can dig the basement and do all exterior grading cuts and fills with maximum efficiency.

Another application of the bulldozer is spreading fill hauled in and dumped by trucks or cast by other excavating equipment.

Tractors are also the ideal power source for hauling sheepsfoot rollers and pneumatic rollers used in compacting fills.

Scraper Application. Scrapers are high production earth moving units which are capable of digging their own load, hauling the load and then spreading the load in controlled layers. Furthermore, the scraper is a precision tool, as a skilled operator can cut a grade to within 0.1 ft. and can spread fill to the same degree of accuracy.

Scrapers may be powered by three types of prime movers: two wheeled or four wheeled rubber-tired units or crawler type tractors.

Scrapers, as a rule, are not used to any great extent in residence building or on any work in congested city areas but for jobs such as suburban and rural schools, hospitals, factories, housing projects, in fact any construction work with a medium to large site grading problem, scrapers are probably the most efficient pieces of excavating equipment for the job.

The scrapers can be used to strip and stockpile top soil, after the site has been cleared, and reclaim the top soil from the stockpile and spread it where required at the end of the job. Many times basement excavation may be carried on in conjunction with site grading cuts and fills by scrapers.

The basic efficiency of scrapers is relatively unaffected by depth of cut, length of haul or type of soil. Quantities being sufficient to warrant the use of scraper equipment, crawler tractor drawn scrapers are economical at haul distances up to 500 feet (152 meters) and with rubber-tired, high speed prime movers can compete with trucks on longer hauls.

Capacities of scrapers range from about 7 cu. yds. (5.35 cu. m) struck measure to approximately 40 cu. yds. (30 cu. m) heaped measure.

Scrapers powered by crawler tractors are capable of self-loading in most soils but generally it is more economical to use pusher assistance from a tractor to obtain heaped loads in the shortest time. With few exceptions, pusher assistance is essential for self-propelled, rubber-tired scrapers. Pusher loading is generally accomplished in an average of about 1 minute in 100 ft. (30.48 m) of travel. To determine the number of scrapers one pusher can handle, divide the scraper cycle by the pusher cycle.

Estimating the Cost of Owning and Operating Tractors and Scrapers. In order to properly prepare an estimate on work to be done by tractors and/or scrapers, the contractor must know what to charge for the use of the equipment. He should work out an hourly charge for the equipment, to

220

be used in pricing his estimates, so that the cost of ownership and operation is carried proportionately by each job and does not have to come out of profit.

Hourly Cost of Ownership. The cost of ownership has as its basis the total delivered cost of the piece of equipment and is composed of several items-depreciation, interest, insurance and taxes. In figuring the hourly depreciation cost, most contractors usually take the full delivered price of the machine, less cost of tires, and divide by the expected economic life of the machine in hours. The life expectancy figure commonly accepted is 5 years of 2,000 hours each or 10,000 hours.

Tandem Power Wheel Tractor-Scraper with Dozer (Courtesy Caterpillar, Inc.)

Interest, insurance, taxes, and storage are computed on the basis of average investment. The average investment in a piece of equipment which is to be fully depreciated in 5 years is 60% of the full delivered cost, as is shown on previous pages under "Power Shovels and Cranes". The assumed rate for interest, insurance, taxes, and storage is a total of 20%, which when applied to the average investment and then divided by 2,000 hours, amounts to 0.006% of the full delivered cost to give the hourly cost of interest, insurance, taxes, and storage.

Total hourly ownership cost is then obtained by totaling the hourly cost of depreciation and the hourly cost of interest, insurance, taxes, and storage.

Hourly Cost of Operation. This cost includes tire replacement, tire repairs, fuel, lubrication, mechanical repairs, maintenance of blades and cable, etc.

The hourly costs of tire replacements and tire repairs are computed in the same manner as previously described, under "Hourly Cost of Ownership and Operation of Trucks", and are based on a tire life of 3,000 hours of operation with repair costs assumed to be 15% of the hourly tire cost.

Fuel costs may be figured on the following basis: Assuming an average machine in good repair, operating under average conditions, diesel fuel consumption of 0.04 gallons per horsepower per hour is used.

Experience has shown that lubricants and lubricating labor amount to approximately 50% of the fuel costs per hour.

Cost of repairs and maintenance, including major overhauls at the end of the working season, should be determined by the contractor from his own experience record.

Many contractors find that total repairs, maintenance and overhauls for the economic life of the machine runs from 70% of the full delivered price for the larger machines to 90% for the smaller machines. This percentage divided by the economic life of the machine in hours (10,000) will give the hourly cost of mechanical repairs, maintenance, etc., and amounts from 0.007% to 0.009% of the total delivered cost.

Approximate Prices of Crawler Type Tractors with Bulldozer Equipment				
Horsepower (HP)	Lbs.	Approx. Cost	Kilowatt (kW)	Kilograms
70	32,000	$73,200	52	14,515
80	16,890	$80,000	60	7,661
90	19,120	$87,000	67	8,673
140	33,010	$147,400	104	14,973
200	44,600	$250,300	149	20,231
335	83,040	$371,700	250	37,667
370	92,550	$478,500	276	41,981

Maintenance and repair as this item will vary depending on the soil conditions. This cost can be estimated by dividing the cost of blades by there expected life in hours, under average conditions can be 450 hrs.

Approximate Prices of Crawler Type Loaders				
Capacity in Cu. Yds	Horsepower (HP)	Approx. Cost	Capacity in Cu. Meters	Kilowatt (kW)
1	70	$67,000	0.76	52.2
1 1/2	90	$98,400	1.15	67.1
2	110	$139,000	1.53	82.1
2 1/4	121	$178,300	1.72	90.3
3	160	$219,000	2.29	119.4
3 3/4	210	$330,000	2.87	156.7

SITE WORK

Approximate Prices of Rubber Tire Type Loaders

Horsepower (HP)	Horsepower (HP)	Approx. Cost	Capacity in Cu. Meters	Kilowatt (kW)
3/4	45	$65,000	0.57	33.6
1	60	$71,800	0.76	44.8
1 1/2	85	$83,800	1.15	63.4
2	98	$92,500	1.53	73.1
2 1/4	105	$113,200	1.72	78.3
3	140	$153,000	2.29	104.4
4 1/4	200	$235,600	3.25	149.2

Approximate Prices of Skid Loader

Horsepower (HP)	Operating Capacity Lbs	Approx. Cost	Kilowatt (kW)	Operating Capacity Kilograms
15	600	$6,200	11.47	272.2
25	950	$16,100	19.12	430.9
43.5	1,300	$20,100	33.26	589.7
56	1,850	$23,100	42.82	839.2
73	1,900	$27,700	55.82	861.8
74	2,400	$35,000	56.58	1,088.6
94	4,000	$60,000	71.87	1,814.4

The hourly cost of maintenance of blades is figured separately from the cost of general maintenance and repair as this item will vary depending on the soil conditions. This cost can be estimated by dividing the replacement cost of blades by their expected life in hours, which under average conditions can be about 450 hours; under favorable conditions, such as digging soft loam, it may be increased to about 800 hours; under unfavorable conditions, where sharp rocks and shale are being loaded, it might be reduced to about 250 hours.

Approximate Prices of Self-Propelled Scrapers

Capacity				Hp	Kw	Approx. Cost	Original Tire Value
Struck		Heaped					
Cu Yd	Cu M	Cu Yd	Cu M				
14	10.70	20	15.29	225	167.9	$441,000	$47,408
21	16.06	31	23.70	250	186.5	$720,000	$77,400
24	18.35	34	26.00	480	358.1	$728,000	$78,260
38	29.05	46	35.17	480	358.1	$735,000	$79,013

223

Approximate Prices of Tractor-Elevating Scrapers

Capacity Cu.Yd.	Capacity Cu.Meter	Hp	Kw	Approx. Cost	Original Tire Value
11.00	8.41	175	130.6	$193,100	$19,310
16.00	12.23	265	197.7	$285,100	$28,510
23.00	17.59	365	272.3	$435,000	$43,500
34.00	26.00	475	354.4	$758,000	$75,800

Ownership and Operation of Bulldozers, Loaders, and Scrapers. The following data is the hourly cost of ownership and operation for various sizes of tractor and scraper equipment and is based on preceding price and horsepower data. It does not include operator's wages, supervision, compensation, and liability insurance or Social Security taxes.

Hourly Cost of Crawler Type Loaders					
Bucket Capacity, CY	1	1 1/2	2 1/4	3	4 1/2
Horsepower	150	180	260	300	390
Approx. Delivered Cost	$67,000	$98,400	$178,300	$219,000	$550,000
Depreciation, 0.01%	$6.70	$9.84	$17.83	$21.90	$55.00
Interest, Insurance and Taxes 0.006%	$4.02	$5.90	$10.70	$13.14	$33.00
Fuel, 0.04 Gal. per hp*	$24.00	$28.80	$41.60	$48.00	$62.40
Lubrication, Including Labor, 50% of Fuel	$12.00	$14.40	$20.80	$24.00	$31.20
Mech. Repairs and Maintenance 0.007%	$4.69	$6.89	$12.48	$15.33	$38.50
Blade Replacement	$0.80	$0.84	$0.88	$0.96	$1.00
Total Hourly Cost**	$52.21	$65.68	$81.87	$142.96	$162.60
Metric					
Bucket Capacity, Cu.M.	0.75	1.12	1.68	2.24	3.36
Kilowatt	111.9	134.3	194.0	223.8	290.9
Fuel, 0.20 liters per kW*	$23.72	$28.47	$41.12	$47.45	$61.68

*Based on $4 per gallon ($1.06 per liter) for diesel fuel.
**Not including Labor, Supervision, Unemployment, Social Security.,
Overhead Expense

Hourly Cost of Crawler Type Bulldozers							
Horsepower	70	80	105	140	200	400	570
Approx. Delivered Cost	$73,200	$80,000	$116,000	$147,400	$250,300	$595,000	$810,000
Depreciation, 0.01%	$7.32	$8.00	$11.60	$14.74	$25.03	$59.50	$81.00
Interest, Insurance and Taxes @ 0.006%	$4.39	$4.80	$6.96	$8.84	$15.02	$35.70	$48.60
Fuel, 0.04 Gal. per HP*	$11.20	$12.80	$16.80	$22.40	$32.00	$64.00	$91.20
Lubrication, Including Labor, 50% of Fuel	$5.60	$6.40	$8.40	$11.20	$16.00	$32.00	$45.60
Mech. Repairs and Maintenance @ 0.007%	$5.12	$5.60	$8.12	$10.32	$17.52	$41.65	$56.70
Blade Repair/Replacement	$0.45	$0.47	$0.51	$0.55	$0.55	$0.60	$1.00
Total Hourly Cost**	$34.09	$38.07	$52.39	$68.05	$106.12	$233.45	$324.10
Metric							
Kilowatt	52	59	78	104	148	296	422
Fuel, 0.20 Liters per kW*	$10.98	$12.55	$16.47	$21.96	$31.38	$62.75	$89.42

*Based on $4 per gallon ($1.06 per liter) for diesel fuel.

**Not including Labor, Supervision, Unemployment, Social Security., Overhead Expense

Hourly Cost of Self-propelled Scrapers				
Capacity, Struck (CY)	14	21	24	38
Capacity, Heaped (CY)	20	31	34	46
Horsepower	225	250	480	480
Approx. Delivered Price	$441,000	$720,000	$728,000	$735,000
Original Tire Value Depreciation 0.01% of Total	$44,100	$72,000	$72,800	$73,500
Depreciation 0.01% of Total Investment Less Tires	$44.10	$72.00	$72.80	$73.50
Interest, Insurance and Taxes, 0.006% of Total Investment	$29.11	$47.52	$48.05	$48.51
Tire Replacement for 3,000 Hr Life	$13.23	$21.60	$21.84	$22.05
Tire Repairs, 15%	$1.98	$3.24	$3.28	$3.31
Fuel, 0.04 Gal. per hp*	$36.00	$40.00	$76.80	$76.80
Lubrication, Incl. Labor, 50% of fuel Mech Repairs & Maintenance 0.007% of Total Investment	$18.00	$20.00	$38.40	$38.40
Blade Replacement	$0.70	$0.80	$1.00	$1.10
Total Hourly Cost**	$143.12	$205.16	$262.16	$263.67
Metric				
Capacity, Struck in Cu.M.	10.70	16.06	18.35	29.05
Capacity, Heaped in Cu.M.	15.29	23.70	26.00	35.17
Kilowatt	167.9	186.5	358.1	358.08
Fuel, 0.20 liter per kW*	$35.58	$39.54	$75.91	$75.91

*Based on $4 per gallon ($1.06 per liter) for diesel fuel.
**Not including Labor, Supervision, Unemployment, Social Security, Overhead Expense

Bulldozer Production

In an average day's work, the bulldozer can be called upon to do several operations. Time will be lost in moving from one task to another. If this time loss is not considered in estimating bulldozer work, the job might not be profitable. Inasmuch as each job has its own peculiar problems, no cost table can reflect the time losses, and the estimator must analyze each job carefully and adjust the production rates to conform to the conditions present. The following table gives the approximate hourly production to be expected from various sized bulldozers and is based on good weather conditions, efficient and well-maintained machinery, and steady operation. It has been established that a well organized and supervised job will result in approximately 50 work-productive minutes each hour, a job efficiency of 83%.

Scraper Production

Relative scraper production is mainly governed by the cycle time of operation; that is, the length of time a scraper will take to load, haul, spread-dump, and return to start over again. In properly organized jobs with good, adequately-powered equipment, a portion of this cycle is usually termed the fixed time, because it is relatively unaffected by the length and grades of the haul and consists of the following: loading, approximately 1.0 to 1.5 minutes, spread-dumping, about 0.5 minutes, two turns of 0.25 minutes each, and shifting and acceleration of 0.5 to 1.0 minute or a total fixed time of 2.5 to 3.5 minutes. The other part of the cycle is the actual time of hauling the load to the fill area plus the time of returning empty to the beginning of a new cut. This traveling time is the big variable in the operation. It is affected by length of haul, road conditions, and grades to be negotiated. It must be estimated accurately. Scrapers powered by rubber-tired tractors can reach a speed of 30 mph (48 kph) when empty and on a good level road, but this will be reduced to 10 to 20 mph (32 kph) when loaded. It must also be remembered that adverse grades greatly reduce the speed, and favorable grades increase the speed. The traveling time plus the fixed time is the cycle time of operation.

Yardage production is figured in terms of bank measure. When estimating scraper production, the struck capacity of the machine is normally used as the pay yardage volume, anticipating the scraper will be heap loaded. The difference between struck and heaped capacity is generally assumed to be the amount the material will swell in being disturbed.

The following tables give the production for various sized machines operating at several different time cycles and are based on ordinary soil conditions, good weather, good equipment, and an 83% job efficiency.

Approximate Hourly Production of Bulldozers Based on an 83% Job Efficiency

Description of Work	Horsepower				
	70	80	90	200	
Simple clearing, vegetation, light brush, tree samplings (SF per hr)	900 - 1,000	1,000 - 1,100	1,250 - 1,350	1,450 - 1,550	
Moderately difficult clearing thick brush & trees saplings (SF per hr)	300 - 350	350 - 400	600 - 650	700 - 750	
Strip top soil or shallow excavation of ordinary soil and placing in piles on premises (CY per hr)	15 - 20	17 - 22	22 - 27	27 - 32	
Backfill ordinary soil placed slow enough to permit compacting (CY per hr)	33 - 38	37 - 42	50 - 55	60 - 65	
Spread material dumped in piles by trucks or cast by other excavating equipment (CY per hr)	42 - 47	45 - 50	60 - 65	70 - 75	

Add extra for trees

Approximate Hourly Production of Bulldozers Based on an 83% Job Efficiency				
Description of Work	Kilowatt			
	51.8	59.2	66.6	148
Simple clearing, vegetation, light brush, tree samplings (SF per hr)	83.6 - 92.9	92.9 - 102.2	116.1 - 125.4	134.7 - 144.0
Moderately difficult clearing thick brush & trees saplings (SF per hr)	27.9 - 32.5	32.5 - 37.2	55.7 - 60.4	65.0 - 69.7
Strip top soil or shallow excavation of ordinary soil and placing in piles on premises (CY per hr)	11.5 - 15.3	13.0 - 16.8	16.8 - 20.6	20.6 - 24.5
Backfill ordinary soil placed slow enough to permit compacting (CY per hr)	25.2 - 29.1	28.3 - 32.1	38.2 - 42.1	45.9 - 49.7
Spread material dumped in piles by trucks or cast by other excavating equipment (CY per hr)	32.1 - 35.9	34.4 - 38.2	45.9 - 49.7	53.5 - 57.3

Add extra for trees

Dozer, D4 Series (Courtesy Caterpillar, Inc.)

Daily Scraper Production for Various Operation Cycles Based on a 50- Minute Hour and an 8-Hour Day

Operation Cycle in Minutes	Struck Capacity of Scraper					
	7 CY	14 CY	24 CY	5.35 Cu.M	10.7 Cu.M	18.35 Cu.M
3	930	1,860	3,200	711	1,422	2,447
4	700	1,400	2,400	535	1,070	1,835
5	560	1,120	1,920	428	856	1,468
6	465	930	1,600	356	711	1,223
7	400	800	1,370	306	612	1,048
8	350	700	1,200	268	535	918
9	310	620	1,070	237	474	818
10	280	560	960	214	428	734

Sample Estimate for Scraper Excavation

Based on a site grading cut containing 8,000 cu. yds. (6116 cu. m), excavated material to be spread-dumped in low areas requiring fill, good conditions permitting a 50-minute hour, an operation cycle of 6 minutes and using two 14 cu. yd. (11 cu. m), struck measure, self-propelled scrapers.

230

Description	Hours	Rate	Total
Bring scrapers to project, Move-in Charge 2 ea.		$ 675.00	$ 1,350.00
Labor Foreman	35.00	$ 30.16	$ 1,055.60
Scraper Operators	70.00	$ 37.89	$ 2,652.30
Scraper Hours	70.00	$ 76.38	$ 5,346.25
Total Direct Costs			$10,404.15
Cost per Cu.Yd			$ 1.30
Cost per Cu. Meter			$ 1.70

If a 105-Hp (78 kW) bulldozer is needed to give pusher assistance in loading, to maintain the operation cycle time of 6 minutes, the additional cost per cu. yd. (cu. m) would be as follows:

Description	Hours	Rate	Total
Bring scrapers to project, Move-in Charge 1 ea.		$ 675.00	$ 675.00
Bulldozer Operator	35.00	$ 37.89	$1,326.15
Bulldozer Charge	35.00	$ 44.88	$1,570.63
Additional Direct Costs			$2,896.78
Cost per Cu.Yd			$ 0.36
Cost per Cu. Meter			$ 0.47

Trenching and Ditching Machine Excavation

Designed strictly for the purpose of cutting ditches, the trenching machine is the fastest earthmover for its weight and horsepower. Before a contractor can reap the benefits of this potential output, he must select a trenching machine of a type and size that fits the work to be done, and even more important, must put an operator on the machine who has both skill and experience.

There are two basic types of machine, the wheel type and the ladder type. The wheel type is generally considered to be the fastest and is by far the most common type in use today. Each type has advantages and disadvantages for performing certain classes of work.

Because most trenching or ditching operations in building construction will be foundations, service trenches, etc., the following data covers the cost and operation of a general purpose, crawler-mounted, wheel machine of the type and size generally used for this work.

This machine is equipped with shifting-tilting boom that permits it to dig close to obstructions, such as foundations or curbs, and to dig a vertical trench even when the machine is on a slope.

231

The cost of excavating with a trenching or ditching machine will vary with the size of the job, width and depth of trenches, kind of soil, accessibility, etc. Most machines are designed to dig certain standard width trenches, so that trenches a trifle narrower or wider will not make any appreciable difference in either output or costs.

While the following table gives the rated capacity of the machine, it will be necessary to allow for factors such as time lost moving machine and bad weather. It is not advisable to figure rated capacity as the average daily output of any machine.

Data on Crawler Mounted Ditching Machine Operation and Cost			
Possible Cutting Widths	Maximum Depth Trench	Weight in Lbs.	Approximate Cost
24 "	7 ' - 0 "	32,000	$ 200,000
30 "	10 ' - 0 "	150,000	$ 590,000
Possible Cutting Widths	Maximum Depth Trench	Weight in Kg	Approximate Cost
610 mm	2.134 M	14,515	$ 200,000
762 mm	2.134 M	68,040	$ 590,000

Maximum digging speed is 34 fpm (10.36 mpm) forward and reverse travel speeds are 0.46, 0.95, 1.7, and 2.9 mph (0.74, 1.53, 2.73 and 4.67 kmph). Digging speeds are controlled by a hydraulic transmission, permitting speed to be varied to match soil conditions. Wheel speeds vary from 0 to 9.2 rpm.

Crawler Mounted Ditching Machine Operation and Cost

When figuring operating cost on a ditching machine, include fuel and oil consumption, depreciation, interest on investment, repairs, replacements, and taxes.

Crawler Mounted Ditching Machine Operation and Cost				
Description	12"	24"	300 mm	600 mm
Approximate Delivered Price	$100,000	$200,000	$100,000	$200,000
Depreciation per year, @ 18%	$ 18,000	$ 36,000	$ 18,000	$ 36,000

Crawler Mounted Ditching Machine Operation and Cost				
Description	12"	24"	300 mm	600 mm
Interest @ 15% per year	$ 15,000	$ 30,000	$ 15,000	$ 30,000
Insurance and Taxes per year, @ 5%	$ 5,000	$ 10,000	$ 5,000	$ 10,000
Repairs and Replacement per year, @ 10%	$ 10,000	$ 20,000	$ 10,000	$ 20,000
Hourly Consumption of Fuel, Gals (Liters)	4	10	$ 15.14	$ 37.85
Daily Cost of Fuel, Oil, Grease @ $4 per Gal. ($1.06 per Liter) + 15%	$ 147.20	$ 368.00	$ 147.65	$ 369.11
Operating Cost per Mile, Based on 75 mi. (120.67 km) Per Year, 200 Working Days per Year Avg., Trench Depth, 5' (1.52m)				
Depreciation	$ 750	$ 1,500	$ 1,207	$ 2,414
Interest	$ 200	$ 400	$ 200	$ 400
Taxes and Insurance	60	120	60	120
Repairs and Replacements	$ 130	$ 260	$ 130	$ 260
Fuel, Oil and Grease	$ 140	$ 280	$ 140	$ 280
Total machine cost per mile (per km)	$ 1,431	$ 2,938	$ 1,900	$ 3,880
Machine cost per lin. ft. (per m.) trench	$ 0.27	$ 0.56	$ 0.08	$ 0.17

Volume of Trench Excavation Cu. Yd. Per 100 Lin. Ft.							
Trench Depth, Inches	Trench width, Inches						
	12 "	18 "	24 "	30 "	36 "	42 "	48 "
12	3.70	5.56	7.41	9.26	11.11	12.96	14.81
18	5.56	8.33	11.11	13.89	16.67	19.44	22.22
24	7.41	11.11	14.81	18.52	22.22	25.93	29.63
30	9.26	13.89	18.52	23.15	27.78	32.41	37.04
36	11.11	16.67	22.22	27.78	33.33	38.89	44.44
42	12.96	19.44	25.93	32.41	38.89	45.37	51.85
48	14.81	22.22	29.63	37.04	44.44	51.85	59.26

Volume of Trench Excavation Cu. Yd. Per 100 Lin. Ft.

Trench Depth, Inches	Trench width, Inches						
	12 "	18 "	24 "	30 "	36 "	42 "	48 "
54	16.67	25.00	33.33	41.67	50.00	58.33	66.67
60	18.52	27.78	37.04	46.30	55.56	64.81	74.07
66	20.37	30.56	40.74	50.93	61.11	71.30	81.48
72	22.22	33.33	44.44	55.56	66.67	77.78	88.89
78	24.07	36.11	48.15	60.19	72.22	84.26	96.30
84	25.93	38.89	51.85	64.81	77.78	90.74	103.70
90	27.78	41.67	55.56	69.44	83.33	97.22	111.11
96	29.63	44.44	59.26	74.07	88.89	103.70	118.52

Volume of Trench Excavation Cu. Meters Per 10 Lin. Meters

Trench Depth, mm	Trench width, mm						
	305 mm	457 mm	610 mm	762 mm	914 mm	1067 mm	1219 mm
305	0.93	1.39	1.86	2.32	2.79	3.25	3.72
457	1.39	2.09	2.79	3.48	4.18	4.88	5.57
610	1.86	2.79	3.72	4.65	5.57	6.50	7.43
762	2.32	3.48	4.65	5.81	6.97	8.13	9.29
914	2.79	4.18	5.57	6.97	8.36	9.76	11.15
1067	3.25	4.88	6.50	8.13	9.76	11.38	13.01
1219	3.72	5.57	7.43	9.29	11.15	13.01	14.87
1372	4.18	6.27	8.36	10.45	12.54	14.63	16.72
1524	4.65	6.97	9.29	11.61	13.94	16.26	18.58
1676	5.11	7.66	10.22	12.77	15.33	17.88	20.44
1829	5.57	8.36	11.15	13.94	16.72	19.51	22.30
1981	6.04	9.06	12.08	15.10	18.12	21.14	24.16
2134	6.50	9.76	13.01	16.26	19.51	22.76	26.01
2286	6.97	10.45	13.94	17.42	20.90	24.39	27.87
2438	7.43	11.15	14.87	18.58	22.30	26.01	29.73

Note: Quantities reflect direct conversion of above english volume.

Cost of Digging 100 L.F. (30.5 M) of Trench 5'-0" (1.52 m) Deep, Using a Crawler Mounted Ditching Machine, Based on 350 Lin.Ft. (106.7 M) per Hour

Description	Hours	Rate	Total
Machine operator	0.29	$ 37.89	$ 10.99
Machine charge per LF, 12" (300mm) wide		$ 0.25	$ 25.00
Cost per 100 LF (30.5 M) x 12" (300 mm) wide		$	35.99

Cost of Digging 100 L.F. (30.5 M) of Trench 5'-0" (1.52 m) Deep, Using a Crawler Mounted Ditching Machine, Based on 350 Lin.Ft. (106.7 M) per Hour			
Description	Hours	Rate	Total
Machine operator	0.29	$ 37.89	$ 10.99
Machine charge per LF, 24" (600mm) wide		$ 0.51	$ 51.00
Cost per 100 LF (30.5 M) x 24" (600 mm) wide			$ 61.99

To the above, add cost of trucking machine to job and removing same at completion, trucking waste material, supervision, compensation and liability insurance, social security and unemployment taxes, overhead expense and profit. Add for backfilling trenches, as required.

Drilling Holes Using an Earth Boring Machine. Where a large number of holes are required for line construction, foundation work, pre-boring, guard rails, piling, fencing, draining, or bridges, an earth boring machine performs the work much faster and at lower cost than most other methods.

These machines will drill holes through loam, clay, hard pan, and shale and are furnished in various sizes by different manufacturers. Some machines will drill up to 8'-0" (2.43 m) diameter holes to a depth of 120 ft. (36.60 m) or more.

Based on 20 holes, 18" (450 mm) diam. and 3'-0" (0.91 m) deep, per hour or 160 holes per 8-hr. day, the daily cost should average as follows:

Description	Hour	Rate	Total
Operator	8.00	$ 37.89	$ 303.12
Machine charge	8.00	$ 25.00	$ 200.00
Truck charge	8.00	$ 19.13	$ 153.00
Total cost 160 holes			$ 656.12
Cost per 3'-0" (0.9 m) hole			$ 4.10
Cost per lin. ft.			$ 1.37
Cost per m			$ 4.48

When drilling holes up to 18" (450 mm) in diameter in ordinary soil, a machine will drill at the rate of 1" (25 mm) depth in 1-1/2 to 2 seconds, or drill a 3'-0" (0.91 m) deep hole in 1 to 1-1/4 minutes. Moving the machine and spotting the holes takes some time, so this will govern the number of holes that may be drilled per day. A machine that will drill holes to 24" (600 mm) diameter and 9'-0" (2.74 m) deep costs approximately $100,000.00

f.o.b. factory. This does not include cost of truck, which will have to be added. The larger machines which are used for drilling caisson foundations may cost as much as $400,000.00.

There are also small portable posthole diggers, mounted on two rubber tired wheels and equipped with a 5 to 6 Hp gasoline engine. This machine can be operated by one person and in ordinary ground it will bore a 9" (225 mm) hole 3'-0" (0.9 m) deep in 1 to 1-1/2 minutes. Such machine costs $2,000.00 to $4,000.00.

Placing And Compacting Fills

Architects and engineers are rigid in their requirements as to the manner of placing and compacting fill materials. The usual specification for this work is to place the fill in from 6" to 12" (150 to 300 mm) layers and to compact each layer with some sort of roller equipment.

Placing and Compacting Site Grading Fills. Fill material in site grading work is placed either by scrapers spread-dumping their loads, trucks dumping their loads more or less in piles, or some sort of excavating equipment casting or bucket loads as they are dug. In the case of the scraper spread-dumping fill material, no further spreading is necessary, because it can be controlled with precision by the scraper operator. In the other cases, however, a bulldozer or loader will be required to spread out the fill material into layers of specified depth. Hourly production for this work can be found in the table under "Bulldozer Production".

When the fill has been placed and spread in the specified depth of layer, the compacting operation follows, and this is usually accomplished by rolling with a tractor-drawn or self-propelled sheepsfoot roller. The degree of compaction may be specified by the number of passes to be made by the roller or by a percentage, such as 95%, of maximum density of the material obtained at optimum moisture content in a soil mechanics laboratory. The latter specification is usually called for in government work and for fills under paving. Obtaining 95% of maximum density compaction may take as many as 12 passes of a sheepsfoot or vibratory roller.

If rolling is not to interfere with production schedules, a rate must be established in cubic yards compacted per hour and applied to the rate of placing and spreading the fill material to determine the size and number of rollers required to keep the job moving. Where frequent turns are made or restricted areas require maneuvering these factors must be considered and the figures adjusted accordingly.

The following table gives rate of compaction, in cu. yds. per hr., for rolling fill with a 5'-0" (1.52 m) wide sheepsfoot roller at 2.5 mph (4.0 kph) with the fill material placed in 12" (300 mm) layers for soil of various compaction factors and for 1 pass of the roller to 12 passes.

236

SITE WORK

No. of Passes of Roller	Hard Tough, Clay 70%		Medium Clay 80%		Loam 90%	
	CY	Cu. M.	CY	Cu. M.	CY	Cu. M.
1	1711	1308	1956	1496	2200	1682
2	856	654	978	748	1100	841
3	570	436	652	499	733	561
4	428	327	489	374	550	421
5	342	262	391	299	440	336
6	285	218	326	249	367	280
7	244	187	279	214	314	240
8	214	164	245	187	275	210
9	190	145	217	166	244	187
10	171	131	196	150	220	168
11	156	119	178	136	200	153
12	143	109	163	125	183	140

Rate of Sheepsfoot Roller Compaction in Cu. Yds (Cu. M.) per Hr Based on 5' (1.52m) Wide Roller, 2-1/2 mph (4kph) Speed, 12" (300mm) Fill Layers and 100% Job Efficiency. Percentage Factor of Pay Yd. (M) to Loose Yd. (M)

For rollers of different widths, adjust the above rates proportionately; for example, for a 10'-0" (3.04 m) wide roller double the rates, and for a 15'-0" (4.57 m) wide roller triple the rates.

For different depths of layers, adjust the above rates proportionately: i.e., for 6" (150 mm) layers reduce the rates to 1/2, and for 9" (225 mm) layers reduce the rates to 3/4.

The above rates must also be adjusted to normal job efficiency and further adjusted to take care of lost time in maneuvering and turning. In small jobs as much as 10% of the time could be lost.

Example: Assume a fill to be placed in 9" (225 mm) lifts and rolled 8 passes with a double drum sheepsfoot roller 10'-0" (3.05 m) wide at 2.5 mph (4.02 kph) in soil with a compaction factor of 80%.

From the table, 8 passes in 80% material will compact 245 cu. yds. (187 cu.m) per hr. Adjusting this rate for a 10'-0" (3.05 m) wide roller gives a rate of 490 cu. yds. (374.6 cu. m) per hr. and further adjusting for 9" fill layers results in a compaction rate of 368 cu. yds. (281 cu.m) per hr. Assuming that a 50-minute hour or a job efficiency of 83% is expected, the rate is reduced to 305 cu. yds. (233 cu. m) per hr. and with a time loss of 10% anticipated for turns and maneuvering the final result is a compaction rate of

275 cu. yds. (210 cu.m) per hr. The approximate cost for compacting fill at rate of 275 cu. yds. (210 cu.m) per hr. is as follows:

Description	Hours	Rate	Total
Hourly charge for 75-Hp (56 kW) bulldozer with double drum sheep foot roller	1	$ 237.38	$ 237.38
Bulldozer operator	1	$ 37.89	$ 37.89
Laborer	1	$ 26.16	$ 26.16
Cost per 275 c.y. (210 cu.m)			$ 301.43
per cu.yd.			$ 1.10
per cu. m.			$ 1.43

Does not include hauling equipment to and from project, workers compensation and liability insurance, social security, taxes and overhead and profits.

After Fill has been placed, Self-powered Compactor is used to Tamp Earth (Courtesy Caterpillar, Inc.)

The sheepsfoot roller method of compaction is applicable to nearly all soils with cohesive qualities, such as loams and clays, and in very dry soils it may be necessary to add moisture to the material to obtain the correct degree of plasticity.

With non-cohesive materials, such as sands and gravels, the sheepsfoot roller usually is ineffective and rubber-tired pneumatic rollers are used instead.

238

SITE WORK

Placing and Compacting Fills for Floors on Grade Inside Buildings. Fill material, necessary to raise the elevation of the subgrade for floors inside buildings, can be placed in a number of ways.

For small jobs this is usually done after the building is enclosed. Fill is brought in through door and window openings, and it is spread and tamped by hand. Three laborers working together should spread and tamp 35 to 40 c.y. (27 to 31 cu.m) of fill per 8-hr. day at the following cost per c.y. (cu.m):

Description	Hours	Rate	Total
Labor per cu.yd.	0.6	$ 26.16	$ 15.70
per cu. m.			$ 20.53

Does not include hauling equipment to and from project, workers compensation and liability insurance, social security, taxes and overhead and profits.

On larger jobs it may be possible to truck the fill material directly into the space and spread-dump it, using a bulldozer to further spread into even layers. Compaction of material in this case can usually be accomplished by using a single drum sheepsfoot roller or small pneumatic roller for the open areas and by hand tamping at walls and columns. Trucking traffic over the various layers of fill is a big help in compacting the material. The roller can be drawn by the same bulldozer that spreads the material and with 2 laborers spreading and tamping at "hard-to-get" places a daily production of 95 to 100 cu. yds. (73 to 76 cu. m) placed and compacted should be realized at the following cost per cu. yd. (cu. m):

Description	Hours	Rate	Total
50Hp (37 kW) Bulldozer and single drum sheepskin roller	0.08	$ 162.38	$ 12.99
Bulldozer operator	0.08	$ 37.89	$ 3.03
Laborer	0.16	$ 26.16	$ 4.19
Cost per Cu.Yd			$ 20.21
per Cu.M.			$ 26.43

Does not include hauling equipment to and from project, workers compensation and liability insurance, social security, taxes and overhead and profits.

Grading

Grading consists of dressing up ground surfaces, either by hand or machine, to conform to specified contours or elevations and is usually preparatory to a subsequent operation such as placing sub-base gravel, a floor slab, walk or drive, or spreading top soil for planting areas.

Grading costs vary according to the degree of accuracy demanded and the method by which it is done. Following are approximate costs on some of the various types of grading usually encountered in construction work.

Rough Grading. After all backfill is in pl ace and the site has been cut and/or filled to the approximate specified contours, rough grading of the site is done, by hand for small jobs and by machine for large areas, usually preparatory to the spreading of top soil. Tolerance for this type of grading is usually plus or minus 0.1 ft. (0.03 m)

On small jobs a laborer should rough grade 800 sq. ft. (74 sq. m) of ground surface per 8-hr. day at the following cost per 100 sq. ft. (9.3 sq. m):

Description	Hours	Rate	Total
Labor	1	$ 26.16	$ 26.16
per sq. ft.			$ 0.26
per sq. m.			$ 2.82

Does not include hauling equipment to and from project, workers compensation and liability insurance, social security, taxes and overhead and profits.

On large jobs a bulldozer is usually employed to do rough grading, and a laborer generally accompanies the machine to check the surface and direct the operation. A 75-Hp (56 kW) bulldozer should rough grade 6,500 to 7,500 sq. ft. (604 to 697 sq. m) per 8-hr. day at the following cost per 1,000 sq. ft. (93 sq. m):

Description	Hours	Rate	Total
75Hp (56 kW) Bulldozer	1.15	$ 159.25	$ 183.14
Bulldozer operator	1.15	$ 37.89	$ 43.57
Laborer	1.15	$ 26.16	$ 30.08
Cost per 1,000 sq. ft.(92.9 sq. m.)			$ 256.80
Cost per sq. ft.			$ 0.26
per sq. m.			$ 2.76

Does not include hauling equipment to and from project, workers compensation and liability insurance, social security, taxes and overhead and profits

Grading for Slabs on Ground. Grading for slabs on ground such as floors, walks, and driveways usually is done by hand unless the job is quite large and can be organized so that some of the work is done by machine. This work generally must be done accurately with tolerances of no more than 1/2" (15 mm). A laborer should grade 500 to 600 sq. ft. (46 to 56 sq. m) per 8-hr. day at the following cost per 100 sq. ft. (9.3 sq. m):

240

SITE WORK

Description	Hours	Rate	Total
Labor	1.5	$ 26.16	$ 39.24
Cost per sq.ft			$ 0.39
per Sq. Meter			$ 4.22

For sloping surfaces, add about 50% to the above cost, depending on steepness of pitch.

Finish Grading of Top Soil. After top soil has been spread over areas specified, a finish grading operation, usually including hand-raking, must be performed prior to seeding or sodding. The tolerance on this work usually is plus or minus 1 inch. A laborer should finish grade and hand-rake 600 to 700 sq. ft. (56 to 65 sq. m) per 8-hr. day at the following cost per 100 sq. ft. (9.3 sq. m):

Description	Hours	Rate	Total
Labor	1.25	$ 26.16	$ 32.70
Cost per sq.ft			$ 0.33
per Sq. Meter			$ 3.52

For sloping surfaces, add about 50% to the above cost, depending on steepness of pitch.

Grading for Footing Bottoms. When footing pits and trenches are dug by machine, there is always cleanup, squaring, and grading work to be done by hand. Many contractors price this work by the sq. ft. of footing bottom. A laborer should clean up, square, and grade 200 sq. ft. (19 sq. m) of footing bottom per 8-hr. day at the following cost per 100 sq. ft. (9.3 sq.m):

Description	Hours	Rate	Total
Labor	4	$ 26.16	$ 104.64
Cost per sq.ft			$ 1.05
per Sq. Meter			$ 11.26

Does not include hauling equipment to and from project, workers compensation and liability insurance, social security, taxes and overhead and profits.

Costs Of Excavating

Digging Fence Post Holes. When digging 250 post holes about 3'-0" (0.91 m) deep in black soil and sand, using an ordinary augur post-hole digger, a worker will dig 4 holes per hour or 32 per 8-hr. day, at the following labor cost per hole:

Description	Hours	Rate	Total
Labor Cost per Hole	0.25	$ 26.16	$ 6.54

Loader Excavating under Favorable Conditions. The following costs are for a theoretical job performed where conditions are favorable, with no time lost because of bad weather or breakdowns.

This is a basement excavation 80'-0"x80'-0" (24 x 24 m) and 13'-0" (4 m) deep. The top 5'-0" (1.5 m) consists of loose clay and loam, while the lower 8'-0" (2.4 m) is blue grey dolomite limestone in beds varying from 6" to 24" (150 to 600 mm).

A 1-1/2 cu. yd. (1.2 cu. m) loader is used, rented at $61.00 per hr. including operator and fuel.

Five 6-cu. yd. (4.60 cu. m) trucks transport the excavated earth and rock to a dump about 1/4-mile (0.4 km) from the job. Trucks are estimated to cost $62.00 per hr. including gasoline and driver.

A 5-1/2"x5" (137.5 x 125 mm) portable air compressor, capacity 900 cu. ft. per minute, and a Type 12 standard I-R jackhammer is used for rock drilling. The rental of the compressor and jackhammer, including gasoline, operator, and repairs to drills, is estimated at $95.00 an hour.

Excavating Loose Clay and Loam. The area of loose clay and loam is 80'-0"x80'-0"x5'-0" (24 x 24 x 1.5 m), a total of 1,185 cu. yds. (864 cu. m) of excavation.

The average on this job is 135 cu. yds. (103 cu. m) an hr. at the following cost per 100 cu. yds. (76 cu. m):

Description	Hours	Rate	Total
Wheel Loader	0.75	$ 62.88	$ 47.16
Operator	0.75	$ 37.89	$ 28.42
Cost per cu.yd.			$ 0.47
per cu. m.			$ 0.62

Does not include hauling equipment to and from project, workers compensation and liability insurance, social security, taxes and overhead and profits

Five 10-cu. yd. (7.6 cu. m) trucks will haul 1,080 cu. yds. (826 cu. m) of excavated earth 1/4 mile (0.40 km) to a dump per day, or an average of 27 cu. yds. (21 cu. m) an hr. per truck, at the following cost per 100 cu. yds.

Description	Hours	Rate	Total
Truck and Driver	3.7	$ 79.77	$ 295.13
Cost per Cu. Yd.			$ 2.95
per Cu. M.			$ 3.86

Four workers at the dump spread the excavated earth after it has been dumped by the trucks. Very little grading is necessary, because the earth dump is a high-terraced fill around a building. A worker will spread about 10 cu. yds. (7.6 cu. m) per hr. at the following labor cost per cu. yd. (cu. m):

242

Description	Hours	Rate	Total
Laborer per Cu. Yd.	0.1	$ 26.16	$ 2.62
Cost per Cu. M.			$ 3.42

The total cost of excavating, hauling and spreading one cu. yd. (cu. m) of loose earth is as follows:

Description	Cu Yds	Cu Meter
Loader Excavation	$ 0.47	$ 0.62
Truck hauling excavated earth	$ 2.95	$ 3.86
Labor spreading loose earth at dump	$ 2.62	$ 3.42
Total Cost	$ 6.04	$ 7.90

Does not include hauling equipment to and from project, workers compensation and liability insurance, social security, taxes and overhead and profits

Rock Excavation

After the top 5'-0" (1.5 m) of soil have been removed, the excavation consists of blue grey dolomite limestone, hard and stratified in beds varying from 6" to 24" (150 to 600 mm). The depth of the cut was 8'-0" (2.4 m), which was taken out in two lifts of 4'-0" (1.2 m) each. The last lift was drilled 5'-0" (1.5 m) deep or 1'-0" (.30 m) below the desired level. Total rock excavation is 80'-0"x80'-0"x8'-0" (24 x 24 x 2.4 m), or 1,900 cu. yds. (1382 cu. m).

Drilling Rock for Blasting. The area drilled is 80'-0"x80'-0" (24 x 24 m), less a 5'-0" (1.5 m) border on 3 sides and a 10'-0" (3.0 m) border on one side. The rows were 5'-0" (1.5 m) apart and the holes were 5'-0" (1.5 m) on centers, staggered, and all of the holes will be 2" (50 mm) in diameter.

There will be 14 rows with 15 holes in a row or a total of 210 holes, as follows: 105 holes 4'-0" (1.2 m) deep, containing 420 lin. ft. (128 m) and 105 holes 5'-0" (1.5 m) deep, containing 525 lin. ft. (160 m), a total of 945 lin. ft. (288 m) of 2" (50 mm) holes.

One worker and a helper will drill 15 lin. ft. (4.6 m), possibly up to 60 lin. ft. (18 m), of 2" (50 mm) hole per hr. or 120 lin. ft. (37 m) per 8-hr. day, at the following cost per 100 lin. ft. (30 m):

Jackhammer Drilling			
Description	Hours	Rate	Total
Compressor 600 CFM	7	$ 38.50	$ 269.50
Jackhammer W/ Drill and bit	7	$ 9.00	$ 63.00
Jackhammer Operator	7	$ 28.16	$ 197.12
Labor helping	7	$ 26.16	$ 183.12
Cost 100 lin. ft.. (30 m)			$ 712.74
Cost per lin. ft.			$ 7.13
Cost per cu. yd.(1,900 cy)			$ 0.38
Cost per lin. meter			$ 23.38
Cost per cu. Meter			$ 0.49

Explosives. The job will require 2.5 cases, 50 lbs. (22.7 kg) each of 40% dynamite, a total of 125 lbs. (57 kg), which is equivalent to 1/8 lb. per lin. ft. (0.4 kg per 0.3 m) of hole or per cu. yd. (0.76 cu. m) of excavation. The cost of explosives is as follows:

Explosives			
Description	Qty	Rate	Total
Lbs dynamite	125	$ 10.50	$ 1,312.50
Number of Electronic detonators with 6' (1.8M) lead wire	220	$ 12.00	$ 2,640.00
Lin. Ft. Cords/Lead-in wires	200	$ 60.96	$12,192.00
Blasting Mats, each	4	$ 2.50	$ 10.00
Total Cost (210 holes)			$16,144.50
Cost per hole			$ 76.88
Cost per Cu.Yd. (1,900 cu.yds.)			$ 8.50
Cost per Cu. M. (1452 cu.m.)			$ 11.11

Note: Due to added security measures, check supplier and local law enforcement agencies for latest security measures and associated costs.

Rock Excavation. There will be 1,900 cu. yds. (1452 cu. m) of rock excavation, taken out in two lifts of 4'-0" (1.2 m) each, and the loader will handle 45 cu. yds. (34 cu. m) per hr. at the following cost per 100 cu. yds. (76 cu. m):

SITE WORK

Rock Excavation			
Description	Hours	Rate	Total
Loader	2.3	$ 62.88	$ 144.61
Loader Operator	2.3	$ 37.89	$ 87.15
Labor with Loader	2.3	$ 26.16	$ 60.17
Powderman	2.0	$ 35.04	$ 291.93
Labor Assisting	5.0	$ 26.16	$ 439.24
Total Cost per 100 cu.yd. (76.46 cu.m)			$ 1,023.10
Cost per cu.yd.			$ 10.23
Cost per cu. Meter			$ 13.38

Two 6-cu. yd. (4.6 cu. m) trucks haul the excavated rock to a dump ¼ mile (0.4 km) from the job, averaging about 23 cu. yds. (18 cu. m) per hr. per truck, at the following cost per 100 cu. yds. (76 cu. m):

Description	Hours	Rate	Total
Truck and Driver (30 Cu.Yd. dump Trailer)	4.4	$ 150.00	$ 652.50
Cost cu.Yd.			$ 6.53
Cost per cu. Meter			$ 8.53

Note: Add Dump or tip fees if appropriate.

Four workers at the dump level 20 cu. yds. (15.3 cu. m) of excavated rock per hr. at the following labor cost per 100 cu. yds. (76 cu. m):

Description	Hours	Rate	Total
Laborer	20.0	$ 26.16	$ 523.20
Cost per Cu.Yd.			$ 5.23
Cost per cu. Meter			$ 6.84

The total cost of rock excavation, including drilling, blasting, excavating, hauling, and spreading, averages as follows per cu. yd. (cu. m):

Total Rock Excavation		
Description	Cu.Yd.	Cu Meter
Drilling	$ 0.38	$ 0.49
Explosives	$ 8.50	$ 11.11
Excavating	$ 10.23	$ 13.38
Hauling	$ 6.53	$ 8.53
Spreading and Leveling	$ 5.23	$ 6.84
Total	$ 30.86	$ 40.36

Rock excavation conditions and costs vary widely. Always obtain information from sources at or near the job site before bidding on this work. Check with one or more of the local powdermen (blasters) they will know what the local rock conditions are and what production may be expected. As a rule of thumb, the harder the rock the less powder that will be required to blast (remove) each cubic yard.

Drilling Rock Using Pneumatic Jackhammer Drills. The hand held jackhammer is a popular tool for rock drilling on all types of construction jobs. They are furnished in sizes from 30 to 80 lbs. (14 to 36 kg), and many rock drilling conditions can be met. Holes up to 20 ft. (6 m) in depth are easily drilled with the heavier self containing rock drilling machines. Most sizes of jackhammer can be furnished in three styles, depending on the conditions of drilling and the depth of holes being drilled (dry style for shallow holes, blower style for deep holes, and wet for jobs where dust must be kept down to a minimum).

Hollow drill steel of various sizes and shapes are used, the most common being 7/8" and 1" hexagon. Bits are sometimes forged on the steel but the detachable bit is by far the most widely used. When drilling holes in rock for work such as trenching and blasting, two workers operating drills should drill 125 to 178 lin. ft. (38 to 53m) of hole per 8-hr. day, at the following cost per 100 lin. ft. (30 m):

Description	Hours	Rate	Total
Jackhammer man	11.0	$ 28.16	$ 309.76
Compressor 900 CFM	5.5	$ 47.00	$ 258.50
Compressor Operator	5.5	$ 37.89	$ 208.40
Cost per 100 lin.ft. (30 m)			$ 776.66
Cost per lin. ft.			$ 7.77
Cost per lin. meter			$ 25.48

Note: Cost do not include bit charges.

Rock Drilling Using Wagon Drills. The wagon drill consists of a one piece tubular frame on pneumatic or steel tracks and a tilting tower, on which is mounted a heavy pneumatic drill. It drills holes downward at any angle from horizontal to vertical, to depths of 25 to 30 feet (7.5-9.0 m). This type of drill is suitable for all kinds of excavation and trenching and usually does the job of at least 2 or 3 heavy handheld jackhammers. It requires only an operator, a helper, and a compressor operator.

Rock Drilling Using Self Contained Crawler Mounted Drills. Rock drilling for blast holes, etc. is the completely mechanized, self-propelled drill unit. Many construction equipment manufacturers make these units, which can do the work of two or three wagon drills. The unit can move over rugged terrain and can tow or carry its own air supply with it. The set-up time for

SITE WORK

each hole is greatly reduced because of the self-propelled features. It is not uncommon for a crawler mounted drill to drill more than 100 lin.ft. or 800 lin. ft. (152 m) per 8-hour shift in hard sedimentary rock. It can be rented with operator for $800 to $1000 per day.

Excavating Using Pneumatic Diggers. These tools can be used to advantage on all classes of excavation that require picking of dirt or clay, such as trench work, tunneling, caisson sinking, and all kinds of building excavation, in fact, on all kinds of work in stiff clay or hard ground where power shovels, trenching machines, etc., cannot be used.

On work of this kind, two workers using air picks and six workers shoveling will loosen and shovel 35 to 40 cu. yds. (27 to 31 cu. m) per 8-hr. day, at the following labor cost per cu. yd. (cu. m):

Description	Hours	Rate	Total
Labor operating picks	0.4	$ 26.16	$ 10.46
Labor Shoveling	1.2	$ 26.16	$ 31.39
Compressor Operator	0.2	$ 37.89	$ 7.58
Compressor 900 CFM	0.2	$ 47.00	$ 9.40
Cost per cu.yd.			$ 58.83
Cost per cu.meter			$ 76.95

Note: Cost do not include bit charges.

On large excavations providing plenty of room for workers, a worker with a pneumatic pick will loosen 25 to 30 cu. yds. (19.1 to 23 cu. m) per 8-hr. day. After the clay or earth is loosened, a worker will shovel 6 to 7 cu. yds. (4.5 to 5.4 cu. m) per 8-hr. day.

Comparative Cost By Hand

Excavating in stiff clay or tough ground, a worker using a hand pick and shovel will loosen 2-1/2 to 3 cu. yds. (1.9 to 2.3 cu. m) of dirt per 8-hr. day. The same worker, picking only, will loosen 3-1/2 to 4 cu. yds. (2.7 to 3.1 cu. m) of dirt per 8-hr. day, and when shoveling only, will remove 6 to 7 cu. yds. (4.6 to 5.4 cu. m) per 8-hr. day.

Using a hand pick, a worker will loosen and shovel one cu. yd. of dirt at the following cost per cu. yd. (cu. m):

Description	Hours	Rate	Total
Labor per cu.yd.	3.0	$ 26.16	$ 78.48
per cu.meter		$ 47.00	$ 102.64

Backfill Tamping Using a Pneumatic Tamper. A pneumatic tamper enables a worker to do nearly 10 times the amount of work he can do by hand. A backfill tamper strikes over 600 hard, snappy blows a minute, rams

247

the fill hard, and works in and around pipe with far greater thoroughness than hand devices are able to do.

In piers, trenches, etc., where the excavated earth can be shoveled directly into the trench, a worker should backfill 16 to 20 cu. yds. (12.2 to 15.3 cu. m) per 8-hr. day.

Where the trenches have to be tamped, one worker with a pneumatic tamper will tamp as much as 2 workers can backfill, or approximately 35 to 40 cu. yds. (26.8 to 30.6 m) per 8-hr. day. The cu. yd. (cu. m) cost should average as follows:

Description	Hours	Rate	Total
Labor backfilling	0.4	$ 26.16	$ 10.46
Labor tampering	0.2	$ 26.16	$ 5.23
Compressor Operator	0.2	$ 37.89	$ 7.58
Compressor 900 CFM	0.2	$ 47.00	$ 9.40
Pneumatic tamper	0.2	$ 51.38	$ 10.28
Cost per cu. Yd.			$ 42.95
Cost per cu. Meter			$ 56.17

Comparative Cost By Hand

A worker tamps by hand 8 to 10 cu. yds. (6.1 to 7.6 cu. m) of backfill per 8-hr. day, at the following cost per cu. yd. (cu. m):

Description	Hours	Rate	Total
Labor backfilling	0.4	$ 26.16	$ 10.46
Tamper	0.8	$ 4.88	$ 3.90
Labor tampering	0.8	$ 26.16	$ 20.93
Cost per cu.yd.			$ 35.29
cost per cu.meter			$ 46.16

02250 SOIL TREATMENT

New regulations are being legislated almost on a daily basis. The estimator should become familiar with local EPA regulations and engage the services of a professional who specializes in this area.

02300 PILE FOUNDATIONS

The estimator needs to become familiar with the many types and models of pile hammers available for installing various pile types, such as wood, steel pipe, steel sheets, "H" piles, precast concrete (both round and square), combinations of concrete and steel, corrugated and step taper piles.

Pile hammers available include air or steam, diesel, hydraulic, and drop hammers.

Drop Hammer. This is simply a heavy weight that is lifted and dropped on the pile. It operates within a guide that directs the impact. In the past the weight was lifted by human or horse power. Occasionally, this is still the case, but usually, a line on a crane or a winch raises the weight to a predetermined height, where a trip releases it to fall freely on the pile.

The weight of a drop hammer should be about one-third the weight of the pile to be driven. The fall must be regulated so that the pile is not damaged, and a suitable cap, with cushion material, is used on the pile to distribute the driving force. Drop hammers are used much more in other parts of the world than the U.S.

Air and Steam Hammers. Single-acting pile hammers, powered by steam or air, are essentially drop hammers with a short stroke and heavy ram. Steam or air is used to raise the weight much more rapidly than for drop hammers. The falling weight opens a port that admits steam or air under the ram to raise it for the next stroke. The rising elements cut off the pressure and open an exhaust port, permitting practically a free fall of the weight on the next stroke.

It is essential that the weight be raised to its full height and fall free to develop the rated capacity. The hammer should be powered at the manufacturer's recommended pressures. It should operate at recommended blows per minute and height of fall. Energy ratings range from 1,000 ft. lbs. up to 60,000 ft. lbs. (454 to 8292 kgm), but they go up to even 1,800,000 ft. lbs. (248,868 kgm) for big offshore hammers.

Double-acting pile hammers use steam or air both to raise the striking hammer and as a force to add energy on the downward stroke. They operate at about twice the speed of single-acting hammers of equal energy but require more steam or air at greater pressure.

A variation is the differential-acting hammer, where non-expansive use of the steam develops more effective pressure. The significant difference between double-acting and differential-acting is the manner and sequence of exhausting on the upward and downward strokes of the cycle. In the differential-acting hammer, there is no drop from the entering pressure to the main effective pressure moving the piston on the downward stroke.

Double-acting and differential-acting hammers usually give better results in granular non-cohesive soils or in soft clays. Used in proper soil conditions with the right pile, almost twice the production can be obtained as with a single-acting hammer.

Diesel Hammers. This type of pile hammer gets its energy from the compression blow of a falling weight and the reaction to controlled instantaneous burning and expansion of fuel, which raises the ram for the next stroke. The diesel hammer does not require steam or air under pressure for operation. The only input is a modest amount of diesel fuel, compressed and atomized to ignition readiness. The hammers have a total weight of one-third to one-half of that of comparable energy-rated single or double-acting steam or air units. Diesel hammers are started by lifting the ram by outside means. In soft overburden there sometimes is a problem where piles move down without developing enough resistance to compress the fuel ignition heat and pressure for the next stroke. Diesels work better in hard driving.

Some diesel hammers have a closed top, designed to give added speed and energy to the down stroke. Contractors like diesel hammers because they are lighter and do not require exterior power.

Vibratory Driver/Extractor. These use rotating weights set eccentric from their centers of rotation. The result is a machine that is a mechanical sine wave oscillator. It is rigidly connected to the pile, usually by clamps, and it oscillates the pile through the soil.

The vibratory driver/extractor incorporates the unique capability of converting from driver to extractor by pulling upon the vibratory hoist line.

There are three basic types of Vibratory hammers

1. Resonance Free Vibratory Hammers, these hammers are designed to eliminate vibration during startup and shutdown. As a result, harmful resonance due to soil's own frequency is avoided
2. Normal Frequency Vibratory Hammers, have high amplitude and high eccentric moment. This type of hammer is suitable for use in clayey soil and soil compaction
3. High Frequency Vibratory Hammers, reduce peak particle velocities, consequently this type of hammer can be used near buildings, sewers, pipes and other buried structures.

Vibratory hammers ordinarily do not operate with leads, so other provisions must be made for holding the piles to plan position for driving. Equipment intended for vibratory or sonic installation or withdrawal of bearing or sheet piles should be capable of adjustment for frequency and amplitude to accommodate to different combinations of soils and piling. For efficient operation the machines must grip the pile firmly, with hydraulic or air actuated rams.

Verification of bearing capacity of piles installed by vibratory machines by load tests or final driving with an impact hammer usually is considered necessary.

When estimating the cost of piles, one must consider overhead expense before the actual cost of driving the piles can be computed. This includes cost of moving the equipment to the job, setting up the pile rig and being ready to operate, and then, dismantling and removing at completion.

This expense will always be pretty much the same, regardless of the size of the job. The equipment move charge is as much on a job with 100 piles as on one containing 1,000 piles. The actual cost of driving the piles will vary with soil conditions, length of piles, and the amount of moving necessary.

250

Types and Specifications of Pile Hammers
Steam and Air Hammers

Rated Energy Ft.Lbs	Manufacturer	Model	Action	Average Blows per Min.	Striking Weight in Lbs.	Rated Energy Kg. Meter	Striking Weight in Kg
1,800,000	Vulcan	6300	Single	42	300,000	248,868	136,080
1,200,000	Conmaco	2000	Single	40	200,000	165,912	90,720
750,000	Vulcan	5150	Single	46	150,000	103,695	68,040
350,000	Conmaco	700E5	Single	43	70,000	48,391	31,752
180,000	Vulcan	360	Single	62	60,000	24,887	27,216
150,000	Vulcan	560	Single	42	30,000	20,739	13,608
120,000	Vulcan	340	Single	60	40,000	16,591	18,144
90,000	Vulcan	030	Single	54	30,000	12,443	13,608
90,000	Conmaco	300	Single	52	30,000	12,443	13,608
60,000	Vulcan	512	Single	41	12,000	8,296	5,443
60,000	Conmaco	200	Single	55	20,000	8,296	9,072
48,750	Vulcan	016	Single	58	16,250	6,740	7,371
48,750	Conmaco	160	Single	50	16,250	6,740	7,371
42,000	Vulcan	014	Single	59	14,000	5,807	6,350
42,000	Conmaco	140	Single	55	14,000	5,807	6,350

Types and Specifications of Pile Hammers

Steam and Air Hammers

Rated Energy Ft.Lbs	Manufacturer	Model	Action	Average Blows per Min.	Striking Weight in Lbs.	Rated Energy Kg. Meter	Striking Weight in Kg
37,375	Conmaco	115	Single	52	11,500	5,167	5,216
32,500	Vulcan	010	Single	50	10,000	4,493	4,536
32,500	Conmaco	65E5	Single	50	6,500	4,493	2,948
26,000	Vulcan	08	Single	50	8,000	3,595	3,629
26,000	Conmaco	80	Single	56	8,000	3,595	3,629
19,500	Vulcan	06	Single	60	6,500	2,696	2,948
19,500	Conmaco	65E3	Single	61	6,500	2,696	2,948
19,150	Mkt Mfg	11B3	Dbl - Act	95	5,000	2,648	2,268
15,000	Vulcan	1	Single	60	5,000	2,074	2,268
15,000	Conmaco	50E3	Single	64	5,000	2,074	2,268
13,100	Mkt Mfg	10E3	Dbl - Act	105	3,000	1,811	1,361
8,750	Mkt Mfg	9E3	Dbl - Act	145	1,600	1,210	726

Diesel-Powered Hammers

Rated Energy in Ft.Lbs.	Manufacturer	Model	Action	Avg Blows per Min	Ram Weight Lbs.	Rated Energy Kg. M.	Ram Weight in Kg.	Est Fuel Consumption	
								Gal/day	L/day
210,000	ICE	205S	Single	41-55	20,000	29,400	9,072	23	87
183,700	ICE	120S-15	Single	41-55	15,000	25,718	6,804	28	106
149,000	ICE	120S	Single	38-55	12,000	20,860	5,443	28	106
120,000	ICE	100S	Single	38-55	10,000	16,800	4,536	22	83
117,000	Delmag	D55	Single	36-47	12,128	16,380	5,501	32	121
105,000	Delmag	D46-23	Single	37-53	10,143	14,700	4,601	26.4	100
99,300	ICE	80S	Single	38-55	8,000	13,902	3,629	18	68
83,100	Delmag	D36-23	Single	36.52	7,936	11,634	3,600	28	106
70,000	Mkt Mfg	DE-70	Single	45-50	7,000	9,800	3,175	26.4	100
72,900	ICE	60S	Single	38-55	8,000	10,206	3,629	18	68
65,600	Mitsubishi	MH-35	Single	42-60	7,720	9,184	3,502	40	151
51,520	Kobe	JK-25	Single	39-60	5,510	7,213	2,499	24	91
48,500	Delmag	D-22-22	Single	38-52	4,850	6,790	2,200	16	61
46,900	Mitsubishi	MH-25	Single	42-60	5,510	6,566	2,499	32	121
42,580	ICE	42S	Single	37-55	4088	5,961	1,854	11	42
32,000	ICE	32S	Single	41-60	3000	4,480	1,361	8.5	32
28,100	Mitsubishi	MH-15	Single	42-60	3,310	3,934	1,501	16	61

Diesel-Powered Hammers

Rated Energy	Manufacturer	Model	Action	Avg Blows	Ram Weight	Rated Energy	Ram Weight	Est Fuel	
								Gal/day	L/day
27,100	Delmag	D-15	Single	40-60	3,300	3,794	1,497	16	61
18,000	Mkt Mfg	DE-20	Single	40-50	2,000	2,520	907	16	61
17,000	Mkt Mfg	DE-20B	Single	40-50	2,000	2,380	907	16	61
9,350	Mkt Mfg	DA-15C	Single	40-50	1,100	1,309	499	8	30
9,100	Delmag	D-5	Single	42-60	1100	1,274	499	8	30
3,630	Delmag	D-4	Single	50-60	836	508	379	4	15
1,815	Delmag	D-2	Single	60-70	484	254	220	2.4	9

Hydraulic Vibratory Drivers/Extractors

Manufacturer	Model	Frequency VPM	Eccentric Moment		Line Pull		Suspended Weight		Engine Power	
			In-lb	kg-m	Tons	Tonnes	Lbs	Kg	HP	kW
ICE	1412BT	1500	22600	260	450	457.20	82,200	37286	2,030	1,523
ICE	1412B	1250	10300	115.2	150	152.40	36,950	16761	800	597
Mkt Mfg	V-20	1650	3000	2.65	60	60.96	1,250	567	-----	-----
Mkt Mfg	V-16	1750	1800	2.65	32	32.51	9,250	4196	-----	-----
Mkt Mfg	V-5	1700	1300	2.65	30	30.48	6,200	2812	-----	-----
ICE	23-40	1600	230	2.65	80	81.28	14,427	6544	40	30
ICE	11-23	1400	1100	12.67	40	40.64	5,775	2620	220	164

Setting Up and Removing Equipment. The cost of moving the pile driving equipment to the job must be taken into consideration, and then after it arrives at the job, there is the task of setting up the pile driver and getting ready to operate.

After the pile driver has been delivered to the project it should take about two to a maximum of 4 hours for a pile driving crew to set it up to be ready to drive piles and this labor should cost as follows:

Setting up Equipment			
Description	Hours	Rate	Total
Foreman	4	$39.04	$ 156.16
Pile driver engineer	4	$35.04	$ 140.16
Oiler	4	$35.89	$ 143.56
4 Pile driver (pile bucks)	16	$37.89	$ 606.24
Total labor cost			$1,046.12

The above item will remain practically the same whether there are 100 or 1,000 piles to be driven.

After the job has been completed, the pile driver must be dismantled, loaded onto trucks and removed from the job, the dismantling should cost as follows:

Dismantling Equipment			
Description	Hours	Rate	Total
Foreman	8	$39.04	$ 312.32
Pile driver engineer	8	$35.04	$ 280.32
Oiler	8	$35.89	$ 287.12
4 Pile driver (pile bucks)	16	$37.89	$ 606.24
Total labor cost			$1,486.00

Various Types of Bearing and Friction Piles

Wood	Concrete	Steel	Sheet Pile
Treated	Cast-in-place	HP shape	Wood
Untreated (plain)	Precast/prestressed	Pipe	Steel
	Shell	Shell	Precast Concrete
	Shell-less		

The above types can be used in various combinations, sometimes known as composite piles. A good example of composite pile would be heavy-walled pipe piles filled with concrete. Normally, the designer indicates on the drawings or in the written specifications, or on both, the type of pile and the size of hammer required to install.

Driving Wood Piles. After the pile driver is set up and ready to operate, the labor cost of driving the piles will vary with the length of the piles and job conditions.

The following table gives the approximate number of piles of various lengths that should be driven per hour and per day:

Length of Piles (Ft)	No. Pile 1-Hour	No. Piles 8-Hour	Length of Piles (m)
22	4.7	38	6.71
24	4.5	36	7.32
26	4.2	34	7.92
28	4.0	32	8.53
30	3.7	30	9.14
32	3.5	28	9.75
34	3.3	26	10.36
36	3.1	25	10.97
38	3.0	24	11.58
40	2.8	22	12.19
45	2.5	20	13.72
50	2.3	18	15.24
55	2.1	17	16.76
60	2.0	16	18.29

Wood piles over 40' are a premium cost.

The labor operating the pile driver per 8-hr. day should cost as follows:

Driving Piles			
Description	Hours	Rate	Total
Foreman	8.00	$ 39.04	$ 312.32
Pile driver engineer	8.00	$ 37.89	$ 303.12
Oiler	8.00	$ 35.89	$ 287.12
4 Pile drivers (pile bucks)	16.00	$ 35.04	$ 560.64
Total labor cost			$ 1,463.20

The above does not include time and costs for pile layout or resurvey, spotting piles, pumping, shoring, excavating, purchasing piles, pile points, or cutting off and disposal of cut-offs for wood piles after they are driven. It only includes the actual time driving the piles. It also does not include the cost of equipment, fuel, oil, and other supplies for the pile driver. These costs should be added.

256

SITE WORK

Cutting Off Wood Piles. After the wood piles have been driven, the tops projecting above the ground will have to be cut off to grade in order to receive the foundation that is to be placed on them.

This cost will vary according to the size of the quarters in which workers must operate, but on the average job, it should require 1/6 to 1/4 hrs. to cut off each wood pile 12" or 14" (300 to 350 mm) in diameter with a chain saw, at the following labor cost per pile:

Cutting Piles			
Description	Hours	Rate	Total
Pile Driver (pile buck)	0.22	$ 35.04	$ 7.71

To the above costs, add the hourly cost of a gas or hydraulic chain saw and the time for removing sawed-off ends, as well as proper removal and disposal.

Computing the Cost of Wood Piles. As described above, the cost of driving wood piles will vary with the number of piles, length, kind of soil, and other job conditions. Pile costs fluctuate so rapidly that current costs must be verified for each project.

The following is a sample estimate, showing a method for arriving at costs for driving 680 treated wood piles 40'-0" (12.2 m) long:

Description	Cost
Trucking pile driver to job	$ 5,000.00
Setting up pile driver, 4-hrs for crew	$ 1,046.12
680 wood piles, 40'-0" long @ $250 Ea.	$ 170,000.00
Driving piles, 22 pile per 8-hr.day = 31 days @$1,157.68 per day	$ 35,888.08
Fuel, oil, misc. supplies, etc. 31 days @ $225 per day	$ 6,975.00
Dismantling pile driver at completion, 1 day	$1,181.92
Trucking pile driver, job to yard	$ 5,000.00
Equipment rental, including leads, hammer and compressor = $2,100 per day	$ 65,100.00
Cost 680 piles	$ 290,191.12
Cost per pile	$ 426.75
Cost per lin. ft.	$ 10.67
Cost per m	$ 35.00

Add cost of cutting off piles after driving, and for overhead & profit.

Driving Steel HP Piles. Once the pile driver is set up and ready to operate, labor costs for actually driving the piles will vary with the length of piles and site conditions.

Rolled structural steel shapes are also used as bearing piles. The shape commonly used for this purpose is the HP. This type of pile has proved especially useful for trestle structures in which the pile extends above the ground and serves not only as a pile but also as a column.

Because of their small cross-sectional area, piles of this type can often be driven through dense soils to point bearing where it would be difficult to drive a pile of solid cross-section, such as a wood, cast-in-place, or precast concrete. But the feature that allows it to penetrate dense soils works to its disadvantage in other soil. Where piles are used to support loads by friction, or where they are used primarily for compaction, a considerably longer HP is required to carry the same load that can be supported by a pile of less length but greater cross-sectional area.

The upper ends of HP piles can be encased in concrete to prevent corrosion. Steel piles weighing more than 200 lbs. per l.f. (303 kg per m) have been driven in lengths up to 130 feet (40 m). Lighter sections have been driven to even greater lengths.

Driving HP Section Piles. The following table gives the approximate number of HP piles of various lengths that can be driven per hour and per day, assuming average job site conditions:

Length of Pile (Ft)	1-Hour	8-Hours	Length of Pile (m)
20	4.00	32	6.10
30	3.75	30	9.14
40	3.00	24	12.19
50	2.50	20	15.24
60	2.25	18	18.29

The labor cost of operating the pile driver per 8-hr. day is as follows:

Labor Costs for Pile Diving 8-Hours			
Description	Hours	Rate	Total
Foreman	8.00	$ 39.04	$ 312.32
Pile driver engineer	8.00	$ 37.89	$ 303.12
Oiler	8.00	$ 32.89	$ 263.12
4 Pile drivers (pile bucks)	16.00	$ 35.04	$ 560.64
Total labor cost			$1,439.20

The above does not include time and costs for pile layout or resurvey, spotting piles, pumping, shoring, excavation, purchase of piles, pile points and pile splices, or cutting off driven piles. It only includes the actual time driving piles. It does not include the cost of equipment, fuel, oil, and other supplies for the pile driver.

SITE WORK

Prices of Steel HP Piles. Prices of steel HP piles vary considerably, mainly due to the material cost of the rolled steel sections. Steel sections commonly used for this purpose may range from an 8-inch (200 mm) HP section weighing 36 lbs. per lin. ft. (55 kg per m) to a 14-inch (350 mm) HP section weighing 117 lbs. per lin. ft. (177 kg per m). Assuming a unit cost of $0.60 per lb. ($1.32 per kg), delivered to job site from warehouse, the HP beam material will cost from $21.60 to $35.10 per l.f. ($70-115 per m). If the job is in an isolated location, additional transportation charges must be figured.

Cutting Off Steel HP Piles. After the steel HP section pile has been driven, the pile tops that project above the ground must be cut off to the required elevation to receive the foundation that will be placed on them.

The cost will vary according to the size of the pile and the quarters in which workers must work. A burning outfit, consisting of gauges, burning torch, hoses, acetylene, and oxygen is normally required to cut HP steel piles at the following cost per pile:

Cutting Steel Piles					
HP		Hours	Rate	Total	HP
8 "	Section	0.17	$35.04	$ 5.96	203 mm
10 "	Section	0.21	$35.04	$ 7.36	254 mm
12 "	Section	0.25	$35.04	$ 8.76	305 mm
14 "	Section	0.31	$35.04	$10.86	356 mm

To the above, add the costs for a complete burning outfit, time for removing the cut off pile sections, and the cost for removal from the site and proper disposal.

Splicing Steel HP Piles. When splicing is necessary, steel HP piles can be spliced in the field. When butt welded, splices should be full penetration butt welds across both flanges and the web. Edges on the ends of the upper section, where welding is to be done, should be beveled for penetration welding. The top of the pile length, where driving will be done, should not be beveled.

Steel HP Piles Butt Welded					
HP		Hours	Rate	Total	HP
8 "	Section	1.33	$35.04	$46.60	203 mm
10 "	Section	1.67	$35.04	$58.52	254 mm
12 "	Section	2.00	$35.04	$70.08	305 mm
14 "	Section	2.50	$35.04	$87.60	356 mm

The costs vary according to the size of the piles and the quarters in which workers are obliged to operate. Add to the above all costs for

259

additional material and equipment (such as plates, welding machine, and pile equipment), and standby time while welding is being performed, and all required testing.

Prefabricated splicers are available from Dougherty Foundation Products, Inc. of Franklin Lakes, New Jersey. These splicers temporarily hold a length of HP pile until the joint can be completed. The driver can set an extension and then move on to other work while the welding takes place, at the following labor cost per pile:

HP	Hours	Rate	Total	HP
8 " to 14 "	0.50	$35.04	$17.52	203 to 356 mm

Material Cost for DFP* Splices		
HP	Cost ea.	HP
8 "	$ 56.00	200 mm
10"	$ 37.00	250 mm
12"	$ 47.00	300 mm
14"	$ 67.00	350 mm

*Dougherty Foundation Products, Inc.,
Franklin Lakes, NJ

The costs vary according to the size of the piles and the quarters in which workers must operate.

CONCRETE PILES

Due to the wide variety of conditions under which concrete piles are installed and used, it is difficult to give dependable cost figures without knowing the details of each installation. The different types of concrete piles vary considerably in cost. When estimating work containing concrete piles, it is always advisable to consult contractors specializing in this class of work. Most of them maintain offices in the principal cities and are usually willing to investigate and quote budget prices on anticipated work. More accurate figures can be obtained than can be given here.

There are two principal types of concrete piles, cast-in-place and precast. The cast-in-place pile is formed in the ground in the position near where it is to be used in the foundation. The precast pile is cast above ground, and after it has been properly cured, it is driven or jetted just like any pile.

In ordinary building foundation work, the cast-in-place pile is most common, because the required length can be readily adjusted in the field as the job progresses. There is no need to predetermine pile lengths, and the required length is installed at each pile location.

In contrast, it is necessary to predetermine the length of precast piles, and to provide for contingencies, it is generally required that piles be ordered longer than the actual anticipated length. Also, it is not possible to determine the pile length at each location, because subsoil conditions at any construction site will show considerable variation. The cost of waste piling for precast piles can be quite high, as is the cost of cutting them off to the proper grade. Precast piles are prestressed and are manufactured at established plants. Job site casting is relatively rare. It is difficult and costly to handle and transport precast or prestressed concrete piles especially in long lengths, and a production plant should be relatively close to the job site to reduce shipping costs.

In marine installations, either in salt water or fresh water, the precast pile is used almost exclusively, because of the difficulty of placing cast-in-place piles in open water. For docks and bulkheads, the cast-in-place pile is sometimes used in the anchorage system. On trestle type structures, such as highway viaducts, the precast pile is commonly used. A portion of the pile often extends above the ground and serves as a column for the superstructure. Precast concrete piles are always reinforced internally so that they resist stress from handling and pile driving. Cast-in-place piles are rarely reinforced, because lateral support of even the poorest of soils is sufficient to overcome any bending moments induced in the pile by column action.

Cast-in-place Concrete Pipe Piles. Pipe can be of any practical diameter. It is almost always filled with concrete, but with adequate wall thickness in non-corrosive ground, it can be used without concrete.

Driving usually produces stresses greater than the working load on the pile. Pipe should conform to *ASTM A252, Specifications for Welded and Seamless Steel Pipe Piles.*

Pipe shell that does not contribute to the strength of the pile as a structural member may be of any metal that adequately resists installation stress and maintains an open shaft to receive the concrete, free from water or other foreign matter. Steel shell should meet the applicable requirements of ASTM Specifications, A-283, A-366, or A-415.

The end closure for pipe or shell may be cast steel meeting requirements of ASTM A-27 or structural steel meeting ASTM A-36 or better.

Concrete for cast-in-place piles should conform to requirements for cement, aggregate, mixing, placing, and protection specified by ACI or PCA for quality concrete.

Cast-in-place concrete piles can be one of four types: 1) closed-end pipe 2) open-end pipe, 3) thin-cased and 4) corrugated shell. The closed-end pipe pile is simply a piece of steel pipe, closed at the bottom with a heavy boot, driven into the ground, and filled with concrete. The uses and allowable loads for these piles are about the same as for other types of cast-in-place piles with driven shells. On some projects the pipe shell is driven all in one piece, but often it is driven in sections that are welded together or fitted together with internal drive fit sleeves as the driving progresses. Pipe wall thickness must be adequate to develop the required stiffness and driveability of the pipe.

Open-end steel pipe piles are usually driven to bearing on rock. Since the pipe is open at the bottom during driving, the interior fills with soil which must be removed before concreting. Cleaning is usually done accomplished by air and water jets, after which the pipe is re-driven to insure proper "seating" in the rock. Remaining water is pumped out and the pipe pile is filled with concrete. Open-end pipe installed in this manner usually carry relatively high working loads. They have been installed up to 60" inches (1500 mm) in diameter and in lengths exceeding 200 feet (61 meters).

Estimating the Cost of Concrete Pipe Piles. The cost of concrete pipe piles will vary greatly, depending upon the number of piles in the project and the length of each pile, splices required, shoes, etc., and because it is the total footage of piles in the project rather than the number of piles that controls the price. For example, the price per foot for 100 piles 35 feet long would be much greater than the price per foot for 1,000 piles 25 feet (7.6 meters) long.

For any type of pile, contractors who specialize in this work are usually willing to submit budget prices, which include all layout, re-survey, labor, materials, pile load tests (if required) and equipment for the complete project. A detailed estimate would only be required when such budget prices are not available, or for some reason the estimator does not desire to seek out a budget price.

As with all pile driving, regardless of the size of the job, there is a fixed charge for moving equipment on and off the project that will cost from $5,000.00 to $25,000.00, provided the equipment is located in the vicinity of the work.

If it is necessary to ship plant and equipment from one city to another, it will be necessary to add additional freight and trucking amounting to $2,000.00 to $4,000.00, making a total fixed charge of $7,000.00 to $29,000.00, before figuring the cost of the piles.

Labor Driving Cast-in-place Concrete Pipe Piles. After the pile driver has been set up and ready to operate, the number of piles driven per day and the labor cost driving them will vary with the length of the piles and job conditions.

For instance, on a job where the piles were closely spaced and driven from a flat surface, a pile driver averaged 100 piles a day (which is exceptional), while on another job where the piles were more widely scattered and had to be driven from different elevations, it was with difficulty that 20 piles were driven per day.

Labor Placing Concrete for Piles. Practically 90% of all concrete used in concrete piles is ready-mix concrete, which is usually discharged directly into the piles. Ordinarily it requires 1/2 to 1 hour time for one worker per cu. yd. (cu. m) of concrete.

| Pipe Sizes Frequently Used for Piling and Concrete Required ||||||||
| Outside Diameter || Thickness Pipe Wall || Weight of Pipe || Concrete ||
Inches	mm	Inches	mm	Lbs. per LF	Kgs. per M.	CY per LF	Cu.M. per M.
8 5/8	219.08	0.125	3.175	11.35	16.89	0.0142	0.0356
		0.188	4.775	16.90	25.15	0.0137	0.0344
		0.219	5.563	19.64	29.23	0.0135	0.0339
		0.250	6.350	22.36	33.28	0.0133	0.0334
		0.312	7.925	27.74	41.28	0.0129	0.0324
		0.375	9.525	33.00	49.11	0.0125	0.0314
10	254.00	0.188	4.775	19.65	29.24	0.0187	0.0469
		0.219	5.563	22.85	34.01	0.0185	0.0464
		0.250	6.350	26.03	38.74	0.0182	0.0457
10 3/4	273.05	0.188	4.775	21.15	31.48	0.0217	0.0544
		0.219	5.563	24.60	36.61	0.0215	0.0539
		0.250	6.350	28.04	41.73	0.0212	0.0532
		0.307	7.798	34.24	50.96	0.0207	0.0519
		0.365	9.271	40.50	60.27	0.0203	0.0509
		0.500	12.700	54.70	81.40	0.0192	0.0482
12	304.80	0.188	4.775	23.65	35.20	0.0273	0.0685
		0.219	5.563	27.62	41.10	0.0270	0.0677
		0.250	6.350	31.40	46.73	0.0267	0.0670
12 3/4	323.85	0.188	4.775	25.16	37.44	0.0309	0.0775
		0.219	5.563	29.28	43.57	0.0306	0.0768
		0.250	6.350	33.40	49.71	0.0303	0.0760
		0.312	7.925	41.50	61.76	0.0297	0.0745
		0.375	9.525	49.60	73.81	0.0291	0.0730
		0.500	12.700	65.40	97.33	0.0279	0.0700
14	355.60	0.188	4.775	27.70	41.22	0.0375	0.0941
		0.219	5.563	32.20	47.92	0.0372	0.0933
		0.250	6.350	36.70	54.62	0.0368	0.0923
		0.312	7.925	45.70	68.01	0.0362	0.0908
		0.375	9.525	54.60	81.26	0.0355	0.0891
		0.500	12.700	72.10	107.30	0.0341	0.0855

Pipe Sizes Frequently Used for Piling and Concrete Required

Outside Diameter		Thickness Pipe Wall		Weight of Pipe		Concrete	
Inches	mm	Inches	mm	Lbs. per LF	Kgs. per M.	CY per LF	Cu.M. per M.
16	406.40	0.188	4.775	31.70	47.18	0.0493	0.1237
		0.219	5.563	36.90	54.91	0.0489	0.1227
		0.250	6.350	42.10	62.65	0.0485	0.1217
		0.312	7.925	52.40	77.98	0.0477	0.1197
		0.375	9.525	62.60	93.16	0.0470	0.1179
		0.500	12.700	82.80	123.22	0.0454	0.1139
18	457.20	0.219	5.563	41.50	61.76	0.0623	0.1563
		0.250	6.350	47.40	70.54	0.0619	0.1553
		0.312	7.925	59.00	87.80	0.0610	0.1530
		0.375	9.525	70.60	105.07	0.0601	0.1508
		0.500	12.700	93.50	139.15	0.0584	0.1465
20	508.00	0.219	5.563	46.20	68.75	0.0773	0.1939
		0.250	6.350	52.70	78.43	0.0768	0.1927
		0.312	7.925	65.70	97.77	0.0758	0.1901
		0.375	9.525	78.60	116.97	0.0749	0.1879
		0.500	12.700	104.10	154.92	0.0729	0.1829
22	558.80	0.250	6.350	58.10	86.46	0.0934	0.2343
		0.375	9.525	86.60	128.88	0.0912	0.2288
		0.500	12.700	114.80	170.84	0.0891	0.2235
24	609.60	0.250	6.350	63.40	94.35	0.1116	0.2800
		0.312	7.925	79.10	117.72	0.1104	0.2769
		0.375	9.525	94.60	140.78	0.1093	0.2742
		0.500	12.700	125.60	186.92	0.1067	0.2677
30	762.00	0.375	9.525	118.70	176.65	0.1728	0.4335
		0.500	12.700	157.50	234.39	0.1700	0.4265
36	914.40	0.375	9.525	142.70	212.36	0.2510	0.6296
		0.500	12.700	189.60	282.16	0.2474	0.6206

264

Length Lin Ft.	12" O.D. x .250 Wall 300mm x 6.25 mm Wall			14" O.D. x .250 Wall 350mm x 6.25 mm Wall			Length Meter
	Cu.Ft	Cu Yd.	Cu.M.	Cu.Ft	Cu Yd.	Cu.M.	
1	0.72	0.0267	0.0204	0.99	0.0367	0.0280	0.30
5	3.60	0.1333	0.1019	4.96	0.1837	0.1405	1.52
10	7.20	0.2667	0.2039	9.93	0.3678	0.2812	3.05
15	10.80	0.4000	0.3058	14.90	0.5519	0.4219	4.57
20	14.40	0.5333	0.4078	19.80	0.7333	0.5607	6.10
22	15.80	0.5852	0.4474	21.80	0.8074	0.6173	6.71
24	17.30	0.6407	0.4899	23.80	0.8815	0.6740	7.32
26	18.70	0.6926	0.5296	25.80	0.9556	0.7306	7.92
28	20.10	0.7444	0.5692	27.80	1.0296	0.7873	8.53
30	21.60	0.8000	0.6117	29.80	1.1037	0.8439	9.14
32	23.00	0.8519	0.6513	31.70	1.1741	0.8977	9.75
34	24.50	0.9074	0.6938	33.70	1.2481	0.9543	10.36
36	10.97	0.4063	0.3107	19.80	0.7333	0.5607	10.97
38	11.58	0.4289	0.3279	20.87	0.7730	0.5910	11.58
40	12.19	0.4515	0.3452	22.02	0.8156	0.6236	12.19
42	12.80	0.4741	0.3625	23.09	0.8552	0.6539	12.80
44	13.41	0.4967	0.3798	24.24	0.8978	0.6864	13.41
46	14.02	0.5193	0.3970	25.31	0.9374	0.7167	14.02
48	14.63	0.5419	0.4143	26.45	0.9796	0.7490	14.63
50	15.24	0.5644	0.4316	27.52	1.0193	0.7793	15.24
52	15.85	0.5870	0.4488	28.59	1.0589	0.8096	15.85
54	16.46	0.6096	0.4661	29.74	1.1015	0.8422	16.46
56	17.07	0.6322	0.4834	30.81	1.1411	0.8725	17.07
58	17.68	0.6548	0.5007	31.96	1.1837	0.9051	17.68
60	18.29	0.6774	0.5179	33.03	1.2233	0.9354	18.29

Quantity of Concrete Required in Various Size Pipe Piles in Cu. Ft., Cu. Yds. and Cu. Meters

Quantity of Concrete Required in Various Size Pipe Piles in Cu. Ft., Cu. Yds. and Cu. Meters

Length Lin Ft.	16" O.D. x .250 Wall 400mm x 6.25 mm Wall			24" O.D. x .250 Wall 600mm x 6.25 mm Wall			Length Meter
	Cu.Ft	Cu Yd.	Cu.M.	Cu.Ft	Cu Yd.	Cu.M.	
1	1.30	0.0481	0.0368	3.01	0.1115	0.0852	0.30
5	6.54	0.2422	0.1852	15.00	0.5556	0.4248	1.52
10	13.00	0.4815	0.3681	30.10	1.1148	0.8524	3.05
15	19.60	0.7259	0.5550	45.10	1.6704	1.2772	4.57
20	26.10	0.9667	0.7391	60.20	2.2296	1.7048	6.10
22	28.80	1.0667	0.8156	66.20	2.4519	1.8747	6.71
24	31.40	1.1630	0.8892	72.30	2.6778	2.0474	7.32
26	34.00	1.2593	0.9628	78.30	2.9000	2.2173	7.92
28	36.60	1.3556	1.0365	84.30	3.1222	2.3873	8.53
30	39.20	1.4519	1.1101	90.30	3.3444	2.5572	9.14
32	41.90	1.5519	1.1865	96.40	3.5704	2.7299	9.75
34	44.50	1.6481	1.2602	102.00	3.7778	2.8885	10.36
36	47.10	1.7444	1.3338	108.00	4.0000	3.0584	10.97
38	49.70	1.8407	1.4074	114.00	4.2222	3.2283	11.58
40	52.30	1.9370	1.4811	120.00	4.4444	3.3982	12.19
42	54.90	2.0333	1.5547	126.00	4.6667	3.5681	12.80
44	57.60	2.1333	1.6311	132.00	4.8889	3.7380	13.41
46	60.20	2.2296	1.7048	138.00	5.1111	3.9080	14.02
48	62.80	2.3259	1.7784	144.00	5.3333	4.0779	14.63
50	65.40	2.4222	1.8520	150.00	5.5556	4.2478	15.24
52	68.00	2.5185	1.9257	156.00	5.7778	4.4177	15.85
54	70.70	2.6185	2.0021	162.00	6.0000	4.5876	16.46
56	73.30	2.7148	2.0757	168.00	6.2222	4.7575	17.07
58	75.90	2.8111	2.1494	174.00	6.4444	4.9274	17.68
60	78.50	2.9074	2.2230	180.00	6.6667	5.0973	18.29

Thin Shell Pipe Piles. Monotube (Monotube Pile Company) piles are the only uniformly tapered steel shell pipe piles available. Monotubes are strictly a friction pile, where the shell is end driven with conventional pile driving equipment plus the use of a mandrel, visually inspected, and filled with concrete.

Monotubes are available in four diameters, four gauges, and three rates of taper, making them suitable for soils with considerable friction resistance, such as sand, silt, and clay. Because of the wedging action of the

266

Specifications for Monotube Piles Tapered Sections

Type	Size — Tip Dia. x Butt Dia. x Length			9 Ga. 0.1495"	7 Ga. 0.1793"	5 Ga. 0.2092"	3 Ga. 0.2391"	Concrete Volume Cu.Yd
"F" Taper 0.14 inch per foot	8 1/2 " x	12 " x	25 '	17	20	24	28	0.43
	8 " x	12 " x	30 '	16	20	23	27	0.55
	8 1/2 " x	14 " x	40 '	19	22	26	31	0.95
	8 " x	16 " x	60 '	20	24	28	33	1.68
	8 " x	18 " x	75 '	-----	26	31	35	2.59
"J" Taper 0.25 Inch per Foot	8 " x	12 " x	17 '	17	20	23	27	0.32
	8 " x	14 " x	25 '	18	22	26	30	0.58
	8 " x	16 " x	33 '	20	24	28	32	0.95
	8 " x	18 " x	40 '	-----	26	30	35	1.37
"Y" Taper 0.40 inch per Foot	8 " x	12 " x	10 '	17	20	24	28	0.18
	8 " x	14 " x	15 '	19	22	26	30	0.34
	8 " x	16 " x	20 '	20	24	28	33	0.56
	8 " x	18 " x	25 '	-----	26	31	35	0.86

Weight Per Foot is given under the four gauge columns (9 Ga., 7 Ga., 5 Ga., 3 Ga.).

Specifications for Monotube Piles Tapered Sections - Metric

Type	Size Tip Dia. x Butt Dia. x Length	Weight per Meter				Concrete Volume Cu.M
		9 Ga. 3.7375 mm	7 Ga. 4.4825 mm	5 Ga. 5.2300 mm	3 Ga. 5.9775 mm	
"F" Taper 3.5 mm	216 mm x 305 mm x 7.62 m	25.30	29.76	35.72	41.67	0.329
	203 mm x 305 mm x 9.14 m	23.81	29.76	34.23	40.18	0.421
	216 mm x 356 mm x 12.19 m	28.28	32.74	38.69	46.13	0.726
	203 mm x 406 mm x 18.29 m	29.76	35.72	41.67	49.11	1.285
per meter	203 mm x 457 mm x 22.86 m	-----	38.69	46.13	52.09	1.980
"J" Taper 6.25 mm	203 mm x 305 mm x 5.18 m	25.30	29.76	34.23	40.18	0.245
	203 mm x 356 mm x 7.62 m	26.79	32.74	38.69	44.65	0.443
	203 mm x 406 mm x 10.06 m	29.76	35.72	41.67	47.62	0.726
per meter	203 mm x 457 mm x 12.19 m	-----	38.69	44.65	52.09	1.048
"Y" Taper 10 mm	203 mm x 305 mm x 3.05 m	25.30	29.76	35.72	41.67	0.138
	203 mm x 356 mm x 4.57 m	28.28	32.74	38.69	44.65	0.260
	203 mm x 406 mm x 6.10 m	29.76	35.72	41.67	49.11	0.428
per meter	203 mm x 457 mm x 7.62 m	-----	38.69	46.13	52.09	0.658

Specifications for Monotube Piles Tapered Sections - Extension Sections

Type	Size Tip Dia. x Butt Dia. x Length	Weight per Foot				Concrete Volume
		9 Ga. 0.1495"	7 Ga. 0.1793"	5 Ga. 0.2092"	3 Ga. 0.2391"	Cu.Yd
N12	12 " x 12 " x 20/40 '	20	24	28	33	0.026
N14	14 " x 14 " x 20/40 '	24	29	33	38	0.035
N16	16 " x 16 " x 20/40 '	27	32	38	44	0.045
N18	18 " x 18 " x 20/40 '	-----	37	43	49	0.058

Type	Size Tip Dia. x Butt Dia. x Length	Weight per Meter				Concrete Volume
		9 Ga. 3.7375 mm	7 Ga. 4.4825 mm	5 Ga. 5.2300 mm	3 Ga. 5.9775 mm	Cu.M
N12	305 mm x 305 mm x 6.1/12.2 m	29.76	35.72	41.67	49.11	0.020
N14	356 mm x 356 mm x 6.1/12.2 m	35.72	43.16	49.11	56.55	0.027
N16	406 mm x 406 mm x 6.1/12.2 m	40.18	47.62	56.55	65.48	0.034
N18	457 mm x 457 mm x 6.1/12.2 m	-----	55.06	63.99	72.92	0.044

uniform taper and the factory attached conical nose, they are commonly driven to the same capacity as straight-sided friction piles with less depth. A combination of cold processing and vertical fluting permits the use of lighter gauges that are equal in strength to pipe piles with thicker walls. Tapered sections can be factory-welded to a consent diameter extension or spliced in the field. Field splices are made after the tapered portion has been driven into the soil. The extension is lifted into the leads and inserted 6 to 8 inches into the taper. Factory-crimped extension sections speed up alignment and field splicing.

Tapertube® Pile. Tapertube piles are made of 50-ksi steel. These piles are 12-sided with lower sections of grade 50 steel and range in thickness from 3/16" to 5/8". Bottom and top diameters can vary from 8" to 24" over lengths from 10 to 35 feet. The circularized top of the polygon conforms to the diameter of the steel pipe extension. A butt weld assures development of the full strength of the piling element. Move in, set-up, driving time and dismantling is the same as the Monotube pile except a mandrel is not required.

Tapertube™ Piles					
Pile Size	Taper Inches	Cu. Yds Conc. per Taper Lgth	Pile Size Metric	Taper mm	Cu. Meter Conc. per Taper Lgth
8 x 12			**200 x 300 mm**		
15 '	0.317	0.25	4.57 m	8.05	0.191
20 '	0.240	0.33	6.10 m	6.10	0.252
25 '	0.190	0.42	7.62 m	4.83	0.321
8 x 14			**200 x 350 mm**		
15 '	0.400	0.35	4.57 m	10.16	0.268
20 '	0.300	0.46	6.10 m	7.62	0.352
25 '	0.240	0.58	7.62 m	6.10	0.443
30 '	0.200	0.70	9.14 m	5.08	0.535
8 x 16			**200 x 400 mm**		
15 '	0.530	0.42	4.57 m	13.46	0.321
20 '	0.100	0.56	6.10 m	2.54	0.428
25 '	0.320	0.70	7.62 m	8.13	0.535
30 '	0.270	0.84	9.14 m	6.86	0.642

Tapertube™ Piles (continued)					
Pile Size	Taper Inches	Cu. Yds Conc. per Taper Lgth	Pile Size Metric	Taper mm	Cu. Meter Conc. per Taper Lgth
8 x 18			**200 x 450 mm**		
15 '	0.670	0.52	4.57 m	17.02	0.398
20 '	0.500	0.69	6.10 m	12.70	0.528
25 '	0.400	0.86	7.62 m	10.16	0.658
30 '	0.330	1.03	9.14 m	8.38	0.788
12 x 18			**300 x 450 mm**		
20 '	0.280	0.83	6.10 m	7.11	0.635
25 '	0.220	1.04	7.62 m	5.59	0.795
30 '	0.180	1.25	9.14 m	4.57	0.956
35 '	0.160	1.46	10.67 m	4.06	1.116
14 x 20			**350 x 500 mm**		
20 '	0.300	1.16	6.10 m	7.62	0.887
25 '	0.240	1.45	7.62 m	6.10	1.109
30 '	0.200	1.75	9.14 m	5.08	1.338

Precast/prestressed Concrete Piles. Precast concrete piles are commonly used in marine structures, such as docks and piers, where they are driven in open water.

Ordinarily they are designed for the particular job and made in a casting yard at or near the job site. The use of precast concrete piles usually delays the starting of actual pile driving because of the time involved in the casting and curing of this type of pile.

Precast concrete piles have the same advantages as other concrete piles over wood and steel piles. They are always reinforced internally to resist stresses produced by handling and driving.

Requirements for precast concrete piles, except for reinforcement requirements, generally apply to prestressed units as well. If prestressed piles are used, the minimum working net prestress in the pile should be 700 psi (5.6 MPa. Prestressing strands are of the ungalvanized, sevenstrand type, conforming to the general requirements of ASTM A-416 and may be either regular or high-strength.

Precast piles may be moved when the compressive strength of concrete has reached 4,000 psi (32 MPa), but are not to be driven until 5,000 psi (40 MPa) strength is attained. Piles are to be lifted or supported only at designated points and handled and driven in such a manner as to avoid excessive bending stresses, cracking, or spalling.

Pile heads must be protected from direct impact of the hammer by a cushion head so arranged that strands or bars projecting from the pile head will not be displaced or deformed during driving. Minimum energy per blow of the hammer is established by the engineer. Jetting may be permitted, or required where necessary, to reach the desired depth.

Driving 20'-0" (6.1 m) Concrete Piles. On a 12-story reinforced concrete building, requiring 280 concrete piles 20'-0" (6.1 m) long, the following costs were obtained.

There was considerable broken time on this job, due to bad weather, and while the job required 17-1/2 days, only 112 hrs. actual driving time was required. This included moving the pile driver from one place to another, because some of the piles were driven from the street level and others from the basement floor level.

The pile driving crew averaged two 20'-0" (6.1 m) concrete piles per hr. or 16 piles—320 lin. ft.(97.5 m)-per 8-hr. day, at the following labor costs:

Setting up Equipment			
Description	Hours	Rate	Total
Foreman	4.00	$ 39.04	$ 156.16
Pile driver engineer	4.00	$ 37.89	$ 151.56
Oiler	4.00	$ 31.93	$ 127.72
4 Pile driver (pile bucks)	16.00	$ 35.04	$ 560.64
Total labor cost			$ 996.08

Driving Piles			
Description	Hours	Rate	Total
Foreman	200.00	$ 39.04	$ 7,808.00
Pile driver engineer	200.00	$ 37.89	$ 7,578.00
Oiler	200.00	$ 32.89	$ 6,578.00
4 Pile driver (pile bucks)	800.00	$ 35.04	$28,032.00
Total labor cost			$49,996.00

Dismantling Equipment			
Description	Hours	Rate	Total
Foreman	8.00	$ 39.04	$ 312.32
Pile driver engineer	8.00	$ 35.04	$ 280.32
Oiler	8.00	$ 32.89	$ 263.12
4 Pile driver (pile bucks)	16.00	$ 35.04	$ 560.64
Total labor cost			$ 1,416.40

Total Costs Pile Driving			
Description	Hours	Rate	Total
Transport Equipment in/Out		$5,000.00	$ 5,000.00
Setting up Equipment			$ 996.08
Driving Piles			$49,996.00
Dismantling Equipment			$ 1,416.40
Total labor cost 280 Piles @ 20'- 0" ea			$57,408.48
Cost per Ea (280 ea.)			$ 205.03
Cost per Foot (5,600 Lin.Ft.)			$ 10.25
Cost per Meter (1,706.88 M)			$ 33.63

Note: Add cost of pile, pile tips, pile driver and hammer plus jetting equipment if required.

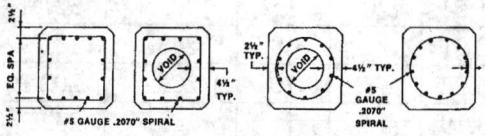

Cross-sections of Square Prestressed Piles

Various Types of Precast Concrete Piles Squared Prestressed Piles						
Pile Size, Inches [1]	Approx. Weight Lbs/Ft. [2]	Strand per Pile Diameter [3]		Design Bearing Capacity in Tons [4]		
		7/16" Strand	1/2" Strand	Perimeter Inch	Concrete 5000 psi	Concrete 6000 psi
10	105	4	4	40	73	90
12	150	6	5	48	105	129
14	205	8	6	56	143	176
16	265	11	8	64	187	229
18	335	13	10	72	237	290
20	415	16	12	80	292	358
22	505	20	15	88	354	433
24	600	23	18	96	421	516
20 HC	320	13	10	80	223	273
22 HC	365	14	11	88	256	314
24 HC	415	16	12	96	291	357

Various Types of Precast Concrete Piles Squared Prestressed Piles - Metric						
Pile Size, mm [1]	Approx. Weight Kg/M [2]	Strand per Pile Diameter [3]		Design Bearing Capacity in Tons [4]		
		10.93 mm Strand	12.50 mm Strand	Perimeter mm	Concrete Strength 34.45 MPa	Capacity Strength 41.34 MPa
254	156.26	4	4	1016	66.22	81.65
305	223.23	6	5	1219	95.25	117.03
356	305.08	8	6	1422	129.73	159.66
406	394.37	11	8	1626	169.64	207.74
457	498.54	13	10	1829	215.00	263.08
508	617.60	16	12	2032	264.90	324.77
559	751.54	20	15	2235	321.14	392.81
610	892.91	23	18	2438	381.92	468.10
508 HC	476.22	13	10	2032	202.30	247.66
559 HC	543.19	14	11	2235	232.24	284.85
610 HC	617.60	16	12	2438	263.99	323.86

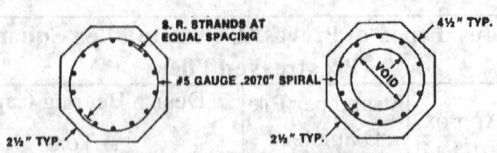

Cross-sections of Octagonal Prestressed Piles

Various Types of Octagonal Prestressed Concrete Piles						
Pile Size, Inches [1]	Approx. Weight Lbs/Ft. [2]	Strand per Pile Diameter [3]		Design Bearing Capacity in Tons [4]		
		7/16" Strand	1/2" Strand	Perimeter Inch	Concrete 5000 psi	Concrete 6000 psi
10	85	4	4	33	61	74
12	125	5	4	40	87	107
14	170	7	5	46	18	145
16	220	9	7	53	155	190
18	280	11	8	60	196	240

SITE WORK

Various Types of Octagonal Prestressed Concrete Piles

Pile Size, Inches [1]	Approx. Weight Lbs/Ft. [2]	Strand per Pile Diameter [3]		Design Bearing Capacity in Tons [4]		
		7/16" Strand	1/2" Strand	Perimeter Inch	Concrete 5000 psi	Concrete 6000 psi
20	345	14	10	66	242	296
22	420	16	12	73	293	359
24	495	19	15	80	348	427
20 HC	245	10	8	66	172	211
22 HC	280	11	8	73	196	240
24 HC	315	12	9	80	219	269

Pile Size, mm [1]	Approx. Weight Kg/M [2]	Strand per Pile Diameter [3]		Design Bearing Capacity in Tons [4]		
		10.93 mm Strand	12.50 mm Strand	Perimeter mm	Concrete Strength 34.45 MPa	Capacity Strength 41.34 MPa
254	126.50	4	4	838	55.34	67.13
305	186.02	5	4	1016	78.92	97.07
356	252.99	7	5	1168	16.33	131.54
406	327.40	9	7	1346	140.61	172.36
457	416.69	11	8	1524	177.81	217.72
508	513.43	14	10	1676	219.54	268.53
559	625.04	16	12	1854	265.80	325.68
610	736.65	19	15	2032	315.70	387.37
508 HC	364.61	10	8	1676	156.03	191.41
559 HC	416.69	11	8	1854	177.81	217.72
610 HC	468.78	12	9	2032	198.67	244.03

Notes (For both square and octagonal piles)
1. Voids in 20", 22", and 24" (500, 550, 600 mm) diameter hollow-core (HC) piles are 11", 13", and 15" (275, 325, 375 mm) diameter respectively, providing a minimum 4-1/2" (112.5 mm) wall thickness.
2. Weights are based on 150 lbs. (68 kg) per cu. ft. of concrete.
3. Based on 7/16" and 1/2" (10.9 and 15 mm) 270 Grade stress-relieved strand with an ultimate strength of 31,000 and 41,300 lbs. (14061 and 18233 kg) respectively.
4. Design bearing capacity based on 5,000 and 6,000 psi (34.5 and 41.4 MPa) concrete and an allowable unit stress on the full section of 0.33 fce - 0.27 fce where fce is the concrete stress in the pile due to prestressing, after all losses. These bearing capacity values may be increased if higher strength concrete is used.

Cutting Off Precast/Prestressed Square or Octagonal Piles. After concrete piles have been driven, it might be necessary to cut off the tops of piles to the proper elevation.

Costs for this work vary according to the size of piles and the space in which workers must operate. There are two methods to cut the pile: 1) Conventional. Before cutting a precast or prestressed pile, make a circumferential cut, with a diamond or carborundum saw, to score the pile and prevent spalling and then jackhammer the portion above the cut off. 2) Use of hydraulic shears. This type of cutting machine normally requires the use of a crane. Though faster, the costs need to be worked out as the most economical method of cut-off.

Size (In.)	Conventional Person-hours		Hydraulic Shears Person-hours		Size mm
	Square	Octagonal	Square Oper Eng. W/ Crane	Octagonal Labor	
10	0.7	1.0	0.25	0.30	254
12	1.0	1.3	0.50	0.60	305
14	1.1	1.4	0.60	0.75	356
16	1.3	1.7	0.80	1.00	406
18	1.5	1.9	1.00	1.10	457
20	1.7	2.1	1.10	1.30	508
22	1.8	2.3	1.30	1.50	559
24	2.0	2.5	1.40	1.75	610

To costs for the above, add the costs of cutting tools and equipment, as well as costs for removal and disposal of cut off sections of pile.

Prestressed Concrete Cylinder Piles. These are centrifugally cast in 16-foot sections, which are assembled end-to-end and post-tensioned, providing a wide range of available pile shapes. These piles are cast with zero-slump concrete, and strengths from 7,000 to 10,000 psi (48 to 69 MPa) are attained. They are made with the Cen-Vi-Ro process, which is a combination of centrifugal casting, vibrating, and rolling. It produces an extremely dense concrete with an absorption factor about 40% less than bed-cast concrete piles. The use of zero-slump concrete prevents migration of the coarse aggregate during the spinning process so that the aggregates are uniformly distributed across the pile wall.

These piles are available in diameters from 36 to 90 inches (900 to 2250 mm), but the standard sizes are 36, 54, and 66 inch (900, 1350, 1650 mm) O.D. Wall thicknesses are 5 inches (125 mm) for the 36- and 54-inch piles (900 and 1350 mm) and 6 inches (150 mm) for the 66-inch (1650 mm) piles. Cylinder Piles are driven either open or closed end. They are capable of

276

carrying extremely high compression loads and resisting large lateral forces or bending moments. They are used for elevated highways, bridges, piers, offshore platforms, and breakwaters. For bridges and other trestletype structures, cylinder piles serve as both foundation pile and pier or column.

Standard Sizes of Cylinder Piles				
Outside Diam. Inch	Inside Diam. Inch	Wall Thickness Inch	Lbs./Ft.	Number of Cables
36	26	5	524	8-16
54	44	5	829	12-24
66	54	6	1217	16-32
Outside Diam. mm	Inside Diam. mm	Wall Thickness mm	Kg/m	Number of Cables
914	660	127	780	8-16
1372	1118	127	1234	12-24
1676	1372	152	1811	16-32

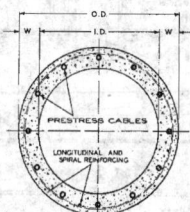

Prestressed Cylinder Piles

Pile Load Testing. When required, load testing might cost anywhere from $5,000 to more than $20,000. Because of the wide variation in piles, and in conditions under which they are tested, it is difficult to give dependable costs without knowing the details of an individual installation. Costs are affected by factors such as the information that is to be obtained and the quarters in which the test is to be undertaken.

There are a number of methods for determining the load-carrying capability of a pile: static load testing with intermittent addition of load; constant rate of penetration; dynamic testing while driving; and others.

Before a pile load test is made, the proposed apparatus and structure for reactions and loading should be fully detailed and costs applied. Load tests are normally carried to twice the design load, and pile tests can be in excess of 500 tons (560 m tons).

For a load test, piles have to be driven. Often, reaction piles are also installed to counteract the test force. This requires the same materials and equipment as planned for production driving. Another pile or two, perhaps with a cast-steel point for comparison, can be added with little increase in the cost.

Typical Pile Load Test Setups. Supports for loads or reaction piles should be at least five pile diameters or 5 feet (1.5 m) from the pile under test. In some soils, this should be as much as 10 diameters. The object is to keep the reaction far enough away that it will not affect the pile under test. Pile load tests vary because of the different types of tests and the various types of piles. The test might vary from $5,000 to $50,000. In some rare cases, the cost has gone as high as $150,000. Not all piles require a load test. Usually, a pile of 20-ton (22 m ton) capacity or less does not require tests.

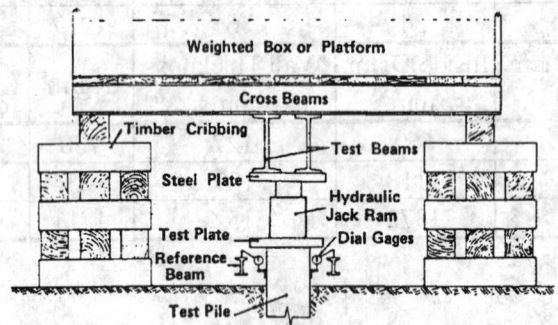

Typical setup for pile load testing in axial compression using a loading platform

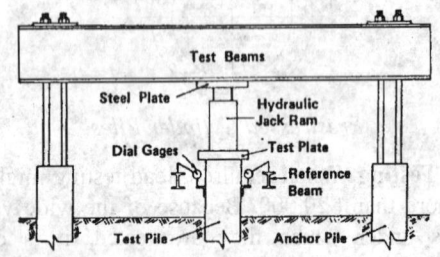

Typical setup for pile load testing in axial compression using anchor piles.

Some local codes mandate a minimum length that must be driven before a pile is considered a pile (for example, 10 feet (3.1 m) of pile driven tip to cut off). Pile driving contractors commonly quote projects giving several different unit costs:

Item Description	Bid Unit
1) Mobilization	Lump Sum
2) Pile Load Test	Each
3) Pile Costs	Each or Lin. Ft. (Meter)
4) Demobilization	Lump Sum

278

SITE WORK

Some pile driving contractors combine the mobilization and demobilization. On small jobs, where tests are not required, they will quote the project as a lump sum.

02350 CAISSONS

Caisson foundations are normally used to support heavy structures where soil conditions near grade cannot support them. Where such conditions exist, the foundation should be carried down through the unsatisfactory soil to material that can carry the imposed load, without causing any detrimental settlement. The size and depth of caissons will vary with the loads to be carried and the distance to material of the required bearing capacity. For example, caissons bearing on rock in downtown Chicago vary from 90 to 190 feet in depth (27 to 58 m).

Caissons can be dug either with straight shafts to the required depth, or they can be belled at the bottom to provide additional bearing area. Most caissons are installed using various types of drill rigs, but under certain conditions hand dug caissons are used. Drilling machines are capable of not only drilling the shaft but mechanically forming a bell. The cost of excavating varies with the diameter and depth of the caisson, site conditions, and the type of soil.

There are contractors who specialize in installing caissons. On jobs of any size, it is advisable to obtain figures from specialty contractors in this work, because they are equipped to perform it much more economically.

Hand Dug Caissons. The hand method of excavation most commonly used is known as the Chicago method. This consists of excavating with hand tools and lining the hole with wood lagging held in place by steel rings. The wood lagging has beveled edges and is tongue grooved. The lagging is milled from either 2"x6" (50 x 150 mm) or 3"x6" (75 x 150 mm) mixed hardwood in 16' (4.9 m) lengths, which are commonly cut into thirds so that a set of lagging is 5'-4" (1.62 m) long. Three-inch steel channels are formed in a half circle with a plate on each end to permit bolting the half sections together. Two complete rings are required for each set of lagging.

To determine the excavation quantity, multiply the cross-sectional area by the length allowing for the increased diameter due to the lagging.

To determine the quantity of concrete required, multiply the cross-sectional area by lengths making proper corrections if the finished concrete surface is to be below the ground surface. If the caisson has a bell, the volume of it must be computed using the formula for a truncated cone: $V = 1/3(A1 + A2 + A1 + A2)$ h, where $A1 + A2$ are the area of the bases and h is the height of the bell.

Excavation quantity for 6'-0" (1.82 m) caisson 106' (32.3 m) deep
Cross-sectional area for 6'-4" (1.9 m) = 31.5 sq. ft. (2.92 sq. m)
31.5 x 106 x 1/27 = 124 cu. yds.
(9.6 x 32.3 x .3057 = 95 cu. m)

Concrete quantity:

Cross-sectional area for 6'-0" (1.82 m) = 28.3 sq. ft. (2.63 sq. m)
28.3 x 106 x 1/27 = 111 cu. yds.
Allow 5% for waste for a total of 117 cu. yds.
(8.63 x 32.3 x .3057 = 85 cu. m
Allow 5% for waste for a total 89 cu. m)

After determining the excavation quantities and the concrete quantities, it is necessary to determine the amount of lagging and rings. By cutting the bevel and tongue and grooves in the lagging, the width of the lagging is reduced, so that it requires 7 pieces of 2"x6" (50 x 150 mm) per foot of diameter of the caisson. For example, a 4' (1.20 m) diameter caisson requires 28 pieces of lagging around the circumference. Therefore, using 2"x6" (50 x 150 mm) lagging, which is one board foot, it would require 28 board feet per foot of depth for a 4' (1.2 m) caisson or 42 board feet (0.08 cu. m) using 3" thick lagging. To determine the amount of lagging required in a 6' (1.8 m) diameter caisson, 106' (32.3 m) deep using 2" (50 mm) lagging, multiply 6 x 7 or 42 board feet per foot (0.08 cu. m) of depth x the depth, which is 106' (32.3 m), or 4,452 board feet (8.9 cu. m) If the lagging had been 3" (75 mm), it would be 6,678. A caisson 106' (32.3 m) deep would require 20 sections of 5'-4" (1.6 m) lagging. Using 2 rings for each section of lagging, the caisson would require 40 rings.

The rings are usually made from 3" (75 mm) channels, weighing 4.2 lbs. per foot (6.3 kg/m). Each ring will weigh approximately 80 lbs. (36 kg) plus 10 lbs. for the plates at the ends of the half circles, for a total of 90 lbs. (40 kg).

Labor Required for Caissons. Labor costs of caissons excavating varies with the diameter and depth of caissons, whether hand or pneumatic tools are used, the type of soil, and whether or not water or sand is encountered. It is generally assumed that one caisson digger can hand dig approximately 4 to 5 yards (3 to 3.8 cu. m) per 8-hour shift and place the lagging and rings.

Only one worker can work in a 4' (1.2 m) diameter shaft, but two workers can work in a 5' (1.5 m) diameter or larger shaft. In a 6' (1.8 m) caisson, two workers each taking out 4 yards (3 cu. m) should be able to excavate approximately 7' (2.1 m) per shift. At the surface, a pneumatic tugger hoist and tripod would be required with two laborers to operate the hoist and dump the buckets. Generally, several caissons are constructed simultaneously, so the cost involved in furnishing compressed air for the hoist is spread over three or more caissons.

In addition to the above, it will be necessary to allow for time of foreman, timekeepers, material clerks, engineers, and cost of setting up and removing equipment, which will cost from $4.00 to $6.00 per cu. yd. ($5.00 to 8.00 per cu. m). It may be estimated separately under overhead expense and an amount allowed to cover the entire job.

280

SITE WORK

To summarize the costs involved for constructing a caisson 6 ft. (1.8 m) in diameter and 106 ft. (32.3 m) deep, with two others being constructed simultaneously, time to excavate and lag:

124 cu. yds. @ 4 cu. yds./shift/digger = 31 shifts.
Using two diggers, this would be 15-1/2 shifts.
Allow 16 shifts.

Excavation			
Description	Hours	Rate	Total
Caisson diggers	256.00	$ 26.16	$ 6,696.96
Hoist operator	128.00	$ 37.89	$ 4,849.92
Top laborers	256.00	$ 26.16	$ 6,696.96
Compressor operator	43.00	$ 35.89	$ 1,543.27
Total Cost Excavation			$ 19,787.11

Material				
Description			Rate	Total
Lagging	4.452	MBF	$760.00	$ 3,383.52
Rings	40.00	Ea.	$ 60.00	$ 2,400.00
Total				$ 5,783.52

Equipment				
Description			Rate	Total
Hoist	16.00	Shifts	$277.00	$ 4,432.00
Air tools	32.00	Shifts	$ 40.00	$ 1,280.00
Compressor 900 CFM	16.00	Shifts	$376.00	$ 6,016.00
Total				$11,728.00

Soil Removal				
Description			Rate	Total
Excavated soil	124.00	Cu.Yd.	$ 16.00	$ 1,984.00

Total	
Excavation	$ 19,787.11
Material	$ 5,783.52
Equipment	$ 11,728.00
Soil Removal	$ 1,984.00
Grand Total -124 Cu.Yd.	$ 39,282.63
Cost per Cu. Yd.	$ 316.80
Cost per Cu. M.	$ 414.33

To the above cost it will be necessary to add for setting up and removing necessary plant and equipment, electric current and wiring, concrete furnishing and placing, overhead expenses, contingencies, miscellaneous expenses, and profit.

Work of this type in the Chicago area generally costs from $200.00 to $275.00 per cu. yd. ($260 to 360 per cu. m) of completed caisson.

Excavating for Caissons with Pneumatic Diggers. When excavating for caissons 5'-0" to 8'-0" (1.5 to 2.4 m) in diameter and up to 100'-0" (30.5 m) deep, a worker with a pneumatic digger will loosen and load into buckets 9 to 11 cu. yds. (6.9 to 8.4 cu. m) per 8-hr. shift, depending upon the toughness of the clay and the working space, at the following cost per cu. yd. (cu. m):

Pneumatic Diggers			
Description	Hours	Rate	Total
Labor Operating Digger and Loading Buckets	1.00	$ 26.16	$ 26.16
Compressor Operator	0.50	$ 35.89	$ 17.95
Compressor 900 CFM	0.50	$ 47.00	$ 23.50
Total Cost per Cu.Yd			$ 67.61
Total Cost per Cu.M			$ 88.42

When excavating caissons by hand, two "diggers" will loosen and load into buckets, 6 to 7 cu. yds. (4.6 to 5.4 cu. m) per 8-hr. shift, at the following labor cost per cu. yd. (cu. m):

Description	Hours	Rate	Total
Labor per Cu.Yd.	2.50	$ 26.16	$ 65.40
Labor per Cu.M			$ 85.53

Drilled Caissons. The use of drilling machines for the installation of caisson foundations has eliminated the hand digging method except in situations where adequate clearances to surrounding structures or utilities cannot be maintained. Drilled caissons have proved more economical than hand-dug caissons for several reasons, the most important of which is the fact that smaller shaft diameters may be used. The minimum size of a handdug caisson is 4' (1.2 m) in diameter, which often provides a greater crosssectional area of concrete than is required for the load imposed upon it.

Another important reason is that drilled caissons can be installed in considerably less time. There are many firms specializing in this type of work throughout the country.

A typical job, consisting of 100 caissons with 2'-0" (.6 m) shaft diameters, 5'-0" (1.5 m) bells, and 25' (7.6 m) deep, might be estimated as follows, assuming normal clay digging. First, compute the total volume of excavation, which in this case would be 418 cu. yds. (320 cu. m). To this figure add about 10% overbreak, making a total of 460 cu. yds. (352 cu. m). Assuming each drill rig can complete 5 caissons per day, the job would take 20 days. Allow approximately 10% extra time for weather and total of 23 days to complete the job.

Description	Hours	Rate	Total
Drill Rig	184.00	$ 186.25	$ 34,270.00
Drill Rig Operator	184.00	$ 37.89	$ 6,971.76
Oiler	184.00	$ 32.89	$ 6,051.76
Caisson Labor	368.00	$ 26.16	$ 9,626.88
Concrete (460 Cu.Yd 351.72 Cu.M)	460.00	$ 100.00	$ 46,000.00
Grand Total			$ 102,920.40
Cost per Cu.Yd. (418 Cu.Yd)			$ 246.22
Cost per Cu.Meter (319.6 CuM)			$ 322.03

These costs are for the complete caisson installation. Allowances must be added for reinforcing steel, additional labor based on union requirements, temporary or permanent casings if necessary, soil and concrete testing, surveying, disposal of excavated materials, and equipment move and set-up charges, if these items are included in the caisson contractor's contract, and overhead and profit.

Churn Drills. In areas where subsurface conditions are such that rotary drilling or augured caissons are difficult or impossible, the use of churn drills is usually effective. This caisson installation is sometimes known as the "drilled in" caisson method.

The cost for a 30" (750 mm) diameter drilled in caisson, complete in place, which includes the concrete, can range from $800 to $1,500 per lineal foot ($2620 to $4920 per m).

If bentonite is used in the installation of the caisson to produce the drillers mud, it is necessary to figure additional cost for the disposal of the bentonite.

This can be an additional $200 to $500 per lin. ft. ($600 to $1640 per m). Labor cost for drilling 4 l.f. (1.2 m) is as follows:

Labor			
Description	Hours	Rate	Total
Driller	8.00	$ 37.89	$ 303.12
Helper	8.00	$ 32.89	$ 263.12
Cost per 4 Lin.Ft. (1.2M)			$ 566.24
Cost per Lin.Ft.			$ 141.56
Cost per Lin. Meter			$ 464.44

Equipment				
Description	Hours	Rate	Total	
Churn Drill (Drill Rig)	8.00	$ 100.00	$	800.00
Cost per Lin.Ft.			$	200.00
Cost per Lin. Meter			$	656.17

02400 SHORING

Wood sheet piling and bracing is usually estimated by the square foot (or square meter or cubic meter taking the number of sq. ft. (sq. m) of bank or trench walls to be braced, plus an additional amount for penetration at the bottom of the excavation and an allowance for extension above the top, and then estimating the cost of the work at a certain price per sq. ft. (sq. m) for labor and material required.

Sheet piling or bracing is required where the soil is not self supporting or where the excavation banksides cannot be sloped back. There is usually considerable vibration adjacent to railroad tracks and roadways that can cause the banks to cave in. Unless there is ample space on all sides of the excavation, the banks should be sheet piled to avoid damage to streets, alleys, or adjacent buildings.

Trenches or foundation piers 5'-0" to 8'-0" (1.5 to 2.4 m) deep and 3'-0" to 4'-0" (.9 to 1.2 m) wide may require bracing or sheet piling, but this is much simpler than is ordinarily required for basement excavations.

The cost of this work will vary with the kind of soil, depth of excavation, amount of bracing or sheet piling required, and the method used to place. In ordinary excavations where no water is encountered, 2" or 3" (50 to 75 mm) square edge planks may be used, but if running sand or water is encountered, tongue and grooved planks should be used, as it is easier to keep them in line, and they are more water tight than square edge planks. It is customary to cut the bottom edge of each plank on a slight angle, so that in driving it is wedged against the preceding plank. The upper corners of the plank should also be cut off so that the effect of the driving will be concentrated along the vertical axis of the plank.

It requires 3 to 3-1/2 b.f.(.007 to .008 Cu.M) of lumber to sheet pile one sq. ft. of trench bank 5'-0" to 8'-0"(1.5-2.4 M)deep, but on large deep excavations it will require 6 to 8 b.f. (.014 to .019 Cu.M.) of lumber per sq. ft. of wall.

Important: The reader needs to know that the conversion from Board Feet to Cubic Meter has been omitted. To do a straight conversion from board feet to cubic meter is not a direct conversion. Since board feet is not really 1/12 of a cubic foot, a dressed piece of softwood lumber 2" x4" x 12' is actually 1 ½" x 3 ½" x 12'. Since metric uses actual sizes and board feet uses nominal sizes, a straight conversion that 424 Board Feet = 1 Cubic Meter will be misleading.

284

SITE WORK

Sheet piling is usually driven with a pneumatic hammer designed especially for this purpose. On very large jobs an ordinary pile driver is used for driving the sheet piling, but this is only on jobs of sufficient size to warrant the cost of a pile driver and crew.

Bracing and Sheet Piling Trenches and Piers. When excavating trenches and piers 5'-0" to 8'-0" (1.5 to 2.4 m) deep, it is not always necessary to sheathe the banks solid but two or three lines of braces placed along the sides of the trench (as illustrated) will often be sufficient.

Where necessary to sheathe the banks solid, it will require about 3 to 3-1/2 b.f. (.007 to .008 cu. m) of lumber per sq. ft. of trench bank. A method of determining the quantity of lumber required for sheet piling a trench similar to the one illustrated is as follows:

Compute the quantity of lumber necessary to sheet pile a trench 50'-0" (15.2 m) long and 7'-0" (2.1 m) deep, sheathed on both sides. To the depth of 7'-0" (2.1 m) must be added an amount for penetration. If 1'-0" (.3 m) will be sufficient, then figure 8'-0" (2.4 m) lumber. If, however, 2'-0" (.6 m) penetration is required, then 10'-0" (3 m) lumber must be figured leaving 1'-0" (.3 m) to extend above grade. Assuming 2'-0" (.6 m) penetration is required and 10'-0" (3 m) lumber must be used, this is equivalent to 100 lin. ft. (30.5 m) of sheathing 10'-0" (3 m) high, or 1,000 sq. ft. (93 sq. m), and the following lumber will be required:

Description			Bd.Ft. of Lumber	Cu. Meters
150 Pcs	2 " x 8 " - 10 '- 0"	Sheeting	2,000	4.717
12 Pcs	4 " x 6 " - 16 '- 0"	Stringers or Wales	384	0.906
26 Pcs	4 " x 6 " - 2 '- 0"	Braces spaced about 4'- 0" apart	104	0.245
Total lumber required			**2,488**	**5.868**
Feet of lumber, bd.ft. per sq.ft. sheathing, divided by 1,000			2.49	0.006
Feet of lumber, bd.ft. per sq.ft. trench bank, divided by 700			3.55	0.008

This is based on using 2" (50 mm) plank for sheathing, 2 rows of 4"x6" (100 x 150 mm) stringers or "wales" placed near the top and bottom as illustrated, and 4"x6" (100 x 150 mm) or 6"x6" (150 x 150 mm) braces. Trenches over 10'-0" (3.0 m) deep will require an additional line of stringers for each additional 4'-0" to 5'-0" (1.2 to 1.5 m) depth.

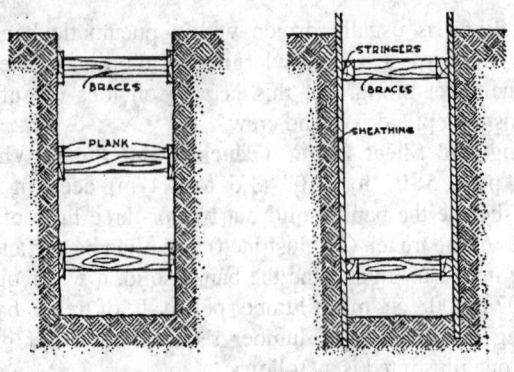

Method of Bracing
Trench Excavation

Method of Sheet Piling
Trench Excavation

Most contractors who do a volume of this work use trench jacks instead of 4"x6" or 6"x6" braces. These jacks cost from $300.00 to $35 .00 each and would not be economical for the contractor who does only an occasional job of this sort. The savings in material and labor is so slight that it is practically negligible.

Sheet Piling for Basements and Deep Foundations. On general basement excavation or large piers 8'-0" to 12'-0" (2.4 to 3.6 m) deep, where it is necessary to brace any or all of the outside banks, the following lumber is required for a bank 50'-0" (15 m) long, 8'-0" (2.4 m) deep, and containing 400 sq. ft. (37 sq. m) of bank to be sheathed:

Description			Bd.Ft. of Lumber	Cu. Meters
80 Pcs	2 " x 8 " - 10 '- 0"	Sheeting	1,067	2.517
6 Pcs	6 " x 8 " - 16 '- 0"	2 Lines of Stringers or Wales	384	0.906
10 Pcs	8 " x 8 " - 12 '- 0"	Top braces spaced about 5'- 0" apart	640	1.509
10 Pcs	8 " x 8 " - 6 '- 0"	Bottom braces spaced about 5'- 0" apart	320	0.755
10 Pcs	8 " x 8 " - 6 '- 0"	Bottom braces spaced about 5' apart	320	0.755
Total lumber required			**2,731**	**6.441**
Bd.Ft. per Sq.Ft. of Sheet Piling (divided by 500)			5.46	0.013
Feet of lumber, bd.ft. per sq.ft. trench bank, (divided by 400)			6.83	0.016

SITE WORK

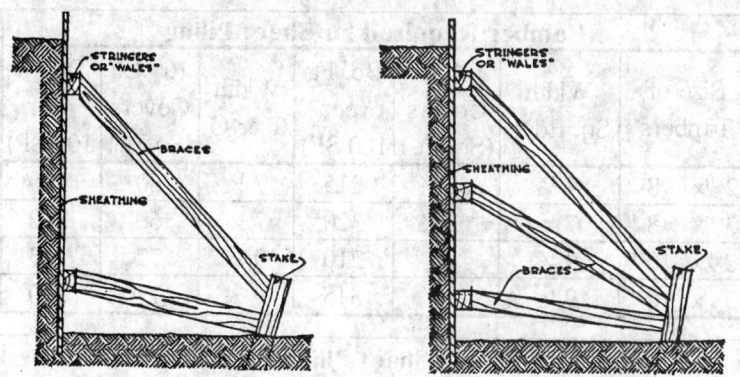

Methods of Bracing Basements or Deep Excavations

On deep excavations and piers 14'-0" to 20'-0" (4.3 to 6.1 m) below grade, 3" (75 mm) plank should be used for sheathing and either 6"x8" (150 x 200 mm) or 8"x8" (200 x 200 mm) lumber for stringers or "wales" and 8"x8" (200 x 200 mm) or 10"x10" (250 x 250 mm) timbers for bracing. Lines of stringers or "wales" should be spaced 4'-0" to 5'-0" (1.2 to 1.5 m) apart, depending upon the earth pressure.

An example of lumber required to sheet pile and brace a bank 50'-0" (15 m) long and 16'-0" (4.9 m) deep, containing 800 sq. ft. (74 sq. m) of bank, is as follows:

Description			Bd.Ft. of Lumber	Cu. Meters
80 Pcs	3 " x 8 " - 18 '- 0"	Sheeting	2,880	6.792
9 Pcs	8 " x 8 " - 16 '- 0"	3 Lines of Stringers or Wales	768	1.811
10 Pcs	8 " x 8 " - 16 '- 0"	Top braces spaced about 5'- 0" apart	853	2.012
10 Pcs	8 " x 8 " - 12 '- 0"	Intermediate braces spaced about 5'- 0" apart	640	1.509
10 Pcs	8 " x 8 " - 6 '- 0"	Bottom braces spaced about 5'- 0" apart	427	1.007
10 Pcs	8 " x 8 " - 6 '- 0"	Stakes spaced about 5'- 0" apart	320	0.755
Total lumber required			**5,888**	**13.887**
Bd.Ft. per Sq.Ft. of Sheet Piling (divided by 900)			6.54	0.015
Feet of lumber, bd.ft. per sq.ft. trench bank, divided by 800			7.36	0.017

Cost of lumber should be computed according to number of times it can be used on the job. Where labor hours are given, either carpenters or laborers may be used, as required.

287

Lumber Required for Sheet Piling

Size of Timbers	Width Sq. Edge	100 LF Covers (Sq. Ft.)	Bd. Ft. (per 100 SF)	Width T & G	100 LF Covers (Sq.Ft.)	Bd.Ft. (per 100 SF)
2 "x 8 "	7 ½	62 ½	215	7 ¼	60	220
3 "x 8 "	7 ½	62 ½	320	7 ¼	60	330
2 "x 10 "	9 ½	79	210	9 ¼	77	215
3 "x 10 "	9 ½	79	315	9 ¼	77	320

Wood Bracing and Sheet Piling Banks by Hand
Quanty and Production per 100 SF of Bank Braced

Class of Work	Depth in Feet	Bd.Ft. Lumber Required 100 Sq.Ft. Bank	No. Sq.Ft. Placed per 8-Hr. Day	Hours Placing 100 Sq.Ft.	Hours Removing 100 Sq.Ft.
Bracing Trenches	5 - 8	100 - 150	150 - 200	4 - 5	1 - 1½
Trench Sheet Piling	5 - 8	300 -350	80 - 90	9 - 10	2 ½ - 3
Trench Sheet Piling	10 - 15	325 - 375	65 - 75	11 - 12	2 ½ - 3
Basement Sheet Piling	8 - 12	675 - 725	40 - 50	17 - 18	5 - 5½
Basement Sheet Piling	14 - 20	750 - 800	35 - 45	20 - 21	5 ½ - 6

Cost of Bracing 100 S.F. (10 Sq.M) of Trench Bank With Plank Bracing

Description	Hours	Rate	Total
125 b.f. lumber *	------	$ 0.75	$ 93.75
Labor placing	4.50	$ 26.16	$ 117.72
Labor removing**	1.25	$ 26.16	$ 32.70
Cost per 100 sq. ft.			$ 244.17
Cost per sq. ft.			$ 2.44
Cost per sq. meter			$ 26.28

Lumber costs can be reduced if lumber is reused
** *Omit Labor removing Bracing or Piling if Left in Place*

Cost of Sheet Piling 100 S.F. (10 Sq.M) of Trench Walls 5' to 8' (1.5 - 2.4m)

Description	Hours	Rate	Total
325 b.f. lumber *	-----	$ 0.75	$ 243.75
Labor placing	9.50	$ 35.04	$ 332.88
Labor removing**	2.75	$ 26.16	$ 71.94
Cost per 100 sq. ft.			$ 648.57
Cost per sq. ft.			$ 6.49
Cost per sq. meter			$ 69.81

* Lumber costs can be reduced if lumber is reused
** Omit Labor removing Bracing or Piling if Left in Place

Cost of Sheet Piling 100 S.F. (10 Sq.M) of Trench Walls 10' to 15' (3.0 - 4.5m)

Description	Hours	Rate	Total
350 b.f. lumber *	-----	$ 0.75	$ 262.50
Labor placing	11.50	$ 35.04	$ 402.96
Labor removing**	2.75	$ 26.16	$ 71.94
Cost per 100 sq. ft.			$ 737.40
Cost per sq. ft.			$ 7.37
Cost per sq. meter			$ 79.38

* Lumber costs can be reduced if lumber is reused
** Omit Labor removing Bracing or Piling if Left in Place

Cost of Sheet 100 S.F. (10 Sq.M) of Sheet Piling for Basements or Foundations 8' to 12' (2.4 - 3.7m)

Description	Hours	Rate	Total
700 b.f. lumber *	-----	$ 0.75	$ 525.00
Labor placing	17.50	$ 35.04	$ 613.20
Labor removing**	5.25	$ 26.16	$ 137.34
Cost per 100 sq. ft.			$1,275.54
Cost per sq. ft.			$ 12.76
Cost per sq. meter			$ 137.30

* Lumber costs can be reduced if lumber is reused
** Omit Labor removing Bracing or Piling if Left in Place

Cost of Sheet 100 S.F. (10 Sq.M) of Sheet Piling for Basements or Foundations 14' to 20' (4.3 - 6.1m)			
Description	Hours	Rate	Total
775 b.f. lumber *	-----	$ 0.75	$ 581.25
Labor placing	20.50	$ 35.04	$ 718.32
Labor removing**	5.75	$ 26.16	$ 150.42
Cost per 100 sq. ft.			$1,449.99
Cost per sq. ft.			$ 14.50
Cost per sq. meter			$ 156.08

Lumber costs can be reduced if lumber is reused

** *Omit Labor removing Bracing or Piling if Left in Place*

Driving Sheet Piling by Compressed Air. The pneumatic pile driver is very effective for driving sheet piling. It is essentially a heavy paving breaker equipped with a special fronthead that adjusts for driving 2" to 3" (50 to 75 mm) piling. These machines weigh about 125 lbs. (57 kg) and will drive wood sheet piling into any ground that the piling can penetrate, such as sand, gravel, or shale, and any gradation between soils.

Driving speed varies from 2'-0" (0.61 m) per minute in hard clay or shale to 9'-0" (2.75 m) per minute in sand or gravel.

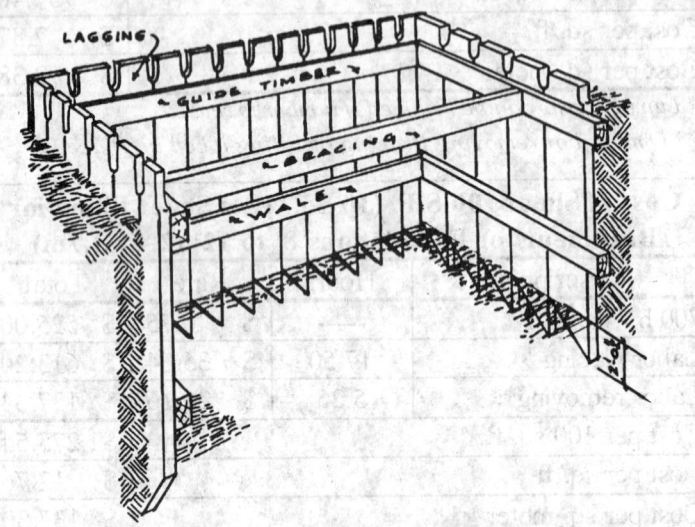

Method of Sheet Piling and Bracing Piers and Pits

Under average conditions, two workers together on one machine should drive 100 l.f. (30.5 m)-60 to 79 s.f. (5.6 to 7.3 sq.m)-per hr. at the following labor cost:

Drive 100 Lin. Ft. (30.48 Lin. M) Sheeting

Description	Hours	Rate	Total
Jack hammer person	1.0	$ 28.16	$ 28.16
Helper	1.0	$ 26.16	$ 26.16
Compressor * & tools	1.0	$ 47.00	$ 47.00
Cost per 100 l.f. (30.48 l.m.)			$ 101.32
per lin.ft.			$ 1.01
per lin.meter			$ 3.32

* Add for compressor engineer if required.

Sheet Piling 100 Sq. Ft. (9.29 Sq. M)
Trench Banks 5'- 0" to 8'- 0" (1.52 M to 2.44 M)

Description	Hours	Rate	Total
325 b.f. Lumber	-----	$ 0.75	$ 243.75
Jack hammer person	2.00	$ 28.16	$ 56.32
Laborer	4.00	$ 26.16	$ 104.64
Labor Removing	2.75	$ 26.16	$ 71.94
Compressor & tools	2.00	$ 47.00	$ 94.00
Compressor Engineer	2.00	$ 35.89	$ 71.78
Cost per 100 Sq.Ft.(9.29 Sq.M)			$ 642.43
per Sq. Ft.			$ 6.42
per Sq.meter			$ 69.15

Sheet Piling 100 Lin. Ft. (9.29 Sq. M)
Trench Banks 10'- 0" to 15'- 0" (3.04 M to 4.57 M)

Description	Hours	Rate	Total
350 b.f. Lumber	-----	$ 0.75	$ 262.50
Jack hammer person	2.00	$ 28.16	$ 56.32
Laborer	4.50	$ 26.16	$ 117.72
Labor Removing	2.75	$ 26.16	$ 71.94
Compressor & tools	2.00	$ 47.00	$ 94.00
Compressor Engineer	2.00	$ 35.89	$ 71.78
Cost per 100 Sq.Ft.(9.29 Sq.M)			$ 674.26
per Sq. Ft.			$ 6.74
per Sq.meter			$ 72.58

STEEL SHEET PILING

Steel sheet piling is used for supporting soil in large excavations, cofferdams, caissons, and deep piers or trenches, where wood sheet piling is impractical. It is driven with a pile driver, the same as wood, steel or concrete piles.

There are a number of different types of steel sheet piling on the market, but the following will give an idea of the common sizes and weights.

Common Size Steel Sheeting				
Section Number	Width (In.)	Flange Thickness (In.)	Weight (Lbs/LF)	Weight (Lbs/SF Wall)
PZ-22	22.00	0.375	40.3	22
PZ-27	18.00	0.375	40.5	27
PZ-35	22.60	0.6	66	35
PZ-40	19.70	0.6	65.5	40
PS-27.5	19.69	0.4	45.1	27.5
PS-31	19.69	0.5	50.9	31
AZ-12	26.38	0.335	44.42	20.22
AZ-17	24.80	0.335	45.96	22.24
AZ-25	24.80	0.472	61.49	29.74
AZ-34	24.80	0.669	77.61	37.54
AZ-46	22.83	0.709	89.1	46.82
Section Number	Width (mm)	Flange Thickness (mm)	Weight (Kg/m)	Weight (Kg/SM Wall)
PZ-22	559	9.5	60	107.4
PZ-27	457	9.5	60.3	131.8
PZ-35	574	15.24	98.2	170.9
PZ-40	500	15.24	97.5	195.3
PS-27.5	500	10.2	67.1	134.3
PS-31	500	12.7	75.7	151.4
AZ-12	670	8.5	66.1	98.7
AZ-17	630	8.5	68.4	108.6
AZ-25	630	12	91.5	145.2
AZ-34	630	17	115.5	183.3
AZ-46	580	18	132.6	228.6

Estimating the Quantity of Steel Sheet Piling. When estimating the quantity of steel sheet piling required for any job, take the entire girth of the basement, cofferdam, or piers to be sheet piled. This gives the number of lineal feet required.

Example: Suppose you have a basement 100' x 125' (30.5 x 38.1 m) to be sheet piled. Adding the four sides of the excavation, 100 + 100 + 125 + 125 = 450 lin. ft. (137.2 m). Using PZ-27 (Skyline Industries) sheet piling 18" (1.5 ft., 450 mm) wide, 450'-0" (147.2 m) divided by 1.50' (0.45 m) = 300. It requires 300 pcs. to sheet pile the excavation. Using sheet piling 30'-0" (9.1 m) long, 300 x 30 = 9,000 lin. ft. (300 x 9.1 m = 2730 m) weighing 40.5 lbs. per lin. ft. = 364,500 lbs. (165,337 kg).

A shorter method is as follows: 450 lin. ft. (137.2 m) around excavation multiplied by 30'-0" (9.1 m) long, equals 13,500 sq. ft. (1248.52 m) of piling at 27 lbs. per sq. ft. equals 364,500 lbs. (165,337 kg).

Either method is satisfactory. But it is necessary to know the number of pieces to be driven in order to compute the labor costs accurately.

Labor Driving Steel Sheet Piling. The labor cost handling and driving steel sheet piling will vary with the size of the job, length of piling, and kind of soil, together with conditions encountered on the job, as described under "Wood Piles".

Example: Find the cost of driving 300 pcs. of PZ-27 sheet piling 30'-0" (9.1 m) long.

Drive 300 Pcs Sheet Piling	
Description	Cost
Trucking pile driver, yard to job	$ 5,000.00
Setting up pile driver, 1/2 day for crew	$ 790.16
300 pcs. PZ-27 piling, 364,500 lbs. (165,337 kg) at $0.60 per Lb	$218,700.00
Driving piling (30 per day) 10 days @ $1,594.96	$ 15,949.60
Fuel, oil, misc. supplies, 10 days @ $150.00	$ 1,500.00
Dismantling pile driver at completion, 1/2 day	$ 790.16
Trucking pile driver, job to yard	$ 5,000.00
Equipment Rental, Leads, Hammer, Fuel. Oil Grease,	
16 days @ $2,100.00 per day	$ 33,600.00
Cost 300 pcs. sheet piling	$281,329.92
per pc.	$ 937.77
per s.f. (13,500 sq. ft. or 1254 sq. m)	$ 20.84
per sq.meter	$ 224.32

Sq. ft. (Sq.Meter) costs will vary with weight of piling

Driving Piles			
Description	Hours	Rate	Total
Foreman	8	$ 39.04	$ 312.32
Pile driver engineer	8	$ 37.89	$ 303.12
Oiler	8	$ 35.89	$ 287.12
4 Pile drivers (pile bucks)	32	$ 35.04	$1,121.28
Total labor cost			$2,023.84

02500 SITE DRAINAGE

Site drainage includes such items as subdrainage, foundation and underslab drainage, drainage structures, sanitary and storm drainage piping, and dewatering and wellpoints. Some of these will be subcontracted directly with specialists. Most will probably be included in the plumbing subcontract, while others may involve local utilities. Some items the contractor may elect to handle directly.

Drainage trenches must have pitch. To find the cubic yards of earth excavation, first determine the average depth by multiplying the pitch by the required length of run, divide by two and add this to the depth at the start of the run. The total length x required width x the average depth = cubic feet (cubic meters). Divide number of cu. ft. by 27 to find cubic yards.

Subsoil drains are usually vitrified clay pipe, set with lowest point at the same elevation as the bottom of the footing, pitched a minimum of 6" (150 mm) in 100 l.f. (30.5 m), laid 1/4" apart and joint covered with 15# (6.8 kg) asphalt felt or #12 (5.4 kg) copper or aluminum mesh. Four-inch (100 mm) pipe is minimum, but 6" (150 mm) pipe is preferred. One laborer can lay approximately 10 l.f. (3.05 m) of 4" (100 mm) pipe per hour, 8 l.f. (2.44 m) of 6" 150 mm) pipe. The 4" (100 mm) pipe costs about $1.00 per l.f. (3.28 per m); the 6" (150 mm) pipe costs $1.40. Excavation is extra, and it is assumed the trench is properly sloped with a firm bottom before pipe is installed.

All sharp turns are formed with fittings, which cost $2.50 for 4" (100 mm) elbows, $4.75 for 6" (150 mm). Wide angles can be made by beveling tile ends into easy radius bends.

The clay drain tile is laid on a thin bed of washed gravel. After the joints have been covered with felt or a filter cloth, more gravel is placed around and on top of the drain tile to allow water filtration into the drain tile itself. Approximately 0.75 to 1.0 c.f. per l.f. (0.069 to 0.09 cu.m per m) of gravel of drain is required for proper filtration.

In recent years, vitrified clay drain tile is being displaced by the use of 4" (100 mm) plastic tubing that is already perforated by the manufacturer. This 4" (100 mm) tubing comes in 25', 50', and 100' (7.6, 15.2, and 30.4 m) rolls. It is very flexible, and since you are not handling small individual pieces as with the vitrified clay, this particular material is installed very rapidly. The joints are made by a coupling sleeve very much like that used with plumbing PVC piping. It is wise to lay a strip of filter cloth above the plastic tubing prior to installing the washed gravel to keep all sandy fines from entering the plastic tubing and creating a stoppage. All other grading preparation and gravels are installed around this particular material the same as for vitrified clay tile.

Subsoil drains, whether perimeter or underslab, can be connected to an existing storm sewer (in some localities, to a sanitary sewer), can be trenched to an outfall at grade level away from the structure (depending on the existing terrain), or connected to a sump pump area within the lower level of the building. If subsoil drainage piping terminates at a sump pump, a submergible pump with an automatic float control switch and discharge piping will be needed.

DRAINAGE STRUCTURES

There are a number of types of drainage structures required in storm drain piping systems. These structures can be of masonry, cast-in-place concrete, or precast concrete.

Curb inlets and yard drains are one type of structure that is used to collect surface water run-off and discharge it into the storm water piping system. The size of the inlet is determined by the area of run-off, as well as the size of the discharge pipe leading away from the inlet structure itself. An average price for a curb inlet drainage structure is $250 to $300 per vertical lineal foot ($820-984 per vertical meter), and the average price of a yard drainage structure is $175 to $210 per vertical lineal foot ($574-689 per vertical meter).

Storm drainage manhole structures are required at certain intervals along a storm pipe line, depending on the local codes, and whenever a storm drainage pipe system changes direction. The majority of manholes are precast concrete. It is much quicker to set an entire manhole in the course of the day instead of keeping the open pit shored while constructing a cast-in-place concrete manhole and bottom chamber. Once the manhole structure is in place and the piping connected, the bottom of the manhole structure is channeled using brick and mortar to aid the water flow from the inlet pipe to the outlet pipe.

Headwalls and endwalls are usually cast-in-place reinforced concrete walls at the ends of pipe, where either water enters the pipe (headwall) or where the water discharges from the pipe (endwall). Quite often at a headwall, there is a concrete flume entrance slab, and there might be a trash rack (a grillage screen) at the entrance to collect debris.

At the endwall an outfall slab or rip-rap area is installed that prevents the discharge water from eroding the soil away. The price of headwalls and endwalls varies tremendously, because each is sized individually depending upon the size of the inlet or discharge pipe and the amount of earth and height of the earth bank above the pipe that is being retained.

STORM PIPING

Concrete, "HOBAS", PVC, Ductile Iron Pipe and corrugated metal are often used for piping other than sewer. The relative material costs per lineal foot (meter) are about as follows:

Storm Piping				
Size of Pipe (In.)	Concrete per LF	Weight of Conc. Pile (Lbs)	Corrugated (per LF) 20' Lengths	PVC Pipe (per LF) 10' Lengths
6	$ 4.20	-----	$ 3.12	$ 2.95
8	$ 4.60	-----	$ 6.93 **	$ 5.24
10	$ 5.10	-----	$ 8.39 **	$ 7.79
12 *	$ 12.20	93	$ 10.12 **	$ 8.75
15 *	$ 15.10	127	$ 12.50 **	$ 12.25
18 *	$ 15.20	168	$ 15.71 **	$ 14.40
24 *	$ 22.20	265	$ 19.28 **	-----
30 *	$ 30.90	384	$ 24.99 **	-----
36 *	$ 43.40	524	$ 36.89 **	-----
42 *	$ 59.50	686	$ 41.06 **	-----
48 *	$ 73.80	867	$ 56.53 **	-----
60 *	$ 118.40	1295	$ 73.19 **	-----
72 *	$ 167.80	1811	$ 111.27 **	-----
84 *	$ 305.80	2409	-----	-----
96 *	$ 368.90	3090	-----	-----

*Reinforced Concrete   ** Coated

Storm Piping				
Size of Pipe (mm)	Concrete per m	Weight of Conc. Pile (Kg)	Corrugated Uncoated (per m) 6m Lengths	PVC Pipe (per m) 3m Lengths
150	$ 13.78	-----	$ 10.23	$ 9.68
200	$ 15.09	-----	$ 22.72	$ 17.18
250	$ 16.73	-----	$ 27.52	$ 25.57
300 *	$ 40.03	42.18	$ 33.19	$ 28.70
375 *	$ 49.54	57.61	$ 40.99	$ 40.19
450 *	$ 49.87	76.20	$ 51.54	$ 47.24

SITE WORK

Storm Piping (continued)				
Size of Pipe (mm)	Concrete per m	Weight of Conc. Pile (Kg)	Corrugated Uncoated (per m) 6m Lengths	PVC Pipe (per m) 3m Lengths
600 *	$ 72.83	120.20	$ 63.25	-----
750 *	$ 101.38	174.18	$ 81.99	-----
900 *	$ 142.39	237.69	$ 121.03	-----
1050 *	$ 195.21	311.17	$ 134.69	-----
1200 *	$ 242.13	393.27	$ 185.45	-----
1500 *	$ 388.45	587.41	$ 240.11	-----
1800 *	$ 550.52	821.47	$ 365.04	-----
2100 *	$ 1,003.28	1092.72	-----	-----
2400 *	$ 1,210.30	1401.62	-----	-----

*Reinforced Concrete

One cannot choose the type of pipe by price alone. While the specifications and job conditions usually dictate the selection, wage costs can also bear heavily. Lighter weight pipe will lay much faster and can be handled on the site much more easily.

For 12" (300 mm) pipe, a 4-worker crew will lay about 100 l.f. (30.5 m) of concrete pipe and 200 l.f. (61 m) of corrugated metal per day. For 36" (900 mm) pipe, a crew will lay about 20 l.f. (6.1 m) of concrete pipe and 60 l.f. (18.3 m) of corrugated metal per day.

WELLPOINT SYSTEM OF DE-WATERING

The following is general information only. For specific applications consult a local de-watering contractor. Wellpoint systems are well known for their ability to de-water water bearing soils, such as sand and gravel, so that excavation and construction of foundations can proceed in the dry.

In addition, wellpoint systems are used extensively in such work as soil stabilization, pressure relief for dams and levees, and for water supply for municipalities and industrial plants.

A wellpoint system in construction work is usually a series of properly sized wellpoints, surrounding or paralleling the area to be de-watered and connected to a header pipe by means of risers and swing piping. Header piping, in turn, is connected to one or more centrifugal pumps depending on the volume of water that must be handled.

Wellpoints are usually self-jetted into place to the correct depth and at the proper spacing to meet requirements. Some soil conditions, however, require pre-drilling or "hole-punching" before the wellpoints can be installed.

The discharge from the pumps should be piped to an area where it will not interfere with construction. The entire installation should be located so that it will interfere as little as possible with other divisions of the work.

This system is usually installed in advance of the excavation, but where the water level is low; it is sometimes possible that a portion of the excavation can be at a cheaper rate when done prior to the installation of the system.

Any de-watering problem that is beyond the scope of ordinary pumping, or which the contractor or engineer thinks can be done more economically with a wellpoint system, should be referred to the engineering department of a wellpoint company. In this way the problem will receive expert analysis, at no cost to the contractor or engineer. It is essential for most contractors to obtain expert advice on the layout and installation of wellpoint systems, because frequently, a relatively small excavation with unusual soil presents greater de-watering difficulties than a larger, deeper job in a well-graded medium sand.

After carefully considering the factors involved, such as soil characteristics, site accessibility, hydrology of area, size and depth of area, total dynamic head involved, excavation and construction schedules, labor and working conditions, and power facilities, the wellpoint company can make a layout recommendation for the job.

Layout recommendations are usually accompanied by a quotation giving rental rates and charges for all equipment required, together with an estimate of the labor hours needed to install and remove the system, fuel and lubrication required per operating day, transportation charges for the equipment to and from the job, and daily rates for salary and expenses of company demonstrator. Companies specializing in this type of work will furnish rental quotations on a "sufficient equipment" basis, i.e., a fixed rental rate regardless of the amount of equipment actually required to do the job. Some of these companies will submit lump sum contract proposals for de-watering projects covering furnishing, installation operation, maintenance and removal of all de-watering equipment with a guaranteed result.

It is sound practice to have a demonstrator from the wellpoint company on the job site to supervise the initial installation of the system. The time and expenses is more than compensated for by his working knowledge of wellpoint systems and their installation.

Operation of a wellpoint system is usually performed on a 24-hr. day, 7 day per week basis and must continue until all work, depending on dry conditions, is completed.

The labor cost of operating the system must be determined by the contractor and is based on the length of time operation is estimated to be required, figuring around-the-clock operation.

From the above information the contractor or estimator should be able to arrive at a lump sum figure for the required wellpoint system to be used in his estimate.

As an example, take the case of a job in which an underground tank was to be constructed. The test borings indicated the soil to be sand with a ground water level at 6'-0" (1.82 m) below the surface. The excavation for

SITE WORK

the circular tank was an average of 84'-0" (25.6 m) in diameter and 18'- 6" (5.6 m) deep. To maintain dry conditions, the water level had to be lowered at least 12'-6" (3.8 m). It was also estimated that a dry condition would have to be maintained for a period of 3 months.

The following sample estimate shows how a detailed de-watering estimate might be put together.

	Hours	Rate	Total
Equipment Rental, First Month			
52 wellpoints, 2" (50 mm)	---	$ 16.00	$ 832.00
260 l.f. (79 m) header pipe, 6" (150 mm)	---	$ 1.80	$ 468.00
50 l.f. (15 m) discharge pipe, 6" (150 mm)	---	$ 1.00	$ 50.00
2 valves, 6" (150 mm)	---	$ 40.00	$ 80.00
1 wellpoint pump	---	$ 950.00	$ 950.00
1 jetwell pump	---	$ 1,025.00	$ 1,025.00
Equipment Rental, Second Month			
52 wellpoints, 2" (50 mm)	---	$ 10.00	$ 520.00
260 l.f. (79 m) header pipe, 6" (150 mm)	---	$ 0.90	$ 234.00
50 l.f. (15 m) discharge pipe, 6" (150 mm)	---	$ 0.50	$ 25.00
2 valves, 6" (150 mm)	---	$ 25.00	$ 50.00
1 wellpoint pump	---	$ 750.00	$ 750.00
1 jetwell pump	---	$ 850.00	$ 850.00
Equipment Rental, Third Month			
52 wellpoints, 2" (50 mm)	---	$ 10.00	$ 520.00
260 l.f. (79 m) header pipe, 6" (150 mm)	---	$ 0.90	$ 234.00
50 l.f. (15 m) discharge pipe, 6" (150 mm)	---	$ 0.50	$ 25.00
2 valves, 6" (150 mm)	---	$ 25.00	$ 50.00
1 wellpoint pump	---	$ 750.00	$ 750.00
1 jetwell pump	---	$ 850.00	$ 850.00
2 weeks rental, 200 l.f. (61 m) jetting hose	---	$ 200.00	$ 400.00
Laborers	160	$ 19.13	$ 3,060.80
Supervisor, 5 days	---	$ 200.00	$ 1,000.00
Supervisor's expenses	---	$ 500.00	$ 500.00
Operating Expense	Hours	Rate	Total
Diesel fuel 91 days at 40 gals. (151 liters) per day	---	$ 4.00	$ 14,560.00
Lubricating oil, 91 days at 1 gal. (3.78 liters) per day	---	$ 5.00	$ 455.00
Pump operator, 3 shifts, regular time	1,560	$ 37.89	$ 59,108.40
Pump operator, 3 shifts, overtime	624	$ 56.84	$ 35,465.04
Removal of Equipment			
Labor	60	$ 26.16	$ 1,569.60
Transportation to and from job, 16,000 lbs. (7258 kg) at $4.00 per 100 lbs. (45 kg)	---		$ 640.00
Total cost of wellpoint system			$125,021.84

Add for insurance, social security, unemployment compensation, overhead and profit.

The jetwell pump is a double-duty unit and is connected into the system as a stand-by wellpoint pump after installation of wellpoints has been completed.

DEEP WELL SYSTEM OF DE-WATERING

A cheaper alternative to the wellpoint system of de-watering is the use of deep wells, which are 18" to 36" (450 to 900 mm) diameter shafts drilled into the ground and lined with a perforated, corrugated metal pipe in which an electric submergible pump is installed. This pump can be discharged directly onto the ground or into another collector pipe discharge system.

The size (diameter and depth) of the drilled shafts, the distance spaced apart, and the size of the electric submergible pump is determined by the depth of the planned excavation, the elevation of the water table, and the subsurface water flow.

When installing the submergible pump, the use of rigid PVC piping is preferred over flexible canvas discharge hoses. Depending on the depth of the shaft, a pump will be required to have sufficient head pressure to pump the water up to ground elevation. Therefore, any restrictions in collapsed discharge piping would create an additional burden for the pump and losing pump efficiency.

Pump costs vary widely depending upon the size of the pump, the head pressure rating, and the GPM capacity. Pumps can be purchased or rented, and the decision should be based on an evaluation of both methods for the amount of time that the pumps will be in operation.

EJECTOR SYSTEM OF DE-WATERING

The ejector method has a distinct advantage over the wellpoint system. Ejectors are not limited by depth, eliminating the need for multiple stages of wellpoints. When conditions are right, ejectors can cost less than deep wells and can be more effective, but to select the right system of de-watering for a specific project, one must have a good knowledge of the project and a thorough understanding of the various methods.

Installation of ejectors is the same as for wellpoints. The system works by supplying water down a pipe to create a vacuum and then pumping up the ejected water and the ground water. The two-pipe ejector system, which is the preferred method, works by pumping water into one pipe and drawing it out the other pipe.

One disadvantage with ejectors is the need for larger pumps with higher operating costs, because one is pumping out more water, both ejected water and ground water.

Ground water quality can create discharge problems. It is always advisable to do a water analysis before deciding on and designing the de-watering system.

Upon completion of the project and termination of the need for the dewatering deep well system, the pump and piping are extracted from the

SITE WORK

hole and sand or gravel material is used to fill the shaft to prevent potential hazards of future cave-in.

The following two charts give the cost per vertical lineal foot for drilled shafts and perforated corrugated metal pipe liner.

Drilled Shafts			
Dia. of shaft, in.	Price per l.f.	Dia. of shaft, mm	Price per m
18	$ 14.00	450	$ 45.93
24	$ 17.00	600	$ 55.77
30	$ 22.00	750	$ 72.18
36	$ 30.00	900	$ 98.43
42	$ 38.00	1,050	$ 124.67

Installation of Perforated Corrugated Pipe			
Dia. of shaft, in.	Price per l.f.	Dia. of shaft, mm	Price per m
18	$ 16.00	450	$ 52.49
24	$ 24.00	600	$ 78.74
30	$ 30.00	750	$ 98.43
36	$ 46.00	900	$ 150.92
42	$ 58.00	1,050	$ 190.29

02550 SITE UTILITIES

Gas, water and electricity are usually brought to the property line or building by the local utility company. From there, the gas and water are normally picked up by the piping contractor doing the building work. Costs on this work are discussed in Chapter 15, Mechanical. The distribution of utilities on new subdivision work is a specialized field subject to many code, health and sanitation requirements.

Installing Service Pipes. Many public utility companies, instead of breaking up sidewalks and digging trenches, are installing service pipes by means of portable compressors and pneumatic tools. With directional drilling as a means and method a contractor can drill in most all soil conditions including rock directly into cellars and basements. In long runs the contractor may elect to go from a starter pit to a receiving pit, thereby reducing a long run.

One method of directional drilling consists of using a drill bit to remove the material (make the opening) and directly behind pull the pipe. The drill bit is then removed and the pipe left in place.

DRILLING WATER WELLS

Pollution of streams, rivers, and lakes has led many municipalities to look elsewhere for sources of water. Many have had to install deep wells as an alternate source of water. Industries sometimes also have to go to wells for their water supply where other sources are not available or are inadequate for their demands.

Large quantities of water are usually found in glacial deposits of sand and gravel. The quality of water from a glacial deposit is usually harder and of lower temperature than water derived from sand rock.

When it is necessary to locate large quantities of water for a town or city, it is customary to explore first by drilling a test hole to locate a water bearing strata, and to determine if there is an ample supply of water at that level. If a strata is found that is determined to be inadequate, blank pipe is then driven to the water bearing strata and a well screen is set in it.

Water traveling through a well screen has a tendency to precipitate the solids from the water on the screening, which eventually will restrict the water flow through it. Voids in the sand and gravel surrounding the pipe and screen may also become filled with the precipitates from the water. Water with a high mineral content may close the screen openings in a relatively short period of time, while water with low mineral content may not restrict the flow of water appreciably for many years. When screens do become clogged up with precipitates they sometimes can be opened by treating them with acid. If this process is not effective, it might be necessary to install a new well.

Rock wells are installed by driving a pipe through the overburden above the rock to the rock surface and continuing by drilling through the rock to the desired depth. It is sometimes possible to increase the quantity of water from a rock well in dense rock by shooting and fracturing the rock in the vicinity of the well. If the water bearing rock is already loose, porous, and creviced, shooting will probably not increase the water quantity.

Wells which extend into glacial sand and gravel deposits will cost approximately $6.00 to $10.00per inch (25 mm) in diameter per lineal foot of vertical depth. For example, a 12" (300 mm) diameter well would cost from $72.00 to $120.00 per lin. ft. ($236 to $394 per m) of depth.

Wells drilled into rock formations generally cost approximately $3.40 to $4.50 per inch (25 mm) of diameter per lin. ft. of depth. A 6" (150 mm) diameter rock well would cost about $40.00 to $54.00per lin. ft. of depth ($131.00 to $177 per m). In the rock itself a pipe is not used, but if there is 100 feet (30.5 m) of earth or sand and gravel above the rock surface, it is necessary to include the cost of the pipe for this distance.

For wells with a capacity of 25 to 250 gallons (95 to 950 liters) per minute, submersible, multistage type pumps are generally used. Turbine type pumps are normally used for wells of more than 250 gallons (950 liters) per minute capacity.

302

SUBMERSIBLE PUMP INSTALLATION

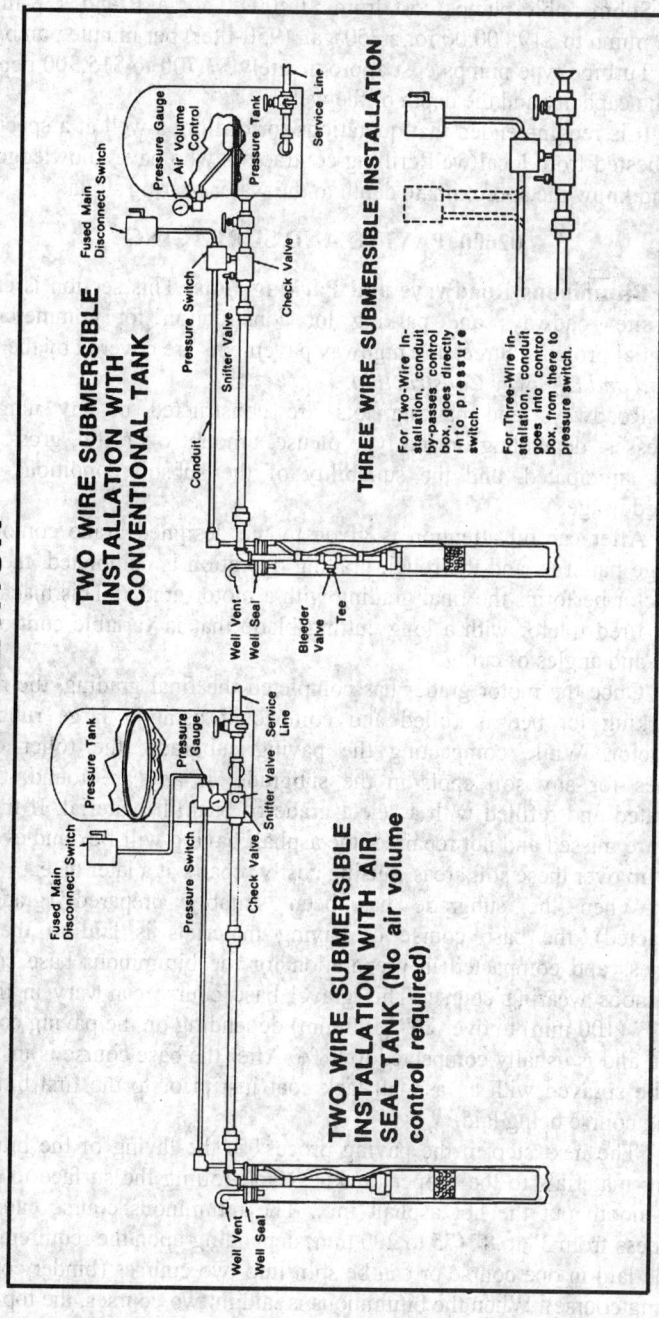

TWO WIRE SUBMERSIBLE INSTALLATION WITH CONVENTIONAL TANK

THREE WIRE SUBMERSIBLE INSTALLATION

For Two-Wire Installation, conduit by-passes control box, goes directly into pressure switch.

For Three-Wire Installation, conduit goes into control box, from there to pressure switch.

TWO WIRE SUBMERSIBLE INSTALLATION WITH AIR SEAL TANK (No air volume control required)

Submersible pumps cost from $1,700.00 for a 10-gal. (38-liter) per minute pump to $19,000.00 for a 250-gal. (950-liter) per minute pump.

Turbine type pumps cost approximately $3.700 to $15,300 depending upon the capacity and the depth of the well.

It is recommended that quotations for drilling a well at a specific site be requested from local well drilling contractors who have knowledge of the area and know the approximate depth to the water bearing strata.

02600 PAVING AND SURFACING

Bituminous Roadways and Parking Lots. This section is confined to on-site roadways and parking lot construction for commercial and residential projects. Street and highway pavements are covered by the chapter on *Road and Highway Construction*.

Roadways and parking lots are constructed in any number of thicknesses, depending on the type of use, amount of traffic, gross vehicle weight anticipated, and the suitability of the subsoil conditions for the intended usage.

After careful attention is given to the designed grade contours and drainage patterns, and the rough grading operation is completed, the paving contractor performs the final grading with a motor grader. This machine is a rubber-tired tractor with a long cutting blade that is variable controlled for depths and angles of cut.

Once the motor grader has completed the final grading, the roadway or parking lot bed is rolled and compacted using a large rubber-tired compactor. While compacting the paving subgrade, the roller operator watches for any soft spots in the subgrade. If any are found, they are excavated and refilled with a select grade of backfill material. If these soft spots are missed and not repaired, the asphalt paving will flex and eventually break up over these soft areas causing costly repairs at a later date.

When the subgrade has been suitably prepared (graded and compacted), the base course of paving materials is laid to the proper thickness and compacted in preparation for the bituminous base course or bituminous wearing course. This gravel base course can vary in thickness from 4" (100 mm) to over 20" (500 mm) depending on the paving composite design and is usually compacted to 98%. After the base course is installed, it may be sprayed with an asphalt tack coat just prior to the first bituminous paving course being laid.

The next step in the paving process is the laying of the bituminous paving materials to the proper thickness and rolling the surface to compact and smooth out the hot asphalt mix. The bituminous course can vary in thickness from 3" to 8" (75 to 200 mm) depending upon the requirements and can be laid in one course or can be split into two courses (binder course and wearing course). When the bituminous is laid in two courses, the top wearing course is usually a finer aggregate mix allowing for a much smoother compacted top surface.

304

SITE WORK

Laying the bituminous paving in two courses will always cost a premium. However, there are some distinct advantages. It allows early paving of a work site without construction abuse of a finished surface. Should soft spots occur, they can be fixed prior to the finish wearing course installation. It allows sufficient time to completely check all areas for low spots (birdbaths) that collect water, and these can then be adjusted in the top wearing course thickness.

When estimating quantities of bituminous paving and base courses for a project, it is customary to figure all quantities in square yards (sq. meters): length in feet x width in feet = area in square feet divided by 9 = square yards; length in meters by width in meters = square meters.

Concrete Curb, Gutter & Paving

This section is not intended to cover road building but only small curb and gutter jobs, and small concrete paving projects. Large jobs of this kind require a vast amount of equipment, and the costs vary widely according to the amount of cut and fill required and the method of receiving and dispersing materials.

The item of plant and equipment should include the cost of transporting equipment to the job and removing it at completion and includes items such as bulldozers, loaders, power rollers, trucks, and wood or metal forms for curb and gutter. These items should be estimated on the basis of total transportation costs both ways and a rental charge for plant and equipment. The total should be divided by the number of sq. yds. (sq.m) of pavement in the job. It is customary to submit bids on paving at a price per sq. yd. (sq.m).

Rolling Streets and Roads. After the excavating and grading has been completed, the subgrade should be rolled to provide a solid foundation to receive the concrete base. A power roller should roll 130 to 150 sq. yds. (109 to 125 sq.m) of subgrade per hr. at the following cost per 100 sq. yds. (84 sq.m):

Compaction Rolling			
Description	Hours	Rate	Total
Operator	0.7	$ 37.89	$ 26.52
Power roller	0.7	$ 74.63	$ 52.24
Cost per 100 sq. yds.			$ 78.76
per sq. yd.			$ 0.79
per 100 sq.m			$ 94.20
per sq. m			$ 0.94

Hauling Sand, Gravel, Crushed Stone, Etc. The cost of hauling paving materials will vary with the length of haul from pit or cars to point of delivery. Tables of hauling time are given under "Excavating".

305

Concrete Curb and Gutter

When estimating the cost of forms for concrete curb or curb and gutter, the size and shape must be considered and the amount of materials estimated.

Combined curb and gutter is usually 12" (300 mm) high at the outside edge (including pavement thickness), 24" to 30" (600 to 750 mm) wide and 6" to 7-1/2" (150 to 187.5 mm) thick.

Separate concrete curb is usually 6" (150 mm) wide at the top, 8" to 9" (200 to 225 mm) wide at the bottom, and 24", 30", or 36" (600, 750, or 900 mm) high.

Forms for Concrete Curb and Gutter. When placing wood forms for separate concrete curb, two mechanics and a helper should set forms for 100 to 110 l.f. (30.5 to 33.5 m) of 24" (600 mm) curb per 8-hr. day, at the following labor cost per 100 l.f. (30 m):

Form Curb and Gutter			
Description	Hours	Rate	Total
Mechanics	15	$ 37.89	$ 568.35
Helpers (laborers)	7.5	$ 26.16	$ 196.20
Cost per 100 l.f.			$ 764.55
per l.f.			$ 7.65
per m			$ 25.08

Two mechanics and a helper should set forms for 90 to 100 l.f. (27.5 to 30.5 m) of 30" (750 mm) concrete curb per 8-hr. day, at the following labor cost per 100 l.f.:

Form Curb and Gutter			
Description	Hours	Rate	Total
Mechanics	17	$ 37.89	$ 644.13
Helpers (laborers)	8.5	$ 26.16	$ 222.36
Cost per 100 lin.ft.			$ 866.49
per lin.ft.			$ 8.66
per m			$ 28.43

When setting wood forms for combined concrete curb and gutter, two mechanics and a helper should set forms for 80 to 90 l.f.(24.4 to 27.5 m) of curb and gutter per 8-hr. day at the following labor cost per 100 l.f. (30 m):

306

Form Curb and Gutter			
Description	Hours	Rate	Total
Mechanics	18.8	$ 37.89	$ 712.33
Helpers (laborers)	9.4	$ 26.16	$ 245.90
Cost per 100 lin.ft.			$ 958.24
per lin.ft.			$ 9.58
per m			$ 31.44

Wood forms for concrete curb and gutter are ordinarily built of 2" (50 mm) lumber. Including all bracing, it requires 2-1/4 to 2-1/2 b.f. of lumber per s.f. of forms (0.004 to 0.005 cu.m per sq.m). This cost will run from $0.18 to 0.22 per s.f. ($1.93 to $2.36 per sq.m) and should be added to the above setting costs, taking into consideration the number of uses expected from the forms. These forms can be used many times, if properly taken care of.

Placing Ready-mix Concrete for Curb and Gutter. A two-worker crew should place about 1 c.y. (0.76 cu.m) of concrete per hr. at the following labor cost per c.y.:

Placing Concrete			
Description	Hours	Rate	Total
Mechanics	2	$ 37.89	$ 75.78
Cost per Cu.Yd.			$ 75.78
Cost per Cu.M			$ 99.11

Concrete Roads and Pavements

Concrete roads and pavements could be thicker in the center than at the sides, so it will be necessary to obtain the average thickness of the pavement when estimating quantities of concrete materials.

Some pavements have uniform thickness. Others have uniform center section and thickened edges. In each case the area of the cross section must be determined and this is multiplied by the length to find the volume of concrete required.

Labor Placing Edge Forms for Concrete Paving. Wood edge forms or concrete paving usually consist of 2" (50 mm) plank of a width equal to the slab thickness, set to line and grade, secured to wood stakes, and braced.

Two mechanics and a helper should set to line and grade from 250 to 300 l.f. (76.2 to 91.4 m) of edge forms per 8-hr day at the following labor cost per 100 l.f. (30.5 m):

Edge Forms			
Description	Hours	Rate	Total
Mechanics	3	$ 37.89	$ 113.67
Helpers (laborers)	1.5	$ 26.16	$ 39.24
Cost per 100 lin.ft.			$ 152.91
per lin.ft.			$ 1.53
per m			$ 5.02

The above costs are based on the forms being straight or set to long radius curves. If sharp curves are required, the form must be either made from 1" (25 mm) boards, or the back of a 2" (50 mm) plank must be kerfed. In either case, the stakes must be set much closer together.

Concrete paving edge forms set to short radius curves will cost 50% to 100% more than straight forms.

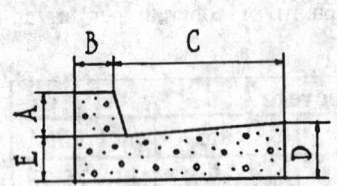

Combined
Curb and
Gutter

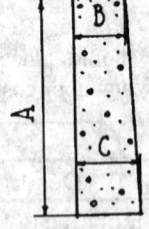

Separate
Concrete
Curb

Approximate Sq. Yds. (Sq. M) of Concrete Pavement Obtainable from One Cu.Yd. of Concrete			
Thickness Inches	No. of Sq.Yds.	Thickness mm	No. of Sq.Meters
3.0	12.0	76.2	10.03
3.5	10.3	88.9	8.61
4.0	9.0	101.6	7.52
4.5	8.0	114.3	6.69
5.0	7.2	127.0	6.02
5.5	6.5	139.7	5.43
6.0	6.0	152.4	5.02
6.5	5.5	165.1	4.60
7.0	5.1	177.8	4.26
7.5	4.8	190.5	4.01
8.0	4.5	203.2	3.76
8.5	4.2	215.9	3.51
9.0	4.0	228.6	3.34
9.5	3.8	241.3	3.18
10.0	3.6	254.0	3.01
10.5	3.4	266.7	2.84
11.0	3.3	279.4	2.76
11.5	3.1	292.1	2.59
12.0	3.0	304.8	2.51

Placing Ready-mix Concrete Pavements. Using ready-mix concrete and wheeling the concrete into place with wheelbarrows, the labor per cu. yd. of concrete should cost as follows:

Wheelbarrow Concrete			
Description	Hours	Rate	Total
Labor	1.2	$ 26.16	$ 31.39
Cost per Cu Yd			$ 31.39
Cost per Cu. Meter			$ 41.06

If ready-mix concrete can be discharged from truck directly into place without wheeling, deduct 0.7-hr. labor time.

Larger jobs, with 100 c.y. (76.5 cu.m) or more of concrete, can realize better organization. Lower costs are the result. Assuming a pavement containing from 100 to 200 c.y. (76.5 to 152.3 cu.m) of ready-mixed concrete wheeled into place, the labor per c.y. (cu.m) of concrete should cost as follows:

Should ready-mix concrete be discharged directly into a paving machine that will spread and finish the concrete, it will take about 96 man-hours. The crew will place approximately 2,500 Sq. Ft. of 12" thick unreinforced concrete pavement. This does not include the setting up running rails for the paving machine nor the time to set up and remove the machine.

Placing Concrete Roadway Pavement with Machine			
Description	Hours	Rate	Total
Paving Machine	8.0	$ 2,031	$2,031.00
Operating Engineer	8	$ 37.89	$ 303.12
Oiler	8	$ 32.89	$ 263.12
Laborers	56	$ 26.16	$1,464.96
Masons	24	$ 36.93	$ 886.32
Total Cost 2,500 Sq. Ft.			$4,948.52
Cost per Cy			$ 51.23
Cost Per Cu. Meter			$ 67.00
Cost per Sq.Ft.			$ 1.98
Cost per Sq.Meter			$ 21.31

Directly from Truck			
Description	Hours	Rate	Total
Labor	0.9	$ 26.16	$ 23.54
Cost per Cu Yd			$ 23.54
Cost per Cu. Meter			$ 30.79

If ready-mix concrete can be discharged from truck directly into place without wheeling, deduct 0.4-hr. labor time.

Finishing Concrete Pavements. Finishing concrete pavements usually consists of a series of operations performed in the following sequence: strike-off and consolidation, straight-edging, floating, brooming, and edging.

Labor Finishing Concrete Pavements. Assuming an 8" (200 mm) concrete pavement being placed at the rate of 25 c.y. (19 cu.m) per hr. or 1,013 s.f. (94 sq.m) per hr., a crew of 10 cement masons should be able to keep up with the pour and perform all the above described finishing operations at the following labor cost per 100 s.f. (9.3 sq.m):

Finish Concrete			
Description	Hours	Rate	Total
Cement Mason Foreman	1.0	$ 40.93	$ 40.93
Cement Mason	1.0	$ 36.93	$ 36.93
Labor	0.5	$ 26.16	$ 13.08
Cost per 100 Sq. Ft.			$ 36.93
Cost per Sq. Ft.			$ 0.37
Cost per Sq. Meter			$ 3.98

The above costs are based on a job that will run from 7,500 to 8,500 s.f. per 8-hr. day. On smaller jobs the costs will increase from 20% to 25%.

02700 SITE IMPROVEMENTS

02710 FENCES & GATES

Industrial chain link fences usually consist of 3" (75 mm) OD tubular end posts and 2" to 2-1/2" (50 to 62.5 mm) OD intermediate posts depending upon the height of the fencing to be installed. Top and bottom rails are not always used, because having a top rail on a chain link fence gives much more stability for intruders to gain a handhold. Therefore, the spring wire is used for top and bottom stability in preventing chain link fabric sag. This spring wire is at least a 6 gauge material. The wire fabric is normally a 9 gauge wire and can be galvanized, aluminized steel, or vinyl coated in a number of colors.

For chain link fencing over 6' (1.8 m) in vertical height, 6 gauge wire fabric will be used in lieu of the 9 gauge fabric.

Wire arms are available either in single or double sided for the installation of 2 to 4 strands of barbed wire per side.

The maximum line post or intermediate post spacing is 10' (3 m) on center and these posts as well as in the corner posts are usually set in a 9" (225 mm) diameter by 3' (900 mm) deep concrete footing.

Industrial Chain Link Fencing Cost per Lin. Ft.				
Material	9 ga.	6 ga.	6 ga.	6 ga.
	4'	5'	6'	8'
Galvanized	$ 13.88	$ 15.49	$ 16.38	$ 18.87
Aluminized Steel	$ 14.58	$ 17.10	$ 18.18	$ 21.24
Vinyl Coated	$ 16.92	$ 18.72	$ 19.98	$ 23.22

Industrial Chain Link Fencing
Cost per Lin. Meter

Material	6 ga. 1.22 M	6 ga. 1.52 M	6 ga. 1.83 M	6 ga. 2.44 M
Galvanized	$ 45.55	$ 50.81	$ 53.73	$ 61.90
Aluminized Steel	$ 47.83	$ 56.10	$ 59.65	$ 69.69
Vinyl Coated	$ 55.51	$ 61.42	$ 65.55	$ 76.18

Add for barbed wire:		
Description	Cost per Lin.Ft	Cost per Lin Meter
Single side - 3 strand	$ 0.56	$ 1.85
Double side - 3 stran	$ 1.00	$ 3.28

Add for corner posts: Diameter		
Height Feet	3" Diam. Cost per Foot	4" Diam. Cost per Foot
4 '	$59.50	$101.15
5 '	$65.45	$119.00
6 '	$74.97	$130.90
8 '	$89.25	$142.80
Height Meter	75 mm Diam. Cost per M	100 mm Diam. Cost per M
1.22 M	$195.21	$331.86
1.52 M	$214.73	$390.42
1.83 M	$245.96	$429.46
2.44 M	$292.81	$468.50

Add for gates:	
3' x 4' (0.91 x 1.22 m)	$ 120.00
3' x 5' (0.91 x 1.52 m)	$ 160.00
3' x 6' (0.91 x 1.83 m)	$ 210.00
3' x 8' (0.91 x 2.44 m)	$ 275.00

SITE WORK

RESIDENTIAL CHAIN LINK FENCING

The standard corner posts are 2" to 2-1/2" (50 to 62.5 mm) diameter. The standard line posts are 1-5/8" (40 mm) diameter set at 10' (3 m) on center. Normally, with residential fencing, a top rail is added for stability.

Residential Chain Link Fencing Cost per Lin. Ft. (Lin. Meter)			
Material	11 ga. 4'	11 ga. 5'	11 ga. 6'
Galvanized	$ 4.80	$ 6.10	$ 7.20
Aluminized Steel	$ 5.20	$ 6.60	$ 7.80
Vinyl Coated	$ 5.70	$ 6.90	$ 8.70
Material	11 ga. 1.22 M	11 ga. 1.52 M	11 ga. 1.83 M
Galvanized	$ 15.75	$ 20.01	$ 23.62
Aluminized Steel	$ 17.06	$ 21.65	$ 25.59
Vinyl Coated	$ 18.70	$ 22.64	$ 28.54

Add for gates:	
3' x 4' (0.91 x 1.22 m)	$ 120.00
3' x 5' (0.91 x 1.52 m)	$ 160.00
3' x 6' (0.91 x 1.83 m)	$ 210.00
3' x 8' (0.91 x 2.44 m)	$ 275.00

SECURITY FENCE

Today there is a new "security" fence that is made of heavy gauge expanded metal mounted on heavy gauge fence posts place into concrete footing. This expanded metal fence is non climbable and normally does not have barbed wire on top. Each expanded metal section is 4' x 8' (1.22 M x 2.44 M) which is overlapped and bolted together to make each section 7'-9" wide x 8' high. (2.36m wide x 2.44 m high). The security fence is sometimes referred to as "Alabama Fence." Cost to install range between $110 to $150 per lin. Foot ($361 to $487 per lin. Meter).

Wood Residential Fences - Cost Per Lin. Ft. (Meter)			
Description - Type of Fence	Per Each	Per Lin. Ft.	Per Lin.Meter
Cedar Stockade Screen Fence, 4' (1.22 m) High		$ 11.50	$ 37.73
Gate 3' (0.91 m) Wide x 4' (1.22 m) High	$ 160.00		

Wood Residential Fences - Cost Per Lin. Ft. (Meter)			
Description - Type of Fence	Per Each	Per Lin. Ft.	Per Lin.Meter
Cedar Stockade Screen Fence 6' (1.83 m) High		$ 13.00	$ 42.65
Gate 3' (0.91 m) Wide x 6' (1.83 m) High	$230.00		
Cedar Split Rail - 2 Rail 3' (0.91 m) High		$ 11.50	$ 37.73
Cedar Split Rail - 2 Rail 3'- 6" (1.07 m) High		$ 13.00	$ 42.65
Redwood or Cedar Board on Board 4' (1.22 m) High		$ 21.00	$ 68.90
Redwood or Cedar Board on Board 6' (1.83 m) High		$ 30.00	$ 98.43
Gates 3' (0.91 m) Wide x 4' (1.22 m) High	$335.00		
Gates 3' (0.91 m) Wide x 6' (1.83 m) High	$475.00		

02720 ROAD AND PARKING APPURTENANCES

Guard rails of formed steel are set on steel, wood, or concrete posts. Rail sections are set 18" (450 mm) to the centerline above grade. Posts are set every 12'-6" (3.8 m) with rails lapped 11-1/2" (450 mm) and attached with 7-5/8" (190 mm) bolts. The 12 ga. sections weigh less than 100 lbs. (45.5 kg), so two workers can handle installation. Special terminal wings are furnished for exposed ends.

A budget price for this type of rail set on steel posts is $35.00 per lin. ft. ($114.83 per m) and about $31.00 lf ($101.70 per m) set on wood posts.

The 4'x8' (1.2 x 2.4 m) timber rails set on wood posts will cost about $19.00 per l.f. ($62.33 per m). They do not offer the protection of the metal rail but may better fit into the landscape.

Wheel stops for parking lots may be either wood or concrete. The concrete ones are shaped on the bottom to allow drainage; wood ones, unless set on standards, tend to collect water. Costs per lin. ft. installed are $6.00 for 6x6 wood dowelled into paving, $6.00 for concrete dowelled into paving and $5.00 for 4x4 wood set on metal brackets.

The 4" striping in parking lots will cost about $0.40 per l.f. (1.31 per m) or $8.00 per stall.

314

02760 SITE FURNISHINGS

Playground Equipment	
Description	Cost Each
Bike Rack, 10'- 0" (3.05 m) long	$ 510.00
Horizontal Monkey Ladder, 12'- 0" (3.66 m)	$ 810.00
Parallel Bars, 10'- 0" (3.05 m) long	$ 475.00
See-saw 2 Units	$ 995.00
See-saw 4 Units	$1,300.00
Slides - Stainless Steel, 12'- 0" (3.66 m) long, 6'- 0" (1.83 m) high	$2,400.00
Swings - 4 seats, 8'- 0" (2.44 m) high	$1,400.00
Swings - 8 seats, 8'- 0" (2.44 m) high	$2,425.00
Whirlers, 10'- 0" (3.05 m) diameter	$2,625.00
Benches	
Precast concrete with wood backs and seats, 8'- 0" (2.44 m) long	$ 825.00
Fiberglass, 8'- 0" (2.44 m) long	$1,000.00
Fiberglass supports with wood backs and seats, 8'- 0" (2.44 m) long	$ 705.00
Wood, 8' (2.44 m) long	$ 500.00
Planters	
Precast concrete, 24" (600 mm) dia. x 30" (760 mm) high	$ 500.00
Precast concrete, 48" (1200 mm) dia. x 36" (900 mm) high	$1,000.00
Fiberglass, 36" (900 mm) dia. x 30" (760 mm) high	$ 450.00
Fiberglass, 60" (1500 mm) dia. x 36" (900 mm) high	$ 800.00
Wood, 48" (1200 mm) dia. x 30" (760 mm) high	$ 800.00

02800 LANDSCAPING

Landscaping requirements are influenced to a considerable degree by the aesthetic appreciation of existing vegetation and the economics of retaining and protection costs during the construction period. Some owners and architects are dedicated to the preservation of the maximum number of trees that do not directly interfere with construction in an effort to blend the new construction into the existing environment. This theory produces exceptional immediate beauty to any new construction project but involves additional costs to construction by requiring extensive protective barriers for trees and vegetation and restricts work areas causing a lowering of productivity for automotive and heavy construction equipment.

In instances where existing trees and vegetation are of low quality, or the owner is unwilling to absorb the cost of retention, the site is cleared of all natural growth and provisions are made to provide new landscaping for the site at the conclusion of construction. The beauty of landscaping is subjective and may be limited to a combination of simple sodding, seeding, and minimal plantings. In other instances, it can progress to an elaborate extent, requiring the special talents of a landscape architect.

Most contractors lack the material sources, experienced personnel, and special equipment required for landscaping, and they rely upon landscape contractors to perform these services. Frequently, specifications require landscaping be performed by a certified contractor who is primarily engaged in this type of work and has the resources and capabilities to provide guarantees and certifications in compliance with specification requirements.

Estimators can develop a general knowledge of the costs involved in landscaping in order to evaluate the proposals received from the subcontractor.

Generally, nurseries providing service to homeowners and transit trade are not competitive in prices for commercial and large scale landscape projects. Soliciting proposals from qualified landscaping subcontractors presents a unique problem for the contractor. It has been found that these specialized subcontractors recognize that their performance will not be required until near the completion of a project, which may be many months in the future, and are reluctant to tender firm price proposals prior to the scheduled bid date.

Estimators must take a special effort to obtain these proposals in a timely manner to allow sufficient time to make an analysis of the bid before it is incorporated into the total project estimate.

The establishment of a work performance schedule to coincide with the accepted local planting season will tend to lower the subcontractor's price and will preserve the guarantee period. It is not unusual to provide for landscaping work slightly out of sequence in order to accommodate the planting season.

For the estimator to develop the proper unit of measure for pricing landscape work, or to determine the completeness of a proposal from a subcontractor, the following are required:

1. Area of seeding and fertilizing in square yards or acres on large projects.
2. Topsoil in cubic yards, with allowance for topsoil available from the site and for the amount that must be purchased from external sources.
3. Area of sodding in square yards.
4. Number, size, and species of shrubs.
5. Number, size and, species of trees.

Description - Item	Unit of Measur	In-Place Cost	Unit of Measure	In-Place Cost
Edging, 1x4 Redwood	Lin. Ft.	$1.30	Lin Meter	$4.27
Edging, Brick (3 per foot, no concrete)	Lin. Ft.	$7.00	Lin Meter	$22.97
Erosion Control - Plastic Netting	Sq. Yd.	$1.00	Sq. Meter	$1.20
Marble Stone Chips - Hand Placed 100 Lbs(45.36 Kg)	Lbs	$9.20	Kg	$20.28
Move Existing Plants - Replanting, 3'-0" (.91 M) high	Each	$50.00	Each	$50.00
Seed, Fertilizer - Small Areas	Sq. Yd.	$1.50	Sq. Meter	$1.79
Seed, Fertilizer - Large Areas, Mechanical Feed 220 Lbs per acres (99.7 Kg/.40 Hectares)	Acre	$1,500.00	Hectares	$3,706.45
Sodding	Sq. Yd.	$3.50	Sq. Meter	$4.19
Trees				
Spruce 8'- 0" (2.4 m)	Each	$175.00	Each	$175.00
Birch 8'- 0" (2.4 m)	Each	$150.00	Each	$150.00
Pine-White 8'- 0" (2.4 m)	Each	$100.00	Each	$100.00
Dogwood 8'- 0" (2.4 m)	Each	$200.00	Each	$200.00
Shrubs				
Honeysuckle 4'- 0" (1.2 m)	Each	$35.00	Each	$35.00
Forsythia 4'- 0" (1.2 m)	Each	$35.00	Each	$35.00
Plant Bed - By Hand	Sq. Ft.	$3.75	Sq.Meter	$40.36

6. Details of any special features, such as rock gardens.
7. Allowance for maintenance for specified periods.
8. Allowance for replacement guarantee, if specified.

After quantities, quality, and features are accurately determined, the estimator is prepared to develop a budget estimate that can be used as a cost comparison with the proposals received from landscaping subcontractors.

For general guidance the table below provides average cost ranges, but due to quantity, cost fluctuation, and geographical conditions, estimators should verify with local landscape subcontractors.

Plant maintenance is necessary until rooting begins. The duration of this activity is determined by local weather conditions and the time of planting. It can extend for many months beyond the conclusion of the actual construction period, so allowances have to be made in the landscape estimate to accommodate this requirement. If the maintenance period extends beyond the conclusion of construction, it will necessitate the dispatch of personnel, including transportation of supplies and equipment, necessary to perform the maintenance work. The per trip cost will be considerably higher than if the work had been performed in the construction phase. These excess costs occur from the additional travel time involved for each trip, which would include the driver's wage cost plus a fair and reasonable allowance for truck rental. Also, the actual work performance would be in an unsupervised, which can mean low productivity.

PLANT GUARANTEE

Should the prime contractor assume the obligation of the guarantee, then the estimate should include an allowance for replanting shrubs and trees that fail to root and grow after transplanting. To arrive at a reasonable allowance for this cost, 20% to 30% of the original planting allowance should be sufficient.

03 CONCRETE

When estimating the total cost of placing concrete for footings, foundations, walls, columns, slabs, and other items, in addition to the material and equipment costs, there are five categories of labor operations to consider:

1. *Form Preparation:* Concrete is placed in a fluid consistency and must be contained or formed to the desired shape before the hardening process. The forming operation includes items such as: earth excavation to the desired shape for footings and foundations, together with supplemental wood forms when necessary for proper elevation, steps (changes in elevation), or unstable soil conditions; proper finish grading of earth surfaces for the placement of concrete slabs on grade to the correct thickness and elevation; and cutting, fabricating, placing, aligning, and bracing wood or metal forms for walls, columns, and slabs.

2. *Embedded Items Placement:* Labor will be required for the proper placement of all items to be cast into the concrete mix. Usually steel reinforcing (mesh or bars) will be necessary to insure the structural integrity of the concrete member. Non-structural items that frequently are installed into cast-in-place concrete include: anchor bolts, plates, or inserts for the future connection of other items of work, corner protection angles, and sleeves for the passage of pipes, conduit, and ductwork.

3. *Placement of Concrete:* The ready-mixed concrete that is delivered to the job site by truck must be physically moved from the truck to the location of placement by the use of a chute, wheelbarrow, buggy, or some mechanical means such as a conveyor, pump, or crane. Any means of moving concrete mix involves a labor operation, whether it be a group of laborers, an equipment operator, or a combination. Once the concrete mix is in place, it is then necessary to tamp or vibrate the concrete for proper consolidation and "strike off" or screed the concrete to its designed level or elevation. In the event that only a small amount of concrete is required, or in remote areas where ready-mixed concrete is unavailable, it will be necessary to estimate the additional labor required to job mix the components (gravel, sand, cement, and water) for the quantity of concrete needed.

4. *Finish, Cure, & Protection:* Once the concrete mix is placed into the forms, vibrated, and screeded to the proper elevation, the next operation is the proper finishing of the exposed surfaces: a steel troweled finish to a slab surface to provide a smooth, hard and consolidated surface, free of voids

319

and unevenness. Concrete requires a chemical action to change it from a fluid consistency to the set or hardened condition, which generates a certain amount of heat. On hot days the heat generated by the setting process and the weather produces a concrete mix that sets too fast and hampers workability for the proper finish. It produces surface cracking, and in some cases, cracking for the full depth of the slab. To help slow the setting process and to seal in moisture, slabs are covered with wet burlap, polyethylene, or a chemical curing compound. Slower curing time produces a stronger product. In cold weather, because of its water content before hardening, concrete must be kept from freezing by either covering with straw or other insulating material to keep in its generated heat, or by a supplemental heating process such as covering the area and installing a temporary heat source. Concrete work, and in particular slabs, must be protected from any form of precipitation prior to the setting process, because the water droplets landing on the slab cause small craters in the finish surface, ruining the quality of finish and diluting the cement ratio of the concrete mix, which weakens at least the top wearing surface of the slab.

5. *Form Removal:* After the concrete has hardened, the forms are removed. This operation entails stripping the nails from the forms, cleaning off the hardened concrete skim particles, preparing the forms for the next forming location by means of applying a form release agent (oiling), and physically moving the forms to the next area of the building for starting the forming process on another scheduled work item. The proper cleaning and oiling (coating the concrete contact surface) of form work is important, because it allows reuses of form materials, reducing the costly expense of new material and fabrication costs.

03100 CONCRETE FORMWORK

Note: *The conversion from Board Feet to Cubic Meter has been rounded off Board Foot, bd. ft. or bf is a unit of volume for measuring lumber. One standard board foot is the volume of twelve inches wide by one inch thick by one-foot length or 1/12 of a cubic foot or approximately 2.360 liters. To do a straight conversion from board feet to cubic meter is not a direct conversion., Since board feet is not really 1/12 of a cubic foot, a dressed piece of softwood lumber 2" x 4" x 12' is actually 1 ½" x 3 ½" x 12 . Since metric uses actual sizes and board feet uses nominal sizes, a straight conversion of 424 Board Feet = 1 Cubic Meter would be misleading.*

Concrete formwork is estimated by computing the actual surface area of the form that comes into contact with the concrete. This is referred to as the "contact surface". The formwork quantities are then categorized into specific groups (wall form, column form, etc.) and in the proper unit of measure-square foot, lineal foot, meter, etc.

Example: Compute the quantity of formwork required for a free standing wall 50'-0" (15.24 m) long, 4'-0" (1.22 m) high, and 1'-0" (0.30 m)

CONCRETE

thick. Forms will be required for both sides of the wall. The formwork and quantity is obtained as follows:

Face Form: 50'-0" (15.24 m) x 4'-0" (1.22 m) x 2 sides = 400 s.f. (37.19 m)
End Form: 1'-0" (0.30 m) x 4'-0" (1.22 m) x 2 ends = 8 s.f. (0.73 sq.m)
Total Contact Area = 400 + 8 = 408 s.f. (37.19 + 0.73 = 37.92 sq.m)

Form Conversion		
Description	English	Metric
Ribbon footing forms	Linear Foot	Linear Meter
Slab edge forms (by height)	Linear Foot	Linear Meter
Foundation edge forms	Square Foot	Square Meter
Wall forms	Square Foot	Square Meter
Pier forms	Square Foot	Square Meter
Column forms	Square Foot	Square Meter
Beam bottom forms	Square Foot	Square Meter
Beam side forms	Square Foot	Square Meter
Supported slab forms	Square Foot	Square Meter

WOOD FORMS

Forms for Concrete Footings. There are two main types of footings in most construction work: column footings and wall footings. Column footings are usually square or rectangular masses of concrete designed to spread the column load over an area to accommodate the soil bearing value and are located at the column centers. Wall footings are usually ribbons of concrete, somewhat wider than the walls they carry, and are similarly designed to spread the loading over a larger area than the actual wall bottom thickness.

Forms for footings that are formed completely are usually simple and rough, using 2" (50 mm) planking for sides and 2"x4" (50 x 100 mm) stakes to hold the sides in place and 2"x4" (50 x 100 mm) struts for braces to the banks. This material may be reused numerous times in the course of the work. Plywood is sometimes substituted for the 2" (50 mm) planking for sides.

Ribbon forms for partly formed footings usually consist of 2" (50 mm) thick lumber in lengths as long as are practical, secured at the correct elevation to 2"x4" (50 x 100 mm) stakes, and tied across the tops at intervals of 4'-0" to 6'-0" (1.22 to 1.83 m) with 1"x2" (25 x 50 mm) spreader-ties, except in the case of column footings, which are usually diagonally tied at the corners with 1" boards and tied transversely, if required, with strap iron.

The following are itemized costs of the column and wall footing forms found in the average simple job. Complicated or "cut-up" jobs require up to 25% more material and up to 50% more labor.

Material Cost of 100 S.F. (9.29 Sq.M) of Column Footing Forms
Footings 5' x 5' x 1'- 6"(1.52 x 1.52 x .46 Meters) Completely Formed

Material	BF	Cu.M.	Rate	Total
Sides, 2" (50 mm) planking	200	0.472	$ 0.13	$ 25.00 *
Stakes and braces, 2" x 4" (50 x 100 mm)	100	0.236	$ 0.13	$ 12.50 *
Nails, etc.			$ 7.50	$ 7.50
Cost per 100 Sq.Ft. (9.29 Sq.M)				$ 45.00
Material cost per s.f.				$ 0.45
Cost per sq.m				$ 4.84

* Based on 6 uses of lumber

Labor Cost of 100 S.F. (9.29 Sq.M) of Column Footing Forms
Footings 5' x 5' x 1'- 6"(1.52 x 1.52 x .46 Meters) Completely Formed

Placing Forms-18-22 Sq. Ft. (1.67- 2.04Sq.M) per Hour	Hours	Rate	Total
Carpenter	5.0	$ 35.04	$ 175.20
Labor helping	2.5	$ 26.16	$ 65.40
Stripping Forms-90-110 Sq.Ft. (8.36-10.22 Sq.M) per Hour	Hours	Rate	Total
Labor	1.0	$ 26.16	$ 26.16
Cost per 100 s.f.			$ 266.76
Labor cost per s.f.			$ 2.67
Cost per sq.m			$ 28.71

Material Cost of 100 L.F. (30.48 M) of Column Footing Ribbon Forms
Footings 5'- 0" x 5'- 0" x 1'- 6" (1.52 x 1.52 x .46 M) Partly Formed

Material	B.F.	Cu.M.	Rate	Total
Ribbons,2"x 4"x 100'- 0" (50x100mm x 30.48M)	67	0.158	$ 0.13	$ 8.38 *
Stakes, 2"x 4" x 3'- 0" long (50x100mm.91M)	80	0.189	$ 0.13	$ 10.00 *
Corner ties, 1"x 6" x 2'-0" long (25x100mm.60M)	20	0.047	$ 0.13	$ 2.50 *
Nails, etc.			$ 4.18	$ 4.18
Cost per 100 LF				$ 25.05
Material cost per LF				$ 0.25
Cost per M				$ 0.82

* Based on 6 uses of lumber

CONCRETE

Labor Cost of 100 L.F. (30.48 M) of Column Footing Ribbon Forms **Footings 5'- 0" x 5'- 0" x 1'- 6" (1.52 x 1.52 x .46 M) Partly Formed**			
Placing Forms-35 l.f. (10.67m) per Hour	Hours	Rate	Total
Carpenter	3.0	$ 35.04	$105.12
Labor helping	1.5	$ 26.16	$ 39.24
Stripping Forms-450-500LF (140-150M) per Hr.	Hours	Rate	Total
Labor	0.21	$ 26.16	$ 5.49
Labor cost per 100 l.f.			$149.85
Cost per l.f.			$ 1.50
Cost per m			$ 4.92

Material Cost of 100 Sq.Ft. (9.29 M) of Wall Footing Forms **Footings 1'- 0" (.30 M) Deep Completely Formed**				
Material	B.F.	Cu.M.	Rate	Total
Sides, 2" x 12" (50 x 300mm), Planks	200	0.472	$ 0.13	$ 25.00 *
Stakes and braces	75	0.177	$ 0.13	$ 9.38 *
Nails, etc.			$ 6.88	$ 6.88
Material Cost per 100 Sq.Ft.				$ 41.25
Cost per Sq.Ft				$ 0.41
Cost per Sq. M				$ 4.44

* Based on 6 uses of lumber

Labor Cost of 100 Sq.Ft. (9.29 M) of Wall Footing Forms **Footings 1'- 0" (.30 M) Deep Completely Formed**			
Placing Forms-25-30 Sq.ft. (2.32-2.79 Sq.M) per Hr.	Hours	Rate	Total
Carpenter	3.5	$ 35.04	$ 122.64
Labor helping	1.8	$ 26.16	$ 45.78
Stripping Forms-450-500LF (140-150M) per Hr.	Hours	Rate	Total
Labor	1.5	$ 26.16	$ 39.24
Labor cost per 100 Sq.Ft.			$ 207.66
Cost per Sq. Ft.			$ 2.08
Cost per Sq. M			$ 22.35

Material Cost of 100 Lin.Ft. (30.48 Lin. M) of Wall Footing Ribbon Forms Footings 1'- 0" (.30 M) Deep Completely Formed					
Material	B.F.	Cu.M.	Rate	Total	
Ribbons, 2" x 4" x 100 LF (50 x 100mm x 30.48 M)	67	0.158	$ 0.13	$ 8.38	*
Stakes, 2" x 4" x 2'- 0" long (50 x 100mm.61M)	33	0.078	$ 0.13	$ 4.13	*
Ties, 1" x 2" x 2'- 0" long (25 x 50mm600mm)	4	0.009	$ 0.13	$ 0.50	
Nails, etc.			$ 2.60	$ 2.60	
Material Cost per 100 Lin. Ft.				$ 15.60	
Cost per Lin.Ft				$ 0.16	
Cost per Lin. M				$ 0.51	

Based on 6 uses of lumber

Material Cost of 100 Lin.Ft. (30.48 Lin. M) of Wall Footing Ribbon Forms Footings 1'- 0" (.30 M) Deep Completely Formed			
Placing Forms-25-30 Sq.ft. (2.32-2.79 Sq.M) per Hr.	Hours	Rate	Total
Carpenter	2.5	$ 35.04	$87.60
Labor helping	1.3	$ 26.16	$32.70
Stripping Forms-450-500LF (140-150M) per Hr.	Hours	Rate	Total
Labor	0.21	$ 26.16	$5.49
Labor cost per 100 Lin.Ft.			$125.79
Cost per Lin. Ft.			$1.26
Cost per Lin. M			$4.13

CONCRETE

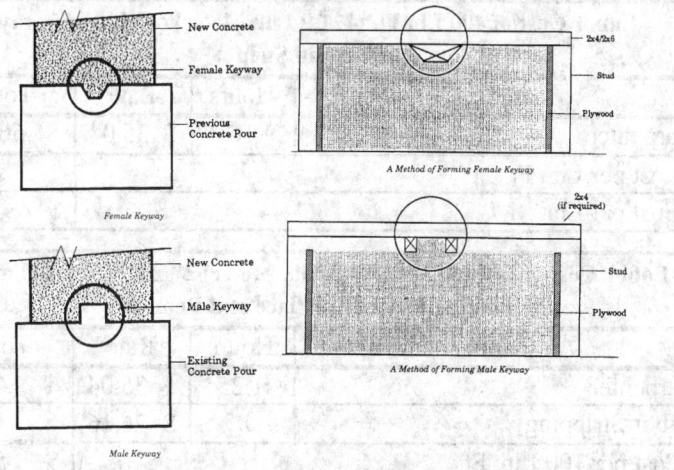

A Method of Forming Female Keyway

Female Keyway

A Method of Forming Male Keyway

Male Keyway

On jobs where the wall footings are stepped down to a lower elevation, a riser form is required at each step. The cost of this riser form is about twice the cost of the regular footing side form.

Grooves are usually formed in the tops of footings and are called "footing keys". They are formed by forcing a beveled 2"x4" (50 x 100 mm) into the surface of the wet concrete. After the concrete has hardened, the "key" forms are pried loose, cleaned, and reused.

Keyways might be required by the contract drawings, specifications, or by local building codes. The two basic types of keyways are female and male. Female keyways are usually either a 2"x 4" (50x100mm) or 2"x 6" (50x150mm), chamfered and placed in the middle of the new concrete pour. Male keyways are pre-formed with the lower pour and placed in the middle of the next pour location.

The following are itemized labor costs for female keyways, based on using 2"x 6" (50 x 150 mm), requiring 1 b.f. per each l.f., plus support of keyway.

Labor Cost of 1,000 Lin.Ft. (304.8 Lin. Meter) of Female Keyway Form and Strip			
Labor	Hours	Rate	Total
Carpenter	8	$ 35.04	$ 280.32
Cost per Lin. Ft.			$0.28
Cost per Lin. M			$ 0.92

The following are itemized labor costs for male keyways, based on using 2"x 4" (50 x 100 mm), requiring 1.33 b.f. (0.20 cu.m) per each l.f. (0.3 m), plus support of keyway.

Labor Cost of 400 Lin.Ft. (121.9 Lin. Meter) of Male Keyway Form and Strip			
Labor	Hours	Rate	Total
Carpenter	16	$ 35.04	$ 560.64
Cost per Lin. Ft.			$1.40
Cost per Lin. M			$ 4.60

Labor Cost of 100 Lin.Ft. (30.48 Lin. Meter) 2" x 4" (20 x 100 mm) Footing Keyways - Place and Remove			
Labor	Hours	Rate	Total
Carpenter	1.25	$ 35.04	$ 43.80
Labor Stripping	0.5	$ 26.16	$ 13.08
Cost per 100 Lin. Ft.			$ 56.88
Cost per Lin. Ft.			$0.57
Cost per Lin. M			$ 1.87

Forms for Pile Caps. After piles are driven and cut off at the required elevation, they are capped with masses of concrete that are designed to transmit the total loading of the structure to the piles. These pile caps vary in size and shape, from simple square caps covering one pile to complex many-sided caps covering clusters of twenty or more piles.

Forms for pile caps are usually prefabricated in panels at the bench and are then set up and put together in the pile location. Side forms are usually made of 3/4" (20 mm) plyform with 2"x 4" (50 x 100 mm) stiffeners nailed to the back face. Five or 6 uses can usually be obtained from these forms.

After panels are set in place, they are secured with 2"x 4" braces and tied with strap iron.

The following is an itemized cost of making average pile cap forms, based on 5'- 6"x 5'-6"x3'-0" (1.68 x 1.68 x 0.9 m), containing 66 s.f. (5.94 sq.m):

Material Costs for Pile Caps				
Making Pile Cap Side Forms	English	Metric	Rate	Total
¾" (18.75 mm) Plyform - 66 Sq.Ft.(6.13 Sq.M)	66 SF	6.13 sq.m.	$ 0.60	$ 39.60
20 pcs. 2" x 4" x 3' long (50 x 100mm x.91 M)	41 BF	0.097 cu.m.	$ 0.75	$ 30.75
8 pcs. 2" x 4" x 5' long (50 x 100mm x 1.52 M)	27 BF	0.064 cu.m.	$ 0.75	$ 20.25
Bracing	English	Metric	Rate	Total
14 pcs. 2" x 4" x 8' long (50 x 100mm x 2.43 M)	75 BF	0.177 cu.m.	$ 0.75	$ 56.25
Nails, etc.			$ 29.37	$ 29.37
Material Cost per 66 Sq. Ft.				$176.22
Cost per Sq.Ft				$ 2.67
Cost per Sq. M				$ 28.74

Divide above costs by number of times forms may be used on the job.

Labor Costs for Making Pile Cap Forms			
Labor	Hours	Rate	Total
Carpenter	4	$ 35.04	$ 140.16
Laborer	2	$ 26.16	$ 52.32
Labor Cost per pile cap - 66 Sq. Ft.			$ 192.48
Cost per Sq. Ft.			$ 2.92
Cost per Sq. M.			$ 31.39

Labor Costs Setting and Removing Pile Cap Forms			
Labor	Hours	Rate	Total
Carpenter	3	$ 35.04	$ 105.12
Labor	1.5	$ 26.16	$ 39.24
Laborer- Stripping, cleaning and moving forms	1	$ 26.16	$ 26.16
Labor Cost per pile cap - 66 Sq. Ft.			$ 170.52
Cost per Sq. Ft.			$ 2.58
Cost per Sq. M.			$ 27.81

Forms for Concrete Piers. When column footing tops are lower than the bottom of the floor slab on the ground, piers are used to carry the column loading to the footings.

Forms for this work are usually made up in side panels at the job bench, hauled to the footings, erected, clamped, and braced.

Side panels are usually made of 3/4" (20 mm) plyform with 2"x 4" (50 x 100 mm) stiffeners. Clamping may be accomplished by using either 2"x 4" (50 x 100 mm) lumber or metal column clamps properly spaced to withstand the pressure developed when placing the concrete.

Bracing is required to hold the form in position and usually consists of 2"x4" (50 x 100 mm) or 2"x 6" (50 x 150 mm) lumber, secured to the form at grade level, spanning the pit, and spiked to stakes driven into the ground. Bracing below this point is done by wedging 2"x4" (50 x 100 mm) struts between the form and the earth banks.

The following are itemized costs of building average pier forms, based on a pier 2'-0"x2'-0"x4'-0" (.60 x .60 x 1.22 m) containing 32 s.f. (2.88 sq.m):

Material Costs				
Making Pier Forms	English	Metric	Rate	Total
¾" (18.75 mm) Plyform, 32 Sq.Ft. (2.97 Sq.M)	32 SF	2.97 sq.m	$ 0.60	$ 19.20 *
12 Pcs. 2"x 4" x 4'- 0" (50 x 100 mm x1.22 M)	32 BF	0.075 cu.m.	$ 0.75	$ 24.00 *
Clamps 8 Pcs. 2"x4"x3' long (50 x 100mm x 0.91m)	16 BF	0.038 cu.m.	$ 0.75	$ 12.00
Nails, etc.			$ 11.04	$ 11.04
Material Cost per 32 Sq. Ft				$ 66.24
Cost per Sq.Ft				$ 2.07
Cost per Sq. M				$ 22.28

* Steel clamps can be used in place of wood.
Divide Sq.F. costs by number of times forms may be used on the project

Labor Costs Making Pier Forms			
Labor	Hours	Rate	Total
Carpenter	1	$ 35.04	$ 35.04
Laborer	0.5	$ 26.16	$ 13.08
Labor Cost per pier, 32 Sq.Ft.			$ 48.12
Cost per Sq. Ft.			$ 1.50
Cost per Sq. M.			$ 16.19

CONCRETE

Labor Costs Erecting Concrete Pier Forms				
Labor	Hours	Rate	Total	
Carpenter	1	$ 35.04	$	35.04
Laborer	0.5	$ 26.16	$	13.08
Labor Stripping, Including cleaning and moving	0.5	$ 26.16	$	13.08
Labor Cost per pier, 32 Sq.Ft.			$	61.20
Cost per Sq. Ft.			$	1.91
Cost per Sq. M.			$	20.59

Forms for Concrete Walls

Forms for concrete walls must be designed and constructed to withstand the horizontal pressures exerted by the fluid concrete. The pressure against any given point of the form varies and is influenced by any one or all of the following factors:

1. Rate of vertical rise in filling the forms in feet (meters) per hour.
2. Temperature of the concrete and the weather (ambient temperature).
3. Proportions of the concrete mix and its consistency.
4. Method of concrete placement and degree of vibration.

When wall forms are filled so rapidly that concrete at the bottom does not have enough time to start its initial setting, then regardless of any other factor, the full height of the fluid concrete must be considered as a column of liquid with a weight of approximately 150 lbs. per c.f. (2403 kg per cu.m) with a head pressure equal to its fluid height.

For example, at 10' (3.04 m) below the top of the concrete, the pressure will amount to 10' x 150 lbs. = 1,500 lbs. per s.f. (3.04 m x 2403 = 7209 kg per sq.m) exerted against the form sides at the bottom. If the wall form has not been adequately tied and braced to withstand this pressure, it will break open.

To avoid the necessity of over designing forms, or form breakage, concrete is placed in layers or "lifts" at the rate of about 4 vertical ft. (1.22 m) per hour. This allows the concrete a chance to start the transformation from the fluid state to a hardened mass that supports itself and relieves pressure on the form surface.

The ambient temperature at which concrete is placed influences the time required for the setting process. Concrete placed at 40°F (4.44°C) takes almost twice the time to set as concrete placed at 80°F (26.67°C). Concrete placed in wall forms during cold weather will require more control over the amount of concrete placed within a given time so that the forms will not be stressed beyond their design capability.

The proportions and consistency of the mix—water/cement ratio per c.y. (cu.m) of concrete—has an effect on the rate of setting time as well as

329

the strength of the concrete mix. The more the cement content and/or the less water needed to provide a workable consistency (slump) per cubic yard, the shorter the time for setting and the stronger the concrete.

When placing concrete into a form, care should be taken to reduce the shock loading of the concrete hitting against the form material. This can be achieved by using a tremie, a flexible enclosed chute in the shape of an elephant's trunk, which confines the concrete until it reaches the bottom of the form or the level of the preceding lift. The tremie will also keep the concrete mix from segregating and should be used whenever placing concrete that could "free fall" a distance more than 4 ft. (1.22 m). This segregation is caused by the large aggregate within the concrete mix hitting the reinforcing steel and shaking the cement/sand matrix coating off, the result being an improperly mixed concrete with inferior strength and possible voids (honeycomb).

After each lift is placed, the concrete mix is mechanically vibrated, necessary for the proper consolidation (elimination of voids) and to ensure that the fine matrix of the mix reaches the form surface for a smooth wall finish. Although vibration of the concrete mix is desirable as well as necessary, it can have some adverse effects on formwork if done improperly. If the vibrator is allowed to pass through the just placed concrete lift into the preceding lift, it will disturb the preceding lift's setting process, causing it to liquefy and increasing the head pressure against the formwork. Over vibrating the concrete will also imbalance the large aggregate to fine aggregate proportion within those areas of the mix.

Concrete Wall Forms Made in Panels or Sections. Many contractors, especially those performing smaller work, use foundation wall forms made up into panels 2'x 6' (0.6 x 1.82 m), 2'x8' (0.6 x 2.43 m), 4'x6' (1.2 x 1.82 m), 4'x8' (1.2 x 2.43 m) and so on. The frame for these sections is usually 2"x4" (50 x 100 m) lumber, with intermediate braces spaced 12" to 16" (300-400 mm) o.c. The corners may be reinforced with galvanized straps or strap iron, and the entire frame is sheathed with 5/8" (15 mm) or 3/4" (18.75 mm) plyform plywood. Forms of this kind can be used a large number of times, especially where the walls are fairly straight.

The labor cost of making panel forms will vary with the method used and the number made up at one time. A carpenter should complete one panel- 28 to 32 s.f. (2.60-2.97 sq.m) in 3/4 to 1 hr. or 8 to 10 panels per 8-hr. day.

The cost of a panel 4'-0"x8'-0" (1.2 x 2.4 sq.m) in size, using 3/4" (18.75 mm) Plyform plywood should cost as follows:

CONCRETE

Material Costs				
Material	BF	Cu.M.	Rate	Total
2 pcs. 2" x 4" x 8' (50 x 100mm x 2.44M)	10.67	0.025	$ 0.75	$ 8.00
7 pcs. 2" x 4" x 4' (50 x 100mm x 1.22M)	18.67	0.044	$ 0.75	$ 14.00
Plyform plywood	SF	Sq.M.	Rate	Total
¾" x 4' x 8' (20mm x 1.2m x 2.4m)	32	2.97	$ 0.60	$ 19.20
Nails, etc.			$ 8.24	$ 8.24
Material cost per panel 32 s.f.				$ 49.45
Cost per S.F.				$ 1.55
Cost per Sq.M.				$ 16.63

Labor Costs			
Labor	Hours	Rate	Total
Carpenter	0.75	$ 35.04	$ 21.19
Labor	0.25	$ 26.16	$ 5.56
Labor cost per panel			$ 26.75
Cost per S.F.			$ 0.84
Cost per Sq.M.			$ 9.00

Many contractors specializing in concrete work, especially in foundations for residences, use panel forms exclusively and get up to 25 uses from the forms, with the usual amount of repairs. These conditions should be given consideration when pricing any wall work using panel forms.

Symons Steel-Ply Forms. Symons Steel-Ply Forms (from Symons Corporation, Des Plaines, IL) are lightweight panels with a rolled steel frame completely enclosing a plastic coated plywood face. The plywood has a high strength face, which is 12% stronger than standard plywood. Due to the plastic coating, effects of moisture are reduced 40%. The welded steel frames are painted to prevent rusting and with reasonable care should last for a number of years.

Panels are 2'-0" (0.60 m) wide and come in lengths ranging from 2'-0" to 8'-0" (0.60-2.40 m) high in one foot increments. Fillers are made in widths 4" to 22" (100-550 mm), from 2'-0" to 8'-0" (0.60-2.40 m) high. All steel fillers are available in 1", 1-1/2", and 2" (25, 38, and 50 mm) widths. The average weight of the Steel-Ply Forms is 6 lbs. per s.f. (30 kg per sq.m). Outside corners are steel angles that lock adjoining forms together. Inside corners are 6"x6" (150 x 150 mm).

331

The side rails of the forms are rolled exclusively for Symons, and have a yield strength in excess of 55,000 psi (379 MPa). "L" shaped cross members on all forms are on 1-ft. centers and have a yield strength in excess of 60,000 psi (414 MPa). The frame is 2-1/2" (64 mm) in width.

In the Symons system, only two basic pieces are necessary for erection—the tie and the one-piece connecting hardware called the wedge-bolt. The same wedge-bolt is also used for connecting any Symons accessories to the form. In erecting forms, the panel tie is placed in the dado slot between the forms.

A wedge-bolt is inserted through the adjoining forms and tie loop. A second wedge-bolt is inserted through the slot of the first, automatically tying and joining the forms in one operation. The Symons system also uses an easy, efficient method for the erection of corners, pilasters, walers, bracing, and scaffolding. Standard panel ties have 1" (25 mm) breakbacks; 1-1/2" (37.5 mm) and 2" (50 mm) are also available with wood cones. Safe load capacity of panel ties is 3,000 lbs. (1357 kg) and 4,000 lbs. (1810 kg); flat ties 4,000 lbs. (1810 kg) and 4,500 lbs. (2036 kg) at 1.5 to 1.0 factor of safety. In stripping, panels may be quickly released from the finished wall. Wedge-bolts are easily loosened with a hammer tap, and panel ties then removed by giving tie loops a half twist, or in the case of flat ties, a vertical hammer blow.

Used panels may be rented for about $1.10 per s.f. ($11.85 per sq.m) for the first month and $1.00 per s.f. ($10.76 per sq.m) for each subsequent month. New panels rent for about $1.35 per s.f. ($14.53 per sq.m) for the first month and $1.15 per s.f. ($12.38 per sq.m) for each subsequent month. Rental costs should be checked with local suppliers. Minimum rental period is one month. Should the contractor wish to purchase the forms, a portion of the paid rentals are applicable to the purchase price. Rentals are charged from

CONCRETE

the date shipped from distribution center and stop on date of return. The contractor pays transportation charges both ways.

A typical foundation wall, 8" (200 mm) thick and 8' (2.4 m) high containing 2500 s.f. (225 sq.m) of plywood faced steel wallforms and using a crew of three carpenters to four laborers would average as follows:

Labor Costs			
Erecting Aligning& Bracing Forms	Hours	Rate	Total
Carpenter Foreman	4	$ 39.04	$ 156.16
Carpenters	16	$ 35.04	$ 560.64
Laborers helping	21	$ 26.16	$ 549.36
Laborers, Helping strip, cleaning, oil & reuse			
Carpenters	8	$ 35.04	$ 280.32
Laborers helping	16	$ 26.16	$ 418.56
Total Cost 2,500 Sq.Ft. (232.25 Sq.M)			$1,965.04
Cost per Sq. Ft.			$ 0.79
Cost per Sq.M.			$ 8.46

To the above add form rental, form tie, and form oil costs.

The Symons Company has made on-the-job time studies of placing, aligning, and stripping its forms. A wall similar to the above example, with a crew of three lead and four supporting workers, should work out to a productivity of 1.355 manhours per 100 s.f. (9 sq.m). In five hours this team should complete some 2500 s.f. (225 sq.m) of contact area.

A study of a curved wall forming a sewage treatment structure and using a crew of one carpenter to one laborer showed a productivity of 2.91 manhours per 100 s.f. (9 sq.m) of contact area. Setting of reinforcement and pouring of concrete are not included in the above figures.

Symons Versiform. This concrete forming system is a steel framed, plywood faced gang forming system for massive concrete structures and architectural concrete. It adapts to usc in walls, columns, piers, culverts, etc., and is available with four different standard facing materials.

Versiform panels are available for rental in 8'x4' (2.4x1.2 m), 8'x2' (2.4x.6 m), 4'x4' (1.2x1.2 m), 4'x2' (1.2x0.6 m), 2'x4' (0.6x1.2 m), and 2'x2' (0.6x0.6 m) panels, with special sizes available for purchase only. A complete line of components and accessory hardware is available for forming virtually any shape or size concrete structure. Versiform panels can be used in conjunction with Symons' other forming systems. A time study of Versiform panels and walers on continuous straight walls showed a productivity of 2.64 manhours per 100 s.f. (9 sq.m) of contact area.

Uni-Form Concrete Wall Forms

Uni-Form Steel Panel Forms. Uni-Forms (from Universal Form Clamp Co., Bellwood, IL) are prefabricated panels consisting of a steel frame 2'x1' (0.6x0.3 m), 2'x2' (0.6x0.6 m), 2'x3' (0.6x0.9 m), 2'x4' (0.6x1.2 m), 2'x5' (0.6x1.5 m), 2'x6' (0.6x1.8 m), 2'x7' (0.6x2.1 m), and 2'x8' (0.6x2.4 m), with a plywood facing. The panels are also furnished in fractional widths. The steel frame is a special rolled tee angle, with an equal legged tee strut to provide additional structural rigidity to the part of frame that acts as a plywood protector.

The plywood face is 5-ply plyform for use in concrete forms. The plywood facing provides a convenient nailing surface. It is advisable to oil the panels after each use. With proper care Uni-Forms have been used as many as 50 times before turning the plywood.

Uni-Form panels require alignment on one side only and employ a simplified method of forming inside corners and fillers.

On a foundation wall job for 47 houses requiring 2,350 s.f. (212 sq.m) of forms per house, using the 7' Uni-Form panels, the following labor costs were obtained:

CONCRETE

Erection Labor Costs			
Labor	Hours	Rate	Total
Carpenter Foreman	6	$ 39.04	$ 234.24
Carpenters	30	$ 35.04	$1,051.20
Laborers, Stripping, cleaning, oil & moving	18	$ 26.16	$ 470.88
Carpenters	10	$ 35.04	$ 350.40
Labor	8	$ 26.16	$ 209.28
Total Cost 2,350 Sq.Ft. (218.32 Sq.M)			$2,316.00
Cost per Sq. Ft.			$ 0.99
Cost per Sq.M.			$ 10.61

To the above add form rental, form tie, and form oil costs.

A carpenter erected and braced 78.3 s.f. (7.27 sq.m) of forms per hr. and required 0.6 hr. labor time per hr. of carpenter time. A carpenter stripped 235 s.f. (21.83 sq.m) of forms per hr. A laborer cleaned, oiled, and moved 294 s.f. (27.31 sq.m) of forms an hr.

On another foundation wall job for 50 houses, requiring 1,550 s.f. (144 sq.m) of forms per house (forms 2'-0"x7'-0" and 1'-0"x7'-0"), the following labor costs were obtained per 100 s.f. (9.29 sq.m) contact area of forms:

Labor Costs for Erecting and Bracing Uni-Form Concrete Wall Forms			
Labor	Hours	Rate	Total
Carpenter Foreman	0.65	$ 39.04	$ 25.38
Carpenters	2.6	$ 35.04	$ 91.10
Laborers, Stripping, cleaning, oil & moving	0.5	$ 26.16	$ 13.08
Carpenters	0.8	$ 35.04	$ 28.03
Labor	0.5	$ 26.16	$ 13.08
Total Cost 100 Sq.Ft. (9.29 Sq.M)			$ 170.67
Cost per Sq. Ft.			$ 1.71
Cost per Sq.M.			$ 18.37

To the above add form rental, form tie, and form oil costs.

A carpenter erected and braced 38.75 s.f. (3.60 sq.m) of forms per hr. A carpenter stripped 194 s.f. (17.50 sq.m) of forms per hr. A laborer oiled 194 s.f. (18.02 sq.m) of forms per hr. A laborer carried from one job to the other or loaded into trucks, 129 s.f. (11.98 sq.m) of forms per hr.

Based on the two preceding examples using Uni-Form prefabricated form panels, one finds that even though the panels, type of construction, and classification of labor are similar, the production time on the first example is considerably faster. There are many factors that affect the production time-weather, adaptability of materials used, accessibility to the work area, layout of the work, efficiency and experience of the workmen, and other factors. The estimator must evaluate all of these variables.

Built-In-Place Wood Forms for Concrete Walls

For those situations where the wall forms are to be built-in-place, the estimator should be familiar with form design parameters so that he can ascertain the correct quantity of form materials necessary for the work.

To order plyform (plywood), state the class, number of pieces, width, length, thickness and grade or veneer. Plywood panel veneer for both front and back of the panel should be specified. Example: For a C-D grade panel, one side conforms to the C grade, and the other side conforms to the D grade.

A. Solid Surface. Shims, circular repair plugs and tight knots to 1" across grain permitted.

B. Tight knots to 1" and 1-1/2". Knotholes to 1" across grain and some to 1-1/2", if the total width of knots and knotholes is within specified limits. Synthetic or wood repairs. Discoloration and sanding defects that do not impair strength permitted. Limited splits allowed. Stitching permitted.

C. Knots and knotholes to 2-1/2" (62.5 mm) width across grain and 1/2" (12.5 mm) larger within specific limits. Limited splits are permitted. Stitching permitted. Limited to exposure or interior panels.

Concrete Pressure For Slab Forms Lbs per Sq.Ft. and Kg per Sq.Meter					
Depth Inches	Non-Motorized Buggies* Lbs/Sq.Ft.	Motorized Buggies** Lbs/Sq.Ft.	Depth mm	Non-Motorized Buggies* Kg/Sq.M.	Motorized Buggies** Kg/Sq.m.
4	100	125	102	488.24	610.30
5	113	138	127	551.71	673.77
6	125	150	152	610.30	732.36
7	138	163	178	673.77	795.83
8	150	175	203	732.36	854.42
9	163	188	229	795.83	917.89
10	175	200	254	854.42	976.48

*Includes 50 psf load for workers, equipment, impact, etc.
**Includes 75 psf load for workers, equipment, impact, etc.

336

CONCRETE

Concrete Pressure For Column and Wall Forms
Pressures for Vibrated Concrete - Lbs. per Sq.Ft. (Kg per Sq.M)

Pour Rate Ft./Hr.	50°F Columns Lbs./Sq.Ft.	50°F Walls Lbs./Sq.Ft.	70°F Columns Lbs./Sq.Ft.	70°F Walls Lbs./Sq.Ft.	Pour Rate mm/Hr.	10°C Columns Kg/Sq.M.	10°C Walls Kg/Sq.M.	21.11°C Columns Kg/Sq.M.	21.11°C Walls Kg/Sq.M.
1	330	330	280	280	305	1611	1611	1367	1367
2	510	510	410	410	610	2490	2490	2002	2002
3	690	690	540	540	914	3369	3369	2637	2637
4	870	870	660	660	1219	4248	4248	3222	3222
5	1050	1050	790	790	1524	5127	5127	3857	3857
6	1230	1230	920	920	1829	6005	6005	4492	4492
7	1410	1410	1050	1050	2134	6884	6884	5127	5127
8	1590	1470	1180	1090	2438	7763	7177	5761	5322
9	1770	1520	1310	1130	2743	8642	7421	6396	5517
10	1950	1580	1440	1170	3048	9521	7714	7031	5712

*Maximum pressure need not exceed 150h, where h is maximum height of pour. Based on concrete with density of 150 pcf (2262 kg/m3) and 4" (100 mm) slump.

Allowable Pressures On Plyform Class I - For Architectural Applications

Allowable Pressure for Face Grain Across Supports - Lbs/Sq.Ft. and Kg./Sq.M.
(deflection limited to 1/360th of the span)

Support Spacing Inches	Plywood Thickness, inches						Support Spacing mm	Plywood Thickness, mm					
	1/2	5/8	3/4	7/8	1	1 1/8	mm	12.70	15.88	19.05	22.23	25.40	28.58
4	2935	3675	4495	4690	5075	5645	102	14330	17943	21946	22899	24778	27561
8	970	1300	1650	1805	1950	2170	203	4736	6347	8056	8813	9521	10595
12	410	575	735	890	1185	1345	305	2002	2807	3589	4345	5786	6567
16	175	270	370	475	645	750	406	854	1318	1806	2319	3149	3662
20	100	160	225	295	410	490	508	488	781	1099	1440	2002	2392
24	-----	-----	120	160	230	280	610	-----	-----	586	781	1123	1367
32	-----	-----	-----	-----	105	130	813	-----	-----	-----	-----	513	635
36	-----	-----	-----	-----	-----	115	914	-----	-----	-----	-----	-----	561

Allowable Pressures for Face Grain Parallel to Supports - Lbs/Sq.Ft. and Kg./Sq.M.

Support Spacing Inches	Plywood Thickness, inches						Support Spacing mm	Plywood Thickness, mm					
	1/2	5/8	3/4	7/8	1	1 1/8	mm	12.70	15.88	19.05	22.23	25.40	28.58
4	1670	2110	2610	3100	4140	4900	102	8154	10302	12743	15135	20213	23924
8	605	810	1005	1190	1595	1885	203	2954	3955	4907	5810	7787	9203
12	215	360	620	740	985	1165	305	1050	1758	3027	3613	4809	5688
16	-----	150	300	480	715	845	406	-----	732	1465	2344	3491	4126
20	-----	105	210	290	400	495	508	-----	513	1025	1416	1953	2417
24	-----	-----	110	180	225	320	610	-----	-----	537	879	1099	1562

*Plywood continuous across two or more spans.

338

Allowable Pressures On Plyform Class I - For Architectural Applications

Allowable Pressure for Face Grain Across Supports - Lbs/Sq.Ft. and Kg./Sq.M. (deflection limited to 1/270th of the span)

Support Spacing Inches	Plywood Thickness, inches						Support Spacing mm	Plywood Thickness, mm					
	1/2	5/8	3/4	7/8	1	1 1/8	mm	12.70	15.88	19.05	22.23	25.40	28.58
4	2935	3675	4495	4690	5075	5645	102	14330	17943	21946	22899	24778	27561
8	970	1300	1650	1805	1950	2170	203	4736	6347	8056	8813	9521	10595
12	430	575	735	890	1185	1345	305	2099	2807	3589	4345	5786	6567
16	235	325	415	500	670	770	406	1147	1587	2026	2441	3271	3759
20	135	210	285	350	465	535	508	659	1025	1391	1709	2270	2612
24	-----	110	160	215	295	340	610	-----	537	781	1050	1440	1660
32	-----	-----	-----	-----	140	170	813	-----	-----	-----	-----	684	830
36	-----	-----	-----	-----	105	120	914	-----	-----	-----	-----	513	586

Allowable Pressures for Face Grain Parallel to Supports - Lbs/Sq.Ft. and Kg./Sq.M.

Support Spacing Inches	Plywood Thickness, inches						Support Spacing mm	Plywood Thickness, mm					
	1/2	5/8	3/4	7/8	1	1 1/8	mm	12.70	15.88	19.05	22.23	25.40	28.58
4	1670	2110	2610	3100	4140	4900	102	8154	10302	12743	15135	20213	23924
8	605	810	1005	1190	1595	1885	203	2954	3955	4907	5810	7787	9203
12	270	405	620	740	985	1165	305	1318	1977	3027	3613	4809	5688
16	115	200	375	525	715	845	406	561	976	1831	2563	3491	4126
20	-----	125	210	290	400	495	508	-----	610	1025	1416	1953	2417
24	-----	-----	135	185	255	320	610	-----	-----	659	903	1245	1562

*Plywood continuous across two or more spans.

339

Maximum Spans for Lumber Framing - Inches*
Douglas Fir- Larch No.2 or Southern Yellow Pine No.2 (KD)

Equivalent Uniform Load Lbs./Ft.	Continuous Over 2 or 3 Supports (1 or 2 spans) Nominal Size - Inches								Continuous Over 4 or More Supports (3 or more spans) Nominal Size - Inches							
	2 x 4	2 x 6	2 x 8	2 x 10	2 x 12	4 X 4	4 X 6	4 X 8	2 x 4	2 x 6	2 x 8	2 x 10	2 x 12	4 X 4	4 X 6	4 X 8
400	36	53	70	90	109	50	79	101	41	60	78	100	122	62	91	118
600	30	43	57	73	89	44	66	88	31	49	64	82	99	51	74	98
800	24	38	50	63	77	39	58	76	25	39	52	66	81	44	64	85
1000	21	33	43	55	67	35	51	58	21	34	44	57	69	39	58	76
1200	19	29	38	49	60	32	47	62	19	30	39	50	61	35	53	69
1400	17	27	35	45	54	30	43	57	17	27	36	46	56	31	49	64
1600	16	25	32	41	50	27	41	54	16	25	33	42	52	28	44	58
1800	15	23	30	39	47	25	38	51	15	24	31	40	48	26	40	53
2000	14	22	29	337	45	23	36	48	14	22	29	38	46	24	37	49
2200	13	21	28	35	43	22	34	45	14	21	28	36	44	22	35	46
2400	13	20	26	34	41	20	32	42	13	20	27	34	42	21	33	44
2600	12	19	26	33	40	19	31	40	13	20	26	33	40	20	31	41
2800	12	19	25	32	38	19	29	38	12	19	25	32	39	19	30	39
3000	12	18	24	31	37	18	28	37	12	19	24	31	38	18	29	38
3200	11	18	23	30	36	17	27	35	12	18	24	30	37	18	28	36
3400	11	17	23	29	36	17	26	34	11	18	23	30	36	17	27	35
3600	11	17	22	29	35	16	25	33	11	17	23	29	35	16	26	34
3800	11	17	22	28	34	16	24	32	11	17	22	29	35	16	25	33
4000	10	16	22	28	34	15	24	31	11	17	22	28	34	15	24	32
4500	10	16	21	27	32	24	22	29	10	16	21	27	33	14	23	30
5000	10	15	20	25	31	13	21	28	10	16	20	26	32	14	22	28

* Spans are based on PS-20 Lumber sizes, single member stress were multiplied by a 1.25 duration-of-load factor for 7-day loads. Deflections limited to 1/360th of the span with a ¼ maximum. Spans are center to center of the supports

CONCRETE

Maximum Spans for Lumber Framing - mm*

Douglas Fir- Larch No.2 or Southern Yellow Pine No.2 (KD)

Equivalent Uniform Load Kg./M.	Continuous Over 2 or 3 Supports (1 or 2 spans) Nominal Size - mm								Continuous Over 4 or More Supports (3 or more spans) Nominal Size - mm							
	50x100	50x150	50x200	50x250	50x300	100x100	100x150	100x200	50x100	50x150	50x200	50x250	50x300	100x100	100x150	100x200
595.3	914	1346	1778	2286	2769	1270	2007	2565	1041	1524	1981	2540	3099	1575	2311	2997
892.9	762	1092	1448	1854	2261	1118	1676	2235	787	1245	1626	2083	2515	1295	1880	2489
1190.5	610	965	1270	1600	1956	991	1473	1930	635	991	1321	1676	2057	1118	1626	2159
1488.2	533	838	1092	1397	1702	889	1295	1473	533	864	1118	1448	1753	991	1473	1930
1785.8	483	737	965	1245	1524	813	1194	1575	483	762	991	1270	1549	889	1346	1753
2083.4	432	686	889	1143	1372	762	1092	1448	432	686	914	1168	1422	787	1245	1626
2381.1	406	635	813	1041	1270	686	1041	1372	406	635	838	1067	1321	711	1118	1473
2678.7	381	584	762	991	1194	635	965	1295	381	610	787	1016	1219	660	1016	1346
2976.3	356	559	737	8560	1143	584	914	1219	356	559	737	965	1168	610	940	1245
3274.0	330	533	711	889	1092	559	864	1143	356	533	711	914	1118	559	889	1168
3571.6	330	508	660	864	1041	508	813	1067	330	508	686	864	1067	533	838	1118
3869.2	305	483	660	838	1016	483	787	1016	330	508	660	838	1016	508	787	1041
4166.8	305	483	635	813	965	483	737	965	305	483	635	813	991	483	762	991
4464.5	305	457	610	787	940	457	711	940	305	483	610	787	965	457	737	965
4762.1	279	457	584	762	914	432	686	889	305	457	610	762	940	457	711	914
5059.7	279	432	584	737	914	432	660	864	279	457	584	762	914	432	686	889
5357.4	279	432	559	737	889	406	635	838	279	432	584	737	889	406	660	864
5655.0	279	432	559	711	864	406	610	813	279	432	559	737	889	406	635	838
5952.6	254	406	559	711	864	381	610	787	279	432	559	711	864	381	610	813
6696.7	254	406	533	686	813	610	559	737	254	406	533	686	838	356	584	762
7440.8	254	381	508	635	787	330	533	711	254	406	508	660	813	356	559	711

* Spans are based on PS-20 Lumber sizes, single member stress were multiplied by a 1.25 duration-of-load factor for 7-day loads. Deflections limited to 1/360th of the span with a ¼ maximum. Spans are center to center of the supports

Maximum Spans for Lumber Framing - Inches*
Hem - Fir No. 2

Equivalent Uniform Load Lbs./Ft.	Continuous Over 2 or 3 Supports (1 or 2 spans) Nominal Size - Inches								Continuous Over 4 or More Supports (3 or more spans) Nominal Size - Inches							
	2 x 4	2 x 6	2 x 8	2 x 10	2 x 12	4 X 4	4 X 6	4 X 8	2 x 4	2 x 6	2 x 8	2 x 10	2 x 12	4 X 4	4 X 6	4 X 8
400	33	48	70	80	97	48	73	96	36	53	70	90	109	56	81	107
600	27	39	57	65	80	41	59	78	27	43	57	72	88	45	66	88
800	22	34	50	57	69	35	51	68	22	35	46	59	72	39	58	76
1000	19	29	43	50	60	31	46	61	19	30	40	51	62	35	51	68
1200	17	26	38	44	54	29	42	55	16	27	36	45	55	31	47	62
1400	15	24	35	41	49	27	39	51	16	25	33	42	51	27	43	57
1600	14	23	32	38	46	24	36	48	15	23	30	39	47	25	39	52
1800	14	21	30	36	43	22	34	45	14	22	29	36	44	23	36	47
2000	13	20	29	34	41	21	33	43	13	21	27	35	42	21	33	44
2200	12	19	28	33	40	19	31	40	13	20	26	33	40	20	31	41
2400	12	19	26	31	38	18	29	38	12	19	25	32	39	19	30	39
2600	12	18	26	30	37	18	28	36	12	18	24	31	38	18	28	37
2800	11	18	25	30	36	17	26	35	11	18	24	30	37	17	27	36
3000	11	17	24	29	35	16	25	33	11	17	23	29	36	17	26	34
3200	11	17	23	28	34	16	24	32	11	17	22	29	35	16	25	33
3400	10	16	23	27	33	15	24	31	11	17	22	28	34	15	24	32
3600	10	16	22	27	32	15	23	30	10	16	22	27	33	15	23	31
3800	10	15	22	26	32	14	22	29	10	16	21	27	33	15	23	30
4000	10	15	22	25	31	14	22	29	10	16	21	27	32	14	22	29
4500	10	14	21	24	29	13	21	27	10	15	20	26	31	13	21	28
5000	9	13	20	23	28	12	20	26	9	15	20	25	30	13	20	26

* Spans are based on PS-20 Lumber sizes, single member stress were multiplied by a 1.25 duration-of-load factor for 7-day loads. Deflections limited to 1/360th of the span with a ¼ maximum. Spans are center to center of the supports

CONCRETE

Maximum Spans for Lumber Framing - mm*

Hem - Fir No. 2

Continuous Over 2 or 3 Supports (1 or 2 spans) — Nominal Size - mm

Equivalent Uniform Load Kg./M.	50x100	50x150	50x200	50x250	50x300	100x100	100x150	100x200
595.3	838	1219	1778	2032	2464	1219	1854	2438
892.9	686	991	1448	1651	2032	1041	1499	1981
1190.5	559	864	1270	1448	1753	889	1295	1727
1488.2	483	737	1092	1270	1524	787	1168	1549
1785.8	432	660	965	1118	1372	737	1067	1397
2083.4	381	610	889	1041	1245	686	991	1295
2381.1	356	584	813	965	1168	610	914	1219
2678.7	356	533	762	914	1092	559	864	1143
2976.3	330	508	737	864	1041	533	838	1092
3274.0	305	483	711	838	1016	483	787	1016
3571.6	305	483	660	787	965	457	737	965
3869.2	305	457	660	762	940	457	711	914
4166.8	279	457	635	762	914	432	660	889
4464.5	279	432	610	737	889	406	635	838
4762.1	279	432	584	711	864	406	610	813
5059.7	254	406	584	686	838	381	610	787
5357.4	254	406	559	686	813	381	584	762
5655.0	254	381	559	660	813	356	559	737
5952.6	254	381	559	635	787	356	559	737
6696.7	254	356	533	610	737	330	533	686
7440.8	229	330	508	584	711	305	508	660

Continuous Over 4 or More Supports (3 or more spans) — Nominal Size - mm

Equivalent Uniform Load Kg./M.	50x100	50x150	50x200	50x250	50x300	100x100	100x150	100x200
595.3	914	1346	1778	2286	2769	1422	2057	2718
892.9	686	1092	1448	1829	2235	1143	1676	2235
1190.5	559	889	1168	1499	1829	991	1473	1930
1488.2	483	762	1016	1295	1575	889	1295	1727
1785.8	406	686	914	1143	1397	787	1194	1575
2083.4	406	635	838	1067	1295	686	1092	1448
2381.1	381	584	762	991	1194	635	991	1321
2678.7	356	559	737	914	1118	584	914	1194
2976.3	330	533	686	889	1067	533	838	1118
3274.0	330	508	660	838	1016	508	787	1041
3571.6	305	483	635	813	991	483	762	991
3869.2	305	457	610	787	965	457	711	940
4166.8	279	457	610	762	940	432	686	914
4464.5	279	432	584	737	914	432	660	864
4762.1	279	432	559	737	889	406	635	838
5059.7	279	432	559	711	864	381	610	813
5357.4	254	406	559	686	838	381	584	787
5655.0	254	406	533	686	838	381	584	762
5952.6	254	406	533	686	813	356	559	737
6696.7	254	381	508	660	787	330	533	711
7440.8	229	381	508	635	762	330	508	660

* Spans are based on PS-20 Lumber sizes, single member stress were multiplied by a 1.25 duration-of-load factor for 7-day loads. Deflections limited to 1/360th of the span with a ¼ maximum. Spans are center to center of the supports

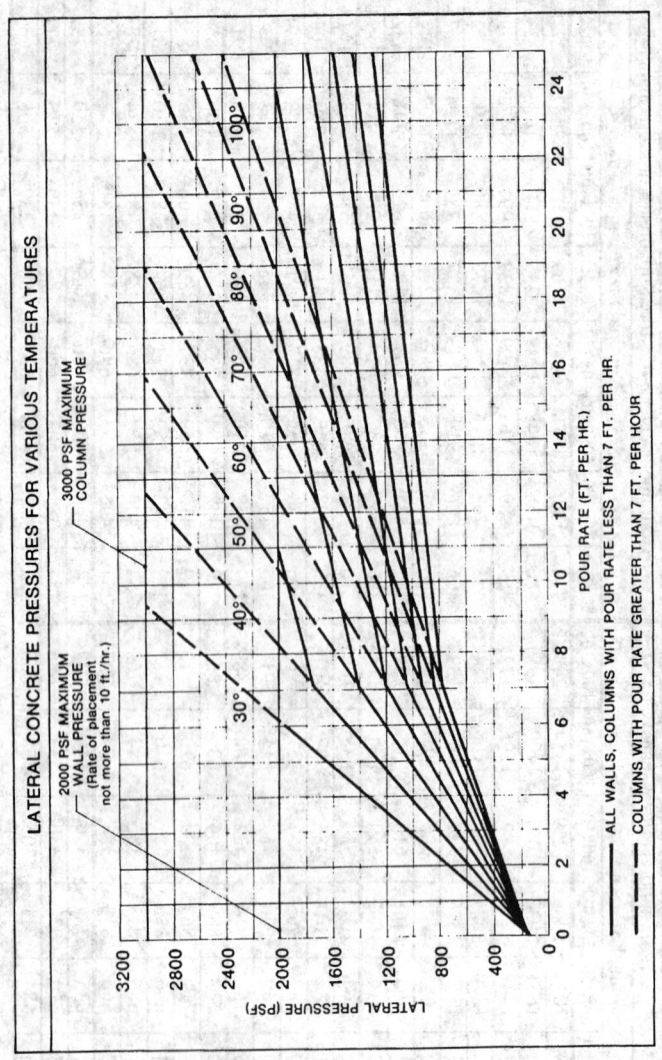

LATERAL CONCRETE PRESSURES FOR VARIOUS TEMPERATURES

3000 PSF MAXIMUM COLUMN PRESSURE

2000 PSF MAXIMUM WALL PRESSURE (Rate of placement not more than 10 ft./hr.)

LATERAL PRESSURE (PSF)

POUR RATE (FT. PER HR.)

—— ALL WALLS, COLUMNS WITH POUR RATE LESS THAN 7 FT. PER HR.

– – – COLUMNS WITH POUR RATE GREATER THAN 7 FT. PER HOUR

Estimating Material Requirements for Wood Forms
Built-In-Place for Concrete Walls

The following examples of wall form material estimates assume concrete at 50°F (10°C) with 4" (100 mm) slump, vertical pour rate of 3 to 4 ft. (0.9 to 1.2 m) per hour, and a maximum form pressure of 750 psf (3770 kg/m2).

CONCRETE

Lumber Required for Concrete Wall Forms 6'-0" (1.8 m) High

Based on a wall 40'-0" (12 m) long and 6'-0" (1.8 m) high, containing 480 s.f. (43 sq.m) of forms.

Lumber required for Concrete Wall Forms 6'- 0" (1.82 M) High				
Description	BF	Cu.M.	Rate	Total
Studs, 82 pcs. 2"x4"x6' (50x100mmx1.83m) @12" (300mm) on Centers	328	0.774	$ 0.75	$ 246.00
Plates, 80 LF (24m) 2"x4" (50x100 mm)	53	0.1250	$ 0.75	$ 39.75
Wales, 3 sets per side 12 pcs. 2"x4"x40' (50x100mmx12.19 m)	320	0.755	$ 0.75	$ 240.00
Bracing and stakes, 2"x4" (50 x100mm)	70	0.165	$ 0.75	$ 52.50
Plyform Plywood	SF	Sq.M.	Rate	Total
15 pcs of 3/4" (18.75mm) 4'x8' (1.22m x 2.44m) Plywood	480	44.59	$ 0.60	$ 288.00
Cost 480 s.f. (44.6 sq.m)				$ 866.25
per s.f.				$ 1.80
per sq.m				$ 19.43
Material not subject to reuse:				
Description	Lb	Kg	Rate	Total
Nails	20	9.07	$ 1.25	$ 25.00
Form ties, 3,000 lbs. (1357 kg) safe load 24" (600mm) o.c. along wales	63	0.149	$ 0.75	$ 47.25
Total Cost 480 s.f. (44.6 sq.m)				$ 72.25
cost per s.f.				$ 0.15
cost per sq.m				$ 1.62

Lumber Required for Concrete Wall Forms 8'-0" (2.43 m) High

Based on a wall 40'-0" (12.19 m) long and 8'-0" (2.43 m) high, containing 640 s.f. (59.46 sq.m) of forms.

Lumber required for Concrete Wall Forms 8'- 0" (2.42 M) High				
Description	BF	Cu.M.	Rate	Total
Studs, 82 pcs. 2"x4"x8' (50x100mmx2.42m) @12" (300mm) on Centers	438	1.033	$ 0.75	$ 328.50
Plates, 80 LF (24m) 2"x4" (50x100 mm)	53	0.1250	$ 0.75	$ 39.75
Wales, 3 sets per side 16 pcs. 2"x4"x40' (50x100mmx12.19 m)	427	1.007	$ 0.75	$ 320.25
Bracing and stakes, 2"x4" (50 x100mm)	80	0.189	$ 0.75	$ 60.00
Plyform Plywood	SF	Sq.M.	Rate	Total
20 pcs of 3/4" (18.75mm) 4'x8' (1.22m x 2.44m) Plywood	640	59.46	$ 0.60	$ 384.00
Cost 640 s.f. (59.46 sq.m)				$1,132.50
per s.f.				$ 1.77
per sq.m				$ 19.05
Material not subject to reuse:				
Description	Lb	Kg	Rate	Total
Nails	30	13.61	$ 1.25	$ 37.50
Form ties, 3,000 lbs. (1357 kg) safe load 24" (600mm) o.c. along wales	84	0.198	$ 0.75	$ 63.00
Total Cost 640 s.f. (59.46 sq.m)				$ 100.50
cost per s.f.				$ 0.16
cost per sq.m				$ 1.69

Total material cost for one use, $1.77 + $0.16 = $1.93 per s.f. ($19.05 + $1.69 = $20.74 per sq.m). Assuming 3 uses of form lumber, $1.77 ÷ 3 = $0.59 ($19.05 ÷ 3 =$6.35) per use of form lumber, plus $0.16 per s.f. for nails and form ties, or a total of$0.75 per s.f. for material. Add for bulkheads or end forms if required.

CONCRETE

Lumber Required for Concrete Wall Forms 12'-0" (3.65 m) High

Based on a wall 40'-0" (12.19 m) long and 12'-0" (3.65 m) high, containing 960 s.f. (89.18 sq.m) of forms.

Lumber required for Concrete Wall Forms 12'- 0" (3.65 M) High				
Description	BF	Cu.M.	Rate	Total
Studs, 82 pcs. 2"x4"x12' (50x100mmx3.65m) @12" (300mm) on Centers	656	1.547	$ 0.75	$ 492.00
Plates, 80 LF (24m) 2"x4" (50x100 mm)	53	0.1250	$ 0.75	$ 39.75
Wales, 3 sets per side 24 pcs. 2"x4"x40' (50x100mmx12.19 m)	640	1.509	$ 0.75	$ 480.00
Bracing and stakes, 2"x4" (50 x100mm)	98	0.231	$ 0.75	$ 73.50
Plyform Plywood	SF	Sq.M.	Rate	Total
30 pcs of 3/4" (18.75mm) 4'x8' (1.22m x 2.44m) Plywood	960	89.18	$ 0.60	$ 576.00
Cost 960 s.f. (89.18 sq.m)				$1,661.25
per s.f.				$ 1.73
per sq.m				$ 18.63
Material not subject to reuse:				
Description	Lb	Kg	Rate	Total
Nails	40	18.14	$ 1.25	$ 50.00
Form ties, 3,000 lbs. (1357 kg) safe load 24" (600mm) o.c. along wales	126	0.297	$ 0.75	$ 94.50
Total Cost 960 s.f. (89.18 sq.m)				$ 144.50
cost per s.f.				$ 0.15
cost per sq.m				$ 1.62

Total material cost for one use, $1.80 + $0.15= $1.95 per s.f. ($19.43 + $1.62 = $21.05 per sq.m). Assuming 3 uses of form lumber, $1.80 ÷ 3 = $0.60 per use for form lumber, plus $.15 for nail and form ties =$ 0.75 per s.f. for material. ($19.43 ÷ 3 = $6.48 +$1.62 = $8.10 per sq.m).

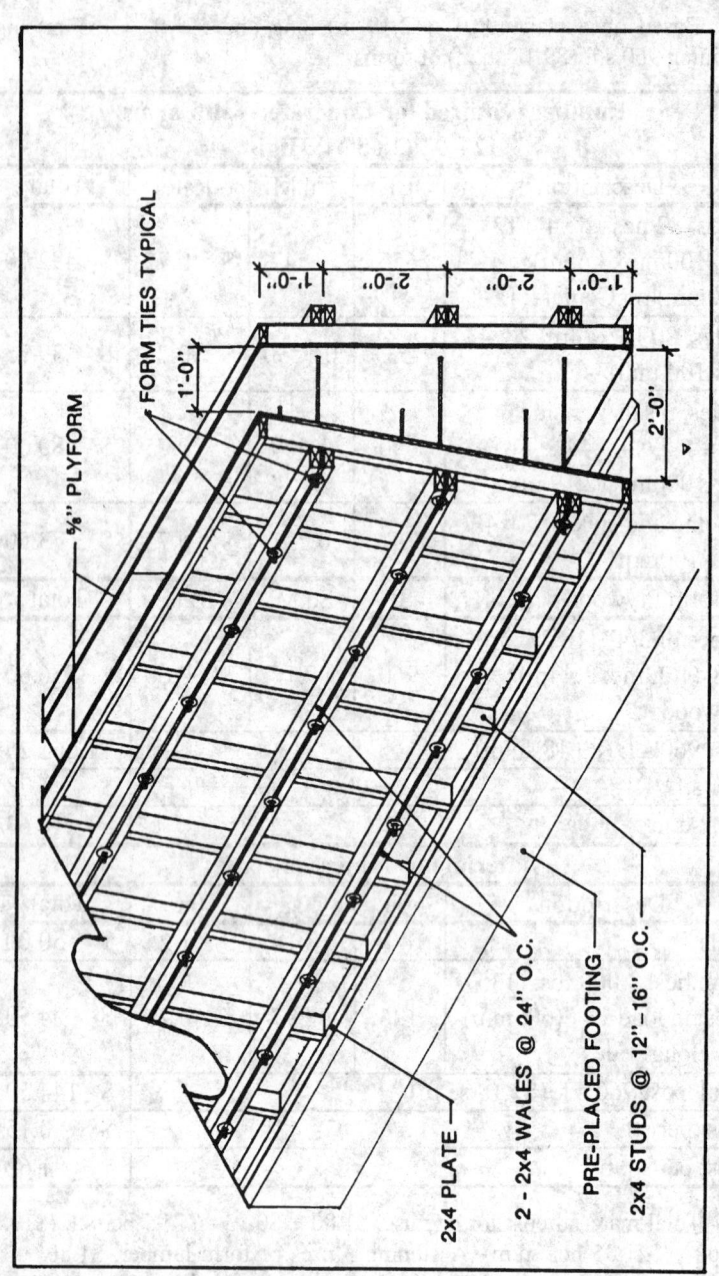

FORM TIES TYPICAL

1'-0"   2'-0"   2'-0"   1'-0"

1'-0"

2'-0"

⅝" PLYFORM

2x4 PLATE

2 - 2x4 WALES @ 24" O.C.

PRE-PLACED FOOTING

2x4 STUDS @ 12"–16" O.C.

In comparing the three preceding sample material estimates for wall forming on 6' (1.82 m), 8' (2.44 m), and 12' (3.65 m) high walls, note that all three are in the same cost range. This holds true for even lower or higher walls, as long as the design parameters for temperature, slump, pour rate, and allowable form pressure remain the same.

If the temperature of the concrete is lower, the mix is wetter, or it is advantageous to place the concrete into the forms at a faster pour rate, then the forms will have to be designed to withstand the increased pressure per square foot on the form surface. The design changes would involve any of the following:

1. Stronger form tie (increased tensile strength)
2. Closer horizontal spacing for form ties
3. Increased stud size to counteract bending under increased loads
4. Closer on-center spacing of stud members to counteract deflection of the plyform, or use of thicker plyform material
5. Increased size of wale members
6. Additional bracing and stakes

Basis for Computing Labor Costs. After checking the labor costs on a large number of jobs over a period of years, it has been found that a carpenter can frame and erect about 325 to 400 b.f. (7.58-9.3 cu.m) of lumber per 8-hr. day and requires 3/8 to 1/2 hr. laborer time (carrying lumber, etc.) for each hour of carpenter time.

If there are no laborers employed on the job and all lumber is handled and carried by carpenters, a carpenter should handle, frame, and erect about 225 to 275 b.f. (5.2-6.4 cu.m) of lumber per 8-hr. day.

When removing or "stripping" concrete forms, a carpenter or laborer should remove 750 to 1,000 b.f. (17.5-23.3 cu.m) of lumber per 8-hr. day.

High walls necessitate working from a scaffold and require 10% to 50% more time per sq. ft. of forms because of the additional handling, hoisting of lumber, and additional bracing. The sq. ft. (sq.m) costs on the following pages have been computed on this basis.

When foundation walls rest directly on earth without a concrete footing, mud sills of 2" (50 mm) or 3" (75 mm) plank are usually employed to serve as layout lines and supports for the wall panel forms required. This will increase the cost of 4'-0" (1.21 m) high panel wall forms about 10% for both material and labor. Similarly, 8'-0" (2.43 m) high wall forms should be increased about 5%.

Production Times for 100 S.F. (9 Sq.M) of Wood Forms for Concrete Walls

Class of Work	Height of Wall in Ft.	BF Lumber per SF of Forms	Place Forms			Removing Forms*	
			SF of Forms per 8-Hr Day	Carpenter Hours	Labor Hours	SF per 8-Hr Day	Labor Hours
Wall Forms Built-in Place	4 to 6	2.6	190 - 210	4.0	2.0	360	2.25
Wall Forms Built-in Place	7 to 8	2.6	165 - 190	4.5	2.3	350	2.2
Wall Forms Built-in Place	9 to 10	2.6	150 - 160	5.3	2.5	325	2.5
Wall Forms Built-in Place	11 to 12	2.6	125 - 135	6.0	3.0	300	2.7

Class of Work	Height of Wall in M.	Lumber Cu.M. per Sq.M. of Forms	Place Forms			Removing Forms*	
			SM of Forms per 8-Hr Day	Carpenter Hours	Labor Hours	Sq.M. per 8-Hr Day	Labor Hours
Wall Forms Built-in Place	1.2 to 1.8	0.066	17.7 - 19.5	4.0	2.0	33.4	2.25
Wall Forms Built-in Place	2.1 to 2.4	0.066	15.3 - 17.7	4.5	2.3	32.5	2.2
Wall Forms Built-in Place	2.7 to 3.0	0.066	13.9 - 14.9	5.3	2.5	30.2	2.5
Wall Forms Built-in Place	3.4 to 3.7	0.066	11.6 - 12.5	6.0	3.0	27.9	2.7

*If Union rules do not permit laborer working with carpenter, figure labor time as additional carpenter time.

Add foreman time as required. Normal practice on large concrete projects is to allow for one foreman when carpenter crew exceeds 3 carpenters.

CONCRETE

Labor Cost of 100 S.F. (9 Sq.M) Built-in-Place Concrete Wall Forms 4'-0" to 6'-0" (1.2 to 1.8 M) High, Requiring 2.6 BF (0.06Cu.M) of Lumber per Sq.Ft. (Sq. M) of Forms

Labor	Hours	Rate	Total
Carpenter	4	$ 35.04	$140.16
Laborer	2	$ 26.16	$52.32
Labor stripping & Cleaning	2.25	$ 26.16	$58.86
Labor cost per 100 s.f. (9.29 Sq.M)			$251.34
cost per s.f.			$2.51
cost per sq.m			$27.05

Labor Cost of 100 S.F. (9 Sq.M) Built-in-Place Concrete Wall Forms 8'-0" (2.43m) High, Requiring 2.6 BF (0.06Cu.M) of Lumber per Sq.Ft. (Sq. M) of Forms

Labor	Hours	Rate	Total
Carpenter	4.5	$ 35.04	$157.68
Laborer	2.3	$ 26.16	$60.17
Labor stripping & Cleaning	2.6	$ 26.16	$68.02
Labor cost per 100 s.f. (9.29 Sq.M)			$285.86
cost per s.f.			$2.86
cost per sq.m			$30.77

Labor Cost of 100 S.F. (9 Sq.M) Built-in-Place Concrete Wall Forms 10'-0" (3 m) High, Requiring 2.6 BF (0.06Cu.M) of Lumber per Sq.Ft. (Sq. M) of Forms

Labor	Hours	Rate	Total
Carpenter	5.3	$ 35.04	$185.71
Laborer	2.5	$ 26.16	$65.40
Labor stripping & Cleaning	2.8	$ 26.16	$73.25
Labor cost per 100 s.f. (9.29 Sq.M)			$324.36
cost per s.f.			$3.24
cost per sq.m			$34.91

Labor Cost of 100 S.F. (9 Sq.M) Built-in-Place Concrete Wall Forms 12'-0" (3.6 m) High, Requiring 2.6 BF (0.06Cu.M) of Lumber per Sq.Ft. (Sq. M) of Forms			
Labor	Hours	Rate	Total
Carpenter	6	$ 35.04	$210.24
Laborer	3	$ 26.16	$78.48
Labor stripping & Cleaning	3.4	$ 26.16	$88.94
Labor cost per 100 s.f. (9.29 Sq.M)			$377.66
cost per s.f.			$3.78
cost per sq.m			$40.65

Pilasters. Walls which are thickened for short intervals of 12" (300 mm) to 36" (900 mm) in length and then return to their former thickness are known as pilasters at the thickened area.

The cost of forming the face and sides of pilasters runs from 25% to 50% more than plain wall forms for both material and labor.

Radial Wall Forms. Wall forms that are built to radii or other curves require closer stud spacing, more bracing, and more labor than straight wall forms. Formwork for radial walls will cost from 50% to 100% more than forms for straight walls.

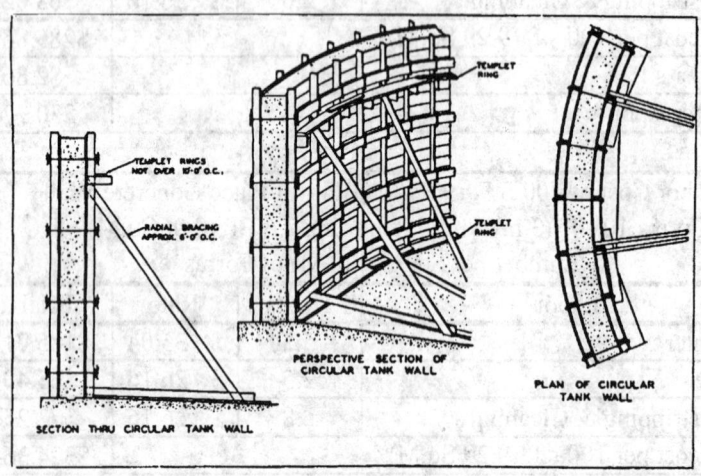

Method of Framing for Radial Concrete Walls

Forming Openings in Concrete Walls. Openings in concrete walls for doors, windows, and louvers are formed by placing bulkheads in the wall forms to form the sills, jambs, and heads of the openings required. These forms usually require from 2-1/4 to 3 b.f. of lumber per s.f. (0.05-0.07 cu.m per sq.m) of contact area, depending upon the size of the opening.

CONCRETE

The following is an itemized labor cost of boxed opening forms, based on 4'-0"x4'-0" (1.21 x 1.21 m) openings in a 1'-0" (300 mm) wall requiring 2-1/2 b.f. of lumber per s.f. (0.06 cu.m per sq.m) of contact area.

Labor Cost of 100 S.F. (9.29 Sq.M) of Boxed Opening Forms			
Labor	Hours	Rate	Total
Carpenter	7	$ 35.04	$245.28
Laborer	3.5	$ 26.16	$91.56
Labor stripping & Cleaning	3	$ 26.16	$78.48
Labor cost per 100 s.f. (9.29 Sq.M)			$415.32
cost per s.f.			$4.15
cost per sq.m			$44.71

Forming Setbacks in Concrete Walls. When concrete foundation walls extend some height above ground level, it is often required that the portion above grade be faced with masonry. This necessitates the portion above grade being reduced in thickness sufficient to accommodate the facing material. This reduction in thickness is accomplished by placing a "setback" form in the regular wall forms.

Setback forms are usually made up of 2" (50 mm) framing lumber faced with 5/8" (15 mm) or 3/4" (19 mm) plyform, and usually require 1-3/4 to 2-1/4 b.f. per s.f. (0.04-0.05 cu.m per sq.m) of contact area, depending mainly upon the thickness of the setback.

The following is an itemized labor cost of 100 s.f. (9.29 sq.m) of setback forms requiring 2 b.f. per s.f. (0.047 cu.m per sq.m) per s.f. (sq.m) of contact area:

Labor Cost of 100 S.F. (9.29 sq.m) of Setback Forms			
Labor	Hours	Rate	Total
Carpenter	6.5	$ 35.04	$227.76
Laborer	3.3	$ 26.16	$86.33
Labor stripping & Cleaning	2.7	$ 26.16	$70.63
Labor cost per 100 s.f. (9.29 Sq.M)			$384.72
cost per s.f.			$3.85
cost per sq.m			$41.41

Forming a Continuous Haunch on Concrete Walls. When an intermediate support on a wall face is required to provide bearing for masonry, slabs, or future construction, and reducing the wall thickness is prohibited by the design, a continuous haunch is commonly used.

A continuous haunch is usually formed of 2" (50 mm) framing lumber faced with 5/8" (15 mm) or 3/4" (19 mm) plyform and requires from 3-1/2 to

4 b.f. (0.08-0.09 m3) of lumber per s.f. (sq.m) of contact area. Following is the itemized labor cost of 100 s.f. (9.29 sq.m) of continuous haunch forms:

Labor Cost of 100 S.F. (9.29 sq.m) of Continuous Haunch Forms			
Labor	Hours	Rate	Total
Carpenter	12	$ 35.04	$420.48
Laborer	6	$ 26.16	$156.96
Labor stripping & Cleaning	2.5	$ 26.16	$65.40
Labor cost per 100 s.f. (9.29 Sq.M)			$642.84
cost per s.f.			$6.43
cost per sq.m			$69.20

Form Coating

For clean, easy stripping of forms and to prolong the life of form facings, the contact surface of all forms should receive a coating, which will prevent concrete from bonding to the form face.

The old method of brushing or spraying form faces with paraffin oil is still used by some contractors and gives adequate results for the average job, such as foundation walls, unexposed columns, beams, and slabs. Forms should always be coated before erection, but where this is impractical, such as for built-in-place wall forms, pan, and joist forms for combination slabs, oiling should be done in advance of setting reinforcing steel and care should be taken not to get any oil on concrete, masonry, or steel bearing surfaces.

On large areas, such as flat slab decks, a laborer can coat 400 to 500 s.f. (37.16-46.45 sq.m) of surface per hour, but production is considerably less when coating forms for columns, beams, combination slabs, etc. A fair overall average for coating forms is about 300 to 350 s.f. (27.87-32.52 sq.m) per hour.

For better class work, when using plywood, plastic-coated plywood, or lined forms, where smooth concrete surfaces are required, there are numerous patented coatings and lacquers available for this purpose. While the initial cost of these coatings is much higher than paraffin oil, better results are obtained and the forms may be used a couple of times without recoating, depending on the care used in applying the coating and in handling the forms when erecting or stripping.

Prices for these coating materials vary from $10.00 to $25.00 per gal. ($2.64-6.60 per liter), but coverage and durability also vary. Some coatings require special thinners before application, while others can be thinned with relatively inexpensive mineral spirits.

Application should be accomplished during dry weather, applied with a low pressure spray using a wide angle, low flow, flat fan type spray nozzle. A brush or roller may also be used in place of a sprayer. New wood forms require a heavy coat of form release agent prior to there first use. Multiple coats of form release agent are usually required to accomplish saturation and

CONCRETE

the coatings should be spaced from 3 to 6 hours apart. A worker can coat about 400 to 500 s.f. (37.16-46.45 sq.m) of form surface per hour.

During application care needs to be taken not to spray the reinforcement steel or other concrte inserts to remain. Symons® Action Kote™ is a mineral oil-based form release agent with a chemically reactive design that minimizes "dusting" and "Bugholes" on concrete surfaces. This material is used mostly for precast and architectural applications. Symons® Thrift Kote™ is a chemically reactive form release, in a solvent or water-based formulation, with optional cherry scent, that provides form stripping characteristics at a moderate price. No-Hold™ (W.R. Grace & Co.®) is an economical alternative to ready to use release agents and form oils. Recommended as a general release agent for steel, sealed concrete or wood forms, and fiberglass forms.

Concrete Column Forms

Reinforced concrete columns are square, rectangular, or round. The forms for square or rectangular columns are usually constructed of wood, while metal or fiber molds are used for round columns.

Forms for reinforced concrete columns should be estimated by the square foot (square meter), obtained by multiplying the girth of the column by the height. Example: Obtain the s.f. (sq.m) of forms in a column 18" (0.45 m) square and 12'-0" (3.65 m) high. Proceed as follows: 18 + 18 + 18 + 18 = 72" or 6'-0" (0.45 + 0.45 + 0.45 + 0.45 = 1.80 m), the girth. Multiply the girth by the height, 6'-0" x 12'-0" = 72 s.f. (1.83 m x 3.65 m = 5.48 sq.m) of forms.

The forms for all square or rectangular columns should be estimated in this manner.

To prevent the column forms from spreading while the concrete is placed and until it hardens, wood or metal clamps should be run around the column at intervals, depending upon the height of the column. In recent years wood clamps have been almost entirely eliminated by metal clamps.

The pressure of the fresh concrete has a greater effect on the design of column forms than on wall forms on account of the smaller amount of concrete to be placed. The contractor usually finds it more convenient and economical to fill a group of medium size columns in as short time as possible, perhaps 30 minutes, when the full liquid pressure of the fresh concrete will be acting, 150 lbs. per s.f. for every foot (755 kg/m2 for every meter) in depth of the concrete.

Column Form Design

There are two principal methods of erecting column forms; one is to unite all the four sides on two or three horses near the site where the form is to be erected and have 4 to 6 laborers hoist the complete form over the column dowels and stand it up temporarily; or in case of heavier columns, a simple 4"x 6" (100 x 150 mm) a small yard crane or a loader with a crane

355

attachment may be used for erection. Sometimes three sides of the column form are united and shoved into place without lifting the heavy box over the dowels, while the other and decidedly more expensive method is to erect each column form side separately.

As an example of the lumber required for a column form, take a concrete column 20" (500 mm) square and 11'-0" (3.35 m) high underneath the lowest beam, and assuming a full liquid pressure of 150 lbs. per c.f. (2400 kg/m3) acting for the entire height of the column. Assume the clear story height as 12'-6" (3.81 m). The area of the column form as usually taken off by the estimator will be 4 sides x 1.67' x 12'-6" high equals 83.50 s.f. (4 x 0.509 m x 3.81 m = 7.76 sq.m).

Using steel column clamps the following lumber would be required for a 20"x 20" (500 x 500 mm) concrete column form 12'-6" (3.81 m) high.

Lumber Required for 20" x 20" (500 mm x 500 mm) Concrete Column 12'-6" (3.81m) High		
Description	BF	Cu.M.
2 pcs. Plywood 3/4" Tk, 1'-9 ½" x 8'-0" (18.75mm Thk, 0.54m x 2.43m)	21.50	0.0507
2 pcs. Plywood 3/4" Tk, 1'-9 ½" x 4'-6" (18.75mm Thk, 0.54m x 1.37m)	12.09	0.0285
2 pcs. Plywood 3/4" Tk, 1'-8" x 8'-0" (18.75mm Thk, 0.51m x 2.43m)	20.00	0.0472
2 pcs. Plywood 3/4" Tk, 1'-8" x 4'-6" (18.75mm Thk, 0.51m x 1.37m)	11.25	0.0265
Total plywood required for 83.33 S.F. (7.74 Sq.M.) forms	64.84	0.1529
20 pcs. 1x6x1'-9" (25mm x 0.15m x 0.53m) cleats	17.50	0.0413
Lumber required for 83.5 S.F. (7.76 Sq.M.) forms	82.34	0.1942
Lumber required per S.F. (Sq.M.)	0.99	0.0250
Bracing lumber per S.F. (Sq.M.) forms	0.58	0.0146
Total lumber required per S.F. (Sq.M.) forms	1.56	0.0397

In order to brace the column form and hold it plumb, it will require 4 pcs. 2"x4" (50 x 100 mm), 16'-0" to 20'-0" (4.93-6.09 m) long, which will add another 48 b.f. to the lumber required. This bracing lumber may be used at least 3 or 4 times, eventually for purposes other than column forms. Bracing lumber is included in the above table.

The number of steel column clamps required may be somewhat larger than usually recommended by the manufacturers, depending on the speed with which the column forms are filled. The spacing in the lower part of the column should be about 9" (225 mm) and about 1'-0" (300 mm) near the middle. The first three spacings from the top should be 1'-3" to 1'-8" (375-500 mm) apart.

An average of one clamp per foot of column height will be safe in this particular instance.

Because plywood is furnished in panels 4'-0" (1.21 m) wide and 8'-0" (2.43 m) long, there is considerable waste unless columns are of sizes that will cut from the plywood sheets without waste, which is unusual.

Another method of stiffening plywood column form panels is by using 2"x4" (50 x 100 mm) vertical cleats. This method permits greater spacing of clamps as shown in the table and the total lumber requirements for a 20" (500 mm) column using 3/4" (18.75 mm) plywood sheathing and 2"x4" (50 x 100 mm) vertical cleats would be as follows:

Description	BF	Cu.M.
Plywood sheathing	100.00	0.24
Cleats	112.00	0.26
Bracing	48.00	0.11
Lumber required for 83.33 s.f. (7.74 sq.m.) forms	260.00	0.61
Lumber per S.F. (Sq.M.) of forms	3.12	0.08

When round concrete columns are used, metal or fiber molds are used for supporting the wet concrete until it has set. Some columns consist of a straight shaft extending from floor to ceiling, while others have a cone head at the top of the column. The forms for round concrete columns should be estimated at a certain price per column, stating whether columns have plain shafts or cone heads, because it costs considerably more to erect the latter. There are several firms in the country who make a specialty of renting these column molds and will quote a lump sum price on the job.

Labor Cost of Wood Forms for Square Concrete Columns. The labor cost of wood forms for square or rectangular concrete columns averages about the same regardless of the type of building. Variations are due to the method used in assembling and building up the column section, their size, method of clamping, etc.

On jobs of any size it is customary to have a mill on the job, where the column forms are built to exact size, including cut-outs for beams or girders, and then sent to the floor on which they are to be used, ready for erection.

Here is an important item when figuring column form costs. How many times can they be used on the job without cutting down or extending their length, or must they be reworked for each floor? This is a matter each estimator or contractor must decide from a study of the plans. For this reason separate labor costs are given for making up the column forms and erection.

Labor Cost of 100SF (9.29 Sq.M.) Concrete Column Forms Using Plywood Sheathing with Adjustable Steel Clamps and 2"x4" (50 x 100mm) Vertical Cleats

Description	Hours	Rate	Total
Making Forms			
Carpenter	4.0	$ 35.04	$140.16
Labor	2.0	$ 26.16	$52.32
Assembling, Erecting, Plumbing and Bracing			
Carpenter	3.0	$ 35.04	$105.12
Labor helping	3.0	$ 26.16	$78.48
Removing Clamps and Forms			
Labor	2.0	$ 26.16	$52.32
Cost per 100 s.f.			$428.40
cost per s.f.			$4.28
cost per sq.m			$46.11

Metal Forms for Round Columns

The cost of metal forms for round concrete columns varies according to the diameter and height of the column, whether shaft is plain or with a conical head, and whether it is necessary to cut out for beams and girders framing into the metal column molds.

Where it is necessary to cut the metal molds or "heads" to frame into wood forms for concrete beams and girders, an extra labor allowance should be made.

When plain round metal forms up to 24" (600 mm) in diameter are used without caps, which do not require cutting and fitting at head for concrete beams and girders, the labor erecting, removing, and oiling one column mold should cost as follows:

Plain Round Column			
Description	Hours	Rate	Total
Carpenter	2.0	$ 35.04	$70.08
Laborer	2.0	$ 26.16	$ 52.32
Total per column			$122.40

Add for scaffolding and/or crane service if required.

If the column has a conical head, necessitating placing a metal cap in addition to the column shaft, the labor erecting, removing, and oiling one column should cost as follows:

CONCRETE

Round Column with Conical Head			
Description	Hours	Rate	Total
Carpenter	2.5	$ 35.04	$87.60
Laborer	2.5	$ 26.16	$ 65.40
Total per column			$153.00

Add for scaffolding and/or crane service if required.

If necessary to cut and fit at column heads for concrete beams and girders, add 1 to 2 hrs. time per column. The above labor costs are for columns up to 16'-0" (4.87 m) in length. For columns 16'-0" to 20'-0" (4.87-6.09 m) long, add 1 to 2 hrs. time.

Leasing or Renting Metal Column Molds. Where metal column molds are leased or rented by the manufacturer to the contractor to be used on a specified job and returned in good condition upon completion of the work, the following prices will prove a fair average for columns up to 24" (600 mm) in diameter.

For a plain metal mold for round concrete columns without cone head, the monthly rental should figure about $9.00 per l.f. ($29.52 per m) of column form to be rented.

For plain round metal forms for concrete columns having cone head, add $70.00 per column per month rental.

To the above prices add delivery charges to the project site.

Adjustable Metal Clamps				
Clamp Size in Inches	Bar Size in Inches	Clamp Opening in Inches	Weight per Clamp in Lbs	Column Size in Inches
36"	5/16 x 2-1/2 x 36	10-1/2 to 28	40	6 to 23-1/2
48"	3/8 x 2-1/2 x 48	14 to 40	56	9-1/2 to 35-1/2
60"	3/8 x 3 x 60	34-1/2 to 50-1/2	85	20 to 46
Clamp Size in mm	Bar Size in mm	Clamp Opening in mm	Weight per Clamp in Kg	Column Size in mm
900	8 x 64 x 914	267 to 711	18	153 to 597
1200	10 x 64 x 1219	356 to 1016	25	242 to 902
1500	10 x 76 x 1524	622 to 1283	39	508 to 1169

C.F. (Cu.M) of Concrete and S.F. (Sq.M) of Forms Required Per L.F. (M) of Height for Round and Square Concrete Columns

Round Columns							
Diam. Inches	Area Sq. In.	Cu.Ft. Concrete	Sq.Ft. of Forms	Diam. mm	Area Sq. M.	Cu.M. of Concrete	Sq.M. of Forms
12	113	0.79	3.14	305	0.073	0.022	0.29
13	133	0.92	3.40	330	0.086	0.026	0.32
14	154	1.07	3.66	356	0.099	0.030	0.34
15	177	1.23	3.93	381	0.114	0.035	0.36
16	201	1.39	4.19	406	0.130	0.039	0.39
17	227	1.57	4.45	432	0.146	0.045	0.41
18	254	1.77	4.71	457	0.164	0.050	0.44
19	283	1.97	4.97	483	0.183	0.056	0.46
20	314	2.18	5.23	508	0.203	0.062	0.49
21	346	2.40	5.50	533	0.223	0.068	0.51
22	380	2.64	5.76	559	0.245	0.075	0.53
23	415	2.88	6.02	584	0.268	0.082	0.56
24	452	3.14	6.28	610	0.292	0.089	0.58
25	491	3.40	6.54	635	0.317	0.096	0.61
26	531	3.66	6.80	660	0.342	0.104	0.63
27	572	3.97	7.07	686	0.369	0.113	0.66
28	615	4.28	7.33	711	0.397	0.121	0.68
29	660	4.58	7.59	737	0.426	0.130	0.70
30	707	4.91	7.85	762	0.456	0.139	0.73
31	754	5.24	8.11	787	0.487	0.148	0.75
32	804	5.58	8.37	813	0.519	0.158	0.78
33	855	5.94	8.64	838	0.552	0.168	0.80
34	907	6.30	8.90	864	0.585	0.179	0.83
35	962	6.67	9.16	889	0.620	0.189	0.85
36	1017	7.07	9.42	914	0.656	0.200	0.88

CONCRETE

Square Columns							
Square Inches	Area Sq. In.	Cu.Ft. of Concrete	Sq.Ft. of Forms	Square Meter	Area Sq. M.	Cu.M. of Concrete	Sq.M. of Forms
12	144	1.00	4.00	305	0.093	0.028	0.37
13	169	1.17	4.33	330	0.109	0.033	0.40
14	196	1.36	4.67	356	0.126	0.039	0.43
15	225	1.56	5.00	381	0.145	0.044	0.46
16	256	1.78	5.33	406	0.165	0.050	0.50
17	289	2.01	5.67	432	0.186	0.057	0.53
18	324	2.25	6.00	457	0.209	0.064	0.56
19	361	2.51	6.33	483	0.233	0.071	0.59
20	400	2.78	6.67	508	0.258	0.079	0.62
21	441	3.06	7.00	533	0.285	0.087	0.65
22	484	3.36	7.33	559	0.312	0.095	0.68
23	529	3.67	7.67	584	0.341	0.104	0.71
24	576	4.00	8.00	610	0.372	0.113	0.74
25	625	4.34	8.33	635	0.403	0.123	0.77
26	676	4.69	8.67	660	0.436	0.133	0.81
27	729	5.06	9.00	686	0.470	0.143	0.84
28	784	5.44	9.33	711	0.506	0.154	0.87
29	841	5.84	9.67	737	0.543	0.165	0.90
30	900	6.25	10.00	762	0.581	0.177	0.93
31	961	6.67	10.33	787	0.620	0.189	0.96
32	1024	7.11	10.67	813	0.661	0.201	0.99
33	1089	7.56	11.00	838	0.703	0.214	1.02
34	1156	8.03	11.33	864	0.746	0.227	1.05
35	1225	8.51	11.67	889	0.790	0.241	1.08
36	1296	9.00	12.00	914	0.836	0.255	1.11

C.F. (Cu.M) of Concrete in Flared Caps For Round Concrete Columns							
Diam. Of Cap at Top Ft.	Diam. Of Cap at Bot. Ft.	Cu.Ft. Conc. Includes Col.Dia	Cu.Ft. Conc. Excludes Col.Dia	Diam. Of Cap at Top mm	Diam. Of Cap at Bot. mm	Cu.M. Conc. Includes Col.Dia	Cu.M. Conc. Excludes Col.Dia
4' - 0"	12"	9.75	8.50	1219	305	0.276	0.241
4' - 0"	14"	9.70	8.00	1219	356	0.275	0.227
4' - 0"	16"	9.60	7.60	1219	406	0.272	0.215
4' - 0"	18"	9.40	7.00	1219	457	0.266	0.198
4' - 0"	20"	9.30	6.50	1219	508	0.263	0.184
4' - 0"	22"	9.00	5.80	1219	559	0.255	0.164
4' - 0"	24"	8.80	5.30	1219	610	0.249	0.150
4' - 0"	26"	8.70	4.90	1219	660	0.246	0.139
4'- 6"	16"	13.60	11.20	1372	406	0.385	0.317
4'- 6"	18"	13.50	10.60	1372	457	0.382	0.300
4'- 6"	20"	13.30	10.00	1372	508	0.377	0.283
4'- 6"	22"	13.10	9.30	1372	559	0.371	0.263
4'- 6"	24"	12.90	8.60	1372	610	0.365	0.244
4'- 6"	26"	12.60	7.80	1372	660	0.357	0.221
4'- 6"	28"	12.30	7.10	1372	711	0.348	0.201
4'- 6"	30"	11.90	6.40	1372	762	0.337	0.181
5'- 0"	20"	18.20	14.30	1524	508	0.515	0.405
5'- 0"	22"	18.00	13.40	1524	559	0.510	0.379
5'- 0"	24"	17.80	12.60	1524	610	0.504	0.357
5'- 0"	26"	17.40	11.70	1524	660	0.493	0.331
5'- 0"	28"	17.10	10.90	1524	711	0.484	0.309
5'- 0"	30"	16.80	10.00	1524	762	0.476	0.283
5'- 0"	32"	16.40	9.10	1524	813	0.464	0.258
5'- 0"	34"	15.80	8.20	1524	864	0.447	0.232
5'- 0"	36"	15.30	7.30	1524	914	0.433	0.207

Sonotube® Fibre Forms. (Sonoco Products Co., Hartsville, SC) are spirally wound, laminated fibre tubes that were developed to provide a fast and economical method of forming for round concrete columns, piers, and the like. They are particularly adaptable as forms for encasing wood or steel piling. They provide a fast, economical method of forming all types of round concrete columns. They permit new design possibilities for architects, provide new structural features for engineers, and save time, labor, and money.

Sonotube fibre forms are available in standard sizes from 4" to 48" (100-1200 mm) inside diameter, with wall thickness ranging from 0.200" to 0.500" (5-12 mm). They are available in lengths up to 48 ft. (14.4 m) and can be easily cut to length on the job with an ordinary hand or power saw.

CONCRETE

There are three types of Sonotube fibre forms to meet virtually all job requirements. The Sonotube smooth or no seam is the premium form, with specially finished inner ply that produces a smoother, continuous concrete surface. There also is the Sonotube that leaves a spiral seam finish and the Sonotube plus that leave a vertical seam finish. Sonotube fibre forms are lightweight, easy to handle, and require few workers for erection, manually or with simple block and tackle. Minimum bracing is required, fixed in place at bottom, with light lumber bracing to keep them erect and plumb. No clamping is necessary.

Under normal conditions for average size columns, 12" to 18" (300-450 mm) diameter, 10 to 14 ft. (3.04-4.27 m) in length, a carpenter and a helper should handle, cut to length, erect, and brace about 12 Sonotube column forms per 8-hr. day at the following labor cost per column:

Erect Sonotube			
Description	Hours	Rate	Total
Carpenter	0.7	$ 35.04	$ 24.53
Labor	0.7	$ 26.16	$ 18.31
Cost per column			$ 42.84

Sonotube fibre forms strip easiest and most economically from 2 to 10 days after concrete is placed. Stripping may be accomplished with the least effort by using an electric hand saw, with blade set to cut slightly less than the wall thickness of Sonotube. Make two or three evenly spaced vertical cuts from bottom to top, after which the form can be pulled free from column. Sonotube forms are for one time use only.

Under normal conditions, a carpenter and a helper should strip 25 to 30 average size Sonotube column forms per 8-hr. day at the following labor cost per column:

Remove Sonotube			
Description	Hours	Rate	Total
Carpenter	0.3	$ 35.04	$ 10.51
Labor	0.3	$ 26.16	$ 7.85
Cost per column			$ 18.36

Approx. Prices of Sonotube Fibre Forms with "A" Coating

Diam. Inches	Wall Thick. In.	Approx. Weight Lbs./Ft.	Price per Lin.Ft.	Diam. mm	Wall Thick. mm	Approx. Weight Kg./M	Price per M.
6	0.125	1.700	$ 1.80	152	3.18	2.53	$ 5.91
8	0.125	1.700	$ 1.90	203	3.18	2.53	$ 6.23
10	0.150	2.300	$ 2.60	254	3.81	3.42	$ 8.53
12	0.150	2.800	$ 3.30	305	3.81	4.17	$ 10.83
14	0.175	3.600	$ 4.15	356	4.45	5.36	$ 13.62
16	0.175	4.800	$ 5.60	406	4.45	7.14	$ 18.37
18	0.200	5.400	$ 6.40	457	5.08	8.04	$ 21.00
20	0.200	7.500	$ 8.20	508	5.08	11.16	$ 26.90
22	0.225	8.200	$ 9.20	559	5.72	12.20	$ 30.18
24	0.225	8.900	$10.50	610	5.72	13.24	$ 34.45
26	0.300	9.600	$12.20	660	7.62	14.29	$ 40.03
28	0.300	10.400	$13.50	711	7.62	15.48	$ 44.29
30	0.300	11.800	$15.10	762	7.62	17.56	$ 49.54
36	0.300	14.500	$19.75	914	7.62	21.58	$ 64.80
42	0.445	18.600	$36.75	1067	11.30	27.68	$120.57
48	0.505	23.500	$47.25	1219	12.83	34.97	$155.02

Sonotube forms are manufactured in plants across the U.S., and distributors maintain stocks in most principal cities. The above prices are f.o.b. the nearest shipping point.

Anchoring Brick, Stone, and Terra Cotta to Concrete Backing

There are a number of methods of anchoring brick, stone, and terra cotta to concrete, some of which consist of metal slots nailed to the column or beam forms, which form a slot in the concrete for placing anchors.

The dovetail anchor slot method of anchoring masonry to concrete consists of a metal slot dovetail in cross section and anchors with dovetail ends for cut stone, brickwork, and terra cotta.

The dovetail anchor slot is nailed to the form with the open side against the wood. The ends are then closed with a wood plug. Or they come with foam filled slots. After the concrete is poured and the forms removed, the slot remains in the concrete, available at any course height to receive the anchors.

The anchors are installed by merely inserting the dovetail end of the anchor edgewise in the slot, then turning the anchor crosswise of the slot with the hand, the other end of the anchor being imbedded in the facing material or mortar joint. The dovetail slot holds the end of anchor securely, making any other means of fastening, drilling or boring unnecessary.

CONCRETE

Forms for Concrete Beams, Girders, Spandrel Beams, and Lintels

Forms for reinforced concrete beams, girders, lintels, etc., should be estimated by the square foot (square meter), and obtained by adding the dimensions of the three sides of the beam and multiplying by the length. For example, a beam or girder may be marked 12"x 24" (300 x 600 mm) on the plans, but maybe 6" (150 mm) of this depth is included in the slab thickness, so the form area should be obtained by taking the dimensions of the beam or girder from the underside of the slab: 18" + 12" + 18" = 48" or 4'-0" (450 + 300 + 450 mm = 1200 mm = 1.23 m) the girth, which multiplied by the length gives the number of s.f. of forms in the beam, 4'-0" x 18'-0" = 72 s.f. (1.23 x 5.48 = 6.74 sq.m).

Spandrel beams, or those projecting above or below the slab, should be measured in the same manner. A good rule to remember when measuring beam forms is to take the area of all forms that come in contact with the concrete.

When preparing an estimate on formwork, beams and girders should always be estimated separately; that is, the columns should be taken off, then the beams, girders and lintels, followed by the floor slabs, stairs.

As an example of the lumber required for beam and girder forms, assume an inside beam 12" (300 mm) wide by 1'-6" (0.45 m) deep by 19'-0" (5.79 m) long.

The contact area of this beam would be 2 sides 1'-6" (0.45 m) deep and a bottom 1'-0" (0.30 m) wide by 19'-0" (5.79 m) long, 1'-6" + 1'-6" + 1'-0" = 4' (0.45 + 0.45 + 0.30 = 1.20 m) girth multiplied by 19'-0" (5.79 m) long, equals 76 s.f. (7.06 sq.m) of forms requiring the following lumber:

Description	Bd.Ft.	Cu.M.
Beam soffit, 1 pc. 5/8" (15.62mm) plyform 12"x19'-0" s.f. (300mm x 1.76sq.m.)	19	0.0448
Beam sides, 2 pcs. 5/8" (15.62mm) plyform 22"x19'-0" s.f. (550mm x 1.76 sq.m.)	57	0.1344
Beam Bottom Joists 4 pcs. 3"x4"x10'-0" (75mm x 100mm x 3.05m)	40	0.0943
Beam Side Template 4 pcs. 2"x4"x10'-0" (50mm x 100mm x 3.05m)	27	0.0637
Beam Side Ledger 4 pcs. 2"x4"x10'-0" (50mm x 100mm x 3.05m)	27	0.0637
Beam Side Studs 40 pcs. 2"x4"x1'-4" (50mm x 100mm x 0.40m)	36	0.0849
Shores 6 pcs. 4"x4"x10'-0" (50mm x 100mm x	80	0.1887
Shores Cross Head (T) 6 pcs. 4"x4"x3'-0" (100mm x 100mm x 0.91m)	24	0.0566
Braces 12 pcs. 1"x4"x2'-6" (25mm x 100mm x	10	0.0236
Lumber required for 76 s.f. (7.06 sq.m.) of forms	320	0.7547
Lumber required per s.f. (sq.m.) of "contact area"	4.21	0.1069

Less than 1" is treated as 1" for the boardfoot calculation.

When beams or girders are more than 2'-0" (0.61 m) deep, the designs given under "Concrete Wall Forms" should be consulted for spacing and size of cleats or studs required, together with bracing and form ties necessary.

Labor Cost of Wood Forms for Concrete Beams and Girders

The labor cost on beam and girder forms will vary considerably on different jobs, due to the amount of duplication, size of beams and girders, and the methods of framing and placing them.

On typical buildings where the same size beams and girders are used on several floors, the forms can be used several times with only a small amount of labor with each reuse. Larger beams are more economical than very small ones, but one of the most important cost factors is the method of fabricating the units and whether they are built up in a job "shop" using power saws. These are factors that cut costs.

The following labor costs are based on efficient management, economical design, and labor-saving tools and equipment. If the old handsaw methods are to be used, increase the costs given 10 to 20%.

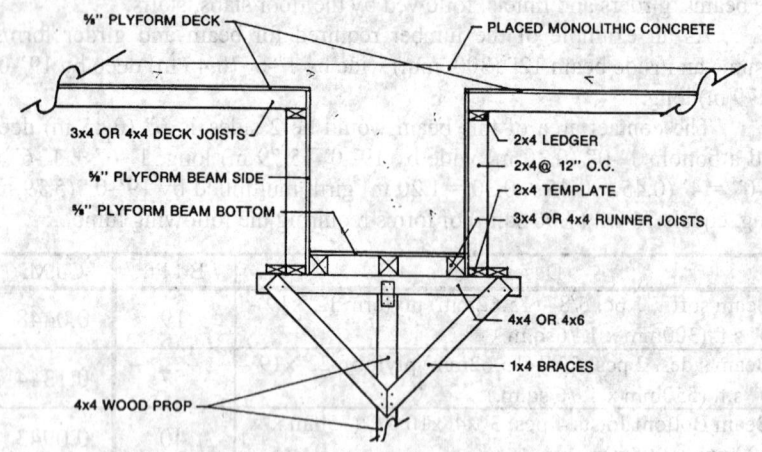

Method of Framing Wood Forms for Concrete Beams

It requires 4 to 4-1/2 b.f. (0.009-0.010 CM) of lumber to complete 1 s.f. (0.09 sq.m) of beam and girder forms, which includes uprights or "shores", beam soffits, bracing, etc.

A carpenter should frame and erect 250 to 275 b.f. (0.58-0.64 CM) of lumber per 8-hr. day, at the following labor cost per 1,000 b.f. (2.35 CM):

Description	Hours	Rate	Total
Carpenter	29.0	$ 35.04	$1,016.16
Labor	14.5	$ 26.16	$ 379.32
Cost per 1,000 b.f.			$1,395.48
Cost per 2.36 cu.m.			$ 591.68

CONCRETE

If laborers are not permitted to carry the lumber and forms for the carpenters, the "helper" time given above should be figured as carpenter time.

A laborer should remove or "strip" 900 to 1,000 b.f. (2.12-2.35 CM) of lumber per 8-hr. day, at the following labor cost per 1,000 b.f. (2.35 CM):

Description	Hours	Rate	Total
Labor	9.0	$ 26.16	$ 235.44

Labor Cost of 100 S.F. (9.29 Sq.M) of Inside Beam and Girder Forms Requiring 4-1/2 B.F. per S.F. (0.10 CM) of Forms, 4"x4" (100 x 100 mm) Wood Shores

Description	Hours	Rate	Total
Making Beam Soffits and Sides			
Carpenter	3.50	$ 35.04	$ 122.64
Labor helping	1.75	$ 26.16	$ 45.78
Assembling, Erecting, Shoring and Bracing			
Carpenter	5.50	$ 35.04	$ 192.72
Labor helping	2.75	$ 26.16	$ 71.94
Removing Forms			
Labor	2.75	$ 26.16	$ 71.94
Cost per 100 s.f.			$ 505.02
cost per s.f.			$ 5.05
cost per sq.m			$ 54.36

Labor Cost of 100 S.F. (9.29 Sq.M) of Spandrel Beam or Lintel Forms Requiring 4 B.F. per S.F. (0.009 CM m3 per sq.m) of Forms, 4"x4" (100 x 100 mm) Wood Shores

Description	Hours	Rate	Total
Spandrel Beams or Lintel Forms			
Carpenter	3.50	$ 35.04	$ 122.64
Labor helping	1.75	$ 26.16	$ 45.78
Assembling, Erecting, Shoring and Bracing			
Carpenter	8.00	$ 35.04	$ 280.32
Labor helping	4.00	$ 26.16	$ 104.64
Removing Forms			
Labor	3.50	$ 26.16	$ 91.56
Cost per 100 s.f.			$ 644.94
cost per s.f.			$ 6.45
cost per sq.m			$ 69.42

If the same forms can be reused without working over and cutting down, "Labor Making Forms" after first use may be reduced or omitted.

Forms for Upturned Concrete Beams. When the top of a beam is higher than the floor slab level, the beam side forms for that portion above the floor must be supported on temporary legs, which are removed after the concrete is poured and is still wet. It is also necessary to use spreaders to maintain the proper beam width.

While it is true that the additional form material required is negligible, the labor cost on this kind of beam side form will run 50 to 100% more than ordinary beam side forms.

Forms for Reinforced Concrete Floors

Forms for all types of reinforced concrete floors should be estimated by the square foot (square meter or millimeter), taking the actual floor area or the area of the wood or metal forms that come in contact with the concrete.

Estimating the Cost of Form Lumber. When computing the lumber cost per s.f. of forms, the estimator should study the plans carefully to determine just how many times it is possible to use the same form lumber in the construction of the building.

On 8- to 12-story buildings, it may be possible to use the same lumber 4 or 5 times, while on a 3-story building, it might be necessary to buy enough lumber to form almost the entire building. These are items that affect the material costs and can only be determined by the estimator or contractor, who should be in a position to know just how soon the forms can be stripped and how fast they will be required to carry on the work.

On many jobs it will be more economical to use high early strength cement instead of ordinary portland cement, which will enable you to strip forms in 3 days instead of 7 to 10 days, saving on forms needed. Keep in mind that form lumber that has been used 3 or 4 times will have little salvage value.

Types of Reinforced Concrete Floors. There are several types of reinforced concrete floor construction that require wood forms, centering, and shores. There are differences in cost. We will describe some of the more commonly used types in detail.

Flat Slab Construction. Mill, factory, warehouse, and industrial buildings designed for heavy floor loads are usually of flat slab construction, which consists of floors without beams or girders, except spandrel beams between columns on outside walls and around areas such as stair wells and elevator shafts. These buildings usually have square or rectangular columns.

Beam and Girder Type Construction. This is the oldest type of concrete structure, having square or round columns with beams and girders running between each row of columns. The floor slab is carried by the beams and girders and then by the columns.

Pan Construction with Concrete Joists. Apartment buildings, hotels, schools, hospitals, department stores, office buildings, and other

368

CONCRETE

structures requiring light floor loads ordinarily use this type of construction, which consists of metal pans in conjunction with concrete joists. The floor is used with round or square columns and beams running at right angles to joists.

Designing Forms for Reinforced Concrete Floors and Estimating Quantity of Lumber Required

The process of designing forms for reinforced concrete floors may be divided into several parts, such as determining the span of the sheathing, the size and spacing of joists and girders or stringers, and the size and spacing of the upright supports or shores.

The spacing of joists for concrete floor forms is governed by the strength of the plywood used. When plywood panels are used for sheathing, the spacing of the joists must coincide with the sizes of the plywood panels, 4'-0" x 8'-0" (1.22 x 2.42.44 m), which means that the joists should be spaced 12", 16", or 19" (300, 400, or 475 mm) on centers, depending on the load to be carried.

Weight of Concrete Floors of Various Thicknesses

In determining the weight per s.f. of floor, wet concrete is figured at 150 lbs. per c.f. (2400 kg/m3). Dead load of form lumber and live load on forms while concrete is being placed is figured at 38 lbs. per s.f. (190 kg/m2), which is sufficient for temporary construction.

Using 2" (50 mm) plank to support the concrete joists 2'-1" (0.63 m) or 2'-11" (0.89 m) o.c., with stringers spaced 4'-0" (1.22 m) apart to support the 2" (50 mm) plank.

The following tables give the sizes of stringers and the distance between supports or shores when supporting various floor loads. The tables are also based on using yellow pine, Douglas fir, or woods of equal strength.

Carrying Capacity of 4" x 4" (100 x 100 mm) Uprights or Shores. The carrying capacity of a given size upright or shore is generally limited by two principal factors. The first is compression that the shore exerts on the cross beam or stringer at right angles to the fibers. The stress per sq. in. (sq. mm) on the bearing area of the stringer on the shore should not greatly exceed 500 lbs. per sq. in. (3.45 MPa) for Yellow Pine, Western Fir, or similar lumber. Otherwise, there is a noticeable impression, often as much as 1/8" (3 mm), of the hore into the fibers of the stringer, especially when the lumber is green or water soaked after a long rain. A 4"x 4" (100 x 100 mm) S4S, 3-1/2"x3-1/2" (87.5 x 87.5 mm) upright should never be loaded to more than 6,000 lbs. (2700 kg), no matter how short it is, and a 4"x 4" (100 x 100 mm) rough shore should not be loaded to more than 8,000 lbs. (3625 kg).

Weight of Concrete Floors

Description	Floor Thickness, inches										
	2 "	3 "	4 "	5 "	6 "	7 "	8 "	9 "	10 "	11 "	12 "
Concrete Weight in Lbs/Sq.Ft.	25.0	37.5	50.0	62.5	75.0	87.5	100.0	112.5	125.0	137.5	150.0
Temp. Live Load in Lbs/Sq.Ft.	40.0	37.5	38.0	37.5	38.0	37.5	38.0	37.5	38.0	37.5	38.0
Total Weight in Lbs/Sq.Ft.	65.0	75.0	88.0	100.0	113.0	125.0	138.0	150.0	163.0	175.0	188.0

(metric)

Description	Floor Thickness, mm										
	# mm	# mm	## mm	## mm	## mm	## mm	## mm	## mm	## mm	## mm	## mm
Concrete Weight in Kg/Sq.M.	122.1	183.1	244.1	305.2	366.2	427.2	488.2	549.3	610.3	671.3	732.4
Temp. Live Load in Kg/Sq.M.	195.3	183.1	185.5	183.1	185.5	183.1	185.5	183.1	185.5	183.1	185.5
Total Weight in Kg/Sq.M.	317.4	366.2	429.7	488.2	551.7	610.3	673.8	732.4	795.8	854.4	917.9

CONCRETE

Table for Designing Slab Forms for Reinforced Concrete Floors Based On Using ¾" Sheathing, with Joists Spaced 2'-0" On Centers

Thickness of Concrete Floor Slab and Total Dead Load and Live Load in Lbs./S.F.									
Thickness Concrete Slab	2"	3"	4"	5"	6"	7"	8"	9"	12"
Total Load in Lbs./S.F.	65	75	88	100	113	125	138	150	188
Distance Between Stringers When Joists are Spaced 2'-0" on Centers									
2"x 4" Joists S4S	6'-0"	5'-6"	5'-6"	5'-0"	5'-0"	4'-6"	4'-6"	4'-0"	4'-0"
2"x 6" Joists S4S	7'-6"	7'-0"	7'-0"	7'-0"	6'-6"	6'-0"	6'-0"	6'-0"	5'-6"
2"x 8" Joists S4S	9'-6"	9'-0"	8'-6"	8'-6"	8'-0"	8'-0"	7'-6"	7'-6"	7'-0"
2"x 10" Joists S4S	11'-0"	10'-6"	10'-0"	10'-0"	9'-6"	9'-6"	9'-0"	9'-0"	8'-6"
Distance Between Shores or Supports Using Stringers of Size and Space Given Below									
4"x 4" S4S Stringers 5' CC	5'-0"	4'-6"	4'-0"	4'-0"	4'-0"	3'-6"	3'-6"	-----	-----
4"x 6" S4S Stringers 6' CC	6'-6"	6'-0"	6'-0"	6'-0"	6'-0"	5'-6"	5'-0"	5'-0"	4'-6"
3"x 8" S4S Stringers 6' CC	8'-6"	7'-6"	7'-0"	7'-0"	7'-0"	6'-6"	6'-0"	6'-0"	5'-6"
2"x 10" S4S Stringers 7' CC	7'-6"	7'-6"	5'-6"	4'-6"	4'-6"	4'-0"	4'-0"	-----	-----

The Distance between shores or Span of Stringers is generally an even Fraction of their Length. Stringers 14'-0" long will permit a spacing of 4'-8" or 7'-0", while 18'-0" stringer will permit spans of 3'-7", 4'-6", 6'-0" and 9'-0".

Based on using yellow pine. Douglas fir or woods of equal strength.

Table for Designing Slab Forms for Reinforced Concrete Floors Based On Using 18.75 mm Sheathing, with Joists Spaced .61 m On Centers

Thickness of Concrete Floor Slab and Total Dead Load and Live Load in Kg/Sq.M.									
Thickness Concrete Slab	50mm	75 mm	100 mm	125 mm	150 mm	175 mm	200 mm	225 mm	300 mm
Total Load inKg/Sq.M.	317.4	366.2	429.7	488.2	551.7	610.3	673.8	732.4	917.9
Distance Between Stringers When Joists are Spaced .61M on Centers									
2"x 4" Joists S4S	1.83	1.68	1.68	1.52	1.52	1.37	1.37	1.22	1.22
2"x 6" Joists S4S	2.29	2.13	2.13	2.13	1.98	1.83	1.83	1.83	1.68
2"x 8" Joists S4S	2.90	2.74	2.59	2.59	2.44	2.44	2.29	2.29	2.13
2"x 10" Joists S4S	3.35	3.20	3.05	3.05	2.90	2.90	2.74	2.74	2.59
Distance Between Shores or Supports Using Stringers of Size and Space Given Below									
4"x 4" S4S Stringers 5' CC	1.52	1.37	1.22	1.22	1.22	1.07	1.07	1.22	1.22
4"x 6" S4S Stringers 6' CC	1.98	1.83	1.83	1.83	1.83	1.68	1.52	1.52	1.37
3"x 8" S4S Stringers 6' CC	2.59	2.29	2.13	2.13	2.13	1.98	1.83	1.83	1.68
2"x 10" S4S Stringers 7' CC	2.29	2.29	1.68	1.37	1.37	1.22	-----	-----	-----

The Distance between shores or Span of Stringers is generally an even Fraction of their Length. Stringers 4.27 m long will permit a spacing of 1.42 M or 2.13 M, while 5.49 stringer will permit spans of 1.09 M, 1.37 M, 1.82 M and 2.74 M.
Based on using yellow pine. Douglas fir or woods of equal strength.

CONCRETE

Table for the Size of Stringers and the Distance Between Supports and Shores when Supporting Various Floor Loads

Size of Stringer		6+2 1/2	8+2 1/2	10+2 1/2	12+2 1/2	14+2 1/2
Thickness of Floor in Inches		6+2 1/2	8+2 1/2	10+2 1/2	12+2 1/2	14+2 1/2
Weight of Floor Lbs per Sq.Ft.		86	92	98	105	112
Distance Between Shores and Supports						
4 " x 4 " S4S	-----	5'- 0"	4'- 9"	4'- 8"	4'- 8"	4'- 8"
3 " x 6 " S4S	-----	6'- 6"	6'- 5"	6'- 3"	6'- 1"	6'- 0"
4 " x 6 " S4S	-----	7'- 0"	7'- 0"	6'- 8"	6'- 7"	6'- 6"
3 " x 8 " S4S	-----	8'- 0"	8'- 0"	7'- 6"	7'- 6"	7'- 6"
2 " x 10 " S4S	-----	8'- 0"	8'- 0"	7'- 6"	7'- 6"	7'- 6"
Thickness of Floor in mm		150+55	200+55	250+55	300+25	350+25
Weight of Floor Kg per Sq.M.		419.89	449.18	478.48	512.65	546.83
Distance Between Shores and Supports						
102 x 102 mm S4S	-----	1.52 M	1.52 M	1.52 M	1.42 M	1.42 M
76 x 152 mm S4S	-----	1.98 M	1.96 M	1.96 M	1.85 M	1.85 M
102 x 152 mm S4S	-----	2.13 M	2.13 M	2.13 M	2.01 M	2.01 M
76 x 203 mm S4S	-----	2.44 M	2.44 M	2.44 M	2.29 M	2.29 M
51 x 254 mm S4S	-----	2.44 M	2.44 M	2.44 M	2.29 M	2.29 M

Based on Yellow Pine, Douglas Fir, or woods of equal strength.

The other principal limitation of strength is due to the length of the shore and the degree of crookedness in its length. Every experienced contractor knows that shores often show bows of 1" to 2" (25-50 mm), and if not cut off exactly square to their length, the carrying capacity of most shores are greatly limited by the eccentric loading impressed on them by the uneven bearing of the stringer and the bow in the shores.

The table on the next page gives the permissible load on shores for various lengths and eccentricities.

Lintels as a rule produce quite large eccentricities on the shores due to the load from the adjoining slab, which comes through on the inside of the lintel forms onto the shore. Never allow a load larger than those given for 3" (75 mm) eccentricity for shores supporting lintels. One frequently sees lintels supported by shores spaced less than 2'-0" (0.61 m) on centers, which is a clear waste of money.

Where the height of the uprights or shores are longer than given in the above table, it is advisable to cross brace the shores, in which case the vertical distance between braces plus 40% may be taken as the post height. Example: You are using a 16'-0" (4.87 m) shore, with 2 rows of cross bracing at 6'-0" (1.8 m) and 12'-0" (3.6 m) above the floor. The distance between braces 6'-0" (1.83 m) plus 40%, or 2.40 ft. (0.73 m), makes the capacity the same as a shore 8.40 or practically 8'-6" (2.59 m) long. Thus an 8'-0" (2.44 m) shore good for 8,000 lbs. (3625 kg) would still be good for that weight when 16'-0" long, if cross braced by two rows of bracing 6'-0" (1.83 m) and 12'-0" (3.65 m) above the floor.

Estimating the Quantity of Lumber Required for Forms for Beam and Girder Type Solid Concrete Floor Slabs. The accompanying illustrations give detailed designs, from which the lumber required per s.f. (sq.m) of forms may be computed. The illustration shows a 4" (100 mm) concrete floor slab with a 7'-6" (2.28 m) clear span between beams. Using a joist spacing of 2'-0" (600 mm) as given in the table for joist spacing, 2"x 8" (50 x 200 mm) joists will carry the required load.

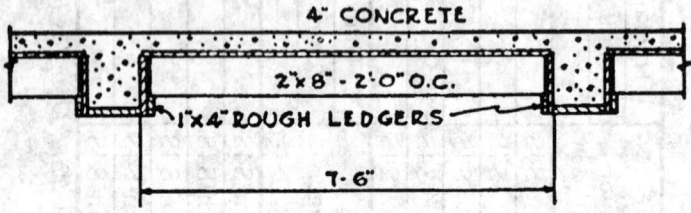

Method of Constructing Forms for Beam and Girder Type
Solid Concrete Floor Slabs

374

Safe Load in Lbs (Kg) on uprights or shores of Various Length in Feet (Meters)

Length of Upright or Shore Eccentricity		4"x 4" S4S YellowPine or Fir Length in Ft.		4"x 4" Rough YellowPine or Fir Length in Ft.		4"x 4" Rough Wisconsin Hemlock Length in Ft.	
		8	10	8	10	8	10
0 "	(Bow+Top)	6,000	6,000	8,000	8,000	8,000	8,000
1 "	(Bow+Top)	5,700	4,800	8,000	7,100	7,300	5,600
1 1/2 "	(Bow+Top)	4,700	3,900	7,300	5,800	6,000	5,100
2 "	(Bow+Top)	4,200	3,400	6,700	5,500	5,200	4,400
2 1/2 "	(Bow+Top)	3,600	3,000	5,300	4,500	4,800	4,100
3 "	(Bow+Top)	3,400	2,800	5,100	4,300	4,500	3,800

Length of Upright or Shore Eccentricity		100 x100 mm S4S YellowPine of Fir Length in M.		100 x100 mm S4S YellowPine of Fir Length in M.		100 x100 mm S4S Wisconsin Hemlock Length in M.	
		2.44	3.05	2.44	3.05	2.44	3.05
0.0 mm	(Bow+Top)	2,722	2,722	3,629	3,629	3,629	3,629
25.4 mm	(Bow+Top)	2,586	2,177	3,629	3,221	3,311	2,540
38.1 mm	(Bow+Top)	2,132	1,769	3,311	2,631	2,722	2,313
50.8 mm	(Bow+Top)	1,905	1,542	3,039	2,495	2,359	1,996
63.5 mm	(Bow+Top)	1,633	1,361	2,404	2,041	2,177	1,860
76.2 mm	(Bow+Top)	1,542	1,270	2,313	1,950	2,041	1,724

375

For a floor area 2'-0"x7'-6" (0.61 x 2.28 m) containing 15 s.f. (1.39 sq.m) the following lumber will be required:

Description	Bd.Ft.	Cu.M.
Sheathing*, 2'-0" x 7'-6" (0.61 x 2.28m)	18	0.0425
Joist, 1 pc. 2" x 8" x 8' (80mm x 200mm x 2.44m) plus 10% for joints	12	0.0283
Ledger boards, 2 pcs. 1" x 4" x 2' (25mm x 100mm x 0.61m)	1	0.0024
Total lumber required for 15 Sq.Ft. (1.39 Sq.M.) forms	31	0.0731
Bd.Ft. (Cu.M.) of lumber required per Sq.Ft. (Sq.M.) of forms	2.07	0.0525

* Includes 20% waste.

SOLID 6" FLOOR SLAB

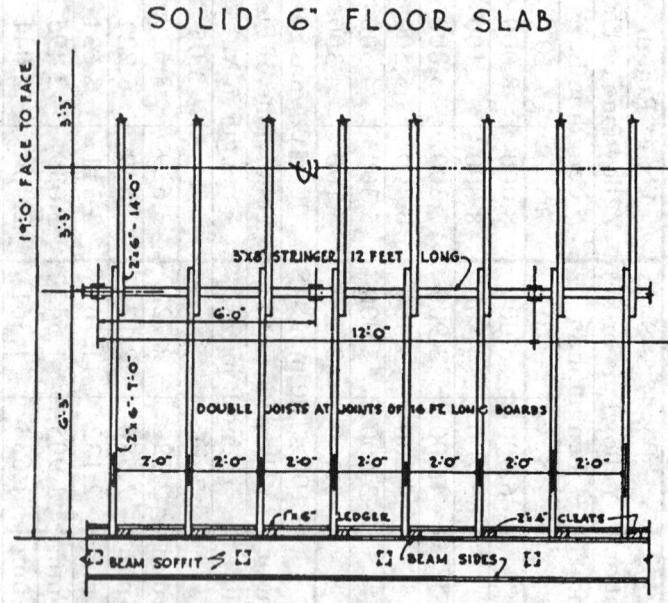

Method of Constructing Forms for Beam and Girder Type Solid Concrete Floors

For a 6" (150 mm) solid concrete slab with a clear span of 11'-0" (3.35 m) between girders, usually it is not economical to support joists with this span directly on the girders. It is better to provide a stringer or girt halfway between the concrete beams so that the joists have a span of only 5'-6" (1.67 m). Using a joist spacing of 2'-0" (600 mm), a 2"x 6" (50 x 150 mm) joist is strong enough, as shown by the table of joist spacings. A 4"x 6" (100 x 150 mm) supported by shores 6'-0" (1.83 m) o.c. will suffice.

CONCRETE

A floor area 2'-0"x11'-0" (0.61 x 3.35 m) containing 22 s.f. (2.04 sq.m) will require the following lumber:

Description	Bd.Ft.	Cu.M.
Sheathing*, 2'-0" x 11'-0" (0.61 x 3.35m)	27	0.0637
Joist, 1 pc. 2" x 6" x 12' (80mm x 150mm x 3.66m) plus 10% for joints	13	0.0307
Stringer, 1 pc. 4" x 6" x 2' (100mm x 150mm x 0.61m)	4	0.0094
Shore, 1/3 or 4" x 4" x 10' (100mm x 100mm x 3.05m)	6	0.0142
Total lumber required for 22 Sq.Ft. (2.04 Sq.M.) forms	50	0.1179
Bd.Ft. (Cu.M.) of lumber required per Sq.Ft. (Sq.M.) of forms	2.27	0.0577

Includes 20% waste.

If shores require bracing, add 0.1 Bd.Ft. per Sq.Ft slab.

Assume a solid concrete floor slab 16'-0" (4.87 m) wide, 19'-0" (5.79 m) long, and 12" (300 mm) thick, containing 304 s.f. (28.24 sq.m) of slab forms. Joists will be 2"x 6" (50 x 100 mm) spaced 24" (600 mm) apart, supported by 3 lines of 3"x 8"-16'-0" (75 x 200 mm - 4.87m) stringers, which in turn are supported by 4"x 4" (100 x 100 mm) shores 5'-4" (1.62 m) apart.

The following lumber will be required for 304 s.f. (28.24 sq.m) of forms:

Description	Bd.Ft.	Cu.M.
Sheathing*, 16'-0" x 19'-0" (4.87 x 5.79m)	365	0.8608
Joist, 9 pc. 2" x 6" x 14' (80mm x 150mm x 4.27m) plus 10% for joints	126	0.2972
Joist, 9 pc. 2" x 6" x 7' (80mm x 150mm x 2.13m) plus 10% for joints	63	0.1486
Stringer, 3 pc. 3" x 8" x 16' (75mm x 200mm x 4.88m)	96	0.2264
Shores, 9 pcs. 4" x 4" x 10'-0" incl. sill & wedges (100mm x 100mm x 3.05m)	153	0.3608
Braces, 6 pcs. 1" x 6" x 16'-0" (25mm x 150mm x 4.88m)	48	0.1132
Joist Support at Girders, 2 pcs. 1" x 6" x 16' (25mm x 150mm x 4.88m)	16	0.0377
Total lumber required for 304 Sq.Ft. (28.24 Sq.M.) forms	867	2.0448
Bd.Ft. (Cu.M.) of lumber required per Sq.Ft. (Sq.M.) of forms	2.85	0.0724

Includes 20% waste.

The above quantities are based on the assumption that the ends of joists and stringers are supported by the beam and girder forms. If a wall bearing job without beams and girders, the following additional lumber is required:

Description	Bd.Ft.	Cu.M.
Stringers, 2 pc. 2" x 8" x 16' at wall (50mm x 200mm x 4.88m)	43	0.1014
Shores, 6 pcs. 4" x 4" x 10'-0" incl. sill & wedges (100mm x 100mm x 3.05m)	102	0.2406
Braces, 2 pcs. 1" x 6" x 16'-0" (25mm x 150mm x 4.88m)	16	0.0377
Total lumber required for 304 Sq.Ft. (28.24 Sq.M.) forms	161	0.3797
Bd.Ft. (Cu.M.) of lumber required per Sq.Ft. (Sq.M.) of forms	0.53	0.0134

* Includes 20% waste.

Estimating the Quantity of Lumber Required for Flat Slab Concrete Floor Forms. Assume a flat slab having a 20'-0" (6.09 m) span, with 7'-0" (2.13 m) square drop heads, 8" (200 mm) thick as shown in illustration.

The maximum span of the joists is 5'-9" (1.75 m). The table of joist spacing shows that 2"x 6" (50 x 150 mm) joists and 3"x 8" (75 x 200 mm) stringers are sufficient. The load on the shores of the center stringer will be 5'-9" (1.75 m) x 6'-0" (1.83 m) = 34.5 s.f. (3.20 sq.m) at 138 lbs. per s.f. (695 kg/m2) = 4,761 lbs. Good practice requires a 4"x 4" (100 x 100 mm) shore rough, though a 4"x 4" (100 x 100 mm) S4S with bracing will do.

The lumber required for a 20'-0"x20'-0" (6.09 x 6.09 m) panel is as follows:

Description	Bd.Ft.	Cu.M.
Sheathing*, 400 Sq.Ft. (37.16 Sq.M.)	480	1.1321
Using 16' (4.88m) long floor sheathing boards requires 9 joist with a double joist being placed at the joint or for 20' (6.10m), it will require 11-1/4 pcs, 2" x 6" x 12' (50mm x 150mm x 3.65m)	135	0.3184
Column Strips, 9 pcs, 2" x 6" x 10' (80mm x 150mm x 3.05m)	90	0.2123
Stringers, 3 pcs 3" x 8" x 20' (75mm x 200mm x 6.10m)	120	0.2830
Stringer, 1 pc 4" x 4" x 12' (100mm x 100mm x 3.65m)	24	0.0566
Shore, 13 pcs 4" x 4" x 10' (100mm x 100mm x 3.05m) plus wedges, etc. @ 17' (5.18m)	221	0.5212

378

CONCRETE

Description	Bd.Ft.	Cu.M.
Horizontal Bracing, 7 pcs 1" x 6" x 20'-0" (25mm x 150mm x 6.09m)	70	0.1651
X Bracing, not less than 1/3 horizontal bracing	23	0.0542
Total lumber required for 400 Sq.Ft. (37.16 Sq.M.) forms	1163	2.7429
For Drop Heads		
Joists, 5 pcs 2" x 6" x 14' (80mm x 150mm x 4.27m)	70	0.1651
Stringers, 2 pcs 2" x 8" x 8' (75mm x 200mm x 2.44m)	21	0.0495
Shores, 9 pcs 4" x 4" x 17 BF (100mm x 100mm x 0.40 cu.m.)	68	0.1604
Total lumber required for 400 Sq.Ft. (37.16 Sq.M.) forms with drop heads	1322	3.1179
Bd.Ft. (Cu.M.) of lumber required per Sq.Ft. (Sq.M.) of forms	3.31	0.0839

* Includes 20% waste.

A 10'-0" (3.04 m) shore is sufficient for a 12'-0" (3.66 m) story height. For a 14'-0" (4.27 m) story height, add 2'-0" (0.61 m) to each of 17 shores or 34 x 1.33 ft. (0.40 m) = 45 b.f. (.000106 cu.m), which adds 0.11 b.f. (0.0002 cu.m) per s.f. of floor. A 16'-0" (4.88 m) story height will require 0.23 b.f. (0.0004 cu.m) per s.f. of floor.

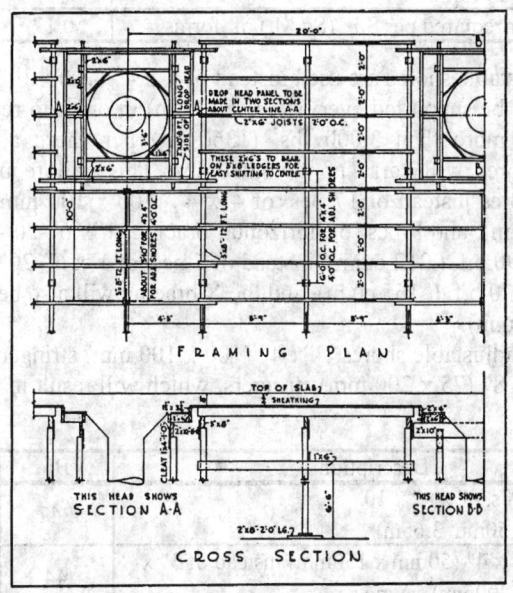

Method of Constructing Forms for Flat Slab Concrete Floors

379

For an 18'-0" (5.48 m) story height, it will be necessary to add for another set of cross bracing in addition to the 0.33 b.f. (0008 cu.m) of lumber required for shores. Another set of cross bracing will add 93 b.f. (.2192 cu.m)of lumber for 400 s.f. (37.16 sq.m) of floor or 0.23 b.f. (0.00054 cu.m) additional, which with the 0.33 ft. required for shores, will increase the lumber (.0004 cu.m) required per s.f. to 3.31 + 0.33 + 0.23 = 3.87 b.f. (.0091 cu.m) per s.f. of forms.

A 20'-0" (6.09 m) story height will add another 0.11 ft. (0.03 m) for extra length of shores, or a total of 3.98 b.f. (0.0093 cu.m) per s.f. of forms.

Where plywood is used for floor sheathing, there is at least a 10% waste in flooring, and while its use saves 480 b.f. (1.13 cu.m) for sheathing, it is necessary to use more joists, because the joists should not be spaced over 19" (475 mm) o.c., and the joists must be doubled every 8'-0" (2.44 m) instead of 16'-0" (4.88 m) where 16'-0" (4.88 m) long sheathing boards are used.

In 400 s.f. (37.16 sq.m) of forms, the following additional lumber would be required:

Description	BF	Cu.M.
Extra joists, 4 pcs. 2"x 6" x 12' (50mm x 150mm 3.65m)	48	0.1132
Extra joists, 2 pcs. 2"x 6" x 10' (50mm x 150mm x 3.04m)	20	0.0472
Total extra lumber required for 400 S.F. (37.16 Sq.M.) forms	68	0.1604
Extra lumber required per S.F. (Sq.M.) of forms	0.17	0.0041

If adjustable shores are used instead of 4"x 4" (100 x 100 mm) shores, stringers must be supported every 4'-0" (1.22 m) in order to reduce the load to not much more than 3,000 lbs. (1360 kg) per shore although some adjustable shores will carry up to 6,000 lbs. This will require 17 to 23 adjustable shores instead of 17 pcs. of 4"x 4" (100 x 100 mm) shores, 247 b.f. (0.582 cu.m), and 9 pcs. of horizontal bracing, 1"x 6"-20'-0" (25 x 150 mm – 6.1m), 90 b.f. (.212 cu.m), instead of 7 pcs. of 1"x 6"-20'-0" (25 x 150 mm – 6.1 m), 70 b.f. (..165 cu.m), and the X bracing will also be increased by 7 b.f. (0.016 cu.m).

Using adjustable shores, 4"x 4" (100 x 100 mm) stringers can be used instead of 3"x 8" (75 x 200 mm) stringers, which will result in the following saving:

Description	BF	Cu.M
Shores, 17 Pcs. 4" x 4" x 10' (50mm x 150mm 3.65m)	247	0.5825
Stringers, 4" x 4" (50mm x 150mm) instead of 3" x 8" (75mm x 200mm) saving s	44	0.1038
Total savings using adjustable shores	291	0.6863

CONCRETE

Description	BF	Cu.M
Extra 1" x 6" (25mm x 150mm) required for bracing	27	0.0637
Extra sills, 6 pcs 2" x 10" x 2' (50mm x 250mm x 0.61m)	20	0.0472
Extra lumber required per S.F. (Sq.M.) of forms	47	0.1108
Description	BF	Cu.M
Total savings on 400 Sq.Ft. (37.16 Sq.M.) forms	244	0.5755
Total savings per Sq.Ft. (Sq.M.) of forms	0.61	0.0146

On the other hand, it will be necessary to handle 17 to 23 shores weighing 1,000 to 1,400 lbs. while the 224-ft. savings weighs only 725 lbs.

For a 12" (300 mm) thick flat slab having a 20'-0" (6.09 m) span, the lumber required can be approximated as follows:

Description	BF	Cu.M
Sheathing per Sq.Ft. (Sq.M.) form	1.20	0.0028
Joists same spacing but using 2" x 8" (50mm x 200mm) joists	0.75	0.0018
Stringers, same as before	0.33	0.0008
Shore, spacing will be 4' to 6" (1.22 to 1.83m) requiring 21 each	0.90	0.0021
Bracing, 9 pcs. 1"x 6" x 20'- 0"	0.23	0.0005
x-bracing, 1/3 of horizontal bracing	0.08	0.0002
Drop Heads same as before	0.25	0.0006
Total Lumber, BF/SF (Cu.M./Sq.M.) for 12'- 6" (3.81m) Story Ht.	3.74	0.0088

Weight of Combination Metal Pan and Concrete Joist Floors

Concrete joists 4", 5", and 6" (100, 125, and 150 mm) wide, with a 2-1/2" (62.5 mm) concrete slab on top, using metal pans of various depths and widths.

For each 1/2" (12.5 mm) variation in slab thickness, add or subtract 6.25 lbs./s.f. (30.5 kg/m2) of floor.

Add 38 lbs./s.f. (185.53 kg/m2) to the above weights to take care of temporary dead and live load on floors while concrete is being poured.

Lumber Required for Floor Forms Using Metal Pans and Concrete Joist Construction. Assume a slab of 19'-0" (5.79 m) clear span and 16'-0" (4.87 m) long, using 8" (200 mm) metal pans with 5" (125 mm) wide ribs spaced 2'-1" (.633 mm) on centers and a 2-1/2" (62.5 mm) concrete top over the entire slab.

According to the table below, this construction weighs 52 lbs./s.f. (253.88 kg/m2) and including a 38-lb. (185.53-kg) live load during construction, makes a total load of 90 lbs./s.f. (439.41 kg/m2).

It is customary to use 2"x8" (50 x 200 mm) S4S planks to support the concrete ribs or joists. If the planks are continuous without patching at the

381

Weight of Combination Metal Pan and Concrete Joist Floors

Conc. Thickness (in)	2 1/2	2 1/2	2 1/2	Conc. Thickness (mm)	63.5	63.5	63.5
Width Conc. Joist In.	4	5	6	Width Conc. Joist mm	102	127	152
Metal Pan Dimesions	Weight Lbs/SF of Slab			Metal Pan Dimesions	Weight Kg/Sq.M. of Slab		
6" deep, 20" wide	45	48	50	150 mm deep, 500 mm wide	219.71	234.36	244.12
6" deep, 30" wide	41	43	45	150 mm deep, 750 mm wide	200.18	209.94	219.71
8" deep, 20" wide	51	54	57	200 mm deep, 500 mm wide	249.00	263.65	278.30
8" deep, 30" wide	45	48	50	200 mm deep, 750 mm wide	219.71	234.36	244.12
10" deep, 20" wide	56	60	64	250 mm deep, 500 mm wide	273.41	292.94	312.47
10" deep, 30" wide	50	52	55	250 mm deep, 750 mm wide	244.12	253.89	268.53
12" deep, 20" wide	63	67	72	300 mm deep, 500 mm wide	307.59	327.12	351.53
12" deep, 30" wide	54	57	61	300 mm deep, 750 mm wide	263.65	278.30	297.83
14" deep, 20" wide	-----	75	80	350 mm deep, 500 mm wide	-----	366.18	390.59
14" deep, 30" wide	-----	62	66	350 mm deep, 750 mm wide	-----	302.71	322.24

end spans, a span of 4'-0" (1.22 m) will support a load of 265 lbs./l.f. (394.36 kg/m) with a deflection of .235" (5.87 mm). In this instance the load is 2.09 x 90 lbs. per s.f. = 188 lbs. per l.f.(279.77 kg/m),which reduces the deflection to 0.166" (4.15 mm) or about 1/6", which is permissible.

It will require a stringer or girt every 4'-0" (1.22 m) to support the planks and the load on the stringers will be 360 lbs./l.f. (53574540 kg/m).

A 2"x10"-16'-0" (50 x 250 mm - 4.87 m) S4S will easily carry this uniform load per l.f. when supported by 4"x4" (100 x 100 mm) shores spaced 8'-0" (2.44 m) apart.

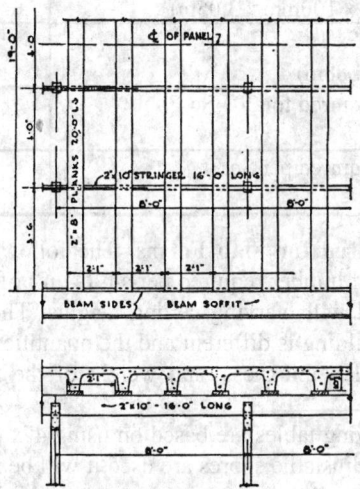

Method of Constructing Forms for Floor of Combination Metal Pans and Concrete Joist Construction

The following lumber will be required for 304 s.f. (28.24 sq.m) of forms:

Description	BF	CM
Planks, 9 pcs. 2" x 8" x 20' (50mm x 200mm x 6.09m)	240	0.5660
Stringers, 4 pcs. 2" x10" x 16' (50mm x 250mm x 4.88m)	107	0.2524
Shores, includes sills and wedges, 8 pcs. 4" x 4" x 10' (100mm x 100mm 3.05m)	136	0.3208
Joist supports at girders, 2 pcs. 1" x 6" x 16' (25mm x 150mm x 4.88m)	16	0.0377
Bracing, 4 pcs. 1" x 6" x 16' (25mm x 150mm x 4.88m)	32	0.0755
Bracing, 2 pcs. 1" x 6" x 20' (25mm x 150mm x 6.09m)	20	0.0472
Total lumber required for 304 s.f. forms	551	1.2995
Lumber required BF/SF (Cu.M./Sq.M.) of forms	1.81	0.0460

The above quantities are based on the assumption that the ends of the joists and stringers are supported by the beam and girder forms. If a wall bearing job without beams or girders, requiring additional stringers and shores at walls, add the following:

Description	BF	CM
Stringers, 2 pcs. 2" x10" x 16' (50mm x 250mm x 4.88m)	54	0.1274
Shores, including sills and wedges, 4 pcs. 4" x 4" x 10' (100mm x 100mm x 3.05m)	68	0.1604
Bracing, 2 pcs. 1" x 6" x 16' (25mm x 150mm x 4.88m)	16	0.0377
Additional lumber required for 304 Sq.Ft. (28.24 Sq.M.) form.	138	0.3255
Additional lumber required per Sq.Ft. (Sq.M.) of form.	0.45	0.0115

Lumber Required for Slab Forms. The following tables give the approximate quantity of lumber required per s.f. (sq.m) of forms for floors of different thickness and with varying ceiling heights. These tables are only approximate. Every building is different and the quantities will vary with the span, ceiling height, slab thickness, and weight of the wet concrete to be supported.

Also, the following tables are based on using 4"x 4" (100 x 100 mm) shores or uprights. If adjustable shores are used it will be necessary to deduct the lumber required for 4"x 4" (100 x 100 mm) and add cost or rental of adjustable shores. Add cost of additional bracing as more shores and bracing are frequently required.

The difference in the quantity of lumber required per s.f. (sq.m) for floors having 10'-0" (3.04m,) ceiling heights and those having 20'-0" (6.09 m) ceiling heights is principally in the shoring and bracing. If adjustable shores are used, deductions can be made on the following basis: floors 3" to 6" (75-150 mm) thick with 10'-0" (3.04 m) ceiling heights require 1/3 to 3/8 ft. (99 to 112.5 mm) of lumber per s.f. (sq.m) for shores; 7" to 9" (175-225 mm) slabs require 1/2 to 5/8 ft. (150-188 mm); and 10" to 12" (250-300 mm) slabs require 5/8 to 3/4 ft. (188-228 mm) of shoring lumber per s.f. (sq.m) of floor.

The above quantities can be used for all 10'-0" (3.04 m) story heights, plus the additional lumber given in the tables for higher ceiling heights.

The forms are designed in yellow pine, Douglas fir, or woods of equal strength, with the loads computed as follows: wet concrete 150 lbs./cu.f. (2.403 g/cu.m), dead load of form lumber, concrete, non-motorized buggies, and workers is an additional 50 lbs./cu.f. (.800 g/cu.m).

The quantities below include lumber required for forming drop heads. If adjustable shores are used, deduct lumber quantities given on previous pages.

CONCRETE

Number of Feet (meters) of Lumber Required per Sq.Ft. (Sq.M.) of Flat Slab Concrete Floor Forms				
Ceiling Height in Ft.	Thickness of Slab in Inches			
	6	8	10	12
	Weight Dead and Live Load, psf			
	125	150	175	200
10	3.0	3.1	3.4	3.6
12	3.1	3.2	3.5	3.7
14	3.3	3.4	3.6	3.9
16	3.4	3.5	3.9	4.0
18	3.4	3.8	4.2	4.4
20	3.5	3.9	4.3	4.6
Ceiling Height in Meters	Thickness of Slab in mm			
	152	203	254	305
	Weight Dead and Live Load, kg/sq.m.			
	610	732	854	976
3.05	9.84	10.17	11.16	11.81
3.66	10.17	10.50	11.48	12.14
4.27	10.83	11.16	11.81	12.80
4.88	11.16	11.48	12.80	13.12
5.49	11.16	12.47	13.78	14.44
6.10	11.48	12.80	14.11	15.09

Note: Dead Load = 150 psf (732.36 kg/sq.m.); Live Load = 50 psf (244.12 kg/sq.m.)

Using Steel Truss Joists Instead of Shoring for Supporting Formwork for Reinforced Concrete Floors. In steel frame buildings where clear floor space is essential, steel truss joists may be used for supporting the wood formwork, as shown in the accompanying illustration.

This is a simple method that is very effective as the salvage value of the truss joists is very high.

Sectional Steel Shoring. Many contractors now use their sectional steel scaffolding units for shoring formwork for reinforced concrete slabs and beams. Prefabricated sectional steel scaffolding sections are as simple to erect for shoring as they are for scaffolding. The same scaffolding components-base plates, adjustable extension legs, welded 5-ft. (1.52 m) wide frames varying in height from 3'-0" to 6'-6" (0.91-1.98 m) and diagonal braces to provide spacing between frames from 2'-6" to 7'-0" (0.76-2.13 m)- are assembled quickly to provide free standing shoring sections suited to the easy placing of stringers, joists, and decking. In most instances, sectional steel shoring can be used wherever conventional shoring methods of adjustable shores or 4"x 4" (100 x 100 mm) lumber are used, but is especially adaptable for shoring forms for large expanses of flat slab construction.

385

Height adjustments are quickly made by the use of 20" (500 mm) adjustable legs, which eliminates the need for cutting, fitting, and wedging. These legs also facilitate stripping operations by the ease with which the shoring units may be lowered. Sectional steel shoring also provides scaffolding within the shoring for forming and stripping, increasing job safety. This method also minimizes the use of wood, reducing the fire hazard and increasing the number of reuses obtainable from form lumber.

Safe working loads for sectional steel shoring depend on the spacing between frames in the same row, distance between rows of frames, and position of the stringers on the frames. Sectional steel shoring will support the largest loads when the stringers are placed directly over the frame legs by means of inverted base plates or U-heads as shown in the illustration. If this is not practical, stringers should be placed as close as possible to frame legs. If required, additional lateral rigidity may be attained by interlacing frames with standard bracing.

It is available in three types: Standard, 20K Heavy Duty for loads up to 10,000 lbs. (4536 kg) on each frame leg, and Extra Heavy Duty for loads up to 40,000 lbs. (18144 kg) on each leg.

The contractor who does not own this equipment may rent sectional steel shoring, engineered for a specific job, from the manufacturer on the same basis as scaffolding. This service is available nationwide.

Since job conditions and requirements vary considerably, it is practically impossible to give unit labor production values on this method of shoring, but various contractors who have used this method report saving up to 30% on the labor and material handling costs over conventional methods. With shoring labor, including bracing, ranging from 15% to 25% of the total form cost, this results in overall savings of 4-1/2% to 7-1/2% on the entire form cost.

Adjustable Shores for Reinforced Concrete Floors

There are a number of adjustable shores on the market to support reinforced concrete beam and slab forms. These shores are a combination of wood and metal, or all metal, which can be raised or lowered within certain limits.

Adjustable Shores. Adjustable shores in a complete unit, as illustrated are made in 3 sizes: 8 ft. (2.44 m) adjustable to 14 ft. (4.27 m); 7 ft. (2.13 m) adjustable to 13 ft. (3.96 m); 5 ft. (1.52 m) adjustable to 9 ft. (2.74 m).

They may be purchased for about $40.00 for the smaller size and $50.00 for the larger sizes. Extension type shores are also available in two sizes: 8 ft. (2.44 m) adjustable to 14 ft. (4.27 m), at about $45.00 each; and 6 ft. (1.83 m) adjustable to 10 ft. 6 in. (3.20 m), at about $43.00 each. They can be easily extended by inserting a 4"x4" (100 x 100 mm) of any length in the top of the shore.

386

CONCRETE

Rental stocks are maintained in most principal cities and may be rented at about $2.50 per month, per section, f.o.b. warehouse and return.

Each shore will support about 3,000 lbs. (1360 kg) with an adequate factor of safety. In some cases, however, shores are spaced to carry a load of not over 2,200 lbs. (1000 kg), as some forming authorities consider it cheaper to use more shores than to use heavier lumber to prevent excessive deflection between supports. For local prices and availability, the estimator should get a firm quote from the local supplier.

Adjustable Shores Versus 4"x 4"s (100 x 100 mm). It is much easier to shore up a floor using adjustable shores instead of 4"x 4"s (100 x 100 mm), which require wedging and cutting or adding to for each change in story height. This additional labor all costs money at present wage scales. However, to offset this, the original cost of the adjustable shores is much more than the 4"x 4"s (100 x 100 mm), but if they can be used often enough they will eventually show a saving. This will have to be considered when contemplating a purchase or rental basis.

Another thing that should be considered is the fact that a 4"x 4" (100 x 100 mm) shore will usually carry a larger load than an adjustable shore, so in many cases it will require more adjustable shores than 4"x 4"s (100 x 100 mm) to carry the same load. However, on some jobs, especially on light constructed floors, the adjustable shores may carry all the load required. This is an item that should be considered by the estimator before pricing the job.

The form costs on the following pages are computed on a rental basis, based on 10 days for each shore or 3 uses a month.

Shoring frames from Symons are used to support plywood decking for a multi-story concrete structure (Courtesy Symons Corp.)

Labor Cost of Flat Slab Forms. Forms for flat slab floors are continuous—without breaks or offsets—except at column heads, stair wells, elevator shafts, etc., where it is necessary to frame for beams and girders.

The s.f. (sq.m) labor cost will vary with the amount of lumber required per s.f. (sq.m) of forms, as the heavier the slab and the higher the ceiling, the more lumber required per s.f. (sq.m).

Forms for exceedingly light floor slabs cost proportionately more per s.f. (sq.m) than heavy forms, because the preliminary work is the same with both types. However, on work of this class, a carpenter should frame and erect 375 to 425 b.f. (.88-1.00 cu.m) of lumber per 8-hr. day, at the following labor cost per 1,000 b.f. (2.35 cu.m):

Description	Hours	Rate	Total
Carpenter	20.0	$ 35.04	$ 700.80
Labor	10.0	$ 26.16	$ 261.60
Cost per 1,000 BF			$ 962.40
Cost per 2.36 Cu.M			$ 408.06

When removing or "stripping" forms, a laborer should remove 1,000 to 1,200 BF (2-2.4 cu.m) of lumber per 8-hr. day, at the following labor cost per 1,000 b.f. (2 cu.m):

Description	Hours	Rate	Total
Labor	7.0	$ 26.16	$ 183.12

Labor Framing for Slab Depressions at Column Heads. On flat slab jobs there may be a depression at each column head 5'-0" to 7'-0" (1.5-2.1 m) square and 4" to 8" (100-200 mm) deep. The labor cost framing for these depressions will cost about 80% more than for straight slab forms.

A satisfactory method of estimating this additional labor is to figure the entire floor area at the regular price and then compute the area of the depressions and figure them at 80% additional labor.

Example: Floor contains 10,000 s.f. (929 sq.m) of slab forms, including 40 column head depressions 5'-0"x5'-0" (1.52 x 1.52 m), containing 25 s.f. (2.32 sq.m) each. Thus 10,000 s.f. (929 sm) of slab forms at $4.00/s.f.($43.05/sq.m) = $40,000. Add for 40 depressions, 5'-0"x5'-0" (1.52 x 1.52 m) 1,000 s.f. (92.9 sq.m.) at 80% more than plain slab forms or $2.60/s.f. ($27.99/sq.m) = $2,600. This makes a total of $6.60/s.f. ($71.04/sq.m.) for labor framing depressions at column heads.

CONCRETE

Labor Cost of 100 S.F. (9.29 Sq.M) of Flat Slab Concrete Floor Forms Requiring 3 BF (.007 cu.m) of Lumber per S.F. of Forms 4"x4" (100 x 100 mm) Shores

4"x4" (100 x 100 mm) Shores			
Description	Hours	Rate	Total
Carpenter	7.00	$ 35.04	$ 245.28
Labor	3.50	$ 26.16	$ 91.56
Labor removing forms	2.25	$ 26.16	$ 58.86
Cost per 100 s.f. (9.29 sq.m)			$ 395.70
cost per s.f.			$ 3.96
cost per sq.m			$ 42.59

For story heights over 16'-0" (4.87 m), add 25% to labor costs. Add labor for framing drop heads

Adjustable Shores			
Description	Hours	Rate	Total
Carpenter	6.00	$ 35.04	$210.24
Labor	3.00	$ 26.16	$78.48
Labor removing forms	2.25	$ 26.16	$58.86
Cost per 100 s.f. (9.29 sq.m)			$347.58
cost per s.f.			$3.48
cost per sq.m			$37.41

For story heights over 16'-0" (4.87 m), add 25% to labor costs. Add labor for framing drop heads

Labor Cost of 100 S.F. (9 Sq.M.) of Flat Slab Forms, 4 B.F. of Lumber Per S.F.

4"x4" (100 x 100 mm) Shores			
Description	Hours	Rate	Total
Carpenter	9.00	$ 35.04	$ 315.36
Labor	4.50	$ 26.16	$ 117.72
Labor removing forms	3.00	$ 26.16	$ 78.48
Cost per 100 s.f. (9.29 sq.m)			$ 511.56
cost per s.f.			$ 5.12
cost per sq.m			$ 55.07

For story heights over 16'-0" (4.87 m), add 25% to labor costs. Add labor for framing drop heads

Adjustable Shores			
Description	Hours	Rate	Total
Carpenter	8.00	$ 35.04	$280.32
Labor	4.00	$ 26.16	$104.64
Labor removing forms	3.00	$ 26.16	$78.48
Cost per 100 s.f. (9.29 sq.m)			$463.44
cost per s.f.			$4.63
cost per sq.m			$49.89

For story heights over 16'-0" (4.87 m), add 25% to labor costs. Add labor for framing drop heads

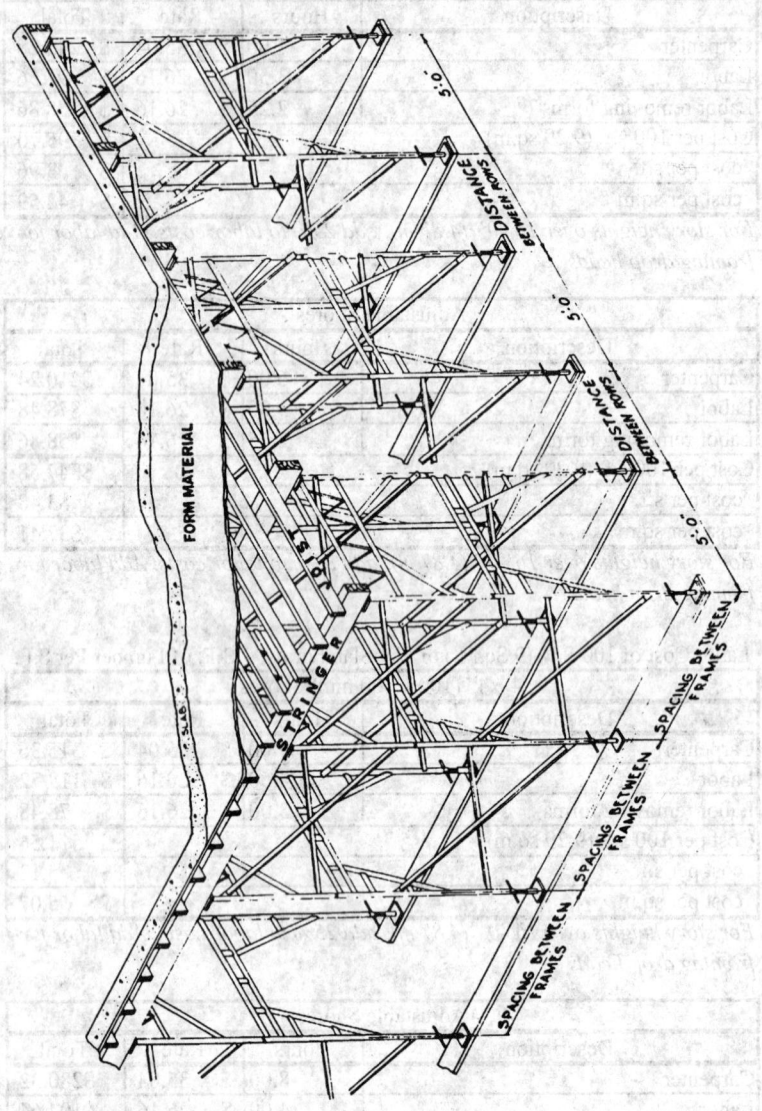

*Method of Shoring for Reinforced Concrete Construction
Using Section Steel Scaffolding Components*

CONCRETE

Forms for Beam and Girder Type Solid Concrete Floors

The labor cost of wood forms for beam and girder type floors will run somewhat higher than for flat slab floors because of the shorter spans, additional framing around beams, girders, etc.

On forms of this class, a carpenter should frame and erect 325 to 375 BF (0.65-0.75 cu.m) of lumber per 8-hr. day, at the following labor cost per 1,000 b.f. (2.3 cu.m):

Description	Hours	Rate	Total
Carpenter	23.00	$ 35.04	$805.92
Labor	11.50	$ 26.16	$300.84
Cost per 1000 bd.ft.(2.35 cu.m)			$1,106.76
cost per Bd. Ft.			$1.11
cost per Cu.M			$469.27

An experienced laborer should remove or "strip" 1,000 to 1,200 b.f. (2.3-2.4 cu.m) of lumber, per 8-hr. day, at the following labor cost per 1,000 b.f. (2.3 cu.m):

Description	Hours	Rate	Total
Labor	7.00	$ 26.16	$183.12

Ceiling Height in Feet	Thickness of Slab in Inches							
	3	4	5	6	7	8	10	12
	Weight Dead and Live Load, psf							
	88	100	113	125	138	150	175	200
10	2.0	2.1	2.4	2.5	2.6	2.6	2.7	2.7
12	2.1	2.2	2.4	2.5	2.6	2.6	2.7	2.8
14	2.1	2.2	2.4	2.5	2.6	2.6	2.7	2.8
16	2.2	2.3	2.5	2.6	2.7	2.7	2.8	2.9
18	2.4	2.5	2.6	2.7	2.8	2.8	2.9	3.0
20	2.5	2.6	2.7	2.8	2.9	2.9	3.0	3.1
Ceiling Height in Meters	Thickness of Slab in mm							
	76	102	127	152	178	203	254	305
	Weight Dead and Live Load, kq/sq.m.							
	430	488	552	610	674	732	854	976
3.0	6.56	6.89	7.87	8.20	8.53	8.53	8.86	8.86
3.7	6.89	7.22	7.87	8.20	8.53	8.53	8.86	9.19
4.3	6.89	7.22	7.87	8.20	8.53	8.53	8.86	9.19
4.9	7.22	7.55	8.20	8.53	8.86	8.86	9.19	9.51
5.5	7.87	8.20	8.53	8.86	9.19	9.19	9.51	9.84
6.1	8.20	8.53	8.86	9.19	9.51	9.51	9.84	10.17

Number of Feet of Lumber Required per Sq.Ft. (Sq.M.) of Beam and Girder Type Solid Concrete Floor Forms

Note: Dead Load = 150 psf; Live Load = 50 psf

Labor Cost of 100 S.F. (9.29 Sq.M) of Slab Forms Between Beam and Girder Forms Requiring 2 Feet of Lumber per Sq.Ft. (Sq.M.) of Floor

4"x4" (100 x 100 mm) Shores			
Description	Hours	Rate	Total
Carpenter	5.00	$ 35.04	$ 175.20
Labor	2.50	$ 26.16	$ 65.40
Labor removing forms	1.50	$ 26.16	$ 39.24
Cost per 100 s.f. (9.29 sq.m)			$ 279.84
per s.f.			$ 2.80
per sq.m			$ 30.12

For story heights over 16'-0" (4.87 m), add 25% to labor costs.

Adjustable Shores			
Description	Hours	Rate	Total
Carpenter	4.60	$ 35.04	$ 161.18
Labor	2.30	$ 26.16	$ 60.17
Labor removing forms	1.50	$ 26.16	$ 39.24
Cost per 100 s.f. (9.29 sq.m)			$ 260.59
per s.f.			$ 2.61
per sq.m			$ 28.05

For story heights over 16'-0" (4.87 m), add 25% to labor costs.

Labor Cost of 100 S.F. (9.29 Sq.M) of Slab Forms Between Beam and Girder Forms Requiring 2 1/2 Feet of Lumber per Sq.Ft. (Sq.M.) of Floor

4"x4" (100 x 100 mm) Shores			
Description	Hours	Rate	Total
Carpenter	6.30	$ 35.04	$ 220.75
Labor	3.15	$ 26.16	$ 82.40
Labor removing forms	1.90	$ 26.16	$ 49.70
Cost per 100 s.f. (9.29 sq.m)			$ 352.86
per s.f.			$ 3.53
per sq.m			$ 37.98

For story heights over 16'-0" (4.87 m), add 25% to labor costs.

Adjustable Shores			
Description	Hours	Rate	Total
Carpenter	5.70	$ 35.04	$ 199.73
Labor	2.85	$ 26.16	$ 74.56
Labor removing forms	1.90	$ 26.16	$ 49.70
Cost per 100 s.f. (9.29 sq.m)			$ 323.99
per s.f.			$ 3.24
per sq.m			$ 34.87

For story heights over 16'-0" (4.87 m), add 25% to labor costs.

CONCRETE

Labor Cost of 100 S.F. (9.29 Sq.M) of Slab Forms Between Beam and Girder Forms
Requiring 3 Feet of Lumber per Sq.Ft. (Sq.M.) of Floor

4"x4" (100 x 100 mm) Shores			
Description	Hours	Rate	Total
Carpenter	7.60	$ 35.04	$ 266.30
Labor	3.80	$ 26.16	$ 99.41
Labor removing forms	2.20	$ 26.16	$ 57.55
Cost per 100 s.f. (9.29 sq.m)			$ 423.26
per s.f.			$ 4.23
per sq.m			$ 45.56

For story heights over 16'-0" (4.87 m), add 25% to labor costs.

Adjustable Shores			
Description	Hours	Rate	Total
Carpenter	6.80	$ 35.04	$ 238.27
Labor	3.40	$ 26.16	$ 88.94
Labor removing forms	2.20	$ 26.16	$ 57.55
Cost per 100 s.f. (9.29 sq.m)			$ 384.77
per s.f.			$ 3.85
per sq.m			$ 41.42

For story heights over 16'-0" (4.87 m), add 25% to labor costs.

Forms for Floors of Metal Pan and Concrete Joist Construction

Wood forms for floors of metal pan and concrete joist construction may be of much lighter construction than solid concrete floors, as the dead load is less. Metal pans are usually furnished 20" to 30" (500-750 mm) wide, with 2"x 6" (50 x 150 mm) or 2"x 8" (50 x 200 mm) plank spaced 24" to 37" (600-925 mm) on centers, depending upon the width of the joists, which vary from 4" to 7" (100-175 mm). Fiberglass pans, sometimes used in place of steel pans, are usually furnished 41" to 52" (1025-1300 mm) wide. The open deck construction is ordinarily used owing to the wide spacing of the floor boards.

In buildings having one-way beams and long floor spans, a carpenter should frame and erect 350 to 400 b.f. (0.825-0.943 cu.m) of lumber per 8-hr. day, at the following labor cost per 1,000 b.f. (2.358 cu.m):

Description	Hours	Rate	Total
Carpenter	21.40	$ 35.04	$ 749.86
Labor	10.70	$ 26.16	$ 279.91
Cost per 1,000 b.f. (2.358 cu.m)			$1,029.77
Cost per b.f.			$1.03
Cost per cu.m.			$436.62

Floors having beams and girders running in both directions with the floor panels averaging 16'-0" x 16'-0" (4.87 x 4.87 m) and smaller, a carpenter should frame and erect 325 to 375 BF (0.766-0.884 cu.m) of lumber per 8-hr. day, at the following labor cost per 1,000 b.f. (2.358 cu.m):

Description	Hours	Rate	Total
Carpenter	22.80	$ 35.04	$ 798.91
Labor	11.40	$ 26.16	$ 298.22
Cost per 1,000 b.f. (2.358 cu.m)			$1,097.14
Cost per b.f.			$1.10
Cost per cu.m.			$465.19

An experienced laborer should remove or strip 1,000 to 1,200 b.f. (2.358-2.829 cu.m) of lumber per 8-hr. day, at the following labor cost per 1,000 b.f. (2.358 cu.m):

Description	Hours	Rate	Total
Labor	7.00	$ 26.16	$ 183.12

Number of Feet of Lumber Required per Sq.Ft. (Sq.M.) of Floor Forms for Metal Pan and Concrete Joist Construction					
Ceiling Height in Feet	Thickness of Floor Slab in Inches				
	8 1/2	10 1/2	12 1/2	14 1/2	16 1/2
	Weight of Dead and Live Load, psf				
	85	90	100	105	115
10	1.7	1.7	1.7	1.8	1.8
12	1.7	1.7	1.8	1.9	1.9
14	1.7	1.7	1.8	1.9	2
16	1.8	1.8	1.9	2	2.1
18	2.1	2.1	2.2	2.3	2.4
20	2.2	2.2	2.3	2.4	2.5
Ceiling Height in Meter	Thickness of Floor Slab in mm				
	216	267	318	368	419
	Weight of Dead and Live Load, kg/sq.m.				
	415.00	439.42	488.24	512.65	561.48
3.05	5.58	5.58	5.58	5.91	5.91
3.66	5.58	5.58	5.91	6.23	6.23
4.27	5.58	5.58	5.91	6.23	6.56
4.88	5.91	5.91	6.23	6.56	6.89
5.49	6.89	6.89	7.22	7.55	7.87
6.10	7.22	7.22	7.55	7.87	8.20

The table assume that ends of joists and stringers are supported by the beam and girder forms.

CONCRETE

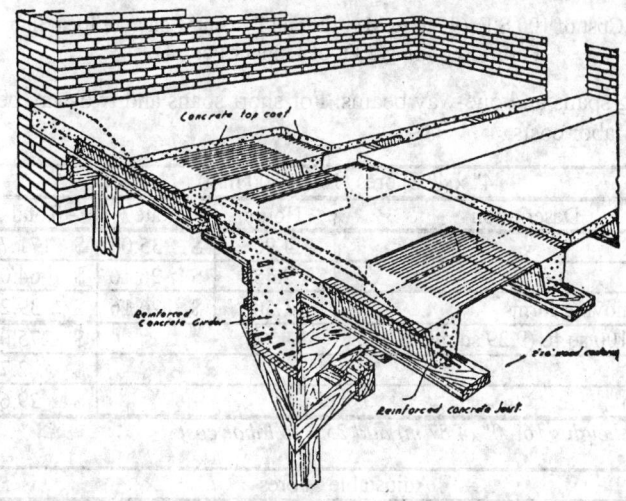

*Method of Constructing Wood Slab Forms for Floor
of Metal Pans and Concrete Joist Construction*

Labor Cost of 100 S.F. (9.29 Sq.M) of Metal Pan and Concrete Joist Slab
Forms Requiring 1-3/4 Ft. of Lumber Per S.F.

Long spans and one-way beams. For short spans and two-way beams,
add 10% to labor costs.

4" x 4" Shores (100 x 100 mm)			
Description	Hours	Rate	Total
Carpenter	4.20	$ 35.04	$ 147.17
Labor	2.10	$ 26.16	$ 54.94
Labor removing forms	1.30	$ 26.16	$ 34.01
Cost per 100 sq.ft. (9.29 sq.m)			$ 236.11
per sq.ft.			$ 2.36
per sq.m			$ 25.42

For story heights 16'- 0" (4.87 m) add 25% to labor costs

Adjustable Shores			
Description	Hours	Rate	Total
Carpenter	3.80	$ 35.04	$ 133.15
Labor	1.90	$ 26.16	$ 49.70
Labor removing forms	1.30	$ 26.16	$ 34.01
Cost per 100 sq.ft. (9.29 sq.m)			$ 216.86
per sq.ft.			$ 2.17
per sq.m			$ 23.34

For story heights 16'- 0" (4.87 m) add 25% to labor costs

Labor Cost of 100 S.F. (9.29 Sq.M) of Metal Pan and Concrete Joist Slab
Forms Requiring 2 Ft. of Lumber Per S.F.

Long spans and one-way beams. For short spans and two-way beams,
add 10% to labor costs.

4" x 4" Shores (100 x 100 mm)			
Description	Hours	Rate	Total
Carpenter	4.90	$ 35.04	$ 171.70
Labor	2.45	$ 26.16	$ 64.09
Labor removing forms	1.50	$ 26.16	$ 39.24
Cost per 100 sq.ft. (9.29 sq.m)			$ 275.03
per sq.ft.			$ 2.75
per sq.m			$ 29.60

For story heights 16'- 0" (4.87 m) add 25% to labor costs

Adjustable Shores			
Description	Hours	Rate	Total
Carpenter	4.30	$ 35.04	$ 150.67
Labor	2.15	$ 26.16	$ 56.24
Labor removing forms	1.50	$ 26.16	$ 39.24
Cost per 100 sq.ft. (9.29 sq.m)			$ 246.16
per sq.ft.			$ 2.46
per sq.m			$ 26.50

For story heights 16'- 0" (4.87 m) add 25% to labor costs

Labor Cost of 100 S.F. (9 Sq.M) of Metal Pan and Concrete Joist Slab
Forms Requiring 2-1/2 Ft. of Lumber Per S.F.

4" x 4" Shores (100 x 100 mm)			
Description	Hours	Rate	Total
Carpenter	6.00	$ 35.04	$ 210.24
Labor	3.00	$ 26.16	$ 78.48
Labor removing forms	1.80	$ 26.16	$ 47.09
Cost per 100 sq.ft. (9.29 sq.m)			$ 335.81
per sq.ft.			$ 3.36
per sq.m			$ 36.15

For story heights 16'- 0" (4.87 m) add 25% to labor costs

Adjustable Shores			
Description	Hours	Rate	Total
Carpenter	5.40	$ 35.04	$ 189.22
Labor	2.70	$ 26.16	$ 70.63
Labor removing forms	1.80	$ 26.16	$ 47.09
Cost per 100 sq.ft. (9.29 sq.m)			$ 306.94
per sq.ft.			$ 3.07
per sq.m			$ 33.04

For story heights 16'- 0" (4.87 m) add 25% to labor costs

CONCRETE

Labor Cost of 100 S.F. (9 Sq.M) of Metal Pan and Concrete Joist Slab
Forms Requiring 3 Ft. of Lumber Per S.F. 4"x4" (100 x 100 mm) Shores

4" x 4" Shores (100 x 100 mm)			
Description	Hours	Rate	Total
Carpenter	7.00	$ 35.04	$ 245.28
Labor	3.50	$ 26.16	$ 91.56
Labor removing forms	2.20	$ 26.16	$ 57.55
Cost per 100 sq.ft. (9.29 sq.m)			$ 394.39
per sq.ft.			$ 3.94
per sq.m			$ 42.45

For story heights 16'- 0" (4.87 m) add 25% to labor costs

Adjustable Shores			
Description	Hours	Rate	Total
Carpenter	6.40	$ 35.04	$ 224.26
Labor	3.20	$ 26.16	$ 83.71
Labor removing forms	2.20	$ 26.16	$ 57.55
Cost per 100 sq.ft. (9.29 sq.m)			$ 365.52
per sq.ft.			$ 3.66
per sq.m			$ 39.35

For story heights 16'- 0" (4.87 m) add 25% to labor costs

Forms for Reinforced Concrete Stairs

The labor cost of framing and erecting concrete stair forms varies with the type of stair, open or box string, straight runs from floor to floor or intermediate landing platforms, straight or winding, square or bull nose treads and risers. The story height also affects the labor costs on account of the shoring and bracing necessary to support the wet concrete.

It is difficult to estimate the cost of stair forms on a s.f. (sq.m) basis, because it costs as much to frame a 3'-0" (0.91 m) stair as one 4'-0" (1.21 m) wide, while the latter contains one-third more forms per l.f. (m). The better method is to allow a certain price per flight of stairs of each type. This will result in more accurate estimates than the square foot (square meter) method. However, where it is desirable to estimate the stair forms by the square foot (sq.m), take the area of the soffit of the stairs and platforms. For example, a stair 4'-0" (1.21 m) wide and 18'-0" (5.48 m) long, contains 72 s.f. (6.68 sq.m) of forms.

Some estimators estimate the cost of stairs at a certain price per l.f. of riser, which includes forms, reinforcing steel, concrete and finishing.

Wood Forms for Straight Concrete Stairs. On straight stairs extending from floor to floor without intermediate landing platforms and having an average story height of 10'-0" to 12'-0" (3.04-3.65 m), two carpenters working together should lay out the stair, place rough stringers,

397

mark off treads and risers on the rough string boards and set them in place in about 8 to 9 hours.

After the rough strings are in place, it will require another 8 to 10 hrs. for 2 carpenters to sheath the stairs, cut, bevel, and place risers, and place all necessary shoring and bracing ready for concrete.

The forms for an average flight of concrete stairs, 4'-0" (1.2 m) wide and 18'-0" (5.4 m) long, containing 72 s.f. (6.69 sq.m) of forms soffit measurement or 72 l.f. (6.69 m) of risers, should cost as follows:

Description	Hours	Rate	Total
Carpenter	36	$ 35.04	$1,261.44
Labor helping and removing forms	9	$ 26.16	$ 235.44
Cost per flight			$1,496.88
per sq.ft. - 72 sq.ft.			$ 20.79
per sq.m			$ 223.79
per lin.ft. of riser			$ 20.79
per meter of riser			$ 68.21
per riser (18)			$ 83.16

The lumber cost is approximately as follows:

Description	B.F.	Cu.M	Rate	Total
Stringers, 2 pcs, 2" x 12" x 20' (50 x 300mm x 6.09m)	80	0.1887	$ 0.40	$ 32.00
Risers, 18 pcs, 2" x 8" x 4' (50 x 200mm x 1.21m)	96	0.2264	$ 0.40	$ 38.40
Soffit sheathing, 4' x 8' (1.2m x 5.44m)	72	0.1698	$ 0.60	$ 43.20
Joists supporting sheathing, 4 pcs 2" x 8" x 18' (50 x 200mm x 5.48m)	96	0.2264	$ 0.40	$ 38.40
Shores or uprights, 6 pcs. 4" x 4" x 10' (100 x 100mm x 3.05m)	80	0.1887	$ 0.40	$ 32.00
Purlins or stringers, 3 pcs. 4" x 6" x 4' (100 x 150mm x 1.21m)	24	0.0566	$ 0.40	$ 9.60
Sills, wedges and bracing	30	0.0708	$ 0.39	$ 11.70
Cost per flight				$ 205.30
per sq.ft. (72)				$ 2.85
per sq.m				$ 30.69

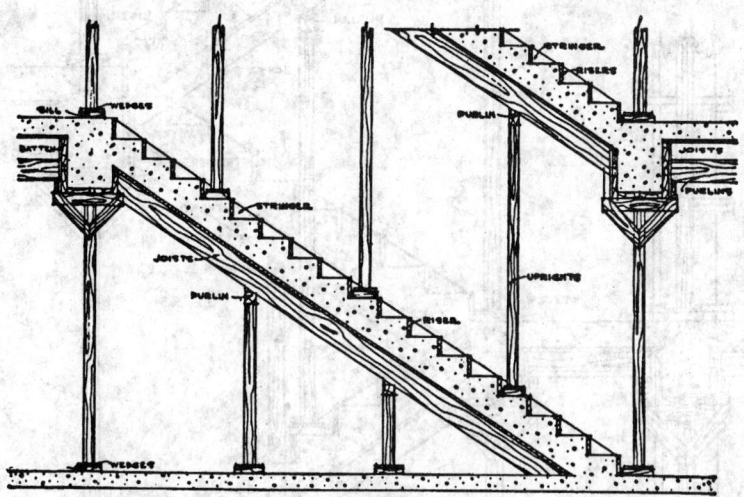

Method of Constructing Wood Forms for Concrete Stairs

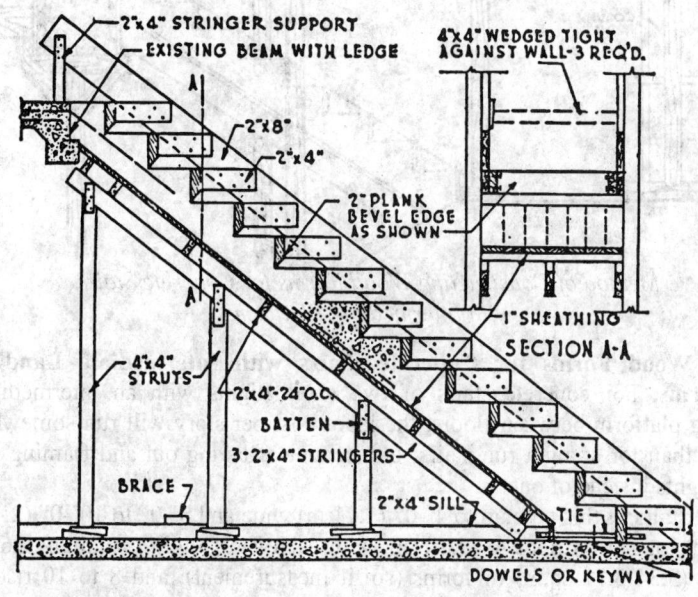

Another Method of Constructing Wood Forms for Concrete Stairs

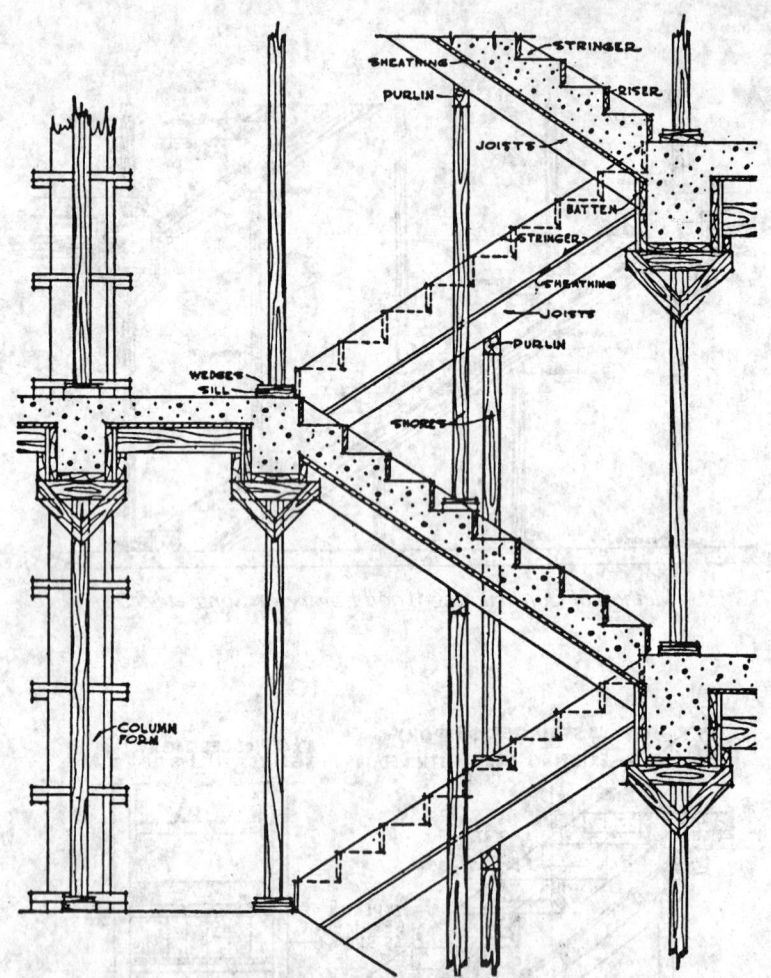

Method of Constructing Wood Forms for Concrete Stairs
with Intermediate Platforms

Wood Forms for Concrete Stairs with Intermediate Landing Platforms. For concrete stairs of two short flights with an intermediate landing platform between floors, the labor cost per story will run somewhat higher than for straight run stairs, on account of laying out and framing for two flights instead of one.

Stairs of this type up to 4'-0" (1.21 m) wide and 8'-0" to 10'-0" (2.44-3.04 m) long from floor to platform, with each short flight containing 36 to 40 s.f. (3.34-3.71 sq.m) of forms (soffit measurement), and 8 to 10 risers, should cost about as follows for labor:

400

CONCRETE

Description	Hours	Rate	Total
Carpenter	23	$ 35.04	$ 805.92
Labor helping and removing forms	7	$ 26.16	$ 183.12
Cost per flight			$ 989.04
per sq.ft. (36)			$ 27.47
per sq.m			$ 295.73
per sq.ft. riser (36)			$ 27.47
per sq.m riser			$ 295.73
per riser (9)			$ 109.89

If the stairs are built between masonry walls and have their bearing in slots or chases left in the wall, it will require about 2 hrs. additional carpenter time cutting out stringers, if using a hand saw, or 1/2 to 3/4 hr., if using a power saw.

Winding stairs, stairs having bull nose treads and risers, and other difficult or complicated construction may cost two or three times as much as given above. The estimator will have to use his judgment, depending upon detail of the stair.

When erecting formwork for concrete stairs over 4'-0" (1.21 m) wide, it will be necessary to place additional "cut-out" stringers, to prevent the risers from bulging or giving way until the wet concrete has set. These "cut-out" stringers should be spaced 3'-0" to 4'-0" (0.91-1.21 m) apart.

When figuring labor costs of landing platform, double the labor cost of plain slab forms.

STEELFORMS FOR JOIST CONSTRUCTED FLOORS AND ROOFS

Where the same size steelforms are used on a number of floors, they are removed after the concrete has attained the desired strength and reused again on the same job until the work is complete. The removable steelforms are available from and can be furnished by companies specializing in this type of work. These companies will erect and remove the steelforms with or without the supporting centering at the stipulated lump sum for the entire job.

Removable steelforms in flange and adjustable type are made from 14, 15, and 16 gauge metal that will stand hard use. The flange type is nailed through the flanges while the adjustables are nailed through the sides. The flange type steelforms are furnished in 20" (500 mm) and 30" (750 mm) standard widths in standard depths of 6", 8", 10", 12", 14", 16", and 20" (150, 200, 250, 300, 350, 400, and 500 mm). The intermediate forms are in lengths of 1', 2', and 3' (.30, .61, and .91 m) with end caps as closures. In addition to the straight ends, there are also single tapered end forms in 3' (0.91 m) lengths for the various depths. Also available are adjustable straight-side forms that can be set to a depth of 8", 10", 12", 14", and 16" (200, 250, 300, 350, and 400 mm). These are furnished in standard 20" and 30" (500 and 750 mm) widths and have the same combination of 1", 2", and 3" (25, 50, and 75 mm) intermediates, end forms, and the 3" (75 mm) single tapered

401

ends. For filling out spaces, filler forms are furnished in 10" and 15" (250 and 375 mm) widths matching the steelform depths as described. Tapered lighter gauge metals do not withstand job abuse, particularly with workers working over them before the concrete is placed. They often dent badly so that it requires considerable extra concrete and gives a poor surface finish, and there is added labor expense reconditioning the forms prior to the next use.

For two-way concrete joist construction, one-piece metal domes are also available. The depths furnished in the 30" x 30" (750 x 750 mm) void are 8", 10", 12", 14", 16", and 20" (200, 250, 300, 350, 400 and 500 mm), and in the 19" x 19" (475 x 475 mm) void are 4", 6", 8",10", 12", 16", and 20" (100, 150, 200, 250, 300, 400, and 500 mm). Flanges for the 30/30 domes are 3" (75 mm) wide to make a 3 ft. (0.91 m) module and for the 19/19 domes are 2-1/2" (62 mm) wide to make a 2 ft. (0.61 m) module. The metal gauge ranges from 14 to 16 gauge.

The steelforms and steeldomes must be nailed to the supporting forms to center and to hold them rigid and in line while the concrete is placed.

It is common practice to oil the forms before installing the reinforcing steel so that they can be removed easier from the ceiling and without concrete adhering to their surfaces.

Estimating Quantities of Steelform. In estimating the area of floor and roof construction requiring removable or permanent forms, the gross floor or roof area is used. No deductions are to be made for beams or for tees of beams or for wide joists. Major openings, 50 s.f. (4.65 sq.m) or more, such as atriums, are deducted when applicable. Typical covering systems generally frame through smaller openings, so no area adjustments are made for these openings.

Concrete joist construction can be supported by any one of three types of structural systems. In addition to the aforementioned general estimating rules, the job area is subject to the following special rules. For a reinforced concrete frame, the areas are to be figured out-to-out of concrete frame. For a structural steel frame, the areas are figured center-to-center spandrel beams. For bearing wall construction, the areas are to be figured clear inside brick walls plus a bearing on all walls of 6" (150 mm).

Labor Handling and Placing Steelforms. The labor cost of placing steelforms will vary with the type of form, type of building, quantity of equipment furnished, amount of horizontal and vertical movement of equipment, depressions, openings in the slabs, headers, etc., and the locality in which the work is being executed.

Steelforms are placed usually by carpenters and laborers. In a few localities, iron workers handle the steelforms from the point of removal from the concrete ceiling to the next point of steelform installation on the job. In most localities, however, laborers assist the carpenters and handle the steelforms to the next point of erection on the job ready for the carpenter to install the steelform to the wood centering. Local practice of work assignment greatly affects the labor costs.

The following costs are based on steelforms being placed by experienced workers, such as ones employed by companies specializing in the erection and removal of steelforms.

On straight run work using flange type steelforms such as factory buildings, garages, warehouses, etc. having typical spans without much cutting for headers, openings, etc., an experienced crew consisting of two carpenters and four laborers should handle and place steelforms over approximately 3,500 s.f. (315.15 sq.m) of floor area in an 8-hr day at the following direct labor cost per 100 s.f. (9.29 sq.m):

Description	Hours	Rate	Total
Mechanic	0.46	$ 37.89	$ 17.43
Labor	0.91	$ 26.16	$ 23.81
Cost per 100 sq.ft. (9 sq.m)			$ 41.24
cost per sq.ft.			$ 0.41
cost per sq.m			$ 4.44

For school buildings, hospitals, apartments, and office buildings, where the concrete joist require fitting and adjusting around pipes, pipe sleeves, chases and conduits, the handling and placing costs of the flange type equipment could increase as much as 40% to 50%, depending on the amount of special work required.

Labor Handling and Removing or Stripping Steelforms

In removing or stripping flange type steelforms after the concrete has set, an experienced crew, usually consisting of two carpenters and three laborers, can remove the steelforms from the ceiling over about 4,000 s.f. (371.60 sq.m) of floor per 8-hr day at the following direct labor cost per 100 s.f. (9.29 sq.m):

Description	Hours	Rate	Total
Mechanic	0.40	$ 37.89	$ 15.16
Labor	0.60	$ 26.16	$ 15.70
Cost per 100 s.f. (9 sq.m)			$ 30.85
cost per s.f.			$ 0.31
cost per sq.m			$ 3.32

It should be noted that the above costs are based on flange type steelforms. The cost for handling and placing the one-piece dome is slightly higher. The cost for handling and placing adjustable type steelform is about 33.33% higher.

Ceco Steelform Construction. Ceco steelform construction is a combination of concrete joist construction and thin top slabs. The steelforms are formed of 14, 15, or 16 gauge steel, depending on the width and depth of the form. Ceco also manufactures a fiberglass for that comes in 3', 4' and 5'

(.91. 1.21 and 1.52 m) modules. These forms are placed on supporting centering and removed from the concrete after the concrete has reached sufficient strength. The type of forms made available to the industry by the Ceco Corporation are flange forms, the most commonly used type of form, followed by the one-piece metal dome, adjustable forms and the one-piece longforms.

The cost of the steelforms installed in a building will depend on a number of factors such as the type of steelform used, the size of the job, the number of reuses of the equipment, location, availability of the steelforms with respect to warehouse stocks, and so on. Local working conditions also play a very important part in job costs.

The Ceco steelform service includes supplying the necessary steelforms and the labor for their erection and removal. Also offered is a complete service of furnishing and erecting all centering for the support of the steelforms. In steeldome construction, centering is usually included for the concrete solid slab areas that are of the same depth as the dome construction and that are within the floor areas containing the steeldomes. The quotations and contracts are generally on a lump sum basis for specific projects. It is always recommended that the steelform prices be obtained from a steelform supplier before finalizing and submitting a bid on this portion of the job costs.

FORMS OTHER THAN WOOD

Lightweight Steel Forming Material for Concrete. High-strength corrugated steel forming material is often used for forming reinforced concrete floor and roof slabs. It comes either galvanized or uncoated and is manufactured from 100,000 psi (689.5 MPa) tough-temper steel. It has a definite reliable structural strength nearly twice that of ordinary steel of equal weight. Used primarily in floor and roof systems having steel joists, junior beams, or purlins, it is also used over pipe tunnels or similar installations where economies of using permanent forms can be realized. Either structural grade concrete or lightweight insulating concrete may be used with corrugated steel forms, but in either case, reinforcing bars or wire mesh should be added to satisfy flexure and temperature steel requirements.

This material offers many advantages over other methods of flexible centering. The light, rigid sheets are quickly placed. No side pull is exerted on joists, and instead, joist top chords are given lateral support by the stiffness of the material. Sheets in place provide a safe working platform. Mesh is easily and effectively placed. Elimination of sag between joists reduces concrete quantity. Uniform thickness of slab over joists and mid span permits monolithic finish. Little cleanup is required underneath.

For exposed construction, it is available as galvanized. For unexposed joist construction, it can be purchased uncoated.

Sheets are placed with corrugations normal to the supporting joists or beams, with the end lap of sheets occurring over the joists or beams. The

CONCRETE

sheets are placed with edge lips up and are lapped a minimum of one corrugation with adjacent sheets. The ends of sheets should lap a minimum of 2" (50 mm). The sheets are fastened to the supporting members by clips, arc-welding, or nailing. For steel joist construction, the most satisfactory method of attachment is arc-welding. The manufacturer can provide special curved washers for use in welding the sheets to the steel framework. These provide fast, high-strength welds.

Concrete Quantities / 30" (750 mm) Widths*							
Depth of LONGform, Inches	Width of Joist, Inches	Cu.Ft. of Concrete per Sq.Ft. for Various Slab Thickness, in Inches*		Depth of LONGform, mm	Width of Joist, mm	Cu.M. of Concrete per Sq.M. for Various Slab Thickness, in mm*	
		3	4 1/2			75	112.5
14	6	0.482	0.607	350	150	0.1469	0.1850
16	6	0.522	0.647	400	150	0.1591	0.1972
20	6	0.605	0.730	500	150	0.1844	0.2225
24	6	0.694	0.819	600	150	0.2115	0.2496
Concrete Quantities / 66" (1650 mm) Widths*							
Depth of LONGform, Inches	Width of Joist, Inches	Cu.Ft. of Concrete per Sq.Ft. for Various Slab Thickness, in Inches*		Depth of LONGform	Width of Joist	Cubic Meter of Concrete per Square Meter for Various Slab Thickness, in mm*	
		3	4 1/2			75	112.5
14	6	0.366	0.491	350	150	0.1116	0.1497
16	6	0.388	0.511	400	150	0.1183	0.1558
20	6	0.427	0.552	500	150	0.1302	0.1683
24	6	0.427	0.597	600	150	0.1302	0.1820

*Apply only for areas over FLANGEforms and joists between them. Bridging joists, special headers, beam tees, etc., not included.

Voids Created by Various Size FLANGEforms							
Depth of LONGform, Inches	Cu.Ft. of Void Created per Lin.Ft. by Width of LONGform, in Inches			Depth of LONGform, mm	Cu.M. of Void Created per Lin.M. by Width of LONGform, in mm		
	66	30	20		1650	750	500
14	6	0.366	0.491	350	150	0.1116	0.1497
16	6	0.388	0.511	400	150	0.1183	0.1558
20	6	0.427	0.552	500	150	0.1302	0.1683
24	6	0.427	0.597	600	150	0.1302	0.1820

Shaded areas above indicate standard filler widths

405

Concrete Quantities / 30" (750 mm) Widths*							
Depth of LONGform, Inches	Width of Joist, Inches	Cu.Ft. of Concrete per Sq.Ft. for Various Slab Thickness, in Inches*		Depth of LONGform, mm	Width of Joist, mm	Cu.M. of Concrete per Sq.M. for Various Slab Thickness, in mm*	
		3	4 1/2			75	112.5
14	5	0.456	0.581	350	125	0.1390	0.1771
	6	0.483	0.608		150	0.1472	0.1853
	7	0.508	0.633		175	0.1548	0.1929
16	6	0.522	0.647	400	150	0.1591	0.1972
	7	0.550	0.675		175	0.1676	0.2057
	8	0.557	0.702		200	0.1698	0.2140
20	6	0.605	0.730	500	150	0.1844	0.2225
	7	0.640	0.765		175	0.1951	0.2332
	8	0.674	0.799		200	0.2054	0.2435
24	6	0.694	0.819	600	150	0.2115	0.2496
	7	0.736	0.861		175	0.2243	0.2624
	8	0.776	0.901		200	0.2365	0.2746

*Apply only for areas over FLANGEforms and joists between them. Bridging joists., special headers, beam tees, etc., not included. For 10" (250mm) and 12" (300mm) depths contact your Ceco dealer

Concrete Quantities / 20" Widths*							
Depth of LONGform, Inches	Width of Joist, Inches	Cu.Ft. of Concrete per Sq.Ft. for Various Slab Thickness, in Inches*		Depth of LONGform, mm	Width of Joist, mm	Cu.M. of Concrete per Sq.M. for Various Slab Thickness, in mm*	
		3	4 1/2			75	112.5
14	5	0.538	0.663	350	125	0.1640	0.2021
	6	0.572	0.697		150	0.1744	0.2125
	7	0.603	0.728		175	0.1838	0.2219
16	6	0.626	0.751	400	150	0.1908	0.2289
	7	0.662	0.787		175	0.2018	0.2399
	8	0.694	0.819		200	0.2115	0.2496
20	6	0.741	0.867	500	150	0.2259	0.2643
	7	0.785	0.910		175	0.2393	0.2774
	8	0.825	0.950		200	0.2515	0.2896

*Apply only for areas over FLANGEforms and joists between them. Bridging joists., special headers, beam tees, etc., not included. For 10" (250mm) and 12" (300mm) depths contact your Ceco dealer

CONCRETE

Voids Created by Various Size FLANGEforms						
Depth of STEELform in Inches	Cu.Ft. of Void Created per Lin.Ft. by Width of FLANGEform in Inches				* Added Cu.Ft. Concrete per Tapered End Condition	
	30	20	15	10	30	20
10	2.023	1.329	0.982	0.634	0.521	0.418
12	2.414	1.581	1.165	0.748	0.625	0.500
14	2.801	1.629	1.343	0.857	0.730	-----
16	3.183	2.027	1.516	0.951	0.834	-----
20	3.933	2.544	1.850	1.155	1.043	-----
24	4.667	3.000	-----	-----	-----	-----
Depth of STEELform in mm	Cu.M. of Void Created per Lin.M. by Width of FLANGEform in mm				* Added Cu.M. Concrete per Tapered End Condition	
	750	500	375	250	750	500
250	0.6166	0.4051	0.2993	0.1933	0.1588	0.1274
300	0.7358	0.4819	0.3551	0.2280	0.1905	0.1524
350	0.8538	0.4965	0.4094	0.2612	0.2225	-----
400	0.9702	0.6179	0.4621	0.2899	0.2542	-----
500	1.1988	0.7754	0.5639	0.3521	0.3179	-----
600	1.4226	0.9144	-----	-----	-----	-----

Shaded areas above indicate standard filler widths

NOTE: LONGforms are available to produce a 6'- 0" and 3'- 0" (1.83 and .91 m) module for one-way joist construction. LONGforms are commonly used with a 6" (150mm) joist resulting in a 3'- 0" or 6'- 0" (1.83 or .91m) module depending on the pan width selected.

NOTE: FLANGEforms are available in standard 2'- 0" and 3'- 0" (0.61 and .91 m) modules. These forms are popular because of their flexibility to accommodate various layouts and joist widths where require.

Estimating. Corrugated steel forming material is sold by the square, with the area determination based on sheet width times actual sheet length.

Erection Costs. Corrugated steel forming material can be rapidly erected in all kinds of weather, providing cover for floors below. The corrugated sheets are large in area but extremely light in weight and can be easily handled by one worker.

A well organized crew, working under favorable conditions, can place and fasten up to 10,000 s.f. (929.0 sq.m) of corrugated steel forming material per 8-hr day, but most jobs will average 3,000 to 5,000 s.f. (278.7-464.50 sq.m) per day.

Using an average figure of 4,000 s.f. (371.60 sq.m) per 8-hr day for a four-worker crew, the labor costs for placing 1,000 s.f. (92.90 sq.m) of corrugated steel forming material would be as follows:

407

Description	Hours	Rate	Total
Ironworker	8.00	$ 40.29	$ 322.32
Cost per sq.ft. (1,000 sq.ft.)			$ 0.32
cost per sq.m			$ 3.47

The above figures are based upon one worker welding or fastening the deck in place and three workers handling and placing the material. Costs do not include equipment, overhead, and profit.

Computing Concrete Quantities. When computing volume of concrete placed over corrugated steel decking, subtract 1/4 in. (6.25 mm) from the slab thickness to allow for corrugations of standard and 3/8 in. (9.37 mm) for heavy duty corrugated steel forming. Size of corrugations: standard, 1/2" depth x 2-3/16" pitch (12.50 x 54.68 mm); heavy duty, 3/4" depth x 3" pitch (18.75 x 75 mm).

ARCHITECTURAL CONCRETE FORMWORK

Appearance is of paramount importance in architectural concrete, and every operation must be planned and executed with this in mind. The architect may design and specify, but regardless of how detailed the specifications, it is the contractor and the superintendent who must have the know-how to detail and construct the forms and be on constant watch during all operations to see that the correct procedures are followed.

Forms for Architectural Concrete

Form design and construction must be approached differently from ordinary structural concrete. In structural concrete some leakage and irregularities in formwork may be tolerated, so long as the forms are strong enough to carry the weight of concrete and imposed loads and are built fairly true to line and grade. In architectural concrete, there must be no leakage through the forms. Chamfer strips at corners and edges are helpful for this purpose. Wherever possible, the forms should be designed so that pressure of the fresh concrete will tighten joints rather than tend to loosen them.

A common source of trouble is bulging of forms at external corners, resulting in both poor alignment and leakage with subsequent ragged corners, sometimes to the extent of exposing the aggregate.

One method of constructing forms to provide locking of the corner is shown. The wales are extended beyond the intersection far enough to permit two vertical strips to be nailed in the corners. Wedges are driven between these strips and the wales to tighten the corner.

Another method of locking the form corners is to use a tie instead of the wood strips and wedges. Since there are no wedges to work loose, a set of wales can be set and tightened on one level around the building without much chance of subsequent loosening when tightening the wales above or below.

CONCRETE

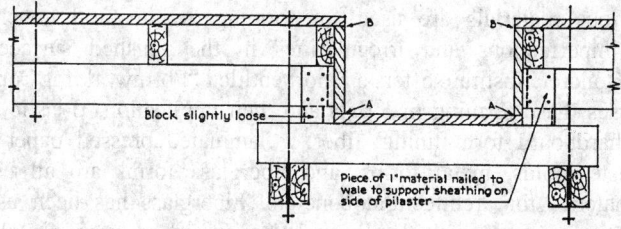

Form Designed so that Pressure of Fresh Concrete Tightens It to Prevent Leakage

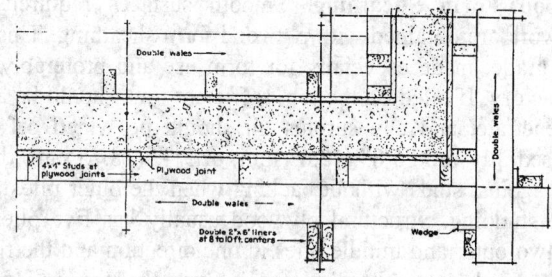

Form Construction to Provide Positive Corner Locking

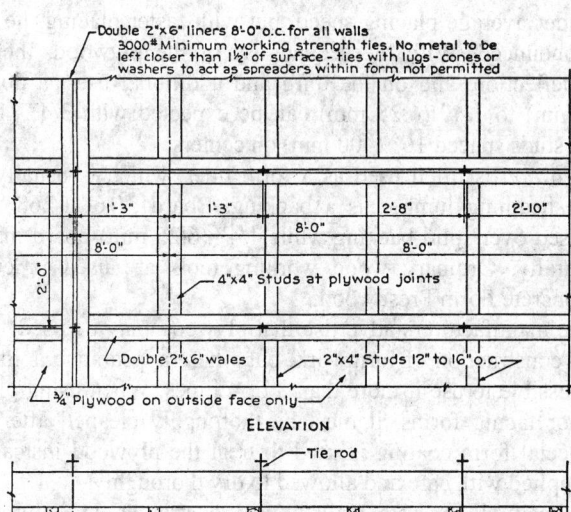

Form Layout for Use of Structural Plywood

409

Various materials are used for forming architectural concrete to eliminate imperfections and irregularities in the finished surface. The American Concrete Institute offers a report entitled "Formwork for Concrete" that discusses these forming materials and their application. Plastic coated plywood, hardboard form lining, fiber or laminated pressed paper tubes, plaster waste molds, metal forms, and fiberglass forms are all used as forming material for architectural concrete. Fiberglass has been used on larger building projects with the forms being reused as many as 60 or 70 times.

Plywood Form Sheathing. Smooth surfaces required on much present day work are obtained with plywood form sheathing. This should be of structural grade, made especially for form use and preferably should be oiled at the factory. If oiled on the job, sufficient time should be allowed for the oil to penetrate, and any surplus oil should be wiped off before the plywood is used. Plywood 5/8" (15.63 mm) or 3/4" (18.75 mm) thick can be used directly against studs without backing when the outer plies are at right angles to the studding. Structural plywood usually has five plies, with the fibers in the two outer and middle plies in one direction and the fibers of the other two plies in the opposite direction, making the plywood stronger and stiffer in one direction. The result may not be serious on the first use of the plywood under average placing speed, but with faster placing the deflection may be pronounced. With the second use of the plywood, there will be noticeable deflection, and on the third and additional uses, a deflection of 1/8" (3.12 mm) to 1/4" (6.25 mm) can be expected with 3/4" (18.75 mm) plywood on studs spaced 16" (400 mm) on centers.

Presdwood is often used as a form liner, which is usually nailed to decking or sheathing lumber as a backing. Board 3/16" (4.68 mm) thick should be used over solid backing while 1/4" (6.25 mm) board may be used semi-structurally. Ordinary wood working tools are used in cutting and applying Concrete Form Presdwood.

Some manufacturers advertise that plywood may be used 10 times or more in the construction of formwork, but where appearance is important, it is seldom possible to use it more than 3 or 4 times. When using plywood for the facade or facing forms, it must be thoroughly cleaned after each use. Where a special form coating is used to treat the plywood instead of oil, it should be applied with care and allowed to dry thoroughly.

Architectural concrete walls must be straight and plumb. This is essential. It means that the 2" x 4" (50 x 100 mm) or preferably 2" x 6" (50 x 150 mm) studs and wales must be of uniform depth and perfectly straight. Otherwise, the walls will be crooked, with the possibility of the work being rejected.

The top of the foundation walls on which the architectural concrete wall rests must be perfectly level and straight, using two 2"x 6" (50 x 150 mm) wales or similar planks, braced at intervals of not less than 4'- 0" (1.21 m) in addition to all other bracing ordinarily required for foundation walls. It

410

CONCRETE

requires about 4-1/2 b.f. (.010 cu.m) of lumber per l.f. of wall for these wales and bracing. In order to hold the studs of the form for the architectural concrete wall in line, it is necessary to insert 5/8" (15.62 mm) greased stud bolts, spaced about 3'-0" (900 mm) on centers, into the top of the foundation wall as illustrated. It requires about 1/8 hr. carpenter and 1/16 hr. labor time per l.f. of wall to brace the top of the foundation wall and to insert the bolts.

As a rule, it is necessary to obtain a perfect horizontal line of concrete at the top of the foundation wall, which will be assured by nailing a 1" x 2" (25 x 50 mm) strip on the outside form of the foundation wall and placing the concrete about 1" higher than the bottom of this strip. About two hours after placing the concrete, the surplus concrete, which generally contains a good deal of laitance, is scraped off to the level of the underside of the strip.

After the forms for the foundation walls have been stripped, the wales should be installed and blocked about 3'- 0" (900 mm) apart against the foundation wall. The 2" x 4" (50 x 100 mm) or 2"x 6" (50 x 150 mm) studs should now be installed between wales and the wall. The studs should have 8d nails, protruding about 1" (25 mm), driven into the outside edge exactly 8" (200 mm) from the lower end and 6d nails, protruding less than 3/4" (18.75 mm) exactly 8" (200 mm) from the lower edge. When inserting the studs between the wales and the foundation wall, the 8d nails form their support on the wales and the 6d nails furnish the support for the 3/4" (18.75 mm) plywood. The studs are usually spaced 12" (300 mm) or 16" (400 mm) on centers but at every vertical joint in the plywood (usually 8'- 0" - 2.44 m) a double stud or a stud 3-1/2" (87.5 mm) wide should be used.

The studs are temporarily braced horizontally by nailing a 1"x 6" (25 x 150 mm) board higher up and are braced vertically by wedging at the bottom between the studs and foundation wall. The first sheet of plywood is then inserted between the studs and the foundation wall and bolts, holding the wale tight until the lowest part of the form is straight and plumb. The wales are spaced 24" (600 mm) vertically and the oiled tie bolts are spaced 2'-0", 2'-8" and 3'-4" (600, 800, and 1000 mm) apart, so that the holes do not interfere with the 16" (400 mm) stud spacing and are in proper relation to the ends of the 8'-0" (2.43 m) long plywood sheets.

The arrangement of the holes for the tie bolts and the arrangement of the 6d box nails by which the sheets of plywood are held to the studs are also shown in the accompanying illustration.

After 2 or 3 of the 4'-0" (1.21 m) wide sheets of 3/4" (18.75 mm) plywood have been fastened to the studs and the proper amount of horizontal walers put in place, the outside form is further strengthened by vertical wales consisting of 2 pcs. of 2" x 6" (50 x 150 mm) spaced not more than 8'-0" (2.44 m, o.c. with about 2'- 0" (0.61 m,) or more projecting over the top of the outside form. These wales are held in place and plumbed on top by two or more No. 9 or 10 wires, secured to column dowels, special bolts, or other means provided in the foundations or in the floor. These vertical wales are bolted to every pair of horizontal wales they cross using 1/2" (12.5 mm) bolts. The form is now ready to support the vertical and horizontal reinforcing.

After the reinforcing steel has been placed, the inside form may be erected. The studs are nailed to a 1"x 6" (25 x 150 mm) sill or plate, which has been nailed to the floor one day after concreting, and if the inside of the wall is to have a finished appearance, requiring the use of plywood, the plywood should be in sheets 24" (600 mm) wide and 8'-0" (2.44 m,) long in order to facilitate the nailing of the plywood to the studs and the insertion of the tie bolts and spreaders.

The joints of the plywood should be filled with a mastic consisting of a mixture of 50% tallow and 50% cement for summer use, while for winter work, the tallow should be replaced by pump grease. The surplus filler should then be scraped off using No. 0 sandpaper.

Estimating Quantities for Architectural Concrete Forms. When estimating the quantities for plain architectural concrete walls, do not make any deductions for ordinary door and window openings, because these must be formed with the wall and then boxes built in to allow for the door or window frame to be inserted later.

The only exception is large display windows in the first story of a commercial building or for skeleton type buildings where the piers and lintels are estimated at a higher rate than plain wall surfaces.

All ornamentation should be estimated separately, either by the lin. ft. or sq. ft. (meter or sq.m) depending upon the class of work. Ornamental work will require the use of wood or plaster waste molds for forming, in addition to the structural backing, so that the entire wall area should first be figured as plain wall surfaces and then ornamentation figured extra.

Water tables, belt courses, cornices, door and window trim, copings, and the like should be estimated by the l.f. (m) when less than 12" (300 mm) wide and by the s.f. when more than 12" (300 mm) wide. Always mention size and detail for each type of ornamentation. This is necessary for pricing the work.

Columns, pilasters, etc. should be estimated separately, giving width, height, and area in s.f. (sq.m), together with a detail of the forming, such as whether fluted.

Sample Wall Form Calculation

This sample shows how to determine square footage of form contact surface area, size and number of studs and wales needed, and the number of steel ties required. The sample data, table, and charts are provided by Richmond Screw Anchor Company, Ft. Worth, Texas.

Given: Footing poured, 56'-0" (17.07 m) long x 5'-0" (1.52 m) wide x
              3'-6" (1.07 m) high

Required: Forms and material required for a concrete wall 50'-0" (15.24 m)
              long x 1'0" (0.30 m) wide x 12'-0" (3.65 m) high

412

CONCRETE

Assumptions:
1. Permanent or reusable formwork, Class "A".
2. 3/4" (18.75 mm) thick Class I plyform with face perpendicular to supports
3. (strong direction).
4. Studs vertical.
5. Wales horizontal.
6. Tie safe working load = 6,000 lbs. (2720 kg)
7. Preferred tie spacing: vertical 2'-0" (0.61 m); horizontal 2'-0" (0.61 m)
8. 2' (0.61m) x 2' (0.61m) = 4 s.f. (0.37 sq.m)

We want to determine stud and wale sizes using 3/4" (18.75 mm) plywood.

Determine Concrete Pressure:

Concrete Pressure = $\dfrac{\text{Tie safe working load}}{\text{Area}}$

Concrete Pressure = $\dfrac{\text{6,000 lbs. (2721.6 kg)}}{\text{4 s.f. (0.37 sq.m)}}$

Concrete Pressure =   1,500 lbs./s.f. (7323.64 kg/sq.m)

Framing Material:

Stud Spacing: Use Table 2, "Spacing of Supports for Class I Plywood, (Strong Direction)-Face Grain Across Supports", or Chart A, "Spacing of Supports for Class I Plywood", which appear at the end of the sample calculation. For Class I plywood with a concrete pressure of 1,500 psf (7323.64kg/cu.m and plywood thickness of 3/4" (18.75 mm), the support (tie) spacing is 7.6" (190 mm), so use 8" (200 mm) center to center (c.c.).

Stud Sizes:

Stud Loading = $\dfrac{\text{Stud Spacing x Design Concrete Pressure}}{\text{12 in./ft. (1000 mm/m)}}$

Stud Loading = $\dfrac{\text{8 in. (200 mm) x 1,500 lbs/ft.2 (7323.64 kg/sq.m)}}{\text{12 in./ft. (1000 mm/m)}}$

Stud Loading =   1,000 lbs./ft. (138.55 kg/m)
Assume studs are continuous or partially continuous.

Using Table 1 "Maximum Wale Spacing Along Stud/Joist Member", and Chart B, "Stud/Joist Loading", stud member sizes can be determined by beginning on Table 1 at 1,000 lbs./ft. (138.55 kg/m) and moving horizontal to 24" (600 mm) vertical tie spacing (wale spacing) and reading lumber sizes above, 2" x 6" or better.

Wale Size:

Wale Loading =   Wale Spacing x Design Concrete Pressure

$$\text{Wale Loading} = \frac{2 \text{ ft. } (0.61 \text{ m}) \times 1,500 \text{ lbs./ft.2 } (7323.64 \text{ kg/sq.m})}{12 \text{ in./ft. } (1000 \text{ mm/m})}$$

Wale Loading =   3,000 lbs./ft. (4536 kg/m)
Assume wales are continuous or partially continuous.

Using Table 2 and Chart C, begin on Table 4 at 3,000 lbs./ft. (4535 kg/m) wale load and progress horizontally to a span of 2'-0" (0.61 m) horizontal tie spacing and read lumber sizes above, double 2x6 (50 x 150 mm) or better.

Summary:
1. Ties: 6,000 lbs. (2720 kg) safe working load
2. Tie Spacing: 2'-0" (0.61 m) horizontal; 2'-0" (0.6 m) vertical
3. Concrete Pressure: 1,500 lbs./ft.2 (7560 kg/m2) (uniform)
4. Plywood: 3/4" (18.75 mm) Class I perpendicular to supports
5. Stud Spacing: 8" (200 mm) c. to c., perpendicular to plywood face grain
6. Stud Member Size: 2" x 6" (50 x 150 mm) or better, continuous or partially continuous
7. Wale Spacing: 2'-0" (0.61 m) c. to c. vertically
8. Wale Member Size: Double 2" x 6" (50 x 150 mm) or better, continuous or partially continuous

Material Calculation List:

3/4" (1.875 mm) plywood, contact surface area, 50' (15.24 m) x 12' (3.65 m) = 60 s.f. (5.57 sq.m) one side x 2 sides = 1,200 s.f. (111.48 sq.m) of contact surface area; add end bulkheads if required

Studs, 50 l.f. (15.24 m) divided by .667 l.f. (0.20 m) c.c. = 75 each side x 2 sides = 150 pcs. 2" x 6" (50 x 150 mm), 12'- 0" (3.65 m) long

Wales, 12' (3.65 m) high divided by 2' (0.61 m) c.c. = 6 each x 2 (doubles) = 12 each side; 12 each x 2 sides = 24 pcs. 2" x 6" (50 x 150 mm), 50' (15.25m) -5 @ 10' = 50' (5 @ 3.04 m = 15.24m) –each

Ties, 50' x 12' ÷ 4 = 150 each (plus hardware)

Notes:
1. Bottom row of ties, maximum 6" (150 mm) from bottom of form
2. Top row of ties, not to exceed 50% of vertical spacing, and 3" (75 mm) concrete cover required
3. First and last vertical row not to exceed 6" (150 mm) from edge of form
4. Concrete placement shall not exceed recommended placement rate of 7 ft. (2.13 m) per hr. in 4-ft. (1.21-m) layers.
5. Consult tie manufacturer for your special requirements.

CONCRETE

TABLE 1 Maximum Wale Spacing Along Stud/Joist Member
Class A - Permanent (Continuous or Partially Continuous Spans)

Load Lbs/Ft	Nominal Lumber Sizes, in Inches										
	2" x 4"	2" x 6"	2" x 8"	2" x 10"	2" x 12"	3" x 4"	3" x 6"	3" x 8"	4" x 4"	4" x 6"	4" x 8"
200	55	78	95	115	133	63	88	108	68	96	118
400	42	65	80	96	122	53	74	91	57	81	99
600	33	53	69	87	101	44	67	82	52	73	90
800	27	42	55	71	86	38	60	77	45	68	83
1,000	23	36	47	60	73	33	52	69	40	63	79
1,200	20	32	42	53	65	29	45	60	37	58	75
1,400	18	29	38	48	59	26	40	53	33	52	69
1,600	17	26	35	45	54	23	37	48	30	47	62
1,800	16	25	33	42	51	22	34	45	27	43	57
2,000	15	23	31	39	48	20	32	42	25	40	53
2,200	14	22	29	37	46	19	30	39	24	37	49
2,400	14	21	38	36	44	18	28	37	22	35	46
2,600	13	21	27	35	42	17	27	35	21	33	44
2,800	13	20	26	33	41	16	26	34	20	32	42
3,000	12	19	25	32	39	16	25	33	19	30	40

Chart from Ramond Screw Anchor Company

TABLE 1 - Metric Maximum Wale Spacing Along Stud/Joist Member
Class A - Permanent (Continuous or Partially Continuous Spans)

Load Kg/M	Nominal Lumber Sizes, in mm										
	50x100	50x150	50x200	50x250	50x300	75x100	75x150	75x200	100x100	100x150	100x200
298	1397	1981	2413	2921	3378	1600	2235	2743	1727	2438	2997
595	1067	1651	2032	2438	3099	1346	1880	2311	1448	2057	2515
893	838	1346	1753	2210	2565	1118	1702	2083	1321	1854	2286
1,191	686	1067	1397	1803	2184	965	1524	1956	1143	1727	2108
1,488	584	914	1194	1524	1854	838	1321	1753	1016	1600	2007
1,786	508	813	1067	1346	1651	737	1143	1524	940	1473	1905
2,083	457	737	965	1219	1499	660	1016	1346	838	1321	1753
2,381	432	660	889	1143	1372	584	940	1219	762	1194	1575
2,679	406	635	838	1067	1295	559	864	1143	686	1092	1448
2,976	381	584	787	991	1219	508	813	1067	635	1016	1346
3,274	356	559	737	940	1168	483	762	991	610	940	1245
3,572	356	533	965	914	1118	457	711	940	559	889	1168
3,869	330	533	686	889	1067	432	686	889	533	838	1118
4,167	330	508	660	838	1041	406	660	864	508	813	1067
4,464	305	483	635	813	991	406	635	838	483	762	1016

TABLE 2 Maximum Mid Tie Spacing Class A (Permanent) Form Design

Simple Spans

Nominal Lumber Sizes, in Inches

Load Lbs/Ft	Single Wale				Double Wales									
	2" x 4"	2" x 6"	2" x 8"	4" x 4"	2" x 4"	2" x 6"	2" x 8"	2" x 10"	2" x 12"	3" x 10"	3" x 12"	4" x 4"	4" x 10"	4" x 12"
200	47	60	81	58	56	79	97	116	135	132	153	69	144	166
400	37	56	68	49	47	66	81	98	113	111	129	58	121	140
600	30	48	62	44	42	60	74	88	102	100	116	53	109	126
800	26	41	54	40	37	56	68	82	95	93	108	50	102	118
1,000	23	37	49	36	33	52	65	78	90	88	102	46	96	111
1,200	21	34	44	33	30	48	62	74	86	84	98	44	92	106
1,400	20	31	41	30	28	44	58	71	83	81	94	42	88	102
1,600	19	29	38	28	26	41	54	69	80	97	91	40	85	99
1,800	18	28	36	27	25	39	51	65	78	76	88	38	83	96
2,000	16	26	34	25	23	37	49	62	75	74	86	36	81	94
2,200	16	25	32	24	22	35	46	59	72	73	84	34	79	91
2,400	15	23	31	23	21	34	44	57	69	71	82	33	77	89
2,600	14	22	30	22	21	32	43	54	66	70	81	31	76	88
2,800	14	22	28	21	20	31	41	52	64	68	79	30	74	86
3,000	13	21	28	21	19	30	40	51	62	65	78	29	73	85

Chart from Ramond Screw Anchor Company

TABLE 2 - Metric Maximum Mid Tie Spacing Class A (Permanent) Form Design

Simple Spans

Nominal Lumber Sizes, in Inches

Load Kg/M	Single Wale				Double Wales									
	50x100	50x150	50x200	100x100	50x100	50x150	50x200	50x250	50x300	75x250	75x300	100x100	100x250	100x300
298	1194	1524	2057	1473	1422	2007	2464	2946	3429	3353	3886	1753	3658	4216
595	940	1422	1727	1245	1194	1676	2057	2489	2870	2819	3277	1473	3073	3556
893	762	1219	1575	1118	1067	1524	1880	2235	2591	2540	2946	1346	2769	3200
1,191	660	1041	1372	1016	940	1422	1727	2083	2413	2362	2743	1270	2591	2997
1,488	584	940	1245	914	838	1321	1651	1981	2286	2235	2591	1168	2438	2819
1,786	533	864	1118	838	762	1219	1575	1880	2184	2134	2489	1118	2337	2692
2,083	508	787	1041	762	711	1118	1473	1803	2108	2057	2388	1067	2235	2591
2,381	483	737	965	711	660	1041	1372	1753	2032	2464	2311	1016	2159	2515
2,679	457	711	914	686	635	991	1295	1651	1981	1930	2235	965	2108	2438
2,976	406	660	864	635	584	940	1245	1575	1905	1880	2184	914	2057	2388
3,274	406	635	813	610	559	889	1168	1499	1829	1854	2134	864	2007	2311
3,572	381	584	787	584	533	864	1118	1448	1753	1803	2083	838	1956	2261
3,869	356	559	762	559	533	813	1092	1372	1676	1778	2057	787	1930	2235
4,167	356	559	711	533	508	787	1041	1321	1626	1727	2007	762	1880	2184
4,464	330	533	711	533	483	762	1016	1295	1575	1651	1981	737	1854	2159

SPACING OF SUPPORTS FOR CLASS I PLYWOOD

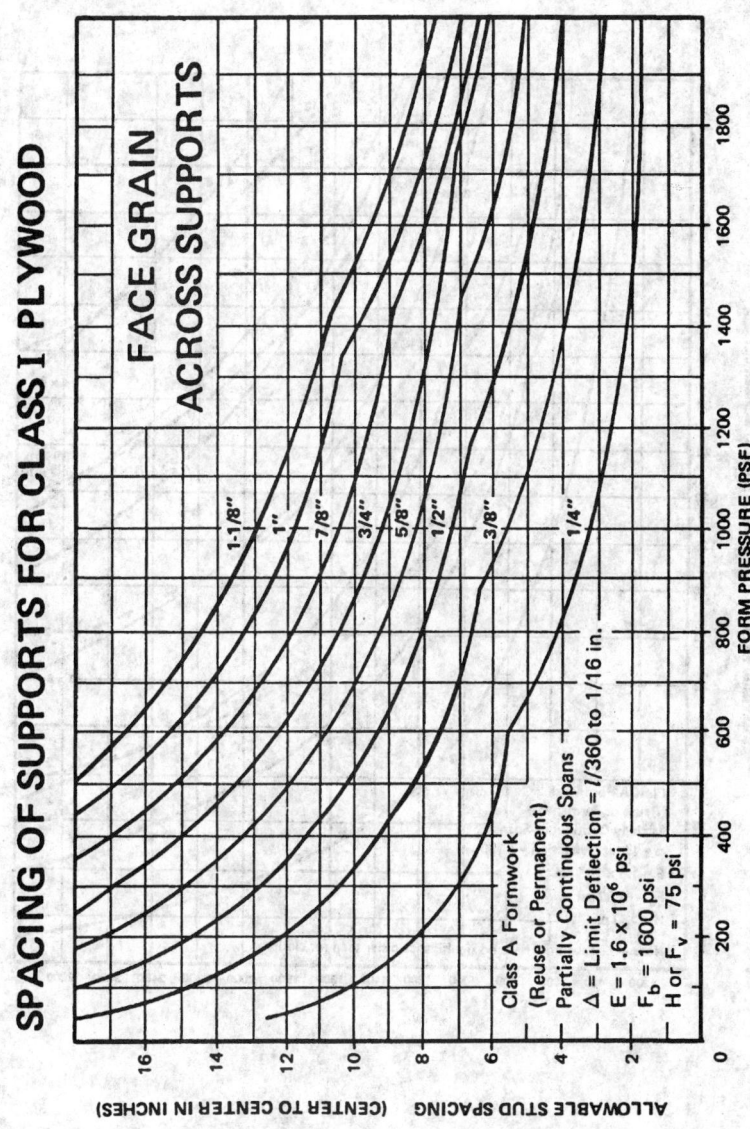

FACE GRAIN ACROSS SUPPORTS

ALLOWABLE STUD SPACING (CENTER TO CENTER IN INCHES)

FORM PRESSURE (PSF)

1-1/8″
1″
7/8″
3/4″
5/8″
1/2″
3/8″
1/4″

Class A Formwork
(Reuse or Permanent)
Partially Continuous Spans
Δ = Limit Deflection = $l/360$ to $1/16$ in.
$E = 1.6 \times 10^6$ psi
$F_b = 1600$ psi
H or $F_v = 75$ psi

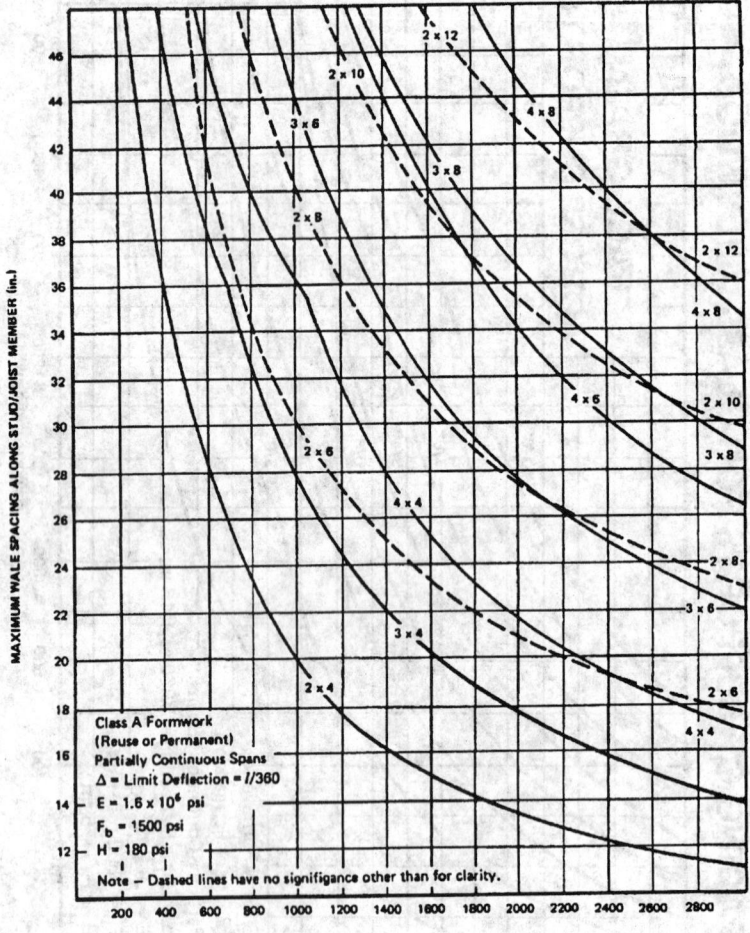

MAXIMUM WALE SPACING ALONG STUD/JOIST MEMBER (in.)

Class A Formwork
(Reuse or Permanent)
Partially Continuous Spans
Δ = Limit Deflection = $l/360$
E = 1.6 x 10⁶ psi
F_b = 1500 psi
H = 180 psi
Note — Dashed lines have no signifigance other than for clarity.

420

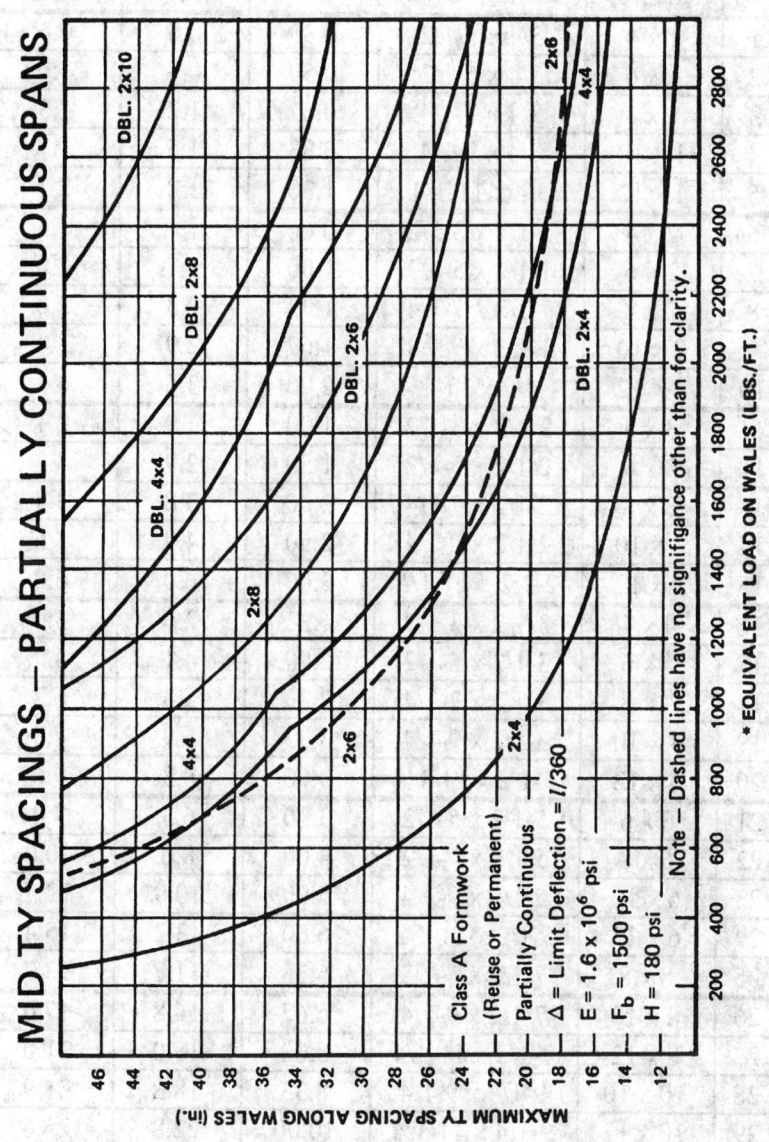

MID TY SPACINGS — PARTIALLY CONTINUOUS SPANS

MAXIMUM TY SPACING ALONG WALES (in.)

* EQUIVALENT LOAD ON WALES (LBS./FT.)

DBL. 2x10

DBL. 2x8

DBL. 2x6

DBL. 4x4

2x8

4x4

2x6

2x4

2x6

2x8

4x4

DBL. 2x4

Class A Formwork
(Reuse or Permanent)
Partially Continuous
Δ = Limit Deflection = $l/360$
E = 1.6×10^6 psi
F_b = 1500 psi
H = 180 psi

Note — Dashed lines have no signifigance other than for clarity.

421

| \multicolumn{7}{c}{Lumber Size and Weight Table} |
Item	Nominal Size, Inches	American Standard, Inches - b x h	B.F. per Lin. Ft.	Weight in Lbs/Lin.Ft.		
				Fir	Pine
1	1 x 4	3/4 x 3-1/2	0.33	0.6	0.6
2	1 x 6	3/4 x 5-1/2	0.50	0.9	1.0
3	1 x 8	3/4 x 7-1/4	0.67	1.2	1.3
4	1 x 10	3/4 x 9-1/4	0.83	1.5	1.7
5	1 x 12	3/4 x 11-1/4	1.00	1.8	2.1
6	2 x 4	1-1/2 x 3-1/2	0.66	1.1	1.3
7	2 x 6	1-1/2 x 5-1/2	1.00	1.7	2.0
8	2 x 8	1-1/2 x 7-1/4	1.33	2.3	2.6
9	2 x 10	1-1/2 x 9-1/4	1.67	2.9	3.4
10	2 x 12	1-1/2 x 11-1/4	2.00	3.6	4.1
11	3 x 4	2-1/2 x 3-1/2	1.00	1.9	2.1
12	3 x 6	2-1/2 x 5-1/2	1.50	2.9	3.3
13	3 x 8	2-1/2 x 7-1/4	2.00	3.8	4.4
14	3 x 10	2-1/2 x 9-1/4	2.50	4.9	5.6
15	3 x 12	2-1/2 x 11-1/4	3.00	6.0	6.8
16	4 x 4	3-1/2 x 3-1/2	1.33	2.6	3.0
17	4 x 6	3-1/2 x 5-1/2	2.00	4.1	4.7
18	4 x 8	3-1/2 x 7-1/4	2.33	5.4	6.2
19	4 x 10	3-1/2 x 9-1/4	3.33	6.9	7.9
20	4 x 12	3-1/2 x 11-1/4	4.00	8.3	9.6
21	6 x 6	5-1/2 x 5-1/2	3.00	6.3	7.4
22	6 x 8	5-1/2 x 7-1/2	4.00	8.6	10.0
23	6 x 10	5-1/2 x 9-1/4	5.00	10.9	12.7
24	6 x 12	5-1/2 x 11-1/2	6.00	13.2	15.4
25	8 x 8	7-1/2 x 7-1/2	5.33	11.7	13.7
26	8 x 10	7-1/2 x 9-1/4	6.67	14.8	17.0
27	8 x 12	7-1/2 x 11-1/4	8.00	18.0	21.0
28	10 x 10	9-1/4 x 9-1/4	8.33	18.8	21.9
29	10 x 12	9-1/4 x 11-1/4	10.00	22.8	26.6
30	12 x 12	11-1/4 x 11-1/4	12.00	27.6	32.1

CONCRETE

				Weight in Kg/Lin. M.	
Lumber Size and Weight Table - Metric					
Item	Nominal Size, mm	American Standard, mm - b x h	Cu.M. per M.	Fir	Pine
1	25 x 100	18.75 x 87.5	0.0026	0.89	0.89
2	25 x 150	18.75 x 137.5	0.0039	1.34	1.49
3	25 x 200	18.75 x 181.3	0.0052	1.79	1.93
4	25 x 250	18.75 x 231.3	0.0064	2.23	2.53
5	25 x 300	18.75 x 281.3	0.0077	2.68	3.13
6	50 x 100	37.50 x 87.5	0.0051	1.64	1.93
7	50 x 150	37.50 x 137.5	0.0077	2.53	2.98
8	50 x 200	37.50 x 181.3	0.0103	3.42	3.87
9	50 x 250	37.50 x 231.3	0.0129	4.32	5.06
10	50 x 300	37.50 x 281.3	0.0155	5.36	6.10
11	75 x 100	62.50 x 87.5	0.0077	2.83	3.13
12	75 x 150	62.50 x 137.5	0.0116	4.32	4.91
13	75 x 200	62.50 x 181.3	0.0155	5.66	6.55
14	75 x 250	62.50 x 231.3	0.0193	7.29	8.33
15	75 x 300	62.50 x 281.3	0.0232	8.93	10.12
16	100 x 100	87.50 x 87.5	0.0103	3.87	4.46
17	100 x 150	87.50 x 137.5	0.0155	6.10	6.99
18	100 x 200	87.50 x 181.3	0.0180	8.04	9.23
19	100 x 250	87.50 x 231.3	0.0258	10.27	11.76
20	100 x 300	87.50 x 281.3	0.0310	12.35	14.29
21	150 x 150	137.5 x 137.5	0.0232	9.38	11.01
22	150 x 200	137.5 x 181.3	0.0310	12.80	14.88
23	150 x 250	137.5 x 231.3	0.0387	16.22	18.90
24	150 x 300	137.5 x 281.3	0.0464	19.64	22.92
25	200 x 200	187.5 x 187.5	0.0412	17.41	20.39
26	200 x 250	187.5 x 231.3	0.0516	22.02	25.30
27	200 x 300	187.5 x 281.3	0.0619	26.79	31.25
28	250 x 250	231.3 x 231.3	0.0645	27.98	32.59
29	250 x 300	231.3 x 281.3	0.0774	33.93	39.59
30	300 x 300	281.3 x 281.3	0.0929	41.07	47.77

Nominal Plywood Weights for Form Plywood				
Nominal Size Tk., Inches	Weight Lbs per Sq.Ft.	Lbs per 4' x 8' Sheet	Nominal Size Tk, mm	Kg per 1.22 x 2.44 Meter Sheet
1/4	0.78	25	6.25	11.32
1/2	1.53	49	12.50	22.21
5/8	1.82	58	15.63	26.42
3/4	2.22	71	18.75	32.22
7/8	2.60	83	21.88	37.74
1 1/8	3.34	107	28.13	48.48

Common Nails					
Sizes	Gauge	Nominal Length in Inches	Approx No. Pcs per Lb.	Nominal Length in mm	Approx No. Pcs per Kg.
2d	12	1	840	25.00	1852
3d	14	1 1/4	543	31.25	1197
4d	12 1/2	1 1/2	294	37.50	648
6d	11 1/2	2	167	50.00	368
7d	11 1/2	2 1/4	149	56.25	328
8d	10 1/4	2 1/2	101	62.50	223
9d	10 1/4	2 3/4	91	68.75	201
10d	9	3	66	75.00	146
12d	9	3 1/4	61	81.25	134
16d	8	3 1/2	47	87.50	104
20d	6	4	30	100.00	66
30d	5	4 1/2	20	112.50	44
40d	4	5	17	125.00	37
50d	3	5 1/2	13	137.50	29
Double Headers/Duplex Nails					
8d	10 1/4	2 1/4	56	88.00	124
16d	8	3	75	44.00	165

Common nails have a thick flat head and used in most phases of construction.
Double header or duplex nails are used in temporary construction as is, form work.

CONCRETE

Masonry/Concrete Nails					
Sizes	Gauge	Nominal Length in Inches	Approx No. Pcs per Lb.	Nominal Length in mm	Approx No. Pcs per Kg.
2d	9	1	185	25.00	408
3d	9	1 1/4	152	31.25	335
4d	9	1 1/2	128	37.50	282
5d	9	1 3/4	112	43.75	247
6d	9	2	99	50.00	218
8d	9	2 1/2	78	62.50	172
9d	9	2 3/4	72	68.75	159
10d	9	3	66	75.00	146
16d	9	3 1/2	55	87.50	121

Masonry/Concrete nails are generally hardened or tempered hardened. The superior strength is required when driving through brick, block or concrete. Hardened nails are resistant to bending when installing.

For examples, see table on next page.

Example: 50'-0" long x 8'-0" high wall, 12" thick, no bulkheads
    50' x 8' = 400 s.f. x 2 sides = 800 s.f. contact surface area.
    50' x 8' x 1' = 400 c.f. ÷ 27 = 14.81 c.y. concrete.
    800 s.f. ÷ 14.81 c.y. = 54 s.f. of forms per c.y.
    14.81 c.y. of concrete ÷ 800 = 0.019 c.y. concrete per s.f. of forms.

Example: 15 m long x 2.4 m high wall, 300 mm thick, no bulkheads
    15 m x 2.4 m = 36 sq.m x 2 sides = 72 sq.m contact surface area.
    15 m x 2.4 m x 0.3 m = 10.8 cu.m of concrete
    72 sq.m ÷ 10.8 cu.m = 6.66 sq.m of forms per cu.m
    10.8 cu.m of concrete ÷ 72 = 0.15 cu.m concrete per sq.m of forms.

Window and Door Openings in Architectural Concrete Walls. It is necessary to form for all door and window openings in concrete walls. The outside wall form is built up solid. Then, boxes or "bucks" are constructed the exact size of the door and window openings. They should be made of 2" (50 mm) lumber and held in place by 1"x 4" (25 x 100 mm) strips nailed to both the outside and inside wall sheathing and braced where necessary using 2"x 4"s (50 x 100 mm) placed horizontally and diagonally. Where the opening is over 3'-0" (0.91 m) wide, it will be necessary to leave an opening in the bottom of the frame so that concrete can be filled to the proper level. After filling to the correct level, the opening is closed by nailing a piece of plank to the bottom of the frame. A few hours after the concrete has set, the plank is removed and the surplus concrete scraped off. To do this, it is necessary to leave an opening in the inside form at every opening.

Concrete Quantities for Wall Forms per Square Foot Contact Area 50'- 0" x 8'- 0" (15.29 x 2.44 m) - No Bulkheads

Wall Thick. in Inches	Contact Area. Less Bulkheads	Cu. Ft. of Concrete	Cu.Yd. of Concrete	Sq.Ft. per Cu.Yd.	Cu.Yd. per Sq.Ft.	Wall Thick. in mm	Contact Area. Less Bulkheads	Cu. M. of Concrete	Sq.M. per Cu.M.	Cu.M. per Sq.M
4	800	132.0	4.89	163.6	0.006	100	74.32	3.74	19.88	0.049
6	800	200.0	7.41	108.0	0.009	150	74.32	5.66	13.12	0.074
8	800	268.0	9.93	80.6	0.012	200	74.32	7.59	9.79	0.099
10	800	332.0	12.30	65.1	0.015	250	74.32	9.40	7.91	0.123
12	800	400.0	14.81	54.0	0.019	300	74.32	11.33	6.56	0.156
14	800	468.0	17.33	46.2	0.022	350	74.32	13.25	5.61	0.181
16	800	532.0	19.70	40.6	0.025	400	74.32	15.06	4.93	0.206
18	800	600.0	22.22	36.0	0.028	450	74.32	16.99	4.37	0.230
20	800	668.0	24.74	32.3	0.031	500	74.32	18.92	3.93	0.255
22	800	732.0	27.11	29.5	0.034	550	74.32	20.73	3.59	0.280
24	800	800.0	29.63	27.0	0.037	600	74.32	22.65	3.28	0.305
26	800	868.0	32.15	24.9	0.040	650	74.32	24.58	3.02	0.329
28	800	932.0	34.52	23.2	0.043	700	74.32	26.39	2.82	0.354
30	800	1000.0	37.04	21.6	0.046	750	74.32	28.32	2.62	0.379
32	800	1068.0	39.56	20.2	0.049	800	74.32	30.24	2.46	0.403
34	800	1132.0	41.93	19.1	0.052	850	74.32	32.05	2.32	0.428
36	800	1200.0	44.44	18.0	0.056	900	74.32	33.98	2.19	0.461

Reveals for door and window openings should be designed so that standard size 2" (50 mm) lumber may be used.

Rustication Strips. Rustication strips should be designed as narrow as possible and only about 3/4" (18.75 mm) deep. When the strips are less than 1" (25 mm) wide, they should have one saw cut. When 2" (50 mm) wide, they should have 2 saw cuts in the back as illustrated. These strips should be nailed to a chalk line, using casing nails long enough to go through the strip and sheathing. It is recommended that the nails be withdrawn by pulling them through the sheathing before the form is removed and allow the wooden strip to remain in place a few days after the form is removed or long enough so that they can be withdrawn without injuring the edges of the concrete.

Ornamentation

Belt courses or other ornamentation where continuous members occur may be formed with wood moldings cut to the proper shape. While some contractors may prefer a waste mold for this particular detail, the wood mold has the advantage of being more easily held in alignment, because the various members are broken at different points, while in a plaster waste mold, the entire form would be cut in two at one point. Soft grained wood that does not warp or split easily should be used. Soft white pine is best, soft-grained Douglas fir is the second choice. In the cornice illustrated, the various members are narrow enough so that common stock sizes can be used throughout. Note the generous use of saw-kerfs in the back side of members to prevent swelling and warping of the lumber. The lumber should be thoroughly oiled on all sides before building the forms as further precaution against swelling and warping.

Large, heavy wood members should be avoided in detailing for ornamental work, because they will swell, and in stripping the corners of the projecting concrete will be fractured. Using several smaller pieces requires less lumber and the members may be arranged so that swelling is away from the concrete, making it easier to strip the form without damage to the concrete corners. Note the illustration where the left-hand detail shows a large member milled from a single wide board in contrast to the smaller pieces shown in the right-hand detail.

In assembling forms, one must constantly keep in mind the steps that can be taken to aid in removing them without injury to the concrete. Boxes, waste molds, wood molds, rustication strips, or anything applied to the face of the forms should be nailed lightly so that when the forms are removed, these members will pull loose from the wall form and remain in the concrete. After these materials have dried out thoroughly and have shrunk, they can be removed without much difficulty and without injury to the concrete corners or edges.

As previously indicated, solid strips of wood, even though well oiled, may swell and result in considerable breakage of concrete corners when the

427

strips are removed. Saw-kerfs on the back of the member will prevent this trouble by relieving any pressure against the concrete. They should be approximately two-thirds the depth of the member, not more than 1-1/2" (37.50 mm) apart, and in general, there should be one kerf within 3/4" (18.75 mm) of each edge.

Waste molds are the only solution for some highly ornamental work. They should be made only by experienced ornamental plaster workers who have been given instructions as to how they are to be fitted into the forms. Ordinarily, the waste molds are about 2" (50 mm) thick, reinforced on the back with 2"x 4"s (50 x 100 mm) that are attached to the mold with burlap dipped in plaster and then wound around the wood. The 2"x 4"s (50 x 100 mm) permit easy handling of the mold and are also used to attach the mold to the form.

When waste molds are large and have deep undercuts, it is usually best to wire them back to the studs or wales with enough wire to be sure that all points will be pressed firmly against the form. Any openings between the form and waste mold should be pointed with plaster of Paris or a patching plaster and nailheads driven through the face and around the edge of the waste molds should be countersunk and similarly pointed.

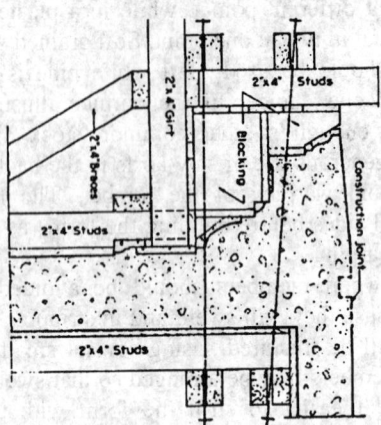

Wood Mold Used in Cornice Form

Ornamental Concrete Cornices. Where it is necessary to use molds for belt courses, cornices, door and window reveals, columns, cornices, entrances, etc., it is more economical to build them of wood, if practicable. However, if plaster waste molds are required, the following will give some idea of their cost, although on this work, it is always advisable to obtain definite prices from a concern specializing in this work.

Plain molds consisting of a few straight lines, $15.00 per s.f. ($161.46 per sq.m). Ordinary molds of stock design will cost from $15.00 to $20 per s.f. ($161.46- $215.28 per sq.m).

Simple molds will cost from $15.00 to $20.00 per s.f. ($161.46 - 215.28 per sq.m) plus the cost of waste molds.

428

CONCRETE

Cornice molds of stock design about 1'-6" (0.45 m) deep and 1'-3" (0.37 m) wide cost about $22.00 to $30.00 l.f. ($72.18 -$98.45 per m).

Capitals for 1'-6" (450 mm) pilasters, stock designs, cost $185 to $295 each.

Joints in plaster waste molds should be patched with non-shrinking patching plaster and painted with shellac.

Individual pieces should not weigh more than 150 lbs. (68.04 kg) to be set without difficulty by 2 workers.

Plaster waste molds are usually given 2 coats of shellac in the shop to make them waterproof and non-absorbent. Before concrete is placed all molds should be greased with a light yellow cup-grease, which may be cut with kerosene if too thick. The grease should be wiped into all angles of the mold and every bit of surplus grease carefully wiped off. Care must be taken not to drop oil, grease or shellac onto hardened concrete or reinforcing.

Construction Joints. Much progress has been made on all types of concrete work in production of satisfactory construction joints, both from the standpoint of good appearance and good bond between successive lifts. On the other hand, some very unsightly joints, for which there is no excuse, occasionally are made on architectural concrete work. It is just as easy to produce a good joint as a poor one. The designer, of course, should indicate on his plans exactly where such joints should be located. Rustications or offsets will mask the joints and are advisable where they can be made to fit into the design.

One requirement that is sometimes neglected is to provide means of holding the form tightly against the hardened concrete. As a result, leakage occurs, discoloring the concrete below, and the upper lift will project slightly beyond the lower lift. The illustration shows the use of 5/8" (15 mm) bolts cast in the concrete for fastening the form for the next lift. Some types of form ties can be used instead of bolts. The bolt or form tie assembly can be used over and over, because only the nut is left in the concrete, if the bolt is used, or the insert, if a form tie is used. In re-erecting the forms for the next lift, the contact surface of the form sheathing should not overlap the hardened concrete more than 1" (25 mm). More overlapping than this only presents more opportunity for leakage due to irregularities in the wall surface against which the sheathing is to be held.

Since sufficient time must be allowed for the concrete between windows or other openings to settle and shrink before the concrete above is placed, the tops of such openings are "must" locations for construction joints. Similarly, it is desirable to locate a joint at the sill line. By so doing, the joints are broken up into short lengths by the opening, making them relatively inconspicuous. Moreover, cracking at corners of openings is thereby minimized.

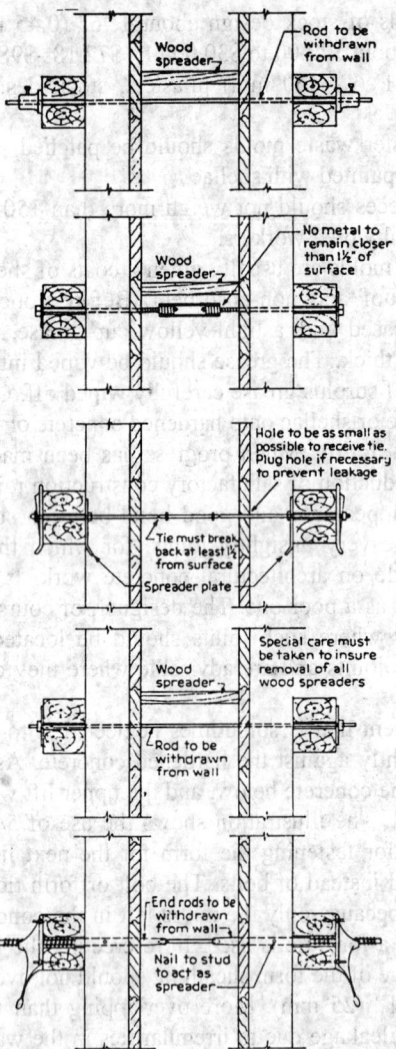

Labels on the figure:
- Wood spreader
- Rod to be withdrawn from wall
- No metal to remain closer than 1½" of surface
- Wood spreader
- Hole to be as small as possible to receive tie. Plug hole if necessary to prevent leakage
- Tie must break back at least 1½" from surface
- Spreader plate
- Special care must be taken to insure removal of all wood spreaders
- Wood spreader
- Rod to be withdrawn from wall
- End rods to be withdrawn from wall
- Nail to stud to act as spreader

Wood Mold Used in Cornice Form

The construction joint should be only a straight thin line at the surface. If a rustication strip is not used at the joint, a 3/4" (18.75 mm) wood strip to a point 1/2" (12.50 mm) above the bottom of this strip and just before the concrete becomes hard, the top should be lightly tamped to be sure that it is tight up against the strip after it has settled or shrunk. All surplus concrete or laitance should then be removed.

In resuming the placing of concrete in the new lift, steps should be taken to get good bond and to avoid honeycomb. The hardened concrete should be clean and thoroughly saturated. A 6" (150 mm) layer of concrete in which the coarse aggregate is reduced about 50%, or a 2" (50 mm) layer of

430

cement-sand grout should then be placed on the hardened concrete followed by the regular concrete.

Control Joints. It is important, both for appearance and performance that control joints be carefully installed so they will be perfectly straight and vertical. Care is required during placing of concrete to avoid knocking the joint out of alignment or otherwise injuring it. When control joints are to be caulked with a plastic compound, this should be done before the final clean down so that any compound smeared onto the surface is removed. Of course, excessive smearing should be avoided.

Form Ties. Various makes of form ties are available for architectural concrete construction. Important features to look for in selecting form ties are: removal of all metal to a depth of at least 1-1/2" (37.50 mm) from the face of the wall, minimum strength of 3,000 lbs. (1360.80 kg) when fully assembled, adjustable length to permit tightening of forms around all openings and inserts and a design that will leave a hole at the surface not more than 7/8" (21.87 mm) in diameter and not less than 3/4" (18.75 mm) deep so that it will hold a patch. Ties should not be fitted with cones or washers acting as spreaders, because tapered and shallow holes cannot be patched properly.

Labor Cost of Architectural Forms. The labor fabricating, erecting, and stripping forms is usually estimated on the basis of sq. ft. (sq. m) of contact area. It is customary on a job having a normal amount of ornamentation to take off the total contact area as though the wall were plain. Window and other openings, unless very large, are figured solid. The area thus obtained is priced as though the wall were unornamented. Separate allowance is made for window frames or "bucks", ornamentation, etc.

The average amount of lumber required for plain wall forms is 2-1/4 to 2-1/2 b.f. (0.005-0.006 cu.m) of lumber per s.f. (sq.m) of contact area, plus the sheathing, plywood, or concrete form Presdwood, used for the facing materials. Figure one tie rod for approximately each 10 s.f. (9.29 sq.m) of form.

On plain walls up to 10 ft. (3 m) story height, it requires about 11 to 13 hrs. carpenter time and 6 to 8 hrs. labor time per 100 s.f. (9.29 sq.m) of contact area. For walls 12' (3.65 m) to 14' (4.26 m) high, add about 10% to these costs. This includes all time necessary for erecting and stripping the forms and hoisting it up for the next use. In addition, figure about 1 to 1-1/4 hrs. labor time for removing hardened concrete from the plywood and coating plywood between each use.

Door and window frame boxes or "bucks" require 5 to 8 hrs. carpenter time for each opening, depending upon size and detail of same.

Rustication strips should be placed at the rate of 25 to 30 l.f. (7.62.5-9.14 m) per hr. for one carpenter.

Ornamental curtain walls, fluted pilasters and piers, reveals, etc., will require about 10 to 12 hrs. carpenter time and 6 hrs. labor time per 100 s.f. (9.29 sq.m) in addition to the cost of plain wall forms.

Ornamental wood cornices built to special detail will require 24 to 28 hrs. carpenter time and 12 to 14 hrs. labor time per 100 sq.ft. (9.29 sq.m) in addition to cost of plain wall forms.

Ornaments projecting from the face of plain walls, such as special sills, balconies, etc., will require about 36 to 40 hrs. carpenter time and 12 to 15 hrs. labor time per 100 sq.ft. (9.29 sq.m) in addition to the cost of plain wall forms.

Where plaster waste molds are used, they may be set either by carpenters or plasterers, depending upon local labor union regulations.

When setting simple belt courses of plaster waste mold, allow 12 to 14 hrs. mechanic time and 6 to 7 hrs. helper time per 100 sq.ft. (9.29 sq.m).

When erecting and blocking cornices and other elaborate moldings using plaster waste molds, figure 20 to 25 hrs. mechanic time and 10 to 12 hrs. helper time per 100 sq.ft. (9.29 sq.m).

When placing letters, a carpenter should set 2 to 3 letters an hr. depending upon size.

Labor Cost of 100 Sq.Ft. (30.48 Sq.M.) of Plain Architectural Concrete Walls Using Plywood for Facing Material					
Description	Hours	Rate		Total	
Carpenter	12.00	$	35.04	$	420.48
Labor helping	6.00	$	26.16	$	156.96
Labor stripping and Cleaning Forms	2.00	$	26.16	$	52.32
Labor coating plywood	0.33	$	26.16	$	8.63
Cost per 100 s.f. (9.29 sq.m)				$	638.39
cost per sq.ft.				$	6.38
cost per sq.m				$	68.72

Details for Architectural Concrete Forms. For the rapid and economical construction of architectural concrete forms, it is necessary to have complete working drawings that show the elevations in large scale, together with every piece of plywood, stud, wale, bolt, lumber for reveals, window boxes or "bucks", and so on, similar to the methods used for detailing cut stone, terra cotta, or structural steel.

INSULATING CONCRETE FORMS

Insulating Concrete Forms (ICF) are a hollow, lightweight form block made of expanded polystyrene. When assembled into a structure or building configuration, the forms represent the exterior wall of the structure and are left in place. The hollow ICF block units are then filled with concrete or reinforced with reinforcement steel to form a structural wall, much like a conventional wall pour, except that the formwork is not removed.

There are some distinct advantages that this type of form material provides to the contractor and in turn to the owner:

432

CONCRETE

1. Ease of installation
2. Energy efficient
3. Interior comfort to the end user
4. Sound attenuation
5. Design flexibility
6. Strength of wall
7. Durability
8. Disaster resistance
9. Pest control
10. No stripping
11. Less waste, less disposal of materials

As with estimating any other specialized material, manufacturer current specifications and guidelines must be used. Specifications for type of form, ties and reinforcement must be adhered to. The ICF form system has been classified into three distinct areas:

Panels. Panels range from 1'- 3" x 8'- 9" to 4'-0" x 8'-0" (.38 x 2.67m to 1.22 x 2.44 m). Edges are flat and require interconnection by either connectors or fasteners, which are made of nonfoam material. The estimator is cautioned to contact supplier of the ICF panel for tie and bracing requirements.

Planks. Planks vary from 1' x 4' to 1' x 8' (.30 x 1.22m to .30 x 2.44m). They look like wood planks and are outfitted with cross pieces, as part of the wall setting sequence requires interconnection by either connector or fastener.

Block. Block units get there name because they look like cement blocks. They range from 8" x 16" (200 x 400 mm) to 16" x 48" (400 x 1200 mm). The blocks are molded with special edges that will interconnect the blocks of the same manufacturer. The most common are tongue and groove interlocking teeth or nubs or raised squares. The cavities inside the ICF block units are in various shapes, which results in the final shape of the concrete. The various final concrete forms fall again into three categories:

Flat. Flat concrete is of consistent thickness, same as conventional flat concrete. As with conventional steel or wood forms, there is a requirement for ties, either of steel or more often plastic, and wall bracing. The estimator is cautioned to contact suppliers for tie requirements.

Grid. Grid concrete is a waffle, with intersecting cylindrical horizontal and vertical concrete members. The size of the waffles varies among manufacturers.

Post and Beam. This consists of more widely and variably-spaced horizontal and vertical members. The name comes from a pattern that is similar to wood post-and-beam construction. As with any wall form, steel or wood there is a requirement for ties, either of steel or more often in ICF systems, plastic and wall bracing if required. Estimator is cautioned contact supplier of the ICF block for tie requirements.

Panels: Labor required to form walls using an ICF forming system (no stripping required). Assuming a carpenter can erect, and brace 125 s.f. (11.61 sq.m) per carpenter per hour.

Description	Hours	Rate	Total
Carpenter Foreman	8.00	$ 39.04	$312.32
Carpenter	8.00	$ 35.04	$280.32
Laborer	2.00	$ 26.16	$52.32
Labor Cost per 2,000 s.f.			$644.96
Cost per sq.f.			$0.32
Cost per sq.m			$3.47

Add material, ties, bracing and scaffolding as required by manufacture.

Planks: Labor required to place planks using an ICF forming system (no stripping required). Planks are normally used as fillers and are more customized forming. Assuming a carpenter can erect, and brace 80 s.f. (7.43 sq.m) per carpenter per hour.

Description	Hours	Rate	Total
Carpenter Foreman	8.00	$ 39.04	$312.32
Carpenter	8.00	$ 35.04	$280.32
Laborer	2.00	$ 26.16	$52.32
Labor Cost per 1,280 s.f.			$644.96
Cost per s.f.			$0.50
Cost per sq.m			$5.42

Planks are normally used as fillers and are more customized forming. Add material, ties, bracing and scaffolding as required by manufacture.

Blocks: Labor required placing ICF forming system using blocks. Blocks are filled with concrete after placement. ICF blocks will vary from manufacture to manufacture, listed below is an average of all manufactures using plastic ties. Add time for specials such as openings or special penetrations. Assuming a standard blocks at 8" x 16", (400 to 800 mm) a carpenter or mason (depending on local area practices) can assemble complete with shoring if required to 8' (2.44 m) high 210 s.f. (19.51 sq.m) per day or 420 s.f. (39.01 sq.m) of contact surface area for concrete.

Description	Hours	Rate	Total
Carpenter Foreman	8.00	$ 39.04	$312.32
Carpenter	8.00	$ 35.04	$280.32
Laborer	4.00	$ 26.16	$104.64
Labor Cost per 420 s.f. (contact surface area)			$697.28
Cost per sq.f. (contact surface area)			$1.66
Cost per sq.m (Contact surface area)			$17.87

Note: Block is filled with concrete. Add material, ties, bracing and scaffolding as required by manufacture.

CONCRETE

Concrete Fill: Labor required to place ICF forming system using 8" x 16" (400 x 800 mm) blocks. After blocks are set, they are filled with concrete, normally with reinforcement. ICF blocks will vary from manufacture to manufacture, so check with the individual manufacturers for quantities of individual units per sq. ft of form area and rate of pour per hour. The costs given below assume:

80 c.y. (61.17 cu.m) per day including setting up and cleaning up.

Description	Hours	Rate	Total
Laborer Foreman	8.00	$ 30.16	$241.28
Laborer	32.00	$ 26.16	$837.12
Concrete Pump Oper.	8.00	$ 37.89	$303.12
Labor Cost per 80 c.y.			$1,381.52
Cost per cu.y.			$17.27
Cost per cu.m			$22.59

Note: Block is filled with concrete. Add material, ties, bracing and scaffolding as required by manufacture.

03200 CONCRETE REINFORCEMENT

Reinforcing steel is estimated by the pound or the ton, obtained by listing all bars of different sizes and lengths and extending the total to pounds. Reinforcing bars may be purchased from warehouse in stock lengths and all cutting and bending done on the job or they may be purchased cut to length and bent, ready to place in the building.

Example: Assume a retaining wall is reinforced with #5 bars spaced 6" (150 mm) on centers both vertically and horizontally. The wall is 40'-0" long and 15'-0" high and with bars spaced 6" (150 mm) on centers it will require the following:

80 #5 bars 15'-0" long = 1,200 l.f. @ 1.043 lbs per ft) = 1,251.60 lbs. (80 #16 @ 4.57 m = 365.76 m x 1.552 kg/m = 567.66 kg)

30 #5 bars 40'-0" long = 1,200 l.f. @ 1.043 lbs. (.47 kg) = 1,251.60 lbs. (30 #16 @ 12.19 m = 365.76 m 1.552 kg/m = 567.66 kg)

Total weight of steel required = 2503.2 lbs. (1135.45 kg)

If reinforcing steel is worth $60.00 per 100 lbs. (45.35 kg) in place, or $0.60 per lb. ($1.32/kg), the cost of the steel in the wall ready for pouring would be 2,504 lbs. x $0.60 = $1502.40 (1135.81 kg x $1.32 per kg = $1499.28)

When purchasing reinforcement steel in stock lengths, labor cost of unloading, sorting, handling, and storing is less than when sent to the job cut to length and bent. The former requires less storage space, because different lengths are placed in compact piles. Steel cut to length and bent requires considerable storage space, because all bars of each size and length must be placed in separate piles.

435

However, shop costs of cutting to length and bending reinforcement steel are much less than the cost to cut and bend on the job, especially where the work is done by ironworkers. Practically all reinforcing steel used today is shop fabricated, cut to length and bent prior to delivery to the job.

Reinforcement steel is now manufactured both in the English and metric systems. Charts below are for direct sizes and manufactured of size, weights and standards as used by the Concrete Reinforcement Steel Institute (CSI) as set by the American Society of Testing Materials (ASTM) Standard Reinforcement Bars. Following the two charts below is a "soft" metric conversion chart.

Standard Sizes and Weights of Concrete Reinforcing Bars							
Nominal Dimensions				Metric - Nominal Dimensions			
Bar Designation (Mark)	Dia. Inch	Area Inch²	Weight Lbs./Ft.	Metric Designation (Mark)	Dia. mm	Area mm²	Weight Kg/m
# 3	0.375	0.11	0.376	# 10	9.53	71	0.560
# 4	0.500	0.20	0.668	# 13	12.70	129	0.994
#5	0.625	0.31	1.043	# 16	15.88	200	1.552
# 6	0.750	0.44	1.502	# 19	19.05	284	2.235
# 7	0.875	0.60	2.044	# 22	22.23	387	3.042
# 8	1.000	0.79	2.670	# 25	25.40	510	3.973
# 9	1.125	1.00	3.400	# 29	28.58	645	5.060
# 10	1.270	1.27	4.303	# 32	32.26	819	6.404
# 11	1.410	1.56	5.313	# 36	35.81	1006	7.907
# 14	1.693	2.25	7.650	# 43	43.00	1452	11.384
# 18	2.257	4.00	13.600	# 57	57.33	2581	20.239

The ASTM Specification for Reinforcement Steel, which has the dimensions for A615/A615M-96a for billet steel and A706/A406M-96b for low-alloy steel, is presented in duel-unit format.

Basis for Estimating Price of Reinforcing Steel. When estimating quantities and weights of reinforcing steel, all bars of each size should be listed separately to obtain the correct weight. No. 6 bars (# 19) and over take the lowest price per lb. (kg), while bars smaller than No. 6 (# 19) increase in price on a graduated scale depending on the size of the bars. The smaller the bars, the higher the price per lb. (kg). On large projects the normal practice is to quote the price per lb. (kg), or ton, based on the entire project as a whole.

Reinforcement steel shipped to a project with a mix of bar sizes an average No.6 (#10) size would run about $0.60 per pound. This price is for stock length. The price excludes, detailing, bending, coating or galvanizining. Due to the low weight of a No.2 (#10) reinforcement bars some manufactures do not manufacture anything less then a No.4 (#13) reinforcement bar. Therefore, one should check with his/her supplier if No. 3 (#10) bars are available if indicted on the drawings. Most detailers will use a no. 4 (#13) bars in place of a No.3 (#3) bar.

436

CONCRETE

Prices on rail steel are theoretically the same as for billet steel but will vary with the scrap market on rails.

There are five specifications covering the use of reinforcement steel:

1. A615/A615M-04b Standard Specification for Deformed and Plain Billet-Steel Bars for Concrete Reinforcement

2. A996/A996M-04 Standard Specification for Rail-Steel and Axle-Steel Deformed Bars for Concrete Reinforcement

3. A706/A706M-04b Standard Specification for Low-Alloy Steel Deformed and Plain Bars for Concrete Reinforcement

4. A184/A184M-05 Standard Specification for Fabricated Deformed Steel Bar Mats for Concrete Reinforcement

5. A934/A934M-04 Standard Specification for Epoxy-Coated Prefabricated Steel Reinforcing Bars

It is of utmost importance that the quantity take-off be correct. Normal practice is to allow between 2% to 5% of the calculated bar weight for waste and splices. Reinforcement steel up to No. 11 bars may be spliced by overlapping them and wiring them together.

Two standard grades of reinforcement steel available are, ASTM A615, Grade 40 and ASTM A615, Grade 60. Bars are of two minimum yield levels: 40,000 psi (275.8 MPa) and 60,000 psi (413.7 MPa), designated as Grade 40 and Grade 60 respectively.

Both grades are standard in building codes and in the design recommendations of the American Concrete Institute.

Reinforcement Bar Supports. Chairs, sometimes referred to as reinforcement bar support, are required in order to accurately support and locate reinforcement steel in the concrete form. The bar support will hold the bar firmly in its proper location during placement of concrete.

Rust Prevention. Reinforcement steel supports are classified with respect to methods to diminish rust spots or similar imperfections from the concrete surface caused by reinforcement steel supports. There are three classes:

Class 1: Maximum protection; plastic protected reinforcement steel support which are intended for use in situations of moderate to severe exposure and/or situations requiring light grinding or sandblasting of concrete surface.

Class 2: Moderate protection; stainless steel protected reinforcement steel supports, which are intended for use in situations of moderate exposure and/or situations requiring light grinding or sandblasting of the concrete surface. Class 2 protection can be obtained by the use of either Type "A" or Type "B" stainless steel protection reinforcement steel supports. The difference between the two types is in the length of the stainless steel tip attached to the bottom of each leg support.

437

Percentages and Weight of standard Steel Reinforcement Spirals (1)

Cure (3) Diam. Inches	% and Wt.	1/4" Ø Spirals (4) Pitch in Inches					3/8" Ø Spirals (4) Pitch in Inches							
		1 3/4	2	2 1/4	2 1/2	F	1 3/4	2	2 1/4	2 1/2	2 3/4	3	3 1/4	F
12	%	0.95	0.83	-----	-----	1.59	2.09	1.83			See Note (1)			3.55
	Wt	3.65	3.19	-----	-----		8.02	7.03						
13	%	0.88	0.77	-----	-----	1.74	1.93	1.69						3.84
	Wt	3.96	3.47	-----	-----		8.71	7.63						
14	%	0.82	0.71	0.64	-----	1.86	1.79	1.57	1.4					4.14
	Wt	4.28	3.71	3.34	-----		9.35	8.21	7.32					
15	%	0.76	0.67	0.598	0.53	2.01	1.67	1.47	1.3	1.17				4.44
	Wt	4.56	4.02	3.54	3.18		10	8.82	7.8	7.02				
16	%	0.71	0.62	0.56	0.5	2.12	1.57	1.37	1.22	1.1				4.73
	Wt	4.85	4.23	3.82	3.41		10.71	9.35	8.33	7.51				
17	%	0.67	0.59	0.52	-----	2.27	1.48	1.29	1.15	1.03	0.94			5.03
	Wt	5.16	4.54	4	-----		11.4	9.93	8.85	7.93	7.24			
18	%	0.64	0.5	-----	-----	2.42	1.4	1.22	1.09	0.98	0.89	0.81		5.32
	Wt	5.53	4.84	-----	-----		12.09	10.52	9.42	8.47	7.68	7		
19	%	0.6	0.53	-----	-----	2.55	1.32	1.16	1.03	0.93	0.84	0.77		5.61
	Wt	5.77	5.1	-----	-----		12.7	11.16	9.91	8.95	8.09	7.41		
20	%	0.57	0.5	-----	-----	2.67	1.25	1.1	0.98	0.8	0.8	0.73	0.68	5.91
	Wt	6.09	5.34	-----	-----		13.34	11.73	10.47	9.4	8.85	7.8	7.26	
21	%	0.54	-----	-----	-----	2.78	1.2	1.05	0.93	0.84	0.76	0.7	0.65	6.21
	Wt	6.35	-----	-----	-----		14.11	12.33	10.92	9.88	8.94	8.22	7.64	
22	%	0.52	-----	-----	-----	2.94	1.14	1	0.89	0.8	0.73	0.67	0.62	6.5
	Wt	6.71	-----	-----	-----		14.7	12.9	11.48	10.32	9.42	8.65	8	
23	%	0.5	-----	-----	-----	3.09	1.09	0.96	0.85	0.77	0.7	0.64	0.59	6.8
	Wt	7.05	-----	-----	-----		15.38	13.53	11.98	10.86	9.88	9.02	8.32	

Percentages and Weight of standard Steel Reinforcement Spirals (1)

Cure (3) Diam. Inches	% and Wt.	1/4" Ø Spirals (4) Pitch in Inches				F	3/8" Ø Spirals (4) Pitch in Inches							F
		1 3/4	2	2 1/4	2 1/2		1 3/4	2	2 1/4	2 1/2	2 3/4	3	3 1/4	
24	%	-----	-----	-----	-----	-----	1.05	0.92	0.81	0.73	0.67	0.61	0.56	7.1
	Wt	-----	-----	-----	-----		16.12	14.13	12.43	11.2	10.29	9.37	8.6	
25	%	-----	-----	-----	-----	-----	1	0.88	0.78	0.71	0.64	0.59	0.54	7.39
	Wt	-----	-----	-----	-----		16.65	14.65	13	11.82	10.67	9.83	9	
26	%	-----	-----	-----	-----	-----	0.97	0.85	0.75	0.68	0.62	0.56	0.52	7.68
	Wt	-----	-----	-----	-----		17.46	15.3	13.5	12.23	11.17	10.08	9.36	
27	%	-----	-----	-----	-----	-----	0.93	0.82	0.72	0.65	0.59	0.54	0.5	7.97
	Wt	-----	-----	-----	-----		18.08	15.93	14	12.63	11.47	10.5	-----	
28	%	-----	-----	-----	-----	-----	0.9	0.79	0.7	0.63	0.57	0.52	-----	8.27
	Wt	-----	-----	-----	-----		18.81	16.52	14.83	13.18	11.92	10.88	-----	
29	%	-----	-----	-----	-----	-----	0.87	0.76	0.67	0.61	0.55	0.51	-----	8.57
	Wt	-----	-----	-----	-----		19.51	17.05	15.03	13.68	12.33	11.43	-----	
30	%	-----	-----	-----	-----	-----	0.84	0.73	0.65	0.59	0.53	-----	-----	8.86
	Wt	-----	-----	-----	-----		20.15	17.52	15.6	14.16	12.72	-----	-----	
31	%	-----	-----	-----	-----	-----	0.81	0.71	0.63	0.57	0.52	-----	-----	9.16
	Wt	-----	-----	-----	-----		20.78	18.2	16.15	14.62	13.33	-----	-----	
32	%	-----	-----	-----	-----	-----	0.79	0.69	0.61	0.55	0.5	-----	-----	9.46
	Wt	-----	-----	-----	-----		21.6	18.86	16.67	15.03	13.67	-----	-----	
33	%	-----	-----	-----	-----	-----	0.76	0.67	0.59	0.53	-----	-----	-----	9.76
	Wt	-----	-----	-----	-----		22.08	19.42	17.11	15.38	-----	-----	-----	

(1) Spirals are based on Theoretical Weights. (2) F equals weight to add for finishing. (includes 1.5 turns at top & 1.5 turns at bottom of spiral. ACI-90-S56). Total weight = Weight/ft. x height + F + weight of spacers. (3) Two spacers required for spirals 20" or less in diameter, 3 spacers for spirals 20" to 30", and 4 spacers for spirals over 30"in diameter and all 3/8" or larger spiral rods, regardless of spiral diameter. Allow 3/4 pound per lin/ft. per spacer. (4) The four spiral-rod sizes shown in the

Percentages and Weight of standard Steel Reinforcement Spirals (1)

Cure (3) Diam. Inches	% and Wt.	1/2" Ø Spirals (4) Pitch in Inches								5/8" Ø Spirals (4) Pitch in Inches							
		2	2 1/4	2 1/2	2 3/4	3	3 1/4	3 1/2	F	2	2 1/4	2 1/2	2 3/4	3	3 1/4	3 1/2	F
12	%	3.33	-----	-----	-----	-----	-----	-----	6.3	5.17	-----	-----	-----	-----	-----	-----	9.84
	Wt	12.79	-----	-----	-----	-----	-----	-----		19.88	-----	-----	-----	-----	-----	-----	
13	%	3.08	-----	-----	-----	-----	-----	-----	6.82	4.77	-----	-----	-----	-----	-----	-----	10.66
	Wt	13.89	-----	-----	-----	-----	-----	-----		21.5	-----	-----	-----	-----	-----	-----	
14	%	2.86	2.54			See Note (1)			7.35	4.43	3.94	-----	-----	-----	-----	-----	11.47
	Wt	14.92	13.26							23.1	20.59	-----	-----	-----	-----	-----	
15	%	2.67	2.37	2.13					7.87	4.14	3.68	3.31	-----	-----	-----	-----	12.29
	Wt	16	14.2	12.78						24.88	22.09	19.85	-----	-----	-----	-----	
16	%	2.5	2.22	2					8.4	3.88	3.45	3.1	-----	-----	-----	-----	13.11
	Wt	17.05	15.14	13.64						26.44	23.53	21.14	-----	-----	-----	-----	
17	%	2.35	2.09	1.88	1.71	-----	-----	-----	8.93	3.65	3.24	2.92	2.65	-----	-----	-----	13.93
	Wt	18.1	16.1	14.48	13.18	-----	-----	-----		28.1	24.98	22.48	20.4	-----	-----	-----	
18	%	2.22	1.97	1.78	1.62	1.48	-----	-----	9.45	3.45	3.06	2.76	2.51	2.3	-----	-----	14.74
	Wt	19.18	17.02	15.38	14	12.79	-----	-----		29.8	26.43	23.83	21.68	19.88	-----	-----	
19	%	2.1	1.87	1.68	1.53	1.4	-----	-----	9.97	3.26	2.9	2.61	2.37	2.17	-----	-----	15.58
	Wt	20.21	18	16.16	14.72	13.47	-----	-----		31.39	27.9	25.1	22.8	20.89	-----	-----	
20	%	2	1.78	1.6	1.45	1.33	1.23	-----	10.5	3.1	2.76	2.48	2.25	2.07	1.91	-----	16.39
	Wt	21.36	19	17.09	15.48	14.2	13.12	-----		33.1	29.45	26.48	24	22.1	20.4	-----	
21	%	1.9	1.69	1.52	1.39	1.27	1.17	1.09	11.02	2.95	2.62	2.36	2.15	1.97	1.82	1.69	7.21
	Wt	22.32	19.85	17.85	16.34	14.92	13.75	12.81		34.66	30.8	27.75	25.29	23.17	21.4	19.85	
22	%	1.82	1.61	1.45	1.32	1.21	1.12	1.04	11.54	2.82	2.51	2.26	2.05	1.88	1.73	1.61	18.03
	Wt	23.48	20.79	18.71	17.02	15.61	14.45	13.42		36.4	32.4	29.19	26.46	24.25	22.3	20.79	
23	%	1.74	1.54	1.39	1.26	1.16	1.07	1	12.08	2.69	2.4	2.16	1.96	1.8	1.66	1.54	18.84
	Wt	24.55	21.7	19.6	17.76	16.35	15.09	14.1		37.98	33.86	30.45	27.62	25.39	23.4	21.7	

Percentages and Weight of standard Steel Reinforcement Spirals (1)

Cure (3) Diam. and Inches	% and Wt.	1/2 " Ø Spirals (4) Pitch in Inches							F	5/8" Ø Spirals (4) Pitch in Inches							F
		2	2 1/4	2 1/2	2 3/4	3	3 1/4	3 1/2		2	2 1/4	2 1/2	2 3/4	3	3 1/4	3 1/2	
24	%	1.67	1.48	1.33	1.21	1.11	1.02	0.95	12.6	2.58	2.3	2.07	1.88	1.72	1.59	1.48	19.68
	Wt	25.64	22.72	20.42	18.59	17.04	15.67	14.59		39.6	35.31	31.8	28.87	26.41	24.41	22.72	
25	%	1.6	1.42	1.28	1.16	1.07	0.98	0.91	13.12	2.48	2.21	1.98	1.8	1.65	1.53	1.42	20.48
	Wt	26.62	23.62	21.3	19.32	17.82	16.31	15.15		41.3	36.8	33	30	27.48	25.48	23.62	
26	%	1.54	1.37	1.23	1.12	1.03	0.95	0.88	13.64	2.38	2.12	1.91	1.74	1.59	1.47	1.36	21.3
	Wt	27.72	24.65	22.14	20.18	18.55	17.1	15.84		42.8	38.2	34.4	31.14	28.62	26.46	24.48	
27	%	1.48	1.31	1.18	1.08	0.99	0.91	0.85	14.17	2.29	2.04	1.84	1.67	1.53	1.41	1.31	22.12
	Wt	28.79	25.42	22.9	21	19.24	17.69	16.52		44.5	39.62	35.79	32.42	29.72	27.4	25.42	
28	%	1.43	1.27	1.14	1.04	0.95	0.88	0.82	14.7	2.21	1.97	1.77	1.61	1.47	1.37	1.27	22.95
	Wt	29.9	26.58	23.82	21.77	19.86	18.4	17.14		46.2	41.2	37	33.68	30.78	28.63	26.58	
29	%	1.38	1.22	1.1	1	0.92	0.85	0.79	15.22	2.14	1.9	1.71	1.55	1.43	1.32	1.22	23.79
	Wt	30.99	27.39	24.65	22.41	20.63	19.08	17.72		48	42.6	38.36	34.79	32.09	29.8	27.39	
30	%	1.33	1.18	1.07	0.97	0.89	0.82	0.765	15.73	2.07	1.84	1.65	1.5	1.38	1.27	1.18	24.6
	Wt	32	28.32	25.7	23.27	21.38	19.69	18.23		49.7	44.2	39.6	36	33.1	30.49	28.32	
31	%	1.29	1.15	1.03	0.94	0.86	0.79	0.74	16.28	2	1.78	1.6	1.49	1.33	1.23	1.14	25.4
	Wt	33.08	29.48	26.39	24.1	22.04	20.25	18.98		51.24	45.6	41	37.41	34.09	31.54	29.21	
32	%	1.25	1.11	1	0.91	0.83	0.77	0.71	16.8	1.94	1.72	1.55	1.41	1.29	1.19	1.11	26.22
	Wt	34.19	30.3	27.33	24.89	22.69	21.05	19.4		53.05	47	42.35	38.52	35.24	32.5	30.3	
33	%	1.21	1.08	0.97	0.88	0.81	0.75	0.69	17.32	1.88	1.67	1.5	1.37	1.25	1.16	1.08	27.08
	Wt	35.1	31.34	28.15	25.51	23.49	21.75	20		54.55	48.49	43.5	39.78	36.25	33.65	31.54	

(1) Spirals are based on Theoretical Weights. (2) F equals weight to add for finishing, (includes 1.5 turns at top & 1.5 turns at bottom of spiral. ACI-90-S56). Total weight = Weight/ft. x height + F + weight of spacers. (3) Two spacers required for spirals 20" or less in diameter, 3 spacers for spirals 20" to 30", and 4 spacers for spirals over 30"in diameter and all 3/8" or larger spiral rods, regardless of spiral diameter. Allow 3/4 pound per lin/ft. per spacer. (4) The four spiral rod sizes shown in the table are In stock, ready for fabrication. (5) Spirals above stepped line are not collapsable for shipment.

Type "A" - Stainless steel protected bar support; a tip of stainless steel is attached to the bottom of each leg so that no portion of the non-stainless steel wire lies closer than 1/4" (6 mm) to the form face.

Type "B" - Stainless steel protected bar support, a tip of stainless steel is attached to the bottom of each leg so that no portion of the non-stainless steel wire lies closer than 3/4" (19 mm) to the form face.

Class 3: No protection; bright basic reinforcement steel supports that have no protection against rusting and which are intended for use in situations where surface blemishes can be tolerated and/or where supports do not come in contact with the exposed concrete.

Costs for reinforcement steel supports can vary widely, from less than 1% to more than 10% of the reinforcement steel costs.

Splicing Reinforcing Steel. Reinforcement shall be spliced only at points shown in the contract document or approved by the engineer when lapped splices are used in reinforcement in which the critical design stress is tensile, splices shall not be used at points of maximum stress. Bars shall be placed in close contact and wired tightly in a manner that the specified distance to the surface of the concrete is maintained. For bar size No.3 (13) to No.11 (36) lap splices are permitted. Bar size No.14 (43) and No. 18 (57) *may not* be lapped.

Lap Splices					
Bar Size	Lap Length Inches		Bar Size	Lap Length Inches	
	Uncoated	Epoxy Coated		Uncoated	Epoxy Coated
4	22	27	13	560	690
5	29	35	16	740	890
6	34	41	19	870	1040
7	43	52	22	1090	1320
8	57	69	25	1450	1750
9	72	87	29	1830	2210
10	92	111	32	2340	2820
11	113	137	36	2870	3480

Bending Reinforcing Steel. The labor cost of bending reinforcing bars will vary with the size of the bars and the amount of bending necessary; the lighter the bars the higher the bending cost per ton.

Type L Bending. Bending all No. 3 (#10) bars; all No. 4 (#13) stirrups and column ties; truss bars and all sizes of bars bent at more than 6 points. It also includes all bars bent in more than one plane and all radius bending with more than one radius in any bar, approximately $90.00 per ton ($99/metric ton).

CONCRETE

Type H Bending. Bending truss bars for beams and slabs, other than No. 3 (#10) and No. 4, (#13) radius bending and types not otherwise described, approximately $40.00 per ton ($44/metric ton).

When cut is made by means other than sawing or milling, $0.05 per end. When cut is made by sawing or milling, $0.15 per end.

Spirals for Reinforced Concrete Columns. Spirals should always be coiled and fabricated in the shop, as it is all machine work.

Prices of spirals vary throughout the country in the same manner and for the same reasons as reinforcing steel prices. Quantity extras are figured independently of reinforcing steel tonnages involved.

Cost of Engineering Service. The cost of making working drawings, bending details, setting plans, etc., will vary with the amount of work involved.

Where the engineers prepare the complete design for a reinforced concrete structure, including working drawings, placing plans, bar lists, and bending details, add $50.00 per ton ($55.55 per tonne) to the detailing charges given below.

For preparing order lists with or without bending details from plans on which all sizes and lengths are shown by others, $0.30 per 100 lbs. (45.35 kg) or $8.50 per ton ($9.44/Tonne).

Hauling Reinforcing Bars From Shops to Job. The cost of hauling reinforcing steel from warehouse to job will vary with the distance of the haul and the size of the city.

For example, in the Chicago area, the trucking charge for deliveries within city limits is about $2.00 per cwt. For jobs located in the Chicago suburbs, the trucking charge is about $2.25 per cwt. Minimum charge for both zones is $250.00.

Unloading Reinforcing Steel From Cars or Trucks. If steel can be unloaded direct from cars or trucks, without sorting or carrying to stock piles, a worker will unload 4-1/2 to 5-1/2 tons (5 – 6 tonne) per 8-hr. day, at the following labor cost per ton:

Description	Hours	Rate	Total
Iron Worker	1.75	$ 40.29	$ 70.51

If it is necessary to carry the bars up to 60 ft. (18.28 m), sort, and place in stock piles, figure 2-1/2 to 3 tons (2.77-3.33 metric tons) per 8-hr. day per worker, at the following labor cost per ton:

Description	Hours	Rate	Total
Iron Worker	3.00	$ 40.29	$ 120.87

Labor Placing Reinforcing Bars. The labor cost of placing reinforcing steel will vary with the average weight of the bars, the manner in which it is placed; i.e., whether bars may be laid loose or whether it is necessary to tie them in place, and upon the class of labor employed.

443

In most of the larger cities there are concerns that make a specialty of setting reinforcing steel. These workers become very adept at this work and under favorable conditions can place steel for 25% less than the average iron worker. For this reason costs are given both ways and the estimator should use his own judgment as to the class of workers obtainable.

Setting Reinforcing Bars. On jobs using No. 5 bars and smaller, where it is not necessary to tie them in place, a worker should handle and place 700 to 900 lbs. (317.52-408.24 kg) per 8-hr. day, at the following labor cost per ton:

Description	Hours	Rate	Total
Iron Worker per ton	20.70	$ 40.29	$ 834.00
per tonne			$ 820.87

Experienced reinforcing steel setters should handle and place 900 to 1,100 lbs. (408.24-498.96 kg) of steel per 8-hr. day, at the following labor cost per ton:

Description	Hours	Rate	Total
Iron worker per ton	16.00	$ 40.29	$ 644.64
per tonne			$ 634.49

On jobs using No. 6 bars and heavier, where it is not necessary to tie the bars in place, a worker should handle and place 900 to 1,100 lbs. (408.24-498.96 kg) of steel per 8-hr. day, at the following labor cost per ton:

Description	Hours	Rate	Total
Iron worker per ton	16.00	$ 40.29	$ 644.64
per tonne			$ 634.49

An experienced reinforcing steel setter should handle and place 1,400 to 1,600 lbs. (635.04-725.76 kg) of steel per 8-hr. day, at the following labor cost per ton:

Description	Hours	Rate	Total
Iron worker per ton	10.60	$ 40.29	$ 427.07
per tonne			$ 420.35

Setting Reinforcing Bars Tied in Place. On jobs having concrete floors where it is necessary to tie the bars in place with tie wire, and where lightweight bars, (No. 5 and under) are used, a worker should handle, place, and tie 600 to 800 lbs. (272.16-362.88 kg) per 8-hr. day, at the following labor cost per ton:

Description	Hours	Rate	Total
Iron worker per ton	23.00	$ 40.29	$ 926.67
per tonne			$ 912.08

444

CONCRETE

An experienced reinforcing steel setter should handle, place, and tie 800 to 1,000 lbs. (362.88-453.60 kg) of steel per 8-hr. day, at the following labor cost per ton:

Description	Hours	Rate	Total
Iron worker per ton	17.80	$ 40.29	$ 717.16
per tonne			$ 705.87

On jobs using heavy bars, (No. 6 and over) a worker should handle, set, and tie 800 to 1,000 lbs. (362.88-453.60 kg) of steel per 8-hr. day, at the following labor cost per ton:

Description	Hours	Rate	Total
Iron worker per ton	17.80	$ 40.29	$ 717.16
per tonne			$ 705.87

An experienced reinforcing steel setter should handle, set, and tie 1,300 to 1,500 lbs. (589.68-680.40 kg) per 8-hr. day, at the following labor cost per ton:

Description	Hours	Rate	Total
Iron worker per ton	11.50	$ 40.29	$ 463.34
per tonne			$ 456.04

Epoxy Coated Reinforcing Bars

Epoxy coated reinforcement steel is used in many concrete structures today. The bars are coated, cut, and bent to the approved configuration of shop drawings at the fabrication and coating shop and then are shipped to the project. Coated reinforcement steel shall meet *ASTM A775/A775M-04a Standard Specification for Epoxy-Coated Steel Reinforcing Bars*.

The estimator should allow as a minimum about 40% additional cost to the base price of reinforcement steel for purchasing epoxy coated reinforcement steel. Additional time in the field is required for the extra care needed to unload and handle the coated steel. Additional care refers to items such as nylon slings for unloading epoxy coated steel bars instead of steel wire slings, not walking on the bars after installation, and possibly covering the steel prior to placement into the form. It is suggested that approximately 5% to 25% additional time be added for the placement of epoxy coated reinforcement steel.

All coated reinforcement bars should be tied with coated tie wire. Coated bar supports should be non-corrosive chairs.

Hot-Dip Galvanized Reinforcement Steel. Galvanized reinforcement steel is effectively and economically used in concrete where unprotected reinforcement will not have adequate durability. The susceptibility of concrete structures to the intrusion of chlorides is the primary incentive for using galvanized steel reinforcement. Galvanized

445

reinforcement steel is especially useful when the reinforcement will be exposed to the weather before construction begins. Galvanizing provides visual assurance that the steel has not rusted

Welding galvanized reinforcement should conform to the requirements of the current edition of the American Welding Society (AWS) standard practice AWS D19.0, welding Zinc Coated Steel.

Detailing of galvanized reinforcement steel should conform to the design specifications for uncoated steel bars and to normal standard practice consistent with the recommendations of the Concrete Reinforcement Steel Institute (CRSI).

The thickness of the galvanized coating is the primary factor in determining the service life of the product. The thicker the coating, the longer it provides corrosion protection. For steel embedded in concrete, the relationship is approximately linear. The minimum coating thickness requirements for reinforcement bars from ASTM A 767, Specifications for Zinc-Coated (Galvanized) Steel Bars for Concrete Reinforcement.

Table 1 – ASTM A 767 Coating Thickness Requirements

Coating Class	Mass of Zinc Coating min oz./ft^2 of surface
Class I	
Bar Designation Size 3	915 (3.00)
Bar Designation Size 4 and Larger	1070 (3.50)
Class II	
Bar Designation Size No.3 and Larger	610 (2.00)

Coated Weight			Coated Thickness	
oz/ft^2	gm/m^2	Mils		Microns
1.00	305.2	1.70		43
1.50	457.8	2.55		65
2.00	610.3	3.40		86
2.50	762.9	4.25		108
3.00	915.5	5.10		130
3.50	1068.1	5.95		153

Welded Steel Fabric or Mesh Reinforcing

Welded steel fabric is a popular and economical reinforcing for concrete work of all kinds, especially driveways and floors. It may also be used for temperature reinforcing, beam and column wrapping, road and pavement reinforcing, etc. It is usually furnished in square or rectangular mesh.

It is usually sold at a certain price per s.f. or s.y. (sq.m), depending on the weight. Prices of mesh vary with quantity ordered and freight from mill to destination. Typical prices for some styles usually stocked, with truck job site delivery are as follows. Approximate prices are per 100 s.f. (9.29 sq.m):

446

Sizes and Weights of Welded Steel Fabric													
Style	Wgt. Per 100 sq.ft. based on 60" width	Wgt. Per 9.3 sq.m. based on 1500mm width	Spacing of Wires				AS&W Co. Steel Wire Guage No.		Sq.in. per ft. Trans.	Sq.in. per ft. Longit.	Sq.cm per M. Trans.	Sq.cm per M. Longit.
			Longit., Inches	Longit., mm	Trans., Inches	Trans., mm	Trans.	Longit.				
22-1616	13	5.90	2	50	2	50	16	16	0.018	0.018	0.38	0.38
22-1414	21	9.53	2	50	2	50	14	14	0.030	0.030	0.64	0.64
22-1313	28	12.70	2	50	2	50	13	13	0.039	0.039	0.83	0.83
22-1212	37	16.78	2	50	2	50	12	12	0.052	0.052	1.10	1.10
22-1111	48	21.77	2	50	2	50	11	11	0.058	0.058	1.23	1.23
22-1010	60	27.22	2	50	2.	50	10	10	0.086	0.086	1.82	1.82
24-1414	16	7.26	4	100	12	300	14	14	0.030	0.015	0.64	0.32
24-1313	19	8.62	4	100	12	300	13	14	0.039	0.015	0.83	0.32
24-1212	28	12.70	4	100	12	300	12	12	0.052	0.026	1.10	0.55
212-38	105	47.63	2	50	12	300	3	8	0.280	0.021	5.93	0.44
212-06	166	75.30	2	50	12	300	0	6	0.443	0.029	9.38	0.61
216-812	46	20.87	2	50	16	400	8	12	0.124	0.007	2.62	0.15
216-711	55	24.95	2	50	16	400	7	11	0.148	0.008	3.13	0.17
216-610	65	29.48	2	50	16	400	6	10	0.174	0.011	3.68	0.23
216-510	75	34.02	2	50	16	400	5	10	0.202	0.011	4.28	0.23
216-49	89	40.37	2	50	16	400	4	9	0.239	0.013	5.06	0.28
216-38	104	47.17	2	50	16	400	3	8	0.280	0.015	5.93	0.32
216-28	119	53.98	2	50	16	400	2	8	0.325	0.015	6.88	0.32
216-17	139	63.05	2	50	16	400	1	7	0.377	0.018	7.98	0.38

Sometime a style such as 66-44 is shown as 6x6-4/4

447

Sizes and Weights of Welded Steel Fabric

Style	Wgt. Per 100 sq.ft. based on 60" width	Wgt. Per 9.3 sq.m. based on 1500mm width	Spacing of Wires				AS&W Co. Steel Wire Guage No.		Sq.in. per ft. Trans.	Sq.in. per ft. Longit.	Sq.cm per M. Trans.	Sq.cm per M. Longit.
			Longit., Inches	Longit., mm	Trans., Inches	Trans., mm	Trans.	Longit.				
33-1414	14	6.35	3	75	3	75	14	14	0.020	0.020	0.42	0.42
33-1212	25	11.34	3	75	3	75	12	12	0.035	0.035	0.74	0.74
33-1111	32	14.52	3	75	3	75	11	11	0.046	0.046	0.97	0.97
33-1010	41	18.60	3	75	3	75	10	10	0.057	0.057	1.21	1.21
33-99	49	22.23	3	75	3	75	9	9	0.069	0.069	1.46	1.46
33-88	58	26.31	3	75	3	75	8	8	0.082	0.082	1.74	1.74
316-812	32	14.52	3	75	16	400	8	12	0.082	0.007	1.74	0.15
316-711	38	17.24	3	75	16	400	7	11	0.098	0.009	2.07	0.19
316-60	45	20.41	3	75	16	400	6	10	0.116	0.011	2.46	0.23
316-510	52	23.59	3	75	16	400	5	10	0.135	0.011	2.86	0.23
316-49	61	27.67	3	75	16	400	4	9	0.159	0.013	3.37	0.28
316-38	72	32.66	3	75	16	400	3	8	0.187	0.015	3.96	0.32
316-28	83	37.65	3	75	16	400	2	8	0.216	0.015	4.57	0.32
316-17	96	43.55	3	75	16	400	1	7	0.252	0.018	5.33	0.38
316-06	113	51.26	3	75	16	400	0	6	0.295	0.022	6.24	0.47
44-1414	11	4.99	4	100	4	100	13	13	0.015	0.015	0.32	0.32
44-1313	14	6.35	4	100	4	100	12	14	0.020	0.020	0.42	0.42
44-1212	19	8.62	4	100	4	100	12	12	0.026	0.026	0.55	0.55
44-1010	31	14.06	4	100	4	100	11	12	0.043	0.043	0.91	0.91

Sometime a style such as 66-44 is shown as 6x6-4/4

CONCRETE

Sizes and Weights of Welded Steel Fabric

Style	Wgt. Per 100 sq.ft. based on 60" width	Wgt. Per 9.3 sq.m. based on 1500mm width	Spacing of Wires				AS&W Co. Steel Wire Guage No.		Sq.in. per ft. Trans.	Sq.in. per ft. Longit.	Sq.cm per M. Trans.	Sq.cm per M. Longit.
			Longit., Inches	Longit., mm	Trans., Inches	Trans., mm	Trans.	Longit.				
44-88	34	15.42	4	100	4	100	10	12	0.062	0.062	1.31	1.31
44-77	53	24.04	4	100	4	100	9	12	0.074	0.074	1.57	1.57
44-66	62	28.12	4	100	4	100	8	12	0.087	0.087	1.84	1.84
44-44	85	38.56	4	100	4	100	7	11	0.120	0.120	2.54	2.54
48-1313	11	4.99	4	100	8	200	13	13	0.020	0.010	0.42	0.21
48-1214	12	5.44	4	100	8	200	12	14	0.026	0.008	0.55	0.17
48-1212	14	6.35	4	100	8	200	12	12	0.026	0.013	0.55	0.28
48-1112	17	7.71	4	100	8	200	11	12	0.034	0.013	0.72	0.28
48-1012	20	9.07	4	100	8	200	10	12	0.043	0.013	0.91	0.28
48-912	23	10.43	4	100	8	200	9	12	0.052	0.013	1.10	0.28
48-812	27	12.25	4	100	8	200	8	12	0.062	0.013	1.31	0.28
48-711	33	14.97	4	100	8	200	7	11	0.074	0.017	1.57	0.36
412-1212	13	5.90	4	100	12	300	12	12	0.026	0.009	0.55	0.19
412-1112	16	7.26	4	100	12	300	11	12	0.034	0.009	0.72	0.19
412-1012	19	8.62	4	100	12	300	10	12	0.043	0.009	0.91	0.19
412-912	22	9.98	4	100	12	300	9	12	0.052	0.009	1.10	0.19
412-812	25	9.98	4	100	12	300	8	12	0.062	0.009	1.31	0.19
412-711	31	11.34	4	100	12	300	7	11	0.040	0.011	0.85	0.23
412-610	66	14.06	4	100	12	300	6	10	0.087	0.014	1.84	0.30

Sometime a style such as 66-44 is shown as 6x6-4/4

Sizes and Weights of Welded Steel Fabric

Style	Wgt. Per 100 sq.ft. based on 60" width	Wgt. Per 9.3 sq.m. based on 1500mm width	Spacing of Wires				AS&W Co. Steel Wire Guage No.		Sq.in. per ft. Trans.	Sq.in. per ft. Longit.	Sq.cm per M. Trans.	Sq.cm per M. Longit.
			Longit., Inches	Longit., mm	Trans., Inches	Trans., mm	Trans.	Longit.				
412-510	42	29.94	4	100	12	300	5	10	0.101	0.014	2.14	0.30
412-57	45	19.05	4	100	12	300	5	7	0.101	0.025	2.14	0.53
412-49	49	20.41	4	100	12	300	4	9	0.120	0.017	2.54	0.36
416-1012	18	8.16	4	100	16	400	10	12	0.043	0.007	0.91	0.15
416-912	21	9.53	4	100	16	400	9	12	0.052	0.007	1.10	0.15
416-812	25	11.34	4	100	16	400	8	12	0.062	0.007	1.31	0.15
416-711	30	13.61	4	100	16	400	7	11	0.074	0.009	1.57	0.19
416-610	35	15.88	4	100	16	400	6	10	0.087	0.011	1.84	0.23
416-510	40	18.14	4	100	16	400	5	10	0.101	0.011	2.14	0.23
416-49	48	21.77	4	100	16	400	4	9	0.120	0.013	2.54	0.28
416-38	56	25.40	4	100	16	400	3	8	0.140	0.015	2.96	0.32
416-28	64	29.03	4	100	16	400	2	8	0.162	0.015	3.43	0.32
66-1212	13	5.90	6	150	6	150	12	12	0.017	0.017	0.36	0.36
66-1010	21	9.53	6	150	6	150	10	10	0.029	0.029	0.61	0.61
66-99	25	11.34	6	150	6	150	9	9	0.035	0.035	0.74	0.74
66-88	30	13.61	6	150	6	150	8	8	0.041	0.041	0.87	0.87
66-77	36	16.33	6	150	6	150	7	7	0.049	0.049	1.04	1.04
66-66	42	19.05	6	150	6	150	6	6	0.058	0.058	1.23	1.23
66-55	49	22.23	6	150	6	150	5	5	0.067	0.067	1.42	1.42

Sometime a style such as 66-44 is shown as 6x6-4/4

Sizes and Weights of Welded Steel Fabric

Style	Wgt. Per 100 sq.ft. based on 60" width	Wgt. Per 9.3 sq.m. based on 1500mm width	Spacing of Wires				AS&W Co. Steel Wire Guage No.		Sq.in. per ft. Trans.	Sq.in. per ft. Longit.	Sq.cm per M. Trans.	Sq.cm per M. Longit.
			Longit., Inches	Longit., mm	Trans., Inches	Trans., mm	Trans.	Longit.				
66-46	50	22.68	6	150	6	150	4	6	0.080	0.058	1.69	1.23
66-44	58	26.31	6	150	6	150	4	4	0.080	0.080	1.69	1.69
66-33	68	30.84	6	150	6	150	3	3	0.093	0.093	1.97	1.97
66-22	78	35.38	6	150	6	150	2	2	0.108	0.108	2.29	2.29
66-11	91	41.28	6	150	6	150	1	1	0.126	0.126	2.67	2.67
66-00	107	48.54	6	150	6	150	0	0	0.148	0.148	3.13	3.13
612-77	27	12.25	6	150	12	300	7	7	0.049	0.025	1.04	0.53
612-66	32	14.52	6	150	12	300	6	6	0.058	0.029	1.23	0.61
612-55	37	16.78	6	150	12	300	5	5	0.067	0.034	1.42	0.72
612-44	44	19.96	6	150	12	300	4	4	0.080	0.040	1.69	0.85
612-33	51	23.13	6	150	12	300	3	3	0.093	0.047	1.97	0.99
612-25	52	23.59	6	150	12	300	2	5	0.108	0.034	2.29	0.72
612-22	59	26.76	6	150	12	300	2	2	0.108	0.054	2.29	1.14
612-17	56	25.40	6	150	12	300	1	7	0.126	0.025	2.67	0.53
612-14	61	27.67	6	150	12	300	1	4	0.126	0.040	2.67	0.85
612-11	69	31.30	6	150	12	300	1	1	0.126	0.063	2.67	1.33
612-06	65	29.48	6	150	12	300	0	6	0.148	0.029	3.13	0.61
612-03	72	32.66	6	150	12	300	0	3	0.148	0.047	3.13	0.99
612-00	81	36.74	6	150	12	300	0	0	0.148	0.074	3.13	1.57

Sometime a style such as 66-44 is shown as 6x6-4/4

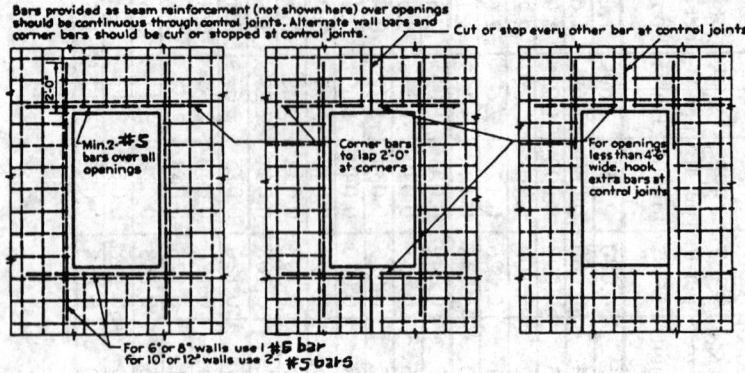

Bars provided as beam reinforcement (not shown here) over openings should be continuous through control joints. Alternate wall bars and corner bars should be cut or stopped at control joints.

Cut or stop every other bar at control joints

Min.2 #5 bars over all openings

Corner bars to lap 2'-0" at corners

For openings less than 4'-0" wide, hook extra bars at control joints

For 6" or 8" walls use 1 #5 bar
for 10" or 12" walls use 2 - #5 bars

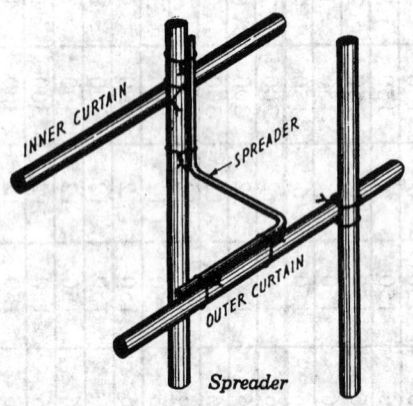

Spreader

Labor Placing Mesh Reinforcing In Walls. When mesh reinforcing is used, furnished in sheets or rolls, a worker should place 700 to 800 s.f. (63-72 sq.m) in walls per 8-hr. day, at the following labor cost per 100 s.f. (9 sq.m):

Description	Hours	Rate	Total
Steel setter	1.00	$ 40.29	$ 40.29
Cost per s.f.			$ 0.40
per sq.m			$ 4.34

Labor Placing Steel Fabric on Floors. When used for reinforcing floor slabs or for temperature reinforcing, a worker should place 1,400 to

452

CONCRETE

1,600 s.f. (126-144 sq.m) of fabric per 8-hr. day, at the following labor cost per 100 s.f. (9 sq.m):

Description	Hours	Rate	Total
Iron worker	0.53	$ 40.29	$ 21.35
Cost per s.f.			$ 0.21
per sq.m			$ 2.30

Reinforcement for Architectural Concrete Walls. It is almost impossible to provide sufficient reinforcement to prevent cracking in a long wall, which is why control joints are used. But it is still necessary to provide adequate reinforcement between joints properly spaced to resist volume change stresses. To be most effective, small bars with a type of deformation to give maximum bond should be placed relatively close together, rather than larger bars at wider spacings. Generally #3 and #4 bars are preferable to larger sizes and the horizontal reinforcement should be at least equal to 0.25% of wall area. In the vertical direction 0.15% reinforcement is sufficient.

Recommended Size and Spacing of Reinforcing Bars in Concrete Walls

Wall Thick. Inches	Horizontal Reinforcement	Vertical Reinforcement
(in outside face of wall)		
6 "	#4 @ 8 "	#4 @ 8 "
8 "	#4 @ 6 "	#4 @ 8 "
(in both faces of wall)		
10 "	#4 @ 10 "	#4 @ 12 "
12 "	#4 @ 8 "	#4 @ 12 "
Wall Thick. mm	Horizontal Reinforcement	Vertical Reinforcement
(in outside face of wall)		
150 mm	#13 @ 200 mm	#13 @ 200 mm
200 mm	#13 @ 150 mm	#13 @ 200 mm
(in both faces of wall)		
250 mm	#13 @ 250 mm	#13 @ 300 mm
300 mm	#13 @ 200 mm	#13 @ 300 mm

The bars should be tied at each intersection with a non-slip tie, and where two curtains of steel are used, it is advisable to include a detail drawing to show how the curtains are to be fastened together with tie spreaders at frequent intervals to make a rigid web of reinforcement.

It is sometimes difficult to place concrete in walls because of a congestion of reinforcing bars. This condition should be avoided insofar as practical.

453

Labor Placing Reinforcing Bars. The labor cost of placing reinforcing steel in architectural concrete walls is expensive on account of the small size of the rods, the fact they run both horizontally and vertically, must be wired together, and the cramped quarters for the work.

On walls 8 to 12 inches (200-300 mm) thick, a reinforcing steel setter will handle, place, and wire 750 to 850 lbs. (340-385 kg) of reinforcing steel per 8-hr. day, at the following labor cost per ton:

Description	Hours	Rate	Total
Iron worker	0.53	$ 40.29	$ 21.35
Cost per s.f.			$ 0.21
per sq.m			$ 2.30

03300 CAST-IN-PLACE CONCRETE

Reinforced and architectural concrete are among our most common building materials and are used on virtually all types of construction work.

They are used in all types of structures, from residential, apartment, hotel and hospital buildings, requiring comparatively light floor loads, to the heaviest constructed warehouses, factory buildings, wharves, and industrial buildings.

Entire buildings are constructed of concrete, including foundations, foundation walls, columns, beams and girders, floor slabs, ornamental exterior walls, cornices, etc.

Regardless of the fact that it is one of our most common building materials, costs on all types of reinforced concrete construction are subject to wide variation. This is due to a number of causes, such as the function and size of building, whether flat slab or beam and girder type, whether solid concrete construction or used in conjunction with steel, building codes, insurance requirements, etc.

The methods used in erecting and moving the formwork, handling, mixing and placing the concrete, and the ability of the superintendent and foreman, all affect the final cost of the work.

Regardless of how carefully or accurately the job has been estimated, if the supervisor and foreman in charge of the work are not proficient, the actual costs are almost certain to run higher than the estimate - not because the work was estimated too low but because it was poorly handled on the project.

All labor costs given here are based on efficient management; economical layout and design of all formwork; use of modern labor-saving methods, such as bench fabrication of column, beam, gang forming and girder forms where possible; and the use of labor saving tools such as power saws, band saws, electric hand saws, etc.

The efficient contractor must use methods and tools that will reduce costs to a minimum, or else be out of the running when it comes to obtaining work and performing it economically. Remember, good management,

454

economical form design, and labor-saving tools and methods are absolutely necessary to a profitable contract.

Symons Steel-Ply forming systems can be moved into place after reinforcing steel is placed (Courtesy Symons)

Items to be Included in the Estimate. Reinforced cast-in-place concrete should be estimated in the same manner and in the same units as the work is constructed on the job. This allows comparisons between estimated and actual costs and enables a contractor to tell where variations occur.

The first operation on any reinforced cast-in-place concrete job is the erection of the temporary wood or metal forms to support the weight of the wet concrete until it has set sufficiently to be self-supporting. M embers that usually require forms are columns, girders, beams, spandrel beams, lintels, floor slabs, stairs, and platforms. Each class of formwork should be estimated separately and in detail.

All types of formwork, including columns, beams, girders, and the various type of floor slabs, have been designed with the best engineering practices-giving lumber requirements, clamp spacing for columns, and beam and girder lumber requirements, including shoring and re-shoring as required. There is a complete set of design tables for all types of concrete floor slabs, giving joist spacing, spans of stringers, and shores required for any floor load.

After the forms are erected, it is necessary to place the reinforcing steel or wire mesh ready to receive the concrete. This may include unloading steel from trucks at the job, cutting steel to length and bending to special shapes for columns, beams, girders and floor slabs, and placing and tying it in position ready to receive the concrete.

After the reinforcing steel is in place and tied off, the job is then ready for concrete placement. This consists set up including the preparing runways, unloading cement and other materials, mixing the cement, sand and gravel into the concrete mass, hoisting, wheeling, placing it in position and vibration of the wet concrete. Then, removing runways, removing old concrete drippings, spillage, cleaning and washing wheelbarrows or buggies, and concrete mixer if ready mix concrete is not used. After the concrete has cured the removal of forms, patching, rubbbing and re-shoring if required especially for slabs.

Each of these items should be estimated in detail to include all material and labor operations necessary to make up the completed unit.

Sample Concrete Pad Estimate

This sample shows how to determine pounds (kg) of rebar, cubic yards (cubic meters) and square feet (square meters) of form work in a concrete pad similar to the drawing below.

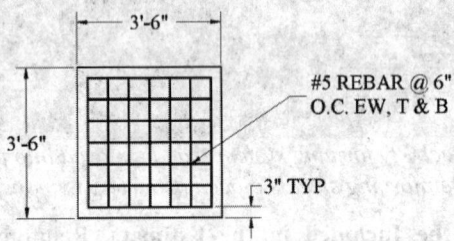

Given: 10 pads each measure 3'-6" (1.07m) x 3'-6" (1.07m) x 1'-4" (0.41m). There are #5 (#16) rebar at 6" (15.2cm) on center layered each way and at the top and bottom of the pad. *Assume 3" (7.6cm) of coverage of concrete is required.*

<u>Number of pounds (kg) of rebar:</u>
    3'-6" – 3" – 3" = 3' (0.91m) length of each piece

    3 ÷ 0.5 spacing + 1 = 7 pcs x 2 directions x 2 layers = 28 pcs per pad
    0.91 ÷ 0.152 + 1 = 7 pcs x 2 x 2 = 28 pcs per pad

    28 pcs x 3' = 84' (25.6m) total length of rebar per pad

    10 x 84' x 1.043 lbs/ft = **876.12 lbs**
    10 x 25.6m x 1.552 kg/m = 397.31 kg

<u>Number of cubic yards (CM) of concrete:</u>
    10 x 3.5' x 3.5' x 1.33' ÷ 27 = **6.03 CY (4.61 CM)**

<u>Number of square feet (SM) of plywood form work:</u>
    10 x (4 x 3.5') x 1.33' = **186.2 SF (17.3 SM)**

    NOTE: If using a ¾"x4'x8' sheet, divide by 32 or whatever square footage (square meter) pieces being used.

CONCRETE

Sample Concrete Pile Cap Estimate

Estimating concrete pile caps is very similar to estimating concrete pads, only the shape may be different. See below for a sample.

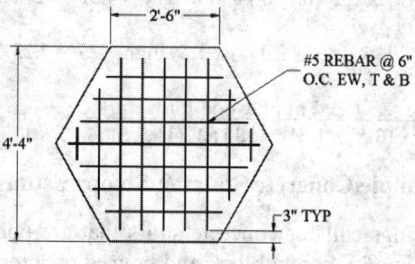

Given: 8 hexagonal pile caps each measure 2'-6" (0.76m) on each side, 4'-4" (1.32m) across parallel sides and is 1'-6" (0.457m) high. There are #5 (#16) rebar at 6" (15.2cm) on center layered each way and at the top and bottom of the pile cap. *Assume 3" (7.6cm) of coverage of concrete is required.*

<u>Number of pounds (kg) of rebar:</u>

For simplicity, the rebar was designed to start at the center of the hex. Let's start with the vertical pieces. Since this is a hex shape, the center horizontal dimension must be twice the side length. Also, there isn't a perimeter piece that follows the hex shape. When counting rebar, the count should result in odd numbers. Use a factor of 1.732' (0.528m) for vertical pieces and 0.577' (0.176m) for horizontal pieces.

$2 \times 2'\text{-}6" = 5'$ (1.52m) max horizontal dimension

$(5' - 0.577') \div 0.5' = 8.8$ or 9 pcs vertically
$2.5' \div 0.5'$ spacing = 5 pcs vertically perpendicular to the side
(1.52 – 0.176) ÷ 0.152 = 9 pcs   and   0.61 ÷ 0.152 + 1 = 5 pcs

$(4.33' - 0.25' - 0.25') \div 0.5' = 7.66$ or 7 pcs horizontally
(1.32 – 0.076 – 0.076) ÷ 0.152 = 7 pcs

5 pcs x (4.33' – 0.25' – 0.25')	= 19.15'
2 pcs x (4.33'– 1.732')	= 5.20'
2 pcs x (4.33'– 1.732' – 1.732')	= 1.73'
1 pcs x (5' – 0.577')	= 4.42'
2 pcs x (5' – 0.577' – 0.577')	= 7.69'
2 pcs x (5' – 0.577' – 0.577' – 0.577')	= 6.54'
2 pcs x (5' – 0.577' – 0.577' – 0.577' – 0.577')	= <u>5.38'</u>
	50.11' (15.3m)

$8 \times 50.11' \times 1.043$ lbs/ft = 418.1 lbs x 2 for top & bottom = **836.2 lbs**
8 x 15.3m x 1.552 kg/m = 190 kg x 2 for top & bottom = 380 kg

NOTE: To be more accurate, the estimator should scale the drawing.

Number of cubic yards (CM) of concrete:
The surface can be divided into six equilateral triangles, one for each side.

Triangle base is 2.5' (0.76m) and height is 4.33' ÷ 2 = 2.17' (0.66m).

8 x 6 sides x ½ x (2.5' x 2.17') x 1.5' high ÷ 27 = **7.23 CY (5.53 CM)**

Number of square feet (SM) of plywood form work:
8 x 6 sides x (2.5' x 1.5') = **180 SF (16.7 SM)**

Sample Concrete Stairs & Stoops Estimate

Many drawings call for concrete stairs that are poured solid concrete or are a combination of concrete block and poured concrete. This example is for solid poured concrete stairs/stoops only. To estimate the poured concrete stairs, a profile or sectional view of the stairs is required.

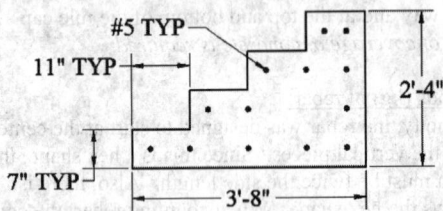

Given: 2 entrance stairs, with a profile as shown above, are 11'-8" (3.56m) wide. Find the lbs (kg) of rebar, CY (CM) of concrete and the SF (SM) of plywood/lumber formwork. *Assume 3" (7.6cm) of coverage and each of the 4 risers are 7" (17.7cm) high and the 3 treads are 11" (27.9cm) deep.*

Number of pounds (kg) of rebar:
11.67' – 0.25' – 0.25' = 11.17' (3.4m)
11.17' x 13 pcs shown x 1.043 lbs/ft = **151.45 lbs**
3.4m x 13 pcs x 1.552 kg/m = 68.6 kg

Number of cubic yards (CM) of concrete:
0.92' x (0.58' + 1.17' + 1.75' + 2.33') x 11.67' ÷ 27 = **2.3 CY**
0.279m x (0.178m + 0.356m + 0.534m + 0.712m) x 3.56m = 1.77 CM

Number of square feet (SM) of plywood/lumber form work:
The lumber is used for each riser only and plywood would be used for the sides and back of the stair.

4 pcs of 2x8 x 12' pcs lumber for the risers, for example

0.92' x (0.58' + 1.17' + 1.75' + 2.33') = 5.36 SF (0.5 SM) sectional area
2 sides x 5.36sf + 2.33' x 11.67' = **37.9 SF (3.52 SM)**

CONCRETE

Sample Concrete Stairs Pans Estimate

Many times in construction of commercial buildings, stair sections will show stair treads as metal pans.

Given: The metal pan size is 1½" x 10¾" x 11'-8". The stairs rise from the 1st thru 4th floors with three concrete landings of 3'-6" (1.07m) x 3'-6" (1.07m) x 3" (7.6cm) between each floor and one at each floor.

Number of cubic yards (CM) of concrete:
Note: 22 equal risers result in 1 less riser in the calculation due to the landings.

1.5" ÷ 12" = 0.13' (3.8cm), 10.75" ÷ 12" = 0.90' (27.3cm)
21 pans x 3 floors x 0.13' x 0.90' x 11.67' ÷ 27 = 3.19 CY (2.44 CM)
3 floors x 4 landings/floor x 3.5' x 3.5' x 0.25' ÷ 27 = 1.36 CY (1.04 CM)
3.19 CY + 1.36 CY = **4.55 total CY (3.48 total CM)**

Equipment Foundations. This type of foundation requires caution. The type of equipment used on these foundations affect the type of foundation, the size of the foundation, the number and size of reinforcing bars needed, etc. Many types of machines transmit live loads to the soil in addition to the dead loads due to the weight of the machine, weight of the foundation, etc.

Some of the machines requiring an equipment foundation are electric motors, turbines, compressors, pumps, etc. Each machine foundation requires the estimator to separate the estimate into a block type foundation for compressors and reciprocating machines.

A caisson type foundation is similar to a column footing or pad. The caisson types are cylinders filled with concrete. The end of the cylinder can be straight or belled as shown below.

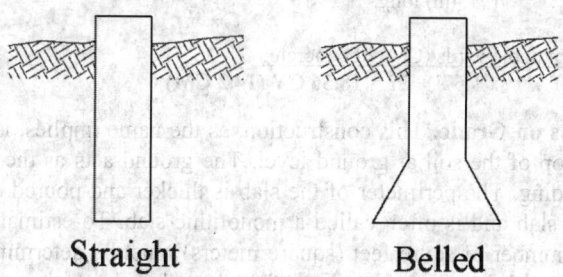

Straight   Belled

The belled caisson increases the bearing capacity of the caisson whereas the straight caisson transfers its bearing capacity by friction between the soil and the sides of the caisson.

CONCRETE

Continuous Footings and Grade Beams. These are commonly found at the base of a wall. Grade beams, as the name implies, are beams at grade where the vertical dimension is greater than the horizontal. See below.

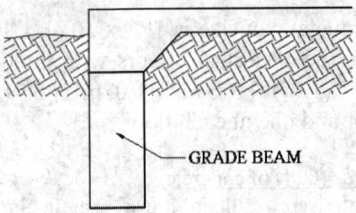

GRADE BEAM

Continuous footers are footers where the horizontal dimension is greater than the vertical. See below.

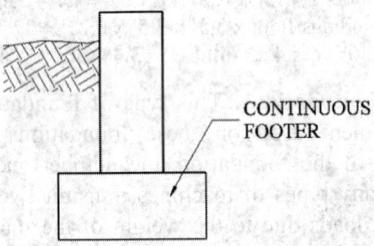

CONTINUOUS FOOTER

To estimate continuous footers or grade beams, determine the centerline lineal footage (lineal meters) of the beam or footer. Then multiply by the cross-sectional area of the footer or grade beam or use the formula $V = \text{length} \times \text{width} \times \text{height} \div 27$.

Sample Continuous Footer Estimate

Given: A continuous footer measuring 12" (0.3m) x 20" (0.51m) is 300' (91:4m) long.

Number of cubic yards (CM) of concrete:
    300' x 1' x 1.67' ÷ 27 = **18.56 CY (14.2 CM)**

Slabs on Grade. This construction, as the name implies, is a flat slab placed on top of the soil at ground level. The ground acts as the foundation for the building. The perimeter of the slab is thicker and poured at the same time as the slab and is often called a monolithic slab. To estimate a slab on grade, the number of squre feet (square meters) must be determined and the number of lineal feet (lineal meters) of thickened perimeter measured at the centerline of this thickness is also required.

Sample Slab on Grade Estimate

For the slab shown below, first get the memory information and then complete the order estimate as illustrated in this example.

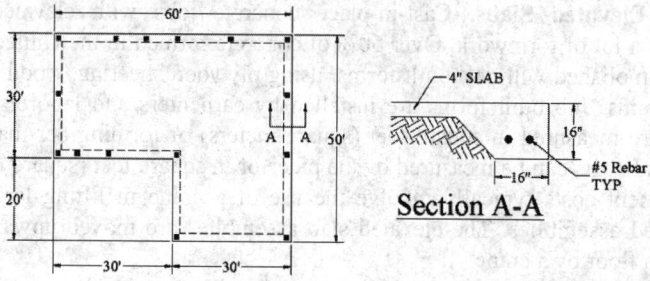

Section A-A

Given: The slab shown above is an overall "L" shape 30' (9.1m) wide
          and is 60' (18.3m) x 50' (15.2m). The section A-A shows the slab
          is 4" (10.2cm) thick and has a monolithic thickened edge 16"
          (40.6cm) deep and 16" (40.6cm) wide at a 45° slope. There are 2
          #5 (#16) rebars in this thickened edge. A visqueen vapor barrier is
          to be placed under the slab and welded wire mesh is placed within
          the slab. *Assume 5% waste for concrete, 4' (1.2m) spacing for
          rebar chairs, use 17' (5.2m) rebar length per 20' piece and use
          825 SF (76.6 SM) per 900 SF roll of welded mesh to allow for lap
          and waste.*

Memory Information:
   60' x 50' – 20' x 30' = **2400 SF (223 SM) of slab**

   16" ÷ 2 = 8" ÷ 12 = 0.67' (20.3cm) to centerline of thickened edge
   Note: Just subtract this as needed to calculate distance at centerline.
   28.66' + 58.66' + 48.66' + 28.66' + 20' + 30' = **214.64' (65.4m) at center**

Order Estimate:
   2400 SF x 0.33' ÷ 27   = 29.33 CY @ slab
   214.64' x 1' x 1.33' ÷ 27   = 10.57 CY @ rectangular footing
   214.64' x 0.5 x 1' x 1' ÷ 27 = 3.97 CY @ triangular "haunch"
                         Total = 43.87 x 1.05 = 46 CY (35.2 CM)
   46 CY (35.2 CM) of concrete with 5% waste

   2400 SF ÷ 825 = 2.91 or 3 rolls
   3 rolls of W.W.M. (6'x150')

   2400 SF ÷ 2000 = 1.2 or 2 rolls
   2 rolls of Visqueen (20' x 100')

   214.64' ÷ 4' = 53.67 or 54 +1 for the end = 55 chairs
   55 each of Rod Chairs 4' o.c.

   214.64' ÷ 17' = 12.63 or 13 pcs x 2 bars = 26 total rebar pcs
   26 each of #5 Rebar of 20' length

   15 vertical drops shown on drawing (count-up item)
   15 each Dowels of 5' length

461

Elevated Slabs. Cast-in-place concrete jobs with elevated slabs require a lot of formwork. Over 50% of concrete formed in the United States is accomplished with job-built forms, using plywood sheeting, wood shoring and beams. Job built forms are installed by carpenters and laborers. Labor costs are measured in square feet (square meters) of forming per man-hour. Material costs can be measured by the piece or in square feet (square meters). Equipment costs typically involve the use of a crane in lifting large truss mounted assemblies. The elevated slab assemblies are moved upward from floor to floor by a crane.

Matted Rebar. For complicated pads, pile caps or other cast-in-place concrete jobs, it may be easier to calculate rebar by using a factor. To estimate the total rebar in a job, use the table below for the size of rebar and spacing required to get the factor. Then using this factor, multiply by the square footage (square meters) per layer and add half the perimeter, if edge pieces are required, as an adjustment.

Matted Rebar Table (lbs/SF)							
	Center-to-Center Spacing of Rebar, Placed Each Way						
	6"	8"	10"	12"	14"	16"	18"
#3	1.504	1.128	0.902	0.752	0.645	0.564	0.501
#4	2.672	2.004	1.603	1.336	1.145	1.002	0.891
#5	4.172	3.129	2.503	2.086	1.788	1.565	1.391
#6	6.008	4.506	3.605	3.004	2.575	2.253	2.003
#7	8.176	6.132	4.906	4.088	3.504	3.066	2.725
#8	10.680	8.010	6.408	5.340	4.577	4.005	3.560
#9	13.600	10.200	8.160	6.800	5.829	5.100	4.533
#10	17.212	12.909	10.327	8.606	7.377	6.455	5.737
#11	21.252	15.939	12.751	10.626	9.108	7.970	7.084
#14	30.600	22.950	18.360	15.300	13.114	11.475	10.200
#18	54.400	40.800	32.640	27.200	23.314	20.400	18.133

Matted Rebar Table (kg/SM)							
	Center-to-Center Spacing of Rebar, Placed Each Way						
	15.2cm	20.3cm	25.4cm	30.5cm	35.6cm	40.6cm	45.7cm
#10	7.343	5.507	4.406	3.672	3.147	2.754	2.448
#13	13.046	9.784	7.827	6.523	5.591	4.892	4.349
#16	20.369	15.277	12.222	10.185	8.730	7.639	6.790
#19	29.333	22.000	17.600	14.667	12.571	11.000	9.778
#22	39.919	29.939	23.951	19.959	17.108	14.969	13.306
#25	52.144	39.108	31.286	26.072	22.347	19.554	17.381
#29	66.401	49.801	39.840	33.200	28.457	24.900	22.134
#32	84.036	63.027	50.422	42.018	36.015	31.513	28.012
#36	103.761	77.821	62.257	51.880	44.469	38.910	34.587
#43	149.402	112.051	89.641	74.701	64.029	56.026	49.801
#57	265.603	199.202	159.362	132.801	113.830	99.601	88.534

Sample Matted Rebar Estimate

Given: A concrete pad measures 8' (2.44m) x 10' (3.05m) and is 8" (20.3cm) thick. This pad is reinforced with #6 (#19) rebar placed at 8" (20.3cm) on center, each way. *Assume 3" (7.6cm) of coverage and the rebar is placed to the edge of the pad.*

462

CONCRETE

Number of pounds (kg) of rebar:
    8' – 0.25' – 0.25' = 7.5' (2.29m) width
    10' – 0.25' – 0.25' = 9.5' (2.90m) long

    7.5' x 9.5' = 71.25 SF (6.62 SM)

    71.25 SF x 4.506 lbs/SF + (7.5'+ 9.5') x 1.502 lbs/ft = **346.6 lbs**
    *6.62 SM x 22 kg/SM + (2.29m + 2.9m) x 2.235 kg/m = **157.2 kg***

Weights of Miscellaneous Building Materials
Weight of Cement, Sand, Gravel, Crushed Stone, Concrete, Etc.

    The following table gives the approximate weights of cement, concrete, and concrete aggregates, and will assist the estimator in figuring freight or when purchasing materials by weight:

Description	Weight, Lbs/CY	Weight, Kg/Cu.M.
Bank sand (dry)	2,500	1,483
Torpedo sand	2,700	1,602
Crushed stone	2,500	1,483
Crushed stone screenings	2,500	1,483
Gravel	2,700	1,602
Roofing gravel	2,700	1,602
Cinder concrete	112	66
Concrete of gravel, limestone, sandstone, trap rock, etc.	150	89

Cement weighs 376 lbs. (1700.55 kg) per bbl.
One sack weighs 94 lbs (42.64 kg) and bulks 1 c.f. (0.028 cu.m.)

    Quantity of Pit Run Sand and Gravel Required for One Cubic Yard of Concrete. Pit run sand and gravel is not recommended for concrete work where strength is a factor, because it is impossible to grade the aggregate. There is either too much gravel and too little sand or too little gravel and too much sand, usually the latter, to make a dependable concrete.

    An aggregate containing approximately 50% sand and 50% gravel (providing it does not contain any humic acid), may be considered acceptable aggregate, and the strength will probably be only 10% less than when using aggregates furnished by washing and grading plants.

    It ordinarily requires about 1.25 cu. yds. of aggregate per cu. yd. of concrete (1.25 cu.m aggregate per cu.m of concrete), depending on the coarseness of the sand.

CONCRETE

Types of Cement

Portland cements are hydraulic cements composed primarily of hydraulic calcium silicates. ASTM C 150-04ae1 Standard Specification for Portland cement and recognizes eight types of portland cement:

Type I - and Type IA* - General purpose cements suitable for all uses where the special properties of other types are not required.

Type II and Type IIA* - Type II cements contain no more than 8% tricalcium aluminate for moderate sulfate resistance. Some type II cements meet the moderate heat of hydration option of ASTM C 150.

Type III and Type IIIA* - chemically and physically similar to Type I cements except they are ground finer to produce higher early strengths.

Type IV - Used in massive concrete structures where the rate and amount of heat generated from hydration must be minimized. It develops strength slower then other types of cement.

Type V - Contains no more then 5% tricalcium aluminate for high sulfate resistance.

air entraining cements

Types of Blended Cement

Blended hydraulic cements are produced by intimately and uniformly integrating or blending two or more types of fine materials. The primary materials are portland cement, ground granulated blast slag, fly ash, silica fume, calcined clay other pozzolans, hydrated lime, and pre-blended combinations of these materials.

ASTM C 595, Specification for Blended Hydraulic Cements, recognizes five primary classes of blended cement

Type IS - Portland blast furnace slag cement

Type IP and tope P - Portland-pozzolan cement

Type I (PM) - Pozzolan-modified portland cement

Type S - Slag Cement

Type I(SM) - Slag-modified portland cement

Types of Hydraulic Cement

All portland and blended cements are hydraulic cements. "Hydraulic cement" is merely a broader term. *ASTM C 1157, Performance specification for Hydraulic Cements*, is a performance specification that includes portland cement, modified portland cement, and blended cements. ASTM C 1157 recognizes six types of hydraulic cements:

Type GU - General use

Type HE - High early strength

Type MS - Moderate sulfate resistance

Type HS - High sulfate resistance

Type MH - Moderate heat of hydration

Type LH - Low heat of hydration

464

CONCRETE

Concrete should be estimated by the cubic foot, or cubic yard containing 27 cu.ft. (for metric, by the cu. m). The cu.yd. is the most convenient unit to use because practically all published tables giving quantities of cement, sand, gravel, or crushed stone are based on the cu.yd. (cu.m) of concrete. The cu.yd. (cu.m) is also a convenient unit to use when estimating labor costs.

It was formerly the practice to specify concrete mixtures as so many volumes of fine and coarse aggregate to each volume of cement, as for example 1:1:2, 1:1.5:3, 1:2:4, 1:3:5, etc. The condition of the aggregate when measured was not specified, nor were the consistency of the concrete or the water content.

As used on the job, sand nearly always contains some surface moisture, usually about 4% or 5% by weight of the sand, although it may be considerably more just after rain or if the sand is used soon after washing it. The moisture causes the sand to bulk up in volume, the amount of bulking depending on the amount of moisture and the grading. Bulking, which might be 20% to 30% or more, reduces the true volume of surface dry sand in a given volume of damp sand. For example, if a 1:2:4 mix is specified and 2 volumes of damp sand are measured, the volume of sand on a surface dry basis would be less. Suppose the sand is bulked 25% due to moisture. The volume of dry sand in each cubic foot of damp sand will be 100 divided by 125, or 0.8 cu.ft. (002 cu.m) of damp sand would give us 2 x 0.8 or 1.6 cu.ft. (.004 cu.m) dry sand.

The coarse aggregate carries very little surface moisture, and the moisture has little effect on the volume. The result is that undersanded mixes were often used. In the above example the 1:2:4 mix if based on damp materials would actually be a 1:1.6:4 mixture of surface dry materials. The amount of sand in this case would be less than 29% of the total volume of aggregate. For average mixtures 35% to 45% is usually desirable for good workability when concrete is placed by hand.

To correct for the bulking in the above example the amount of damp sand should be increased to 1.25 x 2 or 2.5 cu.ft. (.071 cu.m) and the mix would then be 1:2.5:4 based on damp materials as measured in the field.

As a result of the increased knowledge that has been made available, other methods of specifying concrete have been adopted, such as a mix by weight, a minimum cement factor, maximum water content or the strength at a given age, or a combination of these.

The Importance of Mixing Water. The hardening of concrete mixtures is brought about by chemical reactions between cement and water, the aggregates (sand, gravel, stone, etc.) being inactive ingredients used as fillers. Concrete generates heat during hardening as a result of the chemical process by which cement reacts with water to form a hard, stable paste. The heat generated is called heat of hydration; it varies in amount and rate for different Portland cements. Heat generation and buildup are affected by dimensions of the concrete, ambient air temperature, initial concrete temperature, water-cementing materials ratio, cement composition and fineness, amount of cementing materials, and admixtures.

So long as workable mixtures are used, the less water there is in the mix, the stronger, more watertight and more durable will be the concrete. Excess mixing water dilutes the paste made by the cement and water and makes weaker, more porous concrete. For any set of conditions of mixing, placing, and curing and for given materials there is a definite relation between the strength of the concrete and the amount of water used in mixing. This relation can be determined by test and a water-cement ratio-strength curve can be developed representing the actual results on a specific job.

American Concrete Institute (ACI) Standard
for Selecting Proportions for Concrete Mixes

Committee 211 - of the American Concrete Institute develop and report information on the proportioning of concrete mixtures. This committee recommends certain practices for selecting proportions for concrete mixes. Following are extracts from this standard.

Basic Relationship

Concrete proportions must be selected to provide necessary placeability, strength, and durability for the particular application. Well established relationships governing these properties are discussed briefly below.

Placeability (including satisfactory finishing properties) encompasses traits loosely accumulated in the terms "workability" and "consistency". For the purpose of this discussion, workability is considered to be the property of concrete that determines its capacity to be placed and consolidated properly and to be finished without harmful segregation. It embodies such concepts as moldability, cohesiveness, and compactability. It is affected by the grading, particle shape and proportions of aggregate, the amount of cement, the presence of entrained air, admixtures, and the consistency of the mixture. Procedures in the recommended practice take into account these factors.

Consistency, loosely defined, is the wetness of the concrete mixture. It is measured in terms of slump-the higher the slump the wetter the mixture- and it affects the ease with which the concrete will flow during placement. It is related to, but not synonymous with, workability. In properly proportioned concrete, the unit water content required to produce a given slump will depend on several factors. Water requirement increases as aggregates become more angular and rough textured (but this disadvantage may be offset by improvements in other characteristics such as bond to cement paste). Required mixing water decreases as the maximum size of well graded aggregate is increased. It also decreases with the entrainment of air. Mixing water requirement may often be significantly reduced by certain admixtures.

Strength. Strength is an important characteristic of concrete, but other characteristics such as durability, permeability, and wear resistance are often

CONCRETE

equally or more important. These may be related to strength in a general way but are also affected by factors not significantly associated with strength. For a given set of materials and conditions, concrete strength is determined by the net quantity of water used per unit quantity of cement. The net water content excludes water absorbed by the aggregates. Differences in strength, for a given water-cement ratio may result from changes in: maximum size of aggregate; grading, surface texture, shape, strength, and stiffness of aggregate particles; differences in cement types and sources; air content; and the use of admixtures which affect the cement hydration process or develop cementitious properties themselves. To the extent that these effects are predictable in the general sense, they are taken into account in this recommended practice. However, in view of their number and complexity, it should be obvious that accurate predictions of strength must be based on trial batches or experience with the materials to be used.

Durability. Concrete must be able to endure those exposures which may deprive it of its serviceability-freezing and thawing, wetting and drying, heating and cooling, chemicals, de-icing agents, and the like. Resistance to some of these may be enhanced by use of special ingredients: low-alkali cement, pozzolans, or selected aggregate to prevent harmful expansion due to the alkali-aggregate reaction which occurs in some areas when concrete is exposed in a moist environment; sulfate resisting cement or pozzolans for concrete exposed to sea water or sulfate-bearing soils; or aggregate free of excessive soft particles where resistance to surface abrasion is required. Use of a low water-cement ratio will prolong the life of concrete by reducing the penetration of aggressive liquids. Resistance to severe weathering, particularly freezing and thawing, and to salts used for ice removal is greatly improved by incorporation of a proper distribution of entrained air. Entrained air should be used in all exposed concrete in climates where freezing occurs.

Background Data

To the extent possible, selection of concrete proportions should be based on test data or experience with the materials to be used. Where such background is limited or not available, estimates given in this recommended practice may be employed. The following information for available materials will be useful:

1. Sieve analyses of fine and coarse aggregates.
2. Unit weight of coarse aggregate.
3. Bulk specific gravities and absorptions of aggregates.
4. Mixing water requirements of concrete developed from experience with available aggregates.
5. Relationships between strength and water-cement ratio for available combinations of cement and aggregate.

Estimates from Tables 3 and 4 may be used when the last two items of information are not available. As will be shown, proportions can be estimated

467

without the knowledge of aggregate specific gravity and absorption (third item above).

Procedure

Estimating the required batch weights for the concrete involves a sequence of logical, straightforward steps, which in effect fit the characteristics of the available materials into a mixture suitable for the work. The question of suitability is frequently not left to the individual selecting the proportions. The job specifications may dictate some or all of the following:

1. Maximum water-cement ratio
2. Minimum cement content
3. Air content
4. Slump
5. Maximum size of aggregate
6. Strength
7. Other requirements relating to such things as strength over-design, admixtures, and special types of cement or aggregate

Regardless of whether the concrete characteristics are prescribed by the specifications or are left to the individual selecting the proportions, establishment of batch weights per cubic yard of concrete can best be accomplished in the following sequence:

Step 1. Choice of slump. If slump is not specified, a value appropriate for the work can be selected from Table 1. The slump ranges shown apply when vibration is used to consolidate the concrete. Mixes of the stiffest consistency that can be placed efficiently should be used.

Table 1				
	Slump, Inch, (mm)			
Type of Construction	Max.*, Inch	Min., Inch	Max.*, mm	Min., mm
Reinforced foundation wall and footings	3	1	76	25
Plain footings, caissons, pipe piling, and substructure walls	3	1	76	25
Beams and reinforced walls	4	1	102	25
Building columns	4	1	102	25
Pavement and slabs	4	1	102	25
Heavy mass concrete	3	1	76	25

* May increase 1 inch (25 mm) for methods of consolidation other than vibration.

Step 2. Choice of maximum size of aggregate. Large maximum sizes of well graded aggregates have less voids than smaller sizes. Hence, concretes with the larger-sized aggregates require less mortar per unit volume

CONCRETE

of concrete. Generally, the maximum size of aggregate should be the largest that is economically available and consistent with dimensions of the structure. Ordinarily, the ratio of the nominal maximum aggregate size to the minimum dimension within which concrete must be placed should not exceed the value shown in Table 2.

Table 2		
Criterion Dimension		Recommend
Narrowest dimension between sides of forms		1/5
Depth of slab		1/3
Clear spacing between individual reinforcement bars, bundles of bars, or pretensioning strands	In Columns	2/3
	In all other members	3/4

These limitations are sometimes waived if workability and methods of consolidation are such that the concrete can be placed without honeycomb or void. When high strength concrete is desired, best results may be obtained with reduced maximum sizes of aggregate since these produce higher strengths at a given water-cement ratio.

Step 3. Estimation of mixing water and air content. The quantity of water per unit volume of concrete required to produce a given slump is dependent on the maximum size, particle shape and grading of the aggregates, and on the amount of entrained air. It is not greatly affected by the quantity of cement. Table 3 provides estimates of required mixing water for concretes made with various maximum sizes of aggregate, with and without air entrainment. Depending on aggregate texture and shape, mixing water requirements may be somewhat above or below the tabulated values, but they are sufficiently accurate for the first estimate. Such differences in water demand are not necessarily reflected in strength since other compensating factors may be involved. For example, a rounded and an angular coarse aggregate, both well and similarly graded and of good quality, can be expected to produce concrete of about the same compressive strength for the same cement factor in spite of differences in water-cement ratio resulting from the different mixing water requirements. Particle shape per se is not an indicator that an aggregate will be either above or below average in its strength-producing capacity.

Table 3 indicates the approximate amount of entrapped air to be expected in non-air-entrained concrete, and shows the recommended levels of average air content for concrete in which air is to be purposely entrained for durability. Air-entrained concrete should always be used for structures which will be exposed to freezing and thawing, and generally for structures exposed to sea water or sulfates. When severe exposure is not anticipated, beneficial effects of air entrainment on concrete workability and cohesiveness can be achieved at air content levels about half those shown for air-entrained concrete.

469

Table 3
Approximate Water and Air Content Requirements for Different Slumps and Maximum Sizes Aggregates

Water, Lb/Cu.Yd. of Concrete for Indicated Maximum sizes of aggregate (in Inches)*							
Non - Air - entrained concrete							
Slump - In.	3/8	1/2	3/4	1	1 1/2	2	3
1 to 2	360	335	315	300	275	260	240
3 to 4	385	365	340	325	300	285	265
6 to 7	410	385	360	340	315	300	285
Air Content - percent	3	2.5	2	1.5	1	0.5	0.3
Air - entrained concrete							
1 to 2	305	295	280	270	250	240	225
3 to 4	340	325	305	295	275	265	250
6 to 7	365	345	325	310	290	280	270
Air Content - percent	8	7	6	5	4.5	4	3.5
Water, Kg/Cu.M. of Concrete for Indicated Maximum sizes of aggregate (in mm)*							
Slump - mm	9.375	12.5	18.75	25	37.5	50	75
Non - Air - entrained concrete							
25 to 50	213.6	198.7	186.9	178.0	163.1	154.3	142.4
75 to 100	228.4	216.5	201.7	192.8	178.0	169.1	157.2
150 to 175	243.2	228.4	213.6	201.7	186.9	178.0	169.1
Air Content - percent	3	2.5	2	1.5	1	0.5	0.3
Air - entrained concrete							
25 to 50	180.9	175.0	166.1	160.2	148.3	142.4	133.5
75 to 100	201.7	192.8	180.9	175.0	163.1	157.2	148.3
150 to 175	216.5	204.7	192.8	183.9	172.0	166.1	160.2
Air Content - percent	8	7	6	5	4.5	4	3.5

These quantities of mixing water are for use in computing cement factors for trial batches. They are maxima for reasonable well-shaped angular course aggregate within limits of accepted specifications.

When trial batches are used to establish strength relationships or verify strength-producing capability of a mixture, the least favorable combination of mixing water and air content should be used. The air content should be the maximum permitted or likely to occur, and the concrete should be gauged to the highest permissible slump. This will avoid developing an overoptimistic estimate of strength on the assumption that average rather than

CONCRETE

extreme conditions will prevail in the field. For information on air content recommendations, see ACI 201, 301, and 302.

Step 4. Selection of water-cement ratio. The required water-cement ratio is determined not only by strength requirements but also by factors such as durability and finishing properties. Since different aggregates and cements generally produce different strengths at the same water-cement ratio, it is highly desirable to have or develop the relationship between strength and water-cement ratio for the materials actually to be used. In the absence of such data, approximate and relatively conservative values for concrete containing Type I portland cement can be taken from Table 4(a). With typical materials, the tabulated water-cement ratios should produce the strengths shown, based on 28-day tests of specimens cured under standard laboratory conditions. The average strength selected must, of course, exceed the specified strength by a sufficient margin to keep the number of low tests within specified limits.

Table 4(a) Relationships Between Water-Cement Ratio and Compressive Strength of Concrete			
Compressive Strength at 28 Days, psi*	Water-Cement Ratio, by Weight Non Air-entrained concrete	Air-entrained concrete	Compressive Strength at 28 Days, MPa*
6000	0.41	......	41.37
5000	0.48	0.4	34.47
4000	0.57	0.48	27.58
3000	0.68	0.59	20.68
2000	0.82	0.74	13.79

Values are estimated average strengths for concrete containing not more than the percentage of air shown in Table 3. For a consistent water-cement ratio, the strength of concrete is reduced as the air content is increased.

For severe conditions of exposure, the water-cement ratio should be kept low even though strength requirements may be met with a higher value. Table 4(b) gives limiting values.

Strength is based on 6" x 12" (150 x 300 mm) cylinders moist-cured 28 days at 73.4 + 3°F (23 + 1.7°C) in accordance with Section 9(b) of *C31/C31M-03a Standard Practice for Making and Curing Concrete Test Specimens in the Field.*

Relationship assumes maximum size of aggregate about 3/4" to 1" (18.75-25 mm); for a given source, strength produced for a given water-cement ratio will increase as maximum size of aggregate decreases.

Table 4(b) Maximum Permissible Water-Cement Ratios for Concrete in Severe Exposures*		
Types of structures	Structure wet continuously or frequently and exposed to freezing and thawing **	Structure exposed to sea water or sulfates
Thin sections (railings, curbs, sills, ledges, ornamental work) and sections with less than 1 in. cover over steel	0.45	0.40 **
All other structures	0.50	0.45 ***

* Based on reports of ACI Committee 201 "Durability of Concrete in Service", previously cited.

** Concrete should also be air-entrained

*** If sulfate resisting cement (type II or Type V of ASTM C150-04ae1 Standard Specification for Portland Cement is used, permissible water-cement ratio may be increased by 0.05.

Step 5. Calculation of cement content. The amount of cement per unit volume of concrete is fixed by the determinations made in Steps 3 and 4 above. The required cement is equal to the estimated mixing water content (Step 3) divided by the water-cement ratio (Step 4). If, however, the specification includes a separate minimum limit on cement in addition to requirements for strength and durability, the mixture must be based on whichever criterion leads to the larger amount of cement.

Step 6. Estimation of coarse aggregate content. Aggregates of essentially the same maximum size and grading will produce concrete of satisfactory workability when a given volume of coarse aggregate, on a dry-rodded basis, is used per unit volume of concrete. Appropriate values for this aggregate volume are given in Table 6. It can be seen that, for equal workability, the volume of coarse aggregate in a unit volume of concrete is dependent only on its maximum size and the fineness modulus of the fine aggregate. Differences in the amount of mortar required for workability with different aggregates, due to differences in particle shape and grading, are compensated for automatically by differences in dry-rodded void content.

The volume of aggregate, in cubic feet (cu.m), on a dry-rodded basis, for a cubic yard (cubic meter) of concrete is equal to the value from Table 6 multiplied by 27. This volume is converted to dry weight of coarse aggregate required in a cubic yard of concrete by multiplying it by the dry-rodded weight per cubic foot of the coarse aggregate.

These volumes are selected from empirical relationships to produce concrete with a degree of workability suitable for usual reinforced

CONCRETE

construction. For less workable concrete such as required for concrete pavement construction they may be increased about 10%. When placement is to be by pump, they should be reduced about 10%.

Maximum Size Aggregate, Inch	Volume of dry-rodded course aggregate * per unit volume of concrete for different fineness moduli of sand				Maximum Size Aggregate, mm
	2.40	2.60	2.80	3.00	
0.375	0.50	0.48	0.46	0.44	9.38
0.5	0.59	0.57	0.55	0.53	12.50
0.75	0.66	0.64	0.62	0.60	18.75
1	0.71	0.69	0.67	0.65	25.00
1.5	0.75	0.73	0.71	0.69	37.50
2	0.78	0.76	0.74	0.72	50.00
3	0.82	0.80	0.78	0.76	75.00

Table 6
Volume of Course Aggregate Per Unit Volume of Concrete

Volumes are based on aggregates in dry-rodded condition as described in ASTM C29/C29M-97(2003) Standard Test Method for Bulk Density (Unit Weight) and Voids in Aggregate.

Step 7. Estimation of fine aggregate content. At completion of Step 6, all ingredients of the concrete have been estimated except the fine aggregate. Its quantity is determined by difference. Either of two procedures may be employed: the "weight" method or the "absolute volume" method.

If the weight of the concrete per unit volume is assumed or can be estimated from experience, the required weight of fine aggregate is simply the difference between the weight of fresh concrete and the total weight of the other ingredients. Often the unit weight of concrete is known with reasonable accuracy from previous experience with the materials. In the absence of such information, Table 7 can be used to make a first estimate. Even if the estimate of concrete weight per cubic yard (cubic meter) is rough, mixture proportions will be sufficiently accurate to permit easy adjustment on the basis of trial batches as will be shown in the examples.

If a theoretically exact calculation of fresh concrete weight per cubic yard (cubic meter) is desired, the following formula can be used:

$$U = 16.85 \, Ga \, (100 - A)$$
$$+ C \times (1 - Ga/Gc) - W \times (Gc - 1)$$
Where

1. U = weight of fresh concrete per cubic yard, lb. (per cu.m, kg).
2. Ga = weighted average specific gravity of combined fine and coarse aggregate, bulk SSD*
3. Gc = specific gravity of cement (generally 3.15)
4. A = air content, percent

473

5. W = mixing water requirement, lbs. per c.y. (kg per cu.m)
6. C = cement requirement, lbs. per c.y. (kg per cu.m)

*SSD indicates saturated-surface-dry basis used in considering aggregate displacement. The aggregate specific gravity used in calculations must be consistent with the moisture condition assumed in the basic aggregate batch weights-i.e., bulk dry if aggregate weights are stated on a dry basis, and bulk SSD if weights are stated on a saturated-surface-dry basis.

Table 7					
First Estimate of Weight of Fresh Concrete					
Maximum Size of Aggregate Inch	Non-air entrained Concrete Lbs/CY	Air entrained Concrete Lbs/CY	Maximum Size of Aggregate mm	Non-air entrained Concrete Kg/Cu.M.	Air entrained Concrete Kg/Cu.M.
3/8	3,840	3,690	9.38	2,278	2,189
1/2	3,890	3,760	12.50	2,308	2,231
3/4	3,960	3,840	18.75	2,349	2,278
1	4,010	3,900	25.00	2,379	2,314
1 1/2	4,070	3,960	37.50	2,415	2,349
2	4,120	4,000	50.00	2,444	2,373
3	4,160	4,040	75.00	2,468	2,397

A more exact procedure for calculating the required amount of fine aggregate involves the use of volumes displaced by the ingredients. In this case, the total volume displaced by the known ingredients-water, air, cement, and coarse aggregate-is subtracted from the unit volume of concrete to obtain the required volume of fine aggregate. The volume occupied in concrete by any ingredient is equal to its weight divided by the density of that material (the latter being the product of the unit weight of water and the specific gravity of the material).

Step 8. Adjustments for aggregate moisture. The aggregate quantities actually to be weighed out for the concrete must allow for moisture in the aggregates. Generally, the aggregates will be moist and their dry weights should be increased by the percentage of water they contain, both absorbed and surface. The mixing water added to the batch must be reduced by an amount equal to the free moisture contributed by the aggregate-i.e., total moisture minus absorption.

Values calculated by Eq. 7 for concrete of medium richness, 550 lb. of cement per c.y. (326.28 kg/cu.m), and medium slump with aggregate specific gravity of 2.7. Water requirements based on values for 3" to 4" (75-100 mm) slump in Table 3. If desired, the estimated weight may be refined as follows, if necessary information is available: for each 10 lb. (4.53 kg) difference in mixing water from the Table 3 values for 3" to 4" (75-100 mm) slump, correct the weight 15 lbs. per c.y. (6.80 kg/cu.m) in the opposite direction; for each 100 lb. (45.36 kg) difference in cement content from 550 lbs.,(249.48

474

CONCRETE

kg) correct the weight 15 lbs. per c.y. (6.80 kg) in the same direction; for each 0.1 by which aggregate specific gravity deviates from 2.7, correct the concrete weight 100 lbs. (45.36 kg) in the same direction.

Step 9. Trial batch adjustments. The calculated mixture proportions should be checked by means of trial batches prepared and tested in accordance with *ASTM C192/C192M-02 Standard Practice for Making and Curing Concrete Test Specimens in the Laboratory.*

Full-sized field batches. Only sufficient water should be used to produce the required slump, regardless of the amount assumed in selecting the trial proportions. The concrete should be checked for unit weight and yield (ASTM C 138/C138M-01a) and for air content (ASTM C 138/C138M-01a, C 172-04, or C 231-04). It should also be carefully observed for proper workability, freedom from segregation, and finishing properties. Appropriate adjustments should be made in the proportions for subsequent batches in accordance with the following procedure.

Re-estimate the required mixing water per cubic yard of concrete by multiplying the net mixing water content of the trial batch by 27 and dividing the product by the yield of the trial batch in cubic feet (Divide by 0.03 to obtain cubic meters). If the slump of the trial batch was not correct, increase or decrease the re-estimated amount of water by 10 lb. (4.53 kg) for each required increase or decrease of 1" (25 mm) in slump.

If the desired air content (for air-entrained concrete) is not achieved, re-estimate the admixture content required for proper air content and reduce or increase the mixing water content of the above paragraph by 5 lb. (2.26kg) for each 1% by which the air content is to be increased or decreased from that of the previous trial batch.

If estimated weight per cubic yard of fresh concrete is the basis for proportioning, re-estimate that weight by multiplying the unit weight in lbs. per cu. ft. of the trial batch by 27 and reducing or increasing the result by the anticipated percentage increase or decrease in air content of the adjusted batch from the first trial batch.

Calculate new batch weights starting with Step 4, modifying the volume of coarse aggregate from Table 6 if necessary to provide proper workability.

Sample Computations

Two sample problems illustrate application of the proportioning procedures. The following conditions are assumed:

1. Type I non-air-entraining cement will be used, and its specific gravity is assumed to be 3.15. The specific gravity values are not used if proportions are selected to provide a weight of concrete assumed to occupy 1 cu. yd. (.7646 cu.m)
2. Coarse and fine aggregates in each case are of satisfactory quality and are graded within limits of generally accepted specifications, such as *ASTM C33-03 Standard Specification for Concrete Aggregates.*

475

3. The coarse aggregate has a bulk specific gravity of 2.68 and an absorption of 0.5%.
4. The fine aggregate has a bulk specific gravity of 2.64, an absorption of 0.7%, and fineness modulus of 2.8.

Example 1. Concrete is required for a portion of a structure which will be below ground level in a location where it will not be exposed to severe weathering or sulfate attack. Structural considerations require it to have an average 28-day compressive strength of 3500 psi (246.07 kg/cm²).

This is not the specified strength used for structural design but a higher figure expected to be produced on the average. For the method of determining the amount by which average strength should exceed design strength, see *ACI 214-77: Recommended practice for evaluation of strength test results of concrete.*

On the basis of information in Table 1, as well as previous experience, it is determined that under the conditions of placement to be employed, a slump of 3" to 4" (75-100 mm) be used and that the available No. 4 to 1-1/2" (38 mm) coarse aggregate will be suitable. The dry-rodded weight of coarse aggregate is found to be 100 lbs. per c.f. (1511 kg/m3). Employing the sequence outlined in the Procedure section, the quantities of ingredients per cubic yard (cu.m) of concrete are calculated as follows:

Step 1. As indicated, the desired slump is 3" to 4" (75-100 mm)

Step 2. The locally available aggregate graded from No. 4 to 1-1/2" (37.5 mm), has been indicated as suitable.

Step 3. Since the structure will not be exposed to severe weathering, non-air-entrained concrete will be used. The approximate amount of mixing water to produce a 3" to 4" slump (75-100 mm) in non-air-entrained concrete with 1-1/2" (37.50 mm) aggregate is found from Table 3 to be 300 lbs./c.y. (177.98 kg/cu.m). Estimated entrapped air is shown as 1%.

Step 4. From Table 4(a), the water-cement ratio needed to produce a strength of 3500 psi (2.44 kg/mm2) in non-air-entrained concrete is found to be about 0.62.

Step 5. From the information derived in Steps 3 and 4, the required cement content is found to be 300/0.62 = 484 lbs./c.y. (178/0.62 = 287 kg/m3)

Step 6. The quantity of coarse aggregate is estimated from Table 6. For a fine aggregate having a fineness modulus of 2.8 and a 1-1/2" (37.50 mm) maximum size of coarse aggregate, the table indicates that 0.71 cu.ft. (0.20 cu.m) of coarse aggregate, on a dry-rodded basis, may be used in each cu.ft. of concrete. For a cubic yard, the coarse aggregate will be 27 x 0.71 = 19.17 cu.ft. (0.54 cu.m). Since it weighs 100 lb./c.f. (1601.84 kg/m3) the dry weight of coarse aggregate is 1,917 lb. (869.55 kg).

Step 7. With the quantities of water, cement, and coarse aggregate established, the remaining material comprising the cubic yard (cubic meter) of concrete must consist of sand and whatever air will be entrapped. The

CONCRETE

required sand may be determined on the basis of either weight or absolute volume as shown below:

Weight basis. From Table 7, the weight of a cubic yard (cubic meter) of non-air-entrained concrete made with aggregate having a maximum size of 1-1/2" (37.50 mm) is estimated to be 4,070 lbs. (1846.15 kg). For a first trial batch, exact adjustments of this value for usual differences in slump, cement factor, and aggregate specific gravity are not critical. Weights already known are:

Water (net mixing)	300 lbs.	136.08 kg
Cement	484 lbs.	219.54 kg
Coarse aggregate	1,917 lbs. Dry	869.54 kg (dry)*
Total	2,701 lbs.	1,225.15 kg

Aggregate absoption is disregarded since its magnitude is inconsequential in relation to other approximations.

The weight of sand, therefore, is estimated to be 4070 - 2701 = 1369 lbs. (1846.15 – 1225.17 = 620.97 kg) dry*
Aggregate absorption is disregarded since its magnitude is inconsequential in relation to other approximations.

Absolute volume basis. With the quantities of cement, water, and coarse aggregate established, and the approximate entrapped air content (as opposed to purposely entrained air) taken from Table 3, the sand content can be calculated as follows:

Volume of water $= \dfrac{300}{62.4} = 4.81$ c.f.

Solid volume of cement $= \dfrac{484}{3.15 \times 62.4} = 2.46$ c.f.

Solid volume of coarse aggregate $= \dfrac{1917}{2.68 \times 62.4} = 11.46$ c.f.

Volume of entrapped air $= 0.01 \times 27 = 0.27$ c.f.

Total solid volume of ingredients except sand $= 19.00$ c.f.

Solid volume of sand required $= 27 - 19.00 = 8.00$ c.f.

Required weight of dry sand $= 8.00 \times 2.64 \times 62.4 = 1318$ lbs.

Batch weights per cubic yard of concrete calculated on the two bases are compared below:

Concrete Component	Based on estimated concrete weight, Lbs	Based on absolute volume of ingredients Lbs
Water (net mixing)	300	300
Cement	484	484
Coarse aggregate (dry)	1917	1917
Sand (dry)	1369	1318
Concrete Component	Based on estimated concrete weight, Kg	Based on absolute volume of ingredients Kg
Water (net mixing)	136.08	136.08
Cement	219.54	219.54
Coarse aggregate (dry)	869.54	869.54
Sand (dry)	620.97	597.83

These batch weights will require adjustment in the field to take into account moisture on aggregates. Also, some adjustment in proportions may be found desirable on the basis of actual field conditions.

Example 2. Concrete is required for a heavy bridge pier which will be exposed to fresh water in a severe climate. An average 28-day compressive strength of 3000 psi (20.6 MPa) will be required. Placement conditions permit a slump of 1" to 2" (25-50 mm) and the use of large aggregate, but the only economically available coarse aggregate of satisfactory quality is graded from No. 4 to 1" and this will be used. Its dry-rodded weight is found to be 95 lb./cu.f. (1521.75 kg/cu.m) Other characteristics are as indicated in the first part of this section.

The calculations will be shown in skeleton form only. Confusion is avoided by following all steps in the Procedure section, even when they appear to repeat specified requirements.

Step 1. The desired slump is 1" to 2" (25-50 mm).

Step 2. The locally available aggregate, graded from No. 4 to 1" (25 mm), will be used.

Step 3. Since the structure will be exposed to severe weathering, air-entrained concrete will be used. The approximate amount of mixing water to produce a 1" to 2" (25-50 mm) slump in air-entrained concrete with 1" (25 mm) aggregate is found from Table 3 to be 270 lb./c.y. (160 kg/m3). The recommended air content is 5%.

Step 4. From Table 4(a), the water-cement ratio needed to produce a strength of 3000 psi (20.68 MPa) in air-entrained concrete is estimated to be about 0.59. However, reference to Table 4(b) reveals that, for the severe weathering exposure anticipated, the water-cement ratio should not exceed 0.50. This lower figure must govern and will be used in the calculations.

Step 5. From the information derived in Steps 3 and 4, the required cement content is found to be 270/0.50 = 540 lb./cu.yd. (160/0.50 = 320 kg/m3).

478

CONCRETE

Step 6. The quantity of coarse aggregate is estimated from Table 6. With a fine aggregate having a fineness modulus of 2.8 and a 1" (25 mm) maximum size of coarse aggregate, the table indicates that 0.67 c.f. (0.67 cu.m) of coarse aggregate, on a dry-rodded basis, may be used in per cu. ft. (cu.m) of concrete. For a cubic yard, therefore, the coarse aggregate will be 27 x 0.67 = 18.09 c.f. (0.67 cu.m). Because it weighs 95 lb./cu.ft. (1525 kg/m3), the dry weight of coarse aggregate is 18.09 x 95 = 1719 lb. (779.74 kg).

Step 7. With the quantities of water, cement, and coarse aggregate established, the remaining material in the cubic yard of concrete is sand and air. The required sand is determined on the basis of either weight or absolute volume:

Weight basis. From Table 7, the weight of a cubic yard (cubic meter) of air-entrained concrete made with aggregate of 1" (25 mm) maximum size is estimated to be 3900 lbs. (1769.04 kg). For a first trial batch, exact adjustments of this value for differences in slump, cement factor, and aggregate specific gravity are not critical. Weights already known are:

Water (net mixing)	270 Lbs	122.47 Kg
Cement	540 Lbs	244.94 Kg
Coarse aggregate (dry)	1,719 Lbs	779.72 Kg
Total	2,529 Lbs	1,147.13 Kg

The weight of sand, therefore, is estimated to be 3900 - 2529 = 1371 lbs. dry (2310 - 1502 = 808 kg dry)

Absolute volume basis. With the quantities of cement, water, air, and coarse aggregate established, the sand content is calculated as follows:

Volume of water $= \dfrac{270}{62.4} = $ 4.33 c.f.

Solid volume of cement $= \dfrac{540}{3.15 \times 62.4} = $ 2.75 c.f.

Solid volume of coarse aggregate $= \dfrac{1719}{2.68 \times 62.4} = $ 10.28 c.f.

Volume of air $= 0.05 \times 27 = $ 1.35 c.f.

Total solid volume of ingredients except sand $= $ 18.71 c.f.

Solid volume of sand required $= 27 - 18.71 = $ 8.29 c.f.

Required weight of dry sand $= 8.29 \times 2.64 \times 62.4 = $ 1366 lbs.

479

Batch weight per cubic yard of concrete calculated on the two bases are compared below:

Concrete Component	Based on estimated concrete weight, Lbs	Based on absolute volume of ingredients Lbs
Water (net mixing)	270	270
Cement	540	540
Coarse aggregate (dry)	1719	1719
Sand (dry)	1371	1366
Concrete Component	Based on estimated concrete weight, Kg	Based on absolute volume of ingredients Kg
Water (net mixing)	122.47	122.47
Cement	244.94	244.94
Coarse aggregate (dry)	779.72	779.72
Sand (dry)	621.87	619.61

These batch weights will require adjustment in the field to take into account moisture on aggregates. Also, some adjustment in proportions may be found desirable on the basis of actual field conditions.

These batch weights will require adjustment in the field to take into account moisture on aggregates. Also, some adjustment in proportions may be found desirable on the basis of actual field experience.

READY-MIXED CONCRETE

More than 90% of all concrete used in building projects is ready-mixed, which is defined as Portland cement concrete manufactured for delivery to a purchaser in a plastic and unhardened state.

Sometimes, it is mixed completely in a stationary mixer and then is transported to the job in trucks. This is known as central-mixed concrete. In other ready-mixed operations, the materials are dry batched and mixed enroute to the job in truck mixers. This is known as transit-mixed concrete. In another procedure, the concrete is mixed in a stationary mixer at the central plant only enough to intermingle the ingredients, usually about 1/2 minute, and the mixing is completed enroute to the job. This is known as shrink-mixed concrete.

Truck mixers consist essentially of a mixer with a separate water tank and water measuring device mounted on a truck chassis. There are other truck conveyances which are similar but are without provisions for water.

C94/C94M-04a Standard Specification for Ready-Mixed Concrete requires that when a truck mixer is used either for complete mixing or to finish the partial mixing, each batch of concrete is to be mixed not more than 100 revolutions of the drum or blades at the speed of rotation designated by

CONCRETE

the manufacturer as the mixing speed. Any additional mixing is to be done at the agitating speed. The specification also requires that the concrete must be delivered and discharged from the truck mixer or agitator truck within 1-1/2 hours after introduction of the water to the cement and aggregate or the cement to the aggregate.

Quality of Ready-Mixed Concrete. C94/C94M-04a states that in the absence of applicable general specifications, the purchaser shall select one of the alternate bases for specifying the quality of concrete.

Alternate No. 1: When the purchaser assumes responsibility for the proportioning of the concrete mixture, he shall also specify the following:

1. Cement content in bags or pounds per cubic yard of concrete, or equivalent units.
2. Maximum allowable water content in gallons per cubic yard of concrete, or equivalent units, including surface moisture on the aggregates, but excluding water of absorption.
3. If admixtures are required, the type, name, and dosage to be used. The cement content shall not be reduced when admixtures are used under Alternate No. 1 without the written approval of the purchaser.

Alternate No. 2: When the purchaser requires the manufacturer to assume full responsibility for the selection of the proportions for the concrete mixture, the purchaser shall also specify the requirements for compressive strength as determined on samples taken from the transportation unit at the point of discharge. The purchaser shall specify the requirements in terms of the compressive strength of standard specimens cured under standard laboratory conditions for moist curing. Unless otherwise specified the age at test shall be 28 days.

Alternate No. 3: When the purchaser requires the manufacturer to assume responsibility for the selection of the proportions for the concrete mixture with the minimum allowable cement content specified, the purchaser shall also specify the following:

1. Required compressive strength as determined on samples taken from the transportation unit at the point of discharge. The purchaser shall specify the requirements for strength in terms of tests of standard specimens cured under standard laboratory conditions for moist curing. Unless otherwise specified the age at test shall be 28 days.
2. Minimum cement content in bags or pounds per cubic yard of concrete.
3. If admixtures are required, the type, name, and dosage to be used. The cement content shall not be reduced when admixtures are used.

Alternate No. 3 is useful only if the designated minimum cement content is at about the same level that would ordinarily be required for the strength, aggregate size, and slump specified. At the same time, it must be an amount sufficient to assure durability under expected service conditions, as well as satisfactory surface texture and density, in the event specified strength is attained with it.

The proportions arrived at by Alternates 1, 2, or 3 for each class of concrete and approved for use in a project shall be assigned to a designation to facilitate identification of each concrete mixture delivered to the project. This is the designation required and supplies information on concrete proportions when they are not given separately on each delivery ticket. A certified copy of all proportions as established in Alternates 1, 2, and 3 shall be on file at the batch plant.

Curing. The protection of concrete during the early period to prevent loss of moisture at low temperatures is an important factor in the development of both strength and durability in concrete. Specifications require that the concrete be protected to prevent loss of moisture from the surface and to prevent temperatures at the surface from going below 50°F (10°C) for periods of 5 days where normal portland cement is used, and 3 days where high early strength portland cement is used.

These limits recognize that under average conditions curing of concrete does not cease immediately upon removal of the moisture loss protection. Where conditions are extremely severe, such as for thin sections in hot dry air or very low temperatures, it may be desirable to increase somewhat the protection periods specified.

High Early Strength Portland Cement. Most cement companies manufacturing what is designated as high early strength portland cement, which develops practically the same strength in 72 hours as is obtained with normal portland cement in 7 to 10 days.

High early strength cement is used in the same proportions and in exactly the same manner as ordinary portland cement. This cement is advantageous not only in special rush work but in everyday reinforced concrete work. High early strength cement costs $3.50 per barrel more than standard portland cement, or if we assume that approximately 1-1/2 barrels of cement per c.y. (2 per cu.m) will be used, the cost per cu.yd. of concrete is increased by $5.25 ($6.86 per cu.m). If the average amount of concrete per sq.ft. (sq.m) of floor construction (including columns, beams, lintels, etc.) is 8" (0.2 m, 200 mm), it takes 40 sq.ft. of floor for a yard (5 sq.m per cu.m) of concrete. The extra expense of high early strength cement is $0.13 per sq.ft. ($1.44 per sq.m).

Under favorable conditions using high early strength portland cement, it is possible to strip part or an entire floor in three days so it may be possible to save half of the lumber required for a floor. It takes not less than 3 to 4 b.f. of lumber for each sq.ft. (sq.m) of floor construction-including columns, girders, and lintels-in addition to the labor costs per sq.ft. (sq.m) of floor to make up the various sides, panels, shores, wedges, etc. required (see *Concrete Formwork*).

CONCRETE

If it is possible to save one-half of the lumber plus half the labor, the resulting savings per sq.ft. (sq.m) is greater than the additional $0.13 per s.f. ($1.39 sq.m) for high-early cement.

During the winter there is the additional saving of two to three days for cold weather protection.

Air Entrained Concrete. Most agencies are specifying air-entrained concrete for exposed concrete as is, retaining walls, pavements, bridge piers, decks and other structures. In this concrete, 3% to 5 or 6% of air is incorporated in the concrete in the form of minute, separate air bubbles. Such concrete is more resistant to freezing and thawing and to salt action than normal concrete. It is produced by using air-entraining Portland cement or by the addition of an air-entraining agent at the mixer. Originally developed to prevent scaling of pavement where salts are used for ice removal, air-entrained concrete has been adopted for all types of work and in all locations because of its better workability as well as its better resistance to weathering.

In air-entrained concrete the proportion of sand and the amount of water is generally reduced from the normal mix in sufficient quantities to make up for the increased volume of concrete produced by the air. Thus the same cement factor is maintained. In the tables given on the previous pages, the sand can be reduced about 5 lbs. (2.27 kg) for each 1% of air. Thus, if 4% air is incorporated, the quantities of sand shown in these tables would be reduced by 20 lbs. per cu.yd. (12 kg/m3) of concrete. For each 1% of air the amount of water would be reduced by about 6 to 7 lbs. per cu.yd. (3.6-4 kg/m3).

Important Suggestions for Purchasing Materials for Job Mixed Concrete. When purchasing sand, gravel, or crushed stone for concrete, it is important to see that the aggregates and particularly the sand do not contain an abnormal amount of moisture. Otherwise, you will be short on materials, which will reduce profits or may even result in a loss, particularly if there are considerable quantities of aggregates to be purchased. The following suggestions are important when placing orders:

1. Determine whether the sand has a high or low moisture content.
2. Determine whether the sand has a high or low specific gravity.
3. When sand is purchased at a certain price per cu.yd. (cu.m) make certain that you are not paying for water instead of sand, as it is not uncommon to see settlement of 4" (100 mm) in trucks or railroad cars between the time it is shipped and when it arrives at the job. This is especially true where washed sand is loaded immediately after washing or after hard rains.
4. When sand is purchased by weight, also make certain you are not paying for water instead of sand. Ordinarily sand should not contain over 4% to 5% moisture, while wet sand may contain as high as 18% moisture. This makes a difference in the cost of sand.
5. Watch the grading of sand and percentages from fine to coarse.
6. Determine the grading of the gravel and its specific gravity.

Suggested Mixes for Air-entrained Concrete of Medium Consistancy (3" to 4" slump)

Water Cement Ratio Lb/Lb	Max. size of Aggregate in Inch	Air Content Percent	Water, Lb/CY concrete	Cement, Lb/CY of concrete	With Fine Sand fineness modulus = 2.50**			With course Sand fineness modulus = 2.50**		
					Fine aggregate % of total	Fine aggregate lb/cy conc.	Course aggregate lb/cy conc.	Fine aggregate % of total	Fine aggregate lb/cy conc.	Course aggregate lb/cy conc.
0.55	3/8	7.5	340	620	54	1450	1260	58	1560	1150
	1/2	7.5	325	590	45	1250	1520	51	1370	1400
	3/4	6.0	300	545	39	1140	1800	43	1260	1680
	1	6.0	285	520	35	1060	1940	39	1170	1830
	1 1/2	5.0	265	480	33	1030	2110	37	1150	1990
0.60	3/8	7.5	340	656	54	1490	1260	58	1600	1150
	1/2	7.5	325	640	46	1290	1520	50	1410	1400
	3/4	6.0	300	500	40	1180	1800	44	1300	1680
	1	6.0	285	475	36	1100	1940	40	1210	1830
	1 1/2	5.0	265	440	33	1060	2110	37	1180	1990
0.65	3/8	7.5	340	525	55	1530	1260	59	1640	1150
	1/2	7.5	325	500	47	1330	1520	51	1450	1400
	3/4	6.0	300	460	40	1210	1800	44	1330	1680
	1	6.0	285	440	37	1130	1940	40	1240	1830
	1 1/2	5.0	265	110	34	1090	2110	38	1210	1990
0.70	3/8	7.5	340	485	55	1560	1260	59	1670	1150
	1/2	7.5	325	465	47	1360	1520	51	1480	1400
	3/4	6.0	300	430	41	1240	1800	45	1360	1680
	1	6.0	285	405	37	1160	1940	41	1270	1830
	1 1/2	5.0	265	380	34	1110	2110	38	1230	1990

See references bottom of chart 'Suggested Mixes for Non-Air-entrained Concrete of Medium Consistancy (3" to 4" slump)'

484

CONCRETE

Water Cement Ratio Lb/Lb	Max. size of Aggregate in Inch	Air Content Percent (entrapped)	Water, Lb/CY concrete	Cement, Lb/Cu.Yd. of concrete	With Fine Sand fineness modulus = 2.50**			With course Sand fineness modulus = 2.50**		
					Fine aggregate % of total aggregate	Fine aggregate Lb/Cu.Yd. concrete	Course aggregate Lb/Cu.Yd. concrete	Fine aggregate % of total aggregate	Fine aggregate Lb/Cu.Yd. concrete	Course aggregate Lb/Cu.Yd. concrete
0.40	3/8	3.0	385	965	50	1240	1260	54	1350	1150
	1/2	2.5	365	915	42	1100	1520	47	1220	1400
	3/4	2.0	340	850	35	960	1800	39	1080	1680
	1	1.5	325	815	32	910	1940	36	1020	1830
	1 1/2	1.0	300	750	29	880	2110	33	1000	1990
0.45	3/8	3.0	385	855	51	1330	1260	56	1440	1150
	1/2	2.5	365	810	44	1180	1520	48	1300	1400
	3/4	2.0	340	755	37	1040	1800	41	1160	1680
	1	1.5	325	720	34	990	1940	38	1100	1830
	1 1/2	1.0	300	665	31	960	2110	35	1080	1990
0.50	3/8	3	385	770	53	1400	1260	57	1510	1150
	1/2	2.5	365	730	45	1250	1520	49	1370	1400
	3/4	2	340	680	38	1100	1800	42	1220	1680
	1	1.5	325	650	35	1050	1940	39	1160	1830
	1 1/2	1	300	600	32	1010	2110	36	1130	1990
0.55	3/8	3	385	700	54	1460	1260	58	1570	1150
	1/2	2.5	365	665	46	1310	1520	51	1430	1400
	3/4	2	340	620	39	1150	1800	43	1270	1680
	1	1.5	325	590	36	1100	1940	40	1210	1830
	1 1/2	1	300	545	33	1060	2110	37	1180	1990

Suggested Mixes for Non-Air-entrained Concrete of Medium Consistancy (3" to 4" slump)

Suggested Mixes for Non-Air-entrained Concrete of Medium Consistancy (3" to 4" slump) - Continued

Water Cement Ratio Lb/Lb	Max. size of Aggregate in Inch	Air Content Percent (entrapped)	Water, Lb/CY concrete	Cement, Lb/Cu.Yd. of concrete	With Fine Sand fineness modulus = 2.50**			With course Sand fineness modulus = 2.50**		
					Fine aggregate % of total aggregate	Fine aggregate Lb/Cu.Yd. concrete	Course aggregate Lb/Cu.Yd. concrete	Fine aggregate % of total aggregate	Fine aggregate Lb/Cu.Yd. concrete	Course aggregate Lb/Cu.Yd. concrete
0.60	3/8	3	385	640	55	1510	1260	58	1620	1150
	1/2	2.5	365	610	47	1350	1520	51	1470	1400
	3/4	2	340	565	40	1200	1800	44	1320	1680
	1	1.5	325	540	37	1140	1940	41	1250	1830
	1 1/2	1	300	500	34	1090	2110	38	1210	1990
0.65	3/8	3	385	590	55	1550	1260	59	1660	1150
	1/2	2.5	365	560	48	1390	1520	52	1510	1400
	3/4	2	340	525	41	1230	1800	45	1350	1680
	1	1.5	325	500	38	1180	1940	41	1290	1830
	1 1/2	1	300	460	35	1130	2110	39	1250	1990
0.70	3/8	3	385	550	56	1590	1260	60	1700	1150
	1/2	2.5	365	520	48	1430	1520	53	1550	1400
	3/4	2	340	485	41	1270	1800	45	1390	1680
	1	1.5	325	465	38	1210	1940	42	1320	1830
	1 1/2	1	300	430	35	1150	2110	39	1270	1990

*Increase or decrease water per Cu.Yd. by 3% for each increase or decrease of 1 inch in slump, then calculate quantities by absolute volume method. For manufactured fine aggregate, increase % of fine aggregate, by 3 and water by 15 lbs. per Cu.Yd. of concrete. For less workable concrete, as in pavements, decrease % of fine aggregate by 3 and add water at 8 lbs. per Cu.Yd. of concrete.

**For definition of fineness modulus, see Aggregates for Concrete, available from the Portland Cement Association (PCA).

CONCRETE

7. When gravel is purchased by the cu. yd. (cu.m), stipulate measurement at job or destination, if possible.
8. When gravel is purchased by weight, weight at destination should govern instead of weight at loading point, when gravel may contain much moisture.
9. When purchasing crushed stone, obtain the grading and specific gravity.
10. When purchasing crushed stone, determine the weight per cu. yd. (cu.m) also the moisture content.

An easy method of determining the moisture content of sand, gravel, or crushed stone, is to measure out 1 cu.ft. (cu.m) of each and weigh them. Then dry this same material and weigh it. The difference in weight is the moisture content. Example: If one c.f. of loose damp sand weighs 105 lbs. (one cu.m weighs 1690 kg) and after drying it weighs only 90 lbs. (1407 kg/m3) then the moisture content is 15 lbs. (283 kg) or 16.67%.

Quantities of Materials Required for Job-Mixed Concrete. For estimating purposes, material quantities for job mixed concrete given in the following table may be used as a guide. The quantities shown in the table will produce concrete of medium consistency or with a slump of about 3" to 4" (75-100 mm).

Labor Mixing and Placing Concrete

The labor cost of mixing and placing concrete will vary considerably according to the size of the job, the method used in mixing and placing the concrete, the proximity of the stock piles to the mixer, and the distance the concrete has to be transported after mixing.

On large jobs using conveyors and bins for feeding the aggregate to the mixer, the purchase or rental of this equipment, together with the labor cost for placing, should be estimated separately and not included in the cu. yd. price of placing the concrete. This plant and equipment cost may be the same whether the job contains 1,000 or 10,000 cu.yd. (764.6 or 7,646 cu.m.) of concrete, and unless it is kept as a separate item, the resulting labor costs will be misleading and valueless as estimating data for future jobs.

For ready-mixed concrete delivered to the job mixed and ready to place in the forms, it is often possible for the truck mixers to deposit the concrete directly into the trenches or wall forms without any handling on the job, other than spading, vibrating, spreading, and leveling the concrete.

However, on many jobs this is not possible. It is necessary to have a receiving hopper where the truck mixers dump the concrete, which is then hauled from receiving hopper to place of deposit either by wheelbarrows, hand, or power buggies. On large operations the concrete might be pumped from receiving hopper to place of deposit.

Another method that is often used on large operations is to use a crane, which picks up a batch of concrete in a bucket and places it directly over the footings, piers, or forms.

When one of the above methods is used, all mixer labor is eliminated, but it usually requires one or two workers at the receiving hopper or at the truck dump to assist the truck mixer, prevent spillage, cleanup, etc.

Most specifications prohibit the use of long spouts for chuting concrete from the mixer to the forms on account of the great danger of segregation. Cranes with concrete buckets, buggies or buckets are generally desirable. Self-propelled buggies are available and are used on many projects.

Weather conditions also influence the labor costs considerably because placing concrete in freezing weather usually necessitates heating the sand, gravel or stone and water, and protecting the concrete from freezing by the use of tarpaulins, straw, salamanders or other artificial means, all of which add to the overall expense.

With so many small mixers on the market very little concrete is hand mixed today, but occasionally a job presents itself where it is more economical to mix by hand.

Mixes for Small Jobs

For small jobs where time and personnel are not available to determine proportions in accordance with the recommended procedure, mixes in Table 1 will usually provide concrete that is strong and durable

Concrete Mixes for Small Projects						
Maximum size of Aggregate Inch	Mix Designation	Approximate Weights of Solid Ingredienets Per Cu.Ft. Concrete, Lb.				
			Sand*		Course Aggregate	
		Cement	Air-entrained concrete**	Concrete witout air	Gravel or Crushed stone	Iron Blast Furnace Slag
1/2	A	25	48	51	54	47
	B	25	46	49	56	49
	C	25	44	47	58	51
3/4	A	23	45	49	62	54
	B	23	43	47	64	56
	C	23	41	45	66	58
1	A	22	41	45	70	61
	B	22	39	43	72	63
	C	22	37	41	74	65
1 1/2	A	20	41	45	74	65
	B	20	39	43	77	67
	C	20	37	41	79	69
2	A	19	40	45	79	69
	B	19	38	43	81	71
	C	19	36	41	83	72

Concrete Mixes for Small Projects - Metric						
Maximum size of Aggregate mm	Mix Designation	Approximate Weights of Solid Ingredienets Per Cu.m. Concrete, Kg.				
		Sand*			Course Aggregate	
		Cement	Air-entrained concrete**	Concrete witout air	Gravel or Crushed stone	Iron Blast Furnace Slag
12.50	A	400	769	817	865	753
	B	400	737	785	897	785
	C	400	705	753	929	817
18.75	A	368	721	785	993	865
	B	368	689	753	1025	897
	C	368	657	721	1057	929
25.00	A	352	657	721	1121	977
	B	352	625	689	1153	1009
	C	352	593	657	1185	1041
37.50	A	320	657	721	1185	1041
	B	320	625	689	1233	1073
	C	320	593	657	1265	1105
50.00	A	304	641	721	1265	1105
	B	304	609	689	1297	1137
	C	304	577	657	1330	1153

enough, as long as too much water is not added at the mixer. These mixes have been predetermined in conformity with the recommended procedure by assuming conditions applicable to the average small job and for aggregate of medium specific gravity. Three mixes are given for each maximum size of coarse aggregate. For the selected size of coarse aggregate, Mix B is intended for initial use. If this mix proves to be oversanded, change to Mix C; if it is undersanded, change to Mix A. It should be noted that the mixes listed in the table are based on dry or surface-dry sand. If the sand is moist or wet, make the corrections in batch weight prescribed in the footnote.

The approximate cement content per cubic foot of concrete listed in the table will be helpful in estimating cement requirements for the job. These requirements are based on concrete that has just enough water in it to permit ready working into forms without objectionable segregation. Concrete should slide, not run, off a shovel.

Power Buggy. One method of transporting construction materials, such as wet concrete, earth, sand, gravel, cement, etc., is the power-operated buggy, which requires only steering on the part of the operator.

They are furnished in various sizes by different manufacturers and the cost varies according to size and capacity.

The power buggy is mounted on a power driven three-wheel chassis and is steered by means of a rear caster wheel. The maximum speed over the ground conforms to a fast walking pace either on the level or up to a 20% incline.

This machine has a capacity of 1,500 lbs. (680.4 kg); a bucket volume of 10 cu.ft. (0.28 cu.m) and will handle 8 to 9 cu.ft. (0.22-0.25 cu.m) of wet concrete but a larger volume of light, dry materials. This buggy weighs 560 lbs. (254.02 kg), is powered by a 7 Hp. gasoline engine with a fuel capacity of 2-3/4 gals. (10.4 liters), which will operate the machine for eight hours. The machine is 5'-5 1/2" (1.66 m) in length overall, 2'-7 1/2" (0.80 m) wide and is furnished with either single or dual wheels and has a turning radius of 34'-3/4" (10.38m).

Comparison of Hand and Power Buggy Operation. The smaller power buggy having a capacity of 8 to 9 cu.ft. (0.22-0.25 cu.m) of wet concrete will probably do the work of three workers pushing hand powered concrete buggies.

A large machine having a capacity of 10 to 13 cu.ft. (0.28-0.36 cu.m) of wet concrete will probably handle as much concrete as 3 to 4 workers using motor operated buggies, but it may be that heavier runways will be required to handle the larger loads of concrete.

Sizes, Weights and Capacities of Concrete Carts or Buggies				
Turning Radius Inches	Width Overall Inches	Length Overall Inches	Maximum Height Inches	Approx. Weight Lbs.
72	34	87	54	950
82	41	94 1/2	54	1,020
91	54	106 1/2	56	1,400
Turning Radius mm	Width Overall mm	Length Overall mm	Maximum Height mm	Approx. Weight Kg.
1829	864	2210	1372	430.91
2083	1041	2400	1372	462.66
2311	1372	2705	1422	635.03

The following is an approximate comparison of cost between the two methods, based on wheeling 100 cu. yds. (76.5 cu.m) of concrete from a 16-S mixer:

Hand Method	Quant.	Rate	Total
5 Laborers (hours)	40.00	$ 26.16	$1,046.40
Cost per cu.yd.			$ 10.46
per cu.m			$ 13.69

Power Buggy Method	Quant.	Rate	Total
Power buggy, fixed charge per day	1.00	$ 75.00	$ 75.00
Fuel, gals, and oil	5.00	$ 4.00	$ 20.00
Operator (hours)	8.00	$ 37.89	$ 303.12
Total cost per day			$ 398.12
Cost per cu.yd.			$ 3.98
per cu.m			$ 5.21

CONCRETE

On some jobs, it may be that 2 power buggies would be required to keep the mixer operating at capacity, and in such instances would be double that given above. In either case the saving is considerable.

Capacities of Concrete Mixers Based on A.G.C. Standards (in Sacks of Cement per Batch)

Size of mixer denotes number of cu. ft. (cu.m) of mixed concrete, plus 10% excess, when mixer is level.

Capacities of Concrete Mixers Based on AGC Standards (in sacks of Cement per Batch)					
Concrete	Sizes of Standard Mixers, in Cu.Ft.				
Proportions	3-1/2- S	6 - S	11 - S	16 - S	28 - S
1:1-1/2:3	1/2	1	3	4	8
1:1-1/2:3	1/2	1	2	3	7
1:2:3	1/2	1	2	3	7
1:2:3-1/2	1/2	1	2	3	7
1:2:4	1/2	1	2	3	6
1:2-1/2:4	1/2	1	2	3	6
1:3:5	1/2	1	1	2	5
1:3:6	1/2	1	1	2	4
Concrete	Sizes of Standard Mixers, in Cu.M.				
Proportions	3-1/2- S	6 - S	11 - S	16 - S	28 - S
1:1-1/2:3	0.014	0.028	0.085	0.113	0.227
1:1-1/2:3	0.014	0.028	0.057	0.085	0.198
1:2:3	0.014	0.028	0.057	0.085	0.198
1:2:3-1/2	0.014	0.028	0.057	0.085	0.198
1:2:4	0.014	0.028	0.057	0.085	0.170
1:2-1/2:4	0.014	0.028	0.057	0.085	0.170
1:3:5	0.014	0.028	0.028	0.057	0.142
1:3:6	0.014	0.028	0.028	0.057	0.113

Mixing Concrete in a One Sack (6-S) Mixer. Foundation concrete is usually mixed in the proportions 1:2:4, 1:3:5, or 1:3:6, which is a one sack mix using a 6-S mixer.

Often a 1:3:5 mix is used for foundation walls, where watertight concrete is not necessary. This requires 1 sack of cement, 3 cu.ft. (0.8 cu.m) of dry sand and 5 c.f. (0.14 cu.m) of gravel or crushed stone, per batch. Wheelbarrows ordinarily used for dry materials hold 3 to 3-1/2 cu.ft. (0.08-0.10 cu.m) loaded level full, so each batch will require 1 barrow of sand and 2 barrows of gravel or 9 cu.ft. (0.25 cu.m) of dry material.

Where possible to start mixing first thing in the morning and continue all day without delays, a mixer should average 30 batches, 6 to 7 cu.ft. (0.17-0.20 cu.m) of concrete each, per hr. or 240 batches, 54 to 62 cu.yd. (41-47

491

cu.m), per 8-hr day. Runs like this on foundation work are exceptional, because in addition to the actual mixing time, it will be necessary to allow for delays, shutdowns, etc., as maximum output cannot be figured for the entire job. Workers must be paid for time getting ready to start work in the morning and cleaning and washing wheelbarrows, buggies, or concrete mixer at noon and at day; cutting and removing old concrete drippings and spillage; building and removing runways; unloading cement; delays caused by non-delivery of materials, etc.

Based on 6 to 7 cu.yd. (0.17-0.20 cu.m) of concrete per hr., the following crew is required: 2 laborers loading and wheeling gravel or crushed stone; 2 laborers wheeling sand and cement; 1 laborer attending mixer; 3 or 4 laborers wheeling concrete from mixer to place of deposit; 2 or 3 laborers dumping barrows, spreading, leveling, and tamping concrete; 1 mixer engineer if required, and 1 mason foreman in charge of work if required.

Allowing for average conditions, including runways, cleaning and repairing mixer, engine, etc., the labor per cu.yd. (cu.m) of concrete should cost as follows:

Description	Hours	Rate	Total
Labor per cu.yd.	2.50	$ 26.16	$ 65.40
per cu.m			$ 85.53

Under extremely favorable conditions, this cost can be reduced 1/2 to 3/4-hr. per cu.yd. (0.65-1.0 hrs. per cu.m). Add for mason foreman and mixer engineer, if required by union rules.

If a power buggy is used for wheeling concrete from mixer to place of deposit, one power buggy will take away all the concrete a 6-S mixer can turn out. This will eliminate 2 to 3 laborers wheeling concrete and result in a saving of about 1/4-hr. labor time per cu.yd. (0.33 hrs. per cu.m) of concrete.

Mixing Concrete Using a 11-S Mixer. When using a 11-S mixer of 1 to 2 sack mix capacity, the costs will run about the same as given for 6-S or 16-S mixers, under the same conditions.

Mixing Concrete in a 16-S (3 Sack Batch) Mixer. A mixer of this size will turn out about 16 c.f. of mixed concrete at each batch, depending upon the proportions.

Opinions vary as to amount of mixing required. The time of mixing specified by architects and engineers usually varies between 1-1/2 and 2 minutes per batch. The time specified affects the labor costs to a considerable extent. The longer mixing time restricts the day's output.

If the mixer has a side-loader and is fed by wheelbarrows, with the concrete wheeled direct from the mixer to place of deposit in wheelbarrows or concrete buggies, the mixer should average one batch every 2 minutes or 30 batches, 13 to 15 cu.yd. (10-11.5 cu.m) per hr.

Using an average of 2 minutes per batch, the mixer should turn out 104 to 120 cu.yd. (80-92 cu.m) per 8-hr. day, depending on the concrete proportions and the size of each batch. In addition to the actual mixing time,

492

it will be necessary to allow for delays, shutdowns, etc., as maximum output cannot be figured for the entire job. The workers must be paid for time getting ready to start work in the morning, cleaning and washing wheelbarrows and concrete mixer at noon and at night; cutting and removing old concrete drippings and spillage; building and removing runways; delays caused by non-delivery of materials, etc.

Based on an average of 14 cu.yd. (10.75 cu.m) of concrete per hr., it will require a crew of 25 to 28, distributed as follows: 6 laborers loading and wheeling gravel or crushed stone; 3 laborers wheeling sand; 1 or 2 laborers on cement; 1 laborer feeding mixer; 1 laborer dumping concrete from mixer into wheelbarrows or buggies; 5 to 7 laborers wheeling concrete from mixer to place of deposit, depending upon length of haul; 3 to 4 laborers dumping concrete carts or barrows, spreading, spading, and tamping concrete in place; 3 to 4 laborers for general labor, such as building and removing runways, cleaning concrete mixer, wheelbarrows or buggies, picking up or baling cement sacks and other miscellaneous labor; 1 engineer on mixer and 1 mason or concrete foreman in charge.

On the above basis, the labor per cu.yd. (0.765 cu.m) of concrete should cost as follows:

Description	Hours	Rate		Total	
Labor	2.00	$	26.16	$	52.32
Foreman	0.25	$	30.16	$	7.54
Mixer engineer	0.25	$	37.89	$	9.47
Cost per cu.yd.				$	69.33
Cost per cu.m.				$	90.68

In many instances concrete will be placed for 65% to 75% of the above costs, but this will merely represent a day's run and not the average for the entire job.

Using a power buggy to transport the concrete from the mixer or hopper to place of deposit, one or two power buggies should handle all the concrete a 16-S mixer can turn out. This will eliminate 3 to 4 laborers wheeling concrete and result in a saving of 1/4 to 1/3 hr. per cu.yd. (0.33 to 0.50 hr. per cu.m) of concrete.

Mixing Concrete in a 28-S (1 Cu.Yd.) Mixer. If the concrete is mixed in a 28-S mixer fed by wheelbarrows and wheeled to the place of deposit, the labor cost per cu.yd. (cu.m) will run about the same as for 16-S mixers under similar conditions.

While it is true the daily output will be greatly increased, the labor cost feeding the mixer and distributing the concrete will be increased in proportion.

The times in the table below include all lost time in connection with concrete foundations. Add for labor erecting and dismantling hoisting towers, spouts, hoisting engines, and mixers, as this may vary from $750.00 to $7,500.00, depending on the size of job.

Labor Mixing and Placing One Cu.Yd. (Cu.M) of Concrete for Foundations

Description	Hours per Cu.Yd.				Hours per Cu.M.			
Method of Mixing and Placing	Common	Mixer	Hoist.	Labor	Common	Mixer	Hoist.	Labor
	Labor	Engineer	Engineer	Foreman	Labor	Engineer	Engineer	Foreman
Hand mixing using one mixing board *	3.330	-----	-----	0.330	4.355	-----	-----	0.432
Hand mixing using two mixing boards **	3.000	-----	-----	0.250	3.924	-----	-----	0.327
Using one sack (6-S) mixer and Wheeling to place ***	2.500	-----	-----	0.170	3.270	-----	-----	0.222
Using two sack (11-S) mixer and Wheeling to place	2.000	0.125	-----	0.125	2.616	0.163	-----	0.163
Using 16-S mixer and wheeling to place	2.000	0.125	-----	0.125	2.616	0.163	-----	0.163
Using 28-S mixer and wheeling to place	2.000	0.083	-----	0.080	2.616	0.109	-----	0.105
Using conveyors, batch plant, hoist and wheeling to place	1.250	0.060	0.060	0.125	1.635	0.078	0.078	0.163

*In extreme warm weather or southern states, add 1/2 to 2/3 hr.

**If necessary to load concrete into barrows and wheel from mixing board to place of deposit, add 1/2 to 3/4 hr./c.y. (0.65-1.0 hr./m3).

***Under extremely favorable conditions, this may be reduced 1/4 to 1/2 hr.

If power buggies are used for wheeling concrete from mixer or receiving hopper to place of deposit, it usually results in a saving of 1/4 to 1/3 hrs. per cu.yd. (0.33-0.42 hrs. per cu.m) of concrete.

Loading Mixer Using a Front-End Loader

Since labor costs are high, many new methods are being used to reduce costs. One method employs a loader to load the sand and gravel from the stock piles and dump it into the loading skip at the mixer. This eliminates the workers loading wheelbarrows with sand and gravel by hand, but there is usually one worker necessary on the stock pile, plus the small loader and operator. When supplying a 6-S mixer, it would eliminate about 5 workers loading and wheeling sand and gravel but would add one worker for the stock piles, plus the loader and operator. When supplying a 16-S mixer with a loader, it would eliminate about 9 workers loading and wheeling sand and gravel but would add 1 worker on the sand stock pile, 1 worker on the gravel stock pile, the loader, and the operator.

When supplying a 11-S mixer in this manner, it would eliminate 6 to 7 workers loading and wheeling sand and gravel but would add 1 worker on the stock piles, the loader, and the operator.

Mixing and Placing Concrete Where Materials Are Delivered to Job Batch Proportioned. Another method of handling concrete work is to have the concrete aggregates delivered to the job in trucks, correctly proportioned for one batch of concrete, including cement, sand, and gravel or crushed stone.

When this method is used, the concrete aggregates for each batch are proportioned in the material yard and delivered to the job ready to mix. The truck backs up to the mixer and discharges the cement, sand, and gravel or crushed stone into the loading hopper. A paving mixer with a large, flat loading hopper is commonly used for this purpose, because it facilitates dumping the aggregates from the trucks into the loading hopper.

This method eliminates the need for workers loading and wheeling cement, sand, and coarse aggregates and the time required per cu.yd. (cu.m) of concrete should average as given in the following table.

If power buggies are used for wheeling concrete from mixer to place of deposit, it will usually result in a saving of 1/4 to 1/3 hr. labor time per cu.yd. (0.33-0.42 hrs. per cu.m) of concrete.

The above includes all lost time in connection with mixing and placing concrete.

Add for labor erecting and dismantling hoisting towers, hoisting engines, mixers, etc. This might vary from $750.00 to $7,500.00, depending on the size of job.

Placing Concrete with a Crane and Bucket. The crane and bucket method eliminates the need for building and removing runways, labor wheeling concrete, etc., especially when used for ready-mixed concrete. Labor is reduced to a minimum.

	Labor Mixing and Placing 1 CY (0.76 Cu.M.) of Concrete for Foundations When Aggregates are Proportioned and Delivered to the Job a Batch to a Load							
	Hours per Cu.Yd.				Hours per Cu.M.			
Method of Mixing	Common Labor	Mixer Engineer	Hoist. Engineer	Labor Foreman	Common Labor	Mixer Engineer	Hoist. Engineer	Labor Foreman
16-S mixer and wheeling to place	1.250	0.125	-----	0.125	1.635	0.163	-----	0.163
28-S mixer and wheeling to place	1.250	0.083	-----	0.083	1.635	0.109	-----	0.109

	Labor Placing 1 CY (0.76 Cu.M.) of Foundation Concrete Using Crane and Bucket									
	Hrs. per Cu.Yd.					Hrs. per Cu.M				
Method of Mixing	Qty Placed per 8-Hr day	Common Labor	Mixer Engineer	Hoist. Engineer	Labor Foreman	Qty Placed per 8-Hr day	Common Labor	Mixer Engineer	Hoist. Engineer	Labor Foreman
16-S mixer, batched material	70-90	0.70	0.10	0.10	0.10	54-69	0.916	0.131	0.131	0.131
28-S mixer, batched material	150-170	0.50	0.50	0.50	0.50	115-130	0.654	0.654	0.654	0.654
Ready-mix	150-170	0.40	0.05	-----	0.05	115-130	0.523	0.065	-----	0.065

Add for equipment cost or rental of crane and bucket.

*Add only if required.

CONCRETE

Cold Weather Concreting

During the late fall, winter, and early spring, it is necessary to take special precautions to assure proper hardening of the concrete. Careful contractors will use protection whenever the temperature falls below 50°F (10°C). At such time the most important protection required is to heat the sand and mixing water.

During freezing weather an extra having a rating of 50 to 150 Hp according to the size of the job, 10,000 to 25,000 sq.ft. (929-2322.5 sq.m) floor area, with a steam gauge set for a pressure of 2 lbs./sq. in. (0.14/cm2) will be needed with at least 2 perforated pipes for heating the coarse aggregate in addition to that required for heating the sand. On larger jobs or for fast concreting where aggregate bins are used, a grillage of steam pipes 4'-0" (1.2 m) apart at the bottom of the bins will be sufficient. These pipes should have 1/4" (6.25 mm) perforations about 4" (100 mm) on centers. A 1-1/2" (37.5 mm) steam line should be run from the boiler to the water tank and another line run up the side of the building with valves and connections for steam hose to be used for cleaning out ice and snow.

Sufficient steam should be available so the concrete has a temperature of 50 to 70°F (10-21.1°C) when leaving the mixer. This may promote the early setting of the cement and permit earlier removal of the forms, if approved by the engineer of record.

On small jobs the flame of a gas heater is allowed to project into the mixer to heat the materials.

Footings are generally protected by covering the fresh concrete with 8" to 12" (200-300 mm) of hay or straw and with tarpaulins, polyethylene thrown over the hay. The use of winter blankets are recommended and used more and more in place of hay. The economy of the number of re-uses of blankets may be less then salt hay together with its disposal costs.

Foundation walls should be protected in mild weather by covering the top of the wall with hay or winter blankets and the sides of the forms with tarpaulins or winter blankets. During more severe weather, place gas burning salamanders every 30 to 40 ft. (9-12 m) apart and cover with tarpaulins or winter blankets, care needs to be taken to keep the tarpaulins sufficient distance from the salamanders. Some projects will require a "fire watch" to assure the heat is kept and no accidental fire starts.

In skeleton construction it will be necessary to curtain the walls with tarpaulins blankets for the story being concreted, and in very severe weather it may be necessary to curtain two stories with tarpaulins or blankets and to cover the top floor blankets, held down by planks or timbers or weights, which in turn support a layer of boards and planks with tarpaulins or blankets thrown over them. Stair and elevator openings, and openings for steam and plumbing pipes, allow heat to penetrate beneath the protecting layers and to heat the top of the floor.

It will require one salamander for each 200 to 250 s.f. (18-23 sq.m) of floor surface, depending on the severity of the weather. Sufficient fuel should be placed on the floor in the early afternoon to last until next morning. Heat should be supplied for 2 to 5 days or more, depending on whether high-early-strength or normal portland cement is used and on the temperature maintained. The proper fire protection equipment should be available on each floor.

Ready Mixed Concrete

When using ready-mixed concrete for all classes of reinforced construction-floors, beams, girders, columns, etc.-the cost of handling and placing the concrete will run approximately the same for small project as for job-mixed concrete, except for the labor handling materials, supplying and operating mixer, etc., as the trucks dump the ready-mixed concrete into a receiving hopper or bucket, where it is distributed by wheelbarrows, concrete buggies, pumping or hoisting into place.

In addition to saving labor time of about 3/4-hr. per cu.yd. (1 hr. per cu.m) of concrete for charging mixer, the use of ready-mixed concrete saves the cost of cement sheds and provides more room for storing other materials. It permits higher output per hour, because ready-mixed concrete can usually be furnished in any quantity desired. It also saves the cost of installing and dismantling expensive equipment, which may easily cost $15,000 or more on a job of any magnitude. It is necessary to have a fairly large size hopper on the job to receive the concrete from the trucks. This hopper must be of sufficient capacity to supply the hoisting bucket as fast as it can handle the concrete. Otherwise, the work will be slowed down. With the crew idle part of the time, the labor costs go up.

The labor cost of handling and placing ready-mixed concrete will vary with the amount handled and placed per day, the method of handling, hoisting, pumping and placing the concrete, and the class of work. The following tables give the labor hours necessary to handle and place 1 cu.yd. (cu.m) of ready-mixed concrete, using various methods:

PLACING CONCRETE BY PUMPING

The development and perfection of pumps for transporting freshly mixed concrete through pipelines has revolutionized the placing of substantial yardages of concrete in dams, tunnels, retaining walls, bridge decks, mill, factory, warehouse, and apartment buildings, where concrete is required at rates varying from 15 to 150 cu.yd. (11.5-61 cu.m) per hr. Pumps force concrete through a pipeline in much the same manner that a piston pump pumps water. Concrete pumps will transport freshly mixed workable concrete to distances of2500 ft. 762 m) and to heights in excess of 300 plus ft. (91.4 m) directly to the point of placement. Concrete pumps can deliver concrete to forms 2.500 ft (762 m) away or 300 ft. (91.4 m) high, when a well designed concrete pump mix is used.

498

Labor Placing One Cu.Yd.. (Cu.M) of Ready-Mixed Concrete for Concrete Foundations							
Method and Quantity Placed per 8-Hr Day	Cubic Yard			Qty in Metric	Cubic Meter		
	Labor Hrs Cu.Yd.	Hoisting Eng.-Hrs.	Foreman Hours		Labor Hours Cu.M.	Hoisting Eng.-Hrs.	Foreman Hours
Placing concrete directly in forms, 1 truck, 50 – 70 cu.yd.	.35 - .50	-----	0.13	1 truck, 38.2 – 53.5 cu.m.	.46 - .65	-----	0.17
Placing concrete directly in forms 2 trucks, 100 – 150 cu.yd.	.27 - .40	-----	0.08	2 trucks, 76.5-114.7 cu.m.	.35 - .52	-----	0.17
Wheelbarrows, 50-70 cu.yd.	1.25-1.50	.17*	0.17 **	Wheelbarrows, 38.2 - 53.5cu.m	1.63-1.96	0.22 *	0.22 **
Wheelbarrows, 100 cu.yd.	1.00-1.25	.13 *	0.13 **	Wheelbarrows, 76.46 cu.m	1.63-1.96	0.17 *	0.17 **
Concrete buggies, 50-70 cu.yd.	1.00-1.25	.17 *	0.22 **	Concrete buggies, 38.2 - 53.5cu.m	1.63-1.96	0.22 *	0.29 **
Concrete buggies, 100 cu.yd.	0.90-1.10	.13 *	0.17 **	Concrete buggies, 76.46 cu.m	1.20-1.40	0.17 *	0.22 **
13 cu.ft. power concrete	0.65-0.75	.13 *	0.17 **	37 cu.m	0.85- 0.98	0.17 *	0.22 **

*Where the concrete is hoisted and it is necessary to use a hoisting engineer, add as indicated in table.

**If union regulations require a mason or cement finisher foreman in charge of crew, add as indicated in table.

Labor Placing One Cu.Yd. (Cu.M.) of Ready-Mixed Concrete for

Method and Quantity Placed per 8-Hour Day	Cubic Yard			Qty in Metric	Cubic Meter		
	Labor Hrs Cu.Yd.	Hoisting Eng.-Hrs.	Foreman Hours		Labor Hours Cu.M.	Hoisting Eng.-Hrs.	Foreman Hours
Hoisting concrete in wheelbarrows**	2.3	.11*	.11*	....	3.08	0.14*	0.14*
Hoisting concrete in concrete buggies**	1.1	.08*	.08*	....	1.43	0.10*	0.10*
Thin floors.*** Hoisting in wheelbarrow**	2.8	.11*	.11*	....	3.66	0.14*	0.14*
Thin floors.*** Hoisting in concrete buggies**	1.6	.08*	.08*	....	2.09	0.10*	0.10*
Hoisting concrete in concrete bucket, discharging into floor hopper & wheeling into place using wheelbarrows	1.2	.05*	.05*	....	1.57	0.06*	0.06*
Hoisting concrete in concrete bucket, discharging into floor hopper & wheeling into place using concrete buggies	0.8	.04*	.04*	....	1.04	0.05*	0.05*
If power buggies are used for transporting from hopper to place of deposit, deduct per CY	0.25	0.33*	......	Deduct per Cu.M.	0.33	0.43	......
Crane with bucket on ground, hoisting directly to floor or forms & depositing directly into	0.5	......	0.04	....	0.65	......	0.05
Add for rental crane, operator, oiler and comms.	......	.04*	......	....	......	0.05*	....
Add to the above costs***	0.5	......	......	....	0.65	....	....

*Time of hoisting and crane engineer and foreman will vary according to the number of CY of concrete placed per day.

**Assumes using material hoist and wheeling into place.

***Assume for thin concrete floors 2 to 2 1/2" thick (50 - 62.5 mm)

Concrete pumps can handle efficiently concrete of any slump from 1/2" to 7" (12.5-175 mm), but the most dependable slump for general conditions is around 3" to 5" (75-125 mm). However, plastic concrete of about 6" (150 mm) slump can generally be pumped the maximum distance and height. The rate of pumping depends much on the consistency of the concrete.

During the pumping operation, the pipeline is full at all times, which keeps the concrete from segregating, thereby eliminating one of the major objections found in some methods used to transport concrete to the forms. The troweling action of the walls of the pipeline in contact with the concrete improves the quality of the concrete.

The smooth, even flow of concrete into the forms reduces the shock loading on the bracing and ties. Danger of honeycomb is reduced, because there is no "slug" of pebbles dumped in, as in the case of a buggy, at the end of a batch that is not be properly spaded. Walls 120 ft. (3657 m) long and 15 ft. (4.57 m) high have been made at one pour in 3 ft. (.91 m) lifts without signs of longitudinal jointing or uneven surfaces due to the dumping of wheelbarrows or buggies in such a way as to splash the form above the point of placement.

Runways for wheelbarrows or concrete buggies are not required when placing concrete by pump, although it is necessary to provide supports for the pipelines, which usually consists of "horses" or "bridges".

Equipment maintenance varies with the efficiency of the operator, the type of work, abrasiveness of materials, etc., as with all construction machinery.

In addition to replacements for the machine, there will be pipeline maintenance, depending on the abrasiveness of materials. The couplings, which represent approximately three-fourths of the cost of pipe sections, will last indefinitely. The straight pipe will last from 50,000 to 200,000 cu.yd. (38,230 to 152,920 cu.m) and the elbows will last from 30,000 to 100,000 cu.yd. (22,938 to 76,460 cu.m), depending upon the abrasiveness of the concrete. The piping can be replaced as it becomes worn out and the old couplings reused.

Cost of Placing Concrete by Pipeline. The cost of placing concrete by pipeline varies widely as do other methods of distribution, and is governed by the size of the job, area covered, the yardage of concrete that can be placed at one setup of the pipelines, distance from the pump, etc.

When placing concrete by ordinary methods, the cost of rental, erecting, and dismantling the hoisting tower and spouts must be considered, along with the cost of raising the concrete hopper and moving spouts for placing the concrete.

The same conditions must be considered with concrete pump distribution, which includes original cost of equipment, depreciation or salvage value, cost of setting up and removing pumping equipment, pipelines, etc.

Always bear in mind, the cost of mixing the concrete will be the same, regardless of the method of distribution, because where the concrete is mixed and dumped into a hoisting bucket using ordinary methods, it is deposited in the hopper. It is only the method of handling, hoisting, and placing the mixed concrete that should be considered. This also applies where ready-mixed concrete is used.

The crew operating the pumping unit will vary with the size, area, and yardage in the job. An operator is required on the concrete pump. A foreman and 6 to 12 laborers should place and remove pipelines and handle, spread, and level the concrete, and the quantity will vary from 15 to125 cu.yd. (11.46-95.58 cu.m) per hr., depending on the size of the pumping equipment and the class of work.

Placing Concrete on a Commercial Building. Assume a 5-story steel frame commercial building, 140' x 250' (42.67 x 76.20 m) having floors of metal pan construction, with pans from 6" to 14" (150-350 mm) deep and 2-1/2" (62.50 mm) of concrete over the pans, the entire job containing approximately 5,200 cu.yd. (3975.92 cu.m) of concrete.

Assume the concrete is ready-mixed, using 3/4" (18.75 mm) gravel aggregate and having a 5" (125 mm) slump.

The concrete is delivered to the job and deposited directly into the remixing hopper on the pumper and pumped to the various floors to the point of discharge. The labor per cu.yd. (cu.m) cost as follows:

Description	Hour	Rate	Total
1 Labor Foreman	0.08	$ 30.16	$ 2.41
1 Hopper attendant	0.08	$ 26.16	$ 2.09
1 Labor vibrating,	0.08	$ 28.16	$ 2.25
7 laborers*	0.56	$ 26.16	$ 14.65
Pipes (boom)	0.08	$ 80.00	$ 6.40
Concrete Pump	0.08	$ 560.00	$ 44.80
Cost per cu.yd.			$ 72.61
per cu.m			$ 94.96

Labor on pumper includes time placing and removing pipeline.

Average rate of pour 15 cu.yd. (11.46 cu.m) per hr. Maximum rate of pour 100 cu.yd. (22.94 cu.m) per hr. The above costs do not include the cost of the pumper, depreciation, parts and replacements, gas and oil, labor installing and removing equipment, etc.

Placing Concrete by Pipeline on a Sewage Disposal Plant. Assume a sewage disposal plant, size 370' x 380' (112.77 x 115.82 m), containing 15,000 c.y. (11,469 cu.m) of concrete consisting of heavily reinforced concrete in slabs and walls, all concrete to be placed by pipeline.

The mixer will discharge into the remixing hopper of a single pumper, which is placed below grade in the excavation made for the plant addition. From this point the concrete is to be pumped through pipelines to the point of placement.

502

CONCRETE

The floors of the aerating basins, 30 ft. (9.14 m) wide and 120 ft. (3657 m) long are poured by radial spouting from the end of the pipeline, which is supported on horses. The piping is taken up as the concreting progresses and is laid in the next basin.

Staging supported on the forms will permit erection of the pipeline, pouring 112 cu.yd. (85.63 cu.m) of concrete in the wall and removal of the pipe line in one day. The walls of the aerating and settling basins are 15 ft. (4.5 m) high and are poured in 3 ft. (0.91 m) lifts. Valves at 20 ft. (6.09 m) intervals are inserted in the 120 ft. (36.57 m) of distributing line which permits the withdrawal of concrete at these points. Elephant trunks (tremie) direct the concrete into specially constructed chutes leading the concrete down through the maze of steel to the level being poured where it is discharged without spattering the forms above that point. Pouring was started from the valve nearest the pumper. When enough concrete has been withdrawn at a valve, it is closed and the next valve opened for concreting at this point. This procedure is continued until the lift is completed at the end of the pipeline.

With the aid of internal vibration, these high long walls are poured relatively free of blemishes or jointing.

The average labor cost of placing 1 c.y. (0.7646cu.m) of concrete is as follows:

Description	Hour	Rate	Total
1 Labor Foreman	0.08	$ 30.16	$ 2.41
1 Hopper attendant	0.08	$ 26.16	$ 2.09
1 Labor vibrating,	0.08	$ 28.16	$ 2.25
8 laborers*	0.64	$ 26.16	$ 16.74
Pipes (boom)	0.08	$ 80.00	$ 6.40
Concrete Pump	0.08	$ 560.00	$ 44.80
Cost per cu.yd.			$ 74.70
per cu.m			$ 97.70

During the pouring of the concrete in the walls, 2 additional workers are required to operate vibrators. Vibrators would be required regardless of the distribution method used.

CAST IN PLACE CONCRETE

Concrete floors, sidewalks, and lightweight concrete floor fill are estimated by the square foot (square meter), taking the number of sq.ft. (sq.m) of any given thickness - 2", 3", 4", etc. (50, 75, 100 mm, etc.) - to obtain the total cubic yards (cubic meters) for placement.

The quantities are easily computed by obtaining the area of the various spaces to have concrete floors or walks and stating the number of sq.ff. (sq.m) for each thickness.

When preparing an estimate for concrete floors or sidewalks placed on the ground, there are distinct material and labor items to be considered as follows:

1. Labor grading and removing surplus earth under floors or walks or filling where present grade is too low.
2. Cost of sand, gravel, or slag to be placed under concrete floor; labor spreading and tamping same in place.
3. Labor placing wood or metal screeds and forms.
4. Cost of vapor barrier and reinforcing material and placement.
5. Labor mixing and placing concrete.
6. Labor troweling finish surface and blocking off into squares, if required.
7. Labor and materials for curing and hardeners, if required.

Grading for Concrete Floors and Walks. On practically all cement floors and walks placed directly on the ground, it is necessary to do a certain amount of grading before the bed of sand, gravel, or slag can be placed.

If it is not necessary to remove the excavated soil from the premises, a bulldozer or loader is probably the most economical method to use.

A medium size loader and operator should excavate and place in piles, 25 to 30 c.y. (19-23 cu.m) per hour at the following cost per cu.yd. (cu.m):

Description	Hours	Rate	Total
Loader per cu.yd.	1.00	$ 85.00	$ 85.00
Operator per cu.yd.	1.00	$ 37.89	$ 37.89
Cost per Cu.Yd.			$ 122.89
Cost per cu.m			$ 160.72

This cost will vary with the kind of soil encountered, i.e., sand or loam, ordinary black soil, heavy soil, or clay.

Number of Sq.Ft. of Soil of Various Thicknesses Obtainable From One Cu.Yd. (Cu.M)										
Depth in Inches	3	4	5	6	7	8	9	10	11	12
Areas in Sq.Ft.	108	82	64	54	47	40	36	33	29	12
Depth in mm	76.2	101.6	127.0	152.4	177.8	203.2	228.6	254.0	279.4	304.8
Areas in Sq.m	10.03	7.62	5.95	5.02	4.37	3.72	3.34	3.07	2.69	1.11

To obtain the cost of grading 1 sq.ft. of any thickness, divide the cu.yd. cost by the number of sq.ft. obtainable from 1 cu.yd. and the result is the sq.ft. cost for grading. For example:

CONCRETE

1. With excavating at $4.50 per cu.yd., find the sq.ft. cost of removing 5" (125 mm) of top soil. One cu.yd. contains 65 sq.ft. of 5" fill. $4.50 ÷ 65 = $0.069 per sq.ft. for grading.
2. With excavating at $5.88 per cu.m, find the sq.m cost of removing 0.13 m of top soil. One cu.m contains 7.7 sq.m of 0.13 m fill. $5.88 ÷ 7.7 = $0.76 per sq.m for grading.

Hand Grading. If the top soil is loaded by hand into wheelbarrows or power buggies, a worker should loosen, shovel and load 5 to 6 c.y. (3.82-4.58 cu.m) per 8-hr. day, at the following labor cost per cu.yd. (cu.m):

Description	Hours	Rate	Total
Loader per cu.yd.	1.45	$ 26.16	$ 37.93
Cost per cu.m			$ 49.61

Grading and Tamping Sand, Gravel, or Slag Fill Under Concrete Floors or Walks. The cost of this work varies with the thickness of the fill and job conditions.

It costs as much to grade a bed of slag or gravel 3" (75 mm) thick as one 9" (0.23 m, 225 mm) thick, because it is only the surface that is graded and the only additional labor on the thicker fill is for handling, tamping, and spreading a larger quantity of material.

Where the fill varies from 3" to 4" (75-100 mm) thick, a worker will spread, grade, and tamp about 6 cu.yd. (4.58 cu.m) per 8-hr. day, at the following labor cost per c.y. (cu.m):

Description	Hours	Rate	Total
Labor per cu.yd.	1.33	$ 26.16	$ 34.79
per cu.m			$ 45.50

If the fill varies from 5" to 6" (125-150 mm) thick, a worker will spread, grade and tamp about 8-1/2 cu.yd. (6.499 cu.m) per 8-hr. day, at the following labor cost per cu.yd. (cu.m):

Description	Hours	Rate	Total
Labor per cu.yd.	0.94	$ 26.16	$ 24.59
per cu.m			$ 32.16

If the fill varies from 7" to 9" (175-225 mm) thick, a worker will spread, grade, and tamp about 10-1/2 cu.yd. (8.02 cu.m) per 8-hr. day, at the following labor cost per cu.yd. (cu.m):

Description	Hours	Rate	Total
Labor per cu.yd.	0.76	$ 26.16	$ 19.88
per cu.m			$ 26.00

A worker will spread, grade, and tamp about 12-1/2 cu.yd. (9.56 cu.m) of 10" or 12" (250-300 mm) fill per 8-hr. day, at the following labor cost per cu.yd. (cu.m):

Description	Hours	Rate	Total
Labor per cu.yd.	0.64	$ 26.16	$ 16.74
per cu.m			$ 21.90

All fill materials will shrink under compaction, some more than others, and the additional material required to make up for shrinkage, together with the additional labor for handling, must be provided for in the estimate.

The best and easiest way to do this is to increase the quantity before pricing by applying the proper shrinkage factor for the material to the net computed volume.

For example: If an area 100'-0" x 50'-0" (30 x 15 m) is to receive a gravel fill 6" (0.15 m) thick, after tamping the gross volume of gravel would be 100'-0" x 50'-0" x 0'-6" x 1.12 = 2,800 c.f. or approximately 104 cu.yd. (30 x 15 x 0.15 = 75.6 cu.m). In the computation, 1.12 is the shrinkage factor.

The following table gives approximate values for shrinkage percentages and shrinkage factors for various fill materials:

Material	Shrinkage Percentage	Shrinkage Factor
Cinder	30	1.3
Crushed Limestone	20	1.2
Granulated Slag	15	1.15
Gravel	12	1.12
Sand	8	1.08

To obtain the cost of 1 sq.ft. (sq.m) of fill of any thickness, divide the cost per cu.yd. (cu.m) by the number of s.f. (sq.m) obtainable from 1 cu.yd. (cu.m), multiply by the shrinkage factor and the result is the sq.ft. (sq.m) cost.

Forms or Screeds for Concrete Sidewalks. It is customary to use 2"x4" (50x100 mm) or 2"x6" (50x150 mm) lumber for forms at each side of a concrete sidewalk, the depth of the forms depending upon the thickness of the concrete. On ordinary sidewalk work, obtain the number of l.f. (m) of walk and multiply by 2, which will be the number of lin.ft. (m) of forms required. A sidewalk 4" (100 mm) thick will require 2"x4" (50 x 100 mm) lumber; 6" (150 mm) thick, 2"x6" (50 x 150 mm) lumber, etc.

Where concrete floors are placed on the ground it is customary to set wood screeds 6'-0" to 8'-0" (1.8-2.4 m) o.c., so the floor will be level or have a uniform pitch. These screeds are removed as the work progresses and may be used many times. They are usually of 2"x4" (50 x 100 mm) lumber, with stakes placed 3' to 4' (0.9-1.2 m) apart to hold the screeds in place.

506

CONCRETE

Labor Placing Forms and Screeds for Sidewalks and Floors. On all classes of sidewalk work where the forms are either 4", 6", or 8" (100, 150, or 200 mm) wide, a carpenter should place and level 250 to 300 lin.ft. (76.2-91.44 m) per 8-hr. day, at the following labor cost per 100 lin.ft. (30 m):

Description	Hours	Rate	Total
Carpenter	3.00	$ 35.04	$ 105.12
Helper	1.50	$ 26.16	$ 39.24
Cost per 100 lin.ft. (30.48m)			$ 144.36
per lin.ft. of form			$ 1.44
per lin.m. of form			$ 4.74
per lin.ft. of walk			$ 2.89
per lin. m of walk			$ 9.47

The above includes time removing forms.

When placing screeds for concrete floors on the ground, 2 carpenters working together should drive stakes, place and level 650 to 700 lin.ft. of screeds per 8-hr. day at the following labor cost per 100 l.f. (30.48 m):

Description	Hours	Rate	Total
Carpenter	3.00	$ 35.04	$ 105.12
Helper	1.50	$ 26.16	$ 39.24
Cost per 100 lin.ft. (30.48m)			$ 144.36
per lin.ft.			$ 1.44
per lin.m.			$ 4.74

Estimating the Quantity of Concrete Required for Floors and Walks. Compute the number of sq.ft. (sq.m) of concrete floor or sidewalk that may be obtained from one cu.yd. (one cu.m) of concrete, laid out into a cube 3'-0" wide, 3'-0" thick and 3'-0" high (1 x 1 x 1 m). The surface of this cube contains 9 s.f. and it is 36" high.

If the rough concrete is 4" (0.10 m) thick, divide 36" (1 m), the height of the cube, by 4" (0.10 m), the thickness of the slab, and the result is 9 (10), showing that it is possible to obtain 9 slabs, each containing 9 sq.ft. or 81 sq.ft. of 4" floor or walk per cu.yd. of concrete (or 10 slabs, each 1 sq.m, 0.1 m thick).

Materials Required for 1.0 Cu.Yd. (0.76 Cu.M.) of Concrete

This table is based on use of aaverage wet sand and on 3-to4 inchs (75-100 mm) slump. Quantities will vary acording to grading of aggregaate and workability desired. No allowance has been made for waste.
1 Sack cement = 1 Cu.Ft (.02832 Cu.M) / 4 sacks = 1 bbl. (.11328 Cu.M.)

Size of Aggregate, in Inch	Mixing Water* per sack cement, in US Gal	Mixes			Yield of Concrete per Sack of Cement, in Cu.Ft.	Water, in US Gal	Material per Cu.Yd. Concrete		
		Portland Cement, in Sacks	Sand, in Cu.Ft.	Gravel or Crushed Stone, in Cu.Ft.			Portland Cement, in Sacks	Aggrgates	
								Sand, in Cu.Yd.	Gravel or Crushed Stone, in Cu.Yd.
Maximum Size 3/4	5	1	2.50	2.75	4.3	31	6.25	0.58	0.64
Maximum Size 1	5	1	2.50	3.00	4.5	30	6.00	0.50	0.70
Maximum Size 1 1/2	5	1	2.50	3.50	4.7	29	5.75	0.48	0.75

Size of Aggregate, in mm	Mixing Water* per sack cement in Liters	Mixes			Yield of Concrete per Sack of Cement, in Cu.M.	Water, in Liter	Material per Cu.Yd. Concrete		
		Portland Cement, in Sacks	Sand, in Cu.M.	Gravel or Crushed Stone, in Cu.M.			Portland Cement, in Sacks	Aggrgates	
								Sand, in Cu.M.	Gravel or Crushed Stone, in Cu.M.
Maximum Size 18.75	18.93	1	0.07	0.08	0.12	117.34	6.25	0.44	0.49
Maximum Size 25	18.93	1	0.07	0.08	0.13	113.55	6.00	0.38	0.54
Maximum Size 37.50	18.93	1	0.07	0.10	0.13	109.77	5.75	0.37	0.57

*Based on 6 Gallons (22.71 liters) of Water per sack of Cement including water contained in average wet Sand

CONCRETE

Materials Required for 100 Sq.Ft. (9.29 Sq.M.) of Surface Area for Various Thickness of Concrete*

Thickness	Quantity of Concrete	1:2 1/2: 2 3/4			1:2 1/4: 3			1:2 1/4: 3 1/2		
		Cement	Sand	Aggegates Gravel or Crushed Stone	Cement	Sand	Aggegates Gravel or Crushed Stone	Cement	Sand	Aggegates Gravel or Crushed Stone
Inches	Cu.Yd.	Sacks	Cu.Yd.	Cu.Yd.	Sacks	Cu.Yd.	Cu.Yd.	Sacks	Cu.Yd.	Cu.Yd.
3	0.93	5.8	0.54	0.60	5.6	0.47	0.62	5.4	0.45	0.70
4	1.23	7.7	0.71	0.79	7.4	0.61	0.82	7.1	0.59	0.92
5	1.54	9.6	0.89	0.99	9.2	0.77	1.03	8.9	0.74	1.16
6	1.85	11.6	1.07	1.18	11.1	0.93	1.24	10.6	0.89	1.39
8	2.46	15.4	1.43	1.57	14.8	1.23	1.65	14.1	1.18	1.85
10	3.08	19.3	1.79	1.97	18.5	1.54	2.06	17.7	1.48	2.31
12	3.70	23.1	2.14	2.37	22.2	1.85	2.48	21.3	1.78	2.78
mm	Cu.M.	Sacks	Cu.M.	Cu.M.	Sacks	Cu.M.	Cu.M.	Sacks	Cu.M.	Cu.M.
75	0.71	5.8	0.41	0.46	5.6	0.36	0.47	5.4	0.34	0.54
100	0.94	7.7	0.54	0.60	7.4	0.47	0.63	7.1	0.45	0.70
125	1.18	9.6	0.68	0.76	9.2	0.59	0.79	8.9	0.57	0.89
150	1.41	11.6	0.82	0.90	11.1	0.71	0.95	10.6	0.68	1.06
200	1.88	15.4	1.09	1.20	14.8	0.94	1.26	14.1	0.90	1.41
250	2.35	19.3	1.37	1.51	18.5	1.18	1.58	17.7	1.13	1.77
300	2.83	23.1	1.64	1.81	22.2	1.41	1.90	21.3	1.36	2.13

* 100 Sq.Ft. (9.29 Sq.M.) of Portland cement plaster 1/2" (12.5mm) thick requires 1.6 sacks of cement and 0.15 Cu.Yd. (0.11 Cu.M.) of sand

No Allowance has been made for waste

509

Thickness	No. of	Thickness	No. of	Thickness	No. of	Thickness	No. of
Inches	Sq.Ft.	Inches	Sq.Ft.	mm	Sq.M.	mm	Sq.M.
1	324	7	30.1	25.40	30.10	177.80	2.80
1 1/4	259	7 1/4	24.1	31.75	24.06	184.15	2.24
1 1/2	216	7 1/2	20.1	38.10	20.07	190.50	1.87
1 3/4	185	7 3/4	17.2	44.45	17.19	196.85	1.60
2	162	8	15.1	50.80	15.05	203.20	1.40
2 1/4	144	8 1/4	13.4	57.15	13.38	209.55	1.24
2 1/2	130	8 1/2	12.1	63.50	12.08	215.90	1.12
2 3/4	118	8 3/4	11	69.85	10.96	222.25	1.02
3	108	9	10	76.20	10.03	228.60	0.93
3 1/4	100	9 1/4	9.3	82.55	9.29	234.95	0.86
3 1/2	93	9 1/2	8.6	88.90	8.64	241.30	0.80
3 3/4	86	9 3/4	8	95.25	7.99	247.65	0.74
4	81	10	7.5	101.60	7.52	254.00	0.70
4 1/4	76	10 1/4	7.1	107.95	7.06	260.35	0.66
4 1/2	72	10 1/2	6.7	114.30	6.69	266.70	0.62
4 3/4	68	10 3/4	6.3	120.65	6.32	273.05	0.59
5	65	11	6	127.00	6.04	279.40	0.56
5 1/4	62	11 1/4	5.8	133.35	5.76	285.75	0.54
5 1/2	59	11 1/2	5.5	139.70	5.48	292.10	0.51
5 3/4	56	11 3/4	5.2	146.05	5.20	298.45	0.48
6	54	12	5	152.40	5.02	304.80	0.46
6 1/4	52	12 1/4	4.8	158.75	4.83	311.15	0.45
6 1/2	50	12 1/2	4.6	165.10	4.65	317.50	0.43
6 3/4	48	12 3/4	4.5	171.45	4.46	323.85	0.42

Number of S.F. (Sq.M) of Concrete Floor of Any Thickness Obtainable from One Cu.Yd. (Cu.M) of Concrete

Mixing and Placing Concrete for Floors and Walks

Mixing Concrete By Hand. Concrete is seldom, if ever, mixed by hand. It is usually mixed in a small mixer, or ready-mix concrete is used, and discharged direct into the forms or onto the floor. However, for that one-in-a-million job, here is some data on hand mixing.

As with other classes of concrete work, the cost of mixing and placing the concrete depends on the number of times it is turned on the board and the method used in mixing and placing. But on the smaller jobs, it will require 3-1/2 to 4 hrs. labor time to mix and place one cu.yd. of concrete (4.57-5.23 hrs. per cu.m). This class of work does not proceed as fast as foundation work.

On work of this kind, the crew usually consists of 5 or 6 workers, so it is not possible for one individual to work at the same class of work all day but must alternate between wheeling sand, gravel, cement, and mixing concrete, as the job may require.

CONCRETE

Five workers together should handle all materials, mix, place, and tamp 20 to 24 batches-1/2-cu.yd. (0.38 cu.m)-or 10 to 12 cu.yd. (7.65-9.18 cu.m) per 8-hr. day, at the following labor cost per cu.yd. (cu.m):

Description	Hours	Rate	Total
Labor per cu.yd.	3.64	$ 26.16	$95.22
per cu.m			$124.54

Labor Mixing and Placing Concrete for Floors and Walks Using a Small Mixer. A small concrete mixer of 1/2- or 1-sack (3-1/2-S or 6-S) capacity is ordinarily used for floor and sidewalk work.

Where possible to use a full crew and keep the mixer in operation steadily, a mixer should discharge a batch every 2 minutes, but as in all other kinds of construction work there are always delays, such as engine adjustments, repairing mixer, delays caused by non-delivery of sand, gravel or cement, moving and placing runways, cleaning concrete barrows, etc., so the average for a job will be about 20 to 25 batches per hr. or 160 to 200 batches per 8-hr. day. On the above basis, either a 3-1/2-S or 6-S concrete mixer, the labor cost per c.y. (cu.m) should run as follows:

Description	Hours	Rate	Total
Labor per cu.yd.	2.5	$ 26.16	$65.40
per cu.m			$85.53

Using a larger mixer will increase the output but the labor cost will increase in direct proportion. Many contractors have found that a one sack mixer is the most economical unit for floor and sidewalk work.

Labor Placing Ready-Mixed Concrete for Floors. To place ready-mixed concrete for floors with the most economy, it is necessary to have a crew large enough to handle the concrete quickly to eliminate waiting time charges by the material company, and the supply of mixed concrete must be steady and of sufficient quantity to keep the crew busy.

A crew of 14 or 15 laborers, using buggies to wheel concrete, should place 22 to 27 cu.yd. (16.83-20.65 cu.m) of ready-mixed concrete per hr. at the following labor cost per c.y. (cu.m):

Description	Hours	Rate	Total
Labor per cu.yd.	0.60	$ 26.16	$ 15.70
per cu.m			$ 20.53

If the floor is small, such as for a residence, small addition, etc., the work must be organized in a different manner. A job of this sort is too small to keep a number of ready-mix trucks busy. One or two trucks will serve the purpose.

Assuming a floor pour serviced by one 7-cu.yd. (5.36-cu.m) capacity ready-mix truck, delivering one load of concrete per hr., a crew of 6 or 7 laborers should be kept reasonably busy and still handle the ready-mixed load

511

of concrete fast enough to prevent ready-mix concrete truck waiting time penalties.

Based on the above conditions the labor cost of placing one cu.yd. (cu.m) of floor concrete will be as follows:

Description	Hours	Rate	Total
Labor per cu.yd.	1.00	$ 26.16	$ 26.16
per cu.m			$ 34.21

Estimating the Labor Cost Per S.F. (Sq.M) of Floor or Sidewalk. Labor costs given for mixing and placing concrete are based on cubic yards (cu.m), but floors and sidewalks are usually estimated by the square foot (sq.m).

Once the cost per c.y. (cu.m) has been obtained, divide it by the number of sq.ft. (sq.m) of floor obtainable from a c.y. (cu.m) of concrete. The result is the labor cost per sq.ft. (sq.m) of floor. Example: Find the labor cost per sq.ft. of 4" (sq.m of 0.1 m) floor. The table on the previous page shows that 81 sq.ft. of 4" (7.29 sq.m of 0.1 m) floor can be obtained from one cu.yd. (cu.m) of concrete. If the labor cost for concrete placement is $21.04 per cu.yd. (27.35 per cu.m), $21.04 ÷ 81 = $0.26 per sq.ff. (27.35 ÷ 7.29 = $3.75 per sq.m).

Applying a Finish on Reinforced Concrete Floors. The cost of placing a finish on reinforced concrete floors is subject to wide fluctuations. One reason is the custom of placing the concrete over as large an area as possible and then allowing the concrete to stand until the water disappears from the surface before finishing is started. However, the time required for cement to take its initial set depends on the brand of cement used. Some cements are known for being quick setting or slow setting.

The amount of water used in the concrete mix also has considerable bearing on the finishing cost. If "sloppy" concrete is used, it requires much more time for the floor to set sufficiently to receive the finish than if concrete of a "stiff" consistency is used.

Architects and engineers specify the floor must be finished at the same time the concrete is placed, and in some instances it necessitates a large amount of overtime for finishers. When the crew places concrete right up until quitting time, it is a foregone conclusion that the finishers will have to work far into the night.

The item of overtime should be carefully considered by the estimator, because the cement mason often starts the finishing or floating work when all other trades have quit for the day. Provisions should be made and costs added, if required, for lights and for other trades necessary to support the finishers.

While under favorable conditions, some finishers may finish 175 to 200 sq.ft. (15.75-18 sq.m) of floor per hr. on certain classes of work, the general average for a reinforced concrete job will vary from 70 to 80 sq.ft. (6.3-7.2 sq.m) per hr. or 560 to 640 s.f. (50-58 sq.m) per 8-hr. day.

CONCRETE

The labor costs given are for straight time, but the estimator should bear in mind that overtime costs time and a half or double time. The labor per 100 sq.ft. (9 sq.m) of floor should cost as follows:

Description	Hours	Rate	Total
Cement Mason	1.30	$ 36.93	$ 48.01
Cost per sq.ft.			$ 0.48
per sq.m			$ 5.17

Finishing Concrete Floors by Machine. Finishing machines are used for screeding, floating, and troweling. When placing concrete floors, walks, and pavements on the ground, the wood or metal floor screeds are placed in the usual manner and the concrete is struck off by using a power operated screeding machine, which makes 2-1/2" (63 mm) transverse strokes on the header boards, leveling and compacting the mix. A steady pull forward by the operator advances the machine at the rate of 10 ft. (3 m) a minute or 600 lin.ft. (182.88 m) per hour and will strike off a slab 6 to 20 ft. (1.82-6.09 m) in width.

A four-worker crew is usually required with the machine, two laborers spreading the concrete and two cement masons operating the screeding machine. After the concrete slab has been screeded ready for troweling, allow sufficient time for the concrete to set hard enough to walk on before starting finishing.

For floating, the machine is equipped with 12-gauge steel trowels, which revolve on a 46" (1.15 m) diameter. It usually requires 3 cement masons-1 finisher to operate the machine and 2 finishers touching up corners, columns, edging, jointing, etc. On a well prepared slab, one finishing crew can float 4,000 sq.ft. (371.60 sq.m) of floor per hr., but this rate is not usually maintained.

A smaller finishing machine with a 34" (850 mm) diameter will float up to 3,000 s.f. (278.70 sq.m) of floor per hr. under the same conditions described above.

After the floor has been floated, the heavy-gauge steel floating trowels are removed and replaced with steel "finishing" trowels. The cement mason then guides the rotating, adjustable pitch trowels over the slab until a smooth, level surface is obtained.

Using the larger machines, and assuming a "four time over" was necessary to complete the job, it would require 4 hrs. of machine operation and assuming a 3/4 to 1 hour waiting period was necessary between each operation, it would require a total of 6-1/4 to 7 hrs. time to finish 4,000 s.f. (371.60 sq.m) of floor at the following labor cost:

Description	Hours	Rate	Total
Cement Mason	3.00	$ 36.93	$ 110.79
Cost per sq.ft.			$ 0.03
per sq.m			$ 0.30

Add machine rental charges.

After the floor has been screeded, the labor floating and troweling (4 times over), 4,000 s.f. (360 sq.m) of cement finish floor should cost as follows:

Description	Hours	Rate	Total
Cement Mason	32.00	$ 36.93	$1,181.76
Cost per sq.ft.			$0.30
per sq.m			$3.18

Add machine rental charges.

Color for Concrete Floors

Color for concrete is used for sidewalks, driveways, porches, tennis courts, floors, etc., the desired color being mixed into the cement topping which should be either 1/2", 3/4", or 1" (12.50,18.75, or 25 mm) thick, depending on service requirements. The amount required varies with the different manufacturers but the following proportions generally will produce good results.

Description	Inches of Topping			mm of Topping		
	1/2	3/4	1	12.50	18.75	25.00
Gray portland cement, by bag	1.5	2.25	3	1.5	2.25	3
Clean sand, in Lbs (Kg)	150	225	300	68.04	102.06	136.08
Coarse aggregate, 3/8" max. size, in Lbs (Kg)	263	394	525	119.29	178.72	238.14
Conc. cement color, in Lbs (Kg)	13.5	20.3	27	6.12	9.21	12.25

Approximate Quantities Required for 100 Sq.Ft. (9.29 Sq.M) of Floor

For 1" (25 mm) topping, use no more than 5 gals. (18.95 liters) mixing water, including water in aggregates. Smaller mixes in proportion. Sand and coarse aggregate is based on 100 lbs. per c.f. (1601.69 kg/cu.m). Colors are usually packed in 9-lb. (4.08-kg) bags, 100-lb. (45.36-kg) drums and barrels. When mixed in the proportions of 1:2, use 9 lbs. (4.08 kg) of cement color to each bag of portland cement used in the topping.

It is advisable to establish the mixture from samples made under job conditions and allowed to dry. Careful measurement of every ingredient, including water, in each batch is essential to insure uniformity of color.

Final troweling should be done after the concrete has stiffened, as this assures a smooth, polished finish. Integral waterproofing or hardener can be added to mix materials, also polishing or a surface treatment can be applied, if desired.

514

CONCRETE

Colorundum Surface Colorant and Hardener for Concrete. Colorundum (from W.R. Grace & Co.) is a ready-to-use powder containing non-crushable hardeners, dispersing agents, and cementitious binders. It provides a durable, integral concrete color that does not wear off as paint and ordinary surface coatings do. When Colorundum is dusted on and trowelled into freshly poured concrete, it produces bright, decorative surfaces with dense coloration and hard, abrasion resistant surfaces with long life.

Colorundum is for interior and exterior concrete floors and sidewalks. The hard wearing, highly decorative finish of Colorundum floors makes them ideal for use in factories, schools, hospitals, office buildings, showrooms, supermarkets, stores or any public building. Colorundum may be used inside or outside on traffic areas such as sidewalks, ramps, patios, terraces, sun decks and playgrounds. For best results, the slump of concrete monolithic slabs should not exceed 2-1/2" (63 mm) and for toppings 3" (75 mm). Excess water causes laitance to rise to the surface, producing discoloration, and also causes separation, which delays the finishing operation.

Do not use air entraining agents or other admixtures in the mix. When excess water disappears and the topping will hold up knee boards, uniformly dust on 2/3 of specified amount of Colorundum per 100 s.f. (9.29 sq.m) of surface.

After spreading, allow the dry material to wet up. It should then be wood floated and worked into the slab. The first floating should be discontinued as soon as the surface becomes wet. Floating should be resumed when surface moisture has disappeared. Do not steel trowel. Immediately apply the remaining 1/3 of the specified amount per 100 s.f. (9.29 sq.m) and thoroughly wood float. (For heavy duty finish, the quantity of Colorundum may be increased for the second dust coat.) Apply Colorundum exactly as it comes from the container; do not mix with cement or concrete.

Steel trowel to an even plane free from surface marks and voids. When the floor has obtained its initial set, give it a final steel trowel burnishing. Leave exterior areas, such as sidewalks, under wood float finish after the trowelling.

To cure, use an approved curing and sealing compound according to manufacturer's instructions or cover the surface after finish trowelling with polyethylene. This seals against evaporation of the water so vitally necessary to hydrate the cement, thereby developing optimum strength and density throughout the mass. It also protects the colored surface against spillages.

Use approximately 40 lbs. per 100 s.f. (18.14 kg per 9.29 sq.m) for normal traffic areas. For areas subject to extra heavy traffic, the quantity of Colorundum may be increased proportionately.

Gilco Non-Shrink Grouting Compound. Gilco Grout (from Gifford-Hill & Co. Inc.) is a premixed compound containing iron particles, hardening and dispersing agents, binders, and oxidizing agents. When mixed with water, it produces a mass that has controlled expansion to overcome the inherent shrinkage in cementitious mixtures. The metallic aggregate also

515

provides a ductile surface with the "give and take" properties concrete must have to absorb the vibration of machinery, equipment, and columns.

Gilco grout can be used wherever non-shrink concrete is mandatory: setting machinery on concrete foundations; full bearing surface grouting for building columns; and grouting anchor bolts, floor grids, steel sash, and jambs.

Grouting should be continuous once it has started. Do not mix more grout at one time than can be rodded and placed in a period of 20 minutes. Add only enough water to make mix placeable. Avoid an excess of water.

Anchor forms securely to prevent movement during placing and curing. Intersection of form and base slab can be filled with a mix of equal parts of cement and sand before placing grout. Be sure to allow adequate clearance between forms and base plate.

Remove waste material and water from anchor bolt holes. All oil, grease, and paint in contact with grout should be thoroughly removed from base plates before grouting.

It is important to place grout quickly and continuously to avoid all effects of overworking, resulting in segregation, bleeding, and breakdown of initial set. Under no circumstances should the grout be retempered by introducing additional water after it has taken its initial set.

Lightweight Concrete Floor Fill

Concrete made of lightweight aggregate, such as cinders, slag, Haydite, pumice, and vermiculite, is often used for floor fill under marble and tile floors and between wood floor sleepers, screeds, etc.

The labor cost of placing concrete floor fill runs considerably higher than for ordinary concrete, because the fill is usually only 2" or 3" (50 or 75 mm) thick, which necessitates covering a large floor area to place a comparatively small amount of concrete. In many cases it is necessary to pump or hoist the concrete to the upper floors in wheelbarrows or concrete buggies and then wheel it to the different parts of the building where it is to be used.

When placing concrete fill between wood floor sleepers or screeds, it is important that every precaution be taken to prevent the sleepers from being knocked out of level when wheeling over them.

When placing concrete floor fill, it is not advisable to use a large mixer because of the necessity of covering a large floor area to place a small amount of concrete. It usually requires considerable leveling and tamping, making it impractical to handle the concrete in large quantities. A concrete mixer of 1 or 2 sacks capacity is best adapted for this class of work or partial loads of ready-mix can be ordered.

Mixing and Placing Concrete Floor Fill. Because of the additional labor required for spreading, grading, and tamping concrete floor fill, the cu.yd. labor costs run higher than for foundation or sidewalk work.

CONCRETE

There is also more labor required wheeling concrete from the mixer to the hoist and from the hoist to place of deposit, as floor fill is usually placed after the structural portions of the building have been completed, and it is not possible to carry on the work with much speed.

A crew of 13 workers with a 6-S mixer should handle materials, mix, and wheel into place 15 to 20 batches-containing 6 to 7 c.f. (0.17-0.20 cu.m) each or 3-1/2 to 5 c.y. (2.65-3.82 cu.m) of concrete per hour with the crew distributed as follows: 2 workers wheeling coarse aggregate from stock pile to mixer; 2 workers wheeling sand and cement to mixer; 1 worker attending mixer; 3 workers wheeling concrete from mixer to hoist; 3 workers wheeling concrete from hoist to place of deposit; 2 workers dumping barrows and spreading concrete.

The labor per cu.yd. (cu.m) should cost as follows:

Description	Hours	Rate	Total
Labor	3.00	$ 26.16	$ 78.48
Hoisting engineer	0.35	$ 37.89	$ 13.26
Cost per cu.yd.			$ 91.74
Cost per cu.m			$ 119.99

If union rules require a mixer engineer, deduct 0.35 (0.46) hrs. labor time and add 0.35 (0.46) hrs. engineer time per cu.yd. (cu.m).

If concrete is discharged from mixer directly into a hoist concrete bucket and wheeled from a floor hopper to place of deposit, deduct 0.50-hr. labor time.

If no hoisting is required, deduct 0.70-hr. labor time and omit hoisting engineer.

For concrete roof fill on flat roofs, deduct about 10% from the above costs. On pitch or gable roofs, add 25% to 50%, depending on steepness of slope, accessibility, etc.

After the cubic yard (cubic meter) cost has been computed, to obtain the sq.ft. (sq.m) cost, divide the cu.yd. (cu.m) cost by the number of sq.ft. (sq.m) of fill obtainable from one cu.yd. (cu.m) and the result will be the labor cost per sq.ft. (sq.m) of fill. Example: Find the sq.ft. (sq.m) labor cost of 3" (0.075 m) floor fill with a cost of $43.90 per c.y. (57.39 per cu.m). The tables on previous pages give 108 s.f. of 3" (10 sq.m of 0.075 m) fill obtainable from one cu.yd. $43.90 ÷ 108 = $0.406 per sq.ft. (57.39 ÷ 10 = $5.74 per sq.m).

To the above costs must be added the cost of setting screeds and all finishing operations.

Finishing Lightweight Concrete Floor and Roof Fill

All concrete floor and roof fills require some finishing labor, if only setting of screeds and striking off fill at screed height.

Where fill is placed between wood sleepers, screeds are not required, but fill must be struck off and darbied so that no part of it is higher than the top of the sleepers. In addition, the fill should be spaded along both sides of each sleeper so that good anchorage is obtained.

For 2" (50 mm) thick concrete fill between wood floor sleepers, 2 cement masons working with a helper should strike off and darby 5,000 to 6,000 sq.ft. (464.50-557.40 sq.m) per 8-hr. day or about all the fill a small-mixer crew can mix and place in an 8-hr. day. The labor per 100 s.f. (9.29 sq.m) should cost as follows:

Description	Hours	Rate	Total
Cement mason	0.30	$ 36.93	$ 11.08
Helper	0.15	$ 28.16	$ 4.22
Cost per 100 sq.ft. (9.29 sq.m)			$ 15.30
per sq.ft.			$ 0.15
per sq.m			$ 1.65

Finishing Concrete Floor Fill For Resilient Flooring. Where concrete fill is used as a base for cork, asphalt, vinyl, rubber, or wood floor covering, it is necessary to obtain a smooth level finish on the concrete to receive the finish floor material, or all irregularities will show through on the finished surface or have voids between the irregularities.

To obtain the required degree of smoothness, screeds must be placed 4'- 0" to 6'-0" (1.21-1.83 m) o.c. and concrete fill must be struck off, darbied, floated, and troweled.

If the fill is not over 2" (50 mm) thick, 2 cement masons working together should strike off, darby, float, and trowel about 1,600 sq.ft. (148.64 sq.m) of floor per 8-hr. day at the following labor cost per 100 sq.ft. (9.29 sq.m):

Description	Hours	Rate	Total
Cement mason	1.00	$ 36.93	$ 36.93
Cost per sq.ft.			$ 0.37
per sq.m			$ 3.98

Finishing Concrete Roof Fill. Concrete fill for flat roofs are usually struck off, darbied, and floated. If fill is not over 2" to 3" (50 to 75 mm) thick, 2 cement masons with a helper should place screeds, strike off, darby, and float 1,500 to 1,700 s.f. (139.35-157.93 sq.m) of roof fill per 8-hr. day at the following labor cost per 100 sq.ft. (9.29 sq.m):

Description	Hours	Rate	Total
Cement mason	1.00	$ 36.93	$ 36.93
Helper	0.50	$ 28.16	$ 14.08
Cost per 100 Sq,Ft. (9.29 Sq. M)			$ 51.01
per sq.ft.			$ 0.51
per sq.m			$ 5.49

Where concrete fill is placed on pitch or gable roofs, finishing is more difficult. Screeds must be set closer together to help keep the fill from

518

sliding, and an additional helper is usually required. On work of this type, 2 cement masons and 2 helpers should place screeds, strike off, darby, and float 900 to 1,000 sq.ft. (83.61-92.90-90 sq.m) of fill per 8-hr. day at the following labor cost per 100 sq.ft. (9.29 sq.m):

Description	Hours	Rate	Total
Cement mason	1.60	$ 36.93	$ 59.09
Helper	1.60	$ 28.16	$ 45.06
Cost per 100 Sq,Ft. (9.29 Sq. M)			$ 104.14
per sq.ft.			$ 1.04
per sq.m			$ 11.21

CONCRETE SLAB ON GRADE CONSTRUCTION

In the past, this type of floor was built without adequate test data or established construction standards with respect to dampness and floor temperature. Many such floors were cold, especially at the outer edge, and some were damp enough to damage floor coverings and wall construction.

To determine the most satisfactory floor from the standpoint of comfort, the University of Illinois studied the construction of concrete floor slabs that are laid on the ground and designed for use in climates where central heating is necessary. (Small Homes Council, Circular F4.3, "Concrete Floors for Basementless Houses" Printed with permission.)

Nine different types of concrete floors were tested in an effort to determine the proper design for slab on grade houses with respect to heat loss of different types of floors; temperatures at various points throughout the floors for proper placement of floor insulation; and amount of moisture passing from the ground to the top of the concrete slab.

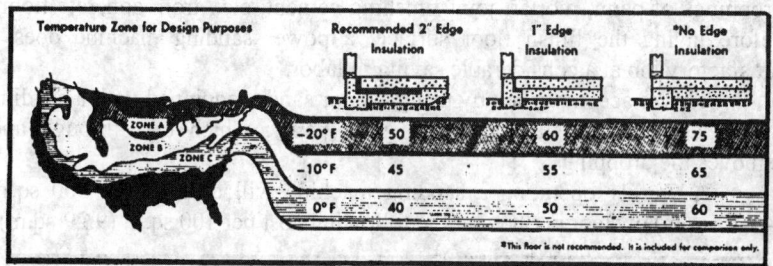

Insulation Requirements for Different Parts of the Country

The floors selected for tests represented standard construction practices and at the same time permitted the use of varying amounts and types of insulation along the edges. The floor slabs were tested simultaneously under similar conditions in a specially constructed laboratory.

When panel heating systems are used with concrete floors, the heat loss to the ground and at the edge is increased due to the higher temperatures maintained in the floors. To prevent excess heat loss in such floors, a minimum of 2" (50 mm) of rigid waterproof insulation should be provided at the edge. In addition, the use of insulation under the entire heated floor area is recommended. Gravel and rock fills are desirable for drainage controls, but they have no insulating value.

The use of concrete floor slabs for basementless homes requires an insulating material around the perimeter of the slab (at the exterior walls) that will not be compressed by imposed loads or soil backfill and that will not be affected by soil acids, moisture, insects, or vermin. Fiberglass perimeter insulation and styrofoam are used successfully for this purpose and come in panels of various widths and thicknesses.

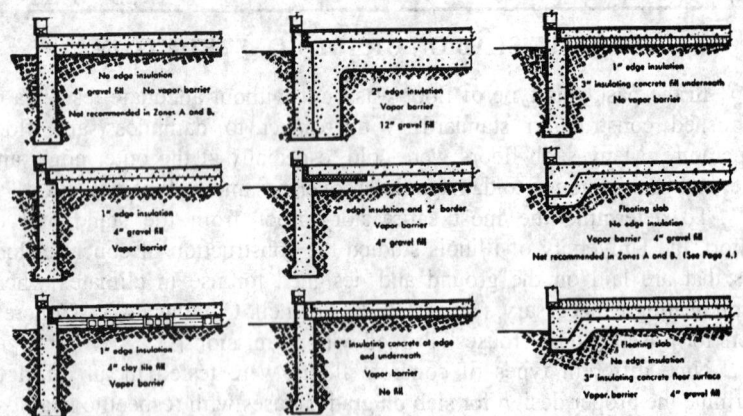

Methods of Constructing Concrete Floors for Basementless Houses

Cleaning Concrete Floors Using a Sanding Machine. To clean droppings of paint, mortar, joint finishing cement, etc., from concrete floors before laying the finish floor surface, a power sanding machine does a satisfactory job at a considerable saving in labor.

This is accomplished by using a floor sander equipped with a sanding disk and open grit silicon carbon grain paper, which cuts through and removes the droppings.

A worker operating a sanding machine will clean about 500 sq.ft. (46.45 sq.m) of floor per hour at the following cost per 100 sq.ft. (9.29 sq.m):

Description	Hours	Rate	Total
Machinery cost	0.20	$ 35.00	$ 7.00
Sanding disk	1.00	$ 5.00	$ 5.00
Labor	0.20	$ 26.16	$ 5.23
Cost per 100 s.f. (9 sq.m)			$ 17.23
per sq.ft.			$ 0.17
per sq.m			$ 1.85

CONCRETE

When this work is performed by hand using a putty knife or steel scraper, a worker will clean and remove droppings from 150 to 175 s.f. (14-16 sq.m) per hour, at the following cost per 100 s.f. (sq.m):

Description	Hours	Rate	Total
Labor	0.60	$ 26.16	$ 15.70
Cost per sq.ft.			$ 0.13
per sq.m			$ 1.40

Lightweight Concrete

There are two types of lightweight concrete. One long-used type is produced by simply combining portland cement and a lightweight aggregate.

The second is known as aerated concrete, which is produced by a process of introducing certain chemicals to generate gases that cause the mass to expand. The weight of aerated concrete can be controlled with a variation from 20 lbs. per cu.ft. (320.21 kg per cu.m) to approximately the weight of ordinary lightweight concrete. For average conditions a weight of 40 to 50 lbs. per c.f. (640.74 – 800.92 kg per cu.m) with a compressive strength of about 500 lbs. per sq. in. (0.35 kg/mm2) is usually selected.

Aerated concrete is fireproof, has a low moisture absorption rate, and provides good insulation against the transmission of heat and sound. It may be sawed or nailed as readily as other lightweight concrete.

All lightweight concretes are of decided advantage for partition walls and fireproofing, whether in monolithic or precast unit form and for floor fill and roof slabs. The estimator should consider the combined advantages of each type of lightweight concrete before selection.

Haydite Concrete. Haydite concrete uses an expanded shale as lightweight aggregate for coarse aggregate instead of gravel or crushed stone. This inert, cellular material weighs less than 50 lbs. per cu.ft. (800.92 kg cu.m) and when mixed with ordinary torpedo sand and portland cement, produces concrete weighing about 98 lbs. per cu.ft. (1569.80 kg per cu.m) instead of 150 lbs. (68 kg) for ordinary concrete.

Full structural strengths can be obtained with Haydite concrete and in many instances a redesign based on lighter weight concrete and reduced dead load more than pays for the slight additional cost of the aggregates.

Perlite Lightweight Concrete. Perlite is a sand or volcanic glass that has been expanded by heat. Perlite concrete aggregate combined with portland cement, air entraining agent, and water produces an ultra lightweight concrete. An air-entraining agent is used to improve the workability and control of water content and insulation value. Perlite concrete may be more accurately defined as concrete containing a minimum of 20 cubic feet (0.56 cu.m) of perlite concrete aggregate per cubic yard (.74 cu.m). For lightweight concrete using Perlite as the aggregate, refer to "Concrete Floors and Walks".

Pumice Lightweight Concrete. A strong, durable porous glass aggregate of volcanic origin. A type of pyroclastic igneous rock. Ideal for

521

lightweight concrete. Using pumice as the aggregate it is possible to get concrete as low as 60 lbs. per cubic foot (966.51 Kg/cu.m). For about 4,000 psi (27.58 MPa) concrete with pumice aggregate the concrete weight will be in the range of 110 Lbs per cubic foot (1762.03 kg/cu.m.), refer to "Concrete Floors and Walks".

Reinforced Concrete

After the forms and reinforcing steel are in place, the job is ready for concrete. This should include the cost of gravel or crushed stone, sand, and cement, as well as the labor cost of handling materials, mixing, hoisting, placing, and curing the concrete.

For many years concrete used for reinforced concrete was specified to be mixed in the proportions of 1:1:2, 1:1-1/2:3, 1:2:4, etc., depending on the class of work. Nowadays, it is specified that concrete used for specific purposes shall develop a compressive strength of 2,500, 3,000, 3,750, 4,000, 6,000 and 10,000 psi or more per sq. in. (17.23, 20.69, 25.85, 27.58, 41.36 and 68.95 or more MPa) at 28 days.

On nearly all types of reinforced concrete building work, such as columns, beams, girders, floor slabs, etc., the coarse aggregate is graded from 1/4" to 1" (6.25-25 mm) in size. Where conditions permit, however, coarse aggregate graded from 1/4" to 1-1/2" (6.25-37.50 mm) is used.

Estimating the Quantity of Concrete Required for Various Types of Concrete Floors

If all concrete floors were merely solid slabs of concrete, it would be an easy matter to compute the quantity of concrete required per sq.ft. (sq.m) of floor but with so many combination floors consisting of metal pans, etc., having joists 4" to 7" (100-175 mm) wide, 4" to 15" (100-375 mm) deep and from 16" to 35" (400-875 mm) o.c., it requires considerable figuring to obtain accurate concrete quantities.

It is practically impossible to estimate accurately the amount of concrete required per s.f. (sq.m) of floor for the various combination joist floors, which invariably have T beam construction at all beams and girders, solid concrete slabs usually extend in 6" to 1'-6" (150-450 mm) from the outside walls, the joists vary in width and are often doubled under partitions, around stair wells, elevator shafts, etc. The only accurate method is to figure the floors as though they were solid concrete and then deduct for the displacement of the tile or pans.

Quantity of Concrete Required for Steelform and Concrete Joist Floor Construction. Obtain the entire slab volume as though it were of solid concrete, and deduct for the displacement of the steelforms.

The deductions for steelform area is not simple. Steelforms are furnished in a number of different sizes, with tapering sides and ends.

Steelforms furnished by different manufacturers vary slightly in shape but not sufficient to materially affect the concrete quantities. Ceco steelforms

CONCRETE

(The Ceco Corporation, Kansas City, Mo.) are used in the tables, because they are typical.

Sizes and Cu. Ft. Displacement of Flange Type Steelforms

Ceco flange type steelforms slope 1" in each 12" (25 in 300 mm) in height. Example: a 12" (300 mm) pan is 20" (500 mm) wide at the bottom and 18" (450 mm) wide at the top, because of the 1" (25 mm) slope on each side. See illustrations on following pages and in the Concrete Formwork section.

The sides of single tapered endforms slope 1" (25 mm) for each 12" (300 mm) in height and one end slopes 1/4" (6.25 mm) for each 1" (25 mm) in height. For estimating purposes their displacement is the same as for standard steelforms in the following tables.

Sizes and Displacement of Tapered End SteelForms						
Depth, in Inches	Width at wide end Inches	Width at narrow end Inches	Avg Width Inches	Length in Feet	Displ. Cu.Ft. per 3 ft. Section	Difference between tapered end forms & 3' length of straight forms Inches
8	20	16	18	3	2.882	0.322
10	20	16	18	3	3.569	0.399
12	20	16	18	3	4.243	0.475
8	30	25	27 1/2	3	4.465	0.406
10	30	25	27 1/2	3	5.548	0.504
12	30	25	27 1/2	3	6.617	0.602
14	30	25	27 1/2	3	7.763	0.698
16	30	25	27 1/2	3	8.715	1.188
20	30	25	27 1/2	3	10.756	-----
Depth, in mm	Width at wide end mm	Width at narrow end mm	Avg Width mm	Length in Meter	Displ. Cu.M. per 1m Section	Difference between tapered end forms & 0.91M length of straight Forms mm
200	500	400	450.0	0.914	0.089	8.05
250	500	400	450.0	0.914	0.111	9.98
300	500	400	450.0	0.914	0.131	11.88
200	750	625	687.5	0.914	0.138	10.15
250	750	625	687.5	0.914	0.172	12.60
300	750	625	687.5	0.914	0.205	15.05
350	750	625	687.5	0.914	0.240	17.45
400	750	625	687.5	0.914	0.270	29.70
500	750	625	687.5	0.914	0.333	-----

Cu.Ft. (Cu.M.) of Concrete Displaced by Adjustable Steelforms and Tapered Endforms

Length of Each Row of Steelforms in lin.ft

Steelform											
Depth	Width	10	11	12	13	14	15	16	17	18	19
8 "	30 "	15.33	16.96	18.58	20.21	21.83	23.46	25.08	26.71	28.33	29.96
10 "	30 "	19.33	21.38	23.42	25.46	27.50	29.54	31.59	33.63	35.67	37.71
12 "	30 "	23.33	25.79	28.25	30.71	33.16	35.62	38.08	40.54	43.00	45.45
14 "	30 "	27.33	30.21	33.08	35.96	38.83	41.71	44.58	47.46	50.33	53.21
15 "	30 "	29.33	32.42	35.50	38.58	41.66	44.75	47.83	50.91	54.00	57.08
Depth	Width	20	21	22	23	24	25	26	27	28	29
8 "	30 "	31.58	33.21	34.83	36.46	38.08	39.71	41.33	42.96	44.58	46.21
10 "	30 "	39.75	41.80	43.84	45.88	47.92	49.96	52.01	54.05	56.09	58.13
12 "	30 "	47.91	50.37	52.83	55.29	57.74	60.20	62.66	65.12	67.58	70.03
14 "	30 "	56.08	58.96	61.83	64.71	67.58	70.46	73.33	76.21	79.08	81.96
15 "	30 "	60.16	63.25	66.33	69.41	72.49	75.58	78.66	81.74	84.83	87.91

Length of Each Row of Steelforms, meters

Steelform											
Depth	Width	3.048	3.353	3.658	3.962	4.267	4.572	4.877	5.182	5.486	5.791
200 mm	750 mm	0.434	0.480	0.526	0.572	0.618	0.664	0.710	0.756	0.802	0.848
250 mm	750 mm	0.547	0.605	0.663	0.721	0.779	0.837	0.895	0.952	1.010	1.068
300 mm	750 mm	0.661	0.730	0.800	0.870	0.939	1.009	1.078	1.148	1.218	1.287
350 mm	750 mm	0.774	0.856	0.937	1.018	1.100	1.181	1.263	1.344	1.425	1.507
375 mm	750 mm	0.831	0.918	1.005	1.093	1.180	1.267	1.355	1.442	1.529	1.617
Depth	Width	6.096	6.401	6.706	7.010	7.315	7.620	7.925	8.230	8.534	8.839
200 mm	750 mm	0.894	0.941	0.986	1.033	1.078	1.125	1.170	1.217	1.263	1.309
250 mm	750 mm	1.126	1.184	1.242	1.299	1.357	1.415	1.473	1.531	1.588	1.646
300 mm	750 mm	1.357	1.426	1.496	1.566	1.635	1.705	1.775	1.844	1.914	1.983
350 mm	750 mm	1.588	1.670	1.751	1.833	1.914	1.995	2.077	2.158	2.240	2.321
375 mm	750 mm	1.704	1.791	1.878	1.966	2.053	2.140	2.228	2.315	2.402	2.490

CONCRETE

Cu. Ft. of Concrete Displaced by Flange Type Steelforms and Straight Endforms

Steelform — Length of Each Row of Steelforms in lin. ft.

Depth	Width	1	2	3	4	5	6	7	8	9	10	11	12	13	14	15	16	17	18	19	20
8 "	10 "	0.52	1.04	1.55	2.07	2.59	3.11	3.63	4.14	4.66	5.18	5.70	6.22	6.73	7.25	7.77	8.29	8.81	9.32	9.84	10.36
10 "	10 "	0.64	1.27	1.91	2.54	3.18	3.82	4.45	5.09	5.72	6.36	7.00	7.62	8.27	8.90	9.54	10.18	10.81	11.45	12.08	12.72
12 "	10 "	0.75	1.50	2.25	3.00	3.75	4.50	5.25	6.00	6.75	7.50	8.25	9.00	9.75	10.50	11.25	12.00	12.75	13.50	14.25	15.00
14 "	10 "	0.86	1.72	2.58	3.44	4.30	5.15	6.01	6.87	7.73	8.59	9.45	10.31	11.17	12.03	12.89	13.74	14.60	15.46	16.32	17.18
16 "	10 "	0.96	1.92	2.88	3.84	4.80	5.75	6.71	7.67	8.63	9.59	10.55	11.51	12.47	13.43	14.39	15.34	16.30	17.26	18.22	19.18

Steelform — Length of Each Row of Steelforms in lin. ft.

Depth	Width	1	2	3	4	5	6	7	8	9	10	11	12	13	14	15	16	17	18	19	20
8 "	15 "	0.80	1.59	2.39	3.18	3.98	4.78	5.57	6.37	7.16	7.96	8.76	9.55	10.35	11.14	11.94	12.74	13.53	14.33	15.12	15.92
10 "	15 "	0.98	1.97	2.95	3.94	4.92	5.90	6.89	7.87	8.86	9.84	10.82	11.81	12.79	13.78	14.76	15.74	16.73	17.71	18.70	19.68
12 "	15 "	1.17	2.33	3.50	4.67	5.84	7.00	8.17	9.34	10.50	11.67	12.84	14.00	15.17	16.34	17.51	18.67	19.84	21.01	22.17	23.34
14 "	15 "	1.35	2.69	4.04	5.38	6.73	8.07	9.42	10.76	12.11	13.45	14.80	16.14	17.49	18.83	20.18	21.52	22.87	24.21	25.56	26.90
16 "	15 "	1.48	2.96	4.44	5.92	7.40	8.88	10.36	11.84	13.32	14.80	16.28	17.76	19.24	20.72	22.20	23.68	25.16	26.64	28.12	29.60

525

Cu. Ft. of Concrete Displaced by Flange Type Steelforms and Straight Endforms (continued)

Steelform

Length of Each Row of Steelforms in lin.ft.

Depth	Width	1	2	3	4	5	6	7	8	9	10
8 "	20 "	1.07	2.15	3.22	4.30	5.37	6.44	7.52	8.59	9.67	10.74
10 "	20 "	1.33	2.66	3.99	5.32	6.66	7.99	9.32	10.65	11.98	13.31
12 "	20 "	1.58	3.17	4.75	6.33	7.92	9.50	11.08	12.66	14.25	15.83
14 "	20 "	1.83	3.66	5.49	7.32	9.16	10.99	12.82	14.65	16.48	18.31
16 "	20 "	2.07	4.14	6.21	8.28	10.35	12.42	14.49	16.56	18.64	20.70

Depth	Width	11	12	13	14	15	16	17	18	19	20
8 "	20 "	11.81	12.89	13.96	15.04	16.11	17.18	18.26	19.33	20.41	21.48
10 "	20 "	14.64	15.97	17.30	18.63	19.97	21.30	22.63	23.96	25.29	26.62
12 "	20 "	17.41	19.00	20.58	22.16	23.75	25.33	26.91	28.49	30.08	31.66
14 "	20 "	20.14	21.97	23.80	25.63	27.47	29.30	31.13	32.96	34.79	36.62
16 "	20 "	22.77	24.84	26.91	28.98	31.05	33.12	35.19	37.26	39.33	41.40

Steelform

Length of Each Row of Steelforms in lin.ft.

Depth	Width	1	2	3	4	5	6	7	8	9	10
8 "	30 "	1.63	3.26	4.89	6.52	8.15	9.77	11.40	13.03	14.66	16.29
10 "	30 "	2.03	4.05	6.08	8.10	10.13	12.15	14.18	16.20	18.23	20.25
12 "	30 "	2.42	4.83	7.25	9.67	12.09	14.50	16.92	19.34	21.75	24.17
14 "	30 "	2.80	5.61	8.41	11.21	14.02	16.82	19.62	22.42	25.23	28.03
16 "	30 "	3.30	6.60	9.90	13.20	16.50	19.80	23.10	26.40	29.70	33.00

Depth	Width	11	12	13	14	15	16	17	18	19	20
8 "	30 "	17.92	19.55	21.18	22.81	24.44	26.06	27.69	29.32	30.95	32.58
10 "	30 "	22.28	24.30	26.33	28.35	30.38	32.40	34.43	36.45	38.48	40.50
12 "	30 "	26.59	29.00	31.42	33.84	36.26	38.67	41.09	43.51	45.92	48.34
14 "	30 "	30.83	33.64	36.44	39.24	42.05	44.85	47.65	50.45	53.26	56.06
16 "	30 "	36.30	39.60	42.90	46.20	49.50	52.80	56.10	59.40	62.70	66.00

CONCRETE

Cu.Meter of Concrete Displaced by Flange Type Steelforms and Straight Endforms

Steelform

Depth	Width	Length of Each Row of Steelforms, meters									
		0.3048	0.6096	0.9144	1.2192	1.524	1.8288	2.1336	2.4384	2.7432	3.048
200 mm	250 mm	0.015	0.029	0.044	0.059	0.073	0.088	0.103	0.117	0.132	0.147
250 mm	250 mm	0.018	0.036	0.054	0.072	0.090	0.108	0.126	0.144	0.162	0.180
300 mm	250 mm	0.021	0.042	0.064	0.085	0.106	0.127	0.149	0.170	0.191	0.212
350 mm	250 mm	0.024	0.049	0.073	0.097	0.122	0.146	0.170	0.195	0.219	0.243
400 mm	250 mm	0.027	0.054	0.082	0.109	0.136	0.163	0.190	0.217	0.244	0.272

Depth	Width	5.462	5.959	6.456	6.952	7.449	7.943	8.440	8.937	9.434	9.930
200 mm	250 mm	0.161	0.176	0.191	0.205	0.220	0.235	0.249	0.264	0.279	0.293
250 mm	250 mm	0.198	0.216	0.234	0.252	0.270	0.288	0.306	0.324	0.342	0.360
300 mm	250 mm	0.234	0.255	0.276	0.297	0.319	0.340	0.361	0.382	0.404	0.425
350 mm	250 mm	0.268	0.292	0.316	0.341	0.365	0.389	0.413	0.438	0.462	0.487
400 mm	250 mm	0.299	0.326	0.353	0.380	0.408	0.434	0.462	0.489	0.516	0.543

Steelform

Depth	Width	Length of Each Row of Steelforms, meters									
		0.3048	0.6096	0.9144	1.2192	1.524	1.8288	2.1336	2.4384	2.7432	3.048
200 mm	375 mm	0.023	0.045	0.068	0.090	0.113	0.135	0.158	0.180	0.203	0.225
250 mm	375 mm	0.028	0.056	0.084	0.112	0.139	0.167	0.195	0.223	0.251	0.279
300 mm	375 mm	0.033	0.066	0.099	0.132	0.165	0.198	0.231	0.265	0.297	0.330
350 mm	375 mm	0.038	0.076	0.114	0.152	0.191	0.229	0.267	0.305	0.343	0.381
400 mm	375 mm	0.042	0.084	0.126	0.168	0.210	0.251	0.293	0.335	0.377	0.419

Depth	Width	3.353	3.658	3.962	4.267	4.572	4.877	5.182	5.486	5.791	6.096
200 mm	375 mm	0.248	0.270	0.293	0.315	0.338	0.361	0.383	0.406	0.428	0.451
250 mm	375 mm	0.306	0.334	0.362	0.390	0.418	0.446	0.474	0.502	0.530	0.557
300 mm	375 mm	0.364	0.396	0.430	0.463	0.496	0.529	0.562	0.595	0.628	0.661
350 mm	375 mm	0.419	0.457	0.495	0.533	0.571	0.609	0.648	0.686	0.724	0.762
400 mm	375 mm	0.461	0.503	0.545	0.587	0.629	0.671	0.713	0.754	0.796	0.838

Cu.Meter of Concrete Displaced by Flange Type Steelforms and Straight Endforms (continued)

Steelform

Depth	Width	Length of Each Row of Steelforms, meters									
		0.3048	0.6096	0.9144	1.2192	1.524	1.8288	2.1336	2.4384	2.7432	3.048
200 mm	500 mm	0.030	0.061	0.091	0.122	0.152	0.182	0.213	0.243	0.274	0.304
250 mm	500 mm	0.038	0.075	0.113	0.151	0.189	0.226	0.264	0.302	0.339	0.377
300 mm	500 mm	0.045	0.090	0.135	0.179	0.224	0.269	0.314	0.359	0.404	0.448
350 mm	500 mm	0.052	0.104	0.155	0.207	0.259	0.311	0.363	0.415	0.467	0.519
400 mm	500 mm	0.059	0.117	0.176	0.234	0.293	0.352	0.410	0.469	0.528	0.586

Depth	Width	3.353	3.658	3.962	4.267	4.572	4.877	5.182	5.486	5.791	6.096
200 mm	500 mm	0.334	0.365	0.395	0.426	0.456	0.487	0.517	0.547	0.578	0.608
250 mm	500 mm	0.415	0.452	0.490	0.528	0.566	0.603	0.641	0.679	0.716	0.754
300 mm	500 mm	0.493	0.538	0.583	0.628	0.673	0.717	0.762	0.807	0.852	0.897
350 mm	500 mm	0.570	0.622	0.674	0.726	0.778	0.830	0.882	0.933	0.985	1.037
400 mm	500 mm	0.645	0.703	0.762	0.821	0.879	0.938	0.997	1.055	1.114	1.172

Steelform

Depth	Width	Length of Each Row of Steelforms, meters									
		0.3048	0.6096	0.9144	1.2192	1.524	1.8288	2.1336	2.4384	2.7432	3.048
200 mm	750 mm	0.046	0.092	0.138	0.185	0.231	0.277	0.323	0.369	0.415	0.461
250 mm	750 mm	0.057	0.115	0.172	0.229	0.287	0.344	0.402	0.459	0.516	0.573
300 mm	750 mm	0.069	0.137	0.205	0.274	0.342	0.411	0.479	0.548	0.616	0.684
350 mm	750 mm	0.079	0.159	0.238	0.317	0.397	0.476	0.556	0.635	0.715	0.794
400 mm	750 mm	0.093	0.187	0.280	0.374	0.467	0.561	0.654	0.748	0.841	0.935

Depth	Width	3.353	3.658	3.962	4.267	4.572	4.877	5.182	5.486	5.791	6.096
200 mm	750 mm	0.507	0.554	0.600	0.646	0.692	0.738	0.784	0.830	0.877	0.923
250 mm	750 mm	0.631	0.688	0.746	0.803	0.860	0.918	0.975	1.032	1.090	1.147
300 mm	750 mm	0.753	0.821	0.890	0.958	1.027	1.095	1.164	1.232	1.300	1.369
350 mm	750 mm	0.873	0.953	1.032	1.111	1.191	1.270	1.349	1.429	1.508	1.588
400 mm	750 mm	1.028	1.121	1.215	1.308	1.402	1.495	1.589	1.682	1.776	1.869

Cu. Ft. of Concrete Displaced by Adjustable Steelforms and Straight Endforms

Steelform — Length of Each Row of Steelforms in lin.ft

Depth	Width	1	2	3	4	5	6	7	8	9	10
8"	10"	0.51	1.03	1.54	2.06	2.57	3.08	3.60	4.11	4.63	5.14
10"	10"	0.65	1.31	1.96	2.61	3.27	3.92	4.57	5.22	5.88	6.53
12"	10"	0.79	1.58	2.38	3.17	3.96	4.75	5.54	6.34	7.13	7.92
14"	10"	0.93	1.86	2.79	3.72	4.66	5.59	6.52	7.45	8.38	9.31
15"	10"	1.00	2.00	3.00	4.00	5.00	6.00	7.00	8.00	9.00	10.00

Depth	Width	11	12	13	14	15	16	17	18	19	20
8"	10"	5.65	6.17	6.68	7.20	7.71	8.22	8.74	9.25	9.77	10.28
10"	10"	7.18	7.84	8.49	9.14	9.80	10.45	11.10	11.75	12.41	13.06
12"	10"	8.71	9.50	10.30	11.09	11.88	12.67	13.46	14.26	15.05	15.84
14"	10"	10.24	11.17	12.10	13.03	13.97	14.90	15.83	16.76	17.69	18.62
15"	10"	11.00	12.00	13.00	14.00	15.00	16.00	17.00	18.00	19.00	20.00

Steelform — Length of Each Row of Steelforms in lin.ft

Depth	Width	1	2	3	4	5	6	7	8	9	10
8"	15"	0.79	1.58	2.38	3.17	3.96	4.75	5.54	6.34	7.13	7.92
10"	15"	1.00	2.00	3.00	4.00	5.00	6.00	7.00	8.00	9.00	10.00
12"	15"	1.21	2.42	3.62	4.83	6.04	7.25	8.46	9.66	10.87	12.08
14"	15"	1.42	2.83	4.25	5.67	7.09	8.50	9.92	11.34	12.75	14.17
15"	15"	1.52	3.04	4.56	6.08	7.61	9.13	10.65	12.17	13.69	15.21

Depth	Width	11	12	13	14	15	16	17	18	19	20
8"	15"	8.71	9.50	10.30	11.09	11.88	12.67	13.46	14.26	15.05	15.84
10"	15"	11.00	12.00	13.00	14.00	15.00	16.00	17.00	18.00	19.00	20.00
12"	15"	13.29	14.50	15.70	16.91	18.12	19.33	20.54	21.74	22.95	24.16
14"	15"	15.59	17.00	18.42	19.84	21.26	22.67	24.09	25.51	26.92	28.34
15"	15"	16.73	18.25	19.77	21.29	22.82	24.34	25.86	27.38	28.90	30.42

529

Cu. Ft. of Concrete Displaced by Adjustable Steelforms and Straight Endforms (continued)

Steelform
Length of Each Row of Steelforms in lin.ft

Depth	Width	1	2	3	4	5	6	7	8	9	10
8 "	20 "	1.07	2.14	3.21	4.28	5.35	6.41	7.48	8.55	9.62	10.69
10 "	20 "	1.35	2.69	4.04	5.39	6.74	8.08	9.43	10.78	12.12	13.47
12 "	20 "	1.63	3.25	4.88	6.50	8.13	9.75	11.38	13.00	14.63	16.25
14 "	20 "	1.90	3.81	5.71	7.61	9.52	11.42	13.32	15.22	17.13	19.03
15 "	20 "	2.04	4.08	6.13	8.17	10.21	12.25	14.29	16.34	18.38	20.42

Depth	Width	11	12	13	14	15	16	17	18	19	20
8 "	20 "	11.76	12.83	13.90	14.97	16.04	17.10	18.17	19.24	20.31	21.38
10 "	20 "	14.82	16.16	17.51	18.86	20.21	21.55	22.90	24.25	25.59	26.94
12 "	20 "	17.88	19.50	21.13	22.75	24.38	26.00	27.63	29.25	30.88	32.50
14 "	20 "	20.93	22.84	24.74	26.64	28.55	30.45	32.35	34.25	36.16	38.06
15 "	20 "	22.46	24.50	26.55	28.59	30.63	32.67	34.71	36.76	38.80	40.84

Steelform
Length of Each Row of Steelforms in lin.ft

Depth	Width	1	2	3	4	5	6	7	8	9	10
8 "	30 "	16.25	17.88	19.50	21.13	22.75	24.38	26.00	27.63	29.25	30.88
10 "	30 "	20.42	22.46	24.50	26.55	28.59	30.63	32.67	34.71	36.76	38.80
12 "	30 "	24.58	27.04	29.50	31.95	34.41	36.87	39.33	41.79	44.24	46.70
14 "	30 "	28.75	31.63	34.50	37.38	40.25	43.13	46.00	48.88	51.75	54.63
15 "	30 "	30.83	33.91	37.00	40.08	43.16	46.25	49.33	52.41	55.49	58.58

Depth	Width	11	12	13	14	15	16	17	18	19	20
8 "	30 "	32.50	34.13	35.75	37.38	39.00	40.63	42.25	43.88	45.50	47.13
10 "	30 "	40.84	42.88	44.92	46.97	49.01	51.05	53.09	55.13	57.18	59.22
12 "	30 "	49.16	51.62	54.08	56.53	58.99	61.45	63.91	66.37	68.82	71.28
14 "	30 "	57.50	60.38	63.25	66.13	69.00	71.88	74.75	77.63	80.50	83.38
15 "	30 "	61.66	64.74	67.83	70.91	73.99	77.08	80.16	83.24	86.32	89.41

CONCRETE

Cu. Meter of Concrete Displaced by Adjustable Steelforms and Straight Endforms

Length of Each Row of Steelforms in lin.M

Steelform Depth	Width	0.305	0.610	0.914	1.219	1.524	1.829	2.134	2.438	2.743	3.048
200 mm	250 mm	0.014	0.029	0.044	0.058	0.073	0.087	0.102	0.116	0.131	0.146
250 mm	250 mm	0.018	0.037	0.056	0.074	0.093	0.111	0.129	0.148	0.167	0.185
300 mm	250 mm	0.022	0.045	0.067	0.090	0.112	0.135	0.157	0.180	0.202	0.224
350 mm	250 mm	0.026	0.053	0.079	0.105	0.132	0.158	0.185	0.211	0.237	0.264
375 mm	250 mm	0.028	0.057	0.085	0.113	0.142	0.170	0.198	0.227	0.255	0.283

Depth	Width	3.353	3.658	3.962	4.267	4.572	4.877	5.182	5.486	5.791	6.096
200 mm	250 mm	0.160	0.175	0.189	0.204	0.218	0.233	0.248	0.262	0.277	0.291
250 mm	250 mm	0.203	0.222	0.240	0.259	0.278	0.296	0.314	0.333	0.351	0.370
300 mm	250 mm	0.247	0.269	0.292	0.314	0.336	0.359	0.381	0.404	0.426	0.449
350 mm	250 mm	0.290	0.316	0.343	0.369	0.396	0.422	0.448	0.475	0.501	0.527
375 mm	250 mm	0.312	0.340	0.368	0.396	0.425	0.453	0.481	0.510	0.538	0.566

Length of Each Row of Steelforms in lin.M

Steelform Depth	Width	0.305	0.610	0.914	1.219	1.524	1.829	2.134	2.438	2.743	3.048
200 mm	375 mm	0.022	0.045	0.067	0.090	0.112	0.135	0.157	0.180	0.202	0.224
250 mm	375 mm	0.028	0.057	0.085	0.113	0.142	0.170	0.198	0.227	0.255	0.283
300 mm	375 mm	0.034	0.069	0.103	0.137	0.171	0.205	0.240	0.274	0.308	0.342
350 mm	375 mm	0.040	0.080	0.120	0.161	0.201	0.241	0.281	0.321	0.361	0.401
375 mm	375 mm	0.043	0.086	0.129	0.172	0.216	0.259	0.302	0.345	0.388	0.431

Depth	Width	3.353	3.658	3.962	4.267	4.572	4.877	5.182	5.486	5.791	6.096
200 mm	375 mm	0.247	0.269	0.292	0.314	0.336	0.359	0.381	0.404	0.426	0.449
250 mm	375 mm	0.312	0.340	0.368	0.396	0.425	0.453	0.481	0.510	0.538	0.566
300 mm	375 mm	0.376	0.411	0.445	0.479	0.513	0.547	0.582	0.616	0.650	0.684
350 mm	375 mm	0.442	0.481	0.522	0.562	0.602	0.642	0.682	0.722	0.762	0.803
375 mm	375 mm	0.474	0.517	0.560	0.603	0.646	0.689	0.732	0.775	0.818	0.861

Cu. Meter of Concrete Displaced by Adjustable Steelforms and Straight Endforms (continued)

Steelform

		Length of Each Row of Steelforms in lin.M									
Depth	Width	0.305	0.820	1.231	1.643	2.054	2.463	2.874	3.286	3.694	4.106
200 mm	500 mm	0.030	0.061	0.091	0.121	0.152	0.182	0.212	0.242	0.272	0.303
250 mm	500 mm	0.038	0.076	0.114	0.153	0.191	0.229	0.267	0.305	0.343	0.381
300 mm	500 mm	0.046	0.092	0.138	0.184	0.230	0.276	0.322	0.368	0.414	0.460
350 mm	500 mm	0.054	0.108	0.162	0.216	0.270	0.323	0.377	0.431	0.485	0.539
375 mm	500 mm	0.058	0.116	0.174	0.231	0.289	0.347	0.405	0.463	0.521	0.578

Depth	Width	3.353	3.658	3.962	4.267	4.572	4.877	5.182	5.486	5.791	6.096
200 mm	500 mm	0.333	0.363	0.394	0.424	0.454	0.484	0.515	0.545	0.575	0.605
250 mm	500 mm	0.420	0.458	0.496	0.534	0.572	0.610	0.649	0.687	0.725	0.763
300 mm	500 mm	0.506	0.552	0.598	0.644	0.690	0.736	0.782	0.828	0.875	0.920
350 mm	500 mm	0.593	0.647	0.701	0.754	0.809	0.862	0.916	0.970	1.024	1.078
375 mm	500 mm	0.636	0.694	0.752	0.810	0.867	0.925	0.983	1.041	1.099	1.157

Steelform

		Length of Each Row of Steelforms in lin.M									
Depth	Width	0.305	10.336	11.278	12.216	13.155	14.097	15.036	15.975	16.913	17.855
200 mm	750 mm	0.460	0.506	0.552	0.598	0.644	0.690	0.736	0.782	0.828	0.875
250 mm	750 mm	0.578	0.636	0.694	0.752	0.810	0.867	0.925	0.983	1.041	1.099
300 mm	750 mm	0.696	0.766	0.835	0.905	0.974	1.044	1.114	1.183	1.253	1.323
350 mm	750 mm	0.814	0.896	0.977	1.059	1.140	1.221	1.303	1.384	1.466	1.547
375 mm	750 mm	0.873	0.960	1.048	1.135	1.222	1.310	1.397	1.484	1.571	1.659

Depth	Width	3.353	15.734	16.484	17.230	17.980	18.730	19.480	20.230	20.976	21.726
200 mm	750 mm	0.920	0.967	1.012	1.059	1.104	1.151	1.197	1.243	1.289	1.335
250 mm	750 mm	1.157	1.214	1.272	1.330	1.388	1.446	1.504	1.561	1.619	1.677
300 mm	750 mm	1.392	1.462	1.532	1.601	1.671	1.740	1.810	1.880	1.949	2.019
350 mm	750 mm	1.628	1.710	1.791	1.873	1.954	2.036	2.117	2.198	2.280	2.361
375 mm	750 mm	1.746	1.833	1.921	2.008	2.095	2.183	2.270	2.357	2.445	2.532

CONCRETE

Sizes and Displacement of Ceco Flange Type Steelforms									
Depth, in Inches	Width at Bot. Inches	Width at Top Inches	Avg Width Inches	Displ. Cu.Ft. per Lin.Ft.	Depth, in mm	Width at Bot. mm	Width at Top mm	Avg Width mm	Displ. Cu.M per Lin.M
8	10	8 2/3	9 1/3	0.516	200	250	216.67	233.33	0.048
10	10	8 1/3	9 1/6	0.634	250	250	208.33	229.17	0.059
12	10	8	9	0.748	300	250	200.00	225.00	0.069
14	10	7 2/3	8 5/6	0.857	350	250	191.67	220.83	0.080
16	10	7 1/3	8 2/3	0.961	400	250	183.33	216.67	0.089
20	10	-----	-----	1.155	500	250	-----	-----	0.107
8	15	13 2/3	14 1/3	0.794	200	375	341.67	358.33	0.074
10	15	13 1/3	14 1/6	0.982	250	375	333.33	354.17	0.091
12	15	13	14	1.165	300	375	325.00	350.00	0.108
14	15	12 2/3	13 5/6	1.343	350	375	316.67	345.83	0.125
16	15	12 1/3	13 2/3	1.516	400	375	308.33	341.67	0.141
20	15	-----	-----	1.85	500	375	-----	-----	0.172
8	20	18 2/3	19 1/3	1.072	200	500	466.67	483.33	0.100
10	20	18 1/3	19 1/6	1.329	250	500	458.33	479.17	0.123
12	20	18	19	1.581	300	500	450.00	475.00	0.147
14	20	17 2/3	18 5/6	1.829	350	500	441.67	470.83	0.170
16	20	17 1/3	18 2/3	2.072	400	500	433.33	466.67	0.192
20	20	-----	-----	2.544	500	500	-----	-----	0.236
8	30	28 2/3	29 1/3	1.628	200	750	716.67	733.33	0.151
10	30	28 1/3	29 1/6	2.023	250	750	708.33	729.17	0.188
12	30	28	29	2.414	300	750	700.00	725.00	0.224
14	30	27 2/3	28 5/6	2.801	350	750	691.67	720.83	0.260
16	30	27 1/3	28 2/3	3.183	400	750	683.33	716.67	0.296

Sizes and Displacement of Adjustable Steelforms							
Depth, in Inches	Width at Bottom Inches	Width at Top Inches	Displ. Cu.Ft. per Lin.Ft.	Depth, in mm	Width at Bottom mm	Width at Top mm	Displ. Cu.M. per Lin.M.
8	10	8	0.516	200	250	200	0.048
10	10	8	0.634	250	250	200	0.059
12	10	8	0.748	300	250	200	0.069
14	10	8	0.857	350	250	200	0.080
15	10	8	0.961	375	250	200	0.089
8	15	13	0.794	200	375	325	0.074
10	15	13	0.982	250	375	325	0.091
12	15	13	1.165	300	375	325	0.108
14	15	13	1.343	350	375	325	0.125
15	15	13	1.516	375	375	325	0.141

Depth, in Inches	Width at Bottom Inches	Width at Top Inches	Displ. Cu.Ft. per Lin.Ft.	Depth, in mm	Width at Bottom mm	Width at Top mm	Displ. Cu.M. per Lin.M.
8	20	18	1.072	200	500	450	0.100
10	20	18	1.329	250	500	450	0.123
12	20	18	1.581	300	500	450	0.147
14	20	18	1.829	350	500	450	0.170
15	20	18	2.072	375	500	450	0.192
8	30	28	1.628	200	750	700	0.151
10	30	28	2.023	250	750	700	0.188
12	30	28	2.414	300	750	700	0.224
14	30	28	2.801	350	750	700	0.260
15	30	28	3.183	375	750	700	0.296

Sizes and Displacement of Adjustable Steelforms (continued)

Sizes and Cu. Ft. (Cu.M) Displacement of Ceco Flange Type Single Tapered Endforms

Endforms are 20" (500 mm) wide at one end tapering to 16" (400 mm) at the other end. The 30" (750 mm) endforms are 30" (750 mm) wide at one end tapering to 25" (625 mm) wide at other end.

Sizes and Cu. Ft. (Cu.M) Displacement of Ceco Adjustable Tapered Steel Endforms

Tapered endforms 3'-0" (0.9 m) long are furnished for 30" (0.75 m, 750 mm) wide steelforms only. Straight endforms 3'-0" (0.9 m, 900 mm) long are furnished for the 30" (0.75 m, 750 mm) standard widths and the 20", 15", and 10" (500, 375, and 250 mm) special widths.

Depth of Steelform	Width Wide End	Width Narrow End	Length in Lin.Ft.	Cu.Ft. displacement per 3' section
8 "	30 "	25 "	3 '	4.317
10 "	30 "	25 "	3 '	5.483
12 "	30 "	25 "	3 '	6.629
14 "	30 "	25 "	3 '	7.775
15 "	30 "	25 "	3 '	8.347
Depth of Stelform	**Width Wide End**	**Width Narrow End**	**Length**	**Cu.M. displacement per .91 M section**
200 mm	750 mm	625 mm	0.914 M	0.1028
250 mm	750 mm	625 mm	0.914 M	0.1306
300 mm	750 mm	625 mm	0.914 M	0.1579
350 mm	750 mm	625 mm	0.914 M	0.1852
375 mm	750 mm	625 mm	0.914 M	0.1988

Another Method of Estimating the Concrete Required for Ceco Steelform Construction. The following method of estimating the concrete requirements for Ceco steelform construction (The Ceco Corporation, Kansas City, Mo.) is that recommended by the manufacturers.

Figure the concrete as though it were a solid slab and then deduct for displacement of a certain number of lin. ft. of straight metal steelforms and tapered endforms.

Ceco steelform construction is a combination of concrete joists and thin slabs. To compute the amount of concrete in this type of floor construction, multiply the floor area by the equivalent thickness of the floor construction. This is better than separately figuring the concrete in the joists, by multiplying the lin. ft. of joists by their unit of volume, and adding the concrete in the thin slabs.

The cu. ft. of concrete per sq. ft. of floor (or roof) area is a function of the thickness of the thin slab, the depth of the joists, and the width and spacing of the joists. The additional concrete in the joists at their ends as formed by tapered endforms cannot be prorated on a square foot basis. The concrete added to the joists by the tapered endforms must be figured per lin. ft. (m) of beam or wall into which the joists frame, and this unit of volume must be multiplied by the number of lin. ft. (m) of beams and walls along which the tapered endforms are set, to determine the total amount of concrete added by the use of the tapered endforms.

For convenience in estimating, an arrangement in tabular form of the quantities of concrete required for Ceco flange type steelform construction is given. Keep in mind that the metal strength of the Ceco steelforms permits sharp angles in the steelforms and an absolutely flat top, allowing a minimum of concrete. These tables should not be used in computing the concrete for other types of metal form construction.

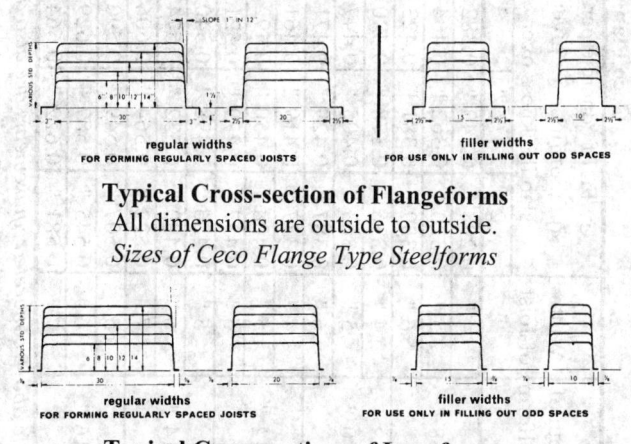

regular widths
FOR FORMING REGULARLY SPACED JOISTS

filler widths
FOR USE ONLY IN FILLING OUT ODD SPACES

Typical Cross-section of Flangeforms
All dimensions are outside to outside.
Sizes of Ceco Flange Type Steelforms

regular widths
FOR FORMING REGULARLY SPACED JOISTS

filler widths
FOR USE ONLY IN FILLING OUT ODD SPACES

Typical Cross-sections of Longforms
All widths and depths are outside dimensions.
Sizes of Ceco Longform Steelforms

535

Concrete Required per Sq. Ft. (Sq.M) of Floor Using Ceco Flange* Type Steelforms 20 Inches (500 mm) Wide

Depth of Steelform	Width of Joist	Cu.Ft. of concrete per Sq.Ft. of floor for given slab thickness over			Add Conc. in CF/LF of beam**	Depth of Steelform	Width of Joist	Cu.M. of concrete per Sq.M. of floor for given slab thickness over steelform			Add Conc. in Cu.M./m of beam**
		2 "	2 1/2 "	3 "				50 mm	#### mm	75 mm	
8 "	4 "	0.298 "	0.340 "	0.382 "	0.17	200 mm	100 mm	0.091 mm	0.104 mm	0.116 mm	0.016
	5 "	0.320 "	0.362 "	0.404 "	0.16		125 mm	0.098 mm	0.110 mm	0.123 mm	0.015
	6 "	0.339 "	0.381 "	0.423 "	0.16		150 mm	0.103 mm	0.116 mm	0.129 mm	0.015
10 "	4 "	0.336 "	0.378 "	0.420 "	0.21	250 mm	100 mm	0.102 mm	0.115 mm	0.128 mm	0.020
	5 "	0.363 "	0.405 "	0.447 "	0.20		125 mm	0.111 mm	0.123 mm	0.136 mm	0.019
	6 "	0.387 "	0.429 "	0.471 "	0.19		150 mm	0.118 mm	0.131 mm	0.144 mm	0.018
12 "	4 "	0.377 "	0.419 "	0.461 "	0.25	300 mm	100 mm	0.115 mm	0.128 mm	0.141 mm	0.023
	5 "	0.408 "	0.450 "	0.492 "	0.24		125 mm	0.124 mm	0.137 mm	0.150 mm	0.022
	6 "	0.437 "	0.479 "	0.521 "	0.23		150 mm	0.133 mm	0.146 mm	0.159 mm	0.021
14 "	5 "	0.456 "	0.498 "	0.540 "	0.28	350 mm	125 mm	0.139 mm	0.152 mm	0.165 mm	0.026
	6 "	0.490 "	0.532 "	0.574 "	0.27		150 mm	0.149 mm	0.162 mm	0.175 mm	0.025
	7 "	0.522 "	0.564 "	0.606 "	0.26		175 mm	0.159 mm	0.172 mm	0.185 mm	0.024
16 "	5 "	0.507 "	0.549 "	0.590 "	0.32	400 mm	125 mm	0.155 mm	0.167 mm	0.180 mm	0.030
	6 "	0.545 "	0.587 "	0.629 "	0.31		150 mm	0.166 mm	0.179 mm	0.192 mm	0.029
	7 "	0.581 "	0.622 "	0.664 "	0.30		175 mm	0.177 mm	0.190 mm	0.202 mm	0.028

* The above concrete requirements are also applicable to one piece longforms

** Amount of concrete given for standard tapered endforms is one side of beam or bearing wall only where std. taper endform is used.

CONCRETE

Concrete Required per Sq. Ft. (Sq.M) of Floor Using Ceco Flange* Adjustable Steelforms 20 Inches (500 mm)Wide

Depth of Steelform	Width of Joist	Cu.Ft. of concrete per Sq.Ft. of floor for given slab thickness over			Add Conc. in CF/LF of beam**
		2 "	2 1/2 "	3 "	
8 "	3.5 "	0.289 "	0.340 "	0.382 "	0.17
	4.5 "	0.309 "	0.362 "	0.404 "	0.16
	5.5 "	0.331 "	0.381 "	0.423 "	0.16
10 "	3.5 "	0.312 "	0.378 "	0.420 "	0.21
	4.5 "	0.340 "	0.405 "	0.447 "	0.20
	5.5 "	0.367 "	0.429 "	0.471 "	0.19
12 "	3.5 "	0.336 "	0.419 "	0.461 "	0.25
	4.5 "	0.371 "	0.450 "	0.492 "	0.24
	5.5 "	0.403 "	0.479 "	0.521 "	0.23
14 "	3.5 "	0.361 "	0.498 "	0.540 "	0.28
	4.5 "	0.402 "	0.532 "	0.574 "	0.27
	5.5 "	0.438 "	0.564 "	0.606 "	0.26

Depth of Steelform	Width of Joist	Cu.M. of concrete per Sq.M. of floor for given slab thickness over steelform			Add Conc. in Cu.M./m of beam**
		50 mm	#### mm	75 mm	
200 mm	88 mm	0.088 mm	0.104 mm	0.116 mm	0.016
	113 mm	0.094 mm	0.110 mm	0.123 mm	0.015
	138 mm	0.101 mm	0.116 mm	0.129 mm	0.015
250 mm	88 mm	0.095 mm	0.115 mm	0.128 mm	0.020
	113 mm	0.104 mm	0.123 mm	0.136 mm	0.019
	138 mm	0.112 mm	0.131 mm	0.144 mm	0.018
300 mm	88 mm	0.102 mm	0.128 mm	0.141 mm	0.023
	113 mm	0.113 mm	0.137 mm	0.150 mm	0.022
	138 mm	0.123 mm	0.146 mm	0.159 mm	0.021
350 mm	88 mm	0.110 mm	0.152 mm	0.165 mm	0.026
	113 mm	0.123 mm	0.162 mm	0.175 mm	0.025
	138 mm	0.134 mm	0.172 mm	0.185 mm	0.024

* The above concrete requirements are also applicable to one piece longforms

** Amount of concrete given for standard tapered endforms is one side of beam or bearing wall only where std. taper endform is used.

Concrete Required per Sq. Ft. (Sq.M) of Floor Using Ceco Flange* Type Steelforms 30 Inches (750 mm)Wide

Depth of Steelform	Width of Joist	Cu.Ft. of concrete per Sq.Ft. of floor for given slab thickness over			Add Conc. in CF/LF of beam**	Depth of Steelform	Width of Joist	Cu.M. of concrete per Sq.M. of floor for given slab thickness over steelform			Add Conc. in Cu.M./m of beam**
		2 "	2 1/2 "	3 "				50 mm	#### mm	75 mm	
8 "	4 "	0.322 "	0.364 "	0.405 "	0.15	200 mm	100 mm	0.098 mm	0.111 mm	0.123 mm	0.014
	5 "	0.338 "	0.380 "	0.421 "	0.15		125 mm	0.103 mm	0.116 mm	0.128 mm	0.014
	6 "	0.510 "	0.392 "	0.434 "	0.14		150 mm	0.155 mm	0.119 mm	0.132 mm	0.013
10 "	4 "	0.353 "	0.395 "	0.436 "	0.19	250 mm	100 mm	0.108 mm	0.120 mm	0.133 mm	0.018
	5 "	0.372 "	0.414 "	0.455 "	0.18		125 mm	0.113 mm	0.126 mm	0.139 mm	0.017
	6 "	0.389 "	0.430 "	0.472 "	0.18		150 mm	0.119 mm	0.131 mm	0.144 mm	0.017
12 "	4 "	0.386 "	0.427 "	0.469 "	0.22	300 mm	100 mm	0.118 mm	0.130 mm	0.143 mm	0.020
	5 "	0.408 "	0.450 "	0.491 "	0.22		125 mm	0.124 mm	0.137 mm	0.150 mm	0.020
	6 "	0.430 "	0.470 "	0.512 "	0.21		150 mm	0.131 mm	0.143 mm	0.156 mm	0.020
14 "	5 "	0.420 "	0.460 "	0.503 "	0.26	350 mm	125 mm	0.128 mm	0.140 mm	0.153 mm	0.024
	6 "	0.447 "	0.488 "	0.530 "	0.25		150 mm	0.136 mm	0.149 mm	0.162 mm	0.023
	7 "	0.470 "	0.511 "	0.553 "	0.24		175 mm	0.143 mm	0.156 mm	0.169 mm	0.022
16 "	5 "	0.451 "	0.493 "	0.535 "	0.29	400 mm	125 mm	0.137 mm	0.150 mm	0.163 mm	0.027
	6 "	0.482 "	0.523 "	0.565 "	0.28		150 mm	0.147 mm	0.159 mm	0.172 mm	0.026
	7 "	0.510 "	0.552 "	0.594 "	0.27		175 mm	0.155 mm	0.168 mm	0.181 mm	0.025

* The above concrete requirements are also applicable to one piece longforms

** Amount of concrete given for standard tapered endforms is one side of beam or bearing wall only where std. taper endform

CONCRETE

Concrete Required per Sq. Ft. (Sq.M) of Floor Using Ceco Flange* Adjustable Steelforms 30 Inches (750 mm)Wide

Depth of Steelform	Width of Joist	Cu.Ft. of concrete per Sq.Ft. of floor for given slab thickness over			Add Conc. in CF/LF of beam**	Depth of Steelform	Width of Joist	Cu.M. of concrete per Sq.M. of floor for given slab thickness over steelform			Add Conc. in Cu.M./m of beam**
		2 "	2 1/2 "	3 "				50 mm	63 mm	75 mm	
8 "	5.5 "	0.326 "	0.368 "	0.408 "	0.15	200 mm	138 mm	0.099 mm	0.112 mm	0.124 mm	0.014
	6.5 "	0.341 "	0.383 "	0.424 "	0.15	200 mm	163 mm	0.104 mm	0.117 mm	0.129 mm	0.014
	7.5 "	0.355 "	0.397 "	0.438 "	0.14	200 mm	188 mm	0.108 mm	0.121 mm	0.134 mm	0.013
10 "	5.5 "	0.352 "	0.394 "	0.434 "	0.19	250 mm	138 mm	0.107 mm	0.120 mm	0.132 mm	0.018
	6.5 "	0.371 "	0.413 "	0.454 "	0.18	250 mm	163 mm	0.113 mm	0.126 mm	0.138 mm	0.017
	7.5 "	0.389 "	0.430 "	0.472 "	0.18	250 mm	188 mm	0.119 mm	0.131 mm	0.144 mm	0.017
12 "	5.5 "	0.378 "	0.420 "	0.461 "	0.22	300 mm	138 mm	0.115 mm	0.128 mm	0.141 mm	0.020
	6.5 "	0.401 "	0.443 "	0.484 "	0.22	300 mm	163 mm	0.122 mm	0.135 mm	0.148 mm	0.020
	7.5 "	0.422 "	0.464 "	0.505 "	0.21	300 mm	188 mm	0.129 mm	0.141 mm	0.154 mm	0.020
14 "	5.5 "	0.404 "	0.446 "	0.486 "	0.26	350 mm	138 mm	0.123 mm	0.136 mm	0.148 mm	0.024
	6.5 "	0.430 "	0.472 "	0.514 "	0.25	350 mm	163 mm	0.131 mm	0.144 mm	0.157 mm	0.023
	7.5 "	0.456 "	0.497 "	0.539 "	0.24	350 mm	188 mm	0.139 mm	0.151 mm	0.164 mm	0.022

* The above concrete requirements are also applicable to one piece longforms

** Amount of concrete given for standard tapered endforms is one side of beam or bearing wall only where std. taper endform

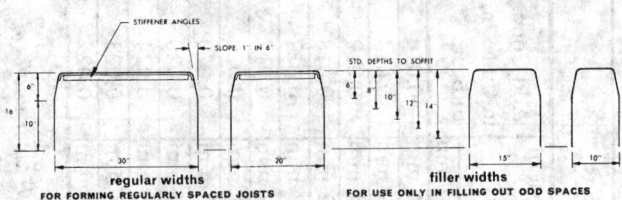

Typical Cross-sections of Adjustable Steelforms
All dimensions are outside to outside.
Sizes of Ceco Adjustable Type Steelforms

Ceco One Piece Steeldomes for Two-Way Dome Slab Construction. To estimate the quantity of concrete required for steeldome construction, figure the concrete required for a solid slab of the depth of the dome plus the top slab, and subtract the voids created for each steeldome as given in the table below.

Table of Concrete Voids				
Voids Created By Steeldome, Cubic Feet				
Depth of Steeldome	19" x 19" System		30" x 30" System	
	Regular 19" x 19"	Regular 30" x 30"	Regular 20" x 30"	Regular 20" x 20"
8 "	1.41	3.85	2.54	1.65
10 "	1.90	4.78	3.13	2.06
12 "	2.14	5.53	3.63	2.41
14 "	-----	6.54	4.27	2.87
16 "	-----	7.44	4.85	3.14
20 "	-----	9.16	5.90	3.81
Voids Created By Steeldome, Cubic Meter				
Depth of Steeldome	475 x 475 mm System		450 x 750 mm System	
	Regular 475 x 475mm	Regular 750 x 750mm	Regular 500 x 750mm	Regular 500 x 500mm
200 mm	0.040	0.109	0.072	0.047
250 mm	0.054	0.135	0.089	0.058
300 mm	0.061	0.157	0.103	0.068
350 mm	-----	0.185	0.121	0.081
400 mm	-----	0.211	0.137	0.089
500 mm	-----	0.259	0.167	0.108

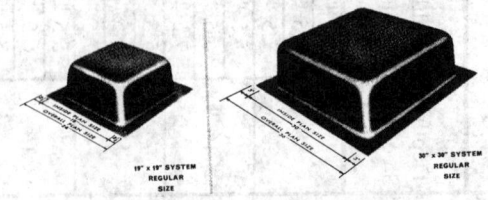

CONCRETE

Item	19" x 19" System	30" x 30" System		
	Standard Size	Standard Size	Filler Size	Filler Size
Plan size of void	19" x 19"	30" x 30"	20" x 30"	20" x 20"
Overall plan size including flanges	24" x 24"	36" x 36"	26" x 36"	26" x 26"
Width of flange	2 1/2"	3"	3"	3"
Depth of domes	4", 6", 8", 10" & 12	4", 6", 8", 10" & 12	8", 10", 12", 14", 16" & 20"	8", 10", 12", 14", 16" & 20"

Item	475 x 475 mm System	750 x 750 mm System		
	Standard Size	Standard Size	Filler Size	Filler Size
Plan size of void	475 x 475 mm	750 x 750 mm	500 x 750 mm	500 x 500 mm
Overall plan size including flanges	600 x 600 mm	900 x 900 mm	650 x 900 mm	650 x 650 mm
Width of flange	62.5 mm	75 mm	75 mm	75 mm
Depth of domes	100, 150, 200, 250, 300 mm	100, 150, 200, 250, 300 mm	200, 250, 300, 350, 400, 500 mm	200, 250, 300, 350, 400, 500 mm

Lightweight Concrete, Insulating Concrete of Pumice, Perlite, Vermiculite and Other Lightweight Aggregates

Lightweight concrete of crushed slag, foamed slag, haydite, pumice, perlite, and vermiculite are used for floor and roof fill, structural roofs, floors in cold storage rooms, and the like. A comparison of concrete aggregates is as follows:

Type of Aggregate	Lbs/Cu.Ft.	Lbs/Cu.Ft. of Concrete	Kg/Cu.M.	Kg/Cu.M. of Concrete
Gravel	120	150	1922	2403
Sand	90 - 100	150	1442 - 1602	2403
Crushed stone	100	145	1602	2323
Crushed Bank Slag	80	100 - 130	1281	1602 - 2082
Haydite	40 - 60	100 - 130	641 - 961	1602 - 2082
Foamed slag	40 - 60	90 - 100	641 - 961	1442 - 1602
Cinders	40 - 50 *	110 - 115	641 - 801 *	1762 - 1842
Pumice	30 - 60	60 - 90	481 - 961	961 - 1442
Diatomite	28 - 40	55 - 70	449 - 641	881 - 1121
Perlite	6 - 16	20 - 50	96 - 256	320 - 801
Vermiculite	6 - 10	20 - 40	96 - 160	320 - 641
Waylite	40 - 60	90 - 100	641 - 961	1442 - 1602
Expanded Shale	44 - 58	92 - 98	705 - 929	1474 - 1570

Plus sand.

Vermiculite Insulating Concrete Aggregate. This is a type of mica which is mined, crushed, and screened to size. When subjected to a temperature of about 2,000°F (1,093°C), it expands into a laminated granule containing millions of tiny dead air cells. This produces a lightweight aggregate of 6 to 10 lbs. per c.f. (97-162 kg/cu.m), compared with sand or crushed rock aggregate weighing about 100 lbs. per c.f. (1620 kg/cu.m). Vermiculite has high insulating value, is fireproof, and will not rot or decay.

Vermiculite insulating concrete is a lightweight building material made like ordinary portland cement concrete, except vermiculite concrete aggregate is used instead of sand, gravel, or crushed stone. Depending on the amount of vermiculite aggregate used, it is possible to make concrete weighing from 20 to 40 lbs. per c.f. (324-648 kg/cu.m) Ordinary concrete weighs 145 to 150 lbs. per c.f. (2350-2430 kg/cu.m).

Vermiculite concrete is used for poured concrete roof decks over a variety of forming materials; for roof insulation over such surfaces as concrete, steel, wood, etc.; and for poured concrete floor slabs.

Because of its high insulation value, vermiculite concrete is an excellent base for radiant heat floor installations and for grade level floors without radiant heat in homes, farm structures, and all types of commercial

542

buildings. At a density of approximately 25 lbs. per c.f. (405 kg/cu.m), 1" (25 mm) of vermiculite concrete has an insulating value equal to approximately 16" (400 mm) of ordinary concrete. The density, insulating value, and strength of vermiculite concrete can be varied to meet a wide range of design requirements.

Vermiculite concrete is incombustible and has earned the highest fire ratings attainable. It carries a 5-hour fire rating for spandrel wall construction and a 1-1/2 hour rating on a steel deck covered with 1-1/2" (37.50mm) of vermiculite concrete and completely unprotected on the underside.

Mixing. Vermiculite concrete should be mixed in a mechanical mixer. The required amount of water and cement is placed in the mixer and then the aggregate. Mixing is limited to the minimum time required to obtain a thorough mix and proper fluidity. (Maximum time recommended, 5 minutes).

When transit-mixed vermiculite concrete is used, the operation is as follows:

1. Introduce water and cement into mixer. (Fill auxiliary water tank before leaving plant.)
2. Rotate the mixer slowly until all aggregate has been added.
3. Continue to rotate mixer for approximately one minute after aggregate is in the mixer.
4. Do not rotate drum on way to job site. Mix concrete at job site at fastest speed until it is uniform and flows freely from the mixer.

Placing. Vermiculite concrete is transported and placed immediately after mixing is completed, and the period between completion of the mixing and placing should be short enough so that the mixture does not appreciably change in consistency.

Vermiculite concrete should not be placed when the temperature is less than 40oF (4.4°C). If the temperature expected after placing of the concrete is near or below 40°F (4.4°C), the mixing water should be heated in the temperature range 75° to 100°F (23.8-37.7°C).

Vermiculite concrete can be placed by the use of standard concrete equipment such as bucket hoists, buggies, or specially designed pumping equipment.

Machine Placement. Pumping or spraying the concrete is especially suited to roof decks with curved, sloped, or irregular surfaces and is just as efficient on flat roof decks.

A pump is fast and simple. It can be set up or taken down in about 30 minutes. The long hose lengths that are possible enable the crew to get over obstructions on the job easily.

The Vermiculite Institute annually approves a national roster of roof deck applicators appointed by institute members to assure the highest quality of finished work.

Vermiculite concrete is covered by the United States of America Standards Institute Specification A122.1-1965.

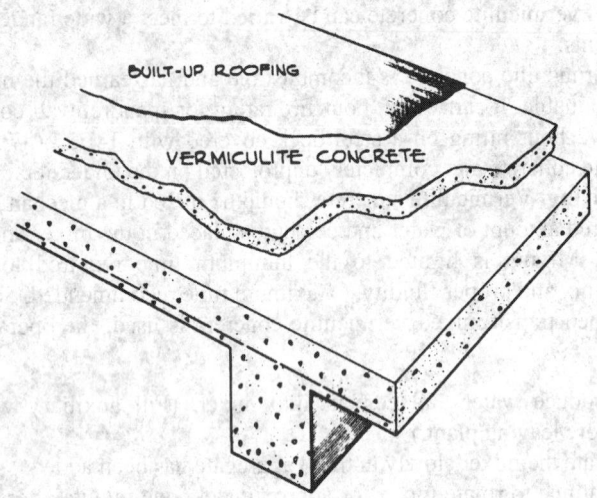

Vermiculite Concrete Roof Insulation Poured Over Structural Concrete Roof Decks

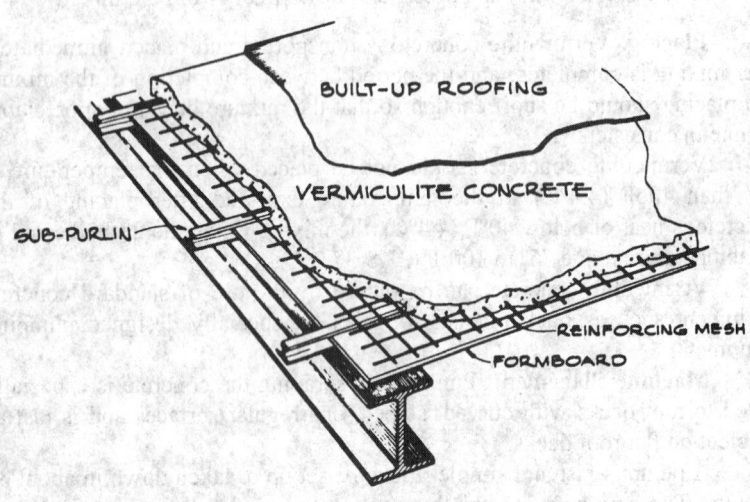

Vermiculite Concrete Roof Deck Poured in Place Over Form Boards

544

CONCRETE

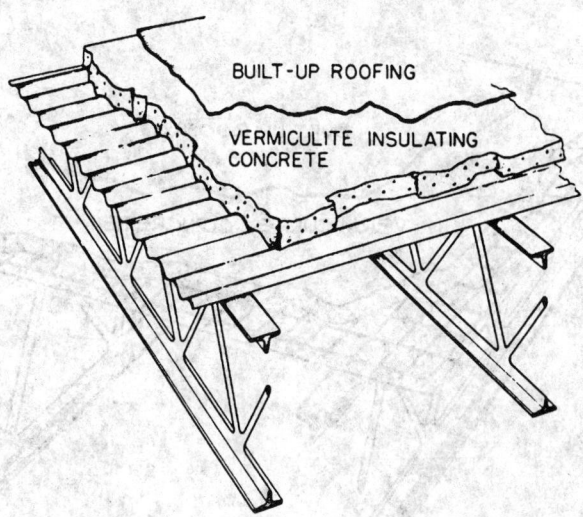

BUILT-UP ROOFING

VERMICULITE INSULATING CONCRETE

Vermiculite Insulating Concrete Over Vented Metal Roof Decks

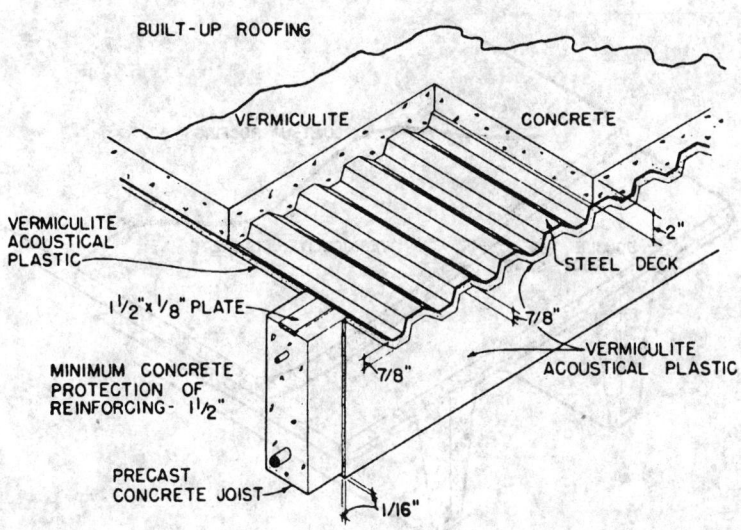

BUILT-UP ROOFING

VERMICULITE   CONCRETE

VERMICULITE ACOUSTICAL PLASTIC

1½"x ⅛" PLATE

MINIMUM CONCRETE PROTECTION OF REINFORCING- 1½"

PRECAST CONCRETE JOIST

2"

STEEL DECK

7/8"

7/8"

VERMICULITE ACOUSTICAL PLASTIC

1/16"

Vermiculite Concrete Insulation Over Vented Galvanized Steel Roof Decks,
Precast Concrete Joists (Two-hour Fire Rated)

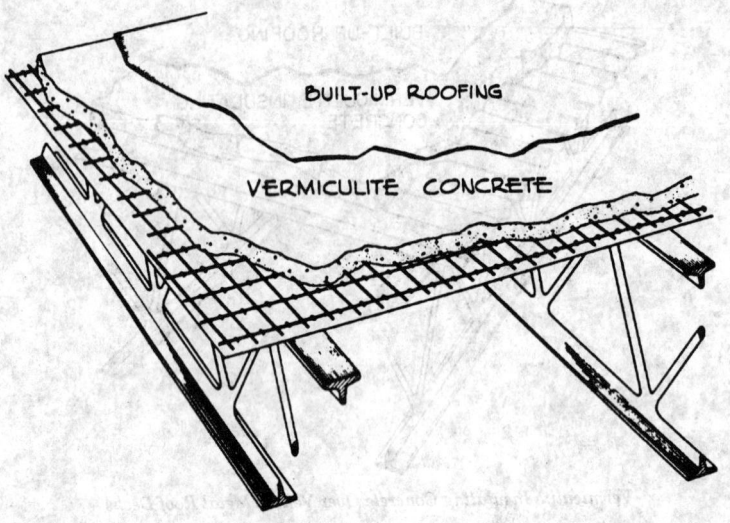

Vermiculite Concrete Roof Deck
Poured-in-Place Over Paper-Backed Wire Lath

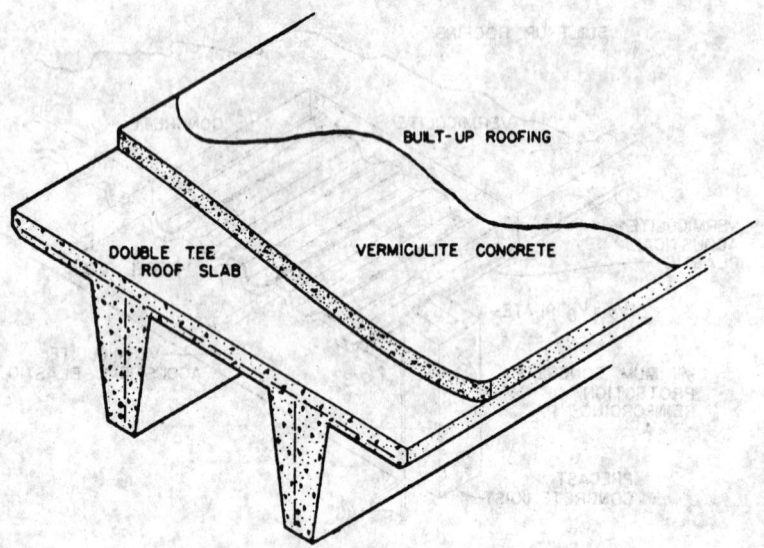

Vermiculite Concrete Insulation Over Precast Prestressed
Concrete Double Tee Roof Slabs

CONCRETE

Vermiculite Insulating Concrete Floors-on-Grade. A 1:4 mix of vermiculite concrete is recommended as an insulating base under a sand-gravel concrete topping for floors on grade. It is particularly effective in reducing heat loss where radiant heating coils or ducts are carried in the floor.

Minimum thickness of the slab should be 3" (75 mm). A reinforcing mesh of 6"x6" (150x150mm) 10/10 ga. welded wire fabric or equal should be placed in the center of the vermiculite concrete, lapping the sides and ends one full mesh.

The sand-gravel topping should be designed to carry the maximum floor load. Minimum thickness should be 1-1/2" (37.50 mm) except when radiant heating coils are used. In that case minimum thickness over the top of the coils should be 1" (25 mm). The radiant heating coils should be supported on chairs to assure a minimum of 1/2" (12.50 mm) of sand-gravel concrete below the coils.

Proportion Table for Vermiculite Concrete				
Description of Work	Vermiculite Aggregate	Portland Cement Cu.Ft.	Vermiculite Aggregate Cu.M.	Portland Cement Cu.M.
Poured concrete roofs	4	1	0.113	0.028
Poured concrete roof insulation	6 or 8	1	0.170 or 0.227	0.028
Poured concrete slabs on grade insulating slab under 1 1/2" ordinary cement topping	4	1	0.113	0.028
Panel or spandrel walls	4	1	0.113	0.028

Mixing Table For Vermiculite Concrete					
Mix by Vol. Vermiculite Cu.Ft.	Portland Cement Sacks	Approx. Density in Lbs/Cu.Ft.		Compressive Strength in psi	
8	1	20 to 25		100 to 125	
6	1	25 to 30		125 to 225	
4	1	35 to 40		350 to 500	
Mix by Vol. Vermiculite Cu.M.	Portland Cement Sacks	Approx. Density in Kg/Cu.M.		Compressive Strength in MPa	
0.227	1	320 to 400		0.69 to 0.86	
0.170	1	400 to 481		0.86 to 1.55	
0.113	1	561 to 641		2.41 to 3.45	

Approximate Quantities Required for One Cu.Yd. of Ready Mixed Vermiculite Concrete			
Mix	4 Cu.Ft. Bags Vermiculite Conc. Aggregate	Sacks (1 cu.ft) Portland Cement	Water Gallons
8-1	7.5	3.75	85 to 95
6-1	7.5	5.00	85 to 95
4-1	7.5	7.50	85 to 95
Mix	0.11 Cu.M. Bags Vermiculite Conc. Aggregate	Sacks (0.028 Cu.M.) Portland Cement	Water Liters
8-1	7.5	3.75	322 to 360
6-1	7.5	5.00	322 to 360
4-1	7.5	7.50	322 to 360

Perlite Concrete Aggregate

Perlite is a siliceous volcanic rock mined in western United States. When crushed and quickly heated to above 1500°F (815.55°C), it expands to form lightweight, non-combustible, glass-like particles of cellular structure. This material, white or light gray in color, weighs 7-1/2 to 15 lbs. per c.f. (121-242 kg/m3).

Perlite aggregate consists of expanded perlite sized for use in lightweight insulating concrete in place of sand or gravel. It is usually packed in 3 or 4 c.f. (0.08 or 0.11 cu.m) paper bags for easy handling.

The many, tiny glass sealed cells in each particle of expanded perlite make it highly insulating as well as comparatively nonabsorptive. Thus perlite mixes with about 30% less water than comparable lightweight aggregates.

Perlite concrete, in combination with portland cement, forms a very lightweight concrete. Where ordinary concrete weighs 150 lbs. per c.f. (2430 kg/cu.m.), the dry weight of perlite concrete can be designed from 20 to 40 lbs. per c.f. (324-648 kg/m3). The extremely light weight and ease of handling make perlite concrete adaptable to practically any shape.

Perlite concrete has received up to 4-hour ratings in Underwriters' Laboratories fire tests. Its light weight makes it suitable for fireproof roof and floor fills, for thin concrete curtain walls (blocks, slabs or monolithic), and for many precast panel and block constructions.

For lightweight structural roof construction, where perlite concrete is placed over galvanized steel forms, paper-backed wire mesh, formboards, structural concrete, or other suitable form materials, use a 25 to 29 lb. per c.f. (405-470 kg/cu.m.) density perlite concrete. This offers an ideal balance of low dead weight, adequate compressive and indentation strengths, and good insulating value. For uses where higher strengths are more important than insulating value, i.e., floor fills and certain lightweight structural roof designs, use a 34 to 40 lb. per c.f. (550-648 kg/cu.m) density perlite concrete.

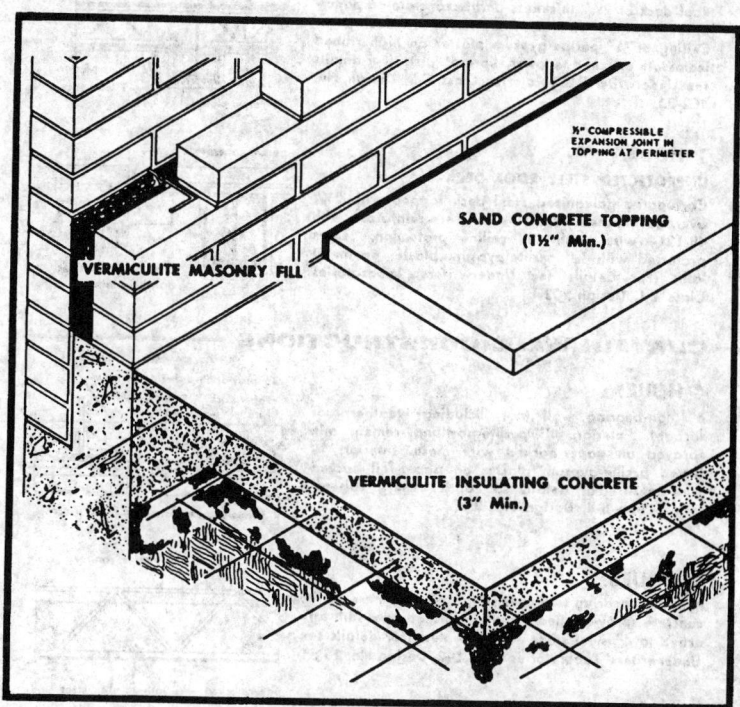

Vermiculite Insulating Concrete Floor

— ROOF CONSTRUCTIONS —

3 HOURS:

Roof deck of 2" minimum thickness perlite concrete on 28 gauge galvanized steel form units supported by steel joists 4' on center. Ceiling of ⅞" perlite-gypsum plaster on expanded metal lath attached to ¾" furring channels wire-tied to lower chord of joists. (For details see Underwriters' Laboratories Class C-3, Design No. RC1-3.)

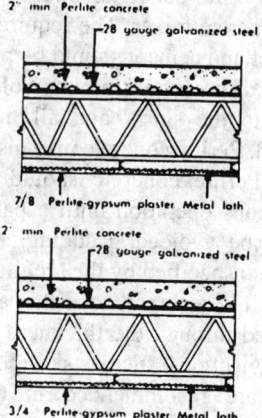

2 HOURS:

Same as 3 hour rated system above except that the perlite-gypsum plaster ceiling shall be ¾" thick. (For details see Underwriters' Laboratories Class D-2, Design No. RC11-2.)

549

3 HOURS:

Roof deck of 2½" thickness perlite concrete on paper-backed welded wire mesh supported by steel joists. Ceiling of ¾" perlite-gypsum plaster on high ribbed metal lath attached to lower chord of joists. (For details see Underwriters' Laboratories Class C-3, Design No. RC2-3.)

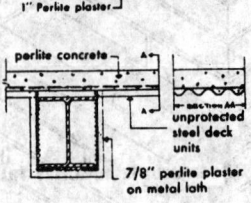

1 HOUR:

UNPROTECTED STEEL ROOF DECK

Corrugated galvanized steel deck topped with 2⅝" average thickness of perlite concrete reinforced with 48-1214 wire mesh. No ceiling protection. Beams protected with ⅞" perlite-gypsum plaster on metal lath. (For details see Underwriters' Laboratories Class E-I, Design RC2-1.)

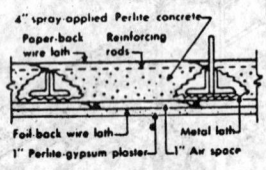

— CURTAIN WALL CONSTRUCTIONS

4 HOURS:

6" non-bearing wall (not including weatherproof facade). Exterior: 4" perlite-portland cement mix sprayed on paper-backed wire mesh. Interior: 1" furred perlite-gypsum plaster on paper-foil backed wire mesh. (For details see Underwriters' Laboratories Class B-4, Design No. 3.)

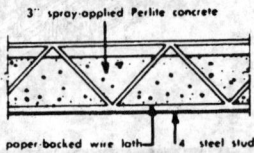

2 HOURS:

Non load-bearing wall consisting of 3" thickness perlite concrete spray-applied to paper-backed wire lath att-ached to 4" steel studs 16" on center. (For details see Underwriters' Laboratories Class D-2, Design No. 18.)

Perlite concrete should be mixed in a paddle type plaster or a drum type concrete mixer. The required amount of water, air entraining admixture, and portland cement should be placed in the mixer and mixed until a slurry is formed. The proper amount of perlite concrete aggregate should then be added to the slurry and all materials mixed until design wet density is reached. Perlite concrete may also be transit mixed.

Perlite concrete should be carefully deposited and screeded in a continuous operation until a panel or section is completed. Steel troweling should be avoided. Rodding, tamping, and vibrating should not be used unless so specified by the architect.

It is recommended that a 1" (25 mm) air space or expansion joint be provided through the thickness of the perlite concrete at the juncture of all roof projections, such as skylights, penthouses, and parapet walls. Expansion joints are recommended in accordance with good concrete and steel construction practice and are based upon the provision that suitable consideration has been given to the use of adequate through-building expansion joints. Other types of insulating concrete may not require expansion joints because of high initial shrinkage.

CONCRETE

Perlite concrete, during the curing period, should be protected for at least the first three days to keep it from drying out too rapidly or freezing. Freshly poured concrete should also be given adequate protection from heavy rain. Allow no traffic until concrete can sustain a man's weight without indentation.

Perlite concrete should not be placed in temperatures under 40°F (4.4°C) nor on frosted surfaces. In near-freezing weather the mixing water should be heated to between 75° and 100°F (23.88-37.77°C).

Perlite concrete is also used as a base for radiant heating coils. For grade level floors a moisture resistant barrier should be used and a suitable wearing surface applied over the perlite insulating concrete.

Considerable savings in dead weight are possible by using perlite concrete as a floor fill over cellular steel or pan-type floors in multiple story buildings. The surface may be covered with concrete topping and other surfacing materials after the perlite concrete has cured several days. Another floor fill material, using perlite blended with other aggregates, is perlite/sand or perlite/expanded shale concrete. These concretes have higher compressive strengths and require only resilient flooring or carpeting as a wearing surface. Following are the results of tests conducted at nationally recognized laboratories. Due to the varying characteristics of naturally occurring aggregates in different geographical areas, trial mixes of aggregate blends are recommended for determination of mix proportions. Perlite concrete can be nailed, sawed, and worked with ordinary carpenter tools.

Mix Proportions By Volume for Lightweight Perlite Insulating Concrete					
Cement (Sacks)	Perlite Cu.Ft.	Water, Gals/sack cement	Air-Entraining Agent-Pints	Density, Lbs/Cu.Ft. Oven-dry	Compressive Srength psi 28 days
1	3	8.5	0.375	72	1000
1	3	8.5	0.25	54	1200
1	4	9	1	36	400
1	5	11	1 1/4	30 1/2	250
1	6	12	1 1/2	27	170
1	7	14	1 3/4	24	120
1	8	16	2	22	95
Cement (Sacks)	Perlite Cu.M.	Water, Liter/sack cement	Air-Entraining Agent-Liters	Density, Kg/Cu.M. Oven-dry	Compressive Srength MPa 28 days
1	0.085	32.18	0.177	1153	6.89
1	0.085	32.18	0.118	865	8.27
1	0.1133	34.07	0.473	577	2.76
1	0.1416	41.64	0.591	489	1.72
1	0.1699	45.43	0.710	432	1.17
1	0.1982	53.00	0.828	384	0.83
1	0.2265	60.57	0.946	352	0.65

Material Required for One Cu. Yd. of Placed Perlite Concrete				
Proportions Cement Perlite	Cement Sacks	Perlite Cu.Ft.	Water Gallons	Air-Entraining Agent - Pints
1:4	6.75	27	61	6 3/4
1:5	5.40	27	59 1/2	6 3/4
1:6	4.50	27	54	6 3/4
1:7	3.85	27	54	6 3/4
1:8	3.38	27	54	6 3/4
Material Required for .7646 Cu. Meter of Placed Perlite Concrete				
Proportions Cement Perlite	Cement Sacks	Perlite Cu.Meter	Water Liters	Air-Entraining Agent - liters
1:4	6.75	0.765	230.89	3 5/6
1:5	7.06	0.765	225.21	3 5/6
1:6	5.89	0.765	204.39	3 5/6
1:7	5.04	0.765	204.39	3 5/6
1:8	4.42	0.765	204.39	3 5/6

Pumice Concrete

Pumice is a natural white or gray glass foam. It is not a volcanic ash. It is mined extensively in the western part of the United States.

Uniform graded pumice is used as aggregate in monolithic concrete, concrete masonry units and precast products. It replaces sand and gravel in any concrete up to strengths of 3,000 psi (20.7 MPa), where thermal insulation, acoustical, and light weight properties are desirable.

All pumice aggregate should be clean, free of all foreign matter, and well graded so as to meet ASTM specifications *C330-04 Standard Specification for Lightweight Aggregates for Structural Concrete.*

All water used in concrete work shall be free from oil, strong acids, alkali, or organic material.

Mixing. Concrete for columns, walls above ground, floors, window sills, and all self-supporting floor and roof slabs, canopies, and other parts of structure not coming in contact with ground shall be of a mixture of water, portland cement, and pumice aggregate, and shall develop at 28 days a strength of not less than 2,500 psi. (17.2 MPa) and with a slump of not more than 2" (50 mm) by standard test.

Water contained in the aggregate shall be deducted from the amount of water used in the mix except that the water needed to saturate the pumice aggregate shall not be deducted. Pumice aggregate shall be thoroughly saturated either in the stockpile or in the mixer prior to the addition of cement and water for mixing.

CONCRETE

Suggested Mix Design Data for Pyramid Pumice Concrete

Compressive Strength. psi	Design Weight Lbs/Cu.Ft.	Sand Fine Cu.Ft.	Sand Course Cu.Ft.	Sand Blend Cu.Ft.	Cement Lbs/Cu.Yd.	Approx. Water Gal	Slump, Inches
150	40	0	36	0	376	81.1	0
250	46	24	12	36	282	81.8	0
500	55	24	12	36	376	82.8	0
1000	60	24	12	36	470	84.8	1/2
1500	65	24	12	36	611	86.8	1/2
2000	70	24	12	36	705	87.8	1
2500	75	24	12	36	752	82.8	2

Compressive Strength, Mpa	Design Weight Kg/Cu.M.	Sand Fine Cu.M.	Sand Course Cu.M.	Sand Blend Cu.M.	Cement Kg/Cu.M.	Approx. Water Liters	Slump, mm
1.034	640.73	0	1.019	0.000	223.07	307.00	0
1.724	736.84	0.680	0.340	1.019	167.30	309.65	0
3.447	881.01	0.680	0.340	1.019	223.07	313.43	0
6.894	961.10	0.680	0.340	1.019	278.84	321.00	12.5
10.342	1041.19	0.680	0.340	1.019	362.49	328.58	12.5
13.789	1121.29	0.680	0.340	1.019	418.26	332.36	25
17.236	1201.38	0.680	0.340	1.019	446.14	313.43	50

Pyramid Pumice Aggregate. This should be saturated prior to the addition of cement. It may be done in the stock pile or in the mixer. A 1" (25 mm) slump in pumice concrete will give a consistency equal to a 3" (75 mm) slump in hard rock concrete. Pumice concrete should be mixed with as little water as possible to get the required workability and should be the consistency of thick mud. No free water should be in evidence. Air entraining agents are recommended to cut water-cement ratio and get greater workability. A combination of fine and coarse aggregate or the blended aggregate should be used. Volume measurements are based on loose volume non-compacted pumice.

Placing. All pumice concrete shall be thoroughly vibrated with a mechanical vibrator. Vibration shall be conducted in such a manner as to assure that the vibrator passes through the pumice mass with not more than 6" (150 mm) of space between the successive positions of the vibrator. In all cases where pumice is placed in layers, the vibrator must penetrate into the underlying concrete layer after passing through the newly placed layer. Ample time should be allowed for the settlement of the concrete before final topping or screeding to finished floor level or grade. The pumice aggregate shall be thoroughly saturated prior to mixing with the cement and water used for mixing and ample time shall be allowed for the saturation to take place.

Application. Pumice must be handled carefully if it is to be successful. Because it will float, it must be pre-saturated before use. The stock pile should be sprayed for at least 48 hours or the pumice should be saturated in the mixer before use. Unless this is done, cement particles will be drawn into the cells of the pumice and excessive shrinkage may occur as a result of loss of water from the mix to the pumice particles. One half of the total water required should be added to the aggregate before the cement is introduced.

The concrete should be water cured for at least seven days, after which time the outer fibers will have become strong and dense enough to protect against evaporation and overly rapid drying.

Gradations Available. Pumice aggregate is graded to meet ASTM specifications for properly graded lightweight aggregates for concrete. Designation C-330 in sizes No. 4 to 0 for fine aggregate; 3/8 (9 mm) and 1/2 (13 mm) to 0 for blended aggregate; 1/2" (12.50 mm) to No. 4 for coarse aggregate.

Shipping Weight. Fine material or 1/4" (6 mm) to 0 aggregate will have an average weight of about 1,200 lbs. per c.y. (711 kg/m3). Blended aggregate or 3/8" (9.37 mm) to 0 material will weigh about 1,100 lbs. per c.y. (625 kg/m3). The coarse aggregate or 1/2" (12.50 mm) to No. 4 material will have an average shipping weight of about 800 lbs. per cu.yd. (474 kg/m3). These weights may increase from 100 to 200 lbs. per cu.yd. (59-119 kg/cu.m) during wet seasons. An open top, side dump, flat bottom (Caswell) car carries approximately 63 cu.yd. (48 cu.m) of material.

CONCRETE

An open top, solid bottom, gondola car will carry approximately 63 cu.yd. (48 cu.m) of material.

An open top, hopper bottom car will vary in capacity from 60 to 100 cu.yd. (46-77 cu.m).

Waylite Lightweight Aggregate

A lightweight, cellular aggregate, refined and expanded by a process of agitating and rapidly cooling in an atmosphere of steam molten hot slag as it comes from the blast furnace at temperatures of 2,500 to 3,000°F. In this process the inherent gases expand and fill the material with minute air cells or bubbles completely sealed. These minute air cells are the active insulation and lightweight principle in all Waylite aggregates.

Waylite aggregate crushed and screened is available in three commercial sizes and average dry loose weights per cubic foot.

Aggregate	Grading, inch	Avg Wgt Lbs	Grading, mm	Avg Wgt Kg
Coarse	7/16 - 3/16	40	10.94 - 4.69	18.14
Acoustical	3/16 - 3/32	45	4.69 - 2.34	20.41
Fine	1/8 - Dust	60	3.13 - Dust	27.22

Waylite concrete requires a higher proportion of fine aggregate to coarse than ordinary concrete. This is due to the lower top sizes of the coarse aggregate – 7/16" (10.93 mm) – and the additional angularity or harshness of the crushed fine and coarse aggregates.

Waylite Concrete Mix Data					
Mix Loose Volume	Gal. Water per Sack	Cement Sacks	Fine Waylite Cu.Yd.	Course Waylite Cu.Yd.	Sand* Cu.Yd.
1 - 2-1/4 - 2-1/2 - 3/4*	9.4	6.4	0.53	0.59	0.18
1 - 2 - 2-1/4 - 3/4*	8.2	7.2	0.53	0.6	0.2
1 - 1-3/4 - 2-1/2*	7.3	8.1	0.54	0.6	0.15
1 - 1-1/2 - 2-1/2*	6.5	8.8	0.49	0.65	0.16
1 - 1-1/2 - 1-3/4 - 1/2*	5.9	9.6	0.53	0.62	0.18
Mix Loose Volume	Liters per Sack	Cement Sacks	Fine Waylite Cu.M.	Course Waylite Cu.M.	Sand* Cu.M
1 - 2-1/4 - 2-1/2 - 3/4*	35.58	6.4	0.405	0.451	0.138
1 - 2 - 2-1/4 - 3/4*	31.04	7.2	0.405	0.459	0.153
1 - 1-3/4 - 2-1/2*	27.63	8.1	0.413	0.459	0.115
1 - 1-1/2 - 2-1/2*	24.61	8.8	0.375	0.497	0.122
1 - 1-1/2 - 1-3/4 - 1/2*	22.33	9.6	0.405	0.474	0.138

*Sand shall be a fine siliceous bank, lake, or mason sand.
The addition of approved concrete air entraining agents materially improve the workability of Waylite concrete mixes.

555

The mixing of Waylite concrete requires a somewhat different procedure than for heavy aggregates. To obtain maximum workability with minimum weight, the aggregate and water should be mixed for one-half minute before the addition of the cement, and then mixed for at least an additional two minutes. A mix that appears harsh at first will improve in plasticity with continued mixing without additional water.

The addition of approved concrete air entraining agents materially improve the workability of Waylite concrete mixes.

PLACING, PATCHING AND
FINISHING ARCHITECTURAL CONCRETE

Architectural concrete demands the use of a plastic, workable mix that can be worked and molded readily in the forms. Surfaces of the hardened concrete should be free of defects such as sand streaking and honeycomb. It is important that sufficient fines be provided in the mixture, and fine aggregate should have at least 15% passing through a 50 mesh sieve and 3% passing the 100-mesh sieve, and the cement factor should not be less than 5-1/2 sacks per cu.yd. (7.2 sacks per cu.m) of concrete. In other words, the concrete should not be richer than 1:2:3 by loose volumes and preferably 1:2-1/3:3-1/3, containing 6-1/2 or 6 sacks of cement per cu.yd. (7.85 or 8 per cu.m) of concrete respectively.

Not more than 6-1/2 gals. (24.6 liters) of water per sack of cement should be allowed so that the concrete will have the durability required to withstand weathering.

Mixing and Placing Concrete. Concreting of architectural concrete walls should proceed at a rather slow rate allowing sufficient time for thorough vibrating at the finished surfaces so that the pressure exerted by the concrete on the forms does not exceed 600 lbs. per s.f. (2929 kg/cu.m).

A reasonable rate of placing the concrete should be considered when locating joints. Generally a rate of not more than 2 ft. (0.61 m) per hr. is most conducive of good workmanship and lifts of 6 to 8 ft. (1.8-2.4 m) between joints are desirable and 12 ft. (3.65 m) should be the absolute maximum.

The concrete should be spaded or vibrated into all corners and along all form surfaces, but even if vibrated, some spading is necessary in the corners and angles. Vibrators will not fill these places at all times. On many jobs small electric hammers moved along the forms at the point of placement have been found helpful. Vibration should be sufficient only for complete compaction, not to the point where segregation occurs and excess water is forced to the form faces. Such over-vibration may result in sand streaking and discoloration.

When concrete is splashed on the forms above the point of deposit and subsequently hardens, the wall in that area is likely to be rough and discolored. Splash boards are sometimes used but are not completely satisfactory, as they do not prevent segregation of the falling concrete. Tremies of various lengths equipped with a hopper at the top through which the concrete is placed will avoid both splash and segregation. The tremies should be spaced not more than about 9 ft. (2.74 m) apart. Sometimes splashing is the result of mortar being thrown from vibrators as they are lifted out of the concrete and raised to the placing platform. This can be prevented by shutting off the power as the vibrator is withdrawn from the concrete.

Labor Placing Concrete. The labor cost of placing architectural concrete will vary considerably, depending on the size of job, size of sections, amount of ornamentation, quantity of reinforcing steel, etc., but on most work of this class, it will require 2 to 3 hrs. labor time per cu.yd. (2.6-4 hrs. per cu.m) of concrete, plus the time for the hoisting engineer and foreman.

Based on a 64-cu.yd. pour per 8-hr. day, the labor cost should average as follows per cu.yd. (cu.m):

Description	Hours	Rate	Total
Foreman	0.13	$ 39.04	$ 5.08
Hoist engineer	0.13	$ 37.89	$ 4.93
Labor	2.50	$ 26.16	$ 65.40
Cost per Cu.yd.			$ 75.40
Cost per Cu.M.			$ 98.61

Curing. On this type of work one of the surest and best ways of curing the concrete is by leaving the forms in place. Whenever it is practical to do so, they should be left in place 4 or 5 days. By that time the concrete is hard enough so that it is not scuffed nor are corners so likely to be broken off during form removal. If the forms are removed earlier, other means must be provided to keep the concrete wet or damp for at least 5 days if a durable concrete is to be obtained.

Patching and Finishing. When the forms are first stripped, the concrete does not always present a pleasing appearance. There may be pock marks, air holes, fins, and some discoloration. Some of the latter will bleach out upon exposure but a simple clean down with cement grout will do wonders for the appearance. The projecting fins and other blemishes extending beyond the face of the wall are removed with a carborundum stone or abrasive hone and then a grout of 1 part portland cement to 1-1/2 parts of very fine sand (sand passing through a fly screen will do) is applied. From 1/3 to 1/2 of the cement should be white cement.

The grout is mixed to the consistency of a heavy paste. After the wall is wet down, the grout is applied uniformly with a stiff fiber brush, completely filling all air voids and holes. Immediately after the grout has been applied, the wall is floated with a wood or cork float. A very fine abrasive hone can also be used. The grout is then allowed to harden partially, the exact time depending upon the weather. When it is hardened sufficiently so it will not be pulled out of the holes, all grout is scraped from the surface of the wall with the edge of a steel trowel. After the wall has dried thoroughly, it is rubbed with a dry piece of burlap to completely remove all

557

of the dried grout or film. There should be no visible film left on the surface. Grout should not be left on the wall overnight. All cleaning of any one panel of concrete should be completed the same day.

Patching mortar used to fill tie holes and imperfections should be of the same mixture of gray and white cements, but should be a stiffer mixture and the cement and sand proportions should be the same as those used in the concrete. Patches should never be steel troweled, but may be finished with wood or cork floats. Allowing the mixed patching mortar to stand for an hour or two before using it reduces the amount of shrinkage but water must not be added in remixing it.

To correct blemishes appearing on the face of the walls, allow about 4 hrs. time for a cement finisher and a helper per 100 s.f. (9.29 sq.m) of wall. The labor cost per 100 s.f. (9.29 sq.m) of wall should average as follows:

Description	Hours	Rate	Total
Cement mason	4.00	$ 36.93	$ 147.72
Helper	4.00	$ 26.16	$ 104.64
Cost per 100 sq.ft			$ 200.40
cost per sq.ft.			$ 2.00
cost per sq.m			$ 21.57

Cutting Concrete with Portable Concrete Saw (Flat Saw/Walk Behind Saw)

Many states specify that control joints in highway paving be cut with a concrete saw, rather than produced by forming. The advantages of this method are improved ridability of joint; less sealing material required to seal joint; better aggregate interlock at joint; and no spalling or cracking at joint, which reduces maintenance costs.

This technique is also widely used for cutting floors and paving for repair work, trenches for electric conduit, pipelines, and sewers, and also for scoring before breaking out concrete for machinery bases and other foundation work.

Target saws from Electrolux located in Kansas City and Diamond Products located in Ohio are a good example. They are available in electric, hydraulic, diesel, and gasoline models. The most popular models are the 18 Hp model, the 37 Hp model, and the 65 Hp model. Approximate prices range from $3,000.00 for the basic model manually propelled to $15,000.00 for the self propelled 65 hp unit. The water pump and spotlight are optional features along with larger blade guard sizes. Many accessories are available on all of these units including a propane option for inside cutting when the proper ventilation is available. The larger 65 Hp engine is generally used for sawing contraction joints in airports, highways, and turnpikes. The smaller units are widely used in patching, trenching, and plant maintenance work. Electrically powered models are available from 3 Hp and up.

Two types of blades are used in cutting concrete with concrete saws as follows: 1) Green-Con wet-dry blades for cutting green concrete 2) diamond blades, of various specifications, for cutting cured or old concrete.

558

CONCRETE

The cost of cutting concrete with a concrete saw varies considerably, depending on age of concrete, size of rebar, type of aggregate, blade specification used, accessibility of work, skill of operator, etc.

Assuming proper blade specification is used for the cutting job, green concrete 24 to 48 hours old may be cut 1" deep at the rate of 12 to 14 l.f. (3.6-4.2 m) per minute and 3" (75 mm) deep at the rate of 6 to 8 l.f. (1.8-2.4 m) per minute. These rates cover only actual cutting time. Overall production will be reduced considerably depending on length of each individual cut, number of machine moves, changes in direction of cuts, etc. Blade wear for a Green-Con blade of proper specification should cost from $0.01-0.02/l.f. (0.033-0.066/m) of cut 1" (25 mm) deep. For a 2" (50 mm) cut, blade cost should be $0.015-0.03/l.f.; 3" (75 mm) cut, $0.02-0.04 per l.f. (0.06-0.13 per m).

When cutting old or cured concrete, using a diamond blade of correct specification, cutting time may be double the amount given above for green concrete and cost of blade wear will probably run 5 to 6 times that given above. Inside cutting of a concrete slab on grade (trench) with rebar may be cut 6" deep at the rate of 40-50 l.f. (12.2-15.3 m) per hour. Trenches 6" deep x 12" wide are typically cut into 2' x 2' pieces for easy removal. Outside cutting of a concrete slab such as bridges and piers 8"-10" deep with rebar may be cut at the rate of 65–75 l.f. (19.9-22.3 m) per hour. Typically bridges and pier decks are cut into 6' x 8'panels and loaded onto truck beds with the aid of a crane or excavator. Roadway concrete with minimal rebar may be cut 8"-10" deep at a rate of 100–150 l.f. (30.5-45.8 m) per hour. Asphalt roadway may be cut 4"-6" deep at a rate of 200–225 l.f. (60.1-68.6 m) per hour. Blade wear considering proper specification when cutting 8-10" deep concrete should cost between $0.05-$0.06 /inch foot. It is best to obtain a firm price quote for any concrete cutting that is required for a project, but if a quick budget estimate is needed, flat sawing concrete is usually priced by the inch foot. For example a slab that is 6" deep would be priced at $0.20-$0.60/inch foot or $1.20-$3.60/l.f. Asphalt 6" deep would be priced at $0.10-$0.20/inch foot or $0.60-$1.20/l.f.

Concrete saws (are available in several electric and gasoline models. The most popular models are the 18 Hp, 30 Hp, 37 Hp, and the 65 Hp models, all gasoline powered. Approximate prices range from $3,000.00 for a basic manually propelled to $20,000.00 a self propelling unit and electric starter, price exclude the blade costs. A water pump and spotlight are optional features. The self propelling unit and other accessories are optional on most models. The larger models with 65 Hp engines are generally used for sawing contraction joints in airports, highways, and turnpikes. The smaller models are widely used in patching, trenching, and plant maintenance work. Electrically powered models are available from 3 Hp and up.

Two types of blades are used in cutting concrete with concrete saws as follows: 1) Green-Con wet-dry abrasive blades for cutting green concrete 2) diamond blades, of various specifications, for cutting cured or old concrete.

CEMENT FLOOR HARDENERS

There are many preparations on the market for waterproofing, dustproofing, and wearproofing cement finish floors. These preparations fall into two classifications, integral treatments in which the entire slab or topping is treated and surface treatments, in which the surface is specially prepared to become waterproof, dustproof and wear resistant.

Integral Treatments. Integral treatments consist of powdered chemicals that are mixed with the cement before sand and water are added and become an integral part of the topping mix throughout its entire thickness. This type of treatment is said to densify and harden the topping and is generally recommended for areas subjected to light pedestrian traffic.

Surface Treatments. There are two types of surface treatments for cement finish floors, dust coat applications which must take place as floors are being finished and liquid chemical applications which may be done at any time after the floor finish is set hard enough for foot traffic.

METALLIC FLOOR TREATMENTS (HARDENERS)

In general, a widespread surface treatment is the metallic dust coat. The term "hardeners" usually applied to metallic floor treatments is somewhat misleading, because iron has a much lower rating on the scale of hardness than even the most common sand or stone aggregate. In this sense iron treatments could be described as "softeners". The basic technique is to provide a floor surface where the hard, brittle sand and stone found in plain concrete is overlaid with a coating of soft, ductile iron. It is reasoned that where hard, brittle aggregates crush under impact and abrasion, iron particles are malleable under these forces and will not fracture or crush out of the floor surface.

Metallic floor treatments consist of a dust coat application of pulverized iron, size graded to pass through mesh screens in the following proportions:

	Percent Passing
4 mesh screen	100%
8 mesh screen	Not less than 90%
14 mesh screen	Not less than 70 nor more than 85%
28 mesh screen	Not less than 35 nor more than 50%
48 mesh screen	Not more than 10%
100 mesh screen	Not more than 5%

Iron particles of the above graduation are mixed with cement and dusted onto freshly floated concrete and then floated and troweled to a smooth even finish. This optimum grading produces a more workable material while plastic and provides a dense surface having great resistance to wear upon hardening.

The iron material should be free from oil and should contain no nonferrous metals. Oil will prevent the cement paste from bonding properly to the iron particles and nonferrous metals might react with the cement to form gases and cause serious surface blistering. Manufacturers, architects, and engineers specify different methods and amounts of iron to be used, which is usually a certain number of pounds of iron mixed with one half that weight of cement. The material is dusted over the floor surface before the finish topping has taken its initial set. These quantities vary from 30 to 120 lbs. (13.60-54.43 kg) of iron per 100 s.f. (9.29 sq.m) of floor, depending on the traffic the floor will have to withstand. When the heavier dust coats are used, water-reducing and plasticizing agents are required to produce a dust coat that can be readily worked onto the surface without using an overly wet concrete mix. In wet mixes there is danger that the iron aggregate will sink into the concrete, where it is of no value as surface armoring.

It is not necessary to add anything to the labor cost of applying the regular cement finish top. It merely means that the finisher dusts the floor with the metallic powder while it is still too wet to trowel. By adding this dust coat, it is possible to trowel the floor sooner, because it absorbs a portion of the surface water.

Durocon Metallic Floor Hardener (Castle Chemical). This is a specially prepared metallic hardener for industrial and heavy duty concrete floors. It contains ocon, which provides concrete of increased strength and resistance to acid, alkali, etc.

For average traffic, 30 to 40 lbs. (13.6-18.14 kg) hardener and 30 lbs. (13.6 kg) portland cement; for heavy traffic, 50 lbs. (22.6 kg) hardener and 30 lbs. (13.6 kg) portland cement; and for extra heavy traffic, 60 lbs. (27.2 kg) hardener and 30 lbs. (13.6 kg) portland cement per 100 s.f. (9 sq.m) floor.

Use 40 lbs. (18.14 kg) colored Durocon with 20 lbs. (9 kg) portland cement per 100 s.f. (9 sq.m).

Standard colors, natural, tile red, linoleum brown, French gray, Persian red, battleship gray, Nile green, black, russet and maroon.

Duroplate (Castle Chemical). A specially prepared, size-graded iron particles combined with dispersing agent, it is applied as a dust coat while concrete is still in a plastic condition and machine floated into the surface. It produces a ductile, high impact- and abrasion-resistant, non-absorbent floor that resists oils and grease. It is easily cleaned and maintained. Floor life is extended up to ten times that of normal concrete floors when Duroplate aggregate is used. Use 60 to 120 lbs. (27-54 kg) per 100 s.f. (9.29 sq.m) of floor depending on degree of duty required.

Durundum Non-Slip Hardener (Castle Chemical). This aggregate floor hardener is composed of powerful coloring mediums and hardening elements plus cementitious binders. It is non-rusting and produces a non-slip surface and is entirely inert, very hard, tough, and resistant to abrasion. Apply as a dust coat, from 30 to 60 lbs. (13.6-27.2 kg) per 100 s.f. (9.29 sq.m), as conditions require.

Ferro-Fax Metallic Floor Hardener (A. C. Horn, Inc.). Ferro-Fax metallic floor hardener meets government specifications. Free from oil and

561

foreign metals, it produces wearproof, waterproof, dustproof concrete floors. Apply as a dust coat, using 30 lbs. (13.6 kg) per 100 s.f. (9.29 sq.m) for light traffic; 40 lbs. (18.14 kg) per 100 s.f. (9.29 sq.m) for moderate traffic; heavy-duty floors, 70 to 100 lbs. (31.7-45 kg) per 100 s.f. (9.29 sq.m).

It comes in two types: Ferro-Fax Standard, to be mixed with portland cement, and Ferro-Fax Ready-Mixed. Immediately following the leveling, deposit upon the surface 2/3 of a uniform dry mixture consisting of 2 sacks Ferro-Fax Standard and 1 sack portland cement.

Allow the Ferro-Fax to absorb the surface water. Float sufficiently to work the Ferro-Fax into the surface. Discontinue floating as soon as the surface becomes wet. Immediately after floating the first shake, apply the final 1/3 uniformly over the surface. Float just until moisture is brought to the surface. Trowel to the desired finish.

Mastercron® (degussa). A ready-to-use colored cementitious dry-shack surface hardener incoproating specially sized and grated mineral aggregate.

The yield, primarily for wear surface 1.0 to 2.0 lbs per Sq.Ft. (4.9 to 9.8 Kg per Sq.m). Primarily for color, 1.5 to 2.0 Lbs per Sq.Ft. (7.3 to 9.8 Kg per Sq.M). The manufaturer recommends a two process: Apply and float ½ to 2/3 of the total amount on the first application. Applying more than 1 lb per per Sq.Ft. (4.88 Kg/Sq.M).

Masterplate® DPS is a highly specialized metallic-aggregate dry-shack surface hardener. It produces spark-resistant, static-disseminating, and wear resistant floors. Material yields 1.0 lbs per Sq.Ft. (4.9 Kg. per Sq.M) of total shake in 1 pass.

Sparkproof Duroplate Pre-Mixed (Castle Chemical). This is static-disseminating and spark-resistant and meets all requirements of the Navy BuDocks Type Specifications, TS-F-15, Metallic Type (Superseding 48Y) for use in hazardous areas. Will not spark from friction with metal as do ordinary concrete aggregates. A special conductive cement binder is interground in this ready-to-use product. Cure with Conductive Kuraseal, 200 s.f. per gal (4.75 sq.m per liter). It is available in French grey, battleship grey, tile red, maroon, seal brown, black, Nile green, tan, and terra cotta.

Harcol R. M. (Sonneborn). A ready to use, rustproof, dry mix of abrasive resistant aggregates, specially selected for their flint-like hardness, portland cement and limeproof pigments of high and uniform tinctorial strength, it is applied as a dust coat at the rate of 30 to 45 lbs. (13.6-20 kg) per 100 s.f. (9 sq.m), floated into the surface, and troweled. Harcol produces an integrally colored surface highly resistant to abrasion, in bright red, dark red, gray, terra cotta, green, and natural.

Ferrolith H. (Sonneborn). A finely ground metallic water absorbent hardener, free from oils and impurities, it requires 60 to 180 lbs. (27-81 kg) per 100 s.f. (9 sq.m) for ordinary traffic, depending on planned use of floor.

Ferrocon Metallic (Castle Chemical). A finely powdered iron mixed with cement and water, it prevents absorption of moisture and seepage of

CONCRETE

water. It is for use on old and new structures. Three to 5 coats are recommended for pressure. Use 10 to 15 lbs. (4.5-7 kg) per 100 s.f. (9.29 sq.m) per coat.

Liquid Chemical Floor Treatments (Hardeners)

As concrete wears down, it is ground into dust, which has an especially detrimental effect in mills and factories where the dust is carried into bearings of machinery, motors, and engines, causing considerable damage and expense.

A number of liquid chemicals that counteract the problem are on the market. They create a chemical action in the cement and cause a complete transformation of the lime so that it solidifies the entire concrete aggregate into a hard, flint-like, homogenous mass that prevents dusting and wearing of the floors.

These liquid chemical compounds are sprayed over the floor until all of the pores in the surface are filled. It is usually necessary to treat the floors two or three times with the liquid to secure satisfactory results. When using these preparations, it is necessary to treat the floors at stated intervals, as the chemical action they create on the cement and lime is not permanent, although some manufacturers claim their product is permanent.

When using these liquid chemicals, care must be exercised to see that the floor is absolutely clean and free from all grease, dust, dirt, oil, or other foreign matter.

Liquid Chemical Floor Hardeners					
Kind of Hardener	No. of Coats	Sq.Ft. per Gal.	Gals. Reqd. per 100 Sq.Ft.	Approx. Price per Gal.	Labor Hours
Granitex, Devoe Paint	2	200	1/2	$15.00	0.80
Hornolith, Tamms	3	150	2/3	$15.00	0.80
Lapidolith, degussa	3	150	2/3	$20.00	1.00
Kind of Hardener	No. of Coats	Sq.M. per Liter	Liters. Reqd. per 10 Sq.M.	Approx. Price per Liter.	Labor Hours
Granitex, Devoe Paint	2	4.91	2.04	$3.96	0.80
Hornolith, Tamms	3	3.68	2.72	$3.96	0.80
Lapidolith, degussa	3	3.68	2.72	$5.28	1.00

Different manufacturers specify varying numbers of applications of their materials but all of them must be applied just as often as the floor shows signs of dusting.

Labor Cleaning Floors. This cost depends entirely on the judgement of the estimator. Floors that are full of oil and grease will require much more time to clean than a floor free of foreign matter.

The labor applying the liquid is also a variable item, but a worker should spray 700 s.f. (63 sq.m) of floor per hr. on the first application and 800 s.f. (72 sq.m) per hr. on subsequent applications.

The labor applying the first coat should cost as follows per 100 s.f. (9.29 sq.m):

Description	Hours	Rate	Total
Labor	0.15	$ 26.16	$ 3.92
Cost per sq.ft.			$ 0.04
per sq.m			$ 0.42

On the second and third applications, the labor per 100 s.f. (9.29 sq.m) should cost as follows:

Description	Hours	Rate	Total
Labor	0.13	$ 26.16	$ 3.40
Cost per sq.ft.			$ 0.03
per sq.m			$ 0.37

Concrete Accelerators and Densifiers

There are numerous preparations on the market for controlling the set of concrete, increasing its early strength, densifying and waterproofing the concrete mass, and for preventing freezing during winter weather by lowering the freezing point of the mixing water.

Anti-Hydro Accelerator and Anti-Freeze. (Anti-Hydro Co., Newark, N.J.) This can be mixed with the water or added to the wet mix. As a general rule, the standard proportion of 1.5 gals. of Anti-Hydro per c.y. (7.43 liters per cu.m) of concrete, with reduced water content to compensate for increased slump, gives protection down to 23°F (-5°C) on concrete deposited in forms except in the most exposed locations.

For cold weather work, be sure that sand and aggregate are free of ice.

To protect concrete or mortar against freezing, use the following table for anticipated outside air temperatures as follows:

1. 32 to 25°F (0 to -3.88°C) use 1 part Anti-Hydro to 12 parts water.
2. 25 to 15°F (-3.88 to -9.44°C) use 1 part Anti-Hydro to 10 parts water.
3. 15°F or less (-9.44°C or less) mechanical heat is necessary.

POZZOLITH-HE (Master Builders). This is an accelerator and water-reducing agent recommended for use in concrete where accelerated set, increased early and ultimate strength and improved performance are required or desired. Concrete made with POZZOLITH-HE provides 3-day normal strength in 1 day, 7-day normal strength in 3 days, 28-day normal strength in 7 days and substantial increases in ultimate strength.

Also, non-chloride high early formulations are available for specialized concreting operations where calcium chloride is not permitted.

564

CONCRETE

POZZOLITH-HE admixture is used at the dosage rate of 16-64 fl. oz. per 100 lbs. (1040 ml to 4160 ml per 100 kg) of cement depending on formulation being used and the amount of set and strength acceleration needed or desired.

Plastiment (Sika). Plastiment is a powder added in the proportion of 1/2 to 1 lb. per sack of cement. It will increase workability, density, and surface hardness and decrease shrinkage. It may be used for concrete or mortar and is also available in a concentrated liquid form for use with automatic dispensing equipment. Use 2 to 4 fl. oz. (59-118 ml) per sack of cement.

Dehydratine No. 80 (W. R. Grace). Liquid integral cement floor hardener, accelerator, and anti-freeze compound, it is used in the proportions of 1 quart per sack of cement.

Sikacrete (Sika). Accelerating and plasticizing liquid for concrete or mortar, used as anti-freeze or rapid hardening compound, it is used in the dilution of 1 part Sikacrete to 3 to 10 parts water. A dilution of 1 to 7 will reduce the hardening rate of mortar or concrete to one-half or protect against freezing to a temperature of 22o F (5.55oC).

Trimix (W. R. Grace). A chemical solution of uniform strength, which when used according to directions functions as an accelerator, integral hardener, or anti-freeze compound, it is used in mortar or concrete work. As an anti-freeze, use it as follows:

1. 25 to 32°F (-3.88 to 0°C), 1-1/2 qts. (1.42 liters). Trimix to each sack of cement.
2. 20 to 24°F (-6.67 to -4.44°C), 2 qts. (1.89 liters). Trimix to each sack of cement.
3. 15 to 19°F (-9.44 to -7.22°C), 2-1/2 qts. (2.37 liters). Trimix to each sack of cement.
4. Below 15°F (Below -9.44°C), heat is necessary.

BONDING OLD AND NEW CONCRETE

The problem of bonding new concrete to old, so the new concrete will remain permanently in place, is one that has received considerable attention among users of concrete. Unless the utmost care is exercised in preparing the old surface, the new concrete will invariably come loose where it is joined to the old concrete.

Different methods have been used to overcome this fault, but the basic condition of all of them is that the old surface must be thoroughly cleaned and washed, and the old aggregate must be exposed, which means that the thin film of cement that covers the surface of the concrete must be removed. One of the methods of preparing the old surface consists of hacking, picking it or hand chipping the surface, and then washing or turning a steam hose under pressure to remove all dust and dirt from the old floor for a perfectly clean surface.

Hacking and Chipping Old Concrete. The cost of hacking and chipping old concrete surfaces, in preparation for bonding new concrete to old, is a variable item. It depends on the hardness of the old concrete and the condition of the surface. Under average conditions a laborer using a small chipping tool should roughen 175 to 200 sq.ft. (16.25-18.58 sq.m) of "green" concrete floor per 8-hr. day, at the following cost per 100 s.f. (9.29 sq.m):

Description	Hours	Rate	Total
Labor	4.00	$ 26.16	$ 104.64
Cost per sq.ft.			$ 1.05
per sq.m			$ 11.26

Where the concrete floors are of old concrete, a worker using hand chipping tools will do well to chip and roughen 8 to 10 sq.ft. (0.74-0.93 sq.m) per hour, and even with a compressor and jackhammer or bush-hammering tool, it is often possible to chip and roughen only 20 to 25 sq.ft. (1.85-2.23 sq.m) per hr. and the cost per 100 sq.ft. (9.29 sq.m) should average as follows:

Description	Hours	Rate	Total
Compressor expense	2.25	$ 44.00	$ 99.00
Labor on jackhammer	4.50	$ 28.16	$ 126.72
Cost per 100 sq.ft.			$ 225.72
per sq.ft.			$ 2.26
per sq.m			$ 24.30

Shotblasting is a methodology used to remove large shallow horizontal areas of concrete to prepare surface overlay's. Shotblasting equipment is a self-contained one piece unit that has an intergral dust collecting system. The shot can be water, steel or ball bearings. The cost of removing approximately 1/8" (3.12 mm) of concrete will range from $0.45 to $1.00 per sq.ft. ($4.84 to $10.76 per sq.m). Cost will vary depending on the existing surface, hardness and reinforment steel. Cost for 18/The cost for removal of the excavated material is not included

Roto-till and hydroblasting can remove up to 6" plus of concretrte or asphalt in several passes. The machines normally have trailing gear attached to them that allow the machine to remove the surface material and scoop the removed material into the trailing gear with its conveyor and deposit the waste material directly into a waiting truck, thus eliminating double handling. The following table is an average for 100 sq.ft. (9.29 sq.m) of concrete material 2" deep deposed into waiting truck(s). The cost of trucking and disposal is separate.

For any demolition of exsiting concrete, asphalt or other material that is adjacent to or interferes with traffic, the cost of maintenance and protection of traffic needs to be addressed and costed.

Description	Hours	Rate	Total
Shot Blasting Machine	1.00	$ 75.00	$ 75.00
Operating Engineer	2.00	$ 37.89	$ 75.78
Labors	8.00	$ 26.16	$ 209.28
Cost per 100 sq.ft.			$ 360.06
per sq.ft.			$ 3.60
per sq.m			$ 38.76

Anti-Hydro Bonding Coat. Rough and clean the old concrete as previously described. Apply a coat of grout composed of 1/2 to 3/4 sack of portland cement added gradually into a solution of 1 gal. of Anti-Hydro in 3 gals. of water (3.79 liters per 11.37 liters of water) until a thick, creamy consistency is obtained. After applying this grout to the prepared surface, the concrete or mortar should be placed while the grout is still wet.

Daraweld-C (W. R. Grace & Co.) Daraweld-C is an emulsion of special internally plasticized high polymer resins uniformly dispersed in water. Mixed with cement mortars or grouts, it forms a durable, highly water-resistant bond. A Daraweld-C bond withstands water immersion without softening or disintegrating, resists the action of heat, cold, oil and gasoline, most acids and other corrosive materials. Its coupling action is so great that grouts containing Daraweld-C will bond not only to concrete and masonry, but to tile, wood, steel, even glass. Ready-to-use, non-settling Daraweld-C will not re-emulsify when subjected to moisture. It is completely compatible with mixes containing calcium chloride.

Daraweld-C is used as a bonding agent admixture in grouts and mortars made from portland or Lumnite cement and assures their lasting adhesion to existing concrete. Formulated for exterior or interior use, it is especially useful for patching concrete floors, repairing cracks in walls, securing an integral bond between successive concrete pours, for waterproofing exteriors below and above grade, in Gunite applications, in bonding concrete to asphalt surfaces. Diluted with water, Daraweld-C is an excellent brush-on sealer for dustproofing concrete floors, and as a primer for painting concrete masonry.

Tremco Floor Bond. A chemically treated metallic powder to use in bonding new concrete toppings to set slabs, where a perfect bond must be secured to prevent cracking.

The old surface must be roughened and cleaned as described previously and the floor bond applied in 2 coats. It requires 25 lbs. (11.33 kg) of Metallic Floor Bond per 100 s.f. (9.29 sq.m) of surface.

When used for bonding vertical surfaces, 2 applications are required, using 15 lbs. (7 kg) of Floor Bond per 100 s.f. (9.29 sq.m).

Weld-Crete. (Larsen Products, Rockville, MD) This product permanently bonds new concrete to old concrete. It also bonds new concrete to many other materials, such as brick, stucco, stone, etc.

The old concrete surface must be structurally sound and free from dust, dirt, loose material, grease, oil, wax, water soluble coatings, etc.

Daraweld-C Bonding Agent Admixture Applications

Description	Volume - Gallons				Yield		Coverage - Sq.Ft.		
	Daraweld C	Cement	Mason's Sand	Water	Cu.Ft.	Gals	1/32 "	1/16 "	1/8 "
Dustproofing	1	----	----	2.0	0.27	2.0	120	----	----
Fine Crakes	1	1	----	1.0	0.32	2.4	260	60	----
Wall Finish	1	3	----	3.0	0.68	5.1	260	130	----
Stucco Patching	1	5	----	1.0	0.52	3.9	200	100	50
Wide Cracks	1	1	1.5	0.5	0.37	2.8	----	70	35
Bonding	1	5	2.0	1.0	0.68	5.1	----	130	65
Floor Finishing	1	5	2.5	1.0	0.72	5.4	----	140	----
Non Skid Surfaces	1	5	5.0	1.5	0.99	7.5	----	190	95
Waterproofing	1	3	6.0	1.0	0.91	6.8	----	175	87
Floor Leveling	1	5	10.0	2.0	1.45	10.9	----	280	140

Description	Volume - Liters				Yield		Coverage - Sq.M		
	Daraweld C	Cement	Mason's Sand	Water	Cu.M	Liter	0.78 mm	1.56 mm	3.13 mm
Dustproofing	3.79	----	----	7.57	0.008	7.57	----	----	----
Fine Crakes	3.79	3.79	----	3.79	0.009	9.09	11.15	5.57	----
Wall Finish	3.79	11.36	----	11.36	0.019	19.31	24.15	12.08	----
Stucco Patching	3.79	18.93	5.68	3.79	0.015	14.76	18.58	9.29	4.65
Wide Cracks	3.79	3.79	5.68	1.89	0.010	10.60	----	6.50	3.25
Bonding	3.79	18.93	7.57	3.79	0.019	19.31	----	12.08	6.04
Floor Finishing	3.79	18.93	9.46	3.79	0.020	20.44	----	13.01	----
Non Skid Surfaces	3.79	18.93	18.93	5.68	0.028	28.39	----	17.65	8.83
Waterproofing	3.79	11.36	22.71	3.79	0.026	25.74	----	16.26	8.08
Floor Leveling	3.79	18.93	37.85	7.57	0.041	41.26	----	26.01	13.01

CONCRETE

It is not necessary to chip, bush hammer or roughen the old surface in any way.

Weld-Crete may be used on either interior or exterior surfaces and may be applied over "green" concrete and damp or dry surfaces. May be applied with a brush, roller or spray. Spraying is most efficient and should be done with heavy industrial spray equipment.

Coverage, when sprayed, from 200 to 300 s.f. per gal. (4.75-7 sq.m per liter). When brushed or rolled, coverage is somewhat less.

Repairing Leaks Against Hydrostatic Pressure

When repairing leaks in basement walls or floors, tunnels, pits, etc., where water pressure exists, the following methods and materials may be used.

Sika 2. Sika 2, red, fast-setting, sealing liquid is mixed with standard portland cement in small quantities. It is used to plug small infiltrations of water against high pressure. Sika 2 mortar will have an initial set of 15 seconds and a final set of 30 seconds. It is used after surrounding leaking concrete has been sealed with Sika 4A. Quantity required varies with degree of leakage.

Sika 4A. Sika 4A is a clear liquid mixed with neat portland cement in small quantities. The resulting mortar will have an initial set of less than 1 minute and a final set of approximately 5 minutes. Sika 4A may be diluted with water and the cement mixed with sand. This mortar is to be applied against damp surfaces or masonry over which water is running. If pressure is too heavy, bleeder pipes may be installed and after smaller leaks are sealed with Sika 4A, the bleeder holes may be plugged with Sika 2, liquid. Quantity required depends upon the degree of leakage.

CONCRETE ADMIXTURES

Since the introduction of the water-cement ratio law governing the strength and durability of concrete, the necessity for using comparatively low water ratio mixes has been generally recognized. Of greatest importance, however, has been the necessity of producing such mixes with a degree of workability that would insure their easy and economical placing and compacting.

In a cubic yard of concrete, approximately 2-1/2 gallons (9.46 liters) of water per sack of cement are required to hydrate the cement. Any water in excess of that amount must be regarded as "placing" water, merely providing sufficient workability to make the mix placeable. This excess water occupies about 10% of the total space in concrete. As it evaporates, it causes the concrete to shrink. Excess water also reduces the strength of the concrete.

The purpose of admixtures is to improve the plasticity, workability, and finish of the concrete, to prevent segregation of the aggregates and to a certain extent provide integral waterproofing qualities by filling the small voids in the concrete mixture.

Pozzolith (Master Builders). This is a water reducing, set-controlling admixture available in several formulations to facilitate optimum performance benefits with the range of cements, sands, and coarse aggregates used for making concrete.

In the plastic concrete, Pozzolith improves finishing characteristics for flat work and cast surfaces, reduces segregation of the mix, and enhances placement of low slump concrete. In the hardened concrete Pozzolith results in reduced cracking, increased compressive and flexural strengths, improved watertightness, and increased resistance of air entrained concrete to damage from freezing and thawing as well as scaling from deicing salts.

Pozzolith Normal and Retarding formulations are generally used at the rate of 5 ±2 fluid ounce per 100 pound (325 +130 ml per 100 kg) of cement.

Plastiment (Sika). Plastiment, concrete densifier and retarding agent, is a liquid to be added to concrete or mortar at the rate of 2 to 4 oz. (59-118 ml) per sack of portland cement. For equal workability the water-cement ratio may be reduced approximately 10%. It results in better workability, density, adhesion to old concrete, surface hardness, and flexural and compressive strength. The delayed set results in elimination of cold joints and reduced shrinkage.

Plastocrete (Sika). A water-reducing admixture for ready-mix and performance concrete to produce higher strengths and workability, it improves structural quality of all concrete and can be used throughout the year.

Trimix (Sonneborn). This is a multi-purpose concrete and mortar admixture that reduces water-cement ratio, produces higher early compression strengths, accelerates set, and generally improves quality and workability of portland cement mixtures.

Used in the proportion of 1 qt. (0.95 l) Trimix to each sack of cement in the mix.

Hydrated Lime as a Concrete Admixture. Hydrated lime is often used as an admixture to increase the plasticity and improve the workability of concrete. For this purpose 5 to 8 lbs. (2-3.5 kg) of hydrated lime are usually added to each bag of portland cement used in the concrete mixture.

Air Entraining Admixtures

In air entrained concrete, 3% to 6% of air is incorporated in the concrete in the form of minute separated air bubbles. Such concrete is more resistant to freezing and thawing and to salt action than normal concrete. Originally developed to prevent scaling of pavement where salts are used for ice removal, air entrained concrete is being widely used for all types of work and in all locations because of its better workability as well as its better resistance to weathering even where salts are not used.

CONCRETE

MB-VR (Master Builders). MB-VR is a neutralized Vinsol resin type air entraining agent for concrete. It is furnished in water solution form and is ready to use as it comes from the drum. MB-VR is compatible for use with other admixtures commonly used in concrete; however, if more than one admixture is used, each should be dispensed into the mix separately. MB-VR is available via bulk tank delivery or 55-gallon (208-liter) steel drums.

Sika Aer. This air entraining resin solution of neutralized wood resin improves various properties of plastic and hardened concrete. Add between 1/2 and 2 fl. oz. (15-59 ml) of Sika Aer solution to entrain between 3% and 6% air as air content is in direct proportion to quantity of Sika Aer added.

Darex AEA (W. R. Grace). Darex AEA is an aqueous solution of highly purified and modified salts of a sulfonated hydrocarbon. It contains a catalyst which promotes more rapid and complete hydration of portland cement. Darex AEA is specifically formulated for use as an air entraining admixture for concrete and is manufactured under rigid control which insures uniform, predictable performance. The addition of 3/4 fluid ounce of Darex AEA per sack of cement will generally entrain 4% to 7% air in a 5-1/2 sack concrete mix.

CONCRETE CURING PROCESS

Efficient methods of curing concrete are of vital importance, because surfaces such as concrete floors, roofs, road, pavements, airport runways, and other surfaces exposed to the sun require protection during the curing period. This used to be accomplished with damp burlap, sand, sawdust, dirt covering, paper, or ponding.

A method that is being used to a large extent involves curing compounds, which provide an air-tight seal over the surface of the concrete and prevent the evaporation of the water from the concrete until it has cured normally. These compounds are usually applied with a brush or spray as soon as the concrete has set sufficiently.

Servicised/Horn Concrete Curing Compounds (A. C. Horn). These are scientifically formulated liquids for spray application to fresh concrete surfaces. The compound quickly forms a vaportight film that seals in 95% or more of the moisture in the concrete for 3 or more days, without changing the concrete color. They can be applied with standard spray equipment. One application should control cure on any size surface area, from mass concrete to a few square feet.

Kure-N-Seal (Sonneborn). Kure-N-Seal cures, seals, and dustproofs in one application, and it increases surface hardness when applied as a curing membrane after final troweling. By locking in essential curing moisture, the hydration process is refined and maximum hardness for the involved mix is assured.

Klearseal (Castle Chemical). This is a clear curing compound for use on surfaces where moderate traffic or corrosive conditions are anticipated, and where a completely clear surface is desired. One gallon covers 300 to 450 s.f. (one liter covers 7-10 sq.m) for smooth trowelled concrete; 250 to 350 s.f. (6-8 sq.m/liter) for broom finished concrete, and 200 to 300 s.f. (4.75-7 sq.m/liter) for rough finished concrete.

03366 POST-TENSIONING CONCRETE

Post-tensioning is the tensioning of concrete after it is placed. It has become increasingly popular in recent years to reduce shrinkage cracks and to enhance the strength, serviceability, and design flexibility of concrete structures. It can provide high quality, watertight floor systems for applications such as parking garage floors, and using corrosion protection systems that are available for post-tensioning tendons protect in corrosive environments.

Many concrete contractors think of post-tensioning as a type of specialty work, but actually, it is just an additional step in producing high-quality reinforced concrete. Of course, the concrete contractor will want the assistance of a qualified post-tensioning engineer and a reliable material supplier to produce a quality job at an economical price.

There are two basic types of post-tensioning: multi-strand and mono-strand. In the multi-strand type several strands of prestressed wire are placed in a duct, and after the strands are tensioned, the duct is grouted. This method is sometimes referred to as a grouted or bonded system.

In the mono-strand type, sometimes referred to as unbonded, individual strands are made up of seven prestressed wires. Strands are coated with a lubricant grease to prevent rusting and reduce friction during the stressing operation. The greased strand is covered with a plastic coat.

Mono-strand post-tensioning. In the estimate the contractor must allow for certain items that are unique to post-tensioning. Stressing access must be provided, usually a platform or floor for extension at the perimeter of the structure.

Forming requirements must be coordinated with the stressing and anchorage of post-tensioning tendons. Horizontal supporting formwork cannot be removed until the stressing operation is completed in a given pour area.

Post-tensioning anchorages are typically designed to permit the stressing of tendons when the concrete reaches a strength of 3,000 psi (2.1 kg/mm2). Check with the post-tensioning material supplier for actual concrete strength requirements for the systems.

The contractor must also make provisions for cutting off the stressing tails of the tendons and patching the stressing pockets.

The post-tensioning material supplier will provide the contractor with the placement drawings for the system and will coordinate the pour sequence with post-tensioning anchorage requirements.

Material for a mono-strand system is estimated by the pound. Contract drawings usually specify effective force requirements. The effective force translates into the number of single strands required by dividing the required

CONCRETE

force by the effective force of each tendon, about 26.5 kips per tendon for 1/2" (13 mm) strand.

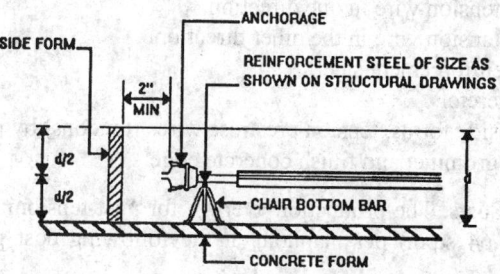

TYPICAL DEAD END POST TENSION ANCHORAGE

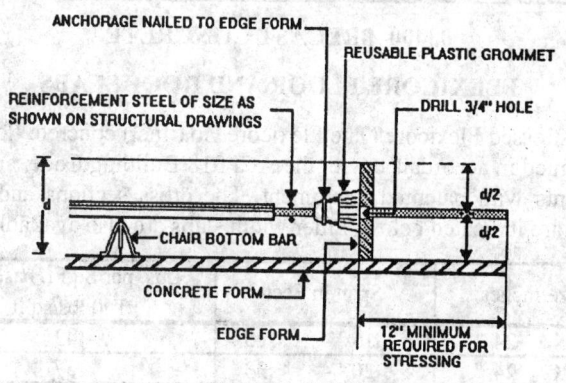

TYPICAL POST TENSION STRESSING ANCHORAGE

The actual effective force provided by each tendon is calculated for each project and must consider seat loss, friction loss, elastic shortening of concrete, and time-dependent losses. The total length of a tendon is obtained by adding stressing tails to the length of each individual tendon, 1'-6" (450 mm) at each stressing end. To obtain nominal strand weight, multiply the length by 0.525 lbs. per ft. (0.79 kg/m) of 1/2" (13 mm) diameter strand.

Material Costs. Post-tensioning material for mono strand averages about $1.1 per lb ($2.42 per Kg). The following are the standard basic steps for placement of post-tensioning tendons:

1. Form concrete slab as usual, but allow about 4'-0" (1.21 m) for post-tensioning set up (jacking) adjacent to the concrete slab side forms.
2. Set side forms, drill holes for post tension wires (tendons), and set grommets.

573

3. Set reinforcement steel chairs about 3'-0" (0.93 m) on center.
4. Set bottom reinforcement steel.
5. Set post-tension wire in one direction.
6. Set post-tension wire in the other direction.
7. Set top reinforcement steel.
8. Pour concrete.
9. Remove side forms, tension prestress wires (tendons) by jacking.
10. Remove grommet and finish concrete edge.

Labor Costs. The production average for post-tensioning is 40 to 50 sq.ft. (3.72 to 4.64 sq.m) per manhour at the following cost per 100 sq.ft. (9.29 sq.m):

Description	Hours	Rate	Total
Ironworker	2.22	$ 40.29	$ 89.44
Cost per sq.ft.			$ 0.89
per sq.m			$9.63

03400 PRECAST CONCRETE

FLEXICORE FLOOR AND ROOF SLABS

Prestressed Flexicore (The Flexicore Co., Inc.) concrete floor and roof slabs, designed in accordance with the ACI 318 Building Code, are precast in central plants with patented equipment. The cross sections and lengths of these units are indicated below. Filler width slabs are also available.

Size Inches	Span in Feet	Lbs. per Sq.Ft. based on 150 lbs/sq.ft.
6 " x 24 "	15 '- 25 '	43
8 " x 24 "	20 '- 33 '	57
10 " x 20 "	28 '- 40 '	61
10 " x 24 "	28 '- 40 '	72
12 " x 24 "	35 '- 50 '	79
Size mm	Span in Meters	Kg. per Sq.M. based on 732.36 Kg/sq.M.
150 x 600 mm	4.57 - 7.62	209.94
200 x 600 mm	6.10 - 10.06	278.30
250 x 500 mm	8.53 - 12.19	297.83
250 x 600 mm	8.53 - 12.19	351.53
300 x 600 mm	10.67 - 15.24	385.71

The amount of deflection allowed shall be consistent with the finish to be applied to the underside of the flexicore units, and shall be within the limits specified by ACI 318 Building Code.

These units are manufactured in a rigid steel form made sufficiently strong to resist the pretensioning force applied by the seven wire strand reinforcing steel. Specially constructed rubber tubes are inflated to form

574

CONCRETE

circular voids so as to constitute approximately 50% of the cross section. The concrete is thoroughly vibrated to assure maximum density and strength.

The slabs are cast of concrete made of gravel, crushed stone, or lightweight aggregate. When lightweight aggregate is used, the unit weight is approximately 25% less. They are cured in a heated kiln.

Not all Flexicore plants are equipped to make all the sections listed in the above table. Availability of sections and lengths must be checked locally.

Erection. Flexicore units are usually delivered to the job by truck. They are hoisted from the truck to the floor location and often placed directly into their final position by the crane. After the slabs are placed side by side, they are aligned and levelled. The keyways in the sides of the slab are then filled with a grout mixed in a ratio of one to three. The erection crew usually consists of six workers, including a crane operator, an oiler, a foreman, and three laborers. A crew will unload, erect, and grout 2,800 sq.ft. (260.12 sq.m) per day on smaller jobs to approximately 6,000 sq.ft. (557.40 sq.m) on some larger jobs.

Cost of Flexicore Floor Slabs. Flexicore slabs are usually quoted on an erected basis by the manufacturer and the price will usually include the cost of grouting and caulking the joints on the ceiling side. The price will vary according to the type of building, size, live load, span, section specified, and location, but usually runs from $5.50 to $7.25 per sq.ft. ($59.20-$78.04 per sq.m) erected, grouted, and caulked.

In localities distant from the manufacturing plant, the slabs are often sold on a "delivered only" basis and erection is performed by the contractor's own personnel.

Caulking the joints underneath is undertaken after the building has been enclosed. A nonstaining type caulking material is used and will average about $0.35 per lin.ft. ($1.15 per m). An underlayment which is not included in the bid is generally used over the Flexicore slabs and is provided by the general contractor.

CONCRETE PLANK

Cantilite (from Concrete Plank Co, Inc., North Arlington, NJ) is a lightweight high-strength concrete plank that is nailable and easily cut in the field. In laying, it is clipped to steel or wood beams. Floor or roof materials are nailed to it. Planks are factory-made in steel forms to give a smooth, even surface on all sides and to assure proper positioning of the reinforcing steel. Side edges are tongued and grooved.

Concrete plank are used for floors and roofs. The plank are 2", 2-3/4" and 3-3/4" (50, 68, 94 mm) thick, 16" (400 mm) wide and made in lengths up to 10'-0" (3.04 m). The 2" (50 mm) plank are used on steel or concrete joists, 4'-0" (1.22 m) o.c. for floors. The 2-3/4" (68.75 mm) plank are placed on steel or concrete joists up to 5'-0" (1.52 m) o.c. for floors. The 3-3/4" (93.75 mm) plank are placed on steel or concrete joists up to 6'-0" (1.83 m) o.c. for floors. A 1" (25 mm) or 2" (50 mm) cement finish top is placed over the plank. Roofs having a span up to 7'-0" (2.1 m) will require 2" (50 mm) plank, while 2-3/4" (69 mm) plank will be required for a roof span up to 8'-0" (2.4 m). The 3-3/4" (94 mm) plank will be required for a roof span up to 10'-0" (3.04 m).

Concrete plank costs $4.00 to $5.00 per sq.ff. ($43.05 to $53.82 per sq.m), f.o.b. factory, depending on quantity in the job. The 2" (50 mm) plank weigh 13 lbs. per sq.ft. (65 kg/m2); the 2-3/4" (68.75 mm) plank, 18 lbs. per sq.ft. (90 kg/m2); the 3-3/4" (93.75mm) plank, 28 lbs. per sq.ft. (141 kg/m2).

Labor Erecting Concrete Plank. A crew consisting of 4 masons and 1 laborer can handle and lay approximately 2,000 sq.ft. (185.80 sq.m) of plank per 8-hr. day, on a flat surface.

Labor Cost of 100 Sq.Ft. (9.29 Sq.M) of Concrete Plank Laid on a Flat Floor or Roof Surface

Description	Hours	Rate	Total
Masons	1.6	$ 36.93	$ 59.09
Labor	0.4	$ 26.16	$ 10.46
Hoisting engineer	0.4	$ 37.89	$ 15.16
Cost per 100 sq.ft. (9.29 sq.m)			$ 84.71
per sq.ft.			$ 0.85
per sq.m			$ 9.12

PRECAST CONCRETE ROOF SLABS

Precast concrete roof slabs are of three general types: rib, flat, and channel. They are adapted to all roof decks, flat or sloping, are fireproof, and with caulked joints, present a smooth surface for the application of built-up roofing. Where slate, ornamental tile, or copper covering is specified, slabs with nailing surfaces can be furnished, permitting direct application.

Precast concrete roof slabs are made with regular concrete aggregates, or where a lightweight slab is desired, they are made of lightweight aggregate.

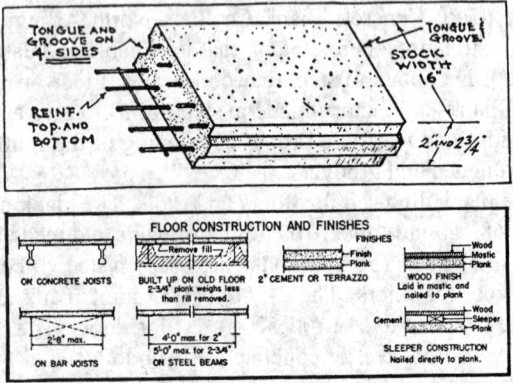

The rib tile is self-weathering. Its attractive red color provides a desirable architectural feature that is often lacking in industrial buildings. Auxiliary pieces are furnished to suit the particular design required, such as ridges, saw-tooth ridges, gable end finishing tile, monitor flashing tile, and other specials required in connection with hip or valley.

TYPE	RIB TILE	FLAT SLAB	2½" CHANNEL TILE	3½" CHANNEL TILE	
WIDTH	24" ⎫ STOCK SIZE	24" ⎫ STOCK SIZE	24"	24"	10"
LENGTH	4'-4" ⎭	3'-0" ⎭	4'-0" MAX.	8'-0" MAX.	3'-0" MAX
THICKNESS	WEB-1½"	1½"	LEG-2½" - WEB-1"	LEG-3½" - WEB-1"	
WEIGHT MAY DIFF. PER SQ. FT. SAND	16 LBS. / 17 LBS.	13.5 LBS. / 19 LBS.	10.5 LBS. / 16 LBS.	12 LBS. / 17.5 LBS.	13.5 LBS. / 19 LBS.
REINFORCEMENT	WELDED GALV. MESH	WELDED GALV. MESH	MESH & 2 DEF. BARS	MESH & 2 DEF. BARS	
ROOF PITCH	⅓ PITCH OR GREATER	FLAT OR PITCHED	FLAT OR PITCHED	FLAT OR PITCHED	
ROOFING	SELF WEATHERING	COMP. ROOFING REQ'D	COMP. ROOFING REQ'D	COMP. ROOFING REQ'D	
PURLIN SPACING	4'-0½" MAX. - 3'-9" MIN.	5'-0" MAX.	4'-0" MAX.	8'-0" MAX.	3'-0" MAX.

Sizes and Weights of Precast Concrete Roof Tile

Tile of the same standard stock sizes but having a glass insert are also furnished where additional light is required. On this type of roof slab, it is necessary to obtain prices from the manufacturer, because the amount of trim varies with each project.

The other two types of precast concrete roof slabs, flat and channel, are used over all roof decks, whether flat or sloping, and present a smooth surface for the application of any type of built-up roofing. When used with a sloping roof and when slate or other types of ornamental covering are to be used, these slabs can have a concrete nailing surface superimposed and manufactured integrally with the structural slab. The illustration gives the sizes and weights of the various types of tile.

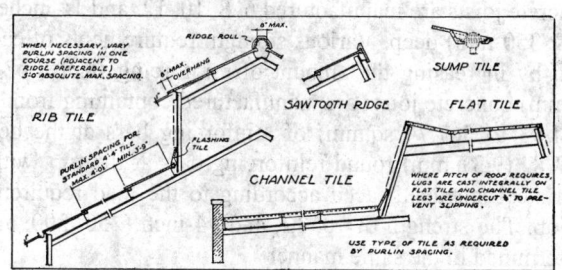

Details of Precast Concrete Roof Tile Construction

Labor Placing Precast Concrete Roof or Floor Slabs. Because of their size and weight, it requires 2 workers to handle precast concrete roof or floor slabs. Each piece weighs from 125 to 200 lbs. (56.70 to 90.72 kg). Ordinarily it requires 2 workers on the ground handling the slabs and getting them to the hoist, 2 more on the floor or roof placing them for the mason, and a mason and helper to lay and caulk the joints. A mason and helper should

577

lay and caulk 960 to 1,050 sq.ft. (89.18 to 97.55 sq.m) of lightweight concrete tile per 8-hr. day on straight roofs.

PRECAST CONCRETE JOIST FLOORS

Precast concrete joists are widely used to support concrete floors in homes, apartment buildings, office buildings, schools, hospitals, and similar structures. Precast concrete joists are also extensively used for supporting concrete roof slabs in all types of buildings.

The design tables are only for floors and roofs that carry relatively light loads and are not intended for designing floors or roofs carrying heavy or concentrated loads or subject to heavy impact loads or to vibration of mounted machinery. These conditions impose special loading problems, and to meet them, floors must be designed accordingly.

Concrete joists are manufactured in a cement products plant and delivered to the job ready to set in place. The joists for each floor of a building should be laid out separately, showing size of joists, spacing, etc., which varies from 20 to 33 inches (500-825 mm), depending on depth of joists, span, floor load, etc.

To develop maximum strength, a reinforced concrete slab 2" to 2-1/2" (50-62.50 mm) thick is placed over the tops of the joists with the joists embedded into the concrete slab to a depth of 1/2" to 3/4" (12.50 to 18.75 mm). Effective bond shall be obtained between the joists and slab. Concrete for both joists and slab shall have an average compressive strength of not less than 3,750 lbs. per sq. in. (2.63 kg/mm2) at age of 28 days. Floors made of precast joists mortised or embedded into a monolithic floor placed or poured on the job are called precast joist cast-in-place concrete slab floors.

Floors made of precast joist over which are laid precast slabs for flooring are called precast joist and slab concrete floors.

Concrete joists are manufactured 6, 8, 10, 12, and 14 inches (150, 200, 250, 300, & 350 mm) deep. Various strength requirements of each size may be obtained by increasing the amount of reinforcing steel. For instance, 8-inch (200-mm) concrete joists are manufactured containing from 0.22 to 0.88 sq. in. (141.93 to 567.74 sq.mm) of reinforcing bars at the bottom of the joists and 3/8" (9.37 mm) round reinforcing bars at the top, with 1/4" (6.25 mm) round steel stirrups spaced according to the load requirements of the particular job. The strength of 10, 12, and 14-inch (250, 300, and 350 mm) joists is determined in the same manner.

Notes for Table Nos. 1, 2 and 3 on following pages. Superimposed loads shown in tables are the loads that the floor or roof is designed to support in addition to the weights shown for the precast joists and concrete slab.

Load values are limited by maximum moment or shear and will produce deflections less than L/360, except in Table No. 3 where load values shown with an asterisk are maximum loads for a deflection of L/360. In Table No. 3 load values in bold face type are limited only by maximum

578

Table 1 - Joists and Slab Designed as T-Beams

No Shoring Used During Construction — Superimposed Loads - psf per clear span - Ft

Joist Size, Inches	Reinforcement Steel Size Top	Bottom	Wgt of Joist and Slab psf	Slab Thk Inch	Joist Spacing C Inch	10	11	12	13	14	15	16	17	18	19	20	21	22	23	24
8	# 3	# 5	34	2	20	--	##	88	69	53	40	30	--	--	--	--	--	--	--	--
			32	2	24	##	90	69	53	40	29	--	--	--	--	--	--	--	--	--
			32	2	27	##	76	57	42	31	--	--	--	--	--	--	--	--	--	--
			37	2 1/2	30	85	61	43	29	--	--	--	--	--	--	--	--	--	--	--
			36	2 1/2	33	74	52	36	23	--	--	--	--	--	--	--	--	--	--	--
	# 3	# 6	34	2	20	--	--	--	##	91	73	59	--	--	--	--	--	--	--	--
			32	2	24	--	--	##	90	72	57	45	--	--	--	--	--	--	--	--
			32	2	27	--	##	96	72	59	46	36	--	--	--	--	--	--	--	--
			37	2 1/2	30	##	##	81	61	46	33	22	--	--	--	--	--	--	--	--
			36	2 1/2	33	##	93	70	53	38	26	--	--	--	--	--	--	--	--	--
	# 3	# 7	34	2	20	--	--	--	--	##	##	94	--	--	--	--	--	--	--	--
			32	2	24	--	--	--	##	##	90	74	--	--	--	--	--	--	--	--
			32	2	27	--	--	##	99	93	76	61	--	--	--	--	--	--	--	--
			37	2 1/2	30	--	##	##	99	78	61	48	--	--	--	--	--	--	--	--
			36	2 1/2	33	--	##	##	87	68	53	40	--	--	--	--	--	--	--	--
10	# 3	# 6	37	2	20	--	--	--	--	##	##	85	70	57	47	38	--	--	--	--
			35	2	24	--	--	--	##	99	81	66	53	43	34	27	--	--	--	--
			34	2	27	--	--	--	##	85	69	55	44	35	27	--	--	--	--	--
			39	2 1/2	30	--	--	--	89	70	54	41	31	22	--	--	--	--	--	--
			38	2 1/2	33	--	--	--	78	60	46	34	25	--	--	--	--	--	--	--
	# 3	# 7	37	2	20	--	--	--	--	--	--	##	##	92	78	66	--	--	--	--
			35	2	24	--	--	--	--	--	##	##	87	72	61	50	--	--	--	--
			34	2	27	--	--	--	--	##	##	89	74	61	51	42	--	--	--	--
			39	2 1/2	30	--	--	--	##	##	90	73	59	47	37	28	--	--	--	--
			38	2 1/2	33	--	--	--	##	98	79	63	50	39	30	22	--	--	--	--

Table 1 - Joists and Slab Designed as T-Beams (continued)

No Shoring Used During Construction

Superimposed Loads - psf per clear span - Ft

Joist Size, Inches	Reinf. Top	Reinf. Bottom	Wgt of Joist and Slab, psf	Slab Thk, Inch	O.C. Joist Spacing, Inch	10	11	12	13	14	15	16	17	18	19	20	21	22	23	24
10	#3	#8	37	2	20	--	--	--	--	--	--	--	--	##	##	99	--	--	--	--
			35	2	24	--	--	--	--	--	--	--	##	##	91	78	--	--	--	--
			34	2	27	--	--	--	--	--	--	##	##	92	78	66	--	--	--	--
			39	2 1/2	30	--	--	--	--	--	--	##	91	76	63	52	--	--	--	--
			38	2 1/2	33	--	--	--	--	--	##	97	80	66	54	44	--	--	--	--
	#3	#6	44	2	20	--	--	--	--	--	--	##	86	71	59	48	39	31	--	--
			41	2	24	--	--	--	--	--	--	83	68	55	44	35	27	--	--	--
			39	2 1/2	27	--	--	--	--	--	87	70	57	46	36	28	--	--	--	--
			43	2 1/2	30	--	--	--	--	90	72	57	44	33	24	--	--	--	--	--
			42	2 1/2	33	--	--	--	99	79	62	48	36	27	--	--	--	--	--	--
12	#3	#7	44	2	20	--	--	--	--	--	--	--	--	##	97	83	70	59	50	42
			41	2	24	--	--	--	--	--	--	--	--	91	77	64	54	45	37	30
			39	2	27	--	--	--	--	--	--	--	93	78	65	54	45	37	30	23
			43	2 1/2	30	--	--	--	--	--	--	95	78	63	51	41	32	25	--	--
			42	2 1/2	33	--	--	--	--	--	--	82	67	54	43	34	26	--	--	--
	#3	#8	44	2	20	--	--	--	--	--	--	--	--	--	--	##	##	93	81	69
			41	2	24	--	--	--	--	--	--	--	--	##	##	98	85	73	62	53
			39	2	27	--	--	--	--	--	--	--	##	##	99	85	72	62	52	44
			43	2 1/2	30	--	--	--	--	--	--	--	##	98	83	69	58	48	39	32
			42	2 1/2	33	--	--	--	--	--	--	--	##	86	72	60	49	40	32	25
	#3	#9	44	2	20	--	--	--	--	--	--	--	--	--	--	--	--	--	--	##
			41	2	24	--	--	--	--	--	--	--	--	--	--	##	##	##	92	80
			39	2	27	--	--	--	--	--	--	--	--	--	--	##	##	90	78	68
			43	2 1/2	30	--	--	--	--	--	--	--	--	--	##	##	87	74	63	54
			42	2 1/2	33	--	--	--	--	--	--	--	--	--	##	89	75	64	54	45

CONCRETE

Table 2 - Joists and Slab Designed as T-Beams

Shoring Used at Mid-Span During Construction
Superimposed Loads - psf per clear span - Ft

Joist Size, Inches	Reinforcement Steel Size Top	Bottom	Wgt of Joist and Slab, psf	Slab Thk, Inch	O.C. Joist Spacing, Inch	10	11	12	13	14	15	16	17	18	19	20	21	22	23	24
8	#3	#5	34	2	20	--	##	99	79	64	51	41	--	--	--	--	--	--	--	--
			32	2	24	##	##	80	63	50	39	30	--	--	--	--	--	--	--	--
			32	2	27	##	86	67	52	41	31	--	--	--	--	--	--	--	--	--
			37	2 1/2	30	##	76	58	44	33	24	--	--	--	--	--	--	--	--	--
			36	2 1/2	33	89	67	50	38	28	--	--	--	--	--	--	--	--	--	--
8	#3	#6	34	2	20	--	--	--	##	##	84	70	--	--	--	--	--	--	--	--
			32	2	24	--	--	##	##	82	67	55	--	--	--	--	--	--	--	--
			32	2	27	--	--	##	86	70	57	46	--	--	--	--	--	--	--	--
			37	2 1/2	30	--	##	96	76	61	48	38	--	--	--	--	--	--	--	--
			36	2 1/2	33	--	##	85	67	53	42	32	--	--	--	--	--	--	--	--
8	#3	#7	34	2	20	--	--	--	--	##	##	85	--	--	--	--	--	--	--	--
			32	2	24	--	--	--	--	##	87	72	--	--	--	--	--	--	--	--
			32	2	27	--	--	--	##	94	77	63	--	--	--	--	--	--	--	--
			37	2 1/2	30	--	--	--	##	83	68	55	--	--	--	--	--	--	--	--
			36	2 1/2	33	--	--	--	90	72	58	46	--	--	--	--	--	--	--	--
10	#3	#6	37	2	20	--	--	--	--	##	##	94	79	66	55	46	--	--	--	--
			35	2	24	--	--	--	--	##	93	75	62	51	42	35	--	--	--	--
			34	2	27	--	--	--	--	##	82	63	52	43	35	28	--	--	--	--
			39	2 1/2	30	--	--	--	--	##	72	53	43	34	26	--	--	--	--	--
			38	2 1/2	33	--	--	--	--	##	58	46	37	29	--	--	--	--	--	--
10	#3	#7	37	2	20	--	--	--	--	--	--	##	##	95	87	75	--	--	--	--
			35	2	24	--	--	--	--	--	--	##	97	82	69	59	--	--	--	--
			34	2	27	--	--	--	--	--	--	##	85	71	59	50	--	--	--	--
			39	2 1/2	30	--	--	--	--	--	##	91	75	62	49	40	--	--	--	--
			38	2 1/2	33	--	--	--	--	--	91	75	62	52	42	35	--	--	--	--

581

Table 2 - Joists and Slab Designed as T-Beams (continued)

Shoring Used at Mid-Span During Construction
Superimposed Loads - psf per clear span - Ft

Joist Size, Inches	Reinf. Top	Reinf. Bottom	Wgt of Joist and Slab, psf	Slab Thk, Inch	O.C. Joist Spacing, Inch	10	11	12	13	14	15	16	17	18	19	20	21	22	23	24
10	#3	#8	37	2	20	--	--	--	--	--	--	--	--	##	##	##	--	--	--	--
	#3	#8	35	2	24	--	--	--	--	--	--	--	##	##	##	87	--	--	--	--
	#3	#8	34	2	27	--	--	--	--	--	##	##	##	88	86	75	--	--	--	--
	#3	#8	39	2 1/2	30	--	--	--	--	##	##	##	92	78	75	64	--	--	--	--
	#3	#8	38	2 1/2	33	--	--	--	--	##	##	##	95	80	66	56	47	39	32	26
	#3	#6	44	2	20	--	--	--	--	##	##	90	75	63	52	43	35	29	--	--
	#3	#6	41	2	24	--	--	--	--	##	##	94	78	65	54	44	36	29	--	--
	#3	#6	39	2	27	--	--	--	##	##	83	67	55	44	38	28	--	--	--	--
	#3	#6	43	2 1/2	30	--	--	##	##	89	72	58	47	37	29	--	--	--	--	--
	#3	#6	42	2 1/2	33	--	--	##	89	72	58	47	37	29	--	--	--	--	--	--
12	#3	#7	44	2	20	--	--	--	--	--	--	--	##	##	##	92	79	68	59	50
	#3	#7	41	2	24	--	--	--	--	--	--	--	##	99	84	72	62	53	45	38
	#3	#7	39	2	27	--	--	--	--	--	--	--	##	86	73	62	53	44	37	31
	#3	#7	43	2 1/2	30	--	--	--	--	--	--	##	89	75	63	52	43	35	29	--
	#3	#7	42	2 1/2	33	--	--	--	--	--	##	93	78	65	54	44	36	29	--	--
	#3	#8	44	2	20	--	--	--	--	--	--	--	--	##	##	##	##	##	90	79
	#3	#8	41	2	24	--	--	--	--	--	--	--	--	##	##	97	92	81	70	61
	#3	#8	39	2	27	--	--	--	--	--	--	--	--	##	##	94	80	69	61	52
	#3	#8	43	2 1/2	30	--	--	--	--	--	--	--	--	##	94	83	69	59	50	43
	#3	#8	42	2 1/2	33	--	--	--	--	--	--	--	##	93	83	71	60	51	43	36
	#3	#9	44	2	20	--	--	--	--	--	--	--	--	##	##	##	##	##	##	##
	#3	#9	41	2	24	--	--	--	--	--	--	--	--	##	##	##	##	##	98	88
	#3	#9	39	2	27	--	--	--	--	--	--	--	--	##	##	##	##	98	86	76
	#3	#9	43	2 1/2	30	--	--	--	--	--	--	--	--	##	##	##	98	85	75	65
	#3	#9	42	2 1/2	33	--	--	--	--	--	--	--	--	##	##	99	86	75	65	58

Table 3 – Joists as Independent Beams

Joist Size, Inches	Reinforcement Steel Size		Wgt of Joist and Slab	O.C. Joist Spacing	No Shoring Used During Construction														
					Superimposed Loads - psf per clear span - Ft														
	Top	Bottom	psf	Inch	10	11	12	13	14	15	16	17	18	19	20	21	22	23	24
8	#3	#5	34	20	112	86	67	52	40	31	--	--	--	--	--	--	--	--	--
			32	24	89	68	52	40	30	22	--	--	--	--	--	--	--	--	--
			32	27	76	57	43	32	23	--	--	--	--	--	--	--	--	--	--
	#4	#6	34	20	--	106	85	69	56	45	--	--	--	--	--	--	--	--	--
			32	24	107	85	67	53	43	34	--	--	--	--	--	--	--	--	--
			32	27	92	72	56	44	35	--	--	--	--	--	--	--	--	--	--
	#6	#7	34	20	--	--	--	--	103	85	71	--	--	--	--	--	--	--	--
			32	24	--	--	--	100	82	67	55	--	--	--	--	--	--	--	--
			32	27	--	--	106	86	69	56	46	--	--	--	--	--	--	--	--
10	#3	#6	37	20	--	--	--	--	100	82	68	56	46	37	--	--	--	--	--
			35	24	--	--	--	98	79	64	52	42	34	--	--	--	--	--	--
			34	27	--	--	104	84	68	54	44	35	--	--	--	--	--	--	--
	#5	#7	37	20	--	--	--	--	--	--	104	88	74	63	53	--	--	--	--
			35	24	--	--	--	--	--	99	82	69	58	48	40	--	--	--	--
			34	27	--	--	--	--	102	85	70	59	49	40	33	--	--	--	--
	#6	#8	37	20	--	--	--	--	--	--	--	--	107	93	80	--	--	--	--
			35	24	--	--	--	--	--	--	--	100	85	73	63	--	--	--	--
			34	27	--	--	--	--	--	--	102	86	73	62	53	--	--	--	--

Table 3 - Joists as Independent Beams

Joist Size, Inches	Reinforcement Steel Size		Wgt of Joist and Slab	O.C. Joist Spacing	No Shoring Used During Construction — Superimposed Loads - psf per clear span - Ft														
	Top	Bottom	psf	Inch	10	11	12	13	14	15	16	17	18	19	20	21	22	23	24
12	#3	#6	44	20	--	--	--	--	--	105	87	72	60	49	40	32	26	--	--
			41	24	--	--	--	--	102	83	68	56	45	36	29	23	--	--	--
			39	27	--	--	--	109	88	72	58	47	38	30	23	--	--	--	--
			43	30	--	--	--	90	71	57	45	35	26	--	--	--	--	--	--
			42	33	--	--	100	79	62	49	38	29	--	--	--	--	--	--	--
	#3	#7	44	20	--	--	--	--	--	--	--	--	96	81	69	58	49	41	34
			41	24	--	--	--	--	--	--	106	89	75	63	53	44	37	30	--
			39	27	--	--	--	--	--	--	92	77	65	54	45	37	30	24	--
			43	30	--	--	--	--	--	91	75	61	50	40	32	25	--	--	--
			42	33	--	--	--	--	98	80	65	53	43	34	--	--	--	--	--
	#4	#8	44	20	--	--	--	--	--	--	--	--	--	--	103	90	78	67	58
			41	24	--	--	--	--	--	--	--	--	--	95	81	70	60	52	44
			39	27	--	--	--	--	--	--	--	--	86	82	70	60	51	44	37
			43	30	--	--	--	--	--	--	--	93	76	66	55	46	38	31	--
			42	33	--	--	--	--	--	--	97	82	68	57	47	39	32	--	--

CONCRETE

moment and shear and will produce deflections greater than L/360.

For Table Nos. 1 and 2, one row of bridging should be used at midspan in residential floor construction having spans over 20 ft. (6.09m) and in all other floor construction having spans over 16 ft. (4.88 m). Bridging is not usually required for roof construction.

For Table No. 3, one row of bridging should be used at midspan for all floor construction having spans over 16 ft. (4.88 m). Bridging is usually not required for roof construction.

For floor or roof construction shown in Table Nos. 1 and 3, no shoring is required during construction.

For floor and roof construction shown in Table No. 2, shoring at midspan is required during construction and must remain in place until the concrete has reached its required strength.

Bar sizes are given in numbers based on the number of eighths of an inch included in the nominal diameter of the bars as specified in A615/A615M-05 *Standard Specification for Deformed and Plain Billet-Steel Bars for Concrete Reinforcement.*

Concrete Joists Designed as Independent Beams. Concrete joists designed to be used with precast concrete slabs, where the slab does not form a part of the beam, use more reinforcing steel in the beams, increasing the cost considerably.

For example, 8" (200 mm) beams with 3/8" (9.37 mm) top bars and 5/8" (15.62 mm) bottom bars are used for spans 10 to 15 ft. (3.05 to4.57 m); 1/2" (12.50 mm) top bars and 3/4" (18.75mm) bottom bars for spans 11 to 16 ft. (3.35 to4.88 m); and joists using 3/4" (18.75 mm) top bars and 7/8" (21.87 mm) bars for spans 12 to 16 ft. (3.65 to 4.88 m) requiring heavier loads.

Ten-inch joists with 3/8" (9.37 mm) top bars and 3/4" (18.75 mm) bottom bars are used for spans 12 to 19 ft. (3.65 to 5.79 m); 5/8" (15.62 mm) top bars and 7/8" (21.87 mm) bottom bars for spans 14 to 20 ft. (4.26 to6.09 m); and 3/4" (18.75 mm) top bars and 1" (25 mm) bottom bars for spans 16 to 20 ft. (4.88 to6.09m) required to carry heavier loads.

Twelve inch joists with 3/8" (9.37 mm) top bars and 3/4" (18.75 mm) bottom bars are used for spans 12 to 22 ft. (3.65 to -6.70 m); 3/8" (18.75 mm) top bars and 7/8" (21.87 mm) bottom bars for spans 14 to 24 ft. (4.26 to 7.31 m); and 1/2" (12.75 mm) top bars and 1" (25 mm) bottom bars for spans 16 to 24 ft. (4.88 to7.31 m) required to carry heavier loads.

These beams require from 1 to 2 lbs. (0.45-0.90 kg) of reinforcing steel per lin.ft. more than the T-beams and cost $0.50 to 0.65 per lin.ft. ($1.64 to $2.13 per m) more than the T-beam joists.

Precast concrete joists are manufactured of Haydite, Waylite, Celocrete, and other lightweight aggregate, also of ordinary aggregate consisting of gravel or crushed stone and sand.

Weight of Precast Concrete Joists, Ordinary Aggregate, Lbs. (Kg.)						
Size of Joist, In.	Length of Joist in Feet					
	10	12	14	16	18	20
8 "	190	228	266	304	-----	-----
10 "	230	276	322	368	414	460
12 "	350	420	490	560	630	700
Size of Joist, mm	Length of Joist in Meters					
	3.048	3.658	4.267	4.877	5.486	6.096
200 mm	86.18	103.41	120.65	137.88	-----	-----
250 mm	104.32	125.18	146.05	166.91	187.77	208.64
300 mm	158.75	190.49	222.24	253.99	285.74	317.49

Cost of Precast Concrete Joists. The cost of precast concrete joists will vary with the size of joists, amount of reinforcing steel required, size of job, locality, etc. Precast concrete joists are manufactured in a concrete products plant. It is important to obtain definite prices for each particular job.

Approximate Prices per Lin.Ft. (M) for Precast Concrete Joists				
6 "	8 "	10 "	12 "	14 "
$ 5.90	$ 7.00	$ 7.75	$ 10.75	$ 12.25
150 mm	200 mm	250 mm	300 mm	350 mm
$ 19.36	$ 22.97	$ 25.43	$ 35.27	$ 40.19

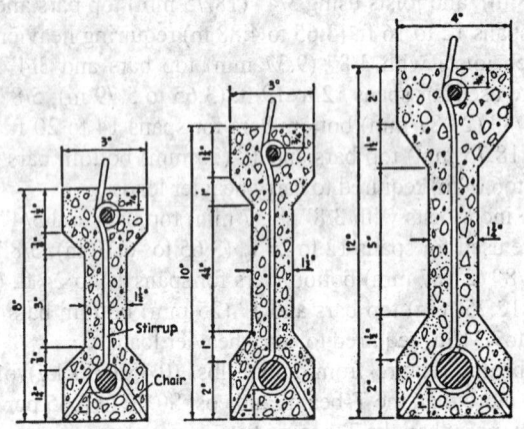

CONCRETE

Description	8" Joist	10" Joist	12" Joist	Description	200 mm Joist	250mm Joist	300 mm Joist
Section Area Sq.In.	18.2	22.1	33.4	Section Area Sq.cm.	117.43	142.59	215.50
Average Wgt. Lb/Lin.Ft.	19	23	35	Average Wgt. Kg/M	28.28	34.23	52.09

Sizes and Weights of Precast Concrete Floor Joists

Where the concrete joists are designed as independent beams for use with precast concrete floor or roof slabs, add from $0.40 to 0.50 per lin.ft. ($1.31 to $1.64per m) to the above prices to allow for the extra reinforcing steel required.

Labor Setting Precast Concrete Joists. As long as the joist is not too heavy to handle by four workers or in some cases by light derricks, the precast joist is more economical than one which is cast in place. They are obtainable to a depth of 14" (0.35 m) and length of 36'-0" (10.95 m). Such a joist weighs approximately 1,150 to 1,330 lbs. (521to 603 kg), if made of lightweight aggregate, falling within the weight where erection by machinery is necessary.

On every job there should be a joist setting plan, prepared by an architect or engineer. This will expedite setting the joists on the job.

Where smaller size joists are used, two workers can carry the joists from the stock pile to the building and set them in place, while with heavier joists; it might require 4 to 6 workers.

Where joists are used on the second floor, two workers will probably be required to carry the joists into the building and 3 workers above to hoist the joists to the second floor. If the joists are not too heavy for two workers to handle, a rope can be slipped around each end of the joist and lifted to the second floor.

After the joists have been placed on the wall, it will require some additional time for exact spacing of the joists, placing headers, etc.

Where the joists can be handled and set by 2 laborers, it will require 6 to 8 minutes to carry one joist from the stock pile, not over 50 ft. (15.24 m), and set it on the wall, at the following labor cost per joist:

Description	Hours	Rate	Total
Labor	0.23	$ 26.16	$ 6.02

If joists are hoisted to the second floor, double the above costs.

When handling and setting joists requiring 4 workers to carry them to the building and set them on the wall, it will require 6 to 8 minutes for 4 workers, at the following labor cost per joist:

Description	Hours	Rate	Total
Labor	0.47	$ 26.16	$ 12.30

If joists are hoisted to the second floor, double the above costs.

When handling and setting 12" (300 mm) joists up to 24 ft. (7.31m) long, requiring 6 workers to handle and set, it will require 6 to 8 minutes at the following labor cost per joist:

Description	Hours	Rate	Total
Labor	0.70	$ 26.16	$ 18.31

If joists are hoisted to the second floor, double the above costs.

Method of Constructing Precast Concrete Joist Floors

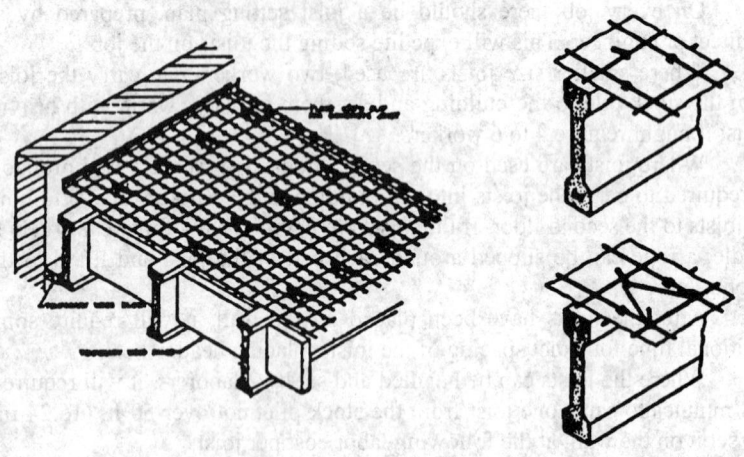

***Method of Constructing Precast Concrete Joist Floors Using Paper-Backed
Floor Lath for Permanent Form Material***

On jobs of sufficient size to warrant the use of a hydraulic crane, and using iron workers to set the joists instead of laborers, the cost of setting one joist should average as follows:

CONCRETE

Description	Hours	Rate	Total
Crane and operator	0.10	$ 125.00	$ 12.50
Crane Operator	0.10	$ 37.89	$ 3.79
Iron workers	0.20	$ 40.29	$ 8.06
Cost per joist			$ 24.35

Wood Forms for Precast Concrete Joist Floors. The reinforced concrete floor slab used with precast concrete joists is usually 2" to 2-1/2" (50 to 62.5 mm) thick, and the joists are spaced from 20" to 33" (500-825 mm) apart, so that very light formwork is sufficient.

The formwork usually consists of short pieces of 2 x 4 (50 x 100 mm) wood spreaders as shown in the illustration, cut to bear on the bottom flange of the concrete joists. These spreaders are placed between the concrete joists, and the 5/8" (15.62 mm) plywood are placed on top of the spreaders to receive the concrete.

On forms of this kind, it will require about 1-1/4 b.f. of lumber per s.f. (0.03 m3 per sq.m) of floor area.

When placing forms for precast concrete joist floors, a carpenter should frame and place forms for 175 to 225 sq.ft. (16.25 to 20.91 sq.m) of floor per 8-hr. day, at the following cost per 100 sq.ft. (9.29 sq.m):

Description	Hours	Rate	Total
Carpenter	4.00	$ 35.04	$ 140.16
Cost per sq.ft.			$ 1.40
per sq.m			$ 15.09

A laborer should remove forms from 600 to 750 sq.ft. (55.74 to 69.67 sq.m) of floor per 8-hr. day, at the following cost per 100 sq.ft. (9.29 sq.m):

Description	Hours	Rate	Total
Labor	1.20	$ 26.16	$ 31.39
Cost per sq.ft.			$ 0.27
per sq.m			$ 2.91

Corrugated Steel Forms for Precast Concrete Joist Floors. There are two methods of using corrugated steel forms for precast concrete joist floors. Corrugated steel forms may be laid continuous over joist tops with end laps occurring only over supports and no adjacent laps over the same joist.

Corrugated steel forms also may be installed between joists, so that tops of joist and protruding stirrups are encased in slab concrete, providing T-beam action. For this method, joists should be precast with ledges on the top edges to furnish support for the corrugated sheets.

Side laps for corrugated steel forms should be one corrugation and end laps a minimum of 2" (50 mm). For estimating purposes, compute area of floor slab placed over corrugated steel sheets and add 5% to 10% for laps and waste, depending upon the method of installation.

589

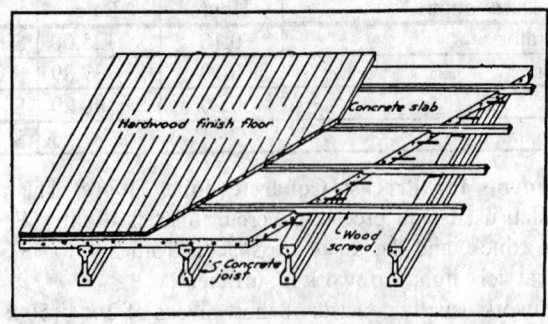

Precast Concrete Joist Floors with Wood Floor Screeds

Where corrugated steel forms are installed in continuous lengths over precast concrete joists for floors at or near grade level, with an area of 5,000 sq.ft. (464.50 sq.m) or more, a 4-worker crew should place and secure 2,500 to 2,700 sq.ft. (232.25 to 250.83 sq.m) per 8-hr. day. For areas less than 5,000 sq.ft. (464.50 sq.m), the crew should install 2,100 to 2,300 sq.ft. (195.09 to 213.67 sq.m) per 8-hr. day. For slabs at higher levels, additional handling labor and hoisting costs must be added.

Labor Cost of 100 Sq.Ft. (9 Sq.M) of Corrugated Steel Slab Forms for Precast Concrete Joist Floors in Residential Construction

Description	Hours	Rate	Total
Ironworker	1.33	$ 40.29	$ 53.59
Cost per sq.ft.			$ 0.54
per sq.m			$ 5.77

Where corrugated steel forms are installed between joists, the above costs should be increased about 10%.

Reinforcing Steel for Precast Concrete Joist Floors. A thin slab of concrete 2" to 2-1/2" (50 to 62.50 mm) thick is used with concrete joists and the reinforcing steel consists of 1/4" (6.25 mm) round rods spaced 10" (250 mm) o.c. at right angles to the joists. It requires about 0.33 lbs. per s.f. (1.66 kg/m2) of steel or 33 lbs. per 100 sq.ft. (488.66 Kg/sq.m) of floor. Steel reinforcing mesh of equal effective steel area may be used instead of reinforcing rods, if desired.

An ironworker should place about 400 lbs. (181.44 kg) of reinforcing steel per 8-hr. day and based on 33 lbs. per 100 sq.ft. (1.66 kg/m2) the labor per 100 sq.ft. (9.29 sq.m) of floor should cost as follows:

Description	Hours	Rate	Total
Ironworker	0.67	$ 40.29	$ 26.99
Cost per sq.ft.			$ 0.27
per sq.m			$ 2.91

590

CONCRETE

Where welded wire mesh is used for reinforcing concrete slabs for precast concrete joist floors, an ironworker should place 1,400 to 1,600 sq.ft. (130.06 to 148.64 sq.m) of mesh per 8-hr. day at the following labor cost per 100 sq.ft. (9.29 sq.m):

Description	Hours	Rate	Total
Ironworker	0.53	$ 40.29	$ 21.35
Cost per sq.ft.			$ 0.21
per sq.m			$ 2.30

Wood Sleepers For Use With Precast Concrete Joist Floors. Where wood floors are used with precast concrete joist floors, it will be necessary to place 2"x2" (50 x 50 mm) or 2"x3" (50 x 75 mm) wood strips or "sleepers" in the concrete for nailing the wood floors. These strips should be spaced about 16" (400 mm) o.c. and should rest on top of the steel stirrups that project above the concrete joist. The sleepers are secured to the joists by wiring around the wood strip and through the top of the steel stirrups. Care must be taken when placing the concrete to see that the strips are level to receive the finish wood floors.

The floor strips should be beveled or nails driven into the sides at short intervals to provide anchorage into the concrete, otherwise after the concrete hardens and the wood strips dry out and shrink they become loose and cause squeaking floors.

Where wood strips are placed over the tops of the joists without leveling, a carpenter should place 350 to 450 lin.ft. (160.68 to 137.16m) per 8-hr. day, at the following cost per 100 lin.ft. (30.48 m):

Description	Hours	Rate	Total
Carpenter	2.00	$ 35.04	$ 70.08
Cost per lin.ft.			$ 0.70
per m			$ 2.30

Cost per 100 s.f. (9.29 sq.m), 16" (400mm) o.c.

Description	Hours	Rate	Total
Carpenter	1.50	$ 35.04	$ 42.38
Cost per sq.ft.			$ 0.42
per sq.m			$ 4.52

Labor Handling and Placing Precast Concrete Slabs. Precast concrete slabs are often used in conjunction with concrete joists instead of cast-in-place concrete slabs.

These slabs are ordinarily used for roofs or for floor slabs in unexcavated areas, where it is either extremely difficult or impossible to remove the forms after the cast-in-place slab has been poured.

When precast slabs are used, it requires joists containing heavier reinforcing steel than where T-beam construction is used. This increases the

591

cost of the concrete joists $0.16 to $0.25 per lin.ft. ($0.52 to $0.85 per m). Precast concrete slabs are furnished in several sizes, including:

2 '- 1 " x 2 '- 6 " x 1 1/4 "	625 x 750 x 31.25 mm
1 '- 2 " x 2 '- 1 " x 1 1/4 "	350 x 625 x 31.25 mm
1 '- 0 " x 2 '- 6 " x 1 1/4 "	300 x 750 x 31.25 mm
1 '- 6 " x 2 '- 1 " x 1 1/4 "	450 x 625 x 31.25 mm

They are reinforced with 4"x4" (100 x 100 mm) No. 10 wire mesh.

When used as roof slabs, they are laid over the concrete joists with about a 1" (25 mm) space between the slabs. After the slabs are placed they are grouted in solid. Insulation board or roofing is applied directly over the slabs.

When used for floors, the precast slabs are placed over the joists and about 1-1/2" (38 mm) of concrete is poured over the slab and either left rough or troweled to a smooth finish to receive the finish flooring.

This method eliminates form cost, but it requires heavier reinforced joists, and it is more difficult for the mechanical trades to place their pipes than where the cast-in-place floor is used.

On smaller jobs, the workers usually work in groups of two carrying and placing the precast slabs over the concrete joists.

On work on the first floor, where the precast slabs are on the floor or convenient to the place where they are to be used, two workers together should handle and set 50 to 60 pcs. an hour, depending upon the size of the slabs used, at the following labor cost per 100 pcs.:

Description	Hours	Rate	Total
Labor	3.60	$ 26.16	$ 94.18
Cost per piece			$ 0.94

Add cost of carrying slabs from stock pile to floor. If slabs are to be hoisted to the second floor, double the above costs. On large jobs where it is possible to use a hydraulic crane to lift the slabs from the ground to the second floor or roof, a crane will hoist 7 slabs at one time and place them on the floor. To the above costs, it will be necessary to add cost of material and labor grouting joints between slabs.

03360 SHOTCRETE

Properly applied shotcrete (under CSI specially placed concrete) is a structurally sound and durable material that exhibits excellent bonding characteristics to existing concrete, rock, steel, and many other materials. Improperly applied shotcrete may create conditions much worse than the untreated condition. Applications for shotcrete include:

1. Repair of damaged surface of concrete, wood or steel structures.
2. Bridges: beam repairs caps, columns, abutments, wingwalls and underdeck.

3. Buildings: repair fire damage, earthquake damage, deterioration, strengthening walls, encasing structural steel for fireproofing.
4. Marine Structures: repair corrosion of steel, impact loading, structural distress, action of waves, sand and gravel.
5. Spillway: repair cavitation erosion or abrasion erosion.
6. Underground excavations: support material, sealing rock surfaces, channeling water flows.
7. Slope and surface protection: temporary protection of exposed rock surfaces.
8. New Structures: Pools, tanks, floors, walls, domes and special shapes.

Dry-Mix Shotcrete. Cementitious material and aggregate are thoroughly mixed and either bagged in a dry condition or mixed and delivered directly to the gun. This is a pneumatically operated gun, which delivers a continuous flow of material through the delivery hose to the nozzle. The interior of the nozzle is fitted with a water ring, which uniformly injects water into the mixture as it is being discharged against the surface being treated.

Wet-Mix Shotcrete. Cement, aggregate, water and admixtures (except accelerators) are thoroughly mixed as for conventional concrete. The mixture is fed into the gun and propelled through the delivery hose to the nozzle by compressed air or pneumatic or mechanical pumping. Air is injected at the nozzle to disperse the stream of concrete and generate the velocity for shotcrete placement.

Comparison of Features of Dry-Mix and Wet-Mix Shotcrete Process

Dry-Mix Process	Wet Mix Process
Mixing water instantaneously controlled at the nozzle by operator to meet variable field conditions	Mixing water controlled at plant and measured at time of batching
Longer hose lengths possible, if necessary	Normal pumping distances
Limited to accelerators as the only practical admixture	Compatible with all ordinary admixtures. Special dispensers for addition of accelerators may be necessary
Use of air-entraining admixture not beneficial.	Air-entrainment possible.
Resistance to freezing and thawing is poor	Acceptable to freezing and thawing
Intermittent use easily accommodated within prescribed time limits	Best suited for continuous application of shotcrete

Exceptional strength performance	Strengths similar to conventional Concrete
Lower production rates	Higher production rates
Higher rebound	Lower rebound
Lower Equip. Maint. costs	Higher Equip. Maint. costs
Higher bond strength	Lower bond strengths, yet often higher then conventional concrete
Dust control	Reduces dust control problems

Fiber Reinforcement. Shotcrete can be reinforced with fibers. The addition of fibers to the shotcrete mixture adds ductility to the material as well as energy absorption capacity and impact resistance. Types of fiber reinforcement: a) steel fibers, b) glass fiber reinforcement (GFRS), and c) synthetic fibers such as nylon, polypropylene, polyethylene, polyester and rayon.

Proportioning of Shotcrete. In general, conventional concrete technology may be applied to shotcrete proportioning. Prior to mixture proportioning, the following should be considered.

1. Types of dry-mix or wet-mix shotcrete appropriate for the work.
2. Specific job constraints on the shotcrete work.
3. Type of specification: performance versus prescription; contractor versus government mixture proportioning.

Shotcrete Application Techniques. The nozzling techniques and procedures used in applying shotcrete greatly affect the quality of the shotcrete and the amount and composition of rebound. Rebound material may become entrapped in succeeding shotcrete if poor nozzling techniques are followed. The entrapment of rebound results in a decrease in the ultimate strength and durability of the shotcrete. Plane surfaces should generally be shotcreted with the nozzle held at 90 degrees to the surface. Areas with reinforcing steel should be gunned at a slight angle from each side.

Rebound. This is aggregate and cement paste that bounce off the surface during the application of shotcrete because of the collision with the hard surface, the reinforcement, or the aggregate particles. The amount of rebound varies with the position of the work, air pressure, cement content, water content, size and grading of aggregate, amount of reinforcement and thickness of the layer.

Rebound from Conventional Cement - Aggregate Shotcrete

Work Surface	Dry-Mix	Wet-Mix
Floor and Slabs	5% - 15%	0 - 5%
Sloping and Vertical Walls	5% - 25%	5% - 10%
Overhead Work	25% - 50%	10% - 20%

CONCRETE

Finishing. The natural gun finish is preferred both from the standpoints of structural soundness and of durability. If desired, the gunned finish or flash coat may be followed by surface finishing using one or more of the following tools:

1. Wood float, giving a granular texture.
2. Rubber floats, giving a course texture and finish.
3. Steel trowel, giving a very smooth finish.

Pre-production phase. The contractor should accomplish or prepare the following:

1. Cementtitous materials: Have the manufacturer-certified test results to verify that the cement and pozzolan meet the contract requirements.
2. Aggregates: Test data should be furnished to verify that the fine and course aggregates meet the quality and grading requirements of the contract.
3. Manufacturer certificates of compliance for admixtures and curing compounds.
4. Manufacture certificate for proposed fibers and reinforcement.
5. Test date to verify contractors proposed mixture proportions produce shotcrete that meets the quality requirements of the specifications.
6. Accelerator test data that the proposed accelerator meets test data ASTM C 1141 (or latest).
7. Nozzle-person holds a current ACI certification.
8. Equipment required is in good working order and meets the designer's specifications.
9. Method of curing and protection is in place.
10. Test panel is scheduled to be shot early to allow sufficient time for evaluation of the panel prior to the start of production.

The quality of the shotcrete depends largely on the skill of the application crew. Crew size will vary depending on the size of the operation and the type and setup of equipment. By far the most important member of the crew is the nozzle-person.

Production of wet placed shotcrete by hand assume placement of 750 sq. ft. per day for 3" thick placement:

Description	Hours	Rate	Total
Labor Foreman	8.00	$ 30.16	$ 241.28
Nozzle-person	8.00	$ 32.16	$ 257.28
Nozzle-helper	8.00	$ 28.16	$ 225.28
Laborer at truck discharge	8.00	$ 26.16	$ 209.28
Operating Engineer (pump)	8.00	$ 37.89	$ 303.12
Laborer(s) (rebound)	8.00	$ 26.16	$ 209.28
Cost per 750 sq.ft.			$1,445.52
per sq.ft.			$ 1.93
per sq.m			$ 20.75

Note: When ordering concrete remember to include waste for rebound.

PRESTRESSED CONCRETE

The use of prestressed concrete began in Europe some 50 years ago. A French engineer, Eugene Freyssinet, is generally regarded as the developer of the prestressing concept. Many spectacular prestressed concrete structures were designed by Mr. Freyssinet and others in Europe, Latin America, and the United States. The first major application of prestressed concrete in the U.S. was the Walnut Lane Bridge, completed in 1950, in Philadelphia, Pa. Since that time thousands of bridges and buildings have been constructed with structural elements of prestressed concrete.

Prestressed concrete was developed partly conventional building materials were scarce and expensive in Europe during and after World War II. The concept of using available materials to their fullest capacities led to techniques that developed prestressing into a workable and practical method of construction.

The American Concrete Institute defines prestressed concrete as "concrete whose stresses resulting from external loadings are counterbalanced by prestressing reinforcement placed in the structure". This is accomplished by precompressing the concrete by means of internal high tensile strands stressed a predetermined amount. This prestressing introduces compressive stresses into the concrete which counteract the stresses induced when the external loading is applied. The high-strength strands are restrained in the member by either bond between the concrete and the strands, or by end bearing devices. These two basic methods of restraining the strands are known as pretensioning and post-tensioning.

In pretensioning, the strands are tensioned prior to casting the concrete, and in post-tensioning the strands are tensioned after the concrete is cast. In general, the post-tensioning process is accomplished at the job site when casting large members. The member is cast with the strands properly positioned in the form, and after the concrete has reached sufficient strength, the tensioning members are stressed and secured by means of end bearing devices.

The pretensioning methods are particularly adaptable to the mass production of building elements such as double and single tees, channel

596

sections, I joists and beams, T joists, hollow slabs and flat planks, among others. These members are usually cast in steel casting beds from 100 to 600 ft. (30-180 m) long. With the use of high quality concrete, low slumps and steam curing, a complete manufacturing cycle may be accomplished in a day.

The prestressed and precast industry is based upon the economics of prefabrication and maximum use of the casting facilities. The larger the project and the number of units of identical span and section, the lower the unit cost.

Building Elements of Prestressed Concrete. Precast/Schokbeton Inc. of Kalamazoo, Michigan has a modern prestressing plant for the manufacture of prestressed concrete building elements. The plant is equipped with a variety of casting beds capable of producing a full range of prestressed products, including both structural and architectural components.

Included in the structural items available in the United States are single and double tee slabs, "F" or Monowing slabs, channel slabs, and Flextee slabs for floor and roof construction. Columns, beams and girders are available in a wide range of sizes and shapes.

Architectural products made at the Precast/Schokbeton plant include precast concrete stadium seats, miscellaneous free form concrete made to order, and Schokbeton products produced under a license from N. V. Schokbeton of Zeist, Holland. These include wall panels, window facades and frames, column shells, stair treads, etc.

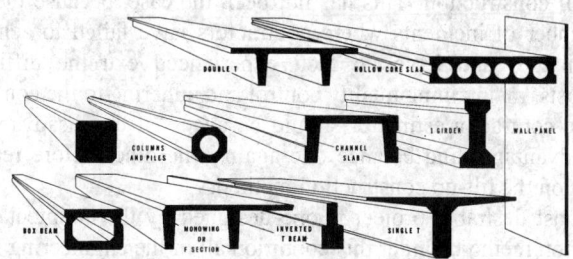

Typical Prestressed Concrete Products

Cost of Prestressed Concrete Structural Elements. As prestressed concrete building elements are now manufactured generally throughout the country, prices may usually be obtained from local sources. As stated previously, casting beds are in most cases long enough to cast any length member desired: length, width, height, and weight must be acceptable to local trucking regulations.

There are, of course, many variables that can influence the cost of the installed product, such as the size of the job; the number of identical units in a job; the distance from the plant to the job site; the length of the units, as extremely long units will probably require special handling at the plant, in transit and at the job site; accessibility at the site; the size and number of

597

cranes or other equipment required for erection; the condition at the ground, etc.

Erection of Prestressed Concrete Building Units. A typical erection setup for an average job would include a crane, crane operator, oiler, five ironworkers, welding machines, and other miscellaneous equipment and tools. With move-in and move-out time for the cranes, dead time due to the weather or breakdowns, contingencies, etc., the daily cost of an erection setup such as this will average about $2,900 per day. On a large job and under good conditions an experienced crew can probably erect 6,000 to 10,000 s.f. (540-900 sq.m) or more per day. On smaller jobs or where working conditions are below average, a crew might only erect 3,000 to 4,000 sq.ft. (278.70 to 371.60 sq.m) per day.

The Prestressed Concrete Institute is a source of the latest reference material on prestressed concrete and can provide assistance to the design profession as well as to the purchaser of prestressed concrete products.

03411 TILT-UP WALL PANELS

The tilt-up construction method can have some distinct advantages in reducing costs through utilization of semi-skilled work forces and producing a completed building project in a shorter number of construction days. These advantages provide a recognizable benefit to the contractor, and it would seem that there would result a major effort to adapt all possible structures to this method of construction. This has not been the case, because there are a sufficient number of incidents where contractors have failed to achieve the several advantages and have, in fact, experienced extreme difficulty in controlling costs, maintaining quality control and adhering to the construction schedule. The contractor/estimator should be aware of the disadvantages as well as the advantages and balance one against the other before reaching a final decision on the tilt-up construction method.

The most desirable project is one designed by the architect to adopt this construction method. Under this condition all of the engineering has been performed during the design stage thus requiring the contractor to cast and erect the panels in a sequence and quality established by the contract plans. All imbedded items, provisions for attachment to other structure members, erection details, allowances for flashing, water sealants and the interrelation with other structure members have been provided for in the design, leaving the contractor unobstructed concentration on scheduling, production, erection and finishes. These are the areas where the contractor has experience and expertise. Both office and field staff are accustomed to these situations and the attendant problems. Under these circumstances, the venture into a tilt-up construction project can be approached with a reasonable degree of assurance the project would succeed, but only after there is a complete understanding of advantages and disadvantages.

Converting Projects to Tilt-up Construction. Attempts to convert or adapt a project to incorporate the tilt-up technique are to be approached

598

with extreme caution. Assuming there is sufficient background and experience to warrant the proposing of a design change, the contractor is exposed to a variety of issues, some of which may be obscure during the period of project planning when the compelling objective is developing methods and techniques to increase production, reduce costs, and thereby enjoy more profitability.

The realities of design change initially demand an in depth exploration of local building codes and restrictions or regulations that may be imposed by the lending institution for construction funding and permanent mortgage money. Some financial institutions have extreme objections to involvement with structures or projects utilizing other than conventional materials and construction methods. This is not because the changes are in any way inferior or detract from the value or stability of the project. More often the objection results from a lack of knowledge or prior experience with the particular construction technique. The same resistance can be encountered in city and county building inspection departments to the point where building permits will not be issued.

Owners usually are more receptive to change, especially in instances where they are the recipients of the ultimate lower cost for the same function level, as this fits with their objective for sponsoring the project. If a reduction in the square foot price is apparent to and understandable by owners, and so long as the project retains the same basic appearance and remains safe for occupancy without addition of excessive maintenance costs, most owners will be receptive to accepting a change in construction methods.

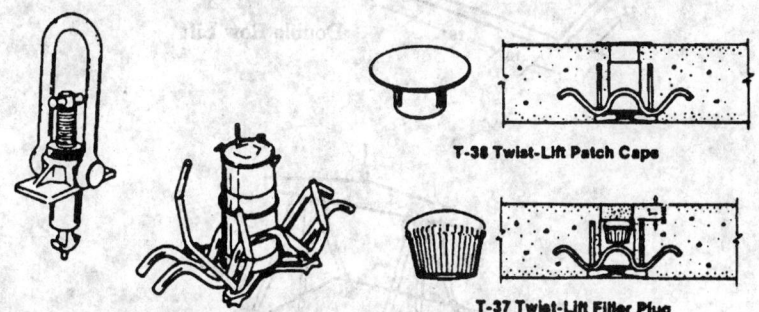

T-38 Twist-Lift Patch Caps

T-37 Twist-Lift Filler Plug

Courtesy of Dayton Superior Corp., Miamisburg, OH
Examples of Hardware Accessories for Lifting Tilt-up Panels

The most serious objection to the conversion process can be expected from the designing architect/engineer who will be reluctant to approve any attempt to modify the design concept which was adopted as the most acceptable solution available. Traditionally, aesthetics and design theory have been entrusted to the professional architect. For outsiders to tamper with these endowed rights is considered an intrusion by persons not meeting the standards, qualifications and professional status enjoyed by the architect. Often such proposed changes are considered a personal affront or even a

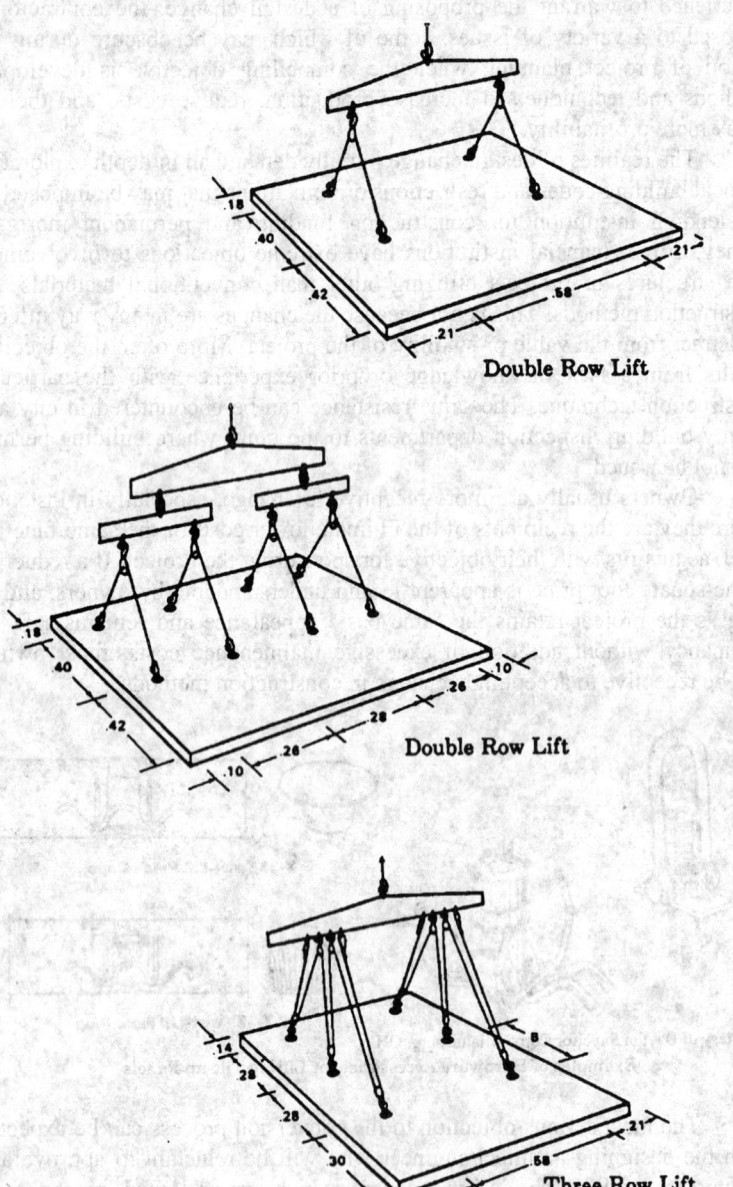

Double Row Lift

Double Row Lift

Three Row Lift

Examples of Rigging Configurations for Tilt-up (Courtesy Dayton Superior)

600

challenge to ability, regardless of the motivation factor which inspired the contractor to initiate the action.

The most serious objection to the conversion process can be expected from the designing architect/engineer who will be reluctant to approve any attempt to modify the design concept which was adopted as the most acceptable solution available. Traditionally, aesthetics and design theory have been entrusted to the professional architect. For outsiders to tamper with these endowed rights is considered an intrusion by persons not meeting the standards, qualifications and professional status enjoyed by the architect. Often such proposed changes are considered a personal affront or even a challenge to ability, regardless of the motivation factor which inspired the contractor to initiate the action.

Should it be found that the designing architect/engineer is inclined to give some degree of consideration to the conversion, the contractor should be prepared to suffer some loss of time in obtaining approval and additional costs in generating engineering data upon which the architect would rely in making the final decision of acceptability. By attempting to modify the original architectural design, the contractor enters an unfamiliar role with many obligations and responsibilities which, if not recognized and accounted for, could transform an otherwise successful venture into a tragedy affecting the financial integrity of the company for many years.

It must be recognized when the contractor assumes even a partial responsibility for the structure design, the services of a registered architect/engineer are necessary to develop and produce the detailed drawings and engineering computations to demonstrate structural integrity and architectural compatibility to the satisfaction of the designing architect. Once design approval is obtained, production and erection drawings have to be provided for use by the field forces.

The loss of construction time resulting from delays in obtaining final approvals and additional engineering costs are strong influences on the feasibility of adapting a structure to the tilt-up method. Some manufacturers of inserts for slab lifting and connection devices offer information on design calculations for their products but this information is not to be relied upon as sufficient or complete in scope to satisfy the engineering responsibility assumed by the contractor. Exposure to design responsibility usually creates the need for a review of insurance coverage to provide adequate protection from the involved risks. Premiums may be exceptionally high or coverage may not exist in instances where this is a first time venture and an experience rating cannot be established.

Once the contractor is aware of the risk element involved, logical solutions can be developed and progress can proceed to accomplish the ultimate objectives of producing a project in a shorter period of time at a lower cost while retaining the functions and quality of the structure.

Evaluation Of The Advantages & Disadvantages

Disadvantages. The disadvantages or special problems provide a base for the contractor/estimator to evaluate the unique situations involved in developing accurate unit costs to support the detailed estimate for tilt-up construction:

1. This type of concrete construction is special and unlike conventional poured-in-place method.
2. Because field personnel will be unfamiliar with the procedure, rejection and opposition undoubtedly will be encountered. To overcome this reaction and resulting low productivity, additional supervision should be assigned during the training period to avoid costly interruption in the work cycle due to indecision. Strong leadership by knowledgeable supervisors will produce coordinated work crews capable of lowering production and erection costs to an acceptable level.
3. Additional lay-out quality is required to insure precise location of all embedded items and openings within panel area.
4. Installation of reinforcing rods demand more time and precision to accommodate a panel type construction containing numerous embedded components and apertures for windows, doors, grilles, and duct, pipe and conduit penetrations.
5. Concrete pouring of horizontal thin 6" (150 mm) panels consisting of 400 s.f. (36 sq.m), 20' x 20' (6 x 6 m) requires the use of crane or chute facilities to economically distribute the concrete to all panel areas.
6. Vibration of the concrete to produce a homogeneous mass within the panel presents additional problems because the operators do not have free and equal access to all areas of the horizontal panel. Obstructions within the panel form such as opening forms, pipes, conduits, reinforcing and embedded items, all contribute to decreasing productivity in vibration of the concrete.
7. Adequate material storage space convenient to the casting area is mandatory to support the orderly flow of materials to the forming and casting area.
8. A well graded and drained access area is to be provided for movement of materials, delivery of pre-mixed concrete and maneuverability of the crane during distribution and erection of the panels.
9. Provisions and location of casting beds for the panels have a direct impact on the final cost of the operation. If sufficient space is available immediately adjacent to the erection location, cost will be lower as distribution and erection time will be less. However, if excessive time is consumed in distribution of the panels, the erection crew will be non-productive during this period thus escalating erection costs.

602

10. Development of a well defined casting schedule and erection drawings will eliminate confusion, delays and interruption to the overall pre-casting and erection operations.

Advantages

1. Pre-casting of concrete panels provides an opportunity to develop work crews having limited prior experience into efficient groups at a reduced cost per hour.
2. This type of construction substantially reduces the investment in large quantities of forming materials and accessories.
3. Using the tilt-up method, most projects can be enclosed more rapidly providing access to interior finishes at an earlier date by avoiding prolonged weather delays.
4. Pre-planning during the casting operation encourages more efficient installation of window and door units in prepared openings.
5. Completion of panel erection supports availability of exterior site areas for storage or site development work.
6. Flexibility of the tilt-up system allows use of precast concrete columns, special wall surface finishes, incorporation of insulation in the wall panels and lightweight aggregate in the concrete.
7. Forming is basically limited to strip side forms at the perimeter of each panel. In instances where openings have to be formed within the panel are for doors and window areas, the cost for forming will be equal to the edge form cost.

Cost Development

Panel size and thickness can vary to accommodate specific job requirements, and engineering design will dictate special features to be found in all non-typical panels.

Cost development for a 20' x 20' x 6" (6.09 x 6.09 x 0.15 m) thick solid panel, 400 sq.ft. (37.16 sq.m), using 3000# (2.10 kg/mm2) pre-mixed concrete, #5 reinforcing rods 6" (150 mm) o.c. each way, would be as follows.

Preparing Pouring Beds. The pouring or casting bed for panels is usually provided by the surface of the concrete slab on grade of the structure. When there is insufficient surface area available to accommodate a reasonable form-pour-cure-erect cycle, a stack method can be employed where panels are cast on previously poured panels.

In both methods care is taken to prevent adherence between the two surfaces. Frequently, the cast bed surface is coated with a brush coat of a commercial release agent or a sheet of poly between pours. Cost for either method is minimal, and the allowances included in the table are adequate.

Edge Forms. Edge forms are constructed of 2"x6"(50 x 150 mm), with a 1"x4" (25 x 100 mm) kick strip secured at the bottom and with the top edge held in alignment by 2"x4" (50 x 100 mm) braces 2'-0" (0.61 m) long at 1'-0" (0.30 m) intervals.

Material Cost		
80 Lin.Ft. =	80 bd.ft.	0.189 Cu.M.
80 Lin.Ft. =	26 bd.ft.	0.061 Cu.M.
84 Pcs 2' X 4" =	112 bd.ft.	0.264 Cu.M.
Total Quantity	218 bd.ft.	0.514 Cu.M.
Forming Materials Cost:	218 Bd.Ft. x $ 0.60	= $ 130.80
Forming Materials Cost:	0.514 Cu.M. x $ 254.40	= $ 130.80
Cost per Sq. Ft		$ 0.33
Cost per Sq. M		$ 3.52

1/2" (12.5mm) strip on edge of 2"x 6" (50x150mm) not considered.
Note: Bd. Ft. is a nominal dimension

Labor Cost

Edge forms for one 20' x 20' (6 x 6 m) panel = 400 s.f. (37.16 sq.m)

Description	Hours	Rate	Total
Carpenter	4.00	$ 35.04	$ 140.16
Helper	4.00	$ 26.16	$ 104.64
Total Labor Cost for 400 SF (37.16 Sq.M.)			$ 244.80
Cost per sq.ft. of panels			$ 0.61
Cost per sq.m. of panels			$ 6.59

Labor cost for openings within the panel area to accommodate doors, windows, etc. would equal the labor cost of the edge forms.

Labor cost for stripping and cleaning forms for re-use would be minimal and are included in the above labor allowances.

Reinforcing. Reinforcing Rods - Plain 42 pcs. x 20'-0" (6.09 m) = 840' (256.03 m) x 1.043# (1.58 kg) = 876 lbs. (396 kg)

Material Costs						
Item	Wgt., Lbs	Price	Cost	Wgt., Kg	Price	Cost
Reinforcing Steel	876	$ 0.60	$ 525.60	397	$ 1.32	$ 525.60
Accessories	876	$ 0.04	$ 35.04	397	$ 0.09	$ 35.04
Tie Wire	876	$ 0.02	$ 17.52	397	$ 0.04	$ 17.52
Total cost			$ 578.16			$ 578.16
Cost per Sq.Ft. (Sq.M.)			$ 1.45			$ 15.56
Cost per Lb. (Kg)			$ 0.66			$ 1.46

Labor Costs			
Description	Hours	Rate	Total
2 Ironworkers	8.00	$ 40.29	$ 322.32
Cost per sq.ft.			$ 0.81
per sq.m			$ 8.67

CONCRETE

Panel Connectors & Inserts. These costs are predicated on the many available alternates. Design requirements and locality will determine the final cost for these items.

For the purpose of developing the average sq. ft. (sq.m) panel cost, an allowance for material cost of $5.30 per lin.ft. ($17.39 per m) of panel edge is considered adequate. Associated labor costs would be $2.40 per lin.ft. ($7.87 per m) or $160.00.

Material Costs			
Description	Mat'l	Rate	Total
80 Lin.Ft. (24.38 m)	80.00	$ 5.30	$ 424.00
Material Cost per sq.ft.			$ 1.06
per sq.m			$ 11.41

Concrete. The cost of concrete includes 5% waste factor. 20' x 20' x 0.5' x 1.05 (waste) = 210 Cu.Ft; 210 cu.ft. ÷ 27 = 7.8 or 8 cu.yd. (6.09 x 6.09 x 0.15 m x 1.05 (waste) = 5.84 cu.m)

Material Costs			
Description	Mat'l	Rate	Total
Concrete, 8 cu.yd. @ $75.00 per cu.yd. (6.11 cu.m @ $98.09)	8.00	$ 55.00	$ 440.00
Material Cost Per sq.ft.			$ 1.10
per sq.m			$ 11.84

Labor Costs			
Description	Hours	Rate	Total
5 Laborers	5.00	$ 26.16	$ 130.80
Cost per sq.ft.			$ 0.33
per sq.m			$ 3.52

Finish. Finish is either float or broom finish.

Labor Costs			
Description	Hours	Rate	Total
2 Laborers @ 4 hrs. ea.	8.00	$ 26.16	$ 209.28
Cost per sq.ft.			$ 0.52
per sq.m			$ 5.63

Special surface finishes can increase finish cost an additional $1.75 to $7.20 per sq.ft. ($18.94 to $77.50 per sq.m). Material unit cost in the table is sufficient to cover the minor material expenditures involved.

Cure. Curing of panel concrete can be accomplished in several ways:

1. Cover with burlap or poly, keeping surface moist and warm;
2. Use of commercial curing compound;
3. Providing a covering of sand or sawdust;

4. Construct canopy to protect from direct sun heat during summer months or to provide enclosure for temporary heat in winter months.

Erection Costs. The labor cost of panel erection is as follows:

Description	Hours	Rate	Total
Crane	8.00	$ 75.00	$ 600.00
Operating Engineer	8.00	$ 37.89	$ 303.12
Carpenter Foreman	8.00	$ 39.04	$ 312.32
4 Carpenters	32.00	$ 35.04	$ 1,121.28
Crew Cost Per Day			$ 2,336.72

Panel erection production includes lifting, distribution, positioning, alignment, and bracing:

Production: 1 completed panel per hour equals
400 sq. ft. x 8 = 3,200 sq.ft. (37.16 sqm x 8 = 297.28 sq.m) panels per day
Erection cost per sq.ft., $1.04
Erection cost per sq.m, $11.19

Frequently erection of panels is assigned to subcontractors having the capability to furnish a complete cost package for equipment and personnel.

Patch & Repair. During the stripping of forms, lifting, distribution, and erection of the panels, minor damage might require remedial work to maintain quality control. An allowance has been included in the panel cost development table.

Tilt panel method is compatible with structural steel, poured-in-place or precast concrete columns. Panels of irregular shape containing an unusual number of openings or requiring substantial modification to accommodate special features will inflict an increase in cost. Proper sequence planning to maximize use of equipment and personnel and work crew efficiency will produce lower cost per square foot.

CONCRETE

Typical Cost Per Sq.Ft. (Sq.M)
For Panels 20'-0"x20'-0"x6" (6.09 x 6.09 x 0.15 m)

Price per Sq. Ft.			
Description	Material	Labor	Total
1. Preparing Pouring Bed (Allowance)	$ 0.06	$ 0.06	$ 0.13
2. Edge Forms	$ 0.35	$ 0.82	$ 1.17
3. Reinforcing	$ 2.21	$ 0.72	$ 2.93
4. Panel Connectors and Inserts	$ 0.96	$ 0.37	$ 1.33
5. Concrete	$ 1.76	$ 0.45	$ 2.21
6. Finish - Float or Broom	$ 0.06	$ 0.72	$ 0.78
7. Cure (Allowance)	$ 0.06	$ 0.10	$ 0.16
8. Erection	-----	$ 1.66	$ 1.66
9. Patch & Repair (Allowance)	$ 0.10	$ 0.19	$ 0.29
Total Panel Cost	$ 5.57	$ 5.09	$ 10.66

Price per Sq. Meter			
Description	Material	Labor	Total
1. Preparing Pouring Bed (Allowance)	$ 0.69	$ 0.69	$ 1.38
2. Edge Forms	$ 3.79	$ 8.78	$ 12.57
3. Reinforcing	$ 23.77	$ 7.75	$ 31.52
4. Panel Connectors and Inserts	$ 10.33	$ 3.96	$ 14.29
5. Concrete	$ 18.95	$ 4.82	$ 23.77
6. Finish - Float or Broom	$ 0.69	$ 7.75	$ 8.44
7. Cure (Allowance)	$ 0.69	$ 1.03	$ 1.72
8. Erection	-----	$ 17.91	$ 17.91
9. Patch & Repair (Allowance)	$ 1.03	$ 2.07	$ 3.10
Total Panel Cost	$ 59.94	$ 54.77	$ 114.70

03500 CEMENTITIOUS DECKS
GYPSUM PLANK

Cementitious deck is lightweight, noncombustible construction for roofs and floors in connection with steel framing. Plank are made in 2"x15"x10'-0" (50 x 375 mm x 3.04 m) units and designed for roofs on spans up to 7'-0" (2.13 m). The plank has a water resistant core.

All plank are molded and have steel tongue and groove edging on sides and ends, firmly anchored into the gypsum. They also have an electrically welded galvanized steel fabric as reinforcement for the gypsum. Weight 13.0 lbs. per sq.ft. (66 kg per sq.m).

On jobs requiring approximately 3,000 sq.ft. (278.70 sq.m) or more, the plank cost about $2.10 per sq.ft. ($22.60 per sq.m) in the eastern zone of the United States.

Labor Laying Gypsum Plank. One plank, 12-1/2 sq.ft. (1.13 sq.m) in area, weighs 160 lbs. (72.5 kg) requiring 2 workers to handle a single plank. Usually, it will require 2 workers on the ground and 2 workers on the roof. This crew should lay approximately 1,200 sq.ft. (111.48sq.m) per 8-hr. day at the following labor costs per 100 sq.ft (9.29 sq.m.):

Description	Hours	Rate	Total
Labor	2.70	$ 26.16	$ 70.63
Cost per sq.ft.			$ 0.71
per sq.m			$ 7.60

The above costs contemplate ordinary flat slab work, on a job containing 3,000 to 4,000 sq.ft. (278.70 to 371.60 sq.m), with a minimum amount of cutting. Add for extra handling and cutting for pitched roof and double the above labor costs for pitched roofs having hips, valleys, dormers, etc. Add extra for hoisting engineer if required.

Monolithic or Poured-in-Place Gypsum Roof Construction

Monolithic or poured-in-place gypsum roof deck, with its continuous reinforcing and securely welded sub-purlins, becomes an integral part of the main steel construction of a building, adding rigidity to the entire structure and can be designed for any required loading condition.

Permanent formboards, which remain as the undersurface of the completed deck, may be 1/2" (13 mm) gypsum formboard; 1/4" (6.25 mm) asbestos cement board; and 1" to 2" (25-50 mm) mineral fiber formboard.

Rolled steel sections designed to carry all loads between supports shall be laid on the top of walls, beams or roof supports at sufficient spacing to receive the formboards on the lower flanges; "reinforcing shall be 48-1214 galvanized welded wire fabric having 12 ga. longitudinal wire spaced 4" (100 mm) on centers and 14 ga. transverse wired 8" (200 mm) on centers or "Keydeck", a galvanized woven mesh as manufactured by Keystone Steel and Wire Co., having 16 ga. longitudinal wires and 19 ga. diagonal wires. The effective cross sectional area of reinforcing shall be not less than 0.026 sq. in. per ft. (56 sq.mm per m) of slab width. Standard (mill mixed) gypsum composition containing at least 87-1/2% calcined gypsum is mixed with water to a medium quaking consistency and poured or pumped in place, forming a solid reinforced slab 2" (50 mm) thick, or more, if desired, plus the board undersurface. The top surface of the slab shall be screeded smooth and left ready to receive the specified roofing.

The density of poured gypsum concrete is approximately 50 lbs. per c.f. (810 kg/m3). When cast in the minimum thickness of 2" (50 mm), the total dead weight of the slab is about 10 lbs. per s.f. (50 kg/m2). The weights, excluding sub purlins and insulating values of poured gypsum roof decks are as follows:

CONCRETE

Description	Wgt, lbs. per Sq.Ft.	"U" Value BTU
2" Gypsum concrete on 1/2" Gypsum Form Boards or plastic faced board	11.0	0.34
2" Gypsum concrete on 1" Mineral fiber formboard	11.2	0.16
2" Gypsum concrete on 2" Mineral fiber formboard	11.2	0.10
Description	Wgt, Kg. per Sq.M.	"U" Value BTU
50 mm gypsum concrete on 12.50 mm Gypsum Form Boards or plastic faced board	53.71	0.34
50 mm gypsum concrete on 25 mm Mineral fiber formboard	54.68	0.16
50 mm gypsum concrete on 50 mm Mineral fiber formboard	54.68	0.10

Practically all gypsum concrete is now mixed mechanically and pumped from ground to roof level as a means of further increasing the speed of operation and uniformity of quality, with corresponding reduction in cost.

The cost of poured-in-place gypsum roof decks will vary with the size and type of structure, availability of materials, freight charges, labor rates, etc., but the simplicity of design, lightweight and speed of erection, noncombustibility and finished undersurface make the overall cost of this construction very economical.

Poured-in-place gypsum roof decks are applied by contractors approved by the manufacturers of the materials. The following are approximate prices only on the various types of poured-in-place gypsum roof decks. These prices may vary locally due to labor and material costs, freight rates or size of specific job. For definite prices on any project it is always advisable to consult a roof deck contractor specializing in this type of construction.

Type of Roof	Approx. price range per Sq.Ft.	
2" Gypsum concrete on 1/2" gypsum formboard	$ 2.00	- $ 2.20
2" Gypsum concrete on 1" mineral fiber formboard	$ 2.30	- $ 2.60
Add for each extra 1/2" in thickness	$ 0.20	- $ 0.20
Type of Roof	Approx. price range per Sq.M.	
50 mm Gypsum concrete on 12.50 mm gypsum formboard	$ 21.53	- $ 23.68
50 mm Gypsum concrete on 25 mm mineral fiber formboard	$ 24.76	- $ 27.99
Add for each extra 12.50 mm in thickness	$ 2.15	- $ 2.15

The illustration that follows shows a poured gypsum roof on steel joists used in combination with a hung ceiling to comply with Underwriters Laboratories design No. P207 for a two hour fire rating. This assembly has a "U" value of 0.19, a noise reduction coefficient of 0.55 to 0.65, and a light reflectance of 85%. Other 2 hr. U.L rated systems are also available.

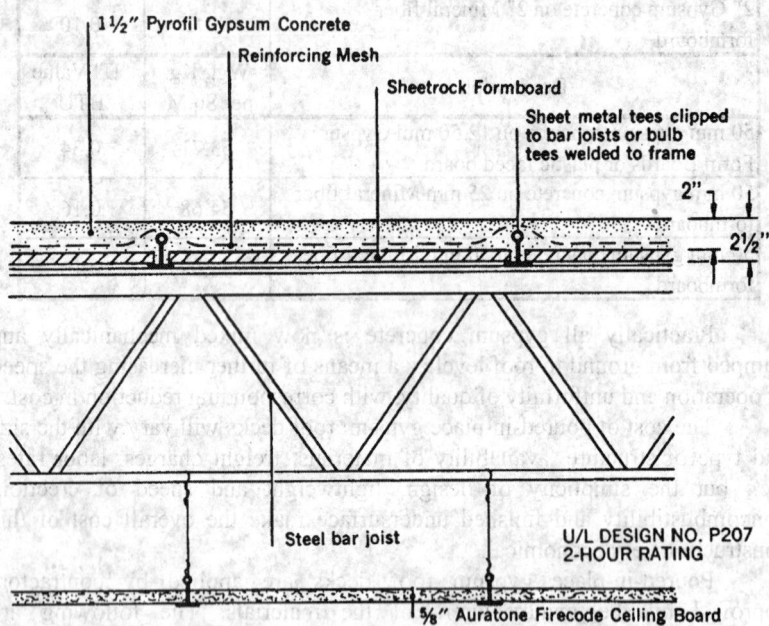

1½" Pyrofil Gypsum Concrete

Reinforcing Mesh

Sheetrock Formboard

Sheet metal tees clipped to bar joists or bulb tees welded to frame

2"

2½"

Steel bar joist

U/L DESIGN NO. P207
2-HOUR RATING

⅝" Auratone Firecode Ceiling Board

Courtesy of U.S. Gypsum Co.
Two Hour Fire Rated Roof and Ceiling Assembly

USG Floor Plank System

04 MASONRY

An accurate estimate of the cost of masonry requires a knowledge of various factors, including type of building, thickness of walls, kind of mortar, style of mortar joint, class of workmanship, ability of workers, and the ability of the foreman.

Estimating masonry is more than a matter of referring to a table or chart to obtain the prevailing wage scale and then determining the labor cost of laying a thousand brick. Accuracy is important in the preparation of your estimates, and all of the above items should be considered when determining the unit costs.

The type of building plays an important part in the costs. A mason can lay far more brick on a factory or warehouse building having long straight walls than on a school or office building that is cut up with numerous windows, which require straight walls, plumb jambs, pilasters, etc.

A mason can lay far more brick on a 16" (400 mm) wall than on an 8" (200 mm) wall and can lay more brick in smooth working lime mortar than in coarse portland cement mortar. He can also lay more brick by merely cutting the mortar joints than by striking them with the point of his trowel.

The class of workmanship plays an important part. Many cheaply constructed, speculative buildings have crooked walls and open mortar joints, presenting a slovenly appearance, while the best grade of workmanship requires full mortar joints, straight walls, plumb jambs, and corners-in other words, good workmanship.

Weather conditions also affect labor costs. A mason can lay more brick on a clear, dry day than when it is cold and wet, or when it is necessary to heat materials during freezing weather.

The ability of the foreman in charge to plan, schedule, and lay out the work affects final costs.

611

Masonry Specifications

In preparing a proposal from plans and specifications, the estimator will usually find masonry units specified according to the American Society for Testing and Materials (ASTM). These specifications set limitations on requirements and properties of the various materials and are generally accepted by manufacturers. ASTM specifications are available for all the following materials:

ASTM Specifications for the following Materials	
C 32*	Sewer and Manhole Brick (made from clay or shale)
C 55*	Concrete Building Brick
C 62*	Building Brick (solid masonry units made from clay or shale)
C 73*	Calcium Silicate Face Brick (sand-lime) brick
C 91*	Masonry Cement
C 105*	Ground Fire Clay or Refractory Mortar for laying up fireclay brickwork
C 126*	Ceramic Glazed Structural Clay Facing Tile, Facing Brick and Solid Masonry Units
C 144*	Aggregate for masonry mortar
C 150*	Portland Cement
C 207*	Hydrated Lime for masonry purposes
C 270*	Mortar for unit masonry
C 216*	Facing Brick (solid masonry units made from clay or shale)
C 279*	Chemical-Resistant Masonry Units
C 287*	Chemical-Resistant Sulphur Mortar
C 331*	Lightweight Aggregates for Concrete Masonry Units
C 476*	Grout for Reinforced and Nonreinforced Masonry
C 652*	Hollow Brick (hollow masonry units made from clay or shale)
C 902*	Paving Brick

Check for latest edition

The estimator should have a general acquaintance with all the above specifications but will deal primarily with face brick, building (formerly called common) brick, and mortar for unit masonry. Building brick is subdivided into 3 grades: SW, MW, and NW, indicating severe, medium, and negligible weathering. Face brick is subdivided into only SE and MW grades, but each of these grades has 3 types—FBX, FBS, and FBA—which set forth requirements for appearance and size.

Mortar specifications cover both portland cement-lime and masonry cement mortars and are subdivided into types M, S, N, and O, based on 28 day compressive strengths of 2,500, 1,800, 750, and 350 lbs./in² (17.24, 12.41, 5.17, and 2.07 MPa). ASTM specifications are also published for all the manufactured ingredients for the mortar.

612

MASONRY

Specifications based on these ASTM designations allow the estimator to figure within well-defined limits, and the ASTM terminology should be adopted in proposals.

04100 MORTAR

Cements. Different types of Portland cement are manufactured to meet different physical and chemical requirements for specific purposes. The American Society for Testing and Materials (ASTM) Designation C 150 provides for eight types of Portland cement:

Type I portland cement is the basic cement used in most mortars. Type II is used where moderate sulphate action is a problem. Type II cement will usually generate less heat at a slower rate than Type I. Type III is a high early portland strength cement that provides high strengths at an early period, usually a week or less. Although richer mixtures of Type I cement can be used to gain high early strength, Type III, high-early-strength portland cement, may provide it more satisfactorily and more economically.

TYPE IA, IIA, IIIA. Specifications for three types of air-entraining Portland cement (Types IA, IIA, and IIIA) are given in ASTM C 150. They correspond in composition to ASTM Types I, II, and III, respectively, except that small quantities of air-entraining materials are interground with the clinker during manufacture to produce minute, well-distributed, and completely separated air bubbles. These cements produce concrete with improved resistance to freeze-thaw action.

TYPE IV. Type IV is a low heat of hydration Portland cement for use where the rate and amount of heat generated must be minimized. It develops strength at a slower rate than Type I cement.

TYPE V. Type V is sulfate-resisting Portland cement used only in concrete exposed to severe sulfate action—principally where soils or groundwater has a high sulfate content.

All are included under ASTM C150-04ae1 Standard Specification for Portland Cement.

Air entraining cements should usually not be substituted, because this affects bond strength. Full information on tests and specifications are available in ASTM publications and in information published by the Portland Cement Association.

Masonry cements are included in ASTM specifications, but their chemical makeup is not regulated and varies widely among the manufacturers. It may therefore take some experimenting to find the product most suited to the job.

Masonry cements do, however, contain additives that increase workability and water retentivity, and the single product saves handling both portland cement and lime on the job, and mortar can be used as soon as mixed. However, where a Type S mortar is selected for bonding strength, care should be taken not to select a masonry cement with air entrainment

613

properties. Masonry cements are marketed under such names as Atlas, Keystone, Lehigh, LaFarge, Alamo Cement Co., etc.

Lime. Lime for construction purposes is available in both the quick (unslaked) and hydrated (slaked) forms, though in recent years hydrated lime has been preferred to quicklime, because it is faster, easier, and safer to use. Building lime is supplied in bulk or packaged in 50-lb. (22.68-kg) multiple-wall, moisture resistant bags. Barrels, once used to package lime, are no longer in use as containers for domestic lime shipments. Lime products sold in bulk or bags are generally quoted on a ton—2000 lbs. (907 kg) net—basis.

Quicklime may be purchased in the following forms, the principal difference being in the size of the particles: pebble lime, crushed lime, ground lime, and pulverized lime. All quicklime must be slaked prior to use and manufacturer's directions for slaking should be followed to secure best results. Due to the greater ease with which complete slaking is secured, the finer divided types of quicklime are preferred. Since quicklime is very reactive with water, care should be exercised at all times during the slaking operation to prevent splattering of the lime and potentially serious burns of the eyes and skin. Slaked quicklime putty should be screened and permitted to age until all the lime particles are completely slaked. The aging period may vary from a few hours to several days, depending on the chemical and physical properties of the quicklime used and on the skill of the operator in slaking the lime.

Hydrated lime for structural use is furnished in the dry powdered form usually packaged in 50 lbs. (22.68 kg) net paper bags and sold by the ton (2000 lbs. or 907.2 kg net). It may also be purchased in bulk, if proper facilities are available for handling and storage. There are several types of hydrated lime 1- Type N, Normal Hydrated Lime. 2- Type NA, Normal Air-entraining hydrated Lime 3- Type S, Special Hydrated Lime and 4- Type SA, Special Air-Entraining Hydrated Lime. Types S and SA are differentiated from types N and NA principally by their ability to develop high early plasticity, higher water retentivity, and by their limitations on unhydrated oxide content. The maximum air content of cement-lime mortar made with types NA and SA is 14%; with types N and S lime 7%

Type "S" or pressure hydrated lime is a highly hydrated product contains a maximum of 8% combined unhydrated calcium and magnesium oxides. In addition, putties made from this lime develop high plasticities instantly upon mixing. Type "S" special hydrated limes meeting all ASTM C 207 requirements are commercially available in both the dolomitic and high calcium varieties. Type "N" normal hydrated limes should be soaked in a paste or putty form for several hours or overnight prior to use in order to enhance their plasticity and workability. The preferred method for soaking is to sift the hydrated lime evenly into a watertight box or vat that has been previously half filled with clean water. Do not stir, mix, or agitate the mass, but allow the lime to settle naturally; continue to add the hydrate slowly and

614

MASONRY

evenly over the entire surface of the water until the thick paste is formed, after which the mass is permitted to soak until required for use.

All building limes should comply with the requirements of the Standard Specifications for Hydrated Lime for Masonry Purposes (ASTM C-207-04) or Quicklime for Structural Purposes (ASTM C-5-03).

Storage of Lime and Lime Putty. Prior to slaking or use, all quicklime and hydrated lime previously delivered to the job site should be stored in a clean, dry place.

Quicklime is shipped in bulk and in waterproofed, multi-wall paper bags. When furnished in bulk, it should be slaked immediately to assure maximum putty yield and to avoid loss. Quicklime furnished in bags may be stored for considerable time, depending on humidity and storage conditions.

Hydrated lime in bulk or packed in bags may be stored for relatively long periods of time provided it is placed in a clean, dry, properly ventilated warehouse.

Lime putty should be aged and stored in clean, tight vats and should be well protected from direct contact with the atmosphere by maintaining a thin film of water over the surface.

Commercial plants for the production of scientifically prepared, well aged putty, are in operation in many markets. The use of commercially prepared lime putty eliminates the need for slaking and aging or soaking of lime on the job. It also has the advantage of providing the estimator with reliable figures for estimating purposes, because purchase order is based on net quantity of lime putty required for the operation. Other advantages of commercially produced lime putty include flexibility during construction, elimination of lost time awaiting preparation of lime putty and admitting of expansion in operations on short notice, reduction in storage space required for materials at the site of the work, and improved quality of putty due to careful supervision during slaking and aging of lime putty.

Lime Putty and Its Preparation. Job specifications generally require mortar to be mixed in definite predetermined proportions by volume. The lime proportion is usually determined in volumes of lime putty, so some knowledge of putty yield of lime is necessary in preparing estimates of quantity of lime required. The putty yield of different limes varies widely, according to chemical and physical properties of the lime, methods of processing, etc. However, the quantities given in the following table may be used as a sound, conservative average for estimating purposes.

Quantity of Lime Putty Obtainable From Various Types of Lime				
Type of Lime	Average C.F. Putty per Ton of Lime	Average Lbs of Lime per C.F. Putty	Average Cu.M. Putty per M Ton of Lime	Average Kg of Lime per Cu.M. Putty
Hydrated Lime	46	44	1.44	704.75
Pebble or Pulverized Quicklime	80	25	2.50	400.42

It takes 109 lbs. (49.44 kg) portland cement to produce 1 cu.ft. (0.028 cu.m) of cement paste of normal working consistency.

It takes 9 cu.ft. (0.25 cu.m) of putty or paste with 27 cu.ft. (0.76 cu.m) of ordinary building sand to produce 1 cu. yd. (0.76 cu.m) of 1:3 mortar.

Quantity of Materials Required for Masonry Using Pebble or Pulverized Quicklime at 80 Cu. Feet Lime Putty per Ton								
Proportion by Volume			Quantity of Materials					
			For One Cu.Yd. Mortar			To Lay 1,000 Brick		
Lime Putty	Portland Cement	Sand	Lime Lbs*	Cement Sacks	Sand CY	Lime Lbs*	Cement Sacks	Sand CY
1	0	3.0	225	0.00	1	150	0.0	0.67
3	1	12.0	169	2.25	1	113	1.5	0.67
2	1	9.0	150	3.00	1	100	2.0	0.67
1 1/2	1	7.5	135	3.60	1	90	2.4	0.67
1	1	6.0	113	4.50	1	75	3.0	0.67
1/2	1	4.5	75	6.00	1	50	4.0	0.67
0	1	3.0	0	9.00	1	0	6.0	0.67
10% **	1	3.0	22.5	9.00	1	15	6.0	0.67
15%	1	3.0	34	9.00	1	23	6.0	0.67

* Dry unslaked Lime
** Based on volume of cement required. Add 5% to 15% for waste.

Quantity of Materials Required for Masonry Using Pebble or Pulverized Quicklime at 2.55 Cu.M. Lime Putty per Ton								
Proportion by Volume			Quantity of Materials					
			For One Cu.M. Mortar			To Lay 1,000 Brick		
Lime Putty	Portland Cement	Sand	Lime Kg*	Cement Sacks	Sand Cu.M.	Lime Kg*	Cement Sacks	Sand Cu.M
1	0	3.0	133	0.00	1.31	68	0.0	0.51
3	1	12.0	100	2.94	1.31	51	1.5	0.51
2	1	9.0	89	3.92	1.31	45	2.0	0.51
1 1/2	1	7.5	80	4.71	1.31	41	2.4	0.51
1	1	6.0	67	5.89	1.31	34	3.0	0.51
1/2	1	4.5	44	7.85	1.31	23	4.0	0.51
0	1	3.0	0	11.77	1.31	0	6.0	0.51
10% **	1	3.0	13	11.77	1.31	7	6.0	0.51
15%	1	3.0	20	11.77	1.31	10	6.0	0.51

* Dry unslaked Lime
** Based on volume of cement required. Add 5% to 15% for waste.

Where hydrated lime is used in dry powdered form, it is usually assumed that a given volume of dry hydrated lime will produce an equivalent

MASONRY

volume of lime putty. A bag of dry hydrated lime - 50 lbs. (22.68 kg) net - equals about 1.15 cu.ft. (0.03 cu.m) of lime putty.

Labor Slaking Lime and Making Mortar. When mixing mortar by hand, a good mortar maker should slake and sand about a ton of quicklime per 8-hr. day, and should make about 4 cu.yd. (3.05 cu.m) of mortar per 8-hr. day.

Quantity of Materials Required for Masonry Mortar Using Hydrated Lime at 46 Cu. Feet Lime per Ton								
Proportion by Volume			Quantity of Materials					
			For One Cu.Yd. Mortar			To Lay 1,000 Brick		
Lime Putty	Portland Cement	Sand	Lime Lbs*	Cement Sacks	Sand CY	Lime Lbs*	Cement Sacks	Sand CY
1	0	3.0	391	0.00	1	96	0.0	0.88
3	1	12.0	294	2.25	1	116	2.0	0.88
2	1	9.0	261	3.00	1	103	2.6	0.88
1 1/2	1	7.5	235	3.60	1	93	3.1	0.88
1	1	6.0	196	4.50	1	77	3.9	0.88
1/2	1	4.5	131	6.00	1	52	5.2	0.88
0	1	3.0	0	9.00	1	0	7.8	0.88
10% **	1	3.0	39	9.00	1	15	7.8	0.88
15%	1	3.0	59	9.00	1	23	7.8	0.88

Dry unslaked Lime
** Based on volume of cement required. Add 5% to 15% for waste.

Quantity of Materials Required for Masonry Mortar Using Hydrated Lime at 1.43 Cu. Meters Lime per Tonne								
Proportion by Volume			Quantity of Materials					
			For One Cu.M. Mortar			To Lay 1,000 Brick		
Lime Putty	Portland Cement	Sand	Lime Kg*	Cement Sacks	Sand Cu.M.	Lime Kg*	Cement Sacks	Sand Cu.M
1	0	3.0	232	0.00	1.31	44	0.0	0.67
3	1	12.0	174	2.94	1.31	53	2.0	0.67
2	1	9.0	155	3.92	1.31	47	2.6	0.67
1 1/2	1	7.5	139	4.71	1.31	42	3.1	0.67
1	1	6.0	116	5.89	1.31	35	3.9	0.67
1/2	1	4.5	77	7.85	1.31	24	5.2	0.67
0	1	3.0	0	11.77	1.31	0	7.8	0.67
10% **	1	3.0	23	11.77	1.31	7	7.8	0.67
15%	1	3.0	35	11.77	1.31	10	7.8	0.67

Dry unslaked Lime
** Based on volume of cement required. Add 5% to 15% for waste.

The actual labor cost of making 18 cu.ft. (0.51 cu.m) of mortar, sufficient for 1,000 common brick, should average as follows:

617

Labor Making 18 Cu.Feet Mortar for 1,000 brick					
Item	Hours	Rate	Total	Rate	Total
Mortar Maker	1.33	$	$	$ 36.93	$ 49.12
Cost per Cu. Foot					**$ 2.73**
Cost per CU. Meter					$ 96.35

On jobs of any size, the mortar is mixed in a mixer, which produces a more uniform and more easily spread mortar. It not only enables a mason to lay more brick but also results in a saving of 10% to 20% in labor required to mix.

Quik-Slake Masons Lime. This is a lime in pulverized form that slakes the instant it hits water. It comes in 50-lb. (22.68-kg) paper sacks, and the manufacturers say that 4 sacks are sufficient to lay, 1,000 brick under average conditions.

Aggregate. Aggregate may be either natural sand that is clean and sharp, or sand manufactured by crushing stone, gravel, or air-cooled iron blast furnace slag. Sand should be properly graded, as set forth in ASTM Specification C 144-04, with all sand passing a No. 4 sieve and approximately 10% passing a No. 200 sieve for manufactured sand. Too coarse a sand decreases workability while too fine a sand decreases water retentivity.

Mortar Colors. Mortar coloring agents may be added, either natural, such as white sand, or pigments. While different color effects will require some experimenting, in general, 4 to 8 lbs. (1.81-3.62 kg) of color to one bag of cement, hydrated lime or c.f. (0.02 cu.m) of putty and 3 cu.ft. (0.08 cu.m) of sand can be figured, or about one 50-lb. (22.68-kg) bag per 1000 bricks with a 3/8" (9.37 mm) joint.

Waterproofing and Shrink proofing Mortar

Leaky masonry walls have aroused considerable concern, and this has led to extensive investigation and tests by the National Bureau of Standards which investigated not only the materials such as brick, tile, concrete blocks, mortar, etc., but also the effects of different grades of workmanship.

After extensive tests, it was determined that workmanship affected the permeability of the walls more than any other factor. Walls with tooled joints were less permeable than similar walls with cut joints. But the quality of the workmanship inside the walls had a greater influence than the kind of surface finish on the joint. In other words, brick as laid on the average contract job, where mortar was spread and furrowed and the joints were not completely filled with mortar, leaked far more than bricks laid in a full bed of mortar and "shoved" into position with full bed and head joints and a complete filling of joints between all brick.

It was also determined that nearly all of the leakage was through the mortar joints regardless of the kind of brick used and that in practically every

618

instance where the "commercial" grade of workmanship was used, leaky walls resulted.

Where brick of high absorptive properties were used, the brick absorbed the water from the mortar, making it difficult to work and resulting in small hair cracks between the mortar joints and the brick. This also caused leakage through the walls.

The use of lime with low plasticity produces mortar with low water retention and greatly increases the permeability of the walls. This effect was more pronounced when the mortar was used with high absorptive brick. Water reduction in cement and lime mortars is highly desirable, because with less water there is less evaporation, and mortar shrinkage is reduced below the critical point.

Since these investigations were made, numerous admixtures have been placed on the market. These are to be added to the mortar mixture and are said to give the mortar greater plasticity, to retain the water in the mortar, to prevent shrinkage and leakage in the mortar joints, and to make a smoother mortar that will work better under the trowel. Some of these products are listed below.

Anti-Hydro. Cement mortar for use with brick, block, or stone shall be of 1:3 mix, gauged and tempered with a solution of 1 gal. (3.78 liters) Anti-Hydro to 15 gals.(56.77 liters) of water, or use 1 qt. (1.13 liters) Anti-Hydro to each sack of cement.

Wet all brick. Push-lay and thoroughly grout all brick at every course. Parge with same mortar between the face brick and back-up masonry.

Castle Aquatite. Mortar to be composed of 1 part portland cement, 3 parts clean and mixed with a solution of 1 part Aquatite to 15 parts water.

Hydratite Plus. A powder in concentrated form, which when mixed with the mortar is dissolved and miscible in the water. It anchors the particle of water intimately to the particle of the cement, sand, and lime with a bond greater than the suction of the brick. With less water it increases the flow or workability of the mortar. Use 2 lbs. (0.90 kg) per sack of cement in 1:1:6 mortar.

Hydrocide Powder. When used for waterproofing masonry mortar, use 1 lb. (0.45 kg) of powder to each sack of cement or per cu.ft. (0.028 cu.m) of lime putty used in the mixture.

Omicron Mortarproofing. A plasticizing and water reducing agent that increases the workability of mortar, increases water retention and minimizes shrinkage. Contains stearate for increased water repellency. Add 1 lb. (0.45 kg) of Omicron for each sack of cement, lime or putty in the mortar mix.

Requires 5 to 6 lbs. (2.27-2.72 kg) per 1,000 brick, depending on mortar proportions and width of joints.

Plastiment (Sika). A powder used in the proportion of up to 1 lb. (0.45 kg) per sack of cement. Increases the workability of mortar, allows a reduction in the water-cement ratio, reduces bleeding and increases the bond between brick and mortar. Of particular advantage in hot weather construction, as it delays the setting time and preserves the workability of the mortar for a longer time.

Toxement IW. Toxement IW powder, paste or concentrated paste may be used for waterproofing and shrinkproofing masonry mortar.

Sikacrete. Accelerating and plasticizing liquid for use in brick mortar. Diluted with the gauging water 1 part Sikacrete to 3 to 7 parts water. Increases the bond and accelerates the hardening rate of mortar so that higher tiers of brick or glass block can be laid and the joints struck sooner.

Types of Mortar

The following mortars are based on proportion set forth in ASTM specifications. These mortars were designated Type A-1, A-2, B, and C prior to 1954.

Type M mortar is a high strength mortar used primarily in foundation masonry, retaining walls, walks, sewers, and manholes. Proportions and costs for a cement-lime mixture are:

Type M Mortar - Proportions and Costs for a Cement - Lime Mixture							
English				Metric			
1 C.F.	Portland Cement	$ 5.75 /C.F.	$ 5.75	1 C.M.	$ 203.04 /C.M.	$ 203.04	
1/4 C.F.	Lime	$ 4.25 /C.F.	$ 1.06	0.25 C.M.	$ 150.07 /C.M.	$ 37.52	
3 C.F.	Sand	$ 0.33 /C.F.	$ 0.99	3 C.M.	$ 11.65 /C.M.	$ 34.96	
Total			$ 7.80	Total		$ 275.51	
Per Cu. Foot			$ 2.60	Per Cu. Meter		$ 91.84	

Proportions and costs for a type M mortar using masonry cement are:

Type M Mortar - Proportions and Costs using Masonry Cement							
English				Metric			
1 C.F.	Portland Cement	$ 5.75 /C.F.	$ 5.75	1 C.M.	$ 203.04 /C.M.	$ 203.04	
1 C.F.	Type II Masonry Cement	$ 5.50 /C.F.	$ 5.50	1 C.M.	$ 194.21 /C.M.	$ 194.21	
6 C.F.	Sand	$ 0.33 /C.F.	$ 1.98	6 C.M.	$ 11.65 /C.M.	$ 69.92	
Total			$ 13.23	Total		$ 467.16	
Per Cu. Foot			$ 2.21	Per Cu. Meter		$ 77.86	

Type S mortar also has a reasonably high compressive strength and develops maximum tensile bond strength between brick and cement-lime mortars. It is recommended for use in reinforced masonry and where flexural strengths are required, such as cavity walls exposed to strong winds, and for maximum bonding power, such as for ceramic veneers. Proportions and costs for cement-lime Type S mixtures are:

620

Type S Mortar - Proportions and Costs for a Cement - Lime Mixture							
English				Metric			
1 C.F.	Portland Cement	$ 5.75 /C.F.	$ 5.75	1 C.M.	$ 203.04 /C.M.		$ 203.04
1/2 C.F.	Lime	$ 4.25 /C.F.	$ 2.13	1/2 C.M.	$ 150.07 /C.M.		$ 75.04
4 1/2 C.F.	Sand	$ 0.33 /C.F.	$ 1.49	4 1/2 C.M.	$ 11.65 /C.M.		$ 52.44
Total			$ 9.36	Total			$ 330.51
Per Cu. Foot			$ 2.08	Per Cu. Meter			$ 73.45

Type S Mortar - Proportions and Costs using Mortar							
English				Metric			
1/2 C.F.	Portland Cement	$ 5.75 /C.F.	$ 2.88	1/2 C.M.	$ 203.04 /C.M.		$ 101.52
1 C.F.	Masonry Cement	$ 5.50 /C.F.	$ 5.50	1 C.M.	$ 194.21 /C.M.		$ 194.21
4 1/2 C.F.	Sand	$ 0.33 /C.F.	$ 1.49	4.5 C.M.	$ 11.65 /C.M.		$ 52.44
Total			$ 9.86	Total			$ 348.16
Per Cu. Foot			$ 2.19	Per Cu. Meter			$ 77.37

Type N mortar is a medium strength mortar, generally used in exposed masonry above grade. Proportions and cost of Type N cement-lime mortar are:

Type N Mortar - Proportions and Costs for a Cement - Lime Mixture							
English				Metric			
1 C.F.	Portland Cement	$ 5.75 /C.F.	$ 5.75	1 C.M.	$ 203.04 /C.M.		$ 203.04
1 C.F.	Lime	$ 4.25 /C.F.	$ 4.25	1 C.M.	$ 150.07 /C.M.		$ 150.07
6 C.F.	Sand	$ 0.33 /C.F.	$ 1.98	6 C.M.	$ 11.65 /C.M.		$ 69.92
Total			$ 11.98	Total			$ 423.02
Per Cu. Foot			$ 2.00	Per Cu. Meter			$ 70.50

Type N Mortar - Proportions and Costs using Masonry Cement							
English				Metric			
1 C.F.	Masonry Cement	$ 5.50 /C.F.	$ 5.50	1 C.M.	$ 194.21 /C.M.		$ 194.21
3 C.F.	Sand	$ 0.33 /C.F.	$ 0.99	3 C.M.	$ 11.65 /C.M.		$ 34.96
Total			$ 6.49	Total			$ 229.17
Per Cu. Foot			$ 2.16	Per Cu. Meter			$ 76.39

Type O mortar is a low strength mixture for general interior use where compressive strengths do not exceed 100 psi (0.68 MPa). It may be used elsewhere where exposures are not severe and no freezing will be encountered. Proportions and costs of Type O cement-lime mortar are:

Type O Mortar - Proportions and Costs for a Cement - Lime Mixture							
English				**Metric**			
1 C.F.	Portland Cement	$ 5.75 /C.F.	$ 5.75	1 C.M.	$ 203.04 /C.M.		$ 203.04
2 C.F.	Lime	$ 4.25 /C.F.	$ 8.50	2 C.M.	$ 150.07 /C.M.		$ 300.14
9 C.F.	Sand	$ 0.33 /C.F.	$ 2.97	9 C.M.	$ 11.65 /C.M.		$ 104.87
Total			$ 17.22	Total			$ 608.05
Per Cu. Foot			$ 1.91	Per Cu. Meter			$ 67.56

Type O Mortar - Proportions and Costs using Mortar							
English				**Metric**			
1 C.F.	Portland Cement	$ 5.50 /C.F.	$ 5.50	1 C.M.	$ 194.21 C.M.		$ 194.21
3 C.F.	Sand	$ 0.33 /C.F.	$ 0.99	3 C.M.	$ 11.65 C.M.		$ 34.96
Total			$ 6.49	Total			$ 229.17
Per Cu. Foot			$ 2.16	Per Cu. Meter			$ 76.39

In the above costs, lime is figured in an exact proportion. As noted under "Quantity of Materials Required for Masonry Mortar Using Hydrated Lime...", a 50-lb. (22.7-kg) bag of hydrated lime will make 1.15 cu.ft. (0.03 cu.m) putty.

The ASTM specifications for Types N and O mortar allow some leeway in lime proportions and a full bag of lime can be figured. Only with very high strength mortars might you have to figure splitting bags to achieve exact proportions.

Cu.Ft. (Cu.M) of Mortar Required to Lay 1,000 Face Brick

Based on standard size brick having 1/4" (6.25 mm) to 3/8" (9.37 mm) end joints and bed joints as follows:

Width of Mortar Joints, inches							
1/8"	1/4"	3/8"	1/2"	5/8"	3/4"	7/8"	1"
4	7	9	12	14	16	18	20
(metric)							
3.1 mm	6.3 mm	9.4 mm	12.5 mm	15.7 mm	18.9 mm	21.9 mm	25.0 mm
0.11	0.2	0.25	0.34	0.4	0.45	0.51	0.57

For Dutch, English, and Flemish Bond, add about 10% to the above quantities on account of additional head joints.

Labor cost to mix mortars is usually figured as part of labor cost setting brick. However, if figured separately, it will be about $0.92 per cu.ft. ($32.48 per cu.m), machine mixed, $1.38 per cu.ft. ($48.72 per cu.m) hand mixed.

Quantity of Mortar Required to Lay 1,000 Brick						
Cubic Feet of Mortar per 1,000 Brick						
Joint Thickness, Inches	Wall Thickness, Inches					
	4"	8"	12"	16"	20"	24"
1/8 "	2.9	5.6	6.5	7.1	7.3	7.5
1/4 "	5.7	8.7	9.7	10.2	10.5	10.7
3/8 "	8.7	11.8	12.9	13.4	13.7	14
1/2 "	11.7	15	16.2	16.8	17.1	17.3
5/8 "	14.8	18.3	19.5	20.1	20.5	20.7
3/4 "	17.9	21.7	23	23.6	24	24.2
7/8 "	21.1	25.1	26.5	27.1	27.5	27.8
1 "	24.4	28.6	30.1	30.8	31.2	31.5
Cubic Meter of Mortar per 1,000 Brick						
Joint Thickness, Inches	Wall Thickness, mm					
	100	200	300	400	500	600
3.1 mm	0.08	0.16	0.18	0.20	0.21	0.21
6.3 mm	0.16	0.25	0.27	0.29	0.30	0.30
9.4 mm	0.25	0.33	0.37	0.38	0.39	0.40
12.5 mm	0.33	0.42	0.46	0.48	0.48	0.49
15.6 mm	0.42	0.52	0.55	0.57	0.58	0.59
18.8 mm	0.51	0.61	0.65	0.67	0.68	0.69
21.9 mm	0.60	0.71	0.75	0.77	0.78	0.79
25.0 mm	0.69	0.81	0.85	0.87	0.88	0.89

For various thicknesses of walls and joints. No allowance for waste.

Selecting Mortars. Often the estimator has no choice in the makeup of mortar, because the architect's specifications set forth exactly what is required. Because replacing masonry that is found not acceptable by an architect or owner is extremely expensive, it is false economy, not to mention the loss of reputation involved, to alter specifications without approval from the architect.

However, many jobs will leave considerable leeway in the selection of mortars, and an informed estimator may even perform a valuable service in suggesting changes to the mortar proportions that will either save money or make a better job. All too often a mortar may be selected on the basis of high compressive strength when the job may require only a 100 psi (0.69 MPa). This not only wastes cement, but such mortars do not have the adhesive and sealing powers of the weaker mortars, and these two characteristics are the primary functions required on most jobs.

Of further importance to the estimator is the workability of the mortar, which will enable the bricklayer to lay more units. A complete discussion of mortars and recommendations is contained in technical notes on brick construction from the Brick Institute of America (BIA) in Reston, VA.

Mortar for Concrete Masonry. For laying masonry walls subject to average conditions of exposure, use a mortar made in the proportions of 1 volume of masonry cement and between 2 and 3 volumes of damp, loose mortar sand; or 1 volume of portland cement and between 1 and 1-1/4 volumes of hydrated lime or lime putty and between 4 and 6 volumes of damp, loose mortar sand.

Walls that will be subject to extremely heavy loads, severe winds, earthquakes, serious frost, or other conditions requiring extra wall strength should be laid with a mortar made of 1 volume masonry cement plus 1 volume portland cement and between 4 and 6 volumes of damp, loose mortar sand; or 1 volume of portland cement, to which may be added up to 1/4 volume of hydrated lime or lime putty, and between 2 and 3 volumes of damp, loose mortar sand.

Cost of One C.Y. (Cu.M.) Cement Mortar for Concrete Masonry

1 Volume Masonry Cement and 3 Volumes Loose Mortar Sand				
	Rate	Total	Rate	Total
9 sacks masonry cement	$....	$....	$ 5.50	$ 49.50
1.00 c.y. sand	$....	$....	$ 8.90	$ 8.90
Cost per c.y.				$ 58.40
per c.f.				$ 2.16
Metric				
11.8 sacks masonry cement	$....	$....	$ 5.50	$ 64.75
1.00 cu.m sand	$....	$....	$ 11.64	$ 11.64
Cost per cu.m				$ 76.39

1 Volume Masonry Cement:1 Volume Hydrated Lime:6 Volumes Loose Mortar Sand				
	Rate	Total	Rate	Total
4.5 sacks portland Cement	$....	$....	$ 5.75	$ 25.88
196 lbs. hydrated lime	$....	$....	$ 0.07	$ 13.72
1.00 c.y. sand	$....	$....	$ 8.90	$ 8.90
Cost per c.y.				$ 48.50
per c.f.				$ 1.80
Metric				
5.9 sacks masonry cement	$....	$....	$ 5.75	$ 33.93
116 kg hydrated lime	$....	$....	$ 0.15	$ 17.90
1.00 cu.m sand	$....	$....	$ 11.64	$ 11.64
Cost per cu.m				$ 63.47

MASONRY

1 Volume Portland Cement:2 Volumes Hydrated Lime:9 Volumes Loose Mortar Sand				
	Rate	Total	Rate	Total
3 sacks masonry cement	$....	$....	$ 5.75	$ 17.25
261 lbs. hydrated lime	$....	$....	$ 0.07	$ 18.27
1.00 c.y. sand	$....	$....	$ 8.90	$ 8.90
Cost per c.y.				$ 44.42
per c.f.				$ 1.65
(metric)				
3.9 sacks masonry cement	$....	$....	$ 5.75	$ 22.43
155 kg hydrated lime	$....	$....	$ 0.15	$ 23.25
1.00 cu.m sand	$....	$....	$ 11.64	$ 11.64
Cost per cu.m				$57.32

1 Volume Masonry Cement:1 Volume Portland Cement:6 Volume Loose Mortar Sand				
	Rate	Total	Rate	Total
4.5 sacks masonry cement	$....	$....	$ 5.50	$ 24.75
4.5 sacks portland cement	$....	$....	$ 5.75	$ 25.88
1.00 c.y. sand	$....	$....	$ 8.90	$ 8.90
Cost per c.y.				$ 59.53
per c.f.				$ 2.20
(metric)				
5.9 sacks masonry cement	$....	$....	$ 5.50	$ 32.45
5.9 sacks portland cement	$....	$....	$ 5.75	$ 33.93
1.00 cu.m sand	$....	$....	$ 11.64	$ 11.64
Cost per cu.m				$ 78.02

1 Volume Portland Cement: 1/4 Volume Hydrated Lime:3 Volumes Loose Mortar Sand				
	Rate	Total	Rate	Total
7.83 sacks masonry cement	$....	$....	$ 5.75	$ 45.02
90 lbs. hydrated lime	$....	$....	$ 0.07	$ 6.30
0.86 c.y. sand	$....	$....	$ 8.90	$ 7.65
Cost per c.y.				$ 58.98
per c.f.				$ 2.18
(metric)				
10.2 sacks masonry cement	$....	$....	$ 5.75	$ 58.65
41 kg hydrated lime	$....	$....	$ 0.15	$ 6.15
1.12 cu.m sand	$....	$....	$ 11.64	$ 13.04
Cost per cu.m				$ 77.84

Quantity of Mortar Required to Lay 1,000 Concrete Brick of Various Sizes				
Width of Bed Joint	Thickness of Wall			
	1 Brick	2 Bricks	2 Bricks	3 Bricks
Inches	4" Wall	8" Wall	8" Backup	12" Wall
mm	100 mm	200 mm	200 mm	300 mm
Modular Size Brick, 2-1/4"x3-5/8"x7-5/8" (56 x 90 x 190 mm)				
5/12"	9.64 c.f.	12.19 c.f.	14.73 c.f.	13.01 c.f.
10.4 mm	0.272 cu.m	0.345 cu.m	0.417 cu.m	0.37 cu.m
Jumbo Brick, 3-5/8" x 3-5/8" x 7-5/8" (90 x 90 x 190 mm)				
3/8"	10.06 c.f.	13.88 c.f.	17.70 c.f.	15.15 c.f.
9.4 mm	0.284 cu.m	0.393 cu.m	0.501 cu.m	0.43 cu.m
Double Brick, 4-7/8" x 3-5/8" x 7-5/8" (122 x 90 x 190 mm)				
11/24"	12.68 c.f.	17.77 c.f.	22.86 c.f.	19.47 c.f.
11.5 mm	0.359 cu.m	0.503 cu.m	0.647 cu.m	0.55 cu.m
Roman Brick, 1-5/8" x 3-5/8" x 11-5/8" (41 x 90 x 290 mm)				
3/8"	11.79 c.f.	14.65 c.f.	—	—
9.4 mm	0.333 cu.m	0.414 cu.m	—	—
Roman Brick, 1-5/8" x 3-5/8" x 15-5/8" (41 x 90 x 390 mm)				
3/8"	15.25 c.f.	19.07 c.f.	—	—
9.4 mm	0.431 cu.m	0.540 cu.m	—	—

All end and back joints figured 3/8" (9.37 mm). Quantities include 10% for waste.

Mortar for glass blocks shall be composed of 1 part waterproof portland cement, 1 part lime, and 4 parts well graded sand, all well mixed to a consistency as stiff and dry as possible.

Cost of Mortar for Laying Glass Blocks				
	Rate	Total	Rate	Total
25 Lbs Quicklime	$....	$....	$ 0.07	$ 1.75
1 Bag W'p'f Cement	$....	$....	$ 7.50	$ 7.50
4 C.F. Sand	$....	$....	$ 0.33	$ 1.32
Cost 4 C.F. Mortar				$ 10.57
Cost per Cubic Foot				$ 2.64
Metric				
14.8 Kg Quicklime	$....	$....	$ 0.15	$ 2.28
1.3 Bag W'p'f Portland Cement	$....	$....	$ 7.50	$ 9.75
4 Cu.M. Sand	$....	$....	$ 11.65	$ 46.60
Cost per Cu.M.				$ 58.63

Mortar Required for Setting Stone. The quantity of mortar required for stone setting will vary with the size of the stone, width of bed, etc., but generally, it will require 4 to 5 cu.ft. (0.11-0.14 cu.m) of mortar per 100 cu.ft. (2.83 cu.m) of stone.

MASONRY

If the stone is to be back-plastered, the quantity of mortar will vary with the thickness of the plaster coat, but as most ashlar is 4" to 8" (100-200 mm) thick, it will require the following cu.ft. of mortar per 100 cu.ft. (2.83 cu.m) of stone:

Cubic Foot of Mortar				.0283 Cubic Meter of Mortar			
Plaster Thick., Inches	Thickness of Stone, Inches			Plaster Thick., mm	Thickness of Stone, mm		
	4 "	6 "	8 "		100.0 mm	150 mm	200 mm
1/4 "	6 1/2 "	4 1/4 "	3 1/4 "	6.25	162.5 mm	106.3 mm	81.3 mm
1/2 "	13 "	8 1/2 "	6 1/2 "	12.5	325.0 mm	212.5 mm	162.5 mm
3/4 "	19 "	13 "	9 1/2 "	18.75	475.0 mm	325.0 mm	237.5 mm

It is customary to use white non-staining portland cement for setting and backplastering limestone or other porous stone to prevent stains from appearing on the face of the stone. Never use ordinary portland cement for setting limestone.

Cost of One C.Y. (0.76 Cu.M) of 1:3 Non-Staining Cement Mortar With 1/5 Part of Hydrated Lime Based on the Cement Volume					
	Rate	Total	Rate		Total
2 bbls. Medusa Stone Sett Cement	$....	$....	$ 10.25	$	20.50
100 lbs. hydrated lime	$....	$....	$ 0.07	$	7.00
1 c.y. sand	$....	$....	$ 8.90	$	8.90
Cost per c.y.				$	36.40
per c.f.				$	1.35
2 bbls. Medusa Stone Sett Cement	$....	$....	$ 10.25	$	20.50
45.36 kg hydrated lime	$....	$....	$ 0.15	$	7.00
0.76 cu.m sand	$....	$....	$ 11.64	$	8.85
Cost per 0.76 cu.m				$	36.35
per cu.m				$	47.82

If white sand is used for mortar, figure $25.50 per ton ($25.91 per tonne) and 3,000 lbs. per cu.yd. (1779.75 Kg/cu.M). If white cement is specified, figure $25.00 per bbl. If white cement (waterproofed) is specified, figure $26.50 per bbl.

Cost of One C.Y. (0.76 Cu.M) of Cement-Lime Mortar Consisting of 1/2 Part Lime, 1/2 Part Stone Set Cement, and 3 Parts Sand						
	Rate	Total	Rate		Total	
1.25 bbls. Stone Set Cement	$....	$....	$	10.25	$	12.81
225 lbs. hydrated lime	$....	$....	$	0.07	$	15.75
1 c.y. sand	$....	$....	$	8.90	$	8.90
Cost per c.y.					$	37.46
per c.f.					$	1.39
2 bbls. Medusa Stone Sett Cement	$....	$....	$	10.25	$	20.50
102.6 kg hydrated lime	$....	$....	$	0.15	$	15.39
0.76 cu.m sand	$....	$....		15.23	$	11.57
Cost per 0.76 cu.m					$	47.46
per cu.m					$	62.45

If white sand is used for mortar, figure $18.25 per ton and 3,000 lbs. per cu.yd. (1361 kg per 0.76 cu.m).

If white cement is specified, figure $17.50 per bbl. If white cement (waterproofed) is specified, figure $18.75 per bbl.

04200 UNIT MASONRY

04210 BRICK

Brick information is provided by the Brick Industry of America (BIA). The BIA separates brick walls into 3 general categories: conventional, bonded walls; drainage type walls; and barrier type walls.

The drainage type walls include brick veneer, the cavity type wall, and the masonry bonded hollow wall. Drainage type walls are recommended for walls subjected to the most severe exposures. The success of this type of wall, of course, depends on the care taken to provide continuous means for water to escape.

The barrier type walls include metal tied walls and reinforced brick masonry. The success of these walls will depend on a solidly filled collar joint. The BIA presently recommends using an 8" (200 mm) metal tied wall in preference to an 8" (200 mm) bonded wall for areas subjected to moderate exposure. Metal tied walls adapt well to brick and block backup.

When estimating the number of brick required for any job, obtain the length, height, and thickness of each wall, stating the totals in cu. ft. (cu.m), or obtain the length and height, stating the totals in sq. ft. (sq.m).

In present day estimating, all openings should be deducted in full, regardless of their size, because the estimate should show as accurately as possible the exact number of brick required for the job. The old method of

MASONRY

counting corners twice, doubling the number of brick required for pilaster, chimney breasts, etc., is not used by modern builders. Knowing actual quantities and costs is the only sure method of getting your share of work at a profit.

Brick walls are usually designated on the plans as 4" (100 mm) or 4-1/2" (112.50 mm); 8" (200 mm) or 9" (225 mm); 12" (300 mm) or 13" (325 mm); 16" (400 mm) or 17" (425 mm), etc., increasing by 4" or 4-1/2" (100 or 112.50 mm) in width. This variation in thickness need not be considered by the estimator because a 4" or 4-1/2" (100 or 112.50 mm) wall is 1 brick thick; and 8" or 9" (200 or 225 mm) wall is 2 bricks thick; a 12" or 13" (300 or 325 mm) wall is 3 bricks thick; a 16" or 17" (400 or 425 mm) wall is 4 bricks thick, etc. The easiest method for the estimator is to mark all walls, 4", 8", 12", or 16" (100, 200, 300, or 400 mm) and then reduce the totals to cu. ft. (cu. meters). After the number of cu. ft. (cu. meters) has been obtained, multiply by the number of brick required per cu. ft. (per cu. meter) of wall and the result will be the number of brick required for the job.

When brickwork is estimated on the above basis, it is advisable to add 1-1/2 to 2% to the total to take care of salmon brick, broken brick, bats, etc. The waste should not exceed this amount unless very poor brick or poor job handling methods are used.

Years ago contractors figured their building brickwork on the basis of 7 or 7-1/2 brick per sq.ft. (0.09 sq.m) of 4" or 4-1/2" (100 or 112.50 mm) wall; 14 or 15 brick per sq.ft. (0.09 sq.m) of 8" or 9" (200 or 225 mm) wall; 21 or 22-1/2 brick per sq.ft. (0.09 sq.m) of 12" or 13" (300 or 325 mm) wall, etc.

This method is no longer used, because modern business demands that actual quantities be figured. Work is being estimated too closely today to permit a 10% to 20% overrun on the brickwork, especially when many contractors are figuring on a 5% to 10% margin of profit.

After the number of cu.ft. (cu.m) of brickwork has been obtained, it will be necessary to determine the number of brick required per cu.ft. (cu.m) of wall. This will vary with the size of the brick and the width of the mortar joints. The standard non-modular size is 8"x2-1/4"x3-3/4" (200 x 56.25 x 93.75 mm). Modular brick is 8"x2-2/3"x4" (200 x 66.50 x 100 mm). Sizes will vary somewhat depending upon the burning, the brick in the center of the kiln usually being more thoroughly burned.

When estimating the number of face brick required for any job, obtain the length and height of all walls to be faced with brick, and the total will be the number of sq.ft. (sq.m) of wall.

Always make deductions in full for all openings, regardless of size, because the estimate should show as accurately as possible the exact number of brick required to complete the job.

When making deductions for door and window openings, always note the depth or width of the brick jambs or "reveals". If they are only 4"

629

(100 mm) deep, deduct the full size of the opening as the 4" (100 mm) end of the brick forms the "reveal" or jamb.

If the jamb is more than 4" (100 mm) wide, the full depth of the jamb or "reveal" should be deducted. Example: Deduct for a 5'-0"x7'-0" (1.52 x 2.13 m) opening in a 12" (300 mm) brick wall where the face brick jamb or "reveal" returns the full thickness of the wall. It will be necessary to deduct for the 12" (300 mm) jamb or reveal on each side of the opening, i.e., 2 x 12" = 24" or 2'-0" (2 x 300 mm = 600 mm or 0.6 m). Deduct this from the width of the opening, i.e., 5' - 2' = 3' (1.52 - 0.60 = 0.92 m). The opening deducted should be 3'-0"x7'-0" (0.91 x 2.13 m) instead of 5'-0"x7'-0" (1.52 x 2.13 m), as shown on the plan.

Standard Brick

A standard non-modular brick is 8" (200 mm) long, 2-1/4" (56.25 mm) high, and contains 18 sq. in. (116.13 sq.cm) on the face. A standard modular brick is 8" (200 mm) long, 2-2/3" (66.50 mm) high, and contains 21.28 sq. in. (137.30 sq.cm) on the face. If using a standard non-modular brick, laid dry without a mortar joint, it would require 8 brick per sq.ft. (0.09 sq.m) of wall: 144 ÷ 18 = 8. However, the thickness of the mortar joint must be added, and this will vary from 3/8" (9.37 mm) to 5/8" (15.62 mm), with 1/2" (12.50 mm) the average width and 3/8" (9.37 mm) the normal thickness. A brick is 8" (200 mm) long, plus 3/8" (9.37 mm) for the vertical or end mortar joint makes a total length of 8-3/8" (209.37 mm). A brick is 2-1/4" (56.25 mm) high, plus 3/8" (9.37 mm) for horizontal or bed mortar joint, making the total height of each brick course 2-5/8" (65.62 mm). 8-3/8" x 2-5/8" = 21.984 or 22 sq. in. (209.37 mm x 65.62 mm = 141.94 sq.cm) on the face.

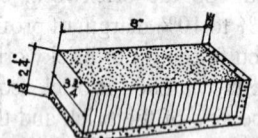

To obtain the number of brick required per s.f. of wall, divide 144 by 22 and the result is 6.54 or 6-1/2 brick per sq.ft. of 4" (100 mm) or 4-1/2" (112.50 mm) wall. If the wall is 8" (200 mm) or 9" (225 mm)—2 brick—thick, 2 x 6-1/2 = 13 brick per sq.ft. of 8" (200 mm) or 9" (225 mm) wall. If 12" (300 mm) or 13" (325 mm)-3 brick-thick, 6-1/2 x 3 = 19.5 brick per sq.ft. of wall, etc.

This same method should be used to obtain the number of brick of any size required for 1 sq.ft. (0.09 sq.m) of wall of any thickness.

MASONRY

Number of Standard Brick (8" x 2-1/4" x 3-3/4") Required for One Square Foot of Brick Wall of any Thickness								
Vertical or End Mortar Joints Figured as 1/4" Wide								
Thickness of Wall	No. of	Width of Horizontal or "Bed" Mortar Joints						
		1/8"	1/4"	3/8"	1/2"	5/8"	3/4"	
4 or 4 1/2 "	1	7.33	7.00	6.67	6.33	6.08	5.80	
8 or 9 "	2	14.67	14.00	13.33	12.67	12.17	11.60	
12 or 13 "*	3	22.00	21.00	20.00	19.00	18.25	17.40	
16 or 17 "	4	29.33	28.00	26.67	25.33	24.33	23.20	
20 or 21 "	5	36.67	35.00	33.33	31.67	30.42	29.00	
24 or 25 "	6	44.00	42.00	40.00	38.00	36.50	34.80	

Number of Standard Brick (200 x 56 x 97 mm) Required for One Square Meter of Brick Wall of any Thickness							
Vertical or End Mortar Joints Figured as 6 mm Wide							
Thickness of Wall.	No. of	Width of Horizontal or "Bed" Mortar Joints					
		3.1mm	6.3mm	9.4mm	12.5mm	15.6mm	18.8mm
100 or 112.5mm	1	78.90	75.35	71.80	68.14	65.44	62.43
200 or 225mm	2	157.91	150.69	143.48	136.38	131.00	124.86
300 or 325mm*	3	236.81	226.04	215.28	204.51	196.44	187.29
400 or 425mm	4	315.71	301.39	287.07	272.65	261.89	249.72
500 or 525mm	5	394.71	376.74	358.76	340.89	327.44	312.15
600 or 625mm	6	473.61	452.08	430.56	409.03	392.88	374.58

Use this column for computing the number of brick required per cu.ft. of wall with any width mortar joint

Variations in Face Brick Quantities. If the brick are laid in running bond without headers, the above quantities will prove sufficient, but if there is a full course of headers every 5th, 6th, or 7th course, it will be necessary to allow for the extra brick required. Also, if the brick are laid in English, Flemish, English Cross, or Dutch Bond, it will be necessary to make an allowance for extra brick where full header courses are required.

Number of Square Inches Occupied by One 8"x2-1/4" Face Brick With Various Width Mortar Joints							
All Vertical Mortar Joints Figured 1/4" Wide							
Width of Horizontal or "Bed" Mortar Joint in Inches							
1/8"	1/4"	3/8"	1/2"	5/8"	3/4"	7/8"	1"
19.30	20.63	21.66	22.69	23.72	25.13	25.78	26.81
Number of Square Millimeters Occupied by One 200 x 56 mm Face Brick with Various Width Mortar Joints in mm							
All Vertical Mortar Joints Figured 6.25 mm Wide							
Width of Horizontal or "Bed" Mortar Joint in mm							
3.1	6.3	9.4	12.5	15.6	18.8	21.9	25.0
12,450	13,306	13,972	14,637	15,302	16,210	16,633	17,298

Number of 8"x 2-1/4"x 3-3/4" Face Brick Required per Sq.Ft. of Wall in Running Bond Without Headers							
All Vertical Mortar Joints Figured 1/4" Wide Width of Horizontal or "Bed" Mortar Joints in Inches							
1/8"	1/4"	3/8"	1/2"	5/8"	3/4"	7/8"	1"
7.46	7.00	6.65	6.35	6.08	5.82	5.60	5.38
Number of 200 x 56 x 94 mm Face Brick Required per Sq.Meter of Wall in Running Bond Without Headers							
All Vertical Mortar Joints Figured 6 mm Wide Width of Horizontal or "Bed" Mortar Joints in mm							
3.1	6.3	9.4	12.5	15.6	18.8	21.9	25.0
80.30	75.35	71.58	68.35	65.44	62.65	60.28	57.91

The above tables do not include any allowance for waste, breakage, or header courses extending into the building brick backing to form a bond. It is based on using all stretchers or blind headers with metal wall ties. On face brick work, 3% to 5% should be added to the net quantity for waste.

For face brick laid with a row of full "headers" every 5th, 6th, or 7th course, add the following percentages to the above quantities:

Percentage to be Added for Various Brick Bonds	
Common (full header course every 5th course)	20% or 1/8
Common (full header course every 6th course)	16.66% or 1/8
Common (full header course every 7th course)	14.33% or 1/7
English or English Cross* (full headers every other course)	50% or 1/2
English or English Cross* (full headers every 6th course)	16.66% or 1/6
Dutch or Dutch Cross* (full headers every other course)	50% or 1/2
Dutch or Dutch Cross* (full headers every 6th course)	16.66% or 1/6
Flemish (full headers every course)	33.33% or 1/3
Flemish (full headers every 6th course)	5.6% or 1/18
Double Header (two headers and a stretcher every 6th course)	8.33% or 1/12
Double Header (two headers and a stretcher every 5th course)	10% or 1/10
Double Flemish (full headers every other course)	10% or 1/10

Add 10% to 15% extra brick for waste in cutting, unless a masonry saw is used.

Percentages To Be Added for Various Brick Bonds	
Double Flemish (full headers every 3rd course)	6.33% or 1/15
3 Stretcher Flemish (full headers every other course)	7.14% or 1/14
3 Stretcher Flemish (full headers every 3rd course)	4.8% or 1/21
4 Stretcher Flemish (full headers every other course)	5.6% or 1/18
4 Stretcher Flemish (full headers every 3rd course)	3.7% or 1/27

For garden walls, porch walls, and other places where an 8" (200 mm) wall is used, with face brick on both sides of the wall, no additional brick are required for any type of bond.

For walks and floors with the brick laid on edge, in any pattern except diagonal, calculate as you would for face brick in common bond

632

without headers. For herringbone or other diagonal work, additional brick will be required because of the waste in chipping or cutting the brick for the borders. The additional number of brick will vary with the width of the walk or floor, as the wider the surface, the smaller the average waste per sq. ft. (sq.m). Walks and floors having the brick laid flat require one-third less brick than where they are laid on edge.

Coursing Tables

The following tables contain vertical coursing dimensions for both non-modular and modular brick. For the non-modular brick, vertical coursing dimensions are shown for both 3/8" (9.37 mm) and 1/2" (12.50 mm) mortar joints. In both tables the bricks are assumed to be positioned in the wall as stretchers.

Vertical Coursing Table for Non-modular Brick*						
No. of Courses	2 1/4" high		2 5/8" high		2 3/4" high	
	Size of Joint					
	3/8" Joint	1/2" Joint	3/8" Joint	1/2" Joint	3/8" Joint	1/2" Joint
1	0'- 2 5/8"	0'- 2 3/4"	0'- 3"	0'- 3 1/8"	0'- 3 1/8"	0'- 3 1/4"
2	0'- 5 1/4"	0'- 5 1/2"	0'- 6"	0'- 6 1/4"	0'- 6 1/4"	0'- 6 1/2"
3	0'- 7 7/8"	0'- 8 1/4"	0'- 9"	0'- 9 3/8"	0'- 9 3/8"	0'- 9 3/4"
4	0'- 10 1/2"	0'- 11"	1'- 0"	1'- 1/2"	1'- 1/2"	1'- 1"
5	1'- 1 1/8"	1'- 1 3/4"	1'- 3"	1'- 3 5/8"	1'- 3 5/8"	1'- 4 1/4"
6	1'- 3 3/4"	1'- 4 1/2"	1'- 6"	1'- 6 3/4"	1'- 6 3/4"	1'- 7 1/2"
7	1'- 6 3/8"	1'- 7 1/4"	1'- 9"	1'- 9 7/8"	1'- 9 7/8"	1'- 10 3/4"
8	1'- 9"	1'- 10"	2'- 0"	2'- 1"	2'- 1"	2'- 2"
9	1'- 11 5/8"	2'- 3/4"	2'- 3"	2'- 4 1/8"	2'- 4 1/8"	2'- 5 1/4"
10	2'- 2 1/4"	2'- 3 1/2"	2'- 6"	2'- 7 1/4"	2'- 7 1/4"	2'- 8 1/2"
11	2'- 4 7/8"	2'- 6 1/4"	2'- 9"	2'- 10 3/8"	2'- 10 3/8"	2'- 11 3/4"
12	2'- 7 1/2"	2'- 9"	3'- 0"	3'- 1 1/2"	3'- 1 1/2"	3'- 3"
13	2'- 10 1/8"	2'- 11 3/4"	3'- 3"	3'- 4 5/8"	3'- 4 5/8"	3'- 6 1/4"
14	3'- 3/4"	3'- 2 1/2"	3'- 6"	3'- 7 3/4"	3'- 7 3/4"	3'- 9 1/2"
15	3'- 3 3/8"	3'- 5 1/4"	3'- 9"	3'- 10 7/8"	3'- 10 7/8"	4'- 3/4"
16	3'- 6"	3'- 8"	4'- 0"	4'- 2"	4'- 2"	4'- 4"
17	3'- 8 5/8"	3'- 10 3/4"	4'- 3"	4'- 5 1/8"	4'- 5 1/8"	4'- 7 1/4"
18	3'- 11 1/4"	4'- 1 1/2"	4'- 6"	4'- 8 1/4"	4'- 8 1/4"	4'- 10 1/2"
19	4'- 1 7/8"	4'- 4 1/4"	4'- 9"	4'- 11 3/8"	4'- 11 3/8"	5'- 1 3/4"
20	4'- 4 1/2"	4'- 7"	5'- 0"	5'- 2 1/2"	5'- 2 1/2"	5'- 5"
21	4'- 7 1/8"	4'- 9 3/4"	5'- 3"	5'- 5 5/8"	5'- 5 5/8"	5'- 8 1/4"
22	4'- 9 3/4"	5'- 1/2"	5'- 6"	5'- 8 3/4"	5'- 8 3/4"	5'- 11 1/2"
23	5'- 3/8"	5'- 3 1/4"	5'- 9"	5'- 11 7/8"	5'- 11 7/8"	6'- 2 3/4"
24	5'- 3"	5'- 6"	6'- 0"	6'- 3"	6'- 3"	6'- 6"
25	5'- 5 5/8"	5'- 8 3/4"	6'- 3"	6'- 6 1/8"	6'- 6 1/8"	6'- 9 1/4"
26	5'- 8 1/4"	5'- 11 1/2"	6'- 6"	6'- 9 1/4"	6'- 9 1/4"	7'- 1/2"
27	5'- 10 7/8"	6'- 2 1/4"	6'- 9"	7'- 3/8"	7'- 3/8"	7'- 3 3/4"
28	6'- 1 1/2"	6'- 5"	7'- 0"	7'- 3 1/2"	7'- 3 1/2"	7'- 7"
29	6'- 4 1/8"	6'- 7 3/4"	7'- 3"	7'- 6 5/8"	7'- 6 5/8"	7'- 11 1/4"
30	6'- 6 3/4"	6'- 10 1/2"	7'- 6"	7'- 9 3/4"	7'- 9 3/4"	8'- 1 1/2"

	Vertical Coursing Table for Non-modular Brick* (continued)					
No. of Courses	2 1/4" high		2 5/8" high		2 3/4" high	
	Size of Joint					
	3/8" Joint	1/2" Joint	3/8" Joint	1/2" Joint	3/8" Joint	1/2" Joint
31	6'- 9 3/8"	7'- 1 1/4"	7'- 9"	8'- 7/8"	8'- 7/8"	8'- 4 3/4"
32	7'- 0"	4'- 4"	8'- 0"	8'- 4"	8'- 4"	8'- 8"
33	7'- 2 5/8"	7'- 6 3/4"	8'- 3"	8'- 7 1/8"	8'- 7 1/8"	8'- 11 1/4"
34	7'- 5 1/4"	7'- 9 1/2"	8'- 6"	8'- 10 1/4"	8'- 10 1/4"	9'- 2 1/2"
35	7'- 7 7/8"	8'- 1/4"	8'- 9"	9'- 1 3/8"	9'- 1 3/8"	9'- 5 3/4"
36	7'- 10 1/2"	8'- 3"	9'- 0"	9'- 4 1/2"	9'- 4 1/2"	9'- 9"
37	8'- 1 1/8"	8'- 5 3/4"	9'- 3"	9'- 7 5/8"	9'- 7 5/8"	10'- 1/4"
38	8'- 3 3/4"	8'- 8 1/2"	9'- 6"	9'- 10 3/4"	9'- 10 3/4"	10'- 3 1/2"
39	8'- 6 3/8"	8'- 11 1/4"	9'- 9"	10'- 1 7/8"	10'- 1 7/8"	10'- 6 3/4"
40	8'- 9"	9'- 2"	10'- 0"	10'- 5"	10'- 5"	10'- 10"
41	8'- 11 5/8"	9'- 4 3/4"	10'- 3"	10'- 8 1/8"	10'- 8 1/8"	11'- 1 1/4"
42	9'- 2 1/4"	9'- 7 1/2"	10'- 6"	10'- 11 1/4"	10'- 11 1/4"	11'- 4 1/2"
43	9'- 4 7/8"	9'- 10 1/4"	10'- 9"	11'- 2 3/8"	11'- 2 3/8"	11'- 7 3/4"
44	9'- 7 1/2"	10'- 1"	11'- 0"	11'- 5 1/2"	11'- 5 1/2"	11'- 11"
45	9'- 10 1/8"	10'- 3 3/4"	11'- 3"	11'- 8 5/8"	11'- 8 5/8"	12'- 2 1/4"
46	10'- 3/4"	10'- 6 1/2"	11'- 6"	11'- 11 3/4"	11'- 11 3/4"	12'- 5 1/2"
47	10'- 3 3/8"	10'- 9 1/4"	11'- 9"	12'- 2 7/8"	12'- 2 7/8"	12'- 8 3/4"
48	10'- 6"	11'- 0"	12'- 0"	12'- 6"	12'- 6"	13'- 0"
49	10'- 8 5/8"	11'- 2 3/4"	12'- 3"	12'- 9 1/8"	12'- 9 1/8"	13'- 3 1/4"
50	10'- 11 1/4"	11'- 5 1/2"	12'- 6"	13'- 1/4"	13'- 1/4"	13'- 6 1/2"
100	21'- 10 1/2"	22'- 11"	25'- 0"	26'- 1/2"	26'- 1/2"	27'- 1"

* Brick positioned in wall as stretchers. Vertical dimensions are from bottom of mortar joint to bottom of mortar joint. (Courtesy of Brick Institute of America)

	Vertical Coursing Table for Non-modular Brick* - Metric					
No. of Courses	56.25 mm high		65.63 mm high		68.75 high	
	Size of Joint					
	9.38 mm Joint	12.5 mm Joint	9.38 mm Joint	12.5 mm Joint	9.38 mm Joint	12.5 mm Joint
1	65.6 mm	68.8 mm	75.0 mm	78.1 mm	78.1 mm	81.25 mm
2	131.3 mm	137.5 mm	150.0 mm	156.3 mm	156.3 mm	162.50 mm
3	196.9 mm	206.3 mm	225.0 mm	234.4 mm	234.4 mm	243.75 mm
4	262.5 mm	275.0 mm	300.0 mm	312.5 mm	312.5 mm	325.00 mm
5	328.1 mm	343.8 mm	375.0 mm	390.6 mm	390.6 mm	406.25 mm
6	393.8 mm	412.5 mm	450.0 mm	468.8 mm	468.8 mm	487.50 mm
7	459.4 mm	481.3 mm	525.0 mm	546.9 mm	546.9 mm	568.75 mm
8	525.0 mm	550.0 mm	600.0 mm	625.0 mm	625.0 mm	650.00 mm
9	590.6 mm	618.8 mm	675.0 mm	703.1 mm	703.1 mm	731.25 mm
10	656.3 mm	687.5 mm	750.0 mm	781.3 mm	781.3 mm	812.50 mm
11	721.9 mm	756.3 mm	825.0 mm	859.4 mm	859.4 mm	893.75 mm
12	787.5 mm	825.0 mm	900.0 mm	937.5 mm	937.5 mm	975.00 mm
13	853.1 mm	893.8 mm	975.0 mm	1015.6 mm	1015.6 mm	1056.25 mm
14	918.8 mm	962.5 mm	1050.0 mm	1093.8 mm	1093.8 mm	1137.50 mm
15	984.4 mm	1031.3 mm	1125.0 mm	1171.9 mm	1171.9 mm	1218.75 mm
16	1050.0 mm	1100.0 mm	1200.0 mm	1250.0 mm	1250.0 mm	1300.00 mm
17	1115.6 mm	1168.8 mm	1275.0 mm	1328.1 mm	1328.1 mm	1381.25 mm
18	1181.3 mm	1237.5 mm	1350.0 mm	1406.3 mm	1406.3 mm	1462.50 mm
19	1246.9 mm	1306.3 mm	1425.0 mm	1484.4 mm	1484.4 mm	1543.75 mm
20	1312.5 mm	1375.0 mm	1500.0 mm	1562.5 mm	1562.5 mm	1625.00 mm

MASONRY

Vertical Coursing Table for Non-modular Brick* - Metric (continued)					
No. of Courses	56.25 mm high		65.63 mm high		68.75 high	
	Size of Joint					
	9.38 mm Joint	12.5 mm Joint	9.38 mm Joint	12.5 mm Joint	9.38 mm Joint	12.5 mm Joint
21	1378.1 mm	1443.8 mm	1575.0 mm	1640.6 mm	1640.6 mm	1706.25 mm
22	1443.8 mm	1512.5 mm	1650.0 mm	1718.8 mm	1718.8 mm	1787.50 mm
23	1509.4 mm	1581.3 mm	1725.0 mm	1796.9 mm	1796.9 mm	1868.75 mm
24	1575.0 mm	1650.0 mm	1800.0 mm	1875.0 mm	1875.0 mm	1950.00 mm
25	1640.6 mm	1718.8 mm	1875.0 mm	1953.1 mm	1953.1 mm	2031.25 mm
26	1706.3 mm	1787.5 mm	1950.0 mm	2031.3 mm	2031.3 mm	2112.50 mm
27	1771.9 mm	1856.3 mm	2025.0 mm	2109.4 mm	2109.4 mm	2193.75 mm
28	1837.5 mm	1925.0 mm	2100.0 mm	2187.5 mm	2181.3 mm	2275.00 mm
29	1903.1 mm	1993.8 mm	2175.0 mm	2265.6 mm	2265.6 mm	2381.25 mm
30	1968.8 mm	2062.5 mm	2250.0 mm	2343.8 mm	2343.8 mm	2437.50 mm
31	2034.4 mm	2131.3 mm	2325.0 mm	2421.9 mm	2421.9 mm	2518.75 mm
32	2100.0 mm	1300.0 mm	2400.0 mm	2500.0 mm	2500.0 mm	2600.00 mm
33	2165.6 mm	2268.8 mm	2475.0 mm	2578.1 mm	2578.1 mm	2681.25 mm
34	2231.3 mm	2337.5 mm	2550.0 mm	2656.3 mm	2656.3 mm	2762.50 mm
35	2296.9 mm	2406.3 mm	2625.0 mm	2734.4 mm	2734.4 mm	2843.75 mm
36	2362.5 mm	2475.0 mm	2700.0 mm	2812.5 mm	2812.5 mm	2925.00 mm
37	2428.1 mm	2543.8 mm	2775.0 mm	2890.6 mm	2890.6 mm	3006.25 mm
38	2493.8 mm	2612.5 mm	2850.0 mm	2968.8 mm	2968.8 mm	3087.50 mm
39	2559.4 mm	2681.3 mm	2925.0 mm	3046.9 mm	3046.9 mm	3168.75 mm
40	2625.0 mm	2750.0 mm	3000.0 mm	3125.0 mm	3125.0 mm	3250.00 mm
41	2690.6 mm	2818.8 mm	3075.0 mm	3203.1 mm	3203.1 mm	3331.25 mm
42	2756.3 mm	2887.5 mm	3150.0 mm	3281.3 mm	3281.3 mm	3412.50 mm
43	2821.9 mm	2956.3 mm	3225.0 mm	3359.4 mm	3359.4 mm	3493.75 mm
44	2887.5 mm	3025.0 mm	3300.0 mm	3437.5 mm	3437.5 mm	3575.00 mm
45	2953.1 mm	3093.8 mm	3375.0 mm	3515.6 mm	3515.6 mm	3656.25 mm
46	3018.8 mm	3162.5 mm	3450.0 mm	3593.8 mm	3593.8 mm	3737.50 mm
47	3084.4 mm	3231.3 mm	3525.0 mm	3671.9 mm	3671.9 mm	3818.75 mm
48	3150.0 mm	3300.0 mm	3600.0 mm	3750.0 mm	3750.0 mm	3900.00 mm
49	3215.6 mm	3368.8 mm	3675.0 mm	3828.1 mm	3828.1 mm	3981.25 mm
50	3281.3 mm	3437.5 mm	3750.0 mm	3906.3 mm	3906.3 mm	4062.50 mm
100	6562.5 mm	6875.0 mm	7500.0 mm	7812.5 mm	7812.5 mm	8125.00 mm

Above based on soft conversion 1" = 25 mm: 1 Foot = 300 mm

Vertical Coursing Table for Non-modular Brick*					
No. of Courses	Nominal Height (h) of Unit				
	2"	2 2/3"	3 1/5"	4"	5 1/3"
1	0 ' - 2 "	0 ' - 2 2/3 "	0 ' - 3 1/5 "	0 ' - 4 "	0 ' - 5 1/3 "
2	0 ' - 4 "	0 ' - 5 1/3 "	0 ' - 6 2/5 "	0 ' - 8 "	0 ' - 10 2/3 "
3	0 ' - 6 "	0 ' - 8 "	0 ' - 9 3/5 "	1 ' - 0 "	1 ' - 4 "
4	0 ' - 8 "	0 ' - 10 2/3 "	1 ' - 4/5 "	1 ' - 4 "	1 ' - 9 1/3 "
5	0 ' - 10 "	1 ' - 1 1/3 "	1 ' - 4 "	1 ' - 8 "	2 ' - 2 2/3 "
6	1 ' - 0 "	1 ' - 4 "	1 ' - 7 1/5 "	2 ' - 0 "	2 ' - 8 "
7	1 ' - 2 "	1 ' - 6 2/3 "	1 ' - 10 2/5 "	2 ' - 4 "	3 ' - 1 1/3 "
8	1 ' - 4 "	1 ' - 9 1/3 "	2 ' - 1 3/5 "	2 ' - 8 "	3 ' - 6 2/3 "
9	1 ' - 6 "	2 ' - 0 "	2 ' - 4 4/5 "	3 ' - 0 "	4 ' - 0 "
10	1 ' - 8 "	2 ' - 2 2/3 "	2 ' - 8 "	3 ' - 4 "	4 ' - 5 1/3 "

No. of Courses	Nominal Height (h) of Unit				
	2"	2 2/3"	3 1/5"	4"	5 1/3"
11	1 ' - 10 "	2 ' - 5 1/3 "	2 ' - 11 1/5 "	3 ' - 8 "	4 ' - 10 2/3 "
12	2 ' - 0 "	2 ' - 8 "	3 ' - 2 2/5 "	4 ' - 0 "	5 ' - 4 "
13	2 ' - 2 "	2 ' - 10 2/3 "	3 ' - 5 3/5 "	4 ' - 4 "	5 ' - 9 1/3 "
14	2 ' - 4 "	3 ' - 1 1/3 "	3 ' - 8 4/5 "	4 '- 8 "	6 ' - 2 2/3 "
15	2 ' - 6 "	3 ' - 4 "	4 '- 0 "	5 ' - 0 "	6 '- 8 "
16	2 ' - 8 "	3 ' - 6 2/3 "	4 ' - 3 1/5 "	5 ' - 4 "	7 '- 1 1/3 "
17	2 ' - 10 "	3 ' - 9 1/3 "	4 '- 6 2/5 "	5 ' - 8 "	7 '- 6 2/3 "
18	3 ' - 0 "	4 ' - 0 "	4 '- 9 3/5 "	6 ' - 0 "	8 '- 0 "
19	3 ' - 2 "	4 ' - 2 2/3 "	5 '- 4/5 "	6 ' - 4 "	8 ' - 5 1/3 "
20	3 ' - 4 "	4 ' - 5 1/3 "	5 '- 4 "	6 ' - 8 "	8 ' - 10 2/3 "
21	3 ' - 6 "	4 '- 8 "	5 '- 7 1/5 "	7 ' - 0 "	9 '- 4 "
22	3 ' - 8 "	4 ' - 10 2/3 "	5 '- 10 2/5 "	7 ' - 4 "	9 '- 9 1/3 "
23	3 ' - 10 "	5 '- 1 1/3 "	6 '- 1 3/5 "	7 ' - 8 "	10 '- 2 2/3 "
24	4 ' - 0 "	5 '- 4 "	6 '- 4 4/5 "	8 '- 0 "	10 '- 8 "
25	4 ' - 2 "	5 '- 6 2/3 "	6 '- 8 "	8 '- 4 "	11 '- 1 1/3 "
26	4 ' - 4 "	5 '- 9 1/3 "	6 '- 11 1/5 "	8 '- 8 "	11 '- 6 2/3 "
27	4 '- 6 "	6 '- 0 "	7 '- 2 2/5 "	9 '- 0 "	12 '- 0 "
28	4 '- 8 "	6 '- 2 2/3 "	7 '- 5 3/5 "	9 '- 4 "	12 '- 5 1/3 "
29	4 '- 10 "	6 '- 5 1/3 "	7 '- 8 4/5 "	9 '- 8 "	12 '- 10 2/3 "
30	5 '- 0 "	6 '- 8 "	8 '- 8 "	10 '- 0 "	13 '- 4 "
31	5 '- 2 "	6 '- 10 2/3 "	8 '- 3 1/5 "	10 '- 4 "	13 '- 9 1/3 "
32	5 '- 4 "	7 '- 1 1/3 "	8 '- 6 2/5 "	10 '- 8 "	14 '- 2 2/3 "
33	5 '- 6 "	7 '- 4 "	8 '- 9 3/5 "	11 '- 0 "	14 '- 8 "
34	5 '- 8 "	7 '- 6 2/3 "	9 '- 4/5 "	11 '- 4 "	15 '- 1 1/3 "
35	5 '- 10 "	7 '- 9 1/3 "	9 '- 4 "	11 '- 8 "	15 '- 6 2/3 "
36	6 '- 0 "	8 '- 0 "	9 '- 7 1/15 "	12 '- 0 "	16 '- 0 "
37	6 '- 2 "	8 '- 2 2/3 "	9 '- 10 2/5 "	12 '- 4 "	16 '- 5 1/3 "
38	6 '- 4 "	8 '- 5 1/3 "	10 '- 1 3/5 "	12 '- 8 "	16 '- 10 2/3 "
39	6 '- 6 "	8 '- 8 "	10 '- 4 4/5 "	13 '- 0 "	17 '- 4 "
40	6 '- 8 "	8 '- 10 2/3 "	10 '- 8 "	13 '- 4 "	17 '- 9 1/3 "
41	6 '- 10 "	9 '- 1 1/3 "	10 '- 11 1/5 "	13 '- 8 "	18 '- 2 2/3 "
42	7 '- 0 "	9 '- 4 "	11 '- 2 2/5 "	14 '- 0 "	18 '- 8 "
43	7 '- 2 "	9 '- 6 2/3 "	11 '- 5 3/5 "	14 '- 4 "	19 '- 1 1/3 "
44	7 '- 4 "	9 '- 9 1/3 "	11 '- 8 4/5 "	14 '- 8 "	19 '- 6 2/3 "
45	7 '- 6 "	10 '- 0 "	12 '- 0 "	15 '- 0 "	20 '- 0 "
46	7 '- 8 "	10 '- 2 2/3 "	12 '- 3 1/5 "	15 '- 4 "	20 '- 5 1/3 "
47	7 '- 10 "	10 '- 5 1/3 "	12 '- 6 2/5 "	15 '- 8 "	20 '- 10 2/3 "
48	8 '- 0 "	10 '- 8 "	12 '- 9 3/5 "	16 '- 0 "	21 '- 4 "
49	8 '- 2 "	10 '- 10 2/3 "	13 '- 4/5 "	16 '- 4 "	21 '- 9 1/3 "
50	8 '- 4 "	11 '- 1 1/3 "	13 '- 4 "	16 '- 8 "	22 '- 2 2/3 "
100	16 '- 8 "	22 '- 2 2/3 "	22 '- 8 "	33 '- 4 "	44 '- 5 1/3 "

Brick positioned in wall as stretchers. (Courtesy of Brick Institute of America)

MASONRY

	Vertical Coursing Table for Modular Brick* - Metric				
No. of	Nominal Height (h) of Unit (mm)				
Courses	50 mm	66.75 mm	80 mm	100 mm	133.25 mm
1	50.0 mm	66.7 mm	80.0 mm	100.0 mm	133.3 mm
2	100.0 mm	133.3 mm	160.0 mm	200.0 mm	266.8 mm
3	150.0 mm	200.0 mm	240.0 mm	300.0 mm	400.0 mm
4	200.0 mm	266.8 mm	320.0 mm	400.0 mm	533.3 mm
5	250.0 mm	333.3 mm	400.0 mm	500.0 mm	666.8 mm
6	300.0 mm	400.0 mm	480.0 mm	600.0 mm	800.0 mm
7	350.0 mm	466.8 mm	560.0 mm	700.0 mm	933.3 mm
8	400.0 mm	533.3 mm	640.0 mm	800.0 mm	1066.8 mm
9	450.0 mm	600.0 mm	720.0 mm	900.0 mm	1200.0 mm
10	500.0 mm	666.8 mm	800.0 mm	1000.0 mm	1333.3 mm
11	550.0 mm	733.3 mm	880.0 mm	1100.0 mm	1466.8 mm
12	600.0 mm	800.0 mm	960.0 mm	1200.0 mm	1600.0 mm
13	650.0 mm	866.8 mm	1040.0 mm	1300.0 mm	1733.3 mm
14	700.0 mm	933.3 mm	1120.0 mm	1400.0 mm	1866.8 mm
15	750.0 mm	1000.0 mm	1200.0 mm	1500.0 mm	2000.0 mm
16	800.0 mm	1066.8 mm	1280.0 mm	1600.0 mm	2133.3 mm
17	850.0 mm	1133.3 mm	1360.0 mm	1700.0 mm	2266.8 mm
18	900.0 mm	1200.0 mm	1440.0 mm	1800.0 mm	2400.0 mm
19	950.0 mm	1266.8 mm	1520.0 mm	1900.0 mm	2533.3 mm
20	1000.0 mm	1333.3 mm	1600.0 mm	2000.0 mm	2666.8 mm
21	1050.0 mm	1400.0 mm	1680.0 mm	2100.0 mm	2800.0 mm
22	1100.0 mm	1466.8 mm	1760.0 mm	2200.0 mm	2933.3 mm
23	1150.0 mm	1533.3 mm	1840.0 mm	2300.0 mm	3066.8 mm
24	1200.0 mm	1600.0 mm	1920.0 mm	2400.0 mm	3200.0 mm
25	1250.0 mm	1666.8 mm	2000.0 mm	2500.0 mm	3333.3 mm
26	1300.0 mm	1733.3 mm	2080.0 mm	2600.0 mm	3466.8 mm
27	1350.0 mm	1800.0 mm	2160.0 mm	2700.0 mm	3600.0 mm
28	1400.0 mm	1866.8 mm	2240.0 mm	2800.0 mm	3733.3 mm
29	1450.0 mm	1933.3 mm	2320.0 mm	2900.0 mm	3866.8 mm
30	1500.0 mm	2000.0 mm	2600.0 mm	3000.0 mm	4000.0 mm
31	1550.0 mm	2066.8 mm	2480.0 mm	3100.0 mm	4133.3 mm
32	1600.0 mm	2133.3 mm	2560.0 mm	3200.0 mm	4266.8 mm
33	1650.0 mm	2200.0 mm	2640.0 mm	3300.0 mm	4400.0 mm
34	1700.0 mm	2266.8 mm	2720.0 mm	3400.0 mm	4533.3 mm
35	1750.0 mm	2333.3 mm	2800.0 mm	3500.0 mm	4666.8 mm
36	1800.0 mm	2400.0 mm	2876.7 mm	3600.0 mm	4800.0 mm
37	1850.0 mm	2466.8 mm	2960.0 mm	3700.0 mm	4933.3 mm
38	1900.0 mm	2533.3 mm	3040.0 mm	3800.0 mm	5066.8 mm
39	1950.0 mm	2600.0 mm	3120.0 mm	3900.0 mm	5200.0 mm
40	2000.0 mm	2666.8 mm	3200.0 mm	4000.0 mm	5333.3 mm

Vertical Coursing Table for Modular Brick* - Metric (continued)					
No. of	Nominal Height (h) of Unit (mm)				
Courses	50 mm	66.75 mm	80 mm	100 mm	133.25 mm
41	2050.0 mm	2733.3 mm	3280.0 mm	4100.0 mm	5466.8 mm
42	2100.0 mm	2800.0 mm	3360.0 mm	4200.0 mm	5600.0 mm
43	2150.0 mm	2866.8 mm	3440.0 mm	4300.0 mm	5733.3 mm
44	2200.0 mm	2933.3 mm	3520.0 mm	4400.0 mm	5866.7 mm
45	2250.0 mm	3000.0 mm	3600.0 mm	4500.0 mm	6000.0 mm
46	2300.0 mm	3066.8 mm	3680.0 mm	4600.0 mm	6133.3 mm
47	2350.0 mm	3133.3 mm	3760.0 mm	4700.0 mm	6266.8 mm
48	2400.0 mm	3200.0 mm	3840.0 mm	4800.0 mm	6400.0 mm
49	2450.0 mm	3266.8 mm	3920.0 mm	4900.0 mm	6533.3 mm
50	2500.0 mm	3333.3 mm	4000.0 mm	5000.0 mm	6666.8 mm
100	5000.0 mm	6666.8 mm	6800.0 mm	10000.0 mm	13333.3 mm

Above based on soft conversion 1" = 25 mm: 1 Foot = 300 mm

Brick Dimensions and Nomenclature

The sizes of brick units shown and discussed below are typical of those currently being produced. However, few manufacturers produce all sizes shown and different manufacturers will vary. It is recommended that the designer consult manufacturers or distributors before proceeding with a design incorporating a specific size of brick that may not be readily available in the particular locality.

The nomenclature indicated, while typical, is not completely standard throughout the industry. Except for the Standard, Roman, and Norman sizes, individual manufacturers may have their own names for certain sizes listed. It is suggested that in order to avoid confusion, the purchaser or specifier first identify the brick by size.

Except for the non-modular Standard, Oversize, and 3-in. (75-mm) units, most bricks are produced in modular sizes. The nominal dimensions of modular brick are equal to the manufactured dimensions plus the thickness of the mortar joint for which the unit is designed. In general, the joint thicknesses used with brick are either 3/8" (9.37 mm) or 1/2" (12.50 mm).

The actual manufactured dimensions of the units may vary from the specified dimensions by not more than the permissible tolerances for variation in dimensions as prescribed in the applicable ASTM specifications (Standard Specification for Facing Brick, ASTM Designation C 216; Standard Specifications for Building Brick, ASTM Designation C 62; Standard Specifications for Hollow Brick, ASTM Designation C 652).

It should be noted that the designated manufacturers heights for the standard brick, the standard modular brick, and all other modular brick designed to be laid 3 courses to 8" (200 mm) are the same—2-1/4" (56.25 mm). There is a very practical reason for this. In 1946-47, with the adoption of modular coordination by the structural clay products industry, brick manufacturers who converted their production completely to modular sizes were faced with a problem in connection with supplying matching brick for additions to existing non-modular buildings. From the standpoint of appearance, most designers required that the vertical coursing in the addition built with modular brick match those in the existing building constructed with non-modular brick. It was agreed that the manufactured face height of the standard modular brick would remain at 2-1/4" (56.25 mm), even though the other dimensions, length and thickness, would become modular. Since that time, the custom has continued. The differences in mortar bed thickness required to maintain the modular coursing of 3 courses in 8" (200 mm) were considered minimal.

Size of Non-Modular Brick						
Unit Designation	**Manufacture Size, Inches**			**Manufacture Size, mm**		
	Thickness Inches	Height Inches	Length Inches	Thickness mm	Height mm	Length mm
Three-Inch (king Size)	3	2 3/4	9 5/8	75.0	68.8	240.6
Three-Inch (Queen Size)	3	2 3/4	8	75.0	68.8	200.0
Three-Inch	3	2 3/4	8 5/8	75.0	68.8	215.6
Standard *	3 5/8	2 1/4	8	90.6	56.3	200.0
Engineered Standard	3 5/8	2 13/16	8	90.6	70.3	200.0
Closure Standard	3 5/8	3 1/2	8	90.6	87.5	200.0

In recent years, the so called "three-inch" brick has gained popularity. The term three-inch designates its thickness or bed depth. The size shown in the table are the ones most commonly produced.
* The manufactured thickness of standard or oversized non-modular brick will vary from 3 1/2 to 3 3/4 inch. If other then a running bond is desired, the designer should check with the manufacture of the brick selected.

Size of Modular Brick in Inches *										
Unit Designation	**Nominal Dimensions, Inches**					**Joint Size, Inches**	**Manufactures Dimensions - Inches**			
	t	x	h	x	l		t	x	h	x	l
Modular	4	x	2 2/3	x	8	3/8	3 5/8	x	2 1/4	x	7 5/8
						1/2	3 1/2	x	2 1/4	x	7 1/2
Engineer, Modular	4	x	3 1/5	x	8	3/8	3 5/8	x	2 13/16	x	7 5/8
						1/2	3 1/2	x	2 11/16	x	7 1/2
Closure Modular	4	x	4	x	8	3/8	3 5/8	x	3 5/8	x	7 5/8
						1/2	3 1/2	x	3 1/2	x	7 1/2
Roman	4	x	2	x	12	3/8	3 5/8	x	1 5/8	x	11 5/8
						1/2	3 1/2	x	1 1/2	x	11 1/2
Norman	4	x	2 2/3	x	13	3/8	3 5/8	x	2 1/4	x	11 5/8
						1/2	3 1/2	x	2 1/4	x	11 1/2
Engineer Norman	4	x	3 1/5	x	12	3/8	3 5/8	x	2 3/4	x	11 5/8
						1/2	3 1/2	x	2 11/16	x	11 1/2
Utility	4	x	4	x	12	3/8	3 5/8	x	3 5/8	x	11 5/8
						1/2	3 1/2	x	3 1/2	x	11 1/2
Triple	4	x	6	x	8	3/8	3 5/8	x	5 5/8	x	7 5/8
						1/2	3 1/2	x	5 1/2	x	7 1/2
Jumbo Utility	4	x	8	x	8	3/8	3 5/8	x	7 5/8	x	7 5/8
						1/2	3 1/2	x	7 1/2	x	7 1/2

Size of Modular Brick in Inches * (continued)							
Unit Designation	**Nominal Dimensions, Inches**			**Joint Size, Inches**	**Manufactures Dimensions - Inches**		
	t x	h x	l		t x	h x	l
6-inch Norwegian	6 x	3 1/5 x	12	3/8	5 5/8 x	2 13/16 x	11 5/8
				1/2	5 1/2 x	2 11/16 x	11 1/2
6-Inch Jumbo	6 x	4 x	12	3/8	5 5/8 x	3 5/8 x	11 5/8
				1/2	5 1/2 x	3 1/2 x	11 1/2
8-Inch Jumbo	8 x	4 x	12	3/8	7 5/8 x	3 1/2 x	11 1/2
				1/2	7 1/2 x	3 1/2 x	11 1/2
8-Inch	8 x	4 x	8	3/8	7 5/8 x	3 5/8 x	7 5/8
15-Inch	4 x	4 x	16	3/8	3 5/8 x	3 5/8 x	15 5/8

* Available as solid units conforming to the latest ASTM requirements, or in a number of cases as hollow brick conforming to the ASTM Specificationsfor hollow brick.

Size of Modular Brick in Metric * - mm							
Unit Designation	**Nominal Dimensions, mm**			**Joint Size, mm**	**Manufactures Dimensions - mm**		
	t x	h x	l		t x	h x	l
Modular	100.0 x	66.7 x	200.0	9.4	90.6 x	56.3 x	190.6
				12.5	87.5 x	56.3 x	187.5
Engineer, Modular	100.0 x	80.0 x	200.0	9.4	90.6 x	70.3 x	190.6
				12.5	87.5 x	67.2 x	187.5
Closure Modular	100.0 x	100.0 x	200.0	9.4	90.6 x	90.6 x	190.6
				12.5	87.5 x	87.5 x	187.5
Roman	100.0 x	50.0 x	300.0	9.4	90.6 x	40.6 x	290.6
				12.5	87.5 x	37.5 x	287.5
Norman	100.0 x	66.7 x	325.0	9.4	90.6 x	56.3 x	290.6
				12.5	87.5 x	56.3 x	287.5
Engineer Norman	100.0 x	80.0 x	300.0	9.4	90.6 x	68.8 x	290.6
				12.5	87.5 x	67.2 x	287.5
Utility	100.0 x	100.0 x	300.0	9.4	90.6 x	90.6 x	290.6
				12.5	87.5 x	87.5 x	287.5
Triple	100.0 x	150.0 x	200.0	9.4	90.6 x	140.6 x	190.6
				12.5	87.5 x	137.5 x	187.5
Jumbo Utility	100.0 x	200.0 x	200.0	9.4	90.6 x	190.6 x	190.6
				12.5	87.5 x	187.5 x	187.5
6-inch Norwegian	150.0 x	80.0 x	300.0	9.4	140.6 x	70.3 x	290.6
				12.5	137.5 x	67.2 x	287.5
6-Inch Jumbo	150.0 x	100.0 x	300.0	9.4	140.6 x	90.6 x	290.6
				12.5	137.5 x	87.5 x	287.5
8-Inch Jumbo	200.0 x	100.0 x	300.0	9.4	190.6 x	87.5 x	287.5
				12.5	187.5 x	87.5 x	287.5
8-Inch	200.0 x	100.0 x	200.0	9.4	190.6 x	90.6 x	190.6
15-Inch	100.0 x	100.0 x	400.0	9.4	90.6 x	90.6 x	390.6

* Available as solid units conforming to the latest ASTM requirements, or in a number of cases as hollow brick conforming to the ASTM Specifications.

MASONRY

Nominal Modular Sizes of Brick*				
Designation	Thickness, Inches	Face Height, Inches	Dimensions Length, Inches	# Unit Courses in 16 Inches
Standard	4	2 2/3	8	6
Engineer	4	3 1/5	8	5
Economy 8 or Jumbo Closure	4	4	8	4
Double	4	5 1/3	8	3
Roman	4	2	12	8
Norman	4	2 2/3	12	6
Norwegian	4	3 1/5	12	5
Economy 12 or Jumbo Utility	4	4	12	4
Triple	4	5 1/3	12	3
6-in. Norwegian	6	3 1/5	12	5
6-in. Jumbo	6	4	12	4
8-in. Jumbo	8	4	12	4
Designation	Thickness, mm	Face Height, mm	Dimensions Length, mm	# Unit Courses in 400 mm
Standard	100	66.67	200	6
Engineer	100	80.00	200	5
Economy 8 or Jumbo Closure	100	100.00	200	4
Double	100	133.33	200	3
Roman	100	50.00	300	8
Norman	100	66.67	300	6
Norwegian	100	80.00	300	5
Economy 12 or Jumbo Utility	100	100.00	300	4
Triple	100	133.33	300	3
6-in. Norwegian	150	80.00	300	5
6-in. Jumbo	150	100.00	300	4
8-in. Jumbo	200	100.00	300	4

Available as solid units conforming to latest ASTM, or in some cases, as hollow brick conforming to ASTM Specifications.

Position of Brick in the Wall. Most brick are laid as stretchers so that the longer of the face dimensions is horizontal. The drawings below also illustrate terms applied to other brick positions as placed in the wall. The shaded areas indicate the surfaces of the brick that are exposed.

Brick Bonds and Patterns

Bond in brickwork is the overlapping of one brick upon the other, either along the length of the wall or through its thickness, in order to bind them together into a secure structural mass. Units are shifted back and forth

641

so that the vertical joints in two successive layers or "courses" do not come into line; in other words, the brick are laid so as to break the joint, the whole forming a natural bond or a structural unit giving strength to the wall.

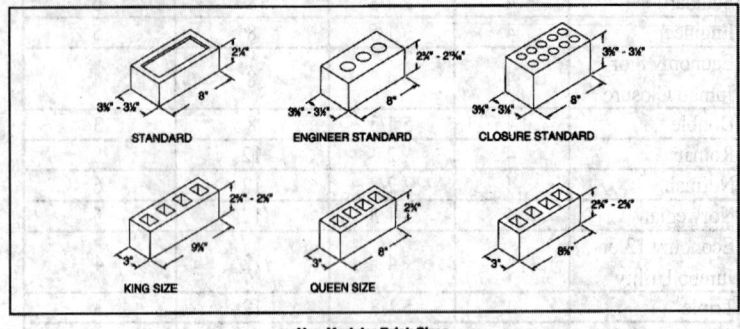

Non-Modular Brick Sizes
(Specified Dimensions)

Bond. The word bond, when used in reference to masonry, has three meanings:

1. *Structural Bond:* The method by which individual masonry units are interlocked or tied together to cause the entire assembly to act as a single structural unit.
2. *Pattern Bond:* The pattern formed by the masonry units and the mortar joints on the face of a wall. The pattern may result from the type of structural bond used or may be purely a decorative one, unrelated to the structural bonding.
3. *Mortar Bond:* The adhesion of mortar to the masonry units or to reinforcing steel.

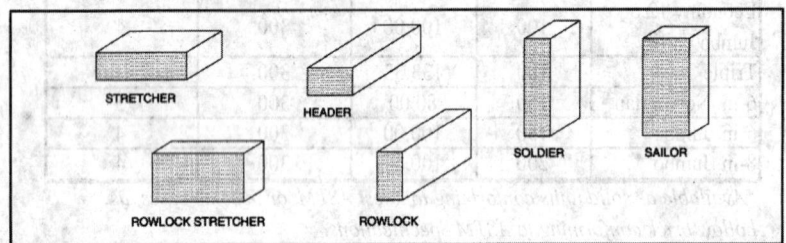

Brick Positions in Wall

Structural Bonds. Structural bonding of masonry walls may be accomplished in three ways: by the overlapping (interlocking) of masonry units, by use of metal ties embedded in connecting joints, and by the adhesion of grout to adjacent wythes of masonry.

The overlapped bond is based on variations of two traditional methods of bonding. The first is known as English bond and consists of alternating courses of headers and stretchers (Figure 1). The second is Flemish bond and consists of alternating headers and stretchers in every

642

course, so arranged that the headers and stretchers in every other course appear in vertical lines (Figure 1).

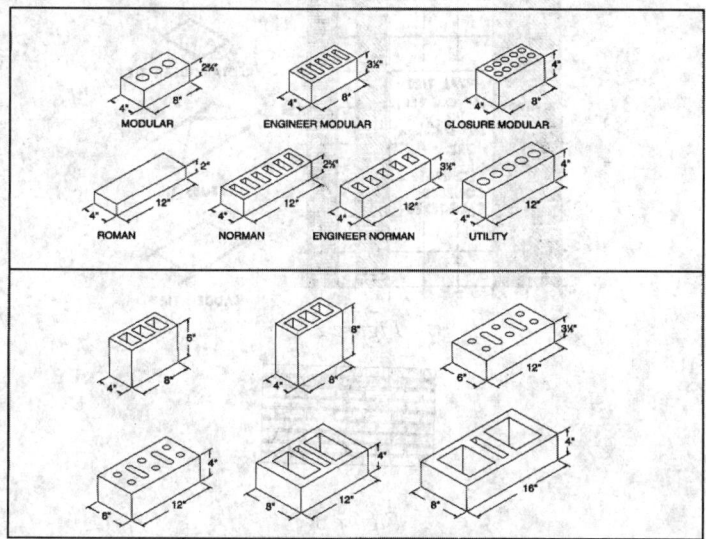

Modular Brick (Nominal Dimensions)

Figure 1

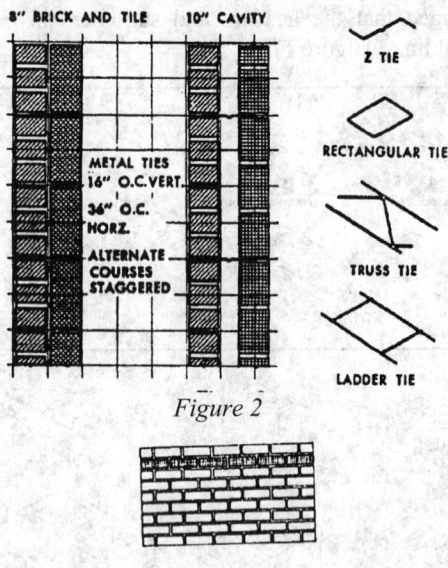

Figure 2

Figure 3

The stretchers, laid with the length of the wall, develop longitudinal bonding strength, while the headers, laid across the width of the wall, bond the wall transversely.

Modern building codes require that masonry-bonded brick walls be bonded so that not less than 4% of the wall surface is composed of headers, with the distance between adjacent headers not exceeding 24" (600 mm), vertically or horizontally.

Structural bonding of masonry walls with metal ties is used in both solid wall and cavity wall construction (Figure 2). Most building codes permit the use of rigid steel bonding ties in solid walls.

At least one metal tie should be used for each 4-1/2 sq.ft. (.41 sq.m) of wall surface. Ties in alternate courses should be staggered. The distance between adjacent ties should not exceed 24" (600 mm) vertically and 36" (900 mm) horizontally. Additional bonding ties, spaced not more than 3' (0.91 m) apart around the perimeter and within 12" (300 mm) of the opening, should be provided at all openings.

If ties less than 3/16" (4.68 mm) in diameter are used, tie spacing should be reduced so that the tie area per sq.ft. (sq.m) of wall is not less than specified above.

Structural bonding of solid and reinforced brick masonry walls is sometimes accomplished by grout, which is poured into the cavity or collar joint between wythes of masonry.

The method of bonding will depend on the use requirements, wall type and other factors. However, the metal tie method is generally recommended for exterior walls. Some of the advantages of this method are

644

greater resistance to rain penetration and ease of construction. Metal ties also allow slight differential movements of the facing and backing, which may relieve stresses and prevent cracking.

Pattern Bonds. Frequently, structural bonds, such as English or Flemish, or variations of these, may be used to create patterns in the face of a wall. However, in the strictest sense of the term, pattern refers to the change or varied arrangement of the brick texture or color used in the face. Therefore, it may be possible to secure many patterns using the same structural bond. Patterns also may be produced by the method of handling the mortar joint or by projecting or recessing certain brick from the plane of the wall, thus creating a distinctive wall texture that is not solely dependent on the texture of the individual brick.

There are five basic structural bonds, which create typical patterns, commonly used today. These are running bond, common or American bond, Flemish bond, English bond, and block or stack bond. Through the use of these bonds and variations of the color and texture of the brick, and of the joint types and color, an unlimited number of patterns can be developed.

Running Bond. The simplest of the basic pattern bonds, the running bond, consists of all stretchers. Because there are no headers in this bond, metal ties are ordinarily used. Running bond is used largely in cavity wall construction and veneered walls of brick, and often in facing tile walls where the bonding may be accomplished by extra width stretcher tile.

Common or American Bond. Common or American bond (Figure 4) is a variation of running bond with a course of full length headers at

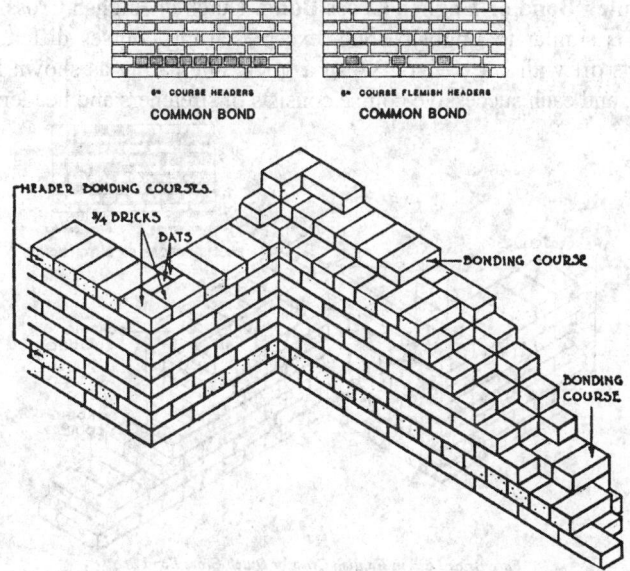

Figure 4 –Face Brick Laid in Common or American Bond

regular intervals. This bond is obtained by laying a course of headers every fifth, sixth, or seventh course. To maintain the effect of the running bond, a special double header bond is sometimes used. This method of using headers, as in Common or American Bond, in order to secure transverse strength of wall, can be treated in a way to produce more pleasing effects, as may be seen in Flemish or English Bonds.

English Bond. English Bond (Figure 5) is made up of alternating courses of stretchers and headers. Ordinarily half brick are used for the header courses, except that every sixth course a full header course is used to tie the face and common brick walls together. The allowance for headers was given previously, under *Percentages to be Added for Various Brick Bonds*. Snap headers are used in courses that are not structural bonding courses.

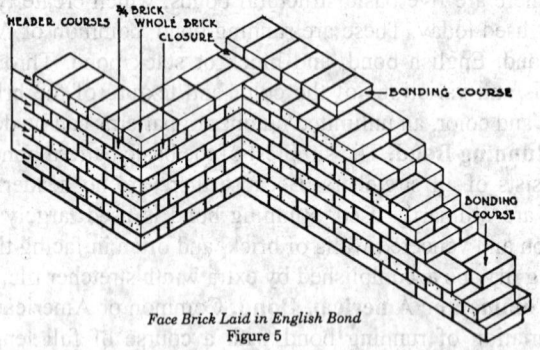

Face Brick Laid in English Bond
Figure 5

Dutch Bond or English Cross Bond. Dutch or English Cross Bond (Figure 6) is similar to English Bond, except starting courses differ. Each course starts off with 1/4, 1/2, or 3/4 of a brick, alternating as shown in the illustration, and each successive course consists of stretchers and headers.

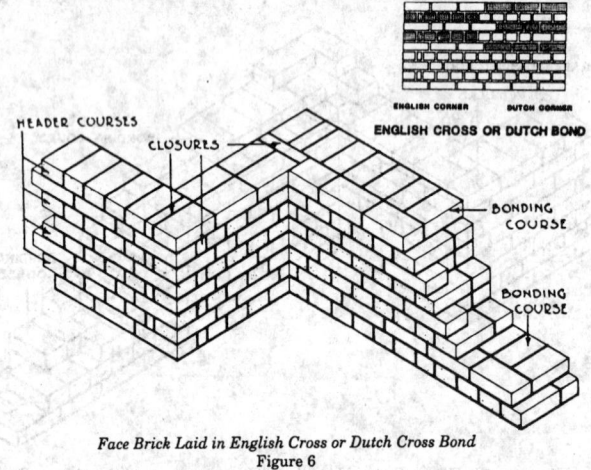

Face Brick Laid in English Cross or Dutch Cross Bond
Figure 6

646

MASONRY

There are two methods used in starting the corners in Flemish and English bonds. In the "Dutch corner", a three-quarter brick closure is used. In the "English corner", a 2" (50 mm) or quarter brick closure, called a "queen closure", is used. The 2" (50 mm) closure should always be placed 4" (100 mm) in from the corner, never at the corner.

On account of the large amount of cutting necessary to obtain the 1/4 and 3/4 pieces of brick, it is necessary to add 10% to 15% additional brick to allow for waste, caused by breakage in cutting.

Block or Stack Bond. Block or stack bond (Figure 7) is purely a pattern bond. There is no overlapping of units. All vertical joints are aligned.

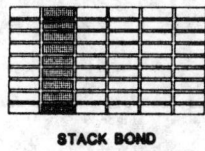

STACK BOND

Figure 7

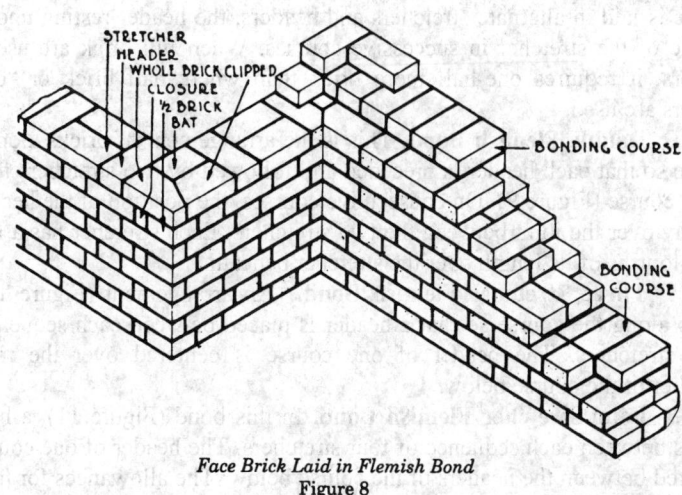

Face Brick Laid in Flemish Bond
Figure 8

Usually this pattern is bonded to the backing with rigid steel ties, but 8" (200 mm) bonder units are often available. In large wall areas and in load bearing construction, it is advisable to reinforce the wall with steel reinforcement placed in the horizontal mortar joints. In stack bond, it is imperative that prematched or dimensionally accurate masonry units be used if the vertical alignment of the head joints is to be maintained. The architect or contractor might want to use a "gauged" brick, allowing for the

647

manufacturer's acceptable production tolerances for brick used in this type of bond. Check with the supplier for additional costs.

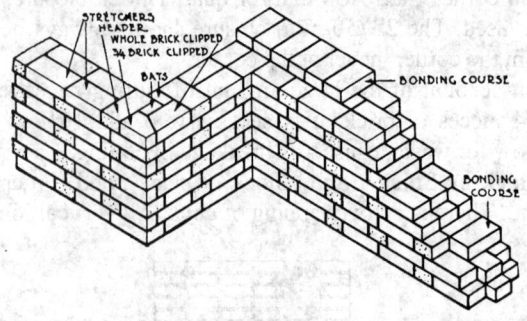

Face Brick Laid in Double Flemish or Garden Wall Bond
Figure 9

Waste in cutting is greatly reduced with a masonry saw, and a brick can be cut in 10 to 15 seconds.

If brick are laid with full header courses every second or sixth course, add the percentages given on the previous pages.

Flemish Bond. Flemish Bond (Figure 8) is effective, because each course is laid in alternate stretchers and headers, the header resting upon the middle of the stretcher in successive courses. When full brick are used for headers, it requires one-half more brick than where half brick or "blind" headers are used.

Double Flemish Bond. This is the arrangement of bricks along the course so that each header is preceded and followed by two stretchers for the entire course (Figure 9). On consecutive courses, the position of the header is directly over the joint between the two stretchers. Each stretcher has a lap of three-fourths of its length over the stretcher beneath.

Three Stretcher Flemish Bond. An arrangement (Figure 10) of bricks along the course so that a header is placed between each sequence of three stretchers. The header of one course is centered over the middle stretcher of the course below.

Four Stretcher Flemish Bond. In this bond (Figure 11), a header occurs between each sequence of four stretchers. The header of one course is centered between the headers of the course below. The allowances for header courses are given on the previous pages.

Garden Wall Bond. Garden Wall Bond (Figure 12) is merely a modification of Flemish Bond, secured by laying courses with two to four stretchers alternating with a header, and requires the same number of brick per square foot as Flemish Bond.

Diamond Bond Patterns. Diamond Bond Patterns (Figure 12) are secured by a modification of the Garden Wall Bond. It is, however, only in case of large wall surfaces that patterns of an elaborate character could be

648

recommended, as any departure from simple bonds adds to the cost of bricklaying

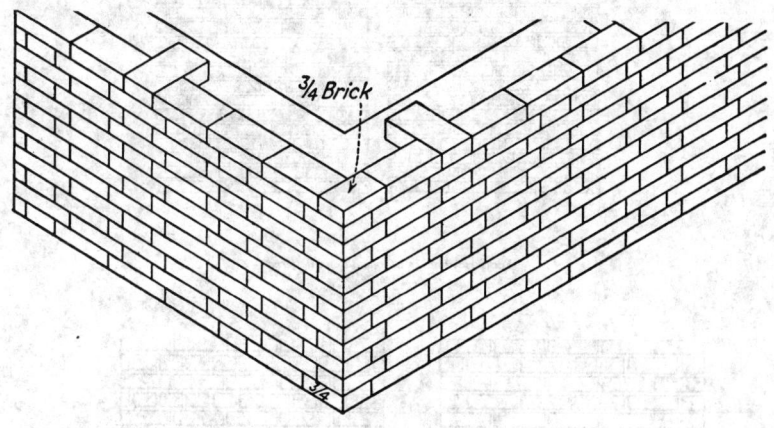

Face Brick Laid in Three Stretcher Flemish Bond
Figure 10

Face Brick Laid in Four Stretcher Flemish Bond
Figure 11

Face Brick Laid in Diamond Bond Patterns
Figure 12

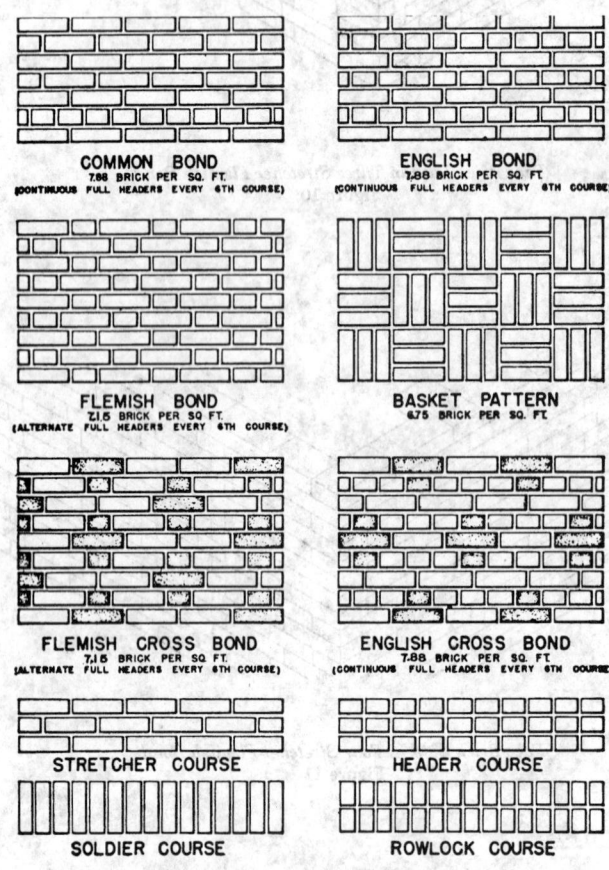

COMMON BOND
7.88 BRICK PER SQ. FT.
(CONTINUOUS FULL HEADERS EVERY 6TH COURSE)

ENGLISH BOND
7.88 BRICK PER SQ. FT.
(CONTINUOUS FULL HEADERS EVERY 6TH COURSE)

FLEMISH BOND
7.15 BRICK PER SQ. FT.
(ALTERNATE FULL HEADERS EVERY 6TH COURSE)

BASKET PATTERN
6.75 BRICK PER SQ. FT.

FLEMISH CROSS BOND
7.15 BRICK PER SQ. FT.
(ALTERNATE FULL HEADERS EVERY 6TH COURSE)

ENGLISH CROSS BOND
7.88 BRICK PER SQ. FT.
(CONTINUOUS FULL HEADERS EVERY 6TH COURSE)

STRETCHER COURSE

HEADER COURSE

SOLDIER COURSE

ROWLOCK COURSE

Elevations of Various Brick Bonds

650

MASONRY
Modular Size Brick

Economy is the prime objective of modular design, which is not a complex and revolutionary plan for coordinating building materials, but rather, a simplified and efficient system of design, layout, and construction resulting in definite economies.

After an exhaustive study, a four-inch (100 mm) module or increment was agreed upon as a unit that would afford the maximum practical standardization and simplification, and at the same time, afford a sufficient flexibility in design.

On the basis of the four-inch (100 mm) module or increment, certain sizes of brick and tile have been recommended. It must be remembered that the four-inch (100 mm) module, governing both the vertical and horizontal dimensions, is used to govern the layout of finished wall areas and to determine heights and widths of doors, windows, and other openings.

Because the module is a unit of wall measurement, it cannot be applied to the individual brick or tile, but must be considered as governing the size of the brick or tile, plus the mortar joint.

In employing the modular system of measurement for masonry walls, the dimension is considered from center to center of the mortar joints. To fit properly into the modular system, the actual sizes of the individual units of either brick or tile are determined by the modular dimension less the mortar joint.

The explanatory details, drawn with 1/2" (12.50 mm) mortar joints, present this matter very clearly. The first illustration indicates a wall section built of standard size brick measuring 2-1/4" x 3-3/4" x 8" (56.25 x 93.75 x 200 mm). It will be noted that a unit of this size does not comply with the modular measurements as indicated by the four-inch (100 mm) modular dimension lines, known as grid lines. Therefore, if standard size brick are used in the construction of a project designed according to the modular system, wasteful cutting and fitting of the brick would be necessary.

Note the difference in the second wall section. In this case a modular brick is employed, having an actual size of 2-1/6" x 3-1/2" x 7-1/2" (54.15 x 87.50 x 187.50 mm). Using a 1/2" (12.50 mm) mortar joint, three courses of brick lay up perfectly to every two horizontal grid lines, and one length of brick fits two vertical grid lines. The same holds true of the other sizes of brick shown.

Exact heights of modular brick will vary, the difference being taken up in the bed joints. It is therefore important to know the exact size of brick. The following table lists standard joint widths.

Modular Face Brick

Three standard modular joint thicknesses are 1/4" (6.25 mm), 3/8" (9.37 mm), and 1/2" (12.5 mm). The number of modular masonry units per square foot (square meter) will remain constant for each nominal size, the difference being taken up in the joint.

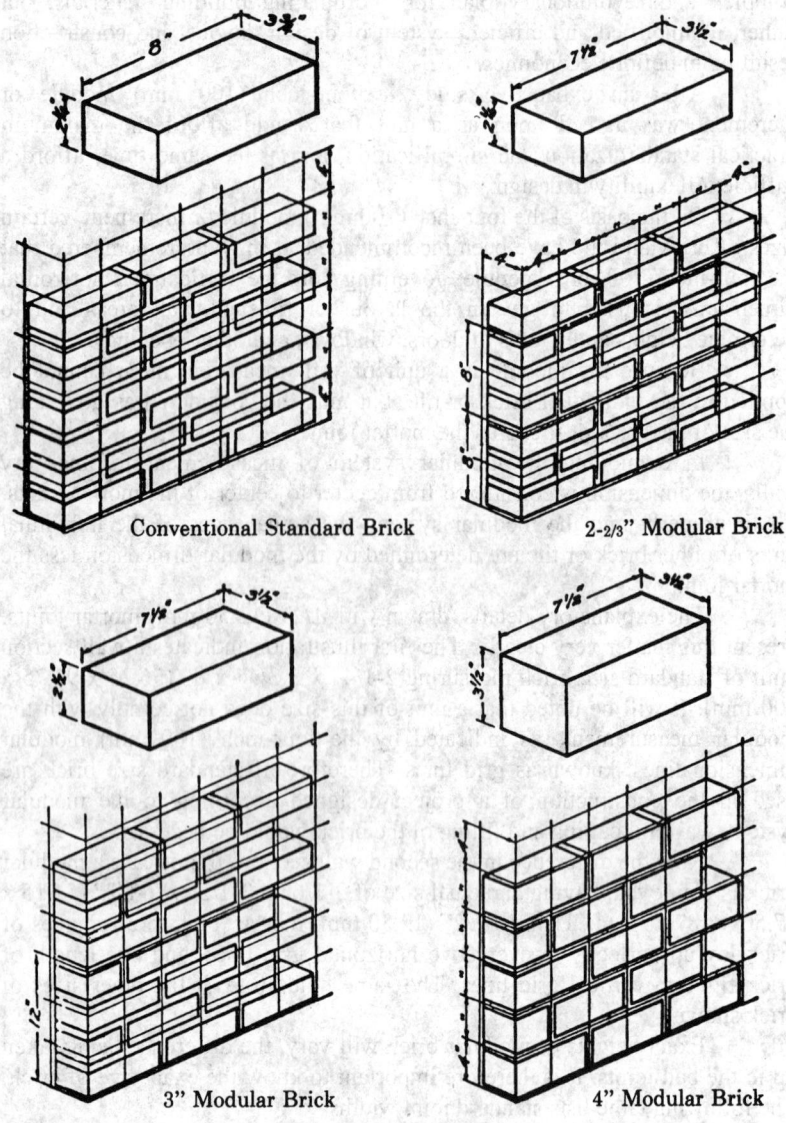

Conventional Standard Brick

2-2/3" Modular Brick

3" Modular Brick

4" Modular Brick

Modular Design

652

MASONRY

Modular Brick Walls Without Headers								
Nominal Size of Brick, Inches	Brick per Square Foot of Wall	Brick per Sq. Ft. of Wall	Cu. Ft. Mortar per 100 Sq.Ft. of Wall		Cu.Ft. of Mortar per 1,000 Brick			
		Joint Thickness				Joint Thickness		
h x t x l		1/4"	3/8"	1/2"	1/4"	3/8"	1/2"	
2 2/3 x 4 x 8	6.750	3.81	5.47	6.95	5.65	8.10	10.30	
3 1/5 x 4 x 8	5.625	3.34	4.79	6.10	5.94	8.52	10.84	
4 x 4 x 8	4.500	-----	4.12	5.24	-----	9.15	11.65	
5 1/3 x 4 x 8	3.375	-----	3.44	4.34	-----	10.19	12.87	
2 x 4 x 12	6.000	-----	6.43	8.20	-----	10.72	13.67	
2 2/3 x 4 x 12	4.500	3.52	5.06	6.46	7.82	11.24	14.35	
3 x 4 x 12	4.000	-----	4.60	5.87	-----	11.51	14.68	
3 1/5 x 4 x 12	3.750	3.04	4.37	5.58	8.11	11.66	14.89	
4 x 4 x 12	3.000	2.56	3.69	4.71	8.54	12.29	15.70	
5 1/3 x 6 x 12	2.250	-----	3.00	3.84	-----	13.34	17.05	
2 2/3 x 6 x 12	4.500	-----	7.85	10.15	-----	17.45	22.55	
3 1/5 x 6 x 12	3.750	-----	6.79	8.77	-----	18.10	23.39	
4 x 6 x 12	3.000	-----	5.72	7.40	-----	19.07	24.67	

Modular Brick Walls Without Headers - Metric								
Nominal Size of Brick, mm.	Brick per Square Meter of Wall	Brick per Sq. M. of Wall	Cu. M. Mortar per 10 Sq.M of Wall		Cu.Meter of Mortar per 1,000 Brick			
		Joint Thickness mm			Joint Thickness mm			
h x t x l		6.25	9.375	12.5	6.25	9.375	12.5	
66.67 x 100.00 x 200.00	72.66	41.01	0.17	0.21	0.16	0.23	0.29	
80.00 x 100.00 x 200.00	60.55	35.95	0.15	0.19	0.17	0.24	0.31	
100.00 x 100.00 x 200.00	48.44	-----	0.13	0.16	-----	0.26	0.33	
133.33 x 100.00 x 200.00	36.33	-----	0.10	0.13	-----	0.29	0.36	
50.00 x 100.00 x 300.00	64.59	-----	0.20	0.25	-----	0.30	0.39	
66.67 x 100.00 x 300.00	48.44	37.89	0.15	0.20	0.22	0.32	0.41	
75.00 x 100.00 x 300.00	43.06	-----	0.14	0.18	-----	0.33	0.42	
80.00 x 100.00 x 300.00	40.37	32.72	0.13	0.17	0.23	0.33	0.42	
100.00 x 100.00 x 300.00	32.29	27.56	0.11	0.14	0.24	0.35	0.44	
133.33 x 150.00 x 300.00	24.22	-----	0.09	0.12	-----	0.38	0.48	
66.67 x 150.00 x 300.00	48.44	-----	0.24	0.31	-----	0.49	0.64	
80.00 x 150.00 x 300.00	40.37	-----	0.21	0.27	-----	0.51	0.66	
100.00 x 150.00 x 300.00	32.29	-----	0.17	0.23	-----	0.54	0.70	

Non-Modular Brick and Mortar Required for Single Wythe Walls in Running Bond (No allowance for Breakage or Waste)						
Brick Size, Inches	# of Brick per 100 Sq.Ft.	Cu.Ft. Mortar per 100 Sq.Ft.	Cu.Ft. Mortar per 1,000 Brick	Number of Brick per 100 Sq.Ft.	Cu.Ft. Mortar per 100 Sq.Ft.	Cu.Ft. Mortar per 1,000 Brick
t x h x l	3/8" Joints			1/2" Joints		
2 3/4 x 2 3/4 x 9 3/4	455	3.2	7.1	432	4.5	10.4
2 5/8 x 2 3/4 x 8 3/4	504	3.4	6.8	470	4.1	8.7
3 3/4 x 2 1/4 x 8	655	5.8	8.8	616	7.2	11.7
3 3/4 x 2 3/4 x 8	551	5.0	9.1	522	6.4	12.2
Brick Size, mm	# of Brick per 10 Sq. Meters	Cu.M. Mortar per 10 Sq.M.	Cu.M. Mortar per 1,000 Brick	Number of Brick per 10 Sq.M.	Cu.Ft. Mortar per 10 Sq.M.	Cu.M. Mortar per 1,000 Brick
t x h x l	9.375 mm Joints			12.5 mm Joints		
68.75 x 68.75 x 243.75	490	0.097	0.20	465	0.137	0.29
65.63 x 68.75 x 218.75	543	0.104	0.19	506	0.125	0.25
93.75 x 56.25 x 200.00	705	0.177	0.25	663	0.219	0.33
93.75 x 68.75 x 200.00	593	0.152	0.26	562	0.195	0.35

The above quantities include only mortar for bed and vertical joints. The following table gives allowances for backing mortar or collar joints.

Cu.Ft. of mortar per 100 Sq.Ft. Wall		
1/4" Joint	3/8" Joint	1/2" Joint
2.08	3.13	4.17
Cu.Meter of Mortar per 10 SQ. Meter Wall		
6.25 mm Joint	9.375mm Joint	12.50 mm Joint
0.06	0.10	0.13

Cubic feet per 1000 units =

10 x cubic feet per 100 sq. ft. of wall
  number of units per sq. ft. of wall

Cubic meters per 1000 units =

10 x cubic meters per 10 sq.m of wall
  number of units per sq.m of wall

Material Quantities per Cubic Foot of Mortar Quantities by Volume				
Material	Mortar Type and Proportions by Volume			
	M	S	N	O
	1:1/4:3	1:1/2:4-1/2	1:01:06	1:02:09
Cement	0.333	0.222	4:00:29	2:39:50
Lime	0.083	0.111	0.167	0.222
Sand	1	1	1	1
Material Quantities per Cubic Meter of Mortar Quantities by Volume				
Material	M	S	N	O
	1:1/4:3	1:1/2:4-1/2	1:01:06	1:02:09
Cement	11.767	7.845	5.901	3.922
Lime	2.933	3.922	5.901	7.845
Sand	35.336	35.336	35.336	35.336

Material Quantities per Cubic Foot of Mortar Quantities by Weight				
Material	Mortar Type and Proportions by Volume			
	M	S	N	O
Cement	31.33	20.89	15.67	10.44
Lime	3.33	4.44	6.67	8.89
Sand	80	80	80	80

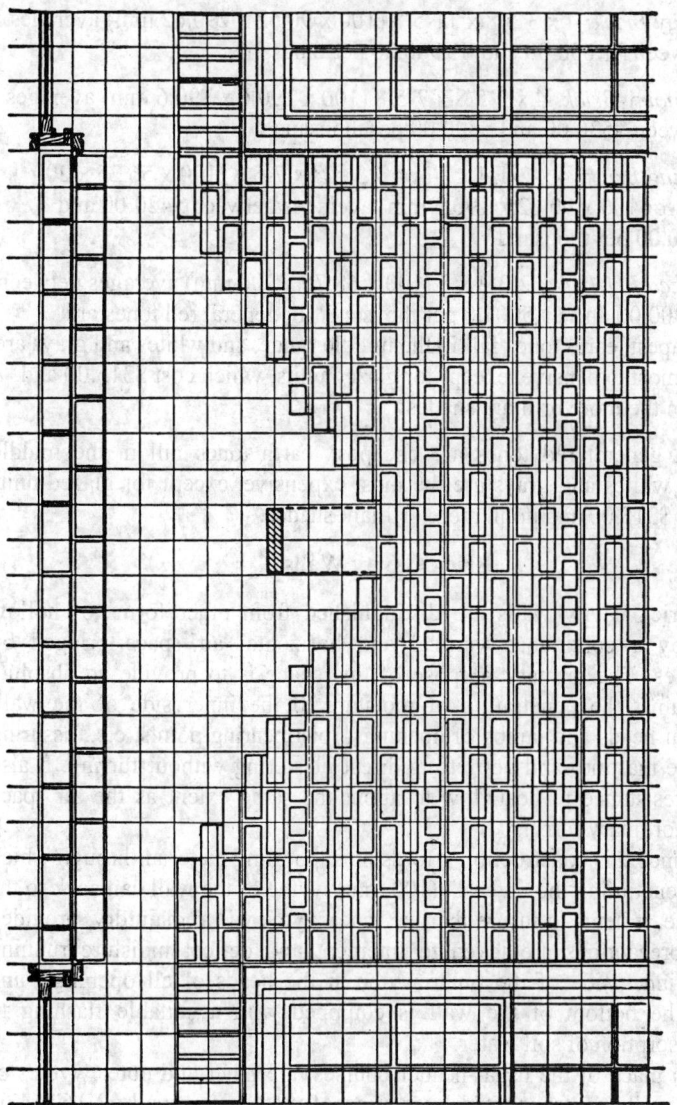

Grid Paper Is Used by the Designer for All Layouts and Details in Modular Design. The Above Plan and Elevation of Brick between Two Window Openings Show How the Brick Fit the Intervening Space Perfectly with No Wasteful Cutting or Fitting.

Material Costs of Brick

Brick costs will vary widely among regions and within color ranges:

1. *Common Brick:* Standard modular—4 x 2-1/4 x 7-5/8 (100 x 56.25 x 190.62 mm)—and standard non-modular—4 x 2-1/4 x 8 (100 x 56.25 x 200 mm)—average between $350.00 and $510.00 per thousand.

2. *Jumbo Brick:* 4 x 2-3/4 x 8 (100 x 68.75 x 200 mm) averages between $500.00 and $670.00 per thousand.

3. *Utility Brick:* 4 x 3-5/8 x 11-5/8 (100 x 90.62 x 290.62 mm) averages between $1,030.00 and $1,900.00 per thousand.

4. *Economy Brick:* 4 x 7-5/8 x 7-5/8 (100 x 190.6 x 190.6 mm) averages between $530.00 and $890.00 per thousand.

5. *Paving Brick:* 4 x 2-1/4 x 8 or 4 x 3-5/8 x 7-5/8 (100 x 56.25 x 200 mm, or 100 x 90.62 x 190.6 mm) averages between $420.00 and $610.00 per thousand.

6. *Glazed Brick:* 4 x 2-3/4 x 8 (100 x 68.75 x 200 mm) averages between $1,300.00 and $1,500.00 per thousand. In general red tones are cheapest, earth tones fall in the middle range, and whites and grays are the most expensive, except for glazed units, which cost $315.00 and up in the more brilliant shades.

In general red tones are cheapest, earth tones fall in the middle range, and whites and grays are the most expensive, except for glazed units which cost $315.00 and up in more brilliant shades.

Brick Cavity Walls

Brick cavity walls are distinguished from other forms of hollow masonry by a continuous vertical and horizontal air space bridged by masonry ties. The primary purpose of this space is to provide an absolute barrier against the penetration of moisture to the inner side of the wall, including at heads and jambs of openings, joist bearing points, etc., assuring trouble-free masonry and permitting direct plastering without furring. It also increases resistance to heat-flow to about the same extent as the air space created by ordinary furring.

Since this construction affords no supporting material through which moisture penetrating the outer 4" (100 mm) wythe of the wall can soak to the inner wythe, it helps to insure that the inner wythe will remain dry, provided adequate precautions are taken to intercept and deflect moisture running down the inner side of the outer wythe at the heads of all openings, and provided the bottom of the wall is equipped with a suitable flashing to prevent absorption of soil water.

In place of the usual header courses, the inner and outer wythes of the cavity walls are connected by rust-proof metal ties-usually 3/16" (4.68 mm) diameter, square-end, Z bars-placed in every sixth course and 3'-0" (0.91 m) on center horizontally.

Labor Laying Building Brick

Many factors influence the labor cost of laying building brick. There are no set quantities of brick that a mason will lay per hour or day. With ten masons laying brick on the same wall and on exactly the same class of work, no two of them will lay the same number of brick. Averages must be taken.

656

MASONRY

Solid brick walls are used much less frequently now than in the past, and where brick is used as a facing, it is most often backed up with cement blocks. Labor costs are much lower for handling and setting cement blocks, because the units are larger than brick. One 8"x8"x16" (200 x 200 x 400 mm) cement block usually displaces 12 building brick in the wall.

The class of work also has a great deal to do with the difference in labor costs. It costs much more to lay brick in an 8" (200 mm) wall with struck joints than in a 16" (400 mm) wall having cut joints on one side of the wall and struck joints on the other. Also, it costs more to lay brick in portland cement mortar than in lime mortar, because the lime mortar spreads far more easily than cement mortar.

Grid paper is used by the designer for all layouts and details in modular design. The above plan and elevation of brick between two window openings show how the brick fit the intervening space perfectly with no wasteful cutting or fitting. It costs more to lay brick in cold, wet weather when the bricks are slippery and the mason's hands cold than in fair weather with all conditions favorable.

It requires more labor to lay brick in the winter, when the brick and mortar must be heated and salamanders placed along the floor or scaffold, requiring additional labor attending salamanders or other heating device(s), removing ice, snow, etc., than in the summer when conditions are favorable

Labor Production for Laying Building Brick						
Class of Work	Motor Joint Style	Kind of Mortar	Avg # of Brick laid per 8-Hour Day	Average Hours 1,000 Brick		
				Mason Hours	Labor Hours	Hoist Engineer*
8" walls, 1 story structure, bungalows, garages, buildings, residences, etc	Cut	Lime	750 - 800	10.5	9.0	-----
	Struck	Lime	700 - 750	11.0	9.0	
12" walls, ordinary construction, apartment buildings, houses, garages, factories, stores, store and apartment buildings, schools, etc.	Cut	Lime	950 - 1,050	8.0	8.0	-----
	Struck	Lime	925 - 975	8.5	8.0	
	Cut	Cement	825 - 900	9.3	8.0	
	Struck	Cement	750 - 850	10.0	8.0	
Hodding Brick to second story add	-----	-----	-----	-----	2.0	-----
16" Walls, heavy warehouse, factory and industrial work. Straight walls	Cut	Lime	1,100 - 1225	7.0	8.0	-----
	Struck	Lime	1,000 - 1,125	7.5	8.0	
	Cut	Cement	950 - 1,050	8.0	8.0	
	Struck	Cement	900 - 1,000	8.4	8.0	
Backing-up face brick, cut stone, terra cotta on ordinary wall bearing buildings, figure same as given above for 8" and 12" walls	-----	-----	-----	-----	-----	-----
Backing-up face brick, cut stone, terra cotta on steel or concrete Skelton frame building. First grade workmanship. Walls 8" to 12" thick.	Cut	Lime	825 - 900	9.3	9.0	-----
	Struck	Lime	750 - 850	10.0	9.0	
	Cut	Cement	725 - 800	10.5	9.0	
	Struck	Cement	700 - 775	10.8	9.0	
Public building, first grade workmanship, schools, colleges and university buildings, courthouses, state capitals, public libraries, etc.. 12" to 20" walls	Cut	Lime	850 - 950	9.0	8.0	-----
	Struck	Lime	800 - 900	9.4	8.0	
	Cut	Cement	750 - 850	10.0	8.0	
	Struck	Cement	700 - 800	10.6	8.0	
Powerhouses and other structures having high walls 12" to 20" thick, without intermediate floors.	Cut	Lime	1,100 - 1,200	7.0	9.0	-----
	Struck	Lime	1,000 - 1,100	7.6	9.0	
	Cut	Cement	925 - 1,050	8.0	9.0	
	Struck	Cement	975 - 950	8.8	9.0	

657

Labor Production for Laying Building Brick (continued)						
Class of Work	Motor Joint Style	Kind of Mortar	Avg # of Brick laid per 8-Hour Day	Average Hours 1,000 Brick		
				Mason Hours	Labor Hours	Hoist Engineer*
Shoved joints. Brick laid in full shoved joints with all vertical joints slushed full of mortar, add to all the above.	-----	-----	-----	1.5	-----	-----
Basement foundation walls, paving brick.	Cut	Cement	900 - 1,000	8.5	8.5	-----
	Struck	Cement	850 - 950	9.0	9.0	
Building brick foundation walls 8" to 12" thick, ordinary workmanship	Cut	Lime	1,000 - 1,100	7.6	8.0	-----
	Struck	Lime	900 - 975	8.5	8.0	
	Cut	Cement	900 - 975	8.5	8.0	
	Struck	Cement	825 - 900	9.3	8.0	
Chimneys and stacks, building brick, 1' -4" to 2'- 0" Sq. 15'- 0" above roof 4" to 8" walls. Hodding brick, extra.	Struck	Lime	500 - 550	16.0	8.0 3.0	-----
Large Chimney and stacks, 3'- 0" to 4'- 0" Sq. 15' to 30' high above roof, 8" to 12" walls.	Struck	Lime	550 - 600	14	16	-----
Large brick stacks 100' to 150' high. Wall from 1'- 8" at base to 12" top. Inside put scaffold (log scaffold). Outside scaffold extra.	Struck	Lime	650 - 750	11.5	14.0	-----
Bricking-in boilers, fire boxes, etc..	Struck	-----	600 - 650	12.5	10	-----

Winter Work: Heating brick, mortar, attending salamanders (heating units) removing snow and ice - add extra for this work.

*** Hoisting:*** add about 1/2 hour hoisting engineer time per 1,000 brick for reasonable size projects using 35,000 or more brick per day; 3/4 hour for projects using 15,000 to 20,000 brick per day; and 1 hour per 1,000 brick for projects using 25,000 to 30,000 brick per day.

If automatic hoists are used, omit all time for hoisting engineer.

The quantities and costs given are based on a good grade of workmanship applicable to the type of building on which it is to be used.

The quantities given are based on the performances of experienced masons working regularly at their trade under present conditions. They will not hold well in small communities where the mason does the cement work and plastering as well as the bricklaying and stone setting of the community.

The costs are based on efficient management and supervision, the use of modern tools and appliances, machine mortar mixers, etc., on jobs large enough to warrant their use. On jobs of sufficient size to require a mason foreman, it will be necessary to add for this time on the basis of a certain price per 1,000 brick or at a certain price per week, based on the estimated duration of the job.

Labor Laying Face Brick

The labor cost of laying face brick will vary with the grade of workmanship, style of bond in which the brick are laid, width and kind of mortar joint, type of building—whether long, straight walls or walls cut up with pilasters, door and window openings—and last but not least, the ability of the foreman in charge.

Common bond is the most economical method of laying face brick. Other styles, such as English, Flemish Dutch, and Garden Wall bond, require more labor on account of the various patterns and the additional cutting

658

required. This is especially so on First Grade work where all vertical mortar joints must be plumb.

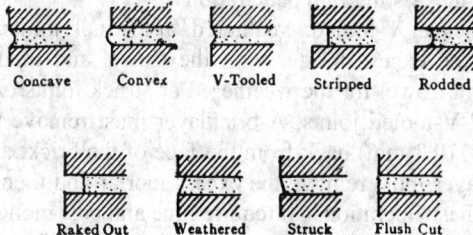

Concave   Convex   V-Tooled   Stripped   Rodded

Raked Out   Weathered   Struck   Flush Cut

Styles of Mortar Joints

For instance, when laying one of the many variations of Dutch, English, or Flemish brick bonds, it is necessary to do a great deal of cutting, especially where half brick are used for headers. It requires just as much time to cut the brick and spread mortar to lay one-half brick as for a full brick. The labor cost of laying 1,000 face brick in Dutch, English, or Flemish bond will be more if half-brick are used for headers than where full brick are used, because it will take just as much time to cut, spread mortar bed, apply mortar to brick, lay brick, and cut the joints.

Flush Cut Joint. The style of mortar joints has considerable bearing on the labor costs. The flush cut joint, sometimes referred to as the rough cut, is the most economical, and because the bricklayer cuts off the excess mortar with his trowel and the joint is finished. This produces an uncompacted joint with a small hairline crack where the mortar is pulled away from the brick by the cutting action. This joint is not always watertight.

Concave and V-Shaped Joints. These joints are normally kept quite small and are formed by the use of a steel jointing tool. These joints are very effective in resisting rain penetration and are recommended for use in areas subjected to heavy rains and high winds.

Weathered Joint. This joint requires care, as it must be worked from below. However, it is the best of the troweled joints, because it is compacted and sheds water readily.

Struck Joint. This is a common joint in ordinary brickwork. American mechanics often work from the inside of the wall, and this is an easy joint to strike with a trowel. Some compaction occurs, but the small ledge does not shed water readily, resulting in a less watertight joint than those discussed above.

Raked Joint. This joint is made by removing the surface of the mortar while it is still soft. While the joint might be compacted, it is difficult to make weathertight and is not recommended where heavy rain, high wind, or freezing is likely. This joint produces marked shadows and tends to darken the overall appearance of the wall.

Grapevine Joint. This is a concave joint with a drip slot placed horizontal into the mortar, approximately midway between bricks. This joint is very effective in resisting rain penetration.

For concave, V-tooled, weathered, and struck joints, it is necessary first to cut off the excess mortar with the trowel and then to do another operation, using a trowel for the weathered or struck joints or a jointing tool for concave and V-tooled joints. A bricklayer must remove the mortar 1/4" (6.2 mm) to 3/8" (9.3 mm) back from the face of the brick, and for a rodded joint, the bricklayer must remove the excess mortar and then strike the joint with a jointer. These operations all require time and cost money.

The stripped joint is the most expensive of any illustrated, because it is necessary to place narrow wood strips in the mortar joints between each course of brick, and these strips must remain until the mortar has set, so it will not squeeze out toward the face of the brick.

If face brick are laid with 7/8" (21.87 mm) or 1" (25 mm) bed mortar joints, the work proceeds very slowly and requires a stiff mortar, because the weight of the top brick courses squeeze the mortar out of the lower mortar joints, producing an uneven appearance. For this reason the work cannot be carried up many courses at a time. Where the joints are struck, it is necessary for the mortar to take its initial set before the joints can be struck to produce a workmanlike job. Brick work of this kind is very expensive.

The quantities and costs given on the following pages are intended to cover all kinds of face brick work, whether laid in veneer in frame buildings, in solid brick or brick and tile walls.

Labor Laying 1,000 Standard Size (8" x 2 1/4" x 3 3/4") or Modular Size (7 5/8" x 2 1/4" x 3 5/8") Face Brick in Common Bond								
Mason Hours	Labor Hours	Hoist Engineer*	# Laid per 8-Hour Day	Style of Mortar Joint	# Laid per 8-Hour Day	Mason Hours	Labor Hours	Hoist Engineer*
2:1:9 Lime - Cement Mortar				Ordinary Workmanship	Cement Mortar			
16.0	12.0	0.5	475 - 525	Flush Cut	425 - 475	17.8	13	0.5
17.8	12.0	0.5	425 - 475	V - Tooled or Concave	400 - 450	19.0	13	0.5
18.0	12.0	0.5	420 - 450	Struck or Weathered	390 - 435	19.5	13	0.5
17.8	12.0	0.5	425 - 475	Raked out	400 - 450	19.0	13	0.5
19.0	12.0	0.5	400 - 450	Rodded	375 - 425	20.0	13	0.5
26.7	15.0	0.5	275 - 325	Stripped	275 - 300	27.8	16	0.5
2:1:9 Lime - Cement Mortar				First Grade Workmanship	Cement Mortar			
19.0	13.0	0.5	400 - 450	Flush Cut	375 - 425	20.0	14	0.5
20.0	13.0	0.5	375 - 425	V - Tooled or Concave	360 - 400	21.0	14	0.5
21.0	13.0	0.5	360 - 400	Struck or Weathered	350 - 390	21.5	14	0.5
20.0	13.0	0.5	375 - 425	Raked out	360 - 400	21.0	14	0.5
23.0	13.0	0.5	325 - 360	Rodded	300 - 340	25.0	14	0.5
32.0	16.0	0.5	225 - 275	Stripped	200 - 240	36.5	18	0.5
32.0	16.0	0.5	225 - 275	7/8" to 1" Flush Joint	200 - 240	36.5	18	0.5
36.0	16.0	0.5	200 - 250	7/8" to 1" Struck	190 - 230	38.0	18	0.5

Add Hoisting Engineer's time only if required.

MASONRY

Labor Laying 1,000 Standard Size (8" x 2 1/4" x 3 3/4") Face Brick in English, Flemish **, Dutch, and Garden Wall Bond. Laid with Full Headers Every 6th Course **

Mason Hours	Labor Hours	Hoist Engineer*	# Laid per 8-Hour Day	Style of Mortar Joint	# Laid per 8-Hour Day	Mason Hours	Labor Hours	Hoist Engineer*
2:1:9 Lime - Cement Mortar				Ordinary Workmanship	Cement Mortar			
20.0	12.0	0.5	380 - 420	Flush Cut	360 - 400	21.0	13	0.5
21.3	12.0	0.5	350 - 400	V - Tooled or Concave	330 - 380	22.5	13	0.5
22.5	12.0	0.5	335 - 375	Struck or Weathered	320 - 360	23.5	13	0.5
21.3	12.0	0.5	350 - 400	Raked out	330 - 380	2.5	13	0.5
22.5	12.0	0.5	335 - 375	Rodded	320 - 360	23.5	13	0.5
29.0	15.0	0.5	250 - 300	Stripped	240 - 285	31.0	15	0.5
2:1:9 Lime - Cement Mortar				First Grade Workmanship	Cement Mortar			
22.5	13.0	0.5	335 - 375	Flush Cut	320 - 355	23.7	14	0.5
24.0	13.0	0.5	315 - 350	V - Tooled or Concave	300 - 335	25.0	14	0.5
25.0	13.0	0.5	300 - 340	Struck or Weathered	285 - 325	26.5	14	0.5
24.0	13.0	0.5	315 - 350	Raked out	300 - 335	25.0	14	0.5
26.7	13.0	0.5	275 - 325	Rodded	260 - 310	29.0	14	0.5
35.5	16.0	0.5	200 - 250	Stripped	190 - 240	37.0	18	0.5
35.5	16.0	0.5	200 - 250	7/8" to 1" Flush Joint	190 - 240	37.0	18	0.5
40.0	16.0	0.5	175 - 225	7/8" to 1" Struck	165 - 215	42.0	18	0.5

Add Hoisting Engineer's time only if required.

**If laid with full headers every course or every 2nd course, increase quantities laid approximately 8.0%.*

Labor Laying 1,000 English Size (8 7/8" x 2 7/8" x 4") Face Brick

Mason Hours	Labor Hours	Hoist Engineer*	# Laid per 8-Hour Day	Style of Mortar Joint	# Laid per 8-Hour Day	Mason Hours	Labor Hours	Hoist Engineer*
2:1:9 Lime - Cement Mortar				Common Bond	Cement Mortar			
22.5	17.0	0.6	335 - 375	Flush Cut	320 - 360	23.5	18	0.6
23.5	17.0	0.6	325 - 360	V - Tooled or Concave	310 - 340	24.6	18	0.6
26.3	17.0	0.6	285 - 325	Struck or Weathered	275 - 315	27.0	18	0.6
23.5	17.0	0.6	325 - 360	Raked out	310 - 340	24.6	18	0.6
28.5	17.0	0.6	260 - 300	Rodded	250 - 290	29.5	18	0.6
2:1:9 Lime - Cement Mortar				English, Flemish, Dutch Bond	Cement Mortar			
24.7	17.0	0.6	300 - 350	Flush Cut	285 - 325	26.3	18	0.6
26.3	17.0	0.6	285 - 325	V - Tooled or Concave	275 - 310	27.0	18	0.6
28.5	17.0	0.6	260 - 300	Struck or Weathered	250 - 290	29.5	18	0.6
26.3	17.0	0.6	285 - 325	Raked out	275 - 310	27.0	18	0.6
31.0	17.0	0.6	240 - 275	Rodded	230 - 265	32.2	18	0.6

Add Hoisting Engineer's time only if required.

Labor Laying 1,000 Standard Size (8" x 2 1/4" x 3 3/4") Vitrified Face Brick

Mason Hours	Labor Hours	Hoist Engineer*	# Laid per 8-Hour Day	Style of Mortar Joint	# Laid per 8-Hour Day	Mason Hours	Labor Hours	Hoist Engineer*
2:1:9 Lime - Cement Mortar				Common Bond	Cement Mortar			
31.0	17.0	0.6	235 - 275	Buttered and Struck	225 - 265	32.7	18	0.6
22.5	16.0	0.6	335 - 375	Flush Cut	320 - 360	23.5	17	0.6
23.5	16.0	0.6	320 - 360	V - Tooled or Concave	300 - 340	25.0	17	0.6
26.3	16.0	0.6	285 - 325	Struck or Weathered	270 - 310	27.5	17	0.6
23.5	16.0	0.6	260 - 300	Rodded	250 - 285	30.0	17	0.6
28.5	16.0	0.6	320 - 360	Rodded Out	300 - 340	25.0	17	0.6
2:1:9 Lime - Cement Mortar				English, Flemish, Dutch Bond**	Cement Mortar			
34.8	17.0	0.6	210 - 250	Buttered and Struck	200 - 240	36.4	17	0.6
25.0	16.0	0.6	300 - 340	Flush Cut	285 - 325	26.3	17	0.6
26.3	16.0	0.6	285 - 325	V - Tooled or Concave	275 - 310	27.3	17	0.6
28.5	16.0	0.6	260 - 300	Struck or Weathered	250 - 290	29.7	17	0.6
26.3	16.0	0.6	285 - 325	Rodded Out	275 - 310	27.3	17	0.6
31.0	17.0	0.6	240 - 275	Rodded	230 - 265	32.7	17	0.6

Add Hoisting Engineer's time only if required.

**If laid with full headers every course or every 2nd course, increase quantities laid approximately 8.0%.*

Roman size face brick, 12" x 1 1/2" x 4" were extensively used years ago for facing many types of buildings. Generally they were laid with thin "buttered" joints and most of the original brick were made by the dry-pressed method and were relatively expensive.

In the past few years the modular Roman brick was introduced having a nominal size of 12" x 2" x 4", including mortar joint, while the actual brick size is 11 5/8" x 1 5/8" x 3 3/8", requiring a 3/8" mortar joint to lay 2" high and 12" long. The bricks are furnished with either a standard brick texture face or a split (rock-like) Texture. The split face is generally preferred as it greatly accentuates the horizontal bonding effect obtainable in any other type of construction. The se units are being produced in all areas of the United States and are in demand, particularly for single story ranch homes, although they are also used for commercial structures, store fronts and for finished masonry interiors.

When laid with a 3/8" mortar joint, it requires 6 bricks per Sq.Ft of wall (6 brick per 0.0929 Sq.Meters) of wall. Add for Headers, breakage and waste.

When laid with a 3/8" mortar joint, it requires 10 3/4 Cu.Ft. (0.304 Cu.Meters) of mortar per 1,000 brick not including any back-up mortar or any allowance for waste.

Labor Laying 1,000 Roman Face Brick (3 5/8" x 1 5/8" x 11 5/8") Face Brick								
Common Bond					Flemish Bond			
Mason Hours	Labor Hours	Hoist Engineer *	Number Laid per 8-Hour Day	Style of Mortar Joint	Number Laid per 8-Hour Day	Mason Hours	Labor Hours	Hoist Engineer *
29.6	17.0	0.5	250 - 290	Buttered and Struck	210 - 240	35.5	17	0.5
20.5	17.0	0.5	370 - 410	Flush Cut	300 - 350	24.6	17	0.5
23.5	17.0	0.5	320 - 360	V - Tooled or Concave	265 - 300	28.3	17	0.5
25.0	17.0	0.5	300 - 340	Struck or Weathered	245 - 285	30.2	17	0.5
23.5	17.0	0.5	320 - 360	Raked out	265 - 300	28.3	17	0.5
26.7	17.0	0.5	280 - 320	Rodded	230 - 270	32.0	17	0.5

Add Hoisting Engineer's time only if required.

Labor Laying Face Brick in English, Flemish, Dutch, Garden Wall, and Diamond Bonds

Face brick laid in any of the above bonds with the courses consisting of various arrangements of headers and stretchers requires considerable extra labor cutting brick on account of the large number of "bats" or headers.

Ordinarily, a full header course extends through the wall every 5th or 6th course, and the other headers are only half brick. On first grade workmanship, it is necessary to plumb all vertical mortar joints in order that the various patterns will work out accurately.

Using the labor laying brick in Common Bond as a base of 100, it will require approximately the following additional time to lay face brick in Dutch, English, and the different variations of Flemish bond, where half brick are used as headers instead of full brick.

662

MASONRY

Kind of Bond	Percent
Common Bond using Full Headers	100
Dutch Bond using Full Headers every other Course	120
Dutch Bond using Full Headers every Sixth Course	129
English Bond using Full Headers every other Course	120
English Bond using Full Headers every Sixth Course	129
Flemish Bond using Full Headers every Course	120
Flemish Bond using Full Headers every Sixth Course	126
Double Flemish Bond using Full Headers every Course	100
Double Flemish Bond using Full Headers every Sixth Course	116
3-Stretcher Flemish Bond using Full Headers every Other Course	107
3-Stretcher Flemish Bond using Full Headers every Sixth Course	112
4-Stretcher Flemish Bond using Full Headers every Other Course	106
4-Stretcher Flemish Bond using Full Headers every Sixth Course	109

Double Size Building Brick

Double size building brick—3-3/4" x 5" x 8" (93.75 x 125 x 200 mm)—are used in different parts of the country. Their use effects a saving in both mortar and labor over standard size building brick—3-3/4" x 2-1/4" x 8" (93.75 x 56.25 x 200 mm).

Each brick contains 40 sq. in. (0.026 sq.m) on the face and when laid with a 1/2" (12.50 mm) bed or horizontal joint and a 1/4" (6.25 mm) vertical or end joint, covers 45.375 or 45-3/8 sq. in. (0.029 sq.m) and requires 3.2 brick per sq.ft. (0.09 sq.m) of 4" (100 mm) wall; 6.4 brick per sq.ft. (0.09 sq.m) of 8" (200 mm) wall and 9.6 brick per sq.ft. (0.09 sq.m) of 12" (300 mm) or 13" (325 mm) wall, not including any allowance for waste.

Mortar Required for Double Size Building Brick

Based on 1/2" (12.50 mm) bed or horizontal joints and 1/4" (6.25 mm) end or vertical joints.

Mortar Required per 1,000 Double Brick		
4" Walls*	8" Walls	12" Walls
12-3/4 c.f.	19-1/4 c.f.	21-1/4 c.f.
Mortar Required per 100 S.F. of Wall		
4" Walls*	8" Walls	12" Walls
4 c.f.	12-1/4 c.f.	20-1/2 c.f.
Mortar Required per 1,000 Double Brick		
100 mm Walls*	200 mm Walls	300 mm Walls
0.36 cu.m	0.55 cu.m	0.60 cu.m
Mortar Required per 9 Sq.M of Wall		
100 mm Walls*	200 mm Walls	300 mm Walls
0.11 cu.m	0.35 cu.m	0.58 cu.m

*Freestanding Walls.
No Allowance For Waste Included In Above Quantities.

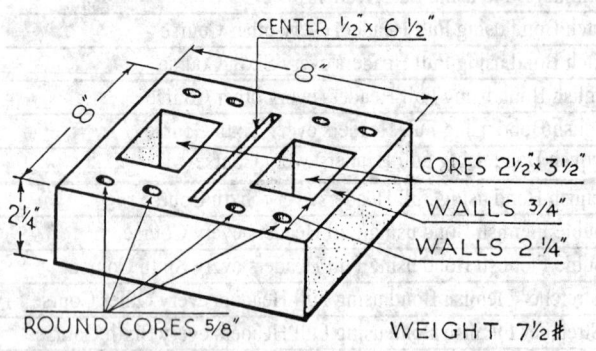

CENTER ½" x 6 ½"

8"

8"

2¼"

CORES 2½" x 3½"

WALLS 3/4"

WALLS 2 ¼"

ROUND CORES 5/8"

WEIGHT 7½#

Labor Laying Double Size Building Brick. A mason will lay about 60% as many double brick—3-3/4" x 5" x 8" (93.75 x 125 x 200 mm)—as standard size building brick—3-3/4" x 2-1/4" x 8" (93.75 x 56.25 x 200 mm). If a mason lays 800 building brick on a certain class of work, he would lay 480 double size building brick on the same class of work. In other words, if it required 10 hrs. mason time per 1,000 building brick, it would require 1-2/3 as many hours per 1,000 double brick, i.e., 10 x 1.667 = 16.67 or 16-2/3 hrs. mason time per 1,000 double brick.

Twin Brick

Twin brick is a double size brick that contains the face brick and a 4" (100 mm) backing brick in one unit, size 8" x 8" x 2-1/4" (200 x 200 x 56.25 mm). It is made with a hollow core so that the twin brick weighs only 7-1/2 lbs. (3.4 kg) and can be picked up and laid in a one-hand operation. It is designed so the wall has no continuous mortar joint from outer to inner wall faces, helping to avoid moisture penetration.

It is intended primarily for residential work having 8" (200 mm) walls that are to be furred and plastered. Extra face brick must be used at openings and corners to insure proper bond.

It requires about 25 cu.ft. (0.70 cu.m) of mortar per 1,000 units.

On four jobs on which these brick were used, requiring 50,000 units, a mason laid 60 twin brick an hour or 480 per 8 hr. day, at the following labor cost per 1,000 units:

Twin Brick					
	Hours	Rate	Total	Rate	Total
Mason	17.00	$....	$....	$36.93	$627.81
Labor	11.50	$....	$....	$ 26.16	$ 300.84
Cost per 1,000 units					$928.65
per brick					$0.93

MASONRY

Skintled Brickwork

Skintled brickwork is used particularly in storefront work and interiors of the "Olde English Pub" variety.

Skintled brickwork consists of laying common brick irregularly, with random projecting brick at intervals and using clinkers or irregular and other off-colored brick to produce a variegated wall.

These brick are laid in a number of different ways to produce the desired effect. In some of them the mortar projects beyond the face of the brick just as it has been squeezed out of the joint; in others the joints are cut flush or raked out to produce shadows, etc.

The labor cost will vary with the design and care used in laying up the brick. When laying building brick without regard to color of wall and with uncut mortar joints or joints cut flush with the face of the wall, a mason should lay 350 to 400 brick per 8-hr. day, at the following labor cost per 1,000:

Skintled Brickwork (uncut joints / cut flush joint)					
	Hours	Rate	Total	Rate	Total
Mason	21.0	$....	$....	$ 36.93	$ 775.53
Labor	15.0	$....	$....	$ 26.16	$ 392.40
Cost per 1,000 brick					$ 1,167.93
per brick					$1.17

If brick joints are raked out, add 2.5 hrs. mason time per 1,000 brick. On skintled brickwork using clinkers and other dark or irregularly shaped brick to produce a textured wall, more care is required in laying the brick, working out the designs and in raking out the joints to produce shadows, etc., and a mason should lay 250 to 300 brick per 8-hr. day, at the following labor cost per 1,000:

Skintled Brickwork (raked out joint)					
	Hours	Rate	Total	Rate	Total
Mason	29.0	$....	$....	$36.93	$1,070.97
Labor	20.0	$....	$....	26.16	523.2
Cost per 1,000 brick					$1,594.17
per brick					$1.59

English Size Face Brick

English size face brick, 8-7/8" x 2-7/8" x 4" (221.87 x 71.87 x 100 mm) in size, are seldom used in this country any more and while the size is not standard, they are still furnished by a few manufacturers.

They are practically half again as large on the face as the standard brick, and a mason is unable to lay as many as the standard size brick. They also require more labor time handling, hoisting, and wheeling to the masons.

Number of English Size 8 7/8" x 2 7/8" x 4" Face Brick Required per Sq.Ft Wall						
All Vertical Mortar Joints Figured 1/4" Wide Width of Horizontal or "Bed" Mortar Joints, inches						
Dry	1/8 "	1/4 "	3/8 "	1/2 "	5/8 "	3/4 "
5.65	5.34	5.06	4.88	4.67	4.52	4.37
All Vertical Mortar Joints Figured 6.25 mm Wide Width of Horizontal or "Bed" Mortar Joints, mm						
Dry	3.13 mm	6.25 mm	9.38 mm	12.50 mm	15.63 mm	18.75 mm
5.65	5.34	5.06	4.88	4.67	4.52	4.37

Add for waste, headers, and the different brick bonds, as given under "Standard Brick"

C.F. of Mortar Required to Lay 1,000 English Size Face Brick Based on 1/4" End Joints and "Bed" Joints as Given Below					
1/8 "	1/4 "	3/8 "	1/2 "	5/8 "	3/4 "
5	9	11	14	17	20
Cu.M. of Mortar Required to Lay 1,000 English Size Face Brick Based on 6.25 mm End Joints and "Bed" Joints as Given Below					
3.13 mm	6.25 mm	9.38 mm	12.50 mm	15.63 mm	18.75 mm
0.142	0.255	0.311	0.396	0.481	0.566

Vitrified Face Brick

There are a number of vitrified face brick on the market. This type is often so hard it is almost impossible to cut. Because of their denseness and brittleness, they do not absorb water from the mortar, and it requires more time for the mortar to take its initial set. It is impossible to lay up many courses of brick at one time, because the weight of the upper brick courses forces the mortar from the joints of the lower courses, causing an unsightly wall. It is also difficult to hold the brick in the wall and keep the wall plumb, because the brick slide out of place very easily.

Brick of this kind are ordinarily laid with "buttered" mortar joints or with 1/4" (6.25 mm) mortar joints struck flush. They are furnished in the standard size, 8" x 2-1/4" x 3-3/4" (200 x 56.25 x 93.75 mm) and require the same number of brick and the same amount of mortar as other standard size face brick.

Brick Floors and Steps

Face Brick Floors Laid in Basket Weave Pattern. Brick floors laid in square or basket weave pattern do not require any cutting and are much easier laid than herringbone designs.

On work of this kind, a mason should lay 180 to 225 brick per 8-hr. day, at the following labor cost per 1,000:

666

BRICK FLOOR PATTERNS

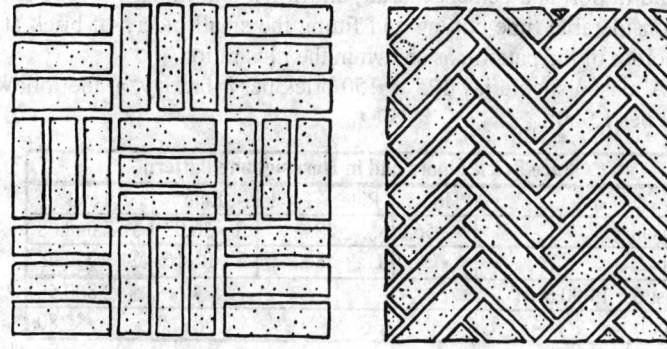

Brick Laid in Basket Weave Pattern.   Brick Laid in Herringbone Pattern.

BRICK STAIR DETAILS

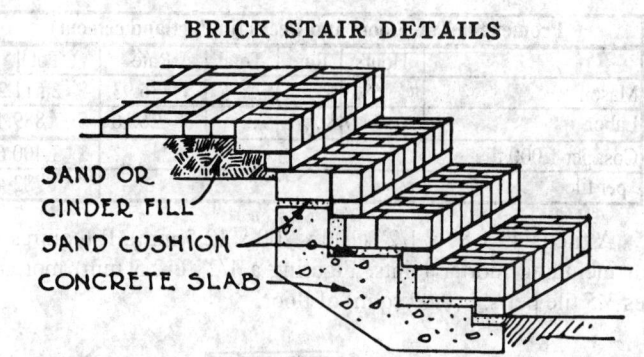

SAND OR
CINDER FILL
SAND CUSHION
CONCRETE SLAB

Brick Steps with End-Set Treads

Face Brick Floors laid in Basket Weave Pattern					
	Hours	Rate	Total	Rate	Total
Mason	40.0	$....	$....	$36.93	$1,477.20
Labor	20.0	$....	$....	$ 26.16	$ 523.20
Cost per 1,000 units					$2,000.40
per brick					$2.00

When using brick for stairs, it is recommended that brick pavers be used instead of standard brick.

Face Brick Floors Laid in Herringbone Pattern. Brick floors are usually laid in portland cement mortar, and when laid in herringbone pattern require considerable time cutting and fitting the small pieces of brick at the edge or border of the pattern, as shown in the illustration.

A mason should lay 125 to 150 brick per 8-hr. day, at the following labor cost per 1,000:

Face Brick Floors laid in Herringbone Pattern					
	Hours	Rate	Total	Rate	Total
Mason	56.0	$....	$....	$ 36.93	$ 2,068.08
Labor	28.0	$....	$....	$ 26.16	$ 732.48
Cost per 1,000 brick					$ 2,800.56
per brick					$2.80

If there is only one mason on the job, it will require a helper for each mason.

Promenade Tile Floor and Decks. When laying 4" x 8" x 1/2" (100 x 200 x 12.5 mm) promenade tile on porch floors and decks, with the tile laid in portland cement mortar and 1/2" (12.5 mm) mortar joints, struck flush, a mason should lay and point 100 to 125 tile per 8-hr. day, at the following labor cost per 1,000:

Promenade Tile Floors and Decks (in Portland cement)					
	Hours	Rate	Total	Rate	Total
Mason	68.0	$....	$....	$ 36.93	$ 2,511.24
Labor	34.0	$....	$....	$ 26.16	$ 889.44
Cost per 1,000 tile					$ 3,400.68
per tile					$3.40

A tile is 4" x 8" x 1/2" (100 x 200 x 12.5 mm). It requires 4 tile per s.f., not including mortar joints. Figuring a 1/2" (12.5 mm) mortar joint, it requires 3.5 tile per s.f. (0.09 sq.m) of floor.

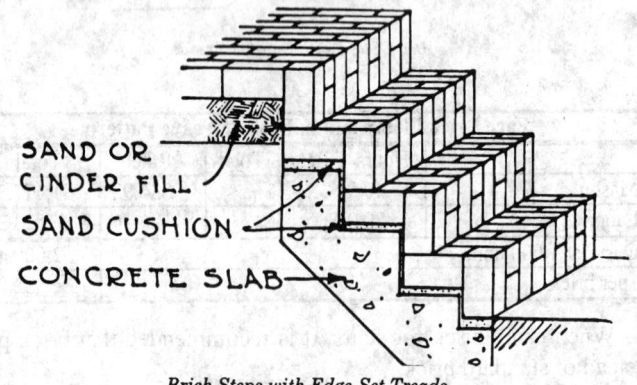

SAND OR CINDER FILL

SAND CUSHION

CONCRETE SLAB

Brick Steps with Edge-Set Treads

MASONRY

A mason will lay 30 to 36 s.f. of floor per 8-hr. day, at the following labor cost per 100 s.f.:

Promenade Tile Floors and Decks (floor)					
	Hours	Rate	Total	Rate	Total
Mason	24.0	$....	$....	$ 36.93	$ 886.32
Labor	12.0	$....	$....	$ 26.16	$ 313.92
Cost per 100 Sq.Ft					$ 1,200.24
Cost per Sq.Ft					$ 12.00
Cost per Sq. Meter					$ 129.19

Brick Steps. When laying brick steps in portland cement mortar, similar to those illustrated, a mason should lay 125 to 150 brick per 8-hr. day, at the following labor cost per 1000:

Brick Steps					
	Hours	Rate	Total	Rate	Total
Mason	56.0	$....	$....	$ 36.93	$ 2,068.08
Labor	28.0	$....	$....	$ 26.16	$ 732.48
Cost per 1,000 brick					$ 2,800.56
Cost per brick					$ 2.80

Turning Face Brick Segmental Arches Over Doors and Windows. Where segmental brick arches are laid over wood centers for 3'-0" (0.9 m) wide door and window openings, it will require about 4 hrs. mason time at the following labor cost per arch:

Turning Face Brick Segmental Arches Over Doors and Windows.					
	Hours	Rate	Total	Rate	Total
Mason	4	$....	$....	$ 36.93	$147.72

For a 5'-0" (1.5 m) arch, figure about 6.5 hrs. mason time, as follows:

For a 5'-0" (1.5 m) arch					
	Hours	Rate	Total	Rate	Total
Mason	6.5	$....	$....	$ 36.93	$240.05

If the brick are cut or chipped to a radius, add cost of chipping as given below.

Laying Face Brick in Flat or "Jack" Arches. When laying face brick in flat or jack arches, a mason should complete one arch up to 3'-6" (1.0 m) wide in about 5 hrs. at the following labor cost:

Laying Face Brick in Flat or "Jack" Arches					
	Hours	Rate	Total	Rate	Total
Mason	5	$....	$....	$ 36.93	$184.65

For a 5'-0" (1.5 m) jack arch, figure about 7-1/2 hrs. mason time, as follows:

For a 5'-0" (1.5 m) arch					
	Hours	Rate	Total	Rate	Total
Mason	7.5	$....	$....	$ 36.93	$276.98

The above costs do not include chipping or rubbing brick to a radius. If this is necessary, add cost of chipping as given below.

Chipping Brick. Where necessary to chip or cut brick for segmental or jack arches, a mason should chip 400 to 500 brick per 8-hr. day, depending on the texture and density of the brick, at the following labor cost per 1,000:

Chipping brick (segmental or jack arch)					
	Hours	Rate	Total	Rate	Total
Mason	17.5	$....	$....	$ 36.93	$646.28

There is considerable lost motion in cutting brick, and unless there is a great deal of duplication in the cutting, where a mason can work continuously on a few special sizes, a mason will cut only 160 to 200 pieces per 8-hr. day at the following labor cost per 100 pieces:

Chipping brick (work continuously)					
Mason	4.5	$....	$....	$ 36.93	$ 166.19
Labor	1.0	$....	$....	$ 26.16	$ 26.16
Cost per 100 cuts					$ 192.35
Cost per cut					$ 1.92

Fire Resistance

Fire Resistance Ratings. Tables and figures below list fire resistance ratings for various brick walls. Ratings listed are for load bearing walls tested with working loads of 160 psi (1.10 MPa) of gross area, except the 4" (100 mm) wall shown was loaded to 92 psi (0.63 MPa). The following table lists the fire resistance ratings for steel columns covered with brick:

Fireproofed Column Fire Ratings

Construction*	Ratings
Steel Columns, 6"x6" (150 x 150 mm) or larger	
3-3/4" (93.75 mm) brick with brick fill	4 hr.
2-1/4" (56.25 mm) brick with brick fill	1 hr.

Column fire resistance varies with the cross-sectional area of solid material; the larger the area, the greater the fire resistance for a given thickness of protection around the structural steel. Column dimensions are outside dimensions. Smaller columns may require more cover to achieve equal ratings. For columns that are not square, protection should equal that of a square column having equal or lesser cross-sectional area. Thickness does not include plaster.

670

MASONRY
Cutting Masonry Products With Portable Masonry Saw

On practically every construction job there is a large amount of cutting necessary on masonry materials, and with the high cost of labor, this becomes an expensive item.

This applies particularly to the cross and diamond bonds in brick masonry, arch brick, radial brick, glazed tile, rotary kiln blocks, concrete blocks, natural stone, roofing tile, and conduits.

By using a masonry saw, this work can be done at a fraction of the cost of hand cutting plus the saving of materials.

Masonry saws are obtainable with 12", 14", 18" and 20" (300, 350, 450, and 500 mm) blade capacities, and are designed for both wet, dustless and for dry cutting. The advantage of wet cutting is in the confined areas where dust would be a problem. An exhaust assembly is available for use with dry cutting masonry saws. This assembly consists of a high velocity fan that draws the dust away from the machine and disposes of it through the nearest opening.

A smaller compact model masonry saw has been developed that is light enough to carry, can be set up on the scaffolding, and can be operated from a pickup truck or the back of a station wagon.

There is also a small, highly portable concrete saw that can do the work of many of the heavier models. There are four types of blades that can be used on masonry saws:

1. wet-dry abrasive blades
2. dry "break-resistant" abrasive blades
3. wet "break-resistant" abrasive blades
4. diamond blades

The wet abrasive and diamond blades should be used only with a masonry saw equipped for wet cutting. There are many specifications of each type of blade; it is very important that the correct specification be used for the particular material to be cut.

Diamond blades for masonry saws vary in price from $340.00 to $660 for 14" (350 mm) diameter blades and $600.00 to $1,275.00 for 18" (450 mm) diameter blades.

The cost of cutting masonry units varies greatly and depends on the type of material, blade specifications, and the operator.

Concrete products vary considerably in hardness, such as sand-gravel blocks compared with Haydite blocks, and blade wear and cutting time will vary accordingly. On average most 8"x8"x16" (200 x 200 x 400 mm) concrete blocks can be sliced in two in 14 to 16 seconds at a cost of less than $0.01 per unit.

Fire brick varies tremendously in hardnesses and types. Ordinary dry press or silica can be cut very rapidly, while Super-duty refractories or Basic, Chrome, and Magnesite require 15 to 18 seconds for each cut

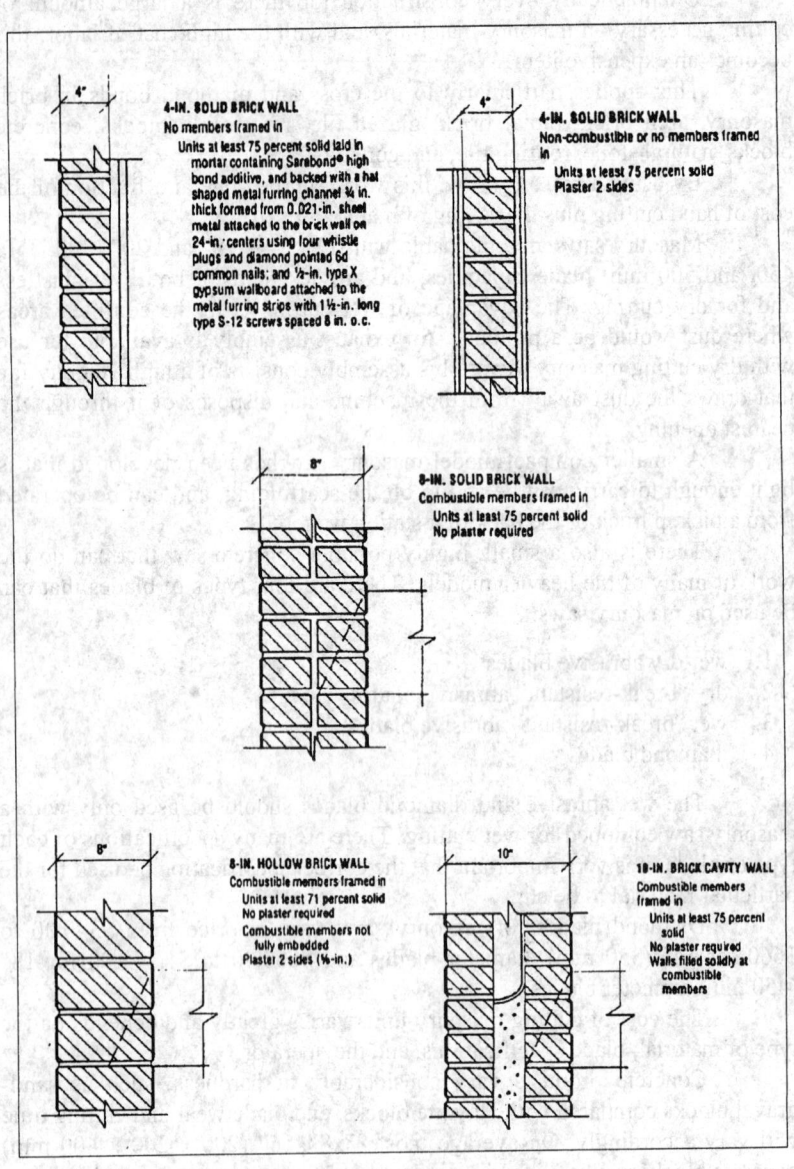

4-IN. SOLID BRICK WALL
No members framed in
  Units at least 75 percent solid laid in mortar containing Sarabond® high bond additive, and backed with a hat shaped metal furring channel ¾ in. thick formed from 0.021-in. sheet metal attached to the brick wall on 24-in. centers using four whistle plugs and diamond pointed 6d common nails; and ½-in. type X gypsum wallboard attached to the metal furring strips with 1½-in. long type S-12 screws spaced 8 in. o.c.

4-IN. SOLID BRICK WALL
Non-combustible or no members framed in
  Units at least 75 percent solid
  Plaster 2 sides

8-IN. SOLID BRICK WALL
Combustible members framed in
  Units at least 75 percent solid
  No plaster required

8-IN. HOLLOW BRICK WALL
Combustible members framed in
  Units at least 71 percent solid
  No plaster required
  Combustible members not fully embedded
  Plaster 2 sides (⅝-in.)

10-IN. BRICK CAVITY WALL
Combustible members framed in
  Units at least 75 percent solid
  No plaster required
  Walls filled solidly at combustible members

Two-Hour Fire Ratings
Constructions Shown are Loadbearing

672

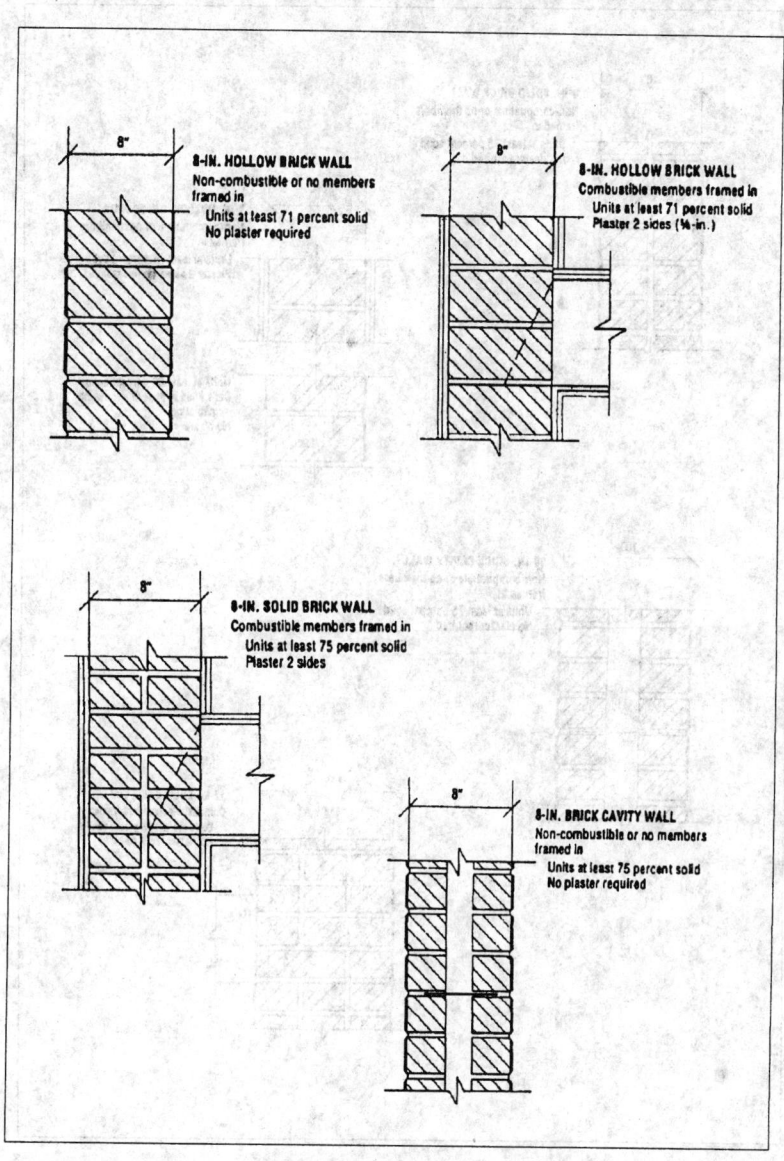

8"
8-IN. HOLLOW BRICK WALL
Non-combustible or no members framed in
Units at least 71 percent solid
No plaster required

8"
8-IN. HOLLOW BRICK WALL
Combustible members framed in
Units at least 71 percent solid
Plaster 2 sides (⅝-in.)

8"
8-IN. SOLID BRICK WALL
Combustible members framed in
Units at least 75 percent solid
Plaster 2 sides

8"
8-IN. BRICK CAVITY WALL
Non-combustible or no members framed in
Units at least 75 percent solid
No plaster required

Three-Hour Fire Ratings
Constructions Shown are Loadbearing

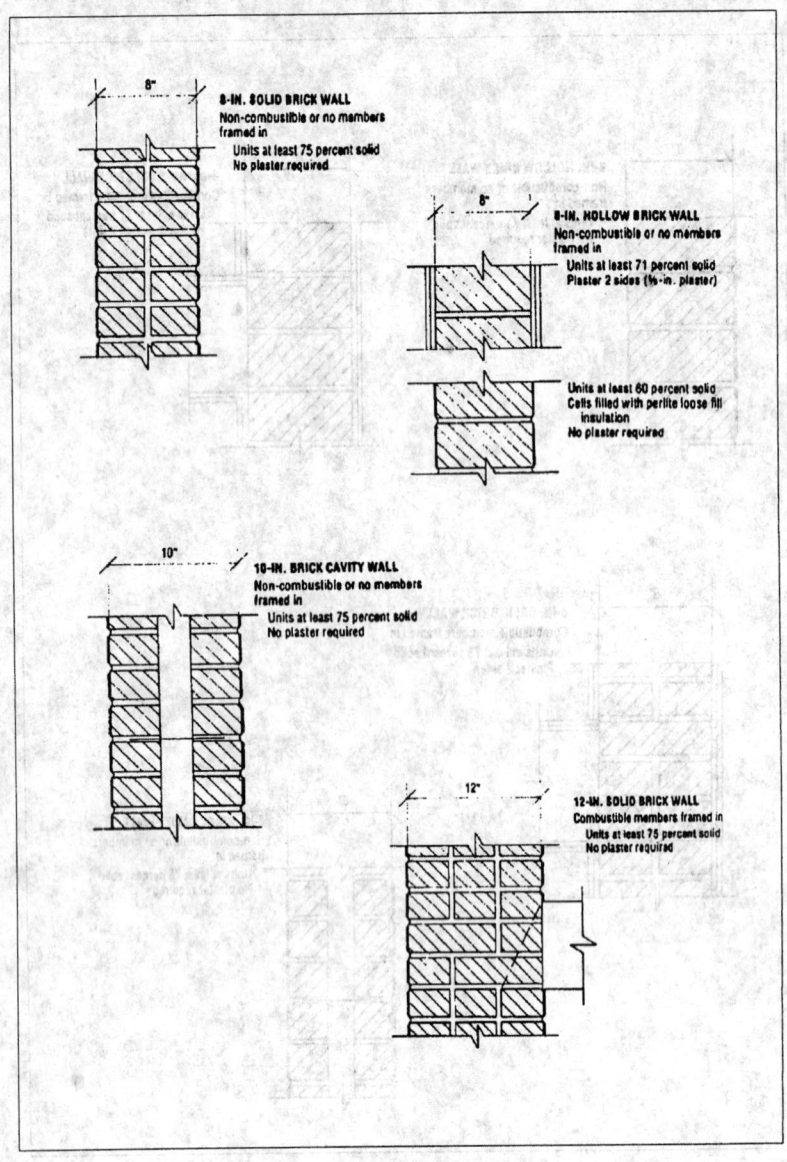

8-IN. SOLID BRICK WALL
Non-combustible or no members
framed in
   Units at least 75 percent solid
   No plaster required

8-IN. HOLLOW BRICK WALL
Non-combustible or no members
framed in
   Units at least 71 percent solid
   Plaster 2 sides (½-in. plaster)

   Units at least 60 percent solid
   Cells filled with perlite loose fill
   insulation
   No plaster required

10-IN. BRICK CAVITY WALL
Non-combustible or no members
framed in
   Units at least 75 percent solid
   No plaster required

12-IN. SOLID BRICK WALL
Combustible members framed in
   Units at least 75 percent solid
   No plaster required

Four-Hour Fire Ratings
Construction Shown are Loadbearing

completely through, at a cost varying from $0.06 to 0.20 depending on actual density.

All the above costs and cutting speeds are based on cross cuts. Longitudinal cuts will be about the same for equivalent area. The following is an example of the approximate time required making cuts in various masonry products:

Material	Type of Cut	Cut Length, Inches	Cut Depth, Inches	Approx. Time, Seconds	Cut Length, mm	Cut Depth, mm
Acid Brick	Miter	4 1/2 "	2 1/4 "	15	112.50 mm	56.25 mm
Concrete Block	Straight	8 "	8 "	19	200.00 mm	200.00 mm
Glazed Brick-Tile	Miter	6 "	4 "	15	150.00 mm	100.00 mm
Magnesite Brick	Miter	2 3/4 "	4 "	15	68.75 mm	100.00 mm
Rotary Kiln Block	Straight	8 "	3 "	11	200.00 mm	75.00 mm
Silica Brick	Straight	8 "	2 1/4 "	10	200.00 mm	56.25 mm
Roofing Tile	Miter	9 "	2 "	14	225.00 mm	50.00 mm
Clay Sewer Pipe	Cross	4" Dia.	—	18	100.00 mm	—
Concrete Cylinder	Cross	4" Dia.	—	21	100.00 mm	—

Cleaning Face Brick Work

The cost of cleaning face brick varies with the kind of brick. It is much easier to clean a smooth face vitrified brick than one having a rough texture. Where smooth face brick are used, and it is not necessary to do any pointing but merely wash the wall, the indiscriminate use of muriatic acid or of the wrong proprietary compound can cause unsightly, hard-to-remove stains. Also, chemical cleaning solutions are generally more effective when the outdoor temperature is 50°F or above.

An experienced mechanic should clean 5,000 to 5,500 brick-750 to 825 sq.ft. (69.67-76.64 sq.m.) of wall-per 8-hr. day at the following labor cost per 1,000 brick, or approximately 150 sq.ft. (13.93 sq.m) of wall:

Clean 150 Sq. Ft. (13.9 Sq.M)					
	Hour	Rate	Total	Rate	Total
Mechanic	1.6	$....	$....	$ 37.89	$ 60.62
Helper	0.8	$....	$....	$ 26.16	$ 20.93
Cost per 1,000 brick				$	81.55
per s.f. of wall				$	0.54
per sq.m				$	5.85

The labor cost of cleaning and washing rough texture face brick will be higher than for smooth brick because of cement and mortar getting into the surface of the brick.

When cleaning and washing rough textured brick, the same crew should clean and wash 3,500 to 4,000 brick-525 to 600 sq.ft. (48.77-55.74

sq.m) of wall-per 8-hr. day at the following labor cost per 1,000 brick, or per 150 sq.ft. (13.93 sq.m) of wall:

Clean (Rough Texture) 150 Sq. Ft. (13.9 Sq.M)						
	Hour	Rate	Total	Rate	Total	
Mechanic	2.2	$....	$....	$ 37.89	$	83.36
Helper	1.1	$....	$....	$ 26.16	$	28.78
Cost per 1,000 brick					$	112.13
per s.f. of wall					$	0.75
per sq.m of wall					$	8.04

The labor cost of cleaning face brick with high-pressure water varies considerably, depending on location and whether pointing and patching are required. When cleaning and washing textured brick, a crew can clean from 6,000 to 8,500 brick—900 to 1,275 sq.ft. (83.61-118.44 sq.m) of wall—per 8-hr. day at the following cost per 1,000 brick or per 150 sq.ft. (13.93 sq.m) of wall:

Clean (High Pressure Water) 150 Sq. Ft. (13.9 Sq.M)						
	Hour	Rate	Total	Rate	Total	
Mechanic	0.55	$....	$....	$ 37.89	$	20.84
Helper	0.55	$....	$....	$ 26.16	$	14.39
Cost per 1,000 brick					$	35.23
per s.f. of wall					$	0.23
per sq.m of wall					$	2.53

When estimating cleaning of masonry units, the following precautions apply, and costs must be added when applicable:

1. Saturation of the brick masonry wall surface with water before and after application of chemical or detergent cleaning solutions, add 0% to 10%.
2. Failure to properly use chemical cleaning solutions. This can result in recleaning or permanent staining.
3. Failure to protect windows, doors and trim. Cost will vary in accordance with the amount of material to be protected.
4. Costs of the chemical or detergents.
5. Classification and safe disposal of the chemicals.
6. Failure to protect adjacent property.

The following table is a guide for the estimator to determine the cleaning method to be employed for various new masonry units.

Cleaning Guide for New Masonry

Brick Category	Cleaning Method
Red and red flashed	Bucket and brush hand cleaning
	High pressure water
	Sandblasting

676

MASONRY

Remarks: Proprietary compounds and emulsifying agents may be used. *Smooth texture:* Mortar stains and smears are generally easier to remove; less surface area exposed; easier to presoak and rinse; unbroken surface, thus more likely to display poor rinsing, acid stain, poor removal of mortar smears. *Rough texture:* Mortar and dirt tend to penetrate deep into textures; additional area for water and acid absorption; essential to use pressurized water during rinsing.

Red, heavy sand finish .... Bucket and brush hand cleaning
.... High pressure water
Remarks: Clean with plain water and scrub brush, or lightly applied high pressure and plain water. Excessive mortar stains may require use of cleaning solutions. Sandblasting is not recommended.

Light-colored units, white, ... Bucket and brush hand cleaning
tan, buff, gray, specks, ... High pressure water
pink, brown and black .... Sandblasting
Remarks: Do not use muriatic acid! Clean with plain water, detergents, emulsifying agents, or suitable proprietary compounds. Manganese colored brick units tend to react to muriatic acid solutions and stain. Light colored brick are more susceptible to "acid burn" and stains, compared to darker units.

Same as light-colored units, ... Bucket and brush hand cleaning
etc., plus sand finish .... High pressure water
Remarks: Lightly apply either method. (See notes for light-colored units, etc.) Sandblasting is not recommended.

Glazed Brick .... Bucket and brush hand cleaning
Remarks: Wiped glazed surface with soft cloth within a few minutes of laying units. Use soft sponge or brush plus ample water supply for final washing. Use detergents where necessary and acid solutions only for very difficult mortar stain. Do not use acid on salt glazed or metallic glazed brick. Do not use abrasive powders.

Colored mortars .... Method generally controlled
.... by the brick unit
Remarks: Many manufacturers of colored mortars do not recommend chemical cleaning solutions. Most acids tend to bleach colored mortars. Mild detergent solutions are generally recommended.

Brick Fireplaces, Mantels, and Hearths

On account of the many sizes, shapes, and styles of brick fireplaces, mantels, and hearths, it is difficult to estimate the cost of the work at a certain price per 1,000 brick. The most satisfactory method is to figure a certain length of time to lay up each fireplace or mantel.

A mason should complete a plain brick fireplace 5'-0" to 6'-0" (1.52-1.83 m) wide and 4'-0" to 5'-0" (1.22-1.52 m) high, requiring 200 to 225 brick, not including hearth, in about 10 hrs. at the following labor cost:

Plain Brick Fireplace					
	Hour	Rate	Total	Rate	Total
Mason	10	$....	$....	$ 36.93	$ 369.30
Labor	5	$....	$....	$ 26.16	$ 130.80
Cost per Fireplace					$ 500.10

On larger and more elaborate brick fireplaces and mantels having raked or rodded mortar joints, a mason should lay about 100 to 125 brick per 8-hr. day, at the following labor cost per 1,000:

Large Fireplace					
	Hour	Rate	Total	Rate	Total
Mason	70	$....	$....	$ 36.93	$ 2,585.10
Labor	35	$....	$....	$ 26.16	$ 915.60
Cost per Fireplace					$ 3,500.70

If there is only one mason working on the fireplace, it will require one laborer with each mason.

Brick Fireplace Linings. A mason should complete the lining for a brick fireplace having an opening 3'-0" to 4'-0" (0.91-1.22 m) wide, 2'-6" to 3'-0" (0.76-0.90 m) high and 1'-4" to 1'-9" (0.40-0.53 m) deep, in 6 to 8 hrs. all depending upon the class of work.

If the fireplace linings are laid in common bond and do not require an excessive amount of cutting and fitting, a mason should complete one fireplace lining in about 5 hrs., but where the back-hearths are of irregular shape and require considerable cutting and fitting at angles and at tops for dampers and smoke chambers, it will require about 8 hrs. mason time for each lining.

Fireplaces having an opening 3'-6" to 4'-0" (1.07 to 1.22M) wide, about 4'-0" (1.22 M) high and 1'-9" (.53M) deep, lined with firebrick laid in herringbone pattern, requiring cutting and fitting at all edges and corners, will require 20 to 24 hrs. mason time to line one fireplace.

Labor Laying Brick Hearths and Back-Hearths. If brick hearths and back-hearths are laid in square or basket weave pattern, a mason should complete one back-hearth in 3 to 4 hrs. and the front hearth will require the same length of time.

If laid in herringbone pattern with cut brick at all edges, figure 4 to 5 hrs. mason time for both hearth and back-hearth.

Figure about 1/2-hr. laborer time to each hour of mason time, except where only one mason is on the job when it will figure hour for hour.

Fire Brick Work

Fire brick and tile are furnished in innumerable sizes and shapes to meet the requirements of all classes of boilers, furnaces, stacks, etc. It is impossible to list them all, but the most common are listed below. Standard

678

MASONRY

fire brick size is 8"x3-1/2"x2-1/4" (200 x 87.50 x 56.25) and will cost about $760.00 per 1,000 for low duty, $1,250.00 for high duty.

Fire Clay. When a close thin joint is desired, the clay should be soaked and mixed thin, and the brick should be dipped and rubbed. For work of this kind, 300 lbs. (135 kg) of finely ground fire clay is required per 1,000 brick.

Sizes and Shapes of Standard Fire Brick Number Required per Sq. Ft.			
Name	**Size**	**Laid Flat**	**Laid on Edge**
8" Straight	8 " x 3 1/2 " x 2 1/4 "	8	5.2
9" Straight	9 " x 4 1/2 " x 2 1/2 "	6.5	3.5
Small 9" Brick	9 " x 3 1/2 " x 2 1/2 "	6.5	4.5
Split Brick	9 " x 4 1/2 " x 1 1/4 "	13	3.5
2" Brick	9 " x 4 1/2 " x 2 "	8	3.5
Soap	9 " x 2 1/2 " x 2 1/2 "	6.5	7
Checker	9 " x 2 3/4 " x 22 3/4 "	6	6

Sizes and Shapes of standard Fire Brick - Metric Number Required per Sq.Meter			
Name	**Size**	**Laid Flat**	**Laid on Edge**
200 mm Straight	200 mm x 87.5 mm x 56.3 mm	86.1	56.0
225 mm Straight	225 mm x 112.5 mm x 62.5 mm	70.0	37.7
Small 225mm Brick	225 mm x 87.5 mm x 62.5 mm	70.0	48.4
Split Brick	225 mm x 112.5 mm x 31.3 mm	139.9	37.7
50 mm Brick	225 mm x 112.5 mm x 50.0 mm	86.1	37.7
Soap	225 mm x 62.5 mm x 62.5 mm	70.0	75.3
Checker	225 mm x 68.8 mm x 568.8 mm	64.6	64.6

Ordinary fire clay costs about $9.25 per 100 lbs. ($20.39 per 45.36 kg) while high-grade refractory fire clay costs about $10.75 per 100 lbs. ($23.69 per 45.36 kg).

Labor Lining Brick Chimneys and Stacks with Fire Brick. When lining small brick chimneys and stacks from 2'-0" to 3'-0" (600-900 mm) square, laid in fire clay, a mason should lay 450 to 525 brick per 8-hr. day, at the following labor cost per 1,000:

Lining Brick Chimneys and Stacks with Fire Brick					
	Hour	Rate	Total	Rate	Total
Mason	16	$....	$....	$ 36.93	$ 590.88
Labor	12	$....	$....	$ 26.16	$ 313.92
Cost per 1,000 brick					$ 904.80

On large brick stacks having an inside diameter of 4'-0" to 6'-0" (1.21-1.83 m), a mason should lay 600 to 750 fire brick per 8-hr. day, at the following labor cost per 1,000:

Lining Brick Chimneys and Stacks with Fire Brick Large Stacks					
	Hour	Rate	Total	Rate	Total
Mason	11.8	$....	$....	$ 36.93	$ 435.77
Labor	10.0	$....	$....	$ 26.16	$ 261.60
Cost per 1,000 brick					$ 697.37

Number of Wedge and Arch Brick Required for Various Circles	
No. 1 Wedge Brick	9 "x 4 1/2 "x (2 1/2 - 1 7/8 ")
No. 2 Wedge Brick	9 "x 4 1/2 "x (2 1/2 - 1 1/2 ")
No. 1 Arch Brick	9 "x 4 1/2 "x (2 1/2 - 2 1/8 ")
No. 2 Arch Brick	9 "x 4 1/2 "x (2 1/2 - 1 3/4 ")

Labor Laying Fire Brick in Fireboxes, Breechings, Etc. When lining fireboxes, breechings, furnaces, etc., with fire brick, requiring arch and radial brick, special shapes, etc., a mason should lay 175 to 225 brick per 8-hr. day, at the following labor cost per 1,000:

Laying Firebrick in Fireboxes, Breechings					
	Hour	Rate	Total	Rate	Total
Mason	40.00	$....	$....	$ 36.93	$ 1,477.20
Labor	40.00	$....	$....	$ 26.16	$ 1,046.40
Cost per 1,000 brick					$ 2,523.60

When estimating brick arches, it is advisable to allow additional labor. It is doubtful that a mason thoroughly experienced in boiler work will average more than 25 brick per hr. for the actual arch and skew-backs, due to adverse conditions and close working space. One contractor specializing in this class of work checks the penetrations through the wall, and regardless of size, adds 4 hours mason time for each one.

Labor Bricking in Boilers. When bricking in steel boilers, fireboxes, breechings, etc., with common or fire brick, where the walls are 8" (200 mm) to 12" (300 mm) thick, a mason should lay 600 to 650 brick per 8-hr. day, at the following labor cost per 1,000:

Bricking in Boilers					
	Hour	Rate	Total	Rate	Total
Mason	13.0	$....	$....	$ 36.93	$ 480.09
Labor	13.0	$....	$....	$ 26.16	$ 340.08
Cost per 1,000 brick					$ 820.17

MASONRY

Table of Wedge Brick Required for Various Circles					Number of Arch Bricks Required for Various Circles				Metric
Diameter Feet -Inch	No.2 Wedges	No.1 Wedges	Square	Total	No.2 Wedges	No.1 Wedges	9-Inch (225 mm)	Total	Dia. MM
2 ' - 0 "	-----	-----	-----	-----	42	-----	-----	42	600
2 ' - 6 "	63.0	-----	-----	63.0	10	40	-----	50	750
3 ' - 0 "	48.0	20.0	-----	68.0	-----	57	-----	57	900
3 ' - 6 "	36.0	39.0	-----	75.0	-----	57	7	64	1050
4 ' - 0 "	24.0	57.0	-----	81.0	-----	57	15	72	1200
4 ' - 6 "	12.0	78.0	-----	90.0	-----	57	22	79	1350
5 ' - 0 "	-----	94.0	-----	94.0	-----	57	29	86	1500
5 ' - 6 "	-----	94.0	7.5	101.5	-----	57	37	94	1650
6 ' - 0 "	-----	94.0	15.0	109.0	-----	57	44	101	1800
6 ' - 6 "	-----	94.0	23.0	117.0	-----	57	52	109	1950
7 ' - 0 "	-----	94.0	30.0	124.0	-----	57	59	116	2100
7 ' - 6 "	-----	94.0	39.0	133.0	-----	57	67	124	2250
8 ' - 0 "	-----	94.0	45.0	139.0	-----	57	74	131	2400
8 ' - 6 "	-----	94.0	53.0	147.0	-----	57	82	139	2550
9 ' - 0 "	-----	94.0	60.0	154.0	-----	57	89	146	2700
9 ' - 6 "	-----	94.0	68.0	162.0	-----	57	97	154	2850
10 ' - 0 "	-----	94.0	75.0	169.0	-----	57	104	161	3000

Fire Clay Tile Flue Lining								
Flue lining is estimated by the lineal foot. It is furnished in 2'-0" lengths in the following sizes:				Flue lining is estimated by the lineal meter. It is furnished in 600 mm lengths in the following sizes:				
Size of Flue Lining	No. of Lin.Ft. per 8-Hour Day	Hrs. To Set 100 Lin.Ft.		Size of Flue Lining	No. of Lin.Meters per 8-Hour Day	Hrs. To Set 30.48 Lin.M.		
		Mason	Labor			Mason	Labor	
4 " x 8 "	165 - 180	4.7	4.7	100 mm x 200 mm	50.3 - 54.9	4.7	4.7	
4 " x 12 "	130 - 150	5.7	5.7	100 mm x 300 mm	39.6 - 45.7	5.7	5.7	
8 " x 8 "	130 - 150	5.7	5.7	200 mm x 200 mm	39.6 - 45.7	5.7	5.7	
8 " x 12 "	105 - 125	7.0	7.0	200 mm x 300 mm	32.0 - 38.1	7.0	7.0	
12 " x 12 "	85 - 100	8.7	8.7	300 mm x 300 mm	25.9 - 30.5	8.7	8.7	
8 " x 18 "	75 - 90	9.7	9.7	200 mm x 450 mm	22.9 - 27.4	9.7	9.7	
12 " x 18 "	70 - 85	10.2	10.2	300 mm x 450 mm	21.3 - 25.9	10.2	10.2	
18 " x 18 "	60 - 70	12.3	12.3	450 mm x 450 mm	18.3 - 21.3	12.3	12.3	
20 " x 20 "	55 - 65	13.3	13.3	500 mm x 500 mm	16.8 - 19.8	13.3	13.3	
20 " x 24 "	50 - 60	14.0	14.0	500 mm x 600 mm	15.2 - 18.3	14.0	14.0	
24 " x 24 "	45 - 55	16.0	16.0	600 mm x 600 mm	13.7 - 16.8	16.0	16.0	
20 " x Round	55 - 65	13.3	13.3	500 mm x Round	16.8 - 19.8	13.3	13.3	

Approximate Prices of Tile Flue Lining

Size, Inches			Price per Lin.Ft.	Size, mm				Price per Meter
8 "	x	8 "	$ 3.30	200 mm	x	200 mm		$ 10.83
8 "	x	12 "	$ 4.45	200 mm	x	300 mm		$ 14.60
8 "	x	18 "	$ 9.10	200 mm	x	450 mm		$ 29.86
12 "	x	12 "	$ 5.60	300 mm	x	300 mm		$ 18.37
13 "	x	13 "	$ 5.60	325 mm	x	325 mm		$ 18.37
12 "	x	16 "	$ 9.45	300 mm	x	400 mm		$ 31.00
12 "	x	18 "	$ 9.95	300 mm	x	450 mm		$ 32.64
13 "	x	18 "	$ 9.95	325 mm	x	450 mm		$ 32.64
16 "	x	16 "	$ 13.25	400 mm	x	400 mm		$ 43.47
16 "	x	20 "	$ 19.15	400 mm	x	500 mm		$ 62.83
18 "	x	18 "	$ 14.70	450 mm	x	450 mm		$ 48.23
20 "	x	20 "	$ 23.00	500 mm	x	500 mm		$ 75.46
20 "	x	24 "	$ 25.10	500 mm	x	600 mm		$ 82.35
24 "	x	24 "	$ 29.60	600 mm	x	600 mm		$ 97.11
6 " Round			$ 4.00	150 mm Round				$ 13.12
6 " Round Thimble			$ 3.10	150 mm Round Thimble				$ 10.17
8 " Round			$ 4.45	200 mm Round				$ 14.60
8 " Round Thimble			$ 6.15	200 mm Round Thimble				$ 20.18
10 " Round			$ 6.25	250 mm Round				$ 20.51
10 " Round Thimble			$ 7.00	250 mm Round Thimble				$ 22.97
12 " Round			$ 6.60	300 mm Round				$ 21.65
12 " Round Thimble			$ 8.50	300 mm Round Thimble				$ 27.89
15 " Round			$ 13.85	375 mm Round				$ 45.44
18 " Round			$ 16.00	450 mm Round				$ 52.49
20 " Round			$ 23.40	500 mm Round				$ 76.77
24 " Round			$ 37.30	600 mm Round				$ 122.38

Labor Setting Terra Cotta Wall Coping Hours Required to Set 100 Lin.Ft.				Labor Setting Terra Cotta Wall Coping Hours Required to Set 30.48 Lin.Meters			
Size of Coping	No. Lin.Ft. Set per 8-Hr.Day	Mason Hours	Labor Hours	Size of Coping	No. Lin.Meters Set per 8-Hr.Day	Mason Hours	Labor Hours
9 "	145 - 160	5.2	5.2	225 mm	44.2 - 48.8	5.2	5.2
12 "	115 - 130	6.5	6.5	300 mm	35.1 - 39.6	6.5	6.5
18 "	100 - 120	7.3	7.3	450 mm	30.5 - 36.6	7.3	7.3

682

MASONRY

	Approx. Prices of Terra Cotta Wall Coping						
Size	Double Slant	Single Slant	Size		Double Slant	Single Slant	
	Price per lineal foot				Price per lineal meter		
9 "	$ 2.50	$ 2.75	225 mm		$8.20	$ 9.02	
13 "	$ 3.15	$ 5.00	325 mm		$10.33	$ 16.40	
18 "	$ 6.30	$ 12.00	450 mm		$20.67	$ 39.37	

Corners, ends, and starters are four times the price of straight coping; angles are six times the price of straight coping.

Brickwork for Incinerators

Incinerators have been used in residences, apartments, apartment hotels, hotels, hospitals, schools, and industrial plants for burning waste. The refuse might be fed into the chimney at each floor, dropped to the incinerator in the basement, or fed direct to the incinerator in the basement.

Practically all incinerators require a separate flue or stack, which should be estimated in the regular way, figuring fire brick, flue lining, and common brick as shown on the plans.

The number of fire and common brick required for the various sizes and types of incinerators will vary with the different manufacturers. However, the following data will assist the estimator in computing the number of brick and labor hours required to construct the brickwork for the different size incinerators, where they are not detailed on the plans.

If 4" (100 mm) outside walls are used, dimensions below will be reduced in length and height by 4" (100 mm) and in width by 8" (200 mm).

	Dimensions			# of Fire Brick	# of Common Brick	Minimum Size of Flue	Mason Hours	Labor Hours
Model	A	B	C					
"A"	6'- 0 "	3 '- 5 "	4 '- 0 "	300	900	12"x12"	20	14
"B"	5 '- 7 "	3 '- 11 "	4 '- 7 1/2 "	350	1100	14"x14"	24	18
"C"	6 '- 0 "	4 '- 3 "	4 '- 7 1/2 "	400	1300	16"x16"	28	20
"D"	6 '- 10 "	4 '- 3 "	5 '- 11 "	525	1500	20"x20"	33	24
"A"	1800 mm	1025 mm	1200 mm	300	900	300x300mm	20	14
"B"	1675 mm	1175 mm	1388 mm	350	1100	350x350mm	24	18
"C"	1800 mm	1275 mm	1388 mm	400	1300	400x400mm	28	20
"D"	2050 mm	1275 mm	1775 mm	525	1500	500x500mm	33	24

Radial Brick Chimneys

Radial brick chimneys are usually constructed by contractors who specialize in chimney construction. These contractors employ workers with years of experience in this class of work and can build them much more economically than the general contractor. Also, most constructors of radial brick chimneys have their own designs and shapes of radial brick,

683

manufactured especially for them, which are not available to the average contractor.

However, some information on this important subject is of value to all contractors. Radial brick chimneys are constructed of perforated radial brick formed to fit the circular and radial lines of the chimney. Bricks are molded with vertical perforations, and should be hard, well burned, acid proof, of necessary refractory powers and crushing strength, and maximum density.

The perforations serve to form a dead air space in the walls of the chimney that tends to prevent rapid heating and cooling of the walls by conserving the heat inside. Naturally, the higher the temperature of the gases the better the draft.

Four sizes of brick are ordinarily used in the construction of radial brick chimneys. All have the same face dimensions, approximately 6-1/2" (162.50 mm) wide by 4-1/2" (112.50 mm) high. The lengths of the blocks vary in order to make possible the breaking of the joints horizontally and vertically in the walls. The combination of bonds with this type of chimney provides for a lighter chimney than can be produced with ordinary building brick.

Radial chimney brick are generally either red or buff in color. Red radial brick are usually the most economical, because they are in greater demand and there are more sources of supply.

In laying the brick, the mortar is worked into the perforations, locking them together on the principle of a mortise and tenon joint. This produces the strongest bonded wall known to brick construction.

Size and Weights of Radial Brick for Radial Brick Chimneys					
Type Brick	Outside Face	Inside Face	Depth Inches	Height Inches	Wt. Lbs
No. 4	6 1/2 "	5 3/4 "	4	4 1/2	7
No. 6	6 1/2 "	5 1/2 "	6	4 1/2	9 1/2
No. 7	6 1/2 "	5 1/2 "	7	4 1/2	11
No. 8	6 1/2 "	5 1/2 "	8	4 1/2	12
Pier	9 "	9 "	4	4 1/2	9
Size and Weights of Radial Brick for Radial Brick Chimneys - Metric					
Type Brick	Outside Face	Inside Face	Depth mm	Height mm	Wt. Kg
No. 4	162.5 mm	143.8 mm	100	112.5	3.18
No. 6	162.5 mm	137.5 mm	150	112.5	4.31
No. 7	162.5 mm	137.5 mm	175	112.5	4.99
No. 8	162.5 mm	137.5 mm	200	112.5	5.44
Pier	225.0 mm	225.0 mm	100	112.5	4.08

Each radial brick with an outside face 6-1/2" (162.50 mm) wide and 4-1/2" (112.50 mm) high covers 29-1/4 sq. in. (18,872 mm2) of surface, or it

684

requires 5 (4.923) blocks per sq.ft. (0.09 sq.m) of wall, not including mortar joints. Allowing 1/2" (12.5 mm) for mortar joints, each brick covers 7"x 5" (175 x 125 mm) or 35 sq. in. (21,875 mm2) of surface, or it requires 4.12 blocks per sq.ft. (0.09 sq.m) of wall. Add about 5% for breakage and waste.

The walls of most radial chimneys taper about 1/4" per 1'-0" (6.25 mm per 300 mm) in height, and the thickness of the walls vary from the bottom to the top, depending upon the height. The wall thickness must be made up of a number of different depth units to provide the proper bond and structural strength. For example, a wall 16-1/4" (406.25 mm) thick might be made up of 1 block 7" (175 mm) deep and 2 blocks 4" (100 mm) deep; a wall 8-5/8" (215.6 mm) thick of 2 blocks 4" (100 mm) deep, etc.

Estimating Quantities of Radial Brick. Radial brick chimney work is frequently estimated by the ton, because wall thicknesses vary, depending on the height. A chimney 125'-0" (38.1 m) high might have walls of the following thicknesses:

Story	Height Feet	Thickness Inches	Height Meter	Thickness mm
First	25 '	16 1/4 "	7.6 M	406.3 mm
Second	20 '	13 1/4 "	6.1 M	331.3 mm
Third	20 '	11 3/4 "	6.1 M	293.8 mm
Fourth	20 '	10 1/4 "	6.1 M	256.3 mm
Fifth	20 '	8 5/8 "	6.1 M	215.6 mm
Top	20 '	7 1/8 "	6.1 M	178.1 mm

These wall thicknesses will be made up of radial brick of different depths to work out the correct wall thickness and proper bond:

Wall Thickness	Size of Units Used	Wgt. 29-1/4 Sq.In. Wall Lbs *	Wgt per s.f., lbs.*
20 " - 21 "	2 Pcs of 8", 1 Pc of 4"	31	128
18 " - 19 "	2 Pcs of 7", 1 Pc of 4"	29	120
16 " - 17 "	2 Pcs of 6", 1 Pc of 4" or 2 Pcs of 8"	24-26	99 - 107
14 " - 15 "	1 Pcs of 6", 2 Pc of 4" or 2 Pcs of 7"	22-23 1/2	91 - 97
12 " - 13 "	1 Pcs of 7", 1 Pc of 4", 1 Pc of 8" or 1 Pc of 4"	18-19	75 - 79
10 " - 11 "	1 Pcs of 6", 1 Pc of 4"	16 1/2	68
8 " - 9 "	2 Pcs of 4" or 1 Pc of 8"	12 1/4	50 - 58
7 " - 8 "	1 Pcs of 7" or 1 Pc of 8"	11-12	45 1/2 - 50

*Net weight of brick does not include mortar.

Wall Thickness	Size of Units Used	Wgt. .019 Sq.M. Wall	Wgt per .0929 Sq.M. Kg.
500 mm - 525 mm	2 Pcs of 200mm, 1 Pc of 100mm	14.1	0.0
450 mm - 475 mm	2 Pcs of 175mm, 1 Pc of 100mm	13.2	0.0
400 mm - 425 mm	2 Pcs of 150mm, 1 Pc of 100mm or 2 Pcs of 200mm	10.9 - 11.8	483.4 - 522.4
350 mm - 375 mm	1 Pcs of 150mm, 2 Pc of 100mm or 2 Pcs of 175mm	10.0 - 10.7	444.3 - 473.6
300 mm - 325 mm	1 Pcs of 175mm, 1 Pc of 100mm, 1 Pc of 200mm or 1 Pc of 100mm	8.2 - 8.6	366.2 - 385.7
250 mm - 275 mm	1 Pcs of 150mm, 1 Pc of 100mm	7.5	0.0
200 mm - 225 mm	2 Pcs of 100mm or 1 Pc of 200mm	5.6	244.1 - 283.2
175 mm - 200 mm	1 Pcs of 175mm or 1 Pc of 200mm	5.0 - 5.4	222.1 - 244.1

Figure radial brick chimney lining at about 30 lbs. per sq.ft. (146.5 kg per sq.m) of wall surface.

Mortar Required Laying Radial Brick. It requires about 2/3 ton or 1,334 lbs. (605.5 kg) of mortar to lay one ton of radial brick. Mortar is usually mixed in the proportions of 1 volume portland cement, 1 volume lime putty, and 4 to 6 volumes clean mortar sand.

Unit	A	B	C	D	Inside Radius	Approx. Ship'g Wt. Lbs.	Approx. % Voids
4 (N)	4 1/2 "	6 1/2 "	4 "	5 5/8 "	3 ' - 4 "	5.5	32
6 (A)	4 1/2 "	6 1/2 "	6 "	5 21/32 "	3 ' - 4 "	8.3	32
7 (B)	4 1/2 "	6 1/2 "	7 "	5 1/2 "	3 ' - 8 "	9.7	32
7- S.R. (B-S.R.)	4 1/2 "	6 1/2 "	7 "	4 15/16 "	2 ' - 0 "	9.7	32
8 (C)	4 1/2 "	6 1/2 "	8 "	5 1/2 "	3 ' - 9 "	10.8	32
8-S.R. (C-S.R.)	4 1/2 "	6 1/2 "	8 "	5 1/16 "	2 ' - 6 "	10.8	32
S (S)	4 1/2 "	8 "	4 "	8 "	Straight	7.0	32

Unit	A	B	C	D	Inside Radius	Approx. Ship'g Wt. Kgs.	Approx. % Voids
4 (N)	112.5 mm	162.5 mm	100 mm	140.6 mm	900 mm - 100 mm	2.5	32
6 (A)	112.5 mm	162.5 mm	150 mm	141.4 mm	900 mm - 100 mm	3.8	32
7 (B)	112.5 mm	162.5 mm	175 mm	137.5 mm	900 mm - 200 mm	4.4	32
7- S.R. (B-S.R.)	112.5 mm	162.5 mm	175 mm	123.4 mm	600 mm	4.4	32
8 (C)	112.5 mm	162.5 mm	200 mm	137.5 mm	900 mm - 225 mm	4.9	32
8-S.R. (C-S.R.)	112.5 mm	162.5 mm	200 mm	126.6 mm	600 mm - 150 mm	4.9	32
S (S)	112.5 mm	200.0 mm	100 mm	200.0 mm	Straight	3.2	32

Radial chimney brick is priced by the ton and will vary with type of brick and locality. Prices are f.o.b. factory, and the estimator must add freight and trucking from factory to job site and cost of unloading and storing brick at job.

Labor Laying Brick in Radial Brick Chimneys. The labor cost of laying radial brick chimneys will vary with the height of the chimney, thickness of walls, etc., but the following are actual production times on six radial brick chimneys constructed.

686

MASONRY

Labor Hours per Ton of Brick				
Description	Tons of Radial Brick	Mason Hours	Labor Hours	Hoist Eng. Hrs.
125'-0" high, Walls 16-1/4" to 7-1/8" thick. 12'-9" diam. at bottom, 7'-0" at top. Lined 50'-0"	157	2.7	4.7	0.9
80'-0" high, walls 11-3/4" to 7-1/8" thick. 8'-10" diam. at bottom, 5'-9" at top. Lined 35'-0"	66.5	3.1	5.3	1
60'-0" high, walls 10-1/4" to 7-1/8" thick. 7'-2" diam. at bottom, 4'-4" at top. Lined 21'-0"	31.1	4.7	8.2	1.9
46'-0" high, walls 10-1/4" to 7-1/8" thick. 6'-6" diam. at bottom, 4'-2" at top. Unlined	24.6	3.3	8	1.5
46'-0" high walls 10-1/4" to 7-1/8" thick. 6'-6" diam. at bottom, 4'-2" at top. Unlined (2)	49.2	3	5	1.5
55'-0" high, walls 10-1/4" to 7-1/8" thick. 6'-11" diam. at bottom, 4'-2" at top. Lined 20'-0"	31.6	3	5	1.5
Labor Hours per Metric Ton of Brick				
38.1 M high, Walls 406 mm to 178.1 mm thick. 3.9 M diam. at bottom, 2.1 M at top. Lined 15.2 M	159.5	2.7	4.7	0.9
24.4M high, walls 293.8 mm to 178.1 mm thick. 2.7 M diam. at bottom, 1.8 M at top. Lined 10.7 M	67.6	3.1	5.3	1
18.3 M high, walls 256.3 mm to 178.1 mm thick. 2.2 M diam. at bottom, 1.3 M at top. Lined 6.4 M	31.6	4.7	8.2	1.9
14.2 M high, walls 256.3 mm to 178.1 mm thick. 2.0 M diam. at bottom, 1.3 M at top. Unlined	25.0	3.3	8	1.5
14.2 M high, walls 256.3 mm to 178.1 mm thick. 2.0 M diam. at bottom, 1.3 M at top. Unlined (2)	50.0	3	5	1.5
16.7 M high, walls 256.3 mm to 178.1 mm thick. 2.1 M diam. at bottom, 1.3 M at top. Lined 6.1 M	32.1	3	5	1.5

The above does not include supervision time.

Labor time includes unloading materials from cars into trucks and unloading trucks at site. All mortar joints on the above stacks were struck flush with point of trowel.

Where stacks are lined, labor includes lining with perforated radial fire clay brick laid in high temperature cement. The lining is laid up from inside after stack is topped out.

Miscellaneous Masonry Costs

Labor Costs on High Buildings. On buildings over 10 to 12 stories high, it will be necessary to add an extra allowance for labor handling and hoisting the bricks and mortar. The higher the building the longer it takes the hoist to make a trip from the street level to the floor on which the bricks are being laid and return, extra work raising and handling scaffolding, etc.

Taking a 12-story building, or 120' (36.57 m) in height, as normal or 100%, add about 1% to the labor cost for each story or each 10' (3.05 m) feet in height above 120' (36.57 m).

For instance, a 12-story building or 120' (36.57 m) in height is 100, the 13th story would be 101; the 14th story, 102; the 15th story, 103, etc. To find the average cost on a 15-story building, figure as follows:

The first 12 stories as 100 each or	1,200
The 13th story as 101	101
The 14th story as 102	102
The 15th story as 103	103
Total for 15 stories or 150' (45.7 m) in height	1,506

Dividing 1,506 by 15 gives 1.004 as the average cost of the masonry labor for a 15-story building. Other story heights would be computed in the same manner. The following table applies to all classes of masonry work on high structures.

Height in Stories	Normal Base	Add Above Normal %	Average for Project
1 to 12	100	None	100.00
13	100	1%	100.08
14	100	2%	100.22
15	100	3%	100.40
16	100	4%	100.63
18	100	6%	101.17
20	100	8%	101.80
22	100	10%	102.50
24	100	12%	103.25
26	100	14%	104.04
28	100	16%	104.86
30	100	18%	105.70
40	100	28%	110.15
50	100	38%	114.82

MASONRY

Brick Catch Basins and Manholes

When laying building or sewer brick in catch basins and manholes from 3'-0" to 5'-0" (0.9-1.5 m) in diameter and having 8" to 12" (200-300 mm) walls, a mason should lay 600 to 750 brick per 8-hr. day, at the following labor cost per 1,000:

Catch Basin - Manholes					
	Hours	Rate	Total	Rate	Total
Mason	11.5	$....	$....	$36.93	$424.70
Labor	11.5	$....	$....	$ 26.16	$ 300.84
Cost per 1,000 brick					$725.54

Pointing Around Steel Sash With Cement Mortar. All of the steel windows in a three story office building were pointed with portland cement mortar between the sash and the brick jambs after they had been set.

There were 39 windows 4'-0"x4'-10" (1.22 x 1.47 m) and 96 windows 4'-0"x7'-0" (1.22 x 2.13 m), making a total of 2,800 lin.ff. (853.44 m) that were pointed at the following labor cost per 100 lin.ft. (30.48 m):

Pointing Around Sash					
	Hours	Rate	Total	Rate	Total
Bricklayer	1.1	$....	$....	$36.93	$40.62
Labor	0.5	$....	$....	$26.16	$13.08
Cost per 100 l.f.					$53.70
per l.f.					$0.54
per m					$1.76

Reinforced Brick Masonry

Reinforced brick masonry has been widely used on the west coast because of its resistance to the lateral forces produced by earthquakes. It is now being used throughout the country, because most areas have been classified as seismic. It is also used where high winds or blast conditions may occur and for retaining walls.

These walls are comparable to reinforced concrete construction with the masonry units serving as a form. The exterior wythes are built up in a conventional manner, the reinforcing is set in the interior joints, and the whole is made homogeneous by filling the interior joints with grout.

No special skills or techniques are needed on the part of the mason, but certain care must be taken, including seeing that there is adequate lime in the mortar; a sound base to start on with the aggregate in concrete exposed and wetted to saturation; that brick is wetted so grout will not dry out too quickly; that grout is sufficiently fluid so none will adhere to a trowel; that grout be poured in a layer of about three bricks high and be allowed to set for 15 minutes before another pour is made, and poured from the inside to avoid splashing the exposed surfaces; and that mortar cuts not be spilled into grout or grout allowed to accumulate on bed joints.

The amount of reinforcing will vary widely with the job. In lightly reinforced walls, the reinforcing will fit within the widths of normal brick walls. The more heavily reinforced walls may require wider inner joints or the use of soap courses to accommodate the steel. Headers are not used in reinforced walls, and the contractor should check the drawings to see if header rows, for aesthetic reasons, are shown, because it will require cutting all these brick. This is especially true when figuring reinforced lintels in conventional walls so that coursing can carry through.

Savings, as well as aesthetic advantages, often can be realized through the use of reinforced brick lintels. These are almost always poured in place with temporary shoring for the soffit. The economies result from savings in steel cost and elimination of painting. In better work it may call for the soffit brick joints to be filled temporarily with sand and then repointed later.

A variation on the above has been developed and is known as high lift grouted reinforced brick masonry. It varies in that it is essentially a cavity wall with a 2" (50 mm) or better inner joint. This allows the reinforcing to be preset and grouting delayed until a full 12' (3.65 m) wall height is reached. Mechanical pumping can thus be used.

Precautions listed for standard reinforced walls also apply here with the following additions. Care must be taken to allow cleaning out of grout space by omitting every other unit on one side; brick must be allowed to set for at least 3 days before grouting; vibrating of grout is necessary; once grout pour is started, it must be continuous; and construction dams may be built every 20 or 25 feet (6.09 or 7.62 meters) to contain the grout laterally.

Handling of Masonry By Palletization

For years, builders considered using palletization for handling masonry materials, and many efforts were made in this direction with mixed results. Now, new systems have overcome most of the problems. Palletized masonry units, whether brick, block, or tile, can be transported from the manufacturing plant, through the job site, up the hoist, and around the scaffold to the mason's station, eliminating all hand handling of units until the mason picks them up for laying. Two companies that manufacture this equipment are the Lull Division of JLG Industries, Inc and Pettibone LLC.

Time studies on over 300 jobs using the one manufacturer's study indicates that with ten or more masons, the conventional ratio of 0.7 to 1.0 labor hour per mason hour can be reduced to 0.3 to 0.5 labor hours per mason hour—a savings of approximately 50% in labor time required for tending masons, or an overall 20% savings on the total masonry labor cost. These ratios include the labor time required to mix mortar, tend mortar, brick, tile, or block, and remove rubbish.

690

The success of these systems is mainly due to the equipment, which consists of properly sized wood pallets, a tractor with specially design forklift mechanism, special brick buggies, and special mortar buggies.

For efficient use of this system and equipment, certain job conditions must be established from the start and maintained throughout the life of the job.

All materials must be purchased on the basis of pallet delivery or palletized as soon as they arrive on the job. In the past, contractors were furnished the wood pallets, but dealers and manufacturers have shown an increasing preference for this type of delivery and in some cases furnish their own pallets. The contractor should specify the quantity of material to be placed on each pallet and how it is arranged. Weight should not exceed 1,000 lbs. (453 kg) per pallet to stay within the capacity of scaffolding and buggies.

The effect on material prices for pallet delivery varies. Some suppliers maintain the same prices, claiming the extra cost of labor and material for palletization is offset by savings in loading and unloading. Other suppliers increase brick prices $3.00 to $5.75 per thousand. Actual cost of a 24"x32" (600 x 800 mm) pallet, for palletization on the job site, is about $5.00 per pallet. One laborer can handle 6 to 7 pallets per hour.

Access roads to the job should be laid out and maintained to facilitate operation of the fork lift tractor. Area surrounding the building should be backfilled, consolidated, and graded before masonry work starts and kept reasonably clear during the work.

Exterior scaffolding must be wide enough to permit maneuvering brick and mortar buggies. Where tubular steel scaffolding is used, this may be accomplished with standard scaffolding components, using a 5' (1.52 m) width tubular scaffold with the addition of a 30" (750 mm) bracket on the inside for the masons and a 20" (500 mm) bracket on the outside for the extra width required for buggies and storage of pallets. No special planking is required, but plank laps should all run in the same direction and a wood easement strip placed at each lap to partially eliminate the irregularity.

Where suspended scaffolding is used, extra width can be obtained by using 8' (2.4 m) wide Wheeling scaffold machines with the addition of 10' (3.05 m) pipe putlogs fastened directly under and through scaffold planking with U-bolts. This method moves the inside drum and cable 2'-0" (0.61 m) out from the wall and creates a clear working space for the mason.

Scaffolding costs will be increased 5% to 15%, depending upon the type of scaffold used.

Standard Pallet 32"x24" (800 x 600 mm). A standard pallet size of 32" x 24" (800 x 600 mm) has been chosen, because it accommodates all common sizes of brick, block and tile, its loaded weight does not exceed safety regulations for scaffolding or light floor construction, and it accommodates just enough material to build 10 lin.ft. (3.05 m) of wall, 4' (1.22 m) high, 4" (100 mm) thick, allowing proper spacing for stacking on the scaffold according to standard masonry practice. In addition, its loaded

691

weight can be handled by one worker using hand powered equipment, and the pallet load can be split into two 16" x 24" (400 x 600 mm) pallet loads, permitting passage through normal doorways, handling on occasional narrow scaffolds and maneuvering in confined spaces.

Pallets should be made to withstand hard use, preferably from a good grade of 1" (25 mm) lumber—1"x8" (25 x 200 mm) boards for the tops and 2"x4" (50 x 100 mm) for skids. Pallets made on the job, using power saws and a production line set-up, should cost about $2.25 per half pallet—16"x24" (400 x 600 mm)—each, including material and labor.

One standard full size pallet is composed of two 16x24s (400 x 600 mm) strapped together with metal banding. When required, the metal banding can be removed, breaking the package into two halves of 16"x24" (400 x 600 mm) for use where operating space is limited.

Forklift Tractors. Manufacturers provide various capacities and sizes of rubber-tired forklifts, with lifting heights from 18'-6" to 30'-6" (5.64-9.30 m) and with capacities from 2,500 to 4,000 lbs. (1134-1814 kg). The 3,000 lbs. (1361 kg) unit can unload and stockpile as many as 4,000 brick in 10 minutes, and can be used to handle concrete planks and roofing materials as well as masonry.

The forklift tractor unloads pallets of material from delivery trucks and places them in stockpiles. From stockpiles, pallets are then transported to the masons' stations at ground level or on the scaffold up to maximum reach of tractor; to construction hoist for work on scaffolds or floors beyond tractor reach; or to building floors within tractor reach for partition work or exterior masonry performed from inside the scaffold.

The forklift can handle as much palletized masonry material as can 20 laborers using conventional hand methods and wheelbarrows. It can also be equipped with other attachments to perform many other tasks on the job.

The fork lift also delivers mortar to the bricklayers from the mixing point by use of mortar buggies, which are mortar containers with a capacity of 7 cu.ft. (0.19 cu.m), equipped with wheels and casters so that they can be readily pushed on floors or scaffolds. The tractor handles these mortar containers to and from the mixer the same as packaged materials.

Brick Buggies. Pallet loads are handled on the scaffold and building floors with brick buggies. These come in a variety of sizes, either hand propelled or power driven, with capacities of 1,000 lbs. (453 kg) for half loads-16"x24" (400 x 600 mm) pallets with 120 bricks—and 1,500 lbs. (680 kg) for full loads—24"x32" (600 x 800 mm) pallets with 240 bricks. Lifting forks can be had for 10" (250 mm), 7'-6" (2.28 m), and 10'-0" (3.05 m) lifts, the latter being able to feed double deck scaffolding. Buggies can be operated under average outdoor conditions as well as on up and down ramps.

Lift trucks for 24"x32" (600 x 800 mm) pallets cost about $6,000 and can service 4 to 6 masons on a medium-rise building, as opposed to one laborer keeping each mason supplied with brick. Two laborers, one on the ground and one delivering, should serve 5 masons. Under good conditions, the daily savings in labor cost should amortize the lift and pallet costs in approximately 5 to 6 weeks. Such handling also cuts down the time the hoist must be held for unloading by hand, which can be critical on larger jobs.

04220 CONCRETE UNIT MASONRY

Concrete masonry units, commonly termed concrete block or concrete brick, are used extensively in all parts of the country for exterior and interior bearing walls, interior partitions, floor and roof fillers, and the like. Much of the information and data in this section comes from the National Concrete Masonry Association.

Concrete masonry units are widely used for backing up veneers. They are used both for bearing walls on wall bearing structures and for curtain walls in steel or reinforced concrete frame structures. The use of concrete masonry has increased greatly. This is partly due to the lower labor cost of handling and setting the larger concrete units in comparison with building brick, when used as a back-up material, but also because of increased use of architectural facing units.

The term concrete masonry is applied to block, brick, or tile building units molded from concrete and laid by masons into a wall. The concrete is made by mixing portland cement with water and other suitable materials, such as sand, gravel, crushed stone, burned clay or shale, blast furnace slag, and pumice.

Concrete masonry units should always be manufactured in a cement products plant where facilities for manufacturing and curing are uniform so as to obtain high quality products. The quality of all concrete masonry units should conform to the standards set by the American Society for Testing and Materials (ASTM) in its standard specifications for concrete masonry units and concrete brick.

Weight of Concrete Masonry. A 7-5/8"x7-5/8"x15-5/8" (190.62 x 190.62 x 390.62 mm) hollow loadbearing concrete block weighs from 35 to 45 lbs. (15.87-20.41 kg) when made from heavyweight aggregate and from 25 to 35 lbs. (11.34-15.87 kg) when made from lightweight aggregate. Heavyweight aggregates are sand, gravel, crushed stone, air-cooled slag, etc. Lightweight aggregates are coal cinders, expanded shale, clay or slag, pumice, etc.

Estimating Quantities of Concrete Brick. Concrete brick are manufactured, throughout the country generally, in the modular size of 2-1/4"x3-5/8"x7-5/8" (56.25 x 90.62 x 190.62 mm). In some localities, additional sizes are available as follows: Jumbo brick, 3-5/8"x3-5/8"x7-5/8" (90.62 x 90.62 x 190.62 mm); double brick, 4-7/8"x3-5/8"x7-5/8" (115.62 x 90.62 x 190.62 mm); and Roman brick, 1-5/8"x3-5/8"x11-5/8" (40.62 x 90.62 x 290.62 mm) and 1-5/8"x3-5/8"x15-5/8" (40.62 x 90.62 x 181.64).

They should be estimated by the sq. ft. (sq. meter) of wall of any thickness or by the cu. ft. (cu. meter), if all wall thicknesses are combined. For instance, if there are only 8" (200 mm) walls in the job, it is acceptable to take the number of sq.ft. (sq.m) of walls, but if the job contains 8" (200 mm), 12" (300 mm), and 16" (400 mm) walls, then it will be easier and more satisfactory to reduce the quantities to cu. ft. (cu.m) and multiply by a certain number of brick per cu. ft. (cu.m) of wall.

The number of concrete brick required per sq. ft. or cu. ft. (sq.m or cu.m) of wall will vary with the size of brick and width of mortar joint used. Concrete brick sizes and widths of mortar joints have been standardized so that they will lay up to an even multiple of the 4-inch (100-mm) module.

All concrete brick units are 3-5/8" (90.62 mm) thick, which together with a 3/8" (9.37 mm) mortar back joint equal the modular thickness of 4 inches (100 mm). Since there is always one less mortar joint than brick, in walls of multiple brick thickness, the actual wall thickness will always be 3/8" (9.6 mm) less than the modular 4 inches; i.e. 3-5/8" (90.62 mm), 7-5/8" (190.62 mm), 11-5/8" (290.62 mm), 15-5/8" (390.62 mm), etc.

In drawing plans, some architects give actual wall thickness dimensions, while others use nominal dimensions. For estimating purposes, nominal dimensions should be used, i.e. 4" (100 mm), 8" (200 mm), 12" (300 mm), 16" (400 mm), etc., to facilitate computations. When measuring the plans and listing the quantities of brick required, always take exact measurements for length and height and deduct all openings in full. Do not count corners twice but have your estimate show as accurately as possible the actual number of brick required to complete the job. For unexposed backup work, add 1% to 2% for waste. For exposed face work, add 3% to 5% for waste.

Modular Size Concrete Brick

Modular size concrete brick, 2-1/4"x3-5/8"x7-5/8" (56.25 x 90.62 x 190.62 mm) are laid so that 3 courses of brick, with 5/12" (10.41 mm) mortar joints, builds 8" (200 mm) in height, and the 7-5/8" (190.6 mm) length, plus 3/8" (9.37 mm) mortar joint, makes a total length of 8" (200 mm) laid in the wall. In other words, 3 brick lay 8"x8" (200 x 200 mm), or 64 sq. in. (40,000 mm2), so it requires 6-3/4 or 6.75 brick per sq.ft. (0.09 sq.m) of wall of nominal 4" (100 mm) thickness.

Jumbo brick are 3-5/8"x3-5/8"x7-5/8" (90.62 x 90.62 x 190.62 mm) in size and are laid so that 3 courses of brick, with 3/8" (9.37 mm) mortar joints, builds 12" (300 mm) in height, and the 7-5/8" (190.62 mm) length, plus 3/8" (15.6 mm) mortar joint, makes a total length of 8" (200 mm) laid in the wall. One jumbo brick lays 32 sq. in. in the wall or it requires 4.5 brick per sq.ft. (0.09 sq.m) of wall of nominal 4" (100 mm) thickness.

Double brick are 4-7/8"x3-5/8"x7-5/8" (121.87 x 90.62 x 190.62 mm) in size, so that 3 courses of brick, with 11/24" (11.18 mm) mortar joint builds 16" (400 mm) high, and the 7-5/8" (190.62 mm) length, plus 3/8" (9.37 mm) mortar joint, makes a total length of 8" (200 mm) in the wall, or one brick lays 42.67 sq. in. in the wall, or it requires 3.4 brick per sq.ft. (0.09 sq.m) of wall of nominal 4" (100 mm) thickness.

MASONRY

Roman size brick are 1-5/8"x3-5/8"x11-5/8" (40.62 x 90.62 x 290.62 mm) and 1-5/8"x3-5/8"x15-5/8" (40.62 x 90.62 x 390.62) and are laid so that 6 courses of brick, with 3/8" (9.37 mm) mortar joints, builds 12" (300 mm) in height, and the 11-5/8" (290.62 mm) or 15-5/8" (390.62 mm) lengths, plus 3/8" (9.37 mm) mortar joints makes a total length of 12" (300 mm) or 16" (400 mm) respectively laid in the wall. One nominal 12" (300 mm) brick lays 24 sq. in. in the wall or it requires 6 brick per sq.ft. (0.09 sq.m) of wall of nominal 4" (100 mm) thickness. One nominal 16" (400 mm) brick lays 32 sq. in. in the wall or it requires 4.5 brick per sq.ft. (0.09 sq.m) of wall of nominal 4" (100 mm) thickness.

The number of modular size brick per sq.ft. (0.09 sq.m) of wall of various thicknesses is as follows:

Number of Modular Size Brick per Sq.Ft. of Wall of Various Thickness					
per s.f.			Thickness of Wall		
			1-Brick (4")	2-Bricks (8")	3-Bricks (12")*
2 1/4 " x	3 5/8 " x	7 5/8 "	6.75	13.50	20.25
3 5/8 " x	3 5/8 " x	7 5/8 "	4.50	9.00	13.50
4 7/8 " x	3 5/8 " x	7 5/8 "	3.40	6.80	10.20
1 5/8 " x	3 5/8 " x	11 5/8 "	6.00	12.00	18.00
1 5/8 " x	3 5/8 " x	15 5/8 "	4.50	9.00	13.50

Number of Modular Size Brick per Sq.Meter of Wall of Various Thickness					
per sq.m			Thickness of Wall		
			1-Brick (100mm)	2-Bricks (200mm)	3-Bricks (300mm)*
56.3 mm x	90.6 mm x	190.6 mm	72.66	145.32	217.98
90.6 mm x	90.6 mm x	190.6 mm	48.44	96.88	145.32
121.9 mm x	90.6 mm x	190.6 mm	36.60	73.20	109.80
40.6 mm x	90.6 mm x	290.6 mm	64.59	129.17	193.76
40.6 mm x	90.6 mm x	390.6 mm	48.44	96.88	145.32

*Use this column for number of brick required per c.f. (0.028 cu.m) of wall.
No allowance is included for waste in above quantities.

Labor Handling and Laying Concrete Brick

The labor cost of handling and laying concrete brick varies, depending on factors such as size of brick, kind of mortar, class of work, thickness of walls, and number of openings.

On buildings having long straight walls without many openings, such as basement walls, garages, and factories, a mason should lay 800 to 900 modular size concrete brick, 2-1/4"x3-5/8"x7-5/8" (56.25 x 90.62 x 190.62), per 8-hr. day, while on dwellings and other structures having 8"

695

(200 mm) walls, cut up with numerous openings and pilasters, a mason should lay 650 to 700 brick per 8-hr. day.

Labor Cost of 1,000 Modular Size Concrete Brick 2 1/4" x 3 5/8' x 7 5/8" Laid in 8" Walls (56.3 mm x 90.6 mm x 190.6 mm laid in 200 mm Walls)					
	Hours	Rate	Total	Rate	Total
Mason	11.8	$....	$....	$36.93	$435.77
Labor	8.8	$....	$....	$ 26.16	$ 230.21
Cost per 1,000 brick					$665.98

The above costs are based on flush cut mortar joints. For struck joints, add 1/2-hr. mason time per 1,000 brick.

Estimate the labor cost of laying concrete face brick the same as given for face brick under Unit Masonry/Brick.

On similar work, using jumbo brick, 3-5/8"x3-5/8"x7-5/8" (90.62 x 90.62 x 190.62 mm), a mason should lay 675 to 775 brick per 8-hr. day in 12" (300 mm) walls and 550 to 600 brick per day in 8" (200 mm) walls. Using double brick, 4-7/8"x3-5/8"x7-5/8" (121.87 x 90.62 x 190.62 mm), a mason should lay 450 to 550 brick per 8-hr. day in 12" (300 mm) walls and 375 to 425 brick per day in 8" (200 mm) walls.

Roman size concrete brick, 1-5/8"x3-5/8"x11-5/8" (40.62 x 90.62 x 290.62 mm) or 1-5/8"x3-5/8"x15-5/8" (40.62 x 90.62 x 390.62 mm), are used only for facing or in 8" (200 mm) walls. The labor cost of handling and laying these brick will vary considerably, depending upon the type of joint treatment, flush cut, V-tooled or concave, struck or weathered, or raked out.

On average work with flush cut joints, a mason should lay 300 to 350 nominal 12" (300 mm) Roman brick per 8-hr. day or 225 to 265 nominal 16" (400 mm) Roman brick per 8-hr. day. For tooled, struck, weathered, or raked out joints, reduce by 8% to 12% the above quantities.

Concrete Blocks and Partition Units

Concrete blocks and partition units are manufactured in weights of concrete. Specific information on density of concrete types and on weights of units should be obtained from local manufacturers. When specific information is not available, assume the following values for products:

Type	Lbs. per Cu.Ft.	Kg per Cu Meter
Pumice	75	1201.3
Expanded shale concrete	85	1361.4
Expanded slag and cinder concrete	95	1521.6
Air-cooled slag concrete	120	1922.0
Crushed stone and gravel concrete	135-145	2162.3 - 2322.5

696

MASONRY

These types of units may be used interchangeably for all purposes, although lightweight units afford a savings of weight to secure economy in design, and they afford increased heat and sound insulation value.

Estimating the Quantity of Concrete Blocks and Partition Units. Concrete blocks and partition units should be estimated by the square foot of wall of any thickness and then multiplied by the number of blocks per 100 sq.ft. (9.29 sq.m) as given in the following tables.

Always take exact measurements and make deductions in full for all openings, regardless of size. The result will be the actual number of sq. ft. (sq.m) or number of blocks required for the job.

Mortar. The same kind of mortar should be used in laying concrete blocks as given for Modular Size Concrete Brick.

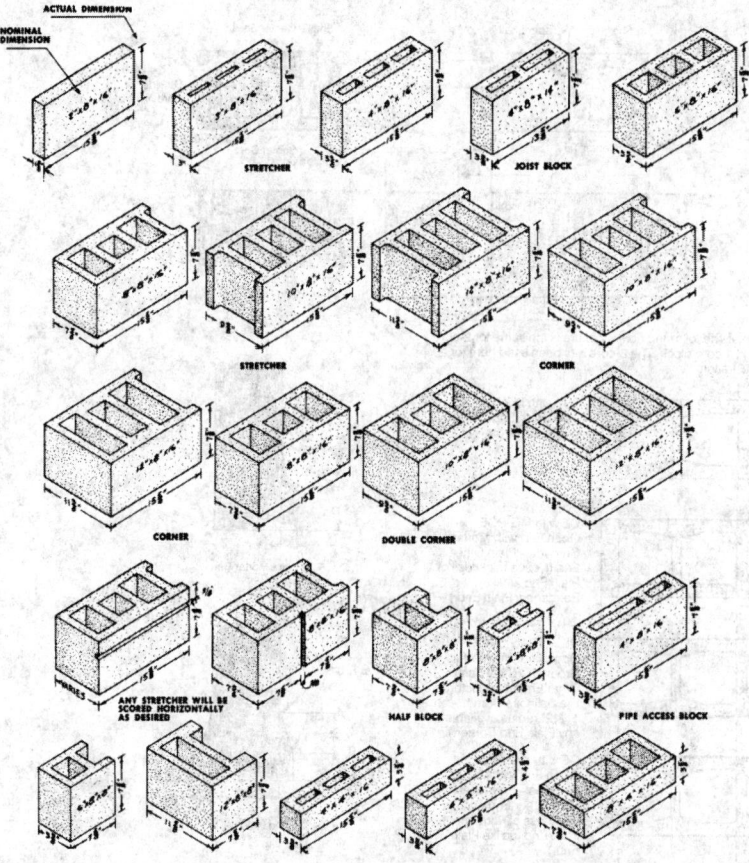

Sizes and Shapes of Concrete Building Units

When laying up exterior walls, face shell bedding shall be used with complete coverage of face shells. Furrowing of the mortar shall not be

697

permitted. Extruded mortar shall be cut off flush with face of wall and the joints firmly compacted, after the mortar has stiffened somewhat.

Tooling is essential in producing tight mortar joints. Mortar has a tendency to shrink slightly and may pull away from the masonry units causing fine, almost invisible cracks at the junction of mortar and masonry units.

Sizes and Weights of Concrete Masonry Units

Concrete masonry units are usually made with standard modular face dimensions 7-5/8" (190.62 mm) high and 15-5/8" (390.62 m) long and are available in thicknesses 3/8" (9.37 mm) less than the nominal 3", 4", 6", 8", 10", and 12" (75, 100, 150, 200, and 300 mm) thickness.

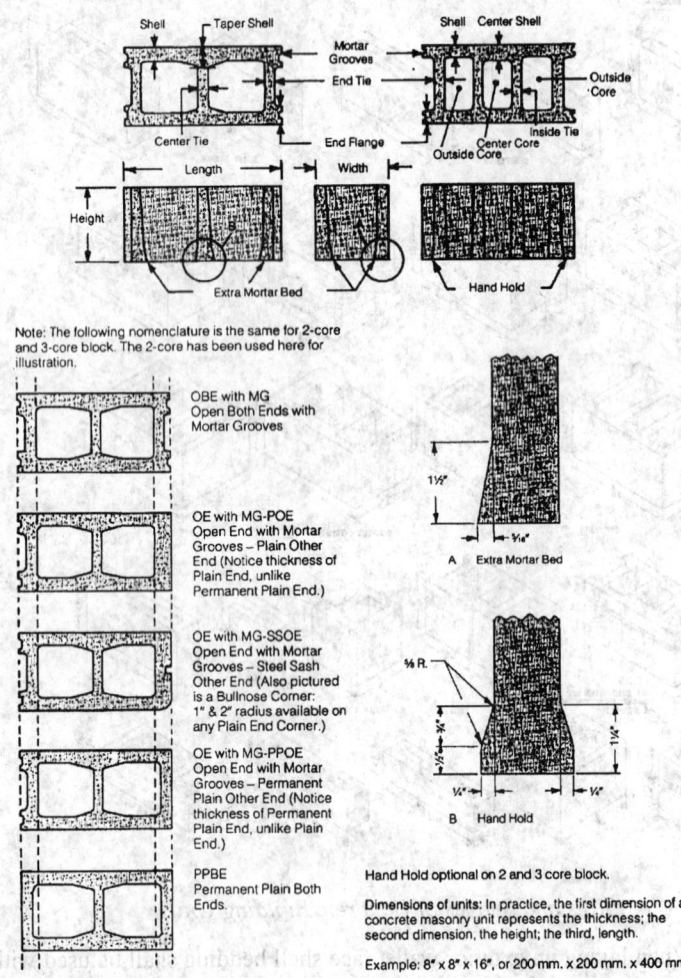

Note: The following nomenclature is the same for 2-core and 3-core block. The 2-core has been used here for illustration.

OBE with MG
Open Both Ends with Mortar Grooves

OE with MG-POE
Open End with Mortar Grooves – Plain Other End (Notice thickness of Plain End, unlike Permanent Plain End.)

OE with MG-SSOE
Open End with Mortar Grooves – Steel Sash Other End (Also pictured is a Bullnose Corner: 1" & 2" radius available on any Plain End Corner.)

OE with MG-PPOE
Open End with Mortar Grooves – Permanent Plain Other End (Notice thickness of Permanent Plain End, unlike Plain End.)

PPBE
Permanent Plain Both Ends.

A   Extra Mortar Bed

B   Hand Hold

Hand Hold optional on 2 and 3 core block.

Dimensions of units: In practice, the first dimension of a concrete masonry unit represents the thickness; the second dimension, the height; the third, length.

Example: 8" x 8" x 16", or 200 mm. x 200 mm. x 400 mm.

698

When laid up with 3/8" (15.6 mm) mortar joints, the units are 8" (200 mm) high and 16" (400 mm) long, requiring 112.5 units per 100 sq.ft. (9.29 sq.m) of wall, not including allowance for waste and breakage.

Hollow and solid partition and furring units have nominal thickness of 2", 3", 4", 6", and 8" (51, 76, 102, 152, and 203 mm).

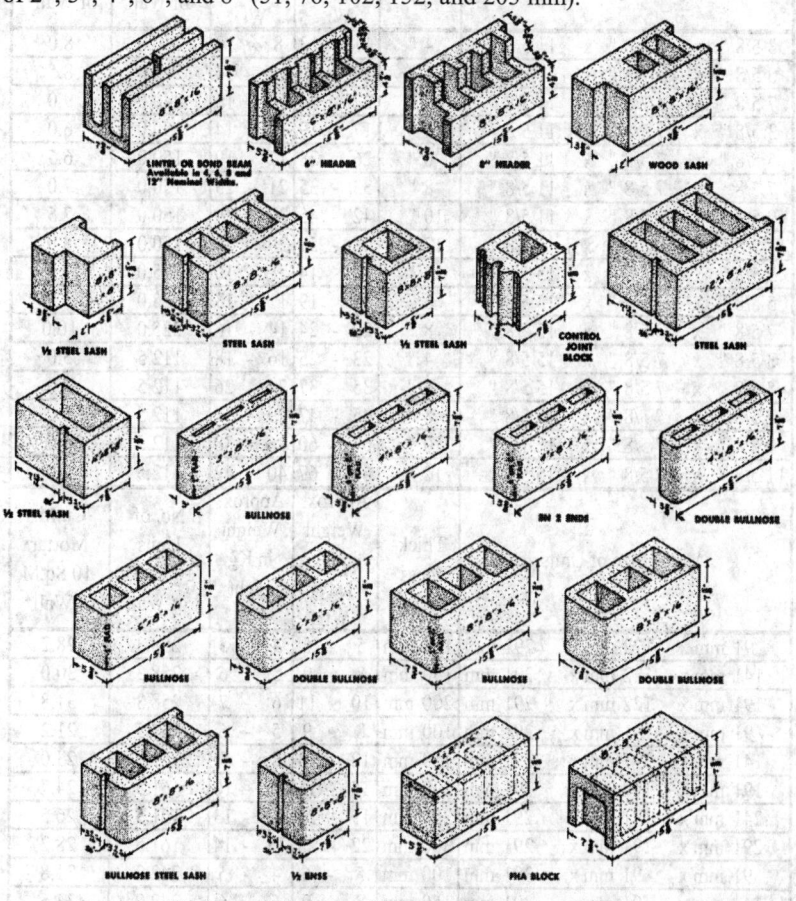

Sizes and Shapes of Concrete Building Units

Hollow and solid loadbearing units have nominal thickness of 4", 6", 8", 10", and 12" (102, 152, 203, 254, and 305 mm) and are also available in half lengths.

Standard specials such as steel and wood sash jambs, bullnose, and closures are also available in full and half lengths. There are also many sizes available on special order.

Sizes, Weights and Quantities of Loadbearing Concrete Blocks and Tile

Actual Size of Units				Thick-ness	Approx. Weight in Lbs. Heavy Wt.	Approx. Wt. in Lbs. Light Wt.	No. of Units 100 Sq.Ft. of Wall	Cu.Ft. Mortar 100 Sq.Ft. of Wall*
3 5/8 "	x	4 7/8 "	x	11 5/8 "	4 "	11 - 13	8 - 10	240.0	8.0
5 5/8 "	x	4 7/8 "	x	11 5/8 "	6 "	17 - 19	12 - 14	240.0	8.5
7 5/8 "	x	4 7/8 "	x	11 5/8 "	8 "	22 - 24	14 - 16	240.0	9.0
3 5/8 "	x	7 5/8 "	x	11 5/8 "	4 "	17 - 19	12 - 14	150.0	6.0
5 5/8 "	x	7 5/8 "	x	11 5/8 "	6 "	26 - 28	17 - 19	150.0	6.5
7 5/8 "	x	7 5/8 "	x	11 5/8 "	8 "	33 - 35	21 - 23	150.0	7.0
9 5/8 "	x	7 5/8 "	x	11 5/8 "	10 "	42 - 45	27 - 29	150.0	7.5
11 5/8 "	x	7 5/8 "	x	11 5/8 "	12 "	48 - 51	29 - 31	150.0	8.0
3 5/8 "	x	3 5/8 "	x	15 5/8 "	4 "	12 - 14	9 - 10	225.0	9.0
5 5/8 "	x	3 5/8 "	x	15 5/8 "	6 "	17 - 19	11 - 13	225.0	9.5
7 5/8 "	x	3 5/8 "	x	15 5/8 "	8 "	22 - 24	14 - 16	225.0	10.0
3 5/8 "	x	7 5/8 "	x	15 5/8 "	4 "	23 - 25	16 - 18	112.5	5.0
5 5/8 "	x	7 5/8 "	x	15 5/8 "	6 "	35 - 37	24 - 26	112.5	5.0
7 5/8 "	x	7 5/8 "	x	15 5/8 "	8 "	45 - 47	29 - 31	112.5	6.0
9 5/8 "	x	7 5/8 "	x	15 1/2 "	10 "	57 - 60	36 - 38	112.5	6.5
11 5/8 "	x	7 5/8 "	x	15 5/8 "	12 "	64 - 67	40 - 42	112.5	7.0

Actual Size of Units				Thick-ness	Approx. Weight in Kg Heavy Wt.	Approx. Weight in Kg Light Wt.	No. of Units 10 Sq.M. of Wall	Cu.M. Mortar 10 Sq.M of Wall*
91 mm x	122 mm x		291 mm	100 mm	5 - 6	4 - 5	258.3	28.2
141 mm x	122 mm x		291 mm	150 mm	8 - 9	5 - 6	258.3	30.0
191 mm x	122 mm x		291 mm	200 mm	10 - 11	6 - 7	258.3	31.8
91 mm x	191 mm x		291 mm	100 mm	8 - 9	5 - 6	161.5	21.2
141 mm x	191 mm x		291 mm	150 mm	12 - 13	8 - 9	161.5	23.0
191 mm x	191 mm x		291 mm	200 mm	15 - 16	10 - 10	161.5	24.7
241 mm x	191 mm x		291 mm	250 mm	19 - 20	12 - 13	161.5	26.5
291 mm x	191 mm x		291 mm	300 mm	22 - 23	13 - 14	161.5	28.2
91 mm x	91 mm x		391 mm	100 mm	5 - 6	4 - 5	242.2	31.8
141 mm x	91 mm x		391 mm	150 mm	8 - 9	5 - 6	242.2	33.5
191 mm x	91 mm x		391 mm	200 mm	10 - 11	6 - 7	242.2	35.3
91 mm x	191 mm x		391 mm	100 mm	10 - 11	7 - 8	121.1	17.7
141 mm x	191 mm x		391 mm	150 mm	16 - 17	11 - 12	121.1	17.7
191 mm x	191 mm x		391 mm	200 mm	20 - 21	13 - 14	121.1	21.2
241 mm x	191 mm x		388 mm	250 mm	26 - 27	16 - 17	121.1	23.0
291 mm x	191 mm x		391 mm	300 mm	29 - 30	18 - 19	121.1	24.7

Actual mortar quantities are about half those given in the table, but experience shows that considerably more mortar is required, due to waste, droppings, etc.

700

Sizes, Weights and Quantities of Concrete Partition Tile						
Actual Size of Units			Thick-ness	Approx. Weight in Lbs. Light Wt.	No. of Units 100 Sq.Ft. of Wall	Cu.Ft. Mortar 100 Sq.Ft. of Wall*
3 5/8 " x	7 5/8 " x	11 5/8 "	4 "	12 - 14	150.0	6.0
3 5/8 " x	7 5/8 " x	11 5/8 "	6 "	17 - 19	150.0	6.5
7 5/8 " x	7 5/8 " x	11 5/8 "	8 "	21 - 23	150.0	7.0
1 5/8 " x	7 5/8 " x	15 5/8 "	2 "	11 - 12	112.5	3.0
3 5/8 " x	7 5/8 " x	15 5/8 "	4 "	16 - 18	112.5	5.0
5 5/8 " x	7 5/8 " x	15 5/8 "	6 "	24 - 26	112.5	5.0
7 5/8 " x	7 5/8 " x	15 5/8 "	8 "	29 - 31	112.5	6.0
Actual Size of Units			Thick-ness	Approx. Weight in Lbs. Light Wt.	No. of Units 10 Sq.M of Wall	Cu.M. Mortar 10 Sq.Ft. of Wall*
91 mm x	191 mm x	291 mm	100 mm	5 - 6	161.5	21.2
91 mm x	191 mm x	291 mm	150 mm	8 - 9	161.5	23.0
191 mm x	191 mm x	291 mm	200 mm	10 - 10	161.5	24.7
41 mm x	191 mm x	391 mm	50 mm	5 - 5	121.1	10.6
91 mm x	191 mm x	391 mm	100 mm	7 - 8	121.1	17.7
141 mm x	191 mm x	391 mm	150 mm	11 - 12	121.1	17.7
191 mm x	191 mm x	391 mm	200 mm	13 - 14	121.1	21.2

Actual mortar quantities are about half those given in the table, but experience shows that considerably more mortar is required, due to waste, droppings, etc.

Labor Laying Concrete Masonry

The labor cost of laying the various types and sizes of concrete blocks and tile will vary with factors such as the size and weight of the blocks, the class of work, and whether there are long straight walls or walls cut up with numerous openings. Labor costs will also vary depending on whether the blocks are laid above or below grade. Basement walls usually proceed faster than exposed work above grade.

It is usually more economical to use lightweight units, even though they cost a few cents more per piece, because a mason can handle and lay them with less effort. In some localities union regulations require two masons to work together where the blocks weigh more than 35 lbs. (15.8 kg) each.

Where concrete blocks are used for exterior or interior facing walls, with the blocks carefully laid to a line and in various patterns and with neatly tooled mortar joints, the labor costs will run considerably higher than on straight structural walls.

The quantities given in the following tables are based on average conditions with blocks laid in 1:1:6 cement-lime mortar. If portland cement mortar is used above grade, reduce daily output about 5 percent.

All quantities are based on the output of one mason and one laborer or hod carrier per 8-hr. day. If hoisting engineer time is required, add about 1/4-hr. per 100 sq.ft. (9.29 sq.m).

Number of Concrete Partition Tile Laid Per 8-Hr. Day By One Mason					
Actual Size of Units			Thick-ness	Light Wt. Units* per 8-Hr. Day	Mason Hrs. per 100 Pieces	Labor Hrs. per 100 Pieces
3 5/8 " x 7 5/8 " x 11 5/8 "			4 "	195 - 215	3.9	3.9
5 5/8 " x 7 5/8 " x 11 5/8 "			6 "	175 - 195	4.3	4.3
7 5/8 " x 7 5/8 " x 11 5/8 "			8 "	155 - 175	4.8	4.8
1 5/8 " x 7 5/8 " x 15 5/8 "			2 "	180 - 200	4.2	4.2
3 5/8 " x 7 5/8 " x 15 5/8 "			4 "	190 - 210	4.0	4.0
5 5/8 " x 7 5/8 " x 15 5/8 "			6 "	170 - 190	4.5	4.5
7 5/8 " x 7 5/8 " x 15 5/8 "			8 "	150 - 170	5.0	5.0
Actual Size			Thick-ness	Light Wt. Units* per 8-Hr. Day	Mason Hrs. per 100 Pieces	Labor Hrs. per 100 Pieces
91 mm x 191 mm x 291 mm			100 mm	195 - 215	3.9	3.9
141 mm x 191 mm x 291 mm			150 mm	175 - 195	4.3	4.3
191 mm x 191 mm x 291 mm			200 mm	155 - 175	4.8	4.8
41 mm x 191 mm x 391 mm			50 mm	180 - 200	4.2	4.2
91 mm x 191 mm x 391 mm			100 mm	190 - 210	4.0	4.0
141 mm x 191 mm x 391 mm			150 mm	170 - 190	4.5	4.5
191 mm x 191 mm x 391 mm			200 mm	150 - 170	5.0	5.0

*For heavyweight concrete units decrease above quantities and increase labor 10 percent.

Number of Concrete Building Units Laid Per 8-Hour Day by One Mason Light Weight					
Actual Size of Units			Thick-ness	Light Wt. Units* per 8-Hr. Day	Mason Hrs. per 100 Pieces	Labor Hrs. per 100 Pieces
3 5/8 " x 4 7/8 " x 11 5/8 "			4 "	215 - 235	3.5	3.5
5 5/8 " x 4 7/8 " x 11 5/8 "			6 "	195 - 215	3.9	3.9
7 5/8 " x 4 7/8 " x 11 5/8 "			8 "	175 - 195	4.3	4.3
3 5/8 " x 7 5/8 " x 11 5/8 "			4 "	195 - 215	3.9	3.9
5 5/8 " x 7 5/8 " x 11 5/8 "			6 "	175 - 195	4.3	4.3
7 5/8 " x 7 5/8 " x 11 5/8 "			8 "	155 - 175	4.8	4.8
9 5/8 " x 7 5/8 " x 11 5/8 "			10 "	135 - 155	5.5	5.5
11 5/8 " x 7 5/8 " x 11 5/8 "			12 "	115 - 135	6.4	6.4

MASONRY

Number of Concrete Building Units Laid Per 8-Hour Day by One Mason Light Weight (continued)					
Actual Size of Units			Thick-ness	Light Wt. Units* per 8-Hr. Day	Mason Hrs. per 100 Pieces	Labor Hrs. per 100 Pieces
3 5/8 " x 3 5/8 " x 15 5/8 "	4 "	215 - 235	3.5	3.5
5 5/8 " x 3 5/8 " x 15 5/8 "	6 "	195 - 215	3.9	3.9
7 5/8 " x 3 5/8 " x 15 5/8 "	8 "	175 - 195	4.3	4.3
3 5/8 " x 7 5/8 " x 15 5/8 "	4 "	190 - 210	4.0	4.0
5 5/8 " x 7 5/8 " x 15 5/8 "	6 "	170 - 190	4.5	4.5
7 5/8 " x 7 5/8 " x 15 5/8 "	8 "	150 - 170	5.0	5.0
9 5/8 " x 7 5/8 " x 15 5/8 "	10 "	130 - 150	5.7	5.7
11 5/8 " x 7 5/8 " x 15 5/8 "	12 "	110 - 130	6.7	6.7
7 5/8 " x 7 5/8 " x 15 5/8 "	8 "	225 - 250*	3.4	3.4
11 5/8 " x 7 5/8 " x 15 5/8 "	12 "	120 - 160*	5.7	5.7

Actual Size of Units			Wall Thick-ness	No. of Light Wt. Units* per 8-Hr.Day	Mason Hrs. Per 100 Pieces	Labor Hrs. per 100 Pieces
91 mm x 122 mm x 291 mm	100 mm	215 - 235	3.5	3.5
141 mm x 122 mm x 291 mm	150 mm	195 - 215	3.9	3.9
191 mm x 122 mm x 291 mm	200 mm	175 - 195	4.3	4.3
91 mm x 191 mm x 291 mm	100 mm	195 - 215	3.9	3.9
141 mm x 191 mm x 291 mm	150 mm	175 - 195	4.3	4.3
191 mm x 191 mm x 291 mm	200 mm	155 - 175	4.8	4.8
241 mm x 191 mm x 291 mm	250 mm	135 - 155	5.5	5.5
291 mm x 191 mm x 291 mm	300 mm	115 - 135	6.4	6.4
91 mm x 91 mm x 391 mm	100 mm	215 - 235	3.5	3.5
141 mm x 91 mm x 391 mm	150 mm	195 - 215	3.9	3.9
191 mm x 91 mm x 391 mm	200 mm	175 - 195	4.3	4.3
91 mm x 191 mm x 391 mm	100 mm	190 - 210	4.0	4.0
141 mm x 191 mm x 391 mm	150 mm	170 - 190	4.5	4.5
191 mm x 191 mm x 391 mm	200 mm	150 - 170	5.0	5.0
241 mm x 191 mm x 391 mm	250 mm	130 - 150	5.7	5.7
291 mm x 191 mm x 391 mm	300 mm	110 - 130	6.7	6.7
191 mm x 191 mm x 391 mm	200 mm	225 - 150*	3.4	3.4
291 mm x 191 mm x 391 mm	300 mm	120 - 160*	5.7	5.7

*For walls below grade.

Number of Concrete Building Units Laid Per 8-Hour Day by One Mason Heavy Weight						
Actual Size of Units				Thick-ness	Light Wt. Units* per 8-Hr. Day	Mason Hrs. per 100 Pieces	Labor Hrs. per 100 Pieces
3 5/8 "	x	4 7/8 "	x	11 5/8 "	4 "	195 - 214	3.9	3.9
5 5/8 "	x	4 7/8 "	x	11 5/8 "	6 "	177 - 195	4.3	4.3
7 5/8 "	x	4 7/8 "	x	11 5/8 "	8 "	159 - 177	4.7	4.7
3 5/8 "	x	7 5/8 "	x	11 5/8 "	4 "	177 - 195	4.3	4.3
5 5/8 "	x	7 5/8 "	x	11 5/8 "	6 "	159 - 177	4.7	4.7
7 5/8 "	x	7 5/8 "	x	11 5/8 "	8 "	141 - 159	5.3	5.3
9 5/8 "	x	7 5/8 "	x	11 5/8 "	10 "	123 - 141	6.1	6.1
11 5/8 "	x	7 5/8 "	x	11 5/8 "	12 "	105 - 123	7.0	7.0
3 5/8 "	x	3 5/8 "	x	15 5/8 "	4 "	195 - 214	3.9	3.9
5 5/8 "	x	3 5/8 "	x	15 5/8 "	6 "	177 - 195	4.3	4.3
7 5/8 "	x	3 5/8 "	x	15 5/8 "	8 "	159 - 177	4.7	4.7
3 5/8 "	x	7 5/8 "	x	15 5/8 "	4 "	173 - 191	4.4	4.4
5 5/8 "	x	7 5/8 "	x	15 5/8 "	6 "	155 - 173	5.0	5.0
7 5/8 "	x	7 5/8 "	x	15 5/8 "	8 "	136 - 155	5.5	5.5
9 5/8 "	x	7 5/8 "	x	15 5/8 "	10 "	118 - 136	6.3	6.3
11 5/8 "	x	7 5/8 "	x	15 5/8 "	12 "	100 - 118	7.4	7.4
7 5/8 "	x	7 5/8 "	x	15 5/8 "	8 "	205 - 227*	3.7	3.7
11 5/8 "	x	7 5/8 "	x	15 5/8 "	12 "	109 - 145*	6.3	6.3
Actual Size of Units				Wall Thick-ness	No. of Light Wt. Units per 8-Hr.Day	Mason Hrs. Per 100 Pieces	Labor Hrs. per 100 Pieces
91 mm	x	122 mm	x	291 mm	100 mm	195 - 214	3.9	3.9
141 mm	x	122 mm	x	291 mm	150 mm	177 - 195	4.3	4.3
191 mm	x	122 mm	x	291 mm	200 mm	159 - 177	4.7	4.7
91 mm	x	191 mm	x	291 mm	100 mm	177 - 195	4.3	4.3
141 mm	x	191 mm	x	291 mm	150 mm	159 - 177	4.7	4.7
191 mm	x	191 mm	x	291 mm	200 mm	141 - 159	5.3	5.3
241 mm	x	191 mm	x	291 mm	250 mm	123 - 141	6.1	6.1
291 mm	x	191 mm	x	291 mm	300 mm	105 - 123	7.0	7.0
91 mm	x	91 mm	x	391 mm	100 mm	195 - 214	3.9	3.9
141 mm	x	91 mm	x	391 mm	150 mm	177 - 195	4.3	4.3
191 mm	x	91 mm	x	391 mm	200 mm	159 - 177	4.7	4.7
91 mm	x	191 mm	x	391 mm	100 mm	173 - 191	4.4	4.4
141 mm	x	191 mm	x	391 mm	150 mm	155 - 173	5.0	5.0
191 mm	x	191 mm	x	391 mm	200 mm	136 - 155	5.5	5.5
241 mm	x	191 mm	x	391 mm	250 mm	118 - 136	6.3	6.3
291 mm	x	191 mm	x	391 mm	300 mm	100 - 118	7.4	7.4
191 mm	x	191 mm	x	391 mm	200 mm	205 - 227*	3.7	3.7
291 mm	x	191 mm	x	391 mm	300 mm	109 - 145*	6.3	6.3

*For walls below grade.

MASONRY

Mortar Joints. Hollow concrete block or tile should be laid with broken joints to secure the maximum resistance to moisture and heat penetration. Sufficient mortar is spread on the inner and outer shell bed and end joints to cover and to provide a joint of the required thickness when pressed into place. Solid units, brick, etc., are laid in solid beds.

Comparison of Concrete Masonry Sizes With Brick. Allowing for the mortar joints in brickwork, the various sizes of concrete masonry units are equivalent to the following number of modular size brick, i.e., 2-1/4"x3-5/8"x7-5/8" (56.25 x 90.62 x 190.62 mm).

Size of Blocks	Brick Equiv.	Metric Size of Blocks	Brick Equiv.
4 " x 8 " x 16 "	6	100 mm x 200 mm x 400 mm	6
8 " x 8 " x 16 "	12	200 mm x 200 mm x 400 mm	12
12 " x 8 " x 16 "	18	300 mm x 200 mm x 400 mm	18

Cost of Concrete Masonry Units. The prices quoted below are for index purposes only and are based on 8"x8"x16" (200 x 200 x 400 mm) block delivered to the job. As concrete masonry is widely manufactured, the estimator is always close to a source of supply. Costs on different size blocks are readily obtainable.

City	Regular (heavy)	Lightweight
Baltimore	$ 0.90	$ 1.20
Boston	$ 0.95	$ 1.25
Chicago	$ 0.80	$ 1.10
Cincinnati	$ 0.80	$ 1.10
Cleveland	$ 0.70	$ 1.00
Dallas	$ 0.80	$ 1.10
Denver	$ 1.05	$ 1.35
Detroit	$ 0.80	$ 1.10
Kansas City	$ 1.05	$ 1.35
Los Angeles	$ 1.00	$ 1.30
Minneapolis	$ 1.50	$ 1.80
New Orleans	$ 0.80	$ 1.10
New York	$ 1.10	$ 1.40
Philadelphia	$ 2.05	$ 2.35
Pittsburgh	$ 0.80	$ 1.10
San Francisco	$ 0.80	$ 1.10
St. Louis	$ 1.30	$ 1.60
Seattle	$ 0.85	$ 1.15

Prices for 8"x 8"x 16" Block
(200 mm x 200 mm x 400 mm ,block)
Based on Truck Load Deliveries

Precast reinforced concrete lintels can usually be obtained from concrete masonry manufacturers.

Estimating Concrete Block (C.M.U.)

Convert inches (mm) to decimal. Use the table in the Mensuration chapter or convert inches by dividing by 12". For example, 4" ÷ 12" = 0.33 feet or 200mm ÷ 1000mm = 0.2m.

Masonry Facts

Block	(SF less openings) x 1.125	= # of Blocks
Mortar	# of Blocks ÷ 55	= # Bags of Mortar
Sand	# Bags ÷ 10	= # CY of Sand
Grout Fill	# Vertical Drops x Wall Height ÷ 125	= # CY of Fill for 8" Block
	# Vertical Drops x Wall Height ÷ 70	= # CY of Fill for 12" Block

Order Estimate

½ Reg Blk	LF of height of openings ÷ 0.66	= # ½ Regular Block
Lintel Blk	LF of opening width ÷ 1.33 + # of openings	= # Lintel or Knock Out Block
	LF of masonry ÷ 1.33	= # of Lintel Block per Course

Lintel block can be used instead of a solid poured concrete beam at the top of a masonry wall. They can also be used over windows, doors and other openings in lieu of a precast concrete lintel beam. In this case, the length of the lintel block will extend a half block on both sides of the opening. No lintel block will be required over these openings if the lintel block forms a continuous beam at the top of the masonry wall.

Header Blk	LF of masonry ÷ 1.33	= # Header Block

Header block are used as part of a stem wall, when the slab will be poured into the block and footing.

¾ Block	Wall Height ÷ 0.66	= # ¾ Block

¾ block is used in a wall that is not a multiple of the block. Blocks that do not need to be cut are a multiple of the block. Even dimensions end on 0" or 8", such as 22'-0" or 48'-0" or 48'-8". Odd dimensions end in 4", such as 13'-4" or 31'-4", are wall dimensions that do not need to be cut or use a special block size. When an even number ends in 4", such as 8'-4", then this wall dimension is 1'-0" longer than a multiple of the block wall. Hence block 12" long (¾ block) are needed.

Inspect Blk	# Vertical Drops	= # Inspection Block
Split Blk	LF of masonry ÷ 1.33	= # Split Block

Split blocks (4" x 8" x 16") may be placed under windows where wall height under window is not a multiple of the block and in any area where wall height is 4" lower than the livable floor slab.

706

MASONRY

Regular Blk   Total block – (0.5 x # ½ Regular Block)
                – # Lintel Block – # Header Block
                – # ¾ Block – # Inspection Block   = # Regular Block

        Cavity Caps   (LF of masonry ÷ 0.66) - # Vertical Drops   = # Cavity Caps

Sample Concrete Block (C.M.U.) Estimate

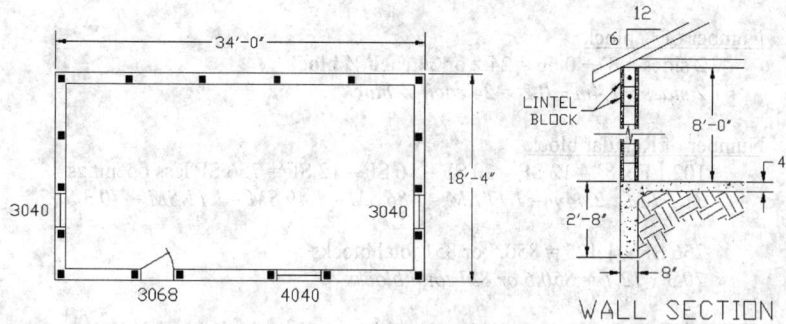

WALL SECTION

Given: Using the drawing shown above, complete the masonry
estimate for the 34'-0" (10.4m) x 18'-4" (5.6m) building.
Note that there are two windows measuring 3'-0" (0.9m) x
4'-0" (1.2m), one window measuring 4'-0" (1.2m) x 4'-0"
(1.2m) and one door measuring 3'-0" (0.9m) x 6'-8" (2m).
The window height (measured at the top of the window) for
all windows is 6'-8" (2m). The masonry wall is 8'-0"
(2.4m) high with vertical drops as shown. There are two
courses of lintel block with one #5 (#16) rebar per course.
The masonry wall uses 8" x 8" x 16" (200mm x 200mm x
400mm) standard CMU block.

Number of LF (LM) of wall at the centerline
     34' – 0.33' – 0.33' = 33.34' (10.2m) wall length at centerline
     18.33' – 0.33' – 0.33' = 17.67' (5.4m) wall width at centerline
     2 x (33.34' + 17.67') = **102 LF (31.1 LM)**

Number of courses of block
     8' ÷ 0.67' = 11.9 or **12 courses of block**
     2.4m ÷ 200mm = 12 courses of block

Number of ½ Regular block
     (4' + 6.67' + 4' + 4') ÷ 0.66 = 28.3 or **28 each ½ Regular block**
     (1.2m + 2m + 1.2m + 1.2m) ÷ 0.2 = 28 each ½ Regular block

Number of Lintel block
     Note that window go up to these lintel courses and so no additional lintel
     block are needed over the windows or door.

     2 courses x 102' ÷ 1.33 = 153.4 or **154 each Lintel block**

*2 courses x 31.1m ÷ 0.406 = 153.2 or **154 each Lintel block***

Number of Header block
There are no header block shown on the wall section.

Number of Inspection block
18 vertical drops = **18 each Inspection block**

Number of ¾ block
2 sides x 8' ÷ 0.66 = 24.2 or **24 each ¾ block**
*2 sides x 2.4m ÷ 0.2 = **24 each ¾ block***

Number of Regular block
102 LF x 8' – 12 SF – 20 SF – 16 SF – 12 SF = 756 SF less openings
31.1 LM x 2.44m – 1.11 SM – 1.86 SM – 1.49 SM – 1.11 SM = 70.3 SM

756 SF x 1.125 = 850.5 or 851 total blocks
70.3 x 12.1 = 850.6 or 851 total blocks

851 – (0.5 x 28 pcs of ½ Reg) – 154 pcs of Lintel – 0 pcs of Header
– 24 pcs of ¾ Block – 18 pcs of Inspection = **641 pcs of Reg block**

Number of bags of mortar
851 blocks ÷ 55 = 15.5 or **16 bags of mortar**

Number of CY (CM) of sand
16 bags ÷ 10 = **1.6 CY (1.1 CM) of sand**

Number of CY (CM) of grout fill
18 drops x 8' ÷ 125 = **1.2 CY (0.88 CM) of grout fill**

Note: Quantities of mortar, sand, and grout fill will vary depending on the crew.

Number of #5 (#16) rebar 10' (3m) long
Each vertical drop will have one rebar vertically for the height of the wall with any remainder bent into the lintel area.

18 drops = **18 each #5 rebar 10' long**

Number of #5 (#16) rebar 20' (6.1m) long
2 courses x 102 LF ÷ 17 = **12 pcs of #5 rebar 20' long**
*2 courses x 31.1 LM ÷ 5.18 = **12 pcs of #16 rebar 6.1m long***

Control Joints in Concrete Masonry

To control movements in concrete masonry walls caused by various kinds of stresses, control joints are used. Control joints are continuous vertical joints built into walls in such locations and in such a manner as to permit slight wall movement without cracking the masonry.

The spacing and location of control joints depends on a number of factors—length of wall, architectural details, and the experience and records

MASONRY

in the particular locality where the structure is to be built. Control joints should be placed at junctions of bearing as well as nonbearing walls; at junctions of walls and columns or pilasters; at construction joints in foundation, roof, and floors; and in walls weakened by chases and openings, columns, and fixtures. In long walls, joints are ordinarily spaced at approximately 20' (6.09 m) intervals, depending on local experience. At return angles in "L", "T", and "U" shaped structures, as a rule, control joint locations are determined by the architect or engineer and are usually indicated on the drawings.

Control joints can be built with regular full- and half-length stretcher block or full- and half-length offset jamb block. With this type of joint construction, a non-corroding metal Z-tiebar, placed in every other horizontal joint across the control joint will provide lateral support to wall sections on each side of the control joint.

In some localities, special control joint block are available. These block have tongue and groove ends, which provide the required lateral support. These block are also made in full- and half-length units.

Other common methods of constructing control joints are illustrated below. The joints permit free longitudinal movement, but they should have sufficient shear and flexural strength to resist lateral loads. They also must be weather tight when located in exterior walls.

Generally, a control joint is placed at one side of an opening less than 6' (1.82 m) in width and at both jambs of openings over 6' (1.82 m) wide. Control joints can be omitted if adequate tensile reinforcement is placed above and below wall openings.

To keep control joints as unnoticeable as possible, care must be taken to build them plumb and of the same thickness as the other mortar joints. If the joint is to be exposed to weather or view, it should be raked out to a depth of at least 3/4" (18.75 mm) and sealed with knife grade caulking compound.

The additional cost of building control joints, over and above the regular wall cost, will be about $0.40 to $0.48 per lin.ft. ($1.32 to $1.57 per m) for material and $0.80 to $96 per lin.ft. ($2.62 to $3.15 per m) for labor.

Reinforced Concrete Masonry

Local conditions frequently demand special construction methods. For example, in earthquake regions it is necessary to provide more than ordinary stability for all types of structures. This is also true in areas where severe wind storms occur or where foundation soils are unstable.

In almost any locality there are structures that are subject to excessive vibrations, very heavy loads, or other stresses. Any of these examples of unusual stress conditions require special attention in design and are often covered by local building codes.

709

In concrete masonry construction, additional strength is obtained by reinforcing the walls with steel reinforcing rods encased in grout. The walls may be reinforced horizontally, vertically, or both.

Bond Beams. To reinforce concrete masonry walls horizontally, bond beams are frequently used at each story height. Under extreme stress conditions, it may be necessary to use them in every second or third course.

Bond beams may be constructed by forming and pouring a continuous ribbon of reinforced concrete at the course height required, but this method breaks up the continuity of the block pattern and may be objectionable.

In many localities, special bond beam blocks are available that are trough-shaped. When laid open side up, they form a continuous trough in which reinforcing steel and concrete encasement may be placed, eliminating the need for wood side forms, and the block pattern of the wall is preserved.

Bond beams serve both as structural elements and as a means of crack control. They are constructed with special-shaped masonry units that are filled with concrete or grout and reinforced with embedded steel. Their value in crack control is due to the increased strength and stiffness they give a masonry wall. Since they are capable of structural function as well as crack control, bond beams will be found serving the following functions:

1. As lintel beams over doors and windows. Lintels may be pre-constructed on the ground and set in place when they have attained sufficient strength or they may be built in place, using wood centers for support until mortar and concrete is strong enough to permit their removal.
2. Below the sill in walls with openings.
3. At the top of walls and at floor level to distribute vertical loads.
4. As horizontal stiffeners incorporated into masonry to transfer flexural stresses to columns and pilasters when unusually high lateral loading is encountered.

As a means of crack control, the bond beam's area of influence normally is presumed to extend 24 inches (600 mm) above and below its location in the wall. In walls without openings, they are spaced four feet apart and may be any length up to 60 feet (18.29 m) maximum.

Reinforcement for bond beams must satisfy structural requirements but should not be less than two no. 4 steel bars. The beams are always discontinuous at expansion joints, and joints should be designed to transfer lateral force along the wall. Beams may be discontinuous at control joints; practice varies depending on structural requirements. Dummy joints are formed when a bond beam is continuous at a control joint.

Vertical Reinforcement in Concrete Masonry Walls. Where vertical reinforcement is required in concrete masonry walls, it is usually located at building corners, jambs of wall openings, and at regular intervals between wall openings. Size and spacing of reinforcement is usually covered

MASONRY

by local building codes. When used in conjunction with bond beams, vertical and horizontal steel should be tied together.

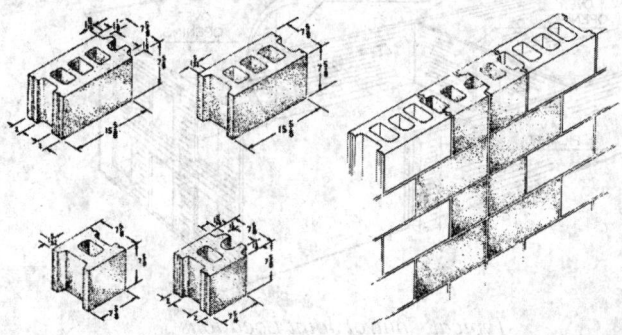

Method of Constructing Control Joints in Concrete Masonry Walls

In placing vertical reinforcement, advantage is taken of the vertical alignment of hollow block cores, which form wells into which the reinforcing bars are placed and grouted solid with poured mortar or concrete. At locations where vertical reinforcement is to occur, the bottom block should be left out for a cleanout hole when wall is laid up. Just prior to setting steel bars in place, the wells should be rodded clean of extruded mortar and debris removed from the cleanout. After cleaning, setting bars, and inspecting, cleanouts are closed with side-forms and the wells are grouted solid.

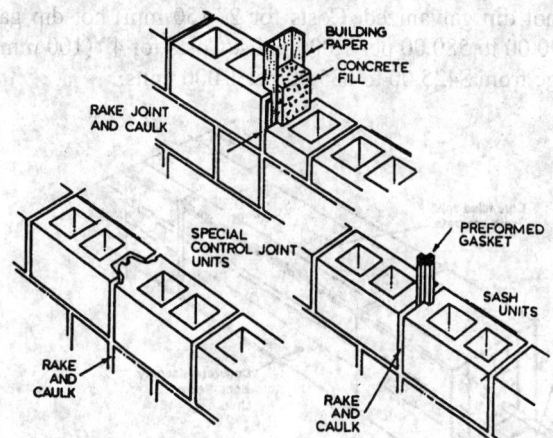

Typical Control Joint Details

711

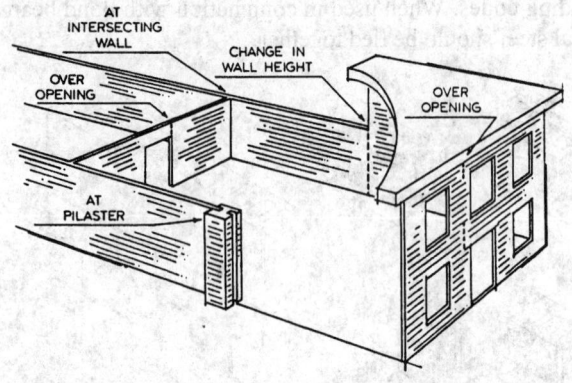

Typical Control Joint Locations

Concrete Masonry Cavity Walls

A cavity wall consists of two walls separated by a continuous air space and securely tied together with non-corroding metal ties of adequate strength. For each 3 sq.ft. (0.27 sq.m) of wall surface, a rectangular tie of 3/16" (4.68 mm) or 1/4" (6.25 mm) wire can be used. The ties are embedded in the horizontal joints of both walls. Additional ties are necessary at all openings, with ties spaced about 3' (900 mm) apart around the perimeter and within 12" (300 mm) of the opening. Rectangular cavity wall ties are manufactured 2" (50 mm) and 4" (100 mm) wide and are either mill galvanized or hot dip galvanized. Costs for 2" (50 mm) hot dip galvanized range from $290.00 to 580.00 per 1,000 units. Costs for 4" (100 mm) hot dip galvanized range from $425.00 to 635.00 per 1,000 units.

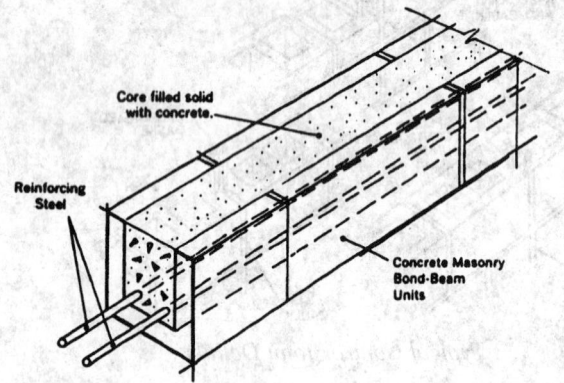

Typical Concrete Masonry Beams

Typical codes require that 10" (250 mm) cavity walls should not exceed 25' (7.6 m) in height. In residential construction the overall thickness of concrete masonry cavity walls is nominally 10" or 12" (250 or 300 mm).

MASONRY

Neither the inner nor outer walls should be less than 4" (100 mm) thick (nominal dimension), and the space between them should not be less than 2" (50 mm) nor more than 3" (75 mm) wide. Usually the outer wall is nominally 4" (100 mm) thick and the remaining wall thickness made up by the air space and the inner wall. For example, a modular concrete masonry cavity wall nominally 12" (300 mm) thick is composed of a 4" (100 mm) outer wall, a 2" (50 mm) air space and a 6" (150 mm) inner wall.

A simple method of preventing the accumulation of mortar droppings between walls and maintaining a clear cavity is to lay a 1"x 2" (25 x 50 mm) wood strip across a level of ties to catch the droppings. As the masonry reaches the next level for placing ties, the strip is raised, cleaned and laid on the ties placed at this level.

The practice of providing special flashing and weep holes in masonry cavity walls primarily presumes that water will enter the wall from the outside. If concrete masonry walls are properly designed and built with well-compacted mortar joints and are painted or stuccoed, the walls should be weathertight, and there should be no need for special flashing and weep holes.

However, in limited areas subject to severe driving rains, or where experience has shown that sufficient water collects in the wall to make flashing and weep holes necessary, the practice is as follows. The heads of windows, doors, and other wall openings, and the bottom course of masonry immediately above any solid belt course or foundation, are flashed so that any moisture entering the wall cavity will be directed toward the outside walls. Only rust-resisting metal or approved materials treated with asphalt or pitch preparations should be used for flashing.

Weep holes are placed every 2 or 3 units apart in the vertical joints of the bottom course of the outside wall immediately above any solid belt course or foundation. In no case should the weep holes be located below grade. Weep holes can be formed by placing well-oiled rubber tubing in the mortar joint and then extracting it after mortar has become hard. The tubing should extend up into the cavity for several inches to provide a drainage channel through any mortar droppings that might have accumulated.

The labor cost of concrete masonry cavity walls varies the same as ordinary block walls according to the size of unit, pattern of laying, and nature of job.

Concrete Masonry Units Used for Exterior Facing

Concrete masonry walls are extensively used for exterior facing the same as face brick or stone. They may be laid up in any of the attractive designs shown in the accompanying illustrations, ranging from straight ashlar to the many variations of random or broken ashlar.

When used for exterior facing, the walls should be true and plumb and laid with full mortar coverage on vertical and horizontal face shells, no furrowing permitted, with all vertical joints shoved tight. Mortar joints should be 3/8" (9.3 mm) thick.

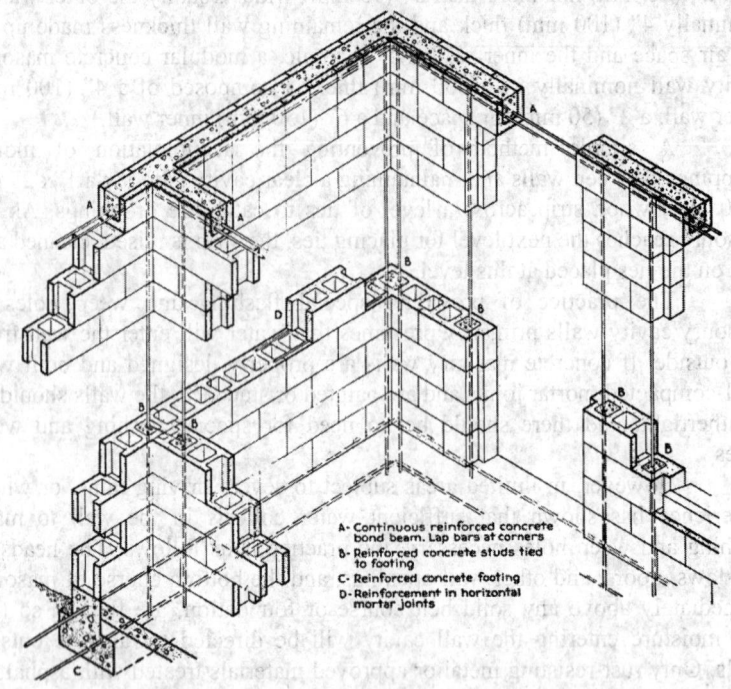

A- Continuous reinforced concrete bond beam. Lap bars at corners
B- Reinforced concrete studs tied to footing
C- Reinforced concrete footing
D- Reinforcement in horizontal mortar joints

Method of Constructing and Reinforcing Concrete

The mortar joints should be struck off flush with wall surface and when partially set shall be compressed and compacted with a rounded or V-shaped tool. This provides a more waterproof mortar joint. An attractive treatment is obtained by emphasizing the horizontal joints and obscuring the vertical joints in concrete walls. This is done by tooling the horizontal joints and striking the vertical joints flush with the wall surface and then rubbing with carpet or burlap to remove the sheen from the troweled mortar surface.

Varying the bond or joint pattern of a concrete masonry wall can create a wide variety of interesting and attractive appearances, using standard units as well as sculptured-face and other architectural facing units. Due to the increased use of concrete masonry as the finished wall surface, the use of bond patterns other than the typical "running bond" has steadily increased for both loadbearing and non-loadbearing.

After running bond construction, the next most widely used bond pattern with concrete masonry units is stacked bond. Lightweight concrete units should be used where obtainable. They provide better insulation than units made of ordinary concrete aggregates.

714

MASONRY

The labor costs given on the following pages are based on using lightweight units. If ordinary concrete units are used, add about 10% to labor costs given.

Labor Cost of 100 S.F. (9.29 Sq.M) of Exterior Facing Concrete Masonry Walls Using 8" x 16" (200 mm x 400 mm) Units, Laid in Regular Ashlar or Running Bond - Walls 8" (200 mm) Thick					
	Hours	Rate	Total	Rate	Total
Mason	7	$....	$....	$ 36.93	$ 258.51
Labor	7	$....	$....	$ 26.16	$ 183.12
Cost per 100 s.f.					$ 441.63
per s.f.					$ 4.42
per sq.m					$ 47.54

Add 10% to labor costs for 10" (250 mm) thick walls, 20% for 12" (300 mm) thick walls. Deduct 25% from labor costs for 4" (100 mm) veneer facing.

Labor Cost of 100 S.F. (9.29 Sq.M) of Exterior Facing Concrete Masonry Walls Using 4" x 16" (100 mm x 400 mm) or half Heights units laid in Regular Ashlar or Running Bond - Walls 8" (200 mm) Thick					
	Hours	Rate	Total	Rate	Total
Mason	10	$....	$....	$ 36.93	$ 369.30
Labor	7	$....	$....	$ 26.16	$ 183.12
Cost per 100 s.f.					$ 552.42
per s.f.					$ 5.52
per sq.m					$ 59.46

Add 10% to labor costs for 10" (250 mm) thick walls, 20% for 12" (300 mm) thick walls. Deduct 25% from labor costs for 4" (100 mm) veneer facing.

Labor Cost of 100 S.F. (9.29 Sq.M) of Exterior Facing Concrete Masonry Walls Using Coursed Ashlar No.1, Using Alternate Courses of 8" x 16" (200 mm x 400 mm) and 4" x 16" (100 mm x 400 mm) Units - Walls 8" (200 mm) Thick					
	Hours	Rate	Total	Rate	Total
Mason	8	$....	$....	$ 36.93	$ 295.44
Labor	6	$....	$....	$ 26.16	$ 156.96
Cost per 100 s.f.					$ 452.40
per s.f.					$ 4.52
per sq.m					$ 48.70

Add 10% to labor costs for 10" (250 mm) thick walls, 20% for 12" (300 mm) thick walls. Deduct 25% from labor costs for 4" (100 mm) veneer facing.

Labor Cost of 100 S.F. (9.29 Sq.M) of Exterior Facing Concrete Masonry Walls Using Coursed Ashlar No.2, Consisting of Two Courses of 8" x 16" (200 mm x 400 mm) Units and One Course of 4" x 16" (100 mm x 400 mm) Units - Walls 8" (200 mm) Thick					
	Hours	Rate	Total	Rate	Total
Mason	7.5	$....	$....	$ 36.93	$ 276.98
Labor	5.5	$....	$....	$ 26.16	$ 143.88
Cost per 100 s.f.					$ 420.86
per s.f.					$ 4.21
per sq.m					$ 45.30

Add 10% to labor costs for 10" (250 mm) thick walls, 20% for 12" (300 mm) thick walls. Deduct 25% from labor costs for 4" (100 mm) veneer facing.

Labor Cost of 100 S.F. (9.29 Sq.M) of Exterior Facing Concrete Masonry Walls Using Coursed Ashlar No.3, Consisting of 8" x 16" (200 mm x 400 mm) and 4" x 16" (100 mm x 400 mm) Units - Walls 8" (200 mm) Thick					
	Hours	Rate	Total	Rate	Total
Mason	8.25	$....	$....	$ 36.93	$ 304.67
Labor	6.5	$....	$....	$ 26.16	$ 170.04
Cost per 100 s.f.					$ 474.71
per s.f.					$ 4.75
per sq.m					$ 51.10

Add 10% to labor costs for 10" (250 mm) thick walls, 20% for 12" (300 mm) thick walls. Deduct 25% from labor costs for 4" (100 mm) veneer facing.

Labor Cost of 100 S.F. (9.29 Sq.M) of Exterior Facing Concrete Masonry Walls Using Coursed Ashlar No.4, Consisting of 8" x 16" (200 mm x 400 mm) and 4" x 16" (100 mm x 400 mm) Units - Walls 8" (200 mm) Thick					
	Hours	Rate	Total	Rate	Total
Mason	7.9	$....	$....	$ 36.93	$ 291.75
Labor	5.7	$....	$....	$ 26.16	$ 149.11
Cost per 100 s.f.					$ 440.86
per s.f.					$ 4.41
per sq.m					$ 47.46

Add 10% to labor costs for 10" (250 mm) thick walls, 20% for 12" (300 mm) thick walls. Deduct 25% from labor costs for 4" (100 mm) veneer facing.

MASONRY

Labor Cost of 100 S.F. (9.29 Sq.M) of Exterior Facing Concrete Masonry Walls Using Vertical Stacking of 8" x 16" (200 mm x 400 mm) Units - Walls 8" (200 mm) Thick

	Hours	Rate	Total	Rate	Total
Mason	9	$....	$....	$ 36.93	$ 332.37
Labor	7	$....	$....	$ 26.16	$ 183.12
Cost per 100 s.f.					$ 515.49
per s.f.					$ 5.15
per sq.m					$ 55.49

Add 10% to labor costs for 10" (250 mm) thick walls, 20% for 12" (300 mm) thick walls. Deduct 25% from labor costs for 4" (100 mm) veneer facing.

Labor Cost of 100 S.F. (9.29 Sq.M) of Exterior Facing Concrete Masonry Walls Using Horizontal Stacking of 8" x 16" (200 mm x 400 mm) Units - Walls 8" (200 mm) Thick

	Hours	Rate	Total	Rate	Total
Mason	7.5	$....	$....	$ 36.93	$ 276.98
Labor	5.5	$....	$....	$ 26.16	$ 143.88
Cost per 100 s.f.					$ 420.86
per s.f.					$ 4.21
per sq.m					$ 45.30

Add 10% to labor costs for 10" (250 mm) thick walls, 20% for 12" (300 mm) thick walls. Deduct 25% from labor costs for 4" (100 mm) veneer facing.

Labor Cost of 100 S.F. (9.29 Sq.M) of Exterior Facing Concrete Masonry Walls Using Square Stacking of 8" x 8" (200 mm x 200 mm) Units - Walls 8" (200 mm) Thick

	Hours	Rate	Total	Rate	Total
Mason	12.5	$....	$....	$ 36.93	$ 461.63
Labor	8.75	$....	$....	$ 26.16	$ 228.90
Cost per 100 s.f.					$ 690.53
per s.f.					$ 6.91
per sq.m					$ 74.33

Add 10% to labor costs for 10" (250 mm) thick walls, 20% for 12" (300 mm) thick walls. Deduct 25% from labor costs for 4" (100 mm) veneer facing.

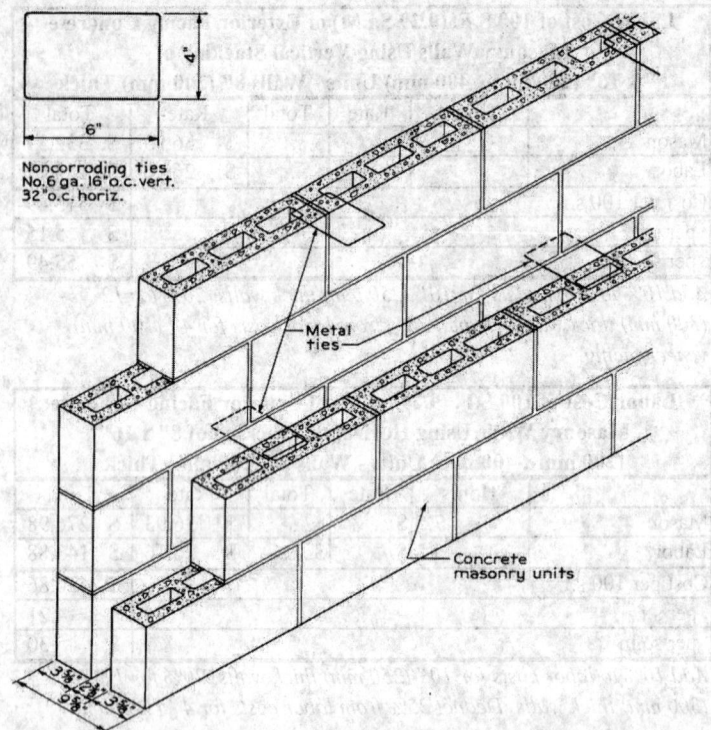

Noncorroding ties
No.6 ga. 16"o.c.vert.
32"o.c.horiz.

Metal ties

Concrete masonry units

Method of constructing Cavity Walls of Concrete Masonry

Labor Cost of 100 S.F. (9.29 Sq.M) of Exterior Facing Concrete Masonry Walls Using Basket Weave Design Consisting of 8" x 16" (200 mm x 400 mm) Units - Walls 8" (200 mm) Thick					
	Hours	Rate	Total	Rate	Total
Mason	7.9	$....	$....	$ 36.93	$ 291.75
Labor	5.7	$....	$....	$ 26.16	$ 149.11
Cost per 100 s.f.					$ 440.86
per s.f.					$ 4.41
per sq.m					$ 47.46

*Add 10% to labor costs for 10" (250 mm) thick walls, 20% for 12"
(300 mm) thick walls. Deduct 25% from labor costs for 4" (100 mm)
veneer facing.*

MASONRY

Labor Cost of 100 S.F. (9.29 Sq.M) of Exterior Facing Concrete Masonry Walls Using Pattern Ashlar Consisting of 8" x 16" (200 mm x 400 mm) and 8" x 8" (200 mm x 200 mm) Units - Walls 8" (200 mm) Thick

	Hours	Rate	Total	Rate	Total
Mason	8.4	$....	$....	$ 36.93	$ 310.21
Labor	6.2	$....	$....	$ 26.16	$ 162.19
Cost per 100 s.f.					$ 472.40
per s.f.					$ 4.72
per sq.m					$ 50.85

Add 10% to labor costs for 10" (250 mm) thick walls, 20% for 12" (300 mm) thick walls. Deduct 25% from labor costs for 4" (100 mm) veneer facing.

Labor Cost of 100 S.F. (9.29 Sq.M) of Exterior Facing Concrete Masonry Walls Using Pattern Ashlar Consisting of 4" x 8" (100 mm x 200 mm), 4" x 12" (100 mm x 300 mm), 4" x 16" (100 mm x 400 mm), 8" x 12" (200 mm x 300 mm), and 8" x 16" (200 mm x 400 mm)

	Hours	Rate	Total	Rate	Total
Mason	14	$....	$....	$ 36.93	$ 517.02
Labor	9	$....	$....	$ 26.16	$ 235.44
Cost per 100 s.f.					$ 752.46
per s.f.					$ 7.52
per sq.m					$ 81.00

Add 10% to labor costs for 10" (250 mm) thick walls, 20% for 12" (300 mm) thick walls. Deduct 25% from labor costs for 4" (100 mm) veneer facing.

Labor Cost of 100 S.F. (9.29 Sq.M) of 8" (200 mm) Brick Faced Concrete Block Wall Bonded Every Seventh Course

	Hours	Rate	Total	Rate	Total
Mason	18.5	$....	$....	$ 36.93	$ 683.21
Labor	13.5	$....	$....	$ 26.16	$ 353.16
Cost per 100 s.f.					$1,036.37
per s.f.					$ 10.36
per sq.m					$ 111.56

Labor Cost of 100 S.F. (9.29 Sq.M) of 12" (300 mm) Brick Faced Concrete Block Wall Bonded Every Sixth Course

	Hours	Rate	Total	Rate	Total
Mason	22	$....	$....	$ 36.93	$ 812.46
Labor	17	$....	$....	$ 26.16	$ 444.72
Cost per 100 s.f.					$1,257.18
per s.f.					$ 12.57
per sq.m					$ 135.33

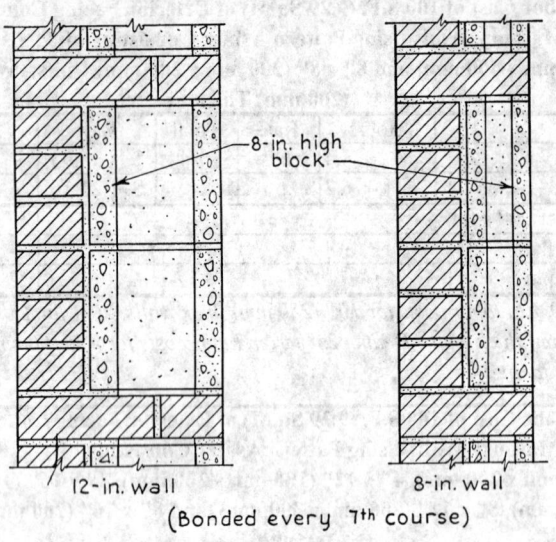

8-in. high block

12-in. wall   8-in. wall

(Bonded every 7ᵗʰ course)

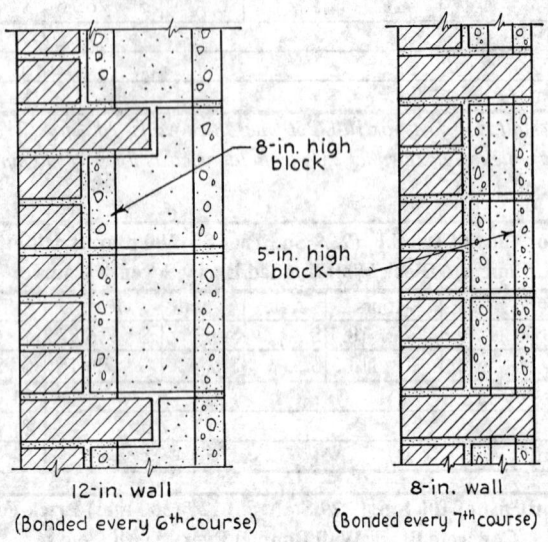

8-in. high block

5-in. high block

12-in. wall

(Bonded every 6ᵗʰ course)

8-in. wall

(Bonded every 7ᵗʰ course)

Examples of Concrete Masonry Used as Backup for Brick

720

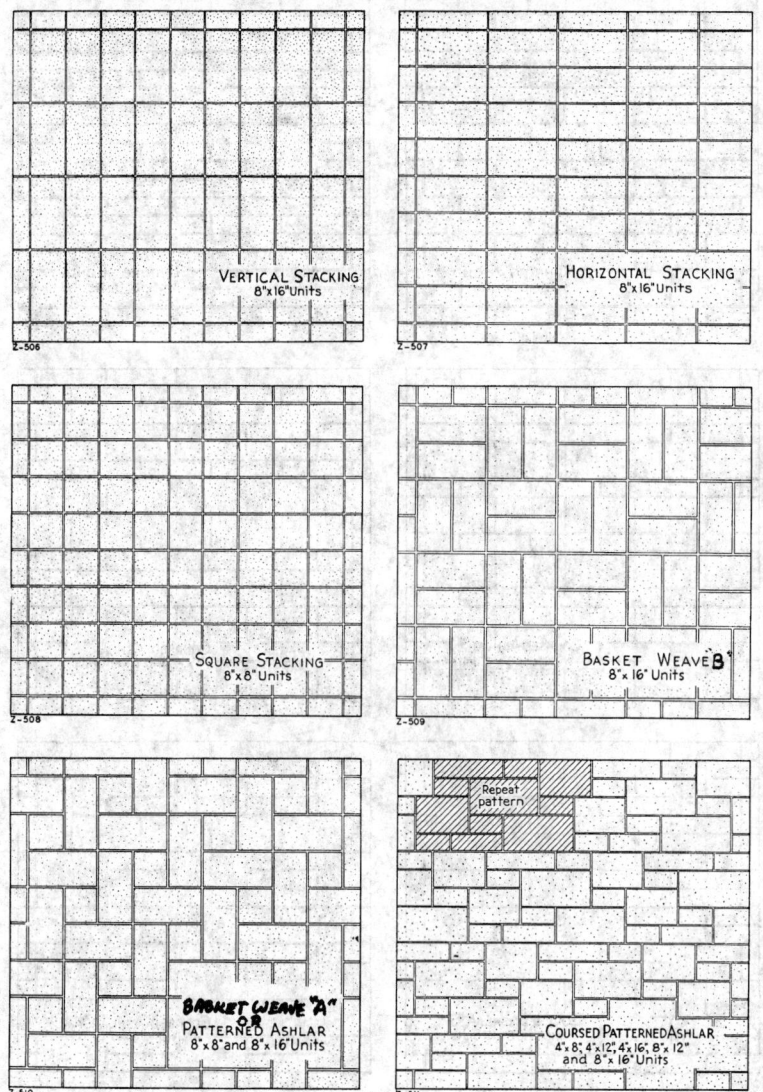

Methods of Laying Concrete Masonry Units for Exterior Facing

721

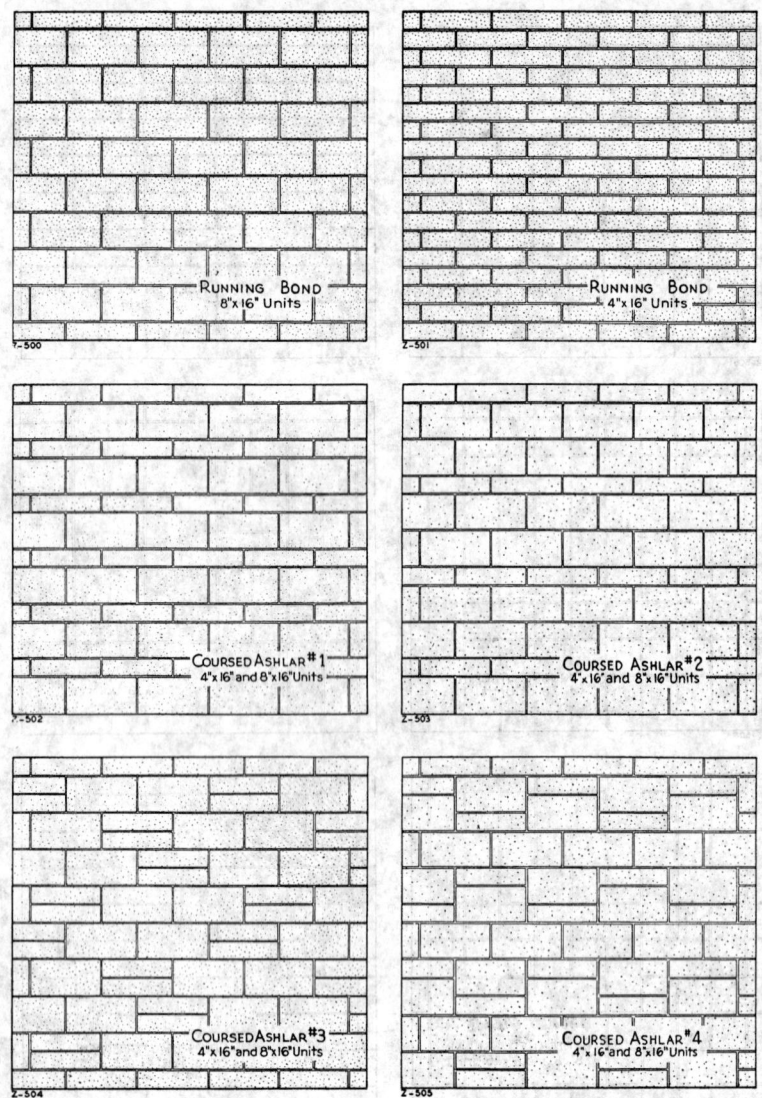

Methods of Laying Concrete Masonry Units for Exterior Facing

Insulating Fill for Concrete Masonry Walls. For insulating masonry walls, a water-repellent vermiculite masonry fill is available. This is a free-flowing granular material, processed to provide a high degree of water-repellency. It is designed for the cores of masonry blocks and the voids of cavity walls. It does not bridge and completely fills cores and cavities. Test installations show that the fill reduces up to 50% of heat loss and 25% of air-

MASONRY

conditioning costs. Masonry fill is also effective sound-deadening in masonry block partitions.

Vermiculite masonry fill is marketed in 4 cu.ft. (0.11 cu.m) bags and costs about $7.00 per bag. Approximate coverage is:

Approximate Coverage					
English		Wall Area Sq. Ft.	Metric		Wall Area Sq.M
2 "	cavity	24	50.0 mm	cavity	2.2
2 1/2 "	cavity	20	62.5 mm	cavity	1.9
4 1/2 "	cavity	11	112.5 mm	cavity	1.0
6 "	block 2 or 3 core	21	150.0 mm	block 2 or 3 core	2.0
8 "	block 2 or 3 core	14 1/2	200.0 mm	block 2 or 3 core	1.3
12 "	block 2 or 3 core	8	300.0 mm	block 2 or 3 core	0.7

Silicone treated perlite loose fill insulation insulates masonry walls to give greater comfort at lower cost. Silicone treated perlite pours readily into the cores of concrete block or cavity type masonry walls. Tests prove it reduces winter and summer heat transfer 50% or more. Silicone treatment of perlite provides lasting water repellency, prevents moisture penetration, and assures constant insulating efficiency in the most severe weather, including wind driven rain. Performance is rated excellent in accordance with test procedures developed by the National Bureau of Standards.

Silicone treated perlite loose fill insulation is manufactured nationally and marketed in 4 cu.ft. (0.11 cu.m) bags. Coverage is approximately the same as shown for vermiculite. Installed costs will also be competitive with other loose fill materials.

In addition to perlite and vermiculite, there are rigid insulating materials on the market. Korfil and Korfil HI-R® are made of preformed expanded polystyrene and designed to fit two core masonry units from 6" to 12" (150-300 mm).

Table A - R Values					
Wall Constructed of 4" (100 mm) Hollow CMU					
Density, lbs./c.f.	60	80	100	120	140
Density, kg/cu.m	961	1281	1602	1922	2242
No insulation	2.8	2.4	2.2	1.9	1.7
Cores filled with vermiculite	4.9	4.1	3.4	2.7	2.2
Cores filled with perlite	5.4	4.4	3.5	2.8	2.3

Table B - R Values					
Walls Construction of 6" (150 mm) Hollow CMU					
Density, lbs./c.f.	60	80	100	120	140
Density, kg/cu.m	961	1281	1602	1922	2242
No insulation	3.2	2.7	2.4	2.1	1.9
Cores filled with vermiculite	7.1	5.7	4.5	3.6	2.8
Cores filled with perlite	7.8	6.1	4.8	3.7	2.9
Rigid premolded insulation for two Core CMU	7.7	6.7	5.9	5	4.2

Table C - R Values					
Wall Constructed of 8" (200 mm) Hollow CMU					
Density lbs./c.f.	60	80	100	120	140
Density kg/cu.m	961	1281	1602	1922	2242
No insulation	3.6	3	2.6	2.3	2
Cores filled with vermiculite	9.6	7.6	6	4.7	3.6
Cores filled with perlite	10.4	8.2	6.3	4.8	3.7
Rigid premolded insulation for two Core CMU	8.3	7.7	6.3	5.3	4.2

Table D - R Values					
Wall Constructed of 10" (250 mm) Hollow CMU					
Density, lbs./c.f.	60	80	100	120	140
Density, kg/cu.m	961	1281	1602	1922	2242
No insulation	3.8	3.2	2.7	2.4	2.1
Cores filled with vermiculite	11.8	9.3	7.2	5.5	4.2
Cores filled with perlite	12.9	10	7.6	5.7	4.3
Rigid premolded insulation for two Core CMU	8.3	7.7	6.7	5.6	4.6

Table E - R Values					
Wall Constructed of 12" (300 mm) Hollow CMU					
Density, lbs./c.f.	60	80	100	120	140
Density, kg/cu.m	961	1281	1602	1922	2242
No insulation	3.9	3.3	2.8	2.5	2.2
Cores filled with vermiculite	14.3	11.3	8.8	6.7	5
Cores filled with perlite	15.7	12.1	9.2	6.9	5.1
Rigid premolded insulation for two core CMU	9.1	8.3	7.1	6.3	5.3

Other products available for insulation of masonry units are extruded and expanded polystyrene, as well as batt insulation. The estimator should be aware of the requirements of the owner and price the material that meets the engineers or owners needs and meets the local building code requirements both for insulation and fire. The following is a list of the common materials:

MASONRY

1. Vermiculite
2. 1" (25 mm) extruded polystyrene
3. 2" (50 mm) expanded polystyrene
4. 2" (50 mm) extruded polystyrene

For Korfil insulation it is suggested that the local masonry unit manufacturers be contacted for availability and costs.

Special Facing Blocks

In many areas, special block are available for the purpose of creating special facing effects. These block range from ordinary cement block, scored across the face to give the appearance of additional joints, to block made or faced with special aggregates which produce various textures and colors.

Scored Block. The simplest departure from ordinary block facing is obtained by scoring regular block one or more times across the face to give the illusion of shallower units and more joints. Most block manufacturers can furnish scored units for about $0.07 per score in addition to the regular block price.

Labor cost for laying up this type of facing is about the same as for a regular running bond facing except when scored grooves are pointed with mortar and tooled to match the actual joints.

Split-Block. Split-blocks offer another variation in wall finish. Split-blocks are made by splitting a hardened concrete unit lengthwise. The units are laid in the wall with the fractured face exposed. Many interesting variations can be obtained by introducing mineral colors and by using aggregates of different grades and colors in the concrete mixture.

Split-block can be laid in simple running bond or in any of the other patterns used in concrete masonry construction. The fractured faces of the units produce a wall of rugged appearance.

Prices for 4" (102 mm) thick split-block veneer units run from $1.25 to $1.40 per sq.ft. ($13.46 to $15.07 per sq.m) of wall depending on the size of units and pattern used.

Labor costs will be about the same as given on previous pages for the various patterns illustrated.

Slump Block. For a weathered stone effect, slump block is used. When manufacturing slump block, a special consistency mixture is used so that the unit will sag or "slump" when released from the molds before complete setting, producing artistic irregularities in height and texture. Many are integrally colored in varying shades. Standard heights vary from 1-5/8" (41 mm) to 3-5/8" (92 mm), and units are usually laid in a random ashlar pattern, although coursed patterns may be used if desired.

The cost of slump block varies with the locality. Prices should be obtained locally when figuring this work. Prices range from $1.20 to $1.60 per sq.ft. ($12.92 to $17.22 per sq.m) of wall facing.

Mortar requirements for slump block facing are high, due to the irregularities of the material and approximate those of random ashlar stone

725

work. On average, it will require about 10 cu.ft. of mortar per 100 sq.ft. (0.28 cu.m per 9.29 sq.m) of 4" (102 mm) veneer facing.

When laying slump block in a random ashlar pattern, a mason and a helper should lay 45 to 55 sq.ft. (1.27 to 1.55 sq.m) of 4" (102 mm) veneer facing per 8-hr. day.

Glazed Concrete Block. Concrete block may be ordered with a factory applied glazed surface on one or more sides in 8"x16" (200 x 400 mm), 8"x8" (200 x 200 mm), and 4"x16" (100 x 400 mm) face sizes and in even thicknesses of 2" (50 mm) through 12" (300 mm). Standard cap, base, and finished end units are also furnished. This glaze is a satin finish, factory applied to a 1/8" (3.12 mm) thickness and is available in a standard range of 48 colors and 11 scored patterns, with special colors and shapes possible. The glazing process is available to local sources through special licensing rather than being centrally manufactured and warehoused, and the local firm should always be consulted for availability and cost.

These units are especially useful where it is desired to have a block finish on one side and glazed on the other, because one unit provides both. This can mean a considerable savings in labor cost and floor space, when compared with building a 4" (100 mm) block partition and adding a 2" (50 mm) glazed furring block. The glazed surface meets both USDA and OSHA requirements for sanitary surfaces. The high performance standards of concrete block for fire resistance and insulation value are retained. Tolerance on face dimensions and face distortions is limited to 1/16" (1.56 mm).

Units are set in standard mortar and struck with a groover but may be pointed or grouted for more finished work. Acid cleaning solutions should not be used, and units must be kept clean as the work progresses. A final cleaning with a masonry cleaning compound should be figured.

Adding the glazed surface to a standard block will add around $1.90 to $2.20 per sq.ft. ($20.45 to $23.68 per sq.m) of glazed surface. Some colors will be more expensive than others, and the amount of scoring within the block surface will also affect the price. Two masons and one laborer will set 240 to 250 sq.ft. (22.29-23.22 sq.m) of 4" (100 mm) partition in running bond per day.

Labor Cost of 100 S.F. (9.29 Sq.M) of 4" (100 mm) Thick, Thick, 8" x 16" (200 mm x 400 mm) Concrete Blocked Glazed One Side Laid in Running Bond					
	Hours	Rate	Total	Rate	Total
Mason	7	$....	$....	$ 36.93	$ 258.51
Labor	3.5	$....	$....	$ 26.16	$ 91.56
Cost per 100 s.f.					$ 350.07
per s.f.					$ 3.50
per sq.m					$ 37.68

Add to the above for pointing, grouting, and cleaning as required.

MASONRY

Number of 6"x7-3/4"x6-3/4" Double Grooved Concrete Manhole Blocks and Quantity of Mortar Required to Lay Up Manholes and Catch Basins of Various Sizes

Depth of Manhole - Ft.		Internal Diameter of Manhole, Inches			
		30 "	36 "	42 "	48 "
4 ft	No. Block (Cone)	8	16	24	32
	No. Block (Bbl.)	35	32	27	20
	Total	43	48	51	52
	Mortar, c.f.	1 1/2	2	2 1/2	3
6 ft	No. Block (Cone)	8	16	24	32
	No. Block (Bbl.)	56	56	54	50
	Total	64	72	78	82
	Mortar, c.f.	2	3	4	5
8 ft	No. Block (Cone)	8	16	24	32
	No. Block (Bbl.)	77	80	81	80
	Total	85	96	105	112
	Mortar, c.f.	2 1/2	3 1/2	4 1/2	5 1/2
10 ft	No. Block (Cone)	8	16	24	32
	No. Block (Bbl.)	99	104	108	110
	Total	107	120	132	142
	Mortar, c.f.	3	4	5	6
12 ft	No. Block (Cone)	8	16	24	32
	No. Block (Bbl.)	119	128	135	140
	Total	127	144	159	172
	Mortar, c.f.	3 1/2	5	6 1/2	8
14 ft	No. Block (Cone)	8	16	24	32
	No. Block (Bbl.)	140	152	162	179
	Total	148	168	186	211
	Mortar, c.f.	4	5 1/2	7 1/2	9 1/2

Number of 152 x 197 x 171 mm Double Grooved Concrete Manhole Blocks and Quantity of Mortar Required to Lay Up Manholes and Catch Basins of Various Sizes

Depth of Manhole - M		Internal Diameter of Manhole, mm			
		750 mm	900 mm	1050 mm	1200 mm
1.22 meter	No. Block (Cone)	8	16	24	32
	No. Block (Bbl.)	35	32	27	20
	Total	43	48	51	52
	Mortar, cu.m	0.042	0.057	0.071	0.085
1.83 meter	No. Block (Cone)	8	16	24	32
	No. Block (Bbl.)	56	56	54	50
	Total	64	72	78	82
	Mortar, cu.m	0.057	0.085	0.113	0.142

Depth of Manhole - M		Internal Diameter of Manhole, mm			
		1925 mm	2000 mm	2025 mm	2000 mm
2.44 meter	No. Block (Cone)	8	16	24	32
	No. Block (Bbl.)	77	80	81	80
	Total	85	96	105	112
	Mortar, c.f.	0.071	0.099	0.127	0.156
3.05 meter	No. Block (Cone)	8	16	24	32
	No. Block (Bbl.)	99	104	108	110
	Total	107	120	132	142
	Mortar, cu.m	0.085	0.113	0.142	0.170
3.66 meter	No. Block (Cone)	8	16	24	32
	No. Block (Bbl.)	119	128	135	140
	Total	127	144	159	172
	Mortar, cu.m	0.099	0.142	0.184	0.227
4.27 meter	No. Block (Cone)	8	16	24	32
	No. Block (Bbl.)	140	152	162	179
	Total	148	168	186	211
	Mortar, cu.m	0.113	0.156	0.212	0.269

Concrete Manhole and Catch Basin Block

Concrete manhole and catch basin block are radial concrete units used for building circular manhole or catch basins in storm or sanitary sewers. The best quality units meet or exceed the specifications of the American Concrete Institute.

Size and Weight of Units. For manholes or catch basins having 48" (1200 mm) inside diameter, 10 units are usually required for each course; for 42" (1050 mm), 9 units; 36" (900 mm), 8 units; and 30" (750 mm), 7 units. However, one popular type of block requires 12 units per course in 48" (1200 mm) structures and 9 units in 36" (900 mm). A block 7-3/4" (193.75 mm) high and 6" (150 mm) thick weighs about 50 lbs. (22.6 kg) and a block 5-3/4" (143.75 mm) high and 6" (150 mm) thick weighs about 35 lbs. (15.8 kg).

Approximate Prices of Manhole Covers and Frames			
Steel / Cast Iron	18 " Diameter	500 mm diameter	$ 250.00
Steel / Cast Iron	24 " Diameter	600 mm diameter	$ 325.00
Steel / Cast Iron	36 " Diameter	1300 mm diameter	$ 720.00

Method of Construction. The concrete base is placed and then the first course of blocks is embedded at least 3" (75 mm) in the plastic concrete, leveling the units at once, insuring a watertight joint at the base.

Labor Costs for Concrete Block Manholes and Catch Basins. The estimated cost of labor for a concrete block manhole, 48" (1200 mm) inside diameter and 8'-0" (2.4 m) deep, is as follows:

MASONRY

Concrete Block Manhole and Catch Basin					
	Hours	Rate	Total	Rate	Total
Mason	8.0	$....	$....	$ 36.93	$ 295.44
Labor	8.0	$....	$....	$ 26.16	$ 209.28
Cost Per Manhole					$ 504.72

Add for Excavation, backfill, concrete base and cover

Precast Manhole 4' - 0" Diameter 8' - 0" Deep					
	Hours	Rate	Total	Rate	Total
Backhoe OP	1.5	$....	$....	$ 37.89	$ 56.84
Labor Foreman	1.5	$....	$....	$ 28.16	$ 42.24
Labor	3.0	$....	$....	$ 26.16	$ 78.48
Mason	2.0	$....	$....	$ 36.93	$ 73.86
Cost Per Manhole					$ 251.42

Add for Excavation, backfill, concrete base and cover

Seismic Concrete Masonry

In some regions of the U.S., there are seismic design requirements, and at times, additional material and labor are necessary to meet these local requirements. Additional design requirements should be included in the drawings and outlined in the specifications by the design engineer.

Reinforcement. In seismic zones 2, 3, and 4 on the accompanying map, the minimum area of reinforcement (in either direction) should not be less than 0.07% of the gross cross-sectional area of the wall. The sum of percents of horizontal and vertical reinforcement should be at least 0.2%. In other areas, where reinforcement is used in selected portions of the wall as needed to resist tensile stresses, there is no minimum percentage requirement for reinforcement. The size of bars used in reinforced masonry wall ranges from No. 4 through No. 11.

The minimum and maximum vertical reinforcement in masonry columns is 0.25 and 4% respectively. Minimum size of bar is No. 4 and the minimum number of bars is four.

When a building experiences earthquake vibrations, its foundation will move back and forth with the ground. These vibrations can be quite intense, creating stresses and deformations throughout the structure. Flexible structures, such as frame buildings without shear wall stiffening elements, will endure large deflections often resulting in extensive non-structural damage. On the other hand, loadbearing concrete masonry buildings are much stiffer than frame buildings and resist deflection. Non-structural damage is minimized, as is hazard from falling debris. Loadbearing buildings are typically designed to resist twice the lateral force as buildings without stiffening elements. While limited research has been conducted on the damping properties of loadbearing buildings, strength and structural field of performance have been demonstrated.

729

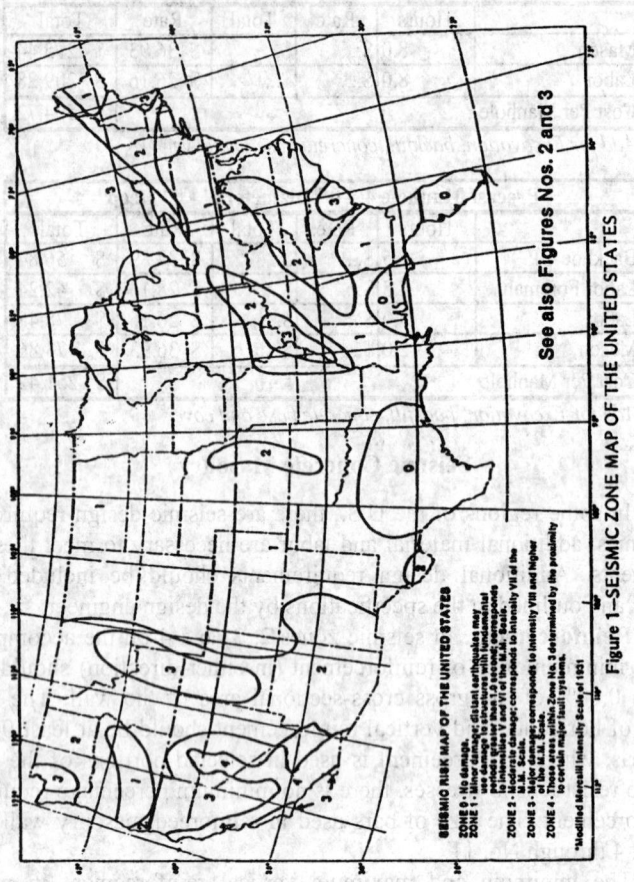

General. The latest edition of the Uniform Building Code requires every structure to be designed and constructed to resist minimum total lateral seismic forces assumed to act non-concurrently in the direction of each of the main axes of the structure.

LOADBEARING TILE

Loadbearing tile is used to carry part of the structural load of the building and is used for both interior and exterior walls. Some units have finished surfaces; some are scored for plaster or other applied finishes.

The American Society for Testing and Materials (ASTM) has set up three specifications covering structural clay loadbearing wall tile. Specification C-34 covers structural clay loadbearing wall tile in general and sets two grades: LBX for exposed locations; LB for units not exposed to frost or where protected with at least 3" (75 mm) of other masonry facing.

730

MASONRY

The tile is manufactured as end construction tile, to be placed in walls with axes of cells vertical and as side construction tile, to be placed with axes of cells horizontal. Average compressive strength of end construction tile, Grade LBX, is 1,400 psi (9.64 MPa), and for Grade LB, 1,000 psi (6.9 MPa). Average compressive strength for side construction tile is 700 psi (4.8 MPa), for both grades, based on gross area.

ASTM Specification C-212 covers structural clay facing tile designed for either interior or exterior exposure, including salt glazed and unglazed units. The two grades are FTX for smoothest faces, low absorption, and high degree of mechanical perfection, and FTS, which provides moderate absorption and mechanical perfection.

The compressive strengths given above for LBX grade tile apply here except that a special duty tile is available with 2,500 psi (17.2 MPa) and 1,200 psi (8.3 MPa) average compressive strength for end and side construction types respectively. The Facing Tile Institute's specifications list considerably higher strengths of 3,000 psi (20.7 MPa) and 2,000 psi (13.8 MPa) respectively, for "select" tiles.

ASTM Specifications C-126 covers ceramic glazed structural clay facing tile, facing brick and solid masonry units. Two grades are given: "S" (select) for use with comparatively narrow joints and "SS for ground edge tile. Grade S is available in all standard sizes except 7-3/4" x 15-3/4" (197 x 400 mm) and "G" (or "SS") is available in 7-3/4" x 15-3/4" (197 x 400 mm) units. These units also carry the designation, Types I and II, which indicates single or double finished faces. The Facing Tile Institute also lists a "B" quality ceramic glaze structural tile. These are not produced to meet a specification, but are units that may be available at the yard which fail to meet one or more of the specifications for "S" grade tile. The estimator should be aware, however, that even when "B" quality is acceptable on a job, it may be impossible to get all the special shapes required to complete the job in anything other than select quality.

04240 INTERIOR STRUCTURAL CLAY FACING TILE

Facing tile discussed below includes those units at least one of whose faces—ceramic glazed, salt glazed, or unglazed—are designed to be left exposed, either as a partition, as backup to another type of facing, or as the inner wythe of a cavity wall.

These units are manufactured in accordance with sizes and specifications adopted by the Facing Tile Institute. The four sizes adopted are modular.

Facing tile are available in such a wide variety of colors and finishes as to be adaptable to most any design requirement, and the estimator must be knowledgeable about the product and grade and the desired end result.

While most facing tile is furnished in 4" (100 mm) thicknesses, the estimator should always check the cost of using 2" (50 mm) soaps in combination with a locally produced backing material against the cost of the full width tile, particularly if the wall carries no superimposed loads. Often, structural tile is used where its loadbearing capacity is never utilized.

Mortar design mixes for tile follow those listed in the previous chapter for brick, except interior non loadbearing partitions may utilize a low strength type "O" mortar that calls for 0.111 cu.ft. (0.003 cu.m) of cement and 0.222 cu.ft. (0.006 cu.m) of lime for each cubic foot of sand. Note, however, that in much tile work white portland cement and white sand may be called for. Most facing tile is figured for 1/4" (6.25 mm) joints, which limits sand to that passing a No. 16 sieve.

In the figures given below, mortar is figured as two separated 1" wide joints, except for the bed joints of horizontal celled units, which are figured as full. Where walls are built two units thick, metal ties should be figured 16" (400 mm) vertically and 36" (900 mm) horizontally, which will be about 25 per 100 sq.ft. (9.29 sq.m). Also, if a solid filled collar joint is desired add 2.6 c.f. of mortar per 100 sq.ft. (9.29 sq.m) of wall.

Material Requirements for Facing Tile - 8W Series			
Size: 8 " x 16 " x 2 " or 4" thick (7 3/4" x 15 3/4") Actual face size			
Tile Per Sq.Ft. Wall = 1.125			
Mortar in cu. ft.:			
For vertical cell tile:			
0.77 per 100 sq. ft.			
6.87 per 1000 units			
For horizontal cell tile:			
1.23 per 100 sq. ft.			
10.92 per 1,000 units			
Cost: 2 " Thick	$ 2,720 per M. or	$ 3.05	Per sq.ft. glazed 1 side
4 " Thick	$ 3,000 per M. or	$ 3.37	Per sq.ft. glazed 1 side
4 " Thick	$ 4,580 per M. or	$ 5.14	Per sq.ft. glazed 2 sides
6 " Thick	$ 4,255 per M. or	$ 4.77	Per sq.ft. glazed 1 side
8 " Thick	$ 5,090 per M. or	$ 5.71	Per sq.ft. glazed 1 side
Size: 200 x 400 x 50 mm tk. 194 x 394 mm Actual face size			
Tile per sq.m wall = 12.11			
Mortar in cu.m:			
For vertical cell tile:			
0.02 per 9.29 sq.M			
0.19 per 1,000 Units			
For horizontal cell tile:			
0.03 per 9.29 sq.M			
0.31 per 1,000 units			
Cost: 50 mm thick =	$ 2,720 per M or	$ 32.86	Per sq.ft. glazed 1 side
100 mm thick =	$ 3,000 per M or	$ 36.24	Per sq.ft. glazed 1 side
100 mm thick =	$ 4,580 per M or	$ 55.33	Per sq.ft. glazed 2 sides
150 mm thick =	$ 4,255 per M or	$ 51.40	Per sq.ft. glazed 1 side
200 mm thick =	$ 5,090 per M or	$ 61.49	Per sq.ft. glazed 1 side

Material Requirements for Facing Tile 6T Series		
Size: 5 1/3 "x 12 "x 2", 4", 6" and 8" Thick		
Tile per sq. ft. wall = 2.250 (5 1/16" x 11 3/4") Actual face size		
Mortar in cu. ft.:		
For vertical cell tile:		
1.11 per 100 sq. ft.		
4.94 per 1,000 units		
For horizontal cell tile:		
1.80 per 100 sq. ft. for 4" wall		
7.98 per 1,000 units for 4" wall		
2.56 per 100 sq. ft. for 6" wall		
11.45 per 1,000 units for 6" wall		
3.36 per 100 sq. ft. for 8" wall		
14.93 per 1,000 units for 8" wall		
Cost: 2" thick $ 1,166 per M. or $ 3.07 Per sq. ft., glazed 1 side		
4" thick $ 1,485 per M. or $ 3.38 Per sq. ft., glazed 1 side		
4" thick $ 2,065 per M. or $ 5.16 Per sq. ft., glazed 2 sides		
6" thick $ 2,175 per M. or $ 4.80 Per sq. ft., glazed 1 side		
8" thick $ 2,600 per M. or $ 5.73 Per sq. ft., glazed 1 side		
Size: 133 x 300 x 50, 100, 150, 200 mm Tk		
(127 x 294 mm actual face size)		
Tile per sq.m wall = 24.2		
Mortar in cu.m:		
For vertical cell tile:		
0.03 per 10 sq.m		
0.14 per 1,000 units		
For horizontal cell tile:		
0.05 cu.m per 9 sq.m for 100 mm wall		
0.23 cu.m per 1,000 units for 100 mm wall		
0.07 cu.m per 9 sq.m for 150 mm wall		
0.32 cu.m per 1,000 units for 150 mm wall		
0.10 cu.m per 9 sq.m for 200 mm wall		
0.42 cu.m per 1,000 units for 200 mm wall		
Cost: 50 mm thick = $ 1,166 per M or $33.05 per sq.m., glazed 1 side		
100 mm thick = $ 1,485 per M or $36.38 per sq.m., glazed 1 side		
100 mm thick = $ 2,065 per M or $55.54 per sq.m., glazed 2 sides		
150 mm thick = $ 2,175 per M or $51.67 per sq.m., glazed 1 side		
200 mm thick = $ 2,600 per M or $61.68 per sq.m., glazed 1 side		

     Special Shapes. In addition to stretcher, or field units, these series also include many special shapes. These are given as belonging to groups; the higher the number, the more expensive the unit.

     In general, the various groups include the following, although not all units will be available in all tile.

733

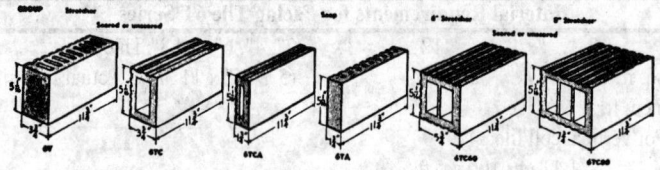

6T Series Structural Facing Tile, 5⅓"x12" Face

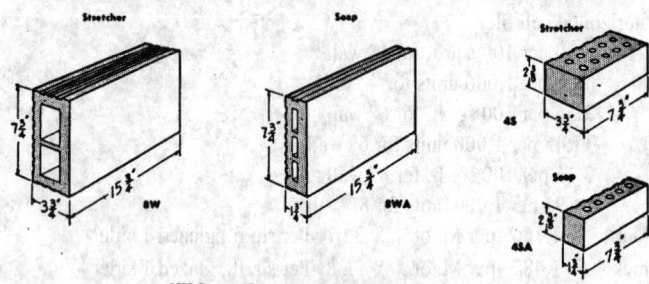

8W Series Structural Facing Tile, 8"x16" Face
4S Series 2⅔"x8" Face

Group I:
4" (100 mm) bed return, bullnose or square
2" (50 mm) and 4" (100 mm) soap return, bullnose or square
coved internal corner
octagonal external corner
octagonal interior corner
cost: 8W $2,960 per M
6T $1,915 per M

Group II:
2" (50 mm) and 4" (100 mm) bed, bullnose cap
2" (50 mm) bed, cove base
cost: 8W $4,980 per M
6T $2,220 per M

Group III:
6" (150 mm) and 8" (200 mm) bed, bullnose cap
4" (100 mm) bed, cove base
2" (50 mm) bed, round top cove base
4" (100 mm) and 8" (200 mm) bed, full and half ends glazed both sides
4" (100 mm) bed, bullnose cap, glazed both sides
4" (100 mm) bed, sloped sill
Standard radial external or internal field units:
2" (50 mm) and 4" (100 mm) bed, 6-3/8" (159 mm) high, bullnose cap unit
2" (50 mm) and 4" (100 mm) bed, 7-3/4" (194 mm) high, bullnose cap unit
cost: 8W $7,060 per M
6T $3,500 per M

734

MASONRY

Group IV:
4" (100 mm) bed, miter unit
4" (100 mm) bed, miter unit with top return
4" (100 mm) bed, miter unit with finished end
2" (50 mm) and 4" (100 mm) bed, bullnose return
4" (100 mm) bed, bullnose return with starter
2" (50 mm) and 4" (100 mm) bed, cove base with return
4" (100 mm) bed, round top cove base
6" (150 mm) bed, bullnose glazed both sides
6" (150 mm) and 8" (200 mm) bed, sloped sill
4" (100 mm) bed, bullnose starter
2" (50 mm) bed, bullnose reveal
2" (50 mm) bed, bullnose coved internal corner
2" (50 mm) bed, bullnose octagonal internal and external corners
4" (100 mm) bed, 6-3/8" (159 mm) high slope sill
4" (100 mm) bed, 7-3/4" (194 mm) high slope sill
cost: 8W $11,250 per M
6T $7,440 per M

Group V:
4" (100 mm) bed, bullnose with 4" (100 mm) return
4" (100 mm) bed, miter 7-3/4" (194 mm) high with face return
4" (100 mm) bed, round top cove base with 4" (100 mm) return
2" (50 mm) bed, round top cove base with 4" (100 mm) return
2" (50 mm) bed, bullnose radial
2" (50 mm) bed, bullnose cove base radial
4" (100 mm) bed, bullnose internal corner with 4" (100 mm) reveal
6" (150 mm) bed, 6-3/8" (159 mm) high slope sill
2" (50 mm) bed, round top cove base with square corner
2" (50 mm) bed, round top cove base with coved interior corner
2" (50 mm) bed, round top cove base with octagonal exterior corner
2" (50 mm) bed, round top cove base with octagonal internal corner
4" (100 mm) bed, field radial starter
cost: 8W $16,215 per M
6T $7,880 per M

Group VI :
4" (100 mm) bed, bullnose full end, or half end glazed both sides
8" (200 mm) bed, bullnose half end, glazed both sides
cost: 8W $20,050 per M
6T $9,725 per M

Group VII :
4" (100 mm) bed, bullnose radial starter
4" (100 mm) bed, cove base radial starter
cost: 8W $20,050 per M
6T $9,725 per M

Estimating Facing Tile Quantities

The most widely used estimating procedure is the wall area method, which is as follows:

Compute gross wall area (height times length). Where coved base or bullnosed caps are specified, do not figure these in overall height.

Determine net area of wall; deduct wall openings. These are computed by using the actual width multiplied by the actual height plus the height of special sill and lintel units. Where openings overlap base or cap units, do not include this in the opening height as it was not included in the gross area. Also deduct for other special shapes by multiplying their height by the following factors:

For exterior bullnose or square corners—1.167 per ft. (3.83 per m) of height.
For Interior coved corners—0.833 per ft. (2.73 per m) of height.
For bullnose or square jambs—0.75 per ft. (2.46 per m) of height.

Determine number of stretchers required by multiplying net area by number of stretchers required per square foot:

	per sq.m
30,58W:1.12	12.06
6T:2.25	24.22
4D:3.37	36.28

Compute number of standard fittings, such as corners and jambs, by the number of courses. Cove bases and caps, which should now be included as well as sills and lintels, are computed by the lineal foot. Miters, cove base corners, etc., should be taken off individually.

Summarize quantities and separate fittings in particular group.
Multiply by the unit cost figures.
Add 2% (or more, depending on job conditions) to cover breakage.

Example:

Step 1	Metric	
Wall Length = 29' - 0"	8.84 Meters	
Wall Height, exclusive of the base course = 8' - 0"	2.44 meters	
Gross Area = 8' 0 " x 29' - 0" = 232 Sq. Ft.	2.44 m x 8.84 m =	21.5696 m

Step 2	Metric	
Deductions		
Window Area 4' x 4.89' = 19.56 SF	1.22 x1.49 =	1.82 Sq.M
Door Area 2' x 6.67' = 13.34 SF	0.61 x 2.03 =	1.24 Sq.M
Exterior Corner 1.167' x 8' = 9.33 SF	0.355x 2.43 =	0.86 Sq.M
Interior Coved Corner = .833' x 8' = 6.66 SF	0.255x 2.43 =	0.62 Sq.M
Jambs (4'+ 4'+ 6.22+ 6.22) x .75' = 15.33 SF	(1.22+1.22+1.9+1.9)x.23 =	1.43 Sq.M
Total 64.21 SF		5.97 Sq.M
Net Wall Area = 232 SF minus 64.21 SF = 167.79 SF	(21.55 - 5.97)	15.58 Sq.M

Step 3
Number of "6T" stretchers required 167.79 Sq.Ft. x 2.25 tile per Sq. Ft. = 378 Ea.
Number of "6T" stretchers required 15.58 Sq.M. x 24.2 tile per Sq.M. = 378 Ea.

Step 4						
	Group					**Metric**
	I	II	III	IV	V	
Exterior Bullnose corner						**Exterior Bullnose corner**
18 courses of 5T4	18					18 courses of 5T4
Interior coved corner						**Interior coved corner**
18 courses of 4T8	18					18 courses of 4T8
Jambs						**Jambs**
24 courses of 6T4	24					24 courses of 6T4
24 courses of 3T4	24					24 courses of 3T4
Lintels						**Lintels**
5 lin. ft. of 6T20		5				1.52 lin.M. of 6T20
1 miter 6T30R (door)				1		1 miter 6T30R (door)
1 miter 6T30L (door)				1		1 miter 6T30L (door)
1 miter 8T31R (window)					1	1 miter 8T31R (window)
1 miter 8T31L (window)					1	1 miter 8T31L (window)
Sill						**Sill**
3 lin. ft. of 6T20		3				.91 lin.M. of 6T20
1 miter 8T31R					1	1 miter 8T31R
1 miter 8T31L					1	1 miter 8T31L
Cove Base						**Cove Base**
25 lin. ft. of 6T50			25			7.62 lin.M. of 6T50
1 outside corner 5T54L				1		1 outside corner 5T54L
1 inside cove corner 4T58L				1		1 inside cove corner 4T58L
1 bullnose jamb 6T504R				1		1 bullnose jamb 6T504R
1 bullnose jamb 6T504L				1		1 bullnose jamb 6T504L
Totals	84	8	25	6	4	**Totals**

Step 5 - Summary		
Group	**Series**	**Qty**
Stretcher	6T	378
I	6T	84
II	6T	8
III	6T	25
IV	6T	6
V	6T	4

Add a percentage for breakage between 2% and 5% depending on job conditions and distance of job from the plant. Chippage can frequently be used for special short lengths or other cuts.

A short cut method, devised by Stark Ceramics, is based on the following formula: (Length x height minus actual openings) x (cost per sq.ft. or sq.m) + (lin.ft. x lin.ft., or m x m, cost factor) = total material cost.

In this method actual total heights are used to figure cross areas, and openings are figured using actual height and width with no allowances for sills, lintels, caps, or bases. Lineal footage includes all special shapes: cove bases, internal coved corners, jambs, sills, lintels, caps, etc. The lineal foot cost factor is an average additional cost to cover these items. It is determined by taking an average of the group costs and deducting the stretcher unit cost already included in the net area figure.

Labor Laying Glazed Structural Facing Tile. To insure the most economical construction, as well as the most attractive appearance in the finished work, all cutting of tile on the job should be done with a power saw using a carborundum blade. The following labor quantities are based on the average job requiring not more than 20% of special shapes, such as bullnose corners, cove base, cap, etc. If the percentage of special pieces vary from the above, increase or decrease labor time proportionately.

Labor Laying Glazed Structural Facing Tile Labor Hours per 100 Pieces					
Nominal Size of Tile Inches	Side Glazed	Pieces per 8-Hr. Day	Mason Hours	Labor Hours	Nominal Size of Tile Metric
2 " x 5 1/3 " x 8 "	1	155 - 175	4.8	4.8	50 mm x 133 mm x 200 mm
4 " x 5 1/3 " x 8 "	1	155 - 175	4.8	4.8	100 mm x 133 mm x 200 mm
4 " x 5 1/3 " x 8 "	2	125 - 155	5.7	5.7	100 mm x 133 mm x 200 mm
6 " x 5 1/3 " x 8 "	1	155 - 175	4.8	4.8	150 mm x 133 mm x 200 mm
6 " x 5 1/3 " x 8 "	2	125 - 155	5.7	5.7	150 mm x 133 mm x 200 mm
8 " x 5 1/3 " x 8 "	1	125 - 155	5.7	5.7	200 mm x 133 mm x 200 mm
8 " x 5 1/3 " x 8 "	2	105 - 120	7.0	7.0	200 mm x 133 mm x 200 mm
2 " x 5 1/3 " x 12 "	1	115 - 135	6.4	6.4	50 mm x 133 mm x 300 mm
4 " x 5 1/3 " x 12 "	1	115 - 135	6.4	6.4	100 mm x 133 mm x 300 mm
4 " x 5 1/3 " x 12 "	2	90 - 110	8.0	8.0	100 mm x 133 mm x 300 mm
6 " x 5 1/3 " x 12 "	1	105 - 125	7.0	7.0	150 mm x 133 mm x 300 mm
6 " x 5 1/3 " x 12 "	2	80 - 100	8.9	8.9	150 mm x 133 mm x 300 mm
8 " x 5 1/3 " x 12 "	1	85 - 100	8.7	8.7	200 mm x 133 mm x 300 mm
8 " x 5 1/3 " x 12 "	2	65 - 85	10.6	10.6	200 mm x 133 mm x 300 mm
2 " x 8 " x 16 "	1	65 - 80	11.0	11.0	50 mm x 200 mm x 400 mm
4 " x 8 " x 16 "	1	55 - 65	13.3	13.3	100 mm x 200 mm x 400 mm

Allow for Hoisting Engineer if Required
Nominal size includes thickness of standard mortar joint (1/4" for glazed Tile) in length, height and thickness

Tile 1-3/4" (43.75 mm) thick is usually used for wall furring or partition wainscoting. Tile glazed one side is often used for partitions that are glazed on one side and plastered on the other side. Tile glazed two sides is often used for partition walls glazed on both sides of wall.

Unglazed Structural Clay Tile

Backup tile is sized for backing up brickwork, but is also used with stone and other facing. These units are available with nominal face sizes of 5"x12" (125 x 300 mm), 8"x12" (200 x 300 mm), and 12"x12" (300 x 300 mm), in thickness of 4" (100 mm), 6" (150 mm), 8" (200 mm), and 12" (300 mm), and are usually furnished scored one side, smooth on the other. These units are often available, especially in larger sizes, in special designs with cast-in handles for greater speed and ease in handling and non-continuous mortar joints to protect against moisture penetration. Generally, backup tile is sized for 1/2" (12.50 mm) joints, the mortar amount for which given below, but certain units will be found to allow for less. Dimensional tolerances are wider in this type than in facing tile. The mortar figures below assume joints consist of two separated 1" (25 mm) wide joints and bed joints are two separated 2" (50 mm) wide joints.

Nominal 5"x12" (125x300mm) tile:
4-7/8" (121.87) x 11-1/2" (287.50 mm) x 3-1/2" (87.50 mm), 5-1/2" (137.50 mm), 7-1/2" (187.50 mm), and 11-1/2" (287.50 mm) actual
Tile per sq.ft. of wall, 2.25 (.21 sq.m)
Mortar per 100 sq.ft. of wall, (9.295 cu.ft. Mortar per 1,000 units, 16.69 cu.ft. (0.47 cu.m)

Nominal 8"x12" (200x300mm) tile:
7-1/2" (187.50 mm) x 11-1/2" (287.50 mm) x 3-1/2" (87.50 mm), 5-1/2" (137.50 mm), 7-1/2" (187.50 mm), and 11-1/2" (287.50 mm) actual
Tile per sq.ft. of wall, 1.50 (0.14 per sq.m)
Mortar per 100 sq.ft. of wall, 2.73 cu.ft. (per 9.29 sq.m of wall, 0.08 cu.m)
Mortar per 1,000 units, 18.23 cu.ft. (0.51 cu.m)

Labor Laying Backup Tile. The following quantities are based on the average job, having the usual number of openings, pilasters, etc.

Following quantities are based on the average project, having the usual number of openings, pilasters, etc.						
Size Tile, Inches	No. Pcs. Laid per 8-Hr. Day		Labor Hrs per 100 Pcs.			Size Tile, Metric
			Mason	Labor	Engr.*	
4 " x 5 " x 12 "	330	- 370	2.3	2.3	0.3	100 mm x 125 mm x 300 mm
8 " x 5 " x 12 "	250	- 290	3.0	3.0	0.3	200 mm x 125 mm x 300 mm
4 " x 8 " x 12 "	200	- 225	3.8	3.8	0.3	100 mm x 200 mm x 300 mm
8 " x 8 " x 12 "	155	- 185	4.7	4.7	0.5	200 mm x 200 mm x 300 mm
Size Tile, Inches	No. Sq. Ft. Laid per 8-Hr. Day		Labor Hrs per 100 Pcs.			Size Tile, Metric
			Mason	Labor	Engr.*	
4 " x 5 " x 12 "	150	- 170	5.0	5.0	0.6	100 mm x 125 mm x 300 mm
8 " x 5 " x 12 "	115	- 130	6.5	6.5	0.7	200 mm x 125 mm x 300 mm
4 " x 8 " x 12 "	135	- 150	5.7	5.7	0.6	100 mm x 200 mm x 300 mm
8 " x 8 " x 12 "	110	- 132	6.6	6.6	0.7	200 mm x 200 mm x 300 mm

Add Operating Engineer if Required

Unit Wall Tile

Some tile units are designed to be used for a complete single unit, loadbearing, nominal 8" (200 mm) wall construction, with finished surfaces inside and out. These are usually manufactured under special patents. Some of these units may have one finished surface only and be used in combination with an inside backup.

Faces are available in buff unglazed, either smooth or Ruggtex finish; salt glazed, red textured (to give the appearance of high quality face brick); and ceramic glaze. Certain units are offered with supplemental full and half jamb corners, and lintel and pilaster units along with interior stretcher units to match the interior face of the unit tile.

Units are also available in 4" (100 mm) widths for use as exterior wythes in cavity wall construction. A wall of this type is highly impervious to water, thoroughly fire, termite, and vermin proof, and cannot rot or decay. Textures offer a broad range including brick-like textures and colors suitable for homes and schools; glazed for situations where a sanitary finish is desired, and smooth faced units in natural range shades suitable for larger structures and areas.

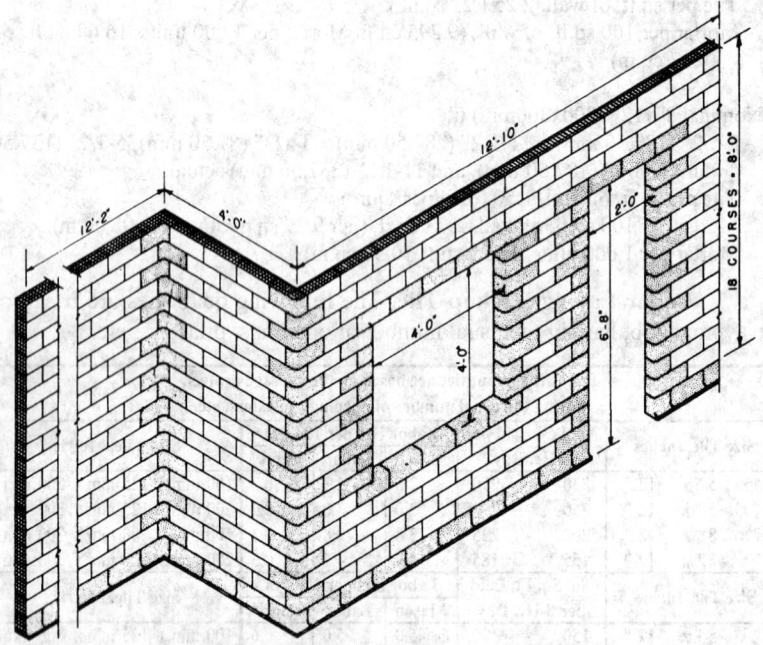

Example of Wall Area Method for Determining Facing Tile Quantities

The estimator must know exactly the quality of interior finish desired. While it is possible to obtain high grade finishes and workmanship both sides, such high quality is not always required, especially in larger warehouse and manufacturing buildings, and large economies can result where "select one face" units can be used. Often the lower wall may be constructed "select two faces" while "select one face" is used above. The estimator must expect a variation of 1/8" (3.12 mm) in thickness, and if this is not acceptable, should plan to use a built-up wall with a separate interior facing.

Unit wall tile comes in two nominal face sizes: 4"x12" (100 x 300 mm), and 5-1/3"x12" (133.25 x 300 mm), in 4" (100 mm) and 8" (200 mm) widths.

The 4"x12" (100 x 300 mm) units, which include Uniwall and Tex Dri-wall, have an actual face dimension of 3-5/8" x11-5/8" (90.62 x 290.62 mm) with 3-5/8" (90.62 mm) and 7-5/8" (190.62 mm) thicknesses. These

740

MASONRY

units will lay up 300 per 100 sq.ft. (9.29 sq.m) of wall and require 7.5 cu.ft. (0.21 cu.m) of mortar for 4" (100 mm) walls; 9.75 cu.ff. (0.28 cu.m) for 8" (200 mm) walls. Labor laying 4"x12" (100 x 300 mm) unit wall tile:

Following quantities are based on the average Project				
Size Tile, Inches	No. Pcs. Laid per 8-Hr. Day	Labor Hrs/100 Pcs.		Size Tile, Metric
		Mason	Labor	
3 5/8 " x 3 5/8 " x 11 5/8 "	170 - 200	4.3	4.3	91 mm x 91 mm x 291 mm
7 3/4 " x 3 5/8 " x 11 5/8 "	150 - 170	5.0	5.0	194 mm x 91 mm x 291 mm
Size Tile, Inches	No. Sq. Ft. Laid per 8-Hr. Day	Labor Hrs/100 Pcs.		Size Tile, Metric
		Mason	Labor	
3 5/8 " x 3 5/8 " x 11 5/8 "	57 - 67	13.0	13.0	91 mm x 91 mm x 291 mm
7 3/4 " x 3 5/8 " x 11 5/8 "	50 - 57	15.0	15.0	194 mm x 91 mm x 291 mm

The 5-1/2"x12" (137.50 x 300 mm) units have actual face dimensions of 5"x11-5/8" (125 x 290.62 mm) with 3-5/8" (90.62 mm) and 7-5/8" (190.62 mm) thicknesses, though they might vary somewhat in various parts of the country from 4-7/8" (121.87 mm) to 5-1/16" (126.56 mm) in height and 11-1/2" (287.50 mm) to 11-3/4" (193.75 mm) in length. Units can be figured to lay up 225 tile per 100 sq.ft. (9.29 sq.m) of wall and require 8.5 cu.ft. (0.24 cu.m) of mortar for 4" (100 mm) walls and 9.25 (0.26 cu.m) for 8" (200 mm) walls, the latter varying as to actual face size. Labor laying 5-1/3"x12" (135 x 305 mm) unit tile:

Following quantities are based on the average Project				
Size Tile, Inches	No. Pcs. Laid per 8-Hr. Day	Labor Hrs/100 Pcs.		Size Tile, Metric
		Mason	Labor	
3 5/8 " x 5 " x 11 5/8 "	175 - 200	4.3	1.9	91 mm x 125 mm x 291 mm
7 5/8 " x 5 " x 11 5/8 "	130 - 155	5.5	2.5	191 mm x 125 mm x 291 mm
Size Tile, Inches	No. Sq. Ft. Laid per 8-Hr. Day	Labor Hrs/100 Pcs.		Size Tile, Metric
		Mason	Labor	
3 5/8 " x 5 " x 11 5/8 "	80 - 90	9.5	4.1	91 mm x 125 mm x 291 mm
7 5/8 " x 5 " x 11 5/8 "	60 - 70	12.3	5.5	191 mm x 125 mm x 291 mm

Clay Masonry Shading Devices

In addition to shades, blinds, and drapes, an effective sun control is the egg crate screen wall design, which baffles the low east and west rays of the sun and the high, direct south rays.

Clay masonry units in glazed and unglazed finishes are available to construct such baffles. They come in a large range of sizes, shapes, and designs. Selection will depend on the degree of shading desired, the wall thickness necessary to provide structural stability, and aesthetic considerations. Glazed solar screen units 4" (100 mm) thick will cost about $9.00 per unit.

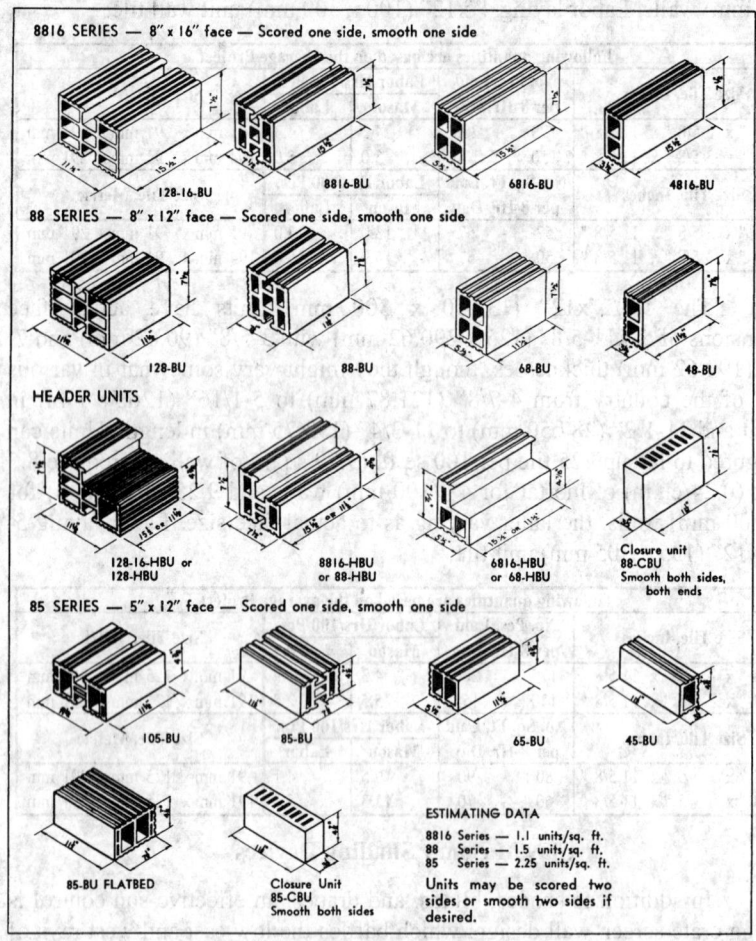

DENISON BACK UP TILE

8816 SERIES — 8" x 16" face — Scored one side, smooth one side

128-16-BU
8816-BU
6816-BU
4816-BU

88 SERIES — 8" x 12" face — Scored one side, smooth one side

128-BU
88-BU
68-BU
48-BU

HEADER UNITS

128-16-HBU or 128-HBU
8816-HBU or 88-HBU
6816-HBU or 68-HBU
Closure unit 88-CBU Smooth both sides, both ends

85 SERIES — 5" x 12" face — Scored one side, smooth one side

105-BU
85-BU
65-BU
45-BU

85-BU FLATBED
Closure Unit 85-CBU Smooth both sides

ESTIMATING DATA

8816 Series — 1.1 units/sq. ft.
88   Series — 1.5 units/sq. ft.
85   Series — 2.25 units/sq. ft.

Units may be scored two sides or smooth two sides if desired.

DENISON STRUCTURAL CLAY TILE
MASON CITY BRICK AND TILE COMPANY, MASON CITY, IOWA

04250 UNIT MASONRY/CERAMIC VENEER

Terra cotta is made of a high grade clay (usually a mixture of several different kinds), to which is added calcined clay to uniformly control the shrinkage and prevent undue warping.

Terra cotta is used for the same purposes as materials like sandstone, limestone, granite, and press brick, for interior or exterior trimmings or for entire fronts. It is also used extensively as glazed terra cotta for light courts, for the interior walls of swimming pools, operating rooms in hospitals, clinics

742

and patient rooms, corridors, vestibules and stair halls (full height of wainscot), and in large kitchens where a ceramic material is required for sanitary purposes as well as for fire resistance. It is used for service stations, power houses, and rooms that house machinery, because it can be kept free from dust, dirt, and grease. Its unlimited possibilities in color effects make it an ideal material for store fronts. It is used as a lining for the interior of restaurants, palm courts, hotels, residences, and railroad stations. In stations and public buildings it is also used in the construction of ceilings and domes. In fact, it is a universal building material with the advantage over other materials that it can be produced in almost any color and can be glazed or unglazed with any surface treatment desired. It furnishes unlimited possibilities to the architect not only in color effect but in the beauty of the original modeling in the plastic clay, an effect that cannot be obtained in carved stone except at tremendous cost.

Method of Manufacturing Architectural Terra Cotta. After shop drawings have been approved, full size models are made of plaster to a shrinkage scale, that is, to an increased size to allow for the shrinkage in drying and burning. A model is made for each different shape on the shop drawings. If a piece is ornamental, such ornament as may be required is modeled in clay and attached to the plaster model. From the models, sectional molds of plaster are cast, from which the required number of pieces of terra cotta is produced. The mixture of clays and fusible materials used in forming terra cotta is carefully selected and proportioned to give the desired degree of plasticity when mixed with water. The composition must be such that when fired at high temperatures, it will produce a homogeneous body amply strong to support the required loads.

Forming terra cotta is a manual operation and consists of pressing the soft clay mixture into the molds. The walls of the pieces should be not less than one-inch thick following the contour of the mold, and the partitions should be properly sized and spaced to produce the strength required. The pressed piece remains in the mold until the clay stiffens, after which it is removed, retouched, and placed in the dryer where most of the moisture in the clay is removed.

Following the drying process the exposed surfaces of each piece are sprayed with the ceramic mixture, which when fired develops into the desired glaze color. The pieces are then placed in kilns and fired by gradually increasing the temperature to 2,000°F (1093°C) or more, depending on the temperature of maturity of the clay and glaze. After proper firing the kiln is allowed to cool slowly to normal temperature.

After firing, the terra cotta is fitted for alignment by cutting or grinding and marked to correspond with the piece numbers on the shop drawings.

Method of Manufacturing Ceramic Veneer. Ceramic veneer is the usual term applied to machine made architectural terra cotta and the process of manufacturing is the same as for handmade pieces, except that the material is extruded instead of hand pressed into molds.

A machine mixes the proportioned clay with water and forces the plastic mixture through a steel die, which forms it into the desired shape. In most cases, before entering the die, the plastic mixture passes through a de-airing chamber, where it is subjected to a partial vacuum, and most of the entrapped air is removed. The pieces are then dried and surface finished to a true plane. Subsequent operations are the same as for handmade pieces and include application of ceramic mixture, firing, fitting, and marking.

Cost of Architectural Terra Cotta and Ceramic Veneer. In estimating the cost of architectural terra cotta and ceramic veneer, there is no unit that can be taken as a definite basis for estimating. All estimates are usually made by the manufacturer in a lump sum price for the entire job.

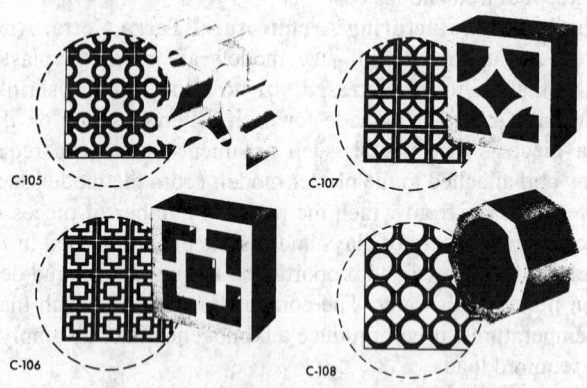

Four Patterns of Solar Screen Tile
7 5/8" x 7 5/8" (194 mm x 194 mm) for 3/8" (9.5 mm) mortar joints
Thickness: 3 5/8" (129 mm) and 7-3/4" (197 mm)

To be able to accurately arrive at the cost of these materials, an estimator must have an intimate knowledge of the manufacturing process. Each job is a custom job. There are no standard sizes, colors, or shapes that are stocked. One manufacturer suggests that the cost of architectural terra cotta will vary from $550.00 to $790.00 and higher per ton ($541.33 to 802.64 per tonne) and that ceramic veneer will cost from $3.90 to $6.25 and higher per sq.ft. ($41.98 to 67.27 per sq.m). The extreme cost range emphasizes the necessity of obtaining firm quotes from manufacturers.

Some of the conditions governing cost of both architectural terra cotta and ceramic veneer are design of work, character and amount of ornamentation, amount of duplication, sizes of pieces, colors of glaze, weight, supply and demand, and shipping conditions.

Estimating Quantities of Architectural Terra Cotta and Ceramic Veneer. When estimating quantities of terra cotta from the plans or details, all measurements should be squared; i.e., all molded courses,

744

cornices, round columns, column caps, bases, coursed ashlar, sills, and copings should be figured from the extreme dimensions and reduced to cubic feet (cubic meters).

The weight of terra cotta varies from about 80 lbs. per cu.ft. (1281 kg per cu.m) for plain 4" (100 mm) thick ashlar to 60 lbs. per cu.ft. (961 kg per cu.m) for large molded courses. For an average job about 72 lbs. per cu.ft. (1153 kg per cu.m), equivalent to 28 cu.ft. per ton of 2,000 lbs. (1153 kg per cu.m) may be used.

After the quantities have been taken off the plans and reduced to cu. ft. (cu.m), they may either be priced in that manner, or multiplying the number of cu. ft. by 72 lbs. (cu.m x 1153 kg) will give the total weight of the terra cotta required for the job.

Ceramic veneer quantities are measured by the square foot area and should include all returns at doors, windows, soffits, etc.

Cost of Handling and Setting Architectural Terra Cotta. The labor cost of handling and setting architectural terra cotta will vary with the sizes of the different pieces, the amount of duplication and the method used in handling and setting same.

Terra cotta weighs about half as much as stone, so it is seldom necessary to use derricks in handling and setting. It is usually furnished in sizes that can be conveniently handled by 2 or 3 workers. Considerable space is required on the job for storing the terra cotta, and as it is sent to the job, each piece is marked to show its exact position in the building and must be carefully handled and placed in convenient piles. When handling the terra cotta on the job, it is customary to load it into wheelbarrows or stone barrows and wheel it to the location where it is to be used. However, if it is necessary to hoist it to the upper floors or roof of the building, the barrows can be placed on the material hoist, raised to the proper floor, and then wheeled to the location.

Because a great deal of terra cotta consists of trimmings or 4" (102 mm) ashlar, it requires considerable handling to set a cubic foot. Most ashlar is 4" (100 mm) thick, and it is necessary to set terra cotta in 3 sq.ft. of wall to set 1 cu.ft. (.28 sq.m to set .02 cu.m). This also applies to sills, lintels, etc.

Based on the average job, consisting of ashlar, sills, lintels, ornamental trimmings, etc., a mason will set about 35 to 40 cu.ft. (.99 to 1.13 cu.m)—105 to 120 sq.ft. (.93 to -11.14 sq.m) of wall—per 8-hr. day, at the following labor cost:

Labor Costs to Set 100 Cu. Ft. of Wall						
	Hours	Rate	Total	Rate	Total	
Terra Cotta Setter	21	$....	$....	$ 36.93	$	775.53
Helper	21	$....	$....	$ 26.16	$	549.36
Labor Sorting, Handling, Wheeling, etc.	44	$....	$....	$ 26.16	$	1,151.04
Cost per 100 Cu.Ft.					$	2,475.93
Cost per Cu. Ft.					$	24.76
Per Sq.Ft. (300)					$	8.25
Per Ton (28 Cu.Ft.)					$	693.26
Cost per 100 Cu.M.					$	87,426.91
Cost per Cu. M.					$	874.27
Per Sq.M. (300)					$	88.84
Per Metric Ton					$	764.18

For labor cost of setting plain terra cotta wall ashlar, use same quantities and costs as given on the previous pages under "Labor Laying Glazed Structural Facing Tile".

On more complicated work, consisting of gothic architecture, balusters and balustrade, glazed and enameled terra cotta, and other intricate work that has a great deal of detail and requires the utmost care in setting, a terra cotta setter will set only 20 to 25 cu.ft. (0.56 to0.71 cu.m)-60 to 75 sq.ft. (5.57 to6.96 sq.m) of wall-per 8-hr. day, at the following labor cost:

Labor Costs to Set 100 Cu. Ft. of Wall						
	Hours	Rate	Total	Rate	Total	
Terra Cotta Setter	36	$....	$....	$ 36.93	$	1,329.48
Helper	36	$....	$....	$ 26.16	$	941.76
Labor Sorting, Handling, Wheeling, etc.	66	$....	$....	$ 26.16	$	1,726.56
Cost per 100 Cu.Ft.					$	3,997.80
Cost per Cu. Ft.					$	39.98
Per Sq.Ft. (300)					$	13.33
Per Ton (28 Cu.Ft.)					$	1,119.38
Cost per 100 Cu.M.					$	141,165.25
Cost per Cu. M.					$	1,411.65
Per Sq.M. (300)					$	143.44
Per Metric Ton					$	1,233.89

Small lintels and similar courses that have to be hung might cost 25% to 50% higher, but these items are exceptional.

In general, the voids in architectural terra cotta should be completely filled with setting mortar or grout to prevent the possible accumulation of moisture in the cavities.

All supporting iron should be firmly embedded in masonry and completely covered with mortar to prevent rusting. Bronze or other non-

MASONRY

corrosive metal or cadmium-coated steel anchors should invariably be used for suspended features or for anchoring or bracing freestanding work.

Mortar Required for Setting Architectural Terra Cotta. Under average conditions it will require about 4 cu.ft. (0.11 cu.m) of mortar to set one ton of terra cotta, or about 1/2 cu.yd. of mortar per 100 cu.ft. (0.343 cu.m of mortar per 2.8 cu.m).

The following proportions are recommended as being suitable for making satisfactory mortar for setting terra cotta: 1 part standard portland cement, 1 part lime putty, and 6 parts sand.

Do not under any circumstances use plaster of Paris or salt in the mortar.

All exposed joints in projecting and overhanging features, parapet work, and work standing free should be raked out and pointed with a good asphaltic elastic cement.

Never use a cement that swells in setting for filling the voids in terra cotta. Use only the best quality of cement.

Handling and Setting Ceramic Veneer. Ceramic veneer is available in two types: adhesion type, commonly called "thin" ceramic veneer, and anchored type.

Adhesion Type Ceramic Veneer. Adhesion type ceramic veneer is not over 1-1/8" (28.12 mm) in thickness, and the maximum face areas of individual slabs do not exceed 540 sq. in. Maximum overall face dimensions are 18"x30" (450 x 750 mm) or 20"x27" (500 x 675 mm). Actual face dimensions will vary with the manufacturer.

This type of ceramic veneer requires no metal anchorage. It is held in place by the adhesion of the mortar to the veneer body and the backing wall. The overall thickness from the face of the veneer to the face of the backing wall is 1-3/4" to 2" (43.75 to 50 mm).

Adhesion type ceramic veneer may be applied to a variety of backings, such as concrete, masonry, wood, and metal lath. Where the backing is wood, a sheet of waterproof building paper should be installed first, followed by a layer of wire mesh. In the case of metal lath backing or wire mesh on wood backing, 9-1/4" (230 mm) scratch coat of mortar should be applied in advance of the setting of veneer slabs.

Mortar Required for Adhesion Type Ceramic Veneer. Under average conditions, it will require 8 to 10 cu.ft. (0.22-0.28 cu.m) of mortar to set 100 sq.ft. (9.29 sq.m) of adhesion type ceramic veneer.

The following proportions are recommended as being suitable for making satisfactory mortar for setting adhesion type ceramic veneer: 2 parts portland cement, 1 part lime putty, and 8 parts sand. A stearate type admixture may be added to the mix in the proportions recommended by the manufacturer.

Setting Adhesion Type Ceramic Veneer. Strict compliance with the manufacturer's recommendations for setting adhesion type ceramic veneer is necessary to assure the development of full shearing strength of veneer to backing. Step-by-step setting directions, recommended by manufacturers are as follows.

All units must be soaked in clean water for at least one hour before setting. The backing wall surface should also be damp at the time of setting. Immediately before setting, the backing wall and the backs of veneer units should both receive a brush coat of portland cement and water. One-half of the setting mortar should be spread on the back of the unit to be set and the other half on the wall surface to be covered. The unit is then set in place, using wood wedges to maintain correct bed joint thickness. Sufficient mortar should be used to create a slight excess, which will be forced out of the joints from the back when the unit is tapped into place, eliminating all air pockets and filling all voids. Face joints may be pointed as soon as mortar has set sufficiently to support the veneer slabs. As a final step, the surface of the veneer should be washed down with clean water.

Cost of Handling and Setting Adhesion Type Ceramic Veneer. On most jobs adhesion type ceramic veneer is delivered to the job in trucks and must be carefully unloaded, stored, and protected in advance of the setting. Two laborers should do this work at the rate of 360 to 400 sq.ft. (33.44 to 37.16 sq.m) of veneer area per 8-hr. day.

Labor Costs Handling and Setting 100 Sq. Ft. Adhesion Type Ceramic Veneer on Concrete or Masonry Backing					
Unloading and Storing	Hours	Rate	Total	Rate	Total
Labor	4.2	$....	$....	$ 26.16	$ 109.87
Setting	Hours	Rate	Total	Rate	Total
Mason	11.5	$....	$....	$ 36.93	$ 424.70
Helper	11.5	$....	$....	$ 26.16	$ 300.84
Cost per 100 Sq. Ft.					$ 835.41
Per Sq.Ft.					$ 8.35
Cost per 10 Sq. M.					$ 89.93
Cost per Sq.M.					$ 8.99

On average jobs, where most veneer consists of typical panels, a mason and a helper should set 65 to 75 sq.ft. (6.03 to 6.96 sq.m) of veneer per 8-hr. day.

If veneer is to be applied to a backing of metal lath, add cost of 1/4" (6.25 mm) scratch coat of mortar.

If veneer is to be applied to a wood backing, add cost of one layer of waterproof building paper, one layer of wire mesh, and 1/4" (6.25 mm) scratch coat of mortar.

Above costs do not include any allowance for scaffolding or hoisting. If hoisting is required, add 1 hr. hoisting engineer and 2 hrs. labor per 100 sq.ft. (9.29 sq.m) of veneer.

Anchored Type Ceramic Veneer. Anchored type ceramic veneer is recommended where a larger size slab is desired. Anchored type slabs are available in face sizes considerably larger than the adhesion type, again

748

depending on the manufacturer. Ribs or scoring are provided on the backs of such units and the overall thickness of the slabs range from 2" (50 mm) to 2-1/2" (62.50 mm). Depending on slab thickness, a total of 3" (75 mm) to 4-1/2" (112.50 mm) is required from rough wall to finished veneer surface to provide adequate grout space between the veneer and the backing. Anchor holes are provided in the bed edges of the slabs for the installation of loose wire anchors, which in turn are fastened to pencil rods anchored to the backing. Once the wire anchoring is in place, the units are bonded to the wall by a poured grout core.

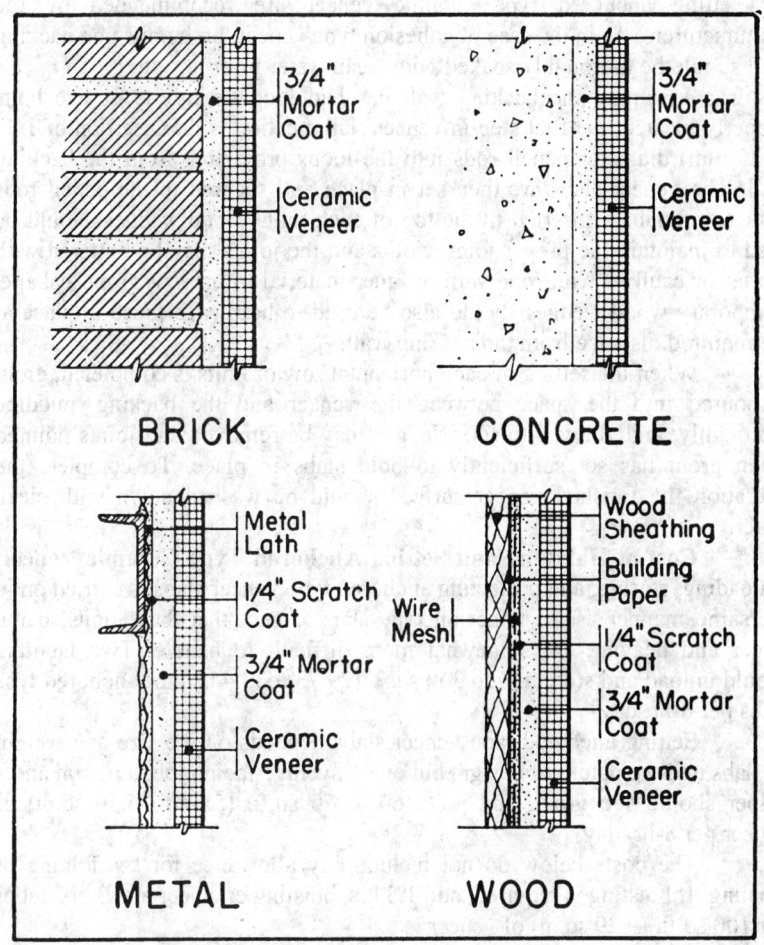

Method of Setting Adhesion Type Ceramic Veneer

Grout Required for Anchored Type Ceramic Veneer. For ordinary work, it will require 20 to 25 cu.ft. (0.56 to .71 cu.m) of grout to fill the space back of 100 sq.ft. (9.29 sq.m) of anchored type ceramic veneer.

749

The following proportions are recommended for a suitable grout mix: 1 part portland cement, 1 part sand, and 5 parts graded pea gravel passing a 1/4" (6.3 mm) sieve. Sufficient water should be added to cause the mixture to flow readily.

In addition to the above grout, it may require 1 to 2 cu.ft. (0.02 to 0.05 cu.m) of mortar per 100 sq.ft. (9.29 sq.m) of veneer for buttering the joints in setting to prevent leakage of grout. Any workable mortar mix may be used.

Setting Anchored Type Ceramic Veneer. The following directions for setting anchored type ceramic veneer are recommended by the manufacturers. As in the case of adhesion type veneer, both units and backing wall should be thoroughly soaked with clean water prior to setting.

Assuming the backing wall has had loop tie inserts or eye bolts properly located, the first step in veneer slab erection is the insertion of 1/4" (6.25 mm) diameter pencil rods into the loops projecting from the backing wall. The veneer slabs are then set in place and secured to the pencil rods with wire anchors inserted in the top of each unit. Wood wedges should be used to maintain the proper joint widths and the joints may be buttered with mortar or caulked with rope yarn or other material to prevent grout leakage. Temporary wood wedges should also be used to hold each piece in place at the required distance from the backing wall.

When the setting of each horizontal row of units is completed, grout is poured into the space between the veneer and the backing, puddled thoroughly, and allowed to set. Wedges may be removed and joints pointed when grout has set sufficiently to hold slabs in place. To complete the operation the finished veneer surface should be washed down with clean water.

Cost of Handling and Setting Anchored Type Ceramic Veneer. Unloading, storing, and protecting anchored type veneer slabs is carried on in the same manner as for adhesion type slabs except that these units, being larger and heavier, are somewhat more difficult to handle. Two laborers should unload and store 320 to 360 sq.ft. (29.73 to 33.44 sq.m) anchored type slabs per 8-hr. day.

Setting anchored type veneer slabs depends on the size and weight of slabs and the nature of design, but on an average facing job, a mason and a helper should set, grout, and point 60 to 70 sq.ft. (5.57 to 6.50 sq.m) of veneer per 8-hr. day.

The costs below do not include any allowance for scaffolding or hoisting. If hoisting is required, add 1.5 hrs. hoisting engineer and 3 hrs. labor per 100 sq.ft. (9.29 sq.m) of veneer.

MASONRY

Labor Costs Handling and Setting 100 Sq. Ft. Anchored Type Ceramic Veneer					
Unloading and Storing	Hours	Rate	Total	Rate	Total
Labor	4.7	$....	$....	$ 26.16	$ 122.95
Setting	Hours	Rate	Total	Rate	Total
Mason	12.3	$....	$....	$ 36.93	$ 454.24
Helper	12.3	$....	$....	$ 26.16	$ 321.77
Cost per 100 Sq. Ft.				$	898.96
Per Sq.Ft.				$	8.99
Cost per 10 Sq. M.				$	96.77
Cost per Sq.M.				$	9.68

Cleaning Down or Washing Terra Cotta. Broadly speaking, there are three different surface finishes in terra cotta: lustrous glazed, matt glazed, and unglazed. The glazed surfaces are the easiest to clean.

For both lustrous and matt glazed finishes a good abrasive soap or washing powder is best. Where required, there can be added to the latter a slight proportion of sharp sand. A stiff lather should be made and the surface scrubbed hard, allowing the lather to remain long enough to soften the dirt and then rinse off with clean water. Acid is not necessary for cleaning glazed surfaces.

In the past, for cleaning unglazed terra cotta, the addition of a slight proportion of commercial muriatic acid was recommended. One quart (0.94 liter) of acid to 4 gallons (15 liters) of water was sufficient in ordinary cases. In no event did it exceed 1-1/2 pints (0.71 liter) of acid to one gallon of water, and care was taken to rinse off the acid wash shortly after applying.

Acid was used with acid-resistant pails and fiber brushes. Metal pails caused a chemical reaction where the solution became a yellow stain instead of a cleaning fluid. Never use hydrofluoric acid. Many cleaners, in order to save hand labor, employ strong solutions of it, which is highly injurious.

Pointing Terra Cotta. Recommended practice is to strike the joints as work progresses. Pointing mortar applied later is usually of different composition and so of questionable permanence.

Cold Weather Protection. Architectural terra cotta or ceramic veneer should not be set in freezing temperatures or when freezing can be expected within 30 days after setting, without providing enclosures and temporary heat.

04270 GLASS BUILDING BLOCKS

Glass blocks are hollow, partially evacuated units of clear or colored glass. They have been engineered to serve several functions, all providing good thermal and sound insulating values.

Clear and other non-light directing units are available in 5-3/4" (143.75 mm), 8" (200 mm), and 12" (300 mm) square units, all 3-7/8" (96.87

751

mm) thick. These units give high light transmission and are available in clear glass for visibility, wide fluted patterns, and wavy, translucent surfaces. These units are often used in partitions and do not screen out sunlight.

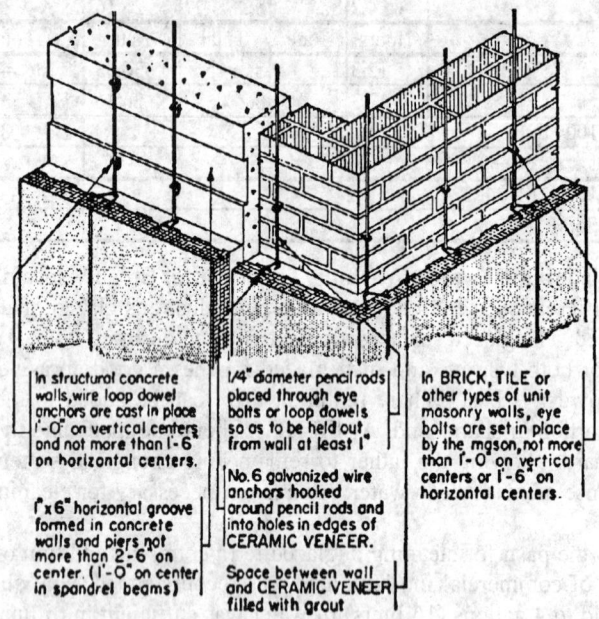

In structural concrete walls,wire loop dowel anchors are cast in place 1'-0" on vertical centers and not more than 1'-6" on horizontal centers.

1"x6" horizontal groove formed in concrete walls and piers not more than 2'-6" on center. (1'-0" on center in spandrel beams)

1/4" diameter pencil rods placed through eye bolts or loop dowels so as to be held out from wall at least 1"

No.6 galvanized wire anchors hooked around pencil rods and into holes in edges of CERAMIC VENEER.

Space between wall and CERAMIC VENEER filled with grout

In BRICK, TILE or other types of unit masonry walls, eye bolts are set in place by the mason, not more than 1'-0" on vertical centers or 1'-6" on horizontal centers.

Method of Setting and Securing Anchored Type Ceramic Veneer Terra Cotta to Brick and Concrete Backing

Light directing units are engineered for use on sun exposures of rooms with the type of finished ceiling that efficiently reflects the directed light to produce indirect daylighting, such as in schools and offices. The basic size is 7-3/4" x 7-3/4" x 3-7/8" (193.75 x 193.75 x 96.87 mm). These units also come with white or green fibrous insert to reduce both glare and solar heat gain. They are also available in 11-3/4" (293.75 mm) square size. The insert reduces the overall light transmission.

Light diffusing units diffuse light in all directions and give maximum light transmission. Size is 7-3/4" (193.75 mm) square. These units are also available with sun control inserts.

Color glass units have a ceramic coating fused to a single face of each block. They are used for decoration, often as accents with other blocks, and are available in 5-3/4" (143.75mm) and 7-3/4" (193.75 mm) square units and 3-3/4" x 11-3/4" (93.75 x 293.75 mm) rectangles.

Sculptured glass modules have various geometric shapes pressed into the glass to a depth of approximately 1-1/2" (37.50 mm) on both sides of the unit. They are 11-3/4" (293.75 mm) square and are available in clear glass and in the lighter hues of the color glass units.

752

MASONRY

Grille wall units are patterned glass with a smooth exterior. They have an outline of grey colored ceramic frit fused at the edges leaving ovals, circles, dots, and hourglass designs exposed in the center. They are also available in solid grey frit. Sizes are 3-3/4"x7-3/4"x3-7/8" (93.75 x 193.75 x 96.8 mm) and 7-3/4" (193.75 mm) square. The finish is scratch, abrasive, and chemical resistant and will not fade from exposure. They make handsome and practical screen walls.

Mortar for Glass Blocks. Materials used in making mortar for laying glass blocks shall be measured by volume. For this purpose 25 lbs. (11.3 kg) of quicklime or 40 lbs. (18.1 kg) of hydrated lime shall equal 1 cu. ft. (0.028 cu.m).

Lime shall be high-calcium or dolomitic hydrated lime. Dolomitic lime must be pressure hydrated so that it does not contain more than 8% by weight of unhydrated oxides, in accordance with ASTM Specification C207-04 Standard Specification for Hydrated Lime Masonry Purposes.

The mortar shall be composed of 1 part waterproof portland cement, 1 part lime, and 4 parts well graded sand. It shall be mixed to a consistency as stiff and dry as possible and still retain good working characteristics. Prepared masonry mortars of high strength and low volume change may be substituted, when approved. Setting accelerators or anti-freeze compounds are not to be used.

Expansion Joints. Expansion joints must be provided at the head and jambs of each glass block panel. This expansion area shall contain expansion strips as designated by the details or specifications. Caulk head and jamb expansion joints and at intermediate reinforcing members with sponge plastic strip or rope and a good grade of non-hardening mastic so that all expansion joints are waterproof.

In residential construction, panels that are not over 5'-0" (1.5 m) in width or 7'-0" (2.1 m) in height, or 25 sq.ft. (2.32 sq.m) in area, may be mortared in solid without the use of expansion joints at the jambs. The expansion joint at the head must still be used to prevent loads from being carried by the panels. Reinforcing wall ties may be omitted in residential panels that do not exceed the above size limits.

Reinforcement. Reinforcing wall ties shall consist of two No. 9 wires spaced 2" (50 mm) apart to which are welded No. 14 gauge cross wires. Ties are to be 8'-0" (2.43 m) long and not more than 0.20" (5 mm) thick at the weld. They shall be galvanized or treated with some other approved corrosion resisting coating. Ties are to run continuously with ends lapped 6" (150 mm) and are to be installed in horizontal mortar joints approximately 24" (600 mm) on centers.

Ties shall not be laid directly on the block, but shall be completely embedded in the center of the mortar joint.

Glass Building Units—Installation (Courtesy Pittsburgh Corning)

1. Sill area to be covered by mortar shall first have a heavy coat of PC Asphalt Emulsion and allowed to dry.
2. Adhere PC Expansion Strips to jambs and head with PC Asphalt Emulsion. Expansion strip must extend to sill.
3. When emulsion on sill is dry, place full mortar bed joint. Do not furrow.
4. Set lower course of block. All mortar joints must be full and not furrowed. Steel tools must not be used to tap blocks into position. Mortar shall not bridge expansion joints. Visible width mortar joint shall be 1/4" (6.25 mm) or as specified.
5. Install PC Panel Reinforcing in horizontal joints where required as follows: a) Place lower half of mortar bed joint. Do not furrow. b) Press panel reinforcing into place. c) Cover panel reinforcing with upper half of mortar bed and trowel smooth. Do not furrow. d) Panel reinforcing must run from end to end of panels and where used continuously must lap 6 inches. Reinforcing must not bridge expansion joints.
6. Place full mortar bed for joints not requiring panel reinforcing. Do not furrow.
7. Follow above instructions for succeeding courses. The number of blocks in successive lifts shall be limited to prevent squeezing out of mortar or movement of blocks.
8. Strike joints smoothly while mortar is still plastic and before final set. At this time rake out all spaces requiring caulking to a depth equal to the width of the spaces. Remove surplus mortar from faces of glass blocks and wipe dry. Tool joints smooth and concave, before mortar sets so that exposed edges of blocks have sharp clean lines.
9. After final mortar set pack PC Oakum tightly between glass block panel and jamb and head construction. Leave space for caulking.
10. Caulk panels as indicated on details.
11. Final cleaning of glass block faces shall not be done until after final mortar set. Surplus mortar shall be removed and block faces wiped dry as joints are tooled. Final cleaning shall be done by others after mortar has attained final set.

Labor Laying Glass Blocks. In laying glass blocks, all mortar joints shall be completely filled with mortar. The joints shall not be less than 3/16" (4.68 mm) thick or more than 3/8" (9.37 mm) thick. Blocks shall not be hit with a metal tool but laid by the method known as "shoving", (working into place with the hands) thereby compressing the vertical joints.

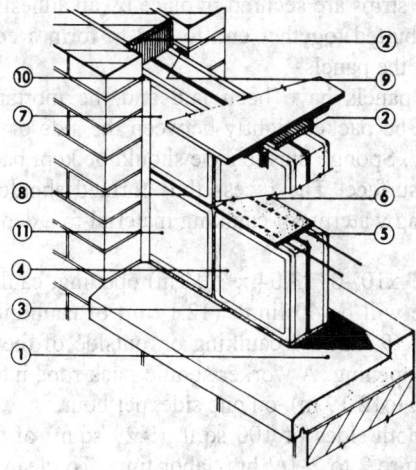

When a bricklayer's trowel handle is capped with a common type of rubber ferule, such as used on the bottom of crutches, it prevents damage to the glass.

After the mortar has passed its initial set, the exposed edges of the joints shall be tooled and thoroughly compressed with a round jointer. The finished surface of the joint shall be slightly concave, smooth, and nonporous. The final cleaning shall not be done until after the mortar has reached its final set.

If blocks are in any way disturbed after laying, during the plumbing, tooling, or cleaning process, such blocks shall be removed and relaid.

Inasmuch as glass blocks do not absorb water, the mortar must be stiff as possible and still be workable, otherwise the blocks will be inclined to slide, and it will be difficult to lay up a straight wall and very few courses can be laid at a time as the weight of the glass blocks will squeeze the mortar out of the joints.

Basis for Estimating Labor Costs. The following labor costs are based on one laborer to two masons, which includes time mixing and supplying mortar, glass blocks, etc. It does not include time building and removing scaffolding, placing expansion joint strips, packing joints with sponge plastic rope and caulking, nor does it include time required cleaning glass blocks after erection. These are given as separate items.

Panels containing more than 144 sq.ft. (13.37 sq.m) and not more than 25'-0" (7.6 m) in length or 20'-0" (6.0 m) in height should be braced by and anchored to structural stiffeners, so a fair average for all classes of work would be panels 10'-0"x10'-0" (3.04 x 3.04 m) in size, containing 100 sq.ft. (9.29 sq.m). Panels of this size have been used as a fair example for determining average costs.

Expansion joint strips 3/8" (9.38 mm) thick, 4-1/8" (103.12 mm) wide, and 25" (625 mm) long are used at the heads, jambs, and mullions of

755

all openings. These strips are secured in place by an adhesive such as asphalt emulsion, and are butted together end to end to form a continuous cushion around the edges of the panels.

After the panels have been laid and the mortar has set, sponge plastic rope should be packed tightly between the side of the block and the sides of the "chase". Sponge plastic rope should be kept back 3/8" (9.38mm) from the finished surface. The recess thus formed should then be caulked with a non-hardening waterproof caulking material to a depth of not less than 3/8" (9.38 mm).

In a 10'-0" x10'-0" (3.04 x 3.04 m) opening, caulked on both sides of the blocks, there will be 40 lin.ft. (12.19 m) of caulking on the inside of the wall and 30 l.f. (9.1 m) of caulking on outside of the wall or 70 lin.ft. (21.33 m) for each opening. A worker should pack rope into joints and caulk 30 to 40 lin.ft. (9.14 to12.19 m), on one side, per hour.

To clean both sides of 100 sq.ft. (9.29 sq.m) of smooth face glass block wall will require 2 to 2-1/2 hrs. labor time. To clean both sides of 100 sq.ft. (9.29 sq.m) of ribbed face glass block (light directional type) will require 3 to 3-1/2 hrs. labor time.

Erecting and removing scaffolding on the average job will require 1-1/2 to 2 hrs. labor time per 100 sq.ft. (9.29 sq.m) of glass area.

Glass blocks must be carefully handled in unloading and handling on the job, as they require the same care as fine enameled brick.

Cost of Glass Blocks. The cost of glass blocks varies with the kind, size, and quantity required. Always obtain delivered prices on blocks, expansion strips, and wall ties for each job figured.

Approximate Prices of Standard Glass Blocks			
Size of Block	Up to 1,000 SF (93 Sq. M)	From 1,000 SF (93 Sq. M)	Size of Block Metric
5 3/4 " x 5 3/4 " x 3 7/8 "	$ 2.20	$ 2.15	144 mm x 144 mm x 97 mm
7 3/4 " x 7 3/4 " x 7 7/8 "	$ 3.00	$ 2.90	194 mm x 194 mm x 197 mm
11 3/4 " x 11 3/4 " x 3 7/8 "	$ 7.00	$ 6.80	294 mm x 294 mm x 97 mm

Note: Special block runs considerably more and should be checked locally for both price and availability.

Estimating Data on Laying Glass Blocks Using 1/4" Mortar Joints - Cost per 100 Sq.Ft. of Wall			
Size of Glass Blocks in Inches	5 3/4" x 5 3/4" x 3 7/8"	7 3/4" x 7 3/4" x 7 7/8"	11 3/4" x 11 3/4" x 3 7/8"
No. Blocks per Sq.Ft. of Wall *	4.00	2.25	1.00
No. Blocks per 100 Sq.Ft. of Wall *	400.00	225.00	100.00
Cu.Ft. Mortar Required **	5.00	3.60	2.33
Mason Hours	21 - 23	14 - 16	10 - 12
Labor Hours	10.5 - 11.5	7 - 8	5 - 6
Labor on Scaffold Hours	2.00	2.00	2.00
Mason Ramming Oakum, Caulking - Hours	1.50	1.50	1.50

MASONRY

Estimating Data on Laying Glass Blocks Using 1/4" Mortar Joints (continued) - Cost per 100 Sq.Ft. of Wall			
Mason Cleaning Block Hours	2.50	2.50	2.50
Expansion Strips 3/8" x 4 1/8", Lin.Ft.	30.00	30.00	30.00
Wall Ties, Lin. Ft.	60.00	60.00	60.00
Sponge Plastic Rope Joints, Lin. Ft.	460.00	60.00	60.00
Caulking Joints, Lin. Ft.	70.00	70.00	70.00
Asphalt Emulsion. 40'- 0" x 4 1/2" pints	0.5	0.5	0.50

Estimating Data on Laying Glass Blocks Using 6.25mm Mortar Joints - Cost per 10 Sq.M. of Wall			
Size of Glass Blocks in mm	143.8 x 143.8 x 96.9 mm	193.8 x 193.8 x 196.9 mm	293.8 x 293.8 x 96.9 mm
No. Blocks per Sq.M. of Wall *	43.00	24.00	11.00
No. Blocks per 10 Sq.M. of Wall *	430.00	240.00	110.00
Cu.M. Mortar Required **	0.14	0.10	0.07
Mason - Hours	21 to 23	14 - 16	10 - 12
Labor - Hours	10.5 to 11.5	7 - 8	5 - 6
Labor on Scaffold - Hours	2.00	2.00	2.00
Mason Ramming Oakum, Caulking - Hours	1.50	1.50	1.50
Mason Cleaning Block - Hours	2.50	2.50	2.50
Expansion Strips 9.38 x 103.13, Lin.M.	9.14	9.14	9.14
Wall Ties, Lin. M.	18.29	18.29	18.29
Sponge Plastic Rope Joints, Lin. M.	140.21	18.29	18.29
Caulking Joints, Lin. M.	21.34	21.34	21.34
Asphalt Emulsion. 12.19M x 112.5 liters	0.28	0.28	0.28

Does not include allowance for breakage
** *Includes 10% for waste*

Laying Glass Blocks Using 1/4" Mortar Joints - Cost per 1,000 Block			
Based on 10' x 10' Panels	5 3/4" x 5 3/4" x 3 7/8"	7 3/4" x 7 3/4" x 7.7/8"	11 3/4" x 11 3/4" x 3 7/8"
Cu. Ft of Mortar Required **	12.50	16.00	23.30
Mason Hours	53 - 57	62 - 71	100 - 120
Labor Hours	27 - 29	31 - 36	50 - 60
Labor on Scaffold Hours	5.00	9.00	20.00
Mason Ramming Oakum, Caulking - Hours	3.00	6.00	15.00
Mason Cleaning Block Hours	6.00	11.00	25.00
Expansion Strips 3/8" x 4 1/8", Lin.Ft.	75.00	135.00	300.00
Wall Ties, Lin. Ft.	110.00	200.00	440.00
Sponge Plastic Rope Joints, Lin. Ft.	150.00	266.00	600.00
Caulking Joints, Lin. Ft.	175.00	310.00	700.00
Asphalt Emulsion. 3 1/2" wide - pints	1.25	2.25	2.50
Number of Pieces laid per 8-Hr day	140 - 150	115 - 125	70 - 80

Laying Glass Blocks Using 6.25 mm Mortar Joints - Cost per 1,000 Block			
Based on 3.05 M x 3.05 M Panels	143.8 x 143.8 x 96.9 mm	193.8 x 193.8 x 196.9 mm	293.8 x 293.8 x 96.9 mm
Cu. M of Mortar Required **	0.35	0.45	0.66
Mason Hours	53 - 57	62 - 71	100 - 120
Labor Hours	27 - 29	31 - 36	50 - 60
Labor on Scaffold Hours	5.00	9.00	20.00
Mason Ramming Oakum, Caulking - Hours	3.00	6.00	15.00
Mason Cleaning Block Hours	6.00	11.00	25.00
Expansion Strips 9.38 x 103.13, Lin.M.	22.86	41.15	91.44
Wall Ties, Lin. M.	33.53	60.96	134.11
Sponge Plastic Rope Joints, Lin. M.	45.72	81.08	182.88
Caulking Joints, Lin. M	53.34	94.49	213.36
Asphalt Emulsion. 87.5 mm wide - liters	0.71	1.28	1.42
Number of Pieces Laid per 8-Hr day	140 - 150	115 - 125	70 - 80

** *Includes 10% for waste*

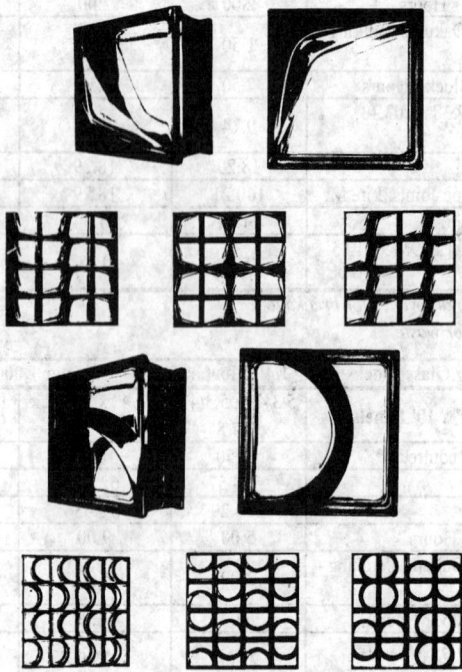

Two Styles of Decorative Glass Block

04280 GYPSUM TILE PARTITIONS AND FIREPROOFING

Gypsum tile are used for partitions and column fireproofing. They are 12"x30" (300 x 750 mm) in face dimensions. The large units are light in weight and require a minimum of labor for erection.

758

MASONRY

Estimating Quantities of Gypsum Tile. Gypsum tile should be estimated by the sq. ft. (sq.m) taking the area of all walls or partitions and making deductions in full for all openings. Gypsum tile are easily handled and may be cut with an ordinary hand saw, reducing waste to a minimum. The average mortar joint is 3/8" (9.37 mm) to 1/2" (12.50 mm) thick.

Sizes and Weights of Gypsum Partition and Furring Tile							
Inches				Metric			
Type of Tile	Size of Tile	For Ceiling Hgt. up to	Wt. Lbs	Type of Tile	Size of Tile	For Ceiling Hgt. up to	Wt. Kg
2" Solid	2" x 12" x 30"	10'	10	50 mm Solid	50 x 300 x 750 mm	3.05 M	###
3" Hollow	3" x 12" x 30"	13'	10	75 mm Hollow	75 x 300 x 750 mm	4.00 M	###
4" Hollow	4" x 12" x 30"	17'	13	100 mm Hollow	100 x 300 x 750 mm	5.18 M	###
6" Hollow	6" x 12" x 30"	30'	19	150 mm Hollow	150 x 300 x 750 mm	9.14 M	###

C.F. of Mortar Required to Lay 100 S.F. of Gypsum Partition Tile									
Width of Mortar Joint	Thickness of Gypsum Tile, in Inches				Width of Mortar Joint	Thickness of Gypsum Tile, in Millimeters			
	2"	3"	4"	6"		50 mm	75 mm	100 mm	150 mm
1/4"	1	1 1/2	2	3	6.25 mm	0.030	0.046	0.061	0.091
3/8"	1 1/2	2	2 1/2	3 1/2	9.38 mm	0.046	0.061	0.076	0.107
1/2"	2*	2.5*	3*	4*	12.50 mm	0.061*	0.076*	0.091*	0.122*

* Quantities ordinarily used. Mortar quantities determined through actual experience and include waste due to dropping, filling around pipe, etc.

Gypsum partition tile cement should always be used for laying up gypsum tile to obtain a secure bond and consequently a rigid wall.

Labor Setting Gypsum Tile. When laying gypsum tile, a worker sets 2-1/2 sq.ft. (0.23 sq.m) at a time, but on account of the greater footage of tile set, it requires more labor time handling and getting them to the mason.

Gypsum tile are set by bricklayers or tile setters (bricklayers who specialize in tile setting), and the labor costs will vary with the experience of the workers setting them. There are firms that specialize in the erection of gypsum tile partitions and fireproofing, and their workers are skilled in this branch of work. Naturally, they set more tile per day than a bricklayer, who lays brick one day, clay tile the next day, and gypsum tile the day after.

Experienced tile setters should set 15% to 25% more tile than inexperienced workers.

Labor Setting Gypsum Partition Tile						
English		Labor Hrs. per 100 S.F.			Metric	
Size of Tile, Inches	Sq.Ft. per 8 Hr Day	Bricklayer	Labor	Engr.*	Size of Tile, MM	Sq.M. per 8-Hr Day
2" Partition	250-300	2.91	2.75	0.25	50 mm Partition	23-28
3" Partition	220-260	3.34	3	0.25	75 mm Partition	20-24
4" Partition	195-235	3.72	3.5	0.25	100 mm Partition	18-22
6" Partition	175-215	4.1	3.75	0.5	150 mm Partition	16-20
2" Column	200-230	3.72	3	0.25	50 mm Column	19-21
3" Column	160-190	4.57	4	0.25	75 mm Column	15-18

* Add hoisting engineer time only if required.

04400 STONE

Being a natural product, stone varies widely in type and in geographical distribution. Being a very heavy material, except for the lava type stones which have very limited use, the availability of stone locally is of utmost importance to the estimator. Transportation costs can become a high percentage of the total cost. Geologically, stones are classified as follows:

1. Igneous: those formed by cooling molten material inside the earth including granite, trap rock, and lava.
2. Sedimentary: those consolidated from particles of decayed rocks deposited along streams and including limestone and sandstone.
3. Metamorphic: those either igneous or sedimentary which have been altered by deep-seated heat and pressure or intrusion of rock materials and including gneiss, marble, quartzite, and slate.

The Building Stone Institute (BSI) classifies stones as follows:

1. Bluestone: a hard sandstone, blue gray and buff in color, quarried in New York and Pennsylvania, used for paving, treads, copings, sills and stools, hearths, and in random ashlar patterns for walls.
2. Granite: a fine to coarse grained igneous rock, grey, red, green, black, or yellow in color, quarried in New England, New York, Virginia, Michigan, Minnesota, and Colorado, used for paving, curbing and rubble, mosaic, ashlar, and veneered wall construction for almost all building conditions.
3. Greenstone: a metamorphosed stone, a distinctive green in color, quarried in Virginia, and used inside and out for flooring, treads, spandrels and curtain walls, sills, copings, hearths.
4. Limestone: a sedimentary rock, oolitic, dolomitic, and crystalline, grey, buff, and pinkish in color, quarried in Indiana, Kentucky, Massachusetts, Minnesota, New York, Alabama, Kansas, and Texas, used for wall facings in all patterns, from rubble through the various ashlars to smooth facings in large sheets and for all trim pieces including carved moldings, panels, and sculpture.
5. Marble: a metamorphic rock, generally recrystallized limestone, available in a wide range of colors and variegated patterns, quarried in Georgia, Alabama, and Vermont, used generally as interior facings and much of which is imported, but are increasingly used as exterior veneers on precast concrete panels and as brick sized and random ashlar units for exterior walls.
6. Quartzite: a compact granular rock composed of quartz crystals in a wide variety of colors, many of which are variegated, quarried in Colorado, Tennessee, and Utah, and used for paving, ashlar, and mosaic wall veneers, copings, sills, hearths, and treads. Quartzite is also highly popular as aggregate for precast panels.
7. Sandstone: a sedimentary rock, in warm earth tones from near white to dark browns and reds, quarried in Colorado, Minnesota, Michigan,

760

Indiana, Ohio, New York, and Massachusetts, used for wall facings of all types rubble and mosaics, ashlars, and smooth sheet facings.

8. Slate: a fine grained metamorphic rock derived from sedimentary rock shale, predominantly gray and blue-black in color but also available in red, green, mottled green, purple, mottled purple, and weathering green, quarried in Pennsylvania, Virginia, New York, and Vermont, and used in slabs for exterior facings, flooring, paving, treads, bases, sills and stools, copings, fireplace trim and, of course, roofing.

The stone pattern and the type of finish also are of utmost importance to the estimator. Patterns used in laying up stone work in walls include:

1. Rubble: including uncoursed field stone, polygonal or mosaic and random coursed.
2. Ashlar: including rubble field mixed with coursed ashlar, random range coursed ashlar, random range interrupted coursed, broken range ashlar, and coursed ashlar.
3. Paneling: including spandrel and curtain wall units of both natural stone and stone aggregate precast with concrete.

Finishes for stone will vary as to the hardness and evenness of the material, and may be generally classified as natural, rough worked, and smooth.

Natural finishes may be weathered as in field stone, or seam or split faced, when the natural quarry texture is exposed.

Rough finishes are obtained mechanically by cutting with pitching chisels; gang sawing with steel shot or rough chat; plain sawn which still leaves saw marks; planed finish including those where small particles are plucked after planing to roughen the texture; hammered, hand, and machine tooled, where 2 to 10 parallel concave grooved cuts per inch are cut into the exposed surface. A newly devised rough finish for granite is obtained by exposing it to flames.

Smooth finishes include those rubbed wet by hand with carborundum block and water; machine rubbed with carborundum disc; and honed, obtained with a polishing machine and which leaves the stone almost without texture. As this last finish is expensive, it should be limited to interior work or the finest exterior within 30' (9.14 m) of eye range. Granite and marble may be polished smooth.

Rough faced stone is used in the construction of churches, college and university buildings, chapels, residences, and retaining walls. It is available in granite, quartzite, and sandstone.

Rubble Stone Work

Rubble stone consists of irregularly shaped pieces only partly trimmed or squared if at all. Finish is usually split faced or natural. Some specifications will restrict the size range. Strip rubble is ledge stone with beds of the natural cleft.

Rubble stone work is estimated by the cu. yd. containing 27 cu.ft. (0.76 cu.m) or by the perch containing 24-3/4 cu.ft (0.70 cu.m). A perch of stone is nominally 16-1/2' (5.02 m) long, 1' (0.30 m) high, and 1-1/2' (0.45 m) wide, but is often computed differently in different locations. In many states, especially west of the Mississippi River, rubble work is figured by the perch containing 16-1/2 cu.ft. (0.46 cu.m). Before submitting prices on rubble work by the perch, be sure to find out the common practice in your locality.

Cost of Rubble Stone. The cost of rubble stone will vary with the locality, kind of stone used, and distance of quarry from the job.

The method of buying and selling rubble stone also varies with the locality. In some places it is sold by the ton containing 2,000 lbs. (907 kg), while in others it is sold by the perch containing 24-3/4 c.f. (0.67 cu.m), or by the cord containing 100 cu.ft. (2.83 cu.m) and weighing 13,000 lbs. (5897 kg), which is about 130 lbs. per cu.ft. (2080 kg/cu.m). In some localities it is sold by the cu. yd., containing 27 c.f., or by the cu.m.

Mortar for Rubble Stone Work. The quantity of mortar required for rubble stone will vary with the size and shape of the stone and the thickness of the walls, but a fair average is 7 to 9 cu.ft. of mortar per c.y. of wall (0.19 to .0.25 cu.m per cu.m of wall). Mortar costs are given under the mortar section early in this chapter.

Cost of Setting Rubble Stone. When handling and setting rubble stone in walls up to 1'-6" (0.45 m) thick, a mason and laborer should handle and set 70 to 85 cu.ft. (1.98 to 2.41 cu.m) per 8-hr. day.

On heavy walls, 2'-0" to 3'-0" (0.61 to 0.91 m) thick, a mason should set 100 cu.ft. (2.83 cu.m) per 8-hr. day, and will require about 1-1/2 hrs. labor time to each hr. mason time.

Labor Costs of Rubble Stone for Walls over 2'- 0" Thick (.61 Meters Thick)						
	Hours	Rate	Total	Rate	Total	
Mason	3.0	$....	$....	$ 36.93	$	110.79
Labor	4.5	$....	$....	$ 26.16	$	117.72
Cost per Cu.Yd.					$	228.51
Cost per Cu. Ft.					$	8.46
Per Cu.M					$	298.86

Ashlar Stone Work

Ashlar Stone has a flat faced surface generally square or rectangular with sawed or dressed beds and joints. Wall patterns are referred to as "coursed" when set to form continuous horizontal joints; "stacked" when set to form continuous vertical joints; and "random" when set with stones of various lengths and heights so that neither horizontal or vertical joints are continuous. The usual restrictions for random ashlar design are that

762

no vertical joint should be higher than the highest course height being used; that no horizontal joint should be more than three stones long; and that no two stones the same height should be placed end to end. Random work will take more planning on the part of the mason than coursed work.

Finish for ashlar work is usually split face, which is the least expensive and brings out the character of the stone. It may also be ordered chat, sand, or shot sawn. The grade of stone for ashlar work is usually a mixture of all the grades available as a full range is part of the desired effect.

Many types of stone are furnished in ashlar stock including granite; limestone in the greys and buffs from Indiana, the creamy pinks from Minnesota and the popular lannon from Wisconsin; both Vermont and Georgia marble; Tennessee quartzite; and the local sandstones, those from Ohio and Pennsylvania being especially popular.

The common heights for ashlar are 2-1/4" (56.25 mm), 5" (125mm), 7-3/4" (193.75mm), and sometimes 10-1/2" (262.50 mm), 13-1/4" (33.12 mm), and 16" (4006 mm). Lengths are shipped from 1' (0.30 m) to 6' (1.82 m) for cutting on the job. Most ashlar work today is set as veneer and comes in thicknesses of 3-1/2" to 4" (87.50 to 100 mm). Some ashlar comes in varying thicknesses to be set to a uniform back and to create deep shadow lines on the face. Some jobs will still specify that the ashlar be built as part of the back-up wall, not veneered to it. Ashlar must then be ordered with bond blocks. These units will be a nominal 8" (200 mm) deep and usually constitute around 10% of the stone furnished, or one stone in each 10 sq.ft. (0.93 sq.m) of wall.

Many producers offer stock patterns under their own trade names. These are less expensive than stone cut to specification, but are limited to veneer thickness and a fixed percentage of heights, usually around 15% of 2-1/4" (56.25 mm), 40% of 5" (125 mm) and 45% of 7-3/4" (193.75 mm).

Veneer ashlar must be tied to the back-up wall with noncorrosive ties not less than 18" (450 mm) apart horizontally and 24" (600 mm) vertically. The stock heights when used with 1/2" (12.50 mm) mortar joints level well with brick and block back-up, but adjustable anchors and stud anchors are also available.

Costs of ashlar stone will vary widely depending on the stone selected and the distance from the quarry as well as whether stock supplies or special cutting and finishes are required.

The cost of setting this stone also varies widely, depending upon the size of the stone and the amount of cutting required. On some of the more intricate jobs it requires as much mason's time trimming and cutting the stone as is required to set it. The entire success of the use of this random ashlar depends upon the artistic manner in which the work is performed, and the labor costs will be governed accordingly.

Labor time handling stone and helping masons varies widely, depending on method of handling stone at job, access of material piles, size of stone, etc.

To show the variations in the cost of setting, six different classes of work are described.

Estimating Quantities of Random Ashlar. Split faced or other finishes of ashlar stone is usually sold by the ton and is furnished with rough beds, backs, and joints. The stone weighs 160 to 165 lbs. per c.f. (2543-2643 kg per cu.m) and there are 12 to 12-1/2 cu.ft. (0.34 to0.34 cu.m) of stone to the ton.

Stone 4" (100 mm) thick produces 36 to 38 sq.ft. of wall to the ton (3.34 to3.53 sq.m per metric ton) and when laid with a 1/2" (12.50 mm) mortar joint should lay 40 to 42 sq.ft. (3.71 to3.90 sq.m) of wall.

Stone 5" (125 mm) thick produces 29 to 30 sq.ft. (2.69 to 2.79 sq.m) of wall to the ton and when laid with a 1/2" (12.50 mm) mortar joint should lay 32 to 33 sq.ft. (2.97 to 3.06 sq.m) of wall.

Where the stone is backed with brick or block, the stone is usually furnished 4" (100 mm) to 8" (200 mm) thick to bond with the brick backing. Based on an average 6" (150 mm) thickness, one ton is sufficient for 24 to 25 sq.ft. (2.23-2.32 sq.m) of wall and when laid with a 1/2" (12.50 mm) mortar joint should lay 26 to 28 sq.ft. (2.41-2.60 sq.m) of wall.

Cost of 100 Sq. Ft. (9.29 Sq.M) of Random Ashlar Used on a Veneer Over Wood Faming and Sheeting, Stone Veneer 4" (100 mm) Thick					
	Hours	Rate	Total	Rate	Total
Mason	24.0	$....	$....	$ 36.93	$ 886.32
Labor	12.0	$....	$....	$ 26.16	$ 313.92
Cost per 100 Sq.Ft.					$ 1,200.24
Cost per Sq.Ft.					$ 12.00
Cost per Sq.M					$ 129.20

Cost per Ton of Stone					
	Hours	Rate	Total	Rate	Total
Mason	10.0	$....	$....	$ 36.93	$ 369.30
Labor	5.0	$....	$....	$ 26.16	$ 130.80
Cost per Ton					$ 500.10
Cost per Tonne					$ 492.22

On work above the second story or where large stones are used, it may require an extra laborer bringing stone to the mason.

Setting Random Ashlar in Straight Work Requiring Backing, Cutting, and Fitting. When setting random ashlar in fairly straight walls as described above, but where the stone must be backed, and end joints cut on the job, an experienced mason should set stone in 17 to 25 sq.ft. (1.57 to 2.32 sq.m) of wall per 8-hr. day, at the following labor cost per 100 sq.ft. (9.29 sq.m):

Setting Random Ashlar in Straight Work Requiring Backing, Cutting and Fitting						
	Hours	Rate	Total	Rate	Total	
Mason	38.0	$....	$....	$ 36.93	$	1,403.34
Labor	38.0	$....	$....	$ 26.16	$	994.08
Cost per 100 Sq.Ft.					$	2,397.42
Cost per Sq.Ft.					$	23.97
Cost per Sq.M					$	258.06

The labor time will vary widely according to job conditions, as some jobs may require only one-half as much labor time as mason time, while others may require twice as much.

Setting Random Ashlar in Churches, Chapels, Etc., Where Stone is Cut to Size in Shops. On churches, chapels, and similar structures having numerous corners, offsets, piers, pilasters, etc., where the stone is all backed in the shop and requires very little cutting and fitting by the masons on the job, an experienced mason should set stone in 25 to 30 sq.ft. (2.32 to 2.79 sq.m) of wall per 8-hr. day, at the following labor cost per 100 sq.ft. (9.29 sq.m):

Setting Random Ashlar in Churches, Chapels, etc., Where Stone is Cut to Size in Shops						
	Hours	Rate	Total	Rate	Total	
Mason	29.0	$....	$....	$ 36.93	$	1,070.97
Labor	29.0	$....	$....	$ 26.16	$	758.64
Cost per 100 Sq.Ft.					$	1,829.61
Cost per Sq.Ft.					$	18.30
Cost per Sq.M					$	196.94

Setting Random Ashlar in Churches, Chapels, Etc., Where Cutting and Fitting is Done on the Job. On churches, chapels, and similar structures having numerous corners, offsets, piers, pilasters, etc., where the rough stone is sent to the job and all backing, jointing, and fitting is done by the masons on the job, an experienced mason should set stone in 14 to 18 sq.ft. (1.30-1.67 sq.m) of wall per 8-hr. day, at the following labor cost per 100 sq.ft. (9.29sq.m):

Setting Random Ashlar in Churches, Chapels, etc., Where Cutting and Fitting is Done on the Project						
	Hours	Rate	Total	Rate	Total	
Mason	50.0	$....	$....	$ 36.93	$	1,846.50
Labor	50.0	$....	$....	$ 26.16	$	1,308.00
Cost per 100 Sq.Ft.					$	3,154.50
Cost per Sq.Ft.					$	31.55
Cost per Sq.M					$	339.56

Random Ashlar in Residence Construction. In a study of a residence faced with 4500 sq.ft. (418.05sq.m) of 5" (125 mm) Lannon stone veneer. The stone varied from 1" (25 mm) to 10" (250 mm) high and consisted of one-third weathered edge, one-third bed face stone, and one-third rock face stone. It was only one to two inches high. Care was taken to produce a natural looking pattern with very little cutting and fitting.

A mason set 34 sq.ft. (3.15 sq.m) of stone per 8-hr. day, at the following labor cost per 100 sq.ft. (9.29 sq.m):

Setting Random Ashlar in Residential Construction						
	Hours	Rate	Total	Rate		Total
Mason	24.0	$....	$....	$ 36.93	$	886.32
Labor	14.0	$....	$....	$ 26.16	$	366.24
Cost per 100 Sq.Ft.					$	1,252.56
Cost per Sq.Ft.					$	12.53
Cost per Sq.M					$	134.83

Random Ashlar in Memorial Chapel. The memorial chapel illustrated is also built of Lannon (Wisconsin) stone, but on account of the numerous corners, pilasters, piers, etc., and the fact that all backing, jointing, and fitting of stone was done on the job, the labor costs ran rather high, as on work of this kind the time required cutting and fitting the stone is almost as much as the setting time.

On this job, a mason averaged 17 sq.ft. (1.58 sq.m) of 8" (200 mm) ashlar per 8-hr. day, at the following labor cost per 100 sq.ft. (9.29 sq.m):

Setting Random Ashlar in Memorial Chapel						
	Hours	Rate	Total	Rate		Total
Mason	47.0	$....	$....	$ 36.93	$	1,735.71
Labor	40.0	$....	$....	$ 26.16	$	1,046.40
Cost per 100 Sq.Ft.					$	2,782.11
Cost per Sq.Ft.					$	27.82
Cost per Sq.M					$	299.47

CUT STONE

The term "cut stone" refers here to stone that is cut to an exact dimension from selected quality stone, given a uniform texture, and delivered to the site ready to set in place. In order to provide such quality control, shop drawings must be made and approved in advance.

Cut stone is used on the highest quality work and demands the highest quality installation. Cut stone may be built up with and bonded to a masonry backup, but it is most often applied as a veneer or a facing panel, sometimes preassembled with an insulating backing and frame. In the past cut stone usually referred to granite or limestone or sometimes sandstone, which was used to face the many "brownstones" that line the older streets of our northern cities. With the shift from slabs to thin veneers, the denser marbles and slate are now often specified.

Shop Drawings. All cut stone work should be cut and set in accordance with approved shop drawings. These should show the bedding, bonding, jointing, and anchoring details. Each stone is dimensioned and given a setting number. The producer will not cut the stone until all details and dimensions are agreed upon. Because jointing in veneer work is generally 1/4" (6.3 mm), extremely accurate measurements are required. Also, the contractor must plan ahead and have cutouts and holes required for lifting large slabs (called Lewis holes), as well as cutouts for posts, door checks, electrical fixtures, etc., as may be required for the work of other trades. Holes for anchorage are also cut by the contractor furnishing the stone.

Mortar for Cut Stone. Three types of mortar are recommended for setting cut stone. The best type for the particular work should be selected.

Non-Staining Cement Mortar. The cement mortar is 1 part non-staining cement to 3 parts sand, with the addition of 1/5 part of hydrated lime based on the cement volume.

Non-Staining Cement-Lime Mortar. Cement-lime mortar shall be composed of 1/2 part non-staining cement, 1/2 part of hydrated lime or quick lime paste, and not to exceed 3 parts of sand.

If quick lime paste is used, first thoroughly mix the quick lime paste and sand and stack to age; the cement shall be added and thoroughly worked into the lime and sand mixture in small batches just prior to use.

Quick lime shall be reduced to a paste by thorough and complete slaking with cold water and screening through mesh screen into a settling box. Lime putty shall stand not less than one week before use.

Sand shall be clean and sharp, free from loam, silt, vegetable matter, salts, and all other injurious substances with grains graded from fine to coarse. Screen through a 6-mesh screen.

Non-Staining Waterproof Cement Mortar. Recommended for use wherever parging is required or when any of the construction materials, such as water, sand, brick, tile, etc., are known to contain salts, which can cause staining or efflorescence; and to walls exposed to excessive amounts of moisture.

Non-staining waterproof cement mortar should be composed of 1 part of non-staining waterproof cement to 3 parts of sand, with the addition of 1/5 part of hydrated lime based on the cement volume.

For quantity and cost of mortar required for setting limestone, refer to the mortar section of this chapter.

Job Conditions. The cut stone contractor must check job conditions carefully. The product is expensive: defective or damaged work is almost always refused and replacements can cause costly job delays.

Carefully unpack and inspect all material to assure it matches samples and is cut to dimensions of shop drawings and tolerances of specifications, and check for shipping damages.

Carefully store all stone on non-staining platforms at least 4" (100 mm) off the ground. Plug all holes during freezing weather.

If hoist and scaffolding are involved, check their condition. Dirty hoists and scaffolding can stain stone.

Check backup construction. Backup must be laid to a true line. Concrete and masonry backup should be dampproofed. If this is not in other contracts, this contractor should provide for it.

If work is to continue alongside or above stone after it is set, add an allowance for protection of exposed surfaces.

Estimating Cut Stone. The cost of cut stone will depend on the kind of stone used (granite, limestone, or marble), the amount of cutting and hand work necessary, and freight from mill to destination.

Cut stone for exterior use is estimated by the cu. ft. (cu.m), taking the cube of the largest dimensions, and any special work such as hand cutting, carving, etc., is estimated separately and added to the cost of the plain stone.

When estimating the cost of plain door and window sills, lintels, steps, platforms, copings, etc., the quantities are sometimes estimated by the lin. ft. (meter) for sills and coping and by the sq. ft. (sq.m) for platforms, but the cu. ft. (cu.m) method is more generally used.

Setting Cut Stone. When estimating the labor cost of handling and setting cut stone, the size of the job should be considered, average size of stone, class of workmanship required, and the method of setting; i.e., whether by hand, hand operated derricks, or by using a mobile crane, stiff leg, guy or other types of derricks.

Setting Cut Stone by Hand. On small cut stone jobs consisting of door and window sills, lintels, wall coping, steps and other light stone that can be handled and set by two workers without the use of derricks, a stone setter should set 50 to 60 cu.ft. (1.41-1.70 cu.m) of stone per 8-hr. day, at the following labor cost per 100 cu.ft. (9.29 cu.m):

Setting Cut Stone by Hand						
	Hours	Rate	Total	Rate		Total
Stone Setter	14.0	$....	$....	$ 36.93	$	517.02
Labor	28.0	$....	$....	$ 26.16	$	732.48
Cost per 100 Cu.Ft.					$	1,249.50
Cost per Cu.Ft.					$	12.50
Cost per Cu.M					$	441.21

Setting Cut Stone Using Hand Derricks. Where the stone consists of coursed ashlar, sills, lintels, lightweight cornices, copings, etc., which can be handled on stone barrows or rollers and set with a breast derrick, gin pole, or other hand operated derricks, a stone setter should set 80 to 100 cu.ft. (2.26

768

to2.83 cu.m) of stone per 8-hr. day, at the following labor cost per 100 cu.ft. (2.83 cu.m):

Setting Cut Stone Using Hand Derricks						
	Hours	Rate	Total	Rate		Total
Stone Setter	9.0	$....	$....	$ 36.93	$	332.37
Helper - Labor	9.0	$....	$....	$ 26.16	$	235.44
Derrick-person	18.0	$....	$....	$ 36.93	$	664.74
Helpers - laborers handling Stone	18.0	$....	$....	$ 26.16	$	470.88
Cost per 100 Cu.Ft.					$	1,703.43
Cost per Cu.Ft.					$	17.03
Cost per Cu.M					$	601.49

Setting Heavy Cut Stone Using Hand Derricks. On jobs requiring heavy stone, such as thick ashlar, heavy molded courses, cornices, copings, etc., where the stone is hoisted and set using hand operated derricks, a stone setter should set 115 to 140 cu.ft. (3.25 to 3.96 cu.m) of stone per 8-hr. day, at the following labor cost per 100 cu.ft. (2.83 cu.m):

Setting Heavy Cut Stone Using Hand Derricks						
	Hours	Rate	Total	Rate		Total
Stone Setter	7.0	$....	$....	$ 36.93	$	258.51
Helper - Labor	7.0	$....	$....	$ 26.16	$	183.12
2- Derrick-person	14.0	$....	$....	$ 36.93	$	517.02
3- Helpers - laborers handling Stone	21.0	$....	$....	$ 26.16	$	549.36
Cost per 100 Cu.Ft.					$	1,508.01
Cost per Cu.Ft.					$	15.08
Cost per Cu.M					$	532.49

Setting Cut Stone Using a Power Crane. On many large jobs it will be possible to set the stone using a mobile power crane to lift the stone from the ground to the wall.

This method is usually less expensive than setting up and dismantling stiff-leg or guy derricks, as the crane can cover the entire perimeter of the building without any setup or dismantling cost to speak of.

Where the stone can be set in this manner, it will require one operator for the crane, 2 laborers on the scaffold helping to land the stone and set it, 4 or 5 laborers on the ground sorting and handling stone from stock pile to derricks, plus the stone setter.

On a job of this kind, a crew should handle and set 200 to 240 cu.ft. (5.66 to 6.80 cu.m) of stone per 8-hr. day, depending on the size of stone, at the following labor cost per 100 cu.ft. (2.83 cu.m), excluding crane and operator costs:

769

Setting Stone Using a Crane						
	Hours	Rate	Total	Rate	Total	
Stone Setter	4.0	$....	$....	$ 36.93	$	147.72
2- Helpers - Labor on Scaffold	8.0	$....	$....	$ 26.16	$	209.28
5- Laborers handling Stone	20.0	$....	$....	$ 26.16	$	523.20
Cost per 100 Cu.Ft.					$	880.20
Cost per Cu.Ft.					$	8.80
Cost per Cu.M					$	310.81

The above does not include any allowance for scaffold. Add for a separate scaffold required for stone setting.

Setting Cut Stone Using Power Derricks. On large jobs where it is necessary to use stiff leg or guy derricks for hoisting and handling the stone, the costs will vary with the size and weight of the stone, quantity, and with the method of handling the stone on the ground.

On work of this kind, it is usually necessary to use stiff leg or guy derricks, or by having the boom of the derrick placed in a boom seat and bolted to the steel frame of the building. On jobs covering a large area, it will be necessary to use a number of derricks. Regardless of the method used in setting the stone, the cost of setting and removing derricks, hoisting engines, etc., should be estimated separately, because this charge will remain the same whether there are 5,000 or 50,000 cu.ft. (141.60 or 1416.00 cu.m) of stone to be set.

After the derricks and equipment are in place, a stone setter should set 200 to 240 cu.ft. (5.66 to 6.80 cu.m) of stone per 8-hr. day. It will require 2 laborers helping to land the stone and set it, 4 or 5 laborers on the ground sorting and handling the stone from stock piles to derricks, 3 derrickmen and one hoisting engineer.

In connection with work of this kind, there is always a certain amount of lost time caused by delays in delivery of stone, repairs to derricks and hoisting engines, oiling and attending derricks, moving and raising derricks, inspecting cables, tightening guy lines, and many other items that require attention and must be charged against the cost of stone setting.

Pointing Cut Stone. Cut stone is often pointed at the time the stone is set. Recommended practice is to strike the joints as the work progresses. Pointing mortar applied afterward is usually of different composition and questionable permanence.

However, on some jobs it may be necessary to point all mortar joints with white or other non-staining cement after all stone has been set. The mortar joints may be struck flush or slightly rounded with a jointer as preferred.

The cost of this work will vary with the size of the individual pieces of stone. The fewer joints the more surface a tuckpointer will be able to point per day.

770

MASONRY

On an average job, a tuckpointer should point 500 to 600 sq.ft. (46.45 to 55.74 sq.m) of surface per 8-hr. day, at the following labor cost per 100 sq.ft. (9.29 sq.m):

Pointing						
	Hours	Rate	Total	Rate	Total	
Stone Setter - Pointer	1.50	$....	$....	$ 36.93	$	55.40
Labor	0.75	$....	$....	$ 26.16	$	19.62
Cost per 100 Cu.Ft.					$	75.02
Cost per Cu.Ft.					$	0.75
Cost per Cu.M					$	26.49

Allow one helper on the ground to each 2 tuckpointers or to each scaffold. Add for mortar and scaffold.

Washing and Cleaning Cut Stone. After the stone work has been pointed, it will be necessary to wash and clean it, using a stiff fiber brush and plenty of clean water or soapy water, if necessary, and then rinsed with clean water, which will remove the construction and mortar stains. A wire brush should never be used, as it tends to produce rust marks in the stone.

On new work or work that is fairly clean, an experienced tuckpointer should wash and clean 600 to 800 sq.ft. (55.74 to 74.32 sq.m) of surface per 8-hr. day, at the following labor cost per 100 sq.ft. (9.29 sq.m):

Wash and Clean Stone						
	Hours	Rate	Total	Rate	Total	
Tuck pointer	1.1	$....	$....	$ 36.93	$	40.62
Labor	0.6	$....	$....	$ 26.16	$	15.70
Cost per 100 Cu.Ft.					$	56.32
Cost per Cu.Ft.					$	0.56
Cost per Cu.M					$	19.89

Allow one helper on the ground to each 2 tuckpointers or to each scaffold. Add for scaffold.

Carving Cut Stone. The cost of carving cut stone is usually figured from full-size details or models furnished by the architect. This cost will vary with the amount of detail and the method used in carving, i.e., whether performed in the shop before setting or after the stone has been set on the job.

Where you are sure of a first class job of setting, it is more economical to have the carving done in the shop. All carved work of a number of pieces is set up with wood strips forming the thickness of mortar joints and accurately fitted before shipping to the job.

Carving Stone by Hand in the Shop. When carving cut stone for medium priced buildings requiring just an average grade of workmanship and no fine detail work, a carver should complete 2-1/2 to 3-1/2 sq.ft. (0.23-0.32 sq.m) per 8-hr. day, at the following labor cost per sq.ft. (0.09 sq.m):

Carving Stone in Shop by Hand - No Fine Detail Work						
	Hours	Rate	Total	Rate	Total	
Stone Carver	2.7	$....	$....	$ 36.93	$	99.71
Cost per Sq. Ft.					$	99.71
Cost per Sq.M					$	1,073.32

On jobs having fine detail work and requiring first class workmanship, a carver should complete 1 to 1-1/2 sq.ft. (0.09-0.14 sq.m) per 8-hr. day, at the following cost per sq.ft. (0.09 sq.m):

Carving Stone in Shop by Hand - Detail Work						
	Hours	Rate	Total	Rate	Total	
Stone Carver	7.0	$....	$....	$ 36.93	$	258.51
Cost per Sq. Ft.					$	258.51
Cost per Sq.M					$	2,782.67

Carving Stone in the Shop Using Pneumatic Tools. The use of air tools will greatly increase the daily output of a stone carver, regardless of the class of workmanship required.

On jobs requiring just an average grade of workmanship and no fine detail work, a carver should complete 5 to 8 sq.ft. (0.46 to 0.74 sq.m) per 8-hr. day, at the following labor cost per sq.ft. (0.09 sq.m):

Carving Stone in Shop with Pneumatic Tool - No Fine Detail Work						
	Hours	Rate	Total	Rate	Total	
Stone Carver	1.5	$....	$....	$ 36.93	$	55.40
Cost per Sq. Ft.					$	55.40
Cost per Sq.M					$	596.29

Where first class workmanship and fine detail is required, a carver should complete 2 to 2-1/2 sq.ft. (0.05 to -0.07 sq.m) per 8-hr. day, at the following labor cost per sq.ft. (0.09 sq.m):

Carving Stone in Shop with Pneumatic Tool - Detail Work						
	Hours	Rate	Total	Rate	Total	
Stone Carver	4.0	$....	$....	$ 36.93	$	147.72
Cost per Sq. Ft.					$	147.72
Cost per Sq.M					$	1,590.10

Carving Stone by Hand After It Is Set in the Building. Where the stone is carved on the job after it has been set in place and where first class workmanship is required, a carver should complete 1 to 1-1/4 sq.ft. (0.02 to -- 0.03 sq.m) per 8-hr. day, at the following labor cost per sq.ft. (0.09 sq.m):

Carving Stone Set in Building - No Fine Detail Work					
	Hours	Rate	Total	Rate	Total
Stone Carver	7.0	$....	$....	$ 36.93	$ 258.51
Cost per Sq. Ft.					$ 258.51
Cost per Sq.M					$ 2,782.67

If the carving contains considerable fine detail, add 3/4 to 1 hr. per sq.ft. (0.06 to 0.09 sq.m). If outside scaffold is required, add labor time to above.

Carving Stone on the Job Using Pneumatic Tools. If air tools are used to carve stone on the job, a carver should complete 1-1/2 to 2 sq.ft. (0.14 to 0.19 sq.m) per 8-hr. day, at the following labor cost per sq.ft. (0.09 sq.m):

Carving Stone Set in Building - Detail Work					
	Hours	Rate	Total	Rate	Total
Stone Carver	4.5	$....	$....	$ 36.93	$ 166.19
Cost per Sq. Ft.					$ 166.19
Cost per Sq.M					$ 1,788.86

If the carving contains considerable fine detail, add 1/2 to 3/4 hr. per sq.ft. (0.06 to 0.09 sq.m). If outside scaffold is required, add labor time to above.

Cost of Models for Stone Carving. On the better class of work, clay or plaster models for all carving are usually furnished by the architect or a stipulated sum is allowed for models.

Clay models run about $60.00 to $150.00 per sq.ft. ($646 to 1615 per sq.m). Plaster models cost about $100.00 to $190.00per sq.ft. ($1076 to 2,045 per sq.m). A small rosette or head cast in plaster will cost about $150.00. The models quite generally cost more than the carving, and often twice as much where models are made for an individual piece. It will often require a week's work for a skilled artisan to produce a carefully made model for a carved plaque or panel, 3'-6" to 4'-0" (0.91 to 1.22 m) in diameter.

The sq. ft. (sq.m) price of models should be based on the sq. ft. (sq.m) of model required and not on the sq. ft. (sq.m) surface of carved stone, as one model may often be used for carving 10 or 15 pieces of stone, or one lin. ft. (meter) model of carved molding in plaster may often be used for carving hundreds of lin. ft. (meters) of stone.

Indiana Limestone

Indiana Limestone is probably the best known and most widely used building stone in the United States. This is doubtless due to its appearance, as well as the ease with which it is worked and its freedom from stratification and cleavage plane, and while it is a sedimentary stone, it is to all intents and purposes a free stone, having about equal strength in all directions.

It is used for the exterior facings of buildings, as well as all kinds of trimmings, such as sills, lintels, copings, steps, platforms, etc.

Indiana limestone can be produced in any reasonable quantity and size and is usually cut ready to place in the building by the mills located adjacent to the quarries. The creamy-pink limestone quarried in Minnesota is another high quality material. Limestone is furnished in a variety of different grades and color tones, a brief summary of which is given below:

Select stock is uniform in color and texture, more so than is required for ordinary building construction. It is recommended for entrance work and those portions of a building within ready range of vision, carving, interior, and other exceptional uses.

Standard stock is sound stone with a range of variation in color, shades, and texture not found in Select but necessarily confined within limits that make it impossible to determine at a distance of a few feet whether it is standard or select stock.

Rustic stock is of more or less coarse grain. Rustic stock is used to a considerable extent for sawed ashlar and various rough textured finishes.

Variegated stock in an irregular mixture of buff and gray produced from the blocks that are quarried where the buff and gray color tones adjoin in the quarry. It shows variation in texture as well as color. Variegated stone, when cut into building units, will exhibit units of each color tone and a small percentage of units with both color tones in one piece. Variegated stone is unusually effective in giving variety to plain surfaces and is a desirable class of material for trim as well as for facing.

There are certain grades of Indiana Oolitic limestone especially adapted for grade courses, steps, buttresses, floor tiling, etc., or any position in the building exposed to abrasion. Some are available in different color tones and may be identified by the following names: special hard gray and special hard buff Indiana limestone.

The following are the trade terms by which the different grades of Indiana limestone are classified: select buff, standard buff, select gray, standard gray, and variegated.

Some of the machine finishes suitable for limestone are sand sawed, chat or shot-sawed finish, smooth machine finish, wet rubbed finish, machine tooled finishes, carborundum finish, and plucked finish. There are also some rough textured finishes. The class of tooling in common practice is understood to cover what is properly termed fluting, that is, concave depressions, four and six parallel, concave grooves to the inch. Convex tooling is sometimes used effectively, but it is never applied unless specifically designated. Tooling and machinery has been developed to produce a variety of rough textured finishes.

It is generally recommended that all molded work be finished smooth. If it is tooled at all, the tooling should run in the direction of the mold and not across it, unless much greater expense is of no consequence. As for tooled ashlar, some architects contend vertical tooling is preferable, others prefer the horizontal.

MASONRY

Indiana Oolitic limestone will take most hand finishes, which under certain conditions are considered preferable to machine finish. Notable examples of this class of work are crandalling and tooth chiselling, but a number of the hand finishes can now be simulated by machine work.

A smooth finish is not commonly furnished other than in the form it comes from the planers. If sand rubbing is desired, it should be specified, and in such event the sand rubbing should be by the wet process or carborundum-and-water rubbed finish.

Indiana limestone weighs 200 lbs. per cu.ft. (3204 kg per cu.m) when furnished in rough quarry blocks and 180 lbs. per cu.ft. (2883 kg per cu.m) in scabbled quarry blocks. These figures include the weight of rough excess stone.

Sawed slabs or other squared material weighs 170 lbs. per cu.ft. (2723 kg per cu.m). The weight of the finished cut stone ready to set in the building is figured at 150 lbs. per cu.ft. (2403 kg per cu.m). If there are any large round columns in the job, they may be figured at 140 lbs. per cu.ft. (2243 kg per cu.m) of stock. The actual net weight of a cubic foot of dry seasoned stone is 144 lbs. (2308 kg per cu.m).

Where the cut stone for any job consisting of ashlar, sills, steps, coping, plain machine moldings, etc., is furnished by any of the mills in the Bedford or Indiana limestone district, the price will vary from $18.00 to $30.00 per cu.ft. ($636 to 1059per cu.m) f.o.b. cars at the mills.

Considerable variation in price is due to the following reasons: Stone consisting of lightweight pieces, such as 4" (100 mm) or 6" (150 mm) ashlar, small molded courses, small turned columns, bases, sills, coping, etc., require a comparatively small amount of stone but involves considerable cutting and finishing expense, and even higher costs will prevail.

On other jobs consisting of fairly large stone, 8" (200 mm) or more thick, large molded pieces, bases, columns, etc., the cubical contents are a deciding factor, and the cu. ft. (cu.m) price of this stone, even for more elaborate work will usually average less in cost than the lighter stone described above.

Jobs containing many large pieces of stone will take the minimum price. Always remember that the cube of the largest dimensions must always be used when computing the quantity, cost and weights of stone of any kind.

Indiana Limestone Sill Stock

Sill stock is available as a standard item and sold by the lineal foot. The designation S4S means stock will be sent in random lengths of from 5' to 12' (1.52 to -3.65 m). The S6S designation means stock will be sawed to exact lengths. Sill stock is sold in rectangular shapes. If slopes, lugs, drips, or moldings are required, profile cuts will be made at additional cost. Sand and other special finishes are also extra.

Sill stock will cost about $14.00 per cu.ft. ($494 per cu.m). If specific lengths are ordered, add $5.00 per cu.ft. ($175 per cu.m). If custom

775

cut to shop drawing details, then sills, copings, steps, and other simple trim will cost about $50.00 per cu.ft. ($1,765 per cu.m). Belt courses, simple cornice molds, and the like cost about $55.00 per cu.ft. ($1,942 per cu.m). Elaborately carved detail work will cost $70.00 and up per cu.ft. ($2,470 or more per cu.m).

Cutting Beds and Joints. When cutting beds and joints, the cost will vary with the size of the stone. The smaller the stone the higher the price per cu. ft. (cu.m). For example, a piece of ashlar 8" (203 mm) thick contains twice as many cu. ft. (cu.m) of stone as a piece 4" (102 mm) thick, while the cost of cutting the joints will not be much more on the larger piece than on the smaller. Where the joints are cut by hand, the cost will vary from $2.00 to $2.50 per cu.ft. ($70.00 to $88.00 per cu.m), depending on the conditions described above.

On the second sawing, where sawed slabs are used, and it is necessary to run them through the saw the second time for top and bottom beds, it will cost $2.50 per cu. ft. ($70 per cu.m) for this work.

Where the jointing is done on the saw, figure $2.50 per cu.ft. ($88.00 per cu.m) for cutting joints.

Smooth or Planer Finish. When Indiana limestone is wanted with a smooth finish as it comes from the planer, the cost will vary from $1.50 to $2.00 per sq.ft. ($16.14 to $21.52 per sq.m) of surface planed.

The slabs are usually furnished sawed to the correct thickness and then the stone is run through the planer to remove the saw marks, which is not a very expensive process on straight work.

On the cheaper classes of work, and on some of the higher grade work where a sawed finish is desired, the sawed slabs are jointed and the stone is sent to the building without having been run through the planer. Work of this class is a little cheaper than stone having a first class finish.

Tooled Finish. If the stone has a tooled finish, having 4, 6, 8, or 10 bats to the inch (0.16, 0.24, 0.31, or 0.39 bats to the mm), the cost of horizontal tooling on plain surfaces is $1.205 to $1.60 per sq.ft. ($12.91 to $17.22 per sq.m) of surface tooled, or about the same as planer finish.

Hand vertical tooling in connection with machine molded work will add $1.20 to $1.60 per sq.ft. ($12.91 to -$ 17.22 per sq.m) of surface finished.

Cross or vertical tooling on curved surfaces will add $2.40 per sq.ft. ($25.83 per sq.m) of surface finished.

Two bat tooling, which is always used vertically, can be done for almost the same cost as given above if it is not made to register. This tooling, however, usually has to be made to register and for that reason is generally more costly than 4, 6, 8, or 10 bat work.

Sand Rubbed Finish. If the stone is to have a sand rubbed finish, making it necessary to rub the surface with sand and water, add $1.20 to $1.60 per sq.ft. ($12.91 to $17.22 per sq.m) for all surface to be rubbed.

Rustication. If the stone is to be furnished with rustication or a "rabbet" along the edges of the stone, the cost will vary according to the

776

depth of the rabbet. This is all planer work, so the width of the rabbet does not make as much difference as the depth to which it must be cut. The following will prove a fair average of the cost of rustication:

Rustication or Rabbet Edges of Stone			
Depth Inches	Price per Lin.Ft.	Depth mm	Price per Lin.Ft.
1/2 "	$ 0.75	12.5 mm	$ 2.46
1 "	$ 1.00	25.0 mm	$ 3.28
1 1/2 "	$ 1.25	37.5 mm	$ 4.10
2 "	$ 1.50	50.0 mm	$ 4.92
2 1/2 "	$ 1.80	62.5 mm	$ 5.91
3 "	$ 1.90	75.0 mm	$ 6.23

Rounded Edges on Stone. Add about $0.60 per lin.ft. ($1.97 per m) for stone with edges slightly rounded.

Stone Moldings. The cost of stone moldings will vary greatly according to the size of the stone, the type of mold, breaks, heads, jointing, etc. The two kinds of moldings most commonly used are plain or Classic moldings and Gothic moldings.

When cutting stone moldings on a planer, the smaller the piece of stone the higher the price per cu.ft. (cu.m), on account of the additional labor involved and the small amount of stone required.

Tools for cutting moldings vary from 3" (75 mm) to 8" (200 mm) wide, so when the moldings are over 8" (203 mm) in width, it is necessary to run them through the planer a second time. The moldings that are the cheapest to run are similar to the illustration of plain moldings, where a large amount of stone is required on which no special work is necessary.

When estimating plain moldings, take the largest dimensions of the stone to obtain the cubical contents and figure at the average price as given on the previous pages, as this will cover most jobs where the molded courses do not form too large a proportion of the total stone in the job.

Gothic Moldings. Gothic moldings are much more expensive than plain or classic moldings, as in almost every instance it is necessary to run the stone through the planer twice or perform a certain amount of hand labor. Gothic moldings cost at least twice as much to run as plain or classic moldings. The cutting of heads for gothic moldings will run much higher than plain moldings, from $5.00 to $24.00 each, depending upon the size and character of the molding.

When an entire job consists of elaborate gothic stone, it frequently will run from $24.00 to $64.00 and more per cu.ft. ($847 to $2,259 and more per cu.m). On work of this kind it is advisable to obtain figures from a reputable concern covering the stone delivered ready to set.

Round Stone Columns. When estimating the cost of plain turned columns, the cubical contents should be obtained by taking the cube of the largest dimension (measured square) and price as follows:

Obtain the cu. ft. cost of the grade of stock to be used, whether select buff, standard buff, select gray, etc., and add seven times the cu. ft. cost of the rough quarry blocks. This will give the cu. ft. price at which the columns should be estimated. For instance, a turned column 3'-0" (0.91 m) in diameter and 36'-0" (10.97 m) long would contain 3'x3'x36' or 324 c.f. of stone (0.91 x 0.91 x 10.97 = 9.07 cu.m of stone). Select buff is worth $25.00 per cu.ft. ($883 per cu.m) in rough blocks. Multiply $24.00 x 7 gives $168.00 per cu. ft.; 324 cu. ft. @ $168.00 equals $54,432.00, the cost of one column.

A stone column 38'-0" (11.58 m) long is the usual limit of present day machine capacity, so when the column shaft is over 36'-0" (10.97 m) long, it is best to get them out in three sections.

Fluted Stone Columns. Fluted stone columns should be figured on the same basis as plain turned columns described above. After the cost of the plain column has been obtained, the cost of the flutes should be estimated at $1.25 per lin.ft. ($4.11 per m) of each flute in the column.

Cutting Raised Letters in Indiana Limestone. The cost of cutting raised letters in Indiana limestone will vary according to the size and style of the letters.

On average work, a stone carver should cut one 6" (150 mm) letter per hr. at the following labor cost per letter:

Cutting Raised Letters 6" (150 mm) High						
	Hours	Rate	Total	Rate	Total	
Stone Carver	1.0	$....	$....	$ 36.93	$	36.93
Cost per letter Ft.					$	36.93

If the letters are 12" (300 mm) high, it will require about 1-1/2 hrs. time per letter, at the following labor cost per letter:

Cutting Raised Letters 12" (300 mm) High						
	Hours	Rate	Total	Rate	Total	
Stone Carver	1.5	$....	$....	$ 36.93	$	55.40
Cost per letter					$	55.40

Cutting Sunk Letters in Indiana Limestone. The cost of cutting sunk letters is considerably less than raised ones.

When cutting letters up to 6" (150 mm) high, a cutter should cut one letter in about 20 minutes at the following labor cost per letter:

Cutting Sunk Letters 6" (150 mm) High						
	Hours	Rate	Total	Rate	Total	
Stone Carver	0.33	$....	$....	$ 36.93	$	12.19
Cost per letter					$	12.19

If the letters are 12" (300 mm) high, a cutter will cut one letter per hr. at the following labor cost per letter:

Cutting Sunk Letters 12" (300 mm) High					
	Hours	Rate	Total	Rate	Total
Stone Carver	1.0	$....	$....	$ 36.93	$ 36.93
Cost per letter					$ 36.93

STANDARD ITEMS

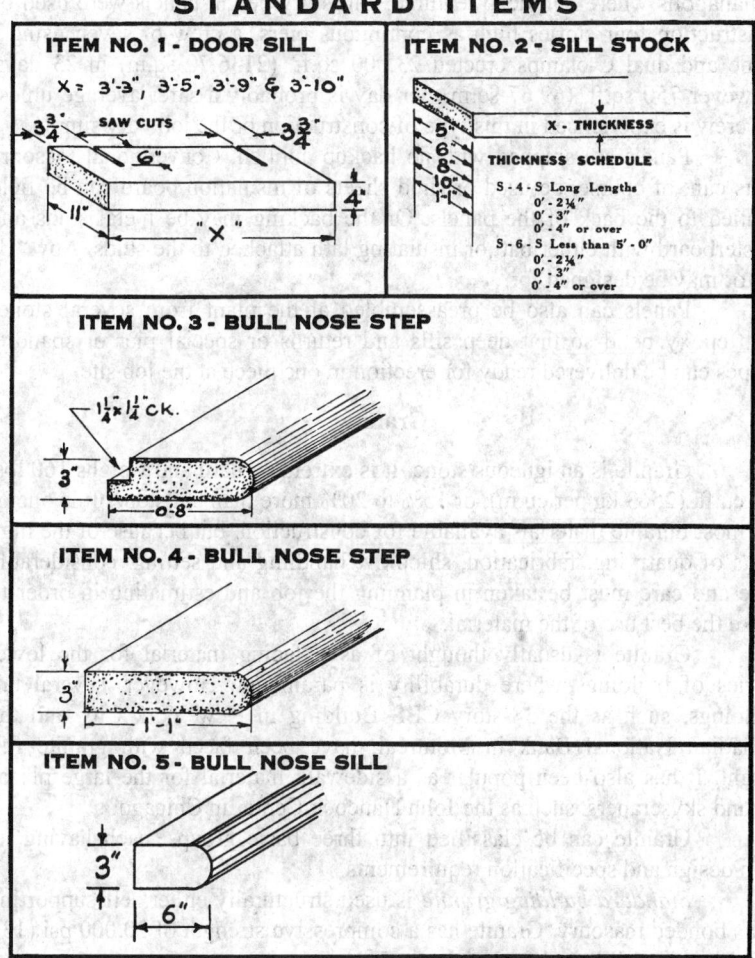

ITEM NO. 1 - DOOR SILL

X = 3'-3", 3'-5", 3'-9" & 3'-10"

ITEM NO. 2 - SILL STOCK

X - THICKNESS

THICKNESS SCHEDULE

S - 4 - S Long Lengths
0' - 2¼"
0' - 3"
0' - 4" or over
S - 6 - S Less than 5' - 0"
0' - 2¼"
0' - 3"
0' - 4" or over

ITEM NO. 3 - BULL NOSE STEP

ITEM NO. 4 - BULL NOSE STEP

ITEM NO. 5 - BULL NOSE SILL

Textured Indiana Limestone Facing Panels

Limestone panels are available as "curtain wall" units, as wall units spanning a full story height, or as spandrels spanning a full bay. Lengths of up to 15' (4.57 m) are possible, but the most practical panels will fall in the range of 40 to 80 sq.ft. (3.71-7.43 sq.m), 120 sq.ft. (11.14 sq.m) being the maximum. Thicknesses will depend on panel size and depth of texture but usually fall in the 4" (100 mm) to 5" (125 mm) range. Nine textures are

779

available. The price range for panels in the 40 to 60 sq.ft. (3.71-5.57 sq.m) size will run from $13.00 to 18.00 per sq.ft. ($139.93 to $ 193.75 per sq.m) plus shipping and erection costs.

Erection costs are estimated to run from $7.00 to $10.00 per sq.ft. ($75.35 to $107.64 per sq.m) including cranes or hoists. On a job in Indianapolis where 346 deep textured, full story height panels were used on construction four stories high as continuous piers, a crew of seven using a crane and dual C-clamps erected 23,000 sq.ft. (2136.70 sq.m) in 23 days. However 750 sq.ft. (69.67 sq.m) per day is probably a safer average unless the crew is experienced in this type of construction or the job very simple.

Panels are shipped with no backup applied. Conventional masonry units can, of course, be used or rigid sheets of insulation board can be field applied to the back of the panels. Or the backing may be metal studs and plasterboard with either batt or insulating lath attached to the studs. Any "U" factor may be designed for.

Panels can also be preassembled at the plant from several stones with epoxy bond so that deep sills and returns or special pier or spandrel shapes can be delivered ready for erection in one piece at the job site.

Granite

Granite is an igneous stone. It is extremely hard and weighs 160 lbs. per cu.ft. (2563 kg per cu.m), or 15% to 20% more than cut stone. It is one of the most durable materials available for construction, but because of the high costs of quarrying, fabrication, shipping, handling and setting, considerable time and care must be taken in planning the job and estimating in order to make the best use of the material.

Granite is usually thought of as a facing material for the lower stories of buildings where durability is paramount. However, several tall buildings, such as the 38-story CBS Building in New York City and the Canadian National Bank in Montreal, have been faced with granite full height. It has also been popular as a sidewalk material for the large plazas around skyscrapers, such as the John Hancock Center in Chicago.

Granite can be classified into three basic groups, each having its own design and specification requirements.

Standard building granite is used structurally either self-supporting or as bonded masonry. Granite has a compressive strength of 20,000 psi (138 MPa) and is well suited for bearing walls.

Granite veneer is a facing used for decoration or protection rather than for its structural value and is applied to masonry backup with stainless steel anchors engineered for the job. Veneers will vary from 7/8" (21.87 mm) in thickness up to 8" (200 mm), the majority being in the 2" (50 mm) to 4" (100 mm) range. Also, precast "curtain wall" panels use granite facing. These are supported from the building framing rather than from the backup and present different construction problems altogether.

780

MASONRY

Masonry granite is granite used as an integral part of a masonry wall in which one of the more rugged finishes are used such as in churches, rambling residences, bridge abutments, and the like.

Granite is available in a wide range of finishes from split faced to mirror finished slabs.

Polished face is a mirror gloss; honed face, a dull gloss; and fine rubbed, a smooth finish with no gloss. Allowable tolerance for these three types is 3/64" (1.17 mm). Next are the rubbed finishes. Sand finish is slightly pebbled with a tolerance of 1/10" (2.50 mm); spot finish has circular markings with an indefinite pattern and 3/32 (2.34 mm) surface tolerance.

Hammered surfaced granite comes in three grades: fine or 8-cut with parallel markings not over 3/32" (2.34 mm); medium or 6-cut with parallel markings not over 1/8" (3.12 mm); and coarse with parallel markings not over 3/32" (2.34 mm). Tolerances are limited to 7/32" (5.46 mm) for the first two, 1/8" (3.12 mm) for the third. These are corrugated finishes which become smoother near arris lines; that is, near the intersections of two dressed faces.

Sawn faces may be had with vertical scorings (VSM), horizontal scorings (HSM), or circular scorings (RSM). The first two are produced by gang saws, the third, by rotary or circular saws. Tolerance is 1/8" (3.12 mm) with scorings 3/32" (2.34 mm) deep. Thermal finishes are produced by flame texturing a sawn surface with torches to flake granite to a dull, uniform, regular rough surface of coarse texture. It is popular for large paving blocks. Tolerance is 1/8" (3.12 mm).

Jet honing is produced by honing previously cut surfaces with a jet of water and compressed air fed with tiny glass spheres under pressure. Tolerance is 1/8" (3.12 mm).

Granite is a natural material and colors are varied, often two to three per material. Color may be selected from samples or specified by quarry. As shipping costs can be a major factor in a material as heavy as granite, local quarries with similar colors can effect major savings.

However, when the architect specifies a color by quarry, any substitution should be made only with his written approval or as an alternate, lest the contractor be held for the difference in shipping costs.

Granites with feldspar produce the red, pink, brown, buff, grey, and off-white varieties; those with horn blende or mica produce the dark greens and blacks.

Graining as well as color is a consideration; a uniformity of texture, whether fine, medium or coarse, being desirable.

The National Building Granite Quarries Assoc., Inc. publishes a list of granite suppliers with colors and graining available from each. Canada and Sweden are also major sources.

Material Cost of Granite. As discussed above, several factors determine the price of granite, including color, graining, and transportation to the site.

The lower priced polished veneers will include the solid blacks and mottled greys. These cost from $15.00 to $20.00 per sq.ft. ($161.46 to $215.28 per sq.m) in 1" (25 mm) thickness, $20.00 to $23.00 per sq.ft. ($215.28 to 247.58 per sq.m) in 2" (50 mm) thickness, and $23.00 to $27.00 per sq.ft. ($247.57 to 290.63 per sq.m) in 4" (100 mm) thickness. The variegated reds, pinks, and bluish and greenish blacks will cost $20.00 to $29.00 per sq.ft. ($215.29 to $312.16 per sq.m). Sawn finish will cost about $2.75 less per sq.ft. ($29.60 per sq.m) and thermal finish about $1.75 less.

Random ashlar granites are considerably cheaper, from $8.00 to $15.00 in 4" (100 mm) thicknesses. This is often priced per cu. ft. (cu.m), the range being $23.00 to $47.00 ($812.14 to $1,659.60 per cu.m).

Dressed granite, machine cut in structural thicknesses, will be $23.00 to $75.00per cu.ft. ($812.14 to $ 2,648.30 per cu.m) depending on the number of openings, washes, and corners. Hand cut granite, like that on a mausoleum, can cost $188.00 or more per sq.ft. ($2,023.68 per sq.m).

Hand cutting, except on the most elaborate work, has been replaced by precision machines. Even surface carving and lettering is done today by sandblasting. Carved letters 6" (150 mm) to 10" (250 mm) high will cost about $30.00 per letter if cut into the granite, $47.00 if raised from the granite.

Cutting of granite into shapes other than square blocks adds quickly to the cost. One source suggests that if a simple square step, which might cost $29.00 per cu.ft. ($1,024.01per cu.m), were to have a wash or 1/8" (3.12 mm) slope on the top surface, the cost would increase 23%; a wash on two surfaces, 30%; a wash on two surfaces plus a drip, 42%; and an ogee molding plus a drip, 55%.

Labor Setting Granite Veneer. One mason and one helper should set from 30 to 40 sq.ft. (2.78 to3.71 sq.m) of veneer per day, polished work with 1/4" (6.25 mm) joints taking more time than straight sawn work. Column enclosures can be figured at around 35 sq.ft. (3.25 sq.m) per day. Straight run work with no openings and no hoisting involved can be figured at 50 sq.ft. (4.64 m) per day.

The following figures are based on one mason and one helper. On large jobs this ratio may be cut down with 3 laborers serving 4 masons, provided scaffolding is arranged so workers can move from position to position. In addition, hoisting costs may have to be added.

On a low-rise building, the square foot labor cost for 100 sq.ft. (9.29 sq.m) of polished granite veneer would be figured as follows:

Low Rise Building - Polish Granite Veneer					
	Hours	Rate	Total	Rate	Total
1 Mason	20.0	$....	$....	$ 36.93	$ 738.60
1 Labor	20.0	$....	$....	$ 26.16	$ 523.20
Cost per 100 Sq.Ft.					$ 1,261.80
Cost per Sq. Ft.					$ 12.62
Cost per Sq. M.					$ 135.82

Work involving special trim at openings, copings, and moldings requires more time:

Special Trim at Openings, Copings and Moldings					
	Hours	Rate	Total	Rate	Total
1 Mason	26.5	$....	$....	$ 36.93	$ 978.65
1 Labor	26.5	$....	$....	$ 26.16	$ 693.24
Cost per 100 Sq.Ft.					$ 1,671.89
Cost per Sq. Ft.					$ 16.72
Cost per Sq. M.					$ 179.97

Artificial Granite. Cast granite is manufactured in slabs 1-1/2" (37.5 mm) to 2-1/2" (62.5 mm) thick as a less expensive substitute for the natural material. It is not as durable and will not take or hold quite the high polish of the true stone. Still, it has been popular on remodeling and store front work. It is set and handled on the job like true granite. The material costs less than half, or in the $5.00 to $8.00 ($53.80-86.11) range. One mason and one helper should set 50 to 60 sq.ft. (4.6-5.6 sq.m) per day.

Labor Setting Granite Ashlar. The labor cost of setting granite ashlar will vary considerably, depending on type of wall, height of rises, and the skill of the mason. Being thin, sawed bed ashlar is light in weight and easy to handle and when finished or semi-finished, a mason should set 45 to 55 sq.ft. (4.18-5.10 sq.m) of wall consisting predominantly of low rises per 8-hr. day, and up to 70 sq.ft. (6.50 sq.m) of wall, where the ashlar consists principally of high rises.

Marble

Marble is classified according to soundness into A, B, C, and D grades. The A grade is in general the cheapest and also the best for exterior use.

Exterior marble veneer varies in thickness from 7/8" (21.87 mm) to 2" (50 mm). Curtain wall material is usually 1-1/4" (31.25 mm) thick, although if set in a metal frame or precast with a concrete backup 7/8" (21.87 mm) material is used.

The finish for exterior marble is either sawn, sand, or honed. A polished surface is not recommended.

Marble will vary widely in cost with the type selected and the distance it must be shipped. In Chicago 7/8" (21.87 mm) Class A marble costs from $14.00 to $23.00 per sq.ft.

Jointing for marble is 1/16" (1.56 mm) to 1/8" (3.12 mm) for thin veneer, 3/16" (4.68 mm) for the thicker stock. All bed and face joints must be solidly filled with either a non-staining mastic or mortar. Most work is either pointed or caulked.

Exterior veneer should be anchored to the backup as follows: two anchors for up to 2 sq.ft. (0.2 sq.m); four for up to 20 sq.ft. (1.85 sq.m); and two for each additional 10 sq.ft. (0.92 sq.m) Anchors may be dowels, straps, wire ties or dovetails, but all must be non-corrosive.

In addition to anchors marble veneer must be supported over all openings and at each story height or not more than 20' (6.09 m) apart vertically.

Marble may be set against a dampproofed, prebuilt wall, in which case it is set out 1" (25 mm) to 1-1/2" (37.50 mm), anchored, and spot mortared 18" (450 mm) on centers; or set along with the backup, in which case it is completely parged on the back with the anchors built into both facing and backup as the work progresses; or it may be solidly grouted to the backup with the grouting carefully paced so as not to dislodge the marble. In all cases the mortar must be non-staining. Soffit marble must have liners. Marble is cleaned with clear water and fibre brushes. Acids will stain the surface.

Two masons and a laborer will set around 8 sq.ft. (0.74 sq.m) of 1" (25 mm) veneer per hour in straight runs. Pointing and caulking and hoisting costs must be added.

Labor Cost of 100 Sq.Ft. (9.29 Sq.M.) of 1/4" (6.25 mm) Marble Veneer Set Against a Prebuilt, Dampproofed Masonry Wall					
	Hours	Rate	Total	Rate	Total
Mason	25.0	$....	$....	$ 36.93	$ 923.25
Labor	12.5	$....	$....	$ 26.16	$ 327.00
Cost per 100 Sq.Ft.					$ 1,250.25
Cost per Sq. Ft.					$ 12.50
Cost per Sq. M.					$ 134.58

Slate

Slate is sometimes used as an exterior veneer. When used in this way, it is usually 1" (25 mm) to 1-1/4" (31.25 mm) thick and furnished with a natural cleft finish. Panel sizes should not exceed 9'-6" (2.85 m) in length, 5' (1.52 m) in width, or about 25 sq.ft. (2.32 sq.m) in area.

Slate veneer is held in place with anchors and spot mortar. Relief angles should be provided at each story height. All angles and anchors must be non-corrosive. Anchors should be provided four to a slab or every 3 sq.ft. (0.27 sq.m). Round anchor holes cut in the edge of the slate to receive wire anchors is the recommended procedure. Mortar should be applied in 6" (150 mm) spots, 18" (200 mm) on center. All joints should be full and are usually 3/8" (9.37 mm) wide.

Slate will run around $15.00per sq.ft. ($161.46 per sq.m) for material. Two masons and one laborer should set around 7 sq.ft. (0.65 sq.m) per hour.

784

MASONRY

Labor Cost of 100 Sq.Ft. (9.29 Sq.M.) of Slate Veneer					
	Hours	Rate	Total	Rate	Total
Mason	28.0	$....	$....	$ 36.93	$ 1,034.04
Labor	14.0	$....	$....	$ 26.16	$ 366.24
Cost per 100 Sq.Ft.					$ 1,400.28
Cost per Sq. Ft.					$ 14.00
Cost per Sq. M.					$ 150.73

04500 MASONRY RESTORATION

Cutting out Mortar Joints in Old Brickwork with an Electric Grinder. On old buildings, where it is necessary to cut out the mortar joints in the old brickwork to a depth of 1/2" (12.50 mm) and repoint them, the most efficient method is to use an electric carborundum wheel, about 5" (125 mm) in diameter and 1/4" (6.25 mm) thick, depending upon the width of the mortar joints. The motor is hung above the scaffolding with a flexible shaft connected to the carborundum wheel.

This method is nearly twice as fast as using an ordinary electric hand saw on account of the weight of the saw, approximately 17 lbs. (7.7 kg), which is difficult and awkward to handle for workers on a swinging scaffold.

Either two or four workers work from each scaffold, depending on the size of the job, and it will require one helper on the ground to each scaffold.

An experienced tuckpointer or operator should grind out lime or lime-cement mortar joints in 350 to 450 sq.ft. (32.51 to 41.80 sq.m) of brick wall per 8-hr. day, at the following labor cost per 100 sq.ft. (9.29 sq.m):

Grind Out Lime or Limestone Joints					
	Hours	Rate	Total	Rate	Total
Tuck pointer	2.0	$....	$....	$ 36.93	$ 73.86
Labor	1.0	$....	$....	$ 26.16	$ 26.16
Cost per 100 Sq.Ft.					$ 100.02
Cost per Sq. Ft.					$ 1.00
Cost per Sq. M.					$ 10.77

Using ordinary methods, with an electric saw or grinder, a mason will cut out about 25 sq.ft. (2.32 sq.m) of joints per hr. at the following labor cost per 100 sq.ft. (9.29 sq.m):

Using Ordinary Methods, Grind Out Lime or Limestone Joints					
	Hours	Rate	Total	Rate	Total
Mason	4.0	$....	$....	$ 36.93	$ 147.72
Cost per 100 Sq.Ft.					$ 1.48
Cost per Sq. Ft.					$ 0.01
Cost per Sq. M.					$ 0.16

785

Repointing Mortar Joints in Old Brickwork. After the mortar joints have been cut out, two tuckpointers working on the scaffold, with one helper on the ground, should point 400 to 500 sq.ft. (37.16-46.45 sq.m) of brick wall per 8-hr. day, at the following labor cost per 100 sq.ft. (9.29 sq.m):

Repoint Mortar Joints in Old Brickwork					
	Hours	Rate	Total	Rate	Total
Tuckpointer	3.5	$....	$....	$ 36.93	$ 129.26
Labor	1.5	$....	$....	$ 26.16	$ 39.24
Cost per 100 Sq.Ft.					$ 168.50
Cost per Sq. Ft.					$ 1.68
Cost per Sq. M.					$ 18.14

The ordinary hanging or swinging scaffold is 25' (7.62 m) long. The tuckpointers working from each scaffold can cover approximately 6' (1.82 m) in height, so that each lift of the scaffold covers 150 sq.ft. (13.93 sq.m).

Cleaning Paving Brick Using Pneumatic Chipping Hammers. A satisfactory method of cleaning brick with pneumatic chipping hammers is to use a 4"x12" (100 x 300 mm) plank about 10'-0" (3.04 m) long, placed on 2 carpenter horses and raised at one end 15 to 18 degrees. This is used to hold the brick, which may be placed lengthwise across the plank, so that one side and one end can be cleaned without moving. A plank will hold about 56 brick. Using a pneumatic chipper with a hexagon nozzle and a 2" (50 mm) chisel bit, a worker should clean 150 to 180 brick per hr. or 1,200 to 1,440 per 8-hr. day.

Sandblast Cleaning of Buildings

The portable compressor and the sandblast are widely used for cleaning buildings, bridges, and other structures, either preparatory to repainting or for the purpose of brightening up the surface.

The cost of this work, however, is contingent entirely upon the condition of the old building, as some of them have nothing more than many years' accumulation of dirt while others have been painted numerous times, making it necessary to remove several coats of paint before the surface of the brick or stone can be touched. Also, it is often necessary to repoint all of the old mortar joints. The estimator should consider all of these items when preparing the estimate.

The process of sandblasting is to force sand against the surface to be cleaned through a small hose connected with the air compressor, and the force of the sand striking the object to be cleaned removes the old surface. Where it requires a maximum pressure of sand and air to remove the old dirt, a 3/4" (18.75 mm) hose is used but where volume is desired more than high pressure a 1" (25 mm) hose is used.

It usually requires four workers to operate a sandblasting machine: one worker attending the machine, one at the nozzle, and two on the ground assisting with hose, scaffold, and sand.

786

MASONRY

When estimating quantities, take the entire area of the surface to be cleaned and make no deductions for door or window openings.

On court houses, state capitols, post offices, public libraries, and other ornamental structures, the quantities are computed as above, and in addition, balustrades, balusters, etc., are measured on both sides, as it is necessary to go around each baluster separately to clean it.

With a portable compressor supplying air, the average operator can cover a strip 25'-0" (7.60 m) wide and 60'-0" (18.25 m) to 75'-0" (22.86 m) high per 8-hr day. The exact area varies with the quality of the stone encountered but the following table shows the speed at which various stones can be cleaned.

Clean Stone					
Material	Sq.Ft. per min.		Sq.M. per min.		
Limestone	7	- 9	0.65	-	0.84
Terra Cotta	8	- 10	0.74	-	0.93
Marble	3	- 5	0.28	-	0.46
Brick or Brownstone	8	- 10	0.74	-	0.93
Granite	6	- 8	0.56	-	0.74
Sand stone	10	- 12	0.93	-	1.11

The quantity of sand required for sandblasting will vary with the method used, but where a canvas screen or covering surrounding the scaffold catches most of the sand, it is then carried back to the tank on the ground. If this is done, about 75% of the sand is saved for use again. This, however, is governed somewhat by the conditions of the work. The following costs are for removing and cleaning only the natural accumulation of dirt and do not include removing paint, oil stains, etc., unless otherwise mentioned.

Maximum quantities are not figured in the tables as average job conditions usually include delays, moving from one job to another, placing and removing scaffold, etc.

Sandblasting 100 Sq.Ft. (9.29) SQ.M.) of Old Brick or Limestone Buildings					
	Hours	Rate	Total	Rate	Total
Compressor Operator	0.33	$....	$....	$ 37.89	$ 12.50
Nozzelman	0.33	$....	$....	$ 36.93	$ 12.19
Labor	0.67	$....	$....	$ 26.16	$ 17.53
Cost per 100 Sq.Ft.					$ 42.22
Cost per Sq. Ft.					$ 0.42
Cost per Sq. M.					$ 4.54

Note: Add for equipment rental, Sand, and Scaffold Cost together with enclosure

Sandblasting Old Brick Work That Has Been Painted. If the old brick work has been painted and the paint can be blasted off without the necessity of repointing the mortar joints, the costs will be about the same as given above for plain work.

Sandblasting Ornamental Buildings, Such as Post Offices, Courthouses, State Capitols, Public Libraries, Etc., Constructed of Granite or Limestone. The cost of sandblasting buildings as enumerated above, having exterior elaborations, such as balustrades, gables, pediments, plain or fluted columns, etc., will be higher than plain buildings on account of the extra labor required for scaffolding, sandblasting balusters, columns, caps, etc., that are not found on the average building.

Repointing Mortar Joints After Sandblasting. On many old buildings that are unusually dirty or that have been painted, it often happens that the mortar is blasted from the joints when removing the paint from the brick. After the job has been cleaned, the tuckpointers must go over the entire surface and repoint all of the brick joints.

05   METALS

05100   STRUCTURAL METAL FRAMING

The preparation of a structural steel estimate involves a great deal of work because of the large amount of detail required in listing the quantities from the plans, computing the weights of the various sections, making the shop drawings and detailing the steel, and estimating the cost of shop fabrication, freight, trucking, and erecting the steel on the job.

The structural steel framework of a building usually consists of anchor bolts, setting plates, base plates, columns, girders, beams, lintels, roof trusses, etc., which are fabricated from standard shapes, such as angles, "S" beams, channels, "W" beams and columns, plates, rods, etc., in combinations designed to give the required strength.

Estimating Quantities of Structural Steel

When estimating the quantity of structural steel required for any job, each class of work (column bases, columns, girders, beams, lintels, trusses) should be estimated separately, because each involves different labor operations in fabrication and erection.

All standard connections can be taken from the AISC Manual of Steel Construction. Be sure to include bolts for connecting steel to steel. In addition to the structural framing of the building, it is often necessary for the structural steel contractor to furnish numerous miscellaneous items, such as bearing plates, loose lintels, and anchor bolts, even though they are set in place by other contractors.

Items to be Included in a Structural Steel Estimate. When preparing an estimate on structural steel, the following items should be included to obtain the erected price:

1. Cost of structural steel shapes at the mill or warehouse.
2. Freight or trucking of structural steel shapes from rolling mill or warehouse to fabricating shop.
3. Cost of making shop drawings and details for shop fabrication.
4. Cost of assembling and fabricating shapes into columns, girders, beams, trusses, etc. including templates in connection with these items.
5. Cost of applying one or more coats of shop paint.
6. Shop overhead expense and profit.

7. Freight on fabricated steel from shop to destination.
8. Cost of erecting steel, including equipment rental, setting up and removing cranes, hoisting equipment, unloading steel at site, erecting steel, and field connections either welded or bolted connections.

All open hot rolled shapes approved by the American Institute of Steel Construction AISC specifications fall under the American Society for Testing and Materials (ASTM).

Standard Steel Shapes and Sections

Shape/Section	ASTM Standard
W	A992
HP	A572, grade 50
M, S, C, MC, L	A36
HSS	A500, grade B
Pipe	A53, grade 53

Structural Plates and Bars

Thickness	ASTM Standard
3/16" up to 4"	A36; A572, grade 50
Greater than 4"	A36

Structural Fasteners

Fastener Type	ASTM Standard
High Strength Bolts	A325; A490
High Strength Control Bolts	F1852; F2280
Common Bolts	A307, grade B
Nuts	A563
Washers	F436
Threaded Rods	A36
Shear Stud Connectors	A108
Anchor Rods	F1554, grade 36

Estimating Cost of Structural Shapes at Mill or Warehouse. After the quantities have been listed from the plans, it will be necessary to compute the cost of the steel sections required.

All prices on structural steel are based on mill shipment from rolling mills in Illinois, Arkansas, Georgia, Texas, etc., depending on the location of the mill where the steel is purchased, plus freight or trucking from rolling mill to fabricating shop.

Where steel is wanted in a hurry, it is sometimes necessary to purchase it from warehouse stocks, which usually costs from $0.05 to $0.75 per lb. ($0.11 to $1.65 per kg) more than when purchased from mills, depending on quantity.

Another item that must be considered when pricing steel work is the base price on standard sections and the additions to it. For example, assume the base price of structural shapes is $26.50 plus surcharge $3.65 for a total

790

METALS

cost of $30.15 per 100 lbs. (CWT) (45.36 kg) or $603.00 per ton ($593.50 per tonne). Find the cost of 4" x 4" x 1/2" (100 x 100 x 12.50 mm) steel angles. The standard classification of extras on the following pages shows that angles of this size cost $1.25 per 100 lbs. (45 kg) more than the base price or $31.40 per 100 lbs. (per 45.36 kg).

Warehouse prices vary from the base price depending on size and quantity ordered. On orders of 30,000 lbs. (13,600 kg) or more, the decrease from base price is $0.70 per 100 lbs. (per 45.36 kg), while on orders from 100 to 399 lbs. (45-180 kg), the increase over base price is about $8.00 per 100 lbs. (per 4536 kg).

Structural steel prices change so often that latest quotations from steel suppliers should always be obtained before figuring jobs of any size.

Cost of Making Working Drawings and Detailing Structural Steel. The cost of making shop or working drawings for steel structures is usually figured by the number of sheets estimated to be required for layout, details, erection diagrams, etc. Drawings, 24"x36" (600 x 900 mm) average about $360.00 to $600.00 per sheet, including drafting room overhead.

If drawing costs are reduced to a cost per ton, the unit costs will vary considerably, depending upon the type of building, weight of steel, amount of duplication, etc. The ton cost will run considerably less where fairly heavy shapes are used than on light structures using tube columns, large lightweight trusses, etc., as the latter type of construction requires considerable detailing without involving much tonnage.

Steel skeleton construction, such as office buildings, hotels, etc., without unusual features can usually be detailed for $120.00 to $150.00 per ton (1.016 tonne).

Ordinary mill and factory buildings consisting of columns, beams, light trusses, and crane runs, requiring considerable detailing without much duplication, will cost from $90.00 to $180.00 per ton ($99.22 to $198.45 per tonne).

Theaters, churches, and other structures of this class will cost from $120.00 to $250.00 per ton ($132.30 to $275.63 per tonne) for details.

Shop Painting of Structural Steel. After the steel has been fabricated, it is customary to give it a coat of paint before it leaves the shop, except where steel is to be encased in concrete.

The cost of painting is usually estimated at a certain price per ton and the quantity of paint required and the surface area to be covered will vary with the class of steel. Ordinary structural beams contain 225 to 275 sq.ft. (20.90 to 25.54 sq.m) of surface per ton; plate girders 125 to 175 sq.ft. (11.61to 16.25 sq.m) of surface per ton, while trusses contain 275 to 325 sq.ft. (25.54 to 30.19 sq.m) of surface per ton. The average for a job will vary from 175 to 225 sq.ft. (16.25-to -20.90 sq.m) per ton of steel.

A gallon of paint will cover 400 to 500 sq.ft. (37.16 to 46.45 sq.m) of surface or about 2 tons (2.03tonnes) of steel, at a material cost of $25.00 per gallon, or $12.50 per ton ($6.60 per liter or $13.88 per tonne).

The labor necessary to paint one ton of structural steel, 200 sq.ft. (18.58 sq.m), with one shop coat of paint should cost as follows:

Description	Hours	Rate	Total	Rate	Total
Labor (spray)	0.75	$	$	$ 26.16	$ 20.12
Cst per Sq.Ft.				$	0.10
Cost per Sq.M.				$	1.08

Basis for Estimating All Classes of Structural Steel Work

Every structural shop has constant or fixed costs that must be considered before estimating the detailed fabricating operations. The following items should be computed before adding the various shop operations required for fabricating the different members.

Approximate Constant or Fixed Costs

Description	Price per 100 Lbs	Price per 50 Kg
Base price of steel, f.o.b. Virginia	$30.15- $112.50	$33.23 - 124.01
Freight or trucking on steel from rolling mill	$2.50*	$2.76*
Extras	$3.60*	$3.97*
Cost of structural steel, f.o.b. fabrication shop	$36.25	$39.96
Shop handling charge which includes unloading steel from cars or trucks upon arrival and loading cars or trucks after fabrication	$1.75	$32.85
Tonnage Charges	......	......
Painting structural steel in shop	$0.95	$1.05
Handling structural steel in shop	$1.00	$1.10
Cutting structural shapes to lengths	$1.40	$1.54
Shop Drawings - cost varies $30-60.00 per ton (metric ton) see explanation on previous pages, average cost	$5.00**	$5.51**
Total Constant or Fixed Costs	$41.35	$76.50

*Varies. **Varies according to type of building. (In rare cases this could be as much as $0.10)

The above constant or fixed costs must be determined in every estimate before adding special fabrications that are needed.

Cost of Fabricating Structural Steel

The cost of fabricating structural steel varies widely, depending upon the type of building, amount of duplication, weight of structural members, number of fabricating operations necessary, etc. The cost of fabricating light steel members usually runs considerably higher than heavier members on a per ton basis. The following gives approximate costs, including shop overhead, of the various shop operations and classifications.

Beam and Channel Punching Only. Where beams and channels are punched only, without any other operation, estimate as follows:

Beams and Channels, Punching Only Inches	Price per 100 Lbs	Beams and Channels, mm				Price per 50.00 Kg
6 " to 10 "	$ 6.66	150	to	250	mm	$ 7.35
12 " to 18 "	$ 5.22	300	to	450	mm	$ 5.75
Over 18 "	$ 4.93	Over		450	mm	$ 5.43

Beams and Channels Framed with Connections. When beam and channel sections are punched and fabricated with the necessary connections, estimate as follows:

Beams and Channels, Framed Conn. Inches	Price per 100 Lbs	Beams and Channels, mm			Price per 50.00 Kg
Up to 10 "	$ 10.43	Up to	250	mm	$ 11.50
12 " to 18 "	$ 8.69	300	to 450	mm	$ 9.58
Over 18 "	$ 6.95	Over	450	mm	$ 7.67

Beam and Angle Framing with Shelf Angles. Where W-beams are fabricated with shelf angles as illustrated, estimate as follows:

Beams and Angle Framing W/shelf Angle, Inches	Price per 100 Lbs	Beams with Shelf angles, mm			Price per 50.00 Kg
Up to 10 "	$ 9.85	Up to	250	mm	$ 10.86
12 " to 18 "	$ 8.69	300	to 450	mm	$ 9.58
Over 18 "	$ 6.95	Over	450	mm	$ 7.67

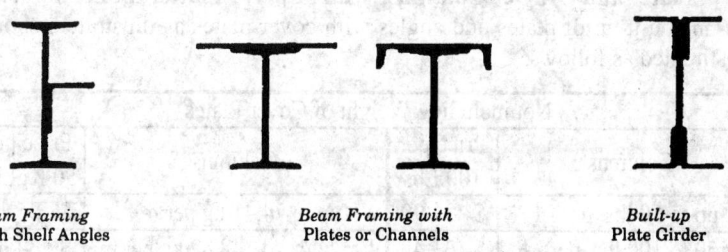

Beam Framing with Shelf Angles   *Beam Framing with Plates or Channels*   *Built-up Plate Girder*

Beam Framing with Plates or Channels. Where beams are fabricated with a plate or channel bolted or welded as illustrated, estimate as follows:

Beam Framing W/Plates or Channels, Inches	Price per 100 Lbs	Beams mm		Price per 50.00 Kg
Up to 10 "	$ 29.40	Up to 250 mm		$ 32.41
12 " to 18 "	$ 30.15	300 to 450 mm		$ 33.23
Over 18 "	$ 30.65	Over 450 mm		$ 33.79

Built-Up Plate Girders. When estimating the cost of built-up plate girders consisting of plates and angles as illustrated, estimate as follows:

Price per 100 lbs.   Price per 45.36 kg
$30.00 $66.13

Built-Up Plate Girders with Cover Plates. Built-up plate girders consisting of plates, angles, and cover plates as illustrated should be estimated per 100 lbs. (per 45.36 kg):

Price per 100 lbs.   Price per 45.36 kg
$36.00 $79.36

Plate and Angle Columns. Structural steel columns built-up of plates and angles as illustrated should be estimated as follows:

Columns	Lbs per Lin. Ft.	Price per 100 Lbs	Kg per Lin. Meter			Price per 50.00 Kg
Weighing up	50	$30.15	74.41			$ 33.23
Weighing	50 to 100	$31.40	74.41	to	148.82	$ 34.61
Weighing	Over 100	$30.90	Over	148.82		$ 34.06

Plate and Angle Columns with Cover Plates. Structural steel columns built-up of plates and angles with cover plates as illustrated, should be estimated as follows:

Not including Weight of Cover Plates			
Columns	Price per 100 Lbs	Columns	Price per 50 KgLbs
Wt. up to 50 lbs. per LF.	$ 31.90	Wt. up to 74.41 kg per m	$ 35.16
Wt. 50 to 100 lbs. per LF.	$ 31.40	Wt. 74.41 to 148.82 kg per m	$ 34.61
Wt. 100 lbs. per LF.	$ 31.40	Wt. 148.82 kg per m	$ 34.61

794

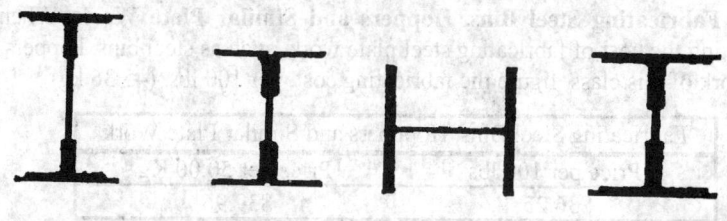

Built-up Plate Girder with Cover Plates   **Built-up Plate and Angle Columns**   **Rolled W Sections**   **Built-up Plate and Angle Columns with Cover Plates**

Rolled W Columns. Structural steel columns consisting of rolled wide flange and similar sections as illustrated should be estimated as follows:

WF Columns	Price per 100 Lbs	WF Columns	Price per 50.00 Kg
Up to 6"	$ 29.90	Up to 150 mm	$32.96
8"	$ 29.15	200 mm	$32.13
Over 8"	$ 30.90	Over 200 mm	$34.06

Crane Columns. Structural steel columns for supporting crane runs, etc.

Crane Columns	Price per 100 Lbs	Crane Columns	Price per 50.00 Kg
Columns	$ 30.90	Columns	$34.06

Steel Roof Trusses. When estimating the cost of structural steel roof trusses, the fabricating cost will vary with the weight of the trusses as follows:

Steel Roof Trusses		
Description	Price per 100 lbs	Price per 50.00 Kg
weighing up to 1,000 lbs. (453.6 Kg) each	$ 45.00	$ 49.60
weighing 1,000 to 2,000 lbs. (453.6 to 907.2 Kg) each	$ 41.25	$ 45.47
weighing from 2,000 to 3,000 lbs. (907.2 to 1,360.8 Kg) each	$ 38.75	$ 42.71
weighing from 3,000 to 4,000 lbs. (1,360.8 to 1,814.4 Kg) each	$ 36.25	$ 39.96
weighing over 4,000 lbs. (1,814.4 Kg) each	$ 35.00	$ 38.58
Knee Bracing. Fabricating cost	$ 28.75	$ 31.69
Angle Struts. Fabricating cost	$ 28.75	$ 31.69
"X" Framing and Bracing. Fabricating cost	$ 17.50	$ 19.29
Tie Rods. Tie rods 5/8" (15.62 mm) rounds, fabricating	$ 56.25	$ 62.00
Bracing Rods, For 3/4" to 1" (18.75 to 25 mm) round, fabricating	$ 37.50	$ 39.68
Knee Bracing, Fabricating cost	$ 28.75	$ 31.69

Fabricating Steel Bins, Hoppers and Similar Plate Work. When estimating the cost of fabricating steel plate work, such as steel bins, hoppers, and work of this class, figure the fabricating cost, per 100 lbs. (45.36 kg)

Fabricating Steel Bins, Hopppers and Similar Plate Work	
Price per 100 lbs	Price per 50.00 Kg
$36.25	$39.96

Floor Plates, Plain or Checkered. Estimate the weight of the material in the usual manner. The material usually costs $0.72 to 0.80 per lb. (0.45 kg), depending on market conditions. The cost of fabricating should be figured at $0.45 to 0.45 per lb. (0.45 kg) additional.

Make Up of a Complete Structural Steel Price

The following items should be included to obtain the total price for structural steel delivered f.o.b. cars or job:

Total Steel Price - Delivered to the Project	Price per 100 Lbs.	Price per 50 Kg.
Total steel price, f.o.b. fabricating shop, including constant or fixed costs given on previous pages	$41.35	$45.58
Shop labor and fabrication cost (varies according to operations required, as given on the previous pages	$14.00**	$15.43**
Total direct shop costs	$55.35	$61.01
Business administration, overhead and sales expense, 20%	$11.07	$12.20
Sub total	$66.42	$73.21
Profit on total cost, including overhead, 10%	$6.64	$7.32
Sub total	$73.06	$80.54
Transportation from shop to job, varies	$2.00	$2.00
Price of structural steel, delivered at job	$75.06	$82.54

This item varies as described on previous pages.

**Fabricating costs vary according to amount of shop labor required as given under cost of the various fabricating operations.*

Mill and Shop Inspection of Structural Steel. Where the structural steel is inspected by one of the testing laboratories, the usual charges for this work are as follows.

Mill inspection is usually figured at $5.00 to $10.00 per ton ($5.51 to $1.02 per tonne) and shop inspection from $6.50 to $7.00 per ton ($7.22 to $7.71 per tonne). Where both mill and shop inspection are let to one firm, a common price is $10.00 to $15.00 per ton ($11.02 to $13.60 per tonne) for both.

Field Inspection of Structural Steel. Where field inspection of the erection of the structural steel is required, figure $2,200 per week for salary and expenses of a field inspector, plus transportation to and from job.

METALS

Other possible added cost items to consider when pricing structural steel are manufacture extras for some extras as not limited to the following items: a)Military Specifications, b) ABS certification, c) steel grade extras, and d) surcharge costs.

Erection of Structural Steel

The field cost of handling and erecting structural steel is difficult to estimate for a number of reasons, such as weather conditions, strikes, delays in receipt of materials, storage facilities, size and type of building, equipment available, labor conditions, number and type of connections.

Any one of the above items may cause a considerable variation in the estimated labor costs and for that reason should be carefully considered when preparing the estimate.

Hauling Structural Steel From Shops to Job. The cost of hauling structural steel from the shop to the job will vary with the facilities for loading trucks, distance of haul, weight of steel members, etc.

The average cost of hauling structural steel is $30 per ton (0.90 metric ton), but varies according to total tonnage.

Plant and Equipment for Steel Erection. The equipment required for steel erection will vary considerably with the size and type of building.

Skeleton frame buildings up to 100'-0" x 100'-0" (30.48 x 30.48 m) in size will require one crane to unload the steel from the trucks and erect it in place, while buildings of irregular shape, such as "L" shape, "U" shape, etc., will probably require additional cranes. One crane can handle a job 100'-0" x 100'-0" (30.48 x 30.48 m), but if the job is 75'-0" x 125'-0" (22.86 x 38.10 m), it will probably require two cranes to set the job, while a "U" shape building may require 3 cranes.

Where the buildings have setbacks at certain story heights, as required by many cities, it will often require 2 cranes on the job, one to unload the steel and hoist it from the street and another to set the steel.

It costs $15,000.00 to $20,000.00 to set up and remove a crane with all its associated equipment for the boom transportation, counterweights, mats, permits, compressor outfit, etc., for a one crane project, with additional cranes at approximately the same cost or less crane depending crane size, rubber tire or track mounted.

On buildings covering considerable ground area, such as high schools, office buildings, auditoriums, mill and factory buildings up to 8 plus stories high, it is often more economical to use, large capacity truck mounted hydraulic cranes, crawler mounted lattice boom cranes or tower cranes depending upon location of the project, the type and number of cranes used must be well planned for any project. Where a movable crane can be used to advantage, that is the most economical method to use, because the crane can be moved about at slight expense. This is in comparison to a tower crane that is stationary but has a large reach. One must remember the larger the reach the less capacity the tower crane can accommodate.

797

Steel Erection Contractors. Most structural steel is erected by specialty contractors, who have an organization and personnel with background and experience. These contractors can erect structural steel much cheaper than the average general contractor, because they have well-organized crews specializing in this type of work. They also have on hand the most modern equipment, such as tower cranes, crawler cranes, truck cranes, lightweight and modern air tools, such as grinders, drills, impact wrenches, welding machines, and other special tools which are furnished to the job as needed and seldom figured in the job cost.

Erecting Structural Steel in Skeleton Frame Buildings. When figuring erection on a job of this type, consider carefully the amount of equipment, size of cranes, etc. required, as described previously.

On a one-crane job, an erection crew should unload, handle, erect, and connect 9 to 10 tons (8.16 to 9.07 tonne) of steel per 8-hr. day, at the following labor cost per ton (metric ton):

Erection Crew					
Description	Hours	Rate	Total	Rate	Total
Crane	0.85	$	$	######	$ 148.75
Operating Engineer	0.85	$	$	$ 37.89	$ 32.21
Oiler	0.85	$	$	$ 32.89	$ 27.96
1 Iron worker foreman	0.85	$	$	$44.29	$ 37.65
2 Ironworkers "hooking on"	1.70	$	$	$40.29	$ 68.49
1 Ironworker "giving signals"	0.85	$	$	$40.29	$ 34.25
4 Ironworkers "connecting"	3.40	$	$	$40.29	$ 136.99
Cost per 10 ton					$ 486.29
per ton					$ 48.63
per tonne					$ 47.86

Add for worker's compensation and liability insurance.

Add for workmen's compensation and liability insurance. If the structural connections are welded instead of bolted, add $36.50 to $43.50 per ton ($40.24 to $47.96 per tonne).

Erecting Structural Steel in Wall Bearing Buildings. On wall bearing buildings, the ends of all beams and channels usually rest on the exterior masonry walls while the interior framing is carried on masonry walls or structural steel columns. Field connections are usually bolted.

Eight workers together should handle and erect (including bolted connections) about 8 tons (7.25 tonnes) of steel per 8-hr. day, at the following labor cost per ton (tonne):

METALS

Wall Bearing Building					
Description	Hours	Rate	Total	Rate	Total
Crane	1.00	$	$	$ 75.00	$ 75.00
Operating Engineer	1.00	$	$	$ 37.89	$ 37.89
Oiler	1.00	$	$	$ 32.89	$ 32.89
1 Ironworker Foreman	1.00	$	$	$ 44.29	$ 44.29
7 Ironworkers	7.00	$	$	$ 40.29	$ 282.03
Cost per 8 ton					$ 472.10
per ton					$ 59.01
per tonne					$ 58.08

Add for worker's compensation and liability insurance.

Erecting Lightweight Steel for Mill Buildings, Foundries, Auditoriums, Etc. On light-constructed steel buildings, such as machine shops, foundries, mill buildings, garages, auditoriums, etc., where the structural steel consists of steel columns and lightweight roof trusses, requiring considerable handling to set only a small tonnage, the labor costs run higher than on average skeleton frame construction.

A rubber tire hydraulic crane or a truck crane should be used for setting the steel. After the crane has been delivered at the job, the labor unloading, handling, erecting, and connecting one ton (0.907 tonne) of steel should cost:

Lightweight Steel					
Description	Hours	Rate	Total	Rate	Total
Crane	1.10	$	$	$ 75.00	$ 82.50
Operating Engineer	1.10	$	$	$ 37.89	$ 41.68
Oiler	1.10	$	$	$ 32.89	$ 36.18
1 Ironworker Foreman	1.10	$	$	$ 44.29	$ 48.72
7 Ironworkers	7.70	$	$	$ 40.29	$ 310.23
Cost per 1 ton					$ 519.31
per tonne					$ 511.13

Add for worker's compensation and liability insurance.

Erecting Steel Bins, Hoppers, and Similar Plate Work. When erecting steel bins, material hoppers, and similar plate work, an erection crew consisting of 5 iron workers and an engineer should erect in place and bolt 4 to 5 tons (3.62 to 4.53 tonne) of steel a day at the following labor cost per ton (metric ton):

Erecting Steel Bins					
Description	Hours	Rate	Total	Rate	Total
Crane	2.00	$	$	$ 75.00	$ 150.00
Operating Engineer	2.00	$	$	$ 37.89	$ 75.78
Oiler	2.00	$	$	$ 32.89	$ 65.78
1 Ironworker Foreman	2.00	$	$	$ 44.29	$ 88.58
5 Ironworkers	10.00	$	$	$ 40.29	$ 402.90
Cost per 5 ton					$ 783.04
per ton					$ 80.58
per tonne					$ 79.31

Add for worker's compensation and liability insurance.

Welding the joints or seams in hopper work runs from $85.00 to $140.00100.00 per ton, additional ($93.71 to $154.35 per tonne). Add for worker's compensation and liability insurance.

Bolting Field Connections. Most structural steel projects have sections bolted together. On work of this kind, an iron worker should place 150 to 200 bolts per 8-hr. day, at the following labor cost per 100:

Bolting Field Connection					
Description	Hours	Rate	Total	Rate	Total
Compressor	2.50	$	$	$ 45.00	$ 112.50
Compressor Operator	2.50	$	$	$ 32.89	$ 82.23
Iron worker	5.00	$	$	$ 40.29	$ 201.45
Total Cost 100 bolts					$ 396.18
Cost per bolt					$ 3.96

Bolted connections normally are completed with a two-worker bolting crew. The cost of bolt units will vary with size, length, and grade of bolt. The most common bolts are ASTM A-325, high strength bolts for structural steel joints, 3/4" diameter x 2" long (18.75 x 50 mm), at an approximate cost per bolt unit of $0.95 each. Each bolt unit consists of bolt, washers, and nut. Other grades sometimes required for connections are A490M-04a Standard Specification for High-Strength Steel Bolts, Classes 10.9 and 10.9.3, for Structural Steel Joints [Metric] heat treated steel structural bolts (add up to 50% to the cost per bolt unit), and for secondary bolted connections, grade *A307-04 Standard Specification for Carbon Steel Bolts and Studs, 60 000 PSI Tensile Strength* carbon steel externally-threaded standard connection (decrease the cost about 15% per bolt unit).

If tension control bolts are used, add additional costs of about 65% per bolt unit.

Instead of counting individual bolts, some estimators use an average count per ton of steel. If one judges that the time involved in making an exact bolt count is not warranted by the cost, then use the following averages.

800

Average Amount of Field Bolts Required per Ton of Steel			
Type of Work	Light	Avg.	Heavy
No. Field Bolts per Ton	30	25	20

Steel Infrastructure Repairs

Repair of steel infrastructure, which usually refers to bridge and elevated highway structures, presents the estimator with many challenging problems that are not found in new construction. One major difficulty is that repairs are often undertaken in areas where the public has access to the work zone or to adjacent areas. There are many items that might not appear on contract drawings and documents, but which the estimator must consider:

1. Maintenance and protection of traffic - For costing and sample traffic patterns in accordance with federal regulations, see Chapter 22, Road and Highway Construction.

2. Maintenance and protection of railroad traffic - Normally, when working within 50 feet of the centerline of a rail track, special precautions must be taken. The estimator should contact local railroad authorities regarding the requirements for flagman, working hours, and the like.

3. Temporary support of existing structures - If temporary supports are not shown on the drawings, the contractor should enlist the services of a professional engineer to design them. The cost of these services will be in direct proportion to the complexity and risk involved. Allow from $65.00 to $150.00 per hour for design services.

4. Activity sequencing - A work sequence must be established. If one has already been set up by the designing engineer, then it is critical that it be followed. In pricing the infrastructure project, the estimator must carefully consider the schedule.

5. Safety - Provisions for netting, rigging, scaffolding, and platforms, as well as environmental conditions, must be considered and priced when required.

6. Work restrictions - Work hours might be restricted, such as during rush hours or holiday periods. In fact, it might be necessary to perform all work outside normal work hours. Area restrictions can result in productivity loss from 10 to 50 %.

7. Hazardous material - Most older steel structures are covered with lead-based paints. Removal of these painted surfaces must comply with regulations set forth by federal, state, and local environmental protection agencies, or other agencies with jurisdiction in this area. For the cost of removing lead-based paints from structural steel, see the "Painting" section in Chapter 9, *Finishes*.

Rivet Removal and Replacement. Rivet heads are snapped off with a hand-held air scaling tool. The use of flame cutting is discouraged and usually not even permitted. The flat end of a chisel point is placed on top of the rivet against the steel member until the rivet head breaks off. The rivet is then hammered out, and a new bolt is placed in the existing hole. Two ironworkers can remove and replace bolts at the rate of about 20 bolts per hr.:

Description	Hours	Rate	Total	Rate	Total
Ironworkers	2.00	$	$	$ 40.29	$ 80.58
Cost per bolt					$ 4.03

Drilling Holes into Existing Steel. The best tool for this purpose is a magnetic base drill with a hole bit. Gaining access to the work location and setting up usually take more time than the actual hole drilling.

The set up will take about 15 minutes per hole. Drilling time is between 1 and 5 minutes for a 3/4" to 1-1/4" (18.75 to 31.25 mm) hole for steel up to 1-1/2" (37.50 mm) thick. Two iron workers can drill about 6 holes per hour:

Description	Hours	Rate	Total	Rate	Total
Ironworkers	2.00	$	$	$ 40.29	$80.58
Cost per bolt					$13.43

Cutting Existing Steel. When rusted or decayed areas need to be cut out and replaced with new steel, allow for two cuts. The first is cut rough, the second more carefully. After the second cut, additional time is needed for grinding. Production time for this work is difficult to estimate. There are so many types of steel and conditions of existing steel. Assume that 3 lin.ft. (0.91 m) of burn can be accomplished in about 10 minutes. If lead paint exists on the steel to be cut, time to remove the lead paint to the bare metal must be added. An area at least 1'-0" each side of the flame cut must cleaned of all lead per local, state and federal requirements in force at the time of the work.

Once repairs are complete, it will be necessary to paint the new steel and bolts, and to touch up areas adjacent to the repair work.

ARC WELDING IN BUILDING CONSTRUCTION

Arc welding structural steel in the construction of buildings and bridges has become an established practice. To aid the structural designer in its use, the American Welding Society, the American Institute of Steel Construction, and the American Association of State Highway Officials have established recommended codes and specifications governing design, fabrication, and erection of welded steel structures and applicable codes of most cities, counties and states are in accordance with them. In addition, most fabricating shops now have the equipment, personnel, and experience required to fabricate steel members for welded construction, and in the field an increasing number of qualified operators are available for welding field joints during erection.

Advantages of Arc Welded Steel Design. Advantages of arc welded steel design in building construction are lower overall construction costs, less construction noise, particularly desirable in hospital zones, residential districts, etc., and smooth, clean appearance of erected members which permits leaving structural steel exposed where desirable or necessary.

Overall lower costs are realized through reduced steel material requirements and savings in foundation and exterior wall construction. When a structure with welded connections is properly designed, savings in steel material are gained from the following:

1. Gross section, rather than effective net section, is used for design, permitting lighter members.
2. Rigid connections provide continuous beam action, permitting a further reduction in member sections.
3. Connection steel, such as splice plates, gussets, etc., are greatly reduced, sometimes eliminated.
4. With beams of less depth, height of building may be reduced, saving column steel and exterior wall construction.
5. In addition, overall costs for detailing, fabrication, freight, trucking, handling and erection are lower due to reduced weight, fewer pieces, and simplified connections.
6. On many structures, where designs and estimates were made for both riveted and welded construction, savings up to 15% on structural steel costs were obtained by using the welded method.

Conditions Affecting Arc Welding Costs

Best results and lowest costs are obtainable only when proper plate preparation and welding procedure is followed. The following conditions have considerable bearing on the cost and quality of welded joints.

Fit-Up. Care in cutting, forming, and handling shapes to be welded to avoid poor fit-up is a major factor in the cost and performance of welded joints. However, a gap of 1/64" to 1/32" (0.39-0.78 mm) is useful in preventing angular distortion and weld cracking. The accompanying graph shows how various sizes of gaps affect welding speeds for square or grooved butt welds. Fillet welds are affected by oversize gaps in a similar manner.

Position of Joints. The position of the joint has considerable effect on the speed and case of welding. Wherever practical, welds should be made in the downhand position with the joint level. Vertical or overhead welds require much more time and skill.

Foreign Matter in Joint. Excessive scale, paint, oil, or rust tend to interfere with welding and should be removed to obtain best speeds and results.

Build-Up or Overwelding. Any amount of weld metal in addition to that actually needed for the specified strength is useless, costly, wasteful, and in some instances, actually harmful. For butt welds, there should be just enough build-up, no more than 1/16" (1.56 mm), to make sure weld is flush with the plate. Excessive build-up not only wastes weld metal but increases welding time.

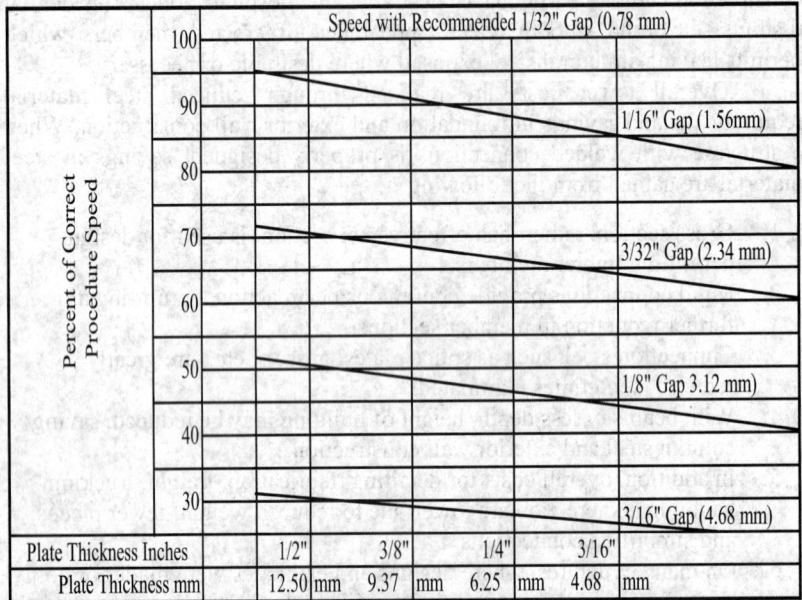

Effect of Gap Size on Welding Speeds for Flat Butt Joints
Welded on Both Sides

Procedures for Making Various Types of Welded Joints

The tables that follow give procedures for making various types of welded joints usually encountered in field erection of structural steel and furnish information for calculating time and material required. In the tables, electrodes specified are those which will make the joints at the lowest cost.

METALS

Horizontal Position   **Flat Position**

Plate Thickness Inches	Gauge Size of Fillet, Inches	Electrode Size, Inches	Current Amps	Electrode Melt-off Rate Inches/Min	Arc Speed, First Pass Inches/Min	Passes or Beads	Joints Welded Ft/Hr (100% Operating Factor)	Lbs, of Electrode per Foot of Weld
3/16	5/32	1/8	170	15	15 - 16	1	75 - 80	0.10
1/4	3/16	5/32	225	14	15 - 16	1	75 - 80	0.15
5/16	1/4	3/16	275	12	14 - 15	1	70 - 75	0.19
3/8	5/16	1/4	350 - 375	9.6 - 10.2	12 - 13	1	60 - 65	0.30
1/2	7/16	1/4	350 - 375	9.6 - 10.2	10 - 11	2	28 - 31	0.57
3/4	5/8	1/4	350 - 375	9.6 - 10.2	10 - 11	3 - 4	16 - 14	1.34
1	3/4	1/4	350 - 375	9.6 - 10.2	10 - 11	5	9 - 10	1.66
Plate Thickness mm	Gauge Size of Fillet, mm	Electrode Size, mm	Current Amps	Electrode Melt-off Rate mm/Min	Arc Speed, First Pass mm/Min	Passes or Beads	Joints Welded m/Hr (100% Operating Factor)	Kg, of Electrode per Meter of Weld
4.69	3.91	3.13	170	375	375 - 400	1	22.9 - 24.4	0.15
6.25	4.69	3.91	225	350	375 - 400	1	22.9 - 24.4	0.22
7.81	6.25	4.69	275	300	350 - 375	1	21.3 - 22.9	0.28
9.38	7.81	6.25	350 - 375	237.5 - 255.0	300 - 325	1	18.3 - 19.8	0.45
12.50	10.94	6.25	350 - 375	237.5 - 255.0	250 - 275	2	8.5 - 9.4	0.85
18.75	15.63	6.25	350 - 375	237.5 - 255.0	250 - 275	3 - 4	4.9 - 4.3	1.99
25.00	18.75	6.25	350 - 375	237.5 - 255.0	250 - 275	5	2.7 - 3.0	2.47

Fillet and Flat Welds in Horizontal Position and Flat Positions

Arc speed in inches per minute is given for single pass welds and for the first pass in multiple pass welds, as the speed of the first pass is important in obtaining proper penetration.

Feet of joints welded per hour is based on actual welding time. No factor has been included for set-up, electrode changing, cleaning, moving scaffold, etc., which will vary with the job. For overall average quantities, these figures should be multiplied by an operating factor which may range from 25% to 50%, depending upon job conditions.

Pounds of electrode per foot of weld is the amount of electrode required to weld the joint with the recommended plate preparation, fit-up, and a build-up of 1/16" (1.56 mm) or less, including normal spatter loss and 2" (50 mm) stub ends. Any increase in gap or build-up will greatly affect amount of electrode required and time for depositing same.

Fillet and Lap Welds in Vertical Position, Welded Up*

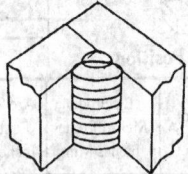

Plate Thickness Inches	Gauge Size of Fillet Inches	Electrode Size, Inches	Current (amp)	Electrode Melt-off Rate, Inches/min	Arc Speed, first pass Inches/min	Passes of Beads	Joints Welded Ft/Hr (100% Oper. Factor)	Lbs. of Electrode per Ft. of Weld
3/16	5/32*	5/32	140	9 1/2	10.0	1	50.0	0.08
1/4	3/16	3/16	150	5	7.0	1	35.0	0.14
5/16	1/4	3/16	170	8 1/2	6.0	1	30.0	0.17
3/8	5/16	3/16	170	8 1/2	3.9	1	19.5	0.26
7/16	3/8	3/16	170	8 1/2	2.7	2	13.5	0.38
1/2	7/16	3/16	170	8 1/2	6.0	2	10.0	0.52
5/8	1/2	3/16	170	8 1/2	4.0	2 or more	7.7	0.67
3/4	5/8	3/16	170	8 1/2	4.0	......	5.0	1.10
1	7/8	3/16	170	8 1/2	4.0	......	2.5	2.10
Plate Thickness mm	Gauge Size of Fillet mm	Electrode Size, mm	Current (amp)	Electrode Melt-off Rate, mm/min	Arc Speed, first pass mm/min	Passes of Beads	M of joint Welded per Hour (100% Oper. Factor)	Kg. of Electrode per M of Weld
4.69	3.91*	3.91	140	237.50	250.0	1	15.24	0.119
6.25	4.69	4.69	150	125.00	175.0	1	10.67	0.208
7.81	6.25	4.69	170	212.50	150.0	1	9.14	0.253
9.38	7.81	4.69	170	212.50	97.5	1	5.94	0.387
10.94	9.38	4.69	170	212.50	67.5	2	4.11	0.566
12.50	10.94	4.69	170	212.50	150.0	2	3.05	0.774
15.63	12.50	4.69	170	212.50	100.0	2 or more	2.35	0.997
18.75	15.63	4.69	170	212.50	100.0	......	1.52	1.637
25.00	21.88	4.69	170	212.50	100.0	......	0.76	3.125

* Weld 5/32" (3.91mm) fillet verticaly down. T-joint welded both sides with this fillet will have plate strength. Total No. of passes varies with operator.

Prepration - Square Edge.
Fit-up: ecommended gap 1/32" (0.78 mm); maximum gap 1/16" (1.56 mm). If Greater gaps must be welded, use same procedure but add width of gap to fillet weld.
Electrode - E6010
Polarity: electrode positive.

METALS
WELDED JOINTS
Standard symbols

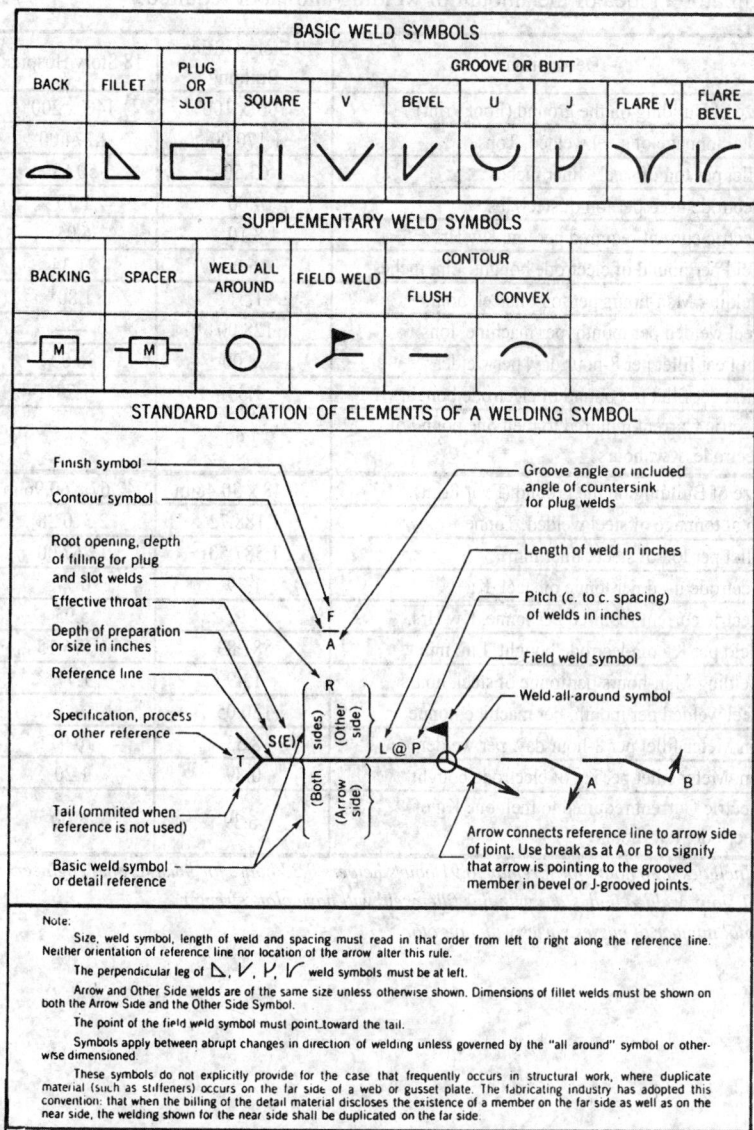

BASIC WELD SYMBOLS

BACK	FILLET	PLUG OR SLOT	GROOVE OR BUTT						
			SQUARE	V	BEVEL	U	J	FLARE V	FLARE BEVEL
⌒	△	▢	‖	∨	⋁	⋃	⊬	⋁	⋁

SUPPLEMENTARY WELD SYMBOLS

BACKING	SPACER	WELD ALL AROUND	FIELD WELD	CONTOUR		
				FLUSH	CONVEX	
▣ M	⊣ M ⊢	○	⚑	—	⌒	

STANDARD LOCATION OF ELEMENTS OF A WELDING SYMBOL

Finish symbol

Contour symbol

Root opening, depth of filling for plug and slot welds

Effective throat

Depth of preparation or size in inches

Reference line

Specification, process or other reference

Tail (ommited when reference is not used)

Basic weld symbol or detail reference

Groove angle or included angle of countersink for plug welds

Length of weld in inches

Pitch (c. to c. spacing) of welds in inches

Field weld symbol

Weld-all-around symbol

Arrow connects reference line to arrow side of joint. Use break as at A or B to signify that arrow is pointing to the grooved member in bevel or J-grooved joints.

F
A
R
(Other side)
(sides)
S(E)
T
L @ P
(Both sides)
(Arrow side)
A
B

Note:
    Size, weld symbol, length of weld and spacing must read in that order from left to right along the reference line. Neither orientation of reference line nor location of the arrow alter this rule.

    The perpendicular leg of △, ∨, ⋎, ⋁ weld symbols must be at left.

    Arrow and Other Side welds are of the same size unless otherwise shown. Dimensions of fillet welds must be shown on both the Arrow Side and the Other Side Symbol.

    The point of the field weld symbol must point toward the tail.

    Symbols apply between abrupt changes in direction of welding unless governed by the "all around" symbol or otherwise dimensioned.

    These symbols do not explicitly provide for the case that frequently occurs in structural work, where duplicate material (such as stiffeners) occurs on the far side of a web or gusset plate. The fabricating industry has adopted this convention: that when the billing of the detail material discloses the existence of a member on the far side as well as on the near side, the welding shown for the near side shall be duplicated on the far side.

Courtesy of American Institute of Steel Construction, Inc.

The following are actual production quantities of field welding on two different jobs located in different parts of the country and will give the estimator an idea of the amount of welding and labor required.

Description	9 Story Offfice Building	18 Story Hospital
Size of Building on the ground (Foot Print)	100' x 100'	140' x 200'
Total tonnage of steel welded, Tons	1,170.00	2,274.00
Fillet per ton of steel , liner inches	63.50	50.68
Electrode used per ton of steel, lbs.	4.00	1.70
Electric current required per ton, Kw Hrs.	8.10	6.05
Weld per pound of electrode bought, Lin. inches	16.00	21.34
Welding Man-hours per ton of steel, hours	1.51	1.80*
Steel welded per month, per machine, tons	128.00	-----
Lin Feet fillet per 8-hour day, per welder	28.00	36.00
Lin. Feet fillet per pound of electrode bought	1.33	1.78
Electric Current required to melt one pound of electrode, Kw. hours	2.90	3.63
Size of Building on the ground (Foot Print)	30.48 x 30.48 m	42.67 x 60.96 m
Total tonnage of steel welded, Tonne	1,188.72	2,310.38
Fillet per ton of steel , liner mm	1,587.50	1,267.00
Electrode used per tonne of steel, Kg	1.79	0.77
Electric current required per tonne, Kw Hrs.	7.97	5.95
Weld per Kg of electrode bought, Lin. mm	881.83	1,176.15
Welding Man-hours per tonne of steel, hours	1.49	1.77*
Steel welded per month, per machine, tonne	130.05	-----
Lin Meter fillet per 8-hour day, per welder	8.53	10.97
Lin. Meter fillet per Kg of electrode bought	0.89	1.20
Electric Current required to melt one Kg of electrode, Kw. hours	6.39	8.00

* Includes 0.26 hours for foreman. 0.91 hours helpers 0.37 hours for hoist (crane) engineer.

* T -joint welded both sides with this fillet weld will have plate strength.

Total number of passes varies with operator

METALS

Data on Fillet and Lap Welds in Overhead Position

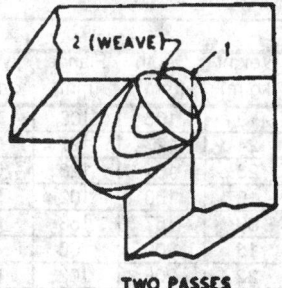

2 (WEAVE)   1

TWO PASSES

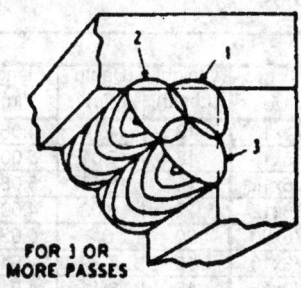

FOR 3 OR MORE PASSES

Data on Fillet and Lap Welds in Overhead Position								
Plate Thickness Inches	Gauge Size of Fillet Inch	Electrode Size Inch	Current (amp)	Electrode Melt-off Rate Inches per Minute	Arc Speed (Inches per Minute for First Pass)	Passes of Beads	Ft. of Joint Welded per Hour (100% Oper. Factor)	Lbs. of Electrode per Ft. of Weld
3/16	5/32	5/32	140	9 1/2	9.00	1	45.0	0.09
1/4	3/16	3/16	160	8 1/4	7.00	1	35.0	0.14
5/16	1/4	3/16	160	8 1/4	5.70	1	28.5	0.18
3/8	5/16	3/16	160	8 1/4	3.70	2	18.5	0.27
7/16	3/8	3/16	160	8 1/4	7.75	3	12.5	0.40
1/2	7/16	3/16	160	8 1/4	7.00	-----	9.2	0.55
5/8	1/2	3/16	160	8 1/4	6.00	-----	7.0	0.72
3/4	5/8	3/16	160	8 1/4	6.00	-----	4.5	1.20
1	7/8	3/16	160	8 1/4	6.00	-----	2.2	2.30
Plate Thickness mm	Gauge Size of Fillet mm	Electrode Size mm	Current (amp)	Electrode Melt-off Rate mm per Minute	Arc Speed (mm per Minute for First Pass)	Passes of Beads	Meter of Joint Welded per Hour (100% Oper. Factor)	Kg. of Electrode per M. of Weld
4.69	3.91	3.91	140	237.50	225.00	1	13.7	0.13
6.25	4.69	4.69	160	206.25	175.00	1	10.7	0.21
7.81	6.25	4.69	160	206.25	142.50	1	8.7	0.27
9.38	7.81	4.69	160	206.25	92.50	2	5.6	0.40
10.94	9.38	4.69	160	206.25	193.75	3	3.8	0.60
12.50	10.94	4.69	160	206.25	175.00	-----	2.8	0.82
15.63	12.50	4.69	160	206.25	150.00	-----	2.1	1.07
18.75	15.63	4.69	160	206.25	150.00	-----	1.4	1.79
25.00	21.88	4.69	160	206.25	150.00	-----	0.7	3.42

Preperation; Square edge.

Fit-up: recommended gap 1/32" (0.78 mm), maximum gap 1/16" (1.56 mm). If greater gaps must be welded, use same procefure but add witdh of gap to fillet size.

Estimating Weights of Standard Steel Shapes

The following tables are only a portion of what is available to be used on jobs.

W Shapes								
Size	Weight (Lbs/ft)	Depth (in)	Flange (in)	Web (in)	Weight (kg/m)	Depth (mm)	Flange (mm)	Web (mm)
W4x13	13	4.16	4.060	0.280	19	106	103	7.1
W5x16	16	5.01	5.000	0.240	24	127	127	6.1
W5x19	19	5.15	5.030	0.270	28	131	128	6.9
W6x12	12	6.03	4.000	0.230	18	153	102	5.8
W6x20	20	6.2	6.020	0.260	30	157	153	6.6
W8x10	10	7.89	3.940	0.170	15	200	100	4.3
W8x15	15	8.11	4.015	0.245	22	206	102	6.2
W8x24	24	7.93	6.495	0.245	36	201	165	6.2
W8x40	40	8.25	8.070	0.360	60	210	205	9.1
W10x12	12	9.87	3.960	0.190	18	251	101	4.8
W10x22	22	10.17	5.750	0.240	33	258	146	6.1
W10x45	45	10.1	8.020	0.350	67	257	204	8.9
W10x88	88	10.84	10.265	0.605	131	275	261	15.4
W12x14	14	11.91	3.970	0.200	21	303	101	5.1
W12x30	30	12.34	6.520	0.260	45	313	166	6.6
W12x50	50	12.19	8.080	0.370	74	310	205	9.4
W12x72	72	12.25	12.040	0.430	107	311	306	10.9
W12x96	96	12.71	12.160	0.550	143	323	309	14.0
W14x22	22	13.74	5.000	0.230	33	349	127	5.8
W14x48	48	13.79	8.030	0.340	71	350	204	8.6
W14x74	74	14.17	10.070	0.450	110	360	256	11.4
W16x40	40	16.01	6.995	0.305	60	407	178	7.7
W16x100	100	16.97	10.425	0.585	149	431	265	14.9
W18x40	40	17.9	6.015	0.315	60	455	153	8.0
W18x76	76	18.21	11.035	0.425	113	463	280	10.8
W21x68	68	21.13	8.270	0.430	101	537	210	10.9
W24x55	55	23.57	7.005	0.395	82	599	178	10.0
W24x104	104	24.06	12.750	0.500	155	611	324	12.7
W27x94	94	26.92	9.990	0.490	140	684	254	12.4
W30x124	124	30.17	10.515	0.585	185	766	267	14.9

Source: American Institute of Steel Construction

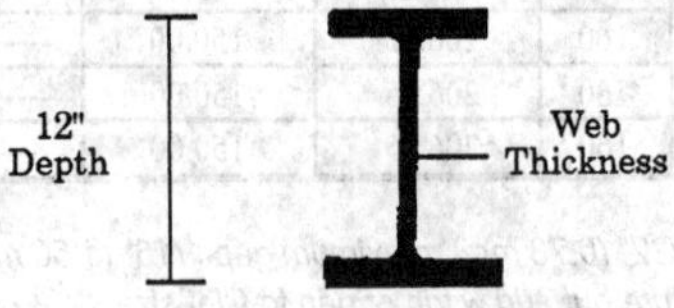

W12x45 (300x67)
W=shape
12=depth in inches (300 mm)
45=weight per l.f. (67 kg per m)

METALS

M Shapes

Size	Weight (Lbs/ft)	Depth (in)	Flange (in)	Web (in)	Weight (kg/m)	Depth (mm)	Flange (mm)	Web (mm)
M4x13	13	4	3.940	0.254	19	102	100	6.5
M5x18.9	18.9	5	5.003	0.316	28	127	127	8.0
M6x4.4	4.4	6	1.844	0.114	7	152	47	2.9
M8x6.5	6.5	8	2.281	0.135	10	203	58	3.4
M10x8	8	9.81	2.690	0.139	12	249	68	3.5
M10x9	9	9.86	2.690	0.157	13	250	68	4.0
M12x10.8	10.8	11.87	3.065	0.162	16	301	78	4.1
M12x11.8	11.8	11.91	3.065	0.177	18	303	78	4.5

Source: American Institute of Steel Construction

S Shapes

Size	Weight (Lbs/ft)	Depth (in)	Flange (in)	Web (in)	Weight (kg/m)	Depth (mm)	Flange (mm)	Web (mm)
S3x5.7	5.7	3	2.330	0.170	8	76	59	4
S3x7.5	7.5	3	2.509	0.349	11	76	64	9
S4x7.7	7.7	4	2.663	0.193	11	102	68	5
S4x9.5	9.5	4	2.796	0.326	14	102	71	8
S5x10	10	5	3.004	0.214	15	127	76	5
S6x12.5	12.5	6	3.332	0.232	19	152	85	6
S8x23	23	8	4.171	0.441	34	203	106	11
S10x35	35	10	4.944	0.594	52	254	126	15
S12x35	35	12	5.078	0.428	52	305	129	11
S12x50	50	12	5.477	0.687	74	305	139	17
S15x50	50	15	5.640	0.550	74	381	143	14
S18x70	70	18	6.251	0.711	104	457	159	18
S20x66	66	20	6.255	0.505	98	508	159	13
S20x75	75	20	6.385	0.635	112	508	162	16
S20x86	86	20.3	7.060	0.660	128	516	179	17
S20x96	96	20.3	7.200	0.800	143	516	183	20
S24x80	80	24	7.000	0.500	119	610	178	13
S24x90	90	24	7.125	0.625	134	610	181	16
S24x100	100	24	7.245	0.745	149	610	184	19

Source: American Institute of Steel Construction

HP Shapes

Size	Weight (Lbs/ft)	Depth (in)	Flange (in)	Web (in)	Weight (kg/m)	Depth (mm)	Flange (mm)	Web (mm)
HP8x36	36	8.02	8.155	0.445	54	204	207	11
HP10x42	42	9.7	10.075	0.415	63	246	256	11
HP10x57	57	9.99	10.225	0.565	85	254	260	14
HP12x53	53	11.78	12.045	0.435	79	299	306	11
HP12x63	63	11.94	12.125	0.515	94	303	308	13
HP12x74	74	12.13	12.215	0.605	110	308	310	15
HP12x84	84	12.28	12.295	0.685	125	312	312	17
HP14x73	73	13.61	14.585	0.505	109	346	370	13
HP14x89	89	13.83	14.695	0.615	132	351	373	16
HP14x102	102	14.01	14.785	0.705	152	356	376	18
HP14x117	117	14.21	14.885	0.805	174	361	378	20

Source: American Institute of Steel Construction

C Shapes

Size	Weight (Lbs/ft)	Depth (in)	Flange (in)	Web (in)	Weight (kg/m)	Depth (mm)	Flange (mm)	Web (mm)
C3X4.1	4.1	3	1.410	0.170	6	76	36	4
C3X5	5	3	1.498	0.258	7	76	38	7
C3X6	6	3	1.596	0.356	9	76	41	9
C4X5.4	5.4	4	1.584	0.184	8	102	40	5
C4X7.25	7.25	4	1.721	0.321	11	102	44	8
C5X6.7	6.7	5	1.750	0.190	10	127	44	5
C5X9	9	5	1.885	0.325	13	127	48	8
C6X10.5	10.5	6	2.034	0.314	16	152	52	8
C6X13	13	6	2.157	0.437	19	152	55	11
C6X8.2	8.2	6	1.920	0.200	12	152	49	5
C7X9.8	9.8	7	2.090	0.210	15	178	53	5
C8X11.5	11.5	8	2.260	0.220	17	203	57	6
C9X15	15	9	2.485	0.285	22	229	63	7
C9X20	20	9	2.648	0.448	30	229	67	11
C10X20	20	10	2.739	0.379	30	254	70	10
C10X25	25	10	2.886	0.526	37	254	73	13
C10X30	30	10	3.033	0.673	45	254	77	17
C12X25	25	12	3.047	0.387	37	305	77	10
C12X30	30	12	3.170	0.510	45	305	81	13
C15X40	40	15	6.520	0.520	60	381	166	13
C15X50	50	15	3.716	0.716	74	381	94	18

Source: American Institute of Steel Construction

MC Shapes

Size	Weight (Lbs/ft)	Depth (in)	Flange (in)	Web (in)	Weight (kg/m)	Depth (mm)	Flange (mm)	Web (mm)
MC6X12	12	6	2.497	0.375	18	152	63	10
MC6X15.1	15.1	6	2.941	0.475	22	152	75	12
MC6X16.3	16.3	6	3.000	0.475	24	152	76	12
MC6X18	18	6	3.504	0.475	27	152	89	12
MC7X19.1	19.1	7	3.452	0.500	28	178	88	13
MC7X22.7	22.7	7	3.603	0.500	34	178	92	13
MC8X8.5	8.5	8	1.874	0.311	13	203	48	8
MC8X20	20	8	3.025	0.500	30	203	77	13
MC8X22.8	22.8	8	3.502	0.525	34	203	89	13
MC9X23.9	23.9	9	3.450	0.550	36	229	88	14
MC9X25.4	25.4	9	3.5	0.55	38	229	89	14
MC10X8.4	8.4	10	1.5	0.28	13	254	38	7
MC10X22	22	10	3.315	0.575	33	254	84	15
MC10X25	25	10	3.405	0.575	37	254	86	15
MC12X31	31	12	3.67	0.7	46	305	93	18
MC12X40	40	12	3.890	0.700	60	305	99	18
MC12X45	45	12	4.012	0.700	67	305	102	18
MC12X50	50	12	4.135	0.700	74	305	105	18
MC13X35	35	13	4.072	0.610	52	330	103	15
MC13X50	50	13	4.412	0.610	74	330	112	15
MC18X58	58	18	4.200	0.625	86	457	107	16

Source: American Institute of Steel Construction

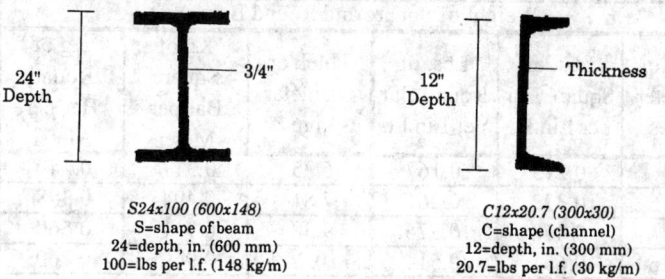

S24x100 (600x148)
S=shape of beam
24=depth, in. (600 mm)
100=lbs per l.f. (148 kg/m)

C12x20.7 (300x30)
C=shape (channel)
12=depth, in. (300 mm)
20.7=lbs per l.f. (30 kg/m)

Example, a job requires 60lf (18.3m) W6x20, 30lf (9.1m) W8x15, 45lf (13.7m) M8x6.5 and 110lf (33.5m) S18x70, then the total number of pounds (kg) of steel is 60 x 20 + 30 x 15 + 45 x 6.5 + 110 x 70 = 9,642.5 lbs (4,373.8 kg). In estimating steel, add all connector pieces, bolts, plates, welding, etc. to get the total number of lbs (kg).

Estimating the Weight of Wrought Iron, Steel, or Cast Iron

When tables of weights are not handy, the following rules will prove of value to the estimator when computing the weights of wrought iron, steel, or cast iron.

Weight of Wrought Iron. One cu. ft. (0.02832) of wrought iron weighs 480 lbs (217.72 kg). One sq. ft. (0.0929 sq.m) of wrought iron 1" (25 mm) thick weighs 40 lbs (18.14 kg). One sq. in. (6.452 sq.cm)of wrought iron one foot (0.3048 m) long weighs 3-1/3 lbs (1.51 kg).

To find the weight of one square foot (0.0929 sq.m) of flat iron of any thickness, multiply the thickness in inches by 40, and the result will be the weight of the iron in lbs.

To find the weight of one lin. ft. (0.3048 m) of wrought iron bar of any size, multiply the cross sectional area in sq. in. by 3-1/3, and the result will be the weight per lin.ft (0.3048 m).

Weight of Steel. One cu. ft. (0.02832 cum) of steel weighs 489.6 lbs (222.08 kg), or 2% more than wrought iron. One sq.ft. (0.0929 sq.m) of 1" (25mm) thick steel weighs 40.8 lbs (18.50 kg). A piece of 1" x 1" x 1' steel weighs 3.4 lbs 25 x 25 x 0.3048 m weighs 1.54 Kg).

To find the weight of one lin.ft. (0.3048 m) of steel bar of any size, multiply the cross sectional area in square inches by 3.4, and the result will be the weight of the steel in lbs. If the weight per lin.ft. is known, the exact sectional area in square inches may be obtained by dividing the weight by 3.4.

Weight of Cast Iron. One cu. ft. of cast iron weighs 450 lbs (0.02832 cu.m = 204.14 kg). One sq.ft. of cast iron, 1" thick, weighs 37-1/2 lbs (0.0929 sq.m weighs 17.01Kg). A piece of cast iron 1" x 1" x 1' weighs 3-1/8 lbs. (25mm x 25mm x 0.3028mweighs 1.41 kg). One cu.in. of cast iron weighs 0.26 lbs. 16.39 cu.cm. weighs 0.12 kg

Weight of Square and Round Bars					
Thick. or Diameter Inches	Lbs. of Square Bar per Lin.Ft.	Lbs. of Round Bar per Lin.Ft.	Thick or Diameter mm	Kg of Square Bar per Meter	Kg of Round Bar per Meter
1/4	0.213	0.167	6.25	0.317	0.249
5/16	0.332	0.261	7.81	0.494	0.388
3/8	0.478	0.376	9.38	0.711	0.560
7/16	0.651	0.511	10.94	0.969	0.760
1/2	0.850	0.668	12.50	1.265	0.994
9/16	1.076	0.845	14.06	1.601	1.258
5/8	1.382	1.040	15.63	2.057	1.548
11/16	1.607	1.260	17.19	2.392	1.875
3/4	1.913	1.500	18.75	2.847	2.232
13/16	2.240	1.760	20.31	3.334	2.619
7/8	2.603	2.040	21.88	3.874	3.036
15/16	......	2.350	23.44	......	3.497
1	3.400	2.670	25.00	5.060	3.973
1 1/16	......	3.010	26.56	......	4.479
1 1/8	4.303	3.380	28.13	6.404	5.030
1 1/4	5.313	4.170	31.25	7.907	6.206
1 5/16	......	4.600	32.81	......	6.846
1 3/8	6.428	5.050	34.38	9.566	7.515
1 7/16	......	5.520	35.94	......	8.215
1 1/2	7.650	6.010	37.50	11.385	8.944
1 5/8	8.928	7.050	40.63	13.287	10.492
1 11/16	......	7.600	42.19	......	11.310
1 3/4	10.410	8.180	43.75	15.492	12.173
1 13/16	......	8.773	45.31	......	13.056
1 7/8	11.950	9.390	46.88	17.784	13.974
2	13.600	10.680	50.00	20.239	15.894
2 1/8	15.350	12.060	53.13	22.844	17.948
2 1/4	17.210	13.520	56.25	25.612	20.120
2 3/8	......	15.060	59.38	......	22.412
2 1/2	21.250	16.690	62.50	31.624	24.838
2 5/8	23.430	18.400	65.63	34.868	27.383
2 3/4	25.710	20.200	68.75	38.261	30.061
2 7/8	......	22.070	71.88	......	32.844
3	30.600	24.030	75.00	45.539	35.761
3 1/8	......	26.080	78.13	......	38.812
3 1/4	35.910	28.210	81.25	53.441	41.982
3 3/8	......	30.420	84.38	......	45.271
3 1/2	......	30.420	87.50	......	45.271
3 5/8	41.650	32.710	90.63	61.983	48.679
3 3/4	47.810	37.550	93.75	71.150	55.881
4	54.400	42.730	100.00	80.957	63.590

Weights of Flat Steel Bars Lbs per foot						
Thick. Inches	Width of Bars					
	1 "	2 "	3 "	4 "	5 "	6 "
3/16	0.64	1.28	1.91	2.55	3.19	3.83
1/4	0.85	1.70	2.55	3.40	4.25	5.10
5/16	1.06	2.12	3.19	4.25	5.31	6.38
3/8	1.28	2.55	3.83	5.10	6.38	7.65
7/16	1.49	2.98	4.46	5.95	7.44	8.95
1/2	1.70	3.40	5.10	6.80	8.40	10.20
9/16	1.92	3.83	5.74	7.65	9.57	11.48
5/8	2.12	4.25	6.38	8.50	10.63	12.75
11/16	2.34	4.67	7.02	9.35	11.69	14.03
3/4	2.55	5.10	7.65	10.20	12.75	15.30
13/16	2.76	5.53	8.29	11.05	13.81	16.58
7/8	2.98	5.98	8.93	11.90	14.87	17.85
15/16	3.19	6.38	9.57	12.75	15.94	19.13
1	3.40	6.80	10.20	13.60	17.00	20.40
1 1/16	3.61	7.22	10.84	14.45	18.06	21.68
1 1/8	3.83	7.65	11.48	15.30	19.13	22.95
1 3/16	4.04	8.08	12.12	16.15	20.19	23.23
1 1/4	4.25	8.50	12.75	17.00	21.25	25.50
1 5/16	4.46	8.93	13.39	17.85	22.32	26.78
1 3/8	4.67	9.35	14.03	18.70	23.38	28.05
1 7/16	4.89	9.78	14.66	19.55	24.44	29.33
1 1/2	5.10	10.20	15.30	20.40	25.50	30.60
1 9/16	5.32	10.63	15.94	21.25	26.57	31.88
1 5/8	5.50	11.05	16.58	22.10	27.63	33.15
1 11/16	5.74	11.47	17.22	22.95	28.69	34.43
1 3/4	5.95	11.90	17.85	23.80	29.75	35.70
1 13/16	6.16	12.33	18.49	24.65	30.81	36.98
1 7/8	6.38	12.75	19.13	25.50	31.87	38.25
1 15/16	6.59	13.18	19.77	26.35	32.94	39.53
2	6.80	13.60	20.40	27.20	34.00	40.80

Weights of Flat Steel Bars Kg. per Meter						
Thick. Mm	Width of Bars					
	25 mm	50 mm	75 mm	100 mm	125 mm	150 mm
4.69	0.95	1.90	2.84	3.79	4.75	5.70
6.25	1.26	2.53	3.79	5.06	6.32	7.59
7.81	1.58	3.15	4.75	6.32	7.90	9.49
9.38	1.90	3.79	5.70	7.59	9.49	11.38
10.94	2.22	4.43	6.64	8.85	11.07	13.32
12.50	2.53	5.06	7.59	10.12	12.50	15.18
14.06	2.86	5.70	8.54	11.38	14.24	17.08
15.63	3.15	6.32	9.49	12.65	15.82	18.97
17.19	3.48	6.95	10.45	13.91	17.40	20.88
18.75	3.79	7.59	11.38	15.18	18.97	22.77
20.31	4.11	8.23	12.34	16.44	20.55	24.67
21.88	4.43	8.90	13.29	17.71	22.13	26.56
23.44	4.75	9.49	14.24	18.97	23.72	28.47
25.00	5.06	10.12	15.18	20.24	25.30	30.36
26.56	5.37	10.74	16.13	21.50	26.88	32.26
28.13	5.70	11.38	17.08	22.77	28.47	34.15
29.69	6.01	12.02	18.04	24.03	30.05	34.57
31.25	6.32	12.65	18.97	25.30	31.62	37.95
32.81	6.64	13.29	19.93	26.56	33.22	39.85
34.38	6.95	13.91	20.88	27.83	34.79	41.74
35.94	7.28	14.55	21.82	29.09	36.37	43.65
37.50	7.59	15.18	22.77	30.36	37.95	45.54
39.06	7.92	15.82	23.72	31.62	39.54	47.44
40.63	8.19	16.44	24.67	32.89	41.12	49.33
42.19	8.54	17.07	25.63	34.15	42.70	51.24
43.75	8.85	17.71	26.56	35.42	44.27	53.13
45.31	9.17	18.35	27.52	36.68	45.85	55.03
46.88	9.49	18.97	28.47	37.95	47.43	56.92
48.44	9.81	19.61	29.42	39.21	49.02	58.83
50.00	10.12	20.24	30.36	40.48	50.60	60.72

METALS

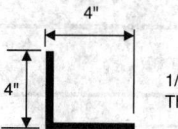

Equal Angles
4" x 4" x 1/2"
(100x100x12.50 mm)
Weight is listed in
AISC manual

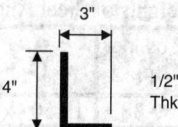

Unequal Angles
4" x 3" x 1/2"
(100x75x12.50 mm)
Weight is listed in
AISC manual

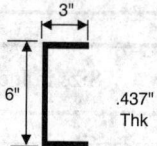

Standard Channel
6 x 13
(150 x 19)
6" (150mm) = channel size
13 Lbs/LF (19 kg/m)

Weights of American Standard Channels							
Depth of Channel, Inches	Weight Lbs per Lin.Ft.	Thickness of Web, Inches	Width of Flange Inches	Depth of Channel, mm	Weight Kg. per Lin. Meter	Thickness of Web, mm	Width of Flange mm
3	6.0	0.362	1.602	75	8.93	9.050	40.050
	5.0	0.264	1.504		7.44	6.600	37.600
	4.1	0.170	1.410		6.10	4.250	35.250
4	7.25	0.325	1.725	100	10.79	8.125	43.125
	5.4	0.180	1.580		8.04	4.500	39.500
5	9.0	0.330	1.890	125	13.39	8.250	47.250
	6.7	0.190	1.750		9.97	4.750	43.750
6	13.0	0.440	2.160	150	19.35	11.000	54.000
	10.5	0.318	2.038		15.63	7.950	50.950
	8.2	0.200	1.920		12.20	5.000	48.000
7	14.75	0.423	2.303	175	21.95	10.575	57.575
	12.25	0.318	2.198		18.23	7.950	54.950
	9.8	0.210	2.090		14.58	5.250	52.250
8	18.75	0.490	2.530	200	27.90	12.250	63.250
	13.75	0.307	2.347		20.46	7.675	58.675
	11.5	0.220	2.260		17.11	5.500	56.500
9	20.0	0.452	2.652	225	29.76	11.300	66.300
	15.0	0.288	2.488		22.32	7.200	62.200
	13.4	0.230	2.430		19.94	5.750	60.750
10	30.0	0.676	3.036	250	44.65	16.900	75.900
	25.0	0.529	2.889		37.20	13.225	72.225
	20.0	0.382	2.742		29.76	9.550	68.550
	15.3	0.240	2.600		22.77	6.000	65.000
12	30.0	0.513	3.173	300	44.65	12.825	79.325
	25.0	0.390	3.050		37.20	9.750	76.250
	20.7	0.280	2.940		30.81	7.000	73.500
15	50.0	0.720	3.720	375	74.41	18.000	93.000
	40.0	0.524	3.524		59.53	13.100	88.100
	33.9	0.400	3.400		50.45	10.000	85.000

817

Weights of Steel Angles										
Size in Inches					Lbs. per Ft.	Size in mm					Kg. per Meter
1/2	x	1/2	x	1/8	0.38	12.50	x	12.50	x	3.13	0.57
5/8	x	5/8	x	1/8	0.48	15.63	x	15.63	x	3.13	0.71
3/4	x	3/4	x	1/8	0.59	18.75	x	18.75	x	3.13	0.88
3/4	x	3/4	x	3/16	0.84	18.75	x	18.75	x	4.69	1.25
7/8	x	7/8	x	1/8	0.70	21.88	x	21.88	x	3.13	1.04
1	x	5/8	x	1/8	0.64	25.00	x	15.63	x	3.13	0.95
1	x	1	x	1/8	0.80	25.00	x	25.00	x	3.13	1.19
1	x	1	x	3/16	1.16	25.00	x	25.00	x	4.69	1.73
1	x	1	x	1/4	1.49	25.00	x	25.00	x	6.25	2.22
1 1/8	x	1 1/8	x	1/8	0.91	28.13	x	28.13	x	3.13	1.35
1 1/4	x	1 1/4	x	1/8	1.01	31.25	x	31.25	x	3.13	1.50
1 1/4	x	1 1/4	x	3/16	1.48	31.25	x	31.25	x	4.69	2.20
1 1/4	x	1 1/4	x	1/2	1.92	31.25	x	31.25	x	12.50	2.86
1 3/8	x	7/8	x	1/8	0.91	34.38	x	21.88	x	3.13	1.35
1 1/2	x	1 1/2	x	1/8	1.23	37.50	x	37.50	x	3.13	1.83
1 1/2	x	1 1/2	x	3/16	1.80	37.50	x	37.50	x	4.69	2.68
1 1/2	x	1 1/2	x	1/4	2.34	37.50	x	37.50	x	6.25	3.48
1 1/2	x	1 1/2	x	5/16	2.86	37.50	x	37.50	x	7.81	4.26
1 1/2	x	1 1/2	x	3/8	3.35	37.50	x	37.50	x	9.38	4.99
1 3/4	x	1 3/4	x	1/8	1.44	43.75	x	43.75	x	3.13	2.14
1 3/4	x	1 3/4	x	3/16	2.12	43.75	x	43.75	x	4.69	3.15
1 3/4	x	1 3/4	x	1/4	2.77	43.75	x	43.75	x	6.25	4.12
1 3/4	x	1 3/4	x	5/16	3.39	43.75	x	43.75	x	7.81	5.04
2	x	1 1/4	x	3/16	1.96	50.00	x	31.25	x	4.69	2.92
2	x	1 1/4	x	1/4	2.55	50.00	x	31.25	x	6.25	3.79
2	x	1 1/2	x	1/8	1.44	50.00	x	37.50	x	3.13	2.14
2	x	1 1/2	x	3/16	2.12	50.00	x	37.50	x	4.69	3.15
2	x	1 1/2	x	1/8	2.77	50.00	x	37.50	x	3.13	4.12
2	x	1 1/2	x	5/16	3.39	50.00	x	37.50	x	7.81	5.04
2	x	2	x	1/8	1.65	50.00	x	50.00	x	3.13	2.46
2	x	2	x	3/16	2.44	50.00	x	50.00	x	4.69	3.63
2	x	2	x	1/4	3.19	50.00	x	50.00	x	6.25	4.75
2	x	2	x	5/16	3.92	50.00	x	50.00	x	7.81	5.83
2	x	2	x	3/8	3.70	50.00	x	50.00	x	9.38	5.51
2	x	2	x	1/2	6.00	50.00	x	50.00	x	12.50	8.93
2 1/4	x	1 1/2	x	3/16	2.28	56.25	x	37.50	x	4.69	3.39
2 1/4	x	2 1/4	x	1/8	1.86	56.25	x	56.25	x	3.13	2.77
2 1/4	x	2 1/4	x	3/16	2.75	56.25	x	56.25	x	4.69	4.09
2 1/4	x	2 1/4	x	1/4	3.62	56.25	x	56.25	x	6.25	5.39
2 1/4	x	2 1/4	x	5/16	4.50	56.25	x	56.25	x	7.81	6.70
2 1/2	x	1 1/2	x	3/16	2.44	62.50	x	37.50	x	4.69	3.63

METALS

Weights of Steel Angles (cont'd)										
Size in Inches					Lbs. per Ft.	Size in mm					Kg. per Meter
2 1/2	x	1 1/2	x	1/4	3.19	62.50	x	37.50	x	6.25	4.75
2 1/2	x	2	x	3/16	2.75	62.50	x	50.00	x	4.69	4.09
2 1/2	x	2	x	1/4	3.62	62.50	x	50.00	x	6.25	5.39
2 1/2	x	2	x	5/16	4.50	62.50	x	50.00	x	7.81	6.70
2 1/2	x	2	x	3/8	5.30	62.50	x	50.00	x	9.38	7.89
2 1/2	x	2	x	1/2	6.74	62.50	x	50.00	x	12.50	10.03
2 1/2	x	2 1/2	x	1/8	2.08	62.50	x	62.50	x	3.13	3.10
2 1/2	x	2 1/2	x	3/16	3.07	62.50	x	62.50	x	4.69	4.57
2 1/2	x	2 1/2	x	1/4	4.10	62.50	x	62.50	x	6.25	6.10
2 1/2	x	2 1/2	x	5/16	5.00	62.50	x	62.50	x	7.81	7.44
2 1/2	x	2 1/2	x	3/8	5.90	62.50	x	62.50	x	9.38	8.78
2 1/2	x	2 1/2	x	1/2	7.70	62.50	x	62.50	x	12.50	11.46
3	x	3	x	1/8	2.50	75.00	x	75.00	x	3.13	3.72
3	x	3	x	5/16	3.71	75.00	x	75.00	x	7.81	5.52

Weights of Structural Steel Angles										
Size in Inches					Lbs. per Ft.	Size in mm					Kg. per Meter
3	x	2	x	1/4	4.10	75.0	x	50.0	x	6.25	6.10
3	x	2	x	5/16	5.00	75.0	x	50.0	x	7.81	7.44
3	x	2	x	3/8	5.90	75.0	x	50.0	x	9.38	8.78
3	x	2	x	1/4	4.50	75.0	x	50.0	x	6.25	6.70
3	x	2	x	5/16	5.60	75.0	x	50.0	x	7.81	8.33
3	x	2	x	3/8	6.60	75.0	x	50.0	x	9.38	9.82
3	x	2	x	1/2	8.50	75.0	x	50.0	x	12.50	12.65
3	x	3	x	1/4	4.90	75.0	x	75.0	x	6.25	7.29
3	x	3	x	5/16	6.10	75.0	x	75.0	x	7.81	9.08
3	x	3	x	3/8	7.20	75.0	x	75.0	x	9.38	10.71
3	x	3	x	7/16	8.30	75.0	x	75.0	x	10.94	12.35
3	x	3	x	1/2	9.40	75.0	x	75.0	x	12.50	13.99
3	x	3	x	5/8	11.50	75.0	x	75.0	x	15.63	17.11
3 1/2	x	2 1/2	x	1/4	4.90	87.5	x	62.5	x	6.25	7.29
3 1/2	x	2 1/2	x	5/16	6.10	87.5	x	62.5	x	7.81	9.08
3 1/2	x	2 1/2	x	3/8	7.20	87.5	x	62.5	x	9.38	10.71
3 1/2	x	2 1/2	x	1/2	9.40	87.5	x	62.5	x	12.50	13.99
3 1/2	x	2 1/2	x	5/8	11.50	87.5	x	62.5	x	15.63	17.11
3 1/2	x	3 1/2	x	7/16	9.80	87.5	x	87.5	x	10.94	14.58
3 1/2	x	3 1/2	x	1/2	11.10	87.5	x	87.5	x	12.50	16.52
3 1/2	x	3 1/2	x	5/8	13.60	87.5	x	87.5	x	15.63	20.24
4	x	3	x	1/4	5.80	100.0	x	75.0	x	6.25	8.63
4	x	3	x	5/16	7.20	100.0	x	75.0	x	7.81	10.71

Weights of Structural Steel Angles (cont'd)									
Size in Inches				Lbs. per Ft.	Size in mm				Kg. per Meter
4	x	3	x	3/8	8.50	100.0 x	75.0	x	9.38	12.65
4	x	3	x	7/16	9.80	100.0 x	75.0	x	10.94	14.58
4	x	3	x	1/2	11.10	100.0 x	75.0	x	12.50	16.52
4	x	3	x	3/4	16.00	100.0 x	75.0	x	18.75	23.81
4	x	3 1/2	x	1/4	6.20	100.0 x	87.5	x	6.25	9.23
4	x	3 1/2	x	5/16	7.70	100.0 x	87.5	x	7.81	11.46
4	x	3 1/2	x	3/8	9.10	100.0 x	87.5	x	9.38	13.54
4	x	3 1/2	x	7/16	10.60	100.0 x	87.5	x	10.94	15.77
4	x	3 1/2	x	1/2	11.90	100.0 x	87.5	x	12.50	17.71
4	x	4	x	1/4	6.60	100.0 x	100.0	x	6.25	9.82
4	x	4	x	5/16	8.20	100.0 x	100.0	x	7.81	12.20
4	x	4	x	3/8	9.80	100.0 x	100.0	x	9.38	14.58
4	x	4	x	7/16	11.30	100.0 x	100.0	x	10.94	16.82
4	x	4	x	1/2	12.80	100.0 x	100.0	x	12.50	19.05
4	x	4	x	5/8	15.70	100.0 x	100.0	x	15.63	23.36
4	x	4	x	3/4	18.50	100.0 x	100.0	x	18.75	27.53
4 1/2	x	3	x	3/8	9.10	112.5 x	75.0	x	9.38	13.54
5	x	3	x	5/16	8.20	125.0 x	75.0	x	7.81	12.20
5	x	3	x	3/8	9.80	125.0 x	75.0	x	9.38	14.58
5	x	3	x	7/16	11.30	125.0 x	75.0	x	10.94	16.82
5	x	3	x	1/2	12.80	125.0 x	75.0	x	12.50	19.05
5	x	3	x	3/4	18.50	125.0 x	75.0	x	18.75	27.53
5	x	3 1/2	x	5/16	8.70	125.0 x	87.5	x	7.81	12.95
5	x	3 1/2	x	3/8	10.40	125.0 x	87.5	x	9.38	15.48
5	x	3 1/2	x	7/16	12.00	125.0 x	87.5	x	10.94	17.86
5	x	3 1/2	x	1/2	13.60	125.0 x	87.5	x	12.50	20.24
5	x	3 1/2	x	5/8	16.80	125.0 x	87.5	x	15.63	25.00
5	x	4	x	3/8	11.00	125.0 x	100.0	x	9.38	16.37
5	x	4	x	1/2	14.50	125.0 x	100.0	x	12.50	21.58
5	x	5	x	3/8	12.30	125.0 x	125.0	x	9.38	18.30
5	x	5	x	7/16	14.30	125.0 x	125.0	x	10.94	21.28
5	x	5	x	1/2	16.20	125.0 x	125.0	x	12.50	24.11
5	x	5	x	5/8	20.00	125.0 x	125.0	x	15.63	29.76
5	x	5	x	3/4	23.60	125.0 x	125.0	x	18.75	35.12
6	x	3 1/2	x	5/16	9.80	150.0 x	87.5	x	7.81	14.58
6	x	3 1/2	x	3/8	11.70	150.0 x	87.5	x	9.38	17.41
6	x	3 1/2	x	7/16	13.50	150.0 x	87.5	x	10.94	20.09
6	x	3 1/2	x	1/2	15.30	150.0 x	87.5	x	12.50	22.77
6	x	3 1/2	x	5/8	18.90	150.0 x	87.5	x	15.63	28.13
6	x	4	x	5/16	10.30	150.0 x	100.0	x	7.81	15.33
6	x	4	x	3/8	12.30	150.0 x	100.0	x	9.38	18.30

METALS

Weights of Structural Steel Angles (cont'd)										
Size in Inches				Lbs. per Ft.	Size in mm					Kg. per Meter
6	x	4	x	7/16	14.30	150.0	x	100.0	x	10.94	21.28
6	x	4	x	1/2	16.20	150.0	x	100.0	x	12.50	24.11
6	x	4	x	9/16	18.10	150.0	x	100.0	x	14.06	26.94
6	x	4	x	5/8	20.00	150.0	x	100.0	x	15.63	29.76
6	x	4	x	3/4	23.60	150.0	x	100.0	x	18.75	35.12
6	x	4	x	7/8	27.20	150.0	x	100.0	x	21.88	40.48
6	x	6	x	3/8	14.90	150.0	x	150.0	x	9.38	22.17
6	x	6	x	7/16	17.20	150.0	x	150.0	x	10.94	25.60
6	x	6	x	1/2	19.60	150.0	x	150.0	x	12.50	29.17
6	x	6	x	9/16	21.90	150.0	x	150.0	x	14.06	32.59
6	x	6	x	5/8	24.20	150.0	x	150.0	x	15.63	36.01
6	x	6	x	3/4	28.70	150.0	x	150.0	x	18.75	42.71
6	x	6	x	7/8	33.10	150.0	x	150.0	x	21.88	49.26
6	x	6	x	1	37.40	150.0	x	150.0	x	25.00	55.66
6	x	6	x	1 3/8	7.90	150.0	x	150.0	x	34.38	11.76
6	x	6	x	1 1/2	10.20	150.0	x	150.0	x	37.50	15.18
7	x	3 1/2	x	3/8	13.00	175.0	x	87.5	x	9.38	19.35
7	x	3 1/2	x	7/16	15.00	175.0	x	87.5	x	10.94	22.32
7	x	3 1/2	x	1/2	17.00	175.0	x	87.5	x	12.50	25.30
7	x	3 1/2	x	5/8	21.00	175.0	x	87.5	x	15.63	31.25
8	x	3 1/2	x	1/2	18.70	200.0	x	87.5	x	12.50	27.83
8	x	6	x	1/2	23.00	200.0	x	150.0	x	12.50	34.23
8	x	6	x	3/4	33.80	200.0	x	150.0	x	18.75	50.30
8	x	8	x	1/2	26.40	200.0	x	200.0	x	12.50	39.29
8	x	8	x	5/8	32.70	200.0	x	200.0	x	15.63	48.66
8	x	8	x	3/4	38.90	200.0	x	200.0	x	18.75	57.89
8	x	8	x	7/8	45.00	200.0	x	200.0	x	21.88	66.97

Weights of Steel Tees						
Size Flange by stem in Inches	Lbs. per lin. ft. Thickness in inches					
	1/8	3/16	1/4	5/16	3/8	1/2
3/4 x 3/4	0.60	-----	-----	-----	-----	-----
7/8 x 7/8	0.73	-----	-----	-----	-----	-----
1 x 1	0.90	1.20	-----	-----	-----	-----
1 1/4 x 1 1/4	-----	1.55	1.98	-----	-----	-----
1 1/2 x 1 1/2	-----	1.88	2.43	-----	-----	-----
1 3/4 x 1 3/4	-----	2.30	2.90	-----	-----	-----
2 x 1 1/2	-----	-----	3.10	-----	-----	-----
2 x 2	-----	-----	3.60	4.40	-----	-----
2 1/4 x 2 1/4	-----	-----	4.10	4.90	-----	-----
2 1/2 x 2 1/2	-----	-----	4.60	5.50	6.40	-----
2 1/2 x 2 3/4	-----	-----	-----	5.90	6.80	-----
2 1/2 x 3	-----	-----	-----	6.10	7.20	-----
3 x 2 1/2	-----	-----	-----	6.10	7.20	-----
3 x 3	-----	-----	-----	6.70	7.80	-----
3 1/2 x 3	-----	-----	-----	-----	8.50	-----
3 x 4	-----	-----	-----	-----	9.30	-----
3 x 3 1/2	-----	-----	-----	-----	8.50	-----
3 1/2 x 3 1/2	-----	-----	-----	-----	9.20	11.70
3 1/2 x 4	-----	-----	-----	-----	9.80	-----
4 x 2 1/2	-----	-----	-----	-----	8.50	-----
4 x 3	-----	-----	-----	-----	9.20	-----
4 x 4	-----	-----	-----	-----	10.50	13.50
4 x 5	-----	-----	-----	-----	11.90	15.30
4 1/2 x 2 1/2	-----	-----	-----	-----	9.20	-----
4 1/2 x 3	-----	-----	-----	8.60	10.00	-----
5 x 3	-----	-----	-----	-----	11.00	13.60
6 1/2 x 6 1/2	-----	-----	-----	-----	-----	19.80

METALS

Weights of Steel Tees - Metric						
Size Flange by stem in Inches	Kg. per lin. Meter Thickness in mm					
	3.13	4.69	6.25	7.81	9.38	12.50
18.75 x 18.75	0.89	-----	-----	-----	-----	-----
21.88 x 21.88	1.09	-----	-----	-----	-----	-----
25.00 x 25.00	1.34	1.79	-----	-----	-----	-----
31.25 x 31.25	-----	2.31	2.95	-----	-----	-----
37.50 x 37.50	-----	2.80	3.62	-----	-----	-----
43.75 x 43.75	-----	3.42	4.32	-----	-----	-----
50.00 x 37.50	-----	-----	4.61	-----	-----	-----
50.00 x 50.00	-----	-----	5.36	6.55	-----	-----
56.25 x 56.25	-----	-----	6.10	7.29	-----	-----
62.50 x 62.50	-----	-----	6.85	8.19	9.52	-----
62.50 x 68.75	-----	-----	-----	8.78	10.12	-----
62.50 x 75.00	-----	-----	-----	9.08	10.71	-----
75.00 x 62.50	-----	-----	-----	9.08	10.71	-----
75.00 x 75.00	-----	-----	-----	9.97	11.61	-----
87.50 x 75.00	-----	-----	-----	-----	12.65	-----
75.00 x 100.00	-----	-----	-----	-----	13.84	-----
75.00 x 87.50	-----	-----	-----	-----	12.65	-----
87.50 x 87.50	-----	-----	-----	-----	13.69	17.41
87.50 x 100.00	-----	-----	-----	-----	14.58	-----
100.00 x 62.50	-----	-----	-----	-----	12.65	-----
100.00 x 75.00	-----	-----	-----	-----	13.69	-----
100.00 x 100.00	-----	-----	-----	-----	15.63	20.09
100.00 x 125.00	-----	-----	-----	-----	17.71	22.77
112.50 x 62.50	-----	-----	-----	-----	13.69	-----
112.50 x 75.00	-----	-----	-----	12.80	14.88	-----
125.00 x 75.00	-----	-----	-----	-----	16.37	20.24
162.50 x 162.50	-----	-----	-----	-----	-----	29.47

Weights of Standard Diamond Steel Floor Plates			
Thickness Inches	Lbs per Sq.Ft.	Thickness	Kg per Sq.Meter
1/8	8.00	3.13	39.06
3/16	8.75	4.69	42.72
1/4	11.25	6.25	54.93
5/16	13.75	7.81	67.14
3/8	16.25	9.38	79.34
1/2	21.50	12.50	104.98

Approximate Weight of Bolts					
Diameter Inches	Length Inches	Weight Lb per 100 units	Diameter mm	Length mm	Weight Kg per 100 Units
5/8	1 1/2	40	15.63	37.50	18.14
	1 3/4	43		43.75	19.50
	2	45		50.00	20.41
	2 1/4	47		56.25	21.32
	2 1/2	49		62.50	22.23
	2 3/4	51		68.75	23.13
	3	54		75.00	24.49
3/4	1 3/4	66	18.75	43.75	29.94
	2	70		50.00	31.75
	2 1/4	73		56.25	33.11
	2 1/2	76		62.50	34.47
	2 3/4	79		68.75	35.83
	3	83		75.00	37.65
	3 3/4	92		93.75	41.73
	4	95		100.00	43.09
	4 1/4	98		106.25	44.45
	4 1/2	102		112.50	46.27
	4 3/4	105		118.75	47.63
	5	108		125.00	48.99
	5 1/2	114		137.50	51.71
	6	118		150.00	53.52
7/8	2	106	21.88	50.00	48.08
	2 1/4	110		56.25	49.90
	2 1/2	114		62.50	51.71
	2 3/4	120		68.75	54.43
	3	124		75.00	56.25
	3 1/4	128		81.25	58.06
	3 1/2	129		87.50	58.51
	3 3/4	131		93.75	59.42
	4	136		100.00	61.69
	4 1/4	140		106.25	63.50
	4 1/2	143		112.50	64.86
	4 3/4	149		118.75	67.59
	5	153		125.00	69.40
	5 1/4	157		131.25	71.22
	5 1/2	162		137.50	73.48
	5 3/4	165		143.75	74.84
	6	170		150.00	77.11
	6 1/2	185		162.50	83.92
	7	194		175.00	88.00
	7 1/2	202		187.50	91.63
	8	209		200.00	94.80

METALS

Approximate Weight of Bolts (cont'd)					
Diameter Inches	Length Inches	Weight Lb per 100 units	Diameter mm	Length mm	Weight Kg per 100 Units
1	2	140	25.00	50.00	63.50
	2 1/4	143		56.25	64.86
	2 1/2	149		62.50	67.59
	2 3/4	154		68.75	69.85
	3	160		75.00	72.58
	3 1/4	166		81.25	75.30
	3 1/2	171		87.50	77.57
	3 3/4	177		93.75	80.29
	4 1/2	194		112.50	88.00
	4 3/4	199		118.75	90.27
	5	205		125.00	92.99
	5 1/2	217		137.50	98.43
	6	228		150.00	103.42
	6 1/2	251		162.50	113.85
	7	262		175.00	118.84
	7 1/2	273		187.50	123.83
	8	288		200.00	130.64
1 1/8	2 1/2	218	28.13	62.50	98.88
	2 3/4	225		68.75	102.06
	3	232		75.00	105.24
	3 1/4	250		81.25	113.40
	3 1/2	252		87.50	114.31
	3 3/4	254		93.75	115.21
	4	260		100.00	117.94
	4 1/4	267		106.25	121.11
	4 1/2	273		112.50	123.83
	4 3/4	280		118.75	127.01
	5	288		125.00	130.64
	5 1/2	302		137.50	136.99
	6	316		150.00	143.34
	7	320		175.00	145.15
	7 1/2	341		187.50	154.68

Each bolt shall be tightened to provide, when all bolts in the joint at tight, at least the minimum tension as indicated in table 1 below:

Typical inspection for bolts: The bolt tension specified in table 1 shall be attained by tightening all bolts in the joint the applicable amount of nut rotation specified in table 2 below by the turn-of-the-nut method

Nut rotation is relative to the bolt, regardless of the element (nut or bolt) being turned. For bolts installed by ½ turn and less, a tolerance of plus or minus 30 is permitted. For bolts installed by 2/3 turn and more, a tolerance of plus or minus 45 is permitted.

Table 1			
Bolt Size, Inches	Bolt Tension* Bolt size Kips, minimum inches A 325	Bolt Size, mm	Bolt Tension* Bolt size kN, minimum inches A 325 M
1/2	12	12.50	-----
5/8	19	15.63	91
3/4	28	18.75	142
7/8	39	21.88	176
1	51	25.00	206
1 1/8	56	28.13	267
1 1/4	71	31.25	327
1 3/8	85	34.38	-----
1 1/2	103	37.50	475

Equal to 70 percent of specified minimum tensile strength of bolts, rounded off to the nearest Kip (kN)

Table 2			
Nut Rotation From Snug Tight Condition			
	Disposition of Outer Faces of Bolted Parts		
Bolt Length (as measured from Underside of head to extreme end of point)	Both Faces normal to bolt axis	One face normal to bolt axis and other face sloped not more than 1:20 (bevel washer not used)	Both Faces sloped not more then 1:20 from normal to bolt axis (bevel washer not used)
Up to an including 4 diameters	1/3 Turn	1/2 Turn	2/3 Turn
Over 4 diameters not exceeding 8 diameters	1/2 Turn	2/3 Turn	5/6 Turn
Over 8 diameters not exceeding 12 diameters	1/3 Turn	5/6 Turn	1 Turn

Bolts shall be inspected by the use of a manual torque wrench normally furnished by the contractor. The inspection wrench shall be calibrated at least each working day. The bolts shall be of sufficient length to project at least ¼" (6.25 mm) beyond the nut when tightened, and the threads on the projecting end shall be burred.

Weights of Steel Plates in Lbs. per Sq. Ft (Kg per Sq.M)			
Thickness Inches	Lbs per Sq. Ft	Thickness mm	Kg per Sq. Meter
3/16	7.65	4.69	37.35
1/4	10.20	6.25	49.80
5/16	12.75	7.81	62.25
3/8	15.30	9.38	74.70
7/16	17.85	10.94	87.16
1/2	20.40	12.50	99.61
5/8	25.50	15.63	124.51
3/4	30.60	18.75	149.41
7/8	35.70	21.88	174.31
1	40.80	25.00	199.21
1 1/4	51.00	31.25	249.02
1 1/2	61.20	37.50	298.82
1 3/4	71.40	43.75	348.62
2	81.60	50.00	398.43

Weight of Corrugated Sheets, Lbs per 100 S.F. (Kg per 10 Sq.M)							
	Black and Painted		Galvanized		Black and Painted		Galvanized	
Gauge Nos.	2, 2-1/2, & 3" corrug.	1 1/4" Corrug.	2, 2-1/2, & 8" corrug.	1 1/4" Corrug.	50, 62.5, & 75 mm corrug.	31.25 mm corrug.	50, 62.5, & 75 mm corrug.	31.25 mm corrug.
16	271	-----	286	-----	132.32	-----	139.64	-----
18	217	-----	232	-----	105.95	-----	113.28	-----
20	163	170	178	185	79.59	83.01	86.91	90.33
22	136	142	151	157	66.40	69.33	73.73	76.66
24	110	114	125	129	53.71	55.66	61.03	62.99
26	83	86	98	101	40.53	41.99	47.85	49.31
27	76	79	91	94	37.11	38.57	44.43	45.90
28	68	72	85	87	33.20	35.16	41.50	42.48

JUNIOR STEEL BEAMS

Junior beams (M shapes) are lightweight, hot-rolled structural beams that are used as secondary floor and roof beams in schools, stores, apartments, hospitals, and other types of light occupancy buildings. They are well adapted for purlins in mill buildings.

They are rolled to ASTM specifications, and are made in 6", 8", 10", 12", and 14" (150, 200, 250, 300, and 350 mm) sizes. Sizes and properties are given below:

Designation			Depth Inches	Lbs. per Lin.Ft.	Web Thick. Inches	Flange Wdth Inches	Area Sq. In.
M	6 x	4.4	6	4.4	0.114	1.844	1.29
M	6 x	20	6	20.0	0.250	5.938	5.89
M	8 x	6.5	8	6.5	0.135	2.281	1.92
M	10 x	9	10	9.0	0.157	2.690	2.65
M	12 x	11.8	12	11.8	0.177	3.065	3.47
M	14 x	18	14	18.0	0.215	4.000	5.10

Designation			Depth mm	Kg per Lin. Meter	Web Thick. mm	Flange Wdth mm	Area Sq. Cm
M	152 x	2.00	152	6.55	2.850	46.100	8.323
M	152 x	9.07	152	29.76	6.250	148.450	38.002
M	203 x	2.95	203	9.67	3.375	57.025	12.388
M	254 x	4.08	254	13.39	3.925	67.250	17.098
M	305 x	5.35	305	17.56	4.425	76.625	22.388
M	356 x	8.16	356	26.79	5.375	100.000	32.905

The cost of cutting, punching holes, and coping junior beams will run approximately as given below, but when much fabricating is required, it is advisable to refer plans to a local fabricator or warehouse for a sub bid.

Top cope, per end $6.00
Bottom cope and seat angles, per end $15.00
Cutting end on diagonal, per end $8.00
Anchor holes in web, per hole $1.75
Conduit holes in web, per hole $1.75

Junior steel beams can be used in conjunction with monolithically finished concrete floors or precast gypsum slabs laid over the top flange of the junior beams. By encasing the top flanges in concrete, the floor is made rigid and does not require bridging in light occupancy buildings.

Where the slab is poured on metal lath or where gypsum plank is used, adjustable steel rigid bridging is furnished.

When a poured-in-place concrete floor is used, the forms are easily constructed by cutting short pieces of lumber to rest on the lower flanges of the beams, which in turn support the floor sheathing.

METALS

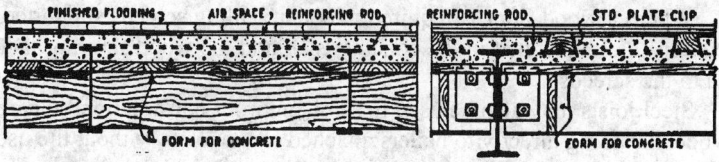

*Method of Constructing Wood Forms for Concrete Floors
Supported by Junior Beams*

Where joists rest on walls or bearing partitions, anchors are provided. Where the tops of junior beams are flush with tops of supporting beams, either clip angles or shelf angles are provided. Where the tops of beams are 2-1/2" (62.50 mm) above the tops of supporting beams, part of the beam is coped out and an angle seat is welded to the junior beam, which rests on the top flange of the supporting beam.

05200 METAL JOISTS

Standard and longspan steel joists were developed primarily to take the place of wood construction. The fire-resistance and structural qualities of this type of floor and roof construction has meant that it has not only replaced wood but other types of fire-resistant floor construction as well. Steel joist construction is recognized and accepted by all building codes.

Standard open web and longspan steel joists are manufactured by welding chord members of hot rolled structural or cold formed sections to round bar or angle web members to form a truss. Information on the types of steel joists can be obtained from the Steel Joist Institute and the joist manufacturers.

Typical Shortspan Steel Joists

When steel joists are used in fireproof construction, it is necessary to place a 2" to 2-1/2" (50-62.50mm) reinforced concrete slab over the joists, and a fire resistant ceiling of gypsum board or plaster beneath the joists to obtain the required fire rating.

829

Where a wood floor is required, wooden nailing screeds supported on screed clips are embedded in the concrete slab and the wood flooring is then nailed to the screeds.

Steel joists may also be used in semi-fireproof construction by nailing the wood sheathing directly to nailers attached to the joists without the use of a protective concrete slab. Some manufacturers of steel joists have met the demand for this type of construction by fabricating a nailer joist with the wood nailer attached in the shops by the use of lag screws or bolts.

Open web joists may be used to great advantage in all types of light occupancy buildings. The open web construction greatly facilitates the placing of plumbing, conduits, etc., making unnecessary the need of a furred ceiling or heavy fills on top of the structural slab.

Steel joists are also used in industrial construction to replace structural steel roof purlins with a resulting economy of materials. Metal roof deck is then usually welded to the joists and lightweight insulating material either poured or rigid board applied to the metal deck, and the roofing is installed.

Estimating Steel Joists. Steel joists and related products are generally sold by the manufacturer directly to the contractor. When preparing general contract estimates the most satisfactory method is to send plans or sketches to the manufacturer or his agent thus assuring a sub-bid based on an economical layout and design and one which includes the necessary accessories.

In its quotation the manufacturer will note the tonnage of joists, information which enables the contractor to figure the necessary handling and labor of erection. The cost of joist erection, including placing of accessories, varies from $1,000 to $2,500 per ton ($1,102.53 to $2,756.33 per tonne), excluding material costs, depending on labor conditions, location, and type of building.

Labor Setting Steel Joists. The labor cost of handling and setting steel joists is usually estimated by the ton, based on the total weight of the joists to be placed.

This cost will vary considerably, depending upon the type of building, size of joists, length of spans, amount of handling and hoisting necessary, etc.

It is much cheaper to set joists on long straight spans and large floor areas, such as garage and factory buildings, office buildings, etc., than on small or other irregularly constructed buildings.

All steel joists must be bridged and connected according to the Steel Joist Institute standards.

On an ordinary job of steel floor joists of regular construction and fairly long spans, five workers together should handle and place about 4-1/2 to 5 tons (4.08 to 4.53 tonnes) of joists, including bridging, per 8-hr. day at the following labor cost per ton:

METALS

Description	Hours	Rate	Total	Rate	Total
Ironworker	40.00	$	$	$ 40.29	$ 1,611.60
Cost per ton					$ 322.32
per metric ton					$ 317.24

On jobs of irregular construction, a five-worker crew together should handle and place about 3-1/2 to 4 tons (3.17 to 3.62 tonnes) of joists, including bridging, per 8-hr. day, at the following labor cost per ton:

Description	Hours	Rate	Total	Rate	Total
Ironworker	40.00	$	$	$ 40.29	$ 1,611.60
Cost per ton					$ 429.76
per metric ton					$ 422.99

Prices of Steel Joists. Costs of standard steel joists vary, depending on several factors, such as size of job, the number of joists of one particular size and length, and the job location. For an average job the cost of standard joists will be in the range of $750.00 to $900.00 per ton ($826.90 to 992.28 per tonne), which would include nominal bridging members and other accessories.

Costs of longspan steel joists will vary the same as those for standard joists. For an average job the cost of the longspan joists will range from $ 800.00 to $1,200.00 per ton ($882.03 to $ 1,323.04 per tonne) including accessories.

05300 METAL DECKING

Steel roof deck is formed from steel sheets in 18, 20, or 22 gauge. Most commercial decks have a standardized cross section with longitudinal ribs spaced 6" (150 mm) on centers. Sections are 30" (750 mm) in width and are manufactured in various lengths to suit job conditions, usually in the 14 to 31-ft. (4.26 to-9.44 m) range, although some can be supplied in greater lengths.

Deck sections are usually 1-1/2" (37.50 mm) deep, but some manufacturers supply a 1-3/4", 2", or 2-1/2" (44.75, 50, or 62.50 mm) deep deck. Sections have interlocking or nesting side laps and telescoping or nesting end laps.

Under ordinary roof live loads, 30 to 40 lbs. per sq.ft. (14648 to 195.31 kg/m^2), 20-gauge deck is normally used for spans from 6'-0" to 7'-6" (1.82-2.28 m) centers of purlins and 18 gauge deck for spans from 7'-0" to 8'-0" (2.13-2.44 m). Specific information on allowable loads can be obtained from the various manufacturers. By extending a single sheet over two or more purlin spaces, structural continuity, increasing load carrying capacity, can be obtained. Dead load of deck, insulation, and built-up roofing is approximately 7 lbs. per sq.ft. (34.18 kg/m^2).

Approximate cost of steel roof deck, delivered to job site in the eastern or northern states, on an average size job, 10,000 sq.ft. (929 sq.m) or more, is as follows:

Gauge of Roof Deck	Price per sq.ft. of 1 1/2" (37.50 mm) Deck
26 Gauge	$ 0.75
24 Gauge	$ 0.85
22 Gauge	$ 0.95
20 Gauge	$ 1.10
18 Gauge	$ 1.40
16 Gauge	$ 1.75

Placing Metal Deck on Top of Steel Joists. When metal deck is placed over the top of steel joists to receive the concrete slab or lightweight insulating materials, the metal deck is usually spot welded or mechanically fastened to the joists every 6" to 8" (150 to 200 mm).

On work of this type two workers together should place and weld about 200 sq.ft. (18 sq.m) of deck per hour at the following labor cost per square of 100 sq.ft. (9.29 sq.m):

Description	Hours	Rate	Total	Rate	Total
Iron Worker	1.00	$	$	$ 40.29	$ 40.29
Cost per sq.ft.					$ 0.40
per sq.m					$ 4.34

Steel roof deck can be erected in almost any kind of weather, permitting other trades to work under cover much more rapidly. Deck must be welded to steel support members in accordance with the Steel Deck Institute recommendations. Two tack welds per 18" (450 mm) width of deck section at end laps and one weld on the outside rib on intermediate supports are recommended.

Longspan Deck

Longspan deck is formed from a single 14, 16, 18, or 20 gauge steel sheet into 4-1/2", 6", or 7-1/2" (112.50, 150, or 187.50 mm) deep pans 12" (300 mm) wide. These are rolled sections with integral stiffening ribs formed into the top flange. Interlocking longitudinal side laps are provided on opposite edges for positive side lap attachment and are fastened by welding. Maximum spans allowable vary from 20'-0" to 30'-0" (6.0-9.0 m).

Longspan deck is easily erected—on an average job, a 5-worker crew should erect 2,200 to 2,400 sq.ft. (204.38 to 222.96 sq.m) per 8-hr. day.

METALS

Approximate cost of longspan steel roof deck, delivered to job sites in the eastern or northern states, is $3.50 to $6.25per sq.ft. ($37.67 to $67.27 sq.m), depending upon gauge and length of sections.

Reinforcing Floor Forms

Steel deck erected in an inverted position (with ribs up) can be used as a reinforcing form for concrete construction. It acts as a form to support the concrete and permits the ribs to act as reinforcing.

Steel deck reinforcing forms can be erected for an entire structure as soon as the steel framework is placed, providing an immediate working platform and protective staging for all trades. Concrete can be placed at any portion of the building without regard for removing, cleaning, and re-setting temporary formwork. Steel deck forms usually provide all necessary reinforcing to satisfy flexure requirements—the only additional reinforcing ordinarily required is temperature mesh to minimize shrinkage cracking. Tight form joints prevent concrete from dripping to lower floors, saving clean-up time.

Steel deck reinforcing forms can also be used in conjunction with composite beam design. Ample area for concrete around stud shear connectors and between deck ribs permits full effectiveness of the connectors. Standard AISC Composite Design Procedure may be followed using the total slab depth in beam property calculations.

Reinforcing floor forms are installed by welding, the same as ordinary steel roof deck, and costs of materials and erection are about the same as given for "Metal Decking".

CELLULAR STEEL DECK AND FLOOR PANELS

Lightgauge cellular steel structural panels are used for longspan structural floors and roofs. Some panels are composed of two identically formed beam sections and a flat plate. Others are formed from two ribbed sections and a flat plate. All panels have interlocking side joints. Components are factory assembled by spot welding. Panels are manufactured in gauges from 20 to 13 and are normally available in 24" (600 mm) widths. Some panels are fabricated in 12" (300 mm) widths. Lengths are available up to 40'-0" (12.19 m), while depths vary from 1-1/2" to 7-1/2" (37.50 to 187.50 mm).

This type of panel is easily handled and erected, because it is deck and joist combined. Concrete floor forms are eliminated and panels provide a working platform for other trades.

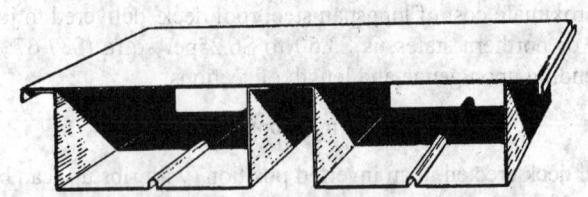

Cellular Steel Deck and Floor Panel

A flat, plate-down installation is often used to provide an attractive ceiling which needs only a finish coat of paint to complete the job. Voids between beam sections may be filled and covered with 2-1/2" (62.50 mm) of concrete for floors or covered with rigid board insulation and built-up roofing for roofs. Installations with beam section forming the ceiling can be merely painted to give a finished fluted ceiling, or a suspended ceiling can be erected to provide a flush type ceiling.

Where panels are covered with a concrete fill, the use of temperature mesh to reduce shrinkage cracks is recommended.

The cost of cellular panels, delivered to job site, ranges from $4.50 per sq.ft. ($48.43 per sq.m) for 20 ga. 2" to 3" (50-75 mm) thick panels, $5.75 to $7.00 per sq.ft. ($61.89 to $75.34 per sq.m) for 3" to 4" (75-100 mm) 18 ga. units on up to $7.50 per sq.ft. $80.73 per sq.m) for 16 ga. deck.

A crew of four structural iron workers plus a foreman should install around 1,600 sq.ft. (148.64 sq.m) of light deck per 8-hr. day, 1200 sq.ft. (111.48 sq.m) of 18 ga. deck, and around 1,000 sq.ft. (92.9 sq.m) of the heavy gauge deck under normal job conditions. For example, 100 sq.ft. (9.29 sq.m) of 18 ga. labor costs would be as follows:

Description	Hours	Rate	Total	Rate	Total
Ironworker Foreman	0.65	$	$	$ 44.29	$ 28.79
Ironworkers	2.60	$	$	$ 40.29	$ 104.75
Cost per 100 s.f.					$ 133.54
per s.f.					$ 1.34
per sq.m					$ 14.37

05400 LIGHTGAGE FRAMING

The components involved are a complete range of studs, joists, and accessories for the steel framing of buildings. Sections are fabricated from structural-grade, high tensile strip steel by cold forming and are designed specifically for strength, light weight and low cost. Yet, structural framing carries all the benefits of conventional steel framing.

834

METALS

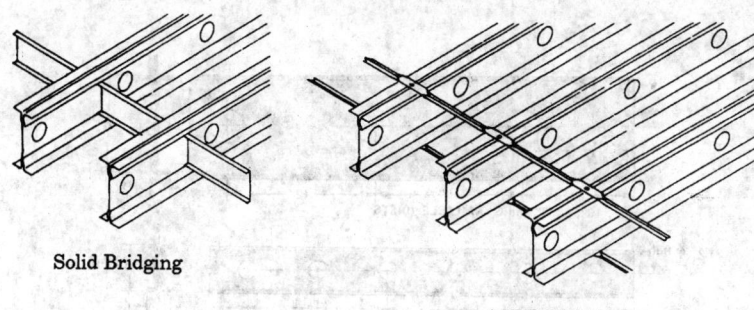

Solid Bridging

"V" Bridging

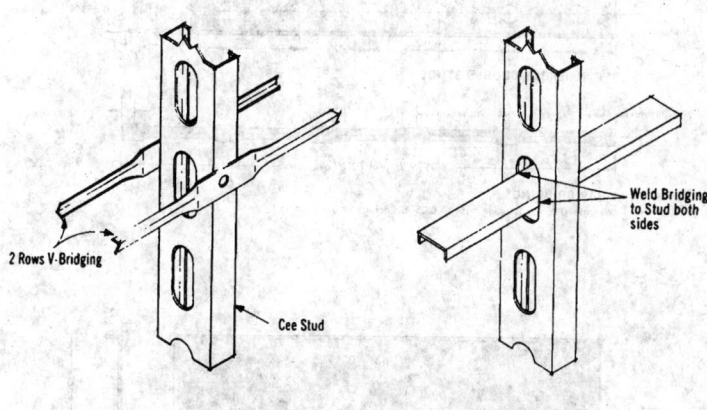

2 Rows V-Bridging

Cee Stud

Weld Bridging
to Stud both
sides

"V" Bridging of Studs

Channel Bridging
of Studs

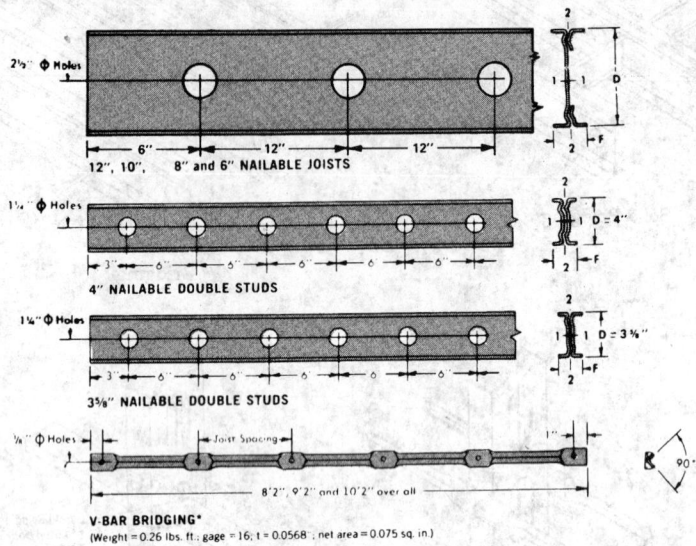

2½" Φ Holes

6" | 12" | 12"

12", 10", 8" and 6" NAILABLE JOISTS

1¼" Φ Holes

3" 6" 6" 6" 6" 6"

D = 4"

4" NAILABLE DOUBLE STUDS

1¼" Φ Holes

3" 6" 6" 6" 6" 6"

D = 3⅝"

3⅝" NAILABLE DOUBLE STUDS

⅛" Φ Holes   Joist Spacing   1"

8'2", 9'2" and 10'2" over all

V-BAR BRIDGING*
(Weight = 0.26 lbs. ft.; gage = 16; t = 0.0568 ; net area = 0.075 sq. in.)

90°

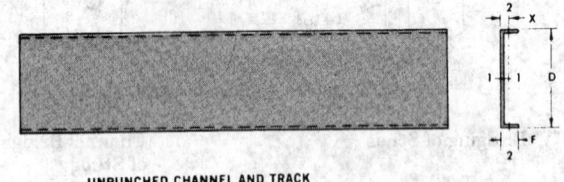

X

D

UNPUNCHED CHANNEL AND TRACK

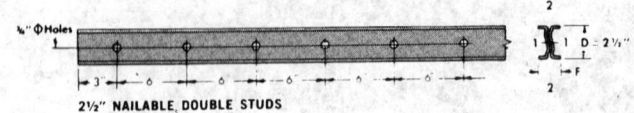

¾" Φ Holes

3" 6" 6" 6" 6" 6"

D = 2½"

2½" NAILABLE DOUBLE STUDS

836

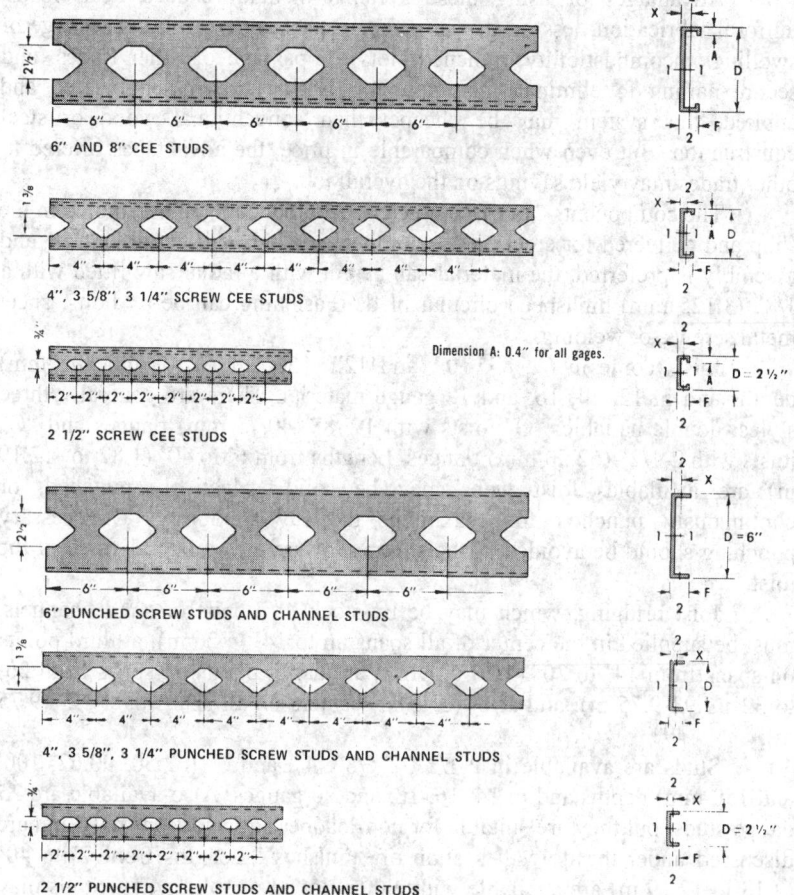

6" AND 8" CEE STUDS

4", 3 5/8", 3 1/4" SCREW CEE STUDS

Dimension A: 0.4" for all gages.

2 1/2" SCREW CEE STUDS

6" PUNCHED SCREW STUDS AND CHANNEL STUDS

4", 3 5/8", 3 1/4" PUNCHED SCREW STUDS AND CHANNEL STUDS

2 1/2" PUNCHED SCREW STUDS AND CHANNEL STUDS

A nailing groove, developed for easy, economical attachment of other materials, is a feature of all double studs and joists. It is obtained by welding two cold-formed steel elements together, so designed that a nail driven into this space is not only held by friction, but is also deformed to provide maximum holding power.

The usual limitations imposed by prefabrication are avoided. Using nailable framing sections architects and engineers have unrestricted freedom in design. All sections are painted at the plant with a coat of oven-dried, rust-resisting red zinc chromate paint, or are galvanized.

Lightgage framing systems can supply complete wall, floor, and roof construction for buildings up to four stories in height, or can be used in combination with other framing systems for interior, load bearing partitions, exterior curtain walls, fire separation walls, parapets, penthouses, trusses, suspended ceilings, and mansard roofs.

837

Advantages of using these systems include reduced dead loads, uniform fabrication, less on-the-job storage space, no warping, shrinkage or swelling, incombustibility, punched slots for passage of other trades, and secure nailing to eliminate nail popping. When properly engineered and applied, this system may be cheaper than conventional wood or steel construction. But even when comparable in price, the advantages offered to other trades may yield savings on the overall job.

The components can be completely detailed, cut, and assembled in the shop and delivered for erection to the job site. Where on-the-job cutting and assembly is preferred, the material can be cut with a radial saw fitted with a 1/8" (3.125 mm) highspeed circular blade. Fastening can be by bolts, sheet metal screws, or welding.

Joists come in 6", 8", 10", and 12" (150, 200, 250, and 300 mm) depths and in 12, 14, 16, and 18 gauge material. There are generally three styles: double nailable, "C" joists with 1-5/8" (40.75 mm) flanges and "C" joists with 2-1/2" (62.50 mm) flanges. Lengths from 6' to 40' (1.82 to -12.19 m) are available. Joist webs may be solid, selectively punched, or continuously punched for maximum raceway flexibility. Unnecessary punching should be avoided as the slotting lowers the structural value of the joist.

Joist bridging, which may be by stock "V" units or solid channels, must be supplied in the center of all spans up to 14' (4.26 m); at third points on spans from 14' to 20' (4.26-6.09 m); at quarter points on spans from 26' to 32' (7.92-9.75 m); and at 8' (2.43 m) centers on all spans over 32' (9.75 m).

Studs are available in 2-1/2", 3-5/8", 4", and 6" (62.50, 90.62, 100, and 150 mm) depths and in 14, 16, 18, and 20 gauges. Also available are 25 gauge studs, but they are suitable for non-loadbearing partitions only and are discussed under the drywall section of "Finishes". Lengths from 7' to 40' (2.13 to 12.19 m) are available with either slotted or solid webs. Studs may be of the double-nailable type, channels, or "C" type with knurled flanges.

Studs mount into standard tracks at floors, ceilings, sills, and fascias. They should be stiffened with lateral bridging at the midpoint of all walls to 10' (3.04 m) in height; at third points on walls from 10' to 14' (3.04 to 4.26 m); and every 4' (1.22 m) on walls above 14' (4.26 m). Stock "V" bars, channels or special clipping systems may be ordered for this.

Costs will vary because of the various options available. Each job should be priced with the supplier when all the design criteria are known. For general budget purposes joists can be figured at around $1.25 per lb. galvanized. A punched, galvanized "C" joist will vary from 2.24 lbs. per ft. (3.33 Kg/m), 6" (150 mm) deep, to 5.85 lbs. (2.65 kg), 12" (300 mm) deep or from around $2.50 to $6.50 per ft ($8.20 to $21.32m). Carload lots will run less.

Studs 3-5/8" (90.62 mm) deep, punched and galvanized, will run $0.75 per ft. ($2.46 per m)) in 20 gauge material in less than carload lots;

$1.25 per ft. ($4.10 per m) in 18 gauge; and $1.35 per ft. ($4.43 per m) in 16 gauge. Tracks will approximate stud costs.

Attachment of Other Materials. Steel or aluminum roof decking, metal siding, or corrugated sheets can be quickly and economically attached to sections with self-tapping screws, or by other conventional methods.

When self-tapping screws are used, a hole slightly smaller than the screw is drilled through the material and the lightgage section. Then, the screw can be placed and tightened by one person. This is faster and more economical than most conventional methods of attachment which require two workers. The 5/8" (15.62 mm) hex-head self-tapping, cadmium-plated screws are recommended.

Attachment of materials is generally with standard nails when the nailable sections are used. When channel studs or other non-nailable sections are used, the self-drilling screws are recommended for metal siding or plywood. When drywall is applied to non-nailable sections, self-drilling screws can be ordered from the drywall manufacturer.

Other types of materials can be readily attached to framing members by the use of bolts, clips, welds, or other methods consistent with the finish desired.

Installation steel track top and bottom 20 gauge plus studs at 24" center to center 4" wide wall x 8'-0" high (100mm x 2.43 m). A team of two carpenters with one labor feeding material should install approximately 150 lin.ft. (45.72 m) per day.

Installation Metal Track and Studs 4" wide x 8' High (100 mm x 2.43 M)						
Description	Hours	Rate	Total	Rate	Total	
Carpenter Foreman	8.00	$	$	$ 39.04	$	312.32
Carpenter	8.00	$	$	$ 35.04	$	280.32
Labor	4.00	$	$	$ 26.16	$	104.64
Total Cost					$	697.28
Cost per foot					$	4.65
Cost per Meter					$	15.25

05500 METAL FABRICATIONS

Metal Stairs. Metal stairs are usually figured at so much per riser with stringers, treads, nosings, and railings included. A standard, 3-ft. (0.91-m) wide simple run metal pan stair can be budgeted at around $85.00 per riser, material cost. For each additional foot of width up to 5 ft. (1.52 m), add 10%. Custom stairs will run from $95.00 per riser to twice that amount where any of the stair components vary from the norm. Simple landings cost about $8.50 per sq. ft. ($91.48 per sq.m) without an allowance for railings.

A four-worker crew should erect around 40 to 50 risers or 150 sq.ft. (13.93 sq.m) of landing per 8 hr. day.

The labor erection cost of a three foot wide steel stair consisting of three flights of sixteen risers and one 3' x 6' (0.93 x 1.82 m) landing each run would figure as follows:

Description	Hours	Rate	Total	Rate	Total
Ironworkers					
Stairs - 48 risers	34.00	$	$	$ 40.29	$ 1,369.86
Landing 54 sq.ft. (5.01 sq.m)	11.10	$	$	$ 40.29	$ 447.22
Total Cost					$ 1,454.48
Cost per stair run (3)					$ 484.83
Cost per riser incl. landings					$ 30.30

Pipe Railings. When estimating plain pipe railing, 3'-6" (1.06 m) high, consisting of two horizontal runs of pipe with uprights 6' to 8' (1.82-2.44 m) on center, figure as follows for the various sizes, erected in place:

Description	Price
1-1/4" (31.25 mm) pipe railing complete as described above per lin.ft. (0.30 m)	$50 per lin.ft. ($164.04 per m)
1-1/2" (37.50 mm) pipe railing complete as described above per lin.ft. (0.30 m)	$55 per lin. Ft. ($180.44 per m) 2"
(50 mm) pipe railing complete as described above per lin.ft. (0.30 m)	$70 per lin.ft. ($229.66 per m)

For each curved section or termination point that is formed to radius, add $45.00 to $60.00.

For following the rake of a stair, add $11.00 per upright.

For each foot (0.30 m) of wall railing, add $10.00.

Aluminum pipe railings can also be ordered from stock and will cost around $80.00 per ft. ($262.47 per m), anodized. Aluminum wall rails will cost around $11.50 per ft. ($37.73 per m). A two-worker crew should erect some 100 lin.ft. (30.48 m) of straight-run railing per day and 130 ft. (39.62 m) of wall railing.

Steel Ladders. Straight steel ladders cost from $60.00 to $90.00 per lin.ft. ($196.85 to $295.27 per m) erected in place. Add $40.00 to $60.00 for curved handles and platform over coping walls. Add $60.00 per lin.ft. ($196.85 per m) if cage is required.

Steel Gratings. Gratings may be welded or of expanded construction and made of steel, aluminum, or stainless steel.

Welded steel gratings weighing 4 lbs run about $5.75 per sq.ft. ($61.89per sq.m); weighing 5 lbs, $5.75; 9 lbs, $8.25; 12 lbs, $10.65; and 16 lbs, $11.60 to $12.80, all depending on the spacing and size of the bars. If the material is galvanized add $1.20 per sq.ft. ($12.91 per sq.m).

Aluminum grating will run from $13.25 per sq.ft. ($142.62 per sq.m), 3/4" (18.75 mm) thick, and weighing 1.5 lbs. (0.68 kg), to $40.0025.00 per

840

sq.ft. ($430.57 sq.m), 2-1/4" (56.25 mm) thick, and weighing 5.9 lbs (2.67 kg) in amounts of 75 sq.ft. (6.96 sq.m) or more.

Checkered plate is sometimes specified in place of or in combination with gratings and can be figured at about $14.50 per sq.ft. ($156.08 per sq.m) for 1/4" (6.25 mm) thickness.

A typical crew laying grating will consist of four workers, who can install about 100 sq.ft. (9.29 sq.m) per hour at the following cost per 100 sq.ft. (9.29 sq.m):

Description	Hours	Rate	Total	Rate	Total
Ironworkers	4.00	$	$	$ 40.29	$ 161.16
Cost per sq.ft.					$ 1.61
per sq.m					$ 17.35

Steel Window Guards. Window guards made with 3/8" x 2" (9.37 x 50 mm) horizontal bars, top and bottom and round or square upright bars spaced 4" (100 mm) on centers. Delivered to job.

Window Guards Made of Crimped Diamond Mesh. Window guards, made with 1" (25 mm) channel iron frames and crimped diamond mesh wire, per sq.ft. (sq.m) erected in place cost $15.00 to $25.00 ($161.46 to $269.10 per sq.m). Do not figure any opening as containing less than 20 sq.ft. (1.85 sq.m).

Curb Angles, Floor Frames, Trench Frames, and Covers. For miscellaneous items, such as curb angles, floor frames, trench frames, and covers, figure $2.25 per lb. ($4.96 per kg) delivered to job.

Window Guards	
Description	Approx. Price per sq.ft.
Window guards with 1/2 " round upright bars	$9.50
Window guards with 5/8 " round upright bars	$12.00
Window guards with 5/8 " square upright bars	$13.00
Window guards with 3/4 " round upright bars	$15.00
Window guards with 3/4 " square upright bars	$18.00

Window Guards - Metric	
Description	Approx. Price per sq.m
Window guards with ##### mm round upright bars	$102.26
Window guards with ##### mm round upright bars	$129.17
Window guards with ##### mm square upright bars	$139.94
Window guards with ##### mm round upright bars	$161.46
Window guards with ##### mm square upright bars	$193.76

Ornamental Iron Entrance Rails. Fabricated from 1-1/4" x 1/4" or 1-1/4" x 3/8" (31.25 x 6.25 or 31.25 x 9.37 mm) bar stock, or 1-1/2" (37.50 mm) bar size channels for rails and 1/2" (12.50 mm) square bar pickets, stock designs for this item, delivered to job, cost about $75.00 per lin.ft. ($246.06 per m) of rail, depending on height and amount of ornamentation. For rails following the rake of stairs, add 50 %. Rails made to special design may cost 2 to 3 times the above prices.

Ornamental Porch Columns. Stock design are scrolled, wrought iron columns, with 3/4" x 3/4" (18.75 x 18.75 mm) bar stock, or 15/16" (23.44 mm) square tubing frames and 3/8" x 1/4" (9.37 x 6.25 mm) or 1/2" x 3/16" (12.55 x 4.68 mm) bar or strip stock scrolls. Flat columns, 8-1/2" (212.50 mm) to 12" (300 mm) wide x 8'-0" (2.43 m) high, cost $110.00 to $190.00 each. Corner columns to match flat columns cost $90.00 to $125.00 each.

Stock design columns, with 3/4" x 3/4" (18.75 x 18.75 mm) bar stock frames and cast iron panels, delivered to job, cost as follows: flat columns, 9-1/2" to 11" (237.50 to -275 mm) wide x 8'-0" (2.43 m) high are $95.00 to $160.00 each; corner columns to match flat columns are $100.00 to $140.00 each. Brackets to match columns are $50.00 each.

Steel Stacks

Steel stack costs vary widely and are dependent on height, diameter, gauge of material, number of joints, transportation, etc. Steel stacks are usually fabricated by companies specializing in steel boiler and tank manufacturing.

When estimating a job that has a steel stack, always obtain a firm figure from a steel stack fabricator before submitting a bid on the work.

Labor Erecting Steel Stacks. When estimating the labor cost of erecting steel stacks for office buildings or other buildings of this type, where the stack extends from the ground to the top of the building and is either built into the building or placed on the outside of one of the rear walls, there are several items to be taken into consideration-the length and weight of each section of stack, the facilities for hoisting and placing, and the amount of rigging necessary.

Where each section of the stack weighs 500 to 1,000 lbs. (226.80 to 453.60 kg), it is customary to place a small gin pole on the roof of the building and lift each section in place.

Four iron workers should set up all necessary rigging in one day, ready to place the sections. After the crane and rigging have been placed to start hoisting, the crew should handle, place, and bolt one 15- to 20-ft. (4.57-6.09 m) section of stack in 4 hours, or 2 sections, 30 to 40 lin.ft. (9.14 to12.19 m), per 8-hour day.

The labor cost placing, rigging and removing cranes for erecting a steel stack should average as follows:

METALS

Description	Hours	Rate	Total	Rate	Total
Ironworkers placing equip.	32.00	$	$	$ 40.29	$ 1,289.28
Ironworkers removing equip.	32.00	$	$	$ 40.29	$ 1,289.28
Labor on equipment					$ 2,578.56

The labor cost erecting a steel stack up to 30" (750 mm) in diam. and 60 lin.ft. (18.28 m) long, should average as follows:

Description	Hours	Rate	Total	Rate	Total
Ironworkers erecting stack	48.00	$	$	$ 40.29	$ 1,933.92
Cost per lin.ft.				$	32.23
per m				$	85.14

To the above costs, add equipment rental and move charges to and from job for equipment.

06   CARPENTRY

06100 ROUGH CARPENTRY

To estimate carpentry quantities and costs accurately, an estimator should know how the various classes of work are constructed, where joists and studs should be doubled, where to place wood furring strips for walls and floors and openings requiring wood bucks, and where to place wood grounds. These may seem like small items, but they are important.

An inexperienced estimator who is unfamiliar with how work is conducted in the field might figure wood bucks for the exterior window frames. He or she must know how a job goes together and what is required for each item of work.

When taking off lumber quantities from the plans, always estimate as closely as possible the exact quantities of each kind required. Some contractors estimate the cost of wood floors at a certain price per sq. ft. including joists, bridging, subfloor, deadening felt, furring strips, and the finish wood flooring. Entering into the construction of such a floor are materials and labor operations of widely varying costs, yet certain contractors are content to estimate their work in this haphazard manner, and at the completion of the job have no more idea of their labor costs than before they took the job.

When estimating labor costs of rough carpentry, bear in mind the costs will vary with the class of work performed, with the ability of the carpenters employed, and with the experience and ability of the foreman or superintendent.

Tools, such as electric hand saws, electric drills, nailers and sanders, play a big part in reducing labor costs. For instance an electric saw can cut a 2"x12" (50 x 300 mm) plank in 4 or 5 seconds, while it takes about a minute to do it by hand; it takes 10 to 12 minutes to rip a 12'-0" (3.65 m) plank 2" (50 mm) thick when sawing by hand, while an electric handsaw will do the work in less than a minute. In other words, an electric saw will cut 10 to 20 times as fast.

Another consideration is the cost of filing saws. On average it will require 1/2 hr. to file a saw by hand, while an automatic saw filer will do the

845

same work in 1/4 hr. The same applies to circular saws used on saw rigs or electric hand saws, as these can be filed in 1/2 to 3/4 hr. by machine, while hand filing requires nearly twice as much time.

The same thing applies to the use of electric drills, electric nailers, electric sanders, etc. The contractor who plans to keep up with the procession must use modern, labor-saving tools and equipment wherever possible.

Here is another thing that should be remembered by every contractor. No matter how carefully the plans have been measured and how much consideration they have been given before pricing, if the work is not executed in an efficient manner, careful estimating will not avert losses.

Estimating Lumber Quantities

When estimating the quantity of lumber required for any job, the only safe method is to take off every piece of lumber required to complete that portion of the work. The following tables can simplify the work as much as possible and at the same time provide accurate material quantities.

Lumber is usually sold by the board foot (bf), sometimes referred to as board foot measure. One board foot is always 144 cubic inches. The formula for figuring the number of board feet in a given length of board or lumber is:

Width (inches) x Thickness (inches) x Length (feet) ÷ 12 = # of BF

12" x 12" x 1 foot = 1' x 1' x 1' = 1 cubic foot or 1728 Cubic Inches
1/12 of a cubic foot = 1/12 of 1728 = 144 Cubic Inches

For example, to figure the board feet measure of a 2" x 4" that is 6'-0" long, (2 x 4 x 6) ÷ 12 = 48 ÷ 12 = 4 board feet (.009 Cu.M).

Note: Conversion from Board Feet to Cubic meter has been rounded off. Board Foot (bd. Ft. or bf) is a unit of volume for measuring lumber. One standard board foot is the volume of twelve inches wide by one inch thick by one-foot length or 1/12 of a cubic foot. To accomplish a straight conversion from board feet to cubic meter is not possible. Since board feet is not really 1/12 of a cubic foot, a dressed piece of softwood lumber is a nominal size of 2" x 4" x 12' which is actually 1 ½" x 3 ½" x 12. Board feet use nominal sizes while metric uses actual sizes and a straight conversion is not possible. Therefore the "soft" conversion that 424 Board Feet = 1 Cubic Meter will not be accurate for converting actual volumes but is the accepted "soft" conversion factor for calculations.

Estimating Wood Joists. When estimating wood joists, always allow 4" to 6" (100 to 150 mm) on each end of the joist for bearing on the wall.

To obtain the number of joists required for any floor, take the length of the floor in feet, divide by the distance the joists are spaced and add 1 to allow for the extra joist required at end of span.

846

CARPENTRY

Example: If the floor is 28 ft. (8.53 m) long and 15 ft. (4.57 m) wide, it will require 16-ft. (4.87-m) joists to allow for wall bearing at each end. Assuming the joists are spaced 16" (400 mm) on centers, one joist will be required every 16" (400 mm) or every 1-1/3 ft. (0.40 m). In other words it will require 3/4 or 0.75 as many joists as the length of the span, plus one. Three-quarters of 28 equal 21, plus 1 extra joist at end, makes 22 joists 16 ft. (4.87 m) long for this space.

The following table gives the number of joists required for any spacing:

Number of Wood Floor Joists Required for any Spacing				
Distance Joists are Placed on Center	Multiply Length of Floor Span by	Add Joists	Distance Joists are Placed On Center	Add Joists
12 "	1	1	300 mm	1
16 "	3/4 or 0.75	1	400 mm	1
20 "	3/5 or 0.60	1	500 mm	1
24 "	1/2 or 0.50	1	600 mm	1
30 "	2/5 or 0.40	1	750 mm	1
36 "	1/3 or 0.33	1	900 mm	1
42 "	2/7 or 0.29	1	1050 mm	1
48 "	1/4 or 0.25	1	1200 mm	1
54 "	2/9 or 0.22	1	1350 mm	1
60 "	1/5 or 0.20	1	1500 mm	1

Board Feet of Lumber Required per 100 Sq.Ft. of Surface When Used for Studs, Joists Rafters, Wall and Floor Furring Strips, etc.				
Size	On Center Spacing			
	12"	16"	20"	24"
1 " x 2 "	16 2/3	12 1/2	10	8 1/3
2 " x 2 "	33 1/3	25	20	16 2/3
2 " x 4 "	66 2/3	50	40	33 1/3
2 " x 5 "	83 1/3	62 1/2	50	41 2/3
2 " x 6 "	100	75	60	50
2 " x 8 "	133 1/3	100	80	66 2/3
2 " x 10 "	166 2/3	125	100	83 1/3
2 " x 12 "	200	150	120	100
2 " x 14 "	233 1/3	175	140	116 2/3
3 " x 6 "	150	112 1/2	90	75
3 " x 8 "	200	150	120	100
3 " x 10 "	250	187 1/2	150	125
3 " x 12 "	300	225	180	150
3 " x 14 "	350	262 1/2	210	175

Metric

Cubic Meter of Lumber Required per 10 Sq.M. of Surface When Used for Studs, Joists Rafters, Wall and Floor Furring Strips, etc.				
Size	On Center Spacing			
	300 mm	400 mm	500 mm	600 mm
25 x 50 mm	0.042	0.032	0.025	0.021
50 x 50 mm	0.085	0.063	0.051	0.042
50 x 100 mm	0.169	0.127	0.102	0.085
50 x 125 mm	0.212	0.159	0.127	0.106
50 x 150 mm	0.254	0.190	0.152	0.127
50 x 200 mm	0.338	0.254	0.203	0.169
50 x 250 mm	0.423	0.317	0.254	0.212
50 x 300 mm	0.508	0.381	0.305	0.254
50 x 350 mm	0.592	0.444	0.355	0.296

CARPENTRY

Cubic Meter of Lumber Required per 10 Sq.M. of Surface When Used for Studs, Joists Rafters, Wall and Floor Furring Strips, etc. (cont'd)				
Size	On Center Spacing			
	300 mm	400 mm	500 mm	600 mm
75 x 150 mm	0.381	0.286	0.228	0.190
75 x 200 mm	0.508	0.381	0.305	0.254
75 x 250 mm	0.635	0.476	0.381	0.317
75 x 300 mm	0.762	0.571	0.457	0.381
75 x 350 mm	0.889	0.666	0.533	0.444

Note: Does not include any waste in cutting, doubling joists under partitions or around stair wells, extra joists at each end of the span, top or bottom plates. Etc. These items vary with each job. Add as required

Estimating Quantity of Bridging. It is customary to place a double row of bridging between joists about 6'-0" to 8'-0" (1.82 to 2.44 m) on centers. Joists 10'-0" to 12'-0" (3.08 to 3.65 m) long will require one double row of bridging or 2 pcs. to each joist.

Joists 14'-0" to 20'-0" (4.26 to 6.09 m) long will require 2 double rows of bridging or 4 pcs. to each joist.

Bridging is usually cut from 1" x 3" (25 x 75 mm), 1" x 4" (25 x 100 mm), 2" x 2" (50 x 50 mm), or 2" x 4" (50 x 100 mm) lumber. The following table gives the approximate number of pcs. and the lin.ft. (m) of bridging required per 100 sq.ft. (9.29 sq.m) of floor.

Joists to 12 feet Long				Joists to 20 feet Long			
12"Centers		16" Centers		12"Centers		16" Centers	
Pcs.	lin.ft.	Pcs.	lin.ft.	Pcs.	lin.ft.	Pcs.	lin.ft.
20	30	16	24	40	60	32	48
Metric							
Joists to 3.65 meters Long				Joists to 6.09 meters Long			
300 mm Cent.		400 mm Cent.		300 mm Cent.		400 mm Cent.	
Pcs.	lin.M.	Pcs.	lin.M.	Pcs.	lin.M.	Pcs.	lin.M.
20	9.14	16	7.32	40	18.29	32	14.63

Number of Wood Joists Required for any Floor and Spacing																					
Length of Flr Lin Ft.	Spacing of Joists, Inches										Length of Flr Lin. M	Spacing of Joists, inches (mm)									
	12	16	20	24	30	36	42	48	54	60		300	400	500	600	750	900	1050	1200	1350	1500
6	7	6	5	4	3	3	3	3	2	2	1.83	7	6	5	4	3	3	3	3	2	2
7	8	6	5	5	4	4	3	3	3	2	2.13	8	6	5	5	4	4	3	3	3	2
8	9	7	6	5	4	4	3	3	3	3	2.44	9	7	6	5	4	4	3	3	3	3
9	10	8	6	6	5	4	4	3	3	3	2.74	10	8	6	6	5	4	4	3	3	3
10	11	9	7	6	5	4	4	4	3	3	3.05	11	9	7	6	5	4	4	4	3	3
11	12	9	8	7	5	5	4	4	3	3	3.35	12	9	8	7	5	5	4	4	3	3
12	13	10	8	7	6	5	4	4	4	3	3.66	13	10	8	7	6	5	4	4	4	3
13	14	11	9	8	6	5	5	4	4	4	3.96	14	11	9	8	6	5	5	4	4	4
14	15	12	9	8	7	6	5	5	4	4	4.27	15	12	9	8	7	6	5	5	4	4
15	16	12	10	9	7	6	5	5	4	4	4.57	16	12	10	9	7	6	5	5	4	4
16	17	13	11	9	7	6	6	5	5	4	4.88	17	13	11	9	7	6	6	5	5	4
17	18	14	11	10	8	7	6	5	5	4	5.18	18	14	11	10	8	7	6	5	5	4
18	19	15	12	10	8	7	6	6	5	4	5.49	19	15	12	10	8	7	6	6	5	4
19	20	15	12	11	9	7	6	6	5	5	5.79	20	15	12	11	9	7	6	6	5	5
20	21	16	13	11	9	8	7	6	5	5	6.10	21	16	13	11	9	8	7	6	5	5
21	22	17	14	12	9	8	7	6	6	5	6.40	22	17	14	12	9	8	7	6	6	5
22	23	18	14	12	10	8	7	7	6	5	6.71	23	18	14	12	10	8	7	7	6	5
23	24	18	15	13	10	9	8	7	6	6	7.01	24	18	15	13	10	9	8	7	6	6
24	25	19	15	13	11	9	8	7	6	6	7.32	25	19	15	13	11	9	8	7	6	6
25	26	20	16	14	11	9	8	7	7	6	7.62	26	20	16	14	11	9	8	7	7	6
26	27	21	17	14	11	10	8	8	7	6	7.92	27	21	17	14	11	10	8	8	7	6
27	28	21	17	15	12	10	9	8	7	6	8.23	28	21	17	15	12	10	9	8	7	6
28	29	22	18	15	12	10	9	8	7	7	8.53	29	22	18	15	12	10	9	8	7	7
29	30	23	18	16	13	11	9	8	7	7	8.84	30	23	18	16	13	11	9	8	7	7
30	31	24	19	16	13	11	10	9	8	7	9.14	31	24	19	16	13	11	10	9	8	7
31	32	24	20	17	13	11	10	9	8	7	9.45	32	24	20	17	13	11	10	9	8	7
32	33	25	20	17	14	12	10	9	8	7	9.75	33	25	20	17	14	12	10	9	8	7
33	34	26	21	18	14	12	10	9	8	8	10.06	34	26	21	18	14	12	10	9	8	8
34	35	27	21	18	15	12	11	10	9	8	10.36	35	27	21	18	15	12	11	10	9	8
35	36	27	22	19	15	13	11	10	9	8	10.67	36	27	22	19	15	13	11	10	9	8
36	37	28	23	19	15	13	11	10	9	8	10.97	37	28	23	19	15	13	11	10	9	8
37	38	29	23	20	16	13	12	10	9	8	11.28	38	29	23	20	16	13	12	10	9	8
38	39	30	24	20	16	14	12	11	9	9	11.58	39	30	24	20	16	14	12	11	9	9
39	40	30	24	21	17	14	12	11	10	9	11.89	40	30	24	21	17	14	12	11	10	9
40	41	31	25	21	17	14	12	11	10	9	12.19	41	31	25	21	17	14	12	11	10	9

One joist has been added to each of the above quantities to take care of extra joist required at end of span. Add doubling joists under all partitions.

Steel Bridging for Wood Joists. Several types of steel bridging are available, fabricated from deformed strip steel varying from 16 to 22 gauge in thickness. Most are galvanized while some others are painted with black asphaltum or have a baked enamel finish. Stock sizes fit all joist depths and spacings encountered in average construction work—special sizes are available made to order.

Prices vary according to type, size, finish, and quantity, but a fair average for galvanized bridging is $2.00 to $5.00 per set.

Installation time is about 50% of that required for wood bridging and with many types can be done after subflooring is laid.

There are two basic types: 1) Nailess cross bridging which is designed for secure grip before the "drive home" blow and deeper prong penetration. One end of the bridging is formed to an angular fork-like projection with two 9/16" (14.06 mm) prongs. This end is installed first by driving the prongs into the wood joist to the bent tab, with hammer blows applied at the opposite end. The opposite end is installed after the other end is

850

CARPENTRY

in place by raising it into position against the joist and driving the prongs into the wood joist with hammer blows to a flat position. 2) Tension type bridging which has maximum nailing flexibility. Each end normally has about seven nails holes but only two nail holes need be used at each end. Recommended nail size should minimum, 11 gauge by 1 ¼" long.

Estimating Number of Wood Studs. When estimating the number of wood partition studs, take the length of each partition and then the total length of all partitions.

If a top and bottom plate is required, take the length of the wood partition and multiply by 2. The result will be the number of lin.ft. of plates required.

If a double plate consisting of two top members and a single bottom plate is used, multiply the length of the wood partitions by 3.

Example: Find the quantity of lumber required to build a stud partition 16'-0" (4.87 m) long, 8'-0" (2.44 m) high, with studs spaced 16" (400 mm) on centers and having single top and bottom plates. 16'-0" = 192" (4.87 m = 4.87m). Next, 192"/16" = 12 studs (4800/400mm = 12), plus 1 extra at the end equals 13 studs 8'-0" (2.44 m) long. Two top and bottom plates 16'-0" (4.87m) long equal 32 lin.ft (9.75).

13 pcs. 2 x 4 @ 8'-0" = 104 '- 0"	13 pcs. 50 x 100 mm @ 2.44 M = 31.72 M	
2 pcs. 2 x 4 @ 16'-0" = 32 '- 0"	2 pcs. 50 x 100 mm @ 4.88 M = 9.76 M	
104' + 32' = 136 lin.ft.	31.72 M + 9.76 M = 41.48M.	
136 lin.ft. x 2/3 = 90.67 bd.ft.	41.48 M = 0.207 Cu M	
	Soft Conversion	
	424bd ft. = 1 Cu M 0.214 Cu M	

Convert the total lineal feet of lumber to board feet. For metric, convert the total lineal meters to cubic meters.

Bd.Ft. of Lumber Required for Wood Stud Partitions 2" x 4" Studs 16" O.C., Single Top and Bottom Plates*						
Length of Partition	No. of Studs**	Height of Partition				
		8'-0"	8'-6"	9'-0"	10'-0"	12'-0"
3 '- 0 "	4	25	27	28	31	36
4 '- 0 "	4	27	28	29	32	37
5 '- 0 "	5	33	35	37	40	47
6 '- 0 "	6	40	42	44	48	56
7 '- 0 "	7	47	49	51	56	65
8 '- 0 "	7	48	50	53	57	67
9 '- 0 "	8	55	57	60	65	76
10 '- 0 "	9	61	64	67	73	85
11 '- 0 "	10	68	71	75	81	95
12 '- 0 "	10	69	73	76	83	96
13 '- 0 "	11	76	80	83	91	105
14 '- 0 "	12	83	87	91	99	115
15 '- 0 "	13	89	94	98	107	124
16 '- 0 "	13	91	95	99	108	125
17 '- 0 "	14	97	102	107	116	135
18 '- 0 "	15	104	109	114	124	144
19 '- 0 "	16	111	116	121	132	153
20 '- 0 "	16	112	117	123	133	155
21 '- 0 "	17	119	124	130	141	164
22 '- 0 "	18	125	131	137	149	173
23 '- 0 "	19	132	138	145	157	183
24 '- 0 "	19	133	140	146	159	184
25 '- 0 "	20	140	147	153	167	193
26 '- 0 "	21	147	154	161	175	203
27 '- 0 "	22	153	161	168	183	212
28 '- 0 "	22	155	162	169	184	213
29 '- 0 "	23	161	169	177	192	223
30 '- 0 "	24	168	176	184	200	232
31 '- 0 "	25	175	183	191	208	241
32 '- 0 "	25	176	184	193	209	243
33 '- 0 "	26	183	191	200	217	252
34 '- 0 "	27	189	198	207	225	261
35 '- 0 "	28	196	205	215	233	271
36 '- 0 "	28	197	207	216	235	272
37 '- 0 "	29	204	214	223	243	281
38 '- 0 "	30	211	221	231	251	291
39 '- 0 "	31	217	228	238	259	300
40 '- 0 "	31	219	229	239	260	301

*Add 2/3 board feet (Bd.Ft.). of lumber for each lin.ft. of double top or bottom plate.
** The formula that precedes the chart is first used to calculate the number of studs (i.e. 2.25 studs are rounded to 3 studs) Bd. Ft. figures are determined based on whole studs then rounded to the nearest Bd.Ft. (2.25 is rounded to 2,2.5 to 3 etc.).

CARPENTRY

Cu.M. of Lumber Required for Wood Stud Partitions 50 x 100 mm Studs 400 mm O.C., Single Top and Bottom Plates*							
Length of Partition		No. of Studs**	Height of Partition				
			2.44 M	2.59 M	2.74 M	3.04 M	3.65 M
0.91	M	4	0.059	0.064	0.066	0.073	0.085
1.22	M	4	0.064	0.066	0.068	0.075	0.087
1.52	M	5	0.078	0.083	0.087	0.094	0.111
1.83	M	6	0.094	0.099	0.104	0.113	0.132
2.13	M	7	0.111	0.116	0.120	0.132	0.153
2.44	M	7	0.113	0.118	0.125	0.134	0.158
2.74	M	8	0.130	0.134	0.142	0.153	0.179
3.05	M	9	0.144	0.151	0.158	0.172	0.200
3.35	M	10	0.160	0.167	0.177	0.191	0.224
3.66	M	10	0.163	0.172	0.179	0.196	0.226
3.96	M	11	0.179	0.189	0.196	0.215	0.248
4.27	M	12	0.196	0.205	0.215	0.233	0.271
4.57	M	13	0.210	0.222	0.231	0.252	0.292
4.88	M	13	0.215	0.224	0.233	0.255	0.295
5.18	M	14	0.229	0.241	0.252	0.274	0.318
5.49	M	15	0.245	0.257	0.269	0.292	0.340
5.79	M	16	0.262	0.274	0.285	0.311	0.361
6.10	M	16	0.264	0.276	0.290	0.314	0.366
6.40	M	17	0.281	0.292	0.307	0.333	0.387
6.71	M	18	0.295	0.309	0.323	0.351	0.408
7.01	M	19	0.311	0.325	0.342	0.370	0.432
7.32	M	19	0.314	0.330	0.344	0.375	0.434
7.62	M	20	0.330	0.347	0.361	0.394	0.455
7.92	M	21	0.347	0.363	0.380	0.413	0.479
8.23	M	22	0.361	0.380	0.396	0.432	0.500
8.53	M	22	0.366	0.382	0.399	0.434	0.502
8.84	M	23	0.380	0.399	0.417	0.453	0.526
9.14	M	24	0.396	0.415	0.434	0.472	0.547
9.45	M	25	0.413	0.432	0.450	0.491	0.568
9.75	M	25	0.415	0.434	0.455	0.493	0.573
10.06	M	26	0.432	0.450	0.472	0.512	0.594
10.36	M	27	0.446	0.467	0.488	0.531	0.616
10.67	M	28	0.462	0.483	0.507	0.550	0.639
10.97	M	28	0.465	0.488	0.509	0.554	0.642
11.28	M	29	0.481	0.505	0.526	0.573	0.663
11.58	M	30	0.498	0.521	0.545	0.592	0.686
11.89	M	31	0.512	0.538	0.561	0.611	0.708
12.19	M	31	0.517	0.540	0.564	0.613	0.710

*Add 0.0008 Cu.M of lumber for each .304 lin.M. of double top or bottom plate.
** The formula that precedes the chart is first used to calculate the number of studs (i.e. 2.25 studs are rounded to 3 studs) Cu.M. figures are determined based on whole studs then rounded to the nearest Cu.M. (2.25 is rounded to 2,2.5 to 3 etc.).

Number of Partition Studs Required for Any Spacing

Distance Apart Studs	Multiply Length of Partition by	Add Wood Studs	Distance Apart Studs	Multiply Length of Partition by	Add Wood Studs
12 "	1.00	1	300 mm	1.00	1
16 "	0.75	1	400 mm	0.75	1
20 "	0.60	1	500 mm	0.60	1
24 "	0.50	1	600 mm	0.50	1

Add for top and bottom plates.

Number of Feet of Lumber Required Per Sq.Ft. (Sq.M) of Wood Stud Partition using 2" x 4" (50 x 100mm) Studs Spaces 16" (400 mm) on Centers With Single Top and Bottom Plate

Length Partition in Feet	No. Studs Req'd	Ceiling Height in Feet				Length Partition in Meters	No. Studs Req'd	Ceiling Height in Meters			
		8'-0"	9'-0"	10'-0"	12'-0"			2.44 M	2.74 M	3.04 M	3.65 M
2	3	1.250	1.167	1.13	1.13	0.610	3	13.46	12.56	12.16	12.16
3	3	0.833	0.812	0.80	0.80	0.914	3	8.97	8.74	8.61	8.61
4	4	0.833	0.812	0.80	0.80	1.219	4	8.97	8.74	8.61	8.61
5	5	0.833	0.812	0.80	0.80	1.524	5	8.97	8.74	8.61	8.61
6	6	0.833	0.812	0.80	0.80	1.829	6	8.97	8.74	8.61	8.61
7	6	0.833	0.75	0.75	0.80	2.134	6	8.97	8.07	8.07	8.61
8	7	0.75	0.75	0.75	0.70	2.438	7	8.07	8.07	8.07	7.53
9	8	0.75	0.75	0.75	0.70	2.743	8	8.07	8.07	8.07	7.53
10	9	0.75	0.75	0.75	0.70	3.048	9	8.07	8.07	8.07	7.53
11	9	0.75	0.70	0.70	0.67	3.353	9	8.07	7.53	7.53	7.21
12	10	0.75	0.70	0.70	0.67	3.658	10	8.07	7.53	7.53	7.21
13	11	0.75	0.70	0.70	0.67	3.962	11	8.07	7.53	7.53	7.21
14	12	0.75	0.70	0.70	0.67	4.267	12	8.07	7.53	7.53	7.21
15	12	0.70	0.70	0.70	0.67	4.572	12	7.53	7.53	7.53	7.21
16	13	0.70	0.70	0.70	0.67	4.877	13	7.53	7.53	7.53	7.21
17	14	0.70	0.70	0.70	0.67	5.182	14	7.53	7.53	7.53	7.21
18	15	0.70	0.70	0.67	0.67	5.486	15	7.53	7.53	7.21	7.21
19	15	0.70	0.70	0.67	0.67	5.791	15	7.53	7.53	7.21	7.21
20	16	0.70	0.70	0.67	0.67	6.096	16	7.53	7.53	7.21	7.21
For Double Plate add per Sq.Ft		0.13	0.11	0.10	0.08	For Double Plate add per Sq.M		1.399	1.184	1.076	0.893

For 2" x 8" studs, double the above quantities. For 2" x 6" studs increase above quantities by 50%

Example: Find the number of bd.ft. of lumber required for a stud partition 18'-0" (5.48 m) long and 9'-0" (2.74 m) high. This partition would contain 18'- 0" x 9'- 0" = 162 sq.ft. (5.48 x 2.74 m = 15.04 sq.m). The table gives 0.70 bd.ft lumber of partition. Multiply 162 sq.ft. x 0.70 bd.ft. = 113.40 bd.ft. (15.04 sq.mx7.53 cu m = 113.25 Sq.m)

854

CARPENTRY

Estimating the number of headers and plywood spacers. In a wood exterior wall or an interior partition wall, where an opening is framed, a header built up of two 2" (50.8mm) members with a ½" (12.7mm) spacer to bring the total thickness out to 3½" (88.9mm) for bearing walls is often used. To estimate the number of pieces, check the specifications of the drawings to determine the size, 2x6, 2x8, 2x10, etc (50x150, 50x200, 50x250). The length of the piece can be estimated by adding 6" (152mm) to each side of the opening then times 2 to get the number of lineal feet (lineal meters) of "two by lumber" for the headers. Add up all the lineal footages (lineal meters) of the same size "two by lumber" and divide the length of a piece to get the total number of 2x6's, 2x8's, 2x10's, etc (50x150's, 50x200's, 50x250's). Figure the number of plywood spacers by using the length of the opening plus 6" (152mm) each side time 1' (0.3m) which equals square footage (square meters). Add up all the square footage (square meters) for each opening, then divide by 32 sq.ft. (2.97 sq.m.) per sheet to equal the number of plywood sheets.

For the example to follow, a sample cross-sectional view is shown below.

Sample Header and Plywood Spacer Estimate

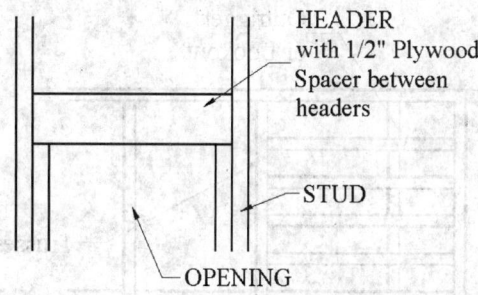

HEADER
with 1/2" Plywood
Spacer between
headers

STUD

OPENING

Given   The example uses 2x12 (40x300) headers and ½"x4'x8' plywood (12.5mm x 1.2m x 2.4m). Assume there are three each 3'x4' (0.9m x 1.2m) windows, three each 3'x7' (0.9m x 2.1m) doors, one each 6'x7' (1.8m x 2.1m) door and one each 8'x4' (2.4m x 1.2m) window.

Number of 2"x12"x16' Headers
    3 windows x (3' + 0.5' + 0.5') = 12' x 2 (for "two by lumber") = 24'
       3 x (0.9m + 0.15m + 0.15m) = 3.65m x 2 = 7.3m
    3 doors x (3' + 0.5' + 0.5') = 12' x 2 (for "two by lumber") = 24'
       3 x (0.9m + 0.15m + 0.15m) = 3.65m x 2 = 7.3m
    1 door x (6' + 0.5' + 0.5') = 7' x 2 (for "two by lumber") = 14'
       1 x (1.8m + 0.15m + 0.15m) = 2.1m x 2 = 4.2m
    1 window x (8' + 0.5' + 0.5') = 9' x 2 (for "two by lumber") = 18'
       1 x (2.4m + 0.15m + 0.15m) = 2.7m x 2 = 5.4m

855

24' + 24' + 14' + 18' = 80' Total lineal feet
7.3m + 7.3m + 4.2m + 5.4m = 21.6m Total lineal meters

80' ÷ 16' lumber length = **5 each of 2"x12"x16' lumber**
24.4m ÷ 4.88m lumber length = 5 each of 50x600x4.88m lumber

Number of ½"x4'x8' Plywood Sheets
 3 windows x (3' + 0.5' + 0.5') x 1' = 12 sq.ft.
  3 x (0.9m + 0.15m + 0.15m) x 0.3m = 1.1 sq.m.
 3 doors x (3' + 0.5' + 0.5') x 1' = 12 sq.ft.
  3 x (0.9m + 0.15m + 0.15m) x 0.3m = 1.1 sq.m.
 1 door x (6' + 0.5' + 0.5') x 1' = 7 sq.ft.
  1 x (1.8m + 0.15m + 0.15m) x 0.3m = 0.6 sq.m.
 1 window x (8' + 0.5' + 0.5') x 1' = 9 sq.ft.
  1 x (2.4m + 0.15m + 0.15m) x 0.3m = 0.8 sq.m.

12sf + 12sf + 7sf + 9sf = 40 sq.ft. Total square feet
1.1sm + 1.1sm + 0.6sm + 0.8sm = 3.7 sq.m. Total square meters

40sf ÷ 32sf plywood area = 1.25 or **2 plywood sheets**
*3.7sm ÷ 2.97sm plywood area = 1.25 or **2 plywood sheets***

Sample Gable End Roof Outrigger or Lookout Estimate

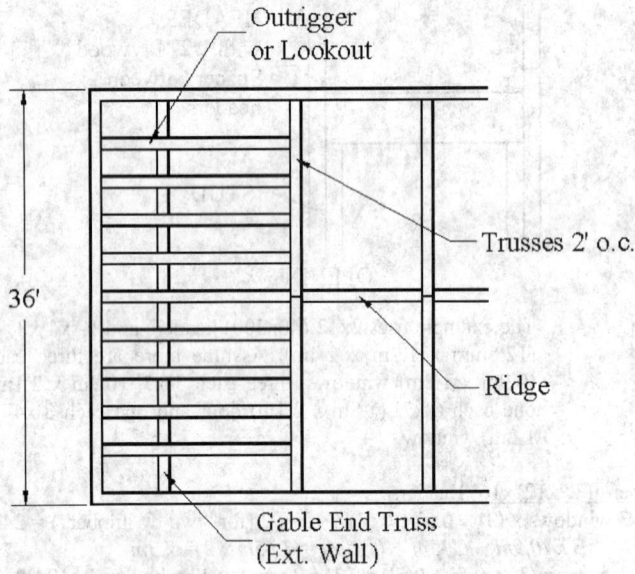

Outrigger
or Lookout

Trusses 2' o.c.

36'

Ridge

Gable End Truss
(Ext. Wall)

Given   An overhang is built with 2x4 (50x100) outriggers using 16'
      (4.9m) long pieces. The overhang is 1' (0.3m) and the outriggers
      are spaced 16" (0.4m) on center. The span of the roof is 36' (11m)
      from overhang to overhang and the slope of the roof is 6 in 12.
      See the drawing above for a plan view. Trusses are spaced at 2'
      (0.6m) on center.

856

CARPENTRY

<u>Number of 2"x4"x16' Outriggers</u>
  36' ÷ 2 = 18' run, or half the span *(11m ÷ 2 = 5.5m run)*

  A 6 in 12 roof factor is 1.118 x run to get length along the roof deck.
  18' x 1.118 = 20.12' ÷ 1.33' + 1 = 16.1 or 17 outriggers per run
  5.5m x 1.118 = 6.15m ÷ 0.4m + 1 = 16.4 or 17 outriggers per run

  Since the outrigger must connect back to the next rafter or truss, the length
    of each outrigger will be 1' + 2' = 3' (0.9m).

  2 runs x 17 outriggers x 3' = 102' Total length of outriggers
  2 x 17 x 0.9m = 31m Total length of outriggers

  102' ÷ 16' lumber length = 6.375 or **7 each of 2"x4"x16' outriggers**
  *31m ÷ 4.88m lumber length = 6.35 or **7 each of 50x100x4.88m outriggers***

Sample Gable End / Vaulted End Wall Framing Estimate

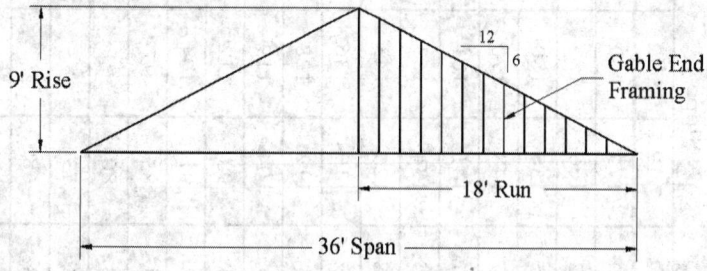

Given   The gable end roof, shown above, where the span is 36' (11m)
             and the slope is 6 in 12. The gable end framing uses 2"x4"x16'
             (50x100x4.88m) is spaced at 16" (0.4m) on center.

<u>Number of 2"x4"x16' Gable End Framing Lumber</u>
  The tables on the following pages gives the total length of lumber.

  121' ÷ 16' = 7.56 or **8 each of 2"x4"x16' lumber**
  *36.9m ÷ 4.88m = 7.56 or **8 each of 50x100x4.88m lumber***

Gable End Framing at 16" o.c. Spacing - Total Length (feet) of Vertical Studs

Span (ft)	Slope									
	3/12	4/12	5/12	6/12	7/12	8/12	9/12	10/12	11/12	12/12
40	75	100	125	150	175	200	225	250	275	300
39	71	95	118	142	166	189	213	237	260	284
38	68	90	113	135	158	180	203	225	248	270
37	64	85	106	127	148	170	191	212	235	257
36	61	81	101	121	141	161	182	202	222	242
35	57	76	96	115	134	153	172	191	210	230
34	54	72	90	108	125	143	161	179	197	217
33	51	68	85	102	119	136	153	170	187	204
32	48	64	80	96	112	128	144	160	176	192
31	45	60	75	89	104	119	134	149	164	179
30	42	56	70	84	98	112	126	140	154	168
29	39	52	65	78	91	104	117	130	145	158
28	37	49	61	73	85	97	110	122	134	146
27	34	45	57	68	80	91	102	114	125	137
26	31	42	52	63	73	83	94	104	115	127
25	29	39	49	58	68	78	87	97	107	117
24	27	36	45	54	63	72	81	90	99	108

Gable End Framing at 40.6mm o.c. Spacing - Total Length (meters) of Vertical Studs										
Span (m)	Slope									
	3/12	4/12	5/12	6/12	7/12	8/12	9/12	10/12	11/12	12/12
12.2	22.9	30.5	38.1	45.7	53.3	61.0	68.6	76.2	83.8	91.4
11.9	21.6	29.0	36.0	43.3	50.6	57.6	64.9	72.2	79.2	86.6
11.6	20.7	27.4	34.4	41.1	48.2	54.9	61.9	68.6	75.6	82.3
11.3	19.5	25.9	32.3	38.7	45.1	51.8	58.2	64.6	71.6	78.3
11.0	18.6	24.7	30.8	36.9	43.0	49.1	55.5	61.6	67.7	73.8
10.7	17.4	23.2	29.3	35.1	40.8	46.6	52.4	58.2	64.0	70.1
10.4	16.5	21.9	27.4	32.9	38.1	43.6	49.1	54.6	60.0	66.1
10.1	15.5	20.7	25.9	31.1	36.3	41.5	46.6	51.8	57.0	62.2
9.8	14.6	19.5	24.4	29.3	34.1	39.0	43.9	48.8	53.6	58.5
9.4	13.7	18.3	22.9	27.1	31.7	36.3	40.8	45.4	50.0	54.6
9.1	12.8	17.1	21.3	25.6	29.9	34.1	38.4	42.7	46.9	51.2
8.8	11.9	15.8	19.8	23.8	27.7	31.7	35.7	39.6	44.2	48.2
8.5	11.3	14.9	18.6	22.3	25.9	29.6	33.5	37.2	40.8	44.5
8.2	10.4	13.7	17.4	20.7	24.4	27.7	31.1	34.7	38.1	41.8
7.9	9.4	12.8	15.8	19.2	22.3	25.3	28.7	31.7	35.1	38.7
7.6	8.8	11.9	14.9	17.7	20.7	23.8	26.5	29.6	32.6	35.7
7.3	8.2	11.0	13.7	16.5	19.2	21.9	24.7	27.4	30.2	32.9

Sample Girders and Joist Hanger Estimate

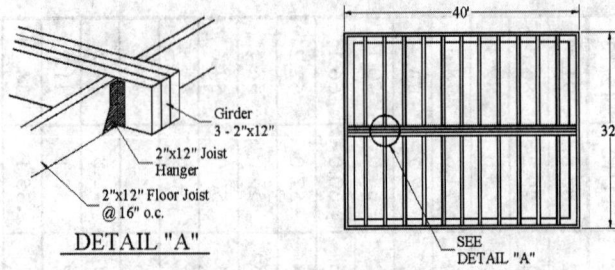

Girder
3 - 2"x12"

2"x12" Joist
Hanger

2"x12" Floor Joist
@ 16" o.c.

DETAIL "A"

40'

32'

SEE
DETAIL "A"

Given   Use 2"x12"x16' (50x300x4.88m) lumber to form the girders,
shown above, where three are nailed together to extend across the
40' (12.2m) length with 2"x12" (50x300) joist hangers spaced 16"
(0.4m) on center.

Number of 2"x12"x16' Girders
40' length ÷ 16' = 2.5 x 3 layers = 7.5 or **8 each of lumber**
*12.2m length ÷ 4.88m = 2.5 x 3 layers = 7.5 or **8 each of lumber**

Number of Joist Hangers or Joists
40' ÷ 1.33 + 1 = 31 x 2 sides = **62 each joist hangers and joists**
*12.2m ÷ 0.4m + 1 = 31 x 2 sides = **62 each joist hangers and joists**

Quantity of Plain End Softwood Flooring Required per 100 Sq.Ft. (9.29 SqM)				
Measured Size Inches	Actual Size Inches	Add for Width Size Inches	Bd.Ft. Req'd per 100 Sq.Ft. Surface	Lbs. Per 1,000 Ft
1 x 3	3/4 x 2-3/8	27%	132	1800
1 x 4	3/4 x 3-1/4	23%	128	1900
(metric)				
Measured Size mm	Actual Size mm	Add for Width Size Inches	Cu.M Req'd per 10.00 Sq.M. Surface	Kg. Per 100 M
25 x 75	18.75 x 59.37	27%	3.46	267.87
25 x 100	18.75 x 81.25	23%	3.57	282.76

The above quantities include 5% for end cutting and waste.

CARPENTRY

Quantity of End Matched Softwood Flooring Required Per 100 Sq.Ft. (9.29 Sq.M) of Floor

Measured Size Inches	Actual Size Inches	Add for Width Size Inches	Bd.Ft. Req'd per 100 Sq.Ft. Surface	Lbs. Per 1,000 Ft
1 x 3	13/16 x 2-3/8	27%	130	1800
1 x 4	13/16 x 3-1/4	23%	126	1900
(metric)				
Measured Size mm	Actual Size mm	Add for Width Size Inches	Cu.M Req'd per 10.00 Sq.M. Surface	Kg. Per 100 M
25 x 75	20.31 x 59.37	27%	3.51	267.87
25 x 100	20.31 x 81.25	23%	3.62	282.76

The above quantities include 3% for end cutting and waste.

Quantity of Square Edged (S4S) Boards Required Per 100 Sq.Ft (9.29 Sq.M). of Surface

Measured Size Inches	Actual Size Inches	Add for Width Size Inches	Bd.Ft. Req'd per 100 Sq.Ft. Surface	Lbs. Per 1,000 Ft
1 x 4	3/4 x 3-1/2	14%	119	2300
1 x 6	3/4 x 5-1/2	9%	114	2300
1 x 8	3/4 x 7-1/4	10%	115	2300
1 x 10	3/4 x 9-1/4	8%	113	2300
1 x 12	3/4 x 11-1/4	7%	112	2400
(metric)				
Measured Size mm	Actual Size mm	Add for Width Size Inches	Cu.M Req'd per 10.00 Sq.M. Surface	Kg. Per 100 M
25 x 100	18.75 x 87.50	14%	3.84	342.28
25 x 150	18.75 x 137.50	9%	4.00	342.28
25 x 200	18.75 x 181.25	10%	3.97	342.28
25 x 250	18.75 x 231.25	27%	4.04	342.28
25 x 300	18.75 x 281.25	23%	4.08	357.17

Lineal Foot Table of Board Measure Number of Board Feet per Liner Foot of Size Indicated					
Lumber Size	Bd.Ft.	Lumber Size	Bd.Ft.	Lumber Size	Bd.Ft.
2 " x 4 " =	0.667	4 " x 4 " =	1.333	8 " x 14 " =	9.333
2 " x 6 " =	1.000	4 " x 6 " =	2.000	8 " x 16 " =	10.667
2 " x 8 " =	1.333	4 " x 8 " =	2.667	10 " x 10 " =	8.333
2 " x 10 " =	1.667	4 " x 10 " =	3.333	10 " x 12 " =	10.000
2 " x 12 " =	2.000	4 " x 12 " =	4.000	10 " x 14 " =	11.667
2 " x 14 " =	2.333	4 " x 14 " =	4.667	10 " x 16 " =	13.333
2 " x 16 " =	2.667	4 " x 16 " =	5.333	10 " x 18 " =	15.000
2.5 " x 12 " =	2.500	6 " x 6 " =	3.000	12 " x 12 " =	12.000
2.5 " x 14 " =	2.917	6 " x 8 " =	4.000	12 " x 14 " =	14.000
2.5 " x 16 " =	3.333	6 " x 10 " =	5.000	12 " x 16 " =	16.000
3 " x 6 " =	1.500	6 " x 12 " =	6.000	12 " x 18 " =	18.000
3 " x 8 " =	2.000	6 " x 14 " =	7.000	14 " x 14 " =	16.333
3 " x 10 " =	2.500	6 " x 16 " =	8.000	14 " x 16 " =	18.667
3 " x 12 " =	3.000	8 " x 8 " =	5.333	14 " x 18 " =	21.000
3 " x 14 " =	3.500	8 " x 10 " =	6.667	16 " x 16 " =	21.333
3 " x 16 " =	4.000	8 " x 12 " =	8.000	16 " x 18 " =	24.000

Metric

Lineal Meter Table of Cubic Feet Measure Number of Cubic Feet per Liner Meter of Size Indicated					
Lumber Size	Cu.M	Lumber Size	Cu.M	Lumber Size	Cu.M
50 x 100 mm =	0.0016	100 x 100 mm =	0.0031	200 x 350 mm =	0.0220
50 x 150 mm =	0.0024	100 x 150 mm =	0.0047	200 x 400 mm =	0.0252
50 x 200 mm =	0.0031	100 x 200 mm =	0.0063	250 x 250 mm =	0.0197
50 x 250 mm =	0.0039	100 x 250 mm =	0.0079	250 x 300 mm =	0.0236
50 x 300 mm =	0.0047	100 x 300 mm =	0.0094	250 x 350 mm =	0.0275
50 x 350 mm =	0.0055	100 x 350 mm =	0.0110	250 x 400 mm =	0.0314
50 x 400 mm =	0.0063	100 x 400 mm =	0.0126	250 x 450 mm =	0.0354
63 x 300 mm =	0.0059	150 x 150 mm =	0.0071	300 x 300 mm =	0.0283
63 x 350 mm =	0.0069	150 x 200 mm =	0.0094	300 x 350 mm =	0.0330
63 x 400 mm =	0.0079	150 x 250 mm =	0.0118	300 x 400 mm =	0.0377
75 x 150 mm =	0.0035	150 x 300 mm =	0.0142	300 x 450 mm =	0.0425
75 x 200 mm =	0.0047	150 x 350 mm =	0.0165	350 x 350 mm =	0.0385
75 x 250 mm =	0.0059	150 x 400 mm =	0.0189	350 x 400 mm =	0.0440
75 x 300 mm =	0.0071	200 x 200 mm =	0.0126	350 x 450 mm =	0.0495
75 x 350 mm =	0.0083	200 x 250 mm =	0.0157	400 x 400 mm =	0.0503
75 x 400 mm =	0.0094	200 x 300 mm =	0.0189	400 x 450 mm =	0.0566

Note: Soft metric conversion

CARPENTRY

Lengths of Common, Hip, and Valley Rafters per 12" (300 mm) of Run						Metric	
1	2	3	4	5*	6**	3	4
Roof Pitch	Rise and Run or Cut	Length in Inches Common Rafter per 12" of Run		% Increase length common Rafter over Run	Length in Inches Hip or Valley Rafters	Length in mm Common Rafter per 12" of Run	
1/12	2 & 12	12.165	0.014	1.014	17.088	304.125	0.350
1/8	3 & 12	12.369	0.031	1.031	17.233	309.225	0.775
1/6	4 & 12	12.649	0.054	1.054	17.433	316.225	1.350
5/24	5 & 12	13.000	0.083	1.083	17.692	325.000	2.075
1/4	6 & 12	13.417	0.118	1.118	18.000	335.425	2.950
7/24	7 & 12	13.892	0.158	1.158	18.358	347.300	3.950
1/3	8 & 12	14.422	0.202	1.202	18.762	360.550	5.050
3/8	9 & 12	15.000	0.250	1.250	19.209	375.000	6.250
5/12	10 & 12	15.620	0.302	1.302	19.698	390.500	7.550
11/24	11 & 12	16.279	0.357	1.357	20.224	406.975	8.925
1/2	12 & 12	16.971	0.413	1.413	20.785	424.275	10.325
13/24	13 & 12	17.692	0.474	1.474	21.378	442.300	11.850
7/12	14 & 12	18.439	0.537	1.537	22.000	460.975	13.425
5/8	15 & 12	19.210	0.601	1.601	22.649	480.250	15.025
2/3	16 & 12	20.000	0.667	1.667	23.324	500.000	16.675
17/24	17 & 12	20.809	0.734	1.734	24.021	520.225	18.350
3/4	18 & 12	21.633	0.803	1.803	24.739	540.825	20.075
19/24	19 & 12	22.500	0.875	1.875	25.475	562.500	21.875
5/6	20 & 12	23.375	0.948	1.948	26.230	584.375	23.700
7/8	21 & 12	24.125	1.010	2.010	27.000	603.125	25.250
11/24	22 & 12	25.000	1.083	2.083	27.785	625.000	27.075
11/12	23 & 12	26.000	1.167	2.167	28.583	650.000	29.175
Full	24 & 12	26.875	1.240	2.240	29.394	671.875	31.000

*Use figures in this column to obtain area of roof for any pitch. See explanation below.

**Figures in last column are length of hip and valley rafters in inches for each 12 inches of common rafter run.

To Obtain Area of Roofs For Any Pitch. To obtain the number of sq.ft. (sq.m) of roof area for any pitch, take the entire flat or horizontal area of the roof and multiply by the figure given in the fifth column (*), and the result will be the area of the roof. Always bear in mind that the width of any overhanging cornice must be added to the building area to obtain the total area to be covered.

Example: Find the area of a roof 26'-0" x 42'-0" (7.92 x 12.80 m), with a 1'-0" (.31 m) overhanging cornice and a 1/4 pitch. To obtain roof area,

26'-0" + 1'-0" + 1'-0" = 28'-0" (7.92+0.31+.031 m = 8.54 m) width
42'-0" + 1'-0" + 1'-0" = 44'-0" (12.80 + .31 + .31 = 13.42 m) length
28 x 44 = 1,232 sq.ft. (8.5 x 13.42 = 114.61 sq.m) flat or horizontal area.

Area at 1/4 pitch: 1,232 x 1.118 = 1377.38 or 1,378 sq.ft. (114.61 x 1.118 = 128.13 or 128 sq.m) roof surface.
Add allowance for overhang on dormer roofs and sides.

Size	Length, liner feet					
Inches	10'	12'	14'	16'	18'	20'
1 " x 2 "	1.67	2.00	2.33	2.67	3.00	3.33
1 " x 3 "	2.50	3.00	3.50	4.00	4.50	5.00
1 " x 4 "	3.33	4.00	4.67	5.33	6.00	6.67
1 " x 6 "	5.00	6.00	7.00	8.00	9.00	10.00
1 " x 8 "	6.67	8.00	9.33	10.67	12.00	13.33
1 " x 10 "	8.33	10.00	11.67	13.33	15.00	16.67
1 " x 12 "	10.00	12.00	14.00	16.00	18.00	20.00
1 1/4 " x 4 "	4.17	5.00	5.83	6.67	7.50	8.33
1 1/4 " x 6 "	6.25	7.50	8.75	10.00	11.25	12.50
1 1/4 " x 8 "	8.33	10.00	11.67	13.33	15.00	16.67
1 1/4 " x 10 "	10.42	12.50	14.58	16.67	18.75	20.83
1 1/4 " x 12 "	12.50	15.00	17.50	20.00	22.50	25.00
1 1/2 " x 4 "	5.00	6.00	7.00	8.00	9.00	10.00
1 1/2 " x 6 "	7.50	9.00	10.50	12.00	13.50	15.00
1 1/2 " x 8 "	10.00	12.00	14.00	16.00	18.00	20.00
1 1/2 " x 10 "	12.50	15.00	17.50	20.00	22.50	25.00
1 1/2 " x 12 "	15.00	18.00	21.00	24.00	27.00	30.00
2 " x 2 "	3.33	4.00	4.67	5.33	6.00	6.67
2 " x 3 "	5.00	6.00	7.00	8.00	9.00	10.00
2 " x 4 "	6.67	8.00	9.33	10.67	12.00	13.33
2 " x 6 "	10.00	12.00	14.00	16.00	18.00	20.00
2 " x 8 "	13.33	16.00	18.67	21.33	24.00	26.67
2 " x 10 "	16.67	20.00	23.33	26.67	30.00	33.33
2 " x 12 "	20.00	24.00	28.00	32.00	36.00	40.00
2 " x 14 "	23.33	28.00	32.67	37.33	42.00	46.67
3 " x 4 "	10.00	12.00	14.00	16.00	18.00	20.00
3 " x 6 "	15.00	18.00	21.00	24.00	27.00	30.00
3 " x 8 "	20.00	24.00	28.00	32.00	36.00	40.00
3 " x 10 "	25.00	30.00	35.00	40.00	45.00	50.00
3 " x 12 "	30.00	36.00	42.00	48.00	54.00	60.00
3 " x 14 "	35.00	42.00	49.00	56.00	63.00	70.00
4 " x 4 "	13.33	16.00	18.67	21.33	24.00	26.67
4 " x 6 "	20.00	24.00	28.00	32.00	36.00	40.00
4 " x 8 "	26.67	32.00	37.33	42.67	48.00	53.33
4 " x 10 "	33.33	40.00	46.67	53.33	60.00	66.67
4 " x 12 "	40.00	48.00	56.00	64.00	72.00	80.00
4 " x 14 "	46.67	56.00	65.33	74.67	84.00	93.33
6 " x 6 "	30.00	36.00	42.00	48.00	54.00	60.00
6 " x 8 "	40.00	48.00	56.00	64.00	72.00	80.00
6 " x 10 "	50.00	60.00	70.00	80.00	90.00	100.00

Board Feet Content of Joists, Scantlings, and Timbers
of Given Lineal Foot Lengths

CARPENTRY

Board Feet Content of Joists, Scantlings, and Timbers of Given Lineal Foot Lengths							
Size		Length, liner feet					
Inches		10'	12'	14'	16'	18'	20'
6 " x	12 "	60.00	72.00	84.00	96.00	108.00	120.00
6 " x	14 "	70.00	84.00	98.00	112.00	126.00	140.00
6 " x	16 "	80.00	96.00	112.00	128.00	144.00	160.00
8 " x	8 "	53.33	64.00	74.67	85.33	96.00	106.67
8 " x	10 "	67.67	80.00	93.33	106.67	120.00	133.33
8 " x	12 "	80.00	96.00	112.00	128.00	144.00	160.00
8 " x	14 "	93.33	112.00	130.67	149.33	168.00	186.67
8 " x	16 "	106.67	128.00	149.33	170.67	192.00	213.33
10 " x	10 "	83.33	100.00	116.67	133.33	150.00	167.67
10 " x	12 "	100.00	120.00	140.00	160.00	180.00	200.00
10 " x	14 "	116.67	140.00	163.33	186.67	210.00	233.33
10 " x	16 "	133.33	160.00	186.67	213.33	240.00	266.67
12 " x	12 "	120.00	144.00	168.00	192.00	216.00	240.00
12 " x	14 "	140.00	168.00	196.00	224.00	252.00	280.00
12 " x	16 "	160.00	192.00	224.00	256.00	288.00	320.00
14 " x	14 "	163.33	196.00	228.67	261.33	294.00	326.67
14 " x	16 "	186.67	224.00	261.33	298.67	336.00	373.33
14 " x	18 "	210.00	252.00	294.00	336.00	378.00	420.00
14 " x	20 "	233.33	280.00	326.67	373.33	420.00	466.67
16 " x	16 "	213.33	256.00	298.67	341.33	384.00	426.67
16 " x	18 "	240.00	288.00	336.00	384.00	432.00	480.00
16 " x	20 "	266.67	320.00	373.33	426.67	480.00	533.33
18 " x	18 "	270.00	324.00	378.00	432.00	486.00	540.00
18 " x	20 "	300.00	360.00	420.00	480.00	540.00	600.00
20 " x	20 "	333.33	400.00	466.67	533.33	600.00	666.67

Metric							
Cubic Meters Content of Joists, Scantlings, and Timbers of Given Lineal Meter Lengths							
Size		Length, liner Meter					
Inches		3.04 M	3.65 M	4.27 M	4.87 M	5.48 M	6.09 M
25 x	50 mm	0.004	0.005	0.005	0.006	0.007	0.008
25 x	75 mm	0.006	0.007	0.008	0.009	0.011	0.012
25 x	100 mm	0.008	0.009	0.011	0.013	0.014	0.016
25 x	150 mm	0.012	0.014	0.017	0.019	0.021	0.024
25 x	200 mm	0.016	0.019	0.022	0.025	0.028	0.031
25 x	250 mm	0.020	0.024	0.028	0.031	0.035	0.039
25 x	300 mm	0.024	0.028	0.033	0.038	0.042	0.047
31.25 x	100 mm	0.010	0.012	0.014	0.016	0.018	0.020
31.25 x	150 mm	0.015	0.018	0.021	0.024	0.027	0.029
31.25 x	200 mm	0.020	0.024	0.028	0.031	0.035	0.039

865

Metric						
Cubic Meters Content of Joists, Scantlings, and Timbers						
of Given Lineal Meter Lengths						
Size		Length, liner Meter					
Inches		3.04 M	3.65 M	4.27 M	4.87 M	5.48 M	6.09 M
31.25 x 250 mm	0.025	0.029	0.034	0.039	0.044	0.049
31.25 x 300 mm	0.029	0.035	0.041	0.047	0.053	0.059
37.5 x 100 mm	0.012	0.014	0.017	0.019	0.021	0.024
37.5 x 150 mm	0.018	0.021	0.025	0.028	0.032	0.035
37.5 x 200 mm	0.024	0.028	0.033	0.038	0.042	0.047
37.5 x 250 mm	0.029	0.035	0.041	0.047	0.053	0.059
37.5 x 300 mm	0.035	0.042	0.050	0.057	0.064	0.071
50 x 50 mm	0.008	0.009	0.011	0.013	0.014	0.016
50 x 75 mm	0.012	0.014	0.017	0.019	0.021	0.024
50 x 100 mm	0.016	0.019	0.022	0.025	0.028	0.031
50 x 150 mm	0.024	0.028	0.033	0.038	0.042	0.047
50 x 200 mm	0.031	0.038	0.044	0.050	0.057	0.063
50 x 250 mm	0.039	0.047	0.055	0.063	0.071	0.079
50 x 300 mm	0.047	0.057	0.066	0.075	0.085	0.094
50 x 350 mm	0.055	0.066	0.077	0.088	0.099	0.110
75 x 100 mm	0.024	0.028	0.033	0.038	0.042	0.047
75 x 150 mm	0.035	0.042	0.050	0.057	0.064	0.071
75 x 200 mm	0.047	0.057	0.066	0.075	0.085	0.094
75 x 250 mm	0.059	0.071	0.083	0.094	0.106	0.118
75 x 300 mm	0.071	0.085	0.099	0.113	0.127	0.142
75 x 350 mm	0.083	0.099	0.116	0.132	0.149	0.165
100 x 100 mm	0.031	0.038	0.044	0.050	0.057	0.063
100 x 150 mm	0.047	0.057	0.066	0.075	0.085	0.094
100 x 200 mm	0.063	0.075	0.088	0.101	0.113	0.126
100 x 250 mm	0.079	0.094	0.110	0.126	0.142	0.157
100 x 300 mm	0.094	0.113	0.132	0.151	0.170	0.189
100 x 350 mm	0.110	0.132	0.154	0.176	0.198	0.220
150 x 150 mm	0.071	0.085	0.099	0.113	0.127	0.142
150 x 200 mm	0.094	0.113	0.132	0.151	0.170	0.189
150 x 250 mm	0.118	0.142	0.165	0.189	0.212	0.236
150 x 300 mm	0.142	0.170	0.198	0.226	0.255	0.283
150 x 350 mm	0.165	0.198	0.231	0.264	0.297	0.330
150 x 400 mm	0.189	0.226	0.264	0.302	0.340	0.377
200 x 200 mm	0.126	0.151	0.176	0.201	0.226	0.252
200 x 250 mm	0.160	0.189	0.220	0.252	0.283	0.314
200 x 300 mm	0.189	0.226	0.264	0.302	0.340	0.377
200 x 350 mm	0.220	0.264	0.308	0.352	0.396	0.440
200 x 400 mm	0.252	0.302	0.352	0.403	0.453	0.503
250 x 250 mm	0.197	0.236	0.275	0.314	0.354	0.395
250 x 300 mm	0.236	0.283	0.330	0.377	0.425	0.472

CARPENTRY

Metric						
Cubic Meters Content of Joists, Scantlings, and Timbers of Given Lineal Meter Lengths						
Size	Length, liner Meter					
Inches	3.04 M	3.65 M	4.27 M	4.87 M	5.48 M	6.09 M
250 x 350 mm	0.275	0.330	0.385	0.440	0.495	0.550
250 x 400 mm	0.314	0.377	0.440	0.503	0.566	0.629
300 x 300 mm	0.283	0.340	0.396	0.453	0.509	0.566
300 x 350 mm	0.330	0.396	0.462	0.528	0.594	0.660
300 x 400 mm	0.377	0.453	0.528	0.604	0.679	0.755
350 x 350 mm	0.385	0.462	0.539	0.616	0.693	0.770
350 x 400 mm	0.440	0.528	0.616	0.704	0.792	0.880
350 x 450 mm	0.495	0.594	0.693	0.792	0.892	0.991
350 x 500 mm	0.550	0.660	0.770	0.880	0.991	1.101
400 x 400 mm	0.503	0.604	0.704	0.805	0.906	1.006
400 x 450 mm	0.566	0.679	0.792	0.906	1.019	1.132
400 x 500 mm	0.629	0.755	0.880	1.006	1.132	1.258
450 x 450 mm	0.637	0.764	0.892	1.019	1.146	1.274
450 x 500 mm	0.708	0.849	0.991	1.132	1.274	1.415
500 x 500 mm	0.786	0.943	1.101	1.258	1.415	1.572

Above from Bd.ft. to Cu.M is based on soft conversion

Nails Required for Carpentry Work

The following tables indicates the number of nails in lbs. (kg) for the various kinds of lumber per 1,000 b.f. (2..36 cu.m), or per 1,000 shingles and lath, or per square (100 sq.ft., 9.29 sq.m) of asphalt slate surfaced shingle, with the number of nails added for loss of material on account of lap or matching of shiplap, flooring, ceiling, and siding of the various widths. The table gives the sizes generally used for certain purposes with the nailing space 16" (400 mm) on centers, and 1 or 2 nails per board for each nailing space.

Splitless Wood Siding Nail Specification

Splitless siding nails have thin shanks and blunt points to reduce splitting. For greater holding power, nails with ring threaded or spiral threaded shanks are suggested. Use only Type 316 Stainless Steel nails for salt water coastal areas.

\multicolumn{8}{c}{Splitless Wood Siding Nail Specifications (Type 304 SS)}							
Size	Gauge	Head Inch	Length Inch	Wt. Lbs	Head mm	Length mm	Wt. Kg
3 d	14	5/32	1 1/4	495	3.91	31.25	224.53
4 d	14	5/32	1 1/2	398	3.91	37.50	180.53
5 d	14	5/32	1 3/4	354	3.91	43.75	160.57
6 d	13	5/32	2	245	3.91	50.00	111.13
7 d	13	5/32	2 1/4	215	3.91	56.25	97.52
8 d	13	5/32	2 1/2	196	3.91	62.50	88.91
10 d	12	7/32	3	120	5.47	75.00	54.43
12 d	12	7/32	3 1/2	110	5.47	87.50	49.90
16 d	11	1/4	3 1/2	88	6.25	87.50	39.92
20 d	10	1/4	4	74	6.25	100.00	33.57

\multicolumn{4}{c}{Recommended Sizes and Quantities Commonly Used}			
Siding Inch	Nail Size	Nail per 1,000 Bd.Ft.	Nail Lbs per 1,000 Bd.Ft.
1/2 " x 4 "	6d	2,280	8
1/2 " x 6 "	7d	1,520	6 1/4
3/4 " x 8 "	8d	1,140	6
3/4 " x 10 "	8d	912	5
3/4 " x 12 "	8d	760	4
Siding Inch	Nail Size	Nail per 3.00 Cu. M.	Nail Lbs per 3.00 Bd.Ft.
12.50 x 100 mm	6d	2,900	5
12.50 x 150 mm	7d	1,933	4
18.75 x 200 mm	8d	1,450	3
18.75 x 250 mm	8d	1,160	3
18.75 x 300 mm	8d	967	2

CARPENTRY

	Recommended Nail Sizes and Quantities Commonly Used				
	12" (300 mm) Horizontal Siding				
Size	Lbs per 1,000 Sq.Ft.	Nails per Sq.Ft	Kg per 100 Sq.M.	Nails per Sq.M	Size
6d	6	1	0.025	11	6d
8d	7	1	0.029	11	8d
10d	10	1	0.042	11	10d
Panel Siding					
6d	17	3	0.072	33	6d
8d	21	3	0.088	33	8d
10d	30	3	0.126	33	10d

Nails are the most common mechanical fasteners used in wood construction. There are many types, sizes and forms of nails manufactured today. The nail fasteners are manufactured singularly for the hand hammer as well as in reels for the automatic "gun" fastener installation.

The diameters of various penny or gauge sizes of bright common nails are indicated below. The penny size designation should be used cautiously. International nail producers sometimes do not adhere to the dimensions indicated in the chart. Thus, penny sizes, although still widely used, are becoming obsolete in the international market. Specifying nail size by length and diameter is recommended together with the penny size, if used.

						Bright Common Nail Specifications					
Size	Gauge	Head Inch	Length Inch	Dia. Inch	Nails per Lb	Gauge	Head mm	Length mm	Dia. mm	Nails per Kg
2d	15	3/16	1	0.072	807	11 1/2	7.03	25.4	1.83	366
3d	14	3/16	1 1/4	0.083	473	11 1/2	7.03	31.8	2.11	215
4d	12	1/4	1 1/2	0.109	228	11 1/2	7.03	38.1	2.77	103
5d	12	1/4	1 3/4	0.109	193	11 1/2	7.03	44.5	2.77	88
6d	11 1/2	9/32	2	0.113	181	11 1/2	7.03	50.8	2.87	82
8d	10 1/4	9/32	2 1/2	0.131	106	10 1/4	7.03	63.5	3.33	48
10d	9	9/32	3	0.148	69	9	7.03	76.2	3.76	31
12d	9	5/16	3 1/4	0.148	73	9	7.81	82.6	3.76	33
16d	8	5/16	3 1/2	0.162	49	8	7.81	88.9	4.11	22
20d	6	3/8	4	0.192	31	6	9.38	101.6	4.88	14
30d	5	3/8	4 1/2	0.207	30	5	9.38	114.3	5.26	14
40d	4	3/8	5	0.225	23	4	9.38	127.0	5.72	10
50d	3	3/8	5 1/2	0.244	21	3	9.38	139.7	6.20	10
60d	2	3/8	6	0.262	22	2	9.38	152.4	6.65	10

Bright box nails are generally of the same length but slightly smaller diameter, see table below.

Bright Box Nail Specifications										
Size	Gauge	Head Inch	Length Inch	Dia. Inch	Nails per lbs	Gauge	Head mm	Length mm	Dia. mm	Nails per Kg
3d	14 1/2	7/32	1 1/4	0.076	491	14 1/2	5.47	31.8	1.93	223
4d	14	7/32	1 1/2	0.08	392	14	5.47	38.1	2.03	178
5d	14	7/32	1 3/4	0.08	293	14	5.47	44.5	2.03	133
6d	12 1/2	17/64	2	0.098	194	12 1/2	6.64	50.8	2.49	88
7d	12 1/2	17/64	2 1/4	0.098	172	12 1/2	6.64	57.2	2.49	78
8d	11 1/2	19/64	2 1/2	0.113	132	11 1/2	7.42	63.5	2.87	60
10d	10 1/2	19/64	3	0.128	103	10 1/2	7.42	76.2	3.25	47
16d	10	5/16	3 1/2	0.135	57	10	7.81	88.9	3.43	26
20d	9	3/8	4	0.148	35	9	9.38	101.6	3.76	16

Recommended Sizes and Quantities Nails Commonly used			
Siding Inch	Nail Size	Nails per 1,000 Bd.Ft.	Nails Lbs per 1,000 Bd.Ft.
1/4 " x 4 "	6d	2,280	11 1/2
1/2 " x 6 "	7d	1,520	9
3/4 " x 8 "	8d	1,140	9 1/2
3/4 " x 10 "	8d	912	7 1/2
3/4 " x 12 "	8d	760	6 1/4
Siding mm	Nail Size	Nails per 3 Cu.M.	Nails Kg per 3 Cu.M.
6.25 x 100 mm	6d	2,900	14.63
12.50 x 150 mm	7d	1,933	11.45
18.75 x 200 mm	8d	1,450	12.08
18.75 x 250 mm	8d	1,160	9.54
18.75 x 300 mm	8d	967	7.95

Note: Use one size larger over insulation

CARPENTRY

Cedar Shake Siding Face Nail Specification

Size	Gauge	Head Inch	Length Inch	Nail per Lb	Gauge	Head mm	Length mm	Nail per Kg
3d	14	1/8	1 1/4	402	14	3.18	31.75	182
4d	14	1/8	1 1/2	402	14	3.18	38.10	182
5d	14	1/8	1 3/4	346	14	3.18	44.45	157
6d	14	1/8	2	310	14	3.18	50.80	141
6d*	12 1/2	3/16	2	209	12 1/2	4.76	50.80	95
8d	13	9/64	2 1/2	196	13	3.57	63.50	89

* For use over high density fiberboard sheeting.

Recommended Sizes and Quantities Commonly Used

Shingles Inch	Size Inch	Nails per Square	Lbs per square	Shingles mm	Size	Nails per 10 Sq.M	Kg per 10 Sq.M
16	2	360	1+	400	50	388	0.045
18	2 1/4	310	1	450	56 1/4	334	0.045

Use only corrosive resistance nails

Casing nails are designed for the secure application of window and door frames, cornices and exterior molding. Normally eliminates need for countersinking if the nail is double dipped in molten zinc.

Casing Nail Specification

Size	Length Inch	Gauge	Head Inch	Nails per Lb	Size mm	Gauge	Head mm	Nails per Kg
6d	2	12 1/2	9/64	212	50.0	12 1/2	3.516	96
8d	2 1/2	11 1/2	9/64	131	62.5	11 1/2	3.516	59
10d	3	10	5/32	85	75.0	10	3.906	39
16d	3 1/2	10	5/32	67	87.5	10	3.906	30
20d	4	9	3/16	48	100.0	9	4.688	22

Finishing nails are designed for trim locations. Brad head sets easily and insures neat finished appearance.

871

Finish Nail Specification								
Size	Length Inch	Gauge	Head Inch	Nails per Lb	Size mm	Gauge	Head mm	Nails per Kg
8d	2 1/2	13	9/64	196	62.5	13	3.516	89
6d	2	14	1/8	110	50.0	14	3.125	50
10d	3	11 1/2	5/32	109	75.0	11 1/2	3.906	49
16d	3 1/2	11 1/2	5/32	95	87.5	11 1/2	3.906	43

Hand-driven nails for roofing–Asphalt and Fiberglass Shingle Nails. These nails are designed for the secure application of asphalt and fiberglass shingles on new construction and re-roofing projects. Ring and spiral shanks allow for extra holding power, especially in high wind-hurricane areas. Roofing nails should be long enough to penetrate ¾" into the wood deck lumber, or completely thru the plywood decking.

Asphalt and Fiberglass Roofing Nail Specifications								
Size	Length Inch	Gauge	Head Inch	Nails per Lb	Size mm	Gauge	Head mm	Nails per Kg
X	7/8	11	3/8	272	21.9	11	9.375	123
X	1	11	3/8	250	25.0	11	9.375	113
3d	1 1/4	11	3/8	202	31.3	11	9.375	92
4d	1 1/2	11	3/8	180	37.5	11	9.375	82
5d	1 3/4	11	3/8	156	43.8	11	9.375	71
6d	2	11	3/8	136	50.0	11	9.375	62
8d	2 1/2	11	3/8	112	62.5	11	9.375	51

Insulation Roof Deck Nails are extra long by design for the application of structural roof deck to wood trusses or rafters. Also important for the application of shingles over rigid foam insulation.

Insulation Roof Deck Nail Specification								
Size	Length Inch	Gauge	Head Inch	Nails per Lb	Size mm	Gauge	Head mm	Nails per Kg
10d	3	10	7/16	72	75.0	11	10.938	33
16d	3 1/2	10	7/16	62	87.5	11	10.938	28
20d	4	9	7/16	46	100.0	11	10.938	21
30d	4 1/2	9	7/16	40	112.5	11	10.938	18
50d	5.50	7	13/32	24	137.5	11	10.156	11

Ideal for aluminum, galvanized and vinyl gutters. Strong steel spikes drive better than aluminum spikes. For best results spikes should be driven into rafters or trusses, not only into the fascia. Minimum spike every 24". Heads should be checked to enhance paint adhesion.

872

		Gutter and Trim Spike Nail Specification						
Size	Gauge	Head Inches	Length Inches	Nails Per Lb	Gauge	Head mm	Length mm	Nails Per Kg
x	3/16	7/16	7	16	3/16	10.94	177.8	7
x	3/16	7/16	8	14	3/16	10.94	203.2	6
x	1/4	9/16	8	8	1/4	14.06	203.2	4
x	1/4	9/16	9	7	1/4	14.06	228.6	3
x	1/4	9/16	10	6.5	1/4	14.06	254.0	3

Originally designed for roofing felt, these nails today are widely used for built up roofing, rigid insulation, attaching sidewall curtains in poultry buildings and for general packing and creating. Large 15/16" domed caps are securely joined to sturdy ring shanks.

		Square Cap Roofing Nail Specification						
Size	Gauge	Sq.Head Inches	Length Inches	Nails Per Lb	Gauge	Head mm	Length mm	Nails Per Kg
x	12	15/16	1/8	80	15/16	23.44	3.2	36
x	12	15/16	3/4	100	15/16	23.44	19.1	45
x	12	15/16	7/8	80	15/16	23.44	22.2	36
x	12	15/16	1	78	15/16	23.44	25.4	35
x	12	15/16	1 1/4	75	15/16	23.44	31.8	34
x	11	15/16	2 1/2	53	15/16	23.44	63.5	24
x	11	15/16	3	48	15/16	23.44	76.2	22
x	9	15/16	3 1/2	45	15/16	23.44	88.9	20

Sharp spiral threads insure maximum holding power. Flat countersunk head is suitable for either hand or machine installation. The Casing head flooring nail is also a sharp spiral thread nail for maximum holding power, but recommended only for hand installation.

		Spiral Nails with Countersunk Heads						
Size	Gauge	Head Inches	Length Inches	Nails Per Lb	Gauge	Head mm	Length mm	Nails Per Kg
x	11 1/2	13/64	2 1/4	152	11 1/2	5.08	57.2	69
x	11 1/2	13/64	2 1/2	138	11 1/2	5.08	63.5	63
		Spiral Nails with Casing Head						
x	11 1/2	13/64	2 1/4	152	11 1/2	5.08	57.2	69
x	11 1/2	13/64	2 1/2	138	11 1/2	5.08	63.5	63

Underlayment nails are flat head nails, sharp ring shanks that will insure extra holding power.

Underlayment Nail Specification								
Size	Gauge	Head Inches	Length Inches	Nails Per Lb	Gauge	Head mm	Length mm	Nails Per Kg
x	14	3/16	1	640	14	4.69	25.4	290
x	14	3/16	1 1/4	528	14	4.69	31.8	240

Plywood nails are recommended by the plywood manufacturers for underlayment sheathing and sub floor applications. Sharp ring-shanks ensure extra holding power.

Plywood Nail Specifications								
Size	Gauge	Head Inches	Length Inches	Nails Per Lb	Gauge	Head mm	Length mm	Nails Per Kg
x	12 1/2	1/4	1 1/2	285	12 1/2	6.25	38.1	129
x	12 1/2	1/4	1 3/4	250	12 1/2	6.25	44.5	113
x	12	1/4	2	190	12	6.25	50.8	86
x	11	9/32	2 1/2	118	11	7.03	63.5	54
x	10	9/32	3	78	10	7.03	76.2	35

These GWB-54 style nails have long sharp diamond points for quick and easy driving. Heads are slightly countersunk with feathered edge to seat properly. Nails in high humidity areas should use a hot-dipped zinc-coated finish.

Drywall Nail Specifications								
Size	Gauge	Head Inches	Length Inches	Nails Per Lb	Gauge	Head mm	Length mm	Nails Per Kg
x	12 1/2	19/64	1 1/4	352	12 1/2	7.42	31.8	160
x	12 1/2	19/64	1 3/8	321	12 1/2	7.42	34.9	146
x	12 1/2	19/64	1 1/2	302	12 1/2	7.42	38.1	137
x	12 1/2	19/64	1 5/8	274	12 1/2	7.42	41.3	124
x	12 1/2	19/64	1 3/4	259	12 1/2	7.42	44.5	117
x	12 1/2	19/64	2	228	12 1/2	7.42	50.8	103

Trim nails are a slim diameter made from high carbon steel for hardwood and other materials to ensure good driving into oak, birch, walnut and other hardwoods. The high carbon steel nail usually costs slightly more per pound but costs less in the end since there are more nails per pound and fewer wood "splits" than conventional finishing nails.

Hardwood Trim Nail Specifications								
Size	Gauge	Head Inches	Length Inches	Nails Per Lb	Gauge	Head mm	Length mm	Nails Per Kg
x	0.072	0.089	1	810	12 1/2	7.42	25.4	367
x	0.072	0.089	1 1/4	685	12 1/2	7.42	31.8	311
x	0.072	0.089	1 1/2	570	12 1/2	7.42	38.1	259
x	0.072	0.089	2	472	12 1/2	7.42	50.8	214
x	0.083	0.083	2 1/2	472	12 1/2	7.42	63.5	214

Duplex nails are also called scaffolding nails. The duplex is like a regular nail, but has one head on top of the other. When driven into the first head, the second head is exposed for pulling the nail out easily. The drywall screw has replaced the duplex nails in some cases since it is easy to remove the screw and prevents tripping hazards. The duplex nail is almost invaluable in nailing temporary concrete forms since the second head makes for easy removal.

Duplex Head Nail Specifications								
Size	Gauge	Length under Lower head, Inch	Dist. between Heads, Inch	Nails Per Lb	Gauge	Length under Lower head, mm	Dist. between Heads, mm	Nails Per Kg
8d	10	2 1/4	1/4	90	10	56.25	6.3	41
10d	9	2 3/4	1/3	59	9	68.75	8.3	27
16d	8	3	3/8	45	8	75.00	9.4	20
20d	6	3 1/2	3/8	25	6	87.50	9.4	11

Cut Flooring Hardened Nails: Hardened steel cut flooring nails are ideal for laying hardwood tongue and groove floors. The slim shanks avoid splitting the groove and have excellent holding power. It is recommended nails be spaced 8" (200 mm) apart on 3/8" (9.37 mm) thick material and 12" (300 mm) apart on ¾" (18.75 mm) flooring.

Cut Flooring Hot-Dipped Galvanized Nails: These nails are good for applications where rust resistance is important. These nails are sometimes referred to as hot-dip galvanized cut finish nails.

Masonry Nail Specifications								
Size	Gauge	Head Inches	Length Inches	Nails Per Lb	Gauge	Head mm	Length mm	Nails Per Kg
Cut Flooring Hot-Dipped Galvanized Nails:								
x	9	x	1 1/2	165	9	x	38.1	75
x	9	x	2	104	9	x	50.8	47
x	9	x	2 1/4	88	9	x	57.2	40
x	9	x	2 1/2	78	9	x	63.5	35
x	9	x	3	53	9	x	76.2	24
x	10	x	3 1/2	35	9	x	88.9	16
Cut Flooring Hardened Nails:								
x	9	x	2	93	9	x	50.8	42
x	9	x	2 1/2	70	9	x	63.5	32
x	9	x	3	48	9	x	76.2	22

Note: Nails should be driven 3" (75 mm) or more from the end or edge of a masonry unit. Normally only 3/4" (18.75 mm) penetration is recommended. Use Safety glasses when using nails, especially hardened nails.

Metal Roof nails with Rubber Washers: Used for fastening corrugated metal roofing and siding as well as aluminum and fiberglass products. Nails should be double dipped in molten zinc after treating so no raw steel is exposed. EDPM rubber washers seal nail holes. Most manufacturers recommend a minimum nail penetration of 1" (25 mm) into a purlin or girt.

Metal Roof Nails with Silicone Washers: The gray silicone washers are normally twice as thick as the black EDPM washers, which allows extra resiliency to the nail so that the washer can seal the nail hole even if the nail is not driven at right angles to the sheet. Nails should be double dipped in molten zinc after treating so no raw steel is exposed. Contractor needs to be careful not to overdrive. The washer should just begin to balloon under the nail head. Most manufactures recommend a minimum nail penetration of 1" (25 mm) into a purlin or girt.

CARPENTRY

Metal Roofing Nail Specifications								
Size	Gauge	Head Inches	Length Inches	Nails Per Lb	Gauge	Head mm	Length mm	Nails Per Kg
Roofing Nails With Flat EDPM Rubber Washer								
x	10	3/8	1 1/2	126	10	9.38	38.1	57
x	10	3/8	1 3/4	111	10	9.38	44.5	50
x	10	3/8	2	106	10	9.38	50.8	48
x	10	3/8	2 1/4	91	10	9.38	57.2	41
x	10	3/8	2 1/2	82	10	9.38	63.5	37
x	10	3/8	3	70	10	9.38	76.2	32
x	10	3/8	3 1/2	61	10	9.38	88.9	28
x	9	7/16	4	45	9	10.94	101.6	20
x	9	7/16	4 1/2	42	9	10.94	114.3	19
Roofing Nails With Silicone Washers								
x	10	7/16	1 1/2	119	10	10.94	38.1	54
x	10	7/16	1 3/4	107	10	10.94	44.5	49
x	10	7/16	2	96	10	10.94	50.8	44
x	10	3/8	2 1/4	91	10	9.38	57.2	41
x	10	7/16	2 1/2	79	10	10.94	63.5	36
x	10	7/16	3	69	10	10.94	76.2	31
x	10	7/16	3 1/2	60	10	10.94	88.9	27
x	9	7/16	4	45	9	10.94	101.6	20

Recommended Sizes and Quantities Commonly used for Roofs			
New Roof			
1 3/4" Nails	100 Nails per Sq.	43.75 mm Nails	108 Nails / 10 Sq.M.
Over Existing Roof			
2 1/2" Nails	100 Nails per Sq.	62.50 mm nails	108 Nails / 10 Sq.M.

Pole Barn nails contain sharp ring and spiral shanks for excellent holding power in fastening timber, trusses, girts, purlins and othr similar types of timber. Oil-quench hardened nails drive well without breaking.

Pole Barn - Post and Framing Nail Specifications								
Size	Gauge	Head Inches	Length Inches	Nails Per Lb	Gauge	Head mm	Length mm	Nails Per Kg
6d	11	9/32	2	115	11	7.03	50.8	52
8d	11	9/32	2 1/2	115	11	7.03	63.5	52
10d	10	9/32	3	78	10	7.03	76.2	35
12d	10	9/32	3 1/4	73	10	7.03	82.6	33
16d	9	5/16	3 1/2	57	9	7.81	88.9	26
20d	7	3/8	4	35	7	9.38	101.6	16
30d	7	15/32	4 1/2	30	7	11.72	114.3	14
40d	5 1/2	15/32	5	23	5 1/2	11.72	127.0	10
60d	5 1/2	15/32	6	19	5 1/2	11.72	152.4	9
70d	5 1/2	15/32	7	16	5 1/2	11.72	177.8	7
80d	5 1/2	15/33	8	14	5 1/2	11.72	203.2	6

Nail Reference Data							
Specifications							
Size	Gauge	Length Inch	Nails Per Lb	Gauge	Length mm	Nails Per Kg
2d	5 1/2	1	876	5 1/2	25.4	397
3d	7	1 1/4	568	7	31.8	258
4d	9	1 1/2	316	9	38.1	143
5d	10	1 3/4	271	10	44.5	123
6d	11	2	181	11	50.8	82
7d	12	2 1/4	161	12	57.2	73
8d	12 1/2	2 1/2	106	12 1/2	63.5	48
9d	13	2 3/4	96	13	69.9	44
10d	14	3	69	14	76.2	31
12d	14 1/2	3 1/4	63	14 1/2	82.6	29
16d	16 1/2	3 1/2	49	16 1/2	88.9	22
20d	x	4	31	x	101.6	14
30d	x	4 1/2	24	x	114.3	11
40d	x	5	18	x	127.0	8
50d	x	5 1/2	14	x	139.7	6
60d	x	6	11	x	152.4	5
70d	x	7	x	x	177.8	x
80d	x	8	x	x	203.2	x

CARPENTRY

Suggested Nailing Schedule	
Connection	**Nailing[1]**
Joist to sill or girder, toenail	3d to 8d
Bridging to joist, toenail each end	2d to 8d
1" x 6" (25x150 mm) Subfloor or less to each joist, face nail	2d to 8d
Wider then 1" x 6" (25x150 mm) subfloor to each joist, face nail	3d to 8d
2" (50 mm) subfloor to joist or girder, blind and face nail	2d to 16d
Sole plate to joist or blocking, face nail	16d at 16" o.c. (400 mm)
Sole plate to joist or blocking, at braced wall panels	3d to 16d per 16" (400 mm)
Top plate to stud, end nail	2d to 16d
Stud to sole plate	4d to 8d toenail or 2d to 16d, end nail
Double studs, face nail	16d at 24" o.c.(600 mm)
Double top plates, typical face nail	16d at 16" o.c.(4000 mm)
Double top plate, lap splice	8d to 16d
Blocking between joists or rafters to top plate, toenail	3d to 8d
Rim joist to top plate, toenail	8d at 6"o.c.
Top plates laps and intersections, face nail	2d to 16d
Continuous header, two pieces	16s at 16"o.c. (400 mm) along each edge
Ceiling joist to plate, toenail	3d to 8d
Continuous header to stud, toenail	4d to 8d
Ceiling Joists, laps over partitions, face nail	3d to 16d
Ceiling Joists to parallel rafters, face nail	3d to 16d
Rafter to plate, toenail	3d to 8d
1" (25 mm) Brace to each stud and plate, face nail	2d to 8d
1" x 8" (25x200 mm) Sheathing or less to each bearing, face nail	2d to 8d
Wider than 1" x 8" sheathing to each bearing, face nail	3d to 8d
Built-up Corner Studs	16d at 24" o.c. (600 mm)
Built-up girder and beams	20d at 32" o.c. (800 mm) at top and bottom and staggered 2-20d at ends and at each splice
2" (50 mm) Planks	2d to 16d at each bearing

Connection		Nailing[1]
2" (50 mm) Planks - Wood structural panels and particleboard [2] Subfloor, roof and wall sheathing to framing:		
1/2" or less	(25 mm or less)	6d [3]
19/32" to 3/4"	(14.84 to 18.75mm)	8d [4] or 6d [5]
7/8" to 1"	(21.87 to 25 mm)	8d [3]
1 1/8" to 1 1/4"	(28.12 to 31.25 mm)	10d [4] or 8d [5]
Combination subfloor-underlayment (to framing)		
3/4" or less	(18.75 mm)	6d [5]
7/8" to 1"	21.87 to 25 mm)	8d [5]
1 1/8" to 1 1/4"	(28.12 to 31.25 mm)	10d [4] or 8d [5]
Panel Siding (to framing)		
1/2" or less	(25 mm or less)	6d [6]
5/8"	(15.62 mm)	8d [6]
Fiberboard Sheathing 7		
1/2"	25 mm	No.11 Ga [8], 6d [4], No. 16 ga [9]
25/32"	(19.53 mm)	No.11 Ga [8], 8d [4], No. 16 ga [9]
Interior Paneling		
1/4"	(6.25 mm)	4d [10]
3/8"	(9.37 mm)	6d [11]

[1] Common or box nails may be used except where otherwise stated.

[2] Nails spaced at 6" (50 mm) on center at edges, 12" (300 mm) at intermediate supports, except 6-inches (150 mm) at all supports where spans are 48-inches (1200 mm) or more. Nails for wall sheathing may be common, box or casing.

[3] Common or deformed shank

[4] Common

[5] Deformed Shank

[6] Corrosion-resistant siding or casing nails

[7] Fasteners spaced 3" (75 mm) on center at exterior edges and 6-inches (150 mm) on center at intermediate supports.

[8] Corrosion-resistant roofing nails with 7/16 inch-diameter(10.93 mm) head and 1 1/2" (37.5 mm) length for 1/2-inch (12.50 mm) sheathing and 1 1/4" (31.25 mm) length for 25/32-inch (19.53 mm) sheathing.

[9] Corrosion-resistant staples with nominal 7/16-inch (10.93 mm) crown and 1 1/8-inch(28.12 mm) length for 1/2-inch (12.50 mm) sheathing and 1 1/2" (37.50 mm) length for 25/32-inch (19.53 mm) sheathing.

[10] Panel supports at 16-inches (20-inches if strength axis in the long direction of the panel, unless otherwise marked). Casing for finishing nails spaced 6-inches on panel edges, 12-inches at intermediate supports.

[11] Panel supports at 24-inches (600 mm) Casing or finish nails spaced 6-inches (150 mm) on panel edges, 12-inches (300 mm) at intermediate supports.

CARPENTRY

Hardware Accessories Used for Wood Framing

The following items may be used to advantage in all types of wood construction. Developed primarily to provide better joints between wood framing members, in many cases their use has resulted in lower overall costs- the additional material cost being more than offset by increased labor efficiency.

Steel Joist Hangers. Used for framing joists to beams and around openings for stair wells, chimneys, hearths, ducts, etc. Made of galvanized. steel, varying from 12 gauge. to 3/16" (4.68 mm) in thickness, with square supporting arms, holes punched for nails and with bearing surfaces proportioned to size of lumber. Approximate prices for sizes most commonly used are as follows:

Joist Size Inch	Gauge	Depth of Seat Inches	Opening in Hanger Inch		Price per 100
2 "x 6 "	12	2 "	1 5/8 " x	5 "	$ 1.22
2 "x 8 "	12	2 "	1 5/8 " x	5 "	$ 1.35
2 "x 10 "	12	2 "	1 5/8 " x	8 1/2 "	$ 1.40
2 "x 12 "	10	2 1/2 "	1 5/8 " x	8 1/2 "	$ 1.73
4 "x 8 "	11	2 "	3 5/8 " x	8 1/4 "	$ 0.13
4 "x 10 "	9	2 "	3 5/8 " x	8 1/4 "	$ 1.94
4 "x 12 "	3/16	2 1/2 "	3 5/8 " x	8 1/2 "	$ 2.99

Joist Size mm	Gauge	Depth of Seat mm	Opening in Hanger mm		Price per 100
50 x 150 mm	12	50.0 mm	40.63 x	125.00 mm	$ 1.22
50 x 200 mm	12	50.0 mm	40.63 x	125.00 mm	$ 1.35
50 x 250 mm	12	50.0 mm	40.63 x	212.50 mm	$ 1.40
50 x 300 mm	10	62.5 mm	40.63 x	212.50 mm	$ 1.73
100 x 200 mm	11	50.0 mm	90.63 x	206.25 mm	$ 0.13
100 x 250 mm	9	50.0 mm	90.63 x	206.25 mm	$ 1.94
100 x 300 mm	3/16	62.5 mm	90.63 x	212.50 mm	$ 2.99

Framing Accessories. These are used in light wood construction to provide face-nailed connections for framing members. Adaptable to most framing connections, they eliminate the uncertainties and weaknesses of toe-nailing. Manufactured of zinc-coated sheet steel in various gauges and styles. Framing accessories are designed to provide nailing on various surfaces. Special nails, approximately equal to 8d common nails, but only 1-1/4" (31.25 mm) long, to prevent complete penetration of standard nominal 2"

(50.00 mm) lumber, are furnished with anchors. Approximate prices are as follows:

Type	Price per 100
Trip-L-Grips	$43.50
Du-Al-Clip	$34.50
Nail-On Plates	$52.50
Post Caps	$113.40
H Clips	$7.50
Angles	$78.00

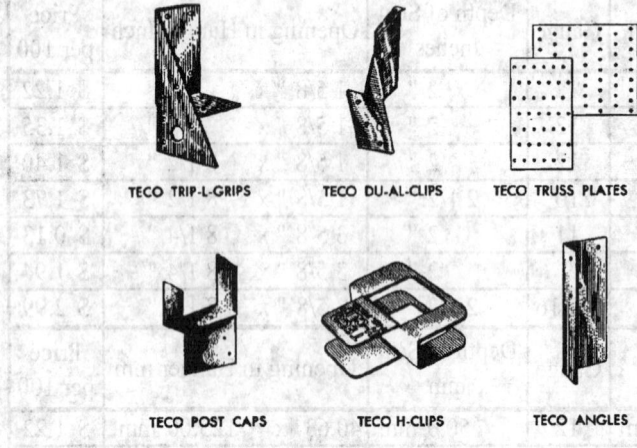

TECO TRIP-L-GRIPS   TECO DU-AL-CLIPS   TECO TRUSS PLATES

TECO POST CAPS   TECO H-CLIPS   TECO ANGLES

Timber Connectors. Timber connectors are devices for increasing the strength of bolted joints in timber construction. They are placed between adjacent faces of overlapping members, embedded to half-depth in each, to increase the bearing area at the connection. Several types of timber connectors are available, each for a particular purpose, on the next page.

Metal Connector Plates (Truss Plates). These are made from structural quality galvanized sheet steel and have integral teeth protruding from one face perpendicular to the plate. They are made in various sizes, thicknesses, and gauges and are designed to laterally transmit load in wood. Metal connector plates are installed by applying uniform pressure over the plate surface, which requires special equipment such as hydraulic or roller presses.

882

CARPENTRY

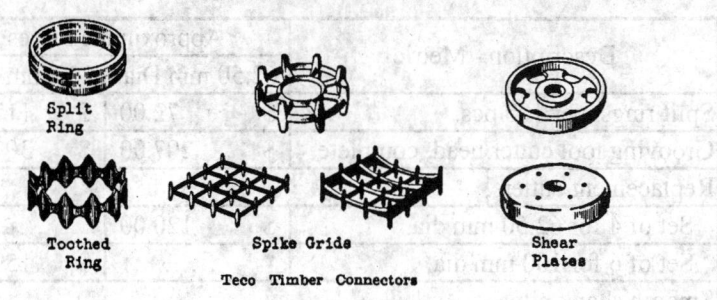

Split Ring

Toothed Ring

Spike Grids

Shear Plates

Teco Timber Connectors

Wedge-Fit Split Rings. Split ring connectors are hot-rolled carbon steel bands, bent to form a ring with a tongue and groove meeting joint. The cross section of the steel band is tapered both ways from center to edges, providing a wedge-fit when connectors are inserted into tapered grooves, conforming to ring section and precut to half-depth in each contact face of overlapping members.

Grooves in wood members must be cut with a special grooving tool, using a power drill press or portable drill. Grooves and bolt holes may be cut in a single operation, using a combination grooving tool and drill bit. If bolt holes are bored first, a smooth shank pilot is used to guide the grooving tool. Split rings are available in two sizes: 2-1/2" (62.50 mm) diameter rings are used for wood framing involving lumber of 2" (50 mm) nominal dimension and are used in trussed rafter construction for spans up to 50' (15.24 m), and 4" (100 mm) diameter rings are used for medium to heavy construction, generally with members 3" (75 mm) or more in nominal thickness. Approximate prices of split rings and grooving tools are as follows:

Description	Approximate Prices	
	2 1/2" Dia.	4" Dia.
Split rings, per 100 pcs.	$ 72.00	$ 117.00
Grooving tool cutter head, complete	$ 207.00	$ 301.50
Replacement cutters,		
Set of 4 for 2-1/2" dia.	$ 120.00	......
Set of 6 for 4" dia.	......	$ 159.00
Smooth shank pilot,		
9/16" dia.	$ 18.00	......
11/16" dia.	$ 18.00	......
13/16" dia.	......	$ 25.50
15/16" dia.	......	$ 30.00

Description - Metric	Approximate Prices	
	62.50 mm Dia.	100 mm Dia.
Split rings, per 100 pcs.	$ 72.00	$ 117.00
Grooving tool cutter head, complete	$ 207.00	$ 301.50
Replacement cutters,		
Set of 4 for 62.50 mm dia.	$ 120.00	......
Set of 6 for 100 mm dia.	......	$ 159.00
Smooth shank pilot,		
14 mm dia.	$ 18.00	......
17 mm dia.	$ 18.00	......
20 mm dia.	......	$ 25.50
23 mm dia.	......	$ 30.00

Shear Plates. Shear plates are used for wood-to-steel connections and for wood structures that are frequently disassembled, where they are used singly for wood-to-steel joints and in pairs for wood-to-wood assemblies. They are also often used in place of split rings for field connected joints. Shear plates are placed in daps precut in the contact faces of wood members with a special grooving tool. Load is transferred from wood member to shear plate, from shear plate to bolt and thence to opposing member, which may be another shear plate in wood, a steel gusset plate, or rolled steel shape. Drilling, grooving, and dapping of wood members are accomplished in the same manner as described for split ring connectors.

Shear plates are available in two sizes-2-5/8" (65.63 mm) diameter made of pressed steel and 4" (100 mm) diameter made of malleable iron. Approximate prices of shear plates and grooving tools are:

Description	Approximate Prices	
	2 5/8" Dia.	4" Dia.
Shear plates, per 100 pcs.	$ 108.00	$ 379.50
Grooving tool cutter head, complete	$ 254.25	$ 405.00
Replacement cutters,		
Set of 3 for 2-5/8" dia.	$ 136.50	......
Set of 5 for 4" dia.	......	$ 169.50
Smooth shank pilot,		
13/16" dia.	$ 21.00	$ 19.50
15/16" dia.	......	$ 25.50

Description - Metric	Approximate Prices	
	65.62 mm Dia.	100 mm Dia.
Shear plates, per 100 pcs.	$ 108.00	$ 379.50
Grooving tool cutter head, complete	$ 254.25	$ 405.00
Replacement cutters,		
Set of 3 for 65.62 mm dia.	$ 136.50	......
Set of 5 for 100 mm dia.	......	$ 169.50
Smooth shank pilot,		
20.31 mm dia.	$ 21.00	$ 19.50
23.43 mm dia.	......	$ 25.50

Toothed Rings. Toothed ring connectors are toothed metal bands with each tooth corrugated or curved along its cross section for greater rigidity. Used for wood-to-wood connections between lighter structural members, they are also used for strengthening structures in place and structural repairs.

The function of toothed rings is similar to that of split rings, but the method of installation is not as efficient and requires more labor. Toothed rings are embedded in wood members by applying pressure. Where a large number of joints are to be assembled, a hydraulic jack setup can be used to advantage. It requires a 5 to 7-1/2 ton (4.5- to 6.8-metric ton) jack for single bolt connections and a 10-ton (9072-kg) jack for joints with two bolts. Where fewer joints are involved, the high strength rod assembly is recommended. The assembly consists of a high strength rod with Acme threads on one end, double depth nuts, ball bearing thrust washer, heavy plate washers, lock washer, and nut. Metal sleeve adapters permit the use of 1/2", 5/8", or 3/4" (12.50, 15.62 or 18.75 mm) diameter rods with a standard ball bearing washer. The length of rod required equals the total thickness of wood members plus 1" (25 mm) for each layer of toothed rings plus 5" (125 mm) for nuts and washers. Ratchet and impact wrenches speed assembly. After the joint is drawn closed, the high-strength rod assembly is replaced with ordinary bolt and washers.

Spike Grids. Spike grid connectors are malleable iron castings of a grid-like structure with teeth protruding from both faces. Available in three styles: square flat type, 4-1/8"x4-1/8" (103.12 x 103.12 mm); circular type, 3-1/4" (81.25 mm) dia., used between sawn timbers; and single curve type, 4-1/8"x4-1/8" (103.12 x 103.12 mm), used between curved faces of poles or piling and sawn timbers. Spike grids are embedded under pressure, using the high strength rod assembly, as previously described for toothed rings. Approximate prices are as follows:

Type of spike grid	Square Flat	Single Curve	Circular
Price per 100 pcs.	$ 421.50	$ 421.50	$ 327.00
Installation tool for spiked grids, complete with adapter			$ 319.50

WOOD FLOOR AND ROOF TRUSSES

Some advantages of wood floor trusses are speed of construction and the ability to install utility lines, such as HVAC ducts, water lines, sprinkler lines, and electrical cables so that they do not interfere with one another. Easy of installation combines with economy.

Floor trusses are used for spans from 12'-0" (3.65 m) to as large as 70'-0" (21.33 m) and are found in all conventional types of construction, as well as in factories, storage facilities, and any other structure where a large, uninterrupted area is required.

Wood roof trusses, either shop or field fabricated, are used in buildings where clear floor space is a requirement and the width of the building exceeds the economical span of roof joists. Some of the advantages of trussed roof construction are fast erection, early availability of unobstructed weather-protected space, simplified installation of ceiling, floor, mechanical and electrical systems, and use of non-loadbearing movable partitions.

Wood roof trusses are used for spans as short as 25'-0" (7.62 m) and can be used up to 200'-0" (60.96 m). They are widely used in garages, factories, hangars, gymnasiums, auditoriums, bowling alleys, dance halls, supermarkets, and other structures of this type.

When using wood roof trusses, it is common to span the shorter dimension of the building. One might use any of several truss types, such as mechanically connected or bolted plywood gusset. The type of truss depends on the degree of economy desired, building use, and the general architectural effect desired. Prices of roof trusses are governed by the following conditions:

1. Cost of material and labor.
2. Loading conditions and spacing.
3. Difficulties of installation.
4. Requirements of local building ordinances.

The last condition causes the price of trusses to vary considerably in different parts of the country. Some cities may require only a 25-lb. (11.25-kg) live load while others require a 50-lb. (22.5-kg) live load, necessitating heavier construction.

The spacing of wood roof trusses that directly support roof sheathing is usually 2'-0" (0.6 m). Where roof loads are light and the

CARPENTRY

installation of a ceiling is not required, spacings of 4'-0" to 4'-10" (1.21 to 1.47 m) are advantageous. Where snow loads are especially heavy, spacings of 16" (400 mm) and even 12" (300 mm) have been used.

Prices given on following pages are only approximate. The estimator must obtain definite prices-for trusses either delivered at the building site or installed in place-from the manufacturer. Truss suppliers are familiar with conditions in all parts of the country and can quote a definite price.

Bowstring Bolted Roof Truss. For industrial and many commercial structures, the bowstring truss is popular. This type of truss has a curved top chord that starts from a point approximately 10" (250 mm) high at the end of the truss, rising to a maximum height at centerline of about 1/8 of the total span. This type of truss gives a curved roof shape. Because of its low height at the ends of the truss, a minimum of masonry is required for parapet or fire walls.

The truss may be left exposed, carrying roof load only, as in garages or industrial plant, or it can be designed to carry monorail, floor systems, or other concentrated loads. The bowstring truss is also designed to carry ceiling loads for use in stores, automobile agency salesrooms, and other commercial structures. Bowstring trusses provide wind bracing for walls through diaphragm action and can be used with "knee braces" to develop portal action in mill type buildings. They are efficiently fabricated with bolts, resulting in little reduction in cross-sectional areas of the pieces connected.

Approximate Prices of Wood Bowstring Truss. They are designed with glued laminated or nailed 2"x3" (50 x 75 mm) or 2"x4" (50 x 100 mm) top chords, built to conform to a parabolic curve, furnished in lots of 5 or more, spaced 16'-0" (4.87 m) on centers, with 25-lb. (11.25-kg) live load and 40-lb. (18.11-kg) total load.

Span Feet	Height Feet	Price	Span Meters	Height Meters
20 '- 0"	2 '- 6 "	$ 586	6.10 M	0.76 M
25 '- 0"	3 '- 1 "	$ 596	7.62 M	0.94 M
30 '- 0"	3 '- 9 "	$ 606	9.14 M	1.14 M
35 '- 0"	4 '- 4 "	$ 633	10.67 M	1.32 M
40 '- 0"	5 '- 0 "	$ 698	12.19 M	1.52 M
45 '- 0"	5 '- 7 "	$ 708	13.72 M	1.70 M
50 '- 0"	6 '- 3 "	$ 782	15.24 M	1.91 M
55 '- 0"	6 '- 10 "	$ 978	16.76 M	2.08 M
60 '- 0"	7 '- 6 "	$1,108	18.29 M	2.29 M
65 '- 0"	8 '- 1 "	$1,268	19.81 M	2.46 M
70 '- 0"	8 '- 9 "	$1,510	21.34 M	2.67 M
75 '- 0"	9 '- 4 "	$1,667	22.86 M	2.84 M
80 '- 0"	10 '- 0 "	$1,947	24.38 M	3.05 M
85 '- 0"	10 '- 7 "	$2,181	25.91 M	3.22 M
90 '- 0"	11 '- 3 "	$2,412	27.43 M	3.43 M
95 '- 0"	11 '- 10 "	$2,735	28.96 M	3.61 M
100 '- 0"	12 '- 6 "	$2,795	30.48 M	3.81 M
Span Feet	Height Feet	Price	Span Meters	Height Meters
105 '- 0"	13 '- 1 "	$3,038	32.00 M	3.99 M
110 '- 0"	13 '- 9 "	$3,167	33.53 M	4.19 M
115 '- 0"	14 '- 4 "	$3,167	35.05 M	4.37 M
120 '- 0"	15 '- 0 "	$3,167	36.58 M	4.57 M
125 '- 0"	15 '- 7 "	$3,167	38.10 M	4.75 M
130 '- 0"	16 '- 3 "	$3,167	39.62 M	4.95 M
135 '- 0"	16 '- 10 "	$3,167	41.15 M	5.13 M
140 '- 0"	17 '- 6 "	$3,167	42.67 M	5.33 M
145 '- 0"	18 '- 2 "	$3,167	44.20 M	5.49 M
150 '- 0"	18 '- 9 "	$3,167	45.72 M	5.72 M

Crescent Type Roof Truss. This is increasingly popular and is identified by its curved lower chord, which affords a higher ceiling at the mid-span of the truss. It produces a curved roof, similar to the bowstring truss. This truss presents a curved ceiling effect and is used in gymnasiums,

888

auditoriums, low-cost churches, and more elaborate stores and restaurants. Recommended span is from 20'-0" to 85'-0" (6.09 to -25.91 m).

Approximate Prices of Crescent Type Wood Roof Trusses. They are furnished in lots of 5 or more, spaced 16'-0" (4.87 m) on centers, 25-lb. (11.25-kg) live load, and 40 lb. (22.5-kg) total load.

Span Feet	Height Feet	Camber Feet	Price Each	Span Meters	Height Meters	Camber Meters
20 - 0"	5 '- 0 "	1 '- 0 "	$ 699.20	6.10 M	1.52 M	0.305 M
25 - 0"	5 '- 6 "	1 '- 3 "	$ 726.40	7.62 M	1.68 M	0.381 M
30 - 0"	6 '- 0 "	1 '- 6 "	$ 772.80	9.14 M	1.83 M	0.457 M
35 - 0"	6 '- 6 "	1 '- 9 "	$ 800.00	10.67 M	1.98 M	0.533 M
40 - 0"	7 '- 0 "	2 '- 0 "	$ 846.40	12.19 M	2.13 M	0.610 M
45 - 0"	7 '- 6 "	2 '- 3 "	$ 920.00	13.72 M	2.29 M	0.686 M
50 - 0"	8 '- 0 "	2 '- 6 "	$ 966.40	15.24 M	2.44 M	0.762 M
55 - 0"	8 '- 6 "	2 '- 9 "	$ 1,168.00	16.76 M	2.59 M	0.838 M
60 - 0"	9 '- 0 "	3 '- 0 "	$ 1,288.00	18.29 M	2.74 M	0.914 M
65 - 0"	9 '- 6 "	3 '- 3 "	$ 1,518.40	19.81 M	2.90 M	0.991 M
70 - 0"	10 '- 0 "	3 '- 6 "	$ 1,638.40	21.34 M	3.05 M	1.067 M
75 - 0"	10 '- 6 "	3 '- 9 "	$ 2,080.00	22.86 M	3.20 M	1.143 M
80 - 0"	11 '- 0 "	4 '- 0 "	$ 2,208.00	24.38 M	3.35 M	1.219 M
85 - 0"	11 '- 6 "	4 '- 3 "	$ 2,438.40	25.91 M	0.46 M	1.295 M

Belgian Roof Truss. This type is used where it is desirable to give a conventional building a more pleasing appearance through the use of a peaked roof. It is recognized by its sloping top chord and horizontal lower chord. In most instances, a ceiling is applied to the lower chord of these trusses, while the top chord may have shingles or slate as a covering. It is used on some higher class store buildings and low-cost churches and is recommended for spans from 20'-0" to 85'-0" (6.09 to25.91 m).

Belgian roof trusses are less efficient than the bowstring type, because the connections generally govern the member sizes. They cost about 50% more than bowstring type trusses.

The Double fink truss is also referred to as a Belgian truss and is used for spans from 36'-0" to 60'-0" (10.97 to 18.29 m).

Flattop Roof Truss. Industrial plants, warehouses, sheds, etc., when built of frame construction, often find use for a truss that absorbs wind loads, in addition to carrying roof and concentrated loads. The flattop truss is ideal for this purpose. It has parallel top and bottom chords, forming a large rectangle. Spans should not exceed 65' (19.81 m) where cost is an important factor.

Types of Trusses

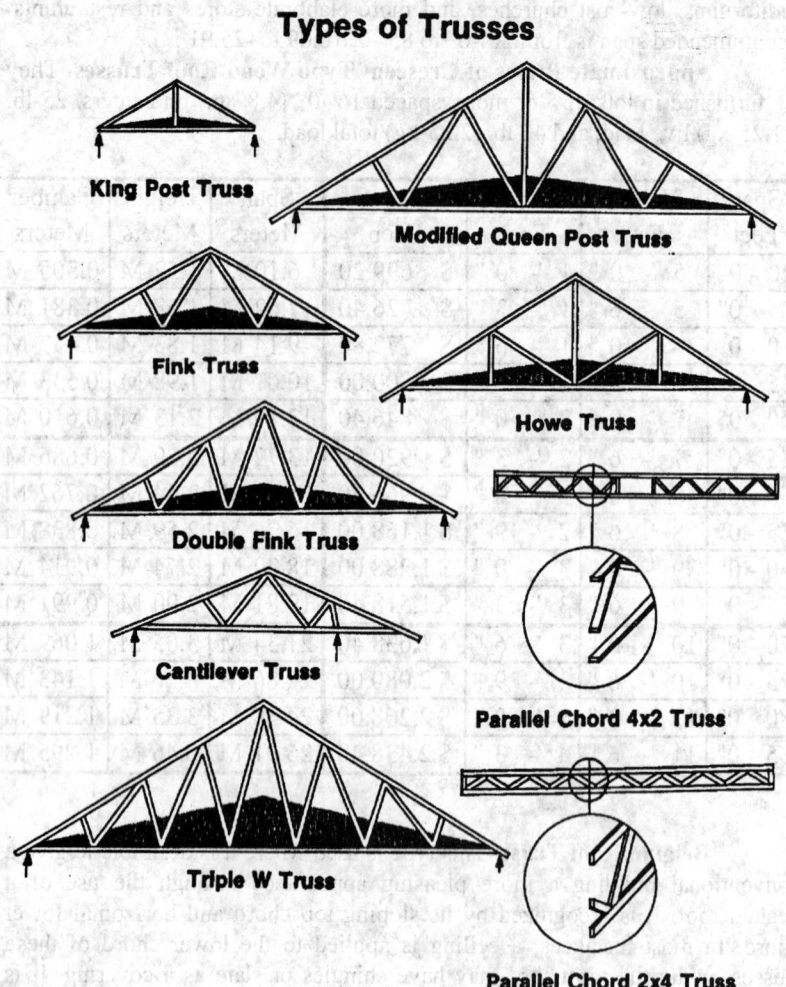

King Post Truss

Modified Queen Post Truss

Fink Truss

Howe Truss

Double Fink Truss

Cantilever Truss

Parallel Chord 4x2 Truss

Triple W Truss

Parallel Chord 2x4 Truss

Howe Truss. The Howe truss can be used for spans from 16' to 18' (4.87 to 5.48 m). It is efficient for loading conditions that balance the top and bottom chords and is often used when the truss is designed as a girder, with loads applied for the bottom chord.

Parallel Chord 4x2 Truss. Normally used in floor truss design, it has parallel chords of stress-related 2"x4" (50 x 100 mm) lumber set flat so that the wider faces are the bearing surface of the truss. The web members are also cut from 2"x4" (50 x 100 mm) and positioned so that they support the chord across their full widths. This type of truss can be manufactured with duct chase openings so that wiring, piping, and ducts can run within the chords.

890

Types of Trusses

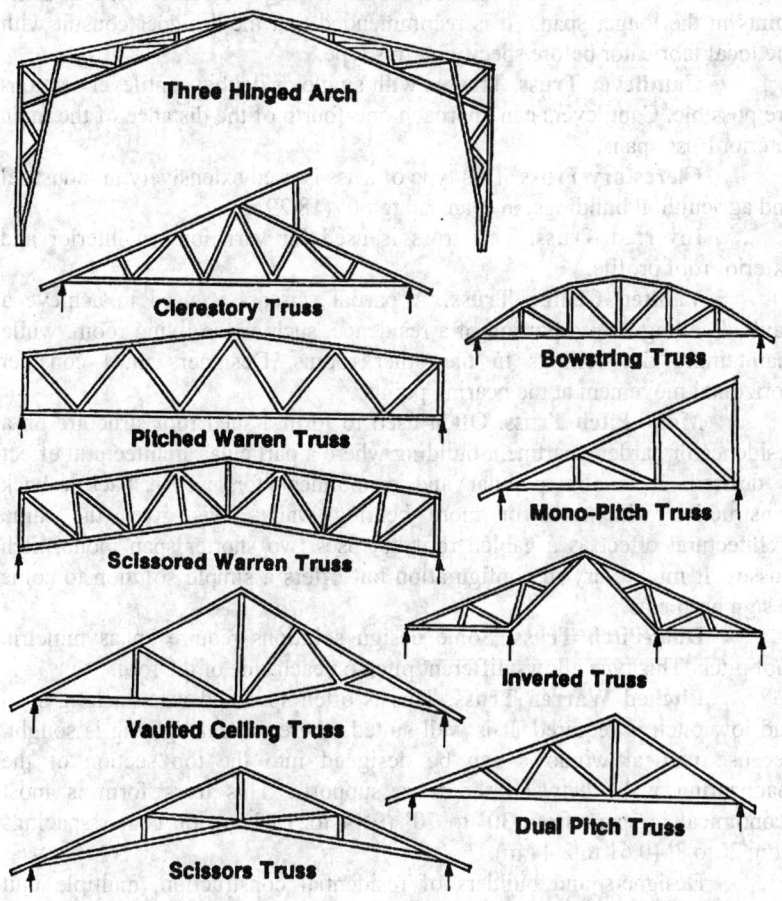

Parallel Chord 2" x 4" Truss. Generally used in roof truss design, it is gaining popularity because of the long, clear spans and shallow depth. The slope can be built into the truss, or drainage can be provided by using different opposite side wall heights to slope the truss. Roof slopes should be at least 1/4" (6.25 mm) per foot of span.

   Modified Queen Post Truss. This type of truss is often used for spans the same as those requiring double W but that have load conditions that require fewer bottom chord panels.

   Fink Truss. The fink truss is generally suitable for spans from 16' (4.87 m) to as long as 46' (14.02 m) and for all classes of construction. It is an efficient and cost effective truss configuration, 50% to 60% the cost of a comparable steel truss.

Three-Hinged Arch. This type of structure provides good overhead clearance and attractive appearance. However, high forces develop in some joints in the longer spans. It is recommended that the designer consult with the local fabricator before specifying this type.

Cantilever Truss. Trusses with single or double cantilever sections are possible. Cantilevers can approach one-fourth of the distance of the main interior truss span.

Clerestory Truss. This type of truss is used extensively in industrial and agricultural buildings, in spans up to 60' (18.29 m).

Inverted Truss. This truss is used for variations in interior and exterior roof profile.

Vaulted Ceiling Truss. A partial scissors is used to achieve a vaulted ceiling over a portion of a residence, such as the living room, while maintaining flat ceilings in the other rooms. Designers must consider horizontal movement at the bearing points.

Mono-Pitch Truss. Often used to form a shed roof structure on a residence or garden apartment building where a particular architectural effect is desired. It is also popular and economical for use in back-to-back construction with three or more bearing walls. This gives the same architectural effect as a gabled roof but uses two shorter span mono-pitch trusses. It might vary in configuration but offers a simple solution to some design problems.

Dual-Pitch Truss. Some design solutions require an asymmetric roof pitch. This type allows different pitch on each side of the roof.

Pitched Warren Truss. This is often specified when a long span and low pitch is required. It is well suited where natural lighting is sought, because vertical windows can be designed into the top section of the loadbearing walls between the truss supports. This truss form is most economical in spans from 30' to 70' (9.14 to 21.33 m), on center spacings from 2' to 8' (0.61 to 2.44 m).

Designers and builders of residential construction, multiple unit housing projects, and light commercial construction use trusses to reduce costs and to shorten construction time. Most small home constructors can benefit from this type of framing. Some of the advantages to using mechanically connected trusses are: roof framing and ceiling framing are accomplished at the same time; members can be made of lighter stock, saving material and reducing weight; trusses can be pre-assembled; trusses can be erected rapidly and the job put under cover quickly; and trusses permit the use of non-bearing partitions for all interior walls.

Trussed rafters used in house building and similar construction vary as to truss pattern and method of construction, depending on design requirements and individual preference of the builder. Some examples of truss patterns are:

The *W-Type* is the most popular type and is adaptable for spans from 18' up to 40' (5.48 to 12.19 m); roof slope from 2 in 12 to 6 in 12 and higher.

892

CARPENTRY

The *Triple-W* is used for spans up to 80' (24.38 m) with slopes of 3 in 12 and higher. Centerline spacings can be from 2' to 20' (0.61 to 6.09 m), depending on requirements.

The *Kingpost* truss is usually recommended for shorter spans. The economical range is up to 26' (7.92 m) under most loading conditions. It has wide application, for residential and commercial garages, carports, and short span storage buildings.

The *Scissors* truss is used where vaulted or sloped ceilings are desired for architectural effect. Slope selection is very important, and a designer can easily overlook the depth in span ratio. The recommended bottom chord slope is one-half the slope of the top chord. Because there is no horizontal bottom chord or tie between the bearing points, the scissor truss develops movement at its bearing points when under load. The designer should anticipate this movement and design the supporting structure accordingly. Economical spans are from 18' to 36' (5.48 to 10.97 m); roof slopes 5 in 12 to 6 in 12; ceiling slopes 2 in 12 to 3 in 12.

The *Scissored Warren* truss has similar depth requirements as the regular chord truss and can achieve many of the advantages of portal frames, as far as internal clearance and appearances are concerned.

Trussed rafter construction varies principally as to the method of assembly, whether members are fastened together at joints by bolting or nailing, with split ring connectors and bolts, or flat, bent, pronged or toothed metal plates, or by gluing.

Using the bolting method, the members at all major joints, i.e., heel joints, apex, splices, etc., are connected by one or more bolts, or by using split rings to transfer the loads. In some cases, one or both ends of diagonal members may be fastened by nailing.

For nailed trussed rafters, all joint connections use metal truss plates, plywood gussets and nails, or ordinary nailed joints.

Other Types of Commonly Used Wood Trusses. The Double Fink truss is generally used for spans from 36' to 60' (10.97 to 18.29 m). It is often used in shorter spans over other types of trusses, because it is easier for workers to handle without special equipment.

The Piggy Back Sectional truss is generally used when the slope (or pitch) exceeds the height limits of the fabricating press or by highway restrictions.

The Long Sectional is a truss that is fabricated in sections for when the overall length exceeds permissible highway length restrictions, the length of the fabricating jig, or if the overall height is too large to be put through the fabricating press. The height is reduced by producing it in two sections.

An extremely long truss with cantilevered ends can be manufactured in three sections. Trusses of this type, 128' (39.01 m) in overall length, have been fabricated in this manner. The overall height would be limited to the width of the fabrication press, but a fourth piggyback section could be added to the center section.

The Sectional Scissors truss is used when overall height exceeds highway regulations or limits of the press. Splices are applied at the job site by general contractor or a subcontractor. Due to its very high center of gravity and unstable configuration, this type requires extreme care in erection.

There are alternate methods of splicing sectional scissors trusses. One such method, when it is feasible, has the advantage of providing manufacturing symmetry.

Roof trusses offer numerous advantages over conventional roof framing: a truss can span greater roof distances; longer clear spans allow complete freedom in the layout of interior partitions; the bottom chord is available for ceiling finish material; and the cost is usually less, when on-site labor is considered.

Special Conditions Associated with Roof Truss. The parallel chord roof truss provides a flat roof that can be sloped to one side to drain by using wedge pieces added to the top chord. An alternate method is a slight variation in the bearing plate heights. This technique is acceptable where a slight variation in the interior room height is acceptable or where a hung ceiling will conceal the variation. When there is significant slope and no horizontal bottom chord, one must allow for horizontal movement at the wall.

Truss Openings. The two most widely used floor truss configurations are Warren and Fan. Each type of truss allows openings for through-truss duct work. The table below shows the allowable openings for each type of truss. Note that the warren and fan trusses both allow a 7" (175 mm) opening when the depth of the truss is 12" (300 mm). However, if the truss is 16" (400 mm) deep, the warren truss allows a 10-1/4" (256.25 mm) opening, the fan truss an 11" (275 mm) opening.

Subtract insulation thickness and working tolerances from these dimensions to determine maximum net duct dimensions.

Installation of Wood Trusses

An estimator must consider the entire process of handling and installing wood trusses. The Truss Plate Institute recommends specific procedures, and the information provided below is excerpted from its publication, *Commentary and Recommendations for Handling Installing & Bracing Metal Plate Connected Wood Trusses (HIB-91)*.

894

Center Duct Depth/Opening					
Truss Depth Inches	Max Opening Height Inches	Width in Inches	Truss Depth mm	Max Opening Height mm	Width in mm
12 "	9 "	18 "	300 mm	225 mm	450 mm
14 "	11 "	22 "	350 mm	275 mm	550 mm
16 "	13 "	24 "	400 mm	325 mm	600 mm
18 "	15 "	24 "	450 mm	375 mm	600 mm
20 "	17 "	24 "	500 mm	425 mm	600 mm
22 "	19 "	24 "	550 mm	475 mm	600 mm
24 "	21 "	24 "	600 mm	525 mm	600 mm
26 "	23 "	24 "	650 mm	575 mm	600 mm
28 "	25 "	24 "	700 mm	625 mm	600 mm
30 "	27 "	24 "	750 mm	675 mm	600 mm

Warren Truss Depth/Opening			
Overall Depth Inches	Max Rnd. Opening Inches	Overall Depth mm	Max Rnd. Opening mm
12 "	7 "	300 mm	175.0 mm
14 "	8 3/4 "	350 mm	218.8 mm
16 "	10 1/4 "	400 mm	256.3 mm
18 "	11 1/2 "	450 mm	287.5 mm
20 "	12 1/2 "	500 mm	312.5 mm
24 "	14 1/2 "	600 mm	362.5 mm
30 "	16 1/2 "	750 mm	412.5 mm

Unloading and Lifting. During the unloading process, and throughout all phases of construction, care should be taken to avoid lateral bending, which can cause joint and lumber damage. Use proper equipment to lift trusses. A crane with spreader bar and cables is strongly recommended for trusses with spans greater than 30' (9.14 m). For spans 30' (9.14 m) or less, bridle and chokers may be used.

Do not lift bundles by the strapping, which is not strong enough to safely support the weight of the trusses. It is only meant for shipping purposes. Do not attach cables, chains, or hooks to web members. Thread cables under the top chord of a bundle of trusses near panel points closest to the quarter or third points of the trusses. Lift the bundle slowly in order to

895

determine the load balance. If it is unbalanced, then lower the bundle, readjust the cable, and lift again.

Lifting with a bridle or spreader bar requires the same care in balancing the load. Never lift an unbalanced bundle of trusses. Do not lift single trusses with spans greater than 30' (9.14 m) by the peak.

The truss fabricator has dimensions of openings for intermediate depths of trusses.			
Overall Depth Inches	Max. Rnd. Opng Inches	Overall Depth mm	Max. Rnd. Opening mm
12 "	7 "	300 mm	175 mm
14 "	9 "	350 mm	225 mm
16 "	11 "	400 mm	275 mm
18 "	13 "	450 mm	325 mm

Storage. Whenever possible, trusses should be unloaded in bundles, on a relatively smooth ground, and picked up by the top chord in a vertical position only. It is common to unload trusses from a trailer one at a time and store them in a stable position to prevent toppling or shifting. When they are stored horizontally, blocking or sleepers should be maximum 10' (3.04 m) centers to prevent lateral bending. Do not store unbraced bundles upright.

If trusses are stored vertically, extreme care should be taken to insure that they are braced and blocked in a stable manner that will prevent toppling. Pitched trusses should be stored with the peak up.

Installation. An open web metal plate connected truss is a manufactured product, comprised of various lumber components that are precisely cut and fitted together with metal connector plates. A metal plate connected wood truss should be installed with greater care than a monolithic item such as sawn wood joist, steel channel, or a beam. Without bracing, or some other type of restraining device, a truss can be unstable.

The installer is responsible for selecting the most suitable method and sequence of installation that is available and which is consistent with the plans and specifications and such information as may be furnished prior to installation. Trusses can be installed by hand or by mechanical means, depending on truss span, installed height above grade, and accessibility by equipment (such as crane or forklift).

The installer must understand truss design drawings, placement plans, and all notes and cautions on them. He should clarify any and all questions with the manufacturer before starting the installation. Use particular care when installing under adverse conditions, such as high winds, uneven or sloping terrain, proximity of high voltage power lines, or constricted job site.

896

Never walk on trusses that are lying flat. Under no circumstances should construction loads of any description be placed on unbraced trusses.

Hand Installation. Installation by hand should be limited to trusses of a size and configuration that can be carried, raised up to bearing support height, and rotated into position without excessive lateral deflection (bow), which produces strain in the lumber or metal connector plates and weakens the joints. Lateral deflection greater than 3" in 10' (75 mm in 3.04 m) of span is excessive. Trusses should be handled so as to ensure support at intervals of 25' (7.62 m) or less.

When installed by hand, trusses are positioned over the side walls and rotated into position using a fork-like lifting pole. The longer the span, the more workers will be necessary to avoid lateral strain on the truss. Depending on length, the truss should be supported at the peak for spans less than or equal to 20' (6.09m) and at quarter points for spans less than or equal to 30' (9.14 m).

Mechanical Installation. Trusses that are installed by mechanical means should be handled so as to ensure support at intervals of 25' (7.62 m) or less. The installer should provide adequate rigging (crane, fork lift, slings, tag lines, and spreader bars) for sufficient control during lifting and placement to assure safety to personnel and to prevent damage to trusses and property. Slings, tag lines, spreader bars, etc. should be used in a manner that will not cause any damage to metal connector plates.

Trusses that are lifted in place in banded bundles should be securely supported by temporary means, permitting the safe removal of banding and the sliding of individual trusses. Do not lift bundled trusses by their strapping.

Take care to position truss bundles so that the supporting structure is not overloaded. The bundle straps (banding) can be broken after the bundle has been placed on the supporting structure and prior to the release of the lifting cables.

Under certain conditions, several trusses may be assembled on the ground into structural subcomponents, complete with temporary bracing members and portions of the roof deck or subfloor, and then carefully lifted into place as a self supporting unit. These units are assembled as needed until the roof or floor structure is complete. Usually some decking is omitted so that the stagger-lap is provided between adjacent units for diaphragm continuity. This method requires substantial planning beforehand and special engineering for the trusses, such as location and design of pick-up or liftpoints. This method has been found to be a safe solution to the installation of especially long spans or complex shapes.

Trusses that are mechanically installed one at a time should be held safely in position with the erection equipment until such time as all temporary bracing has been installed. Lines from the end of the spreader bar should "toe-in". Do not permit lines to "toe-out", which will tend to cause trusses to buckle.

For truss spans greater than 60' (18.29 m), the suggested lift procedure requires the use a strong-back. The strong-back should be attached to the top chord and web members at intervals of approximately 10' (3.0 m) and should be at or above mid height of the truss to prevent overturning. The strong-back should be monolithic with sufficient strength to safely carry the weight of the truss and rigid enough to resist truss bending.

Each truss should be set in position in accordance with the designer's plan and held with the hoisting equipment until the ends of the truss are securely fastened and temporary bracing is installed. Only then should the hoisting equipment be released from the braced truss and used for lifting, placing, and securing the next truss.

Installation tolerances are critical to achieving an acceptable roof or floor line and for effective bracing. A stringline, plumb bob, level, or transit is recommended to attain acceptable installation tolerances.

L Inches	L/200 Inches	L feet	L mm	L/200 mm	L Meter
25 "	1/8 "	2.1 '	625 mm	3.13 mm	0.64 M
50 "	1/4 "	4.2 '	1250 mm	6.25 mm	1.28 M
75 "	3/8 "	6.3 '	1875 mm	9.38 mm	1.92 M
100 "	1/2 "	8.3 '	2500 mm	12.50 mm	2.53 M
125 "	5/8 "	10.4 '	3125 mm	15.63 mm	3.17 M
150 "	3/4 "	12.5 '	3750 mm	18.75 mm	3.81 M
175 "	7/8 "	14.6 '	4375 mm	21.88 mm	4.45 M
200 "	1 "	16.7 '	5000 mm	25.00 mm	5.09 M
225 "	1 1/8 "	18.8 '	5625 mm	28.13 mm	5.73 M
250 "	1 1/4 "	20.8 '	6250 mm	31.25 mm	6.34 M
275 "	1 3/8 "	22.9 '	6875 mm	34.38 mm	6.98 M
300 "	1 1/2 "	25.0 '	7500 mm	37.50 mm	7.62 M
350 "	1 3/4 "	29.2 '	8750 mm	43.75 mm	8.90 M
400 "	2 "	33.3 '	10000 mm	50.00 mm	10.15 M

Trusses should not be installed with an overall bow, or bow in any chord or panel, that exceeds the lesser of L/200 or 2 inches (50 mm), where L is the span of the truss, chord, or panel length in inches.

Trusses should not be installed with a variation from plumb (vertical tolerance) at any point along the length of the truss from top to bottom chords that exceeds 1/50 of the depth of the truss at that point (D/50) or 2" (50 mm), whichever is less. This does not apply to trusses specifically designed to be installed out of plumb.

898

D Inches	D/50 Inches	D Feet	D mm	D/50 mm	D Meters
12 "	1/4 "	1 '	300 mm	6.25 mm	0.305 M
24 "	1/2 "	2 '	600 mm	12.50 mm	0.610 M
36 "	3/4 "	3 '	900 mm	18.75 mm	0.914 M
48 "	1 "	4 '	1200 mm	25.00 mm	1.219 M
60 "	1 1/4 "	5 '	1500 mm	31.25 mm	1.524 M
72 "	1 1/2 "	6 '	1800 mm	37.50 mm	1.829 M
84 "	1 3/4 "	7 '	2100 mm	43.75 mm	2.134 M
96 "	2 "	8 '	2400 mm	50.00 mm	2.438 M
108 "	2 "	9 '	2700 mm	50.00 mm	2.743 M

Location of trusses along the bearing support should be within +1/4" (6.25 mm) of plan dimensions. Special hangers or supports should be located to support trusses within +1/4" (6.25 mm) of plan dimensions. Trusses are to be located at o.c. spacing specified by the design engineer.

Top chord bearing parallel chord trusses should have a maximum gap 1/2" (12.50 mm) between the inside of the bearing and the first diagonal or vertical web.

Correction of Errors in Truss Installation. Errors in building lines or dimensions, or by subcontractors and suppliers, must be corrected before installation begins. Alterations, such as cutting overhangs to proper length, might be required and should be made by the builder of record. Any correction that involves the cutting, drilling, or relocation of a truss member or metal connector plate is considered major and should not be made without notifying the truss manufacturer.

A major correction calls for an engineering analysis by qualified registered engineer prior to the field work. Cutting or drilling trusses in the field without the approval of this engineer should be strictly prohibited. Trusses might be designed by the truss design engineer for specific reasons to be field cut by the installer at the designated locations. Such modifications should be indicated on the engineering drawing.

Field Assembly. Trusses that are too long for delivery to the jobsite in one piece can be designed to be delivered in two or more parts and then spliced together at the jobsite. The installer should carefully follow the splicing specifications shown on the truss drawings. Splicing may be performed on the ground before installation, or the truss sections may be supported by temporary shoring and splices installed by workers on a safe working surface.

Trusses that are too high for delivery to the jobsite in one piece may be manufactured in two or more sections and piggy-backed at the jobsite. The installer should observe the permanent bracing and connection details shown on the truss or piggyback design drawings. The supporting trusses should be completely installed, including the temporary and permanent bracing and sheathing when required, before installing top or supported truss sections.

Installation Sequence for Temporary and Permanent Bracing of Wood Truss. Install the first truss with a ground bracing system, which is constructed as follows:

1. Attach verticals to the end wall at the same spacing as the top chord lateral bracing.
2. Attach diagonals to the vertical and ground stake.
3. Attach horizontal ties or backup ground stakes if required.
4. Attach struts to the diagonals if required.
5. Attach lateral braces to the diagonals if required.
6. Attach end braces to the diagonals and lateral braces.

For the first group of three to six trusses, the sequence is as follows:

1. Install three to six trusses with top chord lateral braces affixed to the first (ground-braced) truss. The number of trusses to be initially installed should be specified in the bracing design.
2. Install top chord diagonal bracing between top chord lateral braces at intervals specified in the bracing design.
3. Install bottom chord lateral braces at no greater than 15' (4.57 m) on center.
4. Install web lateral bracing as specified on the truss design.
5. Install bottom chord diagonal braces between bottom chord lateral braces at the same interval as top chord diagonal bracing.
6. Install web cross bracing on web(s) corresponding to the nearest adjacent panel point(s) to the bottom chord lateral brace(s).

Continue installation of trusses with top chord lateral braces as specified on the bracing design or truss layout plan. The number of trusses to be installed between sets of top chord diagonal braces should be specified on the bracing design or truss layout plan.

Repeat 2) and 4) above for each group of trusses and then install sets of web cross bracing at 20-ft. (6.09-m) intervals. Install sets of bottom chord diagonal braces at the same intervals as top chord diagonal bracing and at each end bay. Top chord diagonal braces may remain as permanent bracing per the building design. Install permanent sheathing materials immediately as the temporary bracing is removed. Ground bracing may be removed only after the top chord plane is completely sheathed. Install any other permanent bracing as specified by the building design.

Note that all bracing lumber should be no less than 2'x4'x10' (0.61 x 1.22 x 3.04 m). A minimum of two 16d double head nails should be used at each connection.

Labor Erecting Wood Roof Trusses. Erection of wood roof trusses in confined areas, or where the supporting structure is not heavy enough to

900

support new loads, has usually been accomplished using a gin pole. However, most truss erection today uses a special hydraulic crane mounted on a truck chassis, with lifting capacity up to 12 tons (10,886 kg).

Before the crane is brought in, area around the building should be clear and all overhead electrical wires removed. Sizes of door openings must be predetermined to permit entrance of the crane.

Where a gin pole is used, the labor cost of erecting will vary with job conditions. Deep excavations, piles of brick, mortar boxes or, lumber piles in the way of derricks prevent erecting crews from making good time.

Mechanical Erection of Trusses. Under average working conditions on jobs requiring 5 or more trusses, a crew consists of 5 workers with an erection machine. Sometimes, the truss fabricator sends trusses with a truck-mounted crane and the cost of unloading and erecting is included in the quoted price. If not, the estimator must add the cost of an erecting machine, such as a rubber-tired cherry picker or a crawler crane. The crew should erect one roof truss of given length at the following rate and costs:

Erecting Roof Trusses Using a Machine							
Length of Truss	Hrs. to Erect	Labor Hrs per Truss	Rate	Total	Rate	Total	Length of Truss
30 '	0.52	2.62	$	$	$35.04	$91.80	9.14 M
40 '	0.75	3.75	$	$	$35.04	$131.40	12.19 M
50 '	1.00	5.00	$	$	$35.04	$175.20	15.24 M
60 '	1.00	5.00	$	$	$35.04	$175.20	18.29 M
75 '	2.00	10.00	$	$	$35.04	$350.40	22.86 M
90 '	3.00	15.00	$	$	$35.04	$525.60	27.43 M
100 '	4.00	20.00	$	$	$35.04	$700.80	30.48 M
120 '	5.00	25.00	$	$	$35.04	$876.00	36.58 M

Remember to add cost of Machine

Erecting Floor Trusses Using a Machine							
Length of Truss	Hrs. to Erect	Labor Hrs per Truss	Rate	Total	Rate	Total	Length of Truss
30 '	0.15	0.75	$	$	$35.04	$26.28	9.14 M
40 '	0.20	1.00	$	$	$35.04	$35.04	12.19 M
50 '	0.25	1.25	$	$	$35.04	$43.80	15.24 M
60 '	0.35	1.75	$	$	$35.04	$61.32	18.29 M
75 '	0.45	2.25	$	$	$35.04	$78.84	22.86 M

Remember to add cost of Machine

Hand Erection of Trusses. Wood roof trusses are erected by hand using a gin pole and normally requires a crew of four workers. The maximum length is 30'-0" (9.14 m), and the contractor must be careful to avoid "bowing" or deflection when erecting trusses by hand.

Erecting Roof Trusses by Hand							
Length of Truss	Hrs. to Erect	Labor Hrs per Truss	Rate	Total	Rate	Total	Length of Truss
15 '	0.25	1.00	$	$	$35.04	$35.04	4.57 M
20 '	0.30	1.20	$	$	$35.04	$42.05	6.10 M
25 '	0.30	1.20	$	$	$35.04	$42.05	7.62 M
30 '	0.45	1.80	$	$	$35.04	$63.07	9.14 M

Erecting Floor Trusses by Hand							
Length of Truss	Hrs. to Erect	Labor Hrs per Truss	Rate	Total	Rate	Total	Length of Truss
15 '	0.15	0.60	$	$	$35.04	$21.02	4.57 M
20 '	0.20	0.80	$	$	$35.04	$28.03	6.10 M
25 '	0.30	1.20	$	$	$35.04	$42.05	7.62 M
30 '	0.45	1.80	$	$	$35.04	$63.07	9.14 M

Fabrication of Trussed Rafter Members. Fabrication of trussed rafter members is ideally suited for mass production methods, due to the large amount of repetition possible. In most cases, trussed rafters are symmetrical about their center lines so that corresponding members of each half are identical.

Since every member of a truss must do its share of the work, each one must be fabricated to exacting specifications. Design dimensions must be rigidly followed and all joints must have a good mechanical fit.

For efficient fabrication, a full-size layout should be made, either in the shop or on the subfloor of the building, after which one trussed rafter is carefully patterned and constructed with temporary connections. The unit may then be disassembled and the members used as templates for the balance of the trusses. On large projects, templates may be made of plywood or sheet metal.

To insure smooth roof and ceiling lines after erection, layout measurements should be made to upper edges of top chords and lower edges of bottom chords to eliminate variations in stock lumber widths.

CARPENTRY

A word of caution-the joints and fabrication of various size trusses are similar. It is important to see that the appropriate drawings are consulted and followed when trusses are fabricated.

Split-Ring Connected Trussed Rafters

Members are connected by means of bolts and split rings at all major joints, as the split rings transfer the loads and the bolts hold the members in contact.

In order that each member may function properly, the bolt holes must be located exactly in accordance with design dimensions, and grooves for split rings must be of correct diameter, width, and depth to provide a snug fit. Grooves must be cut with a grooving tool designed especially for this purpose.

Proper procedure for cutting grooves is as follows: after bolt locations have been established, drill bolt holes 1/16" (1.56 mm) larger in diameter than bolt; grooving tool is then fitted with a pilot mandrel of same diameter as drill and cutters are set to cut to a depth equal to half the width of the split ring; using bolt holes as pilot holes, concentric grooves are then cut in the faces of members which will contact each other. The grooving tool may also be fitted with the correct size drill bit, instead of a pilot mandrel, permitting both drilling and grooving operations to be done at one setup. For best results, drilling and grooving should be done with a bench drill or portable drill mounted in a drill stand.

Material Requirements for Split-Ring Connected Trussed Rafters. The material requirements for split-ring connected trussed rafters are based on the following design conditions.

Trussed rafters are designed to support a dead load plus live load on roof of 35 lbs/sq.ft. (1676 Pa or 171 kgf/sq.m) and a ceiling load of 10 lbs/sq.ft. (0.93 kgf/sq.m) with rafters spaced at 2'-0" (0.61 m) centers.

Lumber shall be good grade of sufficient quality to permit the following unit stresses:

c = 900 lbs/sq.in. (6.20 MPa) Compression parallel to grain
f = 900 lbs/sq.in. (6.20 MPa) Extreme fiber in bending
E = 1,600,000 lbs/sq.in. (11,032 MPa) Modulus of elasticity

All split rings to be 2-1/2" (62.50 mm) diameter. All bolts to be 1/2" (12.50 mm) diameter machine bolts. Washers may be 2"x2"x1/8" (50 x 50 x 3.12 mm) plate washers, 2-1/8" (53.12 mm) diameter cast or malleable iron washers or cut washers.

No allowances have been included for roof overhangs.

Quantities listed for diagonals are most economical lengths to produce a long and short member from a single length.

Material Requirements for One Split-Ring Connected Trussed Rafter of Various Spans for a 4 in 12 Roof Slope					
Span	20'-0"	24'-0"	26'-0"	28'-0"	30'-0"
Top Chords	2-2"x 6"	2-2"x 6"	2-2"x 6"	2-2"x 8"	2-2"x 8"
......	12'-0"	14'-0"	16'-0"	16'-0"	18'-0"
Bottom Chords	2-2" x 4"	2-2" x 4"	2-2" x 4"	2-2" x 4"	2-2" x 4"
......	12'-0"	14'-0"	16'-0"	16'-0"	18'-0"
Diagonals	2-2" x 4"	2-2" x 4"	2-2" x 4"	2-2" x 4"	2-2" x 4"
......	8'-0"	10'-0"	10'-0"	12'-0"	12'-0"
Joint Scabs	1-1" x 4"	1-1" x 4"	1-1" x 4"	1-1" x 4"	1-1" x 4"
......	2'-0"	2'-0"	2'-0"	3'-0"	3'-0"
Total Bd.ft.	51-1/3	60-2/3	67-1/3	81.0	89.0
Split Rings	11	11	11	13	13
Bolts					
1/2"x4"	4	4	4	6	6
1/2"x6"	2	2	2	2	2
1/2"x7-1/2"	1	1	1	1	1
Washers	14	14	14	18	18
Nails, 8d	16	16	16	16	16

Metric					
Span	6.09 M	7.31 M	7.92 M	8.53 M	9.14 M
Top Chords	2-50x150 mm	2-50x150 mm	2-50x150 mm	2-50x200 mm	2-50x200 mm
......	3.65 M	4.26 M	4.87 M	4.87M	5.48 M
Bottom Chords	2-50x100 mm	2-50x100 mm	2-50x100 mm	2-50x100 mm	2-50x100 mm
......	3.65 M	4.26 M	4.87 M	4.87 M	5.48 M
Diagonals	2-50x100 mm	2-50x100 mm	2-50x100 mm	2-50x100 mm	2-50x100 mm
......	2.43 M	3.04 M	3.04 M	3.65 M	3.65 M
Joint Scabs	1-25x100 mm	1-25x100 mm	1-25x100 mm	1-25x100 mm	1-25x100 mm
......	0.61 M	0.61 M	0.61 M	0.91 M	0.91 M
Total Cu.M	0.121	0.143	0.159	0.191	0.210
Split Rings	11	11	11	13	13
Bolts					
12.50x100 mm	4	4	4	6	6
12.50x150 mm	2	2	2	2	2
12.50x187.50mm	1	1	1	1	1
Washers	14	14	14	18	18
Nails, 8d	16	16	16	16	16

Note: Materials requirements same for 9.75M span same as 9.14M span

CARPENTRY

Material Requirements for One Split-Ring Connected Trussed Rafter of Various Spans for a 5 in 12 Roof Slope					
Span	20'-0"	24'-0"	26'-0"	28'-0"	30'-0"
Top Chords	2-2"x 6"	2-2"x 6"	2-2"x 6"	2-2"x 6"	2-2"x 6"
......	12'-0"	14'-0"	16'-0"	18'-0"	18'-0"
Bottom Chords	2-2"x 4"	2-2"x 4"	2-2"x 4"	2-2"x 4"	2-2"x 4"
......	12'-0"	14'-0"	16'-0"	16'-0"	18'-0"
Diagonals	2-2"x 4"	2-2"x 4"	2-2"x 4"	2-2"x 4"	2-2"x 4"
......	10'-0"	12'-0"	12'-0"	12'-0"	14'-0"
Joint Scabs	1-1"x 4"	1-1"x 4"	1-1"x 4"	1-1"x 4"	1-1"x 4"
......	2'-0"	2'-0"	2'-0"	2-0"	2'-0"
Total Bd.ft.	54	63-1/3	70	74	79-1/3
Split Rings	11	11	11	11	11
Bolts					
1/2"x4"	4	4	4	4	4
1/2"x6"	2	2	2	2	2
1/2"x7-1/2"	1	1	1	1	1
Washers	14	14	14	14	14
Nails, 8d	16	16	16	16	16

Metric					
Span	6.09 M	7.31 M	7.92 M	8.53 M	9.14 M
Top Chords	2-50x150mm	2-50x150mm	2-50x150mm	2-50x150mm	2-50x150mm
......	3.65 M	4.26 M	4.87 M	5.48	5.48
Bottom Chords	2-50x100 mm	2-50x100 mm	2-50x100 mm	2-50x100 mm	2-50x100 mm
......	3.65 M	4.26 M	4.87 M	4.87 M	5.48
Diagonals	2-50x100 mm	2-50x100 mm	2-50x100 mm	2-50x100 mm	2-50x100 mm
......	3.04 M	3.65 M	3.65 M	3.65 M	4.26 M
Joint Scabs	1-25x100 mm	1-25x100 mm	1-25x100 mm	1-25x100 mm	1-25x100 mm
......	0.61 M	0.61 M	0.61 M	0.61 M	0.61 M
Total Cu.M.	0.127	0.149	0.165	0.175	0.187
Split Rings	11	11	11	11	11
Bolts					
12.50x 100 mm	4	4	4	4	4
12.50x 150 mm	2	2	2	2	2
12.50x187.5 mm	1	1	1	1	1
Washers	14	14	14	14	14
Nails, 8d	16	16	16	16	16

Note: Materials requirements same for 9.75M span same as 9.14M span

Material Requirements for One Split-Ring Connected Trussed Rafter of Various Spans for a 6 in 12 Roof Slope					
Span	20-0"	24'-0"	26'-0"	28'-0"	30'-0"
Top Chords	2-2"x 6"	2-2"x 6"	2-2"x 6"	2-2"x 6"	2-2"x 6"
......	14'-0"	16'-0"	16'-0"	18'-0"	18'-0"
Bottom Chords	2-2"x 4"	2-2"x 4"	2-2"x 4"	2-2"x 4"	2-2"x 4"
......	12'-0"	14'-0"	16'-0"	18'-0"	18'-0"
Diagonals	2-2"x 4"	2-2"x 4"	2-2"x 4"	2-2"x 4"	2-2"x 4"
......	10'-0"	12'-0"	14'-0"	14'-0"	16'-0"
Joint Scabs	1-1"x 4"	1-1"x 4"	1-1"x 4"	1-1"x 4"	1-1"x 4"
......	2'-0"	2'-0"	2'-0"	2'-0"	2'-0"
Total Bd.ft.	58	67-1/3	72-2/3	76-2/3	82
Bolts					
1/2"x4"	4	4	4	4	4
1/2"x6"	2	2	2	2	2
1/2"x7-1/2"	1	1	1	1	1
Washers	14	14	14	14	14
Nails, 8d	16	16	16	16	16

Metric					
Span	6.09 M	7.31 M	7.92 M	8.53 M	9.14 M
Top Chords	2-50x150 mm	2-50x150 mm	2-50x150 mm	2-50x150 mm	2-50x150 mm
......	4.26 M	4.87 M	4.87 M	5.48 M	5.48 M
Bottom Chords	2-50x100 mm	2-50x100 mm	2-50x100 mm	2-50x100 mm	2-50x100 mm
......	12'-0"	4.26 M	4.87 M	5.48 M	5.48 M
Diagonals	2-50x100 mm	2-50x100 mm	2-50x100 mm	2-50x100 mm	2-50x100 mm
......	10'-0"	12'-0"	14'-0"	14'-0"	16'-0"
Joint Scabs	1-25x100 mm	1-25x100 mm	1-25x100 mm	1-25x100 mm	1-25x100 mm
......	0.61 M	0.61 M	0.61 M	0.61 M	0.61 M
Total Cu.M.	0.137	0.159	0.171	0.181	0.193
Bolts					
12.50x100 mm	4	4	4	4	4
12.50x100 mm	2	2	2	2	2
12.50x187.5mm	1	1	1	1	1
Washers	14	14	14	14	14
Nails, 8d	16	16	16	16	16

Note: Materials requirements same for 9.75M span same as 9.14M span

CARPENTRY

Material Requirements for One Split-Ring Connected Trussed Rafter of Various Spans for a 7 in 12 Roof Slope					
Span	20'-0"	24'-0"	26'-0"	28'-0"	30'-0"
Top Chords	2-2"x 6"	2-2"x 6"	2-2"x 6"	2-2"x 6"	2-2"x 6"
......	14'-0"	16'-0"	16'-0"	18'-0"	20'-0"
Bottom Chords	2-2"x 4"	2-2"x 4"	2-2"x 4"	2-2"x 4"	2-2"x 4"
......	12'-0"	14'-0"	16'-0"	16'-0"	18'-0"
Diagonals	2-2"x 4"	2-2"x 4"	2-2"x 4"	2-2"x 4"	2-2"x 4"
......	12'-0"	14'-0"	14'-0"	16'-0"	16'-0"
Joint Scabs	1-1"x 4"	1-1"x 4"	1-1"x 4"	1-1"x 4"	1-1"x 4"
......	2'-0"	2'-0"	2'-0"	2'-0"	2'-0"
Total Bd.ft.	60-2/3	70	72-2/3	79-1/3	86
Split Rings	11	11	11	11	11
Bolts					
1/2"x4"	4	4	4	4	4
1/2"x6"	2	2	2	2	2
1/2"x7-1/2"	1	1	1	1	1
Washers	14	14	14	14	15
Nails, 8d	16	16	16	16	16

Metric					
Span	6.09 M	7.31 M	7.92 M	8.53 M	9.14 M
Top Chords	2-50x150 mm	2-50x150 mm	2-50x150 mm	2-50x150 mm	2-50x150 mm
......	4.26 M	4.87 M	4.87 M	5.48 M	6.09 M
Bottom Chords	2-50x100 mm	2-50x100 mm	2-50x100 mm	2-50x100 mm	2-50x100 mm
......	12'-0"	14'-0"	4.87 M	4.87 M	5.48 M
Diagonals	2-50x100 mm	2-50x100 mm	2-50x100 mm	2-50x100 mm	2-50x100 mm
......	3.65 M	4.26 M	4.26 M	4.87 M	4.87 M
Joint Scabs	1-25x100 mm	1-25x100 mm	1-25x100 mm	1-25x100 mm	1-25x100 mm
......	0.61 M	0.61 M	0.61 M	0.61 M	0.61 M
Total Cu. M	0.143	0.165	0.171	0.187	0.203
Split Rings	11	11	11	11	11
Bolts					
50x100 mm	4	4	4	4	4
50x100 mm	2	2	2	2	2
50x187.50 mm	1	1	1	1	1
Washers	14	14	14	14	15
Nails, 8d	16	16	16	16	16

Note: Materials requirements same for 9.75M span same as 9.14M span

Labor Framing Lumber in Building Construction

The labor quantities and costs that follow are intended to include all classes of frame construction, such as residences, ranch houses, barns, stables, apartment buildings, combined store and apartment buildings, country clubs, schools, and in fact, all buildings that are constructed entirely of wood or have brick, stone, tile or cement block walls and wood joists, rafters, stud walls, partitions, wood subfloors, wall and roof sheathing, or finish wood floors.

Costs are given on two classes of workmanship. "Ordinary workmanship" is encountered in most buildings, where price is the main factor. "First grade workmanship" is where quality is a factor as well, usually found in high-grade residences, apartments, hotel and office buildings, public buildings, high school, and college and university buildings.

The estimator must use his or her own judgment as to the grade of workmanship required, depending upon the type of building, the reputation of the architect, and the requirements of the specifications.

Another thing that is going to govern labor costs is the kind of equipment used. All costs given are based on using an electric saw for cutting all joists, studs, and rafters to length and for cutting off ends of subflooring, roof sheathing, etc. If an ordinary handsaw were used, you would add 1 to 1-1/2 hrs. carpenter time per 1,000 bd.ft. (2.36 cu.m).

All of the labor quantities and costs are based on laborers or helpers handling and carrying the lumber from the stock piles or benches where the lumber is being cut to length to the building. If the lumber is handled and carried by carpenters, figure labor time at carpenter wages.

Framing and Placing Foundation Wall Plates. Where 2"x4" (50 x 100 mm) or 2"x6" (50 x 150 mm) wood plates are placed on top of foundation walls or concrete slab to receive the floor joists and exterior wall studs, it is customary to place anchor bolts in the walls and floors. The plates are then bored to receive the bolts and placed on the wall (and usually wedged with shingles) ready to receive the joists and exterior wall studs. Special purpose nails may be used for anchoring wall plates to concrete.

Where just an ordinary grade of workmanship is required, two carpenters working together should handle, frame and place 225 to 275 lin.ft. (68.58 to 83.82 m) of 2"x4" (50 x 100 mm) or 2"x6" (50 x 150 mm) plates per 8-hr. day, at the following cost per 100 lin.ft. (30.48 m):

Description	Hours	Rate	Total	Rate	Total
Carpenter	6.40	$	$	$ 35.04	$ 224.26
Cost per lin.ft.					$ 2.24
per m					$ 7.36

908

CARPENTRY

First Grade Workmanship

On jobs where first grade workmanship is required, with the foundation wall plates drilled for bolts, plates set and bedded absolutely level in a bed of cement mortar, two carpenters working together should handle, frame, and place 175 to 225 lin.ft. (53.34 to 68.58 m) of 2"x4" (50 x 100 mm) or 2"x6" (50 x 150 mm) wall plates per 8-hr. day at the following labor cost per 100 lin.ft. (30.48 m):

Description	Hours	Rate	Total	Rate	Total
Carpenter	8.00	$	$	$ 35.04	$ 280.32
Cost per lin.ft.					$ 2.80
per m					$ 7.46

Framing and Placing Box Sills and Plates. On platform framing, where a wood box sill and plate is formed by using a 2" x 4" (50 x 100 mm) or 2"x6" (50 x 150 mm) plate and a 2" x 8" (50 x 200 mm) or 2" x 10" (50 x 250 mm) on the end of the joists to form a box sill, the wall plate should be drilled, leveled, and set the same as described above, and the side or end piece is nailed after the joists are set.

This work should be estimated as given above for "Foundation Wall Plates" and the 2" x 8" (50 x 200 mm) or 2" x 10" (50 x 250 mm) end pieces should be figured in with the floor joists.

Framing and Erecting Exterior Stud Walls for Frame Buildings. The labor cost of framing and erecting stud walls is subject to wide variation, depending upon the type of building, height, regularity of the walls, etc.

The framing on square or rectangular buildings, such as Cape Cod and colonial type houses, costs much less than for English type houses, having walls of irregular shape and height. The average house requires only 1,500 to 2,500 lin.ft. (457 to 762 m) of lumber for outside stud walls.

On square or rectangular shaped buildings, such as colonial, Georgian houses, etc., a carpenter should frame and erect 350 to 400 bd.ft. (0.82 to 0.94 cu.m) of lumber per 8-hr. day, at the following cost per 1,000 bd.ft. (2.36 cu.m):

Description	Hours	Rate	Total	Rate	Total
Carpenter	21.40	$	$	$ 35.04	$ 749.86
Labor	6.00	$	$	$ 26.16	$ 156.96
Cost 1,000 bd.ft.					$ 906.82
per cu.m					$ 384.49

On English type and other buildings having irregular wall construction, a carpenter should frame and erect 250 to 300 bd.ft. (0.59-0.70

909

cu.m) of lumber per 8-hr. day, at the following labor cost per 1,000 bd.ft. (2.36 cu.m):

Description	Hours	Rate	Total	Rate	Total
Carpenter	29.10	$	$	$ 35.04	$ 1,019.66
Labor	6.00	$	$	$ 26.16	$ 156.96
Cost 1,000 bd.ft.					$ 1,176.62
per cu.m					$ 498.89

First Grade Workmanship

In high class wood constructed buildings, where every precaution is taken to prevent settling due to shrinkage, where the wood studs rest on masonry walls or steel I beams and are not set on top of the floor joists, and where it is necessary to bridge or truss between all studs and over door and window openings, a carpenter should frame and erect 225 to 275 bd.ft. (0.53 to 0.65 cu.m) of lumber per 8-hr. day, on square or rectangular type buildings, at the following labor cost per 1,000 bd.ft. (2.36 cu.m):

Description	Hours	Rate	Total	Rate	Total
Carpenter	32.00	$	$	$ 35.04	$ 1,121.28
Labor	10.00	$	$	$ 26.16	$ 261.60
Cost 1,000 bd.ft.					$ 1,382.88
per cu.m					$ 586.34

On English type and other buildings having irregular wall construction, a carpenter should frame and erect 150 to 200 bd.ft. (0.35 to 0.47 cu.m) of lumber per 8-hr day, at the following labor cost per 1,000 bd.ft. (2.36 cu.m):

Description	Hours	Rate	Total	Rate	Total
Carpenter	45.70	$	$	$ 35.04	$ 1,601.33
Labor	10.00	$	$	$ 26.16	$ 261.60
Cost 1,000 bd.ft.					$ 1,862.93
per cu.m					$ 789.88

Framing Interior Stud Partitions. When framing and setting interior stud partitions set on top of rough wood floors that require just the ordinary amount of framing for door openings, a carpenter should frame and erect 375 to 425 bd.ft. (0.88 to 1.00 cu.m) per 8-hr. day, at the following labor cost per 1,000 bd.ft. (2.36 cu.m):

910

Description	Hours	Rate	Total	Rate	Total
Carpenter	20.00	$	$	$ 35.04	$ 700.80
Labor	6.00	$	$	$ 26.16	$ 156.96
Cost 1,000 bd.ft.					$ 857.76
per cu.m					$ 363.69

First Grade Workmanship

Where the wood partition studs are set on masonry walls or steel I beams instead of the wood subfloor, and where it is necessary to brace between studs and truss over all door openings, a carpenter should frame and erect 250 to 300 bd.ft. (0.59 to 0.70 cu.m) of lumber per 8-hr. day, at the following labor cost per 1,000 bd.ft. (2.36 cu.m):

Description	Hours	Rate	Total	Rate	Total
Carpenter	29.10	$	$	$ 35.04	$1,019.66
Labor	8.00	$	$	$ 26.16	$ 209.28
Cost 1,000 bd.ft.					$1,228.94
per cu.m					$ 521.07

Framing and Setting Floor Joists. When framing and placing wood floor joists up to 2"x8" (50 x 200 mm) in buildings of regular construction, a carpenter should frame and place 550 to 600 bd.ft. (1.30 to 1.42 cu.m) per 8-hr. day, at the following labor cost per 1,000 bd.ft. (2.36 cu.m):

Description	Hours	Rate	Total	Rate	Total
Carpenter	13.90	$	$	$ 35.04	$ 487.06
Labor	6.00	$	$	$ 26.16	$ 156.96
Cost 1,000 bd.ft.					$ 644.02
per cu.m					$ 273.06

If 2"x10" (50 x 250 mm) or 2"x12" (50 x 300 mm) joists are used, a carpenter should frame and erect 600 to 650 bd.ft. (1.42-1.53 cu.m) per 8-hr. day, at the following labor cost per 1,000 bd.ft. (2.36 cu.m):

Description	Hours	Rate	Total	Rate	Total
Carpenter	12.80	$	$	$ 35.04	$ 448.51
Labor	6.00	$	$	$ 26.16	$ 156.96
Cost 1,000 bd.ft.					$ 605.47
per cu.m					$ 256.72

First Grade Workmanship

On jobs where the wood joists must be set with the crowning edge up and it is not permissible to block up under the joists with shingles or wood wedges, a carpenter should frame and erect 500 to 550 bd.ft. (1.18 to 1.30 cu.m), in sizes up to 2"x8" (50 x 200 mm), per 8-hr. day, at the following labor cost per 1,000 bd.ft. (2.36 cu.m):

Description	Hours	Rate	Total	Rate	Total
Carpenter	15.20	$	$	$ 35.04	$ 532.61
Labor	4.50	$	$	$ 26.16	$ 117.72
Cost 1,000 bd.ft.					$ 650.33
per cu.m					$ 275.74

Using 2"x10" (50 x 250 mm) or 2"x12" (50 x 300 mm) joists, a carpenter should frame and erect 550 to 650 b.f. (1.30 to 1.53 cu.m) of lumber per 8-hr. day, at the following labor cost per 1,000 bd.ft. (2.36 cu.m):

Description	Hours	Rate	Total	Rate	Total
Carpenter	13.90	$	$	$ 35.04	$ 487.06
Labor	4.00	$	$	$ 26.16	$ 104.64
Cost 1,000 bd.ft.					$ 591.70
per cu.m					$ 250.88

Framing Panel and Girder Floor Systems. When framing and placing 4"x6" (100 x 150 mm) wood beams and 2"x4" (50 x 100 mm) spacers to form 4-ft. (1.22-m) square grids for panel and girder floor systems, 2 carpenters and a helper should frame and place 1,500 to 1,700 bd.ft. (3.54 to 4.0 cu.m) of lumber per 8-hr. day, at the following labor cost per 1,000 bd.ft. (2.36 cu.m):

CARPENTRY

Description	Hours	Rate	Total	Rate	Total
Carpenter	10.00	$	$	$ 35.04	$ 350.40
Labor	5.00	$	$	$ 26.16	$ 130.80
Cost 1,000 bd.ft.					$ 481.20
per cu.m					$ 204.03

Installing Cross Bridging. Under average conditions a carpenter should cut and place 90 to 110 sets (2 pieces) of cross-bridging per 8-hr. day at the following labor cost per 100 sets:

Description	Hours	Rate	Total	Rate	Total
Carpenter	8.00	$	$	$ 35.04	$ 280.32
Cost per Set					$ 28.03

Framing and Erecting Rafters for Gable Roofs. When framing and erecting rafters for plain double pitch or gable roofs without dormers or gables, and where 2"x6" (50 x 150 mm) or 2"x8" (50 x 200 mm) lumber is used, a carpenter should frame and erect 285 to 335 bd.ft. (0.67 to 0.79 cu.m) per 8-hr. day, at the following labor cost per 1,000 bd.ft. (2.36 cu.m):

Description	Hours	Rate	Total	Rate	Total
Carpenter	24.60	$	$	$ 35.04	$ 861.98
Labor	7.00	$	$	$ 26.16	$ 183.12
Cost 1,000 bd.ft.					$1,045.10
per cu.m					$ 443.12

On buildings having double pitch or gable roofs cut up with dormers and gables framing into the main roof, a carpenter should frame and erect 250 to 300 bd.ft. (0.59 to 0.70 cu.m) of lumber per 8-hr. day, at the following labor cost per 1,000 bd.ft. (2.36 cu.m):

Description	Hours	Rate	Total	Rate	Total
Carpenter	29.10	$	$	$ 35.04	$1,019.66
Labor	8.00	$	$	$ 26.16	$ 209.28
Cost 1,000 bd.ft.					$1,228.94
per cu.m					$ 521.07

Framing and Erecting Rafters for Hip Roofs. When framing and erecting plain hip roofs without dormers or gables framing into the main roof,

913

a carpenter should frame and erect 250 to 300 bd.ft. (0.59 to 0.70 cu.m) of lumber per 8-hr. day, at the following labor cost per 1,000 bd.ft. (2.36 cu.m):

Description	Hours	Rate	Total	Rate	Total
Carpenter	29.10	$	$	$ 35.04	$ 1,019.66
Labor	8.00	$	$	$ 26.16	$ 209.28
Cost 1,000 bd.ft.					$ 1,228.94
per cu.m					$ 521.07

On difficult hip roofs, where it is necessary to frame for dormers, gables, and valleys, a carpenter should frame and erect 180 to 220 bd.ft. (0.42 to 0.52 cu.m) of lumber per 8-hr. day, at the following labor cost per 1,000 bd.ft. (2.36 cu.m):

Description	Hours	Rate	Total	Rate	Total
Carpenter	40.00	$	$	$ 35.04	$ 1,401.60
Labor	8.00	$	$	$ 26.16	$ 209.28
Cost 1,000 bd.ft.					$ 1,610.88
per cu.m					$ 683.01

Framing Light Timbers for Exposed Roof Beam Construction, Light timbers, varying from 4"x6" (100 x 150 mm) to 4"x14" (100 x 350 mm) and larger, depending upon span and spacing, are used for exposed beam roof construction. This type of construction is widely used for contemporary design-houses, stores, medical groups, small churches-with flat or low rise sloping roofs.

On work of this type, two carpenters and a helper should frame and place 750 to 850 bd.ft. (1.77-2.0 cu.m) of lumber per 8-hr. day, at the following labor cost per 1,000 bd.ft. (2.36 cu.m):

Description	Hours	Rate	Total	Rate	Total
Carpenter	20.00	$	$	$ 35.04	$ 700.80
Labor	10.00	$	$	$ 26.16	$ 261.60
Cost 1,000 bd.ft.					$ 962.40
per cu.m					$ 408.06

Framing for Roof Saddles on Flat Roofs. On flat roofs where it is necessary to frame saddles to pitch the water toward the drains, and where the saddles are built up of 2"x4" (50 x 100 mm) framing covered with 1" (25 mm) sheathing, a carpenter should frame and erect 400 to 450 bd.ft. (0.94-

914

CARPENTRY

1.06 cu.m) of lumber per 8-hr. day, at the following labor cost per 1,000 bd.ft. (2.36 cu.m):

Description	Hours	Rate	Total	Rate	Total
Carpenter	19.00	$	$	$ 35.04	$ 665.76
Labor	6.00	$	$	$ 26.16	$ 156.96
Cost 1,000 bd.ft.					$ 822.72
per cu.m					$ 348.83

Placing Wood Cant Strips. When cant strips, diagonally cut from 4"x4" (100 x 100 mm) or 6"x6" (150 x 150 mm) lumber, occur in long runs with few breaks, a carpenter should place 475 to 525 lin.ft. (144.78 to 160.02 m) per 8-hr. day at the following labor cost per 100 lin.ft. (30.48 m):

Description	Hours	Rate	Total	Rate	Total
Carpenter	1.60	$	$	$ 35.04	$ 56.06
Cost Per Lin.Ft.					$ 0.56
per Lin. M					$ 1.84

If cant strips occur in short runs, such as around roof opening curbs, or if the runs are cut up with pilasters and other breaks, a carpenter should place 300 to 325 lin.ft. (91.44 to 99.06 m) per 8-hr. day at the following labor cost per 100 lin.ft. (30.04 m):

Description	Hours	Rate	Total	Rate	Total
Carpenter	2.60	$	$	$ 35.04	$ 91.10
Cost Per Lin.Ft.					$ 0.91
per Lin. M					$ 2.99

Laying Rough Wood Floors. When laying 1"x6" (25 x 150 mm) or 1"x8" (25 x 200 mm) rough wood subflooring, a carpenter should lay 825 to 925 bd.ft. (1.95-2.18 cu.m) per 8-hr. day, at the following labor cost per 1,000 bd.ft. (2.36 cu.m):

Description	Hours	Rate	Total	Rate	Total
Carpenter	9.20	$	$	$35.04	$ 322.37
Labor	5.30	$	$	$26.16	$ 138.65
Cost 1,000 bd.ft.					$ 461.02
per cu.m					$ 195.47

915

Laying Rough Wood Floors Diagonally. When laying rough wood subflooring diagonally, making it necessary to cut both ends of the flooring on a bevel, a carpenter should lay 650 to 750 bd.ft. (1.53-1.77 cu.m) of flooring per 8-hr. day, at the following labor cost per 1,000 bd.ft. (2.36 cu.m):

Description	Hours	Rate	Total	Rate	Total
Carpenter	11.50	$	$	$35.04	$ 402.96
Labor	5.50	$	$	$26.16	$ 143.88
Cost 1,000 bd.ft.					$ 546.84
per cu.m					$ 231.86

Laying Wood Sheathing on Flat Roofs. Figure same as given above for Rough Wood Floors.

Labor Laying Plywood Subflooring. The labor cost of handling and laying plywood subflooring can vary widely, depending on the size of the floor area, the regularity of the joist spacing, and the method of laying.

When plywood is used as subflooring, nailed directly to the joists and with the finish floor applied directly to the plywood, all edges (both longitudinal and crosswise) should be nailed about 6" (150 mm) on centers. Otherwise, the finish floor may develop squeaks caused by deflection of the long edges of the sheets when walked upon.

In other words, with the joists spaced 16" (400 mm) o.c., it is necessary to cut pieces of 2"x4" (50 x 100 mm) approximately 14-3/8" (365.12mm) long and the ends of these 2"x4"s (50 x 100 mm) must be nailed into the sides of the joists to provide nailing for the long edges of the plywood sheets. As the plywood is usually 4'-0" (1.21 m) wide, it means a row of 2"x4" (50 x 100 mm) blocking must be run at right angles to the floor joists every 4'-0" (1.21 m) apart to provide nailing for the long edges of the plywood sheets. This requires a lot of extra work and costs a lot of money.

In buildings with regular spans, where an underlayment is to be provided, and where the long edges of the plywood sheets are not nailed, two carpenters working together should handle, fit, lay, and nail 52 to 60 sheets, 1,664 to 1,920 sq.ft. (154.58 to 178.36 sq.m) of 4'-0"x8'-0" (1.21 x 2.42.43 m) plywood 5/8" to 3/4" (15.62 to 18.75 mm) thick per 8-hr. day, at the following labor cost per 100 sq.ft. (9.29 sq.m):

916

CARPENTRY

Description	Hours	Rate	Total	Rate	Total
Carpenter	0.90	$	$	$35.04	$ 31.54
Labor	0.30	$	$	$26.16	$ 7.85
Cost 100 Sq.Ft. (6.29 Sq.M.)					$ 39.38
Cost Sq.Ft.					$ 0.39
per Sq.M					$ 4.24

If 2"x4" (50 x 100 mm) wood blocking is required between the joists, and the plywood sheets must be nailed both longitudinally and crosswise, two carpenters working together should cut and nail blocks, and handle, fit, lay, and nail 26 to 30 sheets of 4'-0"x8'-0" (1.21 x 2.43 m) plywood, 832 to 960 sq.ft. (77.29 to 89.18 sq.m) 5/8" to 3/4" (15.62 to 18.75 mm) thick per 8-hr. day, at the following labor cost per 100 sq.ft. (9.29 sq.m):

Description	Hours	Rate	Total	Rate	Total
Carpenter	1.80	$	$	$35.04	$ 63.07
Labor	0.50	$	$	$26.16	$ 13.08
Cost 100 Sq.Ft. (9.29 Sq.M.)					$ 76.15
Cost Sq.Ft.					$ 0.76
per Sq.M					$ 8.20

Laying Plywood Decking for Panel and Girder Floor Systems. Plywood decking for panel and girder floor construction should be 1" (25 mm) or 1-1/4" (31.25 mm) thick and laid with staggered joints. To minimize waste, inside dimensions of the building should be laid out on a 4-ft. (1.21 m) module, which also reduces cutting of panels to starters in every other row. In some localities, 4'x4' (1.21 x 1.21 m) plywood panels are available as a stock size.

On large projects, where the work may be highly organized, a carpenter and two helpers should lay 4,500 to 4,700 sq.ft. (418.05 to 436.63 sq.m) of plywood decking per 8-hr. day, or enough for 3 to 4 average one-story houses, at the following labor cost per 1,000 sq.ft. (92.90 sq.m):

917

Description	Hours	Rate	Total	Rate	Total
Carpenter	1.75	$	$	$35.04	$ 61.32
Labor	3.50	$	$	$26.16	$ 91.56
Cost 1,000 Sq.Ft. (92.90 Sq.M.)					$ 152.88
Cost Sq.Ft.					$ 0.15
per Sq.M					$ 1.65

For smaller jobs, where only one or two houses are involved, increase the above costs about 25% to absorb lost motion in laying out work and getting started.

Where plywood decking is to be covered with linoleum or other resilient flooring, requiring a fairly smooth surface, a carpenter should spackle and sand joints and surface cracks and other imperfections at the rate of 200 sq.ft. (18.58 sq.m) per hr.

Laying Wood Sheathing on Pitch or Gable Roofs. When laying roof sheathing on plain hip or gable roofs, a carpenter should lay 575 to 615 bd.ft. (1.36-1.45 cu.m) of lumber per 8-hr. day, at the following labor cost per 1,000 bd.ft. (2.36 cu.m):

Description	Hours	Rate	Total	Rate	Total
Carpenter	13.50	$	$	$35.04	$ 473.04
Labor	6.50	$	$	$26.16	$ 170.04
Cost 1,000 bd.ft.					$ 643.08
per cu.m					$ 272.67

On roofs that are very steep or cut up with dormers, hips, and valleys, such as English type houses, a carpenter will lay 275 to 325 bd.ft. (0.65-0.77 cu.m) of sheathing per 8-hr. day at the following labor cost per 1,000 bd.ft. (2.36 cu.m):

Description	Hours	Rate	Total	Rate	Total
Carpenter	26.50	$	$	$35.04	$ 928.56
Labor	7.50	$	$	$26.16	$ 196.20
Cost 1,000 bd.ft.					$ 1,124.76
per cu.m					$ 476.90

Labor Placing Sidewall Sheathing. When sheathing sidewalls of frame buildings with 1"x6" (25 x 150 mm) or 1"x8" (25 x 200 mm) boards laid horizontally, a carpenter should handle and place 625 to 700 bd.ft. (1.47-

918

CARPENTRY

1.65 cu.m) per 8-hr. day, at the following labor cost per 1,000 bd.ft. (2.36 cu.m):

Description	Hours	Rate	Total	Rate	Total
Carpenter	12.00	$	$	$35.04	$ 420.48
Labor	5.00	$	$	$26.16	$ 130.80
Cost 1,000 bd.ft.					$ 551.28
per cu.m					$ 233.74

Labor Placing Diagonal Sidewall Sheathing. When sidewall sheathing is placed diagonally, it usually requires two carpenters working together-one at each end of the board-and usually requires scaffolding the outside walls before the sheathing is placed. The worker at the top must have a scaffold from which to work.

On work of this class, a carpenter should place 450 to 525 bd.ft. (1.06-1.24 cu.m) of 1"x6" (25 x 150 mm) or 1"x8" (25 x 200 mm) sheathing per 8 hr. day, at the following labor cost per 1,000 bd.ft. (2.36 cu.m):

Description	Hours	Rate	Total	Rate	Total
Carpenter	16.50	$	$	$35.04	$ 578.16
Labor	5.50	$	$	$26.16	$ 143.88
Cost 1,000 bd.ft.					$ 722.04
per cu.m					$ 306.14

WOOD BLOCKING, FURRING AND GROUNDS

Placing Wood Furring Strips on Masonry Walls. Where it is necessary to place wood furring strips on brick or tile walls before lathing and plastering, allowing the strips to follow the line of the walls without wedging or blocking out to make them straight or plumb, with the nails driven into dry joints in the brickwork, a carpenter should place 500 to 550 lin.ft. (152.40 to 164.64 m) of furring per 8-hr. day, at the following labor cost per 100 lin.ft. (30.48 m):

Description	Hours	Rate	Total	Rate	Total
Carpenter	1.50	$	$	$35.04	$ 52.56
Cost per Lin.Ft.					$ 0.53
Per Lin. M					$ 1.72

Cost per Square (100 sq.ft. or 9.29 sq.m)

Description	Hours	Rate	Total	Rate	Total
Strips 12" Centers (300 mm)					
Carpenter	1.70	$	$	$35.04	$ 59.57
Strips 16" Centers (400 mm)					
Carpenter	1.30	$	$	$35.04	$ 45.55

First Grade Workmanship

Where it is necessary to plug the masonry walls and place all furring strips absolutely straight and plumb to produce a level surface to receive lath and plaster, a carpenter should plug walls and place 200 to 250 lin.ft. (60.96 to 76.20 m) of strips per 8-hr. day, at the following labor cost per 100 lin.ft. (30.48 m):

Description	Hours	Rate	Total	Rate	Total
Carpenter	3.50	$	$	$35.04	$ 122.64
Cost per Lin.Ft.					$ 1.23
Per Lin. M					$ 4.02

Cost per Square (100 sq.ft. or 9.29 sq.m)

Description	Hours	Rate	Total	Rate	Total
Strips 12" Centers (300 mm)					
Carpenter	3.90	$	$	$35.04	$ 136.66
Strips 16" Centers (400 mm)					
Carpenter	2.90	$	$	$35.04	$ 101.62

Placing Wood Grounds. Where wood grounds are nailed directly to wood furring strips, door and window openings, etc., allowing them to follow the rough furring, partitions and wood bucks, without wedging or blocking to make them absolutely straight, a carpenter should place 540 to 590 lin.ft. (164.59 to 179.83 mm) per 8-hr. day, at the following labor cost per 100 lin.ft. (30.48 m):

920

CARPENTRY

Description	Hours	Rate	Total	Rate	Total
Carpenter	1.40	$	$	$35.04	$ 49.06
Cost per Lin.Ft.					$ 0.49
Per Lin. M					$ 1.61

First Grade Workmanship

Where it is necessary to keep all wood grounds absolutely straight, plumb, and level for the finish plaster and for the interior wood finish, and where it is necessary to plug the masonry walls with wood plugs to hold the nails, a carpenter should place 200 to 250 lin.ft. (60.96 to 76.20 m) of grounds per 8-hr. day, at the following labor cost per 100 lin.ft. (30.48 m).

Description	Hours	Rate	Total	Rate	Total
Carpenter	3.50	$	$	$35.04	$ 122.64
Cost per Lin.Ft.					$ 1.23
Per Lin. M					$ 4.02

Wood Blocking and Grounds Around Window Openings. Where the masonry walls are 12" (300 mm) or more in thickness, requiring a wood jamb lining, a carpenter should place all necessary wood blocking and grounds for a window up to 4'-0"x6'-0" (1.21 x 1.82 m) in size in 3/4 to 1-1/4 hr. at the following labor cost:

Description	Hours	Rate	Total	Rate	Total
Carpenter	1.00	$	$	$35.04	$ 35.04

Windows 4'-0"x7'-0" (1.21 x 2.13 m) to 5'-0"x8'-0" (1.5 2x 2.43 m) in size will require 1-1/8 to 1-3/8 hrs. to place blocking and grounds, at the following labor cost per window:

Description	Hours	Rate	Total	Rate	Total
Carpenter	1.25	$	$	$35.04	$ 43.80

First Grade Workmanship

In the better class of buildings, where all grounds must be absolutely straight and plumb, and it is necessary to plug the brick and block walls for securing the grounds and blocking, a carpenter should complete one window up to 4'-0"x6'-0" (1.21 x 1.82 m) in size in 1-1/4 to 1-3/4 hrs. at the following labor cost:

921

Description	Hours	Rate	Total	Rate	Total
Carpenter	1.50	$	$	$35.04	$ 52.56

A carpenter should plug walls, place blocking and grounds for one window 4'-0"x7'-0" (1.21 x 2.13 m) to 5'-0"x8'-0" (1.52 x 2.43 m) in 1-1/2 to 2 hrs. at the following labor cost per window:

Description	Hours	Rate	Total	Rate	Total
Carpenter	1.75	$	$	$35.04	$ 61.32

Wood Blocking for Cabinet and Case Bases. A carpenter should locate, frame and set 115 to 135 bd.ft. (0.27-0.32 cu.m) of 2"x4" (50 x 100 mm) or 2"x6" (50 x 150 mm) blocking for cabinet or case bases per 8-hr. day at the following labor cost per 1,000 bd.ft. (2.36 cu.m):

Description	Hours	Rate	Total	Rate	Total
Carpenter	64.00	$	$	$35.04	$2,242.56
Cost per 1,000 Bd.Ft.					$2,242.56
Cost per Cu.M					$ 950.85

Wood Furring Strips Over Wood Subfloors. When placing 1"x2" (25 x 50 mm) or 2"x2" (50 x 50 mm) wood furring strips over wood subfloors to receive finish flooring, a carpenter should place 550 to 600 lin.ft. (167.64 to 182.88 m), without wedging or leveling, per 8-hr. day, at the following labor cost per 100 lin.ft. (30.48 m):

Description	Hours	Rate	Total	Rate	Total
Carpenter	1.40	$	$	$35.04	$ 49.06
Labor	0.30	$	$	$26.16	$ 7.85
Cost per 100 Lin.Ft.					$ 56.90
Cost per Lin.Ft.					$ 0.57
Per Lin. M					$ 1.87

CARPENTRY
Cost per Square (100 sq.ft. or 9.29 sq.m)

Description	Hours	Rate	Total	Rate	Total
Strips 12" Centers (300 mm)					
Carpenter	1.90	$	$	$35.04	$ 66.58
Strips 16" Centers (400 mm)					
Carpenter	1.40	$	$	$35.04	$ 49.06

First Grade Workmanship

In the better class of buildings where the wood furring strips must be wedged up and blocked to produce an absolutely level surface to receive finish flooring, a carpenter should place 250 to 300 lin.ft. (76.20 to 91.44 m) of furring strips per 8-hr. day, at the following cost per 100 lin.ft. (30.48 m).

Description	Hours	Rate	Total	Rate	Total
Carpenter	3.00	$	$	$ 35.04	$ 105.12
Labor	0.30	$	$	$ 26.16	$ 7.85
Cost per 100 Lin.Ft.					$ 112.97
Cost per Lin.Ft.					$ 1.13
Per Lin. M					$ 3.71

Cost per Square (100 sq.ft. or 9 sq.m)

Description	Hours	Rate	Total	Rate	Total
Strips 12" Centers (300 mm)					
Carpenter	3.60	$	$	$ 35.04	$ 126.14
Strips 16" Centers (400 mm)					
Carpenter	2.80	$	$	$ 35.04	$ 98.11

Placing Strip Deadening Felt and Wood Furring Strips Over Wood Subfloors. Frequently a narrow strip of deadening felt 1/2"x3" (12.75 x 75 mm) is placed under the wood furring strips and both are nailed to the wood subfloor. This provides better insulation and deadening than where the furring strips are nailed directly to the subfloor.

On work of this kind, a carpenter should place 425 to 475 lin.ft. (129.54 to 144.78 m) of deadening felt and furring strips per 8-hr. day, at the following labor cost per 100 lin.ft. (30.48m):

Description	Hours	Rate	Total	Rate	Total
Carpenter	1.80	$	$	$ 35.04	$ 63.07
Labor	0.40	$	$	$ 26.16	$ 10.46
Cost per 100 Lin.Ft.					$ 73.54
Cost per Lin.Ft.					$ 0.74
Per Lin. M					$ 2.41

Cost per Square (100 sq.ft. or 9.29 sq.m)

Description	Hours	Rate	Total	Rate	Total
Strips 12" Centers (300 mm)					
Carpenter	2.40	$	$	$ 35.04	$ 84.10
Strips 16" Centers (400 mm)					
Carpenter	1.90	$	$	$ 35.04	$ 66.58

First Grade Workmanship

Where it is necessary to wedge and block up the strips to produce an absolutely level surface to receive the finish flooring, a carpenter should place 220 to 250 lin.ft. (67.05 to 76.20 m) of deadening and furring strips per 8-hr. day, at the following labor cost per 100 lin.ft. (30.48 m):

Description	Hours	Rate	Total	Rate	Total
Carpenter	3.50	$	$	$ 35.04	$ 122.64
Labor	0.40	$	$	$ 26.16	$ 10.46
Cost per 100 Lin.Ft.					$ 133.10
Cost per Lin.Ft.					$ 1.33
Per Lin. M					$ 4.37

924

CARPENTRY

Cost per Square (100 sq.ft. or 9.29 sq.m)

Description	Hours	Rate	Total	Rate	Total
Strips 12" Centers (300 mm)					
Carpenter	4.30	$	$	$ 35.04	$ 150.67
Strips 16" Centers (400 mm)					
Carpenter	3.30	$	$	$ 35.04	$ 115.63

Placing Deadening Quilt Over Rough Wood Floors. Where deadening felt or quilt is laid over wood subfloors, a worker should handle and lay 200 to 250 sq.ft. (18.58 to 23.22 sq.m) per hr. at the following cost per 100 sq.ft. (9.29 sq.m):

Description	Hours	Rate	Total	Rate	Total
Labor	0.50	$	$	$ 26.16	$ 13.08
Cost per Sq. Ft .(100 Sq.Ft.)					$ 0.13
Cost per Sq.M					$ 1.41

The above costs are based on deadening felts or quilts in rolls or sheets that are laid on the rough floor with furring strips placed directly over them.

Cost of Placing Deadening Felt and Wood Furring Strips Over 100 Sq.Ft. (9.29 Sq.M) of Floor

Strips 12" (300 mm) O.C.-Ordinary Workmanship

Description	Hours	Rate	Total	Rate	Total
Carpenter	2.00	$	$	$ 35.04	$ 70.08
Labor	0.30	$	$	$ 26.16	$ 7.85
Cost per (100 Sq.Ft.)					$ 77.93
Cost per Sq.Ft.					$ 0.78
Per Lin. Sq.M					$ 8.39

Strips 16" (400 mm) O.C.-Ordinary Workmanship

Description	Hours	Rate	Total	Rate	Total
Carpenter	1.60	$	$	$ 35.04	$ 56.06
Labor	0.30	$	$	$ 26.16	$ 7.85
Cost per 100 Lin.Ft.					$ 63.91
Cost per Lin.Ft.					$ 0.64
Per Lin. M					$ 6.88

Strips 12" (300 mm) O.C.-First Grade Workmanship

Description	Hours	Rate	Total	Rate	Total
Carpenter	3.80	$	$	$ 35.04	$ 133.15
Labor	0.30	$	$	$ 26.16	$ 7.85
Cost per 100 Sq. Ft.					$ 141.00
Cost per Sq.Ft.					$ 1.41
Per Lin. Sq. M					$ 15.18

Strips 16" (400 mm) O.C.-First Grade Workmanship

Description	Hours	Rate	Total	Rate	Total
Carpenter	3.00	$	$	$ 35.04	$ 105.12
Labor	0.30	$	$	$ 26.16	$ 7.85
Cost per (100 Sq.Ft.)					$ 112.97
Cost per Sq.Ft.					$ 1.13
Per Lin. Sq.M					$ 12.16

Placing Wood Floor Sleepers. When placing 2"x3" (50 x 75 mm) or 2"x4" (50 x 100 mm) wood floor screeds or sleepers over rough tile or concrete floors, to receive finish flooring, a carpenter should place 225 to 275 lin.ft. (68-83 m) per 8-hr. day, at the following labor cost per 100 lin.ft. (30 m):

Description	Hours	Rate	Total	Rate	Total
Carpenter	3.20	$	$	$ 35.04	$ 112.13
Labor	0.80	$	$	$ 26.16	$ 20.93
Cost per 100 Lin.Ft.					$ 133.06
Cost per Lin.Ft.					$ 1.33
Per Lin. M					$ 4.37

CARPENTRY

First Grade Workmanship

In the better class of buildings, 2"x3" (50 x 75 mm) or 2"x4" (50 x 100 mm) beveled floor sleepers are placed over the rough concrete floors and wedged or blocked up to provide a perfectly level surface to receive the finish flooring.

The screeds are usually held in place by metal clips placed in the rough concrete, or anchored with special purpose nails.

On work of this class, a carpenter should place, wedge up, and level 130 to 170 lin.ft. (39-51 m) of sleepers per 8-hr. day, including setting sleeper clips, at the following labor cost per 100 lin.ft. (30 m):

Description	Hours	Rate	Total	Rate	Total
Carpenter	5.30	$	$	$ 35.04	$ 185.71
Labor	0.80	$	$	$ 26.16	$ 20.93
Cost per 100 Lin. Ft					$ 206.64
Cost per Lin..Ft.					$ 2.07
Per Lin. Lin. M					$ 6.78

Lumber Required for Wood Door Bucks of 2" x 4" (50x 100 mm) Lumber				
Size of Opening	Lin Ft. of Lumber Required	Bd.Ft of Lumber Req'd.	Transom Add Lin. Ft.	Bd.ft. of Lumber
2 '- 0 " x 4 '- 0 "	12	8	6 '- 0 "	4
2 '- 0 " x 4 '- 6 "	12	8	6 '- 0 "	4
2 '- 0 " x 5 '- 0 "	13 1/2	9	6 '- 0 "	4
2 '- 6 " x 4 '- 0 "	12	8	6 '- 6 "	4 1/3
2 '- 6 " x 4 '- 6 "	13	8 2/3	6 '- 6 "	4 1/3
2 '- 6 " x 5 '- 0 "	14	9 1/3	6 '- 6 "	4 1/3
2 '- 6 " x 5 '- 6 "	15	10	6 '- 6 "	4 1/3
2 '- 6 " x 6 '- 0 "	16	10 2/3	6 '- 6 "	4 1/3
2 '- 6 " x 6 '- 6 "	17	11 1/3	6 '- 6 "	4 1/3
2 '- 6 " x 7 '- 0 "	18	12	6 '- 6 "	4 1/3

2'-8" x 6'-8"	17	11 1/3	7'-0"	4 2/3
2'-8" x 7'-0"	18	12	7'-0"	4 2/3
3'-0" x 7'-0"	19	12 2/3	7'-0"	4 2/3
3'-6" x 7'-0"	19	12 2/3	7'-6"	5
4'-0" x 7'-0"	20	13 1/3	8'-0"	5 1/3
5'-0" x 7'-0"	21	14	9'-0"	6
6'-0" x 7'-0"	22	14 2/3	10'-0"	6 2/3
7'-0" x 7'-0"	24	16	11'-0"	7 1/3
3'-6" x 7'-6"	20	13 1/3	7'-6"	5
4'-0" x 7'-6"	20	13 1/3	8'-0"	5 1/3
4'-6" x 7'-6"	21	14	8'-6"	5 2/3
5'-0" x 7'-6"	21	14	9'-0"	6
6'-0" x 7'-6"	22	14 2/3	10'-0"	6 2/3
7'-0" x 7'-6"	24	16	11'-0"	7 1/3
3'-6" x 8'-0"	21	14	7'-6"	5
4'-0" x 8'-0"	22	14 2/3	8'-0"	5 1/3
5'-0" x 8'-0"	22	14 2/3	9'-0"	6
6'-0" x 8'-0"	23	15 2/3	10'-0"	6 1/3
7'-0" x 8'-0"	24	16	11'-0"	7 1/3
8'-0" x 8'-0"	25	16 2/3	12'-0"	8
9'-0" x 9'-0"	28	18 2/3	9'-0"	6

Lumber quantities given in the table above table are estimated in lengths that cut with the least waste.

Wood Door Bucks

Wood door bucks are made from 2"x4" (50 x 100 mm) or 2"x6" (50 x 150 mm) lumber, depending upon the thickness of the partition in which they are to be used.

Considerable savings are realized by using bucks made from 2"x6" lumber, having the back grooved out the thickness of the partition and 1/2" to 3/4" (12.50 to18.75 mm) deep. For a 5-1/2" (138 mm) finished partition, this allows for a 4" (100 mm) tile or gypsum partition block and 1-1/2" (37.50 mm) for plaster on both sides of the partition. The partition tile fit into the groove in the wood buck and the 3/4" (18.75 mm) extension on each side provides a ground for nailing the trim.

The mills furnishing the lumber usually charge $22.00 to $30.00 per 1,000 bd.ft. (2.36 cu.m) for grooving the backs of 2"x6"s (50 x 150 mm).

CARPENTRY

The lumber quantities given above are estimated in lengths that cut with the least waste.

Labor Making Rough Door Bucks. A carpenter should make and brace a rough door buck for an opening up to 3'-0"x7'-0" (0.91 x 2.13m) in about 1 hour. If the door has a transom, add 1/2 hour carpenter time.

When making rough door bucks for door openings 5'-0"x7'-0" (1.52 x 2.13 m) to 6'-0"x9'-0" (1.82 x 2.74 m), a carpenter should make and brace one door buck in 1-1/4 to 1-1/2 hrs. If the door has a transom, add 1/2 to 3/4 hr. carpenter time.

Rough Wood Bucks for Borrowed Lights. When making rough wood bucks for borrowed lights, a carpenter should make and brace one buck for an average size opening in 1 to 1-1/2 hrs.

Lumber Required for Wood Door Bucks of 2"x 6" (50x 150 mm) Lumber				
Size of Opening	Lin Ft. of Lumber Required	Bd.Ft of Lumber Req'd.	Transom Add Lin. Ft.	Bd.ft. of Lumber
2'- 0"x 4'- 0"	12	12	6'- 0"	6
2'- 0"x 4'- 6"	12	12	6'- 0"	6
2'- 0"x 5'- 0"	13 1/2	13 1/2	6'- 0"	6
2'- 6"x 4'- 0"	12	12	6'- 6"	6 1/2
2'- 6"x 4'- 6"	13	13	6'- 6"	6 1/2
2'- 6"x 5'- 0"	14	14	6'- 6"	6 1/2
2'- 6"x 5'- 6"	15	15	6'- 6"	6 1/2
2'- 6"x 6'- 0"	16	16	6'- 6"	6 1/2
2'- 6"x 6'- 6"	17	17	6'- 6"	6 1/2
2'- 6"x 7'- 0"	18	18	6'- 6"	6 1/2
2'- 8"x 6'- 8"	17	17	7'- 0"	7
3'- 0"x 7'- 0"	19	19	7'- 0"	7
3'- 6"x 7'- 0"	19	19	7'- 6"	7 1/2
4'- 0"x 7'- 0"	20	20	8'- 0"	8
5'- 0"x 7'- 0"	21	21	9'- 0"	9
6'- 0"x 7'- 0"	22	22	10'- 0"	10

7'- 0"x 7'- 0"	24	24	11'- 0"	11
3'- 6"x 7'- 6"	20	20	7'- 6"	7 1/2
4'- 0"x 7'- 6"	20	20	8'- 0"	8
4'- 6"x 7'- 6"	21	21	8'- 6"	8 1/2
5'- 0"x 7'- 6"	21	21	9'- 0"	9
6'- 0"x 7'- 6"	22	22	10'- 0"	10
7'- 0"x 7'- 6"	24	24	11'- 0"	11
3'- 6"x 8'- 0"	21	21	7'- 6"	7 1/2
4'- 0"x 8'- 0"	22	22	8'- 0"	8
5'- 0"x 8'- 0"	22	22	9'- 0"	9
6'- 0"x 8'- 0"	23	23	10'- 0"	10
7'- 0"x 8'- 0"	24	24	11'- 0"	11
8'- 0"x 8'- 0"	25	25	12'- 0"	12
9'- 0"x 9'- 0"	28	28	9'- 0"	9

Lumber quantities given in the table above are estimated in lengths that cut with least waste.

Lumber Required for Wood Door Bucks of 2"x 6" (50x 150 mm) Lumber				
Size of Opening	Lin M. of Lumber Required	Cu.M of Lumber Req'd.	Transom Add Lin. M	Cu.M. of Lumber
0.61 x 1.22 M	3.66	0.03	1.83	0.014
0.61 x 1.37 M	3.66	0.03	1.83	0.014
0.61 x 1.52 M	4.11	0.03	1.83	0.014
0.76 x 1.22 M	3.66	0.03	1.98	0.015
0.76 x 1.37 M	3.96	0.03	1.98	0.015
0.76 x 1.52 M	4.27	0.03	1.98	0.015
0.76 x 1.68 M	4.57	0.04	1.98	0.015
0.76 x 1.83 M	4.88	0.04	1.98	0.015
0.76 x 2.01 M	5.18	0.04	1.98	0.015
0.76 x 2.13 M	5.49	0.04	1.98	0.015

0.81 x 2.03 M	5.18	0.04	2.13	0.017
0.81 x 2.13 M	5.79	0.04	2.13	0.017
0.91 x 2.13 M	5.79	0.04	2.29	0.018
1.07 x 2.13 M	6.10	0.05	2.44	0.019
1.52 x 2.13 M	6.40	0.05	2.74	0.021
1.83 x 2.13 M	6.71	0.05	3.05	0.024
2.13 x 2.13 M	7.32	0.06	3.35	0.026
1.07 x 2.29 M	6.10	0.05	2.29	0.018
1.22 x 2.29 M	6.10	0.05	2.44	0.019
1.52 x 2.29 M	6.40	0.05	2.59	0.020
1.83 x 2.29 M	6.40	0.05	2.74	0.021
2.13 x 2.29 M	6.71	0.05	3.05	0.024
0.76 x 2.13 M	7.32	0.06	3.35	0.026
0.91 x 2.44 M	6.40	0.05	2.29	0.018
1.22 x 2.44 M	6.71	6.71	2.44	0.019
1.52 x 2.44 M	6.71	6.71	2.74	0.021
1.83 x 2.44 M	7.01	7.01	3.05	0.024
2.13 x 2.44 M	7.32	7.32	3.35	0.026
2.44 x 2.44 M	7.62	7.62	3.66	0.028
2.74 x 2.74 M	8.53	8.53	2.74	0.021

Lumber quantities given in the table above are estimated in lengths that cut with least waste.

Setting Rough Door Bucks. After the rough door buck has been made, a carpenter should set, plumb, and brace one door buck for an opening up to 3'-0"x9'-0" (0.91 x 2.74 m) in about l to 1-1/4 hr.

For double doors requiring an opening from 5'-0"x7'-0" (1.52 x 2.13 m) to 6'-0"x9'-0" (1.82 x 2.74 m), a carpenter should set, plumb, and brace one large door buck in 1-1/4 to 1-1/2 hrs.

Setting Rough Door Buck					
Description	Hours	Rate	Total	Rate	Total
Carpenter	1.00	$	$	$ 35.04	$ 35.04
Labor	0.25	$	$	$ 26.16	$ 6.54
Cost Each					$ 41.58

Fireproofing Wood. Where framing lumber and timbers are chemically treated to fireproof the wood, the cost of this treatment almost doubles the cost of wood.

BUILDING AND INSULATING SHEATHING

There are numerous types of insulating sheathing board, which usually consist of felted wood fiber or vegetable fiber products treated to make them moisture resistant.

Some of these sheets are coated with high melting point asphalt or are asphalt impregnated to form a moisture resistant surface which retards moisture penetration.

Insulating sheathing is furnished in sheets 4' (1.22 m) wide and 6' (1.82 m), 7' (2.13 m), 8' (2.43 m), 9' (2.74 m), 10' (3.04 m) and 12' (3.65 m) long, and 1/2" (12.50 mm) and 25/32" (19.53 mm) thick, the same thickness as wood sheathing.

Most manufacturers furnish this type of sheathing 25/32" (19.53 mm) thick, 2' (0.61 m) wide and 8' (2.43 m) long, with the long edges V-jointed or shiplapped.

Insulating sheathing is usually applied to wood studs under wood siding, shingles, stucco or brick veneer and because of the large size sheets used and the asphalt treatment, building paper is not ordinarily used with it, except under stucco.

Use 2" (50 mm) galvanized nails with 3/8" (9.37 mm) or 1/2" (12.50 mm) heads for insulating sheathing.

Place nails 3" (75 mm) apart on all outside edges of board and 6" (150 mm) apart for all intermediate nailing.

Insulating sheathing 25/32" (19.53 mm) thick costs $250 to $300 per 1,000 sq.ft. (92.90 sq.m).

Labor Placing Insulating Sheathing. When placing insulating sheathing on square or rectangular houses of regular construction, a carpenter should place 700 to 900 sq.ft. (65.03 to 83.61 sq.m) per 8-hr. day, at the following labor cost per 100 sq.ft. (9.29 sq.m):

Description	Hours	Rate	Total	Rate	Total
Carpenter	1.00	$	$	$ 35.04	$ 35.04
Labor unloading and carrying Sheets	0.30	$	$	$ 26.16	$ 7.85
Cost per 100 sq.ft. (9 sq.m)					$ 42.89
per sq.ft.					$ 0.43
per sq.m					$ 4.62

CARPENTRY

On buildings of irregular construction, requiring a great deal of cutting and fitting, it is not possible to take advantage of the large size sheets. Too much cutting and fitting is required. A carpenter will place only 350 to 450 sq.ft. (32.51 to 41.80 sq.m) of sheathing per 8-hr. day, at the following labor cost per 100 sq.ft. (9.29 sq.m):

Description	Hours	Rate	Total	Rate	Total
Carpenter	2.00	$	$	$ 35.04	$ 70.08
Labor unloading and carrying Sheets	0.30	$	$	$ 26.16	$ 7.85
Cost per 100 sq.ft. (9 sq.m)					$ 77.93
per sq.ft.					$ 0.78
per sq.m					$ 8.39

Insulating Roof Decking

Several manufacturers produce insulating roof decking made of multiple layers of 1/2" (12.50 mm) insulating board laminated together with vapor resistant cement and fabricated into tongue and groove planks, 2'-0" (0.61 m) wide, 8'-0" (2.43 m) long, and 1-1/2", 2", or 3" (37.50, 50, or 75 mm) thick, with a finish painted undersurface. It provides a roof deck, insulation, vapor barrier, and interior ceiling finish using only one material and is especially adaptable to designs with exposed beam ceilings.

Insulating roof decking can be used for flat, pitched, or monosloped roofs. Flat roofs and surfaces with a slope of 3" in 12" (75 in 300 mm) or less are usually covered with built-up roofing. Steeper roofs may be covered with rigid shingles, slate, or tile roofing, providing wood nailing strips are fastened through decking to supporting beams below.

Decking should be laid so that cross joints are staggered and occur only over supports. Decking should be face nailed to all framing members, spacing nails 4" to 6" (100 to 150 mm) apart and keeping back 3/4" to 1" (18.75 to 25 mm) from edges of plank. Nails should be galvanized common of sufficient length to pass through decking and penetrate supports at least 1-1/2" (37.50 mm) and should be driven flush but not countersunk. Where underside of decking will be exposed, planks must be handled and laid with care to prevent finished ceiling surface from being marred or damaged. Avoid excessive sliding of plank on roof beams.

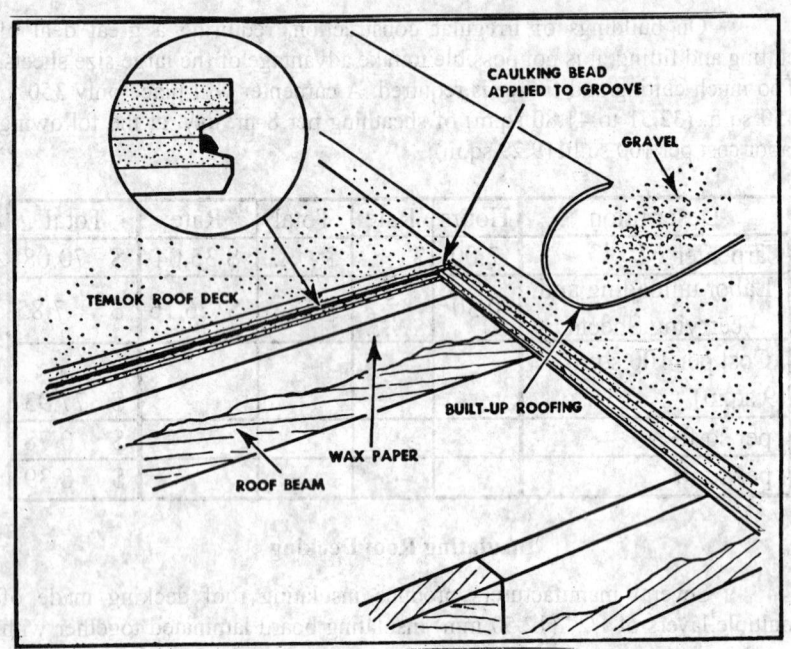

Method of Installing Insulating Roof Decking

Estimating Data on Insulating Roof Decking

Thickness	Max. Distance Between Supports	Size & Type of Nails	Lbs. of Nails Req'd. per 100 Sq.Ft.	Approx. Price per Sq.Ft.
1 1/2 "	24 "	10d Com. Galv.	6.0	$ 1.01
2 "	32 "	16d Com. Galv.	3.5	$ 1.10
3 "	48 "	30d Com. Galv.	3.0	$ 1.30

Thickness mm	Max. Distance Between Supports	Size & Type of Nails	Kg Nails Req'd. per 10 Sq.M	Approx. Price per Sq.M
37.50 mm	600 mm	10d Galv	2.93	$ 10.85
50.00 mm	800 mm	16d Galv	1.71	$ 11.88
75.00 mm	1200 mm	30d Galv	1.46	$ 13.95

Labor Placing Insulating Roof Decking. The labor cost of handling and placing insulating roof decking will be about the same for the various thicknesses, because the savings in handling lighter material is offset by the increased nailing due to closer required spacing of supports.

934

CARPENTRY

On simple, rectangular, flat or low-rise roofs, two carpenters and a helper should place 750 to 850 sq.ft. (69.68 to 78.96 sq.m) of decking per 8-hr. day, including stapling paper strips on roof beams for added protection against marring finished surface, at the following labor cost per 100 sq.ft. (9.29 sq.m):

Description	Hours	Rate	Total	Rate	Total
Carpenter	2.00	$	$	$ 35.04	$ 70.08
Labor	1.00	$	$	$ 26.16	$ 26.16
Cost per 100 sq.ft. (9 sq.m)					$ 96.24
per sq.ft.					$ 0.96
per sq.m					$ 10.36

For steep roofs, with a slope greater than 6" in 12" (150 in 300 mm), add 15% to 25%

PLYWOOD ROOF AND WALL SHEATHING, SUBFLOORING AND UNDERLAYMENT

Roof and Wall Sheathing. Plywood for use in roof and wall sheathing is readily available, easily worked with ordinary tools and skills, and adaptable to almost any light construction application. Because of the large sheet size, installation time is less than for other types of sheathing, and waste is minimal. It may be used under any type of shingle or roofing material, or any type of siding.

Standard size sheets are 4"x 8" (100 x 200 mm), but other sizes are available by special order. Most common sheathing thicknesses are 5/16", 3/8", 1/2", 5/8", and 3/4" (7.81, 9.37, 12.50, 15.62, and 18.75 mm), which are unsanded and are available with interior, intermediate, or exterior glue.

For most roof and wall sheathing installations, where sheathing is to be covered, interior type plywood is used. Where exposed to the weather, such as for roof overhangs or service building siding, use exterior plywood.

Nailing of plywood sheathing should be at 6" (150 mm) o.c. along panel edges and 12" (300 mm) o.c. at intermediate supports. Use 6d common nails for panels of 1/2" (12.50 mm) or less in thickness, and 8d for greater thickness.

Where plywood is used for wall sheathing, corner bracing is not required because of the rigidity of the plywood. Where other types of sheathing are used, plywood of the same thickness as the sheathing is sometimes used at corners to take the place of corner diagonal bracing.

Labor Placing Plywood on a Hip Roof. Two carpenters (or one carpenter and one skilled helper) can place approximately 1,000 to 1,500 sq.ft. (92.90 to 139.35 sq.m) of tongue and grooved plywood on hip or gable roof rafters with few penetrations per day. For other types of roofs, or roofs

935

that are very steep or cut up with dormers, hips and valleys such as English type houses, the estimator should add from 25% to 75% to the rates below. The estimated time is based on a maximum two-story structure from existing ground.

Placing 5/16" and 3/8" (7.81 and 9.37 mm) Plywood on plain or gable roofs a carpenter should place approximately 624 sq.ft. (57.97 sq.m) per day using manual hammer.

Description	Hours	Rate	Total	Rate	Total
Carpenter Foreman	4.00	$	$	$ 39.04	$ 156.16
Carpenter	8.00	$	$	$ 35.04	$ 280.32
Labor - Helper	8.00	$	$	$ 26.16	$ 209.28
Cost 1,248 sq.ft.					$ 645.76
Cost per sq.ft.					$ 0.52
Cost per sq.m					$ 5.57

Placing 1/2" and 5/8" (12.50 and 15.62 mm) plywood on plain or gable roofs a carpenter should place approximately 720 sq.ft. (66.89 sq.m) per day using pneumatic hammer.

Description	Hours	Rate	Total	Rate	Total
Carpenter Foreman	4.00	$	$	$ 39.04	$ 156.16
Carpenter	8.00	$	$	$ 35.04	$ 280.32
Labor - Helper	8.00	$	$	$ 26.16	$ 209.28
Cost 1,440 sq.ft.					$ 645.76
Cost per sq.ft.					$ 0.45
Cost per sq.m					$ 4.83

Placing 3/4" (18.75 mm) plywood on plain or gable roofs a carpenter should place approximately 544 sq.ft. (50.53 sq.m) per day using manual hammer.

Description	Hours	Rate	Total	Rate	Total
Carpenter Foreman	4.00	$	$	$ 39.04	$ 156.16
Carpenter	8.00	$	$	$ 35.04	$ 280.32
Labor - Helper	8.00	$	$	$ 26.16	$ 209.28
Cost 1,088 sq.ft.					$ 645.76
Cost per sq.ft.					$ 0.59
Cost per sq.m					$ 6.39

CARPENTRY

Placing 3/4" (18.75 mm) Plywood on plain or gable roofs a carpenter should place approximately 680 sq.ft. (63.17 sq.m) per day using pneumatic hammer.

Description	Hours	Rate	Total	Rate	Total
Carpenter Foreman	4.00	$	$	$ 39.04	$ 156.16
Carpenter	8.00	$	$	$ 35.04	$ 280.32
Labor - Helper	8.00	$	$	$ 26.16	$ 209.28
Cost 1,360 sq.ft.					$ 645.76
Cost per sq.ft.					$ 0.47
Cost per sq.m					$ 5.11

Subflooring and Underlayment. Plywood subflooring, underlayment, or combined subfloor-underlayment are available in most areas of the country. Some particular panels may not be stocked in all areas, so it is advisable to check with local plywood sources.

Subflooring panels are engineered grades of plywood marked by the American Plywood Association for maximum spans for the various thicknesses. Deflection of the panels under concentrated loads at panel edges is the limiting factor in the design of plywood floors. Panel edges should be supported by solid blocking, or tongue and grooved panels should be used, unless a separate layer of underlayment is installed with its joints offset from the joints in the subfloor.

Nailing of subflooring should be at 6" (150 mm) o.c. along the edges and 10" (250 mm) o.c. at intermediate supports. Use 6d common nails for 1/2" (12.50 mm) plywood, and 8d for 5/8" (15.62 mm) to 7/8" (21.87 mm) thick plywood. For 1-1/8" (28.12 mm) thick panels, use 8d ring shank or 10d common nails spaced 6" (150 mm) o.c. at both panel edges and intermediate supports.

Underlayment grades of plywood are touch sanded panels of 1/4" (6.25 mm) and 3/8" (9.37 mm) thickness with a smooth, solid surface for application of nonstructural flooring finished directly to them. This grade or plywood is made with special inner-ply construction which resists punch-through by concentrated point loading.

Joints in the underlayment should be staggered with respect to the joints in the subfloor. Nail with 3d ring shanked underlayment nails at 6" (150 mm) o.c. along the edges and 8" (200 mm) each way in the interior of the panels. Nails should be countersunk, if subfloor and joists are not dry.

Combined Subfloor-Underlayment Plywood. This is a single application installation combining both subfloor and underlayment, eliminating the expense of installing a separate underlayment. Panel edges perpendicular to the joists must be supported by solid blocking if tongue and grooved panels are not used.

937

Panels should be nailed at 6" (150 mm) o.c. at edges and 10" (250 mm) at intermediate supports with ringshank or spiral-thread nails. Use 6d deformed-shank nails for panel thickness up to 3/4" (18.75 mm) and 8d for thicker panels. Approximate costs of plywood sheathing, subflooring, and underlayment per 1,000 sq.ft. (92.90 sq.m) are as follows:

Types of Plywood - Cost per 1,000 Sq. Ft. (92.9 sq.m)							
Interior Type Standard C-D Interior		Interior Type Structural 1 C-D		Exterior Type C-C Ext		Plywood Underlayment Group I, Interior	
5/16 "	$ 372.60	1/2 "	$ 481.95	5/16 "	$ 450.90	1/2 "	$ 465.75
3/8 "	$ 403.65	5/8 "	$ 606.15	3/8 "	$ 544.05	5/8 "	$ 606.15
1/2 "	$ 346.28	3/4 "	$ 730.35	1/2 "	$ 606.15	3/4 "	$ 776.25
5/8 "	$ 558.90			5/8 "	$ 668.25		
3/4 "	$ 652.05			3/4 "	$ 776.25		
Types of Plywood - Cost per 100 Sq. M.							
Interior Type Standard C-D Interior		Interior Type Structural 1 C-D		Exterior Type C-C Ext		Plywood Underlayment Group I, Interior	
7.81 mm	$ 401.08	12.50 mm	$ 518.78	7.81 mm	$ 485.36	12.50 mm	$ 501.35
9.38 mm	$ 434.50	15.63 mm	$ 652.48	9.38 mm	$ 585.63	15.63 mm	$ 652.48
12.50 mm	$ 372.74	18.75 mm	$ 786.17	12.50 mm	$ 652.48	18.75 mm	$ 835.58
15.63 mm	$ 601.61			15.63 mm	$ 719.32		
18.75 mm	$ 701.88			18.75 mm	$ 835.58		

Prices on all grades and types of plywood should be verified with local suppliers before you submit a proposal for work that includes these materials.

Approximate costs of various materials and labor required for sidewall sheathing are given below.

Labor Cost of 100 Sq.Ft. (9.29 sq.m) of Structural Insulating Sidewall Sheathing Applied to Square or Rectangular Buildings of Regular Construction					
Description	Hours	Rate	Total	Rate	Total
Carpenter sheathing	1.00	$	$	$ 35.04	$ 35.04
Labor unloading and carrying sheets	0.30	$	$	$ 26.16	$ 7.85
Cost per 100 sq.ft.					$ 42.89
per sq.ft.					$ 0.43
per sq.m					$ 4.62

No paper required as insulating sheathing is treated to resist moisture.

On buildings of irregular construction, add 1 hr. carpenter time.

938

CARPENTRY

Labor Cost of 100 Sq.Ft. (9.29 sq.m) of 1"x6" (25 x 150 mm) Wood Sidewall Sheathing Aplied Horizontally to Square or Rectangular Buildings of Regular Construction					
Description	Hours	Rate	Total	Rate	Total
Carpenter sheathing	1.50	$	$	$ 35.04	$ 52.56
Labor unloading and carrying sheets	0.50	$	$	$ 26.16	$ 13.08
Carpenter applying building (wrap) paper	0.40			$ 35.04	$ 14.02
Cost per 100 sq.ft. (9.29 sq.m)					$ 79.66
per sq.ft.					$ 0.80
per sq.m					$ 8.57

If 1"x8" (25 x 200 mm) shiplap is used instead of 1"x6" (25 x 150 mm) D&M lumber, deduct 5 bd.ft. (0.01 cu.m) lumber and 0.1 hr. carpenter time per 100 sq.ft. (9.29 sq.m) of wall.

On buildings of irregular construction, add 1 hr. carpenter time.

Labor Cost of 100 Sq.Ft. (9.29 sq.m) of 1"x6" (25 x 150 mm) Wood Sidewall Sheathing Applied Diagonlly to Square or Rectangulr Buildings of Regular Construction					
Description	Hours	Rate	Total	Rate	Total
Carpenter sheathing	2.00	$	$	$ 35.04	$ 70.08
Labor unloading and Carrying Lumber	0.50	$	$	$ 26.16	$ 13.08
Carpenter applying Paper (wrap)	0.40	$	$	$ 35.04	$ 14.02
Cost per 100 sq.ft.					$ 97.18
per sq.ft.					$ 0.97
per sq.m					$ 10.46

If 1"x8" (25 x 200 mm) shiplap is used instead of 1"x6" (25 x 150 mm) D&M lumber, deduct 5 bd.ft. (0.01 cu.m) lumber and 0.1 hr. carpenter time per 100 sq.ft. (9.29 sq.m) of wall.

On buildings of irregular construction, add 1 hr. carpenter time.

Labor Cost of 100 Sq.Ft. (9 sq.m) of 1/2" (12.50 mm) Plywood Sidewall Sheathing Applied to Square or Rctangular Buildings of Regular Construction

Description	Hours	Rate	Total	Rate	Total
Carpenter sheathing	1.00	$	$	$ 35.04	$ 35.04
Helper unloading and Carrying Sheets	0.30	$	$	$ 26.16	$ 7.85
Carpenter applying paper or wrap	0.40	$	$	$ 35.04	$ 14.02
Cost per 100 sq.ft.					$ 56.90
per sq.ft.					$ 0.57
per sq.m					$ 6.13

Paper optional: plywood comes in sheets 4' x8' (1.22 x 2.44 m)treated with
On buildings of irregular construction add 1 hour carpenter time.

Gypsum Sheathing. Gypsum sheathing is a fireproof solid sheet of gypsum encased in a tough, fibrous, water-resistant covering. The sides and ends are treated to resist moisture. Some manufacturers blend asphalt emulsion into the wet gypsum core mix to provide additional resistance to moisture. It is used for sheathing on frame structures under siding, shingles, stucco, and brick veneer.

Gypsum sheathing has V-joint edges on the long dimension to provide positive assurance of a tight fit at the unsupported joints. It is applied with its length at right angles to the studs.

Gypsum sheathing is 1/2" (12.50 mm) thick, 2' (0.61 m) wide, and 8' (2.43 m) and 10' (3.04 m) long, and 4' (1.22 m) wide and 8' (2.43 m) and 9' (2.74 m) long to fit supports 16" (400 mm) on centers. Nails should be 1-3/4" (43.75 mm) long, No. 10-1/2 gauge, galvanized flat head roofing nails, spaced 4" (100 mm) on centers, except under wood siding and stucco, 8" (200 mm). Requires 14 to 21 lbs. (6.35 to 9.52 kg) per 1,000 sq.ft. (92.9 sq.m). Price $85.00 to $125.00 per 1,000 sq.ft. (92.90 sq.m). Labor cost same as given for insulating or plywood sheathing.

940

Labor Cost of 100 Sq.Ft. (9.29 sq.m) of Gypsum Sidewall Sheathing Applied to Square or Rectangular Buildings of Regular Construction

Description	Hours	Rate	Total	Rate	Total
Carpenter sheathing	1.00	$	$	$ 35.04	$ 35.04
Helper unloading and carrying sheets	0.30	$	$	$ 26.16	$ 7.85
Carpenter applying paper or wrap	0.40	$	$	$ 35.04	$ 14.02
Cost per 100 s.ft.					$ 56.90
per sq.ft.					$ 0.57
per sq.m					$ 6.13

Paper is optional. Gypsum sheathing is waterproofed sides and edges. Some manufacturers mix asphalt emulsion with wet gypsum to provide a waterproof board.

On buildings of irregular construction, add 1 hr. carpenter time.

Paper is optional. Gypsum sheathing is waterproofed sides and edges. Some manufacturers mix asphalt emulsion with wet gypsum to provide a waterproof board. On buildings of irregular construction, add 1 hr. carpenter time.

Insulating Shingle Backer Strips

A number of manufacturers produce a shingle backer strip, of the same material as insulating sheathing board, that is used as the undercourse in double course shingle work. These strips are 4'-0" (1.22 m) long and either 5/16" (7.81 mm) or 3/8" (9.37 mm) thick. They are available in two widths, 13-1/2" (337.50 mm) for 12" (300 mm) shingle exposure and 15-1/2" (287.50 mm) for 14" (350 mm) shingle exposure. They provide added insulation, improve the shadow line appearance of the finished wall, eliminate the necessity of building paper over wood sheathing, and are more economical to use than second grade wood shingle undercoursing. Insulating shingle backer strips cost approximately $0.30 per sq.ft. ($0.09 sq.m).

Labor Placing Insulating Shingle Backer Strips. On straight run jobs, with an average number of openings, a carpenter should place 500 to 600 sq.ft. (46.45 to 55.74 sq.m) of shingle backer strips per 8-hr. day at the following labor cost per 100 sq.ft. (9.29 sq.m):

Description	Hours	Rate	Total	Rate	Total
Carpenter	1.50	$	$	$ 35.04	$ 52.56
Labor	0.50	$	$	$ 26.16	$ 13.08
Cost 100 sq.ft.					$ 65.64
per sq.ft.					$ 0.66
per sq.m					$ 7.07

On complicated jobs, requiring a large amount of cutting and fitting, a carpenter should place 400 to 500 sq.ft. (37.16 to46.45 sq.m) per 8-hr. day at the following labor cost per 100 sq.ft. (9.29 sq.m):

Description	Hours	Rate	Total	Rate	Total
Carpenter	1.80	$	$	$ 35.04	$ 63.07
Labor	0.50	$	$	$ 26.16	$ 13.08
Cost 100 sq.ft.					$ 76.15
per sq.ft.					$ 0.76
per sq.m					$ 8.20

Insulating Lath

Insulating lath is made by several manufacturers of insulating building sheathing. It is made of the same materials as the building boards previously described.

Insulating lath is usually furnished in the 18"x48" (450 x 1200 mm) size, 1/2" (12.50 mm), 3/4" (18.75 mm), and 1" (25 mm) thick, although they are obtainable from some sources in 16"x48" (400 x 1200 mm).

Some manufacturers furnish lath both plain and with the back coated with asphalt to form a vapor barrier.

Most insulating lath are manufactured with the long edges shiplapped and all edges beveled to reinforce the plaster against cracking and eliminate unsightly lath marks. Some of these insulating lath are metal reinforced along the edges to reinforce the unsupported horizontal joint.

Insulating gypsum lath consists of regular gypsum board lath with a thin sheet of aluminum attached to one side to provide insulating value. They are furnished in sheets 16"x32" (400 x 800 mm), 16"x48" (400 x 1200 mm), and 3/8" (9.37 mm) and 1/2" (12.50 mm) thick.

942

CARPENTRY

Estimating Quantities of Insulating Lath					
Size of Lath Inches	No. Sq.Ft. per lath	No. Lath per 100 Sq.Ft.	Size of Lath mm	No Sq.M. per lath	No. Lath per 10 Sq.M.
16" x 48"	5.33	18.76	400 x 1200 mm	57.37	1.88
18" x 48"	6.00	16.67	450 x 1200 mm	64.59	1.67

Approximate prices of insulating lath are as follows per 1,000 sq.ft. (92.9 sq.m): 1/2" (12.50 mm), $320.00; 5/8" (15.62 mm), $400.00.

Labor Applying Insulating Lath. The labor cost of applying insulating lath will vary with the locality and labor restrictions. Where there are no labor restrictions, an experienced worker should place 125 sq. yds. (104.51 sq.m) per 8-hr. day, while a less experienced worker will place only 100 sq. yds. (83.61 sq.m) per 8-hr. day, at the following labor cost per 100 sq. yds. (83.61 sq.m):

Description	Hours	Rate	Total	Rate	Total
Lather	8.00	$	$	$ 40.29	$ 322.32
Cost per sq. yd					$ 3.22
per sq.m					$ 3.86

WOOD AND WIRE FENCE

Setting Wood Fence Posts. When placing wire fence around the lots of 16 houses, it required 218 cedar posts, 5" (125 mm) in diameter and 7'-0" (2.13 m) long, also 32 clothes posts, 7" (175 mm) in diameter and 10'-0" (30.48 m) long.

After the holes had been dug, a worker set 7 posts per hr. at the following labor cost per post:

Description	Hours	Rate	Total	Rate	Total
Carpenter	0.50	$	$	$35.04	$14.13

Treated pine and locust posts may also be used for fencing. Locust and cedar posts should be all heartwood.

Posts can also be made up of a composite material, wood mixed with other fibers such as plastic to prevent insect and decay.

Placing 2"x4" (50 x 100 mm) Top Bottom Fence Rail Ready to Receive Wire. The posts for these fences are spaced about 6'-0" (1.82 m) apart and a 2"x4" (50 x 100 mm) top rail is nailed flatwise to the top of the

943

post and a 2"x4" (50 x 100 mm) bottom rail is notched into the side of the post so that the wire can be stapled to it.

There is a total 1,400 lin.ft. (420 m) of fence or 2,800 lin.ft. (840.44 m) of top and bottom rail. A carpenter places rail for 110 lin.ft. (33.52 m) of fence or 220 lin.ft. (66.06 m) of single rail per 8-hr. day, at the following labor cost per 100 lin.ft. (30.48 m):

Description	Hours	Rate	Total	Rate	Total
Carpenter	7.30	$	$	$35.04	$ 255.79
Labor	0.90	$	$	$26.16	$ 23.54
Cost per 100 lin.ft.					$ 279.34
per lin.ft. of fence					$ 2.79
per lin.m of fence					$ 9.16

Stringing Fence Wire. Cyclone wire fencing in 42" (1050 mm) rolls is used for these fences and is stapled to the top and bottom rails and posts about 4" (100 mm) on centers.

There is a total 1,400 lin.ft. (426.72m) of 42" (1050 mm) wire fencing in the job and a carpenter would string and staple about 280 lin.ft. (85.34 m) per 8-hr. day, at the following labor cost per 100 lin.ft. (30.48 m):

Description	Hours	Rate	Total	Rate	Total
Carpenter	2.90	$	$	$ 35.04	$ 101.62
Cost per lin.ft.					$ 1.02
per lin.m.					$ 3.33

Setting Lally Columns in Basements. In a group of brick houses, 3" (75 mm) iron columns were used to support the 6"x 8" (150 x 200 mm) wood girders carrying the floor joists.

These columns were 3" (75 mm) in diameter and 7'-0" (2.13 m) long, and a carpenter set, plumbed, and braced 1 column per hr. at the following labor cost per column:

Description	Hours	Rate	Total	Rate	Total
Carpenter	1.00	$	$	$ 35.04	$ 35.04
Labor	1.00	$	$	$ 26.16	$ 26.16
Cost per column					$ 61.20

MISCELLANEOUS HARDWARE AND ACCESSORIES

Steel Area Walls and Gratings

These are used around basement windows when first floor is at grade level and are made of 16-ga. and 20-ga. galvanized steel, with stiffening ribs and rounded tops. They are attached to masonry walls by screws or bolts.

Size Width x Depth Ft- In	For Use with Basement Windows		Price Each 16 Ga.	Price Each 20 Ga.	Size Width x Depth	For Use with Basement Windows
3 ' - 2 " x 0 ' - 11 1/2 "	15 " x 12 "	2 lite	$17.40	$12.50	0.966 x 0.293 m	375 x 300 mm
3 ' - 2 " x 1 ' - 5 1/2 "	15 " x 16 "	2 lite	$20.40	$15.20	0.966 x 0.445 m	375 x 400 mm
3 ' - 2 " x 1 ' - 11 1/2 "	15 " x 20 "	2 lite	$27.20	$20.40	0.966 x 0.597 m	375 x 500 mm

Steel gratings for use with above walls. Frames made of 1-1/4"x1/4" (31.25 x 6.25 mm) steel bars with cross bars 1"x13/16" (25 x 20.31 mm) welded into one piece. Supports on the grating hold it flush with the top of the area wall. Price each painted, $35.00.

Metal Foundation Wall Ventilators

Brick Ventilators. Made of aluminum exactly the size of a brick, 2-1/2"x4"x8" (62.50 x 100 x 200 mm). The bottom edge is flanged both front and back to insure a positive mortar lock. Louvered face allows 13 sq. in. (83.87 sq.cm) of free area. Screened back. Price each, $11.00.

Frame Ventilators. Nailed to sheathing just above the foundation or to ends of joists spaced 16" (400 mm) on centers. Made of sheet aluminum, size 16"x8" (400 x 200 mm), providing 89 sq. in. (574.23 sq.cm.) of free area. Screened back. Price each, $50.00.

Concrete Block Ventilators. Made of cast aluminum 16" (400 mm) wide x 8" (200 mm) high, providing 89 sq. in. (574.23 sq.cm.) of free area. Screened back. Price each, $55.00.

The use of stainless steel brick ventilators would add approximately 30% to the material cost.

Metal Roof Ventilators

Metal roof ventilators provide air in the attic space under the roof. They are made of steel or aluminum and usually provide 18 to 75 sq. in. (116.13 to 483.9 sq.cm) of free area.

Roof Opening, Inches	Free Area Sq.In.	Price each Aluminum	Roof Opening mm		Free Area Sq.Cm	
12 "x 18 "	30.0	$ 14.00	300 x 450	mm	193.56	sq.cm.
15 "x 20 "	50.0	$ 18.00	375 x 500	mm	322.60	sq.cm.
16 "x 16 "	65.0	$ 30.00	400 x 400	mm	419.38	sq.cm.

 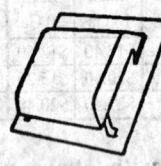 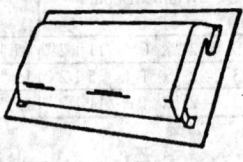

Metal Roof Ventilators

Steel Clothes, Laundry Chute Doors

Price Each

Furnished in steel, white enameled, opening 9"x12" (225x300mm) $24.00
For stainless steel add approximately 25%

Steel Access Doors

For access to piping, wiring installations and mechanical equipment, furnished with a flanged aluminum bottom and aluminum cover with the frame nailed to studs. Size of door 2'-6" x 4'- 6" (.76 x 1.37 m), $1,100.00

ROUGH HARDWARE

While it is possible to include the cost of rough hardware such as nails and screws in the material unit for each carpentry item figured, most contractors and estimators find it more convenient to make an allowance for this item at the end of the carpentry estimate based on the amount of lumber and other material involved. The usual method of arriving at a rough hardware allowance is to total all the material in the carpentry estimate, taking bd.ft. (cu.m) of dimension lumber and boards, sq. ft. (sq.m) area of plywood, paneling, etc., and lin.ft. (meters) of grounds, furring strips, trim, etc., and then figure from 25 to 35 lbs. (11 to 16 kg) of nails per 1,000 bd.ft. (2.36 cu.m). The weight of nails multiplied by the cost per lb. (kg) will give the rough hardware allowance required.

946

CARPENTRY

06130 HEAVY TIMBER CONSTRUCTION

The quantities and costs included here cover types such as heavy constructed mill, warehouse, and factory buildings.

Framing and Erecting Heavy Wood Columns or Posts. The labor cost of erecting and framing heavy wood columns or posts, 10"x10" (250 x 250 mm) or larger, will vary according to the number of operations necessary, such as chamfering corners, cutting out and framing for post caps and bases, boring holes through the center, and upon the distance the posts must be hoisted before placing.

On the average mill constructed job, requiring 10"x10" (250 x 250 mm) or 12"x12" (300 x 300 mm) posts 10'-0" to 16'-0" (3.04 to 4.87 m) long, it will require 3-1/2 to 4 hrs. to chamfer corners, frame for bases and caps, and erect each column, at the following labor cost:

Description	Hours	Rate	Total	Rate	Total
Carpenter	3.70	$	$	$ 35.04	$ 129.65
Labor	0.50	$	$	$ 26.16	$ 13.08
Cost per column					$ 142.73

If hoisted above the second floor, add 1/2 hr. of labor time per column. It is more difficult to estimate the cost of this work by the 1,000 bd.ft. (2.36 cu.m) of lumber, because it costs just as much to chamfer the 4 corners and frame for post caps and bases for an 8"x8" (200 x 200 mm) as a 12"x12" (300 x 300 mm) column, while the former contains less than half as much lumber per lin.ft. as the latter. The following costs are based on an average 10"x10"-12'-0" (250 x 250 mm - 3.65 m) timber and should be increased or reduced accordingly for larger or smaller timbers.

Based on a 10"x10"-12'-0" (250 x 250 mm - 3.65 m) column, a carpenter should chamfer corners, frame for post caps and bases, handle, and erect 200 to 250 bd.ft. (0.47-0.59 cu.m) of lumber per 8-hr. day, at the following labor cost per 1,000 bd.ft. (2.36 cu.m):

Description	Hours	Rate	Total	Rate	Total
Carpenter	35.00	$	$	$ 35.04	$1,226.40
Labor	8.00	$	$	$ 26.16	$ 209.28
Cost 1,000 bd.ft. (2.36 cu.m)					$1,435.68
per cu.m					$ 608.34

Rounding Corners on Large Timbers. Where the corners of wood columns, beams, or girders must be rounded to a small radius, the yards or mills furnishing the lumber usually charge $23.10 to $30.00 per 1,000 bd.ft. (2.36 cu.m) for performing this work.

Boring Holes Through Center of Large Timbers. Where the wood columns or posts have a hole bored through the center for their entire length, the yard or mill furnishing the lumber usually charges $44.00 to $51.00 per 1,000 bd.ft. (2.36 cu.m) for boring these holes.

Framing and Erecting Heavy Wood Beams and Girders for First Floor. The labor cost of framing and erecting heavy wood beams and girders will vary according to the amount of framing necessary and the method used in placing them; i.e., whether they set on top of the wood columns or whether it is necessary to frame for post caps, stirrups, etc.

Placing heavy beams and girders for the first floor of a building does not require as much labor as for the upper floors, because the timbers can be rolled to the place they are to be used without hoisting.

Where heavy wood girders are set on concrete or masonry piers without framing for post caps, joist hangers, etc., a carpenter should frame and erect 1,000 to 1,250 bd.ft. (2.36 to 2.95 cu.m) per 8-hr. day, at the following labor cost per 1,000 bd.ft. (2.36 cu.m):

Description	Hours	Rate	Total	Rate	Total
Carpenter	7.00	$	$	$ 35.04	$ 245.28
Labor	7.00	$	$	$ 26.16	$ 183.12
Cost 1,000 bd.ft. (2.36 cu.m)					$ 428.40
per cu.m					$ 181.53

Where necessary to frame for post caps and bases, stirrups, etc., a carpenter should frame and erect 500 to 600 bd.ft. (1.18-1.42 cu.m) of lumber per 8-hr. day, at the following labor cost per 1,000 bd.ft. (2.36 cu.m):

Description	Hours	Rate	Total	Rate	Total
Carpenter	14.50	$	$	$ 35.04	$ 508.08
Labor	7.00	$	$	$ 26.16	$ 183.12
Cost 1,000 bd.ft. (2.36 cu.m)					$ 691.20
per cu.m					$ 292.88

Framing and Erecting Heavy Wood Girders Above First Floor. Where heavy wood girders must be hoisted above the first floor, using a timber hoist or derrick, the carpenter time framing and placing the timbers will remain practically the same but the labor time handling, hoisting, and skidding the timbers into place will be increased.

On work of this kind, a carpenter should frame and erect 450 to 550 bd.ft. (1.06-1.30 cu.m) of lumber per 8-hr. day, at the following labor cost per 1,000 bd.ft (2.36 cu.m):

Description	Hours	Rate	Total	Rate	Total
Carpenter	16.00	$	$	$ 35.04	$ 560.64
Labor	10.00	$	$	$ 26.16	$ 261.60
Cost 1,000 bd.ft. (2.36 cu.m)					$ 822.24
per cu.m					$ 348.41

Framing and Erecting Heavy Wood Joists. When placing heavy wood joists 4"x12" (100 x 300 mm) or 4"x16" (100 x 400 mm) in size, where the joists are set on top of wood girders or in iron stirrups, a carpenter should frame and erect 1,000 to 1,200 bd.ft. (2.36-2.83 cu.m) per 8-hr. day, at the following labor cost per 1,000 bd.ft. (2.36 cu.m):

Description	Hours	Rate	Total	Rate	Total
Carpenter	7.30	$	$	$ 35.04	$ 255.79
Labor	3.50	$	$	$ 26.16	$ 91.56
Cost 1,000 bd.ft. (2.36 cu.m)					$ 347.35
per cu.m					$ 147.18

Where smaller joists are used, ranging from 3"x10" (75 x 250 mm) to 4"x10" (100 x 250 mm) in size, a carpenter should frame and erect 800 to 1,000 bd.ft. (1.88-2.36 cu.m) per 8-hr. day, at the following labor cost per 1,000 bd.ft. (2.36 cu.m):

Description	Hours	Rate	Total	Rate	Total
Carpenter	9.00	$	$	$ 35.04	$ 315.36
Labor	4.00	$	$	$ 26.16	$ 104.64
Cost 1,000 bd.ft. (2.36 cu.m)					$ 420.00
per cu.m					$ 177.97

Framing and Placing Oak Bumpers at Edges of Loading Docks. Oak timbers, varying from 4"x6" (100 x 150 mm) to 12"x12" (300 x 300 mm), are usually called for at loading dock edges. The timbers generally are secured to the edge of the dock slab by bolts set in the concrete or with cinch anchors.

With bolts already set in place, a carpenter should lay out, drill and countersink holes, and secure in place 150 to 175 bd.ft. (0.35-0.41 cu.m) of oak bumper per 8-hr. day at the following labor cost per 1,000 bd.ft. (2.36 cu.m):

Description	Hours	Rate	Total	Rate	Total
Carpenter	48.00	$	$	$ 35.04	$ 1,681.92
Labor	24.00	$	$	$ 26.16	$ 627.84
Cost 1,000 bd.ft. (2.36 cu.m)					$ 2,309.76
per cu.m					$ 978.71

A carpenter should locate, set, and secure in place in the slab edge or wall form about 25 anchor bolts per 8-hr. day at the following labor cost per bolt:

Description	Hours	Rate	Total	Rate	Total
Carpenter	0.30	$	$	$ 35.04	$10.51

Laying 2"x6" (50 x 150 mm) or 3"x6" (75 x 150 mm) Tongue and Groove Timber Subfloors or Roof Sheathing. When laying 2"x6" (50 x 150 mm) or 3"x6" (75 x 150 mm) tongue and groove flooring or roof decking in large areas without much cutting and fitting, a carpenter should lay 800 to 1,000 bd.ft. (1.88-2.36 cu.m) per 8-hr. day at the following labor cost per 1,000 bd.ft. (2.36 cu.m):

Description	Hours	Rate	Total	Rate	Total
Carpenter	8.90	$	$	$ 35.04	$ 311.86
Labor	6.00	$	$	$ 26.16	$ 156.96
Cost per 1,000 bd.ft. (2.36 cu.m)					$ 468.82
per cu.m					$ 198.65

If tongue and groove flooring or roof sheathing is laid in small spaces requiring considerable cutting and fitting around openings, stair wells, elevator shafts, skylights, etc., a carpenter should lay 700 to 800 bd.ft. (1.65-

950

CARPENTRY

1.88 cu.m) per 8-hr. day at the following labor cost per 1,000 bd.ft. (2.36 cu.m):

Description	Hours	Rate	Total	Rate	Total
Carpenter	10.60	$	$	$ 35.04	$ 371.42
Labor	6.00	$	$	$ 26.16	$ 156.96
Cost per 1,000 bd.ft. (2.36 cu.m)					$ 528.38
per cu.m					$223.89

LAMINATED WOOD FLOORS

Laminated wood floors consisting of 2" (50 mm) or 3" (75 mm) plank set on edge and spiked together are used for floors capable of carrying heavy loads or supporting heavy machinery with a minimum amount of vibration.

Sometimes the floors are designed to use certain length plank reaching from bearing to bearing, while on others they are spiked together with broken joints, regardless of bearing.

Quantity of Lumber Required for Laminated Wood Floors					
Measured Size Inches	Finished Thickness Inches	Add for Waste Percent	To Obtain Qty of Lumber Multiply by Area Sq.Ft.	Lin.Ft. Lumber per 100 Sq.Ft. of Surface	
2 " x 4 "	1 5/8 "	23	4.90	490	74 pcs.
2 " x 6 "	1 5/8 "	23	7.40	740	74 pcs.
2 " x 8 "	1 5/8 "	23	9.84	984	74 pcs.
2 " x 10 "	1 5/8 "	23	12.30	1230	74 pcs.
2 " x 12 "	1 5/8 "	23	14.76	1476	74 pcs.
2 " x 14 "	1 5/8 "	23	17.22	1722	74 pcs.
3 " x 6 "	2 3/4 "	10	6.60	660	44 pcs.
3 " x 8 "	2 3/4 "	10	8.80	880	44 pcs.
3 " x 10 "	2 3/4 "	10	11.10	1110	44 pcs.
3 " x 12 "	2 3/4 "	10	13.20	1320	44 pcs.
3 " x 14 "	2 3/4 "	10	15.40	1540	44 pcs.
4 " x 6 "	3 3/4 "	7	6.42	642	32 pcs.
4 " x 8 "	3 3/4 "	7	8.56	856	32 pcs.
4 " x 10 "	3 3/4 "	7	10.70	1070	32 pcs.
4 " x 12 "	3 3/4 "	7	12.84	1284	32 pcs.
Measured Size mm	Finished Thickness mm	Add for Waste Percent	To Obtain Qty of Lumber Multiply by Area Sq.M.	Lin.Ft. Lumber per 10 Sq.M. of Surface	
50 x 100 mm	40.63 mm	23	0.46	52.823	75 pcs.
50 x 150 mm	40.63 mm	23	0.69	52.823	75 pcs.
50 x 200 mm	40.63 mm	23	0.91	52.823	75 pcs.
50 x 250 mm	40.63 mm	23	1.14	52.823	75 pcs.
50 x 300 mm	40.63 mm	23	1.37	52.823	75 pcs.
50 x 350 mm	40.63 mm	23	1.60	52.823	75 pcs.
75 x 150 mm	68.75 mm	10	0.61	52.823	45 pcs.
75 x 200 mm	68.75 mm	10	0.82	52.823	45 pcs.
75 x 250 mm	68.75 mm	10	1.03	52.823	45 pcs.
75 x 300 mm	68.75 mm	10	1.23	52.823	45 pcs.
75 x 350 mm	68.75 mm	10	1.43	52.823	45 pcs.
100 x 150 mm	93.75 mm	7	0.60	52.823	33 pcs.
100 x 200 mm	93.75 mm	7	0.80	52.823	33 pcs.
100 x 250 mm	93.75 mm	7	0.99	52.823	33 pcs.
100 x 300 mm	93.75 mm	7	1.19	52.823	33 pcs.

CARPENTRY

The above quantities do not include any allowance for waste due to end matching, but are based on spans taking lumber of even lengths or laying flooring with broken joints.

Size of Nails	Nails Required for 100 Sq.Ft. (9.29 Sq.M.) of 2" Laminated Flooring									
	Distance Nails Are Spaced Apart - Inches									
	12		16		20		24		30	
	No. Nails	No. Lbs.	No. Nails	No. Lbs.	No. Nails	No. Lbs.	No. Nails	No. Lbs.	No. Nails	No. Lbs.
16d	740	15	600	12 1/4	518	10 1/2	450	9 1/4	370	7 1/2
20d	740	24	600	19 1/4	518	16 3/4	450	14 1/2	370	12
30d	740	31	600	25	518	21 1/2	450	18 3/4	370	15 1/2
40d	740	41	600	33 1/4	518	28 3/4	450	25	370	20 1/2
50d	740	53	600	43	518	37	450	32	370	26 1/2
60d	740	67	600	55	518	47	450	41	370	33 3/4

Metric

Size of Nails	Nails Required for 10 Sq.M. of 2" Laminated Flooring									
	Distance Nails Are Spaced Apart - mm									
	300		400		500		600		750	
	No. Nails	No. Kg	No. Nails	No. Kg	No. Nails	No. Kg	No. Nails	No. Kg	No. Nails	No. Kg
16d	797	7.32	646	5.98	558	5.13	484	4.52	398	3.66
20d	797	11.72	646	9.40	558	8.18	484	7.08	398	5.86
30d	797	15.14	646	12.21	558	10.50	484	9.16	398	7.57
40d	797	20.02	646	16.23	558	14.04	484	12.21	398	10.01
50d	797	25.88	646	21.00	558	18.07	484	15.62	398	12.94
60d	797	32.71	646	26.85	558	22.95	484	20.02	398	16.48

Size of Nails	Nails Required for 100 Sq.Ft. (9.29 Sq.M) of 3" (75 mm) Laminated Flooring									
	Distance Nails Are Spaced Apart - Inches									
	12		16		20		24		30	
	No. Nails	No. Lbs.	No. Nails	No. Lbs.	No. Nails	No. Lbs.	No. Nails	No. Lbs.	No. Nails	No. Lbs.
16d	440	9	352	7 1/4	308	6 1/4	264	5 1/2	220	4 1/2
20d	440	14 1/4	352	11 1/2	308	10	264	8 1/2	220	7
30d	440	18 1/2	352	14 3/4	308	12 3/4	264	11	220	9 1/4
40d	440	24 1/2	352	19 1/2	308	17	264	14 3/4	220	12 1/4
50d	440	31 1/2	352	25 1/4	308	22	264	18 3/4	220	15 3/4
60d	440	40	352	32	308	28	264	24	220	20

Metric

Size of Nails	Distance Nails Are Spaced Apart - mm									
	300		400		500		600		750	
	No. Nails	No. Kg	No. Nails	No. Kg	No. Nails	No. Kg	No. Nails	No. Kg	No. Nails	No. Kg
	474	4.39	379	3.54	332	3.05	284	2.69	237	2.20
20d	474	6.96	379	5.62	332	4.88	284	4.15	237	3.42
30d	474	9.03	379	7.20	332	6.23	284	5.37	237	4.52
40d	474	11.96	379	9.52	332	8.30	284	7.20	237	5.98
50d	474	15.38	379	12.33	332	10.74	284	9.16	237	7.69
60d	474	19.53	379	15.62	332	13.67	284	11.72	237	9.77

	Nails Required for 100 Sq.Ft. (9.29 Sq.M) of 4" (100 mm) Laminated Flooring									
Size of Nails	Distance Nails are Spaced Apart - inches									
	12		16		20		24		30	
	No. Nails	No. Lbs.	No. Nails	No. Lbs.	No. Nails	No. Lbs.	No. Nails	No. Lbs.	No. Nails	No. Lbs.
30d	320	13 1/2	256	10 3/4	224	9 1/2	192	8	160	6 3/4
40d	320	17 3/4	256	14 1/4	224	12 1/2	192	10 3/4	160	9
50d	320	23	256	18 1/4	224	16	192	13 3/4	160	11 1/2
60d	320	29	256	23 1/4	224	20 1/2	192	17 1/2	160	14 1/2

	Metric									
Size of Nails	Distance Nails are Spaced Apart (mm)									
	300		400		500		600		750	
	No. Nails	No. Kg	No. Nails	No. Kg	No. Nails	No. Kg	No. Nails	No. Kg	No. Nails	No. Kg
30d	344	6.59	276	5.25	241	4.64	207	3.91	172	3.30
40d	0	8.67	276	6.96	241	6.10	207	5.25	172	4.39
50d	344	11.23	276	8.91	241	7.81	207	6.71	172	5.62
60d	344	14.16	276	11.35	241	10.01	207	8.54	172	7.08

06170 PREFABRICATED STRUCTURAL WOOD

Glued Laminated Beam Construction. Beams of glued laminated construction are popular where price is not the controlling factor.

They are made in uniform or tapered shapes. In the beam shape, similar to illustration, they can be used in spans from 20'-0" to 80'-0" (6.09 to 24.38 m).

They are used in schools, auditoriums, churches, stores, and ranch-style homes and are made of kiln-dried structural woods bonded together by glue, applied under controlled conditions of temperature and pressure. They are planed and sanded to a very smooth surface and later treated with a transparent preservative, making their overall appearance very pleasing. With this type of construction, the ceilings are usually omitted, leaving the beams exposed to the room below. Roof insulation is accomplished through the use of standard insulating boards placed upon the roof deck and covered with roofing material. Because of the purlin construction generally used, 2" or 4" decking is recommended.

Span in Feet	Center Hgt. Feet	End Hgt. Feet	Price	Span in Meters	Center Hgt. Meter	End Hgt. Feet
20'-0"	2'-4"	0'-9 1/2"	$ 420	6.10 m	0.71 m	0.24 m
30'-1"	2'-10"	0'-11"	$ 1,010	9.14 m	0.87 m	0.28 m
40'-2"	3'-11"	1'-2 1/2"	$ 1,475	12.19 m	1.19 m	0.37 m
50'-3"	5'-1"	1'-6 1/2"	$ 2,350	15.24 m	1.55 m	0.47 m
60'-4"	6'-1"	1'-9 1/2"	$ 3,350	18.29 m	1.85 m	0.55 m

CARPENTRY

Glued Laminated Three Hinged Arch

Another type of glued laminated construction is the three-hinged arch, which gains its support from floor level, incorporating column and beam in one compact design. Fabricated and finished in the manner of the glued beam, they are used where good appearance and function are the major factors and are used quite extensively in churches and schools.

Purlins are generally used to span the resulting bays and are covered with two inch decking and suitable insulating material.

This is the most expensive of the trusses discussed and costs from 5 to 6 times that of the bowstring type truss.

General Notes Regarding Trusses. Where rafters are used between trusses, 1" (25 mm) roof sheathing or 1/2" (12.50 mm) plywood is generally used. Where purlins are spaced 4'-0" to 6'-0" (1.22 to 1.82 m) apart, 2" (50 mm) sheathing is common practice. Purlins may be eliminated and wood decking of 4" to 6" (100 to 150 mm) may be used to span up to 18'-0" (5.48 m) between trusses.

The use of ceilings supported by the lower chord of the truss should be called to the attention of the truss designer, as it requires a heavier load.

Trusses may bear upon pilasters, steel columns, wood posts and plates, grillages or steel beams. A suitable steel bearing plate should be placed under each end of the truss to distribute the load throughout the masonry. Trusses may afford lateral support through angle irons fastened to truss and column.

Labor Framing Wood Roof Trusses. When framing wood roof trusses, a carpenter experienced in truss construction will handle and frame approximately the following amounts of lumber per hour on the various types of trusses:

Type of Truss	Bd.ft. per Hour	Cu.m per Hour
Bowstring Type Trusses	40.0	0.094
Scissors and Gothic Type Trusses	30.0	0.071
Glued Laminated Type Trusses	15.0	0.035

PREFABRICATED STRUCTURAL TIMBER

Structural timber decking in thicknesses of 2" to 4" (50-100 mm), generally 6" (150 mm) wide and double tongue and grooved is available in hemlock cedar, Douglas fir, and spruce. It can be used for roof decks, structural flooring, and planked walls. Approximate costs are as follows:

Lumber Grade	Size Inches	Hemlock per Sq.Ft.	Cedar or Fir per Sq.Ft.	Spruce per Sq.Ft.
Select	4 " x 6 "	$ 4.00	$ 4.20	$ 5.25
Select	3 " x 6 "	$ 3.25	$ 3.60	$ 4.10
Lumber Grade	Size mm	Hemlock per Sq.M	Cedar or Fir per Sq.M	Spruce per Sq.M
Select	100 x 150 mm	$ 43.06	$ 45.21	$ 56.51
Select	75 x 150 mm	$ 34.98	$ 38.75	$ 44.13

All prices should be verified with local distributors prior to submitting proposals that include these materials.

06200 FINISH CARPENTRY

For estimating finish carpentry work for a job whose specifications are arranged in accordance with the Construction Specifications Institute (CSI) format, you must check at least five divisions to be certain that all of this type of work is included in the bid. Division 6, the subject of this chapter, covers rough and finish carpentry. But wood and related sidings are in Division 7, Thermal and Moisture Protection. Wood doors and windows are in a section of their own, Division 8. Division 10, Specialties, includes such items as tackboards, laminated plastic toilet partitions, signs, folding partitions, telephone booths, and toilet and wardrobe accessories, all of which may be furnished or set by the carpenter. Division 11, Equipment, covers kitchen and laundry cabinets and counters for residential work and such unrelated work as wood bank counters, church pews, bars, lab furniture, dock bumpers, and library shelving. Finally Division 12, Furnishings, includes wood seating, class room furniture, and similar semi-attached items. In addition, it is prudent to check all the allowances set forth in the specifications. Often the furnishing of some of the items in Divisions 10, 11, and 12 will be covered under a cash allowance, but the installation is to be included in the bid submitted by the carpenter.

EXTERIOR FINISH CARPENTRY

Wood siding and wood doors and windows are discussed under their own CSI divisions. The following items cover other adjacent exterior work.

Placing Corner Boards, Fascia Boards, Etc. When placing wood fascia boards, corner boards, etc., a carpenter should place 175 to 225 lin.ft. (53.34 to 68.58 m) per 8-hr. day, at the following labor cost per 100 lin.ft. (30.48 m):

956

Description	Hours	Rate	Total	Rate	Total
Carpenter	4.00	$	$	$ 35.04	$ 140.16
Cost per lin.ft.					$ 1.40
per m					$ 4.60

Placing Exterior Wood Cornices, Verge Boards, Etc. When placing exterior wood cornices, verge boards, fascia, soffits, etc., consisting of two members, two carpenters working together should place 150 to 175 lin.ft. (45.72 to53.34 m) per 8-hr. day, at the following labor cost per 100 lin.ft. (30.48 m):

Description	Hours	Rate	Total	Rate	Total
Carpenter	10.00	$	$	$ 35.04	$ 350.40
Cost per lin.ft.					$ 3.50
per m					$ 11.50

When the exterior cornices consist of three members (crown mold, bed mold, fascia, etc.), two carpenters working together should place 100 to 125 lin.ft. (30.48 to -38.10 m) per 8-hr. day, at the following labor cost per 100 lin.ft. (30.48 m):

Description	Hours	Rate	Total	Rate	Total
Carpenter	14.20	$	$	$ 35.04	$ 497.57
Cost per lin.ft.					$ 4.98
per m					$ 16.32

If a four-member wood cornice is used, two carpenters working together should place 60 to 75 lin.ft. (18.29 to 22.86 m) per 8-hr. day, at the following labor cost per 100 lin.ft. (30.48 m):

Description	Hours	Rate	Total	Rate	Total
Carpenter	23.70	$	$	$ 35.04	$830.45
Cost per lin.ft.					$8.30
per m					$27.25

The above quantities and costs do not include time blocking out for fascia boards, cornices, etc. An extra allowance should be made for all blocking required.

Placing Brick Moldings. A carpenter should fit and set around 32 lin.ft. (7.75 m) of brick molding per hour at the following cost per 100 lin.ft. (30.48 m):

Description	Hours	Rate	Total	Rate	Total
Carpenter	3.20	$	$	$ 35.04	$112.13
Cost per lin.ft.					$1.12
per m					$3.68

Brick moulding costs about $0.80 per lin.ft. (0.3048 m) in random lengths and $1.20 in specified lengths.

Placing Wood Cupolas. One carpenter should set a prefabricated pine cupola in around two hours.

Description	Hours	Rate	Total	Rate	Total
Carpenter	2.00	$	$	$ 35.04	$ 70.08
Labor	2.00	$	$	$ 26.16	$ 52.32
Cost per each					$ 122.40

Porch Work. The labor placing porch work is a variable item owing to the vast difference in the style and construction of porches and the amount of detail involved.

Placing Plain Porch Columns. When placing plain square or turned porch columns, such as commonly used for rear porches and other inexpensive work, a carpenter should place one post in about 1 hr., at the following labor cost:

Description	Hours	Rate	Total	Rate	Total
Carpenter	1.00	$	$	$ 35.04	$ 35.04
Cost per post					$ 28.25

Placing Porch Top and Bottom Rail and Balusters. When placing wood top and bottom rail and wood balusters, such as used on front porches, a carpenter should complete 15 to 20 lin.ft. (4.57 to 6.09 m) of rail per 8-hr. day, at the following labor cost per lin.ft. (0.3048 m):

Description	Hours	Rate	Total	Rate	Total
Carpenter	0.46	$	$	$ 35.04	$ 16.12
Cost per lin.ft.					$ 16.12
per m					$ 52.88

CARPENTRY

When placing top and bottom rails with open balusters or using matched and beaded ceiling, such as is often used for the cheaper grades of work, a carpenter should complete 35 to 45 lin.ft. (10.66 to 13.71 m) of rail per 8-hr. day, at the following labor cost per lin.ft. (0.3048 m):

Description	Hours	Rate	Total	Rate	Total
Carpenter	0.20	$	$	$ 35.04	$ 7.01
Cost per lin.ft.					$ 7.01
per m					$ 22.99

Framing and Erecting Exterior Wood Stairs for Rear Porches. When framing and erecting outside wood stairs for rear porches on apartment buildings, etc., where the stringers are 2"x10" (50 x 250 mm) or 2"x12" (50 x 300 mm) lumber, with treads and risers nailed on the face of the stringers, it will require 18 to 22 hrs. carpenter time per flight of stairs.

This is for an ordinary stair having 14 to 18 risers extending from story to story, and the labor per flight should cost as follows:

Description	Hours	Rate	Total	Rate	Total
Carpenter	20.00	$	$	$ 35.04	$ 700.80

If the stair consists of two short flights with an intermediate landing platform between stories, it will require 12 to 13 hrs. carpenter time per flight or 24 to 26 hrs. per story, including platform, at the following labor cost:

Description	Hours	Rate	Total	Rate	Total
Carpenter	25.00	$	$	$ 35.04	$ 876.00

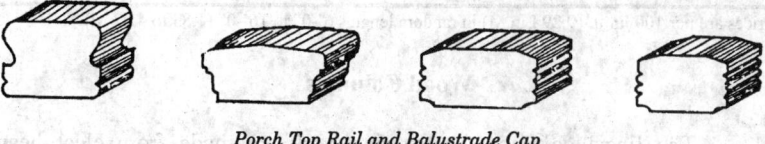

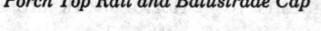

Porch Top Rail and Balustrade Cap

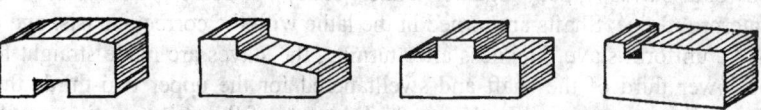

Rabbeted Porch Jamb, Plowed Shoe, and Bottom Rail

Framing and Erecting Wood Stairs Having Winders. Where the wood stairs to rear porches have 4 to 6 winders in each story height, figure 24 to 28 hrs. carpenter time, at the following labor cost per story:

Description	Hours	Rate	Total	Rate	Total
Carpenter	26.00	$	$	$ 35.04	$ 911.04

Clear Ponderosa Pine Porch Material					
Kind of Molding	Size Inches	Price per 100 Lin.Ft.	Size mm	Price per 10 Meters	
Square Baluster Stock	1- 1/8 x 1 1/8	$ 90.00	1- 28.12 x 28.12	$ 96.88	
Square Baluster Stock	1- 3/8 x 1 3/8	$ 118.00	1- 34.37 x 34.37	$ 127.02	
Square Baluster Stock	1- 5/8 x 1 5/8	$ 137.00	1- 40.62 x 40.62	$ 147.47	
Balustrade Cap	3- 5/8 x 1 5/8	$ 240.00	3- 90.62 x 40.62	$ 258.34	
Balustrade Shoe	3- 5/8 x 1 5/8	$ 240.00	3- 90.62 x 40.62	$ 258.34	
Rabbeted Porch Jamb	1- 1/2 x 3 1/2	$ 240.00	1- 37.50 x 87.50	$ 258.34	
Plowed Porch Shoe	1- 1/2 x 3 1/2	$ 240.00	1- 37.50 x 87.50	$ 258.34	
Top Rail	3 x 2 1/4	$ 267.00	75.00 x 56.25	$ 287.41	
Top Rail	3- 3/4 x 1 3/4	$ 240.00	3- 93.75 x 43.75	$ 258.34	
Bottom Rail	3- 1/2 x 1-3/4	$ 244.00	3- 87.50 x 43.75	$ 262.65	

Cornice Boards* Select Grade					
Measured Size Inches	to S4S Inches	Price per 100 Lin. Ft.	Measured Size, mm	to S4S mm	Price per 10 Lin. M.
1 x 3	3/4 x 2 5/8	$ 72.00	25 x 75	18.75 x 65.62	$ 77.50
1 x 4	3/4 x 3 1/2	$ 85.00	25 x 100	18.75 x 87.50	$ 91.50
1 x 6	3/4 x 5 1/2	$ 144.00	25 x 150	18.75 x 137.50	$ 155.01
1 x 8	3/4 x 7 1/2	$ 172.00	25 x 200	18.75 x 187.50	$ 185.15
1 x 10	3/4 x 9 1/2	$ 225.00	25 x 250	18.75 x 237.50	$ 242.20
1 x 12	3/4 x 11 1/2	$ 310.00	25 x 300	18.75 x 287.50	$ 333.69

*Prices are per 100 lin.ft. (9.29 Lin.M) in random lengths 6'-0" to 16'-0" (1.83 to 4.87 M)

Wood Columns

The finest columns for exterior work are made from clear heart redwood staves glued together under clamp pressure with cold water waterproof glue. Shafts are turned in the lathe with the correct entasis, and to insure uniform stave thickness after turning, the staves are made straight for the lower third of the shaft and swelltapered for the upper two-thirds thus giving the correct entasis in the rough. Any style of fluting or reeding may be had.

960

CARPENTRY

The columns receive a prime coat of lead and oil on the exterior surface. Columns over 14" (350 mm) in diameter receive a coat of waterproof asphaltum paint on the hollow core surface.

All columns are furnished with a guarantee as to their workmanship and durability.

No printed price lists are available for these items and each must be priced at the factory. In order of price, the Roman is cheapest followed by the unfluted Ionic, the fluted Attic, the fluted Doric, and finally the fluted Corinthian. A 12" (300 mm) diameter Doric fluted column 10' (3.04 m) high with a plinth and cap is currently quoted at around $500.00 at the factory.

Stock, tapered column shafts in pine can be figured at around $43.00 per vertical lin.ft. (0.30 m) in 12" (300 mm) diameter; $80.00 in 14" (350 mm); $90.00 in 18" (450 mm); and $100.00 in 24" (600 mm).

Fiberglass columns cast in a colonial design will run $9.00 per lin.ft. (0.30 m) in 6" (150 mm) diameter; $10.00 in 8" (200 mm); and $13.00 in 10" (250 mm).

Aluminum columns extruded to a fluted design are stocked in assorted sizes from 6" (150 mm) diameter in 8' (2.43 m) lengths to 12" (300 mm) diameter in 18' (5.48 m) lengths. They may be used as loadbearing members and come factory finished in a high-gloss, white, baked enamel. An 8" (200 mm) diameter shaft 12' (3.65 m) long will cost about $75.00. If no base or cap is used, it may be attached to the slab and ceiling by cutting 2" (50 mm) lumber to fit the column diameter and bolting it to the adjacent construction. If a base and cap is ordered, they will be furnished in aluminum castings and will be about $17.00 each.

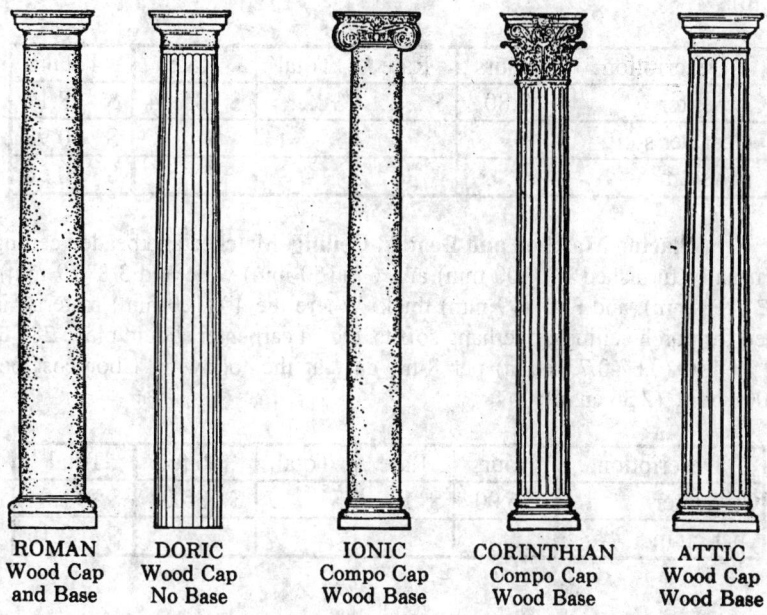

ROMAN	DORIC	IONIC	CORINTHIAN	ATTIC
Wood Cap and Base	Wood Cap No Base	Compo Cap Wood Base	Compo Cap Wood Base	Wood Cap Wood Base

961

Square porch columns costs about $2.00 per lin.ft. (0.30 m) in 4x4 (100 x 100 mm) fir. Turned colonial posts in 8' (2.43 m) lengths will cost about $22.00 in 4x4 (100 x 100 mm) stock, $30.00 in 5x5 (125 x 125 mm) stock. It takes two carpenters to set columns, and they should erect a medium sized column in an hour.

Cast Iron Column Bases. Cast iron column bases keep columns off the floor and permit air to circulate on the inside of the column.

Base Size Inches	Height Inches	Col. Dia.	Approx. Price	Base Size mm	Height mm	Col. Dia. Mm
11 "x 11 "	2 "	8 "	$ 60.00	275.00 x 275.00 mm	50.00 mm	200
13 1/2 "x 13 1/2 "	2 1/2 "	10 "	$ 86.00	337.50 x 337.50 mm	62.50 mm	250
16 "x 16 "	2 1/2 "	12 "	$ 115.00	400.00 x 400.00 mm	62.50 mm	300
18 1/2 "x 18 1/2 "	2 1/2 "	14 "	$ 165.00	462.50 x 462.50 mm	62.50 mm	350
21 1/4 "x 21 1/4 "	2 1/2 "	16 "	$ 220.00	531.25 x 531.25 mm	62.50 mm	400
24 "x 24 "	3 1/4 "	18 "	$ 255.00	600.00 x 600.00 mm	81.25 mm	450
26 1/2 "x 26 1/2 "	3 1/4 "	20 "	$ 295.00	662.50 x 662.50 mm	81.25 mm	500

*For Doric columns without base use one size smaller.

Cast iron bases for pilasters are two-thirds the price of bases for corresponding columns.

Placing Plywood Soffits. Two carpenters should set 75 sq.ft. (6.96 sq.m) of 1/4" (6.25 mm) and 64 sq.ft. (5.94 sq.m) of 1/2" (12.50 mm) plywood on soffits per hour at the following labor cost per 100 sq.ft. (9.29 sq.m):

Description	Hours	Rate	Total	Rate	Total
Carpenter	2.80	$	$	$ 35.04	$ 98.11
Cost per sq.ft.					$ 0.98
per sq.m					$ 10.56

Placing Matched and Beaded Ceiling. Matched and beaded ceiling is usually furnished 4" (100 mm) and 6" (150 mm) wide and 3/8" (9 mm), 1/2" (13 mm), and 3/4" (19 mm) thick. Where the 4" (100 mm) material is used for porch ceilings, overhang soffits, etc., a carpenter should place 285 to 325 bd.ft. (0.67-0.77 cu.m) per 8-hr. day, at the following labor cost per 1,000 bd.ft. (2.36 cu.m):

Description	Hours	Rate	Total	Rate	Total
Carpenter	26.00	$	$	$ 35.04	$ 911.04
per cu.m					$ 2.15

CARPENTRY

The labor cost per square-100 sq.ft. (9.29 sq.m) surface-requiring 128 bd.ft. (0.30 cu.m) of ceiling should run as follows:

Description	Hours	Rate	Total	Rate	Total
Carpenter	3.33	$	$	$ 35.04	$ 116.68

When placing 6" (150 mm) wide matched and beaded ceiling, a carpenter should place 400 to 450 bd.ft. (0.94-1.06 cu.m) of ceiling per 8-hr. day, at the following labor cost per 1,000 bd.ft. (2.36 cu.m):

Description	Hours	Rate	Total	Rate	Total
Carpenter	18.80	$	$	$ 35.04	$ 658.75
per cu.m					$ 1.55

The labor cost per square-100 sq.ft. (9.29 sq.m) surface-requiring 120 bd.ft. (0.28 cu.m) of ceiling should run as follows:

Description	Hours	Rate	Total	Rate	Total
Carpenter	2.25	$	$	$ 35.04	$ 78.84

Placing Exterior Door Trim. Much of the architectural type door trim today is of molded material rather than milled wood. It is pre-engineered to fit together with little or no cutting or fitting, providing the door frame itself is set plumb and true. Stock trim can be ordered for widths of 3' (0.91 m), 5' (1.52 m), and 6' (1.82 m) and for either brick or wood wall openings. Side trim is delivered for 7' (2.13 m) door height but designed so that 4" (100 mm) may be cut to fit a 6'-8" (2.03 m) door without altering the design. Pilasters are nominally 8" (200 mm) in width. Cornices are 6" to 8" (150-200 mm) high and available in several styles. In addition, pediments may be added. A basic cornice piece costs around $80.00 for a 3' (0.91 m) opening, $120.00 for the larger ones. Pediments will run from $80.00 to $120.00 extra, for 3'-0" (0.91 m) openings and up to $180.00 for the double door openings. Pilaster assemblies run $100.00 per pair. Two carpenters should set a cornice in 1/2 hour, the pediment in 1/2 hour, and the pilasters in one hour.

Outside Blinds or Shutters

Shutters are available in wood, aluminum, and molded plastic. Because of the popularity of the latter, it is not always easy to find wood stocked in all sizes. Custom made units will run considerably more than the prices given.

All shutters have stiles, top and middle rails 2-1/4" (56.25 mm) wide; bottom rail 4-1/2" (112.50 mm) wide. All slats are stationary. Shutters

963

with raised panels can be fitted with decorative cutouts in the upper panel at an extra cost of $2.30 per pair. Designs include pine trees, cloverleafs, squirrels, crescents, and roosters.

A carpenter should hang 8 to 10 pair a day depending on size and height off the ground.

Price Per Pair, Hemlock, Preservative or Paint Treated			
Window Size Feet- Inches	With Fixed Slats	With Raised Slats	Window Size Meter
2'- 0 "x 3'- 3"	$20.90	$26.90	0.61 x 0.99 M
2'- 0 "x 3'- 11	$24.70	$29.90	0.61 x 1.19 M
2'- 0 "x 4'- 7"	$28.50	$35.90	0.61 x 1.40 M
2'- 0 "x 5'- 3"	$31.50	$43.40	0.61 x 1.60 M
2'- 4 "x 3'- 3"	$21.70	$26.90	0.71 x 0.99 M
2'- 4 "x 3'- 7"	$22.80	$26.90	0.71 x 1.09 M
2'- 4 "x 3'- 11	$24.70	$26.90	0.71 x 1.19 M
2'- 4 "x 4'- 3"	$26.10	$26.90	0.71 x 1.30 M
2'- 4 "x 4'- 7"	$29.90	$26.90	0.71 x 1.40 M
2'- 4 "x 4'- 11"	$32.10	$26.90	0.71 x 1.50 M
2'- 8 "x 3'- 3"	$22.50	$26.50	0.81 x 0.99 M
2'- 8 "x 3'- 7"	$23.10	$28.50	0.81 x 1.09 M
2'- 8 "x 3'- 11	$25.50	$29.90	0.81 x 1.19 M
2'- 8 "x 4'- 3"	$27.70	$32.90	0.81 x 1.30 M
2'- 8 "x 4'- 7"	$29.90	$34.50	0.81 x 1.40 M
2'- 8 "x 4'- 11"	$31.50	$36.70	0.81 x 1.50 M
2'- 8 "x 5'- 3"	$34.50	$38.90	0.81 x 1.60 M
3'- 0 "x 3'- 3"	$29.90	$32.90	0.91 x 0.99 M
3'- 0 "x 3'- 7"	$31.50	$35.50	0.91 x 1.09 M
3'- 0 "x 3'- 11"	$32.90	$37.40	0.91 x 1.19 M
3'- 0 "x 4'- 3"	$34.50	$40.40	0.91 x 1.30 M
3'- 0 "x 4'- 7"	$35.90	$41.90	0.91 x 1.40 M
3'- 0 "x 4'- 11"	$37.40	$44.10	0.91 x 1.50 M
3'- 0 "x 5'- 3"	$38.90	$44.90	0.91 x 1.60 M
3'- 0 "x 3'- 3"	$31.50	$35.10	0.91 x 0.99 M
3'- 4 "x 3'- 11"	$34.50	$38.90	1.01 x 1.19 M
3'- 4 "x 4'- 7"	$37.40	$43.40	1.01 x 1.40 M
3'- 4 "x 5'- 3"	$41.90	$47.80	1.01 x 1.60 M
3'- 8 "x 4'- 7"	$34.50	$45.60	1.12 x 1.40 M
3'- 8 "x 4'- 11"	$37.40	$47.80	1.12 x 1.50 M
3'- 8 "x 5'- 3"	$40.40	$50.40	1.12 x 1.60 M

Door Blinds - Same as above Except 3 Panels of Slats in Height		
Window Size Feet- Inches	Price per Pair	With Raised Slats
2'- 8" x 6'- 9"	$37.70	0.81 x 2.05 M
3'- 0" x 6'- 9"	$39.00	0.91 x 2.05 M

Aluminum Shutters

Aluminum shutters are designed to screw permanently on the wall. The louver section is stamped out of one piece of metal. They are furnished in white, black, green, or brown and come packaged with mounting screws.

Aluminum Shutters		
Size Feet - Inches	Price per Pair	Size Meter - mm
2 '- 4 "x 34 "	$34.40	0.71 M x 850 mm
2 '- 4 "x 36 "	$35.50	0.71 M x 900 mm
2 '- 4 "x 39 "	$37.20	0.71 M x 975 mm
2 '- 4 "x 43 "	$38.80	0.71 M x 1075 mm
2 '- 4 "x 48 "	$40.00	0.71 M x 1200 mm
2 '- 4 "x 55 "	$42.30	0.71 M x 1375 mm
2 '- 4 "x 64 "	$48.80	0.71 M x 1600 mm
2 '- 4 "x 80 "	$57.70	0.71 M x 2000 mm
2 '- 8 "x 34 "	$36.70	0.81 M x 850 mm
2 '- 8 "x 36 "	$37.80	0.81 M x 900 mm
2 '- 8 "x 39 "	$39.40	0.81 M x 975 mm
2 '- 8 "x 43 "	$41.10	0.81 M x 1075 mm
2 '- 8 "x 48 "	$42.30	0.81 M x 1200 mm
2 '- 8 "x 55 "	$45.50	0.81 M x 1375 mm
2 '- 8 "x 64 "	$51.10	0.81 M x 1600 mm
2 '- 8 "x 80 "	$62.10	0.81 M x 2000 mm

One carpenter should hang one pair in 0.8 hour.

Molded Shutters

Shutters molded of thermo-formed polymer are prefinished in either black or white but may be painted with an exterior latex paint. They are screwed to siding with 2" (50 mm) aluminum screws, 4 per unit to 55" (1375 mm), 6 per unit over 55" (1375 mm).

Molded Shutters		
Panel Inches	Price per pair	Panel, mm
14 "x 35 "	$34.50	350 x 875 mm
14 "x 39 "	$35.50	350 x 975 mm
14 "x 43 "	$37.40	350 x 1075 mm
14 "x 47 "	$39.30	350 x 1175 mm
14 "x 51 "	$40.40	350 x 1275 mm
14 "x 55 "	$42.60	350 x 1375 mm
14 "x 59 "	$43.70	350 x 1475 mm
14 "x 63 "	$49.40	350 x 1575 mm
14 "x 67 "	$50.80	350 x 1675 mm
14 "x 71 "	$52.40	350 x 1775 mm
16 "x 80 "	$56.00	400 x 2000 mm

Labor Costs on Half Timber Work. The following costs on half timber work are taken from a building with timber of white oak 3" (75 mm) thick, 10" (250 mm) and 11" (275 mm) wide, and various lengths.

Nothing was done to the face of the timber and it was left just as received. The backs of all timbers were beveled, the thickness remaining 3" (75 mm). A shop was set up on the job where the timbers were cut, beveled, and fit ready for erection.

There were 14,500 bd.ft. (34.20 cu.m) of lumber used in this half-timber work and the labor cost per 1,000 bd.ft. (2.36 cu.m) was as follows:

Description	Hours	Rate	Total	Rate	Total
Carpenter	66.30	$	$	$ 35.04	$ 2,323.15
Labor helping	12.00	$	$	$ 26.16	$ 313.92
Cost per 1,000 bd.ft.					$ 2,637.07
per cu.m					$ 6.22

After the timbers were cut, beveled, and fit in the shop, the labor cost of placing and bracing the timbers in the building was as follows per 1,000 bd.ft. (2.36 cu.m):

966

Description	Hours	Rate	Total	Rate	Total
Carpenter	41.50	$	$	$ 35.04	$ 1,454.16
Labor helping	2.20	$	$	$ 26.16	$ 57.55
Cost per 1,000 bd.ft.					$ 1,511.71
per cu.m					$ 3.57

INTERIOR FINISH CARPENTRY

When estimating the labor cost of interior finish, determine the grade of workmanship required by the architect or owner, because the quantity of work a carpenter will perform per hour or per day will vary considerably with the grade of workmanship.

Costs are given for two distinct grades of workmanship. "Ordinary work" is by far the most common and is usually found in medium priced residences, cottages, apartment buildings, factory and warehouse buildings, non-fireproof store, and offices and schools.

"First grade workmanship" is required in fine residences, high-class fireproof apartments, hotels, banks and office buildings, high-class stores, fireproof school and university buildings, courthouses, city halls, state capitols, post offices, and other such buildings. The interior finish in buildings of this class is usually selected birch, selected gumwood, plain or quarter-sawed oak, mahogany, walnut or other first-class hardwoods, and is discussed later in this chapter under "Architectural Woodwork".

Placing Wood Base. This cost will vary with the size of the rooms and whether a single-, two- or three-member base is specified.

Where there are 55 to 60 lin.ft. (16.76 to 18.29 m) of two-member base in each room, without an unusually large amount of cutting and fitting, a carpenter should place 125 to 150 lin.ft. (38.10 to 45.72 m) per 8-hr day, at the following labor cost per 100 lin.ft. (30.48 m):

Description	Hours	Rate	Total	Rate	Total
Carpenter	5.80	$	$	$ 35.04	$ 203.23
Labor helping	1.00	$	$	$ 26.16	$ 26.16
Cost per 100 lin.Ft.					$ 229.39
per lin.ft.					$ 2.29
per lin.m.					$ 7.53

If there is an unusually large number of miters, such as required in closets and other small rooms, increase the above costs accordingly.

A carpenter should place almost as many lin.ft. of three-member base (consisting of two base members and a carpet strip), as two-member

967

base (consisting of one member and carpet strip), as it is much easier to fit a small top member against the plastered wall than it is to nail a wide piece of base so that it will fit snug against the wall and follow the irregularities in the plaster.

Where there are 50 to 60 lin.ft. (15.24 to 18.29 m) of base in a room, a carpenter should place 110 to 130 lin.ft. (30.48 to 39.64 m) per 8-hr. day, at the following labor cost per 100 lin.ft. (30.48 m):

Description	Hours	Rate	Total	Rate	Total
Carpenter	6.70	$	$	$ 35.04	$ 234.77
Labor helping	1.00	$	$	$ 26.16	$ 26.16
Cost per 100 lin.Ft.					$ 260.93
per lin.ft.					$ 2.61
per lin.m.					$ 8.56

First Grade Workmanship

In average size rooms, a carpenter should place 100 to 115 lin.ft. (30.48 to 34.05 m) of two-member hardwood base per 8-hr. day at the following labor cost per 100 lin.ft. (30.48 m):

Description	Hours	Rate	Total	Rate	Total
Carpenter	7.40	$	$	$ 35.04	$ 259.30
Labor helping	1.00	$	$	$ 26.16	$ 26.16
Cost per 100 lin.Ft.					$ 285.46
per lin.ft.					$ 2.85
per lin.m.					$ 9.37

Where three-member hardwood base is used in average rooms, a carpenter should place 85 to 100 lin.ft. (25.91 to 30.48 m) two ordinary rooms per 8-hr. day, at the following labor cost per 100 lin.ft. (30.48 m):

Description	Hours	Rate	Total	Rate	Total
Carpenter	8.70	$	$	$ 35.04	$ 304.85
Labor	1.00	$	$	$ 26.16	$ 26.16
Cost per 100 lin.Ft.					$ 331.01
per lin.ft.					$ 3.31
per lin.m.					$ 10.86

CARPENTRY

On work of this class, the wood grounds should be straight, so that it will not be necessary to "force" the wood base to make it fit tight against the finished wall.

Where a single 1"x4" (25 x 100 mm) pine base is to be fitted to straight runs, a single carpenter should set about 200 lin.ft. (60.96 m) per day at the following cost per 100 lin.ft. (30.48 m):

Description	Hours	Rate	Total	Rate	Total
Carpenter	4.00	$	$	$ 35.04	$ 140.16
Cost per 100 lin.Ft.					$ 140.16
per lin.ft.					$ 1.40
per lin.m.					$ 4.60

Placing Wood Picture Molding. Where just an ordinary grade of workmanship is required, a carpenter should place picture molding in 5 or 6 ordinary rooms per 8-hr. day. This is equivalent to 250 to 275 lin.ft. (76.20 to 83.82 m) at the following labor cost per 100 lin.ft. (30.48 m):

Description	Hours	Rate	Total	Rate	Total
Carpenter	3.00	$	$	$ 35.04	$ 105.12
Labor	0.50	$	$	$ 26.16	$ 13.08
Cost per 100 lin.Ft.					$ 118.20
per lin.ft.					$ 1.18
per lin.m.					$ 3.88

First Grade Workmanship

Where the wood picture molding must fit close to the plastered walls with perfect fitting miters, a carpenter should place molding in 4 to 5 ordinary sized rooms per 8-hr. day. This is equivalent to 175 to 200 lin.ft. (53.34 to 60.96 m), at the following cost per 100 lin.ft. (30.48 m):

Description	Hours	Rate	Total	Rate	Total
Carpenter	4.40	$	$	$ 35.04	$ 154.18
Labor	0.50	$	$	$ 26.16	$ 13.08
Cost per 100 lin.Ft.					$ 167.26
per lin.ft.					$ 1.67
per lin.m.					$ 5.49

If the picture molding is placed in fireproof buildings having tile or brick partitions, it will be necessary to place wood grounds for nailing the picture molding, but in non-fireproof buildings the nails may be driven into the plaster so that nails obtain a bearing in the wood studs or wall furring.

Placing Wood Chair or Dado Rail. In large rooms or long, straight corridors, a carpenter should fit and place 275 to 300 lin.ft. (83.82 to 91.44 m) of wood chair rail per 8-hr. day, at the following labor cost per 100 lin.ft. (30.48 m):

Description	Hours	Rate	Total	Rate	Total
Carpenter	2.80	$	$	$ 35.04	$ 98.11
Labor	0.50	$	$	$ 26.16	$ 13.08
Cost per 100 lin.Ft.					$ 111.19
per lin.ft.					$ 1.11
per lin.m.					$ 3.65

In small kitchens, closets, and bathrooms, a carpenter will place only 160 to 180 lin.ft. (48.76 to 54.86 m) of chair rail per 8-hr. day, at the following labor cost per 100 lin.ft. (30.48 m):

Description	Hours	Rate	Total	Rate	Total
Carpenter	4.70	$	$	$ 35.04	$ 164.69
Labor	0.50	$	$	$ 26.16	$ 13.08
Cost per 100 lin.Ft.					$ 177.77
per lin.ft.					$ 1.78
per lin.m.					$ 5.83

First Grade Workmanship

Where first grade workmanship is required, a carpenter should place 200 to 225 lin.ft. (60.96 to 68.58 m) of chair rail per 8-hr. day, at the following labor cost per 100 lin.ft. (30.48 m):

Description	Hours	Rate	Total	Rate	Total
Carpenter	3.80	$	$	$ 35.04	$ 133.15
Labor	0.50	$	$	$ 26.16	$ 13.08
Cost per 100 lin.Ft.					$ 146.23
per lin.ft.					$ 1.46
per lin.m.					$ 4.80

CARPENTRY

In small rooms, such as kitchens, pantries, bathrooms, etc., requiring considerable cutting and fitting around medicine cabinets, wardrobes, kitchen cases, etc., a carpenter should place 120 to 135 lin.ft. (36.57 to 41.14m) of chair rail per 8-hr. day, at the following labor cost per 100 lin.ft. (30.48 m):

Description	Hours	Rate	Total	Rate	Total
Carpenter	6.30	$	$	$ 35.04	$ 220.75
Labor	0.50	$	$	$ 26.16	$ 13.08
Cost per 100 lin.Ft.					$ 233.83
per lin.ft.					$ 2.34
per lin.m.					$ 7.67

Placing Wood Cornices. Where three- or four-member wood cornices are placed in living rooms, reception rooms, dining rooms, etc., a carpenter should place cornice in one average sized room per 8-hr. day, which is equivalent to 50 to 60 lin.ft. (15.24 to 18.29 m), and the labor cost per 100 lin.ft. (30.48 m) would be as follows:

Description	Hours	Rate	Total	Rate	Total
Carpenter	14.50	$	$	$ 35.04	$ 508.08
Labor	2.00	$	$	$ 26.16	$ 52.32
Cost per 100 lin.Ft.					$ 560.40
per lin.ft.					$ 5.60
per lin.m.					$ 18.39

First Grade Workmanship

Where it is necessary that the wood members fit the plastered walls and ceilings closely, with all miters true and even, two carpenters working together should complete 1 to 1-1/4 rooms per day, at the rate of 35 to 40 lin.ft. (10.66-12.19 m) per 8-hr. day for one carpenter and at the following labor cost per 100 lin.ft. (30.48 m):

Description	Hours	Rate	Total	Rate	Total
Carpenter	21.00	$	$	$ 35.04	$ 735.84
Labor	2.00	$	$	$ 26.16	$ 52.32
Cost per 100 lin.Ft.					$ 788.16
per lin.ft.					$ 7.88
per lin.m.					$ 25.86

Placing Vertical Wood Panel Strips. When vertical wood panel strips or "battens" are nailed to plastered walls to produce a paneled effect, a carpenter should place 22 to 28 pcs. 175 to 225 lin.ft. (53.34 to 68.58 m) per 8-hr. day, at the following labor cost per 100 lin.ft. (30 m):

Description	Hours	Rate	Total	Rate	Total
Carpenter	4.00	$	$	$ 35.04	$ 140.16
Labor	0.50	$	$	$ 26.16	$ 13.08
Cost per 100 lin.Ft.					$ 153.24
per lin.ft.					$ 1.53
per lin.m.					$ 5.03

Placing Wood Strip Paneling. Where panels are formed of wood molding 1-1/2" to 2-1/2" (38-63 mm) wide, making it necessary to cut and miter both ends of each panel strip, the lin.ft. (meter) cost will vary with the size of the panels and the amount of cutting and fitting. Almost as much labor is required on a panel 2'-0"x3'-0" (0.61 x 0.91 m) as for one 3'-0"x6'-0" (0.91 x 1.82 m), even though the former is only half as long.

On small panels up to 2'-0"x4'-0" (0.61 x 1.22 m), requiring 12 lin.ft. (3.65 m) of molding, a carpenter should complete 9 to 11 panels, containing 110 to 135 lin.ft. (3352 to 41.14 m) of molding per 8-hr. day, at the following labor cost per 100 lin.ft. (30.48 m):

Description	Hours	Rate	Total	Rate	Total
Carpenter	6.50	$	$	$ 35.04	$ 227.76
Labor	0.50	$	$	$ 26.16	$ 13.08
Cost per 100 lin.Ft.					$ 240.84
per lin.ft.					$ 2.41
per lin.m.					$ 7.90

On larger panels 3'-0"x5'-0" (0.91 x 1.52 m) to 4'-0"x6'-0" (1.22 x 1.82 m), where each panel contains 16 to 20 lin.ft. (4.87-6.09 m) of molding, a carpenter should complete 7 to 9 panels, containing 140 to 180 lin.ft. (42-54 m) of molding per 8-hr. day, at the following labor cost per 100 lin.ft. (30.48 m):

CARPENTRY

Description	Hours	Rate	Total	Rate	Total
Carpenter	5.00	$	$	$ 35.04	$ 175.20
Labor	0.50	$	$	$ 26.16	$ 13.08
Cost per 100 lin.Ft.					$ 188.28
per lin.ft.					$ 1.88
per lin.m.					$ 6.18

First Grade Workmanship

Where wood panel moldings are used over canvassed or burlap walls, with all strips plumb and level, fitting closely to the plastered walls with perfect fitting miters, a carpenter should complete 7 to 9 small panels, requiring 90 to 115 lin.ft. (27.43 to 34.05 m) of molding per 8-hr. day, at the following labor cost per 100 lin.ft. (30.48 m):

Description	Hours	Rate	Total	Rate	Total
Carpenter	7.80	$	$	$ 35.04	$ 273.31
Labor	0.50	$	$	$ 26.16	$ 13.08
Cost per 100 lin.Ft.					$ 286.39
per lin.ft.					$ 2.86
per lin.m.					$ 9.40

On large panels, from 3'-0"x5'-0" (0.9 x 1.5 m) to 4'-0"x6'-0" (1.2 x 1.8 m), where each panel contains 16 to 20 lin.ft. (4.87-6.09 m) of panel molding, a carpenter should complete about 6 to 8 panels, containing 120 to 150 lin.ft. (36.57 to 45.72 m) of molding per 8-hr. day, at the following labor cost per 100 lin.ft. (30.48 m):

Description	Hours	Rate	Total	Rate	Total
Carpenter	6.00	6	$	$ 35.04	$ 210.24
Labor	0.50	$	$	$ 26.16	$ 13.08
Cost per 100 lin.Ft.					$ 223.32
per lin.ft.					$ 2.23
per lin.m.					$ 7.33

Placing Wood Ceiling Beams. In buildings where built-up ceiling beams are used, the labor costs will vary according to the number of intersections of beams in each room and the length of the beams. It is as easy to erect a 12'-0" (3.65 m) built-up beam as an 8'-0" (2.44 m) one.

973

On average work, a carpenter should place 35 to 45 lin.ft. (10.66 to 13.71 m) of built-up wood beams per 8-hr. day, at the following cost per 100 lin.ft. (30.48 m):

Description	Hours	Rate	Total	Rate	Total
Carpenter	20.00	6 $	$	$ 35.04	$ 700.80
Labor	3.00	$	$	$ 26.16	$ 78.48
Cost per 100 lin.Ft.					$ 779.28
per lin.ft.					$ 7.79
per lin.m.					$ 25.57

First Grade Workmanship

In the better class of buildings using wood ceiling beams, a carpenter should place 30 to 35 lin.ft. (9.14 to 10.66 m) per 8-hr. day, at the following labor cost per 100 lin.ft. (30.48 m):

Description	Hours	Rate	Total	Rate	Total
Carpenter	25.00	6 $	$	$ 35.04	$ 876.00
Labor	3.00	$	$	$ 26.16	$ 78.48
Cost per 100 lin.Ft.					$ 954.48
per lin.ft.					$ 9.54
per lin.m.					$ 31.31

Price of Ponderosa Pine Moldings Per 100 Lin.Ft. and 10 M unless otherwise indicated

Description Size in Inches		Price per 100 Lin.Ft. unless noted *	Description Size in mm				Price per 10 Lin.M. unless noted
Aprons							
11/16 "x	2 1/4 "	$ 83.10	17.19	x	56.25	mm	$ 27.26
11/16 "x	2 1/4 "	$ 27.30	17.19	x	56.25	mm	$ 8.96
9/16 "x	3 1/2 "	$ 50.40	14.06	x	87.50	mm	$ 16.54
Astragals							
1 3/8 "x	7'- 0 "	$ 5.70 Ea.	34.38 mm x		2.13 M		$ 18.70 Ea.
1 1/4 "x	7'- 0 "	$ 6.30 Ea.	31.25 mm x		2.13 M		$ 20.67 Ea.
Bases							
9/16 "x	3 "	$ 72.20	14.06	x	75.00	mm	$ 23.69
1/2 "x	3 1/2 "	$ 75.80	12.50	x	87.50	mm	$ 24.87
1/2 "x	4 1/4 "	$ 86.70	12.50	x	106.25	mm	$ 28.44

CARPENTRY

Description Size in Inches		Price per 100 Lin.Ft. unless noted *	Description Size in mm				Price per 10 Lin.M. unless noted
	Base Cap						
11/16 "x	1 3/8 "	$ 36.10	17.19	x	34.38	mm	$ 11.84
	Backband						
11/16 "x	1 1/16 "	$ 28.90	17.19	x	26.56	mm	$ 9.48
	Balusters						
3/4 "x	3/4 "	$ 45.20	18.75	x	18.75	mm	$ 14.83
1 1/16 "x	1 1/16 "	$ 72.20	26.56	x	26.56	mm	$ 23.69
1 5/16 "x	1 5/16 "	$ 117.40	32.81	x	32.81	mm	$ 38.52
1 5/8 "x	1 5/8 "	$ 126.40	40.63	x	40.63	mm	$ 41.47
	Base Shoe Combination						
3/4 "x	2 5/8 "	$ 54.20	18.75	x	65.63	mm	$ 17.78
	Bed Molding						
11/16 "x	1 3/4 "	$ 63.30	17.19	x	43.75	mm	$ 20.77
11/16 "x	2 1/4 "	$ 90.30	17.19	x	56.25	mm	$ 29.63
	Base Shoe						
1/2 "x	3/4 "	$ 36.10	12.50	x	18.75	mm	$ 11.84
	Blind Stop						
3/4 "x	3/8 "	$ 72.20	18.75	x	9.38	mm	$ 23.69
	Brick Molding						
1 5/16 "x	2 "	$ 108.30	32.81	x	50.00	mm	$ 35.53
	Corner Guards						
11/16 "x	2 1/4 "	$ 66.90	17.19	x	56.25	mm	$ 21.95
11/16 "x	3 1/2 "	$ 122.80	17.19	x	87.50	mm	$ 40.29
	Corner Guards						
11/16 "x	2 1/8 "	$ 90.30	17.19	x	53.13	mm	$ 29.63
7/16 "x	2 1/2 "	$ 101.10	10.94	x	62.50	mm	$ 33.17
	Corner Guards						
3/4 "x	3/4 "	$ 38.00	18.75	x	18.75	mm	$ 12.47
1 1/8 "x	1 1/8 "	$ 75.80	28.13	x	28.13	mm	$ 24.87
1 5/16 "x	1 5/16 "	$ 83.10	32.81	x	32.81	mm	$ 27.26
	Corner Guards						
11/16 "x	3 1/4 "	$ 126.40	17.19	x	81.25	mm	$ 41.47
11/16 "x	2 1/4 "	$ 86.70	17.19	x	56.25	mm	$ 28.44
11/16 "x	1 3/4 "	$ 61.40	17.19	x	43.75	mm	$ 20.14
3/4 "x	7/8 "	$ 46.90	18.75	x	21.88	mm	$ 15.39
3/4 "x	3/4 "	$ 43.30	18.75	x	18.75	mm	$ 14.21
1/2 "x	1/2 "	$ 36.10	12.50	x	12.50	mm	$ 11.84
11/16 "x	1 1/8 "	$ 48.80	17.19	x	28.13	mm	$ 16.01
	Crowns						
11/16 "x	4 1/4 "	$ 184.20	17.19	x	106.25	mm	$ 60.43
11/16 "x	3 1/4 "	$ 122.80	17.19	x	81.25	mm	$ 40.29
11/16 "x	2 1/4 "	$ 86.70	17.19	x	56.25	mm	$ 28.44

Description Size in Inches		Price per 100 Lin.Ft. unless noted *	Description Size in mm				Price per 10 Lin.M. unless noted	
Drip Caps								
1 1/16 "x	1 5/8 "	$ 86.70	26.56	x	40.63	mm	$	28.44
3/4 "x	1 5/8 "	$ 75.80	18.75	x	40.63	mm	$	24.87
Full Rounds								
1 5/8 "		$ 135.50	40.63 mm				$	44.46
1 5/16"		$ 90.30	32.81 mm				$	29.63
Glazing Beads								
1/2 "x	9/16 "	$ 32.50	12.50	x	14.06	mm	$	10.66
3/8 "x	3/8 "	$ 27.20	9.38	x	9.38	mm	$	8.92
Half Rounds								
5/16 "x	5/8 "	$ 36.10	7.81	x	15.63	mm	$	11.84
Hand Rails								
1 1/2 "x	1 3/4 "	$ 104.70	37.50	x	43.75	mm	$	34.35
Hook Strips								
9/16 "x	2 7/16 "	$ 75.80	14.06	x	60.94	mm	$	24.87
Jamb Extensions								
3/4 "x	1 15/16 "	$ 66.90	18.75	x	48.44	mm	$	21.95
3/4 "x	7/8 "	$ 50.60	18.75	x	21.88	mm	$	16.60
Lattice								
1/4 "x	1 1/8 "	$ 23.60	6.25	x	28.13	mm	$	7.74
1/4 "x	1 3/8 "	$ 27.20	6.25	x	34.38	mm	$	8.92
1/4 "x	1 3/4 "	$ 34.40	6.25	x	43.75	mm	$	11.29
1/4 "x	2 1/4 "	$ 46.90	6.25	x	56.25	mm	$	15.39
Mullion Casings								
9/16 "x	5 1/2 "	$ 153.50	14.06	x	137.50	mm	$	50.36
3/16 "x	2 "	$ 75.80	4.69	x	50.00	mm	$	24.87
Parting Strip								
1/2 "x	3/4 "	$ 28.90	12.50	x	18.75	mm	$	9.48
Picture Molding								
3/4 "x	1 3/4 "	$ 70.50	18.75	x	43.75	mm	$	23.13
Quarter Rounds								
1/4 "x	1/4 "	$ 23.70	6.25	x	6.25	mm	$	7.78
1/2 "x	1/2 "	$ 21.70	12.50	x	12.50	mm	$	7.12
3/4 "x	3/4 "	$ 36.10	18.75	x	18.75	mm	$	11.84
1 1/16 "x	1 1/16 "	$ 68.60	26.56	x	26.56	mm	$	22.51
Screen Moldings								
5/16 "x	5/8 "	$ 21.70	7.81	x	15.63	mm	$	7.12
1/4 "x	3/4 "	$ 18.10	6.25	x	18.75	mm	$	5.94

CARPENTRY

Description Size in Inches		Price per 100 Lin.Ft. unless noted *	Description Size in mm				Price per 10 Lin.M. unless noted
Screen Stock							
3/4 "x	1 3/4 "	$ 72.20	18.75	x	43.75	mm	$ 23.69
3/4 "x	2 3/4 "	$ 110.20	18.75	x	68.75	mm	$ 36.15
1 1/16 "x	1 3/4 "	$ 113.80	26.56	x	43.75	mm	$ 37.34
1 1/16 "x	2 3/4 "	$ 135.50	26.56	x	68.75	mm	$ 44.46
Shelf Cleat							
11/16 "x	1 1/2 "	$ 36.10	17.19	x	37.50	mm	$ 11.84
Stools							
1 1/16 "x	3 1/4 "	$ 306.90	26.56	x	81.25	mm	$ 100.69
1 1/16 "x	3 5/8 "	$ 334.10	26.56	x	90.63	mm	$ 109.61
1 1/16 "x	3 3/4 "	$ 172.20	26.56	x	93.75	mm	$ 56.50
1 1/16 "x	2 1/4 "	$ 126.40	26.56	x	56.25	mm	$ 41.47
1 1/16 "x	2 5/8 "	$ 135.50	26.56	x	65.63	mm	$ 44.46
1 1/16 "x	2 3/4 "	$ 162.50	26.56	x	68.75	mm	$ 53.31
Stops							
7/16 "x	2 1/8 "	$ 86.70	10.94	x	53.13	mm	$ 28.44
7/16 "x	1 5/8 "	$ 75.80	10.94	x	40.63	mm	$ 24.87
7/16 "x	1 3/8 "	$ 66.90	10.94	x	34.38	mm	$ 21.95
7/16 "x	1 1/8 "	$ 56.00	10.94	x	28.13	mm	$ 18.37
7/16 "x	7/8 "	$ 54.20	10.94	x	21.88	mm	$ 17.78
7/16 "x	15/16 "	$ 56.00	10.94	x	23.44	mm	$ 18.37
7/16 "x	5/8 "	$ 43.30	10.94	x	15.63	mm	$ 14.21

PLASTIC LAMINATE AND STONE COUNTERTOPS

Plastic laminate for countertops will vary in price depending on the number of cutouts, corners, and whether the fabricator must include the cost of taking job measurements. A typical L-shaped kitchen countertop covering 18 lin.ft. (5.48 m) of inside wall, all edges plastic, laminated to 7/8" (21.87 mm) particle board, with 4" (100 mm) splashes back and ends, will cost around $60.00 per lin.ft. (per 0.30 m) in a standard pattern. Rounded edges will cost considerably more; metal banded ones somewhat less. For each cut out add $24.00. For each end splash add $17.00.

There is also engineered stone and products such as Corian® and Granite tops for countertops today. Each has its advantages and disadvantages. The weights of these materials are greater than a laminated countertop and therefore may require additional labor to set in place. Most stones are precut in the shop, transportation and handling may be an added cost in order to prevent the stone from fracturing during transportation. Costs for a solid surface countertop can cost about $100.00 to $150.00 and higher per lin.ft. (0.30 m). Costs for Stone will start at about $80.00 per lin.ft. (0.30 m) and go up based on thickness, size, type and quality stone.

Setting Cabinets and Cases. When setting mill assembled cabinets and cases, such as used in the average residence or apartment building, it will require about 1/6 hr. per sq.ft. of cabinet face area (1.85 hrs. per sq.m) carpenter time at the following labor cost per sq.ft. (sq.m):

Description	Hours	Rate	Total	Rate	Total
Carpenter per sq.ft.	0.16	$	$	$ 35.04	$ 5.61
per sq.m					$ 60.35

A carpenter should fit and hang, complete with pulls and catches, 15 to 17 flush case doors per 8-hr. day, at the following labor cost per door:

Description	Hours	Rate	Total	Rate	Total
Carpenter	0.5	$	$	$ 35.04	$ 17.52

For lip case doors that do not require fitting, a carpenter should install, complete with pulls and catches, 30 to 34 doors per 8-hr. day, at the following cost per door:

Description	Hours	Rate	Total	Rate	Total
Carpenter	0.25	$	$	$ 35.04	$ 8.76

In the better class buildings, where cases, wardrobes, cabinets, etc., are used, and it is necessary to scribe and fit them against the plastered walls or run a small molding to conceal joints between cases and plastered walls, it requires about 1/3 hr. per sq.ft. of cabinet face area (3.67 hrs. per sq.m) at the following labor cost per sq.ft. (sq.m):

Description	Hours	Rate	Total	Rate	Total
Carpenter per sq.ft.	0.33	$	$	$ 35.04	$ 11.56
per sq.m					$ 124.47

A carpenter should fit and hang, complete with pulls and catches, 11 to 13 flush case doors per 8-hr. day, at the following labor cost per door:

Description	Hours	Rate	Total	Rate	Total
Carpenter	0.67	$	$	$ 35.04	$ 23.48

Often, in restricted areas, such as adding cabinets in narrow lavatory areas, it is necessary to have cabinet work delivered in sections so that it may be turned and properly set within the space. Seldom is a job so organized that

978

CARPENTRY

cabinet work is delivered before wall finishes are up, so the fit must be preplanned and exact.

Setting Factory Assembled and Finished Kitchen Cabinets. When setting factory assembled and finished kitchen cabinets, such as used in the average apartment or residence, it will require about 0.1 hr. per sq.ft. (1.11 hrs. per sq.m) of cabinet face area at the following labor cost:

Description	Hours	Rate	Total	Rate	Total
Carpenter per sq.ft.	0.1	$	$	$ 35.04	$ 3.50
per sq.m					$ 37.72

First Grade Workmanship

In better class apartments and residences, where a high class job is required, it takes about 1/6 hr. per sq.ft. (1.85 hrs. per sq.m) of cabinet face area at the following labor cost:

Description	Hours	Rate	Total	Rate	Total
Carpenter per sq.ft.	0.16	$	$	$ 35.04	$ 5.61
per sq.m					$ 60.35

Setting Wood Fireplace Mantels. If wood fireplace mantels are factory assembled, merely requiring fitting and setting, a carpenter should install 2 to 3 mantels per 8-hr. day at the following labor cost per mantel:

Description	Hours	Rate	Total	Rate	Total
Carpenter	3.2	$	$	$ 35.04	$ 112.13
Labor	1	$	$	$ 26.16	$ 26.16
Cost per Mantel					$ 138.29

Fitting and Placing Closet Shelving. A carpenter should fit and place 115 to 135 sq.ft. (10.68 to 12.54 sq.m) of closet shelving per 8-hr. day including setting of shelf cleats at the following labor cost per 100 sq.ft. (9.29 sq.m):

Description	Hours	Rate	Total	Rate	Total
Carpenter	6.4	$	$	$ 35.04	$ 224.26
Cost per sq.ft.					$ 2.24
per sq.m					$ 24.14

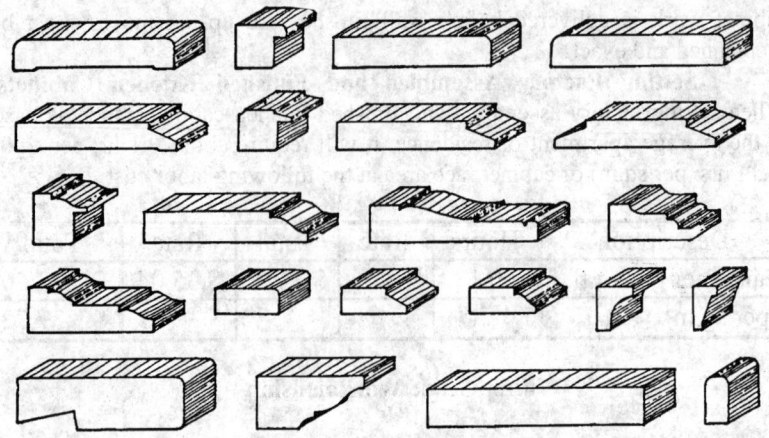

Wood Interior Trim, Casings, Back-band, Stops, Stool, Etc.

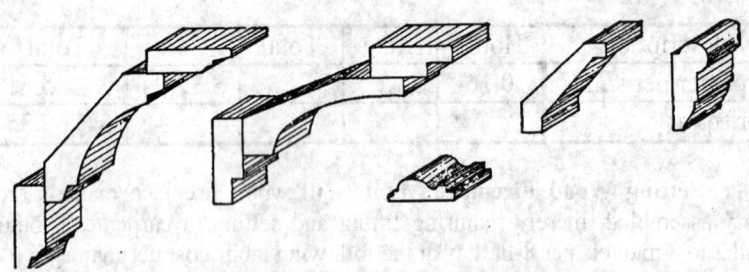

Wood Ceiling Cornice and Picture Mold

Installing Hanging Rods in Closets. A carpenter should install about three hanging rods, including supports, per hr. at the following labor cost per rod:

Description	Hours	Rate	Total	Rate	Total
Carpenter	0.33	$	$	$ 35.04	$ 11.56

Setting Metal Medicine Cabinets. A carpenter should set 10 to 12 average size medicine cabinets per 8-hr. day at the following labor cost per cabinet:

Description	Hours	Rate	Total	Rate	Total
Carpenter	0.75	$	$	$ 35.04	$ 26.28

980

CARPENTRY

Setting Bathroom Accessories. In an 8-hr. day a carpenter should locate and set 30 to 34 bathroom accessories, such as towel bars, soap dishes, paper holders, etc. at the following labor cost per accessory:

Description	Hours	Rate	Total	Rate	Total
Carpenter	0.25	$	$	$ 35.04	$ 8.76

Installing Chalkboards, Corkboards, and Accessories. To install chalkboards, before starting, all grounds should be in place and final coat of finish should be thoroughly dry.

Chalkboard should be set so that all edges will be adequately covered by trim moldings. The writing surface should be plumb and in a true plane with 1/4" (6.25 mm) clearance on all sides. Chalkboard panels should be nailed at the top edges only and the balance secured to the wall with an even coat of spotting cement.

Two carpenters working together should install 125 to 150 sq.ft. (11.61 to 13.93 sq.m) of chalkboard per 8-hr. day at the following labor cost per 100 sq.ft. (9.29 sq.m):

Description	Hours	Rate	Total	Rate	Total
Carpenter	11	$	$	$ 35.04	$ 385.44
Cost per sq.ft.					$ 3.85
per sq.m					$ 41.49

Installation of corkboard is similar to that of chalkboard except nailing is eliminated. The labor cost of installing corkboard is approximately the same as for chalkboard.

Installation of aluminum trim moldings in conjunction with chalkboards and corkboards may be accomplished either with exposed wood screws or with concealed snap-on clips. A carpenter should install 90 to 110 lin.ft. (27.43 to 3352 m) of molding per 8-hr. day at the following labor cost per 100 lin.ft. (30.48 m):

Description	Hours	Rate	Total	Rate	Total
Carpenter	8	$	$	$ 35.04	$ 280.32
Cost per lin. ft.					$ 2.80
per lin.m					$ 9.20

Labor Placing Finish Hardware		
Description of Work	No. Set per 8-Hr. Day	Carp. Hrs. Each
Rim locks or night latches, cheap work	20.0 - 24.0	0.3 - 0.4
Mortised locks in soft or hardwood doors, Ordinary Workmanship	12.0 - 16.0	0.5 - 0.7
Mortised locks in hardwood doors, first grade workmanship	6.0 - 8.0	1.0 - 1.3 *
Cylindrical locks	17.0 - 23.0	0.3 - 0.5
Front entrance cylinder locks in hardware doors, ordinary workmanship	6.0 - 8.0	1.0 - 1.3 **
Front entrance cylinder locks in hardwood doors, first grade workmanship	4.0 - 5.0	1.6 - 2.0 ***
Panic bolts, first grade workmanship	2.0 - 3.0	2.7 - 4.0
Door closers, exposed	7.0 - 9.0	0.9 - 1.1
Door closers, concealed	2.0 - 3.0	2.7 - 4.0
Door holders	20.0 - 24.0	0.3 - 0.4
Sash lifts and locks, no. of windows	20.0 - 24.0	0.3 - 0.4
Cremone bolts for casement doors, windows	20.0 - 24.0	0.3 - 0.4
Kickplates, ordinary workmanship	8.0 - 10.0	0.8 - 1.0
Kickplates, first grade workmanship	6.0 - 8.0	1.0 - 1.3

*On this grade of work, the lock must be flush with the face of the door and must operate perfectly

**On this class of work, the architect or owner can expect some trouble, because time is not sufficient to adjust locks and see that it operates properly

***This includes front door locks having fancy escutcheon plates, knobs, knockers, etc.

982

CARPENTRY

Labor Placing Cabinet Hardware		
Description of Work	No. Set per 8-Hr. Day	Carp. Hrs. Each
Surface butts	180 - 200	0.040 - 0.045
Offset butts	90 - 100	0.080 - 0.090
Elbow catches	90 - 100	0.080 - 0.090
Friction catches	90 - 100	0.080 - 0.090
Drawer and door pulls	140 - 150	0.053 - 0.057
Knobs	180 - 200	0.040 - 0.045
Rim latches	90 - 100	0.080 - 0.090
Mortised locks	20 - 24	0.300 - 0.400

Mantels

There are many stock mantel designs from which to choose. A simple molding 6" (150 mm) wide extending out from the wall 2" (50 mm) on 3 sides will cost about $85.00 for an outside dimension of 4'x7' (1.21 x 2.13 m). A full mantel with 7" (175 mm) shelf and of colonial design will cost from $160.00 to $240.00. More elaborate designs in Georgian or French provincial and with end returns to fit over the brick projection will run from $240.00 to $800.00. These costs are for wood members only. Brick or marble trim around the fireplace opening are, of course, extra. It will take a carpenter 3 to 4 hours to set a prefabricated unit in place.

China Cabinets

Corner cabinets can be purchased from stock to add architectural interest to dining, breakfast, or recreation rooms. Mostly of Ponderosa pine with either Colonial or Georgian design, they will vary from $115.00 for a unit 2'-8" (0.8 m) wide with single doors top and bottom to $375.00 for a unit 3'-6" (1.06 mm) wide with double doors. Cabinets are from 7' to 7'-6" (2.13 to 2.28 m) high. Bottom doors are usually paneled, tops factory glazed. Shipped with front assembled and body knocked down, the unit includes three shelves for the upper portion, one for the lower. Hardware is not included. Allow three hours for one carpenter time to fit and set each unit and apply the hardware.

Ironing Board Cabinet

A built-in feature that appeals to many people is an ironing board cabinet unit complete with adjustable board plus sleeve board, all operating hardware, and flush door which is sized to fit between wall studs 16" (400

983

mm) on centers. It is shipped in a carton and costs around $125.00. The door is available in either birch or Philippine mahogany. No casing is included, but if the unit is set in a matching paneled wall, it is almost inconspicuous.

Louvered Pine Panels

Shuttered panels are available in stock sizes, complete with hardware, or customized to fit exact openings. They are not only used inside window openings, but for cabinet doors and room separators.

Cost of Stock Single Panel Horizontal Slat Unit					
Height	Widths, Inches				
Inches	6"	7"	8"	10"	12"
20	$4.95	$5.55	$6.15	$7.65	$9.15
23	$6.00	$6.45	$7.05	$8.55	$10.05
29	$6.75	$7.35	$8.40	$9.60	$11.70
32	$7.65	$8.25	$9.30	$10.95	$13.05
36	$8.85	$9.15	$10.50	$12.30	$15.15

Cost of Stock Single Panel Horizontal Slat Unit					
Height	Widths, (mm)				
mm	150	175	200	250	300
500	$4.95	$5.55	$6.15	$7.65	$9.15
575	$6.00	$6.45	$7.05	$8.55	$10.05
725	$6.75	$7.35	$8.40	$9.60	$11.70
800	$7.65	$8.25	$9.30	$10.95	$13.05
900	$8.85	$9.15	$10.50	$12.30	$15.15

A four-panel hardware kit costs $2.00. If no trimming is required, one carpenter should set 2 panels per hour or 2 window openings consisting of 8 panel units each in a day.

Custom shutters in lengths of 6" to 18" (150-450 mm), and for openings from 12" to 24" (300-600 mm), 2 panels will cost $40.00 unfinished and $55.00 prefinished; for an opening 24" to 30" (600-750 mm), four panels same height, $65.00 and $85.00; for 30" to 42" (750-1050 mm) opening, $60.00 and $105.00; in lengths of 18" to 24" (450-600 mm), openings 12" to 24" (300-600 mm), 2 panels will run $45.00 and $65.00; openings 24" to 30" (600-750 mm), 4 panels, $65.00 and $110.00; openings 30" to 42" (750-1050 mm), 4 panels, $90.00 and $110.00. Custom door panels to 7' (2.1 m) will run $160.00 unfinished, $157.00 finished for 2 panels for an opening to 24" (600 mm) wide. For openings to 30" (750 mm),

2 panels, $160.00 and $225.00; openings 30" to 42" (750-1050 mm), 4 panels, $315.00 and $440.00.

06300 WOOD TREATMENT

Wood and plywood are often given special treatments at the mill to give additional protection against fire, rot, and termites.

Fire protected or fire retardant wood is pressure-impregnated with mineral salts that will react chemically at a temperature below the ignition point of the wood. An Underwriter's Laboratory Flame Spread Rating of 25 can be obtained, and no periodic maintenance is required to keep the rating. However, the lumber must also be kiln dried to 12% moisture content if it is to be painted. Also, wood is not suitable for direct exposure to weather and ground conditions unless further treatment is given. A fire retardant treatment will add $0.20 to 0.25per bd.ft. (0.002 cu.m) to lumber costs. If it is necessary to kiln dry the lumber, add $0.10 for soft woods, $0.15 for hard woods.

Rot and termite resistance can be obtained by several methods. Creosote, when permitted by local environmental regulations, is impregnated at the rate of around 8 lbs. per cu.ft. (127.11 kg/m^3) at a cost of $0.50to 1.20 per bd.ft. This finish is not readily paintable and can stain adjacent finishes such as plaster and wallboard. Pentachlorophenol is a popular oil treatment and costs $0.35 to $0.60 per sq.ft. and leaves the wood with an oil residue. This residue can be eliminated, and the wood left clean and paintable, if the pentachlorophenol is impregnated into the wood by liquid petroleum. This will add $0.20 to the cost.

06400 ARCHITECTURAL WOODWORK

This section covers the assembly and installation of cabinets, wood and wallboard paneling, wood stairs, and railing from components made up in a shop or factory off the job premises.

If the items under this heading are not stock items, it will be necessary to include the cost of making on-the-job measurements and preparing shop drawings. Most cabinet makers are equipped to do this and will include such costs in their proposals. Often they will also install their work, but usually the general contractor handles this work.

There is considerable difference between the grades of workmanship for architectural woodwork and the estimator must be able to determine the grade required. These same differences apply to the erection and installation of all classes of millwork and interior finish.

The Architectural Woodwork Institute publishes Illustrated Quality Standards, in which the work is divided into three grades: premium, custom, and economy. Premium is the highest grade available in both material and workmanship. It might be used throughout an entire building, but most often, it is for selected spaces or items. Custom, the middle or normal grade in both material and workmanship, is intended for high quality regular work.

985

Economy is the lowest grade and intended for work where price outweighs quality considerations. An estimator may find references to these three grades in an architect's specification.

Methods of Estimating the Labor Cost Millwork and Interior Finish. Different contractors use different methods of estimating the labor cost of erecting and setting millwork and interior finish, but the system generally used is to add a certain percentage of the cost of the millwork to cover the labor cost of erection. While this method is the easiest, it does not produce the best estimates. For example, suppose you are estimating the cost of a building on which you receive four millwork bids, ranging from $7,200 to $9,300. After receiving the bids you add the customary percentage to the lowest acceptable bid to take care of the labor cost of handling and erecting the millwork. This percentage usually runs from 40% to 50% of the cost of the millwork. Taking 40% of $7,200.00 gives $2,880.00 for labor to handle and erect all millwork and interior finish. Suppose a competitor received a bid for $6,400.00 from a mill which did not quote you. If your competitor added 40% to his bid, he would have only $2,560 to erect the same amount of millwork on which you figured $2,280.00.

Why this difference of 12-1/2% on such a small item? While the method saves a great deal of time and is almost universally used, it does not produce the most accurate estimates. The only accurate method is to list each item separately; i.e., the number of lin.ft. of wood base, chair rail, picture mold, etc., the number of door jambs to set, the number of sash to fit and hang, together with the number of window openings to be cased or trimmed; the number of wardrobes, linen cases, vanity and kitchen units to set, etc. Estimating labor costs by this method produces more uniform and accurate estimates, reduces risk to a minimum, and enables the contractor to check actual and estimated costs during construction, but it does require a lot more work.

INTERIOR PANELING

Interior paneling used to mean solid wood with stiles and rails and elaborate mouldings for the mansion or bank job; knotty pine for the suburban retreat; and beaver board for the cottage. Today the selection is much wider.

Solid wood paneling is still with us, but it is more apt to be in the form of planks with face-applied mouldings rather than as elaborate paneling. Plywood veneer with hard wood is often favored over solid wood not only for cost savings but because the veneer exposes the beauty of the wood and offers a more easily cleaned surface.

Most paneling today though is sheet board faced to look like wood grain, marble, tile, stucco, leather, or a decorative pattern all its own. The designs come with a selection of moldings that blend into the overall wall.

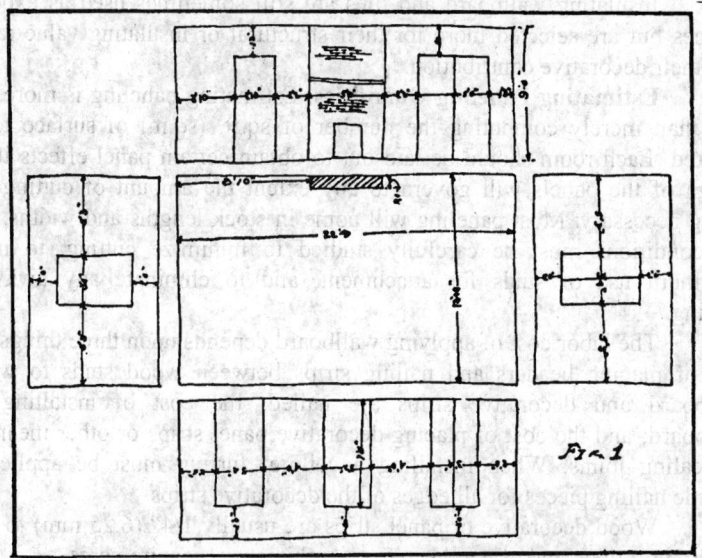

Rough Sketch of Room Showing Location of Doors and Windows

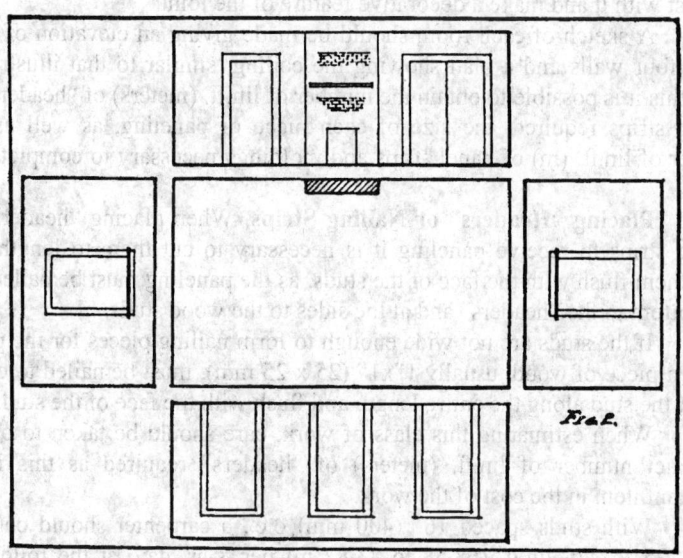

Finished Drawing Showing Location of Doors and Windows

Insulating wallboard and tiles are still sometimes used as exposed finishes but are selected more for their structural or insulating value rather than their decorative contribution.

Estimating Paneling Quantities. Estimating paneling is more of a task than merely computing the number of sq.ft. (sq.m) of surface to be covered. Each room should be laid out to obtain certain panel effects if the design of the panels will govern to any extent the amount of cutting and fitting necessary. Most paneling will come in stock lengths and widths. The job conditions must be carefully studied to minimize cutting, to make maximum use of studs for attachment, and to eliminate any awkward jointing.

The labor cost of applying wallboard depends upon three things: the cost of placing headers and nailing strips between wood studs to which wallboard and decorative strips are nailed; the cost of installing the wallboard; and the cost of placing decorative panel strips or other means of concealing joints. When installing a ceiling, furring must be applied to provide nailing pieces for all edges of the decorative strips.

Wood decorative or panel strips are usually 1/4" (6.25 mm) to 1/2" (12.50 mm) thick and of widths to meet architectural requirements.

There are several ways of covering joints. Today much paneling is grooved so joints are butted and appear as an overall pattern. Most flush panels have cover strips, caps, etc. available to match the basic material or to contrast with it and make a decorative feature of the joint.

A sketch of each room should be made giving an elevation of each of the four walls and a plan showing the ceiling, similar to that illustrated. From this it is possible to obtain the number of lin.ft. (meters) of "headers" or nailing strips required, the size of each piece of paneling, as well as the number of lin.ft. (m) of panel strips and moldings necessary to complete the job.

Placing "Headers" or Nailing Strips. When placing "headers" or nailing strips to receive paneling it is necessary to cut them to length and place them flush with the face of the studs, as the paneling must be nailed top and bottom to the "headers" and at the sides to the wood studs.

If the studs are not wide enough to form nailing pieces for the panel strips, a piece of wood, usually 1"x1" (25 x 25 mm), must be nailed to either side of the stud along the entire length and flush with the face of the stud.

When estimating this class of work, care should be taken to obtain the exact number of lin.ft. (meters) of "headers" required as this is an important item in the cost of the work.

With studs spaced 16" (400 mm) o.c., a carpenter should cut and place 225 to 275 lin.ft. (68.58 to 83.82 m) per 8-hr. day, at the following labor cost per 100 lin.ft. (30.48 m):

988

CARPENTRY

Description	Hours	Rate	Total	Rate	Total
Carpenter	3.20	$	$	$ 35.04	$ 112.13
Cost per lin.ft.					$ 1.12
per m					$ 3.68

Placing Furring Strips for Ceilings. Furring strips should be nailed at right angles to joists. Any convenient width of furring may be used, but it must be wide enough to provide a nailing base to nail all edges of the decorative strips at least every 6" (150 mm). Furring may be any No. 2 material, soft wood being preferable. Make sure that all furring strips are level.

With joists 16" (400 mm) o.c., a carpenter should cut and place 300 to 350 lin.ft. (91.44 to 106.68 m) of furring strips per 8-hr. day, at the following labor cost per 100 lin.ft. (30 m):

Description	Hours	Rate	Total	Rate	Total
Carpenter	2.50	$	$	$ 35.04	$ 87.60
Cost per lin.ft.					$ 0.88
per m					$ 2.87

Placing Sheet Paneling. The actual cost of placing sheet paneling will vary with the size and shape of the room, whether full size sheets may be used or considerable cutting and fitting necessary. It requires practically as much time to place a sheet 16"x96" (400 x 2400 mm) as one 48"x96" (1200 x 2400 mm) on account of the cutting and fitting, while there are three times as many sq. ft. (sq.m) in the latter as in the former.

On straight work, in large rooms, a carpenter should fit and place 450 to 550 sq.ft. (41.80 to 51.09 sq.m) per 8-hr. day, at the following labor cost per 100 sq.ft. (9.29 sq.m):

Description	Hours	Rate	Total	Rate	Total
Carpenter	1.60	$	$	$ 35.04	$ 56.06
Cost per sq.ft.					$ 0.56
per sq.m					$ 6.03

When applying sheets in smaller rooms requiring considerable cutting and fitting, a carpenter should fit and place 275 to 300 sq.ft. (25.54 to 27.87 sq.m) per 8-hr. day, at the following labor cost per 100 sq.ft. (9.29 sq.m):

Description	Hours	Rate	Total	Rate	Total
Carpenter	2.70	$	$	$ 35.04	$ 94.61
Cost per sq.ft.					$ 0.95
per sq.m					$ 10.18

989

As previously mentioned, it usually costs more to place a small piece than a large one, as this generally calls for cutting and fitting above and below windows, around door openings, etc., which requires considerably more work than placing a large sheet where no cutting and fitting is necessary. For this reason, estimating the labor cost by the sq.ft. (sq.m) is not the safest method to use, although it is the most common.

Where time in estimating is not a factor, the safest method is to lay out the work as illustrated, dividing it into panels and then allow 20 to 30 minutes to place each piece of wallboard or plywood depending upon the class of work.

06421 HARDWOOD PLYWOOD PANELS

These plywood panels of woods like birch, oak, and mahogany are usually furnished in sheets 4'-0" (1.22 m) wide and 8'-0" to 10'-0" (2.43-3.04 m) high and are used instead of plaster or other wall finishes.

When using 1/4" (6.25 mm) plywood panels in large sizes for this class of work, it is necessary to have a substantial backing for the thin plywood panels. This may be obtained by applying the plywood direct to a brown or finish plaster coat wall; a 1/2" (12.50 mm) gypsum board backing, or better still, a 1/2" (12.50 mm) thick plywood backing with the sheets running in the opposite direction to that of the finish plywood panels.

These plywood panels are usually applied with the panel running from floor to ceiling in one length, i.e., 4'-0"x8'-0" (1.22 x 2.43 m) or 4'-0"x10'-0" (1.22 x 3.04 m), with the vertical joints butted flush or having a very slight bevel at the edges.

To obtain a first class job, it is usually necessary to apply adhesive to the back of the plywood panels and also use small brads to secure the plywood in place until the adhesive sets.

This is particular work, because the adjoining sheets must be accurately fitted together if a first class job is to result.

If wood panel strips are used to conceal the joints, it is not necessary to use so much care in placing the panels.

Hardwood laminated veneer panels cover a wide variety of products, from the 1/4" (6.25 mm) thick "grooved" veneers to 3/4" (18.75 mm) thick flush panels with or without sequence matching of the flitches on up to special burl veneers suitable for the tops of the finest furniture.

Plywood may have a veneer core with a center and at least two intermediate cross banded veneers, a particleboard core, or a lumber core with strips of lumber edge glued into a solid slab with cross banding veneers either side. The back and face of each sheet then receive additional veneers at right angles to the cross banding.

The face veneers may be rotary cut with the log's annular rings exposing a bold variegated grain; flat sliced, which produces a variegated figure similar to that of plain sawn lumber; quarter sliced with growth rings of the log at right angles to the knife, producing a series of stripes in the

990

veneer; and rift cut, which is used with oak logs to produce fairly uniform grains with a combed look caused by the medullary rays.

There are various grades of face veneers: premium or architectural panels, which have veneers sequence matched and no patching; custom, which allows no patching on face veneers but is not sequence matched; and economy, which allows "good" face veneers, but no matching of grain or colors.

And, finally, there is the type of tree itself which affects the price. The range from natural yellow birch to Brazilian rosewood is around 800%.

Costs should always be determined locally, since not all veneers are available at all times. The following will serve as a general outline of costs. All types are 4'x8'x 1/4" (1.22 m x 2.43 m x 6.25 mm) custom.

Description	Cost per 32 Sq.Ft	Cost per 3 Sq.M
Natural yellow birch	$ 41.40	$ 41.78
Natural hard maple	$ 41.40	$ 41.78
Natural gum	$ 41.40	$ 41.78
Select white birch	$ 53.40	$ 53.89
Poplar	$ 46.80	$ 47.23
Select birch heartwood	$ 48.00	$ 48.44
Select white maple	$ 64.08	$ 64.67
Plain sawn red oak	$ 62.40	$ 62.97
Mahogany - Honduras South / Central America	$ 62.40	$ 62.97
Oak	$ 72.00	$ 72.66
White ash	$ 66.00	$ 66.60
Limba	$ 69.00	$ 69.63
Rift sawn white oak	$ 69.00	$ 69.63
Pecan	$ 73.20	$ 73.87
Cherry / Chilean Cherry	$ 92.40	$ 93.25
Butternut	$ 80.40	$ 81.14
Quarter sawn white oak	$ 90.00	$ 90.82
Walnut	$ 10.50	$ 10.60
Primavera	$ 103.80	$ 104.75
Avodire	$ 111.00	$ 112.02
Teak	$ 165.60	$ 167.12
Brazilian rosewood	$ 414.00	$ 417.79

If matched sequence panels are ordered, the cost will be about 50% more. Add around $17.00 per sheet for 3/4" (18.75 mm) thickness.

The stock V-grooved prefinished 4'x8'x 1/4" (1.22 m x 2.44 m x 6.25 mm) panels are considerably cheaper. Although the range of woods is not so extensive, the finishes vary widely and approximate the colors of more expensive woods still using the grains of the more common species. A random sampling of costs indicates the following range.

Description			Clear	Colonial	Metric
Birch	4' x 8' x 1/4"	V-grooved	$ 22.50	$ 16.25	1.22 m x 2.44 m x 6.25 mm
Oak	4' x 8' x 1/4"	V-grooved	$ 35.00	$ 20.00	1.22 m x 2.44 m x 6.25 mm
Walnut	4' x 8' x 1/4"	V-grooved	$ 40.00	$ 23.75	1.22 m x 2.44 m x 6.25 mm
Pecan	4' x 8' x 1/4"	V-grooved	$ 33.75	$ 31.25	1.22 m x 2.44 m x 6.25 mm

Inlaid paneling in 4'x8'x1/4" (1.22 m x 2.44 m x 6.25 mm) sheets consisting of pecan panels striped with 1-1/4" (31.25 mm) wide walnut stripes costs $55.00 per sheet; with walnut striped in pecan the cost is $64.00 per sheet.

For lower budget jobs, simulated wood grain finishes on 1/4" (6.25 mm) plywood will be $12.00 per 4'x8' (1.22 m x 2.44 m) sheet.

For the lowest budget jobs, simulated wood finishes printed on hardboard are available in the $11.00 to $13.00 range per 4'x8'x 1/4" (1.22 m x 21.44 m x 6.25 mm) sheet.

Labor Placing Flush Plywood Wall Panels. When applying plywood panels 4'-0"x8'-0" (1.22 m x 2.44 m) to 4'-0"x10'-0" (1.22 m x 3.04 m), with the abutting joints fitted tightly together, using adhesive and small brads to secure the plywood to the backing, two carpenters working together should handle, fit, and place one plywood panel in 1-1/2 to 2 hours, or they should place 4 to 5 panels per 8-hr. day.

Many of the panels will have to be cut to half or three-quarter size, and in some instances cut out for doors and windows, so a certain length of time should be allowed for each piece of plywood, rather than figuring on a square foot basis.

06422 SOFTWOOD PLYWOOD PANELING

Softwood plywood consists of an odd number of thin sheets or veneers of Douglas fir of 3, 5, 7, or 9 standard thicknesses. These are laminated with alternating grain direction. Because the grain of each ply composing the material is at right angles to the grain of adjacent plies, the strength of Douglas fir plywood panels is approximately equal in both directions.

All Douglas fir plywood from American Plywood Association mills is made to conform to the moisture-resistance requirements set forth in the revised standards established by the National Bureau of Standards, as U. S. Commercial Standards CS45-55.

Plywall wallboard is the grade of plywood commonly used for interior walls and ceilings instead of lath and plaster or other types of wallboard. It is furnished in sheets 24" to 48" (600-1200 mm) wide and 5' to 12' (1.5-3.6 m) long.

Plywall wallboard is furnished 1/4" (6.25 mm), 3/8" (9.37 mm), 1/2" (12.50 mm), 5/8" (15.62 mm), 3/4" (18.75 mm), 1" (25 mm), and 1-1/8"

CARPENTRY

(28.12 mm) thick. The 1/4" (6.25 mm) thickness is most widely used for walls and ceilings, although for more substantial construction 3/8" (9.37 mm) or 1/2" (12.50 mm) thick is often preferred. Thicker material is used for shelving, wardrobes, cases, etc.

Practically any design may be obtained by the use of plywood as with other types of wallboard. Four penny casing or finishing nails are recommended for paneling up to 1/2" (12.50 mm) thick with the nails spaced 6" to 10" (150-250 mm) apart. One horizontal or fire stop should be cut in between top and bottom plates as a nailing piece.

Plywood panels may be finished by staining, painting, wall papering, or mechanical surfacing.

Labor Applying Plywood Wallboard. The labor cost of applying plywood wallboard will be about the same as for fiber wallboard covering the same class of work.

Where special jointing of the panels is required to work out wall and ceiling panels, allow extra for grooving, beveling, and any other designs worked into the wallboard. Also add extra for applying panel strips as given on previous pages.

Prices of Douglas Fir Plywood

Douglas fir plywood is made with water resisting glue. All panels are sanded two sides to net thickness shown. Plywood prices can be very volatile. Check local sources.

Size of Panel - Inches				Thick Inches	No. of Plies	Price per 1,000 Sq.Ft.	
						Good (A-D) One side	Good Both Sides
24 "x	60 " to	48 "x	96 "	1/4 "	3	$ 386.40	$ 437.00
24 "x	60 " to	48 "x	96 "	3/8 "	3	$ 406.30	$ 687.00
24 "x	60 " to	48 "x	96 "	1/2 "	5	$ 468.80	$ 822.00
24 "x	60 " to	48 "x	96 "	5/8 "	5	$ 625.00	$ 974.40
24 "x	60 " to	48 "x	96 "	3/4 "	5	$ 750.00	$ 1,093.00

Size of Panel - mm				Thick mm		Price per 92.9 Sq.M.	
600 x	1500 mm to 1200 x	2400 mm		0.00 mm	3	$ 386.40	$ 437.00
600 x	1500 mm to 1200 x	2400 mm		6.25 mm	3	$ 406.30	$ 687.00
600 x	1500 mm to 1200 x	2400 mm		9.38 mm	5	$ 468.80	$ 822.00
600 x	1500 mm to 1200 x	2400 mm		12.50 mm	5	$ 625.00	$ 974.40
600 x	1500 mm to 1200 x	2400 mm		15.63 mm	5	$ 750.00	$ 1,093.00

Saw textured plywood siding is popular and relatively inexpensive. It is readily available in Douglas fir and cedar.

Saw Textured Douglas Fir Plywood

Size Thickness Inches	per 1000 sq.ft. (92.9 sq.m)	Size Thickness mm
3/8" Natural	$ 720.00	9.37 mm
3/8" Grooved	$ 804.00	9.37 mm
5/8" Natural	$ 900.00	15.62 mm
5/8" Grooved	$ 1,032.00	15.62 mm

06423 SOLID WOOD PANELING

While most paneling used today is of veneered plywood, nothing matches the beauty of solid wood paneling and its ability to withstand years of use and even improve with age.

Solid paneling is usually quoted in random widths of 4" to 8" (100-200 mm) and random lengths of 6' to 16' (1.82 to 4.87 m). Specified widths and lengths will run 10% or more additional. Traditional thickness is 3/4" (18.75 mm), but much paneling today is available at a considerable savings in 1/2" (12.50 mm) thickness. There are no nationally established grading rules for solid wall paneling, so it is necessary to know the product of the mill quoting to compare prices. Generally, woods are available in two grades: "select", "luxury" or "clear"; and "knotty", "wormy" or "pecky", often designated as "Colonial" grade.

In figuring paneling, add 15% to actual areas for waste and matching.

Paneling is available with V-joint tongue and groove, square joint, shiplap, and Colonial beaded.

The most commonly encountered paneling is knotty pine. Cost is around $1.15 per sq.ft. ($11.11 per sq.m). Clear pine will run around $1.50.

Clear birch in 3/4" (18.75 mm) thickness will be about $2.10 per bd.ft. in lots over 1000 bd.ft. ($900.00 per cu.m), random width and length. Birch is an excellent cabinet wood with a firm texture, but has a rather strong grain pattern.

White ash is an open grained wood, light in color with contrasting light brown heartwood. It is easy to work and somewhat resembles flat grained white oak. It will cost $2.25 for 3/4" (18.75 mm) select grade.

Butternut, often called white walnut, is a soft-textured, easy-to-nail wood which takes a beautiful natural finish. It will run $2.50 per bd.ft. (932.68 per cu.m) select, $1.30 character marked, 3/4" (18.75 mm) thick.

Cherry, cut from the wild black cherry, is a traditional Colonial paneling. It is a firm textured cabinet wood, but is available only in a rather narrow range of sizes. It runs around $2.45 for 3/4" (19 mm) select down to $1.70 for 3/4" (18.75 mm) Colonial.

994

CARPENTRY

Oak is available from either white or red varieties cut on the quarter grain. Oak is a very hard wood and subject to shrinkage and splitting in adverse conditions, but well worth the extra effort for the right place. Oak runs $1.70 for the red and $2.10 for the white in 3/4" (18.75 mm) select, $1.30 and $1.55 in 3/4" (18.75mm) Colonial.

Walnut has become so popular and the sources for it so restricted that it is now a high priced wood and will run $3.75 or better per sq.ft. (0.09 sq.m) for 3/4" (18.75mm) select. Because it is a hard wood, 1/2" (13 mm) thickness is practical and will run $3.65 for select and $2.65 for colonial.

Other domestic woods available are maple, around $1.40 for 3/4" (18.75mm) select; wormy chestnut, at $1.70 for 3/4" (18.75 mm); and cypress at $1.05 for 3/4" (18.75 mm) select. Imported woods are also available. The so-called "exotic" woods are more often found in the architectural grade, book-matched plywood where their veneers can be shown off to better advantage.

Some foreign woods stocked in the Chicago area include paldao, a firm textured striped wood for medium brown from the Philippines; African mahogany and the less costly Philippine mahogany which is more difficult to finish; avodire, a light creamy-colored wood which comes from Africa with a roll figure somewhat on the diagonal; limba, or korina, African wood which is cream or tan in color with a darker heartwood, and tigerwood or African lovea, a dullish brown to golden colored wood with a wide or ribbon stripe.

These are all special woods for special jobs. Prices and availability should always be checked, because they vary considerably, but all are expensive woods.

Placing Wood Wainscot or Paneling. The labor cost of placing wood paneling or wainscoting will vary with the height of the wainscot and the manner in which it is delivered, whether assembled or knocked down.

It takes nearly as long to place wood wainscoting 4'-0" (1.22 m) high as 7'-0" (2.13 m) high, as the top and bottom members are identical and the only difference is the height of the panels.

If the wood wainscot is put together in the mill and the carpenters on the job do what little fitting is necessary, set it against the walls and nail it to the grounds, a carpenter should place 12 to 18 lin.ft. of (3.65 to 5.48 m) wainscot per 8-hr. day, at the following labor costs for the different heights:

Height of Wainscot Feet	No. Sq.Ft. 8-Hr Day	Carp Hrs. per 100 Sq.Ft.	Height of Wainscot Meter	No. Sq.M. 8-Hr Day	Carp Hrs. per 10 Sq.M.
4'-0"	60 - 70	11 - 13	1.22 m	5.57 - 6.50	11.84 - 13.99
5'-0"	70 - 85	9 - 11	1.52 m	6.50 - 7.90	9.69 - 11.84
6'-0"	80 - 100	8 - 10	1.83 m	7.43 - 9.29	8.61 - 10.76
7'-0"	90 - 110	7 - 9	2.13 m	8.36 - 10.22	7.53 - 9.69
8'-0"	95 - 115	7 - 8	2.44 m	8.83 - 10.68	7.53 - 8.61

Placing Wood Wainscot or Paneling (Knocked Down). If wood wainscot or paneling is sent to the job knocked down, making it necessary to assemble the pieces and put them together on the job, a carpenter should place top and bottom rails, cap and intermediate panels for 10 to 15 lin.ft. (3.04 - 4.57 m) of wainscoting per 8-hr. day, at the following costs for the different heights:

Height of Wainscot Feet	No. Sq.Ft. 8-Hr Day	Carp Hrs. per 100 Sq.Ft.	Height of Wainscot Meter	No. Sq.M. 8-Hr Day	Carp Hrs. per 10 Sq.M.
4 '- 0"	50 - 60	13 - 16	1.22 m	4.65 - 5.57	13.99 - 17.22
5 '- 0"	60 - 70	11 - 13	1.52 m	5.57 - 6.50	11.84 - 13.99
6 '- 0"	70 - 80	10 - 11	1.83 m	6.50 - 7.43	10.76 - 11.84
7 '- 0"	80 - 90	9 - 10	2.13 m	7.43 - 8.36	9.69 - 10.76
8 '- 0"	90 - 100	8 - 9	2.44 m	8.36 - 9.29	8.61 - 9.69

06424 SHEET BOARD PANELING

An inexpensive prefinished paneling is plastic coated hardboard. This comes in 1/8" (3.12 mm) and 1/4" (6.25 mm) thick, 4'x8' (1.22 x 2.44 m) panels, and 16" (400 mm) wide, 1/4" (6.25 mm) thick grooved planks. The thinner boards should have solid backing for first class work. These panels have a wide range of printed designs from wood grains, solid colors, scenic designs, and marble textures. A full line of aluminum and presdwood moldings is available colored to match or compliment the selected material.

Vinyl surfaced gypsum wallboard comes in 1/2" (12.50 mm) thickness 8' (2.44 m), 9' (2.74 m), and 10' (3.04 m) lengths. It can be applied with direct nailing using colored nails; glued to furring with a combination adhesive nail-on system, or laminated to a base layer of gypsum backed board. An advantage of this finish is that it can be patched and eventually painted.

Open Work Panels. A popular addition to stock partitioning available today are panels with open work. These vary from 1/8" (3.12 mm) thick prefinished masonite to solid walnut grilles ranging as high as $50.00 per sq.ft. ($538.00 sq.m) The amount of openings can vary from 15% to 50%. These materials make good room dividers, help control sun and glare at windows, and cut air conditioning and drapery costs.

The masonite product is known as Filigree and is available in a series of sizes up to 4'x8' (1.22 x 2.44 m) sheets. Prefinished frame sections, 3/4" (18.75 mm) thick and 1-1/2" (37.50 mm) wide, factory grooved to receive the panel, are available.

Sculptured hardwood panels 3/4" (18.75 mm) thick are available in many designs and woods including walnut, maple, ash, oak, and poplar, in stock sizes from 2'x4' (0.61 x 1.22 m) to 4'x8' (1.22 x 2.44 m). These may be ordered sanded for finishing on the job or factory finished. Woods

996

available listed in order of ascending cost include sycamore or gumwood; poplar, maple, red oak, ash, and walnut. Some designs, nine are available, are also made in 5/8" (15.62 mm) clear acrylic.

The least expensive design in the least expensive wood, sycamore, costs $65.00 for a 2'x6' (0.61 x 1.82 m) unfinished, unframed panel. Framed and finished it will cost $210.00, plus shipping charges. The most expensive designs will run $125.00 unfinished and unframed, $270.00 framed and finished.

Wallboard. There are so many kinds and sizes of wall and insulating board that it is impossible to list all of them.

The thinnest boards are made of wood fiber and are usually furnished in sheets 4' (1.22 m) wide, 4' to 12' (1.22-3.65 m) long, and 1/8" (312 mm), 3/16" (4.68 mm), 1/4" (6.25 mm), and 3/8" (9.37 mm) thick. Prices vary from $0.23 per sq.ft. ($2.55 per sq.m) for the 1/8" (3.25 mm) to $0.40 per sq.ft. ($4.30 per sq.m) for the 3/8" (9.37 mm) board.

Insulating and building board is usually furnished 3/8" (9.37 mm) and 1/2" (12.50 mm) thick, 4' (1.22 m) wide and 6', 7', 8', 9', 10', and 12' (1.82, 2.13, 2.44, 2.74, 3.04, and 3.65 m) long, all having square edges. Board of this type will cost $0.22 per sq.ft. ($2.44 per sq.m) for the 3/8" (9.37 mm) and $0.32 per sq.ft. ($3.45 per sq.m) for the 1/2" (12.50 mm) thickness.

Insulation Board Finish Plank. Used for interior finish, it combines insulation and decoration in one material. Insulating plank are 1/2" (12.50 mm) thick and furnished in random widths 8", 10", 12", and 16" (200, 250, 300, and 400 mm) wide and 8', 10', and 12' (2.44, 3.04, 3.65 m) long. The long edges may be shiplapped with either a plain bevel or bead and bevel design, or have a blind nailing flange.

Plank 1/2" (12.50 mm) thick cost about $0.58 per sq.ft. ($6.24 per sq.m). Wall plank with acoustical properties is also now available in 8' (2.44 m) and 10' (3.04 m) lengths.

Plank Effect Board. The plank effect board is 1/2" (12.50 mm) thick, 4'-0" (1.22 m) wide, and 8'-0" (2.44 m) high. A series of 5 score lines on each edge and through the center of the board gives the effect of 3 planks 16" (400 mm) wide. It can be applied vertically or horizontally either over old plaster or directly to studs. All edges are flush for butt jointing. The board is painted ivory color on one side and costs $0.70 per sq.ft. ($7.53 per sq.m).

Concealed Nailing Clips. These provide an invisible attachment of tile or plank to nailing base. Clips fit into tongue and groove. Usually packed 1,000 to a carton. Four pieces are required for each 16"x16" (400 x 400 mm) tile. Prices per 1,000 pcs., $7.50.

Use 2d common coated nails with clips, with 1 lb. (0.4536 kg) required per 1,000 clips.

Cement or Mastic. Waterproof cement or mastic is used to install insulating plank. One gallon covers 80 to 100 sq.ft. (3.79 liters per 7.2-9.0 sq.m) Price per gal. is $3.75 (per liter is $1.13).

Nails Required for Applying Fiberboard, Tile, Plank, Wainscot and Molding			
For Material 1/2" Thick	Gauge of Nails	Length of Nails	Spacing
Board	No. 17	1 1/4 "	3"
Tile	No. 17	1 1/4 "	6"- 10"
Plank	No. 17	1 1/4 "	6"
Wainscot	No. 17	1 1/4 "	6"
Moldings	No. 15	1 3/4 "	6"

Nails Required for Applying Fiberboard, Tile, Plank, Wainscot and Molding			
For Material 12.50 mm Thick	Gauge of Nails	Length of Nails - mm	Spacing mm
Board	No. 17	31.25 mm	75 mm
Tile	No. 17	31.25 mm	150-250 mm
Plank	No. 17	31.25 mm	150 mm
Wainscot	No. 17	31.25 mm	150 mm
Moldings	No. 15	43.75 mm	150 mm

Board, plank, and wainscot generally require about 6 brads per sq.ft. (0.09 sq.m) or 4 to 5 lbs. (1.8 - 2.25 kg) of 1-1/4" (31.25 mm) brads per 1,000 sq.ft. (92.90 sq.m) Galvanized wire brads are recommended. When they are not available, bright wire brads may be used.

Placing Insulating Plank. The labor cost of placing insulating plank will vary with the surface to which it is applied, length of walls, amount of cutting and fitting necessary, etc. Ceiling work costs more than wall work on account of working on a scaffold and placing the plank overhead.

The most economical method of application is to a solid backing of plywood, wood boards, or other satisfactory nailing surface, because it is necessary to have nailing facilities every 8", 10", 12", or 16" (200, 250, 300, or 400 mm), depending on plank width. If a solid backing is not available, then place wood furring strips, to provide for nailing the boards at edges and ends, and for chair rails, wainscot cap, and other ornamental strips.

For plank without blind nailing flange, nail through face or bead of the plank, never through bevel. Bury nail heads below surface with properly gauged hammer blow or with a nail set.

CARPENTRY

The cost of wood blocking and furring is given on previous pages of this chapter.

Number of Sq.Ft. (Sq.M) of Insulating Plank of Various Widths Placed per 8-Hour Day by One Carpenter, when nailed-in-place				
Plank Width	Height of Wainscoting in Feet			
Inches	4'-0"	6'-0"	8'-0"	10'-0"
8 "	210 - 260	275 - 315	315 - 365	325 - 390
10 "	240 - 300	300 - 360	360 - 420	375 - 450
12 "	275 - 340	350 - 410	410 - 475	425 - 500
16 "	300 - 360	375 - 440	450 - 525	465 - 525
Plank Width	Height of Wainscoting in Meters			
mm	1.22 m	1.82 m	2.44 m	3.04 m
200 mm	19.51 - 24.15	25.55 - 29.26	29.26 - 33.91	30.19 - 36.23
250 mm	22.30 - 27.87	27.87 - 33.44	33.44 - 39.02	34.84 - 41.81
300 mm	25.55 - 31.59	32.52 - 38.09	38.09 - 44.13	39.48 - 46.45
400 mm	27.87 - 33.44	34.84 - 40.88	41.81 - 48.77	43.20 - 48.77

Add for wood furring strips or other backing. When applied to ceilings, reduce above quantities about 25%

Carpenter Hours Required to Place 100 Sq.Ft. of Insulating Plank on Walls When Nailed in Place				
Plank Width	Height of Wainscoting in Feet			
	4'- 0"	6'- 0"	8'- 0"	10'- 0"
8 "	3.50	2.80	2.40	2.30
10 "	3.00	2.50	2.00	1.90
12 "	2.60	2.10	1.80	1.80
16 "	2.50	2.00	1.70	1.70
Metric				
Carpenter Hours Required to Place 10 Sq.M. of Insulating Plank on Walls When Nailed in Place				
Plank Width	Height of Wainscoting in mm			
	1.22 M	1.82 M	2.44 M	3.04 M
200 mm	3.77	3.01	2.58	2.48
250 mm	3.23	2.69	2.15	2.05
300 mm	2.80	2.26	1.94	1.94
400 mm	2.69	2.15	1.83	1.83

Add for wood furring strips or other backing.

Panel Strips and Moldings For Use With Sheet Paneling. Strips for working out decorative effects on walls or ceilings and for concealing the joints may be of the same material as the sheet or they may be wood or metal, depending upon the effect desired.

The labor cost of placing wood or composition decorative strips over the joints will vary with the design of panel strip used, whether it consists of one, two or three members, the size of the panels, and the design of the walls and ceilings, as the more cutting and fitting required the fewer lin.ft. (meters) of strips a worker will place per day.

On straight work, using an ordinary panel strip 1/4"x2" (6.25 x 50 mm) to 1/2"x3" (12.50 x 75 mm), a carpenter should fit and place 350 to 450 lin.ft. (106.68 to 137.15 m) per 8-hr. day, at the following labor cost per 100 lin.ft. (30.48 m):

Description	Hours	Rate	Total	Rate	Total
Carpenter	2.00	$	$	$35.04	$70.08
Cost per lin.ft.					$0.70
per m					$2.30

On more complicated work, using a one-member panel or decorative strip, a carpenter should place 275 to 325 lin.ft. (83.82 to 99.06 m) per 8-hr. day, at the following labor cost per 100 lin.ft. (30.48 m):

Description	Hours	Rate	Total	Rate	Total
Carpenter	2.70	$	$	$35.04	$94.61
Cost per lin.ft.					$0.95
per m					$3.10

Placing Three-Member Decorative or Panel Strips. When the panel strips form a plain wall and ceiling design but consist of three members-a strip 3" (75 mm) or 4" (100 mm) wide with a small molding on each side-a carpenter should fit and place 150 to 175 lin.ft. (45.75 to 53.34 m) per 8-hr. day, at the following labor cost per 100 lin.ft. (30.48 m):

Description	Hours	Rate	Total	Rate	Total
Carpenter	5.00	$	$	$35.04	$175.20
Cost per lin.ft.					$1.75
per m					$5.75

Placing Wood Ceiling Beams, Cornices, Etc. In rooms having beam ceiling effects, wood ceiling cornices, etc., a carpenter should fit and place 175 to 200 lin.ft. (53.34 to 60.96 m) of each member per 8-hr. day, at the following labor cost per 100 lin.ft. (30.48 m):

Description	Hours	Rate	Total	Rate	Total
Carpenter	4.30	$	$	$35.04	$150.67
Cost per lin.ft.					$1.51
per m					$4.94

Three-member cornices would cost three times the above price, four-member cost four times, and so on.

Applying Wallboard Over Existing Plastered Walls and Ceilings. In remodeling work where the wallboard is applied over existing walls and ceilings, it is advisable to remove the existing plaster wherever possible. To obtain a first class job, it is recommended that all interior trim, such as door and window casings, and base, be removed before applying the wallboard. It is also recommended that furring strips be placed over all existing plastered walls, as this will straighten the walls and provide a much better base for securing the wallboard.

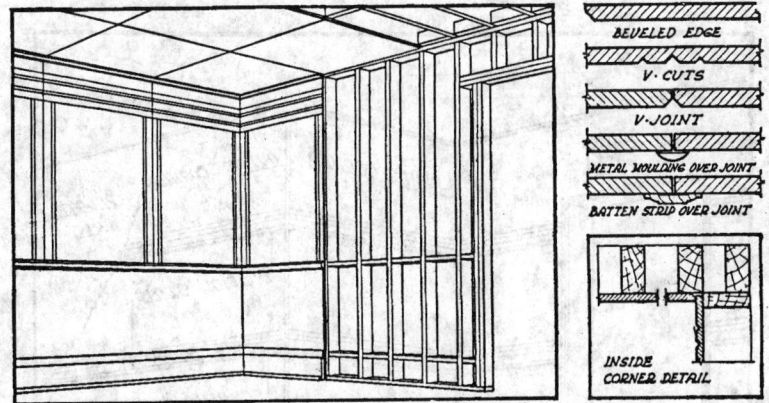

Method of Applying Wallboard and Treating Joints

After the furring strips have been placed, the cost of applying the wallboard will run about the same as given on the previous pages.

The labor cost of placing wood furring strips is given on the previous pages under "Placing Furring Strips for Ceilings".

Prefinished Panels. New developments in wallboard have been most noticeable in prefinishing and the consequent care needed in concealing nailing.

One of the most popular items is V-grooved hardwood, prefinished with a baked-on finish. Grooves are generally random but with a groove every 16" (400 mm) o.c. to hit a stud. Most boards are 1/4" (6 mm) thick in

1001

sheets 4' (1.22 m) wide by 8', 10' or 12' (2.44, 3.04, or 3.65 m) long. These sheets should be applied over a backing board for first class work. Colored nails are available as well as prefinished moldings for cornices, trim, and base. Similar paneling is also available in 7/16" (10.93 mm) thickness for application directly to studs. A wider range of finishes is generally available in the heavier material.

Insulating Tile. Tile made of Celotex, Fir-Tex, Insulite, Nu-Wood, Temlok, and other fiber composition are still sometimes used for wall and ceiling finish. They are furnished with beveled edges either square or with a blind nailing or stapling flange. They come in sizes from 12"x12" (300 x 300 mm) to 16"x32" (400 x 800 mm) and may be used for obtaining ashlar effects on walls and tile effects on ceilings.

They can be applied to wood backing or to furring strips with adhesive, nails, staples, or special clips on the walls or ceilings. When furring strips are used, the strips must be spaced to work out with the tile sizes.

In applying square edged tiles, nails should be driven through the tile at an angle to the surface of approximately 60 degrees. Reverse the angle of half the nails and nail through the face of the tile, never through the bevel. Bury nail heads below the surface with properly gauged hammer blow or with nail set.

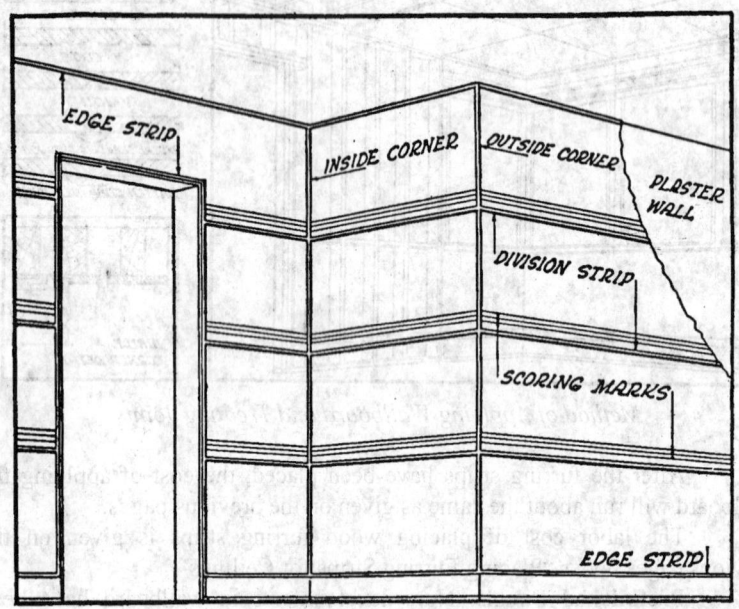

Method of Placing Wallboard with Horizontal Siding Marks and Division Strips

When applying tiles with blind nailing or stapling flanges, drive the nails or staples in the thick part of the flange perpendicular to the surface.

1002

CARPENTRY

Insulating fiber tile 1/2" (12.75 mm) thick costs $10.50 per carton of 64.

Nailing Recommendations for Square Edged Fiber Tile. Where nailing is to be exposed, galvanized wire brads are recommended. When they are not available, bright wire brads may be used. 1-1/4" (31.25 mm) brads or finishing nails should be used for exposed nailing of 1/2" (12.75mm) thick tile.

	Nailing Recommendations for Square Edged Fiber Tile					
Tile Size Inches	Location of Brad	No. Brads per Tile	Approx. Lbs. of Brads per 1,000 Sq.Ft. Tile	Tile Size Inches		Approx. Kg. of Brads per 100 Sq.M. Tile
12 "x 12 "	1 in each corner	4	3 1/2	300 x 300 mm		1.71
12 "x 24 "	4 in each long leg 3 on 2 opposite Edges	8	3 1/2	300 x 600 mm		1.71
16 "x 16 "	3 on center line 5 on each long leg	9	4	400 x 400 mm		1.95
16 "x 32 "	5 on center line	15	3 1/2	400 x 800 mm		1.71

Nailing Recommendations for Flanged Tile. Where attachment of flanged tile is to be by nails, 3d blued lath nails 1-1/8" (28,12 mm) long should be used. Where application is by means of staples, rust resistant staples 1/2" (12.75 mm) or 9/16" (14.06 mm) in length should be used in conjunction with a gun type stapler.

Tile Size in Inches	No. Nails or Staples per Tile	Approx. No. Staples per 1,000 Sq.Ft.	Approx. Lbs. Nails per 1,000 Sq.Ft.	Tile Size in mm	No. Nails or Staples per Tile	Approx. No. Staples per 100 Sq.M	Approx. Lbs. Nails per 100 Sq.Ft.
12 "x 12 "	4	4000	6	300 x 300 mm	4	4305.71	6.46
12 "x 24 "	6	3,000	4.5	300 x 600 mm	6	3229.28	4.84
16 "x 16 "	4	2,250	3.4	400 x 400 mm	4	2421.96	3.66
16 "x 32 "	6	1,690	2.6	400 x 800 mm	6	1819.16	2.80

Labor Placing Fiber Tile When Nailed In Place. The labor cost of handling and placing fiber tile will vary considerably with the size of the tile, the skill of the mechanic, the amount of nailing, and the surface to which the tile is applied. A worker will place more tile on a good solid backing than on furring strips. To the quantities given in the following table, add for blocking and furring where necessary.

1003

Quantities and Labor Costs of Applying Fiber Tile to Walls and Ceilings Where Tile are Nailed In Place

Tile Size Inches	Sq.Ft. per Tile	No. Tile Required per 100 Sq.Ft.	No. Placed per 8-Hr. Day	No. of Sq.Ft. per 8- Hr. Day	Carp. Hrs. Per 100 Sq.Ft
12 "x 12 " *	1.00	100	180 - 200	180 - 200	4.2
12 "x 24 " *	2.00	50	95 - 105	190 - 210	4
16 "x 16 " *	1.77	56*	120 - 130	215 - 235	3.6
16 "x 32 " *	3.54	28*	83 - 90	295 - 315	2.7

(metric)					
Tile Size mm	Sq.M. per Tile	No. Tile Required per 10 Sq.M.	No. Placed per 8-Hr. Day	No. of Sq.M. per 8- Hr. Day	Carp. Hrs. Per 10 Sq.M.
300 x 300 mm*	0.09	107.64	180 - 200	16.72 - 18.58	4.52
300 x 600 mm*	0.19	53.82	95 - 105	17.65 - 19.51	4.31
400 x 400 mm*	0.16	60.28	120 - 130	19.97 - 21.83	3.88
400 x 800 mm*	0.33	30.14	83 - 90	27.41 - 29.26	2.91

*Stock sizes.

Add for wood furring strips or other necessary backing.

Applying Insulating Tile Using Adhesives. Apply the adhesive in spots or dabs about 2" (50 mm) in diameter, 1/4" (6.25 mm) to 3/8" (9.37 mm) thick, one on each corner of the square tileboard and two or more additional spots on the rectangular sizes. Keep adhesive about 1" (25 mm) from edges of fiberboard units and use approximately 4 spots per sq.ft. (0.09 sq.m). Adhesive spots should not be spaced over 10" (250 mm) o.c. in any direction.

After dabs of adhesive have been applied, place the tile in its final position. Slide back and forth for a distance of about 3/4" (18.75 mm) under uniform pressure of the hands to obtain good contact. When working units larger than 16"x16" (400 x 400 mm), two workers should exert pressure simultaneously for best results. If the surface of the unit is below the level of adjoining pieces, do not pull the unit away from the base to bring surface level, but remove it and add more adhesive. Again work into position as described above to obtain correct surface level.

Brad or shore the first units on either walls or ceilings to prevent sliding out of position during application of surrounding units. Do not bend units over high spots.

Tile and plank may be applied to plasterboard, sound plaster, plywood, or lumber. In each case the use of adhesive or a combination of adhesive and brads is recommended.

CARPENTRY

Coverage Chart of Adhesive							
Tile Size Inches	No. Spots per Tile	Sq.Ft. Per Gal.	Gals. Per 1,000 Sq.Ft.	Tile Size mm	No. Spots per Tile	Sq.M. Per Liter.	Liter Per 100 Sq.M.
12 "x 12 "	4	40	13	300 x 300 mm	4	0.98	52.97
12 "x 24 "	6	40	13	300 x 600 mm	6	0.98	52.97
16 "x 16 "	4	60	10	400 x 400 mm	4	1.47	40.74
16 "x 32 "	6	60	10	400 x 800 mm	6	1.47	40.74

Price, $8.00 per gal. ($2.11 per liter)

Labor Placing Fiber Tile Using Adhesive. The labor cost of placing fiber tile using adhesive or adhesive and nails will vary greatly with the skill and experience of the workers, size of tile, etc. For instance, where an experienced mechanic is placing tile and has a helper to build scaffold, unpack tile, and apply adhesive to the backs of tiles, an experienced mechanic and helper should place 40 to 45 sq.ft. (3.71 to 4.18 sq.m) of 12"x12" (300 x 300 mm) tile per hour, but if one person is working alone and must stop to apply the adhesive to each tile, set it in place, and then drive several brads through them, 15 to 20 sq.ft. (1.39 to 1.85 sq.m) per hour would probably be average. Conditions under which the tiles are to be installed must be carefully considered before arriving at the labor cost.

The design in which the tiles are laid also affects the labor costs. For instance when tile are laid in herringbone pattern, it requires about 30% to 40% more time than regular square or common bond. Diagonal patterns require about 20% to 25% more time than regular squares or common bond while tile laid in mixed ashlar designs require about 50% more time than straight squares or common bond.

The quantities and costs given in the following table are based on using two workers-an experienced mechanic to place the tile and a worker to build scaffold, unpack the tile, and place them on the scaffold, and place the adhesive on the back of the tile ready for the mechanic.

If the work is performed by inexperienced workers, the labor costs might be as much as double those given here.

Quantities and Labor Costs of Applying Fiber Wall Tile to Walls and Ceilings Using Adhesive and Nails Where Necessary

Add for cleaning plaster or other work necessary for applying tile. The quantities and costs given in the following table are based on using a 4- or 5-ft. (1.22 or 1.52 m) scaffold, consisting of horses and planks, and laying the tile in plain squares or common bond.

Tile Size Inches	Sq.Ft. per Tile	Tile per 100 Sq.Ft.	No. Placed per 8-Hr Day	Sq.Ft. per 8-Hr Day	Hrs. per 100 Sq.Ft.
12 "x 12 "*	1	100	165 - 185	165 - 185	4.5
12 "x 24 "*	2	50	88 - 98	175 - 195	4.4
16 "x 16 "*	1.77	56*	112 - 125	200 - 220	3.9
16 "x 32 "*	3.54	28*	65 - 70	230 - 250	3.3
(metric)					
Tile Size Inches	Sq.M per Tile	Tile per 10 Sq.M.	No. Placed per 8-Hr Day	Sq.M. per 8-Hr Day	Hrs. per 10 Sq.M
300 x 300 mm*	0.09	92.90	165 - 185	15 - 17	4.18
300 x 600 mm*	0.19	46.45	88 - 98	16 - 18	4.09
400 x 400 mm*	0.16	52.02	112 - 125	19 - 20	3.62
400 x 800 mm*	0.33	26.01	65 - 70	21 - 23	3.07

*Standard sizes

If tile are laid in herringbone pattern, add 30% to 40% to the above labor costs; if laid in diagonal patterns, increase the above labor costs 20% to 25%; and if laid in mixed ashlar designs, add about 50% to the above labor costs.

When placing tile in auditoriums and other spaces requiring high scaffolding, add cost of scaffolding. Large jobs are applied proportionately faster than small jobs.

Presdwood, Hardboard, Panelboard, Temwood. There are a number of makes of wood fiber board that are much thinner and yet possess much greater strength than ordinary wallboard.

They are used for both interior and exterior work-applications such as interior finish of walls and ceilings, cupboards, lining of closets, and clothes chutes.

They are usually furnished 1/8" (3.25 mm), 3/16" (4.68 mm), 1/4" (6.25 mm), and 5/16" (7.81 mm) thick, 4' (1.22 m) wide and 4' to 12' (1.22-3.65 m) long, although they vary somewhat among manufacturers.

Tempered hardboard, Temwood, or Presdwood are the same materials as described above, except they have been subjected to a tempering treatment which further increases the strength, resistance to abrasion, durability, and ease of painting. Tempered hardboards, 1/4" (6.25 mm) and 5/16" (7.81 mm) thick are used for the exterior of homes and other small buildings. It is also used quite extensively for concrete forms.

These hardboards or tempered boards are in no way comparable with ordinary wallboard. It is a far denser material, with greater tensile and transverse strength, greater resistance to abrasion, lower moisture absorption, and when properly applied, no warping or buckling.

CARPENTRY

The labor cost of placing hardboard will run approximately the same as placing other types of board under the same conditions.

Approximate Sq. Ft. (Sq.M) Prices of Hardboard				
Type	Thick. Inch	Price per Sq.Ft.	Thick. mm	Price per Sq.M.
Untempered, Plain	1/8 "	$0.20	3.13 mm	$2.15
Untempered, Plain	1/4 "	$0.28	6.25 mm	$3.01
Untempered, Pegboard	1/8 "	$0.24	3.13 mm	$2.58
Untempered, Pegboard	1/4 "	$0.32	6.25 mm	$3.44
Tempered Plain	1/8 "	$0.23	3.13 mm	$2.43
Tempered Plain	1/4 "	$0.35	6.25 mm	$3.72
Tempered Pegboard	1/8 "	$0.27	3.13 mm	$2.86
Tempered Pegboard	1/4 "	$0.37	6.25 mm	$4.01
Tempered Plastic Face	1/8 "	$1.10	3.13 mm	$11.88
Tempered Plastic Face	1/4 "	$1.22	6.25 mm	$13.17
Tempered Pl. F. Pegb'd	1/8 "	$1.10	3.13 mm	$11.88
Tempered PL. F. Pegb'd	1/4 "	$1.25	6.25 mm	$13.46

Tileboard

Tileboard has score lines impressed in one surface forming 4"x4" (100 x 100 mm) squares. The board is denser than tempered hardboard and has a lower rate of water absorption.

This material is used for walls in bathrooms, kitchens, lavatories, restaurants, barber shops, and similar places. It is furnished in sheets 4'-0" (1.22 m) wide and 2'-0" to 12'-0" (0.61 to 3.65 m) long.

Tempered board or tileboard should be cemented to the walls, using Armstrong's Panelboard Cement or a cement of equal quality. The adhesive should be applied to the back of the boards with a saw-tooth trowel, the boards having first been cut to fit. Use supplementary nailing or shoring to hold the boards firmly until the adhesive takes a partial set. Where it is necessary to nail the board to the walls, small brads should be used where scored lines intersect.

Figure about one gallon waterproof cement per 50 sq.ft. (4.65 sq.m) of wall.

Presdwood Temprtile 1/8" (3.25 mm) thick costs about $0.25 per sq.ft. ($2.77 per sq.m) unpainted.

Prefinished Tileboard. Tileboard 5/32" (4 mm) thick, made of masonite tempered Presdwood, is furnished scored 4"x4" (100 x 100 mm) to represent tile and finished with multiple coats of enamel or lacquer. Tileboard is also furnished without scoring, if desired.

These boards are furnished with dull or velvet finish, or a high gloss finish in a variety of colors with contrasting joints.

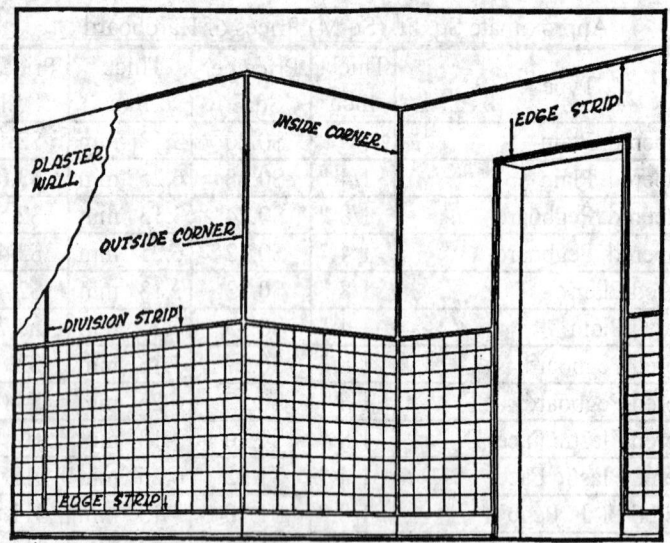

Method of Placing Tileboard in Bathrooms, Kitchens, Etc.

Tileboard is furnished in sheets 4' (1.22 m) wide and 4, 5, 6, 7, 8, 9, 10, and 12 ft. (1.22, 1.52, 1.82, 2.13, 2.43, 2.74, 3.04, and 3.65 m) long.

Tileboard having a semi-gloss or high gloss finish costs about $0.60 to 0.70 per sq.ft. ($6.66-7.77 per sq.m) depending upon size of board.

Cap molding 1-3/4" (44 mm) wide for use at the top of the wainscoting is furnished in strips 4' (1.2 m) and 8' (2.4 m) long, scored to match tile or unscored, costs $0.35 per lin.ft. (1.16 per m).

Base molding 3-3/4" (93.75 mm) high, scored or unscored, in colors to match tileboard costs $0.40 per lin.ft. ($1.33 per m).

Placing Tempered Tileboard. Where tempered tileboard is used in kitchens, for sink splashboards, bathrooms, shower stalls, and other small places, there is usually considerable cutting and fitting required to place a comparatively small square footage (square meter) of board.

The safest method of estimating is to lay out each wall, showing number and sizes of pieces required, as it will cost almost as much to place a small panel as a large one.

Where the tileboard is secured by nails (without the use of adhesive), figure 3/4 to 1 hr. to fit and place each panel. Taking an average of 12 to 16 sq.ft. (1.11 to 1.48 sq.m) per panel, this will equal 125 to 150 sq.ft. (11.61 to 13.93 sq.m) per 8-hr. day.

However, if the tileboard is applied to walls, using an adhesive, a worker should fit and place one panel in 1 to 1-1/4 hrs. or 100 to 125 sq.ft.

CARPENTRY

(9.29 to 11.61 sq.m) per 8-hr. day. This applies to both plain and prefinished tileboard.

One gallon of adhesive will cover about 100 sq.ft. (9.29 sq.m) of surface if "spot" glued or about 65 sq.ft. (6.03 sq.m) of surface if "spread" glued.

Price per Gallon	$10.00
per Liter	$2.64

Labor Cost of 100 Sq.Ft. (9.29 Sq.M) Tempered Tileboard, Secured to Walls With Nails

Description	Hours	Rate	Total	Rate	Total
Carpenter	4.00	$	$	$ 35.04	$ 140.16
Cost per sq.ft.					$ 1.40
per sq.m					$ 15.09

Labor Cost of 100 Sq.Ft. (9.29 Sq.M) Tempered Tileboard, Secured to Walls Using Adhesive

Description	Hours	Rate	Total	Rate	Total
Carpenter	3.50	$	$	$ 35.04	$ 122.64
Cost per sq.ft.					$ 1.23
per sq.m					$ 13.20

Wainscot Trim for Wallboard and Wall Tile

Moldings are made of stainless steel, polished finish in plain designs similar to the illustrations, punched for nails, and furnished in 4'-0" (1.22 m), 6'-0" (1.82 m), and 8'-0" (2.44 m) lengths.

Description	for Wallboard	Cost per lin.ft.	for Wallboard	Cost per lin.M.
Cap Mold	1/8 "	$ 0.33	3.13 mm	$ 1.09
Joint Mold	1/8 "	$ 0.41	3.13 mm	$ 1.35
Outside Corner	1/8 "	$ 0.49	3.13 mm	$ 1.61
Inside Corner	1/8 "	$ 0.49	3.13 mm	$ 1.61
Tub edging or cove base	1/8 "	$ 0.49	3.13 mm	$ 1.61

Metal covered wood cap, base or trim, size 5/16" x 1-1/2" (7.81 x 37.50 mm), chrome zinc, polished finish. Price per lin.ft., $0.35 (per m, 1.17).

Labor Placing Metal Moldings. The labor cost of placing metal moldings will vary with the design and size of the panels, cutting and fitting

1009

necessary, etc., but on the average job, a carpenter should place 100 to 150 lin.ft. (30.48 to 45.72 m) per 8-hr. day, at the following labor cost per 100 lin.ft. (30.48 m):

Description	Hours	Rate	Total	Rate	Total
Carpenter	6.00	$	$	$35.04	$210.24
Cost per lin.ft.					$2.10
per m					$6.90

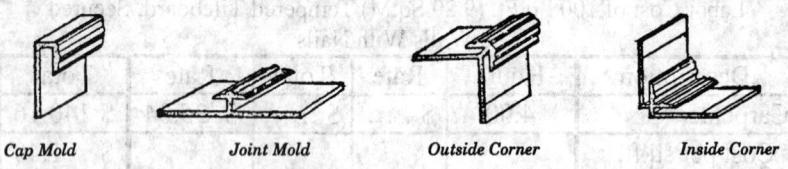

Cap Mold   Joint Mold   Outside Corner   Inside Corner

Items to Include When Making Up Wallboard Estimates. When making up estimates on wallboard construction, the following items should be included to insure a complete and dependable estimate:

1. Headers or nailing strips to receive wallboard.
2. Layout.
3. Furring strips on walls and ceilings where necessary.
4. Wallboard and labor placing same.
5. Wood or metal panel strips and labor placing same.
6. Labor filling nail holes with wood putty or crevice filler.
7. Beam ceilings, ceiling cornices, metal moldings, etc.
8. Painting and decorating wallboard and panel strips.
9. Nails and adhesive.
10. Overhead expense and profit.

06430 STAIR WORK

Framing and Erecting Wood Stairs of Unassembled Stock Material. When framing and erecting ordinary wood stairs with closed stringers (where the treads and risers are not housed out), it will require 3 to 4 hrs. carpenter time to lay out the work, cut out for treads and risers, and place wood stringers for a flight of ordinary wood stairs up to 4'-0" (1.22 m) wide and 11'-0" (3.35 m) story height, at the following labor cost:

1010

CARPENTRY

Description	Hours	Rate	Total	Rate	Total
Carpenter	4.00	$	$	$35.04	$140.16
Labor	2.00	$	$	$26.16	$52.32
Cost per lin.ft.					$192.48

After the wood stringers are in place, it will require 4 hrs. carpenter time to place treads and risers for each flight of stairs (providing it is a plain box stair without nosings to form and without newels, balusters, or handrail), at the following labor cost per flight:

Description	Hours	Rate	Total	Rate	Total
Carpenter	4.00	$	$	$35.04	$140.16

Figure 3 hrs. carpenter time to place 4"x4" 100 x100 mm) newels, 2"x4" (50 x 100 mm) rails and 1"x2" (25 x 50 mm) balusters for one side of this type stairs, at the following labor cost per flight:

Description	Hours	Rate	Total	Rate	Total
Carpenter	3.00	$	$	$35.04	$105.12

To lay out the work and erect the stairs complete, including plain wall hand rail secured by brackets (instead of newels, hand rails, and balusters), figure 8 hrs. carpenter time, at the following labor cost per flight:

Description	Hours	Rate	Total	Rate	Total
Carpenter	8.00	$	$	$35.04	$280.32

For the above stair complete, including newels, handrail, and balusters, one side, figure 10 hrs. carpenter time, at the following labor cost per flight:

Description	Hours	Rate	Total	Rate	Total
Carpenter	10.00	$	$	$35.04	$350.40

Erecting Wood Stairs Made in Shop. Where wood stairs are made in the shop with stringers housed out to receive treads and risers, it will require 8 hrs. stair builder time to lay out the work and to set stringers, treads, and risers (with plain wall rail attached to brackets), at the following labor cost per flight:

1011

Description	Hours	Rate	Total	Rate	Total
Carpenter (stair builder)	8.00	$	$	$35.04	$280.32
Labor	2.00	$	$	$26.16	$52.32
Total Cost					$332.64

If the stair consists of two short flights, having an intermediate landing platform between floors, it will require 10 to 12 hrs. stair builder time to lay out work and to place stringers, treads, risers, and plain wall handrail, at the following labor cost:

Description	Hours	Rate	Total	Rate	Total
Carpenter (stair builder)	11.00	$	$	$35.04	$385.44
Labor	2.00	$	$	$26.16	$52.32
Total Cost					$437.76

Where a plain box stair, 3'-0" to 4'-0" (0.91 to 1.22 m) wide and 9'-0" to 10'-0" (2.74 to 3.04 m) story height, with open stringers is used, it will require 10 to 12 hrs. stair builder time to lay out work and place stringers, treads, and risers (not including newels, balusters, or handrail) for each flight of stairs, at the following labor cost:

Description	Hours	Rate	Total	Rate	Total
Carpenter (stair builder)	11.00	$	$	$35.04	$385.44
Labor	2.00	$	$	$26.16	$52.32
Total Cost					$437.76

If the stairs have newels, wood handrail, and balusters, add 6 to 8 hrs. stair builder time, at the following labor cost per flight:

Description	Hours	Rate	Total	Rate	Total
Carpenter (stair builder)	7.00	$	$	$35.04	$245.28

Plain box stairs consisting of two short flights with an intermediate landing platform between stories will require 6 to 8 hrs. stair builder time to lay out the work and place stringers, treads, and risers for each short flight of stairs at the following labor cost per flight:

1012

CARPENTRY

Description	Hours	Rate	Total	Rate	Total
Carpenter (stair builder)	7.00	$	$	$35.04	$245.28
Labor	2.00	$	$	$26.16	$52.32
Total Cost					$297.60

To place newels, balusters, and handrail on each short flight of stairs will require about 4 hrs. stair builder time, at the following labor cost per flight:

Description	Hours	Rate	Total	Rate	Total
Carpenter (stair builder)	4.00	$	$	$35.04	$140.16

The labor cost complete for two short flights of stairs, including newels, handrails, and balusters, should cost as follows per story height:

Description	Hours	Rate	Total	Rate	Total
Carpenter (stair builder)	22.00	$	$	$35.04	$770.88

Erecting Wood Stairs Having Open Stringers. Wood stairs of the open stringer type, with treads that have a return nosing projecting beyond the face of the stringer, will require about 12 to 14 hrs. stair builder time to lay out the work and place stringers, treads, and risers for one flight of stairs containing 16 to 18 risers, at the following labor cost per flight:

Description	Hours	Rate	Total	Rate	Total
Carpenter (stair builder)	13.00	$	$	$35.04	$455.52
Labor	2.00	$	$	$26.16	$52.32
Total Cost					$507.84

If necessary to place newels, handrails, and baluster, add about 8 to 9 hrs. at the following labor cost per flight:

Description	Hours	Rate	Total	Rate	Total
Carpenter (stair builder)	8.50	$	$	$35.04	$297.84

If the stair consists of two short flights with an intermediate landing platform between stories (each short flight contains 8 to 10 risers), it will require about 8 to 9 hrs. per flight or 16 to 18 hrs. per story, at the following labor cost:

Description	Hours	Rate	Total	Rate	Total
Carpenter (stair builder)	17.00	$	$	$35.04	$595.68
Cost per Short Flight					$297.84

To set starting and landing newels, handrails, and wood balusters will require 7 to 9 hrs. per flight, or 14 to 18 hrs. per story, at the following labor cost:

Description	Hours	Rate	Total	Rate	Total
Carpenter (stair builder)	16.00	$	$	$35.04	$560.64
Cost per Short Flight					$280.32

The above does not include carpenter time framing and erecting intermediate landing platform.

Placing Wood Handrail on Metal Balustrades. A carpenter should place 45 to 55 lin.ft. (13.71 to 16.76 m) of wood handrail on metal balustrades per 8-hr. day at the following labor cost per 100 lin.ft. (30.48 m):

Description	Hours	Rate	Total	Rate	Total
Carpenter (stair builder)	16.00	$	$	$35.04	$560.64
Cost per lin.ft.					$5.61
Cost per lin.m.					$18.39

Placing Wood Handrail on Wall Brackets. A carpenter should set brackets and install 65 to 70 lin.ft. (19.81 to 21.33 m) of wall hung wood handrail per 8-hr. day at the following labor cost per 100 lin.ft. (30.48 m):

CARPENTRY

Description	Hours	Rate	Total	Rate	Total
Carpenter (stair builder)	12.00	$	$	$35.04	$420.48
Cost per lin.ft.					$4.20
Cost per lin.m.					$13.80

Starting Steps	Material	Length Inches	Price	Length Meter
Quarter circle	Red Oak	4'- 0"	$ 57.00	1.22
Half circle	Red Oak	4'- 6"	$ 72.00	1.37
Bull nose	Red Oak	4'- 6"	$ 66.00	1.37
Scroll end	Red Oak	4'- 6"	$ 72.00	1.37

					Price			
Stringers (not housed)								
3/4 "x	11 1/4 "x	8 '- 0 "	Red Oak	$ 48.30	18.75 mm x	281.25 mm x	2.44 M	
3/4 "x	11 1/4 "x	10 '- 0 "	Red Oak	$ 56.93	18.75 mm x	281.25 mm x	2.44 M	
3/4 "x	11 1/4 "x	12 '- 0 "	Red Oak	$ 86.25	18.75 mm x	281.25 mm x	2.44 M	
3/4 "x	11 1/4 "x	14 '- 0 "	Red Oak	$ 96.60	18.75 mm x	281.25 mm x	2.44 M	
3/4 "x	11 1/4 "x	16 '- 0 "	Red Oak	$ 110.40	18.75 mm x	281.25 mm x	2.44 M	
Treads (not returned; for returns add $5.00)								
1 1/16 "x	10 1/2 "x	3 '- 0 "	Red Oak	$ 18.98	26.56 mm x	262.50 mm x	0.91 M	
1 1/16 "x	10 1/2 "x	3 '- 6 "	Red Oak	$ 21.57	26.56 mm x	262.50 mm x	1.07 M	
1 1/16 "x	10 1/2 "x	4 '- 0 "	Red Oak	$ 24.15	26.56 mm x	262.50 mm x	1.22 M	
1 1/16 "x	11 1/2 "x	3 '- 0 "	Red Oak	$ 25.02	26.56 mm x	287.50 mm x	0.91 M	
1 1/16 "x	11 1/2 "x	3 '- 6 "	Red Oak	$ 24.00	26.56 mm x	287.50 mm x	1.07 M	
1 1/16 "x	11 1/2 "x	4 '- 0 "	Red Oak	$ 32.78	26.56 mm x	287.50 mm x	1.22 M	
1 1/16 "x	11 1/2 "x	5 '- 0 "	Red Oak	$ 39.68	26.56 mm x	287.50 mm x	1.52 M	
1 1/16 "x	11 1/2 "x	6 '- 0 "	Red Oak	$ 50.03	26.56 mm x	287.50 mm x	1.83 M	
With 2" (50 mm) Return Nosing								
1 1/8 "x	1 3/4 "x	1 '- 2 "	Red Oak	$ 8.63	28.13 mm x	43.75 mm x	0.36 M	
Tread Nosing								
1 1/8 "x	1 3/4 "x	1 '- 2 "	Red Oak	$ 3.29	28.13 mm x	43.75 mm x	0.36 M	
1 1/16 "x	1 1/8 "x	1 '- 2 "	Red Oak	$ 3.02	26.56 mm x	28.13 mm x	0.36 M	
Landing Tread Nosing								
1 1/16 "x	3 1/2 "x	3 '- 0 "	Red Oak	$ 4.32	26.56 mm x	87.50 mm x	0.91 M	
1 1/16 "x	3 1/2 "x	3 '- 6 "	Red Oak	$ 4.74	26.56 mm x	87.50 mm x	1.07 M	
1 1/16 "x	3 1/2 "x	4 '- 0 "	Red Oak	$ 5.36	26.56 mm x	87.50 mm x	1.22 M	
Tread Brackets								
12 "x	7 3/4 "x	1/4 "	Nat. Birch	$ 3.45	300.0 mm x	193.75 mm x	6.25 mm	
Newel Posts								
			Birch	Red Oak				
3 1/4 "x	3 1/4 "x	3 '- 2 "	$ 24.15	$ 25.02	81.3 mm x	81.25 mm x	0.97 M	
3 1/4 "x	3 1/4 "x	3 '- 10 "	$ 27.60	$ 28.47	81.3 mm x	81.25 mm x	1.17 M	

Tapered Landing						
		Birch	Red Oak			
3 1/4 "x 3 1/4 "x 5 '- 3 "		$ 37.95	$ 39.68	81.3 mm x	81.25 mm x	1.60 M

Tapered Angle						
		Birch	Red Oak			
3 1/4 "x 3 1/4 "x 7 '- 0 "		$ 44.85	$ 46.58	81.3 mm x	81.25 mm x	2.13 M

Balusters						
		Birch	Red Oak			
1 5/16 " Round or Square x 30 "		$ 3.80	$ 3.42	32.81 mm x	750 mm	
1 5/16 " Round or Square x 33 "		$ 3.98	$ 4.74	32.81 mm x	825 mm	
1 5/16 " Round or Square x 36 "		$ 4.14	$ 4.92	32.81 mm x	900 mm	
1 5/16 " Round or Square x 39 "		$ 4.74	$ 5.18	32.81 mm x	975 mm	
1 5/16 " Round or Square x 42 "		$ 5.53	$ 6.04	32.81 mm x	1050 mm	

Railings-Natural birch, random length					
1 3/4 "x 1 5/8 "	$ 2.63	43.75 mm	x	40.63 mm	
1 3/4 "x 1 11/16 "	$ 3.00	43.75 mm	x	42.19 mm	
2 5/8 "x 1 11/16 "	$ 3.23	65.63 mm	x	42.19 mm	
2 1/4 "x 2 3/8 "	$ 6.38	56.25 mm	x	59.38 mm	
Rails Bolts and Plug	$0.75				

For 9'- 0" (2.74 M) finished floors to finished ceiling heights:						
Prefabricated Folding Stairways						
Premium grade	2'- 6"	x	6'- 0"	Opening	$442.90	0.762 M x 1.829 M
Medium grade	2'- 2"	x	4'- 6"	Opening	$156.00	0.762 M x 1.829 M
Economy grade	2'- 2"	x	4'- 6"	Opening	$107.70	0.762 M x 1.829 M

Prefabricated Spiral Stairs. Wood stairs are available in 3'-6" (1.06 m), 4' (1.22 m), 4'-6" (1.37 m), and 5' (1.52 m) diameters with heights made to fit the actual job conditions. The balusters and center column are steel, but the treads, platforms, and railings can be wood to give a finished appearance. Several styles are available and the cost runs from $860.00 for the smallest diameter and the simplest design in an 8' (2.44 m) rise to $1,470.00 for the 5' (1.52 m) diameter. The more elaborate designs will run from $1,350.00 to $1,725.00 for the same sizes. The stairs are factory assembled and finished so that two carpenters should be able to install the unit in one day.

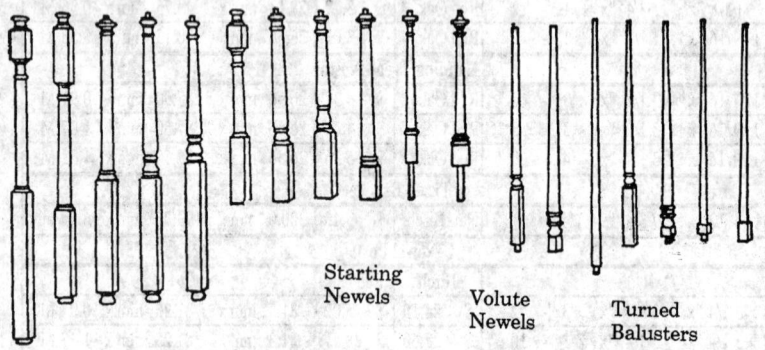

Starting Newels

Volute Newels

Turned Balusters

CARPENTRY

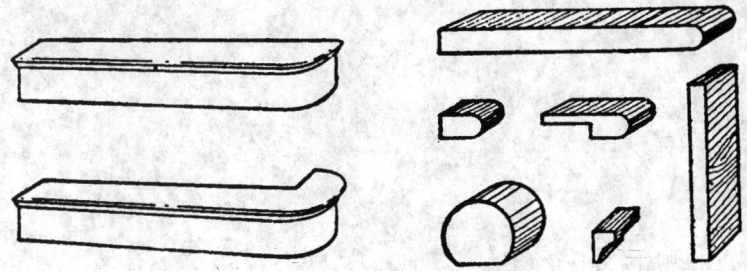

Inside Stair Materials

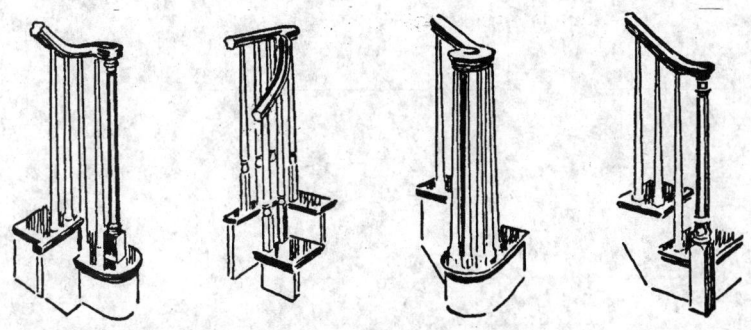

Starting Steps and Colonial Newels rail

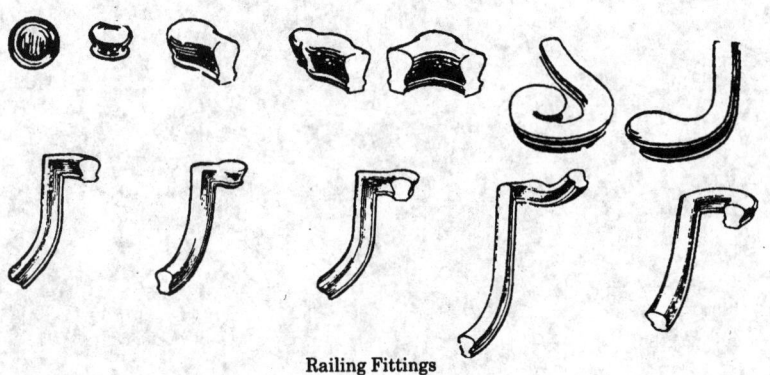

Railing Fittings

1017

07   THERMAL AND MOISTURE PROTECTION

07100 WATERPROOFING

Every building requires water and damp resisting material, plaster bonds, or floor hardeners of one kind or another, and with so many of them on the market, it is impossible even for the best informed contractor to be familiar with all of them. No attempt has been made here to pass on the relative merits of the various preparations. The data is strictly information that is of interest to the estimator—use of the material, covering capacity, and cost of applying. The prices quoted are approximate and may vary considerably in different locations. Always check with local distributors for current prices.

Water and Damp Resisting Methods. There are several methods used by the different manufacturers for mixing and applying their products to produce a water resisting wall or surface. In some instances, the water repelling compounds are incorporated with the cement, sand, and gravel while the concrete is being mixed, and by filling the voids in the cement and sand, they produce a waterproof mixture. Where the various powders or pastes are mixed in the concrete mass, it is called the integral method of waterproofing.

Some water and dampness resisting preparations are furnished in mastic form and are applied with a trowel. Others are mixed with portland cement and sand and applied as a plaster coat to the surfaces to be treated.

There are also numerous water and dampness resisting compounds in the form of heavy paints, which are applied to the concrete or masonry walls, floors, and ceilings with a brush or mop. One or more applications of these materials are supposed to penetrate and fill the pores in the concrete or masonry to such an extent that the treated surfaces will be impervious to dampness or moisture.

Where there is considerable pressure or a large head of water, one of the most satisfactory methods is known as membrane waterproofing. This method consists of applying 2 to 7 hot moppings of pitch or asphalt on 1 to 7 plies of saturated felt or fabric. Waterproofing of this kind is applied in much the same manner as built-up roofing. After the last ply of saturated felt or fabric is applied, the entire surface is mopped with a heavy application of hot pitch or asphalt.

Estimating Quantities of Water and Damp Resisting Preparations. When estimating quantities of the various plaster coats, paints, and membrane waterproofing compounds required for any job, the entire area of the walls or surfaces to be treated should be measured and the quantities stated in sq. ft., sq. yds., sq.m, or by the square (100 sq.ft. or 9.29 sq.m).

Integral water resisting compounds are usually estimated by the cu. ft., cu. yd., or cu.m of concrete to which the liquid, paste, or powder is added.

Water Required for Mixing Concrete. When mixing concrete, sufficient water should be used to produce a plastic mix—no more and no less. A plastic mix is concrete that can be readily molded, and when the mold is removed, will flow sluggishly without segregation of the water or the fine materials from the coarse. Concrete should be mixed at least 2 minutes after all materials have been placed into the mixer. Remember there is nothing more injurious to good concrete than too much mixing water.

Refer to water required for mixing concrete as given in Step 4. "Selection of water-cement ratio" under *Cast-in Place Concrete*.

Waterproofing by the Integral Method

When using the integral method, the admixtures are usually mixed with the cement and aggregate and form a part of the concrete mass. The quantities are based on the specifications of the various manufacturers.

Material prices given are approximate, as it is impossible to quote prices that will apply to every locality. The contractor or estimator should obtain local prices on the materials specified. Prices are based on material being purchased in drum or barrel quantities. When purchased in one gallon or one pound quantities, the cost increases considerably.

Integral Liquid Admixtures

Integral liquid admixtures consist of calcium chloride solutions, oil water repellent preparations, and other integral mixtures that flow freely from a drum.

The cost of hauling the barrel to the mixer, setting it up on blocks, and drawing off the liquid by spigot into a measuring container will cost from $0.15 to $0.20 per gal. $0.04 to $0.05 per liter of waterproofing on small jobs. On large jobs this cost may be halved.

Concrete mixed in the proportions of 1:2:4 or 1:2-1/2:3-1/2 requires 5-1/2 to 6 sacks of cement per cu.yd. (0.76 cu.m) of concrete.

1020

THERMAL AND MOISTURE PROTECTION

Trade Name	Qts. Per sack of Cement	Concrete Proportion	Gals. per Cu. Yd. Concrete	Price per Gal.	Liters Per sack of Cement	Liters per Cu.M. Concrete	Price per Liter
Anti-Hydro*	1	1:2:4	1 1/2	$6.00	1.10	7.43	$1.59
Aquatite	2/3	1:2:4	1	$3.50	0.73	4.95	$0.92
Aquatite**	1	1:2:4	1 1/2	$3.50	1.10	7.43	$0.92
Hydratite	1	1:2:4	1 1/2	$5.00	1.10	7.43	$1.32
Toxement IW	1	1:2:4	1 1/2	$4.75	1.10	7.43	$1.25
Sikacrete	1	1:2:4	1 1/2	$6.75	1.10	7.43	$1.78

*A 5-year maintenance guarantee furnished by manufacturer where
  supervision is supplied at additional cost.
**Use where there is water pressure.

Powdered Integral Admixtures

This material should be added to the portland cement before the
addition of aggregates. Then, it is dry mixed thoroughly before adding
gauging water. Add about $0.45 to 0.60 per lb. ($0.39-0.1.32 per kg) to the
cost of material for labor handling and mixing.

Trade Name	Lbs. per sack of Cement	Concrete Proportion	Lbs. per Cu. Yd. Concrete	Price Per Lb.	Kg. per sack of Cement	Kg. per Cu.M. Concrete	Price Per Kg.
Hydratite Powder	1	1:2:4	6	$0.75	0.45	3.56	$1.65
Hydrocide Powder	1	1:2:4	6	$0.75	0.45	3.56	$1.65
Plastiment	1/2 - 1	1:2:4	3 - 6	$1.25	0.45	1.78 - 3.56	$2.76
Toxement IW	1 1/12	1:2:4	7 1/2	$1.50	0.45	4.45	$3.31

Water and Damp-Resisting Plaster Coats

There are numerous preparations on the market to mix with portland
cement and sand and apply as a plaster coat for waterproofing concrete and
masonry walls. There are other preparations in mastic form that are applied
with a trowel, the same as ordinary cement mortar.

**Estimating Quantities of Water or Damp-Resisting Plaster
Coats**. Where water or damp-resisting plaster coats are applied to brick or
concrete walls, the water resisting ingredient is usually liquid, paste, or
powder.

This compound is usually added to the mortar on the basis of a
certain quantity to each sack of cement used and practically all manufacturers
specify a mortar mixed in the proportions of 1 part portland cement to 2 parts
sand, adding only sufficient water to make a workable mortar.

Quantity of Cement and Sand Required for One Cu.Yd. (0.76 Cu.M) of Cement Mortar

Based on damp loose sand containing 5% of moisture and weighing 2,565 lbs. per cu.yd. (1521069 kg per cu.m).

Proportions by Volume Cement-Sand	Packed Cement Sacks	Loose Sand Cu.Yd.	Loose Sand Cu.M
1:1	18.0	0.70	0.535
1: 1 1/2	15.2	0.84	0.642
1:2	12.5	0.95	0.726
1: 2 1/2	11.0	1.03	0.787
1:3	9.5	1.10	0.841

Above quantities of sand include 5% for waste.

Cement Plaster Coat-Sq.Ft. (sq.m) Obtainable from One Cu.Yd. (Cu.M) of cement Mortar			
Thickness Inches	Square Feet	Thickness mm	Square Meters
1/4	1,296	6.25	120.40
1/2	648	12.50	60.20
3/4	432	18.75	40.13
1	324	25.00	30.10

Cleaning and Washing Old Masonry or Concrete Surfaces. Where old masonry or concrete surfaces are to be given a waterproof plaster coat, it is first necessary to prepare the old surface to receive the cement plaster.

On some jobs it is necessary to rake out the old brick joints to form a key for the plaster, and on others it is necessary to thoroughly wire brush the old surface to remove all surface dirt and then wash the surface with clean water or a mixture of chemical cleaner and water.

Where the surfaces are to be brushed and washed with water, the labor cost of cleaning 100 sq.ft. (9.29 sq.m. of surface should average as follows:

Description	Hours	Rate	Total	Rate	Total
Labor	1.50	$	$	$26.16	$ 39.24
Cost per sq.ft.					$ 0.39
per sq.m					$ 4.22

Roughing and Hacking Old Concrete Walls and Other Vertical Surfaces. Where it is necessary to roughen old, hard concrete surfaces to prepare a suitable bond for the new waterproof plaster coat, this can prove to be very expensive, especially where the old concrete surfaces are very hard

1022

and must be hacked and roughened using hand tools, a jackhammer, or bush-hammering tool. Workers often work in cramped quarters, which prevents the best daily output. On work of this kind, a worker will roughen about 20 sq.ft. (1.8 sq.m) of surface per hour. The following table indicates the labor cost per 100 sq.ft. (9.29 sq.m) to roughen and clean the surface:

Description	Hours	Rate	Total	Rate	Total
Laborer on Jackhammer	5.00	$	$	$28.16	$ 140.80
Labor for Cleaning	2.00	$	$	$26.16	$ 52.32
Cost per 100 sq.ft.					$ 193.12
per sq.ft.					$ 1.93
per sq.m					$ 20.79

Cost of tools (purchase or rental), fuels, etc. has to be estimated, in addition to the labor requirements. These costs will vary, depending on the accessibility, depth to be roughened, and quantity of work. For example, light work or small areas can be accomplished with an electric bush-hammer, while heavier work requires an air compressor, hose, and light jackhammer.

Equipment rental and fuel costs for an air compressor and tools to roughen the 100 sq.ft. (9.29 sq.m) of surface:

Description	Hours	Rate	Total	Rate	Total
Air Compressor and Tools (rental)	5.00	$	$	$60.00	$ 300.00
Fuel Costs 10 Gal (37.85 Liters)		$	$	$ 4.00	$ 40.00
Cost per 100 sq.ft.					$ 340.00
per sq.ft.					$ 3.40
per sq.m					$ 36.60

Applying Slush or Grout Coat of Neat Cement. After the old surfaces are cleaned, it is necessary to give them a slush coat of neat portland cement thoroughly brushed into the pores of the wall, which acts as a bond for the subsequent plaster coats. The labor cost per 100 sq.ft. (9.29 sq.m) of surface should run as follows:

Description	Hours	Rate	Total	Rate	Total
Labor	2.00	$	$	$26.16	$ 52.32
Cost per sq.ft.					$ 0.52
per sq.m					$ 5.63

The material needed for a slush coat of neat portland cement will require one bag of portland cement, mixed with water, per 100 sq.ft. (9.29 sq.m) of surface to be coated. A bag of cement costs about $7.00 each, which amounts to the following material costs: $6.50 per 100 sq.ft. (per 9.29 sq.m), or $0.065 per sq.ft. ($0.70 per sq.m).

1023

Labor Applying Waterproof Plaster Coats to Vertical Surfaces.
When applying waterproof plaster coats to vertical surfaces, the costs run
high for a number of reasons.

1. Interference of other trades.
2. Difficulty of mixing and handling.
3. Delay in setting of coats.
4. Difficulty in finishing vertical surfaces of this type with plaster or
    cement finish.
5. Contingency cost of return to job for breaks and cracks.
6. Waste and spillage of materials.

This work is usually performed by cement masons, plasterers, or
bricklayers.

When applying waterproof plaster coats, it is not necessary to give
the surface a float or brush finish, but the surface must be reasonably smooth
and without uneven projections that would prevent the water from running
down the sides of the walls.

Most manufacturers specify the mortar to be mixed in the
proportions of 1:1-1/2 or 1:2, and the material is usually applied in one or
two coats.

A mason should apply about 350 sq.ft. (32.51 sq.m) of first coat per
8-hr. day, at the following labor cost per 100 sq.ft. (9.29 sq.m):

Description	Hours	Rate	Total	Rate	Total
Plasterer	2.30	$	$	$32.41	$ 74.54
Helper	2.30	$	$	$26.16	$ 60.17
Cost per 100 sq.ft.					$ 134.71
per sq.ft.					$ 1.35
per sq.m					$ 14.50

On the second or finish coat, a mason should apply about 400 sq.ft.
(37.16 sq.m) per 8-hr. day, at the following labor cost per 100 sq.ft. (9.29
sq.m):

Description	Hours	Rate	Total	Rate	Total
Plasterer	2.00	$	$	$32.41	$ 64.82
Helper	2.00	$	$	$26.16	$ 52.32
Cost per 100 sq.ft.					$ 117.14
per sq.ft.					$ 1.17
per sq.m					$ 12.61

If necessary to clean the old masonry or concrete surfaces, or to
roughen the old concrete, add for this work as given on previous page.

1024

THERMAL AND MOISTURE PROTECTION

Water Resisting Plaster Coats Using Powered Intergral Water Repellents Per 100 Sq.Ft. (9.29 Sq.M.)									
Trade Name	Thick Inches	Mix Proportion	# of Coats	Sack Cement	Cu.Ft. Sand	Quan. Water-proof'g	Approx. Price per Lb.	Mech. Hrs	Labor Hrs
Hydratite	1	1:2	2	4	8	8 lbs.	$1.00	8	8
Hydrocide Powder	1	1:2	2	4	8	4 lbs.	$1.00	8	8
Toxement IW	3/4	1:2	2	3	6	4 1/2 lbs	$1.00	7	7
Trade Name	Thick mm	Mix Proportion	# of Coats	Sack Cement	Cu.M. Sand	Quan. Water-proofing	Approx. Price per Lb.	Mech. Hrs	Labor Hrs
Hydratite	25.00	1:2	2	4	0.23	3.62 Kg.	$2.20	8	8
Hydrocide Powder	25.00	1:2	2	4	0.23	1.81 Kg.	$2.20	8	8
Toxement IW	18.75	1:2	2	3	0.17	2.04 Kg.	$2.20	7	7

Water Resisting Palster Coats Using Liquid Integral Admixtures per 100 Sq.Ft. (9.29 Sq.M.)									
Trade	Thick. Inches	Mix Proportion	Number Coats	Sack Cement	Cu.Ft. Sand	Gal. Water proof 'g	Approx. Price per Lb.	Mech. Hrs	Labor Hrs
Anti-Hydro	3/4	1:2	2	3	6	3/4	$6.00	7	7
Anti-Hydro	1	1:2	2	4	8	1	$6.00	8	8
Aquatite	3/4	1:2	2	3	6	3/4	$3.50	7	7
Aquatite	1	1:2	2	4	8	1	$3.50	8	8
Hydratite	3/4	1:2	2	3	6	3/4	$5.00	7	7
Hydratite	1	1:2	2	4	8	1	$5.00	8	8
Toxement IW	3/4	1:2	2	3	6	3/4	$5.00	7	7
Toxement IW	1	1:2	2	4	8	1	$5.00	8	8
Sikacrete	3/4	1:2	2	3	6	3/4	$7.00	7	7
Trimix	3/4	1:2	2	3	6	3/4	$3.50	7	7
(metric)									
Trade	Thick. mm	Mix Proportion	# 0f Coats	Sack Cement	Cu.M. Sand	Liters Water-proof'g	Approx. Price per Kg.	Mech Hrs	Labor Hrs
Anti-Hydro	18.75	1:2	2	3	0.17	2.84	$13.23	7	7
Anti-Hydro	25.00	1:2	2	4	0.23	3.79	$13.23	8	8
Aquatite	18.75	1:2	2	3	0.17	2.84	$7.72	7	7
Aquatite	25.00	1:2	2	4	0.23	3.79	$7.72	8	8
Hydratite	18.75	1:2	2	3	0.17	2.84	$11.02	7	7
Hydratite	25.00	1:2	2	4	0.23	3.79	$11.02	8	8
Toxement IW	18.75	1:2	2	3	0.17	2.84	$11.02	7	7
Toxement IW	25.00	1:2	2	4	0.23	3.79	$11.02	8	8
Sikacrete	18.75	1:2	2	3	0.17	2.84	$15.43	7	7
Trimix	18.75	1:2	2	3	0.17	2.84	$7.72	7	7

Membrane Waterproofing

Membrane waterproofing is constructed in place by building up a strong, waterproof, and impermeable blanket with overlapping plies of tar, asphalt saturated open mesh cotton fabric, or rag felt. The plies are coated and cemented together with hot coal tar pitch or waterproofing asphalt. There is always one or more applications of pitch or asphalt than plies of felt or fabric.

Estimating the Quantity of Felt or Fabric Required for Membrane Waterproofing. When estimating the quantity of felt or fabric required for waterproofing masonry or concrete walls, pits, floors, etc., it is necessary to add a percentage for lap and waste. If the felt is applied to walls, allow about 6" (150 mm) to lap over the footings at the bottom, with a small allowance at the top of the walls above grade. This will average from 7.5% to 10% of the total area to be waterproofed.

If the walls or floors are covered with a single thickness of felt or membrane, lap each strip of felt 4" to 6" (100-150 mm), depending upon the specifications. As most felts are put up in 36" (900 mm) rolls, it is necessary to add about 12% waste for a 4" (100 mm) lap, and 17% for a 6" (150 mm) lap.

If the lap is 4" (100 mm) wide and only a single thickness of felt is used, add about 20% to the actual area to allow for laps and the additional felt required at the top and bottom of the wall. On floors it is customary to run the felt or membrane up the sides of the walls 4" to 6" (100-150 mm). If the lap is 6" (150 mm) and only a single thickness of felt is used, add about 25% to the actual wall or floor area for laps, etc.

When 2, 3, 4, or more plies of felt are used, it is not necessary to allow for laps, except at the ends of the wall or floor, as one thickness of felt overlaps the next without waste. However, extra material will be required at the tops and bottoms of the walls and where the paper runs up the sides of the walls 4" to 6" (100-150 mm). On work of this kind, add about 10% to the actual floor or wall area.

Fabric or Felt Required to Cover 100 Sq.Ft. (9.29 Sq.M.) of Surface							
Class of Work	1 Ply 4" lap	1 Ply 6" lap	2-Ply	3-Ply	4-Ply	5-Ply	Ea. Add. Ply
Add percent	20	25	10	10	10	10	10
Sq.Ft. of felt to cover 100 Sq.Ft.	120	125	220	330	440	550	110
Metric							
Add percent	20	25	10	10	10	10	10
Sq.M of felt to cover 10 Sq.M.	129.17	134.55	236.81	355.22	473.63	592.03	118.41

THERMAL AND MOISTURE PROTECTION

Weight of Tar or Asphalt Felt for Membrane Waterproofing. Tar or asphalt felt for waterproofing is currently furnished in 4 sq. rolls of 432 sq.ft. (40.13 sq.m) weighing 60 lbs. (27.21 kg) per roll.

Double thickness asphalt felt is also furnished in 60 lb. (27.21 kg) per roll containing 216 sq.ft. (20.06 sq.m) and this felt is known as No. 30.

When specifying the grade or weight of felt to be used, it is customary to state that "felt shall weigh not less than 15 lbs. per 108 sq.ft." (6.80 kg per 10.03 sq.m). This is known as No. 15 felt. Felt is furnished in 4 sq. rolls of 432 sq.ft. (40.13 sq.m), so there are 32 sq.ft. (2.97 sq.m) per roll or 8 sq.ft. per 100 sq.ft. (0.74 sq.m per 9.29 sq.m) allowed for laps.

Tar or asphalt saturated fabric is usually sold by the roll containing 50 sq.yds. (41.80 sq.m) or by the sq.yd. (sq.m).

Estimating the Quantity of Pitch or Asphalt Required for Membrane Waterproofing. The quantity of pitch or asphalt required for mopping 100 sq.ft. (9.29 sq.m) of surface with one application is approximately 30 to 35 lbs. (13.60 to 15.87 kg).

Many special coatings are sold. Some are furnished solid, making it necessary to heat and melt them before using, while others are furnished in the form of a heavy black paint that can be either mopped or brushed on cold. The manufacturer usually specifies how the material shall be applied.

Engineers and architects usually specify the quantity of asphalt or compound to be used for each 100 sq.ft. (9.29 sq.m) of surface.

Weight of Pitch or Asphalt Per 100 Sq.Ft. (9.29Sq.M) of Surface			
No. Plies of Saturated Fabric or Tarred Felt	Alternate Moppings of Pitch or Asphalt Required	Lbs. of Pitch or Asphalt Required per 100 Sq.Ft of Waterproofing	Kg. of Pitch or Asphalt Required per 10 Sq.M of Waterproofing
2	3	90 - 105	43.9 - 51.3
3	4	120 - 140	58.6 - 68.4
4	5	150 - 175	73.2 - 85.5
5	6	180 - 210	87.9 - 102.5
6	7	210 - 245	102.5 - 119.6

Applying Membrane Waterproofing. Whether to use saturated fabric or saturated felt in constructing membrane waterproofing depends entirely upon local conditions. Saturated fabric weighing 12 oz. per sq. yd. has approximately three times the tensile strength of 15-lb. (6.80-kg) saturated felt, but it costs considerably more per sq.yd. (sq.m) than saturated felt. On flat surfaces, where the waterproofing will be covered with a protection course immediately after it is applied, and where it will not be subjected to any unusual strain, saturated felts can be used. On any surface where greater strength is required, fabric should be used. It is always desirable to provide a protection course over the waterproofing construction. While the labor operations involved in applying membrane waterproofing are similar to those followed in applying built-up roofing, the work is usually

1027

more expensive than roofing because of the limited working space and the necessity for using scaffolds on vertical surfaces over 8'- 0" (2.43 m) high.

On vertical surfaces the work can be expedited if the plies of felt or fabric are installed up and down the surface like wallpaper rather than across. The pieces of felt or fabric can be cut to the required length and folded so that after the surface has been mopped the top edges can be pushed into the hot bitumen and the remainder of the piece will then fall into place where it can be rubbed into the hot pitch or asphalt. After the required number of plies have been embedded in alternate moppings of hot bitumen all angles at corners, walls, etc., should be reinforced with at least two additional plies of felt or fabric and alternate moppings of bitumen. A protection course of beadboard, fiberboard, or other material should immediately be placed over the waterproofing for best results.

Prior to the application of the first hot bitumen mopping coat, the wall surface is primed, uniformly and completely, with one gallon (3.8 liters) of primer-asphalt primer for asphalt specifications or tarbase primer for pitch specifications-per 100 sq.ft. (9.29 sq.m) of surface area.

For the application of the primer to the surface, one worker should coat about 100 sq.ft. (9.29 sq.m) per hour:

Description	Hours	Rate	Total	Rate	Total
Labor	1.00	$	$	$ 26.16	$ 26.16
Cost per sq.ft.					$ 0.26
per sq.m					$ 2.82

Labor Applying Membrane Waterproofing. When applying membrane waterproofing to walls and floors, it takes three or four workers together. It requires one worker to attend the fire and heat the bitumen, while two or three are mopping the walls and applying the felt or fabric.

If only one application of hot bitumen is applied, two workers together should heat materials and mop 2,000 to 2,200 sq.ft. (185.8 to 204.38 sq.m) per 8-hr. day, at the following labor cost per 100 sq.ft. (9.29sq.m):

Description	Hours	Rate	Total	Rate	Total
Roofer	0.80	$	$	$ 26.16	$ 20.93
Cost per sq.ft.					$ 0.21
per sq.m					$ 2.25

THERMAL AND MOISTURE PROTECTION

Water Pressure Table and Waterproofing Required for Varying Heads of Water					
Hydrostatic Head in Feet	Pressure Lbs. per Sq.In.	Lifting Pressure Lbs per Sq.Ft.	Wall Pressure Lbs. per Sq.Ft.	Plies of Saturated Felt/Fabric	Asphalt Mastic Thickness
0.5	0.21	39.20	15.60	2-Ply	1/8 "
1	0.43	62.50	31.20	2-Ply	1/8 "
2	0.86	125.00	62.50	2-Ply	1/4 "
3	1.30	187.50	93.70	2-Ply	1/4 "
4	1.73	250.00	125.00	3-Ply	5/8 "
5	2.17	312.50	156.20	3-Ply	5/8 "
6	2.60	375.00	187.50	4-Ply	5/8 "
8	3.47	500.00	250.00	4-Ply	5/8 "
10	4.34	625.00	312.50	5-Ply	5/8 "
12	5.21	750.00	375.00	6-Ply	5/8 "
15	6.51	937.50	468.70	6-Ply	3/4 "
20	8.68	1250.00	625.00	7-Ply	3/4 "
25	10.85	1562.50	781.20	8-Ply	3/4 "
30	13.02	1875.00	937.50	10-Ply	3/4 "
40	17.36	2500.00	1250.00	11-Ply	3/4 "
Hydrostatic Head in Meters	Pressure Kg. per Sq. Cm.	Lifting Pressure Kg. per Sq.M.	Wall Pressure Kg. per Sq.M.	Plies of Saturated Felt/Fabric	Asphalt Mastic Thickness
0.15	2.059	191.4	76.2	2-Ply	3.13 mm
0.30	4.217	305.2	152.3	2-Ply	3.13 mm
0.61	8.434	610.3	305.2	2-Ply	6.25 mm
0.91	12.749	915.5	457.5	2-Ply	6.25 mm
1.22	16.966	1220.6	610.3	3-Ply	15.63 mm
1.52	21.280	1525.8	762.6	3-Ply	15.63 mm
1.83	25.497	1830.9	915.5	4-Ply	15.63 mm
2.44	34.029	2441.2	1220.6	4-Ply	15.63 mm
3.05	42.561	3051.5	1525.8	5-Ply	15.63 mm
3.66	51.093	3661.8	1830.9	6-Ply	15.63 mm
4.57	63.841	4577.3	2288.4	6-Ply	18.75 mm
6.10	85.122	6103.0	3051.5	7-Ply	18.75 mm
7.62	106.402	7628.8	3814.2	8-Ply	18.75 mm
9.14	127.683	9154.6	4577.3	10-Ply	18.75 mm
12.19	170.243	12206.1	6103.0	11-Ply	18.75 mm

After the surface has been mopped with hot bitumen, two workers must cut the felt to lengths and place it on the wall. Usually it is cut long enough to lap over the top of the wall a few inches, so that it can be held in place by brick or stone until it is self-supporting.

Two workers placing felt or fabric have to work along with the workers mopping, so they will place felt or fabric on 2,000 to 2,200 sq.ft. (185.80 to 204.38 sq.m) of surface per 8-hr. day, at the following labor cost per 100 sq.ft. (9.29 sq.m):

Description	Hours	Rate	Total	Rate	Total
Roofer	0.80	$	$	$ 30.93	$ 24.74
Cost per sq.ft.					$ 0.25
per sq.m					$ 2.66

Labor Applying 100 Sq.Ft. (9.29 Sq.M) of Membrane Waterproofing				
Description of Work	# of Piles	Mopping Hours	Felt Hours	Total Hours
1-ply fabric, 2 moppings	1	1.6	0.8	2.4
2-ply fabric, 3 moppings	2	2.4	1.6	4
3-ply fabric, 4 moppings	3	3.2	2.4	5.6
4-ply fabric, 5 moppings	4	4	3.2	7.2
Each additional ply of Felt and Mopping	1	0.8	0.8	1.6

Prices of Waterproofing Materials

Prices on waterproofing materials must always be checked, because prices vary considerably due to market conditions, location, and manufacturer.

Asphalt primer costs about $3.50 per gal. ($0.92 per liter) in 55-gal. (208.17-liter) drums; $3.50 per gal. ($0.92 per liter) in 30-gal. (113.55-liter) drums and $3.75 per gal. ($0.99 per liter) in 5-gal (18.92-liter) cans.

Waterproofing pitch or asphalt costs about $400.00 per ton in carload lots; $420.00 per ton in less than carloads.

No. 15 tar or asphalt saturated felt costs $18.00 per roll of 432 sq.ft. (40.13 sq.m) in carloads and $19.00 per roll in less than carloads.

Tar or asphalt saturated cotton fabrics costs $1.50 to $1.75 per sq.yd. ($1.50 to $2.09 per sq.m) in carloads and $1.90 to $2.00 per sq.yd. ($4.31 to $ 2.39 per sq.m) in less than carloads.

THERMAL AND MOISTURE PROTECTION

Material Cost of 100 Sq.Ft. (9.29 Sq.M) of 1-Ply Membrane Waterproofing				
Description	Rate	Total	Rate	Total
1 gal. (3.7 liter) asphalt primer	$	$	$ 3.50	$ 3.50
70 lbs. (31.7 kg) asphalt or pitch	$	$	$ 0.25	$17.50
0.28 rolls No.15 felt, 120 sq.ft. (111.1 sq.m)	$	$	$18.00	$ 5.04
Fuel, mops, etc.	$	$	$ 2.00	$ 2.00
Cost per 100 sq.ft.				$28.04
per sq.ft.				$ 0.28
per sq.m				$ 3.02

Labor Cost of 100 Sq.Ft. (9.29 Sq.M) of 1-Ply Membrane Waterproofing Consisting of 1 ply of felt and 2 moppings of hot bitumen

Description	Hours	Rate	Total	Rate	Total
Roofer**	2.60	$	$	$35.04	$ 91.10
Roofer Placing Felt	0.80	$	$	$35.04	$ 28.03
Cost per 100 sq.ft.					$ 119.14
Cost per sq.ft.					$ 1.19
per sq.m					$ 12.82

** Includes labor costs for primer coat

Material Cost of 100 Sq.Ft. (9.29 Sq.M) of Each Additional Ply of Felt and Hot Bitumen				
Description	Rate	Total	Rate	Total
35 Lbs. (15.8 kg) asphalt or pitch	$	$	$ 0.25	$ 8.75
0.24 rolls No.15 felt, 100 sq.ft. (9.29 sq.m)	$	$	$18.00	$ 4.32
Fuel, mops, etc.	$	$	$ 1.00	$ 1.00
Cost per 100 sq.ft.				$14.07
per sq.ft.				$ 0.14
per sq.m				$ 1.51

Labor Cost per 100 S.F. (9.29 Sq.M) for Each Additional Ply of Felt and Hot Bitumen

Description	Hours	Rate	Total	Rate	Total
Roofer	1.60	$	$	$35.04	$ 56.06
Cost per 100 sq.ft.					$ 56.06
Cost per sq.ft.					$ 0.56
per sq.m					$ 6.03

Material Cost of 100 Sq.Ft. (9.29 Sq.M) of 1-Ply Membrane Waterproofing Using Tar or Asphalt, Saturated Fabric and Hot Pitch or Asphalt Consisting of 1-Ply of Saturated Fabric and 2 Moppings of Hot Bitumen				
Description	Rate	Total	Rate	Total
1 Gal (3.78 Liter) asphalt primer	$	$	$ 3.50	$ 3.50
70 Lbs (32 Kg) asphalt pitch	$	$	$ 0.25	$17.50
13.4 sq.yd. (11.2 sq.m) saturated fabric	$	$	$ 2.09	$28.01
Fuel, mops, etc	$	$	$ 1.50	$ 1.50
Cost per 100 sq.ft.				$50.51
per sq.ft.				$ 0.51
per sq.m				$ 5.44

Material Cost of 100 Sq.Ft. (9.29 Sq.M) Additional Ply of Saturated Fabric and Hot Pitch or Asphalt				
Description	Rate	Total	Rate	Total
35 Lbs. (16 kg) asphalt or pitch	$	$	$ 0.25	$ 8.75
12 Sq.yd. (10 sq.m.) saturated fabric	$	$	$ 1.80	$21.60
Cost per 100 sq.ft.				$30.35
per sq.ft.				$ 0.30
per sq.m				$ 3.27

The labor costs for placing fabric and moppings of hot bitumen are the same as used in placing felts and moppings of hot bitumen.

The Iron Method of Waterproofing

The iron method of waterproofing uses an exceedingly fine metallic powder, containing no grease, asphalt, oil, or other substances subject to disintegration.

The waterproofing is applied to the inside or outside surfaces of walls and to the tops of rough footings and floor slabs, in the form of brush coats, or in a combination of brush and plaster coats.

Applied either upon the inside or outside of walls, this type of waterproofing will resist hydrostatic pressure to a considerable degree.

When preparing old concrete vertical surfaces for iron waterproofing, the vertical surfaces should be given an entirely new bonding exposure by cutting not less than 1/16" (1.56 mm) with bushhammers or other suitable tools. See cost of roughing old surfaces on previous pages. The new surface shall then be thoroughly cleaned by brushing with wire brushes. Surfaces of brick and stone need only thorough cleaning with the wire brush and washing.

1032

THERMAL AND MOISTURE PROTECTION

Holes, cracks, and soft or porous spots in vertical or horizontal surfaces shall be cut out and pointed with one part iron floor bond, 2 parts portland cement and 3 parts sand, all by weight. These materials shall be mixed dry, screened and enough water added to make a stiff mix. Particular care must be taken at the intersections of all walls and floors, in corners, around pipes and other projections through the waterproofing and at construction joints in the concrete.

After the wall surfaces have been cleaned and pointed, they shall be thoroughly cleaned again by hosing. Excess water shall then be removed.

It usually requires 2 to 6 coats of iron waterproofing, depending on the density of the surface and the conditions encountered.

Some of these materials are mixed with sand and water and applied as a brush coat while others are furnished already mixed ready to apply by the addition of water.

Liquid Membrane Waterproofing

Liquid membrane (elastomeric) waterproofing is a cold, fluid applied, synthetic rubber, seamless coating system. It is available in single or two-component form and can be applied by trowel or spray. The material forms a continuous, flexible, water impervious film that bonds tightly to a wide range of surfaces, including concrete, stone, masonry, wood, and metal. On a troweled surface, an application rate of 5 gals. (18.92 liters) per 125 sq.ft. (11.61 sq.m) of surface will produce a cured coating thickness of approximately 55 mils.

Two workers should coat about 650 sq.ft. (60.38 sq.m) per day at the following labor cost:

Description	Hours	Rate	Total	Rate	Total
Labor	16.00	$	$	$26.16	$ 418.56
Cost per 650 sq.ft.					$ 418.56
Cost per sq.ft.					$ 0.64
per sq.m					$ 6.93

Cost of 100 Sq.Ft. (9.29 Sq.M) Mastic Waterproof Coatings

The following materials are furnished in the form of a mastic or plastic cement and are applied with a trowel.

1033

Kind of Waterproofing	Thick. Inches	No. of Coats	Cover Capacity Sq.Ft. per Gal	Quantity Per 100 Sq.Ft., Gal.	Approx. Price per Gallon	Labor Hours
Dehydratine No. 10	1/16	1	26	4 Gal	$ 2.64	3
Hydrocide Mastic	1/16	1	20 - 25	5 Gal	$ 1.81	3
Hydrocide 700	1/16	1	20 - 25	5 Gal	$ 1.60	3
Tremco 103 Mastic Asphalt Base	1/8	1	20	5 Gal	$ 1.60	3
Tar base	1/8	1	20	5 Gal	$ 2.72	3

Kind of Waterproofing	Thick. mm	No. of Coats	Cover Capacity Sq.M. per Liter	Quantity Per 10 Sq.M., liters	Approx. Price per Liter	Labor Hours
Dehydratine No. 10	1.56	1	0.64	16.30 Liter	$ 0.70	3
Hydrocide Mastic	1.56	1	0.49 - 0.61	20.37 Liter	$ 0.48	3
Hydrocide 700	1.56	1	0.49 - 0.61	20.37 Liter	$ 0.42	3
Tremco 103 Mastic Asphalt Base	3.13	1	0.49	20.37 Liter	$ 0.42	3
Tar base	3.13	1	0.49	20.37 Liter	$ 0.72	3

07150 DAMPPROOFING

Dampproof Paints for Exterior Concrete or Masonry Surfaces

Where a heavy head of water is to be overcome, the concrete or masonry should preferably be waterproofed by the membrane method. Where only a dampproofing material is required, there are a number of paints on the market for this purpose. The application of one, two, or more coats of paint is intended to make the walls impervious to dampness.

Number of coats required and covering capacity will vary with porosity and surface of wall.

THERMAL AND MOISTURE PROTECTION

Heavy Dampproof Paints, Per 100 Sq.Ft. (9.29 Sq.M)						
Material	Use on	No. Coats	Sq. Ft. per Gallon	Gals. Required per 100 Sq.Ft.	Price per Gallon	Labor Hours Req.
Dehydratine 4	Fndtn.	2	33	3	$ 2.29	2.2
Dehydratine 10	Masonry	1	30	3 1/3	$ 2.64	1.2
Hydrocide Semi-Mastic	Fndtn.	1	30 - 50	3	$ 1.68	1.2
	Fndtn.	2	15 - 18	6	$ 1.68	2.2
Hydrocide 600	Fndtn.	1	75 - 100	1 1/3	$ 1.81	1.2
Hydrocide 648	Fndtn.	2	50	2	$ 1.81	2.2
Marine Liquid	Blw Grd	1	50 - 75	1 3/4	$ 1.98	1.2
Sikaseal	Fndtn.	2	60 - 80	1 1/2	$ 3.52	2.2
Tremco 110	Fndtn.	1	100	1	$ 1.14	1.2
Material	Use on	No. Coats	Sq. M. per Liter	Liters Required per 10 Sq.M.	Price per Liter	Labor Hours Required
Dehydratine 4	Fndtn.	2	0.81	12.22	$ 0.60	2.2
Dehydratine 10	Masonry	1	0.74	13.58	$ 0.70	1.2
Hydrocide Semi-Mastic	Fndtn.	1	0.74 - 1.23	12.22	$ 0.44	1.2
	Fndtn.	2	0.37 - 0.44	24.45	$ 0.44	2.2
Hydrocide 600	Fndtn.	1	1.84 - 2.45	5.43	$ 0.48	1.2
Hydrocide 648	Fndtn.	2	1.23	8.15	$ 0.48	2.2
Marine Liquid	Blw Grd	1	1.23 - 1.84	7.13	$ 0.52	1.2
Sikaseal	Fndtn.	2	1.47 - 1.96	6.11	$ 0.93	2.2
Tremco 110	Fndtn.	1	2.45	4.07	$ 0.30	1.2

Labor Applying Water and Dampproofing Paint to Exterior Surfaces Below Grade. Where the water or dampproofing consists of applying one or more coats of heavy paint to exterior concrete or masonry surfaces below grade, it is customary to use an ordinary roofer's brush or a mop. On foundation walls below grade, the workers are usually obliged to work in cramped quarters which reduces output.

To make these paints effective, they must be thoroughly brushed into the surface and all voids and pin holes filled, otherwise water and moisture will seep through.

On work of this kind, a worker should apply 600 to 700 sq.ft. (55.74 to 65.03 sq.m) per 8-hr. day, at the following labor cost per sq.ft. (sq.m):

Description	Hours	Rate	Total	Rate	Total
Roofer	1.30	$	$	$ 30.93	$ 40.21
Cost per 100 sq.ft.					$ 40.21
Cost per sq.ft.					$ 0.40
per sq.m					$ 4.33

On the second and third coats, the labor cost will be slightly less than the first coat on account of the smoother surface to work on.

Dampproof and Plaster Bond Paints

Labor Applying Dampproofing and Plaster Bond Paints. Before applying the first coat of plaster to the interior of exterior brick or masonry walls, the walls are often painted with one or more coats of heavy black paint, which acts as a dampproofing and plaster bond.

The labor cost will vary with the consistency of the paint and the condition of the walls, as the paint must be well brushed in and all pin holes and porous places thoroughly covered with paint, otherwise moisture will seep through the walls.

Where the paint is applied by hand, a worker should cover 700 to 800 sq.ft. (65.03 to 74.32 sq.m) per 8-hr. day, at the following labor cost per 100 sq.ft. (9.29 sq.m):

Description	Hours	Rate	Total	Rate	Total
Roofer	1.10	$	$	$ 30.93	$ 34.02
Cost per 100 sq.ft.					$ 34.02
Cost per sq.ft.					$ 0.34
per sq.m					$ 3.66

When dampproof paints are applied with an air spray, a worker should cover 300 to 350 sq.ft. (27.87 to 32.51 sq.m) per hr. at the following labor cost per 100 sq.ft. (9.29 sq.m):

Description	Hours	Rate	Total	Rate	Total
Roofer	0.33	$	$	$ 30.93	$ 10.21
Cost per 100 sq.ft.					$ 10.21
Cost per sq.ft.					$ 0.10
per sq.m					$ 1.10

Dampproof and Plaster Bond Paints, Per 100 Sq.Ft. (9.29 Sq.M)

Heavy black paints are used for dampproofing interior of exterior walls and to form a bond for plaster. Number of coats required will depend upon porosity of wall on which it is used.

1036

THERMAL AND MOISTURE PROTECTION

Labor costs in the following table are based on hand application. If an air spray is used, deduct as given above.

Name of Material	No. Coats	Sq. Ft. per Gallon	Gals. Req'd. 100 Sq.Ft.	Price per Gallon	Labor Hrs. Applying
Dehydratine 4	1	70 - 80	1 1/3	$ 2.29	1
Dehydratine 4	2	45 - 55	2 1/3	$ 2.29	2
Plasterbond 232	1	75 - 80	1 1/3	$ 2.35	1
Plasterbond 232	2	35 - 40	2 1/2	$ 2.35	2
Name of Material	No. Coats	Sq. M. per Liter	Liters Req'd. 10 Sq.M.	Price per Liter	Labor Hrs. Applying
Dehydratine 4	1	1.72 - 1.96	5.43	$ 0.60	1
Dehydratine 4	2	1.10 - 1.35	9.51	$ 0.60	2
Plasterbond 232	1	1.84 - 1.96	5.43	$ 0.62	1
Plasterbond 232	2	0.86 - 0.98	10.19	$ 0.62	2

Transparent Dampproofing of Exterior Masonry Walls

There has been a big change in the use of transparent dampproofings and the method of application due to an investigation by the National Bureau of Standards, Publications BMS7. It was determined that "leaky" walls were mainly caused by poor workmanship. In other words, unless the bricks were laid in a full bed of mortar (not furrowed), with all bed, end, and vertical joints completely filled with mortar, the walls were likely to leak within a short period after the completion of the structure. Hairline cracks between brick and mortar joints is also a cause of leakage.

It was determined that painting the surfaces of masonry walls with transparent "paints" or "waxes" alone does not stop the leaks. One method that was efficient was to cut out the old mortar joints to a depth of 1/2" to 3/4" (12.50 to 18.75 mm) and repoint the joints with new mortar.

This class of work is quite expensive, because the cost depends on the hardness and tenacity of the mortar joints.

The basis for the use of exterior waterproofing, labor and material, is calculated on the ordinary wall constructed with common brick of high absorption with 3/8" to 1/2" (9.37 to 12.50 mm) lime-cement mortar joints.

On lime-cement mortar joints, two masons can rake out and repoint about 12 sq.ft. (1.11 sq.m) of brick masonry surface per hour, removing joints to a depth of 3/4" (18.75 mm). Allow about 1/4 hr. helper's time for moving scaffolds, mixing, etc.

On the above basis, the labor cost of cutting out old mortar joints and repointing mortar joints to 100 sq.ft. (9.29 sq.m) of wall surface will cost:

Description	Hours	Rate	Total	Rate	Total
Mason	16.00	$	$	$36.93	$ 590.88
Helper	4.00	$	$	$26.16	$ 104.64
Cost per 100 sq.ft.					$ 695.52
per sq.ft.					$ 6.96
per sq.m					$ 74.87

On hard Portland cement mortar joints, two masons using a power saw can cut out and repoint joints in about 8 sq.ft. (0.74 sq.m) of wall per hour and requires 1/4 hr. helper time handling tools and mixing mortar.

On the above basis, the labor cost of cutting out old Portland cement mortar joints, and repointing joints to 100 sq.ft. (9.29 sq.m) of wall will cost:

Description	Hours	Rate	Total	Rate	Total
Mason	24.00	$	$	$36.93	$ 886.32
Helper	6.00	$	$	$26.16	$ 156.96
Cost per 100 sq.ft.					$1,043.28
per sq.ft.					$ 10.43
per sq.m					$ 112.30

Colorless or Transparent Liquid Water Repellent Treatments

There are a number of colorless or transparent liquid preparations on the market used on brick, stone, stucco, cement, and concrete surfaces to render them water repellent and help to prevent rain absorption.

These liquids are colorless and do not affect the original color of the surface to which they are applied.

The covering capacities and quantities given in the following table were furnished by the various manufacturers.

It will be noted that some manufacturers specify one coat and others two coats. Regardless of the material used, sufficient liquid must be applied to completely seal the pores of the surface. Otherwise, the results will be unsatisfactory.

Transparent liquids used for dampproofing brick, stone, stucco, concrete, and cement surfaces are:

1038

THERMAL AND MOISTURE PROTECTION

Transparent Liquid Water Repellents, per 100 Sq.Ft. (9.29 Sq. M)					
Name of Material	No. Coats	Sq. Ft. per Gal.	Gal. Reqd. per 100 Sf.	Price per Gal.	Labor Hrs. Applying
Daracone	1	75 - 100	1 1/4	$ 6.48	1.6
Supertox	1	125 - 150	3/4	$ 5.60	0.8
Tremco 141 Invisible	2	75 - 100	1 1/4	$ 5.20	1.6
Name of Material	No. Coats	Sq. M. per Liter	Liter. Reqd. per 10 Sq. M.	Price per Liter	Labor Hrs. Applying
Daracone	1	1.8 - 2.5	5.09	$1.71	1.6
Supertox	1	3.1 - 3.7	3.06	$1.48	0.8
Tremco 141 Invisible	2	1.8 - 2.5	5.09	$1.37	1.6

Colorless or Transparent Silicone Base Liquid Water Repellent Treatments

Silicone base products are truly colorless and permit the masonry to "breathe" while rendering the surface water repellent. They penetrate deeply into the cement or masonry surfaces coating pores, cracks, and fissures with an insoluble, non-oxidizing film of silicone that effectively stops capillary action by which water is absorbed.

Usually one flooding coat is sufficient and can be applied by either brushing or spraying. The covering capacities and quantities given in the following table are furnished by the various manufacturers.

Silicone Base, Transparent Liquid Water Repellents, per 100 Sq.Ft. (9.29 Sq. M.)					
Name of Material	No. Coats	Sq. Ft. per Gal.	Gal. Reqd. per 100 SF	Price per Gallon	Labor Hrs. Applying
Dehydratine 22	1	75 - 100	1 1/4	$ 5.44	0.8
Hydrocide Colorless SX	1	75 - 100	1	$ 7.20	0.8
Sika Transparent	1	80 - 200	2/3	$ 4.32	0.8
Tremco 147-3%	1	100 - 200	2/3	$ 7.52	0.8
Tremco 147-5%	1	100 - 200	2/3	$ 12.16	0.8
Devoe Super-Por-Seal	1	150 - 150	2/3	$ 8.46	0.8
Name of Material	No. Coats	Sq. M. per Liter	Liter. Reqd. per 10 Sq. M.	Price per Liter	Labor Hrs. Applying
Dehydratine 22	1	1.8 - 2.5	5.09	$ 1.44	0.8
Hydrocide Colorless SX	1	1.8 - 2.5	4.07	$ 1.90	0.8
Sika Transparent	1	2.0 - 4.9	2.72	$ 1.14	0.8
Tremco 147-3%	1	2.5 - 4.9	2.72	$ 1.99	0.8
Tremco 147-5%	1	2.5 - 4.9	2.72	$ 3.21	0.8
Devoe Super-Por-Seal	1	3.7 - 3.7	2.72	$ 2.24	0.8

Labor Applying Transparent Liquid Waterproofing. When applying transparent liquid dampproof paints by hand, a worker should cover 900 to 1,100 sq.ft. (83.61 to102.19 sq.m) of surface per 8-hr. day, at the following labor cost per 100 sq.ft. (9.29 sq.m):

Description	Hours	Rate	Total	Rate	Total
Roofer	0.80	$	$	$ 30.93	$ 24.74
Cost per 100 sq.ft.					$ 24.74
per sq.ft.					$ 0.25
per sq.m					$ 2.66

On the second and third coats, a worker should apply 1,000 to 1,200 sq.ft. (92.9 to 111.48 sq.m) per 8-hr. day, at the following labor cost per 100 sq.ft. (9.29 sq.m):

Description	Hours	Rate	Total	Rate	Total
Roofer	0.75	$	$	$ 30.93	$ 23.20
Cost per 100 sq.ft.					$ 23.20
per sq.ft.					$ 0.23
per sq.m					$ 2.50

When applied by spray, a worker should cover 300 to 350 sq.ft. (27.87 to 32.51 sq.m) per hr. at the following labor cost per 100 sq.ft. (9.29 sq.m):

Description	Hours	Rate	Total	Rate	Total
Roofer	0.33	$	$	$ 30.93	$ 10.21
Cost per 100 sq.ft.					$ 10.21
per sq.ft.					$ 0.10
per sq.m					$ 1.10

Add cost of compressor and spray equipment.

Back-Plastering and Back-Painting
Cut Stone and Marble to Prevent Staining

When marble, limestone, brownstone, or sandstone are used in building work, it difficulty is to prevent moisture in the brick or concrete backing from penetrating the surface of the stone or marble and staining or discoloring the face, due to its open grain or porous nature.

To overcome this condition, it is customary to paint the back of the stone, beds, and end joints, up to within 1" (25 mm) of the face, to prevent staining and discoloration. White or non-staining cement is also used for this purpose. The stone is set in non-staining cement mortar, and the back of the stone is plastered or parged with the same mortar.

In many instances the backs, beds, and joints of the stone are painted and then back-plastered with non-staining cement, which acts as a double preventative.

Back-painting of Indiana limestone is not recommended by quarrymen and producers, who say that it tends to create conditions it is supposed to counteract. A full explanation is given under Cut Stone.

Where cut stone comes in contact with concrete, it is recommended that the concrete be painted instead of the stone and a one-inch space left between the concrete and cut stone.

Estimating the Cost of Back-Painting Cut Stone and Marble. It is customary to estimate the cost of cut stone by the cu.ft. (cu.m), so the cost of back-painting should be estimated in the same manner.

Plain stone ashlar is usually 4" to 8" (100-200 mm) thick or in alternating courses of 4" (100 mm) and 8" (200 mm) stone so as to provide a bond with the brick or masonry backing. Where just a veneer of cut stone is required, the stone is usually 4" (100 mm) thick and is bonded into the masonry backing by metal anchors.

If cut stone is furnished in alternating courses of 4" (100 mm) and 8" (200 mm), the stone for the entire job will average 6" (150 mm) thick. It will therefore be necessary to back-paint 2 sq.ft. (0.18 sq.m) of surface to each cu.ft. stone. Including beds and joints, it will double the above quantities or it will be necessary to back-paint 4-1/3 sq.ft. per cu.ft. (0.4 sq.m per cu.m) of stone.

If the stone averages 4" (200 mm) thick, it will be necessary to back-paint about 5 sq.ft. per cu.ft. (0.46 sq.m per cu.m) of stone, including the backs, beds, and joints.

If the stone averages 8" (200 mm) thick, it will be necessary to back-paint about 4 sq.ft. per cu.ft. (0.37 sq.m per cu.m) of stone.

Back-Painting Cut Stone and Marble. When back-painting limestone, sandstone, marble, etc., the stone is often painted in the yard before delivery, although on many jobs it is painted after delivery. It is customary to paint the stone a few days before setting, although the backs are frequently painted after the stone has been set.

Care must be exercised to thoroughly brush the paint into all the pores and not drop any paint on the face of the stone.

When painted at the building site, a worker should paint 600 to 700 sq.ft. (55.74 to 65.03 sq.m) of surface per 8-hr. day, at the following labor cost per 100 sq.ft. (9.29 sq.m):

Description	Hours	Rate	Total	Rate	Total
Mason	1.30	$	$	$ 36.93	$ 48.01
Cost per 100 sq.ft.					$ 48.01
per sq.ft.					$ 0.48
per sq.m					$ 5.17

Heavy black paints are applied to backs and sides of stone to prevent cement and mortar stains. Quantities are based on a single coat:

1042

THERMAL AND MOISTURE PROTECTION

Back-Painting 100 Sq.Ft. (9.3 Sq.M) Cut Stone or Marble to Prevent Staining				
Name of Material	Sq. Ft. per Gal.	Gals. Per Sq.Ft.	Price per Gal.	Labor Hours
Dehydratine 4	100	1.00	$ 2.29	1.3
Hydrocide No. 648	125 - 175	0.75	$ 1.81	1.3
Sikaseal	120	0.80	$ 3.52	1.3
(metric)				
Name of Material	Sq.M per Liter	Liters Per Sq.M.	Price per Liter	Labor Hours
Dehydratine 4	2.45	40.74	$ 0.60	1.3
Hydrocide No. 648	3.07 - 4.30	30.56	$ 0.48	1.3
Sikaseal	2.95	32.59	$ 0.93	1.3

No. of Cu.Ft. (Cu.M) of Stone to be Painted per 100 Sq.Ft. (10 Sq.M) of Surface Includes back, beds and end joints Average Thickness of Stone			
Thickness	Cu. Ft.	Thickness	Cu. M
4 "	20.0	100 mm	0.566
6 "	23.0	150 mm	0.651
8 "	25.0	200 mm	0.708
10 "	28.5	250 mm	0.807
12 "	31.0	300 mm	0.878

No. of Sq.Ft. of Painted Surface in One Cu.Ft. of Stone (No. of Sq.M Painted Surface per .0929 Cu.M of Stone) Includes back, beds and end joints Average Thickness of Stone			
Thickness	Sq. Ft.	Thickness	Sq. M.
4 "	5.0	100 mm	0.465
6 "	4.3	150 mm	0.122
8 "	4.0	200 mm	0.113
10 "	3.5	250 mm	0.099
12 "	3.2	300 mm	0.091

Back-Plastering Cut Stone and Marble. After the stone or marble has been set, it is customary to plaster the back of the stone with non-staining

1043

cement, using an ordinary trowel and applying it 3/16" (4.68 mm) to 1/4" (6.25 mm) thick.

The stone must be back-plastered as the setting progresses, as not to delay the bricklayers in backing up the stone, so the time required is rather indefinite. Working steadily, however, a mason should back-plaster 65 to 75 sq.ft. (6.03 to 6.96 sq.m) per hr. at the following labor cost per 100 sq.ft. (9.29 sq.m):

Description	Hours	Rate	Total	Rate	Total
Mason or stone setter	1.50	$	$	$ 36.93	$ 55.40
Cost per sq.ft.					$ 0.55
per sq.m					$ 5.96

07190 VAPOR BARRIERS/RETARDANTS

The combination of high indoor relative humidities and low outside temperatures causes condensation to form within the structure. Serious damage can result—rotting framing members, paint deterioration, and wet walls and ceilings.

Vapor barriers are recommended where these conditions exist. Vapor seal paper should be installed on the warm side of wall, floor, ceiling, or roof. Paper should be installed with joints running parallel to and over framing members. All joints should be lapped about 2" (50 mm). Where paper is to be exposed, nail wood lath strips over paper along framing members to provide a neat and permanent job.

A satisfactory vapor proof paper consists of a 50-lb. (22.68-kg) continuous asphalt film, faced on both sides with 30-lb. (13.60-kg) basis kraft paper. Other vapor proof papers consist of two 30-lb. (13.60-kg) basis sheets of kraft paper cemented together with asphalt and reinforced with strong jute cords, spaced 1/2" (12.50 mm) to 1" (25 mm) on centers and running in both directions. This same reinforced sheet is furnished with two sheets of asphalt coated paper cemented together and reinforced as described above. Another very satisfactory vapor proof paper consists of a heavy kraft paper coated on one side with a thin sheet of aluminum or copper, which prevents the penetration of moisture. Approximate prices on the various types of vapor seal paper are as follows:

1044

THERMAL AND MOISTURE PROTECTION

Price per Roll			
Description	Price Per Roll 500 Sq.Ft.	Price Per Roll 46.45 Sq.M.	Description
2 sheets 30-lb basis kraft paper, Cemented Together.	$ 8.00	$ 86.11	2 sheets 13.61Kg basis kraft paper.
2 sheets asphalt coated 30-lb kraft paper, Cemented tohether with asphlt and reinforced with jute cords.	$ 14.40	$ 155.01	2 sheets asphalt coated 13.62 Kg.
1 sheet heavy kraft paper coated two sides, with a reflective surface.	$ 12.00	$ 129.17	1 sheet heavy kraft paper coated two sides.

Sisalkraft. A strong, waterproof, windproof building paper consisting of 2 sheets of pretreated kraft paper, cemented together with 2 layers of special asphalt and reinforced with 2 layers of crossed sisal fibers. Costs about $0.20 per sq.ft ($2.15 sq.m).

Copper-Armored Sisalkraft is a combination of Sisalkraft and Anaconda electro-sheet copper bonded under heat and pressure. It is available in 3 weights, 1, 2, or 3 oz. copper per sq.ft. (0.91, 1.8, or 2.7 oz. per sq.m).

It may be used for waterproofing, flashings, ridge roll flashing, etc. The approximate prices are as follows: 1-oz. costs $0.30 per sq.ft. ($3.22 per sq.m); 2-oz. costs $0.50 per sq.ft. ($5.38 per sq.m), and 3-oz. costs $0.70 per sq.ft. ($7.53 per sq.m) Above prices subject to discount on large orders.

Polyethylene. Film comes in 100' (30.48 m) rolls of various widths in thicknesses of 0.002" (0.05 mm), 0.004" (0.1 mm), 0.006" (0.15 mm), and 0.008" (0.2 mm). Costs vary from $0.015 to $0.04 per sq.ft. ($0.16 to $0.43 per sq.m). Tape for sealing joints costs $3.25 for a roll 2" (50 mm) wide by 100' (30.48 m) long in the 0.004" (0.10 mm) thickness.

Labor Placing Vapor Seal Paper. Where vapor seal paper is applied to the interior of exterior stud walls, using a stapling machine, to prevent the penetration of moisture and condensation, a carpenter should handle and place 2,000 to 2,500 sq.ft. (185.8 to to 232.25 sq.m) per 8-hr. day, at the following labor cost per 100 sq.ft. (9.29 sq.m):

Description	Hours	Rate	Total	Rate	Total
Carpenter	0.40	$	$	$ 35.04	$ 14.02
Cost per sq.ft.					$ 0.14
per sq.m					$ 1.51

07200 INSULATION

To insulate is defined as "to make an island of". This "island" not only excludes the outside elements but retains those created within. Today, insulation has become part of a larger, more inclusive concern—energy conservation. Building design must go beyond the casual approach of letting aesthetic and use-function considerations dictate the design and then turning to technology to provide a satisfactory interior climate regardless of the initial and operating cost. The building envelope itself must be modified to contribute in every possible way to interior conditions of comfort-excluding the extremes of weather but welcoming those that pleasantly modify; retaining our manufactured interior weather but also providing means to allow it naturally to renew and maintain itself without resorting exclusively to mechanical means.

There are three stages where crucial decisions are made about the building and its environment. First, in the preliminary planning; second, in material selection; and finally, in the actual construction.

Preliminary planning considers the basics. First the building form should approximate a cube and have a minimum exposure to the elements or sprawl and take maximum advantage of natural light, solar heat, and breezes. Buildings should be sited to work with nature, not fight it. Fenestration should be concentrated to the south and tempered with overhangs engineered to exclude the sun when it is high in the summer sky, but to allow the rays to penetrate and add their warmth when the sun is low in the winter. Overhangs need not always be eaves but can be louvres, screens, balconies, porches, or a few well-spaced deciduous trees. Roof forms should not be chosen solely to exclude the elements but also to shade and vent the space above the top habitable floor. Wind conditions should be considered, doors protected, and windows placed to provide the optimum natural ventilation supplementing them with clear stories and skylights if need be. The earth itself is both a very cheap and a very good insulator and intelligent use of banking can pay large dividends.

Once the general arrangement has been determined, the second consideration is to select the best wall, ceiling, and floor materials for coping with the conditions they will face, and the best methods of installation to assure that the selected materials will perform to their maximum capabilities.

This discussion has concerned itself with new construction. One of the largest opportunities developing for the contractor is bringing existing buildings up to the energy efficient standards acceptable today. Most of the materials and procedures discussed in the following text apply to both new and remodeling work. However, the latter presents its own problems, and for that type of work, the reader is also referred to the chapter on Remodeling Work which deals with work on existing buildings.

THERMAL AND MOISTURE PROTECTION

The exterior and interior faces of a building will be chosen for appearance and wearability, and will by nature be hard and dense and therefore good conductors of heat. Also by nature insulating materials will be full of air pockets and be soft and easily damaged and need protection. Therefore most exterior envelopes will consist of three plies: an outside wearing surface, and intermediate space designed to interrupt the heat flow, and an interior wearing surface.

Even the exterior and interior plies will contribute some insulating quality. Materials are rated for their thermal resistance or their "R" value, the temperature difference between two exposed faces required to cause one BTU to flow through one square foot of the material per hour. "R" values can be added, one to the other, to arrive at a rating for the total wall, ceiling, or floor construction. The following list gives the "R" ratings for some of the more common building materials.

Material		R Value	Material
4 "	Face brick	0.44	100 mm
4 "	Common brick	0.80	100 mm
8 "	Poured concrete	0.64	200 mm
8 "	Cinder block	1.11	200 mm
8 "	Light weight block	2.00	200 mm
4 "	Concrete slab	0.32	100 mm
1/2 "	Plywood	0.47	12.5 mm
1/2 "	Gypsum board	0.45	12.5 mm
1/2 "	Gypsum plaster	0.32	12.5 mm
1" x 8 "	Drop siding	0.79	25 x 200 mm

Even from the above small sampling, one can see that facing materials can be chosen to contribute to the insulating quality of the envelope. In addition, air spaces of over 3/4" (18.75 mm) should also be added at an average of R = 0.91, depending on the position of the air space and whether the air movement is up or down. Also in choosing exterior facings, colors should be considered for their ability to reflect; the darker the color, the more heat absorbed. And finally the jointing of all materials must be carefully studied to eliminate infiltration. A study of one apartment tower found 50% of the heat loss was due to infiltration around the windows!

Heat transfer through the building enclosure is by three means: convection, conduction, and radiation.

Convection is the thermally produced upward and downward movement of air. Warm air rises, cold air falls. The concern here is to construct a total building envelope that blocks air currents. Hollow walls should be solidly blocked top and bottom at each story, and floor and roof construction should be continuously sealed from wall construction so air conditions in the exterior walls will not be able to spread out over the interior. This can be done either by extending the interior facing or by adding insulation or a combination of both.

Conduction is the transmission of heat through a material. The rate of conductance of a material or combination of materials, known as its "U" factor, is the BTU per hour per inch of thickness per square foot per degree temperature difference. The "R" ratings discussed previously are the reciprocals of conductance or 1 divided by the "U" factor. Most common insulating materials are manufactured to be as poor a conductor of heat as possible. However, their positioning within the wall can also make them block convection and when enclosed in reflective coverings and separated from adjacent construction by at least 3/4" (18.75 mm) they also can cut down on the third means of heat transfer, radiation.

Radiation is the emission of energy from a surface. Bright surfaces, such as aluminum foil, are good reflectors and have low emissive coefficients and are therefore poor absorbers of heat. The interior faces of the two outer plies of building envelope tend to be dark, and the heat will radiate from one surface to the other constantly unless interrupted. By inserting one or more layers of reflective surfaces, the heat will be reflected back to the surface from which it escaped. Installation is crucial to the success of a radiant barrier, because the reflective surface will be highly conductive. If allowed to come into contact with either side of the enclosing construction, it will speed heat transfer rather than retard it.

Infiltration, as discussed above, will generally be solved by good detailing at the juncture of one material to another and by caulking. Strips of insulation, such as under sill plates, and loose wools stuffed into place, can also plug air leaks.

There is no simple answer as to how much insulation should be added. Both comfort and the additional costs must be considered. Minimum accepted "R" values in the northern regions have been R19 for ceilings, R13 for floors and R11 for walls. These drop to R13, R9, and R8 in milder areas but should never be less than R9 for ceilings and R7 elsewhere. Increases in fuel costs could always make these figures far too low and R24 for ceilings, R19 for floors and R13 for walls might be recommended. But keep in mind that each additional inch (25 mm) of insulation beyond a certain point reduces the heat loss less than the one before it. There is a point of sharply diminishing returns. To determine the best "R" values for a particular project, one must consider the following:

1. Determine all wall, ceiling, and floor construction types and give each its proper "R" value without insulation.
2. Determine the heat loss (or heat gain for cooling) without insulation through each of the above.
3. Choose an "R" rating for walls, ceiling, and floors based on averages for existing climate and consistent with the type of construction contemplated.

4.   Refigure heat loss with new "R" values.
5.   Figure fuel savings of d) over b). Several alternate fuels might be considered and local utilities can be helpful in determining these costs.
6.   Figure the difference in cost of the insulated construction over the uninsulated, remembering to deduct for savings in size of the mechanical equipment, if any.
7.   Determine how long it will take savings in fuel to pay for insulation costs. This is the payback period and will have to take many things into consideration such as future fuel costs, and the interest on the additional mortgage to finance the insulation. If the payback period looks attractive, the whole procedure can be repeated with a higher "R" value assigned, and this can be repeated until a point of diminishing returns is evident.

Once the desired "R" value is established, the selection of the best type of insulation can begin. The possibilities include:

1.   Blankets and batts, with or without reflective or vapor barriers.
2.   Rigid board type, which may also serve as sheathing or lathing.
3.   Sprayed on or foamed in place, which may also serve as sound or fire retardants.
4.   Poured fill
5.   Reflective barriers

Many of these may be used in combination with the other.

07210 BUILDING INSULATION BLANKET AND BATT INSULATION

Insulation formed into batts or rolls is made in a number of materials, such as stone, slag, glass wools, vegetable and cotton fibers, and suspended pulps. They are all light in weight, easily cut and handled, and need support to stay in place. They are usually encased in either kraft paper or foil and can be used as a vapor barrier. The enclosing material is usually made with flapped edges for easy attachment to studs and joists. Units may be ordered unfaced and fitted between studs by friction. Unfaced units are most often used as a second layer of insulation in attic spaces as they do not form a second vapor barrier if the original layer has one. A narrow form of unfaced blanket 1" (25 mm) thick is made for inserting under sill plates. It will compress to 1/32" (0.78 mm).

Most batt and blanket materials have "R" values of around 3.5 to 3.7 per inch (25 mm) of thickness. The 3" (75 mm) units will provide an "R" of 11, 3-1/2" (87.50 mm) of 13 and 6" (150 mm) of 19. The 6" (150 mm) units are commonly used in attics and fitted between the joists; but more and more, they are used in walls with 2" x 6" (50 x 150 mm) studs placed 24" (600 mm) o.c. in place of the standard 2 x 4 (50 x 100 mm) 16" (400 mm) o.c. If foil faced units are used and an air space of at least 3/4" (18.75 mm) maintained between the foil and the warm side of the room, the "R" rating can be increased some 15%.

Most glass fiber insulation is distributed on a nationwide basis, but other materials are often produced regionally and prices may vary widely. Widths are 15" (375 mm) and 23" (575 mm) to fit normal 16" (400 mm) and 24" (600 mm) stud and joist spacing. Batts come 48" (1200 mm) and 96" (2400 mm) long, blankets in rolls of from 24' (7.31 m) to 40' (12.19 m), often two rolls per package. The square foot cost for foil faced units 3-1/2" (87.50 mm) thick is $0.24; 6" (150 mm) thick is $0.40. Units bound in kraft paper may run a couple of cents less and unfaced units even less. Sill sealer rolls come in 50' (15.24 m) rolls 1" (25 mm) thick and in 3-5/8" (90.62 mm) and 6" (150 mm) widths and cost $0.15 and $0.22 per lin.ft ($0.49 and $0.72 per m).

Because these materials are light and precut to fit standard construction, one carpenter can handle the installation. Friction fit type, with no attachments, will install at the rate of 2000 sq.ft. (185.80 sq.m) per day. Flanged type bound in kraft paper or foil will install at 1800 sq.ft. (167.22 sq.m) per day.

Labor Cost of 100 sq.ft. (9.29 sq.m) Faced Insulation

Description	Hours	Rate	Total	Rate	Total
Carpenter	0.45	$	$	$ 35.04	$ 15.77
Cost per 100 sq.ft.					$ 15.77
per sq.ft.					$ 0.16
per sq.m					$ 1.70

07211 LOOSE FILL INSULATION

Loose fill insulation includes mineral wool, which is molten rock extruded by air and steam into fibers, known as blowing wool for machine applications, or nodules for pouring or spreading by hand; and expanded volcanic rocks such as vermiculite or perlite. The latter are most often specified for filling concrete block or cavity walls.

Loose Insulating Wool. Is suitable for any purpose where insulation can be packed by hand between ceiling joists or side wall studding. For maximum results, it is recommended that wool be applied full thickness of side wall studding and approximately 4" (100 mm) to 16" (400 mm) over ceiling areas.

The covering capacity varies considerably with the density to which it is packed. The National Bureau of Standards conductivity figure for glass or rock wool is 0.27 Btu at a density of 10 lbs. per cu.ft. (162 kg per cu.m).

1050

THERMAL AND MOISTURE PROTECTION

The covering capacity of bulk wool as given by most manufacturers is based on a density of 6 to 8 lbs. per cu.ft. (96.11 to 128.14 kg per cu.m).

No. of Sq.Ft. (Sq.M) of Surface Covered By One Bag of Loose Insulating Wool Weighting 40 Lbs. (18.14 Kg) and Containing 4 Cu.Ft. (.11 CuM.) Not Including Area Covered by Studs or Joists							
Density, Lbs. per Cu.Ft.	Actual Cu. Ft.	Actual Thickness of Loose Insulating Wool in Inches					
		1	2	3	3 1/2	3 5/8	4
6	6.67	80.0	40.0	26.6	22.8	22.0	20.0
7	5.72	68.5	34.4	22.9	19.7	18.8	17.2
8	5.00	60.0	30.0	20.0	17.2	16.6	15.0
9	4.45	53.4	26.7	17.8	15.2	14.7	13.4
10	4.00	48.0	24.0	16.0	13.7	13.3	12.0
Metric							
Density, kg per Cu.M	Actual Cu. M	Actual Thickness of Loose Insulating Wool in mm					
		25.00	50.00	75.00	87.50	90.63	100.00
96.49	0.19	7.43	3.72	2.47	2.12	2.04	1.86
112.57	0.16	6.36	3.20	2.13	1.83	1.75	1.60
128.65	0.14	5.57	2.79	1.86	1.60	1.54	1.39
144.73	0.13	4.96	2.48	1.65	1.41	1.37	1.24
160.81	0.11	4.46	2.23	1.49	1.27	1.24	1.11

No. of S.F. (Sq.M) of Surface Covered By One Bag of Loose Insulating Wool Weighting 40 Lbs (18.14 Kg) and containing 4 Cu.Ft. (.11 Cu.m.) Including Area Covered by Studs or Joists							
Density, Lbs. per Cu.Ft.	Actual Cu. Ft.	Thickness of Loose Insulating Wool in Inches					
		1	2	3	3 1/2	3 5/8	4
6	6.67	85.00	42.50	28.40	24.30	23.50	21.20
7	5.72	73.00	36.50	24.30	20.80	20.00	18.30
8	5.00	63.80	31.90	21.30	18.30	17.60	16.00
9	4.45	56.70	28.40	18.90	16.20	15.70	14.20
10	4.00	51.00	25.50	17.00	14.50	14.10	12.70
Metric							
Density, kg. per Cu.M	Actual Cu. M	Thickness of Loose Insulating Wool in mm					
		25.00	50.00	75.00	87.50	90.63	100.00
96.49	0.19	7.90	3.95	2.64	2.26	2.18	1.97
112.57	0.16	6.78	3.39	2.26	1.93	1.86	1.70
128.65	0.14	5.93	2.96	1.98	1.70	1.64	1.49
144.73	0.13	5.27	2.64	1.76	1.50	1.46	1.32
160.81	0.11	4.74	2.37	1.58	1.35	1.31	1.18

Loose or bulk rock wool is usually in bags weighing 40 lbs. (18.14 kg) and containing 4 cu.ft. (0.11 cu.m). Average price $4.50 to $7.00 per bag.

Granule or Pellet Type Insulating Wool. Granule or pellet type insulating wool is furnished in particles sufficiently large that the material will not sift or dust through wall or ceiling cracks and is used for insulating ceiling areas between supporting joists, or void wall spaces where accessible and which permit pouring the pellets into place, particularly in old house construction.

Granule or pellet type insulating wool is usually furnished in bags weighing 40 lbs. (18.18 kg) and is placed at a density of 7 to 8 lbs. per cu.ft. (112.12 to 128.14 kg per cu.m).

The tables for loose or granular insulating wool, based on 7 to 8 lbs. per cu.ft. (112.12 to 128.14 kg per cu.m) can be used for estimating quantities obtainable per bag.

Granule or pellet type insulating wool in bags or cartons weighing 40 lbs. (18.18 kg) cost about $8.00 each.

Labor Placing Loose or Bulk Insulating Wool. When placing loose or bulk insulating wool between wood studs, a worker should place about 350 to 450 sq.ft. (32.51-41.80 sq.m) per 8-hr. day, at the following labor cost per 100 sq.ft. (9.29 sq.m):

Description	Hours	Rate	Total	Rate	Total
Labor	2.00	$	$	$ 26.16	$ 52.32
Cost per 100 sq.ft.					$ 52.32
per sq.ft.					$ 0.52
per sq.m					$ 5.63

Labor Placing Granule or Pellet Type Insulating Wool Between Ceiling Joists. Where granule or pellet type insulating wool is poured between ceiling joists and other open spaces to a thickness of 3-1/2" (87.50 mm) or 3-5/8" (90.62 mm), a worker should place 700 to 900 sq.ft. (65.03 to 83.61 sq.m) per 8-hr. day, at the following labor cost per 100 sq.ft. (9.29 sq.m):

Description	Hours	Rate	Total	Rate	Total
Carpenter	1.00	$	$	$ 35.04	$ 35.04
Cost per 100 sq.ft.					$ 35.04
per sq.ft.					$ 0.35
per sq.m					$ 3.77

Vermiculite Or Perlite Loose-fill Building Insulation. Vermiculite loose-fill insulation is used to insulate attics, lofts, and side

1052

walls. It is fireproof, not merely "fire resistant". The fusion point is 2200° to 2400°F (1200° to 1315°C). It is completely mineral and does not decompose, decay, or rot. Because of its granular structure, vermiculite is free-flowing, assuring a complete insulation job without joints or seams. Rodents cannot tunnel into it, and it does not attract termites or other vermin. Vermiculite is a non-conductor and is excellent protection around electrical wiring.

Vermiculite loose-fill is marketed in 4-cu.ft. bags (.11 cu.m). Approximate coverage per bag, based on joists spaced 16" (400 mm) o.c., is 14 sq.ft. (1.30 sq.m) for a 3-5/8" (90.62 mm) thickness; 9 sq.ft. (0.83 sq.m) for 5-1/2" (137.50 mm) thickness. A bag will cost about $18.00

Approximate Coverage of Vermiculite Fill in Cavity and Block Walls						
Wall Area Sq.Ft.	No. of Bags Required for Various Wall Types					Wall Area Sq.M.
	1 " cavity	2 " cavity	2 1/2 " cavity	8 " block	12 " block	
100	2	4	5	7	13	9.29
500	10	20	25	34	63	46.45
1,000	21	42	50	69	125	92.90
5,000	104	208	250	545	625	464.50
10,000	208	416	500	1,090	1,250	929.00
	25 mm	50 mm	62.5 mm	200 mm	300 mm	

One mason can pour around 50 bags, 200 cu.ft. (5.66 cu.m) per 8-hr. day, at the following labor cost:

Description	Hours	Rate	Total	Rate	Total
Mason	8.00	$	$	$ 36.93	$ 295.44
Cost per cu.ft.					$ 1.48
per cu.m					$ 52.16

When adequate ventilation is provided in attics or similar spaces, it is not necessary to use a vapor barrier in the ceiling construction below vermiculite fill insulation. A vapor barrier is recommended on the warm side of insulated exterior walls, except when the insulated space is ventilated to permit movement of water vapor and air from the space.

Reinsulation. Where an existing thickness of blanket or other insulation is inadequate, vermiculite can be poured over it to obtain the thickness desired.

Rigid insulation boards are made of expanded polystyrene, often called "beadboard"; extruded polystyrene sold as Styrofoam; urethane; glass fibers; glass foams; and wood and vegetable fibers. Many rigid boards serve dual purposes such as sheathing, lath, and even interior finish. Many are also used for roof decks and perimeter insulation, discussed separately below.

Urethane boards have the highest "R" value per inch (mm) of thickness, 7.14, but are flammable and must be covered over, or treated. The cost is around $0.40 per sq.ft. ($4.30 per sq.m) plain, $0.55 per sq.ft. ($5.92 per sq.m) treated. The sizes available are many, including the standard 4' x 8' (1200 x 2400 mm) sheet and in thicknesses from 1/2" (12.50 mm) to 24" (600 mm). Because of its high insulating value per thickness, it is ideal for use in cavity walls and for applying to the inside face of an exterior masonry wall. In cavity walls it is applied against the outside face of the inside wythe, leaving a clear air space for any penetrating water to drain out.

Where urethane board is used against exterior masonry walls, the surface must be clean and even. Nailers must be applied wherever wood trim will be added, such as at the base and around doors and windows. The board is then applied to the wall with sufficient mastic to cover at least 50% of the back when shoved in place. Some manufacturers supply special channel systems for mechanical attachment as an alternative. The urethane board can then be covered with plasterboard or may be plastered direct. Because of the fire hazard, for whichever method chosen, the interior finish coat must be run up above the finished ceiling to cover all portions of the urethane, as it is the fireproofing for it. Extruded polystyrene, or Styrofoam, under which name it is marketed by Dow Chemical Company, also has a high "R" value, 5.4 per inch (25 mm) of thickness. It costs about $0.45 per sq.ft. ($4.84 per sq.m) in 1" (25 mm) thickness. It is installed in the same locations and by the same methods as urethane board and should also have a continuous interior finish that will protect it from fire.

Polystyrene, which is molded, is known as beadboard. It has a lower "R" rating, 3.85, and is absorbent, but it is much cheaper, about $0.30 per sq.ft. ($3.23 per sq.m) in 1" (25 mm) thickness. It too must be covered with a fireproofing finish.

Glass fiber boards come in various densities. A 1" (25 mm) board in a medium weight gives an "R" of 4.35 and costs about $0.45 per sq.ft. ($4.84 per sq.m).

The stronger rigid boards may also be used for sheathing on wood frame construction. But a 1" (25 mm) board in combination with standard 3-1/2" (87.50 mm) batts will give a wall with the highly desirable rating of R19 or better and may be preferable to using 6" (150 mm) batts and 6" (150 mm) studs to gain such a rating.

Rigid board is light, easily cut, and can be installed by one worker, who should erect about 1,500 sq.ft. (139.35 sq.m) per day depending on the amount of fitting.

Labor Cost of 100 sq.ft. (9.29 sq.m) of 1" (25 mm) Rigid Board
Applied to a Masonry Wall

Description	Hours	Rate	Total	Rate	Total
Carpenter	0.54	$	$	$ 35.04	$ 18.92
Cost per sq.ft.					$ 0.19
per sq.m					$ 2.04

07213 REFLECTIVE INSULATION

Some reflective insulation can be incorporated as the enclosing sheet on batt and blanket insulation and as a backing on rigid insulation. Also available are sheets of insulating material depending entirely on the reflective principle and consisting of alternate layers of trapped air and foil, the outer layers of which are usually laminated to a backing sheet of kraft paper for extra strength and which are extended at the edges to form flaps for stapling. Another type depends on an inner layer or layers of accordion foils. Such insulation must be held away from the inner and outer wall finishes at least 3/4" (18.75 mm) so that there will be an uninterrupted air space which serves the dual purpose of allowing the foil surface to reflect escaping heat (or entering cold) back to where it came from and keeping the highly conductive foil from transmitting the heat by contact.

A typical lamination would consist of three layers: the two outer sheets of foil laminated to a backing sheet and separated from a center sheet of pure foil, usually aluminum, with sealed air spaces both sides. The unit is held together with flanges extended to allow stapling to studs and joists. The "R" value will depend on the placement of the material, because reflective insulation is more effective with the heat flow down than with heat flow up. Where the heat flows down from a hot roof to a ceiling below, 93% of the heat transfer is by radiation and only 7% by conduction. As heated air always travels up, no convection is directed toward the ceiling. Considering the reverse travel, heated interior air travelling through the ceiling to the cold roof above, 50% will be transmitted by radiation, the same 7% by conduction and the remainder by convection. Heat transfer through the side walls will be 65% to 80% by radiation, 15% to 28% by convection, and 5% to 7% by conduction. The "R" rating then on a three-ply, 2" (50 mm) thick reflective sheet will be 26.18 down, 9.52 up and 14.65 through the wall for a mean "R" of 17.85. When we speak of this as 2" (50 mm) thick, it is the distance from foil to foil. Another 3/4" (18.75 mm) minimum must be added to each side to isolate the foil from other wall construction so that a full 3-1/2" (87.50 mm) must be reserved for the reflective insulation.

One-inch (25 mm) thick units with just one inner and one outer layer of reinforced foil separated by a 1" (25 mm) air space will give an "R" value of 21.79 down, 7 up, and 10.72 through the wall for a mean "R" of 14.4.

Single sheets with aluminum foil either side of a reinforcing sheet with no integral air space, if properly installed to preserve air spaces to adjacent construction, will give "R" values of 14.28 down, 3.45 up, and 7 through the wall, for a mean "R" of 9.

The cost of two-ply material is $0.24 per sq.ft. ($2.58 per sq.m); three-ply is $0.30 per sq.ft. ($3.23 per sq.m); and five-ply is $0.40 per sq.ft. ($4.30 per sq.m).

The material is preformed to fit average stud and joist spacing and is very light. One worker can install 2200 sq.ft. (204.38 sq.m) per day if the wall or ceiling space is clear, at the following labor cost per 100 sq.ft. (9.29 sq.m):

Description	Hours	Rate	Total	Rate	Total
Carpenter	0.66	$	$	$ 35.04	$ 23.13
Cost per sq.ft.					$ 0.23
per sq.m					$ 2.49

07214 FOAMED INSULATION

Foamed-in-place insulations are very efficient, and the work is usually subcontracted to an experienced applicator. Foamed insulation will adhere well to most surfaces, will conform to irregular forms, and will set up quickly. The plastic foams flow better into restricted areas than the sprayed-on fibers, which have longer fibers and tend to clog in small spaces.

Application will require special equipment, and the space must be well ventilated. Some foams expand and will set up stresses in restricted areas if care is not taken. Despite these drawbacks, the resulting "R" can be as much as 7.7 per inch (25 mm) and will apply at the rate of 800 sq.ft. (74.32 sq.m) per day with a three-worker crew. Material costs for urethane will run about $0.45 per inch per sq.ft. ($4.88 per 25 mm per sq.m). If subcontracted, a budget figure of $1.00 per sq.ft. ($10.76 per sq.m) for 1" (25 mm) and $2.90 per sq.ft. ($31.21 per sq.m) for 2" (50 mm) should be adequate.

07215 SPRAYED INSULATION

Sprayed fibers of rock wool, glass, gypsum, perlite, and cellulose will have "R" values in the 4 to 5 range. Special binders are required as well as special equipment, so this work, too, will be subcontracted. A two-worker crew will apply some 1200 sq.ft. (111.48 sq.m) of 1" (25 mm) thickness per day, possibly more on wide open ceilings. Material costs will be about $0.40 to 0.50 per sq.ft. ($4.30 to 5.38per sq.m) per 1" (25 mm) of material. If

1056

subcontracted, a budget figure of $2.00 per sq.ft. ($21.52 per sq.m) should be adequate.

07230 HIGH AND LOW TEMPERATURE INSULATION

Rigid foam insulation is composed of expanded polystyrene, glass, or polyurethane, which bound together into rigid board, has largely replaced corkboard for low temperature insulation. It is also used in general building for roofs, because it is quite strong, tough, and resilient; for curtain wall back up because it is lightweight; for core walls as it is self-supporting; and for perimeter and crawl space insulation as it has a low water absorption and a zero rating for capillary action. In addition, it forms a vapor barrier, is dimensionally stable, incombustible, and easily cut. Rigid foam insulation can be placed on concrete and masonry surfaces by adhering units with hot asphalt or with gobs of cement.

Units come in 1" (25 mm), 1-1/2" (37.50 mm), 2" (50 mm), 3" (75 mm), 4" (100 mm), and 6" (150 mm) thicknesses and sizes from 12"x18" (300 x 450 mm) to 36"x36" (900 x 900 mm) in packaged forms and up to 9'-0" (2.74 m) lengths when shipped in carload lots. There are no standard sizes within the industry, and sizes, costs, and insulating values must be carefully checked with the manufacturers.

Back Plaster. Where insulation is to be applied against masonry walls in hot asphalt, the walls must be made true and even with portland cement plaster. The parge coat or back plaster may be omitted if the wall is of poured concrete.

Back plaster should be made by mixing 1 part portland cement and 3 parts clean sharp sand. Back plaster 1/2" (12.50 mm) thick requires 1.5 bags portland cement and 4.6 cu.ft. (0.13 cu.m) of sand per 100 sq.ft. (9.29 sq.m).

Material Cost of One C.Y. (0.76 Cu.M) of 1:3 Portland Cement Back Plaster

Description	Rate	Total	Rate	Total
9.5 Sacks Portland Cement	$	$	$ 6.00	$ 57.00
1.10 cu.yd. (0.84 cu.m) Sand	$	$	$ 9.75	$ 10.73
Cost per cu yd.				$ 67.73
Cost per cu.ft				$ 2.51
per cu.m				$ 88.57

Asphaltic Priming Coat. All concrete or plaster surfaces to which insulation is to be applied in hot asphalt must be primed with asphaltic priming coat made especially for this purpose.

Asphalt. Using the proper grade of asphalt is very important. It must be odorless, as any material with an odor will taint foodstuffs that may be stored in insulated rooms. It should be an oxidized asphalt of 180°F to 200°F (82°C to 93°C) melting point and should preferably be made from a Mexican base oil. The hot asphalt in which corkboard is dipped should be held between 350° and 375°F (176°C and 190°C). Do not heat asphalt to higher temperatures than these. Most manufacturers recommend an asphalt for use with their product.

On dipping floor, wall, and ceiling insulation, allow 0.4 lbs. of asphalt per sq.ft. (1.95 kg per sq.m) of insulation per layer. Mop coating on floors or top of ceiling requires 0.8 lbs. per sq.ft. (3.90 kg per sq.m).

Cold Erection Plastic. Where the use of hot asphalt for the erection of foamboard is not practical, use cold erection plastic on floors, walls, and ceilings. A coat 1/16" (1.56 mm) thick requires 1 gal. (3.785 liters) of cold erection plastic per 25 sq.ft. (2.32 sq.m).

Nails or Impregnated Skewers. Use nails for the attachment of the first layer of insulation to wood and frame construction, for the attachment of the second layer to wood stripped concrete ceilings, or for toenailing the first layer of solid cork walls. All nails should be of galvanized wire and should be 1" (25 mm) longer than the thickness of the insulation. Where two courses of insulation are erected in hot asphalt or cold erection plastic, impregnated hardwood skewers should be used to additionally secure the second course in position. Do not use either nails or skewers when erecting the first course against brick, stone, or concrete walls. Hot asphalt or cold erection plastic is all that is required.

Erection. The requirements (material and labor) for installing one layer of 2" (50 mm) thick insulation, using the hot asphalt dip method of setting, on a masonry wall surface are given in the following tables. Bear in mind that the labor times given are for straight, flat surfaces; and that any irregularities such as columns, pilasters, or beams will require additional labor time due to cutting and setting smaller pieces. If the insulation is thicker than 2" (50 mm), then adjust for the difference in material costs and add about 1/4 hour to both mechanic and helper time per 1" (25 mm) thickness increase over 2" (50 mm).

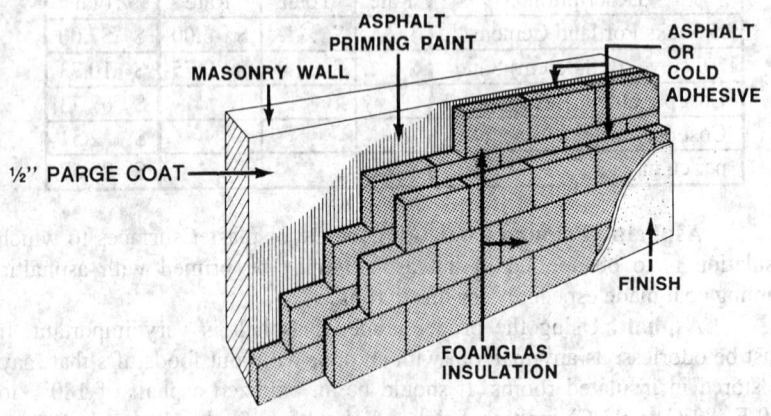

THERMAL AND MOISTURE PROTECTION

Low Temperature Insulation using Two Layers of
Pittsburgh Corning Foamglas

2" (25 mm) Thick Single Insulation Layer, Hot Dip Method Requirements per 100 Sq.Ft. (9.29 Sq. M.)					
Operation Masonry Wall	Material Required	Mechanic Time Hours	Helper Time Hours	Operation Masonry Wall	Material Required
Parge Coat - 1/2"	5.5 cu.ft. mortar	3	3	Parge Coat - 12.50 mm	0.16 cu.m.
Asphalt Prime Coat	1 gal. primer	1	------	Asphalt Prime Coat	3.785 Liter
Hot Asphalt	40 lbs.	------	------	Hot Asphalt	18.14 Kg
Applying Insulation	100 sq.ft.	1.5	1.5	Applying Insulation	9.29 sq.M.

For applications against concrete walls, the parge coat is not required.

Material Cost of 100 S.F. (9.29 Sq.M) of Single Layer Insulation, Set in Hot Asphalt on Masonry Walls					
Description	Rate	Total	Rate	Total	Metric
5.5 cu.ft. cement plaster	$	$	$ 2.67	$ 14.70	0.15 cu.m
1 gal asphalt primer	$	$	$ 3.50	$ 3.50	3.785 Liter
40 Lbs Asphalt	$	$	$ 0.23	$ 9.10	18.14 Kg
100 sq.ft. 2" corkboard	$	$	$ 1.46	$ 145.50	9.29 Sq.M 2" Corkboard
Cost per 100 sq.ft.				$ 172.80	
per sq.ft.				$ 1.73	
per sq.m.					$18.60

Labor Cost of 100 Sq.Ft. (9.3 Sq.M) of Single Layer Insulation, Set in Hot Asphalt on Masonry Walls					
Description	Hours	Rate	Total	Rate	Total
Plaster Coat					
Mechanic	3.00	$	$	$ 37.89	$ 113.67
Helper	3.00	$	$	$ 26.16	$ 78.48
Prime Coat					
Mechanic	1.00	$	$	$ 37.89	$ 37.89
Insulation					
Mechanic	1.50	$	$	$ 37.89	$ 56.84
Helper	1.50	$	$	$ 26.16	$ 39.24
Cost 100 sq.ft.					$ 326.12
per sq.ft.					$ 3.26
per sq.m					$ 35.10

Finishes. Corkboard insulation should receive a finish that will give long service under the temperature and moisture conditions common to cold storage rooms. The finish should bond securely to the corkboard and be reasonably resistant to bumps and abrasion. Cost, good appearance, and availability should also be considered. These requirements are best met by asphalt emulsion finishes or portland cement plaster. Asphalt emulsion finishes are generally used on ceilings and portland cement plaster on walls. Rigid foam insulations may be sprayed or brushed with special white vinyl emulsion paints, aluminum paint, tile set with adhesive, or cement plaster over mesh. Direct plastering should not be considered.

Asphalt emulsion finishes shall be applied directly to corkboard in two coats, each 1/8" (3.1 mm) thick. Before applying the first coat, all voids, open joints, or broken corners must be pointed with emulsion finish. The first coat must be hand dry before the second coat is applied. Trowel the second coat smooth. The final thickness of the finish shall be approximately 1/8" (3.12 mm) when dry. Finish all corners true and clean. The emulsion finishes must be thoroughly dry before any paint is applied, to guard against cracking.

Approximately 5 gals. (18.92 liters) of asphalt emulsion, at a material cost of about $15.00, will be required for the application of the two coats on 100 sq.ft. (9.29 sq.m) of surface, at the following labor costs:

THERMAL AND MOISTURE PROTECTION

Description	Hours	Rate	Total	Rate	Total
First Coat					
Mechanic	2.00	$	$	$ 37.89	$ 75.78
Second Coat					
Mechanic	2.00	$	$	$ 37.89	$ 75.78
Cost 100 sq.ft.					$ 151.56
per sq.ft.					$ 1.52
per sq.m					$ 16.31

The labor cost of 100 sq.ft. (9.3 sq.m) of 1/2" (12.5 mm) of 1:3 portland cement plaster should average as follows:

The labor cost of 100 Sq.Ft. (9.29 sq.m) of 1/2" (12.50 mm) of 1:3 portland cement plaster should average as follows:					
Description	Hours	Rate	Total	Rate	Total
Mechanic	3.00	$	$	$ 37.89	$ 113.67
Helper	3.00	$	$	$ 26.16	$ 78.48
Cost per 100 sq.ft.					$ 192.15
per sq.ft.					$ 1.92
per sq.m					$ 20.68

Portland cement plaster should be mixed in the proportions of 1 part portland cement, 3 parts clean, screened sand, and 5% hydrated lime.

To reduce cracking to a minimum, finish coat of plaster may be scored each way approximately 4'-0" (1200 mm) on centers.

Some manufacturers state that plaster can be applied directly to insulation; others specify wire mesh must be applied, particularly on ceiling work. As rigid foam insulation does not absorb water, plaster coats will dry very slowly. It is also necessary to add the cost of nailer strips for casings, trim, and bases, if any occur. Light trim such as beads may be stapled on.

07240 ROOF AND DECK INSULATION

Roof and deck insulation is often of the same material as rigid board insulation, but the units are generally furnished in smaller sizes more suitable for handling in exposed conditions. In addition to glass fiber, perlite, urethane and polystyrene boards, wood and mineral fiber boards costing as little as $0.50. per sq.ft. per 1" of thickness ($5.38 per sq.m per 25 mm), and foam glass blocks are often specified. Poured decks are also common, but these have been covered at the end of the Concrete chapter. Foamglas (from Pittsburgh Corning) is a unique material that is inorganic, incombustible, dimensionally stable, and water and vapor proof, and weighs only 8.5 lb. per cu.ft. (136.14 kg per cu.m) but is strong enough to be used under parking and promenade decks. It can also be ordered with 1/8" per ft. (3.12 mm per 0.30 m) slope. Minimum thickness is 1-1/2" (37.50 mm), and it is available in 1/2" (12.50 mm) increments up to 4" (100 mm). The 1-1/2" (37.50 mm) material has an "R" value of 4.2 and costs about $1.20 per sq.ft. ($12.91 per sq.m) in 1-1/2" (37.50 mm) thickness. One roofer will install some 800 sq.ft. (74.32 sq.m) per day at the following labor cost:

Labor Cost of 100 s.f. (9.29 sq.m) of 1" (25 mm) Deck Insulation

Description	Hours	Rate	Total	Rate	Total
Roofer	1.00	$	$	$ 30.93	$ 30.93
per sq.ft.					$ 0.31
per sq.m					$ 3.33

07250 PERIMETER INSULATION

Board forms of polystyrene, urethane, fiberglass, and cellular glass are commonly employed as insulation on foundation walls and under slab edges to insulate an on-grade floor construction at the outer wall. Material costs are from $0.25 to $0.50 per sq.ft. ($2.69 to $5.38 per sq.m) in 1" (25 mm) thickness.

As perimeter insulation must fit the outline of the interior face of the outside wall and must be run continuously around offsets, perimeter ducts, column foundations and other irregularities, unit costs will vary widely. In general, insulation is run down 24" (600 mm) below outside grade. In addition, it may be run back 24" (600 mm) under the outer edge of the slab. One carpenter should apply around 800 sq.ft. (74.32 sq.m) per day on straight run work, at the following labor cost per 100 sq.ft. (9.29 sq.m):

Description	Hours	Rate	Total	Rate	Total
Carpenter	1.14	$	$	$ 35.04	$39.95
Cost per sq.ft.					$0.40
per sq.m					$4.30

07300 ROOF SHINGLES AND ROOFING TILES

Roofing is estimated by the square, containing 100 sq.ft. (9.29 sq.m). The method used in computing the quantities will vary with the kind of roofing and the shape of the roof.

The labor cost of applying any type of roofing will be governed by the pitch or slope, size, plan (whether cut up with openings, such as skylights, penthouses, gables, dormers, etc.), and on the distance of the roof from the ground.

Rules for Measuring Plain Double Pitch or Gable Roofs. To obtain the area of a plain double pitch or gable roof as shown in Figure 1,

1062

multiply the length of the ridge (A to B), by the length of the rafter (A to C). This will give the area of one-half the roof. Multiply this by 2 to obtain the total sq.ft. (sq.m) of roof surface.

Example: Assume the length of the ridge (A to B), is 30'-0" (9.14 m) and the length of the rafter (A to C), 20'-0" (6.09 m):

   (A to B) x (A to C)
   30'-0" x 20'-0" = 600 sq.ft.   (9.14 x 6.09 = 55.66 sq.m)
   600 sq.ft. x 2 = 1,200 sq.ft.   55.66 x 2 = 111.32 sq.m

Rules for Measuring Hip Roofs. To obtain the area of a hip roof as shown in Figure 3, multiply the length of the eaves (C to D) by 1/2 the length of the rafter (A to E). This will give the number of sq.ft. (sq.m) of one end of the roof, which multiplied by 2 gives the area of both ends. To obtain the area of the sides of the roof, add the length of the ridge (A to B) to the length of the eaves (D to H). Divide this sum by 2 and multiply by the length of the rafter (F to G). This gives the area of one side of the roof and when multiplied by 2 gives the number of sq.ft. (sq.m) on both sides of the roof.

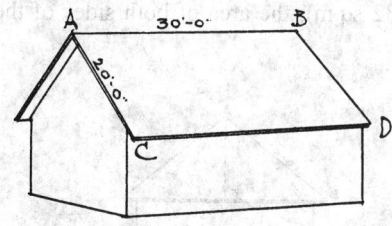

Fig 1. Plain Double Pitch or Gable Roof

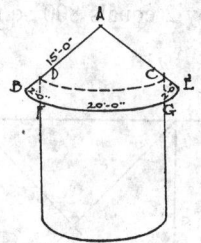

Fig 2. Conical Building and Roof

To obtain the total number of sq.ft. (sq.m) of roof surface, add the area of the two ends to the area of the two sides. This total divided by 100 (or sq.m divided by 9.29) equals the number of squares in the roof.

Example: Assume the length of the eaves (C to D) is 20'-0" (6.09 m) and the length of the rafter (A to E) is 20'-0" (6.09 m):

   (C to D) x 1/2 the length of the rafter (A to E) =
   20'-0" x 10'-0" = 200 sq.ft.   (6.09 x 3.04 = 18.51 sq.m)
   200 x 2 = 400 sq.ft.   (18.51 x 2 = 37.02 sq.m)

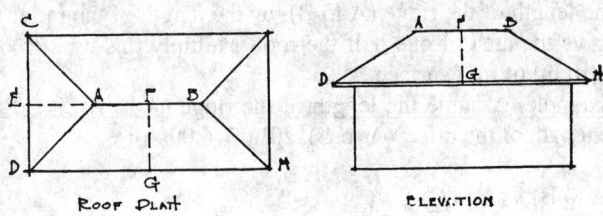

Roof Plan   Elevation

Fig 3. Hip Roof

To obtain the area of the sides of the roof, the length of the ridge (A to B), is 10'-0" (3.04 m) and the length of the eaves (D to H) is 30'-0" (9.14 m):

$$10'\text{-}0" + 30'\text{-}0" = 40'\text{-}0" \quad (9.14 + 3.04 = 12.08 \text{ m})$$

By taking 1/2 the combined length of the ridge and eaves, 1/2 of 40'-0" = 20'-0" (0.5 x 12.19 = 6.09 m), the average length of the roof. Assuming the length of the rafter (F to G) as 20'-0" (6.09 m), 20'-0" x 20'-0" = 400 sq.ft. (6.09 x 6.09 = 37.08 sq.m), the area of one side of the roof, which multiplied by 2 equals 800 sq.ft. (74.32 sq.m), the area of both sides of the roof.

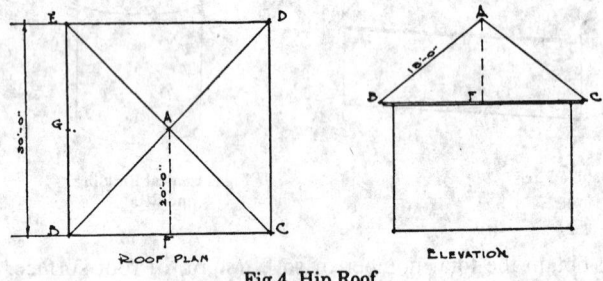

Roof Plan   Elevation

Fig 4. Hip Roof

Adding the area of the two ends to the area of the two sides gives 1,200 sq.ft. (111.48 sq.m) of roof area.

The area of a plain hip roof (Fig. 4) running to a point at the top is obtained by multiplying the length of the eaves (B to E) by 1/2 the length of the rafter (A to F). This gives the area of one end of the roof. To obtain the area of all four sides, multiply by 4. Example: Multiply the length of the eaves (B to E), which is 30'-0" (9.1 m) by 1/2 the length of the rafter (A to F), 10'-0" (3.04 m), and the result is 300 sq.ft. (27.87 sq.m), the area of one end of the roof. 300 x 4 = 1,200 sq.ft. (27.87 x 42 = 11.48 sq.m), the area of the 4 sides.

Rules for Measuring Conical Tower Roofs and Circular Buildings. To obtain the area of a conical tower roof as shown in Figure 2, multiply 1/2 the length of the rafter (A to B) by the distance around the eaves

1064

at B. For example, the length of the rafter (A to B) is 15'-0" (4.57 m) and the diameter of the building at B is 20'-0" (6.09 m).

To obtain the distance around the building, multiply the diameter by 3.1416. If the eaves project beyond the outside walls, and the diameter is given only to the outside walls of the building, add the length of the roof projection on both sides of the building to obtain the correct diameter. Example: If the diameter of the building (C to D) is 20'-0" (6.09 m) and the eaves project 2'-0" (0.61 m) on each side, the diameter of the building at the eaves would be 24'-0" (7.3.15 m).

Multiplying (B to E) 24'-0" (7.3.15 m), the diameter of the building at eaves, by 3.1416, gives 75.3984 (22.9336), or approximately 75'-5" (22.98 m) around the eaves at projection (B to E).

To obtain the area of the roof, multiply 1/2 the length of the rafter (A to B), which is 7-1/2 or 7'-6" (2.28 m), by the distance around the eaves at B and E, which is 75'-5" or 75.42' (22.90 m). The result is 565.5, or 565-1/2 sq.ft. (52.53 sq.m), the area of the roof.

To obtain the wall area of a cylindrical or circular building, multiply the height by the circumference, or the distance around the building, and the result will be the number of sq.ft. (sq.m) to be covered. The circumference is obtained by multiplying the diameter (C to D) of the building by 3.1416. To obtain the area of the outside walls of a cylindrical building whose diameter is 20'-0" (6.09 m) and the height 15'-0" (4.57 m), 20 x 3.1416 = 62.832 or 62'-10" (6.09 x 3.1416 = 19.16 or 19.2 m). Multiply 62.832 (19.2 m), the distance around the building, by 15 (4.6 m), the height of the building = 942.48 or 942-1/2 sq.ft. (87.56 or 87.56 sq.m), the area of the outside walls.

A Short Method of Figuring Roof Areas

To obtain the number of square feet (square meters) of roof area, where the pitch (rise and run) of the roof is known, take the entire flat or horizontal area of the roof and multiply by the factor given below for the roof slope applicable and the result will be the area of the roof.

Always bear in mind that the width of any overhanging cornice must be added to the building area to obtain the total area to be covered.

Example: Find the area of a roof 26'-0" x 42'-0" (7.9 x 12.8 m), with a 12" or 1'-0" (0.3 m) overhanging cornice and a 1/4 pitch or a 6 in 12 rise and run.

To obtain the roof area, 26'-0" + 1'-0" + 1'-0" = 28'-0" width (7.9 + 0.3 + 0.3 = 8.5 m width). 42'-0" + 1'-0" + 1'-0" = 44'-0" length (12.8 + 0.3 + 0.3 = 13.4 m length). 28 x 44 = 1,232 sq.ft. (8.5 x 13.4 = 113.9 sq.m) flat or horizontal area.

To obtain area at 1/4 pitch or 6 in 12 rise and run, 1,232 x 1.118 = 1377.376 or 1,378 sq.ft. of roof surface (113.9 x 1.118 = 128 sq.m of roof surface).

Add allowance for overhang on dormer roofs and sides.

Pitch of Roof	Rise and Run	Multiply Flat Area By	Lin.Ft. Hips or Valleys per Lin.Ft. Common Run
1/12	2 in 12	1.014	1.424
1/8	3 in 12	1.031	1.436
1/6	4 in 12	1.054	1.453
5/24	5 in 12	1.083	1.474
1/4	6 in 12	1.118	1.500
7/24	7 in 12	1.158	1.530
1/3	8 in 12	1.202	1.564
3/8	9 in 12	1.25	1.600
5/12	10 in 12	1.302	1.641
11/24	11 in 12	1.357	1.685
1/2	12 in 12	1.413	1.732
13/24	13 in 12	1.474	1.782
7/12	14 in 12	1.537	1.833
5/8	15 in 12	1.601	1.888
2/3	16 in 12	1.667	1.944
17/24	17 in 12	1.734	2.002
3/4	18 in 12	1.803	2.602
19/24	19 in 12	1.873	2.123
5/6	20 in 12	1.944	2.186
7/8	21 in 12	2.016	2.250
11/12	22 in 12	2.088	2.315
23/24	23 in 12	2.162	2.382
1/1	24 in 12	2.236	2.450

Hips and Valleys. The length of hips and valleys, formed by intersecting roof surfaces that run perpendicular to each other and have the same slope, is also a function of the roof rise and run. For full hips or valleys, where both roofs intersect for their full width, the length is determined by taking the square root of the sum of the rise squared plus twice the run squared.

Using the factors given in the last column of the above table, the length of full hips or valleys is obtained by multiplying the total roof run from eave to ridge, not the hip or valley run, by the factor listed for the roof slope.

THERMAL AND MOISTURE PROTECTION
Sample Roof Factor Estimate

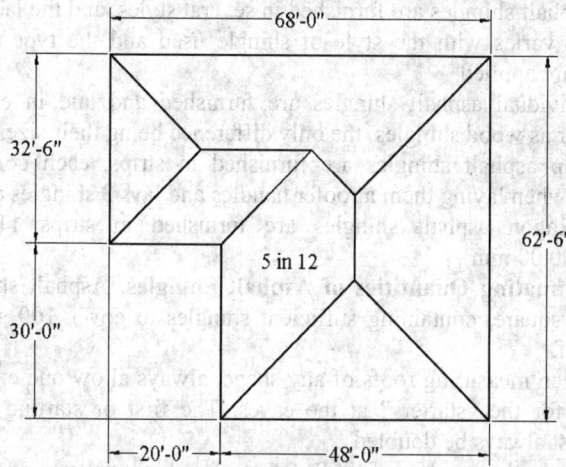

Given

The overall 68'-0" x 62'-6" (20.7m x 19.05m) building above has a uniform 5 in 12 slope. There are two hips on this roof. One hip has a 48'-0" (14.6m) span and the other has a 32'-6" (9.9m) span.

Number of Squares (with 10% waste)

First calculate the flat area and then multiply by the factor in the table.

(68' x 62.5' – 20' x 30') x 1.083 ÷ 100 x 1.1 waste = **43.48 squares**
(20.7m x 19.05m – 6.1m x 9.1m) x 1.083 x 1.1 waste = 403.6 sq.m.

Number of Lineal Feet Drip Edge (with 10% waste)

For hip roofs, the drip edge matches the 2D perimeter shown.

(68' + 62.5' + 48' + 30' + 20' + 32.5') x 1.1 waste = **287.1 lineal feet**
(20.7m + 19.05m + 14.6m + 9.1m + 6.1m + 9.9m) x 1.1 waste = 87 meters

Number of Lineal Feet of Ridge
(62.5' – 24' – 24') + 20' = **34.5 lineal feet**
(19.05m – 7.3m – 7.3m) + 6.1m = 10.5 meters

Number of Lineal Feet of Valley
16.25' x 1.474 = **24 lineal feet**
5m x 1.474 = 7.4 meters

Number of Lineal Feet of Hip Rafters
(16.25' + 4 x 24') x 1.474 = **165.5 lineal feet**
(5m + 4 x 7.3m) x 1.474 = 50.4 meters

07311 ASPHALT SHINGLES

Asphalt shingles are furnished in several styles, and the labor cost of laying them varies with the style of shingle used and the type of roof to which they are applied.

Individual asphalt shingles are furnished and laid in exactly the same manner as wood shingles, the only difference being their size.

Strip asphalt shingles are furnished in strips, each containing 3 shingles, so when laying them a roofer handles and lays 3 shingles at a time.

Hexagon asphalt shingles are furnished in strips 11-1/3"x36" (283.25 x 900.00 mm).

Estimating Quantities of Asphalt Shingles. Asphalt shingles are sold by the square, containing sufficient shingles to cover 100 sq.ft. (9.29 sq.m) of roof.

When measuring roofs of any shape, always allow one extra course of shingles for the "starters" at the eaves. The first or starting course of shingles must always be doubled.

Obtain the number of lin.ft. (meters) of hips, valleys, and ridges to be covered with asphalt shingles and compute as 1'-0" (.30m) wide.

Many roofing contractors do not measure hips, valleys, and ridges, preferring to add a percentage of the roof area to cover these items and for waste. The following percentages are commonly used: Gable roofs, 10%; hip roofs, 15%; hip roofs with dormers and valleys, 20%.

Asphalt shingles must be properly nailed-6 nails to a strip and nailed low enough on the shingle (right at the cut-out); otherwise, they will blow off the roof.

Most manufacturers produce a shingle designed for high wind areas that interlocks in such a manner that all of the shingles are integrated into a single unit. Interlocking shingles are available in single coverage for re-roofing and double coverage for new construction. Self-sealing shingles with adhesive tabs are produced by most manufacturers.

When using asphalt shingles for roofs, roll asphalt roofing of the same materials as the shingles is often used for forming valleys, hips, and roof ridges.

Nails Required for Asphalt Shingles. When laying individual asphalt shingles, use a 12 ga. aluminum nail, 1-1/2" (37.50 mm) long, with a 7/16" (10.93 mm) head. For laying over old roofs, use nails 1-3/4" (43.75 mm) long.

When laying square butt strip shingles, use 11 ga. aluminum nails, 1" (25 mm) long, with a 7/16" (10.93 mm) head. For laying over old roofs, use nails 1-3/4" (43.75 mm) long. It will require about 1 lb. (0.45 kg) of nails per square of shingles laid.

THERMAL AND MOISTURE PROTECTION

Sizes and Estimating Data on Asphalt Shingles											
Kind of Shingle	Nails per Shingle	Size (in)	No. of Shingle per Sq.	Exposure (in)	Nail Length (in)	Lbs Nails per Sq	Size (mm)	No. of Shingle per Sq.M.	Exposure (mm)	Nail Length (mm)	Kg Nails per Sq.M.
3-in-1 strip	4	12" x 36"	80	5	1	1	300 x 900	9	125	25	0.05

Approximate Prices on Mineral Surfaced Asphalt Shingles, Seal Tab						
Kind of Shingle	Size Inches	Lbs. per Square	Price per Square	Size mm	Kg. per Square	Price per Square
3-in-1 strip	12 "x 36 "	235	$ 41.60	300 x 900 mm	107	$ 41.60
3-in-1 strip	12 "x 36 "	340	$ 88.00	300 x 900 mm	154	$ 88.00
3-in-1 strip	12 "x 36 "	350	$ 97.50	300 x 900 mm	159	$ 97.50

Most asphalt shingles carry an Underwriters Class C rating. Incorporating a glass fiber layer results in an "A" rating. Class A shingles weighing 225 lbs. (102.06 kg) cost about $30.00 per square; 300 lbs. (136.08kg), $59.00 per square.

A roofer will lay one square (100 sq.ft. or 9.29 sq.m) of asphalt shingles in 1 hour on double pitched roofs with no hips, valleys, or dormers. Figure about 35 lin.ft. (10.66 m) per hour for fitting hips, valleys, and ridges. Hip and ridge roll material will run about $0.16 per lin.ft. ($0.52 per m); valley, $0.30 per lin.ft. ($0.98 per m.)

Labor Cost of 1 Square (100 sq.ft. or 9.29 sq.m) Strip Asphalt Shingles on Plain Double Pitch or Gable Roofs

Description	Hours	Rate	Total	Rate	Total
Roofer	1.00	$	$	$30.93	$30.93

On roofs with gables, dormers, etc., add 0.2 hr. per square (9.29 sq.m). On difficult constructed hip or English type roofs, add 0.5 hr. per square (9.29 sq.m).

07313 WOOD SHINGLES AND SHAKES

The labor cost of laying wood shingles varies with the type of roof, whether a plain gable roof, a steep roof, or one cut up with gables or dormers, and varies with the manner in which they are laid, whether with regular butts, irregular or staggered butts, or thatched butts.

The costs given on the following pages are based on the actual number of shingles a roofer can lay per day and not upon the number of squares covered, which will vary with the spacing of the shingles. It does not make any difference to the roofer whether the shingles are laid 4" (100 mm), 4-1/2" (112.50 mm), or 5" (125 mm) to the weather. He will lay practically the same number either way. It does make considerable difference in the

1069

number of sq.ft. (sq.m) of surface covered, which will vary from 10% to 40%.

The number of shingles laid will vary with the ability of workers and the class of work. Ordinary carpenters cannot lay as many shingles as carpenters who specialize in shingle laying. On the other hand, experienced shinglers usually demand a higher wage rate than ordinary carpenters.

Some carpenters claim to be able to lay 16 bundles (3,200 shingles) per 8-hr. day, but this is unusual and is generally found only on the cheapest grade of work, where only one nail is driven into each shingle instead of two that are necessary for a craftsman-like job.

Estimating the Quantity of Wood Shingles. Ordinary wood shingles are furnished in random widths, but 1,000 shingles are equivalent to 1,000 shingles 4" (100 mm) wide. Dimension shingles are sawed to a uniform width, being either 4" (100 mm), 5" (125 mm), or 6" (150 mm) wide.

Number of Shingles and Quantity of Nails Required						
Distance Laid to Weather Inch	Area Covered by one shingle Sq. In.	Percent for waste	Actual No. per Square Without Waste	No. per Square with Waste	4-Sq. Bundles Required	Pounds 3d Nails Required
4 "	16	10	900	990	5.0	3.2
4 1/2 "	17	10	850	935	4.7	2.8
4 1/2 "	18	10	800	880	4.4	2.5
5 "	20	10	720	792	4.0	2.0
5 1/2 "	22	10	655	721	3.6	1.6
6 "	24	10	600	660	3.3	1.5
Distance Laid to Weather mm	Area Covered by one shingle Sq. cm	Percent for waste	Actual No. per Square Without Waste	No. per Square with Waste	4-Sq. Bundles Required	Kg 3d Nails Required
100.0 mm	103	10	900	990	5.0	1.45
112.5 mm	110	10	850	935	4.7	1.27
112.5 mm	116	10	800	880	4.4	1.13
125.0 mm	129	10	720	792	4.0	0.91
137.5 mm	142	10	655	721	3.6	0.73
150.0 mm	155	10	600	660	3.3	0.68

Wood shingles are usually sold by the square based on sufficient shingles to lay 100 sq.ft. (9.29 sq.m) of surface, when laid 5" (125 mm) to the weather, 4 bundles to the square.

When estimating the quantity of ordinary wood shingles required to cover any roof, bear in mind that the distance the shingles are laid to the weather makes considerable difference in the actual quantity required.

1070

There are 144 sq. in. in 1 sq. ft. (0.09 sq.m) and an ordinary shingle is 4" (100 mm) wide. When laid with 4" (100 mm) exposed to the weather, each shingle covers 16 sq. in. or it requires 9 shingles per s.f. of surface. There are 100 sq.ft. (9.29 sq.m) in a square. 100 x 9 = 900, and allowing 10% to cover the double row of shingles at the eaves, waste in cutting, narrow shingles, etc., it will require 990 shingles (5 bundles) per 100 sq.ft. of surface.

How to Apply Shingles for Different Roof Slopes. Roof pitches are computed in fractions, such as 1/8, 1/3, 1/2 pitch. In the following cross section, the steepness of distances AB and BC constitutes pitch.

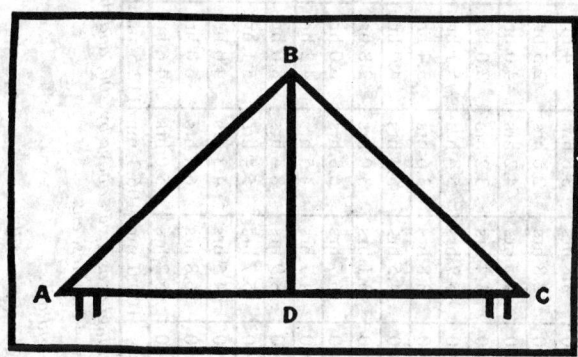

Distance AC, extending from one eave line to the other, is known as the span. One-half of this span, distance AD or DC, is called the run, and distance BD is called the rise. The relationship of the rise to run obviously affects the slope of AB or BD; in fact, roof pitches are computed from the ratio of rise to run. Therefore, the first step is to determine length of the run (AD or DC) and the rise (BD).

Wood shingles are manufactured in three lengths, 16" (400 mm), 18" (450 mm), and 24" (600 mm). The standard weather exposure, the portion of shingle exposed to weather on roof, for 16" (400 mm) shingles is 5" (125 mm), for 18" (450 mm) shingles it is 5-1/2" (137.50 mm), and for 24" (600 mm) shingles is 7-1/2" (187.50 mm). These standard exposures are recommended on all roofs of 1/4 pitch and steeper, 6" (150 mm) rise in 12" (300 mm) run. On flatter roof slopes, the weather exposure should be reduced to 3-3/4" (93.75 mm) for 16" (400 mm) shingles, 4-1/4" (106.25 mm) for 18" (450 mm) shingles, and 5-3/4" (143.75 mm) for 24" (600 mm).

The following diagram shows the weather exposure to be used for various roof pitches. For example, if a roof has a rise of 8" (200 mm) in a run of 12" (300 mm), it is 1/3 pitch and that an exposure of either 5" (125 mm), 5-1/2" (137.50 mm), or 7-1/2" (187.50 mm) should be employed, depending on the length of the shingles used.

Labor Laying One Square 100 Sq.Ft. Wood Shingles

Class of Work	Labor	Number Laid per 8-Hr Day	Distance Shingles are Laid to weather in Hours					
			4 "†	4 1/4 "	4 1/2 "	5 "	5 1/2 "	6 "
Plain Gable or Hip Roof	Carpenter	2,000 – 2,200	3.8 hrs	3.6 hrs	3.4 hrs	3.0 hrs	2.8 hrs	2.5 hrs
	Roofer	2,750 – 3,000	2.8 hrs	2.6 hrs	2.5 hrs	2.3 hrs	2.0 hrs	1.9 hrs
Difficult Gable Roofs, cut up with gables, dormers, hips, valleys, etc.	Carpenter	1,700 – 1,900	4.5 hrs	4.2 hrs	4.0 hrs	3.6 hrs	3.3 hrs	3.0 hrs
	Roofer	2,200 – 2,500	3.4 hrs	3.2 hrs	3.0 hrs	2.7 hrs	2.5 hrs	2.3 hrs
Difficult Hip Roofs, Steep English Roofs, hips, valleys, etc.	Carpenter	1,300 – 1,500	5.7 hrs	5.4 hrs	5.1 hrs	4.6 hrs	4.2 hrs	3.8 hrs
	Roofer	2,000 – 2,200	3.8 hrs	3.6 hrs	3.4 hrs	3.0 hrs	2.8 hrs	2.5 hrs
Shingles laid irregularly or with Staggered Butts on Plain Roofs	Carpenter	1,700 – 1,900	4.5 hrs	4.2 hrs	4.0 hrs	3.6 hrs	3.3 hrs	3.0 hrs
	Roofer	2,400 – 2,700	3.1 hrs	3.0 hrs	2.8 hrs	2.5 hrs	2.3 hrs	2.1 hrs
Shingles laid irregularly or with Staggered Butts on Difficult Constructed Roofs	Carpenter	1,100 – 1,300	6.7 hrs	6.3 hrs	6.0 hrs	5.4 hrs	4.9 hrs	4.5 hrs
	Roofer	1,600 – 1,800	4.7 hrs	4.5 hrs	4.2 hrs	3.8 hrs	3.4 hrs	3.1 hrs
Shingles with Thatched Butts*	Roofer	800 – 1,000	8.9 hrs	8.4 hrs	8.0 hrs	7.1 hrs	6.5 hrs	6.0 hrs
Plan Sidewalls	Carpenter	1,300 – 1,500	5.7 hrs	5.4 hrs	5.1 hrs	4.6 hrs	4.2 hrs	3.8 hrs
	Roofer	1,700 – 1,900	4.5 hrs	4.3 hrs	4.0 hrs	3.6 hrs	3.3 hrs	3.0 hrs
Difficult Sidewalls, having bays window breaks, etc.	Carpenter	1,100 – 1,250	6.8 hrs	6.4 hrs	6.0 hrs	5.4 hrs	5.0 hrs	4.5 hrs
	Roofer	1,400 – 1,650	5.2 hrs	5.0 hrs	4.6 hrs	4.2 hrs	3.8 hrs	3.6 hrs
METRIC			100.0 mm	106.3 mm	112.5 mm	125.0 mm	137.5 mm	150.0 mm

To obtain number of bundles of shingles required, divide number of shingles as given above by 200, and the result will be the number of bundles required, i.e. 2,200 shingles ÷ 200 = 11 bundles: 800 ÷ 200 = 4 bundles, etc.

* Shingles with thatched butts require 25% more shingles than laid regularly.

† Use 4" (100mm) column for carpenter or roofer time per 1,000 shingles (5 bundles).

The above production rate table is based upon using approximately 2 nails each square and 10 percent waste.

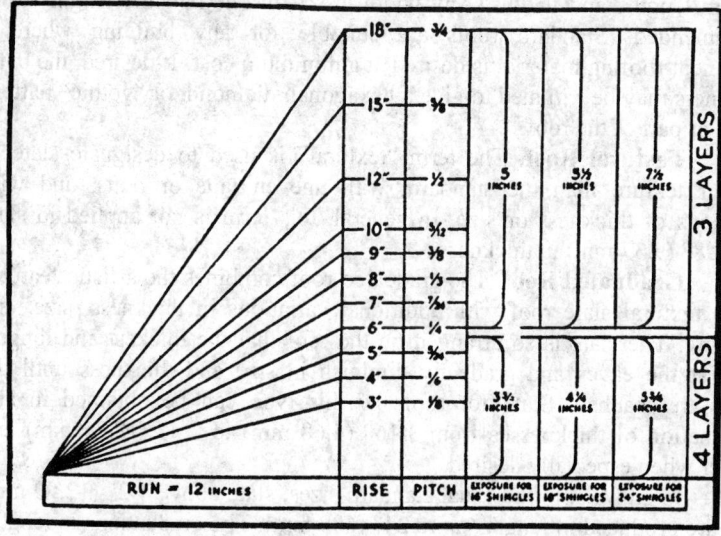

Wood shingles, referred to as shakes, may be laid over open or solid sheathing, with solid preferred where wind driven snow conditions prevail. If open sheathing is selected, use 1"x4"s (25 x 100 mm) at centers equal to the weather exposure at which the shakes are to be laid, but not over 10" (250 mm). Building paper must be added for additional weather protection.

Western Red Cedar Shingles		
144# 16" #1-5X Shingles 4 Bdl. per Sq.	$ 160.00	65.32 Kg per 400 mm
144# 16" #2-5X Shingles 4 Bdl. per Sq.	$ 144.00	65.32 Kg per 400 mm
60# 16" Undercourse 14/14 2 Bdl. per Sq.	$ 40.00	27.22 Kg per 400 mm
158# 18" #1 Perfection 4 Bdl. per Sq.	$ 160.00	71.67 Kg per 450 mm
350# 24" #1 3/4" (18.75 mm) to 5/4" (31.25 mm) Handsplit Shakes, 5 Bdl. per Sq.	$ 176.00	158.76 Kg per 600 mm
For fire retardant shingles add 100%		

Wood shingles are less fire resistant then other types of shingles. For fire retardant shingles add 100%. Due to the less fire resistance, some local jurisdictions restrict the use of wood shingles. The contractor needs to check the local code for compliance.

07314 SLATE ROOFING

Slate roofing is elegant and is furnished in a number of types, sizes, thicknesses, and finishes to meet architectural requirements of the buildings on which it is to be used. Slate roofing while elegant is heavy and the rafters must be designed to carry the additional weight over an asphalt roof. The four principal classifications are standard, textural, graduated and flat slat clay.

Standard Roof. Standard slate roofs are composed of slate approximately 3/16" (4.68 mm) thick (commercial standard slate), of one uniform standard length and width; also one length and random widths or

1073

random lengths and widths. Any width less than one-half the length is not recommended. Standard roofs are suitable for any building where a permanent roofing material is desired at a minimum cost. If desired, the butts or corners may be trimmed to give a hexagonal, diamond, or "gothic" pattern for all or part of the roof.

Textural Roof. The term "textural" is used to designate slate of rougher texture than the standard, with uneven tails or butts, and with variations of thickness or size. In general, this term is not applied to slate over 3/8" (9.37 mm) in thickness.

Graduated Roof. The graduated roof combines the artistic features of the textural slate roof with additional variations in thickness, size, and exposure. The slate is so arranged on the roof that the thickest and longest occur at the eaves and gradually diminish in size and thickness until the ridges are reached. Slate for roofs of this type can be obtained in any combination of thicknesses from 3/16" (4.68 mm) to 3/4" (18.75 mm) and heavier when especially desired.

In addition to the usual standard sizes, slate above 1/2" (12.50 mm) thick are produced in lengths up to 24" (600 mm). The graduations in lengths generally range from 24" (600 mm) to 12" (300 mm). The variations in length will at once provide a graduation in exposure by using the standard 3" (75 mm) lap. To illustrate, a suitable range of lengths might be:

	Thickness Inches	Length Inches	Exposure Inches	Thickness mm	Length mm	Exposure mm
Course Under Eaves	3/8 "	14 "	No exposure	9.38 mm	350 mm	No exposure
First Course	3/4 "	24 "	10 1/2 "	18.75 mm	600 mm	262.50 mm
1	3/4 "	24 "	10 1/2 "	18.75 mm	600 mm	262.50 mm
2	1/2 "	22 "	9 1/2 "	12.50 mm	550 mm	237.50 mm
2	1/2 "	20 "	8 1/2 "	12.50 mm	500 mm	212.50 mm
2	3/8 "	20 "	8 1/2 "	9.38 mm	500 mm	212.50 mm
4	3/8 "	18 "	7 1/2 "	9.38 mm	450 mm	187.50 mm
5	1/4 "	16 "	6 1/2 "	6.25 mm	400 mm	162.50 mm
3	1/4 "	14 "	5 1/2 "	6.25 mm	350 mm	137.50 mm
3	3/16 "	14 "	5 1/2 "	4.69 mm	350 mm	137.50 mm
8	3/16 "	12 "	4 1/2 "	4.69 mm	300 mm	112.50 mm

Random widths should be used and so laid that the vertical joints in each course are broken and covered by the slate of the course above.

Flat Roof. Flat roofs offer a wide field for roofing slate and are designated whether or not they are used for "promenade" purposes.

Slate of any thickness may be used instead of slag or gravel as a surfacing material for the usual built-up type of roof. For ordinary roofs the standard 3/16" (4.68 mm) slate are used, but for promenade or extraordinary service, the slate should be 1/4" (6.25 mm) or 3/8" (9.37 mm) thick.

1074

THERMAL AND MOISTURE PROTECTION

Slate for flat roofs are furnished in the following standard sizes: 6"x6" (150 x 150 mm), 6"x8" (150 x 200 mm), 6"x9" (150 x 225 mm), 10"x6" (250 x 150 mm), 10"x7" (250 x 175 mm), 10"x8" (250 x 200 mm), 12"x6" (300 x 150 mm), 12"x7" (300 x 175 mm), and 12"x8" (300 x 200 mm).

Estimating Quantities of Roofing Slate. Obtain the net area of the roof in sq. ft. as described earlier in this chapter. Add 6" (150 mm) to rafter length to allow for waste of normal roof.

Deduct one-half the area of chimneys and dormers if over 20 sq.ft. (1.85 sq.m) and less than 80 sq.ft. (7.43 sq.m). Make no deductions if less than 20 sq.ft. (1.85 sq.m) and deduct 20 sq.ft. (1.85 sq.m) less than actual area if more than 80 sq.ft. (7.43 sq.m).

Include area of dormer sides, sides of dormers if slated, slate saddles, and other places where slate is used in addition to the main roof area. Include overhanging parts of dormers, etc.

Add 1 sq.ft. for each lin.ft. (0.09 sq m for each 0.30 m) of hips and valleys, for loss in cutting and fitting.

Allow 2% to 15% additional slate, depending on the extent to which the roof is intersected by other roofs, dormers, walls, etc.

Divide the total of the above by 100, which will give the number of "squares" of roofing required.

Slate is always sold at the quarry on the basis of the quantity required to cover 100 sq.ft. (9.29 sq.m) or a "square" of roof when slate is laid with a 3" (75 mm) head lap. If the roof is flat or other than 3" (75 mm) lap is used, the quantity must be corrected to the equivalent amount required as though the 3" (75 mm) lap were used.

Items to be Included in an Estimate for Slate Roofing. The following items should be included to make a complete and accurate estimate on slate roofing:

1. Cost of slate (punched) on cars or trucks at the quarry.
2. Freight or trucking from quarry to destination.
3. Unloading from cars and hauling to job, also unloading and piling at job, unless delivered to the job by truck. Trucking from quarry direct to job within a radius of 500 miles (805 km) is now general practice in the Vermont-New York district.
4. Placing on roof and laying, a) roofing felt, b) elastic cement, c) nails, d) snow guards or snow rails, e) sheet metal, f) labor, g) waste in handling, cutting and fitting.

Size of Slate. The following table gives the different sizes in which roofing slate is furnished, number required per square when laid with 3" (75 mm) lap, and number of nails required per square based on 2 nails to each piece of slate.

Slate Roofing					
Size of Slate Inches	# per Sq.	# Nails req. per Sq.	Size of Slate mm	# per 10 Sq.M.	# req. per 10 Sq. M.
10 " x 6 "	686	1372	250 x 150 mm	738	1477
10 " x 7 "	588	1176	250 x 175 mm	633	1266
10 " x 8 "	515	1030	250 x 200 mm	554	1109
10 " x 10 "	414	828	250 x 250 mm	446	891
12 " x 6 "	533	1066	300 x 150 mm	574	1147
12 " x 7 "	457	914	300 x 175 mm	492	984
12 " x 8 "	400	800	300 x 200 mm	431	861
12 " x 9 "	355	710	300 x 225 mm	382	764
12 " x 10 "	320	640	300 x 250 mm	344	689
12 " x 12 "	267	534	300 x 300 mm	287	575
14 " x 7 "	374	748	350 x 175 mm	403	805
14 " x 8 "	327	654	350 x 200 mm	352	704
14 " x 9 "	290	580	350 x 225 mm	312	624
14 " x 10 "	261	522	350 x 250 mm	281	562
14 " x 12 "	218	436	350 x 300 mm	235	469
14 " x 14 "	188	376	350 x 350 mm	202	405
16 " x 8 "	277	554	400 x 200 mm	298	596
16 " x 9 "	246	492	400 x 225 mm	265	530
16 " x 10 "	221	442	400 x 250 mm	238	476
16 " x 12 "	185	370	400 x 300 mm	199	398
16 " x 14 "	159	318	400 x 350 mm	171	342
18 " x 9 "	213	426	450 x 225 mm	229	459
18 " x 10 "	192	384	450 x 250 mm	207	413
18 " x 11 "	175	350	450 x 275 mm	188	377
18 " x 12 "	160	320	450 x 300 mm	172	344
18 " x 14 "	137	274	450 x 350 mm	147	295
20 " x 10 "	169	338	500 x 250 mm	182	364
20 " x 11 "	154	308	500 x 275 mm	166	332
20 " x 12 "	141	282	500 x 300 mm	152	304
20 " x 14 "	121	242	500 x 350 mm	130	260

THERMAL AND MOISTURE PROTECTION

Size of Slate Inches	# per Sq.	# Nails req. per Sq.	Size of Slate mm	# per 10 Sq.M.	# req. per 10 Sq. M.
20 " x 16 "	106	212	500 x 400 mm	114	228
22 " x 11 "	138	276	550 x 275 mm	149	297
22 " x 12 "	126	252	550 x 300 mm	136	271
22 " x 14 "	109	218	550 x 350 mm	117	235
22 " x 16 "	95	190	550 x 400 mm	102	205
24 " x 12 "	115	230	600 x 300 mm	124	248
24 " x 14 "	98	196	600 x 350 mm	105	211
24 " x 16 "	86	172	600 x 400 mm	93	185

Slate 3/4" (18.75 mm) and thicker and 24" (600 mm) or over in length should have 4 nail holes.

Weight of Slate Roofing. A square of slate roofing, i.e., sufficient slate to cover 100 sq.ft. (9.29 sq.m) of roof surface with a standard 3" (75 mm) lap, will vary from 650 to 8,000 lbs. (294.84 to 295 to 3628.80 kg) for thicknesses from the commercial standard 3/16" (4.68 mm) to 2" (50 mm).

The weight of slate varies with the size of the slate, color, and quarry. Sometimes, it varies even in the same quarry. The variation may be from 10% above to 15% below the weights given in the following table:

Average Weight of Slate per Square (100 sq.ft. or 9.29 sq.m)					
Slate Thickness Inches	Sloping Roof 3" Laps Lbs.	Flat Roof Without Lap, Lbs.	Slate Thickness mm	Sloping Roof 75 mm Lap Kg.	Flat Roof Without Lap, Kg.
Standard	3/16	700	240	4.69	318	109
Selected full	3/16	750	250	4.69	340	113
------	1/4	900	335	6.25	408	152
------	3/8	1,400	500	9.38	635	227
------	1/2	1,800	675	12.50	816	306
------	3/4	2,700	1,000	18.75	1,225	454
------	1	4,000	1,330	25.00	1,814	603
------	1 1/4	5,000	1,670	31.25	2,268	758
------	1 1/2	6,000	2,000	37.50	2,722	907
------	1 3/4	7,000	------	43.75	3,175	------
------	2	8,000	------	50.00	3,629	------

Nails Required for Slate Roofing. The quantity of nails required for slate roofing will vary with size, kind, etc., but the following table gives quantities of those commonly used:

Approximate Number of Nails to the Pound (Kg)					
Length in Inches	"Copperweld" Slating Nails	Copper Wire Slating Nails	Cut Copper Slating Nails	Cut Brass Nails	Cut Yellow-Metal Slate Nails
1	386	270	------	------	------
1 1/4	211	144	190	164	154
1 1/2	176	134	135	140	140
1 3/4	133	112	100	108	------
2	87	104	------	88	------
2 1/4	------	46	------	80	------
2 1/2	------	------	------	64	------
2 3/4	------	------	------	52	------
3	------	------	------	48	------
(metric)					
Length in mm	"Copperweld" Slating Nails	Copper Wire Slating Nails	Cut Copper Slating Nails	Cut Brass Nails	Cut Yellow-Metal Slate Nails
25.00	851	595	------	------	------
31.25	465	317	419	362	340
37.50	388	295	298	309	309
43.75	293	247	220	238	------
50.00	192	229	------	194	------
56.25	------	101	------	176	------
62.50	------	------	------	141	------
68.75	------	------	------	115	------
75.00	------	------	------	106	------

Labor Punching Slate. It is much cheaper to have slate punched at the quarry. The labor cost of punching slate on the job varies with the size of the slate. The larger the slate, the fewer pieces required per square.

If one slater is working on the job with a helper, the helper should punch the slate and carry them to the slater as fast as required. There should be no additional charge for punching. If, however, there is only one worker on the job who must punch the slate and lay them, he should punch one square of slate in 30 to 40 min. or 12 to 16 sqs. (111.48 to 148.64 sq.m) per 8-hr. day, at the following labor cost per square (100 sq.ft. or 9.29 sq.m):

Description	Hours	Rate	Total	Rate	Total
Slater/Roofer	0.60	$	$	$ 30.93	$ 18.56

THERMAL AND MOISTURE PROTECTION

Labor Cost of Laying Roofing Slate. The labor cost of laying roofing slate will vary considerably with the size of the slate, thickness, type of roofs, etc., but when large thick slate are used for graduated roofs, the labor costs will run higher than when the same size slate are used on standard roofs.

The size of slate best adapted for plain roofs are large, wide slates, such as 16"x 12" (400 x 300 mm), 18"x 12" (450 x 300 mm), 20"x 12" (500 x 300 mm), or 24"x 12" (600 x 300 mm). Slate from 16"x 8" (400 x 200 mm) to 20"x10" (500 x 250 mm) are also popular sizes. For roofs cut up into smaller sections, the 14"x 7" (350 x 175 mm) and 16"x 8" (400 x 200 mm) are very popular.

A roofing crew consisting of two slaters and one helper can work to advantage and perform the work more economically than one worker alone. In some localities the helper is considered an apprentice and is permitted to lay slate. In others he can only punch and carry the slate to the slaters.

Laying 16"x 8" (400 x 200 mm) Roofing Slate. When working on plain roofs that do not require much cutting, 2 slaters and a helper should lay felt, handle, and lay 3-1/2 to 4-1/2 sqs. (350-450 sq.ft. or 32.51-41.80 sq.m) of 16"x 8" (400 x 200 mm) slate per 8-hr. day. On more complicated hip or gable roofs, requiring considerable cutting and fitting for hips, valleys, dormers, etc., 2 slaters and a helper should handle and lay 1-3/4 to 2-1/4 sqs. (175-225 sq.ft. or 16.25-20.90 sq.m) of 16"x 8" (400 x 200 mm) slate per 8-hr. day.

Laying 18"x 9" (450 x 225 mm) Roofing Slate. When laying 18"x 9" (450 x 225 mm) slate on plain roofs, 2 slaters and a helper should lay felt, handle, and lay 4 to 5 sqs. (400-500 sq.ft. or 37.16-46.45 sq.m) per 8-hr. day. On more complicated hip or gable roofs, requiring considerable cutting and fitting for hips, valleys, dormers, etc., 2 slaters and a helper should lay 2-1/2 to 3 sqs. (250-300 sq.ft. or 23.22-27.87 sq.m) per 8-hr. day.

Laying 20"x10" (500 x 250 mm) Roofing Slate. When laying 20"x10" (500 x 250 mm) slate on plain roofs, 2 slaters and a helper should lay felt, handle, and lay 5 to 6 sqs. (500-600 sq.ft. or 46.45-55.74 sq.m) of roof per 8-hr. day, but on more complicated roofs, requiring considerable cutting and fitting for hips, valleys, dormers, etc., the same crew would lay only 3 to 3-1/2 sqs. (300-350 sq.ft. or 27.87-32.51 sq.m) per 8-hr. day.

Laying 22"x 12" (550 x 300 mm) Roofing Slate. On straight roofs, 2 slaters and a helper should lay felt, handle, and lay 5-1/2 to 6-1/2 sqs. (550-650 sq.ft. or 51.09-60.38 sq.m) of 22"x12" (550 x 300 mm) slate per 8-hr. day, but on the more complicated roofs, requiring cutting and fitting for hips, valleys, dormers, etc., the same crew would lay only 3-1/2 to 4 sqs. (350-400 sq.ft. or 32.51-37.16 sq.m) per 8-hr. day.

Laying Graduated Roofing Slate. When laying graduated roofing slate from 3/16" (4.68 mm) to 3/4" (18.77 mm) thick and 12" (300 mm) to 24" (600 mm) long, considerable care is required in selecting the right slate for each course and laying them to obtain the desired effect. On work of this

1079

kind 2 slaters and a helper should lay felt, handle, and lay 1-1/2 to 2 sqs.
(150-200 sq.ft. or 13.93-18.58 sq.m) of roof per 8-hr. day.

Labor Cost of One Square (100 sq.ft. or 9.29 sq.m) 16" x 8" (400 x 200 mm) Standard Slate Roof					
Description	Hours	Rate	Total	Rate	Total
Slater/Roofer	4.00	$	$	$ 30.93	$ 123.72
Helper	2.00	$	$	$ 26.16	$ 52.32
Cost per 100 sq.ft.					$ 176.04
per sq.ft.					$ 1.76
per sq.m					$ 18.95

For complicated hip or gable roofs, double the time given above.

Labor Cost of One Square (100 sq.ft. or 9.29 sq.m) 18" x 9" (500 x 225 mm) Standard Roofing Slate					
Description	Hours	Rate	Total	Rate	Total
Slater/Roofer	3.50	$	$	$ 30.93	$ 108.26
Helper	1.75	$	$	$ 26.16	$ 45.78
Cost per 100 sq.ft.					$ 154.04
per sq.ft.					$ 1.54
per sq.m					$ 16.58

For complicated hip or gable roofs, add 2.5 hrs slater & 1.25 hrs helper.

Labor Cost of One Square (100 sq.ft. or 9.29 sq.m) 20" x 10" (500 x 250 mm) Standard Roofing Slate					
Description	Hours	Rate	Total	Rate	Total
Slater/Roofer	3.00	$	$	$ 30.93	$ 92.79
Helper	1.50	$	$	$ 26.16	$ 39.24
Cost per 100 sq.ft.					$ 132.03
per sq.ft.					$ 1.32
per sq.m					$ 14.21

For complicated hip or gable roofs, add 2 hrs. slater & 1 hr. helper time.

Labor Cost of One Square (100 sq.ft. or 9.29 sq.m) 22" x 12" (550 x 300 mm) Standard Roofing Slate					
Description	Hours	Rate	Total	Rate	Total
Slater/Roofer	2.60	$	$	$ 30.93	$ 80.42
Helper	1.30	$	$	$ 26.16	$ 34.01
Cost per 100 sq.ft.					$ 114.43
per sq.ft.					$ 1.14
per sq.m					$ 12.32

For complicated hip or gable roofs, add 1.6 hrs. slater & 0.8 hr. helper.

THERMAL AND MOISTURE PROTECTION

Labor Cost of One Square (100 sq.ft. or 9.3 sq.m) Rough Texture Slate Roof Random Widths and 3/16" (4.68 mm) to 3/8" (9.37 mm) Thick					
Description	Hours	Rate	Total	Rate	Total
Slater/Roofer	7.50	$	$	$ 30.93	$ 231.98
Helper	3.75	$	$	$ 26.16	$ 98.10
Cost per 100 sq.ft.					$ 330.08
per sq.ft.					$ 3.30
per sq.m					$ 35.53

Labor Cost of One Square (100 sq.ft. or 9.29 sq.m) Graduated Slate Roof, Slate 12" (300 mm) to 24" (600 mm) Long 3/16" (4.68 mm) to 3/4" (18.75 mm) Thick					
Description	Hours	Rate	Total	Rate	Total
Slater/Roofer	9.00	$	$	$ 30.93	$ 278.37
Helper	4.50	$	$	$ 26.16	$ 117.72
Cost per 100 sq.ft.					$ 396.09
per sq.ft.					$ 3.96
per sq.m					$ 42.64

Prices on Slate Roofing Material

Slate shingles come from various parts of the country, and the estimator must obtain a quotation from the quarries in the geographical location, not only for price and shipping costs, but for availability of certain sizes and colors.

Vermont slate shingles in 3/16" (4.68 mm) thicknesses will cost as follows: black or gray, $360.00 per square (9.29 sq.m); green and gray, $375.00; and purples, $460.00. The Pennsylvania black #1 clear and the Buckingham, Virginia black slates will cost about $460.00 per square (9.29 sq.m). Slates thicker than 3/16" (4.68 mm) will cost appreciably more.

Copper nails are usually used for nailing the slates. Use 1-1/4" (31.25 mm) long nails for the 3/16" (4.68 mm) thick slates and 1-3/4" (43.75 mm) long nails for the rough texture and graduated slate roofs where thicker slates are encountered.

07315 PORCELAIN ENAMEL SHINGLES

Porcelain enamel shingles are manufactured to have an exposed surface of 10"x10" (250 x 250 mm) with 144 shingles per square (9.29 sq.m). Weight is 225 lbs. per 100 sq.ft. (110 kg per 10 sq.m). Finish is fused on at 1500°F (815°C) and provides a long lasting, self-cleaning finish that will not peel or blister.

1081

In ordering these units, it is important to have complete shop drawings based on actual job conditions so that field cutting can be held to a minimum. Shingles are nailed to the roof deck over 2 layers of 30-lb. (13.6-kg) felt. Special nails, ridge and hip sections are furnished by the manufacturer. Valley flashings and drips should be of non-corrosive metal and furnished by the sheet metal contractor.

Material costs will run in the vicinity of $450.00 per square ($48.43 per sq.m), depending on the number of roof irregularities which run up the cost considerably, as all special fittings should be made up in the shop. One roofer can lay a square in about 6-1/2 hours.

07316 METAL SHINGLES

Metal shingles formed with a wood grain texture and available in a wide range of colors can be used for a lightweight, decorative roof, or false mansard. They can be applied in a conventional way on a solid nailable roof deck over 30-lb. (13.6-kg) felt underlayment or supported and clipped to sub-purlins on steel frames. Weights are 36 lbs. per square (1.75 kg per sq.m) for 0.020 aluminum, 54 lbs. (2.64 kg per sq.m) for 0.030 aluminum, and 88 lbs. (4.3 kg per sq.m) for 30 ga. steel.

The shingles are not individual units, but made in strips either 10" x 60" (250 x 1500 mm), 12" x 36" (300 x 900 mm) or 12" x 48" (300 x 1200 mm).

The shingle manufacturer can provide anchor clips, drip caps, starter strips, trim, ridge, and hip laps. Material Costs will run as follows:

Description	Per square	Per Sq.M.
Aluminum, 0.020", mill finish	$ 144	$ 15.50
Aluminum, 0.030", mill finish	$ 170	$ 18.30
Aluminum, 0.020", anodized color	$ 260	$ 27.99
Aluminum, 0.030", anodized color	$ 300	$ 32.29
Aluminum, 0.020", bonderized	$ 400	$ 43.06
Aluminum, 0.030", bonderized	$ 420	$ 45.21
Steel, 26 ga., galvanized	$ 66	$ 7.10
Steel, 24 ga., galvanized	$ 72	$ 7.75
Steel, 26 ga., color galvanized	$ 150	$ 16.15
Steel, 24 ga., color galvanized	$ 156	$ 16.79

Ridge and cap sections cost $2.50 per lin.ft. ($8.20 per m) for 0.020" aluminum, $4.00 per lin.ft. ($13.12 per m) for 0.030". Valley sections cost $2.75 and $2.50 per lin.ft. ($9.02 and $8.20 per m). One carpenter will take about 4 to 5 hours to lay one square (9.29 sq.m).

THERMAL AND MOISTURE PROTECTION
07321 CLAY ROOFING TILE

The following information on clay roofing tile has been furnished through the courtesy of the Ludowici Roof Tile Company. New Lexington, Ohio. There are six major categories of clay tile manufactured today. 1) S-Tile or sometime referred to as Spanish Tile; 2) Profile tile, sometimes referred to as French Tile; 3) Tower Tile, for steep slopes; 4) Flat Tile; 5) Pans and Cover and 6) Interlocking Flat Tile, which also come in extra large sizes to reduce installation costs.

When estimating the quantities of clay roofing tile, the roof areas are obtained in the same manner as described earlier in this chapter.

After the total number of sq.ft. (sq.m) of surface has been obtained, add about 3% to the net measurement to take care of cutting and waste. Material is sold only in full pallets which vary from 3 to 8.7 squares (28-80 sq.m).

Hip and valley fittings should be figured by the piece. Closure and finishing pieces, such as hip starters, closed ridge ends, terminals, etc., should also be figured by the piece. Fittings are sold only in full boxes.

Where interlocking shingle tiles are used, detached gable rake pieces are required for all gable roofs not intersecting with vertical surfaces. These pieces are measured by the total lineal feet of gable rake, keeping separate quantities for lefts and rights as determined by the roof eave to be covered.

For all shingle tile roofs, end bands or half-tile are required on each course, alternating from one end to the other to break joints. End bands are figured by the lineal feet of gable rake, taking half the length at each gable rake. The following chart lists the several tile types and names together with average weight, width and exposure of different tile sizes.

Laying Interlocking Shingle Tile. On a roof of average difficulty, a crew consisting of 2 roofers and an apprentice or helper should handle and lay 500 to 600 sq.ft. (46.45 to 55.74 sq.m) of interlocking shingle tile per 8-hr. day, at the following labor cost per square (100 sq.ft. or 9.29 sq.m):

Description	Hours	Rate	Total	Rate	Total
Roofer	3.00	$	$	$ 30.93	$ 92.79
Helper	1.50	$	$	$ 26.16	$ 39.24
Cost per 100 sq.ft.					$ 132.03
per sq. ft.					$ 1.32
per sq.m					$ 14.21

On steep or cut-up roofs, add 15% to 25%. For roofs over 3 stories high to eaves, add 1.5 hrs. helper time.

Average Length, Width and Exposures of Tile Styles

Tile Name	Length Inches	Width Inches	Length Exposure Inches	Width Exposure Inches	Pieces per Square	Under Eave Length Inches	Under Eave Width Inches	Avg. Wt. per Sq., Lbs.	Min Slope
Flat Shingle Tiles									
Antique	12 "	7 "	5 "	7 "	412	7 "	7 "	1,650 Lbs	5:12
Brittany	12 "	7 "	5 "	7 "	412	7 "	7 "	1,900 Lbs	5:12
Calais	15 "	7 "	6 1/2 "	7 "	317	8 1/2 "	7 "	1,600 Lbs	5:12
Colonial	14 5/8 "	7 3/8 "	5 5/16 "	7 3/8 "	310	8 5/16 "	7 3/8 "	1,800 Lbs	5:12
Crude	12 "	6 "	5 "	6 "	480	7 "	6 "	1,935 Lbs	5:12
Flat Slab Shingle 3/8"	12 "	6 "	5 "	6 "	480	7 "	6 "	1,300 Lbs	5:12
Flat Slab Shingle 5/8"	12 "	6 "	5 "	6 "	480	7 "	6 "	1,780 Lbs	5:12
Georgian	15 "	8 "	6 1/2 "	8 "	276	8 1/2 "	8 "	1,600 Lbs	5:12
Norman	15 "	7 "	6 1/2 "	7 "	317	8 1/2 "	7 "	1,600 Lbs	5:12
Provincial	15 "	7 "	6 1/2 "	7 "	317	8 1/2 "	7 "	1,575 Lbs	5:12
Interlocking Tiles									
Americana	14 "	9 "	11 "	8 1/4 "	158	4 5/8 "	12 "	800 Lbs	3:12
Americana - XL	16 "	10 3/4 "	13 "	10 1/8 "	109	4 5/8 "	12 "	750 Lbs	3:12
Celadon	16 "	10 3/4 "	13 "	10 1/8 "	109	4 5/8 "	12 "	600 Lbs	3:12
Classic	14 "	9 "	11 "	8 1/4 "	158	4 5/8 "	12 "	800 Lbs	3:12
Classic - XL	16 "	10 3/4 "	13 "	10 1/8 "	109	4 5/8 "	12 "	750 Lbs	3:12
Imperial	15 "	10 "	12 "	9 1/4 "	129	4 5/8 "	12 "	860 Lbs	3:12
Lanai	14 "	9 "	11 "	8 1/4 "	158	4 5/8 "	12 "	800 Lbs	3:12
Lanai - XL	16 "	10 3/4 "	13 "	10 1/8 "	109	4 5/8 "	12 "	750 Lbs	3:12
Williamsburg	14 "	9 "	11 "	8 1/4 "	158	4 5/8 "	12 "	800 Lbs	3:12
Williamsburg - XL	16 "	10 3/4 "	13 "	10 1/8 "	109	4 5/8 "	12 "	750 Lbs	3:12
Interlocking Profile Tiles									
French	16 1/4 "	9 "	13 3/8 "	8 1/8 "	133	N/A	N/A	1,025 Lbs	3:12
S- Tiles									
Spanish 13 1/4"	13 1/4 "	9 3/4 "	10 1/4 "	8 1/4 "	171	N/A	N/A	900 Lbs	4:12
Spanish 18 3/8	18 3/4 "	9 3/4 "	15 3/8 "	8 1/4 "	114	N/A	N/A	800 Lbs	4:12
Pan and Cover Tiles									
16" Tapered Mission and Straight Barrel Mission Tiles	16 "	P = 8" C* = 7"	13 "	P=3.5" C*=7" CC=10.5"	212	N/A	N/A	1,230 Lbs	5:12
16" Tapered Mission Cover and Tapered Pans	16 "	P* = 7" C* = 7"	13 "	P=2" C*=7" CC=9"	246	N/A	N/A	1,300 Lbs	5:12
Greek	12 3/4 "	P* = 9 3/4 C* = 6 1/2	10 "	P=5.5" C*=6.5" CC=12"	240	N/A	N/A	1,550 Lbs	4:12
Roman	12 3/4 "	P* = 9 3/4 C* = 6 1/2	10 "	P=5.5" C*=6.5" CC=12"	240	N/A	N/A	1,550 Lbs	4:12
Straight Barrel Mission	14 1/4 "	P = 8" C = 8"	11 1/4 "	P=3.5" C=8" CC=11.5"	225	N/A	N/A	1,250 Lbs	5:12
Straight Barrel Mission	16 "	P = 8" C = 8"	13 "	P=3.5" C=8" CC=11.5"	192	N/A	N/A	1,190 Lbs	5:12
Straight Barrel Mission	18 3/8 "	P = 8" C = 8"	15 3/8 "	P=3.5" C=8" CC=11.5"	163	N/A	N/A	1,165 Lbs	5:12
Palm Beach	18 3/8 "	P = 8" C* = 8"	15 3/8 "	P=3.5" C*=8" CC=11.5"	163	N/A	N/A	1,100 Lbs	5:12
Italian Pan and Cover	Multiple combinations are possible. Contact Ludowici Roof Tile, New Lexington Ohio for instructions								

P = Pan (*Pan Measurement at Butt of Tile)
C = Cover (* Cover Measurement at Butt of Tile) , CC = Center to Center Dimensions

THERMAL AND MOISTURE PROTECTION

Metric									
Average Length, Width and Exposures of Tile Styles									
Tile Name	Length mm	Width mm	Length Exposure mm	Width Exposure mm	Pieces per 9.29 Sq.M	Under Eave Length mm	Under Eave Width mm	Avg. Wt. per 9.29 Sq. M, Kg.	Min. Slope
Flat Shingle Tiles									
Antique	300 mm	175.0 mm	125.0 mm	175.0 mm	412	175 mm	175 mm	748 Kg	5:12
Brittany	300 mm	175.0 mm	125.0 mm	175.0 mm	412	175 mm	175 mm	862 Kg	5:12
Calais	375 mm	175.0 mm	162.5 mm	175.0 mm	317	212.5 mm	175 mm	726 Kg	5:12
Colonial	365.6 mm	184.4 mm	132.8 mm	184.4 mm	310	207.8 mm	184.4 mm	816 Kg	5:12
Crude	300 mm	150.0 mm	125.0 mm	150.0 mm	480	175 mm	150 mm	878 Kg	5:12
Flat Slab Shingle 3/8"	300 mm	150.0 mm	125.0 mm	150.0 mm	480	175 mm	150 mm	590 Kg	5:12
Flat Slab Shingle 5/8"	300 mm	150.0 mm	125.0 mm	150.0 mm	480	175 mm	150 mm	807 Kg	5:12
Georgian	375 mm	200.0 mm	162.5 mm	200.0 mm	276	212.5 mm	200 mm	726 Kg	5:12
Norman	375 mm	175.0 mm	162.5 mm	175.0 mm	317	212.5 mm	175 mm	726 Kg	5:12
Provincial	375 mm	175.0 mm	162.5 mm	175.0 mm	317	212.5 mm	175 mm	714 Kg	5:12
Interlocking Tiles									
Americana	350 mm	225.0 mm	275.0 mm	206.3 mm	158	4 5/8 "	12 "	363 Kg	3:12
Americana - XL	400 mm	268.8 mm	325.0 mm	253.1 mm	109	4 5/8 "	12 "	340 Kg	3:12
Celadon	400 mm	268.8 mm	325.0 mm	253.1 mm	109	4 5/8 "	12 "	272 Kg	3:12
Classic	350 mm	225.0 mm	275.0 mm	206.3 mm	158	4 5/8 "	12 "	363 Kg	3:12
Classic - XL	400 mm	268.8 mm	325.0 mm	253.1 mm	109	4 5/8 "	12 "	340 Kg	3:12
Imperial	375 mm	250.0 mm	300.0 mm	231.3 mm	129	4 5/8 "	12 "	390 Kg	3:12
Lanai	350 mm	225.0 mm	275.0 mm	206.3 mm	158	4 5/8 "	12 "	363 Kg	3:12
Lanai - XL	400 mm	268.8 mm	325.0 mm	253.1 mm	109	4 5/8 "	12 "	340 Kg	3:12
Williamsburg	350 mm	225.0 mm	275.0 mm	206.3 mm	158	4 5/8 "	12 "	363 Kg	3:12
Williamsburg - XL	400 mm	268.8 mm	325.0 mm	253.1 mm	109	4 5/8 "	12 "	340 Kg	3:12
Interlocking Profile Tiles									
French	406.3 mm	225.0 mm	334.4 mm	203.1 mm	133	N/A	N/A	465 Kg	3:12
S- Tiles									
Spanish 13 1/4"	331.3 mm	243.8 mm	256.3 mm	206.3 mm	171	N/A	N/A	408 Kg	4:12
Spanish 18 3/8	468.8 mm	243.8 mm	384.4 mm	206.3 mm	114	N/A	N/A	363 Kg	4:12
Pan and Cover Tiles									
16" Tapered Mission and Straight Barrel Mission Tiles	400 mm	P = 200mm C* = 175mm	325.0 mm	P = 87.5mm C*= 175mm CC = 263mm	212	N/A	N/A	601 Kg	5:12
16" Tapered Mission Cover and Tapered Pans	400 mm	P* = 175mm C* = 175mm	325.0 mm	P = 50mm C*= 175mm CC = 225mm	246	N/A	N/A	635 Kg	5:12
Greek	318.8 mm	P* = 244mm C* = 163mm	250.0 mm	P = 138mm C*= 163mm CC = 300mm	240	N/A	N/A	757 Kg	4:12
Roman	318.8 mm	P* = 244mm C* = 163mm	250.0 mm	P = 138mm C*= 163mm CC = 300mm	240	N/A	N/A	757 Kg	4:12
Straight Barrel Mission	356.3 mm	P = 200mm C = 200mm	281.3 mm	P = 88mm C = 200mm CC = 288mm	225	N/A	N/A	610 Kg	5:12
Straight Barrel Mission	400 mm	P = 200mm C = 200mm	325.0 mm	P = 88mm C = 200mm CC = 288mm	192	N/A	N/A	581 Kg	5:12
Straight Barrel Mission	459.4 mm	P = 200mm C = 200mm	384.4 mm	P = 88mm C = 200mm CC = 288mm	163	N/A	N/A	569 Kg	5:12
Palm Beach	459.4 mm	P = 200mm C* = 200mm	384.4 mm	P = 88mm C*= 200mm CC =288mm	163	N/A	N/A	537 Kg	5:12
Italian Pan and Cover	Multiple combinations are possible. Contact Ludowici Roof Tile, New Lexington Ohio for instructions								

P = Pan (*Pan Measurement at Butt of Tile)
C = Cover (* Cover Measurement at Butt of Tile), CC= Center to Center dimensions

Laying Tile Shingles (Architectural Patterns). On a roof of average difficulty, 2 roofers and a helper should handle and lay 325 to 375 sq.ft. (30.19 to 34.83 sq.m) of tile shingles per 8-hr. day, at the following labor cost per square (100 sq.ft. or 9.29 sq.m):

Description	Hours	Rate	Total	Rate	Total
Roofer	4.60	$	$	$ 30.93	$ 142.28
Helper	2.30	$	$	$ 26.16	$ 60.17
Cost per 100 sq.ft.					$ 202.45
per sq. ft.					$ 2.02
per sq.m					$ 21.79

On steep or cut-up roofs, add 15% to 25%. For roofs over 3 stories high to eaves, add 2.3 hrs. helper time.

Laying Miscellaneous Tile Pattern Roofs. Where Spanish or French pattern tile is used on new work, figure same labor costs as for interlocking shingle tile.

Labor laying Mission tile will vary according to size and exposure of units. On new work of average difficulty, using 15" x 8" (375 x 200 mm) tile with 12" (300 mm) exposure, 2 roofers and a helper should handle and lay 350 to 400 sq.ft. (32.51 to 37.16 sq.m) per 8-hr. day.

Items to Be Included When Estimating Tile Roofs. The following items should be included when estimating the cost of a clay shingle tile roof.

One roll coated felt or roll roofing weighing 40 lbs. per sq. (1.95 kg per sq.m) and 3 lbs. (1.4 kg) roofing nails, 1-3/4" (43.75 mm) long.

Allow 1 lb. (0.45 kg) colored elastic cement for pointing joints for each 20 lin.ft. (6.09 m) of hips and ridges.

Cost of Clay Tile Roofing Materials. The Luduwici Roof Tile, Inc. New Lexington, Ohio offers a number of different patterns and colors in each of their "Standard" and "Special" series product line. They will also produce custom shapes and colors, if given enough lead time.

Each of the patterns are produced with field tiles and special shapes for the ridge, hip, rake, etc.

Approximate material prices for different patterns and colors are as follows:

Patterns and Colors	per 100 Sq.Ft.	per 10 Sq. M.
Spanish Red	$ 324.00	$ 349.00
Spanish Buff	$ 630.00	$ 678.00
Spanish Blue	$ 990.00	$ 1,066.00
Mission Granada Red	$ 630.00	$ 678.00
Americana Gray or Green	$ 405.00	$ 436.00
Lanai Black or Brown	$ 405.00	$ 436.00
Classic Red	$ 405.00	$ 436.00
French Red	$ 540.00	$ 581.00
French Blue	$ 1,170.00	$ 1,259.00

1086

THERMAL AND MOISTURE PROTECTION

The estimator needs to understand that clay tile roofs normally come with 50 to 75 year warranty. This longer warranty and maintenance free roofing material by life cycle costing may be less expensive then other types of materials.

07400 PREFORMED ROOFING & SIDING

07411 PREFORMED METAL SIDING

Corrugated steel is used extensively for roofing and siding steel mills, manufacturing plants, sheds, grain elevators, and other industrial structures.

It is made with various corrugations, varying in width and depth, but the 2-1/2" (62.50 mm) corrugation width is the most commonly used.

The sheets are usually furnished 26" (650 mm) wide and 6'-0" (1.82 m) to 30'-0" (9.14 m) long. Some manufacturers offer siding protected with vinyl coatings.

When estimating quantities of corrugated siding or roofing, always select lengths that work to best advantage and be sure and allow for both end and side laps.

Corrugated siding or roofing is nailed to the wood framework or siding when used on wood constructed buildings. When used as a wall and roof covering on structural steel framing, the corrugated sheets are fastened to the steel framework, using clip and bolts, which are passed around or under the purlins, which usually consist of channels, angles, or Z bars. When angles are used for purlins, clinch nails are sometimes used for fastening the corrugated sheets.

When used for siding one corrugation lap is usually sufficient, but for roofing two corrugations should be used and if the roof has only a slight pitch, the lap should be three corrugations.

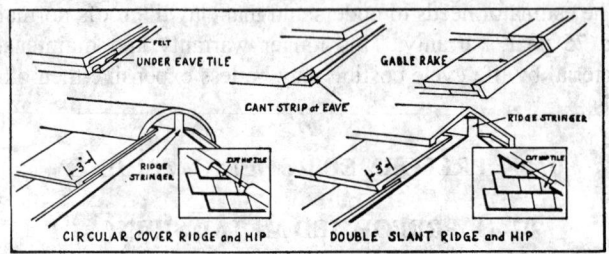

Details of Clay Interlocking Shingle Tile Roofing

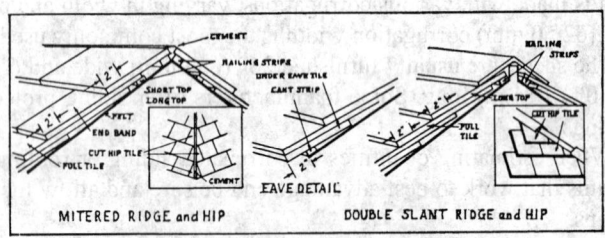

Details of Clay Tile Shingle Roofing

When used for siding, a 1" (25 mm) to 2" (50 mm) end lap is sufficient, but when laid on roofs it should have an end lap of 3" (75 mm) to 6" (150 mm) depending on the pitch of the roof. For a 1/3 pitch, a 3" (75 mm) lap is sufficient; for a 1/4 pitch, a 4" (100 mm) lap should be used; and for a 1/8 pitch, a 5" (125 mm) end lap is recommended.

When applying to wood sheathing or strips, the nails should be spaced about 8" (200 mm) apart at the sides. When applied to steel purlins, the side laps should extend over at least 1-1/2 corrugations, and the sheets should be riveted together every 8" (200 mm) on the sides and at every alternate corrugation on the ends.

Number of Sq.Ft. of Corrugated Sheets Required to Cover 100 Sq.Ft. of surface Using 26" x 96" Sheets						
Description	Length of End Lap in Inches					
2-1/2" Corrugation Width	1 "	2 "	3 "	4 "	5 "	6 "
Side lap, 1 corrugation	110	111	112	113	114	115
Side lap, 1-1/2 corrugations	116	117	118	119	120	121
Side lap, 2 corrugations	123	124	125	126	127	128
Side lap, 2-1/2 corrugations	130	131	132	133	134	135
Side lap, 3 corrugations	138	139	140	141	142	143

THERMAL AND MOISTURE PROTECTION

Number of Sq.M. of Corrugated Sheets Required to Cover 10 Sq.M. of surface Using 650 x 2400 mm Sheets						
Description	Length of End Lap in Inches					
62.50 mm Corrugation Width	25 mm	50 mm	75 mm	100 mm	125 mm	150 mm
Side lap, 1 corrugation	110.0	111.0	112.0	113.0	114.0	115.0
Side lap, 1-1/2 corrugations	116.0	117.0	118.0	119.0	120.0	121.0
Side lap, 2 corrugations	123.0	124.0	125.0	126.0	127.0	128.0
Side lap, 2-1/2 corrugations	130.0	131.0	132.0	133.0	134.0	135.0
Side lap, 3 corrugations	138.0	139.0	140.0	141.0	142.0	143.0

If shorter sheets are used the allowance should be slightly increased

Material Prices. Corrugated steel siding and roofing panels are manufactured in a number of metal gauges, widths, and depths of corrugations, and in galvanized and prefinished painted surfaces. Average material prices, per 100 sq.ft. (9.29 sq.m), for sheeting with a 2-1/2" (62.50 mm) corrugation width are as follows:

Price per 100 Sq.Ft.			Price per 10 Sq.M.		
Metal Gauge	Galvanized	Painted	Metal Gauge	Galvanized	Painted
28	$ 81.00	$ 108.00	28	$ 87.19	$ 116.25
26	$ 99.00	$ 126.00	26	$ 106.57	$ 135.63
24	$ 144.00	$ 180.00	24	$ 155.01	$ 193.76
22	$ 162.00	$ 189.00	22	$ 174.38	$ 203.44
20	$ 171.00	$ 198.00	20	$ 184.07	$ 213.13

Placing Corrugated Steel Roofing or Siding on Wood Framing. If corrugated sheets used for roofing or siding are nailed to wood strips or framing, a carpenter and helper should place one square (100 sq.ft. or 9.29 sq.m) of 26"x 96" (650 x 2400 mm) or larger, in 7/8 to 1-1/8 hr. or 7 to 9 sqs. per 8-hr. day, at the following labor cost per square (100 sq.ft. or 9.29 sq.m):

Labor Cost of One Square (100 sq.ft. or 9.29 sq.m) Corrugated Metal Roofing on Wood Frame					
Description	Hours	Rate	Total	Rate	Total
Carpenter	1.00	$	$	$ 35.04	$ 35.04
Labor	1.00	$	$	$ 26.16	$ 26.16
Cost per 100 sq.ft.					$ 61.20
per sq.ft.					$ 0.61
per sq.m					$ 6.59

Labor Cost of One Square (100 sq.ft. or 9.29 sq.m) Corrugated Metal Siding on Wood Frame					
Description	Hours	Rate	Total	Rate	Total
Carpenter	1.20	$	$	$ 35.04	$ 42.05
Labor	1.20	$	$	$ 26.16	$ 31.39
Cost per 100 sq.ft.					$ 73.44
per sq.ft.					$ 0.73
per sq.m					$ 7.91

Placing Corrugated Steel Roofing on Steel Framing. The labor cost of applying corrugated steel roofing to structural steel framing, where the sheets are fastened with clips or bolts, vary with the size of the roof area, regularity of roof, height above ground, etc., but on a straight job, 2 sheeters and 2 laborers should place 550 to 650 sq.ft. (51.09 to 60.38) per 8-hr. day, at the following labor cost per 100 sq.ft. (9.29 sq.m):

Description	Hours	Rate	Total	Rate	Total
Labor Handling - Hoisting	2.67	$	$	$ 26.16	$ 69.85
Carpenter/Roofer	2.67	$	$	$ 35.04	$ 93.56
Cost per 100 sq.ft.					$ 163.40
per sq.ft.					$ 1.63
per sq.m					$ 17.59

On irregular roofs, above costs may be increased 25% to 50%.

Placing Corrugated Steel Siding on Steel Framing. The labor cost of applying corrugated steel siding to walls varies greatly according to length and height of walls, quantity and regularity of openings, etc., but on a straight job, 2 sheeters and 2 laborers should place 450 to 550 sq.ft. (41.80 to 51.09 sq.m) per 8-hr. day, at the following labor cost per 100 sq.ft. (9.29 sq.m):

Description	Hours	Rate	Total	Rate	Total
Labor Handling - Hoisting	3.20	$	$	$ 26.16	$ 83.71
Carpenter/Roofer	3.20	$	$	$ 35.04	$ 112.13
Cost per 100 sq.ft.					$ 195.84
per sq.ft.					$ 1.96
per sq.m					

On coal tipples, grain elevators and other structures having high and irregular walls, the above costs may be doubled.

THERMAL AND MOISTURE PROTECTION
Protected Metal Roofing and Siding

Protected metal panels consist of a steel core sheet, to which felt is bonded by hot molten zinc. When the zinc has cooled, the felt and the steel core are metallurgically bound together. The felt is then impregnated with asphalt saturant. Finally, a tough, thick, waterproof outer coating is applied. All material is made with protective coating both sides.

There are several colors to choose from. Protected metal panels are manufactured in 4 gauges: 18, 20, 22, and 24, and are available in lengths up to 30 feet (9.14 meters).

Estimating and Cost Data. Panels are available in several profiles. The H.H. Robertson Company, Pittsburgh, Pa., has the following:

Huski-Rib has a deep fluted profile, designed specifically for high strength long span roofing and siding installations. The Huski-Rib sheet is 31-5/32" (779 mm) wide x 1-1/2" (37.5 mm) deep, and provides a coverage of 28" (700 mm). It is manufactured in 18, 20, 22, and 24 gauge metal, a maximum of 30' (9 m) long, and is suitable for spans up to 15'-9" (4725 mm) for siding, and up to 14'-14" (4550 mm) for roofing depending on local design load requirements.

Box-Rib has a strong, precise vertical texture, best suited for sidings only, either in standard or inverted positions. The standard sheet is 29-1/4" (731 mm) wide x 1-1/2" (37.5 mm) deep and provides a coverage of 28". It is manufactured in 18, 20, 22, and 24 gauge metal and is available in lengths up to 30 ft. (9 m).

Sturdi-Rib sheets come 34" (850 mm) wide x 9/16" (14 mm) deep providing a coverage of 29-7/16" (736 mm), and are made in lengths up to 30 ft (9 m). They are made of 18, 20, 22, and 24 gauge metal and are capable of spanning up to 10'-1" (3025 mm) for siding and 8'-9" (2625 mm) for roofing installations, depending upon local design load requirements.

The Magna-Rib profile sheets are 26-5/8" (666 mm) wide x 4" (200 mm) deep providing a coverage of 24" (600 mm) installed. They come in 18, 20, and 22 gauge metal in lengths up to 38 ft. (11.5 m). Magna-Rib sheets are capable of spanning up to 22'-8" (6800 mm) for siding in one span.

07415 CORRUGATED ALUMINUM ROOFING AND SIDING

When using aluminum roofing and siding, it is very important when applying the sheets that they be insulated against electrogalvanic action. Ordinarily aluminum has a high resistance to corrosion, but when it is contiguous to steel or copper and in the presence of moisture, an electrolytic cell is formed, and the resulting flow of current dissolves the aluminum, causing severe pitting. This condition can be avoided by preventing metal-to-metal contact.

Another significant requirement in using aluminum sheet is to seal the openings made for the fastening devices. Methods include the use of special washers and waterproof roofing compounds. Washers should be used with fasteners to distribute stresses and prevent tearing of the sheet.

To provide adequate drainage, the roof surface should never have a slope less than 2-1/2" per ft. (208.13 mm per m) and preferably not less than 3" per ft. (249.75 mm per m).

For roofing, sheets should have a side lap of 1-1/2 corrugations. For siding, sheets should be lapped 1 corrugation.

A 6" (150 mm) overlap at the ends is recommended for roofing and 4" (100 mm) for siding. When corrugated aluminum sheets are used without sheathing on a steel frame, the method shown in (a) is suggested for fastening to purlins and girts. This method utilizes aluminum straps, 12" (300 mm) apart, with aluminum bolts and nuts, aluminum rivets or cadmium-plated steel bolts.

To prevent galvanic action, direct contact between the top of the purlin and roofing should be prevented, preferably by insertion of an aluminum saddle or by painting the flange of the purlin with aluminum or bitumastic paint. Roofing paper may also be used to separate the metals.

Purlin clips (b) are also widely used as a fastening device. The U bend is slipped over the purlin flange and a leg hammered down against the web to lock the clip in place. The roofing is bolted to the clip. If standard steel purlin clips are used, they should be hot-dip galvanized.

Purlin nails, either single (c) or double (d) are also used for anchoring sheeting. A variation of this method employing aluminum washers, nuts, and base plate under the purlin is shown in (e).

The best method of nailing to sheathing is to use galvanized roofing nails about 1-3/4" (43.75 mm) long with a head about 3/8" (9.4 mm) in diameter, spaced at 8" (200 mm) intervals. These nails should have a hot-dip zinc coating.

Corrugated sheets should always be riveted, bolted, or nailed through the top of a corrugation. The reason is that rain running down the roof tends to collect in the bottom of corrugations and would penetrate imperfectly sealed holes there. At the top, water runs away from the holes.

A washer of non-metallic material, such as zinc chromate impregnated fabric, neoprene, or other rubber should be employed under a nailhead. The purpose of this washer is to prevent metal-to-metal contact between the underside of the head and the roofing sheet, to seal the opening made in the sheet and to permit thermal expansion and contraction of the sheeting.

Hot-dip galvanized nails without washers may be used when washers are not available. Roofing nails with cast-on lead heads can be used in other than industrial or seacoast corrosive atmospheres.

1092

THERMAL AND MOISTURE PROTECTION

Sizes, Weights, and Coverage Data on Corrugated Aluminum Sheets Used for Roofing

Lengths 5'-0" (1500 mm) to 12'-0" (3600 mm) in 6" (150 mm) increments. Widths, 35" (875 mm) and 48-3/8" (1209 mm). Thicknesses, 0.024" (0.6 mm) and 0.032" (0.8 mm). Corrugation, 2.67" (66.7 mm) pitch and 7/8" (22 mm) deep. Coverage, 32" (800 mm) for 35" (875 mm) width and 45-3/8" (1134 mm) for 48-3/8" (1209 mm).

Data on 0.032" (1 mm) Corrugated Aluminum Roofing Sheets						
Sheet Length, Feet	Sq.Ft. per Sheet		Weight per Sheet, Lbs.		Approx. No. Sheets per 100 Sq.Ft.*	
	35 "	48 3/8 "	35 "	48 3/8 "	35 "	48 3/8 "
5	14.58	20.16	8.04	11.03	6.86	4.96
5 1/2	16.04	22.17	8.85	12.13	6.23	4.51
6	17.50	24.19	9.65	13.24	5.71	4.13
6 1/2	18.96	26.20	10.46	14.34	5.27	3.82
7	20.42	28.22	11.26	15.44	4.9	3.54
7 1/2	21.88	30.23	12.07	16.55	4.57	3.31
8	23.33	32.25	12.87	17.65	4.29	3.1
8 1/2	24.79	34.27	13.68	18.75	4.03	2.92
9	26.25	36.28	14.48	19.86	3.81	2.76
9 1/2	27.71	38.30	15.29	20.96	3.61	2.61
10	29.17	40.31	16.09	22.06	3.43	2.48
10 1/2	30.63	42.33	16.9	23.16	3.26	2.36
11	32.08	44.34	17.7	24.27	3.12	2.26
11 1/2	33.54	46.36	18.51	25.37	2.98	2.16
12	35.00	48.37	19.31	26.47	2.86	2.07
Sheet Length, Meters	Sq.M. per Sheet		Weight per Sheet, Kg.		Approx. No. Sheets per 10 Sq.M.*	
	875 mm	1209.4 mm	875 mm	1209.4 mm	875 mm	1209.4 mm
1.52	1.35	1.87	3.65	5.00	7.38	5.34
1.68	1.49	2.06	4.01	5.50	6.71	4.85
1.83	1.63	2.25	4.38	6.01	6.15	4.45
1.98	1.76	2.43	4.74	6.50	5.67	4.11
2.13	1.90	2.62	5.11	7.00	5.27	3.81
2.29	2.03	2.81	5.47	7.51	4.92	3.56
2.44	2.17	3.00	5.84	8.01	4.62	3.34
2.59	2.30	3.18	6.21	8.51	4.34	3.14
2.74	2.44	3.37	6.57	9.01	4.10	2.97
2.90	2.57	3.56	6.94	9.51	3.89	2.81
3.05	2.71	3.74	7.30	10.01	3.69	2.67
3.20	2.85	3.93	7.67	10.51	3.51	2.54
3.35	2.98	4.12	8.03	11.01	3.36	2.43
3.51	3.12	4.31	8.40	11.51	3.21	2.33
3.66	3.25	4.49	8.76	12.01	3.08	2.23

For 0.024" (0.6 mm) thickness, reduce weights by 25%.
*For side and end lap allowance, add approximately 16% for 35" (875 mm) width and 12% for 48 3/8" (1209 mm) width.

Sizes, Weight, and Coverage Data on Corrugated Aluminum Sheets Used for Siding

Lengths, 5'-0" (1500 mm) to 12'-0" (3600 mm) in 6" (150 mm) increments. Widths, 33-3/4" (844 mm) and 47-1/8" (1178 mm). Thicknesses, 0.024" (0.6 mm) and 0.032" (0.8 mm). Corrugation, 2.67" (66.7 mm) pitch and 7/8" (22 mm) deep. Coverage, 32" (800 mm) for 33-3/4" (844 mm) width and 45-3/8 (1134 mm) for 47-1/8" (1178 mm) width.

Data on 0.032" (0.8 mm) Corrugated Aluminum Siding Sheets						
Sheet Length Feet	Sq.Ft. per Sheet		Weight per Sheet, Lbs.		Approx. No. Sheets per 100 Sq.Ft.*	
	33 3/4 "	47 1/8 "	33 3/4 "	47 1/8 "	33 3/4 "	47 1/8 "
5	14.06	19.64	7.76	10.75	7.11	5.09
5 1/2	15.47	21.60	8.54	11.82	6.46	4.63
6	16.87	23.56	9.31	12.90	5.93	4.24
6 1/2	18.28	25.53	10.09	13.97	5.47	3.92
7	19.69	27.49	10.87	15.05	5.08	3.64
7 1/2	21.09	29.45	11.65	16.12	4.74	3.40
8	22.50	31.42	12.42	17.20	4.44	3.18
8 1/2	23.91	33.38	13.20	18.27	4.18	3.00
9	25.31	35.34	13.97	19.35	3.95	2.83
9 1/2	26.72	37.31	14.75	20.42	3.74	2.68
10	28.12	39.27	15.52	21.50	3.56	2.55
10 1/2	29.53	41.23	16.30	22.57	3.39	2.43
11	30.94	43.20	17.08	23.65	3.23	2.31
11 1/2	32.34	45.16	17.86	24.72	3.09	2.21
12	33.75	47.13	18.63	25.80	2.96	2.12
Sheet Length Feet	Sq.M. per Sheet		Weight per Sheet, Kg.		Approx. No. Sheets per 10 Sq.M.*	
	843.75 mm	1178.13 mm	843.75 mm	1178.13 mm	843.75 mm	1178.13 mm
1.52	1.31	1.82	3.52	4.88	7.65	5.48
1.68	1.44	2.01	3.87	5.36	6.95	4.98
1.83	1.57	2.19	4.22	5.85	6.38	4.56
1.98	1.70	2.37	4.58	6.34	5.89	4.22
2.13	1.83	2.55	4.93	6.83	5.47	3.92
2.29	1.96	2.74	5.28	7.31	5.10	3.66
2.44	2.09	2.92	5.63	7.80	4.78	3.42
2.59	2.22	3.10	5.99	8.29	4.50	3.23
2.74	2.35	3.28	6.34	8.78	4.25	3.05
2.90	2.48	3.47	6.69	9.26	4.03	2.88
3.05	2.61	3.65	7.04	9.75	3.83	2.74
3.20	2.74	3.83	7.39	10.24	3.65	2.62
3.35	2.87	4.01	7.75	10.73	3.48	2.49
3.51	3.00	4.20	8.10	11.21	3.33	2.38
3.66	3.14	4.38	8.45	11.70	3.19	2.28

For 0.024" (0.6 mm) thickness use data given above, except reduce weights 25%.

*For side and end lap allowance, add approximately 9% for 33-3/4" (800 mm) width and 6% for 47 1/8" (1178 mm) width.

THERMAL AND MOISTURE PROTECTION

Data on Nails Required for Roofing						
Kind of Nail	Length in Inches	# per Lb.	Price per Lb.	Length in mm	# per Kg.	Price per Kg.
Galvanized Needle Point	1 3/4	83	$ 2.00	43.75	182.98	$ 4.41
Galvanized Needle Point	2	77	$ 2.00	50	169.75	$ 4.41
Aluminum Nails-Neoprene Washers	1 3/4	318	$ 6.00	43.75	701.058	$ 13.23
Aluminum Nails-Neoprene Washers	2	285	$ 6.00	50	628.307	$ 13.23

Approximate Prices per Sq.Ft. (Sq.M) of Corrugated Aluminum Roofing and Siding Material				
Thickness Inches	Finish	Price per Sq.Ft.	Thickness mm	Price per Sq.M
0.0175 "	Mill	$ 0.80	0.44	$ 8.61
0.0175 "	Painted	$ 0.90	0.44	$ 9.69
0.0240 "	Mill	$ 1.50	0.60	$ 16.15
0.0240 "	Painted	$ 2.00	0.60	$ 21.53
0.0320 "	Mill	$ 2.00	0.80	$ 21.53
0.0320 "	Painted	$ 2.75	0.80	$ 29.60

After figuring the price of sheets from the above, approximately 20% should be added to cover cost of closures, flashings, fillers, corners, etc.

Labor Placing Corrugated Aluminum Roofing and Siding. Estimate about the same labor cost for erecting aluminum corrugated roofing and siding as given on the previous pages under Corrugated Steel Roofing and Siding.

Aluminum is much lighter than steel, but the labor operations are practically the same.

Labor Cost of One Square (100 sq.ft. or 9.29 sq.m) Corrugated Aluminum Roofing on Wood Framing, Based on using sheets 35" (875 mm) wide, 0.032" (0.8 mm) thick, weighing 0.552 lbs. per sq.ft. (2.69 kg per sq.m)

Description	Hours	Rate	Total	Rate	Total
Carpenter	1.00	$	$	$ 35.04	$ 35.04
Labor	1.00	$	$	$ 26.16	$ 26.16
Cost per 100 sq.ft.					$ 61.20
per sq.ft.					$ 0.61
per sq.m					$ 6.59

Cost of One Square (100 sq.ft. or 9.3 sq.m) Corrugated Aluminum Siding on Wood Framing, Based on using sheets 33-3/4" (844 mm) wide, 0.032" (0.8 mm) thick, weighing 0.552 lbs. per sq.ft. (2.69 kg per sq.m)

Description	Hours	Rate	Total	Rate	Total
Carpenter	1.20	$	$	$ 35.04	$ 42.05
Labor	1.20	$	$	$ 26.16	$ 31.39
Cost per 100 sq.ft.					$ 73.44
per sq.ft.					$ 0.73
per sq.m					$ 7.91

Cost of One Square (100 sq.ft. or 9.3 sq.m) Corrugated Aluminum Siding on Steel Framing, Straight Work, Based on using sheets 33-3/4" (844 mm) wide, 0.032" (0.8 mm) thick, weighing 0.552 lbs. per sq.ft. (2.69 kg per sq.m)

Description	Hours	Rate	Total	Rate	Total
Labor handling, hoisting	3.20	$	$	$ 26.16	$ 83.71
Sheeters	3.20	$	$	$ 26.16	$ 83.71
Cost per 100 sq.ft.					$ 167.42
per sq.ft.					$ 1.67
per sq.m					$ 18.02

Add extra labor for high and irregular walls such as grain elevators, conveyor housing, etc. Add for closures, flashings, etc.

Cost of One Square (100 sq.ft. or 9.3 sq.m) Corrugated Aluminum Roofing on Steel Framing, Straight Work, Based on using sheets 35" (875 mm) wide, 0.032" (0.8 mm) thick, weighing 0.552 lbs. per sq.ft. (2.69 kg per sq.m)

Description	Hours	Rate	Total	Rate	Total
Labor handling, hoisting	2.67	$	$	$ 26.16	$ 69.85
Sheeter	2.67	$	$	$ 26.16	$ 69.85
Cost per 100 sq.ft.					$ 139.69
per sq.ft.					$ 1.40
per 10 sq.m					$ 15.04

Add extra labor for high or irregular roof areas. Add for closures, flashings, etc.

THERMAL AND MOISTURE PROTECTION
07460 CLADDING/SIDING

Wood siding is available in plain bevel, plain shiplap, and tongue and groove patterns to be used horizontally in heights from 4" (100 mm) to 12" (300 mm). Bevel siding is either 1/2" (12.50 mm) or 3/4" (18.75 mm) thick, T&G, and shiplap nominal 1" (25 mm). Cedar is the most common wood for siding but redwood, fir, hemlock, and spruce are also stocked. Wood may also be installed vertically with wood battens at the joints. Bevel siding 1/2" (12.50 mm) thick in clear "A" grade cedar will run about $1.25 per bd.ft., in redwood about $2.40. T&G red cedar 1x 8s (25 x 200 mm) "D" and better will run $2.00 per bd.ft.; T&G redwood, clear and better, runs $3.50.

Hardboard siding 7/16" (10.93 mm) thick and factory primed, costs $1.00 per sq.ft.

Plywood siding in 4' x 8' (1.21 x 2.43 m) sheets, rough textured and grooved to look like individual boards 5/8" (15.75 mm) thick costs about $1.50 per sq.ft. in select grade Douglas fir, $3.00 in redwood.

Aluminum siding, factory finished white in 8" (200 mm) widths, will run $1.50 uninsulated, $2.15 insulated per sq.ft. Flashing strips will cost $0.50 per lin.ft.; inside corners, $1.00; outside corners $2.00 per lin.ft. Colored nails cost $3.75 per lb.

Vinyl siding, factory finished in 8" (200 mm) wide strips, costs about $1.50 per sq.ft. uninsulated, $2.20 with insulating backup. Corner boards run $2.00 for outside positions, $1.00 for inside. Trim moldings run $0.75 per lin.ft., while soffits run $2.50 per sq.ft.

Placing Bevel and Drop Siding. The labor cost of placing bevel or drop siding will vary with the class of work and the method of placing same.

On the less expensive type of construction, it is customary to square only one end of the siding, and the ends at the corners are left rough and are later covered with metal corner pieces. This is the cheapest method of placing siding. Metal corners are used extensively in the South, because the long summers and intense heat cause mitered wood corners to open up, making a very unsightly appearance.

On more expensive buildings, the door and window casings and corner boards are placed, and then it is necessary to cut and fit each piece of siding between the corner boards or between the casings and corner boards. On work of this class, two carpenters usually work together, because it is necessary to square one end of the siding and then measure and cut each board separately to insure a snug fit. This increases labor costs considerably but makes for a better appearance than the metal corners.

This also applies where the siding is mitered at the corners. Two carpenters working together must measure and miter each piece of siding so that it will fit the adjoining corner.

The following quantities and production time are based on using ordinary brackets with plank scaffolding. If a more elaborate scaffold is required for placing siding, add extra.

Labor Placing Drop Siding				
Measured Size Inches	Actual Size Inches	Class of Work	Bd.Ft. Placed per 8-Hr. Day	Carpenter Hours per 1,000 Bd.Ft.
6	5 1/4	Rough Ends	525 - 575	14.5 *
6	5 1/4	Fitted Ends	350 - 400	21.4
6	5 1/4	Mitered Corners	285 - 325	26.3
8	7 1/4	Rough Ends	600 - 650	12.8 *
8	7 1/4	Fitted Ends	415 - 460	18.2
8	7 1/4	Mitered Corners	325 - 375	23.0
(metric)				
Measured Size mm	Actual Size mm	Class of Work	Cu.M. Placed per 8-Hr. Day	Carpenter Hours per 3 Cu.M.
150	131.25	Rough Ends	1.24 - 1.36	18.44 *
150	131.25	Fitted Ends	0.83 - 0.94	27.22
150	131.25	Mitered Corners	0.67 - 0.77	33.45
200	181.25	Rough Ends	1.42 - 1.53	16.28 *
200	181.25	Fitted Ends	0.98 - 1.08	23.15
200	181.25	Mitered Corners	0.77 - 0.88	29.26

*Where an electric saw is used to square both ends of the siding before placing, leaving the exposed corners rough to be covered with metal corner pieces, deduct 1 to 1-1/2 hours time per 1,000 bd.ft. (2.35 Cu.M.)

THERMAL AND MOISTURE PROTECTION

Quantity of Bevel Siding Required Per 100 Sq.Ft. of Wall					
Measured Size Inches	Actual Size Inches	Exposed to Weather	Pattern	Add for Lap	BF Req. per Sq.Ft. Surface
1/2 x 4	1/2 x 3 1/4	2 3/4	Regular	46%	151
1/2 x 5	1/2 x 4 1/4	3 3/4	Regular	33%	138
1/2 x 6	1/2 x 5 1/4	4 3/4	Regular	26%	131
1/2 x 8	1/2 x 7 1/4	6 3/4	Regular	18%	123
5/8 x 8	5/8 x 7 1/4	6 3/4	Regular	18%	123
3/4 x 8	3/4 x 7 1/4	6 3/4	Rabbetted	18%	123
5/8 x 10	5/8 x 9 1/4	8 3/4	Rabbetted	14%	119
3/4 x 10	3/4 x 9 1/4	8 3/4	Rabbetted	14%	119
3/4 x 12	3/4 x 11 1/4	10 3/4	Rabbetted	12%	117
Measured Size mm	Actual Size mm	Exposed to Weather	Pattern	Add for Lap	Cu.M Req. per Sq.M. Surface
12.50 x 100	12.50 x 81.25	68.75	Regular	46%	3.83
12.50 x 125	12.50 x 106.25	93.75	Regular	33%	3.50
12.50 x 150	12.50 x 131.25	118.75	Regular	26%	3.33
12.50 x 200	12.50 x 181.25	168.75	Regular	18%	3.12
15.63 x 200	15.63 x 181.25	168.75	Regular	18%	3.12
18.75 x 200	18.75 x 181.25	168.75	Rabbetted	18%	3.12
15.63 x 250	15.63 x 231.25	218.75	Rabbetted	14%	3.02
18.75 x 250	18.75 x 231.25	218.75	Rabbetted	14%	3.02
18.75 x 300	18.75 x 281.25	268.75	Rabbetted	12%	2.97

The above quantities include 5% for end cutting and waste.

Quantity of Drop Siding Required Per 100 S.F. (10 Sq.M) of Wall				
Measured Size Inches	Actual Size Inches	Exposed to Weather	Add for Lap	BF Required per 1000 Sq.Ft. Surface
1 x 6	3/4 x 5 1/4	5 1/4	14%	119
1 x 8	3/4 x 7 1/4	7 1/4	10%	115
Measured Size MM	Actual Size Inches	Exposed to Weather	Add for Lap	Cu.M Required per 3 Sq.M. Surface
25 x 150	18.75 x 131.25	131.25	14%	0.0091
25 x 200	18.75 x 181.25	181.25	10%	0.0088

The above quantities include 5% for end cutting and waste.

Quantity of Shiplap Required Per 100 S.F. (10 Sq.M) of Surface

Measured Size Inches	Actual Size Inches	Add for Lap	BF Required per 100 Sq.Ft. Surface
1 x 8	3/4 x 7 1/4	10%	115
1 x 10	3/4 x 9 1/4	8%	113
Measured Size MM	Actual Size mm	Add for Lap	Cu.M Required per 10 Sq.M. Surface
25 x 200	18.75 x 181.25	10%	0.2920
25 x 250	18.75 x 231.25	8%	0.2869

The above quantities include 5% for end cutting and waste.

Quantity of bead Ceiling & Partition Required Per 100 S.F. (10 Sq.M) of Surface

Measured Size Inches	Actual Size Inches	Add for Lap	BF Required per 100 Sq.Ft. Surface
1 x 4	5/8 x 3 1/4	23%	128
1 x 4	3/4 x 3 1/4	23%	128
Measured Size MM	Actual Size mm	Add for Lap	Cu.M Required per 10 Sq.M. Surface
25 x 100	15.63 x 81.25	23%	0.3250
25 x 100	18.75 x 81.25	23%	0.3250

The above quantities include 5% for end cutting and waste.

Quantity of Dressed & Matched (D&M) or Tongued & Grooved (T&G) Boards Required Per 100 S.F. (10 Sq.M) of Surface

Measured Size Inches	Actual Size Inches	Add for Lap	BF Required per 100 Sq.Ft. Surface
1 x 6	3/4 x 5 1/4	14%	119
2 x 6	1 5/8 x 5 1/4	14%	238
Measured Size MM	Actual Size mm	Add for Lap	Cu.M Required per 10 Sq.M. Surface
25 x 150	18.75 x 131.25	14%	0.3021
50 x 150	40.63 x 131.25	14%	0.6042

The above quantities include 5% for end cutting and waste.

THERMAL AND MOISTURE PROTECTION

Labor Placing Bevel Siding					
Measured Size Inches	Actual Size Inches	Exposed to Weather Inches	Class of Workmanship	BF placed per 8-Hr Day	Carpenter Hrs per 1,000 Bd.Ft.
4	3 1/4	2 3/4	Rough Ends	350 - 400	21.3 *
4	3 1/4	2 3/4	Fitted Ends	240 - 285	30.5
4	3 1/4	2 3/4	Mitered Corners	200 - 240	36.3
5	4 1/4	3 3/4	Rough Ends	415 - 460	18.2 *
5	4 1/4	3 3/4	Fitted Ends	285 - 330	25.0
5	4 1/4	3 3/4	Mitered Corners	240 - 285	30.5
6	5 1/4	4 3/4	Rough Ends	475 - 525	16.0 *
6	5 1/4	4 3/4	Fitted Ends	325 - 375	23.0
6	5 1/4	4 3/4	Mitered Corners	265 - 310	28.0
8	7 1/4	6 3/4	Rough Ends	570 - 620	13.4 *
8	7 1/4	6 3/4	Fitted Ends	375 - 415	20.0
8	7 1/4	6 3/4	Mitered Corners	300 - 350	24.0
10	9 1/4	8 3/4	Rough Ends	650 - 700	12.0 *
10	9 1/4	8 3/4	Fitted Ends	440 - 480	17.5
10	9 1/4	8 3/4	Mitered Corners	375 - 425	20.0
12	11 1/4	10 3/4	Fitted Ends	475 - 525	16.0
12	11 1/4	10 3/4	Mitered Corners	400 - 450	19.0
Metric					
Measured Size mm	Actual Size mm	Exposed to Weather mm	Class of Workmanship	Cu.M. Placed per 8-Hr Day	Carpenter Hrs per 3 Cu M.
100	81.25	68 3/4	Rough Ends	0.83 - 0.94	27.09 *
100	81.25	68.75	Fitted Ends	0.57 - 0.67	38.80
100	81.25	68.75	Mitered Corners	0.47 - 0.57	46.17
125	106.25	93.75	Rough Ends	0.98 - 1.08	23.15 *
125	106.25	93.75	Fitted Ends	0.67 - 0.78	31.80
125	106.25	93.75	Mitered Corners	0.57 - 0.67	38.80
150	131.25	118.75	Rough Ends	1.12 - 1.24	20.35 *
150	131.25	118.75	Fitted Ends	0.77 - 0.88	29.26
150	131.25	118.75	Mitered Corners	0.63 - 0.73	35.62
200	181.25	168.75	Rough Ends	1.34 - 1.46	17.04 *
200	181.25	168.75	Fitted Ends	0.88 - 0.98	25.44
200	181.25	168.75	Mitered Corners	0.71 - 0.83	30.53
250	231.25	218.75	Rough Ends	1.53 - 1.65	15.26 *
250	231.25	218.75	Fitted Ends	1.04 - 1.13	22.26
250	231.25	218.75	Mitered Corners	0.88 - 1.00	25.44
300	281.25	268.75	Fitted Ends	1.12 - 1.24	20.35
300	281.25	268.75	Mitered Corners	0.94 - 1.06	24.17

*Where an electric handsaw is used to square both ends of the siding before placing, leaving the exposed corners rough to be covered with metal corner pieces, deduct 1 to 1-1/2 hours time per 1,000 b.f.

07500 MEMBRANE ROOFING

Built-up roofing consists of alternate plies of saturated felt and moppings of pitch with tar-saturated felt, or asphalt with asphalt-saturated felt, covered with a top pouring of pitch or asphalt into which slag or gravel is embedded. On flat roofs with slopes of less than 1/4" (6.25 mm) per foot (0.30 m), on which water may collect and stand, coal tar pitch and felt or a low melting point asphalt bitumen and asphalt felt are generally used. Coal tar pitch is not recommended for roofs having an incline in excess of 1" (25 mm) per foot (0.30 m).

On built-up roofs, where slope is over 2" (50 mm) to 4" (100 mm) per foot (0.30 m), steep asphalt, 180°-200°oF (82°-93°C) melting point, is more suitable and slag is embedded in preference to gravel, because it remains embedded better than well rounded gravel. Where slag is not available and where a light gray or white surface is desired, hard limestone chips, angular pieces 1/4" (6.25 mm) to 3/4" (18.75 mm) in size, are embedded in the top pouring of bitumen. On slopes over 2" (50 mm) per foot (0.30 m), double coverage mineral surfaced roll roofing with a 19" (475 mm) selvage edge may be used as the top finish.

Smooth surface built-up roofing consists of alternate plies of felt cemented solid to the base sheet and to each other with asphalt, and a top coating of hot or cold asphalt, uniformly distributed.

The specifications on the types of roofing vary widely, depending on the surface to which the built-up roofing is applied and the service required.

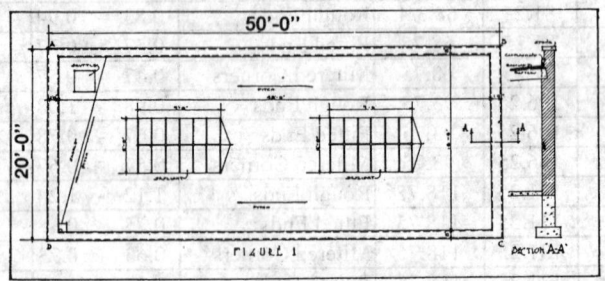

Fig. 1. Flat Roof with Parapet Walls

The cost of built-up roofing is governed by the incline of the roof, size, plan (whether cut up with openings, skylights, penthouses, irregular roof levels, etc.), and the distance of the roof above the ground. The higher the roof, the higher materials must be hoisted, increasing the labor costs.

The term, "built-up roof" will be commonly heard in the trade, but it is more properly "built-up roofing". Technically, the "roof" is the supporting structure over which the "roofing" is applied. Specification writers,

1102

designers, and attorneys will be careful to distinguish between "roof" and "roofing".

Rules for Measuring Flat Roof. When measuring flat roof surfaces that are to be covered with composition, tar and gravel, tin, metal, or prepared roofing, the measurements should be taken from the outside of the walls on all four sides to allow for flashing up the side of each wall. The flashing is usually 8" (200 mm) to 1'- 0" (300 mm) high. This applies particularly to brick, stone, or tile buildings having parapet walls above the roof level.

On flat roofs projecting or overhanging beyond the walls of the building, the measurements should cover the outside dimensions of the roof and not merely to the outside of the walls.

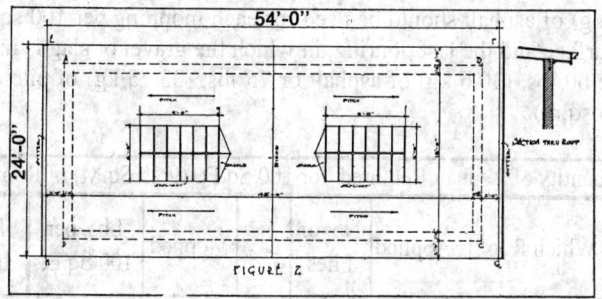

Fig. 2. Flat Roof Overhanging Walls

When estimating the area of flat roof surfaces, do not make deductions for openings containing less than 100 sq.ft. and then deductions should be made for just one-half the size of the opening.

Make deductions in full for all openings having an area of 500 sq.ft. (46 sq.m) or more.

Example: The length of the building, as shown in Figure 1, is 50'-0" (15.24 m). The width of the building is 20'-0" (6.09 m). Multiply the length by the width, 50 x 20 = 1,000 sq.ft. (15.24 x 6.09 = 92.81 sq.m). Referring to the figure, note that the building is only 48'-0" (14.63 m) long and 18'-8" (5.69 m) wide between walls. The method of measuring the full length and width of the building is to allow for flashing up the side of the walls, which usually extends 8" (200 mm) to 12" (300 mm) high.

Figure 2 illustrates a building of the same size with projecting roof. The size of the building proper is 20'-0"x50'-0" (6.09 x 15.24 m), while the roof projects beyond the walls 2'-0" (600 mm) on each side, making the size of the roof 24'-0"x54'-0" (7.31 x 16.45 m). The roof area is obtained by multiplying the length of the roof, 54'-0" (16.45 m), by the width of the roof, 24'-0" (7.31 m), making the total roof area, 54'-0" x 24'-0" = 1,296 sq.ft. (16.45 x 7.31 = 120.24 sq.m).

The skylights on both roofs measure 6'-0"x 9'-0" (1.82 x 2.74 m) and contain only 54 sq.ft. (5.01 sq.m) each. No deduction should be made for these openings, as they contain less than 100 sq.ft. (9.29 sq.m) each and the extra labor flashing up the sides of the walls more than offsets the cost of the small openings.

Quantity of Pitch or Asphalt Required for Built-Up Roofs. The quantity of roofing pitch or asphalt required for any built-up roof will vary with the number of thicknesses or plies of felt used. All manufacturers of roofing materials indicate in their specifications the amount of materials required for first class roof. On every first class roof, plenty of pitch or asphalt is used in each mopping so that each ply of felt is well cemented to the next and in no instance does felt touch felt.

Approximately 25 lbs. (11.3 kg) of coal tar pitch or approximately 20 lbs. (9 kg) of asphalt should be used for each mopping per 100 sq.ft. (9.29 sq.m) of surface and the last pouring, in which the gravel or slag is imbedded, should use 60 lbs. (30.6 kg) of asphalt or 75 lbs. (33.75 kg) of pitch per 100 sq.ft. (9.29 sq.m).

Quantity of Asphalt Required For 100 Sq.Ft. (9.29 Sq.M) of Roof					
Surface to Which Roof is Applied	No. of Plies	Dry	Mopped	Lbs. per 100 Sq.Ft.	Kg. per 10 Sq.M.
Wood, plywood, structural wood fiber	5	2	3	120	58.59
Poured gypsum, lightweight concrete	4	1	3	120	58.59
Concrete, precast concrete or gypsum	4	------	4	140	68.36

For roofing pitch add about 5 lbs. (2 kg) per mopping each ply per 100 sq.ft. (9.29 sq.m)—add 15 lbs. (6.8 kg) for top pouring per 100 sq.ft. (9.29 sq.m).

Estimating the Quantity of Roofing Felt. Asphalt or tarred felt for built-up roofing is furnished in 4 sq. rolls containing 432 sq.ft. (40.13 sq.m) and weighing from 56 to 62 lbs. (25-28 kg) per roll for No. 15 felt. No. 30 felt, which is generally used as the base sheet over wood construction when an asphalt specification is used, is furnished in 2 sq. rolls containing 216 sq.ft. (20.06 sq.m) and weighing 60 lbs. (27 kg). These weights are those used on a bonded job in the best type of building construction.

When specifying the grade or weight of felt to be used, it is customary to state that "felt shall weigh not less than 15 lbs. per 108 sq.ft.", and 15 lbs. per 108 sq.ft. (6.8 kg per 10 sq.m) is standard practice of all roofing manufacturers. Because felt is furnished in 4 sq. rolls of 432 sq.ft.

(40.13 sq.m), 32 sq.ft. (2.97 sq.m) per roll or 8 sq.ft. (0.74 sq.m) per 100 sq.ft. (9.29 sq.m) is allowed for laps.

The following table gives the quantity of roofing felt required to cover 100 sq.ft. (9.29 sq.m) of surface of various thicknesses:

Number of Plies	Add for Waste	Sq.Ft. per 100 Sq.Ft. of Roof	Lbs. per Sq. of Roof Using No. 15 Felt	Sq.M. per 10 Sq.m. of Roof	Kg. per 10 Sq.M. of Roof Using No. 15 Felt
1	8%	108	15	10.80	7.32
2	8%	216	30	21.60	14.65
3	8%	324	45	32.40	21.97
4	8%	432	60	43.20	29.30
5	8%	540	75	54.00	36.62

The general use of No. 30 felt is as a base sheet over wood decks in 1-ply thickness over which subsequent layers of No. 15 felts are mopped in.

Quantity of Roofing Gravel Required for Built-Up Roofs. Roofing gravel should be uniformly embedded into a heavy top pouring of asphalt or pitch so that approximately 400 lbs. (181 kg) of gravel or 300 lbs. (136 kg) of slag is used per 100 sq.ft. (9.29 sq.m) of roof area.

Labor Applying Built-Up Roofing. The labor cost of handling materials and applying built-up roofing will vary with the specification used and the type of building to which the roofing is applied. On the low type of building, 1, 2, or 3 stories high, where there is no great distance from the ground to the roof deck, the labor for hoisting materials, and supplying hot asphalt or pitch from the ground to the roof deck is considerably less than on a high building, where there is a greater distance from the kettle to the roof.

On low type buildings, a 5-worker crew composed of 1 worker at the kettle, 1 worker carrying "hot stuff", 1 worker rolling in felts, 1 worker nailing in and 1 worker mopping in, can lay the following areas per 8-hr. day:

1. 18 sqs. (167 sq.m) of 5-ply asphalt and gravel or pitch and gravel over wood roof deck, consisting of 1 sheathing paper, 2 dry felts and 3 plies of felt mopped in with a top pouring of pitch or asphalt and gravel surfacing;
2. 18 sqs. (20.25 sq.m) of 4-ply asphalt and gravel or pitch and gravel over a concrete deck, all mopped including a top pouring of pitch or asphalt and gravel or slag surfacing;
3. 22 sqs. (29.91 sq.m) of 4-ply asphalt and gravel or pitch and gravel over wood roof deck, consisting of 1 sheathing paper, 2 dry felts and 2 plies of felt mopped in with a top pouring of pitch or asphalt and gravel or slag surfacing;

4. 22 sqs. (39.20 sq.m) of 3-ply asphalt and gravel or pitch and gravel over concrete roof deck, all mopped and including a top pouring of pitch or asphalt and gravel or slag surfacing;

5. 24 sqs. (48.67 sq.m) of 3-ply asphalt and gravel, or pitch and gravel, over wood roof deck, consisting of 1 dry felt, and 2 plies of felt mopped in, with a top pouring of pitch or asphalt, and gravel or slag surfacing.

The above figures apply to buildings up to 3 stories in height.

If the roof surfaces are broken up with skylights, irregular roof level, etc., the above quantities should be reduced about 15% to 20%.

A 5-worker crew should apply the following number of squares (sq.m) of roof per 8-hr. day, based on first class workmanship throughout:

Type of Structure				
Type of Roofing	"A" Squares	"B" Squares	"A" Sq. Meter	"B" Sq.Meter
3-ply roof over wood roof deck	24	18	222.96	167.22
3-ply roof over concrete roof deck or insulation	22	17	204.38	157.93
4-ply roof over wood roof deck	22	17	204.38	157.93
4-ply roof over concrete roof deck or insulation	18	14	167.22	130.06
5-ply roof over wood roof deck	18	14	167.22	130.06

Type "A" buildings are the most convenient type of structure for application of built-up roofing. A low type of building from 1 to 3 stories in height, straight, practically flat area not broken up by very many skylights or variations in the deck elevation.

Type "B" buildings are not convenient type of structures for the application of built-up roofing. These would include high buildings which require considerable handling of materials, and buildings with roofs broken up by skylights or penthouses, sawtooth construction, monitors, or considerable variation in deck levels.

On the average roof where the mopping between layers consists of mopping the width of the lap, a roofer should mop 100 sq.ft. (9.29 sq.m) of surface in 6 to 7 minutes.

On first grade work where the entire roof surface is mopped between each layer of felt, a roofer mops 100 sq.ft. (9.29 sq.m) of roof in about 10 minutes.

After the roof surface has been mopped, a roofer should place roofing felt over 100 sq.ft. (9.29 sq.m) in about 6 minutes.

Where first grade workmanship is required, a roofer should handle pitch and gravel, pour hot pitch or asphalt and spread gravel over one square of roof in 35 to 40 minutes.

Prices on Roofing Materials. Prices on roofing materials such as coal tar pitch, asphalt, felt, etc., vary according to market fluctuations,

THERMAL AND MOISTURE PROTECTION

location of job, etc. Current material prices should always be obtained, before submitting bid.

Description	Hours	Rate	Total	Rate	Total
Roofing Foreman (working)	8.00	$	$	$ 34.93	$ 279.44
Roofers (4)	32.00	$	$	$ 30.93	$ 989.76
Crew cost per 8-hr. day					$1,269.20
Cost per sq. - 14 sqs. per day					$ 90.66
Cost per sq. - 17 sqs. per day					$ 74.66
Cost per sq. - 18 sqs. per day					$ 70.51
Cost per sq. - 22 sqs. per day					$ 57.69
Cost per sq. - 24 sqs. per day					$ 52.88

(metric)

Description	Hours	Rate	Total	Rate	Total
Roofing Foreman (working)	8.00	$	$	$ 34.93	$ 279.44
Roofers (4)	32.00	$	$	$ 30.93	$ 989.76
Crew cost per 8-hr. day					$1,269.20
Cost per sq.m - 130 sq.m per day					$ 9.76
Cost per sq.m - 158 sq.m per day					$ 8.04
Cost per sq.m - 167 sq.m per day					$ 7.59
Cost per sq.m - 204 sq.m per day					$ 6.21
Cost per sq.m - 223 sq.m per day					$ 5.69

Material Cost of One Square (100 sq.ft. or 9.29 sq.m)
3-Ply Tar and Gravel Built-Up Roof
applied Over Wood Roof

Maximum incline 2" per lin.ft. (50 mm per 0.30 m)
One ply No. 30 felt, 2 plies No. 15 felt (1 dry, nailed and
2 mopped), coal tar pitch surfaced with gravel or slag.

Description	Rate	Total	Rate	Total
1.1 sqs. No. 30 felt	$	$	$ 12.00	$ 13.20
2.2 sqs. No. 15 felt	$	$	$ 6.00	$ 13.20
125 lbs. coal tar pitch	$	$	$ 0.30	$ 37.50
400 lbs. roofing gravel	$	$	$ 0.05	$ 20.00
Nails, fuel, mops, etc.	$	$	$ 5.00	$ 5.00
Cost per 100 sq.ft.				$ 88.90
per sq.ft.				$ 0.89

Metric				
10.21 sq.M No. 30 felt	$	$	$ 1.29	$ 13.17
20.43 sq.M. No. 15 felt	$	$	$ 0.64	$ 13.08
56.7 Kg. coal tar pitch	$	$	$ 0.66	$ 37.42
181.44 Kg. roofing gravel	$	$	$ 0.11	$ 19.96
Nails, fuel, mops, etc.	$	$	$ 5.27	$ 5.27
Cost per 9.29 sq.M.				$ 88.90
per sq.m				$ 9.57

Material Cost of One Square (100 sq.ft. or 9.29 sq.m)
3-Ply Tar and Gravel Built-Up Roof Over Poured Concrete
Poured Gypsum Roof Decks, or Insulated board

Maximum incline 2" per lin.ft. (50 mm per 0.30 m)
Roof consisting of 3 plies of felt
coal tar pitch surfaced with gravel or slag.

Description	Rate	Total	Rate	Total	
3.3 sqs. No. 15 felt	$	$	$ 6.00	$	19.80
150 lbs. coal tar pitch	$	$	$ 0.30	$	45.00
400 lbs. roofing gravel	$	$	$ 0.05	$	20.00
Fuel, mops, etc.	$	$	$ 5.00	$	5.00
Cost per 100 sq.ft.				$	89.80
Cost per sq.ft.				$	0.90
Metric					
30.67 Sq.M No. 15 felt	$	$	$ 0.64	$	19.63
68.04 Kg coal tar pitch	$	$	$ 0.66	$	44.91
181.44 Kg. roofing gravel	$	$	$ 0.11	$	19.96
Fuel, mops, etc.	$	$	$ 5.00	$	5.31
Cost per 9.29 sq.M.				$	89.80
per sq.m				$	9.67

THERMAL AND MOISTURE PROTECTION

Material Cost of One Square (100 sq.ft. or 9.29 sq.m) of Each Additional Ply of No. 15 Felt and Hot Mopping of Pitch					
Description	Rate	Total	Rate	Total	
1.1 sqs. No. 15 Felt	$	$	$ 6.00	$ 6.60	
25 lbs. Pitch	$	$	$ 0.30	$ 7.50	
Fuel, Mops, etc.	$	$	$ 0.50	$ 0.50	
Cost per 100 sq.ft.				$ 14.60	
per 10 sq.ft.				$ 0.15	
Metric					
10.21 sq.m No. 15 Felt	$	$	$ 0.64	$ 6.53	
11.34 kg Pitch	$	$	$ 0.66	$ 7.48	
Fuel, Mops, etc.	$	$	$ 0.50	$ 0.50	
Cost per 9.29 sq.m.				$ 14.52	
per sq.m.				$ 1.56	

07520 PREPARED ROLL ROOFING

Ready roofing can be used on small buildings where an inexpensive yet satisfactory roofing is required. It is furnished in various grades and weights and is sold by the roll or in flat sheets packed with the nails and cement necessary for application.

Laying Ready-to-Lay Roll Roofing. When laying ready-to-lay roll roofing, a roofer should handle, lay, and cement or nail in place about 800 sq.ft. (74.32 sq.m) per 8-hr. day, at the following cost per 1 square (100 sq.ft. or 9.29 sq.m):

Description	Hours	Rate	Total	Rate	Total
Carpenter	1.00	$	$	$35.04	$35.04

Asphalt Slate Surfaced Ready-to-Lay Roofing

A ready-to-lay roofing consisting of several plies of asphalt impregnated felt, surfaced with a thick layer of slate granules, either red or green. For roofs having a pitch of 1-1/2" or more per foot (37.50 mm or more per 0.30 m).

Furnished in rolls 36" (900 mm) wide and 36'-0" (10.97 m) long, containing 108 sq.ft. (10.03 sq.m), sufficient for 100 sq.ft. (9.29 sq.m) of roof, at $13.00 per 90-lb. (40.5-kg) roll.

Smooth Surfaced Ready-to-Lay Roofing

Mica surface asphalt roofing consists of good quality felt saturated with pure asphalt. Both sides are covered with flake mica. Rolls are 36" (900 mm) wide and 36'-0" (10.97 m) long.

Description	Lbs. per Roll	Sq.Ft. per Roll	Price per Roll	Kg. per Roll	Sq.M. per Roll	Price per Roll
Mica surfaced	55	108	$ 13.80	24.95	10.03	$ 13.80
Asphalt Roof	65	108	$ 15.90	29.48	10.03	$ 15.90

7530-40 ELASTIC SHEET AND FLUID APPLIED ROOFING

Elastomeric coatings include neoprene, Hypalon, urethane, butyl, and silicone. Some are applied by spraying, others in sheet form, and all are subcontracted to firms licensed by the manufacturer.

Since these coatings are thin, the roof surface to which they are to be applied must be firm, continuous, smooth, clean, and dry. New concrete should be sealed with a primer.

Thinness and ability to conform to any shape, and the fact that they can be had in most any color including white, make them the choice as the roofing surface for decorative and fluid roof forms. Some materials are also used as traffic decks.

A 1/16" (1.56 mm) thick butyl sheet will cost around $0.65 per sq.ft. ($7.00 per sq.m) and one roofer can install some 225 sq.ft. (20.90 sq.m) per day for a total cost of around $1.25 per sq.ft. ($13.45 per sq.m).

A 1/16" (1.56 mm) neoprene sheet will have a material cost of some $1.25 per sq.ft. ($13.55 per sq.m) and will install at the same rate as the butyl for an installed price of around $1.80 per sq.ft. ($19.40 per sq.m).

Fluid applied Hypalon-neoprene 0.02" (0.5 mm) thick will be applied at the rate of some 100-110 sq.ft. (9.29-10.21 sq.m) per day and have a material cost of $0.85 per sq.ft. ($9.00 per sq.m) for a total figure of around $2.00 per sq.ft. ($21.50 per sq.m).

EPDM ROOF SYSTEMS

Ethylene Propylene Diene Methylene Rubber is popularly known in the trade as EPDM. EPDM is formulated as an elastomeric polymer synthesized from ethylene, propylene and diene (a small amount). EPDM sheeting is a vulcanized material; that is, the polymer's molecular structure is set as a result of heating during the manufacturing process. EPDM membranes can be produced in various colors, including black, reinforced or non-reinforced, and in thicknesses ranging from 30 to 90 mils.

THERMAL AND MOISTURE PROTECTION

The material was introduced into the construction industry in the early 1960's and in recent years has become a popular inexpensive flat roof application system. It is estimated that 20% of North America roof installations are EPDM systems. The success of this product during the past several years is due to its ability to withstand a wide range of temperatures, from -50°F to 240°F. (-10°C to 116°C), manufacture and supplier support and because of the stability of the product.

Non-reinforced EPDM is loose-laid over insulation or substrate (roof deck) and attached directly to the substrate using the EPDM manufacturer's approved fasteners and fastening plates, splicing cement and seam sealants. Non-reinforced material normally is available in widths of 7', 10', 10.5', 20' and 30' (2.13, 3.04, 3.20, 6.09 and 9.14 m) by 100' (30.48 m) lengths. Standard or reinforced EPDM, either mechanically fastened, fully-adhered or loose laid, is available in widths up to 10' (30 m) x 200' (60.96 m). Check with individual manufacturers for sizes and color.

Types of EPDM Installment. Mechanically fastened EPDM roofing is secured with manufacturer approved seam fasting plates. The EDPM membrane is rolled into position and overlapping sheets are fastened together with manufacturer approved splicing cement, in-seam sealants.

Fully adhered EPDM roofing is mechanically attached to an approved substrate (roof deck). The substrate is coated with manufacturer approved bonding adhesive, and the EPDM membrane is rolled into place and "brushed" down. Sheets are fastened with approved in-seam sealants or splicing tapes.

Ballasted EPDM roofing is normally installed by first applying approved insulation, which is loose-laid over an approved substrate (roof deck). The EPDM membrane is loose-laid over insulation and secured with a minimum of 10 Lbs. (4.5 Kg) of ballast per square foot.

Warranty. Warranties can vary. The norm is 10 years, and the maximum is 20 years, but 15-year warranties are available. Terms and conditions vary among manufacturers, and the length of a warranty will mandate certain conditions being adhered to during installation. Longer warranty periods will add costs to the installation for items such as certifications, inspections, installation restrictions and applicator certifications.

Materials for EPDM. In addition to the roofing material, the following material supplies are normally required for EDPM roofing installations:

Item	Produced in
Seam Primer	1-gallon containers
Seam Sealer	1/10 gallon cartridges (tubes)
Bonding / Splicing Adhesive	5 gallon pails
Water cut-off mastic	Tubes of 1 gallon cans
Seam tape	3" wide
Wood nailers	to be determined

Item	Produced in
Preformed pipe boots	Sized for pipes
Termination bars	10' x 1"
Corrosion resistant fasteners	to be determined

Because the amounts supplied and used will vary with manufacturers, an estimator should consult a specific manufacturer before estimating a project to assure that all materials required and quantities needed are correct for securing a watertight roof and to determine that the manufacturer offers the warranty period that is mandated by the owner.

EPDM - FULLY ADHERED

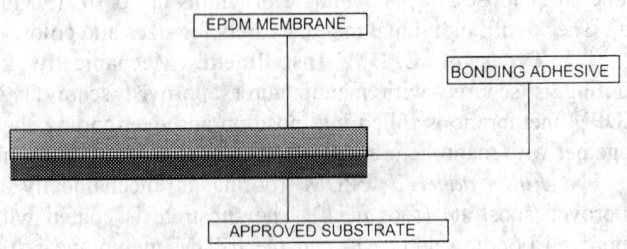

EPDM - FULLY ADHERED
MECHANICAL DESIGN

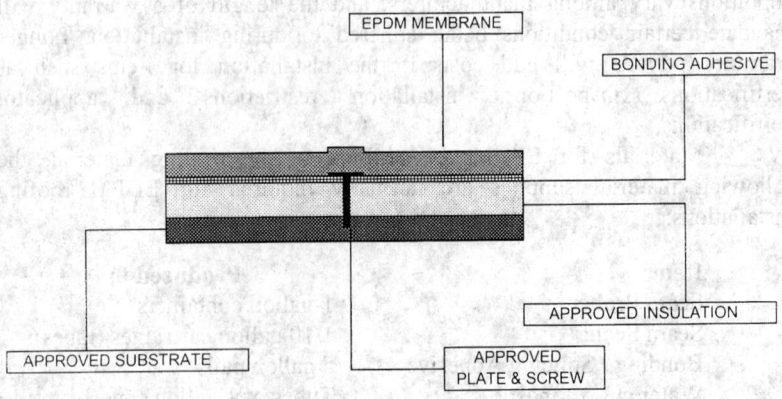

1112

THERMAL AND MOISTURE PROTECTION

Typical EPDM Roof Installation. Position the membrane over the substrate (roof deck) without stretching the sheet. When a new membrane is to be sealed to an abutting membrane, overlap the edge with the new membrane a minimum of 5 inches (125 mm) onto the in-place membrane.

Bond the new sheet to the substrate by folding the sheet lengthwise on to itself so that one half of the underside is exposed. Smooth out all the wrinkles and buckles along the edge. Apply the bonding adhesive evenly at the rate of 120 sq.ft. (11.15 sq.m.) per gallon per surface, using a 9" (225 mm) paint roller (manufacturer approved). Allow the adhesive to become tacky, then roll the membrane onto the substrate avoiding wrinkling by using a wave air motion at the leading edge to roll the sheet into place. Bond the adhered half of the membrane with an 18" to 24" (450 to 600 mm) heavy-duty push broom. Pull back the unbonded half of the sheet and repeat the procedure. The adhesive should not be applied to the top surface of any abutting membrane that is to be used as the seam joint.

All lap joints should be a minimum 4' wide (1.21 m) and shingled toward the drains or roof scuppers. Apply splicing adhesives to both bonding surfaces with a 4" (100 mm) paint brush at the rate of 120 lin.ft. (36.57 m) of 4-foot (1.21 m) lap per gallon, allowing the adhesive to dry until tacky. Join the bonding surfaces, avoiding stretching or fishmouths. Roll the lap joint thoroughly with a roller in a direction perpendicular to one length of the joint. Apply a seam sealer to the exposed edge and tool to assure positive contact with both sheets. Apply seam sealer to all lap joints completed during each day.

Mechanically Fastened. The EPDM membrane is fastened using 2" (50 mm) steel disks (or size as determined by manufacturer). The spacing and the type of fastener used depends on the substrate and approval by the EPDM manufacturer.

A roofing crew should be able to place 3,500 or 35 squares per day of partially adhered (mechanically fastened) EPDM roofing at the following labor cost:

Description	Hours	Rate	Total	Rate	Total
Roofer Foreman	8.00	$	$	$ 34.93	$ 279.44
Roofers	32.00	$	$	$ 30.93	$ 989.76
Laborers	8.00	$	$	$ 26.16	$ 209.28
Total Cost for 35 Squares					$ 1,478.48
per Square					$ 42.24
per sq.ft.					$ 0.42
per sq.m					$ 4.55

A roofing crew should be able to place 2,500 or 25 squares per day of fully adhered EPDM roofing at the following labor cost:

Description	Hours	Rate	Total	Rate	Total
Roofer Foreman	8.00	$	$	$ 34.93	$ 279.44
Roofers	32.00	$	$	$ 30.93	$ 989.76
Laborers	8.00	$	$	$ 26.16	$ 209.28
Total Cost for 25 Squares					$ 1,478.48
per Square					$ 59.14
per sq.ft.					$ 0.59
per sq.m					$ 6.37

07560 ROOF MAINTENANCE

Ease of periodic preventive maintenance is one of the strong points of a flat roof. Good maintenance programs can eliminate the major cause of roof complaints—leaks. Regardless of who does a roof inspection and how they do it, inspections should be held both in the spring and fall, with special inspections after exceptional storm events. Roof inspections include but are not limited to:

1. Interior walls and ceiling for signs of water staining
2. Cap flashing
3. Edge metal
4. Base flashings
5. Penetrations
6. Field of the roof
7. Ballast
8. Roof adhesives
9. Surface coatings

Normal maintenance includes keeping the drainage system clear and functioning and restricting roof access. If traffic patterns develop, to address the situation the owner should consult the roof manufacturer. A professional roofing contractor should accomplish all roof repairs. Taking these steps will help assure an owner (as best as possible) that a flat roof will be free of leak and will assist in the validation of the warranty.

Maintenance costs depend on many factors. A larger roof does not necessarily mean higher maintenance costs. Typical items that add to the cost of maintenance are access, amount of and type of roof penetrations, drainage system, and equipment on roof and foot traffic.

Assuming a roof area of 30,000 sq.ft. (2,787 sq.m), with two annual inspections (one in the spring and one in the fall) inspection cost is as follows:

THERMAL AND MOISTURE PROTECTION

Description	Hours	Rate	Total	Rate	Total
Roofer Foreman	32.00	$	$	$ 34.93	$ 1,117.76
Roofers	32.00	$	$	$ 30.93	$ 989.76
Total Cost - two inspections					$ 2,107.52
per sq.ft.					$ 0.07
per sq.m					$ 0.76

Make allowance for written reports, if required, and for materials for minor and miscellaneous repairs over a 10 to 20 year period (longer if necessary). This allowance can vary greatly (as do opinions concerning roof maintenance) from $0.00 to $1.00 per sq.ft ($0.00 to $ 10.76 sq.m.) per year.

07600 FLASHING AND SHEET METAL

07610 SHEET METAL ROOFING

Metal roofing includes galvanized steel, copper, lead, stainless steel, and aluminum plus the many combinations and alloys of these metals such as lead-coated copper, terne (80% lead, 20% tin over copper-bearing carbon steel) microzinc, and terne coated stainless. Terne and aluminum are the least expensive and copper and lead the costliest. Metal prices tend to fluctuate broadly. Many of the ores are imported and at the mercy of the value of the dollar and the political climate of the country where they are mined.

Metal roofs may be applied in many ways. The simplest is the flat seam roof that may be used on slopes as low as 1/4" (6.25 mm) to the foot. Standing seams and battened seams generally need a slope of at least 2-1/2" (62.50 mm). They are more decorative, and in fact batten designs are often selected solely on their decorative value. Some materials may be ordered with prefabricated battens, others are formed in the traditional way over wood strips. Another decorative roof is the "Bermuda" type, where the metal is applied over wood "steps" provided by the carpenter. The step is based on the width of the metal roll to be used and is sloped at least 2-1/2" per foot (62.50 mm per 0.30 m). The roofer interlocks the rolls at each step edge giving a sharp shadowline.

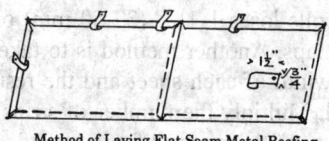

Method of Laying Flat Seam Metal Roofing

1115

For custom work terne and copper, with or without special coatings, are the usual choices. Many batten roofs today have the batten stamped into the metal and are usually factory finished aluminum or steel.

When estimating quantities of metal roofing, the measurements are taken in the same manner as for other types of roofs.

Where the metal roof is applied over flat roofs having parapet walls, an allowance of 1'-0" (0.30 m) on each side and end of the wall should be made for flashing up the sides of the walls.

If metal roofing is applied to a flat surface or pitch roof with overhanging eaves, the measurement is taken to cover the entire roof surface.

Do not make deductions for openings containing less than 25 sq.ft. (2.32 sq.m) as the extra flashing costs as much as the roofing omitted.

Deduct in full for all openings over 50 sq.ft. (4.64 sq.m) and add for flashing.

Flat Seam Metal Roofing. The common sizes of tin plates are 10" x 14" (250 x 350 mm) or multiples of that size. The sizes generally used for roofing are 14" x 20" (350 x 500 mm) and 20" x 28" (500 x 700 mm). The larger sizes are more economical to lay, and for flat roofs the 20" x 28" (500 x 700 mm) size is preferable.

For a flat seam roof, the edges of the sheets are turned in about 1/2" (12.50 mm), locked together and well soaked with solder. The sheets should be fastened to the roof boards by cleats locked in the seams. These cleats are usually spaced about 8" (200 mm) apart and 2 nails are used in each cleat.

For flat seam roofing, a 14" x 20" (350 x 500 mm) sheet of tin, with edges turned up 1/2" (12.50 mm) on each side and end, measures 13" x 19" (325 x 475 mm) and contains 247 sq. in. (1593.64 sq.cm); but the covering capacity when locked to other sheets is only 12-1/2" x18-1/2" (312.50 x 462.50 mm) or 231-1/4 sq. in. (1492.02 sq.cm). It requires 62-1/2 sheets of 14" x 20" (350 x 500 mm) tin to cover 100 sq.ft. (9.29 sq.m) of roof.

Sheets measuring 20" x 28" (500 x 700 mm) with the edges turned for flat seam roofing measure 19" x 27" (475 x 675 mm), and when locked to other sheets, have a covering capacity of 18-1/2" x 26-1/2" (462 x 662 mm) or 490-1/4 sq. in. (3163.09 sq.cm). It requires 20-1/2 sheets to cover 100 sq.ft. (9.29 sq.m) of roof.

Tin for roofs may also be obtained in rolls 14" (350 mm), 20" (500 mm), 24" (600 mm), and 28" (700 mm) wide. When used for flat seam roofing the loss due to turning edges amounts to 1-1/2" (37.50 mm) on the width of any sheet.

Tin in 14" (350 mm) rolls loses 1-1/2" (37.50 mm) or 11.2% of its width due to turning edges and laps. Another method is to take the length of the roof and divide by the net width of each sheet and the result will be the number of strips of tin required. Multiply the number of strips by the length of each strip to obtain the quantity of tin required for any roof.

THERMAL AND MOISTURE PROTECTION

The following table gives the covering capacity and allowances for edging and laps to be figured when using tin roofing in rolls for flat seam roofs:

Width of Inches	Allowance for Edge and Laps-Inches	Actual Width Inches	Add for Loss Due to Edges and Laps	No. Sq.Ft. Required per Sq.
14	1 1/2	12 1/2	12%	112
20	1 1/2	18 1/2	8%	108
24	1 1/2	22 1/2	7%	107
28	1 1/2	26 1/2	6%	106

Width of mm	Allowance for Edge and Laps-mm	Actual Width mm	Add for Loss Due to Edges and Laps	No. Sq.M. Required per Sq.
350	37.50	312.50	12%	10.40
500	37.50	462.50	8%	10.03
600	37.50	562.50	7%	9.94
700	37.50	662.50	6%	9.85

The quantities of galvanized iron or steel sheets required for flat seam roofing will depend upon the size of the sheets, as they may be obtained in sheets from 24" x 96" (600 x 2400 mm) to 48" x 120" (1200 x 3000 mm) in size.

The actual measurements of the sheets will be 1-1/2" (37.50 mm) less in width and length than the size of the sheet used. For instance, the actual covering capacity of a 24" x 96" (600 x 2400 mm) sheet will be only 22-1/2" x 94" = 2,115 sq. in. or 14.68 sq.ft. (562 x 2350 mm = 1,320,700 sq.mm).

The following table gives the approximate covering capacity of one sheet of the different sizes, allowing 1-1/2" (37.50 mm) in width and length for turning edge and lapping, together with the number of sheets of the different sizes required to cover one sq. (100 sq.ft. or 9.29 sq.m) of surface:

Size of Sheet Inches	Allowance for Edge Lap Inches		Actual Size Sheet - Inches	Cover Cap. One Sheet Sq.Ft.	No. Sheets per Square
	Length	Width			
24 "x 96 "	1 1/2	1 1/2	22 1/2 "x 94 1/2 "	14.77	6.8
24 "x 120 "	1 1/2	1 1/2	22 1/2 "x 118 1/2 "	18.52	5.4
26 "x 96 "	1 1/2	1 1/2	24 1/2 "x 94 1/2 "	16.08	6.2
26 "x 120 "	1 1/2	1 1/2	24 1/2 "x 118 1/2 "	20.16	5.0
28 "x 72 "	1 1/2	1 1/2	26 1/2 "x 70 1/2 "	12.97	7.7
28 "x 84 "	1 1/2	1 1/2	26 1/2 "x 82 1/2 "	15.18	6.6
28 "x 96 "	1 1/2	1 1/2	26 1/2 "x 94 1/2 "	17.39	5.8
28 "x 108 "	1 1/2	1 1/2	26 1/2 "x 106 1/2 "	19.60	5.1
28 "x 120 "	1 1/2	1 1/2	26 1/2 "x 118 1/2 "	21.81	4.6
30 "x 96 "	1 1/2	1 1/2	28 1/2 "x 94 1/2 "	18.70	5.3
30 "x 120 "	1 1/2	1 1/2	28 1/2 "x 118 1/2 "	23.45	4.3

Size of Sheet Inches	Allowance for Edge Lap MM		Actual Size Sheet - mm	Cover Cap. One Sheet Sq.M.	No. Sheets per Square
	Length	Width			
600 x 2400 mm	37.50	37.50	562.50 x 2362.50 mm	1.33	6.8
600 x 3000 mm	37.00	37.00	562.50 x 2962.50 mm	1.67	5.4
650 x 2400 mm	37.00	37.00	612.50 x 2362.50 mm	1.45	6.2
650 x 3000 mm	37.00	37.00	612.50 x 2962.50 mm	1.81	5.0
700 x 1800 mm	37.00	37.00	662.50 x 1762.50 mm	1.17	7.7
700 x 2100 mm	37.00	37.00	662.50 x 2062.50 mm	1.37	6.6
700 x 2400 mm	37.00	37.00	662.50 x 2362.50 mm	1.57	5.8
700 x 2700 mm	37.00	37.00	662.50 x 2662.50 mm	1.76	5.1
700 x 3000 mm	37.00	37.00	662.50 x 2962.50 mm	1.96	4.6
750 x 2400 mm	37.00	37.00	712.50 x 2362.50 mm	1.68	5.3
750 x 3000 mm	37.00	37.00	712.50 x 2962.50 mm	2.11	4.3

Standing Seam Metal Roofing. Standing seam roofing requires a larger allowance for waste than flat seam roofing. The standing seam, edged 1-1/4" (31.25 mm) and 1-1/2" (37.50 mm) takes 2-3/4" (68.75 mm) off the width and the flat cross seams edged 3/8" (9.37 mm), take 1-1/8" (28.13 mm) off the length.

If 14" x 20" (350 x 500 mm) tin is used, each sheet covers 11-1/4" x 18-7/8" (281.25 x 471.87 mm) or 212.34 sq. in. (1370.01 sq.cm). It requires 68 sheets of 14" x 20" (350 x 500 mm) tin per 100 sq.ft. (9.29 sq.m) of roof.

If 20" x 28" (500 x 700 mm) tin is used, each sheet covers 17-1/4 x 26-7/8" (431.25 x 671.87 mm) or 463.59 sq. in. (2991.08 sq.cm). It requires 32 sheets of 20" x 28" (500 x 700 mm) tin per 100 sq.ft. (9.29 sq.m).

If roll tin is used, 2-3/4" (68.75 mm) should be allowed for the two standing seams. The end waste will vary with the length of the sheets.

Tin in 14" (350 mm) rolls loses 2-3/4" (68.78 mm) off the width due to the standing seams, making it necessary to add 20% for waste. Another

1118

method is to take the width of the roof and divide by 11-1/4" (281.25 mm), the net width of the sheet, and the result is the number of strips of tin required. Multiply the number of strips by the length of the roof to obtain the quantity of tin required.

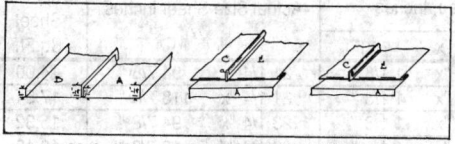

Method of Laying Standing Seam Metal Roofing

The following table gives the covering capacity and allowances for standing seams, when estimating roll roofing:

Width of Sheets Inches	Allowance for Standing Seams - In.	Actual Width Inches	Add for Loss Due to Standing Seams	No. Sq. Ft. Required per Square
14 "	2 3/4	11 1/4	25%	125
20 "	2 3/4	17 1/4	16%	116
24 "	2 3/4	21 1/4	13%	113
28 "	2 3/4	25 1/4	11%	111

(metric)				
Width of Sheets mm	Allowance for Standing Seams -mm	Actual Width mm	Add for Loss Due to Standing Seams	No. Sq. M. Required per Square
350 mm	68.75	281.25	25%	11.61
500 mm	68.75	431.25	16%	10.78
600 mm	68.75	531.25	13%	10.50
700 mm	68.75	631.25	11%	10.31

If galvanized steel or any of the special process metals, such as Armco, Toncan, etc., are used for standing seam roofs, an allowance of 1-1/4" (31.25 mm) and 1-1/2" (37.50 mm) or a total of 2-3/4" (68.75 mm) should be taken from the width of the sheets to allow for standing seams. The cross seams edged 3/8" (9.37 mm) take 1-1/8" (28.12 mm) off the length of each sheet. For instance, where 24" x 96" (600 x 2400 mm) sheets are used, the actual size of the sheets is 21-1/4" x 94-7/8" (531.25 x 2371.87 mm), or each sheet will cover 2,016 sq.in. or 14 sq.ft. (1.30 sq.m). The following table gives the covering capacity of painted or galvanized steel sheets of the different sizes:

Size of Sheet Inches	Standing Seam & End Lap, Inches		Actual Size Sheet Inches	Cover Cap. Per Sheet Sq.Ft.	No. Sheets per Square
	Width x	Length			
24 "x 96 "	2 3/4 "x	1 1/8 "	21 1/4 "x 94 7/8 "	14.00	7.1
24 "x 120 "	2 3/4 "x	1 1/8 "	21 1/4 "x 118 7/8 "	17.54	5.7
26 "x 96 "	2 3/4 "x	1 1/8 "	23 1/4 "x 94 7/8 "	15.32	6.5
26 "x 120 "	2 3/4 "x	1 1/8 "	23 1/4 "x 118 7/8 "	19.19	5.2
30 "x 96 "	2 3/4 "x	1 1/8 "	27 1/4 "x 94 7/8 "	17.95	5.6
30 "x 120 "	2 3/4 "x	1 1/8 "	27 1/4 "x 118 7/8 "	22.50	4.4
36 "x 96 "	2 3/4 "x	1 1/8 "	33 1/4 "x 94 7/8 "	21.91	4.6
36 "x 120 "	2 3/4 "x	1 1/8 "	33 1/4 "x 118 7/8 "	27.45	3.6
42 "x 96 "	2 3/4 "x	1 1/8 "	39 1/4 "x 94 7/8 "	25.86	3.9
42 "x 120 "	2 3/4 "x	1 1/8 "	39 1/4 "x 118 7/8 "	32.40	3.1
48 "x 96 "	2 3/4 "x	1 1/8 "	45 1/4 "x 94 7/8 "	29.81	3.4
48 "x 120 "	2 3/4 "x	1 1/8 "	45 1/4 "x 118 7/8 "	37.35	2.7

Size of Sheet mm	Standing Seam & End Lap, mm		Actual Size Sheet mm	Cover Cap. Per Sheet Sq.M.	No. of Sheets per Sq.M.
	Width x	Length			
600 x 2400 mm	68.75 x	28.13 mm	531.25 x 2371.88 mm	1.26	7.1
600 x 3000 mm	68.75 x	28.13 mm	531.25 x 2971.88 mm	1.58	5.7
650 x 2400 mm	68.75 x	28.13 mm	581.25 x 2371.88 mm	1.38	6.5
650 x 3000 mm	68.75 x	28.13 mm	581.25 x 2971.88 mm	1.73	5.2
750 x 2400 mm	68.75 x	28.13 mm	681.25 x 2371.88 mm	1.62	5.6
750 x 3000 mm	68.75 x	28.13 mm	681.25 x 2971.88 mm	2.02	4.4
900 x 2400 mm	68.75 x	28.13 mm	831.25 x 2371.88 mm	1.97	4.6
900 x 3000 mm	68.75 x	28.13 mm	831.25 x 2971.88 mm	2.47	3.6
1050 x 2400 mm	68.75 x	28.13 mm	981.25 x 2371.88 mm	2.33	3.9
1050 x 3000 mm	68.75 x	28.13 mm	981.25 x 2971.88 mm	2.92	3.1
1200 x 2400 mm	68.75 x	28.13 mm	1131.25 x 2371.88 mm	2.68	3.4
1200 x 3000 mm	68.75 x	28.13 mm	1131.25 x 2971.88 mm	3.36	2.7

V-Crimped Roofing. This type of roofing is used the same as standing seam roofing. It is usually furnished in sheets covering 24" (600 mm) in width and 6'-0" (1.83 m) to 10'-0" (3.04 m) long.

When estimating quantities of V-crimped roofing, allow for the end lap but there is no waste in the width as only the actual covering capacity is charged for by the manufacturer. For instance, a sheet 26" (650 mm) wide before crimping is 24" (600 mm) wide after crimping, but the sheet is sold as 24" (600 mm) wide.

The following table gives the quantity of V-crimp roofing required to cover 100 sq.ft. (9.29 sq.m) of roof with end laps 1" (25 mm) to 6" (150 mm):

1120

THERMAL AND MOISTURE PROTECTION

Length of Sheet - Ft.	End Laps, Inches					
	1	2	3	4	5	6
	Sq.Ft. of V-Crimp Roofing Required					
6	102	103	105	106	108	109
7	102	103	104	105	106	108
8	101	102	103	104	106	107
9	101	102	103	104	105	105
10	101	102	103	104	105	105

Length of Sheet - M.	End Laps, mm					
	25	50	75	100	125	150
	Sq.M of V-Crimp Roofing Required					
1.8	9.48	9.57	9.75	9.85	10.03	10.13
2.1	9.48	9.57	9.66	9.75	9.85	10.03
2.4	9.38	9.48	9.57	9.66	9.85	9.94
2.7	9.38	9.48	9.57	9.66	9.75	9.75
3.0	9.38	9.48	9.57	9.66	9.75	9.75

Material Cost of One Square (100 sq.ft. or 9.29 sq.m) Flat Seam Tin Roofing Using 14" x 20" (350 x 500 mm) Plates

Description	Rate	Total	Rate	Total
1.1 sq.No. 15 felt	$	$	$ 6.00	$ 6.60
63 shts. 14"x20" 40 Lb coated	$	$	$ 3.00	$ 189.00
6 lbs. (2.7 kg) solder	$	$	$ 2.00	$ 12.00
252 cleats	$	$	$ 0.09	$ 22.68
2 lbs. nails	$	$	$ 1.60	$ 3.20
Flux and charcoal	$	$	$ 3.00	$ 3.00
Cost per 100 sq.ft.				$ 236.48 *
per sq.ft.				$ 2.36
Description	Rate	Total	Rate	Total
10.21 sq.m No. 15 felt	$	$	$ 0.65	$ 6.59
63 shts. 350 x 500 mm 18Kg coated	$	$	$ 3.00	$ 189.00
2.72 kg solder	$	$	$ 4.41	$ 11.99
252 cleats	$	$	$ 0.09	$ 22.68
0.90 kg. nails	$	$	$ 3.53	$ 3.17
Flux and charcoal	$	$	$ 3.00	$ 3.00
Cost per 9.29sq.m.				$ 236.44 *
per sq.m				$ 25.45

*Add for ridges, hips, valleys, flashing etc.

Description of Roof	Sq.Ft. per 8-Hr Day	Hours per 100 Sq.Ft (9.29 Sq.M)		Sq.m per 8-Hr Day
		Tinner	Helper	
Flat seam metal, Using 24" x 96" (600 x 2400 mm) or larger	300 - 350	2.5	2.5	27.9 - 32.5
Flat seam tin, using 14" x 20" (350 x 500 mm) plates	175 - 225	4.0	4.0	16.3 - 20.9
Flat seam tin, roofing using 20" x 28" (500 x 700 mm) plates	225 - 275	3.3	3.3	20.9 - 25.5
Flat seam tin, roofing using 20" (500 mm) tin in rolls	325 - 375	2.3	2.3	30.2 - 34.8
Flat seam tin, roofing using 14" (350 mm) tin in rolls	275 - 325	2.8	2.8	25.5 - 30.2
Flat seam tin, roofing using 28" (700 mm) tin in rolls	425 - 475	1.8	1.8	39.5 - 44.1
Standing seam metal roofing using 24" x 96" (600 x 2400 mm) Sheets	250 - 300	3.0	3.0	23.2 - 27.9
Standing seam metal roofing using 14" x 20" (350 x 500 mm) Sheets	125 - 175	5.3	5.3	11.6 - 16.3
Standing seam metal roofing using 20" x 28" (500 x 700 mm) Sheets	175 - 225	4.0	4.0	16.3 - 20.9
Standing seam metal roofing using 14" (350 mm) tin in rolls	200 - 225	3.8	3.8	18.6 - 20.9
Standing seam metal roofing using 20" (500 mm) tin in rolls	250 - 300	3.0	3.0	23.2 - 27.9
Standing seam metal roofing using 24" (600 mm) tin in rolls	275 - 325	2.8	2.8	25.5 - 30.2
Standing seam metal roofing using 28" (700 mm) tin in rolls	300 - 350	2.4	2.4	27.9 - 32.5
V-Clamped metal roofing using 24" x 96" (600 x 2400 mm) sheets or larger	425 - 475	1.8	1.8	39.5 - 44.1

Labor Placing 100 Sq.Ft. (9.29 Sq.M.) Flat and Standing Seam Tin and Metal Roof

Material Cost of One Square (100 sq.ft. or 9.29 sq.m) Flat Seam Tin Roofing Using 20" x28" (350 x 500 mm) Plates				
Description	Rate	Total	Rate	Total
1.1 sq.No. 15 felt	$	$	$ 6.00	$ 6.60
29 shts. 20"x 28" 40 Lb coated	$	$	$ 6.00	$ 174.00
5 lbs. solder	$	$	$ 2.00	$ 10.00
145 cleats	$	$	$ 0.09	$ 13.05
1 lbs. nails	$	$	$ 1.60	$ 1.60
Flux and charcoal	$	$	$ 3.00	$ 3.00
Cost per 100 sq.ft.				$ 208.25 *
per sq.ft.				$ 2.08
Description	Rate	Total	Rate	Total
10.21 sq.m No. 15 felt	$	$	$ 0.65	$ 6.59
29 shts. 350 x 500 mm 18Kg coated	$	$	$ 6.00	$ 174.00
2.26 kg solder	$	$	$ 4.41	$ 9.96
145 cleats	$	$	$ 0.09	$ 13.05
0.45 kg. nails	$	$	$ 3.53	$ 1.59
Flux and charcoal	$	$	$ 3.00	$ 3.00
Cost per 9.29sq.m.				$ 208.20 *
per sq.m				$ 22.41

*Add for ridges, hips, valleys, flashing etc.

Material Cost of One Square (100 sq.ft. or 9.29 sq.m) Flat Seam Tin Roofing Using Roll Roofing				
Description	Rate	Total	Rate	Total
1.1 sq. ft. (10.2 sq.m.) No. 15 felt	$	$	$ 12.00	$ 13.20
108 Sq.Ft 20" roll roofing (10.03 Sq.M. x 500 mm)	$	$	$ 3.00	$ 324.00
4 lbs. (1.81 kg) solder	$	$	$ 2.00	$ 8.00
110 cleats	$	$	$ 0.09	$ 9.90
1 lbs. nails (0.45 Kg.)	$	$	$ 1.60	$ 1.60
Flux and charcoal	$	$	$ 2.25	$ 2.25
Cost per 100 sq.ft.				$ 358.95 *
per sq.ft.				$ 3.59
per sq.m.				$ 38.64

*Add for ridges, hips, valleys, flashing etc.

Material Cost of One Square (100 sq.ft. or 9.29 sq.m) Flat Seam Roofing Using 24" x 96" (600 x 2400 mm) Galvanized Sheets				
Description	Rate	Total	Rate	Total
1.1 sq. ft. (10.2 sq.m.) No. 15 felt	$	$	$ 12.00	$ 13.20
7 Sheets 26 ga. Steel 105 lbs (47.62 Kg)	$	$	$ 2.00	$ 14.00
3 lbs. (1.36 kg) solder	$	$	$ 2.00	$ 6.00
100 cleats	$	$	$ 0.09	$ 9.00
1 lbs. nails (0.45 Kg)	$	$	$ 1.60	$ 1.60
Flux and charcoal	$	$	$ 2.00	$ 2.00
Cost per 100 sq.ft.				$ 45.80 *
per sq.ft.				$ 0.46
per sq.m.				$ 4.93

*Add for ridges, hips, valleys, flashing etc.

THERMAL AND MOISTURE PROTECTION

Material Cost of One Square (100 sq.ft. or 9.29 sq.m) Standing Seam Roofing Using 14" x 20" (350 x 500mm) Plates				
Description	Rate	Total	Rate	Total
1.1 sq. ft. (10.2 sq.m.) No. 15 felt	$	$	$ 12.00	$ 13.20
68-14" x 20" (350 x 500 mm) 40 lbs (18.14 Kg) Coated	$	$	$ 3.00	$ 204.00
3 lbs. (1.36 kg) solder	$	$	$ 2.00	$ 6.00
100 cleats	$	$	$ 0.09	$ 9.00
1 lbs. nails (0.45 Kg)	$	$	$ 1.60	$ 1.60
Flux and charcoal	$	$	$ 1.60	$ 1.60
Cost per 100 sq.ft.				$ 235.40 *
per sq.ft.				$ 2.35
per sq.m.				$ 25.34

*Add for ridges, hips, valleys, flashing etc.

Material Cost of One Square (100 sq.ft. or 9.29 sq.m) Standing Seam Roofing Using 20" x 28" (500 x 700mm) Plates				
Description	Rate	Total	Rate	Total
1.1 sq. ft. (10.2 sq.m.)No. 1	$	$	$ 12.00	$ 13.20
32-20" x 28" (350 x 500 mm) 40 lbs (18.14 Kg) Coated	$	$	$ 3.00	$ 96.00
2 lbs. (0.90 kg) solder	$	$	$ 2.00	$ 4.00
80 cleats	$	$	$ 0.09	$ 7.20
1 lbs. nails 0.45 Kg)	$	$	$ 1.60	$ 1.60
Flux and charcoal	$	$	$ 1.60	$ 1.60
Cost per 100 sq.ft.				$ 123.60 *
per sq.ft.				$ 1.24
per sq.m.				$ 13.30

*Add for ridges, hips, valleys, flashing etc.

Material Cost of One Square (100 sq.ft. or 9.29 sq.m) Standing Seam Roofing Using Roll Roofing				
Description	Rate	Total	Rate	Total
1.1 sq. ft. (10.2 sq.m.) No. 15 felt	$	$	$ 12.00	$ 13.20
116 Sq.Ft. 20" roll Roofing 40 Lbs. (500 mm - 18.14 kg.)	$	$	$ 2.50	$ 290.00
2 lbs. (0.45 kg) solder	$	$	$ 2.00	$ 4.00
80 cleats	$	$	$ 0.09	$ 7.20
1 lbs. nails (0.45 kg)	$	$	$ 1.60	$ 1.60
Flux and charcoal	$	$	$ 1.60	$ 1.60
Cost per 100 sq.ft.				$ 317.60 *
per sq.ft.				$ 3.18
per sq.m.				$ 34.19

*Add for ridges, hips, valleys, flashing etc.

Material Cost of One Square (100 sq.ft. or 9.29 sq.m) Standing Seam Roofing Using 24" x 96" (600 x 2400 mm) Plates Galvanized Steel Sheets				
Description	Rate	Total	Rate	Total
1.1 sq. ft. (10.2 sq.m.) No. 15 felt	$	$	$ 12.00	$ 13.20
8-Sheets 26 Ga. Steel 120 Lbs. (54.43 Kg)	$	$	$ 2.00	$ 16.00
1 lbs. (0.45 kg) solder	$	$	$ 2.00	$ 2.00
60 cleats	$	$	$ 0.09	$ 5.40
1 lbs. nails 0.45 Kg)	$	$	$ 1.60	$ 1.60
Flux and charcoal	$	$	$ 1.60	$ 1.60
Cost per 100 sq.ft.				$ 39.80 *
per sq.ft.				$ 0.40
per sq.m.				$ 4.28

*Add for ridges, hips, valleys, flashing etc.

THERMAL AND MOISTURE PROTECTION

Material Cost of One Square (100 sq.ft. or 9.29 sq.m) V- Crimped Metal Roofing				
Description	Rate	Total	Rate	Total
1.1 sq. ft. (10.2 sq.m.) No. 15 felt	$	$	$ 12.00	$ 13.20
1 Sq (9.29 sq.m) 26 ga. V-crimped Roofing	$	$	$225.00	$ 225.00
5 lbs. (2.27 kg) Special Nails W/washer	$	$	$ 2.00	$ 10.00
60 lin.ft. (18.29 lin.m) wood Strips	$	$	$ 0.50	$ 30.00
Cost per 100 sq.ft.				$ 278.20 *
per sq.ft.				$ 2.78
per sq.m.				$ 29.95

*Add for ridges, hips, valleys, flashing etc.

COPPER ROOFING

Copper is used as roofing for any type of roof. There are various methods of applying copper for roofing, but most frequently used are the flat seam and standing seam. The batten seam method is also used, where durability is of more importance than cost, e.g., cathedrals, government buildings, monumental structures, etc.

Flat Seam Roofing. Flat seam roofing is commonly used on dead level or flat surfaces, but is adaptable to any slope. This method is also adaptable for the construction of water cooling roof panels, permitting water to stand up to a depth of 3" (75 mm).

For flat seam roofing, the use of small sheets of 20-oz (0.56-kg), cold-rolled copper, not larger than 16" x 18" (400 x 450 mm), with 3/4" (18.75 mm) seams flat-locked and soldered, is recommended. The sheets are tinned on all edges for soldering and are then formed to lock 3/4" (18.75 mm) with adjacent sheet. Corners must be clipped to permit folding as shown. Opposite sides of the sheets are folded in opposite directions so they will hook into adjoining sheets.

The sheets are held down by 2" (50 mm) wide copper cleats, two on each of two adjacent sides. The other two sides are held by the edges of adjacent sheets already cleated. For small areas, it is general practice to use only three cleats, two on a long side and one on a short side.

The cross seams are folded in the direction of flow. All seams should be flattened with a mallet and solder sweated to completely fill the seam.

1127

Each 16" x 18" (400 x 450 mm) sheet when folded covers 14-1/2" x 16-1/2" (362.50 x 412.50 mm) or 239-1/4 sq.in. (1543.64 sq.cm). It requires 61 sheets of 16" x 18" (400 x 450 mm) copper per 100 sq.ft. (9.29 sq.m) of roof.

Standing Seam Roofing. Standing seam copper roofing may be used for surfaces with a minimum slope of 3" per foot (75 mm per 0.30 m). Standing seams are usually made to finish 1" (25 mm) high. The spacing of the seams is a matter of scale and architectural effect, but for economy, a choice should be made that will use sheets of stock sizes. The use of 20" x 96" (500 x 2400 mm) sheets is recommended as maximum for ordinary purposes.

For standing seams to finish 1" (25 mm) high, the vertical bends are made on the long edges of the sheets 1-3/4" (43.75 mm) on one side and 1-1/2" (37.50 mm) on the other; the short bend on the one sheet always adjoins the long bend on the adjacent sheet. Recommended weights of material are: 16-oz. (0.45 kg) for widths between seams up to 20" (500 mm); 20-oz. (0.56 kg) for widths over 20" (500 mm).

Formed pans are held to the roof deck with 2" x 3" (50 x 75 mm) cleats spaced 12" (300 mm) apart and locked into the standing seam as shown in the accompanying illustration. Transverse joints should be staggered and each secured with one cleat locked into the flat seam.

Standing seams should not be riveted or soldered. Cross seams should be left unsoldered whenever conditions permit.

A sheet loses 3-1/4" (81.25 mm) in width on account of the standing seam, i.e. a sheet 20" (500 mm) wide would cover only 16-3/4" (418.75 mm). Where 20" (500 mm) sheets are used, add 20% to roof area for standing seams.

The loss at the end of the sheet will depend on the length of the roof and the length of copper sheet used, but allow 1-1/2" (37.50 mm) for each flat seam.

THERMAL AND MOISTURE PROTECTION

Sizes and Weights of Copper Sheets Commonly Used in Building Work				
Size of Sheet	Lbs. per Sheet for Various Sizes of Copper			
	16-Oz.	20-Oz.	24-Oz.	32-Oz.
20 "x 96 "	13.33	16.67	20.00	26.67
24 "x 96 "	16.00	20.00	24.00	32.00
30 "x 96 "	20.00	25.00	30.00	40.00
36 "x 96 "	24.00	30.00	36.00	48.00
20 "x 120 "	16.67	20.83	25.00	33.33
24 "x 120 "	20.00	25.00	30.00	40.00
30 "x 120 "	25.00	31.25	37.50	50.00
36 "x 120 "	30.00	37.50	45.00	60.00
48 "x 120 "	------	------	80.00	96.00
(metric)				
	Kg per Sheet for Various Sizes of Copper			
Size of Sheet	453.6 Grams	567.0 Grams	680.4 Grams	907.2 Grams
500 x 2400 mm	6.05	7.56	9.07	12.10
600 x 2400 mm	7.26	9.07	10.89	14.52
750 x 2400 mm	9.07	11.34	13.61	18.14
900 x 2400 mm	10.89	13.61	16.33	21.77
500 x 3000 mm	7.56	9.45	11.34	15.12
600 x 3000 mm	9.07	11.34	13.61	18.14
750 x 3000 mm	11.34	14.18	17.01	22.68
900 x 3000 mm	13.61	17.01	20.41	27.22
1200 x 3000 mm	------	------	36.29	43.55

Labor Placing Flat Seam Copper Roofing. When using 16" x 18" (400 x 450 mm) copper sheets for flat seam roofing, a tinner and helper should place 175 to 225 sq.ft. (16.25 to 20.90 sq.m) of roof per 8-hr. day, on average size roofs.

Labor Placing Standing Seam Copper Roofing. When using copper sheets 20" (500 mm) wide and up to 8'-0" (2.43 m) long for standing seam roofing, a tinner and helper should place 200 to 250 sq.ft. (18.58 to 23.22 sq.m) per 8-hr. day, on average size roofs.

Material Cost of One Square (100 sq.ft. or 9.29 sq.m) Flat Seam Coppper Roofing Using 16" x 18" (4000 x 450 mm) Sheets				
Description	Rate	Total	Rate	Total
1.1 sq. (10.2 sq.m) No. 15 felt	$	$	$ 6.00	$ 6.60
61 shts. 16"x18" (400x450mm), 153 lbs. (69 kg)	$	$	$ 6.50	$ 396.50
250 copper cleats	$	$	$ 0.50	$ 125.00
2 lbs. (0.9 kg) copper nails	$	$	$ 5.75	$ 11.50
6 lbs. (2.7 kg) solder	$	$	$ 2.00	$ 12.00
Flux and charcoal	$	$	$ 2.50	$ 2.50
Cost per 100 sq.ft.				$ 554.10 *
Cost per Sq.Ft				$ 5.54
per sq.m				$ 59.64

*Add for ridges, hips, valleys, flashing, etc.

Material Cost of One Square (100 s.f. or 9.3 sq.m) Standing Seam Copper Roofing using 20" (500 mm) Copper				
Description	Rate	Total	Rate	Total
1.1 sq. (10.2 sq.m) No. 15 felt	$	$	$ 6.00	$ 6.60
120 sq.ft. 120 lbs. (11.15 sq.m - 54.43 kg)	$	$	$ 4.00	$ 480.00
80 copper cleats	$	$	$ 0.50	$ 40.00
1 lb. (0.45 kg) copper nails	$	$	$ 5.75	$ 5.75
Cost per 100 sq.ft.				$ 532.35 *
Cost per Sq.Ft				$ 5.32
per sq.m				$ 57.30

*Add for ridges, hips, valleys, flashing, etc.

Stainless steel, type 304, monel and lead are also used for flat, standing, and batten seams.

THERMAL AND MOISTURE PROTECTION
07620 SHEET METAL FLASHING AND TRIM

Sheet metal work is a highly specialized business, because the greater part of it is done in the shop, and the fabricated materials are sent to the job ready to erect. An estimate on sheet metal work must take into consideration the cost of the finished materials at the shop or delivered to the building site, plus the labor cost of erection.

Estimating Quantities of Sheet Metal Work

The data given below will assist the estimator in measuring quantities from the plans and listing them on the estimate sheet.

When preparing the quantity survey for the different types, shapes, and sizes of metal items, be sure to identify the type and weight, or gauge, of metal—galvanized steel, aluminum, stainless steel, copper, lead, terne, etc. Each one has its own factors for waste, difficulty in shaping, and method of jointing, and all of these affect the production and installation costs.

Metal Gutters and Eaves Trough. Metal gutters and eave trough are furnished in numerous designs, from 2" (50 mm) to 12" (300 mm) in size. When estimating quantities, obtain the number of pieces of each size gutter and the length. The estimate should state the lin.ft. (m) of each size.

Metal Conductor Pipe or Downspout. Metal conductor pipe is furnished both round and square and from 2" (50 mm) to 6" (150 mm) in size. When estimating quantities, measure the distance from the roof to the ground, which will be the length of each downspout. After the quantities have been obtained in this manner, list the total number of lin.ft. (m) of each size.

Metal Conductor Heads. Metal conductor heads are made in various styles, shapes, and sizes.

Conductor heads are generally used on buildings having flat roofs where a cast iron conductor pipe is placed inside the building and installed by the plumbing contractor. When placed on the outside of a wall at the roof level, an opening is usually left in the masonry wall at the low point of the roof to allow the water to pass through the wall opening into the conductor head and down the conductor pipe.

When estimating conductor heads, always note style and size of each head and the material specified.

Metal Flashing and Counterflashing. Flashing and counterflashing are furnished in tin, stainless steel, galvanized steel, aluminum, or copper. They are measured by the lin.ft. (m) when less than 12" (300 mm) wide and by the sq.ft. (sq.m) when over 12" (300 mm) wide.

Metal flashing used in connection with a flat roof deck covered with felt or built-up roofing usually extends 6" (150 mm) to 12" (300 mm) up the side of the brick fire wall and projects under the roofing the same distance. A counterflashing is then placed over and above this flashing. The upper edge of the counterflashing is inserted in a reglet (slot) in the brick or stone wall.

1131

After the counterflashing has been placed, it is caulked with caulking compound to prevent water from getting in back of the flashing and running under the roofing.

On many jobs using felt or built-up roofing, the felt is run up the wall about 12" (300 mm) and mopped. The metal counterflashing is then placed to extend down over the felt, which prevents water from getting in back of the flashing and running under the roofing.

Where metal counterflashing is used, it is necessary to cut an open joint or reglet in the mortar joint of the masonry wall.

The cost of cutting reglets should be estimated by the lin.ft. (meter).

Metal Valleys and Hips. Metal valleys and hips are frequently used with certain types of shingle and slate roofs. They are either continuous metal strips of a certain size or small shingles of tin, galvanized steel, aluminum, stainless steel, or copper.

They should be estimated by the lin.ft. (m), stating the width of the valley or hip and the kind of metal used.

Metal Ridge Roll, Hip Roll, and Cap. Metal ridge roll, hip roll and cap should be estimated by the lin.ft. (m) giving the width, gauge, and kind of metal used.

Metal Ventilators. Metal ventilators for roofs and skylights should be estimated at a certain price for each ventilator, f.o.b. factory or erected. Always state size and type of ventilators required, as there are numerous kinds on the market at varying prices.

Metal Skylights. When estimating metal skylights, consider the following: size, type of skylight, whether single or double pitch or hip skylight; with or without ventilators or side sash; kind of glass, and whether a metal curb flashing is required.

Labor Placing Sheet Metal Work

The costs given below are for the erection of various kinds of sheet metal work on the job. Add cost of the fabricated material, contractor's overhead expense and profit.

Placing Hanging Gutter or Eaves Trough. This cost will vary with the slope of the roof, distance above the ground, and the method used in hanging the gutter.

On wood constructed buildings having hip or gable roofs, metal hangers or supports are placed over the gutters or eave trough and fastened to the eaves under the shingles or other roofing. The eave trough or gutter must always slope in the direction of the downspout, in order to drain satisfactorily.

On hip or gable roofs having eaves 20'-0" to 25'-0" (6.09 to 7.62 m) above the ground, a tinner and helper should place 180 to 220 lin.ft. (54.86 to 67.05 m) of gutter per 8-hr. day, at the following labor cost per 100 lin.ft. (30.48 m):

Description	Hours	Rate	Total	Rate	Total
Sheet Metal Worker	4.00	$	$	$45.25	$ 181.00
Labor	4.00	$	$	$26.16	$ 104.64
Cost 100 lin.ft.					$ 285.64
per lin.ft.					$ 2.86
per m					$ 9.37

Placing Metal Conductor Pipes or Downspouts. Metal conductor pipes or downspouts are furnished in both round and square and from 1-1/2" (37.50 mm) to 6" (150 mm) diameter. The conductor pipe extends down the side of the building from the eave trough to grade, where it is connected to an elbow that throws the water away from the building or is connected with the drainage system. Metal conductor pipe is usually held in place by hooks fastened to the walls.

On one- or two-story buildings where the conductor pipe is 12'-0" to 25'-0" (3.65 to 7.62 m) long, a tinner and helper should place 200 to 250 lin.ft. (60.96 to 76.20 m) per 8-hr. day, at the following labor cost per 100 lin.ft. (30.48 m):

Description	Hours	Rate	Total	Rate	Total
Sheet Metal Worker	3.50	$	$	$45.25	$ 158.38
Labor	3.50	$	$	$26.16	$ 91.56
Cost 100 lin.ft.					$ 249.94
per lin.ft.					$ 2.50
per m					$ 8.20

Where outside conductor pipe is installed on buildings 3 to 4 stories high, a tinner and helper should place 160 to 200 lin.ft. (48 to 61 m) per 8-hr. day, at the following labor cost per 100 lin.ft. (30 m):

Description	Hours	Rate	Total	Rate	Total
Sheet Metal Worker	4.50	$	$	$45.25	$ 203.63
Labor	4.50	$	$	$26.16	$ 117.72
Cost 100 lin.ft.					$ 321.35
per lin.ft.					$ 3.21
per m					$ 10.54

Placing Metal Flashing. When placing metal roof flashing around parapet walls, etc., it is customary to have the flashing extend under the roofing and up the side of the wall 6" (150 mm) to 12" (300 mm), and then place the counterflashing above to lap over the flashing.

Where it is not necessary to cut reglets in the masonry wall, a tinner should place 140 to 160 lin.ft. (42.67 to 48.76 m) of flashing per 8-hr. day, at the following cost per 100 lin.ft. (30.48 m):

Description	Hours	Rate	Total	Rate	Total
Sheet Metal Worker	5.40	$	$	$45.25	$ 244.35
Cost 100 lin.ft.					$ 244.35
per lin.ft.					$ 2.44
per m					$ 8.02

On frame buildings where the flashing extends under the wood wall siding to protect porch roofs and the like, it is only necessary to place the flashing or metal shingles on the roof and tack them against the sheathing. A tinner should place 165 to 185 lin.ft. (50.29 to 56.39 m) per 8-hr. day, at the following labor cost per 100 lin.ft. (30.48 m):

Description	Hours	Rate	Total	Rate	Total
Sheet Metal Worker	4.60	$	$	$45.25	$ 208.15
Cost 100 lin.ft.					$ 208.15
per lin.ft.					$ 2.08
per m					$ 6.83

Placing Metal Counterflashing. When placing metal counterflashing over felt or metal flashing that extends up on the side of the wall, it will be necessary to place the counterflashing in a reglet or open joint in the masonry that can be caulked or sealed to prevent the water getting in back of the counterflashing and under the roof.

Where reglets have previously been cut, as is usually the case on stone copings, requiring only the sealing of the joint, a tinner should place 140 to 160 lin.ft. (42.67 to 48.76 m) per 8-hr. day, at the following labor cost per 100 lin.ft. (30.48 m):

Description	Hours	Rate	Total	Rate	Total
Sheet Metal Worker	5.40	$	$	$45.25	$ 244.35
Cost 100 lin.ft.					$ 244.35
per lin.ft.					$ 2.44
per m					$ 8.02

THERMAL AND MOISTURE PROTECTION

Cutting Reglets in Masonry Walls. The labor cost of cutting reglets in brick walls will depend entirely upon the kind of mortar used and the condition of the mortar at the time of cutting the reglets.

If the brick are laid in portland cement mortar, it will cost more to cut the reglet than when laid in lime mortar. Also, if the reglets are cut within 24 hrs. after the brick are laid, the labor cost will be much less than cutting them two or three weeks later.

If reglets are cut before the mortar has set, a worker should cut 175 to 225 lin.ft. (53.34 to 68.58 m) per 8-hr. day, at the following labor cost per 100 lin.ft. (30.48 m):

Description	Hours	Rate	Total	Rate	Total
Sheet Metal Worker	4.00	$	$	$45.25	$ 181.00
Cost 100 lin.ft.					$ 181.00
per lin.ft.					$ 1.81
per m					$ 5.94

If the reglets are cut several days after the brick have been laid, giving the mortar time to harden, it will require a hammer and chisel to cut out the joints, and a worker should cut 65 to 80 lin.ft. (19.81 to 24.38 m) per 8-hr. day, at the following labor cost per 100 lin.ft. (30.04 m):

Description	Hours	Rate	Total	Rate	Total
Sheet Metal Worker	11.00	$	$	$45.25	$ 497.75
Cost 100 lin.ft.					$ 497.75
per lin.ft.					$ 4.98
per m					$ 16.33

If a portable electric saw with an abrasive blade or cutting wheel is used, a worker cuts 10 to 15 lin.ft. (3.04 to 4.57 m) of reglet per hour.

Cutting reglets in stone coping or balustrade is usually performed by stone cutters at about the same cost as given above.

Placing Metal Valleys. When placing metal shingles or valleys in connection with gable or hip roofs, dormers, etc., a tinner should place 115 to 135 lin.ft. (35.05 to 41.14 m) per 8-hr. day, at the following labor cost per 100 lin.ft. (30.48 m):

Description	Hours	Rate	Total	Rate	Total
Sheet Metal Worker	6.40	$	$	$45.25	$ 289.60
Cost 100 lin.ft.					$ 289.60
per lin.ft.					$ 2.90
per m					$ 9.50

Placing Metal Ridge Roll of Cap. If metal ridge roll is placed on the ridge of a hip or gable roof, a tinner should place 90 to 110 lin.ft. (27.43 to 33.52 m) per 8-hr. day, at the following labor cost per 100 lin.ft. (30.48 m):

Description	Hours	Rate	Total	Rate	Total
Sheet Metal Worker	8.00	$	$	$45.25	$ 362.00
Cost 100 lin.ft.					$ 362.00
per lin.ft.					$ 3.62
per m					$ 11.88

Placing Metal Conductor Heads. On buildings having flat roofs where the conductor pipe or downspout is placed inside of the building, requiring a conductor head at the low point of the roof which is connected to the downspout, a tinner should place one conductor head (made complete in the shop) in 1-1/2 to 2 hrs. at the following labor cost:

Description	Hours	Rate	Total	Rate	Total
Sheet Metal Worker	1.80	$	$	$45.25	$ 81.45

07630 ROOFING SPECIALTIES

Flashings can be formed from a wide variety of materials of varying permanence:

Material	Thickness or Weight Inches/ Oz./Lbs.	Cost per Sq.Ft.	Sq.Ft. installed per Day	Thickness or Weight mm/ Gram/Kg	Cost per Sq.m	Sq.M installed per Day
Aluminum	0.013 "	$ 0.31	150	0.325 mm	$ 3.34	13.94
(Mill Finish)	0.016 "	$ 0.50	150	0.400 mm	$ 5.38	13.94
	0.019 "	$ 0.50	150	0.475 mm	$ 5.38	13.94
	0.032 "	$ 0.95	145	0.800 mm	$ 10.23	13.47
	0.04 "	$ 1.15	145	1.000 mm	$ 12.38	13.47
	0.05 "	$ 1.40	145	1.250 mm	$ 15.07	13.47
Copper	16 Oz.	$ 2.65	140	453.584 Grams	$ 28.53	13.01
	20 Oz.	$ 3.25	140	566.98 Grams	$ 34.98	13.01
	24 Oz.	$ 3.75	140	680.376 Grams	$ 40.37	13.01
Stainless Steel	26 Ga.	$ 6.40	150	26 Ga.	$ 68.89	13.94
Asph. C't'd Cotton	17 Oz.	$ 0.30	350	481.933 Grams	$ 3.23	32.52
	40 Oz.	$ 0.45	350	1133.96 Grams	$ 4.84	32.52
Lead	2.5 Lb	$ 5.00	140	1.134 Kg	$ 53.82	13.01
Zinc/Copper Alloy		$ 2.40	140		$ 25.83	13.01
Butyl	1/32 "	$ 0.85	275	0.781 mm	$ 9.15	25.55
	1/16 "	$ 1.15	275	1.563 mm	$ 12.38	25.55
Neoprene	1/16 "	$ 1.90	275	1.563 mm	$ 20.45	25.55
Polyvinyl	0.02 "	$ 0.25	275	0.500 mm	$ 2.69	25.55
	0.03 "	$ 0.35	275	0.750 mm	$ 3.77	25.55

THERMAL AND MOISTURE PROTECTION

Gravel stops may be formed on the job or ordered out in extruded form. If extruded be sure to check if any special finish is specified, which will raise costs considerably. A few costs and rates for installation are:

Gravel Stops						
Material	Cost per Lin. Ft. 4" High	Cost Per Lin.Ft. 6" High	Lin.Ft per Day	Cost per Lin.M. 100 mm High	Cost Per Lin.M 150 mm High	Lin.M per Day
Aluminum 0.050" mill finish	$ 2.50	$ 3.25	135	$8.20	$10.66	41.15
Duranodic	$ 4.00	$ 4.50	135	$13.12	$14.76	41.15

Preformed gutters can be installed by one sheet metal worker at the rate of 110 to 120 lin.ft. (33-36 m) per day. Material costs are:

Gutters						
Material	Cost per l.f.			Cost per meter		
	3" dia.	4" dia.	5" dia.	75 mm diam.	100 mm dia.	125 mm dia.
Aluminum 0.025" plain	$ 0.75	$ 1.10	$ 1.50	$2.46	$3.61	$4.92
Enamel	$ 0.80	$ 1.25	$ 1.60	$2.62	$4.10	$5.25
Copper 16 oz.	$ 4.10	$ 5.00	$ 5.90	$13.45	$16.40	$19.36
Stainless 22 Ga.	$ 9.60	$ 12.30	$ 19.00	$31.50	$40.35	$62.34
Vinyl	$ 0.85	$ 1.20	$ 1.40	$2.79	$3.94	$4.59
Galv. Steel 28 ga.	$ 0.65	$ 0.85	$ 1.00	$2.13	$2.79	$3.28

Downspouts can be installed by one sheet metal worker at the rate of 175 lin.ft. (53.34 m) for small sizes, 150 lin.ft. (0.45 m) for medium and 125 lin.ft. (38.1 m) for the larger sizes per day. The material costs will run as follows:

Material	Cost per l.f.				
	3" dia.	4" dia.	5" dia.	2"x3"	3"x4"
Aluminum 0.025" plain	$ 0.68	$ 1.09	$ 1.50	$ 1.02	$ 1.29
Enamel	$ 0.79	$ 1.22	$ 15.05	$ 1.16	$ 1.50
Copper 16 oz.	$ 4.08	$ 4.76	$ 5.85	$ 4.49	$ 5.51
Stainless	$ 9.52	$ 12.24	$ 19.04	$ 9.52	$ 12.24
Vinyl	$ 0.82	$ 0.85	$ 0.90	$ 0.95	$ 1.00
Galv. Steel 28 ga.	$ 0.61	$ 0.82	$ 1.43	$ 0.68	$ 0.88

Material	Cost per meter				
	25 mm dia.	100 mm dia.	125 mm dia.	50 x 75 mm	75 x 100 mm
Aluminum 0.62mm plain	$2.23	$3.57	$4.91	$3.35	$4.24
Enamel	$2.59	$4.02	$49.38	$3.79	$4.91
Copper 0.45kg	$13.39	$15.62	$19.19	$14.72	$18.07
Stainless	$31.23	$40.16	$62.47	$31.23	$40.16
Vinyl	$2.68	$2.79	$2.95	$3.12	$3.28
Galv. Steel 28 ga.	$2.01	$2.68	$4.69	$2.23	$2.90

Elbows	
3" diam (75mm diam) in Plain Aluminum	$ 1.30
3" diam (75mm diam) in Enamel Aluminum	$ 1.50
3" diam (75mm diam) in 16 Oz. Copper	$ 7.00
4" diam (100mm diam) in Plain Aluminum	$ 1.75
3" x 4" (75 x 100 mm) in Plain Aluminum	$ 2.15
3" x 4" (75 x 100 mm) in Enamel Aluminum	$ 2.50
3" x 4" (75 x 100 mm) in 16 Oz. Copper	$ 11.00

07800 ROOF ACCESSORIES

Erecting and Glazing Metal Skylights. Metal skylights are manufactured in the shop and sent to the job "knocked down" or in sections ready to erect and glaze.

An average skylight (single pitch, double pitch, or hip) up to 8'-0" x 12'-0" (2.43 x 3.65 m) in size, containing 100 sq.ft. (9.29 sq.m) of area, should be erected and glazed complete by a sheet metal worker and glazier in 8 to 10 hrs. time, at the following labor cost:

Description	Hours	Rate	Total	Rate	Total
Sheet metal worker	5.00	$	$	$45.25	$ 226.25
Glazier	4.00	$	$	$ 32.50	$ 130.00
Cost per skylight					$ 356.25

When erecting skylights 10'-0" x 10'-0" (3.04 x 3.04 m) to 10'-0" x 15'-0" (3.04 x 4.57 m), containing 100 to 150 sq.ft. (9.29 to 13.93 sq.m), a sheet metal worker and glazier should erect and glaze one skylight in 14 to 16 hrs. at the following labor cost:

THERMAL AND MOISTURE PROTECTION

Description	Hours	Rate	Total	Rate	Total
Sheet metal worker	9.00	$	$	$45.25	$ 407.25
Glazier	6.00	$	$	$ 32.50	$ 195.00
Cost per skylight					$ 602.25

When erecting skylights 10'-0" x 16'-0" (3.04 x 4.87 m) to 10'-0" x 20'-0" (3.04 x 6.09 m) in size, containing 160 to 200 sq.ft. (14.86 to 18.58 sq.m), it will require about 24 hrs. for a sheet metal worker and glazier to erect the skylight and glaze it complete, at the following labor cost:

Description	Hours	Rate	Total	Rate	Total
Sheet metal worker	16.00	$	$	$45.25	$ 724.00
Glazier	8.00	$	$	$ 32.50	$ 260.00
Cost per skylight					$ 984.00

The above costs do not include overhead and profit or the cost of placing a metal curb flashing for skylight. For costs on this work, see below.

Placing Metal Skylight Curb Flashing. Metal skylights are often placed on top of a wood curb projecting 6" (150 mm) or more above the roof level. The space from the top of the curb to the roof level is covered with metal flashing.

When erecting metal skylight curb flashing, a tinner should place 190 to 210 sq.ft. (17.65 to 19.50 sq.m) of metal per 8-hr day, at the following labor cost per 100 sq.ft. (9.29 sq.m):

Description	Hours	Rate	Total	Rate	Total
Sheet metal worker	4.00	$	$	$45.25	$ 181.00
Cost per Sq.Ft.					$ 1.81
Cost per Sq.M.					$ 19.48

Erecting Skylights With Side Sash. If the skylights have side sash, the erection cost will vary with the number of sash in the skylight and whether stationary or pivoted. On an average it will require 1 to 1-1/2 hrs. labor time for each sash in the skylight, at the following labor cost per sash:

Description	Hours	Rate	Total	Rate	Total
Sheet metal worker	1.25	$	$	$45.25	$ 56.56

Erecting Sash Operating Devices for Movable Sash. Skylights furnished with hinged or pivoted sidewall sash require an operating device to open and close them.

1139

An operating device usually consists of a mechanism similar to that illustrated, which is operated from the floor by a worm and gear or a chain attachment, from which all sash on each side of the skylight can be opened and closed.

The labor cost of erecting sash operating devices will vary with the length of each run and the number of runs in each skylight but on an average, a worker should erect 50 to 75 lin.ft. (15.24 to 22.86 m) per 8-hr. day, at the following labor cost per 100 lin.ft. (30.04 m):

Description	Hours	Rate	Total	Rate	Total
Sheet metal worker	13.00	$	$	$45.25	$ 588.25
Cost per Lin.Ft.					$ 5.88
Cost per Lin.M.					$ 19.30

Ventilating Skylights. Ventilating skylights are used in factories, foundries, machine shops, garages, dairies, laundries, and other buildings that require economical light and ventilation. They are furnished in any type of metal desired, galvanized copper-bearing steel, aluminum or copper, depending upon the nature of the building on which they are to be used. Normally, for building of ordinary construction, the standard 18-gauge galvanized copper-bearing steel is furnished. Should the building be of reinforced concrete or steel and concrete construction, the non-corrosive aluminum or copper construction should be furnished.

These skylights are of gable type as illustrated and are furnished in standard widths of 4'-0" (1.21 m) to 20'-0" (6.09 m) and any desired length in multiples of 2'-0" (0.61 m). It is standard practice to interpret the size of a skylight as being outside of curb dimensions.

The length of the ventilating sections depend on the width of the skylights; 4'-0" (1.21 m), 6'-0" (1.83 m) and 8'-0" (2.43 m) widths having maximum ventilating sections of 40'-0" (12.19 m) each controlled from one operating station; 10'-0" (3.04 m) and 12'-0" (3.65 m) wide skylights have maximum ventilation sections of 30'-0" (9.14 m) in length; 14'-0" (4.26 m) and 16'-0" (4.87 m) widths have 40'-0" (12.19 m) maximum lengths, and 18'-0" (5.48 m) and 20'-0" (6.09 m) widths have 30'-0" (9.14 m) maximum ventilating sections.

A mechanical operating device consisting of either loose rod or chain control is furnished as required, so that the skylight can be readily controlled from the floor. All skylights under 30'-0" (9.14 m) in length are arranged for full length ventilation. Longer skylights are arranged with ventilating sections as required.

The cost of ventilating skylights vary with their size and number of ventilating sections in each skylight. Prices should be secured from the manufacturer.

1140

THERMAL AND MOISTURE PROTECTION

In estimating the cost of ventilating skylights, the principal thing to remember is: The smaller the area of the skylight, the higher the price per sq.ft. (sq.m).

Plastic Dome Skylights. This type skylight consists of a thermo-formed acrylic plastic sheet free blown into square, rectangular and circular domes.

This type skylight is furnished as a factory assembled unit; the plastic is mounted in an extruded aluminum frame, and prefabricated curbs are available. The aluminum frame is generally designed to form counterflashing.

Plastic Domes

Size	Cost per Roof Dome	Size
16 " x 16 "	$ 35.00	400 x 400 mm
24 " x 24 "	$ 80.00	600 x 600 mm
36 " x 36 "	$ 180.00	900 x 900 mm
48 " x 48 "	$ 250.00	1200 x 1200 mm
24 " Round	$ 110.00	600 mm Round
48 " Round	$ 250.00	1200 mm Round

Erection cost will vary widely as to the number involved and the preparatory work done by others. As units are vulnerable to breakage and scratching, it takes two workers to handle all but the smallest unit. On a job with several units involved two workers should set 2 units of up to 10 sq.ft. (0.92 sq.m) in a day if the curbing is already in place.

The dome unit is available as an insulating unit, with a secondary acrylic dome mounted under the primary (exterior) dome. This double dome unit provides a dead air space between the domes that reduces heat loss and condensation formation.

ROOF HATCH

A standard size 2'-6" x 3'-0" (750 x 900 mm) steel roof hatch with insulated curb and cover and hardware will cost about $450.00. A mechanic and carpenter should set these in 2 hrs each. A standard size 2'-6" x 8'-0" (750 x 2400 mm) steel roof hatch with insulated curb and cover and hardware will cost about $1,000.00. Setting large 2'-6" x 8'-0" steel roof hatch, a mechanic and carpenter should set in about 4 hrs each.

GRAVITY VENTILATORS

The 12" (300 mm) stationary ventilators of the syphon type with a 24" (600 mm) base will cost about $80.00 and take 2 hours to set. Stationary ventilators of the mushroom type with a 24" (600 mm) base will cost $260.00 with the same labor cost.

07900 SEALANTS

No one sealant can solve the requirements across the full range of construction applications. There are sealants for interior or exterior use, gun or pour grade, or ability to expand or contract. There are variations in service temperature range, paintability, and compatibility with material.

An oil base caulking compound has a life expectancy of about 4 or 5 years. Its restrictions list a joint maximum size of 1/2" (12.50 mm) wide x 3/4" (18.75 mm) deep and is incapable of withstanding any joint movement without rupture. Using this material for exterior applications may not be prudent or cost effective.

Acrylic latex caulk is probably a good starting point for exterior use. It is considered to be the best value among sealants in the middle performance range. It is relatively inexpensive, has a life expectancy of 8 to 12 years, cures quickly, has excellent paintability, is non-staining, and has reasonable elongation and recovery for movement of joints. However, this material has a maximum joint size of 1/2" (12.50mm) wide and 1/2" (12.50 mm) deep and should not be used for expansion joints.

Polysulfide base sealant, available in one or two part compounds, is among the premiere compounds on the market. It is reasonably expensive and requires a little more labor time to install, but its movement capabilities, bonding strength, wide use range (including expansion joints), and a life expectancy of 15 years make it cost effective while providing excellent joint sealing characteristics.

It is important to have a properly designed joint, both in width and depth. Good sealants will expand and contract with the movement of the joint. As the joint moves, the sealant changes shape while the volume remains constant. It is critical that the width-to-depth ratio be designed to withstand the constant elongation and compression cycles over long periods of time.

Deep beads of sealant wastes material and are more prone to failure than shallow sealant beads. Another important factor to consider is that sealant works best when adhering to the two opposing faces and isolated from the third (back) side of the joint to be sealed. This is accomplished by filling the joint with a material such as oakum or polyethylene foam rod to within 3/8" (9.37 mm) or 1/2" (12.50 mm) of the face of the joint. In shallow depth joints, use non-adhering tape on back of the joint to prevent the sealant from bonding.

THERMAL AND MOISTURE PROTECTION

Sealants are manufactured in colors from white to black and some are clear (silicones). This is necessary to be architecturally compatible with the adjoining surfaces. Also, paint does not adhere at all to some sealants.

The following chart provides a guide for the amount of sealant required to fill various size joints; expressed in lin.ft. (lin.m) of joint obtained per gallon of material. Estimator should check with the manufacture for his/her material requirements per gallon (liter).

In using the above chart, if the size of the joint to be sealed is not given on the drawings, use the principle that the depth of the sealant is 1/2 of the width of the joint to be sealed.

When estimating the labor required to caulk a joint, it will depend on the width of the joint, the material used, and the location or accessibility of the joint. One worker should caulk about 600 lin.ft. (182.88 m) per 8 hr. day of joints around doors or window frames when they are close to the ground or accessible without the use of scaffolding. This time may be reduced by half when working off ladders or scaffolding. The same person working on sealing wide expansion joints would only finish about 200 lin.ft. (60.96 m) per 8-hr. day.

Lineal Feet per Gallon

Depth, Inches	#### "	1/8 "	1/4 "	3/8 "	1/2 "	5/8 "	3/4 "	1 "	1 1/4 "	1 1/2 "
					Width, inches					
1/16 "	4,928	2,464	1,232	821	616	493	411	307	-----	-----
1/8 "	-----	1,232	616	411	307	246	205	154	-----	-----
3/16 "	-----	-----	411	275	205	164	137	103	-----	-----
1/4 "	-----	-----	308	205	154	123	103	77	-----	-----
3/8 "	-----	-----	-----	137	103	82	68	51	-----	-----
1/2 "	-----	-----	154	-----	77	-----	-----	-----	-----	-----
3/4 "	-----	-----	103	-----	51.3	-----	34.2	-----	-----	-----
1 "	-----	-----	77	-----	38.5	-----	25.7	19.3	-----	-----
1 1/4 "	-----	-----	62	-----	30.8	-----	20.5	15.4	12.3	-----
1 1/2 "	-----	-----	51	-----	25.7	-----	17.1	12.8	10.3	8.6
1 3/4 "	-----	-----	44	-----	22	-----	14.7	11	8.8	7.3
2 "	-----	-----	39	-----	19.3	-----	12.8	9.6	7.7	6.4
2 1/4 "	-----	-----	31	-----	15.4	-----	10.3	7.7	6.2	5.1
2 1/2 "	-----	-----	26	-----	12.8	-----	8.6	6.4	5.1	4.3

Lineal Meter per Liter

Depth, mm	Width, mm									
	1.56 mm	3.13 mm	6.25 mm	9.38 mm	12.50 mm	15.63 mm	18.75 mm	25.00 mm	31.25 mm	37.50 mm
1.56 mm	396.8	198.4	99.2	66.1	49.6	39.7	33.1	24.7	-----	-----
3.13 mm	-----	99.2	49.6	33.1	24.7	19.8	16.5	12.4	-----	-----
4.69 mm	-----	-----	33.1	22.1	16.5	13.2	11.0	8.3	-----	-----
6.25 mm	-----	-----	24.8	16.5	12.4	9.9	8.3	6.2	-----	-----
9.38 mm	-----	-----	-----	11.0	8.3	6.6	5.5	4.1	-----	-----
12.50 mm	-----	-----	12.4	-----	6.2	-----	-----	-----	-----	-----
18.75 mm	-----	-----	8.3	-----	4.1	-----	2.8	-----	-----	-----
25.00 mm	-----	-----	6.2	-----	3.1	-----	2.1	1.6	1.0	-----
31.25 mm	-----	-----	5.0	-----	2.5	-----	1.7	1.2	0.8	0.7
37.50 mm	-----	-----	4.1	-----	2.1	-----	1.4	1.0	0.8	0.6
43.75 mm	-----	-----	3.5	-----	1.8	-----	1.2	0.9	0.7	0.6
50.00 mm	-----	-----	3.1	-----	1.6	-----	1.0	0.8	0.6	0.5
56.25 mm	-----	-----	2.5	-----	1.2	-----	0.8	0.6	0.5	0.4
62.50 mm	-----	-----	2.1	-----	1.0	-----	0.7	0.5	0.4	0.3

08   DOORS AND WINDOWS

Most of the items covered in this chapter are furnished by material distributors and set by a carpenter on the job; in the estimate, the cost of labor would be included under the proper trade.

Glazing is an exception, and both material and labor are included under a single item. However, sash and doors usually come to the job factory glazed. The contractor must always check this out so as to eliminate any duplication in the estimate and to make certain the pre-glazing complies with both the specifications and local union rules.

Curtainwall material and labor are usually let to a single subcontractor. It is advantageous to have the metal erection, sealants, and glazing all in one contract so that one source will be responsible for the weather tightness of the wall.

The estimator who will perform the quantity survey for these materials should become very familiar with the various schedules that the project architect includes on the plans.

These schedules are usually laid out in block column form, consisting of a listing of the material involved and all of the particulars for each item, and they might show architectural details for each type of item.

Door schedules will normally list the door openings by "mark" numbers or letters, e.g., #1, #2, #3 or A, B, C. After the mark designation, there will be the door size, door material and the frame, a code designation for the type and size of the frame, fire rating requirements (if any), a code designation for any louver or glass panels required, finish hardware set designations, and a "remarks" column for any special instructions. Bear in mind that the "mark" designation refers to that particular design type and may have nothing to do with the total quantity of doors and frames of that design type required for the project.

Window schedules are laid out in much the same fashion, showing: mark designation, size, material, glass, type, etc.

Finish hardware schedules are normally set up in the hardware section of the specifications, rather than on the drawings. The hardware schedule is a listing of the hardware items required to fit a particular door opening, e.g., hinges, latchset or lockset, closer, kickplates, etc.

Each type of opening, such as for toilets, closets, offices, etc., will have its own "Hardware (HW) Set #" designation; and each will vary from the other in some way. For example, a door that leads to a stairway in an office building could list under its HW Set number: "1-1/2 pair hinges, latchset, and closer". A door that leads to an office in the same building could list under its HW Set number: "1-1/2 pair hinges, lockset, and closer". As you can see from the two examples, they are very similar, but because of the lock requirements for the office door opening, a separate HW Set # will be given.

Here again, keep in mind that the HW Set number has nothing to do with the total quantity of finish hardware required. One set number might be applied to 10 door openings and the next set number applied to only one or two door openings.

08100 METAL DOORS AND FRAMES

Each hollow metal manufacturer produces stock design doors and frames, as well as a custom designs to an architect's requirements. Stock lines vary among manufacturers as to size, width of frame, hardware preparation, etc. A hollow metal door from one manufacturer usually cannot be hung in the hollow metal frame from another manufacturer, due to unmatched hinge or lockset locations.

"Stock" will also refer to the cut-out preparation for hinges, and in particular, for locksets. Most manufacturers prepare the frames for 3-hinge locations each sized 3-1/2" (87.50 mm) or 4-1/2" (112.50 mm) in height and for the standard cylindrical lockset. If the finish hardware specified for the project varies from the manufacturer standards, the hollow metal manufacturer will consider the order "custom" and increase the price accordingly, as well as increase the delivery time to the project.

Material prices for hollow metal items are quoted inclusive of all mortising, reinforcing, drilling, and tapping for mounting the door into the frame and preparation for items of finish hardware. Also included are frame anchors and shop prime coat of paint.

Hollow Metal Frames. Frames are made of 18, 16, and 14 gauge metal and formed to receive a 1-3/8" (34.37 mm) or 1-3/4" (43.75 mm) thick door, depending on which side of the stop the hinge and strike cut-outs are placed. The following illustration shows a section through a standard hollow metal frame.

When the manufacturer varies the overall width of the frame to suit the different wall conditions that may be encountered, he simply increases the width of the stop. This allows them to set up any width frame for either thickness of door in lieu of manufacturing frames for 1-3/8" (34.37 mm) doors and 1-3/4" (43.75 mm) doors.

Frames are made in various widths (profile) to suit the thicknesses of the walls to which they are mounted. The standard widths with most

1148

manufacturers are 4-3/4" (118.75 mm), 5-3/4" (143.75 mm), 6-3/4" (168.75 mm), and 8-3/4" (218.75 mm).

In addition to the frame width dimensions, frames are ordered according to the door opening size, the standard being 2'-0" (600 mm), 2'-4" (700 mm), 2'-6" (750 mm), 2'-8" (800 mm), 3'-0" (900 mm), 3'-4" (1000 mm), 3'-6" (1050 mm), 3'-8" (1100 mm), and 4'-0" (1200 mm) wide for single swing and 4'-0" (1200 mm), 4'-8" (1400 mm), 5'-0" (1500 mm), 5'-4" (1600 mm), 6'-0" (1800 mm), 6'-8" (2000 mm), 7'-0" (2100 mm), 7'-4" (2200 mm), and 8'-0" (2400 mm) wide for double swing by 6'-8" (2000 mm), 7'-0" (2100 mm), 7'-2" (2150 mm), and 8'-0" (2400 mm) in height.

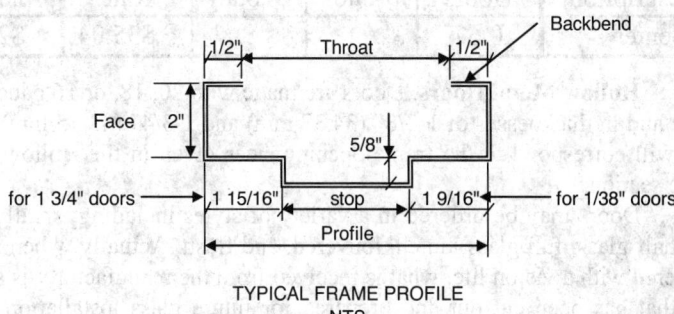

TYPICAL FRAME PROFILE
NTS

Frames are available from the factory in welded joint construction, all three sides put together to form a unit, or in knocked down (K.D.) form for use on drywall partitions. The K.D. frame is considered to be a pressure-fit frame that once in place capping the wall exerts a gripping action on the wall. The welded frame is stronger and must be erected prior to the wall erection, whereas the K.D. frame is erected after the wall is built.

Special frames can be ordered that will incorporate a transom panel above the door or sidelites of various heights and widths beside the door. Average material prices for hollow metal frames of various sizes are as follows:

18 Ga. Welded Hollow Metal Frames					
Size	Inches				Size - mm
	4 3/4	5 3/4	6 3/4	8 3/4	
2'- 6" x 7'- 0"	$ 84.00	$ 88.00	$ 91.00	$ 94.00	750 x 2100 mm
3'- 0" x 7'- 0"	$ 87.00	$ 90.00	$ 92.00	$ 95.00	900 x 2100 mm
5'- 0" x 7'- 0"	$ 108.00	$112.00	$115.00	$118.00	1500 x 2100 mm
6'- 0" x 7'- 0"	$ 112.00	$115.00	$118.00	$120.00	1800 x 2100 mm
Metric	118.75	143.75	168.75	218.75	

Deduct $14.00 each for knocked down (K.D.) frames.
Add $14.00 for "B" Label, $21.00 for "A" Label.

Labor Erecting Hollow Metal Frames. Two carpenters should erect about 16 welded frames per 8-hr. day at the following labor cost per frame:

Description	Hours	Rate	Total	Rate	Total
Carpenter	1.00	$	$	$35.04	$35.04

One carpenter should erect about 12 K.D. frames per 8-hr. day at the following labor cost per frame:

Description	Hours	Rate	Total	Rate	Total
Carpenter	0.67	$	$	$35.04	$23.48

Hollow Metal Doors. Doors are made with 20, 18, or 16 gauge face sheets and in thicknesses of 1-3/8" (34.37 mm) and 1-3/4" (43.75 mm). Door sizes will correspond to the frame opening sizes given in the Hollow Metal Frame section.

Doors may be ordered in a variety of styles including: small vision lites, half glass, full glass, dutch, louvered, and flush. Actually, when a door is ordered with a vision lite, what is received from the manufacturer is a flush door that has been cut out and prepared for future glass installation at the jobsite; the glass is not supplied by the door manufacturer.

The physical construction of the door itself is where manufacturers really differ. Each has its own techniques for assembly-location of seams, inverted channel or flush top and bottom edges, gauge of hinge and closer reinforcements, leading edge reinforcement, core materials, etc. The contractor is obligated to determine the compliance of the product with the requirements contained in the specification. Therefore, be sure of compliance with specification requirements when pricing hollow metal doors, because prices vary widely.

Average material prices for hollow metal doors are as follows:

20 Ga. Hollow Metal Doors					
Size	Inches		Size - mm	mm	
	1 3/8"	1 3/4"		34.38	43.75
2'- 6" x 7"- 0"	$146.00	$156.00	750 x 2100 mm	$146.00	$156.00
3'- 0" x 7"- 0"	$146.00	$157.00	900 x 2100 mm	$146.00	$157.00
3'- 6" x 7"- 0"	$160.00	$181.00	1050 x 2100 mm	$160.00	$181.00

Add $30.00 for "B" Label, $65.00 for "A" Label
Add $85.00 for louver, $70.00 for lite cut-out.
Add $55.00 for mineral core (for temperature rise rating).

Labor Erecting Hollow Metal Doors. Two carpenters should hang, on hinges, about 12 doors per 8-hr. day at the following labor cost per door:

1150

Description	Hours	Rate	Total	Rate	Total
Carpenter	1.34	$	$	$35.04	$46.95
Labor	0.50	$	$	$26.16	$13.08
Cost per Door					$60.03

Fire Label Requirements. According to building codes, certain door locations within a building, such as at stairs, trash rooms, furnace rooms, garage entries, storage rooms, etc., will require a fire rating for the door and frame. This rating is expressed as "C Label" for 3/4-hour rating, for a 1-hour wall rating. The "C" labeled doors are used for openings in corridors and room partitions within the building or in exterior walls subject to moderate to light fire exposure from outside the building. "B Label" for 1-1/2 hour rating in a 2-hour wall rating. Openings in enclosures of vertical communications, such as stairwells or elevator shafts, as well as in exterior walls subject to severe fire exposure from external of the building and "A Label" for 3-hour rating for a 4-hour rated wall. This is for door openings in wall which separate buildings or divide a single building into fire areas. Normally no glass is allowed unless tested for fire rating.

The rating is obtained by the manufacturer subjecting its product to actual fire testing by either Underwriters Laboratory Inc. (UL), Intertek Testing Services/Warnock Hersey (ITS/WHI), or Factory Mutual (FM) testing laboratory. The door and frame is then approved based on the physical construction of the unit. If the manufacturer changes the design after the approval, it must apply for a new test based on the re-designed unit. The approved door and frame then qualifies to have a label affixed, showing that it is a fire rated assembly.

When a door is to be fire rated, it also carries other restrictions, such as that the latch or lockset must be installed with a UL throw bolt, or that glass is either not allowed (as in "A Label" rating) or restricted in size (as in "B or C Label" ratings). If a louver is needed in the door, it must have a fusible link that will melt and allow the louver blades to close when the fire reaches the door, and the door must have a closer device or spring hinges.

Metal Covered Jambs and Trim

Wood jambs and trim covered with No. 24 gauge zinc coated sheet steel, for use with kalamein doors.

Jambs are 1-3/4" x 5-1/2" (43.75 x 137.50 mm) rabbeted for 1-3/4" (43.75 mm) doors. Plain casing is 7/8" x 3-1/2" (21.87 x 87.50 mm). Round edge and brick mold is 7/8" x 7/8" (21.87 x 21.87 mm).

1151

Size Up to	Inches			Size Up to	mm		
	6 3/4	8 3/4	12	mm	168.75	218.75	300.00
3'- 0" x 7'- 0"	$104.00	$112.00	$119.00	900 x 2100 mm	$104.00	$112.00	$119.00
4'- 0" x 7'- 0"	$112.00	$119.00	$126.00	1200 x 2100 mm	$112.00	$119.00	$126.00
5'- 0" x 7'- 0"	$119.00	$126.00	$133.00	1500 x 2100 mm	$119.00	$126.00	$133.00
6'- 0" x 7'- 0"	$126.00	$133.00	$140.00	1800 x 2100 mm	$126.00	$133.00	$140.00

For frames over 7'-0" (2100 mm) add $11.00 for each 6" (150 mm) or less.

Exterior Metal Residential Door Assemblies

Prehung metal doors complete with moldings of the styles usually found on traditional wood· doors and with a frame, threshold, weatherstripping, and a variety of locks are now available in over fifty stock designs. Prices for a 3' (900 mm) wide unit will vary from $250.00 to $1,000.00 depending on the type of lock and the type of glazing.

A plain, flush unit with a double cylinder lock and no glazing runs $300.00; a unit with raised panels runs $320.00; if the upper section is glazed with insulating glass the cost will raise to $350.00. Units similar to the above but in pairs for a 5' (1500 mm) opening will run $560.00, $650.00, and $730.00 respectively. If moldings are ordered for the inside face of the door add $25.00.

08200 WOOD AND PLASTIC DOORS

Flush Veneer Slab Doors for Exterior Use

Furnished with water and weather resistant birch face veneers both sides of the door. All doors 1-3/4" (43.75 mm) thick.

Size	Hollow Core	Solid Core	Size mm
2'- 6" x 6'- 8"	$ 40.00	$ 60.00	750 x 2000 mm
2'- 8" x 6'- 8"	$ 42.00	$ 62.00	801 x 2000 mm
3'- 0" x 6'- 8"	$ 44.00	$ 67.00	900 x 2000 mm
3'- 0" x 7'- 0"	$ 49.00	$ 70.00	900 x 2100 mm
3'- 0" x 8'- 0"	$ 57.00	$ 102.00	900 x 2400 mm

Extras for Altering Solid Core Flush Doors		
For V-grooves, add per groove	$11.00	------
For cutting in one square light, add	$25.00	------
For one circle light, approx. 12" diam., add	$31.00	300 Diam, add
For one circle light, approx. 18" diam., add	$37.00	450 Diam, add
For one peek light, add	$25.00	------
For one diamond light, add	$25.00	------

DOORS AND WINDOWS

White Pine Entrance Doors		
Type	Size	
	2'- 8"x 6'- 8"x 1-3/4"	3'-0"x 6'- 8"x 1-3/4"
6 Panel Colonial	$218.00	$223.00
4 Panel & Light	$220.00	$225.00
X Buck & 9 Light Rect.	$233.00	$238.00
X Buck & 12 Light Diag.	$248.00	$253.00
12 Panel Spanish	$258.00	$263.00
21 Panel Spanish	------	$175.00
Size Metric		
	.81 x 2.03 x .53 M	.91 x 2.03 x .53 M

These doors can usually be purchased prehung in rabbeted frames of standard widths with 1-1/2 pair of butts, without sill, lock or weatherstripping, for about $115.00 extra.

Exterior trim sets for front entrances are available. The price range is great, depending on depth of members, elaborateness, and sharpness of carvings and flutings. One manufacturer offers trim resembling entrance "A" shown at $225.00 in the deluxe model, $130.00 for a modified design; trim for doorways "C", "D", and "E" would cost about $130.00. They may be purchased as complete entrance sets with frames, thresholds, and sidelights if desired.

Western Pine Colonial Front Entrances

Exterior wood doors are usually fitted with combination storm and screen units. These are usually of 1-1/8" (28.12 mm) thick Ponderosa pine with aluminum wire. They run as follows for 6'-9" (2025 mm) height:

Description	2'- 6"	2'- 8"	3'- 0"
4 Light	$120.00	$125.00	$135.00
X Buck & 1 Light	------	$135.00	$150.00
Metric			
	750 mm	800 mm	900 mm

Combination doors are also available in aluminum, 1" (25 mm) thick for either 6'-8" (2000 mm) or 7'-0" (2100 mm) openings in various qualities.

Standard quality, 2'-6", 2'-8"& 3'-0" wide	$168.00	750 mm, 800 mm & 900 mm wide
Deluxe quality, 2'-6", 2'-8" & 3'-0" wide	$175.00	750 mm, 800 mm & 900 mm wide
Deluxe quality with white finish, 3'-0" wide	$193.00	900 mm wide
X-Buck type with white finish, 3'-0" wide	$210.00	900 mm wide

1153

Soft Wood French Doors

These doors are made of clear Ponderosa pine. Stiles and top rails are 4-3/4" (118.75 mm) and bottom rails are 9-5/8" (240.62 mm) overall with bead and cove sticking. Muntins are 1/2" (12.50 mm) between glass. Wood stops or glazing beads are mitred and tacked in place.

Ponderosa Pine						
Description	2'-6"x 6'-8"		2'-8"x 6'-8"		3'-0"x 6'-8"	
Type	1-3/8"	1-3/4"	1-3/8"	1-3/4"	1-3/8"	1-3/4"
1 Light French	$195.00	$210.00	$195.00	$210.00	$215.00	$225.00
5 Light French	$200.00	$205.00	$200.00	$205.00	$220.00	$230.00
15 Light French	$205.00	$215.00	$205.00	$215.00	$225.00	$240.00
(metric)						
Description	750 x 2000mm		800 x 2000mm		900 x 2000mm	
Type	34.37 mm	43.75 mm	34.37 mm	43.75 mm	34.37 mm	43.75 mm

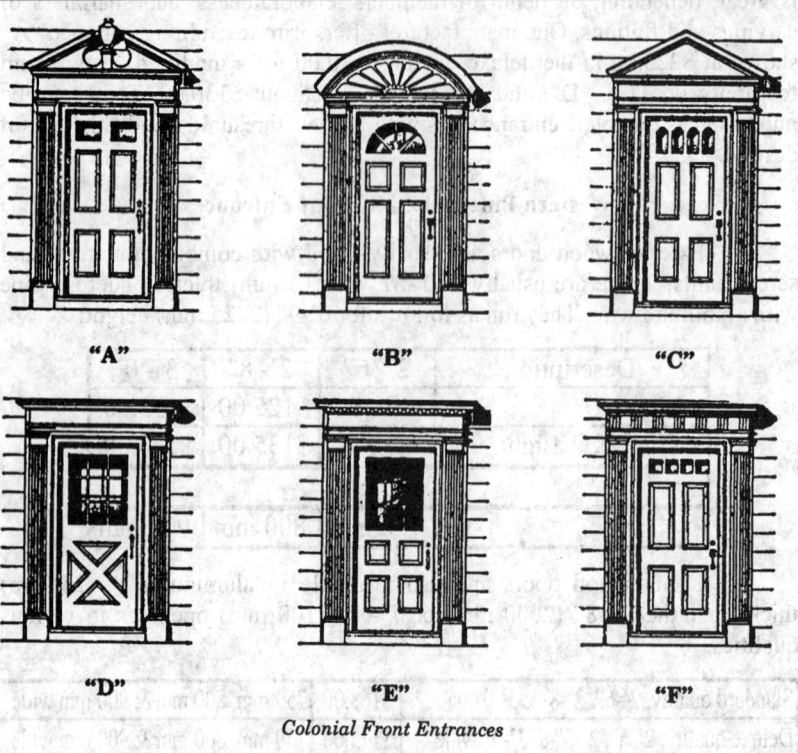

"A"   "B"   "C"

"D"   "E"   "F"

Colonial Front Entrances

DOORS AND WINDOWS

Prices of Western Pine Exterior Door Frames

Size	Drip Cap		Masonry		Size
	4-5/8"	5-1/4"	4-5/8"	5-1/4"	mm
2'-6" x 2'-6"	$110.00	$110.00	$100.00	$100.00	750 x 750 mm
2'-8" x 2'-8"	$115.00	$115.00	$105.00	$105.00	800 x 800 mm
2'-8" x 2'-8"	$115.00	$115.00	$105.00	$105.00	800 x 800 mm
3'-0" x 3'-0"	$120.00	$120.00	$110.00	$110.00	900 x 900 mm
3'-0" x 3'-0"	$120.00	$120.00	$110.00	$110.00	900 x 900 mm
4'-0" x 4'-0"	$125.00	$125.00	$115.00	$115.00	1,200 x 1,200 mm
5'-0" x 5'-0"	$130.00	$130.00	$120.00	$120.00	1,500 x 1,500 mm
6'-0" x 6'-0"	$140.00	$140.00	$130.00	$130.00	1,800 x 1,800 mm
Metric	115.62 mm	131.25 mm	115.62 mm	131.25 mm	

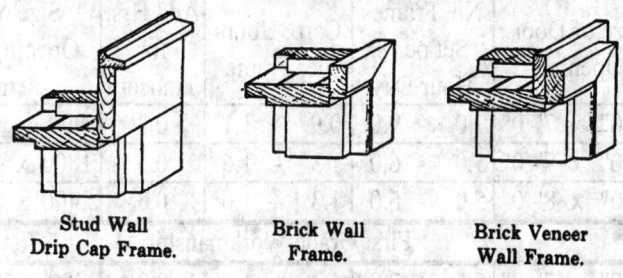

Stud Wall Drip Cap Frame.   **Brick Wall Frame.**   **Brick Veneer Wall Frame.**

Door Frame Prices with Preglazed
Sidelight One Side for 3'-0" x 6'-8" (900 x 2000 mm) Doors

Glass	1-1'-0"x 6'-8"		1-1'-4"x 6'-8"		1-1'-8"x 6'-8"	
	Drip Cap	Masonry	Drip Cap	Masonry	Drip Cap	Masonry
Safety Glass -1 Lite	$240.00	$230.00	$265.00	$255.00	$295.00	$285.00
3 Light Safety Glass	$255.00	$245.00	$270.00	$260.00	$310.00	$300.00
3 Light & Flat Panel	$285.00	$280.00	$300.00	$290.00	$315.00	$305.00
3 Light & Raised Panel .	$310.00	$295.00	$325.00	$310.00	$340.00	$330.00
Metric	1-300 x 2000 mm		1-400 x 2000 mm		1-500 x 2000 mm	

Also available in 4 and 5 light with or without flat or raised panel and with fluted, latticed rondel or loxenged, or insulating glass.

Door Frame Prices with Preglazed Sidelights Both Sides for 3'- 0" x 6' - 8" (900 x 2000 mm) Doors						
	2-1'-0"x 6'-8"		2-1'-4"x 6'-8"		2-1'-8"x 6'-8"	
Glass	Drip Cap	Masonry	Drip Cap	Masonry	Drip Cap	Masonry
Safety Glass -1 Lite	$335.00	$330.00	$365.00	$355.00	$415.00	$400.00
3 Light Safety Glass	$345.00	$340.00	$370.00	$360.00	$430.00	$415.00
3 Light & Flat Panel	$375.00	$370.00	$400.00	$390.00	$435.00	$420.00
3 Light & Raised Panel	$345.00	$340.00	$425.00	$415.00	$460.00	$445.00
Metric	2-300 x 2000 mm		2-400 x 2000 mm		2 -500 x 2000 mm	

Setting Exterior Wood Door Frames				
Size of Door Opening	No. Frames Set per 8 Hour Day	Carp. Hours per Frame	Add Hrs. for Transom	Size of Door Opening Meters mm
3'- 0" x 7'- 0"	7.0 - 9.0	0.9 - 1.1	0.3	900 x 900 mm
6'- 0" x 7'- 0"	5.0 - 6.0	1.3 - 1.6	0.3	1800 x 2100 mm
8'- 0" x 8'- 0"	5.0 - 6.0	1.3 - 1.6	0.3	2400 x 2400 mm
First Grade Workmanship				
3'- 0" x 7'- 0"	5.0 - 7.0	1.1 - 1.6	0.3	900 x 900 mm
6'- 0" x 7'- 0"	4.0 - 5.0	1.6 - 2.0	0.3	1800 x 2100 mm
8'- 0" x 8'- 0"	4.0 - 5.0	1.6 - 2.0	0.3	2400 x 2400 mm

Interior Flush Doors

Interior hollow core (H.C.) flush doors are manufactured with face veneers of tempered hardboard or plywood of various species, e.g., luan mahogany, birch, or oak, walnut, etc. The core, which is not hollow, will have some material such as honeycomb fiber board or wood stripping to give lateral support to the face veneers.

1156

DOORS AND WINDOWS

Interior Hollow Core Flush Doors			
Doors 1-3/8" Thick	Plain Birch	Luan	Doors 34.37 mm Thick
2'-0" x 6'-8"	$38.00	$30.00	600 x 2000 mm
2'-4" x 6'-8"	$39.00	$35.00	700 x 2000 mm
2'-6" x 6'-8"	$42.00	$40.00	750 x 2000 mm
2'-8" x 6'-8"	$45.00	$45.00	800 x 2000 mm
3'-0" x 6'-8"	$46.00	$50.00	900 x 2000 mm

For doors 7'-0" (2100 mm) in height, add $10.00 per door.
For doors 1-3/4" (43.7 mm) thick, add $9.00 per door.

Interior solid core (S.C.) doors have face veneers of plywood in various species, the same as hollow core doors. The cores of the solid door are usually constructed of solid wood blocks (stave) or particleboard.

Interior Solid Core Flush Doors, Particleboard Core				
Size	Rotary Birch	Red Oak	Select Birch	Size mm
2'-8" x 1 3/8"	$100.00	$110.00	$120.00	800 x 34 mm
3'-0" x 1 3/8"	$110.00	$110.00	$135.00	900 x 34 mm
2'-8" x 1 3/4"	$110.00	$115.00	$140.00	800 x 44 mm
3'-0" x 1 3/4"	$115.00	$120.00	$150.00	900 x 44 mm

Pre-Hung Interior Door Units

Pre-hung interior door units consist of the door, a 1-3/8" (34.3 7mm) H.C. frame, stops, face trim, and hinges, all assembled as a unit. The door is bored and mortised for the lockset and bolt, and the frame is mortised for the strike plate.

The frame, called a "split jamb", is in two pieces at the stop for installation in the wall opening.

Prices are for units with adjustable jambs, from 4-3/8" (109.37 mm) to 5-1/2" (137.5 0mm) jambs.

Door Size	Luan Doors	Birch Doors	Door Size mm
1'- 6" x 6'- 8"	$120.00	$150.00	450 x 2000 mm
2'- 0" x 6'- 8"	$130.00	$155.00	600 x 2000 mm
2'- 4" x 6'- 8"	$135.00	$160.00	700 x 2000 mm
2'- 6" x 6'- 8"	$140.00	$165.00	750 x 2000 mm
2'- 8" x 6'- 8"	$145.00	$170.00	800 x 2000 mm

Interior Paneled Doors

Pine panel doors built up of stiles and rails have largely been supplanted by molded doors consisting of a single piece of resin-impregnated wood fibre. The doors are factory primed for finish painting on the job site and are joint-free and shrink and swell resistant. Colonial style 6 panel doors will cost as follows:

Size	Price	Size - mm
2'- 0" x 6'- 8" x 1 3/8 "	$43.00	600 x 2000 x 34 mm
2'- 4" x 6'- 8" x 1 3/8 "	$46.00	700 x 2000 x 34 mm
2'- 6" x 6'- 8" x 1 3/8 "	$47.00	750 x 2000 x 34 mm
2'- 8" x 6'- 8" x 1 3/8 "	$50.00	800 x 2000 x 34 mm
3'- 0" x 6'- 8" x 1 3/8 "	$55.00	900 x 2000 x 34 mm

Sliding and Bi-Folding Closet Doors

With the popularity of sliding doors for closets, many different combinations of doors, materials, and sizes are on the market. With some of them, the doors are the same height as the other doors in the room or building, i.e. 6'-8" (2000 mm) or 7'-0" (2100 mm), while in others the doors extend to the ceiling so that they provide access to the storage shelves at the top.

There are many different types of doors and hardware, but those described below are the type usually found in housing and apartments.

Frames for sliding doors may be obtained in kits containing two side jambs, head assembly with track attached, and necessary hardware for hanging doors. Available with 5-3/8" (134.3 mm) jambs for lath and plaster construction, or with 4-5/8" (115.62 mm) jambs for dry wall construction. Openings may be arranged for two, three, or four bypassing doors. Any 1-3/8" (34.37 mm) thick doors can be used. The following table lists bypassing units, doors, and hardware only for a variety of 1-3/8" (34.37 mm) thick doors.

DOORS AND WINDOWS

2 Panel Opening	H. C. Luan	H. C. Birch	W. P. Louver	Half Louver	6 Pan. W.P. Col	2 Panel Opening mm
4'- 0" x 6'- 8"	$62.00	$75.00	$130.00	$165.00	$135.00	1200 x 2000 mm
5'- 0" x 6'- 8"	$68.00	$82.00	$136.00	$171.00	$240.00	1500 x 2000 mm
6'- 0" x 6'- 8"	$75.00	$88.00	$143.00	$178.00	$245.00	1800 x 2000 mm

Bi-folding doors offer the advantage of full access to the closet. They are available in kits with doors, hinges, pivots, knobs, and guide assembly and track. Jambs and casing are not included. Openings to 3'-2" (950 mm) panels hinged one side; 4'-0" (1200 mm) and over 4 panels hinged two sides.

Opening	Panels per Opening	1-3/8" Hardboard	1-3/8" H.C. Birch	1-1/8" Louver	1-1/8" Half Louver	1-3/8" W. P. Col.	Opening mm
2'- 0" x 6'- 8"	2	$45.00	$60.00	$80.00	$90.00	$98.00	600 x 2000 mm
2'- 6" x 6'- 8"	2	$50.00	$70.00	$90.00	$100.00	$105.00	750 x 2000 mm
3'- 0" x 6'- 8"	2	$55.00	$75.00	$95.00	$105.00	$115.00	900 x 2000 mm
4'- 0" x 6'- 8"	4	$65.00	$120.00	$130.00	$150.00	$200.00	1200 x 2000 mm
5'- 0" x 6'- 8"	4	$75.00	$125.00	$140.00	$160.00	$205.00	1500 x 2000 mm
6'- 0" x 6'- 8"	4	$90.00	$130.00	$145.00	$165.00	$210.00	1800 x 2000 mm
Metric		34.38	34.38	28.13	28.13	34.38	mm

08240 PLASTIC FACED WOOD DOORS

For some uses, especially institutional, some doors are veneered on all surfaces with plastic laminate. The laminate can be used on both solid and hollow core doors as well as fire, acoustical, x-ray, and other special units. Any standard plastic finish may be applied.

Unless otherwise specified, doors are edged as well as faced with plastic. The doors will be shipped in a carton prefinished and pre-mortised for all hardware and cannot be trimmed on the job. Therefore, special care must be taken in checking all details on the shop drawings to see that dimensions are accurately followed on the job.

Since all mortising and fitting is eliminated, hanging of the doors is simpler. One carpenter and one laborer should hang 2 doors per hour.

These doors are fabricated by manufacturers on special order. The veneers are not applied by contact adhesives but are bonded under pressure. Costs should be obtained from the manufacturer or distributor.

For budget purposes a door 2'-8"x 6'-8" (800 x 2000 mm) plastic faced both sides of a 1-3/4" (43.75 mm) solid core will run around $145.00. For one face plastic, one face birch around $110.00.

1159

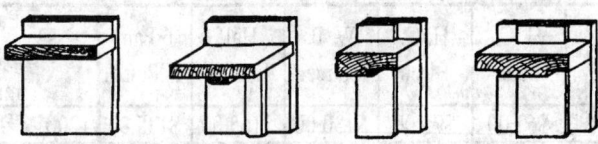

Door Jambs

Interior Door Jambs

The following prices are for plain door jambs 3/4" x 4-1/2" (18.75 x 112.50 mm). Side jambs are dadoed at top for head jambs. The prices are for jambs only, not including stops or trim which must be added. Furnished knocked down.

Price Per Set, Dadoed and Sanded, Not Including Stops			
Size	Ponderosa Pine	Natural Birch	Size mm
2'- 8" x 6'- 8" or smaller	$30.00	$50.00	800 x 2000 mm or smaller
3'- 0" x 7'- 0" or smaller	$35.00	$55.00	900 x 2100 mm or smaller
6'- 0" x 6'- 8" or smaller	$40.00	$60.00	1800 x 2000 mm or smaller

Add for Door Stops 7/16"x1-3/8" (10.93 x 34.37 mm), Cut to Length

Size	Ponderosa Pine	Natural Birch	Size
3'- 0" x 7'-0" or smaller	$2.50	$6.00	900 x 2100 mm or smaller
6'- 0" x 7'-0" or smaller	$1.70	$8.00	1800 x 2100 mm or smaller

If door jambs are put together in the mill, add $3.00 per set.

Setting Interior Wood Door Jambs

Door Size, Description of Jamb	No. Set per 8-Hr. Day	Carp Hours per Jamb	Add for Transom Hours	Door Size mm
3'- 0" x 7'-0" Plain door jambs	8.00 - 10.00	0.80 - 1.00	0.60	900 x 2100 mm
3'- 0" x 7'-0" Paneled door jambs	5.00 - 6.00	1.30 - 1.60	0.60	900 x 2100 mm
6'- 0" x 7'-0" Plain door jambs	5.00 - 6.00	1.30 - 1.60	0.60	1800 x 2100 mm
First Grade Workmanship				
3'- 0" x 7'-0", Plain door jambs	7.00 - 9.00	0.90 - 1.10	0.60	900 x 2100 mm
3'- 0" x 7'-0" Paneled door jambs	4.00 - 5.00	1.60 - 2.00	0.60	900 x 2100 mm
6'- 0" x 7'-0" Plain door jambs	4.00 - 5.00	1.60 - 2.00	0.60	1800 x 2100 mm

Door Jambs

Stock door casings per 100 lin.ft. cost $55.00 ($1.80 m) in Ponderosa pine and $160.00 ($5.25 per m) in birch. Stops will run $19.00 ($0.62 per m) and $36.00 ($1.18 per m).

Oak thresholds, 3-5/8" x 5/8" (90.62 x 15.62 mm) will run $155.00 per 100 lin.ft. ($50.85 per 10 m) and oak sills, 1-5/16" x 7-1/4" (32.81 x

181.25 mm) will run $445.00 per 100 lin.ft. ($146.00 per 10 m). Pre-cut sills will run $4.90 for 2'-8" (800 mm) opening, $5.60 for 3' (900 mm) opening; in the narrow size, $14.00 and $17.50 for the wider.

Interior Door Trim

The labor cost of placing interior door trim varies with the type of trim or casings and the class of workmanship. Production time is given on 4 different types of interior door trim:

Style "A" consists of 1-member casing with either a mitred or square cut head or cap. This is the simplest trim to use.

Style "B" consists of 1-member trim for sides and a built-up cap trim for heads. Caps are either put together in the mill or on the job.

Style "C" is 2-member back-band trim, consisting of a one-piece casing with a back-band of thicker material.

Style "D" is any type of door trim that is assembled and glued in the mill, ready to set in place as a unit.

The class of workmanship has considerable bearing on the labor costs. In ordinary workmanship the casings are nailed to the rough bucks or grounds and the trim might show some hammer marks, but where first grade workmanship is required, the utmost care must be used in cutting and fitting all casings, miters, etc. All nails must be carefully driven and set, the casings must show the same margin on all sides of the jamb and trim must fit close to the walls. In other words, it must be a first class job in every respect. For labor costs for each style of trim, refer to the table on the opposite page.

Fitting and Placing Stationary Transom Sash. Where just an ordinary grade of workmanship is required, a carpenter should fit and place about 12 to 14 stationary transom sash per 8-hr. day, at the following labor cost per sash:

Description	Hours	Rate	Total	Rate	Total
Carpenter	0.60	$	$	$35.04	$21.02

First Grade Workmanship

Figure about 9 to 11 transom sash per 8-hr. day, at the following labor cost per sash:

Description	Hours	Rate	Total	Rate	Total
Carpenter	0.80	$	$	$35.04	$28.03

Labor Erecting Interior Door Casings and Trim

Style Trim	Kind of Trim	Size of Opening	No. Sides Trim per 8-Hr Day	Carpenter Hrs / Side	Carpenter Hrs / Opening	Add for Transom†	Size of Opening, mm
			Ordinary Workmanship				
"A"	Single Casing	3'- 0" x 7'- 0"	14	0.5 - 0.6	1.0 - 1.2	0.6	900 x 2100 mm
"B"	Cap Trim	3'- 0" x 7'- 0"	14	0.5 - 0.6	1.0 - 1.2	0.6	900 x 2100 mm
"B"	Cap Trim*	3'- 0" x 7'- 0"	7	1.0 - 1.1	2.0 - 2.2	0.6	900 x 2100 mm
"C"	Back-Band Trim	3'- 0" x 7'- 0"	9	0.8 - 0.9	1.6 - 1.8	0.6	900 x 2100 mm
"D"	Mill Assembled**	3'- 0" x 7'- 0"	14	0.5 - 0.6	1.0 - 1.2	0.6	900 x 2100 mm
"A"	Single Casing	6'- 0" x 7'- 0"	9	0.8 - 0.9	1.6 - 1.8	0.8	1800 x 2100 mm
"B"	Cap Trim	6'- 0" x 7'- 0"	9	0.8 - 0.9	1.6 - 1.8	0.8	1800 x 2100 mm
"B"	Cap Trim*	6'- 0" x 7'- 0"	5	1.1 - 1.6	2.2 - 3.2	0.8	1800 x 2100 mm
"C"	Back-Band Trim	6'- 0" x 7'- 0"	7	1.0 - 1.1	2.0 - 2.2	0.8	1800 x 2100 mm
"D"	Mill Assembled**	6'- 0" x 7'- 0"	9	0.8 - 0.9	1.6 - 1.8	0.8	1800 x 2100 mm
			Fine Grade Workmanship				
"A"	Single Casing	3'- 0" x 7'- 0"	9	0.8 - 0.9	1.6 - 1.8	0.8	900 x 2100 mm
"B"	Cap Trim	3'- 0" x 7'- 0"	9	0.8 - 0.9	1.6 - 1.8	0.8	900 x 2100 mm
"B"	Cap Trim*	3'- 0" x 7'- 0"	5	1.1 - 1.6	2.2 - 3.2	0.8	900 x 2100 mm
"C"	Back-Band Trim	3'- 0" x 7'- 0"	8	0.9 - 1.0	1.8 - 2.0	0.8	900 x 2100 mm
"D"	Mill Assembled**	3'- 0" x 7'- 0"	11	0.7 - 0.8	1.4 - 1.6	0.8	900 x 2100 mm
"A"	Single Casing	6'- 0" x 7'- 0"	7	1.0 - 1.1	2.0 - 2.2	1.0	1800 x 2100 mm
"B"	Cap Trim	6'- 0" x 7'- 0"	7	1.0 - 1.1	2.0 - 2.2	1.0	1800 x 2100 mm
"B"	Cap Trim*	6'- 0" x 7'- 0"	5	1.3 - 1.6	2.6 - 3.2	1.0	1800 x 2100 mm
"C"	Back-Band Trim	6'- 0" x 7'- 0"	6	1.1 - 1.3	2.2 - 2.6	1.0	1800 x 2100 mm
"D"	Mill Assembled**	6'- 0" x 7'- 0"	7	1.0 - 1.1	2.0 - 2.2	1.0	1800 x 2100 mm

* Cap trim put together by carpenters on the project.

** Trim assembled and glued together in the mill.

† Includes transom jambs, stops and casings but no transom sash.

DOORS AND WINDOWS

Fitting and Hanging Hinged or Pivoted Transom Sash. If the transom sash are hinged or pivoted, a carpenter should fit and hang 7 to 9 sash per 8-hr. day, at the following labor cost per sash:

Description	Hours	Rate	Total	Rate	Total
Carpenter	1.00	$	$	$35.04	$35.04

First Grade Workmanship

Figure about 5 to 7 hinged or pivoted transom sash per 8-hr. day, at the following labor cost per sash:

Description	Hours	Rate	Total	Rate	Total
Carpenter	1.30	$	$	$35.04	$45.55

Fitting and Hanging Wood Doors. The labor cost of fitting and hanging doors varies with the class of workmanship, weight of door, soft or hard wood, and the tools and equipment used. There are power planes, hinge butt routers, and lock mortisers that enable a carpenter to do about three times as much work per day as where the doors are fitted by hand.

Where just an ordinary grade of workmanship is required, the doors do not always show the same margin at the top and sides, the screws are frequently driven instead of placed with a screw driver, and in many instances, the doors are "hinge-bound" so they will not open and close freely.

Ordinary Workmanship

Where doors are fit and hung by hand, a carpenter should fit and hang 8 doors per 8-hr. day, at the following labor cost per door:

Description	Hours	Rate	Total	Rate	Total
Carpenter	1.00	$	$	$35.04	$35.04

On large production line jobs, where power tools can be used to advantage and where the same class of work is performed using a power plane, an electric hinge butt router for door and jambs, and an electric lock mortiser, a carpenter should fit and hang 12 to 16 doors per 8-hr. day, at the following labor cost per door:

Description	Hours	Rate	Total	Rate	Total
Carpenter	0.60	$	$	$35.04	$21.02

The above costs will vary with the ability of the carpenter and the job setup, as a power planer should plane off a door in 5 to 10 minutes; a hinge butt router should do 60 to 75 openings per 8-hr. day, while a lock router should mortise out for a lock in about 1/2 minute, after it has been set on the door. The labor handling the doors is the same for both hand and machine work.

First Grade Workmanship

In the better class of buildings, the doors must show the same margin between the door and jamb at the top and sides, the hinges must be set even and flush with the jambs, and the edge of the door and the doors must be free from "hinge binding" so they will fit well and work easily at the same time.

On this grade of work a carpenter should fit and hang about 4 doors per 8-hr. day, at the following labor cost per door:

Description	Hours	Rate	Total	Rate	Total
Carpenter	2.00	$	$	$35.04	$70.08

Where power planers, hinge butt routers, and lock mortisers are used, a carpenter should fit and hang 7 doors per 8-hr. day, at the following labor cost per door:

Size of Door Opening	Style Trim	Kind of Trim	Carp. Hrs. per Opening	Add for Transom†	Size of Door Opening mm
Labor Setting Jambs, Fitting and Hanging Doors, Placing Stops, Hardware and Casing one Door Opening Complete					
Ordinary Workmanship					
3'- 0" x 7'- 0"	"A"	Single Casing	3.1 - 3.7	2.2	900 x 2100 mm
3'- 0" x 7'- 0"	"B"	Cap Trim	3.1 - 3.7	2.2	900 x 2100 mm
3'- 0" x 7'- 0"	"B"	Cap Trim*	4.1 - 4.7	2.2	900 x 2100 mm
3'- 0" x 7'- 0"	"C"	Back-Band Trim	3.7 - 4.3	2.2	900 x 2100 mm
3'- 0" x 7'- 0"	"D"	Mill Assembled**	3.1 - 3.7	2.2	900 x 2100 mm
6'- 0" x 7'- 0"	"A"	Single Casing	5.6 - 6.5	2.8	1800 x 2100 mm
6'- 0" x 7'- 0"	"B"	Cap Trim	5.6 - 6.5	2.8	1800 x 2100 mm
6'- 0" x 7'- 0"	"B"	Cap Trim*	6.2 - 7.9	2.8	1800 x 2100 mm
6'- 0" x 7'- 0"	"C"	Back-Band Trim	6.0 - 6.9	2.8	1800 x 2100 mm
6'- 0" x 7'- 0"	"D"	Mill Assembled**	5.6 - 6.5	2.8	1800 x 2100 mm

DOORS AND WINDOWS

Size of Door Opening	Style Trim	Kind of Trim	Carp. Hrs. per Opening	Add for Transom†	Size of Door Opening mm
Labor Setting Jambs, Fitting and Hanging Doors, Placing Stops, Hardware and Casing one Door Opening Complete (Cont'd)					
First Grade Workmanship					
3'- 0" x 7'- 0"	"A"	Single Casing	5.5 - 6.2	2.7	900 x 2100 mm
3'- 0" x 7'- 0"	"B"	Cap Trim	5.5 - 6.2	2.7	900 x 2100 mm
3'- 0" x 7'- 0"	"B"	Cap Trim*	6.1 - 7.6	2.7	900 x 2100 mm
3'- 0" x 7'- 0"	"C"	Back-Band Trim	5.7 - 6.4	2.7	900 x 2100 mm
3'- 0" x 7'- 0"	"D"	Mill Assembled**	5.3 - 6.0	2.7	900 x 2100 mm
6'- 0" x 7'- 0"	"A"	Single Casing	9.2 - 10.3	3.4	1800 x 2100 mm
6'- 0" x 7'- 0"	"B"	Cap Trim	9.2 - 10.3	3.4	1800 x 2100 mm
6'- 0" x 7'- 0"	"B"	Cap Trim*	9.8 - 11.3	3.4	1800 x 2100 mm
6'- 0" x 7'- 0"	"C"	Back-Band Trim	9.4 - 10.7	3.4	1800 x 2100 mm
6'- 0" x 7'- 0"	"D"	Mill Assembled**	9.2 - 10.3	3.4	1800 x 2100 mm

* Cap trim put together by carpenters on the project.

** Trim assembled and glued together in the mill.

† Above is for hinged or pivoted transom. For stationary Transom deduct 0.4 - Hour.

Description	Hours	Rate	Total	Rate	Total
Carpenter	1.10	$	$	$35.04	$38.54

For front door hardware, add 0.5-hr. for ordinary work and 1 to 2 hrs. for first grade work.

Where power planers, hinge butt routers, and lock mortisers are used, deduct 0.4 hr. for single door and 0.8-hr. for double door openings on ordinary workmanship; for first grade workmanship, deduct 0.9-hrs. for single door and 1.8 hrs. for double door openings.

Labor Installing Pre-Hung Door Units. Pre-hung door units are used extensively in the construction of houses and apartments ranging from "economy" grade to middle class.

In the completely assembled package, jambs are assembled, the door is hung in place with hinges applied, and stops are mitred and nailed into place.

Where assembled pre-hung door units are used, a carpenter should install 16 units per 8-hr. day, at the following labor cost per unit:

Description	Hours	Rate	Total	Rate	Total
Carpenter	0.50	$	$	$35.04	$17.52

Fitting and Hanging Hardwood Acoustical Doors. The labor cost of fitting and hanging hardwood acoustical doors is always figured on a first grade workmanship basis, because they must be perfectly installed to insure their efficient operation.

1165

Assuming a carpenter, experienced in hanging acoustical doors, is employed to do the work, the following production should be obtained:

Type of Door	Approx. Wgt. Lbs. per Sq.Ft.	Carpenter Hrs per Opening	Helper Hrs per Opening	Approx. Wgt. Kg. per Sq.M.
Hardwood Class 36	4	6	3	19.53
Hardwood Class 41	7	8	4	34.18

Fitting and Hanging Toilet Stall Doors. A carpenter should fit and hang 9 to 11, 7/8" (21.8 mm) or 1-1/8" (28.1 mm) toilet doors per 8-hr. day, at the following labor cost per door:

Description	Hours	Rate	Total	Rate	Total
Carpenter	0.80	$	$	$35.04	$28.03

First Grade Workmanship

Where hardwood stall doors are used and carefully hung, a carpenter should fit and hang 7 to 9 doors per 8-hr. day, at the following cost per door:

Description	Hours	Rate	Total	Rate	Total
Carpenter	1.00	$	$	$35.04	$35.04

Fitting and Hanging Heavy Wood Swinging Doors Up to 4'-0"x 8'-0" (1200 x 2400 mm). When hanging heavy wood swinging doors, such as used in garages, stables, mills and factory buildings, where the doors are 3'-0" (900 mm) to 4'-0" (1200 mm) wide and 7'-0" (2100 mm) to 8'-0" (2400 mm) high, it will require about 4 to 4-1/2 hrs. time to fit and hang one door or 8 to 9 hrs. per pair, at the following labor cost per door:

Description	Hours	Rate	Total	Rate	Total
Carpenter	4.30	$	$	$35.04	$150.67

Fitting and Hanging Heavy Wood Swinging Doors Up to 5'-0" x 10'-0" (1500 x 3000 mm). To fit and hang extra heavy wood swinging doors up to 5'-0" x 10'-0" x 2-1/2" or 3" (1500 x 3000 x 62.5 or 75 mm), will require about 4 to 5 hrs. time for 2 carpenters, at the following labor cost per door:

Description	Hours	Rate	Total	Rate	Total
Carpenter	9.00	$	$	$35.04	$315.36

Fitting and Hanging Wood Sliding Doors Up to 4'-0" x 8'-0" (1200 x 2400 mm). When fitting and hanging heavy wood sliding doors, such as in garages and factories, two carpenters working together should

1166

place tracks and hang one door complete in 3 to 3-1/2 hrs. at the following labor cost per door:

Description	Hours	Rate	Total	Rate	Total
Carpenter	6.50	$	$	$35.04	$227.76

Placing track and hang a pair of double sliding doors for an opening up to 8'-0" x 8'-0" (2400 x 2400 mm) requires 10 to 12 hrs. carpenter time, at the following labor cost per opening:

Description	Hours	Rate	Total	Rate	Total
Carpenter	11.00	$	$	$35.04	$385.44

The above does not include cutting or drilling in masonry walls to place bolts or anchors. It assumes the work is done when the walls are being built.

Fitting and Hanging Heavy Wood Sliding Doors Up to 10'-0" x 10'-0" (3000 x 3000 mm). To place track, fit, and hang one sliding door up to 10'-0" x 10'-0", 2-1/2" to 3" (3000 x 3000, 62.50 to 75 mm) thick, requires about 12 hrs. carpenter time, at the following labor cost per opening:

Description	Hours	Rate	Total	Rate	Total
Carpenter	12.00	$	$	$35.04	$420.48

The above does not include cutting or drilling masonry walls for bolts or anchors.

Fitting and Hanging Heavy Wood Sliding Doors Up to 12'-0" x 18'-0" (3600 x 5400 mm). To place track and hang one large door for an opening up to 12'-0" x 18'-0" (3600 x 5400 mm) requires about 24 to 28 hrs. carpenter time, at the following labor cost per door:

Description	Hours	Rate	Total	Rate	Total
Carpenter	26.00	$	$	$35.04	$911.04

It requires 4 to 6 workers to handle a door of this size.

Making and Setting Sliding Door Pockets. Where sliding door pockets are made on the job for door openings up to 3'-0" x 7'-0" (900 x 2100 mm), a carpenter should make the pocket, fit track, and set the pocket in about 4 hrs. at the following labor cost:

Description	Hours	Rate	Total	Rate	Total
Carpenter	4.00	$	$	$35.04	$140.16

Where double doors are used and it is necessary to make and set 2 pockets, double the above costs.

Setting Sliding Door Pockets. If sliding door pockets are sent to the job assembled, with track in place or steel pocket door T-frames are used, and the carpenters on the job set the pocket or steel frame ready to receive the door, it will require about 1 hr. carpenter time, at the following labor cost per opening:

Description	Hours	Rate	Total	Rate	Total
Carpenter	1.00	$	$	$35.04	$35.04

If double sliding doors are used and it is necessary to set 2 pockets for each opening, double the above costs.

Fitting and Hanging Sliding Doors. For an ordinary sliding door up to 3'-0" x 7'-0" (900 x 2100 mm), to place hangers and hang the door complete, not including finish hardware, requires about 1 hr. carpenter time, at the following labor cost per door:

Description	Hours	Rate	Total	Rate	Total
Carpenter	1.00	$	$	$35.04	$35.04

For double sliding doors, double the above costs.

Labor on Sliding Door Openings Complete with Pockets. Where necessary to make the sliding door pocket, hang the track, set the pocket, and fit and hang the door complete, exclusive of finish hardware, figure about 5 hrs. carpenter time, at the following labor cost per opening:

Description	Hours	Rate	Total	Rate	Total
Carpenter	5.00	$	$	$35.04	$175.20

For double sliding door openings, double the above costs.

If pockets are sent to the job with track fitted in place or steel pocket door T-frames are used, and the job carpenters set the box or steel frame and fit and hang the sliding door, it will require about 2 hrs. carpenter time per opening, exclusive of finish hardware, at the following labor cost per opening:

Description	Hours	Rate	Total	Rate	Total
Carpenter	2.00	$	$	$35.04	$70.08

For double sliding door openings, double the above costs.

Labor Fitting and Hanging Sliding Wood Closet Doors. When hanging and fitting wood closet doors that slide past each other without the necessity for pockets, a carpenter should place track, attach hangers, fit, and hang two doors complete for an opening 3'-0" to 5'-0" (90 to 1500 mm) wide and 6'-8" to 8'-0" (2000 to 2400 mm) high in about 2 hours. This does not include time setting door jambs and casing openings on two sides. The labor cost per opening should average as follows:

1168

Description	Hours	Rate	Total	Rate	Total
Carpenter	2.00	$	$	$35.04	$70.08

08300 SPECIAL DOORS

TIN CLAD FIRE DOORS

Tin clad fire doors are used in factory, mill, and warehouse buildings. They are either sliding or swinging, with fusible links and automatic closing devices.

Sliding doors and lap type swinging doors in brick walls do not ordinarily require a steel frame, but it might be necessary for tile walls. Flush type swinging doors require a frame.

Standard tin clad fire doors are usually manufactured in two thicknesses depending on the fire rating required. For Class "A" rating, the thickness is 2-5/8" (65.6 mm) consisting of 3-plies of tongue-and-groove lumber clad with tin sheeting on all sides. For Class "B" or "C" rating, the thickness is 1-3/4" (43.7 mm) consisting of 2-plies of tongue-and-groove lumber with cladding.

Tin Clad doors shall conform to UL-10. Doors shall have a core made up of layers of boards nailed to each other and encased in terne- or zinc-coated plates that are jointed together at their edges with nails driven through the joints into the core.

Method of Calculating the Area of Fire Doors. For flush doors, multiply the width of the opening by the height of the opening.

Square Top Lap Doors. Obtain size of door by adding 8" (200 mm) to width and 4" (100 mm) to height of opening, then multiply width of door by height of door.

Inclined Top Lap Doors. Obtain width of door by adding 8" (200 mm) to width of opening and obtain height by adding 8" (200 mm) to height of door for doors up to 4'-0" (1200 mm) wide and 12" (300 mm) for doors up to 6'-0" (1800 mm) wide. Multiply width of door by height of door.

Prices of Tin Clad Fire Doors and Shutters

The price includes door only, covered both sides with tin of a quality and in a manner specified by the National Board of Fire Underwriters.

Description	Price per Sq.Ft.
Standard 3-ply, 2-5/8" thick, 13 lbs./sq.ft.	$22.00
Standard 2-ply, 1-3/4" thick, 10 lbs./sq.ft.	$15.00
Standard 2-ply, 1-3/4" thick, shutters, 10 lbs./sq.ft.	$15.00

Above prices include doors only, no hardware or fixtures of any kind or delivery costs.

Description	Price per Sq.M.
65.62 mm thick, 63.47 kg/sq.m	$236.81
43.75 mm thick, 48.82 kg/sq.m	$161.46
43.75 mm thick, shutters, 48.82 kg/sq.m	$161.46

Above prices include doors only, no hardware or fixtures of any kind or delivery costs.

Hardware for Swinging, Automatic Closing Tin Clad Doors

Fixture sets include hinges, pintles, weights, chains, catches, latches, complete automatic device with bolts, and screws for attaching. Installation is not included. Door frames are not included.

Door Width & Height up and Including	Single Swinging Doors		Double Swinging Doors		Door Width & Height up and Including mm
	Flush Type	Lap Type	Flush Type	Lap Type	
3'- 0" x 6'- 6"	$110.00	$150.00	$260.00	$300.00	900 x 1950 mm
3'- 0" x 8'- 6"	$140.00	$190.00	$290.00	$350.00	900 x 2550 mm
3'- 0" x 10'- 0"	$167.00	$225.00	$535.00	$390.00	900 x 3000 mm
4'- 0" x 6'- 6"	$120.00	$160.00	$260.00	$304.00	1200 x 1950 mm
4'- 0" x 8'- 6"	$145.00	$195.00	$295.00	$350.00	1200 x 2550 mm
4'- 0" x 12'- 0"	$175.00	$235.00	$330.00	$400.00	1200 x 3600 mm
5'- 0" x 6'- 6"	$125.00	$160.00	$265.00	$305.00	1500 x 1950 mm
5'- 0" x 8'- 6"	$150.00	$205.00	$300.00	$360.00	1500 x 2550 mm
5'- 0" x 12'- 0"	$180.00	$240.00	$340.00	$410.00	1500 x 3600 mm
6'- 0" x 6'- 6"	$130.00	$165.00	$270.00	$310.00	1800 x 1950 mm
6'- 0" x 8'- 6"	$155.00	$210.00	$310.00	$365.00	1800 x 2550 mm
6'- 0" x 12'- 0"	$190.00	$250.00	$345.00	$410.00	1800 x 3600 mm

If automatic controls are not wanted with swinging tin clad doors, deduct from the above prices, $15.00 for single doors and $20.00 per opening for double swinging

Hardware for Sliding Tin Clad Doors

Prices include non-adjustable hangers, track, weights, and bolts for attaching hardware to doors. Through wall bolts are not included in prices.

Hardware for single sliding door, 3'- 0" (900 mm) wide opening $270.00
For each 6" (150 mm) additional width of opening, add $23.00
Openings over 6'-0" (1800 mm) wide require 3 hangers on each door, add for each additional hanger $16.00

DOORS AND WINDOWS

Hardware for double sliding doors, 6'-0" (1800 mm) wide
opening $455.00
For each 6" (150 mm) additional width of opening, add $23.00

Labor Erecting Tin Clad Doors

The following schedules, are guides for estimating labor to install.

The hours of labor are based on brick walls 13" (325 mm) thick. For thicker walls, add 1/4 hour for one worker for each 4" (100 mm) in thickness per bolt. Concrete and stone walls require slightly more time drilling; terra cotta, slightly less. Where "doors both sides" are noted, this assumes doors back-to-back and tracks bolted together. When tracks cannot be bolted together, use "one side" schedule multiplied by two.

	Width in Inches	Door One Side		Door Both Sides		Width in mm
		Mech.	Helper	Mech.	Helper	
Type Single Pairs	2'- 0" to 4'- 0"	7	7	10	10	600 to 1200 mm
	4'- 0" to 5'- 2"	8	8	12	12	1200 to 1551 mm
	5'- 3" to 7'- 8"	9	9	14	14	1575 to 2301 mm
	7'- 9" to 8'- 8"	11	11	17	17	2325 to 2601 mm
	8'- 9" to 12'- 0"	12	12	18	18	2625 to 3600 mm
	Width in Inches	Door One Side		Door Both Sides		Width in mm
		Mech.	Helper	Mech.	Helper	
Type Slide in Pairs	4'- 0" to 4'- 5"	9	9	14	14	1200 to 1326 mm
	4'- 6" to 9'- 8"	11	11	16	16	1350 to 2901 mm
	9'- 9" to 10'- 6"	13	13	18	18	2925 to 3150 mm
	10'- 7" to 12'- 0"	15	15	20	20	3174 to 3600 mm
	Width in Inches	Door One Side		Door Both Sides		Width in mm
		Mech.	Helper	Mech.	Helper	
Vertical Sliding	to 4'-10"	13	13	21	21	to 1449
	4'-11" to 7'- 6"	15	15	24	24	1476 to 2250 mm
	7'- 7" to 10'- 2"	17	17	26	26	2274 to 3051 mm
	10'- 3" to 12'- 0"	19	19	28	28	3075 to 3600 mm
	Door Height	Door One Side		Door Both Sides		Width in mm
		Mech.	Helper	Mech.	Helper	
Single Swing (Flush or Lap)	to 5'-9"	5	5	7	7	to 1725
	5'-10" to 8'- 9"	5 1/2	5 1/2	8	8	1749 to 2625 mm
	8'-10" to 12'- 0"	6	6	9	9	2649 to 3600 mm
	Door Height	Door One Side		Door Both Sides		Width in mm
		Mech.	Helper	Mech.	Helper	
Swing In Pairs (Flush or Lap)	to 5'-9"	8	8	12	12	to 1725
	5'-10" to 8'- 9"	9	9	14	14	1749 to 2625 mm
	8'-10" to 12'- 0"	10	10	16	16	2649 to 3600 mm

Labor Cost of Installing One Single 3-Ply Sliding Tin Clad Fire Door for 4'-0" x 7'-0" (1200 x 2100 mm) Wall Opening

The following estimate includes door, track, hangers, and necessary hardware installation.

Description	Hours	Rate	Total	Rate	Total
Mechanic	7.00	$	$	$ 37.89	$ 265.23
Helper	7.00	$	$	$ 26.16	$ 183.12
Cost per opening					$ 448.35

Labor Cost of Installing One Pair Double Sliding 3-Ply Tin Clad Fire Doors for 6'-0" x 7'-0" (1800 x 2100 mm) Wall Opening

The following estimate includes doors, track, hangers, and necessary hardware installation.

Description	Hours	Rate	Total	Rate	Total
Mechanic	11.00	$	$	$ 37.89	$ 416.79
Helper	11.00	$	$	$ 26.16	$ 287.76
Cost per opening					$ 704.55

Labor Cost of Installing One Lap Type Single Swinging 2-Ply Tin Clad Fire Door for 3'-0" x 7'-0" (900 x 2100 mm) Wall Opening

The following estimate includes door, hardware, and automatic controls installation.

Description	Hours	Rate	Total	Rate	Total
Mechanic	5.00	$	$	$ 37.89	$ 189.45
Helper	5.00	$	$	$ 26.16	$ 130.80
Cost per opening					$ 320.25

Labor Cost of Installing One Pair Lap Type Double Swinging 2-Ply Tin Clad Fire Doors for 6'-0" x 7'-0" (1800 x 2100 mm) Wall Opening

The following estimate includes doors, hardware, and automatic controls installation.

Description	Hours	Rate	Total	Rate	Total
Mechanic	9.00	$	$	$ 37.89	$ 341.01
Helper	9.00	$	$	$ 26.16	$ 235.44
Cost per opening					$ 576.45

1172

DOORS AND WINDOWS

Labor Cost of Installing One Flush Type Single Swinging 2-Ply Tin Clad
Fire Door for 3'-0" x 7'-0" (900 x 2100 mm) Wall Opening

The following estimate includes door, hardware, and automatic
controls installation.

Description	Hours	Rate	Total	Rate	Total
Mechanic	5.00	$	$	$ 37.89	$ 189.45
Helper	5.00	$	$	$ 26.16	$ 130.80
Cost per opening					$ 320.25

Labor Cost of Installing One Pair Flush Type Double 3-Ply Swinging Tin
Clad Fire Doors for 6'-0" x 7'-0" (1800 x 2100 mm) Wall Opening

The following estimate includes doors, hardware, and automatic
controls installation.

Description	Hours	Rate	Total	Rate	Total
Mechanic	9.00	$	$	$ 37.89	$ 341.01
Helper	9.00	$	$	$ 26.16	$ 235.44
Cost per opening					$ 576.45

Frames for tin clad doors are made of angles and channels usually
furnished under metal fabrications and set by the mason. A 3" x 3" (75 x 75
mm) angle frame costs about $10.00 per lin.ft. ($32.80 per m) installed. An
8" (200 mm) channel frame costs about $21.00 per lin.ft. ($68.90 per m)
installed.

08310 SLIDING GLASS DOORS

Aluminum sliding glass doors are available in several types, as
illustrated, for glazing with crystal glass or 5/8" (15.62 mm) insulated glass.
"X" indicates sliding panel; "O" indicates a fixed panel.

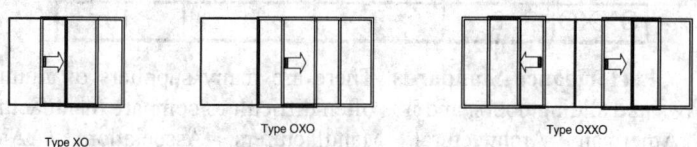

Type XO   Type OXO   Type OXXO

Sliding Glass Doors

Frame, glazing bead and door members are heavy aluminum
extrusions. Door sections move upon bottom mounted, ball bearing, grooved
brass rollers. Frames are designed with provisions for installing bottom roller,
horizontally sliding type screens. Weatherstripping is mohair pile, factory
mounted in channels to form a continuous double seal. Hardware includes

full-grip lucite pulls and touch latch. Cylinder lock sets are available at extra cost.

Screens are made from aluminum extrusions of tubular box type section with bottom rollers and fitted with fiberglass screen.

Prices are for door units set up. Factory assembly includes attachment of all accessory parts, such as weatherstripping, roller mechanisms, latch, corner connecting brackets, and bumpers. Screen units are assembled. Glass and glazing, is included.

Material Prices of Aluminum Sliding Doors				
Type	Frame Size Ft.- In.	Economy Glass	Deluxe Insulating Safety Glass	Frame Size mm
XO	6'- 0" x 7'- 0"	$318.00	$689.00	1800 x 2100
OXO	9'- 0" x 7'- 0"	$498.00	$819.00	2700 x 2100
OXXO	12'- 0" x 7'- 0"	$415.00	$1,154.00	3600 x 2100

For cylinder lock set in lieu of latch, add $15.00.

Labor Erecting Aluminum Sliding Glass Doors. Assembly and erection of aluminum sliding glass doors is simple and does not require special tools. All aluminum members are prefit, with holes drilled and connecting brackets attached. The following table gives average labor hours required for installing aluminum sliding glass doors in ordinary residential work.

Type	Number of Doors	Number of Fixed Panels	Mechanic Hours Required	
			For Assembly	For Erection
XO	1	1	2	6
OXO	1	2	3	8
OXXO	2	2	4	10

Performance Standards. There are many suppliers of aluminum windows and sliding doors, and it is often difficult to compare manufacturers. The American Architectural Manufacturers Association (AAMA), Schaumburg, IL; Window and Door Manufacturers Association (WDMA), Des Plaines, IL.; and the Canadian Standards Association (CSA), Mississauga, Ontario Canada publish quality specifications, and manufacturers belonging to these associations affix their certified labels to their products, attesting that they comply with the designated specification. The AAMA specifications include the following:

1174

DOORS AND WINDOWS

Awing, hopper, projected window	AP
Architectural terrace door	ATD
Basement window	BW
Casement window	C
Duel-action side-hinged door	DASHD
Duel-action window	DAW
Fixed door	FD
Fixed window	FW
Hung window	H
Greenhouse wndow	GH
Hinged Rescue window	HE
Horizontal pivoted window	HP
Horizontal sliding window	HS
Jalousie window	JA
Limited water duel-action side-hinged door	LW DASHD
Limited water side-hinged door	LW SHD
Roof window	RW
Sliding door	SD
Side-hinged door	SHD
Side-hinged (inswinging) window	SHW
Unit skylight - glass glazed	SKG
Unit skylight – plastic glazed	SKP
Side lite	SLT
Specialty product	SP
Tropical awing window	TA
Top-hinged window	TH
ransom	TR
Vertical pivoted window	VP
Vertical sliding window	VS

"Excerpted from AAMA/WDMA/CSA 101/I.S.2/A440-05 used with permission of the American Architectural Manufacturers Association".

The above product designation for windows, doors and unit skylight are covered in the standard specifications of AAMA/WDMA and CSA 101/I.S.2/A440-05 standard/specification for windows, doors and unit skylights. The AAMA also publishes a certified products directory with a current listing of all certified windows and sliding glass doors by company, model and, maximum size approved.

Wood Sliding or Patio Doors

Sliding doors are now available from most of the nationally distributed millwork houses and follow the general designs established for aluminum sliding units. These doors can usually be purchased in two basic widths per panel and in various arrangements including two panels, one fixed, one operating; three panels, two sides fixed and center panel operable; and four panels, two end panels fixed and two center panels operable.

The following price schedule is based on units preglazed with 5/8" safety glass and complete with operating hardware, weatherstripping, sill, and screens for operable sash.

Glass Size Inches	Nominal Door Size	Type	Material Cost	Glass Size mm	Nominal Door Size mm
28 " wide	5'- 0" x 6'- 8"	Double XO	$600.00	700 mm wide	1500 x 2000 mm
34 " wide	6'- 0" x 6'- 8"	Double XO	$646.00	850 mm wide	1800 x 2000 mm
46 " wide	8'- 0" x 6'- 8"	Double XO	$838.00	1150 mm wide	2400 x 2000 mm
34 " wide	9'- 0" x 6'- 8"	Triple OXO	$996.00	850 mm wide	2700 x 2000 mm
46 " wide	12'- 0" x 6'- 8"	Triple OXXO	$1,260.00	1150 mm wide	3600 x 2000 mm

Add for keyed lock, $15.00. Grilles that can be removed for cleaning are available in either diamond or rectangular patterns. Costs are $53.00 each for the 28" (700 mm) width, $54.00 each for the 34" (850 mm) width, $64.00 for the 46" (1150 mm) width in the rectangular pattern; and $54.00, $57.00, and $78.00 each for the diamond patterns.

08330 COILING DOORS

Coiling doors are usually called rolling shutters and are fully factory assembled units made of steel or aluminum. Often, they are labeled fire doors, and installation is governed by Underwriter's specifications. Steel mounting jambs and heads are usually furnished by fabrications contractor.

A manually operated, 20 gauge door up to 12' x 12' (3600 x 3600 mm) costs about $10.00 to $11.00 per sq.ft. ($107.64 to 118.40 per sq.m), including hood, guides, and operating hardware.

Where class A fire ratings is required, doors cost $9.10 per sq.ft. ($97.95 per sq.m) more. If 18-gauge construction, it will add another $1.40 per sq.ft. ($15.07 per sq.m). Motor operation will cost $1,100.00 for smaller sizes, $1,260.00 for medium. Each pass door will add another $1,100.00.

Two mechanics should hang and adjust a rolling shutter, up to 12' x 12' (3600 x 3600 mm), in one 8-hr. day at the following cost per shutter:

Description	Hours	Rate	Total	Rate	Total
Mechanics	16.00	$	$	$ 37.89	$ 606.24

Coiling Grilles

Rolling grilles are used for protection of store fronts, for open fronted stores in enclosed shopping areas, and for closing off counter areas.

Stock commercial grilles will cost about $17.00 per sq.ft. ($182.99 per sq.m) with all hardware. These are often part of a decorative front, made

1176

of anodized aluminum, and cost about $20.00 per sq.ft. ($215.28 per sq.m.). Exact prices can be obtained from the manufacturer, who will also furnish shop drawings upon request to show all the necessary anchoring involved. Steel grilles run about $11.00 per sq.ft. ($118.40 per sq.m). Motorizing and installation costs are the same as for rolling shutters.

08350 FOLDING DOORS

Steel Bi-fold Closet Doors. These may be installed in openings with drywall, wood, or steel jambs and head frame. Each bi-fold door package comes complete with door panels, track, guides, and pulls for a complete installation.

Doors are available in a variety of styles including flush, panel, half louver, or full louver. The doors are shipped "factory finished" with a baked-on paint finish.

A two-panel unit is used for openings up to 3'-0" (900 mm) wide and costs about $70.00. A four-panel unit is used for openings 4'-0" (1200 mm) to 6'-0" (1800 mm) wide, with the 6'-0" (1800 mm) unit costing about $130.00. Units are sold in 6'-8" (2000 mm) and 8'-0" (2400 mm) heights, with the 8'-0" (2400 mm) heights costing an additional $15.00 to $30.00 more per unit.

One carpenter should install a complete bi-fold unit and all hardware in about 45 minutes at the following labor cost per unit:

Description	Hours	Rate	Total	Rate	Total
Carpenter	0.75	$	$	$ 35.04	$ 26.28

Folding Doors. These are often folding partitions that can be ordered completely furnished and with all hardware in wood or vinyl. Folding doors for a 3' x 6'-8" (900 x 2000 mm) opening will cost about $90.00 in vinyl, $130.00 in wood slats. Doors for 6' x 6'-8" (1800 x 2000 mm) openings will run $210.00 for vinyl, $250.00 for wood. Deluxe and custom models will run somewhat more. All the above prices include hardware, but the frame must be furnished by others. When the opening is over 150 sq.ft. (14 sq.m), the units are considered "partitions" rather than doors and cost about $7.00 per sq.ft. ($75.35 per sq.m) for light models to $10.00 per sq.ft. ($107.64 per sq.m) for heavy duty units.

08360 OVERHEAD DOORS

Overhead Garage Doors. Overhead garage doors are furnished in several different styles, including the one-piece overhead door, self-balancing overhead door, and the spring-balanced overhead door.

For an average opening up to 8'-0" x 7'-0" (2400 x 2100 mm), two carpenters should place tracks, fit, and hang doors and apply the necessary hardware in 4 hours, at the following labor cost per door:

Description	Hours	Rate	Total	Rate	Total
Carpenter	8.00	$	$	$ 35.04	$ 280.32

For larger openings up to 16'-0" (4800 mm) wide and 7'-0" (2100 mm) high, where one wide overhead folding door is used, two carpenters should place tracks, fit, and hang doors and apply the necessary hardware complete in 5 hours, at the following labor cost per door:

Description	Hours	Rate	Total	Rate	Total
Carpenter	10.00	$	$	$ 35.04	$ 350.40

GARAGE, FACTORY AND WAREHOUSE DOORS

Overhead Type Garage Doors

Overhead type garage doors are counterbalanced and controlled by oil tempered torsion springs. Cables are secured at the bottom of the door on each side and are directly connected with the pair of balancing springs located immediately over the door.

Ball bearing rollers carry all sections of door over a continuous channel track. The track has a flange that keeps the rollers in place.

Doors have kiln-dried fir or spruce sections, with hardboard panels. All joints are mortised, tenoned, or doweled with steel pins and glued. All sections are fitted to size, squared for opening, and drilled for hardware.

Doors over 12'- 0" (3600 mm) wide have steel struts in intermediate sections to prevent sagging. Doors over 15'- 0" (4500 mm) wide have steel struts in all sections.

Standard Size, Sectional Overhead Type, Garage Doors

Furnished complete with hardware, including cylinder lock, springs, and track. Doors are furnished with all wood panels, or with top section open for glass, which is not included. Approximate prices are:

DOORS AND WINDOWS

Size Opening Feet - Inches	1-3/8" Thick	1-3/4" Thick	Size Opening mm
9'- 0" x 7'- 0"	$290.00	$320.00	2700 x 2100 mm
14'- 0" x 7'- 0"	$600.00	$640.00	4200 x 2100 mm
15'- 0" x 7'- 0"	$640.00	$700.00	4500 x 2100 mm
16'- 0" x 7'- 0"	$670.00	$770.00	4800 x 2100 mm
10'- 0" x 10'- 0"	......	$510.00	3000 x 3000 mm
12'- 0" x 10'- 0"	......	$610.00	3600 x 3000 mm
10'- 0" x 10'- 0"	......	$620.00	3000 x 3000 mm
12'- 0" x 10'- 0"	......	$730.00	3600 x 3000 mm
	34.38 mm Thick	43.75 mm Thick	

Electric Door Operators

Newer garage door openers offer, most importantly, safety and then convenience, since they let you open and close the door remotely. The garage door openers come in three basic types. The most widely used in about 60% to 75% of the installations is the chain-driven type, followed by the belt and screw drive type.

Three types of operators may be used with any type of overhead doors. Motors are suitable for single phase, 60 cycle, 110-120 Volt AC power supply, and are of the reversing type which permit stopping and changing direction at any point of the door travel. Adjustable limit switches stop the door automatically at top and bottom of door openings. A safety mechanism automatically reverses the door if it meets an obstruction while in operation, or the door will stop and the operator will shut off.

Doors are driven by means of a roller chain fastened to a shoe sliding on 2 steel tracks. The shoe is connected to the door by a detachable arm.

Single door residential operator with 1/4-hp. motor for
   door up to 22' (6600mm) wide, 7' (2100mm) high,
   complete with one toggle type control station $200.00
For commercial door operator, with 1/4-hp. motor for
   door not over 140 sq.ft. (13.00 sq.m) nor over 10'
   (3000mm) high $295.00
Commercial door operator, with 1/3-hp. motor for door
   not over 168 sq.ft. (15.61 sq.m) or 12' (3600mm) high $320.00
Commercial door operator, with 1/2-hp. motor for door
   not over 224 sq.ft. (20.80 sq.m) or 14' (4200mm) high $390.00
Extra toggle type control station for commercial operator $20.00
Extra key switch (for residential operator only) $25.00

An experienced mechanic should install an electric door operator in 1.5 to 3.0 hours after the necessary wiring is in place.

Electronic Door Operators

For use where a transmitter is located in each automobile using the garage. A push of the button operates the door from the car.

For residential doors up to 16'- 0"x 8'- 0" (4800 x 2400
   mm) approx. price installed _____ $375.00

Colonial   *Fan Top*   *Vertical Battens*   *Diagonal Battens*

Doors of Special Designs

An experienced mechanic should install an electric door operator in 2 hours after the necessary wiring is in place.

Special Hardware-Add per Door

Low headroom, 6" (150 mm) minimum, for doors not
   over 132 sq.ft. (12.26 q.m) _____ $20.00
Low headroom, 6" (150 mm) minimum, for doors not
   over 133-175 sq.ft. (12.35-16.25 sq.m) _____ $33.00
Low headroom, 6" (150 mm) minimum, for doors over
   16'-0" (4800 mm) wide   or over 12'-0" (3600 mm)
   high _____ $49.00
High lift (for lubritoriums) up to 2'-0" (600 mm) above
   head jamb _____ $29.00
High lift (for lubritoriums) up to 5'-0" (1500 mm) above
   head jamb _____ $57.00

On commercial jobs steel doors are often specified. A 10' x 10' (3000 x 3000 mm) door will cost about $510.00 with manual operator, $900.00 with electric operator.

Fiberglass and aluminum doors are light in weight and let in daylight. A stock door will cost about $288.00 for a single car size, $480.00 for a double car size.

Colonial doors, heavily molded outside with raised panels, add $1.90 per sq.ft. ($20.45 per sq.m).

For fan top on door, add $156.00 extra.

Vertical battens, 2-ply V-jointed or beaded, add $1.90 per sq.ft. ($20.45 per sq.m).

Diagonal battens, 2-ply V-jointed or beaded, add $2.10per sq.ft. ($22.60 per sq.m).

08385 SOUND RETARDANT DOORS

Hardwood Sound Insulating Doors

Acoustical doors are for use in consultation rooms, conference rooms, hospitals, clinics, music rooms, television and radio studios, and other areas where acoustical levels must be closely maintained.

Depending on the desired decibel reduction or acoustical efficiency, doors are identified and furnished in several models.

Units 1-3/4" (43.75 mm) thick are rated in sound transmission class 36. Units 2-1/2" (62.50 mm) thick are rated in sound transmission class 41. A door 3" (75 mm) thick is also made and has superior acoustical qualities.

Acoustical doors are made in a wide selection of face woods. Custom finishing of doors is available. These are veneered doors and can be furnished with veneers to match those of other doors in the building. Any standard hardware may be used for installation.

The following prices include bottom closers, stops, gaskets, and stop adjusters and are for standard rotary cut veneers of natural birch and red oak.

Material Prices of Hardwood Sound Insulating Doors*				
Size	1-3/4"	2-1/2"	3"	Size
Feet - Inches	35 Decibel	40 Decibel	50 Decibel	mm
2'- 6" x 6'- 8"	$265.00	$345.00	$455.00	62.50 x 166.75 mm
2'- 8" x 6'- 8"	$275.00	$355.00	$465.00	66.75 x 166.75 mm
3'- 0" x 6'- 8"	$285.00	$365.00	$475.00	75.00 x 166.75 mm
2'- 6" x 7'- 0"	$275.00	$355.00	$465.00	62.50 x 175.00 mm
2'- 8" x 7'- 0"	$285.00	$365.00	$475.00	66.75 x 175.00 mm
3'- 0" x 7'- 0"	$295.00	$375.00	$485.00	75.00 x 175.00 mm
	43.75 mm	62.5 mm	75 mm	
Light openings: rectangular vision light				
not over 18" square	$35.00	$37.00	$40.00	not over 450 mm square

*f.o.b. factory, uncrated.

Astragals. Astragals for pairs of doors, 2-1/2" x 1" (62.50 x 25 mm) in any wood to match face veneer with 1" x 1/2" (25 x 12.5 mm) rubber gasket, $37.00 per pair. For rounding meeting edges and grooving one edge with Eveleth bumper strip, add $37.00.

Other Woods. For other woods, add to the total price the following percentages:

Sliced natural birch and sliced red oak 10%
Rotary cut red or white birch and rotary cut plain white oak 20%
Sliced red or white birch and sliced plain white oak 25%
Rift sliced white oak and sliced African mahogany 40%
Sliced American black walnut 65%

08400 ENTRANCES AND STORE FRONTS

08410 ALUMINUM STORE FRONT CONSTRUCTION

Aluminum storefront and entrance doors are manufactured by a number of firms in a wide variety of styles to suit the design and function of most any type of building.

The glass and glazing subcontractor is the firm to contact for this type of work, and they typically represent more than one manufacturer of these materials. This subcontractor will handle all the necessary functions of bidding, shop drawing preparation, ordering, and erection of the storefront materials, as well as the glass, for a complete installation.

Materials are made from an aluminum alloy and extruded into many different shapes and sizes of various thicknesses. The extrusions are offered in stock lengths of 24' (7200 mm) long, but longer lengths are available upon special order.

To prevent oxidation of the aluminum surfaces, the extrusions are anodized in a number of colors as well as the clear anodized finish. In recent years this material has become available with a paint coating that offers exceptional durability in addition to a controlled color gradation that is difficult to obtain with the color anodized finishes.

Entrance doors are available in numerous sizes and designs—narrow stile, medium stile, and wide or monumental stile—with various styles of applied panels and finish hardware.

The doors can be hung into the frame by using butt hinges, offset pivots, or center pivots. The controlled automatic closing of the doors can be achieved by installing an exposed overhead closer, a closer concealed in the horizontal head frame member, or by a floor closer that is set into a pocket in the floor under the hinge point of the door.

Installation. Once the subcontractor has a contract agreement to provide the storefront construction, he will prepare shop drawings based on the architectural drawings for the project. The shop drawings clearly define all dimensions and jointing details necessary for the installation, and the materials are ordered based on these drawings.

Prior to fabrication of the exact lengths necessary for erection, the subcontractor will verify the shop drawing dimensions with actual field dimensions taken at the jobsite.

After the extrusions are cut to the proper length for field erection, they are delivered to the jobsite, distributed to the proper location, and installation begins.

The extrusions are fastened to each other with concealed, extruded aluminum clips at all joints. The pre-fabricated frames are then set into the opening and fastened, through the glazing pocket, to the substrate material with screws or expansion anchors. Proper assembly allows the vertical extrusions to run top to bottom, with all horizontal extrusions fitted between the verticals.

DOORS AND WINDOWS

Method of Estimating Materials. Measure the length of each type and size extrusion and list into proper groupings. Since this material is sold in stock lengths, the estimator will have to combine the measured lengths together to arrive at the total number of stock lengths needed with the least amount of waste. Count the joints to ascertain the required number of concealed clips, one per joint. It will also be necessary to measure the total length of vinyl gasket material needed for glazing, because this material is ordered separately from the aluminum extrusions. After all of the material quantities are measured, grouped, and totaled, the appropriate material prices can be applied for the extension of the material costs.

Doors are listed by size and style, and depending on the architectural requirements, they may be stock units or custom units. The pricing manual, supplied by the aluminum manufacturer, will list the material pricing strategy for the stock units as well as for the component parts for the custom units.

Installed Costs of Storefront Construction. Since there are so many different sizes and shapes of extrusions available, it would be difficult to list the material prices and labor time required. The labor hours are directly related to the size and shapes of the extrusions. For instance, a simple storefront and entrance door installation, 40'-0" (12.19m) wide x 10'-0" (3.04 m) high, in one system might take 72 manhours, while a more sophisticated system would require 100 manhours.

For budget purposes only, the following installed costs of aluminum storefront framing, glass not included, are given on a per sq.ft. (per sq.m) basis:

1. Windows without intermediate vertical mullions up to 5'-0"x5'-0" (1500 x 1500 mm) in size, using 1-3/4" x 4-1/2" (43.75 x 112.50 mm) tube - $14.00 per sq.ft. ($150.69 per sq.m).
2. Same as above, but with one vertical mullion - $16.00 per sq.ft. ($172.22 per sq.m).
3. Windows 40'-0" x 10'-0" (12.19 x 3.04 m) in size with vertical mullions spaced 5'-0" (1500 mm) on center, using 1-3/4" x 4-1/2" (43.75 x 112.50 mm) tube - $13.00 per sq.ft. ($139.93 per sq.m).
4. Same as above, but with 3" x 6" (75 x 150 mm) tube - $18.00 per sq.ft. ($193.75 per sq.m).
5. For a narrow stile single door and frame, 3'-0" x 7'-0" (900 x 2100 mm) - $812.00 each.
6. For a pair of narrow stile doors and frame, 6'-0" x 7'-0" (1800 x 2100 mm) - $1,260.00 each.

Add 15% for bronze anodized finish, 20% for black anodized finish. Add for transom over single door, $119.00; over pair of doors, $210.00.

Revolving Doors. Revolving doors are made in a variety of styles and sizes, in aluminum, bronze, and stainless steel. The doors are fitted with a mechanism to control the travel speed and are pivoted so that they may collapse together to provide an unobstructed opening. The semi-circular wings at each side of the four-leaf revolving door provide a surface for the door sweep to ride against. At no point during normal operation is there a clear path to the exterior for the escape of the building heat or air conditioning to the outside atmosphere.

Average installed prices for revolving doors are as follows: Aluminum, $21,000; bronze, $37,100; stainless steel, $30,000.

Storefront Members. The detail drawings that follow show some typical cut sections through aluminum storefront extrusions.

WINDOWS

Manufacturing of windows in the past years has evolved from being mainly a product of steel and wood to incorporate other materials. Recent records by various associations have divided the different types of window material into the following categories and the approximate percentages of each category manufactured.

1. Wood 40%
2. Vinyl 40%
3. Aluminum 18%
4. Steel 2%

All the above materials can be manufactured into almost any shape and size a homeowner or architect may require. Hence, some window manufacturers will price a window based on total inches. For example, a 27" x 43" window is a total of 70" and can be priced between $3.00 to $13.00 per inch or $210.00 to $910.00 per window depending on the material. Steel is the most expensive. But price is affected by glass, type of window, awing, casement, double hung etc. among many other factors, including shipping cost.

08500 METAL WINDOWS

Metal windows today, due to their thermal qualities, are used seldom used and at times do not meet some local thermal codes. The main use of steel windows is to replicate existing conditions and for manufacturing plants and warehouse structures due to the high fire ratings, with few being manufactured for the small homeowner due to price and thermal protection. Today steel windows are manufactured to specific sizes and requirements. Steel and aluminum windows are available in a great variety of types and sizes to meet almost all building requirements. Some of the steel window products have been standardized by the manufacturers to reduce costs and promote quicker deliveries.

The illustrations included in this chapter are for reference when selecting requirements. For full description and specifications of available products refer to manufacturers' catalogs.

All aluminum window and door products are either classified as Standard, Commodity, or Special.

1184

DOORS AND WINDOWS

Commodity are those selected from standards that are subject to the greatest demand and are available from manufacturer warehouse stocks or through dealers engaged in the distribution of these products. Standard and Special units are not available from stock and must be fabricated and shipped from factory. The trend is for distributors to send all orders to the factory for both pricing and fabrication.

The following information deals only with Commodity. For Standard and Special, obtain quotations from manufacturers or dealers.

Labor Installing Steel Windows and Doors. Installation costs vary with the class of labor, type and size of individual units, and job conditions. However, a reasonably accurate estimate of manhours required to install each unit can be made.

Product	Window Opening	Manhours per Unit
Pivoted, Commercial	Single Unit*	2.25
Projected and Architectural	Two or more Units*	
Projected Windows	(a) Precast Sills	1.25
	(b) Poured Sills	1.50
Fixed Industrial	Single Unit	1.75
	Two or More Units	
	(a) Precast Sills	0.75
	(b) Poured Sills	1.00
Fixed Architectural	Single Unit	2.00
	Two or More Units	
	(a) Precast Sills	1.00
	(b) Poured Sills	1.25
Intermediate Windows In Wood (Casement, Combination and Projected)	(a) Single Unit	2.75
	(b) Two or More Units	2.25
In Stone	(a) Single Unit	3.00
	(b) Two or More Units	2.50
In Masonry	(a) Single Unit	2.25
	(b) Two or More Units	1.25
Mullions Only		0.50
Mullions with Covers		0.75
Residence Casements (Roto)	Single Unit	
	(a) In Wood	1.75
	(b) In Masonry or Brick Veneer	1.25
Residence Casements (Roto)	Two or More Units	
	(a) In Wood	1.50
	(b) In Masonry or Brick Veneer	1.00
Residence Casements (Fixed)	Single Unit	
	(a) In Wood	1.50
	(b) In Masonry or Brick Veneer	1.00

Product	Window Opening	Manhours per Unit
Residence Casements (Fixed)	Two or More Units	
	(a) In Wood	1.25
	(b) In Masonry or Brick Veneer	0.75
Casements in Casings	Single or Multiple Units	1.50
Continuous Top Hung, Fixed	per lin.ft. (per.3048 m)	0.50
Continuous Top Hung, Swing	per lin.ft. (per.3048 m)	0.75
Mechanical Operators, Rack and Pinion	per lin.ft. (per.3048 m)	0.75
Mechanical Operator, Lever Arm	per lin.ft. (per.3048 m)	0.50
Mechanical Operators, Tension	per lin.ft. (per.3048 m)	1.50
Industrial Doors and Frames	up to 35 sq.ft. (3.25 sq.m) per sq.ft. (per 0.0929 sq.m)	0.13
	Over 35 sq.ft. (3.25 sq.m) per sq.ft. (3.25 sq.m)	0.10
Double Hung Windows	Single Units	1.50
	Two or More Units	1.25
Screen Installation		
For Projected Windows	Net per Screen	1.10
For Basement and Utility	Net per Screen	0.75
For Security Windows	Net per Screen	1.10
For Casement Windows	Net per Screen	0.75
For Pivoted Windows	Net per Vent	1.50

*Single ventilator units. For each additional ventilator, add $0.50.

Prices of Commercial Steel Steel Windows					
Window Type	Window Size Inches	Window Price	Screen Price	Window Size mm	Window Price
Awning	19 " x 12 "	$775.00	$25.00	475 x 300 mm	$800.00
Awning	30 " x 13 "	$1,075.00	$25.00	750 x 325 mm	$1,100.00
Awning	43 " x 13 "	$1,400.00	$35.00	1075 x 325 mm	$1,435.00
Dbl. Hung	21 " x 36 "	$1,425.00	$30.00	525 x 900 mm	$1,455.00
Dbl. Hung	25 " x 42 "	$1,675.00	$35.00	625 x 1050 mm	$1,710.00
Dbl. Hung	30 " x 48 "	$1,950.00	$35.00	750 x 1200 mm	$1,985.00

Prices of Security Steel Windows					
Window Type	Window Size Inches	Window Price	Screen Price	Window Size mm	Window Price
Security	32 "x 13 "	$1,350.00	$25.00	800 x 325 mm	$1,375.00
Security	33 "x 16 "	$1,470.00	$30.00	825 x 400 mm	$1,500.00
Security	62 "x 16 "	$2,340.00	$45.00	1550 x 400 mm	$2,385.00

1186

DOORS AND WINDOWS

Prices of Residential Steel Casement Windows					
Window Type	Window Size Inches	Window Price	Screen Price	Window Size mm	Window Price
Ventilated	19 "x 12 "	$620.00	$12.00	475 x 300 mm	$632.00
Ventilated	23 "x 12 "	$700.00	$14.00	575 x 300 mm	$714.00
Ventilated	30 "x 12 "	$840.00	$20.00	750 x 300 mm	$860.00
Fixed	19 "x 12 "	$558.00	------	475 x 300 mm	$558.00
Fixed	23 "x 12 "	$630.00	------	575 x 300 mm	$630.00
Fixed	30 "x 12 "	$756.00	------	750 x 300 mm	$756.00

Prices include standard operating hardware, installation, fittings, bonderizing, and one factory coat of paint. Glass, putty, glazing, and glazing clips are not included.

Anchors

Anchors for securing frames to wood or masonry walls are included with window units. Four anchors per opening are furnished for heights up to and including 4'-5-1/2" (1337.50mm). Six anchors per opening to and including 5'-9-1/2" (1737.50 mm). Eight anchors per opening to and including 8'-1-1/2" (2437.50 mm).

Prices of ventilated units include locking handles, Roto underscreen operators (lever underscreen operators also available at reduced price) and standard installation fittings, bonderizing and one factory coat of paint for both ventilated and fixed units. Glass, putty, glazing, glazing clips, and mastic not included. Screen prices include screens for all vents in unit and clips for attachment.

ALUMINUM CASEMENT WINDOWS

Aluminum windows are often overlooked as a viable option. There are a number of advantages to aluminum windows, one being strength. The drawback to aluminum windows is that they are a good conductor of heat and a poor insulator. To overcome this, during the manufacturing of aluminum windows, additional step are required. One of these steps is to place a thermal break for a better U-thermal factor. Prices include aluminum diecast roto operator with aluminum arm. Operator case and crank are brushed to a satin finish and given one coat of clear Duranite phenolic resin baked-on plastic coating. Casements are etched in a hot acid bath and receive a dip coat of methacrylate lacquer in a controlled environment. Also included are wood screws and sill anchors when specified. Prices include 1/2" (12.50 mm) insulating glass and screens.

Prices of Aluminum Residential Casement Windows With Insulating Glass					
Window Type	Window Size Inches	Window Price	Screen Price	Window Size mm	Window Price
Ventilated	12 "x 12 "	$120.00	$12.00	300 x 300 mm	$132.00
Ventilated	22 "x 12 "	$170.00	$14.00	550 x 300 mm	$184.00
Ventilated	32 "x 22 "	$270.00	$14.00	800 x 550 mm	$284.00
Ventilated	44 "x 24 "	$340.00	$20.00	1100 x 600 mm	$360.00
Fixed	19 "x 12 "	$139.50	------	475 x 300 mm	$139.50
Fixed	19 "x 12 "	$139.50	------	475 x 300 mm	$139.50
Fixed	23 "x 12 "	$157.50	------	575 x 300 mm	$157.50
Fixed	30 "x 12 "	$189.00	------	750 x 300 mm	$189.00

The above represent sample sizes. Aluminum can be made to any practical dimension.

Price includes a mechanical operator with crank handle. Windows are glazed by using snap-in aluminum beads. Price includes integral fin, and glazing with 1/2" (12.50 mm) insulating glass.

Windows are weatherstripped and receive a dip coat of methacrylate lacquer.

Screen prices include screens for all vents in unit. Screens have aluminum frames and 18" x 14" (450 x 350 mm) mesh aluminum cloth. Storm sash are factory glazed.

Prices of Aluminum Double Hung Windows

Prices of windows include silicone-treated wool-pile weatherstripping, spiral, sash balances, glazing clips, standard anchors, mullion screws, and window hardware consisting of sash lock, keeper, and pull down for upper sash. Prices include 1/2" (12.50 mm) insulating glass and screens.

All windows are finished with a coat of water clear lacquer.

Screens have extruded or rolled aluminum frame sections and are wired with 18" x 14" (450 x 350 mm) mesh aluminum cloth. Screens consist of full or half panels to allow for screening as desired. Prices include all necessary installation hardware.

Storm sash is furnished glazed with single strength glass set in a plastic seal. All points of contact with window frame are insulated. Prices include all necessary hardware for installation.

Picture windows are shipped knocked down. Picture windows are designed to receive 1/4" (6.2 mm) single glass or 1/2" (12.50 mm), 3/4" (18.75 mm), or 1" (25 mm) insulating glass if specified. Prices include necessary assembly and installation hardware.

1188

DOORS AND WINDOWS

Double Hung	Window Size Inches	Window Price	E - Glass	Window Size mm	Window Price plus E- Galss
\multicolumn{6}{c}{Prices of Aluminum Residential Double Hung Windows With Insulating Glass - Screens Included}

Prices of Aluminum Residential Double Hung Windows
With Insulating Glass - Screens Included

Double Hung	Window Size Inches	Window Price	E - Glass	Window Size mm	Window Price plus E- Galss
DH	20 "x 30 "	$250.00	$20.00	500 x 750 mm	$270.00
DH	28 "x 30 "	$290.00	$23.20	700 x 750 mm	$313.20
DH	30 "x 30 "	$300.00	$24.00	750 x 750 mm	$324.00
DH	34 "x 30 "	$320.00	$25.60	850 x 750 mm	$345.60
DH	38 "x 30 "	$340.00	$27.20	950 x 750 mm	$367.20
DH	20 "x 38 "	$290.00	$23.20	500 x 950 mm	$313.20
DH	28 "x 38 "	$330.00	$26.40	700 x 950 mm	$356.40
DH	34 "x 38 "	$360.00	$28.80	850 x 950 mm	$388.80
DH	38 "x 38 "	$380.00	$30.40	950 x 950 mm	$410.40
DH	20 "x 44 "	$320.00	$25.60	500 x 1,100 mm	$345.60
DH	28 "x 44 "	$360.00	$28.80	700 x 1,100 mm	$388.80
DH	30 "x 44 "	$370.00	$29.60	750 x 1,100 mm	$399.60
DH	34 "x 44 "	$390.00	$31.20	850 x 1,100 mm	$421.20
DH	34 "x 44 "	$390.00	$31.20	850 x 1,100 mm	$421.20
DH	20 "x 50 "	$350.00	$28.00	500 x 1,250 mm	$378.00
DH	28 "x 50 "	$390.00	$31.20	700 x 1,250 mm	$421.20
DH	30 "x 50 "	$400.00	$32.00	750 x 1,250 mm	$432.00
DH	34 "x 50 "	$420.00	$33.60	850 x 1,250 mm	$453.60
DH	50 "x 38 "	$440.00	$35.20	1250 x 950 mm	$475.20
DH	20 "x 60 "	$400.00	$32.00	500 x 1,500 mm	$432.00

The above represent sample sizes. Aluminum can be made to most practical
dimension.

Prices of Aluminum Picture Windows with Horizontal Sliding Flankers
Including Insulating Glass and Screens

Double Hung	Window Size Inches	Window Price	E - Glass	Window Size mm	Window Price plue E- Glass
DH	70 "x 20 "	$450.00	$45.00	1750 x 500 mm	$495.00
DH	80 "x 20 "	$500.00	$50.00	2000 x 500 mm	$550.00
DH	90 "x 20 "	$550.00	$55.00	2250 x 500 mm	$605.00
DH	70 "x 30 "	$500.00	$50.00	1750 x 750 mm	$550.00
DH	80 "x 30 "	$550.00	$55.00	2000 x 750 mm	$605.00
DH	90 "x 30 "	$600.00	$60.00	2250 x 750 mm	$660.00
DH	70 "x 40 "	$600.00	$55.00	1750 x 1,000 mm	$655.00
DH	80 "x 40 "	$600.00	$60.00	2000 x 1,000 mm	$660.00
DH	90 "x 40 "	$650.00	$65.00	2250 x 1,000 mm	$715.00

The above represent sample sizes. Aluminum can be made to any practical
dimension.

08600 WOOD AND PLASTIC WINDOWS

Wood windows generally are sold as complete units with frame, sash, operating hardware, weatherstripping, and glazing assembled at the factory. Frames are usually treated. The exterior of wood windows is generally covered with rigid vinyl (PVC). The interior portion is unfinished, to be finished by painters or others. Standard sizes for each type window may vary slightly with each manufacturer. Most manufacturers now have the ability to manufacture windows of any material to exacting architectural specifications at little to no additional cost. For replacement windows some manufacturers will only manufacture to the specifications of an approved order. The time for manufacture usually is only three weeks from placement of order to delivery. Prices, of course, should be checked locally, because they will vary considerably, depending on geographic location, on size of the order, and on whether delivery is to a local dealer or to job site.

The contractor should add or deduct for items such as insect screens, divided light grilles, hardware, window blinds, or window treatments when comparing individual window costs. The prices quoted below are helpful for comparing the various window and glazing types.

Double Hung Window Units

Frame. Wood members are treated with a water repellent preservative and covered with a rigid vinyl (PVC) sheath. Sill ends are prefinished with polyurea or polyester urethane, depending on whether white or color.

Sash. The exterior of the wood sash is protected with a long-lasting patented polyurea finish or a polyester urethane finish coat, depending on color. Interior face of sash is clear for stain or paint finish.

Glazing. Select quality high-performance or high-performance sun insulating glass. Double-pane insulated glass is available for white only. For high altitude and other special glazings, contact the manufacturer.

Glazing System. Sash has a rigid vinyl snap-in-bead.

Weatherstripping. Foam type weatherstripping is applied to top and bottom rails. Rigid vinyl rib on head jamb liner and sills fits into vinyl covered foam weatherstripping on sash. Rigid vinyl leaf weatherstripping is applied to check rail. Polypropylene leaf weatherstrip with foam inserts contacts side jamb liner ribs.

Anchoring Flange and Windbreak. Factory applied rigid vinyl flanges are at head and sides of outer frame members. A flexible vinyl sill windbreak is factory applied to the bottom of sill as flashing.

Sill. This is wood core treated with a water repellent preservative, covered with PVC sheath.

Jamb Liner. This is white PVC for unit in white and complementary color PVC for unit in color.

Sash Lock and Lift. Stone color decorator finish, sash lock, keeper and lift are factory applied.

1190

DOORS AND WINDOWS

A rough opening normally recommended by the manufactures is ½" larger than the window on the sides for a total of 1" and ¾" larger on top and bottom for a total of 1½" added to the height of the window. Once a rough opening is completed, the new window needs to be placed centered in the opening, leveled with the use of shims if necessary.

Double Hung Units High Performance (H.P.) Insulating Glass, Sun Insulating Glass and Double Pane Insulating Glass						
Window Frame Size, Inches	H.P. Glass	Insect Screen Add	Grille Add	Comb. Screen Storm-Add	Window Frame Size mm	
35 " x 25 "	$224.90	$21.32	$17.29	$93.60	875 x	625 mm
35 " x 29 "	$234.00	$17.30	$13.50	$74.00	875 x	725 mm
35 " x 33 "	$249.60	$18.00	$15.10	$77.00	875 x	825 mm
35 " x 37 "	$269.10	$18.80	$15.60	$80.00	875 x	925 mm
41 " x 25 "	$286.00	$20.50	$16.70	$85.00	1,025 x	625 mm
41 " x 29 "	$302.90	$22.00	$22.60	$104.00	1,025 x	725 mm
41 " x 33 "	$228.80	$14.40	$13.30	$69.00	1,025 x	825 mm
41 " x 37 "	$24.44	$15.30	$13.60	$70.00	1,025 x	925 mm
47 " x 25 "	$261.30	$16.60	$14.20	$75.00	1,175 x	625 mm
47 " x 29 "	$279.50	$19.40	$15.70	$84.00	1,175 x	725 mm
47 " x 33 "	$291.20	$20.40	$17.00	$88.00	1,175 x	825 mm
47 " x 37 "	$314.60	$22.20	$17.80	$94.00	1,175 x	925 mm
53 " x 25 "	$291.70	$20.30	$16.40	$87.40	1,325 x	625 mm
53 " x 29 "	$306.70	$21.30	$17.20	$91.80	1,325 x	725 mm
53 " x 33 "	$321.60	$22.40	$18.10	$96.30	1,325 x	825 mm
53 " x 37 "	$336.60	$23.40	$18.90	$100.80	1,325 x	925 mm
53 " x 41 "	$351.60	$24.40	$19.70	$105.30	1,325 x	1025 mm
57 " x 25 "	$306.70	$21.30	$17.20	$91.80	1,425 x	625 mm
57 " x 29 "	$321.60	$22.40	$18.10	$96.30	1,425 x	725 mm
57 " x 33 "	$336.60	$23.40	$18.90	$100.80	1,425 x	825 mm
57 " x 37 "	$351.60	$24.40	$19.70	$105.30	1,425 x	925 mm
57 " x 41 "	$366.50	$25.50	$20.60	$109.80	1,425 x	1025 mm
59 " x 25 "	$314.20	$21.80	$17.60	$94.10	1,475 x	625 mm
59 " x 29 "	$329.10	$22.90	$18.50	$98.60	1,475 x	725 mm
59 " x 33 "	$344.10	$23.90	$19.30	$103.00	1,475 x	825 mm
59 " x 37 "	$359.00	$25.00	$20.20	$107.50	1,475 x	925 mm
59 " x 41 "	$374.00	$26.00	$21.00	$112.00	1,475 x	1025 mm
65 " x 25 "	$336.60	$23.40	$18.90	$100.80	1,625 x	625 mm
65 " x 29 "	$351.60	$24.40	$19.70	$105.30	1,625 x	725 mm
65 " x 33 "	$366.50	$25.50	$20.60	$109.80	1,625 x	825 mm
65 " x 37 "	$381.50	$26.50	$21.40	$114.20	1,625 x	925 mm
65 " x 41 "	$396.40	$27.60	$22.30	$118.70	1,625 x	1025 mm

Double Hung Units High Performance (H.P.) Insulating Glass, Sun Insulating Glass and Double Pane Insulating Glass (cont'd)					
Window Frame Size, Inches	H.P. Glass	Insect Screen Add	Grille Add	Comb. Screen Storm-Add	Window Frame Size mm
71 " x 25 "	$359.00	$25.00	$20.20	$107.50	1,775 x 625 mm
71 " x 29 "	$374.00	$26.00	$21.00	$112.00	1,775 x 725 mm
71 " x 33 "	$389.00	$27.00	$21.80	$116.50	1,775 x 825 mm
71 " x 37 "	$403.90	$28.10	$22.70	$121.00	1,775 x 925 mm
71 " x 41 "	$418.90	$29.10	$23.50	$125.40	1,775 x 1025 mm

For double hung windows in pairs, triple or quadruple arraignments, multiply costs by two, three or four respectively.

Picture Window Units Combined with 1-8 Flanking Units High Performance (H.P.) Insulating Glass				
Window Frame Size, Inches	HP Insulating Glass	Grille Add	Insect Screen Add	Window Frame Size - Meter
8' - 1" x 4'-5 1/4"	$462.80	$15.33	$81.76	2.46 x 1.35 M
8' - 1" x 4'-9 1/4"	$546.00	$11.97	$63.84	2.46 x 1.45 M
8' - 9" x 4'-5 1/4"	$479.70	$15.33	$81.76	2.67 x 1.35 M
8' - 9" x 4'-9 1/4"	$572.00	$11.97	$63.84	2.67 x 1.45 M
9' - 5" x 4'-9 1/4"	$582.40	$11.97	$63.84	2.87 x 1.45 M
8' - 9" x 4'-5 1/4"	$652.60	$15.33	$81.76	2.67 x 1.35 M
8' - 9" x 4'-9 1/4"	$663.00	$11.97	$63.84	2.67 x 1.45 M
8' - 9" x 6'-5 1/4"	$970.10	$16.17	$86.24	2.67 x 1.96 M
9' - 5" x 4'-5 1/4"	$680.60	$15.33	$81.76	2.87 x 1.35 M
9' - 5" x 4'-9 1/4"	$697.00	$11.97	$63.84	2.87 x 1.45 M
9' - 5" x 5'-5 1/4"	$729.80	$13.65	$72.80	2.87 x 1.66 M
10' - 1" x 4'-9 1/4"	$729.80	$11.97	$63.84	3.07 x 1.45 M

Gliding Window Units

An advantage of high narrow gliding windows is that furniture can be placed under them, making much better interior arrangements possible.

The gliding windows described below are furnished complete with frames, sash, and weatherstrips. Each sash slides to the opposite side of the opening for ventilation but when fully closed is weatherstripped on all four sides. Prices are for complete gliding window units, assembled and carton packed.

Frame. Wood members are treated with water repellent preservative and covered with white, rigid PVC sheath. Retractable inside head stop allows for operating sash removal. Captive screws release stationary sash.

Sash. Wood core treated with a water repellent preservative and completely covered with white PVC sheath.

DOORS AND WINDOWS

Glazing. Select quality double-pane insulating window glass. Installations over 3500' (1066.80 m) altitude require high altitude glass.

Glazing Bead. Rigid vinyl snap-in glazing bead with flexible vinyl tip against glass.

Weatherstripping. Rigid vinyl spring tension type, factory applied for a tight seal between sash and frame.

Flashing. Continuation of the rigid vinyl sheath on frame forms perimeter flashing and anchoring fin.

Corner Seal. Vinyl frame sheath eliminates corner joints. Sash corner joints are welded to form waterproof joints.

Sill Tank. For added tightness under severe conditions of exposure, a molded vinyl tank has been built in to drain any moisture to the exterior.

Glides. Adjustable chrome-plated steel glides, factory applied.

Locking Latch. Center interlock and two-way spring loaded locking mechanism secures sash at head and sill. Steel locking rods and zinc die-cast handle case with stone color decorator finish. Flexible baffles cover meeting stiles.

Combination Unit. A unit for triple glazing. Features aluminum frame and 1/8" (3.1 mm) tempered glass in storm panels and glass fiber screen cloth. Storm panels and screen with baked-on white decorator finish.

Gliding window units may be ordered with a fixed glass in the center and two gliding sash on either end, which when open, pass in front of the fixed panel.

Sash Extras. Glazing with obscure glass, add $4.50 per sq.ft. ($48.44 per sq.m). Rectangular grilles same as double hung. Estimator should verify that it is available.

Gliding Windows, Double Pane Insulating Glass			
Window Frame Size Inches	HP Insulating Glass	Insect Screen	Window Frame Size mm
14 "x 30 "	$264.00	$15.80	350 x 750 mm
14 "x 35 "	$294.00	$17.60	350 x 875 mm
20 "x 35 "	$330.00	$19.80	500 x 875 mm
20 "x 41 "	$366.00	$22.00	500 x 1,025 mm
26 "x 35 "	$366.00	$22.00	650 x 875 mm
26 "x 41 "	$402.00	$24.10	650 x 1,025 mm
26 "x 54 "	$480.00	$28.80	650 x 1,350 mm
32 "x 41 "	$438.00	$26.30	800 x 1,025 mm
32 "x 54 "	$516.00	$31.00	800 x 1,350 mm

Awning Panel Wall Units

Awning units are a system of window panel arrangement ranging from a single individual panel to a multiple 12-panel unit, which is used as a picture frame unit.

Awning units are designed to give a selection of either stationary glass panels, ventilating sash panels, or a combination of both. Both come with high performance (H.P.) sun insulating glass and insulating glass.

Ventilating units have 2" (50 mm) sash, hinged at the top, which can be opened about 35 degrees, closed, or held in intermediate position by rotary operators. Operators have worm gears, twin arms of cadmium plated steel, and removable handles, and they are self-locking.

Insect screens, grilles, and interior blinds are available for all units.

Awning Windows with High Performance (H.P.) Insulating Glass, and Double Pane				
Window Frame Size Inches	HP Insulating Glass	Insect Screen Add	Grille Add	Window Frame Size - mm
19 "x 20 "	$184.80	$12.00	$9.60	475 x 500 mm
31 "x 16 "	$192.00	$13.20	$10.80	775 x 400 mm
31 "x 19 "	$216.00	$14.40	$12.00	775 x 475 mm
36 "x 16 "	$212.40	$15.60	$13.20	900 x 400 mm
36 "x 19 "	$232.80	$19.20	$15.60	900 x 475 mm
43 "x 16 "	$232.80	$26.40	$21.60	1075 x 400 mm
43 "x 19 "	$252.00	$13.20	$10.80	1075 x 475 mm
31 "x 24 "	$226.80	$16.80	$12.00	775 x 600 mm
31 "x 32 "	$235.20	$21.60	$20.40	775 x 800 mm
31 "x 37 "	$333.60	$22.80	$21.60	775 x 925 mm
31 "x 44 "	$367.20	$30.00	$24.00	775 x 1100 mm
31 "x 63 "	$591.60	$44.40	$12.00	775 x 1575 mm
36 "x 36 "	$307.20	$27.60	$25.20	900 x 900 mm
36 "x 44 "	$300.00	$32.40	$30.00	900 x 1100 mm
36 "x 63 "	$537.60	$49.20	$13.20	900 x 1575 mm
43 "x 24 "	$271.20	$19.20	$15.60	1075 x 600 mm
43 "x 44 "	$458.40	$36.00	$21.60	1075 x 1100 mm
43 "x 63 "	$717.60	$48.00	$14.40	1075 x 1575 mm
55 "x 32 "	$434.40	$34.80	$20.40	1375 x 800 mm

DOORS AND WINDOWS

Glass panel wall units come completely set up and glazed. The ventilating units are weatherstripped, and operators are applied.

Frame. Exterior wood members are treated with a water repellent preservative and covered with a pre-formed rigid vinyl (PVC) sheath in white or color.

Sash. This is wood core treated with a water repellent preservative and completely covered with a PVC sheath in white or color.

Glazing. Select quality high-performance, high-performance sun, or double-pane insulating window glass. For high altitude and other special glazings, contact manufacturer.

Glazing Bead. This is rigid vinyl. It has a snap-in feature with flexible vinyl tip against the glass.

Weatherstripping. This is flexible PVC bulb type, factory applied for tight seal between sash and frame.

Flashing. It is a continuation of the pre-formed rigid vinyl sheath on frame forms perimeter flashing and anchoring fin.

Corner Seal. Pre-formed vinyl frame sheath eliminates corner joints. Sash corner joints are welded to form waterproof joint.

Inside Stops. Western cedar pine natural wood, can be finished to match interior decor.

Awning Roto-Lock Operator. This offers a self-locking underscreen operator. Dual arms pull corners in tightly for maximum weathertightness. It has stone color decorator finish. Sash disengages from operator without tools. Picture window units include same materials as ventilating units except no operating hardware. Recommended additional parts include extension jambs of Western clear pine. Sizes are available for 4-9/16" (114 mm), 5-1/4" (131 mm), 6-9/16" (164.06 mm), 7-9/16" (189.06 mm) wall thickness. They are pre-drilled for easy application, except for 7-9/16" (189.06 mm) size.

Optional Accessories. Two different Perma-Fit Divided Light Grilles are available. The white grille is rigid vinyl. Exterior is white and the interior surface has a pine color stainable/paintable polymer finish. The grille in color is polycarbonate. Exterior is color and interior surface has pine color stainable/paintable polymer finish. Perma-Fit grilles without the finish are available in white or color both sides. Grille fits tight against glass and is easily removed for cleaning. Grilles not available for picture window sizes.

Casement Window Units

Casement sash may be ordered in gangs of two, three, four, or five. Prices for each additional unit in a gang arrangement should be quoted at the time of estimate for possible gang arrangement discount. It may be cheaper to use casements in groups than as individual units. Additional savings will also be realized in gang arrangements because of fewer jambs to form and caulk, providing the lintel construction does not become more costly due to the longer span. If some of the sash is ordered as stationary units, a further

1195

savings of $45.00 on the basic price per opening can be made. Casements can also be used in combination with larger fixed units or picture windows.

Casement units swing outward for top to bottom ventilation. They can be combined with other styles of windows and patio doors, or can be arranged in bow, angle bay, and box bay windows.

Frame. Wood members are treated with a water repellent preservative and covered with a rigid PVC sheath in white or color.

Sash. This is wood core treated with a water repellent preservative and completely covered with a rigid PVC sheath in white or color.

Glazing. Select quality high-performance, high-performance sun, or double-pane insulating window glass.

Glazing Bead. This is of rigid vinyl and has snap-in features with flexible vinyl tip against glass.

Weatherstripping. This is flexible PVC bulb type, factory applied for tight seal between sash and frame.

Flashing. This is a continuation of the pre-formed rigid vinyl sheath on frame forms perimeter flashing and anchor fin.

Corner Seal. Pre-formed vinyl frame sheath eliminates corner joints. Sash corner joints are welded to form waterproof joint.

Inside Stops. Western clear pine natural wood can be finished to match interior decor.

Sash Lock. The lock pulls sash firmly closed or ejects for opening. Reach-out action eliminates binding of locks and relieves stress on operator. Stone color decorator finish.

Casement Roto Operator. Under-screen operator has arm that attaches to bracket applied to sash. With either left- or right-hand operation and stone color decor finish, sash disengages from operator without tools. Picture window units include same materials as ventilating units except no operating hardware. Recommended additional parts include extension jambs of Western clear pine. Sizes available from Andersen for 4-9/16" (114.06 mm), 5-1/4" (131.25 mm), 6-9/16" (164.06 mm) or 7-9/16" (189.06 mm) wall thickness. It is pre-drilled for easy application, except for 7-9/16" (189.06 mm) unit.

Optional Accessories. Two different Perma-Fit Divided Light Grilles are available. The white grille is rigid vinyl. Exterior is white and the interior surface has a pine color stainable/paintable polymer finish. The grille in color is polycarbonate. Exterior is color and interior surface has pine color stainable/paintable polymer finish. Perma-Fit grilles without the finish are available in white or color both sides. Grille fits tight against glass and is easily removed for cleaning. Grilles not available for picture window sizes.

DOORS AND WINDOWS

Casement Windows with High Performance Glass (H.P.)., Insulating glass and Double Pane Insulating Glass					
Window Frame Size Inches	HP Insulating Glass	Insect Screen Add	Grille Add	Comb. Storm Add	Window Frame Size - mm
12 "x 19 "	$156.00	$12.00	$9.60	$67.20	300 x 475 mm
12 "x 31 "	$175.20	$13.20	$10.80	$67.20	300 x 775 mm
12 "x 36 "	$190.80	$14.40	$12.00	$69.60	300 x 900 mm
12 "x 43 "	$205.20	$15.60	$13.20	$70.80	300 x 1075 mm
12 "x 55 "	$250.80	$19.20	$3.60	$75.60	300 x 1375 mm
12 "x 67 "	$272.40	$26.40	$21.60	$86.40	300 x 1675 mm
16 "x 31 "	$176.40	$13.20	$10.80	$70.80	400 x 775 mm
16 "x 36 "	$192.00	$15.60	$12.00	$73.20	400 x 900 mm
16 "x 43 "	$207.60	$16.80	$13.20	$75.60	400 x 1075 mm
16 "x 55 "	$254.40	$20.40	$16.80	$80.40	400 x 1375 mm
16 "x 67 "	$276.00	$26.40	$21.60	$88.80	400 x 1675 mm
19 "x 31 "	$188.40	$14.40	$12.00	$74.40	475 x 775 mm
19 "x 36 "	$204.00	$15.60	$13.20	$76.80	475 x 900 mm
19 "x 43 "	$224.40	$18.00	$14.40	$79.20	475 x 1075 mm
19 "x 55 "	$258.00	$20.40	$16.80	$85.20	475 x 1375 mm
19 "x 67 "	$304.80	$26.40	$21.60	$91.20	475 x 1675 mm
24 "x 19 "	$175.20	$14.40	$10.80	$78.00	600 x 475 mm
24 "x 31 "	$206.40	$16.80	$12.00	$81.60	600 x 775 mm
24 "x 36 "	$222.00	$18.00	$13.20	$81.60	600 x 900 mm
24 "x 43 "	$1,452.00	$19.20	$15.60	$85.20	600 x 1075 mm
24 "x 55 "	$300.00	$22.80	$18.00	$92.40	600 x 1375 mm
24 "x 67 "	$360.00	$28.80	$24.00	$99.60	600 x 1675 mm
29 "x 65 "	$385.00	$31.20	$26.00	$102.00	725 x 1625 mm
35 "x 65 "	$403.00	$36.00	$30.00	$108.00	875 x 1625 mm

Circle Top Window Units

Circle top windows are specially designed with a frame/sash profile. The windows are compatible with casement, awning, and double hung units and can be joined to these or can be installed as single units. When combined, the glazing in the circle top window closely aligns with the window unit below it.

Frame. Wood members are treated with a water repellent preservative and covered with a pre-formed rigid PVC sheath in white or color. The vinyl sheath will not chip, flake, peel, blister, rust, pit, or corrode, and it will not need painting every few years.

Flashing. Continuation of the pre-formed rigid vinyl sheath on frame forms a full perimeter flashing and anchoring fin.

Glazing. Select quality high-performance or high-performance sun insulating glass.

Recommended Optional Accessories. Interior arch casing is unfinished wood for painting or staining. Maple or oak is used for laminated portion, which forms the curved piece. Interior casing is available in ranch or colonial style. Divided light grilles are available in two different types. The white grille is rigid vinyl. Exterior is white and the interior surface has a pine color stainable/paintable polymer finish. The grille in color is glass fiber polycarbonate. Exterior is color and interior surface has pine color stainable/paintable polymer finish. Perma-fit grilles without the finish are available in white or color both sides. Grille fits tight against glass and can be removed for cleaning. Grilles not available for picture window sizes.

Plinth Blocks. Solid oak or maple plinth block with decorative radial sunburst pattern or reversible flush face. Plinth blocks come with arch casing or can be purchased separately.

Extension Jambs. Laminated maple curved extension jambs are available for wall thickness of 4-9/16" (114.06 mm), 5-1/4" (131.25 mm), 6-9/16" (164.06 mm) and 7-9/16" (189.06 mm) for casement and awning windows and 5-1/4" (131.25 mm), 6-9/16" (164.06 mm), and 7-9/16" (189.06 mm) for double hung windows. They are pre-drilled, except for casement and awning 7-9/16" (189.06 mm) size. One-piece pine straight extension jamb for sill is available in same sizes as maple parts for casement or awning. Both laminated maple and clear pine parts give uniform appearance when stained.

1198

DOORS AND WINDOWS

Circle Top Windows with High Performance (H.P.) Insulating Glass and Sun Insulating Glass					
Glass Size		H.P. Glass	Grille Add	Glass Size -mm	
Width x	Height			Width x	Height
19 5/8 "x	19 5/8 "	$391.00	$20.00	490.63 x	490.63 mm
19 1/2 "x	9 3/4 "	$351.00	$18.00	487.50 x	243.75 mm
23 3/4 "x	11 7/8 "	$387.00	$23.00	593.75 x	296.88 mm
43 3/8 "x	21 11/16 "	$538.00	$35.00	1084.38 x	542.19 mm
51 7/8 "x	25 15/16 "	$636.00	$44.00	1296.88 x	648.44 mm
67 1/8 "x	35 9/16 "	$1,026.00	$116.00	1678.13 x	889.06 mm
20 3/8 "x	10 3/16 "	$361.00	$20.00	509.38 x	254.69 mm
24 3/8 "x	12 3/16 "	$419.00	$23.00	609.38 x	304.69 mm
28 3/8 "x	14 3/16 "	$452.00	$27.00	709.38 x	354.69 mm
32 3/8 "x	16 3/16 "	$484.00	$30.00	809.38 x	404.69 mm
36 3/8 "x	18 3/16 "	$533.00	$33.00	909.38 x	454.69 mm

Bow, Box Bay and Angle Bay Window Units 30° Angle Bay Unit Sizes with High Performance (H.P.) Insulating Glass, Sun Insulating Glass and Double Pane Insulating Glass					
Size		H.P. Sun Insulating	Insect Screen Add	Grille Add	Size - Meters
Width x	Height				Width x Height
5'- 10"	x 3'-1 15/16"	$820.00	$29.00	$35.00	1.78 x 0.96 M
5'- 10"	x 3'- 6 13/16"	$889.00	$32.00	$38.00	1.78 x 1.09 M
5'- 10"	x 4'- 2"	$954.00	$35.00	$42.00	1.78 x 1.27 M
5'- 10"	x 5'- 1 7/8"	$1,080.00	$42.00	$49.00	1.78 x 1.57 M
5'- 10"	x 6'- 1 7/8"	$1,250.00	$53.00	$66.00	1.78 x 1.88 M
7'- 9 7/8"	x 3'-1 15/16"	$1,051.00	$58.00	$47.00	2.40 x 0.96 M
7'- 9 7/8"	x 3'-6 13/16"	$1,139.00	$64.00	$50.00	2.40 x 1.08 M
7'- 9 7/8"	x 4'- 2"	$1,226.00	$70.00	$55.00	2.40 x 1.27 M

Bow, Box Bay and Angle Bay Window Units 30° Angle Bay Unit Sizes with High Performance (H.P.) Insulating Glass, Sun Insulating Glass and Double Pane Insulating Glass (cont'd)						
Size			H.P. Sun Insulating	Insect Screen Add	Grille Add	Size - Meters
Width	x	Height				Width x Height
7'- 9 7/8"	x	5'-1 7/8"	$1,390.00	$84.00	$66.00	2.40 x 1.57 M
7'- 9 7/8"	x	6'- 1 7/8"	$1,591.00	$106.00	$88.00	2.40 x 1.88 M
9'- 9 3/4"	x	3'- 1 15/16"	$1,248.00	$58.00	$58.00	2.99 x 0.96 M
9'- 9 3/4"	x	3'- 6 13/16"	$1,356.00	$64.00	$64.00	2.99 x 1.08 M
9'- 9 3/4"	x	4'- 2"	$1,454.00	$70.00	$70.00	2.99 x 1.27 M
9'- 9 3/4"	x	5'- 1 7/8"	$1,639.00	$84.00	$83.00	2.99 x 1.57 M

Bow, Box Bay and Angle Bay Window Units 45° Angle Bay Unit Sizes with High Performance (H.P.) Insulating Glass, Sun Insulating Glass and Double Pane Insulating Glass						
Size			H.P. Sun Insulating	Insect Screen Add	Grille Add	Size - Meters
Width x	x	Height				Width x Height
5'- 4 1/8"	x	3'- 1 15/16"	$842.40	$28.80	$34.80	1.63 x 0.96 M
5'- 4 1/8"	x	3'- 6 13/16"	$912.00	$32.40	$38.40	1.63 x 1.08 M
5'- 4 1/8"	x	4'- 2"	$987.60	$34.80	$49.20	1.63 x 1.27 M
5'- 4 1/8"	x	5'- 1 7/8"	$1,003.20	$42.00	$66.00	1.63 x 1.57 M
5'- 4 1/8"	x	6'-1 7/8"	$1,294.80	$52.80	$70.80	1.63 x 1.88 M
7'- 3 7/8"	x	3'- 1 15/16"	$1,087.20	$56.40	$50.40	2.23 x 0.96 M
7'- 3 7/8"	x	3'- 6 13/16"	$956.40	$63.60	$69.60	2.23 x 1.08 M
7'- 3 7/8"	x	4'- 2"	$1,272.00	$75.60	$72.00	2.23 x 1.27 M
7'- 3 7/8"	x	5'- 1 7/8"	$1,446.00	$84.00	$73.20	2.23 x 1.57 M
7'- 3 7/8"	x	6'-1 7/8"	$1,647.60	$105.60	$74.40	2.23 x 1.88 M
9'- 3 3/4"	x	3'- 1 15/16"	$1,288.80	$56.40	$69.60	2.84 x 0.96 M
9'- 3 3/4"	x	3'- 6 13/16"	$1,398.00	$63.60	$82.80	2.84 x 1.08 M
9'- 3 3/4"	x	4'- 2"	$1,506.00	$69.60	$90.00	2.84 x 1.27 M
9'- 3 3/4"	x	5'- 1 7/8"	$1,772.40	$84.00	$96.00	2.84 x 1.57 M

DOORS AND WINDOWS

Bow, Box Bay and Angle Bay Window Units 90° Angle Bay Unit Sizes with High Performance (H.P.) Insulating Glass, Sun Insulating Glass and Double Pane Insulating Glass						
Size		H.P. Sun Insulating	Insect Screen Add	Grille Add	Size - Meters	
Width x	Height				Width x	Height
4'- 8 3/8" x	2'- 11 15/16"	$1,060.00	$55.00	$46.00	1.43 x	0.91 M
4'- 8 3/8" x	3'- 4 13/16"	$1,076.00	$61.00	$49.00	1.43 x	1.03 M
4'- 8 3/8" x	4'- 0"	$1,166.00	$66.00	$54.00	1.43 x	1.22 M
4'- 8 3/8" x	4'- 11 7/8"	$1,362.00	$82.00	$65.00	1.43 x	1.52 M
4'- 8 3/8" x	5'- 11 7/8"	$1,504.00	$106.00	$86.00	1.43 x	1.83 M
6'- 8 1/4" x	2'- 11 15/16"	$1,206.00	$55.00	$56.00	2.05 x	0.91 M
6'- 8 1/4" x	3'- 4 13/16"	$1,302.00	$61.00	$61.00	2.05 x	1.03 M
6'- 8 1/4" x	4'- 0"	$1,403.00	$66.00	$68.00	2.05 x	1.22 M
6'- 8 1/4" x	4'- 11 7/8"	$1,616.00	$82.00	$82.00	2.05 x	1.52 M

Roof Window Units (Skylights)

Roof windows, also called skylights, can enhance the appearance of a living space. However, the estimator must be cautious. Roof window installation is unique and usually requires additional costs. On the exterior, there might be flashing for ice build-up, curbs around the window on a flat roof, or on a sloped roof that is relatively flat, or deflector shield for maximum weather tightness. On the interior, additional finish work might be required.

Roof Windows, Basic and Stationary Unit with High Performance (H.P.) Insulating and Sun Insulating Laminated Tempered Glass								
Size		H.P. Insulating Glass	Set Shingle Flashing	Incline Curb Flashing	Set Tile Flashing	Size - mm		
Width x	Height					Width x		Height
15 1/16 "x	30 5/16 "	$359.00	$82.00	$323.00	$95.00	376.56 x		757.81 mm
15 1/16 "x	40 15/16 "	$420.00	$82.00	$329.00	$107.00	376.56 x		1023.44 mm
15 1/16 "x	68 7/8 "	$613.00	$94.00	$388.00	$167.00	376.56 x		1721.88 mm
26 "x	40 15/16 "	$503.00	$98.00	$348.00	$116.00	650.00 x		1023.44 mm
26 "x	53 15/16 "	$616.00	$98.00	$384.00	$128.00	650.00 x		1348.44 mm
26 "x	68 15/16 "	$716.00	$239.00	$408.00	$160.00	650.00 x		1723.44 mm
38 3/16 "x	40 15/16 "	$631.00	$109.00	$389.00	$133.00	954.69 x		1023.44 mm
38 3/16 "x	53 15/16 "	$768.00	$109.00	$412.00	$143.00	954.69 x		1348.44 mm

Roof Windows, Venting Unit with High Performance (H.P.) Insulating and Sun Insulating Tempered Glass									
Size			H.P. Insulating Glass	Set Shingle Flashing	Incline Curb Flashing	Set Tile Flashing	Size - mm		
Width	x	Height					Width	x	Height
15 1/16 "x		27 2/16 "	$581.00	$82.00	$324.00	$95.00	376.56	x	678.13 mm
15 1/16 "x		37 3/4 "	$668.00	$82.00	$329.00	$107.00	376.56	x	943.75 mm
23 "x		37 3/4 "	$746.00	$98.00	$348.00	$116.00	575.00	x	943.75 mm
23 "x		50 3/4 "	$907.00	$98.00	$384.00	$128.00	575.00	x	1268.75 mm
35 3/16 "x		37 3/4 "	$953.00	$109.00	$388.00	$134.00	879.69	x	943.75 mm
35 3/16 "x		50 3/4 "	$1,158.00	$109.00	$412.00	$143.00	879.69	x	1268.75 mm

Window Screens

Wood window screens have been largely replaced by aluminum and vinyl units. Most prefabricated windows have insect screens available to fit them and can be ordered as part of the window unit or separately later. The following prices are based on prefabricated window glass sizes.

Screen Size, in.	Price	Screen Size, mm		
16 7/16 "x 22 "	$16.40	410.94 x	550 mm
16 7/16 "x 24 "	$17.30	410.94 x	600 mm
20 7/8 "x 22 "	$18.00	521.88 x	550 mm
20 7/8 "x 24 "	$18.80	521.88 x	600 mm
20 7/8 "x 28 "	$20.60	521.88 x	700 mm
20 7/8 "x 68 "	$22.00	521.88 x	1700 mm
24 7/16 "x 22 "	$19.40	610.94 x	550 mm
24 7/16 "x 24 "	$20.50	610.94 x	600 mm
24 7/16 "x 28 "	$22.30	610.94 x	700 mm
28 7/16 "x 22 "	$20.80	710.94 x	550 mm
28 7/16 "x 24 "	$21.80	710.94 x	600 mm
28 7/16 "x 28 "	$24.20	710.94 x	700 mm
28 7/16 "x 60 "	$26.00	710.94 x	1500 mm
28 7/16 "x 68 "	$30.10	710.94 x	1700 mm
32 7/16 "x 22 "	$23.00	810.94 x	550 mm
32 7/16 "x 24 "	$24.20	810.94 x	600 mm
32 7/16 "x 28 "	$26.80	810.94 x	700 mm
32 7/16 "x 60 "	$28.00	810.94 x	1500 mm
32 7/16 "x 68 "	$30.30	810.94 x	1700 mm
36 7/16 "x 22 "	$24.80	910.94 x	550 mm
36 7/16 "x 24 "	$26.00	910.94 x	600 mm
36 7/16 "x 28 "	$28.90	910.94 x	700 mm
36 7/16 "x 60 "	$30.10	910.94 x	1500 mm
40 7/16 "x 24 "	$27.00	1010.94 x	600 mm

DOORS AND WINDOWS

Nailing Wood Window Frames Together. When wood window frames are sent to the job knocked down, it will require 1 to 1-1/4 hr. carpenter time to assemble and nail each frame together, at the following labor cost per frame:

Description	Hours	Rate	Total	Rate	Total
Carpenter	1.20	$	$	$35.04	$42.05

It will require 1-3/4 to 2-1/4 hrs. carpenter time to assemble and nail together a double window frame up to 6'-0" (1800 mm) wide and 5'-6" (1650 mm) or 6'-0" (1800 mm) high, at the following labor cost per frame:

Description	Hours	Rate	Total	Rate	Total
Carpenter	2.00	$	$	$35.04	$70.08

Setting Single Wood Window Frames. The labor cost of setting wood window frames varies with the size of the frame, amount of bracing necessary, and the distance they have to be carried or hoisted.

Once frames have been delivered to the job, it requires 10 to 15 minutes sorting frames and carrying them to the building and 20 to 30 minutes for 2 carpenters to set, plumb, and brace each frame at the rate of 8 to 12 frames per 8-hr. day. The labor handling and setting each frame should cost as follows:

Description	Hours	Rate	Total	Rate	Total
Carpenter	0.80	$	$	$35.04	$28.03
Helper	0.20	$	$	$26.16	$5.23
Cost per Frame					$33.26

Setting Double or Triple Window Frames. It will ordinarily require 10 to 15 minutes for 2 workers to carry a double or triple window frame from the stock pile to place where it is to be used and it will then require 30 to 35 minutes for 2 carpenters to set, plumb, and brace the frame on the wall, at the following labor cost per frame:

Description	Hours	Rate	Total	Rate	Total
Carpenter	1.10	$	$	$35.04	$38.54
Helper	0.40	$	$	$26.16	$10.46
Cost per Frame					$49.01

Labor Handling and Setting Complete Window Units. When handling and setting complete window units, consisting of frame, sash,

1203

balances, and weatherstrips, all installed complete, figure about 1/4 hr. more than for handling and setting window frames of the same size.

Fitting Wood Sash. When fitting wood sash for double hung windows, a carpenter should fit 24 to 28 single sash per 8-hr. day, at the following labor cost per pair:

Description	Hours	Rate	Total	Rate	Total
Carpenter	0.60	$	$	$35.04	$21.02

First Grade Workmanship

Where the sash must fit perfectly, with neither too much nor too little play, a carpenter should fit 18 to 22 single sash per 8-hr. day, at the following labor cost per pair:

Description	Hours	Rate	Total	Rate	Total
Carpenter	0.80	$	$	$35.04	$28.03

Setting Roof Window (Skylight). When placing frame and window, two workers are usually required for ease of installation at the following cost per frame:

Description	Hours	Rate	Total	Rate	Total
Carpenter	3.00	$	$	$35.04	$105.12

It is important that the estimator include any additional roofing material and labor required.

Hanging Wood Sash. Including the time required placing sash cord, chain, and weights or counterbalances, a carpenter should hang 24 to 28 single sash per 8-hr. day, at the following labor cost per pair:

Description	Hours	Rate	Total	Rate	Total
Carpenter	0.60	$	$	$35.04	$21.02

First Grade Workmanship

Where care must be exercised to see that all sash work fits perfectly, and the cord or chain is the exact length with the correct weights or counterbalances, a carpenter should hang 22 to 26 single sash per 8-hr. day, at the following labor cost per pair:

Description	Hours	Rate	Total	Rate	Total
Carpenter	0.70	$	$	$35.04	$24.53

1204

DOORS AND WINDOWS

Fitting and Hanging Wood Sash. A carpenter should fit and hang, complete with sash cord, chain, and weights or counterbalances, 12 to 14 single sash per 8-hr. day, at the following labor cost per pair:

Description	Hours	Rate	Total	Rate	Total
Carpenter	1.20	$	$	$35.04	$42.05

First Grade Workmanship

Where the sash must be fitted to have neither too much nor too little play and hung so they will work with ease and not be too loose, a carpenter should fit and hang, including sash cord, chain, and weights or counterbalances, 10 to 12 single sash per 8-hr. day, at the following labor cost per pair:

Description	Hours	Rate	Total	Rate	Total
Carpenter	1.50	$	$	$35.04	$52.56

Fitting Casement Sash. A carpenter should fit 22 to 26 average size wood casement sash per 8-hr. day, at the following labor cost per pair:

Description	Hours	Rate	Total	Rate	Total
Carpenter	0.70	$	$	$35.04	$24.53

First Grade Workmanship

Where first class workmanship is required with all sash fitted to work easily without binding and hung so that each sash shows the same top, bottom, and side margin, a carpenter should fit 14 to 18 single sash per 8-hr. day, at the following labor cost per pair:

Description	Hours	Rate	Total	Rate	Total
Carpenter	1.00	$	$	$35.04	$35.04

Hanging Casement Sash. The labor cost of hanging casement sash varies, depending on whether they are hung with 2 or 3 butts and style of butts. Where the sash are hung using 2 butts per sash, a carpenter should hang 20 to 24 single sash per 8-hr. day, at the following labor cost per pair:

Description	Hours	Rate	Total	Rate	Total
Carpenter	0.75	$	$	$35.04	$26.28

First Grade Workmanship

In fine residences, apartments, and other buildings where the casement sash must be hung without binding and showing the same margin at the top, bottom, and sides, a carpenter should hang 12 to 14 single sash per 8-hr. day, at the following labor cost per pair:

Description	Hours	Rate	Total	Rate	Total
Carpenter	1.20	$	$	$35.04	$42.05

Fitting and Hanging Casement Sash. Under average conditions, a carpenter should fit and hang 10 to 13 single casement sash per 8-hr. day, at the following labor cost per pair:

Description	Hours	Rate	Total	Rate	Total
Carpenter	1.40	$	$	$35.04	$49.06

First Grade Workmanship

Where casement sash must be fitted and hung to be neither too loose nor so tight as to cause binding when opening and closing and must show the same margin between sash and frame on all sides, a carpenter should fit and hang 7 to 9 single sash per 8-hr. day, at the following labor cost per pair:

Description	Hours	Rate	Total	Rate	Total
Carpenter	2.20	$	$	$35.04	$77.09

Fitting and Hanging Outside Window Shutters. When fitting and hanging outside window shutters on medium priced residences or cottages, a carpenter should fit and hang 7 to 8 pairs per 8-hr. day, at the following labor cost per pair:

Description	Hours	Rate	Total	Rate	Total
Carpenter	1.10	$	$	$35.04	$38.54
Cost per Single Shutter	0.55				$19.27

First Grade Workmanship

On the better class of buildings, where more care must be used in fitting and hanging outside window shutters, a carpenter should fit and hang 5 to 6 pairs per 8-hr. day, at the following labor cost per pair:

DOORS AND WINDOWS

Description	Hours	Rate	Total	Rate	Total
Carpenter	1.40	$	$	$35.04	$49.06
Cost per Single Shutter	0.70				$24.53

Fitting Roof Windows. The labor costs of setting wood, polyvinyl, or steel roof windows varies with the size of the frame, amount of bracing necessary, and the distance that the window must be hoisted or carried.

After frames have been delivered to a project, it will require 10 to 15 minutes for sorting and carrying them to the building, and it will take 20 to 45 minutes for two carpenters to set, plumb, and brace each frame, a rate of 6 to 12 units per day. The most important variable affecting the production rate is the slope of the roof that carpenters are working on.

Description	Hours	Rate	Total	Rate	Total
Carpenter	1.40	$	$	$35.04	$49.06
Laborer	0.70	$	$	$26.16	$18.31
Total Costs					$67.37

If required, add curbs around window unit at approximately 45 minutes per window unit.

Interior Window Trim

The labor cost of placing interior window trim will vary considerably with the type of building, thickness of walls, and style of trim or casing. Frame buildings, stucco buildings, or brick veneer buildings with walls 6" (150 mm) to 8" (200 mm) thick do not require inside wood jamb linings, because the wood window frames usually extend to the inside face of the wall, and the wood casings, window stools, aprons, and stops are nailed directly to the wood frame.

Buildings having walls 10" (250 mm), 12" (300 mm), 16" (400 mm), or more in thickness require a wood jamb lining to extend from the inside of the wood window frame to the face of the finished wall. It will also require wood blocking and grounds to nail the jamb linings, window stools, and aprons.

The style of trim affects labor costs considerably, and for this reason costs are given on four distinct types of interior casings:

Style "A" consists of 1-member trim with either a mitred or square cut head or cap. This is the simplest trim it is possible to use.

Style "B" consists of 1-member trim for sides and a built-up cap trim for window heads. Caps may be put together in the mill or assembled and put together on the job.

Style "C" consists of 2-member back-band trim, consisting of a one-piece casing with a back-band of thicker material.

Style "D" consists of any type of window trim that are assembled and glued together in the mill and set in the building as a unit.

The class of workmanship will also have considerable bearing on the labor costs, as in ordinary structures the casings are nailed to the frames or jamb linings, and the trim may show some hammer marks, but where first grade workmanship is required, the utmost care must be used in cutting and fitting all casings, miters, etc. All nails must be carefully driven and set; casings must show the same margin on all sides and must fit closely to the walls. In other words, it must be a first class job in every respect.

Labor Time Erecting Interior Window Trim			
Style & Kind of Trim	No. Windows 8-Hr. Day	Carp. Hrs. Per Window	Add for Wood Lining
"A" Single Casing	78.0	1.0 - 1.1	0.6
"B" Cap Trim*	57.0	1.1 - 1.6	0.6
"C" Back-band Trim	56.0	1.3 - 1.6	0.6
"D" Mill Assembled**	8.0 - 10.0	0.8 - 1.0	0.6
First Grade Workmanship			
"A" Single Casing	5.0 - 7.0	1.1 - 1.6	0.8
"B" Cap Trim*	4.0 - 6.0	1.3 - 2.0 *	0.8
"C" Back-band Trim	4.0 - 5.0	1.6 - 2.0	0.8
"D" Mill Assembled**	7.0 - 9.0	0.9 - 1.1	0.8

*For cap trim put together on job.

**Trim assembled and glued up in the mill.

On complicated window trim with recessed jamb linings and on trim that has both a casing and a subcasing or panel under the stools, a carpenter will complete only about 1 window per 8-hr. day. This class of work is exceptional and is found only in high grade work.

Labor Setting Window Frames, Fitting and Hanging Wood Sash, And Trimming Interior Window Openings Complete

The following table includes all labor required to set window frame, to fit and hang wood sash, and to place jamb linings, stops, window stool, apron, interior casings, and hardware for one complete single window opening.

1208

DOORS AND WINDOWS

Trim Style and Kind	Windows Per 8-Hr. Day	Hrs. per Window		Add for Wood Jamb Lining
		Carpenter	Labor	
"A" Single Casing	2 1/2 - 3	3.0	0.3	0.6
"B" Cap Trim	2 1/2 - 3	3.0	0.3	0.6
"B" Cap Trim*	2 *	3.5	0.3	0.6
"C" Back-band Trim	2 *	3.5	0.3	0.6
"D" Mill Assembled**	2 1/2 - 3	3.0	0.3	0.6
First Grade Workmanship				
"A" Single Casing	2	3.5	0.3	0.8
"B" Cap Trim	2	3.5	0.3	0.8
"B" Cap Trim*	2	4.1	0.3	0.8
"C" Back-band Trim	2	4.4	0.3	0.8
"D" Mill Assembled**	2 1/4 - 2 1/2	3.3	0.3	0.8

*For cap trim put together on job.

**Trim assembled and glued up in the mill.

08700 HARDWARE AND SPECIALTIES

08710 FINISH HARDWARE

Finish hardware refers to any item that is fitted to a door and frame to perform a specific function-hinges, latch or locksets, closers, stops, and bolts.

Most items of finish hardware can be obtained in various finishes, and each finish has a U.S. Code Symbol Designation and a matching Building Hardware Manufactures Association (BMHA) code as follows:

US Code	Description	BHMA Code
USP	Prime Coat	600
US 1B	Flat Black Coated for Steel Hinges	601
US 3	Polished Brass - Clear Coated	605
US 4	Satin Brass - Clear Coated	606
US 5	Antique Satin Brass, Blackened - Clear Coated	609
US 9	Polished Bronze	-----
US 10	Satin Bronze	612
US 10B	Oil Rubbed Bronze	613
US 14	Polished Nickel	-----

US Code	Description	BHMA Code
US 15	Satin Nickel	619
US 15A	Antique Satin Nickel, Pewter - Clear Coated	620
US 19	Black - Flat - Coated	622
US 26	Polished Chrome	625
US 26D	Satin Chrome	626
US 28	Satin Aluminum - Anodized	-----
US 32	Polished Stainless Steel	629
US 32D	Satin Stainless Steel	630
LT	Lifetime Finish - Clear	-----
BZ	Solid Bronze - Dark Patina	-----
LTBZ	Solid Bronze - Light Patina	-----
TC	Matt Black Powder Coate Bronze	-----

Keep in mind that the finish will affect two areas of concern for the estimator: cost and delivery time. The US 28 finish on finish hardware is one of the cheapest to buy and readily available from most suppliers, compared with US 32, for example, which will be more expensive and a special order from some manufacturers.

Another item that will affect the cost and the delivery time will be the knob and rose trim-ring specified for latch and locksets. There are numerous styles. However, most suppliers only stock the most popular 3 or 4 patterns.

The specification writer will most likely refer to a specific Federal Specification Series Designation when specifying the type of lockset required for the project, and each of these series relates to the construction and design function of the lockset. For example:

ANSI* Series 1000 - Mortise type latchset or lockset and keyed deadbolt housed in one casing.

ANSI Series 2000 - Mortise type latchset or lockset with integral deadbolt operated by the key in the knob.

ANSI* Series 4000 - Cylindrical latch and locksets:

        Grade #1 - Heavy Duty

        Grade #2 - Medium Duty

        Grade #3 - Light Duty

* American National Standards Institute

The following list of finish hardware material prices should be used as a guide only. Always obtain a price quote from a material supplier for the specific type, function, finish, and style of finish hardware specified for the project.

DOORS AND WINDOWS

Material Prices For Finish Hardware	
Cylindrical Locksets	price range
Standard Duty	$25.00 - $125.00
Heavy Duty	$75.00 - $200.00
Mortise Locksets	
Heavy Duty	$140.00 - $225.00
Push-Pull Bars	$60.00 - $225.00
Anti-Panic Device, Single	$300.00 - $500.00
Hinges (per pair)	
Steel Plain	$20.00 - $30.00
Steel B.B.	$30.00 - $100.00
Bronze B.B.	$100.00 - $225.00
Door Closers Types:	
I - Impermeable to the passage of water	$250.00 - $1,000.00
II - Allowing Slight seepage	$175.00 - $500.00
III -Impermeable to the passage of waterborne contamination	$100.00 - $250.00
IV - Barriers to the passage of flood carried debris	$125.00 - $275.00
V- For spaces not meeting requirements of above	$25.00 - $100.00
Surface Bolt	$10.00 - $50.00
Kick Plate	
Aluminum	$15.00 - $30.00
Bronze	$20.00 - $50.00
Door Stops:	
Floor Bumper	$10.00 - $30.00
Wall Bumper	$10.00 - $20.00
Bumper Plus Holder	$25.00 - $50.00
Door Plunger	$20.00 - $35.00
Thresholds	
Aluminum	$20.00 - $35.00
Bronze	$35.00 - $100.00

The following table gives average production time, in carpenter hours, for the installation of finish hardware items:

Items of Finish Hardware	Hrs. to Install
Cylindrical Locksets (door prepared)	0.50 - 0.75
Mortise Locksets (door prepared)	0.70 - 1.00
Push-Pull Bars	0.40 - 0.60
Anti-Panic Device	1.50 - 2.00
Door Closers	0.75 - 8.00
Surface Bolts	0.30 - 0.40
Flush Bolts	1.00 - 1.25
Door Stops - Wall Type	0.30 - 0.40
Floor Type	0.50 - 0.60
Thresholds	0.50 - 0.60

08730 WEATHERSTRIPPING AND SEALS

The cost of metal weather strips varies with the type of window, whether double hung or casement, size of opening, thickness of sash, and the like. These factors all affect the material and labor costs. It will make considerable difference in the labor costs whether the weather strips are installed in an occupied or unoccupied building. Ordinarily a worker must use more care when working in an occupied building, and it requires ¼ to ½ hour more time per opening than when working in new or unoccupied buildings.

The following quantities and costs are based on experienced workers placing weather strips.

Metal Weather Strips for Double Hung Windows. When weatherstripping double hung windows, standard specifications require a track strip at least 3/4" (18.75 mm) wide at the top and bottom of the window, side strips for the lower sash not less than 1/16" (1.5 6mm) wider than the width of the sash runway, and side strips for the upper sash not less than 1/16" (1.56 mm) narrower than the width of the sash runway. Meeting rails to receive two interlocking hook strips or interlocking hook and flat strips.

When estimating the quantities of weather stripping required, take the distance around the window, for the sides, head, and sill strips, plus the width, to take care of the meeting rail between the top and bottom sash. A 4'-0" x 6'-0" (1200 x 1800 mm) double hung window will require 20 lin.ft. (6.09 m) of weather strips for the sides, head, and sill and 4 lin.ft. (1.21 m) of interlocking material for the meeting rails.

The lin.ft. (meter) cost of metal weather strip for double hung windows averages as follows in zinc:

Description	Sash Thickness					
	1 3/8 "	1 3/4 "	2 1/4 "	34.3 mm	43.7 mm	56.2 mm
Side, head and sill strips	$0.60	$0.75	$0.90	$1.97	$2.46	$2.95
Meeting rail strips	$1.97	$2.46	$2.95	$6.46	$8.07	$9.69

The weather strip for a 4'-0" x 6'-0" x 1-3/8" (1200 x 1800 x 34.37 mm) double hung window would cost 20 x $0.45 plus 4 x $1.00 = $13.00. Bronze will run about twice as much.

Labor Installing Metal Weather Strips for Double Hung Windows. When installing metal weather strips on 1-3/8" (34.37 mm) sash for windows up to 3'-0" x 5'-6" (900 x 1650 mm), an experienced carpenter should weatherstrip 1 window in 3/4 to 7/8 hr. or 9 to 10 windows per 8-hr. day, at the following labor cost per window:

DOORS AND WINDOWS

Description	Hours	Rate	Total	Rate	Total
Carpenter	0.90	$	$	$35.04	$31.54

When installing metal weather strips on 1-3/4" (43.75 mm) sash for windows up to 4'-0" x 7'-0" (1200 x 2100 mm), an experienced carpenter should complete one window in 7/8 to 1 hr. or 8 to 9 windows per 8-hr. day, at the following labor cost per window:

Description	Hours	Rate	Total	Rate	Total
Carpenter	1.00	$	$	$35.04	$35.04

If installed in an occupied building, add 1/4 hour per window

Installing Metal Weather Strips for Double Hung Windows Glazed with Plate Glass. When installing metal weather strips on 1-3/4" (43.75 mm) or 2-1/4" (56.25mm) sash, glazed with plate glass, for openings up to 4'-0" x 7'-0" (1200 x 2100 mm), a carpenter should weatherstrip one window in 1-1/8 to 1-1/4 hrs. or 6 to 7 windows per 8-hr. day, at the following labor cost per window:

Description	Hours	Rate	Total	Rate	Total
Carpenter	1.25	$	$	$35.04	$43.80

If installed in an occupied building, add 1/4 hour per window

Weatherstripping In-Swinging Casement Windows. Single or double casement windows that swing in require a 3/4" (18.75 mm) wide track strip on the hinged side, running in a groove in the sash. At the top, on the latch side, and where the sash meet (in the case of double casement windows) interlocking hook and flat strips are generally used. A trough strip and a hook strip are generally used at the bottom of the sash.

Material for jambs, head, and center stiles for in-swinging casement windows costs $0.60 per lin.ft. ($1.96 per m). Brass trough section and interlocking sillstrip, $1.70 per lin.ft. ($5.58 per m).

When estimating the number of lin.ft. of strip required for single casement windows, measure the entire distance around the window, but price the sill section separately.

For double casement windows, take the distance around the window, plus the height, which allows for the strip where the two sash meet. A 4'-0" x 6'-0" (1200 x 1800 mm) double casement window requires 22 lin.ft. (6.70 m) of weather strips for jambs, head, and center stiles and 4'-0" (1200 mm) of sill section.

Labor Weatherstripping In-Swinging Single Casement Windows. An experienced carpenter should weatherstrip one single

1213

casement window in 1 to 1-1/3 hrs. or 6 to 8 windows per 8-hr. day, depending on size, at the following labor cost per window:

Description	Hours	Rate	Total	Rate	Total
Carpenter	1.10	$	$	$35.04	$38.54

If installed in an occupied building, add 1/4 hour per window

Weatherstripping In-Swinging Double Casement Windows. When weatherstripping in-swinging double casement windows, an experienced carpenter should weatherstrip one window in 1-3/4 to 2 hrs. or 4 to 5 windows per 8-hr. day, depending on the size, at the following labor cost per window:

Description	Hours	Rate	Total	Rate	Total
Carpenter	2.00	$	$	$35.04	$70.08

If installed in an occupied building, add 1/4 hour per window

Labor Weatherstripping Out-Swinging Casement Windows. Out-swinging single casement windows should have -3/4" (18.75 mm) wide track strip on the hinged side, running in a groove in the sash. At the top, on the latch side, and where the sash meet in the case of double casement sash, interlocking hook and flat strips are required. A heavy sill section with interlocking hook strip is generally used at the bottom of the sash. The material for jambs, heads, and center stiles will cost about $0.60 per lin.ft. ($1.96 per m), while the sill section will cost about $1.40 per lin.ft. ($4.59 per m).

Labor Weatherstripping Out-Swinging Single Casement Sash. An experienced carpenter should place weather strips on one single casement sash in 3/4 to 1 hr. or 8 to 10 windows per 8-hr. day, at the following labor cost per window:

Description	Hours	Rate	Total	Rate	Total
Carpenter	1.00	$	$	$35.04	$35.04

If installed in an occupied building, add 1/4 hour per window

Labor Weatherstripping Out-Swinging Double Casement Sash. An experienced carpenter should weatherstrip one double casement window in 1-1/4 to 1-1/2 hrs. or 5 to 6 windows per 8-hr. day, at the following labor cost per window:

1214

Description	Hours	Rate	Total	Rate	Total
Carpenter	1.40	$	$	$35.04	$49.06

If installed in an occupied building, add 1/4 hour per window

Weather Strips for Single Wood Doors. Consisting of spring bronze strips for jambs and heads and 1-1/2" (37.50 mm) interlocking brass threshold for sill.

Door Size Up To	Thick	Price	Door Size Up To	Thick
2'- 8" x 6'- 8"	1-3/8"	$12.30	66.75 x 166.75 mm	34.38 mm
3'- 0" x 7'- 0"	1-3/8"	$13.10	75.00 x 175.00 mm	34.38 mm
2'- 8" x 6'- 8"	1-3/4"	$13.20	66.75 x 166.75 mm	43.75 mm
3'- 0" x 7'- 0"	1-3/4"	$14.10	75.00 x 175.00 mm	43.75 mm

If aluminum interlocking thresholds are used instead of brass, deduct $2.50 for 2'-8" (800 mm) openings and $2.80 for 3'-0" x 7'-0" (900 x 2100 mm) openings.

If interlocking bronze is used instead of spring bronze, add $10.00 per door to the above prices.

Extruded aluminum and felt or neoprene sponge 1-1/8" (28.12 mm) wide and 1/4" (6.25 mm) thick to be set on stops at heads and jambs runs $1.55 per lin.ft. ($5.08 per m). Extruded aluminum with neoprene gasket 1-1/4" (31.25 mm) wide by 25/32" (19.53 mm) thick to be set on stops runs $3.10 per lin.ft. ($10.17 per m). A similar gasket for mounting on bottom rail of door costs $1.70 per lin.ft. ($5.58 per m).

A complete extruded aluminum channel with integral drip and vinyl insert at still costs $1.40 per lin.ft. ($4.59 per m). A 6" (12.50 mm) wide extruded aluminum thresholds for use with above rail strips costs $3.15 per lin.ft. ($10.33 per m). A complete extruded aluminum threshold 3-3/4" (93.75 mm) wide by 3/4" (18.75 mm) thick with integral latch track, interlock and hook stop for mounting on outside of door costs $4.20 per lin.ft. ($13.78 per m) complete. A two-piece extruded aluminum with neoprene gasket astragal for surface mounting on double doors runs $3.65 per lin.ft. ($11.97 per m).

A one-piece extruded aluminum and neoprene gasket designed for recessing in meeting rail will cost $2.00 per lin.ft. ($6.56 per m). Adjustable two-piece astragals with pile will run $4.90 per lin.ft. ($16.07 per m) per each door leaf in extruded aluminum, $5.60 per lin.ft. ($18.37 per m) in dull bronze, and $6.75 per lin.ft. ($22.14 per m) in polished bronze or dull chrome.

Labor Weatherstripping Single Wood Doors. An experienced carpenter should weatherstrip one 3'-0" x 7'-0" x 1-3/4" (900 x 2100 x 43.75

mm) door in 1 hour or about 8 doors per 8-hr. day, at the following labor cost per door:

Description	Hours	Rate	Total	Rate	Total
Carpenter	1.00	$	$	$35.04	$35.04

Where interlocking weather strip is used for head and jambs of doors instead of spring bronze, add 1/2 to 3/4 hr. carpenter time to the time given above.

Metal Weatherstripping for French Doors. The material for weatherstripping double doors or French doors consists of spring bronze strips for jambs and head, spring bronze interlocking strip, and threshold for sill. For doors up to 5'-0" x 7'-0" x 1-3/8" (1500 x 2100 x 34.37 mm), the material should cost about $18.50; for a 5'-0" x 7'-0" x 1-3/4" (1500 x 2100 x 43.75 mm) door, about $20.00.

Labor Weatherstripping French Doors. An experienced carpenter should weatherstrip one pair of 1-3/4" (44 mm) French doors in 2 hrs. or 3 to 4 pair per 8-hr. day, at the following labor cost per pair:

Description	Hours	Rate	Total	Rate	Total
Carpenter	2.00	$	$	$35.04	$70.08

Weatherstripping Double Doors. Where double exterior doors are used for openings up to 5'-0" x 7'-0" (1500 x 2100 mm), using spring bronze weather strip for the head, jambs, and between doors, with an interlocking threshold at the bottom of the door, figure 3 to 4 hrs. carpenter time per opening, as follows:

Description	Hours	Rate	Total	Rate	Total
Carpenter	2.00	$	$	$35.04	$70.08

If interlocking weather strip is used for the head, jambs, and meeting rail, add 1 hour per opening.

Metal Thresholds

Price per lin.ft. (m) or fraction thereof. Drilled and countersunk, with screws, ready for installation.

DOORS AND WINDOWS

Metal	Size Inches	Price per Lin. Ft.	Size mm	Price per Lin.M
Brass	1 1/2 " x 1/4 "	$15.00	37.50 x 6.25 mm	$49.00
Brass	3 1/2 " x 3/4 "	$20.00	87.50 x 18.75 mm	$66.00
Brass	4 1/2 " x 3/4 "	$38.00	112.50 x 18.75 mm	$125.00
Brass	5 " x 7/8 "	$45.00	125.00 x 21.88 mm	$148.00
Aluminum	1 1/2 " x 1/4 "	$5.00	37.50 x 6.25 mm	$16.00
Aluminum	3 1/2 " x 3/4 "	$9.00	87.50 x 18.75 mm	$30.00
Aluminum	4 1/2 " x 3/4 "	$12.00	112.50 x 18.75 mm	$39.00
Aluminum	5 " x 7/8 "	$15.00	125.00 x 21.88 mm	$49.00

All the above thresholds are furnished with interlocking Strip

08800 GLAZING

During the late 1970's, the glass manufacturers changed the process for making certain types of flat glass. Before that time, vertically drawn sheet glass was manufactured in various qualities—AA, A, and B—and thicknesses—S.S., D.S., 3/16" (4.68 mm) and 7/32" (5.46 mm). The 3/16" (4.68 mm) and 7/32" (5.46 mm) glass was known as crystal sheet. Plate glass was made by grinding and polishing both surfaces to a level, parallel plane, thereby removing any distortion lines from the glass.

The new process of making glass consists of floating molten glass over large pools of molten tin. Through automated production stages, the floated glass is fire polished to remove all distortion, allowed to cool, and fed onto the tables in a continuous ribbon where it is cut to sheet size for packing into shipping crates. The glass is classified as "float glass" and is made in glazing quality and mirror quality of various thicknesses.

				Clear Float Glass - Glazing Quality				
Product	Thick. Inches	Wt. in lb. per Sq.Ft	Standard Maximum Size	Material Cost per Sq.Ft.	Thick. Mm	Wt. in Kg. per Sq.M	Standard Maximum Size	Material Cost per Sq.M
Float	3/32 "	1.22	40 "x 100 "	$3.50	2.34 mm	5.96	1000 x 2500 mm	$37.67
	1/8 "	1.62	80 "x 120 "	$4.00	3.13 mm	7.91	2000 x 3000 mm	$43.06
	3/16 "	2.43	120 "x 212 "	$4.25 *	4.69 mm	11.86	3000 x 5300 mm	$45.75 *
Float/Plate	1/4 "	3.24	130 "x 212 "	$4.75 *	6.25 mm	15.82	3250 x 5300 mm	$51.13 *
	3/8 "	4.92	124 "x 204 "	$7.50 *	9.38 mm	24.02	3100 x 5100 mm	$80.73 *
	1/2 "	6.56	124 "x 204 "	$15.00	12.50 mm	32.03	3100 x 5100 mm	$161.46
	3/4 "	9.85	124 "x 204 "	$20.00	18.75 mm	48.09	3100 x 5100 mm	$215.29

*Note: Add $2.00 per sq.ft. for gray or bronze tint.

Float glass is also available in glare and heat reducing tints of gray and bronze, heat absorbing (blue-green), tempered, coated with thin metallic coverings for reflective qualities, laminated, and in insulating glass units consisting of two sheets of glass separated by an internal air space.

Tempered Glass. Tempered glass is a heat treated annealed glass that has high mechanical strength. It is 4 to 5 times as strong as annealed glass of the same thickness. When it is broken, it disrupts into innumerable small fragments of a cubical shape. In recent years, tempered glass has become much more reasonable in cost and should be given more consideration in general construction. However, the nature of its manufacture does not allow cutting or drilling on the job and all fabricating must be done prior to heat treatment. It is available in clear, tinted, and heat absorbing. One-quarter inch (6.25 mm) thick units are available in sizes to 72" x 120" (1800 x 3000 mm). Because this glass is all custom ordered, costs should be checked locally, but tempering will add around $2.80 a sq.ft. ($30.14 per sq.m) premium to base glass prices.

Laminated Glass. Laminated or safety glass is composed of two or more lights of glass with a layer or layers of tough, transparent vinyl plastic sandwiched between glass under heat and pressure to form a single unit. Laminated glass will crack but almost always will hold together. It is available in a wide range of thicknesses and sizes from 7/32" (5.46 mm) thick to 1" (25 mm) thick in sheets up to 80" x 120" (2000 x 3000 mm). One-quarter inch (6.25 mm) laminated glass will cost about $6.00 per sq.ft. ($64.58 per sq.m), 1/2" (12.50 mm) will cost about $14.00 per sq.ft. ($150.60 per sq.m).

Insulating Glass. Glass units made from two sheets of glass with a sealed air space between are used to decrease heat loss through windows. The seal can be either metal or glass, and air space will vary from 3/16" (4.68 mm) to 1/2" (12.50 mm) in thickness.

Insulating glass units must be made to order in the exact size to be used, although standard sizes are manufactured.

They require a slightly larger allowance for clearance in setting them into the rabbet than single lights of glass.

Each piece weighs about 2-1/2 times what a single lite of the same thickness would weigh and the glazing cost is about double that of single lights.

The price per square foot varies widely, depending upon kind of glass, size, etc., so it is always advisable to obtain definite prices on the sizes required.

For quick estimates two sheets of 1/8" (3.12 mm) for 1/2" (12.50 mm) thick unit will cost $5.00 per sq.ft. ($53.82 per sq.m).

Two sheets of 1/4" (6.25 mm) for 1" (25 mm) thick unit will cost about $6.50 per sq.ft. ($69.97 per sq.m).

For tinted glass one side, add 15%.

Spandrel Glass. Heat-strengthened glass is used for spandrels and other opaque panels of curtain walls. The glass is coated on the back with a ceramic frit to give opacity and color.

1218

It is manufactured in several standard colors, plus black and white, but can be ordered in almost any color desired.

Spandrel glass is factory cut to size. Job cutting is not possible due to the heat-strengthening process in manufacturing. It is usually 1/4" (6.25 mm) in thickness but may be ordered up to 3/4" (18.75 mm) thick. Sealing of spandrel glass is usually accomplished by the use of liquid polysulfide sealants and neoprene gaskets. Material prices of 1/4" (6.25 mm) spandrel glass are about $5.60 per sq.ft. ($60.28 per sq.m) for standard colors, and about $6.50 per sq.ft. ($69.97 per sq.m) for non-standard colors.

Wired Glass. Wire glass is a protective type glass for use in fire doors and windows, skylights and other places where breakage is a problem. It is available in 1/4" (6.25 mm) thicknesses in sheets 60" x 144" (1500 x 3600 mm) maximum size in hexagonal, square, diamond, and pin stripe patterns, and comes in clear, hammered, and many other patterned glass. Patterned wire glass will cost around $5.25 per sq.ft. ($56.51 per sq.m). Clear wire glass will cost $6.50 per sq.ft. ($69.97 per sq.m) in diamond pattern, $9.45 sq.ft. ($101.72 per sq.m) in pin stripe.

Patterned Glass. Glass can be rolled or figured for decoration and light diffusion. Patterns vary among manufacturers but usually include sandblasted, hammered, ribbed, fluted, pebbled, and rough hammered designs. It is usually available in two thicknesses: 1/8" (3.12 mm) in 48" x 132" (1200 x 3300 mm) maximum size and 7/32" (5.46 mm) in 60" x 132" (1500 x 3300 mm) maximum size.

Some patterns are illustrated on accompanying pages. Cost of material is about $2.35 per sq.ft. ($25.29 per sq.m) for 1/8" (3.12 mm) thicknesses, $2.10 sq.ft. ($22.60 per sq.m) for 7/32" (5.46 mm).

Mirrors. Mirrors are manufactured by taking a sheet of "mirror quality" float glass, in 1/8" (3.12 mm) or 1/4" (6.25 mm) thickness, and hermetically sealing a silver coating with a film of electrolytic copper plating onto one side of the glass. The coating is then given a protective paint coat to seal out moisture and reduce the potential of damaging the silver coating.

Average material prices for mirrors will be about $5.00 per sq.ft. ($53.82 per sq.m).

Any edge work associated with mirrors will cost extra: plain polished edges at $0.35 per inch ($0.01 per mm), bevelled edges at $1.10 per inch ($0.04 per mm).

Estimating The Quantity Of Glass Required. When measuring and listing the quantities of glass for a project, list the glass by:

1. Type of glass and thickness.
2. Size, measuring to the next even inch dimension, listing the width first and then the height.
3. Type of frame the glass is placed in.
4. Note if required to glaze from the inside or outside of the building. If from the outside, ladders or scaffolding may be necessary.

5. Height from the ground, in multi-story buildings, for additional labor requirements of distributing the materials.
6. Be sure not to list any glass for products that will be delivered to the job pre-glazed. It is common practice for sliding door units, aluminum sliding and single or double hung sash, and some wood sash to be glazed at the factory.

Labor Glazing Window Glass. The labor cost of glazing will vary with the size and kind of glass, and whether set in putty or glazing compound, a Thiokol or Silicone sealant, or with wood or metal stops. Putty is now seldom recommended for glazing except on a very small light under 50 united inches (1250 mm), which is length plus width.

Labor Setting Window or Plate Glass in Wood Sash, Using Putty or Glazing Compound			
Approx. Size of Glass Inches	Number Lights Set 8-Hr. Day	Glazier Hours per 100 Lights	Approx. Size of Glass mm
12 "x 14 "	65	12.5	300 x 350 mm
20 "x 28 "	40	20.0	500 x 700 mm
30 "x 40 "	30	26.7	750 x 1000 mm
40 "x 48 "	20	40.0	1000 x 1200 mm

Labor Setting Window or Plate Glass Using Wood Stops			
Approx. Size of Glass Inches	Number Lights Set 8-Hr. Day	Glazier Hours per 100 Lights	Approx. Size of Glass mm
12 "x 14 "	40	20.0	300 x 350 mm
20 "x 28 "	20	40.0	500 x 700 mm
30 "x 40 "	15	53.3	750 x 1000 mm
40 "x 48 "	12	66.7	1000 x 1200 mm

Putty Required Setting Glass in Wood Sash. Window glass set in 1-3/8" (34.37 mm) wood sash requires 1 lb. (0.45 kg) of putty for each 8 to 8-1/2 lin.ft. (2.43 to 2.59 m) of glass or sash rabbet. When set in 1-3/4" (43.75 mm) sash, it requires 1 lb. (0.45 kg) of putty to each 7-1/4 to 7-1/2 lin.ft. (2.21 to 2.28 m) of glass or sash rabbet.

The following table gives the quantity of putty in lbs. (kg) required for glazing various size glass:

1220

Sash	Size of Glass Inches				
	12 "x 14 "	14 "x 20 "	20 "x 28 "	30 "x 36 "	40 "x 48 "
1 3/8 "	1/2 "	2/3 "	1 "	1 3/8 "	1 3/4 "
1 3/4 "	3/5 "	3/4 "	1 1/8 "	1 1/2 "	2 "

Sash	Size of Glass mm				
	300 x 350 mm	350 x 500 mm	500 x 700 mm	750 x 900 mm	1000 x 1200 mm
34.3 mm	12.50 mm	16.67 mm	25.00 mm	34.38 mm	43.75 mm
43.7 mm	15.00 mm	18.75 mm	28.13 mm	37.50 mm	50.00 mm

A good grade of wood sash putty will cost about $0.70 per lb. ($1.54per kg).

Labor Setting Glass in Metal Doors. When glazing metal doors and enclosures, the stops are usually steel and are fastened to the sash with small screws, and a glazier must remove all the screws from the stops, place the glass, and then replace the screws for each small light of glass. On work of this kind a glazier should set about 15 lights per 8-hr. day, at the following labor cost:

Description	Hours	Rate	Total	Rate	Total
Glazer	8.00	$	$	$ 32.50	$ 260.00
Cost per light					$ 17.33

Labor Glazing Steel Sash. The labor cost of glazing steel sash will vary with the size of the glass, amount of putty required per light, and the class of work; i.e., whether sidewall sash, monitor sash, etc., and also whether it is necessary to hoist the glass and erect special scaffolding to set it.

The time of year in which the work is performed will also influence the labor costs, as a glazier can set more glass in warm weather than during the winter months. The labor costs given below are based on summer glazing. For winter work add up to 25% to costs given.

Putty Required for Glazing Steel Sash. Putty used for glazing steel sash is a special mixture, part of which is red or white lead. One lb. (0.45 kg) should be sufficient to glaze 5 to 5-1/2 lin.ft. (1.52 to 1.67 m) of sash rabbet. For instance, a 16" x 20" (400 x 500 mm) light would contain 72" (1800 mm) or 6'-0" (1.82 m) to be puttied.

Elastic glazing compound is displacing the regular steel sash putty to a great extent for use in steel sash and must be used in aluminum sash. This compound costs about $0.85 per lb. ($1.87 per kg). Sufficient putty to glaze 100 lin.ft. (30.48 m) of steel sash should cost as follows:

Description	Lbs.	Rate	Total	Rate	Total
Putty	20.00	$	$	$0.75	$ 15.00
Cost per lin.ft.					$ 0.15
per lin.m					$ 0.49

1221

Description	Lbs.	Rate	Total	Rate	Total
Putty-elastic	20.00	$	$	$0.85	$ 17.00
Cost per lin.ft.					$ 0.17
per lin.m					$ 0.56

Labor Glazing Steel Sash. On jobs where there are no unusual window heights or other abnormal conditions, a glazier should set the following number of lights of glass in the various types of steel sash per 8-hr. day.

In clerestory, monitor, or outside setting (from scaffold or swing stage), it will require one worker on the ground keeping two workers on a scaffold supplied with glass and putty. This adds approximately 50% to the costs given below. Add up to 25% to the costs for winter work.

Glass Size Inches	No. of Lights per 8-Hr day	Glazier Hrs. per 100 Lts	Lbs. Putty per Light	Glass Size mm	Kg. Putty per Light
Pivoted Steel or Commercial Projected Steel Windows*					
22 "x 16 "	50	26.7	1.25	550 x 400 mm	0.57
32 "x 16 "	40	26.7	1.50	800 x 400 mm	0.68
32 "x 22 "	30	32.0	1.67	800 x 550 mm	0.76
Housing and Residence Steel Casements**					
9 "x 12 "	60	20.0	0.75	225 x 300 mm	0.34
16 "x 12 "	60	23.0	1.00	400 x 300 mm	0.45
18 "x 12 "	60	23.0	1.00	450 x 300 mm	0.45
Architectural Projected Steel Windows**					
18 "x 16 "	30	23.0	1.25	450 x 400 mm	0.57
20 "x 16 "	30	23.0	1.25	500 x 400 mm	0.57
38 "x 16 "	25	40.0	1.75	950 x 400 mm	0.79
40 "x 16 "	25	50.0	2.00	1000 x 400 mm	0.91
46 "x 16 "	25	50.0	2.00	1150 x 400 mm	0.91
48 "x 16 "	25	50.0	2.25	1200 x 400 mm	1.02

*Based on setting glass from inside and working from floor.
**Based on setting glass from outside, using glazing compound or steel
sash putty.

Labor Setting Plate Glass in Store Fronts. The labor cost of handling and setting plate glass in store fronts varies with factors such as the size of the glass and the length of the haul.

The average size plate glass used in modern store fronts requires 4 to 6 workers to handle and set each light of glass. No matter how many lights of glass there are in the job, it will be necessary to allow time for all workers

1222

DOORS AND WINDOWS

going to the job and returning to the warehouse after completion, and time for the truck hauling glass from the warehouse to the job and return.

Unless the job is a considerable distance from the warehouse, an allowance of 1 hour for each worker going to the job and returning to the warehouse should prove sufficient.

It is customary for the truck to remain on the job until the glass is set and return to the warehouse with the workers, so truck time should be figured on this basis.

Number of Workers Required for Various Sizes of Plate Glass in Store Fronts				
From	To	No. Workers	From	To
100 "	140 "	2.0	2500 mm	3500 mm
140 "	160 "	3.0	3500 mm	4000 mm
160 "	180 "	4.0	4000 mm	4500 mm
180 "	210 "	5.0	4500 mm	5250 mm
210 "	250 "	6.0	5250 mm	6250 mm
250 "	300 "	8.0	6250 mm	7500 mm

On sizes larger than the above, some thought should be given to a mechanical means of hoisting the glass, in addition to the labor requirement.

It is further required that when glass is set 6' (1.82 m) or more over street level, when weather conditions require it, or when insulated glass, such as Thermopane (a brand name for a window-glass construction that has insulating qualities; the insulating qualities are two layers of glass separated by an air space, sometimes also referred to as double-glazing insulating glass), is installed using additional labor as indicated below:

First size enumerated in the schedule, add. workers 1
Next two sizes enumerated in the schedule, add. workers 2
Next three sizes enumerated in the schedule, add. workers 3

LEADED AND ART GLASS

When estimating leaded or art glass, odd or fractional parts of inches are charged as even inches of the next larger size; for example, a 12-1/4" x 13-1/4" (306.25 x 331.25 mm) lite will be charged as 14" x 14" (350 x 350 mm).

Circles, ovals, or irregular patterns will be charged at the rate of a square or rectangle which will contain them.

The following prices are based on leading, using zinc cames not over 3/8" (9.37 mm) wide:

Description	Price per Sq.Ft.	Price per Sq.M.
Rectangles and squares, Clear sheet glass	$9.20	$99.03
Rectangles and squares, Cathedral glass, white or colored	$9.60 - $40.00	$103.34 - $430.57
Rectangles or squares, Plate glass	$12.80	$137.78
Diamonds, Clear sheet glass	$9.60	$103.34
Diamonds, Cathedral glass, white or colored	$11.20 - $56.00	$120.56 - $602.80
Diamonds, Plate glass	$16.00	$172.23

Colored Art Glass Leaded. Leaded art glass furnished in colors are offered in many designs at $8.40 to $35.00 per sq.ft. ($90.41 to 376.75 per sq.m), which includes only straight work.

Landscape and water scenes in colored art glass will cost from $28.00 to $56.00 per sq.ft. ($301.40 to $302.80 per sq.m).

Very elaborate designs of art glass in colors with flowers, fruits, etc., can be purchased from $70.00 up per sq.ft. ($753.50 per sq.m).

The architect usually specifies an allowance of so much per sq.ft. (sq.m) to cover the cost of all art glass in the building.

08900 WINDOW WALLS/CURTAIN WALLS

INSULATED METAL WALL PANELS

Insulated steel, aluminum, and aluminum-steel wall panels were developed to provide the answer to dry wall construction, enabling the builder to enclose large areas in a minimum of time. They are designed to serve as exterior walls and partitions.

One typical wall panel consists of two members pressed together at the side laps to form a structural unit. This type has a strip of felt inserted the full length of the joint to prevent metal-to-metal contact between the two members. These panels are factory filled with borosilicate glass fiber insulation, 2-1/2 lb. (1.13-kg) density. An end closure is attached to both ends of the panel to insure the proper retention of insulation and to seal the panels. Panels are manufactured in standard widths of 12" (300 mm), 16" (400 mm), 24" (600 mm), or 32" (800 mm) and thicknesses vary from 1-9/16" (39.06 mm) to 3-3/8" (84.37 mm), depending on whether surfaces are flat or fluted and thickness of insulation used.

Lengths vary from 6'-0" (1.83 m) to 30'-0" (9.14 m). Maximum allowable clear span for an all-galvanized steel panel, based on a wind load of 20 lbs. per sq.ft. (97.64 kg per sq.m), is 14'-0" (4.26 m); for an all aluminum panel, 10'-6" (3.20 m); and for an aluminum-galvanized steel panel, 12'-0" (3.65 m).

The interlocking tongue and groove between adjacent panels offers a joint that unites the series of units into a continuous wall surface. When using the interlocking tongue and groove connection the panels can be erected either vertically or horizontally in walls and partitions. This double-acting

1224

tongue and groove joint has three positive bearing surfaces that permit a series of panels to act as a structural unit in resisting loads from either side of a wall. With panels in the horizontal position, the double tongue and groove joint forms a shiplap construction, adding to the weather-resisting properties of the complete wall unit. Where panels are used as semi-permanent constructions enclosing the face of a building to which a future extension is contemplated, details can allow for 100% salvaging of the panels.

Steel Wall Panels. Steel type panels are manufactured of 18 to 26 gauge sheet steel, with either a baked-on coat of shop primer or galvanized finish. The all-steel panel has the advantage of high structural strength and low initial cost. The panels are caulked during erection by buttering a mastic in the female joints prior to putting the panels in place. Welding of panels to supporting steel is the most common method of attachment. The material cost of shop assembled steel wall panels is approximately $3.50 per sq.ft. ($37.67 per sq.m). For two sheets of 26 ga. metal and 1" (25 mm) insulation to $3.90 per sq.ft. ($41.98 per sq.m) for two sheets of 22 ga. If the outer sheet is to have a porcelain enamel finish, add $1.75 per sq.ft. ($18.83 per sq.m).

Aluminum Wall Panels. The extreme light weight of the long length aluminum panels allows easier and faster handling and erecting in the field. A typical aluminum panel, 12'-0" (3.65 m) long, 24" (600 mm) wide, weighs only 60 lbs. (27.21 kg). The erection procedure is the same as for steel panels except that direct contact between the aluminum and steel must be prevented. A coating of bituminous paint is recommended for this purpose. Maintenance costs become negligible with an all-aluminum construction. The original mill finish will eventually darken with weathering. Bolting of panels to supporting steel and use of pre-formed sections are the most common methods of attachments. Aluminum wall panels cost about $5.00 per sq.ft. ($53.82 per sq.m).

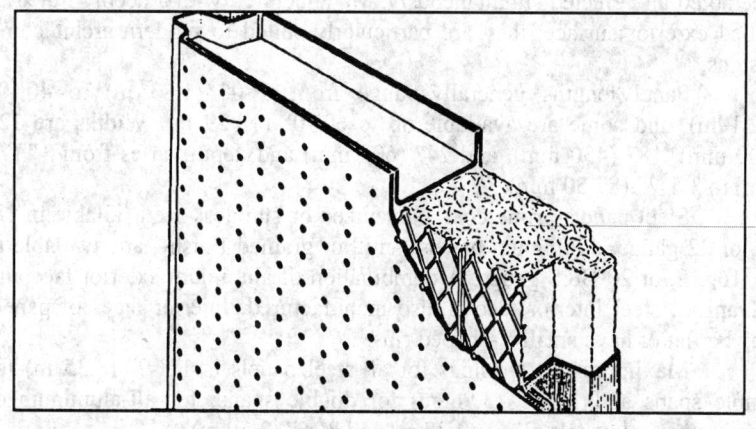

Acoustically Treated Wall Panel

Aluminum-Steel Wall Panels. To supply the need for an economical panel that would have an aluminum exterior surface for high resistance to corrosion and weather but did not require this type surface on the interior, the aluminum-steel panel was developed.

For particular use in conjunction with aluminum-steel panels and all aluminum panels, extrusions have been developed which enable the panel to be placed in position and held through the clamping action of the extrusion at the base of the panel. These extrusions eliminate bolting and accelerate erection. They are also 100% salvageable when placed in buildings that are to be expanded later. At the head of the panels, there is usually no extrusion. Rather, a toggle bolt is used to fasten at this position. The material cost of aluminum-steel wall panels is about $5.00 per sq.ft. ($53.82 per sq.m), depending on the gauge of metals used.

Labor Erecting Metal Wall Panels. A normal wall panel job with panel lengths ranging between 6' (1.82 m) and 8' (2.43 m) should be constructed at the rate of 12 manhours per 100 sq.ft. (9.29 sq.m), which includes handling all work from unloading to actual fastening of the panels in place. A difficult job, involving the use of toggle bolts, aluminum panels, short runs, either long or short panels, will take somewhat longer—up to 20 manhours per 100 sq.ft. (9.29 sq.m). A typical crew would consist of one worker to caulk and hoist, a worker to receive and distribute the panel, a worker to place and weld the bottom, one to plumb and weld the top, and a worker to do the cutting, attach closures, and place flashings and bitumastic. The erection cost for a normal job is about $1.75 per sq.ft. ($18.83 per sq.m) and $3.15 per sq.ft. ($33.90 per sq.m) for a difficult job. On large jobs with no openings, an erection cost of $1.15 per sq.ft. ($12.38 per sq.m) can be achieved, with 180 to 190 sq.ft. (16.72 to 17.65 sq.m) erected per worker per 8-hour day.

Field Assembled Wall Panels. Steel, aluminum, and combination aluminum—steel wall panels are also available for field assembly—to be assembled and erected simultaneously. Manufactured with a deeply fluted or ribbed exterior surface, they are particularly suited to modern architectural designs.

Panel lengths generally range from 6'-0" (1.83 m) to 40'-0" (12.19m), and some are available up to 60'-0" (18.29 m). Widths are 12" (300 mm), 18" (450 mm), and 24" (600 mm) and depth varies from 3" (75 mm) to 3-1/2" (87.50 mm).

Steel panels-prime coat, galvanized or stainless-are available in 18, 20, or 22 gauge. Aluminum panels—mill or grained finish—are available in 14, 16, 18, or 20 B&S gauge. A combination of aluminum exterior face and galvanized steel interior face is also manufactured. Interior faces of panels may be flat or have small V-shaped ribs.

Maximum girt spacings for all-steel panels is 11'-0" (3.35 m) for simple spans and 13'-0" (3.96 m) for double spans; for all-aluminum or aluminum-steel panels, 7'-6" (2.28 m) and 8'-6" (2.59 m).

Material and erection costs for field assembled wall panels total practically the same as for shop assembled panels (given on previous pages) of similar materials.

Movable Fire Partitions. Movable fire partitions, long needed in industrial plants, are available for installation where permanent fire walls are not desired. One such partition has a 2-1/2 hour fire rating.

These lightweight partitions are formed of two 18 gauge galvanized steel, 1-1/2" (37.50 mm) deep fluted sheets of the same type as used for exterior face of field assembled wall panels. One 1/2" (12.50 mm) lamination of gypsum board, containing glass fibers and unexpanded vermiculite, is inserted between the sheets and the build-up is held together by bolts.

Depth of the fire partition is 5" (125 mm) and coverage width per section is 24" (600 mm). Length of section varies with ceiling to floor dimension.

Fire partitions are lightweight and easily erected. Portable attachment to structural steel is by bolting; permanent attachment is by welding.

Cost in place of movable fire partitions is approximately $5.75 per sq.ft. ($61.89 per sq.m).

Curtain Wall Construction. Buildings of all types are now using the curtain wall method of exterior wall construction, in preference to the conventional masonry, because they are lightweight and permit savings in structural framing and foundations.

Units are thin—varying from 1-1/4" (31.25 mm) to 4-1/2" (112.50 mm)—increasing ratio of usable floor area to total building area.

They provide good insulation with the use of efficient insulating filler materials and smooth reflective exterior and interior surfaces.

The large prefabricated and shop assembled panels, containing window units, spandrels, mullions, etc., are erected rapidly, reducing labor costs and permitting earlier completion and occupancy of building.

At the present time, curtain walls must be designed and fabricated to order to fit a particular job. Each job is custom built and costs vary considerably depending on design, materials used, size of units, amount of duplication, type of windows, and secondary framing requirements.

Materials commonly used for curtain wall panels are painted, galvanized or stainless steel, aluminum and porcelain enameled steel, or aluminum. Various combinations of these materials are also used.

Various types of windows are used, including intermediate projected, casement or combination types, awning type, reversible windows, and others.

Panel sizes vary considerably, but most economical designs for multiple story structures have a coverage width of about 4'-0" (1.22 m)and a story to story height of 10'-0" (3.04 m) to 12'-0" (3.65 m). For 2 and 3 story structures, full height panels have been used successfully.

The following average curtain wall costs are given for preliminary estimating purposes only. Always obtain firm figures for curtain walls before submitting bid, unless an allowance has been set up for this work.

Description	Cost per Sq.Ft.	Cost per Sq.M.
1-1/2" (37.5 mm) Steel frames and sash, insulated Panels	$12.00 - $23.00	$129.17 - $247.58
Same in aluminum	$16.00 - $29.00	$172.23 - $312.16
Same in stainless steel	$26.00 - $40.00	$279.87 - $430.57
Same in bronze	$35.00 - $58.00	$376.75 - $624.33

The curtain wall can be priced out after four elements are agreed on: the framing system, the window system, the spandrel material, and the type of glazing.

09   FINISHES

09000 EXTERIOR INSULATION AND FINISH SYSTEM (EIFS)

Exterior Insulation and Finish System is a multilayered exterior wall system that is used both for residential and commercial structures of various sizes and shapes. This system is usually referred to by it acronym, EIFS, on drawings and specifications.

Systems vary among the manufacturers but typically consist of the following components:

1. Insulation board (the thickness can vary), normally made of polystyrene or polyisocyanurate foam. Insulation material is secured to the exterior wall surface with an adhesive approved by the EIFS manufacturer.
2. A tough, durable water-resistant base coat is applied on top of the insulation.
3. The water-resistant base coat is reinforced with a fabric wire mesh of fiber glass for added strength.
4. A finish coat is applied. The finish coat can be an acrylic copolymer or other product that is colorfast and crack-resistant.

The greatest advantages of EIFS material is that it is energy-efficient and offers multifarious applications. Most systems use insulation board with an R-value of R4 to R5 per inch. EIFS can conform to various shapes and can be made to resemble stucco, stone, or other types of expensive finish at a fraction of the cost. EIFS is produced in an almost unlimited selection of colors, finishes and textures.

For warranty purposes the installation of EIFS requires experienced installers who are approved by the manufacturer. Mixing coatings and applying them to the correct thickness is highly specialized work, and certified installers are also necessary to avoid adverse conditions-applying materials while surfaces are wet or just before rain, failing to end a day's work at a joint, not using approved applicators and materials, not caulking properly around openings, flashing and the like.

The following check list of items from a manufacturer's installer will ensure a quality product:

1. Knowledgeable and experienced installers who have been certified by the EIFS manufacturer.
2. Verification from the manufacturer that based on the plans and specifications being used its system will will provide a watertight structure.
3. Validation that all supplied components of the EIFS are approved by the manufacturer.
4. EIFS manufacturer and its approved contractor are aware of any unusual details and conditions.
5. Validation that all materials will come with proper labeling attached for ease of identification.
6. EIFS manufacturer and contractor are aware of and approve the schedule, taking into consideration weather conditions.
7. EIFS manufacturer must examine and approve substrate; if applied by others, substrate must meet the manufacturer's tolerances.
8. EIFS manufacturer must periodically inspect the work while it is being undertaken by the EIFS contractor.
9. At the conclusion of the project, manufacturer must verify that all items have been properly installed and all openings have been caulked correctly, with proper materials around windows, doors, flashing, and penetrations.

The following cost tables show material quantities and labor required for a typical Exterior Insulation and Finish System onto an existing substrate (e.g., plywood). The tables assume a 30,000 sq. ft., (2,787 sq.m.) maximum two-story structure and an R-value of 4.2. For exact material quantities on a job, you would need to check with the manufacturer.

Item 1 - Insulation Board					
Insulation Board for 1,000 Sq. Ft					
Material Costs	Qty	Unit	Rate	Unit	Total
Board (plus 10% waste)	1,100.00	Sq. Ft.	$ 0.90	Sq. Ft.	$990.00
Primer	6.00	Pails	$ 140.00	Pail	$840.00
Cement	3.00	Bags	$ 20.00	Bag	$60.00
Total Material Cost					$1,890.00
per sq.ft.					$1.89
per sq.m					$20.34

FINISHES

Item 1 - Insulation Board (cont'd)					
Installation Costs	Hours	Rate	Total	Rate	Total
Carpenter	16.00	-----	-----	$35.04	$560.64
per sq.ft.					$0.56
per sq.m					$6.03
Total Labor and Material Cost Insulation					$2,450.64
Total per sq. ft..					$2.45
Total per sq. m					$26.38

Item 2 - Fiberglass Reinforcing Mesh					
Material Costs	Qty	Unit	Rate	Unit	Total
Mesh (plus 10% waste - 150+ sq.ft. of Roll)	5.00	Rolls.	$56.00	Roll.	$280.00
Water resistant base coat	6.00	Pails	$117.60	Pail	$705.60
Total Material Cost					$985.60
per sq.ft.					$0.99
per sq.m					$10.61
Installation Costs	Hours	Rate	Total	Rate	Total
Carpenter	16.00	-----	-----	$35.04	$560.64
Labor	16.00	-----	-----	$26.16	$418.56
Total Labor Cost					$979.20
per sq.ft.					$0.98
per sq.m					$10.54
Total Labor and Material Cost Reinforcing Mesh					$1,964.80
Total per sq. ft..					$1.96
Total per sq. m					$21.15

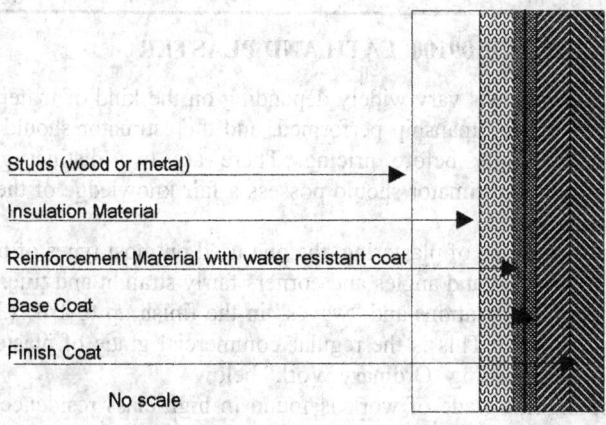

Studs (wood or metal)

Insulation Material

Reinforcement Material with water resistant coat

Base Coat

Finish Coat

No scale

TYPICAL SECTION OF AN EIFS* WALL

* EXTERIOR INSULATION AND FISNISH SYSTEM

1231

Item 3 - Base and Finish Coat					
Material Costs	Qty	Unit	Rate	Unit	Total
Primer	6.00	Pails.	$120.00	Pail.	$720.00
Cement	3.00	Bags	$20.00	Bag	$60.00
Topping Material	9.00	Pails	$125.00	Pail	$1,125.00
Total Material Cost					$1,905.00
per sq.ft.					$1.91
per sq.m					$20.51
Installation Costs	Hours	Rate	Total	Rate	Total
Carpenter	64.00	-----	-----	$35.04	$2,242.56
Labor	32.00	-----	-----	$26.16	$837.12
Total Labor Cost					$3,079.68
per sq.ft.					$3.08
per sq.m					$33.15
Total Labor and Material Cost Reinforcing Mesh					$4,984.68
Total per sq. ft..					$4.98
Total per sq. m					$53.66

Summary - Total Labor & Material Costs	
Item 1 - Insulation Board	$2,450.64
Item 2 - Fiberglass reinforcing Mesh	$1,964.80
Item 3 - Base and Finish Coat	$4,984.68
Total Labor and Material	$9,400.12
Total per sq. ft..	$9.40
Total per sq. m	$101.19

09100 LATH AND PLASTER

Plastering costs vary widely depending on the kind of materials used and the class of workmanship performed, and the estimator should consider these items carefully before pricing. There are two distinct grades of plastering, and the estimator should possess a fair knowledge of the class of work required.

The first grade of plastering, the one used on most types of buildings, has the walls rodded and angles and corners fairly straight and true. Work of this class permits variations and "waves" in the finish up to 1/8" (3.12 mm) to 3/16" (4.68 mm). This is the regular commercial grade of plastering and comes under the heading "Ordinary Work" below.

The highest grade of work is found in high-class residences, hotels, schools, courthouses, federal government buildings, state capitol buildings, public libraries, and the like, where all walls and ceilings, together with all interior and exterior angles, must be absolutely straight and plumb, and not permit variations or "waves" exceeding 1/32" (0.78 mm) to 1/16" (1.56 mm). This is termed "First Grade Workmanship".

Estimating Quantities of Lathing and Plastering. Lathing and plastering are estimated by the square yard (square meter), but the method of deducting openings, etc., varies with the class of work and the individual contractor.

Many plastering contractors make no deductions for door or window openings when estimating new work, while others deduct for one-half of all openings over a certain size.

Both lathing and plastering are estimated by the sq. yd. (sq.m), which is obtained by multiplying the girth of the room by the height, plus the ceiling area. For example: Obtain the number of sq. yds. (sq.m) of plastering in a room 12'-0"x18'-0" (3.6 5x 5.48 m), having 9'-0" (2.74 m) ceilings. 18 + 18 + 12 + 12 = 60 lin.ft. of wall (5.49 + 5.49 + 3.66 + 3.66 = 18.29 m).

60 ' x 9 ' = 540 Sq.Ft. Wall	50.17 Sq.M. Wall
12 ' x 18 ' = 216 Sq.Ft. Ceiling	20.07 Sq.M. Ceiling
540 + 216 = 756 Sq.Ft. Walls and Ceiling	70.23 Sq.M. Walls and Ceiling
756 ÷ 9 = 84 Sq.Yds.	

Openings should be measured, and no deductions should be made for openings less than 2'-0" (0.61 m) wide, and then only one-half the area of each opening should be deducted.

The complete contents of large openings should be deducted, and the contractor should figure 1'-6" (0.45 m) wide for each jamb by the height, when the return or "reveal" is less than 1'-0" (0.30 m).

For all beams or girders projecting below ceiling line, allow 1'-0" (0.30 m) in width by the total length as an extra for each internal or external angle, in addition to the actual area.

Corner Beads, Moldings, Etc. All corner beads, quirks, rule joints, moldings, casings, cornerites, screeds, and other trim accessories necessary to complete a plastering job are measured by the lin. ft. (meter), from their longest extension with 1'-0" (0.30 m) extra added for each stop or miter.

Plaster Cornices. Plaster cornices should be measured by the total length of the walls. If the cornices are 1'-0" (0.30 m) or less in girth, measurement should be made by the lin. ft. (meter), stating the girth of each cornice. If the cornices are over 1'-0" (0.30 m) in girth, measurement may be made either by the sq. ft. (sq.m) of exposed face or by the lin. ft. (meter), stating the girth of each cornice.

Allow 1 lin.ft. (0.30 m) for each internal miter and 3 lin.ft. (0.91 m) for each external miter. Enriched cornices (cast work) should be measured by the lin. ft. (meter) for each enrichment.

Arches, Corbels, Brackets, Ceiling Frieze Plates, Etc. Arches, corbels, brackets, rings, center pieces, pilasters, columns, capitals, bases, rosettes, pendants, and niches are estimated by the piece. Ceiling frieze plates should be estimated by the sq. ft. (sq.m).

Columns. Plain plastered columns should be measured by the lin. ft. (meter) or by the sq. yd. (sq.m), multiplying the girth by the height.

Cement Wainscot. Cement wainscot should be measured by the sq.ft. or sq.yd. (sq.m).

Circular or Elliptical Work. Circular or elliptical work on a small radius should be estimated at double the rate of straight work.

Gypsum Lath

Gypsum lath is manufactured in three types—plain, perforated, and insulating. Each type is made of a core of gypsum with a tough, fibrous surfacing material. For the usual surfaces, either plain or perforated gypsum lath is used. Where maximum fire protection is required or desired, perforated gypsum lath should be employed. For insulation against heat and cold, and as a vapor barrier, insulating gypsum lath having an aluminum foil back should be used on the inside face of exterior walls and for top floor ceilings.

Gypsum lath is furnished 3/8" (9.37 mm) thick and face dimension is approximately 16" x 48" (400 x 1200 mm), 6 laths per bundle. Each bundle of laths contains 32 sq.ft. (2.97 sq.m) and weighs about 50 lbs. (22.68 kg). These sizes fit the usual stud or joist spacings without cutting.

Sizes and Weights of Gypsum Lath					
Size and Thickness, in Inches	Lbs/Sq.Yd.	Approx. Price per 1,000 Sq.Ft.	Size and Thickness, in mm	Kg/Sq.M.	Approx. Price per 100 Sq.M.
16 x 48 x 3/8	14	$175.00	400 x 1200 x 9.38	7,60	$188.37
16 x 48 x 1/2	19	$185.00	400 x 1200 x 12.50	10.31	$199.14

Insulating or aluminum foil backed gypsum lath costs about $0.15 per sq.ft. ($1.61 per sq.m) more than plain lath.

Gypsum lath is also made 1/2" x 16" x 48" (12.50 x 400 x 1200mm) for framing over 16" (400 mm) o.c. but not exceeding 24" (600 mm) on center 3/8" x 24" x job length (9.37 mm x 600 mm x job length) for exterior wall furring with or without foil backing; and 3/8" x 16" x 96" (9.37 x 400 x 2400 mm) plain or perforated.

For use in 2" (50 mm) solid studless partitions, gypsum lath is made 1/2" (12.50 mm) thick, 24" (600 mm) wide, and in lengths up to 12' (3.65 m). These laths are erected with the long dimension vertical.

Nails Required for Gypsum Lath. Nails shall be 1-1/8" (28.12 mm) long for 3/8" (9.37 mm) lath and 1-1/4" (31.25 mm) long for 1/2" (12.50 mm) lath, 13 gauge with 19/64" (7.42 mm) flathead and blued. For 3/8" (9.37 mm) lath on 16" (400 mm) spacing, nails shall be spaced approximately 5" (125 mm) apart using 4 nails per stud and for 1/2" (12.50 mm) lath on 24" (600 mm) spacing, nails shall be spaced approximately 4" (100 mm) apart, using 5 nails per framing member. Lath should be butted lightly together. Gaps over 1/2" (12.50 mm) wide should be reinforced with metal lath.

1234

With studs spaced 16" (400 mm) o.c., it will require 6 to 7 lbs. (2.72 to 3.17 kg) of nails per 100 sq. yds. (83.61 sq.m) of gypsum lath and with studs spaced 24" (600 mm) o.c., it will also require 6 to 7 lbs. per 100 sq.yds. (2.72 to 3.17 kg per 83.61 sq.m) of 1/2" (12.50 mm) lath. Lath may also be attached by staples, screws, or clips.

Labor Placing Gypsum Lath. A lather should place 90 to 105 sq. yds. (75.24 to 87.79 sq.m) of gypsum lath per 8-hr. day on average work, at the following labor cost:

Labor Cost of 100 Sq. Yds. of 3/8" Gypsum Lath (83.61 Sq.M of 9.37 mm Gypsum Lath)			
Description	Hours	Rate	Total
Lather	8.00	$35.04	$280.32
Cost per sq. yd.			$2.80
per sq.m			$3.35

Mechanical Stapling of Gypsum Lath. Increased production in placing gypsum lath can be gained where labor conditions permit the use of air-driven staplers for attaching lath to wood studs or joists.

Air-driven stapler heads, using galvanized frozen staples 7/8" (21.87mm) long with a 3/8" (9.37 mm) crown, cost about $275.00 each. Staples cost $1.00 to $1.10 per thousand depending upon quantity purchased and are packed 12,000 to the carton. An air compressor able to handle one stapler head will cost about $405.00 and a 2-stapler head capacity machine will cost about $815.00.

For this type of attachment, it is general practice to tack lath into place, using 2 nails per piece for wall lath and 4 nails per piece for ceiling lath, until enough area is covered to permit continuous stapling. Attachment is completed by stapling 4" (100 mm) o.c. to all wood studs or joists. Approximately 3,500 staples and 1 lb. (0.45 kg) of nails are required per 100 sq.yds. (83.61 sq.m) of gypsum lath.

Labor Cost of 100 Sq. Yds. (86 Sq.M) of 3/8" (9 mm) Gypsum Lath Applied to Wood Studs or Joists using Air Driven Staples			
Description	Hours	Rate	Total
Lather	5.30	$35.04	$185.71
Cost per sq. yd.			$1.86
per sq.m			$2.22

Two-Inch Solid Gypsum Long Length Lath and Plaster Partition

A studless, non-loadbearing partition consists of 1/2" (12.50 mm) thick, 24" (600 mm) wide, long length, up to 12'-0" (3.6 m), gypsum lath with V'd tongue and groove edges, held vertically in floor and ceiling runners and plastered to 3/4" (18.75 mm) grounds each side.

Temporary bracing consists of 3/4" (18.75 mm) cold rolled channels horizontally, one channel for heights up to 9'-0" (2.74 m) and two channels at the third points for heights up to 12'-0" (3.65 m), is required with adequate vertical braces on partitions over 6'-0" (1.82 m) long, and every 6'-0" (1.82 m) or fraction thereof during initial plastering stages.

A metal ceiling runner is used for attachment at the ceiling and either a 2-1/2" (63 mm) flush metal base or a routed out wood floor runner may be used at the floor to secure the long sheets of gypsum board lath.

On work of this class, it will require 20 to 24 hours lather time to set and place floor and ceiling runners and gypsum lath for 100 sq.yds. (83.61 sq.m) of partition, after the material has been hoisted and stocked by laborers, which will average about 4 hrs. per 100 sq.yds. (83.61 sq.m).

Material Cost per 100 Sq.Yds. (83.61 sq.m.) 2" (50 mm) Solid Gypsum Long Length Lath and Plaster Partition. (Partition only, not including Plaster)		
Description	Rate	Total
115 lin. ft. Ceiling runner	$0.40	$46.00
115 lin.ft. Flush metal base*	$0.70	$80.50
945 sq.ft. 1/2"long length gypsum lath	$0.20	$189.00
50 lin.ft. 3" Strip lath & paper backing	$0.25	$12.50
155 lin.ft. Temp. bracing channels**	$0.15	$23.25
100 Temp. bracing clips	$0.10	$10.00
Cost per 100 sq. yds.		$361.25
per sq. yd.		$3.61
per sq.ft.		$0.40
Description	Rate	Total
35.05 lin. M. Ceiling runner	$1.31	$46.00
35.05 lin. M. Flush metal base*	$2.30	$80.50
87.79 sq. M. 50 mm long length gypsum lath	$2.15	$189.00
15.24 lin. M.75 mm Strip lath & paper backing	$0.82	$12.50
35.05 lin. M. Temp. bracing channels**	$0.49	$17.25
100 Temp. bracing clips	$0.10	$10.00
Cost per 83.61 sq. M.		$355.24
per sq. M.		$4.25

*Floor runner may be a special wood or steel runner channel at correspondingly lower material and installation cost in carpentry labor, but will require wood base at additional cost.
**115 lin.ft. (35.05 m) of horizontal temporary bracing channels, plus five 8-lin.ft (2.44 lin.m.). verticals figured on a minimum of 3 re-uses.

Labor Cost of 100 Sq.Yds. 2" (83.61 Sq.M 50 mm) Solid Gypsum Long Length Lath and Plaster Partition (Partitions only, not including plastering)			
Description	Hours	Rate	Total
Lather	22.00	$35.04	$770.88
Labor	4.00	$26.16	$104.64
Cost per 100 sq. yds			$875.52
Cost per sq. yd.			$8.76
per sq.ft.			$0.97
per sq.m			$10.47

1236

FINISHES

Add for corner beads, cornerites and other accessories required to complete lathing work. Also available are 1" (25 mm) thick gypsum lath either solid or laminated, which reduce the requirements for bracing and grounds.

Floating Systems of Attachment of Gypsum Lath. In order to eliminate sound transmission through walls and ceilings, the principle of resiliently isolating the interior walls and ceilings from their structural elements (studs and joists) has proven extremely effective. It interposes a shock absorber that reduces the energy transmitted to the structural elements of the building. Interior wall and ceiling finish should be completely and thoroughly isolated from all structural members surrounding the source of sound.

Attachment devices for gypsum lath are designed to isolate the lath from framing members. They are made from sheet metal or wire, and generally include field and corner clips for the attachment to wood, metal furring, and masonry.

In addition to reducing sound transmission, plaster cracking is also often eliminated, often caused by movement of the structural elements, which are stronger than the plastered interior. Floating systems of attachment permit a limited movement of the structural members without imposing undue stresses in the lath and plaster interior.

The spacing of framing members should not exceed 16" (400 mm) o.c. for clip attachments. Plain gypsum lath should be used on ceilings, and either plain or perforated lath may be used on walls.

Suspended Gypsum Lath Ceilings. Suspended gypsum lath ceilings consist of metal hangers, runners, and furring channels, which in turn support gypsum lath as a base for plaster, or the adhesive application of acoustical tile.

The hangers, runners, and furring channels for ceilings are about the same as required for metal lath. Furring channel spacing must not exceed 16" (400 mm) o.c., and some fire resistant ratings require 12" (300 mm) spacing. Starter channels along walls should be within 2" (50 mm) of wall. A 1/2" (13 mm) space should be provided between ends of channels and masonry walls. Each lath is supported transversely under the channels by a long wire clip, with not less than 3 such clips per full length of lath.

Stagger the end joints by starting alternative rows with half lengths of lath, allowing end joints to fall between channels.

Material Cost of 100 Sq. Yds. (83.61 Sq.M) Suspended Gypsum Lath Ceilings Using 16" x 48" x 3/8" (400x 1200x 9.37 mm) Gypsum Lath with Main Runners 3'- 0" (.91 m) on Centers and Cross Furring Channels 16" o.c.(400 mm)			
Description	Rate	Rate Metric	Total
77 pcs. 3/16"x 3'-0" (4.68mm x 0.91m) Hangers	$0.06	$0.06	$4.93
364 lin.ft. (110.94m) of 1-1/2" (37.5mm) Channels	$0.40	$1.31	$145.60
700 lin.ft. (213.36m) of 3/4" (18.75mm) Channels	$0.27	$0.89	$190.40
945 sq.ft. (87.79 sq.m.) of 16"x48"x3/8" (400 x 1200 x 9.37mm) gypsum Lath	$0.18	$1.89	$166.32
500 Brace-tite field clips	$0.14	$0.14	$72.00
340 Bridjoint clips, B-1	$0.10	$0.10	$32.64
Cost 100 sq.yds. (83.91 sq.m.)			$611.89
per sq. yd.			$6.12
per sq.ft.			$0.68
per sq.m.			$7.32

Labor Cost of 100 Sq.Yds. (83.61 Sq.M) Suspended Gypsum Lath Ceilings Using 16"x 48" x 3/8" (400x 1200x 9.37 mm) Gypsum Lath with Main Runners 3'- 0" (0.91 M) on Centers and cross Furring Channels 16" (400 mm) on Centers			
Description	Hours	Rate	Total
Lather	20.00	$35.04	$700.80
Labor	4.00	$26.16	$104.64
Cost per 100 sq. yds			$805.44
Cost per sq. yd.			$8.05
per sq.ft.			$0.89
per sq.m			$9.63

The above ceiling has a fire rating of one hour. If the job specification calls for a fire resistance rating of two hours, 14 ga. diagonal wire reinforcement must be added below the lath; to obtain a 4 hour rating, 1" (25 mm) 20 ga. galvanized 1" (25 mm) hex wire mesh must be added. All the above ratings are dependent on the type of floor construction above.

Hollow Steel Stud Partitions Using Gypsum Lath as a Plaster Base. A non-loadbearing partition of trussed steel studs, track, and shoes to which gypsum lath is applied to both sides by means of metal clips, nailing, or stapling.

The construction is similar to hollow steel stud non-loadbearing partitions except that gypsum lath is applied to the studs by means of special metal clips and the studs are spaced 16" (400 mm) o. c. to receive the 16" x 48" (400 x 1200 mm) gypsum lath.

1238

FINISHES

Material Cost of 100 Sq. Yds. (83.61 Sq.M) of Hollow Steel Stud Non-Bearing Plastered Partitions Using Gypsum Lath on Both Sides of Partitions, 2 1/2" (62.50 mm) Studs 16" (400 mm) on Centers, Ready to receive Plaster			
Description	Rate	Rate Metric	Total
105 lin.ft. (32m) top track	$0.40	$1.31	$42.00
105 lin.ft. (32m) bottom track	$0.40	$1.31	$42.00
76 9'-0" (2.74m) studs, Use 750 lin.ft. (228.6m)*	$0.54	$1.78	$408.00
1,900 sq.ft. (176.51 sq.m.) of gypsum lath. 3/8" (0.37mm) thick	$0.18	$1.89	$334.40
1,000 field clips	$0.13	$0.13	$128.00
680 Bridjoint clips	$0.10	$0.10	$65.28
Cost 100 sq.yds. (83.61 sq.m.)			$1,019.68
per sq.yd.			$10.20
per sq.ft.			$1.13
per sq.m			$12.20

Allowance made for opening framing.

Labor Cost of 100 Sq. Yds. (83.61 Sq.M) of Hollow Steel Stud Non-Bearing Plastered Partitions Using Gypsum Lath on Both Sides of Partitions, 2 1/2" (62.50 mm) Studs 16" (400 mm) on Centers, Ready to Receive Plaster			
Description	Hours	Rate	Total
Lather	28.00	$35.04	$981.12
Labor	4.00	$26.16	$104.64
Cost 100 sq. yds.			$1,085.76
per sq. yd.			$10.86
per sq.m			$12.99

Add for corner beads, cornerites, and other accessories required to complete lathing work.

Gypsum lath is also used for exterior wall furring and column fireproofing. Special framing support systems are manufactured for this. Foil backed lath may be used for exterior walls.

Wood Lath

Wood lath for plastering is a thing of the past. Gypsum lath and metal lath are commonly used. The following data is included for that one in a thousand job where wood lath may be used for patching in remodeling work.

Wood lath should be estimated by the sq.yd. (sq.m). Because it is customary to deduct only one-half the area of openings, it is not necessary to add any allowance for waste. When placing wood lath, it is customary to allow 3/8" (9.37 mm) between each lath for a plaster key. A standard wood lath is 1-1/2" (37.50 mm) wide and 48" (1200 mm) long:

Size of Lath	No. Req'd. 100 Sq.Yd.	Size of Lath	No. Req'd. 10 Sq.M.
1 1/2 "x 48 "	1425 - 1450	37.50 x 1200 mm	170.43 - 173.42

Labor Placing Wood Lath. A lather should place 1,400 to 1,600 1-1/2" x 48" (38 x 1200 mm) wood lath, 95 to 110 sq.yds. (79.43 to 91.97 sq.m) per 8-hr. day.

Metal Lath and Furring

Where metal lath is used as a plaster base in partition work, it may be nailed or stapled to wood or steel studs, nailed to masonry, or wire-tied to steel channels and prefabricated steel studs. It may also be used without studs or back-up when attached to floor and ceiling runners and temporarily braced during application and curing of first coat of plaster. Where used as a plaster base in ceiling work, it may be nailed to wood joists, wire-tied to various metal support systems either in contact with or suspended from the building structure, or wire-tied directly to the bottom chords of steel joists. It may also be attached directly to the underside of reinforced concrete floors or joists by means of attachments previously embedded in the concrete.

In addition to its utility as a plaster base, metal lath is also used to reinforce plaster applied to other surfaces. Prefabricated, right-angled strips of metal lath are attached to masonry or other rigid plaster bases, prior to plastering, at the interior angles formed by intersecting walls and along the junction of walls and ceilings. Flat strips of metal lath are attached to solid plaster bases to reinforce the joints in such bases, or to bridge potential crack lines at the corners of door and window openings.

Metal lath is used in the individual fireproofing of beams and columns. It may be adapted to decorative shapes for cornices, coves, curved walls, convoluted ceilings, and various acoustic shapes such as inverted pyramids. In specialty classifications, it is used to form the membrane of hyperbolic paraboloids, or the web section of concrete beams and girders.

Metal Lath Construction. The costs of metal lathing will vary according to the complexity of work, size of the rooms, height above floors, and method of attachment. In the sections below, an attempt is made to cover only the type of lathing that is normal in building construction.

Materials. A number of metal and wire laths and welded wire fabrics are available to suit the various conditions that might be encountered in building construction. In general, there are sheets of slitted and expanded metal, or perforated metal, commonly called metal laths; rolls of netting formed of 19 ga. wire woven into a pattern which provides 2-1/2 strands per inch (25 mm), commonly called wire laths; and sheets of open mesh laths formed of strands of 16 gauge wire welded at right angles to one another and spaced 2" (50 mm) apart, commonly called welded wire fabric. Laths may be of galvanized metal or painted with a rust-inhibitive paint. Some metal laths and all welded wire fabrics are available with a paper backing. Also, some laths may be obtained with a backing of aluminum foil laminated with paper. The backings sometimes perform useful functions insofar as utility and plastering are concerned, but many such laths are often more difficult to install.

1240

FINISHES

Expanded metal lath can be either flat dimpled (self-furring), or ribbed. The flat expanded type is classed as small mesh, having approximately 1000 openings per sq.ft. (0.09 sq.m). The perforated metal lath, commonly called sheet lath, is fabricated either with furring indentations or with channel-shaped ribs. The openings in sheet lath and in rib lath are classified as small. Although they are somewhat larger than those of the expanded metal laths, they are separated by larger areas of unperforated metal. Stucco mesh has diamond-shaped openings, which are approximately 1-1/2" x 3" (37.50 x 75 mm). Wire laths are flat when installed and the openings are classified as medium size. The welded wire fabrics are either flat, or in the crimped, self-furring variety, all the wires laying in one direction are bent approximately 1/4" (6.25 mm) out of a flat plane at their intersection with every third wire lying in the other direction.

Installation Procedures. Rather rigid industry standards govern the installation of metal laths, particularly with respect to ceiling work. Complete specifications are prepared by manufacturer associations, by individual manufacturers, or by sectional committees of the American National Standards Institute. The building construction estimator needs a brief outline of the critical factors of installation in order to do a realistic estimate.

Spans of Lath. In all work, except where laths are supported by solid surfaces, and except in the case of studless solid partitions, the weight of metal lath must be suitable to the distance between studs, joists, or other supporting members. The laths are attached to these members by nails, wire-ties, staples, or clips, at intervals not exceeding 6" (150 mm). The following chart gives maximum spans or distance between supports for various weights of metal and wire laths and welded wire fabric.

Type of Lath	Min Wt. Lath Lbs. per Sq.Yd.	Minimum Allowable Spacing of Supports in Inches				
		Vertical Supports			Horizontal Supports	
		Wood	Metal Solid Partition	Others	Wood or Concrete	Metal
Flat Expanded	2.50	16	16	12	------	------
Metal Lath	3.40	16	16	16	16	13 1/2
Flat Rib	2.75	16	16	16	16	12
Metal Lath	3.40	19	24	24	19	19
3/8" Rib	3.40	24	------	24	24	24
Metal Lath	4.00	24	------	24	24	24
Steel Lath*	4.50	24	------	24	24	24
Wire Lath	2.48	16	16	16	13 1/2	13 1/2
V- Stiffened						
Wire Lath	3.30	24	24	24	19	19
Wire Fabric	------	16	------	16	16	16

| Type of Lath | Min Wt. Lath Kg. per Sq.M. | Minimum Allowable Spacing of Supports in mm | | | | |
| | | Vertical Supports | | | Horizontal Supports | |
		Wood	Metal Solid Partition	Others	Wood or Concrete	Metal
Flat Expanded	1.36	400	400	300	------	------
Metal Lath	1.84	400	400	400	400.0	337.5
Flat Rib	1.49	400	400	400	400.0	300.0
Metal Lath	1.84	475	600	600	475.0	475.0
3/8" Rib	1.84	600	------	600	600.0	600.0
Metal Lath	2.17	600	------	600	600.0	600.0
Steel Lath*	2.44	600	------	600	600.0	600.0
Wire Lath	1.35	400	400	400	337.5	337.5
V- Stiffened						
Wire Lath	1.79	600	600	600	475.0	475.0
Wire Fabric	------	400	------	400	400.0	400.0

Used in studless solid partitions

Laps of Lath. Industry standards require overlapping at ends and sides of sheets of lath. Most sheets are manufactured slightly larger than nominal size and the estimator need not allow extra quantities on this account. A few extra ties are required in all laps, one between each supporting member, and where laps of the ends of sheets fall between supports, such laps should be laced with tie wire. These extra ties are considered part of normal lathing work and no extra allowance need be made for them.

Attachment of Lath. Nail-on attachment of metal lath to horizontal or sloping surfaces (ceilings and soffits) requires the use of No. 11 ga. barbed, galvanized roofing nails with a head diameter of 7/16" (10.93 mm). The length of such nails must permit a minimum penetration into the wood of 1-3/8" (34.37 mm). In nail-on attachment of metal lath to vertical surfaces (walls and partitions), 4d common nails are permitted. They should penetrate the wood at least 3/4" (18.75 mm). However, and the remainder should be bent over the strands of lath. Also, in vertical work, 1" (25 mm) long roofing nails with a head diameter of 7/16" (10.93 mm) are permitted. They are driven home without crushing the strands of metal. In direct attachment to masonry, concrete stub nails with 3/8" (9.37 mm) flat heads are generally used.

Tie-on attachment of metal lath is generally done with 18 ga. galvanized annealed wire, a strand of wire passing through the mesh, around the support member, and back through the mesh with the two ends twisted together approximately three turns. This operation is more difficult for the lather when paper or foilbacked lath is being used, unless adequate openings in the backing have been provided. When estimating hand tying of lath having heavy backing, a reduction in labor productivity should be allowed. In certain cases 18 ga. tie wire is not permitted. Where metal lath is attached directly to structural elements of the building, such as concrete slabs or joists or the bottom chord of open web steel joists, wire of heavier gauge, or more than one loop of 18 ga. wire, is required (16 ga. or 2-18 ga. for steel joist

1242

FINISHES

work; 14 ga. twisted or 10 ga. bent over for concrete work). The extra
difficulty of twisting and cutting the heavier gauge wires has been considered
in the productivity rates suggested below.

Special devices for attachment of laths are permitted, where they
develop a fastening strength equal to the accepted standards. Such devices
often result in noticeable labor savings but the estimator should not neglect to
adjust his material costs as necessary.

Horizontal Support Systems. Horizontal supports for metal lath fall
under the general classifications of contact systems, where the lath is fastened
in direct contact with the structural elements of the building, furred systems,
or suspended systems. In both of the latter systems, the laths are attached in
the usual manner to a framework of pencil rods or channel shapes which have
been fastened directly to, or suspended from, structural elements of the
building. In a furred system, the channels or pencil rod furring are attached to
the structural elements of the building, by means of 16 ga. wire saddle ties—
in open web steel joist construction, or by 18 ga. wire saddle ties supported
by nails driven in or through wood joists. The forming of saddle ties is
considered normal lathing work, and the productivity rates given below
consider this type of tying.

In a suspended system, the channels or pencil rods to which the lath is
attached are called cross-furring. These members are saddle-tied with 16 ga.
wire to the underside of an intersecting framework of heavier channels,
which have been hung from the structural elements of the building by means
of wire, rod, or flat steel hangers. Wire hangers and rods are saddle-tied
around these support channels, which are called main runners, and flats may
be welded or bolted to them.

A combination of the furred and suspended systems is usually found
where furring members are installed in contact with concrete joists. The
furring members, which support the lath, are in contact with the bottoms of
the joists, but they are supported by transverse channels, or runners, which
are suspended between and run parallel to the joists. The furring members are
saddle-tied with 16 ga. wire to the runners, and the runners are suspended by
means of 10 ga. wire hangers embedded in the slab above. The wire hangers
are saddle-tied to the runners.

Hanger Sizes for Suspended Ceilings			
Ceiling Area sq.ft. max.	Hanger Size minimum	Hangers for Supporting up to 25 Sq.Ft. Ceiling	
8	12 ga. wire	6 ga. wire	1 " x 3/16 " flat
12	10 ga. wire	5 ga. wire	1 1/4 " x 1/8 " flat
12 1/2	9 ga. wire	3/16" rod	1 1/4 " x 3/16 " flat
16	8 ga. wire	7/32" rod	1 1/2 " x 1/8 " flat
17 1/2	7 ga. wire	1/4" rod	1 1/2 " x 3/16 " flat

Hanger Sizes for Suspended Ceilings - Metric					
Ceiling Area sq.M. max.	Hanger Size minimum	Hangers for Supporting up to 2.32 Sq.M. Ceiling			
0.74	12 ga. wire	6 ga. wire	25.00	x	4.69 mm flat
1.11	10 ga. wire	5 ga. wire	31.25	x	3.13 mm flat
1.16	9 ga. wire	4.69 mm rod	31.25	x	4.69 mm flat
1.49	8 ga. wire	5.47 mm rod	37.50	x	3.13 mm flat
1.63	7 ga. wire	6.25 mm rod	37.50	x	4.69 mm flat

All wire hangers should be galvanized steel, all flat and rod hangers should be coated with a rust-inhibitor non-toxic paint.

Spans and Spacing for Main Runners in Suspended Ceilings			Max. Span Between Hangers or Supports	Max.CtoC Spacing of Runners
Minimum Size and Type Inches and Lbs.				
3/4 "	channel	0.300 Lbs	3 ' - 0 "	2 ' - 4 "
3/4 "	channel	0.300 Lbs	2 ' - 6 "	2 ' - 6 "
3/4 "	channel	0.475 Lbs	2 ' - 0 "	3 ' - 0 "
1 1/2 "	channel	0.475 Lbs	5 ' - 0 "	2 ' - 0 "
1 1/2 "	channel	0.475 Lbs	4 ' - 0 "	3 ' - 0 "
1 1/2 "	channel	0.475 Lbs	3 ' - 6 "	3 ' - 6 "
1 1/2 "	channel	0.475 Lbs	3 ' - 0 "	4 ' - 0 "
2 "	channel	0.590 Lbs	7 ' - 0 "	2 ' - 0 "
2 "	channel	0.590 Lbs	5 ' - 0 "	3 ' - 6 "
2 "	channel	0.590 Lbs	4 ' - 6 "	4 ' - 0 "
Minimum Size and Type mm and Kg			Max. Span Between Hangers or Supports	Max.C to C Spacing of Runners
18.75 mm channel		0.136 Kg	0.91 M	0.71 M
18.75 mm channel		0.136 Kg	0.76 M	0.76 M
18.75 mm channel		0.215 Kg	0.91 M	0.91 M
37.50 mm channel		0.215 Kg	0.61 M	0.61 M
37.50 mm channel		0.215 Kg	0.91 M	0.91 M
37.50 mm channel		0.215 Kg	1.07 M	1.07 M
37.50 mm channel		0.215 Kg	1.22 M	1.22 M
50.00 mm channel		0.268 Kg	0.61 M	0.61 M
50.00 mm channel		0.268 Kg	1.52 M	1.07 M
50.00 mm channel		0.268 Kg	1.22 M	1.22 M

All weights of channel are for cold-rolled members.

Spans and Spacing for Cross Furring		
Minimum Size and Type Inches & Lbs.	Max. Span Between Runners or Supports	Max. CtoC Spacing of Cross Furring
3/8 " Pencil Rod	2 ' - 6 "	12 "
3/8 " Pencil Rod	2 ' - 0 "	19 "
3/4 " Channel 0.3 Lb	4 ' - 0 "	16 "
3/4 " Channel 0.3 Lb	3 ' - 6 "	19 "
3/4 " Channel 0.3 Lb	3 ' - 0 "	24 "
Minimum Size and Type mm & Kg.	Max. Span Between Runners or Supports	Max. CtoC Spacing of Cross Furring
9.38 mm Pencil Rod	0.762 m	300 mm
9.38 mm Pencil Rod	0.610 m	475 mm
18.75 mm Channel 0.136 Kg	1.219 m	400 mm
18.75 mm Channel 0.136 Kg	1.067 m	475 mm
18.75 mm Channel 0.136 Kg	0.914 m	600 mm

All weights of channel are for cold-rolled members.

In contact, furred or suspended systems, the lath must be supported at spans not exceeding those given in the table of maximum spans, above. In furred and suspended systems, all pencil rods and channels are limited as to distance between supports and center to center spacing. In suspended systems the size of each hanger depends on the area of ceiling that it supports. The following tables list the usual elements and arrangements to be found in furred and suspended ceilings.

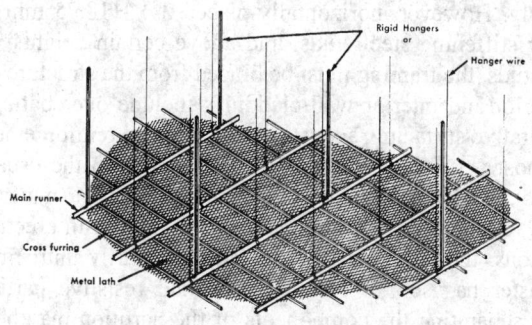

Courtesy Metal Lath Association

Details of Suspended Ceiling

In most horizontal work, the laths, framing members, hangers, and inserts are installed in the usual manner. The estimator should look for the presence of large ducts, which might require additional framing or hangers not detailed on the drawings. He should allow for hung or tied supports at the perimeter of rooms and not plan to set channels into wall pockets. The estimator's checklist of materials should have allowances for lath, furring channels, carrying channels, tie wire, hanger wire, nails, special inserts if any, corner reinforcing, and accessory items, if required, such as expansion bead or casing bead.

Vertical Support Systems. Metal lath may be used in a number of wall and partition framing systems. A generalized classification would include hollow walls and partitions, solid plaster partitions, vertical furring, and wall cladding. Hollow assemblies may have either metal or wood studs. The metal studs are generally used where non-combustible construction is required, where concealed horizontal pipe must be accommodated, or where sound resistive qualities are needed. Although the majority of such studs are rated as non-loadbearing, because they are fabricated from light-gauge steel or rod, tests of completed walls have indicated surprising load carrying capacity and resistance to horizontal impact.

Thin solid plaster partitions may be formed over a core of metal lath alone, or an assembly of metal lath and metal studs. The studless solid partitions are generally 2" (50 mm) thick and may be as much as 10' (3.04 m) in height. The solid partitions with 3/4" (18.75 mm) cold rolled channels may range between 1-1/2" (37.50 mm) thickness and 8-1/2" (212.50 mm) height to 2-1/2" (62.50 mm) thickness and 16' (4.87 m) height. Heavier channels may be used for thicker and higher solid partitions.

Vertical, or wall, furring systems may be braced from exterior walls, pilasters, or interior columns, or may be free-standing. They are sometimes decorative, other times functional. Functional aspects include concealment, fireproofing, acoustic treatment, and condensation and temperature control. The lathing installation is essentially that of a partition with studs, lathed on one side only. However, horizontally-placed 3/4" (18.75 mm) channels are required for stiffening steel studs, and above certain heights, depending on size of the studs, the framing must be braced from the structure behind.

Exterior and interior wall claddings include overcoating (remodeling work), decorative surfacing, acoustic surfacing, and reinforcement.

In most vertical work, the laths are applied in the usual manner and the estimator need not look for hidden items of extra material and labor. However, a few unusual operations are worth noting. In erection of studless solid partitions, the laths must be braced temporarily until first, or scratch, coat of plaster has set. In erection of sound resistive partitions, various systems for separating the components of the partition may be encountered. These might include resilient clips and pencil rod furring between the metal studs and metal lath, staggered stud arrangements that require double the normal number of studs, or double rows of small studs with short pieces of cross-tie channels between them. Note also that most soundproofing partitions extend up to the structure itself, and are not terminated at the level of a suspended ceiling or the bottoms of joists. In wall cladding work, some of the denser back-up materials, such as reinforced concrete or old brick, may be difficult to penetrate with nails, slowing the progress of the lather. In certain areas of the country, the exterior walls of stuccoed dwellings may be constructed without sheathing. In this style of erection, known as line wire construction, the lathing work sometimes includes installation of 18 ga. wires, stretched taut completely around the building, and fastened securely to the studs at 6" (150 mm) intervals vertically. This operation is generally performed only when woven wire or stucco mesh is the plaster base.

The estimator's checklist of materials for vertical work should contain allowances for lath, studs if any, four shoes for each prefabricated metal stud (2 top and 2 bottom), ceiling and floor tracks for holding studs and shoes, nails for attachment of either the tracks or bent-over ends (of small studs) to ceiling and floor, nails for attachment of lath to wood or solid back-up, wire for tying lath and shoes to studs, and shoes to tracks, if required, metal base or screeds, if any, and accessory items such as corner, casing or expansion beads, and picture molds, if required. In estimating requirements for studs, allow for double-studding at door frames where necessary, and where splicing is required, allow for an 8" (200 mm) lap and double wire-tying.

Installation—Labor

Nail-On Attachment of Metal Lath. For either horizontal or vertical work, a lather should apply 90 to 100 sq.yds. (75.24 to 83.61 sq.m) of metal lath per 8-hr. day, depending on size of rooms.

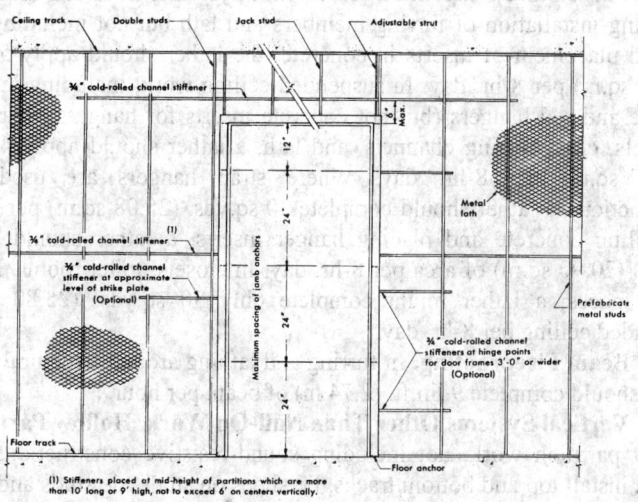

Courtesy Metal Lath Association

Partition Framing At Door Openings

Tie-On Attachment of Metal Lath. Usually, the actual tying of metal lath is not an isolated operation, for the lather must also install the studs or channels to which the lath is tied. However, in contact ceiling work, where the lath is sometimes tied to open web joists or to hangers previously embedded in concrete, the lather should apply 100 sq.yds. (83.61 sq.m) of metal lath per 8-hr. day.

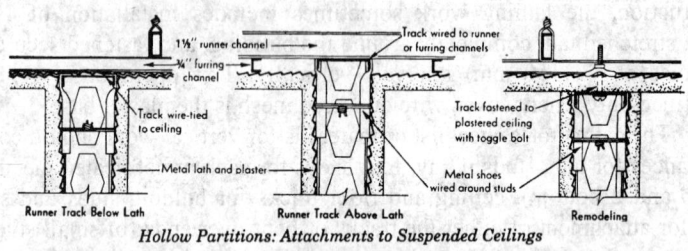

Hollow Partitions: Attachments to Suspended Ceilings

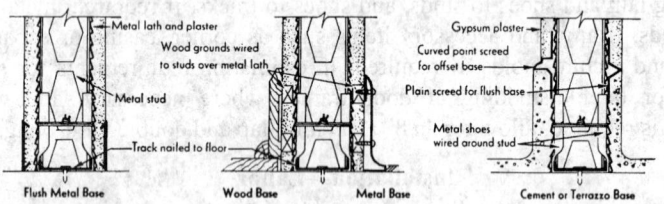

Courtesy Metal Lath Association

Hollow Partitions: Attachments to Floor

Furred and Suspended Horizontal Systems. In furred ceiling work, including installation of furring members and lath but not including drilling for and placement of inserts in concrete, the lather should apply 50 sq.yds. (41.80 sq.m) per 8-hr. day. In suspended ceiling work, including installation of wire and rod hangers (but not concrete inserts for hangers), main runner channels, cross furring channels, and lath, a lather should apply 40 sq.yds. (33.44 sq.m) per 8-hr. day. Where strap hangers are used (bolted construction), a lather should complete 30 sq.yds. (25.08 sq.m) per 8-hr day. In drilling concrete and placing hanger inserts, a lather should cover 25 sq.yds. (20.90 sq.m) of area per 8-hr. day. In closets, toilet rooms, and other small spaces, a lather might complete only 30 sq.yds. (25.80 sq.m) of suspended ceiling per 8-hr. day.

Beam Fireproofing. In furring and lathing around structural beams, a lather should complete 9 lin.ft. (2.74 m) of beam per hour.

Vertical Systems Other Than Nail-On Work, Hollow Partitions. In hollow partition work, not including sound-resistive construction, a lather should install top and bottom tracks, tie shoes on studs, tie studs and shoes to tracks, apply lath, and complete 30 sq.yds. (25.08 sq.m) of partition per 8-hr. day. In 100 sq.yds. (83.61 sq.m) of such partitions, approximately 16 hrs. are required to install runners and studs and the attachment of two surfaces of

1248

lath should require approximately 11 hrs. Another version of the steel stud hollow partition, using a framework of 3/4" (18.75 mm) cold-rolled channels, requires slightly more labor for installation. In this work the lather must erect two walls and tie them together. Productivity is between 24 and 28 sq.yds. (20.06 and 23.41 sq.m) per 8-hr. day. For estimating sound-resistive hollow partitions with prefabricated steel studs, add 5 hrs. per 100 sq.yds. (83.61 sq.m) to cover installation of resilient clips and pencil rods. No extra allowance need be made in the case of hollow partitions with double rows of 3/4" (18.75 mm) channels.

Vertical Systems, Solid Partitions, and Vertical Furring. In erecting solid partitions with 3/4" (18.75 mm) cold rolled channels, a lather should complete 22 sq.yds. (18.39 sq.m) per 8-hr. day, depending on spacing of channels and method of attaching channels at floor and ceiling. Studless solid partitions should be erected at the rate 27 sq.yds. (22.57 sq.m) per 8-hr. day. Labor savings are realized in the studless system due to elimination of wire tying the lath to channel studs. Vertical furring systems should generally be erected at the same rate as solid partitions with channel studs. However, where wall braces are required, add 4 hrs. per 100 sq.yds. (83.61 sq.m).

Vertical Systems, Wall Cladding. In overcoating work, and in other direct-applied lath attachments, a lather should nail metal lath to surfaces of wood and concrete block at the same rate as for nail-on work to wood studs and joists. Lathing should proceed at the rate of 160 sq.yds. (133.77 sq.m) per 8-hr. day. In attachment to old brick or reinforced concrete, however, a lather should be expected to complete only 125 sq.yds. (104.51 sq.m) per 8-hr. day.

Accessories. Many metal lathing accessories are available for a variety of functional or trimming purposes. Only a few of the more common items are covered here. The labor rates given below for erection of some of the functional accessories are inclusive of items of work covered as part of the total assembly. The estimator should use the highest productivity rate for erection of an entire assembly, when he lists these functional accessories as separate items. For example, installation of metal base sometimes includes layout of partition and attachment of clips to floor-items of work which are also contemplated in erection of some partition types.

Flush Metal Base. Double for solid partitions-including layout of partition, attachment of clips to floor, cutting and fitting, attachment of side plates: 5 hrs. per 100 lin.ft. (30.48 m) of partition.

Single for Masonry Trim. Including drilling for and setting masonry clips, cutting and fitting, attachment of side plates: 3 hrs. per 100 lin.ft. (30.48 m).

Single for Wall Furring. Including layout of furring, attachment of clips to floor, cutting and fitting, attachment of side plates: 4 hrs. per 100 lin.ft. (30.48 m).

Flush or Applied Metal Base. Single for snap-on to metal studs or for screwing to wood grounds, including cutting and fitting, attaching side plates, and attaching clips to studs, but not including wood grounds: 3 hrs. per 100 lin.ft. (30.48 m).

Beads and Molding. Including corner and bullnose beads, picture molds, base screeds, casing bead (except when used for door and window trim), and expansion bead (except when lath and/or framing is severed beneath line of bead):

Description	Hours per 100 lin.ft.	Hours per 30.48 M
Wired or stapled to lath bases	2 1/2	2 1/2
Nailed or stapled to wood construction	2	2
Nailed to brick, tile or block construction	3	3
Stub-nailed to concrete construction	3 1/2	3 1/2

If on-the-job mitering is required, add up to 1/2 hr. per 100 lin.ft. (30.48 m) to above rates.

When windows and doors are being trimmed with casing bead, and the operation is separate from the remainder of the lathing work, the lather should complete 1 door or window per hour. When expansion bead is to be installed to a lath base and the base is to be severed beneath the line of bead, add 3 hrs. per 100 lin.ft. (30.48 m) for cutting lath.

Cornerite and Stripite. Placed in interior wall angles and at junctions of walls and ceilings where adjoining constructions are of rigid lath or masonry, and installed along the joints in nonmetallic plaster bases, should be placed at the rate of 500 lin.ft. (152.40 m) per 8-hr. day, depending on whether material is applied in isolated or repetitious patterns.

Material Cost of 100 Sq.Yds. of 3.4 Lb. (83.61 Sq.M 1.54 Kg) Diamond Mesh Metal Lath Applied to wood Studs or Joists, Spaced @ 16" (400 mm) on Centers			
Description	Rate	Rate Metric	Total
105 sq.yds. (87.79 sq.m.) Diamond mesh	$2.54	$3.03	$267.12
9 lbs. (4.08 kg) of 1-1/2" (37.5mm) roofing nails	$1.70	$3.74	$15.26
Cost 100 sq.yds. (83.61 sq.m.)			$282.38
per sq.yd.			$2.82
per sq.m.			$3.38

Labor Cost of 100 Sq.Yds. of 3.4 Lb. (83.61 Sq.M 1.54 Kg) Diamond Mesh Metal Lath Applied to Wood Studs or Joists @ 16" (400mm) on Centers, Ready to receive Plaster			
Description	Hours	Rate	Total
Lather	7.50	$35.04	$262.80
Cost per sq. yd.			$2.63
per sq.m			$3.14

1250

FINISHES

Material Cost of 100 Sq.Yds. of 3.4 Lb. (83.61 Sq.M 1.54 Kg) 3/8" (9.37 mm) Rib Metal Lath Applied to Wood Studs or Joists @ 24" (600 mm) on Centers, Ready to Receive Plaster			
Description	Rate	Rate Metric	Total
105 sq.yds. (87.79 sq.m.) of 3/8" (9.37mm) Rib Metal Lath	$3.73	$4.46	$391.44
6 lbs. (2.72 kg) of 1-1/2" (37.5 mm) roofing nails	$1.70	$3.74	$10.18
Cost 100 sq.yds. (83.61 sq.m.)			$401.62
per sq.yd.			$4.02
per sq.m.			$4.80

Labor Cost of 100 Sq.Yds. of 3.4 Lb. (83.61 Sq.M 1.54 Kg) 3/8" (9.37 mm) Rib Metal Lath Applied to Wood Studs or Joists @ 24" (600 mm) on Centers, Ready to Receive Plaster			
Description	Hours	Rate	Total
Lather	7.00	$35.04	$245.28
Cost per sq. yd.			$2.45
per sq.m			$2.93

Material Cost of 100 Sq.Yds. of 3.4 Lb. (83.61 Sq.M 1.54 Kg) 3/8" (9.37 mm) Rib Metal Lath Applied Directly to Ties Embedded in Concrete or to Steel Joists @ 24" (600 mm) on Centers			
Description	Rate	Rate Metric	Total
105 sq.yds. (87.79 sq.m.) of 3/8" (9.37 mm) Rib Metal Lath	$4.37	$5.22	$458.64
15 lbs. (6.8 kg) of 18 ga. Tie wire	$1.52	$3.35	$22.80
Cost 100 sq.yds. (83.61 sq.m.)			$481.44
per sq.yd.			$4.81
per sq.m.			$5.76

Labor Cost of 100 Sq.Yds. of 3.4 Lb. (83.61 Sq.M 1.54 Kg) 3/8" (9.37 mm) Rib Metal Lath Applied Directly to Ties embedded in Concrete or to Steel Joists @ 24" (600 mm) on Centers			
Description	Hours	Rate	Total
Lather*	9.00	$35.04	$315.36
Cost per sq. yd.			$3.15
per sq.m			$3.77

Lather hours listed are for attachment of lath only. For concrete joist construction, add for placing wire in forms before pour, or if necessary to drill holes in concrete.

Material Cost of 100 Sq.Yds. (83.61 Sq.M) of Furred Ceiling Using 3.4 Lb. (1.54 Kg) Flat Rib Metal Lath Applied to 3/8" (9.37 mm) Pencil Rod 19" (175 mm) on Centers Tied to Structure at 2'- 0" (0.61M) Intervals

Description	Rate	Rate Metric	Total
105 sq.yds. (87.79 sq.m.) of 3/8" (9.37 mm) Flat Rib Metal Lath	$3.73	$4.46	$391.44
660 lin.ft. (201.17 m) of 3/8" (9.37 mm) Pencil Rod	$0.13	$0.42	$84.48
6 lbs. (2.72 kg) of 18 ga. Tie Wire (Lath to Rods)	$1.52	$3.35	$9.12
2 lbs. (0.91 kg) of 16 ga. Wire for splicing rods	$1.36	$3.00	$2.72
4 lbs. (1.81 kg) of 16 ga. Wire for attaching rods	$1.36	$3.00	$5.44
Cost per 100 sq.yds. (83.61 sq.m.)			$493.20
per sq.yd.			$4.93
per sq.m.			$5.90

Labor Cost of 100 Sq.Yds. (83.61 Sq.M) of Furred Ceiling Using 3.4 Lb. (1.54 Kg) Flat Rib Metal Lath Applied to 3/8" (9.37 mm) Pencil Rod 19" (175 mm) on Centers Tied to Structure at 2'- 0" (0.61M) Intervals

Description	Hours	Rate	Total
Lather*	16.00	$35.04	$560.64
Cost per sq. yd.			$5.61
per sq.m			$6.71

Lather hours listed are for attachment of lath only. For concrete joist construction, add for placing wire in forms before pour, or if necessary to drill holes in concrete.

Material Cost of 100 Sq.Yds. (83.61 Sq.M) Suspended Metal Lath Ceiling Using 3.4 Lb. (1.54 Kg) 3/8" (9.37 mm) Rib Metal Lath with Furring Channels 24" on Centers, Runner Channels 36" (900 mm) on Centers, Hangers 48" (1200mm) on Centers

Description	Rate	Rate Metric	Total
100 pcs of 9 ga. x 3'- 0" (0.91 m) hangers, 300'- 0" (91.44 m) or 19 lbs. (8.62 kg)	$1.18	$2.61	$22.50
345 lin.ft. (105.16 m) of 1-1/2" (37.5 mm) c.r. channels	$0.40	$1.31	$138.00
490 lin.ft. (149.35 m) of 3/4" (18.75 mm) c.r.	$0.27	$0.89	$133.28
3 lbs. (1.36 kg) of 16 ga. wire	$1.36	$3.00	$4.08
105 sq.yds. (87.79 sq.m.) of 3/8" (9.37 mm) Rib Metal Lath	$3.73	$4.46	$391.44
5 lbs. (2.27 kg) of 18 ga. wire	$1.52	$3.35	$7.60
Cost per 100 sq.yds. (83.61 sq.m.)			$696.90
per sq.yd.			$6.97
per sq.m.			$8.34

FINISHES

Labor Cost of 100 Sq.Yds. (83.61 Sq.M) Suspended Metal Lath Ceiling Using 3.4 Lb. (1.54 Kg) 3/8" (9.37 mm) Rib Metal Lath with Furring Channels 24" on Centers, Runner Channels 36" (900 mm) on Centers, Hangers 48" (1200mm) on Centers			
Description	Hours	Rate	Total
Lather*	50.00	$35.04	$1,752.00
Cost per sq. yd.			$17.52
per sq.m			$20.95

For erection under concrete, add placing hangers in forms before pour or if necessary to drill holes in concrete or tile, and place inserts.

Material Cost of 100 Sq.Yds. (83.61 Sq.M) Suspended Metal Lath Ceiling Using 3.4 Lb. (1.54 Kg) 3/8" (9.37 mm) Diamond Mesh Metal Lath with Furring Channels 13 1/2" (337.50 mm) on Centers, Runner Channels 48" (1200 mm) on Centers, Hangers 48" (1200mm) on Centers			
Description	Rate	Rate Metric	Total
85 pcs of 8 ga. x 3'- 0" (0.91 m) hangers, 255'- 0" (77.72 m) or 19 lbs. (8.62 kg)	$0.74	$1.63	$14.06
280 lin.ft. (85.34 m) of 1-1/2" (37.5 mm) c.r.	$0.25	$0.82	$70.00
850 lin.ft. (259.08 m) of 3/4" (18.75 mm) c.r.	$0.17	$0.56	$144.50
4 lbs. (1.81 kg) of 16 ga. wire	$0.85	$1.87	$3.40
105 sq.yds. (87.79 sq.m.) of Diamond Mesh	$1.59	$1.90	$166.95
8 lbs. (3.63 kg) of 18 ga. wire	$0.95	$2.09	$7.60
Cost per 100 sq.yds. (83.61 sq.m.)			$406.51
per sq.yd.			$4.07
per sq.m.			$4.86

Note: See above for erection under concrete.

Labor Cost of 100 Sq.Yds. (83.61 Sq.M) Suspended Metal Lath Ceiling Using 3.4 Lb. (1.54 Kg) 3/8" (9.37 mm) Diamond Mesh Metal Lath with Furring Channels 13 1/2" (337.50 mm) on Centers, Runner Channels 48" (1200 mm) on Centers, Hangers 48" (1200mm) on Centers			
Description	Hours	Rate	Total
Lather*	56.00	$35.04	$1,962.24
Cost per sq. yd.			$19.62
per sq.m			$23.47

* See above for erection under concrete.

Material Cost of 100 Lin. Ft. (30.48 Lin. M) Beam or Girder Fireproofing Using 3.4 Lb. (1.54 Kg) Self-Furring Diamond Mesh Metal Lath Tied to Adjacent Ceiling Lath and Wrapped Downwards Around 12" (300 mm) WF Beam			
Description	Rate	Rate Metric	Total
33 sq.yds. (27.59 sq.m.) of 3.4 lbs. (1.54 kg) Self-Furring Diamond Mesh	$2.72	$3.25	$89.76
2.5 lbs. (1.13 kg) of 18 ga. wire	$1.52	$3.35	$3.80
Cost 100 lin.ft. (30.48 m)			$93.56
per lin.ft.			$0.94
per lin.m.			$3.07

Labor Cost of 100 Lin. Ft. (30.48 Lin. M) Beam or Girder Fireproofing Using 3.4 Lb. (1.54 Kg) Self-Furring Diamond Mesh Metal Lath Tied to Adjacent Ceiling Lath and Wrapped Downwards Around 12" (300 mm) WF Beam			
Description	Hours	Rate	Total
Lather	12.00	$35.04	$420.48
Cost per lin.ft.			$4.20
per lin.m.			$13.80

Material Cost of 100 Sq. Yds. (83.61 Sq.M) of Hollow Steel Stud Partitions; Lathed Both Sides, Studs 12" (300 mm) on Centers, Using 2.5 lb (1.12 Kg) Diamond Mesh Metal Lath			
Description	Rate	Rate Metric	Total
200 lin.ft. (60.96 m) of snap-in runner track	$0.40	$1.31	$80.00
Nail or expansion drive allowance	$42.40	$42.40	$42.40
101 pcs of 9'- 0" (2.74 m) x 3-5/8" (90.62 mm) thick galv. Steel Studs, 909 Lin.ft. (277.06 m)*	$0.59	$1.94	$538.13
210 sq.yds. (175.58 sq.m.) of Diamond mesh	$2.03	$2.43	$426.72
17 lbs. (7.71 kg) of 18 ga. Wire	$1.52	$3.35	$25.84
Cost per 100 sq.yds. (83.61 sq.m.)			$1,113.09
per sq.yd.			$11.13
per sq.m.			$13.31

*Add necessary extras for framing around openings, returns, etc.

Labor Cost of 100 Sq. Yds. (83.61 Sq.M) of Hollow Steel Stud Partitions; Lathed Both Sides, Studs 12" (300 mm) on Centers, Using 2.5 lb (1.12 Kg) Diamond Mesh Metal Lath			
Description	Hours	Rate	Total
Lather	27.00	$35.04	$946.08
Cost per sq.yd.			$9.46
per sq.m.			$11.32

1254

FINISHES

Material Cost of 100 Sq. Yds. (83.6 Sq.M) of Sound Insulating Partitions;
Using Prefabricated Steel Studs 24" (600 mm) on Centers, Resilient Clips and
Pencil Rods, Lath Both Sides with 3.4 lb (1.54 kg.) 3/8" (9.37 mm) Rib Metal Lath

Description	Rate	Rate Metric	Total
200 lin.ft. (60.96 m) of plain runner track	$0.40	$1.31	$80.00
Nail or expansion drive allowance	$42.40	$42.40	$42.40
51 pcs iof 9'-0" (2.74 m) x 3-5/8" (90.62 mm) thick Galv Steel Studs, 459 lin.ft. (139.9 m)	$0.59	$1.94	$271.73
204 Shoes	$0.13	$0.13	$26.11
210 sq.yds. (175.58 sq.m.) of 3/8" (9.37 mm) Rib Metal Lath	$3.73	$4.46	$782.88
11 lbs. (4.99 kg) of 18 ga. Wire	$1.26	$2.79	$13.90
Add for Sound Treatment:			
100 lin.ft. (30.48 m) of cork, 3-1/2" (87.5 mm) wide x 1" (25 mm) thick	$0.85	$2.78	$84.80
1020 Resilient clips	$0.13	$0.13	$130.56
918 lin.ft. (279.81 m) of 1/4" (6.25 mm) Rod	$0.11	$0.37	$102.82
Cost per 100 sq.yds. (83.61 sq.m.)			$1,535.19
per sq.yd.			$15.35
per sq.m.			$18.36

Labor Cost of 100 Sq. Yds. (83.61 Sq.M) of Sound Insulating Partitions;
Using Prefabricated Steel Studs 24" (600 mm) on Centers, Resilient Clips and
pencil rods, Lath Both Sides with 3.4 lb (1.54 kg.) 3/8" (9.37 mm) Rib Metal Lath

Description	Hours	Rate	Total
Lather	27.00	$35.04	$946.08
Lather - Added for Sound Treatment	5.00	$35.04	$175.20
Cost per 100 sq.yd. (83.61 sq.m.)			$1,121.28
per sq.yd.			$11.21
per sq.m.			$13.41

Material Cost of 100 Sq. Yds. (83.6 Sq.M) Metal Lath Solid Partitions;
Lathed One Side Only Using Channels 16" (400 mm) on Centers,
Metal Floor and Ceiling Runners and 2.5 lb (1.12 kg) Diamond Mesh Metal Lath

Description	Rate	Rate Metric	Total
690 lin.ft. (210.31 m) of 3/4" (18.75 mm) c.r.	$0.27	$0.89	$187.68
200 lin.ft. (60.96 m) of Z-type runner, slotted	$0.30	$1.00	$60.80
Nails and screws - allowance	$42.40	$42.40	$42.40
105 sq.yds. (87.79 sq.m.) of Diamond mesh	$2.03	$2.43	$213.36
8 lbs. (3.63 kg) of 18 ga wire	$1.52	$3.35	$12.16
Cost per 100 sq.yds. (83.61 sq.m.)			$516.40
per sq.yd.			$5.16
per sq.m.			$6.18

Labor Cost of 100 Sq. Yds. (83.61 Sq.M) Metal Lath Solid Partitions; Lathed One Side Only Using Channels 16" (400 mm) on Centers, Metal Floor and Ceiling Runners and 2.5 lb (1.12 kg) Diamond Mesh Metal Lath			
Description	Hours	Rate	Total
Lather	37.00	$35.04	$1,296.48
Cost per sq.yd.			$12.96
per sq.m.			$15.51

Material Cost of 100 Sq. Yds. (83.61 Sq.M) of Hollow Partitions (Concealment) Using Double Rows of 3/4" (18.75 mm) Channel Studs at 16" (400 mm) on Centers, and 3.4 lb (1.5 kg) Diamond Mesh Metal Lath			
Description	Rate	Rate Metric	Total
152 pcs of 3/4" (18.75 mm) Studs x 9'- 2" (2.8 m), 1,390 lin.ft. (423.67 m) tops of studs tied to ceiling furring channels*	$0.27	$0.89	$378.08
208 lin.ft. (63.40 m) of 3/4" (18.75 mm) stiffener	$0.27	$0.89	$56.58
152 pcs of 3/4" (18.75 mm) spacer, 512'-0" lin.ft. (156.06 m) based on 2'- 0" (0.61 m) between rows of studs	$0.27	$0.89	$139.26
200 lin.ft. (60.96 m) snap-on metal base	$0.77	$2.52	$153.60
152 single base clips, furring clip	$0.13	$0.13	$19.46
Nails or expansions drive-allowance	$42.40	$42.40	$42.40
210 sq.yds. (175.58 sq.m.) of Diamond mesh	$2.54	$3.04	$534.24
17 lbs. (7.71 kg) of 18 ga. Wire	$1.52	$3.35	$25.84
Cost per 100 sq.yds. (83.61 sq.m.)			$1,349.46
per sq.yd.			$13.49
per sq.m.			$16.14

* Add necessary extras for framing around opening, returns, etc.

Labor Cost of 100 Sq. Yds. (83.61 Sq.M) of Hollow Partitions (Concealment) Using Double Rows of 3/4" (18.75 mm) Channel Studs at 16" (400 mm) on Centers, and 3.4 lb (1.5 kg) Diamond Mesh Metal Lath			
Description	Hours	Rate	Total
Lather	32.00	$35.04	$1,121.28
Cost per sq.yd.			$11.21
per sq.m.			$13.41

FINISHES

Material Cost of 100 Sq. Yds. (83.61 Sq.M) of Vertical Furring Using 2.75 lbs. (1.24 kg) Flat Rib Metal Lath and 3/4" (18.75 mm) Channels 16" (400 mm) on Centers			
Description	Rate	Rate Metric	Total
76 pcs of 3/4" (18.75 mm) Studs x 9'- 0" (2.74 m), 684 lin.ft. (208.48 m)*	$0.27	$0.89	$186.05
104 lin.ft. (31.7 m) of 3/4" (18.75 mm) stiffener	$0.27	$0.89	$28.29
51 pcs of 3/4" (18.75 mm) braces x 2'- 0" (0.61 m), 102 lin.ft. (31.09 m)	$0.27	$0.89	$27.74
200 lin.ft. (60.96 m) of double-pronged Ceiling floor runner	$0.40	$1.31	$80.00
Nails or expansions drive-allowance	$42.40	$42.40	$42.40
105 sq.yds. (87.79 sq.m.) of Flat rib metal lath	$2.54	$3.04	$267.12
8 lbs (3.63 kg) of 18 ga. Wire	$1.52	$3.35	$12.16
Cost per 100 sq.yds. (83.61 sq.m.)			$643.76
per sq.yd.			$6.44
per sq.m.			$7.70

* *Add necessary extras for framing around opening, returns, etc.*

Labor Cost of 100 Sq.Yds. (83.61 Sq.M) of Vertical Furring Using 2.75 lbs. (1.24 kg) Flat Rib Metal Lath and 3/4" (18.75 mm) Channels 16" (400 mm) on Centers			
Description	Hours	Rate	Total
Lather	22.00	$35.04	$770.88
Cost per sq.yd.			$7.71
per sq.m.			$9.22

Material Cost of 100 Lin. Ft. (30.48 lin.M) of Metal Corner Bead Attached to lath Bases			
Description	Rate	Rate Metric	Total
100 lin.ft. (30.48 m) of corner bead	$0.16	$0.52	$16.00
Staples or tie wire allowance	$1.05	$1.05	$1.05
Cost per 100 lin.ft. (30.48 m)			$17.05
per lin.ft.			$0.17
per lin.m.			$0.56

Material Cost of 100 Lin. Ft. (30.48 lin.M) of Metal Corner Bead Attached to lath Bases			
Description	Hours	Rate	Total
Lather	2.50	$35.04	$87.60
Cost per lin.ft.			$0.88
per lin.m.			$2.87

In estimating for other beads, molding, or screeds, substitute proper material price, and for other bases, consult discussion of labor on beads and

1257

molding for necessary additions to hours required. Also, if beads or molding are to serve as plaster grounds and all plumbing or straightening work is assigned to the installation of the accessory, add 1/2 hrs. per 100 lin.ft. (30.48 m)

ESTIMATING PLASTER QUANTITIES

The following covering capacities and quantities of plastering materials are based on manufacturer specifications.

Gypsum plasters are sold by the ton and are usually packed in 100-lb. (45-kg) sacks. When sand is mixed with plaster in the proportion of 1:1, 1:2, or 1:3, it means 100, 200, or 300 lbs. (45, 90, or 135 kg) of sand respectively are to be added to each 100-lb. (45-kg) sack of plaster.

Certain fire-resistant ratings may require an increase in the ratio of plaster to aggregate or an increase in the thickness of the plaster over normal grounds, or both, and the estimator must take this into consideration in using the following data.

Covering Capacity. The covering capacity of plaster necessarily depends upon conditions under which the plaster is applied-the mechanic, thickness of grounds, kind and trueness of the surface to be plastered, type of lath, quality of sand, etc.

The covering capacity given for each type of material and for the various surfaces is based on average conditions and may be accepted as a safe basis for calculating the quantity of each class of material required when applied according to the manufacturer's specifications.

Thickness of grounds has considerable bearing on the covering capacity of plaster. The quantities given are based on the following: 3/4" (18.75 mm) grounds for metal lath, or about 5/8" (15.62 mm) of plaster over the face of the lath; 7/8" (21.87 mm) grounds for 3/8" (9.37 mm) gypsum lath or 1/2" (12.50 mm) of plaster; 1" (25 mm) for 1/2" (12.50 mm) insulating board lath or 1/2" (12.50 mm) of plaster and 5/8" (15.62 mm) for unit masonry.

Number of 100 Lb. (45.36 Kg) Sacks of Gypsum Cement Plaster Required per 100 Sq.Yds. (83.61 Sq.M.)		
Kind of Plastering Surface		
Metal Lath	Gypsum Lath	Unit Masonry
18 to 20	9 to 12	10 to 12
(Sanded 1:2;1:3)	(Sanded 1:2;1:3)	(Sanded 1:3)
For each 1/8" (3.12 mm) in Thickness, Proportions 1:3; Add or Deduct Sacks:		
2.5	2.5	2.5

For insulation board lath, use same plaster quantities as given for gypsum lath.

Standard specifications specify that the first (scratch) coat on all types of lath shall be mixed in the proportions of 1 part gypsum neat plaster to not more than 2 parts of sand by weight, or 100 lbs. (45 kg) of gypsum neat plaster to 2 cu.ft. (0.06 cu.m) of perlite or vermiculite aggregate.

First coat on masonry surfaces, except monolithic concrete, and second (brown) coat on all lath bases shall be mixed in the proportions of 1 part gypsum neat plaster to not more than 3 parts of sand by weight, or 100 lbs. (45 kg) of gypsum neat plaster to 2 cu.ft. (0.06 cu.m) of perlite or vermiculite aggregate.

The sacks of gypsum cement plaster shown in the table above per 100 sq.yds. (83.6 sq.m) for various bases are based on these proportions when used with sand aggregate.

Perlite and Vermiculite Plaster

Vermiculite is a mineral that expands when heated at about 2000°F (1093.33°C). Each ore particle expands to about 12 times its original size. It will not rot, decay, burn, or be attacked by vermin or termites.

Perlite is a siliceous volcanic rock mined in western United States. When crushed and quickly heated to above 1500°F (815.55°C), it expands to form lightweight, noncombustible, inert glass-like particles of cellular structure. This material, white in color, is about 1/10th the weight of sand.

Perlite or vermiculite plaster aggregate is a lightweight material. Compared with 100 lbs. per cu.ft. (1589 kg per cu.m) for sand aggregate, perlite ranges from 7-1/2 to 15 lbs. per cu.ft. (119-238 kg per cu.m), and vermiculite 6 to 10 lbs. per cu.ft. (95-159 kg per cu.m). Plaster using this aggregate will weigh about 30 to 45 lbs. per cu.ft. (477-715 kg per cu.m), depending on proportions used, against 100 lbs. per cu.ft. (1589 kg per cu.m), where sand aggregate is used.

When perlite or vermiculite is expanded, each granule has countless small dead air spaces, which in addition to making the material very light, gives it a high insulating value. It has approximately three and one half times as much insulating value as ordinary plaster. It is this insulating quality that provides good fire resistance when combined with gypsum plaster. These aggregates are frequently called for when specific fire ratings are required for the protection of steel framing. As such fire ratings usually require greater plaster thicknesses than normal grounds provide, the estimator is cautioned to determine the proper quantities required per 100 sq.yds. (83.61 sq.m).

Perlite or vermiculite plaster aggregate is usually packed in multi-wall paper bags containing 3 or 4 cu.ft. (0.08-0.11 cu.m). Since there may be some compression or packing in shipping or storage, to insure correct proportioning do not use the original package. Use a cubic foot measuring box.

Machine-applied vermiculite portland cement plaster is used for panel or spandrel walls and exterior columns. The contractor can apply it to exterior columns simultaneously with the wall, since the paper-backed wire fabric used in wall construction is also wrapped around the columns.The protection is also used for interior columns where more than ordinary wear and tear is expected due to the type of occupancy.

The following table gives the approximate quantities of gypsum plaster and perlite or vermiculite for 100 sq.yds. (83.61 sq.m) of various mixtures.

Recommended Proportions and Quantities

In the following table giving recommended proportions for gypsum plaster and perlite or vermiculite, the proportions 1-2, means 100 lbs. (45.35 kg) of gypsum plaster to 2 cu.ft. (0.06 cu.m) of perlite or vermiculite; 1-3 means 100 lbs. (45.35 kg) of gypsum plaster to 3 cu.ft. (0.08 cu.m) of perlite or vermiculite.

Type of Construction	Grounds (see notes)	Actual Plaster Thickness	Coat	Recommended Volume Proportions		Required per 100 Sq.Yds	
						Bags Plaster	Cu. Ft. Aggregate
Metal Lath	3/4 "	1/4 "	Scratch	1	- 2	10	20
		7/16 "	Brown	1	- 2	12 1/2	45
Gypsum Lath	7/8 "	3/16 "	Scratch	1	- 2	5	10
		1/4 "	Brown	1	- 2	10	20
2" Solid Partition	2 "	3/8 "	Scratch (1 Side Only)	1	- 2	15	20
		3/8 "	Brown (1 Side Only)	1	- 2	15	30
		1 1/8 "	Brown (2nd Side)	1	- 3	45	90
Unit Masonry	5/8 "	3/16 "	Scratch	1	- 3	4	12
		3/8 "	Brown	1	- 3	8	24

Type of Construction	Grounds (see notes)	Actual Plaster Thickness	Coat	Recommended Volume Proportions		Required per 83.61 Sq.M.	
						Bags Plaster	Cu. M. Aggregate
Metal Lath	18.75 mm	6.25 mm	Scratch	1	- 2	10	0.57
		10.94 mm	Brown	1	- 2	12 1/2	1.27
Gypsum Lath	21.88 mm	4.69 mm	Scratch	1	- 2	5	0.28
		6.25 mm	Brown	1	- 2	10	0.57
50 mm Solid Partition	50 mm	9.38 mm	Scratch (1 Side Only)	1	- 2	15	0.57
		9.38 mm	Brown (1 Side Only)	1	- 2	15	0.85
		28.13 mm	Brown (2nd Side)	1	- 3	45	2.55
Unit Masonry	15.63 mm	4.69 mm	Scratch	1	- 3	4	0.34
		9.38 mm	Brown	1	- 3	8	0.68

Grounds attached to studs, joists or channels, not the lath. 1/16" (1.56 mm) allowed for finish coat thickness.

FINISHES

Number of 100 Lb. (45.36 Kg) Sacks of Wood Fiber Plaster Required per 100 Sq.Yds. (83.61 Sq.M.)		
Kind of Plastering Surface		
Metal Lath	Gypsum Lath	Unit Masonry
32 to 34	17 to 18	12 to 14
Unsanded	Unsanded	Equal Parts of Sand by
For each 1/8" (3.12 mm) in Thickness, Proportions 1:3; Add or Deduct Sacks:		
5	4 1/2	3

For insulation board lath, use same plaster quantities as given for gypsum lath.

For two and three-coat work on all types of lath, gypsum wood fiber plaster may be used without the addition of aggregate.

For two and three-coat work on masonry surfaces and as scratch coat on lath (except monolithic concrete), gypsum wood fiber plaster shall be mixed in the proportions of 1 part of plaster to 1 part of sand, or perlite by weight.

Gypsum Ready-Mix Plasters Made with Lightweight Aggregate

Gypsum ready-mix plasters are generally made with perlite. The regular formula for lath bases is packed in 80-lb. (36.28-kg) sacks and the masonry formula for application over all masonry bases is packed in 67-lb. (30.39-kg) sacks. The following table gives the number of sacks required per 100 sq.yds. applied over various surfaces:

Kind of Plastering Surface		
Metal Lath	Gypsum Lath	Unit Masonry
32 to 34*	17 to 19*	24 to 26**
For each 1/8" (3.12 mm) in Thickness, Proportions 1:3; Add or Deduct Sacks:		
5	5	5

*80 Lb Sacks   ** 67 Lb Sacks*

Quantity of Putty Obtained from Various Plastering Materials

The quantity of lime putty obtainable from one ton of lime varies with the kind of lime, whether pebble quicklime, pulverized quicklime, or hydrated lime, also upon the quality of the lime, as lime produced in different parts of the country will produce varying amounts of lime putty. Under average conditions the following quantities should be obtained.

Plastering Material	Putty, Cu.Ft. per Ton Mat'l	Lbs. Req'd per Cu.Ft. Putty	Putty, Cu.M. per M.Ton Mat'l	Kg. req'd per Cu.M. Putty
Pebble quicklime	80	25	2.50	400.42
Pulverized quicklime	80	25	2.50	400.42
Hydrated lime	46	43 1/2	1.44	696.74
Gypsum plaster	37	54	1.16	864.92
Keene's cement	30	66 2/3	0.94	1,067.80
Portland cement	18 1/2	109	0.58	1,745.85

1261

Base Coat

Covering Capacity of Lime Plaster. Lime plaster is usually proportioned by volume, using 1 volume of stiff, well-aged lime putty to 2 volumes of dry sand, plus necessary fiber for scratch coat.

For the brown coat, use 1 volume of lime putty to 3 volumes of dry sand, plus fiber as needed.

One cu.ft. (0.028 cu.m) of sand is considered 100 lbs. (45.35 kg) or 2,700 lbs. per cu.yd. (1589 kg per cu.m).

Coat	Covering Capacity of One Cu.Yd. (.7646 Cu.M) of Lime Plaster Based on the Same Thickness of Grounds as given on Previous Pages			
	Plastering Surface Metal Lath	Unit Masonry 5/8" Grounds	Plastering Surface Metal Lath	Unit Masonry 15.63 mm Grounds
Scratch Coat	50 - 60 Sq.Yds.	-----	38.2 - 45.9 Sq.M.	-----
Brown Coat	70 - 80 Sq.Yds.	60 - 75 Sq.Yds.	53.5 - 61.2 Sq.M.	45.9 - 57.3 Sq.M.

Bondcrete For Interior Concrete Surfaces. Bondcrete is used as a bond coat on rough concrete surfaces only. If the concrete surfaces are smooth, they must be roughened by bush-hammering or hacking, or a dash coat of portland cement grout, composed of 1 part portland cement and 1-1/2 parts of fine sand mixed to a mushy consistency, shall be applied. Use a stiff fiber brush forcibly dashing the mix onto the surface and damp cure for two days and then allow to dry.

Bondcrete adheres to dry concrete surfaces because of its cementitious qualities and because it maintains a practically unchanging volume during the process of setting and hardening.

On properly prepared rough concrete ceilings, apply a coat of bond plaster scratched in thoroughly, doubled backed, and filled out to a true, even surface and left rough, ready to receive the finish coat. Total thickness of the plaster should not exceed 3/8" (9.37 mm).

On properly prepared rough concrete columns or walls, scratch in a minimum 1/4" (6.25 mm) coat of bond plaster and then follow with a brown coat of gypsum plaster-one part gypsum neat plaster to not more than three parts sand, by weight-troweled into the scratch coat before it sets. Brown coat shall be brought out to ground, straightened to a true surface, and left rough, ready to receive the finish coat. Total thickness of plaster should not exceed 5/8" (15.62 mm).

It usually requires 1,700 to 2,200 lbs. (771.10 to 997.90 kg) of Bondcrete per 100 sq.yds. (83.61 sq.m) depending upon the regularity of the concrete surface and thickness applied.

Plaster-Weld. Plaster-Weld permanently bonds new gypsum plaster, finish or base coat, or acoustical plaster to most surfaces, but cannot be used over calcimine or other water soluble coatings.

No chipping or roughening of surfaces is required. The surface must be structurally sound and free from dust, dirt, grease, oil, or wax.

FINISHES

It may be applied by brush, roller, or spray equipment. After allowing Plaster-Weld to become dry to the touch (one hour on porous surfaces and two hours on non-porous surfaces), plaster can be applied either immediately or later. A time lag in excess of a week or ten days may result in a need for recoating. Plaster-Weld covers up to 500 sq.ft. per gal. (11.89 sq.m per liter) when sprayed, which gives the most satisfactory results. When applied by brush or roller, coverage is reduced. When sprayed on, the labor cost per 100 sq.yds. (83.61 sq.m) should average as follows:

Description	Hours	Rate	Total
Labor	2.25	$26.16	$58.86
Cost per sq. yd.			$0.59
per sq.ft.			$0.07
per sq.m			$0.70

Plaster-Weld costs approximately $20.00 per gal. ($5.28 per liter) in 1-gal. (3.8-L) cans.

Finishing Lime. Finishing lime is essential to two common types of job mixed finish coats-the white putty coat trowel finish and the sand float finish.

For the trowel finish coat, the proportion is generally 1 part of gypsum gauging plaster to 2 parts of hydrated lime on a dry weight basis, or 100 lbs. (45.35 kg) of gypsum gauging plaster to 4-1/2 cu.ft. (0.13 cu.m) of lime putty. One ton of gypsum gauging plaster and 2 tons of hydrated finishing lime, or 90 cu.ft. (2.55 cu.m) of lime putty will finish approximately 1,200 sq.yds. (1003.32 sq.m).

While a sand float finish may be mixed with 1 part of lime putty and 2-1/2 to 3 parts of sand, it is generally necessary to add Keene's cement or gypsum gauging plaster to give the finish early hardness and strength. A standard mix for a Keene's cement-lime putty sand float finish is 2 parts of lime putty to 1-1/2 parts of Keene's cement and 4-1/2 parts of sand by volume. This is equivalent to 1 ton of hydrated lime, or 45 cu.ft. (1.27 cu.m) of putty, to 1-1/4 tons (1.13 metric tons) of Keene's cement and 4-3/4 tons (4.28 metric tons) of sand, and this quantity will cover approximately 1,600 sq. yds. (1337.76 sq.m).

If type "S" (fully) hydrated limes are used, they must be added to the mix in dry form, as they form immediate plasticity on wetting. If quicklimes are used, they must be slaked and properly aged and added to the mix in the volume indicated above for putty.

Prepared Gypsum Trowel Finishes. There are a number of prepared gypsum trowel finishes, which require only adding water to use. A few of these are Universal White Trowel Finish, Red Top Trowel, and Red Top Imperial (from Larson Products Corp., Jessup, Md.).

Under average conditions, if mixed and applied in accordance with standard specifications, one ton will cover 350 to 400 sq.yds. (293-334 sq.m) of surface, or 500 to 575 lbs. per 100 sq.yds. (2.69-3.10 kg per sq.m).

Prepared Gypsum Sand Float Finishes. There are a number of gypsum sand float finishes. These finishes are ready for use when water is added, and under average conditions one ton is sufficient for 250 to 275 sq.yds. (209.02 to 229.92 sq.m), or it requires 725 to 800 lbs. per 100 sq.yds. (3.9-4.3 kg per sq.m).

Estimating Labor Costs of Plastering

Before it is possible to prepare an accurate estimate on the cost of any plastering job, the quality of work must be considered. Labor costs may vary up to 25 to 30 percent.

The costs given on the following pages cover two grades of workmanship, which are fully described at the beginning of this chapter.

Applying Scratch Coat of Plaster. A plasterer should apply 140 to 165 sq.yds. (11.05 to 137.95 sq.m) of scratch coat per 8-hr. day, where ordinary grade of workmanship is required, and 135 to 155 sq.yds. (112.87 to 129.59 sq.m) per 8-hr. day where first grade workmanship is required.

Applying Brown Coat. Where an ordinary grade of workmanship is required, a plasterer should apply 90 to 110 sq.yds. (75.24 to 91.97 sq.m) of brown coat per 8-hr. day.

Where first grade workmanship is required, a plasterer should brown out 70 to 85 sq.yds. (58.52 to 71.07 sq.m) per 8-hr. day.

Applying Scratch—Double-Back Basecoat. Over masonry bases and in many markets over gypsum lath base, it is common practice to apply a single plaster mix in a "scratch-in" and "double-back" operation to fill out to grounds.

For ordinary work, a plasterer should apply 90 to 110 sq.yds. (75.24 to 91.97 sq.m) per 8-hr. day, and for first grade workmanship, a plasterer should apply 70 to 85 sq.yds. (58.27 to 71.06 sq.m) per 8-hr. day in this manner.

Applying White Finish or "Putty" Coat. On an ordinary job of white coating, a plasterer should apply 90 to 110 sq.yds. (75.24 to 91.97 sq.m) per 8-hr. day.

Where first grade workmanship is specified, a plasterer will apply 60 to 80 sq.yds. (50.16 to 66.89 sq.m) per 8-hr. day.

General Notes Applying to Finish Plaster Coats

All of the tables on the following pages give costs on a finish white coat. If wanted with sand float finish, deduct quantities given in tables for Finish White Coat and add cost of Sand Float Finish as given on the following pages under "Sand Float Finish".

Applying Sand Float Finish. A plasterer should apply 100 to 120 sq.yds. (83.61 to 100.33 sq.m) of simple sand float finish per 8-hr. day, where

1264

FINISHES

an ordinary grade of workmanship is required, or 70 to 90 sq.yds. (58.27 to 75.24 sq.m) when first grade workmanship is required.

Labor Applying 100 Sq.Yds. (83.61 Sq.M) of Plaster to Various Plaster Bases Ordinary Workmanship				
Description of Work	Sq.Yds. Per 8-Hr. Day	Plasterer Hours	Tender Hours	Sq.M per 8- Hr. Day
Gypsum Mortar on Gypsum lath-1/2" (12.50 mm) Plaster-2-Coat, Work 1:2-1/2 Brown Coat	90 - 110	8.0	5.0	75.25 - 91.97
Gypsum Mortar on Unit Masonry-5/8" (15.62 mm) Plaster-2-Coat, Work 1:3 Brown Coat	80 - 100	9.0	6.0	66.89 - 83.61
Lime Mortar on Unit Masonry-5/8" (15.62 mm) Plaster-2-Coat, Work Brown Coat 1:3	90 - 100	8.5	5.5	75.25 - 83.61
Lime Mortar on Metal Lath-3/4"(18.75mm) Grounds 5/8" (15.62 mm) Plaster-3-Coat Work Scratch Coat 1:2	150 - 160	5.0	5.0	125.42 - 133.78
Brown Coat 1:3	90 - 110	8.0	5.0	75.25 - 91.97
Lime Mortar on Unit Masonry 5/8" (15.62 mm) Plaster-3-Coat Work Scratch Coat 1:2	150 - 160	5.0	5.0	75.25 - 75.25
Brown Coat 1:3	90 - 110	8.0	5.0	75.25 - 91.97
Gypsum Mortar on Unit Masonry Tile-5/8" (15.62 mm) Plaster-3-Coat Work Scratch Coat 1:3	150 - 160	5.0	5.0	125.42 - 133.78
Brown Coat 1:3	90 - 110	8.0	5.0	75.25 - 91.97
Gypsum Mortar on Gypsum Lath 1/2" (12.50 mm) Plaster-3-Coat Work Scratch Coat 1:2	150 - 160	5.0	5.0	125.42 - 133.78
Brown Coat 1:3	90 - 110	8.0	5.0	75.25 - 91.97
Gypsum Mortar on Metal Lath-3/4" (18.75 mm) Grounds 5/8" (15.62 mm) Plaster-3-Coat Work Scratch Coat 1:2	150 - 160	5.0	5.0	75.25 - 75.25
Brown Coat 1:3	85 - 100	9.0	6.0	71.07 - 83.61
Two-Inch Solid Plaster Partitions - Metal Lath One Side Back-up Finished 2-Sides				
Gypsum Mortar 3/4" (18.75 mm) Grounds 2" (50 mm) Plaster Scratch Coat 1:2	150 - 160	5.0	5.0	75.25 - 75.25
Back Up Coat 1:2	150 - 160	5.0	5.0	125.42 - 133.78
2 Brown Coats 1:2	40 - 50	18.0	12.0	33.44 - 41.81

Labor Applying 100 Sq. Yds. (83.6 Sq.M) of Plaster to Various Plaster Bases Ordinary Workmanship				
Description of Work	Sq. Yds. Per 8-Hr. Day	Plasterer Hours	Tender Hours	Sq.M per 8-Hr. Day
Hollow Stud Partitions-Lathed 2-Sides - Platstered to 3/4" (18.75 mm)				
Grounds Scratch Coat 1:2 two sides	75 - 80	10.00	10.00	62.71 - 66.89
Brown Coat-2 Sides	40 - 50	18.00	12.00	33.44 - 41.81
Finish White-All Classes of Undercoat	90 - 110	8.00	4.00	75.25 - 91.97
Sand Finish-All Classes of Undercoat	100 - 120	7.25	5.00	83.61 - 100.33
First Grade Workmanship-Where First Grade Workmanship is Required as described on the previous pages, add Hours given below:				
Brown Coat	-----	2.25	1.00	-----
Finish White Coat	-----	3.50	1.00	-----
Sand Finish	-----	2.75	1.00	-----

Applying Finish Coat of Cement Wainscoting. Where cement plaster finishes are used in kitchens, bathrooms, laundries, corridors, etc., using Keene's cement, or similar finishes applied over a brown coat of gypsum cement plaster or portland cement, a plasterer should apply 50 to 60 sq.yds. (41.80 to 50.16 sq.m) per 8-hr. day.

If the wainscoting is blocked off into squares, 3" x 6" (75 x 150 mm), 4" x 4" (100 x 100 mm) or 6" x 6" (150 x 150 mm) to represent tile, a plasterer should block off 45 to 55 sq.yds. (37.62 to 45.98 sq.m) per 8-hr. day.

To apply the finish coat of plaster and block same off into squares to represent tile, a plasterer should complete 25 to 30 sq.yds. (20.90 to 25.08 sq.m) per 8-hr. day.

Where first grade workmanship is required, a plasterer should apply 35 to 45 sq.yds. (29.26 to 37.62 sq.m) of cement wainscoting per 8-hr. day, and block off 40 to 45 sq.yds. (33.44 to 37.62 sq.m) or will apply cement wainscoting and block off 18 to 22 sq.yds. (15.04 to 18.39 sq.m) per 8-hr. day.

Itemized Plastering Costs

The itemized cost given on the following pages are intended to furnish the estimator and contractor detailed information regarding material quantities and labor costs of all kinds of plastering.

All plaster quantities are based on thickness of grounds as given. If the work is "skimped" and the plaster does not fill out the grounds or is applied thinner than specified, then the quantities in the tables will not be correct. If sand is not added in accordance with manufacturer's specifications, the quantities may be more or less than given. Rates given for quantities in lbs. (kg) are usually for a 100-lb. (100-kg) bag. (Note that metric figures are based on a 100-kg bag, which equals about 220 lbs.)

Labor hours include time required for building and removing ordinary scaffolding, but will not cover scaffolding for high work in churches, auditoriums, etc.

1266

FINISHES

2-Coat Plastering

Material Cost of 100 Sq. Yds. (83.6 Sq.M) 2-Coat Gypsum Cement Plaster, White Finish on Gypsum Lath Ordinary Workmanship 3/8" (9.37 mm) Grounds 1/2" (12.50 mm) Plaster		
Brown Coat-1:2.5	Rate	Total
1,000 lbs. (453.6 kg) gypsum plaster	$15.36 *	$ 153.60
25 cu.ft. (0.71 cu.m.) sand	$ 1.15	$ 28.80
Finish White Coat		
340 lbs. (152.22 kg) fin. hydrated lime	$20.40 *	$ 69.36
170 lbs. (77.11 kg) gauging plaster	$22.93 *	$ 38.98
Cost 100 sq.yds. (83.61 sq.m.)		$ 290.74
per sq.yd.		$ 2.91
per sq.ft.		$ 0.32
per sq.m.		$ 3.48

Based on 100 lb. (45.36 kg) bag.
Add for gypsum lath.

Labor Cost of 100 Sq.Yds. (83.61 Sq.M) 2-Coat Gypsum Cement Plaster, White Finish on Gypsum Lath Ordinary Workmanship 3/8" (21.87 mm) Grounds - 1/2" (12.50 mm) Plaster			
Brown Coat-1:2.5	Hours	Rate	Total
Plasterer	8.00	$32.41	$ 259.28
Labor	5.00	$26.16	$ 130.80
Finish White Coat			
Plasterer	8.00	$32.41	$ 259.28
Labor	4.00	$26.16	$ 104.64
Cost 100 sq. yds.			$ 754.00
per sq. yd.			$ 7.54
per sq. ft.			$ 0.84
per sq.m			$ 9.02

Add for gypsum lath.

Material Cost of 100 Sq.Yds. (83.61 Sq.M) 2-Coat Gypsum Cement Plaster, White Finish on Unit Masonry Ordinary Workmanship - 5/8" (15.62 mm) Grounds		
Brown Coat-1:3	Rate	Total
1,100 lbs. (495 kg) gypsum plaster	$10.24 *	$ 112.64
33 cu.ft. (0.93 cu.m.) sand	$ 0.77	$ 25.34
Finish White Coat		
340 lbs. (153 kg) fin. hydrated lime	$13.60 *	$ 46.24
170 lbs. (76.5 kg) gauging plaster	$15.28 *	$ 25.98
Cost 100 sq. yds.		$ 210.20
per sq. yd.		$ 1.31
per sq. ft.		$ 0.15
per sq.m		$ 1.57

Based on 100-lb. bag.
Add for gypsum lath.

Labor Cost of 100 Sq. Yds. (83.6 Sq.M) 2-Coat Gypsum Cement Plaster, White Finish on Unit Masonry Ordinary Workmanship - 5/8" (15.62 mm) Grounds			
Brown Coat-1:3	Hours	Rate	Total
Plasterer	9.00	$32.41	$ 291.69
Labor	6.00	$26.16	$ 156.96
Finish White Coat			
Plasterer	8.00	$32.41	$ 259.28
Labor	4.00	$26.16	$ 104.64
Cost 100 sq. yds.			$ 812.57
per sq. yd.			$ 8.13
per sq. ft.			$ 0.90
per sq.m			$ 9.72

Material Cost of 100 Sq. Yds. (83.6 Sq.M) 2-Coat Wood Fiber Plaster, White Finish on Gypsum Lath Ordinary Workmanship 7/8" (21.87 mm) Grounds 1/2" (12.50 mm) Plaster		
Brown Coat	Rate	Total
1,700 lbs. (765 kg) wood fiber	$10.24 *	$ 174.08
Finish White Coat		
340 lbs. (153 kg) fin. hydrated lime	$13.60 *	$ 46.24
170 lbs. (76.5 kg) gauging plaster	$15.28 *	$ 25.98
Cost 100 sq. yds.		$ 246.30
per sq. yd.		$ 2.46
per sq. yd.		$ 0.27
per sq.m		$ 2.95

Add for gypsum lath.
*Based on 100-lb. bag.

Labor Cost of 100 Sq. Yds. (83.6 Sq.M) 2-Coat Wood Fiber Plaster, White Finish, on Gypsum Lath Ordinary Workmanship 7/8" (21.87 mm) Grounds 1/2" (12.50 mm) Plaster			
Brown Coat	Hours	Rate	Total
Plasterer	9.00	$32.41	$ 291.69
Labor	5.00	$26.16	$ 130.80
Finish White Coat			
Plasterer	8.00	$32.41	$ 259.28
Labor	4.00	$26.16	$ 104.64
Cost 100 sq. yds.			$ 786.41
per sq. yd.			$ 7.86
per sq. ft.			$ 0.87
per sq.m			$ 9.41

Bondcrete plaster can only be applied to rough concrete surfaces. If the concrete surfaces are smooth, they must be roughened by bush-hammering or hacking, or a dash coat of portland cement grout, composed of 1 part portland cement and 1-1/2 parts fine sand, mixed to a mushy

1268

FINISHES

consistency, must be applied. Use a stiff fiber brush forcibly dashing the mix onto the surface and damp cure for two days and then allow to dry. For costs on this work, refer to section on waterproofing.

Material Cost of 100 Sq. Yds. (83.6 Sq.M) 2-Coat Bondcrete Plaster, White Finish on Concrete Ceilings Ordinary Workmanship Plaster 3/8" (9.37 mm) Thick		
Brown Coat	Rate	Total
2,000 lbs. (900 kg) Bondcrete	$11.87 *	$ 237.44
Finish White Coat		
340 lbs. (153 kg) fin. hydrated lime	$13.60 *	$ 46.24
170 lbs. (76.5 kg) gauging plaster	$15.28 *	$ 25.98
Cost 100 sq. yds.		$ 309.66
per sq. yd.		$ 3.10
per sq. ft.		$ 0.34
per sq.m		$ 3.70

Add for gypsum lath.
*Based on 100-lb. bag.

Labor Cost of 100 Sq.Yds. (83.61 Sq.M) 2-Coat Bondcrete Plaster, White Finish on Concrete Ceilings Ordinary Workmanship - Plaster 3/8" (9.37mm) Thick			
Brown Coat	Hours	Rate	Total
Plasterer	8.00	$32.41	$ 259.28
Labor	5.00	$26.16	$ 130.80
Finish White Coat			
Plasterer	8.00	$32.41	$ 259.28
Labor	4.00	$26.16	$ 104.64
Cost 100 sq. yds.			$ 754.00
per sq. yd.			$ 7.54
per sq. ft.			$ 0.84
per sq.m			$ 9.02

3-Coat Plastering

Material Cost of 100 Sq.Yds. (83.61 Sq.M) 3-Coat Gypsum Cement Plaster, White Finish on Gypsum Lath Ordinary Workmanship 7/8" (21.87 mm) Grounds 1/2" (12.50 mm) Plaster		
Scratch Coat-1:2	Rate	Total
400 lbs. (180 kg) gypsum plaster	$10.24 *	$ 40.96
8 cu.ft. (0.22 cu.m.) sand	$ 0.77	$ 6.14
Brown Coat-1:3		
700 lbs. (315 kg) gypsum plaster	$10.24 *	$ 71.68
21 cu.ft. (0.59 cu.m.) sand	$ 0.77	$ 16.13
Finish White Coat		
340 lbs. (153 kg) fin. hydrated lime	$13.60 *	$ 46.24
170 lbs. (76.5 kg) gauging plaster	$15.28 *	$ 25.98
Cost 100 sq. yds.		$ 129.46
per sq. yd.		$ 1.29
per sq. ft.		$ 0.14
per sq.m		$ 1.54

Add for gypsum lath.
*Based on 100-lb. bag.

Labor Cost of 100 Sq.Yds. (83.61 Sq.M) 3-Coat Gypsum Cement Plaster, White Finish on Gypsum Lath Ordinary Workmanship 7/8" (21.87 mm) Grounds 1/2" (12.50 mm) Plaster			
Scratch Coat-1:2	Hours	Rate	Total
Plasterer	5.00	$32.41	$ 162.05
Labor	5.00	$26.16	$ 130.80
Brown Coat-1:3			
Plasterer	8.00	$32.41	$ 259.28
Labor	5.00	$26.16	$ 130.80
Finish White Coat			
Plasterer	8.00	$32.41	$ 259.28
Labor	4.00	$26.16	$ 104.64
Cost 100 sq. yds.			$1,046.85
per sq. yd.			$ 10.47
per sq. yd.			$ 1.16
per sq.m			$ 12.52

If first grade workmanship is required, add 5.75 hrs. plasterer time and 2 hrs. labor time per 100 sq. yds. (83.6 sq.m).

FINISHES

Material Cost of 100 Sq.Yds. (83.61 Sq.M) 3-Coat Gypsum Cement Plaster, White Finish on Unit Masonry Ordinary Workmanship 5/8" (15.62 mm) Grounds		
Scratch Coat-1:3	Rate	Total
500 lbs. (225 kg) gypsum plaster	$10.24 *	$ 51.20
15 cu.ft. (0.42 cu.m.) sand	$ 0.74	$ 11.04
Brown Coat-1 :3		
700 lbs. (315 kg) gypsum plaster	$10.24 *	$ 71.68
21 cu.ft. (0.59 cu.m.) sand	$ 0.77	$ 16.13
Finish White Coat		
340 lbs. (153 kg) fin. hydrated lime	$12.75 *	$ 43.35
170 lbs. (76.5 kg) gauging plaster	$14.33 *	$ 24.35
Cost 100 sq. yds.		$ 217.75
per sq. yd.		$ 2.18
per sq. ft.		$ 0.24
per sq.m		$ 2.60

Based on 100-lb. bag.

Labor Cost of 100 Sq.Yds. (83.61 Sq.M) 3-Coat Gypsum Cement Plaster, White Finish on Unit Masonry Ordinary Workmanship 5/8" (15.62 mm) Grounds			
Scratch Coat-1:3	Hours	Rate	Total
Plasterer	5.00	$32.41	$ 162.05
Labor	5.00	$26.16	$ 130.80
Brown Coat-1:3			
Plasterer	8.00	$32.41	$ 259.28
Labor	5.00	$26.16	$ 130.80
Finish White Coat			
Plasterer	8.00	$32.41	$ 259.28
Labor	4.00	$26.16	$ 104.64
Cost 100 sq. yds.			$1,046.85
per sq. yd.			$ 10.47
per sq. ft			$ 1.16
per sq.m			$ 12.52

If first grade workmanship is required, add 5.75 hrs. plasterer time and 2 hrs. labor time per 100 sq. yds. (83.6 sq.m).

Material Cost of 100 Sq.Yds. (83.61 Sq.M) 3-Coat Gypsum Cement Plaster White Finish on Metal Lath Walls and Ceilings Ordinary Workmanship 3/4"(18.75 mm) Grounds - 5/8" (15.62 mm) Plaster		
Scratch Coat-1:2	Rate	Total
800 lbs. (360 kg) gypsum plaster	$10.24 *	$ 81.92
16 cu.ft. (0.45 cu.m.) sand	$ 0.77	$ 12.29
Brown Coat-1 :3		
1,200 lbs. (540 kg) gypsum plaster	$10.24 *	$ 122.88
36 cu.ft. (1.01 cu.m.) sand	$ 0.77	$ 27.65
Finish White Coat		
340 lbs. (153 kg) fin. hydrated lime	$13.60 *	$ 46.24
170 lbs. (76.5 kg) gauging plaster	$15.28 *	$ 25.98
Cost 100 sq. yds.		$ 316.95
per sq. yd.		$ 3.17
per sq. ft		$ 0.35
per sq.m		$ 3.79

Add for metal lath and furring.
Based on 100-lb. bag.

Labor Cost of 100 Sq.Yds. (83.61 Sq.M) 3-Coat Gypsum Cement Plaster, White Finish on Metal Lath Walls and Ceilings Ordinary Workmanship 3/4" (18.75 mm) Grounds - 5/8" (15.62 mm) Plaster			
Scratch Coat—1:2	Hours	Rate	Total
Plasterer	5.00	$32.41	$ 162.05
Labor	5.00	$26.16	$ 130.80
Brown Coat-1:3			
Plasterer	9.00	$32.41	$ 291.69
Labor	6.00	$26.16	$ 156.96
Finish White Coat			
Plasterer	8.00	$32.41	$ 259.28
Labor	4.00	$26.16	$ 104.64
Cost 100 sq. yds.			$1,105.42
per sq. ft.			$ 11.05
per sq. yd.			$ 1.23
per sq.m			$ 13.22

Add for metal lath and furring.
If first grade workmanship is required, add 5.75 hrs. plasterer time and 2 hrs.
labor time per 100 sq.yds. (83.61 sq.m).

Material Cost of 100 Sq.Yds. (83.61 Sq.M) 2" (50 mm) Solid Plaster Partition on Metal Lath, Ordinary Workmanship		
Scratch Coat-1:2	Rate	Total
2,000 lbs. (900 kg) gypsum plaster	$10.24	$ 204.80
40 cu.ft. (1.13 cu.m.) sand	$ 0.77	$ 30.72
Brown Coat-1:3		
2,800 lbs. (1260 kg) gypsum plaster	$10.24	$ 286.72
84 cu.ft. (2.38 cu.m.) sand	$ 0.77	$ 64.51
Finish White Coat		
680 lbs. (306 kg) fin. hydrated lime	$13.60	$ 92.48
340 lbs. (153 kg) gauging plaster	$15.28	$ 51.95
Cost 100 sq. yds.		$ 731.18
per sq. yd.		$ 7.31
per sq. ft		$ 0.81
per sq.m		$ 8.75

Measurement taken on one side only. Add for metal lath and furring.

Labor Cost of 100 Sq.Yds. (83.61 Sq.M) 2" (50 mm) Solid Plaster Partition on Metal Lath, Ordinary Workmanship			
Scratch Coat-1:2	Hours	Rate	Total
Plasterer	10.00	$32.41	$324.10
Labor	10.00	$26.16	$261.60
Brown Coat-1:3			
Plasterer	18.00	$32.41	$583.38
Labor	12.00	$26.16	$313.92
Finish White Coat			
Plasterer	16.00	$32.41	$518.56
Labor	8.00	$26.16	$209.28
Cost 100 sq. yds.			$2,210.84
per sq. yd.			$22.11
per sq. ft.			$2.46
per sq.m			$26.44

Measurement taken on one side only. Add for metal lath and furring.
If first grade workmanship is required, add 11.5 hrs. plasterer time and 4 hrs.
labor time per 100 sq. yds. (83.6 sq.m).

FINISHES

Material Cost of 100 Sq.Yds. (83.61 Sq.M) 2" (50 mm) Solid Gypsum Long Length Lath and Plaster Partition Ordinary Workmanship Partition 2" (50 mm) Thick; 1/2" (12.250 mm) Gypsum Lath; 3/4" (18.75 mm) of Plaster Each Side of Lath		
Scratch Coat-1:2	Rate	Total
800 lbs. (360 kg) gypsum plaster	$10.24 *	$ 81.92
16 cu.ft. (0.45 cu.m.) sand	$ 0.77	$ 12.29
Brown Coat-1:3		
2,400 lbs. (1080 kg) gypsum plaster	$10.24 *	$ 245.76
72 cu.ft. (2.03 cu.m.) sand	$ 0.77	$ 55.30
Finish White Coat		
680 lbs. (306 kg) fin. hydrated lime	$13.60 *	$ 92.48
340 lbs. (153 kg) gauging plaster	$15.28 *	$ 51.95
Cost 100 sq. yds.		$ 539.70
per sq. yd.		$ 5.40
per sq. ft		$ 0.60
per sq.m		$ 6.45

Add for gypsum lath.
*Based on 100-lb. bag.

Labor Cost of 100 Sq.Yds. (83.61 Sq.M) 2" (50 mm) Solid Gypsum Long Length Lath and Plaster Partition Ordinary Workmanship Partition 2" (50 mm) Thick; 1/2" (12.250 mm) Gypsum Lath; 3/4" (18.75 mm) of Plaster Each Side of Lath			
Scratch Coat-1:2	Hours	Rate	Total
Plasterer	10.00	$32.41	$ 324.10
Labor	10.00	$26.16	$ 261.60
Brown Coat-1:3			
Plasterer	16.00	$32.41	$ 518.56
Labor	10.00	$26.16	$ 261.60
Finish White Coat			
Plasterer	16.00	$32.41	$ 518.56
Labor	8.00	$26.16	$ 209.28
Cost 100 sq. yds.			$2,093.70
per sq. yd.			$ 20.94
per sq. ft			$ 2.33
per sq.m			$ 25.04

Add for gypsum lath.

Material Cost of 100 Sq.Yds. (83.61 Sq.M) 2" (50 mm) Solid Studless Metal Lath and Plaster Partition Ordinary Workmanship Partition 2" (50 mm) Thick; 3.4 Lb (1.53 Kg.) Metal Lath; Plaster 2" (50 mm) Thick		
Scratch Coat-1:2	Rate	Total
2,000 lbs. (900 kg) Gypsum plaster	$10.24 *	$ 204.80
40 cu.ft. (1.13 cu.m.) sand	$ 0.77	$ 30.72
Brown Coat-1 :3		
2,800 lbs. (1260 kg) gypsum plaster	$10.24 *	$ 286.72
84 cu.ft. (2.38 cu.m.) sand	$ 0.77	$ 64.51
Finish White Coat		
680 lbs. (306 kg) fin. hydrated lime	$13.60 *	$ 92.48
340 lbs. (153 kg) gauging plaster	$15.28 *	$ 51.95
Cost 100 sq. yds.		$ 731.18
per sq. yd.		$ 7.31
per sq. ft.		$ 0.81
per sq.m		$ 8.75

Add for metal lath and furring.
*Based on 100-lb. bag.

Labor Cost of 100 Sq.Yds. (83.61 Sq.M) 2" (50 mm) Solid Studless Metal Lath and Plaster Partition Ordinary Workmanship Partition 2" (50 mm) Thick; 3.4 Lb (1.53 Kg.) Metal Lath; Plaster 2" (50 mm) Thick			
Scratch Coat-1:2	Hours	Rate	Total
Plasterer	10.00	$32.41	$ 324.10
Labor	10.00	$26.16	$ 261.60
Brown Coat-1:3			
Plasterer	18.00	$32.41	$ 583.38
Labor	12.00	$26.16	$ 313.92
Finish White Coat			
Plasterer	16.00	$32.41	$ 518.56
Labor	8.00	$26.16	$ 209.28
Cost 100 sq. yds.			$2,210.84
per sq. yd.			$ 22.11
per sq. ft.			$ 2.46
per sq.m			$ 26.44

Add for metal lath and furring.

FINISHES

Material Cost of 100 Sq. Yds. (83.61 Sq.M) 3-Coat Wood Fiber Plaster, White Finish on Metal Lath Walls and Ceilings, Ordinary Workmanship 3/4" (18.75 mm) Grounds - 5/8" (15.62 mm) Plaster		
Scratch Coat	Rate	Total
1,100 lbs. (495 kg) wood fiber	$10.24 *	$ 70.40
Brown Coat		
2,300 lbs. (1035 kg) wood fiber	$10.24 *	$ 235.52
Finish White Coat		
340 lbs. (153 kg) fin. hydrated lime	$13.60 *	$ 46.24
170 lbs. (76.5 kg) gauging plaster	$15.28 *	$ 25.98
Cost 100 sq. yds.		$ 378.14
per sq. yd.		$ 3.78
per sq. ft.		$ 0.42
per sq.m		$ 4.52

Add for metal lath and furring.
*Based on 100-lb. bag.

Labor Cost of 100 Sq. Yds. (83.61 Sq.M) 3-Coat Wood Fiber Plaster, White Finish on Metal Lath Walls and Ceilings, Ordinary Workmanship 3/4" (18.75 mm) Grounds - 5/8" (15.62 mm) Plaster			
Scratch Coat	Hours	Rate	Total
Plasterer	5.50	$32.41	$ 178.26
Labor	5.00	$26.16	$ 130.80
Brown Coat			
Plasterer	10.00	$32.41	$ 324.10
Labor	6.00	$26.16	$ 156.96
Finish White Coat			
Plasterer	8.00	$32.41	$ 259.28
Labor	4.00	$26.16	$ 104.64
Cost 100 sq. yds.			$1,154.04
per sq. yd.			$ 11.54
per sq. ft.			$ 1.28
per sq.m			$ 13.80

If first grade workmanship is required, add 5.75 hrs. plasterer time and 2 hrs. labor time per 100 sq.yds. (83.61 sq.m).

On properly prepared rough concrete columns or walls, scratch in a minimum 1/4" (6.25 mm) coat of bond plaster and then follow with a brown coat of gypsum plaster (1 part gypsum neat plaster to not more than 3 parts sand, by weight) troweled into the scratch coat before it sets. Brown coat shall be brought out to grounds, straightened to a true surface, and left rough ready to receive the finish coat. Total thickness of plaster not to exceed 5/8" (15.62 mm). Add for preparing concrete surfaces, if required.

Material Cost of 100 Sq. Yds. (83.61 Sq.M) 3-Coat Bondcrete Plaster, White Finish on Concrete Walls and Columns, Ordinary Workmanship Plaster 5/8" (15.62 mm) Thick		
Scratch Coat-1/4" (6.25 mm)	Rate	Total
1,500 lbs. (675 kg) Bondcrete	$11.87 *	$ 178.08
Brown Coat-1 :3		
800 lbs. (360 kg) gypsum plaster	$10.24 *	$ 81.92
24 cu.ft. (0.68 cu.m) sand	$ 0.77	$ 18.43
Finish White Coat		
340 lbs. (153 kg) fin. hydrated lime	$13.60 *	$ 46.24
170 lbs. (76.5 kg) gauging plaster	$15.28 *	$ 25.98
Cost 100 sq. yds.		$ 350.65
per sq. yd.		$ 3.51
per sq. ft		$ 0.39
per sq.m		$ 4.19

Based on 100-lb. bag.

Labor Cost of 100 Sq. Yds. (83.61 Sq.M) 3-Coat Bondcrete Plaster, White Finish on Concrete Walls and Columns, Ordinary Workmanship Plaster 5/8" (15.62 mm) Thick			
Scratch Coat-1/4" (6.25 mm)	Hours	Rate	Total
Plasterer	5.00	$32.41	$ 162.05
Labor	5.00	$26.16	$ 130.80
Brown Coat-1:3			
Plasterer	8.00	$32.41	$ 259.28
Labor	5.00	$26.16	$ 130.80
Finish White Coat			
Plasterer	8.00	$32.41	$ 259.28
Labor	4.00	$26.16	$ 104.64
Cost 100 sq. yds.			$1,046.85
per sq. yd.			$ 10.47
per sq. ft.			$ 1.16
per sq.m			$ 12.52

If first grade workmanship is required, add 5.75 hrs. plasterer time and 2 hrs. labor time per 100 sq.yds. (83.61 sq.m).

Plastering Radiant Panel Heat Ceilings. Where radiant panel heating coils are embedded in ceiling plaster, base coat plastering costs will be higher than normal for both material and labor.

For this type of heating, 3/8" (9.37 mm) I.D.—1/2" (12.50 mm) O.D.—copper tubing or electric cable is secured to ceiling framing after lath is installed, in contact with the underside. For proper operation of the heating system, all ceiling coils must be completely embedded in plaster, in perfect contact with plaster (no air pockets permitted) and have a plaster cover of 3/8" (9.37 mm) or more. This means at least 7/8" (21.87 mm) of plaster

1276

FINISHES

below the face of the lath for copper tubing and 1/2" (12.50 mm) for electric cable.

Scratch coat application is more difficult, due to the care which must be taken to obtain proper contact with tubing. Brown coat application will require more labor, due to the increased thickness of material required and the difficulty in making it stick to the ceiling. In some cases, a second scratch coat may be necessary to obtain the required thickness.

There should be no change in the method or cost of applying the various plaster finish coats.

Use sand aggregate only for radiant heat ceilings and walls. Lightweight aggregate is an insulator.

Material Cost of 100 Sq.Yds. (83.61 Sq.M), 3-Coat Gypsum Plaster 7/8" (21.87 mm) Thick, Applied to Gypsum Lath Ceilings with Radiant Heating Coils Embebbed in the Plaster, Ordinary Workmanship - 7/8" (21.87 mm) Plaster		
Scratch Coat-1:2	Rate	Total
400 lbs. (180 kg) gypsum plaster	$10.24 *	$ 40.96
8 cu.ft. (0.22 cu.m) sand	$ 0.77	$ 6.14
Brown Coat-1:3		
1,400 lbs. (630 kg) gypsum plaster	$10.24 *	$ 143.36
42 cu.ft. (1.19 cu.m) sand	$ 0.77	$ 32.26
Finish White Coat		
340 lbs. (153 kg) fin. hydrated lime	$13.60 *	$ 46.24
170 lbs. (76.5 kg) gauging plaster	$15.28 *	$ 25.98
Cost 100 sq. yds.		$ 294.94
per sq. yd.		$ 2.95
per sq. ft.		$ 0.33
per sq.m		$ 3.53

*Based on 100-lb. bag.

Labor Cost of 100 Sq.Yds. (83.61 Sq.M), 3-Coat Gypsum Plaster 7/8" (21.87 mm) Thick, Applied to Gypsum Lath Ceilings with Radiant Heating Coils Embebbed in the Plaster, Ordinary Workmanship - 7/8" (21.87 mm) Plaster			
Scratch Coat-1:2	Hours	Rate	Total
Plasterer	6.00	$32.41	$ 194.46
Labor	6.00	$26.16	$ 156.96
Brown Coat-1:3			
Plasterer	10.00	$32.41	$ 324.10
Labor	6.00	$26.16	$ 156.96
Finish White Coat			
Plasterer	8.00	$32.41	$ 259.28
Labor	4.00	$26.16	$ 104.64
Cost 100 sq. yds.			$1,196.40
per sq. yd.			$ 11.96
per sq. ft.			$ 1.33
per sq.m			$ 14.31

Add for gypsum lath.

Material Cost of 100 Sq.Yds. (83.61 Sq.M) 3-Coat Gypsum Plaster 5/8" (15.62 mm) Thick, Applied to Metal Lath Ceilings with Radiant Electrical Heating Cables Embedded in the Plaster, Ordinary Workmanship - 5/8" (15.62 mm) Plaster		
Scratch Coat-1:2	Rate	Total
800 lbs. (360 kg) gypsum plaster	$10.24 *	$ 81.92
16 cu.ft. (0.45 cu.m) sand	$ 0.77	$ 12.29
Brown Coat-1 :3		
1,200 lbs. (540 kg) gypsum plaster	$10.24 *	$ 122.88
36 cu.ft. (1.01 cu.m) sand	$ 0.77	$ 27.65
Finish White Coat		
340 lbs. (153 kg) fin. hydrated lime	$13.60 *	$ 46.24
170 lbs. (76.5 kg) gauging plaster	$15.28 *	$ 25.98
Cost 100 sq. yds.		$ 316.95
per sq. yd.		$ 3.17
per sq. ft.		$ 0.35
per sq.m		$ 3.79

Add for metal lath.
Based on 100-lb. bag.

Labor Cost of 100 Sq.Yds. (83.61 Sq.M) 3-Coat Gypsum Plaster 5/8" (15.62 mm) Thick, Applied to Metal Lath Ceilings with Radiant Electrical Heating Cables Embedded in the Plaster, Ordinary Workmanship - 5/8" (15.62 mm) Plaster			
Scratch Coat-1:2	Hours	Rate	Total
Plasterer	6.00	$32.41	$ 194.46
Labor	6.00	$26.16	$ 156.96
Brown Coat-1:3			
Plasterer	9.00	$32.41	$ 291.69
Labor	6.00	$26.16	$ 156.96
Finish White Coat			
Plasterer	8.00	$32.41	$ 259.28
Labor	4.00	$26.16	$ 104.64
Cost 100 sq. yds.			$1,163.99
per sq. yd.			$ 11.64
per sq. ft.			$ 1.29
per sq.m			$ 13.92

Add for metal lath.

If it is necessary to apply the plaster in four coats instead of three, add labor for a second scratch coat:

Description	Hours	Rate	Total
Plasterer	6.00	$32.41	$ 194.46
Labor	6.00	$26.16	$ 156.96
Cost 100 sq. yds			$ 351.42
per sq. yd.			$ 3.51
per sq. ft.			$ 0.39
per sq.m			$ 4.20

Labor Applying Perlite or Vermiculite Plaster

When using perlite or vermiculite plaster aggregate, a plasterer is applying material weighing 30 to 45 lbs. per cu.ft. (477-715 kg per cu.m)

1278

FINISHES

against 100 lbs. per cu.ft. (1589 kg per cu.m) for gypsum-sand plaster, so that a plasterer should apply a greater yardage of light aggregate plaster.

Material Cost of 100 Sq. Yds. (83.61 Sq.M) 3-Coat Plaster Applied to Gypsum Lath, Using Perlite or Vermiculite Aggregate Ordinary Workmanship 7/8" (21.87 mm) Grounds - 1 1/2" (37.50 mm) of Plaster		
Scratch Coat-1:2	Rate	Total
500 lbs. (225 kg) gypsum plaster	$ 6.40 *	$ 32.00
10 cu.ft. (0.28 cu.m) aggregate	$ 3.18	$ 31.80
Brown Coat-1:2		
1,000 lbs. (450 kg) gypsum plaster	$ 6.40 *	$ 64.00
20 cu.ft. (0.56 cu.m) aggregate	$ 3.18	$ 63.60
Finish White Coat		
340 lbs. (153 kg) fin. hydrated lime	$ 8.50 *	$ 28.90
170 lbs. (76.5 kg) gauging plaster	$ 9.55 *	$ 16.24
Cost 100 sq. yds.		$ 236.54
per sq. yd.		$ 2.37
per sq. ft		$ 0.26
per sq.m		$ 2.83

Add for gypsum lath.
*Based on 100-lb. bag.

Labor Cost of 100 Sq. Yds. (83.61 Sq.M) 3-Coat Plaster Applied to Gypsum Lath, Using Perlite or Vermiculite Aggregate Ordinary Workmanship 7/8" (21.87 mm) Grounds - 1 1/2" (37.50 mm) of Plaster			
Scratch Coat-1:2	Hours	Rate	Total
Plasterer	4.50	$32.41	$ 145.85
Labor	4.50	$26.16	$ 117.72
Brown Coat-1:2			
Plasterer	7.50	$32.41	$ 243.08
Labor	5.00	$26.16	$ 130.80
Finish White Coat			
Plaster	8.00	$32.41	$ 259.28
Labor	4.00	$26.16	$ 104.64
Cost 100 sq. yds.			$1,001.36
per sq. yd.			$ 10.01
per sq. ft			$ 1.11
per sq.m			$ 11.98

Add for gypsum lath.

Material Cost of 100 Sq. Yds. (83.61 Sq.M) 3-Coat Plaster Applied to Unit Masonry Walls Using Perlite or Vermiculite Aggregate Ordinary Workmanship - 5/8" (15.62 mm) Grounds - 5/8" (15.62 mm) Plaster		
Scratch Coat-1:3	Rate	Total
400 lbs. (180 kg) gypsum plaster	$10.24 *	$ 40.96
12 cu.ft. (0.34 cu.m) aggregate	$ 5.09	$ 61.06
Brown Coat-1:3		
800 lbs. (360 kg) gypsum plaster	$10.24 *	$ 81.92
24 cu.ft. (0.68 cu.m) aggregate	$ 5.09	$ 122.11
Finish White Coat		
340 lbs. (153 kg) fin. gypsum plaster	$13.60 *	$ 46.24
170 lbs. (76.5 kg) gauging plaster	$15.28 *	$ 25.98
Cost 100 sq. yds.		$ 378.27
per sq. yd.		$ 3.78
per sq. yd.		$ 0.42
per sq.m		$ 4.52

*Based on 100-lb. bag.

Labor Cost of 100 Sq. Yds. (83.61 Sq.M) 3-Coat Plaster Applied to Unit Masonry Walls Using Perlite or Vermiculite Aggregate Ordinary Workmanship - 5/8" (15.62 mm) Grounds - 5/8" (15.62 mm) Plaster			
Scratch Coat-1:3	Hours	Rate	Total
Plasterer	4.50	$32.41	$ 145.85
Labor	4.50	$26.16	$ 117.72
Brown Coat-1:3			
Plasterer	7.50	$32.41	$ 243.08
Labor	5.00	$26.16	$ 130.80
Finish White Coat			
Plasterer	8.00	$32.41	$ 259.28
Labor	4.00	$26.16	$ 104.64
Cost 100 sq. yds.			$1,001.36
per sq. yd.			$ 10.01
per sq. ft.			$ 1.11
per sq.m			$ 11.98

Material Cost of 100 Sq. Yds. (83.61 Sq.M) 3-Coat Plaster Applied to Metal Lath, Using Perlite or Vermiculite Aggregate Ordinary Workmanship - 3/4" (18.75 mm) Grounds - 3/4" (18.75 mm) Plaster		
Scratch-1:2	Rate	Total
1,000 lbs. (450 kg) gypsum plaster	$10.24 *	$ 102.40
20 cu.ft. (0.57 cu.m) aggregate	$ 5.09	$ 101.76
Brown Coat-1:2		
1,250 lbs. (562.5 kg) gypsum plaster	$10.24 *	$ 128.00
40 cu.ft. (1.13 cu.m) aggregate	$ 5.09	$ 203.52
Finish White		
340 lbs. (153 kg) fin. hydrated lime	$13.60 *	$ 46.24
170 lbs. (76.5 kg) gauging plaster	$15.28 *	$ 25.98
Cost per 100 sq. yds.		$ 607.90
per sq. yd.		$ 6.08
per sq. ft.		$ 0.68
per sq.m		$ 7.27

Add for metal lath and furring.
*Based on 100-lb. bag.

FINISHES

Labor Cost of 100 Sq. Yds. (83.61 Sq.M) 3-Coat Plaster Applied to Metal Lath, Using Perlite or Vermiculite Aggregate — Ordinary Workmanship - 3/4" (18.75 mm) Grounds - 3/4" (18.75 mm) Plaster			
Scratch Coat-1:2	Hours	Rate	Total
Plasterer	4.50	$32.41	$ 145.85
Labor	4.50	$26.16	$ 117.72
Brown Coat-1:2			
Plasterer	8.00	$32.41	$ 259.28
Labor	6.00	$26.16	$ 156.96
Finish White Coat			
Plasterer	8.00	$32.41	$ 259.28
Labor	4.00	$26.16	$ 104.64
Cost per 100 sq. yds			$1,043.73
per sq. yd.			$ 10.44
per sq. ft.			$ 1.16
per sq.m			$ 12.48

Add for metal lath and furring.

Material Cost of 100 Sq. Yds. (83.61 Sq.M) 2" (50 mm) Solid Plaster Partition Applied to Metal Lath, Using Perlite or Vermiculite Aggregate (Measurement on One Side Only) — Ordinary Workmanship - 2" (50.00 mm) - 2" (50.00 mm)Plaster		
Scratch Coat-1:2	Rate	Total
1,500 lbs. (675 kg) gypsum plaster	$10.24 *	$ 153.60
30 c.f. (0.85 cu.m) aggregate	$ 5.09	$ 152.64
Brown Coat-1:2		
6,000 lbs. (2700 kg) gypsum plaster	$10.24 *	$ 614.40
120 c.f. (3.4 cu.m) aggregate	$ 5.09	$ 610.56
Finish White Coat		
680 lbs. (306 kg) fin. hydrated lime	$13.60 *	$ 92.48
340 lbs. (153 kg) gauging plaster	$15.28 *	$ 51.95
Cost 100 sq. yds.		$1,675.63
per sq. yd.		$ 16.76
per sq. ft.		$ 1.86
per sq.m		$ 20.04

Add for metal lath and furring.
*Based on 100-lb. bag.

Labor Cost of 100 Sq. Yds. (83.61 Sq.M) 2" (50 mm) Solid Plaster Partition Applied to Metal Lath, Using Perlite or Vermiculite Aggregate (Measurement on One Side Only) — Ordinary Workmanship - 2" (50.00 mm) - 2" (50.00 mm) Plaster			
Scratch Coat-1:2	Hours	Rate	Total
Plasterer	5.00	$32.41	$ 162.05
Labor	5.00	$26.16	$ 130.80
Brown Coat-1:2			
Plasterer	23.00	$32.41	$ 745.43
Labor	17.00	$26.16	$ 444.72
Finish White Coat			
Plasterer	16.00	$32.41	$ 518.56
Labor	8.00	$26.16	$ 209.28
Cost 100 sq. yds.			$2,210.84
per sq. yd.			$ 22.11
per sq. ft.			$ 2.46
per sq.m			$ 26.44

Add for metal lath and furring.

Material Cost of 100 Sq. Yds. (83.61 Sq.M) 3-Coat Plaster (Scratch-Double-Back Basecoat) Applied to Gypsum Lath, Using Pre-Mixed Plaster Ordinary Workmanship - 7/8" (21.87 mm) Grounds - 1/2" (12.50 mm) Plaster		
Scratch Coat	Rate	Total
1180-lb. (531 kg) sacks P. plaster	$11.04 *	$ 130.27
Double-Back Brown Coat		
1180-lb. (531 kg) sacks P. plaster	$11.04 *	$ 130.27
Finish White Coat		
340 lbs. (153 kg) fin. hydrated lime	$13.60 *	$ 46.24
170 lbs. (76.5 kg) gauging plaster	$15.28 *	$ 25.98
Cost 100 sq. yds.		$ 332.76
per sq. yd.		$ 3.33
per sq. yd.		$ 0.37
per sq.m		$ 3.98

Add for gypsum lath.
*Based on 100-lb. bag.

Labor Cost of 100 Sq. Yds. (83.61 Sq.M) 3-Coat Plaster (Scratch-Double-Back Basecoat) Applied to Gypsum Lath, Using Pre-Mixed Plaster Ordinary Workmanship - 7/8" (21.87 mm) Grounds - 1/2" (12.50 mm) Plaster			
Scratch Coat	Hours	Rate	Total
Plasterer	4.50	$32.41	$ 145.85
Labor	4.50	$26.16	$ 117.72
Double-Back Brown Coat			
Plasterer	7.50	$32.41	$ 243.08
Labor	5.00	$26.16	$ 130.80
Finish White Coat			
Plasterer	8.00	$32.41	$ 259.28
Labor	4.00	$26.16	$ 104.64
Cost 100 sq. yds.			$1,001.36
per sq. yd.			$ 10.01
per sq. ft.			$ 1.11
per sq.m			$ 11.98

Add for gypsum lath.

Material Cost of 100 Sq. Yds. (83.61 Sq.M) 3-Coat Plaster (Scratch-Double-Back Basecoat) Applied to Unit Masonry Walls Using Premixed Perlite Plaster Ordinary Workmanship—5/8" (16 mm) Grounds—5/8" (16 mm) of Plaster		
Scratch Coat	Rate	Total
1080-lb. (486-kg) sacks P. Plaster	$11.04 *	$ 119.23
Double-Back Brown Coat		
1680-lb. (756-kg) sacks P. Plaster	$11.04 *	$ 185.47
Finish White Coat		
340 lbs. (153 kg) fin. hydrated lime	$13.60 *	$ 46.24
170 lbs. (76.5 kg) gauging plaster	$15.28 *	$ 25.98
Cost 100 sq. yds.		$ 376.92
per sq. yd.		$ 3.77
per sq. ft.		$ 0.42
per sq.m		$ 4.51

Add for gypsum lath.
*Based on 100-lb. bag.

1282

FINISHES

Labor Cost of 100 Sq. Yds. (83.61 Sq.M) 3-Coat Plaster (Scratch-Double-Back Basecoat) Applied to Unit Masonry Walls Using Premixed Perlite Plaster Ordinary Workmanship—5/8" (16 mm) Grounds—5/8" (16 mm) of Plaster			
Scratch Coat	Hours	Rate	Total
Plasterer	4.50	$32.41	$ 145.85
Labor	4.50	$26.16	$ 117.72
Double-Back Brown Coat			
Plasterer	7.50	$32.41	$ 243.08
Labor	5.00	$26.16	$ 130.80
Finish White Coat			
Plasterer	8.00	$32.41	$ 259.28
Labor	4.00	$26.16	$ 104.64
Cost 100 sq. yds.			$1,001.36
per sq. yd.			$ 10.01
per sq. ft.			$ 1.11
per sq.m			$ 11.98

Material Cost of 100 Sq. Yds. (83.61 Sq.M) 3-Coat Plaster Applied to Metal Lath Using Pre-Mixed Perlite Plaster Ordinary Workmanship 3/4" (18.75 mm) Grounds-3/4" (18.75 mm) Plaster		
Scratch Coat	Rate	Total
1580-lb. (711-kg) sacks P. Plaster	$11.04 *	$ 174.43
Brown Coat		
1880-lb. (846-kg) sacks P. Plaster	$11.04 *	$ 207.55
Finish White Coat		
340 lbs. (153 kg) fin. hydrated lime	$13.60 *	$ 46.24
170 lbs. (76.5 kg) gauging plaster	$15.28 *	$ 25.98
Cost 100 sq. yds.		$ 454.20
per sq. yd.		$ 4.54
per sq.ft.		$ 0.50
per sq.m		$ 5.43

Add for metal lath and furring.
*Based on 100-lb. bag.

Labor Cost of 100 Sq. Yds. (83.61 Sq.M) 3-Coat Plaster Applied to Metal Lath Using Pre-Mixed Perlite Plaster Ordinary Workmanship 3/4" (18.75 mm) Grounds-3/4" (18.75 mm) Plaster			
Scratch Coat	Hours	Rate	Total
Plasterer	4.50	$32.41	$145.85
Labor	4.50	$26.16	$117.72
Brown Coat			
Plasterer	8.00	$32.41	$259.28
Labor	6.00	$26.16	$156.96
Finish White Coat			
Plasterer	8.00	$32.41	$259.28
Labor	4.00	$26.16	$104.64
Cost 100 sq. yds.			$1,043.73
per sq. yd.			$10.44
per sq. ft			$1.16
per sq.m			$12.48

Add for metal lath and furring.

Material Cost of 100 Sq. Yds. (83.61 Sq.M) 2" (50 mm) Solid Plaster Partition Applied to Metal Lath Using Pre-Mixed Perlite Plaster (Measurement One Side) Ordinary Workmanship-2" (50 mm) Partition - 2" (50 mm) of Plaster		
Scratch Coat	Rate	Total
1580-lb. (711-kg) sacks P. Plaster	$11.04 *	$ 174.43
Brown Coat		
7080-lb. (3186-kg) sacks P. Plaster	$11.04 *	$ 781.63
Finish White Coat		
680 lbs. (306 kg) fin. hydrated lime	$13.60 *	$ 92.48
340 lbs. (153 kg) gauging plaster	$15.28 *	$ 51.95
Cost 100 sq. yds.		$1,100.50
per sq. yd.		$ 11.00
per sq. ft.		$ 1.22
per sq. m.		$ 13.16

Add for metal lath and furring.
Based on 100-lb. bag.

Labor Cost of 100 Sq. Yds. (83.61 Sq.M) 2" (50 mm) Solid Plaster Partition Applied to Metal Lath Using Pre-Mixed Perlite Plaster (Measurement One Side) Ordinary Workmanship-2" (50 mm) Partition - 2" (50 mm) of Plaster			
Scratch Coat	Hours	Rate	Total
Plasterer	5.00	$32.41	$162.05
Labor	5.00	$26.16	$130.80
Brown Coat			
Plasterer	23.00	$32.41	$745.43
Labor	17.00	$26.16	$444.72
Finish White Coat			
Plasterer	16.00	$32.41	$518.56
Labor	8.00	$26.16	$209.28
Cost 100 sq. yds.			$2,210.84
per sq. yd.			$22.11
per sq. ft.			$2.46
per sq.m			$26.44

Add for metal lath and furring.

Machine Application of Plaster. Various types of machines are available for the spray application of base coat plaster mortar to walls and ceilings where it is then worked in the normal manner. They vary in size from large machines capable of spraying any plaster mortars, including base coat materials for ordinary plastering and membrane fireproofing, to small machines that handle only the finish coat of acoustical plaster or a sand float finish. Most models have self-contained pump units with gasoline or electric power air compressors. Several small hand gun-hopper units are available for attachment to separate air compressors.

When applying plaster mortars by machine, approximately 20% to 25% additional material will be required to provide equal coverage to hawk and trowel application. This is caused by material densification due to machine action and increased material loss from spattering.

1284

FINISHES

Using the machine method, the labor cost of applying the various plaster coats will be considerably less than for the hawk and trowel method, but an allowance must be made for machine cost and maintenance and additional protection and clean-up labor will be required.

Where scratch coat is machine applied over metal lath, an experienced operator can apply 750 to 850 sq.yds. (627.07 to 710.68 sq.m) per 8-hr. day. For this area, an additional 4 hrs. of plasterer time will be required for touching up and scratching surface.

Material Cost of 100 Sq. Yds. (83.61 Sq.M) of Machine Applied Scratch Coat on Metal, Lath Using Perlite or Vermiculite Aggregate		
Description	Rate	Total
1,200 lbs. (540 kg) gypsum plaster	$10.24 *	$ 122.88
24 c.f. (0.68 cu.m) aggregate	$ 5.09	$ 122.11
Machine cost	$ 8.48	$ 8.48
Cost 100 sq. yds.		$ 253.47
per sq. yd.		$ 2.53
per sq. ft		$ 0.28
per sq.m		$ 3.03

Add for metal lath and furring.
*Based on 100-lb. bag.

Labor Cost of 100 Sq. Yds. (83.61 Sq.M) of Machine Applied Scratch Coat on Metal, Lath Using Perlite or Vermiculite Aggregate			
Description	Hours	Rate	Total
Plasterer	1.50	$32.41	$ 48.62
Labor	2.00	$26.16	$ 52.32
Cost 100 sq. yds.			$ 100.94
per sq. yd.			$ 1.01
per sq. ft.			$ 0.11
per sq.m			$ 1.21

Where brown coat is machine applied over scratch coat or gypsum lath, an experienced operator, with another plasterer rodding and leveling, can apply 375 to 425 sq.yds. (313.53 to355.34 sq.m) per 8-hr. day.

Material Cost of 100 Sq. Yds. (83.61 Sq.M) of Machine Applied Brown Coat, 7/16" (10.93 mm) Thick over Scratch Coat or Gypsum Lath Using Perlite or Vermiculite Aggregate		
Description	Rate	Total
1,550 lbs. (697.5 kg) gypsum plaster	$10.24 *	$ 158.72
50 cu.ft. (1.41 cu.m) aggregate	$ 5.09	$ 254.40
Machine cost	$16.96	$ 16.96
Cost 100 sq. yds.		$ 430.08
per sq. yd.		$ 4.30
per sq. ft.		$ 0.48
per sq.m		$ 5.14

*Based on 100-lb. bag.

Labor Cost of 100 Sq. Yds. (83.61 Sq.M) of Machine Applied Brown Coat, 7/16" (10.93 mm) Thick over Scratch Coat or Gypsum Lath Using Perlite or Vermiculite Aggregate			
Description	Hours	Rate	Total
Plasterer	4.00	$32.41	$ 129.64
Labor	4.00	$26.16	$ 104.64
Cost 100 sq. yds.			$ 234.28
per sq. yd.			$ 2.34
per sq. ft.			$ 0.26
per sq.m			$ 2.80

Special Prepared Finishes

There are a number of prepared plaster finishes on the market that are ready for use with the addition of water. These may be applied over a brown coat of gypsum cement plaster.

Smooth trowel finishes such as gypsum gauging-lime putty require a much higher degree of rigidity and strength in the lath and basecoat plaster for good visual performance. When the use of high strength basecoat plasters is impractical, consideration should be given to the elimination of restraint at the perimeter angles or the selection of a more rigid lath, or both. They help to compensate for the lower strength of some basecoat plasters. Where a special finish is specified, deduct the cost of the finish white coat and add the cost of the special finish specified.

Material Cost of 100 Sq. Yds. (83.61 Sq.M) Gypsum Trowel Finish Ordinary Workmanship		
Description	Rate	Total
550 lbs. (247.5 kg) gyp. trowel finish	$25.44 *	$ 139.92
Cost per sq. yd.		$ 1.40
per sq. ft.		$ 0.16
per sq.m		$ 1.67

*Based on 100-lb. bag.

Labor Cost of 100 Sq. Yds. (83.61 Sq.M) Gypsum Trowel Finish Ordinary Workmanship			
Description	Hours	Rate	Total
Plasterer	8.00	$32.41	$ 259.28
Labor	4.00	$26.16	$ 104.64
Cost 100 sq. yds.			$ 363.92
per sq. yd.			$ 3.64
per sq. ft.			$ 0.40
per sq.m			$ 4.35

1286

FINISHES

For first grade workmanship, add 3.5 hrs. plasterer time and 1-hr. labor time per 100 sq.yds. (83.61 sq.m).

If plaster is wanted with a sand float finish, deduct quantities of plaster and labor given under finish white coat and add the following:

Material Cost of 100 Sq. Yds. (83.61 Sq.M) Sand Float Finish Ordinary Workmanship		
Description	Rate	Total
200 lbs. (90 kg) unfibered plaster	$11.04 *	$ 22.08
4 cu.ft. (0.11 cu.m) sand	$ 0.77	$ 3.07
Cost per 100 sq. yds.		$ 25.15
per sq. yd.		$ 0.25
per sq. ft.		$ 0.03
per sq.m		$ 0.30

Based on 100-lb. bag.

Labor Cost of 100 Sq. Yds. (83.6 Sq.M) Sand Float Finish Ordinary Workmanship			
Description	Hours	Rate	Total
Plasterer	7.25	$32.41	$ 234.97
Labor	5.00	$26.16	$ 130.80
Cost per 100 sq. yds			$ 365.77
per sq. yd.			$ 3.66
per sq. ft.			$ 0.41
per sq.m			$ 4.37

If first grade workmanship is required, add 3 hours plasterer time and 1 hour labor time per 100 sq.yds. (83.61 sq.m).

The resistance of gypsum gauging-lime putty trowel finishes to cracking, particularly to check cracking, can be increased by the addition of fine aggregate. The use of not less than 1/2 cu.ft. (0.01 cu.m) of fine silica sand or perlite to each 100 lbs. (45.35 kg) of gauging plaster or Keene's cement increases the factor of safety in trowel finishes.

Acoustical Plaster. Acoustical plaster is occasionally specified instead of trowel putty coat or sand float finishes for noise quieting in classrooms and corridors and for acoustical correction of churches and auditoriums. A number of prepared acoustical plasters are available having sound absorption values of 40% to 45% at 512 cycles and a noise reduction coefficient of 55% to 60%. There are two general types of acoustical plaster. Those with a gypsum or lime binder, such as Gold Bond, are for trowel application only with either a troweled, floated, or stippled and perforated finish. Other acoustical plasters, such as, Sprayolite or Acoustical Plaster, are made with an adhesive type binder that shrinks on drying, providing the porosity for sound absorption, and may be applied by machine or trowel.

In general, acoustical plasters are for application over gypsum base coats (scratch and brown) and should be applied in two coats to a total thickness of 1/2" (12.50 mm)—usually a 3/8" (9.37 mm) first coat and a 1/8" (3 mm) finish coat.

Because of restrictions due to the asbestos fiber content, acoustical plaster is temporarily out of supply but will probably be back on the market with a new mixture. The prices below are for the old formula material.

Where applied by trowel, a plasterer should complete 65 to 75 sq.yds. (54.34 to 62.70 sq.m) of acoustical plaster, two coats to 1/2" (12.50 mm) total thickness, per 8-hr. day.

Material Cost of 100 Sq. Yds. (83.61 Sq.M) of Acoustical Plaster Applied by Trowel		
Description	Rate	Total
20 bags, 5 sq. yds. (4.18 sq.m),	$12.72	$ 254.40
Cost per sq. yd		$ 2.54
per sq.ft.		$ 0.28
per sq.m		$ 3.04

Add for lath and base coats required.

Labor Cost of 100 Sq. Yds. (83.61 Sq.M) of Acoustical Plaster Applied by Trowel			
Description	Hours	Rate	Total
Plasterer	11.50	$32.41	$ 372.72
Laborer	6.00	$26.16	$ 156.96
Cost per 100 sq. yd			$ 529.68
per sq.yd.			$ 5.30
per sq.ft.			$ 0.59
per sq.m			$ 6.34

Add for lath and base coats required.

Where application of the two coat acoustical finish is by machine, a plasterer should complete 100 to 110 sq.yds. (84 to 92 sq.m) per 8-hr. day, but one extra hour of labor time must be figured for the additional cleanup and protection required, and an allowance included for cost and maintenance of equipment.

Material Cost of 100 Sq. Yds. (83.61 Sq.M) of Acoustical Plaster Applied by Machine		
Description	Rate	Total
14 bags, 7.2 sq. yds. (6 sq.m)	$12.72	$ 178.08
Machine cost	$63.60	$ 63.60
Cost per 100 sq. yds		$ 241.68
per sq. yd		$ 2.42
per sq. ft.		$ 0.27
per sq.m		$ 2.89

Add for lath and base coats required.

FINISHES

Labor Cost of 100 Sq. Yds. (83.61 Sq.M) of Acoustical Plaster Applied by Machine			
Description	Hours	Rate	Total
Plasterer	8.00	$32.41	$ 259.28
Laborer	7.00	$26.16	$ 183.12
Laborer - masking	7.00	$26.16	$ 183.12
Cost per 100 sq. yd			$ 625.52
per sq.yd.			$ 6.26
per sq.ft.			$ 0.70
per sq.m			$ 7.48

Add for lath and base coats required.

Vermiculite Direct-to-Steel Fireproofing. Vermiculite Type-MK direct-to-steel fireproofingis a method of protecting steel columns, beams, girders, and trusses, and steel floor and roof sections. It is applied by machine directly to the metal, without a primer, lath, or reinforcing. It presents a tough, fire-protective, and insulating coating that is free of fissures and can be applied in the early stages of construction, when the building is free of other trades.

This is a mill-mixed material that requires only the addition of water. It is packaged in 25- and 50-lb. (11- and 22.5-kg) bags. Coverage varies with specified thickness.

Colored Interior Finish Plaster

There are a number of colored finish plasters on the market used for textured or float plaster finish.

These special finishes are applied over a brown coat of gypsum cement plaster and may be finished in a number of different ways, such as Spanish Palm Finish, Brush Swirl, Float Finish, English Trowel finish, etc., in addition to a simple float finish.

Labor Applying Special Finish Coats. For colored finish plaster with a simple float finish, a plasterer should apply 100 to 120 sq.yds. (84 to 100 sq.m) per 8-hr. day, where an ordinary grade of workmanship is permitted.

Interior Color Finish Plaster, from U.S. Gypsum Co., is used for textured or floated plaster finish. It is made in several shades and white and is ready for use with the addition of water only.

It requires approximately 8 lbs. of finish per sq.yd. depending on the texture used. Price about $20.00 per 100 lbs. ($44.09 per 100 kg).

Material Cost of 100 Sq. Yds. (83.61 Sq.M) of Oriental Interior Colored Finish Plaster, with Simple Float Finish, Applied Over a Brown Coat of Gypsum Cement Plaster - Ordinary Workmanship		
Finish Coat	Rate	Total
800 lbs. (363 kg) Oriental finish	$ 20.35	$ 162.82
Cost per sq. yd		$ 1.63
Cost per sq. ft.		$ 0.18
per sq.m		$ 1.95
Based on 100-lb. bag.		

Labor Cost of 100 Sq. Yds. (83.61 Sq.M) of Oriental Interior Colored Finish Plaster, with Simple Float Finish, Applied Over a Brown Coat of Gypsum Cement Plaster - Ordinary Workmanship			
Finish Coat	Hours	Rate	Total
Plasterer	7.25	$ 32.41	$ 234.97
Labor	4.50	$ 26.16	$ 117.72
Cost 100 sq. yds			$ 352.69
per sq. yd			$ 3.53
per sq. ft.			$ 0.39
per sq.m			$ 4.22

If first grade workmanship is required, add 2.75 hrs. plasterer time and 1 hr. labor time per 100 sq.yds. (83.6 sq.m).

Material Costs Colored Plaster with Smooth or Textured Finish		
Description	Rate	Total
350 lbs.(157.5 kg) cement	$27.14 *	$ 94.98
450 lbs. (202.5 kg) fin. hydrated lime	$13.60 *	$ 61.20
16 lbs. (7.2 kg) color	$ 1.70	$ 27.14
Cost 100 sq. yds		$ 183.31
per sq. yd.		$ 1.83
per sq. ft.		$ 0.20
per sq.m		$ 2.19

*Based on 100-lb. bag.

Material Costs Colored Plaster with Sand Float Finish		
Description	Rate	Total
100 lbs. (45 kg) cement	$16.96 *	$ 16.96
500 lbs. (22.5 kg) fin. hydrated lime	$ 8.50 *	$ 42.50
1,200 lbs. (540 kg) white silica sand	$ 1.06 *	$ 12.72
36 lbs. (16.2 kg) color	$ 1.06	$ 38.16
Cost 100 sq. yds		$ 110.34
per sq. yd.		$ 1.10
per sq. ft		$ 0.12
per sq.m		$ 1.32

FINISHES

The above formulas are standard mixture for permanent colors. Best results are obtained by soaking lime overnight or for 24 hours.

Ornamental Plastering

Ornamental plastering is not used as extensively as it once was, and mechanics who can do this type of work are not always available. It is best to contact a plastering subcontractor for firm prices on intricate ornamental plaster work.

Material Cost of 100 Lin.Ft. (30.48 m) of Plaster Cornice Under 12" (300 mm) in Girth		
Description	Rate	Total
200 lbs. (90 kg) gypsum plaster	$10.18 *	$ 20.35
4 cu.ft. (0.11 cu.m) sand	$ 0.77	$ 3.07
50 lbs. (22.5 kg) molding plaster	$16.96 *	$ 8.48
Cost 100 lin.ft.		$ 31.90
per lin.ft.		$ 0.32
per lin. m		$ 1.05

Based on 100-lb. bag.

Add for metal lath and furring as given on previous pages.

Labor Cost of 100 Lin.Ft. (30.48 m) of Plaster Cornice Under 12" (300 mm) in Girth			
Description	Hours	Rate	Total
Plasterer	18.00	$32.41	$ 583.38
Labor	6.00	$26.16	$ 156.96
Cost 100 lin.ft.			$ 740.34
per lin.ft.			$ 7.40
per lin. M.			$ 24.29

Add for metal lath and furring as given on previous pages.

Material Cost of 100 Sq.Ft. (9.29 Sq.M) of Plaster Cornice Over 1'-0" (0.30 m) in Girth		
Description	Rate	Total
200 lbs. (90 kg) gypsum plaster	$ 6.36 *	$ 12.72
4 cu.ft. (0.11 cu.m) sand	$ 0.48	$ 1.92
50 lbs. (22.5 kg) molding plaster	$10.60 *	$ 5.30
Cost 100 sq.ft.		$ 19.94
per sq.ft.		$ 0.20
per sq.m		$ 2.15

Add for metal lath and furring as given on previous pages.

Labor Cost of 100 Sq.Ft. (9.29 Sq.M) of Plaster Cornice Over 1'-0" (0.30 m) in Girth			
Description	Hours	Rate	Total
Plasterer	17.00	$32.41	$ 550.97
Labor	6.00	$26.16	$ 156.96
Cost 100 sq.ft.			$ 707.93
per sq.ft.			$ 7.08
per sq.m			$ 76.20

Add for metal lath and furring as given on previous pages.

Material Cost of 100 Sq.Ft. (9.29 Sq.M) of Portland Cement Cornice		
Description	Rate	Total
4 sacks Portland cement	$ 6.78	$ 27.14
40 lbs. (18 kg) hydrated lime	$13.60 *	$ 5.44
10 cu.ft. (0.28 cu.m) sand	$ 0.77	$ 7.68
Cost 100 sq.ft.		$ 40.26
per sq.ft.		$ 0.40
per sq.m		$ 4.33

Add for metal lath and furring as given on previous pages.

Labor Cost of 100 Sq.Ft. (9.29 Sq.M) of Portland Cement Cornice			
Description	Hours	Rate	Total
Plasterer	46.00	$32.41	$1,490.86
Labor	10.00	$26.16	$ 261.60
Cost 100 sq.ft.			$1,752.46
per sq.ft.			$ 17.52
per sq.m			$ 188.64

Add for metal lath and furring as given on previous pages.

Running Bull Nose Corners, Ceiling Coves, Etc. When running bull nose corners on plaster columns, pilasters, window jambs, etc., and plain ceiling coves and wall angles, a plasterer should run 80 to 100 lin.ft. (24.38 to 30.48 m) per 8-hr. day, at the following labor cost per 100 lin.ft. (30.48 m):

Description	Hours	Rate	Total
Plasterer	9.00	$32.41	$291.69
Labor	2.00	$26.16	52.32
Cost 100 lin.ft.			$344.01
per lin.ft.			$3.44
per lin. M.			$11.29

1292

FINISHES

Portland Cement Plaster

Portland cement plaster is used for both interior and exterior work. It should not be mixed too rich. A too rich mixture has a tendency to check and shrink.

Proper attention to curing is important. After each coat has set, and before the following coat is applied, the mortar should be kept moist for at least three days and then permitted to dry out gradually.

When using portland cement mortar it is customary to add a small percentage of lime (usually about 10%) to make the mortar work easier, because a straight portland cement mortar is very difficult to apply.

For the cost of portland cement plaster applied over metal lath and the cost of metal lath and furring, see previous pages of this chapter.

Labor Applying 100 Sq.Yds. (83.61 Sq.M) Portland Cement Plaster to Various Plaster Bases				
Description of Work - Ordinary Work*	Sq.Yd. per 8-Hr. Day	Hr. per 100 Plasterer	Tender	Sq.M. per 8-Hr. Day
Interior Work on Brick, Clay Tile or Cement Block - 3/4" (18.75 mm) Thick				
Scratch Coat	120 - 140	6	4	100.3 - 117.1
Brown Coat	75 - 85	10	5	62.7 - 71.1
Float Finish Coat	60 - 70	12	4	50.2 - 58.5
Trowel Finish Coat	45 - 55	16	4	37.6 - 46.0
Interior Portland Cement Plaster on Metal Lath				
Scratch Coat	120 - 140	6	5	100.3 - 117.1
Brown Coat	60 - 70	12	6	50.2 - 58.5
Float Finish Coat	60 - 70	12	4	50.2 - 58.5
Trowel Finish Coat	45 - 55	16	4	37.6 - 46.0
Exterior Portland Cement Stucco on Block, Clay Tile or Cement Block				
Scratch Coat	120 - 140	6	5	100.3 - 117.1
Brown Coat	75 - 85	10	6	62.7 - 71.1
Scaffold	-----	-----	12	-----
Exterior Portland Cement Stucco on Metal Lath on Frame Construction				
Scaffold	-----	-----	12	-----
Scratch Coat	120 - 140	6	6	100.3 - 117.1
Brown Coat	60 - 70	12	8	50.2 - 58.5
Floated Finish Coat	60 - 70	12	6	50.2 - 58.5
Troweled Finish Coat	45 - 55	16	6	37.6 - 46.0
Textured Finish Coat	50 - 60	14	6	41.8 - 50.2
Blocking Off Portland Cement Plaster into Squares to represent Tile	45 - 55	16	4	37.6 - 46.0
Portland Cement Straight Base**	45 - 55	16	4	37.6 - 46.0
Portland Cement Coved Base**	30 - 35	24	6	25.1 - 29.3

For first grade workmanship as described at the beginning of this chapter, add 30% to 40% to plasterer time for brown and finish coats.
**Lineal feet.*

Material Cost of 100 Sq.Yds. (83.61 Sq.M) 2-Coat Portland Cement Plaster Brick, Clay Tile or Cement Block Ordinary Workmanship 5/8" (15.62 mm) Grounds		
Brown Coat	Rate	Total
14 sacks Portland cement	$ 6.78	$ 94.98
140 lbs. (63 kg) hydrated lime	$13.60 *	$ 19.04
42 cu.ft. (1.18 cu.m) sand	$ 0.77	$ 32.26
Finish Coat		
4 sacks Portland cement	$ 6.78	$ 27.14
40 lbs. (18 kg) hydrated lime	$13.60 *	$ 5.44
12 cu.ft. (0.34 cu.m) sand	$ 0.77	$ 9.22
Cost 100 sq. yds.		$ 188.06
per sq. yd.		$ 1.88
per sq. ft.		$ 0.21
per sq.m		$ 2.25

* Based on 100-lb bag.

Labor Cost of 100 Sq.Yds. (83.61 Sq.M) 2-Coat Portland Cement Plaster Brick, Clay Tile or Cement Block Ordinary Workmanship 5/8" (15.62 mm) Grounds			
Brown Coat	Hours	Rate	Total
Plasterer	13.00	$32.41	$ 421.33
Labor	6.00	$26.16	$ 156.96
Finish Coat			
Plasterer	16.00	$32.41	$ 518.56
Labor	4.00	$26.16	$ 104.64
Cost 100 sq. yds			$1,201.49
per sq. yd			$ 12.01
per sq. ft			$ 1.33
per sq.m			$ 14.37

Material Cost of 100 Sq. Yds. (83.61Sq.M) 3-Coat Portland Cement Plaster, on Brick, Clay Tile or Cement Block Ordinary Wormanship - 5/8" (15.62 mm) Grounds		
Scratch Coat	Rate	Total
6 sacks portland cement	$ 6.78	$ 40.70
60 lbs. (27 kg) hydrated lime	$13.60 *	$ 8.16
18 c.f. (0.51 cu.m) sand	$ 0.77	$ 13.82
Brown Coat		
8 sacks portland cement	$ 6.78	$ 54.27
80 lbs. (36 kg) hydrated lime	$13.60 *	$ 10.88
24 cu.ft. (0.68 cu.m) sand	$ 0.77	$ 18.43
Finish Coat		
4 sacks portland cement	$ 6.78	$ 27.14
40 lbs. (18 kg) hydrated lime	$13.60 *	$ 5.44
12 cu.ft. (0.34 cu.m) sand	$ 0.77	$ 9.22
Cost 100 sq. yds		$ 188.06
per sq. yd		$ 1.88
per sq. ft.		$ 0.21
per sq.m		$ 2.25

*Based on 100-lb. bag.

FINISHES

Labor Cost of 100 Sq. Yds. (83.61Sq.M) 3-Coat Portland Cement Plaster, on Brick, Clay Tile or Cement Block Ordinary Workmanship - 5/8" (15.62 mm) Grounds			
Scratch Coat	Hours	Rate	Total
Plasterer	6.00	$32.41	$194.46
Labor	4.00	$26.16	$104.64
Brown Coat			
Plasterer	10.00	$32.41	$324.10
Labor	5.00	$26.16	$130.80
Finish Coat			
Plasterer	16.00	$32.41	$518.56
Labor	4.00	$26.16	$104.64
Cost 100 sq. yds			$1,377.20
per sq. yd			$13.77
per sq. ft.			$1.53
per sq.m			$16.47

If first grade workmanship is required, add 8 hrs. plasterer time and 2 hrs. labor time per 100 sq.yds. (83.61 sq.m).

If blocked off into squares to represent tile, add 16 hrs. plasterer time for ordinary work and 18 hrs. plasterer time for first grade workmanship per 100 sq.yds. (83.61 sq.m).

Material Cost of 100 Sq.Yds. (83.61 Sq.M) 3-Coat Portland Cement Plaster on Metal Lath, Ordinary Workmanship - 3/4" (18.75 mm) Grounds		
Scratch Coat	Rate	Total
7 sacks Portland cement	$ 6.78	$ 47.49
70 lbs. (31.5 kg) hydrated lime	$13.60 *	$ 9.52
21 cu.ft. (0.59 cu.m) sand	$ 0.77	$ 16.13
Brown Coat		
11 sacks Portland cement	$ 6.78	$ 74.62
110 lbs. (49.5 kg) hydrated lime	$13.60 *	$ 14.96
33 cu.ft. (0.93 cu.m) sand	$ 0.77	$ 25.34
Finish Coat		
4 sacks Portland cement	$ 6.78	$ 27.14
40 lbs. (18 kg) hydrated lime	$13.60 *	$ 5.44
12 cu.ft. (0.34 cu.m) sand	$ 0.77	$ 9.22
Cost 100 sq. yds		$ 229.86
per sq. yd.		$ 2.30
per sq. ft.		$ 0.26
per sq.m		$ 2.75

Add for metal lath and furring.
Based on 100-lb. bag.

Labor Cost of 100 Sq.Yds. (83.61 Sq.M) 3-Coat Portland Cement Plaster on Metal Lath, Ordinary Workmanship - 3/4" (18.75 mm) Grounds			
Scratch Coat	Hours	Rate	Total
Plasterer	6.00	$32.41	$ 194.46
Labor	5.00	$26.16	$ 130.80
Brown Coat			
Plasterer	12.00	$32.41	$ 388.92
Labor	6.00	$26.16	$ 156.96
Finish Coat			
Plasterer	16.00	$32.41	$ 518.56
Labor	4.00	$26.16	$ 104.64
Cost 100 sq. yds			$1,494.34
per sq. yd.			$ 14.94
per sq. ft.			$ 1.66
per sq.m			$ 17.87

Add for metal lath and furring.

If blocked off into squares to represent tile, add 16 hrs. plasterer time per 100 sq.yds. (83.61 sq.m).

If 5/8" (15.62 mm) grounds are used instead of 3/4" (18.75 mm), deduct 2 sacks portland cement, 20 lbs. (9.07 kg) lime and 6 cu.ft. (0.17 cu.m) sand.

If first grade workmanship is required, add 8 hrs. plasterer time and 2 hrs. labor time per 100 sq.yds. (83.61 sq.m).

Material Cost of 100 LF. (30.48 M) of 6" (150 mm) Portland Cement Straight Base		
Description	Rate	Total
2 sacks portland cement	$ 6.78	$ 13.57
20 lbs. (9 kg) hydrated lime	$13.60 *	$ 2.72
6 cu.ft. (0.17 cu.m) sand	$ 0.77	$ 4.61
Cost 100 lin.ft.		$ 20.90
per lin.ft.		$ 0.21
per lin. M.		$ 0.69

Based on 100-lb. bag.

Labor Cost of 100 LF (30.48 M) of 6" (150 mm) Portland Cement Straight Base			
Description	Hours	Rate	Total
Plasterer	16.00	$32.41	$ 518.56
Labor	4.00	$26.16	$ 104.64
Cost 100 lin.ft.			$ 623.20
per lin.ft.			$ 6.23
per lin.M.			$ 20.45

For coved base, add 50% to material and labor.

1296

FINISHES

Exterior Stucco

Exterior stucco is generally composed of a portland cement base. The scratch and brown coats should be mixed in the proportions of 1 part of portland cement by weight to 3 parts of clean sand by weight. Do not add more than 8 lbs. (3.62 kg) of lime to each 100 lbs. (45.36 kg) of portland cement used in the mixture. Portland cement plaster should not be applied when the outside temperature is below 32°F (0°C).

The cost of portland cement stucco will vary according to the materials used and the grade of workmanship required.

Applying Scratch Coat of Portland Cement Stucco. Where just an ordinary grade of workmanship is required, a plasterer should apply 120 to 140 sq.yds. (100.33 to 117.05 sq.m) of scratch coat per 8-hr. day on metal lath.

Applying Brown Coat of Portland Cement Stucco. A plasterer should apply 75 to 85 sq.yds. (62.70 to 71.06 sq.m) of brown coat per 8-hr. day, on brick, tile, or concrete block walls and 60 to 70 sq.yds. (50.16 to 58.52 sq.m) per 8-hr. day on metal lath.

Applying Trowel Finish Coat of Portland Cement. A plasterer should apply 45 to 55 sq.yds. (37.620 to 45.99 sq.m) of portland cement trowel finish per 8-hr. day, where just an ordinary grade of workmanship is required and 35 to 40 sq.yds. (29.26 to 33.44 sq.m) per 8-hr. day, when first grade work is required.

Applying Wet Rough Cast Finish to Portland Cement Stucco. If a wet rough cast finish is used and the mortar and aggregate are thrown against the wall with a paddle or similar tool, a plasterer should complete 30 to 40 sq.yds. (25.08 to 33.44 sq.m) per 8-hr. day.

Applying Pebble Dash or Dry Rough Cast. When applying a pebble dash or dry rough cast finish where the aggregate is thrown against the wet cement or "butter" coat, a plasterer should apply 35 to 40 sq.yds. (29.26 to 33.44 sq.m) per 8-hr. day.

Washing Exterior Stucco With Acid to Expose Crystals. Where the finish coat of stucco contains granite, marble, or crystal screenings, it is washed off with a solution of muriatic acid to expose the crystals. A worker should wash 475 to 525 sq.ft. or 53 to 58 sq.yds. (44.31 to 48.49 sq.m) per 8-hr. day.

On some jobs it will be necessary to wash the walls two or three times to bring out the crystals satisfactorily, and in such instances the labor cost should be increased accordingly.

Material Cost of 100 Sq. Yds. (83.61 Sq.M) 3-Coat 1:3 Portland Cement Stucco, Float Finish, Applied to Metal Lath Over Wood Framing Ordinary Workmanship		
Scratch Coat-1:3	Rate	Total
8 sacks Portland cement	$ 6.78	$ 54.27
60 lbs. (27 kg) hydrated lime	$13.60 *	$ 8.16
24 cu.ft. (0.68 cu.m) sand	$ 0.77	$ 18.43
Brown Coat-1:3		
16 sacks Portland cement	$ 6.78	$ 108.54
120 lbs. (54 kg) hydrated lime	$13.60 *	$ 16.32
48 cu.ft. (1.36 cu.m) sand	$ 0.77	$ 36.86
Float Finish-1:3		
5 sacks w'p'f. Portland cement	$ 8.48	$ 42.40
15 cu.ft. (0.42 cu.m) sand	$ 0.77	$ 11.52
Cost 100 sq. yds.		$ 296.51
per sq. yd.		$ 2.97
per sq. ft.		$ 0.33
per sq.m		$ 3.55

Add for metal lath.
*Based on 100-lb. bag.

Labor Cost of 100 Sq. Yds. (83.61 Sq.M) 3-Coat 1:3 Portland Cement Stucco, Float Finish, Applied to Metal Lath Over Wood Framing Ordinary Workmanship			
Scratch Coat-1:3	Hours	Rate	Total
Plasterer	6.00	$32.41	$ 194.46
Labor	6.00	$26.16	$ 156.96
Brown Coat-1:3			
Plasterer	12.00	$32.41	$ 388.92
Labor	8.00	$26.16	$ 209.28
Float Finish-1:3			
Plasterer	12.00	$32.41	$ 388.92
Labor	6.00	$26.16	$ 156.96
Cost 100 sq. yds.			$1,495.50
per sq. yd.			$ 14.96
per sq. ft.			$ 1.66
per sq.m			$ 17.89

Add for metal lath.

If a smooth troweled finish is desired, add 4 hrs. plasterer time. If a textured finish coat is desired, add 2 hrs. plasterer time. If first grade workmanship is required, add 8 hrs. plasterer time to the above. Add for scaffold.

FINISHES

Material Cost of 100 Sq. Yds. (83.61 Sq.M) 3-Coat 1:3 Portland Cement Stucco, Float Finish, Applied Over Brick, Clay Tile or Cement Block Surfaces Ordinary Workmanship		
Scratch Coat-1:3	Rate	Total
6 sacks Portland cement	$ 6.78	$ 40.68
45 lbs. (20.25 kg) hydrated lime	$13.60 *	$ 6.12
18 c.f. (0.51 cu.m) sand	$ 0.77	$ 13.82
Brown Coat-1:3		
10 sacks Portland cement	$ 6.78	$ 67.80
75 lbs. (33.75 kg) hydrated lime	$13.60 *	$ 10.20
30 c.f. (0.85 cu.m) sand	$ 0.77	$ 23.04
Float Finish-1:3		
5 sacks w'p'f. Portland cement	$ 8.48	$ 42.40
15 c.f. (0.42 cu.m) sand	$ 0.77	$ 11.52
Cost 100 sq. yds.		$ 215.58
per sq. yd.		$ 2.16
per sq. ft.		$ 0.24
per sq.m		$ 2.58

Based on 100-lb. bag.

Labor Cost of 100 Sq. Yds. (83.61 Sq.M) 3-Coat 1:3 Portland Cement Stucco, Float Finish, Applied Over Brick, Clay Tile or Cement Block Surfaces Ordinary Workmanship			
Scratch Coat-1:3	Hours	Rate	Total
Plasterer	6.00	$32.41	$ 194.46
Labor	5.00	$26.16	$ 130.80
Brown Coat-1:3			
Plasterer	10.00	$32.41	$ 324.10
Labor	6.00	$26.16	$ 156.96
Float Finish-1:3			
Plasterer	12.00	$32.41	$ 388.92
Labor	6.00	$26.16	$ 156.96
Cost 100 sq. yds.			$1,352.20
per sq. yd.			$ 13.52
per sq. ft.			$ 1.50
per sq.m			$ 16.17

For smooth troweled finish, add 4 hrs. plasterer time. If a textured finish coat is desired, add 2 hrs. plasterer time. If first grade workmanship is required, add 8 hrs. plasterer time to the above. Add for scaffold.

Special Finishes for Portland Cement Stucco

If any of the following special finishes are wanted with portland cement stucco, deduct the finish coat given above and add cost of the finish coat.

Material Cost of 100 Sq. Yds. (83.61 Sq.M) of White Cement Float Finish, Ordinary Workmanship		
Description	Rate	Total
5 sacks w'p'f. white cement	$13.60	$ 68.00
1,500 lbs. (675 kg) white silica sand	$ 1.70 *	$ 25.44
Cost 100 sq. yds.		$ 93.44
per sq. yd.		$ 0.93
per sq. ft.		$ 0.10
per sq.m		$ 1.12

Based on 100-lb. bag.

Labor Cost of 100 Sq. Yds. (83.61 Sq.M) of White Cement Float Finish, Ordinary Workmanship			
Description	Hours	Rate	Total
Plasterer	12.00	$32.41	$ 388.92
Labor	6.00	$26.16	$ 156.96
Cost 100 sq. yds.			$ 545.88
per sq. yd.			$ 5.46
per sq. ft			$ 0.61
per sq.m			$ 6.53

Material Cost of 100 Sq.Yds. (83.611Sq.M) Finish Coat Colored Stucco, Float Finish - Ordinary Workmanship		
Description	Rate	Total
2 sacks w'p'f white cement	$13.60	$ 27.20
100 lbs. (45 kg) fin. hydrated lime	$13.60 *	$ 13.60
600 lbs. (270 kg) white silica sand	$ 1.70 *	$ 10.18
18 lbs. (8.1 kg) color	$ 1.70	$ 30.53
Cost 100 sq. yds.		$ 81.50
per sq. yd.		$ 0.82
per sq. ft.		$ 0.09
per sq.m		$ 0.97

Based on 100-lb. bag.

Labor Cost of 100 Sq.Yds. (83.611Sq.M) Finish Coat Colored Stucco, Float Finish - Ordinary Workmanship			
Description	Hours	Rate	Total
Plasterer	12.00	$32.41	$ 388.92
Labor	6.00	$26.16	$ 156.96
Cost 100 sq. yds.			$ 545.88
per sq. yd.			$ 5.46
per sq. ft.			$ 0.61
per sq.m			$ 6.53

FINISHES

If a smooth troweled finish is desired, add 4 hrs. plasterer time. If a textured finish is desired, add 2 hrs. plasterer time. If first grade workmanship is required, add 8 hrs. plasterer time to the above.

Material Cost of 100 Sq.Yds. (83.61 Sq.M) Wet Rough Cast Finish Coat		
Description	Rate	Total
4 sacks w'p'f. portland cement	$13.60	$ 54.40
12 cu.ft. (0.34 cu.m) sand	$ 0.77	$ 9.22
500 lbs. (225 kg) aggregate	$10.18 *	$ 50.88
Cost 100 sq. yds.		$ 114.50
per sq. yd.		$ 1.14
per sq. ft.		$ 0.13
per sq.m		$ 1.37

Based on 100-lb. bag.

Labor Cost of 100 Sq.Yds. (83.61 Sq.M) Wet Rough Cast Finish Coat			
Description	Hours	Rate	Total
Plasterer	20.00	$32.41	$ 648.20
Labor	10.00	$26.16	$ 261.60
Cost 100 sq. yds.			$ 909.80
per sq. yd.			$ 9.10
per sq. ft.			$ 1.01
per sq.m			$ 10.88

Material Cost of 100 Sq. Yds. (83.61 Sq.M) of Dry Rough Cast Finish		
Description	Rate	Total
4 sacks w'p'f. Portland cement	$13.60	$ 54.40
12 cu.ft. (0.34 cu.m) sand	$ 0.48	$ 5.76
750 lbs. (337.5 kg) aggregate	$ 6.36 *	$ 47.70
Cost 100 sq. yds.		$ 107.86
per sq. yd.		$ 1.08
per sq. ft.		$ 0.12
per sq.m		$ 1.29

Based on 100-lb. bag.

Labor Cost of 100 Sq.Yds. (83.61 Sq.M) of Dry Rough Cast Finish			
Description	Hours	Rate	Total
Plasterer	20.00	$32.41	$ 648.20
Labor	10.00	$26.16	$ 261.60
Cost 100 sq. yds.			$ 909.80
per sq. yd.			$ 9.10
per sq. ft.			$ 1.01
per sq.m			$ 10.88

A dry rough cast finish requires a finish or "butter" coat 1/4" (6 mm) thick applied directly over the brown coat. This is brought to a straight, smooth finish and then the aggregate is thrown onto the "butter" coat dry.

If coarse aggregate is used, it requires about 1,000 lbs. per 100 sq.yds. (453.60 kg per 83.60 sq.m).

If medium size aggregate is used, it requires 750 lbs. per 100 sq.yds. (340.02 kg per 83.61 sq.m). If fine aggregate is used, it requires 500 lbs. per 100 sq.yds. (218.05 kg per 83.61 sq.m).

Approximate Prices on Lathing and Plastering Materials Metal Lath (Expanded Diamond Mesh, Flat)				
Finish	Wgt., Lbs./SY	Price per Sq.Yd.	Wgt., Kg/Sq.M.	Price per Sq.M.
Painted	2.5	$ 2.34	1.3	$ 2.80
Painted	3.4	$ 2.88	1.8	$ 3.44
Galvanized	3.4	$ 3.33	1.8	$ 3.98

Self-furring lath, add $0.25 per sq. yd. to the above prices.

Rib Metal Lath							
Description	Finish	Hgt of Rib, Inch	Wgt., Lbs/SY	Price per Sq.Yd.	Hgt. of Rib, mm	Wgt., Kg/Sq.M.	Price per Sq.M.
Flat rib	Painted	1/8	2.75	$ 2.88	3.13	1.49	$ 3.44
Flat rib	Painted	1/8	3.40	$ 3.06	3.13	1.84	$ 3.66
Rib or	Painted	3/8	3.40	$ 4.14	9.38	1.84	$ 4.95
Rod ribbed	Painted	3/8	4.00	$ 4.05	9.38	2.17	$ 4.84

Hot Rolled Channels							
Width, Inches	Leg, Inches	Lbs. per 1,000 Lin.Ft.	Price per 1,000 Lin.Ft.	Width, mm	Leg, mm	Kg per 100 Lin.M.	Price per 100 Lin.M.
3/4 " Standard	15/32 "	300	$306.25	18.75 mm	11.72 mm	44.65	$100.48
1 " Standard	3/8 "	410	$367.50	25.00 mm	9.38 mm	61.02	$120.57
1 1/2 " Standard	19/32 "	650	$463.75	37.50 mm	14.84 mm	96.73	$152.15
2 " Standard	7/16 "	1,260	$577.50	50.00 mm	10.94 mm	187.51	$189.47

Peforating, add $10.00 per 1,000 l.f. ($33.00 per 1000 m).

Cold Rolled Channels							
Width, Inches	Leg, Inches	Lbs. per 1,000 Lin.Ft	Price per 1,000 Lin.Ft.	Width, mm	Leg, mm	Kg per 100 Lin. M.	Price per 100 Lin.M.
3/4 "	1/2 "	300	$297.50	18.75 mm	12.50 mm	44.65	$ 97.60
1 1/2 "	19/32 "	475	$446.25	37.50 mm	14.84 mm	70.69	$146.41
2 "	19/32 "	590	$560.00	50.00 mm	14.84 mm	87.80	$183.73

1302

FINISHES

Studs & Tracks - 20 Ga.			
Size, Inches	Price per 1,000 Lin. Ft.	Size, mm	Price per 100 Lin.M.
1 5/8 "	$512.00	40.63 mm	$167.98
2 1/2 "	$544.00	62.50 mm	$178.48
3 1/4 "	$592.00	81.25 mm	$194.23
4 "	$624.00	100.00 mm	$204.72
6 "	$800.00	150.00 mm	$262.47
Shoes (per 1,000 pcs.)	$140.00	-----	$45.93

Wire	
Size	Per Cwt.
16 Ga	$148.75
18 Ga	$166.25

Corner Bead		
Type	Price per 1,000 Lin.Ft.	Price per 100 Lin.M.
Regular	$ 280.00	$ 91.86
Expanded	$ 323.75	$ 106.22
Bull Nose	$ 647.50	$ 212.43
Cornerite	$ 183.75	$ 60.29

Rods			
Type, Inches	Price per 1,000 Lin.Ft.	Type, mm	Price per 100 Lin.M.
3/16 " plain	$ 72.00	4.69 mm	$ 23.62
3/16 " galvanized	$ 87.00	4.69 mm	$ 28.54
1/4 " plain	$ 96.00	6.25 mm	$ 31.50
1/4 " galvanized	$ 111.00	6.25 mm	$ 36.42
3/8 " plain	$ 120.00	9.38 mm	$ 39.37

Gypsum Lath				
Type and size, Inches		Price per 1,000 Sq.Ft.	Type and size, mm	Price per 100 Sq.M.
plain or perforated	16 " x 48 " x 3/8 "	$ 192.50	400 x 1200 x 9.38 mm	$ 207.21
foilback	16 " x 48 " x 3/8 "	$ 341.25	400 x 1200 x 9.38 mm	$ 367.33
plain	16 " x 48 " x 1/2 "	$ 201.25	400 x 1200 x 12.50 mm	$ 216.63
foilback	16 " x 48 " x 1/2 "	$ 350.00	400 x 1200 x 12.50 mm	$ 376.75

Nails and Staples				
Type and size, Inches	Price per 100 Lbs	Type and size, mm	Price per 100 Kg
Gypsum lath, 1-1/8" blued	$190.00	Gypsum lath, 28.123 mm blued	$418.87
Gypsum lath, 1-1/4" blued	$190.00	Gypsum lath, 31.25 mm blued	$418.87
Roofing, barbed, 1-1/2" galv	$210.00	Roofing, barbed, 37.50 mm) galv	$462.96
Concrete stub	$240.00	Concrete stub	$529.10
Annular ring nail (blued)	$190.00	Annular ring nail (blued)	$418.87
Brick-plain 2" and 2-1/2"	$210.00	Brick-plain 2" 50 mm & 62.50 mm	$462.96

Lime and Finishing Plaster		
Description	Price per 100 Lbs.	Price per 100 Kg.
Pulverized Quicklime	$ 12.95	$ 28.55
Plasterer's Hydrated Lime	$ 14.88	$ 32.79
Gypsum Gauging Plaster	$ 16.63	$ 36.65
No. 1 Moulding Plaster	$ 18.55	$ 40.90
Keene's Cement, Regular or Fast	$ 29.68	$ 65.43
Portland Cement	$ 7.25	$ 15.98

Plastering Items		
Description	Price per Gallon	Price per Liter
Liquid Bonding Agent	$ 46.38	$ 12.25

Gypsum Plaster		
Description	Price per 100 Lbs	Price per 100 Kg.
Gypsum Cement Plaster	$ 11.20	$ 24.69
Wood Fiber (prepared)	$ 11.20	$ 24.69
Bondcrete	$ 12.95	$ 28.55
Perlite Plaster	$ 12.00	$ 26.46
Gypsum Trowel Finish	$ 12.25	$ 27.01
Silica Sand Float	$ 16.63	$ 36.65

09250 GYPSUM WALLBOARD

Gypsum wallboard, or drywall, is manufactured from a gray-white colored rock called gypsum, which is a nonmetallic mineral composed of calcium sulphate chemically combined with crystallized water.

After the gypsum ore is mined or quarried, it is crushed, dried, and ground into a fine powder. It is then heated to remove most of the chemically combined water. The calcined gypsum is then mixed with other ingredients and water to create a slurry that is sandwiched between two sheets of treated paper to form a smooth gypsum wallboard panel. After the gypsum core has set, the wallboard is cut to length, dried, finished, and packaged for shipment.

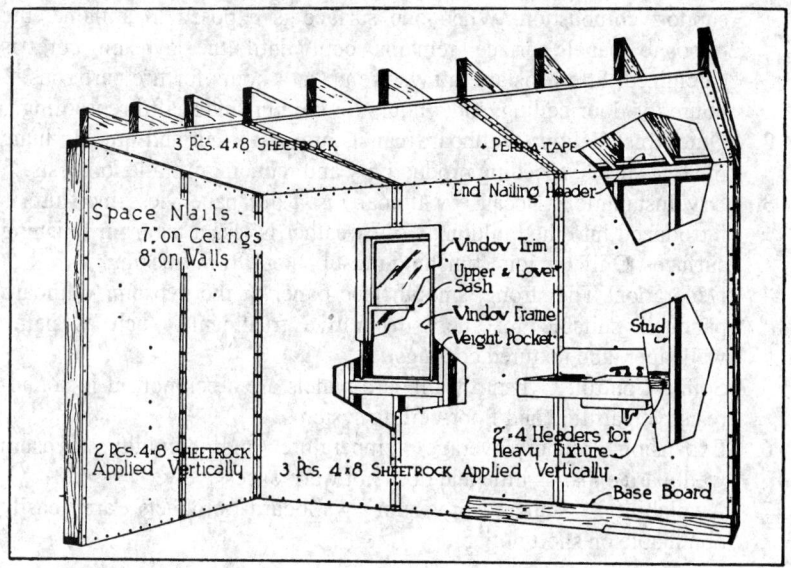

Vertical Application of Gypsum Wallboard

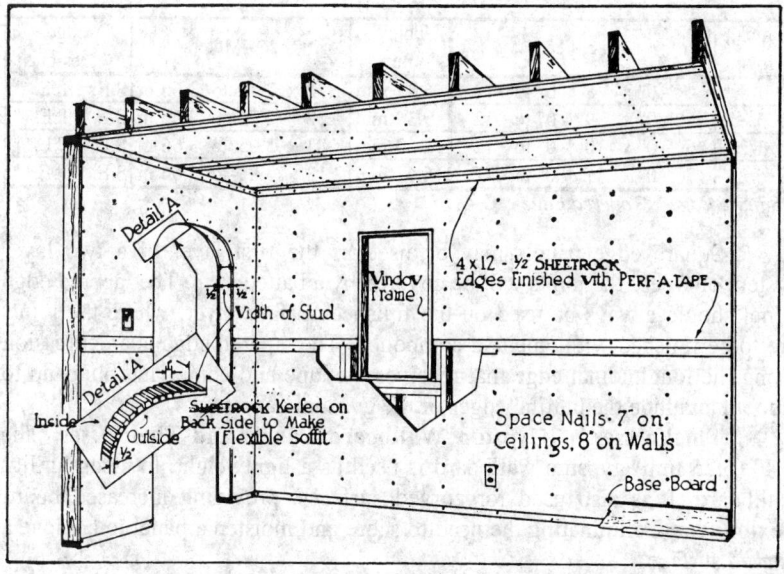

Horizontal Application of Gypsum Wallboard

Advantages. Interior walls and ceilings built with gypsum panels form durable surfaces for most types of decorative treatment. The following are some of the advantages of gypsum wallboard.

1. Fire Protection: Gypsum wallboard is fire resistant and will not support combustion. When one surface is exposed to a flame, the opposite panel surface remains cool until the gypsum core is calcinized. Fire resistant ratings are up to 4 hours for fire partitions, 3 hours for floor-ceiling, and 4 hours for column and shaft fireproofing.
2. Quick Installation: Wallboard panels are easily cut and quickly hung in large sheets, speeding productivity and reducing installation costs.
3. Dry Installation: Because wallboard is a dry panel, less moisture is introduced into the building. Cold weather is less a factor in finishing surfaces. Only the joint finishing must be kept from freezing.
4. Decoration: The strong, smooth face paper of the gypsum wallboard panel is suitable for most decorative treatments, such as paint, wallpaper, and textured coatings.
5. Sound Control: Gypsum wallboard panels are a component in sound-resistive partition and floor-ceiling systems.
6. Expansion: Under average temperature and humidity, gypsum wallboard expands little and does not warp excessively.
7. Availability: Standard gypsum wallboard products are easily obtainable on short notice.

Size and Thickness. Gypsum wallboard panels are manufactured in a number of thicknesses and sizes with various longitudinal edge designs.

Thick., Inch	Size, Feet	Thick. mm	Size - Meters	Edge Design
1/4 "	4 '- 0" x 8 '- 0" to 12 '	6.25 mm	1.22 x 2.44 to 3.66 m	S.E.*, T.E.**
3/8 "	5 '- 0" x 9 '- 0" to 12 '	9.38 mm	1.52 x 2.74 to 3.66 m	S.E.*, T.E.**
1/2 "	6 '- 0" x 10 '- 0" to 12 '	12.50 mm	1.83 x 3.05 to 3.66 m	S.E.*, T.E.**
5/8 "	7 '- 0" x 11 '- 0" to 12 '	15.63 mm	2.13 x 3.35 to 3.66 m	S.E.*, T.E.**

*Square Edge **Tapered Edge*

Square edge panels may be used as the first layer of a two layer system or where the joints will be covered by a batten strip. The tapered edge panel should always be used on the finished surface layer, where the joints are to be finished with tape and compound. The tapered edge is a depression along the longitudinal edge that receives the tape and compound build-up to smooth and hide the butting edges of the gypsum panels.

Thicknesses of Gypsum Wallboard Panels and Their Uses. The 1/4" (6.25 mm) gypsum wallboard is used as a lightweight, low cost utility wallboard. It is also used for curved surfaces, providing increased board flexibility and eliminating the need to score and moisten a panel to achieve a curved surface.

The 3/8" (9.37 mm) gypsum wallboard is used principally for repair and remodeling work or in double wall construction.

The 1/2" (12.50 mm) gypsum wallboard is used in single layer new construction, as well as for remodeling installations.

FINISHES

The 5/8" (15.62 mm) gypsum wallboard is used where a one-hour fire rating is required or where the framing is spaced in excess of 1/2" (12.50 mm) wallboard limitations. The thickness of these panels not only give increased resistance to fire exposure but also additional resistance to sound transmission. The stiffness decreases the likelihood of sagging.

Tape Joint System for Gypsum Wallboard. A drywall joint treatment, using tape and joint compound, must provide a joint that is as strong as the gypsum wallboard itself. The hand method of concealing joints in gypsum wallboard is backed by more than a quarter century of joint reinforcing experience. It conceals the joints between boards and bonds the gypsum wallboard units together into a single, smooth, even wall and ceiling surface unit.

First the hollow or channel at the edges of the wallboard is filled with joint compound, using a flexible 4", 5", or 6" (100, 125, or 150 mm) joint finishing broadknife.

Apply the joint tape immediately, directly over the compound and press it into place with the broadknife, squeezing excess compound out from under the joint tape but leaving enough compound under tape for proper bond. The depth best not exceed 1/32" (0.78 mm). Apply a thin covering coat of joint compound immediately and let dry.

Next, apply another thin coating of joint compound using a 7" to 10" (175 to 250 mm) flexible broadknife so that the tape will be completely hidden. Feather out edges beyond previous coat as smoothly as possible and let dry thoroughly. Intermediate nail heads should also be carefully filled with joint compound and brought flush with the surface of the wallboard.

Apply a third thin finish coat of joint compound. It should be of thinner consistency to even up surface and edges of joint. This is usually accomplished using a wide finish coat applicator, wide enough to extend 2" (50 mm) beyond the edges of the second coat.

Dry sand, as needed, between and after coats to insure a smooth, inconspicuous joint and to eliminate scratches, craters, and nicks.

Mechanical Tape Applicator. There are several types of mechanical and semi-mechanical applicators. Joint compound is applied as before, by hand, and then the joint is taped mechanically. Then joint compound is then wiped down with a flexible broadknife and left to dry.

Apply a second fill coat of joint compound over the tape using a hand finisher tool. Spot finish all nails or screws to bring them flush with the surface of the wallboard.

Apply a third and final finish coat, feathering the edges about 2" (50 mm) beyond the last coat, and allow to dry. Dry sand as needed. Remember, drying time between coat applications will vary according to air temperature and humidity.

Types of Gypsum Panels

Foil-Back Gypsum Panels. These panels are made by laminating a sheet of aluminum foil to the back surface of the gypsum wallboard. The foil reduces outward heat flow in winter and inward heat flow in summer, has a significant thermal insulating value if facing an air space of 3/4" (18.75 mm) or more, and is effective as a vapor barrier.

In addition, foil-backed panels provide a water vapor retarder to help prevent interior moisture from entering wall and ceiling spaces. Foil-backed wallboard has some limitations. It is not recommended over tile, double layer installations, or where high outside temperatures prevail.

Moisture Resistant Gypsum Panels. MR board has a special asphalt composition gypsum core and is covered with chemically treated face papers to prevent moisture penetration. This board is a little harder to cut, has a brownish color core, and is usually covered with a light green finish face paper. These panels were developed for application in bathrooms, kitchens, utility rooms, and other high moisture areas.

Fire Rated Gypsum Panels. A specially formulated mineral gypsum core is used to make panels in 1/2" (12.50 mm) and 5/8" (15.62 mm) thicknesses for application to walls, ceilings, and columns where a fire rated assembly is needed. The formulation must meet or exceed ASTM C36. Based on tests by Underwriters' Laboratories, Inc., certain wall, floor/ceiling, and column assemblies give 45-minute to 4-hour fire resistance ratings.

Cement Boards. These panels are produced with a portland cement core, usually wrapped in a fiber glass mesh and covered with a specially

Fastener Description	Nail Spacing c. to c.	Approx. Lbs. Nails Req'd per MSF Gypsum Panels
1¼″ GWB-54 Annular Ring Nail 12½ ga.; ¼″ dia. head with a slight taper to a small fillet at shank; bright finish; medium diamond point; meets ASTM C380 	7″ ceiling 8″ walls	5¼
1⅜″ Annular Ring Nail (Same as GWB-54 except for length) 	7″ ceiling 8″ walls	5¼

Fastening Application	Fastener Used
GYPSUM PANELS TO STANDARD METAL FRAMING	
½″ single-layer panels to standard studs, runners, channels	 ⅞″ Type S Bugle Head
⅝″ single-layer panels to standard studs, runners, channels	 1″ Type S Bugle Head
½″ double-layer panels to standard studs, runners, channels	 1⁵⁄₁₆″ Type S Bugle Head
⅝″ double-layer panels to standard studs, runners, channels	 1⅝″ Type S Bugle Head
1″ coreboard to metal angle runners in solid partitions	 1¼″ Type S Bugle Head
½″ panels through coreboard to metal angle runners in solid partitions	 1⅝″ Type S Bugle Head
⅝″ panels through coreboard to metal angle runners in solid partitions	 2¼″ Type S Bugle Head
GYPSUM PANELS TO 12-GA. (MAX.) METAL FRAMING	
½″ and ⅝″ panels and gypsum sheathing to 20-ga. studs and runners	 1″ Type S-12 Bugle Head
USG Self-Furring Metal Lath through gypsum sheathing to 20-ga. studs and runners	 1¼″ Type S-12 Bugle Head
½″ and ⅝″ double-layer gypsum panels to 20-ga. studs and runners	 1⅝″ Type S-12 Bugle Head
Multi-layer gypsum panels to 20-ga. studs and runners	 1⅞″ Type S-12 Bugle Head

formulated smooth covering. Cement boards are used for interior and exterior construction, where there are areas subject to water, moisture or high humidity, such as bathroom showers and tubs and for kitchen countertops.

These boards require special screws and tape for proper installation. They are not gypsum boards, and they have different stud spacing requirements.

Gypsum Drywall Accessories

Gypsum wallboard can be attached to framing by several methods, depending on the type of framing and the results desired.

Fasteners. Nails or screws can be used to attach gypsum panels to wood framing. However, screws must be used for attachment to metal framing members. There are four basic ways to fasten panels to framing:

1. Single Nailing
2. Double Nailing
3. Screw Attachment
4. Adhesive Attachment

Nail Application: Nails are applied with a hammer, seating the nail so that the head is in a shallow dimple formed by the last blow of the hammer. The dimple should be no deeper than 1/32" (0.78 mm) for gypsum panels. Drive nails at least 3/8" (9.37 mm) from ends or edges of wallboard.

Screw Application: Screws are applied with an electric power positive-clutch tool called an electric screwgun. Drive screws at least 3/8" (9.37 mm) from ends or edges of wallboard.

Staple Application: Staples can be used to attach base layer boards to wood framing in a double layer application. Staples should be 16-gauge flattened galvanized wire and provide a minimum penetration of 5/8" (15.62 mm) into the wood framing.

Double Nailing or Screwing: In this method perimeter screws are spaced 7" (175 mm) on center for ceilings and 8" (200 mm) on center for walls. Nails or screws are not doubled on the panel perimeter. For nails or screws attached to supports, pairs of nails or screws are attached 2" (50 mm) apart, and pairs are spaced every 12" (300 mm).

Adhesive Application: A continuous bead of adhesive is applied to a wood stud. The drywall panel is pressed onto the adhesive and supplementary screws or nails are attached. The main advantage of using adhesive is that it reduces the number of nails or screws required.

Corner Bead. Metal and plastic corner bead is furnished in 8'-0" (2.43 m) and 10'-0" (3.04 m) lengths and is used on all external corners in drywall construction. Corner bead provides a true and straight finish line to corners.

Stop Bead. Metal or vinyl stop beads are used to cap the edge of a gypsum panel at its termination point, whenever that edge is exposed and not covered by either a tape/compound finish or other trim members.

1310

FINISHES

Joint Tape. A strong fiber tape designed with feathered edges and lightly pre-creased for internal corners is used with the joint compound for finishing the butt joints of gypsum panels. Joint tape is packaged and sold in rolls of 75, 250 and 500 lin.ft. (22.86, 76.2 and 152.40 m) lengths and widths of 1-31/32" (49 mm) 2-1/16" (51.56 mm) wide. Usually, about 0.37 lin.ft. (0.11 m) of tape is required per sq.ft. (.0929 sq.m) of drywall panel.

Joint Compound. This material is available in many different forms, e.g., powdered, ready-mixed, all-purpose, fast setting, exterior application, etc. The ready-mixed all-purpose joint compound is used extensively for all interior applications of taping and joint finishing, and is readily available in 5-gal. (18.93-liter) containers weighing about 61.7 lbs. (27.98 kg) each.

Type of Compound	Approximate Coverage
Ready-mixed joint compound - taping, topping, and all-purpose	0.138 lbs. per square foot 0.674 kg per square meter
Ready-mixed lightweight all-purpose joint compound	0.0094 gal per square foot 0.383 liters per square meter
Powder Joint Compound - taping, topping, and all-purpose	0.083 lbs. per square foot 0.405 kg per square meter
Setting Type Joint Compound, easy sand	0.052 lbs. per square foot 0.253 kg per square meter

Estimating Material Quantities for Gypsum Wallboard and Accessories

The estimator first measures the wall and ceiling surfaces to figure square feet (square meters) of surface to be covered. Do not deduct for openings smaller than 50 sq.ft. (4.65 sq.m).

When doing the quantity takeoff, an estimator separates the square foot (square meter) quantities into proper categories of types, thicknesses, and sizes of panels. From these figures the estimator can easily determine the number of gypsum wallboards and other materials required, including accessories such as fasteners.

Quantities of Fasteners. For nails or screws, the amount required depends on the framing spacing. If framing is spaced 16" (400 mm) on center, and fasteners are spaced 12" (300 mm) apart on a horizontal 8' x 4' (2.43 x 1.21 sq.m) panel, then it would require 1,094 fasteners per 1,000 sq.ft. (92.9 sq.m), or 35 fasteners per panel. "W" is used to denote wood fasteners, and "S" indicates steel fasteners.

Quantities of Joint Tape. Approximately 370 lin. ft. (112.77 m) of joint tape are required per 1,000 sq.ft. (92.9 sq.m) of wallboard.

Quantities of Joint Compound. Approximately 138 lbs. (62.59 kg) of all-purpose joint material is required to complete 1,000 sq.ft. (92.9 sq.m) of wallboard. The amount of compound is reduced if there are very few external corners to be finished.

Drywall Material Prices

Type of Gypsum Panel	Thick., Inches	Price per 32 Sq.Ft.	Price per 1,000 Sq.Ft.	Thick., mm	Price per 3 Sq.M.	Price per 100 Sq.M.
Regular	1/4 "	$ 7.49	$ 233.94	6.25 mm	$ 7.55	$ 251.82
Regular	3/8 "	$ 7.85	$ 245.22	9.38 mm	$ 7.92	$ 263.96
Regular	1/2 "	$ 8.15	$ 254.72	12.50 mm	$ 8.23	$ 274.19
Regular	5/8 "	$ 9.23	$ 288.56	15.63 mm	$ 9.32	$ 310.62
Fire Rated	1/2 "	$ 9.12	$ 285.00	12.50 mm	$ 9.20	$ 306.78
Fire Rated	5/8 "	$ 9.92	$ 309.94	15.63 mm	$ 10.01	$ 333.62
Moisture Resistant	1/2 "	$ 11.80	$ 368.72	12.50 mm	$ 11.91	$ 396.90
Moisture Resistant	5/8 "	$ 12.65	$ 395.44	15.63 mm	$ 12.77	$ 425.66
Add: For Foil Back Panels		$ 5.53	$ 172.78		$ 5.58	$ 185.99

Fasteners

Description	Price per 1,000 Pcs
1 1/4 " (31.25 mm) GWB Ring Nails	$ 5.80
1 1/4 " (31.00 mm) Type S Bugle Head Screws	$ 15.63

Corner Bead

Description	Price per Lin. Ft.	Price per Lin.M.
Corner Bead	$ 4.48	$ 14.70
Stop Bead	$ 12.08	$ 39.62

Joint Tape

Description	Price per Lin. Ft.	Price per Lin.M.
75 ' (22.86 m) Roll	$ 2.75	$ 9.02
250 ' (76.20 m) Roll	$ 5.80	$ 19.03
300 ' (91.44 m) Roll	$ 6.34	$ 20.80

Joint Compound

Description	Price per Container
4.5 gal (17.03 L) Container	$ 16.80

4' x 8' Board - Vertical Attachment — Number of Screws per 1,000 Sq.Ft

Frame Spacing	8 "	12 "	16 "	24 "
8 "	2,844	1,969	1,531	1,094
12 "	2,031	1,406	1,094	781
16 "	1,625	1,125	875	625
24 "	1,219	844	656	469
30 "	914	731	569	406

1.21 M x 2.43 M Board - Vertical Attachment — Number of Screws per 100 Sq.M

Frame Spacing	200 mm	300 mm	400 mm	600 mm
200 mm	3,061	2,119	1,648	1,178
300 mm	2,186	1,513	1,178	841
400 mm	1,749	1,211	942	673
600 mm	1,312	909	706	505
750 mm	984	787	612	437

FINISHES

4' x 10' Board - Vertical Attachment Number of Screws per 1,000 Sq.Ft					1.21 M x 3.04 M Board - Vertical Attachment Number of Screws per 100 Sq.M				
Frame	Screw Spacing				Frame	Screw Spacing - Metric			
Spacing	8 "	12 "	16 "	24 "	Spacing	200 mm	300 mm	400 mm	600 mm
8 "	2,800	1,925	1,488	1,050	200 mm	3,014	2,072	1,602	1,130
12 "	2,000	1,375	1,063	750	300 mm	2,153	1,480	1,144	807
16 "	1,600	1,100	850	600	400 mm	1,722	1,184	915	646
24 "	1,200	825	638	450	600 mm	1,292	888	687	484
30 "	1,040	715	553	390	750 mm	1,119	770	595	420

4' x 12' Board - Vertical Attachment Number of Screws per 1,000 Sq.Ft					1.21 M x 3.66 M Board - Vertical Attachment Number of Screws per 100 Sq.M				
Frame	Screw Spacing				Frame	Screw Spacing - Metric			
Spacing	8 "	12 "	16 "	24 "	Spacing	200 mm	300 mm	400 mm	600 mm
8 "	2,771	1,896	1,458	1,021	200 mm	2,983	2,041	1,569	1,099
12 "	1,979	1,354	1,042	729	300 mm	2,130	1,457	1,122	785
16 "	1,583	1,083	833	583	400 mm	1,704	1,166	897	628
24 "	1,188	813	625	438	600 mm	1,279	875	673	471
30 "	1,029	704	542	379	750 mm	1,108	758	583	408

Typical Weights of Gypsum Board					
Panel Size		4 ' x 8 '	4 ' x 9 '	4 ' x 10 '	4 ' x 12 '
Sq. Ft. per Panel		32 Sq.Ft.	36 Sq.Ft.	40 Sq.Ft.	48 Sq.Ft.
Regular Boards	Lbs per Sq.Ft.	Total Sheet Weight in Lbs			
1/4 "	1.20	38.40	43.20	48.00	57.60
3/8 "	1.40	44.80	50.40	56.00	67.20
1/2 "	1.70	54.40	61.20	68.00	81.60
5/8 "	2.30	73.60	82.80	92.00	110.40
Firecore Boards					
1/2 "	1.90	60.80	68.40	76.00	91.20
5/8 "	2.20	70.40	79.20	88.00	105.60
3/4 "	2.50	80.00	90.00	100.00	120.00
WR Boards					
1/2 "	1.80	57.60	64.80	72.00	86.40
5/8 "	2.20	70.40	79.20	88.00	105.60
Panel Size		1.2 x 2.4 m	1.2 x 2.7 m	1.2 x 3.0 m	1.2 x 3.7 m
Sq.M. per Panel		3.0 Sq.M.	3.3 Sq.M.	3.7 Sq.M.	4.5 Sq.M.
Regular Boards	Kg per Sq.M.	Total Sheet Weight in Kg			
6.25 mm	5.86	17.42	19.60	21.77	26.13
9.38 mm	6.84	20.32	22.86	25.40	30.48
12.50 mm	8.30	24.68	27.76	30.84	37.01
15.63 mm	11.23	33.38	37.56	41.73	50.08
Firecore Boards					
12.50 mm	9.28	27.58	31.03	34.47	41.37
15.63 mm	10.74	31.93	35.93	39.92	47.90
18.75 mm	12.21	36.29	40.82	45.36	54.43
WR Boards					
12.50 mm	8.79	26.13	29.39	32.66	39.19
15.63 mm	10.74	31.93	35.93	39.92	47.90

Example: The following list of material quantities is required for installation and finishing of 300 lin.ft. (91.44 m) of drywall partition, 8'-0" (2.43 m) high, covering both sides of the partition: 300 x 8 x 2 sides = 4,800 sq.ft. (445.92 sq.m.)

Drywall Material Cost		
Item	Rate	Total
3/4" Drywall Regular - 4,800 sq.ft..	$ 0.32	$1,536.00
Screws 4.8M pcs.	$11.04	$ 33.12
Corner Bead 96 lin.ft.	$ 4.48	$ 7.68
Stop Bead 40 lin.ft.	$12.08	$ 4.40
Joint Tape - 8 rolls	$ 2.75	$ 23.20
Joint Compound - 6 five-gallon cans	$16.80	$ 69.00
Cost per 4,800 sq.ft.		$1,673.40
per sq.ft.		$ 0.35
per sq.m		$ 3.75

Labor Placing Gypsum Wallboard-Nailed. When placing 1/2" (12.50 mm) gypsum wallboard in average size rooms, 2 carpenters experienced in wallboard erection should place about 1,700 sq.ft. (157.93 sq.m) of board (one-layer wall and ceiling) per 8-hr. day, at the following labor cost per sq.ft. (sq.m):

Description	Hours	Rate	Total
Carpenter	16.00	$35.04	$ 560.64
Cost per sq.ft.			$ 0.33
per sq.m			$ 3.55

For walls and ceilings higher than 12'-0", decrease production by 10 to 15%.

When placing 5/8" (16 mm) gypsum wallboard in average size rooms, 2 carpenters experienced in wallboard erection should place about 1,500 sq.ft. (139.35 sq.m) of board (one-layer wall and ceiling) per 8-hr. day, at the following labor cost per sq.ft. (sq.m):

Description	Hours	Rate	Total
Carpenter	16.00	$35.04	$ 560.64
Cost per sq.ft.			$ 0.37
per sq.m			$ 4.02

For walls and ceilings higher than 12'-0", decrease production by 10 to 15%.

Labor Placing Gypsum Wallboard-Screwed. When placing 1/2" (12.50 mm) gypsum wallboard in average size rooms, 2 carpenters experienced in wallboard erection should place about 2,000 sq.ft. (185.80 sq.m) of board (one-layer wall and ceiling) per 8-hr. day, at the following labor cost per sq.ft. (sq.m):

FINISHES

Description	Hours	Rate	Total
Carpenter	16.00	$35.04	$ 560.64
Cost per sq.ft.			$ 0.28
per sq.m			$ 3.02

When placing 5/8" (15.62 mm) gypsum wallboard in average size rooms, 2 carpenters experienced in wallboard erection should place about 1,700 sq.ft. (157.93 sq.m) of board (one-layer wall and ceiling) per 8-hr. day, at the following labor cost per sq.ft. (sq.m):

Description	Hours	Rate	Total
Carpenter	16.00	$35.04	$ 560.64
Cost per sq.ft.			$ 0.33
per sq.m			$ 3.55

Labor Placing Gypsum Wallboard in Shafts-Screwed. When placing 1/2" (12.50 mm) wallboard in shafts, 2 carpenters experienced in wallboard erection in shafts should place about 400 sq.ft. (37.16 sq.m) of board per 8-hr. day, at the following labor cost per sq.ft. (sq.m):

Description	Hours	Rate	Total
Carpenter	16.00	$35.04	$ 560.64
Cost per sq.ft.			$ 1.40
per sq.m			$ 15.09

When placing 5/8" (16 mm) gypsum wallboard in shafts, 2 carpenters experienced in this work should place about 350 sq.ft. (32.51 sq.m) of board per 8-hr. day, at the following labor cost per sq.ft. (sq.m):

Description	Hours	Rate	Total
Carpenter	16.00	$35.04	$ 560.64
Cost per sq.ft.			$ 1.60
per sq.m			$ 17.24

When placing 1-1/2" (37.50 mm) core board in shafts, 2 carpenters experienced in this work should place about 400 sq.ft. (37.16 sq.m) of board per 8-hr. day, at the following labor cost per sq.ft. (sq.m):

Description	Hours	Rate	Total
Carpenter	16.00	$35.04	$ 560.64
Cost per sq.ft.			$ 1.40
per sq.m			$ 15.09

Labor Placing Gypsum Wallboard, Column Enclosures-Screwed. When placing 1/2" (12.50 mm) gypsum wallboard around columns, 2 carpenters experienced in wallboard erection should place about 400 sq.ft. (37.16 sq.m) of board per 8-hr. day, at the following labor cost per sq.ft. (sq.m):

1315

Description	Hours	Rate	Total
Carpenter	16.00	$35.04	$ 560.64
Cost per sq.ft.			$ 1.40
per sq.m			$ 15.09

When placing 5/8" (15.62 mm) gypsum wallboard around columns, 2 carpenters experienced in wallboard erection should place about 350 sq.ft. (32.51 sq.m) of board per 8-hr. day, at the following labor cost per sq.ft. (sq.m):

Description	Hours	Rate	Total
Carpenter	16.00	$35.04	$ 560.64
Cost per sq.ft.			$ 1.60
per sq.m			$ 17.24

Labor Placing Gypsum Wallboard, Enclosures Around Beams and Soffits-Screwed. When placing 1/2" (12.50 mm) gypsum wallboard around beams and soffits, 2 carpenters experienced in wallboard erection should place about 400 sq.ft. (37.16 sq.m) of board per 8-hr. day, at the following labor cost per sq.ft. (sq.m):

Description	Hours	Rate	Total
Carpenter	16.00	$35.04	$ 560.64
Cost per sq.ft.			$ 1.40
per sq.m			$ 15.09

When placing 5/8" (15.62 mm) gypsum wallboard around beams and soffits, 2 carpenters experienced in wallboard erection should place about 350 sq.ft. (32.51 sq.m) of board per 8-hr. day, at the following labor cost per sq.ft. (sq.m):

Description	Hours	Rate	Total
Carpenter	16.00	$35.04	$ 560.64
Cost per sq.ft.			$ 1.60
per sq.m			$ 17.24

Labor Placing Gypsum Wallboard, Light Wells-Screwed. When placing 1/2" (12.50 mm) gypsum wallboard in light wells, 2 carpenters experienced in wallboard erection should place about 400 sq.ft. (37.16 sq.m) of board per 8-hr. day, at the following labor cost per sq.ft. (sq.m):

Description	Hours	Rate	Total
Carpenter	16.00	$35.04	$ 560.64
Cost per sq.ft.			$ 1.40
per sq.m			$ 15.09

1316

FINISHES

When placing 5/8" (15.62 mm) gypsum wallboard in light wells, 2 carpenters experienced in wallboard erection should place about 350 sq.ft. (32.51 sq.m) of board per 8-hr. day, at the following labor cost per sq.ft. (sq.m):

Description	Hours	Rate	Total
Carpenter	16.00	$35.04	$ 560.64
Cost per sq.ft.			$ 1.60
per sq.m			$ 17.24

Labor Finishing Gypsum Wallboard. The finishing of gypsum wallboard is a multi-step process of tape coat, block coat, skim coat, and point-up coat. Sanding between the block and skim coats is not required if the finisher is careful and a first class mechanic. The complete finishing labor for 1,000 sq.ft. (92.90 sq.m) of wallboard is as follows:

Description	Hours	Rate	Total
Carpenter - Tape Coat	2.00	$35.04	$ 70.08
Carpenter - Block Coat	1.60	$35.04	$ 56.06
Carpenter -Skim Coat	1.30	$35.04	$ 45.55
Carpenter - Sand	0.40	$35.04	$ 14.02
Carpenter - Point-Up	0.80	$35.04	$ 28.03
Cost per 1,000 sq.ft.			$ 213.74
per sq.ft.			$ 0.21
per sq.m			$ 2.30

Metal Studs and Furring

Stud framing and wall or ceiling furring on commercial projects is normally accomplished with light gauge metal studs, runner channels, and furring members instead of wood framing.

This material is manufactured from cold rolled galvanized metal and is available in a number of sizes, shapes, and gauges.

Metal Stud Partitions. Runner channels are channel shaped members that are positioned at the top and bottom of the studs, as in top and bottom plates for a wood stud partition. The runner channels are attached to the floor and overhead structure with nails, screws, or powder-actuated drive pins.

The metal studs, which are channel shaped with a backbend to give stiffness, are placed within the web of the runner channels, located for on-center spacing, plumbed, and screwed into place with a 3/8" (9.37 mm) panhead screw, top and bottom. The web of the metal studs have cutouts for the passage of conduit and piping.

Runner channels and studs are made in 1-5/8" (40.62 mm), 2-1/2" (62.50 mm), 3-5/8" (91 mm), 4" (100 mm) and 6" (150 mm) widths in 25 ga., 2-1/2" (62.50 mm), 3-5/8" (90.62 mm), 4" (100 mm), 6" (150 mm) widths in 20 ga., and 4" (100 mm), 6" (150 mm) widths in 18 ga. and 16 ga. metals.

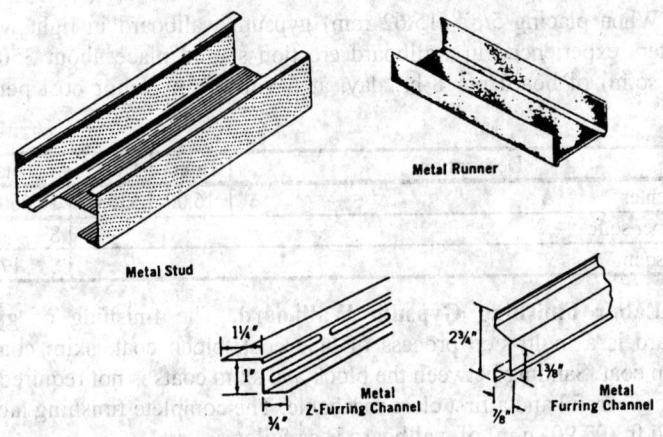

Metal Runner

Metal Stud

1¼"

1"

¾"

Metal
Z-Furring Channel

2¾"

1⅜"

⅞"

Metal
Furring Channel

The 25 ga. runners and studs are usually specified for non-loadbearing partitions up to 14' (4.26 m) to 16' (4.87 m) in height. Over this height and up to 22' (6.70 m), the 20 ga. material is used. The 18 ga. and 16 ga. materials are used for greater heights, loadbearing walls, and for exterior wall construction where lateral pressure (wind loads) will be imposed, requiring greater stiffness and less deflection.

Metal Furring. This material is made from 25 gauge metal in the following shapes:

1. Z furring channel is used in conjunction with rigid insulation board when furring exterior wall surfaces.
2. The hat shaped furring channel is used for furring walls and ceilings when insulation board is not required.
3. The resilient channel is used primarily on ceilings to provide a separation between the gypsum panels and the framing members for the floor above. It is used extensively in wood framed garden apartments for noise dampening between apartment units.

Estimating Material Quantities. To determine the quantity of runner channel, studs, and fasteners for a partition, the estimator would measure the total length of partition to be erected. Note the thickness and gauge of materials specified, ascertain the height of the partition for length of studs, ascertain the required spacing of studs, and add extras for door or window openings and at corners, and determine what type and spacing of fasteners is required for attachment of the runner channels to the floor and overhead structure. Horizontal stud bridging or fire stops are usually not required as in wood frame partition, but be sure to check, because this item in metal is expensive to install.

To determine the quantity of furring channel required, ascertain the on-center spacing of the members for quantity of pieces by the length of the members (height of wall, as furring members are usually vertically installed) to give total lineal feet (meters) to be installed. Determine the fastener type and spacing for attachment to the furring channel.

Material Prices		
Description	Price per 1,000 Lin.Ft.	Price per 100 Lin.M.
25 ga. - 1 5/8 " (40.63 mm) Runner Channel	$ 221.00	$ 72.51
25 ga. - 1 5/8 " (40.63 mm) Stud	$ 237.00	$ 77.76
25 ga. - 2 1/2 " (62.50 mm) Runner Channel	$ 262.00	$ 85.96
25 ga. - 2 1/2 " (62.50 mm) Stud	$ 288.00	$ 94.49
25 ga. - 3 5/8 " (90.63 mm) Runner Channel	$ 322.00	$ 105.64
25 ga. - 3 5/8 " (90.63 mm) Stud	$ 339.00	$ 111.22
25 ga. - 4 " (100.00 mm) Runner Channel	$ 347.00	$ 113.85
25 ga. - 4 " (100.00 mm) Stud	$ 373.00	$ 122.38
25 ga. - 6 " (150.00 mm) Runner Channel	$ 450.00	$ 147.64
25 ga. - 6 " (150.00 mm) Stud	$ 483.00	$ 158.46
25 ga. - 7/8 " (21.88 mm) Hat Furring Channel	$ 237.00	$ 77.76
25 ga. - 1 " (25.00 mm) Z Furring Channel	$ 280.00	$ 91.86
25 ga. - " (0.00 mm) Resilient Channel	$ 246.00	$ 80.71
20 ga. - 2 1/2 " (62.50 mm) Runner Channel	$ 517.00	$ 169.62
20 ga. - 2 1/2 " (62.50 mm) Stud	$ 542.00	$ 177.82
20 ga. - 3 5/8 " (90.63 mm) Runner Channel	$ 576.00	$ 188.98
20 ga. - 3 5/8 " (90.63 mm) Stud	$ 619.00	$ 203.08
20 ga. - 4 " (100.00 mm) Runner Channel	$ 610.00	$ 200.13
20 ga. - 4 " (100.00 mm) Stud	$ 645.00	$ 211.61
20 ga. - 6 " (150.00 mm) Runner Channel	$ 781.00	$ 256.23
20 ga. - 6 " (150.00 mm) Stud	$ 822.00	$ 269.69
	Cost per 1,000	Cost per 1,000
3/8 " (9.38 mm) Pan Head Screws	$ 7.15	$ 7.15
	Cost Ea.	Cost Ea.
Powder Actuated Drive Pins (avg. cost inc. shot)	$ 0.75	$ 0.75

Example. The following list of material quantities would be required for a metal stud partition 200 lin.ft. (60 m) long, 10 lin.ft. (3 m) high, stud at 24" (600 mm) o.c., 25 ga. metal and installed with drive pins at 24" (600 mm) o.c., top and bottom to concrete slabs:

Item	Rate	Total
2-1/2" (62.50 mm) 25 ga. Runner, 400 lin.ft. (121.92 lin.M)	$ 0.26	$ 104.80
2-1/2" (62.50 mm) 25 ga. Stud,	$ 0.29	$ 290.88
1" (25 mm) Drive Pins - 202 ea.	$ 0.75	$ 151.50
3/8" (9 mm) Pan Screws - 0.4 M	$ 7.15	$ 2.86
Cost for 200 lin.ft.		$ 550.04
per lin.ft.		$ 2.75
per lin.M.		$ 9.02

Add extra studs for door and window openings and corners.

Labor to Install Metal Stud Partitions. Using the above example given for materials, 2 carpenters should erect the 200 lin.ft. (60.96 m) of partition at the following labor cost:

Description	Hours	Rate	Total
Layout Wall Line	1.00	$35.04	$ 35.04
Fasten Bottom Runner	4.00	$35.04	$ 140.16
Fasten Top Runner	4.00	$35.04	$ 140.16
Install Studs	3.00	$35.04	$ 105.12
Plumb & Screw Studs	4.00	$35.04	$ 140.16
Cost for 200 lin.ft.			$ 560.64
per lin.ft.			$ 2.80
per lin.M.			$ 9.20

09300 TILE

CERAMIC WALL AND FLOOR TILE

Ceramic tile provides a durable, colorful surface that is virtually maintenance-free. Its applications include interior and exterior finishes for functional and decorative purposes in all types of structures. Special uses include acid resistant and electrically conductive installations. Ceramic tile is available in many sizes, shapes, and finishes.

Glazed Ceramic Wall Tile. This has an impervious facial finish fused onto the body of the tile. The glazed surface comes in a wide variety of colors.

The Porcelain Enamel Institute (P.E.I) has set up a guide to durability. The grades ratings measure the wear resistance of the tile surface. They do not rate the quality or price of tile.

Class I – No foot traffic: Basically for residential bathrooms where softer footwear is worn. Note that many manufactures do not produce this class of tile.

Class II – Light Traffic: For residential traffic, except kitchens, entrance halls, and other heavy traffic areas within a residential dwelling.

Class III - For all residential and light commercial areas such as office and reception areas.

Class IV – Moderate to Heavy Traffic: For medium commercial and light institutional applications such as restaurants hotels, hospital lobbies and corridors. This class of tile is also used in homes

Class V – Heavy to Extra Heavy Traffic: For heavy traffic and wet areas such as walkways, shopping centers, building entrances, food service areas and around Swimming pools. This class of tile usually is not used in homes due to what can be called a commercial look.

Ceramic Mosaic Tile. This tile, either glazed or unglazed, has a facial area of less than 6 sq. in. and is usually mounted on sheets or mesh, about 2' x 1' (0.61 x 0.30 m), to facilitate setting.

FINISHES

Quarry Tile. This is a rugged ceramic tile used primarily as a finish flooring, interior and exterior, where a long wearing, easily cleanable surface is desired.

Recent developments in the ceramic tile industry make it necessary to stress the importance of relating tile costs to each individual job specification. Glazed ceramic wall tile, for example, can be backmounted or unmounted, and can be installed using conventional portland cement mortar, various types of adhesives, or the more recently developed "dry-set" portland cement mortar. All of these variations can affect costs, both material and labor. Review job requirements carefully before doing the estimate.

However, the estimator's job has been simplified in recent years. Virtually the entire ceramic tile industry has adopted a "simplified practice", prepared under the auspices of the Tile Council of America, which has reduced the number of sizes and shapes and which has established generally recognized standards for the industry.

Estimating Quantities. Ceramic tile is estimated by the square foot, with trim pieces such as base, cap, etc. being estimated by the lineal foot. The estimator should deduct door and window openings, but the trim pieces necessary to finish the openings must be added. The quantities should be related to the type and size of ceramic tile, since these items will affect the cost of the ceramic tile when priced.

Estimate Composition. The finished ceramic tile estimate will include the following items:

1. Cost of ceramic tile delivered to the job site.
2. Cost of accessory materials such as wire mesh, sand, and cement for floor fill under ceramic tile.
3. Cost of mixing and placing floor fill.
4. Direct labor cost of laying and cleaning the ceramic tile.
5. The ceramic tile contractor's overhead and profit.

Variable Factors Influencing Costs. As in the other construction trades, estimating the labor costs of setting ceramic tile requires an intimate knowledge of the labor market in which the work is to be performed. Wage rates vary throughout the country, and it is important the estimator determine the rate in his locality.

Ceramic tile is unique in that it is often necessary to install it in very small quantities (for example, in 1 or 2 bathrooms in a house), and the costs can vary greatly. But such small installations are the exception, and the costs given below are based on the assumption that areas involved are large enough to permit the tile contractor to operate efficiently.

These costs are based on 4-1/4" x 4-1/4" (106.25 x 106.25 mm) glazed wall tile, 1" x 1" (25 x 25 mm) ceramic mosaic tile, and 6" x 6" x 1/2" (150 x 150 x 12.50 mm) quarry tile, since these are the sizes most commonly employed. If other sizes are specified, substitute the material price only. For all practical purposes, the labor cost remains constant regardless of size. The

1321

developed costs do not include areas where an unusually large amount of trim is required. Make an allowance for extra trim pieces in such special cases.

Setting Methods. Three common methods of adhering ceramic tile to a subsurface are described below.

Conventional Portland Cement Mortar Method. This method is to bond each ceramic tile with a layer of pure portland cement paste to a portland cement setting bed. This is done while the setting bed is still plastic. Wall tile must be soaked in water so that the water needed for curing is not absorbed from the paste. This is the traditionally accepted method, but it is also the costliest method.

Dry-Set Portland Cement Mortar Method. This method uses a dry curing portland cement mortar, accomplished through the use of water retaining additives, and has made ceramic tile installation cheaper and simpler. "Dry-set" is ideally suitable for use with concrete masonry, brick, poured concrete, and portland cement plaster. It should not be used over wood or gypsum plaster. Labor costs are appreciably reduced when this method is used.

Water-Resistant Organic Adhesive. Organic adhesives can be used over smooth base materials, such as wallboards, plywood, and metal. Labor productivity is comparable to that of "dry-set" mortar.

Description	Labor Productivity per Sq.Ft. (Sq.M) per Team Day Team is 1 Tile Setter and 1 Helper					
	Face mounted Sq.Ft.	Back mounted Sq.Ft.	Un-mounted Sq.Ft.	Face mounted Sq.M.	Back mounted Sq.M.	Un-mounted Sq.M.
Conventional Mortar						
Glazed wall tile	-----	75 - 90	55 - 65	-----	6.97 - 8.36	5.11 - 6.04
Ceramic mosaic						
tile-walls	45 - 50	55 - 65	-----	4.18 - 4.65	5.11 - 6.04	-----
floors	100 - 125	100 - 125	-----	9.29 - 11.61	9.29 - 11.61	-----
Quarry tile	-----	-----	100 - 125	-----	-----	9.29 - 11.61
"Dry-Set" Mortar						
Glazed wall tile	-----	125 - 150	100 - 125	------	11.61 - 13.94	9.29 - 11.61
Ceramic mosaic						
tile-walls	100 - 125	120 - 140	------	9.29 - 11.61	11.15 - 13.01	-----
floors	125 - 150	125 - 150	------	11.61 - 13.94	11.61 - 13.94	-----
Quarry tile floors	-----	-----	125 - 150	-----	-----	11.61 - 13.94
Organic Adhesive						
Glazed wall tile	-----	150 - 175	120 - 140	-----	13.94 - 16.26	11.15 - 13.01
Ceramic mosaic						
tile-walls	120 - 140	125 - 150	-----	11.15 - 13.01	11.61 - 13.94	-----
floors	150 - 175	175 - 200	-----	13.94 - 16.26	16.26 - 18.58	-----
	Lin.Ft			Lin.M.		
Cove or Base	65 - 75			19.8 - 22.9		
Cap	80 - 90			24.4 - 27.4		

FINISHES

Approximate Prices of Ceramic Tile Materials Ceramic Wall Tile				
Description	Size, Inches	Price per Sq.Ft.	Size mm	Price per Sq.M.
Flat tile, all colors, incl. white	4 1/4 " x 4 1/2 "	$ 1.35	106.25 x 112.50 mm	$ 14.53
	6 " x 6 "	$ 1.45	150.00 x 150.00 mm	$ 15.61
	6 " x 4 1/2 "	$ 1.50	150.00 x 112.50 mm	$ 16.15
	6 " x 9 "	$ 2.00	150.00 x 225.00 mm	$ 21.53
	8 1/2 " x 4 1/4 "	$ 2.25	212.50 x 106.25 mm	$ 24.22
For use with 4-1/4"x4-1/4" (106 x 106 mm) tile				
Cove	6 " x 3 1/4 "	$ 0.85	150.00 x 81.25 mm	$ 9.15
Cove	4 1/4 " x 4 1/4 "	$ 0.85	106.25 x 106.25 mm	$ 9.15
Bullnose	2 " x 6 "	$ 0.85	50.00 x 150.00 mm	$ 9.15
Bullnose	4 1/4 " x 4 1/4 "	$ 0.90	106.25 x 106.25 mm	$ 9.69
Double bullnose	4 1/4 " x 4 1/4 "	$ 1.30	106.25 x 106.25 mm	$ 13.99
For use with 6"x6" (150x150mm) tile				
Cove	6 " x 6 "	$ 1.15	150.00 x 150.00 mm	$ 12.38
Bullnose	6 " x 6 "	$ 1.20	150.00 x 150.00 mm	$ 12.92
Double bullnose	6 " x 6 "	$ 1.50	150.00 x 150.00 mm	$ 16.15
Miscellaneous trim shapes				
Bead	6 " x 3/4 "	$ 0.85	150.00 x 18.75 mm	$ 9.15
Bullnose	6 " x 3 3/4 "	$ 1.15	150.00 x 93.75 mm	$ 12.38
Base	6 " x 4 "	$ 1.30	150.00 x 100.00 mm	$ 13.99
Base	6 " x 6 "	$ 1.35	150.00 x 150.00 mm	$ 14.53
Sink Trim	6 " x 2 "	$ 2.00	150.00 x 50.00 mm	$ 21.53
Window Sill	9 " x 5 1/8 "	$ 2.20	225.00 x 128.13 mm	$ 23.68

Prices are for stretchers; for angles, multiply prices by two.

Ceramic Mosaic Tile Description - Unglazed, Modular, Solid Color					
Size, inches	Whites & Grays per Sq.Ft.	Other Colors per Sq.Ft.	Size, mm	Whites & Grays per Sq.M.	Other Colors per Sq.M.
1 " x 1 "	$ 2.08	$ 2.40	25.00 x 25.00 mm	$ 22.39	$ 25.83
2 " x 1 "	$ 2.24	$ 2.40	50.00 x 25.00 mm	$ 24.11	$ 25.83
2 " x 2 "	$ 2.40	$ 2.40	50.00 x 50.00 mm	$ 25.83	$ 25.83
3/4 " x 3/4 "	$ 2.08	$ 2.40	18.75 x 18.75 mm	$ 22.39	$ 25.83
1 9/16 " x 3/4 "	$ 2.08	$ 2.40	39.06 x 18.75 mm	$ 22.39	$ 25.83
1 9/16 " x 1 9/16 "	$ 2.08	$ 2.40	39.06 x 39.06 mm	$ 22.39	$ 25.83
Trim Shapes, price per lin. ft. (meter)					
Bullnose					
1 " x 1 "	-----	$ 2.24	25.00 x 25.00 mm	-----	$ 24.11
2 " x 1 "	-----	$ 2.56	50.00 x 25.00 mm	-----	$ 27.56

3/4" Series trim pieces same price as 1" Series

Various manufacturers furnish, as standard items, patterns made from combinations of modular tile sizes and colors. Prices vary depending on number of different sizes, number of colors, and depth of colors used. Patterns such as these are too numerous to itemize, but the general range is from $1.90 to $2.40 per sq.ft. ($20.45 to 25.83 per sq.m). There are exceptions to this range, and when a specific pattern price is needed, check with the specific supplier.

Description	Price per Sq.Ft.	Price per Sq.M
Abrasive Ceramic Mosaic Tile	$ 2.32 - $ 2.88	$ 24.97 - $ 31.00
Conductive Floor Tile	$ 2.55 - $ 5.12	$ 27.45 - $ 55.11
Granitized Ceramic Mosaic Tile	$ 1.60 - $ 2.48	$ 17.22 - $ 26.70
Abrasive Granitized Ceramic Mosaic Tile	$ 2.48 - $ 2.72	$ 26.70 - $ 29.28
Glazed Satin Finish Ceramic Mosaic Tile	$ 2.24 - $ 2.24	$ 24.11 - $ 24.11
Glazed Textured Finish Ceramic Mosaic Tile	$ 3.20 - $ 2.56	$ 34.45 - $ 27.56
Special Decorative Glazed Ceramic		
Mosaic Tile for Accent Effects	$ 5.12 - $22.08	$ 55.11 - $237.67

Venetian glass mosaics, mounted on sheets 12-1/4" x 12-1/4" (306.25 x 306.25 mm) square, are $3.71 per sq.ft. ($39.93 per sq.m) for 3/4" x 3/4" (18.75 x 18.75 mm), $4.10 per sq.ft. ($44.13 per sq.m) for 1" x 1" (25 x 25 mm) and $4.35 per sq.ft. ($46.82 per sq.m) for 1-1/2" x 1-1/2" (37.50 x 37.50 mm). Metallic finishes run considerably more. Beads and coves for glass mosaics will run $3.00 per lin.ft. ($9.84 per m).

Sculptured tiles which can be used for wall surfacing, murals, screens, and sculpture are available in a wide range of ceramic designs and treatments. Design-Technics of New York have tiles in sizes 4-1/4" x 8-1/2" x 1/4" (106.25 x 212.50 x 6.25 mm) at approximately $6.00 to $7.50 per sq.ft. ($64.58 to $80.73 per sq.m), 12" x 12" x 3/8" (300 x 300 x 9.37 mm) at $7.50 to $9.30 per sq.ft. ($80.73 to $100.10 per sq.m), and 12" x 12" x 1-3/8" (300 x 300 x 34.37 mm) at $9.70 to $11.50 per sq.ft. ($104.41 to $123.79 per sq.m).

Quarry Tile				
Size, Inches	Price per Sq.Ft.	Size, mm		Price per Sq.M.
Deep red, plain surface,				
6 " x 6 " x 1/2 "	$ 2.08	150 x 150.00 x 12.50 mm		$ 22.39
6 " x 2 3/4 " x 1/2 "	$ 2.16	150 x 68.75 x 12.50 mm		$ 23.25
6 " x 6 " x 3/4 "	$ 3.20	150 x 150.00 x 18.75 mm		$ 34.45
9 " x 9 " x 3/4 "	$ 57.78	225 x 225.00 x 18.75 mm		$ 621.92
Deep red, abrasive,				
6 " x 6 " x 3/4 "	$ 3.60	150 x 150.00 x 18.75 mm		$ 38.75
6 " x 6 " x 3/4 "	$ 4.40	150 x 150.00 x 18.75 mm		$ 47.36

Other colors, with the exception of green, are virtually the same price as the red shown above. Green in the 6" x 6" x 1/2" (150x150x13mm) size costs $3.00 per sq.ft. ($32.29 per sq.m).

Extras, when specified, are:

FINISHES

	Price per Sq.Ft.	Price per Sq.M.
Tile ground square after burning	$ 1.11	$ 11.95
Wax coating	$ 2.44	$ 26.26
Floor brick, deep red, plain	Price per 1,000 Pcs.	
tray packed 8 " x 4 " x 1 3/4 " (200 x 100 x 43.75 mm)	$280.00	
in sealed cartons	$329.60	
Floor brick, deep red, abrasive		
tray packed 8 " x 4 " x 1 3/4 " (200 x 100 x 43.75 mm)	$304.00	
in sealed cartons	$355.20	

There are many specially cast heavy floor and wall tiles that compliment the popular provincial and country modes. Many of these are quite expensive and are not regularly stocked. Always check locally for availability and prices, and for similar tiles that might be allowed for substitution.

Ceramic Tile Bathroom Accessories

Numerous ceramic tile bathroom accessories are available in a variety of sizes and qualities. They may be recessed or surface mounted, and they come in the full range of tile colors. Some of the more commonly used items are as follows:

Description	Price per piece
Recessed soap holder	$ 12.75
Recessed glass holder	$ 12.75
Roll paper holder	$ 15.90
Robe hook	$ 4.25
Double robe hook	$ 4.75
Toothbrush holder	$ 6.36
Tile bar brackets (pair)	$ 8.50

Decorative Ceramic Wall Tiles

Glazed ceramic wall tile containing designs as an integral part of each piece of tile is also available. Normally supplied in 4-1/4" x 4-1/4" (106.25 x 106.25 mm) and 6" x 6" (150 x 150 mm) sizes, the price of this material will vary with the design selected and number of colors in the design. If the design is known, the estimator should obtain a firm price from a supplier.

Description Inches	Price per Sq.Ft.	Description mm	Price per Sq.M
One color,			
4 1/4 " x 4 1/4 "	$ 2.65	106.25 x 106.25 mm	$ 28.53
6 " x 6 "	$ 2.50	150.00 x 150.00 mm	$ 26.91
Two colors,			
4 1/4 " x 4 1/4 "	$ 3.20	106.25 x 106.25 mm	$ 34.45
6 " x 6 "	$ 3.00	150.00 x 150.00 mm	$ 32.29
Three colors,			
4 1/4 " x 4 1/4 "	$ 3.70	106.25 x 106.25 mm	$ 39.83
6 " x 6 "	$ 3.45	150.00 x 150.00 mm	$ 37.14

Ceramic Tile Adhesives and Accessory Materials		
Item	Coverage	Price
Organic adhesive	40 - 50 Sq.Ft. per gal	$ 16.15 per gal
Dry-set mortar mix	35 Lbs. per 100 Sq.Ft.	$ 0.95 per Lb
Primer (for use in damp areas)	125 Sq.Ft. per gal	$ 10.40 per gal
Plastic underlayment	100 Sq.Ft. per Unit*	$ 16.15 Per Unit
Wet tile grout	15 - 20 Sq.Ft. pr lb **	$ 0.57 per Lb
Dry tile grout	12 - 15 Sq.Ft. pr lb **	$ 0.67 per Lb

*Unit consists of 1 gallon (3.785 liters) liquid and 34 lbs. (15.3 kg) aggregate.
**When used with 4-1/4" x 4-1/4" (106.25 x 106.25 mm) tile.

Ceramic Tile Adhesives and Accessory Materials - Metric		
Item	Coverage	Price
Organic adhesive	0.98 - 1.23 Sq.M.per liter	$ 4.27 per liter
Dry-set mortar mix	170.9 kg/100 sq.m	$ 2.09 per Kg
Primer (for use in damp areas)	3.07 sq.m/liter	$ 2.75 per Liter
Plastic underlayment	9.29 sq.m/unit	$ 16.15 Per Unit
Wet tile grout	3.07 - 4.1 sq.m/kg	$ 1.26 Per Kg
Dry tile grout	2.46 - 3.1 sq.m/kg	$ 1.47 Per Kg

*Unit consists of 1 gallon (3.785 liters) liquid and 34 lbs. (15.3 kg) aggregate.
**When used with 4-1/4" x 4-1/4" (106.25 x 106.25 mm) tile.

Material Cost of 100 Sq.Ft. (9.29 Sq.M) of 4-1/4"x4-1/4" (106.25 x 106.25 mm) Glazed Ceramic Wall Tile Using Conventional Mortar Method of Installation (Unmounted Tile)		
Description	Rate	Total
2 sacks Portland cement	$ 6.00	$ 12.00
4 cu.ft. (0.11 cu.m) sand	$ 0.72	$ 2.88
100 sq.ft. (9 sq.m) glazed wall tile	$ 2.00	$ 200.00
7 lbs. (3.15 kg) wet tile grout mix	$ 0.41	$ 2.87
Cost per 100 sq.ft.		$ 217.75
per sq.ft.		$ 2.18
per sq.m		$ 23.44

FINISHES

Labor Cost of 100 Sq.Ft. (9.29 Sq.M) of 4-1/4"x4-1/4" (106.25 x 106.25 mm) Glazed Ceramic Wall Tile Using Conventional Mortar Method of Installation (Unmounted Tile)			
Description	Hours	Rate	Total
Tile Setter	13.50	$34.17	$ 461.30
Cost per sq.ft.			$ 4.61
per sq.m			$ 49.66

Material Cost of 100 Sq.Ft. (9.29 Sq.M) of 4-1/4"x4-1/4" (106.25 x 106.25 mm) Glazed Ceramic Wall Tile Using Water Resistant Organic adhesive (Unmounted Tile)		
Description	Rate	Total
2.5 gals. (9.46 liters) adhesive	$12.30	$ 30.75
100 sq.ft. (9 sq.m) glazed wall tile	$ 2.00	$ 200.00
9 lbs. (4.05 kg) dry tile grout mix	$ 0.56	$ 5.04
Cost per 100 sq.ft.		$ 235.79
per sq.ft.		$ 2.36
per sq.m		$ 25.38

Labor Cost of 100 Sq.Ft. (9.29 Sq.M) of 4-1/4"x4-1/4" (106.25 x 106.25 mm) Glazed Ceramic Wall Tile Using Water Resistant Organic adhesive (Unmounted Tile)			
Description	Hours	Rate	Total
Tile Setter	8.00	$34.17	$ 273.36
Cost per sq.ft.			$ 2.73
per sq.m			$ 29.43

Material Cost of 100 Sq.Ft. (9.29 Sq.M) of 4-1/4"x4-1/4" (106.25 x 106.25 mm) Glazed Ceramic Wall Tile Using "Dry-Set" Portland Cement Mortar (Unmounted Tile)		
Description	Rate	Total
35 lbs. (16.75 kg) "dry-set" mortar mix	$ 0.72	$ 25.20
100 sq.ft. (9 sq.m) glazed wall tile	$ 2.00	$ 200.00
9 lbs. (4.05 kg) dry tile grout mix	$ 0.56	$ 5.04
Cost per 100 sq.ft.		$ 230.24
per sq.ft.		$ 2.30
per sq.m		$ 24.78

Labor Cost of 100 Sq.Ft. (9.29 Sq.M) of 4-1/4"x4-1/4" (106.25 x 106.25 mm) Glazed Ceramic Wall Tile Using "Dry-Set" Portland Cement Mortar (Unmounted Tile)

Description	Hours	Rate	Total
Tile Setter	10.00	$34.17	$ 341.70
Cost per sq.ft.			$ 3.42
per sq.m			$ 36.78

Material Cost of 100 Sq.Ft. (9.29 Sq.M) 1"x1" (25 x 25 mm) Ceramic Mosaic Tile Floors Using Conventional Mortar Method of Installation (Face-Mounted Tile)

Description	Rate	Total
2 sacks Portland cement	$ 6.00	$ 12.00
4 cu.ft. (0.11 cu.m) sand	$ 0.72	$ 2.88
100 sq.ft. (9 sq.m) ceramic mosaic tile	$ 1.91	$ 191.00
17 lbs. (7.65 kg) wet tile grout mix	$ 0.41	$ 6.97
Cost per 100 sq.ft.		$ 212.85
per sq.ft.		$ 2.13
per sq.m		$ 22.91

Labor Cost of 100 Sq.Ft. (9.29 Sq.M) 1"x1" (25 x 25 mm) Ceramic Mosaic Tile Floors Using Conventional Mortar Method of Installation (Face-Mounted Tile)

Description	Hours	Rate	Total
Tile Setter	10.50	$34.17	$ 358.79
Cost per sq.ft.			$ 3.59
per sq.m			$ 38.62

Material Cost of 100 Sq.Ft. (9.29 Sq.M) 1"x1" (25 x 25 mm) Ceramic Mosaic Tile Floors Using Water Resistant Organic Adhesive (Face-Mounted Tile)

Description	Rate	Total
2.5 gals. (9.46 liters) adhesive	$12.75	$ 31.88
100 sq.ft. (9 sq.m) ceramic mosaic tile	$ 1.91	$ 191.00
23 lbs. (10.35 kg) dry tile grout mix	$ 0.56	$ 12.88
Cost per 100 sq.ft.		$ 235.76
per sq.ft.		$ 2.36
per sq.m		$ 25.38

FINISHES

Labor Cost of 100 Sq.Ft. (9.29 Sq.M) 1"x1" (25 x 25 mm) Ceramic Mosaic Tile Floors Using Water Resistant Organic Adhesive (Face-Mounted Tile)			
Description	Hours	Rate	Total
Tile Setter	8.00	$34.17	$ 273.36
Cost per sq.ft.			$ 2.73
per sq.m			$ 29.43

Material Cost of 100 Sq.Ft. (9.29 Sq.M) 1"x1" (25 x 25 mm) Ceramic Mosaic Tile Floors Using "Dry-Set" Portland Cement Mortar (Face-Mounted Tile)		
Description	Rate	Total
35 lbs. (15.75 kg) "dry-set" mortar mix	$ 0.72	$ 25.20
1 cu.ft. (0.02 cu.m) of sand	$ 0.72	$ 0.72
100 sq.ft. (9 sq.m) ceramic mosaic tile	$ 1.91	$ 191.00
23 lbs. (10.35 kg) dry tile grout mix	$ 0.56	$ 12.88
Cost per 100 sq.ft.		$ 229.80
per sq.ft.		$ 2.30
per sq.m		$ 24.74

Labor Cost of 100 Sq.Ft. (9.29 Sq.M) 1"x1" (25 x 25 mm) Ceramic Mosaic Tile Floors Using "Dry-Set" Portland Cement Mortar (Face-Mounted Tile)			
Description	Hours	Rate	Total
Tile Setter	9.00	$34.17	$ 307.53
Cost per sq.ft.			$ 3.08
per sq.m			$ 33.10

Material Cost of 100 Sq.Ft. (9.29 Sq.M) 1"x1" (25 x 25 mm) Ceramic Mosaic Tile on Walls Using Conventional Mortar Method of Installation (Face-Mounted Tile)		
Description	Rate	Total
2 sacks portland cement	$ 6.00	$ 12.00
4 cu.ft. (0.11 cu.m) sand	$ 0.72	$ 2.88
100 sq.ft. (9 sq.m) ceramic mosaic tile	$ 1.91	$ 191.00
17 lbs. (7.65 kg) wet tile grout mix	$ 0.41	$ 6.97
Cost per 100 sq.ft.		$ 212.85
per sq.ft.		$ 2.13
per sq.m		$ 22.91

Labor Cost of 100 Sq.Ft. (9.29 Sq.M) 1"x1" (25 x 25 mm) Ceramic Mosaic Tile on Walls Using Conventional Mortar Method of Installation (Face-Mounted Tile)

Description	Hours	Rate	Total
Tile Setter	11.00	$34.17	$ 375.87
Cost per sq.ft.			$ 3.76
per sq.m			$ 40.46

Material Cost of 100 Sq.Ft. (9.29 Sq.M) 1"x1" (25 x 25 mm) Ceramic Mosaic Tile on Walls Using Water Resistant Organic Adhesive (Face-Mounted Tile)

Description	Rate	Total
2.5 gals. (9.46 liters) adhesive	$12.75	$ 31.88
100 sq.ft. (9 sq.m) ceramic mosaic tile	$ 1.91	$ 191.00
23 lbs. (10.35 kg) dry tile grout mix	$ 0.41	$ 9.43
Cost per 100 sq.ft.		$ 232.31
per sq.ft.		$ 2.32
per sq.m		$ 25.01

Labor Cost of 100 Sq.Ft. (9.29 Sq.M) 1"x1" (25 x 25 mm) Ceramic Mosaic Tile on Walls Using Water Resistant Organic Adhesive (Face-Mounted Tile)

Description	Hours	Rate	Total
Tile Setter	9.00	$34.17	$ 307.53
Cost per sq.ft.			$ 3.08
per sq.m			$ 33.10

Material Cost of 100 Sq.Ft. (9.29 Sq.M) 1"x1" (25 x 25 mm) Ceramic Mosaic Tile on Walls Using "Dry-Set" Portland Cement Mortar (Face-Mounted Tile)

Description	Rate	Total
35 lbs. (15.75 kg) "dry-set" mortar mix	$ 0.72	$ 25.20
1 cu.ft. (0.02 cu.m) sand	$ 0.72	$ 0.72
100 sq.ft. (9 sq.m) ceramic mosaic tile	$ 1.91	$ 191.00
23 lbs. (10.35 kg) dry tile grout mix	$ 0.56	$ 12.88
Cost per 100 sq.ft.		$ 229.80
per sq.ft.		$ 2.30
per sq.m		$ 24.74

FINISHES

Labor Cost of 100 Sq.Ft. (9.29 Sq.M) 1"x1" (25 x 25 mm) Ceramic Mosaic Tile on Walls Using "Dry-Set" Portland Cement Mortar (Face-Mounted Tile)			
Description	Hours	Rate	Total
Tile Setter	9.00	$34.17	$ 307.53
Cost per sq.ft.			$ 3.08
per sq.m			$ 33.10

Material Cost of 100 Sq.Ft. (9.29 Sq.M) of 6" x 6"x 1/2" (150 x 150 x 12.50 mm) Quarry Tile Floors Using Conventoional Mortar Method of Installation		
Description	Rate	Total
2 sacks portland cement	$ 6.00	$ 12.00
6 cu.ft. (0.17 cu.m) sand	$ 0.72	$ 4.32
100 sq.ft. (9 sq.m) quarry tile plain Surface	$ 1.91	$ 191.00
35 lbs. (15.75 kg) portland cement grout	$ 0.80	$ 28.00
Cost per 100 sq.ft.		$ 235.32
per sq.ft.		$ 2.35
per sq.m		$ 25.33

Labor Cost of 100 Sq.Ft. (9.29 Sq.M) of 6" x 6"x 1/2" (150 x 150 x 12.50 mm) Quarry Tile Floors Using Conventoional Mortar Method of Installation			
Description	Hours	Rate	Total
Tile Setter	16.00	$34.17	$ 546.72
Cost per sq.ft.			$ 5.47
per sq.m			$ 58.85

Material Cost of 100 Sq.Ft. (9.29 Sq.M) of 6" x 6"x 1/2" (150 x 150 x 12.50 mm) Quarry Tile Floors Using "Dry-Set" Portland Cement Mortar		
Description	Rate	Total
35 lbs. (15.75 kg) "dry-set" mortar Mix	$ 0.72	$ 25.20
2 cu.ft. (0.06 cu.m) of sand	$ 0.72	$ 1.44
100 sq.ft. (9 sq.m) quarry tile plain surface	$ 1.91	$ 191.00
35 lbs. (15.75 kg) Portland cement grout	$ 1.76	$ 61.60
Cost per 100 sq.ft.		$ 279.24
per sq.ft.		$ 2.79
per sq.m		$ 30.06

Labor Cost of 100 Sq.Ft. (9.29 Sq.M) of 6" x 6"x 1/2" (150 x 150 x 12.50 mm) Quarry Tile Floors Using "Dry-Set" Portland Cement Mortar

Description	Hours	Rate	Total
Tile Setter	12.00	$34.17	$ 410.04
Cost per sq.ft.			$ 4.10
per sq.m			$ 44.14

Material Cost of 100 Lin.Ft. (30.48 M) of Cove or Base (Ceramic or Quarry Tile)

Description	Rate	Total
100 lin.ft. (30 m) cove	$ 2.55	$ 255.00
Adhesive or mortar-allow	$ 9.54	$ 9.54
Cost per 100 lin.ft.		$ 264.54
per lin.ft.		$ 2.65
per m		$ 8.68

Labor Cost of 100 Lin.Ft. (30.48 M) of Cove or Base (Ceramic or Quarry Tile)

Description	Hours	Rate	Total
Tile Setter	20.00	$34.17	$ 683.40
Cost per lin.ft.			$ 6.83
per lin.m			$ 22.42

Material Cost of 100 Lin.Ft. (30.48 M) of Wainscot Cap (Glazed Ceramic Wall Tile)

Description	Rate	Total
100 lin.ft. (30 m) bullnose	$ 8.51	$ 851.00
Adhesive or mortar-allow	$ 9.54	$ 9.54
Cost per 100 lin.ft.		$ 860.54
per lin.ft.		$ 8.61
per lin.M.		$ 28.23

Labor Cost of 100 Lin.Ft. (30.48 M) of Wainscot Cap (Glazed Ceramic Wall Tile)

Description	Hours	Rate	Total
Tile Setter	20.00	$34.17	$ 683.40
Cost per lin.ft.			$ 6.83
per lin.M.			$ 22.42

Placing Cement Floor Fill. Cement floor fill under ceramic tile floors is usually placed by tile setters and helpers (one tile setter and one or two helpers working together). The fill is placed one or two days in advance

FINISHES

of the tile if the overall thickness from rough floor to finished tile surface is over 3" (75 mm). For 3" (75 mm) thickness and under, fill and setting bed may be placed in one operation. When such tile floors are to be placed over wood subfloors, it is necessary to first place a layer of waterproof building paper and a layer of wire mesh reinforcing before placing the fill.

A tile setter and two helpers should place 450 to 500 sq.ft. (41.80 to 46.45 sq.m) of fill per 8-hour day in areas large enough to permit efficient operations, at the following cost per 100 sq.ft. (9.29 sq.m):

Material Costs		
Description	Rate	Total
6 bags portland cement	$ 6.00	$ 36.00
1 cu.yd. (0.76 cu.m) sand	$20.00	$ 20.00
Cost per 100 ss.ft.		$ 56.00
per sq.ft.		$ 0.56
per sq.m		$ 6.03

Labor Costs			
Description	Hours	Rate	Total
Tile Setter	1.50	$34.17	$ 51.26
Tile Setter Helper	3.00	$26.16	$ 78.48
Cost per 100 sq.ft.			$ 129.74
Cost per sq.ft.			$ 1.30
per sq.M.			$ 13.97

Metal Wall Tile

Wall tile of aluminum, having a baked enamel finish in various colors, or tile of copper or stainless steel are used for the same purposes.

They require the same type of base and are applied with adhesive. A worker will apply about 100 sq.ft. (9.29 sq.m) per day.

Metal tile mastic weighs about 16 lbs. per gal. (1.92 kg per liter), and it requires 1/2 lb. per sq.ft. (2.5 kg per sq.m), or one gal. (3.78 liters) is sufficient for 32 sq.ft. (2.97 sq.m) of tile. Price about $11.90 per gal. Aluminum metal tile cost about $2.25 per sq.ft. ($24.22 per sq.m). Copper or stainless steel tile cost about $5.55 per sq.ft. ($59.74 per sq.m).

MARBLE TILE

To prepare an accurate estimate of cost on marble work, the estimator should understand the correct method of measuring and listing quantities from the plans, the manner in which the marble should be handled and set on the job, as well as the most economical and practical methods of handling the work.

When estimating marble wainscoting, it is incorrect to take the height of the wainscot from the floor to the cap and multiply by the number of lin. ft. (meters), because in all probability the wainscot is made up of three members, (base, die, and cap) and each member should be estimated separately: the base and cap by the lin. ft. (meters) and the die by the square foot (square meter).

The length and size of the pieces of marble has considerable bearing on the labor cost, because it usually costs as much to set a 6" (150 mm) piece as one 2'-0" (0.60 m) long, so that a job made up of short pieces will cost more per foot than a job having a larger percentage of long pieces of marble.

Trade Practice. Trade practice in the marble industry recognizes all marbles and stones commonly used for interior building purposes, as falling within one of four groups or classifications, according to their respective characteristics and working qualities. Therefore, for purposes of standardization and clearer definition, the Marble Institute of America has officially adopted these four classifications:

Group A: Sound marbles and stones that require no sticking, waxing, or filling; characteristically uniform and favorable working qualities.

Group B: Marbles and stones similar in character to the preceding group but somewhat less favorable working qualities; occasional natural faults and requiring a limited amount of waxing and sticking.

Group C: Marbles of uncertain variation in working qualities, geological flaws, voids, veins, and lines of separation common; common shop practice to repair nature's shortcomings by sticking, waxing, and filling; liners and other forms of reinforcement freely employed when necessary.

Group D: Marbles and stones similar to the preceding group and subject to the same methods of finishing and manufacture but embracing those materials that contain a larger proportion of natural faults and a maximum variation in working qualities, etc. This group comprises many of the highly colored marbles prized for their decorative qualities.

Below is a list of some of the more commonly used varieties of marbles and stones and the groups to which they belong:

Group A

Alabama-usual grades	Napoleon Gray (Vt.)
Blanco P	Tennessee, Gray and Pink
Brocadillo, Vermont	Vermont, Black
Carthage	Vermont, White Grades
Georgia	Vermont Pavonazzo
Italian White	Westland Green Vein Cream
Italian English Vein	

Group B

Alabama Cream Veined A	Imperial Black, Tenn.
Belgian Black	Mankato Buff

FINISHES

Champville	Travertine (Italian)
Cremo Italian	

Group C

Belgian Grand Antique	Red Levanto
Bois Jourdan	Red Verona
Botticino	Rosato
Breche Opal	Tavernelle, all types
Escalette	Verdona
Hauteville	Verona, Yellow
Mankato Pink	Vermont Verde Antique
Ozark Rouge	Westfield Green

Group D

Alps Green	Grand Antique, Italian
Black and Gold	Onyx, Pedrara
Bleu Belge	Rouge Antique
Breche Oriental	Sienna
Forest Green	Verde Antico (Italian)
Grand Antique, French	Vert St. Denis

Estimating Quantities of Marble Work. When measuring and listing quantities of marble work from the plans, there are several general rules that should be followed to insure accuracy.

To estimate accurately the quantity of marble base, obtain the number of lineal feet (meters) of base, listing the various heights separately. All base under 1'-0" (0.30 m) high is extended and priced on a lineal foot (meter) basis. All base 1'-0" (0.30 m) or over is extended and priced on a square foot (square meter) basis. Always bear in mind that no piece of marble should be figured as being less than 1'-0" (0.30 m) long, as it requires just as much time to set a 6" (150 mm) piece as one 2'-0" (0.61 m) long. Short pieces are usually found around pilasters, door and window returns, and the like.

Marble die is estimated by the square foot (square meter), and the labor costs computed in the same manner. When listing quantities, obtain the length of each run and multiply by the height, and the result will be the number of sq. ft. (sq.m) of die. If marble pilasters project beyond the face of the wall, list each projection according to its actual dimensions, and if less than a 6" (150 mm) projection, it should be priced as 6" (150 mm) wide, but if the pilasters are over 6" (150 mm) wide, then use actual dimensions.

Marble wainscot cap under 1'-0" (0.30 m) wide or any marble under 1'-0" (0.30 m) wide, should be measured by the lineal foot (meter).

Marble stair treads should be measured by the square foot (square meter), as they are ordinarily 12" (300 mm) or more in width. Be sure to mention length, width, and thickness of all treads, type of nosing, etc.

Marble stair risers should be estimated by the lin. ft. (meter) giving length, height, and thickness of each riser. It costs practically as much to set a short riser as a long one.

When estimating toilet stall work consisting of toilet backs, partitions, stall fronts, stiles, etc., the fronts should be estimated by the square foot (square meter) if over 12" (300 mm) wide and by the lin. ft. (meter) if less than 12" (300 mm) wide. Partitions and backs should be estimated by the square foot (square meter). Marble cap is usually estimated by the lin. ft. (meter). Some marble contractors figure toilet work at a certain price per stall, including all labor in connection with same, although most contractors recommend pricing on a square or lineal foot (meter) basis.

Marble column bases and caps formed of solid stock over 2" (50 mm) thick, should be estimated at a certain price for each base or cap, mentioning the size and number of cu.ft. (cu.m) in each.

Marble floor tile and border are usually estimated by the square foot (square meter), giving size of tile, such as 6" x 6" (150 x 150 mm), 9" x 9" (225 x 225 mm), 12" x 12" (300 x 300 mm), etc. Small or irregular sized tile, such as dots, etc., are estimated in the same manner, except that a notation should be made of the class of work, as it costs more to lay small and irregular shaped tile.

Marble handrail, moldings, etc., should be estimated by the lin.ft. (meter), giving the size of each, in order that the quantities may also be stated in cu.ft. (cu.m) if over 2" (50 mm) thick.

All classes of circular work, such as base, wainscot, or die, should be estimated in the same units as straight work, except that circular work should be estimated separately, because it usually requires considerably more cutting and fitting than straight work.

All marble work cut and set "on the rake", such as stair wainscoting, wainscoting under stairs, etc., should be estimated by the sq.ft. (sq.m), using the largest dimensions. Work of this kind should be estimated separately as it is more expensive to set than straight work.

Handling and Setting Interior Marble. Estimating the labor cost of handling and setting interior marble will depend much on the conditions under which the work is performed. The amount of labor handling, sorting, and distributing the marble on the job before it is set should be considered, and also whether it is necessary to hoist the marble above the first floor, because this will require considerably more handling. The average size of the pieces should be considered, such as the average length of base, size and thickness of wainscot, length of thresholds, treads, risers, and caps. It costs almost as much to handle and set a short piece of marble as a long one. This is especially true on stock that can be handled by one person.

The labor costs given below are based on the performance of an average crew. If it is necessary to hire temporary workers, daily production may decrease as much as 25%.

On practically all jobs it is necessary to have laborers handle the marble and distribute it ready for the setters, although in some localities, all marble must be handled by marble setters and helpers.

1336

FINISHES

Hoisting Marble. The proposal sheet of practically all marble contractors states that the marble contractor shall have free use of the general contractor's hoisting facilities, runways, etc., so for this reason no allowance has been made for hoisting in the following estimates of cost.

However, if it is necessary for the marble contractor to pay for the use of hoist for hoisting marble to the upper floors of a building, figure about 1/2 hr. hoist time per 100 sq.ft. (9.29 sq.m) of marble.

Trucking on Marble From Mill or Cars to Job. If the marble is furnished by an out of town mill and it is necessary to haul it in trucks from cars to the job, the cost will vary with the length of the haul. If the marble is furnished by a mill located in the same town, the price of the marble will probably include delivery to the job.

Marble Contractor's Foreman Expense. It is necessary to have a foreman on all jobs of any size to supervise the arrival, unloading, and distribution of the stock, as well as to lay out the work and supervise the setting. On average allow 1/16 to 1/8 hr. foreman time to each hour setter's time, depending on the size of the job and the number of setters employed. (Most small jobs will not require a foreman.)

Average Labor Cost Setting Interior Marble. While the labor cost of marble setting varies with the kind of material handled, some of the large marble companies estimate the setting cost of the entire job at a certain price per foot. The quantities for the entire job are taken, which include base, cap, wainscot or die, treads, risers, etc., and the labor is priced per sq.ft. (sq.m); 100 sq.ft. (9.29 sq.m) should cost as follows:

Description	Hours	Rate	Total
Foreman	1.50	$38.17	$ 57.26
Marble setter	12.00	$34.17	$ 410.04
Marble Setter Helper	12.00	$34.17	$ 410.04
Labor or helper	8.00	$26.16	$ 209.28
Cost 100 sq.ft.			$1,086.62
per sq.ft.			$ 10.87
per sq.m			$ 116.97

The above price includes setting on jobs consisting principally of 7/8" (22 mm) and 1-1/4" (31 mm) stock. If 2" (50 mm) marble is used, the handling and setting cost will increase considerably, and the labor per 100 sq.ft. (9.29 sq.m) should cost as follows:

Description	Hours	Rate	Total
Foreman	1.50	$38.17	$ 57.26
Marble setter	16.00	$34.17	$ 546.72
Marble Setter Helper	16.00	$34.17	$ 546.72
Labor or helper	12.00	$26.16	$ 313.92
Cost 100 sq.ft.			$1,464.62
per sq.ft.			$ 14.65
per sq.m			$ 157.66

1337

On jobs having an average amount of base, wainscot or die, cap, toilet stalls, treads and risers, etc., the above prices will prove close enough, but where there are large quantities of any one class of work, it is advisable to refer to that particular class of work to obtain accurate quantities and costs.

Setting Marble Base. The cost of setting marble base varies with the size of the rooms, whether straight walls or broken up with pilasters, piers, etc., and whether the job consists of long or short pieces of base.

In small corridors and other spaces having numerous pilasters, piers, etc., requiring short pieces of base, a setter and helper should set 60 to 75 lin.ft. (18.29 to 22.86 m) per 8-hr. day, at the following labor cost per 100 lin.ft. (30.48 m):

Description	Hours	Rate	Total
Foreman	1.50	$38.17	$ 57.26
Marble setter	12.00	$34.17	$ 410.04
Marble Setter Helper	12.00	$34.17	$ 410.04
Labor or helper	8.00	$26.16	$ 209.28
Cost 100 sq.ft.			$1,086.62
per sq.ft.			$ 10.87
per sq.m			$ 116.97

If the base can be set in reasonably long pieces, a marble setter and helper should set 80 to 100 lin.ft. (24 to 30 m) per 8-hr. day, at the following labor cost per 100 lin.ft. (30.48 m):

Description	Hours	Rate	Total
Foreman	1.00	$38.17	$ 38.17
Marble setter	9.00	$34.17	$ 307.53
Marble Setter Helper	9.00	$34.17	$ 307.53
Labor or helper	8.00	$26.16	$ 209.28
Cost 100 sq.ft.			$ 862.51
per sq.ft.			$ 8.63
per sq.m			$ 92.84

Setting Circular Base. Setting circular base costs about double the cost of straight work of the same kind set under similar conditions. There is usually considerable cutting and fitting on all circular work.

Setting Marble Wainscot or Die. The labor cost of setting marble wainscot or die will vary with the size of the pieces, the height of the wainscot or die, etc. A marble setter will set almost as many lin.ft. (meters) of 5'-0" (1.52 m) wainscot as 3'-0" (0.91 m) wainscot. The labor cost handling stock will be increased somewhat, as it will require 2 to 4 laborers or helpers to handle each piece of marble and place it on the floor ready for the setter, while lighter stock, such as base, risers, treads, etc., can be handled by one worker.

Setting Marble Wainscot Up to 3'-0" (0.91 m) High. When setting marble wainscot up to 3'-0" (0.91 m) high, a marble setter and helper should

1338

set 22 to 26 lin.ft. (6.70 to 7.72 m) containing 66 to 78 sq.ft. (6.13 to 7.24 sq.m) per 8-hr. day, at the following labor cost per 100 sq.ft. (9.29 sq.m):

Description	Hours	Rate	Total
Foreman	1.50	$38.17	$ 57.26
Marble setter	11.00	$34.17	$ 375.87
Marble Setter Helper	11.00	$34.17	$ 375.87
Labor or helper	8.00	$26.16	$ 209.28
Cost 100 sq.ft.			$1,018.28
per sq.ft.			$ 10.18
per sq.m			$ 109.61

The cost of setting circular wainscot or die will run about double the cost of straight work on account of the extra cutting and fitting.

Setting Marble Wainscot 3'-0" (0.9 m) to 4'-0" (1.21 m) High. A marble setter and helper should set 20 to 25 lin.ft. (6.09 to 7.62 m) containing 75 to 90 sq.ft. (6.96 to 8.36 sq.m) of wainscot per 8-hr. day, at the following labor cost per 100 sq.ft. (9.29 sq.m):

Description	Hours	Rate	Total
Foreman	1.20	$38.17	$ 45.80
Marble setter	10.00	$34.17	$ 341.70
Marble Setter Helper	10.00	$34.17	$ 341.70
Labor or helper	8.00	$26.16	$ 209.28
Cost 100 sq.ft.			$ 938.48
per sq.ft.			$ 9.38
per sq.m			$ 101.02

The cost of setting circular wainscot or die will run about double the cost of straight work on account of the extra cutting and fitting.

Setting Marble Wainscot 4'-0" (1.21 m) to 5'-0" (1.52 m) High. Where the marble wainscot varies from 4'-0" to 5'-0" (1.21 to 1.52 m) high, a marble setter and helper should set 20 to 23 lin.ft. (6.09 to 7.01 m) containing 85 to 100 sq.ft. (7.89 to 9.290 sq.m) per 8-hr. day, at the following labor cost per 100 sq.ft. (9.29 sq.m):

Description	Hours	Rate	Total
Foreman	1.00	$38.17	$ 38.17
Marble setter	9.00	$34.17	$ 307.53
Marble Setter Helper	9.00	$34.17	$ 307.53
Labor or helper	8.00	$26.16	$ 209.28
Cost 100 sq.ft.			$ 862.51
per sq.ft.			$ 8.63
per sq.m			$ 92.84

The cost of setting circular wainscot or die will run about double the cost of straight work on account of the extra cutting and fitting.

Setting Marble Wainscot Over 5'-0" (1.52m) High. When setting marble wainscot 5'-0" to 7'-0" (1.52 to 2.13 m) high, a marble setter and helper should set 90 to 110 sq.ft. (8.36 to 10.21 sq.m) per 8-hr. day, at the following labor cost per 100 sq.ft. (9.29 sq.m):

Description	Hours	Rate	Total
Foreman	1.00	$38.17	$ 38.17
Marble setter	8.00	$34.17	$ 273.36
Marble Setter Helper	8.00	$34.17	$ 273.36
Labor or helper	8.00	$26.16	$ 209.28
Cost 100 sq.ft.			$ 794.17
per sq.ft.			$ 7.94
per sq.m			$ 85.49

The cost of setting circular wainscot or die will run about double the cost of straight work on account of the extra cutting and fitting.

Wainscot over 7'-0" (2.13 m) high will require working from a scaffold, which decreases the production considerably. In such cases the labor cost of that portion of wainscot over 7'-0" (2.13 m) high should be increased 50% to 100%, and the cost of erecting and removing steel tubular scaffold should be added.

Setting Marble Wainscot Cap. When setting marble wainscot cap, a marble setter and helper should set 60 to 75 lin.ft. (18.28 to 22.86 m) per 8-hr. day, at the following cost per 100 lin.ft. (30.48 m):

Description	Hours	Rate	Total
Foreman	1.50	$38.17	$ 57.26
Marble setter	12.00	$34.17	$ 410.04
Marble Setter Helper	12.00	$34.17	$ 410.04
Labor or helper	8.00	$26.16	$ 209.28
Cost 100 Lin.ft.			$1,086.62
per lin.ft.			$ 10.87
per lin. M.			$ 116.97

Setting Marble Stair Treads. The cost of setting marble stair treads will vary with the thickness and length of the treads. The shorter each piece of marble, the higher the cost per lin.ft. (meter).

If the treads are 3'-0" (0.91 m) to 3'-6" (1.06 m) long and 1-1/4" (31.25 mm) thick, a marble setter and helper should set 20 to 22 treads containing 60 to 77 lin.ft. (18.29 to 23.46 m) per 8-hr. day, at the following labor cost per 100 lin.ft. (30.48 m):

FINISHES

Description	Hours	Rate	Total
Foreman	1.50	$38.17	$ 57.26
Marble setter	11.00	$34.17	$ 375.87
Marble Setter Helper	11.00	$34.17	$ 375.87
Labor or helper	8.00	$26.16	$ 209.28
Cost 100 lin.ft.			$1,018.28
per lin.ft.			$ 10.18
per lin.M.			$ 109.61

If the treads vary from 4'-0" to 6'-0" (1.21 to 1.82 m) long, a marble setter and helper should set 80 to 90 lin.ft. (24.38 to 27.43 m) per 8-hr. day, at the following labor cost per 100 lin.ft. (30.48 m):

Description	Hours	Rate	Total
Foreman	1.10	$38.17	$ 41.99
Marble setter	9.50	$34.17	$ 324.62
Marble Setter Helper	9.50	$34.17	$ 324.62
Labor or helper	8.00	$26.16	$ 209.28
Cost 100 lin.ft.			$ 900.50
per lin.ft.			$ 9.00
per lin.M.			$ 29.54

Setting Marble Stair Risers. The cost of setting marble stair risers will vary with the length of the pieces of marble, width of stairs, etc.

Where the stairs are 3'-0" to 3'-6" (0.91 to 1.06 m) wide, a marble setter and helper should set 60 to 75 lin.ft. (18.28 to 22.86 m) of risers per 8-hr. day, at the following labor cost per 100 lin.ft. (30.48 m):

Description	Hours	Rate	Total
Foreman	1.50	$38.17	$ 57.26
Marble setter	11.00	$34.17	$ 375.87
Marble Setter Helper	11.00	$34.17	$ 375.87
Labor or helper	8.00	$26.16	$ 209.28
Cost 100 lin.ft.			$1,018.28
per lin.ft.			$ 10.18
per lin.M.			$ 33.41

If the marble stairs are over 4'-0" (1.21 m) wide, a marble setter and helper should set 80 to 90 lin.ft. (24.38 to 27.43 m) of risers per 8-hr. day, at the following labor cost per 100 lin.ft. (30.48 m):

Description	Hours	Rate	Total
Foreman	1.10	$38.17	$ 41.99
Marble setter	9.50	$34.17	$ 324.62
Marble Setter Helper	9.50	$34.17	$ 324.62
Labor or helper	8.00	$26.16	$ 209.28
Cost 100 lin.ft.			$ 900.50
per lin.ft.			$ 9.00
per lin.M.			$ 29.54

Setting Marble Stair Wainscot on the Rake. Where marble wainscoting is set on the rake of the stairs, the labor cost will run considerably higher than straight work on account of the additional cutting and fitting necessary.

If the wainscoting is 3'-0" to 3'-6" (0.91 to 1.06 m) high, a marble setter and helper should set 10 to 15 lin.ft. (3.04 to 4.57 m) containing 35 to 45 sq.ft. (3.25 to 4.18 sq.m) per 8-hr. day, at the following labor cost per 100 sq.ft. (9.29 sq.m):

Description	Hours	Rate	Total
Foreman	2.00	$38.17	$ 76.34
Marble setter	20.00	$34.17	$ 683.40
Marble Setter Helper	20.00	$34.17	$ 683.40
Labor or helper	8.00	$26.16	$ 209.28
Cost 100 sq.ft.			$1,652.42
per sq.ft.			$ 16.52
per sq.M.			$ 177.87

Where the wainscot is 5'-0" to 7'-0" (1.52 to 2.13 m) high and follows the rake of the stairs, a marble setter and helper should set 50 to 65 sq.ft. (4.65 to 6.03 sq.m) per 8-hr. day, at the following labor cost per 100 sq.ft. (9.29 sq.m):

Description	Hours	Rate	Total
Foreman	1.50	$38.17	$ 57.26
Marble setter	14.00	$34.17	$ 478.38
Marble Setter Helper	14.00	$34.17	$ 478.38
Labor or helper	8.00	$26.16	$ 209.28
Cost 100 sq.ft.			$1,223.30
per sq.ft.			$ 12.23
per sq.M.			$ 131.68

Setting Marble Toilet Stalls and Partitions. The cost of handling and setting marble toilet stalls, backs, partitions, stiles, etc., will vary greatly with the individual job. On work of this kind, the marble partition slabs are seldom less than 5'-0" (1.52 m) in height and width and each slab contains 25 to 30 sq.ft. (2.32 to 2.78 m) of marble. The backs are usually the same height as the partitions but not so wide and each back contains 16 to 20 sq.ft. (1.48

1342

to 1.85 sq.m) of marble. The stall fronts or stiles are usually 6" (150 mm) to 8" (200 mm) wide and 6'-0" (1.82 m) to 7'-0" (2.13 m) high and extend to the floor to support the dividing partitions where metal standards are not used.

As an average on toilet stall work, a good marble setter should complete a stall in 3 to 3-1/2 hours, after the marble backs are in place. Since the average partition contains about 25 sq.ft. (2.32 sq.m) and each stile and cap about 7 sq.ft. (0.65 sq.m), a setter and helper should set 75 to 80 sq.ft. (6.96 to 7.43 sq.m) of marble per 8-hr. day, at the following labor cost for one complete stall exclusive of back:

Description	Hours	Rate	Total
Foreman	0.40	$38.17	$ 15.27
Marble setter	3.30	$34.17	$ 112.76
Marble Setter Helper	3.30	$34.17	$ 112.76
Labor or helper	2.60	$26.16	$ 68.02
Cost to Install 1 Stall			$ 308.81

Toilet partition work is estimated by the sq. ft. (sq.m) by most marble contractors.

Setting Marble Toilet Backs. The marble backs for toilet stalls are usually 6'-0" to 7'-0" (1.82 to 2.13 m) high. A marble setter and helper should set 75 to 85 sq.ft. (6.75-7.65 sq.m) per 8-hr. day, at the following labor cost per 100 sq.ft. (9.29 sq.m):

Description	Hours	Rate	Total
Foreman	1.20	$38.17	$ 45.80
Marble setter	10.00	$34.17	$ 341.70
Marble Setter Helper	10.00	$34.17	$ 341.70
Labor or helper	8.00	$26.16	$ 209.28
Cost 100 sq.ft.			$ 938.48
per sq.ft.			$ 9.38
per sq.M.			$ 101.02

Setting Marble Door and Window Trim. A marble setter and helper should set 45 to 55 lin.ft. (13.71 to 16.76 m) of marble door and window trim or casings per 8-hr. day, at the following labor cost per 100 lin.ft. (30.48 m):

Description	Hours	Rate	Total
Foreman	2.00	$38.17	$ 76.34
Marble setter	16.00	$34.17	$ 546.72
Marble Setter Helper	16.00	$34.17	$ 546.72
Labor or helper	8.00	$26.16	$ 209.28
Cost 100 lin.ft.			$1,379.06
per lin.ft.			$ 13.79
per lin.M.			$ 45.24

1343

Setting Marble Countertops. When setting marble countertops, a marble setter and helper should set 75 to 85 sq.ft. (6.96 to 7.89 sq.m) per 8-hr. day, at the following labor cost per 100 sq.ft. (9.29 sq.m):

Description	Hours	Rate	Total
Foreman	1.20	$38.17	$ 45.80
Marble setter	10.00	$34.17	$ 341.70
Marble Setter Helper	10.00	$34.17	$ 341.70
Labor or helper	8.00	$26.16	$ 209.28
Cost 100 sq.ft.			$ 938.48
per sq.ft.			$ 9.38
per sq.M.			$ 101.02

Setting Marble Thresholds. When setting marble thresholds 3'-0" to 3'-6" (0.91 to 1.09 m) long, a marble setter and helper should set 12 to 14 thresholds per 8-hr. day, at the following labor cost each:

Description	Hours	Rate	Total
Marble setter	0.60	$34.17	$ 20.50
Marble Setter Helper	0.60	$34.17	$ 20.50
Cost per Threshold			$ 41.00

Setting Marble Plinths. When setting marble plinths at door openings, a marble setter and helper should set 20 to 25 plinths per 8-hr. day, at the following labor cost per 100 plinths:

Description	Hours	Rate	Total
Foreman	4.50	$38.17	$ 171.77
Marble setter	36.00	$34.17	$1,230.12
Marble Setter Helper	36.00	$34.17	$1,230.12
Labor or helper	8.00	$26.16	$ 209.28
Cost 100 Plinths			$2,841.29
per plinth			$ 28.41

Setting Marble Stair Railings and Balusters. On jobs having marble balusters up to 6" (150 mm) in diameter and 2'-6" (0.76 m) long, a marble setter and helper should set about 20 balusters per 8-hr. day, at the following labor cost per 100 balusters:

Description	Hours	Rate	Total
Foreman	5.00	$38.17	$ 190.85
Marble setter	40.00	$34.17	$1,366.80
Marble Setter Helper	40.00	$34.17	$1,366.80
Labor or helper	16.00	$26.16	$ 418.56
Cost 100 balusters			$3,343.01
per baluster			$ 33.43

Setting Marble Ashlar. Where marble ashlar is used for the interior walls of buildings, it is customary to use 7/8" (21.87 mm) or 1-1/4" (31.25

1344

mm) marble and the method of handling and setting is similar to that used for setting exterior work, as it is often necessary to use a light breast derrick to set the marble.

On work of this kind, a marble setter and helper should handle and set 60 to 80 sq.ft. (5.57 to 7.43 sq.m) of ashlar per 8-hr. day, at the following labor cost per 100 sq.ft. (9.29 sq.m):

Description	Hours	Rate	Total
Foreman	1.50	$38.17	$ 57.26
Marble setter	12.00	$34.17	$ 410.04
Marble Setter Helper	12.00	$34.17	$ 410.04
Labor or helper	12.00	$26.16	$ 313.92
Cost 100 sq.ft.			$1,191.26
per sq.ft.			$ 11.91
per sq.M.			$ 128.23

Setting Marble Moldings, Handrail, Etc. On jobs having light marble moldings, stair handrail, well hole rail, etc., where the marble is about 6" (150 mm) square and furnished in reasonably long pieces, a marble setter and helper should set 75 to 85 lin.ft. (22.86 to 25.91 m) per 8-hr. day, at the following labor cost per 100 lin.ft. (30 m):

Description	Hours	Rate	Total
Foreman	1.20	$38.17	$ 45.80
Marble setter	10.00	$34.17	$ 341.70
Marble Setter Helper	10.00	$34.17	$ 341.70
Labor or helper	8.00	$26.16	$ 209.28
Cost 100 lin.ft.			$ 938.48
per lin.ft.			$ 9.38
per lin.M.			$ 30.79

Setting Marble Column Bases. When setting marble column bases 2'-6" to 3'-0" (0.23 to 0.27 m) square feet and 0'-9" to 1'-3" (0.22 to 0.31 m) high, a marble setter and helper should handle and set about one base per hr. at the following labor cost per base:

Description	Hours	Rate	Total
Foreman	0.10	$38.17	$ 3.82
Marble setter	1.00	$34.17	$ 34.17
Marble Setter Helper	1.00	$34.17	$ 34.17
Labor or helper	0.50	$26.16	$ 13.08
Cost per base			$ 85.24

Laying Marble Floor Tile. There are three labor operations to be considered when estimating the labor cost of setting marble floor tile: the cement bed under the marble floor, labor handling and setting floor tile, and the cost of rubbing or smoothing the floors after they have been laid.

1345

As a general rule, the cement bed under the marble floor is 2-1/2" to 3" (62.50 to 75 mm) thick, composed of cement and sand mixed rather dry. It requires about 1-1/2 bdls. of portland cement and one cu.yd. (0.76 cu.m) of sand to each 100 sq.ft. (9.29 sq.m) of floor.

Material and labor per 100 sq.ft. (9.29 sq.m) of floor fill should cost as follows:

Material Costs			
Description		Rate	Total
6 sacks portland cement		$ 6.00	$ 36.00
1 cu.yd. (0.76 cu.m) sand		$20.00	$ 20.00
Cost 100 sq.ft.			$ 56.00
per sq.ft.			$ 0.56
per sq.m			$ 6.03

Labor Costs			
Description	Hours	Rate	Total
Foreman	0.20	$38.17	$ 7.63
Marble setter	1.50	$34.17	$ 51.26
Marble Helper	3.00	$34.17	$ 102.51
Cost 100 sq.ft.			$ 161.40
per sq.ft.			$ 1.61
per sq.m			$ 17.37

If the marble floor tile vary from 6" x 6" (150 x 150 mm) to 12" x 12" (300 x 300 mm) in size, a marble setter and helper should handle and lay 90 to 110 sq.ft. (8.36 to 10.22 sq.m) of floor per 8-hr. day, at the following labor cost per 100 sq.ft. (9.29 sq.m):

Description	Hours	Rate	Total
Foreman	1.00	$38.17	$ 38.17
Marble setter	8.00	$34.17	$ 273.36
Marble Setter Helper	8.00	$34.17	$ 273.36
Labor or helper	8.00	$26.16	$ 209.28
Cost 100 sq.ft.			$ 794.17
per sq.ft.			$ 7.94
per sq.M.			$ 85.49

When laying marble floors it is customary to place the cement bed just ahead of the tile, and the entire work is performed by the setter and helper laying the floor. If the subfloor is sound and even, marble in residential and light commercial work is only 3/8" (9.37 mm) thick and set in mastic. Setting costs would average about $0.25 per sq.ft. ($3.12 sq.m) less, unless the area was small and irregular, or set with an intricate pattern.

Smoothing Marble Floors. After the marble floors have been laid, it is necessary to rub them with carborundum stone to remove the uneven spots at the joints. It is customary to use a rubbing machine powered by gasoline or electricity, which is operated by one worker and runs back and forth over the floors and grinds them to an even surface. A machine rents for about $7.50 per hour.

The quantity of floor that can be surfaced per day will vary with the kind of marble used, as it is possible to surface considerably more soft marble than hard marble.

When rubbing and surfacing soft marble, a machine should surface 125 to 150 sq.ft. (11.61 to 13.93 sq.m) per 8-hr. day, at the following labor cost per 100 sq.ft. (9.29 sq.m):

Description	Hours	Rate	Total
Machine Operator	6.00	$37.89	$ 227.34
Foreman	0.70	$38.17	$ 26.72
Cost 100 sq.ft.			$ 254.06
per sq.ft.			$ 2.54
per sq.M.			$ 27.35

When rubbing and surfacing hard marble, a rubbing machine should surface 100 to 125 sq.ft. (9.29 to 11.61 sq.m) of floor per 8-hr. day, at the following labor cost per 100 sq.ft. (9.29 sq.m):

Description	Hours	Rate	Total
Machine Operator	7.00	$37.89	$ 265.23
Foreman	0.80	$38.17	$ 30.54
Cost 100 sq.ft.			$ 295.77
per sq.ft.			$ 2.96
per sq.M.			$ 31.84

Setting Marble Floor Border. Where the marble floor has a different colored marble border, the labor setting the border will cost about the same as the balance of the floor.

Setting Structural Slate. The labor cost of setting structural slate work, such as base, treads and risers, toilet stalls and partitions, etc., will run about the same as for the same class of marble work and should be estimated accordingly.

If, however, field cutting should be required, as is often the case with window stools and door thresholds, the labor costs should be increased 20% to 25% as slate is more difficult to work than marble.

Estimating Quantities of Interior Marble

When estimating quantities of interior marble, any fraction of an inch will be treated as a whole inch. No piece of marble will be considered as being less than 6" (150 mm) wide and 1'-0" (0.30 m) long.

Thickness. Thicknesses given are approximate but not exact. For slabs gauged to a given thickness, an extra charge of 10% to the list price is made.

Extra Lengths. On slabs 10'-0" to 12'-0" (3.04 to 3.65 m) long, add 10% to list price; on slabs over 12'-0" (3.65 m) long, add 20%.

Extra Widths. On slabs over 5'-6" (1.67 m) wide, add 10% to list price.

Polished Both Sides. Prices given are based on marble having one side polished unless otherwise noted. If both sides are to be polished, add $2.10 per sq.ft. ($22.60 per sq.m) for marbles in Group A and B and $1.75 per sq.ft. ($18.83 per sq.m) for marbles in Group C and D.

Hone Finish. Hone finish takes the same price as polished work.

Sand Finish. If marble is finished with sand finish on one side only, deduct $0.25 per sq.ft ($2.69 sq.m). If sand finish on both sides, add $0.35 per sq.ft. ($3.76 per sq.m).

Polished Edges. For each polished edge of 7/8" (21.87mm) and 1-1/4" (31.25 mm) thickness, add $0.75 per lin.ft. ($2.46 per m) of edge; on 1-1/2" (37.50 mm) and 2" (50 mm) thickness, add $1.05 per lin.ft. ($3.28 per m).

Beveled or Rounded Edges. For each beveled or rounded edge on 7/8" (21.87 mm) and 1-1/4" (31.25 mm) thickness, add $0.70 per lin.ft. ($2.29 per m); on 1-1/2" (37.50 mm) and 2" (50 mm) thickness, add $0.1.05 per lin.ft. ($3.44 per m).

Ogee Molding. For single member ogee molding $4.60 per lin.ft. ($15.09 per m) of edge on 7/8" (21.87 mm) thick; $4.90 per lin.ft. ($16.07 per m) on 1-1/4" (31.25 mm), and $6.00 per lin.ft. ($19.68 per m) on 1-1/2" (37.50 mm). For each additional member, add $3.15 per lin.ft. ($10.33 per m) of edge on 7/8" (21.87mm) and 1-1/4" (31.25 mm), and $3.50 lin.ft. ($11.48 per m) on 1-1/2" (37.50 mm).

Countersinking. For countersinking 1/4" (6.25 mm) deep or less, $3.00 per sq.ft. ($32.9 per sq.m) of surface. For each additional 1/4" (6.25 mm) deeper, or fraction thereof, add extra $3.50 per sq.ft. ($37.67 per sq.m) of surface.

Grooving. For grooving 1/4" (6.25 mm) deep or less, $2.10 per lin.ft. ($5.89 per m) for each groove. For each additional 1/4" (6.25 mm) deeper, or fraction thereof, add extra $2.10 per lin.ft. ($6.89 per m) of each groove.

Rabbeting. For rabbeting 1/4" (6.25 mm) deep or less, $2.10 per lin.ft. ($6.89 per m) for each rabbet. For each additional 1/4" (6.25 mm) deeper, or fraction thereof, add extra $1.75 per lin.ft. ($5.74 per m) of each rabbet.

Special Cutting, Polishing, Drilling, Etc. All special work other than listed above will be charged for extra.

Shipping Weights of Marble. The approximate shipping weights of marble, boxed for shipment, are as follows per sq.ft. (sq.m): 7/8" (21.87mm) thick, 15 lbs. (75 kg); 1-1/4" (31.25 mm) thick 20 lbs. (100 kg); 1-1/2" (37.50 mm) thick 24 lbs. (120 kg); 2" (50 mm) thick 32 lbs. (160 kg); per cu.ft., 192 lbs. (per cu.m, 3050 kg).

1348

FINISHES

Approximate Material Prices per S.F. (Sq.M) of 7/8" (21.87 mm) Marble, Polished One Side

Always check local sources for price and availability. Foreign marbles are always affected by exchange rates. The trend has been to marbles of even color. Many of the quarries for the more colorful marbles have been shut down, and when a block of such marble is found, the price may be out of line because of scarcity.

Type of Marble	Price per Sq.Ft	Price per Sq.M.
Alabama Clouded "A" or Cream "A"	$ 17.00	$ 182.99
Brocadillo	$ 21.20	$ 228.20
Carthage Gray	$ 15.90	$ 171.15
French Gray	$ 17.00	$ 182.99
Georgia Creole, Cherokee or Mezzotint	$ 17.00	$ 182.99
Gravina	$ 15.90	$ 171.15
Kasota Pink	$ 15.90	$ 171.15
Kasota Yellow	$ 19.10	$ 205.60
Light Cloud Vermont	$ 17.00	$ 182.99
Napoleon Gray	$ 17.00	$ 182.99
Radio Black	$ 19.10	$ 205.60
Tennessee Marbles	$ 17.00	$ 182.99
Verde Antique	$ 21.20	$ 228.20
Westfield Green	$ 23.30	$ 250.81
York Fossil	$ 19.10	$ 205.60
Foreign Marbles		
Alps Green	$ 28.60	$ 307.86
Belgian Black	$ 25.50	$ 274.49
Black and Gold	$ 33.90	$ 364.91
Botticino	$ 25.50	$ 274.49
Bois Jordan	$ 25.50	$ 274.49
Breche Oriental	$ 36.10	$ 388.59
Grand Antique, French	$ 42.40	$ 456.40
Grand Antique, Italian	$ 31.84	$ 342.73

The above prices are for marble in less than carload lots, uncrated, f.o.b. mill. Crating $1.00 to 1.50 per sq.ft. for marble up to 2" (50 mm) thick.

Approximate Material Prices on Sand Finish Marble Floor, Tile and Border 7/8" (21.87 mm) Thick - Tile Size 6" x 6" (150 x 150 mm) to 1'- 0" x 2'- 0" (300 x 600 mm) Prices per Sq.Ft.(Sq.M.)		
Material	Price per Sq.Ft	Price per Sq.M.
Alabama Clouded "A"	$ 11.60	$ 124.87
Carthage Gray	$ 8.00	$ 86.11
Georgia Creole	$ 11.60	$ 124.87
Georgia Cherokee	$ 6.40	$ 68.89
Italian White	$ 3.70	$ 39.83
Italian English Vein	$ 7.40	$ 79.66
Tennessee Pink	$ 9.60	$ 103.34
Tennessee Gray	$ 9.60	$ 103.34
Tennessee Dark	$ 8.50	$ 91.50
Verde Antique	$ 14.90	$ 160.39
Travernelle	$ 11.70	$ 125.94
Travertine	$ 10.60	$ 114.10

Approximate Material Prices on 3/8" (9.37 mm) Thinset Marble Tiles Per Sq.Ft. (Sq. M.)				
Description	Polished per Sq.Ft.	Honed per Sq.Ft.	Polished per Sq.M.	Honed per Sq.M.
Cremo	$ 5.94	$ 5.30	$ 63.94	$ 57.05
Negro-Marquina	$ 7.04	$ 6.30	$ 75.78	$ 67.81
Rose Aurora	$ 5.61	$ 5.00	$ 60.39	$ 53.82
Statuary Vein	$ 4.95	$ 4.40	$ 53.28	$ 47.36
Travertine Filled	$ 4.73	$ 4.20	$ 50.91	$ 45.21
Travertine Unfilled	$ 4.51	$ 4.00	$ 48.55	$ 43.06
White Italian	$ 4.73	$ 4.20	$ 50.91	$ 45.21

12"x12" (300 x 300 mm) tiles, 3/8" (9 mm) thick, 10 tiles per box.

SLATE TILE

Slate is a natural mica granular crystalline stone. Its main sources are Vermont, Pennsylvania and Virginia. Vermont produces colored slates, including greens; purples; mottled green, purple and red; as well as the standard dark grays.

Slate is usually graded into clear stock and ribbon stock. Ribbon stock has bands of a darker color running through it and is cheaper than clear stock, which has no ribbons but will have some spots and veining in it. Finishes available are as follows:

FINISHES

Natural Cleft: Natural split or cleaved face, moderately rough with some textural variations.

Sand Rubbed Finish: Wet sand on a rubbing bed is used to eliminate any natural cleft, and leave slate in an even plane.

Honed Finish: This is a semi-polished surface.

Slate is used for spandrels, sills, stools, treads and risers, toilet stalls, floors and walks, fireplace facings and hearths.

Spandrels are generally clear slate with either cleft, rubbed or honed finish. Thickness is generally 1" (25 mm), 1-1/4" (31 mm), or 1-1/2" (37.50 mm). Sizes are somewhat limited, with a length of no more than 6' (1.8 m) and width of 2'-6" (0.76 m) to 4' (1.21 m) being standard recommendations, although larger sizes may be available for certain conditions. Many spandrels today are set in metal grid frames. Otherwise, they must be anchored with bronze or stainless steel anchors in each corner.

Sills and stools may be either ribbon or clear stock. Exterior sills are generally furnished with a sand rubbed finish in lengths up to 4' (1.21 m) and 1" (25 mm) to 2" (50 mm) in thickness. Sills may have drip grooves cut in at additional cost.

Interior stools and fireplace surrounds are usually 1" (25 mm) thick with honed finish, a maximum length of 4' (1.21 m) being most economical.

Treads and risers on the exterior are generally natural cleft finish. Interior work is usually sand rubbed rather than honed, because it is both cheaper and gives a good non-slip surface, although the front edges may be honed. Thickness is generally 1" (25 mm) or 1-1/4" (31.25 mm) for stairs, 1-1/2" (37.25 mm) to 2" (50 mm) for landings.

The cost of slate is determined by the grade, the size involved and the finish.

Ribbon grade, often used on stairs, will run around $7.00 per sq.ft. ($75.35 per sq.m) in 1" (25 mm) thickness and $11.25 per sq.ft. ($121.10 per sq.m) for 2" (50 mm) thickness, sand finished. Honed finishes will add $0.70 per sq.ft. ($7.53 per sq.m) to the cost.

Clear grade, 1" (25 mm) thick and sand finished, will be about $7.70 per sq.ft. ($82.88 per sq.m) in sizes to 3 sq.ft. (0.27 sq.m), $10.85 to $12.25 in sizes 3 sq.ft. (0.27 sq.m) to 6 sq.ft. (0.55 sq.m), and $15.50 in sizes over 6 sq.ft. (0.55 sq.m). Lineal foot (meter) costs for sills and stools will vary with both length and width. Up to 6" (150 mm) wide stock will cost about $6.65 per lin.ft. ($21.81 per m) for 1" (25 mm) thickness; $10.50 per lin.ft. ($34.45 per m) for 2" (50 mm) thickness. The 10" (250 mm) wide stock will cost $10.50 per lin.ft. ($34.45 per m) for 1" (25 mm) thickness to $13.00 per lin.ft. ($42.65 per m) for 2" (50 mm) thickness.

Slate for flooring is available for either mortar bed or thin set application. This slate is 1/4" (6.25 mm) thick and available in 3" (75 mm) multiples 6" x 6" (150 x 150 mm) to 12" x 12" (300 x 300 mm). It weighs 3-3/4 lbs. per sq.ft. (18.75 kg per sq.m). It is set in 1/4" (6.25 mm) mastic applied with a notched trowel over concrete or plywood. It can have grouted

1351

joints or be ordered for butt joints. Cost will run $2.80 per sq.ft. ($30.14 per sq.m) for material and $3.50 per sq.ft. ($37.67 per sq.m) for labor.

Slate for heavier duty floors and for outside applications is generally 1/2" (12.50 mm), 3/4" (18.75 mm), or 1" (25 mm) thick, weighing 7-1/2 lbs. (3.4 kg), 11-1/4 lbs. (5 kg) and 15 lbs. (6.75 kg) respectively. It is set in 1" (25 mm) mortar beds on concrete slabs or plywood. Plywood should be covered with felt and metal lath, nailed on. Minimum thickness is 1-1/2" (37.50 mm). If plywood is set over joists without a subfloor, it should be at least 3/4" (18.75 mm) thick. Sizes are 6" x 6" (150 x 150 mm) to 24" x 18" (600 x 450 mm), in multiples of 3" (75 mm). Cost runs from $5.50 per sq.ft. ($59.20 per sq.m) for 1/2" (12.50 mm) to $7.00 per sq.ft. ($75.35per sq.m) for 1" (25 mm).

09400 TERRAZZO

The cost of terrazzo work varies with the size of the job, size of rooms or spaces where the floors are to be installed, floor designs, strips of brass or other non-rusting material, and method of laying the floors.

Two methods are used in laying terrazzo floors over concrete construction. One is to bond it to the concrete and the other to separate it from the structural slab.

When the first method is used, the concrete fill is provided for by another contractor and should be left 2" (50 mm) below the finished floor. Before the terrazzo contractor installs his underbed, he must see that this concrete fill is thoroughly cleaned of plaster droppings, wood chips and other debris, and wetted to insure cohesion.

The second method is used in buildings where cracking is anticipated either from settlement, expansion and contraction or vibration. In this case the terrazzo contractor begins his work from the structural floor slab up. This method requires a thickness of at least 3" (75 mm). The concrete slab is covered with a thin bed of dry sand over which a sheet of tar paper is laid. Over the paper the underbed is installed as in the first method except that coarser aggregate can be used in the underbed, such as cinders or fine gravel where its thickness exceeds 2-1/2" (62.50 mm). When this method is used, the cracks originating in the structural slab are not likely to appear on the surface but terminate at the sand bed.

When terrazzo is laid over wood floors a thickness of not less than 2" (50 mm) is required. The floor should first be covered with tarred paper. Over this paper, nail galvanized wire netting no. 14 gauge, 2" (50 mm) mesh. Then lay the concrete underbed as specified under the second method.

The underbed for terrazzo consisting of 1 part portland cement and 4 parts sharp, screened sand shall be spread and brought to a level not less than 1/2" (12.50 mm) nor more than 3/4" (18.75 mm) below the finished floor.

Into this underbed, while still in a semi-plastic state, install metal strips or preassembled decorative units, having proper bonding features.

FINISHES

The terrazzo topping shall be not less than 1/2" (12.50 mm) nor more than 3/4" (183.75 mm) in thickness and shall be of granulated marble of colors selected. The topping shall be uniform in composition and the marble granule that appears on the surface shall be used for its entire thickness. The granule to be composed of such proportions of Nos. 1-2-3 sizes as required.

Composition for terrazzo mixture shall be in the proportion of 200 lbs. (90 kg) of marble granules to 100 lbs. (45 kg) of gray or white portland cement, mixed dry, and water added afterward to make the mix plastic, but, not too wet. The mix shall then be placed in the spaces formed by the metal dividing strips and rolled into a compact mass by means of heavy stone or metal rollers until all superfluous cement and water is extracted, after which it must be hand troweled to an even surface, disclosing the lines of the metal strips on a level with the terrazzo filling.

When the floors have set sufficiently hard, they shall be machine rubbed, using coarse carborundum grit stones for the initial rubbing after which a light grouting of pure gray white portland cement shall be applied to the surface, filling all voids, and allowed to remain until the time of final cleaning.

Floors shall have the grouting coat removed by machine, using a fine carborundum grit, after which it must be thoroughly washed. Do not use acids in cleaning terrazzo floors.

Ramped or other surfaces in terrazzo floors so specified, shall be made anti-slip by the addition of an abrasive aggregate. For heavy duty floors, the proportion shall be 2 parts of abrasive to 3 parts of marble granule and the abrasive shall be mixed in the terrazzo topping for its entire thickness. For light traffic floors, the abrasive shall be sprinkled on the surface only, and show a proportion of 5 parts marble granule and 1 part abrasive aggregate.

Terrazzo Base. Terrazzo base is usually coved at the floor and may be any height desired. If the base is to finish flush with the finish plaster, a metal base bead shall be furnished and set by the plasterer. The base shall be divided approximately every 4 lin.ft. (1.21 m) using metal base dividers. The walls back of the base shall have a scratch coat of cement and sand mortar brought to a line 3/8" (9 mm) back of the finished face of the base, into which the base dividers shall be set. Base shall be finished with a very fine stone, so as to leave the surface at a hone finish.

Estimating the Cost of Terrazzo Work. In addition to the direct cost of labor and materials entering into terrazzo work, an allowance should be made to provide for wood grounds, carborundum stone for rubbing, wiring for electric power, depreciation on machines, use of hoist, cleaning up rubbish, freight and trucking on machines and tools. This will vary with size and location of job.

The costs on the following pages are based on an average job containing about 3,000 sq.ft. (278.70 sq.m). The labor on a small job containing 100 to 200 sq.ft. (9.29 to 18.58 sq.m) will run twice as much per sq.ft. (sq.m) as on a job containing 5,000 sq.ft. (464.50 sq.m).

Metal Strips in Terrazzo Floors and Base. Brass strips or other non-rusting metals are used in practically all terrazzo floors and add to the cost in proportion to the quantity used.

Standard brass strips, B & S gauge No. 20, cost $0.85 per lin.ft. ($2.79 per m) including waste.

Metal strips, B & S gauge No. 14, cost $1.45 per lin.ft. ($4.75 per m) including waste in brass and $0.75 per lin.ft. ($3.29 per m) in zinc.

The cost of metal strips per sq.ft. of floor will vary with the size of the squares or design and should be estimated accordingly.

Terrazzo base is jointed every 4'-0" (1.21 m) with vertical metal joint dividers.

Estimating Quantities of Terrazzo Work. Terrazzo floors are estimated by the sq.ft. (sq.m) and will vary from 2" (50 mm) to 3" (75 mm) thick, depending upon the subfloors.

Terrazzo floor borders are estimated by the sq.ft. (sq.m) when over 12" (300 mm) wide and by the lin.ft. (meter) when less than 12" (300 mm) wide.

Terrazzo base is estimated by the lin.ft. (meter) stating height. An allowance of 1'-0" (0.30 m) should be made for each corner or miter.

Double cove terrazzo base having 2 finished faces, such as ordinarily used under narrow partitions, shower stalls, etc. are estimated by the lin.ft. (meter) and the cost is double that of single cove base.

Labor Placing Terrazzo, Floor and Base. The labor cost of placing terrazzo work will vary with the size of the rooms, design of the floor, size of the squares or pattern requiring brass strips, etc., as the smaller the squares, the more metal strip required and the higher the cost.

The following are approximate quantities of various classes of terrazzo work a crew of 2 mechanics and 3 helpers should install per 8-hr. day:

Description Floor Blocked off into	Sq.Ft per 8-hr. Day	Hrs. per 100 SqFt (9.29 sq.m)		Sq.M per 8-hr. Day	Description Floor Blocked off into
		Mechanic	Helpers		
5'-0" squares	400 - 450	3.75	5.65	37.16 - 41.81	1.52 M. squares
4'-0" squares	375 - 425	4.00	6.00	34.84 - 39.48	1.22 M. squares
3'-0" squares	350 - 400	4.25	6.40	32.52 - 37.16	0.91M squares
2'-0" squares	325 - 375	4.60	6.90	30.19 - 34.84	0.61 M.squares
1'-0" squares	275 - 325	5.33	8.00	25.55 - 30.19	0.30 M squares
Border 12"-24" wide	275 - 325	5.33	8.00	25.55 - 30.19	Border .30-61 M wide
Terrazzo Cove Base		Hrs. per 100 lin.ft. (30 m)			
Terrazzo cove base, 3" high	60 - 65	13.00	13.00	5.57 - 6.04	75 mm high
Terrazzo cove base, 6" high	50 - 55	16.00	16.00	4.65 - 5.11	150 mm high

Double cove base, double time given for single base.

Rubbing and Finishing Terrazzo Work. Terrazzo floors are rubbed by machine and by hand only in corners where the machine cannot reach. On

1354

FINISHES

jobs consisting of large and small rooms, a worker rubs and completes about 100 sq.ft. (9.29 sq.m) of terrazzo floor per 8-hr. day. A worker should rub and finish about 80 lin.ft. (24.38 m) of terrazzo base per 8-hr. day.

Cost of Concrete Subbase Under Terrazzo Floors. There are three methods of applying the concrete underbed for terrazzo floors, as described previously: bonded to the structural concrete slab, separate from the structural slab, or applied over wood floors.

Inasmuch as terrazzo topping is 1/2" (12.50 mm) thick for all types of floors, cost of finished floor varies with type of underbed.

Material Cost of 100 Sq.Ft. (9.29 Sq.M) of Concrete Underbed 1-1/4" (31.25 mm) Thick Bonded to Structural Concrete Slab		
Description	Rate	Total
3 sacks Portland cement	$ 6.00	$ 18.00
11 cu.ft. (0.31 cu.m) sand	$ 0.72	$ 7.92
Cost 100 sq.ft.		$ 25.92
per sq.ft		$ 0.26
per sq.m		$ 2.79

Material Cost of 100 Sq.Ft. (9.29 Sq.M) of Concrete Underbed, Consisting of 1/4" (6.25 mm) Dry Sand, 1 Thickness Tarred Felt and 2" (50 mm) of Concrete, Underbed Separated from Structural Slab		
Description	Rate	Total
2.5 cu.ft. (0.07 cu.m) dry sand	$ 0.72	$ 1.80
1 sq. No. 15 felt	$ 4.77	$ 4.77
4.5 sacks Portland cement	$ 6.00	$ 27.00
18 cu.ft. (0.5 cu.m) sand	$ 0.72	$ 12.96
Cost 100 sq.ft.		$ 46.53
per sq.ft.		$ 0.47
per sq.m		$ 5.01

Material Cost of 100 Sq.Ft. (9.29 Sq.M) of Concrete Underbed 2" (50 mm) Thick, Applied over Wood Floor		
Description	Rate	Total
1 sq. No. 15 felt	$ 4.77	$ 4.77
105 sq.ft. (9.45 sq.m) 14 ga. galv.wire Netting	$ 0.20	$ 20.48
4.5 sacks Portland cement	$ 6.00	$ 27.00
18 cu.ft. (0.50 cu.m) sand	$ 0.72	$ 12.96
Cost 100 sq.ft.		$ 65.21
per sq.ft.		$ 0.65
per sq.m		$ 7.02

Material Cost of 100 Sq.Ft. (9.29 Sq.M) of Terrazzo Floor 1/2" (13 mm) Thick, Applied over Concrete Underbed		
Description	Rate	Total
3 sacks portable cement	$ 6.00	$ 18.00
600 lbs. (270 kg) marble granules *	$ 0.20	$ 117.00
0.5 sack port. cement (grout)	$ 6.00	$ 3.00
Cost 100 sq.ft.		$ 138.00
per sq.ft.		$ 1.38
per sq.m		$ 14.85

Varies according to kind of granules used. Add for metal strips.

Material Cost of 100 Sq.Ft. (9.29 Sq.M) Terrazzo Floors Blocked Into 5'-0" (1.52 m) Squares, Including 1 1/4" (31.25 mm) Concrete Underbed		
Description	Rate	Total
Underbed 1-1/4" (31 mm) thick 3 sacks Portland Cement	$ 6.00	$ 18.00
11 cu.ft. (0.31 cu.m) sand	$ 0.72	$ 7.92
Terrazzo Floor 3 sacks Portland cement	$ 6.00	$ 18.00
600 lbs. (270 kg) marble granules *	$ 0.20	$ 117.00
0.5 sack cement (grout)	$ 6.00	$ 3.00
50 lin.ft. (15 m) metal strips **	$ 0.80	$ 39.75
Cost 100 sq.ft.		$ 203.67
per sq.ft.		$ 2.04
per sq.m		$ 21.92

Varies according to kind of granules used. Add for metal strips.
** *Varies according to kind and gauge of steel*

Labor Cost of 100 Sq.Ft. (9.29 Sq.M) Terrazzo Floors Blocked Into 5'-0" (1.52 m) Squares, Including 1 1/4" (31.25 mm) Concrete Underbed			
Description	Hours	Rate	Total
Mechanic	3.75	$34.17	$ 128.14
Helper	5.65	$26.16	$ 147.80
Labor rub and finish	8.00	$26.16	$ 209.28
Cost 100 sq.ft			$ 485.22
per sq.ft.			$ 4.85
per sq.m			$ 52.23

If terrazzo is placed over wood subfloor or separated from concrete floor, add difference in cost of underbed.

1356

FINISHES

Material Cost of 100 Sq.F.t (9.29 Sq.M) Terrazzo Floors Blocked Into 4'- 0" (1.22 M) Squares Including 1 1/4" (31.25 mm) Concrete Underbed		
Description	Rate	Total
Underbed 1-1/4" (31 mm) Thick, 3 Sacks Portland Cement	$ 6.00	$ 18.00
11 cu.ft. (0.31 cu.m) sand	$ 0.72	$ 7.92
Terrazzo Floor 3 sacks portland cement	$ 6.00	$ 18.00
600 lbs. (270 kg) marble granules *	$ 0.20	$ 117.00
0.5 sack cement (grout)	$ 6.00	$ 3.00
60 lin.ft. (18 m) metal strips **	$ 0.80	$ 47.70
Cost 100 sq.ft.		$ 211.62
per sq.ft.		$ 2.12
per sq.m		$ 22.78

Varies according to kind of granules used. Add for metal strips.
* *Varies according to kind and gauge of steel*

Labor Cost of 100 Sq.Ft. (9.29 Sq.M) Terrazzo Floors Blocked Into 4'- 0" (1.22 mm) Squares Including 1 1/4" (31.25 mm) Concrete Underbed			
Description	Hours	Rate	Total
Mechanic	4.00	$34.17	$ 136.68
Helper	6.00	$26.16	$ 156.96
Labor rub and finish	8.00	$26.16	$ 209.28
Cost 100 sq.ft			$ 502.92
per sq.ft.			$ 5.03
per sq.m			$ 54.14

If terrazzo is placed over wood subfloor or separated from concrete floor, add difference in cost of underbed.

Material Cost of 100 Sq.Ft. (9.29 Sq.M) Terrazzo Floors Blocked Into 3'- 0" (0.90 M) Squares Including 1 1/4" (31.25 mm) Concrete Underbed		
Description	Rate	Total
Underbed 1-1/4" (31 mm) thick 3 Sacks Portland Cement	$ 4.24	$ 12.72
11 cu.ft. (0.31 cu.m) sand	$ 0.48	$ 5.28
Terrazzo Floor 3 sacks Portland cement	$ 4.24	$ 12.72
600 lbs. (270 kg) marble granules *	$ 0.13	$ 78.00
0.5 sacks cement (grout)	$ 4.24	$ 2.12
80 lin.ft. (24 m) metal strips **	$ 0.53	$ 42.40
Cost 100 sq.ft.		$ 153.24
per sq.ft.		$ 1.53
per sq.m		$ 16.50

Varies according to kind of granules used. Add for metal strips.
* *Varies according to kind and gauge of steel*

Labor Cost of 100 Sq.Ft. (9.29 Sq.M) Terrazzo Floors Blocked Into 3'- 0" (0.90 M) Squares Including 1 1/4" (31.25 mm) Concrete Underbed			
Description	Hours	Rate	Total
Mechanic	4.25	$34.17	$ 145.22
Helper	6.40	$26.16	$ 167.42
Labor rub and finish	8.00	$26.16	$ 209.28
Cost 100 sq.ft			$ 521.93
per sq.ft.			$ 5.22
per sq.m			$ 56.18

If terrazzo is placed over wood subfloor or separated from concrete floor, add difference in cost of underbed.

Material Cost of 100 Sq.Ft. (9.29 Sq.M) Terrazzo Floors Blocked Into 2'- 0" (0.90 M) Squares Including 1 1/4" (31.25 mm) Concrete Underbed		
Description	Rate	Total
Underbed 1-1/4" (31 mm) thick, 3 Sacks Portland	$ 6.00	$ 18.00
11 cu.ft. (0.31 m) sand	$ 0.72	$ 7.92
Terrazzo Floor 3 sacks Portland cement	$ 6.00	$ 18.00
600 lbs. (270 kg) marble granules *	$ 0.20	$ 117.00
0.5 sack cement (grout)	$ 6.00	$ 3.00
120 lin.ft. (36 m) metal strips **	$ 0.80	$ 95.40
Cost 100 sq.ft.		$ 259.32
per sq.ft.		$ 2.59
per sq.m		$ 27.91

*Varies according to kind of granules used.
**Varies according to kind and gauge of metal.

Labor Cost of 100 Sq.Ft. (9.29 Sq.M) Terrazzo Floors Blocked Into 2'- 0" (0.60) Squares Including 1 1 /4" (31.25 mm) Concrete Underbed			
Description	Hours	Rate	Total
Mechanic	4.60	$34.17	$ 157.18
Helper	6.90	$26.16	$ 180.50
Labor rub and finish	8.00	$26.16	$ 209.28
Cost 100 sq.ft			$ 546.97
per sq.ft.			$ 5.47
per sq.m			$ 58.88

If terrazzo is placed over wood subfloor or separated from concrete floor, add difference in cost of underbed.

1358

FINISHES

Material Cost of 100 Sq.Ft. (9.29 Sq.M) Terrazzo Floors Blocked Into 1'- 0" (0.30 M) Squares Including 1 1/4" (31.25 mm) Concrete Underbed		
Description	Rate	Total
Underbed 1-1/4" (31 mm) thick 3 sacks Portland cement	$ 6.00	$ 18.00
11 cu.ft. (0.31 m) sand	$ 0.72	$ 7.92
Terrazzo floor 3 sacks Portland cement	$ 6.00	$ 18.00
600 lbs. (272.16 kg) marble granules*	$ 0.20	$ 117.00
0.5 sacks cement (grout)	$ 6.00	$ 3.00
210 lin.ft. (64.08 m) metal strips	$ 0.80	$ 166.95
Cost 100 sq.ft.		$ 330.87
per sq.ft.		$ 3.31
per sq.m		$ 35.62

Varies according to kind of granules used.

Labor Cost of 100 Sq.Ft. (9.29 Sq.M) Terrazzo Floors Blocked Into 1'- 0" (0.30 M) Squares Including 1 1/4" Concrete Underbed			
Description	Hours	Rate	Total
Mechanic	5.35	$34.17	$ 182.81
Helper	8.00	$26.16	$ 209.28
Labor rub and finish	8.00	$26.16	$ 209.28
Cost 100 sq.ft			$ 601.37
per sq.ft.			$ 6.01
per sq.m			$ 64.73

If terrazzo is placed over wood subfloor or separated from concrete floor, add difference in cost of underbed.

Material Cost of 100 Lin.Ft. (30.48 m) of 6" (150 mm) Terrazzo Cove Base		
Description	Rate	Total
Scratch Coat		
1.5 sacks Portland cement	$ 6.00	$ 9.00
3 cu.ft. (0.08 cu.m) sand	$ 0.72	$ 2.16
Terrazzo Base		
1 sack Portland cement	$ 6.00	$ 6.00
200 lbs. (90 kg) marble granules *	$ 0.20	$ 39.00
25 metal dividers**	$ 0.80	$ 19.88
Cost 100 lin.ft.		$ 76.04
per lin.ft.		$ 0.76
per m		$ 2.49

Varies according to kind of granules used.
**Varies according to kind and gauge of metal.*

Labor Cost of 100 Lin.Ft. (30.48 m) of 6" (150 mm) Terrazzo Cove Base			
Description	Hours	Rate	Total
Mechanic	16.00	$34.17	$ 546.72
Helper	16.00	$26.16	$ 418.56
Labor rub and finish	10.00	$26.16	$ 261.60
Cost 100 sq.ft			$1,226.88
per sq.ft.			$ 12.27
per sq.m			$ 132.06

Double cove base, double cost of single cove base.

If terrazzo is placed over wood subfloor or separated from concrete floor, add difference in cost of underbed.

Abrasive Materials for Terrazzo Floors and Stair Treads. Where an anti-slip surface is required on terrazzo floors, platforms, landings, and treads, mix 2 parts abrasive aggregates with 3 parts marble granule. For heavy duty floors, the abrasive aggregate shall be mixed in the terrazzo topping for its entire thickness. For light traffic floors, the abrasive shall be sprinkled on the surface only, to show in the proportion of 3 parts marble granule and 1 part abrasive aggregate.

Abrasive aggregates for 100 sq.ft. (9.29 sq.m) of heavy duty floors should cost as follows:

Description	Rate	Total
240 lbs. (108 kg) abrasive aggregates	$ 0.80	$ 190.80
Cost per sq.ft.		$ 1.91
per sq.m		$ 20.54

Deduct cost of 240 lbs. (108 kg) of marble granules included in itemized estimates. Abrasive aggregates for 100 sq.ft. (9.29 sq.m) of light duty floors should cost as follows:

Description	Rate	Total
10 lbs. (4.5 kg) abrasive aggregates	$ 0.80	$ 7.95
Cost per sq.ft.		$ 0.08
per sq.m		$ 0.86

For prices on special terrazzo floors, such as venetian, mosaic, and conductive, it is advisable to contact a local installation contractor.

There are several products on the market that resemble terrazzo floors and are installed by terrazzo mechanics. These are magnisite, latex, epoxy, and polyester resin type terrazzo floors. These floors can be installed over any sound existing floor in 1/2" (12.50 mm) thickness, with the exception of the epoxy and polyester resin types, which can be installed as thin as the marble granules will allow. The cost of these floors is similar to that of regular terrazzo.

Precast terrazzo is available for base, floor, wainscots, and stair treads. Straight 6" (150 mm) base costs about $6.00 per lin.ft. ($19.68 per m), and

1360

FINISHES

one mechanic sets 125 lin.ft. (38.10 m) per 8 hr. day. Floor tiles, 1" (25 mm) thick cost $6.50 per sq.ft. ($70.00 per sq.m) and one mechanic sets 60 to 70 sq.ft. (5.57 to 6.50 sq.m.) per 8 hr. day. Stair treads cost around $15.75 per lin.ft. ($61.67 per m), because they are heavy. On a stair of any width, it takes a mechanic and helper to set 100 lin.ft. (30.48 m) per 8 hr. day.

09500 ACOUSTICAL TREATMENT

There is a large variety of prefabricated acoustical products available on the market. In general, they fall into three categories: acoustical tiles that are mounted on ceilings, on suspended boards, or clipped or splined to suspension systems; acoustical panels for lay-in systems on exposed runners; and metal pan systems for mechanical suspension. Many systems today are also engineered to distribute air and light as well as to control sound; also, some systems are rated as fire resistive assemblies.

Acoustical tiles are applied to ceilings and walls for the purpose of absorbing and deadening sound in offices, banks, and other spaces where the utmost quiet is desired and in auditoriums to obtain optimum reverberation time.

Acoustical tiles are made of many different materials, such as wood fiber, cane fiber, mineral wool, cork, specially processed mineral filaments, perforated metal units, and other insulating materials.

They are made in a variety of sizes, such as 12" x 12" (300 x 300 mm), 12" x 24" (300 x 600 mm), 24" x 24" (600 x 600 mm), 24" x 48" (600 x 1200 mm), etc., and vary from 1/2" (12.50 mm) to 2" (50 mm) thick, depending upon the materials used.

Acoustical tile may be installed over sound, dry plaster or concrete that is thoroughly seasoned, over gypsum board, or nailed or screwed to 1" x 3" wood furring strips or by means of mechanical suspension systems.

Acoustical tile work is simple to estimate, but because it is almost always factory cut and prefinished, it is extremely important to see that job conditions are the very best. Once installed, there is little opportunity to "touch up" the work. Call backs can quickly eat up profits as well as ruin reputations.

One of the most important conditions to check is temperature and humidity. These should always approximate the interior conditions that will exist when the building is occupied.

All plastering, poured roof decks, concrete and floor work should be complete and dry; all windows and doors in place and glazed. If weather demands it, the building heating, rather than space heating, should be turned on and kept on as though the building were occupied.

The trend today is to square edge tile, which minimizes joint visibility. However, tile work will always have slight unevenness in it, and if the job conditions include cove lights or high windows where light strikes the tile surface at a small angle, tile with beveled edges will help conceal any unevenness.

Budget Prices on Acoustical Tile. Most acoustical tile jobs are sublet to concerns specializing in this work. They have mechanics experienced in applying their particular brands of tile and can complete the job at less cost than a general contractor who purchases the tile and attempts to install them.

The following are approximate prices on the various classes of tile erected in place on fair size jobs as they might be quoted by a subcontractor including his overhead and profit.

Type of Installation	Price per Sq.Ft.	Price per Sq.M.
1/2" (12.50 mm) wood fiber tile surface applied	$ 1.84	$ 19.81
5/8" (15.62 mm) mineral tile suspended*	$ 2.40	$ 25.83
3/4" (18.75 mm) mineral tile suspended*	$ 2.64	$ 28.42
1" (25 mm) pad on steel pan suspended*	$ 4.00	$ 43.06
1" (25 mm) pad on alum. pan suspended*	$ 5.92	$ 63.72
5/8" (15.62 mm) mineral fiberboard suspended*	$ 1.60	$ 17.22
Air distributing ceilings-board* 2'x2' (0.61 x 0.61 m)	$ 2.32	$ 24.97
Air distributing ceilings tile* 12"x12" (300 x 300 mm)	$ 2.80	$ 30.14

Including hangers but not additional channel supports.

Acoustical Tile Directly Applied. When acoustical tile is cemented, four spots of cement or adhesive should be applied to each tile on each corner of 12" x 12" (300 x 300 mm) tiles. The spots should be placed as uniformly as possible with relation to the corners of the tile. A sufficient area on each corner of the tile should be primed prior to the application of the spot to receive the entire spot when it has been pressed to its final diameter. The spots of cement should be applied to these primed areas using a putty knife or trowel. The spot should be approximately the size and shape of a walnut, and should be nearly uniform in size and shape, except where thicker spots are required to level surfaces. They should stand out from the tile about 1/2" (12.50 mm) to 3/4" (18.75 mm) so as to contact the surface at about the same time. The diameter of the spot when pressed in place should be approximately 2-1/2" (62.50mm) to 2-3/4" (68.75mm) and about 1/8" (3.12 mm) thick.

In placing, the tile should be held with both hands so that when the pressure is applied it will be between the spots rather than outside them. The tile should be held as nearly parallel to its final position as possible and pressed to position with a firm pressure. A slight lateral sliding motion of the hands while setting is necessary to settle the cement into the ceiling surface. It is important that the tile be pressed to level when setting.

A crew of at least two workers is required to handle the installation efficiently, one on the scaffold applying the adhesive to the tile and the other placing the tiles on the ceiling.

1362

FINISHES

One gallon (3.79 liters) of acoustical cement is sufficient for approximately 60 sq.ft. (5.57 sq.m) of tile under average conditions, although this varies somewhat with the surface to which it is applied. Adhesive costs about $10.00 per gal. ($2.64 per liter).

The labor cost of placing acoustical tile on ceilings will vary with the size of the rooms or spaces where the tile are to be applied, pattern in which the tile are laid, and height of scaffolding.

The following quantities are based on 2 mechanics working together from a 4' (1.21 m) to 6' (1.82 m) scaffold and laying tile in plain square or ashlar designs:

Size of Acoustical Tile	Pcs. Per 8- Hr. Day	Mechanic Hrs. per 100 Sq.Ft.	Size of Acoustical Tile	Mechanic Hrs. per 10 Sq.M.
12 " x 12 "	600 - 700	2.50	300 x 300 mm	2.69
12 " x 24 "	400 - 450	2.00	300 x 600 mm	2.15

When acoustical tile are laid in herringbone pattern, it requires 30% to 40% more time than regular squares or common bond. Diagonal patterns require 20% to 25% more time than regular squares or common bond, while mixed ashlar designs require about 50% more time than straight squares or common bond.

When placing tile in auditoriums and other spaces requiring high scaffolding, additional scaffolding will have to be added to the cost of installing the tile. Large jobs are applied faster proportionately than small jobs.

Description	Rate	Total
100 sq.ft. (9.29 sq.m) mineral tile, 3/4" (18.75 mm)*	$ 1.05	$ 105.00
1.75-gal. (6.62 liters) adhesive	$11.93	$ 20.87
Cost 100 sq.ft.		$ 125.87
per sq.ft.		$ 1.26
per sq.m		$ 13.55

Cost of tile varies according to kind and thickness.

Labor Cost of 100 Sq.Ft. (9.29 Sq.M) 12" x 12" (300 x 300 mm)			
Description	Hours	Rate	Total
Mechanics	2.50	$35.04	$ 87.60
Cost per sq.ft.			$ 0.88
per sq.m			$ 9.43

Acoustical Tile On Suspended Systems

Lay-in type systems for 2' x 2' (0.61 x 0.61 m) and 2' x 4' (0.61 x 1.22 m) grids will have a material cost of around $0.85 and $0.75 per sq.ft. ($9.14 and $8.07 per sq.m) respectively, not including any channel supports if they are needed. One worker should install 700 sq.ft. (65.03 sq.m) per 8-hr.

1363

day of 2' x 4' (0.61 x 1.22 m) grid, 600 sq.ft. (55.74 sq.m) for 2' x 2' (0.61 x 0.61 m) grid.

Concealed spline systems are more time consuming. Material costs will run about $0.90 per sq.ft. ($9.69 per sq.m), and one worker will install 500 sq.ft. (46.45 sq.m) per day.

If channel carriers are required, the material cost will be about $0.25 per sq.ft. ($2.15 per sq.m), and one worker installs about 500 sq.ft. (46.45 sq.m) per 8-hr. day.

Labor Cost to Install 100 Sq.Ft. (9.29 Sq.M) 2' x 2' (0.61 x 0.61 M.) Lay-in-Grid			
Description	Hours	Rate	Total
Mechanics	1.30	$35.04	$ 45.55
Cost per sq.ft.			$ 0.46
per sq.m			$ 4.90

Labor Cost to Install 100 Sq.Ft. (9.29 Sq.M) 2' x 4' (0.61 x 1.21 M.) Lay-in-Grid			
Description	Hours	Rate	Total
Mechanics	1.20	$35.04	$ 42.05
Cost per sq.ft.			$ 0.42
per sq.m			$ 4.53

Labor Cost to Install 100 Sq.Ft. (9.29 Sq.M) 12 x 12 Z Bar System Grid			
Description	Hours	Rate	Total
Mechanics	1.60	$35.04	$ 56.06
Cost per sq.ft.			$ 0.56
per sq.m			$ 6.03

Labor Cost to Install 100 Sq.Ft. (9.29 Sq.M) Channel Ceiling Supports			
Description	Hours	Rate	Total
Mechanics	1.60	$35.04	$ 56.06
Cost per sq.ft.			$ 0.56
per sq.m			$ 6.03

Acoustical Tile For Suspended Grid Systems. This material is made "cut to size" ready to install into the grid system. The 24" x 24" (600 x 600 mm) and 24" x 48" (600 x 1200 mm) boards cost about $0.75 to 0.90 per sq.ft. ($8.07 to $9.69 per sq.m) depending on the type of tile, and one worker should install about 100 sq.ft. (9.29 sq.m) per hr. The 12" x 12" (300 x 300 mm) tiles for the Z bar system costs about $1.15 to $1.30 per sq.ft. ($12.37 to $14.00 per sq.m), and one worker should install about 100 sq.ft. (9.29 sq.m) in 2 hrs.

Estimating a Suspended Grid Ceiling. To estimate a suspended grid ceiling, it is necessary to determine the number of square feet (square meters) and lineal feet (lineal meters) of the ceiling.

1364

FINISHES

Sample Suspended Ceiling Estimate

Given: A room measures 18' (5.5m) wide by 36' (11m) long and needs a 2x4 (50x100) acoustical ceiling.

Number of Main Runners 12' (3.6m) long:
    Main runners will placed once every 4' (1.2m) in this layout.

    18' x 36' = 648 sq.ft. room *(5.5m x 11m = 60 sq.m.)*

    648 sq.ft. x 0.25 spacing ÷ 12' = 13.5 or **14 each Main Runners**
    *60 sq.m. ÷ 1.2m spacing ÷ 3.6m = 13.8 or **14 each Main Runners***

Number of 4' (1.2m) Tees:
    648 sq.ft. x 0.25 x 0.5 = **81 each 4' Tees**
    *60 sq.m. ÷ 1.2m ÷ 0.61m = **81 each 100mm Tees***

Number of 2'x4' (50mm x 100mm) Tiles:
    648 sq.ft. ÷ 8 sq.ft. = **81 each 2'x4' Tiles**
    *60 sq.m. ÷ 0.74 sq.m. = **81 each 50x100 Tiles***

Number of Wall Angle 12' (3.6m) long:
    (18' + 36') x 2 ÷ 12' = **9 each Wall Angle**
    *(5.5m + 11m) x 2 ÷ 3.6m = **9 each Wall Angle***

Number of Wire:
    648 sq.ft. ÷ 16 sq.ft. = 40.5 or **41 each Wire**
    *60 sq.m. ÷ 1.49 sq.m. = 40.2 or **41 each Wire***

Metal Pan Type Acoustical Tile. This type of acoustical treatment consists of metal units 12" x 24" (300 x 600 mm) in size, having center grooves which result in the appearance of 12" x 12" (300 x 300 mm) units when applied. The edges, which are beveled and returned vertically, are firmly held in place on the 12" (300 mm) sides by special tee runners. Within the metal units and supported on crimped, galvanized wire mesh, or on miniature channels, are mineral wool absorbent pads.

    These systems will install at the rate of around 250 sq.ft. (23.22 sq.m) per day, including sound absorbing pads, but without channel supports if they are required. Materials will cost about $3.00 for steel, $4.50 for aluminum, and $7.00 for stainless.

Labor Cost to Install 100 Sq.Ft. (9.29 Sq.M) Suspended Steel Pan System			
Description	Hours	Rate	Total
Mechanics	3.20	$35.04	$ 112.13
Cost per sq.ft.			$ 1.12
per sq.m			$ 12.07

09550 WOOD FLOORING

Estimating Quantities of Wood Flooring. When estimating the quantity of board feet (b.f.) of wood flooring required for any job, take the actual number of sq.ft. (sq.m) in any room or space to be floored, and add allowances as given in the following tables:

Measured Size, Inches	Finished Size, Inches	No. Pcs.in Bundle	Add for Waste Percentage	To Obtain Quantity Flooring Repaired Multiply area by	No. Bd. Ft. Flooring Required for 100 Sq.Ft. Floor
1 " x 1 "	3/8 " x 7/8 "	24	16-2/3	1 1/16 or 1.170	117
1 " x 2 "	3/8 " x 1 1/2 "	24	33-1/3	1 1/3 or 1.330	133
1 " x 2 1/2 "	3/8 " x 2 "	24	25	1 1/4 or 1.250	125
1 " x 2 1/4 "	25/32 " x 1 1/2 "	12	501/3	1 1/2 or 1.500	150
1 " x 2 3/4 "	25/32 " x 2 "	12	37-1/2	1 3/8 or 1.375	137 1/2
1 " x 3 "	25/32 " x 2 1/4 "	12	33-1/3	1 1/3 or 1.330	133
1 " x 4 "	25/32 " x 3 1/4 "	8	25	1 1/4 or 1.250	125

Measured Size, mm	Finished Size, mm	No. Pcs.in Bundle	Add for Waste Percentage	To Obtain Quantity Flooring Repaired Multiply area by	No. Cu.M. Flooring Required for 10 Sq.M. Floor
25 x 25.0 mm	9.38 x 21.88 mm	24	16-2/3	1 1/16 or 1.170	0.30
25 x 50.0 mm	9.38 x 37.50 mm	24	33-1/3	1 1/3 or 1.330	0.34
25 x 62.5 mm	9.38 x 50.00 mm	24	25	1 1/4 or 1.250	0.32
25 x 56.3 mm	19.53 x 37.50 mm	12	501/3	1 1/2 or 1.500	0.38
25 x 68.8 mm	19.53 x 50.00 mm	12	37-1/2	1 3/8 or 1.375	0.35
25 x 75.0 mm	19.53 x 56.25 mm	12	33-1/3	1 1/3 or 1.330	0.34
25 x 100.0 mm	19.53 x 81.25 mm	8	25	1 1/4 or 1.250	0.32

How Measures	Size Flooring	Covers Sq.Ft. Flooring	Nail Spacing, Inch	Nails Req'd.	
				Cut Nails	Helically Threaded Nails
1 x 2	3/8 x 1-1/2	750	8 "	20 lbs. 4 "d" cut	-----
1 x 2-1/2	3/8 x 2	800	8 "	17 lbs. 4 "d" cut	-----
1 x 2-1/4	25/32 x 1-1/2	667	12 "	70 lbs. 8 "d" cut	60 lbs 7 "d"
1 x 2-3/4	25/32 x 2	727	12 "	56 lbs. 8 "d" cut	47 lbs 7 "d"
1 x 3	25/32 x 2-1/4	750	10 "	64 lbs. 8 "d" cut	54 lbs 7 "d"
1 x 4	25/32 x 3-1/4	800	10 "	29 lbs. 8 "d" cut	24 lbs 7 "d"

Amount of Surface 1,000 Bd.Ft. of Flooring Will Cover and Quantity of Nails Required to Place It

FINISHES

Amount of Surface 2.36 Cu.M. of Flooring Will Cover and Quantity of Nails Required to Place It					
How Measures	Size Flooring	Covers Sq.m.	Nailed Spacing,	Nails Req'd.	
				Cut Nails	Helically
1 x 2	3/8 x 1-1/2	69.68	200 mm	9.07 Kg	-----
1 x 2-1/2	3/8 x 2	74.32	200 mm	7.71 Kg	-----
1 x 2-1/4	25/32 x 1-1/2	61.96	300 mm	31.75 Kg	27.22 Kg
1 x 2-3/4	25/32 x 2	67.54	300 mm	25.40 Kg	21.32 Kg
1 x 3	25/32 x 2-1/4	69.68	250 mm	29.03 Kg	24.49 Kg
1 x 4	25/32 x 3-1/4	74.32	250 mm	13.15 Kg	10.89 Kg

Drill flooring for nails, if possible, for best results. No predrilling is required for helically threaded nails. The above figures are based on laying the flooring straight across a rectangular room without producing any design whatever.

Rules for Grading Hardwood Flooring. In many instances the specifications of the architects and owners are confusing and not in accordance with the standard rules for grading as adopted by the manufacturers. The rules adopted by the National Oak Flooring Manufacturers Association (NOFMA) standardize flooring grades. Below is the NOFMA grading system so that it may be compared with the architect's specifications and the proper grade of flooring figured.

Clear Grade. The highest grade offering the most uniform look and is highly selected in order to obtain uniformity:

Defects: None allowed including knots, splits, checks, wormholes, excessive mineral streaks or contracsting sapwood.

Color: Color selected to disallow any piesces that are too dark/light or wildy grained in order to create a harmonious color/grain selection:

Lengths: In standard nested distribution packaging of 84" to 88" (2100 to 2200 mm) long, the minimum length average will be 36" (900 mm) and typically is slightly longer.

Select Grade. The second highest grade, which differs only from Clear grade in that Select grade allows a greater range of color/grain found in a species and includes the Clear grade that develops in the production run.

Defects: None allowed including knots, splits, checks, wormholes, excessive mineral streaks or contrasting sapwood.

Color: Minimally color selected but does disallow extreme dark/light or wildly grained pieces that contrast too strongly.

Lengths: In standard distribution packaging of 84" to 88" (2100 to 2200 mm) long, the length average will be 36" (900 mm).

Low Select Grade. The third highest grade, which contains all the pieces that are Select Grade in terms of lack of defects, but that have been selected out from the Clear and Select grades as having too much color/grain variation. This grade is available only on an accumulation basis as it develops

1367

as a small percentage of the grading when producing Clear/Select grades of flooring.

Defects: None allowed including knots, splits, checks, wormholes, excessive mineral streaks or contrasting sapwood.

Color: Contains all the dark and light colored pieces as well as wilder grained pieces that have been selected out from the Clear and Select grades.

Lengths: In standard nested distribution packaging of 84" to 88" (2100 to 2200 mm) long, the length average will be approximately 32" (800 mm).

Natural Grade. This grade allows most of a given Species color and grain range along with most of the natural character that develops in a wood.

Defects: Allows small knots, splits, checks, wormholes, mineral streaks and contrasting sapwood.

Color: Contains the full range of Color and Grain to be found in a species.

Lengths: In standard nested distribution packaging of 84" to 88" long, the length average will typically be approximately 26" to 28" (650 to 700 mm) long.

Rustic Grade. This grade allows even more of the natural character that develops in a wood than found in the Natural grade.

Defects: Allows more numerous and larger knots, splits, checks, wormholes, mineral streaks and contrasting sapwood than found in the Natural grade.

Color: Contains the full range of Color and Grain to be found in a species.

Lengths: In standard nested distribution packaging of 84" to 88" (2100 to 2200 mm) long, the length average will typically be approximately 22" to 26" (550 to 650 mm) long.

Graining Selections:

Mixed Grain - is a mixture of straight graining and flat sawn (with the "flower" figure of the wood) and is the typical unselected for graining that most wood flooring is offered in and unless specified otherwise is the default graining offered.

Quartered - is a selection of graining, whereby the flooring is selected for straight graining only. Quartered is also typically used in the oaks to denote the ray / fleck that is visible in some of the straight grained selection. Straight grain provides for more uniform coloring and a more formal, less busy look.

Rift - is a selection of graining whereby the flooring is selected for straight graining only. It is principally used in the oaks to denote the straight grained material without the ray / fleck that is visible in some of the straight grained selection.

Rift and Quartered - is principally used when describing the grain in oaks and is used to describe a straight grain selection which has pieces both with and without rays / flecks.

FINISHES

Flat Sawn w/straight grained pulled - this is the graining left over after all or most of the Quartered and Rift straight grain has been pulled to make that selection leaving mostly the flat sawn graining and a "busier" look.

Color Selections. Occasionally offered, color selections vary species to species but typically are selected light and selected dark selections. See Individual Species for color details.

Clear Grade Light Northern Hard Maple. This grade is selected for light color. The color tones in individual strips will vary somewhat, but after laying, it provides a luxurious "light" floor. It costs about $6.00 per sq.ft. ($64.58 per sq.m) for material.

Selected Clear Grade Amber Northern Hard Maple. This grade is selected for amber color. The color tones in individual strips will vary somewhat, but after laying, this grade provides a luxurious "amber" appearing floor. The cost is $5.75 per sq.ft. ($61.89 per sq.m) for material.

Oak Flooring - Quarter Sawed

Clear. The face shall be practically clear, admitting an average 3/8" (9.37 mm) of bright sap. The question of color shall not be considered. Bundles to be 2' and up. Average length is 3-3/4' (1.14 m).

Select. The face may contain sap, small streaks, pin worm holes, burls, slight imperfections in working, and small tight knots that do not average more than one to every 3' (0.92 m). Bundles to be 2' (0.61 m) and up. Average length is 3-1/4' (0.98 m).

Plain Sawed

Clear. The face shall be practically clear, admitting an average of 3/8" (9.37 mm) bright sap. The question of color shall not be considered. Bundles to be 2' (0.6 m) and up. Average length 3-3/4' (1.14 m).

Select. The face may contain sap, small streaks, pin worm holes, burls, slight imperfections in working, and small tight knots which do not average more than one to every 3' (0.92 m). Bundles to be 2' (0.61 m) and up. Average length 3-1/4' (0.98 m).

Standard Thicknesses and Widths

25/32" (19.53 mm) thickness; widths 1-1/2" (37.50 mm) face, 2" (50 mm) face, 2-1/4" (56.25 mm) face, and 3-1/4" (81.25 mm) face.

3/8" (9.37 mm) thickness; width 1-1/2" (37.50 mm) face and 2" (50 mm) face.

1/2" (12.50 mm) thickness; width 1-1/2" (37.50 mm) face and 2" (50 mm) face.

Above tongue and groove and end matched.

Square Edge Strip Flooring

Grades as shown above but bundling and lineals as follows: 5/16x 2" (7.81 x 50 mm), 5/16x 1-1/2" (7.81 x 37.50 mm), 5/16x 1-1/3" (7.81 x 33.25 mm), 5/16x1-1/4" (7.81 x 31.25 mm), 5/16x 1-1/8" (7.81 x 28.12 mm), 5/16x 1" (7.81 x 25 mm), 5/16x 7/8" (7.81 x 21.87 mm). Also made rough back in 11/32x 2" (8.59 x 50 mm) and 11/32x 11 1/2" (8.59 x 37.50 mm).

Clear. Bundled 2' (0.61 m) and up. Average length 5-1/2' (1.67 m).

Select. Bundled 2' (0.61 m) and up. Average length 4-1/2' (1.37 m).

No. 1 Common. Bundled 2' (0.61 m) and up. Average length 3-1/2' (1.06 m).

No. 2 Common. May contain defects of all characters, but will lay a serviceable floor.

Bundles to be 1-1/4' (0.38 m) and up. Average length 2-1/2' (0.76 m). All faces shown above in 5/16" (7.81 mm) square edge are finished 1/64" (0.39 mm) over face.

Approximate Prices of Oak Flooring

Oak flooring prices will vary widely throughout the country. A recent quote gives 25/32" x 2-1/4 (19.53 x 56.25 mm) oak at $5.00 per sq.ft. ($53.82 per sq.m), select, $4.50 per sq.ft. ($48.44 per sq.m) common. Clear will run 30% more than select, if it is at all available.

Grading Rules on Prefinished Hardwood Flooring

Oak. White and Red Oak to be separated in each grade. Grades are established after the flooring has been sanded and finished.

Prime Grade. Face shall be selected for appearance after finishing, but sapwood and the natural variations of color are permitted. Minimum average length 4' (1.22 m). Bundles 2' (0.61 m) and longer.

Standard Grade. Will contain sound wood characteristics which are even and smooth after filling and finishing and will lay a sound floor without cutting. Minimum average length 3' (0.92 m). Bundles 1-1/4' (0.38 m) and longer.

Standard and Better Grade. A combination of Prime and Standard to contain the full product of the board except that no pieces are to be lower than Standard Grade. Minimum average length 3-1/2' (1.06 m). Bundles 1-1/4' (0.38 m) and longer.

Tavern Grade. Shall be of such nature as will make and lay a serviceable floor without cutting, but purposely containing typical wood characteristics, which are to be properly filled and finished. Minimum average length 2-1/2' (0.76 m). Bundles 1-1/4' (0.38 m) and longer.

Beech and Pecan

(Will be furnished only in a combination grade of Tavern and Better)

FINISHES

Tavern and Better. A combination of Prime, Standard, and Tavern to contain the full product of the board, except that no pieces are to be lower than Tavern Grade. Minimum average length 3' (0.9 m). Bundles 1-1/4' (0.38 m) and longer.

General Rules for All Species

Hardwood flooring is bundled by averaging the lengths. A bundle may include pieces from 6" (150 mm) under to 6" (150 mm) over the nominal length of the bundle. No piece shorter than 9" (225 mm) admitted.

The percentages under 4' (1.22 m) referred to apply on total footage in any one shipment of the item.

A 3/4" (18.75mm) allowance shall be added to the face length when measuring the length of each piece of flooring.

Flooring shall not be considered of standard grade unless the lumber from which the flooring is manufactured has been properly kiln dried.

Laying and Finishing Wood Floors

The following pages contain detailed itemized costs of laying and finishing all kinds of soft and hardwood floors.

Description of Work	Sq.Ft.per 8-Hr. Day	Carp. Hours	Labor Hours	Size in Metric mm	Sq.M.per 8-Hr. Day
Labor Laying 100 Sq.Ft. (9.29 Sq.M) of Hardwood Floors					
Ordinary Workmanship					
25/32 " x 3 1/4 " Face Softwood floors for porches, kitchens, factories, stores, etc.	400 - 500	1.8	0.6	19.53 x 81.25 mm	37.2 - 46.5
25/32 " x 2 1/4 " Face Third Grade Maple, for warehouse, factory and loft building floors	375 - 425	2	0.6	19.53 x 56.25 mm	34.8 - 39.5
25/32 " x 2 1/4 " Face Oak or Birch in residences, apartments, stores, offices, etc	250 - 300	2.9	0.6	19.53 x 56.25 mm	23.2 - 27.9
Same as above laid by experienced floor-layer	300 - 350	2.5	0.6		27.9 - 32.5
25/32 " x 1 1/2 " Face Third Grade Maple for warehouse, factory and loft building floors	300 - 325	2.3	0.7	19.53 x 37.50 mm	27.9 - 30.2
25/32 " x 1 1/2 " Face Oak or Birch in residences, apartments, stores, offices, etc	150 - 200	4.5	0.8	19.53 x 37.50 mm	13.9 - 18.6
Same as above laid by experienced floor-layer	225 - 275	3.2	0.8		20.9 - 25.5
First Grade Workmanship					
25/32 " x 2 1/4 " Face Oak or Birch in fine residences, apartments, hotels, stores and offices	200 - 225	3.8	0.6	19.53 x 56.25 mm	18.6 - 20.9
Same as above laid by experienced floor-layer	225 - 250	3.3	0.6		20.9 - 23.2
25/32 " x 1 1/2 " Face Oak or Birch, same class of work as described above	120 - 150	6	0.8	19.53 x 37.50 mm	11.1 - 13.9
Same as above laid by experienced floor-layer	175 - 200	4.3	0.8		16.3 - 18.6

In many cities there are carpenters who make a specialty of laying and finishing floors, and they seldom do any other kind of work. Due to their long experience, they not only do a better job but accomplish more work per day than the average carpenter, and they expect to be paid accordingly. For this reason, production times are given for floors laid and finished by both carpenters and floorlayers.

Another reason some floorlayers will lay more floors than carpenters is because they do not nail the flooring 8" (200 mm) or 16" (400 mm) o.c. but

1371

will nail at every other bearing unless watched continually. The use of air-activated nailing guns has help to eliminate some of this problem due to the ease of using a nail gun.

The first or Ordinary Workmanship includes the grade of work found in moderate priced homes, apartments, stores, etc., where the floor laying permits some hammer marks. Considerable savings can be obtained by laying the strip flooring all the way through prior to the erection of partitions.

First Grade Workmanship includes all classes of high class buildings where the best in workmanship is required, with all flooring closely driven up, laid free from hammer marks, and with all nails set.

Labor Laying 3/8" (9.37 mm) or 1/2" (12.50 mm) Hardwood Flooring. Figure same costs as given for 25/32" (19.53 mm) flooring.

Setting Sleepers in Mastic on Concrete Floors. A sound, fast, economical way to provide nailing for strip wood flooring installed over concrete subfloors is the sleeper-in-mastic method, used extensively in the construction of slab on grade houses.

Assuming concrete slab is smooth, level, and dampproofed to prevent moisture seepage, 2" x 4" (50 x 100 mm) wood sleepers are laid flat in hot or cold mastic and are immediately ready to receive strip flooring.

After priming concrete slab, hot or cold mastic is spread to a 3/32" (2.34 mm) depth over the entire slab or in 1/4" (6.25 mm) deep rivers along parallel lines in positions where sleepers are to be placed.

Wood sleepers should be straight, flat, dry 2" x 4" (50 x 100 mm) lumber in random lengths from 18" (450 mm) to 48" (1200 mm). For best results, sleepers should be treated with an approved non-creosote wood preservative. Lay sleepers in mastic, flat side down, in staggered rows 12" (300 mm) o.c. at right angles to direction of flooring. Ends should overlap 4" (100 mm) and end joints should be staggered by alternating short and long lengths. Leave 1" (25 mm) clearance between sleepers and all vertical surfaces to allow for normal expansion.

Laying strip wood flooring over sleepers in mastic should follow the same general procedure as for over wood subfloors with the following additional precautions:

1. Use end-matched flooring only.
2. Individual strips of flooring should span at least two sleeper spacings, with no adjacent joints occurring in the same spacing.
3. Flooring strips passing over sleeper laps should be nailed to both sleepers.

Material coverages and approximate prices are as follows:

Asphalt primer, 200 to 400 sq.ft. per gal. (4.90-9.80 liters per sq.m), depending on porosity of concrete surface. Price $7.20 per gal. ($1.90 per liter).

Hot mastic, 25 sq.ft. per gal. (0.61 liters per sq.m). Price $5.60 per gal. ($1.48 per liter).

1372

FINISHES

Cold mastic, 40 to 50 sq.ft. per gal. (0.98-1.23 sq.m per liter). Price $5.20 per gal. ($19.70per liter).

For average size rooms, 100 sq.ft. (9.29 sq.m) of floor area will require 120 to 135 lin.ft. (36.57 to 41.14) or 80 to 90 bd.ft. (0.19-0.21 cu.m) of 2" x 4" (50 x 100 mm) wood sleepers.

Under ordinary conditions, a floorlayer should prime concrete floor, spread mastic, and place wood sleepers at the rate of 100 sq.ft. (9.29 sq.m) per hour.

Labor Laying 1,000 bd.ft (2.25 cu.m) Soft and Hardwood Flooring					
Description of Work	Bd.ft. Lumber per 8-Hr Day	Carp. Hours	Labor Hours	Size in Metric mm	Cu.M. Lumber per 8-Hr Day
Ordinary Workmanship					
25/32 " x 3 1/4 " Face softwood Floors for porches, kitchens, factories, stores, etc.	500 - 600	14.50	4.00	19.53 x 81.25 mm	1.18 - 1.42
25/32 " x 2 1/4 " Face Third Grade maple, used in warehouses, factory and Loft buildings, etc	500 - 575	15	4	19.53 x 56.25 mm	1.18 - 1.36
25/32 " x 2 1/4 " Face Birch or Oak flooring in residences, apartments, stores, etc.	350 - 400	21	4	19.53 x 56.25 mm	0.83 - 0.94
Same as above, laid by experienced floorlayer	400 - 450	19	4		0.94 - 1.06
25/32 " x 1 1/2 " Face Third Grade Maple, used in warehouses, factory and loft buildings,	450 - 500	17	4	19.53 x 37.50 mm	1.06 - 1.18
25/32 " x 1 1/2 " Face Birch or Oak flooring in residences, apartments, store and office buildings, etc	250 - 280	30	5	19.53 x 37.50 mm	0.59 - 0.66
Same as above laid by experienced floorlayer	350 - 400	21	5		0.83 - 0.94
First Grade Workmanship					
25/32 " x 1 1/4 " Face Oak or Birch flooring in fine residences, apartments, hotels, stores and offices	275 - 300	28	4	19.53 x 31.25 mm	0.65 - 0.71
Same as above laid by experienced floorlayer	325 - 375	23	4		0.77 - 0.88
25/32 " x 1 1/2 " Face Oak or Birch, same class of work as described above for 2-1/4" (56.25 mm) face flooring	175 - 225	39	5	19.53 x 37.50 mm	0.41 - 0.53
Same as above laid by experienced floorlayer	275 - 300	28	5		0.65 - 0.71

Sanding Wood Floors by Machine

Practically all wood floors today are sanded and finished by machine. The cost varies, depending on the kind of flooring, whether old or new work, class of workmanship, size of rooms, etc.

The floors are first sanded with a floor sanding machine and then the edges are finished, using a disc type edging machine, which is capable of sanding up to the base shoe or quarter round.

The class of workmanship of number of operations governs production and cost, but where just an ordinary grade of workmanship is required in average size rooms in houses, apartments, offices, etc., a good machine and operator should finish and edge 800 to 900 sq.ft. (74.32 to 83.61 sq.m) per 8-hr. day, at the following labor cost per 100 sq.ft. (9.29 sq.m):

Description	Hours	Rate	Total
Operator-finisher	1.00	$35.04	$35.04
Cost per sq.ft.			$0.35
per sq.m			$3.77

When used in store rooms, auditoriums, and other large floor areas, a machine and operator should finish and edge 1,000 to 1,250 sq.ft. (92.90 to 116.12 sq.m) of new floor per 8-hr. day, at the following labor cost per 100 sq.ft. (9.29 sq.m):

Description	Hours	Rate	Total
Operator-finisher	0.75	$35.04	$26.28
Cost per sq.ft.			$0.26
per sq.m			$2.83

Surfacing Floors by Machine, First Grade Workmanship. In residences, apartments, hotel, store and office buildings, etc., where first grade workmanship is required, eliminating all irregularities and waves in the floor, about four cuts are necessary: first with No. 1-1/2 or 2 grit sandpaper, then with No. 1/2 or 1 grit paper and the last two cuts with No. 0 paper.

On work of this kind an experienced operator should finish and edge 400 to 500 sq.ft. (37.16 to 46.5 sq.m) of floor per 8-hr. day, at the following labor cost per 100 sq.ft. (9.29 sq.m):

Description	Hours	Rate	Total
Operator-finisher	1.75	$35.04	$61.32
Cost per sq.ft.			$0.61
per sq.m			$6.60

Resurfacing Old Floors by Machine. When resurfacing old wood floors, including the removal of old varnish, dirt, grease spots, etc., the cost will vary with the condition of the old floors. This variation will run from 300 to 1,000 sq.ft. (27.87 to 92.90 sq.m) per 8-hr. day and should be determined by the estimator after examining the old floors.

There is little or no difference in resurfacing an old floor or surfacing a new one except that under normal conditions one extra cut with a special open-faced abrasive is usually necessary to remove old varnish or paint.

Sanding Wood Floors

The following specifications list the operations necessary to produce a first class job. Cheaper grades of workmanship will require less sanding and consequently less labor—and lower costs.

Sanding Old Wood Floors Laid in Strips. The first operation is to remove the surface finish by sanding in the direction of the grain of the wood until all surface finish has been removed, using a drum type sanding machine, with extra heavy open coat sandpaper grit No. 3-1/2 or 4.

1374

FINISHES

The second operation is to remove the effects of the first operation and level the floor. Sand as in first operation, using same type machine with sharp sandpaper, grit No. 2 or 2-1/2, to remove the effects of the coarse open coat sandpaper and level the floor surface.

The third operation is to use a spinner along the edges of the floor and hand scrape all corners where the machines could not reach. Use a No. 3-1/2 or 4 grit paper and repeat, using No. 2 or 2-1/2 grit paper.

The fourth operation is to sweep the entire floor to remove the coarse grits and abrasives left by the previous operations.

The fifth operation is to sand as in the first operation using the same type machine with No. 0 or No. 1/2 fine grit.

The sixth operation is the final smoothing of the edges. Sand as in third operation using same type of machine with fine sandpaper, grit 0 or 1/2.

After sweeping the entire floor clean it is ready for any kind of floor finish.

Sanding New Wood Floors Laid in Strips. First, sand to a level surface. Sand in the direction of the grain of the wood until all the high edges or over-wood on the floor has been sanded off to a level surface, using a drum type sanding machine with No. 1-1/2 or 2 grit sandpaper.

The following chart gives the proper abrasives or grits for surfacing all classes of wood floors:

Second, remove the scratch pattern of the first sanding. Sand in the direction of the grain of the wood until the scratch pattern of the first operation is eliminated using the same type sanding machine with No. 1 or 1/2 sandpaper. Frequent changes of sandpaper must be made to insure cutting rather than sliding over the surface.

Third, use a spinner around the edges and scrape by hand where it is impossible to use the machine, without crossing the grain of the wood. Use same grits as on drum sander.

Fourth, sweep the entire floor free from any grit and abrasives and then sand in direction of the grain of the wood, using the drum type machine as in the first operation, with No. 0 sandpaper. Change the sandpaper frequently to insure cutting rather than sliding over the surface.

After sweeping the floor clean it is ready for any kind of floor finish.

Sanding Old Parquet and Herringbone Wood Floors. The first operation is to sand lengthwise of the room or at 45 degrees, using a drum type sanding machine with No. 3-1/2 or 4 open coat grit, using a very light pressure on the drum.

Second, remove the coarse marks of the first operation by sanding crosswise of the room using same type machine with No. 1 or 1-1/2 grit. Change paper frequently to assure cutting rather than sliding over the surface. Then sand lengthwise of the room using No. 1/2 or No. 1 grit.

Third, use a portable edging machine or hand scrape all places where the drum machine and edger cannot reach. Sweep the entire floor to remove coarse grits and abrasives left by the previous operations.

Fourth, use a large diameter rotary sanding machine across and then lengthwise of the room with No. 1/2 sandpaper to remove the cross marks of previous operations.

The fifth and final operation is to sweep the entire floor free from all grits, then sand using the large diameter rotary sander with No. 00 or No. 0 sandpaper, going over the floor once across and once lengthwise of the room. After sweeping the floor clean it is ready for any kind of floor finish.

Finishing Hardwood Floors

The modern and approved method of finishing hardwood floors is to use wood sealers, which are a special compound of ingredients blended so that they penetrate into the fiber and cells of the wood, sealing the pores with tough elastic materials, thereby building up a base resistance to water, dirt, and stains. It wears only as the wood wears down.

Chart for Floor Surfacing Operations			
New Floors			
Kind of floors	Operation	Floor Conditions	Proper Grit
Oak, Maple and Close Grained Hardwood Floors	Roughing	Ordinary Floors	2 - 2 1/2
		Well Laid Floors	1 1/2
		Very Uneven Floors	2 1/2
		Very Hard Floors	2 1/2
	Finishing	Ordinary Finish	1/2
		Fine Finish	0
		Extra Fine finish	0
		Rough Finish	1
Parquet Floor	Roughing	-----	1 1/4
	Finishing	Semi Finish	1/2
		Finishing	0
Old Floors			
Resurfacing Old Floors, Soft and Hard Woods	Removing Varnish, Paints, etc.	Ordinary Conditions	3 1/2
		Extra Heavy Coat Paint or Varnish	4
	Roughing	Ordinary Floors	2 - 2 1/2
	Finishing	Ordinary Finish	1/2
		Fine Finish	0
		For Penetrating Sealer	00

Finishing Hardwood Floors in Natural Color. Before finishing hardwood floors, they should be sanded as described above.

After the floor has been completely sanded and swept clean of all dust, a good grade of permanent penetrating seal should be applied. One such seal on the market is Pentra-Seal (from the American-Lincoln Corp., Toledo, OH). It can be applied quickly with a brush, rag, mop, lamb's wool applicator, or squeegee and does not leave brush or lap marks.

1376

FINISHES

For an extra fast job, Fast-Drying Seal is recommended. Apply a thin, even coat and after it has become tack free remove the excess with a No. 2 steel wool pad. Sweep the floor clean and apply a second coat in the same manner. Where a highly glossy finish is desired, apply a third coat and do not use steel wool. This will remain on the surface, since the preceding coats have thoroughly filled every cell and pore, and result in a glossy surface finish.

The floor should now be given a final protective coat of a good grade of paste wax or Lustre Finish. Paste wax should be applied manually and Lustre Finish can be put on with a mop. Two coats are recommended, followed by buffing with a mixed fiber brush.

Fast-Drying Seal is available in natural as well as seven decorator type colors. Coverage ranges from 400 to 450 sq.ft. per gal. (9.5-10.7 sq.m per liter) for the first coat and 650 to 750 sq.ft. per gal. (15.5-17.8 sq.m per liter) for the second coat. Price, Fast-Drying Sealatural, about $30.00 per gal. ($6.613 per liter); colors, slightly higher.

Coverage of Lustre-Finish is about 1,250 sq.ft. per gal. (29.7 sq.m per liter). Price, about $15.00 per gal. ($3.90 per liter).

Labor Finishing Wood Floors. Assuming the floor has been sanded and is ready for the seal, an experienced worker can finish about 1,000 sq.ft. (92.90 sq.m) of natural finish floor per 8-hr. day. Colored seal takes more time as the excess pigment must be removed by hand wiping.

Labor Cost of 100 Sq.Ft. (9.29 Sq.M) of Fast-Drying Seal and Lustre Finish Floor Finish on Wood Floors, Natural Color			
Description	Hours	Rate	Total
Mechanic	0.80	$30.99	$ 24.79
Cost per sq.ft.			$ 0.25
per sq.m			$ 2.67

Prefinished Hardwood Plank Flooring

Plank or strip flooring in 3/4" (18.75 mm) thickness is available in several widths and finishes, most furnished in random lengths and widths. It is generally of red oak and predrilled for nailing. Widths are in the 3" (75 mm) to 8" (200 mm) range. Special distressed finishes and pegging are also available.

Plain flooring in narrower widths will cost about $4.50 per sq.ft. ($48.44 per sq.m). For pegging, add $0.20 per sq.ft. ($2.15 sq.m). Wider planks will run $5.75 per sq.ft. ($61.89 per sq.m). Premium finishes can raise the cost to as much as $6.00 per sq.ft. ($64.59 per sq.m).

One carpenter should lay at least 150 sq.ft. (13.93 sq.m) per day.

Wood Parquet Flooring

Wood parquet is available in stock designs of oak, walnut, teak, cherry, and maple. Standard thickness is 5/16" (7.81 mm), although heavier parquet is available on custom orders in 11/16" (17.18 mm) and 13/16" (20.31 mm) thicknesses. Sizes will vary with patterns running from 9" x 9" (225 x 225 mm), 12" x 12" (300 x 300 mm), 16" x 16" (400 x 400 mm), on up to 36" x 36" (900 x 900 mm).

Parquet is set in mastic on any level floor above grade. Most parquet is ordered factory finished, but may be unfinished if desired. Many patterns are available in "prime grade" and "character marked", square or bevelled edges.

Costs of all wood flooring are changing rapidly, but cost approximate $2.55 per sq.ft. ($27.44 per sq.m) for a standard 5/16" (7.81 mm) oak, $6.00 per sq.ft. ($64.58 per sq.m) for walnut, and $5.75 per sq.ft. ($61.89 per sq.m) for teak. The prime grades in more elaborate patterns will run much higher. If prefinished, add $0.75 per sq.ft. ($8.07 per sq.m).

Laying Wood Parquet Floors in Mastic Over Concrete. Concrete subfloor shall be primed with one coat of asphalt priming paint and allowed to dry before laying parquet. Average covering capacity of primer 200 to 400 sq.ft. per gal. (4.75-9.5 sq.m per liter) depending on porosity of cement floor finish. Price $8.00 per gal. in 5-gal. cans ($2.11 per liter in 19-liter can).

Where concrete floors are in direct contact with the ground, below grade, or where there is not sufficient ventilation to prevent condensation, a membrane waterproofing shall be applied to the slabs, extending same up the sides of all walls, 4" (100 mm) to 6" (150 mm), according to manufacturer's specifications. Where wood parquet is to be laid in hot mastic, a hot application, two-ply membrane shall be applied. Where parquet is to be laid in cold mastic, a cold application one-ply polyethylene film shall be applied.

Installation of polyethylene film is as follows: after concrete surface is primed, cold mastic is applied at the rate of 80 to 100 sq.ft. per gal. (1.9 to 2.4 sq.m per liter), into which polyethylene film is rolled, lapping joints 4 inches (100 mm). Laps do not have to be sealed.

Parquet shall be laid to a level surface bedded in a suitable mastic without the use of nails or screws. One gallon of mastic will cover 45 to 50 sq.ft. (1 liter covers 1.0 to1.2 sq.m) and costs about $20.000 per gal. ($5.28 per liter). Expansion joints must be left around all edges of rooms according to manufacturer's specifications.

Labor Laying Wood Parquet Flooring. When laying wood parquet floors in mastic in ordinary size rooms such as residences, apartments, etc., an experienced floorlayer with a helper should lay 300 to 350 sq.ft. (27.87 to 32.51 sq.m) of floor per 8 hr. day, while in large spaces such as schools, auditoriums, gymnasiums, etc., an experienced floorlayer should lay 350 to 400 sq.ft. (32.51 to 37.16 sq.m) of floor per 8-hr. day. This applies to either finished or unfinished parquet, as the same care is required in laying either type of flooring. This does not include any time preparing or leveling old concrete floors, as this must be added extra if required.

Where unfinished parquet is used, add for sanding and finishing.

Laying Wood Parquet Floors in Mastic over Other Surfaces. Wood parquet flooring may also be laid in mastic over plywood or board subflooring, present wood floors, asphalt tile, etc., using the same general procedures as given above for over concrete, except subsurface will not require priming before applying mastic.

Subsurfaces must be sound, smooth, clean, and dry-asphalt tile must be firmly bonded and in good condition-wood subflooring or present wood flooring may require re-nailing. Where installation is over square edged board subflooring, one layer of 15-lb. (6.75-kg) asphalt saturated felt should be laid to keep mastic from seeping through joints.

Prices of Wood Parquet Floors. Owing to the variation in freight rates, dealer's markup, etc., it is practically impossible to give prices that will apply to all parts of the country. Dealer's markup will probably vary from 25% on large jobs to 40% on smaller installations, plus freight from factory to destination.

Material Cost of 100 Sq.Ft. (9.29 Sq.M) of 9"x 9"x 5/16" (225 x 225 x 7.81 mm) Oak Parquet Floors Laid in Mastic		
Description	Rate	Total
0.33 gal. (1.25 L) asphalt primer	$ 4.24	$ 1.40
100 sq.ft. (9 sq.m) oak parquet	$ 2.88	$ 288.00
2.5 gal. (9.5 L) hot mastic	$20.40	$ 51.00
Cost per 100 sq.ft		$ 340.40
per sq.ft.		$ 3.40
per sq.m		$ 36.64

Labor Cost of 100 Sq.Ft. (9.29 Sq.M) of 9"x 9"x 5/16" (225 x 225 x 7.81 mm) Oak Parquet Floors Laid in Mastic			
Description	Hours	Rate	Total
Floorlayer	3.00	$35.04	$ 105.12
Helper	3.00	$26.16	$ 78.48
Cost per 100 sq.ft.			$ 183.60
per sq.ft.			$ 1.84
per sq.m			$ 19.76

If wood parquet is laid over wood or asphalt tile surfaces, omit asphalt primer.

If parquet floors are laid in large spaces, such as auditoriums, etc., deduct 1/4 to 1/2 hr. floorlayer time per 100 sq.ft. (9.29 sq.m).

Add for membrane waterproofing, if required.

If unfinished parquet is used, change price of blocks and add for sanding and finish.

Above costs do not include any allowance for preparing subfloor, if necessary.

Laminated Block Flooring. Laminated oak blocks are fabricated from 3-ply all oak plywood. Furnished in 9" x 9"x 1/2" size, factory finished.

A 27 sq.ft. (2.50 sq.m) carton—48 pcs.—will cost about $55.00 in a light or dark finish.

Laminated block flooring is for mastic installation only and may be laid over any sound, smooth, clean and dry subfloor, using hot or cold mastic. General installation procedure is the same as for wood parquet floors given on previous pages, except that no allowance for expansion is necessary.

Labor costs for laying laminated block flooring should be about the same as for 9" x 9" x 5/16" (225 x 225 x 8 mm) wood parquet given above.

Wood Block Industrial Flooring. Creosoted wood blocks 2" (50 mm) to 3" (75 mm) thick will run from $2.60 to $6.00 per sq.ft. ($27.98 to $64.58 per sq.m). One carpenter will install around 200 sq.ft. (18.58 sq.m) per day.

09650 RESILIENT FLOORING

Resilient flooring includes, rubber, cork, vinyl asbestos, pure vinyl tile, and vinyl sheet flooring. Each type has its own qualities and should be selected on these considerations rather than cost alone. If material is to be installed on or below grade, cork tile and certain sheet vinyls are not suitable.

If grease is a problem, rubber, cork tile perform poorly, while if resilience and quietness are important, cork and rubber tile are excellent, but sheet vinyl is poor.

A quick guide for approximate installed costs per sq.ft. (sq. meter) is as follows:

Type	Thickness, Inches	Price per Sq.Ft.	Thickness, mm	Price per Sq.M
Cork tile-Prefinished	1/8"; 3/16"; 5/16"	$3.45 - $5.30	3.13;4.68;7.81	$37.14 - $57.05
Vinyl tile	0.05"	$1.54 - $2.20	1.25 mm	$16.58 - $23.68
Vinyl sheet	0.070" - 0.075"	$3.40 - $5.30	1.75 - 1.87 mm	$36.60 - $57.05
Rubber tile	1/8"; 3/16"	$3.20 - $3.90	3.13;4.68 mm	$34.45 - $41.98

How to Estimate Tile Floors

To Determine and Price the Quantity of Material Required. Compute the actual square footage of surface to be covered, and add a percentage for waste. The percentage will depend on the size of the area and regularity of outline. A fair percentage of waste to add to the actual area is shown in the accompanying table.

After the gross area has been determined, multiply it by the square foot price of tile. The result will be the price of the material.

The square foot (square meter) price of the material will depend on the colors and quantity.

1380

FINISHES

Table of Approximate Percentages of Allowable Waste for Various Areas			
Less then 50 Sq.Ft.	14.00%	Less then 4.65 Sq.M.	
50 - 100 Sq.Ft.	10.00%	4.65 - 9.29 Sq.M.	
100 - 200 Sq.Ft.	7.00% - 9.00%	9.29 - 18.58 Sq.M.	
200 - 300 Sq.Ft.	6.00% - 7.00%	18.58 - 27.87 Sq.M.	
500 - 1,000 Sq.Ft.	4.00% - 5.00%	46.45 - 92.90 Sq.M.	
1,000 - 5,000 Sq.Ft.	4.00%	92.90 - 464.50 Sq.M.	
5,000 - 10,000 Sq.Ft.	3.00%	464.50 - 929.00 Sq.M.	
10,000 Sq. ft. and up	1.50% - 3.00%	929.00 Sq.M. and up	

Quantity of one gallon of adhesive material should lay 175 to 225 sq.ft. (one liter lays 4.29 to 5.5.2 sq.m) over concrete floors. Costs about $18.00 per gal. ($4.75 per liter).

Labor Laying Tile. The cost of laying tile varies with the size and shape of the room or space where the tile are to be laid, size of tile, kind of subfloor, etc., as a tile setter will lay more sq.ft. (sq.m) of tile in a large room than one cut up with pilasters, offsets, etc., also a worker lays more sq.ft. (sq.m) of large size tile than of the smaller sizes.

If felt lining is required, placing felt over wood subfloors, an experienced worker should spread paste, lay, and roll 250 sq.ft. (23.22 sq.m) of felt per hr., depending on the room sizes.

After the lining felt has been laid (over wood floors), an experienced tile setter should lay the following quantities of tile per 8-hr. day:

Size of Tile, Inches	Sq.Ft. Per 8-Hr. Day	Size of Tile, mm	Sq.M. Per 8-Hr. Day
9 " x 9 "	400 - 450	225 x 225 mm	37.16 - 41.81
12 " x 12 "	425 - 575	300 x 300 mm	39.48 - 53.42

The above quantities are based on fairly large rooms. If laid in small spaces, increase the labor costs accordingly.

Vinyl Tile

Vinyl tile is a more expensive resilient flooring material. It is very stable, will not shrink or expand and is highly resistant to oils, fats, greases, acids, alkalis, detergents, soaps, and most solvents.

Furnished in 9" x 9" (225 x 225 mm), 12" x 12" (300 x 300 mm), 18" x 36" (450 x 900 mm) sizes and in 3/32" (1.56mm) and 1/8" (3.12 mm) thicknesses.

Subfloor requirements are the same as for other resilient flooring. May also be laid over concrete floors, or on grade or below grade, using special adhesives.

Vinyl tile is laid in the same manner as other tile and labor costs are about the same.

Material costs vary according to pattern, ranging from $0.75 per sq.ft. ($8.33 per sq.m) for marbleized patterns to $5.00 per sq.ft. ($53.82 per sq.m) for translucent patterns, for 1/8" (3.12 mm) thick tiles. The 3/32" (1.56mm) thick tiles cost about 10% to 15% less.

Rubber Tile Floors and Base

Rubber tile is furnished in 9" x 9" (225 x 225 mm), 12" x 12" (300 x 300 mm), and 18" x 36" (450 x 900 mm) tile, 1/8" (3.12 mm) thick and in a variety of colors.

Rubber floor tile may be laid over wood or above grade concrete floors. It may also be laid over clean, dry, concrete floors on ground or below grade providing a special adhesive, recommended by the manufacturer of the tile, is used.

When laying rubber tile over an old wood floor, all loose, defective, or badly worn boards shall be replaced and face nailed. Any unevenness in the boards shall be planed or sanded smooth and even.

If the wood floors cannot be placed in proper condition by these methods, install Masonite or 1/4" (6 mm) plywood over the original wood floor.

Over the wood base there shall be installed semi-saturated lining felt which shall be securely affixed to the base with mastic. Depending upon the condition of the subfloor, the spreading capacity should be about 140 sq.ft. per gal. (3.43 sq.m per liter). After the felt is laid, it should be thoroughly rolled with a three-section 100-lb. (45-kg) linoleum roller.

The tile should be installed over the felt, using mastic.

The finished floor should be thoroughly rolled with a three-section 100-lb. (45 kg) linoleum roller and any high edges smoothed in place with a flat headed hammer.

Laying Rubber Tile. The cost of laying rubber floor tile varies with the size and shape of the room or space where the tile are to be laid, size of tile, kind of subfloor, etc., as it is easier to lay tile in a large square room than one cut up with offsets, pilasters, bays, etc., and a worker will lay more sq.ft. (sq.m) of large size tile than of the smaller sizes.

When laying lining felt over wood subfloors, an experienced worker should spread paste, lay, and roll 250 sq.ft. (23.22 sq.m) of felt per hr., depending on the room sizes.

After the felt has been laid (over wood floors), an experienced tile setter should lay and finish the following sq.ft. (sq.m) of tile of various sizes per 8-hr. day:

1382

FINISHES

Size of Tile Inches	Sq.Ft. per 8-Hr. Day	Size of Tile mm	Sq.M. per 8-Hr. Day
9 " x 9 "	400 - 450	225 x 225 mm	37.16 - 41.81
12 " x 12 "	425 - 550	300 x 300 mm	39.48 - 51.10
18 " x 36 "	600 - 675	450 x 900 mm	55.74 - 62.71

The above quantities are based on fairly large rooms. If laid in small spaces, increase the labor costs accordingly.

Material Prices of Rubber Floor Tile, All Standard Colors			
Thickness	Price per Sq.Ft.	Thickness	Price per Sq.M.
1/8 "	$ 2.10 - $ 3.60	3.13 mm	$22.60 - $38.75
1/4 "	$ 4.00 - $ 4.75	6.25 mm	$43.06 - $51.13

Rubber tile has lost much of its market to the vinyl's but is still used as a stair tread covering.

Stair Treads						
Size Inches	Plain Marbleized per Sq.Ft.	Plain Black per Sq.Ft.	Size mm	Plain Marbleized per Sq.M.	Plain Black per Sq.M.
12 1/2 " x 36 " x 1/4 "	$ 21.00	$ 20.00	312.5 x 900 x 6.25 mm	$ 226.05	$ 215.29
12 1/2 " x 42 " x 1/4 "	$ 24.50	$ 21.00	312.5 x 1,050 x 6.25 mm	$ 263.72	$ 226.05
12 1/2 " x 48 " x 1/4 "	$ 28.50	$ 26.00	312.5 x 1,200 x 6.25 mm	$ 306.78	$ 279.87
12 1/2 " x 54 " x 1/4 "	$ 30.50	$ 29.00	312.5 x 1,350 x 6.25 mm	$ 328.31	$ 312.16
12 1/2 " x 60 " x 1/4 "	$ 33.00	$ 32.00	312.5 x 1,500 x 6.25 mm	$ 355.22	$ 344.46
12 1/2 " x 72 " x 1/4 "	$ 42.00	$ 40.00	312.5 x 1,800 x 6.25 mm	$ 452.10	$ 430.57

Landing tile will run about $2.20 per sq.ft. ($23.68 per sq.m) and 3/32" (2.34 mm) riser material about $1.30 per lin.ft. ($4.26 per m).

Description	Price per Lin.Ft.	Description	Price per Lin.M.
4 " base	$ 1.00	100 mm base	$ 3.28
6 " base	$ 1.10	150 mm base	$ 3.61

Cork Tile

Cork tile are made from high-grade pure cork shavings compressed and baked in molds. It is used for floors and wainscoting in offices, libraries, hospitals, schools, and other places where footsteps must be muffled.

They are furnished in 6" x 12" (150 x 300 mm), 9" x 9" (225 x 225 mm), and 12" x 12" (300 x 300 mm) size 1/8" (3.12 mm), 3/16" (4.68 mm), and 5/16" (7.81 mm) thick. Cork tile are supplied beveled or unbeveled. With unbeveled cork tile, sanding the finished floor is usually necessary to offset unevenness of the subfloor. With beveled cork tile, the beveling conceals such irregularities, with the result that sanding is unnecessary and the smooth surface given at the tile factory is retained.

Cork tile should never be placed over a single wood subfloor but should be first covered with Masonite or 3/8" (9.37 mm) plywood. Never install on grade or subgrade floors.

The wood or above grade concrete subfloor must be smooth and even before cork tile can be placed. Lining felt is used over wood subfloors but this is not necessary when applied over smooth, level concrete floors.

After the floor has been laid and sanded (if unbeveled tile are used), the floor should be given a coat of cork tile filler, followed by two coats of cork tile finish (glossy or flat) and subsequent waxings with an approved water emulsion wax, or the floor may be finished by merely applying 3 applications of water emulsion wax and vigorously buffed by machine.

Approximate Sq. Ft. (Sq.M) Prices of Cork Tile				
Kind of Tile	Thickness Inches	Price per Sq.Ft.	Thickness mm	Price per Sq.M.
Color according to	1/8"	$ 3.50	3.12 mm	$ 37.67
Manufacturer's Specifications,	3/16"	$ 4.00	4.69 mm	$ 43.06
Regular Sizes	5/16"	$ 4.75	7.81 mm	$ 51.13

Border feature strips: In any width under 3" (75 mm) and any length between 18" (450 mm) and 36" (900 mm) at $0.20 per sq.ft. ($2.15 per sq.m) additional.

Factory finished tile: Add $0.25 per sq.ft. ($2.69 per sq.m) for beveled cork tile finished at factory with sealer, undercoat, and one application of wax. After installation, apply one application of wax and buff.

Material Cost of 100 Sq.Ft. (9.29 Sq.M) 9" x 9" x 1/8" (225 x 225 x 3.12 mm) Cork Floor Tiles Over Wood Flooring		
Description	Rate	Total
0.75-gal. (2.84 liters) adhesive	$ 8.50	$ 6.38
12 sq.yds. (10.03 sq.m.) lining felt	$ 0.32	$ 3.84
105 sq.ft. (9.75 sq.m.) cork tile	$ 0.95	$ 99.75
0.75 gal. (2.84 liters) adhesive finishing floor	$ 8.50	$ 6.38
0.33 gal. (1.25 liters) cork tile sealer	$ 3.18	$ 1.05
0.25 gal. (0.95 liters) cork tile undercoat	$ 3.18	$ 0.80
0.13 gal. (0.49 liters) water emul. wax	$ 3.18	$ 0.41
Cost 100 sq.ft.		$ 118.60
per sq.ft.		$ 1.19
per sq.m		$ 12.77

FINISHES

Labor Cost of 100 Sq.Ft. (9.29 Sq.M) 9" x 9" x 1/8" (225 x 225 x 3.12 mm) Cork Floor Tiles Over Wood Flooring			
Description	Hours	Rate	Total
Mechanic	3.50	$35.04	$ 122.64
Sanding 100 sq.ft finishing Floor	1.00	$35.04	$ 35.04
Clean and wax floor	2.00	$35.04	$ 70.08
Cost 100 sq.ft.			$ 227.76
per sq.ft.			$ 2.28
per sq.m			$ 24.52

Sheet Vinyl Flooring

Sheet vinyl can be installed on concrete floors, below and above grade, and on wooden floors that have a smooth surface. If the wood floor is rough, then install a layer of hardboard or 1/4" (6.25 mm) plywood to provide a good surface for the sheet vinyl.

This material is available in 6' (1.82 m) and 12' (3.65 m) wide rolls up to 120' (36.57 m) in length. Thicknesses vary with the product design, type of backing material, and the manufacturer. However, the most recognized gauges will fall in the 0.065" (1.62 mm), .080" (2.00 mm) and 0.090" (2.25 mm) range for non-cushion backed sheet vinyl.

Most sheet vinyl flooring materials are designed to require a minimum of maintenance, which means that the flooring cleans easily, is resistant to normal abuse, and does not require waxing to maintain the luster appearance.

Installation is by full adhesive coating on the 6' (1.82 m) wide material and adhesive coating on seams only for the 12' (3.65 m) wide rolls.

Material prices will vary from $2.00 to $4.00 per sq.ft. ($21.53 to $43.05 per sq.m) depending on the gauge and backing, but the estimator should be aware that this material is sold by the square yard (square meter) and that proper floor layout should be determined to reduce the waste factor.

Labor for installation will vary depending on the size and shape of the area to be covered and the number of cuts to be made. One worker lays about 600 sq.ft. (55.74 sq.m) in large areas per 8-hr. day, and about 350 sq.ft. (32.51 sq.m) in smaller areas per 8-hr. day.

09680 CARPETING

With the development of man-made fibers special padding and backings, carpeting has become competitive with resilient (vinyl) tile in commercial construction. While almost always subcontracted to a firm specializing in the field, the general contractor should be acquainted with the basic choices available today.

Pure wool has set the high standards most people expect from carpet and no man-made fiber has yet equaled it in overall performance and appearance, although in some respects, such as abrasion resistance, texture retention, and mildew resistance, man-made fibers excel. But wool carpeting is in the high cost range, running $30.00 to $60.00 per sq.yd. ($35.88 to

$71.76 per sq.m) in the 36 to 42 oz. (1021-1191 g) commercial grades, without padding.

Nylon is the most often used of the man-made fibers. It has outstanding wear ability and resiliency, is mildew resistant and mothproof. But there are problems with static electricity unless it is specially treated or has grounding mesh woven into the backing. In the lighter weights, it is extremely economical, about $10.00 per sq.yd. ($11.96 per sq.m) for 15 oz. (425 g); $12.00 ($14.35 per sq.m.) for 22 oz. (624 g), both without padding.

The olefin (polypropylene) fibers are comparable in price to nylon, but they are found mostly in the indoor-outdoor type of carpet as they offer the lowest moisture absorbency and are exceptionally stain-resistant. Their appearance is dull, resiliency low; they have the lowest melting point, so are subject to cigarette burns. Their range is $10.00 to $20.00 per sq.yd. ($11.96 to $23.92 per sq.m).

Acrylics surpass the other man-made fibers in looks and feel, most closely resembling wool, but they do not wear as well as nylon, and will pill some. They are $12.00 to $ 19.00 per sq.yd. ($14.35 to $22.72 per sq.m) for 36 to 42 oz. (1020-1191 g) weights, without padding.

Polyesters have poor resiliency so are seldom used alone, except for shag carpets, seldom encountered in contract work. They fall in a medium cost range. Padding may be laid separately or come with the carpet. It will add $1.60 to $4.50 per sq.yd. ($1.91 to $5.38 per sq.m) to the material cost.

Carpets can be taped down or laid in mastic. Installation cost will be about $4.50 per sq.yd. ($5.38 per sq.m) on open work. If not integral with carpet, add another $2.25 per sq.yd. ($2.69 per sq.m) for laying padding.

Estimating Quantities. Carpet is estimated by the square yard (square meter). There are 9 sq.ft. per sq.yd. Since most carpeting is manufactured in 12' (3.6 m) wide material, the estimator must be careful to determine how the carpet will lay on the floor to minimize the waste factor. Bear in mind that carpet has a definite "lay" to the nap and the "lay" must be in the same direction when joining pieces together, or else the carpet pieces will appear to have a different color.

In commercial use, carpet tiles glued to a concrete floor surface is becoming very popular. It has the ease of placement like vinyl tile and can be replace easily on a tile by tile basis. The most popular size of carpet tile is 18" x 18" (450 x 450 mm). The placement by an experienced tradesperson should range from 125 to 200 sq.yds. (104 to 167.22 sq.m) per day depending upon cuts and pattern.

09900 PAINTING

Estimates and bids on painting and interior decorating vary more than for almost any other construction trade. There are several kinds of painters-good, fair, poor, and just plain daubers. Expect wide variation in figures. When preparing painting estimates, consider the quality of materials and workmanship required.

FINISHES

Methods of Measuring and Listing Painting Quantities

When estimating quantities of painting or interior finishing, the actual surface to be painted should be measured as accurately as possible from the plans or from measurements taken at the building. From these measurements it should be possible to estimate the quantity of materials required, but the labor quantities present a much more difficult problem, owing to the different classes of work and the difficulties encountered in applying the paint. For instance, the covering capacities of paint will be the same for a plain surface or a cornice, but the labor cost painting the cornice will be considerably more owing to the height of the cornice above the ground and the amount of "cutting" or "trimming" necessary. Care must be used when pricing any piece of work, as conditions on each job are different.

Clapboard or Drop Siding Walls. Obtain actual area of all walls and gables. Add 10% to actual surface measurement to allow for painting under edge of boards. Do not deduct for openings less than 10'-0" x 10'-0" (3.04 x 3.04 m).

Shingle Siding. Obtain actual area of all walls and gables and multiply by 1-1/2. Do not deduct for openings less than 10'-0" x 10'-0" (3.04 x 3.04 m).

Brick, Wood, Stucco, Cement, and Stone Walls. Obtain actual area of all walls and gables. Do not deduct for openings less than 10'-0" x 10'-0" (3.04 x 3.04 m).

Eaves. Plain eaves painted same colors as side walls, obtain area and multiply by 1-1/2. If eaves are painted a different color than side walls, obtain area and multiply by 2.

Eaves with rafters running through, obtain area and multiply by 3.

Eaves over brick, stucco or stone walls, obtain area and multiply by 3.

For eaves over 30'-0" (9.14 m) above ground, add 1/2 for each additional 10'-0" (3.04 m) in height.

Cornices, Exterior. Plain cornices, obtain area and multiply by 2. Fancy cornices or cornices containing dentils, etc., obtain area and multiply by 3.

Downspouts and Gutters. Plain downspouts and gutters measure the area and multiply by 2. For fancy downspouts and gutters, obtain area and multiply by 3.

Blinds and Shutters. Plain blinds or shutters, measure one side and multiply by 2. Slatted blinds or shutters, measure one side and multiply by 4.

Columns and Pilasters. Plain columns or pilasters, obtain area in sq.ft. (sq.m); if fluted, obtain area and multiply by 1-1/2; if paneled, obtain area and multiply by 2.

Lattice Work. Measure one side and multiply by 2 if painted one side only; if painted two sides, obtain area and multiply by 4.

Porch Rail, Balustrade, Balusters. If solid balustrade, add one foot to height to cover top and bottom rail. Multiply the length by the height. If painted two sides, multiply by 2.

1387

If balustrade consists of individual balusters and handrail, multiply length by the height, then multiply the result by 4.

Handrail only. If under 1'-0" (0.30 m) girth, figure as 1'-0" (0.30 m) by length and multiply by 2.

Moldings. Cut in on both sides, figure 1 sq.ft. per lin.ft. (0.09 sq.m per m) if under 12" (300 mm) girth. If over 12" (300 mm) girth, take actual measurement.

Doors and Frames, Exterior. Inasmuch as it costs almost as much to paint a small door as a large one, do not figure any door less than 3'-0"x7'-0" (0.91 x 2.13 m). Allowing for frame, add 2'-0" (0.61 m) to the width and 1'-0" (0.30 m) to the height. For instance, a 3'-0"x7'-0" (0.91 x 2.13 m) door would be figured as 5'-0"x8'-0" (1.52 x 2.43 m), or 40 sq.ft. (3.71 sq.m).

If a sash door, containing small lights of glass, add 2 sq.ft. (0.18 sq.m) for each additional light. A 4-lt. door would contain 8 sq.ft. (0.74 sq.m) additional; a 12-lt. door, 24 sq.ft. (2.22 sq.m) additional, etc.

If painted on both sides, obtain sq.ft. (sq.m) area of one side and multiply by 2.

Door frames only, where no door is hung, allow area of opening to take care of both sides.

Windows, Exterior. Inasmuch as it costs almost as much to paint a small window as a large one, do not figure any window less than 3'-0"x6'0" (0.91 x 1.82 m). Add 2'-0" (0.61 m) to both the width and height of the opening, to take care of the sides and head of the frame and the outside casing or brick mold, and multiply to obtain the area. For instance, a window opening 3'-0"x6'-0" (0.91 x 1.82 m), add 2'-0" (0.61 m) to both the width and height, making 5'-0"x8'-0" (1.52 x 2.43 m), containing 40 sq.ft. (3.6 sq.m) of surface.

If sash contain more than one light each, such as casement sash, etc., add 2 sq.ft. (0.18 sq.m) for each additional light. A 6-lt. window would contain 12 sq.ft. (1.11 sq.m) additional; a 12-lt. window, 24 sq.ft. (2.22 sq.m) additional, etc.

Roofs. For flat roofs or nearly flat, measure actual area. For roofs having a quarter pitch, measure actual area and add 25%; roofs having a one-third pitch, measure actual area and add 33-1/3%; roofs having a half pitch, measure actual area and add 50 percent.

Fences. Plain fences, measure one side and multiply by 2; picket fences, measure one side and multiply by 4.

Picture Mold and Chair Rail. On picture mold, chair rail, etc., less than 6" (150 mm) wide, obtain the number of lin.ft. (meters) to be painted or varnished and figure 3/4 sq.ft. per lin.ft. (0.22 sq.m per m).

Wood and Metal Base. Wood or metal base 6" (150 mm) to 1'-0" (0.30 m) high should be figured as 1'-0" (0.30 m) high; under 6" (150 mm), figure at 0.5 sq.ft. per lin.ft. (0.15 sq.m per m).

FINISHES

Wood Panel Strips, Cornices, Etc. When estimating quantities of wood panel strips, wainscot rail, and ceiling cornices, measure the girth of the member and if less than 1'-0" (0.30 m) wide, figure as 1'-0" (0.30 m).

If over 1'-0" (0.30 m) wide, multiply the actual girth by the length: 1'-3" x 200'-0" = 250 sq.ft. (0.38 x 60.96 = 23.16 sq.m).

Refer to rule for exterior cornices.

Interior Doors, Jambs, and Casings. When estimating quantities for interior doors, jambs, and casings, add 2'-0" (0.61 m) to the width and 1'-0" (0.30 m) to the height of the opening. This allows for painting or varnishing the edges of the door, the door jambs which are usually 6" (150 mm) wide, and the casings on each side of the door which average from 4" (100 mm) to 6" (150 mm) wide. Example: on a 3'-0"x7'-0" (0.91 x 2.13 m) door opening, add 2'-0" (0.61 m) to the width and 1'-0" (0.30 m) to the height, which gives an opening 5'-0"x 8'-0" (1.52 x 2.43 m), containing 40 sq.ft. (3.71 sq.m) on each side. Some painters figure all single doors at 40 sq.ft. (3.71 sq.m) per side, or 80 sq.ft. (7.43 sq.m) for both sides, while others figure them at 50 sq.ft. (4.64 sq.m) per side or 100 sq.ft. (9.29 sq.m) for both sides. Do not deduct for glass in doors.

If a sash door, containing small lights of glass, add 2 sq.ft. (0.18 sq.m) for each additional light. A 4-lt. door would contain 8 sq.ft. (0.74 sq.m) additional; a 12-lt. door 24 sq.ft. (2.22 sq.m) additional, etc.

Interior Windows, Jamb Linings, Sash, and Casings. When estimating painting quantities for windows and window trim, add 2'-0" (0.61 m) to the sides and length to allow for jamb linings, casing at the top and sides, and window stool and apron at the bottom. Example: If the window opening is 3'-0"x 6'-0" (0.91 x 1.82 m), adding 2'-0" (0.61 m) to both the width and length gives a window 5'-0"x8'-0" (1.52 x 2.43 m), containing 40 sq.ft. (3.71 sq.m) of surface.

If sash contain more than one light each, such as casement sash, etc., add 2 sq.ft. (0.18 sq.m) for each additional light. A 6-lt. window would contain 12 sq.ft. (1.11 sq.m) additional; a 12-lt. window, 24 sq.ft. (2.22 sq.m) additional, etc.

Stairs. When estimating quantities of paint or varnish for wood stairs, add 2'-0" (0.61 m) to the length of treads and risers to allow for stair strings on each side of the stairs. A stair tread is 10" (250 mm) to 12" (300 mm) wide, riser is about 7" (175 mm) high, and the average cove and underside of a stair tread 2" (50 mm) in girth, so each tread should be figured 2'-0" (0.61 m) wide. Multiply the width by the length to obtain the number of sq.ft. (sq.m) of paint or varnish required for each step. Multiply the area of each step by the number of treads in the stair, and the result will be the number of sq.ft. (sq.m) of finish. Example: A flight of stairs containing 20 treads, 4'-0" (1.21 m) long. Adding 2'-0" (0.61 m) to the length of the treads to allow for the stringers, gives 6 lin.ft. (1.82 m) in each tread, which, multiplied by the girth of 2'-0" (0.61 m) gives 12 sq.ft. (1.11 sq.m) of surface in each step.

Since there are 20 treads in the stairs, 12 x 20 = 240 sq.ft. (3.65 x 6.09 = 22.23).

For painting the soffits of stairs, use same rules of measurement as given above.

Balustrades and Handrails with Balusters. When estimating balustrades around well holes or stair handrail and balusters, measure the distance from the top of the treads to the top of the handrail and add 6" (150 mm) for painting or varnishing the handrail. Multiply the height of the balustrade by the length and the result will be the number of sq.ft. (sq.m) of balustrade. An easy method for computing the length of the stairs is to allow 1 lin.ft. (0.30 m) in length for each tread in the stairs. Example: If the handrail is 2'-6" (0.76 m) above the treads and the stairs contain 20 treads, add 6" (150 mm) to the height of 2'-6" (0.76 m) to take care of the extra work on the handrail proper, making a total height of 3'-0" (0.91 m). 3 x 20 = 60 sq.ft. (0.91 x 6.09 = 5.54 sq.m) of surface for one side of the balustrade or 120 sq.ft. (11.14 sq.m) for two sides. The balustrade around the stair well hole is easily estimated by multiplying the height by the length.

On fancy balustrades having square or turned spindles, which require considerable additional labor, use actual measurements as given above and multiply by 4 to allow for extra labor.

Wood Ceilings. When estimating quantities for wood ceilings, multiply the length by the width. Do not deduct for openings less than 10'-0"x10'-0" (3.04 x 3.04 m).

Wainscoting. Plain wainscoting, obtain actual area. Paneled wainscoting, obtain actual area and multiply by 2.

Floors. To compute the quantities of floors to be finished, multiply the length of each room by the width, and the result will be the number of sq. ft. (sq.m) of floor to be finished.

Plastered Walls and Ceilings. To obtain the area of any ceiling, multiply the length by the breadth and the result will be the number of sq.ft. (sq.m) of ceiling. When estimating walls, measure the entire distance around the room and multiply by the room height. The result will be the number of sq.ft. (sq.m) of wall to be decorated. For instance, a 12'-0"x15'-0" (3.65 x 4.57 m) room has two sides 12'-0" (3.65 m) long and two sides 15'-0" (4.57 m) long, giving a total of 54 lin.ft. (16.452 m) If the ceilings are 9'-0" (2.74 m) high, 54 x 9 = 486 sq.ft. (16.45 x 2.7 4= 45.07 sq.m) area of the walls. Do not deduct for door and window openings.

Cases, Cupboards, Bookcases, Etc. When estimating surfaces of cupboards, wardrobes, bookcases, closets, etc., to be painted or varnished, obtain the area of the front and multiply by 3 if the cases do not have doors.

If the cases have doors, obtain the area of the front and multiply by 5.

This takes care of painting or varnishing doors inside and out, shelves 2 sides, cabinet ends and backs.

Radiators. For each front foot, multiply the face area by 7.

FINISHES

Sanding and Puttying Interior Trim. Sanding and puttying on high grade interior trim should be figured as one coat of paint or varnish; on medium grade trim, figure at 50% of one coat of paint or varnish. For industrial work, figure about 25% of the cost of one coat of paint or varnish.

Painting Concrete and Masonry Surfaces

Hydrocide 600, 700 700b (Sonneborn, degusa), is a cold-applied water base emulsified-asphalt dampproofing and vapor-retarding coatings for use on "green" or slightly damp surfacs Properly applied, speeds dampprooofing of new foundation walls. Minimizes fire hazards during application and withstands normal expansion and contraction, therefore suitable for most climates.

One gallon of material covers 70 to 100 sq.ft. depending on porosity of the surface. (One liter covers 1.43 to 1.66 sq.m). Price, $15.00 per gal. ($3.96 per liter).

SikaGard Hi-bild (Sika). Tough water base coating for concrete, block, steel, wood, and other structural materials. Can be applied to dry, damp, or wet surfaces. Coverage with short nap roller will be approximately 175 to 200 sq.ft. per gal. (4.16-4.75 sq.m per liter); when spraying, 250 to 400 sq.ft. per gal. (5.9-9.5 sq.m per liter). Coverage varies with temperature, substrate, and environment. Available in colors. Price, $35.00 per gal. ($9.24 per liter).

Covering Capacity of Paints

It is difficult to say how many sq. ft. of surface per one gallon (or square meters of surface per one liter) paint will cover, because there are several items that influence the covering capacity. By brushing the paint out very thin, it will cover more surface than when applied thick. Dark paint will hide the surface better than light paint, and for that reason it can be brushed out thinner than the lighter colors. Also, a rough surface will require considerable more paint than a smooth surface.

Soft and porous wood will absorb more oil and require more paint than close grained lumber.

Another thing that must be considered are the ingredients entering into the paint. Different brands of paint vary in covering capacity or hiding power.

Covering Capacity of Oil Base Paint on New Exterior Wood. In figuring the number of square feet a gallon of oil base paint will cover, a great deal depends upon the surface to be painted; that is, the kind of wood, the degree of roughness, etc. Some woods are more porous than others and absorb more paint. Much depends on the way the paint is brushed out as some painters brush the paint out more and thus cover more surface.

The priming coat, properly applied, should cover 450 to 500 sq.ft. per gal. (11.04 to 12.27 sq.m per liter). The second coat should cover 500 to 550 sq.ft. per gal. (12.27 to 13.50 sq.m per liter), and the third coat should cover 575 to 625 sq.ft. per gal. (14.11 to 15.34 sq.m per liter).

When estimating exterior trim painting, with measurements taken in accordance with standard methods, the priming coat should cover 750 to 850 sq.ft. per gal. (18.40 to 20.86 sq.m per liter); the second coat should cover 800 to 900 sq.ft. per gal. (19.63 to 22.08 sq.m per liter), and the third coat should cover 900 to 1,000 sq.ft. per gal. (22.08 to 24.54 sq.m per liter).

Bear in mind the above coverages are based on surfaced lumber, as the covering capacity will be greatly reduced when applied to rough boards.

Painting Old Exterior Wood Surfaces. When estimating repaint work on old exterior wood surfaces, if the existing surface is sound and in good condition, two coats should be sufficient to give a good job and paint coverages should be about the same as for the second and third coats on similar new work.

If the old surface shows cracking, blistering, scaling, and peeling, the old paint should be removed by using either a paste or liquid paint remover of the slow drying type, or a blow torch and scraper.

If the old paint is removed completely, three coats of paint should be applied the same as recommended for new work.

Painting Wood Shingle Siding. Wood shingle siding should receive three coats of paint on new work. Paint coverages should be: first coat, 300 to 325 sq.ft. per gal. (7.36 to 7.97 sq.m per liter); second coat, 400 to 450 sq.ft. per gal. (9.81 to 11.04 sq.m per liter); third coat, 500 to 550 sq.ft. per gal. (12.27 to 13.50 sq.m per liter).

Wood shingles which have been previously painted with an oil paint and are in suitable condition for repainting, should receive two coats of paint. The first coat should cover 400 to 450 sq.ft. per gal. (9.81 to 11.84 sq.m per liter), and the second coat 500 to 550 sq.ft. per gal. (12.27 to 13.50 sq.m per liter).

Painting Exterior Wood Floors and Steps. When painting exterior wood floors and steps with oil base paint, material coverages should be: first coat, 325 to 375 sq.ft. (7.97 to 9.20 sq.m per liter); second coat, 475 to 525 sq.ft. (11.65 to 12.88 sq.m per liter); third coat, 525 to 575 sq.ft. (12.88 to 14.11 sq.m per liter).

Painting Interior Wood. New interior wood should receive three coats of paint, a priming coat, a second coat and a final or finish coat.

Paint coverage for interior wood surfaces will vary with the type of work, such as doors and windows, running trim, paneling, etc., in accordance with the standard method of measuring the quantities.

For doors and windows, coverage should be: first coat, 475 to 500 sq.ft. per gal. (11.3-11.9 sq.m per liter); second coat, 500 to 550 sq.ft. per gal. (12.27 to 13.50 sq.m per liter); third coat, 500 to 550 sq.ft. per gal. (12.27 to 13.50 sq.m per liter).

For running trim up to 6" (150 mm) wide, coverage should be: first coat, 800 to 900 lin.ft. per gal. (64.42 to 72.47 m per liter); second and third coats, 1,100 to 1,200 lin.ft. per gal. (88.58 to 96.63 m per liter).

FINISHES

For ordinary flat wood paneling, coverage should be: first coat, 450 to 500 sq.ft. per gal. (11.84 to 12.27 sq.m per liter); second and third coats, 500 to 550 sq.ft. per gal. (12.27 to 13.50 sq.m per liter).

When painting interior woodwork that has been previously painted and is in good condition, one gallon of paint should cover approximately the same as given for the second and third coats on new wood.

Painting Brick, Stone, Stucco, and Concrete. When painting masonry surfaces that are reasonably smooth, oil base paint should cover 170 to 200 sq.ft. per gal. (4.17 to 4.91 sq.m per liter) for the priming coat; 350 to 400 sq.ft. per gal. (8.59 to 9.81 sq.m per liter) for the second coat and 375 to 425 sq.ft. per gal. (9.20 to 10.43 sq.m per liter) for the third coat.

These quantities are subject to considerable variation due to the porosity of the surfaces to which the paint is applied, smoothness of walls, etc., as a rough porous surface will require considerably more paint than a smooth hard surface.

Painting Metal Work. The area that any paint may be expected to cover on metal work will vary with the surface to be painted. Badly pitted or rough metal will require more paint than a perfectly smooth surface. The covering capacity will also vary with the temperature, consistency of the paint and the effort behind the brush.

However, for ordinary smooth surfaces, one gallon should cover 550 to 650 sq.ft. (one liter covers 13-15.5 sq.m), one coat.

Aluminum Paint. Aluminum paint is used for both priming and finishing coats on metal surfaces. The covering capacity varies with the condition of the surface, but on smooth surfaces, one gallon covers 600 to 700 sq.ft. (one liter covers 14.72 to 17.18 sq.m) with one coat.

Painting Interior Walls. When painting interior walls, the covering capacity of the paint will depend on the surface to be painted, whether smooth or sand finished plaster, porous or hard finished wallboard, etc.

Three coats are recommended for interior plaster that has never been painted: a priming coat, a second or body coat, and a third or finishing coat. However, if a two-coat job on unpainted plaster is desired, use a good wall primer or sealer for the first coat, followed by the second or finishing coat. To make two coats hide better, tint the first to nearly the same color as the second coat. If the surface has been painted before and the old paint is still in good condition, two coats are sufficient, no priming coat being required.

On smooth plaster or hard finished wallboard, wall primer or sealer should cover 575 to 625 sq.ft. per gal. (14.11 to 15.34 sq.m per liter) for the priming coat. For the second coat, figure 500 to 550 sq.ft. per gal. (12.27 to 13.49 sq.m per liter) and for the third coat, 575 to 625 sq.ft. per gal. (14.11 to 15.34 sq.m per liter).

On rough, porous sand finished plaster, wall primer or sealer might cover only 275 to 300 sq.ft. per gal. (6.74 to 7.36 sq.m per liter), while on very porous wallboard, it may be only 150 to 200 sq.ft. per gal. (3.68 to 4.90 sq.m per liter).

For the second and third coats on rough surfaced plaster or wallboard, figure 400 to 475 sq.ft. per gal. (9.81 to 11.65 sq.m per liter).

One Coat Wall Finishes. There are any number of "one-coat" wall finishes on the market, but when applied to new plastered surfaces or over surfaces that have been previously unpainted, it is advisable to use a primer or wall seal coat, which seals the pores in the plaster or other surface and provides a base suitable to receive the one coat finish. Fully described above.

Latex Base Paints. There are any number of latex base paints on the market that are compatible with most surfaces.

When used over smooth plastered surfaces, it should cover 450 to 500 sq.ft. per gal. (11.04 to 12.27-11.9 sq.m per liter) first coat and 550 to 650 sq.ft. per gal. (13.49 to 15.95 sq.m per liter) second coat.

When applied over rough sand finish plaster, it should cover 325 to 350 sq.ft. (7.97 to 8.59 sq.m per liter) per coat.

Covering Capacity of Wall Size. The covering capacity of wall size will vary greatly with the material used.

Wall size consisting of ground or flake glue and water will cover 600 to 700 sq.ft. per gal. (14.72 to 17.18 sq. per liter) of surface.

Varnish size consisting of varnish, benzine, or turpentine and a little paint, will cover 450 to 550 sq.ft. per gal. (11.04 to 13.50 sq.m per liter).

Hard oil or gloss oil size consisting of rosin and benzine will cover 450 to 500 sq.ft. per gal. (11.04 to 12.27 sq.m per liter).

The covering capacity of prepared wall primers and sealers is given on previous pages under Painting Interior Walls.

Shingle Stain. When staining wood shingles after they have been laid, stain should cover 120 to 150 sq.ft. per gal. (2.94 to 3.68 sq.m per liter) of surface, for the first coat.

The second and third coats will go farther, because the wood does not absorb as much stain and should cover 200 to 225 sq.ft. per gal. (4.90 to 5.52 sq.m per liter). Stain should cover about 70 sq.ft. per gal. (1.71 sq.m per liter) of roof with 2 coats, or 50 sq.ft. per gal. (1.22 sq.m per liter) with 3 coats.

If the shingles are dipped in stain before laying, it will require 3 to 3-1/2 gals. (11.35 to 13.24 liters) of stain per 1,000 shingles. When dipping shingles in stain, only two-thirds of the shingle need be dipped, as it is unnecessary to dip the end of the shingle that is not exposed.

Covering Capacity of Oil and Spirit Stains. When staining finishing lumber, such as birch, mahogany, oak, gum, etc., oil stain should cover 700 to 725 sq.ft. per gal. (17.18 to 17.79 sq.m per liter).

One gallon of spirit stain should cover 500 to 600 sq.ft. (12.27 to 14.73 sq.m per liter).

Covering Capacity of Wood Fillers. Liquid filler is normally used with close grained woods, such as pine, birch, etc., Coverage should be 500 to 550 sq.ft. per gal. (12.27 to 13.49 sq.m per liter).

It is necessary to use a paste filler to fill the pores of all open grained wood, such as oak, ash, walnut, mahogany, etc.

FINISHES

If paste filler is used on oak, it should cover about 450 sq.ft. per gal. (11.04 sq.m per liter).

Covering Capacity of Shellac. The covering capacity of shellac varies with its purity. Shellac should cover 550 to 750 sq.ft. per gal. (13.49 to 18.41 sq.m per liter), depending on whether first, second, or third coat.

Covering Capacity of Varnish. On softwood floors, varnish should cover 400 to 450 sq.ft. per gal. (9.81 to 11.04 sq.m per liter) with one coat; 200 to 225 sq.ft. per gal. (4.90 to 5.52 sq.m per liter) with 2 coats, and 135 to 150 sq.ft. per gal. (3.31 to .68 sq.m per liter) with 3 coats.

When applied to hardwood floors, varnish should cover 500 to 550 sq.ft. per gal. (12.27 to 13.49 sq.m per liter) with one coat; 250 to 275 sq.ft. (6.13 to 6.75 sq.m per liter) with 2 coats, and 170 to 185 sq.ft. (4.17 to 4.54 sq.m per liter) with 3 coats.

When applied to softwood interior finish, spar finishing varnish should cover 400 to 425 sq.ft. per gal. (9.81 to 10.43 sq.m per liter) with one coat; 200 to 215 sq.ft. per gal. (4.90 to 5.27 sq.m per liter) with 2 coats and 135 to 145 sq.ft. per gal. (3.31 to 3.56 sq.m per liter) with 3 coats.

When applied to hardwood interior finish, spar varnish should cover 425 to450 sq.ft. per gal. (10.43 to 11.04 sq.m per liter) with one coat; 210 to 225 sq.ft. per gal. (5.15 to 5.52 sq.m per liter) with 2 coats, and 140 to 150 sq.ft. per gal. (3.43 to 3.68 sq.m per liter) with 3 coats.

Covering Capacity of Wax. Liquid wax should cover 1,050 to 1,075 sq.ft. (25.77 to 26.38 sq.m per liter).

Covering Capacity of Varnish Remover. The amount of varnish remover required to remove old varnish from floors and interior finish depends on the condition of the old work, but varnish remover should cover 150 to 180 sq.ft. per gal. (3.68 to 4.42 sq.m per liter).

Enamel Finish. Where interior woodwork is to receive an enamel finish, at least three coats are required: first coat, oil base paint primer and sealer; second coat, prepared enamel undercoat; third coat, enamel finish coat. If a four-coat job is specified, the third coat may be a mixture of 1/2 undercoat and 1/2 enamel, followed by the enamel finish coat.

For flat work, figure material coverages as: paint primer and sealer, 575 to 600 sq.ft. per gal. (14.11 to 14.73 sq.m per liter); undercoat, 375 to 400 sq.ft. (9.20 to 9.82 sq.m per liter); enamel finish, 475 to 500 sq.ft. per gal. (12.69 t 12.27 sq.m per liter).

For running trim, material should cover: paint primer and sealer, 1,100 to 1,200 lin.ft. per gal. (88.58 to 96.63 m per liter); undercoat, 775 to 800 lin.ft. per gal. (62.41 to 64.42 m per liter); enamel finish, 775 to 800 lin.ft. per gal. (62.41to 64.42 m per liter).

Glazed Finish. Glazing liquid, colored with oil colors, should cover 500 to 550 sq.ft. per gal. (12.27 to 13.49 sq.m per liter).

For glazed finish on interior running trim, not over 6" (150 mm) wide, glazing liquid should cover 1,050 to 1,100 lin.ft. per gal. (84.55 to 88.58 m per liter).

Estimating Labor Costs for Painting

Perhaps in no other trade do labor costs vary to the extent they do in painting and decorating. On some classes of work the material costs are almost negligible, while the labor costs make it one of the most expensive kinds of work.

Naturally, the cost of the work will depends on the grade of workmanship specified or upon the class of work the painting contractor figures on doing.

In some locales, there are two classes of painter—those who make a specialty of ordinary or commercial work such as factory buildings, flat or apartment buildings, speculative work and work of this class, and those who specialize in high class work. Naturally, a painter who figures the lower grade of workmanship is bound to be lower in cost. The estimator must consider the class of work required.

Another thing to keep in mind is to be sure that all measurements and quantities are listed and computed in strict accordance with the methods of measurement as given at the beginning of this section. This applies particularly to the allowances for labor when estimating such items as sash doors, windows having divided lights, drop siding and shingle siding, fences, trellis and lattice work where the face and edges of all surfaces must be painted, stairs where it is necessary to paint all four sides or the entire circumference of newels, handrails, and balusters.

These are all exceedingly important, because if the quantities are not computed according to the methods given, then the labor costs given below will be far too low in many cases.

Itemized Costs of Painting

The labor and material tables on the following pages give detailed information on material coverages and production times for various types of paint on different kinds of surfaces.

In preparing an estimate for painting work, after obtaining quantities required for the project, the estimator need only to refer to the proper table to determine the required quantity of material and labor hours necessary to perform the work. Once the gallons of paint and man hours are computed, apply the appropriate unit costs applicable for your area.

Do not overlook the items of protection, scaffolding or other equipment costs in preparing the estimate. These are not given in the following tables.

For example, let's assume that the estimator is performing an estimate for staining shingle siding on a residence, and the quantity of work to be performed amounts to 3,300 sq.ft. (306.57 sq.m) of surface to be covered with two 2 coats of stain. Refer to the chart for exterior residential work and find the proper work classification, "Stain-Shingle Siding".

(Note that there are two sets of charts for *Labor and Materials Required for Painting Operations*. The first set, which is for the English or

1396

FINISHES

Imperial system of measures, is followed by a set of charts giving metric figures for the labor and material for painting.)

The material coverage is stated as 120 to 150 sq.ft. per gal. (2.94 to 3.68 sq.m per liter) for the first coat and 200 to 225 sq.ft. per gal. (4.91 to 5.52 sq.m per liter) for the second coat. The labor production is 1.20 hrs. per 100 sq.ft. (9.29 sq.m) for first coat and 0.82 hours per 100 sq.ft. (9.29 sq.m) for the second coat. The sample estimate for this 3,300 sq.ft. (306.57 sq.m) of surface would be as follows:

Material Costs		
Description	Total	Total
Stain 27 gal. (102.19 liters) - 1st Coat	$15.00	$ 405.00
Stain 16 gal. (60.56 liters) - 2nd Coat	$15.00	$ 240.00
Cost of 3,000 sq.ft.		$ 645.00
per sq.ft.		$ 0.22
per sq.m		$ 2.31

Labor Costs			
Description	Hours	Rate	Total
Painter - 1st Coat	39.60	$30.99	$1,227.20
Painter - 2nd Coat	27.10	$30.99	$ 839.83
Cost of 3,300 sq.ft. (306.57 sq.m)			$2,067.03
per sq.ft.			$ 0.63
per sq.m			$ 6.74

LABOR AND MATERIAL REQUIRED
FOR PAINTING OPERATIONS
English (Imperial) System

Note: Metric figures are given in a separate set of charts following those for the English system.

Exterior Work - Residential - Brush				
Description of Work	Coat	No. of Sq.Ft. per Hour	Hrs. per 100 Sq.ft	Material Coverage Sq.Ft. per Gal.
Sanding & Puttying, Plain Siding & Trim	-----	200 - 210	0.50	-----
Sanding & Puttying, Outside Trim Only	-----	115 - 120	0.86	-----
Burning Paint Off Plain Surfaces	-----	40 - 50	2.20	-----
Burning Paint Off Wood	-----	30 - 40	2.85	-----
Exterior Painting, Plain Siding & Trim	Priming	100 - 110	0.95	400 - 500
	Second	115 - 125	0.85	500 - 550
	Third	125 - 135	0.75	575 - 625
Exterior House Painting, Rubberized Wood Bond	One Coat	125 - 135	0.75	450 - 500
Exterior Trim Only	Priming	65 - 75	1.40	750 - 850
	Second	85 - 95	1.12	800 - 900
	Third	95 - 105	1.00	900 - 1,000
Shingle/Shake Siding* - Oil Paint	First	115 - 125	0.82	250 - 300
	Second	150 - 160	0.65	375 - 425
Shingle/Shake Roofs* - Oil Paint	First	100 - 120	0.90	120 - 150
	Second	155 - 175	0.60	220 - 250
Shingle/Shake Siding* - Stain	First	75 - 85	1.20	120 - 150
	Second	115 - 130	0.82	200 - 225
Shingle/Shake Roofs* - Stain	First	145 - 155	0.67	100 - 120
	Second	190 - 210	0.50	170 - 200
Shingle/Shake Roofs & Siding* - Sealer	First	200 - 210	0.49	125 - 150
	Second	220 - 250	0.43	150 - 200
*If surface is extremely dry, increase labor hours and decrease covering capacity.				
Brick Walls - Oil Paint	First	100 - 120	0.93	170 - 200
	Second	140 - 160	0.67	350 - 400
	Third	155 - 175	0.60	375 - 425
Brick Walls - Latex Paint	First	95 - 110	0.98	180 - 210
	Second	130 - 150	0.71	350 - 410
	Third	150 - 170	0.63	380 - 430
Porch Floors & Steps - Oil Paint	First	235 - 245	0.42	325 - 375
	Second	270 - 280	0.36	475 - 525
	Third	285 - 295	0.35	525 - 575
Smooth Face Brick - Waterproof Cement Paint	First	175 - 185	0.55	90 - 110
	Second	260 - 270	0.38	140 - 160
Asphalt Shingle - Clear Waterproof Paint	First	45 - 55	2.00	120 - 140
	Second	50 - 70	1.67	165 - 185
	Third	60 - 80	1.43	175 - 205
Smooth Face Brick - Clear Waterproof Paint	First	200 - 210	0.50	500 - 510
	Second	210 - 230	0.45	590 - 610
Stucco, Medium Texture - Oil Paint	First	90 - 100	1.10	140 - 160
	Second	150 - 160	0.65	340 - 360
	Third	150 - 160	0.65	340 - 360
Stucco, Medium Texture - Latex Paint	First	110 - 125	0.85	160 - 180
	Second	175 - 100	0.53	180 - 210
	Third	175 - 210	0.53	180 - 220

FINISHES

Description of Work	Coat	No. of Sq.Ft. per Hour	Hrs. per 100 Sq.ft	Material Coverage Sq.Ft. per Gal.
Stucco, Medium Texture - Waterproof	First	130 - 140	0.74	90 - 110
Cement Paint	Second	190 - 210	0.50	125 - 145
Exterior Masonry, Stucco, Shingle Siding	First	100 - 120	0.91	325 - 375
Latex Flat Finish	Second	165 - 175	0.59	375 - 425
Concrete Walls, Smooth - Waterproof	First	175 - 180	0.57	110 - 130
Cement Paint	Second	225 - 285	0.36	150 - 170
Concrete Floors & Steps - Floor Enamel	First	250 - 280	0.38	440 - 460
	Second	190 - 210	0.50	575 - 625
	Third	200 - 220	0.48	575 - 625
Cement Floors - Color Stain & Finish	First	325 - 370	0.28	475 - 525
	Second	280 - 300	0.35	450 - 500
Cement Floors - Epoxy	Primer	100 - 125	0.89	175 - 200
	First	115 - 135	0.80	190 - 220
	Second	150 - 180	0.60	200 - 250
Fences, Plain, Average	First	125 - 135	0.77	530 - 550
	Second	190 - 200	0.51	630 - 650
Fences, Picket, Average	First	140 - 150	0.70	630 - 650
	Second	160 - 170	0.60	650 - 675
Fences, Wire Metal, Average	First	100 - 110	0.95	900 - 1,000
	Second	140 - 150	0.70	750 - 1,125
Shutters, Average, Each Coat	No. Shutter	2 - 3	0.40	11 - 13
Downspout & Gutters - Paint	First	170 - 180	0.57	540 - 560
	Second	185 - 195	0.53	575 - 600
Screens, Wood Only, Average, Each Coat	No. Screens	6 - 8	0.14	45 - 55
Storm Sash, 2 Light, Average, Each Coat	No. Sash	3 - 4	0.29	23 - 27
Sheet Metal Work	First	230 - 240	0.43	500 - 550
	Second	260 - 270	0.38	550 - 600
Steel Factory Sash	First	80 - 90	1.18	951 - 1,000
	Second	100 - 110	0.95	950 - 1,000
	Third	125 - 135	0.77	1,250 - 1,350
Note: For old work, add 10 to 20% to prepare surface				
Doors & Windows - Paint	First	175 - 185	0.55	475 - 500
	Second	130 - 140	0.75	500 - 550
	Third	165 - 175	0.60	500 - 550
Base, Chair Rail, Picture Mold, & other	First	275 - 300	0.35	800 - 900
Trim less than 6" wide (per lin. ft.)	Second	175 - 200	0.55	1,100 - 1,200
	Third	150 - 165	0.65	1,100 - 1,200
Base, Chair Rail, Picture Mold, & other	First	260 - 280	0.37	525 - 575
Trim less than 6" wide - Polyurethane	Second	280 - 300	0.34	550 - 600
(per lin. ft.)	Third	280 - 300	0.34	550 - 600
Masonry Walls - Bonding & Penetrating	First	350 - 360	0.28	225 - 250
Oil Paint	Second	350 - 375	0.28	275 - 300
Common Brick, Smooth Brick, Concrete or Cinder Blocks, Cement Plaster - Waterproofing	First	100 - 110	1.00	90 - 100
	Second	180 - 195	0.53	175 - 200
Wire Cut Tapestry Brick or Very Rough Concrete Block or Stucco - Waterproofing	First	90 - 100	1.05	50 - 60
	Second	150 - 165	0.65	150 - 175
Smooth Plaster Walls - Flat Finish Paint	First	450 - 475	0.22	500 - 550
	Second	475 - 500	0.21	500 - 550
Plaster Walls - Sand Finish	First	400 - 425	0.25	450 - 500
	Second	425 - 450	0.24	400 - 450

Description of Work	Coat	No. of Sq.Ft. per Hour	Hrs. per 100 Sq.ft	Material Coverage Sq.Ft. per Gal.
Smooth Finish Plaster - Industrial Plaster	First	450 - 475	0.22	500 - 550
	Second	500 - 550	0.20	400 - 450
	Third	500 - 550	0.20	450 - 500
Exterior Work - Residential - Roller/Pad				
Shingle/Shake Siding* - Oil Paint	First	130 - 140	0.74	220 - 260
	Second	170 - 185	0.56	330 - 370
Shingle/Shake Roofs* - Oil Paint	First	115 - 135	0.80	105 - 130
	Second	180 - 200	0.53	190 - 215
Shingle/Shake Siding* - Stain	First	85 - 100	0.74	105 - 130
	Second	130 - 150	0.49	170 - 195
Shingle/Shake Roofs* - Stain	First	165 - 180	0.58	85 - 105
	Second	220 - 240	0.43	150 - 170
Shingle/Shake Roofs & Siding* - Sealer	First	230 - 240	0.43	110 - 130
	Second	250 - 285	0.37	130 - 170
*If surface is extremely dry, increase labor hours and decrease covering capacity.				
Brick Walls - Oil Paint	First	115 - 135	0.80	150 - 170
	Second	160 - 180	0.59	305 - 345
	Third	180 - 200	0.53	325 - 370
Brick Walls - Latex Paint	First	110 - 125	0.85	155 - 185
	Second	150 - 170	0.63	305 - 350
	Third	170 - 190	0.55	330 - 370
Porch Floors & Steps - Oil Paint	First	270 - 280	0.36	285 - 325
	Second	310 - 320	0.32	415 - 450
	Third	330 - 340	0.30	460 - 500
Smooth Face Brick - Waterproof Cement Paint	First	200 - 210	0.50	80 - 95
	Second	300 - 310	0.33	120 - 140
Asphalt Shingle - Clear Waterproof Paint	First	50 - 65	1.48	105 - 120
	Second	60 - 80	1.42	140 - 160
	Third	70 - 90	1.25	150 - 180
Smooth Face Brick - Clear Waterproof Paint	First	230 - 240	0.43	430 - 440
	Second	240 - 265	0.50	515 - 530
Stucco, Medium Texture - Oil Paint	First	100 - 115	0.93	120 - 140
	Second	170 - 185	0.55	295 - 310
	Third	170 - 185	0.55	295 - 310
Stucco, Medium Texture - Latex Paint	First	130 - 145	0.72	140 - 155
	Second	200 - 230	0.47	155 - 180
	Third	200 - 240	0.45	155 - 190
Stucco, Medium Texture - Waterproof Cement Paint	First	150 - 160	1.00	75 - 95
	Second	215 - 240	0.44	110 - 125
Exterior Masonry, Stucco, Shingle Siding Latex Flat Finish	First	115 - 135	0.80	280 - 325
	Second	190 - 200	0.51	325 - 370
Concrete Walls, Smooth Waterproof Cement Paint	First	200 - 210	0.49	95 - 115
	Second	260 - 320	0.48	130 - 150
Concrete Floors & Steps - Floor Enamel	First	290 - 320	0.37	380 - 400
	Second	220 - 240	0.43	500 - 540
	Third	230 - 350	0.42	500 - 540
Cement Floors - Color Stain & Finish	First	370 - 420	0.25	410 - 450
	Second	320 - 345	0.30	390 - 430
Cement Floors - Epoxy	Primer	115 - 140	0.78	150 - 170
	First	130 - 155	0.70	165 - 190
	Second	170 - 210	0.53	170 - 210
Fences, Plain, Average	First	145 - 155	0.67	460 - 480
	Second	220 - 230	0.45	550 - 565

FINISHES

Description of Work	Coat	No. of Sq.Ft. per Hour	Hrs. per 100 Sq.ft	Material Coverage Sq.Ft. per Gal.
Fences, Picket, Average	First	160 - 170	0.61	550 - 565
	Second	185 - 195	0.53	565 - 585
Fences, Wire Metal, Average	First	115 - 125	0.83	780 - 870
	Second	160 - 170	0.60	850 - 980
Downspout & Gutters - Paint	First	195 - 205	0.50	470 - 485
	Second	210 - 225	0.46	500 - 520
Masonry Walls - Bonding & Penetrating Oil Paint	First	400 - 410	0.25	200 - 220
	Second	400 - 430	0.24	240 - 260
Common Brick, Smooth Brick, Concrete or Cinder Blocks, Cement Plaster - Waterproofing	First	115 - 125	0.83	85 - 85
	Second	205 - 225	0.53	150 - 170
Smooth Plaster Walls - Flat Finish Paint	First	520 - 550	0.19	435 - 475
	Second	520 - 575	0.18	435 - 475
Plaster Walls - Sand Finish	First	460 - 490	0.21	390 - 430
	Second	490 - 515	0.20	350 - 390
Smooth Finish Plaster - Industrial Plaster	First	515 - 454	0.19	435 - 475
	Second	575 - 630	0.17	350 - 390
	Third	575 - 630	0.17	390 - 430
Spray Painting - Residential & Commercial Work, Brick, Tile & Cement				
Brick, Tile, & Cement - Latex Paint	First	500 - 525	0.20	90 - 110
	Second	525 - 550	0.19	140 - 160
Brick, Tile, & Cement - Oil Paint	First	475 - 500	0.21	250 - 275
	Second	500 - 525	0.20	400 - 450
Brick - Waterproofing	First	550 - 600	0.17	200 - 300
	Second	650 - 700	0.15	225 - 350
Note: For weathered brick, add 15% labor and 35% to 50% for material.				
Rough Brick, Tile, Cement & Stucco - Silicone	One Coat	800 - 900	0.13	80 - 90
Stucco - Exterior Latex Paint	First	400 - 425	0.25	90 - 11
	Second	450 - 500	0.22	125 - 150
Stucco - Oil Paint	First	300 - 325	0.34	200 - 225
	Second	325 - 350	0.30	350 - 400
Stucco - Waterproofing	First	550 - 600	0.17	150 - 180
	Second	650 - 700	0.15	160 - 190
Note: For weathered stucco, add 15% for labor and 35% to 50% for material.				
Shingle Roofs - Oil Paint	First	400 - 450	0.25	125 - 150
	Second	450 - 500	0.22	200 - 225
Shingle Roofs - Stain	First	450 - 475	0.22	125 - 150
	Second	475 - 500	0.21	200 - 225
Shingle Siding - Oil Paint	First	350 - 375	0.29	125 - 150
	Second	400 - 425	0.25	225 - 250
Shingle Siding - Stain	First	400 - 450	0.25	125 - 150
	Second	450 - 500	0.22	200 - 225
Smooth Plaster Walls - Flat Finish Paint	First	450 - 475	0.22	500 - 550
	Second	475 - 500	0.21	500 - 550
Plaster Walls - Sand Finish	First	400 - 425	0.25	450 - 500
	Second	425 - 450	0.24	400 - 450
Smooth Finish Plaster - Industrial Enamel	First	450 - 475	0.22	500 - 55
	Second	500 - 550	0.20	400 - 450
	Third	500 - 550	0.20	450 - 500
Fence, Picket - Oil or Latex Paint	First	475 - 500	0.20	500 - 525
	Second	475 - 500	0.20	550 - 600
Fence, Plain - Oil or Latex Paint	First	450 - 500	0.21	375 - 400
	Second	570 - 600	0.17	400 - 450

Description of Work	Coat	No. of Sq.Ft. per Hour	Hrs. per 100 Sq.ft	Material Coverage Sq.Ft. per Gal.
Doors, Garage - Oil or Latex Paint	First	540 - 550	0.22	550 - 575
	Second	540 - 550	0.22	560 - 600
Blinds or Shutters - Paint	First	120 - 140	0.77	120 - 140
	Second	130 - 170	0.67	140 - 180
	Third	150 - 190	0.59	150 - 190
Note: Trim must be figured separately as hand work.				
Finishing Interior Trim - Residential				
Preparatory Work for Painting	Sanding	290 - 300	0.35	-----
	Sand & Putty	190 - 200	0.50	-----
	Light Sanding	340 - 350	0.30	-----
Back Priming Interior Trim up to 6" Wide	Lin. Ft.	475 - 500	0.21	1,100 - 1,200
Doors & Windows, Interior - Paint	First	140 - 150	0.70	475 - 500
	Second	115 - 125	0.85	500 - 550
	Third	115 - 125	0.85	500 - 550
Wood Sash - Oiling & Priming	No. of Sash	9 - 10	-----	600 - 700
Preparatory Work for Enamel Finish	Sanding	290 - 300	0.35	-----
	Sand & Putty	120 - 130	0.80	-----
	Light Sanding	135 - 145	0.72	-----
Doors & Windows - Enamel Finish	Paint First	140 - 150	0.70	575 - 600
	Undercoat Second	90 - 100	1.10	375 - 400
	Enamel Third	100 - 110	1.00	475 - 500
Four coat work, add 1/2 undercoat, 1/2 enamel	Add'l Coat	90 - 100	1.10	425 - 450
Base, Chair Rail, Picturem Molding and Other Trim up to 6" Wide (lin. ft.)	First	250 - 260	0.40	800 - 900
	Second	145 - 155	0.67	1,100 - 1,200
	Third	125 - 135	0.77	1,100 - 1,200
Base, Chair Rail, Picture Molding and Other Trim up to 6" Wide - Polyurethane (lin. ft.)	First	260 - 280	0.37	525 - 575
	Second	280 - 300	0.34	550 - 560
	Third	280 - 300	0.34	550 - 560
Four coat work, add 1/2 undercoat, 1/2 enamel	Add'l Coat	135 - 145	0.72	1,100 - 1,200
Stain Interior Work	One Coat	215 - 225	0.45	700 - 725
Stain & Fill Interior Woodwork, Wipe Off	One Coat	65 - 70	1.50	400 - 450
Shellac Interior Woodwork	One Coat	215 - 225	0.45	700 - 725
Varnish, Gloss, Interior Woodwork	One Coat	165 - 175	0.60	425 - 450
Varnish, Flat, Interior Woodwork	One Coat	170 - 180	0.57	600 - 625
Standing Trim - Wax & Polish	One Coat	100 - 110	0.95	550 - 600
Standing Trim - Penetrating Stainwax	One Coat	170 - 180	0.57	525 - 550
	Second	190 - 200	0.51	600 - 625
	Polishing Coat	190 - 200	0.51	-----
Sanding for Extra Fine Varnish Finish	Sanding	40 - 45	2.25	-----
Synthetic Resin Finish, Requires Wiping	First	190 - 200	0.50	550 - 600
	Second	210 - 220	0.47	625 - 675
Spackling or Swedish Putty over Flat Trim	One Coat	60 - 65	1.60	140 - 150
Glazing & Wiping over Enamel Trim	One Coat	60 - 65	1.60	1,050 - 1,100
Brush Stippling Interior Trim - Painted	-----	85 - 90	1.15	-----
Flat Varnishing & Brush Stippling Over Glazed Trim	Varnish	170 - 180	0.57	600 - 625
	Stipple	240 - 250	0.42	-----
Finishing Interior Trim - Residential (Old Work)				
Washing Average Enamel Finish	Washing	85 - 90	1.15	-----
Washing Better Grade Enamel Finish	Washing	60 - 65	1.60	-----
Polishing Better Grade Enamel Finish	Polish	170 - 180	0.57	1,700 - 1,800
Removing Varnish with Liquid Remover	Flat Surfaces	30 - 35	3.00	150 - 180

FINISHES

Description of Work	Coat	No. of Sq.Ft. per Hour	Hrs. per 100 Sq.ft	Material Coverage Sq.Ft. per Gal.
Wash, Touch Up, One Coat Varnish	Wash-Touch Up	165 - 175	0.60	-----
	Varnish	100 - 110	0.95	575 - 600
Wash, Touch Up, One Coat Enamel	Wash-Touch Up	140 - 150	0.70	
	Enamel	70 - 75	1.40	375 - 400
Wash, Touch Up, One Coat Undercoat and One Coat Enamel	Wash-Touch Up	140 - 150	0.70	-----
	Undercoat	75 - 85	1.25	475 - 500
	Enamel	75 - 85	1.25	475 - 500
Burning Off Interior Trim	-----	20 - 25	4.00	-----
Burning Off Plain Surfaces	-----	30 - 35	3.00	-----
Finishing Interior Floors				
Wood Floors - Paint	First	290 - 300	0.35	525 - 550
	Second	260 - 270	0.38	425 - 450
Wood Floors - Filling, Wiping	Fill-Wipe	170 - 180	0.57	425 - 450
Hardwood Floors - Penetrating Stainwax	First	390 - 400	0.25	525 - 550
	Second	590 - 600	0.17	1,200 - 1,250
Floor Seal	First	490 - 500	0.20	600 - 625
	Second	590 - 600	0.17	1,200 - 1,250
Shellac	First	390 - 400	0.25	500 - 550
	Second	475 - 500	0.21	650 - 675
Stainfill, 1 Shellac, 1 Varnish	Stainfill	260 - 270	0.38	475 - 500
	Shellac	390 - 400	0.25	650 - 675
	Varnish	300 - 310	0.33	500 - 550
Stainfill, 1 Shellac, Wax & Polish	Stainfill	260 - 270	0.38	475 - 500
	Shellac	390 - 400	0.25	650 - 675
	Wax/Polish	200 - 201	0.50	1,050 - 1,075
Varnish, Each Coat over Shellac	Varnish	300 - 310	0.33	500 - 550
Buffing Floors by Machine	Buffing	390 - 400	0.25	-----
Waxing over 2 Coats of Seal & Polish	Wax-Polish	200 - 210	0.50	1,050 - 1,075
Finishing Interior Floors (Old Work)				
Removing Varnish with Liquid Remover	Flat Surfaces	40 - 45	2.35	170 - 180
Clean, Touch Up, & Varnish	Clean-Touch Up	125 - 135	0.77	-----
	Varnish	300 - 310	0.33	250 - 550
Clean, Touch Up, Wax, & Polish	Clean-Touch Up	125 - 135	0.77	-----
	Wax-Polish	200 - 210	0.50	1,050 - 1,075
Painting Interior Walls				
Taping, Beading, Spotting Nail Heads, & Sanding Gypsum Wallboard	-----	90 - 110	1.00	-----
Texture over Gypsum Wallboard (* per lb.)	One Coat	250 - 270	0.39	8 - 10*
Casein or Resin Emulsion over Textured Gypsum Wallboard	One Coat	300 - 320	0.32	325 - 350
Sizing New Smooth Finish Walls	Sizing	350 - 400	0.27	600 - 700
Wall Sealer or Primer on Smooth Walls	Sealer	200 - 225	0.45	575 - 625
Wall Sealer or Primer on Sand Finish Walls	Sealer	125 - 135	0.80	275 - 300
Smooth Finish Plaster - Flat Finish	Sealer	200 - 225	0.45	575 - 625
	Second	165 - 175	0.60	500 - 550
	Third	175 - 185	0.65	575 - 625
	Stippling	190 - 200	0.52	-----
Sand Finish Plaster - Flat Finish	Sealer	125 - 135	0.80	275 - 300
	Second	135 - 145	0.70	400 - 450
	Third	125 - 135	0.80	425 - 475
Smooth Finish Plaster - Gloss or Semi-Gloss	Sealer	200 - 225	0.45	575 - 625
	Second	170 - 180	0.57	500 - 550
	Third	140 - 150	0.70	500 - 550
	Stippling	85 - 95	1.10	-----

Description of Work	Coat	No. of Sq.Ft. per Hour	Hrs. per 100 Sq.ft	Material Coverage Sq.Ft. per Gal.
Sand Finish Plaster - Gloss or Semi-Gloss	Sealer	125 - 135	0.80	275 - 300
	Second	115 - 125	0.83	275 - 400
	Third	125 - 135	0.80	400 - 450
Texture Plaster, Average Semi-Gloss	Sealer	100 - 110	0.95	250 - 275
	Second	115 - 125	0.83	320 - 350
	Third	125 - 135	0.80	375 - 400
Smooth Finish Plaster - Latex Rubber Paint	First	140 - 160	0.67	300 - 350
	Second	140 - 16	0.67	300 - 350
Sand Finish or Average Texture Plaster - Latex Rubber Paint	First	110 - 130	0.83	250 - 300
	Second	110 - 130	0.83	250 - 300
Glazing & Motting over Smooth Finish Plaster	-----	80 - 90	1.15	1,050 - 1,075
Glazing & Motting over Sand Finish Plaster	-----	60 - 65	1.60	875 - 900
Glazing & Highlighting Textured Plaster	-----	80 - 90	1.15	825 - 850
Starch & Brush Stipple over Painted Glazed Surface	-----	15 - 125	0.83	-----
Flat Varnish & Brush Stipple over Painted Glazed Surface	-----	105 - 115	0.90	500 - 550
Textured Oil Paint over Smooth Finish Plaster	Size	215 - 225	0.45	675 - 725
	Texture	40 - 45	2.30	125 - 150
Water Textured over Smooth Finsh Plaster	Size	215 - 225	0.45	700 - 725
	Texture	50 - 55	1.90	85 - 95
New Smooth Plaster - Latex Base Paint	First	190 - 200	0.50	500 - 550
	Second	210 - 220	0.47	650 - 700
on Rough Sand Finish	One Coat	150 - 160	0.65	325 - 350
on Cement Blocks	One Coat	135 - 145	0.70	300 - 325
on Acoustical Surfaces	One Coat	135 - 145	0.70	200 - 225
Casein Paint over Smooth Finish Plaster	First	270 - 280	0.37	500 - 550
	Second	300 - 310	0.33	500 - 550
over Rough Sand Finish Plaster	One Coat	250 - 260	0.40	325 - 350
over Cement Blocks	One Coat	165 - 175	0.60	300 - 325
over Acoustical Surfaces	One Coat	165 - 175	0.60	200 - 225
over Cinder Concrete Blocks	One Coat	125 - 135	0.70	125 - 150

When applying paint to walls and ceilings using a roller applicator, increase quantities 10% to 15% and reduce painter time about the same amount.

Painting Interior Walls (Old Work)

Description of Work	Coat	No. of Sq.Ft. per Hour	Hrs. per 100 Sq.ft	Material Coverage Sq.Ft. per Gal.
Washing Off Calcimine, Average Surfaces	-----	115 - 125	0.83	-----
Washing Smooth Finish Plaster Walls, Average	-----	145 - 155	0.67	-----
Washing Sand Finish Plaster Walls, Average	-----	100 - 110	1.00	-----
Washing Starched Surfaces & Restarching, Smooth Surfaces	Washing	145 - 155	0.67	-----
	Restarching	130 - 140	0.75	-----
Removing Old Wallpaper, 3 layers or less	-----	65 - 75	1.45	-----
Washing Off Glue after Removing Paper including fixing average cracks & sizing	-----	125 - 135	0.77	-----
Cutting Hard Oil or Varnish Size walls including fixing average cracks	-----	125 - 135	0.77	-----
Cutting Gloss Painted Walls including fixing cracks	-----	125 - 135	0.77	-----
Wash, Touch Up, One Coat Gloss Paint to Smooth Plaster Surfaces	Wash-Touch Up	125 - 135	0.77	-----
	One Coat	125 - 135	0.77	450 - 500
Synthetic-Resin Emulsion Paint over old Painted Walls	One Coat	190 - 200	0.50	400 - 425

FINISHES

Description of Work	Coat	No. of Sq.Ft. per Hour	Hrs. per 100 Sq.ft	Material Coverage Sq.Ft. per Gal.
Wallpaper, Canvas, Coated Fabrics, Paper Hanging, Wood Veneer				
Canvas Sheeting	-----	50 - 60	1.82	-----
Coated Fabrics	Single Roll	2 - 2 1/2	-----	-----
Wallpaper				
One Edge Work	Single Roll	3 1/2 - 4	-----	-----
Wire Edge Work	Single Roll	2 1/2 - 3	-----	-----
Butt Work	Single Roll	2 1/4 - 2 3/4	-----	-----
Scenic Paper 40"x60"	Single Roll	1 - 1 1/4	-----	-----
Wood Veneer Flexwood	Sq. Ft.	18	-----	-----
Lacquer Finish over Wood Veneer	First Coat	100	1.00	275 - 325
	Second	100	1.00	450 - 500
Penetrating Wax or Synthetic Resin	First	100	1.00	550 - 600
Application over Wood Veneer	Second	100	1.00	650 - 675

LABOR AND MATERIAL REQUIRED
FOR PAINTING OPERATIONS
Metric System

Exterior Work - Residential - Brush				
Description of Work	Coat	No. of Sq.M. per Hour	Hrs. per 9.29 Sq.M	Material Coverage Sq.M. per Liter
Sanding & Puttying, Plain Siding & Trim	-----	18.58 - 19.51	0.50	-----
Sanding & Puttying, Outside Trim Only	-----	10.68 - 11.15	0.86	
Burning Paint Off Plain Surfaces	-----	3.72 - 4.65	2.20	-----
Burning Paint Off Wood	-----	2.79 - 3.72	2.85	-----
Exterior Painting, Plain Siding & Trim	Priming	9.29 - 10.22	0.95	9.82 - 12.27
	Second	10.68 - 11.61	0.85	12.27 - 13.50
	Third	11.61 - 12.54	0.75	14.11 - 15.34
Exterior House Painting, Rubberized Wood Bond	One Coat	11.61 - 12.54	0.75	11.04 - 12.27
Exterior Trim Only	Priming	6.04 - 6.97	1.40	18.41 - 20.86
	Second	7.90 - 8.83	1.12	19.64 - 22.09
	Third	8.83 - 9.75	1.00	22.09 - 24.54
Shingle/Shake Siding* - Oil Paint	First	10.68 - 11.61	0.82	6.14 - 7.36
	Second	13.94 - 14.86	0.65	9.20 - 10.43
Shingle/Shake Roofs* - Oil Paint	First	9.29 - 11.15	0.90	2.95 - 3.68
	Second	14.40 - 16.26	0.60	5.40 - 6.14
Shingle/Shake Siding* - Stain	First	6.97 - 7.90	1.20	2.95 - 3.68
	Second	10.68 - 12.08	0.82	4.91 - 5.52
Shingle/Shake Roofs* - Stain	First	13.47 - 14.40	0.67	2.45 - 2.95
	Second	17.65 - 19.51	0.50	4.17 - 4.91
Shingle/Shake Roofs & Siding* - Sealer	First	18.58 - 19.51	0.49	3.07 - 3.68
	Second	20.44 - 23.23	0.43	3.68 - 4.91
*If surface is extremely dry, increase labor hours and decrease covering capacity.				
Brick Walls - Oil Paint	First	9.29 - 11.15	0.93	4.17 - 4.91
	Second	13.01 - 14.86	0.67	8.59 - 9.82
	Third	14.40 - 16.26	0.60	9.20 - 10.43
Brick Walls - Latex Paint	First	8.83 - 10.22	0.98	4.42 - 5.15
	Second	12.08 - 13.94	0.71	8.59 - 10.06
	Third	13.94 - 15.79	0.63	9.33 - 10.55
Porch Floors & Steps - Oil Paint	First	21.83 - 22.76	0.42	7.98 - 9.20
	Second	25.08 - 26.01	0.36	11.66 - 12.89
	Third	26.48 - 27.41	0.35	12.89 - 14.11
Smooth Face Brick - Waterproof Cement Paint	First	16.26 - 17.19	0.55	2.21 - 2.70
	Second	24.15 - 25.08	0.38	3.44 - 3.93
Asphalt Shingle - Clear Waterproof Paint	First	4.18 - 5.11	2.00	2.95 - 3.44
	Second	4.65 - 6.50	1.67	4.05 - 4.54
	Third	5.57 - 7.43	1.43	4.30 - 5.03
Smooth Face Brick - Clear Waterproof Paint	First	18.58 - 19.51	0.50	12.27 - 12.52
	Second	19.51 - 21.37	0.45	14.48 - 14.97
Stucco, Medium Texture - Oil Paint	First	8.36 - 9.29	1.10	3.44 - 3.93
	Second	13.94 - 14.86	0.65	8.35 - 8.84
	Third	13.94 - 14.86	0.65	8.35 - 8.84
Stucco, Medium Texture - Latex Paint	First	10.22 - 11.61	0.85	3.93 - 4.42
	Second	16.26 - 9.29	0.53	4.42 - 5.15
	Third	16.26 - 19.51	0.53	4.42 - 5.40
Stucco, Medium Texture - Waterproof Cement Paint	First	12.08 - 13.01	0.74	2.21 - 2.70
	Second	17.65 - 19.51	0.50	3.07 - 3.56

FINISHES

Description of Work	Coat	No. of Sq.M. per Hour	Hrs. per 9.29 Sq.M	Material Coverage Sq.M. per Liter
Exterior Masonry, Stucco, Shingle Siding	First	9.29 - 11.15	0.91	-----
Latex Flat Finish	Second	15.33 - 16.26	0.59	9.20 - 10.43
Concrete Walls, Smooth - Waterproof	First	16.26 - 16.72	0.57	2.70 - 3.19
Cement Paint	Second	20.90 - 26.48	0.36	3.68 - 4.17
Concrete Floors & Steps - Floor Enamel	First	23.23 - 26.01	0.38	10.80 - 11.29
	Second	17.65 - 19.51	0.50	14.11 - 15.34
	Third	18.58 - 20.44	0.48	14.11 - 15.34
Cement Floors - Color Stain & Finish	First	30.19 - 34.37	0.28	11.66 - 12.89
	Second	26.01 - 27.87	0.35	11.04 - 12.27
Cement Floors - Epoxy	Primer	9.29 - 11.61	0.89	4.30 - 4.91
	First	10.68 - 12.54	0.80	4.66 - 5.40
	Second	13.94 - 16.72	0.60	4.91 - 6.14
Fences, Plain, Average	First	11.61 - 12.54	0.77	13.01 - 13.50
	Second	17.65 - 18.58	0.51	15.46 - 15.95
Fences, Picket, Average	First	13.01 - 13.94	0.70	15.46 - 15.95
	Second	14.86 - 15.79	0.60	15.95 - 16.57
Fences, Wire Metal, Average	First	9.29 - 10.22	0.95	22.09 - 24.54
	Second	13.01 - 13.94	0.70	18.41 - 27.61
Shutters, Average, Each Coat	No. Shutter	2 - 3	0.40	0.27 - 0.32
Downspout & Gutters - Paint	First	15.79 - 16.72	0.57	13.25 - 13.74
	Second	17.19 - 18.12	0.53	14.11 - 14.73
Screens, Wood Only, Average, Each Coat	No. Screens	6 - 8	0.14	1.10 - 1.35
Storm Sash, 2 Light, Average, Each Coat	No. Sash	3 - 4	0.29	0.56 - 0.66
Sheet Metal Work	First	21.37 - 22.30	0.43	12.27 - 13.50
	Second	24.15 - 25.08	0.38	13.50 - 14.73
Steel Factory Sash	First	7.43 - 8.36	1.18	23.34 - 24.54
	Second	9.29 - 10.22	0.95	23.32 - 24.54
	Third	11.61 - 12.54	0.77	30.68 - 33.13
For old work, add 10 to 20% to prepare Surface				
Doors & Windows - Paint	First	16.26 - 17.19	0.55	11.66 - 12.27
	Second	12.08 - 13.01	0.75	12.27 - 13.50
	Third	15.33 - 16.26	0.60	12.27 - 13.50
Base, Chair Rail, Picture Mold, & other	First	83.82 - 91.44	0.35	19.64 - 22.09
Trim less than 150 mm wide (per meter)	Second	53.34 - 60.96	0.55	27.00 - 29.45
	Third	45.72 - 50.29	0.65	27.00 - 29.45
Base, Chair Rail, Picture Mold, &	First	79.25 - 85.34	0.37	12.89 - 14.11
other Trim less than 150 mm wide -	Second	85.34 - 91.44	0.34	13.50 - 14.73
Polyurethane (per meter)	Third	85.34 - 91.44	0.34	13.50 - 14.73
Masonry Walls - Bonding & Penetrating	First	32.52 - 33.44	0.28	5.52 - 6.14
Oil Paint	Second	32.52 - 34.84	0.28	6.75 - 7.36
Common Brick, Smooth Brick, Concrete or Cinder Blocks, Cement Plaster - Waterproofing	First	9.29 - 10.22	1.00	2.21 - 2.45
	Second	16.72 - 18.12	0.53	4.30 - 4.91
Wire Cut Tapestry Brick or Very Rough Concrete Block or Stucco - Waterproofing	First	8.36 - 9.29	1.05	1.23 - 1.47
	Second	13.94 - 15.33	0.65	3.68 - 4.30
Smooth Plaster Walls - Flat Finish Paint	First	41.81 - 44.13	0.22	12.27 - 13.50
	Second	44.13 - 46.45	0.21	12.27 - 13.50
Plaster Walls - Sand Finish	First	37.16 - 39.48	0.25	11.04 - 12.27
	Second	39.48 - 41.81	0.24	9.82 - 11.04
Smooth Finish Plaster - Industrial Plaster	First	41.81 - 44.13	0.22	12.27 - 13.50
	Second	46.45 - 51.10	0.20	9.82 - 11.04
	Third	46.45 - 51.10	0.20	11.04 - 12.27

Description of Work	Coat	No. of Sq.M. per Hour	Hrs. per 9.29 Sq.M	Material Coverage Sq.M. per Liter
Shingle/Shake Siding* - Oil Paint	First	12.08 - 13.01	0.74	5.40 - 6.38
	Second	15.79 - 17.19	0.56	8.10 - 9.08
Shingle/Shake Roofs* - Oil Paint	First	10.68 - 12.54	0.80	2.58 - 3.19
	Second	16.72 - 18.58	0.53	4.66 - 5.28
Shingle/Shake Siding* - Stain	First	7.90 - 9.29	0.74	2.58 - 3.19
	Second	12.08 - 13.94	0.49	4.17 - 4.79
Shingle/Shake Roofs* - Stain	First	15.33 - 16.72	0.58	2.09 - 2.58
	Second	20.44 - 22.30	0.43	3.68 - 4.17
Shingle/Shake Roofs & Siding* - Sealer	First	21.37 - 22.30	0.43	2.70 - 3.19
	Second	23.23 - 26.48	0.37	3.19 - 4.17
*If surface is extremely dry, increase labor hours and decrease covering capacity.				
Brick Walls - Oil Paint	First	10.68 - 12.54	0.80	3.68 - 4.17
	Second	14.86 - 16.72	0.59	7.49 - 8.47
	Third	16.72 - 18.58	0.53	7.98 - 9.08
Brick Walls - Latex Paint	First	10.22 - 11.61	0.85	3.80 - 4.54
	Second	13.94 - 15.79	0.63	7.49 - 8.59
	Third	15.79 - 17.65	0.55	8.10 - 9.08
Porch Floors & Steps - Oil Paint	First	25.08 - 26.01	0.36	7.00 - 7.98
	Second	28.80 - 29.73	0.32	10.19 - 11.04
	Third	30.66 - 31.59	0.30	11.29 - 12.27
Smooth Face Brick - Waterproof Cement Paint	First	18.58 - 19.51	0.50	1.96 - 2.33
	Second	27.87 - 28.80	0.33	2.95 - 3.44
Asphalt Shingle - Clear Waterproof Paint	First	4.65 - 6.04	1.48	2.58 - 2.95
	Second	5.57 - 7.43	1.42	3.44 - 3.93
	Third	6.50 - 8.36	1.25	3.68 - 4.42
Smooth Face Brick - Clear Waterproof Paint	First	21.37 - 22.30	0.43	10.55 - 10.80
	Second	22.30 - 24.62	0.50	12.64 - 13.01
Stucco, Medium Texture - Oil Paint	First	9.29 - 10.68	0.93	2.95 - 3.44
	Second	15.79 - 17.19	0.55	7.24 - 7.61
	Third	15.79 - 17.19	0.55	7.24 - 7.61
Stucco, Medium Texture - Latex Paint	First	12.08 - 13.47	0.72	3.44 - 3.80
	Second	18.58 - 21.37	0.47	3.80 - 4.42
	Third	18.58 - 22.30	0.45	3.80 - 4.66
Stucco, Medium Texture - Waterproof Cement Paint	First	13.94 - 14.86	1.00	1.84 - 2.33
	Second	19.97 - 22.30	0.44	2.70 - 3.07
Exterior Masonry, Stucco, Shingle Siding - Latex Flat Finish	First	10.68 - 12.54	0.80	6.87 - 7.98
	Second	17.65 - 18.58	0.51	7.98 - 9.08
Concrete Walls, Smooth - Waterproof Cement Paint	First	18.58 - 19.51	0.49	2.33 - 2.82
	Second	24.15 - 29.73	0.48	3.19 - 3.68
Concrete Floors & Steps - Floor Enamel	First	26.94 - 29.73	0.37	9.33 - 9.82
	Second	20.44 - 22.30	0.43	12.27 - 13.25
	Third	21.37 - 32.52	0.42	12.27 - 13.25
Cement Floors - Color Stain & Finish	First	34.37 - 39.02	0.25	10.06 - 11.04
	Second	29.73 - 32.05	0.30	9.57 - 10.55
Cement Floors - Epoxy	Primer	10.68 - 13.01	0.78	3.68 - 4.17
	First	12.08 - 14.40	0.70	4.05 - 4.66
	Second	15.79 - 19.51	0.53	4.17 - 5.15
Fences, Plain, Average	First	13.47 - 14.40	0.67	11.29 - 11.78
	Second	20.44 - 21.37	0.45	13.50 - 13.87
Fences, Picket, Average	First	14.86 - 15.79	0.61	13.50 - 13.87
	Second	17.19 - 18.12	0.53	13.87 - 14.36
Fences, Wire Metal, Average	First	10.68 - 11.61	0.83	19.14 - 21.35
	Second	14.86 - 15.79	0.60	20.86 - 24.05

FINISHES

Description of Work	Coat	No. of Sq.M. per Hour	Hrs. per 9.29 Sq.M	Material Coverage Sq.M. per Liter
Downspout & Gutters - Paint	First	18.12 - 19.04	0.50	11.54 - 11.90
	Second	19.51 - 20.90	0.46	12.27 - 12.76
Masonry Walls - Bonding & Penetrating Oil Piant	First	37.16 - 38.09	0.25	4.91 - 5.40
	Second	37.16 - 39.95	0.24	5.89 - 6.38
Common Brick, Smooth Brick, Concrete or Cinder Blocks, Cement Plaster - Waterproofing	First	10.68 - 11.61	0.83	2.09 - 2.09
	Second	19.04 - 20.90	0.53	3.68 - 4.17
Smooth Plaster Walls - Flat Finish Paint	First	48.31 - 51.10	0.19	10.68 - 11.66
	Second	48.31 - 53.42	0.18	10.68 - 11.66
Plaster Walls - Sand Finish	First	42.73 - 45.52	0.21	9.57 - 10.55
	Second	45.52 - 47.84	0.20	8.59 - 9.57
Smooth Finish Plaster - Industrial Plaster	First	47.84 - 42.18	0.19	10.68 - 11.66
	Second	53.42 - 58.53	0.17	8.59 - 9.57
	Third	53.42 - 58.53	0.17	9.57 - 10.55
Spray Painting - Residential & Commercial Work, Brick, Tile & Cement				
Brick, Tile, & Cement - Latex Paint	First	46.45 - 48.77	0.20	2.21 - 2.70
	Second	48.77 - 51.10	0.19	3.44 - 3.93
Brick, Tile, & Cement - Oil Paint	First	44.13 - 46.45	0.21	6.14 - 6.75
	Second	46.45 - 48.77	0.20	9.82 - 11.04
Brick - Waterproofing	First	51.10 - 55.74	0.17	4.91 - 7.36
	Second	60.39 - 65.03	0.15	5.52 - 8.59
Note: For weathered brick, add 15% labor and 35% to 50% for material.				
Rough Brick, Tile, Cement & Stucco - Silicone	One Coat	74.32 - 83.61	0.13	1.96 - 2.21
Stucco - Exterior Latex Paint	First	37.16 - 39.48	0.25	2.21 - 0.27
	Second	41.81 - 46.45	0.22	3.07 - 3.68
Stucco - Oil Paint	First	27.87 - 30.19	0.34	4.91 - 5.52
	Second	30.19 - 32.52	0.30	8.59 - 9.82
Stucco - Waterproofing	First	51.10 - 55.74	0.17	3.68 - 4.42
	Second	60.39 - 65.03	0.15	3.93 - 4.66
Note: For weathered stucco, add 15% for labor and 35% to 50% for material.				
Shingle Roofs - Oil Paint	First	37.16 - 41.81	0.25	3.07 - 3.68
	Second	41.81 - 46.45	0.22	4.91 - 5.52
Shingle Roofs - Stain	First	41.81 - 44.13	0.22	3.07 - 3.68
	Second	44.13 - 46.45	0.21	4.91 - 5.52
Shingle Siding - Oil Paint	First	32.52 - 34.84	0.29	3.07 - 3.68
	Second	37.16 - 39.48	0.25	5.52 - 6.14
Shingle Siding - Stain	First	37.16 - 41.81	0.25	3.07 - 3.68
	Second	41.81 - 46.45	0.22	4.91 - 5.52
Smooth Plaster Walls - Flat Finish Paint	First	41.81 - 44.13	0.22	12.27 - 13.50
	Second	44.13 - 46.45	0.21	12.27 - 13.50
Plaster Walls - Sand Finish	First	37.16 - 39.48	0.25	11.04 - 12.27
	Second	39.48 - 41.81	0.24	9.82 - 11.04
Smooth Finish Plaster - Industrial Enamel	First	41.81 - 44.13	0.22	12.27 - 1.35
	Second	46.45 - 51.10	0.20	9.82 - 11.04
	Third	46.45 - 51.10	0.20	11.04 - 12.27
Fence, Picket - Oil or Latex Paint	First	44.13 - 46.45	0.20	12.27 - 12.89
	Second	44.13 - 46.45	0.20	13.50 - 14.73
Fence, Plain - Oil or Latex Paint	First	41.81 - 46.45	0.21	9.20 - 9.82
	Second	52.95 - 55.74	0.17	9.82 - 11.04
Doors, Garage - Oil or Latex Paint	First	50.17 - 51.10	0.22	13.50 - 14.11
	Second	50.17 - 51.10	0.22	13.74 - 14.73

Description of Work	Coat	No. of Sq.M. per Hour	Hrs. per 9.29 Sq.M	Material Coverage Sq.M. per Liter
Blinds or Shutters - Paint	First	11.15 - 13.01	0.77	2.95 - 3.44
	Second	12.08 - 15.79	0.67	3.44 - 4.42
	Third	13.94 - 17.65	0.59	3.68 - 4.66

Trim must be figured separately as hand work.

Finishing Interior Trim - Residential				
Preparatory Work for Painting	Sanding	26.94 - 27.87	0.35	-----
	Sand & Putty	17.65 - 18.58	0.50	-----
	Light Sanding	31.59 - 32.52	0.30	-----
Back Priming Interior Trim up to 150 mm Wide	Lin. M.	144.78 - 152.40	0.21	27.00 - 29.45
Doors & Windows, Interior - Paint	First	13.01 - 13.94	0.70	11.66 - 12.27
	Second	10.68 - 11.61	0.85	12.27 - 13.50
	Third	10.68 - 11.61	0.85	12.27 - 13.50
Wood Sash - Oiling & Priming	No. of Sash	9 - 10	-----	14.73 - 17.18
Preparatory Work for Enamel Finish	Sanding	26.94 - 27.87	0.35	-----
	Sand & Putty	11.15 - 12.08	0.80	-----
	Light Sanding	12.54 - 13.47	0.72	-----
Doors & Windows - Enamel Finish	Paint First	13.01 - 13.94	0.70	14.11 - 14.73
	Undercoat Second	8.36 - 9.29	1.10	9.20 - 9.82
	Enamel Third	9.29 - 10.22	1.00	11.66 - 12.27
Four coat work, add 1/2 undercoat, 1/2 enamel	Add'l Coat	8.36 - 9.29	1.10	10.43 - 11.04
Base, Chair Rail, Picturem Molding and Other Trim up to 150 mm Wide (lin.M.)	First	76.20 - 79.25	0.40	19.64 - 22.09
	Second	44.20 - 47.24	0.67	27.00 - 29.45
	Third	38.10 - 41.15	0.77	27.00 - 29.45
Base, Chair Rail, Picture Molding and Other Trim up to 150 mm Wide - Polyurethane (lin.M.)	First	79.25 - 85.34	0.37	12.89 - 14.11
	Second	85.34 - 91.44	0.34	13.50 - 13.74
	Third	85.34 - 91.44	0.34	13.50 - 13.74
Four coat work, add 1/2 undercoat, 1/2 enamel	Add'l Coat	12.54 - 13.47	0.72	27.00 - 29.45
Stain Interior Work	One Coat	19.97 - 20.90	0.45	17.18 - 17.79
Stain & Fill Interior Woodwork, Wipe Off	One Coat	6.04 - 6.50	1.50	9.82 - 11.04
Shellac Interior Woodwork	One Coat	19.97 - 20.90	0.45	17.18 - 17.79
Varnish, Gloss, Interior Woodwork	One Coat	15.33 - 16.26	0.60	10.43 - 11.04
Varnish, Flat, Interior Woodwork	One Coat	15.79 - 16.72	0.57	14.73 - 15.34
Standing Trim - Wax & Polish	One Coat	9.29 - 10.22	0.95	13.50 - 14.73
Standing Trim - Penetrating Stainwax	One Coat	15.79 - 16.72	0.57	12.89 - 13.50
	Second	17.65 - 18.58	0.51	14.73 - 15.34
	Polishing Coat	17.65 - 18.58	0.51	-----
Sanding for Extra Fine Varnish Finish	Sanding	3.72 - 4.18	2.25	-----
Synthetic Resin Finish, Requires Wiping	First	17.65 - 18.58	0.50	13.50 - 14.73
	Second	19.51 - 20.44	0.47	15.34 - 16.57
Spackling or Swedish Putty over Flat Trim	One Coat	5.57 - 6.04	1.60	3.44 - 3.68
Glazing & Wiping over Enamel Trim	One Coat	5.57 - 6.04	1.60	25.77 - 27.00
Brush Stippling Interior Trim - Painted	-----	7.90 - 8.36	1.15	-----
Flat Varnishing & Brush Stippling Over Glazed Trim	Varnish	15.79 - 16.72	0.57	14.73 - 15.34
	Stipple	22.30 - 23.23	0.42	-----

Finishing Interior Trim - Residential (Old Work)				
Washing Average Enamel Finish	Washing	7.90 - 8.36	1.15	-----
Washing Better Grade Enamel Finish	Washing	5.57 - 6.04	1.60	-----
Polishing Better Grade Enamel Finish	Polish	15.79 - 16.72	0.57	41.73 - 44.18
Removing Varnish with Liquid Remover	Flat Surfaces	2.79 - 3.25	3.00	3.68 - 4.42

1410

FINISHES

Description of Work	Coat	No. of Sq.M. per Hour	Hrs. per 9.29 Sq.M	Material Coverage Sq.M. per Liter
Wash, Touch Up, One Coat Varnish	Wash-Touch Up	15.33 - 16.26	0.60	-----
	Varnish	9.29 - 10.22	0.95	14.11 - 14.73
Wash, Touch Up, One Coat Enamel	Wash-Touch Up	13.01 - 13.94	0.70	-----
	Enamel	6.50 - 6.97	1.40	9.20 - 9.82
Wash, Touch Up, One Coat Undercoat and One Coat Enamel	Wash-Touch Up	13.01 - 13.94	0.70	-----
	Undercoat	6.97 - 7.90	1.25	11.66 - 12.27
	Enamel	6.97 - 7.90	1.25	11.66 - 12.27
Burning Off Interior Trim	-----	1.86 - 2.32	4.00	-----
Burning Off Plain Surfaces	-----	2.79 - 3.25	3.00	-----
Finishing Interior Floors				
Wood Floors - Paint	First	26.94 - 27.87	0.35	12.89 - 13.50
	Second	24.15 - 25.08	0.38	10.43 - 11.04
Wood Floors - Filling, Wiping	Fill-Wipe	15.79 - 16.72	0.57	10.43 - 11.04
Hardwood Floors - Penetrating Stainwax	First	36.23 - 37.16	0.25	12.89 - 13.50
	Second	54.81 - 55.74	0.17	29.45 - 30.68
Floor Seal	First	45.52 - 46.45	0.20	14.73 - 15.34
	Second	54.81 - 55.74	0.17	29.45 - 30.68
Shellac	First	36.23 - 37.16	0.25	12.27 - 13.50
	Second	44.13 - 46.45	0.21	15.95 - 16.57
Stainfill, 1 Shellac, 1 Varnish	Stainfill	24.15 - 25.08	0.38	11.66 - 12.27
	Shellac	36.23 - 37.16	0.25	15.95 - 16.57
	Varnish	27.87 - 28.80	0.33	12.27 - 13.50
Stainfill, 1 Shellac, Wax & Polish	Stainfill	24.15 - 25.08	0.38	11.66 - 12.27
	Shellac	36.23 - 37.16	0.25	15.95 - 16.57
	Wax/Polish	18.58 - 18.67	0.50	25.77 - 26.39
Varnish, Each Coat over Shellac	Varnish	27.87 - 28.80	0.33	12.27 - 13.50
Buffing Floors by Machine	Buffing	36.23 - 37.16	0.25	-----
Waxing over 2 Coats of Seal & Polish	Wax-Polish	18.58 - 19.51	0.50	25.77 - 26.39
Finishing Interior Floors (Old Work)				
Removing Varnish with Liquid Remover	Flat Surfaces	3.72 - 4.18	2.35	4.17 - 4.42
Clean, Touch Up, & Varnish	Clean-Touch Up	11.61 - 12.54	0.77	-----
	Varnish	27.87 - 28.80	0.33	6.14 - 13.50
Clean, Touch Up, Wax, & Polish	Clean-Touch Up	11.61 - 12.54	0.77	-----
	Wax-Polish	18.58 - 19.51	0.50	25.77 - 26.39
Painting Interior Walls				
Taping, Beading, Spotting Nail Heads, & Sanding Gypsum Wallboard	-----	8.36 - 10.22	1.00	-----
Texture over Gypsum Wallboard (per kg)	One Coat	23.23 - 25.08	0.39	0.20 - 0.25
Casein or Resin Emulsion over Textured Gypsum Wallboard	One Coat	27.87 - 29.73	0.32	7.98 - 8.59
Sizing New Smooth Finish Walls	Sizing	32.52 - 37.16	0.27	14.73 - 17.18
Wall Sealer or Primer on Smooth Walls	Sealer	18.58 - 20.90	0.45	14.11 - 15.34
Wall Sealer or Primer on Sand Finish Walls	Sealer	11.61 - 12.54	0.80	6.75 - 7.36
Smooth Finish Plaster - Flat Finish	Sealer	18.58 - 20.90	0.45	14.11 - 15.34
	Second	15.33 - 16.26	0.60	12.27 - 13.50
	Third	16.26 - 17.19	0.65	14.11 - 15.34
	Stippling	17.65 - 18.58	0.52	-----
Sand Finish Plaster - Flat Finish	Sealer	11.61 - 12.54	0.80	6.75 - 7.36
	Second	12.54 - 13.47	0.70	9.82 - 11.04
	Third	11.61 - 12.54	0.80	10.43 - 11.66

Description of Work	Coat	No. of Sq.M. per Hour	Hrs. per 9.29 Sq.M	Material Coverage Sq.M. per Liter
Smooth Finish Plaster - Gloss or Semi-Gloss	Sealer	18.58 - 20.90	0.45	14.11 - 15.34
	Second	15.79 - 16.72	0.57	12.27 - 13.50
	Third	13.01 - 13.94	0.70	12.27 - 13.50
	Stippling	7.90 - 8.83	1.10	-----
Sand Finish Plaster - Gloss or Semi-Gloss	Sealer	11.61 - 12.54	0.80	6.75 - 7.36
	Second	10.68 - 11.61	0.83	6.75 - 9.82
	Third	11.61 - 12.54	0.80	9.82 - 11.04
Texture Plaster, Average Semi-Gloss	Sealer	9.29 - 10.22	0.95	6.14 - 6.75
	Second	10.68 - 11.61	0.83	7.85 - 8.59
	Third	11.61 - 12.54	0.80	9.20 - 9.82
Smooth Finish Plaster - Latex Rubber Paint	First	13.01 - 14.86	0.67	7.36 - 8.59
	Second	13.01 - 1.49	0.67	7.36 - 8.59
Sand Finish or Average Texture Plaster - Latex Rubber Paint	First	10.22 - 12.08	0.83	6.14 - 7.36
	Second	10.22 - 12.08	0.83	6.14 - 7.36
Glazing & Motting over Smooth Finish Plaster	-----	7.43 - 8.36	1.15	25.77 - 26.39
Glazing & Motting over Sand Finish Plaster	-----	5.57 - 6.04	1.60	21.48 - 22.09
Glazing & Highlighting Textured Plaster	-----	7.43 - 8.36	1.15	20.25 - 20.86
Starch & Brush Stipple over Painted Glazed Surface	-----	1.39 - 11.61	0.83	-----
Flat Varnish & Brush Stipple over Painted Glazed Surface	-----	9.75 - 10.68	0.90	12.27 - 13.50
Textured Oil Paint over Smooth Finish Plaster	Size	19.97 - 20.90	0.45	16.57 - 17.79
	Texture	3.72 - 4.18	2.30	3.07 - 3.68
Water Textured over Smooth Finsh Plaster	Size	19.97 - 20.90	0.45	17.18 - 17.79
	Texture	4.65 - 5.11	1.90	2.09 - 2.33
New Smooth Plaster - Latex Base Paint	First	17.65 - 18.58	0.50	12.27 - 13.50
	Second	19.51 - 20.44	0.47	15.95 - 17.18
on Rough Sand Finish	One Coat	13.94 - 14.86	0.65	7.98 - 8.59
on Cement Blocks	One Coat	12.54 - 13.47	0.70	7.36 - 7.98
on Acoustical Surfaces	One Coat	12.54 - 13.47	0.70	4.91 - 5.52
Casein Paint over Smooth Finish Plaster	First	25.08 - 26.01	0.37	12.27 - 13.50
	Second	27.87 - 28.80	0.33	12.27 - 13.50
over Rough Sand Finish Plaster	One Coat	23.23 - 24.15	0.40	7.98 - 8.59
over Cement Blocks	One Coat	15.33 - 16.26	0.60	7.36 - 7.98
over Acoustical Surfaces	One Coat	15.33 - 16.26	0.60	4.91 - 5.52
over Cinder Concrete Blocks	One Coat	11.61 - 12.54	0.70	3.07 - 3.68

When applying paint to walls and ceilings using a roller applicator, increase quantities 10% to 15% and reduce painter time about the same amount.

Painting Interior Walls (Old Work)

Description of Work	Coat	No. of Sq.M. per Hour	Hrs. per 9.29 Sq.M	Material Coverage Sq.M. per Liter
Washing Off Calcimine, Average Surfaces	-----	10.68 - 11.61	0.83	-----
Washing Smooth Finish Plaster Walls, Average	-----	13.47 - 14.40	0.67	-----
Washing Sand Finish Plaster Walls, Average	-----	9.29 - 10.22	1.00	-----
Washing Starched Surfaces & Restarching, Smooth Surfaces	Washing	13.47 - 14.40	0.67	-----
	Restarching	12.08 - 13.01	0.75	-----
Removing Old Wallpaper, 3 layers or less	-----	6.04 - 6.97	1.45	-----
Washing Off Glue after Removing Paper including fixing average cracks & sizing	-----	11.61 - 12.54	0.77	-----
Cutting Hard Oil or Varnish Size walls including fixing average cracks	-----	11.61 - 12.54	0.77	-----

1412

FINISHES

Description of Work	Coat	No. of Sq.M. per Hour	Hrs. per 9.29 Sq.M	Material Coverage Sq.M. per Liter
Cutting Gloss Painted Walls including fixing cracks	-----	11.61 - 12.54	0.77	-----
Wash, Touch Up, One Coat Gloss Paint to Smooth Plaster Surfaces	Wash-Touch Up	11.61 - 12.54	0.77	-----
	One Coat	11.61 - 12.54	0.77	11.04 - 12.27
Synthetic-Resin Emulsion Paint over old Paited Walls	One Coat	17.65 - 18.58	0.50	9.82 - 10.43
Wallpaper, Canvas, Coated Fabrics, Paper Hanging, Wood Veneer				
Canvas Sheeting	-----	4.65 - 5.57	1.82	-----
Coated Fabrics	Single Roll	0.19 - 0.23	-----	-----
Wallpaper				
One Edge Work	Single Roll	0.33 - 0.37	-----	-----
Wire Edge Work	Single Roll	0.23 - 0.28	-----	-----
Butt Work	Single Roll	0.21 - 0.26	-----	-----
Scenic Paper 1000 x 1500 mm	Single Roll	0.09 - 0.12	-----	-----
Wood Veneer Flexwood	Sq. M.	1.67	-----	-----
Lacquer Finish over Wood Veneer	First Coat	9.29	1.00	6.75 - 7.98
	Second	9.29	1.00	11.04 - 12.27
Penetrating Wax or Synthetic Resin Application over Wood Veneer	First	9.29	1.00	13.50 - 14.73
	Second	9.29	1.00	15.95 - 16.57

Shellac and Varnish		
Kind of Material	Price per Gallon	Price per Liter
Shellac, White, 4-lb. Cut	$23.40 - $28.60	$ 6.18 - $ 7.56
Shellac, Orange, 4-lb. Cut	$20.80 - $26.00	$ 5.50 - $ 6.87
Cabinet Finish Varnish	$26.00 - $33.80	$ 6.87 - $ 8.93
Floor Varnish	$24.70 - $32.50	$ 6.53 - $ 8.59
Flat Varnish	$24.70 - $32.50	$ 6.53 - $ 8.59
Spar Varnish	$24.70 - $37.70	$ 6.53 - $ 9.96
Varnish Remover, Liquid	$19.50 - $32.50	$ 5.15 - $ 8.59
Varnish Remover, Paste	$19.50 - $37.70	$ 5.15 - $ 9.96
Wallpaper Lacquer	$15.60 - $18.20	$ 4.12 - $ 4.81
Wallpaper Lacquer Thinner	$10.40 - $16.90	$ 2.75 - $ 4.46
Lacquer	$26.00 - $32.50	$ 6.87 - $ 8.59
Lacquer Thinner	$13.00 - $15.60	$ 3.43 - $ 4.12

Enamels and Enamel Undercoating		
Kind of Material	Price per Gallon	Price per Liter
Enamel, Non-Yellowing White	$16.90 - $19.50	$ 4.46 - $ 5.15
Architectural Enamel Paint	$16.90 - $23.40	$ 4.46 - $ 6.18
Enamel Undercoater	$14.30 - $19.50	$ 3.78 - $ 5.15
Enamel, White	$14.30 - $19.50	$ 3.78 - $ 5.15
Enamel, Quick Drying	$14.30 - $20.80	$ 3.78 - $ 5.50
Enamel, Industrial	$16.90 - $23.40	$ 4.46 - $ 6.18

Wall Size, Sealer, Paste, Etc.

Kind of Material	Price	Price - Metric
Sizing, prepared per gal (liter)	$10.00 - $18.00	$ 2.64 - $ 4.76
Glue, Dry for Sizing per lb (kg)	$ 1.50 - $ 2.50	$ 3.31 - $ 5.51
Wall Primer and Sealer per gal (liter)	$16.00 - $24.00	$ 4.23 - $ 6.34
Wallpaper Cleaner per lb (kg)	$3.75	$8.27
Starch, gloss per lb (kg)	$5.00	$11.02
Patching Plaster per lb (kg)	$5.00	$11.02
Spackle per lb (kg)	$3.75	$8.27
Glazing Liquid per gal (liter)	$16.00 - $24.00	$ 4.23 - $ 6.34
Pre-mixed Vinyl Adhesive per lb (kg)	$5.00	$11.02

Wood Fillers and Stains

Kind of Material	Price per Gallon	Price per Liter
Oil Stains	$24.00 - $36.00	$ 6.34 - $ 9.51
Acid Stains	$24.00 - $36.00	$ 6.34 - $ 9.51
Penetrating Stainwax	$24.00 - $36.00	$ 6.34 - $ 9.51
Stain, Varnish, all colors	$28.80 - $72.00	$ 7.61 - $19.02
Stain, Fill, and Seal	$24.00 - $28.80	$ 6.34 - $ 7.61
Paste Wood Fillers	$19.20 - $26.40	$ 5.07 - $ 6.97
Floor Seal	$28.80 - $36.00	$ 7.61 - $ 9.51

Floor Wax

Kind of Material	Price per Gallon	Price per Liter
Liquid Floor Wax	$28.80 - $30.80	$ 7.61 - $ 8.14
Self-Polishing Wax	$16.80 - $21.60	$ 4.44 - $ 5.71
Paste Wax per lb. (kg)	$14.40 - $19.20	$ 3.80 - $ 5.07

Miscellaneous Paints

Kind of Material	Price per Gallon	Price per Liter
Ready Mixed House Paint - Latex	$15.00 - $22.50	$ 3.96 - $ 5.94
House Paint-One Coat	$18.00 - $24.00	$ 4.76 - $ 6.34
Porch and Floor Paint	$19.00 - $28.50	$ 5.02 - $ 7.53
Flat Wall Paint - Latex	$15.00 - $22.50	$ 3.96 - $ 5.94
One Coat Wall Paint - Oil Base	$24.75 - $41.25	$ 6.54 - $10.90
Metal Paint for Structural Steel, Bridges, Stacks, etc.	$22.50 - $37.50	$ 5.94 - $ 9.91
Cement Floor Paint	$12.00 - $24.00	$ 3.17 - $ 6.34
Cement Floor Paint - Rubber Base	$12.00 - $24.00	$ 3.17 - $ 6.34
Aluminum Paint	$14.00 - $28.00	$ 3.70 - $ 7.40
Vinyl-Chloride Co-Polymer	$43.75	$11.56
Sand Finish	$10.50 - $15.00	$ 2.77 - $ 3.96
Polyurethane	$17.50 - $31.50	$ 4.62 - $ 8.32
Drywall Primer	$ 9.00 - $15.00	$ 2.38 - $ 3.96
Waterproof Sealer, Clear	$13.50 - $30.00	$ 3.57 - $ 7.93

1414

FINISHES

Oil-based paints are used less and less in today's market place as the price per gallon in some cases has exceeded $40.00. Also clean-up for most oil-based paints requires the use of minerail oils, though some now can be cleaned up with water. Oil paints dry to a rigid coating that in most cases blocks moisture, resulting in cracks appearing, and at times, due to the trapping of moisture, may produce blisters. In its place oil-based paints are acceptable as durable paint. If the new paint being applied is covering an existing oil paint, and the existing paint appears fine, then the use of an oil-based paint can be recommended.

Prices for oil-based paints range from a low of $25.00 to $45.00 per gal ($6.60 to $11.89 per liter). Note that some colors wll cost more. If oil-based paints are purchased in one-half and half pints, the cost per unit will exceed that of the cost per gallon.

Lettering On Glass or Wood

The cost of lettering will vary with the style and size of letters and whether painted or gold leaf. Costs given on the following page include only Roman or Gothic style letters, such as are ordinarily used for signs on doors, transoms, and show windows, and are based on all work being performed by experienced sign writers.

Plain Black Letters 2" to 4" (50-100 mm) High. For plain black Gothic letters and numerals, such as appear on windows, transoms, and office doors, an experienced sign writer should average about 25 letters per hr. at the following labor cost per 100 letters:

Description	Hours	Rate	Total
Sign painter	4.00	$30.99	$123.96
Cost per letter			$1.24

Plain Black Letters 6" to 12" (150-300 mm) High. An experienced sign writer should average 12 to 14 plain Gothic letters 6" to 12" (150-300 mm) high, per hr. at the following labor cost per 100 letters:

Description	Hours	Rate	Total
Sign painter	7.70	$30.99	$238.62
Cost per letter			$2.39

Gold Leaf Letters 2" to 4" (50-100 mm) High. An experienced sign writer should complete about 5 plain Roman or Gothic letters in gold leaf per hr. (this class of work proceeds much slower than painted letters) at the following labor cost per 100 letters:

Description	Hours	Rate	Total
Sign painter	20.00	$30.99	$619.80
Cost per letter			$6.20

1415

Gold Leaf Letters 6" to 12" (150-300 mm) High. An experienced sign writer should complete 2 to 3 plain Roman or Gothic letters 6" to 12" (150-300 mm) high, in gold leaf, per hr. at the following labor cost per 100 letters:

Description	Hours	Rate	Total
Sign painter	40.00	$30.99	$1,239.60
Cost per letter			$12.40

Painting Pipe

The cost of painting pipe varies considerably, depending on accessibility of work, layout of piping, scaffolding required, type of paint, whether one color or more, and in the case of old work, the amount of surface preparation required and whether or not pipes are in service.

Quantities are expressed in square feet of surface. When pipe quantities are measured from drawings, determine whether or not pipes are covered and the covering thickness, which greatly affects the diameter and consequently the surface area to be painted, especially in the smaller sizes. Most estimators convert lin. ft. of pipe to sq. ft. of surface (m to sq. m) according to the following table:

O.D. Pipe or Covering Inches	Sq.Ft. area per Lin.Ft.	O.D. Pipe or Covering mm	Sq.M. area per Lin.M.
3 " and under	1	75 mm and under	0.30
4 " - 7 "	2	100 - 175 mm	0.61
8 " - 11 "	3	200 - 275 mm	0.91
12 " - 15 "	4	300 - 375 mm	1.22
16 " - 19 "	5	400 - 475 mm	1.52
20 " - 22 "	6	500 - 550 mm	1.83
23 " - 26 "	7	575 - 650 mm	2.13
27 " - 30 "	8	675 - 750 mm	2.44
31 " - 34 "	9	775 - 850 mm	2.74
35 " - 38 "	10	875 - 950 mm	3.05
39 " - 42 "	11	975 - 1050 mm	3.35
42 " - 45 "	12	1,050 - 1,125 mm	3.66
46 " - 48 "	13	1,150 - 1,200 mm	3.96

Various types of paint materials are specified for pipes, depending on the function, exposure, whether bare or covered, etc. Different colors may be designated for different pipes, such as blue for cold water, red for hot water, etc. This may require the contractor to have several "set-ups" (scaffolding) to reach the pipe as each color is applied, because different types of paint may be required for different types of pipe. If all colors are applied at the same time, additional set-ups can be eliminated.

FINISHES

Average material coverages and approximate prices for a few of these materials are as follows:

Description	Sq.Ft. per Gallon	Price per Gallon	Sq.M. per Liter	Price per Liter
Bare Pipe				
Black Coal Tar Epoxy paint, one coat	130 - 150	$30.00 - $45.00	3.19 - 3.68	$ 7.93 - $11.89
Aluminum paint, each coat	550 - 600	$30.00 - $36.00	13.50 - 14.73	$ 7.93 - $ 9.51
Radiator enamel, each coat	550 - 600	$36.00 - $50.00	13.50 - 14.73	$ 9.51 - $13.21
Zinc Paint	450 - 500	$25.00 - $50.00	11.04 - 12.27	$ 6.61 - $13.21
Covered Pipe				
Sizing, one coat	350 - 400	$10.00 - $20.00	8.59 - 9.82	$ 2.64 - $ 5.28
Paint (oil base), each coat	450 - 500	$10.00 - $50.00	11.04 - 12.27	$ 2.64 - $13.21
Industrial enamel, 1st Coat	450 - 500	$30.00 - $50.00	11.04 - 12.27	$ 7.93 - $13.21
Industrial enamel, 2nd Coat	550 - 600	$30.00 - $50.00	13.50 - 14.73	$ 7.93 - $13.21
Paint (Latex-water base), each coat	450 - 500	$15.00 - $25.00	11.04 - 12.27	$ 3.96 - $ 6.61
Zinc Paint	350 - 400	$25.00 - $50.00	8.59 - 9.82	$ 6.61 - $13.21

Labor Painting Pipe. Labor costs for painting pipes are extremely variable due to the many conditions that might be encountered.

For simple work, consisting of horizontal runs and risers within stepladder reach, all painted one color, a painter should paint 800 sq.ft. (74.32 sq.m) of pipe, one coat, per 8-hr. day.

For boiler room or steam power plant piping of average difficulty, where some high ladder work is required, a painter should paint 600 sq.ft. (55.74 sq.m) of pipe, one coat, per 8-hr. day.

For difficult industrial process piping or in very congested steam power plants, where most of the work requires high scaffolding, a painter will paint only 350 to 450 sq.ft. (32.51 to 41.81 sq.m) of pipe, one coat, per 8-hr. day.

If pipes are painted different colors for identification, reduce the above production 10% to 15%.

If pipes are coded by color banding at 10' (3.04 m) to 15' (4.57 m) intervals, consider this work as another coat for labor, but practically no material will be required.

If painted flow direction indicators are required after the pipes have been painted, then add $11.50 each.

Painting Structural Steel

The cost of painting structural steel after erection will depend upon a number of factors, i.e., the weight of the steel, whether light or heavy sections; access to work and method of painting, whether brush painting by hand or using a spray.

Lightweight members cost more to paint than heavy members, and the less climbing required, the more paint a worker will apply, either by brush or when using a spray.

Structural steel may be estimated, either by the ton of steel or by the number of square feet of surface to be painted.

The ton method is much easier to compute but the square feet of surface method is much more accurate.

The following tables, giving the approximate number of square feet of area on various size beams, angles, and channels, will assist the estimator in computing accurate quantities where the square foot method is used.

Sq.Ft. Area per Lin.Ft. (Sq.M Area per Meter) of Structural "S" Beams of Various Sizes					
Depth Beam Inches	Lbs. per Lin. Ft.	Sq.Ft. Area per Lin.Ft.	Depth Beam mm	Kg. per Lin.M.	Sq.M. Area per Lin.M
3 "	5 - 8	1.25	75 mm	7.44 - 11.91	0.381
4 "	7 - 10	1.40	100 mm	10.42 - 14.88	0.427
5 "	10 - 15	1.90	125 mm	14.88 - 22.32	0.579
6 "	12 - 17	2.00	150 mm	17.86 - 25.30	0.610
7 "	15 - 20	2.60	175 mm	22.32 - 29.76	0.792
8 "	18 - 25	2.70	200 mm	26.79 - 37.20	0.823
9 "	21 - 35	2.90	225 mm	31.25 - 52.09	0.884
10 "	25 - 40	3.40	250 mm	37.20 - 59.53	1.036
12 "	31 - 55	3.70	300 mm	46.13 - 81.85	1.128
15 "	42 - 75	4.50	375 mm	62.50 - 111.61	1.372
18 "	54 - 70	5.00	450 mm	80.36 - 104.17	1.524
20 "	65 - 100	5.40	500 mm	96.73 - 148.82	1.646
24 "	80 - 120	6.40	600 mm	119.06 - 178.58	1.951
27 "	90 - 135	7.50	675 mm	133.94 - 200.91	2.286

Sq.Ft. Area per Lin.Ft. (Sq.M Area per Meter) of Structural Channels of Various Sizes					
Depth Beam Inches	Lbs. per Lin. Ft.	Sq.Ft. Area per Lin.Ft.	Depth Beam mm	Kg. per Lin.M.	Sq.M. Area per Lin.M
3 "	4 - 6	1.00	75 mm	5.95 - 8.93	0.305
4 "	5 - 7	1.20	100 mm	7.44 - 10.42	0.366
5 "	7 - 11	1.50	125 mm	10.42 - 16.37	0.457
6 "	8 - 16	1.70	150 mm	11.91 - 23.81	0.518
7 "	10 - 20	2.00	175 mm	14.88 - 29.76	0.610
8 "	11 - 21	2.20	200 mm	16.37 - 31.25	0.671
9 "	14 - 25	2.35	225 mm	20.83 - 37.20	0.716
10 "	14 - 35	2.70	250 mm	20.83 - 52.09	0.823
12 "	21 - 40	3.00	300 mm	31.25 - 59.53	0.914
15 "	32 - 50	3.70	375 mm	47.62 - 74.41	1.128
18 "	34 - 55	4.40	450 mm	50.60 - 81.85	1.341
20 "	42 - 58	5.00	500 mm	62.50 - 86.31	1.524

1418

Sq.Ft. Area Per Lin.Ft. (Sq.M Area per Meter) of Structural Angles of Various Sizes						
Depth Chanel Inches	Lbs. per Lin.Ft.	Sq.Ft. Area per Lin.Ft.	Depth Channel mm	Kg. per Lin.M.	Sq.M. Area per Lin.M.
2 " x 2 "	2 - 5	0.70	50 x 50 mm	2.98 - 7.44	0.213
3 " x 3 "	2 - 11	1.00	75 x 75 mm	2.98 - 16.37	0.305
4 " x 4 "	5 - 20	1.35	100 x 100 mm	7.44 - 29.76	0.411
5 " x 5 "	8 - 31	1.70	125 x 125 mm	11.91 - 46.13	0.518
6 " x 6 "	13 - 40	2.00	150 x 150 mm	19.35 - 59.53	0.610
8 " x 8 "	26 - 57	2.70	200 x 200 mm	38.69 - 84.83	0.823
4 " x 3 "	4 - 17	1.20	100 x 75 mm	5.95 - 25.30	0.366
5 " x 3 "	6 - 19	1.40	125 x 75 mm	8.93 - 28.28	0.427
6 " x 4 "	10 - 30	1.70	150 x 100 mm	14.88 - 44.65	0.518
7 " x 4 "	13 - 34	1.90	175 x 100 mm	19.35 - 50.60	0.579
8 " x 4 "	17 - 37	2.00	200 x 100 mm	25.30 - 55.06	0.610
8 " x 6 "	20 - 50	2.40	200 x 150 mm	29.76 - 74.41	0.731

Painting Lightweight Structural Steel After Erection. When painting, by hand and using a brush, lightweight structural steel after erection, a painter should paint 600 to 650 sq.ft. (54 to 60 sq.m) per 8-hr. day on the first coat; 800 to 900 sq.ft. (74.32 to 83.61 sq.m) per 8-hr. day on the second and third coats.

On a tonnage basis, this would be at the rate of 2 to 3 tons (1.32 to 2.73 metric tons) per 8-hr. day on the first coat and 3 to 4 tons (2.73 to 3.64 metric tons) per 8-hr. day on the second and third coats.

When using a spray gun, a painter should paint 3,600 sq.ft. (334.44 sq.m) per 8-hr. day on the first coat and 4,800 sq.ft. (445.92 sq.m) on the second and third coats.

Painting Medium to Heavyweight Structural Steel After Erection. When painting, by hand using a brush, medium to heavyweight structural steel after erection, a painter should paint 750 to 850 sq.ft. (70 to 79 sq.m) on the first coat and 900 to 1,000 sq.ft. (83.61 to 92.90 sq.m) per 8-hr. day on the second and third coats.

On a tonnage basis, this would be at the rate of 4 to 5 tons (3.64 to 4.55 metric tons) per 8-hr. day on the first coat and 5 to 6 tons (4.55 to 5.46 metric tons) per 8-hr. day on the second and third coats.

Using a spray gun, a painter should paint 4,400 sq.ft. (408.76 sq.m) per 8-hr. day on the first coat and 5,600 sq.ft. (520.24 sq.m) per 8-hr. day on the second and third coats.

Labor Cost of 100 Sq.Ft. (9.29 Sq.M) 2-Coat Oil Paint Applied to Lightweight Structural Steel				
Painting by Brush				
Description	Hours	Rate	Total	
Painter - First Coat	1.30	$30.99	$	40.29
Painter - Second Coat	1.00	$30.99	$	30.99
Cost 100 sq.ft. (1 sq.)			$	71.28
per sq.ft.			$	0.71
per sq.m			$	7.67
Spray Painting				
Description	Hours	Rate	Total	
Painter - First Coat	0.23	$30.99	$	7.13
Painter - Second Coat	0.17	$30.99	$	5.27
Cost 100 sq.ft. (1 sq.)			$	12.40
per sq.ft.			$	0.12
per sq.m			$	1.33

For third coat, add same as second coat.
Add for air compressor and engineer, depending upon number of spray guns operated.

Labor Cost of 100 Sq.Ft. (9.29 Sq.M) 2-Coat Oil Paint Applied to Medium to Heavy weight Structural Steel				
Painting by Brush				
Description	Hours	Rate	Total	
Painter - First Coat	1.00	$30.99	$	30.99
Painter - Second Coat	0.90	$30.99	$	27.89
Cost 100 sq.ft. (1 sq.)			$	58.88
per sq.ft.			$	0.59
per sq.m			$	6.34
Spray Painting				
Description	Hours	Rate	Total	
Painter - First Coat	0.18	$30.99	$	5.58
Painter - Second Coat	0.15	$30.99	$	4.65
Cost 100 sq.ft. (1 sq.)			$	10.23
per sq.ft.			$	0.10
per sq.m			$	1.10

For third coat, add same as second coat.
Add for air compressor and engineer, depending upon number of spray guns operated.

Add for air compressor and engineer, depending on number of spray guns operated. Not included in above costs are equipment such as scaffolding, and costs of complying with hazardous waste regulations.

FINISHES
Preparing Steel Surfaces

Proper protection of steel depends upon proper adhesion of the protective paint. When rust, oils, grease, scale, and other dirt are not completely removed, the paint coat or coats will fail no matter how skillfully they are applied.

The Society for Protective Coatings, formally the Steel Structures Painting Council (SSPC), Pittsburgh, PA, publishes detailed specifications for cleaning methods.

The simplest method is hand cleaning using scrapers, sanders, wire brushes, and chipping tools. These remove only loose rust and scale and effectiveness depends largely on the zeal of the worker. Power tool cleaning, using power operated sanders and brushes, is slightly more expensive and slightly more effective.

For a superior job, one of the blasting methods is suggested: commercial, near white or white. The last specifies the metal be blasted down to a uniform "white" surface and will provide a base good for 7 to 8 years. Near white specifies that at least 95% of the metal will meet the "white" surface test, and its durability almost equals white blast.

Commercial blast is difficult to specify, but the results should leave at least two-thirds of the metal in the "white" stage, with durability of from 6 to 7 years.

With all the above methods, it is essential to use a primer with a good "wetting" power.

In figuring all these methods, cleaning up and protection for adjacent surfaces can add materially to the cost.

Surface Preparation by Hand and Abrasive Method. A properly prepared surface can make a good painting job look even better. A poorly prepared one can make an otherwise good job look bad. Remember that paint is not only to make a surface attractive. It protects a material from exposure and prolongs the useful life of the material. Paint protects steel surfaces from oxidation that results in rust.

Estimators need to understand the types of paints and coatings that an architect or engineer might specify for steel. Knowing the proper procedures for applying a specific painting material will give the estimator a competitive edge. The following checklist indicates major items to consider:

1. Type of surface to be painted.
2. Type of surface preparation - scrape or sandblast.
3. Scaffolding and other equipment requirements (OSHA).
4. Federal, state, and local environmental protection agency requirements.
5. Proper applicators - brushes, rollers, or sprayers.
6. Priming requirements - compatible with final material.
7. Number of paint coats required.
8. Disposal and clean up requirements.

The method of cleaning the surface in preparation for painting is usually dictated by the specifications, which might refer to the paint manufacturer's recommendations. Wood surfaces are usually limited to a few choices of methods, but steel and concrete offer a wide range of possible methods, depending on the condition of the existing surface. Scaled or rusted surfaces can be cleaned by hand, with power tools, or using abrasive methods. A required degree of cleanliness is often stated in the contract documents and is defined by SSPC specifications.

Cleaning Structural Steel with Hand Wire Brush (SSPC-SP 2). When cleaning structural steel surfaces by hand with a wire brush, a worker can clean about 300 sq.ft. (28 sq.m) of surface per 8-hr. day at the following labor cost per 100 sq.ft. (9.29 sq.m):

Description	Hours	Rate	Total
Painter	2.67	$30.99	$ 82.74
Cost per sq.ft.			$ 0.83
per sq.m			$ 8.91

Cleaning Piping 14" (350 mm) Diameter or Less with Hand Wire Brush. When cleaning steel pipe surfaces by hand with a wire brush, a worker can clean about 225 sq.ft. (21 sq.m) of surface per 8-hr. day at the following labor cost per 100 sq.ft. (9.29 sq.m):

Description	Hours	Rate	Total
Painter	3.20	$30.99	$ 99.17
Cost per sq.ft.			$ 0.99
per sq.m			$ 10.67

Cleaning Structural Steel with Power Tools (SSPC-SP 3). When cleaning rust and scale from structural steel members, bridges, gas holders, oil tanks, and other metal surfaces before repainting, the number of sq.ft. cleaned per 8-hr. day varies with the type of surface and accessibility. A worker using air-powered wire brushes should clean 1,000 to 1,200 sq.ft. (92.9 to 111.48 sq.m) per day at the following cost per 100 sq.ft. (9.29 sq.m):

Description	Hours	Rate	Total
Painter	0.72	$30.99	$ 22.31
Compressor Operator	0.36	$37.89	$ 13.64
Compressor Expenses and Tools	0.36	$60.00	$ 21.60
Cost per 100 sq.ft.			$ 57.55
per sq.ft.			$ 0.58
per sq.m			$ 6.20

Cleaning Piping 14" (350 mm) Diameter or Less with Power Tools. When cleaning rust and scale off structural steel pipe, the number of sq.ft. cleaned per day will vary with surface and accessibility. A worker should clean 750 to 900 sq.ft. (69.67 to 83.61 sq.m) per day using air-powered wire brushes at the following cost per 100 sq.ft. (9.29 sq.m):

Description	Hours	Rate	Total
Painter	0.97	$30.99	$ 30.06
Compressor Operator	0.49	$37.89	$ 18.57
Compressor Expenses and Tools	0.49	$60.00	$ 29.40
Cost per 100 sq.ft.			$ 78.03
per sq.ft.			$ 0.78
per sq.m			$ 8.40

Add Scaffolding or rigging, clean-up and disposal

Commercial Blast Cleaning of Structural Steel (SSPC-SP 6). A blast cleaned surface should be free of all visible (viewed without magnification) oil, grease, dirt, mill scale, rust, paint, oxides, corrosion products, and other foreign matter, except staining. This should provide a base good for 7 to 8 years when the surface is properly painted.

A worker should clean 500 to 600 sq.ft. (46.45 to 55.74 sq.m) of structural steel members, bridges, etc., per day using abrasive blasting methods.

Description	Hours	Rate	Total
Painter	1.45	$30.99	$ 44.94
Compressor Operator	1.45	$37.89	$ 54.94
Compressor Expenses and Tools	1.45	$60.00	$ 87.00
Sandblasting Equipment	1.45	$50.00	$ 72.50
Cost per 100 sq.ft.			$ 259.38
per sq.ft.			$ 2.59
per sq.m			$ 27.92

Add Scaffolding or rigging, clean-up and disposal

Commercial Blast Cleaning of Piping, 14" (350 mm) Diameter and Less. A worker should clean 300 to 500 sq.ft. (27.87 to 46.45 sq.m) of steel pipe per day using abrasive blasting methods at the following cost:

Description	Hours	Rate	Total
Painter	2.00	$30.99	$ 61.98
Compressor Operator	2.00	$37.89	$ 75.78
Compressor Expenses and Tools	2.00	$60.00	$ 120.00
Sandblasting Equipment	2.00	$50.00	$ 100.00
Cost per 100 sq.ft.			$ 357.76
per sq.ft.			$ 3.58
per sq.m			$ 38.51

Add Scaffolding or rigging, clean-up and disposal

Brush Off Blast Cleaning of Structural Steel (SSPC-SP 7). A brush off blast cleaned surface should be free of all visible (viewed without magnification) oil, grease, dirt, dust, loose mill scale, loose rust, and loose paint. Tightly adherent mill scale, rust, and paint that cannot be removed by lifting with a dull putty knife may remain on the surface.

On structural steel members, bridges, etc., a worker should clean 900 to 1,100 sq.ft. (83.61 to 102.19 sq.m) per day using this method:

Description	Hours	Rate	Total
Painter	0.80	$30.99	$ 24.79
Compressor Operator	0.80	$37.89	$ 30.31
Compressor Expenses and Tools	0.80	$60.00	$ 48.00
Sandblasting Equipment	0.80	$50.00	$ 40.00
Cost per 100 sq.ft.			$ 143.10
per sq.ft.			$ 1.43
per sq.m			$ 15.40

Add Scaffolding or rigging, clean-up and disposal

Brush Off Blast Cleaning Piping 14" (350 mm) Diameter or Less. On steel pipes, a worker should clean 680 to 820 sq.ft. (63.17 to 76.18 sq.m) per day using this method at the following cost:

Description	Hours	Rate	Total
Painter	1.07	$30.99	$ 33.16
Compressor Operator	1.07	$37.89	$ 40.54
Compressor Expenses and Tools	1.07	$60.00	$ 64.20
Sandblasting Equipment	1.07	$50.00	$ 53.50
Cost per 100 sq.ft.			$ 191.40
per sq.ft.			$ 1.91
per sq.m			$ 20.60

Add Scaffolding or rigging, clean-up and disposal

Near White Blast Cleaning of Structural Steel (SSPC-SP 10). A near-white blast cleaned surface, when viewed without magnification, should be free of all visible oil, grease, dirt, dust, mill scale, rust, paint, oxides, corrosion products, and other foreign matter, except for staining.

On structural steel members, bridges, etc., a worker using this method should clean 300 to 400 sq.ft. (27.87 to 37.16 sq.m) per day at the following cost:

Description	Hours	Rate	Total
Painter	2.29	$30.99	$ 70.97
Compressor Operator	2.29	$37.89	$ 86.77
Compressor Expenses and Tools	2.29	$60.00	$ 137.40
Sandblasting Equipment	2.29	$50.00	$ 114.50
Cost per 100 sq.ft.			$ 409.64
per sq.ft.			$ 4.10
per sq.m			$ 44.09

Add Scaffolding or rigging, clean-up and disposal

Near White Blast Cleaning Piping 14" (350 mm) Diameter and Under. On steel pipes, a worker should clean 225 to 300 sq.ft. (20.90 to 27.87 sq.m) per day:

FINISHES

Description	Hours	Rate	Total
Painter	3.05	$30.99	$ 94.52
Compressor Operator	3.05	$37.89	$ 115.56
Compressor Expenses and Tools	3.05	$60.00	$ 183.00
Sandblasting Equipment	3.05	$50.00	$ 152.50
Cost per 100 sq.ft.			$ 545.58
per sq.ft.			$ 5.46
per sq.m			$ 58.73

Add Scaffolding or rigging, clean-up and disposal

White Metal Blast Cleaning Structural Steel (SSPC-SP 5). Viewed without magnification, a white metal blast cleaned surface should be free of all visible oil, grease, dirt, dust, mill scale, rust, paint, oxides, corrosion products, and other foreign matter.

On structural steel members, bridges, etc., a worker should clean 230 to 300 sq.ft. (21.36 to 27.87 sq.m) per day using this method:

Description	Hours	Rate	Total
Painter	3.01	$30.99	$ 93.28
Compressor Operator	3.01	$37.89	$ 114.05
Compressor Expenses and Tools	3.01	$60.00	$ 180.60
Sandblasting Equipment	3.01	$50.00	$ 150.50
Cost per 100 sq.ft.			$ 538.43
per sq.ft.			$ 5.38
per sq.m			$ 57.96

Add Scaffolding or rigging, clean-up and disposal

White Metal Blast Cleaning Piping, 14" (350 mm) Diameter or Less. On steel pipes, a worker should clean 170 to 230 sq.ft. (15.79 to 21.36 sq.m) per day using this method:

Description	Hours	Rate	Total
Painter	4.00	$30.99	$ 123.96
Compressor Operator	4.00	$37.89	$ 151.56
Compressor Expenses and Tools	4.00	$60.00	$ 240.00
Sandblasting Equipment	4.00	$50.00	$ 200.00
Cost per 100 sq.ft.			$ 715.52
per sq.ft.			$ 7.16
per sq.m			$ 77.02

Add Scaffolding or rigging, clean-up and disposal

Spray Painting. Almost any painting job can be done quicker with spray equipment. This is particularly true where large surfaces are to be covered or where there are angles and corners that cannot be reached with brushes or rollers.

Practically all surfaces that are painted with a bristle brush can be coated by spray painting. On rough surfaces such as cement, stucco, rough plaster, brick, tile, shingles, and rough boards, spray painting affects a considerable saving in time and labor.

It is now possible to obtain mechanical painting equipment that will apply any paint, varnish, graphite, mineral oxides, or epoxy.

Painting Oil Tanks and other Large Metal Surfaces by Compressed Air. In painting oil tanks and other large metal surfaces, a worker should cover 1,000 sq.ft. (92.90 sq.m) per hr. at the following cost per 100 sq.ft. (9.29 sq.m):

Description	Hours	Rate	Total
Painter	0.10	$30.99	$ 3.10
Compressor Operator	0.05	$37.89	$ 1.89
Compressor Expenses and Tools	0.05	$60.00	$ 3.00
Cost per 100 sq.ft.			$ 7.99
per sq.ft.			$ 0.08
per sq.m			$ 0.86

Add Scaffolding or rigging, clean-up and disposal

Comparative Cost by Hand

On the same class of work, a painter will brush about 200 sq.ft. (18.58 sq.m) of surface per hr. at the following cost per 100 sq.ft. (9.29 sq.m):

Description	Hours	Rate	Total
Painter	0.50	$30.99	$ 15.50
Cost per sq.ft.			$ 0.15
per sq.m			$ 1.40

Does not include paint, scaffold, or preparation.

09950 WALLCOVERING

Wallpaper is estimated by the roll containing 36 sq.ft. (3.34 sq.m). Most rolls are 18" (450 mm) or 20-1/2" (512.5 mm) wide and contain 36 sq.ft. or 4 sq.yds. (3.34 sq.m) of paper. The double roll contains twice as much as the single roll: 72 sq.ft. or 8 sq.yds. (6.68 sq.m) of paper.

Estimating Quantity of Wallpaper for Any Room. When estimating the quantity of wallpaper required for any room, measure the entire distance around the room in lin.ft. (meters) and multiply by the height of the walls or the distance from the floor to the ceiling. The result will be the number of sq.ft. (sq.m) of surface in the four walls.

Make deductions in full for the area of all openings such as windows, doors, mantels, and built-in bookcases. The difference between the area of the four walls and the area of the openings will be the actual number of sq.ft. (sq.m) of surface to be papered.

When adding for waste in matching, cutting, fitting, etc., paper hangers use different methods and allow different percentages, contingent upon the height of the ceiling, the design of the paper, and the size of the pattern or figure, but an allowance of 15% to 20% is sufficient in nearly all cases.

1426

FINISHES

When estimating the quantity of paper required for walls, bear in mind it is not necessary to run the side wallpaper more than one or two inches (25-50 mm) above the border, when a border is used. The height of the wood base should also be deducted from the total height of the wall.

In rooms having drop ceilings, the depth of the drop should be deducted from the height of the wall, which will decrease the area of side wallpaper in the same amount as it increases the quantity of ceiling paper.

If a border is required, measure the distance around the room and the result will be the number of lin.ft. (meters) of border. Dividing by 3, gives the number of yards of border required.

When estimating the quantity of wallpaper required for ceilings, multiply the width of the room by the length, and the result will be the number of sq.ft. (sq.m) of ceiling to be papered.

In rooms having a drop ceiling, add the depth of the drop or the distance the ceiling paper extends down the side walls to the length and breadth of the room and multiply as described above.

After the areas of both side walls and ceiling have been computed in sq.ft., divide by 36 (sq.m divide by 3.3), and the result will be the number of single rolls of paper required, or by dividing by 72 (sq.m divided by 6.7) gives the number of double rolls of paper required.

Paste Required Hanging Wallpaper. The quantity of paste required for hanging wallpaper will vary with the weight of the paper and the surface to which it is applied.

Where light or medium weight wallpaper is used, one gallon (3.79 liters) of paste should hang 12 single rolls of paper.

If heavy or rough texture paper is used, it will often be necessary to give it two or three applications of paste to obtain satisfactory results. On work of this kind one gallon of paste should hang 4 to 6 single rolls of paper.

There are prepared pastes on the market that require only the addition of cold water to make them ready for use.

One lb. of prepared cold water dry paste should make 1-1/2 to 2 gals. (5.7 to 7.6 liters) of ready to use paste.

Labor Hanging Wallpaper. Different methods of hanging wallpaper are used in different sections of the country. In some localities a paper hanger and helper work together, one worker trimming the paper and pasting while the other hangs the paper. However, in many cities the work is performed by one person, who cuts, trims, fits, pastes, and hangs the paper.

The following quantities and costs are based on the performance of one worker but will prove a fair average for use in any locality.

Hanging Wallpaper on Walls and Ceilings, One Edge Work. When hanging light or medium weight paper on ceilings and drops, a paper hanger should hang 28 to 32 single rolls of paper per 8-hr. day, at the following labor cost per roll:

Description	Hours	Rate	Total
Paper hanger	0.27	$30.99	$ 8.37

Hanging Wallpaper on Walls, Butt Work. Where light or medium weight paper is used for bedrooms, halls, etc., and where a good grade of workmanship is required with all paper trimmed on both edges and hung with butt joints, a paper hanger should trim, fit, and hang 20 to 24 single rolls of paper per 8-hr. day, at the following labor cost per roll:

Description	Hours	Rate	Total
Paper hanger	0.36	$30.99	$ 11.16

Another thing that will slow up the work is where new paper is applied over old rough textured wallpaper, such as oatmeal designs, etc.

Hanging Wallpaper, First Grade Workmanship. Where a good grade of medium or heavy weight wallpaper is used, with all paper hung with butt joints, a paper hanger should trim, fit, and hang 16 to 20 single rolls of paper per 8-hr. day, at the following labor cost per roll:

Description	Hours	Rate	Total
Paper hanger	0.44	$30.99	$ 13.64

Labor Hanging Scenic Paper. Where scenic paper or paper having mural designs are used over the wall or up stairways, a paper hanger will hang only 8 to 10 single rolls per 8-hr. day, at the following labor cost per roll:

Description	Hours	Rate	Total
Paper hanger	0.90	$30.99	$ 27.89

Placing Wallpaper Borders. Wallpaper borders are usually estimated by the lin.ft. or yd. (meter) and as the width varies from 3" to 18" (75 to 450 mm) there is never over one width required.

A paper hanger should place border at the rate of 100 to 125 lin.ft. (30 to 37.5 m) per hr. The labor cost per 100 lin.ft. (30 m) should run as follows:

Description	Hours	Rate	Total
Paper hanger	0.90	$30.99	$ 27.89

Hanging Coated Fabrics. Where coated fabrics, such as Walltex, Sanitas, etc., are used, a paper hanger should hang 16 to 20 single rolls per 8-hr. day, at the following labor cost per roll:

1428

FINISHES

Description	Hours	Rate	Total
Paper hanger	0.44	$30.99	$ 13.64

Canvassing Plastered Walls. When applying canvas to plastered walls, an average mechanic should apply 50 to 60 sq.ft. (4.6 to 5.6 sq.m) per hour.

Description	Hours	Rate	Total
Paper hanger	1.82	$30.99	$ 56.40

Flexwood. Flexwood from U.S. Plywood, Louisville, KY, is a genuine wood veneer cut to 1/85" (0.29 mm) thick and glued under heat and hydraulic pressure to cotton sheeting with a waterproof adhesive. A patented flexing operation breaks the cellular unity of the wood to produce a limp, pliable sheet which may be applied to any smooth surface, flat or curved. Waterproof Flexwood 710 Adhesive, which makes a permanent bond, is used to apply Flexwood. Standard sizes of stock material are 18" (450 mm) and 24" (600 mm) wide and 8' (2.43 m) to 12' (3.65 m) long.

The mechanics of installing Flexwood are as follows: the background is sized with Flexwood cement. Another coating of cement is brushed on the Flexwood, which is then hung in the manner of any sheet wall covering. A stiff, broad knife, used with considerable pressure, smoothes out the Flexwood, removes air spaces and furnishes the necessary contact.

Some of the wood available in Flexwood are mahogany, walnut, oak, prima vera, knotty pine, orientalwood, satinwood, zebrawood, rosewood, English oak, maple, lacewood, etc. The cost of the material varies from $1.90 to $6.00 per sq.ft. ($20.45 to $64.58 per sq.m) depending upon the kind of wood. Most woods run from $2.25 to $3.00 per sq.ft ($24.22 to 32.29 per sq.m). Adhesive costs $15.25 per gal. ($4.03 per liter).

It requires a skilled and thoroughly competent paper hanger to apply this material and the labor costs vary considerably with the design to be obtained.

Considerable time is usually required to lay out any room to obtain the desired effect, spacing of strips, etc. This is especially true where certain designs must be obtained on columns, walls, and pilasters and where narrow strips of contrasting woods are used to obtain inlay effects.

On plain walls without inlays, a paper hanger experienced in this class of work should apply 150 sq.ft. (13.93 sq.m) of Flexwood per 8-hr. day, at the following labor cost per 100 sq.ft. (9.29 sq.m):

Labor Costs of 100 Sq.Ft. Canvassed Walls			
Description	Hours	Rate	Total
Paper hanger	5.40	$30.99	$ 167.35
Cost per sq.ft.			$ 1.67
per sq.m			$ 18.01

1429

On pilasters, columns, and walls inlaid with narrow strips of contrasting woods, the work is considerably slower, due to the additional cutting and fitting required. On work of this kind, a paper hanger, experienced in this class of work should apply 75 sq.ft. (6.96 sq.m) per 8-hr. day, at the following labor cost per 100 sq.ft. (9.29 sq.m):

Description	Hours	Rate	Total
Paper hanger	10.70	$30.99	$ 331.59
Cost per sq.ft.			$ 3.32
per sq.m			$ 35.69

Additional time for paper hanger or foreman will be required laying out the work into patterns, marking off strips, etc. This will vary with the size of the job.

After the Flexwood is applied, the joints should be sanded lightly to remove any imperfections in the joints.

Finishing Flexwood. Flexwood will take any wood finish, but where the natural color of the wood is desired, the Flexwood is given one coat of lacquer sealer, sanded lightly between coats and then given one coat of lacquer for a finish coat.

One gallon (3.79 liters) of lacquer sealer will cover about 300 sq.ft. (27.87 sq.m) of surface, one coat.

After the sealer has been applied, one gallon (3.79 liters) of lacquer should cover 450 to 550 sq.ft. (41.80 to 51.10 sq.m).

An experienced painter should apply lacquer sealer, sand lightly between coats and apply one coat of lacquer to 90 to 110 sq.ft. (8.36 to 10.21 sq.m) of surface per hr. at the following labor cost per 100 sq.ft. (9.29 sq.m):

Description	Hours	Rate	Total
Painter	1.00	$30.99	$ 30.99
Cost per sq.ft.			$ 0.31
per sq.m			$ 3.34

Vinyl Wall Covering. Vinyl wall covering is composed of a woven cotton fabric, to which a compound of vinyl resin, pigment, and plasticizer is electronically fused to one side. It comes 24" (600 mm), 27" (675 mm), and 54" (1350 mm) wide in three weights: heavy, 36 oz. per yd. (1.11 kg per m); medium, 24-33 oz. per yd. (0.74-1.02 kg per m); and light, 22-24 oz. per yd. (0.68-0.74 kg per m).

Some of the patterns available simulate linens, silks, moires, grasses, tweeds, Honduras mahogany, travertine, damasks, and burlaps. Costs range from $0.5635 to $1.45 per sq. ft. ($6.02 to $15.61 per sq.m).

Vinyl fabric is hung using regular wallpaper hanging procedures for hanging fabric baked wallcoverings. A broad knife is used to smooth out any

1430

wrinkles, air pockets and insures a good contact. Wash off any excess paste that remains on the surface of the material. No further finishing is necessary.

A competent paper hanger can apply approximately 600 sq.ft. (55.74 sq.m) per 8 hr. day at the following costs per 100 sq.ft. (9.29 sq.m):

Description	Hours	Rate	Total
Paper hanger	1.34	$30.99	$ 41.53
Cost per sq.ft.			$ 0.42
per sq.m			$ 4.47

If excessive cutting and fitting is required, the above figures should be increased to allow for the above conditions.

10   SPECIALTIES

The Construction Specifications Institute (CSI), whose format we use in the general presentation of this text, follow sections covering the usual building trades with four sections covering allied fields, which may or may not be part of the typical general contract.

CSI Division 10 includes specialties that are prefabricated and added to the building, such as chalkboards, toilet partitions, decorative grilles, freestanding fireplaces, flagpoles, directories and signs, lockers, mailboxes and chutes, movable partitions, sun control devices, telephone booths, and the like. Many of these items will be subcontracted and installed under the general contractor's supervision. Some may be covered by allowances.

10100 CHALKBOARDS AND TACKBOARDS

Chalkboards are made of 1/4" (6.25 mm) or 1/2" (12.50 mm) hardboard with special finishes applied. These include the following which are available in either black or green:

1/4" (6.25 mm) Thick - Material Prices		
Description	Price per Sq.Ft.	Price per Sq.M.
Standard, weighing 1.75 lbs per sq.ft. (8.54 Kg/Sq.M.)	$ 2.24	$ 24.11
Moisture sealed at 1.8 lbs. per sq.ft (8.79 Kg/Sq.M.)	$ 2.59	$ 27.88
Porcelain steel at 2.1 lbs. per sq.ft. (10.25 Kg/Sq.M.)	$ 6.72	$ 72.34

1433

These are available in 4' (1.22 m) widths in sheets to 16' (4.87 m) long and are applied to the wall with mastic at the rate of one gal. per 70 sq.ft (3.875 liter per 6.50 sq.m). They can be set in custom trim, or the following stock aluminum trim, which comes complete with all necessary attaching clips:

Description	Price per Lin. Ft.	Price per Lin. M.
Perimeter trim 1-1/2" (37.5 mm) wide	$ 4.90	$ 16.08
Maprail headtrim	$ 3.15	$ 10.33
Joint strip	$ 1.65	$ 5.42
Chalktrough 2-1/4" (56.25 mm) wide	$ 6.58	$ 21.59

Bulletin board materials are 1/4" (6.25 mm) cork and available as follows		
Description	Price per Sq.Ft.	Price per Sq.M.
Natural cork	$ 2.94	$ 31.65
Vinyl cork	$ 4.34	$ 46.72
Cork on 3/8" (9,37 mm) fiberboard	$ 4.41	$ 47.47
1/8" on 3/8" (3.12 mm on 9.37 mm) fiberboard vinyl f'c'd	$ 4.34	$ 46.72

The cork comes in rolls 4' (1.22 m) wide by up to 90' (27.43 m) long. The panels are 4' (1.22 m) wide up to 16' (4.87 m) long. Stock trim costs $4.00 for 1/4" (6.25 mm) stock, $4.50 for 1/2" (12.50 mm) stock.

10150 COMPARTMENTS AND CUBICLES
Toilet Partitions and Screens

Toilet partitions are most often floor-mounted metal units, with baked enamel finish and chrome hardware. One door and side partition for use in a corner or abutting other stall side partitions cost about $185.00. A porcelain enameled finish adds $600.00 to the cost. Ceiling hung units cost around 10% more, plus of course the cost of ceiling reinforcement if needed. One carpenter and laborer will set a unit in about two hours. Urinal screens hung from the wall cost $330.00 in baked enamel and will take a two-person team half an hour to set. Wall hung toilet stalls will cost $700.00per unit.

10301 PREFABRICATED FIREPLACES

Metal fireplaces are available in a wide range of styles for burning wood or gas. They are made for placing against walls or in corners, and as freestanding units. However, there are requirements for clearances from walls and for floor protection. They can be vented to an existing flue or have a

1434

SPECIALTIES

prefabricated smoke pipe terminating above the roof. A simulated masonry chimney top is available. Finishes are either matte black or porcelain. A 24" (600 mm) wide unit costs $800.00 in matte, $1,000.00 in porcelain. A 45" (1125 mm) wide unit costs $1,150.00 in matte and $1,200.00 in porcelain.

Freestanding units open on all sides will cost $800.00 for matte and $1,200.00 for porcelain in 36" (900 mm) diameter size; $1,100.00and $1,250.00 in 42" (1050 mm) diameter. If they are to rest on the floor, add $100.00 for the metal base. If hung from the ceiling, add $60.00 for chains.

The 8" (200 mm) metal flues cost about $45.00 per lin.ft. ($147.63 per m). Chimney top housings for single flues cost $170.00 to $230.00.

10350 FLAGPOLES

Typical prices for several types of flagpoles are as follows:

The 40'-0" (1219 m) ground type, swaged sectional steel, including fittings, ground tube, and flashing collar, costs $2,000.00 erected in place and painted.

The 40'-0" (12 m) ground type, cone tapered steel, including fittings, ground tube, and flashing collar, costs $2,100.00 erected in place.

To above prices, add cost of concrete foundation and ornamental base if required.

The 20'-0" (6 m) roof type, swaged sectional steel, including fittings, socket plate, tension braces, and flashing collars, costs $1,050.00, erected in place and painted. Add cost of setting anchor bolts.

The 12'-0" (3.65 m) outrigger type, cone tapered aluminum, including fittings and base is priced at $1,100.00 erected in place. Add cost of setting anchor bolts.

10400 IDENTIFYING DEVICES
Directories and Signs

Directory boards are available in glass-enclosed or open-faced models. Glass door units cost $390.00 for 36"x30" (900 x 750 mm), $550.00 for 36"x48" (900 x 1200 mm). Open faced directory boards cost as follows, with 100 letters:

12 "x 18 "	$105.00	300 x 450 mm
24 "x 18 "	$120.00	600 x 450 mm
24 "x 36 "	$160.00	600 x 900 mm

Letters for directory boards are ordered in an assortment, including numbers. An assortment of 200 will cost as follows: 1/2" (12.50 mm), $45.00; 3/4" (18.75 mm), $55.0039.00; 1" (25 mm), $70.00; 1-1/2" (37.50 mm), $84.00.

1435

Plaques

Building committees often will ask the contractor to order a plaque to commemorate a donor, an outstanding citizen, or even themselves. Cast bronze plaques is the ultimate for permanence and costs about $220.00 per sq.ft. ($2,368.14 per sq.m), aluminum is about 20% less, as are built-up plaques.

Signs

Cast letters are often applied to walls and copings to identify the building occupant. Letters are cast from aluminum or bronze and then given a variety of finishes. Prices will cost as follows:

Letter Size	2"	4"	6"	12"	18"	24"
Material & Finish						
Cast Aluminum	$ 14.00	$ 22.75	$ 42.00	$ 64.75	$ 115.50	$ 164.50
Cast Bronze	$ 21.00	$ 42.00	$ 57.75	$ 115.50	$ 175.00	$ 231.00
Metric	50 mm	100 mm	150 mm	300 mm	450 mm	600 mm

A worker should install 3 to 4 letters per hr. Plastic letters will cost $17.00 each in 8" (200 mm) height.

10500 LOCKERS

Lockers are made in single, double, and triple high units totaling 72" (1800 mm) high, single tier also comes 60" (1500 mm) high, in widths 9" (225 mm), 12" (300 mm), 15" (375 mm), and 18" (450 mm) and depths of 12" (300 mm), 15" (375 mm), and 18" (450 mm). There are also box lockers 6 units high.

A 12"x15"x72" (300 x 375 x 1800 mm) unit in baked enamel will cost about $70.00 single tier; $95.00 double tier; $130.00 triple tier. For a sloping top add $3.00 per tier. For closed base, add $5.50 per lin.ft. ($18.04 Lin M). Open mesh or "athletic" types will cost $155.00 to $190.00 per tier 72" (1800 mm) high. Basket racks 8' (2.43 m) high, 5' (1.52 m) wide cost $215.00 plus $11.00 per basket. For a combination lock, add $41.00 per door, $18.00 for a key lock.

Benches for lockers are sold by the lineal foot (meter). Benches made of 1-1/4" (31.25 mm) thick laminated maple cost $35.00 per foot ($114.83 per Lin.M.) plus $44.00 for each pedestal.

10550 POSTAL SPECIALTIES

Mail chutes with glass and aluminum enclosures 9"x3-1/2" (225 x 87.50 mm) cost about $70.00 per lin.ft., and one worker can install two floor heights per day. A lobby collection box will cost from $1,600.00 on up, depending on the size and design.

Keyed mail delivery boxes in gangs will cost around $65.00 per unit rear loading and $75.00 front loading. A worker can set around 30 units per day.

10600 PARTITIONS

Movable partitioning is available in metal in both flush and panel construction, in laminated gypsum board and made of open wire mesh. Solid units can be specially constructed for sound and fire isolation, and can be prefinished with baked enamel, vinyl, or wood veneers or painted in the shop or on the job.

Metal office type partitioning is usually a nominal 2-1/4" (56.25 mm) or 3-1/2" (87.50 mm) thick with a height limit of 12 lin.ft. (3.65 lin.m). Free standing units have a standard cornice height of 7'-1-1/2" (2.17 m). "Banker" types are usually 1-3/4" (43.75 mm) thick and 30" (750 mm) high. A standard glass panel can be mounted on the top rail to bring the height to 42" (1050 mm) or 54" (1350 mm).

The base for movable partitioning is designed to form a raceway for both electrical and telephone lines. Switches can usually be built into the door frame panels, which with doors, glazing panels and louver panels can all be ordered as part of the partitioning.

Mesh partitioning is usually made of 10 ga. wire worked into a 1-1/2" (37.50 mm) diamond mesh set in a channel frame. It will cost about $3.00 per ss.ft. ($32.29 per sq.m) on straight runs using standard sections. A 3'x7' (0.91 x 2.13 m) door will cost $270. A 5'x7' (1.52 x 2.13 m) sliding panel, $420.00. A shop coat of paint is standard. A baked enamel coat will add $0.30 per sq.ft. ($3.23 per sq.m). Two erectors can install some 100 lin.ft. (30.48 m) per day. Add 1.5 hrs. for each door. Ceiling panels will install at the rate of 75 sq.ft. (6.96 sq.m) per hour with two persons working together.

Office partitions come in so many grades and job conditions are so diverse that it is difficult to give any basic costs, but because the field is highly competitive, it is always a simple matter to get prices together. A standard 2-3/8"(59.37 mm) 20 ga. flush steel floor to ceiling partition 8' to 9' (2.43 to 2.74 m) high will cost about $8.00 per sq.ft. ($86.1 per sq.m), without installation. Two workers should erect about 40 lin.ft. (12.19 m) per day. Each door with lock but no closer will cost $230.00. If partitions are glazed add for both labor and material.

Gypsum partitions, 2-1/4" (56.25 mm) thick, will run about $5.20per sq.ft. ($55.97 per sq.m) installed but unfinished. If glazed panels are desired, add $20.00 per sq.ft. ($215.28 per sq.m).

Freestanding "bank" type partitioning 3' (0.91 m) high cost $50.00 to $65.00 per lin.ft. ($163.51 to $213.25 per m) Two workers will set 70 lin.ft. (21.33 m) per day.

10650 SCALES

Built-in scales must be very carefully set, adjusted, and maintained and should be set by a contractor licensed by the scale manufacturer. For general budget allowances the following is a guide:

Platform Size Feet	Capacity Tons	Cost of Mechanical	Cost of Electronic
7 'x 9 '	5	$ 9,870	------
10 'x 24 '	10	$ 11,130	------
10 'x 50 '	30	$ 13,125	------
10 'x 70 '	50	$ 31,500	$ 36,750
10 'x 70 '	60	$ 32,800	$ 39,375
10 'x 70 '	80	$ 37,380	$ 42,000
10 'x 70 '	100	------	$ 47,250

Platform Size Meter	Capacity Metric Ton	Cost of Mechanical	Cost of Electronic
2 x 3 M	4.54	$ 9,870	------
3 x 7 M	9.07	$ 11,130	------
3 x 15 M	27.22	$ 13,125	------
3 x 21 M	45.36	$ 31,500	$ 36,750
3 x 21 M	54.43	$ 32,800	$ 39,375
3 x 21 M	72.57	$ 37,380	$ 42,000
3 x 21 M	90.72	------	$ 47,250

10670 STORAGE SHELVING

Free standing, baked enamel industrial shelving comes knocked down in 3' (0.91 m) sections, complete with all hardware for erection at the job site. Cost will approximate the following:

Price per Lin.Ft. (Lin.M.)					
Depth, inches	4 Shelves per Lin.Ft.	7 Shelves per Lin. Ft.	Depth, mm	4 Shelves per Lin.M	7 Shelves per Lin.M
12	$ 30.45	$ 50.40	300	$ 99.90	$ 165.35
18	$ 35.70	$ 58.80	450	$ 117.13	$ 192.91
24	$ 42.00	$ 68.25	600	$ 137.80	$ 223.92

One worker can erect from 9 to 12 l.f. (2.7-3.6 m) per day.

10700 SUN CONTROL DEVICES

Awnings and canopies of canvas, vinyl, and aluminum are prefabricated in the shop and delivered to the job site ready for erection.

1438

SPECIALTIES

Continuous, wall hung, aluminum canopies will cost from $7.50 to $11.00 per sq.ft ($80.73 to $118.41 per sq.m), and two workers can erect about 250 sq.ft. (23.22 sq.m) per day. Awnings over individual openings 3'-6" (1.06 m) wide will cost $450.00 in aluminum, $300.00 in cloth.

Awnings are often used as decorative features and take forms unrelated to sun control and can easily cost up to $45.00 per sq.ft. ($484.39 per sq.m). Two workers will take two hours to hang an individual awning.

10750 TELEPHONE ENCLOSURES

The trend in telephone noise control has been to rely on sound absorption rather than physical separation. Where booths are still used, a simple steel and wood unit costs $1,500.00 while an extruded metal unit costs twice as much. Two workers can set three units in two days.

Outdoor post types will cost about $900.00 and take one worker 6 hours to set.

Indoor shelf types will average about $500.00 a station and take one worker 3 hours to install. Directory shelves will cost about $375.00 for a size that holds up to four books, and will take one worker one hour to set.

10800 TOILET AND BATH ACCESSORIES

The following list may serve as a guide in purchasing chrome plated bathroom accessories. Sometimes accessories are of ceramic tile, costs of which are discussed under Finishes. Also, some items may be furnished with toilet compartments or will be furnished by service companies who supply the owner. Check to avoid duplication.

Prices of Toilet and Bath Accessories	
Ash Urn	$ 175.00
Clothes hook	
single	$ 5.00
double	$ 8.00
Grab bar	
12" (300 mm)	$ 15.00
18" (450 mm)	$ 21.00
24" (600 mm)	$ 25.00
Hand dryer, electric	$ 600.00
Medicine cabinet, 18"x24" (450 x 600 mm), lighted	$ 150.00
Paper towel dispenser wall-mounted	$ 25.00
recessed with waste receptacle	$ 240.00

Prices of Toilet and Bath Accessories (con't)	
Sanitary napkin dispenser, Coin operated and Surface Mounted	$ 350.00
Shelf, stainless steel, 24" (600 mm) long	$ 40.00
Shower rod	
3' (0.9 m)	$ 20.00
5' (1.5 m)	$ 25.00
Soap dish, recessed	$ 25.00
Soap dish with grab bar	$ 65.00
Soap dispenser	
powder	$ 45.00
liquid	$ 40.00
Toilet tissue dispenser	$ 11.00
Towel bar	
12" (300 mm)	$ 20.00
18" (450 mm)	$ 30.00
24" (600 mm)	$ 40.00
Towel ring	$ 15.00
Towel ladder, 16"x30" (400 x 750 mm)	$ 75.00
Tumbler/toothbrush holder	$ 35.00
Waste receptacle, 10 gal (37.85 liters)	$ 150.00

10900 WARDROBE SPECIALTIES

Hat and coat checking storage units will cost approximately the following:

Price per lin.ft. (lin.M.)				
Material	Single Tier per Lin. Ft.	Double Tier per Lin. Ft.	Single Tier per Lin. M.	Double Tier per Lin. M.
Steel	$ 25.00	$ 27.50	$ 82.02	$ 90.22
Aluminum	$ 36.25	$ 43.75	$ 118.93	$ 143.54
Stainless	$ 43.75	$ 50.00	$ 143.54	$ 164.04

One worker can set 45 lin.ft. (13.71 m) of single tier and 32 lin.ft. (9.75 m) of double tier rack per day.

11   EQUIPMENT

CSI Division 11 includes equipment such as vacuum systems, powered window washers, bank, checkroom, church, school, food service, vending, athletic, lab, laundry, library, medical, parking, waste handling, loading dock, detention, and stage equipment.

These are often purchased by the owner separately and installed by the general contractor. Even if the general contractor is not involved with the installation of such equipment, he should have a list of all equipment going into the building, with cuts and shop drawings, so that he can point out any conflicts between the structure and such equipment. A floor that has to be broken up to conceal a conduit or duct becomes a permanent scar that reflects on the general contractor, even though he might be an innocent partner.

11900 RESIDENTIAL EQUIPMENT
Kitchen Cabinets

The following cabinets are furnished completely assembled and finished. Frames are 3/4" kiln dried hardwood; doors and drawer fronts are 7/8" thick solid lumber; shelves are 1/2" plywood or chipboard; backs of base cabinets are Masonite; backs of wall cabinets and drawer bottoms are plywood; ends of cabinets are flush with hardwood faces.

Cabinets are factory finished with a penetrating sealer coat and clear lacquer. Interiors are finished and drawer bottoms are clear lacquered. Door and drawer hardware come in a variety of styles.

These cabinets are the type found in most residential construction. Economy grade cases for cheaper construction range 25% to 50% less for finished units ready for installation. Economy grade of kitchen cabinets normally starts with the laminates and moves up the scale to oak, maple, birch, hickory, pine and cherry. The wide selection of door styles and finishes, optional storage and decorative enhancements, allows an owner to customize a kitchen.

A contractor pricing a kitchen without the details will be hard pressed to quote a complete price. Top quality kitchen cabinets, the prices will run up to 50% higher for stock sizes and 100% higher or more for custom-made units, depending on the design, materials, hardware, and finish specified. All prices are f.o.b. factory.

Wood Wall Units, 12" (300 mm) Deep						
Width	No.	Height				Width
Inches	Door	12"	15"	24"	30"	mm
12	1	------	------	------	$ 120.00	300
15	1	------	------	------	$ 132.00	375
18	1	------	------	------	$ 140.00	450
21	1	------	------	------	$ 148.00	525
24	2	------	------	$ 134.00	$ 170.00	600
27	2	------	------	$ 140.00	$ 190.00	675
30	2	$ 106.00	$ 116.00	$ 150.00	$ 208.00	750
33	2	$ 116.00	$ 120.00	$ 164.00	$ 218.00	825
36	2	$ 120.00	$ 134.00	$ 166.00	$ 228.00	900
42	2	------	------	$ 178.00	$ 240.00	1050
48	2	------	------	$ 200.00	$ 286.00	1200
		300 mm	375 mm	600 mm	750 mm	
		Height in mm				

Corner Shelf and Wall Units, 30" (750 mm) High

Corner wall unit, 23" (575 mm) deep and wide,
  with 2 shelves $200.00
Corner wall unit, 23" (575 mm) deep and wide, with revolving
  3 shelf "Lazy Susan" $285.00

Utility Cabinet and Built-In Oven Cabinets

Utility cabinet, 84" (2100 mm) high, 24" (600 mm) deep,
  18" (450 mm) wide $445.00
Oven cabinet, 84" (2100 mm) high, 24" (600 mm) deep,
  24" (600 mm) wide $530.00
Oven cabinet, 84" (2100 mm) high, 24" (600 mm) deep,
  27" (675 mm) wide $560.00

Wood Cabinet Base Units			
Base units are 34-1/2" (863 mm) high and 24" (600 mm) deep without tops.			
Description	Width Inches	Price	Width mm
4 drawers	15	$ 254.00	375
4 drawers	18	$ 264.00	450
4 drawers	21	$ 280.00	525
4 drawers	24	$ 296.00	600
1 top drawer, 1 door below	12	$ 182.00	300

Wood Cabinet Base Units (continued)			
Base units are 34-1/2" (863 mm) high and 24" (600 mm) deep without tops.			
Description	Width Inches	Price	Width mm
1 top drawer, 1 door below	15	$ 194.00	375
1 top drawer, 1 door below	18	$ 206.00	450
1 top drawer, 1 door below	21	$ 226.00	525
1 top drawer, 1 door below	24	$ 254.00	600
1 top drawer, 2 doors below	27	$ 276.00	675
2 top drawers, 2 doors below	36	$ 340.00	900
2 top drawers, 2 doors below	42	$ 360.00	1050
2 top drawers, 2 doors below	48	$ 390.00	1200
Corner base cabinet	42	$ 300.00	1050
Corner base cabinet with revolving "Lazy Susan"	35	$ 320.00	875
3" Corner filler	3	$ 12.00	75

Wood Sink Fronts			
Description	Width Inches	Price	Width mm
Sink front only, 1 pair doors	27	$ 144.00	675
	30	$ 148.00	750
	33	$ 154.00	825
	36	$ 166.00	900
	42	$ 178.00	1050
	48	$ 192.00	1200
Sink end	24	$ 24.00	600

Wood Sink Base Units		
Width Inches	Price Without Tops or Sinks	Width mm
27	$ 234.00	675
30	$ 244.00	750
33	$ 254.00	825
36	$ 266.00	900
42	$ 286.00	1,050
48	$ 308.00	1,200

The above charts are a rough guide for prices. Contractor must know the detail information such as material type, storage solutions; trim etc. to price the cabinets correctly including the installation.

Steel Kitchen Units

Prices on steel case units vary considerably with different manufacturers, due to weight of metal used, method of manufacturing, kind of finish, sprayed or baked, and volume of production. Many kinds of steel cases are on a mass production basis, while others are custom made to fit the job. Naturally, the custom cabinets cost more.

The cabinets priced below are completely finished and ready to be installed. They are fabricated from 18, 20, and 22 gauge cold rolled steel and are spot welded, bonderized, and then finished with two coats of enamel baked on at high temperature. They are medium quality.

Base units are mounted on an integral black enameled subbase 4" (100 mm) high and recessed 3" (75 mm) in front for toe space.

Doors are 3/4" (18.75 mm) thick, double-walled, die formed of one piece of steel, and spot welded. Drawers have 3/4" (75 mm) fronts that make them flush with the doors. Drawers operate on roller bearings.

Hardware consists of semi-concealed hinges. Die-cast chromium plated handles are furnished on doors and drawers.

Steel Wall Cabinets						
Wall units are 13" (325 mm) deep, 18" (450 mm) units have 1 removable shelf, and 30" (750 mm) units have 2 shelves.						
Width	No.	Height		Width	Height	
Inches	Doors	18"	30"	mm	450 mm	750 mm
12	1	------	$ 78.00	300	------	$ 78.00
15	1	------	$ 82.00	375	------	$ 82.00
18	2	$ 78.00	$ 92.00	450	$ 78.00	$ 92.00
21	2	$ 84.00	$ 102.00	525	$ 84.00	$ 102.00
24	2	$ 90.00	$ 106.00	600	$ 90.00	$ 106.00
27	2	$ 92.00	$ 114.00	675	$ 92.00	$ 114.00
30	2	$ 96.00	$ 120.00	750	$ 96.00	$ 120.00
36	2	$ 110.00	$ 136.00	900	$ 110.00	$ 136.00

Corner Shelf and Wall Units, 30" (750 mm) High

Description	Price Each
Corner shelf unit, 12-1/4" (306.25 mm) deep, 6" (150 mm) wide, with 2 fixed shelves	$40.00
Corner shelf unit, 12-1/4" (306.25 mm) deep and wide, with 2 fixed shelves	$46.00

1444

EQUIPMENT

Description   Price Each

Corner wall unit, 21" (525 mm) deep and wide, with
  revolving 3-shelf "Lazy Susan" $220.00

Steel Base Cabinets			
Base units—34-1/2" (862.50mm) high and 24" (600 mm) deep without tops			
Description	Width Inches	Price	Width mm
1 drawer above, 1 door below	12	$ 144.00	300
1 drawer above, 1 door below	15	$ 160.00	375
1 drawer above, 1 door below	18	$ 172.00	450
1 drawer above, 2 doors below	21	$ 180.00	525
1 drawer above, 2 doors below	24	$ 200.00	600
1 drawer above, 2 doors below	27	$ 216.00	675
1 drawer above, 2 doors below	30	$ 230.00	750
1 drawer above, 2 doors below	36	$ 258.00	900
3 drawers	18	$ 246.00	450
Corner base cabinet with revolving "Lazy Susan"	33	$ 266.00	825
Steel Sink Fronts			
Description	Width Inches	Price	Width mm
Sink front only-one pair doors	24	$ 88.00	600
Sink front only-one pair doors	30	$ 90.00	750
Sink front only-one pair doors	36	$ 102.00	900

Fillers and End Panels for Steel Cabinets

Description   Price Each

Corner wall filler, 30" (750 mm) high, 15" (375 mm) deep,
  15" (375 mm) wide $30.00

Straight wall filler, 18" (450 mm) high, 13" (325 mm) deep,
  3" (75 mm) wide $14.00

Straight wall filler, 30" (750 mm) high, 13" (325 mm) deep,
  3" (75 mm) wide $18.00

Corner base filler, 34-1/2-" (863 mm) high, 27" (675 mm) deep,
  27" (675 mm) wide $44.00

Straight base filler, 34-1/2" (863 mm) high,
  3" (75 mm) wide $10.00

End panel, 34-1/2" (863 mm) high, 25" (625 mm) deep $24.00

Steel Kitchen Sink Units

Sinks are 36" (900 mm) high, 25" (625 mm) deep. Constructed of 18, 20, or 22 ga. furniture steel with a baked on finish. Sink tops are 14 ga. deep drawn enameling steel to which porcelain enamel is bonded.

Tops have grooved drainboards and 4" (100 mm) backsplash, all insulated to deaden noise. Sink tops are one piece construction, roll rimmed.

Length Inches	Size Bowl Inches						Single or Double Bowl	Price Each
42	20	x	16 1/2	x	7		Single	$ 440.00
54	20	x	16 1/2	x	7		Single	$ 500.00
66 Dbl	14 1/2	x	17 1/2	x	7		Double	$ 600.00

Length mm	Size Bowl mm						Single or Double Bowl	Price Each
1050	500.0	x	412.5	x	175		Single	$ 440.00
1350	500.0	x	412.5	x	175		Single	$ 500.00
1650 Dbl	362.5	x	437.5	x	175		Double	$ 600.00

KITCHEN COUNTERTOP

Most all countertops for kitchen cabinets are plastic laminate over a plywood, particleboard, or Masonite subsurface.

The plastic decorative laminate is offered in a wide variety of colors and patterns including simulated wood grain and butcher-block styles.

The tops are usually made in self-edge design (square edges and 90o angle for the backsplash), however they are available in post-form design (rolled edges and angles) at a slight decrease in cost.

To estimate the total lineal feet of countertop, measure the length of the area to be covered along the longest side (usually the wall surface). Plastic laminate tops, made to size, ready to install will cost from $20.00 to $30.00 per l.f. ($66.67-100.00 per m).

Dupont Corian® is a synthetic countertop material that can be manufactured in one piece. Costs installed range from $110.00 to $200.00 per lin.ft. ($360.89 to $656.17 per M).

Granite counter tops are now becoming popular and can be installed for approximately $110.00 to $250.00 lin.ft. ($360.89 to $ 820.21 per M) depending upon color, size and cut outs. Both manmade material and granite countertops are installed as one piece. So, the bigger the counter, the heaver the piece and more manpower required to install since the installation is by hand.

12   FURNISHINGS

CSI Division 12 covers furnishings. Separate contracts are usually let for this work, but coordination is necessary, and the general contractor should have a clear understanding regarding damages to the building by other parties working on the premises, especially if painting and decorating are part of the general work. Also, furnishings and draperies can play havoc with a carefully balanced air handling system. If the owner is uncomfortable, the general contractor is the one he calls, even though it is the 10-foot sofa that blocks the baseboard heater or the $10,000.00 chandelier that interrupts the ceiling outlet airflow and causes a draft on the boss's neck. The architect should be the coordinator of all this, but an alert contractor can forestall many problems.

Typical furnishings supplied by the contractor and coordinated with the architect and interior decorator are:

Modular furniture
Planters for plants
Benches
Bulletin boards
Cabinets Booths
Trash receptacles
Fire extinguishers
Blinds, shades, or other window decorations
Refrigeration
Shredders
Shop equipment
Gym equipment
Recreation equipment
Mailroom equipment
Special lighting
Special sound system

Local labor practices often determine which trades unload and position special furnishings. Laborers might unload benches to be placed loose in an area, but skilled workers might be needed for benches to be positioned and secured.

13   SPECIAL CONSTRUCTION

CSI Division 13 covers such items as air supported structures, pre-engineered structures, audiometric rooms, hyperbaric rooms and seismic together with sound and vibration control. We have chosen to include in this chapter a section on concrete storage bins and silos.

13030 SPECIAL PURPOSE ROOMS
Sauna Rooms

The luxury residence or office is often fitted with a prefabricated sauna cabinet. These come in a wide variety of sizes and there are several manufacturers. More than one source should be contacted.

The minimum sauna, for sitting position only, is about 6'x4' (1.83 x 1.22m), weighs 800 lbs. (362.87 kg), and needs a separate 120 volt 20 AMP service. The cost complete with door, floor, ceiling, bench, and equipment is about $3,500.00. A 4'x8' (1.22 x 2.43 m) unit, for one lying down or 2 sitting, weighs 1,000 lbs. (453.59 kg) and requires 220 volt 30 AMP service and costs about $4,500.00. To accommodate several people, units up to 8'x10' (2.43 x 3.04 m) weigh 1200 to 1500 lbs. (540 to 675 kg), require 220 to 250 AMP service, and cost $6,500.00 to $9,000.00.

These units are prefabricated and simple to install but requires a plumber, a carpenter, a carpenter helper, and an electrician. Except for the largest units, or when preparatory work is required, the installation should be completed in a day.

CONCRETE STORAGE BINS AND SILOS

Concrete storage bins and silos are used for storing grain, coal, sand, gravel, and the like. In addition there are those used for the processing and preserving of forages for feed on the farm.

Silos are constructed of numerous types of materials such as concrete stave, cast-in-place concrete, concrete block, glazed tile, wood, and steel. The cost of storage silos varies with the storage capacity, the type of structure, and the kind of materials to be stored.

Practically all silos are round or circular, which gives more capacity for the least amount of construction cost, and the stored material packs better.

Farm silos are built for storing corn, sorghum, hay crops, and miscellaneous forages. Research studies reveal that lateral pressures increase with the moisture content of the silage and the size of the silo. All silages exert approximately equal lateral pressures when of the same moisture content. Therefore, the reinforcing requirements are identical. However, hay crop silages are usually stored at excessive moisture content and more reinforcing is recommended.

Excavating and Concrete Foundations for Silos

Footings should be ample to carry the weight of the silo without settlement or tipping. The width and depth are varied according to the height of the silo and the load-carrying capacity of the soil. Recommended footing sizes are given in the accompanying table. These take into consideration the bearing capacities of the different soils as well as the weights of silos of different heights, including the weights of the concrete chutes, roofs, and the weight of the silage carried by the walls.

The depth of the foundation wall below ground level will vary according to the location of the silo. The footings should extend below frost penetration and to firm soil. The bottom of the trench for the footing should be flat and even in width to assure uniform distribution of the load to the soil. Forms will not be required for construction of the footings if care is taken to excavate the trench so that the walls stand vertically. Wall forms are used from the footing up.

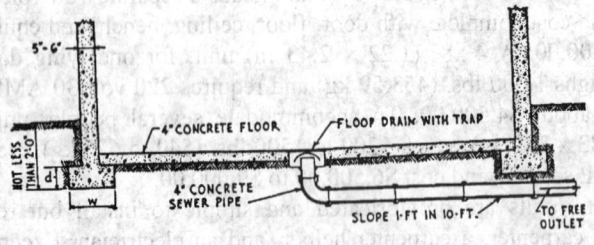

Diameter and Circumference of Standard Size Silos					
Diam. Silo Ft.	Circum. Silo in Ft.	Nominal Circum. in Ft.	Diam. Silo Meter	Circum. Silo in Meter	Nominal Circum. In meter
10	31.42	32	3.05	9.58	9.75
12	37.70	38	3.66	11.49	11.58
14	44.00	44	4.27	13.41	13.41
16	50.27	50	4.88	15.32	15.24
18	56.55	57	5.49	17.24	17.37
20	62.83	63	6.10	19.15	19.20
22	69.10	69	6.71	21.07	21.03
24	75.41	76	7.32	22.98	23.16

SPECIAL CONSTRUCTION

Dimensions of Annular Footings for Silos with Walls 6" Thick						
	Type of Soil*					
Height of Silo feet	Sand & Gravel (Type 1)		Firm clay, wet sand or clay & sand mixture (Type 2)		Soft Clay (Type 3)	
	Width Inches	Depth Inches	Width Inches	Depth Inches	Width Inches	Depth Inches
20	12	8	12	8	17	8
25	12	8	12	8	24	8
30	12	8	16	8	32	11
35	12	8	21	10	------	------
40	13	9	26	13	------	------
45	16	11	32	15	------	------
50	19	13	------	------	------	------
55	23	16	------	------	------	------

Dimensions of Annular Footings for Silos - Metric with Walls 152mm Thick						
	Type of Soil*					
Height of Silo Meter	Sand & Gravel (Type 1)		Firm clay, wet sand or clay & sand mixture (Type 2)		Soft Clay (Type 3)	
	Width mm	Depth mm	Width mm	Depth mm	Width mm	Depth mm
6.10	305	203	305	203	432	203
7.62	305	203	305	203	610	203
9.14	305	203	406	203	813	279
10.67	305	203	533	254	------	------
12.19	330	229	660	330	------	------
13.72	406	279	813	381	------	------
15.24	483	330	------	------	------	------
16.76	584	406	------	------	------	------

Soil pressures in psf Type 1 - 8,000 (40,000 kg per sq.m)
Soil pressures in psf Type 2 - 4,000 (20,000 kg per sq.m)
Soil pressures in psf Type 3 - 2,000 (10,000 kg per sq.m).
Silos higher than 45 ft. (13.71 m) not recommended on Type 2 soil.
Silos higher than 30 ft. (9.14 m) not recommended on Type 3 soil.

To obtain the number of cubic feet or cubic yards of excavation or footings required for any job, proceed as follows. Find the cubic feet of concrete required for a silo 16' (4.87 m) in diameter, requiring a footing 19" (475 mm) wide and 13" (325 mm) thick. Refer to above table: a 16' (4.87 m) diameter silo requiring a footing 19" (475 mm) wide, requires 6.59 cu.ft.(0.18 cu.m) of concrete per inch in thickness. Multiply 6.59 by 13" (thickness of footing) and the result is 85.67 cu.ft. (2.43 cu.m.) of concrete for the footing.

Silo Wall Forms. The use of commercial steel forms is recommended for the construction of cast-in-place silos. Such forms are quickly set and removed and result in true, smooth walls. Several types of commercial steel forms are available. They differ mainly in the size of the sections or panels and in the manner they are fitted together. In all types, the sections are curved

Cu.Ft.(cu.M) of Excavation or Concrete Footing Required for Silo Foundations of Various Sizes per Inch of Depth or Thickness								
Footing Width Inch	Diameter in Feet							
	10"	12"	14"	16"	18"	20"	22"	24"
12	2.67	3.17	3.67	4.17	4.75	5.25	5.75	6.33
13	2.89	3.43	3.97	4.51	5.14	5.69	6.23	6.86
16	3.55	4.22	4.88	5.55	6.33	7.00	7.67	8.44
17	3.77	4.48	5.19	5.90	6.72	7.44	8.14	8.97
19	4.21	5.01	5.80	6.59	7.51	8.31	9.10	10.03
21	4.66	5.53	6.40	7.28	8.30	9.19	10.06	11.08
23	5.10	6.06	7.01	7.97	9.10	10.06	11.02	12.15
24	5.33	6.33	7.33	8.33	9.50	10.50	11.50	12.67
26	5.55	6.86	7.94	9.03	10.29	11.37	12.45	13.72
27	5.77	7.12	8.25	9.37	10.68	11.81	12.95	14.25
32	7.10	8.44	9.77	11.10	12.65	14.00	15.33	16.90
Footing Width mm	Diameter in mm							
	250 mm	300 mm	350 mm	400 mm	450 mm	500 mm	550 mm	600 mm
305	68	81	93	106	121	133	146	161
330	73	87	101	115	131	145	158	174
406	90	107	124	141	161	178	195	214
432	96	114	132	150	171	189	207	228
483	107	127	147	167	191	211	231	255
533	118	140	163	185	211	233	256	281
584	130	154	178	202	231	256	280	309
610	135	161	186	212	241	267	292	322
660	141	174	202	229	261	289	316	348
686	147	181	210	238	271	300	329	362
813	180	214	248	282	321	356	389	429

and joined together snugly with clamps to form a complete circle or course. One course of forms is referred to as a lift. Each course or circle is 2' (0.61 m) or more high, according to the forms used. From 6' to 8' (1.83 to 2.43 m) of wall can be built in a day, depending on the size of the crew and the outdoor temperature. In summer weather, forms can usually be removed in 24 hours. In building 6' or 8' (1.83 to 2.43 m) of wall per day, three or four complete courses of forms, using 2' (0.61 m) high sections, are required. The usual practice is to set and fill one lift at a time. Where three or more courses of forms are in use, it is the usual practice to remove the lower course and set it on top. This procedure is continued until the required height of wall is built. Form faces are cleaned and re-oiled for each resetting.

SPECIAL CONSTRUCTION

A plumb wall is assured when using commercial forms, provided the first course is carefully leveled. Each manufacturer usually furnishes instructions covering the use of its forms. These companies also make forms for casting chutes and roofs and can furnish erection derricks, hoists, and other construction accessories.

Constructing the Walls. Walls of well-built cast-in-place silos are smooth, plumb, and air and watertight. The walls are 4" to 6" (100 to 150 mm) thick. A 5" to 6" (125 to 150 mm) thickness is recommended for large silos.

Concrete for durable silo walls should be of high quality, using clean sharp sand and gravel. Sand should be from very fine up to particles which pass a No. 4 screen (4 openings per inch). Gravel should be clean, hard, and range in size from 1/4" (6.25 mm) up to 1-1/2" (37.50 mm).

Number and Length of Horizontal Reinforcing Bars Required for Cast-in-Place Silos with Continuous Doorway							
Height of Silo, feet	Diameter of Silo, feet						Height of Silo, m
	10	12	14	16	18	20	
	Length of bars, feet*						
	31.83	38.00	44.33	50.50	56.83	63.33	
	Number of bars required						
20	14	17	------	16	------	------	6.10
24	17	21	14	20	------	------	7.32
28	20	25	18	24	------	------	8.53
32	24	29	22	28	20	26	9.75
36	28	33	26	34	24	32	10.97
40	------	39	30	40	30	38	12.19
44	------	------	34	48	36	46	13.41
48	------	------	38	56	42	54	14.63
52	------	------	------	64	50	62	15.85
	No. 3 Bars		No. 4 Bars		No. 5 Bars		
	Diameter of Silo, Meter						
	3.05	3.66	4.27	4.88	5.49	6.10	
	Length of bars, Meter*						
	9.70	11.58	13.51	15.39	17.32	19.30	

Bar lengths do not allow for laps. Add approximately 10% when three lengths of bar are used per circle.

A workable concrete mix is one that is mushy but not soupy. It should be somewhat sticky when worked with a shovel or trowel and should be stiff enough to require some spading in the forms. For machine mixing allow about 2 minutes mixing after all materials are in the mixer. Concrete for footing should be mixed in approximately the proportions of 1 bag air entrained portland cement; 2-3/4 cu.ff. (0.02 cu.m) sand and 4 cu.ft. (0.11 cu.m) gravel. Use about 5 gals. (18.9 liters) water per bag of cement. Concrete for walls, floor, chute, and roof should be mixed in the proportions of 1 bag of air entrained portland cement; 2-1/4 cu.ft. (0.06 cu.m) sand and 3 cu.ft. (0.08 cu.m) gravel. Use about 5 gals. (18.9 liters) water per bag of cement.

Total Weight Lbs. (Kg.) of Horizontal and Vertical Bars Required for Cast-In-Place Concrete Silos						
Height of Silo in Ft.	Diameter of Silo, feet					
	10	12	14	16	18	20
20	547	652	------	------	------	------
24	675	796	968	1,263	------	------
28	787	935	1,174	1,492	------	------
32	910	1,047	1,381	1,721	2,014	2,609
36	1,034	1,213	1,587	2,017	2,351	3,112
40	------	1,400	1,816	2,339	2,836	3,649
44	------	------	2,044	2,702	3,291	4,317
48	------	------	2,351	3,066	3,780	4,986
52	------	------	------	3,430	4,386	5,654
Height of Silo in M.	Diameter of Silo, Meter					
	3.05	3.66	4.27	4.88	5.49	6.10
6.10	166.73	198.73	------	------	------	------
7.32	205.74	242.62	295.05	384.96	------	------
8.53	239.88	284.99	357.84	454.76	------	------
9.75	277.37	319.13	420.93	524.56	613.87	795.22
10.97	315.16	369.72	483.72	614.78	716.58	948.54
12.19	------	426.72	553.52	712.93	864.41	1112.22
13.41	------	------	623.01	823.57	1003.10	1315.82
14.63	------	------	716.58	934.52	1152.14	1519.73
15.85	------	------	------	1045.46	1336.85	1723.34

No allowance made for lapping of bars. Add about 10% when three lengths of bars are used per circle.

Basis for Estimating Labor for Cast-in-Place Concrete Silos. The following data is based on circular type cast-in-place concrete silos.

Labor excavating and placing concrete footings for silos should be estimated separately, because it varies with the depth of excavation and size of silo.

After the footings are in place, an experienced crew of three should build 8 l.f. (2.4 m) of 10'-0" (3 m), 12'-0" (3.6 m), or 14'-0" (4.2 m) diameter silo per 8-hr. day, including the labor on feed room. It will also require about a day to place the roof, and after the concrete is placed, a day will be required to tear down and remove the forms and equipment.

SPECIAL CONSTRUCTION

Number and Length of Vertical Reinforcing Bars Required for Cast-in-Place Concrete Silos

Height of Silo in Ft	Length of Bars	Diameter of Silo in Feet								Height of Silo in Meters
		10	12	14	16	18	20	22	24	
		Number of Bars Required								
20'	20'-0"	22	26	-----	-----	-----	-----	-----	-----	6.10 M
24'	13'-0"	44	52	60	68	-----	-----	-----	-----	7.32 M
28'	15'-0"	44	52	60	68	-----	-----	-----	-----	8.53 M
32'	17'-0"	44	52	60	68	76	86	92	102	9.75 M
36'	19'-0"	44	52	60	68	76	86	92	102	10.97 M
40'	14'-8"	-----	78	90	102	114	129	138	153	12.19 M
44'	16'-0"	-----	78	90	102	114	129	139	153	13.41 M
48'	17'-4"		-----	90	102	114	129	138	153	14.63 M
52'	18'-8"		-----	90	102	114	129	138	153	15.85 M
60'	16'-6"		-----	120	136	152	168	184	204	18.29 M
		3.05	3.658	4.267	4.877	5.486	6.1	6.706	7.315	
		Diameter of Silo in Meters								

All bars are No. 3, spaced 18" (450 mm) apart. Bars are lapped 24" (600 mm) at splices.

1455

Labor Required to Construct Cast-in-Place Concrete Silos Each 8'- 0" Wall Height				
Diam. of Silo Lin.Ft.	Thickness of Wall (in In.)	Size of Crew	No. of Hours for Crew	Labor Hours for Each 8'- 0" (2.44 M) Height
10 '	5 " - 6 "	3	8	24
12 '	5 " - 6 "	3	8	24
14 '	5 " - 6 "	3	8	24
Roof		3	8	24
16 '	5 " - 6 "	4	8	32
18 '	5 " - 6 "	4	8	32
20 '	5 " - 6 "	4	8	32
22 '*	5 " - 6 "	4	8	32
24 '*	5 " - 6 "	4	8	32
Roof		4	8	32
Diam. of Silo Lin.M.	Thickness of Wall (in mm)	Size of Crew	No. of Hours for Crew	Labor Hours for Each 8'- 0" (2.44 M) Height
9.07 M	125 - 150 mm	3	8	24
10.89 M	125 - 150 mm	3	8	24
12.70 M	125 - 150 mm	3	8	24
Roof		3	8	24
14.51 M	125 - 150 mm	4	8	32
16.33 M	125 - 150 mm	4	8	32
18.14 M	125 - 150 mm	4	8	32
19.96 M*	125 - 150 mm	4	8	32
21.77 M*	125 - 150 mm	4	8	32
Roof		4	8	32

Labor required to construct each 6' (1.82 M) of wall height.

A set of metal forms usually includes 4 sets of forms 2'-0" (0.61 m) high. Reinforcing steel may be either rods or mesh but many silo builders prefer mesh reinforcing, because it can be placed at lower cost than rods.

To the above, add labor cost excavating and placing concrete footings.

To the above, add time for 1 day for crew bringing tools and equipment to job and setting up and 1 day tearing down and removing equipment at completion.

Silo Walls Using Slip Forms. For casting monolithic concrete structures of great height, slip forms are often used. This type of form is particularly adaptable to the construction of silos and storage bins. Despite the fact that the sliding forms must be built with more precision and of heavier material than fixed forms, slip forms have proven to be more economical in the long run for high structures. Naturally, the higher the construction the less the cost per square foot of wall.

14 CONVEYING SYSTEMS

14100 DUMBWAITERS

Dumbwaiters must be carefully selected to provide the maximum in convenience and efficiency for the minimum cost. They range from simple hand-operated to computerized models. Standard speeds are 50 fpm, but speeds to 150 fpm are possible on some models. Where speed is not a consideration, lower speeds save initial costs. Car sizes are directly related to capacities.

Size and Capacity of Dumbwaiters					
Capacity Lbs.	Speed Ft. per Min.	Hp	Stops	Cab Size Sq.In.	Price
50	50	3/4	2	472	$6,000
75	50	3/4	2	472	$6,500
100	50	3/4	2	484	$7,000
125	50	3/4	2	484	$8,000
200	50	1	2	540	$10,000
250	50	1	3	540	$13,000
300	50	1	3	900	$16,000
500	50	1	4	1296	$22,000
Capacity Kg	Speed M. per Min.	Hp	Stops	Cab Size Sq.Cm.	Price
22.68	15.24	3/4	2	3045.75	$6,000
34.02	15.24	3/4	2	3045.75	$6,500
45.36	15.24	3/4	2	3122.77	$7,000
56.70	15.24	3/4	2	3122.77	$8,000
90.72	15.24	1	2	3484.08	$10,000
113.40	15.24	1	3	3484.08	$13,000
136.08	15.24	1	3	5806.80	$16,000
226.80	15.24	1	4	8361.79	$22,000

Dumbwaiters can be mounted on channel guides that are attached to the building construction, but a packaged, self-supporting tower is also available. Dumbwaiters are an excellent ways to move merchandise from floor to floor. In residential use dumbwaiters are usually placed between the garage and the kitchen to move groceries, cloths, books and firewood. If dumbwaiters are to be retrofitted into an existing structure, an engineer needs to be consulted to assure the structural capabilities of the structure to accommodate a dumbwaiter. If additional framing is required to accommodate the dumbwaiter, the added costs and time need to be addressed and added as necessary to meet code.

14200 ELEVATORS

Elevator construction is such a highly specialized business that it is out of the question to expect anyone except an experienced elevator contractor to prepare an estimate for the installation of such equipment.

There are many items that affect the cost of elevator construction. All elevator manufacturers submit quotations on an installed basis, and the location of the job must be taken into consideration, as well as the intended use, capacity, car speed, size of car, travel, number of stops and openings, design of the car, and kind of current available.

In addition there are many types of controls, such as car switch, continuous pressure push button, single call push button, up down collective (for operation with an operator or automatic operation) signal control, duplex, and others.

All modern passenger elevator installations are partly or completely computerized. In new office buildings computerized systems are available that recognize traffic demands and automatically program elevator service to these demands. With these programming systems, elevator service is made as efficient as possible during peak or down periods.

Elevator Safety Codes

Many states and cities have their own elevator codes, and all elevator installations must comply with the rules and regulations of that state or city, as well as with American National Standards Institute (ANSI) codes. In general, all municipal or state codes follow closely the Code for Elevators, Dumbwaiters, and Escalators (ANSI Standard A17.1 with supplement ANSI A17-1a) and the Code Rules for Moving Walks (ANSI A17.1.13).

It would be a good idea for anyone endeavoring to estimate the cost of an elevator to become familiar with these codes. For example, the capacity of a passenger elevator must be based upon the effective area of the car platform in accordance with the contact load graphs shown in the latest edition of the ANSI codes.

Essential Information for Estimating the Cost of Elevators. The following information is essential to enable the elevator manufacturer to prepare an intelligent estimate on the elevators required for any building.

CONVEYING SYSTEMS

1. New or renovation project
2. Location of building
3. Number of elevators
4. Use of elevator (passenger, freight, or general purpose)
5. Capacity of elevators
6. Car speed of elevators
7. Controls and operation of elevators
8. Type of motor (single speed, two speed, or variable voltage)
9. Total travel of elevator in feet
10. Number of stops and openings
11. Single or double entrance car
12. Location and size of elevator machinery room or penthouse
13. Kind of current available
14. Size of hatch or platform
15. Design of cab, doors, and frames
16. Car and hatch door power operators
17. Signals
18. Access door location
19. Federal, state, local, and ANSI codes

Costs of Electric and Hydraulic Elevators. The cost of an elevator is a function of its speed, capacity, and type of control, which in turn are determined by type of building to be served, the use of the elevator, and the height of the building or feet of travel and number of stops.

There are so many factors that enter into the cost that an estimator should get definite prices from an elevator contractor whenever possible. The following budget figures can be used for rough estimating only.

Electric Elevators

General Purpose Elevators. There are two basic types of general purpose elevator: gear and gearless. To choose the correct one for a specific need, the contractor should contact a specialist in vertical people moving.

Gearless elevators satisfy most high-rise passenger elevator requirements and are most cost effective for structures of 20 floors. In a gearless elevator system the elevator machine, governor and support elements are combined into a compact, integrated machine and bedplate structure housed within the hoistway eliminating the requirement of a rooftop structure(s). The prices given below represent standard platform sizes, with center opening of doors, for a 20-story building.

1459

Capacity Lbs	Speed Ft. per Min.	No. of Passengers	Cab Size Sq.Ft.	Price
2500	500	9	28.61	$ 175,000
3,000	500	10	31.42	$ 191,000
3,500	700	12	35.82	$ 192,000
4,000	1000	14	41.19	$ 200,000
4,500	1200	15	44.40	$ 210,000
5,000	1200	16	48.20	$ 220,000
Capacity Kg	Speed Meter per Min.	No. of Passengers	Cab Size Sq.M.	Price
1134.00	152.4	9	2.66	$ 175,000
1360.80	152.4	10	2.92	$ 191,000
1587.60	213.36	12	3.33	$ 192,000
1814.40	304.8	14	3.83	$ 200,000
2041.20	365.76	15	4.12	$ 210,000
2268.00	365.76	16	4.48	$ 220,000

For each additional floor above 15 floors, add $7,000.

Many tall building elevators now operate at speeds up to 1,800 fpm (9 mps), but these require special pricing, depending on express and local runs.

Geared elevators satisfy most building requirements to about 27 floors. The conventional geared elevator has a rooftop machine room that houses the elevator machine, governor and controller. The entire structure requires the support of machine beams or a structural concrete slab or combination. The prices given below represent standard platform sizes, with center opening of doors, for an 8-story building.

Capacity Lbs	Speed Ft. per Min.	No. of Passengers	Cab Size Sq.Ft.	Price
2,000	200 - 350	8	24.10	$110,320
2,500	200 - 500	10	28.61	$115,320
3,000	200 - 500	11	31.68	$120,320
3,500	200 - 500	12	36.62	$140,320
4,000	200 - 300	14	41.57	$150,320
4,500	200 - 350	15	44.40	$160,320
5,000	200 - 450	16	48.14	$180,320
Capacity Kg	Speed Meters per Min.	No. of Passengers	Cab Size Sq.M.	Price
907.20	60.96 - 106.68	8	2.24	$110,320
1134.00	60.96 - 152.40	10	2.66	$115,320
1360.80	60.96 - 152.40	11	2.94	$120,320
1587.60	60.96 - 152.40	12	3.40	$140,320
1814.40	60.96 - 91.44	14	3.86	$150,320
2041.20	60.96 - 106.68	15	4.12	$160,320
2268.00	60.96 - 137.16	16	4.47	$180,320

For each additional floor above 8 floors, add $6500.

CONVEYING SYSTEMS

For hospitals, special size elevators are available for increased length requirements and multiple openings necessary for hospitals and modern care facilities.

Capacity Lbs	Speed Ft. per Min.	Cab Size Sq.Ft.	Price
4,000	125	105	$175,000
4,500	200	115	$200,000
Capacity Kg	Speed M. per Min.	Cab Size Sq.M.	Price
1814.40	38.10	9.75	$175,000
2041.20	60.96	10.68	$200,000

The above prices are for 5 stops. For each additional stop, add $8,000

Oil Hydraulic Elevators. There are two types of hydraulic elevators. One type requires the installation of a steel casing into the ground for the piston. Rise is limited to about 60 feet, so they can be used for 6- or 7-story buildings. They are economical to install and require no overhead penthouse. They are available in speeds up to 300 fpm (1.5 mps) and capacity to 5,000 lbs. (2250 kg).

Capacity Lbs.	Speed Ft per Min.	No. of Passengers	Cab Size Sq.Ft	Price
2,000	120	13	42.10	$93,500
2,100	120	13	42.10	$99,000
2,500	120	16	48.00	$104,500
3,000	120	20	52.00	$112,200
3,500	120	23	57.60	$126,500
4,000	200	13	42.00	$132,000
4,500	250	15	45.00	$143,000
5,000	300	16	48.70	$154,000
Capacity Kg.	Speed Meter per Min.	No. of Passengers	Cab Size Sq.M	Price
907.20	36.58	13	3.91	$93,500
952.56	36.58	13	3.91	$99,000
1134.00	36.58	16	4.46	$104,500
1360.80	36.58	20	4.83	$112,200
1587.60	36.58	23	5.35	$126,500
1814.40	60.96	13	3.90	$132,000
2041.20	76.20	15	4.18	$143,000
2268.00	91.44	16	4.52	$154,000

The second type of hydraulic elevator, sometimes referred to as a holeless hydraulic, has limited capacity with a maximum travel of 15'-0" (4.5 m) and maximum two stops. It is available in speeds up to 100 fpm (0.5 mps) and capacity to 2,500 lbs. (1125 kg).

Capacity Lbs.	Speed Ft per Min.	No. of Passengers	Cab Size Sq.Ft	Price
2,000	100	13	42.1	$55,000
2,100	100	13	42.1	$60,500
2,500	100	16	47.9	$66,000

Capacity Kg.	Speed Meter per Min.	No. of Passengers	Cab Size Sq.M	Price
907.20	30.48	13	3.91	$55,000
952.56	30.48	13	3.91	$60,500
1134.00	30.48	16	4.45	$66,000

The cost of an oil hydraulic installation must include the drilling of the hole for the piston. The cost of this hole depends on local conditions such as soil, water, and rock. Elevator prices depend on size, speed, control, and purpose.

Garage Elevator. Freight elevator capacities range from 3,500 to 10,000 lbs. (1575-4500 kg). Speeds vary from 50 to 100 fpm (0.25-0.5 mps), and platforms are from 35 to 112 s.f. (3.15-10.08 sq.m). The following is a typical configuration for this type of equipment:

Freight elevator, single-call push-button control, two-speed motor, automatic leveling, capacity 6,000 lbs. (2700 kg), speed 100 fpm (0.50 mps); total rise 36'-0" (10.8 m) with 4 stops and 4 openings; car platform 10'-0"x20'-0" (3 x 6 m); car built of steel wainscoting on 2 sides 6'-0" (1.8 m) high, a steel car top and two wood or expanded metal lift-up gates with electric contacts. Price installed $75,000. No doors or hatchway gates included.

Residence Type Elevators. During recent years residence type elevators have become quite popular, especially where there are older people or invalids in the house. These elevators are easily installed both in new homes and homes already built. Residential elevator capacities range from 450 to 700 lbs. (203-315 kg), with speeds from 450 to 700 fpm (2.25-3.5 mps).

A typical residential elevator with 450 lbs. (202.5 kg) capacity, 30 fpm (0.15 mps) speed, total travel 11'-0" (3.35 m), 2 stops and 2 openings, 110 or 220 volts, single phase 60 cycles, and platform 36"x42" (900 x 1050 mm), is priced installed approximately $40,000.00. For each additional stop, add $4,000.00.

The same elevator as above with 36"x48" (900 x 1200 mm) platform is priced installed at about $45,000.00. For each additional stop, add $4,000.00.

14400 LIFTS

A wheelchair lift is designed to permit people with physical handicaps to go up and down stairs. The system is usually fastened to a wall but sometimes to the stair treads. Before installing a lift system, the details of walls or stairs must be reviewed to determine whether support members can

1462

CONVEYING SYSTEMS

support a wheelchair lift. A typical wheelchair lift has a travel distance of 27 lin.ft. (8.22 m), a speed of 25 fpm (0.13 mps), and a 1-Hp motor. The chair is supported on 2" (50 mm) OD steel rails, with platform sizes from 2'-6"x3'-8" (0.76 x 1.12 m) to 3'-2"x4'-2" (0.96 x 1.27 m). For a single story lift adjacent to stairway, the cost is about $40,000.00.

14300 MOVING STAIRS

Escalators. Escalators move large volumes of people from one level to another efficiently. They are available in several widths and speeds.

Width Inches	Speed Ft. per Min.	People per Hour	Cost	Width mm	Speed Meters. per Min.
32	90	5,000	$100,000	800	27.43
32	120	6,500	$110,000	800	36.58
40	90	7,000	$104,000	1000	27.43
40	120	9,000	$114,000	1000	36.58
48	90	8,000	$116,000	1200	27.43
48	120	10,000	$120,000	1200	36.58

To the above price, add about $2,000 per lin.ft. ($6,561.68 per m) of rise.

Horizontal People Movers. These are also called power walks, power ramps, or moving walks. Horizontal people movers provide high volume transportation of people within buildings where walking is not advantageous. People movers come in various sizes from 24" to 48" (600-1200 mm) tread width. Larger sizes of 40" (1000 mm) and 48" (1200 mm) accommodate an adult with luggage. For the special requirements for this type of installation, a contractor should consult the manufacturer. For a 40" (1000 mm) wide, flat horizontal people mover with no curves or offsets, prices range from $130,000 to 165,000 per 50 lin.ft. (15.24 m).

15 MECHANICAL

15400 PLUMBING AND SEWERAGE

To prepare an accurate detailed estimate on plumbing and sewerage requires a working knowledge of the trade. All work must be laid out on paper, and the number of lineal feet of each kind of pipe estimated separately, such as tile or cast iron sewers, soil and vent pipe, water pipe, gas pipe, drains, valves, and pipe fittings. Very few, if any, general building contractors or estimators possess this knowledge, so it makes it doubly difficult for them to prepare more than an approximate estimate on the various kinds of plumbing work.

Budget figures for average plumbing installations can be estimated in one of two ways: as a percentage of the total cost of a project, or as the grand total of a predetermined cost allowance for the installation of each fixture.

Plumbing estimates as a percentage of the total job will vary from a minimum of 3% on up to 15%, depending on the number and type of fixtures required and on their disposition within the building. Percentage estimates would be valid for general plumbing only. Process and other special condition piping would of course have to be entered as a separate item.

Buildings that require a high percentage allowance include motels and nursing homes, which are generally spread out over a large area and require plumbing for each room, and apartment high-rises with many studio units, which require full baths and kitchens for relatively small living areas. On the low side would be assembly, mercantile, warehouse, and industrial buildings where plumbing can be concentrated in a central core. Recent OSHA requirements limiting the travel distance to a rest room have raised the costs on some of these types.

The private residence is difficult to peg. Where kitchens and baths can be back to back, the percentage is on the low side. But in a large, rambling house with a central powder room area but with kitchens, family, and master bedrooms all in separate wings and with hose bibs and lawn sprinkling systems on all four sides and sunken bathtubs, shell-shaped lavatories and gold plated fixtures, even 15% may be too low.

However the average building types will fall in the following ranges:

apartments	9%	to	13%	motels	9%	to	12%
assembly	4%	to	7%	office bldgs.	4%	to	10%
banks	3%	to	5%	shops	4%	to	7%
dormitories	7%	to	15%	schools	5%	to	12%
factories	4%	to	8%	stores	3%	to	70%
hospitals	8%	to	15%	warehouses	3%	to	7%

The allowance per fixture estimate is generally more accurate. An experienced plumber can acquire very accurate prices but will have to use some judgment on how much to add for bringing in the main waste and supply lines, depending on how spread out the project.

The per fixture price will be the cost of the fixture itself plus the cost of the immediate piping connections and installation within the room. Some per fixture costs have added to them a prorated cost of the central piping system. It would be more accurate to add it, together with the cost of bringing waste and supply lines from the street to the building and permit costs and overhead and profit, each as separate items, because some buildings require a much more extensive central system to supply fewer fixtures than others. Per fixture allowances are discussed later in this chapter.

PLUMBING SYSTEMS

Estimating Quantities and Costs of Sewer Work. Sewerage is estimated by the lineal foot, obtained by measuring the number of lineal feet (lin. Meter) of each size sewer pipe required, such as 4" (100 mm), 6" (150 mm), 8" (200 mm), and 12" (300 mm) pipe, and all fittings should also be listed in detail, giving the number of ells, Ys, Ts, etc.

Cast-iron and ductile iron sewer pipe is estimated in the same manner as tile sewer pipe. Most plumbers estimate their sewer work at a certain price per lineal foot (per lin.M.), including excavating, sewer pipe, and labor. This is acceptable on ordinary jobs that do not require more than the usual amount of deep excavation, but if there is an exceptionally large amount of deep excavation, from 5 to 15 feet (1.52 to 4.57 m) deep, then the excavating, shoring, and bracing or other type of soil support should be estimated separately.

The best method is to estimate the number of cubic yards of excavating required for sewers, based on a trench 1'-6" (0.45 m) wider than the pipe diameter by the required depth. Refer to chapter on Site Work for labor costs excavating for trenches.

Where 4" to 6" (100 to 150 mm) sewer pipe is used, an experienced sewer layer should lay 12 to 15 lin.ft. (3.65 to 4.57 m) of pipe per hour or 100 to 125 lin.ft. (30.48 to 38.10 m) per 8 hour day, at the following labor cost per 100 lin.ft. (30.48 m):

MECHANICAL

Description	Hours	Rate	Total	Rate	Total
Sewer layer	7.00	$	$	$ 46.90	$ 328.30
per lin.ft.			....		$ 3.28
per lin.m.			....		$ 10.77

Add for excavating and additional time for setting catch basins, gravel basins, triple basins, etc.

Estimating Quantities of Cast Iron or Ductile Iron Soil Pipe, Downspouts, Stacks, and Vents. When estimating soil pipe, stacks, and vents, note the number of stacks and length of each, listing the number of lineal feet of each size pipe, the number of pieces, size, and kind of fittings required, and price them at current market prices.

Estimating Quantity of Pipe and Fittings. When estimating the quantity of black or galvanized pipe, list the number of lin. ft. (meters) of each size pipe required, such as 1/2" (12.5 mm), 3/4" (18.8 mm), 1" (25 mm), 1-1/4" (31.2 mm), 1-1/2" (37.5 mm), 2" (50 mm), 2-1/2" (62.5 mm), 3" (75 mm), and estimate at the current market price.

After the cost of the pipe has been computed, take 60% to 75% of the cost of the pipe to cover the cost of all fittings required.

Brass or copper pipe and fittings should be estimated in the same manner as galvanized pipe.

Estimating the Quantity and Prices of Valves. All valves of the different sizes should be listed on the estimate separately, stating the number of each size required, and pricing them at the current market price.

Estimating the Quantity of Fixtures. Each type of fixture such as sinks, lavatories, laundry trays, water closets, bathtubs, drinking fountains, shower baths, hot water tanks and heaters, house pumps, and bilge pumps, should be listed separately and priced at the current market prices.

Labor Roughing in for Plumbing. The usual practice among plumbers is to allow a certain percentage of the cost of the roughing materials to cover the labor cost of installing them. This will vary with the type of building and grade of work.

The total cost of the roughing-in materials, such as cast-iron soil pipe, downspouts, stacks, vents, and fittings, all black and galvanized pipe and fittings, all valves, increasers, tees, Ys, 1/8 bends, and nipples, in fact, all pipe required to rough in the job, is computed. Then the labor cost is estimated at a certain percentage of the cost of the pipe.

In medium-priced one and two story residences, the labor cost of roughing in the job will vary from 75% to 85% of the cost of the roughing materials, with 80% a fair average.

In apartment buildings of non-fireproof construction, the labor cost of roughing in will average about 90% to 100% of the cost of the roughing in materials, while on high grade fireproof construction, the labor cost roughing in will run from 100% to 120% of the cost of the roughing in materials.

In public garages, the labor cost roughing in the job will average about 90% to 110% of the cost of the roughing in materials. Sewer work is estimated separately from roughing in.

Labor Handling and Placing Plumbing Fixtures. The labor cost handling and placing all kinds of fixtures, such as laundry tubs, kitchen sinks, lavatories, bathtubs, shower baths, water closets, and drinking fountains, is usually estimated at 25% to 30% of the cost of the fixtures.

Preparing Detailed Plumbing Estimates

When preparing detailed estimates on plumbing, it is necessary to list the quantity of each class of work separately, such as the number of lineal feet (meters) of 4" (100 mm), 6" (150 mm), and 8" (200 mm) sewers, and fittings; number of lineal feet of each size soil pipe, vents, and fittings; number of lineal feet of water pipe and fittings of different materials and sizes; number of valves of the various kinds and sizes; number of each type of fixture required, such as hot water heaters, laundry trays, kitchen sinks, slop sinks, bathtubs, lavatories, water closets, shower baths, etc., together with a list of fittings and supplies required for each.

The following is a list of items the plumbing contractor should include in his estimate:

1. Sewerage, including double sewer system
2. House tanks, foundations, etc.
3. Compression tanks, foundations, etc.
4. Sewer ejector and bilge pump and basins
5. Iron, gravel, catch, and other basins
6. Water filters, foundations
7. Water meter
8. Soil and vent pipe and fittings
9. Shower and urinal traps
10. Closet bends
11. Drum traps
12. Lead, solder, sundries
13. Inside downspouts
14. Iron sewer and fittings
15. Floor drains
16. Back water gates
17. Service pipe to building
18. Mason hydrant
19. Stop cock and box
20. Galvanized pipe and fittings
21. Brass or copper pipe and fittings
22. Valves, check valves
23. Pet cocks, sill cocks, hose bibbs
24. Hot water tank and heater
25. Pipe covering, tank covering, heater covering, filter covering, painting

MECHANICAL

26. Gas fitting, mains, ranges, heaters, etc.
27. Detention/Retention ponds
28. Manholes and covers
29. Catch basins and covers
30. Surface drain basins and covers
31. Fire system, incl. pumps, hose, Siamese connection, meter, etc.
32. Roughing in material
33. Fixtures, such as water closets, lavatories, bathtubs, shower baths, sinks, laundry tubs, drinking fountains, N. P. fittings
34. Permits, insurance, trucking, freight, telephone, watchman, etc.
35. Overhead expense and profit

15060 PIPE AND PIPE FITTINGS

Approximate Lineal Foot Prices of Cast-Iron Soil Pipe

Prices on all types of pipe and pipe fittings and supplies should be checked with suppliers at the time of the job, because they are subject to continual change.

Cost per Linear Feet				
Size Inch	Single Hub	Extra Heavy	Double Hub Service	Extra Heavy
2	$ 6.20	$ 9.30	$ 7.13	$ 10.70
3	$ 8.00	$ 12.00	$ 9.20	$ 13.80
4	$ 9.50	$ 14.25	$ 10.93	$ 16.39
5	$ 12.40	$ 18.60	$ 14.26	$ 21.39
6	$ 14.60	$ 21.90	$ 16.79	$ 25.19
8	$ 22.50	$ 33.75	$ 25.88	$ 38.81
10	$ 36.00	$ 54.00	$ 41.40	$ 62.10
12	$ 51.70	$ 77.55	$ 59.46	$ 89.18
Cost per Linear Meter				
Size mm	Single Hub	Extra Heavy	Double Hub Service	Extra Heavy
50	$ 20.34	$ 30.51	$ 23.39	$ 35.09
75	$ 26.25	$ 39.37	$ 30.18	$ 45.28
100	$ 31.17	$ 46.75	$ 35.84	$ 53.76
125	$ 40.68	$ 61.02	$ 46.78	$ 70.18
150	$ 47.90	$ 71.85	$ 55.09	$ 82.63
200	$ 73.82	$110.73	$ 84.89	$127.34
250	$118.11	$177.17	$135.83	$203.74
300	$169.62	$254.43	$195.06	$292.59

Prices of Cast-Iron Soil Pipe Fittings

Description of Fitting	Size in Inches						
	2	3	4	5	6	8	
Quarter bends, service	$ 6.80	$ 6.20	$ 9.00	$ 11.80	$ 15.70	$ 51.70	
Extra heavy	$10.20	$ 9.30	$ 13.50	$ 17.70	$ 23.55	$ 77.55	
Eighth bends, service	....	....	....	....	....	....	
Eighth bends	$ 5.00	$10.10	$ 14.60	$ 20.60	$ 24.70	$ 71.90	
Extra heavy	$ 7.50	$15.15	$ 21.90	$ 30.90	$ 37.05	$107.85	
Sanitary "T" & "Y" Branches	....	....	....	....	....	....	
Extra heavy	$14.00	$ 27.00	$ 33.70	$ 49.40	$ 76.40	$157.30	
Plain Traps S p or 3/4 Service	$19.20	$ 25.80	$ 36.00	$ 69.60	$ 89.80	$204.00	
"T" cleanout Extra Heavy	$60.60	$80.90	$72.00	$206.80	$296.60	$370.80	
Comb. "Y" & eighth bends	....	....	....	....	....	....	
Extra heavy	$20.20	$ 27.00	$ 38.20	$ 74.20	$ 90.00	$209.00	
Size in mm	50	75	100	125	150	200	

Closet bends slip collar type: 4"x4"x12" (100 x 100 x 300 mm), $46.00; 4"x4"x18" (100 x 100 x 450 mm), $54.00.

Lineal Foot Prices of Buttweld Steel Pipe				
Size	Standard Weight		Extra Heavy	
Inches	Black	Galvanized	Black	Galvanized
1/4	$ 0.66	$ 1.00	$ 0.99	$ 1.50
3/8	$ 0.90	$ 1.20	$ 1.35	$ 1.80
1/2	$ 0.94	$ 1.30	$ 1.41	$ 1.95
3/4	$ 1.34	$ 1.60	$ 2.01	$ 2.40
1	$ 2.00	$ 2.50	$ 3.00	$ 3.75
1 1/4	$ 2.40	$ 3.20	$ 3.60	$ 4.80
1 1/2	$ 3.00	$ 3.70	$ 4.50	$ 5.55
2	$ 4.00	$ 5.00	$ 6.00	$ 7.50
2 1/2	$ 5.80	$ 7.30	$ 8.70	$ 10.95
3	$ 7.60	$ 9.50	$ 11.40	$ 14.25
3 1/2	$ 9.40	$ 11.70	$ 14.10	$ 17.55
4	$ 11.20	$ 13.70	$ 16.80	$ 20.55

Lineal Meter Prices of Buttweld Steel Pipe				
Size	Standard Weight		Extra Heavy	
mm	Black	Galvanized	Black	Galvanized
6.25	$ 2.17	$ 3.28	$ 3.25	$ 4.92
9.38	$ 2.95	$ 3.94	$ 4.43	$ 5.91
12.50	$ 3.08	$ 4.27	$ 4.63	$ 6.40
18.75	$ 4.40	$ 5.25	$ 6.59	$ 7.87
25.00	$ 6.56	$ 8.20	$ 9.84	$ 12.30
31.25	$ 7.87	$ 10.50	$ 11.81	$ 15.75
37.50	$ 9.84	$ 12.14	$ 14.76	$ 18.21
50.00	$ 13.12	$ 16.40	$ 19.69	$ 24.61
62.50	$ 19.03	$ 23.95	$ 28.54	$ 35.93
75.00	$ 24.93	$ 31.17	$ 37.40	$ 46.75
87.50	$ 30.84	$ 38.39	$ 46.26	$ 57.58
100.00	$ 36.75	$ 44.95	$ 55.12	$ 67.42

Prices on Standard Malleable Iron Fittings					
Type of Fitting	Size, inches				
	1/2	3/4	1	1 1/4	1 1/2
Tee Black	$ 1.24	$ 1.80	$ 3.20	$ 5.20	$ 6.30
Galv.	$ 1.80	$ 2.60	$ 4.20	$ 6.40	$ 9.00
90° El Black	$ 1.00	$ 1.10	$ 2.10	$ 3.40	$ 4.50
Galv.	$ 1.50	$ 1.60	$ 3.00	$ 4.70	$ 6.30
45° El Black	$ 1.50	$ 1.90	$ 2.50	$ 4.00	$ 5.00
Galv.	$ 2.10	$ 2.50	$ 3.50	$ 5.00	$ 6.80
Size in mm	12.50	18.75	25.00	31.25	37.50

Prices on Standard Malleable Iron Fittings

Type of Fitting	Size, inches			
	2	3	4	6
Tee Black	$ 9.30	$ 23.60	$ 48.30	$ 134.80
Galv.	$ 12.00	$ 32.60	$ 67.40	$ 191.00
90° El Black	$ 6.40	$ 21.40	$ 36.00	$ 107.80
Galv.	$ 9.00	$ 29.20	$ 50.60	$ 146.00
45° El Black	$ 6.80	$ 23.60	$ 38.20	$ 123.60
Galv.	$ 7.20	$ 32.60	$ 53.20	$ 168.60
Size in mm	50.00	75.00	100.00	150.00

Lineal Foot Prices of Copper Water Tubing

Size, Inches					
Type	1/4	3/8	1/2	5/8	3/4
Light Type "M"	$ 1.00	$ 1.68	$ 1.68	$ 2.00	$ 2.38
Medium Type "L"	$ 1.00	$ 2.13	$ 2.13	$ 2.50	$ 3.13
Heavy Type "K"	$ 1.13	$ 2.38	$ 2.38	$ 2.75	$ 4.00

Lineal Foot Prices of Copper Water Tubing

Size, Inches					
Type	1	1 1/4	1 1/2	2	3
Light Type "M"	$ 3.25	$ 4.50	$ 5.88	$ 6.50	$ 16.88
Medium Type "L"	$ 4.25	$ 5.50	$ 7.00	$ 10.50	$ 20.38
Heavy Type "K"	$ 5.00	$ 6.25	$ 8.13	$ 12.25	$ 23.88

Lineal Meter Prices of Copper Water Tubing

Size, mm					
Type	6.25	9.38	12.50	15.63	18.75
Light Type "M"	$ 3.28	$ 5.50	$ 5.50	$ 6.56	$ 7.79
Medium Type "L"	$ 3.28	$ 6.97	$ 6.97	$ 8.20	$ 10.25
Heavy Type "K"	$ 3.69	$ 7.79	$ 7.79	$ 9.02	$ 13.12

Lineal Meter Prices of Copper Water Tubing

Size, mm					
Type	25.00	31.25	37.50	50.00	75.00
Light Type "M"	$ 10.66	$ 14.76	$ 19.27	$ 21.33	$ 55.36
Medium Type "L"	$ 13.94	$ 18.04	$ 22.97	$ 34.45	$ 66.85
Heavy Type "K"	$ 16.40	$ 20.51	$ 26.66	$ 40.19	$ 78.33

MECHANICAL

Prices of Wrought Copper Soldered Joint Fittings				
Size, inches				
Type	1/2	3/4	1	1 1/4
Tee	$ 0.55	$ 1.50	$ 4.25	$ 6.63
90° Elbow	$ 0.43	$ 0.95	$ 2.00	$ 3.13
45° Elbow	$ 0.70	$ 1.25	$ 2.13	$ 3.63
Size in mm	12.50	18.75	25.00	31.25
Prices of Wrought Copper Soldered Joint Fittings				
Size, inches				
Type	1 1/2	2	3	4
Tee	$ 8.43	$ 13.50	$ 44.13	$ 89.88
90° Elbow	$ 4.25	$ 7.50	$ 22.50	$ 50.50
45° Elbow	$ 4.75	$ 7.00	$ 27.38	$ 56.25
Size in mm	37.50	50.00	75.00	100.00

15120 PIPING SPECIALTIES

Roof Drains					
Size in Inches					
Type	3	4	5	6	8
Cast Iron	$ 325.00	$ 338.00	$ 488.00	$538.00	$ 850.00
Galv.	$ 242.00	$ 247.50	$ 308.00	$ 308.00	$ 308.00
Size in mm	75	100	125	150	200

Floor Drains			
Size of Outlet in inches			
Type	3	4	5
Flat round top cast iron	$ 56.00	$ 70.00	$ 119.00
Flat square top cast iron	$ 77.00	$ 77.00	$ 133.00
Funnel type brass	$ 11.20	$ 112.00	......
Drain with bucket cast iron	$ 154.00	$ 154.00	$ 161.00
Trench drain 10"x24"	$ 182.00	$ 210.00	......
Size in mm	75	100	125

Nominal Sizes and Weights of Steel Storage Tanks

No. of Gal.	Tank Size Dia. Feet	Tank Size Length Feet	Approx. Weight in Lbs	No. of Liters	Tank Size Dia. Meters	Tank Size Length Meters	Approx. Weight in Kg
240	3.17	4.00	609	908	0.97	1.22	276
300	3.17	5.00	761	1,136	0.97	1.52	345
500	4.00	5.42	1,042	1,893	1.22	1.65	473
1,000	4.00	10.75	2,066	3,785	1.22	3.28	937
1,500	5.33	9.00	2,305	5,678	1.62	2.74	1,046
2,000	5.33	12.00	3,073	7,570	1.62	3.66	1,394
2,500	5.33	15.00	3,841	9,463	1.62	4.57	1,742
3,000	5.33	18.00	4,609	11,355	1.62	5.49	2,091
4,000	5.33	24.00	6,146	15,140	1.62	7.32	2,788
5,000	6.00	23.83	6,869	18,925	1.83	7.26	3,116
6,000	6.00	28.67	8,264	22,710	1.83	8.74	3,749
8,000	10.00	14.00	6,726	30,280	3.05	4.27	3,051
10,000	10.00	17.00	8,167	37,850	3.05	5.18	3,705
12,000	10.00	20.50	9,849	45,420	3.05	6.25	4,468
15,000	10.00	25.50	12,251	56,775	3.05	7.77	5,557
20,000	10.00	34.00	16,334	75,700	3.05	10.36	7,409
25,000	10.50	38.75	19,547	94,625	3.20	11.81	8,867
30,000	10.50	46.50	23,457	113,550	3.20	14.17	10,640
40,000	12.00	47.50	27,384	151,400	3.66	14.48	12,421
50,000	12.00	59.50	34,302	189,250	3.66	18.14	15,559

Note: weights are empty

Nominal Sizes and Weights of Fiberglass Storage Tanks

No. of Gal.	Tank Size Dia. Feet	Tank Size Length Feet	Approx. Weight in Lbs	No. of Liters	Tank Size Dia. Meters	Tank Size Length Meters	Approx. Weight in Kg
600	4.00	7.29	800	2,271	1.22	2.22	363
1,000	4.00	11.62	1,100	3,785	1.22	3.54	499
2,500	6.00	13.48	1,800	9,463	1.83	4.11	816
3,000	6.00	16.35	2,100	11,355	1.83	4.98	953
4,000	6.00	20.67	2,500	15,140	1.83	6.30	1,134
5,000	6.00	26.42	3,100	18,925	1.83	8.05	1,406
6,000	6.00	30.73	3,600	22,710	1.83	9.37	1,633
8,000	8.00	26.04	3,600	30,280	2.44	7.94	1,633
10,000	8.00	31.54	4,300	37,850	2.44	9.61	1,950
12,000	8.00	37.04	5,000	45,420	2.44	11.29	2,268
15,000	8.00	46.75	6,400	56,775	2.44	14.25	2,903
20,000	10.00	37.73	7,700	75,700	3.05	11.50	3,493
25,000	10.00	47.56	10,000	94,625	3.05	14.50	4,536
30,000	10.00	55.61	11,800	113,550	3.05	16.95	5,352
35,000	10.00	64.06	13,600	132,475	3.05	19.53	6,169
40,000	10.00	73.69	16,000	151,400	3.05	22.46	7,258

Note: weights are empty

MECHANICAL

Septic Sewage Disposal Tanks

Septic tanks are used in rural or suburban districts where running water is available but not sewers and are used for buildings such as homes, hotels, summer resorts.

The capacity of a septic tank should be at least equal to the maximum daily flow of sewage. This is normally estimated at 100 gallons (378 liters) per person per day. For part time service in factories, churches, schools, etc., 50 gallons (189 liters) per person per day is generally satisfactory for estimating tank capacities. In some localities, health authorities require at least a 900-gallon (3407-liter) tank for residential installation.

Tanks are furnished in 12 and 14 gauge copper bearing steel, electrically welded and covered with a thick coating of asphalt to protect the tanks against corrosive action, as well as in plastic/polyethylene. The contractor needs to comply with the local codes and validate the structure for the septic tank that is allowed. If this is different from indicated on the drawings or in the specifications, the contractor needs to notify the architect or engineer for clarification before proceeding with the work.

Size of Tank - Inches	No. of Gallons	Size Connection Inches	Approx. Weight Lbs	Home Capacity No. People	Price Each
46 x 48	300	4 - 6	275	3	$ 170.00
52 x 60	500	4 - 6	380	5	$ 275.00
46 x 120	750	6	575	8	$ 350.00
48 x 144	1,000	6	980	10	$ 500.00

Size of Tank - mm	No. of Liters	Size Connection mm	Approx. Weight kg	Home Capacity # People	Price Each
1150 x 1200	1,136	100 - 150	124.74	3	$ 170.00
1300 x 1500	1,893	100 - 150	172.37	5	$ 275.00
1150 x 3000	2,839	100	260.82	8	$ 350.00
1200 x 3600	3,785	100	444.53	10	$ 500.00

The contractor needs to comply with the local codes and validate the structure for the septic tanks that are allowed. If this is different from that indicated on the drawings or in the specifications, the contractor needs to notify the architect or engineer for clarification before proceeding with the work.

Plastic/polyethylene and precast concrete tanks, not steel tanks, are now the standard in most areas where septic tanks are still required in the U.S. The inner surface of a concrete tank is coated with a tar-base material to prevent gases from deteriorating the concrete. Some precast tank suppliers not only manufacture tanks but also set the tank on delivery into an excavated hole using a delivery truck with boom or other lifting device. Size and weight of the tank limits this service. Consult individual suppliers. Sizes range from 900 to 6,000 gallon capacity.

Plastic / Polyethylene						
Size Gallons	Length Inches	Width Inches	Height Inches	Weight Lbs	Home Capacity No. People	Price Each
300	50.0	50.0	49.5	150	3	$ 570
500	60.0	60.0	63.0	200	5	$ 750
750	80.0	64.0	62.0	250	8	$ 900
1050	100.0	64.0	62.0	300	11	$ 1,160
1250	119.0	64.0	62.0	350	13	$ 1,420
1500	140.0	64.0	62.0	400	15	$ 1,490

Plastic / Polyethylene						
Size Liters	Length mm	Width mm	Height mm	Weight kg	Home Capacity No. People	Price Each
1135.5	1250	1250	1237.5	68.04	3	$ 570
1892.5	1500	1500	1575	90.72	5	$ 750
2838.8	2000	1600	1550	113.40	8	$ 900
3974.3	2500	1600	1550	136.08	11	$ 1,160
4731.3	2975	1600	1550	158.76	13	$ 1,420
5677.5	3500	1600	1550	181.44	15	$ 1,490

Concrete Tanks					
No. of Gallons	Size Feet	Price	Riser per Ft.	Casting	Roof Slab
900	4'-6"x 8'- 0" 4'-6"x 6'- 0"	$580.00	$40.00	$360.00	-----
1,200	8'- 0"	$1,300.00	$40.00	$360.00	$280.00

Concrete Tanks					
No. of Liters	Size Meter	Price	Riser per M	Casting	Roof Slab
3,406.5	1.37 x 2.44 1.37 x 1.83	$580.00	$131.23	$360.00	-----
4,542.0	1.83 M dia.	$1,300.00	$131.23	$360.00	$280.00

For other sizes, consult local suppliers.

MECHANICAL

Prices of Brass Valves					
Description	Size, inches				Pressure
Kind of Valve-Pressure	1/2	3/4	1	1 1/4	kg
Gate valve, 100 lbs.	$16.00	$20.00	$26.00	$33.20	45.36
Gate valve, 125 lbs.	$22.00	$27.50	$34.00	$44.00	56.7
Globe valve, 100 lbs.	$19.00	$25.00	$34.00	$44.00	45.36
Globe valve, 125 lbs.	$20.00	$27.00	$36.00	$50.00	56.7
Check valve, horiz.,125 lbs.	$30.00	$40.00	$56.00	$72.00	56.7
Check valve, 125 lbs. swing	$24.00	$28.00	$36.00	$46.00	56.7
Size, mm	12.50	18.75	25.00	31.25	
Description	Size, inches				Pressure
Kind of Valve-Pressure	1 1/2	2	2 1/2	3	kg
Gate valve, 100 lbs.	$38.50	$60.00	......	......	45.36
Gate valve, 125 lbs.	$56.00	$60.00	$160.00	$240.00	56.7
Globe valve, 100 lbs.	$58.00	$86.00	......	......	45.36
Globe valve, 125 lbs.	$64.00	$96.00	$160.00	$250.00	56.7
Check valve, horiz., 125 lbs.	$90.00	$148.00	$200.00	$280.00	56.7
Check valve, 125 lbs. swing	$54.00	$78.00	$140.00	$200.00	56.7
Size, mm	37.50	50.00	62.50	75.00	

Water - Gas Heaters					
Capacity Gallons	Overall Dia. Inch	Overall Ht. Inches	Approx. Weight Lbs.	Prices 5 Years	Prices 10 Years
20	14 1/2	61	144	$275.00	$300.00
30	16 1/4	64	174	$330.00	$370.00
40	18 1/2	65	195	$475.00	$550.00
50	20 1/4	65	274	$600.00	$725.00
Capacity Liters	Overall Dia. mm	Overall Ht. mm	Approx. Weight kg	Prices 5 Years	Prices 10 Years
75.70	362.5	1525.00	65.32	$275.00	$300.00
113.55	406.3	1600.0	78.9	$330.00	$370.00
151.40	462.5	1625.0	88.5	$475.00	$550.00
189.25	506.3	1625.0	124.3	$600.00	$725.00

Water - Electric Heaters								
# of Gal.	Dia. Inches	Ht. Inches	Approx. Wgt. Lbs	Prices 10 Year	# of Liters	Dia. mm	Ht. mm	Approx. Wgt. Kg
40	20	50	160	$300.00	151.4	500	1250	72.58
52	23	50	181	$350.00	196.8	575	1250	82.10
66	23	62	227	$470.00	249.8	575	1550	102.97
82	25	63	270	$590.00	310.4	625	1575	122.47
110	28	69	430	$700.00	416.4	700	1725	195.05

15440 PLUMBING FIXTURES AND TRIM

Residential kitchens and bathrooms have tended to go full circle. When plumbing first moved indoors, it was generally installed in a full-sized room and was arranged about the walls as so much furniture. Later sanitary concerns became paramount and kitchens and baths were designed as machines—all chrome and tile and porcelain enamel and reduced to minimum dimensions to facilitate cleaning.

Eventually, the kitchen emerged as an entertainment center, where family and friends could gather. Today it is often difficult to tell where the family room ends and the kitchen begins. All this has had its effect on the materials used with appliances and fixtures made as inconspicuous as possible.

The bathroom now seems to be following suit, especially the bathing area which is often in a separate room together with built in equipment for exercise and relaxation and sometimes even a sauna, greenhouse, and fireplace. Often the contractor will be called in to convert a spare bedroom into a bath and dressing room suite.

The following discussion of fixtures and their prices give figures which include the fittings. Also are included averages for the piping hook-ups to the main risers and stacks that might be added to arrive at per fixture installed budget figures.

Bathtubs

The least expensive, standard, recessed tub is made of enamelled steel. This will cost around $740 for the 4'-6" (1350 mm) length and $1,200 for the 5' (1500 mm) length. Units made of cast iron will run $500 and $700 in those sizes in white and $600 and $700 in color. If these units are to be recessed into the floor add another $150. Square recessed tubs 42"x48" (1050 x 1200 mm) will run $1,300 in cast iron.

Units complete with three enclosing walls can be had in fiberglass for around $700.

A plumber with a helper should set a tub with shower in about two hours. To arrive at a per-fixture installed budget figure add around $450.

Whirlpool bathtubs are not only found in luxury homes these days. They are being installed in place of regular units in average homes. Costs of

1478

MECHANICAL

these units vary depending on the number of jets and other options. Prices range from $2,100 to more than $7,500.00. It takes a plumber, helper plus an electrician about one day to install.

Whirlpools usually require greater widths than regular tubs for the water jets and access to the motor must be kept clear at all times. An additional cost is the 1-Hp (1.75 kW) electric motor needed to drive the jets.

Shower Stalls

These can range from simple baked enamel on steel units costing around $450to custom marble enclosures costing six times that. Prefabricated fiberglass units are very popular and can be had with built-in seats and grab bars for the elderly. A 3'x3' (0.9 x 0.9 m) unit will run $850 complete.

Often the plumber will find his work limited to furnishing the receptor and mixing valve and outlet, with other trades providing finished walls. A precast receptor pan in terrazzo together with chrome fittings will cost $350 in 3'x3' (0.9 x 0.9 m) size and $425 in 3'x4' (0.9 x 1.2 m). One worker can install this in half a day.

In industrial, recreational, and educational buildings, group showers with a center outlet column and room for some five persons around it are specified. Such a unit will cost around $1,500 without subdividing partitions, $2,000 with them.

Two plumbers will set the basic unit in 4 hours but will take a full 2 days if partitions are to be set.

Lavatories

Wall hung lavatories are available in porcelain enamelled cast iron or vitreous china. A 19"x17" (475 x 475 mm) unit will run $300 in white, cast iron, $350 colored; $375 china white, $375 china colored. A 22" (550 mm) long unit will run $450 cast iron white, $490 cast iron colored, and $550 china white and $550 china colored.

Allow two workers one hour to install. Add $350 to arrive at a per fixture installed price.

Drop-in, counter top units are mainly self-rimming today. They come in cast iron, china, steel, stainless, and molded plastic. An 18" (450 mm) round unit in cast iron or vitreous will cost about $200in white, $225 in color; in stainless, $250 and in enameled steel, $125. A 20"x18" (500 x 450 mm) unit will cost some $30 more. Two workers will take 1-1/2 to 2 hours to set a unit.

Washfountains

A 3-ft. (900 mm) round, freestanding unit will serve five to six persons at a cost of $600 per unit in precast terrazzo and $725 in fiberglass or stainless. A 4'-6" (1350 mm) unit will serve up to 10 and costs $700 in terrazzo, $920 in fiberglass and $1,100 in stainless.

1479

Half round units are also available. A 4'-6" (1350 mm) unit in terrazzo will cost $600, in stainless $890. Since installation costs will remain almost the same as for circular units, the cost per person served is much more with these units.

While terrazzo units are cheapest, they are also the heaviest, and one should allow another hour to set these units. Otherwise, allow two workers 2-1/2 hours to set a 36" (900 mm) unit, 3 hours to set a 54" (1350 mm) unit.

Also available for vanity units are precast, one piece, bowl and countertops in a variety of marbleized colors and patterns. These are generally locally manufactured and prices should be checked but a 2' (600 mm) long top and bowl should cost about $100 and take two workers 1-1/2 hrs. to set and a 6' (1800 mm) top with double bowl should cost about $225 and take 3 hours to set.

Water Closets

Water closets can be either wall hung or floor mounted and have tank or flush valve water supply. The tank can be a separate piece, or the unit can be cast in one piece with the tank dropped behind the seat. A floor mounted one piece will cost $600 in white, $650 in color.

A floor mounted two piece unit will cost $250 in white, $330 in color. For areas where water supply is a problem, units designed to flush with as little as two quarts (1.9 liters), aided by compressed air, cost about $1,000. Some locales now mandate a flush of not greater than 1.5 gals. (5.67 liters).

Floor mounted units will take two workers 1-1/2 to 2 hours to set a new unit. For a per fixture installed budget figure add $400. In rehabilitation work the old water closet footprint normally does not match the new footprint. The contractor needs to spend some added time to make sure all the old material is not just open. Some tile or carpentry work needs to be added.

In commercial work wall-hung, flush-valve units are the norm. These will cost $375 for the fixture, $300 for a standard carrier and take two workers 2 hours to set.

Urinals

Urinals are available in cast iron or vitreous china, floor, pedestal or wall mounted, single or in gangs or as wall mounted troughs.

Vitreous china, floor mounted units 18" (450 mm) wide and 38" (950 mm) high with 4" (100 mm) recessed into the floor will cost about $400. If set in gangs, add $100 for a spacer-cover 6" (150 mm) wide.

Wall hung china units 18" (450 mm) wide and projecting 12-1/2" (312.50 mm) will cost $400.

A women's pedestal type will run $500; men's type, $450.

Two workers should install a urinal in 2-1/2 to 3 hours each.

For a per fixture installed budget figure, add $400 per unit.

1480

MECHANICAL
Bidets

For the ultimate touch in the luxury bath, install a bidet. It costs about $600 in white, $700 in color, and takes two workers 2 to 3 hours to set.

Sinks

Kitchen sinks are usually self-rimming and set into a cabinet top of plastic or tile. With electric dishwashers, many are only single bowl. Even in luxury kitchens, a single bowl may be set where the dishes are washed, with a second bowl near the refrigerator for washing vegetables and a third near the ice making machine for making drinks.

A single bowl 24"x 21" (600 x 525 mm) will cost $275 in enamelled steel, $325 in white cast iron, $500 in colored cast iron, and $600in stainless. Figure two workers will take 3 to 4 hours to set. Add $800 for a per fixture installed price.

A double unit 42"x 21" (1050 x 525 mm) will cost $400 in steel, $475 in cast iron white, $650 in cast iron color, and $700 in stainless. Two workers will take 3-1/2 to 4 hours to set such a unit.

Cast iron sink tops with drain boards cost $750 for single bowl 42" (1050 mm) unit, $1,000 for double bowl 60" (1500 mm) unit.

Enamelled steel sink and drain board tops cost $300 for a 42" (1050 mm) length, $750 for a double bowl 60" (1500 mm) length.

Laundry sinks in enamelled cast iron cost $225 for a single compartment, $275 for a double; in cast stone, $200 and $225; in fiberglass, $100 and $175; in stainless steel, $400 and $550. Two workers can set a single unit in 1-1/2 to 2 hours; a double unit in 2 to 2 1/2 hours.

Electric Dishwashers

Dishwashers are installed as separate units, or are formed as part of a complete sink. Typical prices for various units are as follows.

Dishwasher unit only with floor cabinet but without top and backsplasher, where used under a continuous countertop, size 24"x25" (600 x 625 mm) and 34-1/2" (862.5 mm) high will cost about $500.00.

Dishwasher unit with porcelain enamel top and backsplasher, cabinets made of electrically welded rust-resistant steel with baked on enamel finish, size 23"x27" (575 x 675 mm) and 39" (975 mm) high will cost about $750.

Add installation charges to all of the above prices.

Garbage Disposal Units. The disposal reduces solid waste but does not do away with the kitchen garbage can. It is a self-contained unit attached to the kitchen sink to form an enlarged drain into which all kitchen food wastes can be placed.

The unit consists of a housing in which is contained a propeller and shredding mechanism. A 1/2-hp (0.37-kW) motor is directly connected to the propeller and shredding mechanism and supplies the power for operating the waste unit.

Waste material can be accumulated in the upper receptacle of the unit until a normal charge is collected, or it may be operated to dispose of the material immediately. The capacity of the waste receptacle is one gallon (3.78 liters).

Waste material passes through a series of shredders where they are reduced to a fine pulp. This pulp then passes into a revolving strainer disc through which it is forced centrifugally into a chamber below and around the flywheel. Fins on the flywheel centrifugally force the pulp into the outlet passage connected to the drain line, which carries the waste to the sewer. Cold water from the faucet flowing through the unit during the grinding operation thoroughly flushes the waste down the drain. The price of this unit averages about $375 Two plumbers should install disposals in about 3 hours each.

15500 HEATING AND AIR CONDITIONING

Over the years methods of heating and cooling buildings have continually improved, especially in residential work, due to the research programs conducted by heating and air conditioning associations, engineering societies, equipment manufacturers, and research departments of various universities.

Types of Conventional Heating Systems. There are four basic types of conventional heating systems—warm air, hot water, steam, and electric, including both resistance types and heat pumps. These types may be used separately or in combination. Solar heating is also now an option.

For small to medium size residences and for stores, churches, and the like of similar size and construction, warm air systems are most widely used. For medium size to large residences and non-residential structures of comparable size, hot water systems are usually selected. For large scale installations, steam systems are generally required. Because there are no definite rules stating just when to select one system in preference to another, there is much overlapping in their application. An owner's preference for a particular system may be governed by initial cost of installation; operation cost; convenience of operation; adaptability of system to other uses, such as summer cooling using ducts of a warm air system, snow melting using hot water piping connected to a hot water system; fuel availability; and aesthetics.

Estimating the Cost of Heating Systems. There is no fast and easy rule-of-thumb method for estimating the cost of a heating system. This requires expert knowledge, because most jobs must be designed and laid out before they can be estimated. In residential and small commercial work, architects seldom give more information than merely specifying the type of system desired, the performance expected, and the kinds of material to use. It is then up to the heating contractor to figure the heating loads and design a system to fit the conditions before estimating the job. On larger work, the heating system is usually designed and laid out diagrammatically, but the

1482

heating contractor must work out details for piping, equipment connections, and controls, all of which influence the cost.

To acquire the necessary technical knowledge for this purpose requires everyday participation in this line of work plus constant study to keep up to date. Heating and air conditioning is too complex a subject to treat in detail in a book of this kind. However, much useful data and information can be gained from manuals published by the HVAC associations (see the List of Associations in Chapter 1). Here, we have provided only the most elementary and approximate data for "roughing" or approximate estimates only.

Boilers are now designed for burning special fuels, such as hard coal, soft coal, oil, and gas, the costs varying with boiler sizes and accessories, such as domestic hot water heaters and thermostatic controls.

The ordinary cast iron radiators standing in the room are pretty much a thing of the past. Heating systems use concealed convection type radiators or baseboard radiation. These are all items that affect costs.

Copper piping is used for both heating and plumbing. The cost of this work depends to a large extent upon the experience of the contractor. Mechanics familiar with sweating joints and installing copper piping can do this work rapidly, while those who are not accustomed will spend considerable time on it.

The estimator should be familiar with the requirements of the trade in order to prepare intelligent estimates.

Determining Heating Loads. The first step in designing and estimating heating work is to determine the maximum heating load. For most residential work, the maximum heating load is the total heat loss of the structure figured at design temperatures. For commercial and industrial structures, special purpose heating loads and pick-up loads may be required which must be added to the total building heat loss to obtain maximum heating load.

Heating loads and heat losses are invariably expressed in British thermal units (Btu) per hour. A Btu is the quantity of heat required to raise the temperature of 1 lb. (0.45 kg) of water 1°F (-17°C).

Building heat losses vary considerably depending upon climatic location of job, exposure, building size, architectural design, purpose of structure, construction materials used, quality of construction workmanship, and other factors.

Heat losses should be figured separately for each room so that the proper amount of radiation may be provided for comfort in all areas. Space limits do not permit giving detailed information and data on figuring building heat losses here. Anything less than a thorough treatment of this subject would be misleading. Complete information and data on this subject are contained in the manuals and guides from the HVAC associations.

Forced Warm Air Heating Systems

Forced warm air heating is the most widely used residential heating method in use at the present time. Generally speaking, the forced warm air heating method differs from the old gravity type by using blower equipment to circulate the air in the system.

It is practically impossible for the average general estimator to compute the cost of a modern forced warm air heating system, unless he has a working knowledge of this subject and is able to calculate building heat losses and design systems which will satisfy heating requirements under design conditions. In addition, the subject has become so complex it cannot be covered in a volume of this kind so as to provide adequate information for the preparation of a cost estimate, other than a rough approximation to be used for preliminary purposes only. Always obtain firm quotations from reputable warm air heating contractors before submitting bids on jobs containing this work.

Winter Air Conditioning or Forced Air Systems. Forced warm air heating systems that are equipped with automatic humidifying devices and air filters are popularly known as winter air conditioning systems. Most forced air installations are of this type.

Each system consists of a direct fired heating unit, a blower, a system of warm air ducts and return air duct system. Air filters are located in the return air duct system just ahead of the blower—humidifying equipment is placed in the warm air plenum chamber over the furnace unit. Air cooling equipment may be added to make this a year-round system.

The unit may be gas, oil, or coal fired. Gas or oil fired units designed for that particular fuel operate at high efficiency, although satisfactory, if less efficient results may be obtained through the use of conversion burners installed in units primarily designed for solid fuel.

Thermostatic control of heating apparatus is highly desirable and invariably used. Automatic controls vary the supply of heat in accordance with demand set up by a thermostat located in the living portion of the home. Actuated by limit controls, the blower commences operation automatically when the heated air in the generating unit has reached a predetermined temperature and cuts out after the source of heat has been shut down and the temperature in the bonnet dropped to a predetermined point.

Design of air conveying systems of this kind presents an engineering problem that can only be solved with a reasonable degree of accuracy by a competent designer. The heat loss of the building, room by room, is established and the volume of heated and conditioned air required for indoor comfort establishes data from which a duct system can be designed with due regard for velocities, static pressures and delivery temperatures.

Items to be Included in the Estimate. Costs vary considerably and methods of estimating also differ, but every estimate for a winter air conditioning system should include the following items:

MECHANICAL

1. Cartage of material and equipment from shop to job.
2. Winter air conditioning unit, complete with humidifier, air filters, and controls.
3. Labor assembling and setting up unit.
4. Oil storage tank installation, including fill and vent piping, oil booster pump if required, oil gauge, piping to unit, oil filter, etc., for oil fired units, or—piping for gas fired units.
5. Installation of smoke pipe, fittings and accessories.
6. Warm air and return air plenum chambers.
7. Warm air and return air duct systems.
8. Diffusers, registers, intakes, grilles, etc.
9. Special insulation.
10. Electrical work.
11. Labor starting plant in operation and balancing system.
12. Service allowance.
13. Miscellaneous costs.
14. Overhead and profit.

Cost of Winter Air or Forced Air Conditioning Units. Sizes, capacities and prices of units vary with the different manufacturers, but the following listings are representative of models available for residential use.

Approximate Net Prices of Winter Air Conditioning Units Complete with Automatic Controls, Humidifier and air Filters				
Bonnet Rating BTU per Hour	Register Rating BTU per Hour	Heating Only	mbh per Hour	Net Price with Cooling
Gas Fired Units				
60,000	48,000	$700.00	24	$2,620.00
76,000	60,800	$740.00	29	$3,210.00
90,000	72,000	$780.00	36	$3,450.00
108,000	86,400	$900.00	47	$4,170.00
120,000	96,000	$1,150.00	47	$4,400.00
150,000	120,000	$1,480.00	58	$5,470.00
180,000	144,000	$1,670.00	-----	-----
Oil Fired Units				
84,000	71,400	$1,250.00	36	$3,450.00
112,000	95,200	$1,430.00	42	$4,170.00
140,000	119,000	$1,680.00	47	$4,760.00
175,000	148,750	$1,900.00	-----	-----
210,000	178,500	$2,140.00	-----	-----

Assembling and Setting up Unit. Most gas fired units can be assembled and set up by a sheet metal worker and a helper in 3 to 4 hours. For an oil fired unit, figure 4 to 5 hours.

Oil Storage Tanks. Common practice for oil storage tanks is the use of one or two inside storage tanks with overflow catch basins with a capacity of 275 gals. (1040 liters) each. Occasionally, outside buried tanks of 550- or 1,000-gal. (2081- or 3785-liter) capacity are used. Buried tanks must be double wall with fuel leak detectors and must be in accordance with local environmental regulations. On a contract basis, these tanks installed, including all piping, cost as follows:

1-275 gallon (1040 liter) inside tank	$800.00
2-275 gallon (1040 liter) inside tanks	$1,560
1-550 gallon (2081 liter) outside tank	$1,200.00
1-1,000 gallon (3875 liter) outside tank	$2,000.00

In the use of an outside tank, a basement wall type pump costing from $240.00 to $320.00, must be used with some types of burner equipment.

Gas Piping. An allowance of $400.00 to $600.00 should cover cost of material and labor for connecting gas to unit within the heating room.

Smoke Pipe. Material and labor for smoke pipe connection from unit to flue should cost $160.00 to $240.00 for gas fired and oil fired units.

Plenum Chambers. Plenum chambers must be fabricated to order in the shop and are usually made from light gauge galvanized steel. They are mounted on top of the furnace and serve to connect the warm air and return air duct systems to the unit. Two plenum chambers are required for each installation.

Plenums cost $16.00 per sq.ft. ($172.23 per sq.m) in 18 ga., $20.00 per sq.ft. ($215.28 per sq.m) in 16 ga., $24.00 per sq.ft. ($258.34 per sq.m) in 14 ga.

Warm Air and Return Air Duct Systems. The cost of duct work will vary with the size of job and type of distribution.

Under ordinary conditions, the warm air heating contractor will measure and list each type and size of duct, fitting, and accessory. These quantities are priced for material, and then an estimate is made of labor required for installation. This is a lengthy operation and requires an experienced heating contractor.

When pressed for time, the duct system cost may be approximated on an outlet basis, using a unit price which covers both material and labor. For ordinary installations, the following costs per outlet are about average where sheet metal workers' wages are from $22.00 to $30.00 per hr.:

Type of System	Cost per Outlet
Conventional system-warm air outlets located on inside walls.	
1st floor outlets	$200.00 to $250.00
2nd floor outlets	$240.00 to $320.00
Capped stacks	$80.00 to $100.00
Radial perimeter system in small 1-story buildings with basement or crawl space.	
Low velocity-6" to 8" (150-200 mm) dia. ducts	$200.00 to $250.00
High velocity-4" and 5" (200-225 mm) dia. ducts	$190.00 to 240.00
Large central return air duct	$260.00 to 280.00

MECHANICAL

Radial perimeter system in small 1-story basementless buildings.

Galv. steel sheet metal pipe and fittings .... $260.00 to $280.00

Fiber duct with galv. steel fittings .... $265.00 to $280.00

Trunk and branch perimeter system in medium-size, 1 story buildings w/basement or crawl space.

Std., galv. steel, rectangular ducts .... $300.00 to $330.00

Add for diffusers, registers, intakes, special insulation, etc. Excavation, backfill and concrete encasement where required, not included.

Diffusers, Registers, and Intakes. Diffusers and registers generally used are of pressed steel construction, with prime coat finish, are fully adjustable and are available in floor, baseboard, or wall types. Return air intakes are of same construction and finish, but are non-adjustable.

Floor type diffusers cost $32.00 to $42.00 for sizes commonly used. Baseboard type diffusers cost $30.00 for 2'-0" (0.60 m) lengths and $42.00 for 4'-0" (1.22 m) lengths. Wall type registers cost $14.00 to $30.00. Wall type intake grilles cost $18.00 to $36.00.

Special Insulation. Ducts in unexcavated spaces or in unheated attic spaces, both supply and return, should be insulated with not less than 1" (25 mm) of adequate insulating material. This material is furnished in flexible form and averages $4.00 to $5.00 per sq.ft. ($43.05 to $53.82 per sq.m) applied.

Electrical Work. Custom varies with respect to the supply and installation of electrical circuit from meter to fused safety switch adjacent to unit. Likewise for the connection of blower motor to 110 volt controls and also for the 22 volt wiring to thermostat when used. Most often, the job electrician runs the 110 volt service to the safety switch adjacent to unit. From this point it is handled as a subcontract under the heating contractor and represents a cost from $180.00 to $500.00 for complete wiring installation of all controls and motors.

Labor Starting and Balancing System. It is customary for the heating contractor to start the plant in operation, test all controls and balance the air distribution system. For jobs completed during summer months, this means a comeback call at the beginning of the heating system.

Labor costs for this work will run $200.00 to $300.00 depending upon size of job. For commercial work, an independent certified testing and balancing contractor might be required. The cost can be up to $7,500.00.

Service Allowance. Most jobs carry a one-year free service warranty, and an allowance should be included in the estimate to cover this contingency. Average costs for service during the first year of operation are $400.00for gas fired and $440.00 for oil fired systems.

Miscellaneous Costs. Under miscellaneous costs are classified such items as federal tax, sales tax, permits for oil burner, or for installation if required.

Approximate Cost of a Complete System Installed. Based on installations in residences of ordinary construction, located in the larger

1487

cities, the cost of average winter air conditioning systems ranges as follows for gas fired units. For oil fired units, add 10% to 15%.

Floor Area Sq.Ft.	Heat Loss BTU per Hour	Price for Complete Gas Fired Installation			Floor Area Sq.M.	Heat Loss Kg-Cal per Hour
Conventional Systems, 1-Sty. Buildings with Basements or Crawl Space						
1,000	65,000	$ 4,080.00	-	$ 5,440.00	92.90	16,380
1,500	80,000	$ 5,440.00	-	$ 6,260.00	139.35	20,160
2,000	100,000	$ 6,260.00	-	$ 6,800.00	185.80	25,200
Conventional Systems, 2-Sty. Buildings with Basements						
1,500	75,000	$ 5,440.00	-	$ 6,260.00	139.35	18,900
2,000	90,000	$ 6,800.00	-	$ 8,160.00	185.80	22,680
Perimeter Systems, 1-Sty. Buildings with Basements or Crawl Space						
1,000	65,000	$ 4,080.00	-	$ 5,440.00	92.90	16,380
1,500	80,000	$ 5,440.00	-	$ 6,260.00	139.35	20,160
2,000	100,000	$ 6,800.00	-	$ 7,620.00	185.80	25,200
Perimeter Systems, 1-Sty. Basement less Buildings						
1,000	65,000	$ 4,080.00	-	$ 5,440.00	92.90	16,380
1,500	80,000	$ 5,440.00	-	$ 6,260.00	139.35	20,160

The above prices are for individual jobs. For multiple housing projects, where large numbers of similar dwelling units are involved, much better prices may be obtained.

As previously mentioned, equipment, design, and practices vary greatly even within localities. No safe guide may be set down for general use. So much depends on correct design, proven equipment, and experienced installation that it behooves the buyer to carefully investigate all of these items before awarding a contract for this important part of the building.

Summer Air Conditioning

Residential summer air conditioning is available at a price the average home owner can afford. This applies to both existing homes and new homes under construction and is especially true where forced warm air heating is used, because the same air distribution duct system can be used for both heating and cooling. Various warm air heating equipment manufacturers make cooling units for this purpose, and installation work is usually done by heating contractors.

The function of summer air conditioning equipment is the reverse of winter air conditioning, circulating air is cooled and dehumidified, but design problems are similar though more complex. A basic requirement for designing a cooling system is calculating the total cooling load or total heat that must be removed from the structure to achieve a predetermined inside temperature and humidity under design conditions.

MECHANICAL

Heat gain calculations are much more involved than those for heat losses, because many additional factors must be considered, such as sun effect on roof, wall, and glass areas; internal heat gains from human occupancy, cooking, lighting, and appliances; and latent heat energy involved in dehumidification. This is a job for an expert. Heat gain calculations are always expressed in Btu per hour. Cooling equipment capacities are usually stated per ton of refrigeration. A ton of refrigeration will remove heat at the rate of 12,000 Btu per hr. In a 24-hr. period, this is equivalent to the heat required to melt a ton of ice. With water-cooled compression refrigeration equipment, one horsepower usually equals about one ton of cooling capacity.

Types of Cooling Equipment for Residential Use. There are several types of cooling units available and in popular use for residential summer air conditioning. In general, equipment may be classified as water-cooled or air-cooled, with each type available for several methods of application.

One method employs a remotely located condensing unit, with refrigerant lines connected to a cooling coil installed in the warm air plenum chamber of a forced warm air heating system. This method uses the heating system blower and ducts for air distribution.

Another method uses a self-contained unit, with the condenser, cooling coil, and blower combined in one cabinet. This type may be cut into the heating ducts, or may have its own duct system where hot water or steam heat is used.

Net Prices of Remote System Cooling Units. Approximate prices are for units factory assembled with heating and cooling thermostat included. Pipe, duct materials, or blower assembly is not included.

# of Tons	Rating BTU per Hour	Net Price Air Cooled	Labor Hrs. to Install	# of Tonns	Rating Kg-cal per Hr.
2	24,000	$ 3,200.00	8	1.81	6,047.85
3	36,000	$ 2,900.00	14	2.72	9,071.77
5	60,000	$ 5,400.00	30	4.54	15,119.62

Where a complete air circulation duct system is required, figure costs the same as for winter air conditioning systems.

Approximate Cost of Complete Cooling Unit Installation. Based on conditions of average difficulty, complete residential cooling unit installations should cost approximately as follows:

# of Tons	Rating BTU per Hr.	Price for Complete Installation		# of Tonns	Rating Kg-cal per Hr.
Remote, Water-cooled System					
2	24,000	$ 7,480.00	- $ 8,500.00	1.81	6,047.85
3	36,000	$ 8,500.00	- $ 9,520.00	2.72	9,071.77
Remote, Air-cooled System					
2	24,000	$ 6,800.00	- $ 7,480.00	1.81	6,047.85
3	36,000	$ 8,500.00	- $ 9,180.00	2.72	9,071.77
5	60,000	$ 15,300.00	- $ 17,000.00	4.54	15,119.62
Self-contained, Water-cooled System					
2	24,000	$ 7,820.00	- $ 8,500.00	1.81	6,047.85
3	36,000	$ 11,900.00	- $ 12,580.00	2.72	9,071.77
Self-contained, Air-cooled System					
2	24,000	$ 5,200.00	- $ 9,520.00	1.81	6,047.85
3	36,000	$ 15,300.00	- $ 16,320.00	2.72	9,071.77
5	60,000	$ 23,800.00	- $ 25,500.00	4.54	15,119.62

Room Size Air Conditioning Units. The cost of portable room size air conditioning units varies according to size, ranging from 1/2-Hp. (0.37-kW), which cools room areas up to 400 s.f. (36 sq.m), to 2-Hp. (1.5 kW) for spaces up to 1,200 sq.ft. (111.48 sq.m). All models are for window installation and include automatic thermostat controls.

Net prices run from $400.00 for a 4,000 Btu (1008 kg-cal) unit to $3,000 for a 20,000 Btu (5040 kg-cal) unit. If electric wiring is necessary, add extra for installation.

Commercial Installations. A number of mechanical equipment manufacturers offer self-contained, one-piece rooftop electric heat pump units for the smaller commercial projects.

These units, which are made in numerous heating and cooling capacities, are pre-wired and precharged at the factory so that when they are delivered to the project site, the installation is relatively simple. The unit is set on a roof curb, power and control wiring connected and duct connections hooked up, and the unit is in operation.

Standard construction of these units includes a manual damper that can be preset to admit up to 25% outdoor ventilation air for year-round fresh air. Some units come equipped with an "economizer cycle" that has a thermostatically-controlled, fully-modulating damper that admits up to 100% outdoor ventilation air to provide low-cost cooling on mild days.

For quick approximation of air cooling costs, cooling units can be estimated from $3,000.00 to $6,000.00 per ton, depending on the ductwork, job conditions, types of grilles, etc. For one ton of air cooling, 12,000 Btus (3024 kg-cal) is required. The following table will give an approximation of final requirements:

1490

MECHANICAL

Type of Space	Floor area, Btu per sq.ft.	Kg-cal per sq.m
Large Stores	40	158.73
Offices	45	178.58
Specialty Shops	50 - 70	198.42 - 277.78
Bars	80	317.47
Restaurants	100 - 120	396.84 - 476.20

Steam and Hot Water Heating

Because of differing conditions, it is impossible to give any set rule that can be accepted without modification for all kinds of buildings. It is necessary to take into consideration all of the conditions in and around a building, and additions or deductions made to suit the requirements.

Methods of Computing Heating. The most advanced method of figuring heating, the one that is generally used by heating engineers, is the Btu Method. This method is based on replacing the heat loss through exposed walls, doors and windows, floors, ceilings, etc., and uses so many factors based on the type of construction that it is impossible to go into detail in a book of this kind. Intricate and elaborate formulas are used for both warm air, steam, and hot water heating. Only approximates are given here.

Flow Control Hot Water Heating Systems. This method of hot water heating has largely superseded the older gravity or "open" type of hot water heating system. It is a complete combined automatic water heating and domestic water supply system for year-round use. Its major advantages are instantaneous heating response, high efficiency, fuel economy, and completely automatic operation.

The flow control valve makes the boiler a reservoir of high temperature water. It prevents circulation through the system when heat is not required. A circulator or pump attached to the main near the boiler forces this hot water through the mains the instant the room thermostat calls for heat. When the thermostat is satisfied, the circulator ceases to operate, and the flow control valve stops all circulation of hot water, thus preventing overheating and needless heat losses.

The flow control hot water heating system requires about 20% more radiation than is figured for steam heat. The mains are smaller, reducing the cost of pipe and fittings and labor to install.

All in all, a flow control hot water heating system costs a trifle less than the older gravity systems requiring large radiators and piping. A gravity hot water heating system can be converted into a flow control system at a nominal expense.

Cost of Special Fittings for Flow Control Hot Water Heating Systems. The cost of fittings varies with the amount of radiation in the job, and whether a complete flow control system is used, or merely a flow control valve and a circulating pump.

1491

Where only the flow control valve and water circulating pump are used, the costs for the special equipment are as follows based on the amount of radiation in the job.

Capacities and Prices of Flow-Control Valves Radiation Capacity in Square Feet (.0929 sq.Meters)					
Valve Size, In.	150 BTU	200 BTU	240 BTU	Weight Lbs	Price Each
1	354	265	221	5	$ 40.00
1 1/4	786	590	492	5	$ 42.00
1 1/2	1,170	875	730	10	$ 48.00
2	2,140	1,610	1,340	21	$ 60.00
2 1/2	3,680	2,750	2,300	33	$ 120.00
3	6,660	5,000	4,160	42	$ 140.00
4	14,000	10,000	8,750	57	$ 340.00

Capacities and Prices of Flow-Control Valves Radiation Capacity in Square Feet (.0929 sq.Meters)					
Valve Size mm	37.80 Kg-Cal	50.40 Kg-Cal	60.48 Kg-Cal	Weight Kg	Price Each
25.00	89.21	66.78	55.69	2.27	$ 40.00
31.25	198.07	148.68	123.98	2.27	$ 42.00
37.50	294.83	220.50	183.96	4.54	$ 48.00
50.00	539.27	405.71	337.67	9.53	$ 60.00
62.50	927.34	692.99	579.59	14.97	$ 120.00
75.00	1,678.29	1,259.98	1,048.30	19.05	$ 140.00
100.00	3,527.93	2,519.95	2,204.96	25.86	$ 340.00

Sizes and Prices of Water Circulators or Booster Pumps

Recommended for use on modern forced circulation systems or for improving gravity hot water systems.

Direct Radiation Capacity in Sq. Ft.						Price Each	
Size In	Motor Hp	Dely Gals. Per Minute	150 Btu	200 Btu	240	Iron Body	All Bronze
1	1/12	31	750	562	469	$320.00	$ 310.00
1 1/4	1/12	33	1,000	750	675	$320.00	$ 360.00
1 1/2	1/12	33	1,000	750	675	$320.00	$ 360.00
2	1/6	75	3,500	2,625	2,187	$560.00	$ 920.00
2 1/2	1/4	120	4,500	3,375	2,820	$810.00	$ 1,470.00
3	1/4	120	4,500	3,375	2,820	$810.00	$ 1,680.00

1492

Direct Radiation Capacity in Sq. Meters							
Size mm	Motor Kw	Dely Liters. Per Min.	37.80 Kg-Cal	50.40 Kg -cal	60.48 Kg-Cal	Price Each	
						Iron Body	All Bronze
25.00	0.06	117.34	8073.20	6049.52	5048.44	$320.00	$ 310.00
31.25	0.06	124.91	10764.26	8073.20	7265.88	$320.00	$ 360.00
37.50	0.06	124.91	10764.26	8073.20	7265.88	$320.00	$ 360.00
50.00	0.12	283.88	37674.92	28256.19	23541.44	$560.00	$ 920.00
62.50	0.18	454.20	48439.18	36329.39	30355.22	$810.00	$ 1,470.00
75.00	0.18	454.20	48439.18	36329.39	30355.22	$810.00	$ 1,680.00

Sizes and Prices of Relief Valves				
ASME Capacity at 30 Lbs in BTU/Hr	Size Inches	Net Price	ASME Capacity at 13.6 Kg in kg-cal/hr	Size mm
175,000	1/4	$24.00	44099.17	6.25
250,000	1/4	$34.00	62998.82	6.25
350,000	1/4	$40.00	88198.34	6.25
480,000	1/4	$42.00	120957.72	6.25
750,000	1	$70.00	188996.45	25.00
1,050,000	1 1/4	$112.00	264595.02	31.25

Sizes and Prices of Air Control Boiler and Tank Fittings

Air control fittings are used for the purpose of removing air from a hot water heating system and preventing its return. Air control boiler fittings consist of a tube inserted into a T-shaped fitting. The tube, in effect, extends the supply main down into the boiler water, thus preventing air, which accumulates at the top of the boiler, from rising into the piping and heating units. This air flows around the tube, up into the T-fitting, and then to the pipe leading to the air control tank fitting.

The air control tank fittings provide two separate passages—one through which air can flow to the top of the tank, and the other for the water displaced by the air to flow back into the pipe connections to the system. These will cost from $44.00 to $90.00 in sizes 1" (25 mm) to 3" (75 mm) for boiler fittings, $44.00 for tank controls up to 24" (600 mm).

Coils for Heating Water. Where coils are inserted in the boiler for heating water for domestic purposes, the size of the boiler should be increased, figuring each gallon of water tank capacity as equivalent to 1 sq.ft. (0.09 sq.m) of steam radiation or 2 sq.ft. (0.18 sq.m) of hot water radiation. For example, a 30-gal. (113.55liter) tank is equivalent to 30 sq.ft. (2.78 sq.m) of steam or 60 sq.ft. (5.57 sq.m) of hot water radiation.

Rule for Computing Boiler Sizes. After the total number of square feet of radiation of total heat loss in Btu (kg-cal) per hour has been obtained, the size of the boiler should be computed.

Practically all boiler manufacturers rate their boilers, giving maximum heat output under full drive and also the recommended load representing the attached load, including piping, which may be placed on the boiler in accordance with accepted installation standards for economical operation.

Regarding Steam Boiler Installation. It is recommended on all installations of steam boilers that drain valves be placed on the returns and that the condensation from such returns be discharged into the sewer for a period of 3 days to one week after starting fire, thereby clearing system of grease and dirt.

At the end of this period the boiler should be thoroughly washed and blown out.

Boiler Capacity. Boiler output is defined as the number of sq. ft. of standard direct column radiation or its equivalent in Btu that a boiler will supply at its outlet in one hour, under standard conditions, with 240 Btu (672 kg-cal) representing 1 sq.ft. (0.09 sq.m) of standard direct column steam radiation and 150 Btu (420 kg-cal) representing 1 sq.ft. (0.09 sq.m) of standard direct column water radiation.

Boiler Horsepower. One boiler horsepower is defined as the evaporation of 34-1/2 lbs. (15.64 kg) of water per hour from a feed water temperature of 212°F (100°C) into steam at 212°F (100°C). This is equal to 34.5 x 970.4 = 33478.8 Btu (8436.65 kg-cal) per hour. One boiler horsepower is equal therefore to 33478.8 ÷ 240 = 139 ss.ft. (8436.5 ÷ by 672 = 12.55 sq.m) of steam radiation. Note that boiler horsepower is not derived from engine horsepower. 1 horsepower hour = 2545 Btu.

To find the Btu output per hour of a steam boiler, multiply the output in sq.ft. by 240 (multiply sq.m by 672 to find kg-cal).

To find the Btu output per hour of a water boiler, multiply the output in sq.ft. by 150. (sq.m x 420 = kg-cal).

To find the output in sq.ft. of steam radiation, divide the output in Btu per hour by 240. (sq.m x 672 = kg-cal).

To find the output in sq.ft. of water radiation, divide the output in Btu per hour by 150. (sq.m x 420 = kg-cal).

Estimating Pipe and Fittings. To prepare an intelligent estimate on the cost of any heating plant, the entire job should be laid out using the quantity of pipe of each size required; connections, such as elbows, tees, nipples, valves, etc., should all be figured separately. This requires considerable labor and an estimator must thoroughly understand the business to figure with any degree of accuracy.

When estimating the quantities of piping required for any heating plant, the risers, mains, and returns, together with all connections to and from the boilers, must be figured. These vary according to the system used and the type of building.

1494

MECHANICAL

A one-pipe steam heating system should have the highest point in the main directly over the boiler, and from this point it pitches down around the circuit until it drops to the floor or enters the boiler. The more pitch the better, but there should be at least one inch (25 mm) in 20 feet (6.09 m). The lowest point in the steam main, where it drops to enter the boiler or to the floor, should not be less than 14 inches (350 mm) above the water line of boiler and the more the better.

Branches for radiators should be taken from top of main, using a nipple and elbow or a nipple and a 45-degree ell, and should pitch up to the radiator.

Connections for steam should never be taken from side of main, as this will cause water hammer and syphoning of the water of condensation in the steam main into the radiator.

Each radiator should be supplied with a good disc radiator valve and a good automatic air valve, and each return main should also have an automatic air valve where it drops to enter the boiler.

If branches are longer than 8 to 10 feet (2.43 to 3.04 m), they should be one size larger than the upright connecting pipes. Where mains drop to floor, a pipe two sizes smaller can be used to return to boiler. Risers for either steam or water should be plumb and straight. If exposed in a room, and it is necessary to use couplings on same, the couplings should be at a uniform height from floor.

The valve on a steam radiator should be either wide open or tightly closed to prevent radiator filling with water.

In a two-pipe gravity hot water heating system the highest return rises from boiler to radiator, the lowest point being at the boiler and the highest point at the end of the main. These pipes may be reduced as the branches are taken off, care being taken not to reduce too rapidly. Each radiator should be fitted with a hot water radiator valve, union elbow and a compression air valve. All pipes should be well supported, so that no pockets or depressions will occur to retard or trap the circulation. All mains, both for steam and hot water, should run about 4 feet (1.22 m) from wall to allow for expansion.

Practically all hot water heating systems installed today are flow control systems where the hot water is circulated from the boiler through the mains and pipes using a circulating or booster pump. When this method is used, the size of the mains can be greatly reduced, as will be noted from the following table. With the water being circulated through the system at 30 to 120 gallons (113.55 to 454.20 liters) per minute, it is very much faster than the old gravity systems. Also the radiators may be placed above or below boiler level.

Sizes of Pipes. When estimating steam heating using a rule of thumb method, the size of the main is determined by taking the total amount of direct radiation and adding 25% for piping. From this total extract the square root and divide by 10, which gives the size of the main to use. This is for one-pipe work.

For two-pipe work, one size less is sufficient, and the return can be one or two sizes less than the supply. A steam main should decrease in size but very little according to the area of its branches.

Example: A job contains 400 sq.ft. (37.16 Sqm) of direct steam radiation. By adding 25% for piping, the result is 500 sq.ft. (49.90 sq,ft.) The square root of 500 is 22.4. Divide 22.4 by 10 equals 2.24; it requires a 2-1/2" (62.50 mm) steam main.

Table Giving Number of Square Feet of Radiation in Exposed Heating Pipe of Various Sizes and lengths						
Length of pipe Feet	Size of pipe - Inches					
	3/4	1	1 1/4	1 1/2	2	2 1/2
1	0.275	0.346	0.434	0.494	0.622	0.753
2	0.5	0.7	0.9	1.0	1.2	1.5
3	0.8	1.0	1.3	1.5	1.9	2.3
4	1.1	1.4	1.7	2.0	2.5	3.0
5	1.4	1.7	2.2	2.4	3.1	3.8
6	1.6	2.1	2.6	2.9	3.7	4.5
7	1.9	2.4	3.0	3.4	4.4	5.3
8	2.2	2.8	3.5	3.9	5.0	6.0
9	2.5	3.1	3.9	4.4	5.6	6.8
10	2.7	3.5	4.3	4.9	6.2	7.5
11	3.0	3.8	4.8	5.4	6.8	8.3
12	3.3	4.1	5.2	5.9	7.5	9.0
13	3.6	4.5	5.6	6.4	8.1	9.8
14	3.8	4.8	6.1	6.9	9.7	10.5
15	4.1	5.2	6.5	7.4	9.3	11.3
16	4.4	5.5	6.9	7.9	10.0	12.0
17	4.7	5.9	7.4	8.4	10.6	12.8
18	5.0	6.2	7.8	8.9	11.2	13.5
19	5.2	6.6	8.3	9.4	11.8	14.3
20	5.5	6.9	8.7	9.9	12.5	15.0
21	5.8	7.3	9.1	10.4	13.0	15.8
22	6.0	7.6	9.6	10.9	13.7	16.5
23	6.3	8.0	10.0	11.3	14.3	17.3
24	6.6	8.3	10.4	11.9	14.9	18.0
25	6.9	8.6	10.9	12.3	15.6	18.8
26	7.1	9.0	11.3	12.8	16.2	19.5
27	7.4	9.4	11.7	13.3	16.8	20.3
28	7.7	9.7	12.2	13.8	17.4	21.0
29	8.0	10.0	12.6	14.3	18.0	21.8
30	8.3	10.4	13.0	14.8	18.7	22.5

Length of pipe Feet	Size of pipe - Inches					
	3	4	5	6	7	8
1	0.916	1.175	1.455	1.739	1.996	2.257
2	1.8	2.4	2.9	3.5	4.0	4.5
3	2.7	3.5	4.4	5.2	6.0	6.8
4	3.6	4.7	5.8	7.0	8.0	9.0
5	4.6	5.8	7.3	7.7	10.0	11.3
6	5.5	7.0	8.7	10.5	12.0	13.5
7	6.4	8.2	10.2	12.1	14.0	15.8
8	7.0	9.4	11.6	13.9	16.0	18.0
9	8.2	10.6	13.1	15.7	18.0	20.3
10	9.1	11.8	14.6	17.4	20.0	22.6
11	10.0	12.9	16.0	19.1	22.0	24.9
12	11.0	14.1	17.4	20.9	24.0	27.1
13	11.9	15.3	18.9	22.6	26.0	29.4
14	12.8	16.5	20.3	24.3	28.0	31.6
15	13.7	17.6	21.8	26.1	30.0	33.9
16	14.6	18.8	23.2	27.8	32.0	36.1
17	15.5	20.0	24.7	29.5	34.0	38.4
18	16.5	21.2	26.2	31.3	36.0	40.6
19	17.4	22.3	27.6	33.1	38.0	42.9
20	18.3	23.5	29.1	34.8	40.0	45.2
21	19.2	24.7	30.5	36.5	42.0	47.4
22	20.2	25.9	32.0	38.3	44.0	49.7
23	21.1	27.0	33.5	40.0	46.0	52.0
24	22.0	28.2	34.9	41.7	48.0	54.2
25	22.9	29.3	36.3	43.5	50.0	56.4
26	23.8	30.5	37.8	45.2	52.0	58.6
27	24.7	31.7	39.3	47.0	54.0	61.0
28	25.6	32.9	40.7	48.7	56.0	63.2
29	26.6	34.1	42.2	50.4	58.0	65.5
30	27.5	35.3	43.6	52.1	60.0	67.7

Table Giving Number of Square Feet of Radiation in Exposed Heating Pipe of Various Sizes and lengths

Length of pipe M	Size of pipe - mm					
	18.75	25.00	31.25	37.50	50.00	62.50
0.30	0.03	0.03	0.04	0.05	0.06	0.07
0.61	0.05	0.07	0.08	0.09	0.11	0.14
0.91	0.07	0.09	0.12	0.14	0.18	0.21
1.22	0.10	0.13	0.16	0.19	0.23	0.28
1.52	0.13	0.16	0.20	0.22	0.29	0.35
1.83	0.15	0.20	0.24	0.27	0.34	0.42
2.13	0.18	0.22	0.28	0.32	0.41	0.49
2.44	0.20	0.26	0.33	0.36	0.46	0.56
2.74	0.23	0.29	0.36	0.41	0.52	0.63
3.05	0.25	0.33	0.40	0.46	0.58	0.70
3.35	0.28	0.35	0.45	0.50	0.63	0.77
3.66	0.31	0.38	0.48	0.55	0.70	0.84
3.96	0.33	0.42	0.52	0.59	0.75	0.91
4.27	0.35	0.45	0.57	0.64	0.90	0.98
4.57	0.38	0.48	0.60	0.69	0.86	1.05
4.88	0.41	0.51	0.64	0.73	0.93	1.11
5.18	0.44	0.55	0.69	0.78	0.98	1.19
5.49	0.46	0.58	0.72	0.83	1.04	1.25
5.79	0.48	0.61	0.77	0.87	1.10	1.33
6.10	0.51	0.64	0.81	0.92	1.16	1.39
6.40	0.54	0.68	0.85	0.97	1.21	1.47
6.71	0.56	0.71	0.89	1.01	1.27	1.53
7.01	0.59	0.74	0.93	1.05	1.33	1.61
7.32	0.61	0.77	0.97	1.11	1.38	1.67
7.62	0.64	0.80	1.01	1.14	1.45	1.75
7.92	0.66	0.84	1.05	1.19	1.50	1.81
8.23	0.69	0.87	1.09	1.24	1.56	1.89
8.53	0.72	0.90	1.13	1.28	1.62	1.95
8.84	0.74	0.93	1.17	1.33	1.67	2.03
9.14	0.77	0.97	1.21	1.37	1.74	2.09

Table Giving Number of Square Feet of Radiation in Exposed Heating Pipe of Various Sizes and lengths

MECHANICAL

Table Giving Number of Square Feet of Radiation in Exposed Heating Pipe of Various Sizes and lengths						
Length of pipe M	Size of pipe - mm					
	75.00	100.00	125.00	150.00	175.00	200.00
0.30	0.09	0.11	0.14	0.16	0.19	0.21
0.61	0.17	0.22	0.27	0.33	0.37	0.42
0.91	0.25	0.33	0.41	0.48	0.56	0.63
1.22	0.33	0.44	0.54	0.65	0.74	0.84
1.52	0.43	0.54	0.68	0.72	0.93	1.05
1.83	0.51	0.65	0.81	0.98	1.11	1.25
2.13	0.59	0.76	0.95	1.12	1.30	1.47
2.44	0.65	0.87	1.08	1.29	1.49	1.67
2.74	0.76	0.98	1.22	1.46	1.67	1.89
3.05	0.85	1.10	1.36	1.62	1.86	2.10
3.35	0.93	1.20	1.49	1.77	2.04	2.31
3.66	1.02	1.31	1.62	1.94	2.23	2.52
3.96	1.11	1.42	1.76	2.10	2.42	2.73
4.27	1.19	1.53	1.89	2.26	2.60	2.94
4.57	1.27	1.64	2.03	2.42	2.79	3.15
4.88	1.36	1.75	2.16	2.58	2.97	3.35
5.18	1.44	1.86	2.29	2.74	3.16	3.57
5.49	1.53	1.97	2.43	2.91	3.34	3.77
5.79	1.62	2.07	2.56	3.07	3.53	3.99
6.10	1.70	2.18	2.70	3.23	3.72	4.20
6.40	1.78	2.29	2.83	3.39	3.90	4.40
6.71	1.88	2.41	2.97	3.56	4.09	4.62
7.01	1.96	2.51	3.11	3.72	4.27	4.83
7.32	2.04	2.62	3.24	3.87	4.46	5.04
7.62	2.13	2.72	3.37	4.04	4.65	5.24
7.92	2.21	2.83	3.51	4.20	4.83	5.44
8.23	2.29	2.94	3.65	4.37	5.02	5.67
8.53	2.38	3.06	3.78	4.52	5.20	5.87
8.84	2.47	3.17	3.92	4.68	5.39	6.08
9.14	2.55	3.28	4.05	4.84	5.57	6.29

List of Sizes of Steam Mains. The following table gives the correct sizes of mains for both one- and two-pipe steam heating systems:

Rad. Sq. Ft.	1-Pipe Work Inches	2-Pipe Work Inches	Rad. Sq.Meter	1-Pipe Work mm	2-Pipe Work mm
125	1 1/2	1 1/4 x 1	11.61	37.5	31.25 x 25.00
250	2	1 1/2 x 1 1/4	23.23	50.0	37.50 x 31.25
400	2 1/2	2 x 1 1/2	37.16	62.5	50.00 x 37.50
650	3	2 1/2 x 2	60.39	75.0	62.50 x 50.00
900	3 1/2	3 x 2 1/2	83.61	87.5	75.00 x 62.50
1,250	4	3 1/2 x 3	116.13	100.0	87.50 x 75.00
1,600	4 1/2	4 x 3 1/2	148.64	112.5	100.00 x 87.50
2,050	5	4 1/2 x 4	190.45	125.0	112.50 x 100.00
2,500	6	5 x 4 1/2	232.25	150.0	125.00 x 112.50
3,600	7	6 x 5	334.44	175.0	150.00 x 125.00
5,000	8	7 x 6	464.50	200.0	175.00 x 150.00
6,500	9	8 x 9	603.85	225.0	200.00 x 225.00
8,100	10	9 x 10	752.49	250.0	225.00 x 250.00

Abbreviation Rad. Equals Radiation

List of Sizes for Hot Water Mains. The following table gives the correct sizes of mains for hot water heating systems:

Radiation Sq.Ft.	Grav. Size Pipe Inches	Forced H.W. 2-Pipe Sys.* Inches	Radiation Sq.Meter	Grav. Size Pipe mm	Forced H.W. 2-Pipe Sys.* mm
75 - 125	1 1/4	1	6.97 - 11.61	31.25	25.00
125 - 175	1 1/2	1	11.61 - 16.26	37.50	25.00
175 - 300	2	1	16.26 - 27.87	50.00	25.00
300 - 475	2 1/2	1	27.87 - 44.13	62.50	25.00
475 - 700	3	1 1/4	44.13 - 65.03	75.00	31.25
700 - 950	3 1/2	1 1/2	65.03 - 88.26	87.50	37.50
950 - 1,200	4	1 1/2	88.26 - 111.48	100.00	37.50
1,200 - 1,575	4 1/2	2	111.48 - 146.32	112.50	50.00
1,575 - 1,975	5	2	146.32 - 183.48	125.00	50.00
1,975 - 2,375	5 1/2	2 1/2	183.48 - 220.64	137.50	62.50
2,375 - 2,850	6	2 1/2	220.64 - 264.77	150.00	62.50

Based on 200 Btu emission and circulators having 1,725 rpm motors.

Hot water flow mains may be reduced in size in proportion to the branches taken off. They should, however, have areas as large as the sum of all branches beyond that point. Returns should be same as flows.

Tables of Mains and Branches. The following table gives the size of mains and the number of branches of different sizes that each main will supply:

MECHANICAL

Mains, Inches	Branches , Inches					
1 will supply	2 - 3/4					
1 1/4 will supply	2 - 1					
1 1/2 will supply	2 - 1 1/4					
2 will supply	2 - 1 1/2					
2 will supply	2 - 1 1/2	1-	1 1/4 or	1- 2	and	1- 1 1/4
3 will supply	1- 2 1/2	1-	2 or	2- 2	and	1- 1 1/2
3 1/2 will supply	2 - 2 1/2	1-	3 or	1- 2	or	3- 2
4 will supply	1- 3 1/2	1-	2 1/2 or	2- 3	and	4- 2
4 1/2 will supply	1- 3 1/2	1-	3 or	1- 4	and	1- 2 1/2
5 will supply	1- 4	1-	3 or	1- 4 1/2	and	1- 2 1/2
6 will supply	2 - 4	1-	3 or	4- 3	or	10- 2
7 will supply	2 - 6	1-	4 or	3- 4	and	1- 2
Mains, mm	Branches, mm					
25 will supply	2 - 18.75					
31.25 will supply	2 - 25.00					
37.5 will supply	2 - 31.25					
50 will supply	2 - 37.50					
50 will supply	2 - 37.50	1-	31.25 or	1- 50.00	and	1- 31.25
75 will supply	1- 62.50	& 1-	50.00 or	2- 50.00	and	1- 37.50
87.5 will supply	2 - 62.50	1-	75.00 or	1- 50.00	or	3- 50.00
100 will supply	1- 87.50	1-	62.50 or	2- 75.00	and	4- 50.00
112.5 will supply	1- 87.50	1-	75.00 or	1- 100.00	and	1- 62.50
125 will supply	1- 100.00	1-	75.00 or	1- 112.50	and	1- 62.50
150 will supply	2 - 100.00	1-	75.00 or	4- 75.00	or	10- 50.00
175 will supply	2 - 150.00	1-	100.00 or	3- 100.00	and	1- 50.00

Size of Expansion Tanks Required for Hot Water Heating Systems

The following table gives the sizes of expansion tanks required for hot water heating systems, complete with trimmings:

# of Gal.	Size Inches			Rad. Sq.Ft.	Price Ea.	# of Liters	Size mm			Rad. Sq. M.
15	12	x	30	350	$126.00	56.78	300	x	750	32.52
18	12	x	36	500	$140.00	68.13	300	x	900	46.45
24	12	x	48	1,000	$154.00	90.84	300	x	1200	92.90
30	12	x	60	2,000	$176.00	113.55	300	x	1500	185.80
40	14	x	60	3,000	$224.00	151.40	350	x	1500	278.70
80	20	x	60	4,500	$360.00	302.80	500	x	1500	418.05

For two gauge glass tapings, add $30.00

Radiation capacities are based on forced circulation and small pipes. On old systems use next larger size tank.

Heights, Sizes and Ratings of Thin Tube Cast Iron Radiators

Cast iron radiations are seldom used these days, but standard measures are 1-3/4" (43.75 mm) in length per section. Add 1/2" (12.50 mm) to the length for each bushing.

Radiant Baseboard Panels

Radiant baseboard panels are produced and marketed by a number of concerns. Some are made of cast iron, others with a finned pipe covered with a steel facing or cover.

They are made in various heights from 7" to 9" (175 to 225 mm) and usually extend out from the wall 1-3/4" to 2-1/2", (43.75 to 62.50 mm) depending on the manufacturer.

Rated Heat Emission of Radiant Baseboard Panels

Rating in sq. ft. is based on standard emission of 240 Btu per hour per sq.ft. (672 kg-cal per sq.m) at average temperature of 215°F (101.67°C).

9-7/8-Inch (246.87-mm) Radiant Panel Baseboard								
Rating of Panel	Water Flow	Heat Output Btu per Hr. per Lin.Ft.						
Sq.ft.	lbs/hr	220°F	210°F	200°F	190°F	180°F	170°F	160°F
1.95	500	460	420	380	350	310	270	240
1.95	2,500	490	450	410	370	330	290	250
Rating of Panel	Water Flow	Heat Output kg-cal per hr per m						
Sq.M.	kg/hr	104.4°C	98.8°C	93.3°C	87.7°C	82.2°C	76.7°C	71.1°C
0.18	226.8	380.31	347.24	314.17	289.36	256.29	223.22	198.42
0.18	1134.0	405.11	372.04	338.97	305.90	272.83	239.76	206.69

9-7/8-Inch (246.87-mm) Radiant Convector Panel Rating based on standards given for baseboard panels.								
Rating of Panel	Water Flow	Heat Output Btu per Hr. per Lin.Ft.						
Sq.ft.	lbs/hr	220°F	210°F	200°F	190°F	180°F	170°F	160°F
3.25	500	780	720	660	600	530	470	420
3.25	2,500	830	760	700	640	570	510	440
Rating of Panel	Water Flow	Heat Output kg-cal per hr. per m						
Sq.M.	kg/hr	104.4°C	98.8°C	93.3°C	87.7°C	82.2°C	76.7°C	71.1°C
0.30	226.8	644.87	595.26	545.66	496.05	438.18	388.58	347.24
0.30	1134	686.21	628.33	578.73	529.12	471.25	421.65	363.77

MECHANICAL

Price of Baseboard Radiation and Accessories		
Description	Weight Lbs.	Approx. Price
Radiant Baseboard Panels 9-7/8", per lin.ft.	12.25	$18.00
Radiant Convector Panels 9-7/8", per lin.ft.	14.00	$36.00
Plain ends, 9-7/8", each	2.50	$12.00
Corner Plate, 9-7/8", each	0.50	$12.00
Angle Radiator Valve, 3/4", each	1.00	$38.00
Description	Weight Kg.	Approx. Price
Panels 246.87 mm per .30 lin.M.	5.56	$18.00
Panels 246.87 mm per .30 lin.M.	6.35	$36.00
Plain Ends 246.87 mm, ea.	1.13	$12.00
Corner Plate 246.87 mm ea.	0.23	$12.00
Valve 18.75 mm	0.45	$38.00

Convector-Radiators

Radiant convectors combine convection heating with quick-acting and comfortable radiant heating. They are formed of cast iron with finned sections, or of copper with or without finned sections. There are so many types of convectors. It is impossible to describe all of them in detail.

Convectors described below are 6-1/4" (156 mm) deep overall and may be installed as freestanding or semi-recessed units. Maximum depth of recess should not exceed 4-3/4" (119 mm).

Size, Capacity, and Approximate Price of Convectors						
Ht. In.	L. In.	Steam SF EDR	Forced Hot Water, 1,000 Btu per			
			180°	200°	220°	Price Ea.
20	16	13	1.74	2.17	2.60	$146.00
20	20	17	2.28	2.84	3.40	$182.00
20	24	21	2.81	3.51	4.20	$218.00
20	28	26	3.48	4.34	5.20	$262.00
20	32	29.5	2.95	4.93	5.90	$308.00
20	36	33.5	4.48	5.59	6.70	$350.00
20	40	38.5	5.15	6.43	7.70	$392.00
20	44	43.5	5.82	7.26	8.70	$436.00
20	48	46.5	6.22	7.76	9.30	$476.00
20	56	55.5	7.43	9.26	11.10	$560.00
20	64	63	8.44	10.5	12.60	$630.00
24	20	18.5	2.48	3.09	3.70	$190.00
24	24	23	3.08	3.84	4.60	$234.00
24	28	28	3.75	4.67	5.60	$164.00
24	32	32.5	4.35	5.42	6.50	$336.00
24	36	36.5	4.88	6.09	7.30	$378.00
24	40	42	5.62	7.01	8.40	$436.00
24	44	47.5	6.36	7.93	9.50	$448.00
24	48	51	6.83	8.51	10.20	$532.00
24	56	60.5	8.18	10.1	12.10	$616.00
24	64	69	9.24	11.52	13.80	$714.00

Size, Capacity, and Approximate Price of Convectors - Metric						
Ht. mm	L. mm	Steam SF EDR	Forced Hot Water, 252 Kg-cal per			
			82.2°	93.3°	104.4°	Price Ea.
500	400	1.17	0.16	0.20	0.23	$146.00
500	500	1.53	0.21	0.26	0.31	$182.00
500	600	1.89	0.25	0.32	0.38	$218.00
500	700	2.34	0.31	0.90	0.47	$262.00
500	800	2.66	0.27	0.44	0.53	$308.00
500	900	3.02	0.40	0.50	0.60	$350.00
500	1000	3.47	0.46	0.58	0.69	$392.00
500	1100	3.92	0.52	0.65	0.78	$436.00
500	1200	4.19	0.56	0.70	0.84	$476.00
500	1400	5.00	0.87	0.83	0.99	$560.00
500	1600	5.67	0.76	0.94	1.13	$630.00
600	500	1.67	0.22	0.25	0.33	$190.00
600	600	2.07	0.27	0.35	0.41	$234.00
600	700	2.52	0.34	0.42	0.50	$164.00
600	800	2.93	0.39	0.48	0.59	$336.00
600	900	3.29	0.44	0.55	0.66	$378.00
600	1000	3.78	0.51	0.63	0.76	$436.00
600	1100	4.28	0.57	0.71	0.85	$448.00
600	1200	4.59	0.61	0.77	0.92	$532.00
600	1400	5.45	0.74	0.91	1.09	$616.00
600	1600	6.21	0.83	1.03	1.24	$714.00

For dampers, add $25.00. For air chambers and accessories less air valve, add $30.55.

Standard Convectors and Cabinets-Non-Ferrous

Convectors listed below are 4" (100 mm), 6" (150 mm), 8" (200 mm), and 10" (250 mm) deep plus 1/4" (6 mm) overall, and may be installed as freestanding or semi-recessed units.

MECHANICAL

Depth Inches	Len. Inches	Height								
		20"			24"			32"		
		EDR	Approx. Price	Wt Lbs	EDR	Approx. Price	Wt Lbs	EDR	Approx. Price	Wt Lbs
4	24	13.1	$134.00	25	15.0	$134.00	25	---	$168.00	---
6	24	19.0	$188.00	30	22.4	$188.00	30	25.5	$230.00	35
	32	27.0	$196.00	40	31.6	$196.00	35	36.0	$252.00	50
	36	30.8	$120.00	40	36.2	$120.00	45	41.2	$266.00	55
	40	34.9	$230.00	45	40.8	$230.00	45	46.5	$300.00	60
	44	38.8	$252.00	50	45.4	$252.00	55	51.7	$336.00	65
	48	42.7	$268.00	55	50.0	$268.00	60	56.9	$346.00	75
	56	50.6	$286.00	60	59.2	$286.00	65	67.4	$386.00	75
	64	58.3	$308.00	65	68.4	$308.00	70	77.9	$400.00	80
8	32	34.0	$230.00	45	37.5	$230.00	45	41.6	$300.00	55
	36	39.1	$246.00	45	43.1	$246.00	45	47.8	$360.00	60
	40	44.2	$286.00	50	48.7	$286.00	50	54.0	$396.00	65
	48	54.3	$336.00	60	59.8	$336.00	65	66.4	$414.00	75
	56	64.4	$364.00	65	70.9	$364.00	70	78.8	$470.00	115
	64	74.4	$406.00	70	82.0	$406.00	70	91.0	$532.00	120
10	36	45.0	$280.00	50	49.0	$280.00	50	54.5	$370.00	65
	40	50.8	$274.00	55	55.3	$274.00	55	61.5	$414.00	70
	48	62.1	$364.00	65	67.7	$364.00	75	75.4	$476.00	85
	56	73.6	$406.00	70	80.2	$406.00	80	89.3	$560.00	120
	64	85.1	$462.00	75	92.7	$462.00	85	103.1	$610.00	130

Dep. mm	Len. mm	Height - Metric								
		500 mm			600 mm			800 mm		
		EDR	Approx. Price	Wt Kg	EDR	Approx. Price	Wt Kg	EDR	Approx. Price	Wt Kg
100	600	1.18	$134.00	11.4	1.35	$134.00	11.4	---	$168.00	---
150	600	1.71	$188.00	13.5	2.02	$188.00	13.5	2.30	$230.00	15.9
	800	2.43	$196.00	18.0	2.84	$196.00	16.0	3.24	$252.00	23.0
	900	2.77	$120.00	18.0	3.26	$120.00	20.0	3.71	$266.00	25.0
	1000	3.14	$230.00	20.0	3.67	$230.00	20.0	4.19	$300.00	27.0
	1100	3.49	$252.00	23.0	4.09	$252.00	25.0	4.65	$336.00	29.0
	1200	3.84	$268.00	25.0	4.50	$268.00	27.0	5.12	$346.00	34.0
	1400	4.55	$286.00	27.0	5.33	$286.00	29.0	6.07	$386.00	34.0
	1600	5.25	$308.00	29.0	6.16	$308.00	31.0	7.01	$400.00	36.0

Height - Metric (continued)											
Dep. mm	Len. mm	500 mm			600 mm			800 mm			
		EDR	Approx. Price	Wt Kg	EDR	Approx. Price	Wt Kg	EDR	Approx. Price	Wt Kg
200	800	3.06	$230.00	20.0	3.38	$230.00	20.0	3.74	$300.00	25.0	
	900	3.52	$246.00	20.0	3.88	$246.00	20.0	4.30	$360.00	27.0	
	1000	3.98	$286.00	23.0	4.38	$286.00	23.0	4.86	$396.00	29.0	
	1200	4.89	$336.00	27.0	5.38	$336.00	29.0	5.98	$414.00	34.0	
	1400	5.80	$364.00	29.0	6.38	$364.00	31.0	7.09	$470.00	52.0	
	1600	6.70	$406.00	31.0	7.38	$406.00	31.0	8.19	$532.00	54.0	
250	900	4.05	$280.00	22.5	4.41	$280.00	22.5	4.91	$370.00	29.0	
	1000	4.57	$274.00	25.0	4.98	$274.00	25.0	5.54	$414.00	32.0	
	1200	5.59	$364.00	29.0	6.09	$364.00	34.0	6.79	$476.00	38.0	
	1400	6.62	$406.00	32.0	7.22	$406.00	36.0	8.04	$560.00	54.0	
	1600	7.66	$462.00	34.0	8.34	$462.00	38.0	9.28	$610.00	59.0	

Stock Model Convector Accessories				
Descrip.	Price	Approx Shpg. Wt., Lbs	Approx Shpg. Wt., Kg	Descrip.
Damper Package, up to 64"	$35.00	5	2.27	up to 1600 mm
Snap-in Inlet Grille for Stock Models Only	$40.00	5	6.54	-----
CM Molding Trim, 24" to 36" long	$50.00	6	13.23	600 to 900 mm long
CM Molding Trim, 40" to 64" long	$65.00	8	8.00	1000 to 1600 mm long
Air Vent-Manual	$30.00	1	0.25	-----
Air Vent-Auto	$30.00	1	0.00	-----

Hot Water Rating in Btu per Sq. Ft. (Kg-Cal per Sq.M.) EDR							
Temp. Drop °F	Average Water Temp. °F			Temp. Drop °C	Average Water Temp. °C		
	180 °	190 °	200 °		82.22 °	87.78 °	93.33 °
10	160 °	181 °	201 °	-12.22	71.11 °	82.78 °	93.89 °
20	141 °	159 °	177 °	-6.67	60.56 °	70.56 °	80.56 °
30	131 °	149 °	166 °	-1.11	55.00 °	65.00 °	74.44 °

Sizes and Approximate Prices of Oil Burning Boiler Units

Price includes boiler complete with oil burner, trombone tank type water heater or tankless type water heater, steel jacket with baked enamel finish, air-cell type insulation furnished with jacket, plain room thermostat, limit control, and combustion control.

MECHANICAL

Steam boiler prices include low water cut-off, pop safety valve, try cocks, water gauge, pressure and vacuum gauge, and flue brush.

Water boiler prices include combination altitude gauge and thermometer and flue brush.

Sizes and Approximate Prices of Cast-Iron Sectional Boilers and Gas or Light Oil Burners with Jacket							Hot Water
	Steam		Hot Water			Steam	Hot Water
Installed Rad. Sq. Ft.*	Price Each	Installed Rad. Sq.Ft.* for Oil Firing	Btu Rating	Price Each	Installed Rad. Sq. M*	Installed Rad. Sq.M.* for Oil Firing	Kg-Cal Rating
845	$1,220	1500	225,000	$1,220	78.50	139.35	56,699
1090	$1,460	1910	287,000	$1,460	101.26	177.44	72,323
1340	$1,675	2320	348,000	$1,675	124.49	215.53	87,694
1600	$1,870	2735	410,000	$1,870	148.64	254.08	103,318
1860	$1,950	3145	472,000	$1,950	172.79	292.17	118,942
2120	$2,050	3560	534,000	$2,050	196.95	330.72	134,565
2380	$2,350	3960	594,000	$2,350	221.10	367.88	149,685
2650	$2,850	4375	656,000	$2,850	246.19	406.44	165,309
2920	$2,900	4785	718,000	$2,900	271.27	444.53	180,933
3200	$3,000	5205	781,000	$3,000	297.28	483.54	196,808
3480	$3,100	5620	843,000	$3,100	323.29	522.10	212,432

Capacities are based on net installed radiation and on the presence of a quantity of net installed radiation sufficient for the requirements of the building. Nothing need be added for normal piping.

All ratings are based on radiation at 240 Btu per sq.ft. for steam and 150 Btu per sq.ft. for water.

Above prices are net to the heating contractor, f.o.b. factory. Prices include boiler trimmings.

Abbreviation Rad. Equals Radiation

Installed Rad. Sq.Ft.*	Steam Boiler Jacked Price with		Net BTU Rating	Installed Rad. Sq.Ft.*	Water Boiler Jacketed price with	
	Trombone Heater	Tankless Heater			Trombone Heater	Tankless Heater
460	$2,600	$2,760	126,000	840	$2,600	$2,760
570	$3,162	$3,300	155,000	1,030	$3,162	1650*2
680	$3,400	$3,530	183,000	1,220	$3,400	$3,570
790	$3,740	$3,910	211,000	1,405	$3,740	$3,910
900	$3,910	$4,100	239,000	1,590	$3,910	$4,100

Installed Rad. Sq.M.*	Steam Boiler Jacked Price with		Net Kg-Cal Rating	Installed Rad. Sq.M.*	Water Boiler Jacketed price with	
	Trombone Heater	Tankless Heater			Trombone Heater	Tankless Heater
42.73	$2,600	$2,760	31,751	78.04	$2,600	$2,760
52.95	$3,162	$3,300	39,059	95.69	$3,162	$3,300
63.17	$3,400	$3,530	46,115	113.34	$3,400	$3,570
73.39	$3,740	$3,910	53,171	130.52	$3,740	$3,910
83.61	$3,910	$4,100	60,227	147.71	$3,910	$4,100

Capacities are based on net installed radiation and on the presence of a quantity of net installed radiation sufficient for the requirements of the building. Nothing need be added for normal piping.
All ratings are based on radiation at 240 Btu per sq.ft. for steam and 150 Btu per sq.ft. for water.

Above prices are net to the heating contractor, f.o.b. factory. Prices include boiler trimmings. If boilers are furnished without tankless heater, deduct $100.00; less trombone tank type heater, deduct $50.00.

Water Heater Capacities

Tankless type   4 gals. (15.14 liters) per minute
Trombone tank type   66 gals. (250249.81 liters) in 3 hrs.

Above capacities are based on 40°F (4.44°C) to 140°F (60°C) rise with boiler water at 180°F (82.2°C). Above prices are f.o.b. factory. Prices include boiler trimmings as listed above.

Approximate Prices of Gas Fired Boilers for Hot Water Heating Systems

Furnished with finned cast iron sections to provide staggered heat travel, draft diverter, heavily insulated steel jacket with baked enamel finish. Boiler price includes gas valve and hydraulic limit control, relay gas valve and recycling manual control switch, transformer, electric safety pilot, gas pressure regulator, manual gas shut-off valve, combination thermometer and

MECHANICAL

altitude gauge, pilot pressure regulator, drain cock, pilot cock, draft diverter, gas manifold, tubing and fittings. Room thermostat not included. Prices are f.o.b. factory.

Rad. Sq.Ft.	Net I-B-R Rating Output BTU/Hr	A.G.A. Gross Output BTU/Hr	Price of Boiler Complete	Rad. Sq.M.	Net I-B-R Rating Output Kg-Cal/Hr	A.G.A. Gross Output Kg-Cal/Hr
515	77,500	103,000	$924	47.84	19,530	25,956
860	129,000	172,000	$1,220	79.89	32,507	43,343
1205	181,000	241,000	$1,595	111.94	45,611	60,731
1550	233,000	310,000	$1,820	144.00	58,715	78,119
1720	258,000	344,000	$2,030	159.79	65,015	86,686
1960	294,000	392,000	$2,240	182.08	74,087	98,782
2240	336,000	448,000	$2,660	208.10	84,670	112,894
2520	378,000	504,000	$2,940	234.11	95,254	127,006
2800	420,000	560,000	$3,150	260.12	105,838	141,117
3080	462,000	616,000	$3,640	286.13	116,422	155,229
3350	502,500	672,000	$4,200	311.22	126,628	169,341
3920	588,000	784,000	$4,760	364.17	148,173	197,564
4480	672,000	896,000	$5,040	416.19	169,341	225,788
5040	756,000	1,008,000	$5,670	468.22	190,508	254,011
5600	840,000	1,120,000	$6,020	520.24	211,676	282,235
6160	924,000	1,232,000	$6,440	572.26	232,844	310,458
6666	1,000,000	1,344,000	$7,000	619.27	251,995	338,682
7222	1,083,350	1,456,000	$7,560	670.92	272,999	366,905
7777	1,166,650	1,568,000	$7,840	722.48	293,990	395,129
8333	1,250,000	1,680,000	$8,120	774.14	314,994	423,352
8888	1,333,300	1,792,000	$8,680	825.70	335,985	451,576

Abbreviation Rad. Equals Radiation

Approximate Prices of Gas Fired Boilers for Steam and Vapor Heating Systems

These are furnished with finned cast iron sections to provide staggered heat travel, draft diverter, and heavily insulated steel jacket with baked enamel finish.

Boiler price includes gas valve, relay gas valve and recycling manual control switch, transformer, electric safety pilot, gas pressure regulator, pilot pressure regulator, manual gas shut-off valve, electric hot water cut-off, steam pressure regulator, retard steam gauge, pop safety valve, water gauge, try cocks, drain cock, pilot and burner cocks, draft diverter, gas manifold, tubing and fittings. Room thermostat not included. Prices are f.o.b. factory.

Rad. Sq.Ft.	Net I-B-R Rating Output BTU/Hr	A.G.A. Gross Output BTU/Hr	Price of Boiler Complete	Rad. Sq.M.	Net I-B-R Rating Output Kg-Cal/Hr	A.G.A. Gross Output Kg-Cal/Hr
280	66,750	103,000	$918	26.01	16,821	25,956
475	113,500	172,000	$1,320	44.13	28,601	43,343
670	161,000	241,000	$1,955	62.24	40,571	60,731
875	210,000	310,000	$2,465	81.29	52,919	78,119
980	235,000	344,000	$2,975	91.04	59,219	86,686
1121	269,100	392,000	$3,400	104.14	67,812	98,782
1293	310,250	448,000	$3,570	120.12	78,182	112,894
1466	351,700	504,000	$3,740	136.19	88,627	127,006
1641	393,800	560,000	$4,080	152.45	99,236	141,117
1996	478,950	672,000	$4,420	185.43	120,693	169,341
2353	565,250	784,000	$5,100	218.59	142,440	197,564
2717	652,100	896,000	$5,780	252.41	164,326	225,788
3081	739,550	1,008,000	$6,120	286.22	186,363	254,011
3449	827,800	1,120,000	$6,460	320.41	208,602	282,235
3814	915,300	1,232,000	$7,140	354.32	230,651	310,458
4167	1,000,000	1,344,000	$7,820	387.11	251,995	338,682
4514	1,083,350	1,456,000	$8,330	419.35	272,999	366,905
4861	1,166,650	1,568,000	$8,840	451.59	293,990	395,129
5208	1,250,000	1,680,000	$9,180	483.82	314,994	423,352
5555	1,333,300	1,792,000	$11,220	516.06	335,985	451,576
5903	1,416,650	1,904,000	$13,600	548.39	356,989	479,799

Abbreviation Rad. Equals Radiation

Net boiler is the amount of actual installed radiation that may be attached to the boiler, based on a heat emission rate from the radiators of 240 Btu per sq.ft. for steam and 150 Btu for gravity hot water. For forced circulation for hot water systems, select the boiler from the net Btu column.

Standard Black and Galvanized Steel Pipe						
ID, Inches	Black Pipe $/LF	Galv. Pipe $/LF	Threads Extra Each End	ID, mm	Black Pipe $/m	Galv. Pipe $/m
1/4	$1.50	$1.75	$2.00	6.25	$4.92	$5.74
3/8	$1.55	$1.90	$2.00	9.38	$5.09	$6.23
1/2	$1.40	$1.65	$2.00	12.50	$4.59	$5.41
3/4	$1.50	$1.95	$2.00	18.75	$4.92	$6.40
1	$2.00	$2.30	$2.00	25.00	$6.56	$7.55
1 1/4	$2.25	$2.90	$2.50	31.25	$7.38	$9.51
1 1/2	$2.60	$3.10	$2.50	37.50	$8.53	$10.17
2	$3.20	$4.15	$2.50	50.00	$10.50	$13.62
2 1/2	$5.10	$5.50	$2.50	62.50	$16.73	$18.04
3	$6.50	$7.00	$3.00	75.00	$21.33	$22.97

Above prices are for pipe cut to order. Add for threads as given.

MECHANICAL

Prices of Steam Radiator Valves						
Descrip.	Sizes in inches					
	1/2	3/4	1	1 1/14	1 1/2	2
Angle steam radiator valve	$58.00	$72.00	$100.00	$138.00	$176.00	$240.00
Gate steam radiator valve	$30.00	$40.00	$48.00	$60.00	$38.00	$110.00
Size mm	12.50	18.75	25.00	26.79	37.50	50.00

Prices of Water Radiator Valves					
Descrip.	Size in inches				
	1/2	3/4	1	1 1/14	1 1/2
Hot water radiator valve	$16.00	$22.00	$30.00	$50.00	$70.00
Hot water radiator union ell	$7.00	$8.00	$13.00	$17.00	$22.00
Size mm	12.50	18.75	25.00	26.79	37.50

Prices on Floor and Ceiling Plates								
Descrip.	Pipe size, inches							
	1/2	3/4	1	1 1/14	1 1/2	2	2 1/2	3
Chrome plated, ea.	$1.92	$1.92	$2.10	$2.43	$2.43	$2.67	$4.11	$5.13
Size mm	12.50	18.75	25.00	26.79	37.50	50.00	62.50	75.00

Floor and ceiling plates made of cold rolled steel, halves securely riveted by a concealed hinge. Can be opened or closed on pipe without effort.

Copper Pipe

The cost of copper pipe on heating work depends to a large extent upon the experience of the steam fitter installing it. The technique of installing copper pipe is entirely different from that of steel pipe. Inexperienced steam fitters figure more for labor than on a steel pipe job, while others figure the same for labor, whether steel or copper. Steam fitters who are thoroughly experienced in installing copper pipe and fittings and some say that they can install copper pipe for 15% to 20% less than steel pipe.

The price of the pipe itself is subject to wide fluctuations, due to the price of copper so that prices on copper pipe and fittings should always be obtained separately for each job. Copper pipe is furnished in 3 weights:

1. Heavy Copper Pipe (Government Type "K") for underground service, such as water, gas, steam, oil lines, industrial uses, also interior plumbing.
2. Standard Copper Pipe (Government Type "L") for interior plumbing and heating.
3. Light Copper Pipe (Government Type "M") for low pressure interior plumbing.

Sizes and Weights of Copper Pipe						
Size Inches	Type "K" Heavy		Type "L" Std.		Type "M" Light	
	Wall Thk. Inch	Wt. Lbs/LF	Wall Thk. Inch	Wt. Lbs/LF	Wall Thk. Inch	Wt. Lbs/LF
3/8	0.049	0.269	0.035	0.198	0.025	0.145
1/2	0.049	0.344	0.040	0.285	0.028	0.204
5/8	0.049	0.418	0.042	0.362	-----	-----
3/4	0.065	0.641	0.045	0.455	0.032	0.328
1	0.065	0.839	0.050	0.655	0.035	0.465
1 1/4	0.065	1.040	0.055	0.884	0.042	0.681
1 1/2	0.072	1.360	0.060	1.140	0.049	0.940
2	0.083	2.060	0.070	1.750	0.058	1.460
2 1/2	0.095	2.930	0.080	2.480	0.065	2.030
3	0.109	4.000	0.090	3.330	0.072	2.680
3 1/2	0.120	5.120	0.100	4.290	0.083	3.580
4	0.134	6.510	0.110	5.380	0.095	4.660
5	0.160	9.670	0.125	7.610	0.109	6.660
6	0.192	13.870	0.140	10.200	0.122	8.910
8	0.271	25.900	0.200	19.300	0.170	16.460
10	0.338	40.300	0.250	30.100	0.212	25.600
12	0.405	57.800	0.280	40.400	0.254	36.700

MECHANICAL

Size mm	Type "K" Heavy		Type "L" Std.		Type "M" Light	
	Wall Thk. mm	Wt. Kg/m	Wall Thk. mm	Wt. Kg/m	Wall Thk. mm	Wt. Kg/m
9.38	1.225	0.400	0.88	0.295	0.63	0.216
12.50	1.225	0.512	1.00	0.424	0.70	0.304
15.63	1.225	0.622	1.05	0.539	-----	-----
18.75	1.625	0.954	1.13	0.677	0.80	0.488
25.00	1.625	1.249	1.25	0.975	0.88	0.692
31.25	1.625	1.548	1.38	1.316	1.05	1.013
37.50	1.800	2.024	1.50	1.697	1.23	1.399
50.00	2.075	3.066	1.75	2.604	1.45	2.173
62.50	2.375	4.360	2.00	3.691	1.63	3.021
75.00	2.725	5.953	2.25	4.956	1.80	3.988
87.50	3.000	7.620	2.50	6.384	2.08	5.328
100.00	3.350	9.688	2.75	8.006	2.38	6.935
125.00	4.000	14.391	3.13	11.325	2.73	9.911
150.00	4.800	20.641	3.50	15.180	3.05	13.260
200.00	6.775	38.544	5.00	28.722	4.25	24.496
250.00	8.450	59.974	6.25	44.794	5.30	38.098
300.00	10.125	86.017	7.00	60.123	6.35	54.617

Table title: Sizes and Weights of Copper Pipe

Because of the smooth inside surface of copper pipe, usually one size smaller pipe can be used than where steel pipe is used. Outside diameter of pipe is 1/8" (3.125) mm more than nominal size of tubing.

Approximate Prices on Copper Tubing

Prices on copper tube and fittings are subject to wide price fluctuations. Always check prices before submitting bids.

Copper Tubing							
Nominal Size, Inch							
1/4	3/8	1/2	5/8	3/4	1	1 1/2	2
Price per lin.ft.							
Type "K" $1.35	$1.75	$2.00	$2.25	$3.25	$3.75	$7.10	$7.75
Type "L" $1.25	$1.35	$1.75	$2.00	$2.25	$3.00	$4.50	$6.50
Nominal Size, mm							
6.25	9.38	12.50	15.63	18.75	25.00	37.50	50.00
Price per Lin.M.							
Type "K" $4.43	$5.73	$6.56	$7.38	$10.66	$12.30	$23.29	$25.43
Type "L" $4.10	$4.43	$5.74	$6.56	$7.38	$9.84	$14.76	$21.33

Prices of Brass Solder - Joint Valves

These valves are made for use with copper tubing and can be used for steam or water service.

Descrip.	Size in Inches					
	1/2	3/4	1	1 1/4	1 1/2	2
Fitting Gate valve	$35.00	$45.00	$49.00	$65.00	$85.00	$120.00
	Size in mm					
	12.50	18.75	25.00	31.25	37.50	50.00

Approximate Quantity of Solder and Flux Required to make 100 Joints							
Size, in.	3/8	1/2	3/4	1	1 1/4	1 1/2	2	2 1/2
Solder, Lbs	1/2	3/4	1	1 1/2	1 3/4	2	2 1/2	3 1/2
Flux, Oz.	1	1 1/2	2	3	3 1/2	4	4	7
Size, mm	9.38	12.50	18.75	25.00	31.25	37.50	50.00	62.50
Solder, kg	0.23	0.34	0.45	0.68	0.79	0.91	1.13	1.59
Flux, grams	28.3	42.5	56.7	85.0	99.2	113.4	113.4	198.4

Approx Qty of Solder and Flux Required to make 100 Joints							
Size, in.	3	3 1/2	4	5	6	8	10
Solder, Lbs	4 1/2	5	6 1/2	9	17	35	45
Flux, Oz.	9	10	13	18	34	70	90
Size, mm	75.00	87.50	100.00	125.00	150.00	200.00	250.00
Solder, kg	2.04	2.27	2.95	4.08	7.71	15.88	20.41
Flux, grams	255.1	283.5	368.5	510.3	963.9	1,984.4	2,551.4

Estimating the Cost of Pipe and Fittings. To estimate the cost of pipe and fittings required for any heating plant, obtain the total number of sq. ft. of radiation in the job, based on radiation, and figure at $4.00 to $5.00 per sq.ft. ($43.06 to $53.82 per sq.m). This is based on prices of pipe as given on the previous pages.

For one story residences having short pipe runs, figure the cost of the pipe and fittings at $4.00 per sq.ft. ($43.06 per sq.m) of radiation, based on standard radiation.

An example of the method used in estimating the cost of pipe and fittings is as follows: For a job containing 1,400 sq.ft. (130.06 sq.m) of standard radiation, take 120% of 1,400 (130) which equals 1,680 (156), or the pipe and fittings would cost approximately $6,800. The above allowances for pipe and fittings do not include radiator valves, floor and ceiling plates, boiler fittings, expansion tanks, etc., but only the pipe and fittings required for roughing-in the job.

Hairpin exchangers: These are a type of water heat exchanger. The hairpin tube takes the condensate and reheats it through the tubes. To estimate the square feet of tube required, the following is a suggested method: Sq.Ft. of each uit = 4.1 x lineal feet of tube.

MECHANICAL

Shell and Tube exchangers: To estimate the square feet of tube required, the following is a suggested method: Sq.Ft. = (outside shell diameter in inches) x bundle length in feet x 0.3142.

Pressure Vessels: To estimate the square feet of tube required, the following is a suggested method:

Sq.Ft. of Shell = (outside shell diameter in inches) + 2(insulation thickness in inches) x (shell length in feet) x 0.2618.

Sq.Ft. per semi-Elliptical Head: (outside head diameter in inches) + 2(insulation thickness in inches) x 2 x 0.00908.

Sq.Ft. per flanged and dished head: (outside head diameter in inches) + 2(insulation thickness in inches) x 2 x 0.00765.

The above formulas can be used for fiberglass board or blanket, urethane or styrene board with vapor barrier, calcium silicate blocks. All the above types of materials can be field-applied.

Pipe Covering

Pipe covering is furnished in several kinds of materials. Fiberglass sectional pipe insulation covering is the most common and cheapest for covering steam and hot water pipes; 85% magnesia pipe covering is for high pressure steam work; and wool felt pipe covering is for protecting pipes against freezing and for general use.

Pipe covering is usually furnished in 3' to 10' (0.91 to 3.04 m) lengths, 3/4" (18.75 mm) to 1" (25 mm) thick and is covered with a canvas or fire safety jacket.

Labor Covering Pipes Up to 3" (75 mm) in Diameter. When placing covering on pipes up to 3" (75 mm) in diameter, a pipe coverer should apply 10 to 15 lin.ft. (3.04 to 4.57 m) of covering per hour, or 80 to 120 lin.ft. (24.38 to 36.57 m) per 8-hr. day, and the labor cost per 100 lin.ft. (30.48 m) should average as follows:

Description	Hours	Rate	Total	Rate	Total
Pipe coverer	8.00	$	$	$ 46.90	$ 375.20
Cost per lin.ft.					$ 3.75
per lin.m.					$ 12.31

Labor Covering Pipes 4" (100 mm) to 6" (150 mm) in Diameter. When placing covering on pipes 4" to 6" (100-150 mm) in diameter, a pipe coverer should apply 9 to 11 l.f. (2.7-3.3 m) of covering per hour or 72 to 88 l.f. (21.6-26.4 m) per 8-hr. day, and the labor cost per 100 l.f. (30 m) should average as follows:

Description	Hours	Rate	Total	Rate	Total
Pipe coverer	10.00	$	$	$ 46.90	$ 469.00
Cost per lin.ft.					$ 4.69
per lin.m.					$ 10.70

When estimating quantities of pipe covering, add 1 lin.ft. (0.30 m) of covering for each fitting for pipe up to 2" (50 mm) and 2 lin.ft. (0.61 m) of covering for each fitting on pipe over 2" (.50 mm) thick.

Temperature Regulators for Heating Plants

Temperature regulators and thermostats are furnished in a variety of different models for controlling the temperature.

There are so many different methods of controlling the inside temperature with straight thermostats, day and night thermostats, inside-outside controls, etc., that it is practically impossible to give even a fraction of them in a book of this kind.

Temperature Regulators With Clock. The thermostat with a clock attachment automatically lowers the temperature at night and raises it in the morning. Electric clocks are now usually furnished with all thermostats of this type.

Models of this kind are furnished with thermometer, clock, electric motor, chain, pulleys, transformer and fittings, at the approximate price of $300.00.

Temperature Regulators Without Clock. Many thermostats or regulators are used without the clock attachment and are either manually operated or the same temperature maintained day and night.

There are a number of different types of regulators on the market and they vary in price from $100.00 to $325.00, depending on the method of wiring and number of controls necessary.

Labor Installing Thermostats. An experienced mechanic should install any of the above thermostat outfits in 1 to 2 hours. If they are installed by the manufacturers in cities where they maintain offices, a charge of $300.00 to $500.00 is usually made.

Making Up a Complete Estimate on a Steam Heating Plant. When making up an estimate on a steam heating system, the following items should always be included in order that the estimate may be complete in every detail.

1. Boiler
2. Oil or gas burner
3. Radiators or convectors
4. Radiator Valves
5. Automatic air valves
6. Pipe and fittings
7. Smoke pipe
8. Special foundation, if required
9. Asbestos boiler covering
10. Pipe covering
11. Floor and ceiling plates
12. Painting radiators
13. Thermostat regulator
14. Freight and trucking
15. Water Connection
16. Labor installing system
17. Bond and insurance
18. Overhead and misc. expense
19. Profit

Making Up a Complete Estimate on a Hot Water Heating Plant. When making up an estimate on a hot water heating plant, the following

MECHANICAL

items should always be included in order that the estimate may be complete in every detail.

1. Special boiler foundation, if required
2. Boiler
3. Oil or gas burner
4. Pipe and fittings
5. Radiators or convectors
6. Radiator valves-union ells
7. Circulating pumps
8. Compression air valves
9. Expansion tank and gauge
10. Altitude gauge
11. Thermometer
12. Smoke pipe
13. Painting radiators
14. Boiler covering
15. Pipe covering
16. Floor/ceiling plates
17. Thermostat Reg
18. Labor install'g Sys.
19. Freight & trucking
20. Water connection
21. Bond and insurance
22. Overhead/misc. exp
23. Profit

Unit Heaters. Unit heaters are often used to heat commercial and industrial areas, garages, and to supplement other heating units at entrances. They may be direct-fired by gas, oil, or electricity, or have hot water or steam piped to them. Direct-fired are cheaper when only a few units are required and no central plant is available. Hot water and steam are the most common units in use where a central plant is available or its cost justified by the number of units it serves.

Units may have propeller or centrifugal fans blowing either horizontally, up or down. Air may be drawn through or blown through, the latter being used for direct-fired units.

The following costs do not include valves, fittings, or piping, costs for which vary widely with job conditions and are based on units served with 2 lbs. (0.9 kg) steam, 60°F (15.5°C) entering air, single phase 115 volts.

Cost of Unit Heaters							
Type	Hp	kW	Btu	1 Speed	2 Speed	Hrs. to Install	kg-cal
Horizontal Delivery	1/40	0.02	25,000	$310.00	-----	1.5	6,299.9
	1/15	0.05	50,000	$400.00	$500.00	2.0	12,599.8
	1/10	0.07	100,000	$600.00	$680.00	3.0	25,199.5
	1/5	0.15	200,000	$960.00	$1,100.00	5.0	50,399.1
Vertical Delivery	1/40	0.02	50,000	$380.00	-----	2.0	12,599.8
	1/12	0.06	100,000	$520.00	$600.00	2.5	25,199.5
	1/3	0.25	200,000	$1,040.00	$1,150.00	6.0	50,399.1
Cabinet	1/2	0.04	25,000	$320.00	-----	2.0	6,299.9
	1/12	0.06	60,000	$460.00	-----	2.5	15,119.7
	1/8	0.09	120,000	$950.00	-----	3.5	30,239.4

For 240 volts, add $150.00. For ceiling suspension, add $100.00.

15300 SPRINKLER SYSTEMS

The construction industry and the general public for many years have been aware of the need for and availability of fire protection systems.

However, for a long time this protection was restricted to areas where local building codes or extreme business risks dictated a demand for their use. In general, there was great insensitivity toward providing fire protection for the public, even in areas of high concentration such as restaurants, theaters, hotels, and sporting events.

This disregard for public safety changed because of a number of nationally publicized fire disasters, inspiring action by federal and state authorities, insurance companies, building owners, and professional architects.

The fire protection industry responded promptly by providing specialized technical design data and producing trained personnel capable of educating the designers and specification writers in selecting the particular system, or combination of systems, most appropriate to provide an adequate degree of protection.

As the need has developed to provide special protection for costly equipment such as computer rooms and tape storage libraries, for irreplaceable documents, and art objects, for sophisticated research projects, and for the aged and the handicapped, new alarm, control, and protection systems have been developed. Some of the systems are mechanically operated, while others are completely electronic. In some cases, alarm and control systems are designed as a combination of mechanical and electrical devices. The estimator must determine with great precision the cost and installation responsibility that each trade will assume in the overall function of the system.

Wet Pipe Sprinkler System

This type of system meets most fire protection requirements for normal hazards, where piping is not subject to freezing. Water is constantly supplied under pressure to each sprinkler head.

Dry Pipe Sprinkler System

This system can be used in areas subject to freezing, because the piping is filled with air under pressure, which contains the water at the supply source until an activated sprinkler head allows the air pressure to disburse and let the water supply move to the head. Sprinkler systems are a specialized type of work, requiring the services of a subcontractor who has access to the necessary equipment, which in many instances is a sole source item and not available on the open market. One usually finds that the contract documents require the services of a specialist who has years of experience in this type of work and who has the capabilities to design the sprinkler system.

For the wet and dry pipe systems, unit costs have been developed on a per head basis. It is assumed that an adequate water supply is available and that special alarm and control work is not included. Different heads are used depending on the degree of hazard, and local fire regulations are consulted for a specific project.

1518

MECHANICAL

For high-rise buildings, booster pumps may be required, which will add from $20,000 to $50,000 to the cost of installation.

Maximum area coverage per head is 130 sq.ft. (12.08 sq.m). The average area covered per head is from 80 sq.ft. to 130 sq.ft. (7.43 to 12.08 sq.m) under normal hazard. Distance between sprinkler heads cannot exceed 15 lin.ft. (4.57 m).

Wet System*	Total Cost per Head
Warehouse	$135.00
Hospital	$225.00
Shopping Center	$150.00
Municipal Buildings	$185.00

For dry system, add $30.00 to $40.00 per head.

From a look at the above costs, it might seem that a fire sprinkler system is costly, but an owner must look at all the benefits. It is insulation against litigation resulting from fire loss and a savings in fire insurance premiums. Most systems return their initial cost in reduced insurance premiums within 3 to 10 years.

16   ELECTRICAL

Even the largest or tallest building may depend completely on electricity for light, power, building services, tenant equipment, and heat. Heating and lighting must be coordinated so that a maximum use of electric energy may be obtained at a minimum of expense.

Obviously such sophisticated designs are well beyond the scope of this book. But even a modest residence will involve electrical planning and power demands well beyond anything thought of 25 years ago.

An electrical estimate, whether commercial or residential, can be divided into four general categories: service, distribution, branch, and devices/loads. The service comprises the outdoor utility-supplied feeders and connection to the building, as well as the revenue metering and main disconnect switches. Distribution consists of the panelboards and switchgear that immediately follow the main switch and the feeders that serve them and emanate from them. The branch work, also referred to as lighting circuitry, is the raceway and wire between the panelboard breakers and the devices/loads. Finally, there are the devices/loads, such as receptacles and switches, lighting outlets, fixtures, and heaters.

The electrical service is installed in two phases: the contractor's portion, consisting of the setting and wiring of the main disconnect switches and the revenue meter pans, followed by the utility company's portion, consisting of the installation of the feeder, either on poles or underground, its connection to the meter pan, and installation of the meter itself.

For residence construction, the typical service size is 200 amperes, 208 volts, single phase; that is, two phase wires and a neutral, making both 120 and 208 volts available. The main disconnect switch is usually incorporated in the main circuit breaker. The typical cost of such a panelboard, 30 circuits and the usual complement of circuit breakers, is $400.00 plus about 24 hours of labor. The matching meter pan would add $100.00 and an additional 5 hours for installation. Where a 100 ampere service is provided, the panelboard costs would be $400.00 plus about 15 hours of labor for installation, with the meter pan not substantially affected.

In commercial work, a separate disconnect switch connected to a distribution panelboard is common. A 200 ampere fused disconnect switch costs $800.00 plus about 10 hours labor for installation. A matching distribution panel with typical breakers, 42 circuits, no main breaker, costs $1,600.00 plus about 36 hours of labor for installation. The wiring between the devices—3/0, 2" (50 mm) emt—costs $12.00 plus about 0.30 hours of labor per l.f. for installation. Where necessary, a 400 ampere switch at $1,3000 plus about 16 hours of labor for installation and a 400 ampere, 42 circuits no main breaker, panelboard at $1,000.00plus about 40 hours of labor for installation may be used. In this case, the wiring between the service switch and the panel would be 500 mom, 3 -1/2" (88 mm) emt, $36.00 plus about 0.43 hours of labor per foot (1.43 per m) of installation.

With the service and distribution completed, the branch work can be analyzed. In residential work, most of the branch work will run as non-metallic sheath cable (romex) or metal sheath cable (bx). Costs are: Romex 2#12 plus ground wire, $0.60 plus about 0.035 hours of labor per lin.ft. (0.12 hrs. per m) of installation; 2#10 plus ground wire, $0.80 plus about 0.038 hours of labor per lin.ft. (0.13 hrs. per m) of installation; BX 3#12, $0.60 plus about 0.05 hours of labor per lin.ft. (0.17 hrs. per m) of installation; BX 3#10, $1.30 plus about 0.06 hours of labor per lin.ft. (0.20 hrs. per m) of installation. In commercial work, emt and building wire is often used. The equivalent costs are: 3/4" (18.75 mm) emt 3#12, $0.90 plus about 0.138 labor hours per lin.ft. (0.46 hrs. per m) of installation; 3/4" (18.75 mm) emt 3#10, $1.25 plus about 0.146 labor hours per lin.ft. (0.49 hrs. per m) of installation.

For very preliminary estimates, electrical work can be assumed to run from 6% to 12% of the total job cost for residences, 8% to 11% for office buildings, and 8% to 12% for industrial buildings.

An estimator should make a detailed electrical take off and must have a thorough knowledge of the working drawings and specifications from which the job will be built, as well as of the building code under which it will be inspected and regulations set by the local utility company.

The electrical take off sheet for residential work should have a column for each of the following:

1. Ceiling outlets
2. Bracket outlets
3. Duplex convenience outlets
4. Triplex convenience outlets
5. Waterproof convenience outlets
6. Range outlets
7. Lighting outlet with lampholder
8. Special purpose outlets (dishwashers, clothes dryers, etc.)
9. Single pole switches
10. Three-way switches
11. Length of wiring

ELECTRICAL

16100 ELECTRICAL GENERAL CONDITIONS

ELECTRICAL PRODUCTIVITY - NORMAL NON-PRODUCTIVE LABOR

Every labor unit in this section includes a prorated amount of the normal non-productive labor. The magnitude of the non-productive labor in each labor unit is known. The normal non-productive labor in the NECA labor units include:

- Personal Comfort Breaks
- Normal Time to obtain drinking water
- Minimum breaks for coffee carried to the work area
- Walking time between the gang shed and the work area
- Minimal walking time to obtain tools and electrical materials
- Minimal lost time due to interference by other crafts
- No time lost waiting for material or personnel hoist or crane
- No time lost due to strikes, work stoppages, or jurisdictional disputes
- No time lost due to egress, job site security, or physical obstructions

Through good project planning and management, electrical contractors generally can keep such things as loitering, interruptions, late starting, early quitting, etc. to a minimum. When these conditions are beyond the control of the electrical contractor, then additional non-productive labor must be considered by the estimator. A factor of no less then 20% must be added to the general conditions.

HOW TO SELECT BASIC LABOR UNITS

Estimating electrical material installation labor is best accomplished by the application of a labor unit to each item on materials prepared from the take-off of the bid contract drawings. Along with the type and size of each item, it is essential that an abbreviated description of the installation condition is noted. The necessity of the type and size of each item is obvious, but of equal necessity are the abbreviated notes regarding the installation conditions so that an appropriate selection may be made from the choice of labor unites available.

If the estimator follows the National Electrical Contractors Association (NECA) labor unit tables, he or she would create three different labor units for each item. The first (lowest) labor unit for each item listed in the take-off would be under **NORMAL INSTALLATION CONDITIONS**. The second (higher) labor unit for each item listed in the take-off would be under **DIFFICULT INSTALLATION CONDITIONS**. The third (highest) labor unit for each item listed in the take-off would be under **VERY DIFFICULT INSTALLATION CONDITIONS**.

NORMAL INSTALLATION CONDITIONS

When all of the conditions associated with the installation of an item will permit the maximum productivity of the electricians on the project these "Normal" labor units are applicable. An example of these normal conditions might include the installation of the surface mounted items up to a height of 8 feet above the floor, the use of a rolling scaffold or 6 foot ladders, a repetitive layout, a minimum of required measurements, a minimum of interference by structural or mechanical obstacles.

DIFFICULT INSTALLATION CONDITIONS

When one or more of the connections associated with the installation of an item will permit less then maximum productivity of the electricians on a typical project, these "Difficult" labor units are applicable. An example of the difficult condition might include the installation of surface mounted items up to a height of 12 feet from the floor, the use of fixed scaffolding installed by others, the use of 10 foot ladders, a non-repetitive layout, a normal amount of interference by the structural or mechanical obstacles, etc. Estimators are also encouraged to consider using labor units whose magnitude are somewhere between the normal labor units and the difficult labor units when conditions justify this consideration.

VERY DIFFICULT INSTALLATION CONDITIONS

When one or more of the conditions associated with the installation of an item will permit substantially less than maximum productivity of the electrician on a typical project, these "Very Difficult" labor units are applicable. An example of the very difficult conditions might include the installation of surface mounted items up to a height of 20 feet above the floor, the use of fixed scaffolding installed by others, the use of 18 foot ladders, a single location of each item, a substantial amount of interference by structural and mechanical items, a difficult fastening method, etc. Estimators are also encouraged to consider using labor units whose magnitude are somewhere between the difficult labor units and the very difficult labor units when the conditions justify this consideration.

INCREASING ESTIMATING ACCURACY

On most typical projects there usually are several identical items whose installation labor units should be determined separately due to significantly different installation conditions. For example, different labor units should be used for estimating the installation labor for conduit located in concrete floor slabs as compared to the identical type and size of conduit installed exposed on walls and ceilings. Conduit in concrete floors slabs is usually installed by the shortest rout possible and therefore requires a lesser amount for point to point distances. Identical exposed conduit installed on walls, on ceilings, or parallel to concrete column lines requires a greater

1524

quantity of conduit, and the labor required per foot of conduit is greater because of the additional effort to maintain the alignment, elevation, and appearances. The labor for installing identical items should be determined separately whenever a significant quantity of the item is required in situations where two or more installation conditions exist.

The best insurance that any construction contractor will continue in business is to bid projects based on the contract documents which include the specifications. BASED ON THE CONTRACT DOCUMENTS is the only justifiable reason for modifying the estimated labor costs in the bid estimates. The adjustment of estimated labor costs based on who the competitors are and how they estimate is not only foolish and unjustifiable but will, without exception, eventually lead to financial disaster.

The estimator then should check off each item and enter the total on the sheet. This is best done floor by floor, or in large one-story residences, wing by wing, so that checking and changes ordered on the job will be simplified.

Enter next the circuit and feeder control centers, listing them by type and size, and then the branch circuits. List separately two-wire and three-wire runs and empty conduit for telephones and such other special items as intercoms, buzzer systems, television outlets, and the like.

A final list of such items as fixture hangers, studs, locknuts, bushings, cable clamps, and outlet covers complete the take off, and then all these items can be transferred to the pricing sheet item by item.

A pricing sheet should include a line for each type of item; the quantity involved; the material cost by unit and in total; the hours estimated for installation per unit and hourly rate; the total labor cost; and finally, the total labor and material cost.

Service Entrance Equipment

The service entrance equipment consists of the service drop, service entrance conductors, the meter, main control center, and the service ground. The local utility will govern the details of this installation and may or may not furnish the meter. Underground service is highly desirable and may be required by local ordinances. Underground service for residential use costs about $16.00 per lin.ft. ($26.67 per m). Overhead lines will run around $1,800.00 per pole. The main control center for residential work consists of a circuit breaker-fuse panel of at least 100 amps. Capacity of 200 amps is recommended.

Circuit Breakers		
Capacity	Cost	Hours to set
100 amp. 3 pole, gen. purpose	$356.00	4
200 amp. 3 pole, gen. purpose	$781.00	7
400 amp. 3 pole, gen. purpose	$1,250.00	11
600 amp. 3 pole, gen. purpose	$2,125.00	14

16110 RACEWAYS

Rigid galvanized steel conduit is the safest system and in some large cities is required by code. Note that in some areas, when reference is made to rigid pipe, it is understood to mean rigid galvanized pipe. Conduit systems may be galvanized steel rigid, aluminum, or electric metallic tubing, generally referred to as thinwall or EMT

Size Inch	Galvanized		Aluminum		EMT		Size mm
	Cost	Hours to Set	Cost	Hours to Set	Cost	Hours to Set	
1/2	$120	15	$150	11	$50	6	12.50
3/4	$140	18	$200	13	$56	11	18.75
1	$202	19	$300	14	$70	12	25.00
1 1/2	$320	26	$480	19	$115	16	37.50
2	$416	30	$700	22	$150	17	50.00
3	$870	38	$1,400	28	$220	19	75.00
4	$1,265	45	$2,200	33	$675	23	100.00

Material and Labor to Set 100 Lin. Ft (30.48 Lin.M). of Conduit (including supports and fittings, drag wire, raceway, terminations, elbows and junction box.)

Conduit Installation

Size Inches	Conduit Cost	Connector Cost	Hours to Set	Size mm
1/2	$60.00	$2.70	5	12.50
1	$150.00	$7.00	8	25.00
1 1/2	$240.00	$12.60	14	37.50
2	$320.00	$20.80	18	50.00
3	$500.00	$64.00	30	75.00

Where codes allow, the least expensive method of wiring is pre-wired armored cable (BX) or non-metallic sheathed cable such as Romex.

Material Cost to Install 100 Lin. Ft. (30.48 Lin. M.) of Armored Cable

Gauge of Wire	No. of Wires	Cost	Hours to Set
#10	2	$126.00	5
#10	3	$165.00	6
#12	2	$75.00	4
#12	3	$105.00	5
#14	2	$65.00	4
#14	3	$85.00	5

ELECTRICAL

Material and Labor to Install 100 Lf (30.48 Lm) of Non-Metallic Sheathed Cable with Ground			
Gauge of Wire	No. of Wires	Cost	Hours to Set
#12	2	$55.00	3.5
#12	3	$85.00	4.0
#14	2	$40.00	3.2
#14	3	$65.00	3.5

Duct System

Underfloor metal duct systems vary widely in their complexity. Some floor systems have raceways built in, others are added in the floor fill and may be a one or two level system. Standard duct is 1-3/8" x 3-1/8" (34.37 x 78.12 mm). Super duct is 1-3/8" x 7-1/4" (34.37 x 181.25 mm). Lengths are 5' (1.52 m), 6' (1.83 m), 10' (3.04 m), and 12' (3.65 m). Detailed take offs on these systems are beyond the scope of this book. The Square D Company catalog gives a suggested take off method with all the required fittings. Unit does not include receptacle, telephone, or other devices outlets or fittings.

Description	Standard		Super	
	Material Cost	Hrs per 100 Lf	Material Cost	Hrs per 100 Lf
Plain Duct per Lf	$7.00	10	$12.00	14
Duct with Inserts per Lf	$7.50	11.5	$12.50	16
Single Box - ea.	$170.00	2.0 ea.	$220.00	3.0 ea.
Double Box - ea.	$200.00	2.5 ea.	$400.00	3.5 ea.
Support - ea.	$10.00	0.25 ea.	$12.00	0.5 ea.
Elbow - ea.	$46.00	0.75 ea.	$88.00	1.0 ea.
Connector - ea.	$8.00	0.30 ea.	$16.00	0.3 ea.

Description	Standard		Super	
	Material Cost	Hrs per 30.48 Lm	Material Cost	Hrs per 30.48 Lm
Plain Duct per Lm	$22.97	10	$39.37	14
Duct with Inserts per Lm	$24.61	11.5	$41.01	16
Single Box	$170.00	2.0 ea.	$220.00	3.0 ea.
Double Box	$200.00	2.5 ea.	$400.00	3.5 ea.
Support	$10.00	0.25 ea.	$12.00	0.5 ea.
Elbow	$46.00	0.75 ea.	$88.00	1.0 ea.
Connector	$8.00	0.30 ea.	$16.00	0.3 ea.

Material and Labor to Pull 1,000 Lin.Ft. (304.80 m) of Wire in Conduit Already in Place (600V Type THHN/THWN)		
Gauge	Material Cost	Hours to Pull
14	$55	6
12	$66	7.25
10	$110	9
8	$198	12
6	$242	13
4	$363	16
2	$616	18
1/0	$968	26
2/0	$1,122	29
4/0	$1,738	38

Note: THW has been replaced by THHN/THWN.

16130 METAL OUTLET BOXES

Installing Metallic Outlet Boxes by Type		
Type	Cost per Unit	Hrs to Install
Switch Box	$3.00	0.30
3 Gang Box	$10.00	0.60
4" (100 mm) Octagon	$1.20	0.50
4" (100 mm) Square	$1.50	0.50
4 Concrete	$2.20	0.40
1 Gang Cast WP	$9.00	0.66
3 Gang Cast	$20.00	1.00

16140 SWITCHES AND RECEPTACLES

Once the box has been set and connected to the conduit and the wiring pulled, the proper switch or receptacle can be installed which will run as follows:

ELECTRICAL

Type	Cost per Unit	Hours to Install
15 amp single pole toggle	$9.00	0.20
20 amp single pole toggle	$12.00	0.30
15 amp 3-way toggle	$15.00	0.35
20 amp 3-way toggle	$18.00	0.50
15 amp 4-way toggle	$42.00	0.60
20amp 4-way toggle	$48.00	0.75
600 watt single pole dimmer	$21.00	0.50
600 watt three pole dimmer	$36.00	0.66
1000 watt single pole dimmer	$150.00	0.50
1000 watt three pole dimmer	$225.00	0.66
15 amp duplex receptacle	$6.00	0.20
20 amp duplex receptacle	$12.00	0.30
30 amp dryer	$24.00	0.50
50 amp range	$33.00	0.75

Cost of Toggle Switch Duplex Receptacle Plates per 100		
Description	Cost in Phenolic	Cost in Stainless
1 Gang Switch	$120.00	$284.00
2 Gang Switch	$240.00	$750.00
3 Gang Switch	$460.00	$1,200.00
4 Gang Switch	$1,000.00	$1,900.00
1 Gang Outlet	$100.00	$284.00
2 Gang Outlet	$240.00	$750.00
3 Gang Outlet	$660.00	$1,200.00
Blank-Single	$120.00	$284.00
Telephone	$120.00	$284.00

Weatherproof switches with grey Hypalon presswitch plates cost $900.00 per hundred for the plate not including the switch or box.

16465 BUSWAY

Busway is a modular power distribution system that consists of solid, insulated copper or aluminum bars encased in a metal housing. The housing is either attached to or mounted to a wall or suspended from the ceiling and provides for power plug-in locations, where one can mount a fusible switch or a circuit breaker.

Busways are an economical solution, providing flexibility to an end user with the amount of electrical power to the appropriate location, very important for manufacturing facilities.

Busway, Feeder Busway, and Riser Busway. "Plug-ins" are normally associated with busways. The busway has frequent provisions for power take-offs. On the other hand, "feeder busway" has no provisions for plug-ins and can be used to bring power from an outdoor transformer to an indoor distribution room or electrical closet. "Riser busways" also normally have no provisions for plug-ins and are used for vertical runs, from floor to floor, but may have plug-ins on one side.

Benefits of using a busway distribution system over wire and conduit distribution system include: quicker installation for a lower installation cost; anticipated lower power losses; and lower voltage drop.

Busway units come in 5' (1500 mm) and 10' (3000 mm) lengths—most common—and 20' (6000-mm) length sections, all requiring bus supports. The type of bracket will depend on whether it is ceiling- or wall-mounted. Types of brackets include: a) twist-in hanger, b) rod-mount hanger, c) T-bar mounting hanger, d) cable suspension assembly and e) hanger with stud. The engineer or contractor needs to assure a rigid installation for both horizontal and vertical alignment. Contractors should consult with and use the manufacturer's recommendations when drawings are absent of such detail. Also, when ordering material, include such accessories that may be required such as 1) end feeder sets, 2) plug-in jacks, 3) coupling sets and 4) dead end caps.

Two workers can install 40 lin.ft.(12.19 lin.m.) per day with 8-ft. (6.15 m) high ceilings. When working from scaffolding, installation rates can be reduced as much as 50%.

Description	Length in Feet	Weight in Pounds	Length in Meters	Weight in Kg
60 amp 2 Pole ground	5	5.0	1.52	2.27
60 amp 2 Pole ground	10	10.0	3.05	4.54
60 amp 2 Pole ground	20	20.0	6.10	9.07
60 amp 3 Pole ground	5	6.0	1.52	2.72
60 amp 3 Pole ground	10	12.0	3.05	5.44
60 amp 3 Pole ground	20	24.0	6.10	10.89
60 amp 4 Pole ground	5	6.2	1.52	2.81
60 amp 4 Pole ground	10	12.5	3.05	5.67
60 amp 4 Pole ground	20	25.0	6.10	11.34
100 amp 2 pole ground	5	6.4	1.52	2.90
100 amp 2 pole ground	10	13.0	3.05	5.90
100 amp 2 pole ground	20	26.0	6.10	11.79
100 amp 3 pole ground	5	7.0	1.52	3.18
100 amp 3 pole ground	10	14.0	3.05	6.35
100 amp 3 pole ground	20	28.0	6.10	12.70
100 amp 4 pole ground	5	8.0	1.52	3.63
100 amp 4 pole ground	10	16.0	3.05	7.26
100 amp 4 pole ground	20	32.0	6.10	14.52

Installing 60 amp straight busways (10' Section)

Description	Hours	Rate	Total
Electrician Foreman	8.00	$47.64	$381.12
Electrician	8.00	$43.64	$349.12
Cost per 40 lin.ft.			$730.24
per lin.ft.			$18.26
per lin.m.			$59.90

Installing 100 amp straight busways (10' Section)

Description	Hours	Rate	Total
Electrician Foreman	8.00	$47.64	$381.12
Electrician	8.00	$43.64	$349.12
Cost per 35 lin.ft.			$730.24
per lin.ft.			$20.86
per lin.m.			$68.45

16500 LIGHTING

The following is a sampling of some fixtures an electrician might encounter on average work.

Type	Wattage	Size	Shielding	Cost	Size
Industrial Fluorescent					
Pendant Mtg.	2-40	4 '	Open	$70	1.22 M
	4-40	8 '	Open	$130	2.44 M
	2-40	4 '	Louvers	$80	1.22 M
	4-40	8 '	Louvers	$160	2.44 M
Commercial Fluorescent	2-40	4 '	35°x35°	$100	1.22 M
Surface Mounted	4-40	4 '	35°x35°	$160	1.22 M
Average Quality	4-40	8 '	35°x35°	$300	2.44 M
Commercial Fluorescent	8-40	8 '	35°x35°	$200	2.44 M
Surface Mounted	2-40	4 '	Acrylic	$130	1.22 M
Premium Quality	4-40	8 '	Acrylic	$200	2.44 M
Commercial Fluorescent	2-40	4 '	Holophane	$160	1.22 M
Recessed Troffers	4-40	4 '	Holophane	$380	1.22 M
Commercial Fluorescent	4-20	2 ' x 2 '	Acrylic	$90	0.61 M x 0.61 M
Square Surface	6-30	3 ' x 3 '	Acrylic	$150	0.91 M x 0.91 M
	6-40	4 ' x 4 '	Acrylic	$350	1.22 M x 1.22 M
Commercial Fluorescent Corridor Unit	1-40	4 '	Acrylic	$260	1.22 M
Incandescent Industrial	1-100	12 "	Diam.- Open	$100	305 mm
Dome Bowl	1-150	14 "	Diam.- Open	$120	356 mm
	1-200	16 "	Diam.- Open	$130	406 mm
	300-500	18 "	Diam.- Open	$150	457 mm
	750-1000	20 "	Diam.- Open	$250	508 mm

1532

Type	Wattage	Size	Shielding	Cost	Size
High Bay Mercury	400	24 "	Diam.- Open	$340	610 mm
	1000	24 "	Diam.- Open	$450	610 mm
Commercial Incandescent					
Recessed Fixed Spot	150	10 "	Diam.- Open	$110	254 mm
Adj. Spot	150	10 "	Diam.- Open	$130	254 mm
Punch Spot	150	8 "	Diam.- Open	$130	203 mm
Recessed Square	100	8 " x 8 "	Diam.- Opal	$160	203 mm
Alum. Wall Bracket	75	5 "	Diam.- Glass	$200	127 mm
Pendant Metal Shade	100	11 "	Diam.- Glass	$100	279 mm
Pendant Sphere	40	16 "	Diam.- Glass	$150	406 mm
Drum	100	18 "	Diam.- Opal	$100	457 mm
For Stems add					
6" Drop	-----	-----	-----	$10	150 mm drop
18"(450 mm) Drop	-----	-----	-----	$20	450 mm drop
30"(750 mm) Drop	-----	-----	-----	$30	750 mm drop
For plaster rings					
for spots add	-----	-----	-----	$14	-----
Track Lighting					
1 Circuit		8 '		$110	2.44 M
3 Circuit		12 '		$140	3.66 M
150 W Spot	-----	-----		$70	-----

16720 ALARM AND DETECTION EQUIPMENT

The largest single threat to property and life is fire. While the primary concern is always potential loss of life, damage to or loss of buildings, equipment, materials, files, and records due to fire can put a company out of business. In the case of a museum or other places where historical data is stored, part of history could be lost forever. Consequently, life safety and fire codes are being enacted across the country for all types of buildings, from institutional to office buildings, from hotels to single family residences.

The threat of property damage and loss of life due to fire can be substantially reduced through the proper use of smoke, fire, and gas detectors. These detectors are either wired into the building's electrical system, battery operated, or both. The detector that is wired to the electrical system should be designed to receive backup power from a battery source in the event of a power failure.

Battery operated detectors normally have a small light that flashes when the battery is operative. Some battery systems have a flashing light, short beeps, or both, periodically indicating the battery is not functioning properly. Battery powered systems do not connect to the main electrical system, which makes this device ideal for use in existing buildings (primarily residences) where exposed wiring is objectionable and concealed wiring means costly "cut and patch" work. The one drawback to the battery operated detector is that while it will warn building occupants, it cannot be connected to an exterior alarm system.

Fire detectors and smoke detectors are different. The former can be adjusted to alarm at a particular temperature and the latter is set to alarm at a certain percentage of smoke in the air surrounding the detector. Because most fire fatalities are due to smoke inhalation, in most cases the smoke detector gives the best protection and the quickest alarm.

Placement of detectors will vary depending on the type of building and the interior partitioning. In institutional and commercial buildings, a detector should be located in every room and storage closet, one for every 900 sq.ft. (83.61 sq.m) in large open areas, and at approximately 60 lin.ft. (18.29 m) on center in corridors. In residential homes a detector should be placed near the kitchen, in hallways leading to bedrooms, and at the top of all stairways at each floor level.

The cost of fire and or smoke detectors varies widely and depends on the amount of sensitivity and sophistication of the alarm system. Residential detectors, either battery or hardwired type, range in price from $20.00 to $110.00 each. Detectors for institutional and commercial buildings will range from $130.00 to $260.00 each.

In addition to the cost of the detectors for non-residential use, there are other associated items of equipment that can be added to the alarm system, such as alarm station pull boxes, test panels, and control console panels, costing several thousand dollars each. Hardwired systems, both commercial and residential, can be provided with an automatic dialers. When

1534

the alarm is activated, the system will automatically notify the proper emergency agencies. The automatic dialer installation costs range from $150.00 to $1,000.00. The wide range of costs is due to the many expenses associated with an automatic dialer. Some emergency units prefer an alarm system wired into a private central station to weed out false alarms. The central control station costs are private and are normally billed directly to the user on an annual basis after installation is completed.

16850 ELECTRICAL HEATING

Heating electrically includes a broad range of methods—radiant units, convector units, baseboard units, electric furnaces with forced air, and heat pumps with forced air. Some of these methods are often included as simple supplemental heaters; others are full plants and involve quite sophisticated design in order to function most efficiently.

Wall heaters are the simplest, most commonly encountered means of heating electrically. Both natural and fan driven units are available and they are usually recessed and have built-in thermostats. A 120 watt ceiling fan (less lighting) unit will cost around $350.00; a 1500 watt ceiling radiant unit around $400.00; a 1000 watt ceiling radiant unit, $450.00. A 1200 watt fan forced unit will run $250.00 with manual control, $225.00 with thermostat.

Baseboard heaters are available in standard and high wattage models in surface mounted and semi-recessed designs. The wiring serving these units may be part of the common wiring system or fed from separate branch control centers.

In some areas electricity for heating is separately metered and must be on a separate system. For preliminary figures, baseboard units will run about $150.00 per lin .ft ($492.12 per lin.m.) installed, not including separate wiring system. Add $50.00 for each thermostat. Baseboards, if used as the sole heating source, are often supplemented by fan units at entrances.

Electric furnaces are similar to conventional warm air furnaces but substitute electric resistance-heating units for flame. They are very compact, require no flue or vent, and can be placed anywhere so that zone heating is simplified. Also, air cooling systems can be added to provide a year-round system. A variation on this are through-the-wall units including both heating and cooling capacities. These are especially suitable for motels, dormitories, and other similar buildings where individual control is important. Electric furnaces will cost about $700.00 for a 35 MBH/hr unit up to $1,500.00 for a 140 MBH model. Assuming wiring is already in the furnace room, two workers can install the unit in half a day.

Heat pumps are more efficient than resistance type heaters, because they are designed to use existing outside conditions to supplement inside heating and cooling. In the winter, the heat pump extracts heat from outside air, ground, or water and conducts it to the inside. In summer, it removes heat from base to the outside. Heat pumps are generally placed outside to eliminate noise. Units are usually designed to handle air cooling loads. In areas of extreme cold where heating capacity of the pump is inadequate, it can be supplemented by electric resistance heaters. The unit is connected to a duct system within the house.

In general, the motor energy used in a residence will vary from home to home, building to building. Some homes require more energy than others, depending on the size of the dwelling, gas or electric heating, automated thermostats, etc.

As indicated below, the energy used by motors is an average. If one could reduce the energy costs one should concentrate on the larger energy consuming items.

The U.S. Green Building Council (USGBC) is the nation's foremost coalition of leaders from across the building industry working to promote buildings that are environmentally responsible, profitable and healthy places to both work and live. USGBC members developed the Leadership in Energy and Environmental Design (LEED) Green Building Rating System to establish a national benchmark for sustainable building performance and a tool for market information. Both nationally and locally, environmentally friendly technologies in building and home construction have been used for a long time on some key resources like energy, water, materials and land. For building owners, such investments can pay for themselves in a short period of time (payback). With electric use being one of the major sources of energy savings, a home or building should be an area for potential savings.

MENSURATION

The information in this chapter will enable contractors and estimators to estimate quantities and costs accurately and efficiently. A number of tables use decimals in the quantities, because by using decimals, it is possible to state the fractional parts of inches, feet, and yards in a smaller space than when regular fractions are used.

For the estimator who is not thoroughly familiar with the decimal system, complete explanations are given covering the use of all classes of decimal fractions so that they can be used rapidly and accurately.

Estimating is nearly all "figures" of one kind or another, so it is essential that the estimator possess a fair working knowledge of arithmetic for estimates to be accurate and dependable.

Most of the estimator's computations involve measurements of surface and volume and are stated in lineal feet (meters), square feet (square meters), square yards, squares, containing 100 sq. ft. (10 sq.m), cubic feet (cubic meters), and cubic yards. These quantities are often further reduced to thousands of brick, board feet of lumber, etc.

The following abbreviations are used throughout the book to make all tables brief and concise and to insert them in the smallest possible space:

		Abbreviation		Symbols
Inches	=	in.	=	"
Lineal feet	=	lin. ft. or l.f.	=	'
Feet-inches	=	ft. in.	=	2'-5"
Square feet	=	sq. ft. or s.f.		
Square yards	=	sq. yds. or s.y.		
Squares	=	sqs.		
Cubic feet	=	cu. ft. or c.f.		
Cubic yards	=	cu. yds. or c.y.		
Board feet	=	b.f.		

		(metric)		
millimeters	=	mm		
meters	=	m		
meters & millimeters	=	m, mm		
square meter	=	sq.m.	=	m^2
cubic meter	=	cu.m.	=	m^3
Cubic yards	=	cu. yds. or c.y.		

Linear Measures

12 inches	= 12 in.	= 12"	= 1 foot	= 1 ft.	= 1'-0"
3 feet	= 3 ft.	= 3'-0"	= 1 yard	= 1 yd.	
16-1/2 feet	= 16-1/2 ft.	= 16'-6"	= 1 rod	= 1 rd.	
40 rods	= 40 rds.	= 1 furlong	= 1 fur.		
8 furlongs	= 8 fur.	= 1 mile	= 1 mi.		
5,280 feet	= 1,760 yds.	= 1 mile	= 1 mi.		

Square Measure or Measure of Surfaces

144 square inches (sq. in.)	=	1 square foot (sq. ft.)
9 square feet (sq. ft.)	=	1 square yard (sq. yd.)
100 square feet (sq. ft.)	=	1 square (sq.)*
30-1/4 square yards (sq. yds.)	=	1 square rod (sq. rd.)
160 square rods	=	1 Acre = 1 A.
43,560 sq. ft.	=	4,840 sq. yds. = 1 Acre = 1 A.
640 acres	=	640 A. = 1 square mile (sq. mi.)

Architects' and builders' measure.

Water Quantities

1 cu. ft.	=	7.4805 gal. = 62.4 lbs.
1 gal.	=	8.34 lbs.
1 acre-foot	=	325,850 gal.
1 gal.	=	3.7854 liters
1 cu. meter	=	264.17 gal.
1 cu. ft. salt water	=	64.4 lbs.

Power Units

1 hp	=	550 ft-lbs. per sec.
1 hp	=	0.746 kw

Cubic Measure or Cubical Contents

1,728 cubic inches (cu. in.)	=	1 cubic foot (cu. ft., c.f.)
27 cu. ft.	=	1 cubic yard (cu. yd., c.y.)
128 cu. ft.	=	1 cord (cd.)
24-3/4 cu. ft.	=	1 perch (P.)*

A perch of stone is nominally 16-1/2' long, 1' high, and 1-1/2' thick, and contains 24-3/4 c.f. However, in some states, especially west of the Mississippi, rubble work is figured by the perch containing 16-1/2 c.f. Before submitting prices on masonry by the perch, find out the common practice in your locality.

MENSURATION
Computing Areas and Volumes

To Compute the Area of a Square, Rectangle, or Parallelogram. Multiply the length by the breadth or height. Example: Obtain the area of a wall 22'-0" (6.6 m) long and 9'-0" (2.7 m) high. 22 x 9 = 198 s.f. (6.6 x 2.7 = 17.82 sq.m).

To Compute the Area of a Triangle. Multiply the base by 1/2 the altitude or perpendicular height. Example: Find the area of the end gable of a house 24'-0" (7.2 m) wide and 12'-0" (3.6 m) high from the base to high point of roof. 24 x 6 (half the height) = 144 s.f. (7.2 x 1.8 = 12.96 sq.m).

To Compute the Circumference of a Circle. Multiply the diameter by 3.1416. The diameter multiplied by 3-1/7 is close enough for all practical purposes. Example: Find the circumference or distance around a circle, the diameter of which is 12'-0". 12 x 3.1416 = 37.6992 ft. the distance around the circle or 12 x 3-1/7 = 37.714 ft. (3.6 x 3.1416 = 11.30 m).

To Compute the Area of a Circle. Multiply the square of the diameter by 0.7854 or multiply the square of the radius by 3.1416. Example: Find the area of a round concrete column 24" or 2'-0" in diameter. The square of the diameter is 2 x 2 = 4. 4 x 0.7854 = 3.1416 s.f. the area of the circle (0.6 x 0.6 = 0.36 x 0.7854 = 0.28 sq.m).

The radius is 1/2 the diameter. If the diameter is 2'-0" (0.6 m), then the radius would be 1'-0" (0.3 m). To obtain the square of the radius, 1 x 1 = 1 (0.3 x 0.3 = 0.09). Multiply the square of the radius, 1 x 3.1416 = 3.1416 (0.09 x 3.1416 = 0.28 sq.m), the area of the circle.

To Compute the Area of a Flare. An inside flare is really one-quarter of a circle. Multiply the square of the radius by 0.785. An outside flair is the turnout of driveways, roads, etc. Multiply the square of the radius by 0.215.

Example: A residential driveway is 10' (3m) wide and 25' (7.6m) long. The driveway flares out toward the street on each side at a radius of 5' (1.5m). The surface area of the driveway will then be 10' x 25' + 2 x $5'^2$ x 0.215 = 250 sq.ft. + 10.75 sq.ft. = 260.75 sq.ft. (24.2 sq.m.)

To Compute the Cubical Contents of a Circular Column. Multiply the area of the circle by the height. Example: Find the cubical contents of a round concrete column 2'-0" (0.6 m) in diameter and 14'-0" (4.2 m) long.

From the previous example, the area of a circle 2'-0" (0.6 m) in diameter is 3.1416 s.f. (0.28 sq.m); 3.1416 x 14'-0" (the height) = 43.9824 c.f. (3.1416 x 4.2 = 13.19 cu.m) or for all practical purposes 44 c.f. of concrete in each column.

To Compute the Cubical Contents of any Solid. Multiply the length by the breadth or height by the thickness. Computations of this kind are used extensively in estimating all classes of building work, such as excavating, concrete foundations, reinforced concrete, brick masonry, cut stone, granite, etc.

Example: Find the cubical contents of a wall 42'-0" (12.6 m) long, 5'-6" (1.65 m) high, and 1'-4" (0.39 m) thick, 42'-0" x 5'-6" x 1'-4" = 308 c.f.

(12.6 x 1.65 x 0.39 = 8.1 cu.m) To reduce cu. ft. to cu. yds., divide 308 by 27. The result is 11-11/27 or 11-1/2 or 11.41 cu. yds.

To Compute the Number of Swollen Cubic Yards (Cubic Meters) from an Excavation. Multiply the cubical contents of the excavation, or bank cubic yards (bank cubic meters), by the appropriate swell factor for the type of soil excavated.

Example: A 10' x 15' (3m x 4.6m) rectangular pit that is 8' (2.4m) deep with vertical sides and a soil swell factor of 15% is to be excavated. The swollen cubic yards will be 10' x 15' x 8' ÷ 27 x 1.15 = 51.1 cu.yds. (39 cu.m.)

To Compute the Number of Swollen Cubic Yards (Cubic Meters) for Backfilling an Excavation. Multiply the cubical contents of the excavation, or bank cubic yards (bank cubic meters), by the appropriate swell factor for the soil and then divide by 1 minus the difference of the compaction percentage and the original density percentage of the soil being backfilled.

Example: A 12' x 12' x 10' deep (3.6m x 3.6m x 3m) excavation that has vertical sides is to be backfilled to a compaction of 98% engineered percentage (compaction percentage) with soil that has a 35% swell and 90% original density. The swollen cubic yards is then 12 x 12 x 10 ÷ 27 x 1.35 ÷ (1 – (0.98-0.9)) = 78.3 cu.yds. (59.8 cu.m.)

To Compute the Slope or Grade. Divide the rise by the run. For grade, it may be necessary to convert to percentage by multiplying by 100.

Example: A parking lot slopes away from the building. If the parking lot is 50' long (15.2m) and drops 8" (203mm), the slope or grade is 0.67' ÷ 50' = 0.0134 slope or 1.34% grade.

The following charts give illustrations and formulas for all the different geometric shapes of area and volume that the estimator is likely to encounter.

AREA (A : Area)		
NAME FORMULA		**SHAPE**
Inside Flair $A = R^2 \times 0.785$		
Outside Flair $A = R^2 \times 0.215$		

AREA (A : Area)

NAME FORMULA	SHAPE
Parallelogram $A = B \times h$	
Trapezoid $A = \dfrac{B + C}{2} \times h$	
Triangle $A = \dfrac{B \times h}{2}$	
Trapezium (Divide into 2 triangles) A = Sum of the 2 triangles (See above)	
Regular Polygon $A = \dfrac{\text{Sum of sides (s)}}{2}$ x Inside Radius (R)	
Circle $A : \begin{cases} (1) \; \pi R^2 \\ (2) \; .7854 \times D^2 \\ (3) \; .0796 \times C^2 \end{cases}$	

AREA (A : Area)

NAME FORMULA	SHAPE
Sector A : (1) $\dfrac{a^{\circ}}{360^{\circ}} \times \pi R^2$ (2) Length of area $\times \dfrac{R}{2}$	
Segment A : Area of sector minus triangle (See above)	
Ellipse $A = M \times m \times .7854$	
Parabola $A = B \times \dfrac{2\ h}{3}$	

1542

VOLUME (V : volume)

NAME FORMULA	SHAPE
Cube $V = a^3$ (in cubic units)	
Rectangular Solids $V = L \times W \times h$	
Prisms $V(1) = \dfrac{B \times A}{2} \times h$ $V(2) = \dfrac{s \times R}{2} \times 6 \times h$ $V = \text{Area of end} \times h$	
Cylinder $V = \pi R^2 \times h$	

VOLUME (V : volume)

NAME FORMULA	SHAPE
Cone $$V = \frac{\pi R^2 \times h}{3}$$	
Pyramide $$V(1) = L \times W \times \frac{h}{3}$$ $$V(2) = \frac{B \times A}{2} \times \frac{h}{3}$$ $$V = \text{Area of Base} \times \frac{h}{3}$$	
Sphere $$V = \frac{1}{6} \pi D^3$$	
Circular Ring (Torus) $$V = 2\pi^2 \times R r^2$$ $$V = \text{Area of section} \times 2\pi R$$	

1544

MENSURATION

How to Use the Decimal System in Estimating

A decimal is a fraction whose denominator is not written and is some power of 10. It is often called a decimal fraction but more often simply a decimal. Example: We know 50 cents is 50/100 or 1/2 of a dollar. Writing the same thing in decimals would be $0.50.

Numerator   50
Denominator   100

When written as 0.50 or 0.5, the denominator of the fraction is not written, but it is understood to be 10 from the fact that 5 occupies the first place to the right of the decimal point. Therefore we can have the following:

0.5 means 5/10, for the 5 extends to the 10th place;
0.25 means 25/100, for the 25 extends to the 100th place;
0.125 means 125/1000, for the 125 extends to the 1000th place.

A whole number and a decimal together, form a *mixed decimal*. Example: 2.25 is the same as 2-25/100 or 2-1/4.

The period written at the left of tenth place is called the decimal point. Example: 0.5 = 5/10; 0.25 = 25/100 = 1/4; 0.375 = 375/1000 = 3/8, etc.

It is not necessary to write a zero at the left of the decimal point in the above examples, because 0.5 means the same as .5. The zero is often written to call attention more quickly to the decimal point.

In the construction business, decimals are used chiefly to denote feet and inches, hours and minutes, the fractional working units of the various kinds of materials, and in money, which is dollars and cents or fractional parts of 100.

Table of Feet and Inches Reduced to Decimals

The following table illustrates how feet and inches may be expressed in four different ways, all meaning the same thing.

1 inch	=	1"	=	1/12th foot	=	0.083
1-1/2 inches	=	1-1/2"	=	1/8th foot	=	0.125
2 inches	=	2"	=	1/6th foot	=	0.1667
2-1/2 inches	=	2-1/2"	=	5/24ths ft	=	0.2087
3 inches	=	3"	=	1/4th foot	=	0.25
3-1/2 inches	=	3-1/2"	=	7/24ths ft	=	0.2917
4 inches	=	4"	=	1/3rd foot	=	0.333
4-1/2 inches	=	4-1/2"	=	3/8ths ft	=	0.375
5 inches	=	5"	=	5/12ths foot	=	0.417
5-1/2 inches	=	5-1/2"	=	11/24ths ft	=	0.458
6 inches	=	6"	=	1/2 foot	=	0.5
6-1/2 inches	=	6-1/2"	=	13/24ths ft	=	0.5417

7 inches	=	7"	=	7/12ths foot	=	0.583
7-1/2 inches	=	7-1/2"	=	5/8ths ft	=	0.625
8 inches	=	8"	=	2/3rds foot	=	0.667
8-1/2 inches	=	8-1/2"	=	17/24ths ft	=	0.708
9 inches	=	9"	=	3/4ths foot	=	0.75
9-1/2 inches	=	9-1/2"	=	19/24ths ft	=	0.792
10 inches	=	10"	=	5/6ths foot	=	0.833
10-1/2 inches	=	10-1/2"	=	7/8ths ft	=	0.875
11 inches	=	11"	=	11/12ths foot	=	0.917
11-1/2 inches	=	11-1/2"	=	23/24ths	=	0.958
12 inches	=	12"	=	1 foot	=	1.0

Example: Write 5 feet, 7-1/2 inches, in decimals. It would be written 5.625, which is equivalent to 5-5/8 feet.

Table of Common Fractions Stated in Decimals

The following table gives the decimal equivalents of common fractions frequently used in estimating:

1/16	=	0.0625		9/16	=	0.5625
1/8	=	0.125		5/8	=	0.625
3/16	=	0.1875		11/16	=	0.6875
1/4	=	0.25		3/4	=	0.75
5/16	=	0.3125		15/16	=	0.9375
3/8	=	0.375		13/16	=	0.8125
7/16	=	0.4375		7/8	=	0.875
1/2	=	0.5		8/8	=	1.0

Annexing zeros to a number does not change its value; for example, 0.5 is the same as 0.500.

MENSURATION

Table of Hours and Minutes Reduced to Decimals

The following table illustrates how minutes may be reduced to fractional parts of hours and to decimal parts of hours.

Number of Minutes		Fractional Part of an Hour		Decimal Part of an Hour	Number of Minutes		Fractional Part of an Hour		Decimal Part of an Hour
1	=	1/60th	=	0.0167	31	=	31/60ths	=	0.5167
2	=	1/30th	=	0.0333	32	=	8/15ths	=	0.5333
3	=	1/20th	=	0.0500	33	=	11/20ths	=	0.5500
4	=	1/15th	=	0.0667	34	=	17/30ths	=	0.5667
5	=	1/12th	=	0.0833	35	=	7/12ths	=	0.5833
6	=	1/10th	=	0.1000	36	=	3/5ths	=	0.6000
7	=	7/60ths	=	0.1167	37	=	37/60ths	=	0.6167
8	=	2/15ths	=	0.1333	38	=	19/30ths	=	0.6333
9	=	3/20ths	=	0.1500	39	=	13/20ths	=	0.6500
10	=	1/6th	=	0.1667	40	=	2/3rds	=	0.6667
11	=	11/60ths	=	0.1833	41	=	41/60ths	=	0.6833
12	=	1/5th	=	0.2000	42	=	7/10ths	=	0.7000
13	=	13/60ths	=	0.2167	43	=	43/60ths	=	0.7167
14	=	7/30ths	=	0.2333	44	=	11/15ths	=	0.7333
15	=	1/4th	=	0.2500	45	=	3/4ths	=	0.7500
16	=	4/15ths	=	0.2667	46	=	23/30ths	=	0.7667
17	=	17/60ths	=	0.2833	47	=	47/60ths	=	0.7833
18	=	3/10ths	=	0.3000	48	=	4/5ths	=	0.8000
19	=	19/60ths	=	0.3167	49	=	49/60ths	=	0.8167
20	=	1/3rd	=	0.3333	50	=	5/6ths	=	0.8333
21	=	7/20ths	=	0.3500	51	=	51/60ths	=	0.8500
22	=	11/30ths	=	0.3667	52	=	13/15ths	=	0.8667
23	=	23/60ths	=	0.3833	53	=	53/60ths	=	0.8833
24	=	2/5ths	=	0.4000	54	=	9/10ths	=	0.9000
25	=	5/12ths	=	0.4167	55	=	11/12ths	=	0.9167
26	=	13/30ths	=	0.4333	56	=	14/15ths	=	0.9333
27	=	9/20ths	=	0.4500	57	=	19/20ths	=	0.9500
28	=	7/15ths	=	0.4667	58	=	29/30ths	=	0.9667
29	=	29/60ths	=	0.4833	59	=	59/60ths	=	0.9833
30	=	1/2	=	0.5000	60	=	1	=	1.0000

Example: Write 4 hours and 37 minutes in decimals. It would be written 4.6167. Now find the labor cost for 4.6167 hrs. at $7.50 an hr.:

Hours	4.6167
X Hourly Rate	$7.50
	2308350
Add	323169
	$34.625250
	or $34.63

Similar Decimals. Decimals that have the same number of decimal places are called similar decimals. Thus, 0.75 and 0.25 are similar decimals and so are 0.150 and 0.275; but 0.15 and 0.275 are dissimilar decimals.

To reduce dissimilar decimals to similar decimals, give them the same number of decimal places by annexing or cutting off zeros. Example, 0.125, 0.25, 0.375, and 0.5 may all be reduced to thousandths as follows: 0.125, 0.250, 0.375, and 0.500.

To Reduce a Decimal to a Common Fraction. Omit the decimal point, write the denominator of the decimal, and then reduce the common fraction to its lowest terms. Example: 0.375 equals 375/1000, reduced to its lowest terms, equals 3/8:

$$375/1000 = 15/40 = 3/8$$

How to Add Decimals. To add numbers containing decimals, write like orders under one another, and then add as with whole numbers. Example, add 0.125, 0.25, 0.375, and 1.0. The sum is 1.750, which equals 1-750/1000, and which may be further reduced to 1-3/4:

```
 0.125
 0.25
 0.375
+1.0
 1.750
```

How to Subtract Decimals. To subtract one number from another, when either or both contain decimals, reduce to similar decimals, write like orders under one another, and subtract as with whole numbers. Example: Subtract 20 hours and 37 minutes from 27 hours and 13 minutes, both being written in decimals. The difference is 6.6000 or 6.6 hours, which reduced to a common fraction is 6-3/5 hours or 6 hours and 36 minutes.

```
 27.2167
-20.6167
  6.6000
```

How to Multiply Decimals. The same method is used as in multiplying other numbers, and the result should contain as many decimal points as there are in both of the numbers multiplied. Example: Multiply 3 feet 6 inches by 4 feet 9 inches.

Multiplicand	4 ft. 9 in.	=	4'-9"	=	4-3/4 ft.	= 4.75	4.75
Multiplier	3 ft. 6 in.	=	3'-6"	=	3-1/2 ft.	= 3.5	3.5
To multiply, proceed as illustrated							2375
							1425
Product							16,625
							or 16.625

In the above example, there are 2 decimals in the multiplicand and 1 decimal in the multiplier. The result or product should contain the same

1548

number of decimals as the multiplicand and multiplier combined, which is three. Starting at the right and counting to the left three places, place the decimal point between the two sixes. The result would be 16.625, which equals 16-625/1000 = 16-5/8 sq. ft.

Practical Examples Using Decimals

The quantities of cement, sand and gravel required for one cu. yd. of concrete are ordinarily stated in decimals. For instance, 1 cu. yd. of concrete mixed in the proportions of 1 part cement:2 parts sand:4 parts gravel is usually stated 1:2:4 and requires the following materials:

1.50 bbls. cement = 1-50/100 bbls. = 1-1/2 bbls. = 6 sacks
0.42 cu. yds. sand = 42/100 cu. yds. = 21/50 cu. yds. = 11-1/3 cu. ft.
0.84 cu. yds. gravel = 84/100 cu. yds. = 22-2/3 cu. ft.

There are 4 sacks of cement to the bbl., and each sack weighs 94 lbs. and contains approximately 1 cu. ft. of cement.

There are 27 cu. ft. in a cu. yd.; to obtain the number of cu. ft. of sand required for a yard of concrete, 0.42 cu. yds = 42/100 cu. yds. = 21/50 cu. yds. and 21/50 of 27 cu. ft. equals:

(21x27) ÷ 50 = 567 ÷ 50 = 11.34 or 11-1/3 cu. ft. sand

To find the cost of a cu. yd. of concrete based on the above quantities, and assuming it requires 2-1/4 hrs. labor time to mix and place one cu. yd. of concrete, proceed as follows:

1-1/2 bbls. cement = 1.50 bbls. @ $9.00 per bbl. $13.5000
11-1/3 cu. ft. sand = 0.42 cu. yd. @ 5.75 per cu. yd.   2.4150
22-2/3 cu. ft. gravel = 0.84 cu. yd. @ 5.75 per cu. yd.   4.8300
2-1/4 hrs. labor = 2.25 hrs. @ 9.80 per hr.   22.0500
Cost per cu. yd. $42.7950

You will note the total is carried in four decimal places. This is to illustrate the actual figures obtained by multiplying. The total would be $42.80 per cu. yd. of concrete.

In making the above multiplications you will note there are 2 decimals in both the multiplicand and the multiplier of all the amounts multiplied, so the result should contain as many decimals as the sum of the multiplicand and multiplier, which is 4. You will note all the totals contain 4 decimal places. Always bear this in mind when making your multiplications, because if the decimal place is wrong it makes a difference of 90 percent in your total. You know the correct result is $42.7950, but suppose by mistake you counted off 5 decimal places. The result would be $4.27950 or $4.28 per cu. yd. or just about 1/10 enough; or if you made a mistake the other way and counted off just 3 decimal places, the result would be $427.950 or $427.95 per cu. yd., or just about 10 times too much. Watch your decimal places.

The same method is used in estimating lumber. All kinds of framing lumber are ordinarily sold by the 1,000 b.f., so 1,000 is the unit or decimal used when estimating lumber.

Suppose you buy 150 pcs. of 2"x8"-16'-0", which contains 3,200 b.f. This is equivalent to 3-200/1000 = 3-2/10 = 3-1/15 thousandths or stated in decimals it may be either 3.2 or 3.200. The cost of this lumber at $230.00 per 1,000 b.f., would be obtained by multiplying 3.2 ft. at $230.00 per 1,000 b.f. as follows: 3.2 mbf of lumber @ $230.00 = $736.00.

Things to Remember When Using Decimals

0.6 indicates tenths, thus 0.6 = 6/10 = 0.60 = 60/100 = 0.600 = 600/1000
0.06 indicates hundredths, thus 0.06 = 6/100 = 0.060 = 60/1000
0.006 indicates thousandths, thus 0.006 = 6/1000
0.0006 indicates ten thousandths, thus 0.0006 = 6/10000

When multiplying decimals, always remember that the result or product must contain as many decimal places as the sum of the decimal places in both the multiplicand and the multiplier. Find the cost of 3 hrs. 20 min. laborer's time at $9.80 an hr.

3 hrs. 20 min. equal 3-1/3 hrs.	3.3333
9 dollars 80 cents an hr.	9.80
	2666640
	299997
	32666340
	32.666340
	or $32.67

There are 3 decimals in the multiplicand and 2 decimals in the multiplier, so the result or total should contain as many decimals as the sum of the multiplicand and multiplier, which is 5. Be sure your decimal point is in the right place and the rest is easy.

Conversion of Feet, Inches to Decimal and Metric Equivalents

Inches	Inches in Decimals	Feet in Decimals	Milli-meters	Centi-meters	Meters
1/16	0.0625	0.005	1.5875	0.15875	0.001588
1/8	0.125	0.010	3.1750	0.31750	0.003175
3/16	0.1875	0.016	4.7625	0.47625	0.004763
1/4	0.25	0.021	6.3500	0.63500	0.006350
5/16	0.3125	0.026	7.9375	0.79375	0.007938
3/8	0.375	0.031	9.5250	0.95250	0.009525
7/16	0.4375	0.036	11.1125	1.11125	0.011113
1/2	0.5	0.042	12.7000	1.27000	0.012700
9/16	0.5625	0.047	14.2875	1.42875	0.014288
5/8	0.625	0.052	15.8750	1.58750	0.015875
11/16	0.6875	0.057	17.4625	1.74625	0.017463
3/4	0.75	0.063	19.0500	1.90500	0.019050

MENSURATION

Inches	Inches in Decimals	Feet in Decimals	Milli-meters	Centi-meters	Meters
13/16	0.8125	0.068	20.6375	2.06375	0.020638
7/8	0.875	0.073	22.2250	2.22250	0.022225
15/16	0.9375	0.078	23.8125	2.38125	0.023813
1	1	0.083	25.400	2.5400	0.025400
1 1/8	1.125	0.094	28.575	2.8575	0.028575
1 1/4	1.25	0.104	31.750	3.1750	0.031750
1 3/8	1.375	0.115	34.925	3.4925	0.034925
1 1/2	1.5	0.125	38.100	3.8100	0.038100
1 5/8	1.625	0.135	41.275	4.1275	0.041275
1 3/4	1.75	0.146	44.450	4.4450	0.044450
1 7/8	1.875	0.156	47.625	4.7625	0.047625
2	2	0.167	50.800	5.0800	0.050800
2 1/8	2.125	0.177	53.975	5.3975	0.053975
2 1/4	2.25	0.188	57.150	5.7150	0.057150
2 3/8	2.375	0.198	60.325	6.0325	0.060325
2 1/2	2.5	0.208	63.500	6.3500	0.063500
2 5/8	2.625	0.219	66.675	6.6675	0.066675
2 3/4	2.75	0.229	69.850	6.9850	0.069850
2 7/8	2.875	0.240	73.025	7.3025	0.073025
3	3	0.250	76.200	7.6200	0.076200
3 1/8	3.125	0.260	79.375	7.9375	0.079375
3 1/4	3.25	0.271	82.550	8.2550	0.082550
3 3/8	3.375	0.281	85.725	8.5725	0.085725
3 1/2	3.5	0.292	88.900	8.8900	0.088900
3 5/8	3.625	0.302	92.075	9.2075	0.092075
3 3/4	3.75	0.313	95.250	9.5250	0.095250
3 7/8	3.875	0.323	98.425	9.8425	0.098425
4	4	0.333	101.60	10.160	0.10160
4 1/4	4.25	0.354	107.95	10.795	0.10795
4 1/2	4.5	0.375	114.30	11.430	0.11430
4 3/4	4.75	0.396	120.65	12.065	0.12065
5	5	0.417	127.00	12.700	0.12700
4 1/4	4.25	0.354	107.95	10.795	0.10795
5 1/4	5.25	0.438	133.35	13.335	0.13335
5 3/4	5.75	0.479	146.05	14.605	0.14605
6	6	0.500	152.40	15.240	0.15240
6 1/4	6.25	0.521	158.75	15.875	0.15875
6 1/2	6.5	0.542	165.10	16.510	0.16510
6 3/4	6.75	0.563	171.45	17.145	0.17145
7	7	0.583	177.80	17.780	0.17780
7 1/4	7.25	0.604	184.15	18.415	0.18415
7 1/2	7.5	0.625	190.50	19.050	0.19050
7 3/4	7.75	0.646	196.85	19.685	0.19685
8	8	0.667	203.20	20.320	0.20320

Inches	Inches in Decimals	Feet in Decimals	Milli-meters	Centi-meters	Meters
8 1/4	8.25	0.688	209.55	20.955	0.20955
8 1/2	8.5	0.708	215.90	21.590	0.21590
8 3/4	8.75	0.729	222.25	22.225	0.22225
9	9	0.750	228.6	22.86	0.2286
9 1/2	9.5	0.792	241.3	24.13	0.2413
10	10	0.833	254.0	25.40	0.2540
10 1/2	10.5	0.875	266.7	26.67	0.2667
11	11	0.917	279.4	27.94	0.2794
11 1/2	11.5	0.958	292.1	29.21	0.2921
12	12	1.000	304.8	30.48	0.3048
18	18	1.500	457.2	45.72	0.4572
24	24	2.000	609.6	60.96	0.6096
30	30	2.500	762.0	76.20	0.7620
36	36	3.000	914.4	91.44	0.9144
42	42	3.500	1066.8	106.68	1.0668
48	48	4.000	1219.2	121.92	1.2192
54	54	4.500	1371.6	137.16	1.3716
60	60	5.000	1524.0	152.40	1.5240
66	66	5.500	1676.4	167.64	1.6764
72	72	6.000	1828.8	182.88	1.8288
78	78	6.500	1981.2	198.12	1.9812
84	84	7.000	2133.6	213.36	2.1336
90	90	7.500	2286.0	228.60	2.2860
96	96	8.000	2438.4	243.84	2.4384
102	102	8.500	2590.8	259.08	2.5908
108	108	9.000	2743.2	274.32	2.7432
114	114	9.500	2895.6	289.56	2.8956
120	120	10.000	3048.0	304.80	3.0480
126	126	10.500	3200.4	320.04	3.2004
132	132	11.000	3352.8	335.28	3.3528
138	138	11.500	3505.2	350.52	3.5052
144	144	12.000	3657.6	365.76	3.6576

MENSURATION

Conversion Factors
S.I. Metric - English Systems

Multiply	by	to obtain
Acres	0.404687	Hectares
	4.04687 x 10^{-3}	Square kilometers
Ares	1076.39	Square feet
Board feet	144 sq.in x 1 in.	Cubic inches
	0.0833	Cubic feet
Bushels	0.3521	Hectoliters
Centimeters	3.28083 x 10^{-2}	Feet
	0.3937	Inches
Cubic centimeters	3.53145 x 10^{-5}	Cubic feet
	6.102 x 10^{-2}	Cubic inches
Cubic feet	2.8317 x 10^4	Cubic milli-meters
	2.8317 x 10^{-2}	Cubic meters
	6.22905	Gallons, imperial
	0.2832	Hectoliters
	28.3170	Liters
	2.38095 x 10^{-2}	Tons, British shipping
	0.025	Tons, U.S. shipping
Cubic inches	16.38716	Cubic centimeters
Cubic meters	35.3145	Cubic feet
	1.30794	Cubic yards
	264.2	Gallons, U.S.
Cubic yards	0.764559	Cubic meters
	7.6336	Hectoliters
Degrees, angular	0.0174533	Radians
Degrees, F (less 32 F)	0.5556	Degrees, C
Degrees, C	1.8	Degrees F (less 32 C)
Foot pounds	0.13826	Kilogram meters
Feet	30.4801	Centimeters
	0.304801	Meters
	304.801	Millimeters
	1.64468 x 10^{-4}	Miles, nautical
Gallons, Canadian	0.004546	Cubic meters
Gallons, Imperial	0.160538	Cubic feet
	1.20091	Gallons, U.S.
	4.54596	Liters
Gallons, U.S.	0.832702	Gallons, Imperial
	0.13368	Cubic feet
	231.0	Cubic inches
	0.003785	Cubic meters
	0.0378	Hectoliters
	3.78543	Liters

Multiply	by	to obtain
Grams, metric	2.20462×10^{-3}	Pounds, avoirdupois
Hectares	2.47104	Acres
	1.076387×10^{5}	Square feet
	3.86101×10^{-3}	Square miles
Hectoliters	3.531	Cubic feet
	2.84	Bushels
	0.131	Cubic yards
	26.42	Gallons
Horsepower, metric	0.98632	Horsepower, U.S.
Horsepower, U.S.	1.01387	Horsepower, metric
Inches	2.54001	Centimeters
	2.54001×10^{-2}	Meters
	25.4001	Millimeters
Kilograms	2.20462	Pounds
	9.84206×10^{-4}	Long tons
	9.807	Newtons
Kilogram meters	7.233	Foot pounds
Kilograms per m	0.671972	Pounds per ft.
Kilograms per sq cm	14.2234	Pounds per sq. in.
Kilograms per sq.m	0.204817	Pounds per sq. ft.
	9.14362×10^{-5}	Long tons per sq. ft.
Kilograms per sq.mm	1422.34	Pounds per sq. in.
	0.634973	Long tons per sq. in.
Kilograms per cu.m	6.24283×10^{-2}	Pounds per cu. ft.
Kilometers	0.62137	Miles, statute
	0.53959	Miles, nautical
	3280.7	Feet
Kips	4448.0	Newtons
	453.8	Kilograms force
Liters	0.219975	Gallons, Imperial
	0.26417	Gallons, U.S.
	3.53145×10^{-2}	Cubic feet
	61.022	Cubic inches
Meters	3.28083	Feet
	39.37	Inches
	1.09361	Yards
Miles, nautical	6080.204	Feet
	1.85325	Kilometers
	1.1516	Miles, statute
Miles, statute	1.60935	Kilometers
	0.8684	Miles, nautical
Millimeters	3.28083×10^{-3}	Feet
	3.937×10^{-2}	Inches

1554

Multiply	by	to obtain
Ounces, U.S. liquid	29.57	Milliliters
Pounds, avoirdupois	453.592	Grams, metric
	0.453592	Kilograms
	4.464×10^{-4}	Tons, long
	4.53592×10^{-4}	Tons, metric
Pounds force	4.448	Newtons
Pounds per foot	1.48816	Kilograms per m
Pounds per sq. ft.	4.88241	Kilograms per sq.m
Pounds per sq. in.	7.031×10^{-2}	Kilograms per sq.cm
	7.031×10^{-4}	Kilograms per sq. mm
Pounds per cu. ft.	16.0184	Kilograms per cu. m
Radians	57.29578	Degrees, angular
Square centimeters	0.1550	Square inches
Square feet	9.29034×10^{-4}	Ares
	9.29034×10^{-6}	Hectares
	0.0929034	Square meters
Square inches	6.45163	Square centimeters
	645.163	Square millimeters
Square kilometers	247.104	Acres
	0.3861	Square miles
Square meters	10.7639	Square feet
	1.19599	Square yards
Square miles	259.0	Hectares
	2.590	Square kilometers
Square millimeters	1.550×10^{-3}	Square inches
Square yards	0.83613	Square meters
Tons, long	1016.05	Kilograms
	2240.0	Pounds
	1.01605	Tons, metric
	1.120	Tons, short
Tons, long, per sq. ft.	1.09366×10^{-4}	Kilograms per sq.m
Tons, long, per sq. in.	1.57494	Kilograms per sq. mm
Tons, metric	2204.62	Pounds
	0.98421	Tons, long
	1.10231	Tons, short
Tons, short	907.185	Kilograms
	0.892857	Tons, long
	0.907185	Tons, metric
Tons, British shipping	42.00	Cubic feet
	0.952381	Tons, U.S. shipping
Tons, U.S. shipping	40.00	Cubic feet
	1.050	Tons, British shipping
Yards	0.914402	Meters

Functions of Numbers, 1 to 49

No.	Square	Cube	Square Root	Cubic Root	Logarithm
1	1	1	1.0000	1.0000	0.00000
2	4	8	1.4142	1.2599	0.30103
3	9	27	1.7321	1.4422	0.47712
4	16	64	2.0000	1.5874	0.60206
5	25	125	2.2361	1.7100	0.69897
6	36	216	2.4495	1.8171	0.77815
7	49	343	2.6458	1.9129	0.84510
8	64	512	2.8284	2.0000	0.90309
9	81	729	3.0000	2.0801	0.95424
10	100	1000	3.1623	2.1544	1.00000
11	121	1331	3.3166	2.2240	1.04139
12	144	1728	3.4641	2.2894	1.07918
13	169	2197	3.6056	2.3513	1.11394
14	196	2744	3.7417	2.4101	1.14613
15	225	3375	3.8730	2.4662	1.17609
16	256	4096	4.0000	2.5198	1.20412
17	289	4913	4.1231	2.5713	1.23045
18	324	5832	4.2426	2.6207	1.25527
19	361	6859	4.3589	2.6684	1.27875
20	400	8000	4.4721	2.7144	1.30103
21	441	9261	4.5826	2.7589	1.32222
22	484	10648	4.6904	2.8020	1.34242
23	529	12167	4.7958	2.8439	1.36173
24	576	13824	4.8990	2.8845	1.38021
25	625	15625	5.0000	2.9240	1.39794
26	676	17576	5.0990	2.9625	1.41497
27	729	19683	5.1962	3.0000	1.43136
28	784	21952	5.2915	3.0366	1.44716
29	841	24389	5.3852	3.0723	1.46240
30	900	27000	5.4772	3.1072	1.47712
31	961	29791	5.5678	3.1414	1.49136
32	1024	32768	5.6569	3.1748	1.50515
33	1089	35937	5.7446	3.2075	1.51851
34	1156	39304	5.8310	3.2396	1.53148
35	1225	42875	5.9161	3.2711	1.54407
36	1296	46656	6.0000	3.3019	1.55630
37	1369	50653	6.0828	3.3322	1.56820
38	1444	54872	6.1644	3.3620	1.57978
39	1521	59319	6.2450	3.3912	1.59106
40	1600	64000	6.3246	3.4200	1.60206
41	1681	68921	6.4031	3.4482	1.61278
42	1764	74088	6.4807	3.4760	1.62325
43	1849	79507	6.5574	3.5034	1.63347
44	1936	85184	6.6332	3.5303	1.64345
45	2025	91125	6.7082	3.5569	1.65321
46	2116	97336	6.7823	3.5830	1.66276
47	2209	103823	6.8557	3.6088	1.67210
48	2304	110592	6.9282	3.6342	1.68124
49	2401	117649	7.0000	3.6593	1.69020

MENSURATION

No.	Square	Cube	Square Root	Cubic Root	Logarithm
50	2500	125000	7.0711	3.6840	1.69897
51	2601	132651	7.1414	3.7084	1.70757
52	2704	140608	7.2111	3.7325	1.71600
53	2809	148877	7.2801	3.7563	1.72428
54	2916	157464	7.3485	3.7798	1.73239
55	3025	166375	7.4162	3.8030	1.74036
56	3136	175616	7.4833	3.8259	1.74819
57	3249	185193	7.5498	3.8485	1.75587
58	3364	195112	7.6158	3.8709	1.76343
59	3481	205379	7.6811	3.8930	1.77085
60	3600	216000	7.7460	3.9149	1.77815
61	3721	226981	7.8102	3.9365	1.78533
62	3844	238328	7.8740	3.9579	1.79239
63	3969	250047	7.9373	3.9791	1.79934
64	4096	262144	8.0000	4.0000	1.80618
65	4225	274625	8.0623	4.0207	1.81291
66	4356	287496	8.1240	4.0412	1.81954
67	4489	300763	8.1854	4.0615	1.82607
68	4624	314432	8.2462	4.0817	1.83251
69	4761	328509	8.3066	4.1016	1.83885
70	4900	343000	8.3666	4.1213	1.84510
71	5041	357911	8.4261	4.1408	1.85126
72	5184	373248	8.4853	4.1602	1.85733
73	5329	389017	8.5440	4.1793	1.86332
74	5476	405224	8.6023	4.1983	1.86923
75	5625	421875	8.6603	4.2172	1.87506
76	5776	438976	8.7178	4.2358	1.88081
77	5929	456533	8.7750	4.2543	1.88649
78	6084	474552	8.8318	4.2727	1.89209
79	6241	493039	8.8882	4.2908	1.89763
80	6400	512000	8.9443	4.3089	1.90309
81	6561	531441	9.0000	4.3267	1.90849
82	6724	551368	9.0554	4.3445	1.91381
83	6889	571787	9.1104	4.3621	1.91908
84	7056	592704	9.1652	4.3795	1.92428
85	7225	614125	9.2195	4.3968	1.92942
86	7396	636056	9.2736	4.4140	1.93450
87	7569	658503	9.3274	4.4310	1.93952
88	7744	681472	9.3808	4.4480	1.94448
89	7921	704969	9.4340	4.4647	1.94939
90	8100	729000	9.4868	4.4814	1.95424
91	8281	753571	9.5394	4.4979	1.95904
92	8464	778688	9.5917	4.5144	1.96379
93	8649	804357	9.6437	4.5307	1.96848
94	8836	830584	9.6954	4.5468	1.97313
95	9025	857375	9.7468	4.5629	1.97772
96	9216	884736	9.7980	4.5789	1.98227
97	9409	912673	9.8489	4.5947	1.98677
98	9604	941192	9.8995	4.6104	1.99123
99	9801	970299	9.9499	4.6261	1.99564

Functions of Numbers, 100 to 149

No.	Square	Cube	Square Root	Cubic Root	Logarithm
100	10000	1000000	10.0000	4.6416	2.00000
101	10201	1030301	10.0499	4.6570	2.00432
102	10404	1061208	10.0995	4.6723	2.00860
103	10609	1092727	10.1489	4.6875	2.01284
104	10816	1124864	10.1980	4.7027	2.01703
105	11025	1157625	10.2470	4.7177	2.02119
106	11236	1191016	10.2956	4.7326	2.02531
107	11449	1225043	10.3441	4.7475	2.02938
108	11664	1259712	10.3923	4.7622	2.03342
109	11881	1295029	10.4403	4.7769	2.03743
110	12100	1331000	10.4881	4.7914	2.04139
111	12321	1367631	10.5357	4.8059	2.04532
112	12544	1404928	10.5830	4.8203	2.04922
113	12769	1442897	10.6301	4.8346	2.05308
114	12996	1481544	10.6771	4.8488	2.05690
115	13225	1520875	10.7238	4.8629	2.06070
116	13456	1560896	10.7703	4.8770	2.06446
117	13689	1601613	10.8167	4.8910	2.06819
118	13924	1643032	10.8628	4.9049	2.07188
119	14161	1685159	10.9087	4.9187	2.07555
120	14400	1728000	10.9545	4.9324	2.07918
121	14641	1771561	11.0000	4.9461	2.08279
122	14884	1815848	11.0454	4.9597	2.08636
123	15129	1860867	11.0905	4.9732	2.08991
124	15376	1906624	11.1355	4.9866	2.09342
125	15625	1953125	11.1803	5.0000	2.09691
126	15876	2000376	11.2250	5.0133	2.10037
127	16129	2048383	11.2694	5.0265	2.10380
128	16384	2097152	11.3137	5.0397	2.10721
129	16641	2146689	11.3578	5.0528	2.11059
130	16900	2197000	11.4018	5.0658	2.11394
131	17161	2248091	11.4455	5.0788	2.11727
132	17424	2299968	11.4891	5.0916	2.12057
133	17689	2352637	11.5326	5.1045	2.12385
134	17956	2406104	11.5758	5.1172	2.12710
135	18225	2460375	11.6190	5.1299	2.13033
136	18496	2515456	11.6619	5.1426	2.13354
137	18769	2571353	11.7047	5.1551	2.13672
138	19044	2628072	11.7473	5.1676	2.13988
139	19321	2685619	11.7898	5.1801	2.14301
140	19600	2744000	11.8322	5.1925	2.14613
141	19881	2803221	11.8743	5.2048	2.14922
142	20164	2863288	11.9164	5.2171	2.15229
143	20449	2924207	11.9583	5.2293	2.15534
144	20736	2985984	12.0000	5.2415	2.15836
145	21025	3048625	12.0416	5.2536	2.16137
146	21316	3112136	12.0830	5.2656	2.16435
147	21609	3176523	12.1244	5.2776	2.16732
148	21904	3241792	12.1655	5.2896	2.17026
149	22201	3307949	12.2066	5.3015	2.17319

MENSURATION

Functions of Numbers, 150 to 199

No.	Square	Cube	Square Root	Cubic Root	Logarithm
150	22500	3375000	12.2474	5.3133	2.17609
151	22801	3442951	12.2882	5.3251	2.17898
152	23104	3511808	12.3288	5.3368	2.18184
153	23409	3581577	12.3693	5.3485	2.18469
154	23716	3652264	12.4097	5.3601	2.18752
155	24025	3723875	12.4499	5.3717	2.19033
156	24336	3796416	12.4900	5.3832	2.19312
157	24649	3869893	12.5300	5.3947	2.19590
158	24964	3944312	12.5698	5.4061	2.19866
159	25281	4019679	12.6095	5.4175	2.20140
160	25600	4096000	12.6491	5.4288	2.20412
161	25921	4173281	12.6886	5.4401	2.20683
162	26244	4251528	12.7279	5.4514	2.20952
163	26569	4330747	12.7671	5.4626	2.21219
164	26896	4410944	12.8062	5.4737	2.21484
165	27225	4492125	12.8452	5.4848	2.21748
166	27556	4574296	12.8841	5.4959	2.22011
167	27889	4657463	12.9228	5.5069	2.22272
168	28224	4741632	12.9615	5.5178	2.22531
169	28561	4826809	13.0000	5.5288	2.22789
170	28900	4913000	13.0384	5.5397	2.23045
171	29241	5000211	13.0767	5.5505	2.23300
172	29584	5088448	13.1149	5.5613	2.23553
173	29929	5177717	13.1529	5.5721	2.23805
174	30276	5268024	13.1909	5.5828	2.24055
175	30625	5359375	13.2288	5.5934	2.24304
176	30976	5451776	13.2665	5.6041	2.24551
177	31329	5545233	13.3041	5.6147	2.24797
178	31684	5639752	13.3417	5.6252	2.25042
179	32041	5735339	13.3791	5.6357	2.25285
180	32400	5832000	13.4164	5.6462	2.25527
181	32761	5929741	13.4536	5.6567	2.25768
182	33124	6028568	13.4907	5.6671	2.26007
183	33489	6128487	13.5277	5.6774	2.26245
184	33856	6229504	13.5647	5.6877	2.26482
185	34225	6331625	13.6015	5.6980	2.26717
186	34596	6434856	13.6382	5.7083	2.26951
187	34969	6539203	13.6748	5.7185	2.27184
188	35344	6644672	13.7113	5.7287	2.27416
189	35721	6751269	13.7477	5.7388	2.27646
190	36100	6859000	13.7840	5.7489	2.27875
191	36481	6967871	13.8203	5.7590	2.28103
192	36864	7077888	13.8564	5.7690	2.28330
193	37249	7189057	13.8924	5.7790	2.28556
194	37636	7301384	13.9284	5.7890	2.28780
195	38025	7414875	13.9642	5.7989	2.29003
196	38416	7529536	14.0000	5.8088	2.29226
197	38809	7645373	14.0357	5.8186	2.29447
198	39204	7762392	14.0712	5.8285	2.29667
199	39601	7880599	14.1067	5.8383	2.29885

No.	Square	Cube	Square Root	Cubic Root	Logarithm
200	40000	8000000	14.1421	5.8480	2.30103
201	40401	8120601	14.1774	5.8578	2.30320
202	40804	8242408	14.2127	5.8675	2.30535
203	41209	8365427	14.2478	5.8771	2.30750
204	41616	8489664	14.2829	5.8868	2.30963
205	42025	8615125	14.3178	5.8964	2.31175
206	42436	8741816	14.3527	5.9059	2.31387
207	42849	8869743	14.3875	5.9155	2.31597
208	43264	8998912	14.4222	5.9250	2.31806
209	43681	9129329	14.4568	5.9345	2.32015
210	44100	9261000	14.4914	5.9439	2.32222
211	44521	9393931	14.5258	5.9533	2.32428
212	44944	9528128	14.5602	5.9627	2.32634
213	45369	9663597	14.5945	5.9721	2.32838
214	45796	9800344	14.6287	5.9814	2.33041
215	46225	9938375	14.6629	5.9907	2.33244
216	46656	10077696	14.6969	6.0000	2.33445
217	47089	10218313	14.7309	6.0092	2.33646
218	47524	10360232	14.7648	6.0185	2.33846
219	47961	10503459	14.7986	6.0277	2.34044
220	48400	10648000	14.8324	6.0368	2.34242
221	48841	10793861	14.8661	6.0459	2.34439
222	49284	10941048	14.8997	6.0550	2.34635
223	49729	11089567	14.9332	6.0641	2.34830
224	50176	11239424	14.9666	6.0732	2.35025
225	50625	11390625	15.0000	6.0822	2.35218
226	51076	11543176	15.0333	6.0912	2.35411
227	51529	11697083	15.0665	6.1002	2.35603
228	51984	11852352	15.0997	6.1091	2.35793
229	52441	12008989	15.1327	6.1180	2.35984
230	52900	12167000	15.1658	6.1269	2.36173
231	53361	12326391	15.1987	6.1358	2.36361
232	53824	12487168	15.2315	6.1446	2.36549
233	54289	12649337	15.2643	6.1534	2.36736
234	54756	12812904	15.2971	6.1622	2.36922
235	55225	12977875	15.3297	6.1710	2.37107
236	55696	13144256	15.3623	6.1797	2.37291
237	56169	13312053	15.3948	6.1885	2.37475
238	56644	13481272	15.4272	6.1972	2.37658
239	57121	13651919	15.4596	6.2058	2.37840
240	57600	13824000	15.4919	6.2145	2.38021
241	58081	13997521	15.5242	6.2231	2.38202
242	58564	14172488	15.5563	6.2317	2.38382
243	59049	14348907	15.5885	6.2403	2.38561
244	59536	14526784	15.6205	6.2488	2.38739
245	60025	14706125	15.6525	6.2573	2.38917
246	60516	14886936	15.6844	6.2658	2.39094
247	61009	15069223	15.7162	6.2743	2.39270
248	61504	15252992	15.7480	6.2828	2.39445
249	62001	15438249	15.7797	6.2912	2.39620

Functions of Numbers, 250 to 299

No.	Square	Cube	Square Root	Cubic Root	Logarithm
250	62500	15625000	15.8114	6.2996	2.39794
251	63001	15813251	15.8430	6.3080	2.39967
252	63504	16003008	15.8745	6.3164	2.40140
253	64009	16194277	15.9060	6.3247	2.40312
254	64516	16387064	15.9374	6.3330	2.40483
255	65025	16581375	15.9687	6.3413	2.40654
256	65536	16777216	16.0000	6.3496	2.40824
257	66049	16974593	16.0312	6.3579	2.40993
258	66564	17173512	16.0624	6.3661	2.41162
259	67081	17373979	16.0935	6.3743	2.41330
260	67600	17576000	16.1245	6.3825	2.41497
261	68121	17779581	16.1555	6.3907	2.41664
262	68644	17984728	16.1864	6.3988	2.41830
263	69169	18191447	16.2173	6.4070	2.41996
264	69696	18399744	16.2481	6.4151	2.42160
265	70225	18609625	16.2788	6.4232	2.42325
266	70756	18821096	16.3095	6.4312	2.42488
267	71289	19034163	16.3401	6.4393	2.42651
268	71824	19248832	16.3707	6.4473	2.42813
269	72361	19465109	16.4012	6.4553	2.42975
270	72900	19683000	16.4317	6.4633	2.43136
271	73441	19902511	16.4621	6.4713	2.43297
272	73984	20123648	16.4924	6.4792	2.43457
273	74529	20346417	16.5227	6.4872	2.43616
274	75076	20570824	16.5529	6.4951	2.43775
275	75625	20796875	16.5831	6.5030	2.43933
276	76176	21024576	16.6132	6.5108	2.44091
277	76729	21253933	16.6433	6.5187	2.44248
278	77284	21484952	16.6733	6.5265	2.44404
279	77841	21717639	16.7033	6.5343	2.44560
280	78400	21952000	16.7332	6.5421	2.44716
281	78961	22188041	16.7631	6.5499	2.44871
282	79524	22425768	16.7929	6.5577	2.45025
283	80089	22665187	16.8226	6.5654	2.45179
284	80656	22906304	16.8523	6.5731	2.45332
285	81225	23149125	16.8819	6.5808	2.45484
286	81796	23393656	16.9115	6.5885	2.45637
287	82369	23639903	16.9411	6.5962	2.45788
288	82944	23887872	16.9706	6.6039	2.45939
289	83521	24137569	17.0000	6.6115	2.46090
290	84100	24389000	17.0294	6.6191	2.46240
291	84681	24642171	17.0587	6.6267	2.46389
292	85264	24897088	17.0880	6.6343	2.46538
293	85849	25153757	17.1172	6.6419	2.46687
294	86436	25412184	17.1464	6.6494	2.46835
295	87025	25672375	17.1756	6.6569	2.46982
296	87616	25934336	17.2047	6.6644	2.47129
297	88209	26198073	17.2337	6.6719	2.47276
298	88804	26463592	17.2627	6.6794	2.47422
299	89401	26730899	17.2916	6.6869	2.47567

GLOSSARY OF CONSTRUCTION TERMS

This glossary is not intended as a complete and comprehensive dictionary of all the specialized terminology that one will encounter in construction estimating. It is to assist the reader in using this text. We have drawn on many sources and gratefully acknowledge the Brick Institute of America, the National Concrete Masonry Association, and the Structural Steel Painting Council for permission to reprint entries from their glossaries.

abatement. Procedures to eliminate fiber release from asbestos containing building materials including encapsulation, enclosure or removal methods.

abrasive. A fine graded (sized) granular or spherical material that is used for cleaning, etc.

absorption. The weight of water a masonry unit absorbs when immersed in either cold or boiling water for a stated length of time, expressed as a percentage of the weight of the dry unit.

accelerator. Material such as calcium chloride and compositions predominately of calcium chloride which accelerate hardening and promote early strength development of concrete or mortar.

acceptance testing. Testing of received products to determine that the quality meets required specifications.

additive. Any substance added in small quantities to another substance, usually to improve properties.

admixtures. Materials other than water, aggregates, and hydraulic cement, used as an ingredient of concrete, mortar, or grout, and added to the batch immediately before or during its mixing.

adsorption. Concentration of a substance at a surface or interface of another substance.

aggregate. Granular material such as natural sand, manufactured sand, expanded clay, shale or slate, pumice, volcanic scoria, bituminous or anthracite cinders, gravel, crushed gravel, crushed stone, heavyweight aggregate such as magnetite or ilmenite, and air-cooled or expanded blast-furnace slag, which when bound together into a conglomerate mass by a matrix forms concrete, mortar, or grout.

1563

air lock. A system for hazardous substance removal and abatement that permits ingress and egress from one room to another while permitting minimal air movement between rooms, typically constructed by placing two overlapping sheets of plastic over an existing or temporarily framed doorway. Two curtained doorways spaced a minimum of 6 ft. apart from an air lock.

air monitoring. The process of measuring asbestos fiber content of a specific volume of air in a stated period of time.

alkyd resins. Synthetic resins formed by the condensation of polyhydric alcohols with polybasic acids.

ambient air quality. Average atmospheric purity, as distinguished from discharge measurements taken at the source of pollution.

amended water. Water to which a surfactant is added.

anchor. A piece or assemblage, usually metal, used to attach building parts (i.e., plates, joists, trusses, etc.) to masonry or masonry materials.

anchor pile. A pile connected to a structure by one or more ties to furnish lateral support or resist uplift.

ANSI. American National Standards Institute

arch. A curved compressive structural member, spanning openings or recesses; also built flat.

back arch. A concealed arch carrying the backing of a wall where the exterior facing is carried by a lintel.

jack arch. One having horizontal or nearly horizontal upper and lower surfaces. Also called flat or straight arch.

major arch. Arch with spans greater than 6 ft. and equivalent to uniform loads greater than 1000 lb. per ft. Typically known as Tudor, semicircular, Gothic, or parabolic arch. Has rise to span ratio greater than 0.15.

minor arch. Arch with maximum span of 6 ft. and loads not exceeding 1,000 lb. per ft. Typically known as jack, segmental, or multicentered arch. Has rise to span ratio less than or equal to 0.15.

relieving arch. One built over a lintel, flat arch, or smaller arch to divert loads, thus relieving the lower member from excessive loading. Also known as discharge or safety arch.

GLOSSARY

trimmer arch. An arch, usually a low rise arch or brick, used for supporting a fireplace hearth.

ashlar masonry. Masonry composed of rectangular units of clay or shale, or stone, generally larger in size than brick and properly bonded, having sawed, dressed or squared beds, and joints laid in mortar. Often the unit size varies to provide a random pattern, random ashlar.

asphalt. Solid or semi-solid mixture of bitumens that are found in natural deposits or created as a petroleum by-product and used mainly for paving and roofing.

natural asphalt (or lake asphalt). Derived from petroleum by the natural process of evaporation leaving the asphalt material in open beds or lakes ready for extraction and processing.

petroleum asphalt. Asphalt refined from crude petroleum.

asphalt cement (AC). Refined asphalt for paving purposes.

asphalt joint filler. Liquid asphalt product used to seal cracks and joints in pavement to prevent moisture penetration.

asphalt joint filler, preformed. Asphalt mixed with various fibrous materials and manufactured in solid form to a specific thickness. Widely used as an expansion joint material.

asphalt primer. Liquid asphalt applied to a non-bituminous surface to promote adhesion between existing surface and new surface application.

ASTM. American Society for Testing and Materials

back filling. 1. Replacing of fill materials in an excavated area 2. Rough masonry built behind a facing or between two faces. 3. Filling over the extrados of an arch. 4. Brickwork in spaces between structural timbers, sometimes called brick nogging.

backhoe. Self-powered excavation equipment that digs by pulling a boom-mounted bucket toward itself.

backup. That part of a masonry wall behind the exterior facing.

backward pass. With the forward pass complete, project calculations are now "turned around" and a second set of calculations are performed to compute the late start (LS) and late finish (LF) times for each activity; right to left.

bat. A piece of brick.

batter. Recessing or sloping masonry back in successive courses; the opposite of corbel.

bed joint. The horizontal layer of mortar on or in which a masonry unit is laid; may cover entire masonry unit or face shell only.

belt course. A narrow horizontal course of masonry, sometimes slightly projected such as window sills, that is made continuous. Sometimes called string or sill course.

benchmark. Point of known elevation from which the surveyors can establish all their grades.

bevel. One side of a solid body which is inclined with respect to the other, with the angle between the two sides either greater or less than a right angle.

blast cleaning. Cleaning or roughing of a surface using metallic or non-metallic grit or shot projected with compressed air, centrifugal force, or water.

block, concrete. A hollow or solid unit consisting of portland cement and suitable aggregates combined with water. Other materials such as lime, fly ash, air-entraining agents, or other admixtures may be permitted.

blocking. A method of bonding two adjoining or intersecting walls, not built at the same time, by means of offsets whose vertical dimensions are not less than 8 in.

blown asphalt. Asphalt specially treated at high temperature levels with blown air to obtain specific characteristics necessary in products for roofing, pipe coating, membrane envelopes, and hydraulic application.

bond. 1. Tying various parts of a masonry wall by lapping units one over another or by connecting with metal ties. 2. Patterns formed by exposed faces of units. 3. Adhesion between mortar or grout and masonry units or reinforcement.

bond, grout. The adhesion to and/or interlocking of grout with the masonry units and reinforcement.

bond, mechanical. Tying masonry units together with metal ties or reinforcing steel or keys.

1566

bond, mortar. The adhesion of mortar to masonry units and reinforcement.

bond beam. Course or courses of a masonry wall grouted and usually reinforced in the horizontal direction. Serves as horizontal tie of wall, bearing course for structural members or as a flexural member itself.

bond breaker. A material used to prevent adhesion between two surfaces.

bond course. The course consisting of units which overlap more than one wythe of masonry.

bond strength. Resistance to separation of mortar from concrete masonry units and of mortar and grout from reinforcing steel and other materials with which it is in contact.

bonder. A bonding unit. See Header.

borrow site. An area from which earth is taken for hauling to a jobsite which is short of earth needed to build an embankment.

bracing. Any type of horizontal or inclined structural member used to increase stability and to resist lateral forces.

breaking joints. Any arrangement of masonry units that prevents continuous vertical joints from occurring in adjacent courses.

brick. A solid masonry unit of clay or shale, formed into a rectangular prism while plastic, and burned or fired in a kiln.

acid-resistant brick. Brick suitable for use in contact with chemicals, usually in conjunction with acid-resistant mortars.

adobe brick. Large roughly-molded, sun-dried clay brick of varying size.

angle brick. Any brick shaped to an oblique angle to fit a salient corner.

arch brick. 1. Wedge-shaped brick for special use in an arch. 2. Extremely hard-burned brick from an arch of a scove kiln.

building brick. Brick for building purposes not especially treated for texture or color. Formerly called common brick.

clinker brick. A very hard-burned brick whose shape is distorted or bloated due to nearly complete vitrification.

common brick. See Building brick.

dry-press brick. Brick formed in molds under high pressure from relatively dry clay (5% to 7% moisture content).

economy brick. Brick whose nominal dimensions are 4"x4"x8".

engineered brick. Brick whose nominal dimensions are 4"x3.2"x8".

facing brick. Brick made especially for facing purposes, often treated to produce surface texture. They are made of selected clay, or treated, to produce desired color.

fire brick. Brick made of refractory ceramic material that resists high temperatures.

floor brick. Smooth dense brick, highly resistant to abrasion, used as finished floor surface.

gauged brick. 1. Brick that have been ground or otherwise produced to accurate dimensions. 2. A tapered arch brick.

hollow brick. A masonry unit of clay or shale whose net cross-sectional area in any plane parallel to the bearing surface is not less than 60% of its gross cross-sectional area measured in the same plane.

jumbo brick. A generic term indicating a brick larger in size than the standard. Some producers use this term to describe oversize brick of specific dimensions manufactured by them.

Norman brick. A brick whose nominal dimensions are 4" x 2-2/3" x 12".

paving brick. Vitrified brick especially suitable for use in pavements where resistance to abrasion is important.

Roman brick. Brick whose nominal dimensions are 4" x 2" x 12".

salmon brick. Generic term for underburned brick that are more porous, slightly larger, and lighter colored than hard-burned brick. Usually pinkish-orange color.

SCR brick. See SCR.

sewer brick. Low absorption, abrasive-resistant brick intended for use in drainage structures.

soft-mud brick. Brick produced by molding relatively wet clay (20% to 30% moisture). Often a hand process. When insides of molds are sanded to prevent sticking of clay, the product is sand-struck brick. When molds are wetted to prevent sticking, the product is water-struck brick.

stiff-mud brick. Brick produced by extruding a stiff but plastic clay (12% to 15% moisture) through a die.

1568

brick and brick. A method of laying brick so that units touch each other with only enough mortar to fill surface irregularities.

brick, concrete. A solid unit having a rectangular prismatic shape, usually not larger than 4" x 4" x 12", made from portland cement and suitable aggregates, with or without the inclusion of other materials.

brick grade. Designation for durability of the unit expressed as SW for severe weathering, MW for moderate weathering, or NW for negligible weathering. See ASTM Specifications C 216, C 62, and C 652.

brick type. Designation for facing brick that controls tolerance, chippage, and distortion. Expressed as FBS, FBX, and FBA for solid brick, and as HBS, HBX, HBA, and HBB for hollow brick. See ASTM Specifications C 216 and C 652.

bridging. Wood or metal members that are placed between trusses and joists in an angled position intended to spread the loads.

british thermal unit (btu). The heat energy required to raise the temperature of one pound of water one degree fahrenheit.

built-up roof. Roofing composed of two or more layers of asphalt. Several layers fastened together.

bullnose block. A unit having one or more rounded exterior corners.

buttress. A masonry pilaster decreasing in area from the base to the top, generally used to give greater lateral strength and stability to a wall.

buttering. Placing mortar on a masonry unit with a trowel.

caisson. Also caisson pile, a large-diameter shaft, hand or machine excavated to bearing stratum inside a protective casing.

capacity insulation. The ability of masonry to store heat as a result of its mass, density and specific heat.

catch basin. A complete drain box made in various depths and sizes. Water drains into pit, then from it through a pipe connected to the box.

cathodic protection. A technique to prevent corrosion of a metal surface by making that surface the cathode of an electrochemical cell.

cavity wall. A wall built of two or more wythes of masonry units separated by a continuous air space (with or without insulating materials) and in which the wythes are securely tied together with rigid corrosion resistant metal ties.

c/b ratio. The ratio of the weight of water absorbed by a masonry unit during immersion in cold water to weight absorbed during immersion in boiling water. An indication of the probable resistance of brick to freezing and thawing. Also called saturation coefficient.

cement paint. Supplied in dry powder form, it is based on portland cement to which pigments are sometimes added for decorative purposes.

centering. Temporary formwork for the support of masonry arches or lintels during construction. Also called center(s).

ceramic color glaze. An opaque colored glaze of satin or gloss finish obtained by spraying the clay body with a compound of metallic oxide, chemicals and clays. It is burned at high temperatures, fusing glaze to body and making them inseparable.

chase. A continuous recess built into a wall to receive pipes, ducts, etc.

class of unit. Distinguishes between masonry units of different grade or type in ASTM specifications, being manufactured from different raw materials, or having different specified compressive strengths.

clay. A natural, mineral aggregate consisting essentially of hydrous aluminum silicate; it is plastic when sufficiently wetted, rigid when dried and vitrified when fired to a sufficiently high temperature.

clay mortar-mix. Finely ground clay used as a plasticizer for masonry mortars.

cleanout. An opening in the first course of masonry for cleaning mortar droppings prior to grout placement in grouted masonry. Required in high lift grouting.

clean room. An uncontaminated area or room that is part of the worker decontamination enclosure system in asbestos abatement and removal, with provisions for storage of street clothes and protective equipment.

clear ceramic glaze. Same as Ceramic Color Glaze except that it is translucent or slightly tinted with a gloss finish.

clip. A portion of a brick cut to length.

1570

GLOSSARY

closer. The last masonry unit laid in a course. It may be whole or a portion of a unit.

closure. Supplementary or short length units used at corners or jambs to maintain bond patterns.

collar beam. Wood member connecting opposite roof rafters.

collar joint. The vertical, longitudinal joint between wythes of masonry.

collar tie. A horizontal member placed between two rafters a specific vertical distance above the very top plate line for the purpose of limiting outward thrust of the rafters.

column. 1. A vertical member whose horizontal dimensions measured at right angles to the thickness does not exceed three times its thickness. 2. (in concrete masonry) a compression member, vertical or nearly vertical, the width of which does not exceed four times its thickness and the height of which exceeds four times its least lateral dimension.

compactor. A machine for compacting soil. Can be pulled or self-powered. The latter has wheels to help compaction.

composite lumber. A group of wood materials that contain wood in whole or fiber form bound together with an adhesive.

composite wall. A multiple wythe masonry wall in which at least one of the wythes is dissimilar to the other wythe with respect to type or grade of unit or mortar.

compressive strength. The maximum compressive load in lbs. that a unit will support divided by the gross cross-sectional area of the unit in square inches.

concrete masonry unit, hollow. A unit whose net cross-sectional area in any plane parallel to the bearing surface is less than 75% of its gross cross-sectional area measured in the same plane.

concrete masonry unit, solid. A unit whose net cross-sectional area in every plane parallel to the bearing surface is 75% or more of its gross cross-sectional area measured in the same plane.

control joint. A continuous unbonded masonry joint to regulate the location and amount of separation resulting from the dimensional change of different parts of a structure so as to avoid the development of excessively high stresses.

coping. The material or masonry units forming a cap or finish on top of a wall, pier, pilaster, chimney, etc. It protects masonry below from penetration of water from above.

corbel. A shelf of ledge formed by projecting successive courses of masonry out from the face of the wall, pier, or column.

core. The molded open space in a masonry concrete unit.

corrosion. The deterioration of a material by direct or electrochemical reaction with its environment.

course. One of the continuous horizontal layers of units bonded with mortar in masonry.

critical activity. An activity with no float time.

cross bracing. Any type of bracing that is constructed at right angles to the main or longitudinal axis.

cross-sectional area, gross. In masonry, the total area of a section perpendicular to the direction of the load, including areas within the cells and within reentrant spaces unless these spaces are to be occupied in the masonry by portions of adjacent masonry. (The gross cross-sectional area of scored units is determined to the outside of the scoring.)

cross-sectional area, net. In masonry, the gross cross-sectional area of a section minus the average area of ungrouted cores or cellular spaces. (The cross-sectional area of grooves in scored units is not deducted from the gross cross-sectional area to obtain the net cross-sectional area.)

culls. Masonry units that do not meet the standards or specifications and have been rejected.

curing. To preserve or finish by chemical or physical process.

curing agent. An additive that promotes curing.

curtain wall. A non-loadbearing, usually prefabricated, wall between columns and piers.

1572

curtained doorway. A device used in asbestos removal to allow ingress or egress from one room to another while permitting minimal air movement between the rooms, typically constructed by placing two overlapping sheets of plastic over an existing or temporarily framed doorway. Two curtained doorways spaced a minimum of 6 ft. apart from an air lock.

customized masonry. Architectural masonry units having textured or sculptured surfaces. Methods used to obtain different surface textures including splitting, grinding, forming vertical striations and causing the units to "slump". Sculptured faces are obtained by forming projecting ribs or flutes, either rounded or angular, as well as vertical and horizontal scoring, recesses and curved faces.

cutback asphalt. Asphalt cement blended with petroleum solvents to produce a more liquid or fluid product thus improving handling and workability qualities. Exposure to atmosphere evaporates the solvents and the remaining asphalt performs the intended function. Variation in the types of solvent used produces the following products: a) Rapid Curing Asphalt (RC), b) Medium Curing Asphalt (MC), and c) Slow Curing Asphalt (SC).

dampcheck. An impervious horizontal layer to prevent vertical penetration of water in a wall consisting of either a course of solid masonry, metal or a thin layer of asphaltic or bituminous material. Generally near grade to prevent upward migration of moisture by capillary action. See damp course below.

damp course. A masonry course or layer of impervious material that prevents capillary entrance of moisture from the ground or a lower course. Same as damp check.

dampproofing. Prevention of moisture penetration by capillary action, usually by addition of one or more coats of a compound that is impervious to water.

decontamination enclosure system. In asbestos abatement and removal, a series of areas or connected rooms, with curtained doorways between them, for decontamination of workers or materials and equipment. The system always contains a minimum of one airlock between two curtained doorways.

direct costs. Costs that are charged directly to a project, such as for labor, taxes, insurance, materials, scaffolding, equipment, and inspection.

disposal. Procedures necessary to transport and deposit hazardous waste containing material removed from a building or site in compliance with regulations.

dog's tooth. Brick laid with their corners projecting from the wall face.

double-walled tank. An underground storage tank in which a rigid secondary container is attached to the primary container and which has an annular space.

dovetail anchor. A splayed tenow shaped like a dove's tail, broader at its end than at its base, which fits into the recess of a corresponding mortise.

dowel. Straight metal bars used to connect two sections of masonry.

drilled pile. Also called an augured pile, it is a cast-in-place concrete pile in an augured hole, possibly belled at the bottom.

drip. A projecting piece of material, shaped to throw off water and prevent its running down the face of the wall or other surface.

dry stack. Masonry work laid without mortar.

EBM. See Engineered brick masonry.

eccentricity. The normal distance between the centroidal axis of a member and the parallel resultant load.

e1/e2. Ratio of virtual eccentricities occurring at the ends of a column or wall under design. The absolute value is always less than or equal to 1.0.

effective area of reinforcement. The area obtained by multiplying the right cross-sectional area of metal reinforcement by the cosine of the angle between its direction and the direction for which the effectiveness of the reinforcement is to be determined.

effective height. The height of a member to be assumed for calculating the slenderness ratio.

effective thickness. The thickness of a member to be assumed for calculating the slenderness ratio.

efflorescence. A powder or stain sometimes found on the surface of masonry, resulting from deposition of water-soluble salts.

1574

emission standard. The maximum amount of a pollutant that is permitted to be discharged from a single source.

encapsulation. Procedures necessary to coat all spray- or trowel-applied asbestos containing materials with an encapsulant to control the release of asbestos fibers into the ambient air.

empirical design. Design based on applying physical limitations based on experience or observations gained through experience without a structural analysis.

engineered brick masonry. Masonry in which design is based on a rational structural analysis.

engineered design. Design based on a rational analysis considering the interrelationships of the various construction materials, their properties and actual design loads in lieu of empirical design procedures.

EPA. Environmental Protection Agency.

epoxy. Thermosetting resin that is mixed with a curing agent to produce adhesives, coatings, etc.

equipment decontamination system. A decontamination enclosure system for materials and equipment, typically consisting of designated area of the work area, a wash room and an uncontaminated area.

equipment decontamination enclosure system. A decontamination enclosure system in asbestos work for materials and equipment, typically consisting of designated area of the work area, a wash room and an uncontaminated area.

expansion joint. A separation between adjoining parts of a structure that is provided to allow small relative movements, such as those caused by thermal changes, to occur without one part affecting an adjacent part.

face. 1. The exposed surface of a wall or masonry unit. 2. The surface of a unit designed to be exposed in the finished structure.

face shell. The side wall of a hollow concrete masonry unit, generally between 3/4" and 1-1/2" thick.

face shell bedding. Mortar is applied only to the horizontal face of the face shells of hollow masonry units and in the head joints to a depth equal to the thickness of the face shell.

facing. Any material, forming a part of a wall, used as a finished surface.

fascia. The flat surface located at the outer end of a roof overhang or cantilever end.

field. The expanse of wall between openings, corners, etc., principally composed of stretchers.

filter block. A hollow, vitrified clay masonry unit, sometimes salt-glazed, designed for trickling filter floors in sewage disposal plants.

finished grade. Any surface, which has been cut to or built to the elevation indicated for that point.

fire clay. A clay that is highly resistant to heat without deforming and used for making brick.

fire-resistive material. See non-combustive material.

fireproofing. Any material or combination protecting structural members to increase their fire resistance.

fire wall. Any wall that subdivides a building so as to resist the spread of fire, by starting at the foundation and extending continuously through all stories to, or above, the roof.

flashing. 1. A thin impervious material placed in mortar joints and through air spaces in masonry to prevent water penetration and/or provide water drainage. 2. Manufacturing methods to produce specific color tones.

flux oil. A thick petroleum product used to soften asphalt to a desired consistency.

foundation wall. A wall below the floor nearest grade serving as a support for a wall, pier, column, or other structural part of a building.

forward pass. The early start (ES) and the early finish (EF) computations that proceed from project start to project finish; left to right.

free float. The amount of time by which an activity can be delayed without affecting the early start (ES) time of any activity in the project.

frog. A depression in the bed surface of a brick. Sometimes called a panel.

furring. Strips of wood or metal used to provide a level surface for finishing or to create an air space, or the installation of such material.

1576

furrowing. The practice of striking a v-shaped trough in a bed of mortar (not recommended).

galvanized. Coating of zinc applied to steel in order to protect it from corrosion.

gambrel. Roof having two slopes on each side, the lower slope usually steeper than the upper one. A curb roof of the same section in all parts with a lower steeper slope and an upper flatter roof.

glove bag system. A portable asbestos abatement system designed for isolated and small areas of pipe, fittings, etc. requiring asbestos removal.

grade. The surface of a road, channel or natural ground area. Usually means the surface level required by plans or specification.

grid pavers. Open type masonry units that allow the growing of grass when employed for soil stabilization in parking areas, along the shoulders of highways and airport runways, embankment erosion control, fire engine lines while providing a base to support vehicular traffic.

grounds. Nailing strips placed in masonry walls as a means of attaching trim or furring.

grout. Mixture of cementitious material and aggregate to which sufficient water is added to produce pouring consistency without segregation of the constituents.

high-lift grouting. The technique of grouting masonry in lifts up to 12 ft.

low-lift grouting. The technique of grouting as the wall is constructed.

grout lift. The height to which grout is placed in a cell, collar joint, or cavity without intermission.

grout pour. The total grouted height between masonry lifts. A grout pour may consist of one or more grout lifts.

grouted masonry. Concrete masonry construction composed of hollow units where hollow cells are filled with grout, or multi-wythe construction in which space between wythes is solidly filled with grout.

hacking. 1. The procedure of stacking brick in a kiln or on a kiln car. 2. Laying brick with the bottom edge set in from the plane surface of the wall.

hand cleaning. Any surface preparation using hand tools such as wire brushes, scrapers, and chipping tools.

hard-burned. Nearly vitrified clay products that have been fired at high temperatures. They have relatively low absorptions and high compressive strengths.

head joint. The vertical mortar joint between ends of masonry units. Often called cross joint.

header. 1. Masonry unit that overlaps two or more adjacent wythes of masonry to tie them together. Often called bonder. 2. A floor or roof beam placed between joists or beams to support the ends.

blind header. A concealed brick header in the interior of the wall, not showing on the face.

clipped header. A bat placed to look like a header for purposes of establishing a pattern. Also called false header.

flare header. A header of darker color than the field of the wall.

heading course. A continuous bonding course of header brick. Also called header course.

HEPA filter. A high efficiency particulate absolute (HEPA) filter capable of trapping and retaining 99.9% of asbestos fibers greater than 0.3 microns in length. These filters are used in vacuum cleaners or in air transfer units (exhaust fans).

HEPA vacuum equipment. High efficiency particulate absolute filtered vacuuming equipment with a filter system capable of collecting and retaining asbestos fibers. Filters should be of a 99.9% efficiency for retaining fibers 0.3 microns (3/1000 of a millimeter) or larger.

hoe. A track mounted, self-powered, shoveling machine that digs by pulling a boom mounted bucket towards itself.

hot-rolled steel. Steel that is hot reduced; that is, it is formed and shaped when it is hot.

hot specs. New and modified federal specifications that are issued every 15 days on microfilm.

indirect costs. Costs that are not directly attributable to just one job, such as office overhead, cost of capital, equipment depreciation.

1578

GLOSSARY

initial rate of absorption. The weight of water absorbed expressed in grams per 30 sq. in. of contact surface when a brick is partially immersed for one minute. Also called suction. See ASTM Specification C 67.

interlocking block paver. Solid masonry units capable of transferring loads and stresses laterally by arching or bridging action between units when subjected to vehicular traffic.

IRA. See initial rate of absorption.

joint reinforcement. Steel wires placed in the mortar joint (over the face shell in hollow masonry) and having cross wires welded between them at regular intervals.

kiln. A furnace oven or heated enclosure used for burning or firing brick or other clay material.

kiln run. Brick from one kiln that have not been sorted or graded for size or color variation.

king closer. A brick cut diagonally to have one 2-in. end and one full-width end.

laitance. A milky white deposit on new concrete.

lateral support. Means whereby walls are braced either vertically or horizontally by columns, pilasters, cross walls, beams, floors, roofs, etc.

latex paint. A paint containing a stable aqueous dispersion of synthetic resin, produced by emulsion polymerization, as the principal constituent of the binder.

lead[1]. The section of a masonry wall built up and racked back on successive courses. A line is attached to leads as a guide for constructing a wall between them.

lead[2]. A heavy metal that can be hazardous to health if breathed or swallowed.

lift. Any layer of material or soil placed upon another.

lime, hydrated. Quicklime to which sufficient water has been added to convert the oxides to hydroxides.

lime putty. Hydrated lime in plastic form ready for addition to mortar.

lintel. A beam placed over an opening in a wall to carry the superimposed weight of the construction and loads above the opening.

loadbearing. A structural system or element designed to carry loads in addition to its own dead weight.

masonry. Brick, stone, concrete, etc. or masonry combinations thereof, bonded with mortar.

masonry cement. 1. A mill-mixed cementitious material to which sand and water must be added. See ASTM C 91. 2. Hydraulic cement produced for use in mortars for masonry construction where greater plasticity and water retention are desired than is obtainable by the use of portland cement alone. Such cements always contain one or more of the following materials: portland cement, portland-pozzolan cement, natural cement, slag cement, hydraulic lime. They usual contain one or more of the following: hydrated lime, pulverized limestone, chalk, talc, pozzolan, clay or gypsum; many masonry cements also include air-entraining and water-repellent additions.

masonry unit. Natural or manufactured building units of burned clay, concrete, stone, glass, gypsum, etc.

hollow masonry unit. One whose net cross-sectional area in any plane parallel to the bearing surface is less than 75% of the gross.

modular masonry unit. One whose nominal dimensions are based on the 4-in. module.

solid masonry unit. One whose net cross-sectional area in every plane parallel to the bearing surface is 75% or more of the gross.

moisture content. The amount of water contained at the time of sampling expressed as a percentage of the total absorption.

mortar. A plastic mixture of cementitious materials, fine aggregate, and water. See ASTM Specifications C 270, C 476 or BIA M1-72.

fat mortar. Mortar containing a high percentage of cementitious components; it is a sticky mortar which adheres to a trowel.

high-bond mortar. Mortar that develops higher bond strengths with masonry units than normally developed with conventional mortar.

lean mortar. Mortar that is deficient in cementitious components; it is usually harsh and difficult to spread.

GLOSSARY

net section. The minimum cross-section of the member under consideration; in masonry, usually the mortar bedded area plus the grouted area.

nominal dimension. A dimension greater than a specified masonry dimension by the thickness of a mortar joint, but not more than 1/2".

non-combustible material. Any material that will neither ignite nor actively support combustion in air at a temperature of 1200°F when exposed to fire.

oil paint. A paint that contains drying oil, oil varnish, or oil-modified resin as the basic vehicle ingredient. Not that the common but technically incorrect definition is any paint soluble in organic solvents.

overhand work. Laying brick from inside a wall by workers standing on a floor or on a scaffold.

pargeting. The process of applying a coat of cement mortar to masonry. Also parging.

particulates. Fine liquid or solid particles, such as dust, smoke, mist, fumes, or smog, found in the air or in emissions.

panel wall. A non-loadbearing wall constructed between columns or piers and wholly supported at each story.

parapet wall. That part of a wall that extends above the roof level.

partition. An interior wall, one story or less in height.

party wall. A wall on an interior lot line, or any wall used to adapt for joint service between two buildings or adjacent living or work spaces.

phenolic resin. Resin made by the condensation of phenols and aldehydes.

pick and dip. A method of laying brick whereby the bricklayer simultaneously picks up a brick with one hand, and with the other hand, enough mortar on a trowel to lay the brick. Sometimes called the eastern or New England method.

pier. Any of various vertical supporting structures; an isolated column of masonry.

pilaster. A wall portion projecting from either or both wall faces and serving as a vertical column or beam.

plasticizer. A substance added to paint, varnish, or plaster to give it flexibility.

plumb rule. This is a combination plumb rule and level. It is used in a horizontal position as a level and in a vertical position as a plumb rule. They are made in lengths of 42" and 48" and short lengths from 12" to 24".

pointing. Troweling mortar into a joint after masonry units are laid.

prefabricated brick masonry. Masonry construction fabricated in a location other than its final inservice location in the structure. Also known as preassembled, panelized, and sectional brick masonry.

prism. 1. Any of a number of polyhedron shapes 2. A small masonry assemblage made with masonry units and mortar. Primarily used to predict the strength of full scale masonry members.

queen closer. A cut brick having a nominal 2" horizontal face dimension.

quoin. A projecting right angle masonry corner.

racking. A method entailing stepping back successive courses of masonry.

raggle. A groove in a joint or special unit to receive roofing or flashing.

RBM. Reinforced brick masonry.

reinforced masonry. Masonry units, reinforcing steel, grout, or mortar combined to act together in resisting forces.

return. Any surface turned back from the face of a principal surface.

reveal. That portion of a jamb or recess that is visible from the face of a wall.

right of way line. A line on the side of a road marking the limit of the construction area and, usually, the beginning of private property.

rippers. Teeth-shaped attachments added to digging equipment for digging through hard pan or rock soil.

rowlock. A brick laid on its face edge so that the normal bedding area is visible in the wall face. Frequently spelled rolok.

salt glaze. A gloss finish obtained by thermochemical reaction between silicates of clay and vapors of salt or chemicals.

GLOSSARY

saturation coefficient: See c/b ratio.

SCR. Structural Clay Research (trademark of the Structural Clay Products Institute, BIA).

SCR acoustile. A side-construction, two-celled facing tile, having a perforated face backed with glass wool for acoustical purposes.

SCR brick. Brick whose nominal dimensions are 6" x 2-2/3" x 12".

SCR building panel. Prefabricated, structural ceramic panels, approximately 2-1/2" thick.

SCR insulated cavity wall. Any cavity wall containing insulation that meets rigid criteria established by the Structural Clay Products Institute (BIA).

SCR masonry process. A construction aid providing greater efficiency, better workmanship, and increased production in masonry construction. It utilizes story poles, marked lines and adjustable scaffolding.

scraper. A digging, hauling and grading machine having a cutting edge, a carrying bowl and a dumping mechanism.

screen block. Open-faced masonry units used for decorative purposes or to partially screen areas from the sun or outside viewers.

sealant. 1. Any type of sealing agent. 2. A material spray applied to the substrate after the wet removal of existing asbestos material. The purpose of this is to prevent the release of residual asbestos fibers left after removal and cleaning operations.

shale. Clay that has been subjected to high pressures until it has hardened.

shear wall. A wall which, in its own plane, carries shear resulting from wind, blast, or seismic forces.

sheepsfoot roller. A compacting roller with feet expanded at their outer tips, used in compacting soil.

shower room. A room constituting an air lock between the clean room and the equipment room in the worker decontamination enclosure system, with hot and cold or warm running water suitably arranged for complete showering of workers during decontamination. Shower room always includes an air lock.

1583

shoved joints. Vertical joints filled by shoving a brick against the next brick when it is being laid in a bed of mortar.

shrinkage. Volume change due to loss of moisture or decrease in temperature.

single wythe wall. A wall of only one masonry unit in thickness.

slenderness ratio. Ratio of the effective height of a member to its effective thickness.

slushed joints. Vertical joints filled, after units are laid, by "throwing" mortar in with the edge of a trowel (generally not recommended).

soap. A masonry unit of normal face dimensions, having a nominal 2" thickness.

soffit. The underside of a beam, lintel, or arch.

soft-burned. Clay products that have been fired at low temperature ranges, producing relatively high absorptions and low compressive strengths.

solar screen. A perforated wall used as a sunshade.

soldier. A stretcher set on end with face showing on the wall surface.

spall. A small fragment removed from the face of a masonry unit by a blow or by action of the elements.

stack. Any structure or part thereof that contains a flue or flues for the discharge of gases.

stacked bond. A bonding pattern where no unit overlaps either the one above or below, all head joints form a continuous vertical line.

station. The point where the worker holds the rod in elevation measurements.

story pole. A marked pole for measuring masonry coursing during construction.

stretcher. A masonry unit laid with its greatest dimension horizontal and its face parallel to the wall face.

stringing mortar. The procedure of spreading enough mortar on a bed to lay several masonry units.

1584

struck joint. Any mortar joint that has been finished with a trowel.

suction. See Initial rate of absorption.

surfactant. In asbestos work, a chemical wetting agent added to water to improve penetration into the asbestos thus reducing the amount of water required for a given operation or area. This surfactant is spray applied to the asbestos to decrease the amount of airborne fibers while removing the asbestos material from the substrate. (Wetting agent-50% polyglycol ether and 50% polyoxyethlene ether or approved equal shall be mixed with water to provide a concentration of 1 oz. surfactant to 5 gallons of water or the manufacturer's recommended concentration.

swale. A shallow dip made to allow the passage of water.

temper. To moisten and mix clay, plaster, or mortar to a proper consistency.

tie. Any unit of material that connects masonry to masonry or other materials. See wall tie.

tooling. Compressing and shaping the face of a mortar joint with a special tool other than a trowel.

toothing. Constructing the temporary end of a masonry wall with the end stretcher of every alternate course projecting. Projecting units are toothers.

total float. The amount of time for an activity, obtained by subtracting its early start (ES) from its late start (LS) or early finish (EF) from its late finish (LF).

traditional masonry. Masonry in which design is based on empirical rules that control minimum thickness, lateral support requirements, and height without a structural analysis.

transformed section. An assumed section of one material having the same elastic properties as the section of two materials.

tuck pointing. The filling in with fresh mortar of cut-out or defective mortar joints in masonry.

veneer. 1. A thin layer of material that is bonded to and serves as the surface for another material. 2. A single wythe of masonry for facing purposes, not structurally bonded.

virtual eccentricity. The eccentricity of a resultant axial load required to produce axial and bending stresses equivalent to those produced by applied axial loads and moments. It is normally found by dividing the moment at a section by the summation of axial loads occurring at that section.

vitrification. The condition resulting when kiln temperatures are sufficient to fuse grains and close pores of a clay product, making the mass impervious.

wall. A vertical member of a structure whose horizontal dimension measured at right angles to the thickness exceeds three times its thickness.

apron wall. That part of a panel wall between window sill and wall support.

area wall. 1. The masonry surrounding or partly surrounding an area. 2. The retaining wall around basement windows below grade.

bearing wall. One that supports a vertical load in addition to its own weight.

cavity wall. A wall built of masonry units so arranged as to provide a continuous air space within the wall (with or without insulating material), and in which the inner and outer wythes of the wall are tied together with metal ties.

composite wall. A multiple-wythe wall in which at least one of the wythes is dissimilar to the other wythe or wythes with respect to type or grade of masonry unit or mortar.

curtain wall. An exterior non-loadbearing wall not wholly supported at each story. Such walls may be anchored to columns, spandrel beams, floors, or bearing walls, but not necessarily built between structural elements.

dwarf wall. A wall or partition that does not extend to the ceiling.

enclosure wall. An exterior non-loadbearing wall in skeleton frame construction. It is anchored to columns, piers, or floors but not necessarily built between columns or piers nor wholly supported at each story.

exterior wall. Any outside wall or vertical enclosure of a building other than a party wall.

faced wall. A composite wall in which the masonry facing and backing are so bonded as to exert a common reaction under load.

fire division wall. Any wall that subdivides a building so as to resist the spread of fire. It is not necessarily continuous through all stories to and above the roof.

fire wall. Any wall that subdivides a building to resist the spread of fire and that extends continuously from the foundation through the roof.

1586

GLOSSARY

foundation wall. That portion of a loadbearing wall below the level of the adjacent grade, or below first floor beams or joists.

hollow wall. A wall built of masonry units arranged to provide an air space within the wall. The separated facing and backing are bonded together with masonry units.

insulated cavity wall. See SCR insulated cavity wall

loadbearing wall. A wall that supports any vertical load in addition to its own weight.

non-loadbearing wall. A wall that supports no vertical load other than its own weight.

panel wall. An exterior, non-loadbearing wall wholly supported at each story.

parapet wall. That part of any wall entirely above the roof line.

party wall. A wall used for joint service by adjoining buildings.

perforated wall. One that contains a considerable number of relatively small openings. Often called pierced wall or screen wall.

shear wall. A wall that resists horizontal forces applied in the plane of the wall.

single wythe wall. A wall containing only one masonry unit in wall thickness.

solid masonry wall. A wall built of solid masonry units, laid contiguously, with joints between units completely filled with mortar or grout.

spandrel wall. That part of a curtain wall above the top of a window in one story and below the sill of the window in the story above.

veneered wall. A wall having a facing of masonry units or other weather-resisting non-combustible materials securely attached to the backing, but not so bonded as to intentionally exert common action under load.

wall plate. A horizontal member anchored to a masonry wall to which other structural elements may be attached. Also called head plate.

wall tie. A bonder or metal piece that connects wythes of masonry to each other or to other materials.

wall tie, cavity. A rigid, corrosion-resistant metal tie that bonds two wythes of a cavity wall. It is usually steel, 3/16" in a diameter and formed in a "Z" shape or a rectangle.

wall tie, veneer. A strip or piece of metal used to tie a facing veneer to the backing.

water retentivity. That property of a material that prevents the rapid loss of water.

water table. A projection of lower masonry on the outside of the wall slightly above the ground. Often a damp course is placed at the level of the water table to prevent upward penetration of ground water.

waterproofing. Prevention of moisture flow through masonry due to water pressure.

web. The cross wall connecting the face shells of a hollow concrete masonry unit.

weep holes. Openings placed in mortar joints of facing material at the level of flashing to permit the escape of moisture.

wet cleaning. The process of eliminating asbestos contamination from building surfaces and objects by using cloths, mops, or other cleaning tools that have been dampened with water and by afterwards disposing of these cleaning tools as asbestos contaminated waste.

with inspection. Masonry designed with the higher stresses allowed under EBM. Requires the establishment of procedures on the job to control mortar mix, workmanship, and protection of masonry materials.

without inspection. Masonry designed with the reduced stresses allowed under EBM.

worker decontamination enclosure system. A decontamination enclosure system for workers typically consisting of a clean room, a locker room, a shower room, and an equipment room.

wythe. 1. Each continuous vertical section of masonry one unit in thickness. 2. The thickness of masonry separating flues in a chimney. Also called withe or tier.

CONSTRUCTION SAFETY

Construction is the largest industry in the U.S. and one of the most dangerous. The issues of accident prevention and safety in construction are in the limelight, partly due to the number of avoidable disasters in recent years.

The most important reason for developing and carefully monitoring a safety program is that employees' lives depend on it, but there are economic advantages as well. Healthy workers are productive ones, and a good safety record can mean lower workers compensation rates, which for some contractors can be 5% to 10% of total contract amount.

There are many agencies and organizations that provide guidelines and regulations for construction safety. The most visible and probably most important is the federal Occupational Safety and Health Administration (OSHA). OSHA regulations are ever changing, and the contractor must keep abreast of the latest revisions in OSHA standards. This point cannot be stressed too much. Listed below are some of the key OSHA publications that concern construction safety, but it is the contractor's responsibility to determine the latest versions of these publications. The easiest way is by contacting the nearest OSHA field office.

At the same time, the contractor must fulfill state and local regulatory requirements regarding the health and safety of workers and the public as well. It might seem like a daunting task, but it is well worth the time and effort.

The list below outlines the subparts of the Code of Federal Regulations (CRF) Volume 29, Chapter XVII - Occupational Safety and Health Administration, Department of Labor. These are the OSHA Construction Industry Standards. The subparts are self-descriptive and cover all major areas of safety concern for the industry. The revisions date indicated below was the latest when the current edition of our book was published. Once again, the contractor must check for more current revisions of these standards:

Occupational Safety and Health Administration
Part Title – Safety and Health Regulations for Construction
Standard Title – 1926
Title – Table of Contents

Revised 2001

Subpart	Topic
A	General
B	General Interpretations
C	General Safety and Health Provisions
D	Occupational Health and Environmental Controls
E	Personal Protective and Life Saving Equipment

F	Fire Protection and Prevention
G	Signs, Signals, and Barricades
H	Materials Handling, Storage, Use, and Disposal
I	Tools – Hand and Power
J	Welding and Cutting
K	Electrical
L	Scaffolds
M	Fall Protection
N	Cranes, Derricks, Hoists, Elevators, and Conveyors
O	Motor Vehicles, Mechanized Equipment, and Marine Operations
P	Excavations
Q	Concrete and Masonry Construction
R	Steel Erection
S	Tunnel and Shafts, Caissons, Cofferdams, and Compressed Air
T	Demolition
U	Blasting and Use of Explosives
V	Power Transmission and Distribution
W	Rollover Protective Structures; Overhead Protection
X	Stairways and Ladders
Y	Commercial Diving Operations
Z	Toxic and Hazardous Substances

In addition to OSHA 2207 - Construction Industry Standards described above, there are a number of other OSHA publications that pertain to the construction industry:

OSHA No.	Title
2098	OSHA Inspection
2201	General Industry Digest
2202	Construction Industry Digest
2209	Handbook for Small Business
2226	Excavations
2254	Training Requirements in OSHA Standards and Training Guidelines
3007	Ground Fault Protection on Construction Sites
3071	Job Hazard Analysis
3072	Sling Safety
3075	Controlling Electrical Hazards
3077	Personal Protective Equipment
3079	Respiratory Protection
3080	Hand and Power Tools
3088	How to Prepare for Workplace Emergencies
3096	Asbestos Standards for Construction Industry
3100	Crane and Derrick Suspended Platforms
3106	Concrete and Masonry Construction
3115	Underground Construction – (Tunneling)

CONSTRUCTION SAFETY

APPENDIX B TO SUBPART P

Sloping and Benching

(a) *Scope and application.* This appendix contains specifications for sloping and benching when used as methods of protecting employees working in excavations from cave-ins. The requirements of this appendix apply when the design of sloping and benching protective systems is to be performed in accordance with the requirements set forth in § 1926.652(b)(2).

(b) *Definitions.*

Actual slope means the slope to which an excavation face is excavated.

Distress means that the soil is in a condition where a cave-in is imminent or is likely to occur. Distress is evidenced by such phenomena as the development of fissures in the face of or adjacent to an open excavation; the subsidence of the edge of an excavation; the slumping of material from the face or the bulging or heaving of material from the bottom of an excavation; the spalling of material from the face of an excavation; and ravelling, i.e., small amounts of material such as pebbles or little clumps of material suddenly separating from the face of an excavation and trickling or rolling down into the excavation.

Maximum allowable slope means the steepest incline of an excavation face that is acceptable for the most favorable site conditions as protection against cave-ins, and is expressed as the ratio of horizontal distance to vertical rise (H:V).

Short term exposure means a period of time less than or equal to 24 hours that an excavation is open.

(c) *Requirements*—(1) *Soil classification.* Soil and rock deposits shall be classified in accordance with appendix A to subpart P of part 1926.

(2) *Maximum allowable slope.* The maximum allowable slope for a soil or rock deposit shall be determined from Table B-1 of this appendix.

(3) *Actual slope.* (i) The actual slope shall not be steeper than the maximum allowable slope.

(ii) The actual slope shall be less steep than the maximum allowable slope, when there are signs of distress. If that situation occurs, the slope shall be cut back to an actual slope which is at least ½ horizontal to one vertical (½H:1V) less steep than the maximum allowable slope.

(iii) When surcharge loads from stored material or equipment, operating equipment, or traffic are present, a competent person shall determine the degree to which the actual slope must be reduced below the maximum allowable slope, and shall assure that such reduction is achieved. Surcharge loads from adjacent structures shall be evaluated in accordance with § 1926.651(i).

(4) *Configurations.* Configurations of sloping and benching systems shall be in accordance with Figure B-1.

1591

TABLE B-1
MAXIMUM ALLOWABLE SLOPES

SOIL OR ROCK TYPE	MAXIMUM ALLOWABLE SLOPES(H:V)[1] FOR EXCAVATIONS LESS THAN 20 FEET DEEP [3]
STABLE ROCK TYPE A [2] TYPE B TYPE C	VERTICAL (90°) 3/4 : 1 (53°) 1:1 (45°) 1¼: 1 (34°)

NOTES:

1. Numbers shown in parentheses next to maximum allowable slopes are angles expressed in degrees from the horizontal. Angles have been rounded off.

2. A short-term maximum allowable slope of 1/2H:1V (63°) is allowed in excavations in Type A soil that are 12 feet (3.67 m) or less in depth. Short-term maximum allowable slopes for excavations greater than 12 feet (3.67 m) in depth shall be 3/4H:1V (53°).

3. Sloping or benching for excavations greater than 20 feet deep shall be designed by a registered professional engineer.

Figure B-1

Slope Configurations

(All slopes stated below are in the horizontal to vertical ratio)

B-1.1 *Excavations made in Type A soil.*

1. All simple slope excavations 20 feet or less in depth shall have a maximum allowable slope of ¾:1.

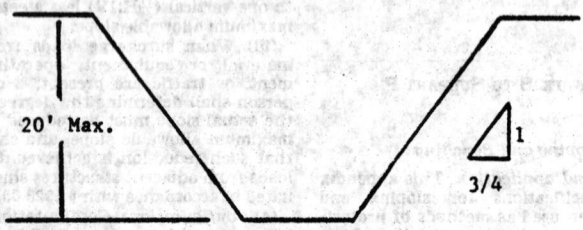

20' Max.

1

3/4

SIMPLE SLOPE—GENERAL

Exception: Simple slope excavations which are open 24 hours or less (short term) and which are 12 feet or less in depth shall have a maximum allowable slope of ½:1.

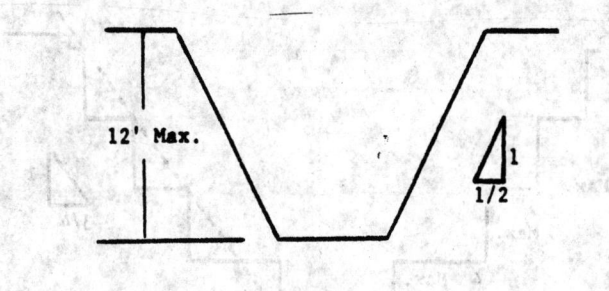

SIMPLE SLOPE—SHORT TERM

2. All benched excavations 20 feet or less in depth shall have a maximum allowable slope of ¾ to 1 and maximum bench dimensions as follows:

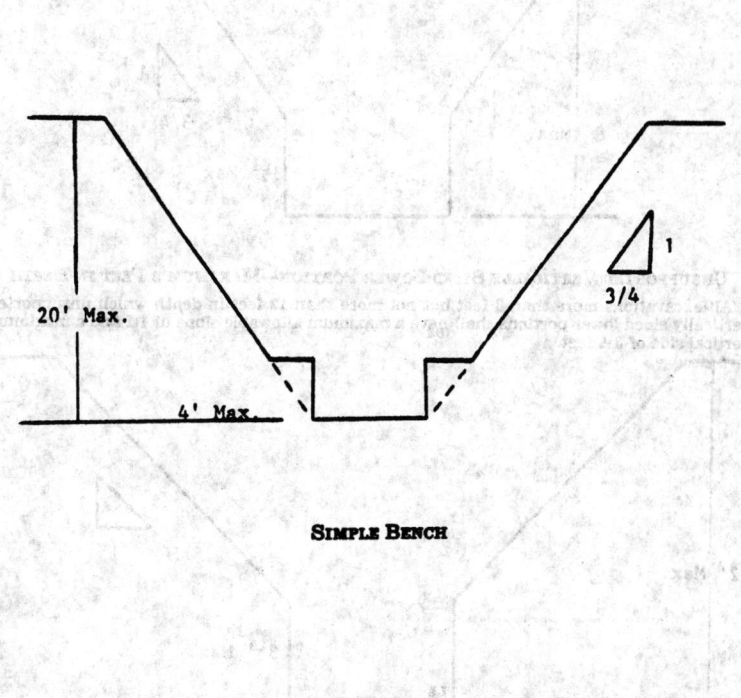

SIMPLE BENCH

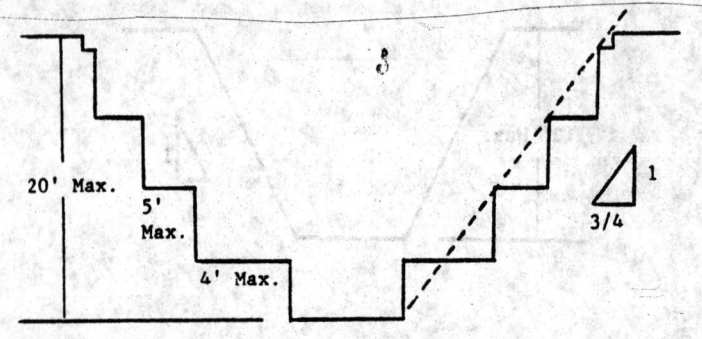

MULTIPLE BENCH

3. All excavations 8 feet or less in depth which have unsupported vertically sided lower portions shall have a maximum vertical side of 3½ feet.

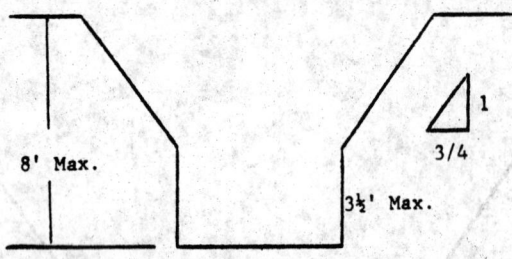

UNSUPPORTED VERTICALLY SIDED LOWER PORTION—MAXIMUM 8 FEET IN DEPTH

All excavations more than 8 feet but not more than 12 feet in depth which unsupported vertically sided lower portions shall have a maximum allowable slope of 1:1 and a maximum vertical side of 3½ feet.

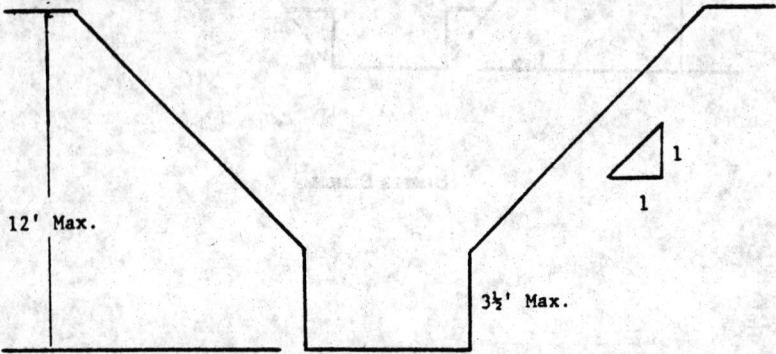

UNSUPPORTED VERTICALLY SIDED LOWER PORTION—MAXIMUM 12 FEET IN DEPTH

All excavations 20 feet or less in depth which have vertically sided lower portions that are supported or shielded shall have a maximum allowable slope of ¾:1. The support or shield system must extend at least 18 inches above the top of the vertical side.

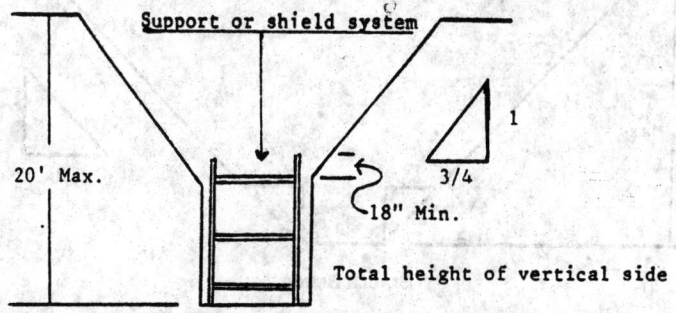

SUPPORTED OR SHIELDED VERTICALLY SIDED LOWER PORTION

4. All other simple slope, compound slope, and vertically sided lower portion excavations shall be in accordance with the other options permitted under § 1926.652(b).

B-1.2 Excavations Made in Type B Soil

1. All simple slope excavations 20 feet or less in depth shall have a maximum allowable slope of 1:1.

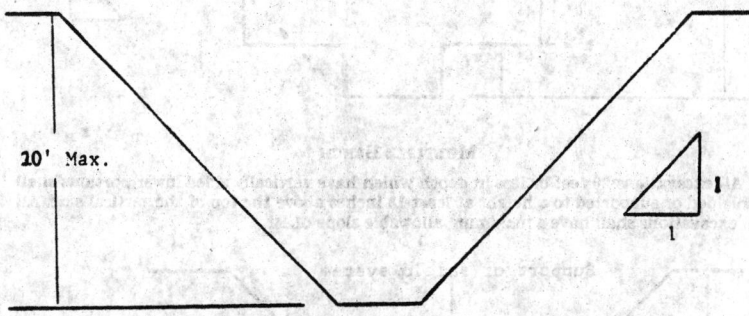

SIMPLE SLOPE

2. All benched excavations 20 feet or less in depth shall have a maximum allowable slope of 1:1 and maximum bench dimensions as follows:

This bench allowed in cohesive soil only.

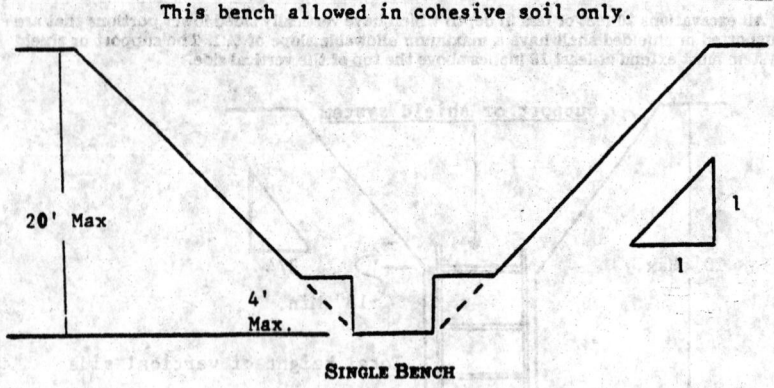

20' Max

4'
Max.

SINGLE BENCH

This bench allowed in cohesive soil only.

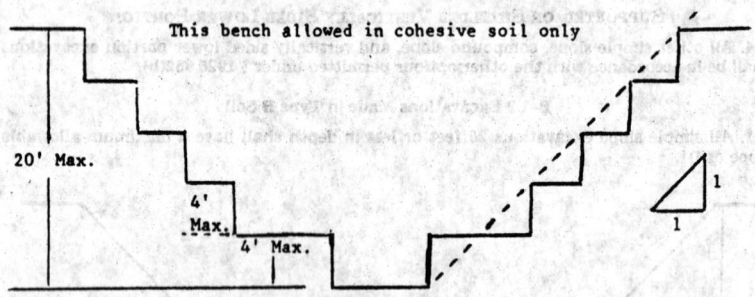

20' Max.

4'
Max.
4' Max.

MULTIPLE BENCH

3. All excavations 20 feet or less in depth which have vertically sided lower portions shall be shielded or supported to a height at least 18 inches above the top of the vertical side. All such excavations shall have a maximum allowable slope of 1:1.

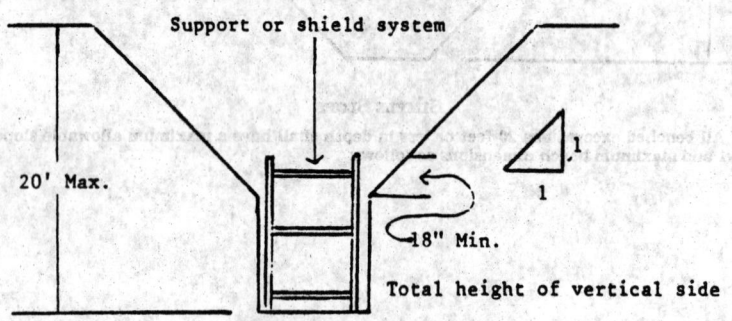

Support or shield system

20' Max.

18" Min.

Total height of vertical side

VERTICALLY SIDED LOWER PORTION

4. All other sloped excavations shall be in accordance with the other options permitted in § 1926.652(b).

B-1.3 Excavations Made in Type C Soil

1. All simple slope excavations 20 feet or less in depth shall have a maximum allowable slope of 1½:1.

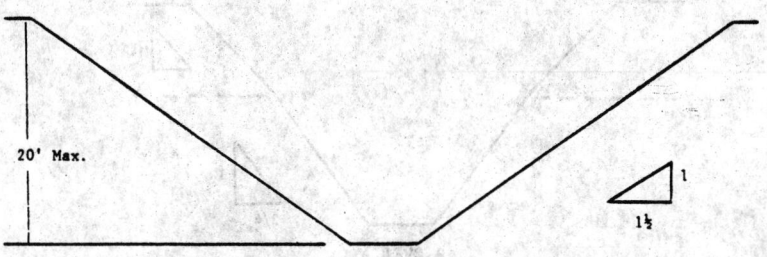

SIMPLE SLOPE

2. All excavations 20 feet or less in depth which have vertically sided lower portions shall be shielded or supported to a height at least 18 inches above the top of the vertical side. All such excavations shall have a maximum allowable slope of 1½:1.

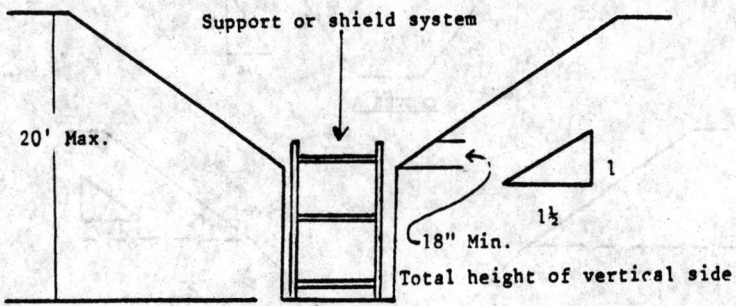

VERTICAL SIDED LOWER PORTION

3. All other sloped excavations shall be in accordance with the other options permitted in § 1926.652(b).

B-1.4 Excavations Made in Layered Soils

1. All excavations 20 feet or less in depth made in layered soils shall have a maximum allowable slope for each layer as set forth below.

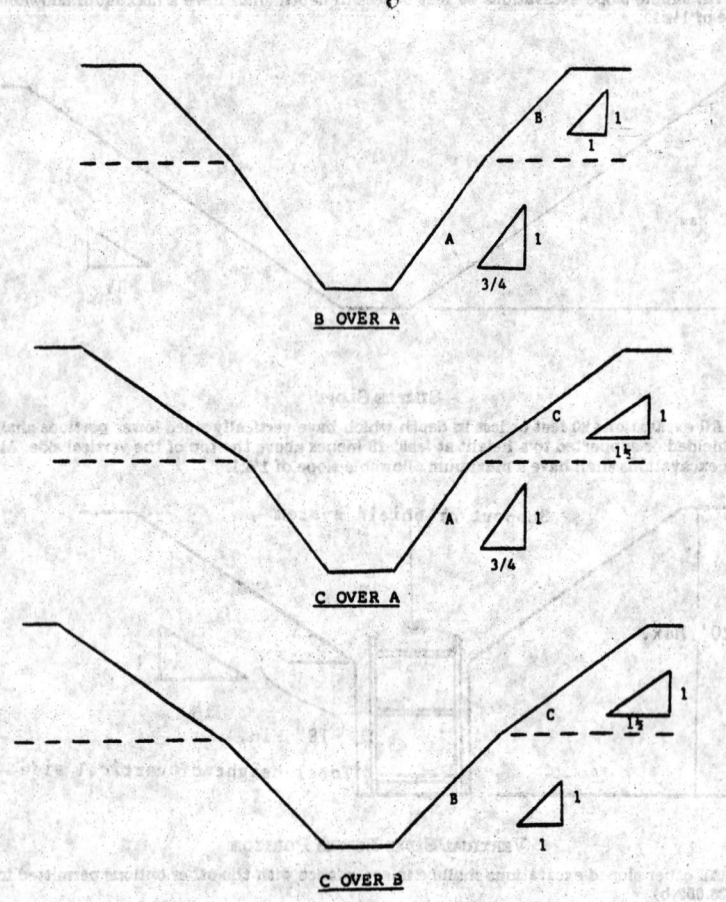

B OVER A

C OVER A

C OVER B

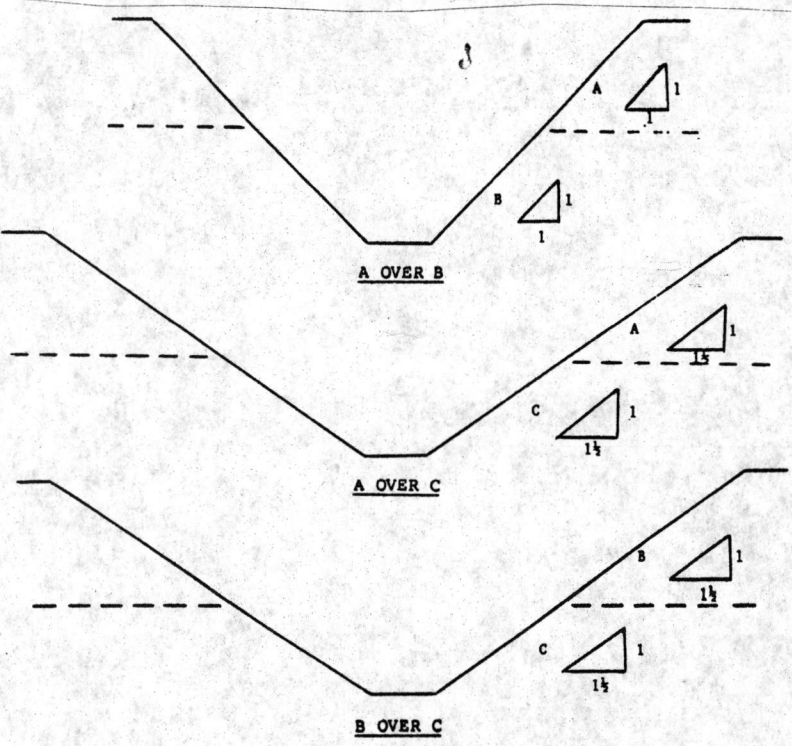

A OVER B

A OVER C

B OVER C

2. All other sloped excavations shall be in accordance with the other options permitted in § 1926.652(b).

INDEX

A

B

1602

INDEX

INDEX

1606

INDEX

INDEX

INDEX

H

INDEX

I

INDEX

K

L

M

INDEX

N

O

P

INDEX

INDEX

INDEX

INDEX

1630

INDEX

INDEX

INDEX

XYZ

NOTES

NOTES